Handbook of Theoretical Computer Science

Volume B

FORMAL MODELS AND SEMANTICS

ELSEVIER
AMSTERDAM • NEW YORK • OXFORD • TOKYO

THE MIT PRESS
CAMBRIDGE, MASSACHUSETTS

Handbook of Theoretical Computer Science

Volume B

FORMAL MODELS AND SEMANTICS

edited by

JAN VAN LEEUWEN,
Utrecht University, The Netherlands

ELSEVIER
AMSTERDAM • NEW YORK • OXFORD • TOKYO

THE MIT PRESS
CAMBRIDGE, MASSACHUSETTS

ELSEVIER SCIENCE PUBLISHERS B.V.
P.O. Box 211,
1000 AE Amsterdam, The Netherlands

Co-publishers for the United States, Canada, and Japan:

The MIT Press
55 Hayward Street
Cambridge, MA 02142, U.S.A.

Library of Congress Cataloging-in-Publication Data

Handbook of theoretical computer science/editor, Jan van Leeuwen.
p. cm.
Includes bibliographical references and indexes.
Contents: v. A. Algorithms and complexity -- v. B. Formal models and semantics.
ISBN 0-444-88075-5 (U.S. : set). -- ISBN 0-444-88071-2 (U.S. : v. A). -- ISBN 0-444-88074-7 (U.S. : v. B). --ISBN 0-262-22040-7 (MIT Press : set). -- ISBN 0-262-22038-5 (MIT Press : v. A). -- ISBN 0-262-22039-3 (MIT Press : v. B).
1. Computer science. I. Leeuwen, J. van (Jan)
QA76.H279 1990
004--dc20 90-3485
CIP

First edition 1990
Second impression 1992

Elsevier Science Publishers B.V.
ISBN: 0 444 88074 7 (Volume B)
ISBN: 0 444 88075 5 (Set of Vols A and B)

The MIT Press
ISBN: 0 262 22039 3 (Volume B)
ISBN: 0 262 22040 7 (Set of Vols A and B)

Printed in The Netherlands

Printed on acid-free paper

Preface

Modern developments in computer and software systems have raised many challenging issues concerning the design and efficiency of complex programming applications. There is an increasing need for "advanced theory", to understand and exploit basic concepts and mechanisms in computing and information processing. The *Handbook of Theoretical Computer Science* is designed to provide a wide audience of professionals and students in Computer Science and related disciplines with an overview of the major results and developments in the theoretical exploration of these issues to date.

There are many different roles for "theory" in Computer Science. On the one hand it provides the necessary mathematical foundations for studying formal systems and algorithms that are needed. On the other hand, it provides concepts and languages to capture the essence, in algorithmic and descriptive terms, of any system from specification to efficient implementation. But the mathematical frameworks that have proved to be invaluable for Computer Science are used increasingly in many other disciplines as well. Wherever notions of information or information processing are identified, questions of representation and computation can be formalized in computer science terms. Theoretical Computer Science concerns itself with all formal models and methods and all techniques of description and analysis that are required in this domain.

As a consequence there are many facets to Theoretical Computer Science. As a discipline it employs advanced techniques from Mathematics and Logic, but at the same time it has established its own formal models and fundamental results in which the original motivations have remained visible. The *Handbook of Theoretical Computer Science* attempts to offer an in-depth view of the field of Theoretical Computer Science as a whole, by a comprehensive exposition of the scientific advances in this area.

In order to keep the Handbook within manageable limits, it was decided to restrict the material to the recognized core areas of Theoretical Computer Science. Even with this restriction imposed, further choices had to be made in the subjects covered and the extent of the material provided on each subject. The current version of the Handbook is presented in two volumes:

Vol. A: Algorithms and Complexity
Vol. B: Formal Models and Semantics

This more or less reflects the division between algorithm-oriented and description-oriented research that can be witnessed in Theoretical Computer Science, and it seemed natural to follow it for these books. Whereas the volumes can be used independently, there are many interesting connections that show that the two areas really are highly intertwined. Together, the volumes give a unique impression of research in Theoretical Computer Science as it is practised today. If not for reference purposes, we hope you will use these books to get a feeling for the theoretical work that is going on in the many fields of Computer Science or to simply satisfy your curiosity.

Each volume consists of close to twenty chapters, with each chapter devoted to a representative subject of current research. Volume A presents the basic material on models of computation, complexity theory, data structures and efficient computation in many recognized subdisciplines of Theoretical Computer Science. Volume B presents a choice of material on the theory of automata and rewriting systems, the foundations of modern programming languages, logics for program specification and verification, and a number of studies aimed at the theoretical modeling of advanced computing applications. Both volumes have been organized to reflect the development of Theoretical Computer Science from its classical roots to the modern complexity-theoretic and logic approaches to, for example, parallel and distributed computing. In most cases an extensive bibliography has been provided to assist in further study of the field and finding specific source documents for further reference. Specific attention has been given to a structured development of the material in the various chapters in order to make the Handbook largely self-contained. However, some familiarity with Computer Science at an undergraduate level is assumed.

The writing of this Handbook was started by an initiative of the publisher, who identified the need for a comprehensive and encyclopedic treatise on the fundamental areas of Theoretical Computer Science. The advisory board consisting of A.R. Meyer (Cambridge, MA), M. Nivat (Paris), M.S. Paterson (Coventry) and D. Perrin (Paris) was instrumental in defining the early version of the Handbook project. The complete project that has now resulted in the present volumes has grown far beyond the limits originally foreseen for the Handbook, and only because of the loyal support of all contributing colleagues to the project has it been possible to complete the gigantic task of presenting the current overview of the core areas of Theoretical Computer Science. I would like to thank all authors, the advisory board and the editorial staff at Elsevier Science Publishers for their invaluable support and contributions to the Handbook project.

J. van Leeuwen
Managing Editor
Utrecht, 1990

List of Contributors to Volume B

K.R. Apt, *CWI, Amsterdam & University of Texas, Austin, TX* (Ch. 10)
H.P. Barendregt, *Katholieke Universiteit Nijmegen* (Ch. 7)
J. Berstel, *Université de Paris VI* (Ch. 2)
L. Boasson, *Université de Paris VII* (Ch. 2)
B. Courcelle, *Université Bordeaux I* (Chs 5, 9)
P. Cousot, *Ecole Polytechnique, Palaiseau* (Ch. 15)
N. Dershowitz, *University of Illinois, Urbana, IL* (Ch. 6)
E.A. Emerson, *University of Texas, Austin, TX* (Ch. 16)
C.A. Gunter, *University of Pennsylvania, Philadelphia, PA* (Ch. 12)
J.P. Jouannaud, *Université de Paris-Sud, Orsay* (Ch. 6)
P. Kanellakis, *Brown University, Providence, RI* (Ch. 17)
D. Kozen, *Cornell University, Ithaca, NY* (Ch. 14)
L. Lamport, *Systems Research Center (Digital), Palo Alto, CA* (Ch. 18)
N.A. Lynch, *Massachusetts Institute of Technology, Cambridge, MA* (Ch. 18)
R. Milner, *University of Edinburgh* (Ch. 19)
J.C. Mitchell, *Stanford University, Palo Alto, CA* (Ch. 8)
P.D. Mosses, *Aarhus Universitet* (Ch. 11)
D. Perrin, *Université de Paris VII* (Ch. 1)
A. Salomaa, *University of Turku* (Ch. 3)
D.S. Scott, *Carnegie-Mellon University, Pittsburgh, PA* (Ch. 12)
W. Thomas, *Universität Kiel* (Ch. 4)
J. Tiuryn, *Warsaw University* (Ch. 14)
M. Wirsing, *Universität Passau* (Ch. 13)

Contents

PREFACE	v
LIST OF CONTRIBUTORS TO VOLUME B	vii
CONTENTS	ix
CHAPTER 1 FINITE AUTOMATA	1
D. Perrin	
1. Introduction	3
2. Finite automata and recognizable sets	4
3. Rational expressions	11
4. Kleene's theorem	14
5. Star-height	19
6. Star-free sets	21
7. Special automata	29
8. Recognizable sets of numbers	35
9. Notes	45
Acknowledgment	52
References	53
CHAPTER 2 CONTEXT-FREE LANGUAGES	59
J. Berstel and L. Boasson	
Introduction	61
1. Languages	62
2. Iteration	72
3. Looking for nongenerators	82
4. Context-free groups	95
Acknowledgment	100
References	100

CHAPTER 3 FORMAL LANGUAGES AND POWER SERIES 103
A. Salomaa
Prelude 105
1. Preliminaries 105
2. Rewriting systems and grammars 109
3. Post canonical systems 113
4. Markov algorithms 114
5. Parallel rewriting systems 116
6. Morphisms and languages 119
7. Rational power series 123
8. Algebraic power series 126
9. What power series do for you 129
10. Final comments 131
References 131

CHAPTER 4 AUTOMATA ON INFINITE OBJECTS 133
W. Thomas
Introduction 135

Part I. Automata on infinite words
Notation 136
1. Büchi automata 136
2. Congruences and complementation 139
3. The sequential calculus 143
4. Determinism and McNaughton's Theorem 147
5. Acceptance conditions and Borel classes 152
6. Star-free ω-languages and temporal logic 156
7. Context-free ω-languages 161

Part II. Automata on infinite trees
Notation 165
8. Tree automata 167
9. Emptiness problem and regular trees 170
10. Complementation and determinacy of games 173
11. Monadic tree theory and decidability results 178
12. Classifications of Rabin recognizable sets 184
Acknowledgment 186
References 186

CHAPTER 5 GRAPH REWRITING: AN ALGEBRAIC AND LOGIC APPROACH 193
B. Courcelle
Introduction 195
1. Logical languages and graph properties 196
2. Graph operations and graph expressions 203
3. Context-free sets of hypergraphs 211
4. Logical properties of context-free sets of hypergraphs 215
5. Sets of finite graphs defined by forbidden minors 222

6. Complexty issues 226
7. Infinite hypergraphs 229
8. Guide to the literature 235
Acknowledgment 238
References 239

CHAPTER 6 REWRITE SYSTEMS 243
N. Dershowitz and J.-P. Jouannaud
1. Introduction 245
2. Syntax 248
3. Semantics 260
4. Church–Rosser properties 266
5. Termination 269
6. Satisfiability 279
7. Critical pairs 285
8. Completion 292
9. Extensions 305
Acknowledgment 309
References 309

CHAPTER 7 FUNCTIONAL PROGRAMMING AND LAMBDA CALCULUS 321
H.P. Barendregt
1. The functional computation model 323
2. Lambda calculus 325
3. Semantics 337
4. Extending the language 347
5. The theory of combinators and implementation issues 355
Acknowledgment 360
References 360

CHAPTER 8 TYPE SYSTEMS FOR PROGRAMMING LANGUAGES 365
J.C. Mitchell
1. Introduction 367
2. Typed lambda calculus with function types 370
3. Logical relations 415
4. Introduction to polymorphism 431
Acknowledgment 452
Bibliography 453

CHAPTER 9 RECURSIVE APPLICATIVE PROGRAM SCHEMES 459
B. Courcelle
Introduction 461
1. Preliminary examples 461
2. Basic definitions 463
3. Operational semantics in discrete interpretations 466
4. Operational semantics in continuous interpretations 471
5 Classes of interpretations 475

6. Least fixed point semantics 481
7. Transformations of program schemes 485
8. Historical notes, other types of program schemes and guide to literature 488
Acknowledgment 490
References 490

CHAPTER 10 LOGIC PROGRAMMING 493
K.R. Apt
1. Introduction 495
2. Syntax and proof theory 496
3. Semantics 511
4. Computability 523
5. Negative information 531
6. General goals 547
7. Stratified programs 555
8. Related topics 566
Appendix 569
Note 570
Acknowledgment 570
References 571

CHAPTER 11 DENOTATIONAL SEMANTICS 575
P.D. Mosses
1. Introduction 577
2. Syntax 578
3. Semantics 585
4. Domains 589
5. Techniques 596
6. Bibliographical notes 623
References 629

CHAPTER 12 SEMANTIC DOMAINS 633
C.A. Gunter and D.S. Scott
1. Introduction 635
2. Recursive definitions of functions 636
3. Effectively presented domains 641
4. Operators and functions 644
5. Powerdomains 653
6. Bifinite domains 660
7. Recursive definitions of domains 663
References 674

CHAPTER 13 ALGEBRAIC SPECIFICATION 675
M. Wirsing
1. Introduction 677
2. Abstract data types 679

3. Algebraic specifications 690
4. Simple specifications 712
5. Specifications with hidden symbols and constructors 718
6. Structured specifications 737
7. Parameterized specifications 752
8. Implementation 759
9. Specification languages 770
Acknowledgment 780
References 780

CHAPTER 14 LOGICS OF PROGRAMS 789
D. Kozen and J. Tiuryn
1. Introduction 791
2. Propositional Dynamic Logic 793
3. First-order Dynamic Logic 811
4. Other approaches 830
Acknowledgment 834
References 834

CHAPTER 15 METHODS AND LOGICS FOR PROVING PROGRAMS 841
P. Cousot
1. Introduction 843
2. Logical, set- and order-theoretic notations 849
3. Syntax and semantics of the programming language 851
4. Partial correctness of a command 858
5. Floyd–Naur partial correctness proof method and some equivalent variants 859
6. Liveness proof methods 875
7. Hoare logic 883
8. Complements to Hoare logic 931
References 978

CHAPTER 16 TEMPORAL AND MODAL LOGIC 995
E.A. Emerson
1. Introduction 997
2. Classification of temporal logics 998
3. The technical framework of Linar Temporal Logic 1000
4. The technical framework of Branching Temporal Logic 1011
5. Concurrent computation: a framework 1017
6. Theoretical aspects of Temporal Logic 1021
7. The application of Temporal Logic to program reasoning 1048
8. Other modal and temporal logics in computer science 1064
Acknowledgment 1067
References 1067

CHAPTER 17 ELEMENTS OF RELATIONAL DATABASE THEORY 1073
P.C. Kanellakis
1. Introduction 1075
2. The relational data model 1079
3. Dependencies and database scheme design 1112
4. Queries and database logic programs 1121
5. Discussion: other issues in relational database theory 1140
6. Conclusion 1144
Acknowledgment 1144
References 1144

CHAPTER 18 DISTRIBUTED COMPUTING: MODELS AND METHODS 1157
L. Lamport and N. Lynch
1. What is distributed computing 1159
2. Models of distributed systems 1160
3. Reasoning about distributed algorithms 1167
4. Some typical distributed algorithms 1178
References 1196

CHAPTER 19 OPERATIONAL AND ALGEBRAIC SEMANTICS OF CONCURRENT PROCESSES 1201
R. Milner
1. Introduction 1203
2. The basic language 1206
3. Strong congruence of processes 1215
4. Observation congruence of processes 1223
5. Analysis of bisimulation equivalences 1234
6. Confluent processes 1238
7. Sources 1240
Note added in proof 1241
References 1241

SUBJECT INDEX 1243

CHAPTER 1

Finite Automata

Dominique PERRIN

LITP, Université Paris 7,
2 Place Jussieu, F-75251 Paris Cedex 05, France

Contents

1. Introduction 3
2. Finite automata and recognizable sets 4
3. Rational expressions 11
4. Kleene's theorem 14
5. Star-height 19
6. Star-free sets 21
7. Special automata 29
8. Recognizable sets of numbers 35
9. Notes 45
Acknowledgment 52
References 53

HANDBOOK OF THEORETICAL COMPUTER SCIENCE
Edited by J. van Leeuwen

1. Introduction

The theory of finite automata has preserved from its origins a great diversity of aspects. From one point of view, it is a branch of mathematics connected with the algebraic theory of semigroups and associative algebras. From another viewpoint, it is a branch of algorithm design concerned with string manipulation and sequence processing. It is perhaps this diversity which has enriched the field to make it presently one with both interesting applications and significant mathematical problems.

The first historical reference to finite automata is a paper of S.C. Kleene of 1954 in which the basic theorem, now known as Kleene's theorem, is already proved [72]. Kleene's paper was actually a mathematical reworking of the ideas of two researchers from the MIT, W. McCulloch and W. Pitts who had presented as early as 1943 a logical model for the behaviour of nervous systems that turned out to be the model of a *finite-state machine* [85]. Indeed a finite automaton can be viewed as a machine model which is as elementary as possible in the sense that the machine has a memory size which is fixed and bounded, independently of the size of the input. The number of possible states of such a machine is itself bounded, whence the notion of a finite-state machine. The historical origin of the finite-state model itself can of course be traced back much earlier to the beginning of this century with the notion of a *Markov chain*. In fact, a Markov chain is the model of a stochastic process in which the probability of an event only depends on the events that happened before at a bounded distance in the past.

Since the origins, the theory of finite automata has developed, stimulated both by its possible application and by its inner mathematical orientations. In an early stage, finite automata appeared as a development of logical circuits obtained by introducing the sequentiality of operations. This lead to the notion of *sequential circuits* which is still of interest in the field of circuit design. But the main present applications of finite automata are related with *text processing*. For example, in the phase of the compiling process, known as *lexical analysis*, the source program is transformed according to simple operations such as comments removal, or parsing of identifiers and keywords. This elementary processing is generally performed by algorithms that are really finite automata and can usefully be designed and handled by the methods of automata theory. In the same way, in natural language processing, finite automata, often called *transition networks*, are used to describe some phases of the lexical analysis. These applications of automata to text processing have a natural extension to subjects as text compression, file manipulation or, more remotely, to the analysis of long sequences of molecules met in molecular biology. Other applications of finite automata are related to the study of parallel processes. In fact most models of concurrency and synchronization of processes use, either explicitly or sometimes implicitly, methods which are close to finite automata.

From the mathematical point of view, the development of automata theory has followed a great number of different paths. A number of them are related with decidability problems, such as the computability of the star-height which has been proved by K. Hashiguchi (see Section 5 below). Another direction is that of studying subfamilies of the family of finite automata. Roughly speaking, one starts with a special kind of finite automaton, defined by a natural condition, and one would like to know under which condition a given finite automaton can be simulated by one from the subfamily. We shall see several examples of this situation in this chapter, the most famous being probably the family of *counter-free automata* introduced by R. McNaughton and characterized by Schützenberger's theorem (see Section 6).

This chapter can serve as an introduction to the field of finite automata. It is to a large extent self-contained and can be read by anyone with a standard mathematical background. It clearly does not cover the entire field but might allow the reader to grasp the flavour of the results and the methods. It also gives pointers to further reading. The related theory of automata working on infinite objects and the relation of finite automata with logic is treated in a separate chapter (see Thomas [124]).

The chapter is organized in nine sections. The first three (Sections 2–4) give the basic definitions and a proof of the fundamental result of Kleene. Section 5 is an introduction to the notion of star-height. As indicated in the notes, this is presently one of the areas of automata theory which is developing very fast. Section 6 deals with star-free sets. This is an important notion especially because of its connection with Logic (see Thomas [124]). We give a complete proof of Schützenberger's theorem, stating the equality between star-free sets and aperiodic sets. We further give, with partial proofs, the syntactic characterization of two important subfamilies of star-free sets namely locally testable and piecewise testable. In Section 7, we introduce the reader to some aspects of the applications of finite automata such as string matching, or file indexing. These aspects are more developed in Aho [4]. Section 8 is an introduction to the field of automata recognizing numbers expanded at some basis. This is an aspect of finite automata related to several fields in classical mathematics such as number theory and ergodic theory. We shall give in this section a complete proof of Cobham's famous theorem. The proof, due to G. Hansel, appears here for the first time in an accessible reference. Section 9 contains notes mentioning extensions of the results presented here in various directions, and indications for further reading. This chapter ends with a bibliography.

2. Finite automata and recognizable sets

In this chapter, we shall describe properties of *words*, that is to say finite sequences of symbols taken out of a given set called the *alphabet*. The nature of this set may vary, according to the application of interest. Its elements may for instance be themselves words over another alphabet as in the case of characters encoded by a fixed-length binary string. They may also be elements of an abstract set of events in the case of an analysis of the behaviour of a system. Let A be an alphabet. We denote a word with a mere juxtaposition of its letters as

$$x = a_1 a_2 \dots a_n$$

when a_i are letters from A for $1 \leqslant i \leqslant n$. The integer n is the *length* of the word x. We denote by ε or 1 the *empty word*, which is the unique word of length zero. We denote by A^* the set of all words on the alphabet A, called the *free monoid* on A.

We recall that a monoid is a set with an associative binary operation and a neutral element 1_M. A *morphism* from a monoid M into a monoid N is a mapping $f: M \to N$ such that $f(1_M) = 1_N$ and $f(xy) = f(x)f(y)$ for all x, y in M.

A *finite automaton* on the alphabet A is given by a finite set Q of *states*, two subsets I, T of Q called the sets of *initial* and *terminal* states and a set $E \subset Q \times A \times Q$ of *edges*. We denote the automaton as a quadruple (Q, I, T, E) or also (Q, I, T) when the set of edges is implicit.

A *path* in the automaton $\mathcal{A} = (Q, I, T, E)$ is a sequence $c = (e_i)_{1 \leqslant i \leqslant n}$ of consecutives edges, i.e. of edges $e_i = (p_i, a_i, p_{i+1})$. The word $w = a_1 a_2 \ldots a_n$ is the *label* of the path, state p_1 its *origin* and state p_{n+1} its *end*. The number n is its *length*. One agrees to define for each state p in Q a unique null path of length 0 with origin and end p. Its label is the empty word 1. A path is *successful* if its origin is in I and its end in T. Finally a word w on the alphabet A is *recognized* by the automaton $\mathcal{A}$ if it is the label of some successful path. The *set recognized* by the automaton, denoted $L(\mathcal{A})$ is the set of all words recognized by $\mathcal{A}$.

A set $X \subset A^*$ is said to be *recognizable* if there exists a finite automaton $\mathcal{A}$ such that $X = L(\mathcal{A})$.

As an example, the automaton depicted on Fig. 1 corresponds to $A = \{a, b\}$, $Q = \{1, 2, 3\}$, $I = \{1\}$, $T = \{3\}$.

It recognizes the set of words ending with *ab*. Note that the automaton of Fig. 2 recognizes the same set.

An automaton $\mathcal{A} = (Q, I, T, E)$ is said to be *deterministic* if $\text{card}(I) = 1$ and if for each pair $(p, a) \in Q \times A$ there is at most one state $q \in Q$ such that $(p, a, q) \in E$. It is said to be *complete* if for each $(p, a) \in Q \times A$ there is at least one $q \in Q$ such that $(p, a, q) \in E$. As an example, the automaton of Fig. 1 is deterministic and complete. The automaton of Fig.

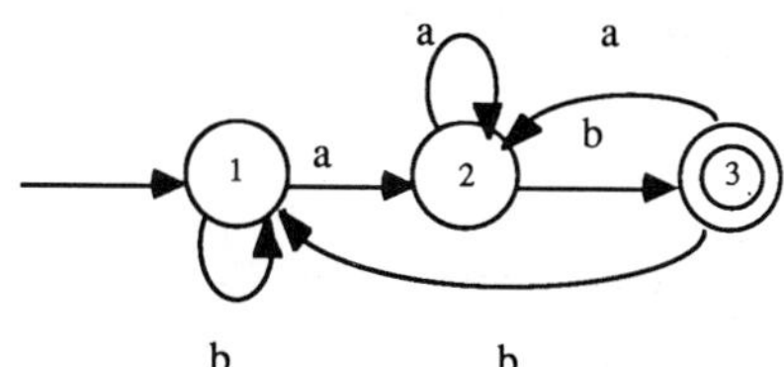

Fig. 1. A deterministic automaton.

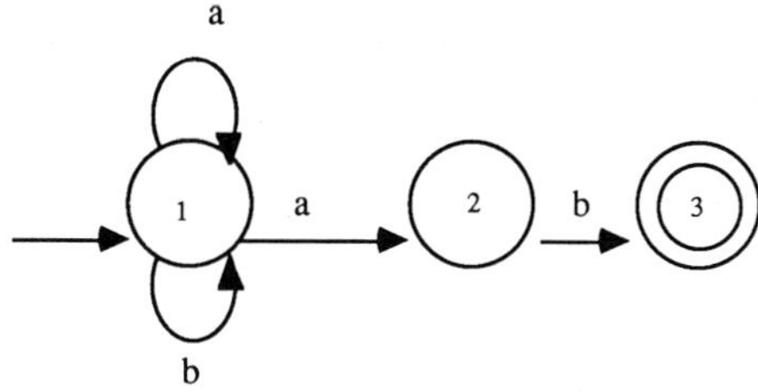

Fig. 2. A nondeterministic automaton.

2 is neither deterministic nor complete. For a deterministic automaton it is convenient to denote $p.a=q$ instead of $(p, a, q) \in E$.

2.1. THEOREM. *For each finite automaton, there exists a deterministic and complete one recognizing the same set.*

PROOF. Starting from $\mathcal{A}=(Q, I, T, E)$, one builds the automaton $\mathcal{B}$ whose states are the subsets of Q, having the set I as initial state, the sets $U \subset Q$ that meet T as terminal states and whose edges are the triples (U, a, V) such that V is the set of all states $q \in Q$ such that there exists an edge $(p, a, q) \in E$ with $p \in U$. □

The above construction is sometimes called the *subset construction.*

For example, starting from the automaton of Fig. 2 one obtains by this construction the automaton of Fig. 3. It has $2^3=8$ states. If we delete the states that cannot be reached from the initial state $\{1\}$, there remain only three states and we obtain the automaton of Fig. 1. We should be careful that it is not the case in general that the "useful" part of the automaton obtained by this construction has a number of states which has the same order of magnitude as the starting one. This is true however in some cases that have practical significance, as we shall see in Section 7.

A deterministic automaton can be easily represented by a $A \times Q$ matrix called its *transition array.* For the automaton of Fig. 1 it is given by the array of Table 1.

A fundamental fact which is a consequence of the subset construction is the following.

2.2. COROLLARY. *The complement of a recognizable set is again recognizable.*

PROOF. Indeed if $\mathcal{A}=(Q, I, T, E)$ is a complete deterministic automaton, changing T into its complement in Q changes $L(\mathcal{A})$ into its complement. □

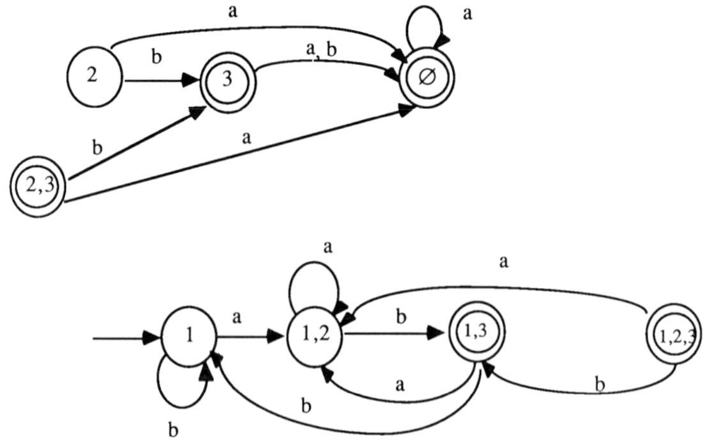

Fig. 3. The subset construction.

Table 1
The transition table

	1	2	3
a	2	2	2
b	1	3	1

This is certainly the basic property of recognizable sets which makes them an object which is relatively easy to handle.

The definition of recognizable sets does not imply a choice in favor of left-to-right reading of words. Indeed, a set X is recognizable iff the set $\tilde{X}$ obtained by reverting words is recognizable. It is only the subset construction which favors left-to-right parsing or words and the automaton obtained in this way for $\tilde{X}$ might look quite different.

It is interesting to realize what happens to the definitions in the case of a finite set X of words. A nondeterministic automaton recognizing X is directly obtained by putting together all the automata obtained for each word of X. Figure 4 shows the resulting automaton for $X = \{\text{the, then, thin, thing, this, sin, sing}\}$.

A deterministic automaton recognizing X is just the same as a *tree* (sometimes called a *trie* as a reference to information re*trie*val) since the subset construction just amounts to collect all elements of X with a common given prefix. Figure 5 gives a deterministic automaton recognizing the seven-element set given above.

The definition of a recognizable set can be extended to a situation which is more general. To do this we have to go to concepts slightly more abstract. To any finite automaton $\mathscr{A} = (Q, I, T, E)$ we may associate a finite monoid, called its *transition monoid*, which is defined as follows. To each word w in A^*, we associate a binary relation $\varphi(w)$ on the set Q of states with $(p, q) \in \varphi(w)$ iff there exists a path from p to

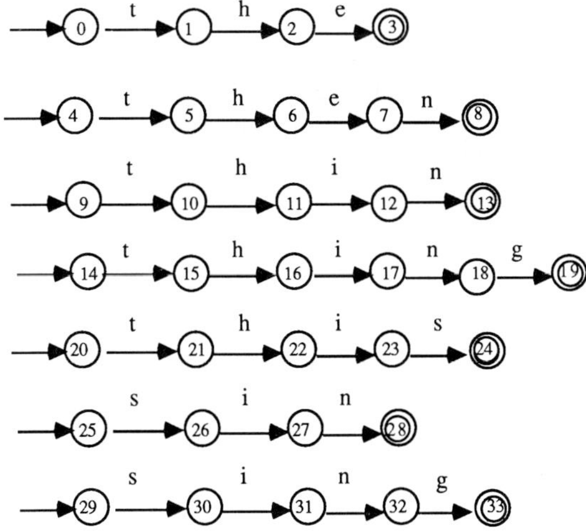

Fig. 4. A nondeterministic automaton.

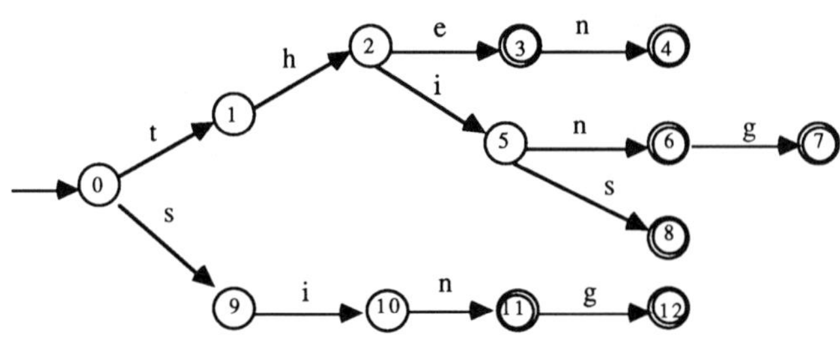

Fig. 5. A trie.

q with label w. Then for two words $u, v \in A^*$ the relation $\varphi(uv)$ is the product under the composition of relations of the relations $\varphi(u)$ and $\varphi(v)$. Therefore, we have thus defined a morphism of monoids $\varphi: A^* \to 2^{Q \times Q}$ from the free monoid A^* into the monoid $N = 2^{Q \times Q}$ of binary relations on the set Q. The morphism φ has the property that whenever two words $u, v \in A^*$ have the same image, i.e. $\varphi(u) = \varphi(v)$, u is accepted by the automaton iff v is. Hence the set $X = L(\mathcal{A})$ recognized by the automaton $\mathcal{A}$ satisfies the equality $\varphi^{-1}\varphi(X) = X$. We shall express this equality by saying that φ *saturates* X.

We can now go to the generalization. Let M be an arbitrary monoid instead of the free monoid A^*. We will say that a set $X \subset M$ is *recognizable* in M if there exists a morphism $\varphi: M \to N$ from M into a finite monoid N that saturates X. What we have to check is that the above definition does coincide with the previous one for $M = A^*$. We have already seen it in one direction. Conversely, let $\varphi: A^* \to N$ be a morphism from A^* into a finite monoid N. Let $X \subset A^*$ be saturated by φ. We construct a finite automaton recognizing X as follows. We take N as set of states, the neutral element 1 of N as initial state, the set $\varphi(X)$ as set of terminal states and we define the edges by

$$n.a = n\varphi(a).$$

The automaton thus constructed recognizes X. Note that it is deterministic and complete.

The above considerations lead naturally to the definition of the *syntactic monoid* of a subset X in a monoid M. To give it, we introduce the notion of the set of *contexts* of an element $m \in M$ with respect to X as the set

$$C_X(m) = \{(u, v) \in M \mid u\,m\,v \in X\}.$$

The equality of the sets of contexts defines an equivalence which is compatible with the multiplication in M. The set of classes has therefore the structure of a monoid. It is called the syntactic monoid of X denoted Synt(X). The natural morphism $\sigma: M \to \text{Synt}(X)$, associating to an element $m \in M$ its class, is called the *syntactic morphism*. This morphism saturates X since $m \in X$ iff $(1, 1) \in C_X(m)$. Also for any surjective morphism $\varphi: M \to N$ that saturates X, there is a unique morphism $\alpha: N \to \text{Synt}(X)$ from N onto Synt(X) such that $\varphi = \alpha\sigma$. This means roughly that Synt(X) is the smallest possible monoid recognizing X. Therefore we have

2.3. PROPOSITION. *A set $X \subset M$ is recognizable iff its syntactic monoid is finite.*

As an example, let us compute the syntactic monoid of the set X of words on the alphabet $A=\{a,b\}$ that end with ab. There are five possible sets of contexts, namely

$$C_X(1)=(A^*\times X)+(A^*a\times b)+(X\times 1),$$
$$C_X(b)=(A^*\times X)+(A^*a\times 1),$$
$$C_X(a)=(A^*\times X)+(A^*\times b),$$
$$C_X(ab)=(A^*\times X)+(A^*\times 1),$$
$$C_X(b^2)=(A^*\times X).$$

The syntactic monoid of X has therefore five elements

$$1=\sigma(1),\quad \alpha=\sigma(a),\quad \beta=\sigma(b),\quad \alpha\beta=\sigma(ab),\quad \beta^2=\sigma(b^2). \tag{2.1}$$

The multiplication in Synt(X) is given by the rules

$$\alpha^2=\beta\alpha=\alpha,\qquad \alpha\beta^2=\beta^3=\beta^2.$$

The computation of the syntactic monoid of a set $X\subset A^*$ can be organized as follows. We suppose that X is given as the set recognized by a deterministic and complete automaton $\mathscr{A}=(Q,i,T,E)$. We first remove all states that are not accessible from the initial state. We then compute an equivalence relation θ on the set Q. This equivalence is defined by $p\equiv q \bmod \theta$ if the set recognized by the automata (Q,p,T,E) and (Q,q,T,E) are equal.

In practice, to compute the equivalence θ one can build a decreasing sequence of equivalence relations

$$\theta_0\geqslant\theta_1\geqslant\theta_2\geqslant\cdots$$

defined as follows. The first one is the partition of the set Q of states into the two classes T and $Q-T$. Further on, θ_{i+1} is defined from θ_i by $p\equiv q \bmod \theta_{i+1}$ iff $p.a\equiv q.a \bmod \theta_i$ for all a in $A\cup 1$. Then $\theta_{n-1}=\theta$ as it can be verified.

By contracting the states belonging to the same θ-class into a unique state and defining the edges in the natural way, one obtains a deterministic and complete automaton whose transition monoid is the syntactic monoid of X. This automaton can also be defined abstractly as follows. For $u\in A^*$ let

$$u^{-1}X=\{v\in A^*\mid uv\in X\}.$$

We build an automaton by taking as states the different sets $u^{-1}X$, as initial state $X=1^{-1}X$, as terminal states the $u^{-1}X$ which contain 1 and with edges defined by

$$(u^{-1}X).a=(ua)^{-1}X.$$

This automaton is identical with the automaton obtained after the reduction procedure described above, whatever was the deterministic automaton recognizing X we started with. This canonical automaton is called the *minimal automaton* of X.

To illustrate this, we treat again the example of the set X of words on the alphabet $\{a,b\}$ that end with ab. The automaton given by Fig. 1 is deterministic and recognizes X. It is minimal, i.e. the equivalence θ is the equality since

$$\theta_0=\{\{1,2\},\{3\}\},\qquad \theta_1=\{\{1\},\{2\},\{3\}\}.$$

Table 2
The syntactic monoid of $X=(a+b)^*ab$

	1	α	β	$\alpha\beta$	β^2
1	1	2	1	3	1
2	2	2	3	3	1
3	3	2	1	3	1

Hence, the syntactic monoid of X is the transition monoid of this automaton. Indeed, this can be checked in Table 2 giving in each column the mapping defined by (2.1), as a generalization of the transition table of the automaton.

As another illustration of the computation of the minimal automata of a set, let us consider again the particular case of a finite set of words, sometimes called a *dictionary* by reference to application in information processing. Whereas a deterministic automaton corresponds in this case to a finite tree as we have seen on Fig. 5, the minimal deterministic automaton gives rise to an acyclic graph sometimes called a *dawg* (for directed acyclic word graph). The minimization procedure can be visualized as a merge of the tree nodes being the root of identical subtrees. The result of the procedure applied to the tree of Fig. 5 is shown on Fig. 6.

One familiar feature of finite automata and recognizable sets is the *iteration property*. Indeed, long enough paths in a finite automaton have to comprise some cycle and this implies the following *iteration lemma*.

2.4. PROPOSITION. *Let $X \subset A^*$ be an infinite recognizable set. There exists an integer n such that any word $x \in X$ of length exceeding n has a factorization $x=uvw$ with $v \neq 1$ and $uv^k w \in X$ for all $k \geqslant 0$.*

Such a statement can be used to prove that some sets are not recognizable. Indeed the set

$$X=\{a^n b^n \mid n \geqslant 0\}$$

obviously does not satisfy the statement. It can also be stated in more refined versions, replacing loops in the automaton by *idempotents* in a finite monoid, that is elements such that $m=m^2$. We shall have the opportunity to use such idempotents in Section 6.

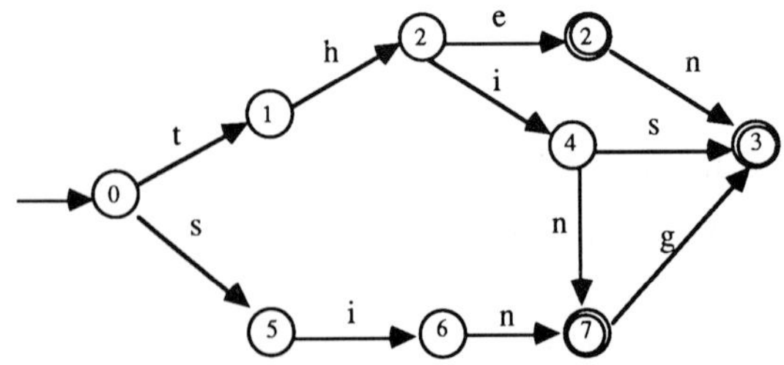

Fig. 6. A directed acyclic word graph.

Another familiar aspect of recognizable sets is the following characterization.

2.5. PROPOSITION. *A set $X \subset A^*$ is recognizable iff there exist a finite alphabet B, two sets U, $V \subset B$, a set $W \subset B^2$ and a substitution $s: B \to A$ such that*

$$X = s(UB^* \cap B^* V - B^* W B^*)$$

(with s extended to words the usual way).

PROOF. Let X be recognized by an automaton $\mathscr{A} = (Q, I, T, F)$. We take $B = F$ and we define s by $s(p, a, q) = a$. We put the edge (p, a, q) in U when p is in I and in V when q is in T. Finally, we choose for W the pairs of consecutive edges. Then the formula above holds by the definition of the set recognized by an automaton.

The converse of the construction is similar. □

The special recognizable sets of the form $UB^* \cap B^* V - B^* WB^*$ are called *local sets.* We shall come back to local sets later on in Section 6 with the notion of locally testable sets.

We end this section with a discussion of decision problems concerning finite automata. The following fact is indeed a fundamental feature of finite automata, which is a corollary of Proposition 2.3.

2.6. COROLLARY. *Given two recognizable sets $X, Y \subset A^*$ it is decidable whether $X \subset Y$.*

PROOF. We compute two morphisms $\varphi: A^* \to M$ and $\psi: A^* \to N$ on finite monoids M and N that saturate X and Y respectively. Let λ be the morphism from A^* into $M \times N$ defined by

$$\lambda(x) = (\varphi(x), \psi(x)).$$

Let P be the submonoid of $M \times N$ defined as $P = \lambda(A^*)$. Let $N' = \varphi(X)$, $M' = \psi(Y)$. We have $X \subset Y$ iff for any (m, n) in P

$$m \in N' \quad \Rightarrow \quad n \in M'.$$

Since M and N are finite monoids this gives a decision procedure to test the inclusion $X \subset Y$. □

3. Rational expressions

Let A be an alphabet. The *rational operations* on the subsets of A^* are the following:

union	$X \cup Y$ also written $X + Y$,
product	$XY = \{xy \mid x \in X, y \in Y\}$,
star	$X^* = \{x_1 x_2 \ldots x_n \mid n \geqslant 0, x_1 \in X\}$

The *rational subsets* of A^* are the sets that can be obtained from finite subsets using a finite number of rational operations. Thus, on the alphabet $A=\{a,b,c\}$ the set

$$X=(a+b)^*c$$

is a rational set, consisting of those words ending with a c.

One may define abstract *rational expressions* by considering the terms of the free algebra over the set $A\cup\{0,1\}$ with function symbols $+,.,^*$. There is a mapping L from this term algebra onto the algebra of rational subsets of A^* defined inductively as follows

$$L(0)=\emptyset,\quad L(1)=\{1\},\quad L(a)=\{a\},$$
$$L(e+e')=L(e)\cup L(e'),\quad L(e.e')=L(e)L(e')$$
$$L(e^*)=L(e)^*$$

We shall say that the expression e *denotes* the set $L(e)$. We shall assume that regular expressions satisfy the axioms making + idempotent and commutative, the product associative and distributive and the usual rules for 0 and 1:

$$e+f=f+e,\quad e+e=e,\quad e.(f+g)=e.f+e.g,\quad (f+g).e=f.e+g.e,$$
$$e.(f.g)=(e.f).g,\quad 0+e=e+0=e,\quad 1.e=e.1=e,\quad 0.e=e.0=0. \tag{3.1}$$

This is of course consistent with the definition of the mapping L. Axioms (3.1) are the axioms of an idempotent semiring. A *semiring* is the same as a ring except that there is no inverse for the + operation. It is said to be *idempotent* because of the axiom $e+e=e$. We shall often omit the dot in rational expression and denote ef instead of $e.f$.

There are of course several expressions denoting the same set and this is the source of *identities* written $e\equiv f$ instead of $L(e)=L(f)$. There are indeed identities relating the star to the other operations, such as

$$(e+f)^*\equiv(e^*f)^*e^*, \tag{3.2}$$
$$(ef)^*\equiv 1+e(fe)^*f. \tag{3.3}$$

The last identity gives for $f=1$ the identity $e^*\equiv 1+ee^*$. One often denotes

$$e^+\equiv ee^*\equiv e^*e.$$

It is an interesting exercise to deduce identities one from the other by means of substitutions. For instance, the identity

$$(e+f)^*\equiv e^*(fe^*)^*$$

can be deduced from the above identities, by the following computation:

$$\begin{aligned}(e+f)^*&\equiv(e^*f)^*e^* && \text{(by (3.2))}\\ &\equiv(1+e^*(fe^*)^*f)e^* && \text{(by (3.3))}\\ &\equiv e^*+e^*(fe^*)^*fe^* && \text{(by (3.1))}\\ &\equiv e^*(1+(fe^*)^*fe^*) && \text{(by (3.1))}\\ &\equiv e^*(fe^*)^* && \text{(by (3.3)).}\end{aligned}$$

There are other identities such as for all $n \geqslant 1$

$$e^* \equiv (1+e+\cdots+e^{n-1})(e^n)^*.$$

It is not very difficult to prove that, on a one-letter alphabet, the identities written up to now form a complete system of identities, i.e. that any other identity is deducible from these by substitution. A much more difficult and somehow disappointing result was proved by J.H. Conway: any complete system of identities on a two-letter alphabet must contain an infinity of identities in two or more variables. (See Section 8 for references.)

The definition of rational subsets of A^* can be readily generalized in the following way. Let M be an arbitrary monoid. The set $\mathscr{P}(M)$ of subsets of M naturally allows the operations of union, denoted $+$, of set product

$$XY = \{xy \in M \mid x \in X, y \in Y\}$$

and star

$$X^* = \{x_1 x_2 \ldots x_n \mid n \geqslant 0, x_i \in X\}.$$

A set $X \subset M$ is then said to be *rational* if it can be obtained from finite subsets of M by a finite number of unions, products and stars.

As an example of this generalization, one may consider the monoid

$$M = A^* \times B^*$$

which is the direct product of A^* and B^*, i.e. whose product is defined by

$$(u, v)(r, s) = (ur, vs).$$

A rational subset of $M = A^* \times B^*$ is called a *rational relation* (or transduction) between A^* and B^*.

For instance if

$$\Lambda = \{a, b, c, \ldots, z\}, \qquad \Gamma = \{A, B, C, \ldots, Z\}$$

are the sets of lower-case and upper-case roman characters, the rational relation, $X \subset \Lambda^* \times \Gamma^*$ defined by

$$X = ((a, A) + (b, B) + \cdots + (z, Z))^*$$

puts in correspondence a word written in lower-case characters with the corresponding one written in upper-case characters.

Rational expressions are used under a more or less complete form in many text processing systems such as text editors or file manipulation languages. For instance, the EMACS text editor uses the ordinary notation for regular expressions plus some useful abbreviations such as

- "." represents the set of all characters except newline,
- [a–z] represents the set of characters between a and z in ASCII ascending order.

These regular expressions are used to find a substring within a file and then possibly operate some modification on it. Special conventions are used when several substrings of a file match the given rational expression. Very often, this convention is to choose the earliest occurrence of a word matching the expression and, for a given position of the beginning of this word to choose the word as long as possible. We shall turn back to applications of automata in Section 7. We now come to the cornerstone of automata theory.

4. Kleene's theorem

We shall prove the following fundamental result. Let A be a finite alphabet.

4.1. THEOREM (Kleene). *A set $X \subset A^*$ is recognizable iff it is rational.*

PROOF. The first half of the proof is easy. Let $X \subset A^*$ be recognizable and let $\mathscr{A} = (Q, I, T)$ be a finite automaton recognizing X. Let $Q = \{1, 2, \ldots, n\}$ to simplify the notation. One may write X as a finite union

$$X = \sum_{i \in I} \sum_{t \in T} X_{it}$$

where X_{it} is the set recognized by the automaton (Q, i, t). It is therefore enough to prove that each X_{it} is rational. To do this, for each $k \geqslant 1$, let $X_{ij}^{(k)}$ denote the set of labels of nonempty paths

$$i \to q_1 \to q_2 \to \cdots \to q_n \to j$$

with $n \geqslant 0$ and $q_i \leqslant k$. Then the set

$$X_{ij}^{(0)} \subset A$$

is rational since the alphabet is supposed to be finite. Further, for each $k \geqslant 0$ one has the following recurrence relation:

$$X_{ij}^{(k+1)} = X_{ij}^{(k)} + X_{i.k+1}^{(k)}(X_{k+1.k+1}^{(k)})^* X_{k+1.j}^{(k)} \tag{4.1}$$

showing that $X_{ij}^{(k+1)}$ is rational if the $X_{ij}^{(k)}$ are rational. Since

$$X_{ij} = \begin{cases} X_{ij}^{(n)} & \text{if } i \neq j, \\ X_{ij}^{(n)} \cup 1 & \text{if } i = j, \end{cases}$$

this proves that the X_{ij} are rational and concludes the proof of the first half of the theorem.

The proof of the second half requires the construction of an automaton recognizing the set described by a regular expression. This can be done recursively as follows. Any set $X \subset A \cup 1$ can be recognized by a two-state automaton.

Furthermore, if $\mathscr{A} = (Q, I, T)$ and $\mathscr{A}' = (Q', I', T')$ are finite automata recognizing sets X, X' respectively, then the automaton $\mathscr{A} + \mathscr{A}' = (Q + Q', I + I', T + T')$, where $+$ denotes the disjoint union of sets, recognizes the set $X + X'$.

To treat the operation of product and star, it is convenient to introduce a definition. An automaton $\mathscr{A} = (Q, I, T)$ is *normalized* if (see also Fig. 7)

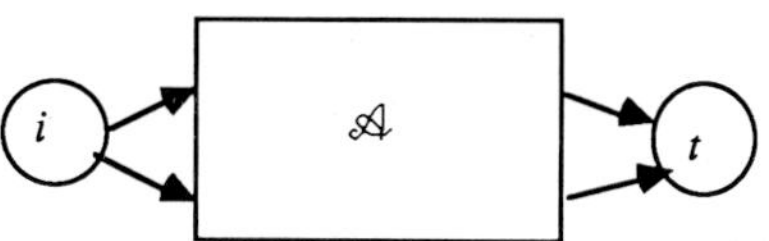

Fig. 7. A normalized automaton.

(i) it has a unique initial state i with no edge entering i,
(ii) it has a unique terminal state t with no edge going out of t.

For any automaton $\mathscr{A}(Q, I, T)$ recognizing a set X, one may build a normalized automaton recognizing the same set X minus possibly the empty word. In fact, it is enough to add two new states i, t and all the edges belonging to the three sets

$$\begin{aligned}
\mathscr{I} &= \{(i, a, q) \mid a \in A \text{ and there is a } p \in I \text{ such that } (p, a, q) \in F\},\\
\mathscr{T} &= \{(p, a, t) \mid a \in A \text{ and there is a } q \in T \text{ such that } (p, a, q) \in F\},\\
\mathscr{O} &= \{(i, a, t) \mid a \in A \text{ and there is a } p \in I, q \in T \text{ such that } (p, a, q) \in F\}.
\end{aligned}$$

To build an automaton recognizing the product XX' of two recognizable sets X, X' one first builds normalized automata $\mathscr{A} = (Q, i, t)$, $\mathscr{A}' = (Q', i', t')$ recognizing $X - 1$, $X' - 1$. We merge t and i' as in Fig. 8. The initial states are i plus possibly $t = i'$ if X contains the empty word 1. The terminal states are t' plus possibly $t = i'$ if X' contains the empty word 1.

Building an automaton for X^* from a normalized automaton $\mathscr{A} = (Q, i, t)$ recognizing $X - 1$ just amounts to merge i and t (see Fig. 9). Since $X^* = (\dot{X} - 1)^*$ this proves that X^* is recognizable whenever X is. □

Given the proof of Kleene's theorem, the time is ripe for some comments on the theorem itself and on the algorithm underlying its proof.

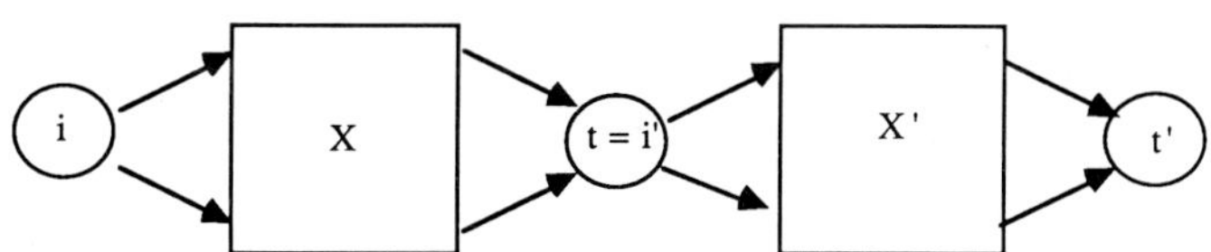

Fig. 8. An automaton for the product XX'.

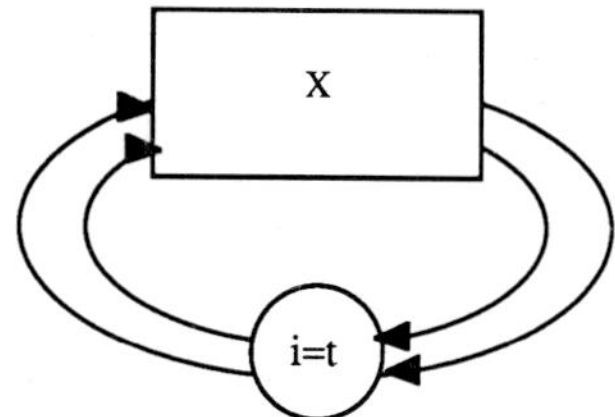

Fig. 9. An automaton for X^*.

The first observation concerns what happens in other monoids than the free monoid A^*, where we have seen that natural definitions of recognizable and rational sets can be given. It is still true that *in any finitely generated monoid M, a recognizable set is rational.* Indeed if $X \subset M$ is recognizable, there is a morphism $\varphi: M \to F$ from M into a finite monoid F such that $\varphi^{-1}\varphi(X)=X$. Since M is finitely generated, we can find a finite alphabet A and a morphism $\alpha: A^* \to M$ from A^* onto M. Now $Y=\alpha^{-1}(X)$ is recognizable in A^* since the morphism $\psi=\varphi\alpha: A^* \to F$ satisfies $\psi^{-1}\psi(Y)=Y$. By Kleene's theorem, Y is rational. It is easy to see that $X=\alpha(Y)$ is also rational (just replace in a rational expression defining Y each letter $a \in A$ by $\alpha(a)$).

However, it is far from true that in an arbitrary monoid any rational set is recognizable. Indeed, in the monoid

$$M = A^* \times A^*$$

the diagonal

$$D = \{(x, x) \mid x \in A^*\}$$

is rational since

$$D = \left[\sum_{a \in A} (a, a)\right]^*.$$

But it is not recognizable since for any morphism $\varphi: M \to F$ such that $\varphi^{-1}\varphi(D)=D$ one must have $\varphi(x, y) \neq \varphi(x', y')$ whenever $x \neq x'$ or $y \neq y'$.

A second observation concerning the theorem is the following. We have seen previously that the complement of a recognizable set is recognizable. Thus Kleene's theorem implies that *the complement of a rational subset of A^* is again rational.* It would certainly be very difficult to prove this directly without using automata. The statement can be false for monoids in which Kleene's theorem does not hold.

We now present two comments concerning the proof of the theorem.

Concerning the first half of the proof, it is useful to see a different presentation of the algorithm allowing to compute rational expressions from automata. The alternative presentation consists in writing a system of equations

$$X_i = \sum_{1 \leqslant j \leqslant n} X_j A_{ji} + \varepsilon_i$$

where A_{ij} is the set of letters a such that (i, a, j) is an edge and $\varepsilon_i = 1$ for $i=1$ and 0 otherwise. It is easy to see that the unique solution to this system is the set X_i denoted previously X_{1i}. To solve this system one uses substitutions according to the following rule: *if $Y, Z \subset A^*$ are such that $1 \notin Y$, then the set ZY^* is the unique solution of the equation $X = XY + Z$.*

EXAMPLE. The system of equations associated with the automaton of Fig. 10 is

$$X_1 = X_1 b + X_3 b + 1, \qquad X_2 = X_1 a + X_2 a, \qquad X_3 = X_2 b.$$

Eliminating X_3 we obtain

$$X_1 = X_1 b + X_2 b^2 + 1, \qquad X_2 = X_1 a + X_2 a.$$

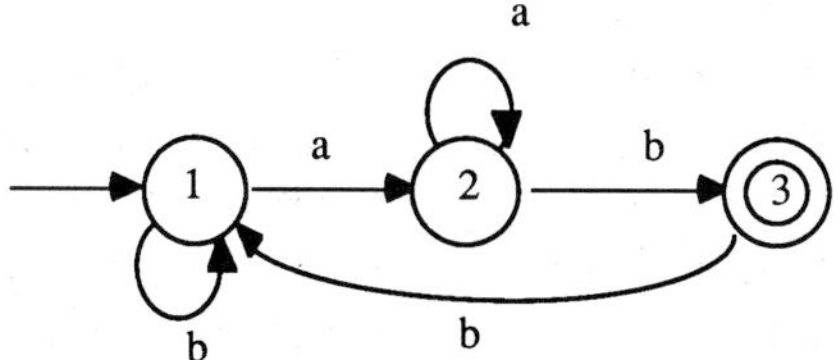

Fig. 10. A deterministic automaton.

Solving the second equation in X_2 gives $X_2 = X_1 a^+$ and substituting in the first equation we get

$$X_1 = X_1 b + X_1 a^+ b^2 + 1.$$

Whence the solution

$$X_1 = (b + a^+ b^2)^*, \qquad X_2 = (b + a^+ b^2)^* a^+, \qquad X_3 = (b + a^+ b^2)^* a^+ b.$$

Concerning the second half of the proof, a useful observation is that there is another completely different method to go from rational expressions to automata. It uses the following notation. For two sets $X, Y \subset A^*$ we denote

$$X^{-1}Y = \{u \in A^* \mid X\, u \cap Y \neq \emptyset\}$$

and call it the *left residual* of Y with respect to X. The following rules recursively allow one to compute the residuals of a rational set. The rules hold for $a \in A$ and $X, Y \subseteq A^*$:

$$a^{-1}(X \cup Y) = a^{-1}X \cup a^{-1}Y, \tag{4.2}$$

$$a^{-1}(XY) = (a^{-1}X)Y \cup (X \cap 1)a^{-1}Y, \tag{4.3}$$

$$a^{-1}(X^*) = (a^{-1}X)X^*. \tag{4.4}$$

The starting point is of course for $b \in A \cup 1$, $a^{-1}b = 1$ if $a = b$ and $\emptyset$ otherwise. To make the system complete one should add the following rules to compute $X \cap 1$:

$$(X + Y) \cap 1 = (X \cap 1) + (Y \cap 1),$$
$$(XY) \cap 1 = (X \cap 1)(Y \cap 1),$$
$$(X^*) \cap 1 = 1$$

and the starting rules $1 \cap 1 = 1$, $a \cap 1 = \emptyset$.

Formulas (4.2)–(4.4) allow one to build a deterministic automaton recognizing a set X from a regular expression e denoting X. Indeed the above rules can be used to define $u^{-1}e$ when u is a word and e a rational expression. Also one may define $\delta(e) = 0$ or 1 according to $L(e) \cap 1 = \emptyset$ or 1. The states of the automaton are the different residuals $u^{-1}e$ obtained (modulo identities (3.1)). The initial state is e and the terminal states are the rational expressions f such that $\delta(f) = 1$.

EXAMPLE. Let $X = (a + b)^* ab$. We then obtain the automaton given in Fig. 11. Indeed, we have $b^{-1}X = X$, $a^{-1}X = X + b$ whence the result.

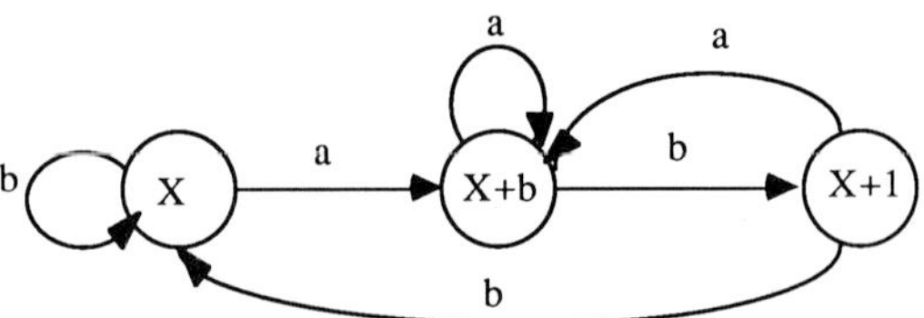

Fig. 11. An automaton recognizing $X=(a+b)^*ab$.

To see that this method really gives an algorithm, we have to prove that rules (1), (2), (3) will only give rise to a finite number of rational expressions. This gives another proof of the second half of Kleene's theorem.

4.2. PROPOSITION. *For any rational expression e, the set*

$$\mathcal{D}(e)=\{u^{-1}e \mid u\in A^*\}$$

is finite.

PROOF. For $w\in A^*$, one has

$$w^{-1}(ef)=(w^{-1}e)f+\sum_{\substack{w=uv\\ v\neq 1}}\delta(u^{-1}e)v^{-1}f$$

and for $w\in A^+$

$$w^{-1}e^*=\sum_{\substack{w=uv\\ v\neq 1}}\delta(u,v)(v^{-1}e)e^*$$

where $\delta(u,v)$ is 0 or 1. This formula shows that the set of rational expressions e for which the set $\mathcal{D}(e)$ is finite is closed under product and star. Since it is closed under sum and contains the basic expressions, the result follows. □

We end this section with a short discussion of the notion of *ambiguity* in rational expressions. One may define the *multiplicity* (e,x) of a word x relatively to a rational expression e as the integer defined by the recursive rules:

$$(e+f,x)=(e,x)+(f,x),$$

$$(ef,x)=\sum_{x=uv}(e,u)(f,v),$$

$$(e^*,x)=\sum_{n\geqslant 0}\ \sum_{x=u_1u_2\ldots u_n}(e,u_1)(e,u_2)\ldots(e,u_n)$$

with the obvious rules $(a,b)=0$ or 1 according to $a\neq b$ or $a=b$, $(1,x)=0$ or 1 according to $x\neq 1$ or $x=1$, $(0,x)=0$.

It is easy to see that $(e,x)\geqslant 1$ iff $x\in L(e)$. We say that e is *unambiguous* if $(e,x)=0$ or 1 for all x in A^*.

In a similar way, we define the multiplicity of a word x relatively to a finite automaton $\mathscr{A}$ as the integer $(\mathscr{A}, x)$ equal to the number of successful paths labeled by x.

One may verify that all the constructions that we have given in the proof of Kleene's theorem are consistent with the notion of multiplicity. More precisely the automaton $\mathscr{A}$ constructed from a rational expression e satisfies

$$(\mathscr{A}, x)=(e, x)$$

for all $x \in A^*$, and the same holds conversely for the rational expression computed by formula (4.1). As a consequence, we have the following result.

4.3. PROPOSITION. *Any rational set has an unambiguous rational expression.*

PROOF. Any rational set can be recognized by a deterministic automaton which, in turn, gives an unambiguous rational expression. □

4.4. EXAMPLE. Let $e=(a+b)^*a(a+b)^*$. The multiplicity (e, x) of a word x in e is equal to the number of occurrences of x in e. A deterministic automaton recognizing $X=L(e)$ is given in Fig. 12. The corresponding unambiguous rational expression is

$$f=b^*a(a+b)^*.$$

5. Star-height

The *star-height* of a rational expression is the maximum number of nested stars in the expression. Formally, it is defined as follows:

$$h(1)=0,$$
$$h(a)=0 \quad \text{if } a \text{ is a letter},$$
$$h(X+Y)=h(XY)=\max\{h(X), h(Y)\},$$
$$h(X^*)=h(X)+1.$$

The star-height of a rational set is the minimal star-height of a rational expression denoting this set. For instance, a set has star-height 0 iff it is finite.

It can be proved that there exist sets having star-height equal to any natural number. The simplest example is the set given for $n \geqslant 1$ by

$$X_n(a, b)=\{x \in (a+b)^* \mid |x|_a \equiv |x|_b \bmod 2^{n-1}\}.$$

It is composed of all words on the alphabet $\{a, b\}$ such that the number of a's is equal mod 2^{n-1} to the number of b's. It can be proved that X_n has exactly star-height n. Let us

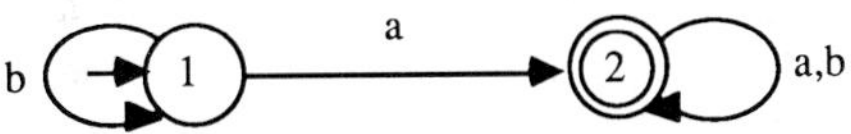

Fig. 12. A deterministic automaton.

verify that $h(X_n)\leqslant n$. Set

$$u=(ab+ba)^*aa, \qquad v=(ab+ba)^*bb.$$

Then we have, for $n\geqslant 1$,

$$X_n(a,b)=X_{n-1}(u,v)(ab+ba)^*,$$

showing that $h(X_n)\leqslant n$ by induction on n since $X_1(a,b)=(a+b)^*$.

The existence of an algorithm to compute the star-height of a given rational set has just been proved by Hashiguchi (see the Notes section). He has previously established that it is effectively decidable whether a set has star-height one. The proof of these results is extremely difficult and will not be presented here. We shall however relate them to some other ones.

A set $X\subset A^*$ is said to have the *finite power property* (f.p.p.) if there exists an integer $n\geqslant 0$ such that

$$X^*=1+X+\cdots+X^{n-1}.$$

For instance $X=a^*+(a+b)^*b$ has f.p.p. with $n=3$ but $X=a^*+a^*b$ does not. This property is closely related to star-height since the star of a set having the f.p.p. can be replaced by sums and products.

K. Hashiguchi and I. Simon proved independently, answering a question of J. Brzozowski, that it is decidable whether a given rational set has f.p.p. Simon's very elegant proof reduces the problem to that of the finiteness of some monoids of matrices. These matrices have coefficients in the *tropical semiring*

$$\mathscr{T}=(\mathscr{N},\min,+)$$

which is the set $\mathscr{N}=\mathbb{N}\cup\infty$ of integers plus infinity equipped with the two operations min (replacing a sum) and $+$ (replacing a product). The infinity ∞ is the neutral element for min and 0 a neutral element for $+$.

Let X be a rational set. Let $\mathscr{A}=(Q,i,t)$ be a normalized automaton recognizing $X-1$ (see Section 3). We merge i and t to obtain an automaton recognizing X^* and we define on the edges output labels in $\mathscr{T}$. All edges are labeled 0 except those entering i (the ones which were entering t before the merge operation) which are labeled 1. We associate in this way to each word $w\in A^*$ an element of $\mathscr{T}$, which is the minimum over all paths (i,w,i) of the sum of the output labels along the path. Clearly X has f.p.p. iff this function is bounded. We associate with each letter a, a $Q\times Q$ matrix with coefficients in $\mathscr{T}$ denoted $f(a)$ defined by

$$(p,f(a),q)=\begin{cases}1 & \text{if } p\xrightarrow{a} q \text{ and } q=i,\\ 0 & \text{if } p\xrightarrow{a} q \text{ and } q\neq i,\\ \infty & \text{otherwise.}\end{cases}$$

The matrices are multiplied according to the operations of $\mathscr{T}$.

One may verify that X has f.p.p. iff the monoid generated by these matrices is finite. Simon has proved that it is decidable whether a finitely generated monoid of matrices with coefficients in $\mathscr{T}$ is finite, thus giving a decision procedure to test whether a given rational set has f.p.p.

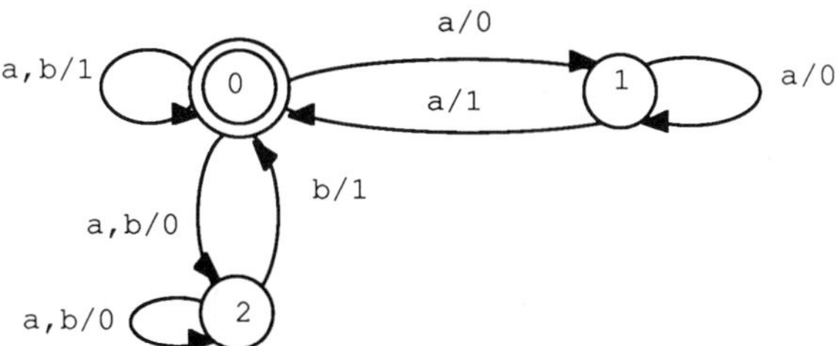

Fig. 13. The automaton with output labels in $\mathcal{T}$.

For example, if $X = a^* + (a+b)^* b$, the automaton with output labels in $\mathcal{T}$ is shown in Fig. 13. The corresponding monoid of matrices with coefficients in $\mathcal{T}$ has the seven elements represented below

$$f(1) = \begin{bmatrix} 0 & \infty & \infty \\ \infty & 0 & \infty \\ \infty & \infty & 0 \end{bmatrix}, \quad f(a) = \begin{bmatrix} 1 & 0 & 0 \\ 1 & 0 & \infty \\ \infty & \infty & 0 \end{bmatrix},$$

$$f(b) = \begin{bmatrix} 1 & \infty & 0 \\ \infty & \infty & \infty \\ 1 & \infty & 0 \end{bmatrix}, \quad f(ab) = \begin{bmatrix} 1 & \infty & 0 \\ 2 & \infty & 1 \\ 1 & \infty & 0 \end{bmatrix}, \quad f(ba) = \begin{bmatrix} 2 & 1 & 0 \\ \infty & \infty & \infty \\ 2 & 1 & 0 \end{bmatrix},$$

$$f(aba) = \begin{bmatrix} 2 & 1 & 0 \\ 3 & 2 & 1 \\ 2 & 1 & 0 \end{bmatrix}, \quad f(a^2) = \begin{bmatrix} 1 & 0 & 0 \\ 1 & 0 & 1 \\ \infty & \infty & 0 \end{bmatrix}.$$

The result of Hashiguchi on star-height one is a generalization of the latter. The main step in the proof is indeed that, given a finite number of rational sets $X, Y, Z, \ldots$, one may effectively decide whether X belongs to the polynomial closure of the family $Y, Z, \ldots$, i.e. can be obtained from these by a finite number of sums and products. A set X having f.p.p. is obviously one such that its star belongs to the polynomial closure of the family reduced to X. The problem can again be formulated in terms of monoids of matrices with coefficients in the tropical semiring also called *automata with a distance function*. One has to prove this time that it is decidable whether, in a given finitely generated monoid of such matrices, the number of values appearing in a given position is or is not finite. The proof of the general result on the computability of star-height again uses automata with distance functions.

6. Star-free sets

The additional use of the Boolean operators in rational expressions leads to the notion of *extended* rational expressions. We limit ourselves to the use of the unary operator $\bar{X}$ denoting the absolute complement of a set X (the alphabet being fixed). For instance

$$\bar{0}(aa+bb)\bar{0}$$

is an extended rational expression. It denotes the set of words having at least two consecutive occurrences of a or b.

The *star-height* of an extended rational expression is defined as the maximum number of nested star operators in the expression. We define it formally as follows:

$$h(0)=h(1)=0,$$
$$h(a)=0 \quad \text{if } a \text{ is a letter},$$
$$h(\bar{X})=h(X),$$
$$h(X+Y)=h(XY)=\max\{h(X), h(Y)\},$$
$$h(X^*)=h(X)+1.$$

The *extended star-height* of a set $X \subset A^*$ is the minimal star-height of an extended rational expression denoting X. It is denoted $eh(X)$. Surprisingly enough, it is not known at present whether or not there exists rational sets whose extended star-height is more than 1. The case of sets of extended star-height zero has received considerable attention especially in connection with formal logic (see [124]). Such a set is called a *star-free* set. To reformulate more precisely the definition, a set $X \subset A^*$ is called star-free if it can be obtained from subsets of the alphabet by a finite number of set-products, unions and absolute complements. The family of star-free sets is therefore a Boolean algebra.

Observe that A^* itself is star-free since it is the complement of the empty set. Also, for instance the set $X=(ab)^*$ is star-free since

$$X=1+(aA^* \cap A^*b)-A^*(a^2+b^2)A^*$$

as one may easily verify. The first question which can be raised is whether or not any rational set is star-free, i.e. whether the extended star-height is always zero. The considerations that we shall develop now will show that the answer is negative and will also provide an algorithm to write a star-free expression when possible.

A set $X \subset A^*$ is said to be *aperiodic* if there exists an integer $n>0$ such that for all $x, y, z \in A^*$ one has

$$xy^nz \in X \Leftrightarrow xy^{n+1}z \in X.$$

The least integer n such that the above equivalence holds is called the *index* of X denoted $i(X)$. The above definition may of course be given more compactly by using the notion introduced previously of contexts. Indeed the index of X is the smallest integer n (when it exists) such that for all $y \in A^*$

$$C_X(y^n)=C_X(y^{n+1})$$

This leads naturally to the notion of an *aperiodic monoid*. It is a monoid M such that identically $m^n=m^{n+1}$ for some integer n. The definition of an aperiodic set X can therefore be reformulated either by requiring that there exists a morphism $\varphi\colon A^* \to M$

into an aperiodic monoid that saturates X or, more simply that the syntactic monoid of X is aperiodic. We shall prove here the following fundamental result.

6.1. THEOREM (Schützenberger). *A rational set $X \subset A^*$ is star-free iff it is aperiodic.*

PROOF. The proof is easy in one direction. Indeed, one has $i(a)=2$ for $a \in A$. Further, it is straightforward to check that

$$i(X+Y) \leqslant \max(i(X), i(Y)), \quad i(\overline{X}) = i(X).$$

Let us finally verify that

$$i(XY) \leqslant i(X) + i(Y).$$

Suppose indeed that $xy^n z \in XY$ with $n \geqslant i(X)+i(Y)$. We may suppose $n>0$. One has decompositions $y=rs$ and $n=h+k+1$ such that $xy^h r \in X, sy^k z \in Y$. Then either $h \geqslant i(X)$ or $k \geqslant i(Y)$. In the first case, one has $xy^{h+1} r \in X$ and in the second case, one has $sy^{k+1} z \in Y$. In both cases $xy^{n+1} z \in XY$. The proof that $xy^{n+1} z \in XY \Rightarrow xy^n z \in XY$ is quite similar.

This proves fairly directly that any star-free set is aperiodic. The proof of the converse implication is more difficult. We shall prove that for any morphism $\varphi: A^* \to M$ from A^* into an aperiodic monoid, the inverse image $\varphi^{-1}(m)$ of any element $m \in M$ is star-free.

To do this, we first prove the following property, later referred to as the *cancellation law* of aperiodic monoids. Let M be an aperiodic monoid. Then for any $p, q, r \in M$ one has

$$q = pqr \;\Rightarrow\; q = pq = qr.$$

In fact, let $n = i(M)$. Then $q = pqr$ implies $q = p^n q r^n$ and since $p^n = p^{n+1}$ we obtain $q = p^{n+1} q r^n = p(p^n q r^n) = pq$. The proof that $q = qr$ is symmetrical.

We need a second property of an aperiodic monoid M. Namely, for any $m \in M$ one has

$$m = (mM \cap Mm) - J \tag{6.1}$$

where $J = \{r \in M \mid m \notin MrM\}$. Clearly m belongs to the set K defined by the expression on the right-hand side. Conversely, let $n \in K$. Then for some $p, q \in M$

$$n = mp = qm$$

and also, since $n \notin J$ there are some $u, v \in M$ such that $m = unv$.

Hence $m = uqmv$. By the cancellation law, this implies $m = uqm$ or also $m = un$. Since $n = mp$, the last equality may be rewritten $m = ump$ and, by the cancellation law, it implies $m = mp = n$.

Now let $\varphi: A^* \to M$ be a morphism from A^* into an aperiodic monoid. We shall prove the following formula. For all $m \in M - 1$, one has

$$\varphi^{-1}(m) = UA^* \cap A^* V - A^* W A^* \tag{6.2}$$

where the sets U, V, W are defined as follows.

First, U is the union of all sets $\varphi^{-1}(n)a$ where $(n, a) \in M \times A$ is such that $n\varphi(a)M = mM$ but $n \notin mM$. Second, V is the union of all sets $a\varphi^{-1}(n)$ where

$(a, n) \in A \times M$ is such that $M\varphi(a)n = Mm$ but $n \notin Mm$. Finally W is the union of all letters $a \in A$ such that $m \notin M\varphi(a)M$ and of all sets $a\varphi^{-1}(n)b$ where $(a, n, b) \in A \times M \times A$ is such that $m \in M\varphi(a)nM \cap Mn\varphi(b)M$ but $m \notin M\varphi(a)n\varphi(b)M$.

Let us first prove the inclusion from left to right in (6.2). Let $w \in \varphi^{-1}(m)$. Let $u \in A^*$ be the shortest prefix of w such that $mM = \varphi(u)M$. We cannot have $u = 1$ since otherwise $mM = M$ whence $mp = 1$ for some $p \in M$ which implies that $m = 1$ by the cancellation law. Let $u = ra$ with $r \in A^*$, $a \in A$ and let $n = \varphi(r)$. Then $n\varphi(a)M = mM$ but $n \notin mM$ since otherwise u would not be of minimal length. Hence $w \in UA^*$. The proof that $w \in A^*V$ is entirely symmetrical. Finally we cannot have $w \in A^*WA^*$ since any word $y \in W$ satisfies $m \notin M\varphi(y)M$. This proves the first inclusion.

Consider now a word x in $UA^* \cap A^*V - A^*WA^*$ and let $n = \varphi(x)$. Since $x \in UA^*$, we have $n \in mM$. Symmetrically, $n \in Mm$. Hence, by (6.1) we will have proved that $n = m$ or equivalently that $x \in \varphi^{-1}(m)$ if we can prove that $m \in MnM$. To do this, we suppose the contrary and let $x = uwv$ with $m \notin M\varphi(w)M$ and w chosen of minimal possible length. It is not possible that $w = 1$. If w is a letter, then $w \in W$, a contradiction. Let then $w = arb$ with $a, b \in A$. Then $m \in M\varphi(a)\varphi(r)M$ since w has been chosen of minimal length. Symmetrically, $m \in M\varphi(r)\varphi(b)M$. Hence $w \in W$, a contradiction. We have thus completed the proof of (6.2).

We are now ready to complete the proof of the theorem. Let $\varphi: A^* \to M$ be a morphism into an aperiodic monoid. We shall prove that $\varphi^{-1}(m)$ is star-free for all $m \in M$ using a descending induction on the number

$$r(m) = \text{card}(MmM).$$

The maximal value of $r(m)$ is reached only for $m = 1$, since by the cancellation law $uv = 1$ implies $u = v = 1$.

Consider first the case $m = 1$. Then

$$\varphi^{-1}(m) = A^* - A^*WA^*$$

where W is the set of letters $a \in A$ such that $\varphi(a) \neq 1$. Indeed, if $\varphi(uv) = 1$, then $\varphi(u) = \varphi(v) = 1$ by the cancellation law.

Let now $m \in M - 1$ and suppose the property proved for all $n \in M$ such that $r(n) > r(m)$. We have by (6.2)

$$\varphi^{-1}(m) = UA^* \cap A^*V - A^*WA^*.$$

We prove that U, V, W are star-free using the induction hypothesis. Let first $(n, a) \in M \times A$ be such that $n\varphi(a)M = mM$ but $n \notin mM$. Clearly $MnM \supset MmM$ whence $r(n) \geqslant r(m)$. Let us suppose ab absurdo that $r(n) = r(m)$. Then $n \in MmM$ hence $n = umw$ for some $u, v \in M$. Since $m \in n\varphi(a)M$, we also have $m = np$ for some $p \in M$. Hence $n = unpv$ which implies by the cancellation law $n = npv = mv$, a contradiction with $n \notin mM$. We have thus proved that $r(n) > r(m)$. Hence $\varphi^{-1}(n)$ is star-free by the induction hypothesis and so is U.

The proof that V is star-free is symmetrical. To prove that W is star-free, we consider $(a, n, b) \in A \times M \times A$ such that $m \in M\varphi(a)nM \cap Mn\varphi(b)M$ and $m \notin M\varphi(a)n\varphi(b)M$. We have $r(n) \geqslant r(m)$. Let us suppose that $r(n) = r(m)$. Then $n = umv$ and also $m = r\varphi(a)ns =$

$pn\varphi(b)t$ for some $p, r, u, v, s, t \in M$. Then $n = ur\varphi(a)nsv$ implies $n = ur\varphi(a)n$ whence

$$m = pur\varphi(a)n\varphi(b)t,$$

a contradiction. This completes the proof of Theorem 6.1. □

As an example, we consider on $A = \{a, b\}$ the set

$$X = (ab)^*.$$

Its syntactic monoid $S = \sigma(A^*)$ has six elements. Indeed, denoting α, β the images of a and b we have the equalities

$$\alpha\beta\alpha = \alpha, \qquad \beta\alpha\beta = \beta, \qquad \alpha^2 = \beta^2 = \alpha^3 = \beta^3$$

giving $S = \{1, \alpha, \beta, \alpha\beta, \beta\alpha, \alpha^2\}$. We have $X = \sigma^{-1}(1) + \sigma^{-1}(\alpha\beta)$, and the above algorithm gives

$$\sigma^{-1}(\alpha\beta) = (aA^* \cap A^*b) - (A^*(a^2 + b^2)A^*).$$

A more complicated example is the set

$$Y = (ab + ba)^*.$$

Its syntactic monoid has 15 elements and the completed form of the computation gives

$$Y = R + S + T$$

with $R = (ab)^+$, $S = (ba)^+$ that we have already written in star-free form. Further, with $U = a(ba)^*$, $V = b(ab)^*$ we have $T = (Sa + Rb)A^* \cap A^*(aR + bS) - A^*(aUa + bVb)A^*$.

There are several interesting subclasses of aperiodic sets that one may naturally consider. We shall mention here two of them and state without a complete proof in each case a characterization in terms of the structure of their syntactic monoid.

The first one is that of *locally testable* sets. A set $X \subset A^*$ is said to be locally testable if it is a finite Boolean combination of sets of the type UA^* or VA^* or A^*WA^* where U, V, W are finite sets. Such sets are certainly all aperiodic but we shall give later on an example of an aperiodic set that is not locally testable.

To formulate the definition in equivalent terms let us denote, for an integer n, and a word w of length at least n,

$l_n(w)$ = the prefix of length n of w,
$F_n(w)$ = the set of factors of length $n+1$ of w,
$r_n(w)$ = the suffix of length n of w.

Recall that x is called a *factor* of w if x has an occurrence in w, that is, if w can be written $w = pxs$ for some words p, s.

A set X is locally testable iff there exists an integer $n \geqslant 1$ such that for two words u, v of length $\geqslant n$, whenever $l_n(u) = l_n(v)$, $F_n(u) = F_n(v)$, $r_n(u) = r_n(v)$, $u \in X$ *iff* $v \in X$. This can be visualized as in Fig. 14 where the status of a word with respect to the set X is tested by a machine using a sliding window of length $n+1$.

The decision that the word processed is or is not in X only depends on the

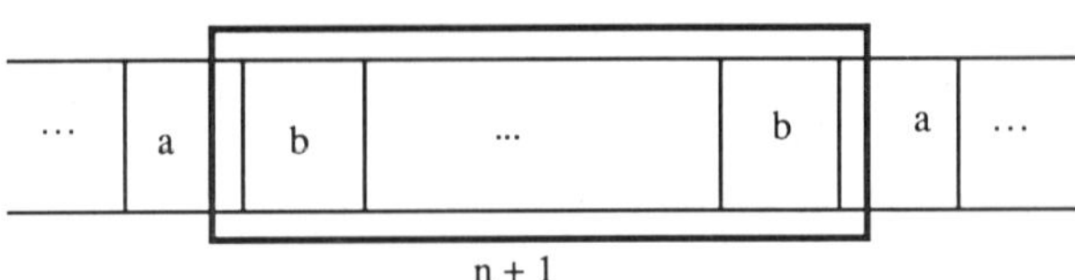

Fig. 14. The sliding window.

(unordered) set of successive contents of the window, with the convention that special markers have been added at both ends of the input word.

For example, the set $X=(a+b)^*ab$ considered in Section 1 is locally testable. Indeed, one has

$$x \in X \text{ iff } r_2(x)=ab.$$

It is of course not obvious at all whether a given recognizable or even aperiodic set is or is not locally testable. The answer relies on the choice of an integer n making the window wide enough.

It is not difficult to state a necessary condition. Let $X \subset A^*$ be a locally testable set. Let us denote $u \equiv v$ iff the words u, v have the same sets of contexts with respect to X, i.e. iff $C_X(u)=C_X(v)$. Let $x \in A^+$ be a nonempty word such that

$$x \equiv x^2$$

in such a way that x may be repeated an arbitrary number of times with the same contexts. Since these repetitions allow to extend beyond a window of any size, for any y, z having x both as a prefix and a suffix we must have the identities

$$y \equiv y^2, \tag{6.3}$$

$$yz \equiv zy; \tag{6.4}$$

the first one because two occurrences of y separated by a very long repetition of x's just look the same as one occurrence and the second one because no window will be able to account for an occurrence of y before z rather than the converse.

These necessary conditions allow us to see an example of an aperiodic set which is not locally testable. We have already given a star-free expression of the set

$$X=(ab+ba)^*.$$

The word $x=ab$ satisfies the requirement that $x \equiv x^2$ since

$$C_X(x)=C_X(x^2)=(X \times X)+(Xb \times aX).$$

Choosing $y=abbab$ we obtain

$$C_X(y)=X \times aX, \qquad C_X(y^2)=\emptyset$$

and therefore the identity (6.3) is not satisfied. Hence X is not locally testable.

The wonderful thing is that identities (6.3) and (6.4) are enough to characterize locally testable sets. More precisely we have the following theorem.

6.2. THEOREM (Brzozowski–Simon, McNaughton, Zalcstein). *A recognizable set $X \subset A^*$ is locally testable iff it satisfies identities* (6.3) *and* (6.4) *for all $x \in A^+$ such that $x \equiv x^2$ and all $y, z \in xA^* \cap A^*x$.*

The above characterization of locally testable sets may of course be checked on the syntactic monoid of X, therefore giving a decision algorithm to check whether a recognizable set is locally testable. Identities (6.3) and (6.4) moreover have a natural algebraic interpretation.

We shall obtain a proof of Theorem 6.2 via the use of a nice combinatorial result which is the following theorem.

6.3. THEOREM (Simon). *Let M be a monoid and let $\mathscr{A}$ be a deterministic automaton on a set P of states and with an output function*

$$(p, a) \in P \times A \to p{*}a \in M$$

with the following property. For any state p and any cycles around p

$$p \xrightarrow{y} p, \qquad p \xrightarrow{z} p,$$

one has

$$p{*}y = p{*}y^2, \qquad p{*}yz = p{*}zy.$$

Then, for any words y, z and any states (p, q) such that the paths

$$p \xrightarrow{y} q, \qquad p \xrightarrow{z} q$$

both use the same set of edges, one has $p{*}y = p{*}z$.

We shall not prove Theorem 6.3 here (see the Notes section for references). Let us show how one can deduce Theorem 6.2 from Theorem 6.3.

PROOF OF THEOREM 6.2. Let $X \subset A^*$ be a set satisfying the hypotheses of Theorem 6.2. Let M be the syntactic monoid of X and let $f: A^* \to M$ be the corresponding morphism. Let $n = \text{card}(M)$ and let us prove that X is n-locally testable, that is, locally testable with $n+1$ as window size.

Let $\mathscr{A}$ be the deterministic automaton obtained by taking $Q = A^n$ and defining for a in A the transitions by the rule

$$(a_1, a_2, \ldots, a_n).a = (a_2, \ldots, a_n, a).$$

This is the automaton that "remembers the n last symbols" called the *n-local universal automaton* (see Fig. 15 for a picture of this automaton when $n = 2$, $A = \{a, b\}$).

We define an output function by

$$p{*}y = f(py).$$

We now verify that the hypotheses of Theorem 6.3. are satisfied. Let $p = (a_1, a_2, \ldots, a_n)$. Since $n = \text{card}(M)$, the elements

$$1, f(a_1), f(a_1 a_2), \ldots, f(a_1 a_2 \ldots a_n)$$

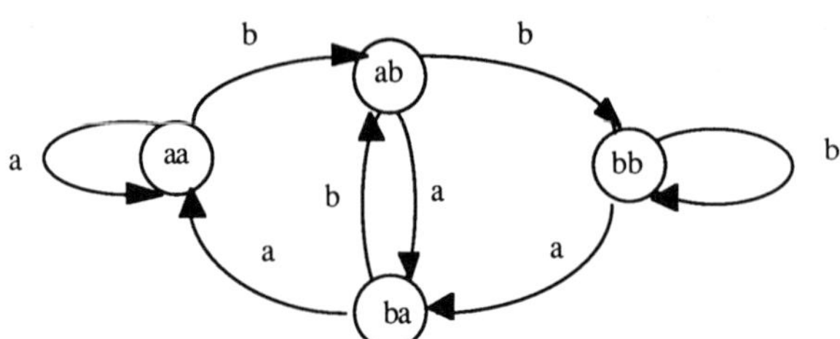

Fig. 15. The 2-local universal automaton.

cannot all be different. Hence we can find a factorization $p=rst$ with $s\neq 1$ such that $rs\equiv r$. Since M is finite, there is an integer $k\geqslant 1$ such that $x=s^k$ satisfies $x\equiv x^2$.

Let y be the label of a cycle $p\to p$ around p. Then the word py ends with p and we can write

$$sty=y't$$

for some y' such that ry' ends with rs. In particular we have $ry's\equiv ry'$, hence

$$py^2=ry'ty\equiv ry'sty=ry'y't\equiv rxy'xy'xt\equiv rxy'xt\equiv ry't=py.$$

Hence $p*y=p*y^2$. In the same way, if z is the label of another loop around p, let $stz=z't$. Then

$$pyz=ry'tz\equiv ry'stz=ry'z't\equiv rxy'xz'xt\equiv rxz'xy'xt\equiv pzy$$

and therefore $p*yz=p*zy$.

By Theorem 6.3 we have $py\equiv pz$ for any pair y, z of words such that the paths $p\xrightarrow{y}q$ and $p\xrightarrow{z}q$ exist and use the same set of edges. Let then $u, v\in A^*$ be such that

$$l_n(u)=l_n(v),\qquad F_n(u)=F_n(v),\qquad r_n(u)=r_n(v).$$

Let $p=l_n(u), q=r_n(u)$ and let $u=py, v=pz$. Then there are paths $p\xrightarrow{y}q$ and $p\xrightarrow{z}q$. Moreover, they both use the set of edges of the form

$$r\xrightarrow{a}t$$

for all $ra\in F_n(u)$. Hence by the above argument $u\equiv v$. This proves that X is locally testable and completes the proof of Theorem 6.2. □

The second special class of star-free sets that we shall mention is that of *piecewise testable* sets. By definition, a set $X\subset A^*$ is piecewise testable iff it is a finite Boolean combination of sets of the form

$$A^*a_1A^*a_2A^*\ldots A^*a_mA^*.$$

The notion of piecewise testable sets is related to the notion of a *subword* of a word in the same way as locally testable sets are related to factors. To see this, we introduce a partial order on words by denoting

$$uv<uav$$

for any $u, v\in A^*$ and $a\in A$ and by taking the reflexive and transitive closure of the above relation. Whenever $u<v$, we say that u is a subword of v. Now for an integer $n\geqslant 1$ and a word $w\in A^*$ we denote by $S_n(w)$ the set of subwords of w of length n. For instance

$$S_3(abca)=\{abc, aba, aca, bca\}.$$

Then a set X is piecewise testable iff there is an integer $n\geqslant 1$ such that for two words u, v of length $\geqslant n$,

$$S_n(u)=S_n(v) \quad\Rightarrow\quad u\equiv v. \tag{6.5}$$

As for locally testable sets, it is far from obvious to decide whether a given recognizable set is or is not piecewise testable. Again a necessary condition can be easily formulated. Let indeed $X\subset A^*$ be piecewise testable. Let $x, y\in A^*$ be arbitrary words and let $n>0$ be an integer such that (6.5) holds. Then, for any $x, y\in A^+$ one has the equivalences

$$(xy)^n x\equiv(xy)^n\equiv y(xy)^n \tag{6.6}$$

as one may easily verify.

The miracle is that identities (6.6) are enough to characterize piecewise testable sets.

6.4. THEOREM (Simon). *A recognizable set is piecewise testable iff Equations* (6.6) *are satisfied for every $x, y\in A^+$, for some large enough n.*

Observe that (6.6) may be checked easily on the syntactic monoid $M=\mathrm{Synt}(X)$ of the set X. Indeed, (6.6) are satisfied for all large enough n iff for any idempotent $e\in M$, and any $u, v\in M$ such that $uv=e$ one has

$$eu=ve=e.$$

The proof of the above theorem is difficult. It relies on an interesting combinatorial result on words which is the following. For two distinct words $u, v\in A^*$, let us define

$$\delta(u, v)$$

to be the minimal length of a word x which is a subword of one of them but not of the other. Said otherwise, $\delta(u, v)-1$ is the maximal integer n such that $S_n(u)=S_n(v)$.

The key lemma in the proof of the above result is the following theorem.

6.5. THEOREM (Simon). *For any words $x, y\in A^*$ there exists a word $z\in A^*$ such that $x\leqslant z$, $y\leqslant z$ and $\delta(x, y)=\delta(x, z)=\delta(y, z)$.*

As an example, let $x=a^3b^3a^3b^3$ and $y=a^2b^4a^4b^2$. We have $\delta(x, y)=5$ since $S_4(x)=S_4(y)=(a+b)^4-baba$. The shortest possible z satisfying the conditions stated in the theorem is $z=a^3b^4a^4b^3$.

7. Special automata

The standard algorithms described in Section 4 allow one to associate to each rational expression a finite automaton. This automaton is in general nondeterministic but its size is relatively small. More precisely, the number of edges of the automaton grows as a linear function in the length of the rational expression, and so does the number of states. However, applying the subset construction may lead to an automaton with a number of states which is an exponential in the length of the rational expression that we started with, even if the automaton has been reduced to the minimal one by the minimization algorithm. For instance, if the rational expression is

$$(a+b)^* a(a+b)^n,$$

a nondeterministic automaton with $n+2$ states is shown on Fig. 16.

The corresponding minimal deterministic automaton is shown in Fig. 17. It has 2^{n+1} states, a fact which could have been anticipated since the automaton has to keep in memory the $n+1$ last letters of the input word (the automaton is in fact the $(n+1)$-local universal automaton introduced in Section 6).

It is the aim of this section to study some special cases where the "exponential blow-up" does not occur between the rational expression and the deterministic minimal automaton. These special cases moreover have a practical significance in text processing algorithms as we shall see.

The first case is that of a rational expression of the form

$$X = A^* x$$

where $x = a_1 a_2 \ldots a_n$ is a word of length n. A nondeterministic automaton with $n+1$ states is shown in Fig. 18.

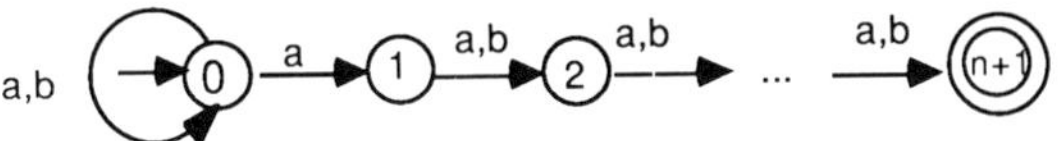

Fig. 16. A nondeterministic automaton.

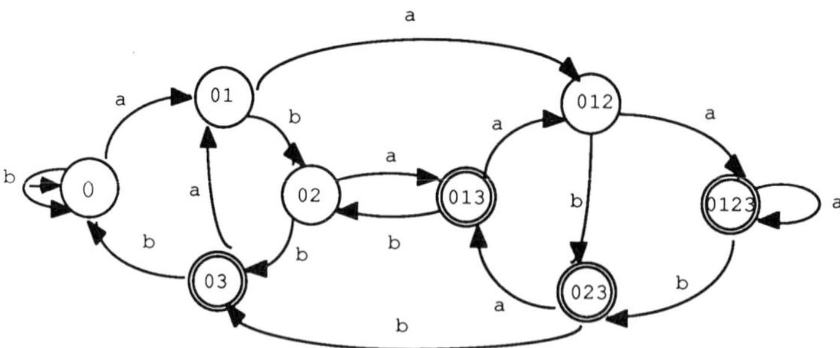

Fig. 17. The corresponding deterministic automaton ($n=2$).

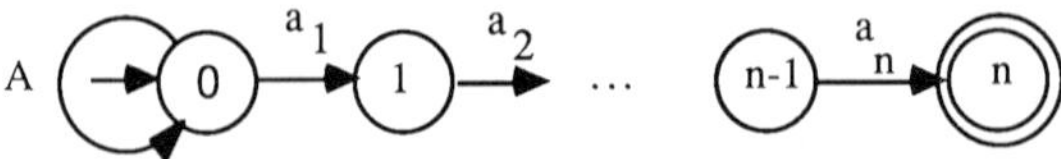

Fig. 18. A nondeterministic automaton.

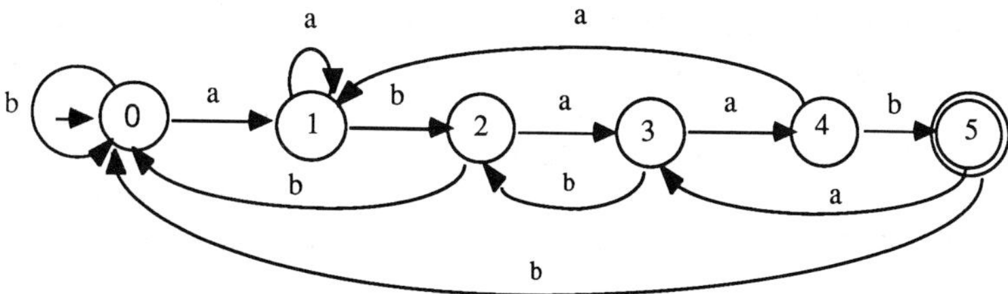

Fig. 19. The case $x = abaab$.

It is a remarkable fact that the accessible part of the corresponding deterministic automaton has also $n+1$ states. Figure 1 gives the result for $x = ab$. Figure 19 gives the result for $x = abaab$.

To analyze the construction of the deterministic automaton, we proceed as follows. Let P be the set of prefixes of the word x

$$P = \{\varepsilon, a_1, a_1 a_2, \ldots, x\}.$$

The set of states of the nondeterministic automaton of Fig. 18 may be identified by P with the number i standing for $a_1 a_2 \ldots a_i$. Now, the state of the deterministic automaton reached from the initial state by a word y is the set of elements of P that are suffixes of y. This set is entirely defined by the longest of its elements, say p, since the others are the suffixes of p that belong to P. Hence the set of states of the deterministic automaton can also be identified with elements of P and this accounts for the fact that it still has $n+1$ states.

Now if we want to describe directly the transitions of the deterministic automaton, we need a way to run through the elements of P making up a state. This is obviously achieved by considering the function f assigning to a nonempty prefix p the longest proper suffix of p that is in P. In this way, the elements of the state represented by p are

$$p, f(p), f^2(p), \ldots, \varepsilon.$$

The values of the function f for the set of prefixes of $x = abaab$ are given in Table 3, where the prefix p is represented by its length.

If the function f is previously computed, it is then easy to compute the transitions of the deterministic automaton by the recursive rule

$$p.a = \begin{cases} pa & \text{if } pa \in P, \\ f(p).a & \text{otherwise} \end{cases}$$

which for each nonempty prefix p gives the transition from the state represented by

Table 3
The failure function for $x = abaab$

p	1	2	3	4	5
f	0	0	1	1	2

p under the input letter a. For $p=\varepsilon$, we have the obvious transitions $\varepsilon.a=a$ if a is in P and $\varepsilon.a=\varepsilon$ otherwise.

Such a function f allowing to run through the components of a state of an automaton obtained by the determinization algorithm is called a *failure function* and can be used in other cases. It presents the advantage of reducing the space required for the storage of the automaton.

The algorithm implementing the above transition rule is known as the *Morris and Pratt's algorithm*. It can be used to search in a text the occurrences of a word x called a *pattern*. The implementation of the transition rule allows to process a text of length m applying the rule at most $2m$ times. Let indeed y be the part of the text already processed and consider the number

$$2|y|-|p|.$$

At each application of the recursive rule, this number increases strictly since either $pa \in P$ and y is changed in ya and p in pa or otherwise p is changed in $f(p)$ which is strictly shorter. The precomputation of the failure function f obeys itself the same rule since indeed we have $f(a)=\varepsilon$ for a in A and

$$f(p\,a)=\begin{cases} f(p)a & \text{if } f(p)a \in P,\\ f(f(p)a) & \text{otherwise,}\end{cases}$$

for $p\neq\varepsilon$, a in A.

We add to this presentation two observations. First, one may apply the same construction to any finite set S instead of the word x. Nothing needs to be changed in the above formulas. The set of states of the deterministic automaton obtained for $X=A^*S$ can be identified with the set P of prefixes of S. The example given at the beginning of this section does not bring a contradiction since the corresponding set P has precisely 2^{n+1} elements. What happens in this case is only that the minimal automaton of S can be itself much smaller.

The second observation is that, in the case of a single word x, one may use a different failure function resulting in a faster simulation of the automaton. Let indeed l be the function from P to A defined by

$$l(a_1a_2\ldots a_i)=a_{i+1}.$$

By convention, we consider $l(x)$ to be a letter that does not appear in x. Let then g be the function assigning to each nonempty prefix p of x the longest suffix q of p that belongs to P and such that $l(p)\neq l(q)$, with $g(p)=\varepsilon$ if no such suffix exists. The values of the failure function g are represented in Table 4 in the case of $x=abaab$. We may then

Table 4
The values of g for $x=abaab$

p	1	2	3	4	5
g	0	0	1	0	2

define the transitions of the automaton as follows for $p \neq \varepsilon$:

$$p.a = \begin{cases} pa & \text{if } a = l(p), \\ g(p).a & \text{otherwise.} \end{cases}$$

The string-matching algorithm obtained by using this modified failure function is known as Knuth, Morris and Pratt's algorithm.

The precomputation of the new failure function g is given by the following recursive scheme. For $p \neq \varepsilon$ we have

$$g(p) = \begin{cases} f(p) & \text{if } l(p) \neq l(f(p)), \\ g(f(p)) & \text{otherwise} \end{cases}$$

with the convention $g(\varepsilon) = \varepsilon$.

We now come to another interesting case where the exponential blow-up does not occur in the determinization procedure. It is also related to string-searching algorithms.

Consider the set S of suffixes of a given word of length n

$$x = a_1 a_2 \dots a_n.$$

A nondeterministic automaton for S is directly obtained as in Fig. 20, by taking all the $n+1$ states as initial states. Now we might expect the minimal deterministic automaton to have a number of states whose order of magnitude is the sum of the lengths of the words in S, that is to say the square of n. We shall see however that this is not the case and that the minimal automaton of S has at most $2n$ states. We represent in Fig. 21 the automaton corresponding to $x = abbc$.

Let Q be the set of states of the minimal deterministic automaton of S. The elements of Q can be identified with the nonempty sets

$$p^{-1}S = \{s \in S \mid ps \in S\}$$

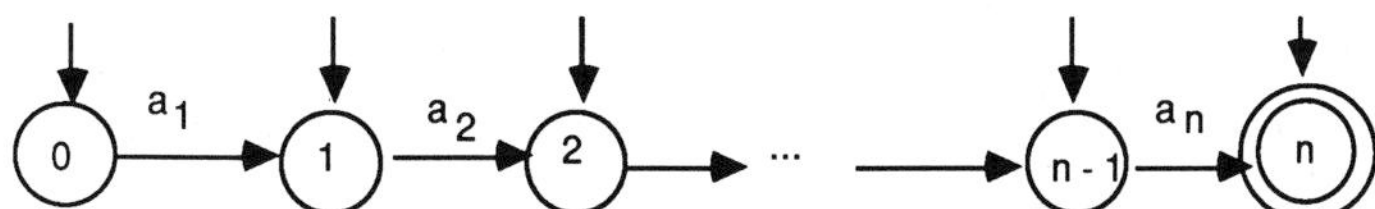

Fig. 20. A nondeterministic automaton for S.

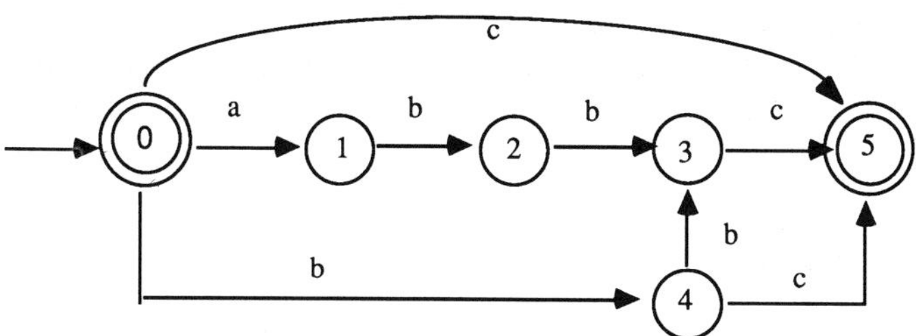

Fig. 21. The automaton of suffixes of $x = abbc$.

varying p in A^*. In fact, p runs through the factors of the word x. We order the set Q according to the inclusion of the corresponding sets $p^{-1}S$. The fundamental remark is that two sets $p^{-1}S$, $q^{-1}S$ are either disjoint or comparable. Indeed if a suffix s is both in $p^{-1}S$ and $q^{-1}S$ this implies that p is itself a suffix of q or conversely. Hence the order defined on Q corresponds to a tree $\mathcal{T}(S)$. It is represented in Fig. 22 for the example of $x = abbc$. In this tree all nodes either are leaves or have at least two sons by the definition of the minimal automaton.

This construction implies that Q has at most $2n$ elements since a tree with $n+1$ leaves and such that each interior node has at least two sons can have at most $2n$ nodes.

The automaton recognizing the suffixes of a word x of length n has therefore a size which is linear in n, since its number of states is at most $2n$ and its number of edges at most $3n$. It can also be computed by an algorithm operating in time $O(n)$.

This automaton can be used to perform several algorithms. It can be used for string matching to look for a pattern x in a text y in two different ways. First one may compute the automaton of suffixes of the text y. This automaton can be transformed to give a transducer computing the position of a given word x within the text y. The result is therefore a kind of index of the text y. It is called the *suffix transducer* of y. The suffix transducer of $y = abbc$ is represented on Fig. 23. Each edge is labeled by a pair of a letter and a number. Along a path, one concatenates the first components and adds the second ones. The result is a pair (p, i) where i is the minimal length of a word l which is a prefix of y and has p as a suffix.

There is a second way to use the suffix automaton for string matching. Instead of

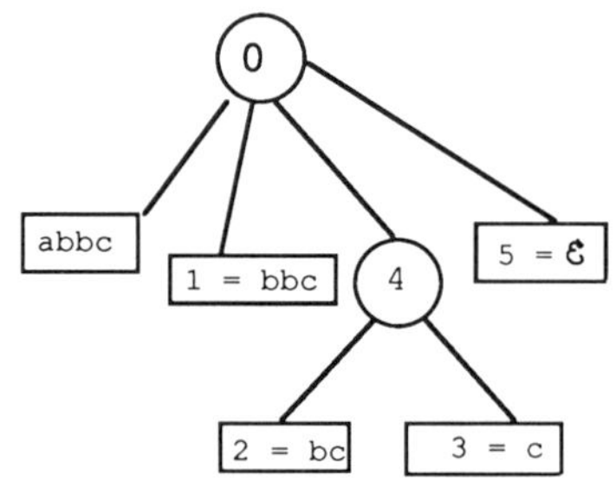

Fig. 22. The tree $\mathcal{T}(abbc)$.

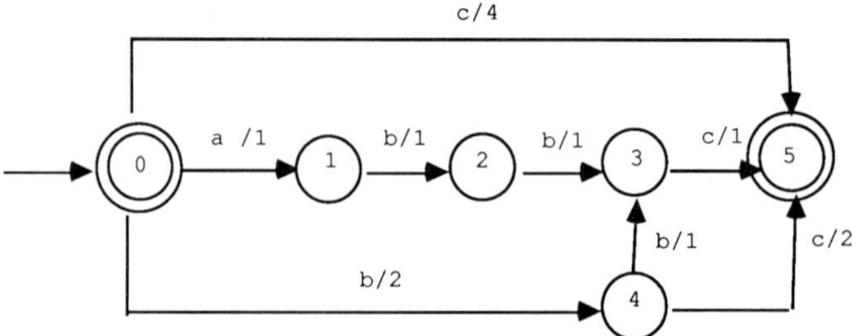

Fig. 23. The suffix transducer of $y = abbc$.

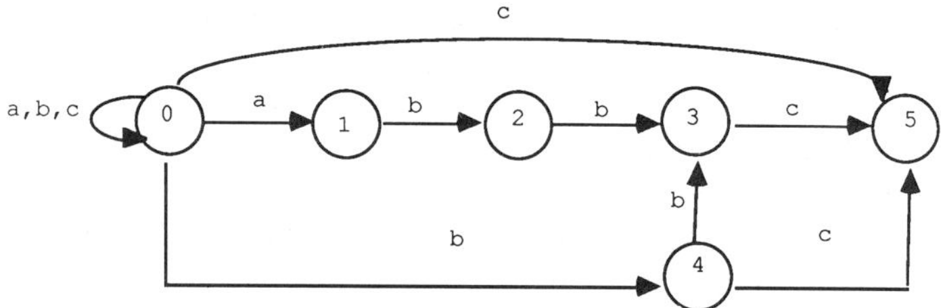

Fig. 24. Completion of the suffix automaton of $x = abbc$.

computing the suffix automaton of the text y, one computes the suffix automaton of the pattern x. One then considers the automaton obtained by adding on the initial state a loop for each letter in the alphabet as we did in Fig. 24 on the simple automaton scanning the word x.

It happens that the corresponding deterministic automaton still has the same number of states. In fact one may identify its states with the states of the suffix automaton and define the transitions using the failure function

$$f: Q \rightarrow Q$$

assigning to a state q its father in the tree $\mathscr{T}(S)$; see Table 5.

The automaton thus obtained can be used in a string matching algorithm as follows. When processing the text y by a left-right scanning one updates variables (p, i). The first one is the state reached thus far in the automaton constructed above. The second one is the length of the longest factor or x which is a suffix of the part of text processed so far. One may show that i can be computed by a simple rule as a function of the transitions in the automaton, as shown by Crochemore [44].

8. Recognizable sets of numbers

Finite automata can be used to specify simple properties of integers written at some basis. A well-known example of this situation is the recognition of multiples of a given number on their decimal representation. We shall see here some interesting aspects of this idea.

Let us first recall the basic definitions of the representation of integers at some basis. Let $k \geqslant 2$ be an integer called the *basis* and let $\boldsymbol{k} = \{0, 1, \ldots, k-1\}$. We associate to each

Table 5
The transitions of the deterministic version of Fig. 24

	0	1	2	3	4	5
a	1	1	1	1	1	1
b	4	2	3	3	3	4
c	5	5	5	5	5	5

word x in $\boldsymbol{k}^*$ its *value* denoted $[x]_k$ or simply $[x]$ defined inductively by $[\varepsilon]=0$ and for x in $\boldsymbol{k}^*$, a in $\boldsymbol{k}$ by

$$[xa]=[x]k+a.$$

The mapping $x \mapsto [x]$ cannot be inverted because of the ambiguity introduced by the leading zeros. However its restriction to the set $X=\boldsymbol{k}^+ - 0\boldsymbol{k}^+$ of *normalized* words is one-to-one. Given an integer n, the unique x in X such that $[x]=n$ is denoted $(n)_k$ or simply (n). It is called the *expansion of n at base k*.

A set $X \subset \mathbb{N}$ of integers is called *k-recognizable* if the set $(X)_k = \{(x)_k \mid x \in X\}$ is a recognizable subset of $\boldsymbol{k}^*$. Note that $(X)_k$ is recognizable iff the set $\{x \in \boldsymbol{k}^* \mid [x] \in X\}$ is recognizable since they only differ by leading zeros.

The following statement provides a family of sets k-recognizable for all k. It is a precise formulation of the fact that there are simple rules to test the divisibility by some fixed number on the expansion at some basis of natural numbers.

8.1. PROPOSITION. *Let $m, p \geqslant 0$. The set $E_{m,p} = \{m+rp \mid r \geqslant 0\}$ is k-recognizable for all $k \geqslant 2$.*

PROOF. We restrict ourselves to the case $0 \leqslant m < p$. The other cases can be easily reduced to this one. We define a deterministic automaton on the alphabet $\boldsymbol{k}$ with a set of states $Q=\{0, 1, \ldots, p-1\}$ and transitions $q.a=r$ where r is defined by

$$r \equiv qk+a \bmod p.$$

The initial state is 0 and the terminal state is m. It is a direct consequence of the definition, that this automaton recognizes the set $(E_{m,p})$. □

We observe that the sets of integers which are finite unions of sets $E_{m,p}$ are precisely the recognizable subsets of the monoid $\mathbb{N}$. Indeed such sets are obviously recognizable in $\mathbb{N}$ since recognizability in $\mathbb{N}$ corresponds to 1-recognizability (which we did not introduce for technical reasons). Conversely any recognizable subset of $\mathbb{N}$ is an ultimately periodic set, that is a finite union of sets $E_{m,p}$.

8.2. EXAMPLE. The automaton of Fig. 25 recognizes the binary expansion of multiples of 3.

The following result contains Proposition 8.1 as a particular case. It shows that multiplication, division, addition and substraction of a constant are operations which preserve the notion of k-recognizability.

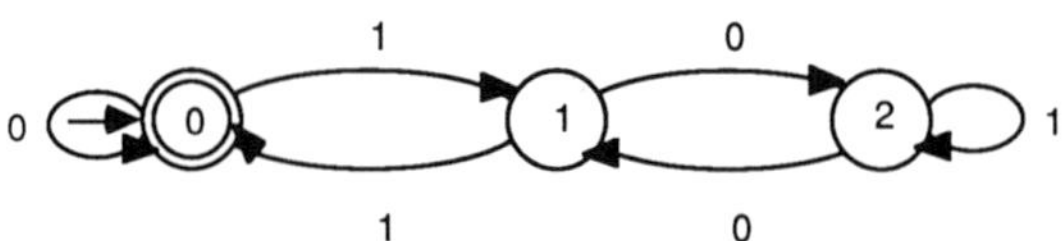

Fig. 25. The multiples of 3 written at base 2.

8.3. PROPOSITION. *Let $m, p \geqslant 0$ and $k \geqslant 2$ be integers. If X is k-recognizable, then*

$$Y = \{m + rp \mid r \in X\}$$

is k-recognizable. If Y is k-recognizable, then

$$X = \{r \geqslant 0 \mid m + rp \in Y\}$$

is k-recognizable.

PROOF (*sketch*). The statement is a direct consequence of the fact for fixed m and p the relations $\{(r, r+m) \mid r \in \mathbb{N}\}$ and $\{(r, rp) \mid r \in \mathbb{N}\}$ are rational relations. The first relation is clearly rational since the addition of a constant only modifies a finite number of symbols on the right. That the division by a fixed number can be performed by a finite automaton is also easy. It is in fact possible to realize it by adding an output to the automaton built in the proof of Proposition 8.1. □

8.4. EXAMPLE. The automaton of Fig. 26 realizes the division by 3 of a number written in binary notation.

The automaton is obtained by adding an output to the automaton of Fig. 25. Note that the automaton is nondeterministic in its output. This is very natural since multiplication by 3 cannot be easily performed from left to right. It is however codeterministic in its output, and this corresponds to the fact that the same automaton used backwards with input and output exchanged performs the multiplication by 3.

The notion of a k-recognizable set of numbers has a fundamental link with the notion of *iterated morphism* or *substitution*, which we introduce now. Let $k \geqslant 1$ be an integer, let A be a finite alphabet and let

$$\alpha : A \rightarrow A^k$$

be a mapping often called a *substitution* or also a k-uniform morphism. We extend α to words, finite or infinite in the obvious way by the rule

$$\alpha(w_0 w_1 \ldots) = \alpha(w_0)\alpha(w_1)\ldots.$$

For each symbol a in A such that a is the first letter of $\alpha(a)$, one may verify by induction on $n \geqslant 1$ that $\alpha^n(a)$ is a prefix of $\alpha^{n+1}(a)$. Thus there is a unique infinite word $w = w_0 w_1 w_2 \ldots$ which has all $\alpha^n(a)$ as prefixes. It is is denoted $\alpha^\omega(a)$. One has the equality $\alpha(w) = w$ and the infinite word w is called a *fixpoint* of the subtitution α.

8.5. PROPOSITION. *A set X of integers is k-recognizable iff there exists a finite alphabet A,*

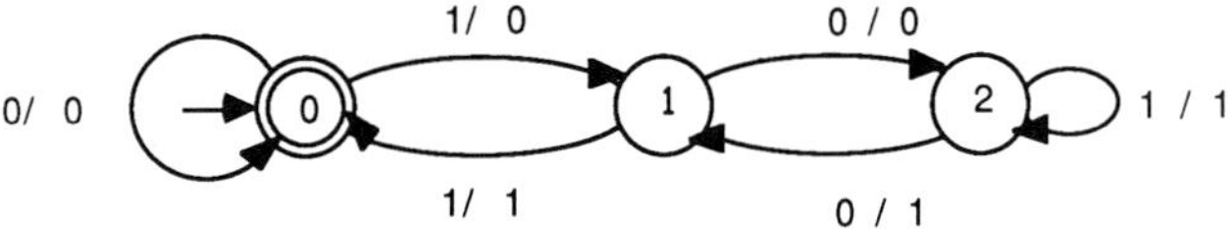

Fig. 26. The division by 3.

a substitution $\alpha: A \to A^k$ *and a fixpoint* w *of* α *such that* X *is a union of sets*

$$X_a = \{n \geqslant 0 \mid w_n = a\}$$

for a *in* A.

PROOF (*sketch*). Let $(X)_k$ be recognized by a finite deterministic automaton (Q, i, T) on the alphabet $\{0, 1, \ldots, k-1\}$. We take $A = Q$ and define the substitution α by

$$\alpha(a) = s_0 s_1 \ldots s_{k-1}$$

where s_i is the state reached from a on input i. We consider the fixpoint $w = \alpha^{\omega}(i)$. It is then possible to verify that for all $n \geqslant 0$ one has

$$w_n \in T \Leftrightarrow n \in X.$$

The proof of the reversed implication of the theorem uses the same construction backwards. □

8.6. EXAMPLE. We consider the following substitution:

$$\alpha(a) = ab, \qquad \alpha(b) = ba.$$

The fixpoint $m = \alpha^{\omega}(a)$ is known as the *Thue–Morse sequence* and it has several interesting combinatorial properties. The set X_a of positions of a in w is a set recognizable in base 2 according to Proposition 8.5. Indeed X_a is the set of numbers whose base 2 expansion has an even number of 1's (see Fig. 27).

There is an alternative characterization of sets of numbers recognizable at base k when k is a prime number. It uses the field of integers modulo k, denoted $\mathbb{F}_k$. One associates with a set X of numbers a formal series $\sigma_X(z)$ with coefficients in $\mathbb{F}_k$ in the variable z

$$\sigma_X(z) = \sum_{n \in X} z^n.$$

We denote by $\mathbb{F}_k(z)$ the field of fractions $p(z)/q(z)$ where p and $q \neq 0$ are polynomials with coefficients in $\mathbb{F}_k$.

8.7. PROPOSITION. *Let* $k > 1$ *be a prime number. A set* X *of numbers is* k*-recognizable iff there exists a nonzero polynomial* p *with coefficients in* $\mathbb{F}_k(z)$ *such that* $p(\sigma_X) = 0$.

We do not prove Proposition 8.7 here (see the Notes section for references). We show

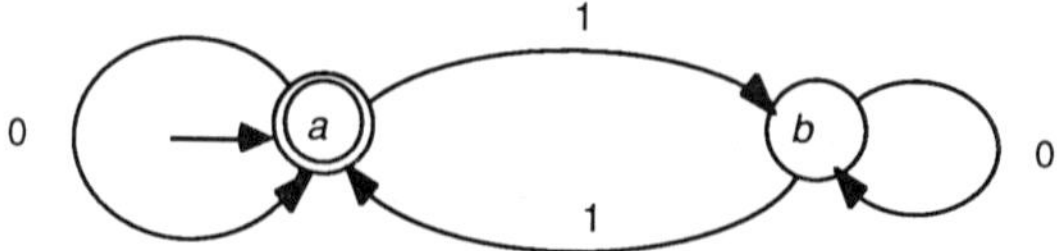

Fig. 27. Even numbers of ones.

however in the following example the main idea of the construction. It relies on the well-known fact that for a polynomial p with coefficients in $\mathbb{F}_k$ one has

$$p(z^k)=p(z)^k.$$

8.8. EXAMPLE. Let X be the set of numbers whose base 2 expansion has an even number of ones (cf. Example 8.6). Let

$$\sigma_X(z)=\sum_{n\in X} z^n=\sum_{n\geqslant 0} u_n z^n.$$

It is easy to verify that the sequence u_n satisfies the following recurrence relation:

$$\begin{cases} u_{2n}=u_n, \\ u_{2n+1}=1-u_n. \end{cases}$$

Hence

$$\sigma_X(z)=\sum_{n\geqslant 0} u_{2n}z^{2n}+\sum_{n\geqslant 0} u_{2n+1}z^{2n+1}$$

$$=\sum_{n\geqslant 0} u_n z^{2n}+\sum_{n\geqslant 0}(1+u_n)z^{2n+1}$$

$$=(1+z)\sigma_X(z^2)+z/(1+z^2).$$

Since $\sigma_X(z^2)=\sigma_X(z)^2$ by the observation made earlier, we obtain that $\sigma_X(z)$ is a root of the polynomial

$$p(t)=(1+z)^3t^2+(1+z)^2t+z.$$

We now come to the main result concerning automata and numbers. Two integers $k, l\geqslant 2$ are said to be *multiplicatively dependent* if the exist two integers $p, q>0$ such that $k^p=l^q$. Otherwise k, l are said to be multiplicatively independent.

8.9. THEOREM (Cobham). *If k and l are multiplicatively independent, then any set which is both k- and l-recognizable is recognizable.*

This theorem expresses the fact that the notion of k-recognizability heavily depends on the base k. As an illustration, we may deduce from it that the set

$$E=\{1, 2, 4, 8, 16, 32, 64, 128, 256, 512, 1024, \ldots\}$$

of decimal expansions of powers of two is not recognizable in $\{0, 1, \ldots, 9\}^*$. This can of course be proved directly (using an iteration lemma) by observing that if E were 10-recognizable, it would contain an infinite subset $E'=\{n_0, n_1, \ldots\}$ such that the quotients n_{i+1}/n_i converge to a power of 10 when $i\to\infty$. This argument will enter in the proof of the general result, as we shall see.

Also, it is not difficult to see that, for any $k>1$ and $p>0$, a set of numbers is k-recognizable iff it is k^p-recognizable. Hence, when k and l are multiplicatively dependent, the sets of numbers which are k-recognizable or l-recognizable are the same.

We shall now prove a series of statements, most of them interesting in their own, which, altogether, give a proof of Theorem 8.9 due to G. Hansel.

We shall need the following elementary result from number theory. If k, l are multiplicatively independent, any interval of the positive real line contains some number of the form k^p/l^q with $p, q \geqslant 0$ (see the Notes section for references).

We say that a set $X \subset A^*$ is *right dense* if for any word u in A^* there exists a word v in A^* such that $uv \in X$. Hence X is right dense iff any word appears as a prefix of a word from X.

The first result accounts for the familiar fact that one cannot determine much concerning the initial digits of a decimal number by reading the initial digits of the corresponding binary number.

8.10. PROPOSITION. *Let X be an infinite k-recognizable set of integers. For any integer l multiplicatively independent of k, the set $0^*(X)_l$ of (unnormalized) expansions of X at base l is right dense.*

PROOF. Since X is infinite, there exists $t, u, v \in \boldsymbol{k}^*$ with $u \neq \varepsilon$ such that $tu^*v \subset (X)_k$. Let $x \in \boldsymbol{l}^*$. Since k and l are multiplicatively independent, there exist $p, q > 0$, arbitrarily large, such that

$$[x]_l + \tfrac{1}{4} < \left([t]_k + \frac{[u]_k}{k^g - 1}\right) k^{gp+h}/l^q < [x]_l + \tfrac{1}{2}$$

where $g = |u|$, $h = |v|$. For q large enough we have

$$-\tfrac{1}{4} < \left([v]_k - \frac{k^h [u]_k}{k^g - 1}\right) \Big/ l^q < \tfrac{1}{2}.$$

Adding these inequalities term by term, we obtain

$$[x]_l l^q < [t]_k k^{gp+h} + [u]_k \frac{k^{gp} - 1}{k^g - 1} k^h + [v]_k < ([x]_l + 1) l^q,$$

whence, rewriting the central term,

$$[x]_l l^q < [tu^p v]_k < ([x]_l + 1) l^q.$$

This proves the existence of an integer j with $0 < j < l^q$ such that $[x]_l l^q + j = [tu^p v]_k$. Hence there exists a word $y \in l^*$ such that $[xy]_l = [tu^p v]_k$ whence $xy \in X$. □

Let $d \geqslant 1$ be an integer. A set X of integers is called *d-syndetic*, or just syndetic if for all x in X there is a y in X such that $x < y \leqslant x + d$.

8.11. PROPOSITION. *A k-recognizable set X is syndetic iff the set $0^*(X)_k$ of (unnormalized) expansions of X at base k is right dense.*

PROOF. Let us suppose that $0^*(X)_k$ is right-dense. By the definition, there exists for each integer n two integers p and $t<k^p$ such that

$$nk^p+t\in X.$$

Since X is k-recognizable, the integer p can be bounded uniformly. This shows that X contains an element in each interval of length k^p and hence is k^p-syndetic. The converse is true independently of the hypothesis that X is k-recognizable. □

We now obtain easily from Propositions 8.10 and 8.11 the following weak version of Theorem 8.9.

8.12. COROLLARY. *If k and l are multiplicatively independent, any infinite set of integers which is both k- and l-recognizable is syndetic.*

We further prove the following technical approximation lemma.

8.13. LEMMA. *Let X be a d-syndetic set of integers. For all integers K, L, h and each $\eta>0$ such that*

$$K<L<K+\eta,$$

there exists x in X and an integer y such that

$$yL\leqslant xK+h\leqslant yL+\eta d.$$

PROOF. Let r be the smallest integer such that $rK+h<rL$. Then we have for all $i\geqslant 1$

$$(r-i)L<(r-i)K+h$$

by minimality of r and moreover

$$(r-i)K+h=rK+h-iK<rL-iK<rL-iL+i\eta=(r-i)L+i\eta.$$

We obtain thus for $1\leqslant i\leqslant d$ the inequalities

$$(r-i)L<(r-i)K+h<(r-i)L+\eta d.$$

Let j be an integer such that $jL+r-d\geqslant x$ for some x in X and $jK+r-d\geqslant 0$. We add jKL to the three terms in the above inequalities to obtain for all i with $1\leqslant i\leqslant d$ the inequalities

$$(jK+r-i)L<(jL+r-i)K+h<(jK+r-i)L+\eta d.$$

Since X is d- syndetic, there is an x in X of the form $x=jL+r-i$ whence the desired result. □

We shall finally need a combinatorial result on infinite words. Let $x=a_0a_1a_2\ldots$ be a right-infinite word, that is an infinite sequence of letters a_i. For $n<m$, we denote $x[n,m]=a_na_{n+1}\ldots a_{m-1}$. A word w is said to be a *factor* of x if there exist n,m with $n<m$ such that $w=x[n,m]$ and w is said to be a *recurrent* factor if there exists an infinity of such pairs n,m.

The infinite word x is said to be *ultimately periodic* if there exists an integer $p \leqslant 1$ such that, for large enough n one has $a_{n+p} = a_n$. Such an integer p is called an *ultimate period* of x.

8.14. PROPOSITION. *An infinite word x is ultimately periodic iff there exists an integer m such that the number of recurrent factors of length m is at most m.*

PROOF. The condition is clearly necessary since, for large enough m, the number of recurrent factors of an ultimately periodic word is constant. Conversely, let m be the smallest integer such that the condition is satisfied. The number of recurrent factors of length $m-1$ is therefore equal to m. Hence, for large enough n, the factor $x[n, n+m-1]$ determines the letter a_{n+m-1} that follows it. Let $x[n, n+m-1]$ and $x[n', n'+m-1]$ with $n<n'$ be two occurrences of the same recurrent factor of length $m-1$. Then $p = n'-n$ is an ultimate period of x. $\square$

We are now able to prove Theorem 8.9. Let $X \subset \mathbb{N}$ be an infinite set of integers which is both k- and l-recognizable with k, l multiplicatively independent. For all $t, j \geqslant 0$, the set

$$E_{tj} = \{ y \mid yk^j + t \in X \}$$

is l-recognizable. But for all $u \in \boldsymbol{k}^*$ we have $v \in (X)_k u^{-1}$ iff $[v]_k \in E_{t,j}$ with $t \in [u]_k, j = |u|$. Hence all sets $(X)_k u^{-1}$ are l-recognizable. Let ρ_k be the equivalence on $\mathbb{N}$ corresponding to the minimal automaton of $(X)_k$ that is to say defined by $x \sim y \bmod \rho_k$ iff $(x)_k^{-1}(X)_k = (y)_k^{-1}(X)_k$. The classes of ρ_k are Boolean combinations of sets $(X)_k u^{-1}$ for $u \in \boldsymbol{k}^*$ since a class of ρ_k is defined by its right contexts. Hence all classes of ρ_k are l-recognizable.

Therefore, there exists an equivalence of finite index θ which is a refinement of ρ_k and which is *l-stable*, that is to say

$$x \sim y \bmod \theta \;\Rightarrow\; xl^j + t \sim yl^j + t \bmod \theta$$

for all $t, j \geqslant 0$ such that $t < l^j$. Let c be the number of classes of θ. We denote by u_n the class of $n \bmod \theta$ and we define

$$u = u_0 u_1 u_2 \ldots .$$

Let also r be the right-infinite word

$$r = r_0 r_1 r_2 \ldots$$

where r_n is the class of $n \bmod \rho_k$. For all recurrent factors w of length 2 or r, the set of indices n such that $r[n, n+2] = w$ is k- and l-recognizable and therefore syndetic. Hence there is an integer d such that any recurrent factor w of length 2 of r has a second occurrence at distance at most d. We choose a real number ε such that

$$0 < \varepsilon < 1 \quad \text{and} \quad c\,\varepsilon/(1-\varepsilon) < \tfrac{1}{2}$$

and we choose integers $p, q \geqslant 0$ such that

$$1 < l^q/k^p < 1 + \varepsilon/d.$$

Let $K=k^p$, $L=l^q$ and m be the integer part of $K(1-\varepsilon)$. We are going to prove that *for any recurrent factor w of length m of r there exists an integer y such that $r[yL,(y+1)L]=swt$ with $|s|\leqslant \varepsilon K$.*

For this, we first observe that w, being a recurrent factor of length at most K, appears infinitely often in factors of the form $r[xK,(x+2)K]$ and always in the same position within that factor. Since ρ_k is k-stable, $r[xK,(x+2)K]$ is determined by $r[x,x+2]$. Since, by the definition of d, every factor of length 2 of r has a second occurrence at distance at most d, there exists a strictly increasing sequence of integers (x_n) such that $x_{n+1}-x_n\leqslant d$ and

$$r[x_nK,(x_n+2)K]=w'ww''.$$

Let $h=|w'|$. We apply Lemma 8.13 with $\eta=K\varepsilon/d$ and $X=\{x_1,x_2,\ldots\}$. We obtain the existence of some integer y such that $r[yL,(y+1)L]=swt$ with $|s|\leqslant d\eta$. This proves the claim; see also Fig. 28.

The number of factors of r of the form $r[yL,(y+1)L]$ is at most equal to c since θ is an l-stable refinement of ρ_k of index c. The number of recurrent factors of length m of r is therefore at most equal to

$$K\varepsilon c\leqslant\tfrac{1}{2}K(1-\varepsilon)\leqslant\tfrac{1}{2}(m+1)\leqslant m.$$

By Proposition 8.14, the word r is ultimately periodic and this concludes the proof of Theorem 8.9. □

Cobham's theorem can be considered as a negative result although it can also be viewed as a uniqueness theorem. It raises indeed the natural problem of computing, given a set X of integers, the essentially unique basis k, if it exists, such that X is k-recognizable. We do no know presently how such a computation can be done.

Besides the representation of numbers at base k that we have discussed so far, there exist other representations, somtimes called *nonstandard*.

One of them is the Fibonacci representation discussed in the following example. (See the Notes section for further indications.)

8.15. EXAMPLE. Let $(\varphi_k)_{k\geqslant 0}$ be the sequence defined by $\varphi_0=1$, $\varphi_1=2$ and inductively by the recurrence relation

$$\varphi_{k+1}=\varphi_k+\varphi_{k-1},\quad k\geqslant 1.$$

The sequence (φ_k) is known as the *Fibonacci sequence*. It is not difficult to prove that

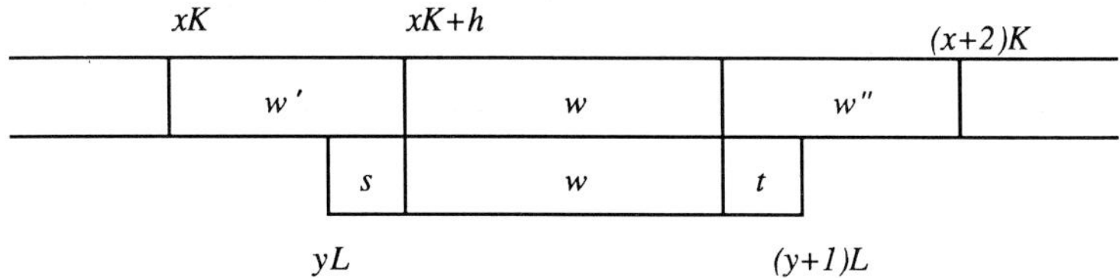

Fig. 28. Localizing the occurrences of w.

any integer $n \geqslant 0$ has a representation as

$$n = \sum_{k=0}^{K} \varepsilon_k \varphi_k$$

with $\varepsilon_k = 0$ or 1. Indeed, if $\varphi_k \leqslant n < \varphi_{k+1}$, then $0 \leqslant n - \varphi_k < \varphi_{k-1}$ whence the existence of the representation by induction on K. Moreover the representation is unique provided no consecutive ε_k are equal to 1 and $\varepsilon_K = 1$. This leads to a representation of numbers by words over the alphabet $\{0, 1\}$. Thus, for example, the words

$$101101, 110001, 1000001$$

all represent the number 22. It is amusing that the relation between a word representing a number n and the normalized representation representing the same number is a rational relation represented in Fig. 29. To make the relation length-preserving, we have assumed that the input begins with 0.

The proof that the transducer of Fig. 29 correctly normalizes any word beginning with 0 can be made easily by applying the subset construction to the automaton reading the input symbols, as on Fig. 30.

The proof of completeness is done by checking that the final state 1 appears in any state reachable from the initial state $\{1, 2\}$. The proof of correctness can be obtained as follows. First no two consecutive edges have output labels equal to 1. Hence the output is normalized. Second, if (x, y) is the label of a path $1 \rightarrow p$ or $2 \rightarrow p$, then the difference $[x] - [y]$ between the numbers represented by x and y is given in Table 6 as a function on p, as may be checked by induction on the length of x.

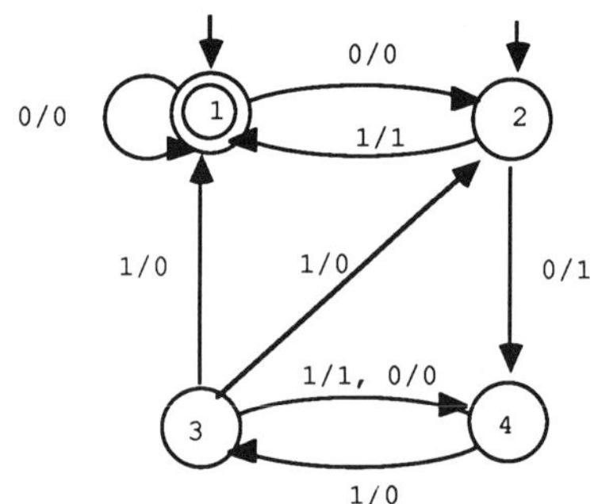

Fig. 29. The Fibonacci normalizer.

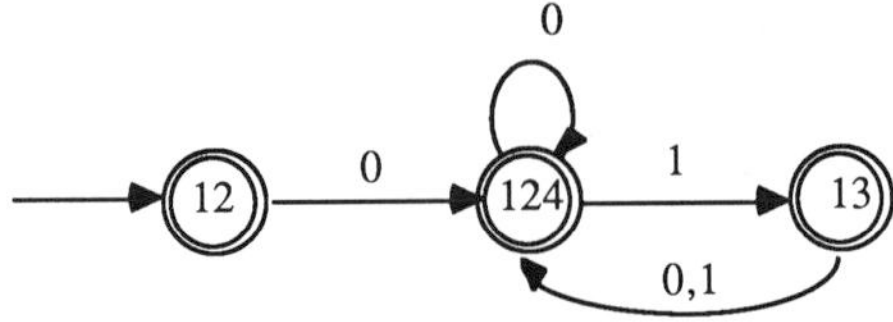

Fig. 30. Verifying the completeness.

Table 6
The difference of values [input] − [output]

p	1	2	3	4
$[x]-[y]$	0	0	−1	−1

9. Notes

This section contains notes concerning the material presented in the chapter. The notes are grouped according to the section to which they refer.

Section 2

The terminology on finite automata is not presently standard. It is very common that rational expressions are called *regular expressions* and correspondingly that rational sets are called *regular sets*. The terms *rational* and *recognizable* have been systematically used in Eilenberg's book [49, 50]. The adjective rational emphasizes the analogy between rational expressions and rational power series or fractions as used in classical algebra. It is M.P. Schützenberger's point of view on automata theory that finite automata and rational expressions correspond to rational formal series whereas context-free languages correspond to algebraic series, i.e. solution of a system of polynomial equations [108]. This analogy can be made extremely precise at the cost of introducing finite automata with multiplicities and to replace sets of words by functions from words into a semiring of coefficients. The case of sets of words corresponds to the Boolean semiring. For a treatment of this extension of automata theory see [49] or [16] or [107]. When the coefficients are real numbers, they can be interpreted as probabilities, giving rise to the notion of a probabilistic automaton [94].

The construction of the minimal automaton in Section 1 uses a minimization algorithm consisting in a stepwise refinement of a partition of the states set. This algorithm is known as *Moore's algorithm*. It works in time $O(n^2)$ on an n-state automaton. A refinement due to J. Hopcroft gives an $O(n \log n)$ running time [6].

The exposition of decision problems concerning finite automata was done for the first time in the classical paper of Rabin and Scott [100].

We have not touched here the subject of the hardware implementation of finite automata, in particular by the so-called technique of *programmable logical arrays* (PLA). See [54, 89]. The software implementation, in particular using *failure functions* is discussed in [7]; see also Section 7.

The definition of recognizable sets can be given by making use of formal grammars instead of finite automata. Indeed, finite automata are equivalent to a special type of context-free grammars in which all rules are of the form

$$x \rightarrow ay \quad \text{or} \quad x \rightarrow 1.$$

These grammars are a particular case of the context-free grammars dealt with in [14].

There are several natural extensions of the notion of a finite automaton. One is the notion of automaton with multiplicities mentioned above. Another one is the notion of a *two-way automaton.* This is an automaton which can scan its input both ways instead of a left-to-right way as we have assumed. It is an interesting result that such an automaton is equivalent to an ordinary one [114]; see [70]. Another extension is that of *multitape finite automata.* These automata have several input words instead of just one. This notion is closely related with the notion of a rational relation introduced at the end of Section 3. It has recently been proved that the equivalence of multitape deterministic finite automata is decidable (T. Harju and J. Karhumäki, The equivalence problem of multitape finite automata, to appear in *Theoret. Comput. Sci.* **78** (January 1991). This had been a longstanding open problem, proved for the case of two tapes in [18].

Another aspect of finite automata which has not been touched here is the notion of *synchronizing word.* Let $\mathcal{A}$ be a complete deterministic automaton with n states. A word w is said to be *synchronizing* if, starting from any state one reaches the same fixed state after reading w. An automaton is called *synchronizing* if there exists a synchronizing word. It is not difficult to prove that if there is a synchronizing word, then there is one of length at most n^3. It has been conjectured that n^3 could be replaced by n^2 but this is still unproved. See [15] for a bibliography and [28] for a recent result. Another conjecture on the same notion appears in [3] and it is known as the *road coloring problem.* The conjecture says that, except for a trivial case due to periodicity, it is always possible to transform an automaton into a synchronizing one just by exchanging the labels of edges with a common origin.

Section 3

Identities on rational expressions have received considerable attention. The proof that any complete system of identities on a two-letter alphabet is infinite is due to Redko and Salomaa [38]. Conway has developed the theory in a decisive way. He has shown that one may associate to each monoid morphism

$$f: A^* \to M$$

from A^* onto a finite monoid M an identity of the form

$$A^* = \sum_{m \in M} f^{-1}(m).$$

He has shown that when M is a group, the corresponding identity is deducible from the cyclic identities

$$a^* = (1 + a + \cdots + a^{n-1})(a^n)^*,$$

iff M is a solvable group. For recent results in this area, see [77].

The notion of a rational relation introduced at the end of Section 3 is a fundamental one. There are several characterizations of rational relations among which one known as Nivat's theorem (see [13]) asserting that $X \subset A^* \times B^*$ is rational iff there exists an alphabet C, two morphisms $f: C^* \to A^*$, $g: C^* \to B^*$ and a rational set $R \subset C^*$ such that

$$X = \{(f(r), g(r)) \mid r \in R\}.$$

Another important result is Eilenberg's *cross-section theorem* concerning rational functions, i.e. functions $f: A^* \to B^*$ such that the set $(x, f(x))$ is a rational relation. According to this theorem any such function has a rational cross-section, i.e., a rational set $X \subset A^*$ such that f is one-to-one from X into B^*. Further results on rational functions appear in [30, 32, 33, 71]. The simplest nontrivial example of a rational function is obtained by considering the inverse of an injective morphism $f: A^* \to B^*$. This relates rational function with the theory of *codes*, a subject not treated here (see [15]). Rational functions and codes are also closely related to the notion of an *unambiguous automaton.* An automaton is said to be unambiguous if for any word x and any pair p, q of states, there is at most one path from p to q with label x. Thus a deterministic automaton is unambiguous and so is a reverse deterministic automaton. It can be proved that any rational function can be realized by an automaton with output which is unambiguous with respect to its input (see [13]). The notion of unambiguous automata is related with the concept of a *finite-to-one* map in symbolic dynamics (see [2, 11]). For further results on unambiguous automata see, [13, 21, 27].

Section 4

The original reference of Kleene's theorem is [72]. The algorithm going from automata to rational expressions is commonly referred to as McNaughton and Yamada's algorithm. It can be shown to be related with Gauss's algorithm to solve a system of linear equations by substitutions. The algorithm going from rational expressions to automata can be performed in several slightly different ways. One of them uses automata with ε-moves and is known as Thomson's construction. It is suitable for practical implementation and it is used in the Lex software building lexical analyzers (see [7]). The method using left residuals mentioned at the end of the section is originally due to J. Brzozowski [23] who called them *derivatives.* A recent progress in this direction appears in [12].

There is an aspect of the theory of rational sets which has not been treated here. It concerns the case of a *commutative* alphabet, the free monoid A^* on a k-letter alphabet being replaced by the free commutative monoid N^k. It is then preferable to use the additive notation instead of the multiplicative one. Hence we start with the set operations

$$\cup, +, {}^*$$

of *union, sum*

$$X+Y=\{x+y \mid x \in X, y \in Y\}$$

and *star*

$$X^* = \{x_1 + x_2 + \cdots + x_n \mid n \geqslant 0, x_j \in X\}.$$

Starting from singleton sets and using these operations, one obtains the *rational subsets* of N^k. The main results are as follows: the family of rational sets is a Boolean algebra (Ginsburg and Spanier), see [59] or [38] or [104]. The family of rational identities has a finite basis (Redko), see [38]. Any rational set is unambiguous (Eilenberg and Schützenberger), see [51]. There is a close connection between rational subsets of free commutative monoids and *Presburger's Arithmetic*. (See [124]), and also with the theory of *Petri Nets* (see [104]).

Recently the case of *partially commutative* alphabets has been investigated. One starts with a symmetric relation $\theta \subset A \times A$ and considers the monoid $M(A, \theta)$ generated by A with the relators (ab, ba) for all pairs $(a, b) \in \theta$. The study of these monoids originated in the work of Cartier and Foata [29] and was followed by [53]. It has further been studied as a model of concurrency and several results have been obtained about the rational subsets of $M(A, \theta)$. See [1, 17, 31, 41, 47, 52, 83, 90, 93, 106]. W. Zielonka has introduced a new model of finite automata that he calls *asynchronous*. The states of these automata are vectors

$$(p_1, p_2, \ldots, p_n).$$

A subset of the index set is associated to each letter a and the action of a takes into account and modifies only the components corresponding to these indices. Two letters a, b that commute correspond to disjoint sets of indices. Zielonka's theorem says that *any recognizable set $X \subset A^*$ which is a union of commutation classes can be recognized by an asynchronous automaton* [40, 42, 130].

The study of *normal forms* in partially commutative monoids has motivated several investigations [46, 96].

Section 5

The notion of star-height of a rational set is a very natural one and the problem of its effective computability has been raised since the beginning of automata theory, see [48]. Reference [45] gives a proof that the star-height of the set X_n is exactly n and [86] shows how to compute the star-height of a set whose syntactic monoid is a group, as X_n^* for example.

The proof that it is decidable whether a given rational set is limited was obtained by Simon [116] and Hashiguchi [65] independently. The decidability of star-height one is proved in the three papers [66–68]. The solution of the general problem is in [69]. The proof is however very difficult to understand and a lot remains to be done to make it a tutorial presentation. New developments improving the understanding on star-height one are due to Hing Lung and to Imre Simon; see [81, 117–119].

Section 6

The notion of star-free sets was introduced by McNaughton in connection with logic [88], see [124].

The original reference of Schützenberger's theorem is [109]. Several proofs of this theorem have been proposed. Our proof here is a slightly simplified version of the original one. It incorporates some improvements from [80]. We have also constructed the induction without changing the underlying monoid. A reader familiar with monoids will remark that we are at each time working in the 0-minimal ideal of a monoid which shrinks at each step of the induction. Also, a crucial fact used in the proof without stating it explicitly is that, in an aperiodic monoid, the Boolean closure of the ideals is the whole powerset of the monoid.

Another completely different proof can be given using the *decomposition* of finite automata. We have not touched this matter here. It can be presented very roughly as follows. If (P, M) and (Q, N) are monoids of transformations on finite sets P and Q, then one may define a new monoid of transformations

$$(P \times Q, M \circ N)$$

called the *wreath product* of M and N. It is obtained by considering the set $M^Q \times N$ of pairs (f, n) with $f: Q \rightarrow M$ a mapping from Q into M and $n \in N$. The action is defined by

$$(p, q)(f, n) = (p(qf), qn)$$

where all mappings are written on the right of their argument. The wreath product of monoids is the operation corresponding to the series composition of automata equipped with output functions, hence to the composition of rational functions.

A fundamental theorem known as Krohn–Rhodes theorem states that any finite monoid is an image by a morphism of a submonoid of a wreath product

$$M_1 \circ M_2 \circ \cdots \circ M_n$$

in which each component M_i is either a finite group or a monoid of constant mappings (the so-called *reset* monoids). Moreover the groups used in the decomposition appear as semigroups of the original one. For a proof of this result see the expositions in book form in [50, 60, 78].

When M is an aperiodic monoid, it is covered in the above sense by a wreath product of reset monoids. When M is the syntactic monoid of a set $X \subset A^*$, this leads easily to an expression of X using products and Boolean operation. This proof of Schützenberger's theorem is due to Meyer [91], and appears in [50]. The original proof of Theorem 6.2 appears as an independent work of McNaughton and Zalcstein in [87, 129], and of Brzozowski and Simon in [25]. Our proof here follows [50] where the reader can find a proof of Theorem 6.3 (Chapter VIII, A theorem on graphs, pp. 222–228). We have simplified the presentation, avoiding the use of Tilson's trace-delay theorem [50, p. 85]).

The proof of Theorem 6.3 appeared originally in [115]. It is reproduced in [50, 98]. The proof of Theorem 6.5 appears in Lothaire's book [79] where it is related with other combinatorial properties of words and their subwords. A new proof of Theorem 6.3 appears in [122].

The presentation in this section of three families of recognizable sets (star-free, locally testable, piecewise testable) is the visible part of an iceberg. A general framework was set by Schützenberger in [110] and Eilenberg in [50]. It involves, the notion of a (pseudo)-*variety* of monoids, which is a family of finite monoids closed under morphisms, submonoids and (finite) direct products. Thus aperiodic monoids form a pseudo variety and so do the syntactic semigroups S such that for each idempotent e, the semigroup eSe is idempotent and commutative. The study of varieties of monoids, in particular in connection with the operation of wreath-product, contains many other results. See in particular the books of Eilenberg, Lallement and Pin [50, 78, 98].

The construction of a star-free set makes use of two kinds of operations. Products on the one hand and Boolean operations on the other hand. Let $\mathscr{B}_0$ be the family of subsets of the alphabet and let inductively $\mathscr{B}_{n+1}$ be the Boolean closure of the family of sets of the form

$$X_1 X_2 \ldots X_k$$

with $X_i \in \mathscr{B}_n$. Then, by definition, any star-free set falls into some $\mathscr{B}_n$. The least integer n such that $X \in \mathscr{B}_n$ is called the *dot-depth* of X. It has been proved by Brzozowski and Knast in [24] that there exist star-free sets of arbitrary high dot-depth. Also Knast [73] has found a characterization of the syntactic monoids of sets of dot-depth one. Further results on dot-depth two can be found in [126]. On the connection between dot-depth, varieties and the complexity of logical formulas see [97, 123].

A recent result characterizing a natural family of automata by identities in the same way as we have seen for piecewise testable sets and locally testable sets has been obtained by Pin [99]. It characterizes by identities the syntactic monoids of sets that may be recognized by finite *reversible automata.*, i.e. finite automata in which no two edges with the same label lead to the same state. This family of automata has a natural connection with notions met in the *free group* [14] and also with the problem of automata *inference* [8].

The processing of an input string by an automaton can be viewed as a computation taking place in a finite monoid. One would like to extend this algebra-combinatorics relation to richer classes of languages. A recent approach has provided such a link for families of functions arising naturally in the study of parallel algorithms. It is due to D. Barrington and D. Thérien [10]. We shall very briefly describe this research.

The model of computation to be analyzed is that of Boolean circuits, essentially as described in [39], but presented here in its nonuniform version. A circuit C is given by a family $(C_n)_{n\in\mathbb{N}}$ of directed acyclic graphs. Each C_n contains $2n$ nodes of indegree 0, labeled $x_1, \ldots, x_n, \ldots, \bar{x}_1, \ldots, \bar{x}_n$; the other nodes are labeled by AND or OR. Given an input string $x \in \{0, 1\}^n$, each node of C_n computes a Boolean value as follows: a node labeled x_i ($\bar{x}_i$) returns the (complement of the) value of the ith bit of x; a node labeled AND (OR) returns 1 iff all (one of) its incoming edges are leaving nodes that return a value 1. One vertex is distinguished as the output node of the graph C_n. The circuit thus computes a function from $\{0, 1\}^n$ into $\{0, 1\}$. The language $L \subset \{0, 1\}^*$ recognized by C is the set $\{x \mid x \in \{0, 1\}^n$ and $C_n(x) = 1\}$.

We isolate two classes of subsets of $\{0, 1\}^*$. Let NC^1 denote the family of languages accepted by circuits in which internal nodes are of indegree 2, and for which C_n has $O(\log n)$ depth (and hence $n^{O(1)}$ vertices). Let AC^0 represents the languages accepted by circuits of depth $O(1)$ and size $n^{O(1)}$, with no restriction on the indegree. It is a nontrivial result, first proved in [56] that AC^0 is strictly contained in NC^1.

The link with finite monoids is obtained by considering automata that can access the input string in any order, possibly interrogating the same position several times. *A random access automaton* $\mathscr{A}$ over a monoid M is a family $(\mathscr{A}_n)_{n\in N}$, where $\mathscr{A}_n$ is a finite sequence of instructions, each of the form

$$(i\colon m_0, m_1) \quad \text{where } 1 \leqslant i \leqslant n, \text{ and } m_0, m_1 \in M.$$

Each $\mathscr{A}_n$ computes a function from $\{0, 1\}^n$ into M as follows: an instruction $(i\colon m_0, m_1)$

yields, on input x, the value m_j iff the ith bit of x has value j. A sequence of instructions returns the product in M of the values produced by each instruction in the order in which they occur. The set $L \subset \{0, 1\}^*$ can be recognized by $\mathscr{A}$ iff for each n we can choose $X_n \subset M$ such that $L \cap \{0, 1\}^n$ is the set of inputs x for which $\mathscr{A}_n(x) \in X_n$. We then have the following two results.

THEOREM (Barrington [8]). *The set L is in* NC^1 *iff it can be recognized by a random-access automaton $\mathscr{A} = \langle \mathscr{A}_n \rangle$ over some finite monoid, where the length of $\mathscr{A}_n$ is $n^{O(1)}$.*

THEOREM (Barrington & Thérien [9]). *The set L is in* AC^0 *iff it can be recognized by a random access automaton $\mathscr{A} = (\mathscr{A}_n)$ over some finite aperiodic monoid, where the length of $\mathscr{A}_n$ is $n^{O(1)}$.*

Reference [10] contains several more results in the same vein.

Section 7. Special automata

As mentioned in Section 3, finite automata and regular expressions are used in several text manipulation systems including lexical analyzers for compilers, text editors, file manipulation systems and others. The idea that regular expressions are a powerful enough but still tractable query language was supported by Kleene's theorem. See [37] for references or [125] for an early step in that direction.

The special automata presented in Section 7 are those which, informally speaking, do not blow up in the determinization algorithm thus providing efficient string processing algorithms. The first papers on this kind of automata are [95] and [111]. The first one studies sets of the form A^*S with S finite and characterizes the associated automata, called *definite*. The second one is a systematic study of sets of the form

$$X = A^*S - A^*SA^+$$

called *semaphore codes* in [15]. The main result of [111] is that for a semaphore code which is not of the form X^n with $n \geqslant 2$, the set X^* can be recognized by a synchronizing automaton.

The original reference to Knuth, Morris and Pratt's algorithm is the paper [76] which appeared in 1977 after circulating several years as a preprint. Other references to string matching algorithms are given in [5, 22, 113].

The suffix automaton was invented by Blumer et al. [20]. It can be considered an heir of the *suffix tree* construction of [127] and [84]. The construction of the *suffix transducer* is due to Crochemore [43]. Crochemore has applied his construction to several problems in [44]. See also [121]. One is the string-matching method mentioned at the end of the section. It gives a real-time string-matching algorithm competing with that of [120] and [58].

For further references and a more complete treatment of the subject matter of this section, see [4]. See [62] for a presentation of other applications of automata to text processing and, in particular, to data compression.

Section 8. Recognizable sets of numbers

The idea of automata recognizing numbers written at some basis appeared at the beginning of automata theory. It was already proved in [26] that the set of powers of an integer k is l-recognizable only when k and l are multiplicatively dependent. A number of results on k-recognizable sets tend to prove that some naturally defined sets of numbers, such as prime numbers, for instance, are not k-recognizable for any k [92] or even context-free [112].

Proposition 8.5 gives an alternative definition of k-recognizable sets using a substitution. It is due to Cobham [36]. Proposition 8.7 is due to Christol et al. [34]. The original reference to Cobham's theorem is [35]. We have reproduced here the proof invented by Hansel [63] following the presentation subsequently made by Reutenauer [103]. It appears here for the first time in an easily accessible form with the kind permission of G. Hansel and C. Reutenauer.

This proof is inspired by concepts borrowed from the field of *topological dynamics*, such as the notion of a syndetic set or the study of infinite words in relation with the growth rate of the number of factors of given length. This is a field which has some important interactions with automata theory. See for example [128] where the notion of a *sofic system* is introduced, which is equivalent to a finite automaton. We mention the historical references of [61] and the more recent book of Furstenberg [57]. Recent applications of these concepts to coding problems appear in [2, 11, 19, 82].

We have used the following lemma: *if k, l are multiplicatively independent, any interval of the positive real line contains some quotient k^p/l^q for some integers $p, q \geqslant 0$.* This can be proved as follows. Two integers k, l are multiplicatively independent iff $\alpha = \log k/\log l$ is irrational.

Let x be a positive real number. One has $k^p/l^q \sim x$ iff

$$p \log k - q \log l \sim \log x.$$

Since we may change k, l for some powers K, L we may assume $0 < \log x < \log l$. Let $y = \log x/\log l$. Then we obtain the above approximation iff we can find integers $p, q \geqslant 0$ such that

$$p\alpha - q \sim y.$$

It is a well-known fact that for any irrational α, the fractional parts of its multiples are dense in the interval $[0, 1]$ (see [64, p. 375] for instance). Thus for any y in $[0, 1]$ there are some $p, q \geqslant 0$ such that $|p\alpha - q - y|$ is as small as we wish whence the desired approximation.

The Fibonacci representation appears in [74, 75]. The nonstandard representations of numbers have been studied in [55] with a relevant bibliography. New developments in this subject in relation with dynamical systems appear in [101, 102].

Acknowledgment

I would like to acknowledge the help received during the preparation of this work. Marie-Pierre Béal, Jean Berstel, Maxime Crochemore, Jean-Eric Pin, Andreas

Podelski, Wolfgang Thomas, Imre Simon and Pascal Weil have brought improvements to the presentation by their hints and comments on successive versions of the manuscript. Denis Thérien has kindly provided the part of the Notes section dedicated to circuits and random access automata. M.P. Schützenberger has brought a "touche finale" to the whole thing. Many thanks are due to Arlette Dupont for her patience.

References

[1] AALBERSBERG, I. and G. ROZENBERG, Theory of traces, *Theoret. Comput. Sci.* **60** (1988) 1–82.

[2] ADLER, R.L., D. COPPERSMITH and M. HASSNER, Algorithms for sliding block codes, *IEEE Trans. Inform. Theory*, (1983) 5–22.

[3] ADLER, R.L., L.W. GOODWIN and B. WEISS, Equivalence of topological Markow shifts, *Israël J. Math.* **27** (1977) 49–63.

[4] AHO, A., Algorithms for finding patterns in strings, in: J. van Leeuwen, ed., *Handbook of Theoretical Computer Science, Vol. A* (North-Holland, Amsterdam, 1990) 255–300.

[5] AHO, A.V. and M.J. CORASIK, Efficient string matching: An aid to bibliographic search, *Comm. Assoc. Comput. Mach.* **18** (1975) 333–340.

[6] AHO, A., J. HOPCROFT and J. ULLMAN, *The Design and Analysis of Computer Algorithms* (Addison-Wesley Reading, MA, 1974).

[7] AHO, A., R. SETHI and J. ULLMAN, 1986, *Compilers, Principles, Techniques and Tools* (Addison-Wesley, Reading, MA, 1986).

[8] ANGLUIN, D., Inference of reversible languages, *J. Assoc. Comput. Mach.* **29** (1982) 741–765.

[9] BARRINGTON, D.A., Bounded-width polynomial size branching programs recognize exactly those languages in NC^1, in: *Proc. 18th Ann. ACM Symp. Theory of Computing* (1986) 1–5.

[10] BARRINGTON, D.A. and D. THÉRIEN, Finite monoids and the fine structure of NC^1, in: *Proc. 19th Ann. ACM Symp. Theory of Computing* (1987) 101–109.

[11] BÉAL, M.P., Codes circulaires, automates locaux et entropie, *Theoret. Comput. Sci.* **57** (1988) 283–302.

[12] BERRY, G. and R. SETHI, From regular expressions to deterministic automata, *Theoret. Comput. Sci.* **48** (1986) 117–126.

[13] BERSTEL, *Transductions and Context-Free Languages* (Teubner, Stuttgart, 1979).

[14] BERSTEL, J. and L. BOASSON, Context-free grammars, in: J. van Leeuwen, ed., *Handbook of Theoretical Computer Science, Vol. B* (North-Holland, Amsterdam, 1990) 59–102.

[15] BERSTEL, J. and D. PERRIN, *Theory of Codes* (Academic Press, New York, 1984).

[16] BERSTEL, J. and C. REUTENAUER, 1988, *Rational Series and Their Languages* (Springer, Berlin, 1988).

[17] BERTONI, A., G. MAURI and N. SABADINI, Equivalence and membership problems for regular trace languages, in: *Automata, Languages and Programming*, Lecture Notes in Computer Science **140** (Springer, Berlin, 1982) 61–71.

[18] BIRD, M., The equivalence problem for deterministic two-tape automata, *J. Comput. System Sci.* **7** (1973) 218–236.

[19] BLANCHARD, P. and G. HANSEL., Systèmes codés, *Theoret. Comput. Sci.* **44** (1986) 17–49.

[20] BLUMER, A., J. BLUMER, A. EHRENFEUCHT, D. HAUSSLER, M.T. CHEN and J. SEIFERAS, The smallest automaton recognizing the subwords of a text, *Theoret. Comput. Sci.* **40** (1985) 31–56.

[21] BOË, J.M., Les boîtes, *Theoret. Comput. Sci.*, to appear.

[22] BOYER, R.S. and J.S. MOORE, A fast string searching algorithm, *Comm. Assoc. Comput. Mach.* **20** (1977) 762–772.

[23] BRZOZOWSKI, J., Derivatives of regular expressions, *J. Assoc. Comput. Mach.* **11** (1964) 481–494.

[24] BRZOZOWSKI, J.A. and R. KNAST, The dot-depth hierarchy of star-free languages is infinite, *J. Comput. System Sci.* **20** (1980) 32–49.

[25] BRZOZOWSKI, J.A. and I. SIMON, Characterization of locally testable events, *Discrete Math.* **4** (1973) 243–271.

[26] BÜCHI, R., Weak second-order arithmetic and finite automata, *Z. Math. Logik Grundlagen Math.* **6** (1960) 66–92.

[27] CARPI, A., On unambiguous reductions of monoids of unambiguous relations, *Theoret. Comput. Sci.* **51** (1987) 215–220.

[28] CARPI, A., On synchronizing unambiguous automata, *Theoret. Comput. Sci.* **60** (1988) 285–296.

[29] CARTIER, P. and D.FOATA, *Problèmes Combinatoires de Commutation et Réarrangements*, Lecture Notes in Mathematics **85** (Springer, Berlin, 1969).

[30] CHOFFRUT, C., Une caractérisation des fonctions séquentielles et des fonctions sous-séquentielles en tant que relations rationnelles, *Theoret. Comput. Sci.* **5** (1977) 325–338.

[31] CHOFFRUT, C. and C. DUBOC, 1987, A star-height problem in free monoids with partial commutations, in: *Automata Languages and Programming*, Lecture Notes in Computer Science **267** (Springer, Berlin, 1987) 190–201.

[32] CHOFFRUT, C. and M.P. SCHÜTZENBERGER, Décomposition de fonctions rationnelles, in: *STACS 86*, Lecture Notes in Computer Science **210** (Springer, Berlin, 1986) 213–226.

[33] CHOFFRUT, C. and M.P. SCHÜTZENBERGER, 1986b, Counting with rational functions, in: *Automata, Languages and Programming*, Lecture Notes in Computer Science **226** (Springer, Berlin, 1986) 79–88.

[34] CHRISTOL, G., T. KAMAE, M. MENDES-FRANCE and G. RAUZY Suites algébriques, automates et substitutions, *Bull. Soc. Math. de France* **108** (1980) 401–419.

[35] COBHAM, A., On the base-dependance of sets of numbers recognizable by finite automata, *Math. Systems Theor.* **3** (1969) 186–192.

[36] COBHAM, A., Uniform tag sequences, *Math. Systems Theor.* **6** (1972) 164–192.

[37] CONSTABLE, R.L., The role of finite automata in the development of modern computing theory, in: J. Barwise et al. eds., *The Kleene Symposium* (North-Holland) 61–83.

[38] CONWAY, J., *Regular Algebra and Finite Machines* (Chapman an Hall, London, 1971).

[39] COOK, S.A., A taxonomy of problems with fast parallel algorithms, *Inform. and Control* **64** (1985) 2–22.

[40] CORI, R. and Y. MÉTIVIER, Approximation of a trace, asynchronous automata and the ordering of events in a distributed system, in: *Automata, Languages and Programming*, Lecture Notes in Computer Science **317** (Springer, Berlin, 1988) 147–161.

[41] CORI, R. and D. PERRIN, Automates et commutations partielles, *RAIRO Inform. Théor.* **9** (1985) 21–32.

[42] CORI, R., E. SOPENA, M. LATTEUX and Y. ROOS, 2-asynchronous automata, *Theoret. Comput. Sci.* **61** (1988) 93–102.

[43] CROCHEMORE, M., Transducers and repetitions, *Theoret. Comput. Sci.* **45** (1986) 63–86.

[44] CROCHEMORE, M., 1987, Longest common factor of two words, in: *TAPSOFT'87*, Lecture Notes in Computer Science **249** (Springer Verlag, Berlin, 1987) 26–36.

[45] DEJEAN, M. and M.P. SCHÜTZENBERGER, On a question of Eggan, *Inform. and Control* **9** (1966) 23–25.

[46] DIEKERT, V., Transitive orientations, Möbius functions and complete semi-Thue systems for free partially commutative monoids, in: *Automata, Languages and Programming*, Lecture Notes in Computer Science **317** (Springer, Berlin, 1988) 176–187.

[47] DUBOC, C., On some equations in free partially commutative monoids, *Theoret. Comput. Sci.* **46** (1986) 159–174.

[48] EGGAN, L.C., (1963) Transition graphs and the star height of regular events, *Michigan Math. J.* **10**, 385–397.

[49] EILENBERG, S., *Automata, Languages and Machines, Vol. A* (Academic Press, New York, 1974).

[50] EILENBERG, S., *Automata, Languages and Machines, Vol. B* (Academic Press, New York, 1976).

[51] EILENBERG, S. and M.P. SCHÜTZENBERGER, Rational sets in commutative monoids, *J. Algebra* **13** (1969) 173–191.

[52] FLÉ, M.P. and G. ROUCAIROL, Maximal serializability of iterated transactions, *Theoret. Comput. Sci.* **38** (1985) 1–16.

[53] FLIESS, M., Sur divers produits de séries formelles, *Bull. Soc. Math. France* **102** (1974) 181–191.

[54] FLOYD, R. and J.D. ULLMAN, The compilation of regular expressions into integrated circuits, *J. Assoc. Comput. Mach.* **29** (1982) 603–622.

[55] FROUGNY, C., Linear numeration systems of order two, *Inform. Comput.* **77**(1988) 233–259.

[56] FURST, M., SAXE, J.B. and M. SIPSER, Parity circuits and the polynomial time hierarchy, in: *Proc. 22nd Ann. IEEE Symp. Foundations of Computer Science* (1981) 260–270.

[57] FURSTENBERG, H., *Recurrence in Ergodic Theory and Combinatorial Number Theory* (Princeton Univ. Press, Princeton, NJ, 1981).

[58] GALIL, Z., String matching in real time, *J. Assoc. Comput. Mach.* **28** (1981) 134–149.

[59] GINSBURG, S. and E. SPANIER, Bounded ALGOL-Like languages, *Trans. Amer. Math. Soc.* **113** (1964) 333–368.

[60] GINZBURG, A., *Algebraic Theory of Automata* (Academic Press, New York, 1969).

[61] GOTTSCHALK, W.H. and G.A. HEDLUND, Topological dynamics, *Amer. Math. Soc. Coll. Publ.* **36** (Amer. Math. Soc., Princeton, NJ, 1955).

[62] GROSS, M. and D. PERRIN, *Electronic Dictionaries and Automata in Computational Linguistics*, Lecture Notes in Computer Science **377** (Springer, Berlin, 1989).

[63] HANSEL, G., A propos d'un théorème de Cobham, in: D. Perrin, ed., *Actes de la Fête des Mots*, Greco de Programmation, CNRS, Rouen (1982).

[64] HARDY, G.H. and E.M. WRIGHT, *An Introduction to the Theory of Numbers* (Oxford University Press, Oxford, 5th ed., 1979).

[65] HASHIGUCHI, K., A decision procedure for the order of regular events, *Theoret. Comput. Sci.* **8** (1979) 69–72.

[66] HASHIGUCHI, K., Regular languages of star height one, *Inform. and Control* **53** (1982) 199–210.

[67] HASHIGUCHI, K., Limitedness theorem on automata with distance functions, *J. Comput. System Sci.* **24** (1982) 233–244.

[68] HASHIGUCHI, K., Representation theorems on regular languages, *J. Comput. System Sci.* **27** (1983) 101–115.

[69] HASHIGUCHI, K., Algorithms for determining relative star height and star height, *Inform. Comput.* **78** (1987) 124–169.

[70] HOPCROFT, J.E. and J.D. ULLMANN., *Introduction to Automata Theory, Languages and Computation* (Addison-Wesley, Reading, MA, 1979).

[71] JOHNSON, J.H., Rational equivalence relations, *Theoret. Comput. Sci.* **47** (1986) 39–60.

[72] KLEENE, S.C., Representation of events in nerve nets and finite automata, in: C. Shannon and J. McCarthy, eds. *Automata Studies* (Princeton Univ. Press, Princeton, NJ, 1956) 3–41.

[73] KNAST, R., A semigroup characterization of dot-depth one languages, *RAIRO Inform. Theor.* **17** (1983) 321–330.

[74] KNUTH, D.E., *The Art of Computer Programming, Vol. 1: Fundamental Algorithms* (Addison-Wesley, Reading, MA, 1968).

[75] KNUTH, D.E., *The Art of Computer Programming, Vol. 2: Seminumerical Algorithms* (Addison-Wesley, Reading, MA, 1969).

[76] KNUTH, D.E., J.H. MORRIS and V.R. PRATT, Fast pattern matching in strings, *SIAM J. Comput.* **6** (1977) 323–350.

[77] KROB, D., Expressions k-rationnelles, Thèse, Université Paris 7, Paris, 1988.

[78] LALLEMENT, G., *Semigroups and Combinatorial Applications* (Wiley, New York, 1979).

[79] LOTHAIRE, M., *Combinatorics on Words*, Encyclopedia of Mathematics (Cambridge Univ. Press, Cambridge, 1983).

[80] LUCCHESI, C.L., I. SIMON, J. SIMON and T. KOWALTOWSKI, *Aspectos Téoricos da Computaçao* (IMPA, Sao Paulo, 1979).

[81] LUNG, H., An algebraic method for solving decision problems in finite automata theory, Ph.D. Thesis, Penn. State Univ., University Park, PA, 1987.

[82] MARCUS, B., Sofic systems and encoding data, *IEEE Trans. Inform. Theory* **31** (1985) 366–377.

[83] MAZURKIEWICZ, A., 1984, Traces histories and graphs: instances of a process monoid, in: *Mathematical Foundations of Computer Science*, Lecture Notes in Computer Science **176** (Springer Berlin, 1984) 115–133.

[84] MCCREIGHT, E.M., A space-economical suffix-tree construction algorithm, *J. Assoc. Comput. Mach* **23** (1976) 262–272.

[85] MCCULLOCH, W.S. and W. PITTS, A logical calculus of ideas immanent in nervous activity, *Bull. Math. Biophys.* **5** (1943) 115–133.

[86] MCNAUGHTON, R., The loop complexity of pure group events, *Inform. and Control* **11** (1967) 167–176.

[87] McNAUGHTON, R., Algebraic decision procedures for local testability, *Math. Systems Theor.* **8** (1974) 60–76.

[88] McNAUGHTON, R. and S. PAPERT, *Counter-free Automata* (MIT Press, Cambridge, MA, 1971).

[89] MEAD, C. and L. CONWAY, *Introduction to VLSI Systems* (Addison-Wesley, Reading, MA, 1980).

[90] MÉTIVIER, Y., On recognizable subsets of free partially commutative monoids, in: *Automata, Languages and Programming*, Lecture Notes in Computer Science **226** (Springer, Berlin, 1986) 254–264.

[91] MEYER, A.R., A note on star-free events, *J. Assoc. Comput. Mach.* **16** (1969) 220–225.

[92] MINSKY, M. and S. PAPERT, Unrecognizable sets of numbers, *J. Assoc. Comput. Mach.* **13** (1966) 281–286.

[93] OCHMANSKI, E., Regular behaviour of concurrent systems, *Bull. EATCS* **27** (1985) 56–67.

[94] PAZ, A., *Introduction to Probabilistic Automata* (Academic Press, New York, 1971).

[95] PERLES, M., M.O. RABIN and E. SHAMIR, The theory of definite automata, *IEEE Trans. Elect. Comput.* **12** (1963) 233–243.

[96] PERRIN, D., 1985, Words over a partially commutative alphabet, in: A. Apostolico and Z. Galil, eds.: *Combinatorial Algorithms on Words*, NATO-ASI Series (Springer, Berlin, 1985) 329–340.

[97] PERRIN, D. and J.-E. PIN, First-order logic and star-free sets, *J. Comput. System Sci.* **32** (1985) 393–406.

[98] PIN, J.-E., *Variétés de Langages Formels* (Masson, Paris, 1984); *Varieties of Formal Languages* (North Oxford/Plenum, London/New York, 1986).

[99] PIN, J.-E., On languages accepted by finite reversible automata, in: *Automata, Languages and Programming, 14th ICALP*, Lecture Notes in Computer Science **267** (Springer, Berlin, 1987) 237–249.

[100] RABIN, M.O. and D. SCOTT, Finite automata and their decision problems, *IBMJ. Res.* **3** (1959) 115–124.

[101] RAUZY, G., Nombres algébriques et subtitutions, *Bull. Soc. Math. France* **110** (1982) 147–178.

[102] RAUZY, G., Sequences defined by iterated morphisms, in: *Workshop on Sequences*, Lecture Notes in Computer Science (Springer, Berlin) to appear.

[103] REUTENAUER, C., Démonstration du théorème de Cobham sur les ensembles de nombres reconnaissables par automate fini, d'après Hansel, in: *Séminaire d'Informatique Théorique*, Année 1983–84, LITP, Université Paris 7, Paris, 1983.

[104] REUTENAUER, C., 1988, *Aspects Mathématiques des Réseaux de Petri* (Masson, Paris, 1988).

[105] RITCHIE, R.W., Finite automata and the set of squares, *J. Assoc. Comput. Mach.* **10** (1963) 528–531.

[106] SAKAROVITCH, J., On regular trace languages, *Theoret. Comput. Sci.* **52** (1987) 59–75.

[107] SALOMAA, A. and M. SOITTOLA, *Automata Theoretic Aspects of Formal Power Series* (Springer, Berlin, 1978).

[108] SCHÜTZENBERGER, M.P., Certain elementary families of automata, in: *Proc. Symp. Mathematical Theory of Automata*, Polytechnic Institute of Brooklin (1962) 139–153.

[109] SCHÜTZENBERGER, M.P., On finite monoids having only trivial subgroups, *Inform. and Control* **8** (1965) 190–194.

[110] SCHÜTZENBERGER, M.P., Sur certaines variétés de monoids finis, in: E.R. Caianiello, ed., *Automata Theory* (Academic Press, New York, 1966).

[111] SCHÜTZENBERGER, M.P., On synchronizing prefix codes, *Inform. and Control* **11** (1967) 396–401.

[112] SCHÜTZENBERGER, M.P., A remark on acceptable sets of numbers, *J. Assoc. Comput. Mach.* **15** (1968) 300–303.

[113] SEDGEWICK, R., *Algorithms* (Addison-Wesley, Reading, MA, 1983).

[114] SHEPHERDSON, J.C., The reduction of two-way automata to one-way automata, *IBMJ. Res. Develop.* **3** (1959) 198–200; reprinted in: E.F. Moore, ed., *Sequential Machines* (Addison Wesley, Reading, MA, 1965) 92–97.

[115] SIMON, I., Piecewise testable events,in: *Proc. 2nd GI Conf.*, Lecture Notes in Computer Science **33** (Springer, Berlin, 1975) 214–222.

[116] SIMON, I., Limited subsets of a free monoid, in: *Proc. 19th Ann. IEEE Symp. Foundations of Computer Science* (1978) 143–150.

[117] SIMON, I., The nondeterministic complexity of a finite automaton, Preprint, 1987.

[118] SIMON, I., Factorization forests of finite height, *Theoret. Comput. Sci.* **6** (1987) 151–167.

[119] SIMON, I., Recognizable sets with multiplicities in the tropical semiring, in: *Mathematical Foundations*

of Computer Science, Lecture Notes in Computer Science **324** (Springer, Berlin, 1988) 107–120.

[120] SLISENKO, A.O., Detection of periodicities and string-matching in real time, *J. Soviet. Math.* **22** (1983) 1316–1387.

[121] SPEHNER, J.C., La reconnaissance des facteurs d'un langage fini dans un texte en temps linéaire, in: *Automata, Languages and Programming*, Lecture Notes in Computer Science **317** (Springer, Berlin, 1988) 547–560.

[122] STERN, J., Characterizations of some classes of regular events, *Theoret. Comput. Sci.* **35** (1985) 17–42.

[123] STRAUBING, H., D. THÉRIEN and W. THOMAS, Regular languages defined with generalized quantifiers, in: *Automata, Languages and Programming*, Lecture Notes in Computer Science **317** (Springer, Berlin, 1988) 561–575.

[124] THOMAS, W., Automata on infinite objects, in: J. van Leeuwen, ed., *Handbook of Theoretical Computer Science, Vol. B* (North-Holland, Amsterdam, 1990) 133–191.

[125] THOMSON, K., Regular expression search algorithm, *Comm. Assoc. Comput. Mach.* **11** (1968) 419–422.

[126] WEIL, P., Inverse monoids and the dot-depth hierarchy, Ph.D. Thesis, University of Nebraska, Lincoln, 1988.

[127] WEINER, P., Linear pattern-matching algorithms, in: *Proc. 14th Ann. IEEE Symp. on Switching and Automata Theory* (1973) 1–11.

[128] WEIS, B., Subshifts of finite type and sofic systems, *Monatsh. Math.* **77** (1973) 462–474.

[129] ZALCSTEIN, Y., Locally testable languages, *J. Comput. System Sci.* **6** (1972) 151–167.

[130] ZIELONKA, W., Notes on finite asynchronous automata, *RAIRO Inform. Theor.* **21** (1987) 99–135.

CHAPTER 2

Context-Free Languages

J. BERSTEL

LITP, Université de Paris VI, 4 Place Jussieu, F-75252 Paris Cedex, France

L. BOASSON

Université de Paris VII, 2 Place Jussieu, F-75221 Paris Cedex, France

Contents

Introduction 61
1. Languages 62
2. Iteration 72
3. Looking for nongenerators 82
4. Context-free groups 95
Acknowledgment 100
References 100

HANDBOOK OF THEORETICAL COMPUTER SCIENCE
Edited by J. van Leeuwen

Introduction

This chapter is devoted to the presentation of some recent results about context-free languages. There have been many results that appeared in the last years, both on classical topics and in new directions of research. The choice of the material to be presented was guided by the idea to emphasize on results which are not presented in any of the available textbooks [2, 12, 32, 33, 38, 55].

Context-free languages and grammars were designed initially to formalize grammatical properties of natural languages. They subsequently appeared to be most suitable for the formal description of the syntax of programming languages. This led to a considerable development of the theory. The recent research is oriented toward a more algebraic treatment of the main topics, in connection with mathematical theories; it also pursues investigations about famous open problems, such as the equivalence of deterministic pushdown automata, or the existence of principal cones with a principal cone of nongenerators.

Most of the theorems given in this chapter have been proved in the second half of the 1980s. It appears (as systematically indicated in the text) that some of them constitute answers to questions listed in [3]. It should be observed that nearly all the conjectures of [3] are now solved (even if some of the solutions are not given here). As usual, these answers raise new questions, some of which are mentioned below.

Even when restricted to recent results, we had to make a choice for the material to be presented. In the first section, we first illustrate the algebraic development of the theory by showing the existence of an invariant for context-free languages, namely the Hotz group. Then we give an account of recent refinements to the proof of inherent ambiguity by a clever investigation of generating functions.

Section 2 is devoted to iteration. We first prove the iteration lemma of Bader and Moura; we then discuss the interchange lemma and some of its applications, mainly to square-free words. Finally we prove that a context-free language which has only degenerated iterative pairs is in fact regular.

Section 3 describes the state of our knowledge concerning generators in cones of context-free languages. The main conjecture, namely that the cone of nongenerators of the context-free languages is not principal, is still open. New facts are: this cone is not the substitution closure of any strict subcone, and it is not generated by any family of deterministic context-free languages. New results concerning the "geography" of the context-free cones are also reported.

In the final section, we give an account of the theory of context-free groups, that is, the groups for which the word problem is context-free. We give a global characterization, a description in terms of Caley graphs, and a relation to the theory of "ends".

There are major topics not presented here. Among them, the theory of rewriting systems has been treated in a monograph by Jantzen [41]. Connections to infinite words are only scarcely sketched in the text. For an overview, see [13] and [61]. The decidability of equivalence of deterministic context-free languages has made considerable progress in the late 1980s. However, it seems not yet ripe for a systematic treatment, and the interested reader is referred to [51, 59, 62].

1. Languages

1.1. Notation and examples

1.1.1. Grammars

A *context-free grammar* $G=(V, A, P, S)$ is composed of a finite alphabet V, a subset A of V called the *terminal alphabet*, a finite set $P\subset(V-A)\times V^*$ of *productions*, and a distinguished element $S\in V-A$ called the *axiom*. A letter in $V-A$ is a *nonterminal* or *variable*.

Given words $u, v\in V^*$, we write $u\to v$ (sometimes subscripted by G or by P) whenever there exist factorizations $u=xXy$, $v=x\alpha y$, with (X,α) a production. A *derivation* of length $k\geqslant 0$ from u to v is a sequence $(u_0, u_1, \ldots, u_k)$ of words in V^* such that $u_{i-1}\to u_i$ for $i=1,\ldots,k$, and $u=u_0, v=u_k$. If this hold, we write $u\overset{k}{\to}v$. The existence of some derivation from u to v is denoted by $u\overset{*}{\to}v$. If there is a proper derivation (i.e. of length $\geqslant 1$), we use the notation $u\overset{+}{\to}v$. The *language generated by* G is the set

$$L(G)=\{w\in A^*\,|\,S\overset{*}{\to}w\}.$$

If X is a variable in G, we write

$$L_G(X)=\{w\in A^*\,|\,X\overset{*}{\to}w\}.$$

Thus $L(G)=L_G(S)$. A language L is called *context-free* if it is the language generated by some context-free grammar.

Consider a derivation $u=u_0\to u_1\to\cdots\to u_k=u$. It is a *leftmost* derivation if, for any derivation step $u_i\to u_{i+1}$, the variable in u_i that is replaced is the leftmost occurrence of a variable in u_i. Rightmost derivations are defined symmetrically. A *derivation tree* in G is a rooted, planted labelled tree. The internal nodes of the tree are labelled with variables. Leaves are labelled with elements in $A\cup\{1\}$, subject to the following condition. Let s be a node, and let $s_1, s_2, \ldots, s_n$ be the children of s, ordered from left to right. If X is the label of s, and if Y_i is the label of s_i, then $(X, Y_1\ldots Y_n)\in P$.

Moreover, if $n\geqslant 2$, then none of the Y_i's is the empty word. It is well-known that there is a bijection between derivation trees with root X and leaves in $A\cup\{1\}$, and leftmost derivations (rightmost derivations) from X into words over A.

Two grammars are *equivalent* if they generate the same language.

1.1.2. Examples

There are several convenient shorthands to describe context-free grammars. Usually, a production (X,α) is written $X\to\alpha$, and productions with same left-hand side are grouped together, the corresponding right-hand sides being separated by a "+". Usually, the variables and terminal letters are clear from the context. The language generated by a context-free grammar is denoted by the list of productions enclosed in a pair of brackets, the axiom being the first left-hand side of a production.

(i) The *Dyck languages*. Let $A=\{a_1,\ldots,a_n\}$, $\bar{A}=\{\bar{a}_1,\ldots,\bar{a}_n\}$ be two disjoint alphabets. The *Dyck language* over $A\cup\bar{A}$ is the language

$$D_n^*=\langle S\to TS+1;\ T\to a_1S\bar{a}_1+\cdots+a_nS\bar{a}_n\rangle.$$

The notation is justified by the fact that D_n^* is indeed a submonoid of $(A\cup\bar{A})^*$. It is even a free submonoid, generated by the language of *Dyck primes*

$$D_n=\langle T\to a_1S\bar{a}_1+\cdots+a_nS\bar{a}_n;\ S\to TS+1\rangle.$$

If $n=1$, we write D^* and D instead of D_1^* and D_1.

The *two-sided Dyck language* over $A\cup\bar{A}$ is the language

$$\hat{D}_n^*=\left\langle S\to TS+1;\ T\to\sum_{i=1}^{n}a_iS\bar{a}_i+\sum_{i=1}^{n}\bar{a}_iSa_i\right\rangle.$$

Again $\hat{D}_n^*$ is a free submonoid of $(A\cup\bar{A})^*$ generated by the set $\hat{D}_n$ of *two-sided Dyck primes*. This set is also context-free, and is generated by T in the following grammar

$$T\to\ \sum_{i=1}^{n}T_i+\sum_{i=1}^{n}\bar{T}_i,$$

$$T_i\to\ a_iS_i\bar{a}_i;\qquad \bar{T}_i\to\bar{a}_i\bar{S}_ia_i\quad(i=1,\ldots,n),$$

$$S_i\to\ 1+\sum_{j=1}^{n}T_jS_i+\sum_{j\neq i}\bar{T}_jS_j\quad(i=1,\ldots,n),$$

$$\bar{S}_i\to\ 1+\sum_{j\neq i}T_j\bar{S}_i+\sum_{j=1}^{n}\bar{T}_j\bar{S}_i\quad(i=1,\ldots,n).$$

Again, we write $\hat{D}^*$ and $\hat{D}$ instead of $\hat{D}_1^*$ and $\hat{D}_1$.

There is an alternative way to define these languages as follows. Consider the congruence δ (resp. $\hat{\delta}$) over $A\cup\bar{A}$ defined by

$$a_i\bar{a}_i\equiv 1 \bmod \delta\quad(i=1,\ldots,n),\qquad a_i\bar{a}_i\equiv\bar{a}_ia_i\equiv 1 \bmod \hat{\delta}\quad(i=1,\ldots,n).$$

Then $D_n^*=\{w\in(A\cup\bar{A})^*\mid w\equiv 1 \bmod \delta\}$ and $\hat{D}_n^*=\{w\in(A\cup\bar{A})^*\mid w\equiv 1 \bmod \hat{\delta}\}$. Moreover, the quotient $(A\cup\bar{A})^*/\hat{\delta}$ is a group, called *the free group generated by A* and denoted by $F(A)$.

(ii) The *Lukasiewicz language* over a set $A=A_0\cup A_1\cup\cdots\cup A_n$ partitioned into subsets A_i of symbols of "arity" i is the language

$$\langle S\to A_0+A_1S+\cdots+A_nS^n\rangle.$$

The most well-known case is when $A_0=\{b\}$, $A_2=\{a\}$, and the other sets are empty. This gives the language

$$Ł=\langle S\to b+aSS\rangle.$$

(iii) The languages of *completely parenthesized expressions*

$$E_n=\left\langle S\to\sum_{k=1}^{n}a_kSbSc_k+d\right\rangle.$$

For $n=1$, we write E instead of E_1: $E=\langle S\to aSbSc+d\rangle$.

(iv) The set of *palindromes* over an alphabet A

$$Pal = \left\langle S \to \sum_{a \in A} aSa + \sum_{a \in A} a + 1 \right\rangle$$

is the set of words $w \in A^*$ with $w = w^\sim$ where $w^\sim$ denotes the *reversal* of w. Related to this set are the *symmetric languages* Sym_n defined over the alphabet $\{a_1, \ldots, a_n, \bar{a}_1, \ldots, \bar{a}_n\}$ by

$$Sym_n = \left\langle S \to \sum_{i=1}^{n} a_i S \bar{a}_i + 1 \right\rangle.$$

Contrary to previous conventions, Sym will denote the language Sym_2.

It is interesting to observe that the languages

$$\{w \# w^\sim \mid w \in A^*\} \quad \text{and} \quad \{w \# w' \mid w' \neq w^\sim\}$$

(with $\#$ not in A) are both context-free. On the contrary, the language $Copy = \{w \# w \mid w \in A^*\}$ is not context-free (as can be shown by one of the pumping lemmas given below); however, the language $\{w \# w' \mid w' \neq w\}$ *is* context-free.

(v) The *Goldstine language* G over $\{a, b\}$ is the set G of words

$$a^{n_1} b a^{n_2} b \ldots a^{n_p} b$$

with $p \geqslant 1$, $n_i \geqslant 0$, and $n_j \neq j$ for some j, $1 \leqslant j \leqslant p$. To see that this language is context-free we start with the context-free language

$$\{a^p b^q c \mid q \neq p+1,\ q, p \geqslant 0\}$$

and then apply the substitution

$$a \mapsto a^* b, \qquad b \mapsto a, \qquad c \mapsto b(a^* b)^*.$$

The language G is the result of this substitution. Since rational (and even context-free) substitution preserves context-freeness, the language G is context-free.

Observe that G is related to the infinite word

$$\boldsymbol{x} = aba^2ba^3b \ldots a^n ba^{n+1} b \ldots$$

Let indeed

$$\text{Co-Pref}(\boldsymbol{x}) = \{w \mid w \text{ is not a prefix of } \boldsymbol{x}\}.$$

Then G is just composed of those words in Co-Pref($\boldsymbol{x}$) which end with the letter b. Further, consider the context-free language

$$\{a^n b^p \mid p > n+1,\ n \geqslant 0\}$$

and then apply the substitution

$$a \mapsto a^* b, \qquad b \mapsto a.$$

Let H be the resulting language. Then

$$\text{Co-Pref}(\boldsymbol{x}) = Ga^* \cup H.$$

This shows that Co-Pref($\boldsymbol{x}$) is a context-free language.

1.1.3. Ambiguity

A grammar $G=(V, A, P, S)$ is *unambiguous* if every word in $L(G)$ has exactly one leftmost derivation. It is equivalent to saying that there is only one derivation tree for each word, whence only one right derivation. A language is *unambiguous* if there is an unambiguous grammar to generate it, otherwise it is called *inherently ambiguous*.

Ambiguity is undecidable. However there are techniques that work in special cases to prove inherent ambiguity. One method is by using iteration lemmas in a clever way (see e.g. [38]). This can be used, for instance, to prove that the language $\{a^n b^p c^q \mid n=p$ or $p=q\}$ is inherently ambiguous. The same method applies sometimes to unambiguity relatively to a subclass. Let us just give one example. A grammar is *linear* if, for every production, the right-hand side contains at most one occurrence of a variable. A language is linear if there exists a linear grammar that generates it.

Consider the following language over $\{a, b, \#\}$:

$$M=\{a^n b^n \# a^p b^q \mid n, p, q \geqslant 1\} \cup \{a^p b^q \# a^n b^n \mid n, p, q \geqslant 1\}.$$

This language is linear. However, it can be shown that every linear grammar generating M is ambiguous. On the other hand,

$$M=\{a^n b^n \# a^p b^q \mid n, q, p \geqslant 1\} \cup \{a^n b^m \# a^p b^p \mid n, m, p \geqslant 1,\ n \neq m\}$$

is the disjoint union of two (nonlinear) languages, which both are unambiguous, thus M is unambiguous.

We shall see another way to attack ambiguity below.

1.1.4. Reduced grammars, normal forms

There exist a great number of normal forms for grammars. These normal forms have mainly theoretical interest and are of little help in practical applications such as parsing. Reduction is a first step toward these normal forms.

A grammar $G=(V, A, P, S)$ is *reduced* if the following three conditions are fulfilled:

(i) for every nonterminal X, the language $L_G(X)$ is nonempty;

(ii) for every $X \in V-A$, there exist $u, v \in A^*$ such that $S \xrightarrow{*} uXv$;

(iii) the axiom S appears in no right-hand side of a production.

It is not difficult to see that for every grammar G with $L(G) \neq \emptyset$, an equivalent reduced grammar can effectively be constructed. A variation of this construction which is sometimes useful requires that $L_G(X)$ is infinite for every variable X. A production is called an *ε-production* if its right-hand side is the empty word. At least one ε-production is necessary if the language generated by the grammar contains the empty word. It is not too difficult to construct, for every context-free grammar G, an equivalent grammar with no ε-production except a production $S \to 1$ if $1 \in L(G)$. The final special kind of grammars we want to mention is the class of proper grammars. A grammar G is *proper* if it has neither ε-productions nor any production of the form $X \to Y$, with Y a variable. Again, an equivalent proper grammar can effectively be constructed for any grammar G if $L(G) \not\ni 1$. The two most common normal forms for context-free grammars are the so called Chomsky normal form and the Greibach normal form. A grammar $G=(V, A, P, S)$ is in *Chomsky normal form* if every production $X \to \alpha$ satisfies $\alpha \in A \cup (V-A)^2$. It is in *Greibach normal form* if $\alpha \in A \cup A(V-A) \cup (V-A)^2$. For every context-free grammar G with $1 \notin L(G)$ equivalent grammars in Chomsky normal form and in

Greibach normal form can effectively be constructed. A less usual normal form is the *double Greibach normal form* where every production $X \to \alpha$ satisfies

$$\alpha \in A \cup A^2 \cup A[(V-A) \cup (V-A)^2]A.$$

There again, for every context-free grammar G with $1 \notin L(G)$, an equivalent grammar in double Greibach normal form can be effectively constructed [39]. A very large variety of such normal forms exists [14].

1.1.5. Systems of equations

Let $G=(V, A, P, S)$ be a context-free grammar. For each variable X, let P_X be the set of right-hand sides of productions having X as left-hand side. With our notation, the set of productions can be written as

$$X \to \sum_{p \in P_X} p \quad (X \in V-A)$$

or simply as

$$X \to P_X \quad (X \in V-A).$$

The *system of equations* associated with the grammar G is the set of equations

$$X = P_X \quad (X \in V-A).$$

A solution of this system is a family $L=(L_X)_{X \in V-A}$ of subsets of A^* such that

$$L_X = P_X(L) \quad (X \in V-A)$$

with the notation

$$P_X(L) = \bigcup_{p \in P_X} p(L)$$

and $p(L)$ being the product of languages obtained by replacing, in p, each occurrence of a variable Y by the language L_Y. Solutions of a system of equations are ordered by component-wise set inclusion. Then one has the following theorem.

1.1. THEOREM (Schützenberger [57]). *Let $G=(V, A, P, S)$ be a context-free grammar. The family $L=(L_X)$ with $L_X = L_G(X)$ is the least solution of the associated set of equations. If the grammar G is proper, then the associated system has a unique solution.*

1.2. EXAMPLE. The grammar $S \to aSS + b$ is proper. Thus the Lukasiewicz language $\L$ is the unique language satisfying the equation $\L = a\L\L \cup b$.

1.3. EXAMPLE. The grammar $S \to S$ generates the empty set which is the least solution of the equation $X = X$. Every language is indeed solution of this equation.

For more details along these lines see [43, 56].

1.1.6. Pushdown automata

A *pushdown automaton* (pda) $\mathscr{A}=(A, V, Q, \delta, v^0, q^0, Q')$ is composed of a finite terminal alphabet A, a finite nonterminal alphabet V, a (nondeterministic) transition function δ from $(A\cup\{\varepsilon\})\times Q\times V$ into the finite subsets of $Q\times V^*$, an initial pushdownstore symbol v^0 in V, an initial state q^0 in Q and a set Q' of terminal states, a subset of Q. A *configuration* of $\mathscr{A}$ is a triple $c=(\gamma, q, x)$ in $V^*\times Q\times A^*$. The automaton moves directly from configuration $c=(\gamma, q, x)$ into configuration $c'=(\gamma', q', x')$, denoted $c\vdash c'$ iff

- *either* $\gamma=\gamma_1 v$ $(v\in V)$, $x=ax'$ $(a\in A)$, $\gamma'=\gamma_1 m$ $(m\in V^*)$ and $\delta(a, q, v)\ni(q', m)$; this is an "$a$-move";
- *or* $\gamma=\gamma_1 v (v\in V)$, $x=x'$, $\gamma'=\gamma_1 m$ $(m\in V^*)$ and $\delta(\varepsilon, q, v)\ni(q', m)$; this is an "$\varepsilon$-move".

We denote by $\overset{*}{\vdash}$ the reflexive and transitive closure of the relation $\vdash$, and we define the *language recognized by empty store* by $\mathscr{A}$ as

$$Null(\mathscr{A})=\{x\in A^*\mid (v^0, q^0, x)\overset{*}{\vdash}(1, q, 1),\ q\in Q\}$$

and the *language recognized by terminal state by* $\mathscr{A}$ as

$$T(\mathscr{A})=\{x\in A^*\mid (v^0, q^0, x)\overset{*}{\vdash}(\gamma, q', 1),\ \gamma\in V^*, q'\in Q'\}.$$

The context-free languages are then characterized in terms of pda's: $L\subset A^*$ is context-free iff there exists a pda $\mathscr{A}$ such that $L=T(\mathscr{A})$ (resp. $L=Null(\mathscr{A})$). Moreover, this result holds even if $\mathscr{A}$ is restricted to be *real-time* (i.e., involves no ε-moves in its transition function).

A pda is *deterministic* (is a dpda) iff, for each q in Q and each v in V,

- *either* $\delta(\varepsilon, q, v)$ is a singleton in $Q\times V^*$ and, for each a in A, $\delta(a, q, v)=\emptyset$;
- *or* $\delta(\varepsilon, q, v)=\emptyset$ and, for each a in A, the set $\delta(a, q, v)$ is either empty or a singleton.

A context-free language L is *deterministic* iff there exists a dpda $\mathscr{A}$ such $L=T(\mathscr{A})$. It should be noted that, contrarily to what happens for nondeterministic pda's, the family of languages recognized by empty store by a dpda $\mathscr{A}$ forms a strict subfamily of the deterministic languages. Similarly, the family of real-time dpda's gives rise to a strict subfamily of deterministic languages. (See [38] for all these classical results.)

1.2. Hotz group

One of the most interesting questions concerning the relation between grammars and languages is whether two grammars are equivalent, i.e., generate the same language. Since this question is undecidable in general, one may look for weaker formulations of this question, i.e., properties which are implied by the equivalence of context-free grammars. One such invariant has been discovered by Hotz [40] and is the topic of this section.

Consider any set A. The *free group* $F(A)$ over A is the quotient monoid

$$F(A)=(A\cup\bar{A})^*/\hat{\delta}$$

where $\bar{A}=\{\bar{a}\mid a\in A\}$ is a disjoint copy of A and where $\hat{\delta}$ is the congruence generated by the relations $a\bar{a}\equiv\bar{a}a\equiv 1$ $(a\in A)$(see also Example (i) in Section 1.1.2).

Let $G=(V, A, P, S)$ be a context-free grammar. The *Hotz group* of G is the group

$$\mathscr{H}(G)=F(V)/[P]$$

where $[P]$ is the group congruence generated by P, that is $u\equiv v \bmod[P]$ iff u and v can be obtained from each other by successive application of productions or their inverses in *both* directions: $u\equiv v \bmod[P]$ iff there exist $k\geqslant 0$ and $w_0,\ldots,w_k$ such that $u=w_0$, $v=w_k$, and for $i=0,\ldots,k-1$.

$$w_i\rightarrow w_{i+1} \quad \text{or} \quad w_{i+1}\rightarrow w_i \quad \text{or} \quad \bar{w}_i\rightarrow\bar{w}_{i+1} \quad \text{or} \quad \bar{w}_{i+1}\rightarrow\bar{w}_i.$$

1.4. EXAMPLE. For the grammar $G=\langle S\rightarrow 1+aSb\rangle$, the congruence is generated by $S\equiv 1$, $S\equiv aSb$; clearly, $a^nb^p\equiv a^{n-p}$ for n, $p\in\mathbf{Z}$. Thus $\mathscr{H}(Q)=\mathbf{Z}$.

1.5. THEOREM. *Let G_1 and G_2 be two reduced context-free grammars. If $L(G_1)=L(G_2)$, then $\mathscr{H}(G_1)\cong\mathscr{H}(G_2)$.*

This result means that the Hotz group, defined for a grammar, is in fact a property of the language generated by the grammar. It is well-known that other algebraic objects are associated with a formal language. The most frequently quoted is the syntactic monoid (for a discussion, see [52]).

Theorem 1.5 is an immediate consequence of the following intrinsic characterization of the Hotz group. Let $L\subset A^*$ be a language. The *collapsing group* of L is the quotient

$$\mathscr{C}(L)=F(A)/[L\times L]$$

of the free group over A by the finest (group) congruence such that L is contained in a single class.

1.6. THEOREM. *Let L be a context-free language, and let G be a reduced context-free grammar generating L; then the collapsing group of L and the Hotz group of G are isomorphic, i.e. $\mathscr{C}(L)\cong\mathscr{H}(G)$.*

PROOF. Let $G=(V, A, P, S)$. For convenience, we denote the congruence mod $[P]$ by $\sim$, and the congruence mod $[L\times L]$ by $\equiv$. Observe that $\sim$ is defined over $F(V)$, and $\equiv$ is only defined over $F(A)$.

We first show that if $u, v\in F(A)$, then

$$u\equiv v \quad\Leftrightarrow\quad u\sim v.$$

For this, consider words $w, w'\in L$. Then $S\xrightarrow{*}w'$. Consequently, $w\sim S$, and $S\sim w'$, whence $w\sim w'$. By induction, it follows that, for $u, v\in F(A)$,

$$u\equiv v \quad\Rightarrow\quad u\sim v.$$

Conversely, consider words $w, w'\in L_G(X)$ for some variable X. Since the grammar is reduced, $uwv, uw'v\in L$ for some words $u, v\in A^*$. Thus $uwv\equiv uw'v$, and since we have a group congruence, it follows that $w\equiv w'$. Thus each $L_G(X)$ is contained in a single class of the collapsing congruence.

Consider now, for each variable X, a fixed word $\varphi(X) \in L_G(X)$, and extend φ to a morphism from V^* into A^*, and then from $F(V)$ into $F(A)$, by setting $\varphi(a) = a$ for $a \in A$. In view of the previous discussion, given any production $X \to \alpha$ in P, one has $\varphi(X) \equiv \varphi(\alpha)$. Assume now that $u, v \in F(V)$, and $u \sim v$. Then there are $k \geqslant 0$, and $w_0, \ldots, w_k \in F(V)$, such that $u = w_0$, $v = v_k$, and $\varphi(w_i) \equiv \varphi(w_{i+1})$ for $i = 0, \ldots, k-1$. Consequently $\varphi(u) \equiv \varphi(v)$, and since $\varphi(u) = u$, $\varphi(v) = v$, this shows that $u \equiv v$. Thus, our claim is proved.

Let p be the canonical morphism

$$p: F(V) \to F(V)/\sim \quad (\cong \mathscr{H}(G)).$$

In order to complete the proof, it suffices to observe that $\mathscr{H}(G) = p(F(A))$ since, for each $w \in V^*$, one has $w \sim \varphi(w)$, and $\varphi(w) \in F(A)$. □

The concept of collapsing group appears in [64]. A more systematic formulation of Theorem 1.6 is given by Frougny et al. in [30]. For recent developments on these lines, see [23, 24].

The relation between the congruence $[P]$ and the so-called NTS-languages is the following: rather than considering the group $F(V)/[P]$, we may consider the quotient monoid $V^*/[P]$ also called the *Hotz monoid*. Obviously, this is no more an invariant. However, it may happen, for some grammar $G = (V, A, P, S)$ that the congruence class of each variable X is exactly the set of sentential forms generated by X:

$$\{x \in V^* \mid x \equiv X \bmod [P]\} = \{x \in V^* \mid X \xrightarrow{*} x\}.$$

In that case, the grammar G is called an *NTS-grammar*. Some striking results concerning these grammars are [58]:

(i) languages generated by NTS-grammars are deterministic context-free languages;

(ii) the equivalence of NTS-grammars is decidable;

(iii) given a context-free grammar G, it is decidable whether G is an NTS-grammar.

There remains an interesting open problem: is the family of NTS-languages (i.e., languages generated by NTS-grammars) closed under inverse morphism? This question seems to be related to another one concerning a weakening of the NTS-condition. A grammar $G = (V, A, P, S)$ is called *pre-NTS* if, for each variable X, the restriction of its congruence class to terminal words is exactly the language generated by X, thus if $L_G(X) = \{x \in A^* \mid X \equiv x \bmod [P]\}$. Clearly, any NTS-grammar is pre-NTS. The question is whether the converse holds for languages.

1.3. *Ambiguity and transcendence*

As already mentioned above, a proof that a given context-free language is inherently ambiguous by means of combinational arguments is rather delicate. The reason for this is that one has to show that *every* grammar is ambiguous. Another reason is that pumping lemmas like those described in the next section only concern local structure.

A fundamental technique for proving ambiguity is based on the use on generating functions. This technique has recently been refined and successfully employed by

Flajolet [28, 29]. Let $L \subset A^*$ be any language over a finite alphabet A. The *generating function* of L is given by the series

$$f_L(z) = \sum a_n z^n$$

where

$$a_n = \mathrm{card}\{w \in L \mid |w| = n\}.$$

Since $a_n \leqslant \mathrm{card}(A)^n$ for $n \geqslant 0$, the series $f_L(z)$ is an analytic function in the neighbourhood of the origin, and its radius of convergence p satisfies $p \geqslant 1/\mathrm{card}(A)$. The basic result for the study of ambiguity is the following classical theorem of Chomsky and Schützenberger.

1.7. THEOREM. *Let $f_L(z)$ be the generating function of a context-free language L. If L is unambiguous, then $f_L(z)$ is an algebraic function.*

This result means of course that if $f_L(z)$ is transcendental and L is context-free, then L is inherently ambiguous. In order to use this statement, it suffices to rely on well-known classical families of transcendental functions or to use more or less easy criteria for transcendence. We just give some examples.

1.8. PROPOSITION (Flajolet [28]). *The Goldstine language G is inherently ambiguous.*

PROOF. Consider indeed a word which is not in G. Then either it ends with the letter a, or it is one of the words in the set

$$F = \{1, ab, aba^2b, aba^2ba^3b, \ldots\}.$$

The generating function of the words over $A = \{a, b\}$ ending with the letter a is $z/(1-2z)$. The generating function of the set F is

$$1 + z^2 + z^5 + z^9 + \cdots = \sum_{n \geqslant 1} z^{n(n+1)/2-1}.$$

Thus

$$f_G(z) = \frac{1-z}{1-2z} - \frac{1}{z} \sum_{n \geqslant 1} z^{n(n+1)/2} = \frac{1-z}{1-2z} - \frac{g(z)}{z}.$$

Now $f_G(z)$ is transcendental iff $g(z)$ is so. And indeed, $g(z)$ is transcendental. To see this, recall that an algebraic function has a finite number of singularities. On the other hand, a powerful "gap theorem" [54] states that if for some series $h(z) = \Sigma_{n \geqslant 0} a_n z^{c_n}$ the exponents satisfy $\sup(c_{n+1} - c_n) = \infty$, then $h(z)$ admits its circle of convergence as a natural boundary. This holds for our function $g(z)$ which thus has infinitely many singularities and is not algebraic. □

Two points should be observed. First, the fact that the generating function is algebraic holds for a language as well as for its complement. Thus, there is no way to get any converse of the Chomsky–Schützenberger Theorem. Next, the technique of natural boundaries is quite general for proving transcendence of functions.

1.9. PROPOSITION. *Let $A=\{a, b\}$, and let C be the language of products of two palindromes:*

$$C=\{w_1 w_2 \mid w_1 w_2 \in A^*,\ w_1=\tilde{w_1},\ w_2=\tilde{w_2}\}.$$

The language C is inherently ambiguous.

A first proof of this proposition is due to Crestin [21] who has proved that its ambiguity is unbounded, i.e., that there exists no bound for the number of leftmost derivations for some word in any grammar. Kemp [42] computed the generating function of C which is

$$f_C(z)=1+2\sum_{m\geqslant 1}\mu(m)\frac{z^m(1+z^m)(1+2z^m)}{(1-2z^{2m})^2}$$

where $\mu(m)=\Pi_{p|m}(1-p)$, the product being over all prime divisors of m. A delicate analysis by Flajolet [29] shows that $f_C(z)$ has singularities at the points $2^{-1/2m}e^{ij/m\pi}$. Thus there are infinitely many singularities and $f_C(z)$ is not algebraic. □

1.10. PROPOSITION. *The language*

$$L=\{w\in A^* \mid |w|_a\neq|w|_b \text{ or } |w|_b\neq|w|_c\}$$

over $A=\{a, b, c\}$ is inherently ambiguous.

PROOF. The complement M of this language is

$$M=\{w\in A^* \mid |w|_a=|w|_b=|w|_c\}$$

and its generating function is

$$f_M(z)=\sum_{n\geqslant 0}\frac{(3n)!}{(n!)^3}z^{3n}=\sum a_n z^{3n}.$$

In order to show that $f_M(z)$ is not algebraic, we observe that by Stirling's formula

$$a_n\sim 3^{3n}\frac{\sqrt{3}}{2\pi n}$$

On the other hand, Flajolet [29] shows that an asymptotic equivalent of the form β^n/n is characteristic for transcendental functions. Thus $f_M(z)$ and $f_L(z)$ are transcendental. □

Flajolet's paper [29] contains many other examples of inherently ambiguous languages, and develops a systematic classification of languages which can be handled by analytic methods (see also [6]).

2. Iteration

Iteration is the most direct method for proving that a formal language is not in a given family of languages. In general, the iteration is on the words of the language and reflects some property of the way languages in the considered family are constructed. The iteration lemmas in fact give a property of "regularity", which expresses the finiteness of the construction mechanism.

There exist numerous results concerning iteration, depending on the type of languages studied. For a fixed family of languages, there may exist several variations. A bibliography on this topic has been compiled by Nijholt [49].

We are concerned here with iteration lemmas for context-free languages. We give three recent results of different nature. The first (Theorem 2.4) is the conclusion of several statements of increasing precision concerning the constraints for iterative pairs in a context-free language. The second (Theorem 2.7) states a different kind of property: if a language is "rich", i.e., if there are "many" words of given length, then factors of words may be interchanged without leaving the language.

The third result (Theorem 2.10) expresses to what amount an iteration property may characterize languages: this is interesting because iteration is only a "local" property of words, and is therefore difficult to relate to global properties of grammars.

2.1. *Iteration lemmas*

Consider a fixed language $L \subset A^*$. An *iterative pair* in L is a tuple of words $\eta = (x, u, y, v, z)$ with $uv \neq 1$, such that for all $n \geqslant 0$

$$xu^n yv^n z \in L.$$

Given a word $w \in L$, the iterative pair η is a pair *for* w provided $w = xuyvz$; a word w is said to *admit* an iterative pair if there exists an iterative pair for w.

2.1. EXAMPLE. For the language D of Dyck primes over $\{a, \bar{a}\}$, the tuple $(a, a, 1, \bar{a}, \bar{a})$ is an iterative pair. However, $(1, a, 1, \bar{a}, 1)$ is not because $1 \notin D$. The latter is of course an iterative pair in D^*.

An easy way to construct iterative pairs for an infinite algebraic language L generated by some (reduced) grammar $G = (V, A, P, S)$ is the following. Since L is infinite, there exists in G some variable X and some derivation $X \xrightarrow{*} uXv$ for some words $u, v \in A^*$. Since G is reduced,

$$S \xrightarrow{*} xXz, \qquad X \xrightarrow{*} y$$

for some words x, y, z in A^*. But then

$$S \xrightarrow{*} xu^n yv^n z \quad \text{for all } n \geqslant 0.$$

Historically the first iteration result concerning context-free languages is the following theorem.

2.2. THEOREM (Bar-Hillel, Perles and Shamir [38]). *Let $L \subset A^*$ be a context-free language. There exists an integer N such that any word $w \in L$ of length $|w| \geqslant N$ admits an iterative pair in L.*

This theorem is a special case of Ogden's refinement, the result stated below. In order to formulate it, we need a definition.

Let $w \in A^*$ be a word of length n. A *position* in w is any integer in $\{1, 2, \ldots, n\}$. Choosing a set of distinguished position in w is thus equivalent to choosing a subset of $\{1, \ldots, n\}$.

2.3. THEOREM (Ogden [38]). *Let $L \subset A^*$ be a context-free language. There exists an integer N such that, for any word $w \in L$ and for any choice of at least N distinguished positions in w, there exists an iterative pair $\eta = (x, u, y, v, z)$ for w in L such that, additionally,*

(i) *either x, u, y each contain at least one distinguished position, or y, v, z each contain at least one distinguished position;*

(ii) *the word uyv contains at most N distinguished positions.*

Observe that in (i), both conditions may be satisfied. Theorem 2.2 is a consequence of this result obtained by considering that all positions in a word are distinguished.

This theorem means that the place of the iterating group uyv within the factorization $w = xuyvz$ can be chosen to some extent: assume that the distinguished positions are chosen to be consecutive in w. Then either u or v is entirely composed of letters at distinguished positions.

One cannot expect to have a much more precise information on the position of u and v within the word w, and in particular it is impossible to force both u and v to be at distinguished places. However, the following result of Bader and Moura indicates that, to some extent, both u and v cannot be at certain places.

2.4. THEOREM (Bader and Moura [7]). *Let L be a context-free language. There exists an integer N such that, for any word w and for any choice of at least d distinguished and e excluded positions in w with $d > N^{1+e}$, there exists an iterative pair $\eta = (x, u, y, v, z)$ for w in L such that, additionally,*

(i) *either x, u, y each contain at least one distinguished position, or y, v, z each contain at least one distinguished position;*

(ii) *the word uv contains no excluded position;*

(iii) *if uyv contains r distinguished and s excluded positions, then $r \leqslant N^{1+s}$.*

Before going into the proof, let us first give an example of the use of the theorem, to show how it works.

2.5. EXAMPLE. Let $A = \{a, b\}$, and let

$$L = b^* \cup aa^+b^* \cup \{ab^p \mid p \text{ prime}\}.$$

This language is of course not context-free. However Ogden's iteration lemma does not prove it, because there is no way to get rid of pumping the initial a. On the other hand, when the initial a is considered as in an excluded position, then Bader and Moura's iteration lemma ensures that $uv \in b^*$, and then it is easily seen that L is not context-free.

PROOF OF THEOREM 2.4. Let $G=(V, A, P, S)$ be a context-free grammar generating L. Let t be the maximal length of the right-hand side of the productions, and let $N=t^{2k+6}$ with k being the number of nonterminals in G.

Let $w \in L$ be a word with e excluded and $d > N^{1+e}$ distinguished positions. Consider a derivation tree for w. A node in the tree is a *branch point* if it has at least two children which have distinguished descendants. Let P be a path with the greatest number of branch points. Since w has at least $t^{2(k+3)(e+1)}$ distinguished positions, the path P has at least $2(k+3)(e+1)$ branch points. The branch points in path P are grouped into two sets. A *left* branch point is a branch point having a child with a distinguished descendant to the left of path P. Right branch points are defined symmetrically. Observe that a node may be both a left and a right branch point. Clearly, P has at least $(k+3)(e+1)$ left or at least $(k+3)(e+1)$ right branch points.

Assume that P has at least $(k+3)(e+1)$ left branch points and divide the lowermost part of P in $e+3$ subpaths. The subpath nearest to the bottom contains $e+1$ left branch points, each of the following $e+1$ "internal paths" has $k+1$ left branch points and the topmost again $e+1$ left branch points (see Fig. 1). In each of the internal subpaths P_i $(0 \leqslant i \leqslant e)$, there must be two branch points with the same label, say X_i. Thus there exist words $u_i, v_i \in A^*$, such that $X_i \xrightarrow{*} u_i X_i v_i$. Moreover, since the node is a left branch point, the word u_i contains at least one distinguished position.

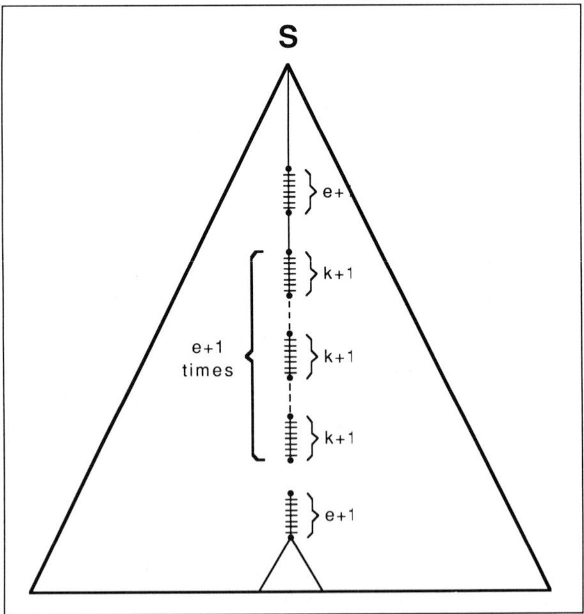

Fig. 1. Factorization of a path.

Now observe that there are $e+1$ pairs of words (u_i, v_i) but only e excluded positions. Therefore one of the words $u_i v_i$ contains no excluded position.

The remaining claims are easily verified. □

2.6. EXAMPLE. As an application of the previous iteration lemma, we consider the following situation. Given an infinite word

$$\boldsymbol{x}=a_0 a_1 a_2 \ldots a_n \ldots$$

over some alphabet A, the language

$$\text{Co-Pref}(\boldsymbol{x})=\{w \mid w \neq a_0 a_1 \ldots \mathrm{a}_{|w|-1}\} \subset A^*$$

is the set of all words over A which are not initial segments of $\boldsymbol{x}$. We consider infinite words of the form

$$x=a^{f(1)} b a^{f(2)} b \ldots a^{f(n)} b \ldots \tag{$*$}$$

where $f: \mathbf{N} \rightarrow \mathbf{N}$ is some function, and we prove that Co-Pref(x) is not context-free provided f grows rapidly enough. Assume indeed that for some function f, the language Co-Pref(x) is context-free, let N be the integer associated with this language in Theorem 2.4, and consider the word

$$w=a^{f(1)} b^{f(2)} b \ldots b a^{1+f(n)} \in \text{Co-Pref}(\boldsymbol{x}).$$

Declare that all positions in $a^{f(1)} b \ldots a^{f(n-1)} b$ are excluded, and that the last $1+f(n)$ are distinguished. In order to be allowed to apply the theorem, it is required that

$$1+f(n)>N^{n+\sum_{i=1}^{n-1} f(i)}.$$

It is easily seen that this holds for large enough n whenever

$$f(n) \geqslant 2^{2^{\cdot^{\cdot^{2}}}}$$

where there are $2n$ stacked exponents. In this case, w has an iterative pair $\eta=(\alpha, u, \beta, v, \gamma)$ with $u \beta v \gamma$ a right factor of the factor $a^{1+f(n)}$. Consequently, $\alpha \beta \gamma$ is an initial segment of $\boldsymbol{x}$, contradicting the theorem.

Let us mention that this brute-force technique gives weaker results than Grazon's cleaver iterated use of Ogden's lemma [35]. It appears indeed to be rather involved to prove that a language of the form Co-Pref($\boldsymbol{x}$) is not context-free. It has been shown by Grazon that all these languages satisfy Ogden's lemma. M.G. Main (personal communication) observed that they also satisfy the interchange lemma given in the next section. Grazon shows that for a language Co-Pref($\boldsymbol{x}$) with $\boldsymbol{x}$ of the form ($*$) to be not context-free, it suffices that $f(n) \geqslant 2^n$ for $n \geqslant 1$.

2.2. *Interchange lemma*

The interchange lemma we describe below gives a different kind of constraint for words of a context-free language. Instead of iteration, the unavoidable property described concerns the possibility of exchanging factors of words in some positions without leaving the language. One interesting aspect of this lemma is that it holds for languages which have "many" words of given length. (Observe that this is precisely the case where the classical iteration lemmas are difficult to use.)

Let L be a language over some alphabet A, and let $n \geqslant 0$ be an integer. A subset $R \subset L \cap A^n$ is an *interchange* set for L if there exist integers p, q with $0 \leqslant p+q \leqslant n$ such that for all u, v, w, u', v', w' with uvw, $u'v'w' \in R$, $|u|=|u'|=p$, $|v|=|v'|=q$ implies $uv'w$, $u'vw' \in L$. If this holds, the integer q is called the *span* of R.

2.7. THEOREM ("interchange lemma", Ogden, Ross, Winklmann [50]). *Let $L \subset A^*$ be a context-free language. There exists a number C, such that, for any integers n, m with $2 \leqslant m \leqslant n$ and for any set $Q \subset L \cap A^n$, there exists an interchange set $R \subset Q$ for L of size*

$$\operatorname{card}(R) \geqslant \frac{\operatorname{card}(Q)}{Cn^2}$$

and of span q with $m/2 \leqslant q \leqslant m$.

Clearly, the interchange set R may be empty unless $\operatorname{card}(Q) \geqslant Cn^2$, which means in practice that the number of words in $L \cap A^n$ should grow faster than n^2.

PROOF. Let L be generated by a context-free grammar $G=(V, A, P, S)$ in Chomsky normal form. Let n be an integer, and let $Q \subset L \cap A^n$. For all $X \in V-A$, and integers n_1, n_2 with $0 \leqslant n_1+n_2 \leqslant n$, we denote by $Q(n_1, X, n_2)$ the set of words $w \in Q$ such that there is a derivation

$$S \xrightarrow{*} uXv \xrightarrow{*} uxv = w$$

with $|u|=n_1$, $|v|=n_2$. Clearly, each set $Q(n_1, X, n_2)$ is an interchange set with span $n-(n_1+n_2)$. It is also clear that $Q(n_1, X, n_2) \subset Q$.

Let now m be any integer that satisfies $2 \leqslant m \leqslant n$. We claim that

$$Q \subset \bigcup Q(n_1, X, n_2) \tag{2.1}$$

where the union is over all sets with span $q=n-(n_1+n_2)$ satisfying the relation $m/2 < q \leqslant m$.

Let indeed $w \in Q$. Then clearly $w \in Q(n_1, X, n_2)$ for some n_1, X, n_2. It remains to show that the parameters n_1, X, n_2 can be chosen in such a way that the span of $Q(n_1, X, n_2)$ is in the desired interval. It is clear that the span can always be chosen greater than $m/2$ (take $X=S$). Assume now that span q is strictly greater than m. Then the derivation $X \xrightarrow{*} x$ where $|x|=q$, may be factorized into

$$X \to YZ \xrightarrow{*} x = yz$$

for some y, z with $Y \xrightarrow{*} y$ and $Z \xrightarrow{*} z$. Clearly, one of the words y or z has length strictly greater than $m/2$. Assume it is y. Then w is in $Q(n_1, Y, n_2+|z|)$ which has span $q-|z|$. The conclusion follows by induction.

Now observe that, in (2.1), the union is over at most $\operatorname{card}(V-A) \cdot n^2$ terms.

Consequently, there is at least one set $R=Q(n_1, X_1, n_2)$ with

$$\operatorname{card}(R) \geqslant \frac{\operatorname{card}(Q)}{\operatorname{card}(V-A)\cdot n^2}. \qquad \square$$

We now apply the interchange lemma to prove that a special language is not context-free. For this, we call a *square* any word of the form uu, with u nonempty. A word is *square-free* if none of its factors is a square. It is easily seen that there are only finitely many square-free words over 1 or 2 letters. We quote without proof the following result.

2.8. THEOREM (Thue [44]). *The set of square-free words over a three-letter alphabet is infinite.*

For a proof and a systematic exposition of the topic, see [44]. It is easily seen that the set of square-free words over at least three letters is not context-free. The same question concerning its complement, i.e., the set of words containing squares, was stated as a conjecture in [3] and was open for a long time. Two different proofs were given, the first by Ehrenfeucht and Rozenberg [26] based on growth considerations for E0L systems, the second by Ross and Winklmann [53] contains some preliminary version of the interchange lemma. The proof given here is from [50].

2.9. THEOREM. *The language of words containing a square over an alphabet with at least three letters is not context-free.*

Denote by $\mathscr{C}_n$ the language of words over n letters containing a square. The proof is in two steps. Only the first one is sketched below. The first step is to show that $\mathscr{C}_6$ is not context-free. Then a so-called square-free morphism (i.e., a morphism preserving square-free words) is applied to show that $\mathscr{C}_3$ also is not context-free. (For more details about square-free morphisms, see [8, 18, 22]).

PROOF OF STEP 1 OF THEOREM 2.9. Let us consider the six-letter alphabet $A=\{\$, 0, 1, a, b, c\}$ and assume that $\mathscr{C}_6 \subset A^*$ is context-free. We choose a large enough integer N and fix a square-free word over $B=\{a, b, c\}$ of length N, say $v=c_1c_2\ldots c_N$, $c_i \in B$. Next we consider the set

$$Q=\{\$t\$t \mid t=d_0c_1d_1c_2\ldots c_Nd_N,\ d_i\in\{0,1\}\}.$$

Any word in Q is a square and contains no other square. Each word in Q has $4(N+1)$ letters, and $\operatorname{card}(Q)=2^{N+1}$. Choose $n=4(N+1)$ and $m=n/2=2(N+1)$ in the interchange lemma. Then there exists an interchange set $R\subset Q$ of size

$$\operatorname{card}(R) \geqslant \frac{\operatorname{card}(Q)}{Cn^2} = \frac{2^{N+1}}{16C(N+1)^2} \tag{2.2}$$

for some constant C. Moreover, the span q of R satisfies $(N+1) < q \leqslant 2(N+1)$.

Using the notation of the beginning of the paragraph, consider words uxv, $u'x'v'$ in R, such that $ux'v$ is in $\mathscr{C}_6$. Since $|u|=|u'|$, $|x|=|x'|=q$, the words x and x' have the same letters in B at the same places. But since v is square-free, uxv and $ux'v$ both contain a square only if $x=x'$. This shows that if uxv, $u'x'v' \in R$ interchange, then $x=x'$. Consequently, the number of words in R is bounded by

$$\operatorname{card}(R) \leqslant 2^{N+1-(1+q/2)} \leqslant 2^{(N+1)/2}$$

since $|x|=q \geqslant N+1$. From (2.2), it follows that

$$2^{N+1}16C(N+1)^2 \leqslant \operatorname{card}(R) \leqslant 2^{(N+1)/2}$$

which is impossible for large enough N. □

There have been several developments around the topic of context-freeness related to square-free words. In [3], a conjecture claims that any context-free language containing the language $\mathscr{C}_n$ for $n \geqslant 3$ must be the complement of a finite language. This conjecture has been disproved by Main [45]. He shows that the set Co-Pref($\boldsymbol{x}$) is context-free where $\boldsymbol{x}$ is an infinite word which is square-free (and even overlap-free). An analogue of Theorem 2.9 for words with overlaps (an overlap is a word of the form $uxuxu$ with u nonempty) has been given by Gabarró [31]. For other developments along these lines, see [44].

2.3. *Degeneracy*

In Section 2.1, we have defined an iterative pair of a word w in a language L as a tuple (x, u, y, v, z) such that

(i) $w = xuyvz$;

(ii) $xu^n yv^n z \in L$ for all integers $n \geqslant 0$

Such a pair is called *degenerated* if $xu^*yv^*z \subset L$ (for more details, see Section 3.1). The aim of this section is to prove the following theorem.

2.10. THEOREM. *If all the iterative pairs in a given context-free language are degenerated, then the language is regular.*

It should be observed that this result does not characterize regular sets. For instance, the language $R = \{a^n b^p \mid n, p \geqslant 0,\ n \equiv p \bmod 2\}$ is regular. It has the following pair $(a, a, 1, b, b)$ which is not degenerated because $aa^2bb \notin R$. On the other hand, there do exist nonregular languages having all their iterative pairs degenerated. (Obviously, they cannot be context-free!) Such an example is

$$\{a^n b^n c^n \mid n \geqslant 1\} \cup a^+ \cup b^+ \cup c^+.$$

The following definitions and proofs are from [27]. A *type* over A is a word y such that $|y|_a \leqslant 1$ for any $a \in A$. Note that the number of different types is finite.

A word x over the alphabet A has type y iff there exists a morphism h such that $h(y) = x$ and such that $h(a) \in aA^*a \cup a$ for all letters $a \in A$.

2.11. EXAMPLE. Let $A = \{a, b, c\}$. The word $x = acabc$ is of type abc; just set $h(a) = aca$, $h(b) = b$, $h(c) = c$. It is also of type ac with $h'(a) = a$, $h'(b) = b$, $h'(c) = cabc$.

As seen in the example, a word may have several types. On the other hand, we have the following lemma.

2.12. LEMMA. *Every word x over A has a type.*

PROOF. The proof is by induction on the number k of different letters in x. If $k=0$ or $k=1$, the result is obvious. Assume the result holds for some k and let x be a word using $k+1$ letters. Let a be the first letter of x and let x_1 be the longest prefix of x ending with a. Then $x=x_1x_2$. Since x_2 uses at most k letters, by induction it has a type y and, clearly, ay is a type for x. □

2.11. EXAMPLE (*continued*). Applied to $x=acabc$, this method gives rise to a factorization $x=aca.bc$ and to the morphism $h(a)=aca$, $h(b)=b$, $h(c)=c$. This computation thus gives the type abc.

Given two words x and x' of some type y, we define the *interleaving* x'' of x and x' (according to the associated morphisms h and h') as follows. Let $y=a_1a_2\ldots a_k$, $x=x_1x_2\ldots x_k$, $x'=x'_1x'_2\ldots x'_k$ with $h(a_i)=x_i=\bar{x}_ia_i$ and $h'(a_i)=x'_i$; then, the interleaving is $x''=\bar{x}_1x'_1\bar{x}_2x'_2\ldots\bar{x}_kx'_k$.

2.13. EXAMPLE. Let $x=acabc$ be of type abc with $h(a)=aca$, $h(b)=b$, $h(c)=c$. Let $x'=ababcbcbc$ be of type abc with $h'(a)=aba$, $h'(b)=bcbcb$, $h'(c)=c$. Then, their interleaving x'' is equal to $acaba.bcbcb.c$. If we were to change h' in $h''(a)=a$, $h''(b)=babcbcb$, $h''(c)=c$, the (new) interleaving would be $aca.babcbcb.c$.

We now turn to some constructs on derivation trees for a given grammar G in Chomsky normal form generating a language $L(G)$ having all its iterative pairs degenerated.

Pruned trees: Given a derivation tree, a fixed path in this tree and two nodes of the path having the same label, we define the *left directly pruned* tree as the tree obtained by erasing all subtrees to the left of the path between the two chosen nodes including the first one. Clearly, in general, the new tree obtained is no more a derivation tree. However, if L has all its pairs degenerated, the produced word still is in L. We define similarly the *right directly pruned* tree (see Fig. 2).

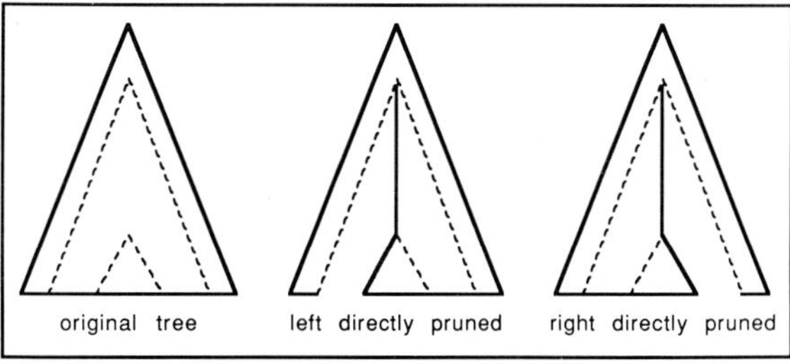

Fig. 2. Pruning a tree.

We now extend this construct to several pairs of nodes on the fixed path. To do so, we choose k pairs of nodes $(r_1, s_1), \ldots, (r_k, s_k)$ such that

(i) for each i, r_i and s_i have the same label;

(ii) for each i, s_i is a descendant of r_i;

(iii) for each i, $r_{i+1} = s_i$ or r_{i+1} is a descendant of s_i.

With each pair (r_i, s_i), we associate an *index* L or R. Then the corresponding pruned tree is, by definition, the tree obtained by successively left directly or right directly pruning the original tree on the pair (r_i, s_i) according to the fact of the index being L or R. Again, the resulting tree may not be a derivation tree. However, as soon as L has all its pairs degenerated, the produced word will belong to the language.

Relabelled trees: We define a new set of labels for the nodes, by

$$\bar{V} = \{(X, Y, Z, i) \mid X, Y, Z \in (V - A),\ i \in \{0, 1\}\}$$
$$\cup \{(X, a) \mid x \in (V, A),\ a \in A\}.$$

Given a derivation tree and a fixed path in it, we relabel the nodes of the path by elements of $\bar{V}$ according to the following rules:

(i) if the node r is labelled by a variable X which is derived in YZ in the tree, the new label of r will be (X, Y, Z, i) with $i = 0$ if Y is on the path and $i = 1$ if Z is on the path;

(ii) if the node r is labelled by the variable X which is derived in $a \in A$ in the tree, the new label of r will be (X, a).

The word of new labels collected on the path from the root to the leaf is called the *spine* of the marked tree.

Further definitions: First, let μ be the function from $V^+ \times \bar{V}^+$ into the subsets of V^* defined as follows: $\mu(\alpha, \bar{\beta})$ is the set of words γ such that $\alpha\gamma$ is generated in the grammar G by a derivation tree where the path from the root to the last letter of α gives rise to a marked tree with spine $\bar{\beta}$. Now, let δ be the function from V^+ into the subsets of $\bar{V}^+$ defined by

$$\delta(\alpha) = \{\bar{\beta} \in \bar{V}^+ \mid \mu(\alpha, \bar{\beta}) \neq \emptyset\}.$$

Finally, for each nonempty left factor α of a word in $L(G)$, we define $\Theta(\alpha) = \{\bar{\beta}_0 \in \bar{V}^+ \mid \bar{\beta}_0$ is a type of some $\bar{\beta}$ in $\delta(\alpha)\}$. Clearly, for each α, $\Theta(\alpha)$ is a finite set and the number of such possible sets is finite. We are now ready to prove the following crucial lemma needed for the theorem.

2.14. LEMMA. *Let α and α' be two nonempty left factors of $L(G)$ and assume $\Theta(\alpha) = \Theta(\alpha')$; then*

$$\{\gamma \mid \alpha\gamma \in L(G)\} = \{\gamma \mid \alpha'\gamma \in L(G)\}.$$

PROOF. Clearly, it suffices to prove

$$\Theta(\alpha) = \Theta(\alpha') \quad \text{and} \quad \alpha\gamma \in L(G) \ \Rightarrow\ \alpha\gamma' \in L(G). \tag{$*$}$$

Sketch of the proof of $(*)$: As $\alpha\gamma$ is in $L(G)$, there exists a derivation tree T producing $\alpha\gamma$. In T, choose the path from the root to the last letter of α and build the marked version of T giving rise to its spine $\bar{\beta}$. If $\bar{\beta}$ has type $\bar{\beta}_0$, then $\bar{\beta}$ is in $\Theta(\alpha')$ and there exists a word $\bar{\beta}'$ of type $\bar{\beta}_0$ such that $\bar{\beta}' \in \delta(\alpha')$. Hence, we can build a derivation tree T' which has a marked version with spine $\bar{\beta}'$ producing a word $\alpha'\gamma'$ in $L(G)$. So now we have two derivation trees T and T' producing $\alpha\gamma$ and $\alpha'\gamma'$ respectively, with in each a selected path whose marked versions are of same type $\bar{\beta}_0$. The idea is then to produce a new tree T'' looking like an interleaving of T and T' along the selected paths. Namely, let $\bar{\beta}_0 = \bar{v}_1 v_2 \ldots \bar{v}_k$ and $h(\bar{\beta}_0) = \bar{\beta}$, $h'(\bar{\beta}_0) = \bar{\beta}'$. Let $\bar{\beta}''$ be the interleaving of $\bar{\beta}$ and β' according to h and h'.

We shall now build the tree T'' by completing this path into a derivation tree. At the same time that we indicate how to build up T'', we indicate on the path $\bar{\beta}''$ some pairs of nodes with an index L or R. This is done in such a way that when we prune T'' according to these pairs of nodes, we get a new $\hat{T}''$ producing $\alpha\gamma'$. Then, because the language has all its pairs degenerated, $\hat{T}''$ will produce a word in the language and (∗) will be proved.

We now describe the construction of T'' together with the pairs of nodes to be pruned. For this, we go through $\bar{\beta}''$ by segments, each of which is the contribution of a letter $\bar{v}_i$ of $\bar{\beta}_0$. We shall picture what to do by indicating which subtrees (from T or T') have to be added to the right and left of the path $\bar{\beta}''$. The dotted arrows (in Figs. 3–6) will show what is dropped out after the pruning. The reader will check that, in each case, the pruned tree obtained $\hat{T}''$ produces a part of γ' on the right of the path and a part of α on the left of it. Hence $\hat{T}''$ produces $\alpha\gamma'$ as already announced.

There are four cases according to the fact of the images through h and h' of a letter $\bar{v}_i$

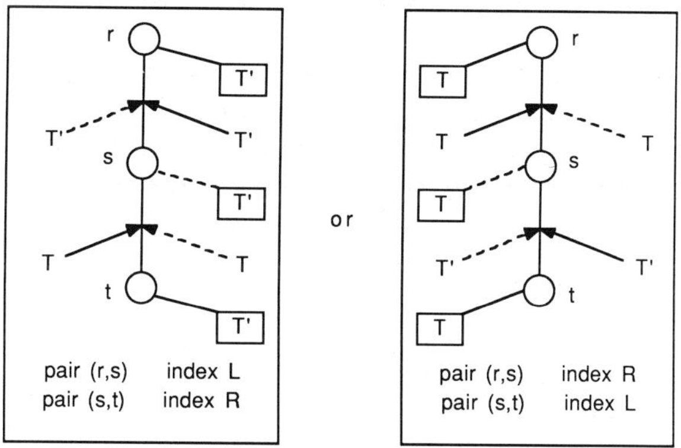

Fig. 3. Case 1: $h(\bar{v}_i) \neq \bar{v}_i \neq h'(\bar{v}_i)$, $\bar{\beta}'' = \bar{v}_i \bar{\beta} \bar{v}_i \bar{\beta}' \bar{v}_i$.

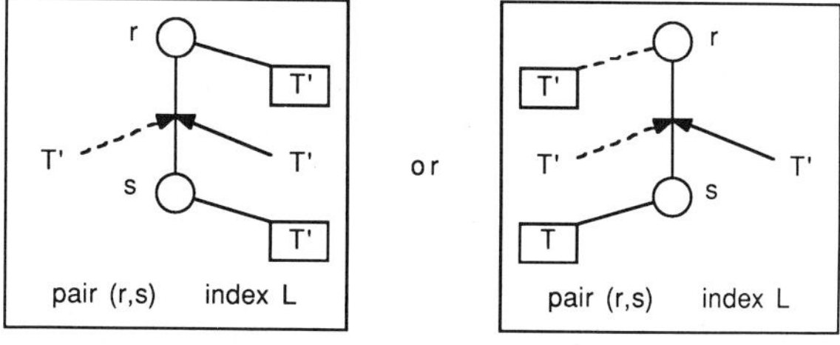

Fig. 4. Case 2: $h(\bar{v}_i) = \bar{v}_i \neq h'(\bar{v}_i)$, $\bar{\beta}'' = \bar{v}_i \bar{\beta}' \bar{v}_i$.

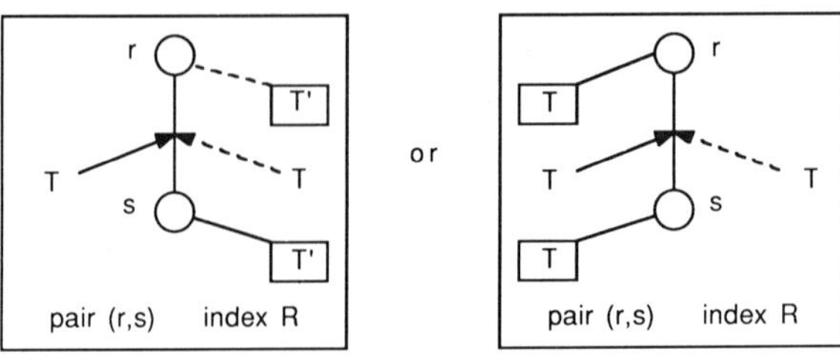

Fig. 5. Case 3: $h(\bar{v}_i) \neq \bar{v}_i = h'(\bar{v}_i)$, $\bar{\beta}'' = \bar{v}_i \bar{\beta} \bar{v}_i$.

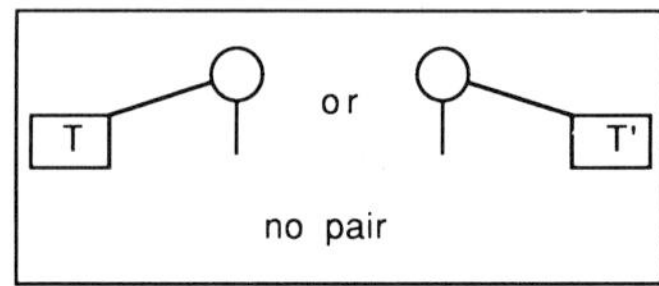

Fig. 6. Case 4: $h(\bar{v}_i) = \bar{v}_i = h'(\bar{v}_i)$, $\bar{\beta}'' = \bar{v}_i$.

being equal to $\bar{v}_i$ or not. In each case, we have two symmetric situations, according to the fact of the selected path leaving the node through its left or right son (see Fig. 3–6). □

Now, we can prove the announced result.

PROOF OF THEOREM. 2.10. By the lemma, the nonempty left factors of L fall into a finite number of classes modulo L: there are at most as many classes as the number of different sets $\Theta(\alpha)$. On the other hand, there is at most one new class containing the empty word. As all the words which are not left factors of L fall in one specific class, the equivalence modulo L has at most two more classes than the number of possible sets $\Theta(\alpha)$. Hence, this equivalence is of finite index which shows that L is regular. □

3. Looking for nongenerators

3.1. Preliminaries

A transduction from A^* to B^* is any function from A^* to the set of subsets of B^*. Such a mapping is a *rational transduction* iff its graph is rational. Many characterizations of rational transductions have been given (see [12]). We will here use the following theorem.

3.1. THEOREM. *A transduction τ from A^* to B^* is rational iff there exist an alphabet C, a regular set R over C and two alphabetic morphisms g and h from C^* to A^* and B^**

respectively, such that for all $x \in A^*$,

$$\tau(x) = h(g^{-1}(x) \cap R).$$

This is known as Nivat's Theorem. (A morphism is alphabetic or length-decreasing if it maps a letter into either a letter, or the empty word.) As usual, we will use transductions on languages: the image of a language L is the union of the images of each word in L. From Theorem 3.1, it is clear that any context-free language will have a context-free image through a rational transduction. One of the basic properties of rational transductions is the following theorem.

3.2. THEOREM. *Given two rational transductions* $\tau_1: A^* \to B^*$ *and* $\tau_2: B^* \to C^*$, *their composition* $\tau_2 \circ \tau_1$ *is a rational transduction from* A^* *to* C^*.

Usually, this is stated as "rational transductions are closed under composition" and is known as Elgot and Mezei's Theorem. Now, given two languages L_1 and L_2, it may happen that there exists a rational transduction τ such that $L_2 = \tau(L_1)$. Then we say that L_1 *dominates* L_2 and denote this by $L_1 \geqslant L_2$. Note that because of Theorem 3.2, this relation is transitive. If it happens then that $L_1 \geqslant L_2$ and $L_2 \geqslant L_1$, we say that L_1 and L_2 are *rationally equivalent*; in symbols: $L_1 \approx L_2$. Again, because of Theorem 3.2, this relation is transitive. Hence, it is an equivalence relation. Two languages are *incomparable* if no relation holds between them.

3.3. EXAMPLE. The Dyck languages D_2^* and $\hat{D}_2^*$ are equivalent. The Dyck language D_2 dominates D_1^*. These languages are not equivalent: D_1^* does not dominate D_2^*. The languages D_1^* and $\hat{D}_1^*$ are incomparable.

The main motivation in using rational transductions for comparing context-free languages comes from the idea that if $L_1 \geqslant L_2$, then L_1 should be at least as "complicated" as L_2. This idea is more or less formalized in the framework of iterative pairs.

In Section 2.1, an *iterative pair* of a word w in a language L is defined as a tuple (x, u, y, v, z) such that $w = xuyvz$ and $xu^n yv^n z \in L$ for all $n \geqslant 0$. The classical pumping lemma for context-free languages (see Section 2.1) ensures that if L is context-free, any word in L long enough will have such a pair. In general the set of possible exponents

$$\{(n, m) \in \mathbf{N} \times \mathbf{N} \mid xu^n yv^m z \in L\}$$

contains, by definition, the diagonal $\{(n, n) \mid n \geqslant 0\}$. If it reduces to that set, the iterative pair is called *strict*; if, to the contrary, the set of exponents is the whole set $\mathbf{N} \times \mathbf{N}$, the iterative pair is *degenerated*. Arbitrary intermediate situations may arise.

3.4. EXAMPLE. (i) Let D^* be the Dyck language over $A = \{a, \bar{a}\}$. Then $x = aa\bar{a}a\bar{a}\bar{a}$ admits the following pair $\pi = (a, a, \bar{a}a\bar{a}, \bar{a}, 1)$. It is a strict iterative pair.

(ii) Let $S_< = \{a^n b^m \mid 1 \leqslant m \leqslant n\}$. The word $x = aaabb$ admits $\pi = (a, a, ab, b, 1)$ as an iterative pair. Note that π is neither strict nor degenerated. On the other hand, $(a, aa, 1,$

$bb, 1)$ is *not* a pair because exponent 0 yields $a \notin S_<$. We leave to reader to check that $S_<$ has no strict iterative pair.

3.5. THEOREM. *Let L_1 and L_2 be two context-free languages such that $L_1 \geqslant L_2$; if L_2 has a strict iterative pair then so does L_1.*

This result shows that strict pairs cannot be created by rational transductions. This is extensively used to show that $L_1 \geqslant L_2$ does not hold. For instance, going back to the above example, Theorem 3.5 shows that $S_<$ does not dominate D^*.

Theorem 3.5 has been considerably strengthened by the consideration of systems of pairs. A system consists in several iterative pairs in the same word, each of which can be pumped up independently. We will not give here any details on this interesting extension (see [12]).

The notion of rational transduction naturally leads to considering families of languages closed under this operation. Such a family is called a *rational cone*. We have already remarked that the image of a context-free language through any rational transduction is context-free, so we can state: the family **Alg** of context-free languages is a rational cone. The same holds, for instance, for **Rat** and **Lin**, which are respectively the families of regular and linear languages. If we look at some family $\mathscr{L}$ (such as the deterministic languages) which fails to be a rational cone, we can define the least rational cone containing $\mathscr{L}$. It is denoted by $\mathscr{T}(\mathscr{L})$ and called the rational cone generated by $\mathscr{L}$. If $\mathscr{L}$ is the family of deterministic languages, then $\mathscr{T}(\mathscr{L})$ will be **Alg** again.

On the other hand, given a rational cone $\mathscr{L}$, we may look for the smallest possible set of languages $\mathscr{G}$ such that $\mathscr{L} = \mathscr{T}(\mathscr{G})$. Whenever $\mathscr{G}$ can be chosen to be one single language G, the cone $\mathscr{L}$ is said to be a *principal rational cone* with *generator G*; it is denoted $\mathscr{T}(G)$. From the Chomsky–Schützenberger Theorem, we derive that the cone **Alg** is principal and that $D_2'^*$ is a generator. Note then that, as two equivalent languages generate the same cone, the following languages are generators of **Alg** as well: $D_2, \hat{D}_2'^*, D_n^*, D_n, \hat{D}_n^*, \hat{D}_n$ $(n \geqslant 2)$, E, E_n. So, for instance, we may write $\mathbf{Alg} = \mathscr{T}(E)$. The family of linear languages turns out to be a principal cone too. The languages *Pal* or *Sym* are generators of **Lin**. The family **Rat** is principal too. Any nonempty regular language is a generator of it. Moreover, **Rat** is included in any rational cone. Besides all this, as we have mentioned that $D_1'^*$ does not dominate $D_2'^*$, we know that $\mathscr{T}(D_1'^*)$ is a strict subfamily of **Alg**. It is the family **Rocl** of restricted one-counter languages, i.e., those languages which can be recognized by a pushdown automaton with one symbol in the pushdown alphabet and no zero test (see [12]). Now the question of building up some nonprincipal cones is raised. If we do not accept "trivial" constructions such as the union of two principal cones with incomparable generators, we get that any set of generators of a nonprincipal cone must be infinite. To prove that such cones exist and to build them, the easiest method is to use substitution.

Given a language L over the alphabet A and, for each letter a of A, a language L_a, we define the substitution σ as the morphism from A^* into its subsets given by $\sigma(a) = L_a$ for each $a \in A$. If all the languages L_a are in some family $\mathscr{L}$, then σ is an $\mathscr{L}$-substitution. So, a rational substitution is a substitution such that each L_a is regular. (Note that in this case, a substitution is a rational transduction). Given two families of languages $\mathscr{L}$ and $\mathscr{M}$, we can define the substitution of $\mathscr{M}$ in $\mathscr{L}$ by $\mathscr{L}\square\mathscr{M}=\{\sigma(L) \mid L\in\mathscr{L}, \sigma$ is an $\mathscr{M}$-substitution$\}$. Again, substitution and rational transductions nicely mix together.

3.6. THEOREM. *If $\mathscr{L}$ and $\mathscr{M}$ are two rational cones, then so is $\mathscr{L}\square\mathscr{M}$. Moreover, if $\mathscr{L}$ and $\mathscr{M}$ are principal, so is $\mathscr{L}\square\mathscr{M}$.*

The proof of Theorem 3.6 uses a very special substitution called the *syntactic substitution*: given a language L over A and a language M over B with $A\cap B=\emptyset$, the syntactic substitution of M in L, denoted by $L\uparrow M$, is the substitution defined by $\sigma(a)=aM$ for $a\in A$. This special substitution gives rise to the following crucial lemma.

3.7. LEMMA ("Syntactic lemma"). *Given two languages L and M and two rational cones $\mathscr{L}$ and $\mathscr{M}$, then $L\uparrow M\in\mathscr{L}\square\mathscr{M}$ implies either $L\in\mathscr{L}$ or $M\in\mathscr{M}$.*

EXAMPLE (*application*). We want to prove that if $\mathbf{Alg}=\mathscr{L}\square\mathscr{M}$, then either $\mathscr{L}$ or $\mathscr{M}$ is equal to **Alg**. For this we use the syntactic lemma with two copies of E:

$$E\uparrow E'\in\mathscr{L}\square\mathscr{M} \quad\Rightarrow\quad E\in\mathscr{L} \text{ or } E'\in\mathscr{M}$$

and the result is proved.

Another application allows us to build nonprincipal cones.

3.8. LEMMA. *Given a cone $\mathscr{L}$, either $\mathscr{L}\square\mathscr{L}=\mathscr{L}$ or the smallest substitution closed rational cone $\mathscr{T}^{\sigma}(\mathscr{L})$ containing $\mathscr{L}$ is not principal.*

Take now $\mathscr{L}=\mathbf{Lin}$. As $\mathbf{Lin}\square\mathbf{Lin}$ is not included in **Lin**, the cone $\mathscr{T}^{\sigma}(\mathbf{Lin})=\mathscr{T}^{\sigma}(Sym)$ is not principal. This cone is the family of quasi rational languages (or of nonexpansive languages) and is denoted by **Qrt**. The same conclusion holds with $\mathscr{L}=\mathbf{Rocl}$ giving rise to the nonprincipal cone **Ict** of iterated-counter languages. We may even take $\mathscr{L}=\mathbf{Lin}\cup\mathbf{Rocl}$ and get a nonprincipal cone of so-called *Greibach languages*, denoted by **Gre**. For each of these cones, we get a generating family using the syntactic substitution. For instance, **Qrt** will be generated by $\{Sym, Sym\uparrow Sym, \ldots, Sym\uparrow (Sym\uparrow\cdots\uparrow Sym), \ldots\}$. Up to now, we have used two methods to get rational cones. We choose a family $\mathscr{L}$ and look either at the cone $\mathscr{T}(\mathscr{L})$ it generates, or at the substitution-closed cone $\mathscr{T}^{\sigma}(\mathscr{L})$ it generates. However, there exist other methods. We shall present here two new ways of getting cones, each of them raising more questions than it answers!

Given a principal cone $\mathscr{L}$, we may distinguish between the family of those languages which are generators of $\mathscr{L}$ and the family of those languages in $\mathscr{L}$ which are not. It is easy to see that this second family is a rational cone $\mathscr{N}\mathscr{L}$ which is the largest subcone of $\mathscr{L}$. In the particular case of $\mathscr{L}=\mathbf{Alg}$, this largest subcone is rather denoted by **Nge**. One of the most popular conjectures in this framework is the following one.

CONJECTURE. **Nge** *is a nonprincipal rational cone.*

Several attempts to prove this conjecture have been made. A great number of them tried to use the substitution operation to build up **Nge** in a similar way as we built **Qrt**. We now know that such an attempt cannot succeed (see Section 3.3). Besides, in the sequel, we will show that **Nge** differs from all the already presented cones in the sense that there is no family of deterministic languages which can generate **Nge** (see Section 3.4). It should be noted that, for any principal rational cone $\mathscr{L}$ (except the cone **Rat**), the question of the principality of $\mathscr{N}\mathscr{L}$ is open ($\mathscr{N}\mathbf{Rat}=\{\emptyset\}$). Along the same lines, we can mention that nobody knows whether or not there exists a rational cone $\mathscr{L}$ which contains only **Rat** as a strict subcone. Such a cone has to be principal. So we have the following question.

QUESTION. Does there exist a principal rational cone $\mathscr{L}$ such that $\mathscr{N}\mathscr{L}=\mathbf{Rat}$?

A second method for getting new rational cones comes from the following obvious observation: given two rational cones $\mathscr{L}$ and $\mathscr{M}$, their intersection $\mathscr{L}\cap\mathscr{M}$ is a rational cone. Here again we do not know much about such cones and we mainly have open questions rather than results. For instance, if we take $\mathscr{L}=\mathbf{Lin}$ and $\mathscr{M}=\mathbf{Rocl}$, we only have that $\mathbf{Lin}\cap\mathbf{Rocl}\supseteq\mathscr{T}(S)$, where $S=\{a^nb^n \mid n\geqslant 0\}$. This inclusion has recently been proved to be strict by Brandenburg [19]. However, we still do not know whether this intersection is a principal cone. In Section 3.5, we present some results in this area showing that such intersections seem to be, in general, larger than was thought in the beginning. Note that here again some attempts have been made to describe **Nge** as the intersection of two rational cones (see [36]).

3.2. Generators

We state here one of the most important characterizations of generators of the cone of context-free languages. The proofs of the following results are very technical and will not be given here (they can be found in [10]). Over the fixed alphabet $A=\{\#_1, \#_2, a, b, c, d\}$, we define the language E' by $E'=\#_1 E \#_2$. Then we have the following theorem.

3.9. THEOREM. *A language L over B is a generator of the cone of context-free languages iff there exist a morphism h from A^* to B^* and two regular sets R over A and K over B such that*

(i) $h(E')=L\cap K$;
(ii) $h^{-1}(L)\cap R=E'$;
(iii) $|h^{-1}(w)\cap R|=1$ *for all words* $w\in L$.

This is known as Beauquier's Theorem. From this result we get the following corollary.

3.10. COROLLARY *For any generator L, there exists a regular set K such that $L\cap K$ is an unambiguous generator.*

Theorem 3.9 can be stated in a slightly different way as follows.

3.11. THEOREM. *A language $L \subset B^*$ is a generator iff there exist six words $x, y, \alpha, \beta, \gamma, \delta \in B^*$ and a regular set $K \subset B^*$ such that*

$$L \cap K = \langle S \to xTy,\ T \to \alpha T \beta T \gamma + \delta \rangle.$$

A recent improvement shows that $\alpha, \beta, \gamma, \delta$ can be chosen to be a (biprefix) code [11].
Essentially, these results show that, in any generator, there is an encoded version of the language E. We now present an application.

3.12. PROPOSITION. *The cone* **Alg** *of context-free languages has no commutative generator.*

PROOF. Let L be a commutative language over B. Suppose that L is a generator. Then Theorem 3.9 holds. Set $h(\#_1)=x$, $h(\#_2)=y$, $h(a)=\alpha$, $h(b)=\beta$, $h(c)=\gamma$, $h(d)=\delta$.
Since $u = \#_1 aa^n d(bdc)^n ba^n d(bdc)^n c \#_2$ is in E' for all integers n, we can choose n large enough to find in each block of a's an iterative factor a^λ in the regular set R. Then

$$z = \#_1 aa^{n+\lambda} d(bdc)^n ba^{n-\lambda} d(bdc)^n c \#_2$$

is in R. Moreover, it has the same commutative image as u. Hence $h(z)$ is in L and $h^{-1}(h(z)) \in h^{-1}(L) \cap R$. Now z is in $h^{-1}(h(z))$ and by (ii) of Theorem 3.9 it should also be in E' which is not true. □

COMMENT. Corollary 3.10 stated above naturally leads to the following notion: a language L is *strongly ambiguous* if, for any regular set K such that $L \cap K$ and L are equivalent, $L \cap K$ is ambiguous. So, we know that no generator of **Alg** is strongly ambiguous. We can even extend this notion as follows: a language L is *intrinsically ambiguous* if there is no language equivalent to L which is unambiguous. Such languages do exist. For instance, the language

$$\{a^n ba^m ba^p ba^q \mid (n \geqslant q \text{ and } m \geqslant p) \text{ or } (n \geqslant m \text{ and } p \geqslant q)\}$$

is intrinsically ambiguous (see [9]). We leave it to the reader to check that, generally, the classical examples of ambiguous languages are not intrinsically ambiguous (not even strongly ambiguous).

3.3. Nongenerators and substitution

The aim of this section is to prove the following theorem.

3.13. THEOREM. *The cone* **Nge** *of nongenerators of context-free languages is not the substitution closure of any strict subcone.*

This result puts an end to any attempt towards proving that **Nge** is nonprincipal by showing that this cone is the substitution closure of simpler subcones. It implies, for instance, that the family of Greibach languages is strictly included in **Nge**. Theorem 3.13 will follow from the following more general result.

3.14. THEOREM. *For any given context-free language L, there exists a context-free language $L^{\uparrow}$ such that*

(i) *$L^{\uparrow}$ is a nongenerator if L is so;*

(ii) *the rational cone generated by $L^{\uparrow}$ contains the substitution closure of the cone generated by L, i.e., $\mathscr{T}(L^{\uparrow})\supset\mathscr{T}^{\sigma}(L)$.*

Theorem 3.14 has other consequences, such as the following corollary which answers Conjecture 8 in [3].

3.15. COROLLARY. *There does exist a principal rational cone of nongenerators containing the family* **Qrt** *of nonexpansive languages.*

The proofs given here come from [15]. Let us turn first to the proof of Theorem 3.14. We start by some definitions necessary to construct the language $L^{\uparrow}$ associated with L.

Given a word x in D over the alphabet $A=\{a,\bar{a}\}$, we define the *height* of an occurrence (x_1,α,x_2) in x (with $x=x_1\alpha x_2$, $\alpha\in A$) by $|x_1\alpha|_a-|x_1\alpha|_{\bar{a}}$. Then the height of the word x is the maximum of all heights over the occurrences in x. It is easy to check that we can compute the heights of the occurrences from left to right in a sequential manner.

Namely, if the current height is k, add 1 if you read a letter a and substract 1 if you read a letter $\bar{a}$. On the other hand, the set of words in D of height at most k is a regular set. Hence, there exists a gsm-mapping which, when reading a word x in D of height at most k, produces the word y obtained by indexing the letters with their heights. It will be denoted by num_k.

3.16. EXAMPLE. Consider the word $x=aa\bar{a}aa\bar{a}\bar{a}a\bar{a}\bar{a}$. The height of the third a is 2. The height of x is 3. Then $num_2(x)=\emptyset$ and

$$num_3(x)=a_1a_2\bar{a}_1a_2a_3\bar{a}_2\bar{a}_1a_2\bar{a}_1\bar{a}_0.$$

Note that, in $num_k(x)$, a letter a_i matches a letter $\bar{a}_{i-1}$.

Given the alphabet $C=A\cup B$, with $B\cap A=\emptyset$, we consider the projection p from C^* into A^*. A word x over C is called a *D-word* if $p(x)$ is in D and x begins and ends with a letter in A. The height of an occurrence in such a D-word is then defined in the same way as for a word in D. Again, we define the gsm-mapping num_k producing from a D-word of height at most k the word y obtained by indexing the letters of x with their height.

3.17. EXAMPLE. Let $B=\{b,c\}$. The word $x=aabc\bar{a}bababc\bar{a}bc\bar{a}cabc\bar{a}\bar{a}$ is a D-word; further, $num_2(x)=\emptyset$ and

$$num_3(x)=a_1a_2b_2c_2\bar{a}_1b_1a_2b_2a_3b_3c_3\bar{a}_2b_2c_2\bar{a}_1c_1a_2b_2c_2\bar{a}_1\bar{a}_0.$$

Given a context-free language L over B, we consider first the marked version of L which is the language $M=aL\bar{a}\cup\{1\}$ over $C=A\cup B$. We then denote by M_i the copy of M obtained by indexing all the letters of L with i, the letter a with i and the letter

$\bar{a}$ with $i-1$. Thus, in M_i the letter a_i will match $\bar{a}_{i-1}$. We use these copies to define

$$M_{(1)}=M_1,\ M_{(2)}=M_1 \Uparrow M_2, \ldots,\ M_{(k)}=M_1 \Uparrow (M_2(\cdots \Uparrow M_k\cdots)), \ldots,$$

where $\Uparrow$ stands for a substitution very similar to the syntactic substitution. The substitution $\Uparrow$ is defined as follows: the image of any b_i is b_iM_{i+1} when b is in B; the image of a_i is a_iM_{i+1}; and the image of $\bar{a}_{i-1}$ is just $\bar{a}_{i-1}$. Clearly, $\Uparrow$ is so near to the usual syntactic substitution that the family $\{M_{(1)}, M_{(2)}, \ldots, M_{(k)}, \ldots\}$ generates the substitution-closed rational cone $\mathcal{T}^\sigma(M)$ generated by M, which is the same as $\mathcal{T}^\sigma(L)$. Using the morphism h from $(\bigcup_{i\geqslant 1} C_i)^*$ onto C^* which just erases the indices, we define $M^{(k)}=h(M_{(k)})$ and $M^{(\infty)}=\bigcup_{k\geqslant 1}M^{(k)}$. (For instance, if $L=\{bc\}$, the word x of Example 3.17 will be in $M^{(3)}$.) We then get our first proposition.

3.18. PROPOSITION. *The language $M^{(\infty)}$ is context-free.*

To prove the proposition, note first the following observation.

OBSERVATION. *$x \in M^{(\infty)}$ iff $x \in M=M^{(1)}$ or*

$$x=ay_1az_1\bar{a}y_2az_2\bar{a}\ldots y_{n-1}az_{n-1}\bar{a}y_n\bar{a}$$

with $y_1, y_n \in B^*$, $y_i \in B^+$ (for $2\leqslant i\leqslant n-1$), $az_i\bar{a} \in M^{(\infty)}$ and $ay_1y_2\ldots y_{n-1}y_n\bar{a} \in M$.

This observation is proved by a straightforward induction on the integer k for which $x \in M^{(k)}$.

PROOF OF PROPOSITION 3.18. Using the observation, we get that $M^{(\infty)}$ can be generated by the following generalized context-free grammar which has infinitely many productions: $\langle S\rightarrow ay_1Sy_2S\ldots y_{n-1}Sy_n\bar{a}+1 \mid ay_1y_2\ldots y_{n-1}y_n\bar{a} \in M\rangle$. Clearly, the set of right members is context-free and it is well known that such a generalized grammar does generate a context-free language. □

Still using the above observation, it is easy to prove that any word x in $M^{(\infty)}$ is a D-word over C. Moreover, if x is of height at most k, then $num_k(x)$ is in $M_{(k)}$. So, we get $num_k(M^{(\infty)})=num_k(M^{(k)})=M_{(k)}-\{1\}$.

Now define $M^{(+)}$ to be the set of those words over C which are not D-words; we note that $M^{(+)}$ is context-free. Set $L^\uparrow=M^{(+)}\cup M^{(\infty)}$. Clearly, $L^\uparrow$ is context-free. Moreover, since num_k equals $\emptyset$ on $M^{(+)}$, we get $num_k(L^\uparrow)=num_k(M^{(\infty)})=M_{(k)}-\{1\}$. This shows that the cone of $L^\uparrow$ does contain all the $M_{(k)}$. So, we can state the following proposition.

3.19. PROPOSITION. *$L^\uparrow$ is a context-free language such that $\mathcal{T}(L^\uparrow)\supseteq\mathcal{T}^\sigma(L)$.*

Note that this proposition is the condition (ii) of Theorem 3.14. So, we now turn to condition (i) and prove this proposition.

3.20. PROPOSITION. *$L^\uparrow$ is a generator of the cone* **Alg** *of context-free languages iff L is.*

Before starting the proof, we introduce a notation. Given a language P over an alphabet disjoint from A, let $\lhd P \rhd$ denote the language

$$\lhd P \rhd = \{a^n y \bar{a}^n \mid n \geqslant 1,\ y \in P\}.$$

Let us first show the following lemma.

3.21. LEMMA. *Given a language P, if $\lhd P \rhd$ is in $\mathscr{T}(L^\uparrow)$, then there exists an integer k such that P is in $\mathscr{T}(M^{(k)})$.*

PROOF (*outline*). The detailed proof is rather technical. We just outline it. As $\lhd P \rhd$ is in $\mathscr{T}(L^\uparrow)$, we may write $\lhd P \rhd = g(f^{-1}(L^\uparrow) \cap R)$.

Let k be the number of states of a finite automaton recognizing R; we want to show that all the words $a^n y \bar{a}^n$ with $n > k$ are images of words in $M^{(k)}$. For this, we choose a word z of minimal length in $f^{-1}(L^\uparrow) \cap R$ such that $g(z) = a^n y \bar{a}^n$ for some $n > k$. Then z naturally factorizes into $z_1 z_2 z_3$ such that $g(z_2) = y$ and z_2 is maximal with this property. Moreover, as n is larger than k, z_1 can be factorized in $z'_1 u z''_1$ such that $z'_1 u^* z''_1 z_2 z_3 \subseteq R$ and $g(u) \in a^+$. So, for all integers $m \neq 1$, $f(z'_1 u^m z''_1 z_2 z_3)$ is not in $L^\uparrow$ because $g(z_i u^m z''_1 z_2 z_3)$ is not in $\lhd P \rhd$. As z is of minimal length, $f(u)$ is nonempty which implies that $f(z)$ is not in $M^{(+)}$; otherwise all its iterated versions with at most one exception would be in $M^{(+)}$, hence in $L^\uparrow$. So $f(z)$ is in $M^{(\infty)}$. Assume then that $f(z)$ is not in $M^{(k)}$. In $x = f(z)$, there exist occurrences of letters of height larger than k. This implies that z can be factorized into $z = \hat{z}_1 \hat{z}_2 \hat{z}_3$ such that $\hat{z}_1 \hat{z}_2^* \hat{z} \subseteq R$ and $p(f(\hat{z}_2)) = a^s$ for some $s \neq 0$. Hence $f(\hat{z}_1 \hat{z}_2^r \hat{z}_3)$ is in $M^{(+)}$ and $g(\hat{z}_1) g(\hat{z}_2)^* g(\hat{z}_3)$ is in $\lhd P \rhd$. As z is of minimal length, $g(\hat{z}_2)$ is not empty. Then it has to be a factor of y. So, we know that $\hat{z}_2$ is a factor of z_2 and we may write $z = z'_1 u z''_1 z'_2 \hat{z}_2 z''_2 z_3$ with $f(z'_1 u z''_1 z'_2 \hat{z}_2^2 z''_2 z_3) \in M^{(+)}$. Then, for all m but at most one we have $f(z'_1 u^m z''_1 z'_2 \hat{z}_2^2 z''_2 z_3) \in M^{(+)}$. This contradicts the fact that the image of such a word is in $\lhd P \rhd$ only for $m = 1$. $\square$

We are now ready for the following proof.

PROOF OF PROPOSITION 3.20. Clearly, if L is a generator, so is $L^\uparrow$. So, only the converse has to be proved. Assume that $L^\uparrow$ is a generator. Then, for any context-free language P, $\lhd P \rhd \in \mathscr{T}(L^\uparrow)$. In particular, $\lhd E \rhd \in \mathscr{T}(L^\uparrow)$. By Lemma 3.21, there exists an integer k such that $E \in \mathscr{T}(M^{(k)})$ and thus $M^{(k)}$ is a generator. As we know that $M^{(k)}$ and $M_{(k)}$ are equivalent, we have that $M_{(k)}$ is a generator. A simple application of the syntactic lemma then shows that L is generator. $\square$

Clearly, Propositions 3.19 and 3.20 prove Theorem 3.2. We now show how we can derive Theorem 3.13.

PROOF OF THEOREM 3.13. It is known that if L is a rational cone, its closure under union $\mathscr{L}_\cup$ satisfies $\mathscr{T}^\sigma(\mathscr{L}) = \mathscr{T}^\sigma(\mathscr{L}_\cup)$. On the other hand, $\mathscr{L}_\cup = \mathbf{Nge}$ implies $\mathscr{L} = \mathbf{Nge}$. Assume that **Nge** is the substitution closure of some subcone $\mathscr{L}$. Then, we may assume that $\mathscr{L}$ is closed under union. Let L be any language in **Nge**. Then, by (i) of Theorem 3.14, $L^{\uparrow}$ is in **Nge**. It follows that $\lhd L^{\uparrow} \rhd \in \mathbf{Nge} = \mathscr{T}^{\sigma}(\mathscr{L})$. So there exists a finite number of languages in $\mathscr{L}$, say $L_1, L_2, \ldots, L_k$ such that $\lhd L^{\uparrow} \rhd \in \mathscr{T}^{\sigma}(\{L_1, L_2, \ldots, L_k\})$. If we now consider L_0 to be the union of disjoint copies of $L_1, L_2, \ldots, L_k$, we get $\lhd L^{\uparrow} \rhd \in \mathscr{T}^{\sigma}(L_0)$ with $L_0 \in \mathscr{L}$. As $\mathscr{T}^{\sigma}(L_0)$ is contained in $\mathscr{T}(L_0^{\uparrow})$, we have $\lhd L^{\uparrow} \rhd \in \mathscr{T}(L_0^{\updownarrow})$.

By the lemma, there exists an integer k such that $L^{\uparrow} \in \mathscr{T}(L_0^{(k)})$ and by Theorem 3.14(ii), $\mathscr{T}^{\sigma}(L)$ is included in $\mathscr{T}(L^{\uparrow})$. Hence $\mathscr{T}^{\sigma}(L) \subset \mathscr{T}(L^{\uparrow}) \subset \mathscr{T}(L_0^{(k)})$. Then, by the syntactic lemma, $\mathscr{T}(L) \subset \mathscr{T}(L_0)$. As L_0 is in $\mathscr{L}$, we have that any language L in **Nge** is in $\mathscr{L}$, which means $\mathscr{L} \supset \mathbf{Nge}$. Since the reverse inclusion is obvious, we have $\mathscr{L} = \mathbf{Nge}$. □

CONCLUDING REMARKS. It is worthwhile pointing out that if L is deterministic, so is the language $L^{\uparrow}$ constructed here. Thus, for instance, there does exist a principal subcone of **Alg** containing **Qrt** which has a deterministic generator. However, the proposed construction leaves the following questions open.

QUESTION **1**. Does there exist a principal substitution-closed cone strictly included in **Alg**, larger than **Rat**?

QUESTION **2**. Does there exist a nongenerator L such that $\mathscr{T}(L) = \mathscr{T}(L^{\uparrow})$?

Note that if we can answer Question 2 positively, we will have a positive answer to Question 1. Note also that if **Nge** is principal, both questions can be answered positively.

3.4. Nongenerators and determinism

The aim of this section is to prove the following fact.

3.22. THEOREM. *The cone* **Nge** *is not generated by any family of deterministic context-free languages.*

This result shows that the cone **Nge** is very different from all the classical cones of context-free languages (principal or not) which all have deterministic generators. The theorem will follow from the existence of a particular language L which has the following two properties:

(i) L is a nongenerator;

(ii) any deterministic language which dominates L is a generator

The proof proposed here is an adaptation of the one given in [16]. We will start with some general results on dpda's before defining the language L. Then, we will show that L satisfies (i) and (ii) and, finally, prove Theorem 3.22.

Given a dpda (deterministic pushdown automaton) $\mathscr{A}$ with input alphabet A, nonterminal alphabet V and set of states Q, recall (see Section 1.1.6) that a *configuration* of $\mathscr{A}$ is a triple $c = (\gamma, q, x)$ in $V^* \times Q \times A^*$. The rightmost symbol of γ is the top of the pushdown store. The dpda is *normalized* if any ε-move of $\mathscr{A}$ erases the top of the pushdown store. It is well known that, for any dpda, there exists an equivalent normalized dpda (see [38]). So, from now on, we will assume that $\mathscr{A}$ is normalized. As usual, a *computation* of $\mathscr{A}$ is a sequence of configuration $c_i=(\gamma_i, q_i, x_i)$ such that $\mathscr{A}$ goes from c_i to c_{i+1} in one move and where $c_0=(S_0, q_0, x_0)$ is initial (i.e., S_0 is the initial pushdown and q_0 the initial state). The computation $c_0, c_1, \ldots, c_n$ is *maximal* if $c_n=(\gamma_n, q_n, 1)$ (which means that the input x_0 has been completely read) and there is no configuration which can possibly follow c_n. It is well known that, as soon as $\mathscr{A}$ is normalized for each input x over A, there is exactly one maximal associated computation. It will be called *the* computation of $\mathscr{A}$ over x, the configuration c_n is said to be its *result*.

A computation $c_0, c_1, \ldots, c_n$, with $c_i=(\gamma_i, q_i, x_i)$ is said to *contain an iterative pair* if there exist four integers i, j, k, l with $0 \leqslant i < j < k < l \leqslant n$, such that the following holds for all m:

(i) $\gamma_i=\gamma S$ for some $\gamma \in V^*$, $S \in V$;
(ii) $\gamma_m \in \gamma V^+$ for $i \leqslant m \leqslant j$;
(iii) $\gamma_j=\gamma\mu S$ for some $\mu \in V^*$ and $q_i=q_j$;
(iv) $\gamma_m \in \gamma\mu V^*$ for $j<m<k$;
(v) $\gamma_k=\gamma\mu$;
(vi) $\gamma_m \in \gamma V^+$ for $k<m<l$;
(vii) $\gamma_l=\gamma$ and $q_l=q_k$.

If we look at the words $x_0=yx_i$, $x_i=ux_j$, $x_j=zx_k$, $x_k=vx_l$, we get that the computations of $\mathscr{A}$ over the words $yu^nzv^nx_l$ $(n \geqslant 0)$ all lead to the same result. Moreover, as $\mathscr{A}$ is normalized, we known that u is not empty. So, for any suffix t such that x_0t is accepted by $\mathscr{A}$, the word x_0t will have (y, u, z, v, x_lt) as an iterative pair. Conversely, if x is a recognized word containing a nondegenerated iterative pair, there exists an iteration of x for which the associated computation contains an iterative pair.

A *prime computation* is a computation containing no iterative pair. A *prime word* is a word on which the computation of $\mathscr{A}$ is prime. We then leave, the proof of the following facts to the reader.

FACT 1: *Given a dpda $\mathscr{A}$, the set $P(\mathscr{A})$ of prime words is regular.*

FACT 2: *For any word x in the regular set $Q(\mathscr{A})=A^*-P(\mathscr{A})$, there exist infinitely many words whose computations have the same result as the computation over x.*

We are now ready to define our special language L. Let $A=\{a,b,c,d\}$. Over $A \cup \{\#\}$, consider the languages

$$S_< = \{a^nb^p \mid n \geqslant p\}, \qquad L_1 = S_< \# E, \qquad L_2=\{x\#y \mid x, y \in A^+,\ |x|<|y|\}.$$

Then the language L is defined by $L=L_1 \cup L_2$. Clearly, L is context-free. We first prove the following proposition.

3.23. PROPOSITION. *L is not a generator of the cone* **Alg**.

PROOF. We show that L has no strict iterative pair. Let (x, u, y, v, z) be an iterative pair of $w = xuyvz$ in L. We then look for the occurrence of # in w. Clearly, it can be neither in u nor in v. So the following situations may arise:

(i) # lies in x. Then there exists an integer k such that $|u^k yv^k z| > |x|$. Hence, $xu^k u^+ yv^+ v^k z \subset L_2 \subset L$ and (x, u, y, v, z) *is not strict.*

(ii) # lies in z. Then $z = z_1 \# z_2$ and, as $xu^n yv^n z_1$ becomes longer than z_2, the word z_2 is in E. Thus (x, u, y, v, z_1) must be a pair of $S_<$ and it cannot be strict.

(iii) # lies in y. Then $y = y_1 \# y_2$. Clearly, if u or v is empty, the pair is not strict. So we assume $v \neq 1$. Then, there is at most one integer k such that $y_2 v^k z$ is in E. Hence, for all n but possibly one, $xu^n y_1 \# y_2 v^n z$ is in L_2, which implies $xu^n y_1 \# y_2 v^n v^+ z$ is in L_2 and the pair is not strict. □

Before proving that no deterministic nongenerator dominates L, we need a preliminary result on the pairs of L.

3.24. PROPOSITION. *For any word y in E, there exists a word x such that $x \# y \in L$ and such that $x \# y$ has a nondegenerated pair whose iterative elements are within x.*

PROOF. Let N be the integer associated with L by Bader and Moura's Theorem 2.4. Given y in E, choose $x = a^d b^d$ with $d > N^{2+|y|}$. In the word $x \# y$, distinguish the d occurrences of the letter a, and exclude the $|y|+1$ letters of $\# y$. Bader and Moura's Theorem guarantees that $x \# y$ contains an iterative pair within x. Clearly, it is not degenerated. □

We can now prove the following crucial result needed for Theorem 3.22.

3.25. PROPOSITION. *Let T be a deterministic language and h an alphabetic morphism such that $L = h(T)$; then T is a generator of* **Alg.**

PROOF (*sketch*). Let T be recognized by a dpda $\mathscr{A}$ over B. Let B' be the letters θ such that $h(\theta) = \#$. Recall that $P(\mathscr{A})$ is the regular set of those words over B which have a prime computation and that $Q(\mathscr{A})$ is its complement. We then define the regular set $B^* - h^{-1}(h(P(\mathscr{A})) = K$. This is the set of words x' on B such that there exists no prime word x'' satisfying $h(x') = h(x'')$. We then define $R = KB'$ and we claim that $E = h(R^{-1}T)$.

(i): $E \subset h(R^{-1}T)$. By Proposition 3.24, for any word y in E, we can find a word x such that $x \# y$ is in L and has an iterative pair within x that is not degenerated. Then, for all words x' and y' and any θ in B' such that $h(x') = x$, $h(y') = y$ and $x'\theta y' \in T$, the word $x'\theta y'$ has a nondegenerated pair within x'. Hence x' is in K and $x'\theta$ is in R. It then follows that y' is in $R^{-1}T$ and then that $y \in h(R^{-1}T)$.

(ii): $h(R^{-1}T) \subset E$. Let y' be in $R^{-1}T$. Then, there exists $x'\theta$ in R such that $x'\theta y'$ is in T. Choose the shortest such $x'\theta$. The word x' is in K, so it is not prime. Moreover, the iterative pair that it contains has an image by h which is a iterative pair of $h(x'\theta y')$. So, we can find infinitely many words leading to the same result as x'. These words have images under h which are of increasing length because, as x' was chosen as short as possible, the iterative elements in x' have a nonempty image by h. Then, there will exist a word x'' leading to the same result as x' such that $|h(x'')|>|h(y')|$. This implies that $h(y')$ is in E.

So, putting together (i) and (ii), we have $E=h(R^{-1}T)$ and thus T is a generator. □

We can now prove Theorem 3.22.

PROOF OF THEOREM 3.22. Suppose that there exists a family $\{T_1, T_2, \ldots, T_n, \ldots\}$ of deterministic languages generating **Nge**. Then, there would exist an integer k such that the language L is in $\mathcal{T}(T_k)$. Thus, for some alphabetic morphisms g, h and some regular language R, one would have $L=h(g^{-1}(T_k)\cap R)$. Set $T=g^{-1}(T_k)\cap R$; then $L=h(T)$ with T a deterministic language. By Proposition 3.25, T is a generator of **Alg** and obviously, so is T_k. Then $\{T_1, \ldots, T_n, \ldots\}$ will generate **Alg.** □

Observe that this result shows in particular that if **Nge** is principal, it has no deterministic generator. On the other hand, this proof leaves the following question open.

QUESTION. May **Nge** be generated by a family of unambiguous languages?

3.5. *Intersection of principal cones*

It is easily verified that the intersection of two cones is again a cone. However, it is not known whether the intersection of two principal cones of context-free languages is again principal. (There is a counterexample by Ullian [63] concerning noncontext-free cones. In his definition, morphisms are required to be nonerasing.) In fact, for no pair of context-free principal cones, the status of the intersection is known, except in trivial cases.

Among the various conjectures concerning these problems, two were disproved by Brandenburg [19] ad Wagner [65]. Consider the languages

$$S=\{a^nb^n \mid n\geqslant 1\}, \qquad Copy=\{w\#w \mid w\in\{a,b\}^+\}.$$

3.26. THEOREM (Brandenburg [19]). *Let* $B=\{a^ib^jc^md^n \mid i\neq m \text{ or } j\leqslant m\leqslant n\}$. *Then* $B\in$ **Lin**$\cap$**Rocl**, *but* $B\notin\mathcal{T}(S)$.

This result shows that **Lin**$\cap$**Rocl** strictly contains $\mathcal{T}(S)$, disproving the conjecture that these two cones coincide. It leaves the question open whether **Lin**$\cap$**Rocl** is principal. The proof is delicate, and we omit it here.

3.27. THEOREM (Wagner [65]). *Let* $W=\{u\#v\#w \mid v\neq u^{\sim} \text{ or } v=w\}$. *Then* $W\in$ **Lin**$\cap$**Reset**, *but* $W\notin$ **Ocl.**

PROOF (*sketch*). Clearly, $\{u\#v\#w \mid v=w\}$ is in **Reset**. On the other hand, as

$\{u\#v \mid v \neq u^{\sim}\} = \{fag\#g'bf' \mid a,b \in A, a \neq b, |f| = |f'|\}$, we see that $\{u\#v\#w \mid v \neq u^{\sim}\}$ is in **Reset**. Hence $W \in$ **Reset**. Similarly, as W can be written $\{u\#v\#w \mid v \neq u^{\sim}$ or $w = u^{\sim}\}$, we have that $W \in$ **Lin** and, consequently, $W \in$ **Lin**$\cap$**Reset**.

The fact that $W \notin$ **Ocl** is more delicate. The argument can be sketched as follows. It is known that any counter pda may be assumed to be real-time [34]. Hence, the height of the pushdown store after reading an input of length n is bounded by kn for some fixed k. It follows that the number of different possible configurations then reached is bounded by $k'n$ for some k' (remember that the pda is just a counter). On the other hand, after reading $u\#u^{\sim}\#$, the only possible suffix is $u^{\sim}$. So, any configuration reached after reading $u\#u^{\sim}\#$ will accept $u^{\sim}$ or nothing. This shows that no successful configuration can be simultaneously reached by $u\#u^{\sim}\#$ and $u'\#u'^{\sim}\#$ for $u \neq u'$. But, the number of different such words is 2^n whence the number of different configurations is at most $k'n$. Hence the contradiction. □

Note the Brandenburg [19] shows the same result using, instead of W, the language

$$\{a^i b^j c^m d^n a^r b^s c^p d^q \mid i \neq n \text{ or } j \neq m \text{ or } (i = n = r = q \text{ and } j = m = s = p)\}.$$

In the same direction, we mention the following result.

3.28. THEOREM (Brandenburg [19]). *Let C be the language defined by*

$$C = \{a^i b^j c^m d^n \mid i \neq n \text{ or } (i \geqslant j \text{ and } j = m)\}.$$

*Then $C \in$ **Rocl**$\cap$**Reset** and $C \notin$ **Lin**.*

Finaly, let **Queue** be the cone of languages accepted by queue machines (FIFO languages or simple Post languages).

3.29. THEOREM. *Let $S^{(2)} = \{a^n b^m c^m d^n \mid n, m \geqslant 0\}$. Then $S^{(2)} \in$ **Lin**$\cap$**Queue**, but $S^{(2)} \notin$ **Ocl** and $S^{(2)} \notin$ **Reset**.*

3.30. COROLLARY. *The language $S = \{a^n b^n \mid n \geqslant 1\}$ is not a generator of any of the following cones:* **Lin**$\cap$**Rocl**, **Lin**$\cap$**Reset**, **Rocl**$\cap$**Reset**, *and* **Lin**$\cap$**Queue**.

4. Context-free groups

There exist several relations between context-free languages and groups, such as the Hotz group described in Section 1, or the definition of the two-sided Dyck language as the set of words equivalent to the empty word in the morphism of the free monoid onto the free group. This section describes recent results, mainly by Muller and Schupp [46, 47, 48] on the reverse problem raised by Anisimov [1]: Consider a presentation of a finitely generated group, and consider the language of all words which are equivalent to the empty word. This language may or may not be context-free. What does it mean about the group that the language *is* context-free? Before we give some recent answers concerning this question, let us recall some basic concepts.

4.1. Context-free groups

4.1.1. Presentation

A *presentation* of a group G is defined by an alphabet A, and a set R of *relators*. A new alphabet $\bar{A}$, disjoint from A, is chosen with a bijection $a \mapsto \bar{a}$ between A and $\bar{A}$. This bijection is extended to $A \cup \bar{A}$ by setting $\bar{\bar{a}}=a$. For $w \in (A \cup \bar{A})^*$, $w=a_1a_2\ldots a_n$, the word $\bar{w}$ is defined by $\bar{w}=\bar{a}_n\ldots\bar{a}_1$. The relators are pairs (u, v) of words. Since G is assumed to be a group, the *trivial relations* $(a\bar{a}, 1)$, for $a \in A \cup \bar{A}$ are tacitly required to be in R and are omitted in the notation. Then any relation (u, v) is equivalent to $(u\bar{v}, 1)$ and therefore R may also be considered as a subset of $(A \cup \bar{A})^*$. The pair $\langle A;R\rangle$ is a presentation of G if $G \simeq (A \cup \bar{A})^*/[R]$, where $[R]$ is the congruence generated by R. The group is finitely generated if A is finite.

4.1. EXAMPLE. The free commutative group over $A=\{a, b\}$ is defined by $R=\{ab=ba\}$ (it is easier to read equations than pairs). In the free monoid over the alphabet $A \cup \bar{A}=\{a, b, \bar{a}, \bar{b}\}$, this relator, together with $a\bar{a}=\bar{a}a=b\bar{b}=\bar{b}b=1$, induces other relators, such as $\bar{a}\bar{b}=\bar{b}\bar{a}$, $a\bar{b}=\bar{b}a$, etc. In fact, this free abelian group is isomorphic to $\mathbf{Z}^2$.

4.1.2. Word problem

Let G be a finitely generated group with presentation $\langle A;R\rangle$. The *word problem* of G (more accurately one should say the word problem of the presentation) is the set $W(G)$ of words over $A \cup \bar{A}$ that are equivalent to the empty word.

4.2. EXAMPLE. For the free group $F(A)$ over an n-letter alphabet, the word problem is the two-sided Dyck language $\hat{D}_n^*$ which is context-free.

4.3. EXAMPLE. It is easily seen that the word problem for the free abelian group $\langle a, b; ab=ba\rangle$ is the set of words $w \in (A \cup \bar{A})^*$ such that $|w|_a=|w|_{\bar{a}}$ and $|w|_b=|w|_{\bar{b}}$. This language is not context-free.

A finitely generated group G is called *context-free* if there exists a presentation $\langle A;R\rangle$ of G for which the word problem is a context-free language. The following observation states that context-freeness is a property of the group and not of the presentation.

4.4. PROPOSITION. *Let G be a context-free group. Then the word problem of any finitely generated presentation of G is a context-free language.*

Thus, the free group is context-free, and the free abelian group with two generators is not.

4.1.3. A global characterization

The question to determine the context-free groups was raised by Anisimov. It was solved by Muller and Schupp [47] up to a special argument contributed by Dunwoody [25]. It uses also a theorem of [60] concerning the structure of groups with more than one end (in the sense of Section 4.3 below).

First, we recall that a group G is *virtually free* if G has a free subgroup of finite index.

4.5. THEOREM. *Let G be a finitely generated group. Then G is context-free iff G is virtually free.*

It can be expected that there is a relation between special groups and special context-free languages. One such result has been given by Haring-Smith [37]: the word problem of a finitely presented group is freely generated by a simple context-free language [38] if and only if the group is a free product of a free group of finite rank and of a finite number of finite groups.

4.2. *Cayley graphs*

In this section, we describe a characterization of context-free groups by a triangulation property of the Cayley graph associated to one of its presentation.

Let G be a finitely generated group and let $\langle A;R\rangle$ be a presentation of G with A finite. The *Cayley graph* $\Gamma(G)$ of the presentation has as vertices the elements of G. The edges are labelled by elements of the alphabet $A \cup \bar{A}$, There is an edge from g to g' labelled by c iff $gc \equiv g' \bmod [R]$. For each edge, there is an inverse edge, from g' to g and labelled by $\bar{c}$.

4.6. EXAMPLE. Figures 7–9 represent the Cayley graphs of the free group with one generator, the free abelian group with two generators and the free group with two generators. In drawing graphs, we represent only one of the pairs composed of an edge and of its inverse. Each edge being labelled (by a letter), the label of a path is the word

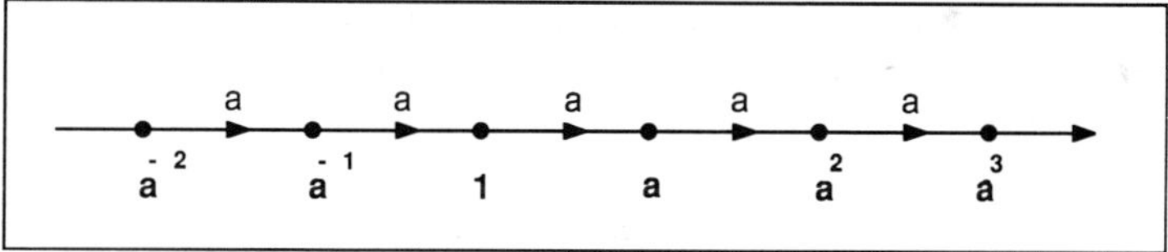

Fig. 7. Cayley graph of **Z**.

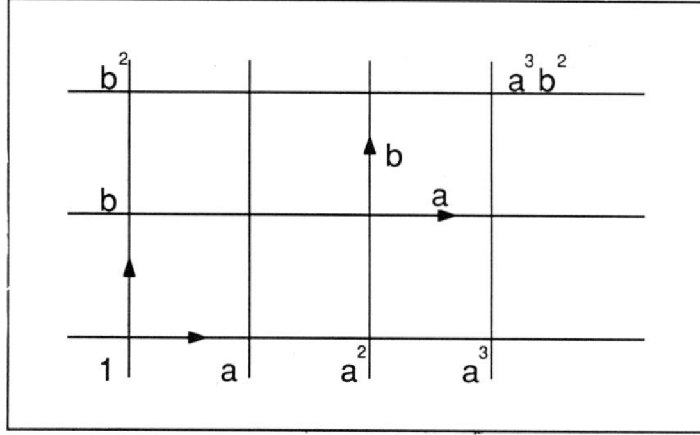

Fig. 8. Cayley graph of $\mathbf{Z}^2$.

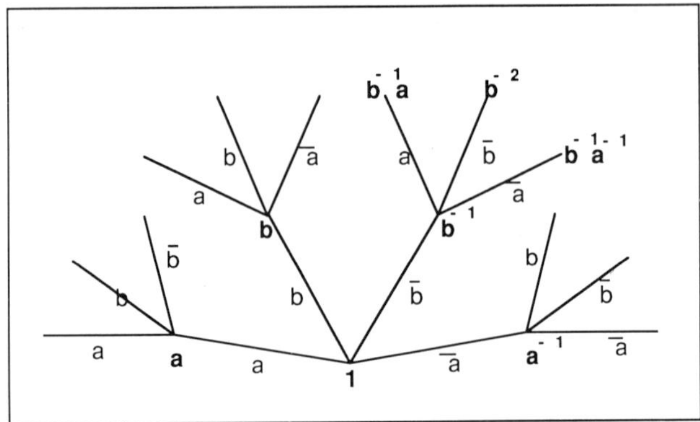

Fig. 9. Cayley graph of the free group with two generators.

composed of the labels of its edges. Clearly, any label of a closed path (a cycle) is in the word problem of the representation, and conversely. A cycle is *simple* if nonconsecutive edges do not share a common vertex.

A *diagonal* triangulation T of a simple cycle P is a triangulation of P with the following properties:

(1) The vertices of the triangles are among those of P.

(2) Each new edge has a label over $(A \cup \bar{A})^*$; these labels are such that reading around the boundary of each triangle gives a relation which holds in the group G.

A *bound* for the triangulation is an upper bound to the length of the new labels.

4.7. EXAMPLE. In the free abelian group over $\{a, b\}$, a closed path is given in fat lines in Fig. 10, the additional edges are drawn in thin lines. A bound for the triangulation is 3. There is also a triangulation of bound 2, but none of bound 1.

The Cayley graph can be *uniformly triangulated* if there is an integer K such that every simple cycle admits a diagonal triangulation with bound K.

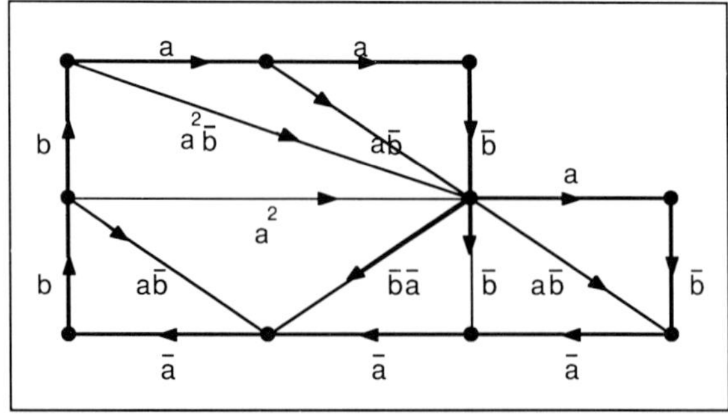

Fig. 10. Triangulations in $\mathbf{Z}^2$.

4.8. THEOREM. *A finitely generated group G given by a presentation* $\langle A;R\rangle$ *is context-free if and only if its Cayley graph can be uniformly triangulated.*

PROOF (*sketch*). Take a reduced context-free grammar $G=(V, A\cup\bar{A}, P, S)$ generating the word problem, in Chomsky normal form (see Section 1.1). For each variable X, let $u_X\in(A\cup\bar{A})^*$ be a fixed word generated by X. Let $w=a_1a_2\ldots a_n$ $(a_i\in A\cup\bar{A})$ be the label of a simple cycle, and assume $n\geqslant 4$. Since $S\xrightarrow{*}w$, one has

$$S\to XY\xrightarrow{*}xy \qquad x=a_1\ldots a_i,\quad y=a_{i+1}\ldots a_n.$$

Introduce a new edge from the starting point of the first edge to the endpoint of the ith edge with label u_X $(=u_Y^{-1})$. Then the relations $u_X=a_1\ldots a_i$, $u_Y=a_{i+1}\ldots a_n$ hold in G and the cycles are shorter.

The reverse construction is similar. First, a variable X_u is introduced for any word $u\in(A\cup\bar{A})^*$ of length less or equal to the uniform bound. Then productions $A_u\to u$, and $A_u\to A_vA_w$, are introduced whenever $u=vw$ in the group G. Finally, one shows by induction on the length of the cycles that if $u_1,\ldots,u_n$ are the labels of the edges of a cycle, each of bounded length, then $A_1\xrightarrow{*}A_{u_1}A_{u_2}\ldots A_{u_n}$. □

This proof shows that in a context-free group G there always exists a finite subset H with the following property: any word w in the word problem of G can be split into two parts w_1w_2 such that each w_i is a product of elements of H. In some sense, the constructed grammar just takes this set H as set of variables. It is then easily seen that the grammar generates a pre-NTS language. On the other hand, a very similar construction has been used to show that some special languages are NTS [4]. So, it was natural to ask if the above proof could be improved to get the following theorem.

4.9. THEOREM. *Any context-free group has an NTS word-problem.*

No proof of this result is known which uses the triangulation result. However, using a constructive characterization of virtually free groups, the above theorem was recently shown to be a consequence of Theorem 4.5 [5]. Note that Theorem 4.9 may be used to construct a set of generators in a context-free group in such a way that the uniform bound of Theorem 4.8 is 1.

4.3. Ends

The triangulation result of the previous section gives a "local" characterization. We now quote a more global one.

Let Γ be the Cayley graph of a finitely generated presentation of a group G. Denote by $\Gamma^{(n)}$ the subgraph with vertices at distance at most n from the origin (the vertex with label 1), and let k_n be the number of *connected components* in $\Gamma-\Gamma^{(n)}$. The *number of ends* of Γ is $\lim_{n\to\infty}k_n$.

4.10. EXAMPLE. For the Cayley graph of the infinite cyclic group, the number of ends is

2. For the free abelian group, the number of ends is 1; for the free group with two generators, the number of ends is infinite.

4.11. PROPOSITION. *If G is an infinite context-free group, then the Cayley graph of one of its finitely generated presentations has more than one end.*

Stallings [60] gives a complete description of the structure of groups with more than one end. This result, showing that there must be several ends for a context-free group, will now be opposed to a statement which claims that there must be only few non-isomorphic subgraphs. For this, we need some definitions.

Let $n \geqslant 0$, and let C be a connected component of $\Gamma - \Gamma^{(n)}$. A *frontier point* of C is a vertex u of C at distance $n+1$ from the origin. If v is a vertex of Γ at distance say $n+1$, we denote by $\Gamma(v)$ the component of $\Gamma - \Gamma^{(n)}$ containing v, and by $\Delta(v)$ the set of all frontier points of $\Gamma(v)$.

An *end-isomorphism* between two subgraphs $\Gamma(u)$ and $\Gamma(v)$ is a label-preserving graph isomorphism that maps $\Delta(u)$ onto $\Delta(v)$. The Cayley graph is *context-free* if the set $\{\Gamma(v) \mid v \text{ a vertex of } \Gamma\}$ has only finitely many isomorphism classes under end-isomorphism.

4.12. THEOREM. *A group is context-free if and only if its Cayley graph is context-free.*

4.13. EXAMPLE. Consider first the infinite cycle group. There are exactly two isomorphism classes. For the free abelian group with two generators the number of frontier points of the unique connected component of $\Gamma - \Gamma^{(n)}$ grows quadratically and therefore all these components are nonisomorphic. Thus the Cayley graph is not context-free. For the free group over two generators, again there are four isomorphism classes.

There is a strong relation between context-free graphs and pushdown automata. Muller, Schupp [48] proved this last theorem.

4.14. THEOREM. *A graph is context-free if and only if it is the complete transition graph of a pushdown automaton.*

Acknowledgment

We acknowledge helpful discussions with J.-M. Autebert, C. Reutenauer and P. Schupp.

References

[1] ANISIMOV, A.V., Group languages, *Kibernetika (Kiev)* **4** (1971) 18–24.

[2] AUTEBERT, J.M., *Langages Algébriques* (Masson, Paris, 1987).

[3] AUTEBERT, J.M., J. BEAUQUIER, L. BOASSON and M. NIVAT, Quelques problèmes ouverts en théorie des langages, *RAIRO Inform. Théor. Appl.* **13** (1979) 363–379.

[4] AUTEBERT, J.M., J. BEAUQUIER, L. BOASSON and G. SÉNIZERGUES, Langages de parenthèses, langages NTS et homomorphismes inverses, *RAIRO Inform. Théor. Appl.* **18** (1984) 327–344.

[5] AUTEBERT, J.M., L. BOASSON and G. SÉNIZERGUES, Groups and NTS languages, *J. Comput. System Sci.* **35** (1987) 213–267.

[6] AUTEBERT, J.M., P. FLAJOLET and J. GABARRÓ, Prefixes of infinite words and ambiguous context-free languages, *Inform. Process. Lett.* **25** (1987) 211–216.

[7] BADER, C. and A. MOURA., A generalization of Ogden's lemma, *J. Assoc. Comput. Mach.* **29** (1982) 404–407.

[8] BEAN, D.R., A. EHRENFEUCHT and G.F. MCNULTY, Avoidable patterns in strings of symbols, *Pacific J. Math.* **85** (1979) 261–294.

[9] BEAUQUIER, J., Ambiguïté forte, in: *Proc. 5th Internat. Coll. on Automata, Languages and Programming*, Lecture Notes in Computer Science, Vol. 62 (Springer, Berlin, 1978) 52–62.

[10] BEAUQUIER, J., Générateurs algébriques et systèmes de paires itérantes, *Theoret. Comput. Sci.* **8** (1979) 293–323.

[11] BEAUQUIER, J., and F. GIRE, On context-free generators, *Theoret. Comput. Sci.* **51** (1987) 117–127.

[12] BERSTEL, J., *Transductions and Context-Free Languages* (Teubner, Stuttgart, 1979).

[13] BERSTEL, J., Properties of infinite words: recent results, in: R. Cori, B. Monien, eds., *Proc. 6th Ann. Symp. STACS*, Lecture Notes in Computer Science, Vol. 349 (Springer, Berlin, 1989) 36–46.

[14] BLATTNER, M. and S. GINSBURG, Position restricted grammar forms and grammars, *Theoret. Comput. Sci.* **17** (1982) 1–27.

[15] BOASSON, L., Non-générateurs algébriques et substitution, *RAIRO Inform. Théor. Appl.* **19** (1985) 125–136.

[16] BOASSON, L. and A. PETIT, Deterministic languages and nongenerators, *RAIRO Inform. Théor. Appl.* **21** (1987) 41–57.

[17] BOOK, R.V., S.A. GREIBACH and C. WRATHALL, Reset machines, *J. Comput. System Sci.* **19** (1973) 256–276.

[18] BRANDENBURG, F.J., Uniformely growing k-th powerfree homomorphisms, *Theoret. Comput. Sci.* **23** (1983) 69–82.

[19] BRANDENBURG, F.J., On the intersection of stacks and queues, *Theoret. Comput. Sci.* **58** (1988) 69–80.

[20] COHEN, D.E., *Groups of Cohomological Dimension One*, Lecture Notes in Mathematics, Vol. 245 (Springer, Berlin, 1972).

[21] CRESTIN, J.P., Un langage non ambigu dont le carré est d'ambiguïté non bornée, in: *Proc. 1st Internat. Coll. on Automata, Languages and Programming* (North-Holland, Amsterdam, 1973) 377–390.

[22] CROCHEMORE, M., A sharp characterization of square-free morphisms, *Theoret. Comput. Sci.* **18** (1982) 221–226.

[23] DIEKERT, V., Investigations on Hotz groups for arbitrary grammars, *Acta Inform.* **22** (1986) 679–698.

[24] DIEKERT, V. and A. MÖBUS, Hotz-isomorphism theorems in formal language theory, in: *Proc. 5th Ann. Symp. STACS*, Lecture Notes in Computer Science, Vol. 294 (Springer, Berlin, 1988) 125–135.

[25] DUNWOODY, M.J., The accessibility of finitely presented groups, *Invent. Math.* **81** (1985) 449–457.

[26] EHRENFEUCHT, A. and G. ROZENBERG, On the separative power of E0L systems, *RAIRO Inform. Théor. Appl.* **17** (1983) 13–32.

[27] EHRENFEUCHT, A. and G. ROZENBERG, Strong iterative pairs and the regularity of context-free languages, *RAIRO Inform. Théor. Appl.* **19** (1985) 43–56.

[28] FLAJOLET, P., Ambiguity and transcendence, in: *Proc. 12th Internat. Coll. on Automata, Languages and Programming*, Lecture Notes in Computer Science, Vol. 194 (Springer, Berlin, 1985) 179–188.

[29] FLAJOLET, P., Analytic models and ambiguity of context-free languages, *Theoret. Comput. Sci.* **49** (1987) 283–309.

[30] FROUGNY, C., J. SAKAROVITCH and E. VALKEMA, On the Hotz group of a context-free grammar, *Acta Inform.* **18** (1982) 109–115.

[31] GABARRÓ, J., Some applications of the interchange lemma, *Bull. EATCS* **25** (1985) 19–21.

[32] GINSBURG, S., *The Mathematical Theory of Context-Free Languages* (McGraw-Hill, New York, 1966).

[33] GINSBURG, S., *Algebraic and Automata-Theoretic Properties of Formal Languages* (North-Holland, Amsterdam, 1975).

[34] GINSBURG, S., J. GOLDSTINE and S.A. GREIBACH, Uniformly erasable families of languages, *J. Comput. System Sci.* **10** (1975) 165–182.

[35] GRAZON, A., An infinite word language which is not co-CFL, *Inform. Process. Lett.* **24** (1987) 81–86.

[36] GREIBACH, S.A., Checking automata and one-way stack languages, *J. Comput. System Sci.* **3** (1969) 196–217.

[37] HARING-SMITH, R.H., Groups and simple languages, Ph.D. Thesis, Univ. of Illinois at Urbana-Champaign, 1981.

[38] HARRISON, M., *Introduction to Formal Language Theory* (Addison-Wesley, Reading, MA, 1978).

[39] HOTZ, G., Normal form transformations of context-free grammars, *Acta Cybernet.* **4** (1978) 65–84.

[40] HOTZ, G., Eine neue Invariante für kontextfreie Sprachen, *Theoret. Comput. Sci.* **11** (1980) 107–116.

[41] JANTZEN, M., *Confluent String Rewriting*, EATCS Monographs on Theoretical Computer Science, Vol. 14 (Springer, Berlin, 1988).

[42] KEMP, R., On the number of words in the language $\{w \in \Sigma^* \mid w = w^{\sim}\}$, *Discrete Math.* **40** (1980) 225–234.

[43] KUICH, W. and A. SALOMAA, *Formal Power Series and Languages*, EATCS Monographs on Theoretical Computer Science, Vol. 5 (Springer, Berlin, 1986).

[44] LOTHAIRE, M., *Combinatorics on Words* (Addison-Wesley, Reading, MA, 1983).

[45] MAIN, M.G., An infinite square-free co-CFL, *Inform. Process. Lett.* **20** (1985) 105–107.

[46] MULLER, D.E. and P.E. SCHUPP, Pushdown automata, ends, second-order logic and reachability problems, in: *Proc. 13th Ann. ACM Symp. on the Theory of Computing* (1981) 46–54.

[47] MULLER, D.E. and P.E. SCHUPP, Groups, the theory of ends, and context-free languages, *J. Comput. System Sci.* **26**, (1983) 295–310.

[48] MULLER, D.E. and P.E. SCHUPP, The theory of ends, pushdown automata, and second-order logic, *Theoret. Comput. Sci.* **37** (1985) 51–75.

[49] NIJHOLT, A., An annotated bibliography of pumping, *Bull. EATCS* **17** (1982) 34–52.

[50] OGDEN, W., R. ROSS and K. WINKLMANN, An "interchange lemma" for context-free languages, *SIAM J. Comput.* **14** (1985) 410–415.

[51] OYAMAGUCHI, M., The equivalence problem for realtime dpda's, *J. Assoc. Comp. Mach.* **34** (3) (1987) 731–760.

[52] PERRIN, D., Finite automata, in: J. van Leeuwen, ed., *Handbook of Theoretical Computer Science, Vol. B* (North-Holland Amsterdam, 1990) 1–57.

[53] ROSS, R. and K. WINKLMANN, Repetitive strings are not context-free, *RAIRO Inform. Théor. Appl.* **16** (1982) 191–199.

[54] RUDIN, W., *Real and Complex Analysis* (McGraw-Hill, New York, 1974).

[55] SALOMAA, A., *Formal Languages* (Academic Press, New York, 1973).

[56] SALOMAA, A., Formal languages and power series, in: J. van Leeuwen, ed., *Handbook of Theoretical Computer Science, Vol. B* (North-Holland, Amsterdam, 1990) 59–102.

[57] SCHÜTZENBERGER, M.P., On a theorem of R. Jungen, *Proc. Amer. Math. Soc.* **13** (1962) 885–889.

[58] SÉNIZERGUES, G., The equivalence and inclusion problems for NTS languages, *J. Comput. System Sci.* **31** (1985) 303–331.

[59] SÉNIZERGUES, G., Some decision problems about controlled rewriting systems, *Theoret. Comput. Sci.* **71** (1990) 281–346.

[60] STALLINGS, J., Groups of cohomological dimension one, *Proceedings of Symposia in Pure Mathematics, Vol. B* (North-Holland, Amsterdam, 1990) 133–191.

[61] THOMAS, W., Automata on infinite objects, in: J. van Leeuwen, ed., *Handbook of Theoretical Computer Science, Vol. B* (North-Holland, Amsterdam, 1990) 133–191.

[62] TOMITA, E. and K. SEINO, A direct branching algorithm for checking the equivalence of two deterministic pushdown transducers, one of which is realtime strict, *Theoret. Comput. Sci.* **64** (1989) 39–53.

[63] ULLIAN, J.S., Three theorems concerning principal AFL's, *J. Comput. System Sci.* **5** (1971) 304–314.

[64] VALKEMA, E., On some relations between formal languages and groups, *Proc. Categorical and Algebraic Methods in Computer Science and System Theory* (1978) 116–123.

[65] WAGNER, K., On the intersection of the class of linear context-free languages and the class of single-reset languages, *Inform. Process. Lett.* **23** (1986) 143–146.

CHAPTER 3

Formal Languages and Power Series

A. SALOMAA

Department of Mathematics, University of Turku, Turku, Finland

Contents

Prelude . . . 105
1. Preliminaries . . . 105
2. Rewriting systems and grammars . . . 109
3. Post canonical systems . . . 113
4. Markov algorithms . . . 114
5. Parallel rewriting systems . . . 116
6. Morphisms and languages . . . 119
7. Rational power series . . . 123
8. Algebraic power series . . . 126
9. What power series do for you . . . 129
10. Final comments . . . 131

References . . . 131

HANDBOOK OF THEORETICAL COMPUTER SCIENCE
Edited by J. van Leeuwen

Prelude

Both natural and programming languages can be viewed as sets of sentences—that is, finite strings of elements of some basic vocabulary. The notion of a language introduced below is very general. It certainly includes both natural and programming languages and also all kinds of nonsense one might think of. Traditionally formal language theory is concerned with the syntactic specification of a language rather than with any semantic issues. A syntactic specification of a language with finitely many sentences can be given, at least in principle, by listing the sentences. This is not possible for languages with infinitely many sentences. The main task of formal language theory is the study of finitary specifications of infinite languages.

The basic theory of computation, as well as of its various branches, is inseparably connected with language theory. The input and output sets of a computational device can be viewed as languages, and—more profoundly—models of computation can be identified with classes of language specifications, in a sense to be made more precise. Thus, for instance, Turing machines can be identified with phrase-structure grammars and finite automata with regular grammars.

The purpose of this chapter is to discuss formal language theory with the emphasis on a specific tool: *formal power series*.

Formal language theory is—together with automata theory (which is really inseparable from language theory)—the oldest branch of theoretical computer science. In some sense, the role of language and automata theory in computer science is analogous to that of philosophy in general science: it constitutes the stem from which the individual branches of knowledge emerge.

In view of this fact it is obvious that our presentation cannot penetrate very deep. Many of the topics discussed below could fill a respectable book and, indeed, many books have been written about some of the topics.

As regards literature, we have tried to make the list of references rather short. This means that we have not tried to trace each individual result to its very origin. The interested reader will find more bibliographical information from the books cited. A detailed exposition of the theory of formal power series in its relation to automata and formal languages was given in [33] (see also [16]).

1. Preliminaries

This section contains some fundamental notions about formal power series and languages. The knowledgeable reader may skip the section and consult it later on if need arises.

A *monoid* consists of a set M, an associative binary operation on M and of a neutral element 1 such that $1a=a1=a$ for every a. A monoid is called *commutative* iff $ab=ba$ for every a and b. The binary operation is usually (as we did here) denoted by juxtaposition and often called *product*. If the operation and the neutral of M are understood, then we denote the monoid simply by M.

The most important type of a monoid in our considerations is the *free monoid* Σ^*

generated by a nonempty finite set Σ. It has all the finite *strings*, also referred to as *words*,

$$x_1 \ldots x_n, \quad x_i \in \Sigma,$$

as its elements and the product $w_1 w_2$ is formed by writing the string w_2 immediately after the string w_1. The neutral element of Σ^*, also referred to as the *empty word*, is denoted by λ.

The members of Σ are called *letters* or symbols. The set Σ itself is called an *alphabet*. The length of a word w, in symbols $|w|$, is defined to be the number of letters of Σ occurring in w, whereby each letter is counted as many times as it occurs in w. By definition, the length of the empty word equals 0.

A *morphism* h of a monoid M into a monoid M' is a mapping $h: M \to M'$ compatible with the neutral elements and operations in M and M', i.e.,

$$h(1) = 1', \quad \text{where 1 and } 1' \text{ are the neutral elements of } M \text{ and } M',$$

respectively, and

$$h(m_1 m_2) = h(m_1)h(m_2) \quad \text{for all } m_1, m_2 \in M.$$

Since Σ^* is the free monoid generated by Σ, every mapping of Σ into a monoid M can be uniquely extended to a morphism of Σ^* into M.

The mapping $h(w) = |w|$ provides an example of a morphism of Σ^* into N, the monoid of nonnegative integers with addition as the operation.

By a *semiring* we mean a set A together with two binary operations $+$ and $\cdot$ and two constant elements 0 and 1 such that

(i) A forms a commutative monoid with respect to $+$ and 0,
(ii) A forms a monoid with respect to $\cdot$ and 1,
(iii) the distribution laws $a \cdot (b+c) = a \cdot b + a \cdot c$ and $(a+b) \cdot c = a \cdot c + b \cdot c$ hold for every a, b, c,
(iv) $0 \cdot a = a \cdot 0 = 0$ for every a.

A semiring is called *commutative* iff $a \cdot b = b \cdot a$ for every a and b. Intuitively, a semiring is a ring (with unity) without subtraction. A very typical example of a semiring that is also very important in the sequel is the semiring of nonnegative integers N. The most important semiring in connection with language theory is the Boolean semiring $B = \{0, 1\}$ where $1 + 1 = 1 \cdot 1 = 1$. Clearly, all rings (with unity), as well as all fields, are semirings, e.g., integers Z, rationals Q, reals R, complex numbers C etc. The reader might want to verify that each of the following is a semiring:

$$\langle R \cup \{\infty\}, \min, +, \infty, 0\rangle,$$
$$\langle R \cup \{-\infty\}, \max, +, -\infty, 0\rangle,$$
$$\langle \{x \in R \mid 0 \leqslant x \leqslant 1\}, \max, \cdot, 0, 1\rangle,$$
$$\langle \{x \in R \mid x \geqslant 0\} \cup \{\infty\}, \max, \min, 0, \infty\rangle,$$

where min and max are defined in the obvious fashion.

Consider a monoid Σ^* as defined above and a set S. Mappings r of Σ^* into S are called *formal power series*. The values of r are denoted by (r, w) where $w \in \Sigma^*$, and r itself is written as a formal sum

$$r = \sum_{w \in \Sigma^*} (r, w)w.$$

The values (r, w) are also referred to as the *coefficients* of the series. We say also that r is a series with (noncommuting) *variables* in Σ. The collection of all power series r as defined above is denoted by $S\langle\langle \Sigma^* \rangle\rangle$.

This terminology reflects the intuitive ideas connected with power series. We call the power series "formal" to indicate that we are not interested in summing up the series but rather, for instance, in various operations defined for series. The power series notation makes it very convenient to discuss such operations in case S has enough structure, for instance S is a semiring. The difference between our formal power series and the ones studied in combinatorics is that in the latter the variables commute. Both approaches have in common the basic idea to view the collection of all power series as an algebraic structure. In what follows we only consider the cases where S is a monoid or a semiring.

Given $r \in A\langle\langle \Sigma^* \rangle\rangle$, where A is a semiring, the subset of Σ^* defined by

$$\{w \mid (r, w) \neq 0\}$$

is termed the *support* of r and denoted by supp(r). The subset of $A\langle\langle \Sigma^* \rangle\rangle$ consisting of all series with a finite support is denoted by $A\langle \Sigma^* \rangle$. Series of $A\langle \Sigma^* \rangle$ are referred to as *polynomials*.

Subsets of Σ^* are called *languages* over Σ. Thus the consideration of supports brings about a natural interconnection between the theory of formal languages and the theory of formal power series.

Examples of polynomials belonging to $A\langle \Sigma^* \rangle$ for every A are 0, w, aw defined by

$$\begin{array}{ll} (0, w) = 0 & \text{for all } w, \\ (w, w) = 1 \text{ and } (w, w') = 0 & \text{for } w' \neq w, \\ (aw, w) = a \text{ and } (aw, w') = 0 & \text{for } w' \neq w. \end{array}$$

Note that w equals $1w$.

We now introduce some operations inducing a monoid or semiring structure to power series. For $r_1, r_2 \in S\langle\langle \Sigma^* \rangle\rangle$, S being a monoid or a semiring, we define the *sum* $r_1 + r_2 \in S\langle\langle \Sigma^* \rangle\rangle$ by $(r_1 + r_2, w) = (r_1, w) + (r_2, w)$ for all $w \in \Sigma^*$.

For r_1, $r_2 \in A\langle\langle \Sigma^* \rangle\rangle$, A being a semiring, we define the (Cauchy) *product* $r_1 r_2 \in A\langle\langle \Sigma^* \rangle\rangle$ by $(r_1 r_2, w) = \Sigma_{w_1 w_2 = w}(r_1, w_1)(r_2, w_2)$ for all $w \in \Sigma^*$. Clearly, $\langle A\langle\langle \Sigma^* \rangle\rangle, +, \cdot, 0, \lambda\rangle$ and $\langle A\langle \Sigma^* \rangle, +, \cdot, 0, \lambda\rangle$ are semirings again.

A series r in $A\langle\langle \Sigma^* \rangle\rangle$, where every coefficient equals 0 or 1, is termed the *characteristic series* of its support L, in symbols,

$$r = \text{char}(L).$$

We now introduce similar operations on formal languages and in this fashion make the interconnection with formal power series referred to above more explicit. *Boolean operations* are defined for languages over Σ in the natural fashion. The *product* or *catenation* of two languages L_1 and L_2 is defined by

$$L_1L_2 = \{w_1w_2 \mid w_1 \in L_1, w_2 \in L_2\}.$$

The set of all subsets of a set S, i.e., the *power set* of S, is denoted by $\mathscr{P}(S)$. The power set includes also the *empty set* $\emptyset$.

Clearly, $\langle \mathscr{P}(\Sigma^*), \cup, \cdot, \emptyset, \{\varepsilon\}\rangle$ is a semiring isomorphic to $\langle B\langle\langle\Sigma^*\rangle\rangle, +, \cdot, 0, \varepsilon\rangle$. Essentially a transition from $\mathscr{P}(\Sigma^*)$ to $B\langle\langle\Sigma^*\rangle\rangle$, and vice versa, means a transition from L to char(L) and from r to supp(r), respectively.

Observe that we have now already defined most of the basic operations customary in language theory. A few additional remarks are in order. The morphic image of a language is defined additively:

$$h(L) = \{h(w) \mid w \text{ in } L\}.$$

The *catenation closure* L^* (also called Kleene star) of a language L is the union of all powers of L, including the power $L^0 = \{\lambda\}$. The *λ-free catenation closure* L^+ consists of all positive powers of L. Corresponding notions can be defined for formal power series but convergence has to be taken into consideration.

A *finite substitution* is a generalization of a morphism: each letter a can be replaced by an arbitrary word from a finite language $L(a)$. Otherwise the definitions are exactly as for morphisms. Thus a morphism is a finite substitution where each of the languages $L(a)$ is a singleton.

The following purely language-theoretic example can also be considered as an introduction to the next section.

The language L over the alphabet $\{a, b\}$ consists of all words that can be obtained from the empty word λ by finitely many applications of the following three rules.

Rule 1. If w is in L, then so is awb.

Rule 2. If w is in L, then so is bwa.

Rule 3. If w_1 and w_2 are in L, then w_1w_2 is also in L.

For instance, Rule 1 shows that *ab* is in L. Hence, by Rule 2, *baba* is in L and, by Rule 3, *babaab* is also in L. This same word can also be derived by first applying Rules 2 and 1 to get *ba* and *ab* and then applying Rule 3 twice. Hence the derivation of a word in L is by no means unique.

We claim that L consists of all words with an equal number of a's and b's. Indeed, since λ has this property and Rules 1–3 preserve it, we conclude that L contains only words with this property. We still have to show that, conversely, every word with this property is in L.

Clearly L contains all words of length 0 with equally many a's and b's. Proceeding inductively, we assume that L contains all words of length no greater than $2i$ with this property. Let w be an arbitrary word of length $2(i+1)$ with this property. Clearly $w \neq \lambda$. We assume without loss of generality that the first letter of w is a. (If this is not the case originally, we may interchange the roles of a and b because Rules 1–3 are symmetric with respect to a and b.) If the last letter of w is b, we obtain $w = aw_1b$, where w_1 is a word of length $2i$ with equally many a's and b's. By our inductive hypothesis, w_1 is in L. Hence by Rule 1, w is in L. If the last letter of w is a, w must have a nonempty prefix $w_1 \neq w$ containing equally many a's and b's. Consequently $w=w_1w_2$, where both w_1 and w_2 are words of length no greater than $2i$ with equally many a's and b's. By our inductive hypothesis, both w_1 and w_2 are in L. Hence by Rule 3, w is in L also in this case.

2. Rewriting systems and grammars

The notion of a *rewriting rule* or *production* is fundamental in language theory: a certain string can be replaced by another string. A rewriting rule can also be viewed as a (rather primitive) rule of inference in the sense customary in logic. Similarly a rewriting system, consisting of finitely many such rules, can be viewed as a system of logic.

Rewriting systems are used to obtain new words from given ones. By the addition of a suitable input–output bus, a rewriting system is converted to a language-defining device. Basically there are two ways for doing this.

Some *start words* (also called axioms) are chosen. The language defined by the system consists of all words obtainable from some start word by successive applications of rewriting rules. Some additional conditions may be imposed: for instance, words in the defined language have to be over a specific alphabet. The number of start words is finite; in case of grammars usually one. The procedure converts a rewriting system into a *generative* device.

Conversely, some *final words* (or target words) can be chosen. The language defined by the system consists of all words w such that a final word is obtainable from w by successive applications of rewriting rules. This procedure converts a rewriting system into an *accepting* device—various types of automata are examples of such devices.

We now give the formal definitions.

A *rewriting system* is a pair (Σ, P), where Σ is an alphabet and P is a finite set of ordered pairs of words over Σ. The elements (w, u) of P are referred to as *rewriting rules* or *productions* and denoted by $w \rightarrow u$. Given a rewriting system, the *yield relation* $\Rightarrow$ on the set Σ^* is defined as follows. For any words x and y, $x \Rightarrow y$ holds if and only if there are words x_1, x_2, w, u such that

$$x=x_1wx_2, \qquad y=x_1ux_2$$

and $w \rightarrow u$ is a production in the system. The reflexive transitive closure of the relation $\Rightarrow$ is denoted by $\Rightarrow^*$. (Thus $x \Rightarrow^* y$ holds if and only if there are $n \geqslant 1$ words $x_1, \ldots, x_n$ such that $x=x_1, y=x_n$ and $x_i \Rightarrow x_{i+1}$ holds for every $i=1, \ldots, n-1$.)

A (*phrase structure*) *grammar* is a quadruple $G=(R, \Sigma_N, \Sigma_T, \Sigma_S)$ where $R=(\Sigma, P)$ is a rewriting system, Σ_N and Σ_T are disjoint alphabets such that $\Sigma=\Sigma_N \cup \Sigma_T$, and Σ_S is a subset of Σ_N. The elements of Σ_N, Σ_T and Σ_S are referred to as *nonterminal*, *terminal* and *start* letters, respectively. The language *generated* by the grammar G is defined by

$$L(G)=\{w \in \Sigma_T^* \mid S \Rightarrow^* w \text{ for some } S \in \Sigma_S\}.$$

(Here $\Rightarrow^*$ is the relation determined by the rewriting system R.)

For $i=0, 1, 2, 3$, the grammar G is of *type i* if the restrictions (i) on the production set

P, as given below, are satisfied:

(0) No restrictions.

(1) Each production in P is of the form $w_1 A w_2 \rightarrow w_1 w w_2$, where w_1 and w_2 are arbitrary words over Σ, w is a nonempty word over Σ, and A belongs to Σ_N (with the possible exception of the production $S \rightarrow \lambda$, $S \in \Sigma_S$, whose occurrence in P implies, however, that S does not occur on the right side of any production).

(2) Each production is of the form $A \rightarrow w$ with A in Σ_N.

(3) Each production is of the form $A \rightarrow aB$, where $A, B \in \Sigma_N$ and $a \in \Sigma_T$, or of the form $A \rightarrow \lambda$, where $A \in \Sigma_N$.

A language is of *type i*, $i = 0, 1, 2, 3$, if it is generated by a grammar of type i.

It can be shown that the resulting four families of languages constitute a strictly decreasing hierarchy of language families, usually referred to as the *Chomsky hierarchy*. Type 0 languages are also called *recursively enumerable*. Type 1 (resp. type 2, type 3) grammars and languages are also called *context-sensitive* (resp. *context-free*, *right-linear* or *regular*).

For instance, the language considered at the end of Section 1 is generated by the context-free grammar with the productions

$$S \rightarrow aSb, \qquad S \rightarrow bSa, \qquad S \rightarrow SS, \qquad S \rightarrow \lambda.$$

Here the start letter S is the only nonterminal. In general, we apply the convention that capital letters are nonterminals and lower case letters terminals.

Derivations can be depicted as *derivation trees* or *parse trees* in case of context-free grammars. For instance, the derivation

$$S \Rightarrow SS \Rightarrow aSbS \Rightarrow aaSbbS \Rightarrow aabbS \Rightarrow aabbbSa \Rightarrow aabbba$$

according to the preceding grammar can be depicted as

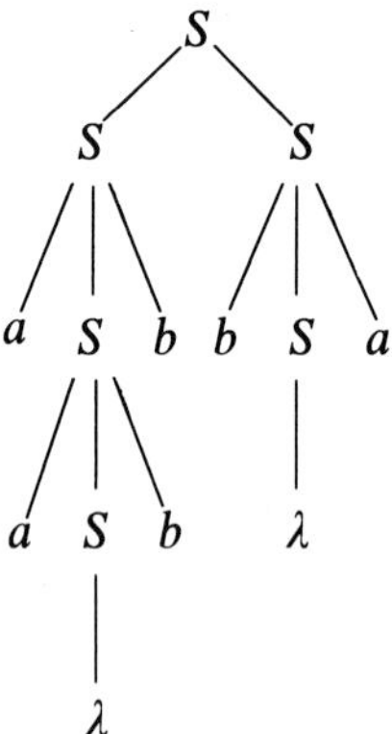

It is often very difficult to show that a specific grammar generates the intended language. Logically this is an undecidable problem. It can also be viewed as a rudimentary version of the problem of showing that a computer program works as wanted. Below are a couple of examples of nontrivial grammars where such a proof is easy.

Consider the grammar G determined by the productions

$$S \to abc, \quad S \to aAbc, \quad Ab \to bA, \quad Ac \to Bbcc, \quad bB \to Bb, \quad aB \to aaA, \quad aB \to aa.$$

We claim that

$$L(G) = \{a^n b^n c^n \mid n \geqslant 1\}.$$

In fact, any derivation according to G begins with an application of the first or second production, the first production directly yielding the terminal word abc. Consider any derivation D from a word $a^i A b^i c^i$, where $i \geqslant 1$, leading to a word over the terminal alphabet. D must begin with i applications of the third production (A travels to the right) and then continue with an application of the fourth production (one further occurrence of b and c is deposited). Now we have derived the word $a^i b^i B b c^{i+1}$. For this the only possibility is to apply the fifth production i times (B travels to the left), after which we have the word $a^i B b^{i+1} c^{i+1}$. This word directly yields one of the two words

$$a^{i+1} A b^{i+1} c^{i+1} \quad \text{or} \quad a^{i+1} b^{i+1} c^{i+1}$$

by the last two productions (one further a is deposited, and either a new cycle is entered or the derivation is terminated). This argument shows that G generates all words belonging to the language and nothing else; every step in a derivation is uniquely determined, the only exception being that there is a choice between termination and entrance into a new cycle.

The reader may verify that the grammar

$$G_1: \quad S \to aSAB, \quad S \to abB, \quad BA \to AB, \quad bA \to bb, \quad bB \to bc, \quad cB \to cc,$$

and the grammar

$$G_2: \quad S \to aSA, \quad S \to aB, \quad BA \to bBc, \quad cA \to Ac, \quad B \to bc$$

are both equivalent to G, that is, generate the same language. The language generated is perhaps the most typical non-context-free language.

For the following example G_3, due to [28], it is also easy to show that the generated language is

$$L(G_3) = \{a^{n^2} \mid n \geqslant 1\}.$$

The productions of G_3 are

$$\begin{array}{llllll} S \to a, & S \to CD, & C \to ACB, & C \to AB, & AB \to aBA, & Aa \to aA, \\ Ba \to aB, & AD \to Da, & BD \to Ea, & BE \to Ea, & E \to a. & \end{array}$$

First words of the form $A^n B^n D$ are derived, and from these $a^{n^2} B^n A^n D$. The procedure is deterministic enough not to allow any "parasite" derivations.

The reader is referred to [30] or [31] for a discussion of various types of automata as rewriting systems. As an example we give here the definition of a pushdown automaton.

A rewriting system (Σ, P) is called a *pushdown automaton* PDA iff the following conditions are satisfied:

(i) Σ is divided into two disjoint alphabets Q and $\Sigma_t \cup \Sigma_p$. The sets Q, Σ_t and Σ_p are called the *state*, *tape* and *pushdown* alphabet, respectively. The sets Σ_t and Σ_p are nonempty but not necessarily disjoint.

(ii) Elements $q_0 \in Q, z_0 \in \Sigma_p$, and a subset $Q_1 \subseteq Q$, are specified, called the *initial state*, *start letter* and *final state set*.

(iii) The productions in P are of the two forms

$$zq_i \rightarrow wq_j, \quad z \in \Sigma_p,\ w \in \Sigma_p^*,\ q_i, q_j \in Q,$$

$$zq_i a \rightarrow wq_j, \quad z \in \Sigma_p,\ w \in \Sigma_p^*,\ a \in \Sigma_t,\ q_i, q_j \in Q.$$

The language *accepted* by PDA is defined by

$$L(\text{PDA}) = \{u \in \Sigma_t^* \mid z_0 q_0 u \Rightarrow^* w' q_1, \text{ for some } q_1 \in Q_1,\ w' \in \Sigma_p^*\}.$$

The mode of acceptance used here is by final state. Acceptance by the empty stack (as well as by both the empty stack and final state) is defined similarly.

The family of languages acceptable by PDAs equals the family of context-free languages. Automata corresponding to the remaining three classes in the Chomsky hierarchy are Turing machines, linear bounded automata and finite automata, respectively.

A further discussion about the Chomsky hierarchy lies beyond the scope of this chapter. A good starting point is to consider various *normal forms* for grammars. For instance, every context-free language (not containing the empty word) is generated by a grammar in the *Greibach normal form*: all productions are of the three types

$$A \rightarrow aBC, \qquad A \rightarrow aB, \qquad A \rightarrow a.$$

Thus every application of a production "deposits" a terminal. This is useful, for instance, in parsing considerations.

Every recursively enumerable language is generated, [34], by a grammar whose productions are of the three types

$$AB \rightarrow \lambda, \qquad A \rightarrow BC, \qquad A \rightarrow a.$$

Rewriting systems and grammars have been extended to include *regulated rewriting*. This means that the application of productions is not arbitrary but is (at least to some extent) controlled by some device. *Matrix grammars* constitute the oldest form of regulated rewriting. Finitely many finite sequences of productions ("matrices") are given. The whole sequence of productions has to be applied before the next matrix can be chosen. For instance, the context-free matrix grammar with the three matrices

$$[S \rightarrow ABC], \qquad [A \rightarrow aA,\ B \rightarrow bB,\ C \rightarrow cC], \qquad [A \rightarrow a,\ B \rightarrow b,\ C \rightarrow c]$$

generates the non-context-free language

$$\{a^n b^n c^n \mid n \geqslant 1\}.$$

Very little is known about the generative capacity of matrix grammars. Characterization results are difficult to obtain because, for instance, power series techniques do not seem to be applicable to regulated rewriting. Some interesting interconnections with other research areas have been observed. For instance, the emptiness problem of languages generated by context-free matrix grammars is equivalent, from the point of view of decidability, to the famous reachability problem in Petri nets. The reader is referred to [7] as the most comprehensive exposition concerning regulated rewriting.

3. Post canonical systems

We now introduce a class of generating devices somewhat different from grammars but still equivalent to grammars in the sense that the family of languages generated by our new devices equals the family of recursively enumerable languages. The new devices—called *Post canonical systems*, or *Post systems*, after Emil Post—very closely resemble formal systems in logic. The productions have the shape of inference rules: given some premises, a conclusion can be made. The formal definition is as follows.

A *Post system* is a quadruple

$$\mathrm{PS}=(\Sigma_{\mathrm{V}}, \Sigma_{\mathrm{C}}, P, S),$$

where Σ_{V} and Σ_{C} are disjoint aphabets, referred to as the *alphabet of variables* and the *alphabet of constants*, respectively; S is a finite set of words over Σ_{C}, the words being referred to as *axioms*, or *primitive assertions*; and P is a finite set of *productions* where a production is a $(k+1)$-tuple, $k \geqslant 1$,

$$(\alpha_1, \ldots, \alpha_k, \alpha)$$

of words over the alphabet $\Sigma=\Sigma_{\mathrm{V}} \cup \Sigma_{\mathrm{C}}$. For different productions, the number k may be different.

Productions are usually denoted by $(\alpha_1, \ldots, \alpha_k) \rightarrow \alpha$ (for $k=1$ we omit the parentheses) and read "the conclusion α can be made from the premises $\alpha_1, \ldots, \alpha_k$". More specifically, the conclusion is made in the following fashion. For all variables appearing in the words $\alpha_1, \ldots, \alpha_k$, α, we uniformly substitute some words over the constant alphabet Σ_{C}. Here *uniform substitution* means that all occurrences of the same variable A have to be replaced by the same word over the constant alphabet Σ_{C}. The constant word (meaning: word over Σ_{C}) $\bar{\alpha}$ obtained in this fashion can be inferred from the corresponding constant words $\bar{\alpha}_1, \ldots, \bar{\alpha}_k$. Initially we have the constant words in S, and we can use them as premises. Whenever we have inferred some new constant word, we can use it as a premise in later inferences. The language $L(\mathrm{PS}, \Sigma_{\mathrm{T}})$ *generated* by the Post system PS with respect to the *terminal* alphabet $\Sigma_{\mathrm{T}} \subseteq \Sigma_{\mathrm{C}}$ consists of all words over Σ_{T} obtained in this fashion. Thus we may decompose the alphabet Σ_{C} in two parts, $\Sigma_{\mathrm{C}}=\Sigma_{\mathrm{N}} \cup \Sigma_{\mathrm{T}}$, such that an inferred word has to be over the alphabet Σ_{T} in order to be included in the language generated by the Post system. Here Σ_{N} may be empty.

The Post system with λ as the only axiom, $\Sigma_{\mathrm{V}}=\{A, A_1, A_2\}$, $\Sigma_{\mathrm{C}}=\{a, b\}$, and with the productions

$$A \rightarrow aAb, \qquad A \rightarrow bAa, \qquad (A_1, A_2) \rightarrow A_1 A_2$$

generates, with respect to the terminal alphabet $\Sigma_{\mathrm{T}}=\Sigma_{\mathrm{C}}$, the language considered at the end of Section 1.

The family of languages generated by Post systems equals the family of recursively enumerable languages. Furthermore the following normal form result holds. For every Post system an equivalent one can be constructed such that the productions of the latter are of the form

$$uA \rightarrow Aw,$$

where A is a variable and u and w are words over the alphabet of constants.

A somewhat modified aspect of this result has popped up in several contexts lately. Consider Post systems whose productions are of the form

$$uA \rightarrow wA,$$

where A, u and w are as above. According to a result of [3], see also [32], a language is regular iff it is generated by such a Post system.

4. Markov algorithms

The model studied in this section resembles the intuitive notion of an algorithm: it gives rise to an effective procedure for processing an input. At each step of the procedure, the next step is uniquely determined. This property of *monogenicity* is possessed neither by grammars nor by Post systems, where in the course of a derivation we may have several choices. However viewed as language-defining devices, Markov algorithms are equivalent to grammars and Post systems.

A *Markov algorithm* is a rewriting system (Σ, P), where the elements of P are given in a linear order

$$\alpha_1 \rightarrow \beta_1, \ldots, \alpha_k \rightarrow \beta_k,$$

and, furthermore, a subset P_1 of P is given. The elements (α, β) of P_1 are called *final* productions and are denoted by $\alpha \rightarrow . \beta$.

At each step of the rewriting process, the first applicable production must be chosen and, furthermore, the leftmost occurrence of its left side must be rewritten. The rewriting process terminates when some final production is applied or when none of the productions is any more applicable. These notions are formalized as follows.

Given a Markov algorithm MA with productions as above, a binary relation $\Rightarrow_{\mathrm{MA}}$—or, briefly, $\Rightarrow$ (yields directly)—on the set of all words over the alphabet of MA is defined as follows. For any words u and w, $u \Rightarrow w$ holds iff each of the conditions (i)–(iv) is satisfied.

(i) There is a number i, $1 \leqslant i \leqslant k$, and words u_1, u_2 such that $u = u_1 \alpha_i u_2$, $w = u_1 \beta_i u_2$.

(ii) None of the words α_j with $j < i$ is a subword of u.

(iii) α_i occurs as a subword of $u_1 \alpha_i$ only once.

(iv) One of the words $\alpha_1, \ldots, \alpha_k$ is a subword of w, and $\alpha_i \rightarrow \beta_i$ is not a final production.

We may, of course, consider the reflexive, transitive closure $\Rightarrow^*$ of the relation $\Rightarrow$ introduced above. However we do this differently, implementing the applications of final productions.

For any words u and w over the alphabet of MA, $u \Rightarrow . w$ holds iff conditions (i)–(iii) are satisfied but condition (iv) is not satisfied (that is, either $\alpha_i \to \beta_i$ is a final production, or none of the words $\alpha_1, \ldots, \alpha_k$ is a subword of w). Furthermore $u \Rightarrow\!\!.^{*}\, w$ holds iff there is a finite sequence of words

$$u = x_0, x_1, \ldots, x_n, w, \quad n \geqslant 0,$$

such that $x_j \Rightarrow x_{j+1}$ holds for $0 \leqslant j \leqslant n-1$ and $x_n \Rightarrow . w$, or else $u = w$ and none of the words $\alpha_1, \ldots, \alpha_k$ is a subword of u.

It is an immediate consequence of the definitions that for any word u, there is at most one word w with the property $u \Rightarrow\!\!.^{*}\, w$. If such a w exists, we say that our Markov algorithm MA *halts* with u and *translates* u into w. Otherwise, we say that MA *loops* with u.

For instance, the Markov algorithm with the alphabet $\{a, b\}$ and the list of productions

$$a \to a^2, \qquad b^2 \to \lambda, \qquad b \to . ab$$

loops with every word containing an occurrence of a and translates, for every $i \geqslant 0$, the word b^{2i} (resp. b^{2i+1}) into λ (resp. ab).

The Markov algorithm with the alphabet $\{a, A, B, \#\}$ and the list of productions

$$\begin{array}{llll} Ba \to aB, & Aa \to aBA, & A \to \lambda, & a\# \to \# A, \\ \# a \to \#, & \# \to \lambda, & B \to a & \end{array}$$

translates every word of the form

$$a^i \# a^j, \quad i, j \geqslant 0,$$

into the word a^{ij}. Hence in this notation, this algorithm multiplies two numbers.

The Markov algorithm with the alphabet $\{a, A, B, C, \#\}$ and the list of productions

$$\begin{array}{llll} aA \to Aa, & a\#a \to A\#, & a\# \to \# B, & B \to a, \\ A \to C, & C \to a, & \# \to \lambda & \end{array}$$

translates every word $a^i \# a^j$ into the word a^k, where k is the greatest common divisor of i and j.

The verification of these facts is left to the reader. For both of the latter Markov algorithms, the set of final productions is empty.

We discussed in Section 2 the general principles of converting a rewriting system into a generative or accepting device for languages. As regards Markov algorithms, the former alternative is not satisfactory: only finitely many words can be generated from a finite set of axioms. Therefore, we convert a Markov algorithm into an accepting device as follows.

Let MA be a Markov algorithm and Σ_{T} a subset of its alphabet. Then we define

$$L(\mathrm{MA}, \Sigma_{\mathrm{T}}) = \{ w \in \Sigma_{\mathrm{T}}^{*} \mid w \Rightarrow\!\!.^{*}\, \lambda \}.$$

A language is recursively enumerable iff it is of the form $L(\mathrm{MA}, \Sigma_{\mathrm{T}})$.

From this it immediately follows (for instance, by a reduction to the halting problem of Turing machines) that there is no algorithm for deciding whether or not a given MA halts with a given word. However, the decidability status of some interesting subproblems is open. We conjecture that the problem is decidable for Markov algorithms with only one production. (The production should be strictly length-increasing; otherwise, the decision method is trivial.) If the conjecture holds, the next step is to study Markov algorithms with two productions, and so forth.

5. Parallel rewriting systems

All the models considered in the preceding sections have the property that at each step of the rewriting process, only some specific part of the word considered is actually rewritten. This is to be contrasted with *parallel* rewriting: at each step of the process, all letters of the word considered must be rewritten.

The most-widely studied parallel rewriting systems are referred to as *L systems*; they were introduced by Lindenmayer [18] to provide a model for the development of filamentous organisms. In this connection parallel rewriting is most natural, since the development takes place simultaneously everywhere in the organism rather than in some specific part only. Of course words as such represent one-dimensional organisms, but by suitable conventions one may use words to describe the many-dimensional case as well [29].

To take a very simple example, consider the production $a \rightarrow aa$. When it is viewed as a production in a rewriting system, all words a^i, $i \geqslant 1$, are derivable from the start word a. On the other hand, if parallel rewriting is considered, then only words of the form a^{2^i}, $i \geqslant 0$, are derivable from the start word a. This follows because, for instance, the word a^2 directly yields only the word a^4 because the production $a \rightarrow aa$ has to be applied for *both* occurrences of the letter a. If we have the productions $a \rightarrow aa$ and $a \rightarrow a$ available, then all words a^i, $i \geqslant 1$, are derivable from the start word a, even if parallel rewriting is considered. Thus, the presence of the production $a \rightarrow a$, which is immaterial for rewriting systems, makes a big difference if parallel rewriting is considered: it makes possible for us to keep some occurrences of a unaltered while the derivation is carried on in some other parts of the word.

We now define the most-widely studied types of L systems, 0L and D0L systems. We use the notations customary in the theory of L systems: D stands for *deterministic* and 0 stands for *context-free*, the latter abbreviation being based on the original biological connotation: the communication between different cells in the development process is 0-sided—that is, there is no communication at all.

A *0L-system* is a triple, $G = (\Sigma, P, w)$, where Σ is an alphabet, w is a word over Σ (start word or axiom), and P is a finite set of ordered pairs (a, α), where a is in Σ and α is in Σ^*, such that for each a in Σ, there is at least one pair (a, α) in P. The elements (a, α) of P are called *productions* and are denoted by $a \rightarrow \alpha$. If for each a in Σ, there is exactly one production $a \rightarrow \alpha$ in P, then G is called a *D0L system.*

The above definition defines 0L systems as rewriting systems of a special kind. However we do not want to call them rewriting systems because the binary relation $\Rightarrow$ (yields directly) is defined differently for 0L systems and for rewriting systems. The definition for 0L systems, which takes care of the parallelism, will now be given.

Given a 0L system $G=(\Sigma, P, w)$, a binary relation $\Rightarrow_G$ — or, briefly, $\Rightarrow$ — on the set Σ^* is defined as follows:

$$a_1 \ldots a_n \Rightarrow \alpha_1 \ldots \alpha_n, \quad n \geqslant 1, \text{ each } a_i \in \Sigma$$

holds iff $a_i \rightarrow \alpha_i$ is in P for each $i=1, \ldots, n$. The *language generated by* G is defined by

$$L(G)=\{u \mid w \Rightarrow^* u\},$$

where $\Rightarrow^*$ is the reflexive, transitive closure of $\Rightarrow$. Languages of this form, where G is a 0L (resp. D0L) system, are referred to as *0L* (resp. *D0L*) *languages.*

As in case of grammars, we may specify a terminal alphabet $\Sigma_T \subseteq \Sigma$ for 0L systems. The generated language is defined as above except that now u must be a word over Σ_T. The resulting languages are referred to as *E0L languages* (short for "extended 0L"). Thus, every E0L language is of the form

$$L(G) \cap \Sigma_T^*,$$

for some 0L system G and a subset Σ_T of its alphabet. It is easy to see that context-free languages constitute a subfamily of the family of E0L languages. That the subfamily is proper is seen from the following example.

Consider the E0L system G with the axiom S and productions $S \rightarrow ABC$, $A \rightarrow A\bar{A}$, $B \rightarrow B\bar{B}$, $C \rightarrow C\bar{C}$, $\bar{A} \rightarrow \bar{A}$, $\bar{B} \rightarrow \bar{B}$, $\bar{C} \rightarrow \bar{C}$, $\bar{A} \rightarrow a$, $\bar{B} \rightarrow b$, $\bar{C} \rightarrow c$, $A \rightarrow a$, $B \rightarrow b$, $C \rightarrow c$, $a \rightarrow N$, $b \rightarrow N$, $c \rightarrow N$ and $N \rightarrow N$, where $\{a, b, c\}$ is the terminal alphabet. It is not difficult to verify that

$$L(G)=\{a^n b^n c^n \mid n \geqslant 1\}.$$

The system G is based on the *synchronization* of terminals: the terminals have to be introduced simultaneously in the whole word. If terminals are introduced only in some parts of the word, then the derivation cannot be continued to yield a word over the terminal alphabet. This is due to the fact that terminals give rise only to the "garbage" letter N, which cannot be eliminated once it is introduced.

The L systems considered so far are based on context-free productions: each letter is rewritten independently of its neighbors. This is indicated by the number 0 in the name of the system. We do not define formally the notion of an *IL system* (I meaning interactions). The productions of an IL system have the form

$$(\beta_1, a, \beta_2) \rightarrow \alpha,$$

meaning that the letter a can be rewritten as the word α *between* the words β_1 and β_2. Productions are again applied *in parallel*: each letter has to be rewritten during one step of the derivation process. Languages obtained in this fashion are called *IL languages. EIL* languages are defined analogously. It is not surprising that a language is recursively enumerable iff it is an EIL language.

An area quite essential in the study of L systems is the theory of *growth functions.* From the biological point of view, the growth function measures the size of the organism modeled. From the theoretical point of view, the theory of growth functions reflects a fundamental aspect of L systems: generating languages as *sequences* of words. The theory is closely linked with the theory of formal power series.

For a D0L system G, we associate the function f_G mapping the set of nonnegative integers into itself as follows. For $i \geqslant 0$, $f_G(i)$ is defined to be length of the ith word in the sequence $E(G)$, whereby the axiom of G is understood to be the 0th word. The function f_G is called the *growth function* of G.

For instance, for the D0L system G with the axiom a and the only production $a \rightarrow aa$, we have $f_G(i) = 2^i$ for all $i \geqslant 0$.

For the D0L system G with the axiom a and the productions $a \rightarrow ab$ and $b \rightarrow b$, we have $f_G(i) = i + 1$ for all $i \geqslant 0$.

For the D0L system G with the axiom a and productions $a \rightarrow abc^2$, $b \rightarrow bc^2$, and $c \rightarrow c$ (resp. $a \rightarrow abd^6$, $b \rightarrow bcd^{11}$, $c \rightarrow cd^6$ and $d \rightarrow d$) we have $f_G(i) = (i+1)^2$ (resp. $f_G(i) = (i+1)^3$).

To a given D0L system G with the alphabet $\{a_1, \ldots, a_n\}$, we associate the *axiom vector* $\pi_G = (v_1, \ldots, v_n)$, where v_i equals the number of occurrences of a_i in the axiom of G for each $i = 1, \ldots, n$. We also associate with G the $n \times n$ *growth matrix* M_G whose (i, j)th entry equals the number of occurrences of a_j in the production for a_i for all i and j. Finally, η is a column vector of dimension n consisting entirely of 1s.

It is easy to show by induction that

$$f_G(i) = \pi_G M^i \eta \quad \text{for all } i \geqslant 0.$$

This indicates the interconnection with formal power series: growth functions constitute a subclass of N-rational functions.

Systems with *tables* (abbreviated T) constitute a widely studied class of L systems. Finitely many sets of productions ("tables") are specified. At each derivation step productions belonging to the same table have to be used. Rewriting is parallel as before. Tables are not necessarily disjoint. It should now be clear that what one means by T0L and ET0L systems and languages.

According to a result of [17], ET0L systems constitute perhaps the simplest grammatical device for defining an NP-complete language. (This means that the ET0L membership problem is NP-complete, since it is easy to see that the problem is in NP.) We now give the construction for the language SAT_3, consisting of satisfiable formulas of propositional calculus in 3-conjunctive normal form and with the unary notation for variables.

We define an ET0L system G as follows. The nonterminal alphabet consists of the letters S (initial), R, T and F. The terminal alphabet consists of the letters A (intuitively, disjunction), N (negation), 1, (, and). The system G has three tables.

The first table consists of the productions

$$S \rightarrow (xAyAz)S, \qquad S \rightarrow (xAyAz), \qquad a \rightarrow a \quad \text{for } a \neq S,$$

where (x, y, z) ranges over all combinations of (T, NT, F, NF) which do not consist entirely of NTs and Fs. The second table consists of the productions

$$S \rightarrow R, \qquad T \rightarrow 1T, \qquad F \rightarrow 1F, \qquad T \rightarrow 1, \qquad a \rightarrow a,$$

for $a \neq S, T, F$. The third table is obtained from the second by replacing $T \rightarrow 1$ with $F \rightarrow 1$.

It is easy to verify that $L(G) = \text{SAT}_3$. The first table generates conjunctions of disjunctions with three terms in a form already indicating the truth-value assignments. The terminating productions in the other two tables guarantee that the truth-value assignment is consistent.

6. Morphisms and languages

The operation of a morphism has turned out to be of particular importance in language theory. Also some central issues about power series are involved. Some typical results, with emphasis on recent work, are outlined in this section.

A fundamental notion is defined as follows. Consider two morphisms

$$g, h : \Sigma^* \to \Sigma_1^*.$$

Their *equality set* is defined by

$$E(g, h) = \{w \in \Sigma^* \mid g(w) = h(w)\}.$$

For instance, $E(g, h)$ is the language considered at the end of Section 1 if g and h are defined by

$$g(a) = h(b) = a, \qquad g(b) = h(a) = aa.$$

For the morphisms g and h defined by

$$g(a) = aab, \qquad g(b) = a,$$

$$h(a) = a, \qquad h(b) = baa,$$

we have $E(g, h) = \{a^2b^2\}^*$.

An instance of the *Post correspondence problem* can be viewed as given by two such morphisms g and h. The instance possesses a solution iff $E(g, h)$ contains a nonempty word. Hence the problem of whether or not $E(g, h)$ contains a nonempty word is undecidable.

Every recursively enumerable language L can be represented in the form

$$L = f(E(g, h) \cap R),$$

where R is a regular language and f is a morphism erasing some letters and leaving the others unchanged. The proof of this result is based on derivations according to the grammar G generating L. Essentially, the idea in the proof is the following. Every word in L appears as the last word in some derivation D according to G. D can be viewed as a single word, where the individual derivation steps are separated by a marker. For this word, the morphisms g and h satisfy

$$g(D) = h(D).$$

However g "runs faster" than h on prefixes of D, and h "catches up" only at the end. The morphism f erases everything except the end result. Finally R is used to make sure that only words of the proper form are taken into account. The reader is referred to [31] and to [4] for related stronger results.

The result mentioned is only one of the numerous results concerning morphic representations of language families. Another such result (for details see [31]) is that there is a context-free language L_0 with the following property. For every context-free language L (not containing the empty word), there is a morphism h with the property

$$L = h^{-1}(L_0).$$

Indeed, morphisms play an important role (at least implicitly) in the earliest work in language theory: the work done by Axel Thue before 1910. In our terminology, the main issues of this work can be described as follows.

Consider the D0L system G with the axiom a and productions

$$a \rightarrow ab, \qquad b \rightarrow ba.$$

The first few words in the sequence are

$$a, \quad ab, \quad abba, \quad abbabaab, \quad abbabaabbaababba.$$

In general, it is easy to verify inductively that, for every $i \geqslant 1$, the $(i+1)$st word w_{i+1} in the sequence satisfies

$$w_{i+1} = w_i w_i',$$

where w_i' results from w_i by interchanging a and b. This also shows that every word in the sequence is a prefix of the next one. (This property is, in general, a consequence of the fact that the axiom is a prefix of the next word.) Consequently G defines an "infinite word" e_G — that is, an infinite sequence of letters of the alphabet $\{a, b\}$ in a natural fashion. To find the jth letter of e_G, we just look at the jth letter in some w_i with $|w_i| \geqslant j$.

The sequence e_G is the celebrated *Thue sequence*, due to Thue [36]. It has the property of *cube-freeness*: it has no subsequence of the form x^3, $x \neq \lambda$, The proof should be not too difficult, following the subsequent guidelines: (1) Neither a^3 nor b^3 occurs as a subword in e_G because e_G is obtained by catenating the two words ab and ba. (2) Consequently, neither $ababa$ nor $babab$ occurs as a subword in e_G. (3) Whenever a^2 or b^2 occurs as a subword in e_G, starting with the jth letter of e_G, then j is even. By (2) and (3), it is now easy to establish inductively the cube-freeness of e_G, recalling the fact that e_G is obtained by iterating a morphism h.

It is clear that no word of length greater than or equal to 4 over a two-letter alphabet $\{a, b\}$ is square-free in the sense that it has no subword x^2, $x \neq \lambda$; beginning with a, the next letter has to be b and the following a, after which a square is unavoidable. On the other hand, it is possible to construct an infinite square-free word over a three-letter alphabet, using e_G.

There has been quite much activity dealing with *Ehrenfeucht's conjecture* during recent years (cf. [15]). According to the conjecture, every language L whatsoever possesses a *finite* subset $F \subseteq L$, referred to as a *test set* of L, with the following property: whenever g and h are morphisms such that the equation

$$g(w) = h(w)$$

holds for every word w in F, it holds also for every word w in L.

Thus if we want to test of two arbitrary morphisms whether or not they coincide pointwise on L, it suffices to do this for the words of F. Since L is an arbitrary language (not necessarily even recursively enumerable), the construction of F is not in general effective. Otherwise, many well-known undecidable problems would become decidable!

Ehrenfeucht's conjecture was proven in 1985, the first proofs being due to Albert and Lawrence [1] and Guba [11]. Several other proofs have been presented later on (see, e.g. [15]). Essentially all proofs are based on two reduction steps. We now outline an argument.

The given language L is first replaced by a system S of word equations as follows. For each letter a in the alphabet of L, also its "barred version" $\bar{a}$ is considered. The system S consists of all equations $w = \bar{w}$, where w is in L and $\bar{w}$ is obtained from w by replacing every letter with its barred version. Thus, S is an infinite system iff L is infinite. For instance, if we have

$$L = \{a^n b^n \mid n \geqslant 1\},$$

then S consists of the equations

$$a^n b^n = \bar{a}^n \bar{b}^n, \quad n = 1, 2, \ldots$$

The original letters, as well as their barred versions, are viewed as variables in S. Thus a solution for S is an assignment of words (perhaps over an entirely different alphabet) for the variables making all equations in S valid. If the morphisms g and h coincide on L, then the values

$$g(a), g(b), \ldots \quad \text{and} \quad h(a), h(b), \ldots$$

constitute such an assignment. This means that Ehrenfeucht's conjecture holds if every system S of word equations possesses a finite equivalent subsystem.

The second reduction step consists of replacing S by a system S' of *numerical* equations. If a solution for S involves words over the alphabet $\{a_0, a_1, \ldots, a_{m-1}\}$, then all words are viewed as integers in m-ary notation, and the equations in S' express the fact that the corresponding integers are the same. For instance, to the equation $uvu = v$ in S there correspond two equations

$$u_1 v_2 u_2 + v_1 u_2 + u_1 = v_1, \qquad u_2 v_2 u_2 = v_2$$

in S'. Intuitively, u_1 is the number represented by the word u, and u_2 is the length of this word.

By Hilbert's Basissatz (which is not difficult to prove directly), the system S' has a finite equivalent subsystem S''. A finite equivalent subsystem of S is now obtained by collecting all equations giving rise to an equation in S''. Observe finally that S' may have solutions not leading to any solution of S.

We have already emphasized that the construction of a test set $F \subseteq L$ cannot be effective in general. However, for some families of languages L it is indeed effective. Many interesting results have been recently obtained along these lines. We now mention one example.

The *DOL equivalence problem* consists of deciding whether or not two given DOL systems generate the same sequence of words. The decidability of this problem was for many years maybe the most celebrated open problem in language theory. We omit the history and the different interconnections and implications, and only outline a particularly simple recent solution due to [6].

It is an easy consequence of a result of Makanin [19] that the equivalence problem for two finite systems of word equations is decidable. Consequently it is decidable of two finite languages F_1 and F_2, $F_1 \subseteq F_2$, whether or not F_1 is a test set of F_2.

It follows immediately by the definition of a test set that, whenever F is a test set of L and g is a morphism, then $g(F)$ is a test set of $g(L)$.

Let us now return to the DOL equivalence problem. We may assume that the two given D0L systems G and H have the same axiom w. (Otherwise, the sequences are certainly not the same.) Let g and h be the morphisms of the two systems, respectively. Clearly the word sequences defined by G and H are the same iff the morphisms g and h coincide on the language $L(G)$, that is, $g(x)=h(x)$ holds for every x in $L(G)$.

Let L_i be the finite language consisting of the first i words in the sequence of G. Thus $L_1=\{w\}$. For $i=1,2,\ldots$, we now test (which can be done effectively) whether or not L_i is a test set for L_{i+1}. We must find an integer j such that L_j is a test set of L_{j+1}. (Otherwise $L(G)$ would possess no finite test set, contradicting Ehrenfeucht's conjecture that is by now established.) Then $g(L_j)$ is a test set of $g(L_{j+1})$, which implies inductively that L_j is a test set of $L(G)$. Hence, we can decide whether g and h coincide on $L(G)$.

It remains an open problem whether the use of Makanin's result (whose proof is rather complicated) can be eliminated from the proof given above.

Also *grammar forms* and *grammatical families* are defined using morphisms. For simplicity we restrict ourselves to the context-free case.

Let G be a context-free grammar, determined by the alphabets Σ_N, Σ_T, Σ_S and the production set P. Let

$$h:\Sigma_\mathrm{N}^*\to\Delta_\mathrm{N}^*,\quad h:\Sigma_\mathrm{T}^*\to\Delta_\mathrm{T}^*,\qquad \Delta_\mathrm{N}\cap\Delta_\mathrm{T}=\emptyset$$

be a *letter-to-letter* morphism. Then

$$G_1=h(G)=(h(\Sigma_\mathrm{N}),h(\Sigma_\mathrm{T}),h(\Sigma_\mathrm{S}),h(P))$$

is also a context-free grammar, referred to as a morphic image of G. If $G_1=h(G)$ holds for a subgrammar G_1 of a grammar G_1' (that is, each item of G_1 is a subset of the corresponding item of G_1'), we say that G is an inverse morphic image of G_1', in symbols,

$$G\in h^{-1}(G_1').$$

The *language family* generated by a grammar G is defined by

$$\mathscr{L}(G)=\{L(G_1)\mid G_1\in h^{-1}(G)\text{ for some }h\}.$$

A family $\mathscr{L}$ of languages is *grammatical* iff $\mathscr{L}=\mathscr{L}(G)$, for some G.

A grammar is often referred to as a *grammar form* if its language family is considered. A grammar G is *complete* iff $\mathscr{L}(G)$ equals the whole family of context-free languages. (Here the omission of λ from a language is disregarded.) For instance, the two grammars determined by the productions

$$S\to SS,\quad S\to a\qquad\text{and}\qquad S\to aSS,\quad S\to aS,\quad S\to a$$

are both complete. The grammars correspond to the Chomsky and Greibach normal forms, respectively. Indeed a characterization of completeness also gives a characterization of all possible *normal forms* for context-free grammars. Such a characterization was given in [22] and [24].

Two grammars are called *family equivalent* iff their language families coincide. The undecidability of family equivalence for (context-free) grammars was established in [25].

Grammatical families possess the following property of *density*. Under certain conditions (that are by now pretty well understood) one is able to "squeeze in" a third grammatical family $\mathscr{L}_3$ between any two grammatical families $\mathscr{L}_1$ and $\mathscr{L}_2$ satisfying $\mathscr{L}_1 \subsetneq \mathscr{L}_2$. This means that $\mathscr{L}_1 \subsetneq \mathscr{L}_3 \subsetneq \mathscr{L}_2$. Basic facts about density are contained in [23].

A generalization of the customary notion of graph coloring is obtained by considering grammar forms of a certain type. This leads to a rather natural classification of graphs [21, 31].

7. Rational power series

We now return to a direct discussion concerning formal power series. It is often stated that rational series correspond to regular and algebraic series to context-free languages. However this is true only to some extent. Some details will be presented below.

A detailed theory of *convergence* in semirings was developed in [16]. Such a theory lies beyond the scope of this article. For our purposes it suffices to consider a special type of convergence (called *discrete convergence* in [16]). The reader is referred to [16] for cases where other types of convergence become necessary.

A sequence $r_1, r_2, \ldots$ of elements of $A\langle\langle\Sigma^*\rangle\rangle$ *converges* to the limit r, in symbols,

$$\lim_{n\to\infty} r_n = r$$

iff for all k there exists an m such that the conditions $|w| \leqslant k$ and $j > m$ imply the condition

$$(r_j, w) = (r, w).$$

Thus, the coefficient of an arbitrary word w in the limit series r is seen from the coefficient of w in r_j, provided j is sufficiently large with respect to $|w|$.

A series r in $A\langle\langle\Sigma^*\rangle\rangle$ is termed *quasi regular* iff $(r, \lambda) = 0$. For a quasi regular series r, the sequence $r, r^2, r^3, \ldots$ converges to the limit 0.

If r is a quasi regular, then

$$\lim_{m\to\infty} \sum_{k=1}^{m} r^k$$

exists. It is called the *quasi inverse* of r and denoted by r^+. The series $r^+ = s$ is the only series satisfying the equations

$$r + rs = r + sr = s.$$

Note also that a nonerasing homomorphism h satisfies the equation $(hr)^+ = h(r^+)$.

A subsemiring of $A\langle\langle\Sigma^*\rangle\rangle$ is *rationally closed* iff it contains the quasi inverse of every quasi regular element. The family of *A-rational* series over Σ, in symbols $A^{\mathrm{rat}}\langle\langle\Sigma^*\rangle\rangle$, is the smallest rationally closed subset of $A\langle\langle\Sigma^*\rangle\rangle$ which contains all polynomials. Thus, every A-rational series can be constructed from polynomials by a finite number of applications of the operations of sum, (Cauchy) product and quasi inversion.

A series r of $A\langle\langle\Sigma^*\rangle\rangle$ is termed *A-recognizable* (in symbols $r \in A^{\text{rec}}\langle\langle\Sigma^*\rangle\rangle$) iff

$$r = \sum_{w \neq \lambda} p(\mu w)w,$$

where $\mu: \Sigma^* \to A^{m \times m}$, $m \geqslant 1$, is a morphism and $p: A^{m \times m} \to A$ is a mapping such that, for an $m \times m$ matrix (a_{ij}), the value $p(a_{ij})$ can be expressed as a linear combination of the entries a_{ij} with coefficients in A:

$$p(a_{ij}) = \sum_{i,j} a_{ij} p_{ij}, \quad p_{ij} \in A.$$

We now present some examples. We consider series in $N\langle\langle\Sigma^*\rangle\rangle$ with $\Sigma = \{x, \bar{x}\}$. The series

$$r = \sum_{n=1}^{\infty} 2^n (x\bar{x})^n x + 3x = (2x\bar{x})^+ x + 3x$$

is N-rational. Clearly supp(r) is the regular language

$$(x\bar{x})^+ x \cup x.$$

Consider next the morphism μ defined by

$$\mu(x) = \begin{pmatrix} 0 & 1 \\ 1 & 1 \end{pmatrix}, \qquad \mu(\bar{x}) = \begin{pmatrix} 0 & 0 \\ 0 & 0 \end{pmatrix},$$

and let p be the mapping defined by

$$p(a_{ij}) = a_{11} + a_{12}.$$

Then the N-recognizable series

$$r' = \sum_{w \in \Sigma^*} p(\mu w)w$$

can be written in the form

$$r' = \sum_{n=0}^{\infty} a_n x^n,$$

where the sequence $a_0, a_1, a_2, \ldots$ constitutes the Fibonacci sequence. The same sequence is defined by the growth function of the D0L system with the axiom a and productions

$$a \to b, \qquad b \to ab.$$

According to the celebrated *Schützenberger's Representation Theorem* the families of A-rational and A-recognizable formal power series coincide, that is,

$$A^{\text{rat}}\langle\langle\Sigma^*\rangle\rangle = A^{\text{rec}}\langle\langle\Sigma^*\rangle\rangle.$$

(For a proof, see [33].) This can be viewed as a generalization of the well-known characterization of languages acceptable by finite automata: they are exactly the languages obtainable from finite languages by the operations of union, catenation and catenation closure.

Let us call a semiring A *positive* iff the mapping $h: A \to B$ (Boolean semiring) defined by

$$h(0)=0, \qquad h(a)=1 \quad \text{for } a \neq 0,$$

is a morphism. Then the interconnection between regular languages and supports of rational series can be stated as follows.

For all regular languages L and semirings A, the characteristic series of L is A-rational. If A is a positive semiring, then the support of any series in $A^{\text{rat}}\langle\langle \Sigma^* \rangle\rangle$ is regular. Consequently the following three conditions are equivalent:

(i) L is a regular language over Σ,
(ii) The characteristic series of L is in $B^{\text{rat}}\langle\langle \Sigma^* \rangle\rangle$,
(iii) The characteristic series of L is in $N^{\text{rat}}\langle\langle \Sigma^* \rangle\rangle$.

A series r is N-rational iff there are a regular language L and a letter-to-letter morphism h such that

$$(r, w) = \text{card}(h^{-1} w \cap L)$$

holds for all $w \neq \lambda$.

On the other hand, the support of a Z-rational series need not even be context-free.

We now mention some decidability properties. It is clear that the well-known decidability results for finite automata can be deduced as corollaries. Decidability status is often different for sequences. An *A-rational sequence* consists of the coefficients of $x^i, i=0, 1, \ldots$, in a series belonging to

$$A^{\text{rat}}\langle\langle \Sigma^* \rangle\rangle \text{ where } \Sigma = \{x\}.$$

More information about these and related matters can be found in [33] and [16].

It is undecidable whether or not a given series in $Z^{\text{rat}}\langle\langle \Sigma^* \rangle\rangle$

(i) has a zero coefficient,
(ii) has infinitely many zero coefficients,
(iii) has a positive coefficient,
(iv) has infinitely many positive coefficients,
(v) has its coefficients ultimately nonnegative,
(vi) has two equal coefficients.

On the other hand, it is decidable whether or not such a series equals 0, as well as whether or not two such series are identical.

It is decidable whether or not a given Z-rational sequence

(i) is identically zero,
(ii) is a polynomial,
(iii) has infinitely many zero coefficients.

Point (iii) (or the number-theoretic result leading to it) is usually referred to as the *Skolem–Mahler–Lech Theorem.* A particularly simple proof for this theorem was recently given in [12].

A famous open problem is to decide whether or not a Z-rational sequence has a zero coefficient. (As mentioned above, this problem is undecidable for Z-rational series.) The problem has many important implications to language theory, see [5].

It is also decidable whether or not a given Z-rational sequence is N-rational [33]. Indeed there are Z-rational sequences of nonnegative integers that are not N-rational. One example is the sequence a_n defined by

$$a_{2n} = 30^n, \qquad a_{2n+1} = 25^n \cos^2 2\pi n\alpha,$$

where the angle α is defined by $\cos 2\pi\alpha = \frac{3}{5}$, $\sin 2\pi\alpha = \frac{4}{5}$.

8. Algebraic power series

This section gives a survey about the interconnection between algebraic power series and context-free languages. The first extensive treatment of this topic was given in [26]. Apart from the language being the support of a certain series, the coefficients indicate the degrees of ambiguity of the individual words.

The *degree of ambiguity* of a word w (according to context-free grammar G) is defined to be the number of leftmost derivations of w from the initial variable y of G. The number of such derivations can also be infinite, in which case we say that w has an infinite degree of ambiguity. The degree 0 means that w is not in $L(G)$. A word having the degree 1 is said to be *unambiguous* (according to G). The context-free grammar G is termed *unambiguous* iff every word in $L(G)$ is unambiguous.

We are now ready to define the basic notions concerning algebraic systems. An *$A\langle\langle\Sigma^*\rangle\rangle$-algebraic system* (briefly *algebraic system*) with *variables* in $Y = \{y_1, \ldots, y_n\}$, $Y \cap \Sigma = \emptyset$, is a system of equations

$$y_i = p_i, \quad 1 \leqslant i \leqslant n,$$

where each p_i is a polynomial in $A\langle(\Sigma \cup Y)^*\rangle$.

Defining the two column vectors

$$Y = \begin{pmatrix} y_1 \\ \vdots \\ y_n \end{pmatrix} \quad \text{and} \quad C = \begin{pmatrix} p_1 \\ \vdots \\ p_n \end{pmatrix},$$

we can write our algebraic system in the matrix notation

$$Y = C.$$

No confusion will arise with the fact that Y also stands for the alphabet of variables because the meaning will always be clear from the context. When not stated otherwise, Y consists of the variables $y_1, \ldots, y_n$.

Intuitively a solution of the algebraic system $Y = C$ is given by n power series $\sigma_1, \ldots, \sigma_n$ in $A\langle\langle\Sigma^*\rangle\rangle$ "satisfying" the algebraic system in the sense that if each variable y_i is replaced by the series σ_i, then valid equations result.

More formally, consider

$$\sigma=\begin{pmatrix}\sigma_1\\ \vdots\\ \sigma_n\end{pmatrix}\in(A\langle\langle(\Sigma\cup Y)^*\rangle\rangle)^{n\times 1}.$$

Then we can define a morphism

$$\sigma:(\Sigma\cup Y)^*\to A\langle\langle(\Sigma\cup Y)^*\rangle\rangle$$

by $\sigma(y_i)=\sigma_i$, $1\leqslant i\leqslant n$, and $\sigma(x)=x$, $x\in\Sigma$. As usual, extend σ to a mapping

$$\sigma: A\langle(\Sigma\cup Y)^*\rangle\to A\langle\langle(\Sigma\cup Y)^*\rangle\rangle$$

by the definition

$$\sigma(p)=\sum_{\gamma\in(\Sigma\cup Y)^*}(p,\gamma)\sigma(\gamma),$$

where p is in $A\langle(\Sigma\cup Y)^*\rangle$. Observe that $\sigma(ap)=a\sigma(p)$ and $\sigma(p+p')=\sigma(p)+\sigma(p')$ for all $p, p'\in A\langle(\Sigma\cup Y)^*\rangle$ and $a\in A$. Furthermore because p is a polynomial, we will have no difficulties with infinite sums.

A *solution* to the algebraic system $y_i=p_i$, $1\leqslant i\leqslant n$, is given by a column vector $\sigma\in(A\langle\langle\Sigma^*\rangle\rangle)^{n\times 1}$ such that $\sigma_i=\sigma(p_i)$, $1\leqslant i\leqslant n$.

It has been customary in literature, when dealing with algebraic power series, to assume that the basic semiring A is commutative. However we want to emphazise that commutativity is not needed for some of the really basic results, e.g., the connection of algebraic systems with pushdown automata.

Matrix notation can be used by extending the mapping σ entrywise to vectors and matrices. In this fashion, a solution to the algebraic system $Y=C$ is given by a column vector σ such that $\sigma=\sigma(C)$.

The *approximation sequence*

$$\sigma^0,\sigma^1,\sigma^2,\ldots,\sigma^j,\ldots \quad \text{where each } \sigma^j\in(A\langle\Sigma^*\rangle)^{n\times 1},$$

associated to an algebraic system $Y=C$ is defined as follows:

$$\sigma^0=0,\qquad \sigma^{j+1}=\sigma^j(C),\quad j\geqslant 0.$$

If the approximation sequence converges, that is,

$$\lim_{j\to\infty}\sigma^j=\sigma,$$

then σ is referred to as the *strong solution*. Clearly, by definition, the strong solution is unique whenever it exists. It can be shown below that the strong solution is, in fact, a solution.

Assume that A is partially ordered. A solution σ of the $A\langle\langle\Sigma^*\rangle\rangle$-algebraic system $Y=C$ is termed *minimal* iff $\sigma\geqslant\tau$ holds for all solutions τ of $Y=C$. Minimal solutions (when they exist) are always unique. The strong solution (when it exists) coincides with the minimal solution. Every $B\langle\langle\Sigma^*\rangle\rangle$-algebraic system has a strong solution.

Every context-free grammar with the terminal alphabet Σ and every semiring A give rise to an $A\langle\langle\Sigma^*\rangle\rangle$-algebraic system. Conversely every $A\langle\langle\Sigma^*\rangle\rangle$-algebraic system gives rise to a context-free grammar. More explicitly, this interrelation is defined as follows.

Consider a context-free grammar G with Y and Σ as the nonterminal and terminal alphabets, R the set of productions and y_1 the only start symbol. Then define the $A\langle\langle\Sigma^*\rangle\rangle$-algebraic system $y_i = p_i$, $1 \leqslant i \leqslant n$, by

$$(p_i, \gamma) = 1 \quad \text{if } y_i \to \gamma \in R,$$

and

$$(p_i, \gamma) = 0 \quad \text{otherwise},$$

where γ is in $(\Sigma \cup Y)^*$. Conversely give an $A\langle\langle\Sigma^*\rangle\rangle$-algebraic system $y_i = p_i$, $1 \leqslant i \leqslant n$, define the context-free grammar $G = (Y, \Sigma, R, y_1)$ by

$$y_i \to \gamma \in R \quad \text{iff} \quad (p_i, \gamma) \neq 0,$$

where γ is in $(\Sigma \cup Y)$.

If we begin with an algebraic system, form the corresponding context-free grammar and, finally, form the corresponding algebraic system then the resulting algebraic system is not necessarily the same as the original one. However if attention is restricted to algebraic systems with coefficients 0 and 1, then our correspondence is indeed one-to-one. Observe also that if we begin with a context-free grammar, form the corresponding algebraic system and, finally, form the corresponding context-free grammar, then the resulting context-free grammar always coincides with the original one.

Assume that G is a context-free grammar and $y_i = p_i$, $1 \leqslant i \leqslant n$, is the corresponding $B\langle\langle\Sigma^*\rangle\rangle$-algebraic system with the strong solution σ. Then

$$L(G) = \text{supp}(\sigma_1).$$

Assume that $G = (Y, \Sigma, R, y_1)$ is a context-free grammar such that, for all $w \in \Sigma^*$ and all context-free grammars $G_i = (Y, \Sigma, R, y_i)$, $1 \leqslant i \leqslant n$, the degree of ambiguity $d_i(w)$ of w (according to G_i) is finite. Let $y_i = p_i$, $1 \leqslant i \leqslant n$, be the $N\langle\langle\Sigma^*\rangle\rangle$-algebraic system corresponding to G. Then the strong solution σ for the algebraic system exists and, moreover, for all $w \in \Sigma^*$ and $1 \leqslant i \leqslant n$,

$$(\sigma_i, w) = d_i(w).$$

Under the same assumptions, G is unambiguous iff, for all $w \in \Sigma^*$,

$$(\sigma_1, w) \leqslant 1.$$

In considerations dealing with context-free grammars, the choice of the semiring of the corresponding algebraic system reflects the particular point of view we want to emphasize. If we are interested only in the language $L(G)$ we choose the semiring B. The semiring N is chosen if we want to discuss ambiguity. Also modifications of context-free grammars, such as weighted and probabilistic grammars, can be taken into account.

For probabilistic grammars, the natural choice of the semiring is R_+, the semiring of nonnegative reals.

We now define special classes of algebraic systems for which strong solutions always exist.

An $A\langle\langle\Sigma^*\rangle\rangle$-algebraic system $y_i=p_i$, $1\leqslant i\leqslant n$, is termed *proper* iff $\mathrm{supp}(p_i)\subseteq(\Sigma\cup Y)^+-Y$ for all $1\leqslant i\leqslant n$. It is termed *weakly strict* iff $\mathrm{supp}(p_i)\subseteq\lambda\cup(\Sigma\cup Y)^*\Sigma(\Sigma\cup Y)^*$ for all $1\leqslant i\leqslant n$. It is termed *weakly strict* iff $\mathrm{supp}(p_i)\subseteq\lambda\cup(\Sigma\cup Y)^*\Sigma(\Sigma\cup Y)^*$ for all $1\leqslant i\leqslant n$. Finally it is termed *strict* iff $\mathrm{supp}(p_i)\subseteq\lambda\cup\Sigma(\Sigma\cup Y)^*$ for all $1\leqslant i\leqslant n$.

Strong solutions exist for all proper and weakly strict algebraic systems. Moreover, the strong solution is the only quasi regular solution of a proper algebraic system and the unique solution of a weakly strict algebraic system.

If r is a component of the strong solution of a $B\langle\langle\Sigma^*\rangle\rangle$-algebraic system, then the quasi regular part of r is a component of the strong solution of a proper $B\langle\langle\Sigma^*\rangle\rangle$-algebraic system.

Many open problems remain in this area. There is no characterization for the supports of Z-algebraic series. The difference between Z-algebraic and Z-rational sequences is not properly understood.

Pushdown automata can be generalized to accept power series. Then all proofs (such as the equivalence of PDAs and context-free grammars) become computational in nature. The reader is referred to [16] for a theory about the interconnection between PDA-series and algebraic series.

9. What power series do for you

The technique of formal power series leads to an arithmetization of the theory. Customary treatments of automata and languages are often unsatisfactory in the sense that entirely different ad hoc proofs are given in very similar situations and, moreover, many of the proofs still remain inadequate from the mathematical point of view. It is obvious that many different proofs can be unified and generalized by the power series methods and, moreover, that many proofs become more satisfactory in this fashion. However we are also convinced that these and analogous methods yield new interesting results that are difficult, if not impossible, to obtain by other means. We give a few examples.

We denote by $c(\Sigma^*)$ the *free commutative monoid* generated by the alphabet Σ. Thus the order of letters is irrelevant when elements of $c(\Sigma^*)$ are considered. Hence for $\Sigma=\{x_1,\ldots,x_m\}$, each element of $c(\Sigma^*)$ can be represented in the form

$$x_1^{n_1}\ldots x_m^{n_m},\quad n_i\geqslant 0.$$

In analogy with our original definition of formal power series, we define a formal power series r in *commuting variables* to be a mapping of $c(\Sigma^*)$ into A. As before, the series is written as a formal sum

$$r=\sum_{v\in c(\Sigma^*)}(r,v)v.$$

Consequently every power series $r \in A\langle\langle c(\Sigma^*)\rangle\rangle$ can be written in the form

$$r = \sum_{n_1,\ldots,n_m \geq 0} (r, x_1^{n_1} \ldots x_m^{n_m}) x_1^{n_1} \ldots x_m^{n_m}.$$

Consider the power series $r = (x_1 + x_2)^*$ in $N\langle\langle \{x_1, x_2\}^*\rangle\rangle$. Then

$$c(r) = \sum_{n_1,n_2 \geq 0} \binom{n_1+n_2}{n_1} x_1^{n_1} x_2^{n_2}.$$

The introduction of commuting variables makes it possible to consider power series as a method of defining functions similarly as in classical analysis. This method or classical elimination theory can be used to establish the following results [35, 16, 33].

It is decidable whether or not a given formal power series in $Q^{\text{alg}}\langle\langle c(\Sigma^*)\rangle\rangle$ is in $Q^{\text{rat}}\langle\langle c(\Sigma^*)\rangle\rangle$, as well as whether or not it is in $Z^{\text{rat}}\langle\langle c(\Sigma^*)\rangle\rangle$.

It is decidable whether or not a given Q-algebraic sequence is N-algebraic.

It is decidable whether or not two given formal power series in $Q^{\text{alg}}\langle\langle c(\Sigma^*)\rangle\rangle$ are equal.

It is decidable whether or not two given formal power series r_1, r_2 in $N^{\text{alg}}\langle\langle \Sigma^*\rangle\rangle$ satisfying

$$(r_1, w) \geq (r_2, w) \quad \text{for all } w$$

are equal.

It is decidable whether or not two given formal power series

$$r_1 \in N^{\text{alg}}\langle\langle \Sigma^*\rangle\rangle \quad \text{and} \quad r_2 \in N^{\text{rat}}\langle\langle \Sigma^*\rangle\rangle$$

with coefficients in $\{0, 1\}$ are equal.

Assume that G and G_1 are given unambiguous context-free grammars such that $L(G_1) \supseteq L(G)$. Then it is decidable whether or not $L(G_1) = L(G)$.

Assume that G and G_1 are given context-free grammars such that $L(G_1) = L(G)$ and G is unambiguous. Then it is decidable whether or not G_1 is unambiguous.

Given an unambiguous context-free grammar G and a regular language R, it is decidable whether or not $L(G) = R$.

Another line of examples comes from the theory of growth functions of L systems. The role of *erasing* (biologically, the role of *cell death*) is very important in growth considerations. In particular, erasing productions $a \to \lambda$ make it possible that $f(i+1) < f(i)$, for the growth function f and for some value i.

But can erasing productions do more? Indeed they can. Let us call a D0L system *propagating* (in symbols, a *PD0L system*) if it has no erasing productions. There are strictly increasing D0L growth functions (that is, $f(i) < f(i+1)$) that are not PD0L growth functions. It is hardly conceivable that such an example could be given without relying on the theory of formal power series. One example can be based on the example given at the end of Section 7. The reader is referred to [33] for a general characterization of PD0L growth functions.

It is usually very hard to show that a given context-free language is inherently ambiguous. (Of course, in its general formulation, this problem is undecidable.) Again formal power series constitute an indispensable tool. For instance, lacunary series from classical analysis are very important in this respect. By the methods of [9], one is able to show that each of the following languages L_1–L_4 over $\Sigma=\{a,b\}$ is inherently ambiguous.

$$L_1=\{w_1w_2 \mid w_i=\mathrm{mi}(w_i),\ w_i\in\Sigma^*,\ i=1,2\},\quad \text{(mi stands for mirror image)}$$

$$L_2=\{a^nbw_1a^nw_2 \mid w_i\in\Sigma^*, n\geqslant 0\},$$

$$L_3=\{a^{n_1}ba^{n_2}b\ldots ba^{n_k}b \mid k\geqslant 1 \text{ and there is a } j,\ 1\leqslant j\leqslant k, \text{ such that } n_j\neq j\},$$

$$L_4=\{a^{n_1}ba^{n_2}b\ldots ba^{n_k}b \mid k\geqslant 1 \text{ and there is a } j,\ 1\leqslant j\leqslant k-1, \text{ with } n_{j+1}\neq n_j\}.$$

10. Final comments

The reader is referred to [2, 13, 14, 30] for further aspects of language theory. References [20] and [27] are the original references for Markov algorithms and Post canonical systems, respectively. More information about L systems can be found in [29]. Reference [8] presents finite automata from essentially the power series point of view.

As emphasized above, this chapter gives only a selection of topics. Many topics such as AFL-theory (see [10] and [16] for its generalization to power series) have been entirely omitted.

References

[1] ALBERT, M.H. and J. LAWRENCE, A proof of Ehrenfeucht's Conjecture, *Theoret. Comput. Sci.* **41** (1985) 121–123.

[2] BUCHER, W. and H.A. MAURER, *Theoretische Grundlagen der Programmiersprachen* (Bibliographisches Institut, Mannheim, 1984).

[3] BÜCHI, J.R., Regular canonical systems, *Arch. Math. Logik Grundlag.* **6** (1964) 91–111.

[4] CULIK, K., A purely homomorphic characterization of recursively enumerable sets, *J. Assoc. Comput. Mach.* **26** (1979) 345–350.

[5] CULIK, K., Homomorphisms: decidability, equality and test sets, in: R. Book, ed., *Formal Language Theory, Perspectives, and Open Problems* (Academic Press, New York, 1980) 167–194.

[6] CULIK, K. and J. KARHUMÄKI, A new proof for the D0L sequence equivalence problem and its implications, in: G. Rozenberg and A. Salomaa, eds., *The Book of L* (Springer, Berlin 1986) 63–74.

[7] DASSOW, J. and G. PAŬN, *Regulated Rewriting in Formal Language Theory* (Akademie, Berlin, 1986).

[8] EILENBERG, S., *Automata, Languages and Machines, Vol A* (Academic Press, New York, 1974).

[9] FLAJOLET, P., Ambiguity and trascendence, in: *Proc. 12th Internat. Coll. on Automata, Languages and Programming*, Lecture Notes in Computer Science, Vol. 194 (Springer, Berlin, 1985) 179–188.

[10] GINSBURG, S., *Algebraic and Automata-Theoretic Properties of Formal Languages* (North-Holland, Amsterdam, 1975).

[11] GUBA, V.S., A proof of Ehrenfeucht's Conjecture, Personal communication by A.L. Semenov.

[12] HANSEL, G., Une démonstration simple du théorème de Skolem–Mahler–Lech, *Theoret. Comput. Sci.* **43** (1986) 91–98.

[13] HARRISON, M., *Introduction to Formal Language Theory* (Addison-Wesley, Reading, MA, 1978).

[14] HOPCROFT, J. and J. ULLMAN, *Introduction to Automata Theory, Languages and Computation* (Addison-Wesley, Reading, MA, 1979).

[15] KARHUMÄKI, J., On recent trends in formal language theory, in: *Proc. 14th Internat. Coll. on Automata,*

Languages and Programming, Lecture Notes in Computer Science, Vol. 261 (Springer, Berlin, 1987) 136–162.

[16] KUICH, W. and A. SALOMAA, *Semirings, Automata, Languages* (Springer, Berlin, 1986).

[17] VAN LEEUWEN, J., The membership question for ET0L languages is polynomially complete, *Inform. Process. Lett.* **3** (1975) 138–143.

[18] LINDENMAYER, A., Mathematical models for cellular interaction in development I–II, *J. Theoret. Biol.* **18** (1986) 280–315.

[19] MAKANIN, G.S., The problem of the solvability of equations in a free semigroup, *Mat. Sb* (*N.S.*) **103** (1977) 147–236 (in Russian).

[20] MARKOV, A.A., *The Theory of Algorithms* (Israel Program for Scientific Translations, Jerusalem, 1961).

[21] MAURER, H.A., A. SALOMAA and D. WOOD, Colorings and interpretations: a connection between graphs and grammar forms, *Discrete Appl. Math.* **3** (1981) 289–299.

[22] MAURER, H.A., A. SALOMAA and D. WOOD, Completeness of context-free grammar forms, *J. Comput. System Sci.* **23** (1981) 1–10.

[23] MAURER, H.A., A. SALOMAA and D. WOOD, Dense hierarchies of grammatical families, *J. Assoc. Comput. Mach.* **29** (1982) 118–126.

[24] MAURER, H.A., A. SALOMAA and D. WOOD, A supernormal-form theorem for context-free grammars, *J. Assoc. Comput. Mach.* **30** (1983) 95–102.

[25] NIEMI, V., The undecidability of form equivalence for context-free and E0L forms, *Theoret. Comput. Sci.* **32** (1984) 261–277.

[26] NIVAT, M., Transductions des langages de Chomsky, *Ann. Inst. Fourier* **18** (1968) 339–455.

[27] POST, E.L., Formal reductions of the general combinatorial decision problem, *Amer. J. Math.* **65** (1943) 197–215.

[28] RÉVÉSZ, G., *Introduction to Formal Languages* (McGraw-Hill, New York, 1983).

[29] ROZENBERG, G. and A. SALOMAA, *The Mathematical Theory of L Systems* (Academic Press, New York, 1980).

[30] SALOMAA, A., *Formal Languages* (Academic Press, New York, 1973).

[31] SALOMAA, A., *Jewels of Formal Language Theory* (Computer Science Press, Rockville, MD, 1981).

[32] SALOMAA, A., *Computation and Automata*, Encyclopedia of Mathematics and its Applications, Vol. 25 (Cambridge Univ. Press, Cambridge, 1985).

[33] SALOMAA, A. and M. SOITTOLA, *Automata-Theoretic Aspects of Formal Power Series* (Springer, Berlin, 1978).

[34] SAVITCH, W., How to make arbitrary grammars look like context-free grammars, *SIAM J. Comput.* **2** (1973) 174–182.

[35] SEMENOV, A.L., Algorithmic problems for power series and context-free grammars, *Dokl. Akad. Nauk SSSR* **212** (1973) 50–52 (in Russian).

[36] THUE, A., Über unendliche Zeichenreihen, *Videnskapsselskapets Skrifter. I. Mat.-naturv. Klasse, Kristiania* (1906) 1–22.

CHAPTER 4

Automata on Infinite Objects

Wolfgang THOMAS

Institut für Informatik und Praktische Mathematik, Universität Kiel, Olshausenstrasse 40, D-2300 Kiel, FRG

Contents

Introduction 135

Part I. Automata on infinite words

Notation 136
1. Büchi automata 136
2. Congruences and complementation 139
3. The sequential calculus 143
4. Determinism and McNaughton's Theorem 147
5. Acceptance conditions and Borel classes 152
6. Star-free ω-languages and temporal logic 156
7. Context-free ω-languages 161

Part II. Automata on infinite trees

Notation 165
8. Tree automata 167
9. Emptiness problem and regular trees 170
10. Complementation and determinacy of games 173
11. Monadic tree theory and decidability results 178
12. Classifications of Rabin recognizable sets 184

Acknowledgment 186

References 186

HANDBOOK OF THEORETICAL COMPUTER SCIENCE
Edited by J. van Leeuwen

Introduction

The subject of finite automata on infinite sequences and infinite trees was established in the sixties by Büchi [9], McNaughton [62], and Rabin [93]. Their work introduced intricate automaton constructions, opened connections between automata theory and other fields (for example, logic and set-theoretic topology), and resulted in a theory which is fundamental for those areas in computer science where nonterminating computations are studied.

The early papers were motivated by decision problems in mathematical logic. Büchi showed that finite automata provide a normal form for certain monadic second-order theories. Rabin'sTree Theorem (which states that the monadic second-order theory of the infinite binary tree is decidable) turned out to be a powerful result to which a large number of other decision problems could be reduced.

From this core the theory has developed into many directions. Today, a rapidly growing literature is concerned with applications in modal logics of programs and in the specification and verification of concurrent programs. Many of the logical specification formalisms for programs (systems of program logic or temporal logic) are embeddable in the monadic second-order theories studied by Büchi and Rabin; indeed these theories can be considered as "universal process logics" for linear, respectively branching computations, as far as programs with finite state spaces are concerned.

Besides these applications (see also [30, 52]), the following lines of research should be mentioned:

- the investigation of other (usually more general) models of computation than finite automata: grammars, pushdown automata, Turing machines, Petri nets, etc.,
- the classification theory of sequence (or tree) properties, e.g. by different acceptance modes of automata or by topological conditions,
- algebraic aspects, in particular the semigroup-theoretical analysis of automata over ω-sequences,
- the connection with fixed point calculi,
- the study of more general structures than ω-sequences and ω-trees in connection with automata (e.g. sequences over the integers or certain graphs), and the extension of Rabin's decidability result to stronger theories.

The aim of this chapter is to provide the reader with a more detailed overview of these and related developments of the field, based on a self-contained exposition of the central results.

Some topics are only treated in brief remarks or had to be skipped, among them combinatorics on infinite words, connections with semantics of program schemes, and the discussion of related calculi for concurrency (such as CCS).

Also the references listed at the end of the chapter do not cover the subject. However, we expect that few papers will be missed when the reader traces the articles which are cited within the given references.

PART I. AUTOMATA ON INFINITE WORDS

Notation

Throughout this chapter A denotes a finite alphabet and A^*, respectively A^ω, denote the set of finite words, resp. the set of ω-sequences (or: *ω-words*) over A. An ω-word over A is written in the form $\alpha=\alpha(0)\alpha(1)\ldots$ with $\alpha(i)\in A$. Let $A^\infty=A^*\cup A^\omega$. Finite words are indicated by $u, v, w, \ldots$, the empty word by ε, and sets of finite words by $U, V, W, \ldots$ Letters $\alpha, \beta, \ldots$ are used for ω-words and $L, L', \ldots$ for sets of ω-words (i.e., *ω-languages*). Notations for segments of ω-words are

$$\alpha(m,n):=\alpha(m)\ldots\alpha(n-1) \quad (\text{for } m\leqslant n), \quad \text{and} \quad \alpha(m,\omega):=\alpha(m)\alpha(m+1)\ldots$$

The logical connectives are written $\neg, \vee, \wedge, \rightarrow, \exists, \forall$. As a shorthand for the quantifiers "there exist infinitely many n" and "there are only finitely many n" we use "$\exists^\omega n$", respectively "$\exists^{<\omega} n$".

The following operations on sets of finite words are basic: for $W\subseteq A^*$ let

$$\text{pref } W:=\{u\in A^* \mid \exists v\, uv\in W\},$$
$$W^\omega:=\{\alpha\in A^\omega \mid \alpha=w_0w_1\ldots \text{ with } w_i\in W \text{ for } i\geqslant 0\},$$
$$\vec{W}:=\{\alpha\in A^\omega \mid \exists^\omega n\, \alpha(0,n)\in W\}.$$

Other notations for $\vec{W}$ found in the literature are lim W and W^δ. Finally, for an ω-sequence $\sigma=\sigma(0)\sigma(1)\ldots$ from S^ω, the "infinity set" of σ is

$$\text{In}(\sigma):=\{s\in S \mid \exists^\omega n\, \sigma(n)=s\}.$$

1. Büchi automata

Büchi automata are nondeterministic finite automata equipped with an acceptance condition that is appropriate for ω-words: an ω-word is accepted if the automaton can read it from left to right while assuming a sequence of states in which some final state ocurs infinitely often (*Büchi acceptance*).

DEFINITION. A *Büchi automaton* over the alphabet A is of the form $\mathcal{A}=(Q, q_0, \Delta, F)$ with finite state set Q, initial state $q_0\in Q$, transition relation $\Delta\subseteq Q\times A\times Q$, and a set $F\subseteq Q$ of final states. A *run* of $\mathcal{A}$ on an ω-word $\alpha=\alpha(0)\alpha(1)\ldots$ from A^ω is a sequence $\sigma=\sigma(0)\sigma(1)\ldots$ such that $\sigma(0)=q_0$ and $(\sigma(i), \alpha(i), \sigma(i+1))\in\Delta$ for $i\geqslant 0$; the run is called *successful* if $\text{In}(\sigma)\cap F\neq\emptyset$, i.e. some state of F occurs infinitely often in it. *$\mathcal{A}$ accepts* α if there is a successful run of $\mathcal{A}$ on α. Let

$$L(\mathcal{A})=\{\alpha\in A^\omega \mid \mathcal{A} \text{ accepts } \alpha\}$$

be the ω-language *recognized* by $\mathcal{A}$. If $L=L(\mathcal{A})$ for some Büchi automaton $\mathcal{A}$, L is said to be *Büchi recognizable.*

EXAMPLE. Consider the alphabet $A=\{a,b,c\}$. Define $L_1\subseteq A^\omega$ by

$\alpha\in L_1$ iff after any occurrence of letter a there is some occurrence of letter b in α.

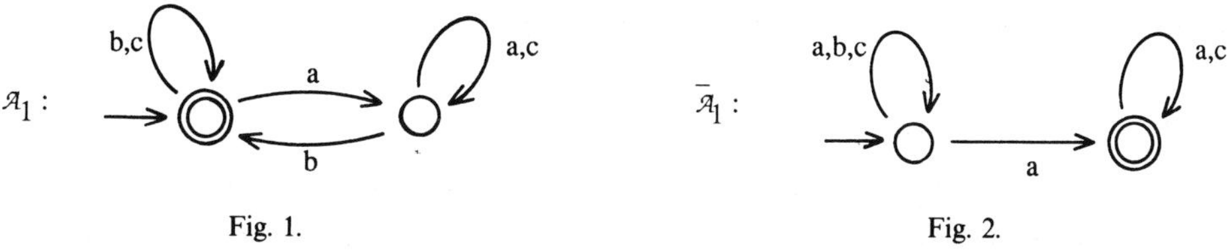

Fig. 1. Fig. 2.

A Büchi automaton recognizing L_1 is (in state graph representation) shown in Fig. 1. The complement $A^\omega - L_1$ is recognized by the Büchi automaton in Fig. 2. Finally, the ω-language $L_2 \subseteq A^\omega$ with

$\alpha \in L_2$ iff between any two occurrences of letter a in α there is an even number of letters b, c

is recognized by the automaton shown in Fig. 3.

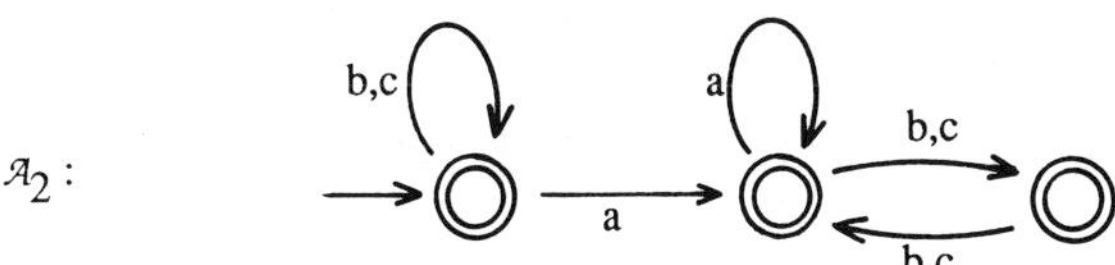

Fig. 3.

For a closer analysis of Büchi recognizable ω-languages we use the following notations, given some fixed Büchi automaton $\mathcal{A} = (Q, q_0, \Delta, F)$: if $w = a_0 \ldots a_{n-1}$ is a finite word over A, write $s \underset{w}{\rightarrow} s'$ if there is a state sequence $s_0, \ldots, s_n$ such that $s_0 = s$, $(s_i, a_i, s_{i+1}) \in \Delta$ for $i < n$, and $s_n = s'$. Let

$$W_{ss'} = \{w \in A^* \mid s \underset{w}{\rightarrow} s'\}.$$

Each of the finitely many languages $W_{ss'}$ is regular. By definition of Büchi acceptance, the ω-language recognized by $\mathcal{A}$ is

$$L(\mathcal{A}) = \bigcup_{s \in F} W_{q_0 s} \cdot (W_{ss})^\omega. \tag{1.1}$$

1.1. THEOREM (Büchi [9]). *An ω-language $L \subseteq A^\omega$ is Büchi recognizable iff L is a finite union of sets $U.V^\omega$ where U, $V \subseteq A^*$ are regular sets of finite words (and where, moreover, one may assume $V.V \subseteq V$).*

In the proof, the direction from left to right is clear from equation (1.1); note that $W_{ss} . W_{ss} \subseteq W_{ss}$. For the converse, we verify the following closure properties of Büchi recognizable sets.

1.2. LEMMA. (a) *If $V \subseteq A^*$ is regular, then V^ω is Büchi recognizable.*

(b) *If $U \subseteq A^*$ is regular and $L \subseteq A^\omega$ is Büchi recognizable, then $U.L$ is Büchi recognizable.*

(c) *If* L_1, $L_2 \subseteq A^\omega$ *are Büchi recognizable, then* $L_1 \cup L_2$ *and* $L_1 \cap L_2$ *are Büchi recognizable.*

PROOF. (*a*) Since $V^\omega = (V - \{\varepsilon\})^\omega$, assume that V does not contain the empty word; further suppose that there is no transition into the initial state q_0 of the finite automaton which recognizes V. A Büchi automaton $\mathcal{A}'$ recognizing V^ω is obtained from the given automaton $\mathcal{A}$ by adding a transition (s, a, q_0) for any transition (s, a, s') with $s' \in F$, and by declaring q_0 as single final state of $\mathcal{A}'$.

The claims in (b) and (c) concerning concatenation and union are proved in the same way as for regular sets of finite words. For later use we show closure of Büchi recognizable sets under intersection. Suppose L_1 is recognized by $\mathcal{A}_1 = (Q_1, q_1, \Delta_1, F_1)$ and L_2 by $\mathcal{A}_2 = (Q_2, q_2, \Delta_2, F_2)$. A Büchi automaton recognizing $L_1 \cap L_2$ is of the form $\mathcal{A} = (Q_1 \times Q_2 \times \{0, 1, 2\}, (q_1, q_2, 0), \Delta, F)$, where the transition relation Δ copies Δ_1 and Δ_2 in the first two components of states, and changes the third component from 0 to 1 when an F_1-state occurs in the first component, from 1 to 2 when subsequently an F_2-state occurs in the second component and back to 0 immediately afterwards. Then 2 occurs infinitely often as third component in a run iff some F_1-state and some F_2-state occur infinitely often in the first two components. Hence with $F := Q_1 \times Q_2 \times \{2\}$ we obtain a Büchi automaton as desired. □

A representation of an ω-language in the form $L = \bigcup_{i=1}^{n} U_i . V_i^\omega$, where the U_i, V_i are given by regular expressions, is called an *ω-regular expression.* Since the constructions in Lemma 1.2 are effective, the conversion of ω-regular expressions into Büchi automata and vice versa can be carried out effectively. Hence, Büchi recognizable ω-languages are called *regular ω-languages*; other terms used in the literature are *ω-regular, rational, ω-rational.*

The formalism of ω-regular expressions can also be applied to *relations* over finite words instead of languages. In the case of binary relations, say for $S_1, S_2 \subseteq A^* \times B^*$, one defines $S_1 . S_2^\omega \subseteq A^\omega \times B^\omega$ by applying the operations of product and ω-power componentwise. The *rational ω-relations* are obtained as finite unions of relations $S_1 . S_2^\omega$ where S_1 and S_2 are rational relations over finite words (i.e., generated via usual regular expressions, cf. [5] or [27]). The rational ω-relations are characterized by a multitape version of Büchi automata with one reading head for each tape (Gire and Nivat [38]). However, as for rational relations over finite words, certain closure and decidability results cannot be transferred from regular ω-languages to the case of ω-relations (for instance, Theorems 2.1 and 2.3 below). In the sequel we restrict ourselves to ω-languages.

As is clear from equation (1.1), a Büchi automaton $\mathcal{A}$ accepts some ω-word iff $\mathcal{A}$ reaches some final state (say via the word u) which can then be revisited by a loop (say via the word v). The existence of a reachable final state which is located in a loop of $\mathcal{A}$ can be checked by an effective procedure. Hence, we obtain the following theorem.

1.3. THEOREM. (a) *Any nonempty regular ω-language contains an ultimately periodic ω-word (i.e., an ω-word of the form $uvvv\ldots$).*

(b) *The emptiness problem for Büchi automata is decidable.*

Vardi and Wolper [131] show that the nonemptiness-problem for Büchi automata ("$L(\mathcal{A}) \neq \emptyset$?") is logspace complete for NLOGSPACE, and Sistla, Vardi and Wolper [107] prove that the nonuniversality problem for Büchi automata ("$L(\mathcal{A}) \neq A^\omega$?") is logspace complete for PSPACE.

2. Congruences and complementation

Closure of Büchi recognizable ω-languages under complement is nontrivial and involves an interesting combinatorial argument. As will be seen in Section 4, it is not possible to work with a reduction to deterministic Büchi automata.

2.1. THEOREM (Büchi [9]). *If $L \subseteq A^\omega$ is Büchi recognizable, so is $A^\omega - L$. Moreover, from a Büchi automaton recognizing L one can construct one recognizing $A^\omega - L$.*

For the proof we shall represent both L and $A^\omega - L$ as finite unions of sets $U.V^\omega$ where U and V are regular sets of a special kind, namely classes of a certain congruence relation over A^* of finite index. (A congruence is here an equivalence relation compatible with concatenation.) Let $L = L(\mathcal{A})$ where $\mathcal{A} = (Q, q_0, \Delta, F)$ is a Büchi automaton. Write $s \xrightarrow[w]{F} s'$ if there is a run of $\mathcal{A}$ on w from state s to state s' such that at least one of the states in the run (including s and s') belongs to F. The set

$$W^F_{ss'} := \{w \in A^* \mid s \xrightarrow[w]{F} s'\}$$

is regular. Now define the equivalence relation $\sim_{\mathcal{A}}$ over A^* as follows:

$$u \sim_{\mathcal{A}} v \text{ iff } \forall s, s' \in Q\,(s \xrightarrow[u]{} s' \Leftrightarrow s \xrightarrow[v]{} s' \text{ and } s \xrightarrow[u]{F} s' \Leftrightarrow s \xrightarrow[v]{F} s').$$

The relation $\sim_{\mathcal{A}}$ is a congruence over A^*, which is of finite index by finiteness of Q. Hence each $\sim_{\mathcal{A}}$-class is regular. (The $\sim_{\mathcal{A}}$-class $[w]$ containing the word w is the intersection of the sets $W_{ss'}$ and $W^F_{ss'}$ and $A^* - W_{ss'}$ and $A^* - W^F_{ss'}$ containing w.)

A representation of $L(\mathcal{A})$ and $A^\omega - L(\mathcal{A})$ in terms of the $\sim_{\mathcal{A}}$-classes will be provided by the following lemma:

2.2. LEMMA. (a) *Let $\mathcal{A}$ be a Büchi automaton. For any $\sim_{\mathcal{A}}$-classes U, V: If $U.V^\omega \cap L(\mathcal{A}) \neq \emptyset$, then $U.V^\omega \subseteq L(\mathcal{A})$. (Hence: if $U.V^\omega \cap (A^\omega - L(\mathcal{A})) \neq \emptyset$, then $U.V^\omega \subseteq A^\omega - L(\mathcal{A})$.)*

(b) *Let $\sim$ be a congruence over A^* of finite index. For any ω-word $\alpha \in A^\omega$ there are $\sim$-classes U, V (even with $V.V \subseteq V$) such that $\alpha \in U.V^\omega$.*

Before the proof let us apply the lemma. Part (a) states a "saturation" property of $\sim_{\mathcal{A}}$ with respect to $L(\mathcal{A})$ and $A^\omega - L(\mathcal{A})$. By definition, a congruence $\sim$ over A^* *saturates* an ω-language $L \subseteq A^\omega$ if

$$U.V^\omega \cap L \neq \emptyset \text{ implies } U.V^\omega \subseteq L \quad \text{for all } \sim\text{-classes } U, V.$$

If $\sim$ saturates L and also is of finite index, then

$$L = \bigcup\{U.V^\omega \mid U, V \ \sim\text{-classes}, U.V^\omega \cap L \neq \emptyset\};$$

the inclusion "$\supseteq$" holds by saturation, and "$\subseteq$" follows from Lemma 2.2(b). Moreover, by the assumption that $\sim$ has finite index the $\sim$-classes are regular and the union is finite; so L is a regular ω-language. For the congruence $\sim_{\mathscr{A}}$, which saturates $A^\omega - L(\mathscr{A})$ by Lemma 2.2(a) and which is of finite index, we obtain that $A^\omega - L(\mathscr{A})$ is regular. Note that emptiness of $U.V^\omega \cap L(\mathscr{A})$ (and hence also nonemptiness of $U.V^\omega \cap (A^\omega - L(\mathscr{A}))$ can be decided effectively by Lemma 1.2(c) and Theorem 1.3. Hence, using Theorem 1.1, a Büchi automaton recognizing $A^\omega - L(\mathscr{A})$ can be constructed effectively from the given automaton $\mathscr{A}$. So Lemma 2.2 suffices to prove Theorem 2.1.

PROOF OF LEMMA 2.2. Let $\mathscr{A} = (Q, q_0, \Delta, F)$. Suppose $\alpha = uv_1v_2\ldots$ where $u \in U$ and $v_i \in V$ for $i > 0$, and assume further that there is a successful run of $\mathscr{A}$ on α. From this run we obtain states $s_1, s_2, \ldots$ such that

$$q_0 \underset{u}{\rightarrow} s_1 \underset{v_1}{\rightarrow} s_2 \underset{v_2}{\rightarrow} s_3 \rightarrow \cdots$$

where we even have

$$s_i \overset{F}{\underset{v_1}{\rightarrow}} s_{i+1} \text{ for infinitely many } i.$$

Let $\beta \in U.V^\omega$ be arbitrary. We show that $\beta \in L(\mathscr{A})$. We have $\beta = u'v'_1v_2\ldots$ where $u' \in U$ and $v'_i \in V$ for $i > 0$. Since U, V are $\sim_{\mathscr{A}}$-classes and hence $u \sim_{\mathscr{A}} u'$, $v_i \sim_{\mathscr{A}} v'_i$, we obtain

$$q_0 \underset{u'}{\rightarrow} s_1 \underset{v'_1}{\rightarrow} s_2 \underset{v'_2}{\rightarrow} s_3 \rightarrow \cdots$$

and

$$s_i \overset{F}{\underset{v'_1}{\rightarrow}} s_{i+1} \text{ for infinitely many } i.$$

This yields a run of $\mathscr{A}$ on β in which some F-state occurs infinitely often. Hence, $\beta \in L(\mathscr{A})$.

(b) Let $\sim$ be a congruence of finite index over A^*. Given $\alpha \in A^\omega$, two positions k, k' are said to *merge at position m* (where $m > k, k'$) if $\alpha(k, m) \sim \alpha(k', m)$. In this case write $k \cong_\alpha k'(m)$. Note that then also $k \cong_\alpha k'(m')$ for any $m' > m$ (because $\alpha(k, m) \sim \alpha(k', m)$ implies $\alpha(k, m)\alpha(m, m') \sim \alpha(k', m)\alpha(m, m')$). Write $k \cong_\alpha k'$ if $k \cong_\alpha k'(m)$ for some m. The relation $\cong_\alpha$ is an equivalence relation of finite index over ω (because $\sim$ is of finite index). Hence there is an infinite sequence $k_0, k_1, \ldots$ of positions which all belong to the same $\cong_\alpha$-class. By passing to a subsequence (if necessary), we can assume $k_0 > 0$ and that for $i > 0$ the segments $\alpha(k_0, k_i)$ all belong to the same $\sim$-class V. Let U be the $\sim$-class of $\alpha(0, k_0)$. We obtain

$$\exists k_0\,(\alpha(0, k_0) \in U \wedge \exists^\omega k\,(\alpha(k_0, k) \in V \wedge \exists m\, k_0 \cong_\alpha k\,(m))). \tag{2.1}$$

We shall show that (2.1) implies $\alpha \in U.V^\omega$ and $V.V \subseteq V$ (which completes the proof of (b)). Suppose that k_0 and a sequence $k_1, k_2, \ldots$ are given as guaranteed by (2.1). Again by passing to an infinite subsequence, we may assume that for all $i \geqslant 0$, the positions $k_0, \ldots, k_i$ merge at some $m < k_{i+1}$ and hence at k_{i+1}. We show $\alpha(k_i, k_{i+1}) \in V$ for $i \geqslant 0$. From (2.1) it is clear that $\alpha(k_0, k_1) \in V$. By induction assume that $\alpha(k_j, k_{j+1}) \in V$ for $j < i$. We know $\alpha(k_0, k_{i+1}) \in V$ and that k_0, k_i merge at k_{i+1}. Thus $\alpha(k_i, k_{i+1}) \in V$ and hence $\alpha \in U.V$.

Finally, in order to verify the claim $V.V \subseteq V$, it suffices to show $V.V \cap V \neq \emptyset$ (since V is a class of a congruence). But this is clear since $\alpha(k_0, k_i)$, $\alpha(k_i, k_{i+1})$ and $\alpha(k_0, k_{i+1})$ belong to V for any $i>0$. □

The use of the merging relation $\cong_\alpha$ in the preceding proof can be avoided if Ramsey's Theorem is invoked (as done in the original proof by Büchi [9]): One notes that $\sim$ induces a finite partition of the set $\{(i,j) \mid i<j\}$ by defining that (i,j) and (i',j') belong to the same class iff $\alpha(i,j) \sim \alpha(i',j')$. Now Ramsey's Theorem (in the version for countable sets) states that there is an infinite "homogeneous" set, i.e., a set $\{i_0, i_1, \ldots\}$ such that all segments $\alpha(i_k, i_l)$ with $k<l$ are in one $\sim$-class, in particular all $\alpha(i_k, i_{k+1})$ are in this class. Define V to be this $\sim$-class and let U be the $\sim$-class of $\alpha(0, i_0)$. Then $\alpha \in U.V^\omega$. We gave the above self-contained proof because condition (2.1) will be used again in Section 4.

By Lemma 1.2 and Theorem 2.1, the regular ω-languages are effectively closed under Boolean operations. Hence the inclusion test and the equivalence test for Büchi automata reduce to the emptiness test (using $L_1 \subseteq L_2$ iff $L_1 \cap \sim L_2 = \emptyset$).

2.3. THEOREM. *The inclusion problem and the equivalence problem for Büchi automata are decidable.*

Let us consider the complexity of the complementation process and the equivalence test. Given a Büchi automaton with n states, there are n^2 different pairs (s, s') and hence $O(2^{2n^2})$ different $\sim_{\mathcal{A}}$-classes. This leads to a size bound of $O(2^{4n^2})$ states for the complement automaton [83, 107]. An improved and essentially optimal bound of $2^{O(n \log n)}$ is given in [99]. In [107] it is shown that the equivalence problem for Büchi automata is logspace complete for PSPACE.

The equivalence problem has also been studied in terms of equations between ω-regular expressions, building on work of Salomaa for classical regular expressions. A sound and complete axiom system for deriving these equations is given in [133]; for further references see [110].

Lemma 2.2 above not only shows complementation for regular ω-languages but also serves as a starting point for an investigation of these ω-languages in terms of finite semigroups. Recall that a language $W \subseteq A^*$ is regular iff there is a finite monoid M and a monoid homomorphism $f: A^* \to M$ such that W is a union of sets $f^{-1}(m)$ where $m \in M$. Since $A^*/\sim_{\mathcal{A}}$ is a finite monoid (for any Büchi automaton $\mathcal{A}$), we obtain, from Theorem 1.1 and Lemma 2.2, the following theorem.

2.4. THEOREM. *An ω-language $L \subseteq A^\omega$ is regular iff there is a finite monoid M and a monoid homomorphism $f: A^* \to M$ such that L is a union of sets $f^{-1}(m).(f^{-1}(e))^\omega$ with $m, e \in M$ and where e can be assumed to be idempotent (i.e., satisfying $e.e=e$).*

As will be shown in Theorem 2.6 below, there is a canonical minimal monoid with this property. In other words, there is a coarsest congruence $\approx_L$ over A^* which saturates L. We introduce $\approx_L$ here together with another natural congruence associated with an ω-language L. Given $L \subseteq A^\omega$, define for $u, v \in A^*$

$$u \sim_L v \quad \text{iff} \quad \forall \alpha \in A^\omega \, (u\alpha \in L \Leftrightarrow v\alpha \in L)$$

(a right congruence, cf. [124, 108]);

$$u \approx_L v \quad \text{iff} \quad \forall x, y, z \in A^* \, (xuyz^\omega \in L \Leftrightarrow xvyz^\omega \in L$$
$$\text{and } x(yuz)^\omega \in L \Leftrightarrow x(yvz)^\omega \in L)$$

(cf. [2]). The congruence $\approx_L$ regards two finite words as equivalent iff they cannot be distinguished by L as corresponding segments of ultimately periodic ω-words. Regularity of L implies that $\sim_L$ and $\approx_L$ are of finite index (since they are refined by the finite congruence $\sim_{\mathscr{A}}$ if $\mathscr{A}$ is a Büchi automaton that recognizes L). We note that the converse fails.

2.5. REMARK (Trakhtenbrot [124]). *There are nonregular sets $L \subseteq A^\omega$ such that $\sim_L$ and $\approx_L$ are of finite index.*

PROOF. For given $\beta \in A^\omega$, let $L(\beta)$ contain all ω-words that have a common suffix with β. Then any two words u, v are $\sim_{L(\beta)}$-equivalent since, for two ω-words $u\alpha$, $v\alpha$, membership in $L(\beta)$ does not depend on u, v. So there is only one $\sim_{L(\beta)}$-class. If we choose β to be not ultimately periodic, $L(\beta)$ is not regular (by Theorem 1.3). Furthermore, in this latter case also $\approx_{L(\beta)}$ has only one congruence class. □

We now show the mentioned maximality property of $\approx_L$.

2.6. THEOREM (Arnold [2]). *An ω-language L is regular iff $\approx_L$ is of finite index and saturates L; moreover, $\approx_L$ is the coarsest congruence saturating L.*

PROOF. If $\approx_L$ is of finite index and saturates L, then $L = \bigcup\{U.V^\omega \mid U, V$ are $\approx_L$-classes, $U.V^\omega \cap L \neq \emptyset\}$ and L hence is regular (see the remark following Lemma 2.2). Conversely, suppose L is regular; then (as seen before Remark 2.5) $\approx_L$ is of finite index. We show that $\approx_L$ saturates L, i.e. $U.V^\omega \cap L \neq \emptyset$ implies $U.V^\omega \subseteq L$ for any $\approx_L$-classes U, V. Since $U.V^\omega \cap L$ is regular, we can assume that there is an ultimately periodic ω-word xy^ω in $U.V^\omega \cap L$. In a decomposition of xy^ω into a U-segment and a sequence of V-segments, we find two V-segments which start after the same prefix y_1 of period y; so we obtain $w := xy^m y_1 \in U.V^r$ and $z := y_2 y^n y_1 \in V^s$ for some m, n, r, s and $y_1 y_2 = y$, so that $xy^\omega = wz^\omega$. Denote by $[w]$ and $[z]$ the $\approx_L$-classes of w and z. Since $[w] \cap U.V^r \neq \emptyset$ we have $U.V^r \subseteq [w]$; similarly, $V^s \subseteq [z]$, and hence $U.V^\omega \subseteq [w].[z]^\omega$. It remains to prove $[w].[z]^\omega \subseteq L$. For a contradiction, assume there is $\alpha \in [w].[z]^\omega - L$, say $\alpha = w_0 z_1 z_2 \ldots$ where $w_0 \approx_L w$, $z_i \approx_L z$. Since α may be assumed again to be ultimately periodic, we obtain p, q with

$$\alpha = w_0 z_1 \ldots z_p (z_{p+1} \ldots z_{p+q})^\omega.$$

But then, from $wz^\omega = xy^\omega \in L$, we know $wz^p(z^q)^\omega \in L$, so

$$w_0 z_1 \ldots z_p (z_{p+1} \ldots z_{p+q})^\omega \in L$$

by definition of $\approx_L$ and thus $\alpha \in L$, a contradiction.

It remains to show that $\approx_L$ is the coarsest among the congruences $\sim$ saturating L. So assume $\sim$ is such a congruence and suppose $u \sim v$ (or: $\langle u \rangle = \langle v \rangle$ for the $\sim$-classes of

u and v). We verify $u \approx_L v$. We have $xuyz^\omega \in L$ iff $\langle xuy \rangle \langle z \rangle^\omega \subseteq L$ (since $\sim$ saturates L) iff $\langle xvy \rangle \langle z \rangle^\omega \subseteq L$ (since $u \sim v$) iff $xvyz^\omega \in L$. Similarly, one obtains $x(yuz)^\omega \in L$ iff $x(yvz)^\omega \in L$; thus $u \approx_L v$. □

The preceding result justifies calling $A^*/\approx_L$ the *syntactic monoid of L*, with concatenation of classes as the product. It allows us to classify the regular ω-languages by reference to selected varieties of monoids, extending the classification theory for regular sets of finite words (cf. [87]). Examples will be mentioned in Section 6.

3. The sequential calculus

One motivation for considering automata on infinite sequences was the analysis of the "sequential calculus", a system of monadic second-order logic for the formalization of properties of sequences. Büchi [9] showed the surprising fact that any condition on sequences that is written in this calculus can be reformulated as a statement about acceptance of sequences by an automaton.

For questions of logical definability, an ω-word $\alpha \in A^\omega$ is represented as a model-theoretic structure of the form $\underline{\alpha} = (\omega, 0, +1, <, (Q_a)_{a \in A})$, where $(\omega, 0, +1, <)$ is the structure of the natural numbers with zero, successor function, and the usual ordering, and where $Q_a = \{i \in \omega \mid \alpha(i) = a\}$ (for $a \in A$). The corresponding first-order language contains variables $x, y, \ldots$ for natural numbers, i.e. for the positions in ω-words. Typical atomic formulas are "$x+1<y$" ("the position following x comes before y") or "$x \in Q_a$" ("position x carries letter a"). In this framework, the example set $L_1 \subseteq \{a, b, c\}^\omega$ of Section 1 (containing the ω-words where after any letter a there is eventually a letter b) can be defined by the sentence

p. 11

We shall also allow variables $X, Y, \ldots$ for *sets* of natural numbers and quantifiers ranging over them. For example, they occur in a definition of the ω-language L_2 of Section 1 (containing the ω-words where between any two succeeding occurrences of letter a there is an even number of letters b, c):

$$\begin{aligned}\varphi_2\colon\quad & \forall x\, \forall y\, (x \in Q_a \wedge y \in Q_a \wedge x<y \wedge \neg\exists z\, (x<z \wedge z<y \wedge z \in Q_a) \\ & \rightarrow \exists X\, (x \in X \wedge \forall z\, (z \in X \leftrightarrow \neg(z+1 \in X)) \wedge \neg(y \in X))).\end{aligned}$$

Note that the set quantifier postulates a set containing every second position starting with position x; this ensures that the number of letters between positions x and y is even. The sequential calculus consists of all the conditions on ω-words which can be written in this logical language.

One also calls this framework S1S for "second-order theory of *one* successor". (Below, in Theorem 3.1, it will be seen that $<$ is second-order definable in terms of successor and hence inessential). It is a system of *monadic second-order logic*, due to the quantification over sets, which are unary relations and hence "monadic second-order objects".

DEFINITION. The interpreted system $S1S_A$ (*second-order theory of one successor* over A) is built up as follows: *Terms* are constructed from the constant 0 and the variables $x, y, \ldots$ by applications of "$+1$" (successor function). *Atomic formulas* are of the form $t=t'$, $t<t'$, $t \in X$, $t \in Q_a$ (for $a \in A$) where t, t' are terms and X is a set variable, and $S1S_A$-*formulas* are constructed from atomic formulas using the connectives $\neg, \vee, \rightarrow, \leftrightarrow$ and the quantifiers $\exists, \forall$ acting on either kind of variables. We write $\varphi(X_1, \ldots, X_n)$ to indicate that at most the variables $X_1, \ldots, X_n$ occur free in φ (i.e., are not in the scope of a quantifier). Formulas without free variables are called *sentences*. Given $\alpha \in A^\omega$ and an $S1S_A$-sentence φ, we write $\underline{\alpha} \models \varphi$ if φ is satisfied in $\underline{\alpha}$ under the canonical interpretation (over the domain ω of $\underline{\alpha}$, as described above). The ω-language defined by an $S1S_A$-sentence φ is $L(\varphi) = \{\alpha \in A^\omega \mid \underline{\alpha} \models \varphi\}$. □

For instance, if $\alpha = abcaabcaaabc\ldots$ and φ_1 is as before, we have $\underline{\alpha} \models \varphi_1$, and $L(\varphi_1)$ is the ω-language L_1 of the example in Section 1.

Sometimes it is convenient to cancel the predicate symbols Q_a and use free set variables X_k in their place. We denote the resulting formalism by S1S. In this case we consider formulas $\varphi(X_1, \ldots, X_n)$ without the symbols Q_a and interpret them in ω-words over the special alphabet $\{0, 1\}^n$. In $\alpha \in (\{0, 1\}^n)^\omega$, the formula $x \in X_k$ says that the xth letter of α has 1 in its kth component. As an example, consider the following sequence $\alpha \in (\{0, 1\}^2)^\omega$, where the letters from $\{0, 1\}^2$ are written as columns:

$$\alpha: \begin{matrix} 1 & 0 & 1 & 0 & 1 & 0 & 1 & 0 & 1 \\ 1 & 1 & 1 & 0 & 0 & 0 & 0 & 0 & 0 \end{matrix} \cdots$$

Since in α there are infinitely many letters with first component 1 and second component 0, we have

$$\underline{\alpha} \models \forall x\, \exists y\, (x<y \wedge y \in X_1 \wedge \neg(y \in X_2)).$$

Formally, we represent $\alpha \in (\{0, 1\}^n)^\omega$ by the structure $\underline{\alpha} = (\omega, 0, +1, <, P_1, \ldots, P_n)$ where $P_k = \{i \mid (\alpha(i))_k = 1\}$, writing $\underline{\alpha} \models \varphi(X_1, \ldots, X_n)$ iff φ holds in $\underline{\alpha}$ with P_k as interpretation of X_k. For an S1S-formula $\varphi = \varphi(X_1, \ldots, X_n)$ define $L(\varphi) = \{\alpha \in (\{0, 1\}^n)^\omega \mid \underline{\alpha} \models \varphi(X_1, \ldots, X_n)\}$.

By embedding a given alphabet A into a set $\{0, 1\}^n$ for suitable n, $S1S_A$-sentences can be reformulated as S1S-formulas $\varphi(X_1, \ldots, X_n)$. (Instead of "$x \in Q_a$", write the corresponding conjunction consisting of formulas "$x \in X_k$" and "$\neg\, x \in X_k$".) Depending on the alphabet under consideration, we shall call an ω-language L simply *definable in* S1S if for some sentence φ with the symbols Q_a, respectively for some formula $\varphi = \varphi(X_1, \ldots, X_n)$ without the Q_a, we have $L = L(\varphi)$.

3.1. BÜCHI'S THEOREM (cf. [9]). *An ω-language is definable in* S1S *iff it is regular.*

PROOF. The direction from right to left is easy: Let $\mathscr{A} = (Q, q_0, \Delta, F)$ be a Büchi automaton over the alphabet A, and assume $Q = \{0, \ldots, m\}$ and $q_0 = 0$. The existence of a successful run on an ω-word $\alpha \in A^\omega$ can be expressed by the existence of a suitable $(m+1)$-tuple of sets $Y_0, \ldots, Y_m$; Y_i contains the positions where the run assumes state i

$L(\mathcal{A})$ is defined by the sentence

$$\exists Y_0 \ldots \exists Y_m \Big(\bigwedge_{i \neq j} \neg \exists y (y \in Y_i \wedge y \in Y_j) \wedge 0 \in Y_0 \\ \wedge \forall x \bigvee_{(i,a,j) \in \Delta} (x \in Y_i \wedge x \in Q_a \wedge x+1 \in Y_j) \\ \wedge \bigvee_{i \in F} \forall x \exists y (x < y \wedge y \in Y_i) \Big). \tag{3.1}$$

We shall prove the converse for S1S-formulas $\varphi(X_1, \ldots, X_n)$ interpreted in ω-words over $\{0, 1\}^n$ as explained above. We proceed in two steps: First S1S is reduced to a simpler formalism $S1S_0$ where *only* second-order variables X_i occur and the atomic formulas are of the form $X_i \subseteq X_j$("X_i is a subset of X_j") and $\text{Succ}(X_i, X_j)$ ("X_i, X_j are singletons $\{x\}, \{y\}$ where $x+1=y$"). In a second step an induction over $S1S_0$-formulas $\varphi(X_1, \ldots, X_n)$ shows that $L(\varphi)$ is Büchi recognizable.

(1) *Reduction of* S1S *to* $S1S_0$. Carry out the following steps, starting with a given S1S-formula:

(i) Eliminate superpositions of "+1" by rewriting, e.g.,

"$(x+1)+1 \in X$" as "$\exists y \exists z (x+1=y \wedge y+1=z \wedge z \in X)$".

(ii) Eliminate the symbol 0 and then the symbol < by rewriting, e.g.,

"$0 \in X$" as "$\exists x (x \in X \wedge \neg \exists y (y < x))$",

"$x < y$" as "$\forall X (x+1 \in X \wedge \forall z (z \in X \rightarrow z+1 \in X) \rightarrow y \in X)$".

We arrive at a formula with atomic formulas of type $x=y$, $x+1=y$, and $x \in X$ only. For the remaining step we use the shorthands

"$X=Y$" for "$X \subseteq Y \wedge Y \subseteq X$", "$X \neq Y$" for "$\neg X=Y$",

"Sing X" ("X is a singleton")

for "$\exists Y (Y \subseteq X \wedge Y \neq X \wedge \neg \exists Z (Z \subseteq X \wedge Z \neq X \wedge Z \neq Y))$"

("there is exactly one proper subset of X").

(iii) Eliminate first-order variables, by rewriting, e.g.,

"$\forall x \exists y (x+1=y \wedge y \in Z)$"

as "$\forall X (\text{Sing } X \rightarrow \exists Y (\text{Sing } Y \wedge \text{Succ}(X, Y) \wedge Y \subseteq Z))$".

We obtain an $S1S_0$-formula equivalent to the given S1S-formula.

(2) *Büchi recognizability of* $S1S_0$*-definable sets.* By induction over $S1S_0$-formulas we show that for any $S1S_0$-formula $\varphi(X_1, \ldots, X_n)$ there is a Büchi automaton $\mathcal{A}$ over $\{0, 1\}^n$ with $L(\mathcal{A}) = L(\varphi)$. As typical examples of atomic formulas, consider $X_1 \subseteq X_2$ and $\text{Succ}(X_1, X_2)$. Corresponding Büchi automata are given by the state graphs shown in Fig. 4.

For the induction step it suffices to treat $\neg$, $\vee$, and $\exists$ (since $\wedge$, $\rightarrow$, $\leftrightarrow$, $\forall$ are expressible in terms of $\neg$, $\vee$, $\exists$). Cases $\neg$ and $\vee$ are clear by closure of the regular ω-languages under complement and union. Concerning $\exists$, we have to show closure of the regular

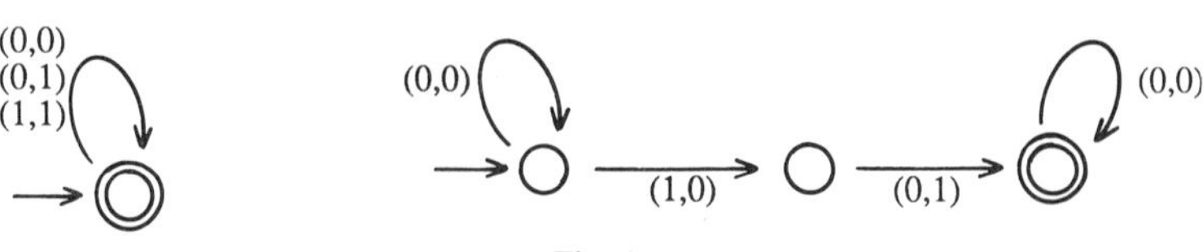

Fig. 4.

ω-languages under *projection*: Assume, for example, that, for the $S1S_0$-formula $\varphi(X_1, X_2)$, a Büchi automaton $\mathcal{A}$ over $\{0, 1\}^2$ exists with $L(\mathcal{A}) = L(\varphi)$; we have to find an automaton $\mathcal{A}'$ over $\{0, 1\}$ for the formula $\varphi'(X_1) = \exists X_2\ \varphi(X_1, X_2)$. We obtain $\mathcal{A}'$ by changing the letters of the transitions of $\mathcal{A}$ from $(1, 0)$, $(1, 1)$ to 1, respectively from $(0, 0)$, $(0, 1)$ to 0. A successful run of $\mathcal{A}'$ thus "guesses" a second component for the given input $\alpha \in \{0, 1\}^\omega$, and for the resulting sequence from $(\{0, 1\}^2)^\omega$ it is a successful run of $\mathcal{A}$. Thus $\mathcal{A}'$ recognizes $L(\varphi')$. $\square$

Büchi's Theorem shows that "global" properties of sequences, as formulated in S1S-conditions by means of quantifiers over arbitrary elements and sets, can be given a strictly "operational" meaning, represented by the stepwise working of a Büchi automaton. From a dual point of view, Büchi automata are, despite their conceptual simplicity, a very powerful formalism for the specification of sequence properties. In this connection we note that the strength of Büchi automata is also reflected by their equivalence with extended models, e.g. the *two-way Büchi automaton* [82, 128] and *alternating ω-automata* [64].

With minor modifications Büchi's Theorem also holds for sets of *finite* words. Thus "regular" is equivalent to "monadic second-order definable" for languages as well as for ω-languages. In a finite word w of length k, the variables $x, y, \ldots$ refer to the positions of w (from 1 to k) and the variables $X, Y, \ldots$ to subsets of $\{1, \ldots, k\}$. Moreover, the successor function has to be redefined for the maximal element k; we may set, for instance, $k + 1 = k$. Also a suitable convention for the empty word is adopted (it should satisfy universal sentences but not existential sentences). With the resulting notion of monadic second-order definability for sets $W \subseteq A^*$ one obtains the following equivalence result.

3.2. THEOREM (Büchi [8], Elgot [28]). *A set $W \subseteq A^*$ is regular iff it is definable in* S1S *(interpreted over finite word models).*

Looking back at the projection step of Theorem 3.1, one notes that it works also for a formula $\varphi' = \exists X_1\ \varphi(X_1)$, which is a *sentence* of S1S, i.e. a formula without free variables. The resulting automaton $\mathcal{A}'$ has unlabelled transitions and in this sense works "input-free". $\mathcal{A}'$ admits a successful run iff the existential sentence φ' is true over the domain ω of the natural numbers. Since the existence of a successful run is effective (see Theorem 1.3), it can be decided whether an S1S-sentence holds in $(\omega, 0, +1, <)$:

3.3. THEOREM (Büchi [9]). *Truth of sentences of* S1S *is decidable.*

The decision problem for S1S was a main motivation in Büchi's investigations. Automata represented a normal form of S1S-formulas (namely, (3.1) in the proof of Theorem 3.1) which was simple enough to be decided effectively.

If set quantification refers to finite sets only, one speaks of the *weak monadic theory of successor*, denoted WS1S. Decidability was shown earlier for WS1S than for S1S (in connection with the characterization of regular sets of finite words, by Büchi [8] and Elgot [28]); it also follows from Theorem 3.3 by an interpretation of WS1S in S1S (since finiteness of sets is definable in S1S).

The decidability result on S1S has been extended in many directions. One of them, Rabin's Tree Theorem (stating decidability of the monadic theory S2S of two successor functions, i.e. of the binary tree) will be discussed in Section 11 below. Another kind of extension is concerned with addition of further number-theoretic relations or functions to the theory S1S (besides successor and order). Elgot and Rabin [29] showed that one may add the unary predicates "is a factorial" or "is a power of k" (for $k \geqslant 2$) without destroying decidability. However, S1S enriched by the function $x \mapsto 2x$ already allows to interpret full first-order arithmetic and thus is undecidable. More recent results are discussed in [102].

Büchi [10] extended his proof of Theorem 3.3 to transfinite ordinals, using automata over ordinal numbers, and showed that the monadic second-order theory of the ordinal $(\omega_1, <)$ is decidable. Siefkes [104], and Büchi and Siefkes [14] presented axiomatizations of the monadic theories of $(\omega, <)$ and $(\omega_1, <)$. In subsequent work of Gurevich, Magidor and Shelah it became clear that from ω_2 onwards the monadic second-order theory of an ordinal depends on set-theoretic hypotheses (cf. [41]).

A fundamental progress in the study of monadic theories was made by Shelah [103]. He developed a model-theoretic technique for obtaining decidability results, which does not refer to automata and is applicable to a larger class of structures. A central idea in this approach is to "compose" a finite fragment of the theory of an ordering (given, say, by the formulas up to some quantifier depth) from the corresponding theory fragments of suborderings and the way these suborderings are arranged. In a series of papers Gurevich and Shelah applied the method to show strong decidability results, in particular concerning dense orderings and trees. However, again based on [103], they also showed that the monadic second-order theory of $(R, <)$, the ordering of the real numbers, is undecidable. For an exhaustive presentation of the subject, see Gurevich's survey [41].

Recently, the automata-theoretic aspects of the monadic theory of the *integers*, i.e. of the ordering $(Z, <)$, have been investigated. This theory is closely related to problems in ergodic theory and symbolic dynamics. Words over Z ("*bi-infinite words*") lack a distinguished position, like the first position of an ω-word. Thus two Z-words are identified if they can be transformed into each other by a finite shift; and a natural automaton model does not refer to some "start position" but works on Z-words from left to right, "coming from infinity" and "going to infinity". A development of the theory of regular Z-languages, establishing results analogous to the case of ω-languages, is given in [76, 89].

4. Determinism and McNaughton's Theorem

A simple argument, given in Example 4.2 below, shows that deterministic Büchi automata are not closed under complement and hence strictly weaker than Büchi automata in general. Nevertheless, by refining the notion of acceptance, it is possible to define a type of deterministic automaton which characterizes the regular ω-languages. In the present section we discuss this fundamental determinization theorem, due to McNaughton [62].

If a deterministic finite automaton $\mathscr{A}$, which recognizes the set $W \subseteq A^*$, is used as a Büchi automaton, it accepts an ω-word α iff infinitely many prefixes of α lead $\mathscr{A}$ to a final state, i.e. belong to W. Collecting these α, we obtain the set

$$\overrightarrow{W} := \{\alpha \in A^\omega \mid \exists^\omega n\, \alpha(0, n) \in W\}.$$

By the mentioned conversion of finite automata into deterministic Büchi automata and vice versa we immediately have the following result.

4.1. REMARK. *An ω-language $L \subseteq A^\omega$ is recognized by a deterministic Büchi automaton iff $L = \overrightarrow{W}$ for some regular set $W \subseteq A^*$.*

The following example shows that not every regular ω-language is of this form; at the same time we see that closure under complement fails for deterministic Büchi automata.

4.2. EXAMPLE. *Let $A = \{a, b\}$ and $L := \{\alpha \in A^\omega \mid \exists^{<\omega} n\, \alpha(n) = a\}$* (i.e., $L = A^\omega - \overrightarrow{(b^*a)^*}$). *Then L is not of the form $\overrightarrow{W}$ with $W \subseteq A^*$.*

PROOF. Assuming $L = \overrightarrow{W}$ one obtains a contradiction as follows: For some n_1, $b^{n_1} \in W$ (because $b^\omega \in L = \overrightarrow{W}$). For this n_1 there is some n_2 such that $b^{n_1}ab^{n_2} \in W$ (because $b^{n_1}ab^\omega \in L = \overrightarrow{W}$). Proceeding in this way, one obtains a sequence of words $b^{n_1}ab^{n_2} \ldots ab^{n_k} \in W$ $(k = 1, 2, \ldots)$. Hence the ω-word $b^{n_1}ab^{n_2} \ldots$ is in $\overrightarrow{W}$ and thus in L, contradicting the definition of L. □

A suitably generalized acceptance condition for deterministic automata on ω-words which captures the power of Büchi automata was defined by Muller [71] (in connection with a problem in asynchronous switching theory).

DEFINITION. A *Muller automaton* over the alphabet A is of the form $\mathscr{A} = (Q, q_0, \delta, \mathscr{F})$ with finite state set Q, initial state $q_0 \in Q$, transition function $\delta: Q \times A \to Q$, and a collection $\mathscr{F} \subseteq 2^Q$ of final state sets. A run σ of $\mathscr{A}$ is *successful* if $\text{In}(\sigma) \in \mathscr{F}$, i.e. the states that $\mathscr{A}$ assumes infinitely often in σ form a set from $\mathscr{F}$. $\mathscr{A}$ *accepts* an ω-word α if the unique run σ of $\mathscr{A}$ on α is successful. An ω-language $L \subseteq A^\omega$ is called *Muller recognizable* if it consists of all ω-words accepted by some Muller automaton over A.

In the nondeterministic version of Muller automaton, a transition relation $\Delta \subseteq Q \times A \times Q$ replaces δ, and acceptance of α means existence of a run σ on α with $\text{In}(\sigma) \in \mathscr{F}$. The ω-languages recognized by nondeterministic Muller automata are definable in S1S (by the same idea as in Theorem 3.1) and hence, obviously coincide with the regular ω-languages. In the sequel we consider only the deterministic version.

Each deterministic Büchi automaton $\mathscr{A} = (Q, q_0, \delta, F)$ is equivalent to a Muller automaton, namely to the automaton $\mathscr{A}' = (Q, q_0, \delta, \mathscr{F})$ where $\mathscr{F}$ consists of all subsets of Q having a nonempty intersection with F. Furthermore, the Muller recognizable ω-languages are closed under Boolean operations: If $\mathscr{A}=(Q, q_0, \delta, \mathscr{F})$ recognizes $L \subseteq A^\omega$, then $(Q, q_0, \delta, 2^Q - \mathscr{F})$ recognizes $A^\omega - L$. Given $\mathscr{A}=(Q, q_0, \delta, \mathscr{F})$ and $\mathscr{A}'=(Q', q'_0, \delta', \mathscr{F}')$ recognizing L and L' respectively, $L \cup L'$ is recognized by the product automaton of $\mathscr{A}$ and $\mathscr{A}'$, where the collection of final state sets contains $\{(q_1, q'_1), \ldots, (q_n, q'_n)\}$ iff $\{(q_1, \ldots, q_n\} \in \mathscr{F}$ or $\{q'_1, \ldots, q'_n\} \in \mathscr{F}'$. These observations, together with Remark 4.1, yield the direction from right to left of the following lemma.

4.3. LEMMA. *An ω-language $L \subseteq A^\omega$ is Muller recognizable iff L is a Boolean combination (over A^ω) of sets $\overrightarrow{W}$ with regular $W \subseteq A^*$.*

PROOF. For the direction from left to right, consider a Muller automaton $\mathscr{A}=(Q, q_0, \delta, \mathscr{F})$ recognizing $L \subseteq A^\omega$; write W_q for the set of finite words recognized by the finite automaton $(Q, q_0, \delta, \{q\})$. By definition, $\mathscr{A}$ accepts α iff, for some $F \in \mathscr{F}$, α belongs to $\overrightarrow{W_q}$ for all $q \in F$ and α does not belong to $\overrightarrow{W_q}$ for all $q \in Q-F$. Thus we get the desired representation (writing $\sim$ for the complement w.r.t. A^ω):

$$L = \bigcup_{F \in \mathscr{F}} \left(\bigcap_{q \in F} \overrightarrow{W_q} \cap \bigcap_{q \in Q-F} \sim \overrightarrow{W_q} \right). \quad \square$$

We now show that deterministic Muller automata and nondeterministic Büchi automata are equivalent in recognition power. Note that this reproves closure of Büchi recognizable sets under complement. The main difficulty lies in the fact that a membership test "$\alpha \in U.V^\omega$?" as performed by a Büchi automaton, i.e. the test whether α can be split into segments $uv_1v_2v_3\ldots$ with $u \in U$ and $v_i \in V$, involves "unlimited guessing" in choosing the segments v_i; this has to be reduced to a deterministic procedure which should depend only on uniformly bounded information about finite prefixes of α.

4.4. MCNAUGHTON'S THEOREM (cf. [62]). *An ω-language is regular (i.e., Büchi recognizable) iff it is Muller recognizable.*

The direction from right to left follows from Lemma 4.3 and the closure properties for Büchi automata. For the converse, it will suffice, again by Lemma 4.3, to show the following theorem.

4.5. THEOREM. *Every Büchi recognizable ω-language $L \subseteq A^\omega$ is a finite union of sets of the form $\overrightarrow{W} \cap \sim\overrightarrow{W'}$ where $W, W' \subseteq A^*$ are regular.*

Before turning to the proof, we mention that Theorem 4.5 motivates also the following modified version of Muller automaton, introduced by Rabin [95].

DEFINITION. A (sequential) *Rabin automaton* over the alphabet A is of the form $\mathscr{A}=(Q, q_0, \delta, \Omega)$ where Q, q_0, δ are as for Muller automata and $\Omega=\{(L_1, U_1), \ldots, (L_n, U_n)\}$ is a collection of *accepting pairs* (L_i, U_i) with $L_i, U_i \subseteq Q$. A run σ of $\mathscr{A}$ is *successful* if for some $i \in \{1, \ldots, n\}$ we have $\mathrm{In}(\sigma) \cap L_i = \emptyset$ and $\mathrm{In}(\sigma) \cap U_i \neq \emptyset$, i.e. the

L_i-states occur only finitely often but some U_i-state occurs infinitely often in σ. $\mathcal{A}$ *accepts* an ω-word $\alpha \in A^\omega$ if the unique run σ of $\mathcal{A}$ on α is successful in this sense.

Let $\mathcal{A}$ be a Rabin automaton as in the above definition. If W_i (respectively W'_i) is the regular language recognized by the finite automaton (Q, q_0, δ, U_i) (resp. (Q, q_0, δ, L_i)), then the Rabin automaton (Q, q_0, δ, Ω) recognizes the ω-language

$$\bigcup_{i=1}^{n} (\overrightarrow{W_i} \cap \sim \overrightarrow{W'_i}),$$

hence a language as required in Theorem 4.5. From Lemma 4.3 and Theorem 4.5, it follows that any Rabin automaton is equivalent to a Muller automaton and vice versa.

PROOF OF THEOREM 4.5. Let $\mathcal{A}$ be a Büchi automaton. By Lemma 2.2 it suffices to consider the case $L = U.V^\omega$ with $\sim_\mathcal{A}$-classes U, V and $V.V \subseteq V$. We use the notation introduced in the proof of Lemma 2.2, in particular the merging relation $\cong_\alpha$. (By $k \cong_\alpha k'\,(m)$ we mean here that $\alpha(k, m) \sim_\mathcal{A} \alpha(k', m)$.) In Lemma 2.2, the condition $\alpha \in U.V^\omega$ was shown to be equivalent to (cf. Fig. 5)

$$\exists k_0\,(\alpha(0, k_0) \in U \wedge \exists^\omega k\,(\alpha(k_0, k) \in V \wedge \exists m\, k_0 \cong_\alpha k\,(m))). \tag{4.1}$$

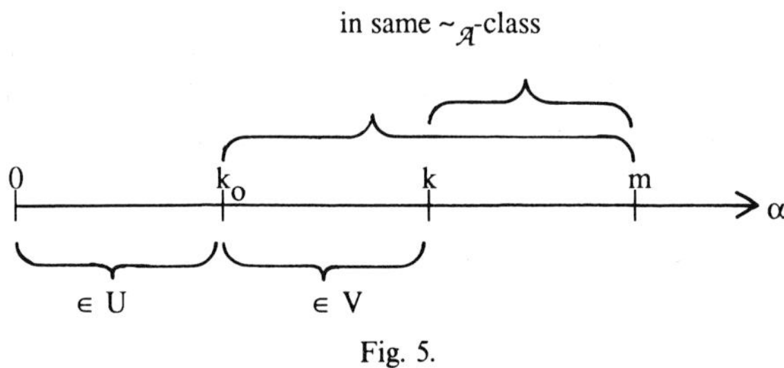

Fig. 5.

Call a segment $\alpha(k_0, m)$ a *V-witness* if, for some k with $k_0 < k < m$, m is the *smallest* position such that $\alpha(k_0, k) \in V$ and $k_0 \cong_\alpha k\,(m)$. It is not difficult to verify that the set $W_V \subseteq A^*$ of V-witnesses is regular. Since for any k as described in (4.1) there is a unique V-witness, we obtain that (4.1) is equivalent to

$$\exists k_0\,(\alpha(0, k_0) \in U \wedge \exists^\omega m\; \alpha(k_0, m) \text{ is a } V\text{-witness}), \tag{4.2}$$

in other words, $\alpha \in U.\overrightarrow{W_V}$.

The aim is to rewrite (4.2) as a Boolean combination of conditions "$\exists^\omega m\, \alpha(0, m) \in W$" (with regular W); so we want to exchange the quantifiers $\exists k_0$ and $\exists^\omega m$ in (4.2). In the desired condition of the form "$\exists^\omega m \exists k_0 \ldots$" we have to ensure that k_0 may be chosen fixed (independent of the m). The idea is to postulate infinitely many prefixes $\alpha(0, m)$ which admit a decomposition $\alpha(0, m) = \alpha(0, k_0)\alpha(k_0, m)$ with $\alpha(0, k_0) \in U$ and $\alpha(k_0, m) \in W_V$, and to guarantee that only finitely many choices of k_0 occur while m increases. A simple approach would be to refer always to the smallest $k_0 < m$ with $\alpha(0, k_0) \in U$.

However, this k_0 might not have the property that there are infinitely many m with $\alpha(k_0, m) \in W_V$: it may happen that only some greater k_1 with $\alpha(0, k_1) \in U$, which does not merge with k_0, has this property. Suppose there are exactly r $\cong_\alpha$-classes with elements k such that $\alpha(0, k) \in U$, i.e. there are r positions, say $k_1, \ldots, k_r$, in α which do not merge pairwise and satisfy $\alpha(0, k_i) \in U$ ("case r"). Then k_0 may be chosen as one of these positions. So in "case r" we require for infinitely many m the existence of $k_1, \ldots, k_r < m$ which do not merge pairwise at m and satisfy $\alpha(0, k_i) \in U$. We express that for increasing m these k_i stay below a fixed finite bound (so that only finitely many choices of k_0 are possible) by saying that for only finitely many m a new choice of $k_1, \ldots, k_r$ (or just of the maximal position k_r) is necessary.

To give a precise formulation, let us call k *r-appropriate for m* iff k is the smallest number less than m such that k is the last element k_r of an r-tuple $(k_1, \ldots, k_r)$ with $0 < k_1 < \cdots < k_r$, $\alpha(0, k_i) \in U$ for all k_i, and not $k_i \cong_\alpha k_j (m)$ for $k_i \neq k_j$. If in addition $\alpha(k_i, m) \in W_V$ for some k_i, we say that k is r-appropriate for m *via* W_V. Finally, a *new* k r-appropriate for m is a number which is r-appropriate for m and fails to be r-appropriate for any $m' < m$. Then, assuming "case r", (4.2) amounts to the following:

$$\begin{aligned}&\exists^\omega m\,[\text{there exists a } k \text{ } r\text{-appropriate for } m \text{ via } W_V]\\ &\wedge \exists^{<\omega} m\,\{\text{there exists a new } k \text{ } r\text{-appropriate for } m\}.\end{aligned} \qquad (4.3)_r$$

Both conditions [...] and {...} above depend only on the segment $\alpha(0, m)$; moreover, one can construct two finite automata accepting exactly the words $\alpha(0, m)$ satisfying these requirements. (As an alternative, one may note definability of [...] and {...} in WS1S and apply Theorem 3.2.) Denote by W_r and W_r' the (regular) sets of words $\alpha(0, m)$ satisfying [...], respectively {...}. Then $(4.3)_r$ says that $\alpha \in \overrightarrow{W_r} \cap \sim\overrightarrow{W_r'}$.

The disjunction of the conditions $(4.3)_r$ over all possible r is equivalent to (4.2). Since r cannot exceed the finite number n of different $\sim_{\mathscr{A}}$-classes, we obtain

$$\bigcup_{r=1}^{n} (\overrightarrow{W_r} \cap \sim\overrightarrow{W_r'})$$

as the desired representation of $U.V^\omega$. □

McNaughton's Theorem has many interesting consequences, among them further characterizations of the regular ω-languages. For example, parts (c) and (d) of the following result give a simple inductive construction of the regular sequence sets, and show the surprising fact that S1S and the weak theory WS1S have the same expressive power.

4.6. THEOREM. *For an ω-language $L \subseteq A^\omega$ the following conditions are equivalent:*

(a) *L is regular.*

(b) *L is a finite union of sets $U.\overrightarrow{W}$ where $U, W \subseteq A^*$ are regular.*

(c) *L can be obtained from A^ω by finitely many applications of union, complement (w.r.t. A^ω) and concatenation with a regular set $W \subseteq A^*$ on the left.*

(d) *L is definable in* WS1S.

PROOF. Implication (a) → (b) is clear from the proof of Theorem 4.5 (cf. condition (4.2)).

For (b) → (c) it suffices to represent $\vec{W}$ as stated in (c), supposing that W is regular. Let (Q, q_0, δ, F) be a finite automaton recognizing W; set $W_{pq} := \{w \in A^* \mid \delta(p, w) = q\}$. We have $\vec{W} = \sim L$ where L contains all ω-words which have only finitely many prefixes in W, i.e. no prefix in W or a last prefix in W. Note that $\alpha(0, m)$ is a last prefix of α in W iff for some state $p \in F$, $\delta(q_0, \alpha(0, m)) = p$ and there is no $n > m$ and no $q \in F$ with $\delta(p, \alpha(m, n)) = q$. This yields the equation

$$L = \sim(W.A^\omega) \cup \bigcup_{p \in F}\left(W_{q_0 p}.\sim\left(\bigcup_{q \in F}(W_{pq} \cap A^+).A^\omega\right)\right).$$

So L and hence $\vec{W}$ are represented as desired.

Implications (c) → (a) and (d) → (a) are obvious; so it remains to show (a) → (d). By McNaughton's Theorem, a regular ω-language L is a Boolean combination of sets $\vec{W}$ with regular W. From a WS1S-formula $\varphi(\bar{X})$ which defines W by Theorem 3.2, we obtain a WS1S-formula $\psi(\bar{X}, y)$ which expresses over ω-words that the prefix up to position y satisfies $\varphi(\bar{X})$ (use relativization to the positions $<y$). Now L is definable in WS1S by a Boolean combination of formulas $\forall x\, \exists y\, (x < y \wedge \psi(\bar{X}, y))$. □

There are several fully worked-out expositions of McNaughton's Theorem; we mention [95, 10, 125, 15, 27]. The above proof of Theorem 4.5 follows [119]. An elegant automaton construction, giving an essentially optimal size bound for the constructed deterministic automaton, was recently found by Safra [99]: for any nondeterministic Büchi automaton with n states there is an equivalent deterministic Rabin automaton with $2^{O(n \log n)}$ states and n accepting pairs. This result has applications in obtaining good upper complexity bounds for several logics of programs [32, 23, 137].

5. Acceptance conditions and Borel classes

Definability of sequence sets is a traditional subject in set-theoretic topology and descriptive set theory. Automata on ω-words constitute a special type of "finitary" definability of sequence sets and lead to combinatorial problems not considered in the classical mathematical literature. (McNaughton's Theorem is an example.) Conversely, several basic topological notions, in particular in connection with the Borel hierarchy, have a natural meaning in the context of automata and help to systematize the acceptance conditions. In the present section we introduce the classification of acceptance modes for automata in a topological setting, and indicate applications to sequence properties which arise in distributed systems. For a more complete discussion, see the surveys [49, 110].

We refer to the *Cantor topology* on the set A^ω, where A is a finite alphabet. This topology may be characterized in several equivalent ways:

- as the product topology of the discrete topology on A,
- by declaring as *open sets* all ω-languages $W.A^\omega$ with $W \subseteq A^*$, or the sets $\{w\}.A^\omega$, as a basis,

– by the *metric* $d\colon A^\omega \times A^\omega \to R$, given by

$$d(\alpha, \beta) = \begin{cases} 0 & \text{if } \alpha = \beta, \\ 1/2^n \text{ with } n = \min\{i \mid \alpha(i) \neq \beta(i)\} & \text{else.} \end{cases}$$

Taking the example $A = \{0, 1\}$, the elements of the space A^ω may be considered as paths through the infinite binary tree (where "0" means "branch left" and "1" means "branch right"). Two paths α, β are close to each other if they have a long common initial segment. The so-called Cantor discontinuum is a linearly ordered representation of the space, corresponding to the left to right ordering of the paths in the binary tree.

Following classical terminology we denote by G the class of open ω-languages and by F ("fermé") the class of closed ω-languages. The *Borel hierarchy* is obtained by taking alternately countable intersections and unions starting with ω-languages in G or F. By G_δ (F_σ) one denotes the countable intersections (resp. unions) of sets in G (resp. F), by $G_{\delta\sigma}$ ($F_{\sigma\delta}$) the countable unions (resp. intersections) of sets in G_δ (resp. F_σ), etc. Then a hierarchy of the form

open $= G$ —— G_δ —— $G_{\delta\sigma}$ —— …

closed $= F$ —— F_σ —— $F_{\sigma\delta}$ —— …

results, where each line indicates a proper inclusion. A set L belongs to some class in the hierarchy (say $L \in G_\delta$) iff its complement is in the dual class (in the example, ${\sim}L \in F_\sigma$).

Our aim is to locate the regular ω-languages in the Borel hierarchy. As a preparation we note two useful facts about open, respectively closed sets. First, an open set $W.A^\omega$ is equal to $W_0.A^\omega$ where W_0 contains all words from W with no proper prefix in W. Any two words in W_0 are then incomparable w.r.t. the prefix relation; if this holds and $W.A^\omega = W_0.A^\omega$, we say that W_0 is a *minimal basis* for $W.A^\omega$. The property of being a minimal basis will not be destroyed if we replace, for some n, a word $w \in W_0$ by all words in $\{w\}.A^n$. Using thus longer and longer words in minimal bases, it is possible to represent a sequence $W_1.A^\omega, W_2.A^\omega, \ldots$ of open sets by a sequence of pairwise disjoint minimal bases W'_i.

Secondly, note that the closed sets in the Cantor topology do not coincide with the sets $\overrightarrow{W} \subseteq A^\omega$. The topological *closure* of a set L consists of all sequences that have arbitrary long common prefixes with ω-words in L. Thus the set $L = \overrightarrow{a^*bb^*}$ is not closed since a^ω is in the closure of L but does not belong to L itself. The sets $\overrightarrow{W}$ are more general than closed sets in the following sense.

5.1. REMARK. *$L = \overrightarrow{W}$ for some $W \subseteq A^*$ iff L is in G_δ.*

PROOF. Assume first $L \in G_\delta$, i.e. $L = \bigcap_{n \geq 0} W_n.A^\omega$. Without loss of generality, we have $W_0 \supseteq W_1 \supseteq \cdots$ (if not, replace W_n by $\bigcap_{i \leq n} W_i$). Choose a sequence $W'_0, W'_1, \ldots$ of disjoint minimal bases (as explained above) of $W_0.A^\omega, W_1.A^\omega, \ldots$ and set $W := \bigcup_{n \geq 0} W'_n$. Then $\alpha \in \overrightarrow{W}$ iff α has a prefix in W'_n for infinitely many n iff α has a prefix in W_n for infinitely many n iff (by monotonicity) $\alpha \in \bigcap_{n \geq 0} W_n.A^\omega$, i.e. $\alpha \in L$. Conversely, suppose that $L=\overrightarrow{W}$. Denote by $A^{\geqslant n}$ the set of words of length $\geqslant n$ over A. We have

$$\overrightarrow{W}=\bigcap_{n\geqslant 0}(W\cap A^{\geqslant n}).A^{\omega}$$

and hence $L\in G_{\delta}$. □

For a discussion of the topology induced by the operation $\overrightarrow{\ }$ as topological closure see [97]. By Remark 5.1 and McNaugthon's Theorem, the regular ω-languages occur very low in the Borel hierarchy, as expressed in the following theorem.

5.2. THEOREM. *Every regular ω-language L is a Boolean combination of G_{δ}-(and/or F_{σ}-)sets; in particular, $L\in G_{\delta\sigma}\cap F_{\sigma\delta}$.*

We now discuss refinements of this result due to Landweber [55], which yield a characterization of the regular sets $L\in A^{\omega}$ occurring on the levels G, F, G_{δ}, F_{σ} of the Borel hierarchy, and establish effective procedures for deciding whether a regular set is of one of these types.

DEFINITION. Let $\mathscr{A}=(Q,q_0,\delta,F)$ be a deterministic finite automaton over A and $\alpha\in A^{\omega}$. $\mathscr{A}$ *1-accepts* α iff $\exists n\ \delta(q_0,\alpha(0,n))\in F$, and $\mathscr{A}$ *2-accepts* α iff $\exists^{\omega} n\ \delta(q_0,\alpha(0,n))\in F$. An ω-language $L\subseteq A^{\omega}$ is called 1-(resp. 2-)*recognizable* if L consists of the ω-words 1-(resp. 2-)accepted by some deterministic finite automaton.

So 2-recognizability is Büchi recognizability by deterministic automata. Since L is 1-(resp. 2-)recognizable iff $L=W.A^{\omega}$ (resp. $L=\overrightarrow{W}$) for some regular set W, one calls such ω-languages regular-open (resp. regular-G_{δ}). The following result shows in particular that regular ω-languages which are open are in fact regular-open, and similarly for G_{δ}.

5.3. THEOREM (Landweber [55]). (a) *A regular ω-language is in G iff it is 1-recognizable.*

(b) *A regular ω-language is in G_{δ} iff it is 2-recognizable.*

(c) *It is decidable whether a regular ω-language (represented, say, by a Muller automaton) is 1-recognizable, resp. 2-recognizable.*

For the theorem it suffices to show the following lemma.

5.4. LEMMA. *There are effective procedures transforming any Muller automaton $\mathscr{A}$ into a Muller automaton $\mathscr{A}_1$, resp. $\mathscr{A}_2$, such that $L(\mathscr{A}_1)$ is 1-recognizable, $L(\mathscr{A}_2)$ is 2-recognizable, and*

$$L(\mathscr{A})\in G \text{ iff } \mathscr{A}=\mathscr{A}_1; \qquad L(\mathscr{A})\in G_{\delta} \text{ iff } \mathscr{A}=\mathscr{A}_2.$$

PROOF. Assume $\mathscr{A}=(Q,q_0,\delta,\mathscr{F})$ is a Muller automaton, w.l.o.g., such that each $q\in Q$ is reachable from q_0 (i.e., there is $w\in A^*$ with $\delta(q_0,w)=q$). By a "loop of $\mathscr{A}$" we shall mean a strongly connected subset of $\mathscr{A}$ (considering $\mathscr{A}$ as a directed graph).

(a) Call a state q of $\mathcal{A}$ an $\mathcal{F}$-state iff q is located in a loop of $\mathcal{A}$ which forms a set from $\mathcal{F}$. Define $\mathcal{A}_1 := (Q, q_0, \delta, \mathcal{F}_1)$ by extending $\mathcal{F}$ to $\mathcal{F}_1$ such that $\mathcal{F}_1$ contains all loops of $\mathcal{A}$ which are reachable from some $\mathcal{F}$-state of $\mathcal{A}$. Let F_1 be the union of the sets $F \in \mathcal{F}_1$. $\mathcal{A}_1$ is equivalent to (Q, q_0, δ, F_1) with 1-acceptance. Namely, some state $q \in F_1$ is reached by (Q, q_0, δ, F_1) on input α iff on α $\mathcal{A}_1$ assumes ultimately the states of a loop from $\mathcal{F}_1$. It remains to show that $L(\mathcal{A}) \in G$ implies $\mathcal{A} = \mathcal{A}_1$. For this we have to verify $\mathcal{F}_1 \subseteq \mathcal{F}$. Let $F \in \mathcal{F}_1$; so there is an $\mathcal{F}$-state q of $\mathcal{A}$ from which the loop F can be reached. Since q is an $\mathcal{F}$-state, we can choose a sequence α inducing $\mathcal{A}$ to assume a loop from $\mathcal{F}$ in which q is located. By assumption, $L(\mathcal{A}) = W . A^{\omega}$ for suitable W; so some prefix w of α is in W. Since *all* sequences $w\beta$ are in $L(\mathcal{A})$, any loop reachable from q, and hence F, is realized via some ω-word in $L(\mathcal{A})$. Thus $F \in \mathcal{F}$.

(b) $\mathcal{A}_2$ is constructed from $\mathcal{A}$ by enlarging $\mathcal{F}$ to $\mathcal{F}_2$ as follows: for any state q in a loop forming a set $F \in \mathcal{F}$, and any set E forming some loop containing q, add $F \cup E$ to $\mathcal{F}_2$. $L(\mathcal{A}_2)$ is 2-recognizable: for this we construct a Büchi automaton working as $\mathcal{A}$ does, but further equipped with a "state memory" S which collects the visited states and is set back to $\emptyset$ whenever S includes some $\mathcal{F}$-set. So the states are of form $(q, S) \in Q \times 2^Q$, and from (q, S) the Büchi automaton passes to state $(\delta(q, a), S \cup \{\delta(q, a)\})$ by letter a if $S \cup \{\delta(q, a)\}$ does not include a set from $\mathcal{F}$; otherwise, the transition goes to $(\delta(q, a), \emptyset)$. The initial state is $(q_0, \emptyset)$, and the states $(q, \emptyset)$ are the final ones. 2-Acceptance for this automaton means that some loop extending a loop forming an $\mathcal{F}$-set is ultimately assumed on the given input, i.e. that $\mathcal{A}_2$ accepts. Suppose now $L(\mathcal{A}) \in G_\delta$, say $L(\mathcal{A}) = \overrightarrow{W}$, and consider two loops F and E of $\mathcal{A}$ as above with common state q. We show $F \cup E \in \mathcal{F}$ (and hence $\mathcal{F} = \mathcal{F}_2$). Pick w such that $\delta(q_0, w) = q$. Via the loop F we can extend w to a sequence α of $L(\mathcal{A})$ $(= \overrightarrow{W})$ and hence reach some finite prefix of α in W, say $wu_1 \in W$. From wu_1 we may complete the loop F back to state q (say via v_1) and continue further through the loop E, again back to q (say via w_1). Repeating the process, we obtain a sequence $wu_1v_1w_1u_2v_2w_2\ldots$ which is in $\overrightarrow{W}$ and causes $\mathcal{A}$ to assume ultimately the states in $F \cup E$. Hence $F \cup E \in \mathcal{F}$. $\square$

A characterization similar to Theorem 5.3 can also be given for the regular ω-languages in the Borel class F (closed sets), respectively F_σ: in these cases one refers to nondeterministic finite automata with the acceptance condition that in some run all states (resp. almost all states) are final. The boolean closures of F and F_σ are characterized by automata with systems $\mathcal{F}$ of final state sets (as in Muller automata).

A more refined system of structural invariants for regular ω-languages is given in [133]. It allows for example, to estimate the length of the Boolean expressions arising in McNaughton's Theorem: Consider representations of a regular set $L \subseteq A^{\omega}$ by Boolean combinations $\bigcup_{i=1}^{n} (\overrightarrow{W_i} \cap \sim \overrightarrow{W_i'})$ of G_δ-sets, and define the *Rabin index* of L to be the smallest n such that L is obtained in this form with regular W_i, W_i'. Wagner [133] (and independently Kaminski [50]) show that this index induces a strict hierarchy; moreover, the Rabin index of a regular ω-language is effectively computable. For a rather complete survey of further results in this field see [110].

Restricted acceptance conditions have been investigated for various other machine models (besides finite automata), for example, pushdown automata [17], deterministic pushdown automata [18, 58], Petri nets [126], and Turing machines [19, 109]. A unified treatment (referring to the general notion of storage type) is given by Engelfriet and Hoogeboom [138]. Generalized acceptance conditions for finite automata that lead beyond the regular ω-languages are considered in [134].

Recently, the topological approach has also been applied in the classification of sequence properties that arise in distributed systems. Intuitively, one calls a property of state sequences of a system a *safety property* if it ensures that nothing "bad" (like deadlock) happens at any time instance. Similarly, a *liveness property* guarantees that, given any time instance, something "good" (like entering a critical section) will eventually happen. For a systematic treatment of such sequence properties (leading to specific verification strategies), several exact definitions for describing these intuitive notions have been proposed, e.g. by Alpern and Schneider [1], Lichtenstein, Pnueli and Zuck [57], and Manna and Pnueli [60]. These proposals agree in identifying safety properties with closed sets (in the Cantor topology). Liveness properties, however, have been defined in different says. Alpern and Schneider [1] suggest to consider them as *dense* sets in the Cantor space. (A set $L \subseteq A^{\omega}$ is dense iff every $w \in A^*$ is extendible to a sequence $w\alpha \in L$.) In this case, a topological fact ("every set of the space is the intersection of a closed set with a dense set") can be applied to obtain a decomposition of correctness proofs into two parts, one establishing a safety property and the other showing a liveness property.

In [57, 60], liveness properties are connected with G_{δ}-sets. The latter paper sets up a correspondence between the Borel hierarchy levels $F, G, F_{\sigma}, G_{\delta}$ and the sequence properties of type "safety", "guarantee", "persistence", and "recurrence" respectively. Proof principles for the verification of such properties are presented, using the framework of temporal logic.

An ω-language of type $(\sim \vec{U}) \cup \vec{V}$ (which is in the Boolean closure of G_{δ}) can be regarded as a so-called *fairness condition*: it represents a sequence property which states that infinitely many instances of event U (as prefix of a sequence) imply infinitely many instances of V. For example, this can mean that (in a state sequence of a system) an action which is "enabled" again and again in fact happens again and again. In this way, fairness conditions may serve to ensure the absence of "starvation" in concurrent programs (where a process is enabled infinitely often but does not continue). A corresponding language-theoretical operation ("fair merge" of two ω-languages) is considered by Park [81]. "Fair" acceptance conditions for finite automata over ω-sequences are studied by Priese, Rehrmann and Willecke-Klemme [92]. It is shown there that nonregular ω-languages can be defined when the fairness constraints refer to unbounded future segments of ω-sequences (instead of initial segments that refer to the past). The study of fairness notions remains an important topic of current research. For a general exposition, see the monograph by Francez [34]; as a recent contribution which characterizes regular ω-languages by fairness conditions in the calculus SCCS (synchronous CCS) we mention [40].

6. Star-free ω-languages and temporal logic

An interesting class of regular ω-languages is obtained when the defining monadic second-order formalism S1S is restricted to first-order logic; in this case quantification is allowed only over elements (i.e., positions in sequences). The first-order definable ω-languages are closely related to the class of *star-free languages* and to the *propositional temporal logic of linear time*. In this section we discuss both aspects. For a more detailed treatment of temporal logic see Emerson's survey [30] (in this Handbook).

Recall that a language $W \subseteq A^*$ is *star-free* if it can be generated from finite sets of words by repeated application of the Boolean operations and concatenation (see [87]). The resulting star-free expressions are very similar to first-order formulas, by the close correspondence between the operations $\sim$, $\cup$, . and the connectives $\neg$, $\vee$, $\exists$. For example, the star-free language $A^*.(a\cup c).\sim(A^*.b.A^*)$ over $A=\{a,b,c\}$ is defined by the first-order sentence

$$\exists x\,((x \in Q_a \vee x \in Q_c) \wedge \neg\exists y\,(x<y \wedge y \in Q_b)).$$

An easy induction shows that each star-free language is first-order definable. In particular, if U, V are defined by the first-order formulas φ, ψ, then $U.V$ is defined by $\exists x\,(\varphi'(x) \wedge \psi'(x))$ where $\varphi'(x)$ and $\psi'(x)$ are the relativizations of φ, ψ to the elements $\leqslant x$ and $>x$ respectively (assuming here for simplicity that $\varepsilon \notin U$). More difficult is the converse translation, which yields the following result.

6.1. THEOREM. (McNaughton and Papert [63]). *A language $W \subseteq A^*$ is star-free iff it is first-order definable (in the signature with $<$ and the unary predicates Q_a for $a \in A$).*

PROOF. For the translation from first-order logic into star-free expressions we use induction over quantifier depth of formulas (i.e., the maximum number of nested quantifiers). In the induction step, we consider the case of the existential quantifier, here for a formula $\exists x\,\varphi(x)$ of quantifier depth $n+1$, assuming that sentences of quantifier depth n define star-free sets. (The general situation $\exists x\,\varphi(x,\bar{y})$, with free variables $\bar{y}$ as parameters, is a little more technical.) We shall reduce the statement $\exists x\,\varphi(x)$ to statements of the form $\exists x\,(\varphi_{<x} \wedge x \in Q_a \wedge \varphi_{>x})$ where $\varphi_{<x}, \varphi_{>x}$ speak only about the elements $<x$, resp. $>x$. If $\varphi_{<x}$ and $\varphi_{>x}$ are of quantifier depth n, then such a formula describes a language $U.a.V$ where U, V are star-free by induction hypothesis.

As a preparation we introduce an equivalence relation $\equiv_n$ over A^*: define, for $u, v \in A^*$,

$$u \equiv_n v \text{ iff } u \text{ and } v \text{ satisfy the same sentences of quantifier depth } n.$$

Also an extended version of this definition is needed for formulas with free variables, in our case for formulas $\varphi(x)$ with one free variable x. The corresponding models are words with some distinguished position, of the form (u,r) where $1 \leqslant r \leqslant |u|$. We write $(u,r) \equiv_{n,1} (v,s)$ if (u,r) and (v,s) satisfy the same formulas $\varphi(x)$ of quantifier depth n. An induction over quantifier depth shows the basic fact that there are, for any $n \geqslant 1$, only finitely many equivalence classes of $\equiv_n$ and $\equiv_{n,1}$, and that any such class W of words u, resp. class $\underline{W}$ of word models (u,r), can be defined by a sentence φ_W, resp. formula $\varphi_{\underline{W}}(x)$, of quantifier depth n. It follows that

> any formula $\varphi(x)$ of quantifier depth n is equivalent to a finite disjunction of formulas $\varphi_{\underline{W}}(x)$ (namely those $\varphi_{\underline{W}}(x)$ where $\underline{W}$ contains some (u,r) satisfying $\varphi(x)$). (6.1)

Next we note a fact which (unlike (6.1)) depends on the present choice of signature with ordering and unary predicates only: The relations $\equiv_n$ and $\equiv_{n,1}$ are congruences. Namely,

$$\text{if } u \equiv_n v \text{ and } u' \equiv_n v', \quad \text{then } uu' \equiv_n vv'; \tag{6.2}$$

$$\text{if } u \equiv_n v, a \in A, \text{ and } u' \equiv_n v', \text{ then } (uav, |u|+1) \equiv_{n,1} (u'av', |u'|+1). \qquad (6.2)_1$$

(A convenient way of showing (6.2) and $(6.2)_1$ uses a characterization of $\equiv_n$ and $\equiv_{n,1}$ by the Ehrenfeucht–Fraissé Game. For more details on this game, see [98].) It turns out that (6.2) and $(6.2)_1$ do not depend on the fact that u, v are finite; indeed these statements hold for any linear orderings expanded by unary predicates.

We now can find the desired star-free expression for $\exists x\, \varphi(x)$: by (6.1) it suffices to treat the case of a formula $\exists x\, \varphi_{\underline{W}}(x)$ since

$$\exists x\, \varphi(x) \leftrightarrow \exists x \bigvee_{\underline{W}} \varphi_{\underline{W}}(x) \leftrightarrow \bigvee_{\underline{W}} \exists x\, \varphi_{\underline{W}}(x).$$

Consider now a triple (U, a, V) where U, V are $\equiv_n$-classes and $a \in A$. If there are $u_0 \in U$, $v_0 \in V$ such that $(u_0 a v_0, |u_0|+1) \in \underline{W}$, then, by $(6.2)_1$, for *all* words uav with $u \in U$, $v \in V$, we have that $(uav, |u|+1) \in \underline{W}$. Hence all words from $U.a.V$ satisfy $\exists x\, \varphi_{\underline{W}}(x)$. Thus $\exists x\, \varphi_{\underline{W}}(x)$ defines the union of the sets $U.a.V$ taken over all triples (U, a, V) where U, V are $\equiv_n$-classes and contain words u_0, v_0 as above. Since U, V are star-free by induction hypothesis, $\exists x\, \varphi_{\underline{W}}(x)$ defines a star-free set. □

The correspondence between star-free expressions and first-order formulas is even tighter than expressed in the statement of Theorem 6.1: the classification of star-free languages by dot-depth (= number of alternations between concatenation and Boolean operations) coincides with the classification of first-order definable word-sets in terms of quantifier alternation depth (cf. [120, 116]).

The proof of Theorem 6.1 provides the main prerequisites which are needed for a development of a theory of star-free (or first-order) ω-languages in close analogy to the regular case. It suffices essentially to repeat the proofs in Sections 2 and 4 (in particular, Lemma 2.2, and Theorems 4.5 and 4.6), replacing the congruences $\sim_{\mathscr{A}}$ by the congruences $\equiv_n$ and using the fact that $\equiv_n$-classes are star-free.

6.2 THEOREM (Ladner [53], Thomas [118, 119]). *For an ω-language $L \subseteq A^\omega$, the following conditions are equivalent:*

(a) *L is first-order definable (in the signature $<$, Q_a for $a \in A$).*
(b) *L is a finite union of sets $U.V^\omega$ where $U, V \subseteq A^*$ are star-free and $V.V \subseteq V$.*
(c) *L is a finite union of sets $\vec{U} \cap {\sim}\vec{V}$ where $U, V \subseteq A^*$ are star-free.*
(d) *L is obtained from A^ω by repeated application of Boolean operations and concatenation with star-free sets $W \subseteq A^*$ on the left.*

An ω-language satisfying one of the conditions above is called *star-free*. The notion was introduced by Ladner [53], who referred to condition (d) and showed that the star-free ω-languages form a proper subclass of the regular ones. A short and self-contained proof of Theorems 6.1 and 6.2(a)$\leftrightarrow$(d) is given in [88]. Perrin [86] shows that the equivalence between (b) and (c) remains true for any class of regular languages which is associated with a variety of semigroups that is closed under the Schützenberger product.

Further aspects of the star-free ω-languages are revealed when one considers their syntactic monoid as introduced in Theorem 2.6. Referring to this monoid, it is possible to extend Schützenberger's Theorem from languages to ω-languages. Recall that this theorem states that a regular language $W \subseteq A^*$ is star-free iff its syntactic monoid is group-free. Via condition (b) of Theorem 6.2, this implies the following equivalence.

6.3 COROLLARY (Perrin [85]). *A regular ω-language $L \subseteq A^\omega$ is star-free iff its syntactic monoid $A^*/\approx_L$ is group-free.*

This result yields an effective test deciding whether a given regular ω-language is star-free. Also one can use this characterization to exhibit regular ω-languages that are not star-free. For this purpose one observes that a nontrivial group exists in $A^*/\approx_L$ iff there are words $u, x, y, z \in A^*$ such that, for infinitely many n, $xu^nyz^\omega \in L$ and, for infinitely many n, $xu^nyz^\omega \notin L$ (or analogously for $x(yu^nz)^\omega$). So the example language L_2 of Section 1 ("between any two a's there is an even number of b, c's") is not star-free, as can be seen by taking $x, y, z = a$ and $u = b$.

As in the theory of regular languages of finite words, the group-free monoids are just a first example of a variety of semigroups that characterizes an interesting language class. Further cases in the domain of ω-languages have been studied by Pécuchet [84].

Much of the recent interest in the star-free ω-languages rests on their connection with propositional temporal logic of linear time (interpreted over ω-models).

DEFINITION. The system PLTL of *propositional linear-time logic* is built up from atomic propositions $p_1, p_2, \ldots$ by means of the Boolean connectives, the unary temporal operators X ("ne*x*t"), F ("eventually", "*f*inally"), G ("henceforth", "*g*lobally"), and the binary operator U ("*u*ntil").

PLTL-formulas φ with atomic propositions $p_1, \ldots, p_n$ are interpreted in ω-sequences $\alpha \in (\{0,1\}^n)^\omega$. The satisfaction relation $\alpha \models \varphi$ is defined inductively by the following clauses (we skip the Boolean connectives):

$\alpha \models p_i$	iff	$\alpha(0)$ has a 1 in its ith component,	
$\alpha \models \mathrm{X}\varphi$	iff	$\alpha(1,\omega) \models \varphi$	("φ holds next time"),
$\alpha \models \mathrm{F}\varphi$	iff	there is an $i \geqslant 0$ s.t. $\alpha(i,\omega) \models \varphi$	("φ holds eventually"),
$\alpha \models \mathrm{G}\varphi$	iff	for all $i \geqslant 0$, $\alpha(i,\omega) \models \varphi$	("φ holds henceforth"),
$\alpha \models \varphi \mathrm{U} \psi$	iff	there is an $i \geqslant 0$ s.t. $\alpha(i,\omega) \models \psi$ and $\alpha(j,\omega) \models \varphi$ for $0 \leqslant j < i$	("φ holds until eventually ψ holds").

It is straightforward to translate PLTL-formulas (over atomic propositions $p_1, \ldots, p_n$) into formulas of first-order logic (with unary predicate symbols $X_1, \ldots, X_n$ and using the interpretation described in Section 3). For example, the PLTL-formula

$$G(p_1 \to X((\neg p_2) \text{ U } p_1))$$

is equivalent (over ω-sequences $\alpha \in (\{0,1\}^2)^\omega$) to the first-order formula

$$\forall x (x \in X_1 \to \exists y (x < y \wedge y \in X_1 \wedge \forall z (x < z \wedge z < y \to \neg z \in X_2))).$$

Thus PLTL may be considered as a system of first-order logic with only implicit use of variables. Quantification in PLTL-formulas refers to segments that are unbounded to the right, with the only exception of the bounded quantification involved in the until-operator. Hence in PLTL it tends to be hard to express statements about finite segments of sequences. Nevertheless, first-order sentences can be written as PLTL-formulas as stated in the following theorem.

6.4. THEOREM (Kamp [51], Gabbay, Pnueli, Shelah and Stavi [35]). *Propositional temporal logic PLTL (with atomic propositions $p_1, \ldots, p_n$) is expressively equivalent to first-order logic over ω-sequences (in the signature with $<$ and n unary predicates).*

The translation from first-order logic to PLTL is difficult and will not be described here. It involves a nonelementary blow-up in the length of the formulas, as can be seen from the fact that satisfiability of PLTL-formulas is PSPACE-complete [106], while satisfiability over ω-words is nonelementary for star-free expressions or for first-order formulas in the given signature [111].

An extension of PLTL which allows to define exactly the regular sets of ω-sequences was proposed by Wolper [135]; it is called ETL ("extended temporal logic"). The idea is to admit an infinity of temporal operators, each of them associated with a regular grammar (or an automaton). The standard operators of PLTL are included in this set-up as simple examples. Complexity issues for several versions of ETL are studied by Vardi and Wolper [131], Safra and Vardi [141]. Vardi [128] presents another formalism equivalent to ETL, in which the temporal operators (except "next" and "previous") are defined in terms of least and greatest fixed points.

Temporal logic has attracted attention as a framework for the specification and verification of concurrent programs. Temporal logic formulas are well-suited for this purpose since arguing about concurrent programs is primarily concerned with their ongoing behavior in time and not so much with a relation between initial input and final output (to which the formulas of Hoare logic correspond more directly). From the extensive literature on the subject, we select a few aspects connected with Büchi automata and ω-language theory. (The reader will find more on the topic in the surveys [30, 60, 90].)

We refer to a model of computation with shared variables (cf. [30]). A *concurrent program P* consists of, say, n modules $P_1, \ldots, P_n$ where each of the P_i is a sequential program composed of labelled instructions; both shared variables and variables private to the P_i are admitted. The possible computations of P are defined as the interleavings of the computations of the P_i. A state of the program is identified with an n-tuple of instruction labels from $P_1, \ldots, P_n$ and a tuple of values for the program variables. Since the desired properties of such a program (like deadlock freedom, fairness etc.) are typically concerned with the flow of control and not so much with an infinity of possible values for variables, it is often appropriate to assume that the number of essentially different states is *finite*. In this case one speaks of a *finite-state program*.

Suppose that for a specification of the program P the properties $p_1, \ldots, p_m$ of states are relevant. Then the program can be represented as an annotated directed graph, where nodes are states and arrows represent transitions between states in one step. The state s is annotated by those p_i which are true in s. Formally, the graph is a *Kripke structure* $\mathcal{M}_P = (S, R, \Phi)$ where S is the set of states, $R \subseteq S \times S$ the transition relation between states, and $\Phi: S \to 2^{\{p_1, \ldots, p_m\}}$ a truth valuation. Note that each $\Phi(s)$ can be regarded as an m-bit vector from $\{0,1\}^m$; thus any computation $\sigma = s_0 s_1 \ldots \in S^\omega$ induces a corresponding sequence $\alpha = \Phi(s_0)\Phi(s_1)\ldots$ from $(\{0,1\}^m)^\omega$, containing the relevant information about σ with respect to the properties $p_1, \ldots, p_m$.

In this framework, the *correctness problem* (whether a program P is in accordance with a PLTL-specification φ) is the following question: do all sequences $\alpha \in (\{0,1\}^m)^\omega$ that are given by paths through $\mathcal{M}_P$ satisfy the PLTL-formula φ? In this case one says that "$\mathcal{M}_P$ is a model of φ"; so the correctness problem has been rephrased as a *model checking* problem.

For finite-state programs and PLTL- (or ETL-) specifications, model checking can be carried out effectively. There are several approaches to obtain efficient model checking procedures. Lichtenstein and Pnueli [56] use tableaux (which are extensions of the annotations of $\mathcal{M}_P$ by arbitrary subformulas of the specification) and they obtain a procedure using linear time in the size of the state space and exponential time in the length of the formula. Vardi and Wolper [130] suggest to apply the theory of Büchi automata for the programs *and* for the specifications: First it is possible to view the Kripke structure $\mathcal{M}_P$ as a Büchi automaton $\mathcal{A}_P$ such that $L(\mathcal{A}_P)$ contains the sequences $\alpha \in (\{0,1\}^m)^\omega$ given by $\mathcal{M}_P$. Secondly a PLTL-formula φ defines a regular ω-language (by Theorems 6.2, 6.3) and thus is also representable by a Büchi automaton $\mathcal{A}_\varphi$. Hence the correctness problem reduces to the containment problem for ω-languages "$L(\mathcal{A}_P) \subseteq L(\mathcal{A}_\varphi)$?" or, in negated form, to the intersection problem "$L(\mathcal{A}_P) \cap L(\mathcal{A}_{\neg\varphi}) \neq \emptyset$?". Alpern and Schneider [1], and Manna and Pnueli [59] suggest using Büchi automata as a genuine specification formalism, taking advantage of the fact that a pictoral graph representation can be more transparent than logical formulas. In [59] a variant of Büchi automata is used (called "∀-automata", equivalent to Büchi automata), which involves a "universal" acceptance condition on runs instead of an "existential" one, and is hence better suited for the question whether *all* $\mathcal{A}_P$-runs satisfy the specification.

Further applications of ω-language theory are based on normal form theorems, in particular the representation of regular ω-languages as unions of sets $\vec{U}_i \cap \sim \vec{V}_i$ in McNaughton's Theorem and in Theorem 6.2 (see, e.g., [57, 90, 60]). The determinization of Büchi automata is applied to the verification of probabilistic finite-state programs by Vardi [127], and Courcoubetis and Yannakakis [23].

7. Context-free ω-languages

In this section we give a short account on the use of grammars for the generation of ω-words. A natural approach is to allow infinite leftmost derivations. As an example, consider the following grammar:

$$G_0: \qquad x_1 \to x_2 x_1, \qquad x_2 \to a x_2 b \mid ab.$$

The infinite derivation

$$x_1 \vdash x_2 x_1 \vdash abx_1 \vdash abx_2x_1 \vdash ababx_1 \vdash ababx_2x_1 \vdash \cdots \tag{7.1}$$

generates the ω-word $(ab)^\omega$ from left to right, and

$$x_1 \vdash x_2x_1 \vdash abx_1 \vdash abx_2x_1 \vdash abax_2bx_1 \vdash abaax_2bbx_1 \vdash abaaax_2bbbx_1 \vdash \cdots \tag{7.2}$$

yields from left to right the ω-word aba^ω.

Derivation (7.2) will be excluded if we impose the condition that both variables x_1, x_2 be used infinitely often (as left-hand side of applied rules). Thus there are two variants of context-free generation of ω-languages, depending on whether arbitrary (leftmost) derivations are admitted or the variables used infinitely often are also specified.

In the sequel, a context-free grammar G over the alphabet A and with variables (nonterminals) $x_1, \ldots, x_n$ is given by an n-tuple $(G_1, \ldots, G_n)$ of finite sets $G_i \subseteq (A \cup \{x_1, \ldots, x_n\})^*$, where $w \in G_i$ means that $x_i \to w$ is a rule of G. As start symbol the variable x_1 is used. A leftmost derivation

$$y_0 \vdash u_1 y_1 v_1 \vdash u_2 y_2 v_2 \vdash \cdots$$

with $y_0 = x_1, u_i \in A^*$, and $y_i \in \{x_1, \ldots, x_n\}$ generates either an ω-word (the unique common extension of the u_i) or a finite word (if for some n, $u_n = u_{n+1} = \cdots$). Of course, a finite word can also be generated by a terminating derivation. Hence, an appropriate domain for the present discussion is A^∞ instead of A^ω.

DEFINITION. (a) A ∞-language $L \subseteq A^\infty$ is *algebraic* if there is a context-free grammar G such that L consists of all words of A^∞ that are generated from x_1 by a leftmost derivation of G.

(b) A ∞-language $L \subseteq A^\infty$ is *context-free* if there is a context-free grammar G and a system $\mathcal{F}$ of sets of variables of G such that L consists of the words of A^∞ that are generated from x_1 by a leftmost derivation of G where the variables used infinitely often form a set in $\mathcal{F}$.

Algebraic and context-free *ω-languages* are defined analogously (using A^ω instead of A^∞). When part (b) of the definition is applied to right-linear grammars, one obtains the *regular ∞-languages*. Note that in this case the derivations can be viewed as (possibly terminating) runs of nondeterministic Muller automata. Similarly, the ∞-languages defined by right-linear grammars in the sense of (a) are characterized by Büchi automata in which each state is final.

Let us first note that the regular, algebraic, and context-free ω-languages (and hence ∞-languages) form a proper hierarchy.

7.1 THEOREM (Cohen, Gold [17]). *The class of regular ω-languages is properly contained in the class of algebraic ω-languages which itself is properly contained in the class of context-free ω-languages.*

PROOF. It is obvious that any algebraic ω-language is context-free. To show that regular ω-languages are algebraic, we refer to the representation of regular ω-languages in the form $L=\bigcup_{i=1}^{n} U_i . V_i^{\omega}$ where U_i, V_i are regular. The proof will be clear when the case $L=V^{\omega}$ (assuming $\varepsilon \notin V$) is settled. Let G_V be a left-linear grammar which generates the regular set V (say with start symbol y_1). Then G_V extended by the rule $x_1 \rightarrow y_1 x_1$ generates ω-words by leftmost derivations of the form

$$x_1 \vdash y_1 x_1 \vdash^* v_1 x_1 \vdash v_1 y_1 x_1 \vdash^* v_1 v_2 x_1 \vdash \cdots$$

and hence defines the ω-language V^{ω}. (Note that if G_V were right-linear, one would obtain the ω-words in $V^* . \overrightarrow{\text{pref } V}$ which in general is different from V^{ω}.)

We indicate properness of the inclusions by definition of suitable example languages: The algebraic ω-language

$$\{a^{n_1}b^{n_1}a^{n_2}b^{n_2} \ldots \mid n_i \geqslant 1\} \cup \{a^{n_1}b^{n_1} \ldots a^{n_r}b^{n_r}a^{\omega} \mid r \geqslant 1, n_1, \ldots, n_r \geqslant 1\}$$

is generated by the example grammar G_0 above and easily shown to be nonregular (by an adaptation of the pumping lemma for nondeterministic finite automata). On the other hand, the first part of the above union is context-free since it is generated by G_0 with the system $\mathscr{F}=\{\{x_1, x_2\}\}$ of one designated variable set (see the above derivation examples (7.1,2)). It turns out that this ω-language is indeed not algebraic. $\square$

The above proof for regular ω-languages can be extended to a characterization of the context-free ω-languages. If $\mathscr{L}$ is a class of languages ($\subseteq A^*$), the *ω-Kleene closure* of $\mathscr{L}$ is the class of all ω-languages which are finite unions of sets $U . V^{\omega}$ with $U, V \in \mathscr{L}$.

7.2. THEOREM (Cohen, Gold [17]). *An ω-language is context-free iff it belongs to the ω-Kleene closure of the class of context-free languages.*

Some standard results on context-free grammars fail when considered in the domain of ∞-languages. We discuss an interesting example of this kind, the reduction to Greibach normal form (i.e., to grammars $G=(G_1, \ldots, G_n)$ where each G_i is contained in $A.(A \cup \{x_1, \ldots, x_n\})^*$). Call an ∞-language *Greibach-algebraic* if it is given as in the definition of "algebraic" but referring to grammars in Greibach form. We shall show that the Greibach algebraic ∞-languages do not cover the class of regular ∞-languages. For the proof one extends the topology over A^{ω} (as introduced in Section 5) to a topology over A^{∞}, by introduction of a new symbol $\Omega \notin A$ and representation of finite words w as sequences $w . \Omega^{\omega}$. We shall verify that Greibach algebraic ∞-languages are closed in this topology. Since there are nonclosed regular ∞-languages (for example, $\overrightarrow{a^*bb^*}$, as shown before Remark 5.1), the Greibach algebraic ∞-languages form a proper subclass of the algebraic ones which is incomparable with the class of regular ∞-languages.

The topological closure $\text{cl}(L)$ of a set $L \subseteq A^{\infty}$ contains the sequences that have arbitrary long common prefixes with sequences from L. In other words, the set

$$\text{adh}(L) := \overrightarrow{\text{pref } L},$$

the *adherence of* L, is the set of accumulation points of L, and we have $\text{cl}(L) = L \cup \text{adh}(L)$.

In order to show that Greibach algebraic ∞-languages are closed, let $\alpha \in A^\omega$ be an accumulation point of the Greibach algebraic set L, i.e., for infinitely many prefixes u of α, there is a $\beta \in A^\infty$ with $u\beta \in L$. We have to verify $\alpha \in L$. Consider all leftmost derivations (by a Greibach grammar G for L) which generate a sequence $u\beta \in A^\infty$ where u is prefix of α. We organize the finite initial parts of such derivations in tree form. We obtain a finitely branching tree of finite derivations which has the zero-step derivation x_1 as root and is infinite by assumption on α. By König's Lemma there is an infinite path, i.e. an infinite derivation. This derivation must describe a generation of α (and not only of a finite prefix of α) because G is Greibach.

Boasson and Nivat [7] proved a sharpened form of this result, including also a converse statement.

7.3. THEOREM (Boasson, Nivat [7]). *A context-free ∞-language $L \subseteq A^\infty$ is Greibach algebraic iff it is the (topological) closure of a context-free language $W \subseteq A^*$.*

A common framework for characterizations of the Greibach algebraic, algebraic and context-free ∞-languages has been developed by Niwinski [77], continuing previous work of Nivat, Park, and others. Here a grammar $G=(G_1, \ldots, G_n)$ is regarded as a *fixed point operator*. The operator maps n-tuples of ∞-languages to n-tuples of ∞-languages. The first component of its (well-defined) greatest fixed point is, by definition, the ∞-language defined by G. More precisely, any word w from $(A \cup \{x_1, \ldots, x_n\})^*$ defines a map F_w that sends an n-tuple $(L_1, \ldots, L_n)$, where $L_i \subseteq A^\infty$, to the ∞-language which results from w by substituting L_i for x_i. (For cases like $w=x_1x_2$, certain conventions concerning concatenation are used, such as $L_1 . L_2 = L_1$ for $L_1 \subseteq A^\omega, L_2 \subseteq A^\infty$.) Now with a set $G_i \subseteq (A \cup \{x_1, \ldots, x_n\})^*$ one associates the map which for $(L_1, \ldots, L_n)$ yields the union of the sets $F_w(L_1, \ldots, L_n)$ with $w \in G_i$.

In this way G induces an operator $\bar{G}$ mapping n-tuples to n-tuples of ∞-languages. Since $\bar{G}$ is monotone (w.r.t. set inclusion taken componentwise), the Knaster–Tarski Theorem guarantees a greatest fixed point $(K_1, \ldots, K_n)$. (The n-tuple $(K_1, \ldots, K_n)$ is obtained from $(A^\infty, \ldots, A^\infty)$ by a β-fold iteration of $\bar{G}$ for some ordinal β, possibly greater than ω.) Note that in this set-up the sets G_i need not be finite. For a class $\mathscr{L}$ of languages $L \subseteq A^*$, denote by GFP($\mathscr{L}$) the class of ∞-languages that are obtained from an operator $\bar{G}$ as first component of its greatest fixed point, where the components G_i of G are in $\mathscr{L}$. For $\mathscr{L}=$ FIN (the finite languages), resp. REG (the regular languages), and CF (the context-free languages), we arrive at the mentioned characterization.

7.4. THEOREM (Niwinski [77]). *For any ∞-language L:*

(a) *L is Greibach algebraic iff* $L \in$ GFP(FIN),

(b) *L is algebraic iff* $L \in$ GFP(REG),

(c) *L is context-free iff* $L \in$ GFP(CF).

A much more complicated classification of context-free ω- (or ∞-)languages arises when we consider definitions by deterministic or nondeterministic *pushdown automata* with various modes of acceptance. Cohen and Gold [17] characterize the context-free ω-languages by pushdown automata with a Muller acceptance condition, and the algebraic ω-languages by pushdown automata with a weakened acceptance mode, requiring a run such that all its states are in one of the designated state sets. A detailed analysis of ∞-language classes induced by deterministic pushdown automata is given in [18, 58].

In the above treatment of grammars, a certain asymmetry is manifested in the convention that only left to right generation of ω-words is considered and terminal symbols are ignored when not reached from the left within ω steps. Dropping this restriction, the derivation example (7.2) of the beginning of this section would be regarded as producing the *generalized word abaaa...... bbb*. Formally, generalized words over the alphabet A are identified with A-labelled countable orderings (where the ordering $(\omega, <)$ occurs as a special case). Recent references on derivations of generalized words and finite expressions for their description are [24, 122].

There are further devices of computation that have been investigated in connection with ω-languages but cannot be treated in detail here. We mention two such models: *Turing machines* (studied in [19, 109]), and *Petri nets*. From a Petri net, a language and an ω-language can be extracted essentially by collecting all sequences of transitions that describe admissible firing sequences. In [126] a language-theoretical characterization of Petri net ω-languages is established close to the representation of regular ω-sets as unions of sets $U.V^{\omega}$ with regular U, V. Parigot and Pelz [80] describe a logical formalism (extending Büchi's theory S1S) which characterizes Petri net (ω-)languages; they refer to existential formulas in a signature where a primitive for comparison of finite cardinalities has been added.

PART II. AUTOMATA ON INFINITE TREES

Notation

Given an alphabet A, an A-valued tree t is specified by its set of nodes (the "domain" $\mathrm{dom}(t)$) and a valuation of the nodes in the alphabet. A. Formally, a *k-ary A-valued tree* is a map $t:\mathrm{dom}(t)\to A$ where $\mathrm{dom}(t)\subseteq\{0,\ldots,k-1\}^*$ is a nonempty set, closed under prefixes, which satisfies

$$wj\in\mathrm{dom}(t),\ i<j \ \Rightarrow\ wi\in\mathrm{dom}(t).$$

As an example over $A=\{f, g, c\}$ consider a finite tree:

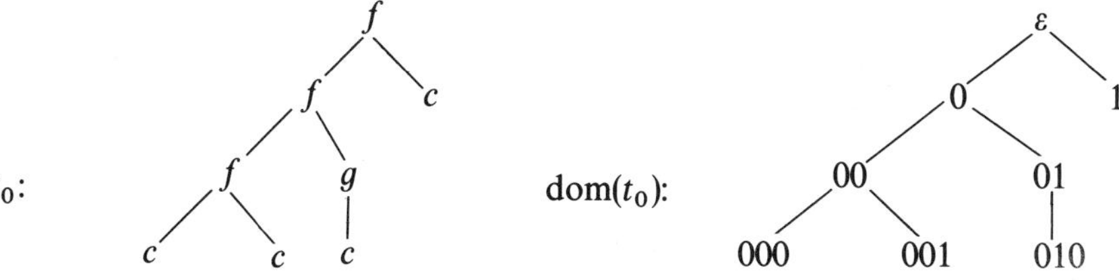

The *frontier* of t is the set

$$\mathrm{fr}(t)=\{w\in\mathrm{dom}(t)\mid \neg\exists i\ wi\in\mathrm{dom}(t)\},$$

and the *outer frontier* $\mathrm{fr}^+(t)$ contains the points $wi \notin \mathrm{dom}(t)$ where $w \in \mathrm{dom}(t)$ and $i<k$.

We set $\mathrm{dom}^+(t) = \mathrm{dom}(t) \cup \mathrm{fr}^+(t)$. The (proper) prefix relation over $\{0,\ldots,k-1\}^*$ is written $<$. A *path* through t is a maximal subset of $\mathrm{dom}(t)$ linearly ordered by $<$. If π is a path through t, then $t|\pi$ denotes the restriction of the function t to the set π. The *subtree* t_w of t at node $w \in \mathrm{dom}(t)$ is given by $\mathrm{dom}(t_w) = \{v \in \{0,\ldots,k-1\}^* \mid wv \in \mathrm{dom}(t)\}$ *and* $t_w(v) = t(wv)$ for $v \in \mathrm{dom}(t_w)$.

Often trees arise as terms (possibly infinite terms). In this case one refers to a *ranked alphabet* $A = A_0 \cup \cdots \cup A_k$ where A_i contains i-ary function symbols. The example tree t_0 above represents the term $f(f(f(c,c),g(c)),c)$ over the ranked alphabet $A = A_0 \cup A_1 \cup A_2$ with $A_0 = \{c\}$, $A_1 = \{g\}$, $A_2 = \{f\}$.

For easier exposition we shall restrict to *binary trees* in the sequel: By T_A we denote the set of finite binary A-valued trees (where a node has either two sons or no sons), and by T_A^ω the set of infinite A-valued trees with domain $\{0,1\}^*$. Set $T_A^\infty = T_A \cup T_A^\omega$. Subsets T of T_A or T_A^ω will sometimes be called *tree languages*. Binary trees represent the typical case from which all notions and results to be discussed below are easily transferred to the general situation. For instance, arbitrary k-ary A-valued trees can be represented within binary trees from $T_{A \cup \{\Omega\}}^\omega$, where Ω is a new symbol, via the coding which maps node $n_1 \ldots n_r$ $(n_i < k)$ to node $1^{n_1}0 \ldots 1^{n_r}0$ and associates the value Ω with all nodes of $\{0,1\}^*$ which are outside the range of this map. In a similar way it is possible to handle countably branching trees in the framework of binary trees.

We now introduce notation concerning *tree concatenation* (defined here in terms of tree substitution). Let $T \subseteq T_A$, $T' \subseteq T_A^\infty$, and $c \in A$. Then $T \cdot_c T'$ contains all trees which result from some $t \in T$ by replacing each occurrence of c on $\mathrm{fr}(t)$ by a tree from T', where different trees are admitted for different occurrences of c. Define a corresponding star operation ^{*c} by

$$T^{*c} = \bigcup_{n \geq 0} T^{nc}$$

where

$$T^{0c} = \{c\}, \qquad T^{(n+1)c} = T^{nc} \cup (T \cdot_c T^{nc}).$$

A tree language $T \subseteq T_A$ is called *regular* iff for some finite set C disjoint from A, T can be obtained from finite subsets of $T_{A \cup C}$ by applications of union, concatenations $\cdot_c$, and star operations ^{*c} where $c \in C$. Note that this notion of regularity generalizes the one for sets of finite words if a word $w = a_1 a_2 \ldots a_r$ over A is considered as a unary tree over $A \cup \{c\}$ of the form $a_1(a_2(\ldots a_r(c)\ldots))$.

We shall also refer to tuples $\boldsymbol{c} = (c_1, \ldots, c_m)$ of concatenation symbols instead of single symbols c. For $T, T_1, \ldots, T_m \subseteq T_A$ let $T \cdot_{\boldsymbol{c}} (T_1, \ldots, T_m)$ be the set of trees obtained from trees $t \in T$ by substituting, for $i = 1, \ldots, m$, each occurrence of c_i on $\mathrm{fr}(t)$ by some tree in T_i. Furthermore, the set $(T_1, \ldots, T_m)^{\omega \boldsymbol{c}}$ is defined to consist of all infinite trees obtained by ω-fold iteration of this tree concatenation; more precisely, it contains all trees $t \in T_A^\omega$ for which there are trees $t_0, t_1, \ldots$ such that $t_0 \in \{c_1, \ldots, c_m\}$, $t_{m+1} \in \{t_m\} \cdot_{\boldsymbol{c}} (T_1, \ldots, T_m)$, and t is the common extension of the trees t'_m which result from the t_m by deleting the symbols c_i at their frontiers.

We shall use expressions like $t_1 \cdot_c t_2$ or $t_0 \cdot_{\boldsymbol{c}} (t_1, \ldots, t_m)^{\omega \boldsymbol{c}}$ as shorthands for $\{t_1\} \cdot_c \{t_2\}$, $\{t_0\} \cdot_{\boldsymbol{c}} (\{t_1\}, \ldots, \{t_m\})^{\omega \boldsymbol{c}}$ respectively. This notation is extended to *infinite* trees t_i with domain $\{0, 1\}^*$ by the convention that instead of frontier occurrences of the c_i the "first" occurrences of the c_i are used for replacement, i.e. the occurrences at nodes w such that no $v < w$ exists with a value c_j. We write $t_1 . t_2$ and $t_0 .(t_1, \ldots, t_m)^\omega$ if the symbols c, $\boldsymbol{c}$ are clear from the context.

8. Tree automata

Tree automata generalize sequential automata in the following way: On a given A-valued binary tree, the automaton starts its computation at the root in an initial state and then simultaneously works down the paths of the tree level by level. The transition relation specifies which pairs (q_1, q_2) of states can be assumed at the two sons of a node, given the node's value in A and the state assumed there. The tree automaton accepts the tree if there is a run built up in this fashion which is "successful". A run is successful if all its paths are successful in a sense given by an acceptance condition for sequential automata. It turns out that for infinite trees the reference to Büchi and Muller acceptance leads to nonequivalent types of tree automata. In this section we introduce these tree automata, first studied by Rabin [93, 94].

As a preparation we collect some notions and facts concerning automata over finite trees.

DEFINITION. A (nondeterministic top-down) *tree automaton* over A is of the form $\mathcal{A} = (Q, Q_0, \Delta, F)$, where Q is a finite set of states, $Q_0, F \subseteq Q$ are the sets of initial, respectively final states, and $\Delta \subseteq Q \times A \times Q \times Q$ is the transition relation. A *run* of $\mathcal{A}$ on the finite binary tree t is a tree $r: \mathrm{dom}^+(t) \to Q$ where $r(\varepsilon) \in Q_0$ and $(r(w), t(w), r(w0), r(w1)) \in \Delta$ for each $w \in \mathrm{dom}(t)$; it is *successful* if $r(w) \in F$ for all $w \in \mathrm{fr}^+(t)$. The tree language $T(\mathcal{A})$ recognized by $\mathcal{A}$ consists of all trees t for which there is a successful run of $\mathcal{A}$ on t, and a set $T \subseteq T_A$ is said to be *recognizable* if $T = T(\mathcal{A})$ for some tree automaton $\mathcal{A}$.

Most of the basic results on regular word languages can be reproved for recognizable tree languages, including a Kleene theorem and closure under Boolean operations (cf. Theorem 8.1 below). However, an important difference between the sequential and the tree case appears in the question of *determinism* since deterministic top-down tree automata, where a function $\delta: Q \times A \to Q \times Q$ replaces the transition relation, are strictly weaker than nondeterministic ones. (For instance, any deterministic top-down tree automaton accepting the trees $f(a, b)$ and $f(b, a)$ would have to accept $f(a, a)$ as well; so it does not recognize the finite set $\{f(a, b), f(b, a)\}$, which is clearly recognized by a nondeterministic top-down tree automaton.) Intuitively, tree properties specified by deterministic top-down automata can depend only on path properties. A reduction to determinism is possible when the working direction of tree automata is reversed from "top-down" to "bottom-up". Nondeterministic *bottom-up tree automata* are of the form (Q, Q_0, Δ, F) with $\Delta \subseteq Q \times Q \times A \times Q$ and Q_0, F as before. A successful run on a tree t should have a state from Q_0 at each point of $\mathrm{fr}^+(t)$ and a state from F at the root ε. By an obvious correspondence, nondeterministic bottom-up tree automata are equivalent to nondeterministic top-down tree automata. However, for the bottom-up version it is possible to carry out the "subset construction" (as for usual finite automata) to obtain equivalent deterministic bottom-up tree automata. Note that the computation of such an automaton, (say, with state set Q), on a term t as input tree may be viewed as an evaluation of t in the finite domain Q. For a detailed treatment, see [37]. In the sequel we refer to the nondeterministic top-down version.

Let us summarize the properties of recognizable tree languages that are needed in the sequel:

8.1. THEOREM (Doner [26], Thatcher and Wright [117]). (a) *The emptiness problem for tree automata over finite trees is decidable* (*in polynomial time*).

(b) *A tree language* $T \subseteq T_A$ *is recognizable iff* T *is regular.*

(c) *The class of recognizable tree languages* $T \subseteq T_A$ *is closed under Boolean operations and projection.*

PROOF. For (a), the decidability claim is clear from the fact that a tree automaton $\mathcal{A}$, say with n states, accepts some tree iff $\mathcal{A}$ accepts one of the finitely many trees of height $\leqslant n$ (eliminate state repetitions on paths). A polynomial algorithm results from the observation that for the decision it even suffices to build up partial run trees in which each transition of the automaton is used at most once.

Part (b) is shown in close analogy to the proof of Kleene's Theorem for sets of finite words. For details, see [37].

In (c), the step concerning union is straightforward. Closure under complement is shown using the equivalence between nondeterministic top-down and deterministic bottom-up tree automata: for the latter, complementation simply means to change the nonfinal states into final ones and vice versa. For projection, assume $T \subseteq T_{A\times B}$ is recognizable and consider its projection to the A-component, i.e. the set

$$T' = \{s \in T_A \mid \exists t \in T_B\, s\hat{\ }t \in T\}$$

where $s\hat{\ }t$ is given by $s\hat{\ }t(w) = (s(w), t(w))$. If T is recognized by the tree automaton $\mathcal{A}$, then T' is recognized by an automaton which on a tree $s \in T_A$ guesses the B-component and works on the resulting tree as $\mathcal{A}$. □

For nondeterministic automata it is possible to restrict the sets of initial, respectively final states to singletons. For technical reasons we shall henceforth assume that tree automata have a single initial state, which moreover occurs only at the root of run trees (i.e., it does not appear in the third and fourth component of transitions).

We now turn to tree automata over infinite trees. There are two basic types, the Büchi tree automaton (called "special automaton" in [94]) and the Rabin tree automaton, which inherit their acceptance modes from sequential Büchi automata, respectively sequential Rabin automata.

DEFINITION. A *Büchi tree automaton* over the alphabet A is of the form $\mathcal{A} = (Q, q_0, \Delta, F)$ with finite state set Q, initial state $q_0 \in Q$, transition relation $\Delta \subseteq Q \times A \times Q \times Q$, and a set $F \subseteq Q$ of final states. A *run* of $\mathcal{A}$ on a tree $t \in T_A^\omega$ is a map $r: \{0, 1\}^* \rightarrow Q$ with $r(\varepsilon) = q_0$

and $(r(w), t(w), r(w0), r(w1)) \in \Delta$ for $w \in \{0, 1\}^*$. The run r is *successful* if on each path some final state occurs infinitely often, i.e.

$$\text{for all paths } \pi, \quad \text{In}(r|\pi) \cap F \neq \emptyset.$$

A *Rabin tree automaton* over A has the form $\mathcal{A} = (Q, q_0, \Delta, \Omega)$, where Q, q_0, Δ are as before, and $\Omega = \{(L_1, U_1), \ldots, (L_n, U_n)\}$ is a collection of "accepting pairs" of state sets L_i, $U_i \subseteq Q$. A run r of the Rabin automaton $\mathcal{A}$ is *successful* if

$$\text{for all paths } \pi \text{ there exists an } i \in \{1, \ldots, n\} \text{ with}$$

$$\text{In}(r|\pi) \cap L_i = \emptyset \quad \text{and} \quad \text{In}(r|\pi) \cap U_i \neq \emptyset.$$

A tree $t \in T_A^\omega$ is *accepted* by the Büchi, respectively Rabin tree automaton $\mathcal{A}$ if some run of $\mathcal{A}$ on t is successful in the respective sense. A set $T \subseteq T_A^\omega$ is *Büchi recognizable*, respectively *Rabin recognizable* if it consists of the trees accepted by a Büchi, resp. Rabin tree automaton.

Since any Büchi tree automaton may be regarded as a Rabin tree automaton (set $\Omega = \{(\emptyset, F)\}$), any Büchi recognizable set of infinite trees is Rabin recognizable.

We mention two equivalent variants of Rabin tree automata. In the first, the *Muller tree automaton*, the collection Ω is replaced by a system $\mathcal{F}$ of state sets, and acceptance is defined via existence of a run r such that for each path π, $\text{In}(r|\pi) \in \mathcal{F}$. The second variant, the *Streett automaton* as introduced by Streett [112], is specified as a Rabin tree automaton but uses the negation of the Rabin condition on a given path of a run: a run r of a Streett automaton (Q, q_0, Δ, Ω) is successful if

$$\text{for all paths } \pi \text{ and for all } i \in \{1, \ldots, n\},$$

$$\text{In}(r|\pi) \cap U_i \neq \emptyset \text{ implies } \text{In}(r|\pi) \cap L_i \neq \emptyset.$$

To illustrate the function of Büchi and Rabin tree automata consider an example tree language T_0 over $A = \{a, b\}$:

$$T_0 = \{t \in T_A^\omega \mid \text{some path through } t \text{ carries infinitely many } a\}.$$

A Büchi tree automaton $\mathcal{A}$ which recognizes T_0 may work as follows: by nondeterministic choice, $\mathcal{A}$ guesses a path down the tree and on this path assumes a final state iff letter a is met; on the other parts of the tree only a fixed final state is computed. Then the existence of a successful run amounts to existence of a path in t with infinitely many values a. Thus T_0 is Büchi recognizable (and of course Rabin recognizable). The complement language

$$T_1 = \{t \in T_A^\omega \mid \text{all paths through } t \text{ carry only finitely many } a\}$$

is recognized by a Rabin automaton with one accepting pair $(\{q_a\}, Q)$ where Q is the state set and q_a is computed iff letter a is encountered. Büchi tree automata, however, do not offer in their acceptance condition such a "finiteness test" along paths. Indeed, they cannot recognize T_1, as stated in the following theorem.

8.2. THEOREM (Rabin [94]). *The set T_1 is a tree language which is Rabin recognizable but not Büchi recognizable.*

PROOF. Assume for a contradiction that T_1 is recognized by the Büchi tree automaton $\mathscr{A}=(Q, q_0, \Delta, F)$, say with $n-1$ states. $\mathscr{A}$ accepts all trees t_i which have letter a at the positions $\varepsilon, 1^{m_1}0, \ldots, 1^{m_1}0 \ldots 1^{m_i}0$ where $m_1, \ldots, m_i > 0$, and letter b elsewhere. The figure on the left indicates the nodes with value a in t_2 (see Fig. 6).

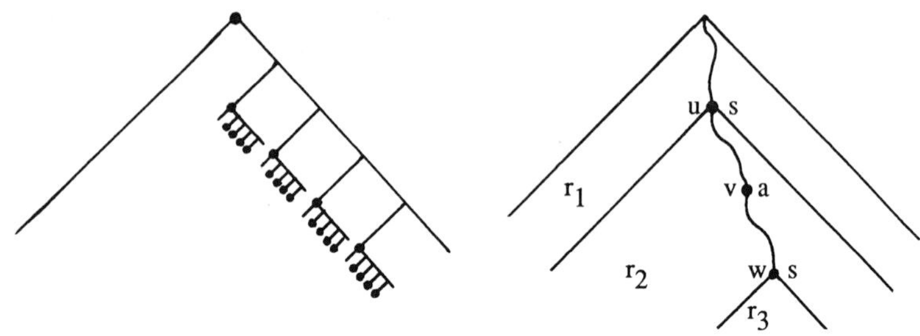

Fig. 6.

Consider a successful run r of $\mathscr{A}$ on t_n. By induction on n one shows that there must be a path in t_n with three nodes $u<v<w$ such that $r(u)=r(w)=s\in F$ and $t_n(v)=a$. Nodes u and w induce a decomposition of r (and t_n) into three parts as follows: Obtain r_1 from r by deleting r_u and setting $r_1(u)=c$. Obtain r_2 similarly from r_u by deleting r_w, and let $r_3=r_w$. Then $r=r_1 . r_2 . r_3$. In the analogous way a decomposition of the underlying tree t_n in the form $t_n=s_1 . s_2 . s_3$ is defined. Now consider $r_1 . r_2^{\omega}$. It is a successful run of $\mathscr{A}$ on the tree $s_1 . s_2^{\omega}$. Since by choice of u, v, w this tree contains a path with infinitely many letters a, we obtain a contradiction. □

The following sections present results analogous to Theorem 8.1 for Büchi and Rabin tree automata, excepting only a Kleene-type characterization for Rabin recognizable sets and complementation for Büchi tree automata. For Rabin tree automata, however, closure under Boolean operations and projection holds, which leads to an equivalence with monadic second-order logic as in Theorem 3.1.

9. Emptiness problem and regular trees

In this section the structure of successful runs of Büchi and Rabin tree automata is analyzed.

First we consider Büchi tree automata and show a representation of Büchi recognizable sets in terms of recognizable sets of finite trees.

Let $\mathscr{A}=(Q, q_0, \Delta, F)$ be a Büchi tree automaton, where $F=\{q_1, \ldots, q_m\}$, and let $r: \{0,1\}^* \to Q$ be a successful run of $\mathscr{A}$ on the tree $t\in T_A^{\omega}$. We claim that r can be built up from finite run trees of $\mathscr{A}$ which are delimited by final states of $\mathscr{A}$. Indeed, the following observation shows that starting from any node u of r the next occurrences of final states enclose a finite tree: Let

$$D_u := \{w \in u.\{0,1\}^* \mid r(v) \notin F \text{ for all } v \text{ with } u<v\leqslant w\}.$$

D_u defines a finitely branching tree which does not contain an infinite path (since r is a successful run). So by König's Lemma D_u is finite, and the nodes of its outer frontier have r-values which are final states. This argument allows to decompose r into "layers" consisting of finite trees: the first layer consists of D_ε and, given the outer frontier F_n of the nth layer, the (n+1)st layer is the union of all D_u with $u \in F_n$.

As a consequence we can represent $T(\mathscr{A})$ using recognizable sets of finite trees. We refer to trees from the set $T_{A \cup F}$ which have values from F exactly at the frontier and are otherwise valued in A. (If t is such a tree, $\bar{t}$ results from t by deleting the F-valued frontier.) Now for $q \in Q$, let T_q consist of all such trees t where there is a run of $\mathscr{A}$ on $\bar{t}$ which starts in q and reaches on $\mathrm{fr}^+(\bar{t})$ exactly the states of $\mathrm{fr}(t)$. Each set T_q is recognizable. The argument above shows that an infinite tree is accepted by the Büchi tree automaton $\mathscr{A}$ iff it belongs to

$$T_{q_0} \cdot_{\boldsymbol{q}} (T_{q_1}, \ldots, T_{q_m})^{\omega \boldsymbol{q}} \tag{9.1}$$

where $\boldsymbol{q} = (q_1, \ldots, q_m)$ is the sequence of all final states in F. Conversely, it is easy to see that for any $(m+1)$-tuple $(T_0, T_1, \ldots, T_m)$ of recognizable sets of finite trees, the expression corresponding to (9.1) defines a Büchi recognizable set of infinite trees:

9.1. THEOREM. *A set $T \subseteq T_A^\omega$ is Büchi recognizable iff there are recognizable sets $T_0, T_1, \ldots, T_m \subseteq T_{A \cup C}$ (where $C = \{c_1, \ldots, c_m\}$) such that $T = T_0 \cdot_c (T_1, \ldots, T_m)^{\omega c}$.*

The representation is implicit in Rabin's solution of the emptiness problem [94]; for a stronger statement see [115] (where also rational expressions for tree languages are introduced).

Using the sets T_q above, the emptiness problem for Büchi tree automata is shown to be decidable. For this, we set up an algorithm which eliminates step by step those states of a given Büchi tree automaton $\mathscr{A} = (Q, q_0, \Delta, F)$ that are useless for successful runs. Certainly, a state q cannot appear in a successful run if the set T_q is empty. So eliminate successively those states q from $\mathscr{A}$ where T_q is empty and update the transition relation of $\mathscr{A}$ accordingly. (Note that each T_q is recognizable, so its emptiness can be checked in polynomial time by Theorem 8.1(a).) The elimination procedure stops after at most $|Q|$ steps, delivering a state set Q_0. We claim that $\mathscr{A}$ accepts some infinite tree iff Q_0 still contains the initial state q_0 (which establishes the desired algorithm). To prove this, assume q_0 is not eliminated and let $q_1, \ldots, q_m$ be the final states remaining in Q_0; note that $m \geqslant 1$ by nonemptiness of T_{q_0} and that also $T_{q_1}, \ldots, T_{q_m}$ are nonempty. So the set (9.1) is nonempty, which is a subset of $T(\mathscr{A})$.

Conversely, it is clear that, in case $T(\mathscr{A}) \neq \emptyset$, the state q_0 will not be eliminated. A closer analysis of the algorithm yields also a polynomial complexity bound:

9.2. THEOREM (Rabin [94], Vardi and Wolper [129]). *The emptiness problem for Büchi tree automata is decidable; moreover, it is logspace complete for* PTIME.

An example tree t in a nonempty Büchi recognizable set T can be obtained by choosing finite trees $t_0, \ldots, t_m$ from the sets $T_{q_0}, \ldots, T_{q_m}$ which the above algorithm produces, and setting $t = t_0 \cdot (t_1, \ldots, t_m)^\omega$. In this infinite tree t the number of distinct subtrees t_u is bounded by $\Sigma_i\,|\text{dom}(t_i)|$ and hence finite. Trees satisfying this finiteness condition are called "regular".

DEFINITION. An infinite tree $t \in T_A^\omega$ is said to be *regular* if there are only finitely distinct subtrees t_u in t (where $u \in \{0, 1\}^*$).

An equivalent definition states that $t: \{0, 1\}^* \to A$ is regular iff there is a finite automaton $\mathscr{A}$ over finite words which "generates t", i.e. whose state set is partitioned into sets Q_a ($a \in A$) such that $\mathscr{A}$ reaches a state in Q_a via input u iff $t(u) = a$. In terms of regular expressions this means that, for each letter $a \in A$, there is a regular expression r_a which defines the language $\{u \in \{0, 1\}^* \mid t(u) = a\}$.

Regular trees represent the simplest infinite terms; they are useful in several other areas of computer science, for instance in semantics of program schemes (where they appear as unravellings of finite flowcharts) and in the foundations of logic programming. Reference [20] is a survey which covers the basic theory and applications in semantics.

Extending the above consideration on Büchi recognizable sets, we shall show that also any nonempty Rabin recognizable set contains a regular tree, and thereby see that the emptiness problem for Rabin tree automata is decidable. Since unary regular trees are ultimately periodic ω-words, this generalizes Theorem 1.3 on Büchi automata in a natural way.

9.3. THEOREM (Rabin [95]). (a) *Any nonempty Rabin recognizable set of trees contains a regular tree.*

(b) *The emptiness problem for Rabin tree automata is decidable.*

PROOF. (a) First reduce the problem to "input-free" tree automata with a transition relation $\Delta \subseteq Q \times Q \times Q$. For this, transform a given Rabin tree automaton $\mathscr{A} = (Q, q_0, \Delta, \Omega)$ over A into $\mathscr{A}' = (Q \times A, Q_0, \Delta', \Omega')$ where $\Delta' \subseteq (Q \times A) \times (Q \times A) \times (Q \times A)$ contains a transition $((q, a), (q', a'), (q'', a''))$ iff $(q, a, q', q'') \in \Delta$. Q_0 contains all states (q_0, a), and Ω' contains those pairs of sets of states from $Q \times A$ where the Q-components yield a pair from Ω. Then the successful runs r' of $\mathscr{A}'$ are the pairs $r\,\hat{}\,t$ where r is a successful run of $\mathscr{A}$ on t; and in this case r is regular provided r' is regular. A corresponding statement holds if $\mathscr{A}'$ has been reduced to an automaton with a single initial state (as mentioned after Theorem 8.1).

So it suffices to show that an imput-free Rabin tree automaton with some successful run admits also a regular sucessful run. Let $\mathscr{A} = (Q, q_0, \Delta, \Omega)$ be a Rabin tree automaton with $\Delta \subseteq Q \times Q \times Q$. Call a state $q \in Q$ *live* if $q \neq q_0$ and the automaton is not forced to stay in q by the single available transition (q, q, q). Using induction on the number of live states of $\mathscr{A}$ we transform a given successful run r into a regular successful run.

If there are no live states, the run r will be stationary from the sons of its root onwards and hence be regular.

In the induction step, distinguish three cases. First, assume that in r some live state q of $\mathscr{A}$ is missing. Then the induction hypothesis can be applied to the automaton where q has been cancelled, and we obtain the desired regular run of $\mathscr{A}$. Secondly, suppose in r there is a node u such that $r(u)=q$ is live but some live state q' does not appear beyond node u. We find two regular runs r_1, r_2, replacing the r-parts "up to first occurrences of q" and "from first occurrences of q onwards", such that the regular run $r_1 \cdot_q r_2$ is successful for $\mathscr{A}$. Run r_1 is obtained by declaring q as nonlive in $\mathscr{A}$, i.e. we consider the modified automaton where only transition (q, q, q) is available for q, and apply the induction hypothesis. Run r_2 is found from the modified automaton where q is taken as initial state and q' is deleted, again by induction hypothesis.

It remains to treat the case that all live states appear in r beyond any given node. Then we may choose a path π_0 through r where all live states appear again and again (and hence no nonlive states can occur). Since r is successful, there is an accepting pair, say (L_1, U_1), such that $\text{In}(r|\pi_0) \cap L_1 = \emptyset$ and $\text{In}(r|\pi_0) \cap U_1 \neq \emptyset$. Note that L_1 contains only nonlive states since $\text{In}(r|\pi_0)$ is the set of live states. Pick q from $\text{In}(r|\pi_0) \cap U_1$. Again we find two regular runs r_1, r_2, now aiming at the property that the regular run $r_1 \cdot_q r_2^{\omega q}$ is successful for $\mathscr{A}$. Run r_1 is given as in the second case above. Run r_2 is obtained from a modification of $\mathscr{A}$, where q is used in two copies: as initial state and as a nonlive state when revisited the first time; to this modified automaton the induction hypothesis can be applied. In the verification that $r_1 \cdot_q r_2^{\omega q}$ is indeed successful, the interesting case concerns those paths π where q appears infinitely often. We show for such a π that (L_1, U_1) is an appropriate accepting pair: Concerning U_1 it suffices to note that $q \in U_1$. Suppose some L_1-state q' occurs infinitely often on π. Since L_1 contains only nonlive states, q' is nonlive and hence must be the only state occurring infinitely often on π; but this contradicts the fact that already the (live) state q occurs infinitely often.

(b) As in the preceding proof, it is enough to consider input-free automata. Suppose an input-free Rabin automaton $\mathscr{A}$ has to be checked for existence of a successful run. Assume $\mathscr{A}$ has n live states. Only finitely many automata can be obtained from $\mathscr{A}$ by the above-mentioned modifications which reduce the number of live states by 1. Iterating these reductions, we obtain in a constructive way finitely many automata derived from $\mathscr{A}$, where the number i of live states ranges from n to 0. In case $i=0$ (nonlive states only), it is trivial to decide whether a successful run exists. For an automaton with $i+1$ live states, the proof above shows how to decide existence of a successful run, given this information for the automata with i live states. Using n such steps, the answer concerning $\mathscr{A}$ is computed. □

The analysis of the algorithm yields an exponential time bound for its execution. Emerson and Jutla [32] show that the nonemptiness problem for Rabin tree automata is NP-complete.

10. Complementation and determinacy of games

In this survey it is not possible to give a proof of the complementation theorem for Rabin tree automata, the most intricate part of Rabin's paper [93]. We shall explain, however, a treatment of this problem in the framework of "infinite games". This approach was proposed by Büchi [11, 12] and Gurevich, Harrington [42]. It allows to study the complementation problem in the context of descriptive set theory and clarifies the connections between tree automata and sequential automata. Moreover, some of the earliest problems on ω-automata, the questions of Church [16] on "solvability of sequential conditions", can be settled via these connections.

We consider infinite games as studied by Gale and Stewart [36], and Davis [25].

DEFINITION. Let A and B be alphabets (each with at least two letters) and let $\Gamma \subseteq (A \times B)^\omega$ be an ω-language. The associated *Gale–Stewart game*, again denoted Γ, is played between two players I and II. A single *play* is performed as follows: First I picks some $a_0 \in A$, then II picks some $b_0 \in B$, then I some $a_1 \in A$, and so on in turns. Player I wins the play if the resulting ω-word $(a_0, b_0)(a_1, b_1)\ldots$ is in Γ, otherwise II wins. (If $\alpha = a_0 a_1 \ldots$ and $\beta = b_0 b_1 \ldots$ we shall denote the sequence $(a_0, b_0)(a_1, b_1)\ldots$ by $\alpha^\wedge\beta$.) A *strategy* for I is a function $f: B^* \to A$, telling I to choose $a_n = f(b_0 \ldots b_{n-1})$ if II has chosen $b_0, \ldots, b_{n-1}$. The strategy f induces a transformation $\bar{f}: B^\omega \to A^\omega$; if II builds up β and I plays strategy f, the play $\bar{f}(\beta)^\wedge\beta$ will emerge. We say that f is a *winning strategy* for I if, for all $\beta \in B^\omega$, $\bar{f}(\beta)^\wedge\beta \in \Gamma$. If there is such a strategy for I, we say that I *wins* Γ. Analogous definitions apply to player II (where a strategy for II is a map $g: A^+ \to B$, inducing a transformation $\bar{g}: A^\omega \to B^\omega$).

A game Γ is called *determined* if player I or player II wins Γ.

Determinacy of Γ amounts to an infinitary version of a quantifier law:

$$\neg\, \exists a_0 \forall b_0 \exists a_1 \forall b_1 \ldots (a_0, b_0)(a_1, b_1)\ldots \in \Gamma \qquad \text{("I does not win } \Gamma\text{")}$$
$$\text{iff } \forall a_0 \exists b_0 \forall a_1 \exists b_1 \ldots (a_0, b_0)(a_1, b_1)\ldots \notin \Gamma \quad \text{("II wins } \Gamma\text{").}$$

But the assumption that all games Γ are determined is a strong set-theoretic hypothesis which contradicts the Axiom of Choice (cf. [66]). Under certain restrictions, however, determinacy has been shown: Martin [61] proved that a game Γ is determined provided Γ belongs to the Borel hierarchy. (See Gurevich [139] for a lucid discussion of the cases that Γ is open or in the Borel class F_σ.)

In the sequel, the relevance of determinacy lies in the fact that it allows to transform the statement

$$\neg\, \exists \text{ strategy } f\, \forall \beta \;\; \bar{f}(\beta)^\wedge\beta \in \Gamma \quad \text{("I does not win } \Gamma\text{")}$$

into the form

$$\exists \text{ strategy } g\, \forall \alpha \;\; \alpha^\wedge\bar{g}(\alpha) \notin \Gamma \quad \text{("II wins } \Gamma\text{")}$$

and hence to write the negation of an existential statement again as an existential statement. Complementation of Rabin tree automata is a natural application, since it requires to express nonexistence of a successful run by one automaton as the existence of a successful run by the complement automaton.

For this purpose the following observation is crucial: *runs of tree automata (and, more generally, valued trees) are strategies.* Just view a strategy $f: B^* \to A$ as a $|B|$-ary A-valued tree: its nodes represent the words from B^* (the root corresponding to the empty word), and the value $a \in A$ at node w indicates that $f(w) = a$. Similarly, strategies $g: A^+ \to B$ are $|A|$-ary B-valued trees with a default value at the root. If such a tree is *regular* (cf. Section 9), it codes a special kind of strategy: since in this case the value of the tree, say at a node w, is computable by a finite automaton (determined by the state reached after reading w), the corresponding strategy is "executable by a finite automaton", or shorter: a *finite-state strategy*. Thus for a finite-state strategy f, the choice $f(w)$ depends only on uniformly bounded finite information in w.

We now specialize the games Γ in two ways: First Γ is defined in terms of automata. The key example are the *regular games* $\Gamma \subseteq (A \times B)^{\omega}$, i.e. games which are recognized by Büchi (or Muller) automata when considered as ω-languages over $A \times B$. Note that in this case Γ belongs to the Borel hierarchy; hence by Martin's result stated above [61] regular games are determined. (Since, by Theorem 5.2, a regular game Γ is even in the Boolean closure of the Borel class F_{σ}, determinacy may be inferred from an easier result of Davis [25].) Secondly, we impose also corresponding restrictions on the two players: their strategies are now required to be *finite-state*. So the following sharpened questions on determinacy arise:

SOLVABILITY: Given (a presentation of) a regular game Γ, can one decide effectively who wins Γ?

SYNTHESIS: Can one exhibit a finite-state winning strategy for the winner of a regular game Γ?

Both problems were posed by Church [16], referring however to different motivation and terminology. Γ was considered as a "sequential condition" on pairs of sequences (i.e., $\Gamma \subseteq A^{\omega} \times B^{\omega}$), expressed in the sequential calculus as a "synthesis requirement" for digital circuits. The circuits should realize a transformation producing for any input sequence β an output sequence α such that $(\alpha, \beta) \in \Gamma$. In recent literature (e.g., [91, 136, 140]) this view is also applied to arbitrary *reactive systems*, i.e. to nonterminating programs (like operating systems), whose purpose is to interact with their environment and to maintain a certain behavior under any influence effected by the environment.

Church's problems were solved positively by Landweber in his thesis (cf. [13]). We sketch here a short proof due to Rabin [95] which exploits the correspondence between strategies and trees.

10.1. THEOREM (Büchi, Landweber [13], see also Trakhtenbrot and Barzdin [125]). *Regular games are determined in the following strong sense: it can be decided effectively who wins, and the winner has a finite-state winning strategy.*

PROOF (Rabin [95]). As a typical example consider the case $A = \{0, 1\}$, $B = \{0, 1\}$. Let $\Gamma \subseteq (\{0, 1\} \times \{0, 1\})^{\omega}$ be regular, say recognized by the Muller automaton $\mathcal{M} = (Q, q_0, \delta, \mathcal{F})$ over $\{0, 1\} \times \{0, 1\}$. We transform $\mathcal{M}$ into a (deterministic) tree automaton $\mathcal{R} = (Q, q_0, \bar{\delta}, \mathcal{F})$ over $\{0, 1\}$, by defining

$$\bar{\delta}(q, a) = (q', q'') \text{ iff } \delta(q, (a, 0)) = q' \text{ and } \delta(q, (a, 1)) = q'' \quad (a \in \{0, 1\}).$$

$\mathcal{R}$ accepts a tree $t \in T_A^{\omega}$ iff along all paths β the states assumed infinitely often by $\mathcal{R}$ form a set in $\mathcal{F}$. This means that for all $\beta = d_1 d_2 \ldots$ (where $d_i \in \{0, 1\}$) the sequence $(t(\varepsilon), d_1)(t(d_1), d_2)(t(d_1 d_2), d_3) \ldots$ is accepted by $\mathcal{M}$ and thus in Γ. Hence $\mathcal{R}$ accepts t iff t is a winning strategy for I in Γ. We may assume that $\mathcal{R}$ is (redefined as) a Rabin tree automaton. The existence of a winning strategy for I can now be decided effectively, by deciding nonemptiness of $T(\mathscr{R})$ (see Theorem 9.3(b)), and a finite-state strategy is guaranteed in this case by Theorem 9.3(a). The case that II wins (which is the only other possibility by the determinacy result of [25]) is handled similarly. □

The complementation problem for Rabin tree automata requires a more general type of game: With any Rabin tree automaton $\mathscr{A}=(Q, q_0, \Delta, \Omega)$ (accepting A-valued trees) and any tree $t \in T_A^\omega$ we associate a game $\Gamma_{\mathscr{A},t} \subseteq (\Delta \times \{0,1\})^\omega$. Thus, player I picks transitions from Δ, and player II picks elements from $\{0,1\}$, i.e. directions building up a path through the tree t. $\Gamma_{\mathscr{A},t}$ contains all sequences $\alpha \hat{} \beta \in (\Delta \times \{0,1\})^\omega$ which "describe a successful path for $\mathscr{A}$ on t". Formally,

$$\alpha \hat{} \beta = ((s_0, a_0, s'_0, s''_0), d_1)((s_1, a_1, s'_1, s''_1), d_2)\ldots$$

should satisfy $s_0 = q_0$, $a_i = t(d_1 \ldots d_{i-1})$, $s_{i+1} = s'_i$ if $d_{i+1}=0$, $s_{i+1}=s''_i$ if $d_{i+1}=1$, and the state sequence $s_0 s_1 s_2 \ldots$ should fulfil the acceptance condition Ω. Then the winning strategies $f:\{0,1\}^* \to \Delta$ for I are in one-to-one correspondence with the successful runs of $\mathscr{A}$ on t, and we have

$$\mathscr{A} \text{ accepts } t \text{ iff I wins } \Gamma_{\mathscr{A},t}. \tag{10.1}$$

Note that the underlying tree t is completely arbitrary and one can no more expect that the winning strategies are finite-state. However, it turns out that *relativized finite-state strategies* (as we call them) can be guaranteed. Such a strategy, say for player II, is executed by a finite automaton $\mathscr{C}$ which is allowed to use an auxiliary tree $t' \in T_{A'}^\omega$ over an alphabet A' in addition to the given tree $t \in T_A^\omega$ (and the transitions from Δ picked by player I). $\mathscr{C}$ works over $\Delta \times A \times A'$ and outputs directions from $\{0,1\}$ (via a partition of its state set into "0-states" and "1-states"). More precisely, consider the game situation

$$\begin{array}{ll} \text{I:} & \bar{\tau} = \tau_0 \quad \tau_1 \ldots \quad \tau_n \\ \text{II:} & w = d_1 \quad \ldots d_n \end{array}$$

and denote by $(t|w)\hat{}(t'|w)$ the sequence of $A \times A'$-values of $t \hat{} t'$ for the nodes visited along w (namely $\varepsilon, d_1, d_1 d_2, \ldots, d_1 \ldots d_n$). Then the state reached by $\mathscr{C}$ after reading the word $\bar{\tau}\hat{}(t|w)\hat{}(t'|w) \in (\Delta \times A \times A')^{n+1}$ fixes the next choice of player II.

Determinacy for the games $\Gamma_{\mathscr{A},t}$ by such relativized finite-state strategies was stated by Büchi [11] with a short proof hint and it was given a detailed exposition in Büchi [12]. A different and simpler proof was given by Gurevich and Harrington in [42]. We state this difficult result here without proof and from it conclude the complementation for Rabin tree automata.

10.2. THEOREM (Büchi [11, 12], Gurevich and Harrington [42]). *Let $\mathscr{A}$ be a Rabin tree automaton over A and $t \in T_A^\omega$. The game $\Gamma_{\mathscr{A},t}$ is determined, and the winner has a relativized finite-state strategy.*

10.3. COROLLARY. *For any Rabin tree automaton $\mathscr{A}$ (by effective construction) there is a Rabin tree automaton $\mathscr{A}'$ recognizing $T_A^\omega - T(\mathscr{A})$.*

PROOF. Let $\mathcal{A} = (Q, q_0, \Delta, \Omega)$ be a Rabin tree automaton over A. We have to find a Rabin tree automaton $\mathcal{A}'$ such that, for all $t \in T_A^\omega$,

$$\mathcal{A} \text{ does not accept } t \text{ iff } \mathcal{A}' \text{ accepts } t.$$

By (10.1), $\mathcal{A}$ does not accept t iff I does not win $\Gamma_{\mathcal{A},t}$. By Theorem 10.2, this holds iff II wins $\Gamma_{\mathcal{A},t}$ in the following sense, using a finite automaton $\mathscr{C}$.

> for some tree $t' \in T_A^\omega$, over an auxiliary alphabet A',
> if $\alpha \in \Delta^\omega$ is chosen by I and $\beta \in \{0, 1\}^\omega$ is the response by II (computed by $\mathscr{C}$ from α, t, t'),
> then $\alpha\wedge\beta$ does not describe a successful path for $\mathcal{A}$ on t.

This can be formulated as follows:

> (1) for some $t' \in T_{A'}^\omega$,
> (2) for all $\beta \in \{0, 1\}^\omega$,
> (3) for all $\alpha \in \Delta^\omega$, if β results from α, t, t' by output of $\mathscr{C}$, then $\alpha\wedge\beta$ does not describe a successful path for $\mathcal{A}$ on t.

Note that (3) is a condition on sequences of the form $\beta\wedge(t|\beta)\wedge(t'|\beta) \in (\{0, 1\} \times A \times A')^\omega$, that (2) is a condition on trees from $T_{A \times A'}^\omega$, and that (1) is a condition on trees from T_A^ω. Since (3) is easily expressed in S1S, it can be defined by a Muller automaton $\mathcal{M}$. Now construct from $\mathcal{M}$ a deterministic tree automaton $\mathcal{R}$ as in the poof of Theorem 10.1. Then $\mathcal{R}$ accepts the trees satisfying (3) on each path β, i.e. all trees $t \in T_{A \times A'}^\omega$ with property (2). Projection to T_A^ω yields a (nondeterministic) Rabin automaton $\mathcal{A}'$, accepting the trees $t \in T_A^\omega$ with property (1). Since these trees were just the trees for which II wins $\Gamma_{\mathcal{A},t}$, $\mathcal{A}'$ is a Rabin tree automaton as desired. □

The complementation problem for Rabin tree automata has been (and continues to be) investigated by several authors. Muchnik [70] presents an elegant proof using an induction over the number of states in the automata. An interesting approach has recently been developed by Muller and Schupp [75], based on the idea of *alternating automata* over trees. In addition to performing nondeterministic choice, these automata are able to pursue several computations simultaneously. The operation of an alternating tree automaton with state set Q is described by the elements of the free lattice over $\{0, 1\} \times Q$. The intended meaning of such an element, say

$$((0, q_1) \wedge (1, q_2)) \vee ((1, q_1) \wedge (1, q_2)),$$

is that the automaton proceeds from a given node with q_1 to the left and q_2 to the right, or proceeds with q_1 to the right and also with q_2 to the right. The second possibility is missing in Rabin tree automata. One can now collect all possible *histories* of the alternating automaton $\mathcal{A}$ over t in a *computation tree* $C(\mathcal{A}, t)$, where one history follows one simultaneous realization of states through the levels of the tree. (So a history is a kind of "multirun".) $\mathcal{A}$ accepts t if for some infinite history, i.e. some path in $C(\mathcal{A}, t)$, all state sequences along paths described by it are successful in the sense of Muller acceptance. Complementation for these alternating automata is easy, performed by dualizing the given automaton (exchange $\wedge$ and $\vee$ in the transitions, and complement the system of final state sets). The hard closure property is projection, which is shown again by an application of Theorem 10.2. An advantage of this approach is that fragments of the monadic theory of the tree which are defined in terms of restricted second-order quantifiers may be handled by suitable restrictions of the projection operation for alternating automata and hence by weakened versions of Theorem 10.2 (cf. [72, 73]).

11. Monadic tree theory and decidability results

For a transfer of the preceding results from automata theory to logic, trees are represented as model-theoretic structures. If A is the alphabet $\{0,1\}^n$, a tree $t \in T_A^\omega$ is coded by a model of the form

$$\underline{t} = (\{0,1\}^*, \varepsilon, \mathrm{succ}_0, \mathrm{succ}_1, <, P_1, \ldots, P_n),$$

where succ_0, succ_1 are the two successor functions over $\{0,1\}^*$ with $\mathrm{succ}_0(w) = w0$, $\mathrm{succ}_1(w) = w1$, $<$ is the proper prefix relation over $\{0,1\}^*$, and $P_1, \ldots, P_n$ are subsets of $\{0,1\}^*$ with $w \in P_i$ iff the ith component of $t(w)$ is 1.

DEFINITION. The interpreted formalism S2S ("*second-order theory of two successors*") contains variables $x, y, \ldots$ and $X, Y, \ldots$ (ranging over elements, respectively subsets of $\{0,1\}^*$). *Terms* are obtained from the individual variables $x, y, \ldots$ and the constant ε by applications of succ_0, succ_1; we write $x0$ and $x1$ instead of $\mathrm{succ}_0(x)$ and $\mathrm{succ}_1(x)$. *Atomic formulas* are of the form $t = t'$, $t < t'$, $t \in X$ where t, t' are terms and X is a set variable; and arbitrary *formulas* are generated from atomic formulas by Boolean connectives and the quantifiers $\exists$, $\forall$ (ranging over either kind of variables). If $\varphi(X_1, \ldots, X_n)$ is an S2S-formula and $\underline{t}$ a tree model as above, write $\underline{t} \models \varphi(X_1, \ldots, X_n)$ if φ is satisfied in $\underline{t}$ with P_i as interpretation for X_i. Let $T(\varphi) = \{t \in T_A^\omega \mid \underline{t} \models \varphi(\bar{X})\}$. If $T = T(\varphi)$ for some S2S-formula φ, T is called *definable in* S2S.

The "weak" system WS2S is obtained when the set quantifiers range over finite subsets of $\{0,1\}^*$ only. If $T = T(\varphi)$ for some WS2S-formula φ (i.e., some S2S-formula using this "weak interpretation"), T is *definable in* WS2S (or simply: "weakly definable").

The above definitions are analogously applied to finite tree models $\underline{t}$ (where $t \in T_A$). In this case there is no difference between the weak and the strong interpretation.

Note that S1S (as introduced in Section 3) results from S2S by deleting the successor function succ_1 and by restriction of the underlying models to the domain 0^*. Similarly to the case of S1S, the primitive ε and $<$ are definable in terms of succ_0, succ_1 and hence could be cancelled; we use them for easier formalizations. By "the infinite binary tree" (as a model-theoretic structure) we shall mean the structure $(\{0,1\}^*, \mathrm{succ}_0, \mathrm{succ}_1)$.

We list some examples of S2S-formulas (using freely abbreviations such as $x \leqslant y$, $X \subseteq Y$, etc.):

$$\mathrm{Chain}(X): \quad \forall x\, \forall y\, (x \in X \wedge y \in X \rightarrow x < y \vee x = y \vee y < x),$$

Path(X): Chain(X) $\wedge \neg\exists Y(X \subseteq Y \wedge X \neq Y \wedge \text{Chain}(Y))$,

$x \preccurlyeq y$: $x \leqslant y \vee \exists z\,(z0 \leqslant x \wedge z1 \leqslant y)$
(this is the total lexicographical ordering of $\{0,1\}^*$),

Fin(X): $\forall Y(Y \subseteq X \wedge Y \neq \emptyset \rightarrow (\exists y$ "y is $\preccurlyeq$-minimal in Y" $\wedge \exists y$ "y is $\preccurlyeq$-maximal in Y"))
(this shows definability of finiteness in S2S; hence WS2S can be interpreted in S2S).

As an example tree language definable in S2S consider the set T_0 of Section 8; it is defined as follows (identifying letters a, b with $1, 0$):

$$\varphi(X_1): \qquad \exists Y(\text{Path}(Y) \wedge \forall x\,(x \in Y \rightarrow \exists y\,(y \in Y \wedge x < y \wedge y \in X_1))).$$

We now can state the analog of Büchi's Theorem 3.1 for sets of trees.

11.1. THEOREM. (a) (Doner [26], Thatcher and Wright [117]) *A set $T \subseteq T_A$ of finite trees is definable in* (W)S2S iff *T is recognizable.*

(b) (Rabin [93]) *A set $T \subseteq T_A^\omega$ is definable in* S2S *iff T is Rabin recognizable.*

PROOF. Assume $A = \{0,1\}^n$. For the implications from right to left, formalize the acceptance condition for the given tree automaton $\mathscr{A}$: suppose, for (b), that the Rabin tree automaton $\mathscr{A}$ has the states $0, \ldots, m$ and the accepting pairs $(L_1, U_1), \ldots, (L_r, U_r)$. $T(\mathscr{A})$ is defined by an S2S-formula which says (see proof of Theorem 3.1)

$$\exists Y_0 \ldots \exists Y_m \Bigg(\text{“}Y_0, \ldots, Y_m \text{ represent a run of } \mathscr{A} \text{ on } X_1, \ldots, X_n\text{”}$$
$$\wedge \forall Z \Bigg(\text{Path}(Z) \rightarrow \bigvee_{1 \leqslant i \leqslant r} \Bigg(\bigwedge_{j \in L_i} \text{“}\exists^{<\omega} x\,(x \in Z \wedge x \in Y_j)\text{”}$$
$$\wedge \bigvee_{j \in U_i} \text{“}\exists^{\omega} x (x \in Z \wedge x \in Y_j)\text{”} \Bigg)\Bigg)\Bigg).$$

The converse is also shown similarly to Theorem 3.1: First S2S is reduced to a pure second-order formalism S2S_0 with atomic formulas of the form $\text{Succ}_0(X_i, X_j)$, $\text{Succ}_1(X_i, X_j)$, $X_i \subseteq X_j$ only. Induction over S2S_0-formulas $\varphi(X_1, \ldots, X_n)$ shows recognizability, respectively Rabin recognizability of $T(\varphi)$. The steps for $\vee$ and $\exists$ are easy since the (nondeterministic) automata are closed under union and projection. (For part (a) apply Theorem 8.1(c); the proof for Rabin tree automata is similar.) Concerning negation, use again Theorem 8.1(c), respectively the complementation result in Corollary 10.3. □

Büchi tree automata correspond to a proper fragment of S2S which still allows us to express many interesting tree properties; they are also used for a beautiful characterization of WS2S. We state the result here without proof.

11.2. Theorem (Rabin [94]). *Let* $A=\{0,1\}^n$.

(a) *A set* $T\subseteq T_A^\omega$ *is Büchi recognizable iff* T *is definable by an* S2S-*formula* $\exists Y_1 \ldots \exists Y_m\, \varphi(Y_1,\ldots,Y_m,X_1,\ldots,X_n)$ *where* φ *is a* WS2S-*formula.*

(b) *A set* $T\subseteq T_A^\omega$ *is definable in* WS2S *iff* T *and* $T_A^\omega - T$ *are Büchi recognizable.*

A direct automata-theoretic characterization of WS2S (in terms of alternating automata over trees) is given by Muller, Saoudi and Schupp [72].

In Theorem 4.6 it was shown that S1S and WS1S are expressively equivalent. From Theorem 11.2(a) and the failure of complementation for Büchi tree automata (cf. Theorem 8.2) it follows that WS2S is strictly less expressive than S2S and even than Büchi tree automata. However, this applies only to formulas $\varphi(X_1,\ldots,X_n)$ which speak about arbitrary sets: Läuchli and Savioz [54] show that any S2S-formula $\varphi(x_1,\ldots,x_n)$, where only individual variables occur free, can be expressed as a WS2S-formula, and Thomas [144] proves an analog for formulas $\varphi(X_1,\ldots,X_n)$ where $X_1,\ldots,X_n$ stand for paths.

An application of the above equivalence theorems is the analysis of variants of Büchi, respectively Rabin tree automata. Let us mention one such automaton model, the subtree automaton of Vardi and Wolper [129].

A *subtree automaton* over A is of the form $\mathscr{A}=(Q,\Delta,f,F)$ where Q, Δ, F are as for Büchi tree automata and $f\colon A\rightarrow Q$. A tree $t\in T_A^\omega$ is accepted by $\mathscr{A}$ if all its nodes x are roots of finite trees accepted by the tree automaton $\mathscr{A}'=(Q,f(x),\Delta,F)$. Hence a set that is recognized by the subtree automaton $\mathscr{A}$ is definable by a formula $\psi=\forall x\,\exists X\,\varphi(x,X)$ where $\varphi(x,X)$ expresses that the finite tree with root x and (finite) frontier X is recognized by $\mathscr{A}'$. Since ψ is a WS2S-formula, subtree automata recognize only WS2S-definable sets and hence are a proper specialization of Büchi tree automata. Subtree automata are tailored for obtaining good upper complexity bounds for program logics. Vardi and Wolper [129] show that for several progam logics the satisfiability problem amounts to a test on existence of certain trees ("Hintikka-trees") which are defined in terms of eventuality properties as defined by ψ above.

We turn to applications of Theorem 11.1 in decision problems. The starting point is the following fundamental result.

11.3. Rabin's Tree Theorem (cf. [93]). *The monadic second-order theory of the infinite binary tree is decidable.*

Proof. For any S2S-sentence φ one can construct, by the proof of Theorem 11.1, an input-free Rabin tree automaton $\mathscr{A}_\varphi$ such that φ is true in the infinite binary tree iff $\mathscr{A}_\varphi$ has a successful run. The latter condition is decided effectively by Theorem 9.3. □

The result is easily generalized to the monadic second-order theory of the full n-ary tree, where n successor functions $\mathrm{succ}_0,\ldots\ \mathrm{succ}_{n-1}$ are allowed in the formulas. Similarly, the monadic second-order theory SωS of the countably branching tree is proved decidable; here one usually refers to the signature with $<$ (prefix relation over ω^*) and $\preccurlyeq$ (lexicographic order over ω^*), because each of the infinitely many successor functions succ_i is definable in terms of $\leqslant$ and $\preccurlyeq$.

Many theories of mathematical logic have been shown to be decidable via Rabin's

Tree Theorem; for some examples see [96]. In the following, we outline a standard application in dynamic logic: the solution of the *satisfiability problem for modal logics of programs*. The method (and refinements of it) have been used for several logics, for example propositional dynamic logic and extensions [112], process logic [47], the calculus L_μ [113], and computation tree logic CTL* [31, 33]. In all instances the satisfiability question: "is there a model $\mathcal{M}$ satisfying the formula φ?" is effectively transformed to a question "is there a tree t satisfying the S2S-sentence φ'?". (Logics allowing this reduction are said to share the *tree model property*.) We present the conceptually simplest form of this translation for the example CTL* of computation tree logic. Further developments of the method (with much better complexity bounds) are surveyed by Emerson [30]. Muller, Saoudi and Schupp [73] present a uniform method to obtain exponential time bounds for logics that involve only quantifications over finite computation paths, based on the alternating automata mentioned at the end of Section 10.

Computation tree logic CTL* is a system of modal logic which allows to specify properties of paths through Kripke structures. From atomic propositions, say $p_1, \ldots, p_n$ for the following discussion, the CTL*-formulas are built up using Boolean connectives, the linear time temporal operators, X, F, G, U (cf. Section 6) and the additional unary operator E. Recall (from Section 6) that a Kripke structure is of the form $\mathcal{M}=(S, R, \Phi)$ where S is a (here at most countable) set of states, $R \subseteq S \times S$ the transition relation, and Φ: $S \rightarrow 2^{\{p_1,\ldots,p_n\}}$ a truth valuation. For simplicity we assume that there is a distinguished start state s_0. The semantics of CTL*-formulas in Kripke structures is based on the usual meaning of X, F, G, U over given state paths and the interpretation of E by "there is an infinite state path". Consider an example: the CTL*-formula

$$p_1 \wedge \mathrm{E}(\mathrm{G}p_2 \wedge \mathrm{FEF}p_3)$$

says that

> "p_1 is true in s_0, and starting from s_0 there is an infinite path π through the model such that all states on π satisfy p_2, and in some state of π a path π' starts with some state satisfying p_3".

The decision procedure for satisfiability of CTL*-formulas is based on the *unravelling of Kripke structures* in tree form: given the Kripke structure $\mathcal{M}=(S, R, \Phi)$, define the structure $\mathcal{M}'=(S', R', \Phi')$ by

$$S' = s_0 S^*,$$

$$(r_1 \ldots r_m) R'(s_1 \ldots s_k s) \text{ iff } k=m,\ r_i = s_i \text{ for } 1 \leqslant i \leqslant k,\ s_k R s,$$

$$\Phi'(s_1 \ldots s_k) = \Phi(s_k).$$

$\mathcal{M}'$ can be considered as an (at most countably branching) $\{0,1\}^n$-valued tree in which the nodes are finite histories of states from S, the relation "is-father-of" is represented by R, and the valuation by Φ. $\mathcal{M}'$ is encoded over the binary tree by the map $n_1 \ldots n_r \mapsto 10^{n_1} \ldots 10^{n_r}$. We obtain a $\{0,1\}^{n+1}$-valued binary tree $t_{\mathcal{M}}$ where the additional component of the valuation indicates the range of $\mathcal{M}$ under this map. Let $\mu(X, Y_1, \ldots, Y_n)$ be an S2S-formula which says that $X \subseteq (10^*)^*$ is a range of a tree under the coding and that $Y_1, \ldots, Y_n \subseteq X$. Any tree of form $t_{\mathcal{M}}$ satisfies μ; conversely, any tree satisfying μ induces a Kripke model (over the set $(10^*)^*$).

It is straightforward to reformulate a given CTL*-formula φ as an S2S-formula

$\varphi'(X, Y_1, \ldots, Y_n)$ such that it is true over a tree $t_{\mathcal{M}}$ iff φ holds in $\mathcal{M}$. Hence φ is satisfiable iff the S2S-sentence

$$\exists X \exists Y_1 \ldots \exists Y_n (\mu(X, Y_1, \ldots, Y_n) \wedge \varphi'(X, Y_1, \ldots, Y_n))$$

is true over the infinite binary tree. So by Rabin's Tree Theorem, satisfiability of CTL*-formulas is decidable.

The decision procedure induced by the above transformation is nonelementary. (Each level of negation in the given formula requires a corresponding complementation of a Rabin automaton and hence an at least exponential blow-up in the size of the automata.) Better procedures are obtained by incorporating more information than just for the atomic formulas in the tree model $t_{\mathcal{M}}$, e.g. by including the "Fischer–Ladner closure" of the given formula. For more details see [30, 32].

We end this section with the formulation of two interesting generalizations of Rabin's Tree Theorem and some remarks on undecidable extensions of the monadic theory of the binary tree.

Stupp [114], continuing work of Shelah [103], extended Rabin's techniques (in particular concerning the complementation theorem for Rabin tree automata) to "higher-dimensional trees" and similar structures.

11.4. THEOREM (Shelah [103], Stupp [114]). *Let $\mathcal{M} = (M, (R_i)_{i<k})$ be a relational structure (say with binary relations $R_i \subseteq M \times M$). Define the structure $\mathcal{M}^* = (M^*, <, (R_i^*)_{i<k})$ by*

$$u < v \text{ iff } u \text{ is a proper prefix of } v \text{ (over } M^*)$$
$$u R_i^* v \text{ iff } \exists m_1, \ldots, m_k, m, m' \in M \text{ such that}$$
$$u = m_1 \ldots m_k m,\ v = m_1 \ldots m_k m',\ m R_i m'.$$

If the monadic second-order theory of $\mathcal{M}$ is decidable, so is the monadic second-order theory of $\mathcal{M}^*$.

The special case of Theorem 11.4 which yields Rabin's Tree Theorem concerns the finite structure $\mathcal{M}_0 = (\{0, 1\}, <_0)$ where $<_0$ is the usual order on $\{0, 1\}$ (and which of course has a decidable monadic second-order theory). We have $\mathcal{M}_0^* = (\{0, 1\}^*, <_1, <_0)$ where

$$u <_1 v \text{ iff } u \text{ is a proper prefix of } v,$$
$$u <_0 v \text{ iff } u = w0,\ v = w1 \text{ for some } w \in \{0, 1\}^*.$$

The functions succ_0, succ_1 are (monadic second-order) definable in terms of $<_1$, $<_0$ and vice versa. So $\mathcal{M}_0^*$ is essentially the infinite binary tree.

Another extension of Rabin's Tree Theorem is due to Muller and Schupp [74]. They consider infinite directed graphs with labelled edges. Such a graph is said to be "finitely generated" if it has a distinguished vertex v_0 ("origin"), a finite label alphabet A and a fixed finite bound on the degrees of the vertices. The model-theoretic structures which represent these graphs are of the form $\underline{G} = (G, v_0, (R_a)_{a \in A})$ such that $u R_a v$ iff there is an edge labelled a from u to v. A finitely generated graph is called a *context-free graph* if it is "finitely behaved at infinity" in the following sense: one obtains only finitely many distinct isomorphism types by collecting, for all vertices v, the substructures $\Gamma(v)$ which remain as connected components when the points are deleted with distance to v_0 smaller than between v and v_0. For more information on these graphs and the background from group theory, see the survey of Berstel and Boasson [6] in this Handbook.

11.5. THEOREM (Muller, Schupp [74]). *The monadic second-order theory of any context-free graph is decidable.*

The proof of Theorem 11.5 is based on Rabin's Tree Theorem. The general problem of reducing monadic theories of graphs to theories of trees is further investigated by Seese [101]; he considers the conjecture that *any* decidable monadic theory of a class of graphs has an interpretation in the monadic theory of a class of trees, and shows that this is true for the class of planar graphs. Other results in this direction are given in [21], where "equational graphs" (defined by certain graph rewriting systems) are shown to have a decidable monadic second-order theory.

The theories in Theorems 11.4 and 11.5 seem close to the margin of undecidability. We mention some variants, respectively extensions of the monadic theory of the binary tree which are undecidable. (These results have been shown by several authors; a recent reference is [54]).

The most basic example is the monadic theory of the "grid" $(\omega \times \omega, s_0, s_1)$ where $s_0(m,n)=(m+1,n)$ and $s_1(m,n)=(m,n+1)$. This structure is obtained from the free algebra $(\{0,1\}^*, \mathrm{succ}_0, \mathrm{succ}_1)$ by adding the relation $\mathrm{succ}_0 \circ \mathrm{succ}_1 = \mathrm{succ}_1 \circ \mathrm{succ}_0$.

11.6. THEOREM (a) *The (weak) monadic second-order theory of the grid* $(\omega \times \omega, s_0, s_1)$ *is undecidable.*

(b) *The (weak) monadic second-order theory of the infinite binary tree extended by the function s with* $s(w)=0w$ (*for* $w \in \{0,1\}^*$) *is undecidable.*

(c) *The (weak) monadic second-order theory of the infinite binary tree extended by the "equal-level predicate" E, given by* $u\,E\,v$ *iff* $|u|=|v|$ (*for* $u, v \in \{0,1\}^*$), *is undecidable.*

PROOF (a) The idea is similar as in the undecidability proof for the origin constrained domino problem. For any Turing machine $\mathscr{A}$, construct a sentence $\varphi_{\mathscr{A}}$ in the weak monadic second-order language of the grid which expresses existence of a halting computation of $\mathscr{A}$ when $\mathscr{A}$ is started on the empty tape (the tape is assumed here left-bounded and right-infinite). As in the domino problem, the ith cell of the jth configuration is represented by the point $(i,j) \in \omega \times \omega$. Using existential quantification over auxiliary unary predicates (which code the letters and states of $\mathscr{A}$), it is easy to formalize that a halting configuration is reached. Since only finitely many steps and a finite portion of the tape are involved, weak second-order quantification suffices.

(b) Identify the (weakly definable) subset 0^*1^* of the binary tree with $\omega \times \omega$. Note that $(\omega \times \omega, s_0, s_1)$ is isomorphic to $(0^*1^*, s, \mathrm{succ}_1)$; so (a) can be applied.

(c) Using the predicate E, the function s is weakly definable on 0^*1^* since we have

(for $u, v \in 0^*1^*$)

$$s(u)=v \text{ iff } (u \in 0^* \wedge v=u0)$$
$$\vee \exists w \in 0^* \, \exists u' \quad (u \in w1^* \wedge u' \in w01^* \wedge u\,E\,u' \wedge u'1=v).$$

Now the claim follows from (b). □

By decidability of S2S one obtains as a corollary that the function s and the relation E over the binary tree are not definable in S2S. In contrast to part (c) of Theorem 11.6, a decidable theory is obtained when the relation E is adjoined to a restricted version of S2S where the second-order variables range over paths only (cf. [144]).

12. Classifications of Rabin recognizable sets

In this final section we give a short overview of the (mostly ongoing) work which studies the "fine structure" of the class of Rabin recognizable sets of trees. The results presented here fall in three categories, depending on the formalism in which tree properties are classified: monadic second-order logic, tree automata, and fixed point calculi.

12.1 Restrictions of monadic second-order logic

Natural subsystems of the monadic second-order formalism S2S are obtained when the range of set quantification is narrowed to special subsets of $\{0,1\}^*$.

First we consider the question for which restricted set quantifiers the same sentences are true (over the infinite binary tree) as in the case of quantification over arbitrary subsets. Rabin [95] proved (as a corollary to his result given in Theorem 9.3(a)) that the regular subsets of $\{0,1\}^*$ constitute such a restriction. Siefkes [105] showed that this fails for the recursive subsets of $\{0,1\}^*$. A related question is the uniformization problem: is there, for any given formula $\varphi(X, Y)$ such that $\forall X\, \exists Y\, \varphi(X, Y)$ holds, a "definable choice function", i.e. a (set) function $X \mapsto Y$ defined by a formula $\psi(X, Y)$ such that $\forall X\, \forall Y(\psi(X, Y) \rightarrow \varphi(X, Y))$? Siefkes [105] proves this for S1S, and Gurevich and Shelah [43] showed that it fails for S2S using a very intricate proof method.

We discuss two further restricted set quantifiers: those ranging over finite sets ("weak quantifiers"), and those ranging over paths through the infinite binary tree.

The weakly definable sets of trees, already considered in the preceding section, have been further classified in [121] (see also [68]): An infinite hierarchy is induced by the alternation depth of "unbounded" quantifiers over finite sets and elements. ("Bounded" set quantifiers are of the form $\exists X \leqslant Y\, \varphi(X, Y)$, where $X \leqslant Y$ is an abbreviation for $\forall x(x \in X \rightarrow \exists y\,(y \in Y \wedge x \leqslant y))$.) Over ω-words the analogous hierarchy is finite, because, by McNaughton's Theorem, two unbounded quantifiers suffice for the definition of regular ω-languages (cf. Theorem 4.6).

When set quantifiers refer to chains in trees (i.e., sets linearly ordered by the prefix relation $<$) or to paths (i.e. maximal chains), one obtains *chain logic*, respectively *path logic* over the binary tree. In [123], these systems are characterized in terms of the regular and star-free ω-languages. A close connection between path quantifiers and systems of branching time logic is set up in [46], where computation tree logic CTL* and path logic are shown expressively equivalent, provided that binary tree models in the signature $<$ are considered.

A general decidability result on path quantifiers over trees was shown by Gurevich and Shelah [44]. They prove that in the language of path logic the theory of arbitrary trees (considered as any partial order where each set $\{y \mid y \leqslant x\}$ is totally ordered) is decidable.

12.2. *Restrictions in Rabin tree automata*

Several authors investigated the possibility of extending Landweber's Theorem 5.3 to Rabin recognizable sets of trees. The notions of 1- and 2-acceptance can be transferred from sequential automata to tree automata (namely, as conditions for all paths of a run). Also the Cantor topology is extended canonically from A^ω to the space T_A^ω (the sets $t.T_A^\omega$, where t is a finite tree, form an open basis). Results on inclusion relations for these acceptance conditions and their topological meaning are presented in [48, 65]. Skurczyński showed in [143] that the Borel hierarchy of Rabin recognizable tree sets is infinite. Mostowski, Skurczyński and Wagner [69] obtain a partial transfer of Theorem 5.3 (including decidability results) to tree languages recognized by deterministic Rabin tree automata. Further results on deterministic tree automata and the power of several acceptance conditions are given in [100].

Niwinski [78] showed that the Rabin index (the number of accepting pairs in Rabin tree automata) defines an infinite hierarchy of sets of trees: for each n there is a set $T_n \subseteq T_A^\omega$ which is recognized only by Rabin tree automata with at least n accepting pairs. The presented example languages T_n belong to the Boolean closure of the Büchi recognizable sets of trees; Hafer [45] proves that this Boolean closure is still properly contained in the class of Rabin recognizable sets. Mostowski [67] presents a "standard form" of Rabin tree automata in which an ordering of the state set enters the acceptance condition. As an application, a calculus of regular-like expressions is set up which allows to define the Rabin recognizable sets.

12.3. *Fixed point calculi*

Niwinski [78] and Takahashi [115] studied the specification of tree properties by a fixed point calculus in which least and greatest fixed points are included. Following [78], we define the *μ-terms* over an alphabet A and with variables $x_1, x_2, \ldots$ by the clauses

- each variable x_i is a μ-term,
- if τ_1, τ_2 are μ-terms and $a \in A$, then $a(\tau_1, \tau_2)$ and $\tau_1 \cup \tau_2$ are μ-terms,
- if τ is a μ-term and x is a variable, then $\mu x \tau$ and $\nu x \tau$ are μ-terms.

Any μ-term $\tau(x_1, \ldots, x_n)$ with free variables $x_1, \ldots, x_n$ defines a function $F_\tau : (T_A^\omega)^n \to T_A^\omega$. For $\tau = x_i$ it is the ith projection. If $\tau(x_1, \ldots, x_n)$ has the form $\tau = \tau_1 \cup \tau_2$, let

$$F_\tau(T_1, \ldots, T_n) = F_{\tau_1}(T_1, \ldots, T_n) \cup F_{\tau_2}(T_1, \ldots, T_n);$$

similarly for $\tau = a(\tau_1, \tau_2)$,

$$F_\tau(T_1, \ldots, T_n) = \{t \in T_A^\omega \mid t(\varepsilon) = a,\ t_0 \in F_{\tau_1}(T_1, \ldots, T_n),\ t_1 \in F_{\tau_2}(T_1, \ldots, T_n)\}.$$

Finally, for $\tau = \mu y \tau_0(y, x_1, \ldots, x_n)$, respectively $\tau = \nu y \tau_0(y, x_1, \ldots, x_n)$, let $F_\tau(T_1, \ldots, T_n)$ be the least, resp. greatest fixed point of the function $T \mapsto F_{\tau_0}(T, T_1, \ldots, T_n)$. (These fixed points exist by the Knaster–Tarski Theorem.)

Each μ-term τ without free variables defines a set $T \subseteq T_A^\omega$, denoted here by $T(\tau)$. As an example over the alphabet $A = \{a, b\}$, consider the μ-term

$$\mu x_1 \nu x_0 (b(x_0, x_0) \cup a(x_1, x_1));$$

it defines the set T_1 of Section 8, containing all trees such that on each path there are only finitely many letters a.

By induction over the μ-terms τ one verifies that the functions F_τ are definable in S2S. It follows that the sets $T(\tau)$ are Rabin recognizable. Niwinski [79] proves also the converse; so the μ-terms have the same expressive power as Rabin tree automata (or S2S). A strict hierarchy of sets of trees is generated by increasing the number of alternations between the least and greatest fixed point operators μ and ν in the defining terms. Moreover, the second level of the hierarchy (given by the terms where all ν-operators precede all μ-operators) characterizes the Büchi recognizable sets of trees (cf. [78, 115, 3]).

Greatest fixed points consisting of finite and infinite trees also arise naturally in the theory of nondeterministic *recursive program schemes* (cf. [4] or [22, Section 8.3]. These schemes and the associated fixed point operators can be considered as tree replacement systems (tree grammars). Since in their derivations a context-free substitution mechanism is involved (which is not present in the μ-terms above), the tree languages generated by recursive program schemes cannot be described by finite-state tree automata. An extended model of "pushdown tree automaton" that is appropriate for this purpose has been introduced in [39, 142].

Acknowledgment

I would like to thank all who contributed comments and corrections to a draft version of this paper; in particular, A. Arnold, H.D. Ebbinghaus, J. Flum, Y. Gurevich, A.W. Mostowski, D. Niwinski, D. Perrin, J.E. Pin, D. Seese, D. Siefkes, L. Staiger and P. Wolper, as well as students and colleagues from the RWTH Aachen (where this paper was written). I also thank Carla Meckler for her endless work in typing and arranging the manuscript.

References

[1] ALPERN, B. and F.B. SCHNEIDER, Recognizing safety and liveness, *Distributed Comput.* **2** (1987) 117–126.

[2] ARNOLD, A., A syntactic congruence for rational ω-languages, *Theoret. Comput. Sci.* **39** (1985) 333–335.

[3] ARNOLD, A., Logical definability of fixed points, *Theoret. Comput. Sci.* **61** (1988) 289–297.

[4] ARNOLD, A. and M. NIVAT, Formal computations of nondeterministic recursive program schemes, *Math. Systems Theory* **13** (1980) 219–236.

[5] BERSTEL, J., *Transductions and Context-Free Languages* (Teubner, Stuttgart, 1979).

[6] BERSTEL, J. and L. BOASSON, Context-free languages, in: J. van Leeuwen, ed., *Handbook of Theoretical Computer Science, Vol. B* (North-Holland, Amsterdam, 1990).

[7] BOASSON, L. and M. NIVAT, Adherences of languages, *J. Comput. System Sci.* **20** (1980) 285–309.

[8] BÜCHI, J.R., Weak second-order arithmetic and finite automata, *Z. Math Logik Grundlag. Math.* **6** (1960) 66–92.

[9] BÜCHI, J.R., On a decision method in restricted second order arithmetic, in: E. Nagel et al., eds., *Proc. Internat. Congr. on Logic, Methodology and Philosophy of Science* (Stanford Univ. Press, Stanford, CA, 1960) 1–11.

[10] BÜCHI, J.R., The monadic theory of ω_1, in: *Decidable Theories II*, Lecture Notes in Mathematics, Vol. 328 (Springer, Berlin, 1973) 1–127.

[11] BÜCHI, J.R., Using determinacy to eliminate quantifiers, in: M. Karpinski, ed., *Fundamentals of Computation Theory*, Lecture Notes in Computer Science, Vol. 56 (Springer, Berlin, 1977) 367–378.

[12] BÜCHI, J.R., State-strategies for games in $F_{\sigma\delta}\cap G_{\delta\sigma}$, *J. Symbolic Logic* **48** (1983) 1171–1198.

[13] BÜCHI, J.R. and L.H. LANDWEBER, Solving sequential conditions by finite-state strategies, *Trans. Amer. Math. Soc.* **138** (1969) 295–311.

[14] BÜCHI, J.R. and D. SIEFKES, Axiomatization of the monadic second order theory of ω_1, in: *Decidable Theories II*, Lecture Notes in Mathematics, Vol. 328 (Springer, Berlin, 1973) 129–217.

[15] CHOUEKA, Y., Theories of automata on ω-tapes: a simplified approach, *J. Comput. System Sci.* **8** (1974) 117–141.

[16] CHURCH, A., Logic, arithmetic and automata, in: *Proc. Internat. Congress Math.* (1963) 23–35.

[17] COHEN, R.S. and A.Y. GOLD, Theory of ω-languages, *J. Comput. System Sci.* **15** (1977) 169–184 and 185–203.

[18] COHEN, R.S. and A.Y. GOLD, ω-Computations of deterministic pushdown machines, *J. Comput. System Sci.* **16** (1978) 275–300.

[19] COHEN, R.S. and A.Y. GOLD, ω-Computations on Turing machines, *Theoret. Comput. Sci.* **6** (1978) 1–23.

[20] COURCELLE, B., Fundamental properties of infinite trees, *Theoret. Comput. Sci.* **25** (1983) 95–169.

[21] COURCELLE, B., The monadic second-order logic of graphs II: infinite graphs of bounded width, *Math. Systems Theory* **21** (1989) 187–221.

[22] COURCELLE, B., Recursive applicative program schemes, in: J. van Leeuwen, ed., *Handbook of Theoretical Computer Science, Vol. B* (North-Holland, Amsterdam, 1990).

[23] COURCOUBETIS, C. and M. YANNAKAKIS, Verifying properties of finite-state probabilistic programs, in: *Proc. 29th Ann. IEEE Symp. on Foundations of Computer Science* (1988) 338–345.

[24] DAUCHET, M. and E. TIMMERMAN, Continuous monoids and yields of infinite trees, *RAIRO Inform. Théor. Appl.* **20** (1986) 251–274.

[25] DAVIS, M., Infinite games of perfect information, in: *Advances in Game Theory* (Princeton Univ. Press, Princeton, NJ, 1964) 85–101.

[26] DONER, J., Tree acceptors and some of their applications, *J. Comput. System Sci.* **4** (1970) 406–451.

[27] EILENBERG, S., *Automata, Languages and Machines, Vol. A* (Academic Press, New York, 1974).

[28] ELGOT, C.C., Decision problems of finite automata design and related arithmetics, *Trans. Amer. Math. Soc.* **98** (1961) 21–52.

[29] ELGOT, C.C. and M.O. RABIN, Decidability and undecidability of second (first) order theory of (generalized) successor, *J. Symbolic Logic* **31** (1966) 169–181.

[30] EMERSON, E.A., Temporal and modal logic, in: J. van Leeuwen, ed., *Handbook of Theoretical Computer Science, Vol. B* (North-Holland, Amsterdam, 1990).

[31] EMERSON, E.A. and J.Y. HALPERN, "Sometimes" and "Not Never" revisited: On branching time versus linear time, *J. Assoc. Comput. Mach.* **33** (1986) 151–178.

[32] EMERSON, E.A. and C.S. JUTLA, The complexity of tree automata and logics of programs, in: *Proc. 29th Ann. IEEE Symp. on Foundations of Computer Science* (1988) 328–337.

[33] EMERSON, E.A. and A.P. SISTLA, Deciding full branching time logic, *Inform. and Control* **61** (1984) 175–201.

[34] FRANCEZ, N., *Fairness* (Springer, Berlin, 1987).

[35] GABBAY, A., A. PNUELI, S. SHELAH, and J. STAVI, On the temporal analysis of fairness, in: *Proc. 7th Ann. ACM Symp. on Principles of Programming Languages* (1980) 163–173.

[36] GALE, D. and F.M. STEWART, Infinite games with perfect information, in: *Contributions to the Theory of Games* (Princeton Univ. Press, Princeton, NJ, 1953) 245–266.

[37] GÉCSEG, F. and M. STEINBY, *Tree Automata* (Akadémiai Kiadó, Budapest, 1984).

[38] GIRE, F. and M. NIVAT, Relations rationelles infinitaires, *Calcolo* **21** (1984) 91–125.

[39] GUESSARIAN, I., Pushdown tree automata, *Math. Systems Theory* **16** (1983) 237–264.

[40] GUESSARIAN, I. and W. NIAR-DINEDANE, Fairness and regularity for SCCS processes, *Inform. Théor. Appl.* **23** (1989) 59–86.

[41] GUREVICH, Y., Monadic second-order theories in: J. Barwise and S. Feferman, eds., *Model-theoretic Logics* (Springer, Berlin, 1985), 479–506.

[42] GUREVICH, Y. and L.A. HARRINGTON, Automata, trees, and games, in: *Proc. 14th Ann. ACM Symp. on the Theory of Computing* (1982) 60–65.

[43] GUREVICH, Y. and S. SHELAH, Rabin's uniformization problem, *J. Symbolic Logic* **48** (1983) 1105–1119.

[44] GUREVICH, Y. and S. SHELAH, The decision problem for branching time logic, *J. Symbolic Logic* **50** (1985) 668–681.

[45] HAFER, T., On the boolean closure of Büchi tree automaton definable sets of ω-trees, Aachener Inform. Ber. Nr. 87–16, RWTH Aachen, 1987.

[46] HAFER, T. and W. THOMAS, Computation tree logic CTL* and path quantifiers in the monadic theory of the binary tree, in: T. Ottmann, ed., *Proc. 14th Internat. Coll. on Automata, Languages and Programming*, Lecture Notes in Computer Science, Vol. 267 (Springer, Berlin, 1987) 269–279.

[47] HAREL, D., D. KOZEN and R. PARIKH, Process logic: expressiveness, decidability, completeness, *J. Comput. System Sci.* **25** (1982) 144–170.

[48] HAYASHI, T. and S. MIYANO, Finite tree automata on infinite trees, *Bull. Inform. Cybernet.* **21** (1985) 71–82.

[49] HOOGEBOOM, H.J. and G. ROZENBERG, Infinitary languages: basic theory and applications to concurrent systems, in: J. W. de Bakker et al., eds., *Current Trends in Concurrency*, Lecture Notes in Computer Science, Vol. 224 (Springer, Berlin, 1986) 266–342.

[50] KAMINSKI, M., A classification of ω-regular languages, *Theoret. Comput. Sci.* **36** (1985) 217–239.

[51] KAMP, H.W., Tense logic and the theory of linear order, Ph.D. Thesis, Univ. of California, Los Angeles, CA, 1968.

[52] KOZEN, D. and J. TIURYN, Logics of programs, in: J. van Leeuwen, ed., *Handbook of Theoretical Computer Science, Vol. B* (North-Holland, Amsterdam, 1990) 789–840.

[53] LADNER, R.E., Application of model-theoretic games to discrete linear orders and finite automata, *Inform. and Control* **33** (1977) 281–303.

[54] LÄUCHLI, H. and C. SAVOIZ, C. Monadic second order definable relations on the binary tree, *J. Symbolic Logic* **52** (1987) 219–226.

[55] LANDWEBER, L.H., Decision problems for ω-automata, *Math. Systems Theory* **3** (1969) 376–384.

[56] LICHTENSTEIN, O. and A. PNUELI, Checking that finite-state concurrent programs satisfy their specification, in: *Proc. 12th Ann. ACM Symp. on Principles of Programming Languages* (1985) 97–107.

[57] LICHTENSTEIN, O., A. PNUELI and L. ZUCK, The glory of the past, in: R. Parikh, ed., *Logics of Programs*, Lecture Notes in Computer Science, Vol. 193 (Springer, Berlin, 1985) 196–218.

[58] LINNA, M., On ω-sets associated with context-free languages, *Inform. and Control* **31** (1976) 272–293.

[59] MANNA, Z., and A. PNUELI, Specification and verification of concurrent programs by $\forall$-automata, in: *Proc. 14th Ann. ACM Symp. on Principles of Programming Languages* (1987) 1–12.

[60] MANNA, Z. and A. PNUELI, The anchored version of the temporal framework, in: J.W. de Bakker et al., eds. *Linear Time, Branching Time and Partial Order in Logics and Models for Concurrency*, Lecture Notes in Computer Science, Vol. 345 (Springer, Berlin, 1989) 201–284.

[61] MARTIN, D.A., Borel determinacy, *Ann. Math.* **102** (1975) 363–371.

[62] MCNAUGHTON, R., Testing and generating infinite sequences by a finite automaton, *Inform. and Control* **9** (1966) 521–530.

[63] MCNAUGHTON, R. and S. PAPERT, *Counter-Free Automata* (MIT Press, Cambridge, MA, 1971).

[64] MIYANO, S. and T. HAYASHI, Alternating automata on ω-words, *Theoret. Comput. Sci.* **32** (1984) 321–330.

[65] MORIYA, T., Topological characterizations of infinite tree languages, *Theoret. Comput. Sci.* **52** (1987) 165–171.

[66] MOSCHOVAKIS, Y.N., *Descriptive Set Theory* (North-Holland, Amsterdam, 1980).

[67] MOSTOWSKI, A.W., Regular expressions for infinite trees and a standard form of automata, in: A. Skowron, ed., *Computation Theory*, Lecture Notes in Computer Science, Vol. 208 (Springer, Berlin, 1984) 157–168.

[68] MOSTOWSKI, A.W., Hierarchies of weak monadic formulas for two successors arithmetic, *J. Inform. Process. Cybernet.* **23** (1987) 509–515.

[69] MOSTOWSKI, A.W., J. SKURCZYŃSKI, and K. WAGNER, Deterministic automata on infinite trees and the Borel hierarchy, in: M. Arato, I. Kátai and L. Varga, eds., *Proc. 4th Hungarian Conf. on Computer Science* (1985) 103–115.

[70] MUCHNIK, A.A.,Games on infinite trees and automata with dead-ends: a new proof of the decidability of the monadic theory of two successors, *Semiotics and Information* **24** (1984) 17–40 (in Russian).

[71] MULLER, D.E., Infinite sequences and finite machines, in: *Proc. 4th Ann. IEEE Symp. on Switching Circuit Theory and Logical Design* (1963) 3–16.

[72] MULLER, D.E., A. SAOUDI and P.E. SCHUPP, Alternating automata, the weak monadic theory of the tree, and its complexity, in: L. Kott, ed., *Proc. 13th Internat. Coll. on Automata, Languages and Programming*, Lecture Notes in Computer Science, Vol. 226 (Springer, Berlin, 1986) 275–283.

[73] MULLER, D.E., A. SAOUDI and P.E. SCHUPP, Weak alternating automata give a simple explanation of why most temporal and dynamic logics are decidable in exponential time, in: *Proc. 3rd IEEE Ann. Symp. on Logic in Computer Science* (1988) 422–427.

[74] MULLER, D.E. and P.E. SCHUPP, The theory of ends, pushdown automata, and second-order logic, *Theoret. Comput. Sci.* **37** (1985) 51–75.

[75] MULLER, D.E. and P.E. SCHUPP, Alternating automata on infinite trees, *Theoret. Comput. Sci.* **54** (1987) 267–276.

[76] NIVAT, M. and D. PERRIN, Ensembles reconnaissables de mots biinfinis, *Canad. J. Math.* **38** (1986) 513–537.

[77] NIWINSKI, D., Fixed-point characterization of context-free ∞-languages, *Inform. and Control* **61** (1984) 247–276.

[78] NIWINSKI, D., On fixed-point clones, in L. Kott, ed., *Proc. 13th Internat. Coll. on Automata, Languages and Programming*, Lecture Notes in Computer Science Vol. 226 (Springer, Berlin, 1986) 464–473.

[79] NIWINSKI, D., Fixed points vs. infinite generation, in: *Proc. 3rd Ann. IEEE Symp. on Logic in Computer Science* (1988) 402–409.

[80] PARIGOT, M. and E. PELZ, A logical approach of Petri net languages, *Theoret. Comput. Sci.* **39** (1985) 155–169.

[81] PARK, D., Concurrency and automata on infinite sequences, in P. Deussen, ed., *Theoretical Computer Science*, Lecture Notes in Computer Science, Vol. 104 (Springer, Berlin, 1981) 167–183.

[82] PÉCUCHET, J.P., Automates boustrophédons et mots infinis, *Theoret. Comput. Sci.* **35** (1985) 115–122.

[83] PÉCUCHET, J.P., On the complementation of Büchi automata, *Theoret. Comput. Sci.* **47** (1986) 95–98.

[84] PÉCUCHET, J.P., Etude syntaxique des parties reconnaissables de mots infinis, in: L. Kott, ed., *Proc. 13th Internat. Coll. on Automata, Languages and Programming*, Lecture Notes in Computer Science, Vol. 226 (Springer, Berlin, 1986) 294–303.

[85] PERRIN, D., Recent results on automata and infinite words, in: M.P. Chytil and V. Koubek, eds., *Mathematical Foundations of Computer Science '84*, Lecture Notes in Computer Science, Vol. 176 (Springer, Berlin, 1984) 134–148.

[86] PERRIN, D., Variétés de semigroupes et mots infinis, *C.R. Acad. Sci. Paris* **295** (1985) 595–598.

[87] PERRIN, D., Finite automata, in: J. van Leeuwen, ed., *Handbook of Theoretical Computer Science, Vol. B* (North-Holland, Amsterdam, 1990) 1–57.

[88] PERRIN, D. and J.E. PIN, First order logic and star-free sets, *J. Comput. System Sci.* **32** (1986) 393–406.

[89] PERRIN, D. and P.E. SCHUPP, Automata on the integers, recurrence distinguishability, and the equivalence and decidability of monadic theories, in: *Proc. 1st IEEE Symp. on Logic in Computer Science* (1986) 301–304.

[90] PNUELI, A., Applications of temporal logic to the specification and verification of reactive systems: a survey of current trends, in: J.W. de Bakker et al., eds., *Current Trends in Concurrency*, Lecture Notes in Computer Science, Vol. 224 (Springer, Berlin, 1986) 510–584.

[91] PNUELI, A. and R. ROSNER, On the synthesis of a reactive module, in: *Proc. 16th Symp. Princ. of Prog. Lang.* (1989) 179–190.
[92] PRIESE, L., R. REHRMANN, and U. WILLECKE-KLEMME, An introduction to the regular theory of fairness, *Theoret. Comput. Sci.* **54** (1987) 139–163.
[93] RABIN, M.O., Decidability of second-order theories and automata on infinite trees, *Trans. Amer. Math. Soc.* **141** (1969) 1–35.
[94] RABIN, M.O., Weakly definable relations and special automata in: Y. Bar-Hillel, ed., *Mathematical Logic and Foundations of Set Theory* (North-Holland, Amsterdam, 1970) 1–23.
[95] RABIN, M.O., *Automata on Infinite Objects and Church's Problem* (Amer. Mathematical Soc., Providence, RI, 1972).
[96] RABIN, M.O., Decidable theories, in J. Barwise, ed., *Handbook of Mathematical Logic* (North-Holland, Amsterdam, 1977) 595–629.
[97] REDZIEJOWSKI, R.R., Infinite-word languages and continuous mappings, *Theoret. Comput. Sci.* **43** (1985) 59–79.
[98] ROSENSTEIN, J.G., *Linear Orderings* (Academic Press, New York, 1982).
[99] SAFRA, S., On the complexity of ω-automata, in: *Proc. 29th Ann. IEEE Symp. on Foundations of Computer Science* (1988) 319–327.
[100] SAOUDI, A., Variétés d'automates descendants d'arbres infinis, *Theoret. Comput. Sci.* **43** (1986) 315–335.
[101] SEESE, D., The structure of the models of decidable monadic theories of graphs, Akad. d. Wiss. der DDR, Inst. f. Math. Berlin, 1988.
[102] SEMENOV, A.L., Decidability of monadic theories, in: M.P. Chytil and V. Koubek, eds., *Mathematical Foundations of Computer Science*, Lecture Notes in Computer Science, Vol. 176 (Springer, Berlin, 1984) 162–175.
[103] SHELAH, S., The monadic theory of order, *Ann. of Math.* **102** (1975) 379–419.
[104] SIEFKES, D., *Decidable Theories I: Büchi's Monadic Second Order Successor Arithmetic*, Lecture Notes in Mathematics, Vol. 120 (Springer, Berlin, 1970).
[105] SIEFKES, D., The recursive sets in certain monadic second order fragments of arithmetic, *Arch. Math. Logik* **17** (1975) 71–80.
[106] SISTLA, A.P. and E.M. CLARKE, The complexity of propositional linear time logics, *J. Assoc. Comput. Mach.* **32** (1985) 733–749.
[107] SISTLA, A.P., M.Y. VARDI and P. WOLPER, The complementation problem for Büchi automata with applications to temporal logic, *Theoret. Comput. Sci.* **49** (1987) 217–237.
[108] STAIGER, L., Finite-state ω-languages, *J. Comput. System Sci.* **27** (1983) 434–448.
[109] STAIGER, L., Hierarchies of recursive ω-languages, *J. Inform. Process. Cybernet.* **22** (1986) 219–241.
[110] STAIGER, L., Research in the theory of ω-languages, *J. Inform. Process. Cybernet.* **23** (1987) 415–439.
[111] STOCKMEYER, L.J. and A.R. MEYER, Word problems requiring exponential time: preliminary report, in: *Proc. 5th Ann. ACM Symp. on the Theory of Computing* (1973) 1–9.
[112] STREETT, R.S., Propositional dynamic logic of looping and converse, *Inform. and Control* **54** (1982) 121–141.
[113] STREETT, R.S. and E.A. EMERSON, The propositional Mu-calculus is elementary, in: J. Paredaens, ed., *Proc. 11th Internat. Coll. on Automata, Languages and Programming*, Lecture Notes in Computer Science, Vol. 172 (Springer, Berlin, 1984) 465–472.
[114] STUPP, J., The lattice model is recursive in the original model, manuscript (The Hebrew University, Jerusalem, 1975).
[115] TAKAHASHI, M., The greatest fixed-points and rational omega-tree languages, *Theoret. Comput. Sci.* **44** (1986) 259–274.
[116] TAKAHASHI, M., Brzozowski hierarchy of ω-languages, *Theoret. Comput. Sci.* **49** (1987) 1–12.
[117] THATCHER, J.W. and J.B. WRIGHT, Generalized finite automata with an application to a decision problem of second-order logic, *Math. Systems Theory* **2** (1968) 57–82.
[118] THOMAS, W., Star-free regular sets of ω-sequences, *Inform. and Control* **42** (1979) 148–156.
[119] THOMAS, W., A combinatorial approach to the theory of ω-automata, *Inform. and Control* **48** (1981) 261–283.
[120] THOMAS, W., Classifying regular events in symbolic logic, *J. Comput. System Sci.* **25** (1982) 360–376.

[121] THOMAS, W., A hierarchy of sets of infinite trees, in: A.B. Cremers and H.P. Kriegel, eds., *Theoretical Computer Science*, Lecture Notes in Computer Science, Vol. 145 (Springer, Berlin, 1982) 335–342.

[122] THOMAS, W., On frontiers of regular trees, *RAIRO Inform. Théor. Appl.* **20** (1986) 371–381.

[123] THOMAS, W., On chain logic, path logic, and first-order logic over infinite trees, in: *Proc. 2nd Ann. IEEE Symp. on Logic in Computer Science* (1987) 245–256.

[124] TRAKHTENBROT, B.A., Finite automata and the logic of one-place predicates, *Siberian Math. J.* **3**, 103–131; English translation in: *AMS Transl.* **59** (1966) 23–55.

[125] TRAKHTENBROT, B.A. and Y.M. BARZDIN, *Finite Automata* (North-Holland, Amsterdam, 1973).

[126] VALK, R., Infinite behaviour of Petri nets, *Theoret. Comput. Sci.* **25** (1983) 311–341.

[127] VARDI, M.Y., Automatic verification of probabilistic concurrent finite-state programs, in: *Proc. 26th Ann. IEEE Symp. on Foundations of Computer Science* (1985) 327–338.

[128] VARDI, M.Y., A temporal fixed point calculus, in: *Proc. 15th Ann. ACM Symp. on Principles of Programming Languages* (1988) 250–259.

[129] VARDI, M.Y. and P. WOLPER, Automata-theoretic techniques for modal logics of programs, *J. Comput. System Sci.* **32** (1986) 183–221.

[130] VARDI, M.Y. and P. WOLPER, An automata theoretic approach to automatic program verification, in: *Proc. 1st Ann. IEEE Symp. on Logic in Computer Science* (1986) 332–334.

[131] VARDI, M.Y. and P. WOLPER, Reasoning about infinite computation paths, to appear.

[132] WAGNER, K., Eine Axiomatisierung der Theorie der regulären Folgenmengen, *J. Inform. Process. Cybernet.* **12** (1976) 337–354.

[133] WAGNER, K., On ω-regular sets, *Inform. and Control* **43** (1979) 123–177.

[134] WISNIEWSKI, K., A generalization of finite automata, *Fund. Inform.* **10** (1987) 415–436.

[135] WOLPER, P., Temporal logic can be more expressive, *Inform. and Control* **56** (1983) 72–99.

[136] ABADI, M., L. LAMPORT and P. WOLPER, Realizable and unrealizable specifications of reactive systems, in: G. Ausiello et al., eds., *Proc. 16th Internat. Coll. on Automata, Languages, and Programming*, Lecture Notes in Computer Science, Vol. 372 (Springer, Berlin, 1989) 1–17.

[137] EMERSON, E.A. and C.S. JUTLA, On simultaneously determinizing and complementing ω-automata, in: *Proc. 4th Ann. Symp. on Logic in Computer Science* (1989) 333–342.

[138] ENGELFRIET, J. and H.J. HOOGEBOOM, Automata with storage on infinite words, in: G. Ausiello et al., eds., *Proc. 16th Intern. Coll. on Automata, Languages, and Programmimg*, Lecture Notes in Computer Science, Vol. 372 (Springer, Berlin, 1989) 289–303.

[139] GUREVICH, Y., Games People Play, in: S. MacLane and D. Siefkes, eds., *The Collected Works of J.R. Büchi* (Springer, Berlin, 1990) 518–524.

[140] PNUELI, A. and R. ROSNER, On the synthesis of an asynchronous reactive module, in: D. Ausiello et al., eds., *Proc. 16th Internat. Coll. on Automata, Languages, and Programing*, Lecture Notes in Computer Science, Vol. 372 (Springer, Berlin, 1989) 652–671.

[141] SAFRA, S. and M.Y. VARDI, On ω-automata and temporal logic, in: *Proc. 21st Ann. Symp. on Theory of Computing* (1989) 127–137.

[142] SAOUDI, A., Pushdown automata on infinite trees and omega-Kleene closure of context-free tree sets, in: A. Kreczmar and G. Mirkowka, eds., *Math. Found. of Comput. Sci. 1989*, Lecture Notes in Computer Science, Vol. 379 (Springer, Berlin, 1989) 445–457.

[143] SKURCZYŃSKI, J., The Borel Hierarchy is infinite in the class of regular sets of trees, in: J. Csirik et al., eds., *Fundamentals of Computation Theory*, Lecture Notes in Computer Science, Vol. 380 (Springer, Berlin, 1989) 416–423.

[144] THOMAS, W., Infinite trees and automaton definable relations over ω-words, in: C. Choffrut and T. Lengauer, eds., *Proc. 7th Ann. Symp. STACS 90*, Lecture Notes in Computer Science, Vol. 415 (Springer, Berlin, 1990) 263–277.

CHAPTER 5

Graph Rewriting: An Algebraic and Logic Approach

Bruno COURCELLE

Laboratoire d'Informatique, Université Bordeaux I, 351 Cours de la Libération, F-33405 Talence, France

Contents

Introduction 195
1. Logical languages and graph properties 196
2. Graph operations and graph expressions 203
3. Context-free sets of hypergraphs 211
4. Logical properties of context-free sets of hypergraphs 215
5. Sets of finite graphs defined by forbidden minors 222
6. Complexity issues 226
7. Infinite hypergraphs 229
8. Guide to the literature. 235
Acknowledgment 238
References 239

HANDBOOK OF THEORETICAL COMPUTER SCIENCE
Edited by J. van Leeuwen

Introduction

The theory of formal languages, i.e., the study of sets of words (finite strings of symbols), of binary relations on these sets, and of devices defining such sets and relations, has been extended so as to deal with infinite strings and with finite and infinite trees. We refer the reader to the volumes edited by Book [16], by Nivat and Perrin [74] and to the chapter by Thomas in this handbook [88].

In a natural way, finite and infinite graphs form the next step. Grammars that generate sets of finite graphs have been considered for a long time. We refer the reader to the proceedings of three international workshops edited by Ehrig et al. [18, 38, 39] for motivations, applications and results. By the name "graph rewriting", we refer to the following topics: graph rewriting rules, context-free graph-grammars, and descriptions of infinite graphs and sets of infinite graphs by rewriting rules and by systems of equations.

Graphs are intrinsically more complicated than strings and trees. The basic notions are not as firmly established as in the case of strings: several types of graph-grammars can be considered as context-free; no satisfactory notion of a regular set of graphs, satisfying the nice properties of regular sets of strings and trees, has yet been found, and finally, no satisfactory notion of a graph automaton has been proposed. In this chapter, we present three mathematical tools that can help to remedy this situation: mathematical logic, universal algebra and category theory. As the reader will see in the course of this chapter, these tools already *provide mathematical results*, and are not only useful to *describe* graph-grammars and the sets they generate.

In a few words we now show how these mathematical notions appear in studies concerning graphs, which is not necessarily obvious for the reader, even when he is familiar with graph theory. Graphs can be considered as logical structures and logical formulas can express their properties. A logical formula defines a set of graphs, namely the set of graphs satisfying the corresponding property. Hence, each logical language defines a class of graph properties and, equivalently, a class of sets of graphs. We shall mainly discuss first-order logic, second-order logic, and monadic second-order logic, together with a few variants and restrictions of these three languages.

Category theory will be used for two purposes: for specifying graph rewriting rules in a precise and concise way, and for properly defining the least (or rather the initial) solution of a system of graph equations.

Universal algebra will be useful in the following way. A suitable class of graphs can be equipped with operations that make it possible to build larger graphs from smaller ones. It follows that every finite graph can be represented by a finite algebraic expression called a *graph expression*. The algebra of all finite graphs is the free R-algebra generated by a set of symbols (the edge labels), where R is a set of equational axioms providing a complete description of the properties of the graph operations. Graph rewriting systems can be considered as term rewriting systems that rewrite graph expressions. Context-free graph-grammars can be considered as (*polynomial*) *systems of equations* of which least solutions are taken in the algebra of all sets of finite graphs. *Recognizable* sets of graphs can also be defined in terms of congruences. Furthermore, *infinite graph expressions* can be defined. (They are comparable to formal power series or continued fractions.) They define countable graphs and every countable graph can be defined by such an expression. Since infinite graph expressions can be considered as infinite trees, automata and systems of equations defining infinite trees can be used to define infinite graphs. Hence, we shall be able to treat certain aspects of graph transformations with the techniques of universal algebra, language theory and term rewriting systems.

In the subfield of language theory dealing with strings and trees, some of these theories are already useful: the theory of regular languages can be based on the study of semigroups, and monadic second-order logic is essential in the study of sets of infinite trees. This chapter will show that they are much more important when graphs and sets of graphs are considered. In addition, we present results establishing close links between the notions yielded by these three theories that are relevant to graph transformations. Let us finally mention that we deal with one specific class of graphs (oriented hypergraphs of a certain type), but that the methodology extends to other classes (see Section 8).

This chapter is organized as follows. Section 1 is devoted to expressing graph properties by logical formulas: we shall compare the powers of several logical languages. Sections 2 and 3 present algebraic techniques that are useful for defining and studying graph rewriting ruies and context-free graph-grammars. Section 4 presents some links between these context-free graph-grammars and monadic second-order logic. Some applications to the definition of sets of graphs by forbidden configurations, and to the theory of NP-completeness are given in Sections 5 and 6 respectively. Section 7 extends some of the previous results to infinite graphs. The last section is a bibliographical review where most of the references can be found. The references in the main text are kept to a minimum, in order to facilitate a continuous reading.

The reader is assumed to be acquainted with some basic notions in logic (first-order formulas and their validity), in universal algebra (many-sorted algebras, terms), in category theory (push-outs, colimits), and in language theory (context-free grammars, least solutions of systems of equations). These basic notions can be found in, e.g., [6, 92, 68, 11] respectively.

1. Logical languages and graph properties

Graphs can be represented by first-order relational structures, and their properties can be written as logical formulas. There are many notions of graphs: a graph can be directed or not; it can be a hypergraph; it can have labels. Many of the results we present hold for all these variants by means of straightforward modifications. In order to state precise results, we introduce a few classes of graphs, together with their representations by classes of logical structures.

1.1. The class of simple, directed graphs $\boldsymbol{S}$

We denote by $\boldsymbol{S}$ the class of finite or infinite simple, directed graphs. With G in $\boldsymbol{S}$, we associate the logical structure $|G| := \langle V_G, \mathbf{edg}_G \rangle$, the domain of which is V_G, the set of vertices of G. We let **edg** be a binary relation symbol. Its meaning in $|G|$ is the relation $\mathbf{edg}_G$ such that

$$\mathbf{edg}_G(x, y) \quad \Leftrightarrow \quad \text{there is an edge from } x \text{ to } y \text{ in } G.$$

If φ is a closed, first-order formula written with $=$ and **edg** as basic relation symbols, then it is either true or false in the structure $|G|$. We write $|G| \models \varphi$ iff it is true in $|G|$, and in this case we say that φ *holds in* G, or that φ is *true in* G, or that G *satisfies* φ.

Consider, for example, the formula φ:

$$\forall x\, \exists y\, [(\mathbf{edg}(x, y) \vee \mathbf{edg}(y, x)) \wedge \neg(x = y)].$$

It is true in G iff G has no isolated vertex, i.e., iff every vertex in G is linked (by an edge) to some other vertex (irrespective of orientation).

In the above definition, we have chosen to represent a graph by a structure, the domain of which is the set of vertices of the graph. Another possibility would be to take a domain consisting of vertices and edges, together with a unary predicate that distinguishes vertices from edges. This alternative representation brings something new, namely the possibility of quantifying over vertices *and* edges, whereas, in the first representation, quantifications are restricted to vertices. We shall see below that certain graph properties cannot be expressed without quantifications over edges. Rather than structures with a single domain of vertices and edges, we shall use two-sorted structures, i.e., many-sorted structures with two domains.

1.2. *The class of graphs* $\boldsymbol{D}(A)$

Let A be a finite alphabet. We denote by $\boldsymbol{D}(A)$ the class of finite or infinite directed graphs, the edges of which are labelled with labels from A. These graphs may have multiple edges and loops. A graph G in $\boldsymbol{D}(A)$ can be represented by a logical structure in different ways. Let $|G|_1$, $|G|_2$, $|G|_3$ be the following logical structures:

$$|G|_1 := \langle \boldsymbol{V}_G, \boldsymbol{E}_G, (\mathbf{edg}_{aG})_{a \in A} \rangle,$$
$$|G|_2 := \langle \boldsymbol{V}_G, \boldsymbol{E}_G, (\mathbf{lab}_{aG})_{a \in A}, \mathbf{edg}_G \rangle,$$
$$|G|_3 := \langle \boldsymbol{V}_G, \boldsymbol{E}_G, (\mathbf{lab}_{aG})_{a \in A}, (\mathbf{edg}_{aG})_{a \in A}, \mathbf{edg}_G \rangle.$$

In each case, $\boldsymbol{V}_G$ is the set of vertices of G and is the domain of sort $\boldsymbol{v}$. The set $\boldsymbol{E}_G$ is its set of edges and is the domain of sort $\boldsymbol{e}$. These structures are defined in terms of the following relations:

$$\begin{aligned} &\mathbf{lab}_{aG}(x) &&:\Leftrightarrow \quad x \text{ is an edge with label } a,\\ &\mathbf{edg}_G(x, y, z) &&:\Leftrightarrow \quad x \text{ is an edge linking } y \text{ to } z,\\ &\mathbf{edg}_{aG}(x, y, x) &&:\Leftrightarrow \quad \mathbf{lab}_{aG}(x) \wedge \mathbf{edg}_G(x, y, z). \end{aligned}$$

Each of these structures describes G completely, and $|G|_3$ does that redundantly since $\mathbf{edg}_{aG}$ is expressible in terms of the other relations.

We can now write down the graph property considered in Subsection 1.1. Let ψ be the formula

$$\forall x\, \exists y\, \exists z \left[\bigvee \{\mathbf{edg}_a(z, x, y) \vee \mathbf{edg}_a(z, y, x) \mid a \in A\} \wedge \neg(x = y) \right].$$

(We denote by $\bigvee$ the disjunction of a set of formulas. We assume that x and y are two variables of sort $\boldsymbol{v}$ and that the variable z is of sort $\boldsymbol{e}$. This hypothesis implies that the quantifications over x and y are restricted to $\boldsymbol{V}_G$, and that the quantifications over z are restricted to $\boldsymbol{E}_G$.) For a graph G in $\boldsymbol{D}(A)$, we have $|G|_1 \models \psi$ iff G has no isolated vertex. With appropriate modifications of ψ, one can easily express the same property of a graph G in $\boldsymbol{D}(A)$ when it is represented by the structures $|G|_2$ or $|G|_3$. It is clear that the same properties of the graphs in $\boldsymbol{D}(A)$ can be written in first-order logic, for all three representations defined above.

The following definitions concern not only the graphs and the graph representations of Subsections 1.1 and 1.2, but also the general notions of a graph, of a graph property, and of the representation of a graph by a logical structure. These definitions apply in particular to other notions of graphs we shall consider below.

1.3. Graph properties

Graphs are defined up to isomorphism: two graphs G and G' are *isomorphic* if there exist two bijections: $\boldsymbol{V}_G \rightarrow \boldsymbol{V}_{G'}$ and $\boldsymbol{E}_G \rightarrow \boldsymbol{E}_{G'}$ that preserve incidence relations and labels. By a *graph property*, we mean a predicate on a class of graphs that is stable under isomorphism. (If we need to consider a class of graphs whose vertices are specific objects like integers, then we let these integers be labels attached to vertices. Hence, there is no loss of generality when considering two isomorphic graphs as equal.)

When representing a graph G by a logical structure $|G|$ we put into $|G|$ some relations concerning the labelling of vertices and edges, and incidences. Hence, if two graphs G and G' are isomorphic, the associated structures $|G|$ and $|G'|$ are also isomorphic.

Let $\mathscr{C}$ be a given class of graphs. We say that $|G|$ *represents* a graph G belonging to $\mathscr{C}$ if, for every graph G' in $\mathscr{C}$, if $|G|$ is isomorphic to $|G'|$ then G is isomorphic (hence equal) to G'. Since logical formulas do not distinguish between isomorphic structures, they define graph properties.

Hence, given a class of graphs $\mathscr{C}$, a representation of its graphs by logical structures and a logical language $\mathscr{L}$, the natural question is: *what graph properties of the graphs of $\mathscr{C}$ can be expressed in $\mathscr{L}$ via the considered representation*? We shall see that, for certain logical languages $\mathscr{L}$, the expressibility of a graph property in $\mathscr{L}$ has consequences for its decidability on certain classes of graphs and its complexity.

A set of graphs L is *$\mathscr{L}$-definable*, where $\mathscr{L}$ is some logical language, iff there is a formula φ in $\mathscr{L}$, such that $L=\{G \mid |G| \models \varphi\}$. This definition assumes that a logical structure $|G|$ has been chosen to represent a graph G of the considered class. It is usually known from the context. We shall now describe three main logical languages and a few graph properties that they are able to express (or not).

1.4. First-order graph properties

We first consider *first-order logic*, denoted by **FOL**. The graph property considered in Subsections 1.1 and 1.2 is written in first-order logic. Here are a few examples of properties of a graph G in the class $\boldsymbol{D}(A)$ that are first-order expressible:

- G is simple;

- G is k-regular for some fixed integer k;
- G is of degree at most k for some fixed integer k;
- G is edge-colored by A, i.e., any two edges of G having the same label have no vertex in common.

The following properties are not expressible in **FOL**:

- G is connected;
- G is planar;
- G is Hamiltonian.

(See also Table 1 in Subsection 1.6.) First-order logic is rather weak, and perhaps of little use for graph theory. Intuitively, first-order formulas can only express *local properties of graphs*. This notion has been made precise by Gaifman [45] in the following result.

1.1. THEOREM. *A property of the graphs of* $\boldsymbol{S}$ *is expressible by a closed first-order formula iff it is equivalent to a Boolean combination of properties of the form*:

$$\exists v_1,\ldots,v_s\left[\bigwedge\{P(N(v_i,r))\mid i\in[s]\}\wedge\bigwedge\{d(v_i,v_j)>2r\mid 1\leqslant i<j\leqslant s\}\right].$$

In this statement, $v_1,\ldots,v_2$ denote vertices, $d(x, y)$ denotes the distance of two vertices, i.e., the minimal length of a path from x to y (where the edges can be traversed in either direction), r denotes an integer, $N(x, r)$ denotes the r-neighborhood of x, i.e., the subgraph of the graph under consideration consisting of all vertices of distance at most r of x and of all the edges linking these vertices, and P is a first-order expressible property. Since the condition $d(x, y)\leqslant r$ is first-order expressible, the "iff" direction is rather obvious. The other direction is proved in [45], in a more general form applicable to arbitrary relational structures and to graphs in $\boldsymbol{D}(A)$.

Theorem 1.1 is useful for proving that certain properties of finite graphs, such as the ones mentioned above, are not expressible in first-order logic. (That "connectedness" of a finite *or infinite* graph is not first-order expressible is provable by a standard compactness argument.)

1.5. Monadic second-order graph properties

The second logical language we wish to consider is *monadic second-order logic*, denoted by **MSOL**. This language is an extension of first-order logic and a restriction of second-order logic, to be discussed below. It uses set variables, denoting sets of vertices or sets of edges and, of course, quantifications over them. The atomic formulas concerning these new variables (always denoted by upper case letters, whereas the other variables are denoted by lower case letters) are of the form $x\in X$.

The following formula expresses 2-colorability of a graph G in $\boldsymbol{D}(A)$ represented by the structure $|G|_2$ or $|G|_3$ (the variables X, Y, x, y, are of sort $\boldsymbol{v}$ and the variable z is of sort $\boldsymbol{e}$):

$$\begin{aligned}\exists X,\ Y[\forall x\{x\in X\vee x\in Y\}\wedge\forall x\{\neg(x\in X\wedge x\in Y)\}\\ \wedge\forall x\forall y\forall z\{\mathbf{edg}(z,x,y)\Rightarrow\neg(x\in X\wedge y\in X)\wedge\neg(x\in Y\wedge y\in Y)\}].\end{aligned}$$

This formula expresses the existence of a vertex-coloring of a graph using two colors: if two sets X and Y exist as demanded, then a vertex in X is said to have color 1, and a vertex in Y is said to have color 2.

No result comparable to Gaifman's Theorem (Theorem 1.1) is known for monadic second-order logic. The main tool for showing a graph property to be expressible in monadic second-order logic is the following lemma.

1.2. LEMMA. *If a binary relation is definable in monadic second-order logic, then so is its reflexive and transitive closure.*

PROOF. Let R be a binary relation on a set D (typically V_G, the set of vertices of a graph G). We say that a subset X of D is *R-closed* iff, for every t in X and u in D, if $(t, u) \in R$ then $u \in X$. For every $x \in D$ there exists a smallest R-closed subset of D containing x. Let us denote it by Y. It is clear that (x, y) belongs to R^* (the reflexive and transitive closure of R) iff $y \in Y$.

From a monadic second-order formula defining R it is not hard to construct a monadic second-order formula ψ, with free variables x, y, and Y, expressing the following:

$$x \in Y \wedge \text{“}Y \text{ is } R\text{-closed”} \wedge \forall X[(x \in X \wedge \text{“}X \text{ is } R\text{-closed”}) \Rightarrow \text{“}Y \subseteq X\text{”}].$$

Then the formula $\exists Y[y \in Y \wedge \psi]$ expresses that $(x, y) \in R^*$. □

From this lemma, it follows that many properties concerning paths can be expressed in monadic second-order logic, in particular:

- the existence of a path from x to y, all the edges of which are in a set U;
- connectedness, strong connectedness;
- the existence of circuits;
- the property of being a tree;
- the existence of a Hamiltonian path linking x to y, where x and y are two distinct vertices.

The last property can be written as follows: *there exists a set of edges U such that* [{*there exists a path from x to y, all edges of which are in U, and U is minimal for inclusion with this property*} *and* {*every vertex belongs to some edge of U*}].

Planarity is expressible in monadic second-order logic via Kuratowski's Theorem (or a variant of it established by Wagner [90]; see Section 5). Some other useful graph properties are not expressible in monadic second-order logic [26]:

- the existence of a nontrivial graph automorphism;
- the property that a graph has as many edges labelled by a as by b (the reason being that one cannot "count" in monadic second-order logic).

One cannot even count modulo an integer: it cannot be expressed that a graph has an even number of vertices [26].

Quantifying over edges is essential for expressing certain properties. For example, in monadic second-order logic one cannot express that a graph G in **S**, represented as in Subsection 1.1, has a Hamiltonian circuit. To see this, consider the family of bipartite graphs $K_{n,m}$ having n "blue" vertices, m "red" ones, edges from any blue vertex to any red one, edges from any red one to any blue one, and no other edges. This family is definable by an **MSOL** formula. $K_{n,m}$ has a Hamiltonian circuit iff $n=m$. If the property of having a Halmiltonian circuit would be definable in **MSOL** (without edge quantifications), then it could be expressed that two sets have the same cardinality. We know that this is not the case.

1.6. *Second-order graph properties*

The third logical language we wish to present is *second-order logic*, denoted by **SOL**. This language uses variables denoting n-ary relations, $n \geqslant 2$, in addition to variables denoting vertices or edges and sets of edges or vertices. Note that in second-order logic, the definability of the transitive closure of a definable binary relation R (i.e., the analog of Lemma 1.2) is straightforward: it suffices to express its definition, being the smallest transitive relation containing R.

We first show how the existence of bijections can be expressed in **SOL**. By first-order quantifications it can be expressed that a given binary relation is a one-to-one function. In **SOL**, one can quantify over binary relations; hence, one can express that there *exists a bijection having a certain property*.

Let us construct a formula expressing that a graph G of $\boldsymbol{D}(A)$ has as many edges labelled with a as with b. We first introduce an auxiliary formula φ_1 expressing that a binary relation X is a one-to-one partial mapping. Let φ_1 be

$$\forall x, y, z\, [(x, y) \in X \wedge (x, z) \in X \Rightarrow y = z] \wedge \forall x, y, z\, [(y, x) \in X \wedge (z, x) \in X \Rightarrow y = z].$$

The desired formula is then

$$\exists X\, [\varphi_1 \wedge \forall x \{\mathbf{lab}_a(x) \Leftrightarrow \exists y\, (x, y) \in X\} \wedge \forall y\, \{\mathbf{lab}_b(y) \Leftrightarrow \exists x\, (x, y) \in X\}].$$

One can also express that a graph has a nontrivial automorphism or that a graph is finite. For the latter property, one writes that $\boldsymbol{V}_G$ has no proper subset in bijection with itself, and similarily for $\boldsymbol{E}_G$.

To take a more complicated property, let us consider the notion of a *perfect graph*. For every graph G, let $\omega(G)$ be the *chromatic number* of G (i.e., the minimum number of colors that is necessary to color the vertices of G) and $\kappa(G)$ the maximum number of vertices of a clique in G. It is clear that $\kappa(G) \leqslant \omega(G)$. Equality holds iff $\omega(G) \leqslant \kappa(G)$, i.e., iff G satisfies the following: there is a binary relation $R \subseteq \boldsymbol{V}_G \times \boldsymbol{V}_G$ such that

- the induced subgraph with set of vertices $\mathbf{Im}(R)$ is a clique,
- the relation R is functional and defines a coloring of $\boldsymbol{V}_G$ with set of colors $\mathbf{Im}(R)$ (x has color y iff $(x, y) \in R$).

By $\mathbf{Im}(R)$ we denote the set $\{y \in \boldsymbol{V}_G \mid (x, y) \in R \text{ for some } x \text{ in } \boldsymbol{V}_G\}$. From this formulation, it is easy to construct a second-order formula expressing that $\kappa(G) = \omega(G)$. A graph G is *perfect* iff, for every induced subgraph G' of G, one has $\kappa(G') = \omega(G')$. Hence, this property is expressible in **SOL**. If the strong perfect graph conjecture holds, then the perfectness of a graph becomes expressible in **MSOL**.

The expressive powers of the three languages discussed thus far can be compared as follows: **FOL** $\subset$ **MSOL** $\subset$ **SOL**. Table 1 gives examples establishing that the inclusions are strict.

Table 1

Properties of a graph G in $\mathbf{D}(A)$	**FOL**	**MSOL**	**SOL**
– simple – k-regular – degree $\leqslant k$	yes	yes	yes
– connected – planar – k-colorable – Hamiltonian – is a tree	no	yes	yes
– has an even number of vertices – has as many a's as b's – has a nontrivial automorphism – finite	no	no	yes
– card (V_G) belongs to a given nonrecursive set	no	no	no

1.7. *Theories*

If $\mathscr{L}$ is a logical language, and $\mathscr{C}$ is a class of graphs, or more precisely, a class of structures representing the graphs of a class, then we let $\mathbf{Th}_{\mathscr{L}}(\mathscr{C})$ denote the *$\mathscr{L}$-theory* of $\mathscr{C}$, i.e., the set of closed formulas φ of $\mathscr{L}$ such that $G \models \varphi$ for all G in $\mathscr{C}$. Graph theorists might hope to solve conjectures by logical techniques. In particular, the 4-color conjecture can be expressed by a monadic second-order formula φ saying that "if G is planar, then it is 4-colorable". Hence, it is equivalent to the statement that φ belongs to $\mathbf{Th}_{\mathbf{MSOL}}(\mathbf{D}(A))$. Unfortunately, undecidability is the common situation. (By $\mathbf{D}_{\mathrm{f}}(A)$ we denote the set of finite graphs in $\mathbf{D}(A)$.)

1.3. THEOREM (Trakhtenbrot [89]). $\mathbf{Th}_{\mathbf{FOL}}(\mathbf{D}_{\mathrm{f}}(A))$ *is undecidable.*

Since, as we have seen above, **FOL** is a rather weak language for expressing graph properties, it is not very interesting to define restrictions $\boldsymbol{L}$ of **FOL** for which $\mathbf{Th}_{\mathrm{L}}(\mathbf{D}_{\mathrm{f}}(A))$ becomes decidable. It is more interesting to restrict the classes of graphs under consideration:

1.4. THEOREM (Courcelle [23, 24, 26]). $\mathbf{Th}_{\mathbf{MSOL}}(\mathscr{C})$ *is decidable if* $\mathscr{C}$ *is a subset of* $\mathbf{D}_{\mathrm{f}}(A)$ *defined by a hyperedge-replacement graph-grammar.*

(The hyperedge-replacement graph-grammars will be introduced below in Section 3.) Theorem 1.4 cannot be applied to the 4-color conjecture because neither the set $\mathbf{D}_{\mathrm{f}}(A)$ nor the set of finite planar graphs are generated by appropriate grammars. Furthermore, Theorem 1.5 below will show that this failure cannot be remedied by the discovery of a more powerful notion of graph grammar.

1.5. THEOREM (Seese [86]). *Let $\mathscr{C} \subseteq \boldsymbol{D}(A)$ be a class of (finite or infinite) graphs having a decidable* **MSOL**-*theory. The graphs in $\mathscr{C}$ are of bounded tree-width.*

The tree-width of a graph will be defined in Section 5. Let us mention here that the smaller the tree-width of a graph G is, the closer G is to a tree. Finite planar graphs are of unbounded tree-width. Hence, conjectures of the form: "φ holds for all finite planar graphs", with φ written in monadic second-order logic, cannot be decided by any result as in Theorem 1.4, even for more powerful graph-generating mechanisms.

2. Graph operations and graph expressions

By a *graph operation* we mean any mapping that associates a graph with one or several graphs. Typical examples are the addition of an edge to a graph at some specified place, the contraction of a certain edge, or the gluing together of two graphs in some precise way. Graphs can be constructed by finitely (or even infinitely) many applications of certain graph operations, starting from finitely many given finite graphs. For example the set of series-parallel graphs can be defined in this way by means of two operations called *series-composition* and *parallel-composition* (see Example 2.4).

Whenever graph operations are defined on a set $\mathscr{C}$ of graphs, an *algebra* (in the sense of universal algebra) is obtained. This aspect of graph operations has not been considered much. Before starting our exposition, let us stress one important point: In order to apply the techniques of universal algebra, we need *deterministic operations*; hence, the addition of an edge "anywhere" in a graph is not an operation because an edge can be added to a graph in many different ways. It becomes an operation only if the vertices to be linked by this new edge are specified. This necessity motivates the introduction of *sources*, i.e., of distinguished vertices to which new edges can be "glued".

This section will be devoted to the algebraic structure introduced in [9, 22]. It is intimately related to the class of hyperedge-replacement graph-grammars in a precise sense (see Theorem 3.4). Another algebraic structure is presented in [31] which is similarly related to another class of graph-grammars. For the reasons explained above, we need to work with graphs with sources. We give a two-step generalization of the class $\boldsymbol{D}(A)$ introduced in Subsection 1.2.

2.1. *Graphs with sources*

For every $k \geqslant 1$, we denote by $\boldsymbol{D}(A)_k$ the set of pairs $H = \langle G,s \rangle$ consisting of a graph G in $\boldsymbol{D}(A)$ and a sequence s of k vertices of G. This sequence may have repetitions. Its elements are called the *sources* of H. The term "source" is just a convenient word for "distinguished vertex". There is no notion of flow involved. The sources of a graph must rather be understood as "gluing vertices". The graph operations described in Subsection 2.3 make it possible to build a graph by "gluing" two graphs at their sources.

An element H of $\boldsymbol{D}(A)_k$ is called a *k-graph* or a *graph of type k*. It can also be defined as a five tuple $H = \langle V_H, E_H, \mathbf{lab}_H, \mathbf{vert}_H, \mathbf{src}_H \rangle$ where $\mathbf{lab}_H : E_H \to A$ defines the label of an edge, $\mathbf{vert}_H: \boldsymbol{E}_H \rightarrow \boldsymbol{V}_H \times \boldsymbol{V}_H$ defines the pair of vertices of an edge, and $\mathbf{src}_H$ is the sequence of sources of H. The set of finite k-graphs is denoted by $\mathbf{FD}(A)_k$.

A word $w = a_1 a_2 \ldots a_k$ in A^* can be represented by the following 2-graph denoted by $G(w)$:

$$1 \bullet \xrightarrow{a_1} \bullet \xrightarrow{a_2} \bullet \cdots \bullet \xrightarrow{a_k} \bullet 2$$

The empty word corresponds to a graph with no edges and a single vertex, which is simultaneously the first and the second source.

We now describe a second extension of $\boldsymbol{D}(A)$.

2.2. *Hypergraphs with sources*

In a directed graph, an edge is a sequence of two (possibly identical) vertices. In a directed hypergraph, an edge has a sequence of k vertices, where k may be any nonnegative integer, called the *type* of the hyperedge. A hyperedge of type 2 is an edge; a hyperedge of type 1 is a label attached to a vertex. We may also have hyperedges of type 0, having no vertex.

The hypergraphs we shall define below have *labelled hyperedges*. They also have *sources*, like the graphs of Subsection 2.1. The alphabet of hyperedge labels is a *ranked alphabet* B, i.e., an alphabet that is given with a type mapping $\tau: B \rightarrow \mathbb{N}$. The *type* of a hyperedge must be equal to the type of its label.

A *hypergraph over B of type n* is a five tuple $H = \langle \boldsymbol{V}_H, \boldsymbol{E}_H, \mathbf{lab}_H, \mathbf{vert}_H, \mathbf{src}_H \rangle$ where $\boldsymbol{V}_H$ is the set of vertices, $\boldsymbol{E}_H$ is the set of hyperedges, $\mathbf{lab}_H$ is a mapping: $\boldsymbol{E}_H \rightarrow B$ defining the label of a hyperedge, $\mathbf{vert}_H$ is a mapping: $\boldsymbol{E}_H \rightarrow \boldsymbol{V}_H^*$ defining the sequence of vertices of a hyperedge, and $\mathbf{src}_H$ is a sequence of n vertices of H. We impose the condition that the length of $\mathbf{vert}_H(e)$ is equal to $\tau(\mathbf{lab}_H(e))$ for all e in $\boldsymbol{E}_H$. One may also have labels of type 0 labelling hyperedges with no vertex. An element of $\mathbf{src}_H$ is called a *source* of H. We denote by $\mathbf{G}(B)_n$ the set of all hypergraphs over B of type n and by $\mathbf{FG}(B)_n$ the set of finite ones. A hypergraph of type n is also called an *n-hypergraph*.

We introduce hypergraphs for the following reasons. Hypergraphs are useful as right-hand sides of production rules in graph-grammars, even if one only wants to generate subsets of $\boldsymbol{D}(A)_0$, i.e., sets of graphs. A second reason is that hypergraphs are useful by themselves: they offer natural representations of relational databases [25]; of flowcharts [85]; of terms and terms with sharing, i.e., dag's [25]; of networks of constraints [71]; and of VLSI layouts [53], to mention only a few examples.

2.3. *Operations on hypergraphs*

If G is an n-hypergraph and G' is an m-hypergraph, then $G \oplus G'$ denotes their *disjoint union* equipped with the concatenation of $\mathbf{src}_G$ and $\mathbf{src}_{G'}$ as the sequence of sources. (This operation is not commutative.)

If δ is an equivalence relation on $[n]$ and G is an n-hypergraph, then $\theta_\delta(G)$ is the n-hypergraph obtained from G by *fusing its ith and jth sources* for every (i,j) in δ. (By $[n]$ we denote the set $\{1, 2, \ldots n\}$.)

If $\alpha: [p] \rightarrow [n]$ is a total mapping and $G \in \boldsymbol{G}(B)_n$, then $\sigma_\alpha(G)$ is the hypergraph in $\boldsymbol{G}(B)_p$

consisting of G equipped with $(\mathbf{src}_G(\alpha(1)), \ldots, \mathbf{src}_G(\alpha(p)))$ as the sequence of sources. (By $\mathbf{src}_G(i)$ we denote the ith element of $\mathbf{src}_G$.)

Note that we write $\theta_\delta(G)$ and $\sigma_\alpha(G)$, and not $\theta(\delta, G)$ and $\sigma(\alpha, G)$: we do not consider δ and α as arguments of binary functions θ and σ, but rather as indices for families of operations (θ_δ), (σ_α). Our three operations $\oplus$, θ_δ, σ_α are actually operation schemes, defining infinitely many operations. This is a key feature of our algebraic treatment.

The three operation schemes define a *many-sorted algebra of hypergraphs* $\mathbf{G}(B)$: $\mathbb{N}$ is the (infinite) set of sorts, $\mathbf{G}(B)_n$ is the domain (or carrier) of sort n, and the three operation schemes define an infinite $\mathbb{N}$-signature $\mathbf{H}$, consisting of

$$\oplus_{n,m} \text{ of profile } (n,m)\rightarrow n+m,$$
$$\theta_{\delta,n} \text{ of profile } n\rightarrow n,$$
$$\sigma_{\alpha,p,n} \text{ of profile } n\rightarrow p.$$

(We refer the reader to [92, 37] for many-sorted algebras.)

For every $b \in B$, we also denote by b the $\tau(b)$-hypergraph H consisting of one edge e, labelled by b and with $\mathbf{src}_H = \mathbf{vert}_H(e)$. We denote by $\mathbf{1}$ the 1-graph consisting of one vertex that is its unique source. We denote the empty graph by $\mathbf{0}$. We let $\boldsymbol{H}_B := \boldsymbol{H} \cup B \cup \{\mathbf{1}, \mathbf{0}\}$.

2.1. THEOREM (Bauderon and Courcelle [9], Courcelle [22]). *Every finite hypergraph over B (can be) denoted by an algebraic expression constructed over* $\boldsymbol{H}_B$.

We denote by $\mathbf{FE}(B)_n$ the set of all such expressions that are of sort n. We call them *hypergraph expressions* (or *expressions* for short when there is no risk of confusion) *of sort n*. We denote by $\mathbf{val}(g)$ the finite hypergraph defined by an expression g.

When writing expressions, we omit the subscripts n, m, in the operation symbols $\oplus_{n,m}$, $\sigma_{\alpha,n,m}$ and $\theta_{\delta,n}$. Provided the sorts of the variables appearing in an expression are known, its sort can be computed and its well-formedness can be checked. We also consider $\oplus$ as an infix associative operation symbol, and we omit parentheses. When δ is generated by a single pair (i, j), we write $\theta_{i,j}$ for θ_δ. We also write $\sigma_{i_1,i_2,\ldots,i_p}$ for σ_α if $\alpha: [p]\rightarrow[n]$ and $\alpha(1)=i_1, \ldots, \alpha(p)=i_p$.

2.2. EXAMPLE. Figure 1 shows a 3-hypergraph G such that $V_G = \{u, w, x, y, z\}$, E_G

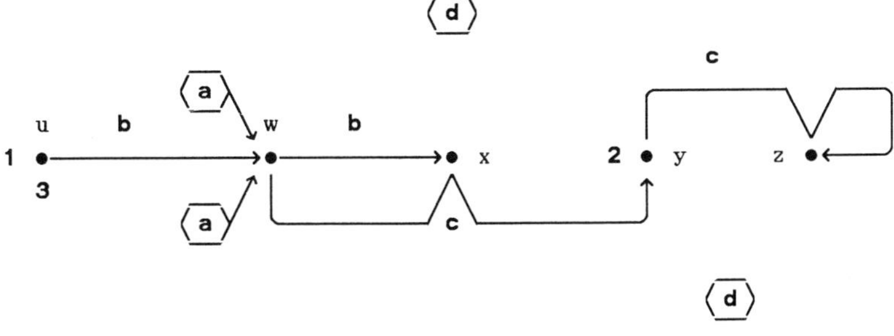

Fig. 1.

consists of eight hyperedges labelled by a, b, c, d, such that $\tau(a)=1$, $\tau(b)=2$, $\tau(c)=3$, $\tau(d)=0$. The sequence $\mathbf{src}_G$ is (u, y, u). When drawing G, we use the following conventions:

- edges are represented as usual;
- unary hyperedges are labels attached to vertices (the same vertex may be tagged with several labels or several occurrences of the same label);
- hyperedges of rank greater than 2 are represented as edges with intermediate nodes (in Fig. 1, there is one hyperedge e such that $\mathbf{vert}_G(e)=(w, x, y)$ and another one, e', with $\mathbf{vert}_G(e')=(y, z, z)$);
- nullary hyperedges are represented as circles floating around.

The positions of the source vertices in $\mathbf{src}_G$ are indicated by integers. The hypergraph is the value of the expression

$$g=\sigma_\alpha(\theta_\delta(a \oplus b \oplus a \oplus b \oplus c \oplus c)) \oplus d \oplus d$$

where δ is the equivalence relation on [12] generated by $\{(1,3), (3,4), (3,5), (3,7), (6,8), (9,10), (11,12)\}$, and α maps 1 to 2, 2 to 10 and 3 to 2.

2.4. *The width of a hypergraph*

The *width* $\mathbf{wd}(g)$ of g in $\mathbf{FE}(B)_n$ is the maximal sort of a symbol of $\boldsymbol{H}_B$ occurring in g. The *width* of a finite n-hypergraph G is $\mathbf{wd}(G) := \min\{\mathbf{wd}(g) \mid g \in \mathbf{FE}(B)_n, \mathbf{val}(g)=G\}$.

The signature $\boldsymbol{H}$ is infinite. From the following lemma, one can see that no finite subset of $\boldsymbol{H}$ can generate the set of all finite hypergraphs of a given type.

2.3. Lemma (Bauderon and Courcelle [9]). *If G is a finite clique with n vertices, then* $\mathbf{wd}(G) \geqslant n$.

In the sequel, we will omit the mention of B, except when necessary. Hence, $\boldsymbol{G}_n$ denotes $\boldsymbol{G}(B)_n$. By $\mathbf{FG}_n^k$ we denote the set $\{G \in \mathbf{FG}_n \mid \mathbf{wd}(G) \leqslant k\}$. By $\mathbf{FG}$ we denote the subalgebra of $\boldsymbol{G}$ consisting of finite hypergraphs.

Hypergraph expressions as in Example 2.2 are not very readable. They are necessarily long because the three operations $\oplus$, σ_α, θ_δ are very elementary. These operations, however, can be used to build new operations, called *derived operations*, appropriate for various purposes.

2.5. *Derived operations*

Let X be an $\mathbb{N}$-sorted set of variables, i.e., a set X wih a mapping $\tau: X \to \mathbb{N}$. A variable x will take its values in $\boldsymbol{G}(B)_{\tau(x)}$. By $\mathbf{FE}(B, X)_n$ we denote the set of finite well-formed terms of sort n, constructed with $\boldsymbol{H}_B$ and X. The set X is always finite and enumerated in a fixed way as $\{x_1, \ldots, x_m\}$ (or also, to avoid subscripts, as $\{x, y, z\}$ if $m=3$). It follows that an expression e in $\mathbf{FE}(B, X)_n$ denotes a mapping, called a *derived operation*:

$$\mathbf{derop}_{\boldsymbol{G}(B)}(e): \boldsymbol{G}(B)_{\tau(x_1)} \times \cdots \times \boldsymbol{G}(B)_{\tau(x_m)} \to \boldsymbol{G}(B)_n.$$

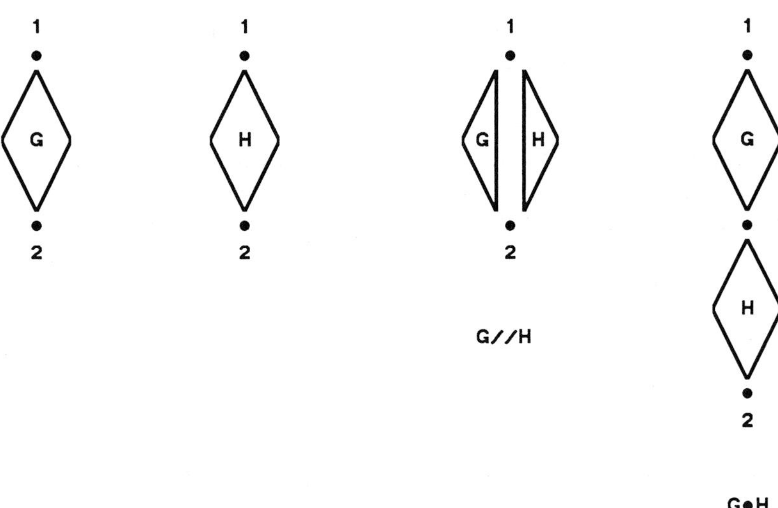

Fig. 2.

2.4. EXAMPLE (*series- and parallel-composition*). Consider the two operations $\boldsymbol{G}(B)_2 \times \boldsymbol{G}(B)_2 \rightarrow \boldsymbol{G}(B)_2$ illustrated in Fig. 2 and defined as follows:

$$G \,||\, G' = \sigma_{1,2}(\theta_{1,3}(\theta_{2,4}(G \oplus G'))), \qquad G \cdot G' = \sigma_{1,4}(\theta_{2,3}(G \oplus G')).$$

They are the classical parallel- and series-composition of 2-graphs. They are the derived operations associated with the graph expressions with variables $e_p := \sigma_{1,2}(\sigma_{1,3}(\theta_{2,4}(x \oplus y)))$ and $e_s := \sigma_{1,4}(\theta_{2,3}(x \oplus y))$, where x and y are variables of sort 2, i.e., ranging on $\boldsymbol{G}(B)_2$.

The graph operations satisfy some algebraic laws that can be described by eleven axiom schemes which we do not list here (they are given in [9]). Each scheme defines an infinite set of equational axioms. Let R be the set of all equational axioms defined by the equation schemes, and $\overset{*}{\underset{R}{\leftrightarrow}}$ the associated (symmetric) rewriting relation (see [32] on term rewriting systems).

2.5. THEOREM (Bauderon and Courcelle [9]). (1) *Two hypergraph expressions g and g' define the same hypergraph iff $g \overset{*}{\underset{R}{\leftrightarrow}} g'$.*

(2) *Two expressions with variables g and g' define the same derived operations iff $g \overset{*}{\underset{R}{\leftrightarrow}} g'$.*

The first assertion means that $\mathbf{FG}(B)$ is the initial $\langle \mathbb{N}, \boldsymbol{H}_B, R \rangle$-algebra. (By an $\langle \mathbb{N}, \boldsymbol{H}_B, R \rangle$-algebra we mean an $\boldsymbol{H}_B$-algebra in which the axioms of R are valid; the triple $\langle \mathbb{N}, \boldsymbol{H}_B, R \rangle$ is an *equational specification* as defined in [37, 92]. The second assertion means that R provides a *complete equational axiomatization* of the algebraic properties of the operations of $\boldsymbol{G}(B)$.

As we do not give R, we do not prove Theorem 2.5. The proof, given in [9], is quite long and technical. But we shall show how the second assertion of Theorem 2.5 reduces to the first one. For this purpose, we introduce *hyperedge replacements*. This notion is also essential for the class of grammars that we shall consider in Section 3.

2.6. *Hyperedge replacements*

Let $G \in \boldsymbol{G}(B)$, $e \in \boldsymbol{E}_G$ and let $H \in \boldsymbol{G}(B)$ be a hypergraph of type $\tau(e)$. We denote by $G[H/e]$ the result of the *replacement* (or *substitution*) *of H for e in G*. This hypergraph can be constructed as follows:

- first construct G' by deleting e from G (but keep the vertices of e);
- add to G' a copy $\bar{H}$ of H, disjoint from G';
- fuse the vertex $\mathbf{vert}_G(e, i)$ (which is still a vertex of G') with the ith source of $\bar{H}$; do this for all $i=1, \ldots, \tau(e)$.

The sequence of sources of $G[H/e]$ is that of G'.

If $e_1, \ldots, e_k$ are pairwise distinct hyperedges of G and if $H_1, \ldots, H_k$ are hypergraphs of respective types $\tau(e_1), \ldots, \tau(e_k)$, then the replacements in G of H_1 for $e_1, \ldots, H_k$ for e_k can be done in any order: the result is the same and it is denoted by $G[H_1/e_1, \ldots, H_k/e_k]$. This operation is the *simultaneous* substitution (or replacement) of H_1 for $e_1, \ldots, H_k$ for e_k in G.

Finally, if $b_1, \ldots, b_k$ are pairwise distinct elements of B, if $H_1, \ldots, H_k$ are of respective types $\tau(b_1), \ldots, \tau(b_k)$, and if G is finite, then we denote by $G[H_1/b_1, \ldots, H_k/b_k]$ the result of the simultaneous substitution of H_i for all hyperedges labelled by b_i for all $i=1, \ldots, k$.

Let us now consider an expression with variables g in $\mathbf{FE}(B, X)_n$. It also belongs to $\mathbf{FE}(B \cup X)_n$. This means that the variables can also be considered as hyperedge labels, hence that $\mathbf{val}(g)$ is a hypergraph in $\mathbf{FG}(B \cup X)_n$. If $X=\{x_1, \ldots, x_m\}$ we have the following lemma.

2.6. LEMMA. *Let* $g \in \mathbf{FE}(B, X)_n$.

(1) *If* $H_1, \ldots, H_m \in \mathbf{G}(B)$ *with* $\tau(H_i)=\tau(x_i)$ *for all* $i=1, \ldots, m$, *then*

$$\mathbf{derop}(g)(H_1, \ldots, H_m)=\mathbf{val}(g)[H_1/x_1, \ldots, H_m/x_m].$$

(2) *If* $h_1, \ldots, h_m \in \mathbf{FE}(B)$ *with* $\tau(h_i)=\tau(x_i)$ *for all* $i=1, \ldots, m$ *then*

$$\mathbf{val}(g[h_1/x_1, \ldots, h_m/x_m])=\mathbf{val}(g)[\mathbf{val}(h_1)/x_1, \ldots, \mathbf{val}(h_m)/x_m].$$

(The proof of this lemma uses an induction on the structure of g.)

2.7. EXAMPLE (*series- and parallel-composition* (continued)). As an illustration of Lemma 2.6, we have $G \| G' = P[G/x, G'/y]$ and $G \cdot G' = S[G/x, G'/y]$, where $P=\mathbf{val}(e_p)$ and $S=\mathbf{val}(e_s)$. These graphs are

$$P = 1\bullet \underset{y}{\overset{x}{\rightrightarrows}} \bullet 2 \quad \text{and} \quad S = 1\bullet \overset{x}{\longrightarrow} \bullet \overset{y}{\longrightarrow} \bullet 2.$$

PROOF OF THEOREM 2.5((1)⇒(2)). Let g, $g' \in \mathbf{FE}(B,X)_n = \mathbf{FE}(B \cup X)_n$ be such that $\mathbf{derop}(g) = \mathbf{derop}(g')$. It follows from Lemma 2.6 that $\mathbf{val}(g) = \mathbf{derop}(g)(x_1, \ldots, x_m) = \mathbf{derop}(g')(x_1, \ldots, x_m) = \mathbf{val}(g')$. Hence, $g \overset{*}{\leftrightarrow} g'$ by Theorem 2.5(1).

Conversely, assume that $g \overset{*}{\underset{R}{\leftrightarrow}} g'$. We have $\mathbf{val}(g) = \mathbf{val}(g')$ and it follows immediately from Lemma 2.6 that $\mathbf{derop}(g) = \mathbf{derop}(g')$. □

We conclude this section by showing that graph and hypergraph rewriting rules defined in terms of push-out diagrams can also be expressed as ground rewriting systems of graph and hypergraph expressions. We first present the use of category theory for the formalization of graph rewriting rules, as introduced in [40].

2.7. *Rewriting rules in a category*

Let $\mathscr{C}$ be a category. A *$\mathscr{C}$-rewriting rule* is a pair of morphisms $p = (b_1, b_2)$ having the same domain. Actually, it is convenient to write such a pair as a five-tuple

$$p = \left(B_1 \xleftarrow{b_1} K \xrightarrow{b_2} B_2 \right)$$

that also indicates the domain K and the codomains B_1 and B_2 of b_1 and b_2.

If G and G' are objects in $\mathscr{C}$, we say that *G rewrites into G' via p* iff there exists a commutative diagram

$$\begin{array}{ccccc} B_1 & \xleftarrow{b_1} & K & \xrightarrow{b_2} & B_2 \\ g\downarrow & & \downarrow d & & \downarrow g' \\ G & \longleftarrow & D & \longrightarrow & G' \end{array}$$

where both squares are push-outs. We write this as $G \underset{p}{\Rightarrow} G'$. (We refer the reader to [68, 34, 35] for the definition of push-outs.)

This definition can be used by letting $\mathscr{C}$ be any category of hypergraphs. By a *category of hypergraphs*, we mean a category formed as follows: its objects are the elements of a certain set of hypergraphs, its homsets are certain sets of hypergraph homomorphisms. A homomorphism $g: B_1 \to G$ represents an occurrence of B_1 in G, and G' is the result of the substitution in G of B_2 for B_1. In [40] the extra condition that d is injective is also imposed. A major concern is clearly the existence and unicity of G' for p, g, G given as above. Another consideration is the efficiency of an algorithm constructing G'. These questions are dealt with in [36, 82] in particular.

For $\mathscr{C}$ we shall take the category of 0-hypergraphs over some ranked alphabet B with *hyperedge-injective homomorphisms*. We shall denote it by $\mathscr{E}$ in the sequel, assuming that B is known from the context. Let us specify that a homomorphism $h: G \to H$, where G and H are finite k-hypergraphs for some k, is a pair of mappings (h_V, h_E) such that

$$\begin{array}{ll} h_V: V_G \to V_H, \quad h_E: E_G \to E_H, & \\ \mathbf{lab}_H(h_E(e)) = \mathbf{lab}_G(e) & \text{for all } e \text{ in } E_G, \\ \mathbf{vert}_H(h_E(e), i) = h_V(\mathbf{vert}_G(e), i) & \text{for all } e \text{ in } E_G, \text{ all } i \text{ in } [\tau(e)], \\ \mathbf{src}_H(i) = h_V(\mathbf{src}_G(i)) & \text{for all } i \text{ in } [k]. \end{array}$$

The homomorphism is *hyperedge-injective* if h_E is injective. By $\mathbf{vert}_G(e,i)$ we denote the ith vertex of $\mathbf{vert}_G(e)$.) The $\mathscr{E}$-rewriting rules can be characterized in terms of ground rewriting systems of hypergraph expressions.

2.8. *Hypergraph rewriting rules*

We define a *hypergraph rewriting rule* as a pair $p=(g,g')$ of expressions of the same sort. The elementary rewriting step associated with a rewriting rule $p=(g,g')$ is defined as follows: $H \xrightarrow[p]{} H'$ iff there exist expressions h and h', defining respectively H and H', such that g appears as a subexpression of h and h' is the result of the substitution of g' in h for this occurrence of g.

A *semi-Thue system over hypergraphs* is a finite set P of rewriting rules. In such a case, we have

$$\{H' \in \mathbf{FG}(B) \mid H \xrightarrow[P]{*} H'\} = \{\mathbf{val}(h') \mid h (\xrightarrow[P]{} \cup \xleftrightarrow[R]{})^* h'\}$$

where h defines H. In other words, the theory of semi-Thue graph rewriting systems is reducible to the theory of ground rewriting systems on $\mathbf{FE}(B)$ modulo $\xleftrightarrow[R]{*}$. (We refer the reader to [32] for rewriting systems modulo equivalence.)

2.8. THEOREM (Bauderon and Courcelle [9]). *Every hypergraph rewriting rule is equivalent to an $\mathscr{E}$-rewriting rule and vice versa.*

PROOF (*sketch*). For every hypergraph rewriting rule $p=(g,g')$, let

$$\bar{p}=(\mathbf{val}(g) \xleftarrow{b} \boldsymbol{n} \xrightarrow{b'} \mathbf{val}(g'))$$

be the $\mathscr{E}$-rewriting rule, where n is the common sort of g and g', $\boldsymbol{n}$ is the graph $\mathbf{1}\oplus\mathbf{1}\oplus\cdots\oplus\mathbf{1}$ (with n vertices, n pairwise distinct sources and no edges), b is the unique graph homomorphism: $\boldsymbol{n}\rightarrow\mathbf{val}(g)$, and similarly for b'. It can be proved that, for all hypergraphs H and H' in $\mathbf{FG}(B)_0$,

$$H \underset{\bar{p}}{\Rightarrow} H' \quad \text{iff} \quad H \xrightarrow[p]{} H.$$

Conversely, if $q=(D \xleftarrow{d} K \xrightarrow{d'} D')$ is an $\mathscr{E}$-rewriting rule, one first proves that $q'=(D \xleftarrow{d} K' \xrightarrow{d'} D')$ defines the same rewriting relation as q, where K' is obtained from K by the deletion of its edges. It is then easy to find a graph rewriting rule $p=(g,g')$ such that $\bar{p}=q'$. This completes the proof. □

We can comment on the restriction to hyperedge-injective homomorphisms as follows. We consider a hypergraph as a set of hyperedges "glued" together by means of vertices, and not as a set of vertices connected by hyperedges. Hence, we consider that G "appears in H" if the hyperedges of G can be mapped injectively into $\boldsymbol{E}_H$. Of course, the "gluing" of the hyperedges and their labels must be preserved in this mapping. New gluings, however, can occur in the "embedding" of G into H. Our operations on hypergraphs are able to fuse vertices, but they do not fuse hyperedges. Hence, rewriting rules defined in a category with homomorphisms that are not hyperedge-injective cannot be expressed in terms of the expressions of Subsection 2.3, i.e., as hypergraph rewriting rules, according to Subsection 2.8. Perhaps they can be in terms of different (more powerful) expressions.

3. Context-free sets of hypergraphs

Hyperedge-replacement grammars, generate finite hypergraphs, as context-free grammars generate words. Their basic rewriting step is the replacement of a hypergraph for a hyperedge (as defined in Subsection 2.6), the replacement of a word for a letter thus extending to hypergraphs. These grammars are context-free in a precise sense to be discussed in Section 8. The sets they generate are the *context-free hyperedge replacements sets of hypergraphs*, or, more simply, the *context-free sets of hypergraphs*. Other notions of context-free grammars can be defined (see [31]), yielding other types of sets of hypergraphs that can also be called context-free. We characterize the context-free sets of hypergraphs as the *equational sets of hypergraphs*, i.e., as the components of least solutions of certain systems of equations to be solved in $\mathcal{P}(\mathbf{G}(B))$.

We shall examine a few easy closure properties of the family of context-free sets of hypergraphs. In the next section, we shall see that the intersection of a context-free set and an **MSOL**-definable set is context-free.

3.1. *Hyperedge replacement graph grammars*

A *hyperedge-replacement grammar* (an *HR grammar* for short) is a four-tuple $\Gamma = \langle B, U, P, Z\rangle$ where B is a finite *terminal ranked* alphabet, U is a finite *nonterminal ranked* alphabet, P is a finite set of *production rules*, i.e., a finite set of pairs of the form (u, D) where $D \in \mathbf{FG}(B \cup U)_{\tau(u)}$ and $u \in U$, and Z is the *axiom*, i.e., a hypergraph in $\mathbf{FG}(B \cup U)$. The set of hypergraphs defined by Γ is $L(\Gamma) := L(\Gamma, Z)$ where, for every hypergraph $K \in \mathbf{FG}(B \cup U)_n$,

$$L(\Gamma, K) := \{H \in \mathbf{FG}(B)_n \mid K \xrightarrow[P]{*} H\},$$

and $\xrightarrow[P]{}$ is the elementary rewriting step defined as follows: $K \xrightarrow[P]{} H$ iff there exist a hyperedge e in K whose label is some u in U and a production rule (u, D) in P such that $H = K[D/e]$, i.e, H is the result of the replacement of D for e in K (see Subsection 2.6).

A set of hypergraphs is *context-free* if it is defined by an HR grammar.

3.1. EXAMPLE (*Series-parallel graphs*). Let B be reduced to a single symbol a of type 2. The set SP of directed series-parallel graphs whose edges are labelled by a is the subset of $\mathbf{FG}(B)_2$ generated by the HR grammar Γ, the set of production rules of which is shown in Fig. 3. An example of a graph belonging to $L(\Gamma)$ is also shown in Fig. 3.

3.2. *The regular tree-grammars associated with an HR grammar*

The right-hand sides of the production rules of a grammar as in Subsection 3.1 can be written as expressions over $B \cup U$, where U is the set of nonterminal symbols. We

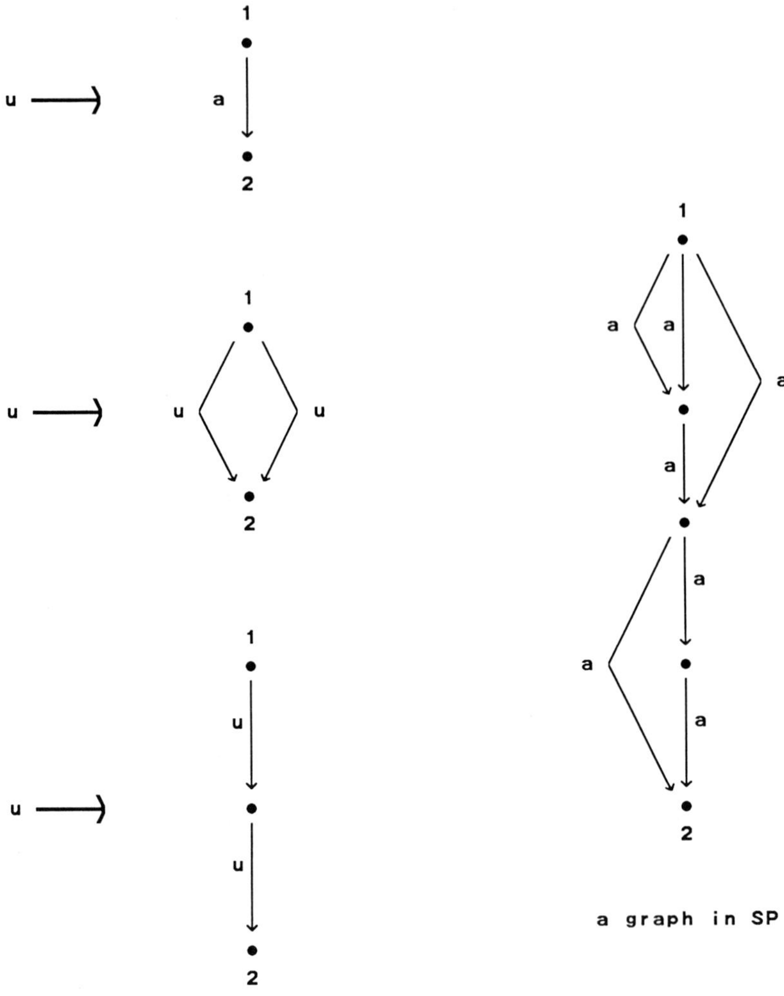

Fig. 3.

denote by $\bar{P}$ this new set of production rules. We say that $\bar{\Gamma} := \langle B, U, \bar{P}, e_Z \rangle$ (where $\mathbf{val}(e_Z) = Z$) is a *regular tree grammar associated with* Γ (see [47] on regular tree grammars). It generates the following set of expressions:

$$L(\bar{\Gamma}, e_Z) := \{e \in \mathbf{FE}(B)_n \mid e_Z \xrightarrow[\bar{P}]{*} e\}$$

where $\bar{P}$ is considered as a ground term rewriting system over $\mathbf{FE}(B \cup U)$. Note that $\bar{\Gamma}$ is not uniquely defined (since different expressions generate the same graph).

The grammar Γ of Example 3.1 can thus be rewritten as

$$\bar{P}\quad\begin{cases} u\to a, \\ u\to\sigma_{1,2}(\theta_{1,3}(\theta_{2,4}(u\oplus u))), \\ u\to\sigma_{1,4}(\theta_{2,3}(u\oplus u)); \end{cases}$$

or even more compactly as

$$\bar{P}\quad\begin{cases} u\to a, \\ u\to u \parallel u, \\ u\to u\cdot u \end{cases}$$

with the help of the derived operations introduced in Example 2.4. The same can be done in the general case.

3.2. LEMMA. *For every hypergraph expression k over $B\cup U$,*

$$L(\Gamma,\mathbf{val}(k))=\{\mathbf{val}(h)\mid h\in\mathbf{FE}(B),\ k\overset{*}{\underset{\bar{P}}{\to}} h\}.$$

PROOF. We first prove "$\supseteq$". Let $k=k_0\to k_1\to k_2\to\cdots\to k_m=h$ be a rewriting sequence w.r.t. $\bar{P}$, where $k=k_0,k_1,k_2,\ldots\in\mathbf{FE}(B\cup U)$ and $k_m=h\in\mathbf{FE}(B)$. Let $K_i=\mathbf{val}(k_i)$ for all i, and let us establish that $K_0=\mathbf{val}(k)\to K_1\to K_2\to\cdots\to K_m$ is a derivation sequence w.r.t. Γ, from which it follows that $K_m=\mathbf{val}(h)\in L(\Gamma,\mathbf{val}(k))$.

Let us assume, by induction, that $K_0\overset{*}{\underset{\bar{P}}{\to}} K_i$. Let us consider the rewriting step $k_i\to k_{i+1}$. It consists of the replacement of some occurrence of u in k_i by d in $\mathbf{FE}(B\cup U)$ such that $(u,\mathbf{val}(d))$ is a production of Γ. This occurrence of u in k_i corresponds to a hyperedge e of $\mathbf{val}(k_i)$ labelled by u. We have

$$\begin{array}{ccc} k_i & \xrightarrow[\bar{P}]{} & k_{i+1} \\ \downarrow & & \downarrow \\ \mathbf{val}(k_i)=K_i & \xrightarrow[\Gamma]{} & K_{i+1}=K_i[\mathbf{val}(d)/e]=\mathbf{val}(k_{i+1}) \end{array}$$

The fact that $\mathbf{val}(k_{i+1})=\mathbf{val}(k_i)[\mathbf{val}(d)/e]$ follows from Lemma 2.6(2).

Conversely, by a similar argument, for every derivation sequence $K_0\overset{*}{\underset{\Gamma}{\to}} K_m$ one can find a derivation sequence $k_0\overset{*}{\underset{\bar{P}}{\to}} k_m$ such that $\mathbf{val}(k_0)=K_0$ and $\mathbf{val}(k_m)=K_m$. This proves the other inclusion and completes the proof (omitting only a few technical details). $\square$

By this lemma, every hypergraph belonging to $L(\mathrm{G},\mathbf{val}(k))$ can be defined by an expression built with finitely many operations occurring in k or in $\bar{P}$. Hence, we have the following proposition.

3.3. PROPOSITION. *If L is context-free set of hypergraphs, then $\mathbf{wd}(L):=\mathbf{max}\{\mathbf{wd}(H)\mid H\in L\}$ is finite.*

It is not hard to construct a grammar generating $\mathbf{FG}(B)_n^k$. But it follows from Lemma

2.3 that the set of all finite hypergraphs is not context-free. This is a fundamental difference with the case of words.

3.3. *Equational set of hypergraphs*

The *equational* subsets of **FG** can be defined as in any algebra [69, 20]. We say that $L \subseteq \mathbf{FG}_n$ is equational, and write $L \in \mathbf{Equat(FG)}_n$ iff L is a component of the least solution in $\mathscr{P}(\mathbf{FG})$ (the powerset algebra of **FG**) of a system of equations written with $\cup$ (set union), and the operations of $\boldsymbol{H}$ (canonically extended to sets of hypergraphs). As for words, one has the following theorem.

3.4. THEOREM (Bauderon and Courcelle [9], Courcelle [23]). *A set of hypergraphs is equational iff it is context-free.*

More precisely, if $U = \{u_1, \ldots, u_m\}$ is the set of nonterminals of an HR grammar $\Gamma = \langle B, U, P, Z\rangle$, then the m-tuple of sets of finite hypergraphs, $(L(\Gamma, u_1), \ldots, L(\Gamma, u_m))$ is the least solution in $\mathscr{P}(\mathbf{FG}(B))$ of a system associated with Γ, exactly as, in the case of words, a system of equations in languages is associated with a context-free grammar. Using again the example of series-parallel graphs, we can characterize their set SP as the least solution L in $\mathscr{P}(\mathbf{FG}(B)_2)$ of the equation

$$L = \{a\} \cup L \parallel L \cup L \cdot L$$

where $L \parallel L' := \{G \parallel G' \mid G \in L,\ G' \in L'\}$ for $L, L' \subset \mathbf{FG}(B)_2$, and similarly for $L \cdot L'$.

PROOF OF THEOREM 3.4 (*sketch*). Let $\Gamma = \langle B, U, P, Z\rangle$ be an HR grammar. Let $\bar{\Gamma} = \langle B, U, \bar{P}, e_Z\rangle$ be an associated regular tree grammar. Let S be the corresponding system of equations: $S = \langle u_1 = p_1, \ldots, u_n = p_n\rangle$ where $U = \{u_1, \ldots, u_n\}$ and each p_i is a sum of the form $d_{i,1} \cup \cdots \cup d_{i,m_i}$ where $\{d_{i,1}, \ldots, d_{i,m_i}\}$ is the set of right-hand sides of the rules in $\bar{P}$ having left-hand side u_i. Then S has a least solution in $\mathscr{P}(\mathbf{FE}(B))$, and this solution is the n-tuple $(L(\bar{\Gamma}, u_1), \ldots, L(\bar{\Gamma}, u_n))$ by the Theorem of Ginsburg and Rice [48] (or its formulation given in [20, Proposition 13.5]). Since **val** is a homomorphism: $\mathbf{FE}(B) \to \mathbf{FE}(B)$, from a fundamental lemma of Mezei and Wright [69, Lemma 5.3] (see also [20, Proposition 13.1]) saying that least solutions of systems of equations are preserved under homomorphisms, it follows that $(\mathbf{val}(L(\bar{\Gamma}, u_1)), \ldots, \mathbf{val}(L(\bar{\Gamma}, u_n)))$ is the least solution of S in $\mathscr{P}(\mathbf{FG}(B))$. But this n-tuple is equal to $(L(\Gamma, u_1), \ldots, L(\Gamma, u_n))$ by Lemma 3.2. This concludes the proof. $\square$

3.4. *Properties of context-free sets of hypergraphs*

3.5. PROPOSITION. *For every HR grammar Γ, one can construct a parsing algorithm, i.e., an algorithm that decides whether a given hypergraph belongs to $L(\Gamma)$ and produces a derivation sequence if the answer is positive.*

PROOF. Let us define the size of a hypergraph G as $\mathbf{size}(G) := \mathbf{card}(\mathbf{V}_G) + \mathbf{card}(\mathbf{E}_G)$. From

Γ one can compute an integer h such that G belongs to $L(\Gamma)$ iff it is derivable from the axiom of G by a derivation of length at most $h \cdot \mathbf{size}(G)$. See [28, Lemma 1.7]. Since there are finitely many derivations of given length, the hypergraphs defined by derivations of length at most $h \cdot \mathbf{size}(G)$ can all be constructed. The hypergraph G belongs to $L(\Gamma)$ iff it is equal to one of them. □

This algorithm is clearly exponential. Some grammars have polynomial parsing algorithms. We shall discuss this question in Section 6.

3.6. PROPOSITION. (1) *The empty set is context-free. One can decide whether the set generated by an HR grammar is empty.*

(2) *The union of two context-free subsets of* $\mathbf{FG}(B)_n$ *is context-free.*

(3) *The intersection of two context-free subsets of* $\mathbf{FG}(B)_n$ *is not context-free in general.*

PROOF. (1) It is easy to construct an HR grammar generating no hypergraph. By Lemma 3.2, deciding whether $L(\Gamma)=\emptyset$ reduces to deciding whether a regular tree grammar generates the empty set, which is decidable.

(2) Easy construction, as for context-free languages.

(3) We have seen in Subsection 2.1 that every word w in A^* corresponds to a graph $G(w)$ in $\mathbf{FD}(A)_2$. It is not hard to construct, for every context-free (word) grammar generating $L \subseteq A^*$, an HR grammar generating $G(L) := \{G(w) \mid w \in L\}$. Example 3.7 will show that one may have $G(L)$ context-free as a set of graphs, with L non-context-free as a language. It follows that one cannot deduce the result from the corresponding one for context-free languages. Here is another proof.

Let $L \subseteq A^*$ be a recursively enumerable nonrecursive language equal to $h(L_1 \cap L_2)$, where $L_1, L_2 \subseteq A'^*$ are context-free languages, and h is a homomorphism: $A'^* \to A^*$. (The existence of such L, A, A', h is a classical result; see [41, Theorem 11].) Then $G(L_1 \cap L_2) = G(L_1) \cap G(L_2)$, and $G(L_1)$ and $G(L_2)$ are context-free subsets of $\mathbf{FD}(A')_2$. If $G(L_1 \cap L_2)$ would be context-free, so would $G(L) = G(h(L_1 \cap L_2))$ (easy construction). Hence $G(L)$ would be recursive and L would be recursive. This contradicts the initial assumption. □

3.7. EXAMPLE. $L = \{a^n b^n c^n \mid n \geqslant 1\}$ is a context-free set of graphs. The set of graphs $G(L) \subseteq \mathbf{FD}(A)_2$ is generated by the HR grammar shown in Fig. 4, with axiom u and nonterminal symbols u and w of respective types 2 and 4.

4. Logical properties of context-free sets of hypergraphs

In this section, we prove that the intersection of a context-free set of hypergraphs and the set of hypergraphs satisfying a formula of monadic second-order logic is context-free. Since every **MSOL**-definable set of hypergraphs is recognizable (in the sense of Mezei and Wright [69]) this result can be considered as an extension of the well-known fact that the intersection of a context-free language and a regular one is context-free. It provides a link between logical definitions of sets of hypergraphs on the

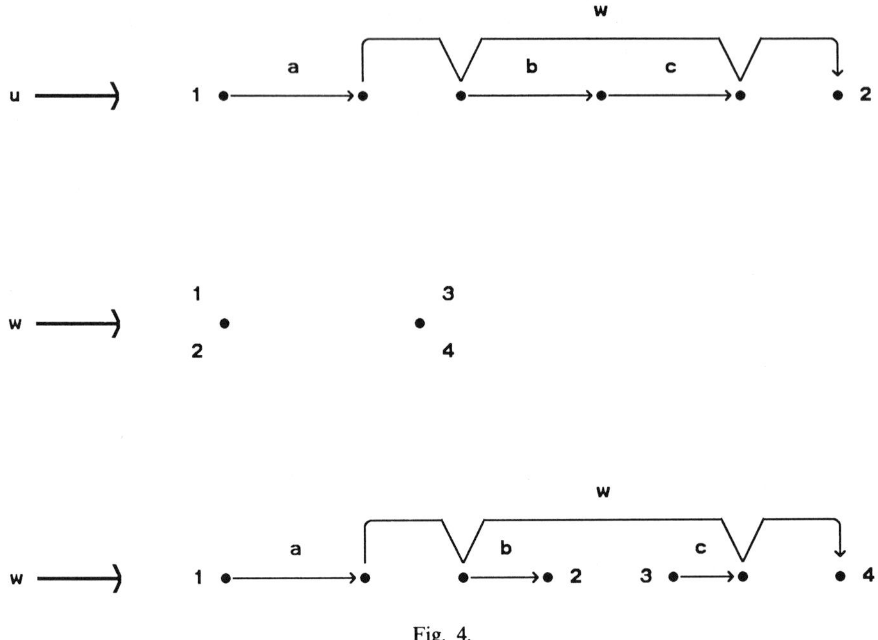

Fig. 4.

one hand, and algebraic ones on the other. "Algebraic" refers to systems of equations, hence to grammars by Proposition 3.3 and to the notion of a recognizable set of hypergraphs, defined (below) in terms of congruences. The link between recognizability and monadic second-order definability is known for words and trees from the works of Büchi and Rabin (see also [88]). The key concept for establishing the forthcoming results is the notion of an inductive set of predicates.

4.1. *Inductive sets of predicates*

Let $\mathbb{M}=\langle \boldsymbol{M}, (f_M)_{f\in F}\rangle$ be an F-algebra. For simplicity, we assume that it is one-sorted, and the extension to a many-sorted algebra is straightforward. By a *predicate* φ on $\boldsymbol{M}$, we mean a total mapping, $\varphi \mapsto \varphi_m : \boldsymbol{M} \to \{\textbf{true}, \textbf{false}\}$. A set Φ of predicates on $\boldsymbol{M}$ is F-*inductive* if, for every φ in Φ and for every f in F of arity k, one can find k sequences $(\varphi^{1,1}, \ldots, \varphi^{1,n_m}), \ldots, (\varphi^{k,1}, \ldots, \varphi^{k,n_k})$ of predicates of Φ, and an $(n_1+n_2+\cdots+n_k)$-place Boolean expression B such that, for all $m_1, \ldots, m_k$ in $\boldsymbol{M}$, if $m=f_M(m_1, \ldots, m_k)$, then

$$\varphi_m = B[\varphi_{m_1}^{1,1}, \ldots, \varphi_{m_1}^{1,n_1}, \ldots, \varphi_{m_2}^{2,1}, \ldots, \varphi_{m_k}^{k,n_k}].$$

This condition means that the validity of φ at $f_M(m_1, \ldots, m_k)$ can be determined from the truth, values of $\varphi^{1,1}, \ldots, \varphi^{1,n_1}$ at $m_1, \ldots,$ and of $\varphi^{k,1}, \ldots, \varphi^{k,n_k}$ at m_k. Every property of hypergraphs defines a predicate on $\mathbf{FG}(B)$. We shall also use the notion of an *inductive set of properties*. "Inductive" here means $\boldsymbol{H}_B$-inductive.

If S is a polynomial system of equations, $\mathbb{M}$ an F-algebra and u an unknown of S, then we denote by $L((S, \mathbb{M}), u)$ the component corresponding to u of the least solution of S in $\mathscr{P}(\mathbb{M})$.

4.1. THEOREM. *Let $\mathbb{M}$ be an F-algebra. Let Φ be a finite F-inductive set of predicates on $\boldsymbol{M}$. Let S be a polynomial system of equations and L a component of its least solution in $\mathscr{P}(\mathbb{M})$. One can construct a polynomial system that defines $\{m \in L \mid \varphi_m = \mathbf{true}\}$ where φ is some predicate in Φ.*

PROOF. Let $U = \{u_1, \ldots, u_n\}$ be the set of unknowns of S. The ith equation of S will be taken in the general form

$$u_i = t_{i,1} \cup \cdots \cup t_{i,n_i} \tag{4.1}$$

where the $t_{i,j}$'s are *monomials*, i.e., terms written with F and U.

We can assume that S is *uniform*, i.e., that the monomials are all of the form $f(u_{j_1}, \ldots, u_{j_k})$. If S is not uniform, one can transform it into a uniform system S_1, with set of unknowns $U \cup U_1$ and such that for every F-algebra $\mathbb{P}$, $L((S', \mathbb{P}), u) = L((S, \mathbb{P}), u)$ for every u in U. This transformation is a generalization of the one that puts a context-free grammar in Chomsky Normal Form (see [69, 20]).

Let Θ be the set of mappings: $\Phi \rightarrow \{\mathbf{true}, \mathbf{false}\}$. For every u_i in U and θ in Θ, we let $[u_i, \theta]$ be a new unknown. We shall construct a polynomial system S' in the new set of unknowns such that $L((S', \mathbb{M}), [u_i, \theta]) = \{m \in L((S, \mathbb{M}), u_i) \mid \varphi_m = \theta(\varphi) \text{ for all } \varphi \in \Phi\}$.

The equation of S' defining $[u_i, \theta]$ is constructed as follows from the equation of S defining u_i. We assume that this equation is of the form (4.1). Consider one of its monomials, say $t_{i,j}$, of the form $f(u_{j_1}, \ldots, u_{j_k})$. Let $T_{i,j,\theta}$ be the set of all monomials of the form $f([u_{j_1}, \theta_1], \ldots, [u_{j_k}, \theta_k])$ where $\theta_1, \ldots, \theta_k$ in Θ are such that, for all $m_1, \ldots, m_k$ in $\mathbb{M}$ such that θ_i is the mapping $\varphi \mapsto \varphi_{m_i}$ for all $i = 1, \ldots, k$, θ is the mapping $\varphi \rightarrow \varphi_m$, where $m = f_M(m_1, \ldots, m_k)$. From the hypothesis that Φ is F-inductive (and from the knowledge of the associated Boolean combinations), one can compute the above set of k-tuples $(\theta_1, \ldots, \theta_k)$, hence the set $T_{i,j,\theta}$. We then let $t_{i,j,\theta}$ be the formal union of the members of $T_{i,j,\theta}$, and

$$[u_i, \theta] = t_{i,1,\theta} \cup t_{i,2,\theta} \cup \cdots \cup t_{i,n_i,\theta}.$$

We now prove that

$$L((S', \mathbb{M}), [u_i, \theta]) = L((S, \mathbb{M}), u_i) \cap L_\theta \tag{4.2}$$

where $L_\theta := \{m \in \mathbb{M} \mid \theta(\varphi) = \varphi_m \text{ for all } \varphi \in \Phi\}$. From the construction of S', it is easy to see that the sets $(L((S, \mathbb{M}), u_i) \cap L_\theta)_{[u_i, \theta] \in U'}$ form a solution of S'. This proves "$\subseteq$" in (4.2). Similarily, the sets $L_i := \bigcup \{L((S', \mathbb{M}), [u_i, \theta]) \mid \theta \in \Theta\}$, $i = 1, \ldots, n$, form a solution of S. Hence, $L_i \supseteq L((S, \mathbb{M}), u_i)$ for all i. It follows that

$$L_i \cap L_\theta \supseteq L((S, \mathbb{M}), u_i) \cap L_\theta,$$

but

$$L_i \cap L_\theta = L((S', \mathbb{M}), [u_i, \theta]) \cap L_\theta$$

since $L((S', \mathbb{M}), [u_i, \theta]) \subseteq L_\theta$ and the sets L_θ are pairwise disjoint. Hence we have

$$L((S', \mathbb{M}), [u_i, \theta]) \supseteq L((S, \mathbb{M}), u_i) \cap L_\theta,$$

which establishes (4.2). It follows that

$$\{m \in L((S, \mathbb{M}), u_i) \mid \varphi_m = \mathbf{true}\} = \cup\{L((S', \mathbb{M}), [u_i, \theta]) \mid \theta(\varphi) = \mathbf{true}\}.$$

In order to define this set, it suffices to add a new unknown u to S', with defining equation

$$u = [u_i, \theta_1] \cup \cdots \cup [u_i, \theta_k]$$

where $\{\theta_1, \ldots, \theta_k\} = \{\theta \mid \theta(\varphi) = \mathbf{true}\}$, and the desired set is $L((S', \mathbb{M}), u)$. □

4.2. EXAMPLE (*2-colorability of series-parallel graphs*). Let B be reduced to a single symbol a of type 2 (as in Example 3.1). Let γ, δ, σ be the following predicates on $\mathbf{FG}_2 = \mathbf{FG}(B)_2$:

$\gamma_G \Leftrightarrow G$ is 2-colorable;

$\delta_G \Leftrightarrow G$ is 2-colorable in such a way that its two sources have different colors;

$\sigma_G \Leftrightarrow G$ is 2-colorable in such a way that its two sources have the same color.

It is clear that $\gamma_G \Leftrightarrow \delta_G \vee \sigma_G$ for all G in $\mathbf{FG}_2$. We now show that $\{\delta, \sigma\}$ is $\{\|, \cdot\}$-inductive. (It then follows that $\{\gamma, \delta, \sigma\}$ is $\{\|, \cdot\}$-inductive.) It suffices to remark that

(1) if $G = H \cdot K$, then

$$\sigma_G \Leftrightarrow (\sigma_H \wedge \sigma_K) \vee (\delta_H \wedge \delta_K), \qquad \delta_G \Leftrightarrow (\sigma_H \wedge \delta_K) \vee (\delta_H \wedge \sigma_K);$$

(2) if $G = H \| K$ then

$$\sigma_G \Leftrightarrow (\sigma_H \wedge \sigma_K), \qquad \delta_G \Leftrightarrow \delta_H \wedge \delta_K.$$

We now give the result of the construction of Theorem 4.1. Let S be reduced to the equation $u = u \| u \cup u \cdot u \cup a$ whose least solution is the set SP of series-parallel graphs (see Theorem 3.4). We now construct S'. Its set of unknowns U' has four elements denoted by $[\sigma\delta]$, $[\bar{\sigma}\delta]$, $[\sigma\bar{\delta}]$, $[\bar{\sigma}\bar{\delta}]$ (where $[\sigma\bar{\delta}]$ stands for $[u, \theta]$ where $\theta(\sigma) = \mathbf{true}$, $\theta(\delta) = \mathbf{false}$, and similarily for the other ones.)

Its least solution in $\mathscr{P}(\mathbf{FG})$ is the four-tuple $(L(S', [\sigma\delta]), L(S', [\bar{\sigma}\delta]), L(S', [\sigma\bar{\delta}]), L(S', [\bar{\sigma}\bar{\delta}]))$. Hence, $\mathrm{SP} = \cup\{L(S', w) \mid w \in U'\}$ and $L(S', [\sigma\bar{\delta}])$ is the set of graphs in SP that satisfy σ but not δ, i.e., that are 2-colorable with their two sources necessarily of the same color. An example of such a graph is

$$1 \bullet \overset{a}{\rightarrow} \bullet \overset{a}{\rightarrow} \bullet 2.$$

We give only three equations of S', the last one of which is more complicated (sixteen monomials):

$$[\sigma\delta] = [\sigma\delta] \| [\sigma\delta] \cup [\sigma\delta] \cdot [\sigma\delta] \cup [\sigma\bar{\delta}] \cdot [\sigma\delta] \cup [\sigma\delta] \cdot [\sigma\bar{\delta}] \cup [\bar{\sigma}\delta] \cdot [\sigma\delta] \cup [\sigma\delta] \cdot [\bar{\sigma}\delta],$$

$$[\sigma\bar{\delta}] = [\sigma\delta] \| [\sigma\bar{\delta}] \cup [\sigma\bar{\delta}] \| [\sigma\delta] \cup [\sigma\bar{\delta}] \| [\sigma\bar{\delta}] \cup [\sigma\bar{\delta}] \cdot [\sigma\bar{\delta}] \cup [\bar{\sigma}\delta] \cdot [\bar{\sigma}\delta],$$

$$[\bar{\sigma}\delta] = a \cup [\sigma\delta] \| [\bar{\sigma}\delta] \cup [\bar{\sigma}\delta] \| [\sigma\delta] \cup [\bar{\sigma}\delta] \| [\bar{\sigma}\delta] \cup [\bar{\sigma}\delta] \cdot [\sigma\bar{\delta}] \cup [\sigma\bar{\delta}] \cdot [\bar{\sigma}\delta].$$

The monomial a in the third equation stems from the fact that for $G=a$, σ_G is **false** and δ_G is **true.** The first monomial of the second equation stems from the fact that if σ_K, δ_K and σ_H hold and δ_H does not, then σ_G holds and δ_G does not (where $G=K \parallel H$).

From these equations one can see that $L(S', [\sigma\delta])=\emptyset$ (proving that no graph in SP satisfies both σ and δ), and that the other sets are nonempty. The graph shown in Fig. 3 in Example 3.1 belongs to $L(S', [\bar{\sigma}\bar{\delta}])$ and is hence not 2-colorable.

It can be observed that $\{\gamma\}$ is not $\{\parallel, \cdot\}$-inductive. We had to introduce auxiliary predicates, namely σ and δ, forming an inductive set with γ. In the general case of Theorem 4.1, the size of S' is exponential in the number of auxiliary predicates. For practical use, one must find "small" inductive sets of predicates.

Since the systems of equations associated with HR grammars have unknowns of different sorts, we need to generalize the notion of an inductive family of predicates to many-sorted algebras. The generalization is straightforward: each predicate has a *fixed* sort, i.e., "says something" on the objects of this sort. A family of predicates is *locally finite* if it contains finitely many predicates of each sort. Theorem 4.1 actually holds for a many-sorted F-algebra $\mathbb{M}$, a locally finite F-inductive family of predicates, and a polynomial system over $\mathbb{M}$.

Our aim is to apply Theorem 4.1 to predicates on hypergraphs expressed in monadic second-order logic. We will need to specify the logical structures representing hypergraphs.

4.2. *Hypergraphs as logical structures*

Let $G \in \boldsymbol{G}(B)_n$ be a hypergraph of the general form $\langle V_G, E_G, \mathbf{lab}_G, \mathbf{vert}_G, \mathbf{src}_G \rangle$. For every b in B, let us introduce a relation symbol $\mathbf{edg}_b$ of rank $\tau(b)+1$ and of arity $(\boldsymbol{e}, \boldsymbol{v}, \ldots, \boldsymbol{v})$. By this we mean that its first argument is a hyperedge and that the $\tau(b)$ others are vertices. For every i in $[n]$, we let $\boldsymbol{s}_i$ be a constant of sort $\boldsymbol{v}$.

Let $|G|$ be the $\{\boldsymbol{v}, \boldsymbol{e}\}$-sorted structure

$$|G| := \langle V_G, E_G, (\mathbf{edg}_{bG})_{b \in B}, (s_{iG})_{i \in [n]} \rangle$$

where

$$\mathbf{edg}_{bG}(e, v_1, \ldots, v_k) :\Leftrightarrow \mathbf{lab}_G(e)=b \wedge \mathbf{vert}_G(e)=(v_1, \ldots, v_k),$$
$$s_{iG} := \mathbf{src}_G(i).$$

Let $\mathscr{L}_{B,n}$ denote the set of closed monadic second-order formulas written with $\{\mathbf{edg}_b \mid b \in B\}$ and $\{s_i \mid i=1, \ldots, n\}$. They express properties of hypergraphs in $\boldsymbol{G}(B)_n$, i.e., they define predicates on $\boldsymbol{G}(B)_n$.

For every $h \in \mathbb{N}$, we denote by $\mathscr{L}^{(h)}_{B,n}$ the set of formulas of $\mathscr{L}_{B,n}$ having at most h levels of nested quantifications. The basic result is the following theorem.

4.3. THEOREM. *For every h, the formulas of $(\mathscr{L}^{(h)}_{B,n})_{n \in \mathbb{N}}$ define a locally finite, $\boldsymbol{H}_B$-inductive family of predicates over* $\mathbf{FG}(B)$.

Since the formulas defining graph properties do not use function symbols, there are only finitely many of them, up to tautological equivalence. Hence, the family of associated predicates is locally finite. The $\boldsymbol{H}_B$-inductivity is a long technical lemma (see [26, 21] for the proof of a similar result).

4.4. THEOREM. *Consider an HR grammar Γ and a monadic second-order formula φ.*

(1) *One can construct an HR grammar Γ' generating $\{G \in L(G) \mid G \models \varphi\}$.*

(2) *One can decide whether there exists a hypergraph G in $L(\Gamma)$ such that $G \models \varphi$.*

(3) *One can decide whether every hypergraph G in $L(\Gamma)$ satisfies φ, i.e. whether $\varphi \in \mathbf{Th}_{\mathbf{MSOL}}(L(\Gamma))$.*

PROOF. Let n and B be such that $L(\Gamma) \subseteq \mathbf{FG}(B)_n$. Let h be such that $\varphi \in \mathscr{L}^{(h)}_{B,n}$.

(1) By Theorems 4.3 and 4.1, one can construct a system of equations, one component of the least solution of which is $L = \{G \in L(\Gamma) \mid G \models \varphi\}$. By Theorem 3.4, an HR grammar Γ' can be obtained that generates L. Note that to construct Γ', one need not consider the infinite family of sets of predicates associated with the sets of formulas $\mathscr{L}^{(h)}_{B,n}$ for all $n \in \mathbb{N}$, but only the finitely many ones for those integers n that are the sorts of symbols occurring in a regular tree grammar $\bar{\Gamma}$ associated with Γ according to Subsection 3.2. This ensures that the construction of Γ' is effective (although practically intractable).

(2) There exists G in $L(\Gamma)$ satisfying φ iff $L(\Gamma') \neq \emptyset$, and this is decidable by Proposition 3.6.

(3) If we similarly construct Γ'' such that $L(\Gamma'') = \{G \in L(\Gamma) \mid G \models \neg\varphi\}$, then $\varphi \in \mathbf{Th}_{\mathbf{MSOL}}(L(\Gamma))$, i.e., $G \models \varphi$ for all G in $L(\Gamma)$ iff $L(\Gamma'') \neq \emptyset$. This is decidable. □

Hence we have also proved Theorem 1.4 now.

4.5. COROLLARY. *Let $L \subseteq \mathbf{FG}(B)_n$ be* **MSOL**-*definable. The following properties are equivalent:*

(1) *L is context-free;*

(2) *$L \subseteq L'$ for some context-free set L';*

(3) $\mathbf{Th}_{\mathbf{MSOL}}(L)$ *is decidable.*

PROOF. (1)⇒(2) is trivial.

(2)⇒(1) is an immediate consequence of Theorem 4.4(1).

(1)⇒(3) is Theorem 4.4(3).

(3)⇒(2) If L has a decidable monadic theory, then $L \subseteq \mathbf{FG}(B)^k_n$ for some k. This is a consequence of Theorem 1.5, mainly because there exists an integer k' such that, for every G in $\mathbf{FG}(B)_n$, $\mathbf{wd}(G) \leqslant 2\mathbf{twd}(G) + k'$ where $\mathbf{twd}(G)$ denotes the tree-width of G. The technical details can be found in [28]. The result follows since $\mathbf{FG}(B)^k_n$ is context-free. □

Theorem 4.4 can be reformulated algebraically in terms of the notion of a recognizable set of finite hypergraphs. The recognizable subsets of **FG** can be defined as in any algebra [69].

4.3. *Recognizable sets of finite hypergraphs*

A subset L of $\mathbf{FG}_n$ is *recognizable* iff it is a union of classes of a congruence over **FG** that has a finitely many classes of each sort. The set of recognizable subsets of $\mathbf{FG}_n$ is denoted by $\mathbf{Rec(FG)}_n$. (A *congruence over* **FG** is a family $\sim\ =(\sim_n)_{n\in\mathbb{N}}$ where $\sim_n$ is an equivalence relation on $\mathbf{FG}_n$, and these relations are preserved by the operations of **FG**.) The notion of a congruence can be reformulated independently of the operations on hypergraphs introduced in Subsection 2.3.

4.6. LEMMA. *A family of equivalence relations on* **FG**, $\sim\ =(\sim_n)_{n\in\mathbb{N}}$, *is a congruence iff, for every hypergraph K in $\mathbf{FG}_n$, for every integer p, for every edge e of K of type p, and for every two hypergraphs G and G' in $\mathbf{FG}_p$,*

$$\text{if } G\sim_p G' \quad \text{then } K[G/e]\ \sim_n K[G'/e].$$

PROOF. We have seen that every finite hypergraph can be defined by an expression. It follows from Lemma 2.6 that the mapping $G\mapsto K[G/e]$ is a derived operation: $\mathbf{FG}_p\to\mathbf{FG}_n$. This proves the "only-if" direction.

Conversely, every linear derived operation $f\colon \mathbf{FG}_p\to\mathbf{FG}_n$ can be written in the form $f(G)=K[G/e]$ for some K and e. The result follows. □

4.7. LEMMA. *A set of finite hypergraphs, all of type m, is recognizable iff it is the set of hypergraphs of type m satisfying a property that belongs to a locally finite inductive set of properties.*

PROOF. For each equivalence class C of a congruence $\sim$ on **FG** having finitely many classes of each sort n in $\mathbb{N}$, we let p_C be the predicate such that $p_C(G):\Leftrightarrow G\in C$. Let P be the set of all such predicates; it is locally finite. It is H-inductive since $\sim$ is a congruence. Let L be a recognizable set of the form $C_1\cup\cdots\cup C_k$. The set $P\cup\{p_L\}$, where

$$p_L(G)\quad :\Leftrightarrow\quad P_{C_1}(G)\vee\cdots\vee P_{C_k}(G),$$

is also locally finite and H-inductive, and $L=\{G\mid p_L(G) \text{ holds}\}$. The proof of the other direction is similar. □

We conclude this section by listing a few properties of the family of recognizable sets of finite hypergraphs.

4.8. THEOREM. (1) $\mathbf{Rec(FG)}_n$ *is closed under Boolean operations.*

(2) *If $L\in\mathbf{Equat(FG)}_n$ and $K\in\mathbf{Rec(FG)}_n$, then $L\cap K\in\mathbf{Equat(FG)}_n$.*

This theorem actually holds in any many-sorted algebra. Its second part is the extension to sets of finite hypergraphs of the theorem that states that the intersection of a context-free language and a regular one is context-free. It follows immediately from Theorem 4.1 and Lemma 4.7.

The following result is (perhaps) unexpected.

4.9. LEMMA (Courcelle [26]). *There are uncountably many recognizable sets of finite hypergraphs.*

One can establish that every set of planar square grids is recognizable by looking at appropriately defined "syntactical" congruences. This proves the lemma. We also have the following theorem.

4.10. THEOREM. *The families* **Equat(FG)**$_n$ *and* **Rec(FG)**$_n$ *are incomparable. The family* **Rec(FG**^{k}**)**$_n$ *is strictly included in* **Equat(FG**^{k}**)**$_n$.

In this theorem, we denote by $\mathbf{FG}^k$ the $\boldsymbol{H}^k$-algebra of finite graphs of type and width at most k, where $\boldsymbol{H}^k$ is the $[0, k]$-signature, defined as the restriction of $\boldsymbol{H}$ to symbols of sort at most k and or arity in $[0, k]^*$ (where $[0, k] = \{0, 1, \ldots, k\}$).

The following result is an immediate consequence of Lemma 4.7 and Theorem 4.3.

4.11. THEOREM (Courcelle [26]). *Every* **MSOL**-*definable set of finite hypergraphs is recognizable.*

A comparability diagram for various families of sets of finite graphs will be given at the end of the next section. The inclusion and incomparability results hold for the corresponding families of sets of finite hypergraphs.

5. Sets of finite graphs defined by forbidden minors

In this section we consider subsets of $\mathbf{FD}(A)_0$ and compare their possible definitions by HR grammars, by logical formulas, and by forbidden minors.

From the characterization of planar graphs given by Kuratowski (saying that a graph is planar iff none of its subgraphs is homeomorphic to K_5 or to $K_{3,3}$), it is easy to deduce an expression of planarity in monadic second-order logic. *Minor containment* is an alternative to subgraph homeomorphism from which one can also characterize the set of planar graphs: *a graph is planar iff it contains neither* K_5 *nor* $K_{3,3}$ *as a minor* [90]. The notion of minors has been investigated in depth by Robertson and Seymour [79–81]. Some of their results will find applications here (see also Chapter 10 on graph algorithms in Vol. A of this Handbook).

In this section, we fix a finite (unranked) alphabet A, and we let $\mathbf{FD}_k = \mathbf{FD}(A)_k$ for all k (see Subsection 2.1) We denote by $\mathbf{FD}$ the corresponding $\boldsymbol{H}_A$-algebra (see Subsection 2.3) By a *graph* we mean an element of $\mathbf{FD}_0$.

5.1. *Minor containment*

Two graphs are *quasi isomorphic* if they are isomorphic up to edge labels and orientations. In particular, two quasi isomorphic graphs have the same numbers of vertices and edges. They also have the same width since every graph expression defining one of them can be easily transformed into a graph expression of the same width defining the other.

Let $G, H \in \mathbf{FD}_0$. One says that *H is contained as a minor in G* (or that *H is a minor of G*) if there exists a graph H' such that G can be transformed into H' by a finite sequence of edge deletions, edge contractions and deletions of isolated vertices, and if H' is quasi isomorphic to H. This is written $H \ll G$. For every set of finite graphs K, we denote by $\mathbf{FORB}(K)$ the set of graphs G such that $H \ll G$ for no H in K. (The set K is also called an *obstruction set* of $\mathbf{FORB}(K)$).

A subset L of $\mathbf{FD}_0$ is *minor-closed* if it contains all the minors of all its elements. If L is minor-closed, then clearly, $L = \mathbf{FORB}(\mathbf{FD}_0 - L)$. Robertson and Seymour have proved a conjecture by Wagner saying that, in such a case, there exists a finite subset K of $\mathbf{FD}_0 - L$ such that $L = \mathbf{FORB}(K)$ (see [81]).

An equivalent definition of minor containment is the following. $H \ll G$ iff there exist three mappings $f: \boldsymbol{V}_H \to \mathscr{P}(\boldsymbol{V}_G)$, $f': \boldsymbol{V}_H \to \mathscr{P}(\boldsymbol{E}_G)$ and $g: \boldsymbol{E}_H \to \boldsymbol{E}_G$ satisfying the following conditions:

(1) for every v in $\boldsymbol{V}_H$, there is a connected subgraph K of G such that $\boldsymbol{V}_K = f(v)$ and $\boldsymbol{E}_K = f'(v)$;

(2) for every v and v' in $\boldsymbol{V}_H$, if $v \neq v'$ then $f(v) \cap f(v') = \emptyset$;

(3) g is injective, and $g(e) \notin f'(v)$ for every e in $\boldsymbol{E}_H$ and v in $\boldsymbol{V}_H$;

(4) if e is an edge of $\boldsymbol{E}_H$ linking v to v', then $g(e)$ is an edge of $\boldsymbol{E}_H$ linking a vertex of $f(v)$ and a vertex of $f(v')$ (we may have $v = v'$).

Intuitively, H is a minor of G iff

- every vertex v of H corresponds to a connected subgraph $\bar{f}(v)$ of G and disjoint subgraphs of G correspond to distinct vertices of H;
- every edge e of H corresponds to an edge $g(e)$ of G that is not in subgraphs corresponding to vertices of H, and if e links v and v' then $g(e)$ links a vertex of $\bar{f}(v)$ and a vertex of $\bar{f}(v')$.

From this characterization (adapted from [78]) we have the following lemma.

5.1. LEMMA. *For every finite graph H*, the set $\{G \in \mathbf{FD}_0 \mid H \ll G\}$ *is* **MSOL**-*definable.*

The following theorem then immediately follows from the theorem of Robertson and Seymour [81] and from Theorem 4.11.

5.2. THEOREM. *If a set of finite graphs is minor-closed, then it is defined by a finite set of forbidden minors. It is* **MSOL**-*definable and recognizable.*

5.3. PROPOSITION. (1) *The set* $\mathbf{FD}_0^k$ *of graphs of width at most k is minor-closed*, **MSOL**-*definable and recognizable.* (*It is also context-free.*)

(2) *The set of graphs of width* (*exactly*) *k is* **MSOL**-*definable, recognizable and context-free.*

PROOF. (1) That $\mathbf{FD}_0^k$ is minor-closed is not hard to see from the first definition of a minor (see [28]). We know that $\mathbf{FD}_0^k$ is context-free. The other assertions follow from Theorem 5.2.

(2) Since the difference of two recognizable sets is recognizable, the result follows from (1). The context-freeness follows from Theorem 4.8(2). □

Since the theorem in [81] is not effective, we do not know the finite set K of forbidden minors that defines $\mathbf{FD}_0^k$. Hence, we know that this set is recognizable, but we do not know the congruence defining it. As a consequence, we do not know the grammar defining the set of graphs of width k, but it does exist. Hence, we know that certain problems are decidable (for example, whether all graphs of width k satisfy a monadic second-order formula), without knowing the algorithm. Johnson has already pointed out a similar situation for polynomial decidability in [58].

We now define the *tree-width* of a graph. This notion has been introduced by Robertson and Seymour in connection with forbidden minors. It is also essential for the study of sets of graphs having a decidable monadic theory (see Theorem 1.5) and, as we shall see below, in investigations concerning the border between NP-complete problems and polynomial ones. (See also Chapter 10 on graph algorithms in Vol. A of this Handbook.)

5.2. *Tree-width and tree decompositions*

Let $G \in \mathbf{FD}_0$. A *tree decomposition* of G is a pair (T, f) consisting of an unrooted unoriented tree T and a mapping $f: V_T \rightarrow \mathscr{P}(V_G)$ such that

(1) $V_G = \bigcup \{f(i) \mid i \in V_T\}$;
(2) every edge of G has its two ends in some set $f(i)$;
(3) if $v \in f(i) \cap f(j)$, then $v \in f(k)$ for every k belonging to the unique loop-free path in T linking i to j.

The width of a tree decomposition is defined as $\mathbf{max}\{\mathbf{card}(f(i)) \mid i \in V_T\} - 1$, and the *tree-width* of G, denoted by $\mathbf{twd}(G)$, is the minimum width of a tree decomposition of G. It is clear that trees are of tree-width 1 and that $\mathbf{twd}(G) \leqslant \mathbf{card}(V_G) - 1$. This upper bound is reached if G is a clique.

From a tree decomposition of G one can construct an expression defining G, and vice versa. This allows us to establish the following lemma.

5.4. LEMMA (Courcelle [28]). *For every G in* $\mathbf{FD}_0$, *one has* $\mathbf{twd}(G) + 1 \leqslant \mathbf{wd}(G) \leqslant 2\mathbf{twd}(G) + 3$.

It follows from Proposition 3.3 that every context-free set of graphs is of bounded tree-width. Furthermore, one can construct an HR grammar generating TW($\leqslant k$), the set of graphs of tree-width at most k.

Many results of the complexity of graph algorithms refer to partial k-trees. A finite simple unoriented graph is a *partial k-tree* iff it is a subgraph of a k-tree. A *k-tree* is a graph constructed as follows: start with a clique with k vertices. Add successively vertices and edges as follows: choose a clique with k vertices in the k-tree already constructed, add a new vertex and link this new vertex to the k vertices of the chosen clique. This addition of vertices and edges is repeated finitely many times. Every k-tree is of tree-width exactly k. It follows that a graph has tree-width at most k iff its underlying simple unoriented graph is a partial k-tree.

The set TW($\leqslant k$) is minor-closed. It follows that Proposition 5.3 holds with "tree-width" instead of "width". The forbidden-minor characterizations of TW($\leqslant k$) is known for $k=0$, 1, 2, 3 (see [5]). Hence, the results of Proposition 5.3 are (presently) ineffective for $k>3$. We go back to the characterization of sets of graphs defined by forbidden minors.

5.5. THEOREM (Courcelle [28]). *If K is a finite set of graphs, one of which is planar, then* **FORB**(K) *is a set of graphs of bounded width. It follows that* **FORB**(K) *is context-free,* **MSOL**-*definable and recognizable. Conversely, if L is context-free and is of the form* **FORB**(K) *for some finite set of graphs K, then* $L=$ **FORB**$(K\cup\{Q\})$ *for some planar graph* Q.

PROOF. Seymour and Robertson have established in [79] that **FORB**$(\{H\})$ is included in the set of graphs of tree-width at most k (an integer computable from H) if H is a planar graph. The result follows from Theorem 4.8(2) and Lemma 5.4. To establish the converse, take a sufficiently large grid Q. □

5.3. *A comparison diagram*

The results can be summarized in the diagram shown in Fig. 5, where several families of sets of finite graphs are considered. (In this diagram, the scope of a family name is the largest rectangle in the upper left corner of which it has been written).

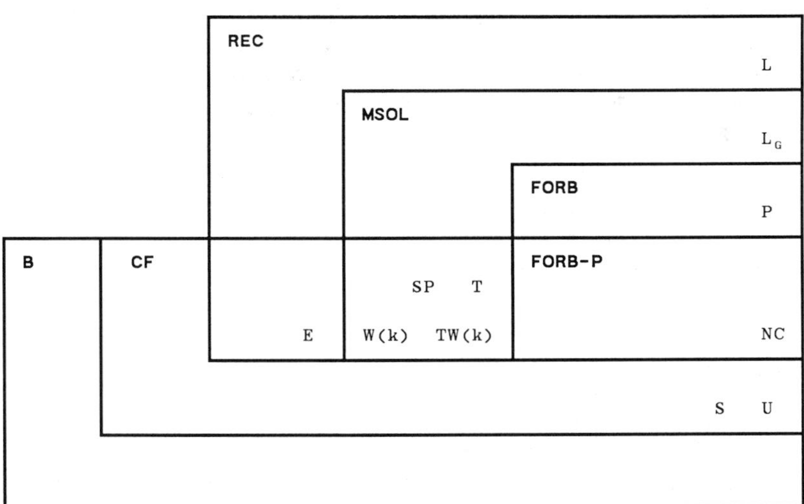

Fig. 5.

REC:	the family of recognizable sets of graphs,
MSOL:	the family of **MSOL**-definable sets of graphs,
CF:	the family of context-free (or equational) sets of graphs,
FORB:	the family of sets defined by a finite set of forbidden minors,
FORB-P:	the family of sets defined by a finite set of forbidden minors one of which is planar,
B:	the family of all subsets of $\mathbf{FD}_0^k$ for all $k\geqslant 0$.

5.6. REMARKS. (1) The families **B** and **REC** are uncountable.

(2) Theorem 5.5 says that **CF**$\cap$**FORB**=**FORP-P**.

(3) By Theorem 4.8(2), the equality of two sets of graphs in **CF**$\cap$**REC** is decidable, provided each of them is effectively given as a context-free set (by a grammar) and as a recognizable set (by a tree automaton on graph expressions).

(4) All inclusions shown in Fig. 5 are proper. All boxes are nonempty.

The diagram also locates the following sets of graphs:

- L_G, the set of square grids;
- L, the set of all square grids of size $n \times n$ where n is an integer, belonging to some nonrecursive set B;
- P, the set of planar graphs;
- T, the set of undirected, unrooted trees;
- NC, the set of cycle-free graphs;
- SP, the set of series-parallel graphs;
- S, the language $\{a^n b^n \mid n>0\}$, considered as a set of graphs;
- U, the language $\{a^n b^n c^n \mid n>0\}$, considered as a set of graphs (see Example 3.7);
- $W(k)$, the set of graphs of width exactly k;
- TW(k), the set of graphs of tree-width exactly k;
- E, the set of discrete graphs (all vertices of which are isolated) having an even number of vertices.

It has been proved in [26] that L_G is **MSOL**-definable, and that every subset of L_G is recognizable. The set L is such a set. If it were defined by a formula φ, then one would have $G_n \models \varphi$ iff $n \in B$, where G_n is the square $n \times n$ grid. Since one can decide whether a finite graph satisfies a formula, one would then be able to decide whether an integer n belongs to B. This would contradict the choice of B. It has also been proved in [26] that E is not **MSOL**-definable. The reason is that in **MSOL** one cannot express whether the number of elements of a set is even. The set P is equal to **FORB**($\{K_{3,3}, K_5\}$) by a result of Wagner [90]. It is not in **B** since it contains all grids, and there exist grids of arbitrarily large width. The set NC of cycle-free graphs if **FORB**($\{G\}$) where G is the graph with one vertex and one edge forming a loop. It has been proved in [26] that S is not recognizable. The same holds for U.

6. Complexity issues

Many papers have been published relating the complexity of a property of finite graphs or of finite relational structures (i.e., of finite hypergraphs) to its logical form. We refer the reader to [59, 54]. Let us quote the first result in this field of research.

6.1. THEOREM (Fagin [44]). *A graph property is in the class* NP *iff it is expressible by an existential second-order formula.*

An existential second-order formula is a formula of the form $\exists R_1, \ldots, R_n \varphi$ where

$R_1, \ldots, R_n$ are set or relation variables, and φ is first-order. On the other hand, we also have the following theorem.

6.2. THEOREM. *A graph property is in the class* P *if it is first-order.*

This result is not that interesting since, as we have already observed, first-order logic is rather weak for expressing graph properties. We shall see that polynomial algorithms exist for **MSOL**-graph properties *restricted to the class* of (finite) graphs or hypergraphs of bounded width (or tree-width). Since certain **MSOL**-graph properties, such as 3-colorability are NP-complete, one cannot hope to have polynomial algorithms for **MSOL**-graph properties without such a restriction (unless P = NP). Our main tool is the following result.

6.3. PROPOSITION. *Let φ be a closed monadic second-order formula in $\mathscr{L}_{B,n}$. One can decide in time* $O(\mathbf{size}(e))$ *whether the hypergraph defined by an expression e in* $\mathbf{FE}(B)_n$ *satisfies φ.*

PROOF. By Theorem 4.3, one can construct a finite-state deterministic bottom-up tree automaton recognizing $\{g \in \mathbf{FE}(B)_n \mid \mathbf{val}(g) \models \varphi\}$ (see [47, or 88] on tree automata). Hence one can test in linear time whether e belongs to this set. □

Here is the first application of this result.

6.4. PROPOSITION. *Given φ,k and n such that $\varphi \in \mathscr{L}_{B,n}$ and $k \geqslant n$, one can construct an algorithm that says for every G in* $\mathbf{FG}(B)_n$, *in time* $O((\mathbf{card}(E_G)+1)\cdot\mathbf{card}(V_G))$, *that* $\mathbf{wd}(G) > k+1$, *or* $\mathbf{wd}(G) \leqslant k'$ *and* $G \models \varphi$, *or* $\mathbf{wd}(G) \leqslant k'$ *and* $G \models \neg\varphi$, *where* $k' = 9k+20+\mathbf{max}\{\tau(b) \mid b \in B\}$.

PROOF (*sketch*). Seymour and Robertson [80] have given an algorithm that, for every finite graph G in S, either constructs a so-called *branch decomposition* of G of width at most $3m$ or reports that G has no branch decomposition of width at most m, where m is a fixed integer. This algorithm runs in time $O((\mathbf{card}(E_G)\cdot\mathbf{card}(V_G))$; it can be adapted to hypergraphs. A branch decomposition is close to a tree decomposition and can be transformed into an expression e denoting G. The algorithm of Proposition 6.3 can be applied to e. The remaining technical details can be found in [28]. □

Several NP-complete problems are expressible in monadic second-order logic. A few examples are 3-COLORABILITY [GT4], MONOCHROMATIC TRIANGLE [GT6], PARTITION INTO TRIANGLES [GT11], CUBIC SUBGRAPH [GT32], HAMILTONIAN CIRCUIT [GT38]. (The numbering is from [46], where the definitions of these problems can be found.) Other NP-complete problems can be expressed in a more powerful languages called *extended monadic second-order logic*, introduced by Arnborg et al. [3]. (This extension allows a few numerical computations; integer values can be part of the input.) Propositions 6.3 and 6.4 also hold for extended monadic second-order formulas φ. The following

NP-complete problems can be expressed in this language: VERTEX COVER [GT1] and SIMPLE MAXCUT [N16]. They take as input a graph and an integer.

Finally, the SUBGRAPH ISOMORPHISM problem for finite connected graphs of degree at most some fixed $d \geqslant 3$, is NP-complete. It can be shown to be not expressible in monadic second-order logic (by an adaptation of the proof that the set of graphs S of Subsection 5.3 is not recognizable). It does not seem to be expressible in extended monadic second-order logic either. But it belongs to the class LCC of problems, defined by Bodlaender in [13], and it can thus be decided in polynomial time for graphs of tree-width at most a fixed number.

6.5. PROPOSITION. *Let $\varphi \in \mathscr{L}_{B,n}$ and let Γ be an HR grammar such that $L(\Gamma) \subseteq \mathbf{FB}(B)_n$. It can be decided in time $O(|d|)$ whether the hypergraph defined by a derivation sequence d of Γ satisfies φ.*

PROOF. By Lemma 3.2, a derivation sequence d' corresponding to d can be constructed of the same length and relative to the regular tree grammar $\bar{\Gamma}$. It produces an expression defining the hypergraph generated by d. This expression is of size $O(|d'|) = O(|d|)$. The result then follows from Proposition 6.3. □

6.6. COROLLARY. *Let B, n, φ, Γ be as in Proposition 6.5. Assume that Γ has a polynomial parsing algorithm. It can be decided in time polynomial in $\mathbf{size}(G)$ which of the following three cases holds (where $G \in \mathbf{FG}(B)_n$):*

(1) $G \notin L(\Gamma)$,

(2) $G \in L(\Gamma)$ *and* $G \models \varphi$,

(3) $G \in L(\Gamma)$ *and* $G \models \neg\varphi$.

PROOF. There exists an integer k such that, for every G in $\mathbf{FG}(B)_n$, one can decide in time $O(\mathbf{size}(G)^k)$ whether $G \in L(\Gamma)$ and one can obtain a derivation sequence d of G if the answer is positive. This sequence can be reduced to d', of length $h \cdot \mathbf{size}(G)$ for some h depending only on G (by the proof of Proposition 3.5), and this transformation can be done in time $O(|d|)$. The result then follows from Proposition 6.5. □

Lautemann has proved in [64] that the HR grammars that generate sets of graphs of bounded s-separability have polynomial parsing algorithms. (A set of graphs is of *bounded s-separability* (with s an integer) if there is a constant k such that for every graph G in the set and for every subset V of V_G of cardinality at most s, the graph obtained from G by the deletion of the vertices of V and of their incident edges has at most k connected components.) On the other hand, there exist HR grammars having an NP-complete membership problem. An example is the grammar generating the set of graphs of cyclic bandwidth at most 2 [67].

But, since every context-free set of graphs (or hypergraphs) is of bounded width, Proposition 6.4 can be used. It suffices to choose k such that all hypergraphs in $L(\Gamma)$ are of width at most $k+1$. Then the quadratic algorithm of Proposition 6.4 yields the following possible answers:

(1) $\mathbf{wd}(G) > k+1$ and $G \notin L(\Gamma)$,

(2) $\mathbf{wd}(G) \leqslant k'$ and $G \models \varphi$,
(3) $\mathbf{wd}(G) \leqslant k'$ and $G \models \neg\varphi$.
In the last two cases, one does not know whether G belongs to $L(\Gamma)$.

7. Infinite hypergraphs

Countably infinite hypergraphs can also be defined by infinite expressions, analogous to formal power series in analysis. These expressions can be manipulated as infinite trees in terms of systems of equations and of tree automata. The corresponding hypergraphs can also be defined by monadic second-order formulas. We fix a finite ranked alphabet B. All hypergraphs are finite or countable.

7.1. *Infinite hypergraph expressions*

An expression in $\mathbf{FE}(B)$ can be considered as a finite tree. Infinite trees constructed over signatures exist (see [19, 49]). They can be considered as *infinite expressions.* We shall consider $\boldsymbol{E}(B)_n$, the set of infinite expressions of sort n. By $\mathbf{FE}(B)$ and $\boldsymbol{E}(B)$ we denote the corresponding $\boldsymbol{H}$-algebras of expressions. In order to simplify notation, we let $\boldsymbol{G} := \boldsymbol{G}(B)$, $\mathbf{GF} := \mathbf{FG}(B)$, $\boldsymbol{E} := \boldsymbol{E}(B)$ and $\mathbf{FE} := \mathbf{FE}(B)$.

7.1. PROPOSITION (Bauderon [8]). *The homomorphism* $\mathbf{val}: \mathbf{FE} \to \mathbf{FG}$ *can be extended to a homomorphism* $\mathbf{val}: \boldsymbol{E} \to \boldsymbol{G}$.

PROOF (*sketch*). The value of an infinite hypergraph expression g is a finite or infinite hypergraph $\mathbf{val}(g)$. Roughly speaking, we can define $\mathbf{val}(g)$ as a quotient of a hypergraph G' obtained from g (considered as a tree) by associating a set of k vertices with every node of g of type k, and by letting hyperedges correspond to the occurrences of the symbols of B in g. The graph $\mathbf{val}(g)$ is then defined as the quotient of G' by an equivalence relation on vertices. □

7.2. PROPOSITION (Bauderon [8]). *Every hypergraph in* $\boldsymbol{G}_n$ *is the value of an expression in* $\boldsymbol{E}_n$.

The width $\mathbf{wd}(g)$ of g in $\boldsymbol{E}_n$ is defined as the least upper bound of the sorts of the symbols occurring in g. It may be infinite. If $G \in \boldsymbol{G}_n$, then $\mathbf{wd}(G) := \min\{\mathbf{wd}(g) \mid g \in \boldsymbol{E}_n, \mathbf{val}(g) = G\}$.

7.3. LEMMA (Courcelle [27]). *A complete graph with countably many vertices has infinite width.*

The next theorem is in some sense an extension of Theorem 4.11. It relates the monadic second-order theory of a hypergraph with the structure of the finite and infinite expressions defining it. The monadic second-order theory of infinite binary trees has been considered by Rabin (see [88]). He has given a characterization in terms of finite-state tree automata of the set of infinite binary trees with nodes labelled in a finite alphabet A that satisfy a formula of **MSOL**. A difficult algorithm on tree automata gives a decidability result for the **MSOL**-theory of the set of infinite trees.

7.2. *Monadic properties of infinite hypergraphs*

In order to express the logical properties of infinite hypergraphs, we shall add to our language **MSOL** new atomic formulas, written **fin**(U). The formula **fin**(U) means that the set U is finite. This extension is denoted by $\mathbf{MSOL}^{\mathrm{f}}$. Since in trees the formulas **fin**(U) are expressible in **MSOL** [75], the languages **MSOL** and $\mathbf{MSOL}^{\mathrm{f}}$ have equivalent power to express tree properties. For graphs however, $\mathbf{MSOL}^{\mathrm{f}}$ is strictly more powerful than **MSOL**. Recall that a set of infinite trees over a finite signature is definable in monadic second-order logic iff it is definable by a Rabin automaton. In the sequel, $\boldsymbol{E}_n$ will be considered as a set of finite and infinite trees.

7.4. THEOREM (Courcelle [27]). *For every closed formula φ in* $\mathbf{MSOL}^{\mathrm{f}}$, *the set of finite and infinite expressions g in $\boldsymbol{E}_n$ of width at most k and such that* $\mathbf{val}(g) \models \varphi$ *is* **MSOL**-*definable.*

Since emptiness of sets defined by Rabin automata is decidable, and by the constructions of [75], we obtain the following theorem.

7.5. THEOREM (Courcelle [27]). *The* $\mathbf{MSOL}^{\mathrm{f}}$-*theory of $\boldsymbol{G}_n^k$ is decidable.*

Theorem 1.5 says that every set of graphs having a decidable monadic second-order theory is included in $\boldsymbol{G}^k$ for some k. Hence there is no hope of relieving the limitation to finite-width hypergraphs.

Let us now clarify the relations between Theorem 7.4 and Theorem 4.11. Recall that Theorem 4.11 states that a set of finite hypergraphs is recognizable if it is **MSOL**-definable. Let $\varphi \in \mathbf{MSOL}$ be closed and let L_φ be the set of finite graphs satisfying it. A consequence of Theorem 7.4 is that the set of *finite* expressions g in $\mathbf{E}_n$ of width at most k and such that $\mathbf{val}(g) \models \varphi$ is **MSOL**-definable as a set of trees. It follows from a theorem by Doner [33] (cf. [88, Theorem 11.1]) that this set of trees is recognizable in $\mathbf{M}(\mathbf{H}_B^k)$. This implies that $L_\varphi \cap \mathbf{FG}_n^k$ is recognizable in the $\mathbf{H}_B^k$-algebra $\mathbf{FG}^k$. It is clear that if L is recognizable in **FG**, then $L \cap \mathbf{FG}_n^k$ is recognizable in $\mathbf{FG}^k$ for all k, but the converse does not hold. Hence, one cannot conclude that L_φ is recognizable as a corollary of Theorem 7.4.

We now introduce the notion of an *equational hypergraph*; it is a hypergraph that is a component of the *canonical solution* of a system of equations in hypergraphs (*not in sets of hypergraphs*). Lengauer [65] considered HR grammars defining sets of graphs reduced to a single finite graph. This notion is interesting because it provides a short description of a graph and efficient algorithms following the techniques of Propositions 6.3 and 6.4. (Such descriptions can be exponentially shorter.) Our systems of equations can be similarly considered as HR graph-grammars generating single infinite graphs.

7.3. *Systems of equations in hypergraphs*

A *regular system of equations over* $\mathbf{G}(B)$ is a system of the form $S=\langle u_1=H_1,\ldots,u_n=H_n\rangle$ where $U=\{u_1,\ldots,u_n\}$ is a ranked alphabet called the *set of unknowns* of S, and where, for every $i=1,\ldots,n$, H_i is a hypergraph in $\mathbf{FG}(B\cup U)_{\tau(u_i)}$. A *solution of* S is an n-tuple of hypergraphs $(G_1,\ldots,G_n)$ with G_i in $\mathbf{G}(B)_{\tau(u_i)}$, and such that $G_i=H_i[G_1/u_1,\ldots,G_n/u_n]$ for all i. (We denote by $H[G_1/u_1,\ldots,G_n/u_n]$ the simultaneous substitution of G_j in H for all edges of H labelled by u_j, $j=1,\ldots,n$; see Subsection 2.6). Solutions of systems can be constructed by means of the following notions of category theory: *the colimit of a diagram* (which generalizes the usual notion of least upper bound of an increasing sequence in a complete partial order), and *the initial fixed point of a functor* (which generalizes the least fixed point of a continuous function in a complete partial order.) We review a few definitions from [1, 7, 8].

7.4. *The initial fixed point of a functor*

Let K be a category, let $F:K\to K$ be a functor. A *fixed point of* F is a pair (X,h) where X is an object of K, and h is an isomorphism $FX\to X$. An *initial fixed point* of F is a fixed point (X_0,h_0) of F such that, for every fixed point (X,h) of F, there is a unique morphism $f:X_0\to X$ making the following diagram commutative:

$$\begin{array}{ccc} FX_0 & \xrightarrow{h_0} & X_0 \\ {\scriptstyle Ff}\downarrow & & \downarrow{\scriptstyle f} \\ FX & \xrightarrow{h} & X \end{array}$$

If F has an initial fixed point, then it is unique up to isomorphism. We also call it the *initial solution* of the equation $X=FX$. The following lemma is a special case of Proposition 5 of Adamek and Koubek [1].

7.6. LEMMA. *Let K be a category having an initial object $\mathbb{1}$. Let $F:K\to K$ be a functor. Let w be the unique morphism: $\mathbb{1}\to F\mathbb{1}$. If the diagram*

$$\mathbb{1}\xrightarrow{w} F\mathbb{1}\xrightarrow{fw} F^2\mathbb{1}\xrightarrow{F^2w}\ \cdots\to F^n\mathbb{1}\xrightarrow{F^nw}\ \cdots$$

has a colimit X and if the canonical morphism $h:X\to FX$ is an isomorphism, then (X,h^{-1}) is the initial fixed point of F.

The canonical morphism h is defined as follows. Since X is the colimit, one has a morphism $w_n:F^n\mathbb{1}\to X$ such that $w_n=F^nw\cdot w_{n+1}$. Hence, one has a family of morphisms $Fw_n:F^{n+1}\mathbb{1}\to FX$, such that $Fw_n=F^{n+1}w\cdot Fw_{n+1}$, and a unique morphism $h:X\to FX$ by the universal property of the colimit. This is illustrated in the diagram in Fig. 6.

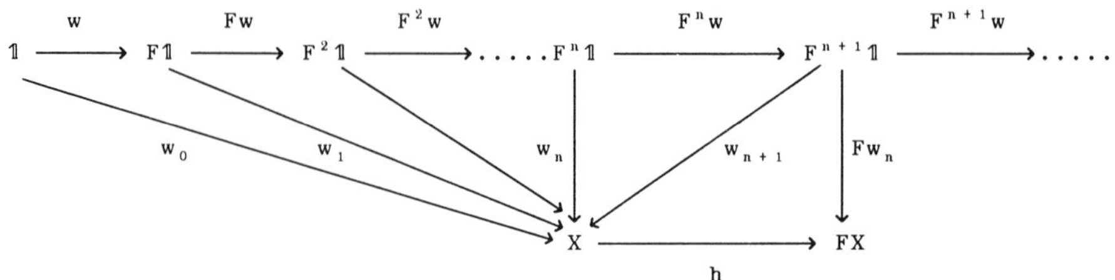

Fig. 6.

7.5. *The initial solution of a system of equations in hypergraphs*

Let $S=\langle u_1=H_1,\ldots,u_n=H_n\rangle$ be a system of equations in the general form of Subsection 7.3. We let $p_i=\tau(u_i)=\tau(H_i)$ for all i. With S we associate a category K_S, and a functor $F_S\colon K_S\to K_S$. The objects of K_S are n-tuples of hypergraphs $(K_1,\ldots,K_n)$ in $\boldsymbol{G}(B)_{p_1}\times\cdots\times\boldsymbol{G}(B)_{p_n}$. A morphism $h\colon(K_1,\ldots,K_n)\to(K'_1,\ldots,K'_n)$ is an n-tuple $h=(h_1,\ldots,h_n)$ of hyperedge-injective graph homomorphisms where $h_i\colon K_i\to K'_i$ for all i. This category has an initial object, denoted by 𝟙, equal to the n-tuple of graphs $(\boldsymbol{p}_1,\boldsymbol{p}_2,\ldots,\boldsymbol{p}_n)$.

We now define a functor $F_S\colon K_S\to K_S$ as follows:

- $F_S(K_1,\ldots,K_n)=(K'_1,\ldots,K'_n)$, where $K'_i=H_i[K_1/u_1,\ldots,K_n/u_n]$ for all i;
- if $h=(h_1,\ldots,h_n)\colon(K_1,\ldots,K_n)\to(K'_1,\ldots,K'_n)$, then $F_S(h)$ is the morphism $(h'_1,\ldots,h'_n)$ where h'_i is the morphism extending $(h_1,\ldots,h_n)$ to

$$H_i[K_1/u_1,\ldots,K_n/u_n]\to H_i[K'_1/u_1,\ldots,K'_n/u_n].$$

It is proved in [7, 8] that K_S and F_S satisfy the conditions of Lemma 7.6. Hence, F_S has an initial fixed point that is, by definition, the *initial solution of S*.

A regular system of equations can also be *solved formally* in the domain $\boldsymbol{E}(B)$ of finite and infinite expressions. It suffices to replace each H_i by an expression denoting it and to solve the associated system $\bar{S}$ in the appropriate algebra of infinite trees. (See [19] for regular systems of equations and their solutions in infinite trees.)

7.7. THEOREM. *Let S be a regular system of equations in hypergraphs, let $(G_1,\ldots,G_n)$ be its initial solution and let $(g_1,\ldots,g_n)$ be the formal solution of any one of the associated systems $\bar{S}$. Then $(G_1,\ldots,G_n)=(\mathbf{val}(g_1),\ldots,\mathbf{val}(g_n))$.*

Hence, the same tuple of graphs can be obtained in two different ways: either by a "least fixed-point" construction in an appropriate category, or by a formal resolution of the system following by an evaluation. (It follows in particular that the tuple $(\mathbf{val}(g_1),\ldots,\mathbf{val}(g_n))$ does not depend on the choice of $\bar{S}$.) A similar situation occurs in the theory of recursive applicative program schemes: the least solution of a system of mutually recursive definitions is the tuple of functions canonically associated with the least solution of the system in the domain of infinite trees (see [30, 49] for other, similar situations).

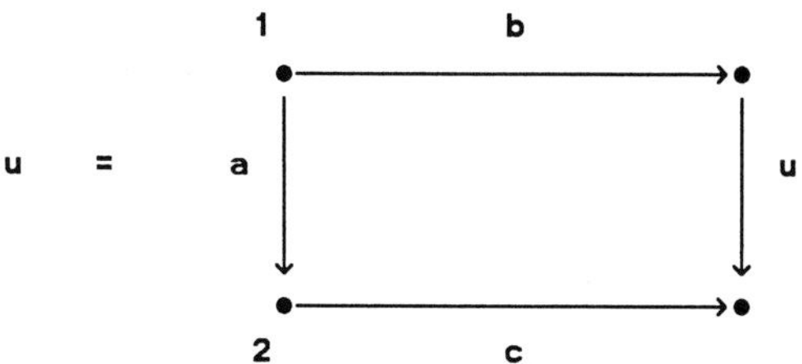

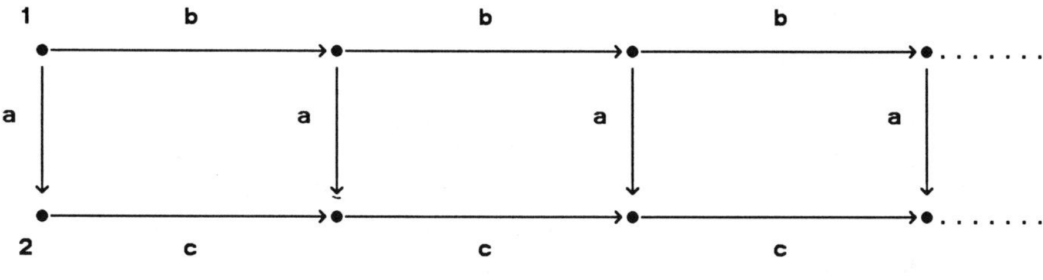

Fig. 7.

7.8. EXAMPLE. Consider the system S reduced to the equation shown in Fig. 7, where u is an unknown of type 2. Its solution is the infinite graph partially shown in the lower part of Fig. 7.

In terms of graph expressions, this equation can be written as $u = \sigma_\alpha(\theta_\delta(a \oplus b \oplus u \oplus c))$ where α is the identity mapping and δ is the equivalence relation on [8] generated by the set of pairs $\{(1,3), (4,5), (6,7), (2,8)\}$. Its solution in $\boldsymbol{E}_2$ is the regular tree of Fig. 8, the value of which is the "ladder" graph in Fig. 7.

7.9. COROLLARY. *A hypergraph is equational iff it is the value of an expression in* **E** *that is regular when considered as a tree. Hence, an equational graph is of finite width.*

Since by a theorem of Rabin [76], every nonempty **MSOL**-definable set of infinite trees contains a regular tree, the following result is a consequence of Theorem 7.4 and Corollary 7.9.

7.10. COROLLARY. *If a closed monadic second-order formula is satisfied by a hypergraph of finite width, then it is satisfied by an equational one.*

7.6. *Monadic properties of equational hypergraphs*

7.11. THEOREM. *The* $\mathbf{MSOL}^{\mathrm{f}}$*-theory of an equational hypergraph is decidable.*

PROOF. Let G be an equational n-hypergraph, given by a system of equations. One can construct k and a regular tree g in $\boldsymbol{E}_n^k$ such that $\mathbf{val}(g) = G$. One can also construct

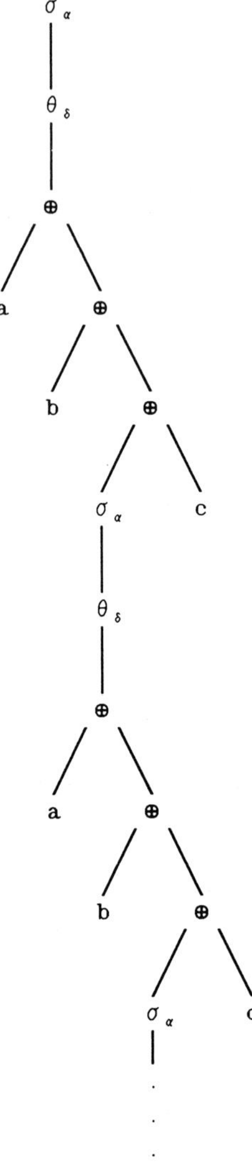

Fig. 8.

a Rabin automaton defining $\{g' \in \boldsymbol{E}_n^k \mid \mathbf{val}(g') \models \varphi\}$, where φ is a given formula in $\mathbf{MSOL}^{\mathrm{f}}$. By the main theorem of [75] one can decide whether g belongs to this set, i.e., whether $G \models \varphi$. $\square$

We say that a graph G is *$\mathscr{L}$-definable* for some logical language $\mathscr{L}$ if the set $\{G\}$ is definable in this language, i.e., if G can be characterized up to isomorphism by a formula of $\mathscr{L}$.

7.12. THEOREM (Courcelle [27, 29]). *A hypergraph is equational iff it is* **MSOL**-*definable and of finite width.*

PROOF (*outline*). Let $G \in \boldsymbol{G}_n^k$ be defined by φ. From the definable set $\{t \in \boldsymbol{E}_n^k \mid \mathbf{val}(t) \models \varphi\}$ one can extract a regular tree t_0. Hence, $\mathbf{val}(t_0)$ is an equational graph G_0 such that $G_0 \models \varphi$. Since φ defines G, we have $G = G_0$,

The other direction is a long and technical proof that constructs, from a regular system of equations over $\boldsymbol{G}(B)$, a formula φ that defines the first component of the solution of the system. □

As a consequence of Theorem 1.5, if an $\mathbf{MSOL}^{\mathrm{f}}$-definable graph has a decidable monadic second-order theory, then it is of finite width and hence, it is equational. Here is a consequence of Theorems 7.11 and 7.12.

7.13. COROLLARY. *The equality problem for equational graphs is decidable.*

8. Guide to the literature

This section describes references and related work not cited in the main text.

Section 1

Finite first-order relational structures can be considered as finite relational databases. This fact has motivated a lot of work on the characterization of the expressive powers of logical languages. We refer the reader to the chapter in this Handbook by Kanellakis [59] for a survey of these results and a bibliography. A systematic classification of graph properties according to their logical definitions remains to be done. Theorem 1.4 and the results of Section 6 motivate such a classification.

Section 2

This section is based on the work by Bauderon and Courcelle [9]. An informal presentation can be found in [22]. The idea of using a sequence of distinguished vertices as "gluing points" in order to combine graphs into larger ones also appears in [52, 66].

Algebraic structures on graphs have also been proposed by Schmeck [85] in order to formalize flowcharts, and by Hotz et al. [53] in order to describe VLSI layouts algebraically. Whether these various structures are equivalent to the one presented in Section 2 and in what precise sense they can be remains to be investigated in depth.

This idea of formalizing graph rewriting rules with push-out diagrams was first introduced by Ehrig, Pfender and Schneider in [40]. It has been developed by Ehrig and other authors in several papers, among which we quote only [15, 36, 82]. A bibliography has been established by Nagl [73], and two tutorial papers have been written by Ehrig [34, 35].

We have considered only the relations between hypergraphs defined by single productions. The simultaneous application of several rules to the same hypergraph has been considered in several papers. We refer the reader to Boehm et al. [15] and to the survey papers of Kreowski and Wilharm [62, 63], where full lists of references can be found. These results can certainly be reformulated (and perhaps also proved) in the framework of hypergraph rewriting rules. This reformulation remains an open research topic.

Let us finally mention two related studies on graph rewriting. Raoult [77] and Kennaway [61] consider categorical graph rewriting rules defined in terms of *single push-out diagrams*, in contrast with the double push-out diagrams of Subsection 2. This is convenient for their purpose, namely the rewriting of graphs representing terms with sharing. A precise comparison of the two approaches of Subsections 2.7 and 2.8 in a more general case remains to be done. (Work in this direction has been announced in [61]).

Section 3

Edge-replacement grammars dealing only with binary graphs (the edges of which have two vertices) have been investigated by Habel and Kreowski in [51]. Hyperedge-replacement grammars have been considered in [52] and by Montanari and Rossi in [71], with the restriction that all hypergraphs have pairwise distinct sources. This restriction is not present in [9].

We have said that hyperedge-replacement grammars are *context-free*. An analysis of the notion of context-freeness for general graph- (and hypergraph-) grammars has been done by Courcelle [21]. The basic aspects of context-freeness are the following ones:

(1) A rewriting step is performed at a certain *occurrence of a nonterminal component*: vertex, edge, hyperedge or subgraph (called a "handle" in [31]), depending on the case.

(2) Such an occurrence has a *context*, and the graph (or hypergraph) is totally defined by this occurrence and its context.

(3) Whether a production rule is applicable at an occurrence depends *only on the nonterminal symbol of the occurrence*, and not on its context.

(4) The context is *not modified* in the rewriting step.

(5) The derivation sequences can be described by *derivation trees*. Different traversals of a derivation tree correspond to different rewriting sequences yielding the same graph. Every derivation sequence is thus equivalent to a *leftmost* (or a *rightmost*) one.

(6) A *system of equations* in sets of graphs (or hypergraphs) can be associated with a context-free grammar, and its *least solution* is the tuple of sets generated by the nonterminals of the grammar.

The last requirement establishes links between graph rewritings and algebras of graphs (or hypergraphs). The so-called *context-free tree-grammars* [43] are *not* context-free as graph grammars in the above sense: they do not satisfy requirement (4) (subtrees of the context can be duplicated, or can disappear), and the IO-grammars do not satisfy requirement (3) (the subtrees below the derived occurrence must be terminal). The context-free-like grammars of [60, 56, 83, 84] satisfy requirements (1)–(5). We say that they are *confluent*. Requirement (6) is not always satisfied for these classes. It is nevertheless useful because proof techniques can be based on it and decidability results can be obtained: see [20, Section 16.6] on "iterative algorithms" and Theorem 4.1.

We now explain how a system of equations can be associated with a context-free grammar Γ. Let D be a hypergraph with finitely many occurrences of nonterminals, numbered from 1 to k. We can make D into an operation $\bar{D}$ that associates with $H_1, \ldots, H_k$ a hypergraph $\bar{D}(H_1, \ldots, H_k)$ constructed by substituting successively H_i for the ith nonterminal occurrence in D for $i=1, \ldots, k$. This defines an algebra of hypergraphs with infinitely many operations. With Γ, we associate a system of equations as follows. Each nonterminal is considered as an unknown. If $D_1, \ldots, D_m$ are the right-hand sides of the production rules with left-hand side u_i, then the corresponding equation is

$$u_i = \bar{D}_1(u_{1,1}, \ldots, u_{1,k_1}) \cup \cdots \cup \bar{D}_m(u_{m,1}, \ldots, u_{m,k_m})$$

where $(u_{i,1}, \ldots, u_{i,k_i})$ is the sequence of nonterminals of D_i. It can be proved that the n-tuple of generated sets is the least solution of the associated system. This gives a general version of Theorem 3.4 but the corresponding algebra is rather ad hoc: it is associated with each grammar. In Proposition 3.3 the algebraic structure is defined in terms of three primitive operations.

Section 4

Janssens, Rozenberg and Welzl [55, 56, 83, 84] have proved that certain classes of graph-grammars satisfy Theorem 4.4 for specific graph properties like k-colorability, planarity or connectedness, as opposed to monadic second-order ones. Their results have been extended by Courcelle [21] to all graph properties expressible in **MSOL** with vertex quantifications only. Similar results have been obtained by Habel [50], and by Lengauer and Wanke [66] for HR grammars and for graph properties that are "compatible with context-free graph derivations" [50], or that are "finite" [66].

The compatibility of a graph property with an HR grammar introduced by Habel [50] consists in saying that this property belongs to a finite F'-inductive family of properties, where F' is the set of operations occurring in the system of equations associated with the given grammar. Lengauer and Wanke [66] have introduced the notion of a finite graph property. It follows from Lemma 4.6 that a graph property is finite iff it is decidable and the set of graphs satisfying it is recognizable. Corresponding versions of Theorem 4.4, where "monadic second-order" is replaced by "compatible" or "finite", can be found in [50, 66].

Theorem 4.4(2) also holds for *controlled grammars*. Let Γ be a context-free grammar of any kind (generating words, trees, graphs; see [21]), let $T(\Gamma)$ be its set of derivation trees. A controlled grammar is then a pair (Γ, C) consisting of Γ as above and $C \subseteq T(\Gamma)$. The set generated by (Γ, C) is the set of objects generated by Γ, with derivation trees in C. It is denoted by $L(\Gamma, C)$. If Γ is an HR grammar, it follows from Theorem 4.3 that the set of trees in Γ that define graphs satisfying a given formula φ in **MSOL** is recognizable. If one can decide whether $C \cap T = \emptyset$ for every recognizable subset of $T(\Gamma)$, then one can decide whether some G in $L(\Gamma, C)$ satisfies φ. Controlled grammars are considered in this way in [66].

Theorem 4.4 has been proved in [26] for an extension of **MSOL** called *counting* monadic second-order logic and denoted by **CMSOL**. In this language, additional atomic formulas of the form $\mathbf{card}_{p,q}(U)$ can say whether the cardinality of a set is even.

However, for expressing properties of strings and, more generally, of trees of bounded degree, **CMSOL** is not more powerful than **MSOL**.

Section 5.

In the comparison diagram of Section 5, there could have been an intermediate box between **REC** and **MSOL**, namely the class of sets of finite graphs definable in counting monadic second-order logic. We conjecture that if $L \in \mathbf{REC} \cap \mathbf{CF}$, then it is definable in **CMSOL**.

Section 6

In [57] Johnson discusses a few NP-complete problems that become polynomial when restricted to special classes of graphs like trees, planar graphs or perfect graphs. Many papers establish results of this kind for special classes of graphs of bounded width, and most of the problems they consider are expressible in the extended monadic second-order logic of Arnborg et al. [3] (see also the remarks following Proposition 6.4 and Chapter 10 on graph algorithms in Volume A of this Handbook).

Examples of such classes of graphs are the classes of graphs of bandwidth at most k considered by Monien and Sudborough [70], the class of partial k-trees considered by Arnborg et al. [3, 4] and by Bodlaender [12–14] and the k-terminal recursive classes of Wimer et al. [91]. All these results use bottom-up evaluations on the finite trees corresponding to tree decompositions or to derivation trees with respect to graph grammars. The first paper using this idea was by Takamizawa et al. [87], dealing with series-parallel grammars. Another one was by Bern et al. [10] where subgraphs, and not only truth values, are computed.

The grammatical version of these results, namely Proposition 6.5, can be applied to confluent NLC grammars and to monadic second-order formulas with quantifications on vertices or sets of vertices (excluding quatifications on edges and sets of edges). This generalizes some results of Rozenberg and Welzl [84], since the boundary NLC grammars they consider are confluent.

Section 7

Applications to recursive program schemes and to sets of finite and infinite graphs defined by forbidden minors can be found respectively in [27, 28]. Muller and Schupp [72] consider infinite graphs called *context-free graphs*. They prove that the monadic theory of a context-free graph is decidable. A context-free graph in their sense is equational (see [8]), hence that their result is a special case of Theorem 7.11.

Acknowledgment

Many references used in this survey have been communicated to me by W. Thomas and D. Seese. I thank them, together with J. Engelfriet and H. Bodlaender for useful and pleasant discussions. I also thank K. Callaway and F. Clairand for their help in producing the manuscript. A first version was published as a survey in the Bulletin of EATCS, Volume 36, October 1988.

This work has been supported by the "Programme de Recherches Coordonnées: Mathématiques et Informatique".

References

[1] ADAMEK, J. and V. KOUBEK, Least fixed point of a functor, *J. Comput. System Sci.* **19** (1979) 163–178.
[2] ARNBORG, S., D. CORNEIL and A. PROSKUROWSKI, Complexity of finding an embedding in a *k*-tree, *SIAM J. of Algebraic Discrete Methods* 8 (1987) 277–284.
[3] ARNBORG, S., J. LAGERGREN and D. SEESE, Problems easy for tree decomposable graphs, in: T. Lepistö and A. Salomaa, eds., *Proc. 15th Internat. Coll. on Automata, Languages and Programming*, Lecture Notes in Computer Science, Vol. 317 (Springer, Berlin, 1988) 38–51.
[4] ARNBORG, S. and A. PROSKUROWSKI, Linear time algorithms for NP-hard problems on graphs embedded in *k*-trees, Tech. Report TRITA-NA 8404, Stockholm, 1984.
[5] ARNBORG, S., A. PROSKUROWSKI and D. CORNEIL, Forbidden minors characterization of partial 3-trees, Report CIS-TR-8607, Univ. of Oregon, 1986.
[6] BARWISE, J., ed., *Handbook of Mathematical Logic* (North-Holland, Amsterdam, 1977).
[7] BAUDERON, M., One systems of equations defining infinite graphs, in: J. van Leeuwen, ed., *Internat. Workshop on Graph-Theoretic Concepts in Computer Science*, Lecture Notes in Computer Science, Vol. 344 (Springer, Berlin, 1989) 54–73.
[8] BAUDERON, M., Infinite hypergraphs, *Theoret. Comput. Sci.*, to appear.
[9] BAUDERON, M. and B. COURCELLE, Graph expressions and graph rewritings, *Math. Systems Theory* **20** (1987) 83–127.
[10] BERN, M., E. LAWLER and A. WONG, Linear time computation of optimal subgraphs of decomposable graphs, *J. Algorithms* **8** (1987) 216–235.
[11] BERSTEL, J. and L. BOASSON, Context-free languages, in: J. van Leeuwen, ed., *Handbook of Theoretical Computer Science, Vol. B* (North-Holland, Amsterdam, 1990) 59–102.
[12] BODLAENDER, H., Polynomial algorithms for chromatic index and graph isomorphism on partial *k*-trees, in: R. Karlsson and A. Lingas, eds., *Proc. 1st Scandanavian Workshop on Algorithmic Theory*, Lecture Notes on Computer Science, Vol. 318 (Springer, Berlin, 1988) 223–232.
[13] BODLAENDER, H., Dynamic programming on graphs with bounded tree-width, in: T. Lepistö and A. Salomaa, eds., *Proc. 15th Internat. Coll. on Antomata, Languages and Programming*, Lecture Notes in Computer Science, Vol. 317 (Springer, Berlin, 1988) 105–118.
[14] BODLAENDER, H., NC-algorithms for graphs with small tree-width, in: J. van Leeuwen, ed., *Internat. Workshop on Graph-Theoretic Concepts in Computer Science*, Lecture Notes in Computer Science, Vol. 344 (Springer, Berlin, 1989) 1–10.
[15] BOEHM, P., H.-R. FONIO and A. HABEL, Amalgamation of graph transformations: a synchronization mechanism, *J. Comput. System Sci.* **34** (1987) 377–408.
[16] BOOK, R., ed., *Formal Languages, Perspectives and open Problems* (Academic Press, New York, 1980).
[17] BÜCHI, J., Weak second order logic and finite automata, *Z. Math. Logik Grundlag. Math* **5** (1960) 66–92.
[18] CLAUS, V., H. EHRIG and G. ROZENBERG, eds., *1st International Workshop, on Graph-Grammars and their Application to Computer Science and Biology*, Lecture Notes in Computer Science, Vol. 73 (Springer, Berlin, 1979).
[19] COURCELLE, B., Fundamental properties of infinite trees, *Theoret. Comput. Sci.* **25** (1983) 95–109.
[20] COURCELLE, B., Equivalence and transformations of regular systems; applications to recursive program schemes and grammars, *Theoret. Comput. Sci.* **42** (1986) 1–122.
[21] COURCELLE, B., An axiomatic definition of context-free rewriting and its application to NLC graph grammars, *Theoret. Comput. Sci.* **55** (1987) 141–181.
[22] COURCELLE, B., A representation of graphs by algebraic expressions and its use for graph rewriting systems, in: [39] 112–132.
[23] COURCELLE, B., On context-free sets of graphs and their monadic second-order theory, in: [39] 133–146.

[24] COURCELLE, B., The monadic second-order logic of graphs: definable sets of finite graphs, in: J. van Leeuwen, ed., *Internat. Workshop on Graph-Theoretic Concepts in Computer Science*, Lecture Notes in Computer Science, Vol. 344 (Springer, Berlin, 1989) 30–53.

[25] COURCELLE, B., On using context-free graph grammars for analyzing recursive definitions, in: K. Fuchi and L. Kott, eds., *Programming of Future-generation Computers II* (North-Holland, Amsterdam, 1988) 83–122.

[26] COURCELLE, B., The monadic second-order theory of graphs I: recognizable sets of finite graphs, *Inform. and Comput.* **85** (1990) 12–75.

[27] COURCELLE, B., The monadic second-order logic of graphs II: infinite graphs of bounded width, *Math. Systems Theory* **21** (1990) 187–221.

[28] COURCELLE, B., The monadic second-order logic of graphs III: tree-width, forbidden minors, and complexity issues, Report 8852, Lab. d'Informatique, Univ. Bordeaux I, 1988.

[29] COURCELLE, B., The monadic second-order logic of graphs IV: every equational graph is definable, Research Report 8830, Lab. d'Informatique, Univ. Bordeaux I, 1988.

[30] COURCELLE, B., Recursive applicative program schemes, in: J. van Leeuwen, ed., *Handbook of Theoretical Computer Science, Vol. B* (North-Holland, Amsterdam, 1990) 459–492.

[31] COURCELLE, B., J. ENGELFRIET and G. ROZENBERG, Handle rewriting hypergraph grammars, in: *Proc. 4th Internat. Workshop on Graph Grammars*, Bremen, Lecture Notes in Computer Science (Springer, Berlin, 1990) to appear.

[32] DERSHOWITZ, N. and J.P. JOUANNAUD, Rewrite systems, in: J. van Leeuwen, ed., *Handbook of Theoretical Computer Science, Vol. B* (North-Holland, Amsterdam, 1990) 243–320.

[33] DONER, J., Tree acceptors and some of their applications, *J. Comput. System Sci.* **4** (1970) 406–451.

[34] EHRIG, H., Introduction to the algebraic theory of graph grammars, in: [18] 1–69.

[35] EHRIG H., Tutorial introduction to the algebraic approach of graph grammars, in: [39] 3–14.

[36] EHRIG H. H.J. KREOWSKI, A. MAGGIOLO-SCHETTINI, B. ROSEN and J. WINKOWSKI, Transformations of structures: an algebraic approach, *Math. Systems Theory* **14** (1981) 305–334.

[37] EHRIG H. and B. MAHR, *Fundamentals of Algebraic Specification I* (Springer, Berlin, 1985).

[38] EHRIG, H., M. NAGL and G. ROZENBERG, eds., *2nd Internat. Workshop on Graph-Grammars and their Application to Computer Science*, Lecture Notes in Computer Science, Vol. 153 (Springer, Berlin, 1983).

[39] EHRIG, H., M. NAGL, G. ROZENBERG and A. ROSENFELD, eds., *3rd Internat. Workshop on Graph-Grammars and their Application to Computer Science*, Lecture Notes in Computer Science, Vol. 291, (Springer, Berlin, 1987).

[40] EHRIG, H., M. PFENDER and H. SCHNEIDER, Graph grammars: an algebraic approach, in: *Proc. 14th IEEE Symp. on Switching and Automata Theory*, (1973) 167–180.

[41] ENGELFRIET, J. and G. ROZENBERG, Fixed point languages, equality languages, and representations of recursively enumerable languages, *J. ACM* **27** (1980) 499–518.

[42] ENGELFRIET, J. and G. ROZENBERG, A comparison of boundary graph grammars and context-free hypergraph grammars, *Inform. and Comput.*, **84** (1990) 163–206.

[43] ENGELFRIET, J. and E. SCHMIDT, IO and OI, *J. Comput. System Sci.* **15** (1977) 328–353 and 16 (1978) 67–99.

[44] FAGIN, R., Generalized first-order spectra and polynomial-time recognizable sets, in: R. Karp, ed., *Complexity of Computation*, SIAM-AMS Proceedings, Vol.7 (1974) 43–73.

[45] GAIFMAN, H., On local and non-local properties, J. Stern, ed., *Proc. of the Herbrand Symposium, Logic Colloqium '81* (North-Holland, Amsterdam, 1982) 105–135.

[46] GAREY, M. and D. JOHNSON, *Computers and Intractability: A Guide to the Theory of NP-Completeness* (Freeman, San Francisco, CA 1979).

[47] GECSEG, F. and M. STEINBY, *Tree-automata* (Akademiai Kiado, Budapest, 1984).

[48] GINSBURG, S. and H. RICE, Two families of languages related to ALGOL, *J. ACM* **9** (1962) 350–371.

[49] GOGUEN, J., J. THATCHER, E. WAGNER and J. WRIGHT, Initial algebra semantics and continuous algebras, *J.ACM* **24** (1977) 68–95.

[50] HABEL A., Graph-theoretic properties compatible with graph derivations, in: J. van Leeuwen, ed., *Internat. Workshop on Graph-Theoretic Concepts in Computer Science*, Lecture Notes in Computer Science, Vol. 344 (Springer, Berlin, 1989). 11–29.

[51] HABEL, A. and H.-J. KREOWSKI, Characteristics of graph languages generated by edge replacements, *Theoret. Comput. Sci.* **51** (1987) 81–115.

[52] HABEL, A. and H.-J. KREOWSKI, May we introduce to you: hyperedge replacement, in: [39] 15–26.

[53] HOTZ, G., R. KOLLA and P. MOLITOR, On network algebras and recursive equations, in: [39] 250–261.

[54] IMMERMAN, N., Languages that capture complexity classes, *SIAM J. Comput.* **16** (1987) 760–777.

[55] JANSSENS, D., G. ROZENBERG, A survey of NLC grammars, in: G. Ausiello and M. Protasi, eds., *Proc. 8th Coll. CAAP'83*, Lecture Notes in Computer Science, Vol. 159 (Springer, Berlin, 1983) 114–128.

[56] JANSSENS, D. and G. ROZENBERG, Neighborhood uniform NLC grammars, *Computer Vision, Graphics and Image Processing* **35** (1986) 131–151.

[57] JOHNSON, D., The NP-completeness column: an ongoing guide (16th edition), *J. Algorithms* **6** (1985) 434–451.

[58] JOHNSON, D., The NP-completeness column: an on going guide (19th edition), *J. Algorithms* **8** (1987) 285–303.

[59] KANELLAKIS, P., Elements of relational database theory, in: J. van Leeuwen, ed., *Handbook of Theoretical Computer Science, Vol. B* (North-Holland, Amsterdam, 1990) 1073–1156.

[60] KAUL, M., Syntaxanalyse von Graphen bei Praezedenz-Graph-Grammatiken, Ph.D. Thesis, Univ. Passau, Passau, 1986.

[61] KENNAWAY, R., On "On graph rewritings", *Theoret. Comput. Sci.* **52** (1987) 37–58; correction, *Theoret. Comput. Sci.* **61** (1988) 317–320.

[62] KREOWSKI, H-J., Is parallelism already concurrency? Part 1: Derivations in graph grammars, in: [39] 343–360.

[63] KREOWSKI, H-J. and A. WILHARM, Is parallelism already concurrency? Part 2: Nonsequential processes in graph grammars, in: [39] 361–377.

[64] LAUTEMANN, C., Efficient algorithms on context-free graph languages, in: T. Lepistö and A. Salomaa, eds., *Proc. 15th Internat. Coll. on Automata, Languages and Programming*, Lecture Notes in Computer Science, Vol. 317 (Springer, Berlin, 1988) 362–378.

[65] LENGAUER, T., Efficient algorithms for finding minimum spanning forests of hierarchically defined graphs, *J. Algorithms* **8** (1987) 260–284.

[66] LENGAUER, T., E. WANKE, Efficient analysis of graph properties on context-free graph languages, in: T. Lepistö and A. Salomaa, eds., *Proc. 15th Internat. Coll. on Automata, Languages and Programming*, Lecture Notes in Computer Science, Vol. 317 (Springer, Berlin, 1988) 379–393.

[67] LEUNG, J., J. WITTHOF and O. VORNBERGER, On some variations on the bandwidth minimization problem, *SIAM J. Comput.* **13** (1984) 650–667.

[68] MCLANE, S., *Category Theory for the Working Mathematician* (Springer, New York, 1971).

[69] MEZEI, J. and J. WRIGHT, Algebraic automata and context-free sets, *Inform. and Control* **11** (1967) 3–29.

[70] MONIEN, B. and I. SUDBOROUGH, Bandwidth constrained NP-complete problems, in: *Proc. 13th Ann. ACM Symp. on Theory of Computing* (1981) 207–217.

[71] MONTANARI, U. and F. ROSI, An efficient algorithm for the solution of hierarchical networks of constraints, in: [39] 440–457.

[72] MULLER, D. and P. SCHUPP, The theory of ends, pushdown automata, and second-order logic, *Theoret. Comput. Sci.* **37** (1985) 51–75.

[73] NAGL, M., Bibliography on graph-rewriting systems (graph grammars), in: [38] 415–448.

[74] NIVAT, M. and D. PERRIN, eds., *Automata on Infinite Words*, Lecture Notes in Computer Science, Vol. 192 (Springer, Berlin, 1985).

[75] RABIN, M., Decidability of second-order theories and automata on infinite trees, *Trans. Amer. Math. Soc* **141** (1969) 1–35.

[76] RABIN, M., *Automata on Infinite Objects and Church's Problem*, A.M.S. Regional Conference Series in Mathematics, Vol. 13 (Amer. Mathematical Soc., Providence, RI, 1972).

[77] RAOULT, J.C., On graph rewritings, *Theoret. Comput. Sci.* **32** (1984) 1–24.

[78] ROBERTSON, N. and P. SEYMOUR, Some new results on the well-quasi-ordering of graphs, in: Annals of Discrete Mathematics, Vol. 23 (Elsevier, Amsterdam, 1984) 343–354.

[79] ROBERTSON, N. and P. SEYMOUR, Graph minors V: excluding a planar graph, *J. Combin. Theory Ser. B* **41** (1986) 92–114.

[80] ROBERTSON, N. and P. SEYMOUR, Graph minors XIII: the disjoint paths problem, Preprint, September 1986.

[81] ROBERTSON, N. and P. SEYMOUR, Graph minors XV: Wagner's conjecture, Preprint, March 1988.

[82] ROSEN, B., Deriving graphs from graphs by applying a production, *Acta Inform.* **4** (1975) 337—357.

[83] ROZENBERG, G. and E. WELZL, Graph-theoretic closure properties of the family of boundary NLC graph languages, *Acta Inform.* **23** (1986) 289–309.

[84] ROZENBERG, G. and E. WELZL, Boundary NLC grammars: basic definitions, normal forms and complexity, *Inform. and Control* **69** (1986) 136–167.

[85] SCHMECK, H., Algebraic characterization of reducible flowcharts, *J. Comput. System Sci.* **27** (1983) 165–199.

[86] SEESE, D., The structure of the models of decidable monadic theories of graphs, *Annals of Pure and Applied Logic*, to appear.

[87] TAKAMIZAWA, K., T. NISHIZEKI and N. SAITO, Linear time computability of combinatorial problems on series-parallel graphs, *J.ACM* **29** (1982) 623–641.

[88] THOMAS, W., Automata on infinite objects, in: J. van Leeuwen, ed., *Handbook of Theoretical Computer Science, Vol. B* (North-Holland, Amsterdam, 1990) 133–191.

[89] TRAKHTENBROT, B., Impossibility of an algorithm for the decision problem on finite classes, *Dokl. Akad. Nauk SSSR* **70** (1950) 569–572.

[90] WAGNER, K., Über eine Eigenshaft der ebenen Komplexe, *Math. Ann.* **114** (1937) 570–590.

[91] WIMER, T., S. HEDETNIEMI and R. LASKAR, A methodology for constructing linear graph algorithms, *Congressus Numerantium* **50** (1985) 43–60.

[92] WIRSING M., Algebraic specification, in: J. van Leeuwen, ed., *Handbook of Theoretical Computer Science, Vol. B* (North-Holland, Amsterdam, 1990) 1201–1242.

CHAPTER 6

Rewrite Systems

Nachum DERSHOWITZ

Department of Computer Science, University of Illinois, Urbana, IL 61801, USA

Jean-Pierre JOUANNAUD

Laboratoire de Recherche en Informatique, Université de Paris-Sud, F-91405 Orsay, France

Contents

1. Introduction 245
2. Syntax 248
3. Semantics 260
4. Church–Rosser properties 266
5. Termination 269
6. Satisfiability 279
7. Critical pairs 285
8. Completion 292
9. Extensions 305
Acknowledgment 309
References. 309

HANDBOOK OF THEORETICAL COMPUTER SCIENCE
Edited by J. van Leeuwen

1. Introduction

Equations are ubiquitous in mathematics and the sciences. Sometimes one tries to determine if an identity follows logically from given axioms; other times, one looks for solutions to a given equation. These reasoning abilities are also important in many computer applications, including symbolic algebraic computation, automated theorem proving, program specification and verification, and high-level programming languages and environments.

Rewrite systems are directed equations used to compute by repeatedly replacing subterms of a given formula with equal terms until the simplest form possible is obtained. The idea of simplifying expressions has been around as long as algebra has. As a form of computer program, rewrite systems made their debut in Gorn [94]; many modern programs for symbolic manipulation continue to use rewrite rules for simplification in an ad hoc manner. As a formalism, rewrite systems have the full power of Turing machines and may be thought of as nondeterministic Markov algorithms over terms, rather than strings. (Regarding Markov algorithms, see, e.g. Tourlakis [233].) The theory of rewriting is in essence a theory of normal forms; to some extent it is an outgrowth of the study of Church's Lambda Calculus and Curry's Combinatory Logic.

To introduce some of the central ideas in rewriting, we consider several variations on the "Coffee Can Problem" (attributed to C.S. Scholten in Gries [95]). Imagine a can containing coffee beans of two varieties, *white* and *black*, arranged in some order. Representing the contents of the can as a sequence of bean colors, e.g.

white white black black white white black black,

the rules of our first game are as follows:

black white	$\rightarrow$	*black*
white black	$\rightarrow$	*black*
black black	$\rightarrow$	*white*

This set of rules is an example of a rewrite system. Each rule describes a legal move: the first two state that, at any stage of the game, the white bean of any adjacent pair of different beans may be discarded; the last rule states that two adjacent black beans may be replaced by one white one (an unlimited supply of white beans is on hand). For example, the following is a possible sequence of moves (the underlined beans participate in the current move):

white white black <u>black white</u> white black black
white white <u>black black</u> white black black
white white white <u>white black</u> black
white white <u>white black</u> black
white <u>white black</u> black
<u>white black</u> black
<u>black black</u>
white

The object of this game is to end up with as few beans as possible. It is not hard to see that with an odd number of black beans the game will always end with one black bean, since the "parity" of black beans is unchanging. Played right (keeping at least one black bean around until the end), an even (non-zero) number of black beans leads to one white bean, but other pure-white results are also possible. For instance, applying the third rule right off to both pairs of black beans leaves six white beans.

By adding an additional rule, the above game may be modified to always end in a single bean:

black white	→	*black*
white black	→	*black*
black black	→	*white*
white white	→	*white*

What is different about this new game is that one of the rules applies to any can containing more than one bean. What is interesting is that the outcome of the game is completely independent of the choice of which move is made when. To establish this, an analysis of the different possible divergences helps. The order in which moves at nonoverlapping locations are made is immaterial, since whichever moves were not taken can still be taken later. The critical cases occur when the possible moves overlap and making one precludes making the other. For example, from *white black black*, either *black black* or *white white* can result. The point is that these two situations can both lead to the same single-white state. The same is true for other overlapping divergences. With this independence in mind, it is a trivial matter to predict the deterministic outcome of any game, by picking a sequence amenable to a simple analysis. Indeed, any initial state with an even number of black beans must now end in one white one. (The "semantic" argument given in Gries [95], based on the invarying parity of black beans, requires some insight, whereas the above analysis is entirely mechanical, as we will see later in this chapter.)

It is obvious that neither of the above games can go on forever, since the number of beans is reduced with each move. A potentially much longer, but still finite, game is

black white	→	*white white white black*
white black	→	*black*
black black	→	*white white white white*
white white	→	*white*

The new rules have the same end-effect as the original ones: no matter how often a bean-increasing move is made, in the final analysis the can must be emptied down to one bean.

Finally, we consider a variant (Scholten's original problem) in which the rules apply to *any* two (not necessarily adjacent) beans. The new rules are

black ... white	$\to$	*black ...*
white ... black	$\to$	*black ...*
black ... black	$\to$	*white ...*
white ... white	$\to$	*white ...*

where an ellipsis on the right refers to the same beans as covered by its counterpart on the left. The new rules, in effect, allow a player to "shake" the can prior to making a move. Again, it can be shown that the outcome is uniquely determined by the initial setup and is, consequently, the same as that of the previous two games.

The final result of an unextendible sequence of rule applications is called a "normal form". Rewrite systems defining at most one normal form for any input term can serve as functional programs or as interpreters for equational programs (O'Donnell [191]). When computations for equal terms always terminate in a unique normal form, a rewrite system may be used as a nondeterministic functional program (Goguen & Tardo [92]). Such a system also serves as a procedure for deciding whether two terms are equal in the equational theory defined by the rules, and, in particular, solves the "word problem" for that theory. Knuth [149] devised an effective test (based on critical overlaps) to determine for any given terminating system if, in fact, all computations converge to a canonical form, regardless of the nondeterministic choices made. In that seminal paper, it was also demonstrated how failure of the test (as transpires for the first Coffee Can game) often suggests additional rules that can be used to "complete" a nonconvergent system into a convergent one. The discovery (Fay [76]) that convergent rewrite systems can also be used to enumerate answers to satisfiability questions for equational theories led to their application (Dershowitz [56]) within the logic programming paradigm.

Rewriting methods have turned out to be among the more successful approaches to equational theorem proving. In this context, completion is used for "forward reasoning", while rewriting is a form of "backward reasoning". Completion utilizes an ordering on terms to provide strong guidance during forward reasoning and to direct the simplification of equations. Besides the use of convergent systems as decision procedures, Lankford [156, 157] proposed that completion-like methods supplant paramodulation for equational deduction within resolution-based theorem provers; later, Hsiang [106] showed how a variant of completion can be used in place of resolution for (refutational) theorem proving in first-order predicate calculus. Although completion often generates an infinite number of additional rules, and—at the same time—deletes many old rules, Huet [114] demonstrated that "fairly" implemented completion serves as a semi-decision procedure for the equational theory defined by the given equations when it does not abort (something it might be forced to do on account of equations that cannot be directed without loss of termination). Lankford's procedure paramodulates to circumvent failure of completion. Rewriting techniques have also been applied (Musser [186]) to proving inductive theorems by showing that no contradiction can result from assuming the validity of the theorem in question.

In the next two sections, we take a quick look at the syntax and semantics of equations from the algebraic, logical, and operational points of view. To use a rewrite system as a decision procedure, it must be convergent; this fundamental concept is studied in Section 4 as an abstract property of binary relations. To use a rewrite system for computation or as a decision procedure for validity of identities, the termination property is crucial; basic methods for proving termination are presented in Section 5. Section 6 is devoted to the question of satisfiability of equations. Then, in Section 7, we return to the convergence property as applied to rewriting. The completion procedure, its extensions, refinements, and main uses, are examined in Section 8. Brief mention of variations on the rewriting theme is made in the final section.

1.1. Further reading

Previous surveys of term rewriting include Huet & Oppen [119], Buchberger & Loos [36], Jouannaud & Lescanne [126], and Klop [147]. Suggestions regarding notation and terminology may be found in Dershowitz & Jouannaud [252].

2. Syntax

Algebraic data types are an important application area for rewrite-based equational reasoning. In the abstract approach to data specification, data are treated as abstract objects and the semantics of functions operating on data are described by a set of constraints. When constraints are given in the form of equations, a specification is called *algebraic* (Guttag [97]). In this section, we talk about the syntax of equations and of equational proofs. As we will see, by turning equations into left-to-right "rules", a useful concept of "direct" proof is obtained.

2.1. Terms

Suppose we wish to define the standard stack operations, *top* and *pop*, as well as an operation *alternate* that combines two stacks. Stacks of natural numbers can be represented by terms of the form $push(s_1, push(s_2, \ldots, push(s_n, \Lambda)\ldots))$, where Λ is the empty stack and the s_i denote representations of natural numbers, 0, *succ*(0), *succ*(*succ*(0)), and so on. The precise syntax of these representations can be given in the following inductive way:

$$\begin{array}{|lll|} Zero & = & \{0\} \\ Nat & = & Zero \cup succ(Nat) \\ Empty & = & \{\Lambda\} \\ Stack & = & Empty \cup push(Nat, Stack) \end{array}$$

The left sides of these equations name sets of different kinds of terms. An expression like *succ*(*Nat*) denotes the set of all terms *succ*(*s*), with $s \in Nat$. The symbols 0, *succ*, Λ, and *push*, used to build data, are called "constructors"; any term built according to these rules is a *constructor term.*

To specify the desired stack operations, we must also define the syntax of non-constructor terms:

top:	*Stack*	$\rightarrow$	*Nat*
pop:	*Stack*	$\rightarrow$	*Stack*
alternate:	*Stack* $\times$ *Stack*	$\rightarrow$	*Stack*

Then we give semantics to the new functions by constraining them to satisfy the following set of equations:

$$\begin{aligned} top(push(x,y)) &= x\\ pop(push(x,y)) &= y\\ alternate(\Lambda,z) &= z\\ alternate(push(x,y),z) &= push(x, alternate(z,y)) \end{aligned}$$

(where x, y, and z are variables ranging over all data of the appropriate type). Inverses of constructors, like *top* and *pop*, are called *selectors.* With these equations, it can be shown, for example, that

$$alternate(push(top(push(0,\Lambda)),\Lambda), pop(push(succ(0),\Lambda))) = push(0,\Lambda).$$

A specification is said to be "sufficiently complete" if, according to the semantics, every term is equal to a term built only from constructors; the above operations are not well-defined in this sense, since terms like $pop(\Lambda)$ are not equal to any constructor term. (See Section 3.2.)

In general, given a set $\mathcal{F} = \bigcup_{n \geqslant 0} \mathcal{F}_n$ of function symbols—called a (finitary) *vocabulary* or *signature*—and a (denumerable) set $\mathcal{X}$ of variable symbols, the set of (first-order) *terms* $\mathcal{T}(\mathcal{F}, \mathcal{X})$ over $\mathcal{F}$ and $\mathcal{X}$ is the smallest set containing $\mathcal{X}$ such that $f(t_1, \ldots, t_n)$ is in $\mathcal{T}(\mathcal{F}, \mathcal{X})$ whenever $f \in \mathcal{F}_n$ and $t_i \in \mathcal{T}(\mathcal{F}, \mathcal{X})$ for $i = 1, \ldots, n$. The stack example uses $\mathcal{F}_0 = \{0, \Lambda\}$, $\mathcal{F}_1 = \{top, pop, succ\}$, $\mathcal{F}_2 = \{push, alternate\}$, and $\mathcal{X} = \{x, y, z\}$. The syntax of the stack example also differentiates between terms of type *Nat* and of type *Stack*. Categorizing function symbols, variables, and terms into classes, called *sorts*, can be very helpful in practice; from now on, however, we will suppose that there is only one, all-inclusive sort. All concepts developed here can be carried over to the many-sorted case, as will be sketched in Section 9.

Each symbol f in $\mathcal{F}$ has an *arity* (*rank*) which is the index n of the set $\mathcal{F}_n$ to which it belongs. (We will assume that the $\mathcal{F}_n$ are disjoint, though "varyadic" vocabularies pose little problem.) In a well-formed term, each symbol of arity n has n immediate subterms. Elements of arity zero are called *constants*, of which we will always make the (rarely critical) assumption that there is at least one. Terms in $\mathcal{T}(\mathcal{F}_0 \cup \mathcal{F}_1, \mathcal{X})$ are called *monadic*; they are "words" spelled with unary symbols (from $\mathscr{F}_1$) and ending in a constant (from $\mathscr{F}_0$) or variable (from $\mathscr{X}$). Variable-free terms are called *ground* (or *closed*); the set $\mathscr{T}(\mathscr{F}, \emptyset)$ of ground terms will be denoted by $\mathscr{G}(\mathscr{F})$. Note that $\mathscr{G}(\mathscr{F})$ is nonempty by the previous assumption regarding $\mathscr{F}_0$. We will often use $\mathscr{T}$ to refer to a set of terms $\mathscr{T}(\mathscr{F}, \mathscr{X})$, with $\mathscr{F}$ and $\mathscr{X}$ left unspecified, and $\mathscr{G}$ to refer to the corresponding set of ground terms. In examples, we occasionally use prefix or postfix notation for $\mathscr{F}_1$ and infix for $\mathscr{F}_2$.

A term t in $\mathscr{T}(\mathscr{F}, \mathscr{X})$ may be viewed as a finite ordered tree, the leaves of which are labeled with variables (from $\mathscr{X}$) or constants (from $\mathscr{F}_0$) and the internal nodes of which are labeled with function symbols (from $\mathscr{F}_1 \cup \mathscr{F}_2 \cup \cdots$) of positive arity, with outdegree equal to the arity of the label. A *position* within a term may be represented—in Dewey decimal notation—as a sequence of positive integers, describing the path from the outermost, "root" symbol to the head of the subterm at that position. By $t|_p$, we denote the *subterm* of t rooted at position p. For example, if $t = push(0, pop(push(y, z)))$, then $t|_{2.1}$ is the first subterm of t's second subterm, which is $push(y, z)$. Positions are often called *occurrences*; we will use this latter denomination to refer, instead, to the subterm $t|_p$. We write $t \trianglerighteq s$ to mean that s is a subterm of t. We speak of position p as being *above* position q in some term t if p (represented as a sequence of numbers) is a prefix of q, i.e. if occurrence $t|_q$ is within $t|_p$. A subterm of t is called *proper* if it is distinct from t.

Reasoning with equations requires replacing subterms by other terms. The term t with its subterm $t|_p$ replaced by a term s is denoted by $t[s]_p$. We refer to any term u that is the same as t everywhere except below p, i.e. such that $u[s]_p = t$, as the *context* within which the replacement takes place; more precisely, a context is a term u with a "hole" (i.e. a lambda-bound variable) at a distinguished position p.

A *substitution* is a special kind of replacement operation, uniquely defined by a mapping from variables to terms, and written out as $\{x_1 \mapsto s_1, \ldots, x_m \mapsto s_m\}$ when there are only finitely many variables x_i not mapped to themselves. Formally, a substitution σ is a function from $\mathscr{X}$ to $\mathscr{T}(\mathscr{F}, \mathscr{X})$, extended to a function from $\mathscr{T}$ to itself (also denoted σ and for which we use postfix notation) in such a way that $f(t_1, \ldots, t_n)\sigma = f(t_1\sigma, \ldots, t_n\sigma)$, for each f (of arity n) in $\mathscr{F}$ and for all terms $t_i \in \mathscr{T}$. A term t *matches* a term s if $s\sigma = t$ for some substitution σ; in that case we write $s \leqslant t$ and also say that t is an *instance* of s or that s *subsumes* t. The relation $\leqslant$ is a quasi-ordering on terms, called *subsumption*.[1] For example, $f(z) < f(a)$, and $f(z) < f(f(a))$, since z is "less specific" than a or $f(a)$. On the other hand, $f(x)$ and $f(z)$ are equally general; we write $f(x) \doteq f(z)$, where $\doteq$ is the equivalence relation associated with $\leqslant$, called *literal similarity* (*α-conversion* in λ-calculus parlance; *renaming*, in other circles). Subsumption and the subterm ordering are special cases of the *encompassment* quasi-ordering (called *containment* in Huet [114]), in which $s \trianglelefteq t$ if a subterm of t is an instance of s. For example, $f(z) \triangleleft g(f(a))$, since $f(a)$ is an instance of $f(z)$.

The *composition* of two substitutions, denoted by juxtaposition, is just the composition of the two functions; thus, if $x\sigma = s$ for some variable x, then $x\sigma\tau = s\tau$. We say that substitution σ is *at least as general* as substitution ρ (with respect to a subset $\mathscr{X}'$

[1] A *quasi-ordering* $\preccurlyeq$ is any reflexive and transitive binary relation; the associated equivalence relation $\sim$ is the intersection of $\preccurlyeq$ with its inverse; the associated "strict" (i.e. irreflexive) partial order $\prec$ is their difference.

of $\mathscr{X}$) if there exists a substitution τ such that $\sigma\tau=\rho$ (when σ and ρ are restricted to $\mathscr{X}'$); we use the same symbols to denote this quasi-ordering on substitutions as we used for subsumption of terms. For example, $\{x\mapsto a, y\mapsto f(a)\}\gtrdot\{x\mapsto y, y\mapsto f(z)\}\doteq\{x\mapsto z, y\mapsto f(x)\}$ (with respect to x and y only). Here, and everywhere, we use the mirror image of a binary relation symbol like $\lessdot$ for its inverse.

In this survey, we will mainly be dealing with binary relations on terms that possess the following fundamental properties:

DEFINITION. A binary relation $\rightarrow$ over a set of terms $\mathscr{T}$ is a *rewrite relation* if it is closed both under context application (the "replacement" property) and under substitutions (the "fully invariant property"). A rewrite relation is called a *rewrite ordering* if it is transitive and irreflexive.

In other words, $\rightarrow$ is a rewrite relation if $s\rightarrow t$ implies $u[s\sigma]_p\rightarrow u[t\sigma]_p$, for all terms s and t in $\mathscr{T}$, contexts u, positions p, and substitutions σ. The inverse, symmetric closure, reflexive closure, and transitive closure of any rewrite relation are also rewrite relations as are the union and intersection of two rewrite relations.

To fix nomenclature, the letters a through h will be used for function symbols; l, r, and s through w will denote arbitrary terms; x, y, and z will be reserved for variables; p and q, for positions; lower case Greek letters, for substitutions. Binary relations will frequently be denoted by arrows of one kind or another. If $\rightarrow$ is a binary relation, then $\leftarrow$ is its inverse, $\leftrightarrow$ is its symmetric closure ($\leftarrow\cup\rightarrow$), $\overset{=}{\rightarrow}$ is its reflexive closure ($\rightarrow\cup=$), $\overset{*}{\rightarrow}$ is its reflexive-transitive closure ($\rightarrow\circ\cdots\circ\rightarrow$) and $\overset{+}{\rightarrow}$ is its transitive closure ($\rightarrow\circ\overset{*}{\rightarrow}$).

2.2. Equations

Replacement leads to the important notion of "congruence": an equivalence relation $\sim$ on a set of terms is a *congruence* if $f(s_1,\ldots,s_n)\sim f(t_1,\ldots,t_n)$ whenever $s_i\sim t_i$ for $i=1,\ldots,n$. In particular, the reflexive-symmetric-transitive closure $\overset{*}{\leftrightarrow}$ of any rewrite relation $\rightarrow$ is a congruence. Note that rewrite relations and congruences form a complete lattice with respect to intersection.

Our primary interest is in congruences generated by instances of equations. For our purposes, an *equation* is an unordered pair $\{s,t\}$ of terms. (For other purposes, it is preferable to regard equations as ordered pairs.) Equations will be written in the form $s=t$.[2] The two terms may contain variables; these are understood as being universally quantified. Given a (finite or infinite) set of equations E over a set of terms $\mathscr{T}$, the *equational theory* of E, $\mathscr{Th}(E)$, is the set of equations that can be obtained by taking reflexivity, symmetry, transitivity, and context application (or functional reflexivity) as inference rules and all instances of equations in E as axioms. Thus, if E is recursively-enumerable, so are its theorems $\mathscr{Th}(E)$. We write $E\vdash s=t$ if $s=t\in\mathscr{Th}(E)$.

A more compact inference system is based on the familiar notion of "replacement of equals for equals" (a.k.a. Leibniz's Law). We write $s\underset{E}{\leftrightarrow}t$, for terms s and t in $\mathscr{T}$, if s has

[2] To avoid confusion, authors are sometimes compelled to use a different symbol in the syntax of equations, instead of the heavily overloaded "equals sign"—a precaution we choose not to take in this survey.

a subterm that is an instance of one side of an equation in E and t is the result of replacing that subterm with the corresponding instance of the other side of the equation. Formally, $s \underset{E}{\leftrightarrow} t$ if $s=u[l\sigma]_p$ and $t=u[r\sigma]_p$ for some context u, position p in u, equation $l=r$ (or $r=l$) in E, and substitution σ. It is folk knowledge that $E \vdash s=t$ iff $s \overset{*}{\underset{E}{\leftrightarrow}} t$, where $\overset{*}{\underset{E}{\leftrightarrow}}$ is the reflexive-transitive closure of $\underset{E}{\leftrightarrow}$; in other words, two terms are provably equal in predicate calculus with equality if one may be obtained from the other by a finite number of replacements of equal subterms. The relation $\underset{E}{\leftrightarrow}$ is the "rewrite" closure of E, when the latter is viewed as a symmetric relation, and $\overset{*}{\underset{E}{\leftrightarrow}}$ is the congruence closure of $\underset{E}{\leftrightarrow}$, i.e. $\overset{*}{\underset{E}{\leftrightarrow}}$ is the smallest congruence over $\mathcal{T}$ such that $l\sigma \overset{*}{\underset{E}{\leftrightarrow}} r\sigma$ for all equations $l=r$ in E and substitutions σ over $\mathcal{T}$. We will write $[s]_E$ for the congruence class of a term s, and denote by $\mathcal{T}/E$ the set of all congruence classes, i.e. the quotient of the set $\mathcal{T}(\mathcal{F}, \mathcal{X})$ of terms and the *provability* relation $\overset{*}{\underset{E}{\leftrightarrow}}$.

A *derivation* in E is any sequence $s_0 \underset{E}{\leftrightarrow} s_1 \underset{E}{\leftrightarrow} \cdots \underset{E}{\leftrightarrow} s_i \underset{E}{\leftrightarrow} \cdots$ of applications of equational axioms in E. A *proof* in E of an equation $s=t$ is a "justified" finite derivation $s=s_0 \underset{E}{\leftrightarrow} \cdots \underset{E}{\leftrightarrow} s_n = t$ $(n \geqslant 0)$, each step $s_i \underset{E}{\leftrightarrow} s_{i+1}$ of which is justified by reference to an axiom $l=r$ in E, a position p in s_i, and a substitution σ, such that $s_i|_p = l\sigma$ and $s_{i+1} = s_i[r\sigma]_p$. Returning to our stack specification, and letting E be its axioms, the following is an example of a derivation:

$$\begin{aligned}&alternate(push(top(push(0, z)), z), \Lambda)\\ &\underset{E}{\leftrightarrow}\ alternate(push(0, z), \Lambda)\\ &\underset{E}{\leftrightarrow}\ alternate(push(0, pop(push(succ(y), z))), \Lambda).\end{aligned}$$

The first step may be justified by the axiom $top(push(x, y))=x$, position 1.1, and substitution $\{x \mapsto 0, y \mapsto z\}$; the second step, by the axiom $pop(push(x, y))=y$ (used from right to left), position 1.2, and substitution $\{x \mapsto succ(y), y \mapsto z\}$.

2.3. *Rewrite rules*

The central idea of rewriting is to impose directionality on the use of equations in proofs. Unlike equations which are unordered, a *rule* over a set of terms $\mathcal{T}$ is an ordered pair $\langle l, r\rangle$ of terms, which we write as $l \rightarrow r$. Rules differ from equations by their use. Like equations, rules are used to replace instances of l by corresponding instances of r; unlike equations, rules are not used to replace instances of the right-hand side r. A (finite or infinite) set of rules R over $\mathcal{T}$ is called a *rewrite system*, or (more specifically) a *term-rewriting system*. A system R may be thought of as a (nonsymmetric) binary relation on $\mathcal{T}$; the rewrite closure $\underset{R}{\leftrightarrow}$ of this relation describes the effect of a left-to-right application of a rule in R.

DEFINITION. For given rewrite system R, a term s in $\mathcal{T}$ *rewrites* to a term t in $\mathcal{T}$, written $s \underset{R}{\rightarrow} t$, if $s|_p = l\sigma$ and $t = s[r\sigma]_p$, for some rule $l \rightarrow r$ in R, position p in s, and substitution σ.

This is the same as saying that $s=u[l\sigma]_p$ and $t=u[r\sigma]_p$, for some context u and position p in u. A subterm $s|_p$ at which a rewrite can take place is called a *redex*; we say that s is *irreducible*, or in *normal form*, if it has no redex, i.e. if there is no t in $\mathcal{T}$ such that $s \underset{R}{\rightarrow} t$.

Systems of rules are used to compute by rewriting repeatedly, until, perhaps, a normal form is reached. A *derivation* in R is any (finite or infinite) sequence $t_0 \underset{R}{\rightarrow} t_1 \underset{R}{\rightarrow} \cdots \underset{R}{\rightarrow} t_i \underset{R}{\rightarrow} \cdots$ of applications of rewrite rules in R. The *reducibility*, or *derivability*, relation is the quasi-ordering $\overset{*}{\underset{R}{\rightarrow}}$, i.e. the reflexive-transitive closure of $\underset{R}{\rightarrow}$. We write $s \overset{!}{\underset{R}{\rightarrow}} t$ if $s \overset{*}{\underset{R}{\rightarrow}} t$ and t is irreducible, in which case we say that t is a normal form of s. One says that a rewrite system is *normalizing* if every term has at least one normal form. This *normalizability* relation $\overset{!}{\underset{R}{\rightarrow}}$ is not a rewrite relation however, since normalizing a subterm does not mean that its superterm is in normal form.

A *ground rewriting-system* is one all the rules of which are ground (i.e. elements of $\mathcal{G} \times \mathcal{G}$); an important early paper on ground rewriting is Rosen [212]. A *string-rewriting system*, or *semi-Thue system*, is one that has monadic words ending in the same variable (i.e. strings of elements of $\mathcal{T}(\mathcal{F}_1, \{x\})$) as left- and right-hand side terms; Book [29] is a survey of string rewriting. The (first three) Coffee Can Games can be formulated as string-rewriting systems, with *white* and *black* as monadic symbols. A *left-linear* system is one in which no variable occurs more than once on any left-hand side. (Ground- and string-rewriting systems are special cases of left-linear systems, with no variable and one variable per term, respectively.)

For our purposes, one of the most essential properties a rewrite system R can enjoy is *unique normalization*, by which is meant that every term t in $\mathcal{T}$ possesses exactly one normal form. The normalizability relation $\overset{!}{\underset{R}{\rightarrow}}$ for uniquely-normalizing systems defines a function, and we denote by $R(t)$ the value of that function for a term t in $\mathcal{T}$. If *all* sequences of rewrites lead to a unique normal form, the system will be called *convergent*. A rewrite system R is said to be *(inter-)reduced* if, for each rule $l \rightarrow r$ in R, the right-hand side r is irreducible under R and no term s less than l in the encompassment ordering $\trianglerighteq$ is reducible. (For convergent R, this is equivalent to the standard definition (Huet [114]) which requires that the left-hand side l not be rewritable by any other rule.) As we will see this, too, is a convenient property. We will reserve the adjective *canonical* for reduced convergent systems, though, in the literature, "canonical" is usually synonymous with "convergent".

Orienting the equations of the stack example gives a canonical rewrite system R:

$top(push(x,y))$	$\rightarrow$	x
$pop(push(x,y))$	$\rightarrow$	y
$alternate(\Lambda, z)$	$\rightarrow$	z
$alternate(push(x,y),z)$	$\rightarrow$	$push(x, alternate(z,y))$

(That every term has at least one normal form will be shown in Section 5.3; that there can be no more than one normal form will be shown in Section 7.2.) An example of a derivation is

$$\begin{aligned} alternate(push(top(push(0,z)),z),\Lambda) &\underset{R}{\rightarrow} alternate(push(0,z),\Lambda) \\ &\underset{R}{\rightarrow} push(0, alternate(\Lambda,z)) \\ &\underset{R}{\rightarrow} push(0,z). \end{aligned}$$

The first step is an application of the *top-push* rule at the occurrence *top*(*push*(0, *z*)); the second is an application of the *alternate-push* rule, with 0 for *x*, *z* for *y*, and Λ for *z*; the third, of $alternate(\Lambda, z) \rightarrow z$. Note that an alternative derivation is possible, leading to the same normal form:

$$
\begin{aligned}
& alternate(push(top(push(0, z)), z), \Lambda) \\
& \underset{R}{\rightarrow} push(top(push(0, z)), alternate(\Lambda, z)) \\
& \underset{R}{\rightarrow} push(top(push(0, z)), z) \underset{R}{\rightarrow} push(0, z).
\end{aligned}
$$

Operationally, rewriting is a nondeterministic computation, with the choice of rule and position left open. For convergent systems, the choice among possible rewrites at each step does not affect the normal form computed for any given input term.

DEFINITION. A binary relation $\rightarrow$ on a set *T* is *terminating* if there exists no endless chain $t_1 \rightarrow t_2 \rightarrow t_3 \rightarrow \cdots$ of elements of *T*.

Termination demands more than that the system be normalizing, since the latter allows some derivations to be infinite. A partial (irreflexive) ordering $\succ$ of a set *T* is *well-founded* if there exists no infinite descending chain $t_1 \succ t_2 \succ \cdots$ of elements of *T*. Thus, a relation $\rightarrow$ is terminating if its transitive closure $\overset{+}{\rightarrow}$ is a well-founded ordering. The importance of terminating relations lies in the possibility of inductive proofs in which the hypothesis is assumed to hold for all elements *t* such that $s \overset{+}{\rightarrow} t$ when proving it for arbitrary *s*. Induction on terminating relations, called "Noetherian induction", is essentially well-founded induction (i.e. transfinite induction extended to partial orderings); see, for example, Cohn [42]. (For this reason, terminating relations have sometimes been called *Noetherian* in the term-rewriting literature—after the algebraicist, Emily Noether—though the adjective is ordinarily used to exclude infinite ascending chains.) We will have occasion to employ this induction technique in Sections 4.1 and 8.2.

A rewrite system *R* is *terminating* for a set of terms $\mathscr{T}$ if the rewrite relation $\underset{R}{\rightarrow}$ over $\mathscr{T}$ is terminating, i.e. if there are no infinite derivations $t_1 \underset{R}{\rightarrow} t_2 \underset{R}{\rightarrow} \cdots$ of terms in $\mathscr{T}$. When a system is terminating, every term has at least one normal form. Note that a terminating (untyped) system cannot have any rule, like $alternate(y, \Lambda) \rightarrow pop(push(x, y))$, with a variable on the right that is not also on the left (since *x* could, for example, be $top(alternate(y, \Lambda))$), nor can a left-hand side be just a variable, like $z \rightarrow alternate(\Lambda, z)$. These two restrictions are often placed a priori on rewrite rules (cf. Huet [113]), something we recommend against. Methods for establishing termination are described in Section 5. For nonterminating systems the choice of which rewrite to perform can be crucial; see Section 7.2.

A *valley*, or *rewrite*, proof of an equation $s = t$ for a system *R* takes the form $s \overset{*}{\underset{R}{\rightarrow}} v \overset{*}{\underset{R}{\leftarrow}} t$, in which the same term *v* is reached by rewriting *s* and *t*. Here is an example, using the above system:

$$
\begin{array}{ll}
alternate(push(top(push(0, z)), z), \Lambda) & alternate(push(0, pop(push(y, z))), \Lambda) \\
\underset{R}{\rightarrow} alternate(push(0, z), \Lambda) & push(0, alternate(\Lambda, pop(push(y, z)))) \underset{R}{\leftarrow} \\
\end{array}
$$

$$\underset{R}{\rightarrow} push(0, alternate(\Lambda, z)) \underset{R}{\leftarrow}$$

With a terminating system of only a finite number of rules, the search space for rewrite proofs is finite. Of course, there is—in general—no guarantee that such a "direct" proof exists for a particular consequence of the equations represented by R. When it is the case, i.e. when the relation $\overset{*}{\underset{R}{\leftrightarrow}}$ is contained in $\overset{*}{\underset{R}{\rightarrow}} \circ \overset{*}{\underset{R}{\leftarrow}}$, the system is called *Church–Rosser*, after a property in Church & Rosser [41]. See Fig. 1(a). Equivalent properties are defined in Section 4 and methods of establishing them are described in Section 7.

2.4. Decision procedures

One of our main concerns is in decision procedures for equational theories; an early example of such a procedure, for groups, is Dehn [52]. Terminating, Church–Rosser

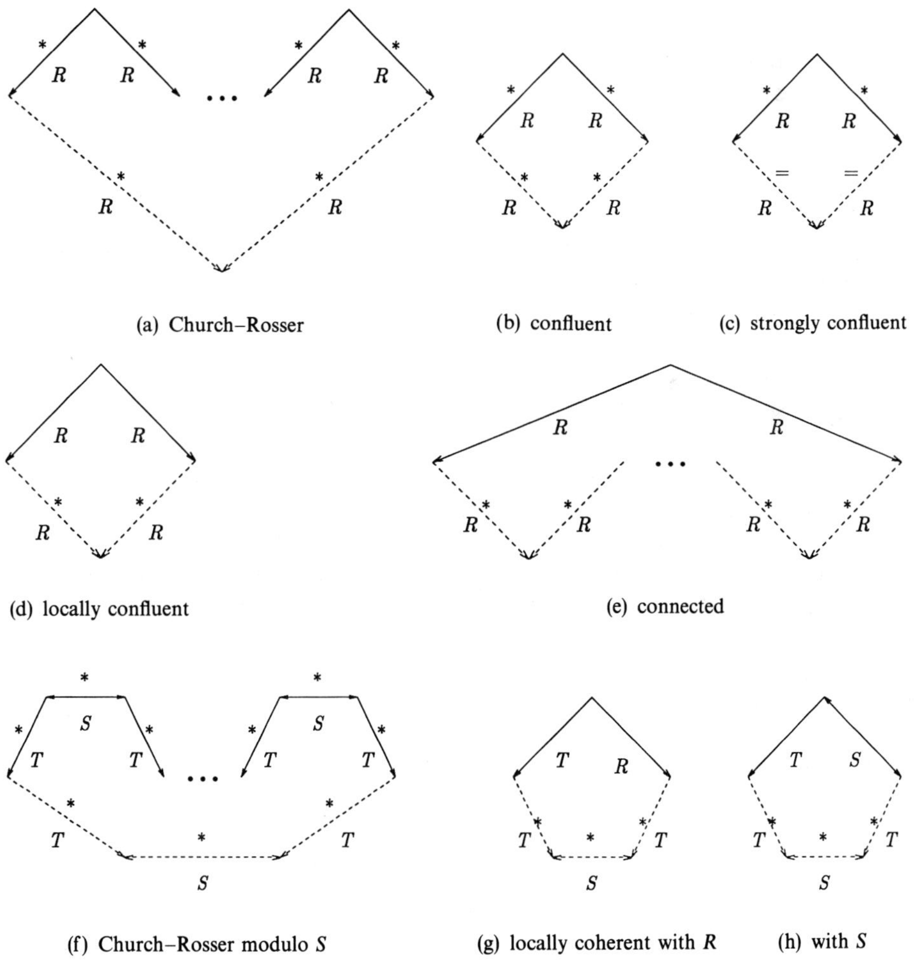

Fig. 1. Joinability properties of relations.

rewrite systems are convergent and define unique normal forms. For a convergent system R to determine provability in its underlying equational theory (treating its rules as equational axioms), it should have only a finite number of rules. Then, to decide if $s \underset{R}{=} t$, one can test if computing the normal forms $R(s)$ and $R(t)$ results in the same term. The Church–Rosser property means that $\overset{*}{\underset{R}{\leftrightarrow}} = \overset{*}{\underset{R}{\rightarrow}} \circ \overset{*}{\underset{R}{\leftarrow}}$; termination ensures that $\overset{*}{\underset{R}{\rightarrow}} \circ \overset{*}{\underset{R}{\leftarrow}} = \overset{!}{\underset{R}{\rightarrow}} \circ \overset{!}{\underset{R}{\leftarrow}}$; finiteness of R makes $\overset{!}{\underset{R}{\rightarrow}}$ decidable; and Church–Rosser implies that $\overset{!}{\underset{R}{\rightarrow}}$ defines a function $R(\cdot)$. Thus, a system R provides a decision procedure for the equational theory of a set of axioms E if R is (a) finite, (b) terminating, (c) Church–Rosser, and (d) sound and adequate for E. Here, *soundness* means that the rules of R are contained in the relation $\overset{*}{\underset{E}{\leftrightarrow}}$, and *adequacy* means that E is contained in $\overset{*}{\underset{R}{\leftrightarrow}}$; together, they imply that $\overset{*}{\underset{E}{\leftrightarrow}} = \overset{*}{\underset{R}{\leftrightarrow}}$. In our stack example, R is sound and adequate for E; hence, R decides equality in E. In general, a system R with properties (a)–(d) is said to be *complete* for E.

Given a set of equations E, its *word problem* is the question whether an arbitrary equation $s=t$ between two *ground* terms follows from E. The word problem is, thus, a special case of provability in E for arbitrary equations. If R is a convergent system for E, its word problem is decidable by reducing s and t to their R-normal form. Actually, it is enough if R is *ground-convergent*, that is, if every ground term rewrites to a unique normal form. Many rewriting-system decision procedures are known; perhaps the first rewriting-based decision procedure for a word problem is the one in Evans [72] for "loops". When a rewriting decision procedure exists, it can be very effective. In Section 8, we will elaborate on systematic methods used to generate convergent systems from given equational axioms.

Of course, not all equational theories can be decided by rewriting—for a variety of reasons. First, some theories (classes of equations) are not *finitely based*; for such theories there exists no finite set of axioms from which all other equations in the theory follow. An example (Taylor [230]) is the intersection of the theories (consequences) of the following two semigroups (abbreviating products by juxtaposition and exponentiation):

$$\begin{array}{lcl} x(yz) & = & (xy)z \\ (xyz)^2 & = & x^2y^2z^2 \\ x^3y^3z_1^2z_2^3 & = & y^3x^3z_1^2z_2^3 \end{array} \qquad \begin{array}{lcl} x(yz) & = & (xy)z \\ x^3y^3 & = & y^3x^3 \end{array}$$

Nor are all finitely-based equational theories decidable, the first counter-examples having been given by Markov [174] and Post [201], in a slightly different context. (See the interesting historical comments at the end of Tarski & Givant [229].) A prime example of an undecidable equational theory is Combinatory Logic (with binary infix symbol "·" and constants S, K, and I):

$$\begin{array}{lcl} I\cdot x & = & x \\ (K\cdot x)\cdot y & = & x \\ ((S\cdot x)\cdot y)\cdot z & = & (x\cdot z)\cdot(y\cdot z) \end{array}$$

(see Curry & Feys [47]). Most disconcertingly, there are finitely-based, decidable theories for which there can be no rewriting-system decision procedure. For example, no finite system—even over an enlarged vocabulary—can rewrite any two terms, equal by commutativity, to the same term (Dershowitz et al. [66]). In other words, no term-rewriting system can decide validity in the decidable theory defined by the commutativity axiom, $x \cdot y = y \cdot x$, since that equation, oriented in either direction, gives a nonterminating rewrite system. Deciding word problems is a somewhat different question, since then one looks only at ground equations over a finite vocabulary. But this same example (let $\mathscr{F} = \{0, succ, \cdot\}$) demonstrates that not all decidable word problems can be decided by a finite rewrite system over the same vocabulary; cf. Klop [147]. The following is one of the simple (semigroup) theories with an undecidable word problem given in Matijasevic [178] ($\mathscr{F}_0 = \{a, b\}$, $\mathscr{F}_2 = \{\cdot\}$, and products are abbreviated as before):

$$\begin{array}{lcl} x(yz) & = & (xy)z \\ aba^2b^2 & = & b^2a^2ba \\ a^2bab^2a & = & b^2a^3ba \\ aba^3b^2 & = & ab^2aba^2 \\ b^3a^2b^2a^2ba & = & b^3a^2b^2a^4 \\ a^4b^2a^2ba & = & b^2a^4 \end{array}$$

2.5. *Extensions*

Happily, rewriting techniques can be adapted to handle some of the more important cases in which any orientation of the axioms yields a nonterminating system. The simplest example of a "structural" axiom requiring special treatment is commutativity: any system containing either $x \cdot y \rightarrow y \cdot x$ or $y \cdot x \rightarrow x \cdot y$ is perforce nonterminating. We describe two techniques for dealing with such axioms, "class rewriting" and "ordered rewriting".

A (*congruence-*) *class-rewriting system* comes in two parts: rules and equations. By R/S we denote the class system composed of a set $R = \{l_i \rightarrow r_i\}$ of rewrite rules and a set $S = \{u_i \leftrightarrow v_i\}$ of equations, the latter written with double-headed arrows to stress their symmetrical usage. Generalizing the notion of term rewriting, we say that s rewrites to t *modulo* S, denoted $s \xrightarrow[R/S]{} t$, if $s \overset{*}{\underset{S}{\leftrightarrow}} u[l\sigma]_p$ and $u[r\sigma]_p \overset{*}{\underset{S}{\leftrightarrow}} t$, for some context u, position p in u, rule $l \rightarrow r$ in R, and substitution σ. Thus, R is essentially computing in the quotient set $\mathscr{T}/S = \{[t]_S \mid t \in \mathscr{T}\}$ of S-congruence classes (more precisely, $\overset{*}{\underset{S}{\leftrightarrow}}$-congruence classes), rewriting a term by rewriting any S-equivalent term. Class-rewriting systems were introduced in Lankford & Ballantyne [160] for *permutative* congruences, that is, congruences for which each congruence class is finite. Of great practical importance are associative-commutative (*AC*) rewrite systems, where S is an equational system consisting of associativity and commutativity axioms for a subset of the binary symbols (in $\mathscr{F}_2$). The last Coffee Can Game can be formulated as an AC system, by using an associative-commutative operator for adjacency with the four rules of the second game.

Then, *black white black* can rewrite directly to *black black* or, via *black black white*, to *white white*.

The notions of derivation and normal form extend naturally to class-rewriting systems. We say that R/S is terminating if $\xrightarrow[R/S]{}$ is terminating and that it is *Church–Rosser modulo S* if $\xleftrightarrow[R/S]{*}$ is contained in $\xrightarrow[R/S]{*} \circ \overset{*}{\underset{S}{\leftrightarrow}} \circ \xleftarrow[R/S]{*}$. More generally, a relation $\underset{T}{\rightarrow}$ is Church–Rosser modulo S if $\xleftrightarrow[T \cup S]{*} \subseteq \overset{*}{\underset{T}{\rightarrow}} \circ \overset{*}{\underset{S}{\leftrightarrow}} \circ \overset{*}{\underset{T}{\leftarrow}}$. See Fig. 1(f). For instance, let BA/AC be the following class system (over vocabulary $\mathcal{F}_2 = \{and, xor\}$ and $\mathcal{F}_0 = \{F, T\}$):

BA		
$and(x, T)$	$\rightarrow$	x
$and(x, F)$	$\rightarrow$	F
$and(x, x)$	$\rightarrow$	x
$xor(x, F)$	$\rightarrow$	x
$xor(x, x)$	$\rightarrow$	F
$and(xor(x, y), z)$	$\rightarrow$	$xor(and(x, z), and(y, z))$

AC		
$and(x, y)$	$\leftrightarrow$	$and(y, x)$
$and(x, and(y, z))$	$\leftrightarrow$	$and(and(x, y), z)$
$xor(x, y)$	$\leftrightarrow$	$xor(y, x)$
$xor(x, xor(y, z))$	$\leftrightarrow$	$xor(xor(x, y), z)$

This system is convergent, i.e. terminating and Church–Rosser modulo AC (Hsiang [106]); it computes normal forms of Boolean-ring expressions that are unique up to permutations under associativity and commutativity. (The exclusive-or normal form is due to Zhegalkin [242] and Stone [224].)

The idea, then, is to put equations that cannot be handled by rewriting into S, placing in R only rules that preserve termination. If R/S is also Church–Rosser modulo S, then $s \xleftrightarrow[R \cup S]{*} t$ iff their normal forms are S-equivalent. For this to work, there are two additional considerations: S-equivalence must be decidable, and R/S normal forms must be effectively computable. The latter requirement does not, however, come automatically, even if R is finite and S-equivalence is decidable, since a rule in R is applicable to a term when any S-equivalent term contains an instance of a left-hand side, whereas S-equivalence classes need be neither finite nor computable. Note that a class system R/S cannot be terminating if (R is nonempty and) S contains an equation with a variable on one side not also on the other, or if it contains an axiom like idempotency, $x \cdot x = x$, with a lone variable on one side and more than one occurrence of it on the other.

Even if S-equivalence classes are computable, they may be impractically large, making class-rewriting prohibitively expensive. These difficulties with R/S are usually circumvented by using a weaker rewrite relation, introduced in Peterson & Stickel [197]. We denote this relation by $\xrightarrow[S \backslash R]{}$ (others have used $\xrightarrow[R,S]{}$); under it, a term is rewritten only if it has a subterm that is equivalent to an instance of a left-hand side. We call $\xrightarrow[S \backslash R]{}$ the "extended" rewrite relation for R; our notation is meant to suggest that S-steps are not applied above the R-step. Formally:

DEFINITION. For given rewrite system R and congruence relation S, the *S-extended* rewrite relation $\xrightarrow[S \backslash R]{}$ is defined by $s \xrightarrow[S \backslash R]{} t$, for two terms s, t in $\mathcal{T}$, iff $s|_p \overset{*}{\underset{S}{\leftrightarrow}} l\sigma$ and $t = s[r\sigma]_p$, for some rule $l \rightarrow r$ in R, position p in s, and substitution σ.

The notions of normal-form, etc., are analogous to the previous definitions. We say that $S\backslash R$ is Church–Rosser modulo S if $\xrightarrow[S\backslash R]{}$ is, i.e. if any two terms, equal in $R\cup S$, lead to S-equivalent terms via extended rewriting with $S\backslash R$.

Extended rewriting avoids the need to compute S-congruence classes, requiring instead an S-matching algorithm: We say that a term t *S-matches* l, if there exists a substitution σ such that $l\sigma \overset{*}{\underset{S}{\leftrightarrow}} t$. Matching algorithms are known for many theories, including associativity, commutativity, and associativity with commutativity; see Section 6.2. If R is finite and S-matching substitutions are computable, then $\xrightarrow[S\backslash R]{}$ is computable, too.

The relation $\xrightarrow[S\backslash R]{}$ is a subset of $\xrightarrow[R/S]{}$, and hence does not necessarily render the same normal forms. For the above system BA/AC, we have $and(a, and(a, b)) \xrightarrow[BA/AC]{} and(a, b)$, but $and(a, and(a, b)) \not\xrightarrow[BA\backslash AC]{} and(a, b)$. However, it is often the case that by adding certain consequences as new rules the two relations can be made to coincide, as shown in Peterson & Stickel [197] for AC, and in Jouannaud & Kirchner [124] for the general case. When R/S is terminating and $S\backslash R$ is Church–Rosser modulo S, the theory $R\cup S$ can be decided by computing $S\backslash R$-normal forms and testing for S-equivalence. For example, if BA is augmented with the two rules, $and(x, and(x, y)) \rightarrow and(x, y)$ and $xor(x, xor(x, y)) \rightarrow y$, then AC-extended rewriting suffices to compute the normal forms of BA/AC. In Section 4.2, conditions for equivalence of normal forms are discussed.

An *ordered-rewriting systems* also comes in two parts: a set of equations and an ordering. Ordered rewriting does not require that a particular equation always be used from left-to-right. Instead, an equation may be used in whichever direction agrees with the given ordering on terms. Suppose, for example, that $x \cdot y = y \cdot x$ is an equation and that $a \cdot b$ is greater than $b \cdot a$ in the ordering. Then, we would use commutativity to rewrite $a \cdot b$ to the normal form $b \cdot a$, but not vice versa. This idea dates back to Brown [35] and Lankford [156]. (A form of ordered-string-rewriting was used in Book [245].) We use a stronger, more "context-sensitive" relation than has been used heretofore in the literature.

DEFINITION. Given a set E of equations over a set of terms $\mathscr{T}$ and a rewrite ordering (transitive and irreflexive rewrite relation) $\succ$ over $\mathscr{T}$, a term s in $\mathscr{T}$ *rewrites* to a term t in $\mathscr{T}$ *according* to $\succ$, denoted $s \underset{\succ}{\rightarrow} t$, if $s = u[l\sigma]_p$, $t = u[l\sigma]_p$ and $s \succ t$, for some context u, position p in u, equation $l = r$ in E, and substitution σ.

This corresponds to considering each use of an equation as a rewrite rule going one way or the other. Thus, the ordered-rewriting relation $\underset{\succ}{\rightarrow}$, is just the intersection of the two rewrite relations, $\underset{E}{\leftrightarrow}$ and $\succ$.

2.6. *Further reading*

The standard work on the Lambda Calculus is Barendregt [17]; its role in the semantics of functional programming is discussed in Barendregt [18]. Word problems are covered in Börger [246]. Evans [72] and Knuth [149] pioneered the use of rewrite systems as decision procedures for validity in equational theories. Bledsoe [27] was an early advocate of incorporating rewriting techniques in general-purpose theorem provers.

3. Semantics

As is usual in logic, models give meaning to syntactic constructs. In our case, the models of equational theories are just algebras, that is, sets with operations on their elements, and provability by equational reasoning coincides with truth in all algebras. When a rewrite system has the unique normalization property, a term's normal form can be its "meaning". It turns out that the set of irreducible terms of a convergent system yields an agebra that is "free" among all the algebras satisfying the axioms expressed by its rules. The free algebra is that model in which the only equalities are those that are valid in all models.

3.1. *Algebras*

Let $\mathcal{F} = \bigcup_n \mathcal{F}_n$ be a signature. An *$\mathcal{F}$-algebra* $\boldsymbol{A}$ consists of a nonempty domain of values, called the *universe* (or *carrier*, or *underlying set*), which we also denote A (when feasible, we will use boldface for the algebra and italics for the corresponding universe), and a family $\mathcal{F}_A$ of $\mathcal{F}$-indexed (*finitary*) *fundamental operations*, such that for every symbol f of arity n in $\mathcal{F}_n$, the corresponding operation f_A in $\mathcal{F}_A$ maps A^n to A. Since we presume the existence of at least one constant (in $\mathcal{F}_0$), universes will always be nonempty. Given an assignment $\theta: \mathcal{X} \to A$ of values to each of the variables in $\mathcal{X}$, the $\mathcal{F}$-algebra $\boldsymbol{A}$ attaches a meaning to each term t of $\mathcal{T}(\mathcal{F}, \mathcal{X})$, which is the result of applying the operations corresponding to each function symbol in t using the values assigned by θ for the variables. Let E be a set of equations; an algebra $\boldsymbol{A}$ is a *model* of E if, for every equation $s = t$ in E, and for every assignment of values to variables in s and t, the meanings of s and t are identical. (From now on, we generally omit reference to $\mathcal{F}$ and just speak of "algebras".) By $\mathcal{Mod}(E)$ we denote the class of all models of E, each of which is an algebra.

Consider the algebra $\boldsymbol{A}$ the universe of which contains two elements, a black coffee bean and a white one: $A = \{black, white\}$. The algebra has two operations: a unary operation "invert" which turns a white bean into a black one and vice versa, and a binary operation "move" which takes two beans and returns a bean according to the rules of the second Coffee Can Game of Section 1. It is easy to see that this algebra is a model of the associative axiom $(x \cdot y) \cdot z = x \cdot (y \cdot z)$, interpreting "$\cdot$" as "move", since the result of a game does not depend upon the orders of moves. However, interpreting the unary symbol "$^-$" as "invert" and the identity constant 1 as *white* does not yield a group, since a *black white* move gives a black bean, and not the identity element. To obtain a group (as might be preferred by the mathematically inclined), we must change the rules of the game slightly:

black white	$\to$	*black*
white black	$\to$	*black*
black black	$\to$	*white*
white white	$\to$	*white*

This algebra is a model of all three group axioms.

A class $\mathscr{K}$ of algebras is a *variety* if there exists a set E of equations such that $\mathscr{K}=\mathscr{M}od(E)$. For example, though groups are axiomatizable nonequationally by giving one associative operator "$\cdot$" and a constant 1 satisfying $\forall x \exists y(x\cdot y=1)$, they may also be axiomatized in the following way:

$$
\begin{array}{|lcl|}
\hline
1\cdot x & = & x \\
x^{-}\cdot x & = & 1 \\
(x\cdot y)\cdot z & = & x\cdot(y\cdot z) \\
\hline
\end{array}
$$

Groups are actually "one-based", with the following axiom providing a basis:

$$x/(((x/x)/y)/z)/(((x/x)/x)/z) \;=\; y$$

(Higman & Neumann [103]). Note that groups defined in the latter way give a different variety than the previous axiomatization, since their signatures differ; nevertheless, the two equational theories are essentially the same, since the operations of one are definable in terms of the other (in particular, $x/y=x\cdot y^{-}$ and $x\cdot y=x/((x/x)/y)$). Rings, commutative rings, and lattices are also varieties; fields are not. Tarski has endeavored to equationally axiomatize the foundations of mathematics; see Tarski & Givant [229]. Huet [115] has shown that much of category theory is equational.

A mapping φ is a *homomorphism* from algebra $\boldsymbol{A}$ to algebra $\boldsymbol{B}$ if $f_A(a_1,\ldots,a_n)\varphi = f_B(a_1\varphi,\ldots,a_n\varphi)$, for all $f\in\mathscr{F}_n$ and $a_i\in A$. An *isomorphism* is a bijective homomorphism. Any assignment $\sigma:\mathscr{X}\to\boldsymbol{B}$ of values to variables extends in this way to a homomorphism $\sigma:\mathscr{T}\to\boldsymbol{B}$ by letting $f(t_1,\ldots,t_n)\sigma=f_B(t_1\sigma,\ldots,t_n\sigma)$. An equation $s=t$ is *valid* (or *true*) in a specific algebra $\boldsymbol{B}$ if, for all assignments σ of values in B to variables in s and t, $s\sigma$ and $t\sigma$ represent the same element of B. "Satisfiability" is the dual of validity: an equation is *satisfiable* in an algebra if it has a solution in that algebra, that is, if there is an assignment of values to variables for which both sides yield the same value. Validity of an equation $s=t$ (that is, validity in *all* models) is expressed as $\mathscr{M}od(E)\models s=t$, or $s\underset{E}{=}t$ for short.

Varieties are characterized in the following algebraic way (Birkhoff [25]): A class of algebras $\mathscr{K}$ is a variety iff it is closed under Cartesian products, subalgebras, and homomorphic images. That is, a class $\mathscr{K}$ of algebras is a variety if

(a) for any $\boldsymbol{A}_1,\ldots,\boldsymbol{A}_n$ in $\mathscr{K}$ $(n\geqslant 0)$, their product $\boldsymbol{A}_1\times\cdots\times\boldsymbol{A}_n$ is also in $\mathscr{K}$, where $f_{A_1\times\cdots\times A_n}(\ldots\langle a_1,\ldots,a_n\rangle\ldots)=\langle f_{A_1}(\ldots a_1\ldots),\ldots,f_{A_n}(\ldots a_n\ldots)\rangle$;

(b) for any subset B of A for algebra $\boldsymbol{A}$ in $\mathscr{K}$, the subalgebra obtained by restricting f_A to B for each f in $\mathscr{F}$ is also in $\mathscr{K}$; and

(c) for any homomorphism $\theta: A\to B$ between universes, if A is in $\mathscr{K}$, then so is the algebra $\boldsymbol{B}$ wherein $f_B(\ldots a_i\theta\ldots)=f_A(\ldots a_i\ldots)\theta$.

This result of Birkhoff's can be used to show that an operation is not equationally axiomatizable. For instance, the models of strict[3] *if*$\cdot$*then*$\cdot$*else*$\cdot$ are not closed under products, hence, no set of equations can characterize that operation. Still, it is remarkable

[3]An operation is *strict*, or *naturally extended*, if it yields the undefined value whenever one of its arguments is undefined.

that equational axioms E can be given for *if·then·else·* such that an equation is valid for $\mathcal{M}od(E)$ iff it is valid in the "if-then-else" models (Bloom & Tindell [31], Guessarian & Meseguer [96]).

Let E be a set of equations. Clearly, replacement of equals for equals is sound, i.e. $E \vdash s=t$ implies $\mathcal{M}od(E) \models s=t$ for all s and t. For the other direction, consider the quotient algebra $\mathcal{T}/E$ (described in Section 2.2). It is one of the models of E. Since classes $\mathcal{T}/E$ are defined by a congruence $\overset{*}{\underset{E}{\leftrightarrow}}$, which is just replacement of equals, we have $\mathcal{T}/E \models s=t$ implies $E \vdash s=t$. Together, we get the following:

COMPLETENESS THEOREM (Birkhoff [25]). *For any set of equations E and terms s and t in $\mathcal{T}$, $\mathcal{M}od(E) \models s=t$ iff $\mathcal{T}/E \models s=t$ iff $E \vdash s=t$.*

Accordingly, we may use the semantic notion $\underset{E}{=}$ and syntactic notion $\overset{*}{\underset{E}{\leftrightarrow}}$ interchangeably. It follows that a convergent rewrite system R decides validity for the models of its rules, since $s \overset{*}{\underset{R}{\leftrightarrow}} t$ iff $R(s)=R(t)$.

A substitution is a homomorphism from $\mathcal{T}$ to itself. If there exists a substitution $\sigma: \mathcal{T} \to \mathcal{T}$ such that $s\sigma$ and $t\sigma$ are identical, then for any algebra $\boldsymbol{B}$ there exists a homomorphism $\theta: \mathcal{T} \to \boldsymbol{B}$ such that $s\theta = t\theta$. In other words, if an equation is satisfiable in the term algebra $\mathcal{T}$, then it is satisfiable in all algebras. Similarly, any equation satisfiable in the quotient algebra $\mathcal{T}/E$ is satisfiable in all algebras in $\mathcal{M}od(E)$. Satisfiability in $\mathcal{T}$ is called *unifiability*; satisfiability in $\mathcal{T}/E$ is called *E-unifiability*. The unification problem is the subject of Section 6.

3.2. *Initial algebras*

For many purposes, not all models are of equal interest. One generally asks whether an equation is valid in a specific model. For example, one might ask whether the equation $alternate(y, \Lambda)=y$ is true for all stacks y. (Of course, all equations are valid, let alone satisfiable, in a trivial algebra having only one element in its universe.) For applications like abstract data types, attention is often focused on those "standard" models that are (finitely) generated from the signature itself, in which every element of the universe is the interpretation of some term.

An algebra $\boldsymbol{A}$ in a class $\mathcal{K}$ of algebras is *free* over a set $\mathcal{X}$ of variables if $\mathcal{X}$ is a subset of A and, for any algebra $\boldsymbol{B} \in \mathcal{K}$ and assignment $\theta: \mathcal{X} \to B$, there exists a unique homomorphism $\varphi: \boldsymbol{A} \to \boldsymbol{B}$ such that φ and θ agree on $\mathcal{X}$. A free algebra is unique up to isomorphism, whenever it exists. The (*absolutely*) free algebra over $\mathcal{X}$ among all algebras is just (isomorphic to) the *term algebra* $\mathcal{T}(\mathcal{F}, \mathcal{X})$ with the symbol $f \in \mathcal{F}$ itself as the operator $f_{\mathcal{T}}$. An algebra $\boldsymbol{A}$ in a class $\mathcal{K}$ of algebras is *initial* if, for any algebra $\boldsymbol{B}$ in $\mathcal{K}$ there exists a unique homomorphism $\varphi: \boldsymbol{A} \to \boldsymbol{B}$. The initial object among all $\mathcal{F}$-algebras is (isomorphic to) the *ground-term algebra* $\mathcal{G}(\mathcal{F})$, again with the function symbol itself as operator, and corresponds to the Herbrand universe over the symbols in $\mathcal{F}$. The importance of the initial algebra lies in its uniqueness (it is the free algebra for empty $\mathcal{X}$), and in the fact that the class $\mathcal{K}$ consists of its homomorphic images, making it the most "abstract" amongst them.

Among all models of a set of equations E, the prototypical one is the initial algebra

$I(E)$ of E. Its universe consists of one element for each E-congruence class of ground terms. In other words, $I(E)$ is (isomorphic to) the quotient $\mathcal{G}/E$ of the ground-term algebra $\mathcal{G}$ and the congruence $\overset{*}{\underset{E}{\leftrightarrow}}$ (restricted to $\mathcal{G}$). This algebra can be realized if R is a ground-convergent rewrite system for E, since, then, E-equivalent ground terms have the same R-normal form. Accordingly, the *normal-form algebra* of R has the set of ground R-normal forms as its universe and operations f_R defined by $f_R(t_1, \ldots, t_n) = R(f(t_1, \ldots, t_n))$ for all normal forms t_i. This algebra is (isomorphic to) the initial algebra $I(R)$ of the variety defined by the rules in R considered as equations (Goguen [87]). Thus, rewriting computes ground normal forms that are representatives of their congruence classes. It is in this sense that rewriting is a "correct" implementation of initial-algebra semantics. Specification languages based on abstract data types, such as OBJ (Futatsugi et al. [80]), follow this implementation scheme: equations are used as rewrite rules, and unique normalization is needed for the operational and initial-algebra semantics to coincide.

Exactly those variable-free equations that follow necessarily from E hold in the initial algebra. Thus, the word problem for E, i.e. deciding, for *ground* terms s and t, whether $s=t$ holds in every model of E, is the same as determining if $I(E) \models s=t$. More generally, one may ask if an equation $s=t$ (possibly containing variables) is valid in the initial algebra $I(E)$, which is the case iff all of its ground instances hold for $\mathcal{Mod}(E)$. We will write $s \underset{I(E)}{=} t$ as an abbreviation for $I(E) \models s=t$ and call the class $\mathcal{Ind}(E)$ of equations $s=t$ valid in $I(E)$ the *inductive theory* of E. It is easy to verify that the relation $\underset{I(E)}{=}$ is a congruence over $\mathcal{T}(\mathcal{F}, \mathcal{X})$. Unlike equational theories, inductive theories are not necessarily recursively enumerable (even for finite E). The inductive theory includes all the equations in the equational theory; on the other hand, an equation that holds in the initial model need not hold in all models, i.e. the inclusion $\mathcal{Th}(E) \subseteq \mathcal{Ind}(E)$ may be strict. Tarski [228] dubbed *ω-complete* those equational theories that coincide with the associated inductive theory, but ω-completeness is not possible in general (Henkin [99]). Finally, note that if an equation $s=t$ is not valid in $I(E)$, that means that some ground equation $u=v$ which does not hold for $\mathcal{Mod}(E)$ does hold for $\mathcal{Mod}(E \cup \{s=t\})$.

For example, let $\mathcal{F} = \{0, succ, \Lambda, push, alternate\}$ and let E be

$$\begin{array}{|lcl|} \hline alternate(\Lambda, z) & = & z \\ alternate(push(x, y), z) & = & push(x, alternate(z, y)) \\ \hline \end{array}$$

The equation $alternate(y, \Lambda) = y$ is valid in $I(E)$, since it is provable for all ground terms of the form $push(s_1, push(s_2, \ldots, push(s_n, \Lambda) \ldots))$, and all other ground terms (entailing *alternate*) are provably equal to one of this form. It is not, however, valid in a model A that, besides the usual stacks, includes stacks built on top of another empty-stack value, Λ_0, and for which $alternate_A(\Lambda_0, \Lambda_A) = \Lambda_A$. Thus, by Birkhoff's Completeness Theorem, $alternate(y, \Lambda) = y$ is not an equational consequence of the given axioms.

For a system R to correctly implement an algebraic specification E, it is enough that their inductive theories are the same, i.e. that $\mathcal{Ind}(E) = \mathcal{Ind}(R)$, when the rules of R are considered as equations. For example, the convergent three-rule system

$alternate(\Lambda, z)$	$\rightarrow$	z
$alternate(y, \Lambda)$	$\rightarrow$	y
$alternate(push(x, y), z)$	$\rightarrow$	$push(x, alternate(z, y))$

is a correct implementation of the above specification E of *alternate*, since all equations in E are deductive theorems of R and all rules in R are inductive theorems of E.

The notion of sufficient completeness of function definitions (and its relation to software specification) was introduced in Guttag [97]:

DEFINITION. Let the set $\mathcal{F}$ of function symbols be split into a set $\mathcal{C}$ of *constructors* and a set $\mathcal{F}-\mathcal{C}$ of other symbols. Let E be a set of equations in $\mathcal{T}(\mathcal{F}, \mathcal{X})$. The specification E is *sufficiently complete* (or "has no junk") with respect to $\mathcal{C}$, if every ground term t in $\mathcal{G}(\mathcal{F})$ is provably equal to a constructor term s in $\mathcal{G}(\mathcal{C})$.

DEFINITION. Let $\mathcal{C}$ be a set of constructors and let E be a set of equations split into a set $E_{\mathcal{C}}$ of equations in $\mathcal{T}(\mathcal{C}, \mathcal{X})$ and a set $E_{\mathcal{F}-\mathcal{C}}$ of other equations. The constructors $\mathcal{C}$ are said to be *free* when $E_{\mathcal{C}}$ is empty. The specification E is *consistent* (or "has no confusion") with respect to $\mathcal{C}$, if, for arbitrary ground constructor terms s and t in $\mathcal{G}(\mathcal{C})$, $s \underset{E}{=} t$ iff $s \underset{E_{\mathcal{C}}}{=} t$.

This generalizes the standard notion of consistency, which is with respect to the Boolean values T and F. For example, the previous specification for stacks becomes inconsistent with respect to the free constructors $\{0, succ, \Lambda, push\}$, if it is "enriched" with the equation $alternate(y, \Lambda) = alternate(y, y)$, since that implies $push(0, \Lambda) = push(0, push(0, \Lambda))$. Adding an equation $push(x, push(y, z)) = push(y, push(x, z))$ to $E_{\mathcal{C}}$ makes the constructors nonfree and constructor terms represent unordered multisets ("bags"). When a set of equations is both consistent and sufficiently complete, it is reasonable to consider it a "specification" of the functions in $\mathcal{F}-\mathcal{C}$. In this case, the algebra $\mathcal{G}(\mathcal{F})/E$, considered as a $\mathcal{C}$-algebra, is isomorphic to $\mathcal{G}(\mathcal{C})/E_{\mathcal{C}}$. This allows one to build complex specifications from simpler ones. Unfortunately, both properties are undecidable in general (see Guttag [97]).

As mentioned above, term-rewriting is used to compute in the initial algebra. More generally, if R/S is a ground-convergent class-rewriting (or ordered-rewriting) system, then the normal-form algebra, with universe $R(\mathcal{G}) = \{[R(t)]_S \mid t \in \mathcal{G}\}$ and operations defined by $f_R(a_1, \ldots, a_n) = [R(f(a_1, \ldots, a_n))]_S$ for all $a_i \in R(\mathcal{G})$, is initial for the variety defined by $R \cup S$, and $R(\mathcal{G})$ is isomorphic to $I(R \cup S)$. In implementing a sufficiently complete specification, one would want all ground normal forms to be constructor terms, i.e. that $R(\mathcal{G}(\mathcal{F})) \subseteq \mathcal{G}(\mathcal{C})$.

In certain cases, sufficient completeness can be related (Kounalis [150]) to the following more tractable property:

DEFINITION. For any rewrite relation $\underset{T}{\rightarrow}$, a term s in $\mathcal{T}$ is *ground T-reducible*, if all its ground instances $s\gamma \in \mathcal{G}$ are rewritable by $\underset{T}{\rightarrow}$.

Suppose that

(a) a ground-convergent class-rewriting system R/S is complete for an equational specification E,

(b) the R/S-normal form of any ground constructor term in $\mathcal{G}(\mathcal{C})$ is a ground constructor term, and

(c) S does not equate any ground constructor term with a ground non-constructor term in $\mathcal{G}(\mathcal{F})-\mathcal{G}(\mathcal{C})$.

For a system R/S satisfying these properties, E is sufficiently complete iff all terms $f(x_1,\ldots,x_n)$ are ground R/S-reducible, when f is in $\mathcal{F}-\mathcal{C}$ and x_i are distinct variables in $\mathcal{X}$. The rationale is that, by ground reducibility, any non-constructor term t must contain a reducible subterm, and, since the system is terminating and sound for E, t must be equal to a constructor term. This connection between sufficient completeness and ground reducibility is implicit in Plaisted [199]. If each left-hand side of a rule in R and each side of an equation in S contains a non-constructor symbol, then property (b) is ensured; cf. Huet & Hullot [116].

Ground reducibility is decidable for finite R and empty S (Plaisted [199], Kapur et al. [139]). A faster decision method is obtained by reducing ground reducibility to the emptiness problem of the language produced by a "conditional tree grammar" describing the system's ground normal forms (Comon [45]). Testing for ground R-reducibility, however, requires exponential time, even for left-linear R (Kapur et al. [138]). In the special case where all constructors are free, ground reducibility is more easily testable. This case had been considered in Nipkow & Weikum [189] for left-linear systems. The general case was considered in Dershowitz [57] and Kounalis [150]. The former defines a "test set" for ground-reducibility by instantiating $f(x_1,\ldots,x_n)$ in all possible ways up to a bound that depends on the maximal depth of a left-hand side; the latter constructs a smaller test set, computed by repeated unification of $f(x_1,\ldots,x_n)$ with left-hand sides, and improves on Thiel [231]. Ground R/S-reducibility is undecidable when S is a set of associative-commutative axioms (Kapur et al. [138]), but is decidable when R is left-linear (Jouannaud & Kounalis [125]).

For ground-convergent systems R, any equation between distinct R-normal forms is considered to be inconsistent with R (considering all symbols in $\mathcal{F}$ as constructors). The observation that an equation $s=t$ is valid in the initial algebra $I(R)$ iff no inconsistency follows from $R\cup\{s=t\}$ is the basis of the *proof by consistency* method of inductive theorem proving (for proving theorems in $\mathcal{Ind}(R)$), pioneered by Musser [186] (and so named in Kapur & Musser [131]). If there exists a ground-convergent system R', with the same ground normal forms as R, and which presents the same equational theory as $R\cup\{s=t\}$, then inconsistency is precluded (Lankford [159]). It can readily be shown that $R(\mathcal{G})\subseteq R'(\mathcal{G})$, for any two systems R and R', iff every left-hand side of R' is ground R-reducible (Dershowitz [55], Jouannaud & Kounalis [125]). It follows that R and R' have the same inductive theory if they are both ground convergent and every left-hand side of one system is ground reducible by the other. This method, relating validity in the initial algebra to ground-reducibility, extends to class-rewriting, with ground R/S-reducibility replacing its ordinary counterpart (Goguen [87], Lankford [159], Jouannaud & Kounalis [125]). In Section 8, we will consider how to search for an appropriate R'.

Note that the equation *alternate*$(y, \Lambda)=y$ is *not* an inductive theorem of the earlier stack specification, given at the beginning of Section 2.1, even though it holds for all stacks Λ, *push*(s_1, Λ), etc. The problem is that the full vocabulary has a richer set of ground terms, involving *top* and *pop*, but their specification is not sufficiently complete. In particular, the equation in question does not hold true for "error" terms like *pop*(Λ): *alternate*(*pop*(Λ), Λ)$=$*pop*(Λ) does not follow from the axioms. Regarding proofs in the (noninitial) constructor model, see Zhang [241] and Kapur & Musser [264].

3.3. Computable algebras

When a system R is not terminating, rewriting will not necessarily compute a representative for the congruence class of a term. However, as long as R is Church–Rosser, one knows that *if* a normal form is obtained, it is unique. Of course, one can always turn a finite set of equations into a Church–Rosser system by turning each equation into a symmetric pair of rules, but then no term at all has a normal form. More interesting is the ability to code interpreters for functional languages as Church–Rosser systems that are normalizing for input programs that terminate for the given input values; see O'Donnell [192]. Furthermore, for certain systems, there are computational strategies (that is, specific choices of where to rewrite next), such as not forever ignoring an "outermost" redex (one that is not a subterm of another redex), that are guaranteed to result in a normal form whenever there is one; see Section 7.2.

Turing machine computations can be simulated by rewrite systems in at least two different ways: by systems of monadic rules that rewrite instantaneous descriptions according to the machine's transitions (Huet & Lankford [117]), and by a (nonmonadic) one-rule system in which the transitions appear as part of the terms (Dauchet [49], refining Dershowitz [58]). Thus, rewrite systems provide a fully general programming paradigm (to the extent that Church's Thesis defines "fully general"). These constructions also imply that most interesting properties, including convergence, are in general undecidable. On the other hand, equality (the word problem) is decidable in what are called "computable" algebras (Meseguer & Goguen [182]); see Wirsing [239].

3.4. Further reading

For a survey of equational logic, see Taylor [230]. A comprehensive multi-volume work on varieties is McKenzie et al. [180] and Freese et al. [78]. Some relevant recent results are summarized in McNulty [181]. A detailed exposition of algebraic aspects of rewriting is Meseguer & Goguen [182]; algebraic semantics are the subject of Wirsing [239].

4. Church–Rosser properties

Newman [188] developed a general theory of "sets of moves", that is, of arbitrary binary relations. It has since become customary to deal separately with properties of such abstract binary relations and with those of relations on terms. In our discussion of the Church–Rosser property, we continue in that tradition, putting off almost all mention of rewrite systems to later sections.

4.1. *Confluence*

In Section 2, we defined the Church–Rosser property for rewrite systems. The analogous property can hold for any binary relation:

DEFINITION. A binary relation $\rightarrow$ on any set T is *Church–Rosser* if its reflexive-symmetric-transitive closure $\overset{*}{\leftrightarrow}$ is contained in the *joinability* relation $\overset{*}{\rightarrow} \circ \overset{*}{\leftarrow}$.

See Fig. 1(a). This is equivalent to the following simpler property:

DEFINITION. A binary relation $\rightarrow$ on any set T is *confluent* if the relation $\overset{*}{\leftarrow} \circ \overset{*}{\rightarrow}$ is contained in the joinability relation $\overset{*}{\rightarrow} \circ \overset{*}{\leftarrow}$.

Confluence says that no matter how one diverges from a common ancestor, there are paths joining at a common descendent. Sometimes the notation $\uparrow$ is used for the common ancestor relation and $\downarrow$ for joinability (common descendent); then confluence boils down to $\uparrow \subseteq \downarrow$. See Fig. 1(b). The equivalence with the Church–Rosser property (Newman [188]) can be shown by a simple inductive argument on the number of divergences $\overset{*}{\leftarrow} \circ \overset{*}{\rightarrow}$ making up $\overset{*}{\leftrightarrow}$.

For arbitrary $\rightarrow$, define $s \overset{!}{\rightarrow} t$ iff $s \overset{*}{\rightarrow} t$ and there is no u such that $t \rightarrow u$ and call t the *normal form* of s. Confluence implies the impossibility of more than one normal form. A binary relation $\rightarrow$ on a set is *strongly confluent* if any local divergence $\leftarrow \circ \rightarrow$ is contained in the immediate descendent relation $\overset{=}{\rightarrow} \circ \overset{=}{\leftarrow}$, i.e. if for any *peak* $s \leftarrow u \rightarrow t$ of elements s, t, and u, one of the following four cases holds: $s = t$, $s \leftarrow t$, $s \rightarrow t$, or $s \rightarrow v \leftarrow t$ (for some element v). See Fig. 1(c). (A slightly weaker definition is given in Huet [113], namely $\leftarrow \circ \rightarrow \subseteq \overset{*}{\rightarrow} \circ \overset{=}{\leftarrow}$, which also allows for circumstances like $a \rightarrow b$, $a \rightarrow c$, $b \rightarrow d \rightarrow c$, and $c \rightarrow d \rightarrow b$.) Strong confluence implies confluence (Newman [188]) by a "tiling" argument. Strong confluence is used in the classical proofs of the Church–Rosser property for the λ-calculus, since confluence of $\rightarrow$ is exactly strong confluence of $\overset{*}{\rightarrow}$ (see Barendregt [17]).

DEFINITION. A binary relation $\rightarrow$ on any set T is *locally confluent* if any local divergence $\leftarrow \circ \rightarrow$ is contained in the joinability relation $\overset{*}{\rightarrow} \circ \overset{*}{\leftarrow}$.

See Fig. 1(d). Local confluence does not generally imply confluence; see the counterexamples in Fig. 2, due to Newman [188] and Hindley [104]. However:

DIAMOND LEMMA (Newman [188]). *A terminating relation is confluent iff it is locally confluent.*

The name derives from the pictorial proof in Fig. 3, due to Huet [113], which uses induction with respect to the terminating relation. When $\rightarrow$ is terminating, it follows from the above results that it is Church–Rosser iff $\leftarrow \circ \rightarrow$ is contained in $\overset{!}{\rightarrow} \circ \overset{!}{\leftarrow}$.

Confluence is sometimes established by well-founded induction in the following way:

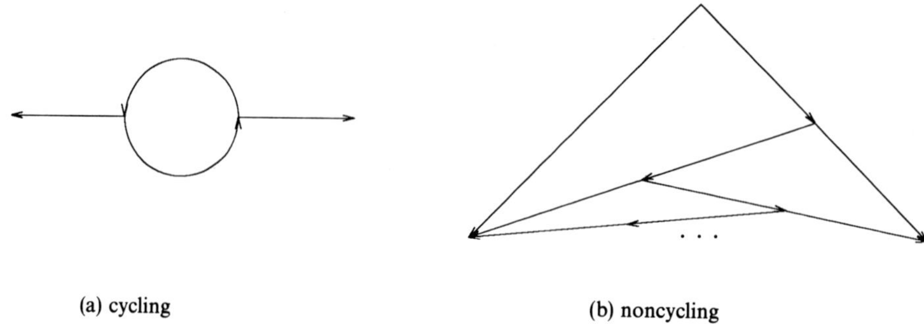

Fig. 2. Two locally-confluent relations.

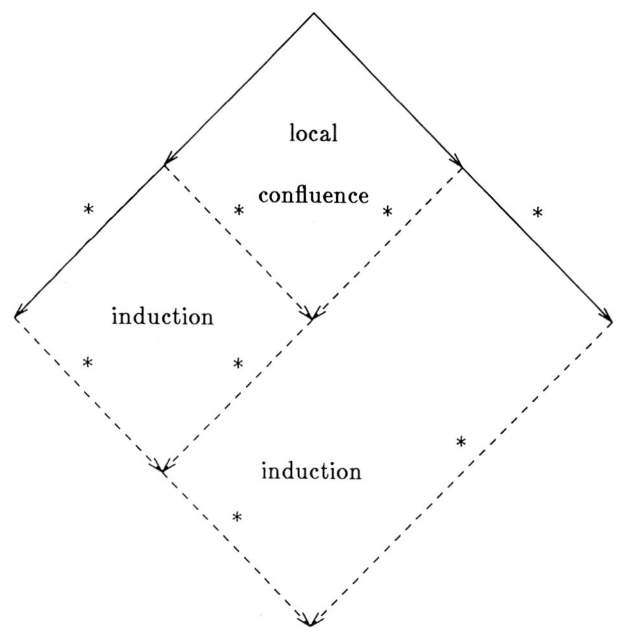

Fig. 3. Proof of Diamond Lemma.

Let $\succ$ be a well-founded ordering on the elements and suppose that for every "peak" $s \leftarrow u \rightarrow t$ there exists an (undirected) path $s = w_0 \leftrightarrow w_1 \leftrightarrow \cdots \leftrightarrow w_n = t$ $(n \geqslant 0)$ such that $u \succ w_1, \ldots, w_{n-1}$. Then it can be shown (by induction on multisets of elements; see Section 5.1) that $\rightarrow$ is Church–Rosser (Winkler & Buchberger [238]). See Fig. 1(e).

4.2. Coherence

As preparation for the study, in Section 7.3, of Church–Rosser properties of extended-rewriting with a congruence, we consider here abstract properties of combinations of an arbitrary binary relation $\underset{R}{\rightarrow}$ and a symmetric binary relation $\underset{S}{\leftrightarrow}$, both on the same set. Let $\xrightarrow[T]{}$ be a relation lying anywhere between $\xrightarrow[R]{}$ and the quotient relation $\xrightarrow[R/S]{}$. Note that $\xleftrightarrow[R\cup S]{*} = \xleftrightarrow[S\cup T]{*}$. In what follows, we will also use R, S, T, and R/S to refer to the relations $\xrightarrow[R]{}$, $\xleftrightarrow[S]{}$, $\xrightarrow[T]{}$, and $\xrightarrow[R/S]{}$, respectively. At one extreme, T can be R, a case partially dealt with in Sethi [216], and Huet [113]; at the other extreme, T can be R/S; the general case—considered here—was first studied in Jouannaud & Kirchner [124].

If one is going to compute normal forms with T instead of with R, the natural question to ask is if their normal forms are equivalent. That is, under what conditions is $\xleftarrow[R]{!} \circ \xrightarrow[T]{!}$ contained in $\xleftrightarrow[S]{*}$? The relation T is said to be *Church–Rosser modulo S* if $\xleftrightarrow[S\cup T]{*}$ is contained in $\xrightarrow[T]{*} \circ \xleftrightarrow[S]{*} \circ \xleftarrow[T]{*}$; see Fig. 1(f). When T is terminating and Church–Rosser modulo S, one can determine if $s \xleftrightarrow[R\cup S]{*} t$, by finding T-normal forms of s and t and testing them for S-equivalence, since $\xleftrightarrow[S\cup T]{*}$ is contained in $\xrightarrow[T]{!} \circ \xleftrightarrow[S]{*} \circ \xleftarrow[T]{!}$. In particular, R/S- and T-normal forms must be equivalent.

When R/S is terminating, the Church–Rosser property may be decomposed into two local ones: T is *locally coherent modulo S with R* if $\xleftarrow[T]{} \circ \xrightarrow[R]{}$ is contained in $\xrightarrow[T]{*} \circ \xleftrightarrow[S]{*} \circ \xleftarrow[T]{*}$; T is *locally coherent modulo S with S* if $\xleftarrow[T]{} \circ \xleftrightarrow[S]{}$ is contained in $\xrightarrow[T]{*} \circ \xleftrightarrow[S]{*} \circ \xleftarrow[T]{*}$. Compare Figs. 2(g) and (h) with 2(d). The concept of coherence was developed by Jouannaud & Kirchner [124] and generalizes *compatibility*, as in Peterson & Stickel [197].

COHERENCE LEMMA (Jouannaud & Kirchner [124]). *Let $R \subseteq T \subseteq R/S$. If R/S is terminating, then T is Church–Rosser modulo S iff it is locally coherent modulo S with both R and S.*

One proves that $\xleftrightarrow[R\cup S]{*}$ is contained in $\xrightarrow[T]{!} \circ \xleftrightarrow[S]{*} \circ \xleftarrow[T]{!}$ by induction on the multiset of elements in a path $t_0 \xleftrightarrow[R\cup S]{} t_1 \xleftrightarrow[R\cup S]{} \cdots \xleftrightarrow[R\cup S]{} t_n$. The induction is with respect to the well-founded ordering on multisets (see Section 5.1) induced by $\xrightarrow[R/S]{+}$. It then follows that $\xleftrightarrow[S\cup T]{*}$ is contained in $\xrightarrow[T]{*} \circ \xleftrightarrow[S]{*} \circ \xleftarrow[T]{*}$.

4.3. *Further reading*

Newman's 1942 paper [188] defined the basic notions. Huet [113] introduced the use of (Noetherian) induction on terminating relations for studying these notions. Confluence (the vanilla-flavored kind) and many related properties of relations are discussed in Klop [147].

5. Termination

Recall that a rewrite system R is terminating for a set of terms $\mathcal{T}$ if there are no infinite derivations $t_1 \xrightarrow[R]{} t_2 \xrightarrow[R]{} \cdots$ of terms in $\mathcal{T}$. The following is one example of a nonterminating system (Toyama [235]):

$$\begin{array}{|lcl|} \hline f(a,b,x) & \rightarrow & f(x,x,x) \\ g(x,y) & \rightarrow & x \\ g(x,y) & \rightarrow & y \\ \hline \end{array}$$

The depth (i.e. the maximum nesting of symbols) of a term in any of its derivations is bounded by the depth of the initial term, but there is a cycling derivation starting from $f(g(a,b), g(a,b), g(a,b))$. If $\underset{R}{\rightarrow}$ is contained in some well-founded partial ordering $\succ$ on $\mathcal{T}$, then R is obviously terminating. The rule $f(f(x)) \rightarrow f(g(f(x)))$, for instance, is terminating, since the number of adjacent f's is reduced with each application. The rule $f(f(x)) \rightarrow g(g(f(f(x))))$ is nonterminating. In general, it is undecidable whether a system is terminating, even if both sides of all rules are monadic (Huet & Lankford [117]) or if it has only one left-linear rule (Dauchet [49]). For finite ground systems, however, termination is decidable (Huet & Lankford [117]). In particular, reduced ground systems are terminating as a corollary of the proof. The decidability of termination of non-length-increasing string-rewriting systems is open.

5.1. *Reduction orderings*

The above method of establishing termination requires one to reason about the global effect of applying a rule at a subterm. To avoid consideration of the infinite number of possible contexts, one can use rewrite orderings on terms:

DEFINITION. A *reduction ordering* on a set of terms $\mathcal{T}$ is any well-founded rewrite ordering of $\mathcal{T}$.

Termination is assured if each of the *rules* in R is contained in a reduction ordering; conversely, if R is terminating, then the relation $\overset{+}{\underset{R}{\rightarrow}}$ itself is a reduction ordering. As suggested in Manna & Ness [173], it is oftentimes convenient to separate reduction orderings into a homomorphism τ from the ground terms $\mathcal{G}(\mathcal{F})$ to an $\mathcal{F}$-algebra $\boldsymbol{W}$ and a "standard" well-founded ordering $\succ$ on $\boldsymbol{W}$. The homomorphism and ordering are constrained to satisfy the following monotonicity condition: $f_\tau(\ldots x \ldots) \succ f_\tau(\ldots y \ldots)$ whenever $x \succ y$, for all f in $\mathcal{F}$ and all x, y, etc. in W. Then, the ordering $\succ_\tau$, under which $s \succ_\tau t$ if $\tau(s) \succ \tau(t)$, for s and t in $\mathcal{G}$, is well-founded. To compare *free* (*open*) terms s and t in $\mathcal{T}(\mathcal{F}, \mathcal{X})$, one can demand $s\sigma \succ_\tau t\sigma$ for all ground substitutions σ, which might be undecidable. Instead, it is common to compare by adding variables to W, extending τ to map variables in terms to distinct variables in $W(\mathcal{X})$, and define $s \succ_\tau t$ only if $\tau(s) \succ \tau(t)$ for all assignments of values in W to the variables in $\tau(s)$ and $\tau(t)$. A system is terminating iff such $\boldsymbol{W}$, τ, and $\succ$ exist. For example, the system below, which computes the disjunctive normal form of formulae, can be shown terminating (Filman [253]) with an exponential mapping into the natural numbers:

$not(not(x))$	$\rightarrow$	x
$not(or(x, y))$	$\rightarrow$	$and(not(x), not(y))$
$not(and(x, y))$	$\rightarrow$	$or(not(x), not(y))$
$and(x, or(y, z))$	$\rightarrow$	$or(and(x, y), and(x, z))$
$and(or(y, z), x)$	$\rightarrow$	$or(and(y, x), and(z, x))$

The mapping $\tau: \mathcal{T} \rightarrow \{2, 3, \ldots\}$ is defined by

$$or_\tau(a, b) = a + b + 1, \qquad not_\tau(a) = 2^a,$$
$$and_\tau(a, b) = a \cdot b, \qquad c_\tau = 2$$

for all a and b in $\{2, 3, \ldots\}$ and constant c in $\mathscr{F}_0$, with numbers compared under their natural ordering $>$. The second rule, for instance, always decreases the interpretation of a term, since $\tau(not(or(x, y))) = 2^{x+y+1}$ is greater than $\tau(and(not(x), not(y))) = 2^{x+y}$, for all x and y. A class of exponential interpretations was used for termination arguments in Iturriaga [121].

The use, in particular, of *polynomial interpretations* was developed in Lankford [156–158]. Here a multivariate integer polynomial $f_\tau(x_1, \ldots, x_n)$ of n variables is associated with each n-ary symbol f in $\mathscr{F}_n$, for all n. The choice of coefficients must satisfy the monotonicity condition and ensure that terms are mapped into nonnegative integers only, as is the case, for example, when all coefficients are positive. Each rule must be shown to be reducing; that is, for each rule $l \rightarrow r$, the polynomial $\tau(l) - \tau(r)$ must be positive for all values of variables greater than the minimal value of a ground term (τ interprets variables in $\mathscr{X}$ as variables ranging over the naturals).

To prove termination of an associative-commutative class-rewriting system, the interpretation of an associative-commutative operator ought to be an associative-commutative polynomial. In general, such polynomials must be of either the quadratic form $f_\tau(x, y) = \alpha xy + \beta(x+y) + \beta(\beta-1)/\alpha$ ($\alpha \neq 0$) or the linear form $f_\tau(x, y) = x + y + \gamma$, where α, β, and γ are natural numbers (Ben Cherifa & Lescanne [22]). For example, one can use the following polynomial interpretation to prove termination of the BA/AC system of Section 2.5

$$\begin{aligned} xor_\tau(a, b) &= a + b + 1 \\ and_\tau(a, b) &= a \cdot b \\ c_\tau &= 2. \end{aligned}$$

Primitive-recursive interpretations cannot suffice for termination proofs in general, since they would place a primitive-recursive bound on the length of computations (Stickel [221]). In particular, integer polynomials place a double-exponential bound on the length of a derivation (Lautemann [165]). The following system, based on the "Battle of Hydra and Hercules" in Kirby & Paris [141], is terminating, but not provably so in Peano Arithmetic:

$$\begin{array}{lcl} h(z, e(x)) & \rightarrow & h(c(z), d(z, x)) \\ d(z, g(0, 0)) & \rightarrow & e(0) \\ d(z, g(x, y)) & \rightarrow & g(e(x), d(z, y)) \\ d(c(z), g(g(x, y), 0)) & \rightarrow & g(d(c(z), g(x, y)), d(z, g(x, y))) \\ g(e(x), e(y)) & \rightarrow & e(g(x, y)) \end{array}$$

Think of $g(x, y)$ as the ordinal $\omega^x + y$, of $d(c^n(0), x)$ as any of the kth predecessors of x, $k \leqslant n$, and of $e(x)$ as x (e is just a place marker). Transfinite (ε_0-) induction is required for a proof of termination.

Nor do total reduction orderings suffice in termination arguments, as can be seen from the terminating system $\{f(a) \rightarrow f(b),\ g(b) \rightarrow g(a)\}$, for which a and b must be incomparable. Nevertheless, most of the orderings used in practice do extend to total reduction orderings. For a *total* rewrite ordering to be well-founded, it is necessary that it contain the proper subterm relation $\rhd$, since if $t|_p \succ t$ for some term t and position p, then there is an infinite descending sequence $t \succ t[t]_p \succ t[t[t]_p]_p \succ \cdots$ (Plaisted [198]). As we will see shortly, for *finite* vocabularies, this "subterm" condition is also sufficient for well-foundedness of a rewrite ordering.

For termination of an ordered rewrite relation $\underset{>}{\rightarrow}$, the ordering $\succ$ according to which rewriting is performed must be a (well-founded) reduction ordering. Note that the variables occurring on the two sides of an equation need not coincide for termination, since the terms substituted for the variables will be such that the rewritten term is smaller vis-à-vis the reduction ordering. If the ordering $\succ$ is total on ground terms $\mathcal{G}$, then each ground instance of an equation in E can be oriented one way or another. As we will see, such ground orderings do exist. Of course, no rewrite ordering can be total on $\mathcal{T}$, since no two distinct variables x and y can be ordered. (Were we to have $x \succ y$, then we would have to have $y \succ x$, as well, the latter being an instance of the former.)

5.2. *Simplification orderings*

Termination arguments are often facilitated by the observation that all rewrite orderings containing the subterm relation $\trianglerighteq$ are well-founded (when $\mathcal{F}$ is finite). To see why the subterm property suffices, we need a stronger notion than well-foundedness:

DEFINITION. A quasi-ordering $\succcurlyeq$ on a set T is a *well-quasi-ordering* if every infinite sequence $t_1, t_2, \ldots$ of elements of T contains a pair of elements t_j and $t_k, j<k$, such that $t_j \preccurlyeq t_k$.

By the Pigeon-Hole Principle, an equivalence relation is a well-quasi-ordering iff there are only a finite number of equivalence classes.

The partial ordering $\succ$ associated with a quasi-ordering $\succcurlyeq$ is well-founded iff from some point on all elements in any infinite "quasi-descending" chain $t_1 \succcurlyeq t_2 \succcurlyeq t_3 \succcurlyeq \cdots$ are equivalent. Thus, any well-quasi-ordered set is well-founded (by the associated partial order), while a set is well-quasi-ordered if it is well-founded and has only a finite number of pairwise incomparable elements (the "finite antichain property"). It is also important to note that any (transitive) extension of a well-quasi-ordering is a well-quasi-ordering and any restriction (that is still a quasi-ordering) is well-founded (though it need not be well-quasi-ordered). This is what makes well-quasi-orderings convenient.

Our interest focuses on well-quasi-orderings of terms. Any well-quasi-ordering $\succcurlyeq$ on a vocabulary $\mathcal{F}$ induces a well-quasi-ordering $\succcurlyeq_{emb}$ on the terms $\mathcal{T}$ by means of the following set of schematic rules and equations:

$f(s_1, \ldots, s_n)$	$\rightarrow$	s_i	$1 \leqslant i \leqslant n$
$f(s_1, \ldots, s_n)$	$\rightarrow$	$g(s_1, \ldots, s_n)$	if $f \succ g$
$f(\ldots, s_i, \ldots)$	$\rightarrow$	$g(\ldots, s_{i-1}, s_{i+1}, \ldots)$	if $f \succcurlyeq g$
$f(s_1, \ldots, s_n)$	$\leftrightarrow$	$g(s_1, \ldots, s_n)$	if $f \sim g$

These schemata apply to all f and g in $\mathscr{F}$. The first deletes context; the second decreases a function symbol; the third deletes subterms; the last replaces symbols with equivalents. We write $s \succcurlyeq_{emb} t$ if t is derivable from s using the above rules. The way the rules have been written, the strict part $\succ_{emb}$ of embedding is derivability by at least one application of the (manifestly terminating) first three rules. The equivalence part of $\succcurlyeq_{emb}$ is just renaming symbols with equivalents under $\sim$ using the last rule. Viewing terms as (ordered) trees: $s \succcurlyeq_{emb} t$ if there is a mapping from the nodes in t into the nodes in s such that the function symbol labeling a node in t is less than or equivalent to (under $\succcurlyeq$) the label of the corresponding node in s, and such that distinct edges in t map to disjoint paths of s.

Taking the transitive closure of the third rule avoids the violaton of arity in which the above schema indulges (the arity of g might not be $n-1$). Combining the new rule with the second and fourth (and blurring the distnction between $\succcurlyeq_{emb}$ and $\succ_{emb}$ thereby), gives the following, alternative schema for $\succ_{emb}$:

$$\begin{array}{|l|}\hline f(s_1,\ldots,s_n) \;\rightarrow\; s_i \quad 1 \leqslant i \leqslant n \\ f(s_1,\ldots,s_n) \;\rightarrow\; g(s_{i_1},\ldots,s_{i_k}) \\ \quad \text{if } f \succcurlyeq g,\ 1 \leqslant i_1 < \cdots < i_k \leqslant n,\ k \leqslant n \\ \hline\end{array}$$

The following deep result is at the heart of the argument:

TREE THEOREM (Kruskal [153]). *If $\succcurlyeq$ is a well-quasi-ordering of a vocabulary $\mathscr{F}$, then the embedding relation $\succcurlyeq_{emb}$ is a well-quasi-ordering of the terms $\mathscr{T}(\mathscr{F})$.*

For finite $\mathscr{F}$, this theorem is due Higman [102]. The general case has a beautiful proof, due to Nash-Williams [187]:

PROOF. Note that, by the infinite version of Ramsey's Theorem, any infinite sequence of elements of a well-quasi-ordered set must contain a subsequence that constitutes an infinite quasi-ascending chain.

Suppose, now, that the theorem were false. Then, there would exist (by the Axiom of Choice) a "minimal counter-example" sequence $t_1, t_2, \ldots, t_i, \ldots$, of which each element t_i is chosen so that it is smallest (in number of symbols) among all sequences of terms beginning with $t_1, t_2, \ldots, t_{i-1}$ and having no embedding $t_j \preccurlyeq_{emb} t_k$ for $j<k$. By the minimality hypothesis, the set of proper subterms of the elements of the minimal counter-example must be well-quasi-ordered (or else $t_1, t_2, \ldots, t_{l-1}, s_1, s_2, \ldots$ would be a smaller counter-example, where $s_1, s_2, \ldots$ is a counter-example of subterms of $t_l, t_{l+1}, \ldots$, such that s_1 is a subterm of t_l).

Since $\mathscr{F}$ is well-quasi-ordered by $\succcurlyeq$, there must exist an infinite subsequence $t_{i_1}, t_{i_2}, \ldots$ of the minimal counter-example such that their roots are a quasi-ascending chain. If any of these terms t_{i_j} are elements of $\mathscr{F}$, the original sequence could not have been a counter-example, because then $t_{i_j} \preccurlyeq_{emb} t_{i_{j+1}}$. Consider, then, the immediate subterms $w_{i_1}, w_{i_2}, \ldots$ of that subsequence. For example, if t_{i_j} is $f(g(a), b, g(b))$, then w_{i_j} is the word $g(a)b\,g(b)$. As noted above, the set of all these words must be well-quasi-ordered.

Using an auxiliary minimal counter-example argument, it can be shown that any infinite sequence of words over a well-quasi-ordered set contains a pair of words such that the first is a (not necessarily contiguous) subword of the second. (This result is known as "Higman's Lemma".) In our case, this means that the infinite sequence of words composed of the immediate subterms of $t_{i_1}, t_{i_2}, \ldots$ must contain a pair w_{i_j} and w_{i_k} $(k>j)$ such that w_{i_j} is a subword of w_{i_k}. That, however, would imply that $t_{i_j} \preccurlyeq_{emb} t_{i_k}$, a contradiction. □

The *(pure) homeomorphic embedding* relation is the special case of embedding induced by simple equality of symbols for $\succcurlyeq$. It is derivability using only the first rule $f(s_1, \ldots, s_n) \rightarrow s_i$ of the previous system. It follows from the above theorem that any extension of homeomorphic embedding is a well-quasi-ordering of terms over a finite vocabulary. Since any rewrite ordering containing the subterm relation $\trianglerighteq$ also contains homeomorphic embedding, the subterm condition suffices for well-foundedness of term orderings over finite vocabularies, as claimed. Such orderings are the main tool for proving termination of rewriting:

DEFINITION. A transitive and reflexive rewrite relation $\succcurlyeq$ is a *simplification ordering* if it contains the subterm ordering $\trianglerighteq$.

Simplification orderings (called "quasi-simplification orderings" in Dershowitz [54]) are quasi-orderings and are what Higman [102] called "divisibility orders". For finite R, only a finite number of function symbols can appear in any derivation $t_1 \underset{R}{\rightarrow} t_2 \underset{R}{\rightarrow} \cdots$. Thus, a finite R over $\mathcal{T}$ is terminating if there exists any simplification ordering $\succcurlyeq$ of $\mathcal{T}$ such that R is contained in its strict part $\succ$ (Dershowitz [54]). The existence of such a simplification ordering means that $t_j \not\preccurlyeq t_k$ for all $k>j$, which precludes any t_j from being homeomorphically embedded in a subsequent t_k, as would necessarily be the case for any infinite derivation.

Virtually all reduction orderings used in rewriting-system termination proofs are simplification orderings. For instance, integer polynomial interpretations with non-negative coefficients are. One can even associate polynomials over the reals with function symbols and interpret terms as before (Dershowitz [53]). For a given choice τ of real polynomials to define a simplification ordering, $f_\tau(\cdots a \cdots) \geqslant a$ must always hold and $a \geqslant b$ must always imply $f_\tau(\cdots a \cdots) \geqslant f_\tau(\cdots b \cdots)$. For termination, $\tau(l)$ must be greater than $\tau(r)$ for each rule $l \rightarrow r$. All these inequalities need hold only when their variables are assigned values at least as large as the minimal interpretation of a constant, and are decidable (Tarski [227]).

In difficult termination proofs, it is frequently useful to build more complicated orderings on top of simpler ones. For example, if $\succ_1$ and $\succ_2$ are partial orderings of S_1 and S_2, respectively, then we say that the pair $\langle s_1, s_2 \rangle$ is *lexicographically greater* than a pair $\langle s_1', s_2' \rangle$ (for s_1, s_1' in S_1 and s_2, s_2' in S_2), if $s_1 \succ_1 s_1'$, or else $s_1 = s_1'$ and $s_2 \succ_2 s_2'$. If $\succ_1$ and $\succ_2$ are well-founded, then the lexicographic ordering of the cross-product $S_1 \times S_2$ is also well-founded. In the same way, well-founded lexicographic orderings are defined on n-tuples of elements of well-founded sets.

Lexicographic orderings work for tuples of fixed length. For collections of arbitrary size, another tool is needed. A (*finite*) *multiset* (or *bag*) is a finite unordered collection in which the number of occurrences of each element is significant. Formally, a multiset is a function from an element set S to the natural numbers giving the multiplicity of each element. In general, if $\succ$ is a partial ordering on S, then the ordering $\succ_{mul}$ on multisets of elements of S is defined as the transitive closure of the replacement of an element with any finite number (including zero) of elements that are smaller under $\succ$. If $\succ$ is well-founded, the induced ordering $\succ_{mul}$ also is, as a consequence of König's Lemma for infinite trees (Dershowitz & Manna [61]). A geometric interpretation of orderings on multisets is given in Martin [267].

As a somewhat contrived example of the application of lexicographic and multiset orderings to termination proofs, consider the rule:

$$x \cdot (y+z) \rightarrow (x \cdot y)+(x \cdot z)$$

We define a reduction ordering on terms as follows: Working our way from each innermost dot to the enclosing outermost dot, we construct a tuple of numbers, listing the size (total number of symbols) of the subterm headed by each dot encountered along the way. Each term is measured by the multiset of all its tuples (one for each innermost dot), with multisets compared in the ordering induced by the lexicographic ordering on tuples. The term $a \cdot ((b \cdot c) \cdot (d+(e \cdot f)))$, for example, is represented by $\{\langle 3,9,11\rangle, \langle 3,9,11\rangle\}$, while the term $a \cdot (((b \cdot c) \cdot d)+(b \cdot c) \cdot (e \cdot f))$ (after rewriting) is represented by $\{\langle 3,5,15\rangle, \langle 3,7,15\rangle, \langle 3,7,15\rangle\}$. The latter multiset is smaller, since each of its elements is lexicographically smaller than $\langle 3,9,11\rangle$, which appears in the former multiset (but not in the latter).

Multiset orderings will play an important role in Section 8.1.

5.3. *Path orderings*

The above termination proof of the single distributivity rule is a complicated way of capturing the intuition that "$\cdot$" is, in some sense, the most significant function symbol. This suggests the possibility of constructing simplification orderings directly from well-founded orderings of vocabularies, or *precedences*. The idea (Plaisted [198], Dershowitz [54]) is that a term s should be bigger than any term that is built from terms smaller than s which are connected together by a structure of function symbols smaller, in the precedence, than the root of s. By "smaller than s", we mean recursively in the same ordering. In particular, s is bigger than all terms containing only symbols smaller than its root. Since simplification orderings contain the subterm ordering, if a proper subterm of s is greater than or equivalent to a term t, then s is strictly greater than t. This means that two terms s and t can be compared by first comparing their roots in the precedence. If the root of s is smaller, then the only way s can dominate t is if one of the subterms of s does. If the root of t is smaller, then s must dominate each of the subterms of t. When the roots are of the same precedence, their immediate subterms need to be compared in some recursive fashion. Different ways of comparing these subterms lead to different classes of "path orderings".

One such ordering is the "multiset path ordering" introduced in Dershowitz [54]:

DEFINITION. For any given precedence $\succcurlyeq$, the *multiset path ordering* $\succcurlyeq_{mpo}$ is defined on ground terms as derivability using the following schematic system *mpo*:

$f(s_1 \ldots, s_n)$	$\to s_i$	$1 \leqslant i \leqslant n$
$f(s_1, \ldots, s_n)$	$\to g(t_1, \ldots, t_m)$	if $f \succ g, f(s_1, \ldots, s_n) \xrightarrow[mpo]{+} t_1, \ldots, t_m$
$f(\ldots, s_i, \ldots)$	$\to g(\ldots t_1, \ldots, t_k, \ldots, s_n)$	if $f \preccurlyeq g, s_i \xrightarrow[mpo]{+} t_1, \ldots, t_k, k \geqslant 0$
$f(s_1, \ldots, s_n)$	$\leftrightarrow g(s_{\pi_1}, \ldots, s_{\pi_n})$	if $f \sim g$, π *is a permutation*

The second rule replaces a term with one having a smaller root symbol. The third replaces a subterm with any number of smaller ones; in particular, it allows deletion of subterms ($k=0$). Actually, the third rule must also permit any number of immediate subterms to be replaced by smaller terms at the same time using $(\xrightarrow[mpo]{+})_{mul}$ (to avoid violating the arity of symbols in the vocabulary).

The multiset path ordering contains the homeomorphic embedding relation and is, therefore, a simplification ordering. (That derivability with only the one-way rules is irreflexive is true, but not self-evident.) Moreover, if $\succ$ is a well-founded ordering of (possibly infinite) $\mathscr{F}$, then $\succ_{mpo}$ is a well-founded ordering of $\mathscr{F}$. To see this (Dershowitz [54]), note that (by Zorn's Lemma) a given precedence $\succ$ may be extended to an ordering $>$ such that the quasi-ordering $> \cup \sim$, call it $\gtrsim$, is a total well-quasi-ordering of $\mathscr{F}$. By Kruskal's Tree Theorem, the induced embedding relation $\gtrsim_{emb}$ well-quasi-orders $\mathscr{F}$, as does the total multiset path ordering $\gtrsim_{mpo}$ induced by the total precedence $\gtrsim$. Thus $>_{mpo}$ is well-founded. Since the mapping from precedence $\succcurlyeq$ to term ordering $\succcurlyeq_{mpo}$ is *incremental*, in the sense that extending the precedence extends the corresponding ordering on terms, the smaller ordering $\succ_{mpo}$ must also be well-founded. Though in principle we say $s \succ_{mpo} t$, for free terms s and t, if $s\sigma \succ_{mpo} t\sigma$ for all ground substitutions σ, one computes in practice with $\xrightarrow[mpo]{}$, since the latter is contained in the former and is more easily computed.

The multiset path ordering establishes termination of our stack and disjunctive normal form examples (as well as the bean-increasing Coffee Can Game). For the four-rule stack system (of Section 2.3), take the precedence *alternate* > *push*. The first three rules are contained in $>_{mpo}$ by the first rule of *mpo*. For example, *pop* (*push*(*x*, *y*)) $\xrightarrow[mpo]{}$ *push*(*x*, *y*) $\xrightarrow[mpo]{}$ *y*. For the remaining stack rule, we have *alternate*(*push* (*x*, *y*), *z*) $>_{mpo}$ *push*(*x*, *alternate*(*z*, *y*)), since *alternate* > *push* and *alternate*(*push* (*x*, *y*), *z*) $\xrightarrow[mpo]{+}$ *x*, *alternate*(*z*, *y*), the latter since *push*(*x*, *y*) $\xrightarrow[mpo]{}$ *y* and *alternate*(*y*, *z*) $\xleftrightarrow[mpo]{}$ *alternate*(*z*, *y*). Termination of the disjunctive normal form system (Section 5.1) may be shown using the precedence *not* > *and* > *or*.

One can think of the multiset path ordering as a functional, mapping an ordering on function symbols (the precedence) to an ordering on terms. A related class of orderings (Kamin & Lévy [129]) compares subterms lexicographically,

DEFINITION. For any given precedence $\succcurlyeq$, the *lexicographic path ordering* $\succcurlyeq_{lpo}$ is defined as derivability by the following schematic system *lpo*:

$f(s_1, \ldots, s_n)$	$\rightarrow s_i$	$1 \leqslant i \leqslant n$
$f(s_1, \ldots, s_n)$	$\rightarrow g(t_1, \ldots, t_m)$	if $f \succ g, f(s_1, \ldots, s_n) \xrightarrow[lpo]{+} t_1, \ldots, t_m$
$f(s_1, \ldots, s_i, \ldots, s_n)$	$\rightarrow f(s_1, \ldots, s_{i-1}, t_i, \ldots, t_n)$	if $s_i \xrightarrow[lpo]{+} t_i, f(s_1, \ldots, s_n) \xrightarrow[lpo]{+} t_{i+1}, \ldots, t_n$
$f(s_1, \ldots, s_n)$	$\rightarrow g(s_1, \ldots, s_m)$	if $f \sim g, m < n$
$f(s_1, \ldots, s_n)$	$\leftrightarrow g(s_1, \ldots, s_n)$	if $f \sim g$

As in the multiset path ordering, the precedence $\succ$ induces an ordering on terms, but, here, subterms of the same function symbol are compared left-to-right, lexicographically. (They could just as well be compared right-to-left, or in any fixed order.)

The following traditional example—for Ackermann's function—illustrates its use with a precedence $ack > succ$:

$ack(0, y)$	$\rightarrow$	$succ(y)$
$ack(succ(x), 0)$	$\rightarrow$	$ack(x, succ(0))$
$ack(succ(x), succ(y))$	$\rightarrow$	$ack(x, ack(succ(x), y))$

For example, the third rule is contained in $>_{lpo}$ since x occurs in $succ(x)$ and $ack(succ(x), succ(y))$ is lexicographically greater than $ack(succ(x), y)$.

If the strict part of a precedence is of order type α, then the multiset path ordering on the set of terms is of order type $\varphi^{\alpha}(0)$ in the notation of Feferman [77]. Combining multiset and lexicographic path orderings into one (Kamin & Lévy [129]), gives a more powerful ordering, which we call the *recursive path ordering* and which is related to Ackermann's ordinal notation (Dershowitz & Okada [63]) sufficing for any simplification ordering (Okada & Steele [269]). (The original "recursive path ordering" (Dershowitz [54]) was of the multiset variety.) Determining if a precedence exists that makes two ground terms comparable in *mpo* is NP-complete (Krishnamoorthy & Narendran [152]); determining whether $s \succ_{lpo} t$ for open terms, in the sense of $s\sigma \succ_{lpo} t\sigma$ for all ground σ is decidable (Comon [250]).

These precedence-based orderings are "syntactic", looking at function symbols one at a time. Similar semantically-oriented orderings have been devised; they replace the condition $f \succ g$ by $f(s_1, \ldots, s_n) \succ g(t_1, \ldots, t_m)$, where $\succ$ is now a well-founded quasi-ordering of terms, not function symbols. Rewriting should not make a term bigger under $\succ$. For example, the *Knuth–Bendix ordering* [149] assigns a weight to a term which is the sum of the weights of its constituent function symbols. Terms of equal weight have their subterms compared lexicographically. Methods for choosing weights are described in Lankford [158] and Martin [177].

None of these orderings, however, can directly prove termination of the rule $f(f(x)) \rightarrow f(g(f(x)))$, since the right-hand side is embedded in the left. To overcome this problem, Puel [204] compares "unavoidable patterns" instead of function symbols in the definition of $\succ_{mpo}$. By *unavoidable*, we mean that any sufficiently large term in $\mathcal{T}$ must be greater, under the encompassment ordering $\trianglerighteq$, than one of the unavoidable patterns. For example, any term constructed from a constant a and three or more f's and g's must contain an occurrence of one of the three patterns: $f(f(x))$, $f(g(x))$, or $g(g(x))$. The well-foundedness of this ordering is based on a powerful extension of Kruskal's Tree Theorem (Puel [203]), analogous to a similar theorem on strings in Ehrenfeucht et al. [70].

5.4. *Combined systems*

We saw above that polynomial orderings are applicable to associative-commutative systems, but are severely restrictive. The multiset path ordering, though compatible with commutativity axioms, is not well-founded when associativity is added as a bi-directional rule to *mpo*. For example, let the precedence be $c \succ b$. Then, $(c+b)+b \underset{mpo}{\longrightarrow} c+(b+b) \underset{S}{\longleftrightarrow} (c+b)+b$, where S is associativity of $+$. To overcome this problem, the multiset path ordering has been adapted to handle associative-commutative operators by flattening and also transforming terms (distributing symbols that are bigger in the precedence over smaller ones) before comparing (Bachmair & Plaisted [15]), i.e. s is greater than t iff the T-normal form $T(s)$ of s is greater under $\succ_{mpo}$ than the T-normal form $T(t)$ of t, for some convergent "transform" system T and precedence $\succ$. The general use of rewrite systems as transforms and the formulation of abstract conditions of the resultant reduction ordering are explored in Bachmair & Dershowitz [9], Bellegarde & Lescanne [21], and Geser [256].

Termination of the union of term- or class-rewriting systems can be reduced to the termination of each: Let R and S be two binary relations contained in well-founded orderings $\succ_R$ and $\succ_S$, respectively. If $\succ_S$ *commutes* over $\succ_R$, i.e. if $\succ_R \circ \succ_S$ is contained in $\succ_S \circ \succ_R$, then the union $R \cup S$ is terminating. Actually, it suffices if $\succ_S$ *quasi commutes* over $\succ_R$, by which we mean that $\succ_R \circ \succ_S$ is contained in $\succ_S \circ \succcurlyeq_R$. For example, if R is any terminating rewrite system, then the union of $\underset{R}{\rightarrow}$ with the proper subterm relation $\rhd$ is well-founded, since taking subterms is well-founded and if a subterm rewrites then so does the superterm. More generally, the union of any terminating rewrite relation with the (proper) encompassment ordering $\trianglerighteq$ is also well-founded, since encompassment is just $>$ (subsumption) and/or $\rhd$, both of which are well-founded, and any rewrite relation commutes over both. Similarly, if R is a binary relation contained in a well-founded ordering $\succ_R$, S is a symmetric binary relation contained in a congruence $\sim_S$, and $\succ_R$ commutes over $\sim_S$, then the composite relation $R \circ S$ is terminating. To prove termination of a combined term-rewriting system $R \cup S$, it is necessary and sufficient that R and S be contained in reduction orderings that commute as above; to prove termination of a class-rewriting system R/S, it is necessary and sufficient that R be contained in a reduction ordering that commutes over a symmetric and transitive rewrite relation that contains S. These ideas generalize results in Bachmair & Dershowitz [9] and Jouannaud & Muñoz [127]. Note that commutation of $\underset{R}{\rightarrow}$ and $\underset{S}{\rightarrow}$ is *not* ensured by R and S having disjoint vocabularies, the system at the beginning of this section being a counter-example (Toyama [235]); see, however, Section 7.2.

5.5. *Further reading*

Martin Gardner [83] talks about multiset orderings and the Hydra battle. For a survey of the history and applications of well-quasi-orderings, see Kruskal [154]. For a comprehensive survey of termination, see Dershowitz [58]. The multiset and lexicographic path orderings, and their variants (see Rusinowitch [213]), have been implemented in many rewriting-rule based theorem provers (e.g. Lescanne [169]). Some results on the complexity of derivations appear in Choppy et al. [40].

6. Satisfiability

We turn out attention now to the determination of satisfiability. If an equation $s=t$ is satisfiable in the (free-) term algebra $\mathcal{T}$, that is, if $s\sigma$ and $t\sigma$ are identical for some substitution σ, then s and t are said to be *unifiable*. The unification problem, per se, is to determine if two terms are unifiable. More particularly, we are interested in determining the set of all unifying substitutions σ. Though unifiable terms may have an infinite number of unifiers, there is—as we will see—a unique substitution (unique up to literal similarity) that "subsumes" all others under the subsumption ordering. More generally, we are interested in solving equations in the presence of equational axioms specifying properties of the operators. For a given equational theory E, we say that s and t are *E-unifiable* if $s=t$ is satisfiable in the free quotient algebra $\mathcal{T}/E$, in which case it is satisfiable in all models of E. In general, there may be no minimal solution to a given E-unification problem.

6.1. *Syntactic unification*

Let $\doteq$ be literal similarity of terms, under which two terms s and t are equivalent if each is an instance of the other. We show that the quotient $\mathcal{T}/\doteq$, ordered by the subsumption relation $\gtrdot$, is a lower semilattice, by showing that every pair of terms, s and t, has a greatest lower bound, $glb(s, t)$, called their *least* (*general*) *generalization* (Plotkin [200]). (Note that $\mathcal{T}/\doteq$ is not an algebra, because literal similarity is not a congruence.) Let LG be the following set of "transformation" rules, operating on pairs $(P; w)$, where w is a term containing the partial solution, and P contains the pairs yet to be solved:

Decompose:	$(\{f(s_1,\ldots,s_m) \sqcap_x f(t_1,\ldots,t_m)\}\cup P;\ w)$
	$\Rightarrow (\{s_1 \sqcap_{x_1} t_1,\ldots,s_n \sqcap_{x_n} t_n\}\cup P;\ w\sigma)$
	where σ is $\{x \mapsto f(x_1,\ldots,x_n)\}$
	and $x_1,\ldots,x_n$ are distinct new variable symbols
Coalesce:	$(\{s \sqcap_x t, s \sqcap_y t\}\cup P;\ w) \Rightarrow (\{s \sqcap_y t\}\cup P;\ w\sigma)$
	where σ is $\{x \mapsto y\}$

Each pair is written as $s \sqcap_x t$, where x is a variable of w. Applying these rules to $(\{s \sqcap_x t\}; x)$, where x is not a variable of s or t, until none is applicable, results in $(\{u_i \sqcap_{x_i} v_i\}; glb(s,t))$. To prove that repeated applications of LG always terminate, note that each application of a rule decreases the number of function symbols (not including $\sqcap$) in P. Since the system $\underset{LG}{\Rightarrow}$ is actually Church–Rosser (on $\mathcal{T}/\doteq$), least generalizations are unique up to literal similarity. For example, the least generalization of $f(g(a), g(b), a)$ and $f(g(b), g(a), b)$ is $f(g(x), g(y), x)$. It follows from properties of well-founded lattices (Birkhoff [26]) that every pair of terms s and t that are bounded from above (i.e., there exists a term that is an instance of both) has a least (i.e. most general) upper bound, deoted $lub(s, t)$.

An equation $s=t$ has a *solution* σ if $s\sigma = t\sigma$. Here, s and t may share variables and we demand that applying a single substitution σ (mapping all occurrences of the same variable to the same term) result in identical terms. The least solution with respect to subsumption (and the variables of s and t), $mgu(s,t)$, is called their *most general unifier*. For example, the most general common instance of $alternate(y', \Lambda)$ and $alternate(push(x, y), z)$ is $alternate(push(x, y), \Lambda)$, or anything literally similar; the *mgu* of $alternate(y, \Lambda)$ and $alternate(\Lambda, z)$ *is* $\{y \mapsto \Lambda,\ z \mapsto \Lambda\}$; there is no solution to $alternate(z, \Lambda) = alternate(push(x, y), z)$. Most general unifiers and least upper bounds of terms are closely related: by varying s and t so that the two terms have disjoint variables, we get $lub(s,t) = s\mu$, where $\mu = mgu(s', t')$, for literally similar terms s' and t' of s and t, respectively; in the other direction, $s\mu$ and $t\mu$ are both equal to $lub(eq(x,x), eq(s,t))$, for $\mu = mgu(s,t)$, where eq is any binary symbol and x is any variable. As a consequence, the most general unifier is unique up to literal similarity, but need not always exist. This fundamental uniqueness result for first-order terms does not hold true for higher-order languages with function variables (Huet [112]).

Robinson [211] was the first to give an algorithm for finding most general unifiers. Following Herbrand [100] and Martelli & Montanari [175], we view unification as a step-by-step process of transforming multisets of equations until a "solved form" is obtained from which the most general unifier can be extracted. A *solved form* is any set of equations $\{x_1 = s_1, \ldots, x_n = s_n\}$ such that the x_i are distinct and no x_i is a variable in any s_j. Then, the most general unifier is the substitution $\{x \mapsto s_1, \ldots, x_n \mapsto s_n\}$. Let $|t|$ denote the *size* of the term t, that is the total number of its function symbols. (The size of a variable is zero.) Equations to be solved will be written $s \stackrel{?}{=} t$. Define a well-founded ordering $\rhd$ on equations as follows: $u \stackrel{?}{=} v \rhd s \stackrel{?}{=} t$ if $\max(|u|,|v|) > \max(|s|,|t|)$, or else $\max(|u|,|v|) = \max(|s|,|t|)$ and $\max(|u|,|v|) - \min(|u|,|v|)$ is greater than $\max(|s|,|t|) - \min(|s|,|t|)$. We also use a constant F to denote the absence of a solution, and make it smaller than any equation. Let MM be the following set of transformation rules operating on pairs $(P; S)$ of sets of equations, with P containing the equations yet to be solved and S, the partial solution:

Delete:	$(\{s \stackrel{?}{=} s\} \cup P; S)$	$\Rightarrow$	$(P; S)$
Decompose:	$(\{f(s_1, \ldots, s_m) \stackrel{?}{=} f(t_1, \ldots, t_m)\} \cup P; S)$	$\Rightarrow$	$(\{s_1 \stackrel{?}{=} t_1, \ldots, s_m \stackrel{?}{=} t_m\} \cup P; S)$

Conflict:	$(\{f(s_1,\ldots,s_m)\stackrel{?}{=}g(t_1,\ldots,t_n)\}\cup P;S)$ if $f\neq g$	$\Rightarrow (\emptyset;\{F\})$
Merge:	$(\{x\stackrel{?}{=}s, x\stackrel{?}{=}t\}\cup P;S)$ if $x\in\mathscr{X}$ and $x\stackrel{?}{=}t \rhd s\stackrel{?}{=}t$	$\Rightarrow (\{x\stackrel{?}{=}s, s\stackrel{?}{=}t\}\cup P;S)$
Check:	$(\{x\stackrel{?}{=}s\}\cup P;S)$ if $x\in\mathscr{X}$, x occurs in s, and $x\neq s$	$\Rightarrow (\emptyset;\{F\})$
Eliminate:	$(\{x\stackrel{?}{=}s\}\cup P;S)$ if $x\in\mathscr{X}$, $s\notin\mathscr{X}$, and x does not occur in s, where $\sigma=\{x\mapsto s\}$	$\Rightarrow (P\sigma;S\sigma\cup\{x=s\})$

DEFINITION. A (*syntactic*) *unification procedure* is any program that takes a finite set P_0 of equations, and uses the above rules MM to generate a sequence of problems from $(P_0;\emptyset)$.

Starting with $(\{s\stackrel{?}{=}t\};\emptyset)$ and using the unification rules repeatedly until none is applicable results in $(\emptyset;\{F\})$ iff $s\stackrel{?}{=}t$ has no solution, or else it results in a solved form $(\emptyset;\{x_1=s_1,\ldots,x_n=s_n\})$. The application of any of these rules does not change the set of solutions. Hence, the former situation signifies failure, and in the latter case, $\sigma=\{x_1\mapsto s_1,\ldots,x_n\mapsto s_n\}$ is a most general unifier of s and t. That σ is most general follows from the fact that the rules preserve all solutions.

For example, the most general unifier of $f(x,x,a)$ and $f(g(y),g(a),y)$ is $\{x\mapsto g(a), y\mapsto a\}$, since

$$
\begin{aligned}
&(\{f(x,x,a)\stackrel{?}{=}f(g(y),g(a),y)\};\emptyset) \underset{MM}{\Rightarrow} (\{x\stackrel{?}{=}g(y), x\stackrel{?}{=}g(a), a\stackrel{?}{=}y\};\emptyset) \underset{MM}{\Rightarrow}\\
&(\{x\stackrel{?}{=}g(y), g(y)\stackrel{?}{=}g(a), a\stackrel{?}{=}y\};\emptyset) \underset{MM}{\Rightarrow} (\{x\stackrel{?}{=}g(a), g(a)\stackrel{?}{=}g(a)\};\{y=a\}) \underset{MM}{\Rightarrow}\\
&(\{x\stackrel{?}{=}g(a)\};\{y=a\}) \underset{MM}{\Rightarrow} (\emptyset;\{x=g(a), y=a\}).
\end{aligned}
$$

On the other hand, $f(x,x,x)$ and $f(g(y),g(a),y)$ are not unifiable, since

$$
\begin{aligned}
&(\{f(x,x,x)\stackrel{?}{=}f(g(y),g(a),y)\};\emptyset) \underset{MM}{\Rightarrow} (\{x\stackrel{?}{=}g(y), x\stackrel{?}{=}g(a), x\stackrel{?}{=}y\};\emptyset) \underset{MM}{\Rightarrow}\\
&(\{y\stackrel{?}{=}g(y), y\stackrel{?}{=}g(a)\};\{x=y\}) \underset{MM}{\Rightarrow} (\emptyset;\{F\})
\end{aligned}
$$

on account of an "occur check."

To prove that repeated applications of MM always terminate, we can use a lexicographic combination of an ordering on numbers and the multiset extension of the ordering $\rhd$ on equations. With each application of a rule, $(P;S)\underset{MM}{\Rightarrow}(P';S')$, either the solved set S is enlarged, or the problem set P is reduced under $\rhd_{mul}$. Since the solved set cannot increase without bound (it can have at most one equation per variable), nor can the unsolved set decrease without limit (since $\rhd_{mul}$ is well-founded), there can be no infinite MM-derivations.

Uncontrolled use of **eliminate** leads to exponential time complexity. With appropriate data structures and control strategies, an efficient algorithm is obtained, which is quasi linear in the worst case (e.g. Baxter [20]); and Ružička & Prívara [272]);

more careful implementations provide for truly linear, but less practical, algorithms (e.g. Paterson & Wegman [194]). Eliminating the **check** and **eliminate** rules produces solutions over the domain of (infinite) "rational" trees (Huet [112]), and has ramifications for the semantics of some Prolog implementations (Colmerauer [44]).

6.2. Semantic unification

When it comes to E-unification, the situation is much more complex. A substitution σ is a *solution in E* to an equation $s=t$ if $s\sigma \underset{E}{=} t\sigma$, in which case we say that σ is an *E-unifier* of s and t; we say that t *E-matches* s if there exists a substitution σ such that $s\sigma \underset{E}{=} t$. Whenever the word problem is undecidable E-unifiability is, and in many other cases as well. For example, the solvability of Diophantine equations, that is, polynomial equations over the integers, is undecidable (Matijasevic [179]), as is unifiability under associativity and distributivity alone (Szabo [225]). Satisfiability may be undecidable even when congruence classes are finite (as for associativity, commutativity, and distributivity; see Siekmann [218]. Second-order unifiability (equivalence of function definitions) is also undecidable, in general (Goldfarb [93]). On the brighter side, many other theories have decidable unification problems, including Presburger arithmetic (Presburger [202], Shostak [217]), real closed fields (Tarski [227], Collins [43]) and monoids (Makanin [172]).

When more than one solution may exist for a theory E, we define a solution σ of $s \underset{E}{=} t$ to be *more general* than a solution ρ if $\sigma \lessdot_E \rho$ in the E-subsumption ordering $\lessdot_E$, i.e. if there exists a substitution τ such that $x\sigma\tau \underset{E}{=} x\rho$, for all variables x in s and t, but not vice versa. An E-unifier is *most general* if no more general unifier exists. Note that E-subsumption is not well-founded for all E. There are decidable theories with infinite sets of most general unifiers (an example is the set of solutions $\{a^i | i \geqslant 1\}$ to $x \cdot a = a \cdot x$, where "$\cdot$" is associative (Plotkin [200])), and there are some for which there are solutions, but no most general one (Fages & Huet [75]) (an example is associativity plus idempotence (Baader [5])). A procedure that generates all most general unifiers for associativity (monoids) is presented in Jaffar [260]. We say that a set S of E-unifiers is *complete* if for every E-unifier there is one in S that is more general with respect to E-subsumption. For example, a complete unification algorithm exists for associativity and commutativity (AC) (Stickel [222], Herold & Siekmann [101]); alternative algorithms with better performance are given in Kirchner [145] and Boudet et al. [32]. Other theories for which algorithms are available that compute finite, complete sets of most general E-unifiers include commutativity, AC with identity and/or idempotency (see Fages [74]), as well as Boolean rings (see Boudet et al. [33]). For many of these theories, unification is believed intractable from the time-complexity point of view (Kapur & Narendran [134]).

Of course, E-unifiability is semidecidable for recursively-enumerable E. Paramodulation (without the functional reflexivity axioms) (Robinson & Wos [210]) is one improvement over the obvious "British-museum" method of interleaving the production of substitutions with the search for equational proofs.

Paramodulation may be improved upon by a more goal-oriented process. The following set of rules, inspired by Gallier & Snyder [81] does the trick:

Decompose:	$(\{f(s_1,\ldots,s_n) \overset{?}{\leftrightarrow} f(t_1,\ldots,t_n)\}\cup P; S)$	$\leadsto (\{s_1 \overset{?}{\leftrightarrow} t_1,\ldots,s_n \overset{?}{\leftrightarrow} t_n\}\cup P; S)$
Eliminate:	$(\{x \overset{?}{\leftrightarrow} s\}\cup P; S)$	$\leadsto (P\sigma; S\sigma\cup\{x=s\})$
	if $x\in\mathscr{X}$, and x does not occur in s, where $\sigma=\{x\mapsto s\}$	
Mutate:	$(\{f(s_1,\ldots,s_n) \overset{?}{\leftrightarrow} t\}\cup P; S)$	$\leadsto (\{s_1 \overset{?}{\leftrightarrow} u_1,\ldots,s_n \overset{?}{\leftrightarrow} u_n,$ $r \overset{?}{\leftrightarrow} t\}; S)$
	if $f(u_1,\ldots,u_n)=r$ is literally similar to an equation in E but has no variables in common with S, P, t, or the s_i	
Splice:	$(\{s \overset{?}{\leftrightarrow} t\}\cup P; S)$	$\leadsto (\{s \overset{?}{\leftrightarrow} x,\ r \overset{?}{\leftrightarrow} t\}\cup P; S)$
	if $x\in\mathscr{X}$ and if $x=r$ is literally similar to an equation in E but has no variables in common with S, P, t, or the s_i	
Imitate:	$(\{f(s_1,\ldots,s_n) \overset{?}{\leftrightarrow} y\}\cup P; S)$	$\leadsto (\{s_1 \overset{?}{\leftrightarrow} y_1,\ldots,s_n \overset{?}{\leftrightarrow} y_n,$ $y \overset{?}{\leftrightarrow} f(y_1,\ldots,y_n)\}\cup P; S)$
	if $y\in\mathscr{X}$ and the y_i are new variables	

We call this set of rules *EU*. These rules *nondeterministically* compute solved forms (which is why we use a different relation symbol $\leadsto$), each of which represents an E-unifier.

DEFINITION. An *E-unification procedure* is any (nondeterministic) program that takes a finite set P_0 of equations, and uses the above rules *EU* to generate sequences of problems from $(P_0; \emptyset)$.

To generate a complete set of unifiers may not require computing all possible sequences. In particular, the use of **imitate** can perhaps be severely restricted (Hsiang & Jouannaud [109]). The set of rules *EU* can be improved for particular classes of equations. Two special cases have been investigated:

(i) when E is "simple" in the sense that for any valid equation $s=t$, there exists a proof with at most one proof step taking place at the top position (Kirchner [143]);

(ii) when there exists a ground-convergent set of rules complete for E.

In the first case, axioms of the form $x=t$, for variable x, are disallowed; hence **splice** is superfluous. Moreover, **decompose** must always apply after any necessary application of **mutate**; hence, the two can be compiled into a single rule. For example, commutativity uses the following "mutate and decompose" rule:

$$(\{f(s_1,s_2) \overset{?}{\leftrightarrow} f(t_1,t_2)\}\cup P; S) \leadsto (\{s_1 \overset{?}{\leftrightarrow} t_2, s_2 \overset{?}{\leftrightarrow} t_1\}\cup P; S).$$

Case (ii) is dealt with in Section 6.3.

Methods of combining unification algorithms for well-behaved theories that do not share symbols have been given in Yelick [240], Kirchner [145] and Boudet et al.

[33]. The general case was solved in Boudet [247] and Schmidt-Schauß [215]. Note that a unification algorithm that generates a complete set of most general unifiers (for terms without free constants) does not automatically work for matching (one cannot just treat variables as constants, since that changes the algebra and may introduce unsound solutions); see Bürckert et al. [38].

6.3. *Narrowing*

Even when a convergent system R exists for a theory E, the E-unification problem remains only semidecidable. For example, the system

$x+0 \to x$	$x+succ(y) \to succ(x+y)$	$x+pred(y) \to pred(x+y)$
$x-0 \to x$	$x-succ(y) \to pred(x-y)$	$x-pred(y) \to succ(x-y)$
$x*0 \to 0$	$x*succ(y) \to (x*y)+x$	$x*pred(y) \to (x*y)-x$
	$succ(pred(x)) \to x$	$pred(succ(x)) \to x$

for addition and multiplication of integers is canonical, but were R-unification (or R-matching) decidable, then the existence of integer solutions to Diophantine equations, such as

$$x \cdot x + y \cdot y - succ(succ(succ(0))) = 0,$$

would also be decidable. The latter is Hilbert's Tenth Problem, shown to be undecidable in Matijasevic [179]. (Cf. Bockmayr [28], Heilbrunner & Hölldobler [98].)

When a convergent R is available, a one-way sort of paramodulation suffices, due to the existence of a rewrite proof for an arbitary valid equation (Dershowitz & Sivakumar [68], Martelli et al. [176]). The following set of rules, RU, restricts uses of equations to left-hand sides of rules:

Decompose: $(\{f(s_1,\ldots,s_n)\overset{?}{\to}f(t_1,\ldots,t_n)\}\cup P;S) \rightsquigarrow (\{s_1\overset{?}{\to}t_1,\ldots,s_n\overset{?}{\to}t_n\}\cup P;S)$

Eliminate: $(\{x\overset{?}{\to}s\}\cup P;S) \rightsquigarrow (P\sigma;S\sigma\cup\{x=s\})$
if $x\in\mathscr{X}$ and x does not occur in s, where $\sigma=\{x\mapsto s\}$

Mutate: $(\{f(s_1,\ldots,s_n)\overset{?}{\to}t\}\cup P;S) \rightsquigarrow (\{s_1\overset{?}{\to}u_1,\ldots,s_n\overset{?}{\to}u_n,r\overset{?}{\to}t\};S)$
if $f(u_1,\ldots,u_n)\to r$ is literally similar to a rule in R but has no variables in common with S, P, t, or the s_i

Imitate: $(\{f(s_1,\ldots,s_n)\overset{?}{\to}y\}\cup P;S) \rightsquigarrow (\{s_1\overset{?}{\to}y_1,\ldots,s_n\overset{?}{\to}y_n, y\overset{?}{\to}f(y_1,\ldots,y_n)\}\cup P;S)$
if $y\in\mathscr{X}$ and the y_i are new variables

This set of rules subsumes "narrowing", as used for this purpose in Fay [76]:

DEFINITION. A term s *narrows* (in one step) to a term t, via substitution μ, symbolized $s\underset{R}{\rightsquigarrow}t$, if t is $s\mu[r\mu]_p$, for some nonvariable position p in s, rule $l\to r$ in R (the variables of which have been renamed so that they are distinct from those in s), and most general unifier μ of $s|_p$ and l.

The verb "narrow" perhaps carries the wrong connotation: it is the set of R-congruence classes of instances of the term that is being narrowed. It is easy to see that RU mimics narrowing by using **decompose, imitate**, and **mutate**. For example, if R is $\{f(x,x)\to c(x), a\to b\}$, then the problem $\{f(a,y)\xrightarrow{?}z, f(y,b)\xrightarrow{?}z\}$ mutates to $\{a\xrightarrow{?}x, y\xrightarrow{?}x, c(x)\xrightarrow{?}z, f(y,b)\xrightarrow{?}z\}$. Then a is imitated by x, and x and y are eliminated by substituting a for them: $\{c(a)\xrightarrow{?}z, f(a,b)\xrightarrow{?}z\}$. This corresponds to narrowing of $f(a,y)$ to $c(a)$ by instantiating $y\mapsto a$, and eventually yields the solution $\{y\mapsto a, z\mapsto c(b)\}$.

Narrowing has the following property:

NARROWING LEMMA. *If R is a ground convergent rewrite system and $s\sigma \xrightarrow[R]{*} t$, then there exist terms u and v such that $s \overset{*}{\underset{R}{\leadsto}} u$, $t \xrightarrow[R]{*} v$, and $v \geqslant u$.*

If σ is irreducible (that is, if $x\sigma$ is irreducible for all variables x), then $v=t$, and the lemma holds even for nonconvergent R (Hullot [120]). Without convergence, reducible solutions are lost. For example, if R is $\{f(a,b)\to c, a\to b\}$ or $\{f(x,g(x))\to c, a\to g(a)\}$, then $f(y,y)$ cannot be narrowed, and $f(y,y)\xrightarrow{?}c$ fails to lead to a solved form, despite the fact that there is a solution $\{y\mapsto a\}$.

Variations on narrowing include: *normal narrowing* (Fay [76]) (in which terms are normalized via $\xrightarrow[R]{!}$ before narrowing), *basic narrowing* (Hullot [120]) (in which the substitution part of prior narrowings is not subsequently narrowed), and their combination (Nutt et al. [190]). All are semicomplete for convergent R. Class-rewriting yields similar results (Jouannaud et al. [123]).

6.4. *Further reading*

For a survey regarding syntactic unification, see Lassez et al. [164]; for unification in general, see Huet [112]. For a survey of theory and applications of syntactic and semantic unification, see Knight [148]. Questions of decidability of unification in equational theories are summarized in Siekmann [218]; a summary of complexity results for some of the decidable cases is Kapur & Narendran [135]. A popular exposition on the undecidability of the existence of solutions to Diophantine equations is Davis & Hersh [51]. For a comprehensive treatment of narrowing and E-unification, see Kirchner [142]. For the satisfiability problem of arbitrary first-order formulae with equality as the only predicate, see Comon & Lescanne [46] and Maher [171]; for a survey of relevant work, see Comon [250]. For an expanded survey of the topics in this section, see Jouannaud & Kirchner [261]; see also Gallier and Snyder [255].

7. Critical pairs

In this section, we continue our study of the Church–Rosser property for rewrite systems. In particular, we will see that confluence is decidable for finite, terminating systems. Confluence, in general, is undecidable (Huet [113]), even if all rules are monadic (Book et al. [30]). For finite ground systems—even if they are nonterminating—decision procedures exist (see Dauchet et al. [50] or Oyamaguchi [193]). Ground confluence, on the other hand, is undecidable, even if the system is terminating (Kapur et al. [137]). Even for convergent systems R, the questions whether congruence classes defined by $\overset{*}{\underset{R}{\leftrightarrow}}$ are finite in number, or are all finite in size, are undecidable, unless R is ground (Raoult [205]).

7.1. *Term rewriting*

Let $l \rightarrow r$ and $s \rightarrow t$ be two rules. We say that the left-hand side l *overlaps* the left-hand side s if there is a *nonvariable* subterm $s|_p$ of s such that l and $s|_p$ have a common upper bound with respect to subsumption. To determine overlap, the variables in the two (not necessarily distinct) rules are renamed, if necessary, so that they are disjoint. Then, l overlaps s if there exists a unifying substitution σ such that $l\sigma = s\sigma|_p$. When there is an overlap, the overlapped term $s\sigma$ can be rewritten to either $t\sigma$ or $s\sigma[r\sigma]_p$. The two-step proof $t\sigma \underset{R}{\leftarrow} s\sigma[l\sigma]_p \underset{R}{\rightarrow} s\sigma[r\sigma]_p$ is called a *critical peak.*

DEFINITION. If $l \rightarrow r$ and $s \rightarrow t$ are two rewrite rules (with variables distinct), p is the position of a nonvariable subterm of s, and μ is a most general unifier of $s|_p$ and l, then the equation $t\mu = s\mu[r\mu]_p$ is a *critical pair* formed from those rules.

Thus, a critical pair is the equation arising from a most general nonvariable overlap between two left-hand sides. For example, $push(x, alternate(\Lambda, y)) = push(x, y)$ is the critical pair obtained from $alternate(push(x, y), z) \rightarrow push(x, alternate(z, y))$ and $alternate(y, \Lambda) \rightarrow y$.

Let $\mathbf{cp}(R)$ denote the set of all critical pairs between (not necessarily distinct, but perhaps renamed) rules in R and let $\underset{\mathbf{cp}(R)}{\longleftrightarrow}$ denote its symmetric rewrite closure.

CRITICAL PAIR LEMMA (Knuth & Bendix [149]). *For any rewrite system R and peak $s \underset{R}{\leftarrow} u \underset{R}{\rightarrow} t$, there either exists a rewrite proof $s \overset{*}{\underset{R}{\rightarrow}} v \overset{*}{\underset{R}{\leftarrow}} t$ or a critical-pair proof $s \underset{\mathbf{cp}(R)}{\longleftrightarrow} t$.*

The proof (Knuth & Bendix [149]), depicted in Fig. 4(a,b,c) considers all relative positions of the two redexes. As stressed in Huet [113], no assumption of termination is necessary for this lemma. It follows from this and the Diamond Lemma, that a terminating system R is confluent iff $\mathbf{cp}(R)$, regarded as a relation, is a subset of the joinability relation $\overset{*}{\underset{R}{\rightarrow}} \circ \overset{*}{\underset{R}{\leftarrow}}$. This holds, for instance, for the stack interleaving example of Section 3.2. Since finite systems have a finite number of critical pairs, their confluence is decidable, provided they are terminating. This criterion for confluence is called the *superposition test* (Knuth & Bendix [149]). In particular, reduced ground systems (which are terminating and have no critical pairs) are confluent.

Without termination, a system may have no critical pairs (hence be locally confluent), and still be nonconfluent. A system without critical pairs (except trivial ones of the form $t = t$) is called *nonoverlapping*, or *nonambiguous*. The following example of a nonoverlapping, but nonconfluent system (Huet [113]), is based on the "ladder"

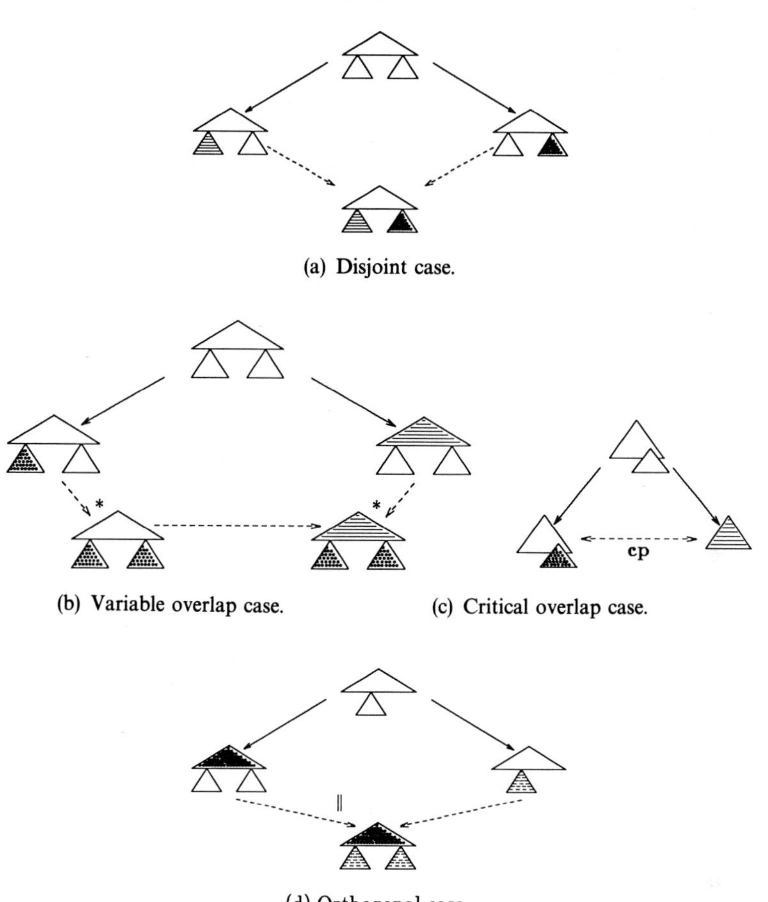

(a) Disjoint case.

(b) Variable overlap case.

(c) Critical overlap case.

(d) Orthogonal case.

Fig. 4. Proof of Critical Pair Lemma.

in Fig. 2(b):

$f(x,x)$	$\rightarrow$	a
$f(x,g(x))$	$\rightarrow$	b
c	$\rightarrow$	$g(c)$

The term c has no normal form, but $f(c,c)$ has two, a and b. This example can be modified so that the system is normalizing (Sivakumar [274]):

$f(x,x)$	$\rightarrow$	$g(x)$
$f(x,g(x))$	$\rightarrow$	b
$h(c,y)$	$\rightarrow$	$f(h(y,c),h(y,y))$

Chew [249] gives stringent conditions under which non-left-linear nonoverlapping systems are uniquely normalizing.

The Critical Pair Lemma can be weakened so that not all pairs need be considered. Various such *critical pair criteria* have been investigated, all revolving around the case of a critical peak that is rewritable in additional ways, making it possible to replace the peak with an alternate proof that is in some sense smaller. The *connectedness* criterion is based on the well-founded method of establishing local confluence (mentioned in Section 4.1) and ignores any critical pair $s=t$ derived from an overlap $s \underset{R}{\leftarrow} u \underset{R}{\rightarrow} t$ such that there exists another proof $s \underset{R}{\overset{*}{\leftrightarrow}} t$, each term of which is derivable from u by $\underset{R}{\overset{+}{\rightarrow}}$ (Winkler & Buchberger [238]); the *compositeness* criterion ignores critical pairs for which the overlapped term can be rewritten at a position strictly below the point of overlap (Kapur et al. [132]). See also Küchlin [265] and Bachmair & Dershowitz [12].

7.2. *Orthogonal systems*

By enforcing strong restrictions on the form of left-hand sides, confluence can be ensured even for nonterminating systems.

DEFINITION. A rewrite system is *orthogonal* if it is both left-linear (no multiple occurrences of a variable on the left) and nonoverlapping (no nontrivial critical pairs).

Examples of orthogonal systems are the stack and Combinatory Logic systems (of Sections 2.3 and 2.4, respectively). Mutually recursive function definitions, with one equation (employing *if* · *then* · *else* ·) per defined function, are orthogonal. Orthogonality (formerly called "regularity") imparts local confluence in the shape of Fig. 4(d) or its mirror image, depending on the redex positions. Moreover, orthogonal systems are always confluent, by the following result:

PARALLEL MOVES LEMMA (Huet [113]). *If R is orthogonal then* $\underset{R}{\overset{\parallel}{\rightarrow}}$ *is strongly confluent.*

The symbol $\underset{R}{\overset{\parallel}{\rightarrow}}$ denotes one "parallel" application of a rule in R at (zero or more) disjoint redexes. (For the name and inspiration of this lemma, cf. Curry & Feys [47].) The confluence of orthogonal systems implies that normal forms can be computed by rewriting. It also establishes the consistency of the operational semantics of recursive programming languages; see Raoult & Vuillemin [207]. The above lemma may be weakened to allow critical pairs that join in one parallel step (Huet [113]); the ground case was considered in Rosen [212].

For orthogonal systems, normal forms can be computed by a "parallel-outermost" redex evaluation scheme (O'Donnell [191]), but not by a "leftmost-outermost" scheme (Huet & Lévy [118]); with additional (decidable) "sequentiality" requirements, one can efficiently compute normal forms, without lookahead (Hoffmann & O'Donnell [105], Huet & Lévy [118]). How to avoid all unnecessary rewrites and obtain an optimal strategy (i.e. one with normalizing derivations of minimal length) is, however, an undecidable problem, in general (Huet & Lévy [118]).How to do as best as feasible is developed in Sekar & Ramakrishnan [273].

For modularity of programming with rewrite systems, one would have wished that the union of two convergent systems over disjoint vocabularies be convergent. Unfortunately, though the union of two confluent systems sharing no function symbols is confluent (Toyama [234]), and the disjoint union of systems with no more than one normal form per term also has that characteristic (Middeldorp [268]), the termination property (as we saw in Section 5.4) is not preserved under disjoint union. This is true even for confluent systems (Toyama [235]). If, however, the two convergent systems are also left-linear, then their union is convergent (Toyama et al. [236]).

7.3. *Class rewriting*

Critical pairs also provide a necessary and sufficient condition for a left-linear terminating system R to be Church–Rosser modulo a congruence S. That is, if $\mathbf{cp}(R\cup S)$ is a subset of $\xrightarrow[R]{!}\circ\overset{*}{\underset{S}{\leftrightarrow}}\circ\xleftarrow[R]{!}$, then $\xleftrightarrow[R\cup S]{*}$ is also contained therein (Huet [113]). Then, R-normal forms may be used to decide validity, provided S-equivalence is decidable. For example, if S is commutativity and R includes all commutativity variants of its rules, then R-normal forms are unique up to permutations of operands.

To handle rewriting modulo a congruence in the presence of non-left-linear rules, Peterson & Stickel [197] suggested using the extended rewrite relation $S\backslash R$ to compute normal forms. The set of critical peaks of the form $t\mu \xleftarrow[S\backslash R]{} s\mu \xrightarrow[S\backslash R]{} s\mu[r\mu]_p$ is in general infinite, so the Critical Pair Lemma is of little practical help. Instead, we consider peaks $t\mu \xleftarrow[R]{} s\mu \xrightarrow[S\backslash R]{} s\mu[r\mu]_p$ and *cliffs* $t\mu \underset{S}{\leftrightarrow} s\mu \xrightarrow[S\backslash R]{} s\mu[r\mu]_p$ separately.

DEFINITION. Let S be a set of equations. If $s\rightarrow t$ and $l\rightarrow r$ are two rewrite rules with distinct variables, p is the position of a nonvariable subterm of s, and μ is a most general substitution (most general, with respect to subsumption modulo S) such that $s\mu|_p \underset{S}{=} l\mu$, then $t\mu = s\mu[r\mu]_p$ is an *S-critical pair* of the two rules. If $s\leftrightarrow t$ is an equation in S, $l\rightarrow r$ is a rewrite rule (renamed as necessary), p is the position of a nonvariable proper subterm of s, and μ is a most general substitution such that $s\mu|_p \underset{S}{=} l\mu$, then $t\mu \rightarrow s\mu[r\mu]_p$ is an *S-extended rule* of $l\rightarrow r$.

In the set $\mathbf{cp}_S(R)$, we include all critical pairs obtained by overlapping S-variants of rules in R on (renamed) rules in R. We also need the set $\mathbf{ex}_S(R)$ of extended rules obtained by overlapping variants of rules in R on (renamed) equations in S.

EXTENDED CRITICAL PAIR LEMMA (Jouannaud [122]). *For any rewrite system R, equational system S, and peak $s \xleftarrow[R]{} \circ \xrightarrow[S\backslash R]{} t$, there exists a rewrite proof $s \xrightarrow[S\backslash R]{*} \circ \overset{*}{\underset{S}{\leftrightarrow}} \circ \xleftarrow[S\backslash R]{*} t$, or a critical-pair proof $s \overset{*}{\underset{S}{\leftrightarrow}} \circ \xleftrightarrow[\mathbf{cp}_S(R)]{} \circ \overset{*}{\underset{S}{\leftrightarrow}} t$ that involves S-steps within the critical pair's variable part only. Similarly, for any cliff $s \overset{*}{\underset{S}{\leftrightarrow}} \circ \xrightarrow[S\backslash R]{} t$, there exists a rewrite proof $s \xrightarrow[S\backslash R]{*} \circ \overset{*}{\underset{S}{\leftrightarrow}} \circ \xleftarrow[S\backslash R]{*} t$, or an extended-rule proof $s \overset{*}{\underset{S}{\leftrightarrow}} \circ \xrightarrow[\mathbf{ex}_S(R)]{} \circ \overset{*}{\underset{S}{\leftrightarrow}} t$ that involves S-steps within the extended rule's variable part only.*

The point is that in the absence of a rewrite proof, there must be a proof that is an application of an "S-instance" of a critical pair. The possible need for S-steps in the variable part is illustrated by the equation $a \leftrightarrow b$ and rules $f(x) \rightarrow g(x)$ and $f(x) \rightarrow h(x)$. A critical peak $g(a) \underset{R}{\leftarrow} f(a) \underset{S}{\leftrightarrow} f(b) \underset{R}{\rightarrow} h(b)$ lends itself to the critical pair proof $g(a) \underset{S}{\leftrightarrow} g(b) \underset{\mathbf{cp}(R)}{\longleftrightarrow} h(b)$.

Using this lemma, it can be shown that if R/S is terminating and the subterm relation modulo S is well-founded, then $S \backslash R$ is Church–Rosser modulo S iff $\mathbf{cp}_S(R)$ and $\mathbf{ex}_S(R)$ are contained in $\overset{!}{\underset{S\backslash R}{\rightarrow}} \circ \overset{*}{\underset{S}{\leftrightarrow}} \circ \overset{!}{\underset{S\backslash R}{\leftarrow}}$ (Jouannaud & Kirchner [124]). If these conditions are satisfied, and an S-matching procedure is available, then validity in $R \cup S$ can be decided. Note that subterm modulo S is well-founded when S-congruence classes are finite.

It is possible to combine the above results by partitioning rules into left-linear and not necessarily left-linear subsets. The critical pair condition can then be tailored to the different kinds of rules, with term rewriting used for the left-linear subset and extended rewriting for the rest (Jouannaud & Kirchner [124]). Additional improvements are provided by critical pair criteria for extended rewriting, as described in Bachmair & Dershowitz [10].

7.4. *Ordered rewriting*

Ordered rewriting systems enjoy a similar critical pair condition for confluence, but only for certain classes of orderings and only for ground terms. An ordering $>$ is called a *complete simplification ordering* if it is a simplification ordering that is total on $\mathcal{G}$, i.e. for any two distinct ground terms u and v, either $u > v$ or $v > u$. For example, if a precedence is total, it is easy to show that the induced lexicographic path ordering $>_{lpo}$ is a complete simplification ordering (a property not shared by the multiset path ordering).

DEFINITION. Let $l = r$ and $s = t$ be two equations in E (with disjoint variables) such that l overlaps s at nonvariable position p with most general unifier μ. The equation $t\mu = s\mu[r\mu]_p$ is a *critical pair* if the participating steps $t\mu \underset{E}{\leftrightarrow} s\mu \underset{E}{\leftrightarrow} s\mu[r\mu]_p$ can form a peak; in other words, if $t\mu\gamma \underset{<}{\leftarrow} s\mu\gamma \underset{>}{\rightarrow} s\mu\gamma[r\mu\gamma]_p$, for some substitution γ.

Let $\mathbf{cp}_>(E)$ denote the set of all such critical pairs between equations in E.

ORDERED CRITICAL PAIR LEMMA (after Lankford [157]). *For any set of equations E, complete simplification ordering $>$, and peak $s \underset{<}{\leftarrow} u \underset{>}{\rightarrow} t$ between ground terms s, t, u, there either exists an ordered-rewrite proof $s \overset{*}{\underset{>}{\rightarrow}} \circ \overset{*}{\underset{<}{\leftarrow}} t$ or a critical-pair proof $s \underset{\mathbf{cp}_>(E)}{\longleftrightarrow} t$.*

The hypotheses that $>$ is total on ground terms and has the replacement property are crucial for proving the variable overlap case displayed in Fig. 4(b); unlike the original Critical Pair Lemma, in this case, the bottom step may go one way or the other.

It may seem that local confluence is not guaranteed by the critical pair condition when equations in E have variables on one side that do not appear on the other. For example, with $E = \{x + g(y) = x + f(z)\}$, and $>$ the lexicographic path ordering induced by the precedence $g > f > b > a$, there is a peak $a + f(b) \underset{<}{\leftarrow} a + g(a) \underset{>}{\rightarrow} a + f(a)$, whereas there is no rewrite proof $a+f(b) \underset{>}{\rightarrow} a+f(a)$. But, in fact, there is a critical pair hidden here: overlapping the left-hand side of the equation on a literally similar instance $x+g(y')=x+f(z')$ gives the pair $x+f(z)=x+f(z')$. Thus, $a+f(b) \underset{\mathrm{cp}>(E)}{\longleftrightarrow} a+f(a)$.

With local confluence established, we have a ground convergent rewrite relation $\underset{>}{\rightarrow}$. Such an ordered-rewriting system—when finite—may be used to decide validity: if $s=t$ is valid for ground convergent E, then by *Skolemizing* the variables in s and t, that is by treating those variables as constants, and extending the ordering $>$ to include them, both s and t will have the same normal form. For example, if E is the commutativity axiom $x\cdot y=y\cdot x$ and $\cdot>z>y>x$ in a lexicographic path ordering, then $(y\cdot x)\cdot z$ and $z\cdot(y\cdot x)$ have the same normal form, $z\cdot(x\cdot y)$.

As a more interesting example, consider the following system for entropic groupoids (Hsiang & Rusinowitch [111]):

$(x\cdot y)\cdot x$	$\rightarrow$	x
$x\cdot(y\cdot z)$	$\rightarrow$	$x\cdot z$
$((x\cdot y_1)\cdot y_2)\cdot z)$	$\rightarrow$	$x\cdot z$
$(x\cdot y_1)\cdot z$	$\leftrightarrow$	$(x\cdot y_2)\cdot z$

and suppose we wish to decide validity of an equation $s=t$. First, any variables $x_1,\ldots,x_n$ appearing in s and t are replaced by Skolem constants $c_1,\ldots,c_n$. Then, a lexicographic path ordering is used with a precedence, say, in which "$\cdot$" is larger than the constants, and the constants are linearly ordered: $c_n>\cdots>c_1$. The equation is used to rewrite any product of the form $(x\cdot y_1)\cdot z$ to the same term with the occurrence of y_1 replaced by a sufficiently small term (viz. c_1) under $>_{lpo}$.

7.5. *Reduced systems*

By reducing right-hand sides and deleting rules with rewritable left-hand sides, a convergent system can always be converted into a canonical, i.e. reduced and convergent, one (see, e.g., Metivier [184]). One of the nice things about reduced systems is that, for any given equational theory, there can be only one (finite or infinite) canonical system contained in a particular reduction ordering (Butler & Lankford [39], Metivier [184]). This uniqueness result is up to literal similarity (renaming of variables). Uniqueness does not, however, hold for arbitrary canonical class-rewriting systems (Dershowitz et al. [62]), but does for associative-commutative systems (Lankford & Ballantyne [163]).

7.6. *Further reading*

A detailed study of the Church-Rosser property of nonoverlapping systems is Klop [146]. Computing normal forms with orthogonal systems is the subject of O'Donnell [191].

8. Completion

In the previous section, we saw that confluence of finite terminating systems can be decided using the superposition test. Suppose a given system fails that test because some critical pair has no rewrite proof. Building on ideas of Evans [72], Knuth and Bendix [149] suggested extending such a system with a new rule designed to cover the offending critical pair. Of course, new rules mean new critical pairs, some of which may also not pass the test. But, often enough, repeating this process eventually leads to a convergent system, with all critical pairs having rewrite proofs. This procedure is called *completion.* Interestingly, the critical pairs generated along the way are frequently the kind of lemmata a mathematician would come up with (Knuth & Bendix [149]).

Starting with a finite set of equations and a reduction ordering on terms, the completion procedure attempts to find a finite canonical system for the theory presented by the equations by generating critical pairs and orienting them as necessary. If reducing the two sides of a critical pair $s=t$ yields an equation $u=v$, where u and v are not identical, then adding a new rule $u \to v$ or $v \to u$ supplies a rewrite proof for $s=t$. To decide between the two orientations, the given reduction ordering is employed: if $u \succ v$ then $u \to v$ is added, while if $v \succ u$ then $v \to u$ is chosen. The new rule, $u \to v$ or $v \to u$, is then used to form new critical pairs. Running the procedure can have one of three outcomes: success in finding a canonical system, failure in finding anything, or looping and generating an infinite number of rules (forming an infinite canonical system).

8.1. *Abstract completion*

Completion has recently been put in a more abstract framework (Bachmair et al. [13]), an approach we adopt here. As in traditional proof theory (cf. Takeuti [226]), proofs are reduced, in some well-founded sense, by replacing locally maximal subproofs with smaller ones, until a normal-form proof is obtained. In completion, the axioms used are in a constant state of flux; these changes are expressed as inference rules, which add a dynamic character to establishing the existence of reducible subproofs. This view of completion, then, has two main components: an inference system, used in the completion process to generate new rewrite rules, and a rewrite relation that shows how any proof can be normalized, as long as the appropriate rules have been generated.

An *inference rule* (for our purposes) is a binary relation between pairs $(E; R)$, where E is a set of (unordered) equations and R is a set of (oriented) rewrite rules. (Rules or equations that differ only in the names of their variable are, for all intents and purposes, treated as identical.) Let $\succ$ be a reduction ordering, and $\blacktriangleright$ the well-founded ordering on rules defined as follows: $s \to t \blacktriangleright l \to r$ if (i) $s \rhd l$ under the encompassment ordering, or else (ii) $s \doteq l$ (s and l are literally similar) and $t \succ r$.

We define the following set KB of six inference rules:

Delete:	$(E\cup\{s=s\}; R)$	$\vdash$	$(E; R)$	
Compose:	$(E; R\cup\{s\to t\})$	$\vdash$	$(E; R\cup\{s\to u\})$	if $t \underset{R}{\to} u$
Simplify:	$(E\cup\{s=t\}; R)$	$\vdash$	$(E\cup\{s=u\}; R)$	if $t \underset{R}{\to} u$
Orient:	$(E\cup\{s=t\}; R)$	$\vdash$	$(E; R\cup\{s\to t\})$	if $s\succ t$
Collapse:	$(E; R\cup\{s\to t\})$	$\vdash$	$(E\cup\{u=t\}; R)$	if $s \underset{R}{\to} u$ by $l\to r$ with $s\to t \rhd l\to r$
Deduce:	$(E; R)$	$\vdash$	$(E\cup\{s=t\}; R)$	if $s=t\in \mathbf{cp}(R)$

We write $(E; R)\vdash_{KB}(E'; R')$ if the latter may be obtained from the former by one application of a rule in KB. **Delete** removes a trivial equation $s=s$. **Compose** rewrites the right-hand side t of a rule $s\to t$, if possible. **Simplify** rewrites either side of an equation $s=t$. **Orient** turns an equation $s=t$ that is orientable ($s\succ t$ or $t\succ s$) into a rewrite rule. **Collapse** rewrites the left-hand side of a rule $s\to t$ and turns the result into an equation $u=t$, but only when the rule $l\to r$ being applied to s is smaller under the rule ordering $\rhd$ than the rule being removed. **Deduce** adds equational consequences to E, but only those that follow from critical overlaps $s \underset{R}{\leftarrow} u \underset{R}{\to} t$.

DEFINITION. A (*standard*) *completion procedure* is any program that takes a finite set E_0 of equations and a reduction ordering $\succ$, and uses the above rules KB to generate a sequence of inferences from $(E_0; \emptyset)$.

In practice, the completion rules are usually applied in roughly the given order, saving space by preserving only reduced rules and equations. The (*inductive*) *limit* of a possibly infinite completion sequence $(E_0; \emptyset)\vdash_{KB}(E_1; R_1)\vdash_{KB}\cdots$ is the pair $(E_\infty; R_\infty)$, where E_∞ is the set $\bigcup_{i\geqslant 0}\bigcap_{j\geqslant i}E_j$ of *persisting* equations and R_∞ is the set $\bigcup_{i\geqslant 0}\bigcap_{j\geqslant i}R_j$ of *persisting* rules. For a finite sequence $(E_0; \emptyset)\vdash_{KB}\cdots\vdash_{KB}(E_n; R_n)$, we let $(E_\infty, R_\infty)=(E_n; R_n)$. We say that a completion sequence is *successful*, if E_∞ is empty and R_∞ is canonical.

When success occurs after a finite number of steps, the resultant system R_∞ is a decision procedure for E_0. But completion may "loop", producing an infinitely large set of persisting rules. A simple example (Ardis [3]) of looping is provided by the equation $f(g(f(x)))=g(f(x))$. Oriented the only way possible, the rule $f(g(f(x)))\to g(f(x))$ overlaps itself, generating the critical pair $g(f(g(f(x))))=f(g(g(f(x))))$, which simplifies to $g(g(f(x)))=f(g(g(f(x))))$. Continuing in the same manner, an infinite set of rules $\{f(g^i(f(x)))\to g^i(f(x)) \mid i\geqslant 1\}$ is produced.

The rules in KB are evidently *sound*, in that the class of provable theorems is unchanged by an inference step. Furthermore, only rules contained in $\succ$ are added. We are thus assured that the limit R_∞ of any (finite or infinite) successful completion sequence is terminating and presents the same equational theory as did E_0. Table 1

Table 1
A successful completion sequence for a fragment of group theory

<table>
<tr><th>i</th><th>R_i</th><th>E_i</th><th>inference</th></tr>
<tr><td>0</td><td></td><td>$1\cdot x = x$
$x\cdot 1 = x$
$x^{-}\cdot(x\cdot y) = y$</td><td></td></tr>
<tr><td>1</td><td>$x\cdot 1 \to x$</td><td>$1\cdot x = x$
$x^{-}\cdot(x\cdot y) = y$</td><td>orient</td></tr>
<tr><td>2</td><td>R_1
$1\cdot x \to x$</td><td>$x^{-}\cdot(x\cdot y) = y$</td><td>orient</td></tr>
<tr><td>3</td><td rowspan="2">R_2
$x^{-}\cdot(x\cdot y) \to y$</td><td></td><td>orient</td></tr>
<tr><td>4</td><td>$x^{-}\cdot x = 1$</td><td>deduce(1, 3)</td></tr>
<tr><td>5</td><td rowspan="2">R_3
$x^{-}\cdot x \to 1$</td><td></td><td>orient</td></tr>
<tr><td>6</td><td>$1^{-} = 1$</td><td>deduce(1, 5)</td></tr>
<tr><td>7</td><td rowspan="2">R_5
$1^{-} \to 1$</td><td></td><td>orient</td></tr>
<tr><td>8</td><td>$(x^{-})^{-}\cdot y = x\cdot y$</td><td>deduce(3, 3)</td></tr>
<tr><td>9</td><td rowspan="2">R_7
$(x^{-})^{-}\cdot y \to x\cdot y$</td><td></td><td>orient</td></tr>
<tr><td>10</td><td>$1^{-}\cdot y = 1\cdot y$</td><td>deduce(7, 9)</td></tr>
<tr><td>11</td><td>$1^{-}\cdot y \to 1\cdot y$</td><td rowspan="2"></td><td>orient</td></tr>
<tr><td>12</td><td>R_9
$1^{-}\cdot y \to y$</td><td>compose(11, 2)</td></tr>
<tr><td>13</td><td rowspan="4">R_9</td><td>$1\cdot y = y$</td><td>collapse(12, 7)</td></tr>
<tr><td>14</td><td>$y = y$</td><td>simplify(2)</td></tr>
<tr><td>15</td><td></td><td>delete</td></tr>
<tr><td>16</td><td>$(x^{-})^{-} = x$</td><td>deduce(1, 9)</td></tr>
<tr><td>17</td><td>R_9
$(x^{-})^{-} \to x$</td><td></td><td>orient</td></tr>
<tr><td>18</td><td rowspan="3">R_7
$1^{-}\cdot y \to y$
$(x^{-})^{-} \to x$</td><td>$x\cdot y = x\cdot y$</td><td>collapse(9, 17)</td></tr>
<tr><td>19</td><td></td><td>delete</td></tr>
<tr><td>20</td><td>$x\cdot(x^{-}\cdot y) = y$</td><td>deduce(3, 17)</td></tr>
<tr><td>21</td><td rowspan="2">R_{18}
$x\cdot(x^{-}\cdot y) \to y$</td><td></td><td>orient</td></tr>
<tr><td>22</td><td>$x\cdot x^{-} = 1$</td><td>deduce(1, 21)</td></tr>
<tr><td>23</td><td>R_{21}
$x\cdot x^{-} \to 1$</td><td></td><td>orient</td></tr>
</table>

shows an example of a successful completion sequence. Starting with the three axioms

$x \cdot 1$	$=$	x
$1 \cdot x$	$=$	x
$x^{-} \cdot (x \cdot y)$	$=$	y

over a vocabulary containing a constant 1, postfix unary symbol "$^{-}$", and infix binary symbol "$\cdot$", it generates the eight-rule canonical system

$1 \cdot x$	$\to$	x	$x \cdot 1$	$\to$	x
$x^{-} \cdot x$	$\to$	1	$x \cdot x^{-}$	$\to$	1
1^{-}	$\to$	1	$(x^{-})^{-}$	$\to$	x
$x^{-} \cdot (x \cdot y)$	$\to$	y	$x \cdot (x^{-} \cdot y)$	$\to$	y

using size as the reduction ordering.

For a given reduction ordering $\succ$, a (not necessarily finite) convergent system R contained in $\succ$ exists for an equational theory E, iff each E-congruence class of terms has a unique minimal element vis-à-vis $\succ$ (Avenhaus [4]). Nonetheless, completion may fail to find R, even when given $\xrightarrow[R]{+}$ as the reduction ordering (Dershowitz et al. [66]). For example, despite the existence of $\{f(a) \to a, c \to a, b \to a\}$, no successful sequence exists for $\{f(b)=a, f(c)=c, b=c\}$, as long as b and c are incomparable under the given ordering. In fact, on account of the partialness of the ordering, some sequences may fail while others may succeed (Avenhaus [4], Dershowitz et al. [66]). For example, let $\succ$ be a recursive path ordering with precedence $f \succ d \succ c \succ a$ and $d \succ b \succ a$ (but b and c are incomparable), and let $E_0 = \{f(c)=c, b=d, c=d, f(d)=a\}$. There is a successful sequence:

$$\begin{aligned}(E_0; \emptyset) &\vdash^{+}_{KB} (\{b=d, f(d)=a\}; \{d \to c, f(c) \to c\}) \\ &\vdash^{+}_{KB} (\emptyset; \{f(a) \to a, b \to a, c \to a, d \to a\})\end{aligned}$$

as well as a failing one:

$$\begin{aligned}(E_0; \emptyset) &\vdash^{+}_{KB} (\{c=d, f(d)=a\}; \{d \to b, f(c) \to c\}) \\ &\vdash^{+}_{KB} (\{b=c\}; \{f(b) \to a, d \to b, f(c) \to c\}).\end{aligned}$$

The latter sequence cannot be extended further.

As pointed out already in Knuth & Bendix [149], such failures can be circumvented by incorporating an inference rule that adds $s \to k(x_1, \ldots, x_n)$ and $t \to k(x_1, \ldots, x_n)$ to R_i if $s=t$ is an unorientable equation in E_i, where k is a new function symbol not in the original vocabulary and $x_1, \ldots, x_n$ are those variables appearing in s and t. Though this inference is not sound (it constitutes a conservative extension), it results in a decision procedure if ultimately successful. In the above failing example, replacing $b=c$ with $b \to k$ and $c \to k$ leads directly to $\{a \to k, b \to k, c \to k, d \to k, f(k) \to k\}$. Two terms

s and t in $\mathcal{T}(\{a, b, c, d, f\})$ are equal in the original theory iff they have the same normal form in this system. Unfortunately, this process can, in general, degenerate into unsuccessful sequences that add infinitely many new symbols. As we will see below, completion has been extended in various other ways, in particular to handle the associative-commutative axioms, which cannot be handled as rules.

Completion can also be used as a mechanical theorem prover. The idea is that, even when the procedure loops, any valid theorem should eventually have a rewrite proof using rules already on hand. This is not actually the case, since a procedure might abort a sequence when no equation is orientable. (See Section 8.4, however.) For implementations that are fair in their choice of inferences, one can show that—barring abortion—all provable equations eventually lend themselves to a direct proof by rewriting. This perspective on completion was taken by Huet [114], and by Lankford [156] from the refutational point of view. The difficult part is the need to show that deleting simplifiable rules (to which **compose** or **collapse** apply) does not—in the long run—shrink the class of equations having rewrite proofs. Once we have established that completion acts as a semi-decision procedure for validity when it loops, we will be ready to apply it to theorem-proving in inductive theories and in first-order predicate calculus.

A proof in $E \cup R$ is a sequence of E-steps and R-steps. By applying the above inference rules, it may be possible to simplify a given proof, replacing some steps with alternate ones from $E' \cup R'$, whenever $(E; R) \vdash^*_{KB} (E'; R')$. By a *proof pattern* we intend a schema describing a class of subproofs; e.g. to characterize rewrite proofs, we use the pattern $s \xrightarrow[R]{*} v \xleftarrow[R]{*} t$, where s, t, and v denote arbitrary terms. If a proof contains no peaks $s \xleftarrow[R]{} u \xrightarrow[R]{} t$ nor applications $s \xleftrightarrow[E]{} t$ of equations, it must be a rewrite proof. Let E^* be the set of all equations logically deducible from E_0 and R^* be the orientable subset of E^* that intersects with the reduction ordering $\succ$. The following set C of proof-pattern rules captures the elimination of the undesirable patterns and the simplification of proofs which takes place during completion:

$s \xleftrightarrow[E^*]{} s$	$\Rightarrow$	s	
$s \xrightarrow[R^*]{} t$	$\Rightarrow$	$s \xrightarrow[R^*]{} v \xleftarrow[R^*]{} t$	where $s \xrightarrow[R^*]{} t$ by $l \to r$ and $s \xrightarrow[R^*]{} v$ by $l' \to r'$ and $l \to r \rhd l' \to r'$
$s \xleftrightarrow[E^*]{} t$	$\Rightarrow$	$s \xrightarrow[R^*]{} v \xleftrightarrow[E^*]{} t$	
$s \xleftrightarrow[E^*]{} t$	$\Rightarrow$	$s \xrightarrow[R^*]{} t$	
$s \xrightarrow[R^*]{} t$	$\Rightarrow$	$s \xrightarrow[R^*]{} v \xleftrightarrow[E^*]{} t$	where $s \xrightarrow[R^*]{} t$ by $l \to r$ and $s \xrightarrow[R^*]{} v$ by $l' \to r'$ and $l \to r \rhd l' \to r'$
$s \xleftarrow[R^*]{} u \xrightarrow[R^*]{} t$	$\Rightarrow$	$s \xleftrightarrow[E^*]{} t$	
$s \xleftarrow[R^*]{} u \xrightarrow[R^*]{} t$	$\Rightarrow$	$s \xrightarrow[R^*]{*} v \xleftarrow[R^*]{*} t$	

Symmetric rules, with the proof patterns on both sides of $\Rightarrow$ inverted, are also needed.

Note how these rules correspond exactly to the effect of the inference rules: Any proof step involving a deleted equation can be omitted; when a rule is replaced with a new composed rule, any rewrite step using the original can be replaced by a two-step valley; when an equation is simplified, its use in a proof is replaced by a rewrite step and an equational step; when an equation is oriented, the corresponding proof step becomes a rewrite step; when a rule is collapsed into an equation, a combination of rewrite and equational steps may be used instead; when a critical pair is deduced, the corresponding critical peak in a proof may be replaced by a new equational step. Finally the last proof-pattern rule corresponds to a noncritical peak, which can always be converted into a valley.

We assume that the ordering is such that for any equation $s=t$ in E^* there is a term v for which $s \xrightarrow[R^*]{*} v \xleftarrow[R^*]{*} t$. (With a reduction ordering that does not satisfy this condition there is no chance of a successful completion.) Then, at least one of the rules of C can be applied to any nonrewrite proof or to any proof employing a nonreduced rule. Thus, C-normal forms are R^*-rewrite proofs that use only reduced rules. Furthermore, we can apply in $E^* \cup R^*$ techniques of Section 5 and show that the proof-normalization relation $\Rightarrow$ is terminating: Consider the ordering $\succ_c$ which compares proofs by comparing multisets containing the pair $\langle \{s\}, l \rightarrow r \rangle$ for each application $s \xrightarrow[R^*]{} t$ of a rule $l \rightarrow r$ and $\langle \{s,t\}, l \rightarrow r \rangle$ for each application $s \xleftrightarrow[E^*]{} t$ of an equation $l=r$. Pairs are compared lexicographically, using the multiset ordering $\succ_{mul}$ induced by the given reduction ordering $\succ$ for the first component, and the ordering $\rhd$ on rules for the second. Multisets of pairs, measuring the complexity of proofs, are compared by $\succ_c$, the multiset ordering induced by this lexicographic ordering. Since $\succ$ and $\rhd$ are both well-founded, $\succ_c$ is a reduction ordering on *proofs*. Since it can be verified that C is contained in $\succ_c$, the former is terminating. Note how this *proof ordering* considers the justification of a proof and not just the terms in it. For further details, consult Bachmair [14].

8.2. *Fairness*

For any given completion sequence $(E_0; \emptyset) \vdash_{KB} (E_1; R_1) \vdash_{KB} \cdots$, let $\overset{*}{\underset{i}{\leftrightarrow}}$ stand for $\xleftrightarrow[E_i \cup R_i]{*}$ that is, for a proof at the ith stage, using rewriting with R_i in either direction or equational steps with E_i. The proof normalization relation C mirrors the inference system KB in that for any completion sequence and for any proof $s \overset{*}{\underset{i}{\leftrightarrow}} t$ at stage i there exists a proof $s \overset{*}{\underset{j}{\leftrightarrow}} t$ at each subsequent stage j such that $s \overset{*}{\underset{i}{\leftrightarrow}} t \overset{*}{\underset{C}{\Rightarrow}} s \overset{*}{\underset{j}{\leftrightarrow}} t$. In this way, inference rules are used to generate rules needed for proofs to be C-reducible. A (possibly infinite) sequence $(E_0; \emptyset) \vdash_{KB} (E_1; R_1) \vdash_{KB} \cdots$ is deemed *fair* if for any proof $s \overset{*}{\underset{i}{\leftrightarrow}} t$ that is C-reducible, there exists a step j, such that $s \overset{*}{\underset{i}{\leftrightarrow}} t \overset{+}{\underset{C}{\Rightarrow}} s \overset{*}{\underset{j}{\leftrightarrow}} t$. Imposing fairness, we have:

PROOF NORMALIZATION THEOREM (Huet [114]). *If a completion sequence* $(E_0; \emptyset) \vdash_{KB} (E_1; R_1) \vdash_{KB} \cdots$ *is fair, then for any proof* $s \overset{*}{\underset{i}{\leftrightarrow}} t$ *there exists a proof* $s \xrightarrow[R_\infty]{*} \circ \xleftarrow[R_\infty]{*} t$ *using reduced rules only.*

Huet introduced the notion of fairness of completion and proved this theorem for a specific fair implementation; the following proof (Bachmair et al. [13]) builds on the above development and holds for any implementation of inference system KB:

PROOF. The proof is by induction with respect to $\underset{C}{\Rightarrow}$. Suppose that $s \overset{*}{\underset{i}{\leftrightarrow}} t$ is not a rewrite proof $s \xrightarrow[R_\infty]{*} \circ \xleftarrow[R_\infty]{*} t$. Then it must contain a peak or equational step that is reducible by C. Similarly, if $s \overset{*}{\underset{i}{\leftrightarrow}} t$ involves a nonpersistent step, then it is C-reducible. By fairness, $s \overset{*}{\underset{i}{\leftrightarrow}} t \overset{+}{\underset{C}{\Rightarrow}} s \overset{*}{\underset{j}{\leftrightarrow}} t$ for some step j and by induction there is a proof $s \xrightarrow[R_\infty]{*} \circ \xleftarrow[R_\infty]{*} t$ with only reduced rules. □

By the Critical Pair Lemma, noncritical peaks are C-reducible. Thus, it can be shown that a completion sequence is fair if all persistent critical pairs are accounted for ($\mathbf{cp}(R_\infty)$ is a subset of $\bigcup E_i$), no simplifiable rule persists (R_∞ is reduced), and no equation persists (E_∞ is empty). Practically speaking, fairness means that critical pairs are generated for all new rules, and need eventually to be simplified or oriented, unless the new rule itself is later simplified. A marking scheme is generally used to keep track of which rules still need to be overlapped with which; see, for instance, Huet [114]. By generating the remaining critical pairs and then eliminating them, Table 1 turns fair.

An n-step completion sequence $(E_0; \emptyset) \vdash_{KB} (E_1; R_1) \vdash_{KB} \cdots \vdash_{KB} (E_n; R_n)$ succeeds if E_n is empty, R_n is reduced, and each of the latter's critical pairs already appeared in some E_i. The sequence *fails* if no fair sequence has it as a prefix; in that case there is little point continuing. A completion procedure is considered *correct* if it only generates fair, successful sequences—when it does not abort. Assuming the procedure never discriminates against any critical pair or simplifiable rule or equation, the only good reason to abort (and produce a failing sequence) is if all equations are unorientable. The critical pair criteria mentioned in Section 7.1 may be used to improve on the above requirements for fairness, by necessitating consideration of fewer critical pairs; see Bachmair & Dershowitz [12].

It follows from the above theorem that the limit R_∞ of a fair sequence is canonical. Furthermore, since $\overset{*}{\underset{\infty}{\leftrightarrow}} = \overset{*}{\underset{0}{\leftrightarrow}}$, completion, correctly implemented, provides a semi-decision procedure for validity, or else aborts. That is, if $\mathcal{M}od(E_0) \models s=t$, and if completion does not give up in the middle, then, at some step k, it will be possible to check that $s \xrightarrow[R_k]{!} \circ \xleftarrow[R_k]{!} t$. With additional inferences for unorientable rules, completion need not give up; see Section 8.4.

Note that two successful completions, given the same starting set E_0 and reduction ordering $\succ$, must output the same canonical systems, be they finite or infinite, since there can be but one reduced, canonical system (up to literal similarity) contained in $\succ$ (see Section 7.5). Thus, if there exists a complete system R for E_0 contained in $\succ$, then a correct procedure, given E_0 and $\succ$, cannot succeed with any system but R—though it may abort without finding it. Furthermore, if R is finite, then an infinite, looping completion sequence is likewise impossible, since R_∞ must be of the same size as R.

8.3. Extended completion

Before we consider other theorem-proving applications of completion, we adapt it to handle extended rewriting. Let S be an equational system and $\succ$ a reduction ordering such that $\succ$ commutes over S. Rules are compared using the following ordering: $s \to t \rhd l \to r$ if $s \unrhd l' \underset{S}{=} l$, for some l' (i.e. if s properly encompasses a term that is S-equivalent to l), and the S-steps are below the top of s, or else $s \doteq l$ and $t \succ r$.

We define the following set KB/S of inference rules:

Delete:	$(E\cup\{s=t\}; R)$	$\vdash$	$(E; R)$	if $s \overset{*}{\underset{S}{\leftrightarrow}} t$
Compose:	$(E; R\cup\{s\to t\})$	$\vdash$	$(E; R\cup\{s\to v\})$	if $t \underset{R/S}{\to} v$
Simplify:	$(E\cup\{s=t\}; R)$	$\vdash$	$(E\cup\{u=t\}; R)$	if $s \underset{R/S}{\to} u$
Orient:	$(E\cup\{s=t\}; R)$	$\vdash$	$(E; R\cup\{s\to t\})$	if $s \succ t$
Collapse:	$(E; R\cup\{s\to t\})$	$\vdash$	$(E\cup\{v=t\}; R)$	if $s \underset{R/S}{\to} v$ by $l\to r$ with $s\to t \blacktriangleright l\to r$
Extend:	$(E; R)$	$\vdash$	$(E; R\cup\{s\to t\})$	if $s\to t \in \mathbf{ex}_S(R)$
Deduce:	$(E; R)$	$\vdash$	$(E\cup\{s=t\}; R)$	if $s=t \in \mathbf{cp}_S(R)$

As before, we write $(E; R) \vdash_{\overline{KB/S}} (E'; R')$ if the latter may be obtained from the former by one application of a rule in KB/S. With this inference system, **delete** removes equations between S-equivalent terms; **collapse** simplifies left-hand sides; **extend** adds extended rules; **deduce** generates extended critical pairs. Extended rewriting requires S-matching; S-completion requires S-unification to generate critical pairs and extended rules. The set S is unchanging throughout.

DEFINITION. An *S-completion procedure* is any program that takes a finite set E_0 of equations, an S-unification procedure, and a reduction ordering $\succ$ that commutes over S, and uses the above rules KB/S to generate a sequence of inferences from $(E_0; \emptyset)$.

The most important case, in practice, is when S is a set of commutativity (C) or associativity-commutativity (AC) axioms for some binary function symbols (Lankford & Ballantyne [161], Peterson & Stickel [197]). For the purposes of AC-completion, rules are commonly *flattened* by removing nested occurrences of associative-commutative symbols. An associative-commutative unification algorithm is employed—in place of the standard (syntactic) unification algorithm—to generate the critical pairs in $\mathbf{cp}_{AC}(R_i)$, and associative-commutative matching is used to apply rules. For each rule $f(s,t)\to r$ headed by an associative-commutative symbol f, an extended rule $f(s,f(t,z))\to f(r,z)$ is added and flattened out to $f(s,t,z)\to f(r,z)$; extensions of AC-extended rules are redundant.

For example, consider the same set E_0 as in Table 1, along with the following set S of AC axioms:

$x\cdot y$	$\leftrightarrow$	$y\cdot x$
$x\cdot(y\cdot z)$	$\leftrightarrow$	$(x\cdot y)\cdot z$

Extend uses associativity to create two extended rules, $(1\cdot x)\cdot z\to x\cdot z$ and $x^-\cdot(x\cdot z)\to 1\cdot z$, the first of which collapses away, and the second of which composes to yield an existing rule. **Deduce** generates $(x\cdot y)^-\cdot x\to y^-$ from an S-overlap of

$x^- \cdot (x \cdot y) \to y$ on itself (at position 2). The resultant rule is extended to $(x \cdot y)^- \cdot (x \cdot z) \to y^- \cdot z$, which forms an S-critical pair with $x^- \cdot x \to 1$ and generates $x^- \cdot y^- \to (x \cdot y)^-$. Extending as necessary, cleaning up, and flattening products, the final result is

$1^- \to 1$	$x \cdot 1 \to x$
$x \cdot x \to 1$	$x \cdot x^- \cdot z \to z$
$(x^-)^- \to x$	$(x \cdot y^-)^- \to x^- \cdot y$
$(x \cdot y)^- \cdot x \to y^-$	$(x \cdot y)^- \cdot x \cdot z \to y^- \cdot z$
$x^- \cdot y^- \to (y \cdot x)^-$	$x^- \cdot y^- \cdot z \to (y \cdot x)^- \cdot z$

A better choice of ordering, one that would make $(y \cdot x)^-$ greater than $x^- \cdot y^-$, would result in the following neater system G/AC for Abelian (commutative) groups:

$1^- \to 1$	$x \cdot 1 \to x$
$x \cdot x^- \to 1$	$x \cdot x^- \cdot z \to z$
$(x^-)^- \to x$	$(y \cdot x)^- \to x^- \cdot y^-$

A proof in $S \cup E \cup R$ is a sequence of S-steps, E-steps, and R-steps. Analogous to the standard case, a relation $\underset{C/S}{\Rightarrow}$ can be defined that describes the simplifying effect of S-completion at the proof level. Using the Extended Critical Pair Lemma, it can then be shown that a completion sequence (in $\vdash_{KB/S}$) is fair (with respect to $\underset{C/S}{\Rightarrow}$) if all persistent critical pairs are accounted for ($\mathbf{cp}_S(R_\infty)$ is a subset of the S-variants of $\bigcup E_i$), all persistent extended rules are accounted for ($\mathbf{ex}_S(R_\infty)$ is a subset of the S-variants of $\bigcup R_i$), and no equation persists (E_∞ is empty). With fairness, we get that an extended-rewrite proof $s \xrightarrow[S \backslash R_\infty]{*} \circ \overset{*}{\underset{S}{\leftrightarrow}} \circ \xleftarrow[S \backslash R_\infty]{*} t$ will eventually be generated if $s \xleftrightarrow[S \cup E_0]{} t$ (Jouannaud & Kirchner [124]). However, the limit $S \backslash R_\infty$ need not be reduced.

Additional aspects of completion modulo equational theories have been considered: Huet [113] deals with the left-linear case; Jouannaud & Kirchner [124] analyze exactly which critical pairs are necessary when some rules are left-linear and others not; Bachmair & Dershowitz [10] take the inference rule approach and generalize previous results.

8.4. *Ordered completion*

We have seen that completion can have any one of three outcomes: it may succeed in finding a decision procedure for validity after a finite number of steps; it may loop and generate more and more rules until—at some point—any particular valid equation has a rewrite proof; or it may abort with unorientable equations before finding any proof.

Since there are total reduction orderings for any set of ground terms (the lexicographic path ordering with total precedence is one such), completion of ground terms—given such an ordering—will not abort. Moreover, ground completion need never apply the **deduce** inference rule, since the **collapse** rule always applies to one of the rules contributing to a critical pair. And without **deduce**, completion will not loop. Thus, for any finite set of ground equations, completion is sure to generate a decision procedure (Lankford [156]), which is not surprising, since all such theories are decidable (Ackermann [1]). In fact, various $O(n \lg n)$ congruence-closure algorithms exist for the purpose (e.g. Downey et al. [69]; see also Snyder [220]). More interesting are those cases where there are nonground rules for which a canonical rewrite system is available, and all critical pairs between a ground rule and a nonground one are either ground or simplify to a trivial rule. By supplying completion with a complete simplification ordering, these critical pairs can always be oriented. (The complete ordering must be compatible with the canonical system for the non-ground rules.) For example, AC-completion can be used in this way to generate decision procedures for finitely-presented Abelian groups starting from G/AC (Lankford et al. [266]).

We now turn our attention to completion of ordered rewriting systems, and call the process "ordered" (or "unfailing") completion. Ordered completion either returns a (finite) ordered rewriting system in finite time, or else loops and generates an infinite system. With a finite system, validity can be decided by ordered rewriting; an infinite system can only serve as a semi-decision procedure. Since all orientable instances of equations are used to rewrite, there will be no need to explicitly distinguish between rewrite rules and other equations in the inference rules. Let $\succ$ be a reduction ordering that can be extended to a complete simplification ordering, and let $\trianglerighteq$ be the encompassment ordering. Consider the following set OC of inference rules, operating on set of equations E (cf. Bachmair et al. [14]):

Delete:	$E \cup \{s=s\}$	$\vdash$	E	
Simplify:	$E \cup \{s=t\}$	$\vdash$	$E \cup \{s=u\}$	if $t \underset{\succ}{\rightarrow} u$ and $s \succ u$
Collapse:	$E \cup \{s=t\}$	$\vdash$	$E \cup \{s=u\}$	if $t \underset{\succ}{\rightarrow} u$ by $l=r$ with $t \rhd l$
Deduce:	E	$\vdash$	$E \cup \{s=t\}$	if $s=t \in \mathbf{cp}_{\succ}(E)$

We write $E \vdash_{OC} E'$ if the latter may be obtained from the former by one application of a rule in OC. With this inference system, **deduce** generates ordered critical pairs, and the other rules simplify them.

DEFINITION. An *ordered completion procedure* is any program that takes a finite set E_0 of equations and a reduction ordering $\succ$ that can be extended to a complete simplification ordering, and uses the above rules OC to generate a sequence of inferences from E_0.

For example, consider the following axioms for entropic groupoids:

$$\begin{array}{|lcl|}\hline (x\cdot y)\cdot x & = & x \\ (x\cdot y_1)\cdot(y_2\cdot z) & = & (x\cdot y_2)\cdot(y_1\cdot z) \\ \hline\end{array}$$

The second equation is permutative and cannot be oriented by any reduction ordering. Completion will therefore fail. Ordered completion, on the other hand, yields the ground-convergent ordered-rewriting system shown in Section 7.4. As another example, given equational axioms for Abelian groups and a suitable ordering, ordered completion generates the system G/AC shown at the end of Section 8.3.

Analogous to the standard case, a relation $\underset{OC}{\Rightarrow}$ can be defined that describes the simplifying effect of ordered completion at the proof level. Fairness is defined accordingly. Using the Ordered Critical Pair Lemma, it can then be shown that a completion sequence is fair (with respect to $\underset{OC}{\Rightarrow}$) if all persistent critical pairs are accounted for, i.e. if $\mathbf{cp}_{\succ}(E_\infty)$ is a subset of $\bigcup E_i$. With fairness, we get that a rewrite proof between two ground terms s and t will eventually be generated iff $s \overset{*}{\underset{E_0}{\leftrightarrow}}$ (Hsiang & Rusinowitch [111]). Thus, the word problem in arbitrary equational theories can always be semidecided by ordered completion. (See Boudet et al. [33] for an interesting application.)

It is not hard to see that OC can mimic KB for any given equational theory E and reduction ordering $\succ$ (not necessarily total on $\mathcal{G}$). The natural question is whether ordered completion must succeed in generating a canonical set of rules whenever one exists for the given reduction ordering $\succ$. The answer is affirmative (Bachmair et al. [14]), provided $\succ$ can be extended to a complete reduction ordering. For example, if $f(b)\succ a$, but $f(b)$ and $f(c)$ are incomparable, ordered completion infers

$$\begin{aligned} &\{b=c,\ f(b)=a,\ f(c)\to c\} \\ &\quad \vdash_{OC} \{b=c,\ f(c)=a,\ f(b)=a, f(c)=c\}. \end{aligned}$$

With the recursive path ordering in which $f\succ c\succ a$, this sequence continues until success:

$$\begin{aligned} &\vdash_{OC} \{b=c,\ c=a,\ f(b)=a,\ f(c)=c,\ f(c)=a\} \\ &\quad \vdash^{+}_{OC} \{f(a)=a,\ b=a,\ c=a,\ f(b)=a\}. \end{aligned}$$

Ordered completion can also be modified to act as a refutationally complete inference system for validity in equational theories. To prove $s=t$, its Skolemized negation $eq(s',t')=F$ is added to the initial set $E\cup\{eq(x,x)=T\}$ of equations. With a completable reduction ordering (the Skolem constants are added to the algebra, hence must be comparable), the outcome $T=F$ characterizes validity of $s=t$ in the theory of E (Hsiang & Rusinowitch [111]).

8.5. *Inductive theorem proving*

An inductive theorem-proving method based on completion was first proposed in Musser [186]. Recall from Section 3.2 that an inductive theorem $s \underset{I(E)}{=} t$ holds iff there is no equation $u=v$ between ground terms that follows from $E\cup\{s=t\}$, but not from E alone. Let H be a set of equational hypotheses. Given a ground-convergent system R for E, we aim to find a ground-convergent system R' for $E\cup H$ with the same ground normal forms. If R' is the result of a successful completion sequence starting from $(H; R)$ and using an ordering containing R, then, by the nature of completion, R' is convergent and every term reducible by R is reducible by R'. To check that every ground term reducible by R' is also R-reducible, it is necessary and sufficient that every left-hand side of R' be ground R-reducible. If R' passes this test, then the two systems have the same ground normal forms and $s \underset{I(R)}{=} t$ for each $s=t$ in H (Dershowitz [55], Jouannaud & Kounalis [125]). If, on the other hand, $s \underset{I(R)}{\neq} t$, then fair completion will uncover an inconsistency, if it does not fail. This approach is also valid when extended-completion (Goguen [87]), or ordered-completion (Jouannaud & Kounalis [125], Bachmair [6]) is used.

For example, let $\mathcal{F}$ be $\{\Lambda, push, alternate\}$ and R be the following canonical system for interleaving stacks:

$$\boxed{\begin{array}{lll} alternate(\Lambda, z) & \rightarrow & z \\ alternate(push(x, y), z) & \rightarrow & push(x, alternate(z, y)) \end{array}}$$

and suppose we wish to prove that $alternate(y, \Lambda)= y$ is an inductive theorem. This equation can be oriented from left to right. Since the critical pairs $\Lambda=\Lambda$ and $push(x, y)=push(x, alternate(\Lambda, y))$ are provable by rewriting, completion ends with the system $R' = R\cup\{alternate(y, \Lambda)\rightarrow y\}$. Since the left-hand side $alternate(y, \Lambda)$ of the new rule is ground R-reducible, for all ground terms y in $\mathcal{G}(\{\Lambda, push, alternate\})$, the theorem is valid in $I(R)$. In more complicated cases, additional lemmata may be generated along the way.

Finding an R' that is complete for $H\cup R$ (as completion does) is actually much more than needed to preclude inconsistency. One need only show that some R' provides a valley proof for all ground consequences of $H\cup R$ (and that $R(\mathcal{G})\subseteq R'(\mathcal{G})$). Thus, for each inference step $(H_i; R_i) \vdash (H_{i+1}; R_{i+1})$, the equational hypotheses in H_i should be inductive consequences of $H_{i+1}\cup R_i$, and the rules in R_{i+1} should be inductive consequences of R_i (Bachmair [6]). Then, at each state n, $H_0\cup R_0$ follows from $H_n\cup R_0$; the original hypotheses H_0 are proved as soon as an empty H_n suffices. Assuming R_i is always the original R, it is enough if any ground cliff $v \underset{R}{\leftarrow} u \underset{H_i}{\leftrightarrow} w$ can be reduced to a smaller proof in $H_{i+1}\cup R$ (smaller, in some well-founded sense), and no instance of a hypothesis in H_i is itself an inconsistency (that is, an equation between two distinct ground R-normal forms). Consequently, there is no need (though it may help) to generate critical pairs between two rules derived from H_0; it suffices to consider critical pairs obtained by narrowing with R at a set of "covering" positions in H_i (Fribourg [79]). When, and if, these pairs all simplify away, the inductive theorems are proved. On the other hand, if any of the hypotheses are false, a contradiction must eventually surface in a fair derivation. The use of critical pair criteria in this connection is explored in Göbel [84] and Küchlin [155]. How to handle inductive equations that cannot be oriented is discussed in Bachmair [6] and Jouannaud & Kounalis [125].

8.6. *First-order theorem proving*

Rewriting techniques have been applied to refutational first-order theorem proving in two ways. One approach is to use resolution for non-equality literals together with some kind of superposition of left-hand sides of equality literals within clauses (Bachmair & Ganzinger [243], Lankford & Ballantyne [162], Peterson [196], Zhang & Kapur [276] and others). The notions of rewriting permit the incorporation of simplification (demodulation) without loss of completeness.

An alternative approach (Hsiang & Dershowitz [108]) is to use the Boolean ring system BA of Section 2 and treat logical connectives equationally. Let R be BA, S be AC for *xor* and *and*, and E be $\{s=T\}$, where $\forall x_1, \ldots, x_n\ s$ is a logically unsatisfiable closed Boolean formula. It follows from Herbrand's Theorem (Herbrand [100]) that a finite conjunction of instances $s\sigma_i$ of s has normal form F under BA/AC. Thus, ordered AC-completion, starting with R and E, will reduce the proof

$$T \overset{*}{\underset{BA}{\leftarrow}} and(\cdots and(T, T), \ldots, T) \overset{*}{\underset{E}{\leftrightarrow}} and(\cdots and(s\sigma_1, s\sigma_2), \ldots, s\sigma_n) \overset{*}{\underset{BA/AC}{\longrightarrow}} F$$

to a critical pair $T=F$ (see also Paul [271]). As with resolution theorem-proving, one can prove validity of a closed formula in first-order predicate calculus by deriving a contradiction from its Skolemized negation.

There are several problems with such an approach:

(a) AC-unification is required to compute critical pairs;

(b) failure is possible unless the more expensive ordered procedure is used; and

(c) critical pairs with the distributivity axiom in BA are very costly.

Hsiang [107] showed that if each equation in E is of the form $xor(s, T)=F$, where s is the exclusive-or normal form of a clause, then only a subset of the critical pairs with BA must be computed. (See Müller & Socher-Ambrosius [185] for clarifications regarding the need for "factoring" if terms are simplified via BA.) This approach allows the integration of convergent systems for relevant equational theories, when such are available, as Slagle [275] did in the context of resolution. A different completion-like procedure for first-order theorem proving, incorporating simplification, has been proved correct by Bachmair & Dershowitz [11]. A first-order method, using Boolean rings and based on polynomial ideals, is proposed in Kapur & Narendran [133].

8.7. *Further reading*

A survey of completion and its manifold applications may be found in Dershowitz [59]. For a book focusing on the abstract view of completion and its variants, see Bachmair [8]. The similarity between completion and algorithms for finding canonical bases of polynomial ideals was noted in Loos [170] and Buchberger [36]; see Kandri-Rody et al. [262]. REVE (Lescanne [168]), KB (Fages [73]), KADS (Stickel [223]), and RRL (Kapur & Zhang [140] are four current implementations of

AC-completion. Rewrite-based decision procedures for semigroups, monoids, and groups are investigated in Benninghofen et al. [23]; experiments with the completion of finitely-presented algebras are described in Lankford et al. [266] and Le Chenadec [166]; some new classes of decidable monadic word problems were found in Pedersen [195]. The use of completion and its relation to Dehn's method for deciding word problems and small cancellation theory is explored in Le Chenadec [167].

Early forerunners of ordered completion were Lankford [156] and Brown [35]. Ordered completion modulo a congruence has been implemented by Anantharaman & Hsiang [2]. One of the first implementations of inductive theorem proving by completion was by Huet & Hullot [116]. A recent book on using rewrite techniques for theorem proving in first-order predicate calculus with equality is Rusinowitch [214].

9. Extensions

In this section, we briefly consider four variations on the rewriting theme: "order-sorted rewriting", "conditional rewriting", "priority rewriting", and "graph rewriting".

9.1. Order-sorted rewriting

In ordered rewriting, replacement is constrained to make terms smaller in a given ordering; more syntactic means of limiting rewriting are obtained by taxonomies of term types. For example, some data may be of type Boolean, some may represent natural numbers, others, stacks of reals. The appropriate semantic notion in this case is the many-sorted (heterogeneous) algebra. Under reasonable assumptions, virtually everything we have said extends to the multisorted case. Sorted (i.e. typed) rewriting has been dealt with, for example, in Huet & Oppen [119] and Goguen & Meseguer [89]. Recently, issues that arise in higher-order rewriting, with some terms representing functions, have been attached (e.g., in Breazu–Tannen & Gallier [248] and Halpern et al. [258]).

Order-sorted algebras, introduced in Goguen [86] and developed in Gogolla [85], Cunningham & Dick [48], Goguen et al. [88], Goguen & Meseguer [91], Smolka et al. [219], and others, extend sorted algebras by imposing a partial ordering on the sorts, intended to capture the subset relation between them. For example, one can distinguish between stacks, in general, and nonempty stacks by declaring *NonEmptyStack* to be a subsort of *Stack* and specifying their operations to have the following types:

Λ :	$\rightarrow$	*Stack*
push: *Nat* $\times$ *Stack*	$\rightarrow$	*NonEmptyStack*
top : *NonEmptyStack*	$\rightarrow$	*Nat*
pop : *NonEmptyStack*	$\rightarrow$	*Stack*

The main advantage is that definitions can be sufficiently complete without introducing error elements for functions applied outside their intended domains (like *top*(Λ)).

Free algebras can be constructed in the order-sorted case (Goguen & Meseguer [91]), but the algebra obtained is, in general, an amalgamation of the term algebra. To avoid this complication, a syntactic "regularity" condition (namely, that every term has a least sort) may be imposed on the signature (Goguen & Meseguer [91]). Subsorts also require run-time checking for syntactic validity. For example, the term $pop(push(e, push(e, \Lambda)))$ has sort *Stack*, so its *top* cannot be taken, yet that term is equal to the nonempty stack $push(e, \Lambda)$. Fortunately, the run-time checks do not require additional overhead (Goguen et al. [88]).

Deduction in order-sorted algebras also presents some difficulties. For example, suppose that a sort S' contains two constants a and b, and S is its supersort $S' \cup \{c\}$. Let E consist of two identities: $a=c$ and $b=c$, and consider the equation $f(a)=f(b)$, where f is of type $S' \to S$. Clearly, the equation holds in all models, but this cannot be shown by replacement of equals, since $f(c)$ is not well-formed. Additional problems are caused when a sort is allowed to be empty, as pointed out in Huet & Oppen [119]. A sound and complete set of inference rules for order-sorted equality is given in Goguen & Meseguer [91].

For order-sorted rewriting, confluence does not imply that every theorem has a rewrite proof. Consider the same example as in the previous paragraph. The system $\{a \to c, b \to c\}$ is confluent, but there is no rewrite proof for $a=b$, since a term is not rewritten outside its sort. One way to preclude such anomalies is to insist that rules be "sort decreasing", i.e. that any instance of a right-hand side have a sort no larger than that of the corresponding left-hand side (Goguen et al. [88]). Order-sorted unification is investigated in Walther [237] and Meseguer et al. [183], and order-sorted completion in Gnaedig et al. [257].

9.2. *Conditional rewriting*

Another way to restrict applicability of equations is to add enabling conditions. A *conditional equation* is an equational implication $u_1=v_1 \wedge \cdots \wedge u_n=v_n \supset s=t$, for $n \geqslant 0$ (that is, a universal Horn clause with equality literals only). An example of a conditional equation with only one premiss is $empty?(x)=no \supset push(top(x), pop(x))=x$. Initial algebras exist for classes of algebras presented by conditional equations. A *conditional rule* is an equational implication in which the equation in the conclusion is oriented, for which we use the format $u_1=v_1 \wedge \cdots \wedge u_n=v_n | l \to r$. The following is an example of a system of both conditional and unconditional rules:

$$
\begin{array}{rcl}
top(push(x, y)) & \to & x \\
pop(push(x, y)) & \to & y \\
empty?(\Lambda) & \to & yes \\
empty?(push(x, y)) & \to & no \\
empty?(x)=no \,|\, push(top(x), pop(x)) & \to & x
\end{array}
$$

To give operational semantics to conditional systems, the conditions under which a rewrite may be performed need to be made precise. In the above example, is a term $push(top(s), pop(s))$ rewritten whenever the subterm s is such that the condition $empty?(s) = no$ can be proved, whenever a rewrite proof exists, or only when $empty?(s)$ rewrites to no? The ramifications of various choices are discussed in Brand et al. [34], Bergstra & Klop [24] and Dershowitz et al. [66]. The most popular convention is that each condition admit a rewrite proof; in other words, for a given system R, a rule $u_1 = v_1 \wedge \cdots \wedge u_n = v_n \mid l \rightarrow r$ is applied to a term $t[l\sigma]_p$ if $u_i\sigma \overset{*}{\underset{R}{\rightarrow}} \circ \overset{*}{\underset{R}{\leftarrow}} v_i\sigma$, for each condition $u_i = v_i$, in which case $t[l\sigma]_p \underset{R}{\rightarrow} t[r\sigma]_p$. Recent proposals for logic programming languages, incorporating equality, have been based on conditional rewriting and narrowing (e.g. Dershowitz & Plaisted [67] and Goguen & Meseguer [90]; see Reddy [208] and Padawitz [270]).

For the above recursive definition of $\underset{R}{\rightarrow}$ to yield a decidable relation, restrictions must be made on the rules. The most general well-behaved proposal is *decreasing* systems, terminating systems with conditions that are smaller (in a well-defined sense) than left-hand sides; hence, recursively evaluating the conditions always terminates. More precisely, a rule $u_1 = v_1 \wedge \cdots \wedge u_n = v_n \mid l \rightarrow r$ is decreasing if there exists a well-founded extension $\succ$ of the proper subterm ordering such that $\succ$ contains $\underset{R}{\rightarrow}$ and $l\sigma \succ u_i\sigma, v_i\sigma$ for $i = 1, \ldots, n$. Decreasing systems (Dershowitz et al. [66]) generalize the concept of "hierarchy" in the work of Rémy [209], and are slightly more general than the systems considered in Kaplan [130] and Jouannaud & Waldmann [128]; they have been extended in Dershowitz & Okada [64] and Bertling & Ganzinger [244] to cover systems (important in logic programming) with variables in conditions that do not also appear in the left-hand side, e.g. $g(x) = z \mid f(x) \rightarrow h(z)$.

The rewrite relation $\underset{R}{\rightarrow}$ for decreasing systems is terminating, and—when there are only a finite number of rules—is decidable. For confluence, a suitable notion of "conditional critical pair", which is just the conditional equation derived from overlapping left-hand sides, is defined. The Critical Pair Lemma holds for decreasing systems (Kaplan [130]): a decreasing system R is locally confluent (and hence convergent) iff there is a rewrite proof $s\sigma \overset{*}{\underset{R}{\rightarrow}} \circ \overset{*}{\underset{R}{\leftarrow}} t\sigma$ for each critical pair $u_1 = v_1 \wedge \cdots \wedge u_n = v_n \supset s = t$ and substitution σ such that $u_i\sigma \underset{R}{=} v_i\sigma$ for $i = 1, \ldots, n$. Nevertheless, confluence is only semidecidable, on account of the semidecidability of satisfiability of the conditions $u_i = v_i$. For nondecreasing systems, even ones for which the rewrite relation is terminating, the rewrite relation may be undecidable (Kaplan [130]), and the Critical Pair Lemma does not hold, though terminating systems having no critical pairs are confluent (see Dershowitz et al. [65]). To handle more general systems of conditional rules, rules must be overlapped on conditions, extending ordered completion to what has been called "oriented paramodulation". This is an active area of research (Ganzinger [82], Dershowitz [60], Kounalis & Rusinowitch [151]).

9.3. *Priority rewriting*

In priority rewriting, the choice among several possible redexes is constrained to meet, a priori, given priorities on the rules. Priorities, then, are just a partial ordering of rules. The original Markov algorithms were priority string rewriting systems, in which the written order of the rules determined their priority. The following definition of subtraction of natural numbers and divisibility of integers by naturals illustrates the conciseness made possible by using subsorts and priorities:

$Zero$	$=$	$\{0\}$
Pos	$=$	$succ(Zero) \cup succ(Pos)$
Nat	$=$	$Zero \cup Pos$
Neg	$=$	$neg(Pos)$
Int	$=$	$Nat \cup Neg$
$-: Nat \times Nat$	$\rightarrow$	Int
$\vert: Nat \times Int$	$\rightarrow$	$\{T, F, error\}$
$x - 0$	$\rightarrow$	x
$0 - y$	$\rightarrow$	$neg(y)$
$succ(x) - succ(y)$	$\rightarrow$	$x - y$
$0 \vert x$	$\rightarrow$	$error$
$z \vert n: Neg$	$\rightarrow$	F
$z \vert 0$	$\rightarrow$	T
$z \vert x$	$\rightarrow$	$z \vert (x - z)$

Note how priority systems (with a total priority ordering) can have no "critical" overlaps between two left-hand sides at the top. Priority term-rewriting systems were first formally studied in Baeten et al. [16]. Their definition is subtle, since an outermost redex is rewritten only if no possible derivation of its proper subterms enables a higher-priority rule at the outermost position. A condition can be given under which this definition of rewriting is computable. Priority systems cannot, in general, be expressed as term-rewriting systems.

9.4. *Graph rewriting*

Rewriting has also been generalized to apply to graphs, instead of terms. In fact, the idea of rewriting all kinds of objects is already in Thue [232]. In graph rewriting, subgraphs are replaced according to rules, containing variables, themselves referring to subgraphs. For example, a rule $f(g(x)) \rightarrow h(x)$, when applied to a directed acyclic graph $k(g(a), f(g(a)))$, where $g(a)$ is shared by k's two subterms, should rewrite the graph to $k(g(a), h(a))$, with the a still shared. Unlike trees, graphs do not have a simple structure lending itself to inductive definitions and proofs, for which reason the graph-rewriting definitions, as introduced in Ehrig [71] and simplified in Raoult [206], have a global flavor. A categorical framework is used to precisely define matching and replacement; a rewrite is then a pushout in a suitable category. Though the categorical apparatus leads to apparently complicated definitions, many proofs, notably the Critical Pair Lemma, become nothing more than commutativity of diagrams.

A completely different approach to graph rewriting is taken in Bauderon & Courcelle [19], where finite graphs are treated as algebraic expressions. Finitely-oriented labeled hypergraphs are considered as a set of hyperedges glued together by means of vertices. This generalizes the situation of words, with hyperedge labels as the constants, gluing for concatenation, and a set S of equational laws defining equivalence of expressions (instead of associativity). Class-rewriting in R/S is, then, tantamount to graph-rewriting with a system R.

9.5. *Further reading*

Polymorphic order-sorted Horn clauses are treated in depth in [277]. Research on conditional rewriting is summarized in [263]. A detailed survey of graph rewriting is [251].

Acknowledgment

We thank Leo Bachmair, Hubert Comon, Kokichi Futatsugi, Steve Garland, Rob Hasker, Jieh Hsiang, Gérard Huet, Stéphane Kaplan, Deepak Kapur, Claude Kirchner, Hélène Kirchner, Emmanuel Kounalis, Pierre Lescanne, Joseph Loyall, George McNulty, José Meseguer, Mike Mitchell, David Plaisted, Paul Purdom, and Hantao Zhang for their help with aspects of this survey. It goes without saying that our work in this field would not have been possible without the collaboration of many other colleagues and students over the years.

The first author's research was supported in part by the National Science Foundation under Grant DCR 85-13417, and by the Center for Advanced Study of the University of Illinois at Urbana–Champaign. The second author's research was supported in part by the Greco Programmation du Centre National de la Recherche Scientifique, the ESPRIT project METEOR, and a CNRS/NSF grant.

References

[1] ACKERMANN, W., *Solvable Cases of the Decision Problem* (North-Holland, Amsterdam, 1954).

[2] ANANTHARAMAN, S. and J. HSIANG, Identities in alternative rings, *J. Automat. Reasoning* **6** (1990) 79–109.

[3] ARDIS, M.A., Data abstraction transformations, Ph.D. thesis, Report TR-925, Dept. of Computer Science, Univ. of Maryland, August 1980.

[4] AVENHAUS, J., On the termination of the Knuth–Bendix completion algorithm, Report 120/84, Univ. Kaiserslautern, Kaiserslautern, West Germany, 1985.

[5] BAADER, F., Unification in idempotent semigroups is of type zero, *J. Automat. Reasoning* **2** (3) (1986).

[6] BACHMAIR, L., Proof by consistency in equational theories, in: *Proc. Third IEEE Symposium on Logic in Computer Science*, Edinburgh, Scotland (1988) 228–233.

[7] BACHMAIR, L., Proof normalization for resolution and paramodulation, in: *Proc. Third International Conference on Rewriting Techniques and Applications*, Chapel Hill, NC, Lecture Notes in Computer Science **355** (Springer, Berlin, 1989).

[8] BACHMAIR, L., *Canonical Equational Proofs* (Pitman–Wiley, London, 1990).

[9] BACHMAIR, L. and N. DERSHOWITZ, Commutation, transformation, and termination, in: *Proc. Eighth International Conference on Automated Deduction*, Oxford, England, Lecture Notes in Computer Science **230** (Springer, Berlin, 1986) 5–20.

[10] BACHMAIR, L. and N. DERSHOWITZ, Completion for rewriting modulo a congruence, *Theoret. Comput. Sci.* **67**(2 & 3) (1989) 173–201.

[11] BACHMAIR, L. and N. DERSHOWITZ, Inference rules for rewrite-based first-order theorem proving, in: *Proc. Second IEEE Symposium on Logic in Computer Science*, Ithaca, NY (1987) 331–337.

[12] BACHMAIR, L. and N. DERSHOWITZ, Critical pair criteria for completion, *J. Symbolic Comput.* **6**(1) (1988) 1–18.

[13] BACHMAIR, L., N. DERSHOWITZ and J. HSIANG, Orderings for equational proofs, in: *Proc. IEEE Symposium on Logic in Computer Science*, Cambridge, MA (1986) 346–357.

[14] BACHMAIR, L., N. DERSHOWITZ and D.A. PLAISTED, Completion without failure, in: H. Ait-Kaci and M. Nivat, eds., *Resolution of Equations in Algebraic Structures, Vol. II: Rewriting Techniques* (Academic Press, New York, 1989) 1–30.

[15] BACHMAIR, L. and D.A. PLAISTED, Associative path ordering, *J. Symbolic Comput.* **1** (4) (1985) 329–349.

[16] BAETEN, J.C.M., J.A. BERGSTRA and J.W. KLOP, Priority rewrite systems, in: *Proc. Second International Conference on Rewriting Techniques and Applications*, Bordeaux, France, Lecture Notes in Computer Science **256** (Springer, Berlin, 1987) 83–94.

[17] BARENDREGT, H.P., *The Lambda Calculus, its Syntax and Semantics* (North-Holland, Amsterdam, 2nd ed., 1984).

[18] BARENDREGT, H.P., Functional programming and lambda calculus, in: J. van Leeuwen, ed., *Handbook of Theoretical Computer Science, Vol. B* (North-Holland, Amsterdam, 1990) 321–363.

[19] BAUDERON, M., and B. COURCELLE, Graph expressions and graph rewritings, *Math. Systems Theory* **20** (1987) 83–127.

[20] BAXTER, L.D., A practically linear unification algorithm, Report CS-76-13, Univ. of Waterloo, Waterloo, Canada, 1976.

[21] BELLEGARDE, F. and P. LESCANNE, Transformation orderings, in: *Proc. Twelfth Colloquium on Trees in Algebra and Programming*, Pisa, Italy, Lecture Notes in Computer Science **249** (Springer, Berlin, 1987) 69–80.

[22] BEN, CHERIFA A. and P. LESCANNE, Termination of rewriting systems by polynomial interpretations and its implementation, *Sci. Comput. Programming* **9**(2) (1987) 137–159.

[23] BENNINGHOFEN, B., S. KEMMERICH and M.M. RICHTER, *Systems of Reductions*, Lecture Notes in Computer Science **277** (Springer, Berlin, 1987).

[24] BERGSTRA, J.A. and J.W. KLOP, Conditional rewrite rules: Confluence and termination, *J. Comput. System Sci.* **32** (1986) 323–362.

[25] BIRKHOFF, G., On the structure of abstract algebras, *Proc. Cambridge Philos. Soc.* **31** (1935) 433–454.

[26] BIRKHOFF, G., *Lattice Theory* (American Mathematical Society, New York, 3rd ed., 1967).

[27] BLEDSOE, W., Non-resolution theorem proving, *Artificial Intelligence* **9** (1977) 1–35.

[28] BOCKMAYR, A., A note on a canonical theory with undecidable unification and matching problem, *J. Automat. Reasoning* **3** (1987) 379–381.

[29] BOOK, R.V., Thue systems as rewriting systems, *J. Symbolic Comput.* **3** (1987) 39–68.

[30] BOOK, R.V., M. JANTZEN, and C. WRATHALL, Monadic Thue systems, *Theoret. Comput. Sci.* **19**(3) (1982) 321–251.

[31] BLOOM, S.L. and R. TINDELL, Varieties of "if-then-else", *SIAM J. Comput.* **12**(4) (1983) 677–707.

[32] BOUDET, A., E. CONTEJEAN and H. DEVIE, A new AC unification algorithm with a new algorithm for solving Diophantine equations, Tech. Report, Univ. de Paris-Sud, Orsay, France, 1990.

[33] BOUDET, A., J.-P. JOUANNAUD, and M. SCHMIDT-SCHAUß, Unification of Boolean rings and Abelian groups, *J. Symbolic Comput.* **8** (1989) 449–477.

[34] BRAND, D., J.A. DARRINGER and W.J. JOYNER, JR., Completeness of conditional reductions, in: *Proc. Fourth Workshop on Automated Deduction*, Austin, TX (1979).

[35] BROWN, T.C., JR., A structured design-method for specialized proof procedures, Ph.D. Thesis, California Institute of Technology, Pasadena, CA, 1975.

[36] BUCHBERGER, B., History and basic features of the critical-pair/completion procedure, *J. Symbolic Comput.* **3**(1&2) (1987) 3–38.

[37] BUCHBERGER, B., and R. LOOS, Algebraic simplification, in: *Computer Algebra* (Springer, Berlin, 1982) 11–43.

[38] BÜRCKERT, H.-J. J., A. HEROLD and M. SCHMIDT-SCHAUß, On equational theories, unification and decidability, in: *Proc. Second International Conference on Rewriting Techniques and Applications*, Bordeaux, France, Lecture Notes in Computer Science **256** (Springer, Berlin, 1987) 204–215.

[39] BUTLER, G. and D.S. LANKFORD, Experiments with computer implementations of procedures which often derive decision algorithms for the word problem in abstract algebras, Memo MTP-7, Dept. of Mathematics, Louisiana Tech. Univ., Ruston, LA, 1980.

[40] CHOPPY, C., S. KAPLAN and M. SORIA, Algorithmic complexity of term rewriting systems, in: *Proc. Second International Conference on Rewriting Techniques and Applications*, Bordeaux, France, Lecture Notes in Computer Science **256** (Springer, Berlin, 1987) 256–285.

[41] CHURCH, A. and J.B. ROSSER, Some properties of conversion, *Trans. Amer. Math. Soc.* **39** (1936) 472–482.

[42] COHN, P.M., *Universal Algebra* (Reidel, Dordrecht, Holland, 2nd ed., 1981).

[43] COLLINS, G., Quantifier elimination for real closed fields by cylindrical algebraic decomposition, in: *Proc. Second GI Conference on Automata Theory and Formal Languages*, Lecture Notes in Computer Science **33** (Springer, Berlin, 1975) 134–183.

[44] COLMERAUER, A., Equations and inequations on finite and infinite trees, in: *Proc. Second International Conference on Fifth Generation Computer Systems*, Tokyo, Japan (1984) 85–99.

[45] COMON, H., Unification et disunification: théorie et applications, Thèse de Doctorat, Institut Polytechnique de Grenoble, Grenoble, France, 1988.

[46] COMON, H. and P. LESCANNE, Equational problems and disunification, *J. Symbolic Comput.* **7** (1989) 371–425.

[47] CURRY, H.B. and R. FEYS, *Combinatory Logic* (North-Holland, Amsterdam, 1958).

[48] CUNNINGHAM, R.J. and A.J.J. DICK, Rewrite systems on a lattice of types, *Acta Informat.* **22** (1985) 149–169.

[49] DAUCHET, M., Simulation of Turing machines by a left-linear rewrite rule, in: *Proc. Third International Conference on Rewriting Techniques and Applications*, Chapel Hill, NC, Lecture Notes in Computer Science **355** (Springer, Berlin, 1989) 109–120.

[50] DAUCHERT, M., T. HEUILLARD, P. LESCANNE and S. TISON, Decidability of the confluence of finite ground term rewriting systems and of other related term rewriting systems, *Inform. and Comput.*, to appear.

[51] DAVIS, M. and R. HERSH, Hilbert's 10th problem, *Sci. Amer.* **229**(5) (1973) 84–91.

[52] DEHN, M., Uber unendliche diskontinuierliche Gruppen, *Math. Ann.* **71** (1911) 116–144.

[53] DERSHOWITZ, N., A note on simplification orderings, *Inform. Process. Lett.* **9**(5) (1979) 212–215.

[54] DERSHOWITZ, N., Orderings for term-rewriting systems, *Theoret. Comput. Sci.* **17**(3) (1982) 279–301.

[55] DERSHOWITZ, N., Applications of the Knuth–Bendix completion procedure, in: *Proc. Seminaire d'Informatizue Theorique*, Paris, France (1982) 95–111.

[56] DERSHOWITZ, N., Equations as programming language, in: *Proc. Fourth Jerusalem Conference on Information Technology* (IEEE Computer Society, Silver Spring, MD, 1984) 114–124.

[57] DERSHOWITZ, N. Computing with rewrite systems, *Inform. and Control* **64**(2/3) (1985) 122–157.

[58] DERSHOWITZ, N., Termination of rewriting, *J. Symbolic Comput.* **3**(1&2) (1987) 69–115, Corrigendum: **4**(3) (1987) 409–410.

[59] DERSHOWITZ, N., Completion and its applications, in: H. Ait-Kaci and M. Nivar, eds., *Resolution of Equations in Algebraic Structures, Vol. II: Rewriting Techniques* (Academic Press, New York, 1989) 31–86.

[60] DERSHOWITZ, N., A maximal-literal unit strategy for Horn clauses, in: *Proc. Second International Workshop on Conditional and Typed Rewriting Systems*, Montreal, Canada, Lecture Notes in Computer Science (Springer, Berlin, to appear).

[61] DERSHOWITZ, N. and Z. MANNA. Proving termination with multiset orderings, *Comm. AM* **22**(8) (1979) 465–476.

[62] DERSHOWITZ, N., L., MARCUS and A. TARLECKI, Existence, uniqueness, and construction of rewrite systems, *SIAM J. Comput.* **17**(4) (1988) 629–639.

[63] DERSHOWITZ, N. and M. OKADA, Proof-theoretic techniques and the theory of rewriting, in: *Proc. Third IEEE Symposium on Logic in Computer Science*, Edinburgh, Scotland (1988) 104–111.

[64] DERSHOWITZ, N. and M. OKADA, A rationale for conditional equational rewriting, *Theoret. Comput. Sci.* **75**(1/2) (1990) 111–137.

[65] DERSHOWITZ, N., M. OKADA and G. SIVAKUMAR, Confluence of conditional rewrite systems, in: *Proc. First International Workshop on Conditional Term Rewriting Systems*, Orsay, France, Lecture Notes in Computer Science **308** (Springer, Berlin, 1988) 31–44.

[66] DERSHOWITZ, N., M. OKADA and G. SIVAKUMAR, Canonical conditional rewrite systems, in: *Proc. Ninth Conference on Automated Deduction*, Argonne, IL, Lecture Notes in Computer Science **310** (Springer, Berlin, 1988) 538–549.

[67] DERSHOWITZ, N. and D.A. PLAISTED, Logic programming *cum* applicative programming, in: *Proc. IEEE Symposium on Logic Programming*, Boston, MA (1985) 54–66.

[68] DERSHOWITZ, N. and G. SIVAKUMAR, Solving goals in equational languages, in: *Proc. First International Workshop on Conditional Term Rewriting Systems*, Orsay, France, Lecture Notes in Computer Science **308** (Springer, Berlin, 1988) 45–55.

[69] DOWNEY, P.J., R. SETHI and R.E. TARJAN, Variations on the common subexpressions problem, *J. Assoc. Computing. Mach.* **27**(4) (1980) 785–771.

[70] EHRENFEUCHT, A., D. HAUSSLER and G. ROZENBERG, On regularity of context-free languages, *Theoret. Comput. Sci.* **27**(3), 311–332.

[71] EHRIG, H., Introduction to the algebraic theory of graph grammars, in: *Proc. International Conference on the Fundamentals of Complexity Theory*, Poznán-Kórnik, Poland, Lecture Notes in Computer Science **56** (Springer, Berlin, 1977) 245–255.

[72] EVANS, T., On multiplicative systems defined by generators and relations, I, *Proc. Cambridge Philos. Soc.* **47** (1951) 637–649.

[73] FAGES, F., Le système KB: manuel de référence: présentation et bibliographie, mise en oeuvre, Report R.G. 10.84, Greco de Programmation, Bordeaux, France, 1984.

[74] FAGES, F., Associative-commutative unification, *J. Symbolic Comput.* **3**(3) (1987) 257–275.

[75] FAGES, F. and G. HUET, Unification and matching in equational theories, in: *Proc. Eighth Colloquium on Trees in Algebra and Programming*, l'Aquilla, Italy, Lecture Notes in Computer Science **159** (Springer, Berlin, 1983) 205–220.

[76] FAY, M., First-order unification in an equational theory, in: *Proc. Fourth Workshop on Automated Deduction*, Austin, TX (1979) 161–167.

[77] FEFERMAN, S., Systems of predicative analysis II: Representation of ordinals, *J. Symbolic Logic* **33**(2) (1968) 193–220.

[78] FREESE, R., R.N. MCKENZIE, G.F. MCNULTY and W. TAYLOR, *Algebras, Lattices, Varieties, Vols. 2–4* (Wadsworth, Monterey, CA, to appear).

[79] FRIBOURG, L., A strong restricton of the inductive completion procedure, *J. Symbolic Comput.* **8**(3) (1989) 253–276.

[80] FUTATSUGI, K., J.A. GOGUEN, J.-P. JOUANNAUD and J. MESEGUER, Principles of OBJ2, in: *Conference Record Twelfth ACM Symposium on Principles of Programming Languages*, New Orleans, LA (1985) 52–66.

[81] GALLIER, J.H. and W. SNYDER, A general complete E-unification procedure, in: *Proc. Second International Conference on Rewriting Techniques and Applications*, Bordeaux, France, Lecture Notes in Computer Science **256** (Springer, Berlin, 1987) 192–203.

[82] GANZINGER, H., A completion procedure for conditional equations, in: *Proc. First International Workshop on Conditional Term Rewriting Systems*, Orsay, France, Lecture Notes in Computer Science **308** (Springer, Berlin, 1988) 62–83.

[83] GARDNER, M., Mathematical games: Tasks you cannot help finishing no matter how hard you try to block finishing them, *Sci. Amer.* **24**(2) (1983) 12–21.

[84] GÖBEL, R., Ground confluence, in: *Proc. Second International Conference on Rewriting Techniques and Applications*, Bordeaux, France, Lecture Notes in Computer Science **256** (Springer, Berlin, 1987) 156–167.

[85] GOGOLLA, M., Algebraic specifications with partially ordered sets and declarations, *Forschungsbericht Informatik 169*, Univ. Dortmund, Dortmund, West Germany, 1983.

[86] GOGUEN, J.A., Exception and error sorts, coercion and overloading operators, Research Report, Stanford Research Institute, 1978.

[87] GOGUEN, J.A., How to prove algebraic inductive hypotheses without induction, with applications to the correctness of data type implementations, in: *Proc. Fifth International Conference on Automated Deduction*, Les Arcs, France, Lecture Notes in Computer Science **87** (Springer, Berlin, 1980) 356–373.

[88] GOGUEN, J.A., J.-P. JOUANNAUD and J. MESEGUER, Operational semantics of order-sorted algebra, in: *Proc. Twelfth International EATCS Colloquium on Automata, Languages and Programming*, Nafplion, Greece. Lecture Notes in Computer Science **194** (Springer, Berlin, 1985) 221–231.

[89] GOGUEN, J.A. and J. MESEGUER, Completeness of many sorted equational deduction, *Houston J. Math.* **11**(3) (1985) 307–334.

[90] GOGUEN, J.A. and J. MESEGUER, EQLOG: Equality, types, and generic modules for logic programming, in: D. De Groot and G. Lindstrom, eds., *Logic Programming: Functions, Relations, and Equations* (Prentince-Hall, Englewood Cliffs, NJ, 1986) 295–363.

[91] GOGUEN, J.A. and J. MESEGUER, Order-sorted algebra I: Partial and overloaded operators, errors and inheritance, Unpublished Report, SRI International, Menlo Park, CA, 1987.

[92] GOGUEN, J.A. and J.J. TARDO, An introduction to OBJ: A language for writing and testing formal algebraic specifications, in: *Proc. Specification of Reliable Software Conference* (1979) 170–189.

[93] GOLDFARB, W.D., Note on the undecidability of the second-order unification problem, *Theoret. Comput. Sci.* **13**(2) (1981) 225–230.

[94] GORN, S., Explicit definitions and linguistic dominoes, in: J. Hart and S. Takasu, eds., *Systems and Computer Science* (Univ. of Toronto Press, Toronto, 1967) 77–115.

[95] GRIES, D., *The Science of Programming* (Springer, New York, 1981).

[96] GUESSARIAN, I. and J. MESEGUER, On the axiomatization of "if-then-else", *SIAM J. Comput.* **16**(2) (1987) 332–357.

[97] GUTTAG, J.V., Abstract data types and the development of data structures, *Comm. ACM* **20**(6) (1977) 396–404.

[98] HEILBRUNNER, S. and S. HÖLLDOBLER, The undecidability of the unification and matching problem for canonical theories, *Acta Informat.* **24**(2) (1987) 157–171.

[99] HENKIN, L., The logic of equality, *Amer. Math. Monthly* **84**(8) (1977) 597–612.

[100] HERBRAND, J., Recherches sur la théorie de la démonstration, Thèse de Doctorat, Univ. de Paris, Paris, France, 1930.

[101] HEROLD, A. and J. SIEKMANN, Unification in Abelian semi-groups, *J. Automat. Reasoning* **3**(3) (1987) 247–283.

[102] HIGMAN, G., Ordering by divisibility in abstract algebras, *Proc. London Math. Soc. (3)* **2**(7) (1952) 326–336.

[103] HIGMAN, G. and B.H. NEUMANN, Groups as groupoids with one law, *Publ. Math. Debrecen* **2** (1952) 215–221.

[104] HINDLEY, J.R., The Church–Rosser property and a result in combinatory logic, Ph.D. Thesis, Univ. of Newcastle-upon-Tyne, 1964.

[105] HOFFMANN, C.M. and M.J. O'DONNELL, Interpreter generation using tree pattern matching, in: *Conf. Rec. Sixth ACM Symposium on Principles of Programming Languages*, San Antonio, TX (1979) 169–179.

[106] HSIANG, J., Topics in automated theorem proving and program generation, Ph.D. Thesis, Report R-82-1113, Dept. of Computer Science, Univ. of Illinois, Urbana, IL, 1982.

[107] HSIANG, J., Refutational theorem proving using term-rewriting systems, *Artificial Intelligence* **25** (1985) 255–300.

[108] HSIANG, J. and N. DERSHOWITZ, Rewrite methods for clausal and non-clausal theorem proving, in: *Proc. Tenth International Colloquium on Automata, Languages and Programming*, Barcelona, Spain, Lecture Notes in Computer Science **154** (Springer, Berlin, 1983) 331–346.

[109] HSIANG, J. and J.-P. JOUANNAUD, Complete sets of inference rules for E-unification, in: *Proc. Second Unification Workshop*, Val d'Ajol, France (1988); available as: CRIN Report, Univ. de Nancy, France.

[110] HSIANG, J. and M. RUSINOWITCH, A new method for establishing refutational completeness in theorem proving, in: *Proc. Eight International Conference on Automated Deduction*, Oxford, England, Lecture Notes in Computer Science **230** (Springer, Berlin, 1986) 141–152.

[111] HSIANG, J. and M. RUSINOWITCH, On word problems in equational theories, in: *Proc. Fourteenth International Colloquium on Automata, Languages and Programming*, Karlsruhe, West Germany, Lecture Notes in Computer Science **267** (Springer, Berlin, 1987) 54–71.

[112] HUET, G., Résolution d'équations dans les langages d'ordre $1, 2, \ldots \omega$, Thèse de Doctorat, Univ. de Paris VII, Paris, France, 1976.

[113] HUET, G., Confluent reductions: Abstract properties and applications to term rewriting systems, *J. Assoc. Comput. Mach.* **27**(4) (1980) 797–821.

[114] HUET, G., A complete proof of correctness of the Knuth–Bendix completion algorithm, *J. Comput. System Sci.* **23**(1) (1981) 11–21.

[115] HUET, G., Cartesian closed categories and lambda-calculus, in: *Proc. Spring School on Combinators and Functional Programming Languages*, Val d'Ajol, France, Lecture Notes in Computer Science (Springer, Berlin) 123–135.

[116] HUET, G. and J.-M. HULLOT, Proofs by induction in equational theories with constructors, in: *Proc. Twenty-First IEEE Symposium on Foundations of Computer Science*, Lake Placid, NY (1980) 96–107.

[117] HUET, G. and D.S. LANKFORD, On the uniform halting problem for term rewriting systems, Rapport Laboria 283, Institut de Recherche d'Informatique et d'Automatique, Le Chesnay, France, 1978.

[118] HUET, G. and J.-J. LÉVY, Call by need computations in non-ambiguous linear term rewriting systems, in: J. Lassez and G. Plotkin, eds., *Computational Logic: Essays in Honour of Alan Robinson* (MIT Press, Cambridge, MA, to appear).

[119] HUET, G. and D.C. OPPEN, Equations and rewrite rules: A survey, in: R. Book, ed., *Formal Language Theory: Perspectives and Open Problems* (Academic Press, New York, 1980) 349–405.

[120] HULLOT, J.-M., Canonical forms and unification, in: *Proc. Fifth International Conference on Automated Deduction*, Les Arcs, France, Lecture Notes in Computer Science **87** (Springer, Berlin, 1980) 318–334.

[121] ITURRIAGA, R., Contributions to mechanical mathematics, Ph.D. Thesis, Dept. of Computer Science, Carnegie-Mellon Univ., Pittsburgh, PA, 1967.

[122] JOUANNAUD, J.-P., Confluent and coherent sets of reductions with equations: Application to proofs in abstract data types, in: *Proc. Eighth Colloquium on Trees in Algebra and Programming*, Lecture Notes in Computer Science **59** (Springer, Berlin, 1983) 269–283.

[123] JOUANNAUD, J.-P., C. KIRCHNER and H. KIRCHNER, Incremental construction of unification algorithms in equational theories, in: *Proc. Tenth International Colloquium on Automata, Languages and Programming*, Barcelona, Spain, Lecture Notes in Computer Science **154** (Springer, Berlin, 1983) 361–373.

[124] JOUANNAUD, J.-P. and H. KIRCHNER, Completion of a set of rules modulo a set of equations, *SIAM J. Comput.* **15** (1986) 1155–1194.

[125] JOUANNAUD, J.-P. and E. KOUNALIS, Automatic proofs by induction in equational theories without constructors, *Inform. and Comput.* **82**(1) (1989) 1–330.

[126] JOUANNAUD, J.-P. and P. LESCANNE, Rewriting systems, *Tech. Sci. Inform.* **6**(3) (1987) 181–199; translation from French "La réécriture", *Tech. Sci. Inform.* **5**(6) (1986) 433–452.

[127] JOUANNAUD, J.-P. and M. MUÑOZ, Termination of a set of rules modulo a set of equations, in: *Proc. Seventh International Conference on Automated Deduction*, Napa, CA, Lecture Notes in Computer Science **170** (Springer, Berlin, 1984) 175–193.

[128] JOUANNAUD, J.-P. and B. WALDMANN, Reductive conditional term rewriting systems, in: *Proc. Third IFIP Working Conference on Formal Description of Programming Concepts*, Ebberup, Denmark (1986).

[129] KAMIN, S. and J.-J. LÉVY, Two generalizations of the recursive path ordering, Unpublished note, Dept. of Computer Science, Univ. of Illinois, Urbana, IL, 1980.

[130] KAPLAN, S., Simplifying conditional term rewriting systems: Unification, termination and confluence, *J. Symbolic Comput.* **4**(3) (1987) 295–334.

[131] KAPUR, D. and D.R. MUSSER, Proof by consistency, *Artificial Intelligence* **31**(2) (1987) 125–157.

[132] KAPUR, D., D.R. MUSSER and P. NARENDRAN, Only prime superpositions need be considered for the Knuth–Bendix procedure, *J. Symbolic Comput.* **4** (1988) 19–36.

[133] KAPUR, D. and P. NARENDRAN, An equational approach to theorem proving in first-order predicate calculus, in: *Proc. Ninth International Joint Conference on Artificial Intelligence*, Los Angeles, CA (1985) 1146–1153.

[134] KAPUR, D. and P. NARENDRAN, NP-completeness of set unification and matching problems, in: *Proc. Eighth International Conference on Automated Deduction*, Oxford, England, Lecture Notes in Computer Science **230** (Springer, Berlin, 1986) 489–495.

[135] KAPUR, D. and P. NARENDRAN, Matching, unification and complexity (A preliminary note), *SIGSAM Bull.* **21**(4) (1987) 6–9.

[136] KAPUR, D., P. NARENDRAN and H. ZHANG, Proof by induction using test sets, in: *Proc. Eighth International Conference on Automated Deduction*, Oxford, England, Lecture Notes in Computer Science **230** (Springer-Verlag, Berlin, 1986) 99–117.

[137] KAPUR, D., P. NARENDRAN and F. OTTO, On ground confluence of term rewriting systems, *Inform. Comput.* **86**(1) (1989) 14–31.

[138] KAPUR, D., P. NARENDRAN, D.J. ROSENKRANTZ and H. ZHANG, Sufficient-completeness, quasi-reducibility and their complexity, Technical Report, Dept. of Computer Science, State Univ. of New York, Albany, NY, 1987.

[139] KAPUR, D., P. NARENDRAN and H. ZHANG, On sufficient completeness and related properties of term rewriting systems, *Acta Informat.* **24**(4) (1987) 395–415.

[140] KAPUR, D. and H. ZHANG, An overview of Rewrite Rule Laboratory (RRL), in: *Proc. Third International Conference on Rewriting Techniques and Applications*, Chapel Hill, NC, Lecture Notes in Computer Science **355** (Springer, Berlin, 1989) 559–563.

[141] KIRBY, L. and J. PARIS, Accessible independence results for Peano arithmetic, *Bull. London Math. Soc.* **14** (1982) 285–293.

[142] KIRCHNER, C., Méthodes et outils de conception systématique d'algorithmes d'unification dans les théories équationnelles, Thèse d'Etat, Univ. de Nancy, Nancy, France, 1985.

[143] KIRCHNER, C., Computing unification algorithms, in: *Proc. First IEEE Symposium on Logic in Computer Science*, Cambridge, MA. (1986) 206–216.

[144] KIRCHNER, C., Order-sorted equational unification, Report 954, Institut National de Recherche d'Informatique et d'Automatique, Le Chesnay, France, 1988.

[145] KIRCHNER, C., From unification in combination of equational theories to a new AC-unification algorithm, in: H. Ait-Kaci and M. Nivat, eds., *Resolution of Equations in Algebraic Structures, Vol. II: Rewriting Techniques* (Academic Press, New York, 1989) 171–210.

[146] KLOP, J.-W., *Combinatory Reduction Systems*, Mathematical Centre Tracts **127** (Mathematisch Centrum, Amsterdam, 1980).

[147] KLOP, J.-W., Term rewriting systems: A tutorial, *Bull. European Assoc. Theoret. Comput. Sci.* **32** (1987) 143–183.

[148] KNIGHT, K., Unification: A multidisciplinary survey, *Comput. Surveys* **21**(1) (1989) 93–124.

[149] KNUTH, D.E. and P.B. BENDIX, Simple word problems in universal algebras, in: J. Leech, ed., *Computational Problems in Abstract Algebra* (Pergamon Press, Oxford, 1970) 263–297; reprinted in: *Automation of Reasoning 2* (Springer, Berlin, 1983) 342–376.

[150] KOUNALIS, E., Completeness in data type specifications, in: *Proc. EUROCAL Conference*, Linz, Austria (1985).

[151] KOUNALIS, E. and M. RUSINOWITCH, On word problems in Horn theories, in: *Proc. Ninth International Conference on Automated Deduction*, Argonne, IL, Lecture Notes in Computer Science **310** (Springer-Verlag, Berlin, 1988) 527–537.

[152] KRISHNAMOORTHY, M.S. and P. NARENDRAN, A note on recursive path ordering, Unpublished Note, General Electric Corporate Research and Development, Schenectady, NY, 1984.

[153] KRUSKAL, J.B., Well-quasi-ordering, the Tree Theorem, and Vazsonyi's conjecture, *Trans. Amer. Math. Society* **95** (1960) 210 – 225.

[154] KRUSKAL, J.B., The theory of well-quasi-ordering: A frequently discovered concept, *J. Combin. Theory Ser. A* **13**(3) (1972) 297–305.

[155] KÜCHLIN, W., Inductive completion by ground proof transformation, in: H. Ait-Kaci and M. Nivat, eds., *Resolution of Equations in Algebraic Structures, Vol. II: Rewriting Techniques* (Academic Press, New York, 1989) 211–244.

[156] LANKFORD, D.S., Canonical inference, Memo ATP-32, Automatic Theorem Proving Project, Univ. of Texas, Austin, TX, 1975.

[157] LANKFORD, D.S., Canonical algebraic simplification in computational logic, Memo ATP-25, Automatic Theorem Proving Project, Univ. of Texas, Austin, TX, 1975.

[158] LANKFORD, D.S., On proving term rewriting systems are Noetherian, Memo MTP-3, Dept. of Mathematics, Louisiana Tech. Univ., Ruston, LA, 1979.

[159] LANKFORD, D.S., A simple explanation of inductionless induction, Memo MTP-14, Dept. of Mathematics, Louisiana Tech. Univ., Ruston, LA, 1981.

[160] LANKFORD, D.S. and A.M. BALLANTYNE, Decision procedures for simple equational theories with permutative axioms: complete sets of permutative reductions, ATP-37, Dept. of Mathematics and Computer Sciences, Univ. of Texas, Austin, TX, 1977.

[161] LANKFORD, D.S. and A.M. BALLANTYNE, Decision procedures for simple equational theories with commutative-associative axioms: Complete sets of commutative-associative reductions, Memo ATP-39, Dept. of Mathematics and Computer Sciences, Univ. of Texas, Austin, TX, 1977.

[162] LANKFORD, D.S. and A.M. BALLANTYNE, The refutation completeness of blocked permutative narrowing and resolution, in: *Proc. Fourth Workshop on Automated Deduction*, Austin, TX (1979) 53–59.

[163] LANKFORD, D.S. and A.M. BALLANTYNE, On the uniqueness of term rewriting systems, Unpublished Note, Dept. of Mathematics, Louisiana Tech. Univ., Ruston, LA, 1983.

[164] LASSEZ, J.-L., M.J. MAHER and K. MARRIOTT, Unification revisited, in: J. Minker, ed., *Foundations of Deductive Databases and Logic Programming* (Morgan Kaufmann, Los Altos, CA, 1988) 587–625.

[165] LAUTEMANN, C., A note on polynomial interpretation, *Bull. European Assoc. Theoret. Comput. Sci.* **36** (1988) 129–131.

[166] LE CHENADEC, P., *Canonical Forms in Finitely Presented Algebras* (Pitman–Wiley, London, 1985).

[167] LE CHENADEC, P., Analysis of Dehn's algorithm by critical pairs, *Theoret. Comput. Sci.* **51**(1, 2) (1987) 27–52.

[168] LESCANNE, P., Computer experiments with the REVE term rewriting system generator, in: *Proc. Tenth ACM Symposium on Principles of Programming Languages*, Austin, TX (1983) 99–108.

[169] LESCANNE, P., Uniform termination of term-rewriting systems with status, in: *Proc. Ninth Colloquium on Trees in Algebra and Programming* (Cambridge University Press, Cambridge, 1984).

[170] LOOS, R., Term reduction systems and algebraic algorithms, in: *Proc. Fifth GI Workshop on Artificial Intelligence*, Bad Honnef, West Germany, Informatik Fachberichte (Springer, Berlin, 1981) 214–234.

[171] MAHER, M.J., Complete axiomatizations of the algebras of the finite, rational and infinite trees, in: *Proc. Third IEEE Symposium on Logic in Computer Science*, Edinburgh, Scotland (1988) 348–357.

[172] MAKANIN, J., The problem of solvability of equations in a free semi-group, *Akad. Nauk SSSR* **233**(2) (1977); also: *Math. USSR Sb.* **32**(2) (1977).

[173] MANNA, Z. and S. NESS, On the termination of Markov algorithms, in: *Proc. Third Hawaii International Conference on System Science*, Honolulu, HI (1970) 789–792.

[174] MARKOV, A.A., The impossibility of some algorithms in the theory of associative systems, *Dokl. Akad. Nauk SSSR* **55**(7) (1947) 587–590 (in Russian).

[175] MARTELLI, A. and U. MONTANARI, An efficient unification algorithm, *ACM Trans. Programm. Languages and Systems* **4**(2) (1982) 258–282.

[176] MARTELLI, A., G.F. ROSSI and C. MOISO, Lazy unification algorithms for canonical rewrite systems, in: H. Ait-Kaci and M. Nivat, eds., *Resolution of Equations in Algebraic Structures, Vol II: Rewriting Techniques* (Academic Press, New York, 1989) 245–274.

[177] MARTIN, U., How to choose the weights in the Knuth–Bendix ordering, in: *Proc. Second International Conference on Rewriting Techniques and Applications*, Bordeaux, France, Lecture Notes in Computer Science **256** (Springer, Berlin, 1987) 42–53.

[178] MATIJASEVIC, J.V., Simple examples of undecidable associative calculi, *Soviet Math. (Dokl.)* **8**(2) (1967) 555–557.

[179] MATIJASEVIC, J.V., Enumerable sets are Diophantine, *Soviet Math. (Dokl.)* **11** (1970) 354–357.

[180] MCKENZIE, R.N., G.F. MCNULTY and W. TAYLOR, *Algebras, Lattices, Varieties, Vol. 1* (Wadsworth, Monterey, CA, 1987).

[181] MCNULTY, G.F., An equational logic sampler, in: *Proc. Third International Conference on Rewriting Techniques and Applications*, Chapel Hill, NC, Lecture Notes in Computer Science **355** (Springer, Berlin, 1989) 234–262.

[182] MESEGUER, J. and J.A. GOGUEN, Initiality, induction and computability, in: M. Nivat and J. Reynolds, eds., *Algebraic Methods in Semantics* (Cambridge University Press, Cambridge, 1985) 459–540.

[183] MESEGUER, J., J.A. GOGUEN and G. SMOLKA, Order-sorted unification, *J. Symbolic Comput.* 8(4) (1989) 383–414.

[184] METIVIER, Y., About the rewriting systems produced by the Knuth–Bendix completion algorithm, *Inform. Process. Lett.* **16**(1) (1983) 31–34.

[185] MÜLLER, J. and R. SOCHER-AMBROSIUS, Topics in completion theorem proving, SEKI-Report SR-88-13, Fachbereich Informatik, Univ. Kaiserslautern, Kaiserslautern, West Germany, 1988.

[186] MUSSER, D.R., On proving inductive properties of abstract data types, in: *Proc. Seventh ACM Symposium on Principles of Programming Languages*, Las Vegas, NV (1980) 154–162.

[187] NASH-WILLIAMS, C. ST. J.A., On well-quasi-ordering finite trees, *Proc. Cambridge Philos. Soc.* **59**(4) (1963) 833–835.

[188] NEWMAN, M.H.A., On theories with a combinatorial definition of 'equivalence', *Ann. Math.* **43**(2) (1942) 223–243.

[189] NIPKOW, T. and G. WEIKUM, A decidability result about sufficient-completeness of axiomatically specified abstract data types, in: *Proc. Sixth GI Conference on Theoretical Computer Science*, Lecture Notes in Computer Science **145** (Springer, Berlin, 1982) 257–268.

[190] NUTT, W., P. RÉTY and G. SMOLKA, Basic narrowing revisited, *J. Symbolic Comput.* **7**(3/4) (1989) 295–318.

[191] O'DONNELL, M.J., *Computing in Systems Described by Equations*, Lecture Notes in Computer Science **58** (Springer, Berlin, 1977).

[192] O'DONNELL, M.J., Subtree replacement systems: A unifying theory for recursive equations, LISP, Lucid and Combinatory Logic, in: *Proc. Ninth ACM Symposium on Theory of Computing* (1977) 295–305.

[193] OYAMAGUCHI, M., The Church–Rosser property for ground term rewriting systems is decidable, *Theoret. Comput. Sci.* **49**(1) (1987) 43–79.

[194] PATERSON, M.S. and M.N. WEGMAN, Linear unification, *J. Comput. System Sci.* **16** (1978) 158–167.

[195] PEDERSEN, J., Morphocompletion for one-relation monoids, in: *Proc. Third International Conference on Rewriting Techniques and Applications*, Chapel Hill, NC, Lecture Notes in Computer Science **355** (Springer, Berlin, 1989).

[196] PETERSON, G.E., A technique for establishing completeness results in theorem proving with equality, *SIAM J. Comput.* **12**(1) (1983) 82–100.

[197] PETERSON, G.E. and M.E. STICKEL, Complete sets of reductions for some equational theories, *J. Assoc. Comput. Mach.* **28**(2) (1981) 233–264.

[198] PLAISTED, D.A., A recursively defined ordering for proving termination of term rewriting systems, Report R-78-943, Dept. of Computer Science, Univ. of Illinois, Urbana, IL, 1978.

[199] PLAISTED, D.A., Semantic confluence tests and completion methods, *Inform. and Control* **65**(2/3), (1985) 182–215.

[200] PLOTKIN, G., Lattice theoretic properties of subsumption, Tech. Report MIP-R-77, Univ. of Edinburgh, Edinburgh, Scotland, 1970.

[201] POST, E.L., Recursive unsolvability of a problem of Thue, *J. Symbolic Logic* **13** (1947) 1–11.

[202] PRESBURGER, M., Über der Vollständigkeit eines gewissen Systems der Arithmetik ganzer Zahlen, in welchen die Addition als einzige Operation hervortritt, in: *Comptes Rendus Premier Congrès des Mathématicienes des Pays Slaves*, Warsaw (1927) 92–101, 395.

[203] PUEL, L., Using unavoidable sets of trees to generalize Kruskal's theorem, *J. Symbolic Comput.* **8**(4) (1989) 335–382.

[204] PUEL, L., Embedding with patterns and associated recursive ordering, in: *Proc. Third International*

Conference on Rewriting Techniques and Applications, Chapel Hill, NC, Lecture Notes in Computer Science **355** (Springer, Berlin, 1989) 371–387.

[205] RAOULT, J.-C., Finiteness results on rewriting systems, *R.A.I.R.O. Theoret. Inform.* **15**(4) (1981) 373–391.

[206] RAOULT, J.-C., On graph rewritings, *Theoret. Comput. Sci.* **32**(1, 2) (1984) 1–24.

[207] RAOULT, J.-C. and J. VUILLEMIN, Operational and semantic equivalence between recursive programs, *J. Assoc. Comput. Mach.* **27**(4) (1980) 772–796.

[208] REDDY, U.S., On the relationship between logic and functional languages, in: D. De Groot and G. Lindstrom, eds., *Logic Programming: Functions, Relations, and Equations* (Prentice-Hall, Englewood Cliffs, NJ, 1986) 3–36.

[209] RÉMY, J.-L., Etude des systèmes de réécriture conditionnels et applications aux types abstraits algébriques, Thèse, Institut National Polytechnique de Lorraine, Nancy, France, 1982.

[210] ROBINSON, G. and L. WOS, Paramodulation and theorem-proving in first order theories with equality, in: *Machine Intelligence 4* (Edinburgh University Press, Edinburgh, Scotland, 1969) 135–150.

[211] ROBINSON, J.A., A machine-oriented logic based on the resolution principle, *J. Assoc. Comput. Mach.* **12**(1) January (1965) 23–41.

[212] ROSEN, B.K., Tree-manipulating systems and Church–Rosser theorems, *J. Assoc. Comput. Mach.* **20**(1) (1973) 160–187.

[213] RUSINOWITCH, M., Path of subterms ordering and recursive decomposition ordering revisited, *J. Symbolic Comput.* **3**(1&2) (1987).

[214] RUSINOWITCH, M., *Démonstration Automatique: Techniques de Réécriture* (InterEditions, Paris, France, 1989).

[215] SCHMIDT-SCHAUß, M. Unification in a combination of arbitrary disjoint equational theories, in: *Proc. Ninth International Conference on Automated Deduction*, Argonne, IL, Lecture Notes in Computer Science **310** (Springer, Berlin, 1988) 378–396.

[216] SETHI, R., Testing for the Church–Rosser property, *J. Assoc. Comput. Mach.* **21** (1974) 671–679; **22** (1974) 424.

[217] SHOSTAK, R.E., An efficient decision procedure for arithmetic with function symbols, *J. Assoc. Comput. Mach.* **26**(2) (1979) 351–360.

[218] SIEKMANN, J., Universal unification, in: *Proc. Seventh International Conference on Automated Deduction*, Napa, CA, Lecture Notes in Computer Science **170** (Springer, Berlin, 1984) 1–42.

[219] SMOLKA, G., W. NUTT, J.A. GOGUEN and J. MESEGUER, Order-sorted equational computation, in: H. Ait-Kaci and M. Nivat, eds., *Resolution of Equations in Algebraic Structures, Vol II: Rewriting Techniques* (Academic Press, New York, 1989) 299–369.

[220] SNYDER, W., Efficient ground completion: An $O(n \log n)$ algorithm for generating reduced sets of ground rewrite rules equivalent to a set of ground equations E, in: *Proc. Third International Conference on Rewriting Techniques and Applications*, Chapel Hill, NC, Lecture Notes in Computer Science **355** (Springer, Berlin, 1989) 419–433.

[221] STICKEL, M.E., The inadequacy of primitive recursive complexity measures for determining finite termination of sets of reductions, Unpublished manuscript, Univ. of Arizona, Tuscon, AZ, 1976.

[222] STICKEL, M.E., A unification algorithm for associative-commutative functions, *J. Assoc. Comput. Mach.* **28**(3) (1981) 423–434.

[223] STICKEL, M.E., The KLAUS automated deduction system, in: *Proc. Eighth International Conference on Automated Deduction*, Oxford, England, Lecture Notes in Computer Science **230** (Springer, Berlin, 1986) 703–704.

[224] STONE, M., The theory of representations for Boolean algebra, *Trans. Amer. Math. Soc.* **40** (1936) 37–111.

[225] SZABO, P., Unifikationstheorie erster Ordnung, Thesis, Fakultät für Informatik, Univ. Karlsruhe, Karlsruhe, West Germany, 1982.

[226] TAKEUTI, G., *Proof Theory* (North-Holland, Amsterdam, rev. ed., 1987).

[227] TARSKI, A., *A Decision Method for Elementary Algebra and Geometry* (Univ. of California Press, Berkeley, CA, 1951).

[228] TARSKI, A., Equational logic and equational theories of algebras, in: K. Schutte, ed., *Contributions to Mathematical Logic* (North-Holland, Amsterdam, 1968).

[229] TARSKI, A. and S. GIVANT, A formalization of set theory without variables, in: *Colloquium Publications* **41** (American Mathematical Society, Providence, RI, 1985).

[230] TAYLOR, W., Equational logic, in: G. Grätzer, ed., *Universal Algebra* (Springer, New York, 2nd ed., 1979) 378–400.

[231] THIEL, J.J., Stop losing sleep over incomplete data type specifications, in: *Conf. Rec. ACM Symposium on Principles of Programming Languages*, Salt Lake City, UT (1984) 76–82.

[232] THUE, A., Probleme über Veranderungen von Zeichenreihen nach gegeben Regeln, *Skr. Vid. Kristianaia I. Mat. Naturv. Klasse* **10/34** (1914).

[233] TOURLAKIS, G.J., *Computability* (Reston, Reston, VA, 1984).

[234] TOYAMA, Y., On the Church–Rosser property for the direct sum of term rewriting systems, *J. Assoc. Comput. Mach.* **34**(1) (1987) 128–143.

[235] TOYAMA, Y., Counterexamples to termination for the direct sum of term rewriting systems, *Inform. Process. Lett.* **25** (1987) 141–143.

[236] TOYAMA, Y., J.W. KLOP and H.P. BARENDREGT, Termination for the direct sum of left-linear term rewriting systems, in: *Proc. Third International Conference on Rewriting Techniques and Applications*, Chapel Hill, NC, Lecture Notes in Computer Science **355** (Springer, Berlin, 1989) 477–491.

[237] WALTHER, C., Many-sorted unification, *J. Assoc. Comput. Mach.* **35**(1) (1988) 1–17.

[238] WINKLER, F. and B. BUCHBERGER, A criterion for eliminating unnecessary reductions in the Knuth–Bendix algorithm, in: *Proc. Colloquium on Algebra, Combinatorics and Logic in Computer Science*, Györ, Hungary (1983).

[239] WIRSING, M., Algebraic specification, in: J. van Leeuwen, ed., *Handbook of Theoretical Computer Science, Vol. B* (North-Holland, Amsterdam, 1990) 675–788.

[240] YELICK, K.A., Unification in combinations of collapse-free regular theories, *J. Symbolic Comput.* **3**(1&2) (1987) 153–181.

[241] ZHANG, H., Reduction, superposition & induction: automated reasoning in an equational logic, Ph.D. Thesis, Rensselaer Polytechnic Institute, Troy, NY, 1988.

[242] ZHEGALKIN, I.I., On a technique of evaluation of propositions in symbolic logic, *Mat. Sb.* **34**(1) (1927) 9–27.

[243] BACHMAIR, L. and H. GANZINGER, On restrictions of ordered paramodulation with simplification, in: *Proc. Second International Workshop on Conditional and Typed Rewriting Systems*, Montreal, Canada, Lecture Notes in Computer Science (Springer, Berlin, to appear).

[244] BERTLING, H. and H. GANZINGER, Completion-time optimization of rewrite-time goal solving, in: *Proc. Third International Conference on Rewriting Techniques and Applications*, Chapel Hill, NC, Lecture Notes in Computer Science **355** (Springer, Berlin, 1989) 45–58.

[245] BOOK, R.V., Confluent and other types of Thue systems, *J. Assoc. Comput. Machin.* **29**(1) (1982) 171–182.

[246] BÖRGER, E., *Computability, Complexity, Logic*, Studies in Logic and the Foundations of Mathematics **128** (North-Holland, Amsterdam, 1990).

[247] BOUDET, A., Unification dans les mélanges de théories équationnelles, Thèse de Doctorat, Univ. de Paris-Sud, Orsay, Paris, 1990.

[248] BREAZU-TANNEN, V. and J. GALLIER, Polymorphic rewriting conserves algebraic strong normalization, *Theoret, Comput. Sci.*, to appear.

[249] CHEW, P., An improved algorithm for computing with equations, in: *Proc. Twenty-First IEEE Symposium on Foundations of Computer Science*, Lake Placid, NY (1980) 108–117.

[250] COMON, H., Disunification: A survey, in: J. Lassez and G. Plotkin, eds., *Computational Logic: Essays in Honour of Alan Robinson* (MIT Press, Cambridge, MA, to appear).

[251] COURCELLE, D., Graph rewriting: an algebraic and logic approach, in: J. van Leeuwen, ed., *Handbook of Theoretical Computer Science, Vol. B* (North-Holland, Amsterdam, 1990) 193–242.

[252] DERSHOWITZ, N. and J.-P. JOUANNAUD, Notations for rewriting, *Bull. European Assoc. Theoret. Comput. Sci.* (Spring, 1990).

[253] FILMAN, R.E., Personal communication, 1978.

[254] GALLIER, J. and W. SNYDER, Complete sets of transformations for general E-Unification, *Theoret.* Comput. Sci. **67** (1989) 203–260.

[255] GALLIER, J. and W. SNYDER, Designing unification procedures using transformations: A survey, in: *Proc. Workshop on Logic from Computer Science*, Berkeley, CA (1990).

[256] GESER, A., *Termination Relative*, Doctoral Dissertation, Univ. Passau, Passau, 1989.

[257] GNAEDIG, I., C. KIRCHNER and H. KIRCHNER, Equational completion in order-sorted agebras, in: *Proc. Thirteenth Colloquium on Trees in Algebra and Programming*, Nancy, France (1988) 165–184.

[258] HALPERN, J.Y., J.H. WILLIAMS and E.L. WIMMERS, Completeness of rewrite rules and rewrite strategies for FP, *J. Assoc. Comput. Mach.* **37**(1) (1990) 86–143.

[259] HUET, G. and J.-J. LÉVY, Computations in orthogonal term rewriting systems, in: J. Lassez and G. Plotkin, eds., *Computational Logic: Essays in Honour of Alan Robinson* (MIT Press, Cambridge, MA, to appear).

[260] JAFFAR, J., Minimal and complete word unification, *J. Assoc. Comput. Mach.* **37**(1) (1990) 47–85.

[261] JOUANNAUD, J. and C. KIRCHNER, Solving equations in abstract algebras: A rule-based survey of unification, in: J. Lassez and G. Plotkin, eds., *Computational Logic: Essays in Honour of Alan Robinson* (MIT Press, Cambridge, MA, to appear).

[262] KANDRI–RODY, A., D. MAPUR and F. WINKLER, Knuth-Bendix procedure and Buchberger algorithm—A synthesis, Tech. Report 89–18, Dept. of Computer Science, State Univ. of New York, Albany, NY, 1989.

[263] KAPLAN, S. and J.-L. RÉMY, Completion algorithms for conditional rewriting systems, in: H. Ait-Kaci and M. Nivat, eds., *Resolution of Equations in Algebraic Structures* (Academic Press, Boston, 1989) 141–170.

[264] KAPUR, D. and D.R. MUSSER, Inductive reasoning for incomplete specifications, in: *Proc. IEEE Symposium on Logic in Computer Science*, Cambridge, MA (1986) 367–377.

[265] KÜCHLIN, W., A confluence criterion based on the generalized Newman lemma, in: *Proc. EUROCAL Conference*, Linz, Austria (1985).

[266] LANKFORD, D., G. BUTLER and A. BALLANTYNE, A progress report on new decision algorithms for finitely presented Abelian groups, in: *Proc. Seventh International Conference on Automated Deduction*, Napa, CA, Lecture Notes in Computer Science **170** (Springer, Berlin, 1984) 128–141.

[267] MARTIN, U., A geometrical approach to multiset orderings, *Theoret. Comput. Sci.* **67** (1989) 37–54.

[268] MIDDELDORP, A., Modular aspects of properties of term rewriting systems related to normal forms, in: *Proc. Third International Conference on Rewriting Techniques and Applications*, Chapel Hill, NC, Lecture Notes in Computer Science **355** (Springer, Berlin, 1989) 263–277.

[269] OKADA, M. and A. STEELE, Ordering structures and the Knuth–Bendix completon algorithm, Unpublished manuscript, Dept. Computer Science, Concordia Univ., Montreal, Canada.

[270] PADAWITZ, P., Horn logic and rewriting for functional and logic program design, Report MIP-9002, Fakultät für Mathematik und Informatik, Univ. Passau, Passau, West Germany, 1990.

[271] PAUL, E., Equational methods in first order predicate calculus, *J. Symbolic Comput.* **1**(1) (1985) 7–29.

[272] RUŽIČKA, P. and I. PRÍVARA, An almost linear Robinson unification algorithm, *Acta Informat.* **27** (1989) 61–71.

[273] SEKAR, R.C. and I.V. RAMAKRISHNAN, Programming in equational logic: Beyond strong sequentiality, in: *Proc. Fifth IEEE Symposium on Logic in Computer Science*, Philadelphia, PA (1990).

[274] SIVAKUMAR, G., Personal communication, 1986.

[275] SLAGLE, J.R., Automated theorem-proving for theories with simplifiers, commutativity, and associativity, *J. Assoc. Comput. Mach.* **21**(4) (1974) 622–642.

[276] ZHANG, H. and D. KAPUR, First-order theorem proving using conditional equations, in: *Proc. Ninth International Conference on Automated Deduction*, Argonne, IL, Lecture Notes in Computer Science **310** (Springer, Berlin, 1988) 1–20.

[277] SMOLKA, G., Logic programming over polymorphically order-sorted types, Ph.D. Thesis, Univ. Kaiserslautern, 1989.

CHAPTER 7

Functional Programming and Lambda Calculus

H. P. BARENDREGT

Faculty of Mathematics and Computer Science, Catholic University Nijmegen, Toernooiveld 1, 6525 ED Nijmegen, Netherlands

Contents

1. The functional computation model . . . 323
2. Lambda calculus . . . 325
3. Semantics . . . 337
4. Extending the language . . . 347
5. The theory of combinators and implementation issues . . . 355
Acknowledgment . . . 360
References . . . 360

HANDBOOK OF THEORETICAL COMPUTER SCIENCE
Edited by J. van Leeuwen

1. The functional computation model

Some history

In 1936 two computation models were introduced:

(1) Alan Turing invented a class of machines (later to be called *Turing machines*) and defined the notion of computable function via these machines (see [87]).

(2) Alonzo Church invented a formal system called the *lambda calculus* and defined the notion of computable function via this system, see [22].

In [88] it was proved that both models are equally strong in the sense that they define the same class of computable functions.

Based on the concept of a Turing machine are the present-day *von Neumann computers*. Conceptually these are Turing machines with random-access registers. *Imperative programming languages* such as FORTRAN, Pascal etcetera, as well as all the assembler languages are based on the way a Turing machine is instructed: by a sequence of statements. *Functional programming languages*, like ML, MIRANDA, HOPE etcetera, are related to the lambda calculus. An early (although somewhat hybrid) example of such a language is LISP. *Reduction machines* are specifically designed for the execution of these functional languages.

Böhm and Gross [13] and Landin [58, 59, 60] were the first to suggest the importance of lambda calculus for programming. In fact, [59] already describes several aspects of functional languages developed much later. In [57] a reduction machine for lambda calculus is described. Because this resulted in rather slow actual implementations people did not believe very much in the possibility of functional programming. Wadsworth [93] introduced the notion of graph reduction that, at least in theory, could improve the performance of implementations of functional languages. Turner [90] made an implementation using graph reduction that was reasonably fast. The period 1965–1979 is a long one, but matters were going slowly because technology had provided fast and cheap von Neumann computers and little need was felt for new kinds of languages or machines. In the meantime Backus [4] had emphasised the limitations of imperative programming languages. In the early eighties things went faster and rather good sequential implementations of functional languages were developed, e.g. the G-machine [3, 49] or the compiler described in [15]. Presently, several research groups are working towards fast parallel reduction machines.

Reduction and functional programming

A *functional program* consists of an expression E (representing both the algorithm and the input). This expression E is subject to some rewrite rules. *Reduction* consists of replacing a part P of E by another expression P' according to the given rewrite rules. In schematic notation

$$E[P] \to E[P'],$$

provided that $P \to P'$ is according to the rules. This process of reduction will be repeated until the resulting expression has no more parts that can be rewritten. The expression

E^* thus obtained is called the *normal form* of E and constitutes the output of the given functional program.

EXAMPLE

$$\begin{aligned}(7+4)*(8+5*3) &\to 11*(8+5*3)\\ &\to 11*(8+15)\\ &\to 11*23\\ &\to 253.\end{aligned}$$

In this example the reduction rules consist of the "tables" of addition and of multiplication on the numerals (as usual, $*$ takes priority over $+$). Note that the "meaning" of an expression is preserved after reduction: $7+4$ and 11 have the same interpretation. This feature of the evaluation of functional programs is called *referential transparancy*.

Also symbolic computations can be done by reduction. For example,

first-of(sort(append("dog", "rabbit") (sort("mouse", "cat"))))
→first-of(sort(append("dog", "rabbit") ("cat", "mouse")))
→first-of(sort("dog", "rabbit", "cat", "mouse"))
→first-of("cat", "dog", "mouse", "rabbit")
→"cat".

The necessary rewrite rules for *append* and *sort* can be programmed easily by a few lines in terms of more primitive rewrite rules.

Reduction systems for functional languages usually satisfy the *Church–Rosser property*, which implies that the normal form obtained is independent of the order of evaluation of subterms. Indeed, the first example may be reduced as follows:

$$\begin{aligned}(7+4)*(8+5*3) &\to (7+4)*(8+15)\\ &\to 11*(8+15)\\ &\to 11*23\\ &\to 253.\end{aligned}$$

Or even by evaluating several expressions at the same time:

$$\begin{aligned}(7+4)*(8+5*3) &\Rightarrow 11*(8+15)\\ &\to 11*23\\ &\to 253.\end{aligned}$$

($\Rightarrow$ indicates that several reduction steps are done in parallel). This gives the possibility of parallel execution of functional languages.

Functional programming and process control

A functional program transforms data into other data according to a certain algorithm. Functional programs cannot deal with input/output, cannot switch on and off lamps depending on the value of a certain parameter; in general they cannot deal with *process control*. These points are sometimes held as arguments against functional programming. However a reduction machine can produce *code* for process control, code that is executed by some interface. A von Neumann computer also needs interfaces for I/O and other controls. Therefore, a reduction machine with environment will consist of a pure reduction machine (dealing with algorithms that transform data) together with interfaces for process control (like I/O), see Fig. 1. This is of course a logical picture. The hardware of the two parts may be geometrically interleaved or may be even the same.

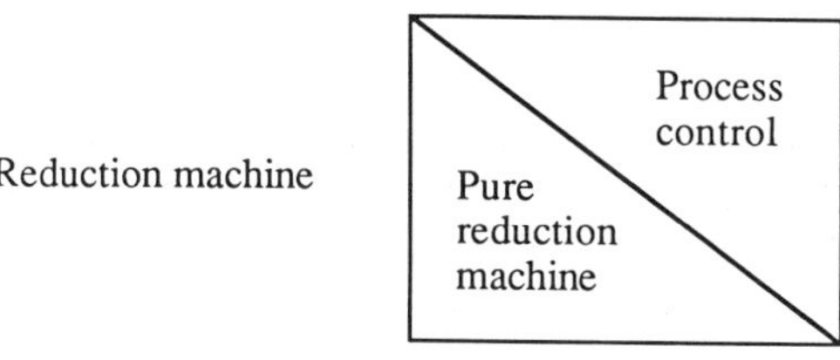

Fig. 1.

Process control via streams

In spite of the previous remarks, there is an approach for functional languages that incorporates I/O. Suppose that we want to compute a function F on several arguments $A_0, A_1, A_2, \ldots$ appearing consecutively in time. One can view this so-called *stream* $(A_0, A_1, A_2, \ldots)$ as a potentially infinite list $A=(A_0:A_1:A_2:\cdots:A_n:\bot)$ where $\bot$ stands for "unspecified" and is interactively updated to $A_{n+1}:\bot$ each time the user has a new argument A_{n+1}. Then on this list A, the system applies the function F^* defined by

$$F^*(A:B)=(FA):(F^*B)$$

obtaining

$$F^*A=(FA_0:FA_1:FA_2:\cdots:FA_n:F^*\bot)$$

also appearing consecutively in time.

This idea of using streams is used in some (implementations of) functional languages, e.g. in the Miranda implementation of Turner [92]. We prefer not to use this mechanism in a pure reduction machine. The reason is that, although F^* is purely functional, the use of streams is not. The way $\bot$ is treated depends essentially on process control. Moreover, it is not always natural to simulate process-like actions using streams as above. For example, this is the case with "control-C", the statement to stop a process of computation that is not yet finished. Therefore we view process control as a necessary interface.

2. Lambda calculus

In this section we introduce lambda calculus and show how this system is able to capture all computable functions. For more information, see [7, 42].

2.1. *Conversion*

We start with an informal description of the system.

Application and abstraction

The lambda calculus has two basic operations. The first one is *application*. The expression

$$F.A$$

(usually written as A) denotes the data F considered as an algorithm applied to A considered as input. The theory is *type-free*: it is allowed to consider expressions like FF, that is, F applied to itself. This will be useful to simulate recursion.

The other basic operation is *abstraction*: if $M = M[x]$ is an expression containing ("depending on") x, then $\lambda x.M[x]$ denotes the map

$$x \mapsto M[x].$$

Application and abstraction work together in the following intuitive formula:

$$(\lambda x.2*x+1)3 = 2*3+1 \ (=7).$$

That is, $(\lambda x.2*x+1)3$ denotes the function $x \mapsto 2*x+1$ applied to the argument 3, giving $2*3+1$ (which is 7). In general we have

$$(\lambda x.M[x])\,N = M[N].$$

This last equation is preferably written as

$$(\beta) \qquad (\lambda x.M[x])N = M[N].$$

where $[x := N]$ denotes substitution of N for x. This equation is called β-conversion. It is remarkable that although it is the only essential axiom of lambda calculus, the resulting theory is rather involved.

Free and bound variables

Abstraction is said to *bind* the *free* variable x in M. For instance, we say that $\lambda x.yx$ has x as bound and y as free variable. Substitution $[x := N]$ is only performed in the free occurrences of x:

$$yx(\lambda x.x)[x := N] = yN(\lambda x.x).$$

In calculus there is a similar variable binding. In $\int_a^b f(x, y)\,dx$ the variable x is bound and y is free. It does not make sense to substitute 7 for x: $\int_a^b f(7, y)\,d7$; but substitution for y does make sense: $\int_a^b f(x, 7)\,dx$.

For reasons of hygiene it will always be assumed that the bound variables occurring in a certain expression are different from the free ones. This can be fulfilled by renaming bound variables. For instance, $\lambda x.x$ becomes $\lambda y.y$. Indeed, these expressions act the same way:

$$(\lambda x.x)\,a = a = (\lambda y.y)\,a$$

and in fact they denote the same intended algorithm. Therefore expressions that differ only in the names of bound variables are identified. Equations like $\lambda x.x = \lambda y.y$ are usually called *α-conversion*.

Functions of several arguments

Functions of several arguments can be obtained by iteration of application. This is due to Schönfinkel [79] but is often called "currying", after H.B. Curry who made the method popular. Intuitively, if $f(x, y)$ depends on two arguments, one can define

$$F_x = \lambda y.f(x, y), \qquad F = \lambda x.F_x.$$

Then

$$(Fx)y = F_x y = f(x, y). \tag{2.1}$$

This last equation shows that it is convenient to use *association to the left* for iterated application:

$$FM_1 \ldots M_n \text{ denotes } (\ldots((FM_1)M_2)\ldots M_n).$$

Equation (2.1) then becomes $Fxy = f(x, y)$.

Dually, iterated abstraction uses *association to the right*:

$$\lambda x_1 \ldots x_n.f(x_1 \ldots x_n) \text{ denotes } \lambda x_1.(\lambda x_2.(\ldots(\lambda x_n.f(x_1 \ldots x_n))\ldots)).$$

Then we have, for F defined above, $F = \lambda xy.f(x, y)$ and (2.1) becomes

$$(\lambda xy.f(x, y))xy = f(x, y).$$

For n arguments we have

$$(\lambda x_1 \ldots x_n . f(x_1 \ldots x_n))x_1 \ldots x_n = f(x_1 \ldots x_n)$$

by using n times (β). This last equation becomes, in convenient vector notation,

$$(\lambda \vec{x}.f(\vec{x}))\vec{x} = f(\vec{x});$$

more generally, one has for $\vec{N} = N_1, \ldots, N_n$

$$(\lambda \vec{x}.f(\vec{x}))\vec{N} = f(\vec{N}).$$

Now we give the formal description of lambda calculus.

2.1.1. DEFINITION. The set of *λ-terms* (notation Λ) is built up from a infinite set of constants $C = \{c, c', c'', \ldots\}$ and of variables $V = \{v, v', v'', \ldots\}$ using application and (function) abstraction.

$$c \in C \Rightarrow c \in \Lambda, \qquad x \in V \Rightarrow x \in \Lambda,$$
$$M, N \in \Lambda \Rightarrow (MN) \in \Lambda, \qquad M \in \Lambda, x \in V \Rightarrow (\lambda x M) \in \Lambda.$$

In BN-form this is

constant ::= "c" | constant "′"
variable ::= "v" | variable "′",
λ-term ::= constant | variable | "(" λ-term λ-term ")" | "(λ" variable λ-term ")".

Or, using abstract syntax, (see the chapter by Mosses in this Handbook), one may write

$$\Lambda = C \mid V \mid \Lambda\Lambda \mid \lambda V \Lambda.$$

2.1.2. EXAMPLE. The following are λ-terms:

$$v, \qquad (vc), \qquad (\lambda v(vc)), \qquad (v'(\lambda v(vc))), \qquad (\lambda v(vc))v').$$

2.1.3. CONVENTION. (i) $c, d, e, \ldots$ denote arbitrary constants, $x, y, z, \ldots,$ denote arbitrary variables, $M, N, L, \ldots$ denote arbitrary λ-terms. Outermost parentheses are not written.

(ii) As already mentioned informally, the following abbreviations are used:

$$FM_1 \ldots M_n \text{ stands for } (\ldots((FM_1)M_2)\ldots M_n)$$

and

$$\lambda x_1 \ldots x_n.M \text{ stands for } \lambda x_1(\lambda x_2(\ldots(\lambda x_n(M))\ldots)).$$

The examples in 2.1.2 now may be written as follows:

$$x, \qquad xc, \qquad \lambda x.xc, \qquad y(\lambda x.xc), \qquad (\lambda x.xc)y.$$

Note that $\lambda x.yx$ is $(\lambda x(yx))$ and not $((\lambda xy)x)$.

2.1.4. DEFINITION. (i) The set of *free variables* of M (notation $\mathrm{FV}(M)$) is inductively defined as follows:

$$\mathrm{FV}(x) = \{x\}; \qquad \mathrm{FV}(MN) = \mathrm{FV}(M) \cup \mathrm{FV}(N);$$
$$\mathrm{FV}(\lambda x.M) = \mathrm{FV}(M) - \{x\}.$$

(ii) M is a *closed λ-term* (or *combinator*) if $\mathrm{FV}(M) = \emptyset$. The set of closed λ-terms is denoted by Λ^0.

In the λ-term $y(\lambda xy.xyz)$, y and z occur as free variables; x and y occur as bound variables. The term $\lambda xy.xxy$ is closed.

Now we introduce lambda calculus as a formal theory of equations between λ-terms.

2.1.5. DEFINITION. (i) The principal axiom scheme of lambda calculus is

$$(\beta) \qquad (\lambda x.M)N = M[x := N] \quad \text{for all } M, N \in \Lambda.$$

(ii) There are also "logical" axioms and rules:

$$M = M,$$
$$M = N \Rightarrow N = M, \qquad M = N, N = L \Rightarrow M = L,$$
$$M = M' \Rightarrow MZ = M'Z, \qquad M = M' \Rightarrow ZM = ZM',$$
$$(\text{rule } \xi) \qquad M = M' \Rightarrow \lambda x.M = \lambda x.M'.$$

(iii) If $M = N$ is provable in the lambda calculus, then we write $\lambda \vdash M = N$ or often just $M = N$ and say that M and N are (β-)*convertible*. $M \equiv N$ denotes that M and N are the same term or can be obtained from each other by renaming bound variables. For instance,

$$(\lambda x.y)z \equiv (\lambda x.y)z, \qquad (\lambda x.x)z \equiv (\lambda y.y)z, \qquad (\lambda x.x)z \not\equiv (\lambda x.y)z.$$

REMARK. We have identified terms that differ only in the names of bound variables. An alternative is to add to the lambda calculus the following axiom scheme:

$$(\alpha) \qquad \lambda x.M = \lambda y.M[x := y],$$

provided that y does not occur in M. The axiom β above was originally the second axiom; hence its name. We prefer our version of the theory in which the identifications are made on a syntactic level. These identifications are done in our mind and not on paper. For implementations of the lambda calculus, the machine has to deal with this so-called α-conversion. A good way of doing this is provided by the name-free notation of de Bruijn (see [7, Appendix C]). In this notation $\lambda x.(\lambda y.xy)$ is denoted by $\lambda(\lambda 21)$, the 2 denoting a variable bound "two lambdas above".

Development of the theory

2.1.6. EXAMPLES (*standard combinators*). Define the combinators

$$\mathbf{I} \equiv \lambda x.x, \qquad \mathbf{K} \equiv \lambda xy.x,$$
$$\mathbf{K}_* \equiv \lambda xy.y, \qquad \mathbf{S} \equiv \lambda xyz.xz(yz).$$

Then the following equations are provable:

$$\mathbf{I}M = M, \qquad \mathbf{K}\mathrm{MN} = \mathrm{M},$$
$$\mathbf{K}_*MN = N, \qquad \mathbf{S}\mathrm{MNL} = \mathrm{ML(NL)}.$$

The following result provides one way to represent recursion in the lambda-calculus.

2.1.7. FIXED POINT THEOREM. (i) $\forall F\, \exists X\; FX = X$. (*This means that for all* $F \in \Lambda$ *there is an* $X \in \Lambda$ *such that* $\lambda \vdash FX = X$.)

(ii) *There is a fixed point combinator*

$$\mathbf{Y} \equiv \lambda f.(\lambda x.f(xx))(\lambda x.f(xx))$$

such that

$$\forall F\; F(\mathbf{Y}F) = \mathbf{Y}F.$$

PROOF. (i) Define $W \equiv \lambda x.F(xx)$ and $X \equiv WW$. Then

$$X \equiv WW \equiv (\lambda x.F(xx))W = F(WW) \equiv FX.$$

(ii) By the proof of (i).

2.1.8. APPLICATION. Given a context $C[f, x]$ (that is, a term possibly containing the displayed free variables), then

$$\exists F\, \forall X\; FX = C[F, X].$$

Here $C[F, X]$ is of course the sbustitution result $C[f, x][f := F][x := X]$. Indeed,

$$\begin{aligned}\forall X\, FX = C[F, X] &\Leftarrow Fx = C[F, x]\\ &\Leftarrow F = \lambda x.C[F, x]\\ &\Leftarrow F = (\lambda fx.C(f, x])F\\ &\Leftarrow F \equiv \mathbf{Y}(\lambda fx.C[f, x]).\end{aligned}$$

This also holds for more arguments: $\exists F\, \forall \vec{x}\, F\vec{x} = C[F, \vec{x}]$.

In lambda calculus one can define numerals and represent numeric functions on them.

2.1.9. DEFINITION. (i) $F^n(M)$ with $n \in \mathbf{N}$ (the set of natural numbers) is defined inductively as follows:

$$F^0(M) \equiv M; \qquad F^{n+1}(M) \equiv F(F^n(M)).$$

(ii) The *Church numerals* $\boldsymbol{c}_0, \boldsymbol{c}_1, \boldsymbol{c}_2, \ldots$ are defined by

$$\boldsymbol{c}_n \equiv \lambda fx.f^n(x).$$

2.1.10. PROPOSITION (Rosser). *Define*

$$\mathbf{A}_+ \equiv \lambda xypq.xp(ypq); \qquad \mathbf{A}_* \equiv \lambda xyz.x(yz); \qquad \mathbf{A}_{\exp} \equiv \lambda xy.yx.$$

Then one has, for all $n, m \in \mathbf{N}$,

(i) $\mathbf{A}_+ \boldsymbol{c}_n \boldsymbol{c}_m = \boldsymbol{c}_{n+m}$;
(ii) $\mathbf{A}_* \boldsymbol{c}_n \boldsymbol{c}_m = \boldsymbol{c}_{n*m}$;
(iii) $\mathbf{A}_{\exp} \boldsymbol{c}_n \boldsymbol{c}_m = \boldsymbol{c}_{(n^m)}$, *except for* $m = 0$ (*Rosser started counting from* 1).

PROOF. *Lemma*: (i) $(\boldsymbol{c}_n x)^m(y) = x^{n*m}(y)$;
(ii) $(\boldsymbol{c}_n)^m(x) = \boldsymbol{c}_{(n^m)}(x)$ *for* $m > 0$.

Proof of lemma: (i) By induction on m: If $m = 0$, then LHS $= y =$ RHS. Assume (i) is correct for m (induction hypothesis, abbreviated IH). Then

$$\begin{aligned}(\boldsymbol{c}_n x)^{m+1}(y) &= \boldsymbol{c}_n x((\boldsymbol{c}_n x)^m(y)) =_{\mathrm{IH}} \boldsymbol{c}_n x(x^{n*m} y) = x^n(x^{n*m} y)\\ &\equiv x^{n+n*m}(y) \equiv x^{n*(m+1)}(y).\end{aligned}$$

(ii) By induction on $m > 0$: If $m = 1$, then LHS $\equiv \boldsymbol{c}_n x \equiv$ RHS. If (ii) is correct for m, then

$$\begin{aligned}\boldsymbol{c}_n^{m+1}(x) &= \boldsymbol{c}_n(\boldsymbol{c}_n^m(x)) =_{\mathrm{IH}} \boldsymbol{c}_n(\boldsymbol{c}_{(n^m)}(x)) = \lambda y.(\boldsymbol{c}_{(n^m)}(x))^n(y)\\ &=_{(\mathrm{i})} \lambda y.x^{n^m * n}(y) = \boldsymbol{c}_{(n^{m+1})} x.\end{aligned}$$

Now the proof of the proposition easily follows.

(i) By induction on m.
(ii) Use Lemma (i).
(iii) By Lemma (ii) we have, for $n > 0$,

$$\mathbf{A}_{\exp} \boldsymbol{c}_n \boldsymbol{c}_m = \boldsymbol{c}_m \boldsymbol{c}_n = \lambda x.\boldsymbol{c}_n^m(x) = \lambda x.\boldsymbol{c}_{(n^m)} x = \boldsymbol{c}_{(n^m)},$$

since $\lambda x.Mx = M$ if $M \equiv \lambda y.M'[y]$. Indeed,

$$\lambda x.Mx \equiv \lambda x.(\lambda y.M'[y])x = \lambda x.M'[x] \equiv \lambda y.M'[y] \equiv M. \qquad \square$$

2.2. *Representing the computable functions*

We have seen that the functions plus, times and exponentiation on $\mathbf{N}$ can be represented in the lambda calculus using Church numerals. In this section we will show that all computable (recursive) functions can be represented. In order to do this we will first introduce Booleans, pairs and a different system of numerals.

2.2.1. DEFINITION. (i) $\mathbf{true} \equiv \mathbf{K}$, $\mathbf{false} \equiv \mathbf{K}_*$.

(ii) If B is a Boolean, i.e. a term that is either **true**, or **false**, then

$$\textbf{if } B \textbf{ then } P \textbf{ else } Q \text{ can be represented by } BPQ.$$

Indeed, $\mathbf{true}PQ \equiv \mathbf{K}PQ = P$ and $\mathbf{false}PQ \equiv \mathbf{K}_*PQ = Q$ (see Example 2.1.6).

2.2.2. DEFINITION (*ordered pairs*). For $M, N \in L$ write

$$[M, N] = \lambda z.zMN.$$

Then

$$[\mathrm{M}, \mathrm{N}]\mathbf{true} = \mathrm{M}, \qquad [M, N]\mathbf{false} = N$$

and hence $[M, N]$ can serve as an ordered pair.

2.2.3. DEFINITION (*numerals*). $\ulcorner 0 \urcorner \equiv \mathbf{I}$, $\ulcorner n+1 \urcorner \equiv [\mathbf{false}, \ulcorner n \urcorner]$.

2.2.4. LEMMA (successor, predecessor, test for zero). *There are combinators* $\mathbf{S}^+, \mathbf{P}^-$, **zero** *such that for all* $n \in \mathbf{N}$ *one has*

$$\mathbf{S}^+ \ulcorner n \urcorner = \ulcorner n+1 \urcorner, \qquad \mathbf{P}^- \ulcorner n+1 \urcorner = \ulcorner n \urcorner,$$

$$\mathbf{Zero} \ulcorner 0 \urcorner = \mathbf{true}, \qquad \mathbf{Zero} \ulcorner n+1 \urcorner = \mathbf{false}.$$

PROOF. Take

$$\mathbf{S}^+ \equiv \lambda x.[\mathbf{false}, x], \qquad \mathbf{P}^- \equiv \lambda x.x\mathbf{false}, \qquad \mathbf{Zero} \equiv \lambda x.x\mathbf{true}. \qquad \square$$

2.2.5. DEFINITION. (i) A *numeric function* is a map $f: \mathbf{N}^p \to \mathbf{N}$ for some p.

(ii) A numeric function f with p arguments is called *λ-definable* if one has for some combinator F

$$F \ulcorner n_1 \urcorner \ldots \ulcorner n_p \urcorner = \ulcorner f(n_1, \ldots, n_p) \urcorner \tag{2.2}$$

for all $n_1, \ldots, n_p \in \mathbf{N}$. If (2.2) holds, then f is said to be *λ-defined* by F.

2.2.6. DEFINITION. (i) The *initial functions* are the numeric functions U_n^i, S^+, Z defined by

$$U_n^i(x_1, \ldots, x_n) = x_i, \quad 1 \leqslant i \leqslant n,$$
$$S^+(n) = n+1, \qquad Z(n) = 0.$$

(ii) Let $P(n)$ be a numeric relation. As usual, $\mu m[P(m)]$ denotes the least number m such that $P(m)$ holds if there is such a number; otherwise it is undefined.

2.2.7. DEFINITION. Let $\mathscr{A}$ be a class of numeric functions.

(i) $\mathscr{A}$ is *closed under composition* if, for all f defined by

$$f(\vec{n}) = g(h_1(\vec{n}), \ldots, h_m(\vec{n})) \quad \text{with } g, h_1, \ldots, h_m \in \mathscr{A},$$

one has $f \in \mathscr{A}$.

(ii) $\mathscr{A}$ is *closed under primitive recursion* if, for all f defined by

$$f(0, \vec{n}) = g(\vec{n}), \qquad f(k+1, \vec{n}) = h(f(k, \vec{n}), k, \vec{n}) \qquad \text{with } g, h \in \mathscr{A},$$

one has $f \in \mathscr{A}$.

(iii) $\mathscr{A}$ is *closed under minimalization* if, for all f defined by

$$f(\vec{n}) = \mu m[g(\vec{n}, m) = 0] \quad \text{with } g \in \mathscr{A} \text{ such that } \forall \vec{n}\, \exists m\, g(\vec{n}, m) = 0,$$

one has $f \in \mathscr{A}$.

2.2.8. DEFINITION. The class $\mathscr{R}$ of *recursive functions* is the smallest class of numeric functions that contains all initial functions and is closed under composition, primitive recursion and minimalization.

So $\mathscr{R}$ is an inductively defined class. The proof that all recursive functions are λ-definable is by a corresponding induction argument. The result is originally due to Kleene [54].

2.2.9. LEMMA. *The initial functions are λ-definable.*

PROOF. Take as defining terms

$$\mathbf{U}_p^i \equiv \lambda x_1 \ldots x_p.x_i, \qquad \mathbf{S}^+ \equiv \lambda x.[\mathbf{false}, x], \qquad \mathbf{Z} \equiv \lambda x.\ulcorner 0 \urcorner. \qquad \square$$

2.2.10. LEMMA. *The λ-definable functions are closed under composition.*

PROOF. Let $g, h_1, \ldots, h_m$ be λ-*defined by* $G, H_1, \ldots, H_m$ respectively. Then $f(\vec{n}) = g(h_1(\vec{n}), \ldots, h_m(\vec{n}))$ is λ-defined by

$$F \equiv \lambda \vec{x}.G(H_1\vec{x}) \ldots (H_m\vec{x}). \qquad \square$$

2.2.11. LEMMA. *The λ-definable functions are closed under primitive recursion.*

PROOF. Let f be defined by

$$f(0, \vec{n}) = g(\vec{n}), \qquad f(k+1, \vec{n}) = h(f(k, \vec{n}), k, \vec{n})$$

where g, h are λ-defined by G, H respectively. An intuitive way to compute $f(k, \vec{n})$ is the following:

- test whether $k=0$;
- if yes, then give output $g(\vec{n})$;
- if no, them compute $h(f(k-1, \vec{n}), k-1, \vec{n})$.

Therefore we want a term F such that

$$\begin{aligned} F x \vec{y} &= \textit{if } (\mathbf{Zero}\ x) \textit{ then } G\vec{y} \textit{ else } H(F(\mathbf{P}^- x)\vec{y})(\mathbf{P}^- x)\vec{y} \\ &\equiv D(F, x, \vec{y}). \end{aligned}$$

Now such an F can be found by Application 2.1.8 of the Fixed Point Theorem. □

2.2.12. LEMMA. *The λ-definable functions are closed under minimalization.*

PROOF. Let f be defined by $f(\vec{n}) = \mu m[g(\vec{n}, m) = 0]$, where g is λ-defined by G. Again, by the Fixed Point Theorem, there is a term H such that

$$H\vec{x}y = \textit{if } (\mathbf{zero}(G\vec{x}y)) \textit{ then } y \textit{ else } H\vec{x}(\mathbf{S}^+ y).$$

Set $F = \lambda\vec{x}.H\vec{x}[0]$. Then F λ-defines f:

$$\begin{aligned} F[\vec{n}] &= H\ulcorner\vec{n}\urcorner\ulcorner 0\urcorner \\ &= \begin{cases} \ulcorner 0\urcorner & \text{if } G\ulcorner\vec{n}\urcorner\ulcorner 0\urcorner = \ulcorner 0\urcorner \\ H\ulcorner\vec{n}\urcorner\ulcorner 1\urcorner & \text{else} \end{cases} \\ &= \begin{cases} \ulcorner 1\urcorner & \text{if } G\ulcorner\vec{n}\urcorner\ulcorner 1\urcorner = \ulcorner 0\urcorner \\ H\ulcorner\vec{n}\urcorner\ulcorner 2\urcorner & \text{else} \end{cases} \\ &= \begin{cases} \ulcorner 2\urcorner & \text{if } \ldots \\ \ldots & \end{cases} \end{aligned}$$

□

2.2.13. THEOREM. *All recursive functions are λ-definable.*

PROOF. By Lemmas 2.2.9–12. □

The converse also holds. The idea is that if a function is λ-definable, then its graph is recursively enumerable, because equations derivable in the lambda calculus can be enumerated. It then follows that the function is recursive. So, for numeric functions, we have that f is recursive iff f is λ-definable. Moreover, also for partial functions, a notion of λ-definability exists. If ψ is a partial numeric function, then we have

$$\psi \text{ is partial recursive iff } \psi \text{ is } \lambda\text{-definable.}$$

The notions λ-definable and recursive both are intended to be formalisations of the intuitive concept of computability. Another formalisation was proposed by Turing in the form of Turing computability. The equivalence of the notions recursive, λ-definable and Turing computable (see, besides the original [88], also [30]) provides evidence for the Church–Turing thesis that states that "recursive" is the proper formalisation of the intuitive notion "computable".

Now we show the λ-definability of recursive functions with respect to the Church numerals.

2.2.14. THEOREM. *With respect to the Church numericals* $\mathbf{c}_n$ *all recursive functions can be* λ*-defined.*

PROOF. Define

$$\mathbf{S}_c^+ \equiv \lambda xyz.y(xyz),$$
$$\mathbf{P}_c^- \equiv \lambda xyz.x(\lambda pq.q(py))(\mathbf{K}z)\mathbf{I} \quad \text{(this term was found by J. Velmans)},$$
$$\mathbf{zero}_c \equiv \lambda x.x\mathbf{S}^+ \ulcorner 0 \urcorner \mathbf{true}.$$

Then these terms represent the successor, predecessor and test for zero. Then, as before, all recursive functions can be λ-defined. □

An alternative proof uses "translators" between the numerals $\ulcorner n \urcorner$ and c_n.

2.2.15. PROPOSITION. *There exist terms* $\mathbf{T}$ *and* $\mathbf{T}^{-1}$ *such that for all* n
(i) $\mathbf{T}c_n = \ulcorner n \urcorner$;
(ii) $\mathbf{T}^{-1}\ulcorner n \urcorner = c_n$.

PROOF. (i) Take $\mathbf{T} \equiv \lambda x.x\mathbf{S}^+\ulcorner 0 \urcorner$.
(ii) Take $\mathbf{T}^{-1} \equiv \lambda x.\mathit{if}\ \mathbf{Zero}\ x\ \mathit{then}\ c_0\ \mathit{else}\ \mathbf{S}_c^+(\mathbf{T}^{-1}(\mathbf{P}^-x))$. □

2.2.16. COROLLARY. *Second proof of Theorem* 2.2.14.

PROOF. Let f be a recursive function (of arity 2, say). Let F represent f with respect to the numerals $\ulcorner n \urcorner$. Define

$$F_c = \lambda xy.\mathbf{T}^{-1}(F(\mathbf{T}x)(\mathbf{T}y)).$$

Then F_c represents f with respect to the Church numerals. □

We end this section with some results showing the flexibility of lambda calculus. First, we give the double Fixed Point Theorem.

2.2.17. THEOREM. $\forall A, B\, \exists X, Y\ X = AXY\ \&\ Y = BXY.$

PROOF. Define $F \equiv \lambda x.[A(x\mathbf{true})(x\mathbf{false}), B(x\mathbf{true})(x\mathbf{false})]$. By the simple Fixed Point Theorem 2.1.7 there exists a Z such that $FZ = Z$. Take $X \equiv Z\mathbf{true}$, $Y \equiv Z\mathbf{false}$. Then

$$X \equiv Z\mathbf{true} = FZ\mathbf{true} = A(Z\mathbf{true})(Z\mathbf{false}) \equiv AXY$$

and, similarly, $Y = BXY$. □

ALTERNATIVE PROOF (Smullyan). By the ordinary Fixed Point Theorem, we can construct a term N such that

$$Nxyz = x(Nyyz)(Nzyz).$$

Now take $X = NAAB$ and $Y = NBAB$. □

2.2.18. APPLICATION. *Given context $C_i \equiv C_i[f, g, \vec{x}]$, $i = 1, 2$, there exist F_1, F_2 such that*

$$F_1\vec{x} = C_1[F_1, F_2, \vec{x}] \quad \textit{and} \quad F_2\vec{x} = C_2[F_1, F_2, \vec{x}].$$

PROOF. Define $A_i \equiv \lambda f_1 f_2 x. C_i[f_1, f_2, \vec{x}]$ and apply Theorem 2.2.17 to A_1, A_2. □

Of course, Theorem 2.2.17 (both proofs) and Application 2.2.18 generalise to a k-fold fixed point theorem and k-fold recursion.

Now it will be shown that there is a "self-interpreter" in the lambda calculus, i.e. a term E such that for closed terms M (without constants) we have the E applied to the "code" of M yields M itself (which is an executable on a reduction machine). First , we give a definition of coding.

2.2.19. DEFINITION (*coding λ-terms without constants*)

(i) $v^{(0)} = v$; $v^{(n+1)} = v^{(n)\prime}$; similarly, one defines $c^{(n)}$.

(ii) $\#v^{(n)} = \langle 0, n\rangle$; $\#c^{(n)} = \langle 1, n\rangle$;
$\#(MN) = \langle 2, \langle \#M, \#N\rangle\rangle$; $\#(\lambda x.M) = \langle 3, \langle \#x, \#M\rangle\rangle$.

(iii) Notation: $\ulcorner M \urcorner = \ulcorner \#M \urcorner$.

2.2.20. THEOREM (Kleene). *There is an "interpreter" $E \in \Lambda^0$ for closed λ-terms without constants. That is, for all closed M without constants one has $E\ulcorner M\urcorner = M$.*

PROOF (P. de Bruin). Construct an E_1 such that

$$E_1 F\ulcorner x\urcorner = F\ulcorner x\urcorner,$$
$$E_1 F\ulcorner MN\urcorner = (E_1 F\ulcorner M\urcorner)\,(E_1 F\ulcorner N\urcorner),$$
$$E_1 F\ulcorner \lambda x.M\urcorner = \lambda z.(E_1 F_{[x:=z]}\ulcorner M\urcorner)$$

where

$$F_{[x:=z]}\ulcorner n\urcorner = \begin{cases} F\ulcorner n\urcorner & \text{if } n \neq \#x, \\ z & \text{if } n = \#x. \end{cases}$$

(Note that $F_{[x:=z]}$ can be written in the form $G\ulcorner x\urcorner zF$.) By induction on M if follows that

$$E_1 F\ulcorner M\urcorner = M[x_1 := F\ulcorner x_1\urcorner, \ldots, x_n := F\ulcorner x_n\urcorner][c_1 := F\ulcorner c_1\urcorner, \ldots, c_m := F\ulcorner c_m\urcorner]$$

where $\{x_1, \ldots, x_n\} = \mathrm{FV}(M)$ and $\{c_1, \ldots, c_m\}$ are the constants in M. Hence, for closed M without constants, one has $E_1\mathsf{I}\ulcorner M\urcorner = M$. Now take $E \equiv \lambda x.E_1\mathsf{I}x = E_1\mathsf{I}$. □

For terms possibly containing a fixed finite set of constants $\vec{c}$ and variables $\vec{x}$, one can also construct an interpreter $E_{\vec{c},\vec{x}}$.

2.2.21. APPLICATION. Remember $\mathbf{U}_n^n = \lambda x_1 \ldots x_n.x_n$. Construct an $F \in \Lambda^0$ such that $F\ulcorner M\urcorner = \mathbf{U}_n^n$.

SOLUTION. Clearly, $\#\mathbf{U}_n^n$ depends recursively on n. Therefore, for some recursive g with λ-defining term G, one has

$$\#\mathbf{U}_n^n = g(n) \Rightarrow \ulcorner\mathbf{U}_n^n\urcorner = G\ulcorner n\urcorner.$$

Now take $F = \lambda x.E(Gx)$. Then

$$F\ulcorner n\urcorner = E(G\ulcorner n\urcorner) = E\ulcorner\mathbf{U}_n^n\urcorner = \mathbf{U}_n^n.$$

EXERCISE (Vree). Find a relatively simple F such that $F\ulcorner n\urcorner = \mathbf{U}_n^n$.

Now we prove the so-called second Fixed Point Theorem.

2.2.22. THEOREM. $\forall F\, \exists X\; F\ulcorner X\urcorner = X$.

PROOF. By the effectiveness of $\#$, there are recursive functions Ap and Num such that $\text{Ap}(\#M, \#N) = \#MN$ and $\text{Num}(n) = \#\ulcorner n\urcorner$. Let Ap and Num be λ-defined by **Ap** and $\mathbf{Num} \in \Lambda^0$. Then

$$\mathbf{Ap}\ulcorner M\urcorner\ulcorner N\urcorner = \ulcorner MN\urcorner, \qquad \mathbf{Num}\ulcorner n\urcorner = \ulcorner\ulcorner n\urcorner\urcorner;$$

hence, in particular, $\mathbf{Num}\ulcorner M\urcorner = \ulcorner\ulcorner M\urcorner\urcorner$. Now define

$$W \equiv \lambda x.F(\mathbf{Ap}x(\mathbf{Num}x)), \qquad X \equiv W\ulcorner W\urcorner.$$

Then

$$\begin{aligned} X \equiv W\ulcorner W\urcorner &= F(\mathbf{Ap}\ulcorner W\urcorner(\mathbf{Num}\ulcorner W\urcorner)) \\ &= F\ulcorner W\ulcorner W\urcorner\urcorner \equiv F\ulcorner X\urcorner. \qquad \square \end{aligned}$$

An application will be given in Section 5.2. Another application is the following result, due to Scott, which is quite useful for proving undecidability results.

2.2.23. THEOREM. *Let $A \subseteq \Lambda$ such that A is nontrivial, i.e. $A \neq \emptyset$, $A \neq \Lambda$. Suppose that A is closed under $=$, that is,*

$$M \in A,\ M = N \Rightarrow N \in A.$$

Then A is not recursive; that is $\#A = \{\#M \mid M \in A\}$ is not recursive.

PROOF. Suppose A is recursive. It follows that there is an $F \in \Lambda^0$ with

$$M \in A \Leftrightarrow F\ulcorner M\urcorner = \ulcorner 0\urcorner, \qquad M \notin A \Leftrightarrow F\ulcorner M\urcorner = \ulcorner 1\urcorner.$$

Let $M_0 \in A$, $M_1 \notin A$. We can find a $G \in \Lambda$ such that

$$M \in A \Leftrightarrow G\ulcorner M\urcorner = M_1 \notin A, \qquad M \notin A \Leftrightarrow G\ulcorner M\urcorner = M_0 \in A.$$

(Take $Gx = $ *if* $\mathbf{Zero}(Fx)$ *then* M_1 *else* M_0.) By the second Fixed Point Theorem, there is a term M such that $G\ulcorner M\urcorner = M$. Hence,

$$M \in A \Leftrightarrow M = G\ulcorner M\urcorner \notin A,$$

a contradiction. □

The following application, first proved in [22] using another method, was in fact historically the first example of a noncomputable property. We say that a term M is in *normal form* (*nf*) if M has no part of the form $(\lambda x.P)Q$. A term M *has* a normal form if $M = M'$ and M' is in nf. For example **I** is in nf and **IK** has a nf.

2.2.24. COROLLARY. *The set* $\mathrm{NF} = \{M \mid M \text{ has a nf}\}$ *is not recursive.*

PROOF. NF is closed under $=$ and is nontrivial. □

3. Semantics

Semantics of a (formal) language L gives a "meaning" to the expressions in L. This meaning can be given in two ways: (1) By providing a way in which expressions of L are *used.* (2) By *translating* the expressions of L into expressions of another language that is already known.

In Section 3.1 the meaning of a λ-term is explained through the notions reduction and strategy. Once a reduction strategy is chosen, the behaviour of a term is determined. This gives a so-called *operational semantics.*

In Section 3.2 the meaning of a λ-term is given by translating it to (an expression denoting) a set $[\![M]\!]$. This set is an element of a mathematical structure in which application and abstraction are well-defined operations and the map $[\![\ ]\!]$ preserves these operations. In this way we obtain a so-called *denotational semantics.*

In both cases the semantics gives a particular view on terms and classifies them. For example, in the operational semantics two terms may act the same on the numerals. The equivalence relation on terms induced by a denotational semantics, i.e. $M \approx N$ iff $[\![\mathrm{M}]\!] = [\![N]\!]$, is often of considerable interest and can settle questions of purely syntactical character; see for example the proof of Theorem 21.2.21 in [7].

3.1. Operational semantics: reduction and strategies

There is a certain asymmetry in the basic scheme (β). The statement

$$(\lambda x.x^2+1)\,3 = 3^2+1$$

can be interpreted as "3^2+1 is the result of computing $(\lambda x.x^2+1)\,3$", but not vice versa. This computational aspect will be expressed by writing

$$(\lambda x.x^2+1)\,3 \twoheadrightarrow 3^2+1$$

which reads "$(\lambda x.x^2+1)\,3$ *reduces to* 3^2+1".

Apart from this conceptual aspect, reduction is also useful for an analysis of convertibility. The Church–Rosser Theorem says that if two terms are convertible, then there is a term to which they both reduce. In many cases the inconvertibility of two terms can be proved by showing that they do not reduce to a common term.

3.1.1. DEFINITION. (i) A binary relation R on Λ is called *compatible* (with the operations) if

$$M\,R\,N \;\Rightarrow\; ZM\,R\,ZN,\ MZ\ R\ NZ \text{ and } \lambda x.M\,R\,\lambda x.N.$$

(ii) A *congruence* relation on Λ is a compatible equivalence relation.
(iii) A *reduction* relation on Λ is a compatible reflexive and transitive relation.

3.1.2. DEFINITION. The binary relations $\rightarrow_\beta$, $\twoheadrightarrow_\beta$ and $=_\beta$ on Λ are defined inductively as follows:

(i) (1) $(\lambda x.M)N \rightarrow_\beta M[x:=N]$;
(2) $M \rightarrow_\beta N \Rightarrow ZM \rightarrow_\beta ZN$,
$M \rightarrow_\beta N \Rightarrow MZ \rightarrow_\beta NZ$,
$M \rightarrow_\beta N \Rightarrow \lambda x.M \rightarrow_\beta \lambda x.M$.

(ii) (1) $M \twoheadrightarrow_\beta M$;
(2) $M \rightarrow_\beta N \Rightarrow M \twoheadrightarrow_\beta N$;
(3) $M \twoheadrightarrow_\beta N,\ N \twoheadrightarrow_\beta L \Rightarrow M \twoheadrightarrow_\beta L$.

(iii) (1) $M \twoheadrightarrow_\beta N \Rightarrow M =_\beta N$;
(2) $M =_\beta N \Rightarrow N =_\beta M$;
(3) $M =_\beta N,\ N =_\beta L \Rightarrow M =_\beta L$.

These relations are pronounced as follows:

- $M \twoheadrightarrow_\beta N$ reads "M β-*reduces* to N";
- $M \rightarrow_\beta N$ reads "M β-reduces to N *in one step*";
- $M =_\beta N$ reads "M is β-*convertible* to N".

By definition we have that $\rightarrow_\beta$ is compatible, $\twoheadrightarrow_\beta$ is a reduction relation and $=_\beta$ is a congruence relation; $\twoheadrightarrow_\beta$ is the transitive reflexive closure of $\rightarrow_\beta$ and $=_\beta$ is the equivalence relation generated by $\rightarrow_\beta$.

3.1.3. PROPOSITION. $M =_\beta N \Leftrightarrow \lambda \vdash M = N$.

PROOF. ($\Rightarrow$): By induction on the generation of $=_\beta$.
($\Leftarrow$): By induction on the length of proof. □

3.1.4. DEFINITION. (i) A β-*redex* is a term of the form $(\lambda x.M)N$ and $M[x:=N]$ is its *contractum*.

(ii) A λ-term M *is* a β-*normal form* (β-nf) if it does not have a β-redex as subexpression.

(iii) A term M *has* a β-normal form if $M =_\beta N$ and N is a β-nf, for some N.

EXAMPLE. $(\lambda x.xx)y$ is not a β-nf, but has as β-nf the term yy.

An immediate property of nf's is the following.

3.1.5. LEMMA. *Let M be a β-nf. Then*

$$M \twoheadrightarrow_\beta N \;\Rightarrow\; N \equiv M.$$

PROOF. This is true if $\twoheadrightarrow_\beta$ is $\rightarrow_\beta$. The result follows by transitivity. □

3.1.6. THEOREM (Church–Rosser). *If $M \twoheadrightarrow_\beta N_1$, $M \twoheadrightarrow_\beta N_2$, then for some* N_3 *one has $N_1 \twoheadrightarrow_\beta M_3$ and $N_2 \twoheadrightarrow_\beta N_3$ (see Fig. 2).*

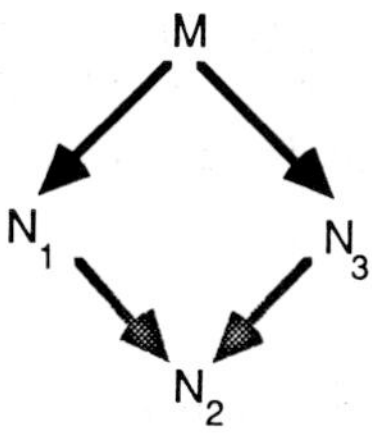

Fig. 2.

The proof will not be given, but can be found in, e.g., [7, Section 11.1].

3.1.7. COROLLARY. *If $M =_\beta N$, then there is an L such that $M \twoheadrightarrow_\beta L$ and $N \twoheadrightarrow_\beta L$.*

PROOF. Induction on the generation of $=_\beta$.

Case 1: $M =_\beta N$ because $M \twoheadrightarrow_\beta N$. Take $L \equiv N$.

Case 2: $M =_\beta N$ because $N =_\beta M$. By the IH there is a common β-reduct L_1 of N and M. Take $L \equiv L_1$.

Case 3: $M =_\beta N$ because $M =_\beta N'$, $N' =_\beta N$. Apply the IH to find L_1, L_2 as in Fig. 3 and then use Theorem 3.1.6 to find L. □

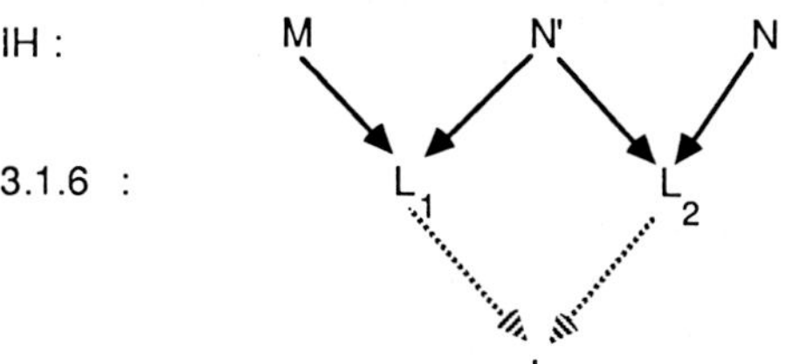

Fig. 3.

3.1.8. COROLLARY. (i) *If M has N as β-nf, then $M \twoheadrightarrow_\beta N$.*
(ii) *A λ-term has at most one β-nf.*

PROOF. (i) Suppose $M =_\beta N$ with N in β-nf. By Corollary 3.1.7, $M \twoheadrightarrow_\beta L$, $N \twoheadrightarrow_\beta L$ for some L. But then $N \equiv L$ by Lemma 3.1.5, so $M \twoheadrightarrow_\beta N$.
(ii) Suppose M has β-nf's N_1, N_2. Then $N_1 =_\beta N_2 (=_\beta M)$. By Corollary 3.1.7, $N_1 \twoheadrightarrow_\beta L$, $N_2 \twoheadrightarrow_\beta L$ for some L. But then $N_1 \equiv L \equiv N_2$ by Lemma 3.1.5. □

Some consequences are the following:
(1) The λ-calculus is consistent, i.e. $\lambda \nvdash \mathbf{true} = \mathbf{false}$. Otherwise $\mathbf{true} =_\beta \mathbf{false}$ by Proposition 3.1.3, which is impossible by Corollary 3.1.8 since **true** and **false** are distinct β-nf's. This is a syntactic consistency proof.
(2) $\mathbf{\Omega} \equiv (\lambda x.xx)(\lambda x.xx)$ has no β-nf. Otherwise $\mathbf{\Omega} \twoheadrightarrow_\beta N$ with N in β-nf. But $\mathbf{\Omega}$ only reduces to itself and is not in β-nf.
(3) In order to find the β-nf of a term M (if it exists), the various subexpressions of M may be reduced in different orders. By Corollary 3.1.8(ii) the β-nf is unique.
The combinator $\mathbf{Y}$ introduced in Theorem 2.1.7. finds fixed points: $\mathbf{Y}F = F(\mathbf{Y}F)$. One does not have

$$\mathbf{Y}F \twoheadrightarrow F(\mathbf{Y}F),$$

although this is often desirable. Turing introduced another fixed point operator that does have the desired property.

3.1.9. THEOREM (Turings fixed point combinator). *Let $\mathbf{\Theta} = AA$, with $A = \lambda xy.y(xxy)$. Then $\mathbf{\Theta}F \twoheadrightarrow F(\mathbf{\Theta F})$.*

PROOF. $\mathbf{\Theta}F \equiv AAF \to F(AAF) \equiv F(\mathbf{\Theta}F)$. □

Similarly, one can find solutions for the double and for the second fixed point theorem that do reduce in an analogous manner.

Strategies
There are terms having a nf but that are such that not all reduction paths lead to the nf. For example $A \equiv \mathbf{K}\mathbf{I}B$, with B a term without a nf, has as normal form $\mathbf{I}$; but A also has an infinite reduction path (by reducing within B).
A *reduction strategy* chooses among the various possible redexes which one(s) to contract, and thereby it determines how to reduce a term. The following theorem states that the *leftmost* (or also called *normal*) reduction strategy always normalises terms that do have a nf.

3.1.10. DEFINITION. (i) The *main symbol* of a redex $(\lambda x.M)N$ is the first λ.
(ii) Let R_1, R_2 be redex occurrences in a term M. Then R_1 is *to the left of* R_2 if the main symbol of R_1 is to the left of that of R_2.
(iii) We write $M \to_\ell N$ if N results from N by contracting the leftmost redex in M. $\twoheadrightarrow_\ell$ is the transitive reflexive closure of $\to_\ell$.

The following theorem, due to Curry, states that if a term has a nf, then that nf can be found by leftmost reduction. For a proof see [7, Theorem 13.2.2]. A more direct and simpler proof is in [86, Theorem 4.7].

3.1.11. THEOREM. *If M has a nf N, then $M \twoheadrightarrow_{\ell} N$.*

The leftmost reduction strategy is sometimes called *lazy reduction*, because in an expression like $(\lambda ab.C[a, b])AB$ it does not first evaluate A and B to nf, but substitutes these subterms directly into $C[a, b]$. *Eager reduction* is such that first A and B are reduced to nf before these are substituted in C.

For the lambda calculus it is not possible to have an eager evaluation mechanism. (For this reason the lambda calculus is sometimes called a *lazy language*.) This is due to the possibility of so-called nonstrict functions like $\lambda x.[0]$. It is an advantage for the expressive power of the lambda calculus, but a complication when implementing the language.

3.2. Denotational semantics: lambda models

Trying to construct a model for the lambda calculus one would like a space D such that D is isomorphic to D^D. For cardinality reasons this is impossible. In 1969 Scott solved this problem by restricting D^D to the *continuous* functions on D provided with a proper topology. Because of Schönfinkel's identification of $D^{D \times D}$ with $(D^D)^D$, it is natural to use a class of topological spaces that form a Cartesian closed category. For this reason Scott worked with complete lattices with an induced topology and constructed a D such that $D^D \cong D$ in this category. It turned out that for a model of the lambda calculus it is sufficent to find a D such that D^D is a retract of D. Such a model will be constructed in this subsection.

After Scott introduced the first set-theoretic models of the lambda calculus, various categories of domains have been studied in which these models "live". For a survey of this domain theory, see the chapter of Gunter and Scott in this Handbook.

3.2.1. DEFINITION. A *complete lattice* is a partially ordered set $D = (D, \sqsubseteq)$ such that for all $X \subseteq D$ the supremum $\sup X \in D$ exists.

$D, D', \ldots$ range over complete lattices. Each D has a largest element *top* $\top = \sup D$ and a least element *bottom* $\bot = \sup \emptyset$ and every $X \subseteq D$ has an *infimum* $\inf X = \sup\{y \mid \forall x \in X\ y \sqsubseteq x\}$. A subset $X \subseteq D$ is *directed* if $X \neq \emptyset$ and $\forall x,y \in X\ \exists z \in X\ [x \sqsubseteq z$ and $y \sqsubseteq z]$.

3.2.2. DEFINITION. A map $f: D \to D'$ is *continuous* if for all directed $X \subseteq D$ one has $f(\sup X) = \sup f(X)$ $(= \sup\{f(x) \mid x \in X\})$.

There is a topology on complete lattices such that "continuous" in Definition 3.2.2 coincides with the usual notion.

Note that a continuous function is automatically monotonic.

$$\begin{aligned} x \sqsubseteq y \;&\Rightarrow\; y = \sup\{x, y\} \\ &\Rightarrow\; f(y) = f(\sup\{x, y\}) = \sup\{f(x), f(y)\} \\ &\Rightarrow\; f(x) \sqsubseteq f(y). \end{aligned}$$

3.2.3. DEFINITION. Let $D = (D, \sqsubseteq)$, $D' = (D', \sqsubseteq')$.

(i) $D \times D' = \{(d, d') \mid d \in D, d' \in D'\}$ is the Cartesian product of D, D' and is partially ordered by

$$(d_1, d_1') \sqsubseteq (d_2, d_2') \;\Leftrightarrow\; d_1 \sqsubseteq d_2 \text{ and } d_1' \sqsubseteq' d_2'.$$

(ii) $[D \to D'] = \{f : D \to D' \mid f \text{ is continuous}\}$ is a function space partially ordered by

$$f \sqsubseteq g \;\Leftrightarrow\; \forall d \in D\; f(d) \sqsubseteq' g(d).$$

3.2.4. PROPOSITION. (i) *$D \times D'$ is a complete lattice with for $X \subseteq D \times D'$*

$$\sup X = (\sup(X)_0, \sup(X)_1),$$

where

$$\begin{aligned} (X)_0 &= \{d \in D \mid \exists d' \in D'\,(d, d') \in X\}, \\ (X)_1 &= \{d' \in D' \mid \exists d \in D\,(d, d') \in X\}. \end{aligned}$$

(ii) *Let $f_i : D \to D'$, $i \in I$, be a collection of continuous maps. Define $f(x) = \sup_i(f_i(x))$. Then f is continuous and in $[D \to D']$ one has $f = \sup f_i$. Therefore $[D \to D']$ is a complete lattice.*

PROOF. (i) Easy.

(ii) Let $X \subseteq D$ be directed. Then

$$\begin{aligned} f(\sup X) &= \sup_i f_i(\sup X) \\ &= \sup_i \sup_{x \in X} f_i(x) \quad \text{by continuity of } f_i, \\ &= \sup_{x \in X} \sup_i f_i(x) \\ &= \sup_{x \in X} f(x). \end{aligned}$$

Therefore f is continuous. Hence $f = \sup f_i$ in $[D \to D']$. □

If Λx denotes λ-abstraction in set theory, then we have, as a consequence of Proposition 3.2.4(ii)

$$\sup_i \Lambda x. f_i(x) = \Lambda x. \sup_i(f_i(x)),$$

i.e., sup commutes with Λ.

3.2.5. PROPOSITION. *Let $f \in [D \to D]$. Then f has a least fixed point defined by*

$$\mathrm{Fix}(f) = \sup_n f^n(\bot).$$

PROOF. Note that the set $\{f^n(\bot) \mid n \in \mathbf{N}\}$ is directed: $\bot \sqsubseteq f(\bot)$; hence, by monotonicity, $f(\bot) \sqsubseteq f^2(\bot)$ etcetera, so $\bot \sqsubseteq f(\bot) \sqsubseteq f^2(\bot) \sqsubseteq \cdots$. Therefore

$$f(\mathit{fix}(f)) = \sup_n f(f^n(\bot)) = \sup_n f^{n+1}(\bot) = \mathrm{Fix}(f).$$

If x is another fixed point of f, then $f(x) = x$ and $\bot \sqsubseteq x$; so, by monotonicity, $f^n(\bot) \sqsubseteq f^n(x) = x$. Therefore $\mathrm{Fix}(f) \sqsubseteq x$. □

3.2.6. LEMMA. *Let $f\colon D \times D' \to D''$. Then f is continuous iff f is continuous in each of its variables separately (i.e. $\lambda x.f(x, x'_0)$ and $\lambda x'.f(x_0, x')$ are continuous for all x_0, x'_0).*

PROOF. ($\Rightarrow$): As usual.

($\Leftarrow$): Let $X \subseteq D \times D'$ be directed. Then

$$\begin{aligned} f(\sup X) &= f(\sup(X)_0, \sup(X)_1) \\ &= \sup_{x \in (X)_0} f(x, \sup(X)_1) \\ &= \sup_{x \in (X)_0} \sup_{x' \in (X)_1} f(x, x') \\ &= \sup_{(x,x') \in X} f(x, x'). \end{aligned}$$

The last equality holds because X is directed. Therefore f is continuous. □

3.2.7. PROPOSITION (continuity of application). *Define* $\mathrm{Ap}\colon [D \to D'] \times D \to D'$ *by* $\mathrm{Ap}(f, x) = f(x)$. *Then* Ap *is continuous.*

PROOF. Apply Lemma 3.2.6: $\lambda x.\mathrm{Ap}(f, x) = \lambda x.f(x) = f$ is continuous since $f \in [D \to D']$. Let $H = \lambda f.\mathrm{Ap}(f, x_0) = \lambda f.f(x_0)$. Then, for $f_i, i \in I$, directed,

$$\begin{aligned} H\Big(\sup_i f_i\Big) &= \Big(\sup_i f_i\Big)(x_0) \\ &= \sup_i (f_i(x_0)) \quad \text{by Proposition 3.2.4(ii),} \\ &= \sup_i H(f_i). \quad \square \end{aligned}$$

3.2.8. PROPOSITION (continuity of abstraction). *Let $f \in [D \times D' \to D'']$. Then $\lambda y.f(x, y) \in [D' \to D'']$ and depends continuously on x.*

PROOF. By Lemma 3.2.6 it follows that $\lambda y.f(x, y) \in [D' \to D'']$. Moreover, let $X \subseteq D$

be directed. Then

$$\Lambda y.f(\sup X, y) = \Lambda y.\sup_x f(x, y)$$
$$= \sup_x \Lambda y.f(x, y)$$

by continuity of f and the remark following Proposition 3.2.4. □

It now follows that the category of complete lattices with continuous maps forms a Cartesian closed category. We will not use this terminology however.

3.2.9. DEFINITION. (i) D is a *retract* of D' (notation $D < D'$) if there are continuous maps $F: D' \to D, G: D \to D'$ such that $F \circ G = \mathrm{id}_D$.

(ii) D is called *reflexive* if $[D \to D] < D$.

REMARK. If $D < D'$ via the maps F, G, then F is surjective and G injective. We may identify D with its image $G(D) \subseteq D'$. Then F "retracts" the larger space D' to the subspace D.

Now it will be shown how a reflexive D can be turned into a model of the lambda-calculus.

3.2.10. DEFINITION. Let D be reflexive via F, G.

(i) F retracts D to its function space $[D \to D] \subseteq D$. So for $x \in D$ one has $F(x) \in [D \to D]$. In this way elements of D become functions on D and one may write

$$x \cdot_F y = F(x)(y) (\in D).$$

(ii) Conversely, every continuous function on D becomes via G an element of D. Now one may write

$$\lambda^G x.f(x) = G(f) (\in D).$$

for f continuous.

A *valuation* in D is a map ρ: variables$\to D$.

3.2.11. DEFINITION. Let D be reflexive via F,G.

(i) Given a valuation ρ in D and $M \in \Lambda$ the interpretation of M *in* D *under the valuation* ρ (notation $[\![M]\!]_\rho^D$) is defined as shown in Table 1, where $\rho(x := d)$ is the valuation ρ' with

$$\rho'(y) = \begin{cases} \rho(y) & \text{if } y \not\equiv x \\ d & \text{if } y \equiv x. \end{cases}$$

This definition is correct: by induction on P one can show the continuity of $\Lambda d.[\![P]\!]_{\rho(x := d)}$.

(ii) $M = N$ is *true in* D (notation $D \models M = N$) if for all ρ one has $[\![M]\!]_\rho^D = [\![N]\!]_\rho^D$.

Intuitively $[\![M]\!]_\rho^D$ is M interpreted in D where each lambda calculus application .

Table 1

M	$[\![M]\!]_\rho^D$
x	$\rho(x)$
PQ	$[\![P]\!]^D \cdot_F [\![Q]\!]_\rho^D$
$\lambda x.P$	$\lambda^G d.[\![P]\!]^D_{\rho(x:=d)}$

is interpreted as $\cdot_F$ and each λ as λ^G. For instance,

$$[\![\lambda x.xy]\!]_\rho^D = \lambda^G d.d\,\rho(y) = \lambda^G x.x\,\rho(y).$$

INFORMAL NOTATION. If a reflexive D is given and $\rho(y)=d$, then we will loosely write $\lambda x.xd$ to denote the more formal $[\![\lambda x.xy]\!]_\rho^D$.

Clearly $[\![M]\!]_\rho^D$ depends only on the values of ρ on FV(M). That is,

$$\rho \upharpoonright \mathrm{FV}(M) = \rho' \upharpoonright \mathrm{FV}(M) \;\Rightarrow\; [\![M]\!]_\rho^D = [\![M]\!]_{\rho'}^D,$$

where $\upharpoonright$ denotes function restriction. In particular for combinators, $[\![M]\!]_\rho^D$ does not depend on ρ and may be written $[\![M]\!]^D$. If D is clear from the context we write $[\![M]\!]_\rho$ or $[\![M]\!]$.

3.2.12. THEOREM. *If D is reflexive, then D is a sound model for the lambda calculus, i.e.*

$$\lambda \vdash M = N \;\Rightarrow\; D \models M = N.$$

PROOF. Induction on the proof of $M=N$. The only two interesting cases are the axioms (β) and the rule (ξ).

As to (β). This was the scheme $(\lambda x.M)N = M[x:=N]$. Now

$$\begin{aligned}[\![(\lambda x.M)N]\!]_\rho &= (\lambda^G d.[\![M]\!]_{\rho(x:=d)}) \cdot_F [\![N]\!]_\rho \\ &= F(G(\Lambda d.[\![M]\!]_{\rho(x:=d)}))([\![N]\!]_\rho) \\ &= (\Lambda d.[\![M]\!]_{\rho(x:=d)})([\![N]\!]_\rho) \quad \text{since } F \circ G = \mathrm{id}, \\ &= [\![M]\!]_{\rho(x:=[\![N]\!]_\rho)}.\end{aligned}$$

Sublemma: $[\![M[x:=N]]\!]_\rho = [\![M]\!]_{\rho(x:=[\![N]\!]_\rho)}$.

Subproof: Induction on the structure of M. Write $P^* \equiv P[x:=N]$, $\rho^* = \rho(x:=[\![N]\!]_\rho)$ and check Table 2.

Table 2

M	$[\![M^*]\!]_\rho$	$[\![M]\!]_{\rho^*}$	Comment
x	$[\![N]\!]_\rho$	$[\![N]\!]_\rho$	OK
y	$\rho(y)$	$\rho(y)$	OK
PQ	$[\![P^*]\!]_\rho \cdot [\![Q^*]\!]_\rho$	$[\![P]\!]_{\rho^*} \cdot_F [\![Q]\!]_{\rho^*}$	IH
$\lambda y.P$	$\lambda^G d.[\![P^*]\!]_{\rho(y:=d)}$	$\lambda^G d.[\![P]\!]_{\rho^*(y:=d)}$	$(\rho(y:=d))^* = \rho^*(y:=d)$

By the Sublemma, the proof of soundness of (β) is complete. As to (ξ), this was $M=N \Rightarrow \lambda x.M=\lambda x.M$. We have to show

$$D\models M=N \;\Rightarrow\; D\models \lambda x.M=\lambda x.M.$$

Now

$$\begin{aligned} & D\models M=N \\ & \quad\Rightarrow [\![M]\!]_\rho=[\![N]\!]_\rho && \text{for all } \rho \\ & \quad\Rightarrow [\![M]\!]_{\rho(x:=d)}=[\![N]\!]_{\rho(x:=d)} && \text{for all } \rho, d \\ & \quad\Rightarrow \lambda\!\!\lambda d.[\![M]\!]_{\rho(x:=d)}=\lambda\!\!\lambda d.[\![N]\!]_{\rho(x:=d)} && \text{for all } \rho \\ & \quad\Rightarrow \lambda^G d.[\![M]\!]_{\rho(x:=d)}=\lambda^G d.[\![N]\!]_{\rho(x:=d)} && \text{for all } \rho \\ & \quad\Rightarrow [\![\lambda x.M]\!]_\rho=[\![\lambda x.N]\!]_\rho && \text{for all } \rho \\ & \quad\Rightarrow D\models \lambda x.M=\lambda x.N. \quad \square \end{aligned}$$

Now we will give an example of a reflexive complete lattice called D_A. The method is due to Engeler [32] and is a code-free variant of the graph model $P\omega$ due to Plotkin [75] and Scott [80].

3.2.13. DEFINITION. (i) Let A be a set. Define

$$B_0=A, \qquad B_{n+1}=B_n\cup\{(\beta,b)\mid b\in B_n \text{ and } \beta\subseteq B_n, \beta \text{ finite}\},$$

$$B=\bigcup_n B_n.$$

$D_A=P(B)=\{x\mid x\subseteq B\}$, considered as complete lattice under inclusion ($\subseteq$). The set B is just the closure of A under the operation of forming ordered pairs (β,b). It is assumed that A consists of urelements, that is, A does not contain pairs $(\beta,b)\in B$.

(ii) Define $F:D_A\to[D_A\to D_A]$, $G:[D_A\to D_A]\to D_A$ by

$$F(x)(y)=\{b\mid \exists\beta\subseteq y\,(\beta,b)\in x\},$$

$$G(f)=\{(\beta,b)\mid b\in f(\beta)\}.$$

3.2.14. THEOREM. *D_A is reflexive via the maps F, G.*

PROOF. F and G are clearly continuous (use that the β's are finite). Moreover, for continuous f,

$$\begin{aligned} F\circ G(f)(y) &= F(\{\beta,b)\mid b\in f(\beta)\})(y) \\ &= \{b\mid \exists\beta\subseteq y\; b\in f(\beta)\} \\ &= \bigcup_{\beta\subseteq y} f(\beta) \\ &= f(y). \end{aligned}$$

since $\sup=\bigcup$ in D_A and $y=\bigcup_{\beta\subseteq y}\beta$ is a directed supremum. Therefore, $F\circ G(f)=f$ and hence, $F\circ G=\mathrm{id}_{[D_A\to D_A]}$. $\square$

Now a semantic proof of the consistency of the lambda calculus can be given.

3.2.15. COROLLARY. *The lambda-calculus is consistent, i.e.* $\lambda \nvdash$ *true* $=$ *false.*

PROOF. Otherwise, $\lambda \vdash x = y$; but then $D_A \models x = y$. This is not so; take $\rho(x) \neq \rho(y)$ in a D_A with $A \neq \emptyset$. $\square$

4. Extending the language

In Section 2 we have seen that all computable functions can be expressed in the lambda calculus. For reasons of *efficiency, reliability* and *convenience* this language will be extended.

In Section 4.1 some of the constants are selected to represent primitive data (such as numbers) and operations on these (such as addition). Some new reduction rules (the so-called *δ-rules*) are introduced to express the operational semantics of these operations. Even if these constants and operations can be implemented in the lambda calculus, it is worthwhile to have primitive symbols for them. The reason is that in an implementation of the lambda calculus addition of the Church numerals runs less efficiently than the usual implementation in hardware of addition of binary represented numbers. Having numerals and addition as primitives therefore creates the possibility to interprete these efficiently.

In Section 4.2 types are introduced as a tool towards writing correct software. Types are assigned to terms in various ways using several formal systems. For some of these systems it is decidable whether a term can be typed correctly. This ensures partial correctness of a program.

Before we start Section 4.1, we introduce the following language constructs that are quite useful and are often used in functional programming languages. It allows the programmer to introduce abbreviations for (recursively) defined functions.

4.0. DEFINITION. (i) The expression

"Let $x = M$ in E"

stands for "$(\lambda x.E)M$" or "$E[x := M]$". (The latter has some advantages if we want to type expressions: the various occurrences of M may need to be typed differently.)

(ii) The expression

"Letrec $fx = C[f, \vec{x}]$ in E"

stands for "Let $f = \boldsymbol{\Theta}(\lambda f\vec{x}.C[f, \vec{x}])$ in E", where $\boldsymbol{\Theta}$ is Turings fixed point operator (Theorem 3.1.9).

(iii) Similarly, one can have a letrec depending on the double fixed point theorem:

$$\text{"Letrec } \begin{aligned} f\vec{x} &= C_1[f, g, \vec{x}] \\ g\vec{x} &= C_2[f, g, \vec{x}]\text{".} \end{aligned}$$

These language constructs occur in languages like ML [66], Miranda [92] and TALE [11].

4.1. Delta rules

Let $X \subseteq \Lambda^0$ be a set of closed normal forms. Usually we take $X \subseteq C$. Let $f: X^k \to \Lambda$ be an "externally defined" function. In order to represent f, a so-called **δ**-rule may be added to the lambda calculus. This is done as follows:

(1) A special constant in C is selected and is given some name, say $\boldsymbol{\delta}$ $(=\boldsymbol{\delta}_f)$.
(2) The following contraction rules are added to those of the lambda calculus:

$$\boldsymbol{\delta} M_1 \ldots M_k \to f(M_1, \ldots, M_k), \quad M_1, \ldots, M_k \in X.$$

Note that, for a given function f, this is not one contraction rule but in fact a rule *scheme*. The resulting extension is called the **λδ**-calculus. The corresponding notion of (one-step) reduction is denoted by $(\to_{\beta\delta})$ $\twoheadrightarrow_{\beta\delta}$. So δ-reduction is not an absolute notion, but depends on the choice of f.

4.1.1. THEOREM (Mitschke). *Let f be a function on closed normal forms. Then the resulting notion of reduction $\twoheadrightarrow_{\beta\delta}$ satisfies the Church–Rosser Theorem.*

PROOF. This follows from Theorem 15.3.3 in [7]. □

The notion of normal form generalises to βδ-normal form. So does the concept of leftmost reduction. The βδ-normal forms can be found by a leftmost reduction.

4.1.2. THEOREM. *If $M \twoheadrightarrow_{\beta\delta} N$ and N is in βδ-nf, then $M \twoheadrightarrow_{\ell\beta\delta} N$.*

PROOF. Analogous to the proof of Theorem 3.1.10 for β-normal forms. □

4.1.3. EXAMPLE. One of the first versions of a **δ**-rule can be found in [23]. Here X is the set of all closed normal forms and, for $M, N \in X$, we have

$$\boldsymbol{\delta}_C MN \to \lambda xy.x \quad \text{if } M \equiv N,$$
$$\boldsymbol{\delta}_C MN \to \lambda xy.y \quad \text{if } M \not\equiv N.$$

4.1.4. EXERCISE. Let k_n be defined by $k_0 = k$ and $k_{n+1} = \mathbf{K}(k_n)$. Show that on the k_n the recursive functions can be represented by terms in $\lambda\boldsymbol{\delta}_C$.

Another possible set of **δ**-rules is for the Booleans.

4.1.5. EXAMPLE. The following constants are selected in C.

$\underline{\text{true}}$, $\underline{\text{false}}$, $\underline{\text{not}}$, $\underline{\text{and}}$, $\underline{\text{ite}}$ (for *if then else*).

The following **δ**-rules are introduced:

$$\begin{array}{ll} \underline{\text{not}}\ \underline{\text{true}} \to \underline{\text{false}}, & \underline{\text{not}}\ \underline{\text{false}} \to \underline{\text{true}}, \\ \underline{\text{and}}\ \underline{\text{true}}\ \underline{\text{true}} \to \underline{\text{true}}, & \underline{\text{and}}\ \underline{\text{true}}\ \underline{\text{false}} \to \underline{\text{false}}, \\ \underline{\text{and}}\ \underline{\text{false}}\ \underline{\text{true}} \to \underline{\text{false}}, & \underline{\text{and}}\ \underline{\text{false}}\ \underline{\text{false}} \to \underline{\text{false}}, \\ \underline{\text{ite}}\ \underline{\text{true}} \to \textbf{true} \equiv \lambda xy.x, & \underline{\text{ite}}\ \underline{\text{false}} \to \textbf{false} \equiv \lambda xy.y. \end{array}$$

It follows that

$$\underline{\text{ite}}\ \underline{\text{true}}\ x\ y \twoheadrightarrow x, \qquad \underline{\text{ite}}\ \underline{\text{false}}\ x\ y \twoheadrightarrow y.$$

Now we introduce as **δ**-rules some operations on the set of integers $\mathbf{Z}=\{\ldots,-2,-1,0,1,2,\ldots\}$.

4.1.6. EXAMPLE. For each $n \in \mathbf{Z}$ a constant in C is selected and given the name $\underline{n}$. (We will express this as follows: for each $n \in \mathbf{Z}$ a constant $n \in C$ is choosen.) Moreover, the following constants in C are selected:

$$\underline{\text{plus}}, \quad \underline{\text{minus}}, \quad \underline{\text{times}}, \quad \underline{\text{divide}}, \quad \underline{\text{error}}, \quad \underline{\text{equal}}.$$

Then we introduce the following **δ**-rules (schemes). For m, $n \in \mathbf{Z}$,

$$\begin{array}{lll} \underline{\text{plus}}\ \underline{n}\ \underline{m} \to \underline{n+m}, & \underline{\text{minus}}\ \underline{n}\ \underline{m} \to \underline{n-m}, & \underline{\text{times}}\ \underline{n}\ \underline{m} \to \underline{n*m}, \\ \underline{\text{divide}}\ \underline{n}\ \underline{m} \to \underline{n:m} \quad \text{if } m \neq 0; & \underline{\text{divide}}\ \underline{n}\ 0 \to \underline{\text{error}}, & \\ \underline{\text{equal}}\ \underline{n}\ \underline{n} \to \underline{\text{true}}, & \underline{\text{equal}}\ \underline{n}\ \underline{m} \to \underline{\text{false}} \quad \text{if } n \neq m. & \end{array}$$

We may add rules like $\underline{\text{plus}}\ \underline{n}\ \underline{\text{error}} \to \underline{\text{error}}$.

4.1.7. EXERCISE. Write down a **λδ**-term F such that $F\ \underline{n} \twoheadrightarrow \underline{n!+n}$.

Similar **δ**-rules can be introduced for the set of reals.
Again, another set of **δ**-rules is concerned with characters.

4.1.8. EXAMPLE. Let Σ be some linearly ordered alphabet. For each symbol $s \in \Sigma$ we choose a constant 's' $\in C$. Moreover, we choose two constants $\boldsymbol{\delta}_{\leqslant}$ and $\boldsymbol{\delta}_{=}$ in C and formulate the following **δ**-rules:

$$\begin{array}{ll} \boldsymbol{\delta}_{\leqslant}\,`s_1\text{'}\,`s_2\text{'} \to \underline{\text{true}} & \text{if } s_1 \text{ precedes } s_2 \text{ in the ordering of } \Sigma, \\ \boldsymbol{\delta}_{\leqslant}\,`s_1\text{'}\,`s_2\text{'} \to \underline{\text{false}} & \text{otherwise.} \\ \boldsymbol{\delta}_{=}\,`s_1\text{'}\,`s_2\text{'} \to \underline{\text{true}} & \text{if } s_1 = s_2, \\ \boldsymbol{\delta}_{=}\,`s_1\text{'}\,`s_2\text{'} \to \underline{\text{false}} & \text{otherwise.} \end{array}$$

4.1.9. EXERCISE. Write down a **λδ**-term F such that for $s_1, s_2, t_1, t_2 \in \Sigma$ we have

$$\begin{array}{ll} F[`s_1\text{'}, `t_1\text{'}][`s_2\text{'}, `t_2\text{'}] \twoheadrightarrow \underline{\text{true}} & \text{if } [s_1, t_1] \text{ precedes } [s_2, t_2] \text{ in the} \\ & \text{lexicographical ordering of } \Sigma \times \Sigma, \\ F[`s_1\text{'}, `t_1\text{'}][`s_2\text{'}, `t_2\text{'}] \twoheadrightarrow \underline{\text{false}} & \text{otherwise.} \end{array}$$

The following is a less orthodox **δ**-rule.

4.1.10. EXAMPLE. For each integer function P in Pascal (say with two integer arguments) choose a constant $\#P$ in C. Add as **δ**-rule

$$\boldsymbol{\delta}_{\textbf{Pascal}}\,\#P\ \underline{n}\ \underline{m} \to \underline{k},$$

provided that k is the value of P with input n and m.

It is also possible to represent "multiple value" functions F by putting as δ-rule

$$\delta\,\underline{n} \to \underline{m}, \quad \text{provided that } F(n)=m.$$

Of course, the resulting $\lambda\delta$-calculus does not satisfy the Church–Rosser Theorem and can be used to deal with nondeterministic computations. We will not consider this.

4.2. *Types*

Types are certain objects, usually syntactic expressions, that may be assigned to terms denoting programs. Types serve as a classification of the (objects denoted by the) terms. Each type σ has as semantics a set D_σ of "objects of type σ". There are several systems of type assignment with different collections of types. We will consider a few of these. (For the more complicated type systems the semantics, D_σ will in general be not a set, but an object in some category.)

Type assignment is done for the following reasons. Firstly, the type of a term F gives a *partial specification* of what the function denoted by F is supposed to do. Usually, this type a specification is given before the term as program is constructed. The verification whether this term, once it has been constructed, is indeed of the required type provides a *partial correctness* proof for the program. Secondly, types play a role in *efficiency*. If it is known that a subterm S of a program has a certain type, then S may be executed more efficiently by making use of the type information.

To explain the idea of type assignment, we present type systems of various strengths. We start with a simple set of types Type_1, inductively defined as follows.

4.2.1. DEFINITION

$$\mathbf{B} \in \text{Type}_1, \qquad \mathbf{Z} \in \text{Type}_1, \qquad \mathbf{Char} \in \text{Type}_1 \qquad \text{(type constants)},$$
$$\sigma \in \text{Type}_1,\ \tau \in \text{Type}_1 \ \Rightarrow\ (\sigma \to \tau) \in \text{Type}_1.$$

CONVENTION. $\sigma_1 \to \sigma_2 \to \cdots \to \sigma_n$ stands for $(\sigma_1 \to (\sigma_2 \to \cdots (\sigma_{n-1} \to \sigma_n)))$.

The intuitive semantics for these types is as follows. The types **Z**, **Char** and **B** are used to denote respectively the sets of integers, characters and Booleans. That is, $D_{\mathbf{Z}} = \{\ldots, -2, -1, 0, 1, 2, \ldots\}$ etcetera. $D_{\sigma \to \tau} = D_\tau^{D_\sigma}$, i.e. the set of functions from D_σ to D_τ. (For more complicated systems $D_{\sigma \to \tau}$ is only a subset or subobject of $D_\tau^{D_\sigma}$.)

Now we will see how types are used in the form of so called type assignment.

4.2.2. DEFINITON. Let Type be a set of types.

(i) A *statement* is of the form $M : \sigma$, where $\sigma \in \text{Type}$ and M is a $\lambda\delta$-term. M is called the *subject* and σ the *predicate* of the statement.

(ii) A *theory of type assignment* consist of a set of statements axiomatised in some way.

4.2.3. DEFINITION. Consider the $\lambda\delta$-terms for Booleans, integers and characters introduced in Examples 4.1.3–5. Consider the set Type_1 introduced in Definition 4.2.1. The theory of type assignment T_1 is defined by the following set of axioms and rules.

Axioms:

(1) $\underline{\text{true}}$: **B**, $\underline{\text{false}}$: **B**, $\underline{\text{not}}$: **B**→**B**, $\underline{\text{and}}$: **B**→**B**→**B**,

(2) $\underline{n}$: **Z**; $\underline{\text{error}}$: **Z**,
$\underline{\text{plus}}$, $\underline{\text{minus}}$, $\underline{\text{times}}$, $\underline{\text{divide}}$: **Z**→**Z**→**Z**,
$\underline{\text{equal}}$: **Z**→**Z**→**B**,

(3) $\boldsymbol{\delta}_{\leqslant}, \boldsymbol{\delta}_{=}$: **Char**→**Char**→**B**.

Rule:

(4) If $M: \sigma \to \tau$ and $N: \sigma$, then $MN: \tau$.

The system T_1 is very weak. We can only derive statements like $\underline{\text{equal}}\,(\underline{\text{plus}}\ \underline{3}\ \underline{4})\ \underline{12}$: **B**. No types more complex than the ones given in the axioms are possible in type assignment. In the following system T_2, essentially due to Curry, more statements can be derived.

4.2.4. DEFINITION The set of types Type_2 is inductively defined as follows:

$$\mathbf{B} \in \text{Type}_2, \qquad \mathbf{Z} \in \text{Type}_2, \qquad \mathbf{Char} \in \text{Type}_2,$$
$$\alpha_0, \alpha_1, \ldots \in \text{Type}_2 \quad \text{(type variables)},$$
$$\sigma \in \text{Type}_2, \tau \in \text{Type}_2 \ \Rightarrow\ (\sigma \to \tau) \in \text{Type}_2.$$

The $\alpha_0, \alpha_1, \ldots$ are called type variables and are often denoted by $\alpha, \beta, \gamma, \ldots$.

The interpretation of types now becomes dependent on a valuation ξ that assigns to type variables some set (or object in some category) $\xi(\alpha)$:

$$D^{\xi}_{\alpha} = \xi(\alpha), \qquad D^{\xi}_{\mathbf{Z}} = \{\ldots, -2, -1, 0, 1, 2, \ldots\} \quad \text{etcetera},$$
$$D^{\xi}_{\sigma \to \tau} = D^{\xi(D_\sigma{}^\xi)}.$$

In order to give the system for type assignment we need some more concepts.

4.2.5. DEFINITION. A *basis* is a set B of statements in which the subjects are distinct term variables. The notion that a statement $M: \sigma$ is *derivable* from a basis B, notation $B \vdash M: \sigma$, is defined inductively as follows:
Axioms:

(0) $x: \sigma \in B \ \Rightarrow\ B \vdash x: \sigma$,

(1) $B \vdash \underline{\text{true}}$: **B**, $B \vdash \underline{\text{false}}$: **B**,
$B \vdash \underline{\text{not}}$: **B**→**B**, $B \vdash \underline{\text{and}}$: **B**→**B**→**B**,
$B \vdash \underline{n}$: **Z**; $B \vdash \underline{\text{error}}$: **Z**,
$B \vdash \underline{\text{plus}}$, $\underline{\text{minus}}$, $\underline{\text{times}}$, $\underline{\text{divide}}$: **Z**→**Z**→**Z**,
$B \vdash \underline{\text{equal}}$: **Z**→**Z**→**B**
$B \vdash \boldsymbol{\delta}_{\leqslant}, \boldsymbol{\delta}_{=}$: **Char**→**Char**→**B**.

Rules:

(2) $B \vdash M: \sigma \to \tau,\ B \vdash N: \sigma \ \Rightarrow\ B \vdash MN: \tau$,

(3) $\quad B \cup \{x:\sigma\} \vdash M:\tau \;\Rightarrow\; B \vdash \lambda x.M:\sigma \to \tau.$

The axioms and rules can be summarised in a convenient natural deduction notation as follows:

Axioms:

(1) $\quad \underline{\text{true}}: \mathbf{B} \qquad \underline{\text{false}}: \mathbf{B} \qquad$ etcetera.

(The axiom (0) is not written down in this notation, but implicitly understood.)

Rules:

(2)
$$\frac{M:\sigma\to\tau \qquad N:\sigma}{MN:\tau}$$

(3)
$$\frac{\begin{array}{c}[x:\sigma]\\ \vdots \\ M:\tau\end{array}}{\lambda x.M:\sigma\to\tau}$$

EXAMPLE. In T_2 we can derive the following statements. We write $\vdash M:\sigma$ for $\emptyset \vdash M:\sigma$.

$$\vdash \lambda xy.x:\sigma\to(\tau\to\sigma), \qquad y:\tau \vdash \lambda x.y:\sigma\to\tau,$$
$$\vdash (\lambda xyz.\underline{\text{times}}\,x\,(\underline{\text{plus}}\;y\,z)): \mathbf{Z}\to\mathbf{Z}\to\mathbf{Z}\to\mathbf{Z}.$$

A derivation for the first statement looks like

$$\frac{\dfrac{\dfrac{[x:\sigma] \qquad [y:\tau]}{x:\sigma}}{\lambda y.x:\tau\to\sigma}}{\lambda xy.x:\sigma\to(\tau\to\sigma)}$$

Note that in this derived statement σ and τ are arbitrary types.

If T is a system of type assignment, then a term M is called *typable* if for some B and σ one has $B \vdash M:\sigma$. The following result is independently due to Curry [28] and Hindley [41] and was rediscovered by Milner [65].

4.2.6. THEOREM. (i) *It is decidable whether a term is typable in* T_2.

(ii) *If a term M has a type in* T_2, *then M has a unique* principal type scheme, *i.e. a type* σ *such that every possible type for M is a substitution instance of* σ. *Moreover,* σ *is computable from M.*

For example the principal type scheme of $\lambda xy.x$ is $\alpha\to(\beta\to\alpha)$.

The following result states that terms typable in T_2 all have a normal form.

4.2.7. THEOREM. *In* T_2 *we have* $B \vdash M:\sigma \;\Rightarrow\;$ *M has a normal form.*

PROOF. This follows from the same property for T_3, see Theorem 4.2.13. □

The following result is called the "subject reduction theorem" by Curry and follows inductively from the rules of type assignment.

4.2.8 THEOREM. *The following holds for* T_2

$$B \vdash M : \sigma \quad \text{and} \quad M \twoheadrightarrow M' \;\Rightarrow\; B \vdash M' : \sigma.$$

The type assignment system T_2 is still rather poor. For example, the term ite representing the conditional has no natural type. For this reason we will extend the set of types and the rules for type assignment. The resulting system is a variant (the "Curry version" in the terminology of [8]) of the polymorphic second-order lambda calculus independently due to [37] and [78].

4.2.9. DEFINITION. Type_3 is the set of types inductively defined by

$$\mathbf{B} \in \text{Type}_3, \qquad \mathbf{Z} \in \text{Type}_3, \qquad \mathbf{Char} \in \text{Type}_3, \qquad \alpha_0, \alpha_1, \ldots \in \text{Type}_3,$$
$$\sigma \in \text{Type}_3, \tau \in \text{Type}_3 \;\Rightarrow\; (\sigma \to \tau) \in \text{Type}_3,$$
$$\sigma \in \text{Type}_3 \;\Rightarrow\; \forall \alpha_i \sigma \in \text{Type}_3.$$

4.2.10. DEFINITION. T_3 is the system of type assignment with types in Type_3 defined by the following axioms and rules (in a natural deduction formulation).

(1) $\underline{\text{true}}: \mathbf{B}$, $\underline{\text{false}}: \mathbf{B}$, $\underline{\text{not}}: \mathbf{B} \to \mathbf{B}$, $\underline{\text{and}}: \mathbf{B} \to \mathbf{B} \to \mathbf{B}$,
$\underline{n}: \mathbf{Z}$: $\underline{\text{error}}: \forall \alpha . \alpha$,
$\underline{\text{plus}}, \underline{\text{minus}}, \underline{\text{times}}, \underline{\text{divide}}; \mathbf{Z} \to \mathbf{Z} \to \mathbf{Z}$,
$\underline{\text{equal}}: \mathbf{Z} \to \mathbf{Z} \to \mathbf{B}$,
$\boldsymbol{\delta}_{\leqslant}, \boldsymbol{\delta}_{=}: \mathbf{Char} \to \mathbf{Char} \to \mathbf{B}$,
$\underline{\text{ite}}: \forall \alpha . \mathbf{B} \to \alpha \to \alpha \to \alpha$.

(2) $$\frac{M : \sigma \to \tau \qquad N : \sigma}{MN : \tau}$$

(3) $$\frac{\begin{matrix}[x : \sigma] \\ \vdots \\ M : \tau\end{matrix}}{\lambda x . M : \sigma \to \tau}$$

(4) $$\frac{M : \forall \alpha . \sigma}{M : \sigma[\alpha := \tau]}$$

(5) $$\frac{M : \sigma}{M : \forall \alpha . \sigma}$$

In rule (5) the type variable may not occur in the assumptions on which $M : \sigma$ depends. That is, rule (5) should be read as follows:

$$B \vdash M : \sigma, \ \alpha \text{ not free in } B \;\Rightarrow\; B \vdash M : \forall \alpha . \sigma.$$

4.2.11. EXAMPLES. In T_3 the following statements can be derived:

(i) $\vdash \mathbf{I} : \forall \alpha . \alpha \to \alpha$;

(ii) Define $\text{Nat} = \forall \alpha . \alpha \to \alpha \to \alpha \to \alpha$. Then for the Church numerals $c_n \equiv \lambda f x . f^n x$ we have $\vdash_{c_n}$: Nat.

This example shows the intended meaning of the types $\forall \alpha . \sigma$. For example in T_2 we had $\vdash \mathbf{I} : \sigma \to \sigma$ for all σ. Now in T_3 we have $\vdash \mathbf{I} : \forall \alpha . \alpha \to \alpha$. A formal semantics for T_3 gives rise to interesting (categorical) notions and will not be discussed here (see [37, 82, 48].

On the Church numerals, many computable functions can be represented by terms of type Nat→Nat. The next result is due to [36]; see also [34].

4.2.12. THEOREM. *A numeric function f is representable by a term of type* Nat→Nat *iff f is an in analysis provable total recursive function.*

The following normalisation result is due to [36]. For a proof see also [85]. The subject reduction theorem is also valid.

4.2.13. THEOREM. *In T_3 we have*

(i) $B \vdash M : \sigma \;\Rightarrow\; M$ *has a normal form*;

(ii) $B \vdash M : \sigma$ *and* $M \twoheadrightarrow M' \;\Rightarrow\; B \vdash M' : \sigma$.

4.2.14. OPEN PROBLEM. Is it decidable whether a term has a type in T_3?

So far all systems of type assignment are such that typable terms have a normal form. It will now be shown that type assignment systems with this property have as limitation that not all computable functions are representable by a typed term.

4.2.15. THEOREM. *Let $\boldsymbol{\lambda\delta}$ be an extension of the lambda calculus by means of effective $\boldsymbol{\delta}$-rules ($M \twoheadrightarrow_{\beta\delta} N$ should be an r.e. relation in M, N). Let T be an effective system of type assignment ($\vdash M : \sigma$ should be an r.e. relation in M) such that every typable term has a normal form. Then not all computable functions are representable in $\boldsymbol{\lambda\delta}$ by a term typable in T.*

PROOF. Recall that a numeric function $F(n, m)$ is called *universal* for a class of unary numeric functions $\mathscr{A}$ if $\mathscr{A} = \{f_0, f_1, \ldots\}$, where f_n is defined by $f_n(m) = F(n, m)$.

Let $\mathscr{A}^k$ be the class of total numeric computable functions with k arguments. It is well-known that there is no $F \in \mathscr{A}^2$ that is universal for $\mathscr{A}^1$. (Otherwise the function $g(n) = F(n, n) + 1$ will yield a contradiction.) It follows that not all computable functions are representable by a typable term. (Otherwise one could obtain a computable function universal for $\mathscr{A}^1$ by enumerating terms: define

$$F(n, m) = k \Leftrightarrow \text{the } n\text{th typed term (say with type } \mathbf{Z} \to \mathbf{Z}) \text{ applied to the argument } \underline{m} \text{ has as normal form } \underline{k}.$$

In order to construct this F, we need that reduction and typing are effective.) □

The following system of type assignment is such that all computable functions are representable by a typed term. Indeed, the system also assigns types to nonnormalising terms by introducing a primitive fixed point combinator $\mathbf{Y}$ having type $\forall\alpha.(\alpha\to\alpha)\to\alpha$.

4.2.16. DEFINITION. (i) The $\boldsymbol{\lambda Y\delta}$-calculus is an extension of the $\boldsymbol{\lambda\delta}$-calculus in which there is a constant $\mathbf{Y}$ with reduction rule $\mathbf{Y}f \to f(\mathbf{Y}f)$.

(ii) T_4 is the system that assigns types in Type_3 to $\boldsymbol{\lambda Y\delta}$-terms. T_4 is defined by adding to the system T_3 the axiom

$$\mathbf{Y} : \forall\alpha.(\alpha\to\alpha)\to\alpha.$$

Because of the presence of **Y**, not all terms have a normal form. Without proof we state the following theorem.

4.2.17. THEOREM. (i) *The* $\lambda\mathbf{Y}\delta$*-calculus satisfies the Church–Rosser property.*

(ii) *If a term in the* $\lambda\mathbf{Y}\delta$*-calculus has a normal form, then it can be found using leftmost reduction.*

(iii) *The subject reduction theorem holds for* T_4.

4.2.18. THEOREM. *All computable functions can be represented in* $\lambda\mathbf{y}\delta$ *by a term typable in* T_4.

PROOF. The construction uses the primitive numerals $\underline{n}$. If we take $\underline{\text{zero}} \equiv \lambda x.\underline{\text{equal}}\,\underline{0}\,x$ and $P^- \equiv \lambda x.\underline{\text{minus}}\,x\,\underline{1}$, then the proof of Theorem 2.2.13 can be imitated using **Y** instead of the fixed point combinator. The types for the functions defined using **Y** are natural. □

Because T_4 is an extension of T_3, it is not known whether typability in T_4 is decidable. The language ML [66], has a system of type assignment that is a subsystem of T_4 in which the $\forall\alpha$ construction on types may only occur on the outside (i.e., $\forall\alpha.\alpha \to \alpha$ is allowed, but not $\beta \to \forall\alpha.\alpha$). For this system typability is decidable and all computable functions are representable by typable terms.

Many aspects of types have been omitted in this section, notably

(1) types for (Cartesian) products, (disjoint) unions and intersections,
(2) recursive types,
(3) dependent types,
(4) type inclusion,
(5) abstract data types,
(6) formulae as types,
(7) semantics: for simple, polymorphic and dependent types, fully abstract models.

Some relevant articles are: [6, 12, 14, 16, 20, 21, 24, 25, 35, 43, 48, 61, 62, 64, 70, 81, 82]. The following volumes contain many articles on typed lambda calculus: [51] and the Proceedings of the Conferences on Logic in Computer Science (IEEE, 1986, 1987 and 1988). See also the chapter on type systems for programming languages by Mitchell in this Handbook, [8] and [38].

5. The theory of combinators and implementation issues

5.1. The theory of combinators

In the lambda calculus the notation of bound variable is used. This does not only cause some complications in the theory (avoiding clashes of bound and free variables) but also in implementations of the lambda calculus. On way to avoid these problems is to translate the lambda calculus into systems without free variables, the so-called *combinatory systems*.

5.1.1. DEFINITION. The set of *combinatory terms*, notation $\mathscr{C}$, is inductively defined as follows.

$$c \in C \Rightarrow c \in \mathscr{C}, \qquad x \in V \Rightarrow x \in \mathscr{C},$$
$$P, Q \in \mathscr{C} \Rightarrow (PQ) \in \mathscr{C}.$$

where C and V are the sets of constants and variables respectively, as in Definition 2.1.1.

5.1.2. DEFINITION. (i) The theory $\mathbf{CL}(\boldsymbol{I}, \boldsymbol{K}, \boldsymbol{S})$ has as terms the set $\mathscr{C}$. Among the constant of $\mathscr{C}$ three are selected and given the names $\boldsymbol{I}$, $\boldsymbol{K}$ and $\boldsymbol{S}$ respectively. The theory is axiomatised by the following set of axioms:

$$\boldsymbol{I}P = P, \qquad \boldsymbol{K}PQ = P, \qquad \boldsymbol{S}PQR = PR(QR).$$

The rules are just the rules of equality and are the same as in Definition 2.1.5(ii), except that the rule (ξ) is not present. We still usually write **CL** for $\mathbf{CL}(\boldsymbol{I}, \boldsymbol{K}, \boldsymbol{S})$.

(ii) The axioms in (i) correspond to a relation *of weak reduction* by setting

$$\boldsymbol{I}P \rightarrow_{\mathrm{w}} P, \qquad \boldsymbol{K}PQ \rightarrow_{\mathrm{w}} P, \qquad \boldsymbol{S}PQR \rightarrow_{\mathrm{w}} PR(QR)$$

The relation $\rightarrow_{\mathrm{w}}$ which is compatible with application generates just as in Definition 3.1.2 the reduction relation $\twoheadrightarrow_{\mathrm{w}}$ and conversion relation $=_{\mathrm{w}}$. Similarly to Proposition 3.1.3 we have

$$\mathbf{CL} \vdash \mathrm{P} = \mathrm{Q} \Leftrightarrow P =_{\mathrm{w}} Q.$$

(iii) The $\boldsymbol{\delta}$-rules discussed in Section 4.1 can be added to the theory $\mathbf{CL}(\boldsymbol{I}, \boldsymbol{K}, \boldsymbol{S})$, obtaining $\mathbf{CL}(\boldsymbol{I}, \boldsymbol{K}, \boldsymbol{S}, \boldsymbol{\delta})$ or simply $\mathbf{CL}\boldsymbol{\delta}$.

5.1.3. DEFINITION. (i) For $P \in \mathscr{C}$ we define $\lambda^* x.P \in \mathscr{C}$ to simulate abstraction in **CL**.

$$\lambda^* x.x \equiv \boldsymbol{I},$$
$$\lambda^* x.P \equiv \boldsymbol{K}P \quad \text{if } x \text{ does not occur in } P,$$
$$\lambda^* x.PQ \equiv \boldsymbol{S}(\lambda^* x.P)(\lambda^* x.Q).$$

(ii) Lambda terms can be translated into combinatory terms: for $M \in \Lambda$ we define $M_{\mathrm{CL}} \in \mathscr{C}$ as follows:

$$c_{\mathrm{CL}} \equiv c; \qquad x_{\mathrm{CL}} \equiv x,$$
$$(MN)_{\mathrm{CL}} \equiv M_{\mathrm{CL}} N_{\mathrm{CL}},$$
$$(\lambda x.M)_{\mathrm{CL}} \equiv \lambda^* x.M_{\mathrm{CL}}.$$

5.1.4. LEMMA. *For $P, Q, \in \mathscr{C}$ and $M, N \in \Lambda$ we have*

(i) $(\lambda^* x.P)Q \twoheadrightarrow_{\mathrm{w}} P[x := Q]$,

(ii) $(\lambda^* x.P)[y := Q] \equiv \lambda^* x.(P[y := Q])$,

(iii) $(M[x := N]_{\mathrm{CL}} \equiv M_{\mathrm{CL}}[x := N_{\mathrm{CL}}]$

PROOF. (i), (ii) By induction on the structure of P.

(iii) By induction on the structure of M, using (ii). $\square$

There is a difference between β-reduction and w-reduction. M_{CL} may be a nf for w-reduction but M itself not for β-reduction. Take, e.g., $M \equiv \lambda x.\boldsymbol{I}x$ with $M_{\mathrm{CL}} \equiv \boldsymbol{S}(\boldsymbol{KI})\boldsymbol{I}$. Therefore **CL** does not seem to be a good intermediate language for evaluating λ-terms. We will show now that for terms having a ground type the normal form can be found via translation and reduction in **CL**.

5.1.5. LEMMA. (i) *Let* $M \rightarrow_{\ell\beta\delta} N$ *with* M *closed and not of the form* $\lambda x.M$; *then* $M_{\mathrm{CL}} \rightarrow_{\mathrm{w}\delta} N_{\mathrm{CL}}$

(ii) *Let* $\twoheadrightarrow_{\ell\beta\delta} N$ *with* M *closed and typable in* T_4 *with a ground type* (**Z**, **B** *or* **Char**); *then* $M_{\mathrm{CL}} \twoheadrightarrow_{\mathrm{w}\delta} N_{\mathrm{CL}}$.

PROOF. (i) Induction on the shape of M. If $M \equiv (\lambda x.M_0)M_1\vec{L}$, then the result follows from Lemma 5.1.4. If $M \equiv \boldsymbol{c}L_1, \ldots, L_k$ and $N \equiv \boldsymbol{c}L'_1 \ldots L_k$, then the induction hypothesis applies to L_1. If $M \equiv \boldsymbol{c}\vec{L}$ is a **δ**-redex, then its **CL**-translation is a **δ**-redex too.

(ii) If M has a ground type, then so does each reduct M' of M and hence $M' \not\equiv \lambda x.M''$. The result follows by repeated application of (i). □

5.1.6. COROLLARY. *Let* $\boldsymbol{\lambda\delta} \vdash M = c$ *with* $\boldsymbol{c}$ *a constant and* M *typable in* T_4 *with a ground type. Then* $\mathbf{CL\delta} \vdash M_{\mathrm{CL}} = \boldsymbol{c}$.

PROOF

$$\begin{aligned} \boldsymbol{\lambda\delta} \vdash M = \boldsymbol{c} &\Rightarrow M =_{\beta\delta} \boldsymbol{c} \\ &\Rightarrow M \twoheadrightarrow_{\ell\beta\delta} \boldsymbol{c} \\ &\Rightarrow M_{CL} \twoheadrightarrow_{\mathrm{w}\delta} \boldsymbol{c}_{\mathrm{CL}} \equiv \boldsymbol{c} \quad \text{by Lemma 5.1.5(ii)} \\ &\Rightarrow \mathbf{CL\delta} \vdash M_{\mathrm{CL}} = \boldsymbol{c}. \quad \Box \end{aligned}$$

An important consequence is that in order to compute the λ-terms it is sufficient to build an implementation for combinatory reduction as long as the output is of a basic type. Indeed, many reduction machines are based on a combinator implementation (see, e.g., [90]).

5.2. *Implementation issues*

Implementing functional languages has become an art in itself. In this subsection we briefly discuss some issues and refer the reader to the literature. In particular to the comprehensive [74] and also to the conference proceedings [26, 50, 5, 52].

LISP

LISP is a language that is not exactly functional but has several features of it. Because implementations of several variants of LISP are quite successfully used, we want to indicate why we prefer functional languages. This preference explains why research is being done to make also fast implementations for these languages.

The following features of LISP 1.5 [63] make it different from functional languages: (1) There are assignment and goto statements. (2) There is "dynamic binding", which is a wrong kind of substitution in which a free variable may become bound. (3) There is a "quote" operator that "freezes" the evaluation of subterms. This quote is not referentially transparent, since quote $(Ic) \neq$ quote c.

In modern versions of LISP, e.g. SCHEME [84], there are no more goto statements and dynamic binding has been replaced by "static binding", i.e. the proper way to do substitution. However in SCHEME one still uses assignments and the quote operator.

Since there is no assignment in functional languages, there is a more clear semantics and therefore we have easier correctness proofs. For the same reason parallel computing is more natural using functional languages. Also functional languages have strong typing, something that is missing in LISP. The quote is sometimes thought to be essential to construct the "metacircular" interpreter for LISP written in LISP. However, self-interpretation is also possible in functional programming, see Theorem 2.2.20.

Another aspect of LISP is that the quote may be used in order to write "self-modifying" programs. This is also possible in the lambda calculus, even if the quote is not λ-definable. Suppose that we want a typically self-modifying function F, e.g.,

$$Fx = \textit{if } Px \textit{ then } G_1 x \textit{ else } G_2 \ulcorner F \urcorner x,$$

where P, G_1 and G_2 are given and $\ulcorner F \urcorner$ is the code of F represented as numeral in the lambda calculus. Then such an F can be found as a λ-term using the second Fixed Point Theorem 2.2.24:

$$F = (\lambda fx.\ \textit{if } Px \textit{ then } G_1 x \textit{ else } G_2 fx) \ulcorner F \urcorner.$$

So with lambda calculus we can do in a more hygienic way most things that can be done with LISP. Nevertheless, LISP is an important language. Experience gained by using and implementing LISP has been useful in functional languages.

The SECD-machine

The first proposal to implement a functional language was done by Landin [57]. He introduced the so called SECD-machine. It uses the idea of an environment, i.e. a list of pairs of variable an associated value that is dynamically updated. The SECD-machine can be used to support both eager and lazy evaluation [17]. A modern version of the SECD-machine is the CAM (Categorical Abstract Machine) [27].

Other combinators

The λ^*-algorithm to imitate abstraction in the combinatory system is rather inefficient. If M if of length n, then the length of M_{CL} is $O(n^3)$. Turner [90] gives an improved algorithm in which M_{CL} is $O(n^2)$. Statman [83] gives an algorithm with M_{CL} of length $O(n \log n)$ and shows that this is indeed best possible. However, his constant factor is such that the algorithm of Turner seems to be better in practical cases. In [53, 69] the complexity of various translations into fixed sets or families of combinators are compared.

Another possibility is to translate λ-terms to "user-defined combinators". For example a λ-term like

$$\lambda xyz\,.\,x(\lambda ab\,.\,ayz)w$$

may be translated into Aw with the ad hoc rules

$$Awxyz \to x(Byz)w, \qquad Byzab \to ayz.$$

This approach is taken in [47] (supercombinators), [44] (serial combinators) and [49] (lambda lifting).

Term rewrite systems

In the lambda calculus we have ordered pairs and projections. Also other constructors and selectors are definable. Introducing primitive constants for these allows a more efficient implementation. For example we may select some constants p, p_1, p_2 and add the rules

$$p_1(pxy) \to x, \qquad p_2(pxy) \to y.$$

These kind of rules are typical members of so called term rewrite systems (TRS). These TRS's are quite flexible to describe what a functional program should do. For example, a (double) fixed point combinator does not require a theorem, but can be written down immediately. See [46, 55, 56] for an exposition of the subject and [72] for the use of TRS's in programming.

Fig. 4.

Graph reduction

If a variable occurs more than once in a term M ($\equiv M[x, x]$, say), then the reduction

$$(\lambda x . M[x, x])N \to M[N, N]$$

with the actual substitution of N in M results in an increase of complexity. In [93] it was proposed that instead of performing the actual substitution, the contraction may be represented as a graph (see Fig. 4) with two pointers from within M towards one occurrence of N. This saves not only space but also execution time, since contraction within N now needs to be done only once, not twice. In [90] another use of graphs is proposed. For example, the **Y** combinator with contraction rule $\mathbf{Y}f \to f(\mathbf{Y}f)$ can be presented using a cyclic graph (see Fig. 5). Cyclic graphs do speed up performance of implemented functional languages; however, garbage collection becomes more complicated.

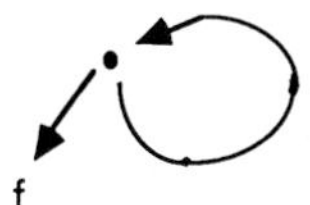

Fig. 5.

In [9] theorems are proved relating ordinary reduction to graph reduction. In [15] the language Clean based on term rewrite systems and graph reduction is proposed as an intermediate between functional languages and reduction machines.

Strictness analysis

A function F is said to be *strict* if for the evaluation of FA the argument A is always needed. For instance $\mathbf{I} \equiv \lambda x.x$ is strict, but $\mathbf{K}\underline{0} = \lambda x.\underline{0}$ is not. Knowing which parts of a functional program are strict is useful for a compiler, since then eager evaluation may be used for those parts, hence obtaining a performance improvement. Although the notion of strictness is in general undecidable, there are some interesting decidable approximations of this notion. See [45, 71, 18, 10, 77] for work in this direction.

Information about types and strictness are important for obtaining speed optimalisations. This is explained in [74] and exploited in e.g. the implementations described in [3, 49, 15].

Modifications of the lambda-calculus

Many implementations of functional languages are such that expressions like $\lambda x.M$ are not reduced any further. For this reason Abramsky introduced the notion of "lazy lambda calculus" in order to make the theory correspond better to this practice, see [73]. Also [76] (call-by-name lambda calculus) and [67] (partial lambda calculus) are relevant in this respect.

Acknowledgment

I would like to thank Mariangiola Dezani-Ciancaglini, Roger Hindley, Bart Jacobs and Pieter Koopman for useful suggestions on a draft version of this chapter.

References

[1] ABELSON, H. and G.J. SUSSMAN, *Structure and Interpretation of Computer Programs* (MIT Press, Cambridge, MA, 1985).

[2] ABRAMSKY, S., D.M. GABBAI and T.S.E. MAIBAUM, eds., *Handbook of Logic in Computer Science* (Oxford University Press, Oxford, to appear).

[3] AUGUSTSSON, L., A compiler for lazy ML, in: *Proc. ACM Symp. on LISP and Functional Programming* (1984) 218–227.

[4] BACKUS, J., Can programming be liberated from the von Neumann style? A functional style and its algebra of proofs, *Comm. ACM* **21** (8) (1978) 613–641.

[5] BAKKER, J.W. DE, A.J. NIJMAN and P.C. TRELEAVEN, eds., *PARLE, Parallel Architectures and Languages Europe*, Lecture Notes in Computer Science, Vol. 258 (Springer, Berlin, 1987).

[6] BARENDREGT, H.P., M. COPPO and M. DEZANI-CIANCAGLINI, A filter lambda model and the completeness of type assignment, *J. Symbolic Logic* **48** (1983) 931–940.

[7] BARENDREGT, H.P., *The Lambda Calculus, its Syntax and Semantics* (North-Holland, Amsterdam, 1984).

[8] BARENDREGT, H.P., Lambda calculi with types, to appear in [2].

[9] BARENDREGT, H.P., M.C.J.D. VAN EEKELEN, J.R.W. GLAUERT, J.R. KENNAWAY, M.J. PLASMEIJER and

M.R. SLEEP, Term graph rewriting, in: PARLE, *Parallel Architectures and Languages Europe, Vol. II: Parallel Languages*, Lecture Notes in Computer Science, Vol. 259 (Springer, Berlin, 1987) 141–158.

[10] BARENDREGT, H.P., J.R. KENNAWAY, J.W. KLOP and M.R. SLEEP, Needed reduction and spine strategies for the lambda calculus, *Inform. and Comput.* **75**(3) (1987) 191–231.

[11] BARENDREGT, H.P. and M. VAN LEEUWEN, Functional programming and the language TALE, in: J.W. de Bakker, W.-P. de Roever and G. Rozenberg, eds., *Current Trends in Concurrency*, Lecture Notes in Computer Science, Vol. 224 (Springer, Berlin, 1986) 122–208.

[12] BÖHM, C. and A. BERARDUCCI, Automatic synthesis of typed Λ-programs on term algebras, *Theoret. Comput. Sci.* **39** (1985) 135–154.

[13] BÖHM, C. and W. GROSS, Introduction to the Cuch, in: E. Caianiello, ed., *Theory of Automata* (Academic Press, London, 1965) 35–65.

[14] BRUCE, K.B. and A.R. MEYER, The semantics of second order polymorphic lambda calculus, in: [51] 131–144.

[15] BRUS, T.H., M.C.J.D. VAN EEKELEN, M.O. VAN LEER and M.J. PLASMEIJER, Clean, a language for functional graph rewriting, in: [52] 364–384.

[16] BRUIJN, N.G. DE, A survey of the project AUTOMATH, in: J.R. Hindley and J.P. Seldin, eds., *To H.B. Curry: Essays on Combinatory Logic, Lambda Calculus and Formalism* (Academic Press, London, 1980) 579–606.

[17] BURGE, W., *Recursive Programming Techniques* (Addison-Wesley, Reading, MA, 1975).

[18] BURN, G., C.L. HANKIN and S. ABRAMSKY, Strictness analysis for higher order functions, *Sci. Comput. Programming* **7** (1986) 249–278.

[19] BURSTALL, T., D. MACQUEEN and D. SANELLA, Hope: an experimental applicative language: in: *Proc. 1980 LISP Conf.* (ACM, New York, 1980) 136–143.

[20] CARDELLI, L., A semantics of multiple inheritance, in: *Proc. on Semantics of Data Types*, Lecture Notes in Computer Science, Vol. 173 (Springer Berlin, 1984) 51–68.

[21] CARDELLI, L. and P. WEGNER, On understanding types, data abstraction and polymorphism, *Comput. Surv.* **17** (4) (1985) 471–522.

[22] CHURCH, A., An unsolvable problem of elementary number theory, *Amer. J. Math.* **58** (1936) 354–363.

[23] CHURCH, A., *The Calculi of Lambda Conversion* (Princeton Univ. Press, Princeton, NJ, 1941).

[24] COPPO, M., A completeness result for recursively defined types, in: *Proc. 12th Internat. Coll. on Automata, Languages and Programming*, Lecture Notes in Computer Science, Vol. 194 (Springer, Berlin, 1985) 120–129.

[25] COQUAND, T. and G. HUET, Constructions: a higher-order proof system for mechanizing mathematics, in: B. Buchberger ed., *EUROCAL'85*, Lecture Notes in Computer Science, Vol. 203 (Springer, Berlin, 1985) 151–184.

[26] COUSINEAU, G., P.-L. CURIEN and B. ROBINET, eds., *Proceedings on Combinators and Functional Programming Languages*, Lecture Notes in Computer Science, Vol. 242 (Springer, Berlin, 1985).

[27] CURIEN, P.-L., *Categorical Combinators, Sequential Algorithms and Functional Programming* (Pitman, London, 1985).

[28] CURY, H.B., Modified basic functionality in combinatory logic, *Dialectica* **23** (1969) 83–92.

[29] DARLINGTON, J., P. HENDERSON and D.A. TURNER, eds., *Functional Programming and its Applications* (Cambridge Univ. Press, Cambridge, 1982).

[30] DAVIS, M., *Computability and Unsolvability* (McGraw-Hill, New York, 1958).

[31] EISENBACH, S., *Functional Programming* (J. Wiley, Chichester, UK, 1987).

[32] ENGELER, E., Algebras and combinators, *Algebra Universalis* **13**(3) (1981) 389–392.

[33] FIELD, A.J. and P.G. HARRISON, *Functional Programming* (Addison-Wesley, New York, 1987).

[34] FORTUNE, S., D. LEIVANT and M. O'DONNELL, The expressiveness of simple and second-order type structures, *J. ACM* **30** (1) (1983) 151–185.

[35] FRIEDMAN, H., Equality between functionals, in; R. Parikh, ed., *Logic Colloquium*, Lecture Notes in Mathematics, Vol. 453 (Springer, Berlin, 1975) 22–37.

[36] GIRARD, J.-Y., Intepretation fonctionelle et élimination des coupures dans l'arithmetique d'ordre supérieur, Ph.D. Thesis, Univ. de Paris VIII, 1972.

[37] GIRARD, J.-Y., The system *F* of variable types, fifteen years later. *Theoret. Comput. Sci.* **45** (1986) 159–192.

[38] GIRARD, J.-Y., Y. LAFONT and P. TAYLOR, *Proofs and Types* (Cambridge Univ. Press, Cambridge, 1989).
[39] GORDON, M., R. MILNER and C. WADSWORTH, *Edinburgh LCF*, Lecture Notes in Computer Science, Vol. 78 (Springer, Berlin, 1978).
[40] HENDERSON, P., *Functional Programming, Application and Implementation* (Prentice-Hall, London, 1980).
[41] HINDLEY, J.R., The principal type-scheme of an object in combinatory logic, *Trans. Amer. Math. Soc.* **146** (1969) 29–60.
[42] HINDLEY, J.R. and J.P. SELDIN, *Introduction to Combinators and λ-calculus* (Cambridge Univ. Press, Cambridge, 1986).
[43] HOWARD, W., The formulae-as-types notion of construction, in: J.R. Hindley and J.P. Seldin, eds., *To H.B. Curry: Essays on Combinatory Logic, Lambda Calculus and Formalism* (Academic Press, London, 1980) 479–490.
[44] HUDAK, P. and B. GOLDBERG, Serial combinators: optimal grains of parallelism, in: [50] 382–399.
[45] HUET, G. and J.-J. LÉVY, Call by need computations in non-ambiguous term rewriting systems, Tech. Report 359, INRIA, Le Chesnay, France, 1979.
[46] HUET, G. and D.C. OPPEN, Equations and rewrite rules: a survey, in: R. Book, ed., *Formal Language Theory: Perspectives and Open Problems* (Academic Press, New York, 1980) 349–405.
[47] HUGHES, J., Supercombinators, a new implementation method for applicative languages, in: *Proc. ACM Symp. on Lisp and Functional Programming* (1982) 1–10.
[48] HYLAND, E. and A. PITTS, The theory of constructions: categorical semantics and topos-theory, in: J. Gray and A. Scredov, eds., *Contemporary Mathematics, vol. 92* (AMS, Providence, RI, 1989).
[49] JOHNSSON, T., Efficient compilation of lazy evaluation, in: *Proc. ACM Conf. on Compiler Construction* (1984) 58–69.
[50] JOUANNOUD, J.-P., ed., *Functional Programming Languages and Computer Architecture*, Lecture Notes in Computer Science, Vol. 201 (Springer, Berlin, 1985).
[51] KAHN, G., D.B. MACQUEEN and G. PLOTKIN, eds., *Semantics of Data Types*, Lecture Notes in Computer Science, Vol. 173 (Springer, Berlin, 1984).
[52] KAHN, G. et al., eds., *Proc. 3th Internat. Conf. on Functional Programming Languages and Computer Architecture*, Lecture Notes in Computer Science, Vol. 274 (Springer, Berlin, 1987).
[53] KENNAWAY, R., The complexity of a translation of lambda calculus to combinators, Preprint, Dept. of Computer Science, Univ. of East Anglia, Norwich, England, 1982.
[54] KLEENE, S.C., λ-definability and recursiveness, *Duke Math. J.* **2** (1936) 340–353.
[55] KLOP, J.W., *Combinatory Reduction Systems*, Mathematical Centre Tracts, Vol. 127 (Centre for Mathematics and Computer Science, Amsterdam, 1980).
[56] KLOP, J.W., Term rewrite systems, to appear in [2].
[57] LANDIN, P., The mechanical evaluation of expressions, *Comput. J.* **6** (1963) 308–320.
[58] LANDIN, P., A correspondence between ALGOL 60 and Church's lambda notation, *Comm. ACM* **8** (1965) 89–101, and 158–165.
[59] LANDIN, P., The next 700 programming languages, *Comm. ACM* **9**(3) (1966) 157–164.
[60] LANDIN, P., A λ-calculus approach, in: L. Fox, ed., *Advances in Programming and Nonnumerical Computation*, (Pergamon Press, New York, 1966) 97–141.
[61] MACQUEEN, D., G. PLOTKIN and R. SETHI, An ideal model for recursive polymorphic types, in: *Proc. 11th Ann. ACM Symp. on Principles of Programming Languages* (1984) 165–174.
[62] MARTIN-LÖF, P., *Intuitionistic Type Theory* (Bibliopolis, Napoli, 1980).
[63] MCCARTHY, J., P.W. ABRAHAMS, D.J. EDWARDS, T.P. HART and M.I. LEVIN, *The Lisp 1.5 Programmers Manual* (MIT Press, Cambridge, MA, 1962).
[64] MILNER, R., Fully abstract models of typed λ-calculi, *Theoret. Comput. Sci.* **4** (1977) 1–22.
[65] MILNER, R., A theory of type polymorphism in programming. *J. Comput. System Sci.* **17** (1978) 348–375.
[66] MILNER, R., A proposal for standard ML, in: *Proc. ACM Symp. on Lisp and Functional Programming* (1984) 184–197.
[67] MOGGI, E., The partial lambda calculus, Ph.D. Thesis, *Dept. of Comput. Sci.*, Univ. of Edinburgh, 1988.
[68] MORRIS, J.H., Lambda calculus models of programming languages, Ph.D. Thesis, MIT, Cambridge, MA, 1968.

[69] MULDER, H., Complexity of combinatory code, Preprint 389, Dept. of Mathematics, Univ. of Utrecht, Utrecht, 1985.

[70] MULMULEY, K., Fully abstract submodels of typed lambda calculus, *J. Comput. System Sci.* **33** (1986) 2–46.

[71] MYCROFT, A., Abstract interpretation and optimising transformations for applicative programs, Ph.D. Thesis, Dept. of Computer Science, Univ. of Edinburgh, 1981.

[72] O'DONNELL, M., *Equational Logic as a Programming Language* (MIT Press, Cambridge, MA, 1985).

[73] ONG, L., Lazy lambda calculus: an investigation into the foundation of functional programming, Ph.D. Thesis, Imperial College, Univ. of London, London, 1988.

[74] PEYTON-JONES, S., *The Implementation of Functional Programming Languages* (Prentice-Hall, London, 1986).

[75] PLOTKIN, G., A set-theoretical definition of application, Memo MIP-R-95, School of Artificial Intelligence, Univ. of Edinburgh, 1972.

[76] PLOTKIN, G., Call-by-name and call-by-value and the lambda calculus, *Theoret. Comput. Sci.* **1** (1975) 125–159.

[77] RENARDEL, G., Strictness Analysis for a language with polymorphic and recursive types, Tech. Report, Dept. of Philosophy, Univ. of Utrecht, Utrecht, 1988.

[78] REYNOLDS, J., Towards a theory of type structure, in: *Proc. Programming Symp.*, Lecture Notes in Computer Science, Vol. 19 (Springer, Berlin, 1974) 408–425.

[79] SCHÖNFINKEL, M., Über die Bausteine der mathematische Logik, *Math. Ann.* **92** (1924) 305–316.

[80] SCOTT, D.S., Data types as lattices, *SIAM J. Comput.* **5** (1975) 522–587.

[81] SEELY, R.A.G., Locally cartesian closed categories and type theory, *Math. Proc. Cambridge Philos. Soc.* **95** (1984) 33–48.

[82] SEELY, R.A.G., Categorical semantics for higher-order polymorphic lambda calculus, *J. Symbolic Logic* **52** (1987) 969–989.

[83] STATMAN, R., An optimal translation of λ-terms into combinators, Manuscript, Dept. of Computer Science, Carnegie-Mellon Univ., Pittsburgh, PA, 1983.

[84] STEELE, G.L.J. and G.J. SUSSMAN, SCHEME: An interpreter for extended lambda calculus, AI Memo 349, MIT AI Lab, Cambridge, MA, 1975.

[85] Tait, W.W., A realizability interpretation of the theory of species, in: *Logic Colloquium*, Lecture Notes in Mathematics, Vol. 453 (Springer, Berlin, 1975) 240–251.

[86] TAKAHASHI, M., Parallel reductions in λ-calculus, Research reports on Information Sciences No. C-82, Dept. of Information Sciences, Tokyo Institute of Technology, Ookayama, Meguro, Tokyo 152, Japan, 1987.

[87] TURING, A., On computable numbers with an application to the Entscheidungsproblem, *Proc. London Math. Soc.* **42** (1936) 230–265.

[88] TURING, A., Computability and λ-definability, *J. Symbolic Logic* **2** (1937) 153–163.

[89] TURNER, D.A., *SASL Language Manual* (Univ. of Kent, Canterbury, 1979).

[90] TURNER, D.A., A new implementation technique for applicative languages, *Software–Practice and Experience* **9** (1979) 31–49.

[91] TURNER, D.A., The semantic elegance of applicative languages, in: *Proc. ACM Conf. on Functional Programming, Languages and Computer Architecture* (1981) 85–92.

[92] TURNER, D.A., Miranda: a nonstrict functional language with polymorphic types, in [50] 1–16.

[93] WADSWORTH, C.P., Semantics and pragmatics of the lambda-calculus, Dissertation, Oxford Univ., Oxford, 1971.

CHAPTER 8

Type Systems for Programming Languages

John C. MITCHELL

Department of Computer Science, Stanford University, Stanford, CA, USA

Contents

1. Introduction 367
2. Typed lambda calculus with function types 370
3. Logical relations 415
4. Introduction to polymorphism 431
Acknowledgment 452
Bibliography 453

HANDBOOK OF THEORETICAL COMPUTER SCIENCE
Edited by J. van Leeuwen

1. Introduction

This chapter illustrates some basic issues in the type-theoretic study of programming language concepts. The larger part of the chapter uses Church's typed lambda calculus as an illustrative example. The remainder surveys typing features such as polymorphism, data abstraction and automatic inference of type information. The main topics in the study of typed lambda calculus are context-sensitive syntax and typing rules, a reduction system modeling program execution, an equational axiom system, and semantic models. Henkin models are used as the primary semantic framework, with emphasis on three examples: set-theoretic, recursion-theoretic and domain-theoretic hierarchies of functions. Completeness theorems for arbitrary Henkin models and models without empty types are described, followed by an introduction to Cartesian closed categories and Kripke-style models. The technique of logical relations is used to prove a semantic completeness theorem and Church–Rosser and strong normalization properties of reduction. Logical relations also encompass basic model-theoretic constructions such as quotients. The study of simple typed lambda calculus concludes with brief discussions of representation independence, a property related to programming language implementation, and speculation on generalizing logical relations to categorical settings.

1.1. Overview

Many studies of types in programming languages have appeared in recent years. Most of the theoretical studies have centered on typed function calculi, which may be regarded either as simple functional programming languages or as metalanguages for understanding other (e.g., imperative languages). In general, a typed function (or lambda) calculus consists of a syntax of expressions, an equational proof system, a set of reduction rules which allow us to simplify or "execute" expressions, and some form of semantics. For many systems, there is substantial mathematical difficulty in constructing intuitive and rigorous semantics.

Although typed function calculi do not reflect all of the intricacies of practical programming languages, they are useful in studying many fundamental and practical issues. Typed function calculi often provide useful "models" of programming languages, much the way Turing machines or PRAMs provide useful models of machine computation. In studies of syntax, function calculi allow a number of typing problems to be stated in simple terms. For example, algorithmic problems in type-checking are often stated using some form of lambda calculus. In semantics, typed function calculi provide natural "intermediate languages" which simplify and structure the mathematical interpretation of programming languages, in much the way that intermediate code may be used to simplify and structure compilation. As in compilation, an intermediate semantic language also allows us to develop basic techniques that are applicable to a family of related programming languages. When typing considerations are relevant, it is natural to use a calculus which reflects the type structure of the programming language. For further discussion of the relationship between programming languages and function (or lambda) calculi, the reader is referred to [69, 92, 111, 131].

This chapter is divided into four sections, including this introduction. The second and third parts attempt to illustrate common properties of type systems using the most basic function calculus: Church's simple typed lambda calculus. The fourth part surveys type systems with polymorphism and data abstraction, along with algorithms for automatically inferring type information. While the simple typed lambda calculus is occasionally misleadingly simple in comparison with other typed lambda calculi, it is sufficient to illustrate a number of general points. Moreover, since the technical machinery is reasonably well developed, simple typed lambda calculus often provides a motivating example when exploring other type systems. It is hoped that this chapter will convey some of the mathematical depth of type-theoretic analysis and provide a basis for further research in the area.

The presentation of typed lambda calculus in Section 2 begins with a discussion of context-sensitive syntax and defining languages by typing rules. Section 2 continues with an equational proof system and reduction rules that model program execution. The main semantic emphasis is on Henkin models, and three examples: set-theoretic, recursion-theoretic and domain-theoretic hierarchies of functions. Two general completeness theorems for Henkin models are described, one assuming all types are nonempty and another for arbitrary models. An introduction to categorical models (Cartesian closed categories) and a brief overview Kripke models are also included.

Section 3 is concerned with the framework of "logical relations", which seems to generalize to a variety of type systems. For the simple typed lambda calculus, logical relations may be used to prove virtually all of the central, basic theorems. For example, we describe a completeness proof analogous to the standard proof from universal algebra, using the congruence relation of a theory to form a quotient structure. Church–Rosser and strong normalization proofs are also outlined, followed by a discussion of representation independence. Intuitively, representation independence is the property of a typed programming language that the behavior of a program does not depend on the precise way that basic values are represented in a computer, only the behavior of these values under the operations provided. The Basic Lemma for logical relations may be used to show that typed lambda calculus has a representation independence property. The discussion of logical relations concludes with some speculative generalizations to categorical settings.

Section 4 of the chapter contains an overview of polymorphic extensions to simple typed lambda calculus, data abstraction, and algorithms for automatically inferring type information.

In general, this chapter is a mix of results from the literature, previously unpublished folklore, and research results of the author, with an unfortunate bias towards my past research areas. While every attempt has been made to give credit where credit is due, there are undoubtably some omissions. For these I would like to apologize in advance.

1.2. Pure and applied lambda calculus

While it is often insightful to view Pascal, ADA or LISP program phrases as expressions from some function calculus of (cf. [69, 111]), this generally involves extending the basic mathematical calculus in some way. For example, one striking difference between lambda calculus and many programming languages is the absence of assignment. Nonetheless, typed function calculi may be used to study conventional programming languages. To do so, we regard side-effect-producing operations as functions which manipulate something called the "store". This technique is central to the Scott–Strachey approach to denotational semantics. Since the functional treatment of side-effects is discussed at length in standard treatments of denotational semantics [47, 131, 119], we will not go into it here.

The common thread through many versions of lambda calculus is a basic calculus of functions. The main constructs are *lambda abstraction*, which we use to write function expressions, and *application*, which allows us to make use of the functions we define. The domain of a function is specified by giving a type to the *formal parameter*. If M is some expression, possibly containing a variable x of type σ, then $\lambda x{:}\sigma.M$ is the function defined by treating M as a function of x. Since "$x{:}\sigma$" explicitly specifies that the formal parameter x has type σ, the domain of $\lambda x{:}\sigma.M$ is σ. The range of $\lambda x{:}\sigma.M$ is determined from the form of M using the typing rules of the language. A simple example is the typed lambda expression

$$\lambda x{:}int.x$$

for the identity function on integers. Since declaring $x{:}int$ guarantees that the function body x has type *int*, the range of this function is *int*. To say that the integer identity function is a mapping from integers to integers, we write

$$\lambda x{:}int.x : int \rightarrow int.$$

Another mathematical notation for $\lambda x{:}int.x$ is $x \mapsto x$, or perhaps $x{:}int \mapsto x$. A more familiar way of defining the identity function in programming languages is to write something like

$$Id(x{:}int) = x.$$

This alternative is perfectly rigorous, but more complicated than necessary since the basic idea of function is combined with the separate notion of binding a name (in this case *Id*) to a value. In addition, this syntax makes it difficult to define a function in the middle of an expression, something that lambda calculus makes particularly easy.

In lambda calculus, a function M is applied to an argument N simply by writing MN. An application MN is considered well-formed (by which we mean *well-typed*) whenever M has a function type, and N has the appropriate type to be an argument to M. For example, we can apply the integer identity function to the number 3 by writing

$$(\lambda x{:}int.x)3.$$

The value of this expression is the identity function, applied to the number 3, which just ends up to be 3. Thus we have

$$(\lambda x{:}int.x)3 = 3.$$

An important part of lambda calculus is a proof system for deriving equations. The equational proof system also gives rise to a set of reduction rules, which constitute a simple model of a programming language interpreter.

The basic calculus of typed lambda abstraction and application is called the *pure typed lambda calculus with function types*. We will refer to this system by the symbol $\lambda^{\rightarrow}$, since $\rightarrow$ is used to write the type of a function. The calculus $\lambda^{\rightarrow}$ may be extended by adding Cartesian product types $\sigma \times \tau$, for example, with the associated pairing operation $\langle \cdot, \cdot \rangle$ and projection functions $\mathbf{Proj}_1$, $\mathbf{Proj}_2$. The resulting system, which we will denote by $\lambda^{\times,\rightarrow}$, is a pure typed lambda calculus with function and product types. Products are very useful in mathematics and programming, but in many ways the general properties of $\lambda^{\times,\rightarrow}$ and $\lambda^{\rightarrow}$ are quite similar. For this reason, we will tend to emphasize the simpler $\lambda^{\rightarrow}$, except when Cartesian products are required. In Section 4, we will consider extensions of $\lambda^{\rightarrow}$ with polymorphism, data abstraction, and other typing features.

Another way to extend $\lambda^{\rightarrow}$ is to add a "basic" type, like the integers. To do this, we might add a type symbol (type constant) *int* to the type expressions, and add numerals 0, 1, 2,... and arithmetic operations such as $+$, $-$ and $\cdot$ (multiplication) to value expressions. To make a natural calculus, we must extend the equational proof system with axioms like $3+2=5$. We will call a typed lambda system with an added basic data type an *applied lambda calculus*. A fundamental idea underlying the relation between lambda calculus and programming languages is the "slogan"

$$\begin{aligned}\text{programming language} &= \text{applied lambda calculus}\\ &= \text{pure lambda system} + \text{basic data types.}\end{aligned}$$

The distinction between pure and applied lambda calculus is often a bit vague, since many useful "basic" types can be defined in more expressive pure systems. For example, the integers may be defined using recursive type declarations. However, constructs such as *stores*, with operations for assignment and storage allocation, do not seem to arise naturally in any pure lambda calculus; it is necessary to add stores when we need them. The main point of the slogan is to suggest that most programming languages begin with some basic data types, like integers, booleans and strings, and provide various ways of building functions and more complicated data structures over them. Many programming languages may be summarized by describing the ways of defining new values from old, which is essentially a description of the "pure" type system used to define the language.

2. Typed lambda calculus with function types

2.1. *Types*

2.1.1. *Syntax*

Like many-sorted algebra and first-order logic, the typed lambda calculus $\lambda^{\rightarrow}$ may be defined using an arbitrary collection of base types (sorts) and constant symbols. Typical base types in programming languages are numeric types and booleans. In typed lambda calculus, constant symbols may have basic types, like constant symbols in universal algebra, or may have function types. Typical constants of programming languages include 3, + and if ... then ... else ... Using b to stand for any base type, the type expressions of $\lambda^{\rightarrow}$ are defined by the grammar

$$\sigma ::= b \mid \sigma_1 \rightarrow \sigma_2.$$

For example, if *nat* is a base type, then *nat*→*nat* and *nat*→(*nat*→*nat*) are types. We can avoid writing too many parentheses by adopting the convention that → associates to the right. Thus *nat*→*nat*→*nat* is the type of functions which, given a natural number argument, return a numeric function.

There are several similar versions of lambda calculus with function types. However, once we have the basic calculus of functions, extensions such as Cartesian products and disjoint unions (sums) may be added without altering the general framework significantly. To add Cartesian products, for example, we would extend the type expressions by adding types of the form $\sigma_1 \times \sigma_2$, and appropriate term-formation rules. Since products and other extensions complicate the technical arguments by adding more cases, we will illustrate the main ideas by focusing on $\lambda^{\rightarrow}$, the pure calculus with function types.

2.1.2. Interpretation of types

Intuitively, $\sigma\rightarrow\tau$ is the type, or collection, of all functions from σ to τ. However, in different settings, we may adopt different formal notions of function. We will see later that the introduction and elimination rules for → types characterize what we expect of "functions" precisely. For the most part, however, it is sufficient to think of a function f from σ to τ as either a rule that assigns a value $f(x){:}\tau$ to each $x{:}\sigma$, or some kind of "code" for such a rule. Intuitively, we think of $\sigma\rightarrow\tau$ as containing all functions from σ to τ that "make sense" in some context, and regard two functions as equal if they take the same value at every point in their domain. This equality principle, which may be written out as

$$f = g : \sigma\rightarrow\tau \text{ iff } \forall x{:}\sigma.\, f(x) = g(x)$$

is commonly referred to as *extensional* equality of functions.

There are many reasonable interpretations of $\lambda^{\rightarrow}$, involving many different technical formalizations of function. To emphasize from the outset that many views are possible, we will discuss three representative interpretations: classical set-theoretic functions, recursive functions coded by Gödel numbers, and continuous functions on complete partial orders (domains). Further details on models in general, and recursive and continuous functions in particular, will be given later on. In the *full set-theoretic interpretation*,

(i) a base type b may denote any set A^b, and

(ii) the function type $\sigma\rightarrow\tau$ denotes the set $A^{\sigma\rightarrow\tau}$ of all functions from A^σ to A^τ.

Note that $A^{\sigma\rightarrow\tau}$ is determined by the sets A^σ and A^τ.

There are several forms of recursion-theoretic interpretation for $\lambda^{\rightarrow}$, each involving sets with some kind of enumeration function or relation. We will consider a representative framework based on partial enumeration functions. A *modest set* $\langle A, e_A\rangle$ is a set A together with a surjective partial function $e_A : \mathcal{N} \rightarrow A$ from the natural numbers to A.

Intuitively, the natural number n is the "code" for $e_A(n) \in A$. Since e_A is partial, $e_A(n)$ may not be defined, in which case n is not a code for any element of A. The importance of having codes for every element of A is that we use coding to define the recursive functions on A. Specifically, $f: A \to B$ is *recursive* if there is a recursive map on natural numbers taking any code for $a \in A$ to some code for $f(a) \in B$. In the *recursive interpretation*,

(i) a base type b may denote any modest set $\langle A^b, e_b \rangle$, and
(ii) the function type $\sigma \to \tau$ denotes the modest set $\langle A^{\sigma\to\tau}, e_{\sigma\to\tau} \rangle$ of all total *recursive* functions from A^σ to A^τ.

Since every total recursive function from A^σ to A^τ has a Gödel number, we can use any standard numbering of recursive functions to enumerate the elements of the function type $A^{\sigma\to\tau}$. Thus, if we begin with modest sets as base types, we can associate a modest set with every functional type.

The *continuous* or *domain-theoretic interpretation* uses complete partial orders. The primary motivations for this interpretation are to give semantics to recursively defined functions and recursive type definitions. However, recursive type definitions are beyond the scope of this article. A *complete partial order*, or *cpo*, $\langle D, \leqslant \rangle$ is a set D partially ordered by $\leqslant$ in such a way that every directed[1] subset $E \subseteq D$ has a least upper bound $\bigvee E$. In the domain-theoretic interpretation,

(i) a base type b may denote any cpo $\langle A^b, \leqslant_b \rangle$, and
(ii) the function type $\sigma \to \tau$ denotes the cpo $\langle A^{\sigma\to\tau}, \leqslant_{\sigma\to\tau} \rangle$ of all *continuous* functions from A^σ to A^τ.

A function $f: A^\sigma \to A^\tau$ is continuous if it is monotonic and preserves limits (least upper bounds) of directed sets. Note that in order to make sense of higher types like $(b \to b) \to (b \to b)$, we need to make sure that every function space $b \to b$ is a cpo. This may be accomplished by considering $f \leqslant_{\sigma\to\tau} g$ whenever we have $f(x) \leqslant_\tau g(x)$ for every $x \in A^\sigma$. Further details will be given in Section 2.6.

2.2. *Terms*

2.2.1. *Context-sensitive syntax*

Programming languages are often defined using a context-free grammar, as in the definition of type expressions above. However, a context-free description of a statically typed language tells only half the story, since typing constraints are typically context-sensitive. For example, we would generally consider $x + 3$ well-typed only if x is a numeric variable. In most languages, the declaration for x may precede the expression $x + 3$ by an arbitrary number of symbols, and so there can be no context-free grammar generating precisely the well-typed terms (cf. [55]).

We will define typed languages using a formalism based on logic. Specifically, we will define expressions and their types simultaneously using axioms and inference rules. The

[1] A subset E of a cpo is *directed* if every finite subset $S \subseteq E$ has an upper bound in E. Further details are given later in this chapter, and in the chapter by Scott and Gunter.

atomic expressions of a language are given by *typing axioms*. Informally, a typing axiom $c{:}\tau$ means that the symbol c has type τ. Put another way, $c{:}\tau$ is an axiom about the type membership relation, which we write as ":". The axiom says that the type membership relation holds between c and τ. An example axiom for a language with natural numbers is 3:*nat*, which specifies that 3 is a natural number.

The compound expressions and their types are defined simultaneously using inference rules. These are rules that let us derive more complicated facts about the ":" relation. One form of inference rule is

$$\frac{M_1{:}\sigma_1, \ldots, M_k{:}\sigma_k}{N{:}\tau}.$$

Intuitively, this rule says that if $M_1, \ldots, M_k$ are well-formed terms of types $\sigma_1, \ldots, \sigma_k$ respectively, then N is a well-formed term of type τ. Typically, $M_1, \ldots, M_k$ are the subterms of N, since the type of a term will generally depend on the types of its subterms. This rule is similar to a production rule: it tells us how to produce a well-formed term N of type τ from well-formed terms $M_1, \ldots, M_k$.

The inference rule above is not powerful enough to capture the context-sensitive syntax of expressions which declare the types of variables. To take the types of variables into account, we will work with typing assertions

$$\Gamma \rhd M{:}\tau,$$

where Γ is a *type assignment* of the form

$$\Gamma = \{x_1{:}\sigma_1, \ldots, x_k{:}\sigma_k\},$$

with no x_i occurring twice. Intuitively, the assertion $\Gamma \rhd M{:}\tau$ says that if variables $x_1, \ldots, x_k$ have types $\sigma_1, \ldots, \sigma_k$ (respectively), then M is a well-formed term of type τ. A type assignment is also called a *typing context* so that $\Gamma \rhd M{:}\tau$ may be read; "in context Γ, the term M has type τ."

The general form of a typing rule with contexts is

$$\frac{\Gamma_1 \rhd M_1{:}\sigma_1, \ldots, \Gamma_k \rhd M_k{:}\sigma_k}{\Gamma \rhd N{:}\tau},$$

which says that if each M_i has type σ_i in typing context Γ_i, then N has type τ in context Γ. Again, the terms $M_1, \ldots, M_k$ are usually the subterms of N.

If Γ is any type assignment, we will write Γ, $x{:}\sigma$ for the type assignment

$$\Gamma, x{:}\sigma = \Gamma \cup \{x{:}\sigma\}.$$

In doing so, we always assume that x does not appear in Γ.

In the literature on type systems, several different formulations of syntax are used. One common approach is to use natural deduction proof systems, writing $\Gamma \vdash M{:}\sigma$ to indicate that the typing assertion $M{:}\sigma$ is provable from the set Γ of typing assumptions. Sequent calculus formulations, with sequents written as $\Gamma \vdash M{:}\sigma$, are also used. The main reason for using $\rhd$ instead of $\vdash$ is to reserve $\vdash$ for provability in the equational proof system.

2.2.2. Syntax of terms

The syntax of terms depends on the choice of base types and constant symbols. A $\lambda^{\rightarrow}$ *signature* $\Sigma = \langle B, C \rangle$ consists of

- a set B whose elements are called *base types* or *type constants*, and
- a collection C of disjoint sets C^{σ} indexed by type expressions over B; the elements of C^{σ} are called *constant symbols of type* σ, or *term constants*.

Note that the base types and term constants must be consistent, in that the type of each constant may only contain the given base types. For example, it only makes sense to have a natural number constant $3 \in C^{nat}$ when we have *nat* as a base type.

The $\lambda^{\rightarrow}$ expressions over Σ and their types are defined simultaneously using axioms and inference rules. For each constant c of type σ, we have the axiom

(cst) $\quad \emptyset \rhd c : \sigma.$

The typing context here is empty, since the type of a constant is fixed, and therefore independent of the context in which it occurs. It is common to leave out the empty context, so that in an applied calculus with natural numbers and booleans, we might have typing axioms like

$$0, 1, 2, \ldots : nat, \qquad true, false : bool,$$
$$+ : nat \rightarrow nat \rightarrow nat, \qquad cond : bool \rightarrow nat \rightarrow nat \rightarrow nat$$

giving us names for elements of *nat* and *bool*, and functions over these types.

We assume some countably infinite set *Var* of variables $\{v_0, v_1, \ldots\}$. Variables are given types by the axiom

(var) $\quad x : \sigma \rhd x : \sigma,$

which says that a variable x has whatever type it is declared to have. Some authors assume each variable $v \in Var$ has a fixed type, and therefore do not mention typing contexts explicitly. However, this seemingly simpler presentation does not generalize to lambda calculi with polymorphic functions or abstract data type declarations. In addition, when we wish to reason about types that may be empty, it is essential to keep track of the free variables used in proofs. This is easily taken care of in the formalism we use.

Compound expressions and their types are specified using inference rules. A straightforward inference rule is the following "structural" rule that applies to terms of any form. The rule

(add hyp) $$\frac{\Gamma \rhd M : \sigma}{\Gamma, x : \tau \rhd M : \sigma}$$

allows us to add an additional hypothesis to the typing context. In words rule (add hyp) says that if M has type σ in context Γ, then M has type σ in the context $\Gamma, x : \tau$ which also gives a type to x. Recall that in writing Γ, $x : \tau$ we assume x does not appear in Γ. Consequently, the type of M could not have depended on the type of x. In fact, after we have seen all of the rules, it will be easy to prove that if $\Gamma \rhd M : \sigma$ is derivable, then every free variable of M must appear in Γ. Therefore, in rule (add hyp), we can be sure that x does not occur free in M.

While the set-theoretic view of functions as sets of ordered pairs has proven useful in classical mathematics, the syntax of $\lambda^{\rightarrow}$ seems closer to the view of functions as rules. Intuitively, then, a function f from σ to τ is a rule that assigns a value $f(x):\tau$ to each $x:\sigma$. If M is a well-typed term with a free variable $x:\sigma$, then M is not just a rule for computing a single value, but a rule which makes sense for every value of the variable x. Therefore, we should expect a term and free variable to determine a function.

In lambda calculus, lambda abstraction is used to distinguish between variables used as function arguments, and variables used as "parameters", in the sense of "variables held fixed while computing the function value". This is made precise by the following term formation rule.

$$(\rightarrow\text{Intro}) \qquad \frac{\Gamma,\ x:\sigma \rhd M:\tau}{\Gamma \rhd (\lambda x:\sigma.M):\sigma\rightarrow\tau}.$$

Intuitively, the rule says that if M specifies a result of type τ for every $x:\sigma$, then the expression $\lambda x:\sigma.M$ defines a function of type $\sigma\rightarrow\tau$. (Other free variables of M are unaffected by lambda abstraction, and must be given types in Γ.) Note that while the type of M may depend on the type of variable x, the type of $\lambda x:\sigma.M$ does not, since the type of x is explicitly declared in $\lambda x:\sigma$. This rule is called ($\rightarrow$Intro), since it "introduces" a term of functional type. An important aspect of lambda abstraction is that the variable x is *bound* in $\lambda x:\sigma$, which means that x is used as a place holder within M, and we could uniformly replace x with any other variable y without changing the meaning of the term, except in the special case that y is already used in some other way in M.

Another way to read the ($\rightarrow$Intro) rule may make it seem more familiar to computer scientists. Suppose we want to type-check a function declaration $\lambda x:\sigma.M$, and we have a "symbol table" Γ associating types with variables that might occur free in M. Then reading from bottom to top, the ($\rightarrow$Intro) rule says that in order to check that $\Gamma \rhd \lambda x:\sigma.M:\sigma\rightarrow\tau$, we modify Γ by incorporating the specification $x:\sigma$, and then check that the function body M has type τ. This should be familiar to anyone who has considered how to type-check a function declaration

```
function f(x:σ);
begin
  ⟨function-body⟩
end;
```

in any ALGOL-like programming language.

Function applications are written according to the rule

$$(\rightarrow\text{Elim}) \qquad \frac{\Gamma \rhd M:\sigma\rightarrow\tau,\ \ \Gamma \rhd N:\sigma}{\Gamma \rhd MN:\tau}$$

which says that we may apply any function with type $\sigma\rightarrow\tau$ to an argument of type σ to produce a result of type τ. Note that while $\rightarrow$ appears in the antecedent, this symbol has been "eliminated" in the consequent, hence the name ($\rightarrow$Elim).

We say *M is a $\lambda^{\rightarrow}$ term over signature Σ with type τ in context Γ if* $\Gamma \rhd M:\tau$ is either a typing axiom for Σ, or follows from axioms by rules (add hyp), ($\rightarrow$Intro) and ($\rightarrow$Elim). As an expository convenience, we will often write $\Gamma \rhd M:\tau$ to mean that "$\Gamma \rhd M:\tau$ is derivable", in much the same way as one often writes a formula $\forall x.P(x)$ in logic as a way of saying "$\forall x.P(x)$ is true".

The free and bound occurrences of a variable x in term M have the usual inductive definition. A variable x occurs free δ unless it is within the scope of λx, in which case it becomes bound. An occurrence of a variable immediately following a λ is called a *binding occurrence.* Since the name of a bound variable is not important, we will generally identify terms that differ only in the names of bound variables. This will be discussed more precisely in the next section. The following lemmas are proved by straightforward inductions on typing derivations.

2.2.1. LEMMA. *If $\Gamma \rhd M:\sigma$, then every free variable of M appears in Γ.*

2.2.2. LEMMA. *If $\Gamma \rhd M:\sigma$ and $\Gamma' \subseteq \Gamma$ contains all the free variables of M, then $\Gamma' \rhd M:\sigma$.*

These lemmas generalize to most type systems, including all of the systems mentioned in this article.

We will write $[N/x]M$ for the result of substituting N for free occurrences of x in M. In defining $[N/x]M$, we must be careful to rename bound variables of M to avoid capture of free variables in N. Specifically, if y occurs free in N, then

$$[N/x](\lambda y{:}\sigma.M) = \lambda z{:}\sigma.[N/x][z/y]M$$

where z is a fresh variable not occurring free in M or N. Two useful facts about $\lambda^{\rightarrow}$, which hold for other typed lambda systems, are the following lemmas about typing and substitution.

2.2.3. LEMMA. *If $\Gamma \rhd M:\sigma$ and y does not occur in Γ, then $[y/x]\Gamma \rhd [y/x]M:\sigma$.*

2.2.4. LEMMA. *If $\Gamma, x{:}\sigma \rhd M:\tau$ and $\Gamma \rhd N:\sigma$ are terms of $\lambda^{\rightarrow}$, then so is the substitution instance $\Gamma \rhd [N/x]M:\tau$.*

2.3. *Proof system*

2.3.1. *Equations*

Typed equations have the form

$$\Gamma \rhd M = N : \tau$$

where we assume that M and N have type τ in context Γ. Intuitively, the equation $\{x_1{:}\sigma_1, \ldots, x_k{:}\sigma_k\} \rhd M = N : \tau$ means that for all type-correct values of the variables $x_1{:}\sigma_1, \ldots, x_k{:}\sigma_k$, expressions M and N denote the same element of type τ. Another way of writing this equation might be

$$\forall x_1{:}\sigma_1 \ldots \forall x_k{:}\sigma_k.M = N : \tau.$$

Because the variables listed in the type assignment are universally quantified, an equation may hold vacuously if some type is empty. Specifically, if σ is empty, then an equation $\forall x{:}\sigma.M = N : \tau$ will be true simply because there is no possible value for x.

Since we include type assignments in equations, we have an equational version of the structural rule

$$\text{(add hyp)} \quad \frac{\Gamma \rhd M = N{:}\sigma}{\Gamma, x{:}\tau \rhd M = N : \sigma}$$

which lets us add an additional typing hypothesis.

The next group of axioms and inference rules make provable equality an equivalence relation, and a congruence with respect to the term-formation operations. To make equality an equivalence relation, we have the axiom and rules

$$\text{(ref)} \qquad \Gamma \rhd M = M{:}\sigma,$$

$$\text{(sym)} \qquad \frac{\Gamma \rhd M = N{:}\sigma}{\Gamma \rhd N = M{:}\sigma},$$

$$\text{(trans)} \qquad \frac{\Gamma \rhd M = N{:}\sigma, \quad \Gamma \rhd N = P{:}\sigma}{\Gamma \rhd M = P{:}\sigma}.$$

The two term-formation operations of $\lambda^{\rightarrow}$ are abstraction and application, both of which preserve equality. The rule

$$(\xi) \qquad \frac{\Gamma, x{:}\sigma \rhd M = N{:}\tau}{\Gamma \rhd \lambda x{:}\sigma.M = \lambda x{:}\sigma : \sigma \rightarrow \tau}$$

says that if M and N are equal for all values of x, then the two functions $\lambda x{:}\sigma.M$ and $\lambda{:}\sigma.N$ are equal. For application, we have the rule

$$(\mu) \qquad \frac{\Gamma \rhd M_1 = M_2{:}\sigma \rightarrow \tau, \quad \Gamma \rhd N_1 = N_2{:}\sigma}{\Gamma \rhd M_1 N_1 = M_2 N_2{:}\tau}$$

saying that equals applied to equals yield equals. Rules (ξ) and (μ) may be explained using the equality principle associated with $\rightarrow$ types, stated earlier. The equality principle is that two functions are equal iff they map equal arguments to equal results. One direction of this "iff" gives rule (μ), and the other (ξ). It is interesting to note that the two congruence rules have the same form as the introduction and elimination rules for $\rightarrow$.

Three axioms remain. The first describes renaming of bound variables, while the other two specify that the introduction and elimination rules are "inverses" of each other (in categorical terminology, they form an adjoint situation). Since λ is a binding operator, the name of a lambda bound variable is not important. This is formalized in the axiom

$$(\alpha) \qquad \Gamma \rhd \lambda x{:}\sigma.M = \lambda y{:}\sigma.[y/x]M : \sigma \rightarrow \tau, \quad \text{provided } y \notin FV(M).$$

We say two terms $\Gamma \rhd M{:}\sigma$ and $\Gamma \rhd N{:}\sigma$ of the same type are *α-equivalent*, and write $\Gamma \rhd M =_{\alpha} N : \sigma$, if the equation $\Gamma \rhd M = N : \sigma$ is provable from the axioms and inference rules given so far. What this boils down to is that $\Gamma \rhd M =_{\alpha} N : \sigma$ if M and N differ only in the names of bound variables. Since substitution is only defined up to α-equivalance, it is conventional not to distinguish between α-equivalent terms.

Symbolically, function application may be evaluated by substituting the function argument for the lambda-bound variable. This "copy rule," as it is called in ALGOL 60, is written as follows:

$$(\beta) \qquad \Gamma \rhd (\lambda x{:}\,\sigma.M)N = [N/x]M : \tau.$$

In the special case that the argument N is a variable, this axiom says that introduction (lambda abstraction) composed with elimination (application) is the identity operation. The other composition of elimination and introduction is also the identity.

$$(\eta) \qquad \Gamma \rhd \lambda x{:}\,\sigma.(Mx) = M : \sigma \rightarrow \tau, \quad \text{provided } x \notin FV(M).$$

The axiom of η-equivalence makes good sense in the light of extensional equality of functions. If $x \notin FV(M)$, then by (β) we have $(\lambda x{:}\,\sigma.Mx)y = My$ for any argument $y{:}\,\sigma$. Therefore M and $\lambda x{:}\,\sigma.Mx$ define the same function.

A *typed lambda theory* (or *$\lambda^{\rightarrow}$ theory*) over signature Σ is a set of well-typed equations between Σ-terms which includes all instances of the axioms and is closed under the inference rules. If $\mathscr{E}$ is any set of well-typed equations, we write $\mathscr{E} \vdash \Gamma \rhd M = N{:}\,\sigma$ to mean that the equation $\Gamma \rhd M = N{:}\,\sigma$ is provable from the axioms and equations of $\mathscr{E}$. It is worth mentioning that including type assignments in equations is unnecessary for pure typed lambda calculus, but not so in the presence of nonlogical axioms.

2.3.1. PROPOSITION. *If* $\vdash \Gamma \rhd M = N : \sigma$ *and* $\Gamma \cap \Gamma'$ *contains all free variables of* M *and* N, *then* $\vdash \Gamma' \rhd M = N : \sigma$.

2.3.2. PROPOSITION. *There exist a typed lambda theory* $\mathscr{E}$ *and terms* M *and* N *without* x *free such that* $\mathscr{E} \vdash \Gamma, x{:}\,\sigma \rhd M = N : \tau$ *but* $\mathscr{E} \nvdash \Gamma \rhd M = N : \tau$.

One example of a theory $\mathscr{E}$ satisfying the conditions of Proposition 2.3.2 is discussed in [82, 94]. The main idea is that $\Gamma \rhd M = N : \tau$ must hold in any model of the theory with σ nonempty, but not necessarily otherwise. Since the typing hypothesis $x{:}\,\sigma$ is a way of assuming that σ is nonempty, we may prove $\Gamma, x{:}\,\sigma \rhd M = N : \tau$ from $\mathscr{E}$. However, $\mathscr{E} \nvdash \Gamma \rhd M = N : \tau$.

2.3.2. Reduction rules

Reduction is a "directed" form of equational reasoning that resembles symbolic execution of programs. We will see that every provable equation $\Gamma \rhd M = N : \sigma$ may be derived by reducing M and N to a common term. This is a consequence of the Church–Rosser property of typed lambda calculus, also called the confluence or the diamond property. In addition, reduction of $\lambda^{\rightarrow}$ terms always leads to a *normal form*, a term which can no longer be reduced. Both of these properties are proved in Section 3 using logical relations.

Technically, reduction is a relation on α-equivalence classes of terms. While we are only interested in reducing typed terms, we will define reduction without mentioning types. Since reduction models program execution, this is a way of emphasizing that $\lambda^{\rightarrow}$ execution may be done without examining the types of terms. We will also see that the type of a term does not change as it is reduced. Together, these two facts imply that type-independent execution of $\lambda^{\rightarrow}$ terms is type-correct, or "$\lambda^{\rightarrow}$ does not require run time type checking".

The rule of *β-reduction* is

$$(\beta)_{\text{red}} \qquad (\lambda x{:}\sigma.M)N \xrightarrow{\beta} [N/x]M,$$

where $[N/x]M$ is the result of substituting N for free occurrences of x in M. For example, the value of $\lambda f{:}\sigma\rightarrow\sigma.fx$ applied to $\lambda y{:}\sigma.y$ may be calculated by substituting the argument $\lambda y{:}\sigma.y$ for the bound variable f:

$$(\lambda f{:}\sigma\rightarrow\sigma.fx)(\lambda y{:}\sigma.y) \xrightarrow{\beta} (\lambda y{:}\sigma.y)x.$$

Of course, $(\lambda y{:}\sigma.y)x$ may be reduced again, and so we have

$$(\lambda f{:}\sigma\rightarrow\sigma.fx)\lambda y{:}\sigma.y \xrightarrow{\beta} (\lambda y{:}\sigma.y)x \xrightarrow{\beta} x.$$

Corresponding to η-equivalence, we also have *η-reduction*:

$$(\eta)_{\text{red}} \qquad \lambda x{:}\sigma.Mx \xrightarrow{\eta} M, \quad \text{provided } x \notin FV(M).$$

A term of the form $(\lambda x{:}\sigma.M)N$ is called a *β-redex* and $\lambda x{:}\sigma.Mx$ and *η-redex*. We say *M reduces to N in one step*, written $M\rightarrow N$, if N can be obtained by applying (β) or (η) to some subterm of M. The reduction relation $\twoheadrightarrow$ is the reflexive and transitive closure of one-step reduction. Using Lemma 2.2.4 and inspection of an η-redex, it is easy to show that one-step reduction preserves type.

2.3.3. LEMMA. *If $\Gamma \rhd M{:}\sigma$, and $M\rightarrow N$, then $\Gamma \rhd N{:}\sigma$.*

It follows by an easy induction that $\twoheadrightarrow$ also preserves types. Since we will only be interested in reduction on well-typed terms, it is useful to write $\Gamma \rhd M \twoheadrightarrow N{:}\sigma$ when $\Gamma \rhd M{:}\sigma$ is well-typed and $M\twoheadrightarrow N$. We know by the lemma above that in this case, we also have $\Gamma \rhd N{:}\sigma$. A term M is in *normal form* if there is no N with $M\rightarrow N$.

Reduction on typed lambda terms is *confluent*, a property that may be drawn graphically as follows:

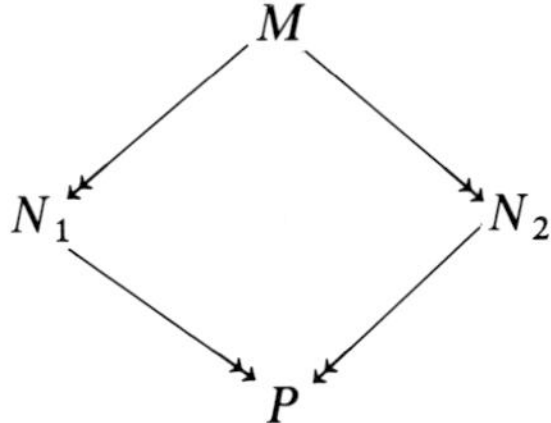

In this picture, the two top arrows are universally quantified, and the bottom two existentially, so the picture "says" that whenever $M\twoheadrightarrow N_1$ and $M\twoheadrightarrow N_2$, there exists a term P such that $N_1\twoheadrightarrow$P and $N_2\twoheadrightarrow P$. Confluence is also referred to as the *diamond property*, because of the diamond shape of the picture given above. It is traditional to say that a confluent notion of reduction is *Church–Rosser*, since confluence for untyped lambda calculus was first proved by Church and Rosser [17].

2.3.4. THEOREM (confluence). *If* $\Gamma \rhd M \twoheadrightarrow N_1 : \sigma$ *and* $\Gamma \rhd M \twoheadrightarrow N_2 : \sigma$, *then there is a term* $\Gamma \rhd P : \sigma$ *such that* $\Gamma \rhd N_1 \twoheadrightarrow P : \sigma$ *and* $\Gamma \rhd N_2 \twoheadrightarrow P : \sigma$.

For terms that reduce to a normal form, confluence guarantees that there are no "wrong" reductions. More precisely, if we want to simplify M as much as possible, we may have several reduction paths. Suppose M reduces to normal form N_1, and $M \twoheadrightarrow N_2$ is an alternate reduction with N_2 not in normal form. Although $M \twoheadrightarrow N_2$ may not lead to the shortest reduction to normal form, there is no irreparable harm in reducing M to N_2. The reason is that for some P, both $N_1 \twoheadrightarrow P$ and $N_2 \twoheadrightarrow P$. But since N_1 cannot be reduced (it is a normal form), it must be that $N_2 \twoheadrightarrow N_1$. Similar reasoning shows that the normal form of any term is unique.

2.3.5. EXERCISE. Use confluence to show that if M reduces to a normal form, then the normal form of M is unique.

Another important consequence of confluence is the following connection between reduction and provable equality.

2.3.6. COROLLARY. *An equation* $\Gamma \rhd M = N : \tau$ *is provable iff there is some term* P *with* $M \twoheadrightarrow P$ *and* $N \twoheadrightarrow P$.

The proof is suggested in Exercise 2.3.8 below.

It is worth emphasizing that reduction is not confluent on what we might call *pre-terms*, which are strings that look like terms but are not necessarily well-typed. To see this, consider the pre-term

$$\lambda x : \sigma . (\lambda y : \tau . y) x.$$

Using β-reduction, we may simplify this to $\lambda x : \sigma . x$, while η-reduction gives us $\lambda y : \tau . y$. Since these normal forms differ by more than names of bound variables when $\sigma \neq \tau$, confluence fails for pre-terms. The importance of this example is that the simple "proof" of confluence for typed lambda calculus by appeal to the Church–Rosser Theorem for untyped lambda calculus applies to pre-terms as well as typed terms. Since this argument leads to an incorrect conclusion for pre-terms, it is not a correct proof for typed terms. More generally, it seems that no proof of confluence for β, η-reduction on $\lambda^{\rightarrow}$ can ignore types[2]. We will give a direct proof of Church–Rosserness for $\lambda^{\rightarrow}$ in Section 3.

[2]It seems to be a "folk theorem" that confluence for $\lambda^{\rightarrow}$ follows immediately from the Church–Rosser Theorem (confluence) for untyped lambda calculus (cf. [6, Appendix A]). The example given above, which is essentially from [138, 100], disproves the folk theorem for the formulation of $\lambda^{\rightarrow}$ used here. The folk theorem also fails for the somewhat more common formulation of $\lambda^{\rightarrow}$ using variables that are each given a fixed type. In the alternate presentation of $\lambda^{\rightarrow}$, α-conversion must be restricted so that we only replace one bound variable by another with the same type. With this restriction on α-conversion, the example demonstrating failure of confluence still applies. Thus, confluence for $\lambda^{\rightarrow}$ does not seem to follow from the Church–Rosser Theorem for untyped β, η-reduction directly. It is worth noting, however, that if we drop η-reduction, then we *do* have confluence for β-reduction on $\lambda^{\rightarrow}$ pre-terms.

The convertibility relation $\leftrightarrow$ on typed terms is the least type-respecting equivalence relation containing reduction $\twoheadrightarrow$. This can be visualized by saying that $\Gamma \rhd M \leftrightarrow N : \sigma$ iff there is a sequence of terms $M_0, \ldots, M_k$ with $\Gamma \rhd M_i : \sigma$ such that

$$M \equiv M_0 \twoheadrightarrow M_1 \twoheadleftarrow \cdots \twoheadrightarrow M_k \equiv N.$$

In this picture, the directions of $\twoheadrightarrow$ and $\twoheadleftarrow$ should not be regarded as significant. However, by reflexivity and transitivity of $\twoheadrightarrow$, this order of reduction and "backward reduction" is completely general. A few words are in order regarding the assumption that $\Gamma \rhd M_i : \sigma$ for each i. For pure typed lambda calculus, this assumption is not necessary; if $\Gamma \rhd M \leftrightarrow N : \sigma$ and $\Gamma \cap \Gamma'$ mentions all free variables of M and N, then $\Gamma' \rhd M \leftrightarrow N : \sigma$. However, for extensions of pure typed lambda calculus obtained by adding algebraic rewrite rules for basic types commonly found in programming languages, this fails (see Proposition 2.3.2). Since the main applications of typed lambda calculus to the study of programming languages involve extensions of the pure calculus, we have chosen definitions and notation which generalize easily.

2.3.7. EXERCISE. Show that $\Gamma \rhd M \leftrightarrow N : \sigma$ iff $\vdash \Gamma \rhd M = N : \sigma$. Note that confluence is not required.

2.3.8. EXERCISE. Use Exercise 2.3.7 and confluence to prove Corollary 2.3.6.

2.3.9. EXERCISE. Prove that if $\Gamma \rhd M : \sigma$ and $\Gamma \rhd N : \sigma$ are normal forms that differ by more than the names of bound variables, then the equation $\Gamma \rhd M = N : \sigma$ is not provable (except from nonlogical axioms).

The reader familiar with untyped lambda calculus may be surprised that reduction on typed terms always terminates. This is a nontrivial property of typed lambda calculus that accounts for many technical differences between untyped and typed systems.

2.3.10. THEOREM (strong normalization). *There is no infinite reduction sequence* $M_0 \to M_1 \to M_2 \to \cdots$ *of* $\lambda^{\to}$ *terms.*

Since reduction is effective, strong normalization implies that, for any reasonable encoding of the natural numbers, we cannot encode all partial recursive functions in pure $\lambda^{\to}$. Every function we can encode in pure $\lambda^{\to}$ terminates on all input. However, adding constants for basic numeric functions and recursion operators gives us a typed lambda calculus with universal computing power (cf. [104]).

2.4. Semantics and soundness

2.4.1. General models and the meanings of terms

We have already seen three notions of function space: classical set-theoretic functions, recursive functions on modest sets, and continuous functions on cpo's. Each of these interpretations gives us a *model* of $\lambda^{\rightarrow}$, which means that every term may be interpreted as a function of the appropriate type, and all of the provable equations between terms are true. A common feature of these interpretations is that $\sigma \rightarrow \tau$ denotes a set of functions from A^{σ} to A^{τ}, possibly enhanced with some extra structure. While the extra structure (enumeration functions or partial orders) may be used to characterize the class of functions in $\sigma \rightarrow \tau$, the extra structure plays no role in giving meaning to pure lambda terms. Therefore, from the point of view of pure lambda calculus, it makes sense to concentrate on the sets of functions involved. This allows us to see all three examples as instances of a general notion of model.

There are several equivalent definitions of "general models", which we will call *Henkin models* after [51]. The definition we will use has three parts. We first define typed applicative structures, and then specify two additional conditions that are required of models. A typed applicative structure consists of families of sets and mappings indexed by types, or pairs of types, of the form necessary to give meanings to terms. To be a model, an applicative structure must be extensional, which guarantees that the meaning of a term is unique, and there must be enough elements so that every well-typed term in fact has a meaning. Two common ways of formalizing that there must be "enough elements" are the environment model definition, and the combinatory model definition. Since the two definitions are equivalent, we will call a structure satisfying either definition a Henkin model. If we add Cartesian products and a terminal ("one-point") type 1 to the language, then an equivalent definition is easily stated using category theory. In categorical terms, Henkin models for $\lambda^{\rightarrow, \times, 1}$ are the same as Cartesian closed categories C such that the functor $Hom(1, \cdot)\colon C \rightarrow Set$ is faithful.

The equational proof system is easily shown to be sound for Henkin models. However, additional rules are needed for deductive completeness. With a relatively minor addition, we have completeness over models without empty types. To achieve completeness for models that may have empty types requires a more substantial extension of the proof system. By generalizing from Henkin models to Cartesian closed categories, we can also obtain completeness theorems for the proof system exactly as presented in Section 2.3.1 (see [68, 97] for further discussion).

2.4.2. *Applicative structures, extensionality and frames*

A typed applicative structure $\mathscr{A}$ for signature Σ is a tuple

$$\langle \{A^{\sigma}\}, \{\mathbf{App}^{\sigma,\tau}\}, Const \rangle$$

of families of sets and mappings indexed by type expressions over the type constants from Σ. For each σ and τ we assume the following conditions:

- A^{σ} is some set,
- $\mathbf{App}^{\sigma,\tau}$ is a set-theoretic map $\mathbf{App}^{\sigma,\tau}\colon A^{(\sigma\rightarrow\tau)} \rightarrow A^{\sigma} \rightarrow A^{\tau}$,
- *Const* is a mapping from term constants of Σ to elements of the appropriate A^{σ}'s.

Note that applicative structures are ordinary algebras for a particular kind of signature with infinitely many sorts (one for each type expression). Therefore, we may appeal to standard algebraic definitions of *homomorphism* and *isomorphism.*

An applicative structure is *extensional* if it satisfies the condition

- for all $f, g \in A^{\sigma\rightarrow\tau}$, if $\mathbf{App}\, f\, d = \mathbf{App}\, g\, d$ for all $d \in A^{\sigma}$, then $f = g$.

The set-theoretic, recursion-theoretic and continuous interpretations all yield extensional applicative structures in which $\mathbf{App}^{\sigma,\tau} f\,x = f(x)$ is simply function application. The set of terms over any signature Σ may also be viewed as an applicative structure, as follows.

2.4.1. EXAMPLE. Let Σ be any signature and let $\mathscr{H}$ be any finite or infinite type assignment $\mathscr{H} = \{x_1 : \sigma_1, x_2 : \sigma_2, \ldots\}$. An applicative structure

$$\mathscr{T} = \langle \{T^\sigma\}, \{\mathbf{App}^{\sigma,\tau}\}, Const \rangle$$

may be defined by letting

$$T^\sigma = \{M \mid \Gamma \rhd M : \sigma \text{ some finite } \Gamma \subseteq \mathscr{H}\}$$

and defining $\mathbf{App}^{\sigma,\tau}\, M\, N = MN$ for every $M \in T^{\sigma\to\tau}$ and $N \in T^\sigma$. For each term constant from Σ, we take $Const(c) = c$. If $\mathscr{H}$ provides variables of each type, then $\mathscr{T}$ is extensional. However, if $\mathscr{H}$ does not then we may have some T^σ empty. This allows two distinct elements of $T^{\sigma\to\tau}$ to be extensionally equal (vacuously), so extensionality fails in general.

In place of applicative structures, many authors use a definition that assumes extensionality. Instead of letting $A^{\sigma\to\tau}$ be any set, and requiring an application map to make elements of $A^{\sigma\to\tau}$ behave like functions, we could require that $A^{\sigma\to\tau}$ actually be some set of functions from A^σ to A^τ. To be precise, a *type frame* is an applicative structure $\langle \{A^\sigma\}, \{\mathbf{App}^{\sigma,\tau}\}, Const \rangle$ such that

$$A^{\sigma\to\tau} \subseteq (A^\tau)^{(A\sigma)}$$

and $\mathbf{App}^{\sigma,\tau} f\, d = f(d)$. In the formula above, exponentiation A^B denotes the usual set-theoretic collection of all functions from B to A.

2.4.2. LEMMA. *An applicative structure $\mathscr{A}$ is extensional iff there is an isomorphic type frame $\mathscr{B} \cong \mathscr{A}$.*

The proof is a straightforward induction on types. If we generalize $\lambda^\to$ to allowing certain types to be identified, then this equivalence fails. For example, it is possible to have an extensional applicative structure with $A^\sigma = A^{\sigma\to\sigma}$, but in this case $A^{\sigma\to\sigma}$ cannot be a set of functions from A^σ to A^σ.

Since **App** is always function application in any type frame, it is common to omit **App**, writing $\mathscr{A} = \langle \{A^\sigma\}, Const \rangle$. If Σ has no constants, then we may also drop *Const*, and think of a type frame $\mathscr{A} = \{A^\sigma\}$ as simply an indexed family of sets.

2.4.3. Environment models

One characterization of Henkin model is the environment model definition, which uses the subsidiary notion of environment. An *environment* η for applicative structure $\mathscr{A}$ is a mapping from variables to the union of all A^σ. If Γ is a type assignment, then we say η *satisfies* Γ, written $\eta \models \Gamma$, if $\eta(x) \in A^\sigma$ for every $x : \sigma \in \Gamma$. If η is any environment for $\mathscr{A}$ and $d \in A^\sigma$, then $\eta[d/x]$ is the environment mapping y to $\eta(y)$ for y different from x and $\eta[d/x](x) = d$.

An *environment model* is an extensional applicative structure such that the clauses below define a total meaning function $[\![\cdot]\!]\cdot$ on terms $\Gamma \rhd M{:}\sigma$ and environments η such that $\eta \models \Gamma$. Recall that if $\Gamma \rhd M{:}\sigma$ is well-typed, then there is proof of this using the typing rules. To make the definition of meaning as simple as possible, we will use induction on typing derivations, later showing that the meaning is independent of which derivation we choose. Specifically, we define the meaning of a well-typed term $\Gamma \rhd M{:}\sigma$ in environment $\eta \models \Gamma$ by five inductive clauses, corresponding to typing axioms (var), (cst), and typing rules (add hyp), (→Elim) and (→Intro). In computer science jargon, the "abstract syntax" of lambda terms allows us to determine the type and free variables of a term, and the typing derivation used to establish that the term is well-typed. All of these are used in defining the meaning of a term.

$$
\begin{aligned}
[\![x{:}\sigma \rhd x{:}\sigma]\!]\eta &= \eta(x),\\
[\![\emptyset \rhd c{:}\sigma]\!]\eta &= \mathit{Const}(c),\\
[\![\Gamma, x{:}\sigma \rhd M{:}\tau]\!]\eta &= [\![\Gamma \rhd M{:}\tau]\!]\eta,\\
[\![\Gamma \rhd MN{:}\tau]\!]\eta &= \mathbf{App}^{\sigma,\tau}\,[\![\Gamma \rhd M{:}\sigma\to\tau]\!]\eta\,[\![\Gamma \rhd N{:}\sigma]\!]\eta,\\
[\![\Gamma \rhd \lambda x{:}\sigma.M{:}\sigma\to\tau]\!]\eta &= \text{the unique } f \in A^{\sigma\to\tau} \text{ such that}\\
&\qquad \forall d \in A^{\sigma}.\ \mathbf{App}\, f\, d = [\![\Gamma, x{:}\sigma \rhd M{:}\tau]\!]\eta[d/x].
\end{aligned}
$$

The main reason for using induction on typing derivations is that in defining the meaning of a lambda abstraction $\Gamma \rhd \lambda x{:}\sigma.M{:}\sigma\to\tau$, we need to refer to the meaning of M in typing context Γ, $x{:}\sigma$. If we know that $\Gamma \rhd \lambda x{:}\sigma.M{:}\sigma\to\tau$ is typed according to rule (→Intro), then we are guaranteed that Γ, $x{:}\sigma$ is well-formed, since this must have occurred in the hypothesis of the rule. There are other ways to solve this minor technical problem, but using induction on typing derivations seems as simple as any of the alternatives, overall, and also generalizes most easily to other type systems. Since two derivations of typing statement $\Gamma \rhd M{:}\sigma$ differ only in the places that (add hyp) is used, it is an easy matter to prove the following lemma.

2.4.3. LEMMA. *Let A be an environment model, $\Gamma \rhd M{:}\sigma$ a well-typed term, and $\eta \models \Gamma$. Then $[\![\Gamma \rhd M{:}\sigma]\!]\eta$ does not depend on which typing derivation we use to define the meaning of the term.*

An extensional applicative structure might fail to be an environment model if $A^{\sigma\to\tau}$ does not contain any f satisfying the conditions given in the clause for $[\![\Gamma \rhd \lambda x{:}\sigma.M{:}\sigma\to\tau]\!]\eta$. This is the only potential problem since extensionality guarantees that if any f exists, it is unique. It is easy to show that any $\mathscr{A}$ containing all set-theoretic functions is an environment model.

2.4.4. EXERCISE. A full set-theoretic function hierarchy is a frame $\mathscr{A} = \langle \{A^{\sigma}\}, \mathit{Const}\rangle$ such that $A^{\sigma\to\tau}$ contains all set-theoretic functions from A^{σ} *to* A^{τ}. Show, by induction on terms, that for any full set-theoretic hierarchy $\mathscr{A}$, typed term $\Gamma \rhd M{:}\sigma$ and any environment $\eta \models \Gamma$, the meaning $[\![\Gamma \rhd M{:}\sigma]\!]\eta$ is a well-defined element of A^{σ}.

We will show later that frames of recursion-theoretic or continuous functions are also environment models, and hence Henkin models.

2.4.4. Type soundness

Type soundness is described by the following lemma, which is part of saying that the environment model definition makes sense.

2.4.5. LEMMA. *Let $\mathscr{A}$ be any environment model, and $\eta \models \Gamma$ an environment for $\mathscr{A}$. Then $[\![\Gamma \rhd M:\sigma]\!]\eta \in \mathscr{A}^\sigma$.*

2.4.5. Equational soundness

A simple fact about meaning is that the meaning of $\Gamma \rhd M:\sigma$ in environment η does not depend on $\eta(y)$ if y is not free in M. We also have a substitution lemma for $\lambda^{\rightarrow}$.

2.4.6. LEMMA (free variables). *Suppose $\eta_1, \eta_2 \models \Gamma$ are environments for $\mathscr{A}$ such that $\eta_1(x) = \eta_2(x)$ for every $x \in FV(M)$. Then $[\![\Gamma \rhd M:\sigma]\!]\eta_1 = [\![\Gamma \rhd M:\sigma]\!]\eta_2$.*

2.4.7. LEMMA (substitution). *Let Γ, $x{:}\sigma \rhd M:\tau$ and $\Gamma \rhd N:\sigma$ be terms, $\eta \models \Gamma$, and $d = [\![\Gamma \rhd M:\sigma]\!]\eta$. Then*

$$[\![\Gamma \rhd [N/x]M:\tau]\!]\eta = [\![\Gamma, x:\sigma \rhd M:\tau]\!](\eta[d/x]).$$

Intuitively, the Substitution Lemma says that the effect of substituting an expression N for x is the same as letting x denote the meaning of N.

The standard notions of satisfaction and validity for equations between typed lambda terms have routine definitions. The only minor point which may require some clarification is that we only define satisfaction

$$\mathscr{A}, \eta \models \Gamma \rhd M = N:\sigma$$

by a model $\mathscr{A}$ and environment η when $\eta \models \Gamma$. We say a model $\mathscr{A}$ *satisfies* an equation $\Gamma \rhd M = N:\sigma$ if $\mathscr{A}$ and environment η satisfy this equation, for every η satisfying Γ. Therefore, if Γ is unsatisfiable (some type is empty in $\mathscr{A}$), an equation may hold vacuously.

2.4.8. THEOREM (soundness). *If $\mathscr{E} \vdash \Gamma \rhd M = N:\tau$, for any set $\mathscr{E}$ of typed equations, then every environment model satisfying $\mathscr{E}$ also satisfies $\Gamma \rhd M = N:\tau$.*

2.4.6. Completeness for Henkin models without empty types

Although the proof system is sound, the proof rules are not complete for Henkin models. The reason has to do with type emptiness (see [82, 94, 97] for further discussion). With the additional rule

$$\text{(nonempty)} \quad \frac{\Gamma,\ x:\sigma \rhd M = N:\tau}{\Gamma \rhd M = N:\tau} \quad x \text{ not free in } M, N$$

we have completeness over models without empty types. If we know σ is not empty, then the Free Variable Lemma may be used to show that (nonempty) is sound. However, if a type σ is empty (*i.e.*, $A^\sigma = \emptyset$), then Γ, $x{:}\sigma \rhd M = N:\tau$ may hold solely because no environment can give x a value of type σ. Therefore, it is incorrect to apply rule (nonempty).

2.4.9. THEOREM (completeness without empty types). *Let $\mathscr{E}$ be any lambda theory closed under the rule* (nonempty). *Then there is a Henkin model $\mathscr{A}$, with no $A^\sigma = \emptyset$, satisfying precisely the equations belonging to $\mathscr{E}$.*

This theorem may be proved directly using a "term model" construction similar to the applicative structure $\mathscr{T}$ in Example 2.4.1. We will sketch another way of constructing term models in Section 3 using logical relations.

An important special case of Theorem 2.4.9 is the pure theory $\mathscr{E}$ closed under the inference rules of Section 2.3.1, with no nonlogical axioms. It follows from Proposition 2.3.1 that although rule (nonempty) is not among the rules in Section 2.3.1, the pure theory is closed under (nonempty). Therefore, we have the following corollary.

2.4.10. COROLLARY. *Let $\mathscr{E}$ be the pure theory of β,η-conversion. There is a Henkin model $\mathscr{A}$ without empty types satisfying precisely the equations of $\mathscr{E}$.*

A model $\mathscr{A}$ is *nontrivial* if there is some equation that is not satisfied by $\mathscr{A}$. If a signature Σ has only one base type b, then a simple argument shows that any nontrivial Henkin model must satisfy (nonempty): if A^b is empty, then by induction on types no A^σ may have more than one element. Therefore, if $\mathscr{A}$ is nontrivial, A^b must not be empty. But then no type is empty, as is easily verified by induction on types. For this reason, when only one base type is used, it has been common practice to write equations without specifying the set of free variables explicitly.

2.4.7. Completeness with empty types

In general, we may be interested in typed lambda calculus over an arbitrary collection of base types. Since some of these may not have any definable elements, or may naturally be considered empty (as may arise when types are given by specifications), it is important to be able to reason about terms over possibly empty types (see [82, 94] for further discussion).

For Henkin models that may have empty types, we may achieve completeness using additional rules presented in [82]. The main purpose of these additional rules is to capture reasoning of the form "if $M = N$ whenever σ is empty, and $M = N$ whenever σ is nonempty, then we must have $M = N$". To facilitate reasoning about empty types, it is convenient to add assumptions of the form *empty*(σ) to type assignments. An *extended equation* will be a formula $\Gamma \rhd M = N : \sigma$ with Γ the union of a type assignment Γ_1 and a set Γ_2 of formulas *empty*(τ). We require that $\Gamma_1 \rhd M : \sigma$ and $\Gamma_1 \rhd N : \sigma$, so that emptiness assertions do not affect the syntactic types of terms.

The proof system for reasoning about empty types uses an axiom scheme for introducing equations with emptiness assertions

(empty I) $\quad \Gamma, \mathit{empty}(\sigma), x{:}\sigma \rhd M = N : \tau$

and an inference rule which lets us use emptiness assertions to reason by cases

(empty E) $$\frac{\Gamma, x{:}\sigma \rhd M = N : \tau, \quad \Gamma, \mathit{empty}(\sigma) \rhd M = N : \tau}{\Gamma \rhd M = N : \tau} \quad x \notin FV(M, N).$$

Technically speaking, the side condition of this rule is redundant, since the second equation in the antecedent can be well-formed only if $x \notin FV(M, N)$. We write $\vdash^{(\text{empty I,E})}$ for provability using the proof rules of Section 2.3.1 and the additional axiom and inference rule for empty types.

2.4.11. THEOREM (completeness for Henkin models [82]). *Let $\mathscr{E}$ be a set of extended equations, possibly containing emptiness assertions, and $\Gamma \rhd M = N : \sigma$ an extended equation. Then $\mathscr{E} \vdash^{(\text{empty I,E})} \Gamma \rhd M = N : \sigma$ iff every Henkin model satisfying $\mathscr{E}$ also satisfies $\Gamma \rhd M = N : \sigma$.*

Since (empty E) is not an equational inference, the rules for reasoning about empty types have a different flavor from the proof systems without (empty E). In particular, we give up the property that every set of equations closed under semantic implication is the theory of a single "minimal" model (see [94] for further discussion).

2.4.8. Combinatory models

The environment model condition that every term has a meaning is equivalent to the existence of certain elements called combinators, with each combinator characterized by an equational axiom. One advantage of the combinatory characterization of Henkin models is that it does not refer to syntax or the meaning function on typed lambda terms. We simply require that certain equational conditions are satisfied.

We say an applicative structure $\mathscr{A}$ *has combinators* if, for all ρ, σ, τ, there exist elements

$$K_{\sigma,\tau} \in A^{\sigma \to (\tau \to \sigma)}, \qquad S_{\rho,\sigma,\tau} \in A^{(\rho \to \sigma \to \tau) \to (\rho \to \sigma) \to \rho \to \tau}$$

satisfying the equational conditions

$$K_{\sigma,\tau} xy = x, \qquad S_{\rho,\sigma,\tau} xyz = (xz)(yz)$$

for all x, y, z of the appropriate types.

2.4.12. LEMMA. *An extensional applicative structure $\mathscr{A}$ is an environment model iff $\mathscr{A}$ has combinators.*

The lemma is proved by translating lambda terms into applicative combinations of combinators, as in [6, 81].

In logical terms, a typed applicative structure may be viewed as any first-order model of a many-sorted signature with one sort for each type, and containing a function symbol

$$\mathbf{App}^{\sigma,\tau} : ((\sigma \to \tau) \times \sigma) \to \tau$$

for each σ and τ. Extensionality may be written as a first-order formula (over the same signature), while each combinator is defined by an equational axiom. Therefore, environment models may be characterized as models of a multisorted first-order theory comprising the extensionality and combinator axioms. This is often called the *combinatory model definition*, since combinators play an important role.

2.4.9. Combinatory and lambda algebras

Two nonextensional structures are occasionally of interest. Combinatory algebras are typed applicative structures that have combinators (as described above), but are not necessarily extensional. In a sense, every term can be given a meaning in a combinatory algebra, since every term can be translated into combinators, and every applicative combination of combinators has a straightforward interpretation. However, many natural equations between lambda terms may fail. For example, $SKK = SKS$ holds in every lambda model (when these combinators are each typed appropriately), since these functions are extensionally equal. However, both are in combinatory normal form, and so by the combinatory analog of Corollary 2.3.6 may have distinct interpretations in a combinatory algebra. Consequently, combinatory algebras are not models of the equational theory of typed lambda calculus.

There exists an equationally axiomatized class of combinatory algebras satisfying the pure equational theory of typed lambda calculus (β, η-conversion). These structures are called *lambda algebras*, and the unmemorable axioms may be found in [6, 81], for example. The main difference between lambda algebras and models is that lambda algebras do not satisfy (ξ) in the standard sense. As a consequence, the set of equations satisfied by an arbitrary lambda algebra is not necessarily closed under (ξ). We generally interpret

$$(\xi) \qquad \frac{\Gamma, x{:}\sigma \rhd M = N{:}\tau,}{\Gamma \rhd \lambda x{:}\sigma.M = \lambda x{:}\sigma.N{:}\sigma \to \tau}$$

as saying that whenever M and N have the same meaning for all values of x, we have $\lambda x{:}\sigma.M = \lambda x{:}\sigma.N$. This semantic interpretation of (ξ) is guaranteed by extensionality, but fails in arbitrary lambda algebras.

2.5. Recursion-theoretic models

2.5.1. Introduction

We have already seen that every full classical function hierarchy is a Henkin model. In this section and the next, we will consider two other kinds of models, "recursion-theoretic" models based on Gödel numbering of partial recursive functions, and "domain-theoretic" models of continuous functions on complete partial orders.

Throughout this section, it will be useful to let $\varphi_0, \varphi_1, \ldots$ be some enumeration of the set $\mathcal{PR}$ of all partial recursive functions. We assume that this enumeration has all of the convenient, standard properties assumed in recursive function theory. In particular, we will require recursive pairing and projection functions, and the *s-m-n* theorem. (Essentially, the *s-m-n* theorem says that given any recursive function $f(x_1, \ldots, x_m, y_1, \ldots, y_n)$ of $m+n$ arguments and natural numbers $a_1, \ldots, a_m$, we may effectively find a recursive function g of n arguments such that $g(y_1, \ldots, y_n) = f(a_1, \ldots, a_m, y_1, \ldots, y_n)$; see [116], for example, for further discussion.)

2.5.2. Modest sets

We will use numeric codes to characterize recursive functions on arbitrary values. A simple and fairly general framework involves sets with partial enumeration functions. These are essentially the "effective objects" of [58], which have been dubbed the "modest sets" by Dana Scott (see [59]). The main ideas go back to Kleene realizability [63, 64], recursion in higher types [66], and the model HEO described in [135]. Formally, a *modest set* is a pair $\langle A, e_A \rangle$, where A is any set and $e_A\colon \mathcal{N} \to A$ is a surjective partial function from the natural numbers to A. We often refer to e_A as the *partial enumeration function for A*. Informally, we think of n as the "code" for the element $e_A(n)$ of A. Since e_A is surjective, every element has a code. This forces A to be countable. However, because e_A may be partial, some natural numbers might not code any element of A. Since e_A will be used simply to define a class of recursive functions on A, we will not require effective procedures for determining whether given $n, m \in \mathcal{N}$ code the same (or any) element of A.

2.5.1. EXAMPLE. The pair $\langle \mathcal{PR}, \lambda n.\varphi_n \rangle$ is a modest set.

2.5.2. EXAMPLE. If $\mathcal{T}$ is the set of total recursive functions, and e_T is the map

$$e_T(n) = \begin{cases} \varphi_n & \text{if } \varphi_n \text{ is total,} \\ \text{undefined} & \text{otherwise,} \end{cases}$$

then $\langle \mathcal{T}, e_T \rangle$ is a modest set.

2.5.3. EXAMPLE. The set of typed lambda terms forms a modest set, with enumeration e_{terms} mapping n to the term M with Gödel code $\ulcorner M \urcorner = n$.

We will often want to refer to the set of codes of an element. For any modest set $\langle A, e_A \rangle$ and $a \in A$, we will write $|a|_A$ for the set of natural numbers: $|a|_A = \{n \mid e_A(n) = a\}$. We will omit the subscript when it is clear from the context. Note that since e_A is a function, the sets $|a|$ and $|b|$ are disjoint whenever $a \neq b$. Furthermore, it is easy to see that if $|\cdot|\colon A \to \mathcal{P}(\mathcal{N})$ is some function assigning disjoint, nonempty sets to distinct elements of A, then $|\cdot|$ determines a unique partial surjection $e\colon \mathcal{N} \to A$. Thus a modest set may be defined by giving $|\cdot|$ instead of e.

Intuitively, the recursive functions on modest sets are the functions which correspond to recursive functions on codes. To be more precise, let $f\colon A \to B$ be a total function on the underlying values of modest sets $\langle A, e_A \rangle$ and $\langle B, e_B \rangle$. A partial function $g\colon \mathcal{N} \rightharpoonup \mathcal{N}$ *tracks* f if

$$n \in |a| \text{ implies } g(n)\!\downarrow \text{ and } g(n) \in |f(a)|,$$

where $g(n)\!\downarrow$ means the function g is defined on n. We say f is *recursive* (or *computable*) if there is a partial recursive g tracking f. Note that while f is total, g may be partial. When f is recursive, we write $|f|$ for the set of codes of partial recursive functions tracking f, i.e., $|f| = \{n \mid \varphi_n \text{ tracks } f\}$. If $A = \langle A, e_A \rangle$ and $B = \langle B, e_B \rangle$ are modest sets, then we let $A \to B = \langle A \to B, e_{A\to B} \rangle$ be the collection of all total recursive functions from A to B, enumerated by $e_{A\to B}(n) = f$ iff $n \in |f|$.

2.5.4. LEMMA. *If A and B are modest sets, then $A \to B$ is a modest set.*

2.5.3. Full recursive hierarchy

We define a type frame called the *full recursive hierarchy* $\mathscr{A}$ *over modest sets* $\langle A^{b_0}, e_0\rangle, \ldots, \langle A^{b_k}, e_k\rangle$ by taking $A^{b_0}, \ldots, A^{b_k}$ as base types, and

$$A^{\sigma\to\tau} = \text{all recursive } f: \langle A^\sigma, e_\sigma\rangle \to \langle A^\tau, e_\tau\rangle,$$

enumerated by $e_{A^\sigma\to A^\tau}$. Note that, technically speaking, a type frame is a collection of sets, rather than a collection of modest sets. In other words, while partial enumerations are used to define function types, they are not part of the resulting type frame. This is sometimes awkward, since many discussions of recursive hierarchies will refer to the partial enumeration functions. However, as mentioned earlier, the meaning of a pure lambda term does not depend on the way any modest set is enumerated.

2.5.5. THEOREM. *Any full recursive hierarchy is a Henkin model.*

PROOF. We will prove that the environment model condition is satisfied, using a slightly stronger induction hypothesis than the theorem. Specifically, we will show that the meaning of every term exists and is computable from a coding of the environment. We will use the assumption that meanings are computable to show that every function defined by lambda abstraction is computable.

To state the induction hypothesis precisely, we assume that all variables are numbered in some standard way. To simplify the notation a bit, we will assume that the variables used in any type assignment are numbered consecutively. For an environment $\eta \models \Gamma$, we code η on Γ by a coded sequence of natural numbers $\langle n_1, \ldots, n_k\rangle$, where $x_1, \ldots, x_k$ is a list of variables in Γ in numeric order, and $n_i \in |(\eta x_i)|$. We say $[\![\Gamma, x{:}\sigma \rhd M{:}\sigma]\!]$ is *computable* if, whenever $\eta \models \Gamma$, a code $m \in |[\![\Gamma \rhd M{:}\sigma]\!]\eta|$ is uniformly computable from any coding of η on Γ. The main steps of the inductive argument are given below, with additional details left to the reader.

$$[\![x_1{:}\sigma_1, \ldots, x_k{:}\sigma_k \rhd x_i{:}\sigma_i]\!]\eta = \eta x_i$$

is computable since $n_i \in |\eta x_i|$ may be computed from $\langle n_1, \ldots, n_k\rangle$;

$$[\![\Gamma \rhd MN{:}\tau]\!]\eta = [\![\Gamma \rhd M{:}\sigma\to\tau]\!]\eta\,([\![\Gamma \rhd N{:}\sigma]\!]\eta)$$

is computed by $\varphi_m(n)$, where $m \in |[\![\Gamma \rhd M{:}\sigma\to\tau]\!]\eta|$ and $n \in |[\![\Gamma \rhd N{:}\sigma]\!]\eta|$;

$$[\![\Gamma \rhd \lambda x{:}\,\sigma.M{:}\sigma\to\tau]\!]\eta = f$$

with $f(a) = [\![\Gamma, x{:}\sigma \rhd M{:}\tau]\!]\eta[a/x]$ is computable since $m \in |[\![\Gamma, x{:}\,\sigma \rhd M{:}\tau]\!]\eta[a/x]|$ is computable from $n \in |a|$.

In the lambda abstraction case, we use the *s-m-n* theorem to show that if

$$m \in |[\![\Gamma, x{:}\sigma \rhd M{:}\,\tau]\!]\eta[a/x]|$$

is computable from a coding of $\eta[a/x]$ on $\Gamma, x{:}\sigma$, then the function mapping

$$n \in |a| \mapsto m \in |[\![\Gamma, x{:}\sigma \rhd M{:}\tau]\!]\eta[a/x]|$$

is computable from a coding of η on Γ. This completes the proof. $\square$

An alternative formulation of recursive models uses partial equivalence relations (*per's*) instead of partial enumeration functions. For example, the model HEO of [135] is a full recursive hierarchy formulated using per's over the natural numbers. A generalization of that construction is discussed in [13, 90], for example. To explain the basic idea, we need a few preliminary definitions. A *partial equivalence relation*, or *per*, on a set A is an equivalence relation on a subset of A, or equivalently, a symmetric and transitive relation on A. If $\sim$ is a per on A, then we can form the *subquotient* $A/\sim$ by taking the collection $A/\sim = \{[a] \mid a \sim a\}$ of nonempty equivalence classes $[a] = \{a' \mid a' \sim a\}$. It is relatively easy to show that any modest set $A = \langle A, e_A \rangle$ is recursively isomorphic to a modest set $B = \langle \mathcal{N}/\sim, e_B \rangle$ whose underlying set is a subquotient of $\mathcal{N}$. Specifically, we define the per $\sim$ on $\mathcal{N}$ by

$$n \sim m \text{ iff } e_A(n) = e_A(m)$$

and define the partial enumeration e_B by $e_B(n) = [n]$. Since the identity on $\mathcal{N}$ tracks the isomorphism $a \mapsto |a|$, every modest set is given by a per on $\mathcal{N}$, up to recursive isomorphism.

2.6. Domain-theoretic models

2.6.1. Recursive definitions and fixed point operators

Domain theory provides a number of model constructions that are very useful in studying typed and untyped languages. From a $\lambda^{\rightarrow}$ point of view, domain-theoretic models may be motivated by considering recursive function definitions. For example, suppose we wish to add a definition form

$$\mathsf{letrec}\, f{:}\sigma = M \;\mathsf{in}\; N$$

with the intended meaning that within N, the variable f denotes a solution to the equation $f = M$. In general, f may occur in M. For the declaration to make sense, M must have type σ under the hypothesis that $f{:}\sigma$. To see how recursive declarations may be used, consider the following example. Using a conditional, written out as a three-place function if ... then ... else ..., and other constants for common natural number and Boolean operations, the expression

$$\mathsf{letrec}\; f{:}nat \rightarrow nat = \lambda y{:}nat.(\mathsf{if}\; zero?\; y\; \mathsf{then}\; 1\; \mathsf{else}\; y * f(y-1))\; \mathsf{in}\; f\; 5$$

defines the familiar factorial function by recursion and applies this function to the natural number 5. Essentially, f is *defined* to be a solution to the equation

$$f = \lambda y{:}nat.\, \mathsf{if}\; zero\,?\; y\; \mathsf{then}\; 1\; \mathsf{else}\; y * f(y-1),$$

or equivalently, a fixed point of the operator

$$F \;::=\; \lambda f{:}nat \rightarrow nat.\lambda y{:}nat.\, \mathsf{if}\; zero?\; y\; \mathsf{then}\; 1\; \mathsf{else}\; y * f(y-1).$$

Therefore, using lambda abstraction and an operator fix_σ which returns a fixed point of any function from σ to σ, we can regard the recursive declaration form above as an abbreviation

$$\text{letrec } f{:}\sigma = M \text{ in } N \ ::= \ (\lambda f{:}\sigma.N)(\mathit{fix}_\sigma \lambda f{:}\sigma.M)$$

for an ordinary $\lambda^{\rightarrow}$ expression involving a fixed-point constant.

There are several properties we might reasonably expect of $\mathit{fix}_\sigma\,(\sigma\rightarrow\sigma)\rightarrow\sigma$. The first is that it indeed produce a fixed point, i.e.,

(fix) $\qquad \mathit{fix}_\sigma = \lambda f{:}\sigma\rightarrow\sigma.f(\mathit{fix}_\sigma f).$

We can use the axiom (fix) in equational reasoning, or define (fix) reduction by directing this equation from left to right. To see how (fix) reduction works, we will continue the factorial example. Since there are other constants besides fix, we also assume reduction rules for conditional, zero test, subtraction and multiplication.

Using $\mathit{fix}_{nat\rightarrow nat}$, the factorial function may be written $\mathit{fact} ::= \mathit{fix}_{nat\rightarrow nat} F$, where F is the expression defined above. Since the type is clear from context, we will drop the subscript from *fix*. To compute *fact n*, we may expand the definition, and use reduction to obtain the following.

$$\begin{aligned} \mathit{fact}\,n &= (\lambda f{:}nat\rightarrow nat.f(\mathit{fix}\,f))\,F\,n \\ &= F\,(\mathit{fix}\ F)\,n \\ &= (\lambda f{:}nat\rightarrow nat.\lambda y{:}nat.\text{ if } \mathit{zero?}\ y \text{ then } 1 \text{ else } y*f(y-1))(\mathit{fix}\,F)\,n \\ &= \text{if } \mathit{zero?}\,n \text{ then } 1 \text{ else } n*(\mathit{fix}\,F)(n-1). \end{aligned}$$

Note that in addition to (fix) reduction, we have only used ordinary β-reduction. When $n=0$, we can use the obvious axioms for conditional tests to simplify *fact* 0 to 1. For $n>0$, we can simplify the test to obtain $n*(\mathit{fix}\,F)(n-1)$, and continue as above. For any natural number n, it is clear that we will eventually compute $\mathit{fact}\,n = n!$. Put more formally, we may use ordinary induction to prove the metatheorem that for every natural number n the expression *fact n* may be reduced to the numeral for $n!$.

An alternative approach to understanding *fact* is to consider the finite expansions of $\mathit{fix}\,F$. To make this as intuitive as possible, let us temporarily think of $nat\rightarrow nat$ as a collection of partial functions on the natural numbers, represented by sets of ordered pairs. Using a constant *diverge* for the "nowhere defined" function (the empty set of ordered pairs), we let the "zeroth expansion" $(\mathit{fix}\,F)^{[0]} = \mathit{diverge}$ and define

$$(\mathit{fix}\,F)^{[n+1]} = F(\mathit{fix}\,F)^{[n]}.$$

In computational terms, $(\mathit{fix}\,F)^{[n]}$ describes the recursive function computed using at most n evaluations of the body of F. Or, put another way, $(\mathit{fix}\,F)^{[n]}$ is the best we could do with a machine having such limited memory that allocating space for more than n function calls would overflow the run-time stack.

Viewed as sets of ordered pairs, the finite expansions of $\mathit{fix}\,F$ are linearly ordered by set-theoretic containment. Specifically, $(\mathit{fix}\ \ F)^{[0]} = \emptyset$ is the least element in this ordering, and $(\mathit{fix}\,F)^{[n+1]} = (\mathit{fix}\,F)^{[n]} \cup \langle n, n!\rangle$ properly contains all $(\mathit{fix}\,F)^{[i]}$ for $i \leqslant n$. This reflects the fact that if we are allowed more recursive calls, we may compute factorial for larger natural numbers. In addition, since every terminating computation involving factorial uses only a finite number of recursive calls, it would make intuitive sense to let *fact* $=\bigcup_n (\mathit{fix}\ F)^{[n]}$. A priori, there is no good reason to believe that $\bigcup_n (\mathit{fix}\ F)^{[n]}$ will be a fixed point of F. However, by imposing relatively natural conditions on F (or the basic functions used to define F), we can guarantee that this countable union is a fixed point. In fact, since any fixed point of F must contain the functions defined by finite expansions of F, this will be the least fixed point of F.

In the domain-theoretic sematics, we will return to viewing $nat \to nat$ as a collection of total functions. However, we will alter the interpretation of *nat* so that $nat \to nat$ corresponds to the partial functions on the ordinary natural numbers in a straightforward way. This will allow us to define an ordering on total functions that reflects the set-theoretic containment $(\mathit{fix}\,F)^{[n]} \subseteq (\mathit{fix}\,F)^{[n+1]}$ of partial functions. In addition, all functions in domain-theoretic models will be continuous, in a certain sense. This will imply that least fixed points can be characterized as least upper bounds of countable sets like $\{(\mathit{fix}\,F)^{[n]}\}$.

2.6.2. Domains and continuous functions

In domain-theoretic models of typed lambda calculus, types denote partially ordered sets of values called domains. Although it is possible to develop domains using partial functions [106], we will follow the more standard approach of total functions. Since recursion allows us write expressions that describe nonterminating computations, we must give meaning to expressions that do not seem to define any standard value. Rather than saying an expression $M{:}nat$ that does not simplify to any standard natural number is undefined, the domain of "natural numbers" will include an additional value $\bot_{nat}{:}nat$ to represent all nonterminating computations of "type" *nat* (i.e., computations which we expect to produce a natural number, but which do not produce any standard value). This gives us a way to represent partial functions as total ones, since we may view any partial numeric function as a function into the domain of natural numbers with $\bot_{nat}$ added.

The ordering of a domain is intended to characterize what might be called "information content" or "degree of definedness". Since a nonterminating computation is less informative than any terminating computation, $\bot_{nat}$ will be the least element in the ordering on the domain of natural numbers. We order $nat \to nat$ pointwise, which gives rise to an ordering that strongly resembles the containment ordering on partial functions. For example, since the constant function $\lambda x{:}nat.\bot_{nat}$ produces the least element from any argument, it will be the least element of $nat \to nat$. Functions such as $(\mathit{fix}\,F)^{[k]}$ which are defined on some arguments, and intuitively "undefined" elsewhere, will be greater than the least element of the domain, but less than $(\mathit{fix}\,F)^{[j]}$ for $j>k$. By requiring that every function be continuous with respect to the ordering, we may interpret *fix* as the least fixed-point functional. For continuous F, the least fixed point will be the least upper bound of all finite expansions $(\mathit{fix}\,F)^{[k]}$.

Many families of ordered structures with these basic properties have been called *domains*. We will focus on the complete partial orders, since most families of domains are obtained from these by imposing additional conditions. Formally, a *partial order* $\langle D, \leqslant \rangle$ is a set D wih a reflexive, antisymmetric and transitive relation $\leqslant$. A subset $S \subseteq D$ is *directed* if every finite $S_0 \subseteq S$ has an upper bound in S. Since even the empty subset must have a least upper bound, every directed set is nonempty. A *complete partial order*, or *cpo* for short, is a partial order $\langle D, \leqslant\rangle$ with a least element $\perp_D \in D$ such that every directed $S \subseteq D$ has a least upper bound $\bigvee S$. When the cpo is clear from the context, we will omit the subscript from $\perp_D$.

The continuous functions are the functions which respect the order structure and least upper bounds of directed sets. Suppose $D = \langle D, \leqslant_D\rangle$ and $E = \langle E, \leqslant_E\rangle$ are cpo's, and $f: D \to E$ is a function on the underlying sets of these cpo's. We say f is *monotonic* if $d \leqslant d'$ implies $f(d) \leqslant f(d')$. It is easy to show that if f is monotonic and S is directed, then $f(S) = \{f(d) \mid d \in S\}$ is directed. A monotonic function f is *continuous* if, for every directed $S \subseteq D$, we have $f(\bigvee S) = \bigvee f(S)$. If $f(\perp) = \perp$, then we say f is *strict*. Since cpo's are required to have least elements, it might seem natural to focus on functions that preserve least elements. However, we will not be particularly concerned with strict functions. The reason is that we can define constant functions using lambda terms, and constant functions do not necessarily map $\perp$ to $\perp$.

One common point of confusion is that since "nontermination" is considered a value, we have functions which are defined on what might intuitively appear to be "undefined" arguments. For example, the constant function $\lambda v{:}nat.5$ maps $\perp_{nat}$ to 5, even though a program such as

$$\text{letrec } f{:}nat \to nat = \lambda x{:}nat.f(x+1) \text{ in } (\lambda x{:}nat.5)(f\ 3)$$

might diverge in many programming languages. However, termination and divergence depend on the order in which programs are "executed", or simplified using reduction rules. Cpo's give semantics to an evaluation order in which a function may ignore its arguments (so-called "lazy" or "call-by-name" evaluation). This is consistent with the equational proof system for $\lambda^{\to}$. In contrast, domains with partial functions, or domains with strict functions only, correspond to socalled "call-by-value" order, in which the program above would diverge.

To interpret every type as a cpo, we must be able to view the collection of continuous functions from one cpo to another as a cpo. We will partially order functions pointwise, as follows. Suppose $D = \langle D, \leqslant_D\rangle$ and $E = \langle E, \leqslant_E\rangle$ are cpo's. For continuous $f, g: D \to E$, we say $f \leqslant_{D\to E} g$ if, for every $d \in D$, we have $f(d) \leqslant_E g(d)$. We will write $D \to E = \langle D \to E, \leqslant_{D\to E}\rangle$ for the collection of continuous functions from D to E, ordered pointwise.

2.6.1. LEMMA. *For any cpo's D and E, the collection $D \to E$ of continuous functions ordered pointwise is a cpo.*

An n-ary function $f: D_1 \times \cdots \times D_n \to E$ is *continuous* if it is continuous in each argument, varied separately. The following two lemmas identify useful continuity properties involving multiary functions. The first is similar to the *s-m-n* theorem for recursive functions.

2.6.2. LEMMA. *If $f: D_1 \times \cdots \times D_k \times D \to E$ is continuous, then there is a unique continuous function (Curry f): $D_1 \times \cdots \times D_k \to (D \to E)$ such that for all $d_1, \ldots, d_k$ with $d_i \in D_i$ and*

$d \in D$, we have

$$f\langle d_1, \ldots, d_k, d\rangle = (Curry\ f)\langle d_1, \ldots, d_k\rangle(d).$$

In addition, the map Curry is continuous.

2.6.3. LEMMA. *If $f: D_1 \times \cdots \times D_k \to (D \to E)$ and $g: D_1 \times \cdots \times D_k \to D$ are continuous, then so is the function App f g given by*

$$(App\ f\ g)\langle d_1, \ldots, d_k\rangle = (f\langle d_1, \ldots, d_k\rangle)(g\langle d_1, \ldots, d_k\rangle)$$

for all $d_1, \ldots, d_k$ with $d_i \in D_i$ and *$d \in D$. In addition, the map App is continuous.*

2.6.3. Full continuous type hierarchy

The type frame called the *full continuous hierarchy $\mathscr{A}$ over* cpo's $\langle A^{b_0}, \leqslant_0\rangle, \ldots, \langle A^{b_k}, \leqslant_k\rangle$ is defined by taking $A^{b_0}, \ldots, A^{b_k}$ as base types, and

$$A^{\sigma\to\tau} = \text{all continuous } f: \langle A^\sigma, \leqslant_\sigma\rangle \to \langle A^\tau, \leqslant_\tau\rangle$$

with $A^{\sigma\to\tau}$ ordered pointwise by $\leqslant_{\sigma\to\tau}$. By Lemma 2.6.1, each $\langle A^\sigma, \leqslant_\sigma\rangle$is a cpo. Technically speaking, a type frame is a collection of sets, and so the full continuous hierarchy is the collection $\mathscr{A} = \{A^\sigma\}$ of base types and sets of functions. As with enumeration functions, partial orderings are used to define functional types, but are not part of the resulting type. This is somewhat awkward, since many discussions of continuous hierarchies refer to partial orders. The best that may be said is that the meaning of a pure lambda term is determined by the type frame, and does not depend on any order properties.

2.6.4. THEOREM. *The full continuous hierarchy over any collection of cpo's is a Henkin model.*

PROOF. The proof is essentially similar to the proof of Theorem 2.5.5. We prove that the environment model condition is satisfied using an induction hypothesis that is slightly stronger than the theorem. Specifically, in addition to showing that the meaning of every term exists, we show that the meaning is a continuous function of the values assigned by the environment. Like computability in the case of recursive models, the continuity hypothesis is used to show that a function defined by lambda abstraction is continuous.

We say $[\![\Gamma, \rhd M:\tau]\!]$ is *continuous* if, for every $x:\sigma \in \Gamma$ and $\eta \models \Gamma$, the map

$$a \in A^\sigma \mapsto [\![\Gamma, \rhd M:\tau]\!]\eta[a/x]$$

is a continuous function from A^σ to A^τ. This is the same as saying $[\![\Gamma, \rhd M:\tau]\!]$ is continuous as an n-ary function $A^{\sigma_1} \times \cdots \times A^{\sigma_n} \to A^\tau$, where $\Gamma = \{x_1:\sigma_1, \ldots, x_n:\sigma_n\}$.

Using Lemmas 2.6.2 and 2.6.3, the inductive proof is essentially straightforward. It is easy to see that

$$[\![x_1:\sigma_1, \ldots, x_k:\sigma_k \rhd x_i:\sigma_i]\!]\eta = \eta x_i$$

is continuous, since identity and constant functions are continuous. For application, we use Lemma 2.6.3, and for lambda abstraction, Lemma 2.6.2. □

Of course, a major motivation for domain-theoretic models is to interpret recursive definitions. In particular, cpo's provide a natural interpretation of *fix* constants as least-fixed-point operators.

2.6.5. LEMMA. *Every continuous* $f: D \to D$ *has a least fixed point*

$$fix_D\, f = \bigvee \{f^n(\bot_D)\},$$

where $f^0(\bot) = \bot$ *and* $f^{n+1}(\bot) = f(f^n(\bot))$. *Furthermore, the map* fix_D *is continuous.*

2.6.6. THEOREM. *Let* $\mathscr{A}$ *be a full continuous hierarchy over any collection of cpo's. By interpreting fixed-point constants appropriately,* $\mathscr{A}$ *may be extended to a Henkin model for a* $\lambda^{\to}$ *signature with fixed-point constants in such a way that each* fix_σ *denotes the least-fixed-point operator on* $A^{\sigma \to \sigma}$.

2.7. *Cartesian closed categories*

2.7.1. *Categories*

An alternative to Henkin models may be formulated using category theory. Cartesian closed categories (*ccc*'s) correspond closely to typed lambda calculus with pairing and a one-element type **1**. Since this calculus is a relatively mild extension of $\lambda^{\to}$, we may regard ccc's as an alternative semantics for $\lambda^{\to}$. To a first approximation, we may read the definition of ccc as a summary of the properties of domains and continuous functions (or modest sets and recursive functions) which we use to demonstrate that every hierarchy of continuous functions forms a Henkin model. However, because of the way these properties are formulated, ccc's turn out to be more general than Henkin models. In fact, the traditional proof rules for typed lambda calculus are sound and complete for Cartesian closed categories. Since we only achieve completeness for Henkin models by adding extra axioms, Cartesian closed categories give us a more natural completeness theorem than Henkin models.

One useful aspect of the categorical view of typed lambda calculus is that ccc's seem as much an alternative formulation of syntax as they do a semantics. This allows us to treat various syntactic and semantic notions uniformly. For example, an interpretation of terms in a model has the same character as a map between "models". However, it also becomes difficult to distinguish between syntax and semantics. In particular, arbitrary Cartesian closed categories do not seem as "semantic" as Henkin models. An attempt to identify certain classes of categories as "models" may be found in [68, Part II, Chapter 17]. One class of Cartesian closed categories that are more general than Henkin models, yet seem semantic rather than syntactic, are the Kripke lambda models discussed in Section 2.8.

The full correspondence between type theory and category theory is too extensive to survey here. For further information, the reader is referred to [68, 121] and proceedings from recent symposia on category theory and computer science. One technique that seems likely to prove fruitful in computer science is the categorical interpretation of logic, as described in [76, 68, 65]. By interpreting logical formulas in categories, we may "relativize" classical constructions to computational settings. One interesting and readable example is the development of fixed-point theory in an arbitrary ccc with finite limits and natural numbers object [7]. In this chapter, we will merely sketch the definition of Cartesian closed category and survey some connections with typed lambda calculus.

In basic terms, category theory is concerned with structures and the maps between them. In algebra (and many category theory books), the usual examples involve groups, rings or fields and the algebraic homomorphisms between them. In lambda calculus, we may regard categories as either an alternative formulation of syntax or semantics. From a syntactic point of view, the "structures" of interest are the type expressions of some calculus and the maps are terms, often modulo provable equality. From a semantic point of view, the usual structures are interpretations of type expressions (such as sets, domains or modest sets), and the maps are the functions we might expect to denote by terms (arbitrary functions, continuous functions or recursive functions).

Formally, a *category*

$$C = \langle C^o, C^a, dom, cod, id, comp \rangle$$

consists of two collections, C^o and C^a, and four maps,

$$dom: C^a \to C^o, \qquad cod: C^a \to C^o,$$
$$id: C^o \to C^a, \qquad comp: C^a \times C^a \rightharpoonup C^a$$

where "$\rightharpoonup$" indicates that *comp* is a partial function. Elements of C^o are called *objects* and elements of C^a are called *arrows* or *morphisms*. In many categories, the object are sets with some kind of structure, and the morphisms are functions that preserve this structure. However, many other mathematical objects satisfy the category axioms.

Intuitively, the maps *dom* and *cod* give the *domain* and *codomain* (or range) of a morphism, *id*(a) is the identity map on object a, and *comp* is function composition. Formally, we require that, for each object $a \in C^o$, the morphism *id*(a) have domain and codomain a. In addition, *id*(a) must satisfy certain axioms given below. Although *comp* is generally partial, *comp*(f, g) must be defined on any pair of morphisms $f, g \in C^a$ with *cod*(g) = *dom*(f). In this case, the domain and codomain of *comp*(f, g) are the domain of g and the codomain of f respectively. The map *comp* must also satisfy the associativity axiom given below. It is conventional to write id_a for *id*(a) and $f \circ g$ for *comp*(f,g). There are several notational conventions involving domains and codomains of morphisms. One is to write $f: a \to b$ for the pair of conditions *dom*(f) = a and *cod*(f) = b. The collection of all arrows $f: a \to b$ is often called a *hom-set*, since this is a set of homomorphisms in many common categories, and written *Hom*(a, b). Hom-sets play a central role in category theory; in fact, a category is determined by its hom-sets. If a and b are objects of category C, then we may also write $C(a, b)$ for *Hom*(a, b). This is particularly useful in discussions referring to more than one category.

The remaining category axioms are that each identity is a unit with respect to composition, and composition is associative. More precisely,

$$f = id_b \circ f = f \circ id_a$$

for every $f: a \to b$, and

$$f \circ (g \circ h) = (f \circ g) \circ h.$$

for every $h: a \to b$, $g: b \to c$ and $f: c \to d$. It is easy to check that ordinary sets (with set-theoretic functions as arrows), groups (with group homomorphisms), modest sets (with recursive functions), and domains (with continuous functions) all form categories.

As presented, categories "almost" form an equational class. A category C is a two-sorted algebra, with four function symbols satisfying a set of equational axioms, *except* that *comp* is not required to be a total function. This exception notwithstanding, it is often useful to consider categories as two-sorted algebras, as we have presented them. (Our presentation of categories may be formalized using sketches, as elaborated in [8]. This definition is useful in developing *internal categories*, which are categories whose collections of objects and arrows are represented by objects of some external category.) An alternative and often useful presentation of categories *is* equational. Instead of using two sorts, one for objects and one for arrows we may represent a category using one sort for each hom-set. In this alternative, we need a family of constants id_a, one for each object, and a family of composition maps $comp_{a,b,c}$, one for each triple of objects. While the category axioms become equational, the alternative presentation does not suggest the definition of functor as clearly as the definition above. In addition, our definition provides better guidance when it comes to studying functors which respect additional structure, such as Cartesian closedness.

Two important notions in category theory are functors and natural transformations. If we view categories as algebras, then functors are ordinary algebraic homomorphisms. More precisely, a *functor* from C to D is a pair of maps $F = \langle F^{\mathrm{o}}, F^{\mathrm{a}} \rangle$ with

- $F^{\mathrm{o}}: C^{\mathrm{o}} \to D^{\mathrm{o}}$ mapping objects to objects;
- $F^{\mathrm{a}}: C^{\mathrm{a}} \to D^{\mathrm{a}}$ mapping arrows to arrows, such that
- if $f: a \to b$, then $F^{\mathrm{a}}(f): F^{\mathrm{o}}(a) \to F^{\mathrm{o}}(b)$,
- $F^{\mathrm{a}}(id_{\mathrm{a}}) = id_{F^{\mathrm{o}}(a)}$,
- $F^{\mathrm{a}}(f \circ g) = F^{\mathrm{a}}(f) \circ F^{\mathrm{a}}(g)$.

It is common to omit superscripts from the maps F^{o} and F^{a}. Following this convention, we may write $F(f): F(a) \to F(b)$, for example. Part of what makes category theory "work" is that categories and functors form a category: we have an identity functor for each category (mapping each object or arrow to itself), and the composition of two functors (defined by composing object and arrow maps separately) is a functor. It is easy to check that all of the conditions are satisfied.

A natural transformation is a "map" between functors with the same domain and range. Since a functor is not a collection, this idea takes a bit of explanation. A vague but helpful intuition is to think of a functor as providing a "view" of a category. To simplify the picture, let us assume that F and G are functors $F, G: C \to D$ which are both injective (one-to-one) on objects and arrows. We may think of drawing F by drawing its image (a subcategory of D), with each object and arrow labeled by its preimage in C. We may draw G similarly. This gives us two subcategories of D labeled in two different ways. A natural transformation from F to G is a "translation" from one picture to another, given by a family of arrows in D. Specifically, for each object a of C, a natural transformation selects an arrow from $F(a)$ to $G(a)$. The conditions of the definition below guarantee that this collection of arrows respect the structure of "F's picture of C".

Formally, a *natural transformation* $v: F \to G$ between functors $F, G: C \to D$ is a map from objects of C to arrows of D satisfying the following two conditions:

- for every $a \in C^{\circ}$, we have $v(a): F(a) \to G(a)$,
- for every $f: a \to b$ in C, the following diagram commutes:

$$\begin{array}{ccc} G(a) & \xrightarrow{G(f)} & G(b) \\ {\scriptstyle v(a)}\uparrow & & \uparrow{\scriptstyle v(b)} \\ F(a) & \xrightarrow[F(f)]{} & F(b) \end{array}$$

The commutative diagram "says" that v translates F's "picture" $F(f): F(a) \to F(b)$ of the arrow $f: a \to b$ in C to G's "picture" $G(f): G(a) \to G(b)$. Since functors F and G must preserve identities and composition, v may be regarded as a translation of any diagram (collection of objects and arrows) in the image of F. For any categories C and D, the collection of functors from C to D forms a category, with natural transformations as arrows.

2.7.2. Typed lambda calculus with pairing

In Sections 2.5 and 2.6, we used products to show that full recursive hierarchies and full continuous hierarchies are Henkin models. The reason was that we needed to show that the meaning functions associated with these models were recursive or continuous, and meaning is a function of several arguments when terms have several free variables. By adding Cartesian products to the language, we may simplify the meaning function by changing the syntax so that each term has only one free variable. The main idea is to write a term with free variables $x_1: \sigma_1, \ldots, x_k: \sigma_k$ as a term with single free variable $x: \sigma_1 \times \cdots \times \sigma_k$ representing the k-tuple of values of $x_1, \ldots, x_k$. To gloss over the special case of closed terms, we will add the "empty product," or one-element type $\mathbf{1}$. This allows us to write a closed term as a term with a single "trivial" free variable of type $\mathbf{1}$.

The typed lambda calculus $\lambda^{\mathbf{1}, \times, \to}$ with $\mathbf{1}$, product and function types is defined using types

$$\sigma ::= b \mid \mathbf{1} \mid \sigma_1 \times \sigma_2 \mid \sigma_1 \to \sigma_2$$

closed under $\times$. The terms are defined by taking $*$ as a constant symbol of type $\mathbf{1}$, and adding the following to the term formation rules of $\lambda^{\to}$.

$$(\times\,\text{Intro}) \qquad \frac{\Gamma \rhd M: \sigma,\ \Gamma \rhd N: \tau}{\Gamma \rhd \langle M, N \rangle : \sigma \times \tau},$$

$$(\times\,\text{Elim}) \qquad \frac{\Gamma \rhd M: \sigma \times \tau}{\Gamma \rhd \mathbf{Proj}_1^{\sigma,\tau}\, M: \sigma,\quad \Gamma \rhd \mathbf{Proj}_2^{\sigma,\tau}\, M: \tau}.$$

In addition to the axioms and inference rules of $\lambda^{\rightarrow}$, we have the axioms

$$\text{(one)} \qquad \Gamma \rhd x = *: \mathbf{1},$$

$$(\mathbf{Proj}_1) \qquad \Gamma \rhd \mathbf{Proj}_1^{\sigma,\tau}\langle M, N\rangle = M: \sigma,$$

$$(\mathbf{Proj}_2) \qquad \Gamma \rhd \mathbf{Proj}_2^{\sigma,\tau}\langle M, N\rangle = N: \tau,$$

$$(\mathbf{Pair}) \qquad \Gamma \rhd \langle \mathbf{Proj}_1^{\sigma,\tau} M,\ \mathbf{Proj}_2^{\sigma,\tau} M\rangle = M: \sigma \times \tau$$

for **1**, pairing and projection. It is common to drop typing superscripts from $\mathbf{Proj}_1$ and $\mathbf{Proj}_2$ when they are irrelevant or determined by context.

It is possible to show that $\lambda^{\mathbf{1},\times,\rightarrow}$ is conservative over $\lambda^{\rightarrow}$, by proof-theoretic analysis.

2.7.1. LEMMA. *Let $\mathscr{E}$ be a set of equations between $\lambda^{\rightarrow}$ terms, and E another such equation. If $\mathscr{E} \vdash E$ using the equational proof rules of $\lambda^{\mathbf{1},\times,\rightarrow}$, then we can prove E from $\mathscr{E}$ using only the rules of $\lambda^{\rightarrow}$.*

Since we do not have a completeness theorem for pure $\lambda^{\rightarrow}$ without extra rules for empty or nonempty types, it does not seem easy to prove this lemma usng Henkin models. However, there is a semantic proof using Kripke lambda models [94].

The remainder of this section will be concerned with reformulating $\lambda^{\mathbf{1},\times,\rightarrow}$ so that each term has exactly one free variable. To see the connection between arbitrary terms and terms with one free variable, let us write $x^{(i,k)}$ for the expression

$$x^{(i,k)} = \begin{cases} \mathbf{Proj}_1(\mathbf{Proj}_2^{(i-1)} x), & i<k, \\ \mathbf{Proj}_2^{(k-1)} x, & i=k \end{cases}$$

where $\mathbf{Proj}_2^{(j)}\ x$ is the result of applying j projection functions to x. Intuitively, if $x{:}(\ldots(\sigma_1 \times \sigma_2)\ldots) \times \sigma_k$, then $x^{(i,k)}$ refers to the ith component of the tuple x. Using this notation, we may see that any term $\{x_1{:}\sigma_1, \ldots, x_k{:}\sigma_k\} \rhd M: \tau$ is essentially equivalent to a term $x{:}(\sigma_1 \times \cdots \times \sigma_k) \rhd M': \tau$, where M' is obtained by replacing x_i with $x^{(i,k)}$. These two terms are semantically different in the framework of Henkin models, since one depends on the values of variables $x_1, \ldots, x_k$ and the other depends on x. However, there is a natural correspondence between the two meanings. If we consider the meaning of $\{x_1{:}\sigma_1, \ldots, x_k{:}\sigma_k\} \rhd M: \tau$ as a function of the tuple of values of $x_1, \ldots, x_k$, this is the same function as given by the meaning of $x{:}(\sigma_1 \times \cdots \times \sigma_k) \rhd M': \tau$, regarding x as a variable ranging over k-tuples. For example, $\{x{:}\ a,\ y{:}\ b\} \rhd (\lambda z{:}\ c.x){:}c \rightarrow a$ defines the same binary function from $a \times b$ to $c \rightarrow a$ as the term $u{:}\ (a \times b) \rhd \lambda z{:}\ c.(\mathbf{Proj}_1^{a,b} u){:}c \rightarrow a$. Since the semantic correspondence between these two terms is easily stated, we leave the precise statement as an exercise for the reader. One special case that deserves noticing is the treatment of a closed term. Since elements of any set A are in one-to-one correspondence with functions from a one-element set into A, we may identify the meaning of a closed term $\emptyset \rhd M: \sigma$ with the meaning of $x{:}\ \mathbf{1} \rhd M: \sigma$ with vacuous free variable of type **1**. Thus we may translate any term into an "essentially equivalent" term with exactly one free variable.

Although the resulting system may seem a bit awkward, the typing rules of $\lambda^{\mathbf{1},\times,\rightarrow}$ may be modified so that we never need to mention any term with more than one free variable. To accomplish this, we introduce a substitution typing rule

$$\text{(subst)} \qquad \frac{x{:}\,\sigma \rhd M{:}\,\tau, \;\; y{:}\,\tau \rhd N{:}\,\rho}{x{:}\,\sigma \rhd [M/y]N{:}\,\rho}.$$

A special case of this rule allows us to rename the free variable of any term. Specifically, if $y{:}\,\tau \rhd N{:}\,\rho$ is any typable term, we have typing axiom $x{:}\,\tau \rhd x{:}\,\tau$, and so by (subst) we may derive $x{:}\,\tau \rhd [x/y]N{:}\,\rho$. For this reason, the names of variables are no longer important.

The only rule that requires substantial revision is ($\to$Intro), since this rule changes the number of free variables of a term. If we have a term $x{:}\,(\sigma\times\tau) \rhd M{:}\,\rho$ in the modified presentation of $\lambda^{1,\times,\to}$, this may correspond to a term $\{y{:}\,\sigma, z{:}\,\tau\} \rhd [\langle y, z\rangle/x]M{:}\,\rho$ in the ordinary syntax, and so we may wish to lambda-abstract only over "part" of the free variable x, i.e., the component of the pair $x{:}\,(\sigma\times\tau)$ which we would ordinarily name by a separate free variable y. To account for this, we will use the slightly strange rule

$$(\to\text{Intro})_1 \quad \frac{x{:}(\sigma\times\tau) \rhd M{:}\rho}{y{:}\sigma \rhd \lambda z{:}\tau.[\langle y, z\rangle/x]N{:}\tau\to\rho}$$

where we allow $\mathbf{Proj}_i\langle y, z\rangle$ in M to be simplified to y or z. The net effect of this rule is to separate a single variable denoting a pair into a pair of variables. Another change that is possible, in the presence of (subst), is that we may drop (add hyp) in favor of a stronger form of constant axiom

$$(\text{cst})_1 \qquad x{:}\,\sigma \rhd c{:}\,\tau \quad (c \text{ a constant of type } \tau).$$

The reason we do not need (add hyp) is that, for any term with a free variable $x{:}\,\sigma$, we may replace x by a variable $y{:}\,\sigma\times\tau$ using substitution. More precisely, if we begin with a term $x{:}\,\sigma \rhd M{:}\,\tau$, then rule (add hyp) lets us write $\{x{:}\,\sigma, y{:}\,\rho\} \rhd M{:}\,\tau$, which corresponds to $z{:}(\sigma\times\rho) \rhd [\mathbf{Proj}_1 z/x]M{:}\,\tau$ if we use only one free variable. We may achieve the same result using the new rule (subst) to substitute $\mathbf{Proj}_1 z$ for x. Of course, this requires that every term have some "free variable" to substitute for, which is why we need the stronger typing axiom for constants.

In summary, the terms of $\lambda^{1,\times,\to}$ with at most one free variable may be defined using typing axiom (var) as before, strengthened constant axioms of the form $(\text{cst})_1$, typing rules ($\times$ Intro), ($\times$ Elim), ($\to$Elim) restricted to contexts with only one free variable, the substitution rule (subst), and the modified lambda abstraction rule $(\to\text{Intro})_1$.

2.7.2. LEMMA. *Let $x{:}\,\sigma \rhd M{:}\,\tau$ be any term of $\lambda^{1,\times,\to}$ with one free variable. There is a typing derivation for this term, using the modified typing rules, such that every term appearing in the typing derivation has precisely one free variable.*

2.7.3. Cartesian closure

There are several ways of defining Cartesian closed categories, differing primarily in the amount of category theory that they require. We will mention some technical aspects of the definition, and then discuss some of the intuition involved. A category theorist might simply say that a *Cartesian closed category* (ccc) is a category with specified terminal object, products and exponentials. The standard categorical concepts of terminal object, products and exponents are all defined using adjoint situations, as outlined in [73]. The word “specified” means that a ccc is a structure consisting of a category and a specific choice of terminal object, products and exponentials, much the way a partial order is a set together with a specific order relation. In particular, there may be more than one way to view a category as a ccc[3].

Although adjoints are one of the most useful concepts in category theory, the general definition of adjoint is too abstract to serve as a useful introduction to ccc’s. A more accessible definition is to say that a ccc is a category with a specified object **1** and binary maps $\times$ and $\rightarrow$ on objects such that, for all objects $a, b,$ and c,

- $Hom(a, \mathbf{1})$ has only one element,
- $Hom(a, b) \times Hom(a, c) \cong Hom(a, b \times c)$,
- $Hom(a \times b, c) \cong Hom(a, b \rightarrow c)$

where the first “$\times$” in the second line is the ordinary Cartesian product of sets. A technical point is that the isomorphisms involving products and function spaces must be natural (in the sense of natural transformation) in all of the objects at hand. The naturality conditions are discussed precisely in [73] and in Section 2.7.4.

The intuitive idea behind ccc’s is that we have a one-element type and, given any two types, we also have their Cartesian product and the collection of functions from one to the other. Each of these properties is stated axiomatically, by referring to collections of arrows. Since objects are not really sets, we cannot say “**1** has only one element” directly. However, the condition that $Hom(a, \mathbf{1})$ has only one element is equivalent in many categories. For example, in the category of sets, if there is only one map from B to A, for all B, then A must have only one element. In a similar way, the axiom that $Hom(a, b \times c)$ is isomorphic to the set of pairs of arrows $Hom(a, b) \times Hom(a, c)$ says that $b \times c$ is the collection of pairs of elements from b and c. The final axiom, $Hom(a \times b, c) \cong Hom(a, b \rightarrow c)$, says that the object $b \rightarrow c$ is essentially the collection of all arrows from b to c. A special case of the axiom which illustrates this point is $Hom(\mathbf{1} \times b, c) \cong Hom(\mathbf{1}, b \rightarrow c)$. Since **1** is intended to be a one-element collection, it should not be surprising that we may prove $b \cong \mathbf{1} \times b$ from the ccc axioms. Hence $Hom(\mathbf{1} \times b, c)$ is isomorphic to $Hom(b, c)$. From this, the reader may see that the hom-set $Hom(b, c)$ is isomorphic to $Hom(\mathbf{1}, b \rightarrow c)$. Since any set A is in one-to-one correspondence with the collection of functions from a one-element set into A, the isomorphism $Hom(b, c) \cong Hom(\mathbf{1}, b \rightarrow c)$ is a way of saying that the object $b \rightarrow c$ is a representation of $Hom(b, c)$ inside the category.

The isomorphisms giving us products and functions spaces must be natural in a, b and c. This is important since the natural transformations involved are needed to give meaning to typed lambda terms. Naturality for the terminal object **1** is largely degenerate, except that we must have a map $\mathcal{O}$ from objects to arrows such that $\mathcal{O}^a$ is the unique element of $Hom(a, \mathbf{1})$. This is only slightly different from saying that for every a there exists a unique arrow $\mathcal{O}^a$, but the distinction is important in certain contexts.

[3]However, the situation is not as arbitrary as the comparison with partial orders may suggest. The choice of terminal object, products and exponentials are all determined up to isomorphism, as pointed out in any book on category theory.

For products, we must have a map sending any triple of objects a, b and c to a "pairing function"

$$\langle\cdot,\cdot\rangle^{a,b,c}: Hom(a,b) \times Hom(a,c) \to Hom(a, b \times c),$$

and a corresponding inverse function from $Hom(a, b \times c)$ to pairs of arrows. Similarly, a ccc must have a map sending every triple of objects a, b and c to a currying function

$$\mathbf{Curry}^{a,b,c}: Hom(a \times b, c) \to Hom(a, b \to c),$$

and a map giving an inverse for $\mathbf{Curry}^{a,b,c}$.

In summary, the requirement of a terminal object ($\mathbf{1}$), products and exponentials boils down to the following extra structure.

- an object $\mathbf{1}$ with arrow $\mathcal{O}^a: a \to 1$ for each object a,
- a binary object map $\times$, function $\langle\cdot,\cdot\rangle: Hom(a,b_1) \times Hom(a,b_2) \to Hom(a, b_1 \times b_2)$, and arrows $\mathbf{Proj}_1^{b_1,b_2}, \mathbf{Proj}_2^{b_1,b_2}$, with $\mathbf{Proj}_i^{b_1,b_2}: b_1 \times b_2 \to b_i$ for all a, b_1, b_2,
- a binary object map $\to$ with function $\mathbf{Curry}^{a,b,c}: Hom(a \times b, c) \to Hom(a, b \to c)$ and arrow $\mathbf{App}^{a,b}: (a \to b) \times a \to b$ for all objects a, b, c

These maps and arrows must satisfy the following conditions, for all appropriate f, h and k (for simplicity, most superscripts are omitted):

- $\mathcal{O}^a$ is the only arrow $a \to 1$,
- $\mathbf{Proj}_i \circ \langle f_1, f_2 \rangle = f_i$ and $\langle \mathbf{Proj}_1 \circ f, \mathbf{Proj}_2 \circ f \rangle = f$,
- $\mathbf{App} \circ \langle (\mathbf{Curry}\ h) \circ \mathbf{Proj}_1, \mathbf{Proj}_2 \rangle = h$ and $\mathbf{Curry}(\mathbf{App} \circ \langle k \circ \mathbf{Proj}_1, \mathbf{Proj}_2 \rangle) = k$.

If we view categories as two-sorted algebras, then the definition of ccc may be written usng equational axioms over the expanded signature with function symbols $\mathcal{O}$, $\times$, $\langle\cdot,\cdot\rangle$, and so on.

It is relatively easy to give each term of $\lambda^{1,\times,\to}$ with one free variable a meaning in any Cartesian closed category. To begin with, we must choose an object for each base type of the signature. Using $\mathbf{1}$ and object maps $\times$ and $\to$, this allows us to interpret every type expression as some object of the category. We may interpret a term $x: \sigma \rhd M: \tau$ as an arrow from the object named by σ to the object named by τ. Intuitively, we interpret a lambda term as the "function" from the value of its free variable to the corresponding value of the term. To simplify the discussion, we will not distinguish between a type expression σ and the object it names.

The easiest way to assign meaning to a term is to use induction on a typing derivation in which every term has exactly one free variable. A term given by axiom (var) denotes an identity morphism, and a term $x: \sigma \rhd *: \mathbf{1}$ given by an instance of the typing axiom for $*$ denotes the unique morphism $\mathcal{O}^\sigma$ from σ to $\mathbf{1}$. For any other constant of the signature, we must have a given map from $\mathbf{1}$ to the appropriate type. For ($\times$ Intro), we apply pairing to the pair of arrows, and for ($\times$ Elim) we compose with one of the projection functions $\mathbf{Proj}_i$. If $y: \sigma \rhd \lambda z: \tau.[\langle y, z \rangle / x]M: \tau \to \rho$ follows from $x: (\sigma \times \tau) \rhd M: \rho$ by rule ($\to$Intro), then the meaning of the lambda term is obtained by applying **Curry** to the meaning of the term with free variable $x: (\sigma \times \tau)$. For ($\to$Elim), we use pairing (on arrows) and composition with an application map **App.** Further discussion may be found in [68, 121, 97]. The typing rule (subst) simply corresponds to composition: if $x: \sigma \rhd M: \tau$ denotes the arrow $f: \sigma \to \tau$ and $y: \tau \rhd N: \rho$ denotes the arrow $g: \tau \to \rho$, then the meaning of $x: \sigma \rhd [M/y]N: \rho$ is the arrow $g \circ f$.

If we say a ccc *satisfies* an equation whenever both terms denote the same arrow, then we have the following soundness theorem for Cartesian closed categories.

2.7.3. THEOREM. *If $\mathscr{E} \vdash \Gamma \rhd M : \sigma$, then every ccc satisfying $\mathscr{E}$ also satisfies $\Gamma \rhd M : \sigma$.*

2.7.4. Semialgebraic formulation of lambda calculus

In this section, we will consider the pedagogic exercise of translating the syntax and proof rules of $\lambda^{1,\times,\to}$ into a form of algebra with partial functions, along the lines discussed in [3] for example. The sentences we use to write typing rules and equational proof rules will be implicational formulas with antecedents of a particularly simple nature. The primary purpose of this exercise is to give a lambda calculus motivation for the definition of Cartesian closed category, and to make the connection between lambda terms and ccc's as straightforward as possible. The basic plan seems implicit in discussions of ccc's as "an alternate formulation of lambda calculus" in [67, 68] (see also [65]), and a similar presentation of syntax is used in [18] for a different purpose. While the properties of algebras with partial functions merit investigation, space limitations prohibit further discussion. However, it is worth saying that an equation $M = N$ will mean that both expressions are defined and have equal value.

Let us begin by discussing the algebraic formulation of some typed language in which each term has exactly one free variable. (This applies to the alternate formulation of $\lambda^{1,\times,\to}$ developed in Section 2.7.2). Since types and terms are distinct syntactic classes, our algebraic "metalanguage" will have two sorts, *typ* and *trm*. We describe the typing of each term by two maps, $tp: trm \to typ$ which tells us the type of each term, and $fv: trm \to typ$, which gives the type of the free variable. In other words, our "intended model" is the two-sorted algebra with sort *type* containing all type expressions of some language (such as $\lambda^{1,\times,\to}$), and sort *trm* containing all well-typed terms $x: \sigma \rhd M : \tau$ with one free variable. In this algebra, *tp* is the function mapping $(x: \sigma \rhd M : \tau) \mapsto \tau$ and *fv* is the function mapping $(x: \sigma \rhd M : \tau) \mapsto \sigma$. An alternative to calling *fv* the "free-variable" function might be to call this the "context" function, since the typing context for term $x: \sigma \rhd\ M : \tau$ is the variable of type σ. Since each term has precisely one free variable, we need not distinguish variables by name.

Since we assume that variables are terms, we have a term $x: \sigma \rhd x : \sigma$ for each type σ. The collection of all such terms may be represented by a function $var: typ \to trm$, with $tp(var(\sigma)) = fv(var(\sigma)) = \sigma$. It is also useful to have substitution as a basic operation on terms. Substitution is a partial map $subst: trm \times trm \to trm$ which is defined whenever the free variable of the first argument agrees with the type of the second. In this case, substitution $subst(M, N)$ produces a term whose free variable has type $fv(N)$ and type is $typ(M)$. Since variables are nameless, substitution involving a variable (either substituting a variable into a term or substituting a term into a variable) has no effect. In addition, substitution is associative. These properties are easily axiomatized by implicational formulas with typing constraints in the antecedents.

Before considering the type constructors and term-building operations specific to $\lambda^{1,\times,\to}$, let us summarize the framework so far. We have developed a two-sorted signature with four maps

$$tp: trm \to typ, \qquad fv: trm \to typ,$$
$$var: typ \to trm, \qquad subst: trm \times trm \rightharpoonup trm$$

where "$\rightharpoonup$" indicates that *subst* is a partial function. The reader may find it amusing to look back at the definition of category. It should be clear that the obvious axioms for these four maps correspond exactly to the conditions imposed on *dom*, *cod*, *id* and *comp* in the definition of category. This illustrates part of the connection between categories and logical theories, explained in more detail in [65], for example.

In the rest of this section, we will focus on $\lambda^{\mathbf{1},\times,\to}$. To incorporate **1**, products and function spaces, we extend our signature with constant $\mathbf{1}: typ$ and binary functions

$$cross: typ \times typ \to typ, \qquad arrow: typ \times typ \to typ.$$

These allow us to write an algebraic expression of sort *typ* for each type expression of $\lambda^{\mathbf{1},\times,\to}$. We will write a term of $\lambda^{\mathbf{1},\times,\to}$ by writing an algebraic expression for its typing derivation. In computer science jargon, our algebraic formulation of syntax may be viewed as a king of "abstract syntax": we describe terms by giving a set of "abstract" operations are based on typing considerations.

To write typing derivations as algebraic expressions, we need functions for each typing axiom and term-building operation. The functions *var* and *subst* give us typing axiom (var) and rule (subst), so it remains to add functions for the constant $*$ and the "introduction" and "elimination" rules. Since ($\times$ E) has two conclusions, we will have two corresponding function symbols.

$$*: typ \to trm,$$
$$arrowI: trm \to trm, \qquad arrowE: trm \times trm \to trm,$$
$$crossI: trm \times trm \to trm \qquad crossE_i: trm \to trm \quad (i = 1, 2).$$

These functions allow us to write an algebraic expression corresponding to the typing derivation of any lambda term, provided we use the modified typing rules discussed at the end of the last section. For example, we write $*(\sigma)$ for the lambda term $x: \sigma \rhd\ *: \mathbf{1}$ given by the axiom scheme $(\text{cst})_1$. In $\lambda^{\mathbf{1},\times,\to}$, a closed identity function may be written $y: \mathbf{1} \rhd\ (\lambda Z: \sigma . z): \sigma \to \sigma$, with, one "vacuous" free variable. We may translate this into an algebraic expression by considering the obvious typing derivation. The derivation for this term begins with axiom $x: (\mathbf{1} \times \sigma) \rhd x: (\mathbf{1} \times \sigma)$, followed by rule ($\times$ Elim) to obtain $x: (\mathbf{1} \times \sigma) \rhd \mathbf{Proj}_2 x: \sigma$. Lambda-abstracting the second component of $x: (\mathbf{1} \times \sigma)$ according to the revised version of ($\to$Intro) gives us $y: \mathbf{1} \rhd\ (\lambda z: \sigma . z): \sigma \to \sigma$. Thus we translate the lambda term to the algebraic expression $arrowI(crossE_2(var(\mathbf{1} \times \sigma)))$.

To give algebraic expressions the appropriate typing properties, we use the following axioms. To make the axioms easier to read, let us write $\sigma, \tau, \ldots$ for variables of sort *typ* and $M, N, \ldots$ for variables of sort *trm*. It is also helpful to write $\times$ and $\to$ for *cross* and *arrow*, and to use $M: \sigma \Rightarrow \tau$ for the conjunction $fv(M) = \sigma \wedge tp(M) = \tau$. The axioms below are chosen so that if $x: \sigma \rhd\ M: \tau$ is a typed lambda term, and N is the algebraic expression obtained from a typing derivation for this term, then we may prove $N: \sigma \Rightarrow \tau$.

The typing axiom for $*$ and the term-formation rules may be written as the following implications:

$$
\begin{array}{lll}
 & & *(\sigma):\sigma\Rightarrow 1, \\
M:\sigma\Rightarrow\tau \wedge N:\sigma\Rightarrow\rho & \supset & crossI(M,N):\sigma\Rightarrow(\tau\times\rho), \\
M:\sigma\Rightarrow(\tau_1\times\tau_2) & \supset & crossE_i(M):\sigma\Rightarrow\tau_i \quad (i=1,2), \\
M:(\sigma\times\tau)\Rightarrow\rho & \supset & arrowI(M):\sigma\Rightarrow(\tau\rightarrow\rho), \\
M:\sigma\Rightarrow(\tau\rightarrow\rho)\wedge N:\sigma\Rightarrow\tau & \supset & arrowE(M,N):\sigma\Rightarrow\rho.
\end{array}
$$

The following lemma shows that we now have an adequate presentation of the syntax of $\lambda^{1,\times,\rightarrow}$.

2.7.4. LEMMA. *Let $x:\sigma \rhd M:\tau$ be a typed term of $\lambda^{1,\times,\rightarrow}$ and let N be the algebraic expression constructed according to a typing derivation for $x:\sigma \rhd M:\tau$, using the modified typing rules that only allow a single free variable. Then we may prove $N:\sigma\Rightarrow\tau$ from the axioms given in this section, using propositional reasoning and ordinary quantifier-free rules of algebra.*

The inference system for equations between typed lambda terms may be adapted to this setting relatively easily. To begin with, ordinary algebraic reasoning guarantees that provable equality is a congruence relation on terms, so axiom (ref) and proof rules (sym), (trans) (ξ) and (μ) are all subsumed by the logical framework. In addition, we do not need (add hyp) for equations (by the reasoning given for the (add hyp) typing rule), and axiom (α) is not necessary since we no longer have named variables. However, our abstract treatment of substitution means that we must axiomatize the essential properties of the substitution operator on terms. Up until this point, we have only specified that substitution is an associative operation, and that substitution involving a variable has no effect. We must also specify that if we substitute into a pair, the result is also a pair, and similarly for the other term-building operations. These axioms are relatively straightforward, except in the case of lambda abstraction (→Intro).

Intuitively, the substitution axiom for pairing

$$
\begin{array}{l}
M:\sigma\Rightarrow\tau\wedge N:\sigma\Rightarrow\rho\wedge P:\nu\Rightarrow\sigma \\
\quad \supset subst(crossI(M,N),P) = crossI(subst(M,P)subst(N,P))
\end{array}
$$

says that, subject to the relevant typing constraints, substituting into a pair is equivalent to forming a pair from the results of substitution. The substitution axioms for projection and application are similar.

$$
\begin{array}{l}
M:\sigma\Rightarrow\tau_1\times\tau_2\wedge P:\nu\Rightarrow\sigma \\
\quad \supset subst(crossE_i(M),P) = crossE_i(subst(M,P)), \\
M:\sigma\Rightarrow(\tau\rightarrow\rho)\wedge N:\sigma\Rightarrow\tau\wedge P:\nu\Rightarrow\sigma \\
\quad \supset \quad subst(crossE(M,N),P) \quad = \quad arrowE(subst(M,P)subst(N,P)).
\end{array}
$$

The substitution axiom for lambda abstraction is somewhat more complicated, since we must take variable-binding into account. The rule may be expressed more simply by adopting the abbreviation

$$M \times N ::= crossI((\text{subst}(M, crossE_1(var(\sigma \times \tau)))), (subst(N, crossE_2(var(\sigma \times \tau))))),$$

where $M : \sigma \Rightarrow \mu$ and $N : \tau \Rightarrow \nu$. Essentially, the "function" $\mathrm{M} \times \mathrm{N} : \sigma \times \tau \Rightarrow \mu \times \nu$ produces a pair of type $\mu \times \nu$ from a pair of type $\sigma \times \tau$ by applying M to the first component and N to the second. Using this notation, the substitution axiom for *arrowI* is written as follows.

$$\begin{aligned} &M : (\sigma \times \tau) \Rightarrow \rho \wedge P : \nu \Rightarrow \sigma \\ &\quad \supset \quad subst(arrowI(M), P) = arrowI(subst(M, (P \times var(\tau)))). \end{aligned}$$

Intuitively, if we think of M as having free variables $x : \sigma$ and $y : \tau$, then $subst(M, (P \times var(\tau)))$ is the result of substituting P for x and leaving y free. Unfortunately, this is a bit cumbersome to express in the present formalism.

The axioms for equations involving $*$ and Cartesian products are straightforward translations of the equational axioms (one), (**Pair**) and (**Proj**$_i$) from Section 2.7.2.

$$\begin{aligned} &M : \sigma \Rightarrow 1 && \supset \quad M = *(\sigma), \\ &M : \sigma \Rightarrow (\tau \times \rho) && \supset \quad M = crossI(crossE_1(M), crossE_2(M)), \\ &N_1 : \sigma \Rightarrow \tau_1 \wedge N_2 : \sigma \Rightarrow \tau_2 && \supset \quad N_i = crossE_i(crossI(N_1, N_2)) \quad (i = 1, 2). \end{aligned}$$

However, the axioms for lambda abstraction (arrowI) and application (arrowE) require a bit of explanation. Intuitively, we would expect one axiom for (β) and another for (η). Let us consider β-equivalence first. Suppose we apply the lambda term $x : \sigma \rhd \lambda y : \tau . M : \tau \to \rho$ to $x : \sigma \rhd N : \tau$. The result, by (β), is $x : \sigma \rhd [N/y]M : \rho$. Let us abuse notation and write M and N for the algebraic expressions corresponding to lambda terms M and N. In the algebraic notation, the body of the lambda abstraction is $M : \sigma \times \tau \Rightarrow \rho$, and the lambda abstraction itself is $arrowI(M) : \sigma \Rightarrow (\tau \to \rho)$. We apply this lambda abstraction to $N : \sigma \Rightarrow \tau$. The application is formed using (→Elim), so the algebraic expression for the whole term is *arrowE(arrowI(M), N)*. It remains to write the substitution $x : \sigma \rhd [N/y]M : \rho$ as an algebraic expression. The complication here is that the function $subst : trm \times trm \to trm$ only allows us to substitute a term for the unique free variable of another term. However, our intent is to substitute N for "y" in M, leaving "x" still free. The way to accomplish this is to substitute the pair $crossI(var(\sigma), N)$ into M, effectively substituting a variable for x and N for y. This gives us the following statement of (β):

$$\begin{aligned} M : \sigma \times \tau \Rightarrow \rho \wedge N : \sigma \Rightarrow \tau \quad \supset \quad & arrowE(arrowI(M), N) \\ & = subst(M, crossI(var(\sigma), N)) \end{aligned}$$

The translation of (η) requires us to "simulate" rule (add hyp) using substitution and pairing, as discussed in Section 2.7.2. We begin with an algebraic expression $M : \sigma \Rightarrow (\tau \to \rho)$ for a lambda expression of functional type, with free variable of type σ. We want to apply this function to a "fresh" variable not appearing in the lambda term represented by M. A straightforward approach that does not involve looking at the structure of the lambda term is to use (add hyp) to add another variable to the context.

As described in Section 2.7.2, this may be accomplished by substituting a "part" of a variable into the term. Specifically, $subst(M, crossE_1(var(\sigma \times \tau)))$ is identical to M, except that it depends vacuously on the second component of the new free variable of type $\sigma \times \tau$. We may apply this to the second component of the free variable using an expression of the form $arrowE(\ldots, crossE_2(var(\sigma \times \tau)))$. This leads us to the following statement of (η).

$$\begin{aligned} &M: \sigma \Rightarrow (\tau \rightarrow \rho) \\ &\quad \supset \quad M = arrowI(arrowE(subst(M, crossE_1(var(\sigma \times \tau))), \\ &\qquad\qquad\qquad\qquad crossE_2(var(\sigma \times \tau)))). \end{aligned}$$

The equations between lambda terms may be summarized using three statements about collections of terms. To make the connection with Cartesian closed categories, let us write $Hom(\sigma, \tau)$ for the collection of all $M: \sigma \Rightarrow \tau$. Then the axioms we have just given are sufficient to prove that

- $Hom(\sigma, \mathbf{1})$ has only one element,
- $Hom(\sigma, \tau) \times Hom(\sigma, \rho) \cong Hom(\sigma, \tau \times \rho)$, and
- $Hom(\sigma \times \tau, \rho) \cong Hom(\sigma, \tau \rightarrow \rho)$.

In other words, we may prove $x: \sigma \Rightarrow \mathbf{1}$ equal to $\mathcal{O}^\sigma$, and for each isomorphism we may give a pair of expressions with one free variable each and prove that this pair defines an isomorphism. In general (see [73, p. 79]), there are two naturality conditions associated with each adjoint situation. For **1**, these are trivial. For products, one of the conditions is precisely the substitution condition for *crossI*, which is equivalent to the substitution condition for *crossE*. The second naturality condition for products turns out to be redundant. The situation is similar for function spaces: the only nontrivial naturality condition is the substitution condition for *arrowI*, which is equivalent to the substitution condition for *arrowE*. Thus all aspects of the definition of Cartesian closed category are covered by the axioms we have formulated in this section. In fact, except for the difference between the statement of β-equivalence and one half of the isomorphism $Hom(\sigma \times \tau, \rho) \cong Hom(\sigma, \tau \rightarrow \rho)$ all of the axioms are identical. It is worth mentioning that projection arrows may be defined by

$$\mathbf{Proj}_i^{\sigma, \tau} ::= crossE_i(var(\sigma \times \tau))$$

and application by

$$\mathbf{App}^{\sigma, \tau} ::= arrowE(crossE_1(var((\sigma \rightarrow \tau) \times \sigma)), crossE_2(var((\sigma \rightarrow \tau) \times \sigma))).$$

The reader who tries to verify any of the identities presented in this section will likely discover that the standard category-theoretic notation is far more convenient than the algebraic notation used here. However, it is often useful to formulate category-theoretic concepts in lambda calculus syntax. The two formalisms are sufficiently different that certain properties that are obvious in one setting may not be readily apparent in the other.

2.7.5. Connections between ccc's and Henkin models

Any Henkin model $\mathcal{A}$ determines a Cartesian closed category $\mathcal{C}_{\mathcal{A}}$ whose objects are the type expressions of the signature and whose arrows from σ to τ are the elements of

$A^{\sigma\to\tau}$. We may form an applicative structure $\mathscr{A}_C$ from any ccc C by taking the objects as types and letting $A^\sigma = C(\mathbf{1}, \sigma)$. However, this does not always produce a Henkin model. The reason is that extensionality may fail for the applicative structure $\mathscr{A}_C$. It may seem reasonable to conclude that Cartesian closed categories are "less extensional" than Henkin models, and in some ways this is correct. However, there are several subtle points in this conclusion. To begin with, it is important to emphasize that by the soundness theorem for ccc's, we know that ccc's support all of the axioms and inference rules of typed lambda calculus, including extensionality (axiom (η) and rule (ξ)). Moreover, any conclusion about the extensionality of C that is based on the interpretation of terms in $\mathscr{A}_C$ necessarily assumes some relationship between the interpretation of terms in C and their interpretation in $\mathscr{A}_C$. It is argued in [97] that extensionality of ccc's holds in a very general sense, and that the point of view implicit in [5, Sections 5.5, 5.6] deserves further scrutiny.

We may distinguish ccc's that correspond to Henkin models from those that do not, using a concept that goes by many names. A relatively neutral term was suggested by Peter Freyd. We will say a category C is *well-pointed* if the following condition holds:

(well-pointed) if $f \circ h = g \circ h$ for all $h: \mathbf{1} \to \sigma$, then $f = g$.

The name comes from the fact that arrows $h: \mathbf{1} \to \sigma$ are often called the *points* of σ. An alternative word for point is *global element.* In the literature, several synonyms for well-pointed are used, such as "having enough points" or "having **1** as a generator." A term that is used in some of the computer science literature is "concrete", which is problematic since concrete does not always mean well-pointed in the category-theoretic literature. The importance of the concept is illustrated by the following proposition.

2.7.5. PROPOSITION. *Every Henkin model for $\lambda^{1,\times,\to}$ determines a well-pointed Cartesian closed category, and conversely, every well-pointed ccc determines a Henkin model. The maps between Henkin models and well-pointed ccc's are inverses, and preserve the meanings of terms.*

2.7.6. Completeness for Cartesian closed categories

We have already seen that the axioms and inference rules of $\lambda^{1,\times,\to}$ are sound for Cartesian closed categories. This implies that the set of equations satisfied by any ccc is a typed lambda theory. We also have the following converse.

2.7.6. THEOREM. *Let $\mathscr{E}$ be any lambda theory, not necessarily closed under* (nonempty) *or* (empty I) *and* (empty E). *There is a ccc satisfying precisely the equations belonging to $\mathscr{E}$.*

By Lemma 2.71, completeness for $\lambda^\to$ follows.

This theorem clearly illustrates the difference between ccc's and Henkin models. We do not have deductive completeness for Henkin models unless we add additional axioms or rules of inference. However, we have a straightforward correspondence between lambda theories and Cartesian closed categories.

2.8. *Kripke lambda models*

2.8.1. *Motivation*

Kripke lambda models provide a general and relatively natural class of Cartesian closed categories that are not well-pointed. For simplicity, we will focus on Kripke models for $\lambda^{\rightarrow}$ without products or **1**. However, the framework is easily extended to $\lambda^{1,\times,\rightarrow}$. The main results in this section are soundness and completeness theorems for Kripke lambda models and a brief discussion of connections with Cartesian closed categories.

2.8.2. *Possible worlds*

As with other Kripke-style semantics, a Kripke lambda model will include a partially ordered set $\mathscr{W}$ of "possible worlds". Instead of having a set of elements of each type, a Kripke lambda model will have a set of elements of each type at each possible world $w \in \mathscr{W}$. The relationship between elements of type σ at worlds $w \leqslant w'$ is that every $a{:}\,\sigma$ *at* w is associated with some unique $a'{:}\,\sigma$ *at* w'. Informally, using the common metaphor of $\leqslant$ as relation in time, this means that every element of σ at w will continue to be an element of σ in every possible $w' \geqslant w$. As we move from w to a possible future world w', two things might happen: we may acquire more elements, and distinct elements may become identified. These changes may be explained by saying that as time progresses, we may become aware of (or construct) more elements of our universe, and we may come to know more "properties" of elements. In the case of lambda calculus, the properties of interest are equations, and so we may have more equations in future worlds. Since a type σ may be empty at some world w and then become nonempty at $w' \geqslant w$, some types may be neither "globally" empty nor nonempty.

2.8.3. *Applicative structures*

Kripke lambda models are defined using a Kripke form of applicative structure. A *Kripke applicative structure*

$$\mathscr{A} = \langle \mathscr{W}, \leqslant, \{A_w^\sigma\}, \{\mathbf{App}_w^{\sigma,\tau}\}, \{i_{w,w'}^\sigma\}, \mathit{Const}\rangle$$

for $\lambda^{\rightarrow}$ signature Σ consists of

- a set $\mathscr{W}$ of "possible worlds" partially-ordered by $\leqslant$,
- a family $\{A_w^\sigma\}$ of sets indexed by type expressions σ over signature Σ and worlds $w \in \mathscr{W}$,
- a family $\{\mathbf{App}_w^{\sigma,\tau}\}$ of "application maps" $\mathbf{App}_w^{\sigma,\tau}: A_w^{\sigma\rightarrow\tau} \times A_w^\sigma \rightarrow A_w^\tau$ indexed by pairs of type expressions σ, τ and worlds $w \in \mathscr{W}$,
- a family $\{i_{w,w'}^\sigma\}$ of "transition functions" $i_{w,w'}^\sigma: A_w^\sigma \rightarrow A_{w'}^\sigma$ indexed by type expressions σ and pairs of worlds $w \leqslant w'$,
- a function *Const* from term constants of the signature Σ to *global elements* (defined below) of the appropriate type,

subject to the following conditions: We want the transition from A_w^σ to A_w^σ to be the identity:

(id) $\quad i_{w,w}^\sigma: A_w^\sigma \rightarrow A_w^\sigma$ is the identity, and other transition functions to compose:

(comp) $\quad i^{\sigma}_{w',w''} \circ i^{\sigma}_{w,w'} = i^{\sigma}_{w,w''}$ for all $w \leqslant w' \leqslant w''$

so that there is exactly one mapping of A^{σ}_{w} into $A^{\sigma}_{w'}$ given for $w \leqslant w'$. We also require that application and transition commute in a natural way

(nat) $\quad \forall f \in A^{\sigma\to\tau}_{w}.\forall a \in A^{\sigma}_{w}.$

$$i^{\tau}_{w,w'}(\mathbf{App}^{\sigma,\tau}_{w}(f,a)) = \mathbf{App}^{\sigma,\tau}_{w'}((i^{\sigma\to\tau}_{w,w'}f),(i^{\sigma}_{w,w'}a)),$$

which may be drawn

$$\begin{array}{ccc} A^{\sigma\to\tau}_{w'} \times A^{\sigma}_{w'} & \xrightarrow{\mathbf{App}^{\sigma,\tau}_{w'}} & A^{\tau}_{w'} \\ \uparrow{\scriptstyle \langle i^{\sigma\to\tau}, i^{\sigma}\rangle} & & \uparrow{\scriptstyle i^{\tau}} \\ A^{\sigma\to\tau}_{w} \times A^{\sigma}_{w} & \xrightarrow[\mathbf{App}^{\sigma,\tau}_{w}]{} & A^{\tau}_{w} \end{array}$$

and will be described informally below.

Intuitively, a global element $\boldsymbol{a}$ of type σ is a "consistent" family of elements $a_w \in A^{\sigma}_{w}$, one for each $w \in \mathscr{W}$. More precisely, a *global element of type* σ is a map $\boldsymbol{a}\colon \mathscr{W} \to \bigcup_{w\in\mathscr{W}} A^{\sigma}_{w}$ such that whenever $w \leqslant w'$, we have $\boldsymbol{a}(w') = i^{\sigma}_{w,w'}\boldsymbol{a}(w)$. It is conventional to write a_w for $\boldsymbol{a}(w)$. This completes the definition.

If $a \in A^{\sigma}_{w}$ and $w \leqslant w'$, then we can read $i^{\sigma}_{w,w'}a \in A^{\sigma}_{w'}$ as "a viewed at world w'". The purpose of the application map $\mathbf{App}^{\sigma,\tau}_{w}$ is to associate a function $\mathbf{App}^{\sigma,\tau}_{w}(f,\cdot)$ from A^{σ}_{w} to A^{τ}_{w} with each element $f \in A^{\sigma\to\tau}_{w}$. Since we can view $f \in A^{\sigma\to\tau}_{w}$ as an element at any future world $w' \geqslant w$, the application map at world w' also associates a function with $i^{\sigma\to\tau}_{w,w'}f$ at w'. The condition (nat) is intended to give a degree of coherence to the functions associated with different views of f. Basically, (nat) says that if we apply f to argument a at world w, and then view the result at a later world $w' \geqslant w$, we see the same value as when we view f and a as elements of world w', and apply f to a there.

Kripke applicative structures can also be defined using category-theoretic concepts. The usual definition of applicative structure may be understood in any Cartesian category, as follows: Recall that an ordinary applicative structure is a structure $\langle\{A^{\sigma}\}, \{\mathbf{App}^{\sigma,\tau}\}, \mathit{Const}\rangle$, where $\{A^{\sigma}\}$ is a family of sets indexed by type expressions over the signature, $\{\mathbf{App}^{\sigma,\tau}\}$ is a family of application functions $\mathbf{App}^{\sigma,\tau}\colon A^{\sigma\to\tau} \times A^{\sigma} \to A^{\tau}$ indexed by pairs of type expressions, and *Const* is a map from constants of the signature to elements of the model. We may interpret this definition "in" a category $\mathscr{C}$ with products and terminal object by regarding the word "set" as meaning "object from $\mathscr{C}$", the word "function" as meaning "morphism from $\mathscr{C}$" and substituting "global elements" for "elements" in the range of *Const*. (We need products to give $\mathbf{App}^{\sigma,\tau}$ the appropriate domain, and a terminal object to define global element.) Thus an *applicative structure in a category* $\mathscr{C}$ *with products and terminal object* is a collection of objects $\{C^{\sigma}\}$ indexed by type expressions of the signature, a collection of morphisms $\{\mathbf{App}^{\sigma,\tau}\}$ indexed by pairs of type expressions (such that $\mathbf{App}^{\sigma,\tau}$ has the domain and codomain given above) and a map *Const* from to global elements of the appropriate types. To derive our definition of Kripke applicative structure from this general idea, we regard a poset $\langle\mathscr{W}, \leqslant\rangle$ as a category in the usual way (see Section 2.8.7), and consider the category $Set^{<\mathscr{W},\leqslant>}$ of functors from $\langle\mathscr{W},\leqslant\rangle$ to sets. If we work out what *applicative structure* means in a category of the form $Set^{<\mathscr{W},\leqslant>}$, then "sets" will be functors, "functions" natural transformations and "elements" will determine global elements in the sense of Kripke models. In short, we end up with the definition of applicative structure spelled out explicitly above.

It is often convenient to omit the application map **App**, writing fx for $\mathbf{App}_w^{\sigma,\tau}(f,x)$ when this does not seem confusing.

2.8.4. Extensionality and combinators

A classical applicative structure may fail to be a model for two reasons, and these reasons apply to Kripke applicative structures as well. The first possibility is that we may not have enough elements. For example, $\sigma\to\sigma$ might be empty, making it impossible to give meaning to the identity function $\lambda x\!:\!\sigma.x$. The second problem is that application may not be extensional, i.e., we may have two distinct elements of functional type which have the same functional behavior. Consequently, the meaning of a lambda term $\lambda x\!:\!\sigma.M$ may not be determined uniquely.

The usual statement of extensionality is that $f=g$ whenever $fx=gx$ for all x of the appropriate type. In Kripke applicative structures, we are concerned not only with the behavior of elements $f,g\in A_w^{\sigma\to\tau}$ as functions from A_w^{σ} to A_w^{q}, but also as functions from $A_{w'}^{\sigma}$ to $A_{w'}^{\tau}$ for all $w'\geqslant w$. Therefore, we must specify that, for all $f,g\in A_w^{\sigma\to\tau}$,

$$f=g \quad \text{whenever} \quad \forall w'\geqslant w.\forall a\in A_{w'}^{\sigma}.(i_{w,w'}^{\sigma\to\tau}f)a=(i_{w,w'}^{\sigma\to\tau}g)a.$$

This can also be phrased using the usual interpretation of predicate logic in Kripke structures, as explained in [94].

There are two ways to identify the extensional applicative structures that have enough elements to be models. As with ordinary applicative structures, the environment and combinatory model definitions are equivalent. Since the combinatory condition is simpler, we will use that. For all types σ, τ and ρ, we need global elements $\boldsymbol{K}_{\sigma,\tau}$ of type $\sigma\to\tau\to\sigma$ and $\boldsymbol{S}_{\sigma,\tau,\rho}$ of type $(\rho\to\sigma\to\tau)\to(\rho\to\sigma)\to\rho\to\tau$ (cf Section 2.4.8). The equational condition on $\boldsymbol{K}_{\sigma,\tau}$ is that for $a\in A_w^{\sigma}$, $b\in A_w^{\tau}$, we must have $(\boldsymbol{K}_{\sigma,\tau})_w ab=a$. The condition for S is derived from the equational axiom in Section 2.4.8 similarly.

We define a *Kripke lambda model* to be a Kripke applicative structure $\mathscr{A}$ which is extensional and has combinators.

2.8.5. Environments and meanings of terms

An *environment* η for a Kripke applicative structure $\mathscr{A}$ is a partial mapping from variables and worlds to elements of $\mathscr{A}$ such that

(env) if $\eta xw\in A_w^{\sigma}$ and $w'\geqslant w$, then $\eta xw'=i_{w,w'}^{\sigma}(\eta xw)$.

Intuitively, an environment η maps a variable x to a "partial element" ηx which may exist (or be defined) at some worlds, but not necessarily all worlds. Since a type may be empty at one world and then nonempty later, we need to have environments such that ηxw is undefined at some w, and then "becomes" defined at a later $w'\geqslant w$. We will return to this point after defining the meanings of terms.

If η is an environment and $a \in A^\sigma_w$, we write $\eta[a/x]$ for the environment identical to η on variables other than x, and with

$$(\eta[a/x])xw' = i^{p}_{w,w'}a$$

for all $w' \geqslant w$. We take $(\eta[a/x])xw'$ to be undefined form w' not $\geqslant w$.

If η is an environment for applicative structure $\mathscr{A}$, and Γ is a type assignment, we say w *satisfies* Γ *at* η, written $w \Vdash \Gamma[\eta]$ if

$$\eta xw \in A^\sigma_w \quad \text{for all } x{:}\sigma \in \Gamma.$$

Note that if $\Vdash \Gamma[\eta]$ and $w' \geqslant w$, then $w' \Vdash \Gamma[\eta]$.

For any Kriple model $\mathscr{A}$ and environment $w \Vdash \Gamma[\eta]$, we define the meaning $[\![\Gamma \rhd M{:}\sigma]\!]\eta w$ of term $\Gamma \rhd M{:}\sigma$ in environment η at world w by induction on the structure of terms:

$$[\![\Gamma \rhd x{:}\sigma]\!]\eta w = \eta xw,$$
$$[\![\Gamma \rhd c{:}\sigma]\!]\eta w = \mathit{Const}(c)_w,$$
$$[\![\Gamma \rhd MN{:}\tau]\!]\eta w = \mathbf{App}^{\sigma,\tau}_w([\![\Gamma \rhd M{:}\sigma\to\tau]\!]\eta w)([\![\Gamma \rhd N{:}\sigma]\!]\eta w),$$
$$[\![\Gamma \rhd \lambda x{:}\sigma.M{:}\sigma\to\tau]\!]\eta w =$$
the unique $d \in A^{\sigma\to\tau}_w$ such that, for all $w' \geqslant w$ and $a \in A^\sigma_{w'}$,
$\mathbf{App}^{\sigma,\tau}_{w'}(i^{\sigma\to\tau}_{w,w'}d)a = [\![\Gamma, x{:}\sigma \rhd M{:}\tau]\!]\eta[a/x]w'$.

Combinators and extensionality guarantee that in the $\Gamma \rhd \lambda x{:}\sigma.M{:}\sigma\to\tau$ case, d exists and is unique. This is proved as in the standard set-theoretic setting, using translation into combinators [5, 54, 81] for existence, and extensionality for uniqueness.

We say an equation $\Gamma \rhd M = N{:}\sigma$ *holds at* w *and* η, written

$$w \Vdash (\Gamma \rhd M = N{:}\sigma)[\eta]$$

if, whenever $w \Vdash \Gamma[\eta]$, we have

$$[\![\Gamma \rhd M{:}\sigma]\!]\eta w = [\![\Gamma \rhd N{:}\sigma]\!]\eta w.$$

A model $\mathscr{A}$ *satisfies* $\Gamma \rhd M = N{:}\sigma$, written $\mathscr{A} \Vdash \Gamma \rhd M = N{:}\sigma$, if every w and η for $\mathscr{A}$ satisfy the equation.

2.8.6. Soundness and completeness

2.8.1. LEMMA (soundness). *Let $\mathscr{E}$ be a set of well-typed equations. If $\mathscr{E} \vdash \Gamma \rhd M = N{:}\sigma$, then every model satisfying $\mathscr{E}$ also satisfies $\Gamma \rhd \mathrm{M} = \mathrm{N}{:}\sigma$.*

2.8.2. THEOREM (completeness). *For every lambda theory $\mathscr{E}$, there is a Kripke lambda model satisfying precisely the equations belonging to $\mathscr{E}$.*

PROOF (sketch). The completeness theorem is proved by constructing a term model $\mathscr{A} = \langle \mathscr{W}, \leqslant, \{A^\sigma_w\}, \{\mathbf{App}^{\sigma,\tau}_w\}, \{i^\sigma_{w,w'}\}, \mathit{Const}\rangle$ of the following form:

- $\mathscr{W}$ is the poset of finite type assignments Γ ordered by inclusion; in what follows, we will write Γ for an arbitrary element of $\mathscr{W}$.

- A_Γ^σ is the set of all $[\Gamma \rhd M : \sigma]$, where $\Gamma \rhd M : \sigma$ is well-typed, and

$$[\Gamma \rhd M : \sigma] = \{\Gamma \rhd N : \sigma \mid \mathscr{E} \vdash \Gamma \rhd \ M = N : \sigma\}$$

is the equivalence class of $\Gamma \rhd M : \sigma$ with respect to $\mathscr{E}$.
- $\mathbf{App}_\Gamma^{\sigma,\tau}([\Gamma \rhd M : \sigma \to \tau], [\Gamma \rhd N : \sigma]) = [\Gamma \rhd \ MN : \tau]$.
- $i_{\Gamma,\Gamma'}^\sigma([\Gamma \rhd M : \sigma]) = [\Gamma' \rhd M : \sigma]$ for $\Gamma \subseteq \Gamma'$.

It is easy to check that the definition makes sense, and that we have global elements $\boldsymbol{K}$ and $\boldsymbol{S}$ at all appropriate types. For example,

$$\boldsymbol{K}_\Gamma = [\lambda x : \sigma . \lambda y : \tau . x].$$

The proof of extensionality illustrates the difference between Kripke models and ordinary Henkin models. Suppose that $[\Gamma \rhd M : \sigma \to \tau]$ and $[\Gamma \rhd N : \sigma \to \tau]$ have the same functional behavior, i.e., for all $\Gamma' \geqslant \Gamma$ and $\Gamma' \rhd P : \sigma$, we have $[\Gamma' \rhd MP : \tau] = [\Gamma' \rhd NP : \tau]$. Then, in particular, for $\Gamma' \equiv \Gamma, x : \sigma$ with x not in Γ, we have

$$[\Gamma, x : \sigma \rhd Mx : \tau] = [\Gamma, x : \sigma \rhd Nx : \tau]$$

and so by rule (ξ) and axiom (η), we have $[\Gamma \rhd M : \sigma \to \tau] = [\Gamma \rhd N : \sigma \to \tau]$. Thus $\mathscr{A}$ is a Kripke lambda model. The remaining parts of the proof are omitted. □

2.8.7. Kripke lambda models as Cartesian closed categories

It is easy to extend the definitions of Kripke applicative structure and lambda model to include Cartesian product types $\sigma \times \tau$ and a terminal type 1 with one element at each world. In this section, we will see than any Kripke model $\mathscr{A}$ with products and a terminal type determines a Cartesian closed category $\mathscr{C}_{\mathscr{A}}$. As one would hope, the categorical interpretation of a term $\Gamma \rhd M : \sigma$ in $\mathscr{C}_{\mathscr{A}}$ coincides with the meaning of $\Gamma \rhd M : \sigma$ in $\mathscr{A}$ given above.

We regard a partially ordered set $\langle \mathscr{W}, \leqslant \rangle$ as a category in the usual way. Specifically, the objects of this category are the elements of $\mathscr{W}$ and there is a unique "less-than-or-equal-to" arrow $l_{w,w'}$ from w to w' iff $w \leqslant w'$. Since a category must have identities and be closed under composition, we let $l_{w,w}$ be the identity on w and define composition by $l_{w',w''} \circ l_{w,w'} = l_{w,w''}$.

Given a Kripke applicative structure $\mathscr{A}$, it is easy to see that each type σ determines a functor Φ_σ from $\langle \mathscr{W}, \leqslant \rangle$ to sets. Specifically, we take

$$\Phi_\sigma(w) = A_w^\sigma, \qquad \Phi_\sigma(l_{w,w'}) = i_{w,w'}^\sigma$$

and use conditions (id) and (comp) in the definition of Kripke applicative structure to show that this map is functorial. While it may seem simplest to use functors Φ_σ as objects of $\mathscr{C}_{\mathscr{A}}$, this may identify types in the case where $\sigma \neq \tau$ syntactically, but $A_w^\sigma = A_w^\tau$ happen to be the same set. Since we would not necessarily want to identify application functions on the two types, this could lead to unnecessary confusion. Therefore, we will use the type expressions as the objects of $\mathscr{C}_{\mathscr{A}}$.

With each type determining a functor, we will use natural transformations as the morphisms of $\mathscr{C}_{\mathscr{A}}$. For each pair of types σ and τ, condition (nat) in the definition of Kripke applicative structure says that the map $w \mapsto \mathbf{App}_w^{\sigma,\tau}$, which we shall write simply as $\mathbf{App}^{\sigma,\tau}$, is a natural transformation from $\Phi_{\sigma\to\tau} \times \Phi_\sigma$ to Φ_τ. Using $\mathbf{App}^{\sigma,\tau}$, we can see that every global element $\boldsymbol{a}$ of type $\sigma\rightarrow\tau$ induces a natural transformation ν from Φ_σ to Φ_τ, namely $\nu_w = \mathbf{App}_w^{\sigma,\tau}(a_w, \cdot)$. For extensional applicative structures (and hence models), it is easy to see that if two global elements a and b determine the same natural transformation, then $a_w = b_w$ at every world w. We let the morphisms from σ to τ in $\mathscr{C}_{\mathscr{A}}$ be all natural transformations $\nu: \Phi_\sigma \rightarrow \Phi_\tau$ induced by global elements of $\mathscr{A}$ of type $\sigma\rightarrow\tau$. Composition of morphisms is simply composition of natural transformations in $Set^{\langle \mathscr{W}, \leqslant \rangle}$.

A routine calculation shows that if $\mathscr{A}$ is a Kripke lambda model, then $\mathscr{C}_{\mathscr{A}}$ is a category with an object for each type, and there is a one-one correspondence between global elements of type $\sigma\rightarrow\tau$ in $\mathscr{A}$ and morphisms from σ to τ in $\mathscr{C}_{\mathscr{A}}$. In addition, it is easy to show that $\mathscr{C}_{\mathscr{A}}$ is Cartesian closed if $\mathscr{A}$ has products and a terminal object.

3. Logical relations

3.1. Introduction

Logical relations are an important tool in the study of typed lambda calculus. In fact, virtually all of the basic results about $\lambda^{\rightarrow}$ may be proved using logical relations. Essentially, a logical relation $\mathscr{R}$ is a family $\{R^\sigma\}$ of typed relations with the relation $R^{\sigma\rightarrow\tau}$ for type $\sigma\rightarrow\tau$ determined from the relations R^σ and R^τ in a way that guarantees closure under application and lambda abstraction. Since logical relations are very general, some important algebraic concepts like homomorphism and congruence relation lead to special classes of logical relations. However, logical relations lack many familiar properties of their algebraic analogs. For example, the composition of two logical relations may not be a logical relation, even in the special cases where both logical relations seem to act as homomorphisms or congruence relations.

The main topics in our brief study of logical relations will be the following:

- The "Basic Lemma", which says, intuitively, that logical relations are closed under application and lambda abstraction. In particular, if $\mathscr{R}$ is a logical relation between models $\mathscr{A}$ and $\mathscr{B}$, and M is a closed term, then $\mathscr{R}$ relates the meaning of M in $\mathscr{A}$ to the meaning of M in $\mathscr{B}$.
- Logical partial functions and partial equivalence relations, which are roughly similar to homomorphisms and congruence relations over algebraic structures.
- Completeness, confluence and strong normalization theorems.
- Representation independence, a property saying that implementors of programming languages based on $\lambda^{\rightarrow}$ have some freedom in the way they implement data types. For example, in implementing a version of $\lambda^{\rightarrow}$ with base type *bool*, it does not matter whether *true* is represented by 1 and *false* by 0, or vice versa, as long as operations like conditional behave properly.

Logical relations for typed lambda calculus were developed by Howard, Tait, Friedman, Statman, Plotkin, and others. In particular, see [56, 133, 38, 105] and the three papers by Statman [128, 130, 129]. Many of the results presented in this section are proved in [130], using a slightly different framework. The uniform treatment of relations over models and terms, and the revised definition of 'admissible," are based on

[90]. Some of the basic theory is generalized to polymorphic lambda calculus in [93].

There are many interesting properties of logical relations which we will not have sufficient space to go into. Among other things, logical relations have application to strictness analysis [12, 1], full abstraction [99], and realizability semantics of constructive logics [88, 95].

3.2. *Logical relations over applicative structures*

3.2.1. *Logical relations*

We will formulate the basic definitions using typed applicative structures. Since the set of terms, as well as any model, forms a typed applicative structure, this will allow us to use logical relations to prove syntactic as well as semantic results about typed lambda calculus.

Let $\mathscr{A}=\langle\{A^\sigma\}, \{\mathbf{App}_{\mathscr{A}}^{\sigma,\tau}\}, Const_{\mathscr{A}}\rangle$ and $\mathscr{B}=\langle\{B^\sigma\}, \{\mathbf{App}_{\mathscr{B}}^{\sigma,\tau}\}, Const_{\mathscr{B}}\rangle$ be applicative structures for some signature Σ. A *logical relation* $\mathscr{R}=\{R^\sigma\}$ *over* $\mathscr{A}$ *and* $\mathscr{B}$ is a family of relations indexed by the type expressions over Σ such that

- $R^\sigma \subseteq A^\sigma \times B^\sigma$ for each type σ,
- $R^{\sigma\to\tau}(f, g)$ iff $\forall x\in A^\sigma . \forall y\in B^\sigma . R^\sigma(x, y) \supset R^\tau(\mathbf{App}_{\mathscr{A}}\, fx, \mathbf{App}_{\mathscr{B}}\, g\, y)$,
- $R^\sigma(Const_{\mathscr{A}}(c), Const_{\mathscr{B}}(c))$ for every typed constant $c:\sigma$ of Σ.

The central property is that two functions f, g are logically related iff they map related arguments to related results. Given R^σ and R^τ, this determines $R^{\sigma\to\tau}$ uniquely. We will often write $R(x, y)$ for $R^\sigma(x, y)$ when σ is either clear from context or irrelevant. We also write $\mathscr{R}\subseteq\mathscr{A}\times\mathscr{B}$ to indicate that $\mathscr{R}$ is a relation over $\mathscr{A}$ and $\mathscr{B}$.

Some trivial examples of logical relations are the identity and "true everywhere" relations:

(1) If $\mathscr{A}$ is extensional, then the identity relation $I\subseteq\mathscr{A}\times\mathscr{A}$ with $I^\sigma(x, y)$ iff $x=y\in A^\sigma$ is logical.

(2) For any applicative structures $\mathscr{A}$ and $\mathscr{B}$ of the same signature, the relation $R\subseteq\mathscr{A}\times\mathscr{B}$ that relates every element of A^σ to every element of B^σ is logical.

When the signature Σ has no constants, logical relations may be constructed by choosing arbitrary $R^b\subseteq A^b\times B^b$ for each base type b, and extending to higher types inductively. For signatures with constants, it is sometimes more difficult to construct logical relations. However, if Σ has no constants of functional type (i.e., if $c:\sigma$ is a constant of Σ, then σ does not have the form $\sigma_1\to\sigma_2$), we may construct logical relations by choosing arbitrary relations that respect constants at base types. It is also relatively easy to accommodate first-order function constants.

Binary logical relations illustrate the general properties of logical relations quite nicely, and will be sufficient for most of our purposes. However, it is worth noting that the definition above generalizes easily to *k-ary* logical relations $\mathscr{R}$ over applicative structures $\mathscr{A}_1,\ldots,\mathscr{A}_k$, including the case $k=1$. In addition, all of the results of this chapter generalize easily to relations of arbitrary arity; only the notation becomes more complicated. The reason is that logical relations of any arity may be viewed as unary logical relations. The reader may enjoy formulating the definition of logical predicate (unary logical relation) and proving that a logical relation $\mathscr{R}\subseteq\mathscr{A}\times\mathscr{B}$ is just a logical predicate over the straightforwardly defined product applicative structure $\mathscr{A}\times\mathscr{B}$.

3.2.2. The Basic Lemma

The Basic Lemma establishes that the meaning of a term in one model is always logically related to its meaning in any other model. On the face of it, the Basic Lemma only seems to apply to models, since the meaning of a term is not necessarily defined in an arbitrary applicative structure. However, with a little extra work, we can also state a version of the lemma which will be useful in proving properties of applicative structures that are not models. We will take this up after stating the Basic Lemma for models.

An auxiliary definition will make the lemma easier to write down. A logical relation $\mathscr{R} \subseteq \mathscr{A} \times \mathscr{B}$ may be regarded as a relation on environments for $\mathscr{A}$ and $\mathscr{B}$, as follows. We say that *environments η_a for $\mathscr{A}$ and η_b for $\mathscr{B}$ are related by $\mathscr{R}$*, and write $\mathscr{R}(\eta_a, \eta_b)$, if $\mathscr{R}(\eta_a(x), \eta_b(x))$ for every variable x. In [130], the Basic Lemma is called the "Fundamental Theorem of Logical Relations".

3.2.1. LEMMA (Basic Lemma, for models). *Let $\mathscr{R} \subseteq \mathscr{A} \times \mathscr{B}$ be a logical relation and let η_a and η_b be related environments for models $\mathscr{A}$ and $\mathscr{B}$ with $\eta_a, \eta_b \models \Gamma$. Then $\mathscr{R}(\mathscr{A}[\![\Gamma \rhd M:\sigma]\!]\eta_a, \mathscr{B}[\![\Gamma \rhd M:\sigma]\!]\eta_b)$ for every typed $\Gamma \rhd M:\sigma$.*

The proof is an easy induction on terms.

The Basic Lemma may be generalized to applicative structures that are not models by including some hypotheses about the way terms are interpreted. Let $\mathscr{A}$ be an applicative structure. A partial mapping $\mathscr{A}[\![\ ]\!]$ from terms and environments to $\mathscr{A}$ is *an acceptable meaning function* if

$$\mathscr{A}[\![\Gamma \rhd M:\sigma]\!]\eta \in A^\sigma \quad \text{whenever } \eta \models \Gamma$$

and the following conditions are satisfied:

$$\begin{array}{lll}
\mathscr{A}[\![\Gamma \rhd x:\sigma]\!]\eta & = \eta(x), & \\
\mathscr{A}[\![\Gamma \rhd c:\sigma]\!]\eta & = Const(c), & \\
\mathscr{A}[\![\Gamma \rhd MN:\tau]\!]\eta & = \mathbf{App}\, \mathscr{A}[\![\Gamma \rhd M:\sigma\to\tau]\!]\eta\, \mathscr{A}[\![\Gamma \rhd N:\sigma]\!]\eta, & \\
\mathscr{A}[\![\Gamma \rhd \lambda x:\sigma.M:\sigma\to\tau]\!]\eta & = \mathscr{A}[\![\Gamma \rhd \lambda y:\sigma.[y/x]M:\sigma\to\tau]\!]\eta, & \\
\mathscr{A}[\![\Gamma \rhd M:\sigma]\!]\eta_1 & = \mathscr{A}[\![\Gamma \rhd M:\sigma]\!]\eta_2 & \text{whenever } \eta_1(x)=\eta_2(x) \\
 & & \text{for all } x \in FV(M), \\
\mathscr{A}[\![\Gamma, x:\sigma \rhd M:\tau]\!]\eta & = \mathscr{A}[\![\Gamma \rhd M:\tau]\!] & \text{for } x \text{ not in } \Gamma.
\end{array}$$

Two important examples of acceptable meaning functions are meaning functions for models and substitution on applicative structures of terms.

3.2.2. EXAMPLE. *If $\mathscr{A}$ is a model, then the ordinary meaning function $\mathscr{A}[\![\ ]\!]$ is acceptable.*

3.2.3. EXAMPLE. Let $\mathscr{T}$ be an applicative structure of terms M such that $\Gamma \rhd M:\sigma$ for some $\Gamma \subseteq \mathscr{H}$ as in Example 2.4.1. An environment η for $\mathscr{T}$ is a mapping from variables to terms, which we may regard as a substitution. Let us write ηM for the result of substituting terms for free variables in M, and define a meaning function on $\mathscr{T}$ by

$\mathscr{T}[\![\Gamma \rhd M:\sigma]\!]\eta = \eta M$. It is a worthwhile exercise to verify that $\mathscr{T}[\![\ ]\!]$ is an acceptable meaning function.

In addition to acceptable meaning functions, we will need some assumptions about the behavior of a logical relation on lambda abstractions. A logical relation over models is necessarily closed under lambda abstraction, as a consequence of the way that lambda abstraction and application interact. However, in an arbitrary applicative structure, abstraction and application may not be "inverses", and so we need an additional assumption to prove the Basic Lemma.

If $\mathscr{R} \subseteq \mathscr{A} \times \mathscr{B}$ and $\mathscr{A}[\![\ ]\!], \mathscr{B}[\![\ ]\!]$ are acceptable meaning functions, we say $\mathscr{R}$ is *admissible for* $\mathscr{A}[\![\ ]\!]$ *and* $\mathscr{B}[\![\ ]\!]$ if, for all related environments $\eta_a, \eta_b \models \Gamma$ and terms $\Gamma, x:\sigma \rhd M:\tau$ and $\Gamma, x:\sigma \rhd N:\tau$,

$$\forall a, b.\, R^\sigma(a, b) \supset R^\tau(A[\![\Gamma, x:\sigma \rhd M:\tau]\!]\eta_a[a/x], \mathscr{B}[\![\Gamma, x:\sigma \rhd N:\tau]\!]\eta_b[b/x])$$

implies

$$\forall a, b.\, R^\sigma(a, b) \supset R^\tau(\mathbf{App}\,(A[\![\Gamma \rhd \lambda x:\sigma.M:\sigma \to \tau]\!]\eta_a)\, a, \\ \mathbf{App}\,(\mathscr{B}[\![\Gamma \rhd \lambda x:\sigma.N:\sigma \to \tau]\!]\eta_b) b).$$

It is clear that if $\mathscr{A}$ and $\mathscr{B}$ are models, then any logical relation is admissible, since then

$$\mathscr{A}[\![\Gamma, x:\sigma \rhd M:\tau]\!]\eta_a[a/x] = \mathbf{App}\,(\mathscr{A}[\![\Gamma \rhd \lambda x:\sigma.M:\sigma \to \tau]\!]\eta_a)\, a,$$

and similarly for $\Gamma, x:\sigma \rhd N:\tau$. This definition of "admissible" is similar to the definition given in [130], but slightly weaker. Using our definition, we may prove strong normalization directly, by a construction that does not seem possible in Statman's framework (see [130, Example 4]). Using acceptable meaning functions and admissible relations, we have the following general form of the Basic Lemma.

3.2.4. LEMMA (Basic Lemma, general version). *Let $\mathscr{A}$ and $\mathscr{B}$ be applicative structures with acceptable meaning functions $\mathscr{A}[\![\ ]\!]$ and $\mathscr{B}[\![\ ]\!]$ and let $\mathscr{R} \subseteq \mathscr{A} \times \mathscr{B}$ be an admissible logical relation. Suppose η_a and η_b are related environments satisfying the type assignment Γ. Then*

$$\mathscr{R}(\mathscr{A}[\![\Gamma \rhd M:\sigma]\!]\eta_a, \mathscr{B}[\![\Gamma \rhd M:\sigma]\!]\eta_b)$$

for every typed term $\Gamma \rhd M:\sigma$.

The proof is similar to the proof of the Basic Lemma for models, since the acceptability of $\mathscr{A}[\![\ ]\!]$ and $\mathscr{B}[\![\ ]\!]$, and the admissibility of $\mathscr{R}$, are enough to make the original line of reasoning go through. However, it is a worthwhile exercise to check the details.

The following lemma is useful for establishing that logical relations over term applicative structures are admissible. For simplicity, we state the lemma for logical predicates only.

3.2.5. LEMMA. *Let $\mathscr{T}$ be an applicative structure of terms with meaning function defined by substitution, as in Example 3.2.3. Let $\mathscr{P} \subseteq \mathscr{T}$ be a logical predicate. If*

$$\forall N \in P^\sigma.\, P^b(([N/x]M)N_1 \ldots N_k) \quad \textit{implies} \\ \forall N \in P^\sigma.\, P^b((\lambda x:\sigma.M)NN_1 \ldots N_k)$$

for any base type b, term $\lambda x{:}\sigma.M$ in $T^{\sigma\to\tau_1\to\cdots\to\tau_k\to b}$ and terms $N_1 \in T^{\tau_1}, \ldots, N_k \in T^{\tau_k}$, then $\mathscr{P}$ is admissible.

3.3. Logical partial functions and equivalence relations

3.3.1. Partial functions and theories of models

A logical relation $\mathscr{R} \subseteq \mathscr{A} \times \mathscr{B}$ is called a *logical partial function* if each R^σ is a partial function from A^σ to B^σ, i.e.,

$$R^\sigma(a, b_1) \text{ and } R^\sigma(a, b_2) \quad \text{implies} \quad b_1 = b_2.$$

Logical partial functions are related to the theories of typed lambda models by the following lemma.

3.3.1. LEMMA. *If $\mathscr{R} \subseteq \mathscr{A} \times \mathscr{B}$ is a logical partial function from model $\mathscr{A}$ to model $\mathscr{B}$, then $Th(\mathscr{A}) \subseteq Th(\mathscr{B})$.*

By Corollary 2.4.10 (completeness for the pure theory of β, η-conversion) we know there is a model $\mathscr{A}$ satisfying $\Gamma \rhd M = N : \sigma$ iff $\vdash \Gamma \rhd M = N : \sigma$. Since the theory of this model is contained in the theory of every other model, we may show that a model $\mathscr{B}$ has the same theory by constructing a logical partial function $\mathscr{R} \subseteq \mathscr{B} \times \mathscr{A}$. This technique was first used by Friedman to show that β, η-conversion is complete for the full set-theoretic type hierarchy over any infinite ground types [38, 129].

3.3.2. COROLLARY. *Let $\mathscr{A}$ be a model of the pure theory of β, η-conversion. If $\mathscr{R} \subseteq \mathscr{B} \times \mathscr{A}$ is a logical partial function and $\mathscr{B}$ is a model, then $Th(\mathscr{B}) = Th(\mathscr{A})$.*

Since logical partial functions are structure-preserving mappings, it is tempting to think of logical partial functions as the natural generalization of homomorphisms. This often provides useful intuition. However, the composition of two logical partial functions need not be a logical relation. It is interesting to note that for a fixed model $\mathscr{A}$ and environment η, the meaning function $\mathscr{A}[\![\ ]\!]\eta$ from $\mathscr{T}$ to $\mathscr{A}$ is not a logical relation unless, for every a in $\mathscr{A}$, there are infinitely many variables y with $\eta(y) = a$. In contrast, meaning functions in algebra are always homomorphisms.

3.3.2. Logical partial equivalence relations

Another useful class of logical relations are the logical partial equivalence relations (logical *per*'s), which have many properties of congruence relations. If we regard typed applicative structures as multisorted algebras, we may see that logical partial equivalence relations are somewhat more than congruence relations. Specifically, any logical per $\mathscr{R}$ is a congruence, since $\mathscr{R}$ is an equivalence relation closed under application, but $\mathscr{R}$ must also be closed under lambda abstraction.

Before looking at the definition of logical partial equivalence relation, it is worth digressing briefly to discuss partial equivalence relations on any set. There are two equivalent definitions. The more intuitive one is that a *partial equivalence relation on set* A is an equivalence relation on some subset of A, i.e., a pair $\langle A', R \rangle$ with $A' \subseteq A$ and

R an equivalence relation on A'. However, because R determines $A' = \{a \mid R(a, a)\}$, the subset A' is technically redundant. Moreover, R is an equivalence relation on the subset $A' = \{a \mid R(a, a)\}$ iff R is symmetric and transitive on A. Therefore, we will define partial equivalence relations as symmetric and transitive relations.

A *logical partial equivalence relation on* $\mathscr{A}$ is a logical relation $\mathscr{R} \subseteq \mathscr{A} \times \mathscr{A}$ such that each R^σ is symmetric and transitive. It is useful to have a notation for the field of a partial equivalence relation. We will write $|R^\sigma|$ for the set

$$|R^\sigma| = \{a \in A^\sigma \mid R^\sigma(a, a)\}.$$

We say a partial equivalence relation is *total* if each R^σ is reflexive, i.e., $|R^\sigma| = A^\sigma$. From the following lemma, we can see that there are many examples of partial equivalence relations.

3.3.3. LEMMA. *Let $\mathscr{R} \subseteq \mathscr{A} \times \mathscr{A}$ be a logical relation. If R^b is symmetric and transitive, for each base type b, then every R^σ is symmetric and transitive.*

Intuitively, this lemma shows that being a partial equivalence relation is a "hereditary" property; it is inherited at higher types. In contrast, being a total equivalence relation is not. Specifically, there exist logical relations which are total equivalence relations at base type, but not total equivalence relations at every type. This is one reason for concentrating on partial instead of total equivalence relations.

3.3.3. Quotients and extensionality

If $\mathscr{R}$ is a partial equivalence relation over applicative structure $\mathscr{A}$, then we may form a quotient structure $\mathscr{A}/\mathscr{R}$ of equivalence classes of $\mathscr{A}$. This is actually a "partial" quotient, since some elements of $\mathscr{A}$ may not have equivalence classes with respect to $\mathscr{R}$. However, we will just call $\mathscr{A}/\mathscr{R}$ a "quotient". One important fact about the structure $\mathscr{A}/\mathscr{R}$ is that it is always extensional. Another is that under certain reasonable assumptions, $\mathscr{A}/\mathscr{R}$ will be model with the meaning of each term in $\mathscr{A}/\mathscr{R}$ equal to the equivalence class of its meaning in $\mathscr{A}$.

Let $\mathscr{R} \subseteq \mathscr{A} \times \mathscr{A}$ be a partial equivalence relation on the applicative structure $\mathscr{A} = \langle \{A^\sigma\}, \{\mathbf{App}^{\sigma,\tau}\}, Const\}\rangle$. The quotient applicative structure

$$\mathscr{A}/\mathscr{R} = \langle \{A^\sigma/\mathscr{R}\}, \{\mathbf{App}^{\sigma,\tau}/\mathscr{R}\}, Const/\mathscr{R}\rangle$$

is defined as follows. For any $a \in |R^\sigma|$, the *equivalence class* $[a]_{\mathscr{R}}$, *of a with respect to* $\mathscr{R}$ is defined by

$$[a]_{\mathscr{R}} = \{a' \mid \mathscr{R}(a, a')\}.$$

Note that by definition of $|R^\sigma|$, we have $a \in [a]_{\mathscr{R}}$. We let $A^\sigma/\mathscr{R} = \{[a]_{\mathscr{R}} \mid a \in |R^\sigma|\}$ be the collection of all equivalence classes from $|R^\sigma|$, and define application on equivalence classes by

$$(\mathbf{App}^{\sigma,\tau}/\mathscr{R})\,[a]\,[b] = [\mathbf{App}^{\sigma,\tau}\, a\, b].$$

The map $Const/\mathscr{R}$ interprets each constant c as the equivalence class $[Const(c)]_{\mathscr{R}}$. This completes the definition of $\mathscr{A}/\mathscr{R}$. It is easy to see that application is well-defined: if $R^{\sigma\to\tau}(a,a')$ and $R^{\sigma}(b,b')$, then, since $\mathscr{R}$ is logical, we have $R^{\tau}(\mathbf{App}^{\sigma,\tau}\,a\,b, \mathbf{App}^{\sigma,\tau}\,a'\,b')$.

3.3.4. LEMMA. *If $\mathscr{R}\subseteq\mathscr{A}\times\mathscr{A}$ is a partial equivalence relation, then $\mathscr{A}/\mathscr{R}$ is an extensional applicative structure.*

A special case of this lemma is the *extensional collapse*, $\mathscr{A}^E$, of an applicative structure $\mathscr{A}$; this is also called the *Gandy hull of* $\mathscr{A}$, and appears to have been discovered independently by Zucker [135] and Gandy [40]. The extensional collapse $\mathscr{A}^E$ is defined as follows. Given $\mathscr{A}$, there is at most one logical relation $\mathscr{R}$ with

$$R^b(a_1,a_2) \quad \text{iff} \quad a_1=a_2\in A^b$$

for each base type b. (Such an $\mathscr{R}$ is guaranteed to exist when the type of each constant is either a base type or first-order function type.) If there is a logical relation $\mathscr{R}$ which is the identity relation on all base types, then we define $\mathscr{A}^E=\mathscr{A}/\mathscr{R}$.

3.3.5. COROLLARY (extensional collapse). *Let $\mathscr{A}$ be an applicative structure for signature Σ. If there is a logical relation $\mathscr{R}$ which is the identity relation on all base types, then the extensional collapse $\mathscr{A}^E$ is an extensional applicative structure for Σ.*

In some cases, $\mathscr{A}/\mathscr{R}$ will not only be extensional, but also a model. This will be true if $\mathscr{A}$ is a model to begin with. In addition, since $\mathscr{A}/\mathscr{R}$ is always extensional, we would expect to have a model whenever $\mathscr{A}$ has "enough functions", and β-equivalent terms are equivalent modulo $\mathscr{R}$. If $\mathscr{A}[\![\ ]\!]$ is an acceptable meaning function, then this will guarantee that A has enough functions. When we have an acceptable meaning function, we can also express the following condition involving β-equivalence. We say *$\mathscr{R}$ satisfies* (β) *with respect to $\mathscr{A}[\![\ ]\!]$* if, for every pair of terms $\Gamma, x{:}\sigma\rhd M{:}\tau$ and $\Gamma\rhd N{:}\sigma$, and environment $\eta\models\Gamma$, we have

$$\mathscr{R}(\mathscr{A}[\![\Gamma\rhd(\lambda x{:}\sigma.M)N{:}\tau]\!]\eta, \mathscr{A}[\![\Gamma\rhd[N/x]M{:}\tau]\!]\eta),$$

To describe the meanings of terms in $\mathscr{A}/\mathscr{R}$, we will associate an environment η_R for $\mathscr{A}/\mathscr{R}$ with each environment η for $\mathscr{A}$ with $\mathscr{R}(\eta,\eta)$. We define $\eta_{\mathscr{R}}$ by

$$\eta_{\mathscr{R}}(x)=[\eta(x)]_{\mathscr{R}},$$

and note that $\eta\models\Gamma$ iff $\eta_{\mathscr{R}}\models\Gamma$. Using these definitions, we can give useful conditions which imply that $\mathscr{A}/\mathscr{R}$ is a model, and characterize the meanings of terms in $\mathscr{A}/\mathscr{R}$.

3.3.6. LEMMA (quotient models). *Let $\mathscr{R}\subseteq\mathscr{A}\times\mathscr{A}$ be an admissible partial equivalence relation over applicative structure $\mathscr{A}$ with acceptable meaning function $\mathscr{A}[\![\ ]\!]$, and suppose that $\mathscr{R}$ satisfies* (β). *Then $\mathscr{A}/\mathscr{R}$ is a model such that for any $\mathscr{A}$-environment $\eta\models\Gamma$ with $\mathscr{R}(\eta,\eta)$, we have*

$$(\mathscr{A}/\mathscr{R})[\![\Gamma\rhd M{:}\sigma]\!]\eta_{\mathscr{R}}=[\mathscr{A}[\![\Gamma\rhd M{:}\sigma]\!]\eta]_{\mathscr{R}}.$$

In other words, under the assumptions specified in the lemma, the meaning of a term

$\Gamma \rhd M:\sigma$ in $\mathscr{A}/\mathscr{R}$ is the equivalence class of the meaning of $\Gamma \rhd M:\sigma$ in $\mathscr{A}$, modulo $\mathscr{R}$. Lemma 3.3.6 is similar to the "Characterization Theorem" of [90], which seems to be the first use of this idea.

3.4. *Proof-theoretic results*

3.4.1. *Completeness without empty types*

In universal algebra, we may prove equational completeness (assuming no sort is empty) by showing that any theory $\mathscr{E}$ is a congruence relation on the term algebra $\mathscr{T}$, and noting that $\mathscr{T}/\mathscr{E}$ satisfies an equation $M=N$ iff $M=N\in\mathscr{E}$. We may give similar completeness proofs for typed lambda calculus, using logical equivalence relations in place of congruence relations. In this subsection, we will illustrate the main ideas by proving completeness for models without empty types. Similar proofs may be given for Henkin models with possibly empty types, or Kripke lambda models (using Kripke-style logical relations) [94].

Throughout this section, we let $\mathscr{T}$ be the applicative structure of terms typed using infinite type assignment $\mathscr{H}$, with meaning function $\mathscr{T}[\![\ ]\!]$ defined by substitution. (These were defined in Example 3.2.3). We assume $\mathscr{H}$ provides infinitely many variables of each type, so each T^σ is infinite. If $\mathscr{E}$ is any set of equations, we define the relation $\mathscr{R}_\mathscr{E}=\{R^\sigma\}$ over $\mathscr{T}\times\mathscr{T}$ by taking

$$R^\sigma(M,N) \quad \text{iff} \quad \mathscr{E}\vdash\Gamma\rhd M=N:\sigma \quad \text{for some finite } \Gamma\subseteq\mathscr{H}.$$

3.4.1. Lemma. *For any theory $\mathscr{E}$ closed under rule* (nonempty), *the relation $\mathscr{R}_\mathscr{E}\subseteq\mathscr{T}\times\mathscr{T}$ is an admissible logical equivalence relation.*

From Lemma 3.3.6, and the fact that any theory $\mathscr{E}$ satisfies (β), we can form a quotient model from any theory.

3.4.2. Lemma. *For any theory $\mathscr{E}$ closed under* (nonempty), *the quotient structure $\mathscr{A}=\mathscr{T}/\mathscr{R}_\mathscr{E}$ is a model. Furthermore, if η is any substitution (environment for $\mathscr{T}$) with $\eta\models\Gamma$, we have $\mathscr{A}[\![\Gamma\rhd M:\sigma]\!]\eta_\mathscr{E}=[\eta M]_\mathscr{E}$.*

This allows us to prove strong completeness for Henkin models without empty types.

3.4.3. Theorem. *Let $\mathscr{E}$ be any theory closed under* (nonempty), *and $\mathscr{A}=\mathscr{T}/\mathscr{R}_\mathscr{E}$. Then $\mathscr{A}$ is a Henkin model without empty types satisfying precisely the equations belonging to $\mathscr{E}$.*

3.4.2. *Normalization*

Logical relations may also be used to prove various properties of reduction on typed terms. In this section, we sketch the proof that every typed term is strongly normalizing. More precisely, we show that for every typed term $\Gamma\rhd M:\sigma$, we have $SN(M)$, where SN is defined by

$$SN(M) \quad \text{iff} \quad \text{there is no infinite sequence of reductions from M.}$$

We will generalize the argument to other reduction properties in the next section. The proof has three main parts:

(1) define a logical predicate $\mathscr{P}$;
(2) show that $P^\sigma(M)$ implies $SN(M)$;
(3) show that $\mathscr{P}$ is admissible, so that we may use the Basic Lemma to conclude that $P^\sigma(M)$ holds for every well-typed term $\Gamma \rhd M:\sigma$.

Since the Basic Lemma is proved by induction on terms, the logical relation $\mathscr{P}$ is a kind of "induction hypothesis" for proving that all terms are strongly normalizing.

We define the logical predicate (unary logical relation) $\mathscr{P} \subseteq \mathscr{T}$ by taking P^b to be all strongly normalizing terms of base type, and extending to higher types by

$$P^{\sigma\to\tau}(M) \quad \text{iff} \quad \forall N \in P^\sigma . P^\tau(MN).$$

It is easy to see that this predicate is logical when the signature has no term constants. If we are interested in terms with constants, then it follows from Lemma 3.4.4 below that $\mathscr{P}$ is logical. The next step is to show that $\mathscr{P}$ contains only strongly normalizing terms. This is a relatively easy induction on types, as long as we take the correct induction hypothesis. The hypothesis we use is the conjunction of (i) and (ii) of the following lemma.

3.4.4. LEMMA. *For each type σ, the predicate P^σ satisfies the following two conditions:*
(i) *If $\Gamma \rhd xM_1 \dots M_k : \sigma$ and $SN(M_1), \dots, SN(M_k)$, then $P^\sigma(xM_1 \dots M_k)$. Similarly for any constant c in place of variable x.*
(ii) *If $P^\sigma(M)$, then $SN(M)$.*

The main reason condition (i) is included in the lemma is to guarantee that $\mathscr{P}$ contains variables of each type, since variables are useful in the proof of condition (ii). Condition (i) is slightly stronger than "every variable belongs to $\mathscr{P}$" since the stronger statement is easier to prove inductively. Using Lemma 3.2.5, we may show that the logical predicate $\mathscr{P}$ is admissible.

3.4.5. LEMMA. *The logical relation $\mathscr{P} \subseteq \mathscr{T}$ is admissible.*

From the two lemmas above, and the Basic Lemma for logical relations, we have the following theorem.

3.4.6. THEOREM (strong normalization). *For every well-typed term $\Gamma \rhd M:\sigma$, we have $SN(M)$.*

3.4.3. Confluence and other reduction properties

We may also prove confluence and other properties of typed reduction using logical relations. While it is well-known that β,η-reduction is confluent on untyped lambda terms, this does not immediately carry over to typed reduction; some proof is necessary. In particular, with variables distinguished by type, α-conversion on typed terms is slightly different from untyped α-conversion. For this reason, β,η-reduction is *not* Church–Rosser on un-typechecked terms with typed variables, as discussed in Section

2.3.2. We will prove confluence using logical relations and discuss alternative proofs at the end of this section. The original logical relations proof of confluence is due to Statman [130].

Many parts of the Church–Rosser proof are identical to the proof of strong normalization. Rather than simply repeat the lemmas of the last section, we will state a general theorem about properties of typed terms, based on [90]. To be precise, let $\mathcal{S}$ be a property of typed lambda terms, which we may regard as a typed predicate $\mathcal{S}=\{S^\sigma\}$ on a term applicative structure $\mathcal{T}$. (We do not assume that S is logical.) We say $\mathcal{S}$ is *type-closed* if the following three conditions are satisfied:

(a) $\mathcal{S}(M_1)\wedge\cdots\wedge\mathcal{S}(M_k) \supset \mathcal{S}(xM_1\ldots M_k)$ and similarly for constant c,

(b) $(\forall x\in S^\sigma)S^\tau(Mx) \supset S^{\sigma\to\tau}(M)$,

(c) $\mathcal{S}(N)\wedge\mathcal{S}(([N/x]M)N_1\ldots N_k) \supset \mathcal{S}((\lambda x{:}\sigma.M)NN_1\ldots N_k)$.

In verifying that a particular predicate satisfies (b), we generally assume $S(Mx)$ for x not free in M and show $S(M)$. The strong normalization proof given in the last section also proves the following theorem.

3.4.7. THEOREM. *Let $\mathcal{S}$ be any type-closed property of typed lambda terms. Then $\mathcal{S}(M)$ for every typed lambda term $\Gamma\rhd M{:}\sigma$.*

We may obtain strong normalization and confluence as relatively straightforward corollaries. In addition, as pointed out in [130], we may also prove termination of standard reduction, completeness of standard reduction and η-postponement using logical relations. In fact, these results all follow as corollaries of Theorem 3.4.7.

The rest of this section will be concerned with confluence. We may regard confluence as a predicate on typed lambda terms by defining

$$CR(M)\quad\text{iff}\quad \forall M_1,M_2[M\twoheadrightarrow M_1\wedge M\twoheadrightarrow M_2\supset\exists N.\,M_1\twoheadrightarrow N\wedge M_2\twoheadrightarrow N].$$

In verifying that CR is type-closed, the most difficult condition is (b). While it is easy to see that $SN(Mx)\supset SN(M)$, it is not immediate that $CR(Mx)\supset CR(M)$, since there may be reductions of Mx that do not apply to M. Condition (a) is trivial and condition (c) is essentially straightforward.

3.4.8. LEMMA. *The predicate CR is type-closed.*

3.4.9. COROLLARY. *β, η-Reduction is confluent on typed lambda terms.*

An alternative proof of confuence involves verifying local confluence directly. The *weak Church–Rosser property*, also called *local confluence*, is

$$M\to M_1\wedge M\to M_2\quad\text{implies}\quad\exists N.\,M_1\twoheadrightarrow N\wedge M_2\twoheadrightarrow N.$$

which we regard as a predicate by defining

$$WCR(M)\quad\text{iff}\quad\forall M_1,M_2.\,M\to M_1\wedge M\to M_2\supset\exists N.\,M_1\twoheadrightarrow N\wedge M_2\twoheadrightarrow N.$$

By the following lemma, due to a number of researchers (see [5]), it suffices to prove $WCR(M)$ for every well-typed term M.

3.4.10. LEMMA. *If* $\twoheadrightarrow$ *is strongly normalizing and weakly Church–Rosser, then* $\twoheadrightarrow$ *is strongly Church–Rosser.*

Somewhat surprisingly, it seems difficult to show directly that WCR is type-closed. The problem lies in condition (c), where the hypothesis that $WCR(([N/x]M)N_1 \ldots N_k)$ does not seem strong enough to show $WCR((\lambda z\colon\sigma.M)NN_1 \ldots N_k)$. In contrast, as noted above, this is straightforward for CR.

3.5. *Representation independence*

3.5.1. *Motivation*

In addition to proving basic results about $\lambda^{\rightarrow}$, logical relations have useful applications in analyzing typed programming languages. One pragmatic motive for type-checking is to guarantee a degree of "representation independence", which is useful in compiler design. Intuitively, representation independence is the property that programs should not depend on the way data types are represented, only on the behavior of data types with respect to operations provided. For example, it should not matter whether stacks are implemented using arrays or linked lists, as long as the *push* and *pop* operations give the correct results.

The intent of a representation independence theorem is to characterize the kinds of implementation decisions that do not affect the meanings of programs. Various formal statements of the property have been proposed by Reynolds, Donahue, Haynes and Mitchell and Meyer [30, 50, 112, 93, 89]. Essentially, all of these have the following general form:

> If two interpretations $\mathscr{A}$ and $\mathscr{B}$ are related in a certain way, then the meaning $\mathscr{A}[\![\Gamma \rhd M\colon\sigma]\!]$ of any closed term $\Gamma \rhd M\colon\sigma$ in $\mathscr{A}$ is related to the meaning $\mathscr{B}[\![\Gamma \rhd M\colon\sigma]\!]$ in $\mathscr{B}$ in the same certain way.

The pragmatic consequence of this sort of theorem is that if two programming language interpreters are related in this "certain way", then the result of executing any program using one interpreter will correspond to the result of executing the same program using the other interpreter. Thus the precise statement of the theorem describes the kind of implementation decisions that do not affect the meanings of programs. Of course, the kinds of relations we are interested in will turn out to be logical relations.

3.5.2. *Example language*

The main ideas seem best illustrated by example. We will consider multisets of natural numbers, assuming that we observe the behavior of programs by computing natural numbers. Since we do not have type declarations in $\lambda^{\rightarrow}$, we will compare implementations of multisets by assuming the implementations are provided as part of the semantics of a simple programming language. However, the main ideas may also be applied to languages with abstract data type declarations, as sketched out in [89].

To study programs with multisets, we let Σ be the $\lambda^{\rightarrow}$ signature with base types *nat* (for natural numbers) and *s* (for multisets), term constants

$$0, 1\colon nat, \qquad +\colon nat \rightarrow nat \rightarrow nat$$

to provide arithmetic, and constants

$$empty\colon s, \qquad insert\colon nat \rightarrow s \rightarrow s, \qquad count\colon s \rightarrow nat \rightarrow nat$$

for multiset operations. Informally, *empty* is the empty multiset, *insert* adds an element to a multiset, and *count* returns the multiplicity of a multiset element.

In this language, we intend to write programs with natural-number inputs and outputs. Multisets and function expressions may occur in programs, but are used only at "intermediate stages". Our goal is to describe the conditions under which different representations (or implementations) of multisets are indistinguishable by programs. Since our definition will involve quantifying over the collection of all programs, it suffices to consider programs with no input, and only one output. Therefore, we let the *programs over* Σ be the closed terms $M\colon nat$ of type natural number.

We will compare different representations of multisets by examining the values of programs over appropriate models. Since we consider natural numbers as the "printable output", we will assume that each model interprets *nat* as the usual natural numbers 0, 1, 2, ... For the purpose of this discussion, we define an *implementation for* Σ to be a model

$$\mathcal{A} = \langle \{A^\sigma\}, 0, 1, +, empty^{\mathcal{A}}, insert^{\mathcal{A}}, count^{\mathcal{A}} \rangle$$

with A^{nat} and 0, 1, + standard, but A^s and operations *empty*, *insert*, *count* arbitrary.

Two compilers or interpreters for a programming language would generally be considered equivalent if every program produces identical results on either one. We say that implementations $\mathcal{A}$ and $\mathcal{B}$ for Σ are *observationally equivalent with respect to the natural numbers* if, for any program $M\colon nat$, we have

$$\mathcal{A}[\![M\colon nat]\!] = \mathcal{B}[\![M\colon nat]\!],$$

i.e., the meaning of $M\colon nat$ in A is the same as the meaning of $M\colon nat$ in $\mathcal{B}$. The rest of this section is devoted to characterizing observational equivalence of implementations.

A first guess might be that $\mathcal{A}$ and $\mathcal{B}$ are observationally equivalent iff there exists some kind of mapping between multisorted first-order structures

$$\langle A^{nat}, A^s, 0, 1, +, empty^{\mathcal{A}}, insert^{\mathcal{A}}, count^{\mathcal{A}} \rangle$$

and

$$\langle B^{nat}, B^s, 0, 1, +, empty^{\mathcal{B}}, insert^{\mathcal{B}}, count^{\mathcal{B}} \rangle$$

say a homomorphism preserving 0, 1 and +. This is partly correct, since a first-order homomorphism lifts to a logical relation, and therefore the Basic Lemma guarantees equivalent results for any program. However, there are two reasons why homomorphisms are not an exact characterization. The first has to do with the fact that elements of A^s and B^s which are not definable by terms are irrelevant. For example, suppose $\mathcal{A}$ is derived from $\mathcal{B}$ by adding some "nonstandard" multiset a with the property

$$insert\ x\ a = a \quad \text{for all } x.$$

Then $\mathcal{A}$ and $\mathcal{B}$ will be observationally equivalent. (Intuitively, this is because the multiset a is not definable in any program.) However, there is no homomorphism h since there is no reasonable choice for $h(a)$.

Another shortcoming of homomorphisms may be explained by describing straightforward computer implementations of multisets. One way of representing a multiset is as a linked list of pairs of the form $\langle element, count\rangle$ where the natural number *count* is the number of times *element* has been inserted. The *empty* multiset is then the empty linked list, *insert* adds a new pair or increments the appropriate count, and the function *count* searches the list and returns the appropriate count. An alternative representation makes sense if we assume that small numbers $1, \ldots, 10$ will occur quite frequently, with other numbers much less likely. In this case, we might use an array of length 10 to count the number of times $1, \ldots, 10$ are inserted, together with a simple list of other insertions in the order they occur. Note that repeatedly inserting 12, for example, will result in 12 appearing several times in the list. With *insert* and *count* implemented properly, these two representations will be observationally equivalent. However, there can be no homomorphism from the first implementation to the second. To see why this is so, consider the result of inserting three elements into the empty multiset. To simplify notation, we will use the abbreviation

$$insert^3\, x\, y\, z \quad \text{for} \quad insert\, x\, (insert\, y\, (insert\, z\, empty)).$$

In the first representation, assuming $x, y > 10$, we have

$$insert^3\, x\, y\, x = insert^3\, x\, x\, y = \langle\langle x, 2\rangle, \langle y, 1\rangle\rangle,$$

whereas in the second representation.

$$insert^3\, x\, y\, x = Array; \qquad \langle x, y, x\rangle \neq Array; \qquad \langle x, x, y\rangle = insert^3\, x\, x\, y.$$

Any homomorphism h from the first representation to the second would have to map the list of pairs $\langle\langle x, 2\rangle, \langle y, 1\rangle\rangle$ to two values, but this is impossible. Conversely, $insert^3\, 1\, 2\, 3 = insert^3\, 3\, 2\, 1$ in the second representation, since both set $Array[1] = Array[2] = Array[3] = 1$. However, in the first representation using lists of $\langle element, count\rangle$ pairs, $insert^3\, 1\, 2\, 3$ and $insert^3\, 3\, 2\, 1$ yield lists in different orders. So there is no homomorphism in either direction.

As the discussion illustrates, the correct chracterization must be a combination of homomorphism and quotient of a subset, which is exactly what a logical relation provides.

3.5.1. PROPOSITION. *Implementations $\mathcal{A}$ and $\mathcal{B}$ for signature Σ with natural number multisets are observationally equivalent with respect to the natural numbers iff there is a logical relation $\mathcal{R} \subseteq \mathcal{A} \times \mathcal{B}$ such that R^{nat} is the identity relation on the natural numbers.*

The generalization of this proposition to structures other than multisets is essentially straightforward.

3.6. *Other kinds of logical relations*

3.6.1. *Introduction*

We have seen that logical relations are useful for proving a variety of theorems about typed lambda calculus and for carrying out certain model-theoretic constructions.

However, there are several reasons for considering variations on the basic definition. Several speculative ideas are outlined in this section. One aesthetic reason to generalize the definition is that since logical relations do not compose to form logical relations, we might wonder whether a more general notion could be closed under composition, without losing essential properties such as the Basic Lemma. Another reason to generalize logical relations is that for Cartesian closed categories that are not Henkin models, we cannot make sense of the definition of logical relation. It is not clear what to quantify over in deciding whether two arrows are logically related. A final reason to consider variations on the general theory is that for languages with function constants, we need methods for constructing logical relations that respect the constants. For example, typed lambda calculus with fixed-point constants is often interpreted over partially ordered *domains*, with fix_σ denoting the least-fixed-point operator. Since the fixed-point constants are interpreted systematically, it is possible to develop a specialized theory of logical relations over domains with the property that every specialized logical relation respects fixed-point operators. We will discuss generalizations and specializations of logical relations in the next four subsections.

3.6.2. Algebraic relations

It is relatively straightforward to generalize homomorphisms to relations. This was suggested to the author by Gordon Plotkin and Samson Abramsky independently. Specifically, if $\mathscr{A}$ and $\mathscr{B}$ are multisorted algebras with the same signature, we may define an *algebraic relation* over $\mathscr{A}$ and $\mathscr{B}$ to be a family of relations, one for each sort, satisfying a pair of properties similar to homomorphisms. The first is that if c is a constant symbol, then $c^{\mathscr{A}}$ and $c^{\mathscr{B}}$ must be related. The second condition is that, for every function symbol f, the functions $f^{\mathscr{A}}$ and $f^{\mathscr{B}}$ must map related arguments to related results: if $\mathscr{R}(a_i, b_i)$, then $\mathscr{R}(f^{\mathscr{A}}(a_1, \ldots, a_k), f^{\mathscr{B}}(b_1, \ldots, b_k))$. Algebraic relations seem quite natural, and have been used to study behavioral equivalence of algebraic data types in [120]. It is easy to prove a version of the Basic Lemma, by the usual argument for homomorphisms. In addition, the composition of two algebraic relations is always an algebraic relation. We may compare algebraic relations to logical relations by considering Henkin models as multisorted algebras.

Recall that an applicative structure is simply a multisorted algebra for a signature with one sort for each type and an application function for each pair of types $\sigma \rightarrow \tau$ and σ. By the combinatory model characterization discussed in Section 2.4.8, a Henkin model may be regarded as an algebra for such a signature, with combinator constants added. Since combinators are lambda-definable, it is easy to see that every logical relation is an algebraic relation for this signature. We may also show that algebraic relations respect that meaning of lambda terms, as follows. If M is any lambda term, then, as mentioned in Section 2.4.8, we may translate M into an algebraic term with combinator constants and application functions. Since any algebraic relation $\mathscr{R}$ must respect algebraic terms, $\mathscr{R}$ must respect the meaning of M. Thus algebraic relations generalize logical relations, but not so much that the Basic Lemma fails.

Since algebraic relations are more general than logical relations, yet satisfy the Basic Lemma, it would seem reasonable to consider algebraic relations more basic. However, since logical relations are easily constructed by induction on types, logical relations seem to be the important special case for proving basic properties of typed lambda calculus.

3.6.3. Kripke logical relations

Before considering a generalization of logical relations to arbitrary Cartesian closed categories, we will consider logical relations for Kripke models. While Kripke structures are not ordinary applicative structures, it is relatively easy to generalize logical relations to Kripke structures. The reason is that Kripke applicative structures are not only models of typed lambda calculus, but also provide a natural interpretation of quantifiers (the one familiar from Kripke models of first-order logic).

The main ideas for Kriple logical relations over ordinary Henkin models were first outlined in [105], and then adapted to Kripke lambda models in [94]. Instead of having a relation for each type, we have a relation for each type and possible world, with the main condition quantifying over all possible future worlds. Specifically, a *Kripke logical relation* over Kripke applicative structures $\mathscr{A}$ and $\mathscr{B}$ sharing the same structure $\langle \mathscr{W}, \leqslant \rangle$ of possible worlds is a family $\mathscr{R} = \{R^\sigma_w\}$ of relations $R^\sigma_w \subseteq A^\sigma_w \times B^\sigma_w$ indexed by types σ and worlds $w \in \mathscr{W}$ satisfying the following three conditions. The first is a "monotonicity" condition:

$$R^\sigma_w(a, b) \quad \text{implies} \quad R^\sigma_{w'}(i^\sigma_{w,w'}\, a, i^\sigma_{w,w'}\, b) \quad \text{for all } w' \geqslant w,$$

which says that when $w \leqslant w'$, the relation R^σ_w is contained in $R^\sigma_{w'}$, modulo the transition functions. The second condition

$$R^{\sigma\to\tau}_w(f, g) \quad \text{iff} \quad \forall w' \geqslant w. \forall a, b.\, R^\sigma_{w'}(a, b) \quad \text{implies} \quad R^\tau_{w'}((i^{\sigma\to\tau}_{w,w'} f)a, (i^{\sigma\to\tau}_{w,w'} g)b),$$

says that the relation $R^{\sigma\to\tau}_w$ contains all functions mapping related arguments to related results. The third condition is that constants of the language must be related at every possible world. This completes the definition. The Basic Lemma for Kripke logical relations is proved by a straightforward argument given in [105]. Additional properties are investigated in [94].

3.6.4. Logical relations for arbitrary ccc's

There are two reasonable generalizations that make sense for arbitrary Cartesian closed categories, one resembling algebraic relations and the other akin to Kripke logical relations. The former seems more natural, but the latter may also have applications.

As suggested by Peter Freyd, relations between Cartesian closed categories arise from the standard categorical notion of relation, taken in the category of ccc's. More precisely, the category $\mathscr{CCC}$ of ccc's has Cartesian closed categories as objects, and functors that preserve the Cartesian closed structure as arrows. A predicate (subobject) in a category is an equivalence class of monomorphisms, which are arrows satisfying a cancellation law [73, 76]. A relation in a category (with products) is simply a predicate on the product of two objects. Putting these pieces together, we may define a *ccc-relation* to be a relation in the category $\mathscr{CCC}$, or a subobject of the product of two ccc's. For the subcategory of Henkin models (well-pointed ccc's), ccc-relations turn out be algebraic relations. And for arbitrary ccc's, a version of the

Basic Lemma may be proved. Thus algebraic relations generalize to arbitrary ccc's using standard categorical concepts.

An alternative is to interpret the usual definition of logical relation in a nonstandard logic. For example, the definition of Kripke logical relation described here may be obtained by interpreting the standard definition in the predicate logic of Kripke structures. While this approach requires more than Cartesian closedness, since the definition of logical relation involves quantifiers, it may be applied to arbitrary ccc's using the Yoneda embedding [73, 68, 121]. At present, it is not clear how useful this approach might be.

3.6.5. Logical relations for domains

The final topic in our brief survey of variations on logical relations will be relations respecting least-fixed-point operators. Since the interpretation of *fix* is based on the order structure of a domain, it is not too surprising that relations with certain order properties respect least fixed points. The main ideas may be developed directly, or derived by adapting ccc-relations to the category of domain models of typed lambda calculus.

Returning to the standard framework of logical relations, we would like to find a class of logical relations over domains which respect terms with least-fixed-point operators. By the Basic Lemma, it suffices to find logical relations which relate fixed-point operators fix_σ for each σ. Since we often construct logical relations by choosing relations at base type, and then extending to all types inductively, we would like a condition which may be applied easily at base type.

Since least fixed points are least upper bounds of directed sets, the natural restriction of logical relation involves least upper bounds. Following [84, 109], we say a logical predicate $\mathscr{R} \subseteq \mathscr{A}$ over an applicative structure of domains is *directed complete* if each R^σ is closed under limits of directed sets, i.e.,

$$\text{if } S \subseteq R^\sigma \text{ is directed,} \quad \text{then } \bigvee S \in R^\sigma.$$

This extends to binary logical relations simply by considering $\mathscr{R} \subseteq \mathscr{A} \times \mathscr{B}$ as a unary logical predicate over the product structure $\mathscr{A} \times \mathscr{B}$. Since the domain $A^\sigma \times B^\sigma$ is ordered coordinatewise, $S \subseteq A^\sigma \times B^\sigma$ is directed if, for any $\langle a_1, b_1 \rangle, \langle a_2, b_2 \rangle \in S$, there exists some $\langle a, b \rangle \in S$ with $a_1, a_2 \leqslant a$ and $b_1, b_2 \leqslant b$. The least upper bound of such a set is a pair $\langle a, b \rangle = \bigvee S$ also determined by the coordinatewise ordering.

An easy argument using $fix\, f = \bigvee \{f^n(\bot)\}$ shows that every directed complete logical relation preserves least-fixed-point operators.

3.6.1. LEMMA. *Let $\mathscr{R} \subseteq \mathscr{A} \times \mathscr{B}$ be a directed complete logical relation over cpo models. Then for every type σ, we have $R^{(\sigma \to \sigma) \to \sigma}(fix_\sigma^{\mathscr{A}}, fix_\sigma^{\mathscr{B}})$, where $fix_\sigma^{\mathscr{B}}$ is the least-fixed-point operator on $A^{\sigma \to \sigma}$, and similarly for $fix_\sigma^{\mathscr{B}}$.*

As a consequence, directed complete logical relations relate the meanings of terms with *fix*. The situation is summarized by the following proposition.

3.6.2. PROPOSITION. *Let $\mathscr{A}$ and $\mathscr{B}$ be Henkin models with each A^σ and B^σ a domain.*

Suppose $\mathcal{R} \subseteq \mathcal{A} \times \mathcal{B}$ *is a directed complete logical relation for a signature* Σ *without least-fixed-point constants, and let* Σ' *be* Σ *with* fix_σ *added. Let* $\mathcal{A}'$ *and* $\mathcal{B}'$ *be the models obtained by interpreting fixed-point constants as least-fixed-point operators in* $\mathcal{A}$ *and* $\mathcal{B}$. *Then* $\mathcal{R} \subseteq \mathcal{A}' \times \mathcal{B}'$ *is a logical relation for the extended applicative structures interpreting* Σ'. *Consequently, by the Basic Lemma,* $\mathcal{R}$ *preserves the meaning of any term of typed lambda calculus with fixed point constants.*

A method for constructing directed complete logical relations is described in [109]. A similar class of relations may be obtained by considering the category of Cartesian closed categories whose objects are domains. The morphisms of this category are functors preserving Cartesian closed structure that are continuous on arrows. The relations in this category are directed complete algebraic relations, which preserve the meaning of any typed lambda term with fixed-point operators.

4. Introduction to polymorphism

4.1. *Types as function arguments*

In the typed lambda calculus $\lambda^{\rightarrow}$, and in programming languages with similar typing constraints, types serve a number of purposes. For example, the type soundness theorem (Lemma 2.4.5) shows that every expression of type $\sigma \rightarrow \tau$ determines a function from type σ to τ. Since every term has meaning in every Henkin model, it follows that no typed term contains a function of type $nat \rightarrow nat$, say, applied to an argument of type *bool*. This may be regarded as a precise way of saying that $\lambda^{\rightarrow}$ terms do not contain type errors. The language $\lambda^{\rightarrow}$ also enjoys a degree of representation independence, as discussed in Section 3.5. However, the typing constraints of $\lambda^{\rightarrow}$ have some obvious drawbacks: there are many useful and computationally meaningful programs that are not well-typed. In this section, we will consider ways of making $\lambda^{\rightarrow}$ more flexible without giving up the advantages of typing.

Some of the inconvenience of $\lambda^{\rightarrow}$ may be illustrated by considering sorting functions in a language like Pascal. The reader familiar with a variety of sorting algorithms will know that most may be explained without referring to the kind of data at hand. We typically assume that we are given an array of pointers to records, and that each record has an associated *key*. It does not really matter what type of value the key is, as long as we have a "comparison" procedure which finds the minimum of any pair of keys. Most sorting algorithms simply compare keys using the comparison procedure and swap records by resetting pointers accordingly. Since this form of sorting algorithm does not depend on the type of data, we should be able to write a procedure *Sort* which sorts any type of records, given a comparison procedure as a parameter. However, in a language such as Pascal, every *Sort* procedure must have a type of the form

$$Sort: (t \times t \rightarrow bool) \times Array[ptr\ t] \rightarrow Array[ptr\ t]$$

for some fixed type t. (This is the type of functions which, given a binary relation on t and an array of pointers to t, return an array of pointers to t.) This forces us to write a separate procedure for each type of data, even though there is nothing inherent in the sorting algorithm to require this. From a programmer's point of view, there are several obvious disadvantages of this inflexibility. The first is that if we wish to build a program library, we must anticipate in advance the types of data we wish to sort. This is unreasonably restrictive. Second, even if we are only sorting a few types of data, it is a bothersome chore to duplicate a procedure declaration several times. Not only does this consume programming time, but having several copies of a procedure may increase debugging and maintenance time. It is far more convenient to use a language which allows one procedure to be applied to many types of data.

One family of solutions to this problem may illustrated using function composition, which is slightly simpler than sorting. The $\lambda^{\rightarrow}$ function

$$compose_{nat,nat,nat} ::= \mathrm{G}\lambda f\colon nat\rightarrow nat\,.\,\lambda g\colon nat\rightarrow nat\,.\,\lambda x\colon nat\,.\,f(gx)$$

composes two natural number functions. Higher-order functions of type $nat\rightarrow(nat\rightarrow nat)$ and $(nat\rightarrow nat)\rightarrow nat$ may be composed using the function

$$compose_{nat,nat\rightarrow nat,nat} ::= \mathrm{G}\lambda f\colon (nat\rightarrow nat)\rightarrow nat\,.\,\lambda g\colon nat\rightarrow(nat\rightarrow nat)\,.\,\lambda x\colon nat\,.\,f(g\,x)$$

It is easy to see that these two composition functions differ only in the types given to the formal parameters. Since these functions compute their results in precisely the same way, we might expect both to be "instances" of some general composition function.

A *polymorphic function* is a function that may be applied to many types of arguments[4]. We may extent $\lambda^{\rightarrow}$ to include polymorphic functions by extending lambda abstraction to type variables. Using type variables r, s and t, we may write a "generic" composition function

$$compose_{r,s,t,} ::= \mathrm{G}\lambda f\colon s\rightarrow t\,.\,\lambda g\colon r\rightarrow s\,.\,\lambda x\colon r\,.\,f(gx).$$

It should be clear that $compose_{r,s,t}$ denotes a composition function of type $(s\rightarrow t)\rightarrow(r\rightarrow s)\rightarrow(r\rightarrow t)$, independent of the semantic values assigned to variables r, s and t. In particular, when all three type variables are interpreted as the type of natural numbers, the meaning of $compose_{r,s,t}$ is the same as $compose_{nat,nat,nat}$. By lambda abstracting r, s and t, we may write a function that produces $compose_{nat,nat,nat}$ when applied to type argruments *nat*, *nat* and *nat*. To distinguish type variables from ordinary variables, we will use the symbol T for some collection of types. As we shall see shortly, there are several possible interpretations for T. Since these are more easily considered after presenting abstraction and application for type variables, we will try to think of T as some collection of types that at least includes all the $\lambda^{\rightarrow}$ types, and perhaps other types not yet specified. There is no need to think of T itself as a "type", in the sense of "element of T". In other words, we do not need T to include itself as a member.

Using lambda abstraction over types, which we will call *type abstraction*, we may define a polymorphic composition function

$$compose ::= \lambda r\colon T\,.\,\lambda s\colon T\,.\,\lambda t\colon T\,.\,compose_{r,s,t}.$$

[4]This is not intended to be a precise definition of polymorphism. Instead of defining polymorphism in general, we will give precise definitions of several polymorphic lambda calculi.

This function requires three type parameters before it may be applied to an ordinary function. We will refer to the application of a polymorphic function to a type parameter as *type application*. Type applications may be simplified using a version of β-reduction, with the type argument substituted in place of the bound type variable. If we want to apply *compose* to two numeric functions $f, g: nat \rightarrow nat$, then we first apply *compose* to type arguments *nat*, *nat* and *nat*. Using β-reduction for type application, this gives us

$$\begin{aligned} compose\ nat\ nat\ nat &= (\lambda r\colon T.\lambda s\colon T.\lambda t\colon T.\lambda f\colon s \rightarrow t.\lambda g\colon r \rightarrow s.\lambda x\colon r.f(gx))\ nat\ nat\ nat \\ &= \lambda f\colon nat \rightarrow nat.\lambda g\colon nat \rightarrow nat.\lambda x\colon nat.f(gx). \end{aligned}$$

In other words, $compose\ nat\ nat\ nat = compose_{nat,nat,nat}$. By applying *compose* to other arguments, we may obtain composition functions of other types.

In a typed function calculus with type abstraction and type application, we must give types to polymorphic functions. A polymorphic function that illustrates most of the typing issues is the *polymorphic identity*

$$Id ::= \lambda t\colon T.\lambda x\colon t.x.$$

It should be clear from the syntax that the domain of *Id* is *T*. The range is somewhat more difficult to describe. Two common notations for the type of *Id* are $\Pi t\colon T.t \rightarrow t$ and $\forall t\colon T.t \rightarrow t$. In these expressions, Π and $\forall$ bind t. The $\forall$ notation allows informal readings such as "the type of functions which, *for all t*, give us a map from t to t". The Π notation is closer to standard mathematical usage, since a product $\Pi_{i \in I} A_i$ over an infinite index set I consists of all functions $f\colon I \rightarrow \bigcup_{i \in I} A_i$ such that $f(i) \in A_i$. Since $Id\, t\colon t \rightarrow t$ for every $t \in T$, the polymorphic identity has the correct functional behavior to belong to the infinite product $\Pi_{t:T} t \rightarrow t$. We may compute the type of a type application $Id\ \tau$ by substituting τ for the bound variable t in the type expression $(t \rightarrow t)$. We will use Π notation in the first polymorphic lambda calculus we consider and $\forall$ in a later system, primarily as way of distinguishing the two systems.

Having introduced types for polymorphic functions, we must decide how these types fit into the language. At the point we introduced T, we had only defned nonpolymorphic types. With the addition of polymorphic types, there are three natural choices for T. We may choose to let T contain only the $\lambda^{\rightarrow}$ types over some collection of base types. This leads us to a semantically straightforward calculus of what is called *predicative polymorphism*. The word "predicative" refers to the fact that T is introduced only after we have defined all of the members of T (see [31, 33, 34]). In other words, the collection T does not contain any types that rely on T for their definition. An influential predicative calculus is Martin-Löf's constructive type theory [77, 78, 79], which provides the basis for the Nuprl proof-development system [19]. A second form of polymorphism is obtained by letting T also contain all the polymorphic types (such as $\Pi t\colon T.t \rightarrow t$), but not consider T itself a type. This T is "impredicative", since we assume that T is closed under type definitions that refer to the entire collection T. The resulting language, formulated independently by Girard [42, 43] and Reynolds [110], is variously called the *second-order lambda calculus*, *system F*, or the calculus of *impredicative polymorphism*. (It is also referred to as the *polymorphic lambda calculus*, but we will consider this phrase ambiguous.) The final alternative is to consider T a type, and let T contain all types, including itself. This may seem strange, since assumptions like "there is a set of all sets" are well-known to lead to foundational problems. However, from a computational point of view, it is not immediately clear that there is anything wrong with introducing a "type of all types". Discussions of this language appear in [14, 24, 83].

Some simple differences between these three kinds of polymorphism may be illustrated by considering uses of *Id*, since the domain of *Id* is T. If we limit T to the collection of all $\lambda^{\rightarrow}$ types, we may only apply *Id* to a nonpolymorphic type such as *nat* or $(nat \to nat) \to (nat \to bool)$. If we are only interested in manipulating functions with $\lambda^{\rightarrow}$ types, then this seems sufficient. A technical advantage of this restricted language is that any model of $\lambda^{\rightarrow}$ may be extended to a model of predicative polymorphism. In particular, there are classical set-theoretic models of predicative polymorphism.

If we let T be the collection of all types, then the polymorphic identity may be applied to any type, including its own. By β-reduction on types, the result of this application is an identity function

$$Id\,(\Pi t\colon T.t \to t) = \lambda x\colon (\Pi t\colon T.t \to t).x$$

which we may apply to *Id* itself. In this calculus, it is impossible to interpret every polymorphic lambda term as a set-theoretic function and every type as the set of all functions it describes. The reason is that the interpretation of *Id* would have to be a function whose domain contains a set (the meaning of $\Pi t\colon T.t \to t$) which contains *Id*. The reader may find it amusing to draw a Venn diagram of this situation. If we recall that a function is a set, then we begin with a circle for the set *Id* of ordered pairs $Id = \{\langle \tau, \lambda x\colon \tau.x\rangle \mid \tau \in T\}$. But one of the elements of *Id* is the pair $\langle (\Pi t\colon T.t \to t), \lambda x\colon (\Pi t\colon T.t \to t).x\rangle$ allowing the function to be applied to its own type. One of the components of this ordered pair is a set which contains *Id*, so clearly we do not have a sensible picture. Similar reasoning may be used to show that a naive interpretation of the impredicative second-order calculus conflicts with basic axioms of set theory. The semantics of impredicative polymorphism has been a popular research topic in the last few years, and several appealing models have been found [11, 13, 25, 44, 59, 90].

In the final form of polymorphism, we let T be a type which contains all types, including itself. This design is often referred to as *type*: *type*. One effect of *type*: *type* is to allow polymorphic functions such as *Id* to act as functions from types to types, as well as functions from types to elements of types. For example, the application $Id\,T$ is considered meaningful, since $T\colon T$. By β-reduction on types, we have $Id\,T = \lambda x\colon T.x$ and so $(Id\,T)\colon T \to T$ is a function from types to types. We may also write more complicated functions from T to T and use these in type expressions. While languages based on the other two choices for T are strongly normalizing (without fixed-point operators) and have efficiently decidable typing rules, strong normalization fails for most languages with *type*:*type*. In fact, it may be shown [83] that the set of provable equations in a minimally powerful language with *type*: *type* is undecidable (see also [77, 24]). Since arbitrarily complex expressions may be incorporated into type expressions, the set of well-typed terms becomes undecidable. This presents an obstacle to practical implementation.

4.2 *Predicative polymorphic calculus*

4.2.1 *Predicative type abstraction*

In this section, we define a simple predicative polymorphic lambda calculus $\lambda^{\rightarrow,\Pi}$. As it stands, pure $\lambda^{\rightarrow,\Pi}$ is not a very significant extension of $\lambda^{\rightarrow}$ since the only operations we may perform on an expression of polymorphic type are type abstraction and application. Since we do not have lambda abstraction over polymorphic variables, we cannot pass polymorphic functions as arguments or model polymorphic function declarations. However, $\lambda^{\rightarrow,\Pi}$ illustrates some of the essential features common to all languages with type abstraction. In addition, $\lambda^{\rightarrow,\Pi}$ appears as a fragment or restriction of several useful languages. By adding a declaration form for polymorphic functions, we obtain a language resembling a core fragment of the programming language ML [47, 86]. When distinctions imposed by predicativity are eliminated, we obtain the Girard–Reynolds calculus of impredicative polymorphism.

The types of $\lambda^{\rightarrow,\Pi}$ fall into two classes, corresponding to the $\lambda^{\rightarrow}$ types and the polymorphic types constructed using Π. Following the terminology of Martin-Löf's type theory [77], we will call these classes *universes*. To simplify the presentation of $\lambda^{\rightarrow,\Pi}$, we will use two sorts of variables, ordinary variables $x, y, z, \ldots$ and type variables $r, s, t, \ldots$. The first universe is essentially the predicative T of the last section, while the second universe contains all the polymorphic types. To introduce some terminology and notation, we will say that τ is a *type of the first universe*, and write $\tau\colon U_1$, if τ is built up from base types and type variables using the function-space constructor $\rightarrow$. All of the type variables of $\lambda^{\rightarrow,\Pi}$ will range over U_1. For this reason, we will not have to specify $t\colon U_1$ or $t\colon U_2$ in any typing context. To be precise, we define the U_1 types by the grammar

$$\tau ::= t \mid b \mid \tau \rightarrow \tau$$

where t may be any type variable and b may be any base type (from some signature). The polymorphic type expressions of $\lambda^{\rightarrow,\Pi}$ are defined by forming products over U_1, so it is natural to regard these types as being of a "higher" second universe. We will say that σ is a *type of the second universe*, and write $\sigma\colon U_2$, if σ has the form $\Pi t_1, \ldots \Pi t_n . \tau$ for type variables $t_1, \ldots, t_n$ and $\tau\colon U_1$. Note that, by design, U_2 is not closed under $\rightarrow$. This places severe restrictions on the language, as we shall see in Section 4.2.2.

The typing rules for $\lambda^{\rightarrow,\Pi}$ will be simplified using several metalinguistic conventions. We will use $\tau, \tau_1, \ldots$ as metavariables for U_1 types, $\sigma, \sigma_1, \ldots$ for U_2 types, and $\rho, \rho_1, \ldots$ for types belonging to U_1 or U_2. All type variables $r, s, t, \ldots$ range implicitly over U_1, so we will not declare the universes of type variables explicitly in type assignments. The type assignments in our presentation of $\lambda^{\rightarrow,\Pi}$ will only contain assertions of the form $s\colon \tau$ assigning U_1 types to ordinary variables. While we have expressions for U_2 types, we do not have variables ranging over U_2. The reason is that since we do not have binding operators for U_2 variables, it seems impossible to use U_2 variables for any significant purpose. If we were to add variables and lambda abstraction over U_2, this would lead us to introduce a third universe U_3. Continuing in this way, we might also add universes U_4, U_5, U_6 and so on. The reader may consult [19, 77, 78, 79] for type systems of this form.

The terms of $\lambda^{\to,\Pi}$ may be summarized by defining the set of "preterms", or expressions which look like terms but are not guaranteed to satisfy all the typing constraints. The unchecked *preterms* of $\lambda^{\to,\Pi}$ are given by the grammar

$$M ::= x \mid \lambda x\colon\tau.M \mid MM \mid \lambda t.M \mid M\tau.$$

In other words, every term of $\lambda^{\to,\Pi}$ must have one of these forms, but not all expressions generated by this grammar are well-typed $\lambda^{\to,\Pi}$ terms. The precise typing rules include axioms (var) and (cst) for variables and constants, rule (add hyp) to add hypotheses about the types of free variables, and rules ($\to$Intro) and ($\to$Elim), restricted to U_1 types. Consequently, any $\lambda^{\to}$ term, perhaps containing type variables in place of base types, is a term of $\lambda^{\to,\Pi}$. In addition, the terms of $\lambda^{\to,\Pi}$ include type abstraction and application formed according to the following rules.

$$(\Pi\ \text{Intro})\qquad \frac{\Gamma \rhd M\colon\rho}{\Gamma \rhd \lambda t.M\colon \Pi t.\rho}\quad (t \text{ not free in } \Gamma),$$

$$(\Pi\ \text{Elim})\qquad \frac{\Gamma \rhd M\colon \Pi t.\rho}{\Gamma \rhd M\tau\colon [\tau/t]\rho}.$$

The restriction "t not free in Γ" in (Π Intro) prevents meaningless terms such as $\{x\colon t\} \rhd \lambda t.x\colon \Pi t.t$, in which it is not clear whether t is free or bound. We say *M is a term of* $\lambda^{\to,\Pi}$ *with type* ρ *in context* Γ if $\Gamma \rhd M\colon\rho$ is derivable from the axioms and $\lambda^{\to,\Pi}$ typing rules.

The equational inference system for $\lambda^{\to,\Pi}$ is an extension of the $\lambda^{\to}$ proof system, with additional axioms for type abstraction and application:

$$(\alpha_\Pi)\qquad \Gamma \rhd \lambda t.M = \lambda s.[s/t]M\colon \Pi t.\rho,$$

$$(\beta_\Pi)\qquad \Gamma \rhd (\lambda t.M)\tau = [\tau/t]M\colon [\tau/t]\rho,$$

$$(\eta_\Pi)\qquad \Gamma \rhd \lambda t.Mt = M\colon \Pi t.\rho \quad t \text{ not free in } M$$

and inference rules making equality a congruence with respect to the additional term formation rules:

$$\frac{\Gamma \rhd M = N\colon\rho}{\Gamma \rhd \lambda t.M = \lambda t.N\colon \Pi t.\rho},$$

$$\frac{\Gamma \rhd M = N\colon \Pi t.\rho}{\Gamma \rhd M\tau = N\tau\colon [\tau/t]\rho}.$$

We consider type expressions that only differ in the names of bound variables to be equivalent.

Most of the basic theorems about $\lambda^{\to}$ generalize to $\lambda^{\to,\Pi}$ for example, if we define reduction by orienting the equational axioms, we may prove confluence and strong normalization for polymorphic $\lambda^{\to,\Pi}$ terms. Extending applicative structures and Henkin models to $\lambda^{\to,\Pi}$ gives us type soundness and the same forms of equational soundness and completeness theorems. Categorical connections may be established with *hyperdoctrines*, *relatively Cartesain closed categories* [134], or *locally Cartesain closed* categories [122], although none of these connections is as straightforward as the categorical equivalence between $\lambda^{\rightarrow}$ and ccc's described in [68].

4.2.2. ML-style declarations

The type system of the programming language ML may be viewed as an extension of $\lambda^{\rightarrow,\Pi}$, as discussed in [75, 92]. The primary typing difference between $\lambda^{\rightarrow,\Pi}$ and the core expression language of ML is that ML includes a polymorphic let declaration. The form of this declaration may be motivated by considering the use of polymorphic functions in $\lambda^{\rightarrow,\Pi}$.

In $\lambda^{\rightarrow,\Pi}$, we may write a polymorphic identity function

$$Id ::= \lambda t.\lambda x{:}\, t.x$$

of type $\Pi t.t \rightarrow t$. Using type application, we may apply Id to any argument belonging to a U_1 type. For example, the expressions $Id\ nat\ 3$ and $Id\ bool\ true$ are well-typed in any signature with $3{:}\ nat$ and $true{:}\ bool$. However, we cannot lambda-bind variables of type $\Pi t.t \rightarrow t$, since the ($\rightarrow$Intro) rule is restricted to variables with U_1 types. As a consequence, we cannot write terms of the form

$$(\lambda f{:}(\Pi t.t \rightarrow t) \ldots f\ nat\ 3 \ldots f\ bool\ true)\lambda t.\lambda x{:}\, t.x$$

in which the polymorphic identity occurs only once, but will be applied to values of different types when the term is reduced. Introducing lambda abstraction over variables with U_2 types would allow us to write the function above, but would also complicate the ML typechecker (see Section 4.7). In fact, since the implementation of ML uses type inference to simplify program syntax, it is not known whether allowing lambda-abstraction over U_2 types would be algorithmically tractable. However, it is possible to allow a limited form of variable-binding for U_2 types which has an acceptably efficient type inference algorithm (cf. [61]) and which allows us to declare polymorphic functions.

ML polymorphism may be illustrated using the language $\lambda^{\rightarrow,\Pi,\mathsf{let}}$, which extends $\lambda^{\rightarrow,\Pi}$ with polymorphic declarations. Intuitively, the value of expression

$$\mathsf{let}\ x{:}\,\rho = N\ \mathsf{in}\ M$$

is the value of M when x is given the value of N. The typing rule for this expression is based on the presentation of ML given in [28]. Since we will declare functions with polymorphic types, we must extend the definition of type assignment to allow assumptions of the form $x{:}\,\sigma$, where $\sigma{:}\, U_2$ is a type from the second universe. Using type assignments that may contain polymorphic variables, the let rule may be written as follows.

$$\text{(let)} \qquad \frac{\Gamma \rhd N{:}\,\rho,\ \ \Gamma, x{:}\,\rho \rhd M{:}\,\tau}{\Gamma \rhd (\mathsf{let}\ x{:}\,\rho = N\ \mathsf{in}\ M){:}\,\tau}.$$

This rule follows [28] in restricting the type of a let expression to U_1. As noted in [124, Exercise 12.8], there is little significance to this restriction.

The intuitive description of the value of a let expression (given above) is reflected in the equational axiom

$$\Gamma \rhd (\mathsf{let}\ x{:}\,\rho = N\ \mathsf{in}\ M) = [N/x]M : \tau.$$

We may derive a reduction rule from this axiom by reading the equation from left to right.

One distinction between $\lambda^{\rightarrow,\Pi}$ and $\lambda^{\rightarrow,\Pi,\mathsf{let}}$ may be illustrated using any $\lambda^{\rightarrow,\Pi}$ term in which a polymorphic subterm occurs several times. For concreteness, let us consider a specific polymorphic function, say $compose{:}\Pi r.\Pi s.\Pi t.\tau$, where $\tau = (s \rightarrow t) \rightarrow (r \rightarrow s) \rightarrow r \rightarrow t$. If we use *compose* several times in one expression, we might prefer to follow standard programming practice and write the function body only once. Given any expression $[compose/x]M$ of $\lambda^{\rightarrow,\Pi}$, we may use let to write a $\lambda^{\rightarrow,\Pi,\mathsf{let}}$ expression

$$\mathsf{let}\ x{:}\,(\Pi r.\Pi s.\Pi t.\tau) = compose\ \mathsf{in}\ M$$

with only one occurrence of *compose*. The equational axiom for let specifies that

$$\Gamma \rhd (\mathsf{let}\ x{:}\,(\Pi r.\Pi s.\Pi t.\tau) = compose\ \mathsf{in}\ M) = [compose/x]M : \rho$$

so the let expression may be viewed as a more concise way of writing $[compose/x]M$. However, since $\Gamma, x{:}\,(\Pi r.\Pi s.\Pi t.\tau) \rhd M : \rho$ has a free polymorphic variable, we cannot even type this term in $\lambda^{\rightarrow,\Pi}$. Therefore, due to typing constraints, we cannot hope to "simulate" let in $\lambda^{\rightarrow,\Pi}$ in any general way. This shows that $\lambda^{\rightarrow,\Pi,\mathsf{let}}$ is indeed more flexible than $\lambda^{\rightarrow,\Pi}$. Of course, $\lambda^{\rightarrow,\Pi,\mathsf{let}}$ also has its limitations. For example, since we may write expressions of the form $\Gamma, x{:}\,(\Pi r.\Pi s.\Pi t.\tau) \rhd M : \rho$, we might wish to lambda-abstract the polymorphic variable $x{:}\,\Pi r.\Pi s.\Pi t.\tau$. However, since the resulting function has a type that does not belong to U_2, this is not possible in $\lambda^{\rightarrow,\Pi,\mathsf{let}}$.

For further discussion of $\lambda^{\rightarrow,\Pi,\mathsf{let}}$, see [92, 124], where the language is called *Core-XML* since it is an explicitly typed version of the core expression language of ML. An alternate explanation of let is given in Section 4.7.

4.3. *Impredicative polymorphism*

The Girard–Reynolds calculus of impredicative polymorphism was developed independently by Girard [42, 43] and Reynolds [110]. Girard used the caculus to prove the consistency of second-order Peano arithmetic while Reynolds proposed the calculus as a model of type structure in programming languages. In the terminology of this chapter, the Girard–Reynolds calculus, which we shall denote by $\lambda^{\rightarrow,\forall}$, may be obtained from the predicative $\lambda^{\rightarrow,\Pi}$ by dropping the distinction between universes U_1 and U_2. More precisely, the usual syntactic presentation of $\lambda^{\rightarrow,\forall}$ may be obtained from our definition of $\lambda^{\rightarrow,\Pi}$ by replacing U_1 and U_2 by a single collection T of all types. In particular, types are formed according to the grammar

$$\sigma\ ::=\ b \mid t \mid \sigma \rightarrow \sigma \mid \Pi t.\sigma$$

and terms are formed axioms (var) and (cst) for variables and constants, rule (add hyp) to add hypotheses about the types of free variables, and rules ($\rightarrow$Intro), ($\rightarrow$Elim), (Π Intro), (Π Elim) without restrictions related to type universes. To distinguish the impredicative system from the predicative one, we will use $\forall$ in place of Π, hence the name $\lambda^{\rightarrow,\forall}$. Thus the type of the impredicative polymorphic identity is written $\forall t.t \rightarrow t$. To keep the distinction between systems straight, the reader may think of Π as standing for "p" in "predicative". Another justification for the choice of notation is the formulas-as-types analogy discussed below.

The Girard–Reynolds calculus is probably the most widely studied of the three forms of polymorphism outlined in Section 4.1, primarily because of its syntactic simplicity, programming flexibility, and semantic complexity. The language is syntactically simpler than $\lambda^{\rightarrow,\Pi}$, since there are no universe restrictions, but it is substantially more difficult to prove strong normalization or to provide intuitive and mathematically rigorous semantics.

One important idea that has not been given its due in this chapter is the "formulas-as-types" principle explained in [57]. This principle, sometimes called the *Curry–Howard isomorphism*, is a general syntactic connection between typed lambda calculi and natural deduction proof systems for intuitionistic logics. In the analogy, $\lambda^{\rightarrow}$ corresponds to the propositional intuitionistic logic of implication. One aspect of the correspondence is that there is a straightforward bijection between typed lambda terms and natural deduction proofs in the logic. Moreover, if we read the function-space constructor $\rightarrow$ as implication, then the type of a $\lambda^{\rightarrow}$ term is precisely the formula proved by the corresponding natural deduction proof. What makes the correspondence useful in proof theory is that reduction steps in the calculus correspond precisely to normalization steps in the proof system (see [107]). Consequently, the normalization proof for $\lambda^{\rightarrow}$ implies that every natural deduction proof in the corresponding logic may be put in normal form. It follows that the logic is consistent. In the formulas-as-types correspondence, the calculus $\lambda^{\rightarrow,\forall}$ corresponds to propositional intuitionistic logic with implication and universal quantification (over propositions). This logic is undecidable [39, 72] and sufficient to express intuitionistic existential quantification and the other propositional connectives [107]. Moreover, the normalization theorem for $\lambda^{\rightarrow,\forall}$ implies normalization for second-order Peano arithmetic and therefore implies consistency. It follows from Gödel's incompleteness theorem that the normalization theorem for $\lambda^{\rightarrow,\forall}$, first proved by Girard [42, 43], cannot be formalized in second-order arithmetic. It is also shown in [43] that the numeric functions definable in pure $\lambda^{\rightarrow,\forall}$ are precisely the partial recursive functions that may be proved total in second-order arithmetic (see also [35, 127]). This representability theorem may be extended to higher-order functionals (see [37, 118]). A good general reference on these topics is [45].

From a programming point of view, $\lambda^{\rightarrow,\forall}$ provides a very flexible polymorphic type system. The polymorphic features of languages such as ADA, CLU and ML [137, 71, 47, 86] may all be regarded as restrictions of $\lambda^{\rightarrow,\forall}$ polymorphism. For example, we may mimic ML let using lambda abstraction in $\lambda^{\rightarrow,\forall}$, as follows. Suppose $\Gamma \rhd M: \forall t.\sigma$ is a polymorphic expression and $\Gamma,\ x: \forall t.\sigma \rhd N: \tau$ is a term which may have several occurrences of a polymorphic variable. We might consider extending $\lambda^{\rightarrow,\forall}$ with a let declaration form which would allow us to write

$$\Gamma \rhd \mathsf{let}\ x: \forall t.\sigma = M\ \mathsf{in}\ N: \tau.$$

However, this is not necessary, since the let expression may be considered an abbreviation for the term

$$\Gamma \rhd (\lambda x\colon \forall t.\sigma.N)M\colon\tau$$

with the same type, same meaning, and same immediate subterms. This illustrates some of the programming flexibility that results from dropping the universe restrictions of $\lambda^{\rightarrow,\Pi}$. However, as suggested in Section 4.1, the semantics of $\lambda^{\rightarrow,\forall}$ is rather complicated. Since the polymorphic identity may be applied to its own type, we cannot interpret types as sets and functions as elements of these sets. For a general discussion of the semantics of $\lambda^{\rightarrow,\forall}$, soundness and completeness for a form of Henkin models, and some model examples, see [11]. Some beautiful domain-theoretic models are developed in [44] and [25]. See also [37, 102, 80, 123] for a discussion of categorical approaches to semantics.

One particularly interesting semantic property of $\lambda^{\rightarrow,\forall}$ is that it is impossible to interpret $\rightarrow$ as full set-theoretic function space, regardless of how we interpret $\forall$. This was first proved by Reynolds [113]. The proof is clarified in [114] and generalized in [102, 103]. Another issue that has received considerable attention is "parametricity", a term coined by Strachey [132] and taken up in earnest by Reynolds [110, 112, 113]. Put in the simplest possible terms, the basic idea of parametricity is that the polymorphic functions that we can define by terms all operate "uniformly" over all types. Therefore, it makes sense to impose corresponding uniformity conditions on semantic models. Parametricity is put in a natural categorical framework in [4, 36].

4.4. Data abstraction and existential types

Abstract data type declarations are used in a number of contemporary programming languages [137, 71, 86]. In fact, data abstraction and associated notions of program specification and modularity are among the most influential programming language developments of the 1970s. While space considerations prohibit a full discussion of data abstraction and its applications, we will describe a general form of data type declaration that may be incorporated into any language with type variables, including the three versions of polymorphic typed lambda calculus mentioned in Section 4.1. For further information, the reader is referred to [71, 96, 98, 112, 124].

The declaration form

$$\textsf{abstype}\ t\ \textsf{with}\ x_1\colon\sigma_1,\ldots,x_k\colon\sigma_k\ \textsf{is}\ \langle\tau;\ M_1,\ldots,M_k\rangle\ \textsf{in}\ N$$

declares an abstract type t with "operations" $x_1,\ldots,x_k$ and implementation $\langle\tau;\ M_1,\ldots,M_k\rangle$. The scope of this declaration is N. For example, the expression

$$\begin{array}{ll}
\textsf{abstype} & \textit{stream}\ \textsf{with} \\
 & s\colon \textit{stream}, \\
 & \textit{first}\colon \textit{stream}\rightarrow\textit{nat}, \\
 & \textit{rest}\colon \textit{stream}\rightarrow\textit{stream} \\
\textsf{is} & \\
 & \langle\tau; M_1, M_2, M_3\rangle \\
\textsf{in} & \\
 & N
\end{array}$$

declares an abstract data type *stream* with distinguished element s: *stream* and functions *first* and *rest* for operating on streams. Within the scope N of the declaration, the stream s and operations *first* and *rest* may be used to compute natural numbers or other results. However, the type of N may not be *stream*, since this type is local to the expression. In computational terms, the elements of the abstract type *stream* are represented by values of the type τ given in the implementation. Operations s, *first* and *rest* are implemented by expressions M_1, M_2, M_3. Since the value of s must be a stream, the expression M_1 must have type τ, the type of values used to represent streams. Similarly, we must have $M_2 : \tau \rightarrow nat$ and $M_3 : \tau \rightarrow \tau$. Using Cartesian products, we may put any abstract data type declaration in the form abstype t with $x : \sigma$ is M in N. For example, the *stream* declaration may be put in this form by combining the three operations s, *first* and *rest* into a single operation of type $stream \times (stream \rightarrow nat) \times (stream \rightarrow stream)$. There is no loss in doing so, since we may recover s, *first* and *rest* using projection functions.

Abstract data type declarations may be added to either predicative or impredicative languages, using the general form described in [96]. While it is possible to formulate a typing rule and equational axiom for abstype as described above (as in the original formulation of ML [47]), some useful flexibility is gained by considering abstract data type declarations and data type implementations separately. An implementation for the abstract type *stream* mentioned above consists of a type τ, used to represent *stream*'s, together with expressions for the specified stream s and the stream operations *first* and *rest*. If we want to describe implementations of streams in general, we might say that in any implementation "there exists a type t with elements of types t, $t \rightarrow nat$, and $t \rightarrow t$". This description would give just enough information about an arbitrary implementation to determine that an abstype declaration makes sense, without giving any information about how streams are represented. This fits the general goals of data abstraction, as discussed in [98], for example.

We may add abstract data type implementations and abstype declarations to a language with type variables as follows. The first step is to extend the syntax of type expressions

$$\sigma ::= \ldots \mid \exists t . \sigma$$

to include *existential types* of the form $\exists t . \sigma$. In a predicative language, $\exists t_1 \ldots \exists t_k . \tau$ would belong to U_2, assuming $\tau : U_1$. Existential types are used to type implementations of abstract data types. Intuitively, each element of an existential type $\exists t . \sigma$ consists of a type τ and an element of $[\tau/t]\sigma$. Using products to combine s, *first*, and *rest*, an implementation of *stream* would have type $\exists t . [t \times (t \rightarrow nat) \times (t \rightarrow t)]$, for example. Note that if we read "$\times$" as "and", this type expression may be read; "there exists a type t with elements of type t, $t \rightarrow nat$, and $t \rightarrow t$", which matches the informal description given above. Existential types, which were part of Girard's system F[42, 43] (but not linked to abstract data types), may also be explained using the formulas-as-types analogy.

When we write abstract data type implementations apart from the declarations that use them, it is necessary to included some type information which might at first appear redundant. We will write an implementation in the form $\langle t = \tau, M : \sigma \rangle$, where $t = \tau$ binds t in the remainder of the expression. The reader may think of $\langle t = \tau, M : \sigma \rangle$ as a "pair" $\langle \tau, M \rangle$ in which access to the representation type τ has been restricted. The bound type variable t and the type expression σ serve to disambiguate the type of the expression. The type of a well-formed expression $\langle t=\tau, M:\sigma\rangle$ is $\exists t.\sigma$, according to the following rule:

$$(\exists\,\text{Intro}) \qquad \frac{\Gamma \rhd M:[\tau/t]\sigma}{\Gamma \rhd \langle t=\tau; M:\sigma\rangle : \exists t.\sigma}.$$

We may read this rule as saying that if $M:[\tau/t]\sigma$, the in the "pair" $\langle t=\tau, M:\sigma\rangle$, there exists a type t together with a value of type σ. Since a type σ is not determined by the form of a substitution instance $[\tau/t]\sigma$, the type of a simpler implementation form $\langle \tau, M\rangle$ would not be determined uniquely. The following rule for abstype declarations allows us to bind names to the type and value components of a data type implementation:

$$(\exists\,\text{Elim}) \qquad \frac{\Gamma, x:\tau \rhd N:\rho, \;\; \Gamma \rhd M:\exists t.\tau}{\Gamma \rhd (\mathsf{abstype}\ t\ \mathsf{with}\ x:\tau\ \mathsf{is}\ M\ \mathsf{in}\ N):\rho} \quad t \text{ not free in } \Gamma \text{ or } \rho.$$

Informally, this rule binds type variable t and ordinary variable x to the type and value part of the implementation M, with scope N. However, there is no syntactic restriction on the form of the implementation. In particular, the data type implementation may be let-bound, or a formal parameter to a function containing this declaration. The equational axiom for abstype is

$$\Gamma \rhd (\mathsf{abstype}\ t\ \mathsf{with}\ x:\sigma\ \mathsf{is}\ \langle t=\tau, M:\sigma\rangle\ \mathsf{in}\ N) = [M/x][\tau/t]N:\rho$$

which allows us to simplify an abstype expression by substituting the two parts of an implementation into the body of the declaration.

As argued in [96], the form of abstract data type declaration presented here is a natural generalization of the constructs used in many programming languages. In addition to capturing the "essence" of abstract data type declarations, this formulation allows data type implementations to be passed as function arguments, returned as function values, or manipulated in any other way provided by the language (see also [16]). The side condition "t not free in ρ" in ($\exists$ Elim) has been questioned in the literature [75]. For the languages discussed in this chapter, it is a necessary scoping condition. However, there are languages with more expressive type systems that allow us to drop this restriction (with some loss of abstraction), notably Martin-Löf's type theory [79] (see Section 4.8.1).

4.5. *Introduction to type inference*

Type inference is the general problem of transforming untyped or partially typed syntax into well-typed terms. One motivation for type inference is pragmatic. If we wish to program in a typed language, with type errors detected at compile time, but find the process of declaring the type of every variable tedious, then we might use a type inference algorithm to insert type designations automatically. This is particularly useful in polymorphic languages, since polymorphic terms involve quite a bit of type information. We will see that type inference for $\lambda^{\rightarrow}$ (with type variables added, for technical reasons) is essentially straightforward. In addition, the type inference algorithm for $\lambda^{\to}$ may be extended to $\lambda^{\to,\Pi,\text{let}}$ with polymorphic declarations. The latter algorithm is used in the programming languages ML [47, 86] and Miranda [136]. The algorithm for $\lambda^{\to}$ was developed by Hindley [52] and the algorithm for polymorphic declarations is due to Milner, independently [85]. While type inference problems have been considered for more flexible type systems [10, 41, 91, 74], relatively few algorithms have been found [60, 87, 101, 108, 125, 139]. In particular, the algorithmic claims of [62, 70] have not been substantiated.

Since type inference may be studied for a variety of languages, it may be useful to present the problem in a general way. A type inference problem is determined by choosing some function *Erase* from a typed language $\mathscr{L}$ to some other language $\mathscr{L}'$. Usually, $\mathscr{L}'$ will be a related "untyped" language with context-free syntax. For $\lambda^{\to}$, we may define *Erase* as follows:

$$\begin{aligned} Erase(x) &= x, \\ Erase(MN) &= Erase(M)Erase(N), \\ Erase(\lambda x{:}\sigma.M) &= \lambda x.Erase(M). \end{aligned}$$

This function simply erases the type designations from lambda bindings, producing a untyped lambda term from any $\lambda^{\to}$ term. For extensions or variants of $\lambda^{\to}$, there may be one or more analogous *Erase* functions. For languages with several kinds of type designations, we may make type inference problem easier or harder by varying the definition of *Erase*.

Given a language and erasure function, type inference is the problem of recovering a term from its erasure. More precisely, the *type inference problem* for language $\mathscr{L}$ and erasure function $Erase\colon \mathscr{L}\to\mathscr{L}'$ is

> given a term $U\in\mathscr{L}'$, find a typed term $\Gamma\rhd M{:}\sigma$ from $\mathscr{L}$ such that $Erase(M)=U$.

One reason we may be interested in finding a fully typed term is that types of subterms may be useful in compilation or optimization. However, many variations on this problem are possible. For example, instead of supplying U and asking for $\Gamma\rhd M{:}\sigma$, we might supply U and σ, and ask only whether a suitable Γ and M exist. Or, given U, we might ask whether Γ, M and σ exist. This turns type inference into a recognition problem. From a certain standpoint, specifying more information does not seem to make the problem any easier. The reason is that in order to decide any typing property of U, it generally seems that we must take into account all possible ways of typing subterms of U. If $\Gamma\rhd M{:}\sigma$ is a well-formed typed term with $Erase(M)=U$, we say the pair Γ, σ *is a typing for* U, and that U *is typable.*

Most type inference algorithms use some form of principal typings. Let us say that a typing Γ, σ *subsumes* Γ', σ' if every term with typing Γ, σ also has typing Γ',σ'. A *principal typing for term* U is a typing for U that subsumes all other typings for U. We say a typed lambda calculus *has principal types* if every typable term has a principal typing. If a language has principal types, and subsumption is decidable, then the problem of deciding whether Γ,σ is a typing for U reduces to computing a principal typing. Since subsumption is not generally antisymmetric, principal typings are generally not unique.

4.6. *Type inference for $\lambda^{\rightarrow}$ with type variables*

4.6.1. *The language $\lambda_t^{\rightarrow}$*

In this section, we will discuss type inference for the language $\lambda_t^{\rightarrow}$, which is the result of adding type variables to $\lambda^{\rightarrow}$. To be precise, the types of $\lambda_t^{\rightarrow}$ are defined by the grammar

$$\sigma ::= b \mid t \mid \sigma \rightarrow \tau$$

where b may be any type constant and t may be any type variable. The terms of $\lambda_t^{\rightarrow}$ are defined by the same axioms and inference rules as $\lambda^{\rightarrow}$, but with type expressions allowed to contain type variables. Put another way, $\lambda_t^{\rightarrow}$ is the Π-free fragment of $\lambda^{\rightarrow,\Pi}$.

The *Erase* function of Section 4.5 may be regarded as a function from $\lambda_t^{\rightarrow}$ terms to untyped lambda terms. We will consider the type inference problem determined by this function, which we will call the *Curry type inference problem*, after H.B. Curry [27].

4.6.2. *Implicit typing*

Historically, type inference problems have been formulated using proof rules for untyped terms. These proof systems are often referred to as systems for *implicit typing*, since the type of a term is not explicitly given by the syntax of terms. The following rules may be used to deduce typing assertions of the form $\Gamma \rhd U : \sigma$, where U is any *untyped* lambda term, σ is a $\lambda_t^{\rightarrow}$ type expression, and Γ is a $\lambda_t^{\rightarrow}$ type assignment.

(var) $\quad x : \sigma \rhd x : \sigma,$

(abs) $$\frac{\Gamma, x : \sigma \rhd U : \tau}{\Gamma \rhd (\lambda x . U) : \sigma \rightarrow \tau}$$

(app) $$\frac{\Gamma \rhd U : \sigma \rightarrow \tau, \quad \Gamma \rhd V : \sigma}{\Gamma \rhd UV : \tau},$$

(add hyp) $$\frac{\Gamma \rhd U : \sigma}{\Gamma, x : \tau \rhd U : \sigma}, \quad x \text{ not in } \Gamma.$$

These rules are often called the *Curry typing rules*. Using C as an abbreviation for Curry, we will write $\vdash_C \Gamma \rhd U : \sigma$ if this assertion is provable using the axiom and rules above.

There is a recipe for obtaining the proof system above from the definition of $\lambda_t^{\rightarrow}$: we simply apply *Erase* to all of the terms appearing in the terms appearing in the antecedent and consequent of every rule. Given this characterization of the inference rules, the following lemma should not be too surprising.

4.6.1. LEMMA. *If $\Gamma \rhd M : \sigma$ is a well-typed term of $\lambda_t^{\rightarrow}$, then $\vdash_C \Gamma \rhd Erase(M) : \sigma$. Conversely, if $\vdash_C \Gamma \rhd U : \sigma$, then there is a typed term $\Gamma \rhd M : \sigma$ of $\lambda_t^{\rightarrow}$ with $Erase(M) = U$.*

What may be more surprising is that this lemma fails for certain type systems, as shown by Anne Salveson [117].

The strong normalization property of $\lambda_t^{\rightarrow}$ (which follows from strong normalization for $\lambda^{\rightarrow}$) may be used to show that certain typing assertions are not derivable. Specifically, if U is not strongly normalizing, then no typing of the form $\Gamma \rhd U{:}\,\sigma$ is provable. A specific example is the untyped term

$$\Omega ::= (\lambda x\,.\,xx)(\lambda x\,.\,xx)$$

which has no normal form. Since $(\lambda x\,.\,\lambda y\,.\,y)\Omega$ has a nonnormalizing subterm, we cannot type this term either, even though it reduces to a typable term in one step. This illustrates the fact that while the typable terms are closed under reduction (cf. [27, 53, 91]), the Curry typable terms are not closed under conversion.

It is possible to study the Curry typing rules, and similar systems, by interpreting them semantically. As in other logical systems, a semantic model may make it easy to see that certain typing assertions are not provable. To interpret $\Gamma \rhd U{:}\,\sigma$ as an assertion about U, we need an untyped lambda model to make sense of U, and additional machinery to interpret type expressions as subsets of the lambda model. One straightforward interpretation of type expressions is to map type variables to arbitrary subsets, and interpret $\sigma \rightarrow \tau$ as all elements of the lambda model which map elements of σ to elements of τ (via application). This is the so-called *simple semantics of types*, discussed in [5, 53], for example. It is easy to prove soundness of the rules in this framework, showing that if $\vdash_C \Gamma \rhd U{:}\,\sigma$, then the meaning of U belongs to the collection designated by σ. With an additional inference rule

$$\text{(term eq)}\ \frac{\Gamma \rhd U{:}\sigma,\ U = V}{\Gamma \rhd V{:}\sigma}$$

for giving equal untyped terms the same types, we also have semantic completeness. Proofs of this are given in [5, 53]. Some generalizations and related studies may be found in [6, 21, 20, 23, 22, 74, 91], for example.

4.6.3. *Instances and principal typings*

As mentioned earlier, the provable $\lambda_t^{\rightarrow}$ typings are closed under substitution. For completeness, we will review substitution and unification briefly. In the next subsection, we will see that a unification-based algorithm computes principal typings.

A *substitution* will be a function from type variables to type expressions. If σ is a type expression and S a substitution, then $S\sigma$ is the type expression obtained by replacing each variable t in σ with $S(t)$. A substitution S applied to a type assignment Γ is the assignment $S\Gamma$ with

$$S\Gamma = \{x{:}\, S\sigma \mid x{:}\,\sigma \in \Gamma\}.$$

For technical convenience, we will use "instance" to refer to a slightly more general relation than substitution. Because of rule (add hyp), typings are preserved by adding hypotheses to type assignments. Combining (add hyp) with substitution leads to the following definition. A typing Γ', σ' *is an instance of* Γ, σ if there exists a substitution S with

$$\Gamma' \supseteq S\Gamma \quad \text{and} \quad \sigma' = S\sigma.$$

In this case we say Γ', σ' is an instance of Γ, σ by substitution S. We say typing statement

$\Gamma' \rhd U:\sigma'$ *is an instance of* $\Gamma \rhd U:\sigma$ if Γ', σ' is an instance of Γ, σ. One important fact about instances is that every instance of a provable typing statement is also provable.

4.6.2. LEMMA. *Let* $\Gamma' \rhd U:\sigma'$ *be an instance of* $\Gamma \rhd U:\sigma$. *If* $\vdash_C \Gamma \rhd U:\sigma$, *then* $\vdash_C \Gamma' \rhd U:\sigma'$.

In the terminology of Section 4.5, this lemma shows that if Γ', σ' is an instance of Γ, σ then Γ, σ subsumes Γ', σ'. Lemma 4.6.2 is easily proved by induction on typing derivations.

Recall that a principal typing for U is a typing for U that subsumes all others. Assuming that the "instance" relation coincides with subsumption, which we will eventually prove for "satisfiable" typings, this means that $\Gamma \rhd U:\sigma$ is principal for U if $\vdash \Gamma \rhd U:\sigma$ and, whenever $\vdash \Gamma' \rhd U:\sigma'$, there is a substitution S with $\Gamma' \supset S\Gamma$ and $\sigma' = S\sigma$. It is easy to see the principal Curry typing is unique, up to renaming of type variables. In addition, since a substitution cannot decrease the size of an expression, a principal typing for U is necessarily a minimum-length typing for U.

An important part of the algorithm for computing principal typings is the way that unification is used to combine typing statements about subterms. A unifier is a substitution which makes two expressions syntactically equal. More generally, if E is a set of equations, then a substitution S *unifies* E if $S\sigma = S\tau$ for every equation $\sigma = \tau \in E$. The unification algorithm computes a most general unifying substitution, where S *is more general than* R if there is a substitution T with $R = T \circ S$.

4.6.3. LEMMA (cf. [115]). *Let E be any set of equations between type expressions. There is an algorithm UNIFY such that if E is unifiable, then UNIFY(E) computes a most general unifier. If E is not unifiable, then UNIFY(E) fails.*

If Γ_1 and Γ_2 are type assignments, then unification can be used to find a most general substitution S such that $S\Gamma_1 \cup S\Gamma_2$ is well-formed, which means that no variable x is assigned two types by $S\Gamma_1 \cup S\Gamma_2$. To find such a substitution, we simply unify the set of all equations $\sigma = \tau$ with $x:\sigma \in \Gamma_1$ and $x:\tau \in \Gamma_2$.

4.6.4. An algorithm for principal curry typings

Given any untyped term U, *the algorithm* $PT(U)$ below either produces a provable typing $\Gamma \rhd U:\sigma$, or *fails*. In fact, we may prove that when $PT(U)$ succeeds, a principal typing is produced, and if $PT(U)$ fails, there is no provable typing for U. It follows that every typable term has a principal typing. The algorithm is written using an applicative, pattern-matching notation resembling the programming language standard ML. The reader may consult [2, 15, 140] for alternative presentations.

$PT(x) = \{x:t\} \rhd x:t;$

$PT(UV) =$ **let**

$\Gamma_1 \rhd U:\sigma = PT(U)$

$\Gamma_2 \rhd V:\tau = PT(V)$,

with type variables renamed to be disjoint from those in $PT(U)$

$$S = UNIFY(\{\alpha = \beta \mid x{:}\,\alpha \in \Gamma_1 \text{ and } x{:}\,\beta \in \Gamma_2\} \cup \{\sigma = \tau \to t\}).$$
where t is a fresh type variable

in

$S\Gamma_1 \cup S\Gamma_2 \rhd UV{:}\,St;$

$PT(\lambda x.U) =$ **let** $\Gamma \rhd U{:}\,\tau = PT(U)$

in

if $x{:}\,\sigma \in \Gamma$ for some σ

then $\Gamma - \{x{:}\,\sigma\} \rhd \lambda x.U{:}\,\sigma \to \tau$

else $\Gamma \rhd \lambda x.U{:}\,s \to \tau$

where s is a fresh type variable.

The variable case is straightforward: the most general statement we can make about a variable x is that if we assume $x{:}\,t$, then the variable x has type t. The lambda abstraction case is also relatively simple. If we know that the body U of a lambda abstraction has typing $\Gamma \rhd U{:}\,\tau$, then we would expect to give $\lambda x.U$ the typing $\Gamma' \rhd \lambda x.U{:}\,\sigma \to \tau$, where $\Gamma = \Gamma', x{:}\,\sigma$. The test in the algorithm takes care of the possibility that possibility that Γ might not have form $\Gamma', x{:}\,\sigma$. An easy inductive argument shows that if x does not appear in Γ, then x must not occur free in U. In this case, $\lambda x.U$ may accept any type of argument.

The application case will succeed in producing a typing only if unification succeeds. Given typings $\Gamma_1 \rhd U{:}\,\sigma$ and $\Gamma_2 \rhd V{:}\,\tau$, we must find instances which match the antecedent of the (app) typing rule. In general, we must find a substitution which allows us to combine type assignments, and which gives the terms appropriate types. Since type variables in the typing for V have been renamed, the type assignments may be combined by finding a substitution S such that $S\Gamma_1 \cup S\Gamma_2$ gives each variable a single type. We must also give U a functional type and V the type of the domain of U. These constraints may be satisfied simultaneously iff there exists a substitution

$$S = UNIFY(\{\alpha = \beta \mid x{:}\,\alpha \in \Gamma_1 \text{ and } x{:}\,\beta \in \Gamma_2\} \cup \{\sigma = \tau \to t\})$$

for some fresh type variable t not occurring in either typing. It is implicit in our notation (as in Standard ML) that if there is no unifier, the result of unification is undefined (or results in raising an exception) and the entire algorithm *fails*. In the case of failure, the term UV does not have a Curry typing, as demonstrated in Theorem 4.6.5 below.

We may prove that if $PT(U)$ succeeds, then it produces a provable typing for U.

4.6.4. THEOREM. *If* $PT(U) = \Gamma \rhd U{:}\,\tau$, *then* $\vdash_C \Gamma \rhd U{:}\,\tau$.

It follows, by Lemma 4.6.2, that every instance of $PT(U)$ is provable. Conversely, we can also prove that every provable typing for U is an instance of $PT(U)$.

4.6.5. THEOREM. *Suppose* $\vdash_C \Gamma \rhd U{:}\,\tau$ *is a provable typing assertion. Then* $PT(U)$ *succeeds and produces a typing with* $\Gamma \rhd U{:}\,\tau$ *as an instance.*

This form of typing algorithm seems to have originated with [70]. The form of typing algorithm given in [85] differs slightly in that it takes a term U and type assignment

Γ as input, producing a substitution S and type σ such that $\vdash_C S\Gamma \rhd U{:}\,\sigma$, or failing if no such S and σ exist.

An interesting property of Curry typing is that if Γ, σ is a typing for any term, then Γ, σ is a principal typing for some term [52]. This gives us the following corollary to Theorem 4.6.5.

4.6.6. COROLLARY. *If Γ, σ is a typing for some term, then Γ', σ' is an instance of Γ, σ iff Γ, σ subsumes Γ', σ'.*

As shown in [126], it is PSPACE-complete to determine whether a given Γ, σ is a typing for some (untyped) lambda term.

4.7. *Type inference with polymorphic declarations*

4.7.1. *ML and ML_1 type inference*

In this section, we will consider type inference for the language $\lambda^{\rightarrow,\Pi,\mathsf{let}}$ with polymorphic declarations. Since this language exhibits the kind of polymorphism provided by the programming language ML, we will call the resulting type inference problem *ML type inference.*

The ML type inference problem may be defined precisely by extending the *Erase* function of Section 4.5 to type abstraction, type application and let as follows:

$$\begin{aligned}
&\mathit{Erase}(\lambda t.M) &&= \mathit{Erase}(M),\\
&\mathit{Erase}(M\tau) &&= \mathit{Erase}(M),\\
&\mathit{Erase}(\mathsf{let}\ x{:}\,\sigma = M\ \mathsf{in}\ N) &&= \mathsf{let}\ x = \mathit{Erase}(M)\ \mathsf{in}\ \mathit{Erase}(N).
\end{aligned}$$

Given any $\lambda^{\rightarrow,\Pi,\mathsf{let}}$ pre-term, *Erase* produces a term of an untyped lambda calculus extended with untyped let declarations. We say that an untyped lambda term U, possibly containing untyped let's, *has the ML typing Γ, ρ* if there is some well-typed $\lambda^{\rightarrow,\Pi,\mathsf{let}}$ term $\Gamma \rhd M{:}\,\rho$ with $\mathit{Erase}(M) = U$. The *ML type inference problem* is to find an ML typing for any given untyped U, or determine that none exists.

A special case of the ML type inference problem might be called ML_1 *type inference,* since it only involves the U_1 types of $\lambda^{\rightarrow,\Pi,\mathsf{let}}$. We will say that Γ, τ is an ML_1 *typing for* V if Γ, τ is an ML typing for V with type assignment Γ containing only U_1 types and τ a U_1 type. It is easy to see that many untyped terms have ML typings, but no ML_1 typings. For example, the term xx has the ML typing $\{x{:}\,\Pi t.t\} \rhd (xx){:}\,t$, since $\{x{:}\,\Pi t.t\} \rhd (x\,t \rightarrow t)(x\,t){:}\,t$ is a well-typed term of $\lambda^{\rightarrow,\Pi,\mathsf{let}}$. However, xx has no ML_1 typing, since no U_1 type for x will allow us to apply x to itself[5].

The purpose of introducing ML_1 typing is to simplify the presentation of Milner's typing algorithm and its correctness properties. From a theoretical point of view, the

[5]It seems difficult to give a simple proof of this. However, the untypability of xx follows from the correctness of the typing algorithm *PTL* given in Section 4.7.3 and the fact that the algorithm fails on this term (see also [85]).

reason it suffices to consider ML_1 typing is that a closed term has an ML typing iff it has a ML_1 typing.

4.7.1. LEMMA. *Let V be a closed, untyped lambda term, possible containing* let's. *There is an ML typing for V iff V has an* ML_1 *typing.*

The algorithm *PTL* given in Section 4.7.3 will find the most general ML_1 typing for any ML_1-typable term. It follows that *PTL* finds a typing for any closed, ML-typable term. This is sufficient in practice, since any executable program must give every identifier a value, and therefore must be closed. A technical detail is that ML programs contain constants for predefined operations. These may be accounted for in the algorithm, as will become clear in Section 4.7.3.

The property of $\lambda^{\rightarrow,\Pi,\mathrm{let}}$ which complicates type inference for open terms is that there are two distinct type universes. This means that there there are two classes of types that may be used in type assignments. If we attempt to infer ML typings "bottom-up", as in the typing algorithm *PT* of Section 4.6.4, then it is not clear whether we should assume that a term variable x has a U_1 or U_2 type. If we choose arbitrarily, then it seems difficult to reverse this decision without recomputing the entire typing. However, the problem may be avoided if we only consider closed terms. The reason is that each variable in a closed term must either be lambda-bound or let-bound. The lambda-bound variables will have U_1 types, while each let-bound variable must be declared equal to some term. In a subterm let $x=U$ in V, the type of x must be a type of U. We will take advantage of this constraint in the typing algorithm by using a principal typing for U to choose typings for x in V.

4.7.2. *Implicit typing rules*

An implicit typing rule for let may be obtained from the $\lambda^{\rightarrow,\Pi,\mathrm{let}}$ rule by applying *Erase* to the antecedent and consequent. This gives us the typing rule

$$(\mathrm{let})_2 \qquad \frac{\Gamma \rhd U\colon\rho,\ \Gamma, x\colon\rho \rhd V\colon\tau}{\Gamma \rhd (\mathrm{let}\ \ x=U\ \ \mathrm{in}\ \ V)\colon\tau}$$

of [28]. With this rule, we have the straightforward analog of Lemma 4.6.1 for $\lambda^{\rightarrow,\Pi,\mathrm{let}}$, proved in [92]. Since we often want let-bound variables to have polymorphic types, this rule requires U_2 types, hence the name $(\mathrm{let})_2$.

An alternative rule for let, first suggested to the author by Albert Meyer, may be formulated using substitution on terms. Since we use substitution to account for multiple occurrences of a let-bound variable, this formulation eliminates the need for U_2 types. The following rule is based on the fact that let $x=U$ in V has precisely the same ML typings as $[U/x]V$, provided $\Gamma \rhd U\colon\rho$ for some type ρ:

$$(\mathrm{let})_1 \qquad \frac{\Gamma \rhd U\colon\rho,\ \Gamma \rhd [U/x]V\colon\tau}{\Gamma \rhd\ \mathrm{let}\ \ x=U\ \ \mathrm{in}\ \ V\colon\tau}.$$

It is not hard to show that if x occurs in V, and $\Gamma \rhd [U/x]V\colon\tau$ is an ML typing, then we must have $\Gamma \rhd U:\rho$ for some ρ. Therefore, the assumption about U only prevents us from typing let $x=U$ in V when x does not occur in V and U is untypable. We will write $\vdash_{ML_1} \Gamma \rhd U:\tau$ if the typing assertion $\Gamma \rhd U:\tau$ is provable from the Curry rules and $(\text{let})_1$. We have the following correspondence between $\vdash_{ML_1}$ and ML_1 typing.

4.7.2. LEMMA. *Let τ be any U_1 type of $\lambda^{\to,\Pi,\text{let}}$ and Γ any type assignment containing only U_1 types. If $\Gamma \rhd M:\tau$ is a well-typed term of $\lambda^{\to,\Pi,\text{let}}$, then $\vdash_{ML_1} \Gamma \rhd Erase(M):\tau$. Conversely, if $\vdash_{ML_1} \Gamma \rhd V:\tau$, then there is a typed term $\Gamma \rhd M:\tau$ of $\lambda^{\to,\Pi,\text{let}}$ with $Erase(M)=V$.*

Since ML_1 typings use the same form of typing assertions as Curry typing, we may adopt the definition of instance given in Section 4.6.3. It is not hard to see that Lemma 4.6.2 extends immediately to $\vdash_{ML_1}$.

4.7.3. Type inference algorithm

The algorithm *PTL* given in Fig. 1 computes a principal ML_1 typing for any ML_1-typable term. In particular, the algorithm will find a typing for any closed ML-typable term. The algorithm has two arguments, a term to be typed, and an environment mapping variables to typing assertions. The purpose of the environment is to handle let-bound variables. In a "top-level" function call, with a closed expression,

$PTL(x, A) =$ **if** $A(x) = \Gamma \rhd U:\sigma$ **then** $\Gamma \rhd x:\sigma$
  else $\{x:t\} \rhd x:t$

$PTL(UV, A) =$ **let**
  $\Gamma_1 \rhd U:\sigma = PTL(U, A)$
  $\Gamma_2 \rhd V:\tau = PTL(U, A)$,
   with type variables renamed to be disjoint from those in $PTL(U, A)$
  $S = UNIFY(\{\alpha=\beta \mid x:\alpha \in \Gamma_1 \text{ and } x:\beta \in \Gamma_2\} \cup \{\sigma = \tau \to t\})$
   where t is a fresh type variable
 in
  $S\Gamma_1 \cup S\Gamma_2 \rhd UV: St$

$PTL(\lambda x.U, U, A) =$ **let** $\Gamma \rhd U:\tau = PTL(U, A)$
  in
   if $x:\sigma \in$ for some σ
   then $\Gamma - \{x:\sigma\} \rhd \lambda x.U:\sigma\to\tau$
   else $\Gamma \rhd \lambda x.U:s\to\tau$
   where s is a fresh type variable

$PTL(\text{let } x=U \text{ in } V, A)$
 $=$ **let** $\Gamma_1 \rhd U:\sigma = PTL(U, A)$
  $A' = A \cup \{x \mapsto \Gamma_1 \rhd U:\sigma\}$
  $\Gamma_2 \rhd V:\tau = PTL(V, A')$
  $S = UNIFY(\{\alpha=\beta \mid y:\alpha \in \Gamma_1 \text{ and } y:\beta \in \Gamma_2\})$
 in $S\Gamma_1 \cup S\Gamma_2 \rhd$ **let** $x=U$ **in** $V:S\tau$

Fig. 1. Algorithm *PTL* to compute principal typing.

this environment would be empty. Polymorphic constants (cf. [85]) would be treated exactly as if they were given typings in the initial environment. It is assumed that the input to PTL is an expression with all bound variables renamed to be distinct, an operation that is commonly done in lexical analysis. This guarantees that if a variable is let-bound, it is not also lambda-bound.

Algorithm PTL may *fail* in the application or let case if the call to $UNIFY$ fails. We can prove that if $PTL(U, \emptyset)$ succeeds, then it produces an ML_1 typing for U.

4.7.3. THEOREM. *If* $PTL(U, \emptyset) = \Gamma \rhd U : \tau$, *then* $\vdash_{ML_1} \Gamma \rhd U : \tau$.

It follows, by Lemma 4.6.2, that every instance of $PTL(U, \emptyset)$ is provable. Conversely, every provable typing for U is an instance of $PTL(U, \emptyset)$.

4.7.4. THEOREM. *Suppose* $\vdash_{ML_1} \Gamma \rhd U : \tau$ *is an* ML_1 *typing for* U. *Then* $PTL(U, \emptyset)$ *succeeds and produces a typing with* $\Gamma \rhd U : \tau$ *as an instance.*

4.8. *Additional typing concepts*

4.8.1. *General products and sums*

There are a number of variations and extensions of the type systems described in this chapter. One historically important series of typed lambda calculi are the Automath languages, summarized in [29]. Some conceptual descendants of the Automath languages are Martin-Löf's intuitionistic type theory [77, 78, 79] and the closely related Nuprl system for program verification and formal mathematics [19] (see also [9]). Two important constructs in these predicative type systems are the "general" product and sum types, written using Π and Σ respectively.

Intuitively, general sums and products may be viewed as straightforward set-theoretic constructions. If A is an expression defining some collection (either a type or universe, for example), and B is an expression with free variable x which defines a collection for each x in A, then $\Sigma x: A.B$ and $\Pi x: A.B$ are called the *sum* and *product of the family B over the index set A* respectively. In set-theoretic terms, the product $\Pi x: A.B$ is the Cartesian product of the family of sets $\{B(x) \mid x \in A\}$. The elements of this product are functions f such that $f(a) \in [a/x]B$ for each $a \in A$. The sum $\Sigma x: A.B$ is the disjoint union of the family $\{B(x) \mid x \in A\}$. Its members are ordered pairs $\langle a, b \rangle$ with $a \in A$ and $b \in [a/x]B$. Since the elements of sum types are pairs, general sums have projection functions *first* and *second* for first and second components. We may regard the polymorphic types $\Pi t.\rho$ and $\forall t.\sigma$ of $\lambda^{\rightarrow,\Pi}$ and $\lambda^{\rightarrow,\forall}$ as particular uses of general products. The existential types of Section 4.4 are similar to general sums, but restricted in a significant way. More precisely, if $\langle a, b \rangle : \Sigma x: A.B$, then $first\langle a, b \rangle$ is an expression for an element of A with $first\langle a, b \rangle = a$. However, if $\langle t = \tau, M : \sigma \rangle : \exists t.\sigma$, we cannot use abstype to retrieve the first component (τ). Since the typing rules specify that, for any expression of the form abstype t with $x : \sigma$ is $\langle t = \tau, M : \sigma \rangle$ in N, the type of N cannot have a free occurrence of t, the type component of any element of $\exists t.\rho$ is hidden in a certain way. This is consistent with the goals of data abstraction (see Section 4.4), but may seem unnecessarily restrictive from other points of view. Some debate on this point may be found in [75, 92, 96].

An interesting impredicative language with general products is the Calculus of Constructions [26], which extends Girard's system F.

4.8.2. Types as specifications

The more expressive type systems mentioned in Section 4.8.1 all serve dual purposes, following the formulas-as-types analogy discussed in Section 4.3. This is explained in some detail in [57, 78, 19]. In brief, a type system with function types, Cartesian products, polymorphism and general sums provides a logic with implication, conjunction, universal and existential quantification. If we introduce types that correspond to atomic formulas, together with term formation rules that correspond to the usual logical axioms and inference rules, then we obtain a system that is both a function calculus and an expressive logic. To give some idea of how atomic formulas may be added, we will briefly summarize a simple encoding of equality between terms. We may introduce types of the form $Eq_A(x, y)$ to represent the atomic formula $x = y$, where $x, y\colon A$. Since equality is reflexive, we might have a typing rule so that, for any $M\colon A$, we have $\mathit{reflexivity}_A(M)\colon Eq_A(M, M)$. In other words, for $M{:}A$, we have the proof $\mathit{reflexivity}_A(M)$ of $Eq_A(M, M)$. To account for symmetry, we might have a rule such that for any $M\colon Eq_A(N, P)$ we have $\mathit{symmetry}_A(M)\colon Eq_A(P, N)$. In addition, we would expect typing rules corresponding to transitivity and other properties of equality (e.g., substitutivity of equivalents). In such a language, we could write a term of type $Eq_A(M, N)$ whenever M and N are provably equal terms of type A.

In a rich typed function calculus with appropriate type operations, we may regard types as "program specifications" and well-typed terms as "verified programs". To give a concrete example, suppose $\mathit{prime}(x)$ is a type corresponding to the assertion that $x\colon \mathit{nat}$ is prime, and $\mathit{divides}(x, y)$ is a type "saying" that x divides y. Then we would expect a term of type

$$\Pi x\colon \mathit{nat}.(x>1 \rightarrow \Sigma y\colon \mathit{nat}.(\mathit{prime}(y) \times \mathit{divides}(y, x))$$

to define a function which, given any natural number x, takes a proof of $x>1$ and returns a pair $\langle N, M\rangle$ with $N\colon \mathit{nat}$ and M proving that N is a prime number dividing x. Based on the general idea of types as logical assertions, Automath, Nuprl and the Calculus of Constructions have all been proposed as systems for verifying programs or mechanically checking mathematical proofs. Some related systems for checking proofs via type checking are the Edinburgh Logical Framework [48] and PX [49], which is based on Feferman's logical system [32].

Acknowledgment

For comments on this chapter I would like to thank S. Abramsky, V. Breazu-Tannen, R. Casley, P.-L. Curien, C. Gunter, R. Harper, I. Mason, A. Meyer, E. Moggi, N. Marti-Oliet, and P. Scott.

Bibliography

[1] ABRAMSKY, S., Abstract interpretation, logical relations and Kan extensions, *J. Logic Comput.*, to appear.

[2] AHO, A.V., R. SETHI and J.D. ULLMAN, *Compilers: Principles, Techniques, Tools* (Addison-Wesley, Reading, MA, 1986).

[3] ANDRÉKA, H., W. CRAIG and I. NÉMETI, A system of logic for partial functions under existence-dependent Kleene equality, *J. Symbolic Logic* **53** (3) (1988) 834–839.

[4] BAINBRIDGE, E.S., P.J. FREYD, A. SCEDROV and P.J. SCOTT, Functional polymorphism, in: G. Huet, ed. *Logical Foundations of Functional Programming* (Addison-Wesley, Reading, MA, 1989).

[5] BARENDREGT, H.P., *The Lambda Calculus: Its Syntax and Semantics* (North-Holland, Amsterdam, 1984).

[6] BARENDREGT, H., M. COPPO and M. DEZANI-CIANCAGLINI, A filter lambda model and the completeness of type assignment, *J. Symbolic Logic* **48** (4) (1983) 931–940.

[7] BARR, M., Fixed points in cartesian closed categories, *Theoret. Comput. Sci.* **70** (1990).

[8] BARR, M. and C. WELLS, *Toposes, Triples and Theories* (Springer, New York, 1985).

[9] BEESON, M., *Foundations of Constructive Mathematics* (Springer, New York, 1985).

[10] BOEHM, H.-J., Partial polymorphic type inference is undecidable, in: *Proc. 26th Ann. IEEE Symp. on Foundations of Computer Science* (1985) 339–345.

[11] BRUCE, K.B., A.R. MEYER and J.C. MITCHELL, The semantics of second-order lambda calculus, *Inform. and Comput.* **85**(1) (1990) 76–134.

[12] BURN, G.L., C. HANKIN and S. ABRAMSKY, Strictness analysis for higher-order functions. *Sci. Comput. Programming* **7** (1986) 249–278.

[13] CARBONI, A., P.J. FREYD and A. SCEDROV, A categorical approach to realizability and polymorphic types, in: M. Main et al., eds., *Proc. 3rd ACM Workshop on Mathematical Foundations of Programming Language Semantics*, Lecture Notes in Computer Science, Vol. 298 (Springer, Berlin, 1988) 23–42.

[14] CARDELLI, L., A polymorphic lambda calculus with type:type, Tech. Report 10, DEC Systems Research Center, 1986.

[15] CARDELLI, L., Basic polymorphic typechecking, *Sci. Comput. Programming* **8** (2) (1987) 147–172.

[16] CARDELLI, L. and P. WEGNER, On understanding types, data abstraction, and polymorphism, *Computing Surveys* **17** (4) (1985) 471–522.

[17] CHURCH, A., *The Calculi of Lambda Conversion* (Princeton Univ. Press, Princeton, NJ, 1941; reprinted by University Microfilms Inc., Ann Arbor, MI, 1963).

[18] CLEMENT, D., J. DESPEYROUX, T. DESPEYROUX and G. KAHN, A simple applicative language: Mini-ML, in: *Proc. ACM Conf. on LISP and Functional Programming* (1986) 13–27.

[19] CONSTABLE, R.L. et al., *Implementing Mathematics with the Nuprl Proof Development System*, Graduate Texts in Mathematics, Vol. 37 (Prentice-Hall, Englewood Cliffs, NJ, 1986).

[20] COPPO, M., On the semantics of polymorphism, *Acta Inform.* **20** (1983) 159–170.

[21] COPPO, M., M. DEZANI-CIANCAGLINI and B. VENNERI, Principal type schemes and lambda calculus semantics, in: *To H.B. Curry: Essays on Combinatory Logic, Lambda Calculus and Formalism* (Academic Press, New York, 1980) 535–560.

[22] COPPO, M., M. DEZANI-CIANCAGLINI and M. ZACCHI, Type theories, normal forms and D_∞ lambda models, *Inform. and Comput.* **72** (1987) 85–116.

[23] COPPO, M. and M. ZACCHI, Type inference and logical relations, in: *Proc. IEEE Symp. on Logic in Computer Science* (1986) 218–226.

[24] COQUAND, T., An analysis of Girard's paradox, in: *Proc. IEEE Symp. on Logic in Computer Science* (1986) 227–236.

[25] COQUAND, T., C.A. GUNTER and G. WINSKEL, Domain-theoretic models of polymorphism, *Inform. and Comput.* **81** (2) (1989) 123–167.

[26] COQUAND, T. and G. HUET, The calculus of constructions, *Inform. and Comput.* **76** (2, 3) (1988) 95–120.

[27] CURRY, H.B. and R. FEYS, *Combinatory Logic I* (North-Holland, Amsterdam, 1958).

[28] DAMAS, L., and R. MILNER, Principal type schemes for functional programs, in: *Proc. 9th ACM Symp. on Principles of Programming Languages* (1982) 207–212.

[29] DE BRUIJN, N.G., A survey of the project Automath, in: *To H.B. Curry: Essays on Combinatory Logic, Lambda Calculus and Formalism* (Academic Press, New York, 1980) 579–607.
[30] DONAHUE, J., On the semantics of data type, *SIAM J. Comput.* **8** (1979) 546–560.
[31] FEFERMAN, S., Systems of predicative analysis, in: *Algebra and Logic*, Lecture Notes in Mathematics, Vol. 450 (Springer, Berlin, 1975) 87–139.
[32] FEFERMAN, S., Constructive theories of functions and classes, in: *Logic Colloquium '78* (North-Holland, Amsterdam, 1979) 159–224.
[33] FEFERMAN, S., A theory of variable types, *Rev. Columbiana Mat.* **19** (1985) 95–105.
[34] FEFERMAN, S., Polymorphic typed lambda-calculi in a type-free axiomatic framework, in: *Proc. Workshop on Logic and Computation*, Contemporary Mathematics (Amer. Mathematical Soc., Providence, RI, 1989).
[35] FORTUNE, S., D. LEIVANT and M. O'DONNELL, The expressiveness of simple and second order type structures, *J. ACM* **30** (1) (1983) 151–185.
[36] FREYD, P., J.-Y. GIRARD, A. SCEDROV and P.J. SCOTT, Semantic parametricity in polymorphic lambda calculus, in: *Proc. 3rd Ann. IEEE Symp. on Logic in Computer Science* (1988) 274–279.
[37] FREYD, P. and A. SCEDROV, Some semantic aspects of polymorphic lambda calculus, in: *Proc. 2nd Ann. IEEE Symp. on Logic in Computer Science* (1987) 315–319.
[38] FRIEDMAN, H., Equality between functionals, in: R. Parikh, ed., *Logic Colloquium* (Springer, New York, 1975) 22–37.
[39] GABBAY, D.M., *Semantical Investigations in Heyting's Intuitionistic Logic* (Reidel, Dordrecht, 1981).
[40] GANDY, R.O., On the axiom of extensionality — part I, *J. Symbolic Logic* **21** (1956).
[41] GIANNINI, P. and S. RONCHI DELLA ROCCA, Characterization of typings in polymorphic type discipline, in: *Proc. 3rd Ann. IEEE Symp. on Logic in Computer Science* (1988) 61–71.
[42] GIRARD, J.-Y, Une extension de l'interprétation de Gödel à l'analyse, et son application à l'élimination des coupures dans l'analyse et la théorie des types, in: J.E. Fenstad, ed., *Proc. 2nd Scandinavian Logic Symposium* (North-Holland, Amsterdam, 1971) 63–92.
[43] GIRARD, J.-Y., Interprétation fonctionelle et élimination des coupures de l'arithmétique d'ordre supérieur, Thèse D'Etat, Université de Paris VII, Paris, 1972.
[44] GIRARD, J.-Y., The system *F* of variable types, fifteen years later, *Theoret Comput Sci.* **45** (2) (1986) 159–192.
[45] GIRARD, J.-Y., Y. LAFONT and P. TAYLOR, *Proofs and Types*, Cambridge Tracts in Theoretical Computer Science, Vol. 7 (Cambridge Univ. Press, Cambridge, UK, 1989).
[46] GORDON, M.J.C., *The Denotational Description of Programming Languages* (Springer, Berlin, 1979).
[47] GORDON, M.J., R. MILNER and C.P. WADSWORTH, *Edinburgh LCF*, Lecture Notes in Computer Science, Vol. 78 (Springer, Berlin, 1979).
[48] HARPER, R., F. HONSELL and G. PLOTKIN, A framework for defining logics, in: *Proc. 2nd Ann. IEEE Symp. on Logic in Computer Science* (1987) 194–204.
[49] HAYASHI, S. and H. NAKANO, *PX—a Computational Logic* (MIT Press, Cambridge, MA, 1988).
[50] HAYNES, C.T., A theory of data type representation independence, in: *Proc. Internat. Symp. on Semantics of Data Types*, Lecture Notes in Computer Science. Vol. 173 (Springer, Berlin, 1984) 157–176.
[51] HENKIN, L., Completeness in the theory of types, *J. Symbolic Logic* **15** (2) (1950) 81–91.
[52] HINDLEY, R., The principal type-scheme of an object in combinatory logic, *Trans. AMS*, **146** (1969) 29–60.
[53] HINDLEY, R., The completeness theorem for typing lambda terms, *Theoret. Comput. Sci.* **22** (1983) 1–17.
[54] HINDLEY, J.R. and J.P. SELDIN, *Introduction or Combinators and Lambda Calculus* (London Mathematical Society, London, 1986).
[55] HOPCROFT, J.E. and J.D. ULLMAN, *Introduction to Automata Theory, Languages and Computation* (Addison-Wesley, Reading, MA, 1979).
[56] HOWARD, W., Hereditarily majorizable functionals, in: *Mathematical Investigation of Intuitionistic Arithmetic and Analysis*, Lecture Notes in Mathematics Vol. 344 (Springer, Berlin, 1973) 454–461.
[57] HOWARD, W., The formulas-as-types notion of construction, in: *To H.B. Curry: Essays on Combinatory Logic, Lambda-Calculus and Formalism* (Academic Press, New-York, 1980) 479–490.

[58] HYLAND, J.M.E., The effective topos, in: *The L.E.J. Brouwer Centenary Symposium* (North-Holland, Amsterdam, 1982) 165–216.

[59] HYLAND, J.M.E., A small complete category, in: *Proc. Conf. on Church's Thesis: Fifty Years Later* (1987).

[60] JATEGAONKAR, L. and J.C. MITCHELL, ML with extended pattern matching and subtypes, in: *Proc. ACM Symp. on Lisp and Functional Programming Languages* (1988) 198–212.

[61] KANELLAKIS, P.C. and J.C. MITCHELL, Polymorphic unification and ML typing, in: *Proc. 16th ACM Symp. on Principles of Programming Languages* (1989) 105–115.

[62] KFOURY, A.J., J. TIURYN and P. URZYCZYN, A proper extension of ML with effective type assignment, in: *Proc. 15th ACM Symp. on Principles of Programming Languages* (1988) 58–69.

[63] KLEENE, S.C., On the interpretation of intuitionistic number theory, *J. Symbolic Logic*, **10** (1945) 109–124.

[64] KLEENE, S.C., Realizability: a retrospective survey, in: *Cambridge Summer School in Mathematical Logic*, Lecture Notes in Mathematics, Vol. 337 (Springer, Berlin, 1971) 95–112.

[65] KOCK, A. and G.E. REYES, Doctrines in categorical logic, in: *Handbook of Mathematical Logic* (North-Holland, Amsterdam, 1977) 283–316.

[66] KREISEL, G., Interpretation of analysis by means of constructive functionals of finite types, in: A. Heyting, ed., *Constructivity in Mathematics*, (North-Holland, Amsterdam, 1959) 101–128.

[67] LAMBEK, J., From lambda calculus to cartesian closed categories, in: *To H.B. Curry: Essays on Combinatory Logic, Lambda Calculus and Formalism* (Academic Press, New York, 1980) 375–402.

[68] LAMBEK, J. and P.J. SCOTT, *Introduction to Higher-Order Categorical Logic*, Cambridge Studies in Advanced Mathematics, Vol. 7. (Cambridge Univ. Press, Cambridge, UK, 1986).

[69] LANDIN, P.J., A correspondence between ALGOL 60 and Church's lambda notation, *Comm. ACM* **8** (1965) 89–101 and 158–165.

[70] LEIVANT, D., Polymorphic type inference, in: *Proc. 10th ACM Symp. on Principles of Programming Languages*, (1983) 88–98.

[71] LISKOV, B. et al., *CLU Reference Manual*, Lecture Notes in Computer Science, Vol. 114 (Springer, Berlin, 1981).

[72] LÖB, M.H., Embedding first-order predicate logic in fragments of intuitionistic logic, Tech. Report 75-8, Mathematisch Institut, Amsterdam, 1975.

[73] MACLANE, S., *Categories for the Working Mathematician*, Graduate Texts in Mathematics, Vol. 5 (Springer, Berlin, 1971).

[74] MACQUEEN, D., G. PLOTKIN and R. SETHI, An ideal model for recursive polymorphic types. *Inform. and Control* **71** (1, 2) (1986) 95–130.

[75] MACQUEEN, D.B., Using dependent types to express modular structure, in: *Proc. 13th ACM Symp. on Principles of Programming Languages* (1986) 277–286.

[76] MAKKAI, M., and G.E. REYES, *First-order Categorical Logic*, Lecture Notes in Mathematics, Vol. 611 (Springer, Berlin, 1977).

[77] MARTIN-LÖF, P., An intuitionistic theory of types: Predicative part, in: H.E. Rose and J.C. Shepherdson, eds., *Logic Colloquium, '73* (North-Holland, Amsterdam, 1973) 73–118.

[78] MARTIN-LÖF, P., Constructive mathematics and computer programming, in: *Proc. 6th Internat. Congr. for Logic, Methodology, and Philosophy of Science* (North-Holland, Amsterdam, 1982) 153–175.

[79] MARTIN-LÖF, P., *Intuitionistic Type Theory* (Bibliopolis, Naples, 1984).

[80] MESEGUER, J., Relating models of polymorphism, in: *Proc. 16th ACM Symp. on Principles of Programming Languages* (1989) 228–241.

[81] MEYER, A.R., What is a model of the lambda calculus? *Inform. and Control* **52** (1) (1982) 87–122.

[82] MEYER, A.R., J.C. MITCHELL, E. MOGGI and R. STATMAN, Empty types in polymorphic lambda calculus, in: *Proc. 14th ACM Symp. on Principles of Programming Languages* (1987) 253–262; revised version in: G. Huet, eds., *Logical Foundations of Functional Programming*, (Addison-Wesley, Reading, MA, 1990) 273–284.

[83] MEYER, A.R. and M.B. REINHOLD, Type is not a type, in: *Proc. 13th ACM Symp. on Principles of Programming Languages* (1986) 287–295.

[84] MILNE, R.E. and C. STRACHEY, *A Theory of Programming Language Semantics* (Chapman and Hall, London, and Wiley, New York, 1976).

[85] MILNER, R., A theory of type polymorphism in programming, *J. Comput. System Sci.* **17** (1978) 348–375.

[86] MILNER, R., The Standard ML core language, *Polymorphism* **2** (2) (1985); an earlier version appeared in: *Proc. 1984 ACM Symp. on Lisp and Functional Programming.*

[87] MITCHELL, J.C., Coercion and type inference (summary), in: *Proc. 11th ACM Symp. on Principles of Programming Languages* (1984) 175–185.

[88] MITCHELL, J.C., Abstract realizability for intuitionistic and relevant implication (abstract), *J. Symbolic Logic* **51** (3) (1986) 851–852.

[89] MITCHELL, J.C., Representation independence and data abstraction, in: *Proc. 13th ACM Symp. on Principles of Programming Languages* (1986) 263–276.

[90] MITCHELL, J.C., A Type-inference approach to reduction properties and semantics of polymorphic expressions, in: *ACM Conf. on LISP and Functional Programming* (1986) 308–319; revised version in: G. Huet, ed., *Logical Foundations of Functional Programming* (Addison-Wesley, Reading, MA, 1990) 195–212.

[91] MITCHELL, J.C., Polymorphic type inference and containment, *Inform. and Comput.* **76** (2, 3) (1988) 211–249.

[92] MITCHELL, J.C. and R. HARPER, The essence of ML, in: *Proc. 15th ACM Symp. on Principles of Programming Laguages* (1988) 28–46.

[93] MITCHELL, J.C. and A.R. MEYER, Second-order logical relations (extended abstract), in: *Logics of Programs*, Lecture Notes in Computer Science, Vol. 193 (Springer, Berlin, 1985) 225–236.

[94] MITCHELL, J.C. and E. MOGGI, Kripke-style models for typed lambda calculus, in: *Proc. 2nd Ann. IEEE Symp. on Logic in Computer Science*, (1987) 303–314; revised version in: *J. Pure and Applied Logic*, to appear.

[95] MITCHELL, J.C. and M.J. O'DONNELL, Realizability semantics for error-tolerant logics, in: *Theoretical Aspects of Reasoning About Knowledge* (Morgan Kaufman, Los Altos, CA, 1986).

[96] MITCHELL, J.C. and G.D. PLOTKIN, Abstract types have existential types, *ACM Trans. on Programming Languages and Systems* **10** (3) (1988) 470–502; preliminary version appeared in: *Proc. 12th ACM Symp. on Principles of Programming Languages* (1985).

[97] MITCHELL, J.C. and P.J. SCOTT, Typed lambda calculus and cartesian closed categories, in: *Proc. Conf. on Computer Science and Logic* Contemporary Mathematics (Amer. Mathematical Soc., Providence, RI, 1989) 301–316.

[98] MORRIS, J.H., Types are not sets. in: *Proc. 1st. ACM Symp. on Principles of Programming Languages* (1973) 120–124.

[99] MULMULEY, K., A semantic characterization of full abstraction for typed lambda calculus, in: *Proc. 25th Ann. IEEE Symp. on Foundations of Computer Science* (1984) 279–288.

[100] NEDERPELT, R.P., Strong normalization in a typed lambda calculus with lambda structured types. Ph.D. Thesis, Technological Univ. Eindhoven, 1973.

[101] OHORI, A. and P. BUNEMAN, Type inference in a database language, in: *Proc. ACM Symp. on LISP and Functional Programming Languages*, (1988) 174–183.

[102] PITTS, A.M., Polymorphism is set-theoretic, constructively, in: *Proc. Summer Conf. on Category Theory and Computer Science*, Lecture Notes in Computer Science, Vol. 283 (Springer, Berlin, 1987) 12–39.

[103] PITTS, A.M., Non-trivial power types can't be subtypes of polymorphic types, in: *Proc. 4th IEEE Symp. on Logic in Computer Science* (1989) 6–13.

[104] PLOTKIN, G.D., LCF considered as a programming language, *Theoret. Comput. Sci.* **5** (1977) 223–255.

[105] PLOTKIN, G.D., Lambda definability in the full type hierarchy, in: *To H.B. Curry: Essays on Combinatory Logic, Lambda Calculus and Formalism* (Academic Press, New York, 1980) 363–373.

[106] PLOTKIN, G.D., *Denotational Semantics with Partial Functions*, Lecture notes, C.S.L.I. Summer School, Stanford, 1985.

[107] PRAWITZ, D., *Natural Deduction* (Almquist and Wiksell, Stockholm, 1965).

[108] RÉMY, D. Typechecking records and variants in a natural extension of ML, in: *Proc. 16th ACM Symp. on Principles of Programming Languages* (1989) 60–76.

[109] REYNOLDS, J.C., On the relation between direct and continuation semantics, in: *Proc. 2nd Internat. Call. on Automata, Languages and Programming*, Lecture Notes in Computer Science, Vol. 14 (Springer, Berlin, 1974) 141–156.
[110] REYNOLDS, J.C., Towards a theory of type structure, in: *Proc. Paris Symp. on Programming*, Lecture Notes in Computer Science, Vol. 19 (Springer, Berlin, 1974) 408–425.
[111] REYNOLDS, J.C., The essence of *ALGOL*, in: J.W. De Bakker and J.C. Van Vliet, eds., *Algorithmic Languages*, IFIP Congress Series (North-Holland, Amsterdam, 1981) 345–372.
[112] REYNOLDS, J.C., Types, abstraction, and parametric polymorphism, in: *Information Processing '83*, (North-Holland, Amsterdam, 1983) 513–523.
[113] REYNOLDS, J.C., Polymorphism is not set-theoretic, in: *Proc. Internat. Symp. on Semantics of Data Types*, Lecture Notes in Computer Science, Vol. 173 (Springer, Berlin, 1984) 145–156.
[114] REYNOLDS, J.C. and G.D. PLOTKIN, On functors expressible in the polymorphic lambda calculus, *Inform. and Comput.*, to appear.
[115] ROBINSON, J.A., A machine-oriented logic based on the resolution principle, *J. ACM* **12** (1) (1965) 23–41.
[116] ROGERS, H., *Theory of Recursive Functions and Effective Computability* (McGraw-Hill, New York, 1967).
[117] SALVESON, A., Polymorphism and monomorphism in Martin-Löf's type theory, Ph.D. Thesis, Institutt for Informatikk, University of Oslo, 1989.
[118] SCEDROV, A., Kleene computable functionals and the higher-order existence property, *J. Pure Appl. Algebra* **52** (1988) 313–320.
[119] SCHMIDT, D.A., *Denotational Semantics* (Allyn and Bacon, Newton, MA, 1986).
[120] SCHOETT, O., Data abstraction and the correctness of modular programs, Tech. Report CST-42-87, Univ. of Edinburgh, 1987.
[121] SCOTT, D.S., Relating theories of the lambda calculus, in: *To H.B. Curry: Essays on Combinatory Logic, Lambda Calculus and Formalism* (Academic Press, New York, 1980) 403–450.
[122] SEELY, R.A.G., Locally cartesian closed categories and type theory, *Math. Proc. Cambridge Philos. Soc* **95** (1984) 33–48.
[123] SEELY, R.A.G., Categorical semantics for higher-order polymorphic lambda calculus, *J. Symbolic Logic* **52** (1987) 969–989.
[124] SETHI, R., *Programming Languages: Concepts and Constructs* (Addison-Wesley, Reading, MA, 1989).
[125] STANSIFER, R., Type inference with subtypes, in: *Proc. 15th ACM Symp. on Principles of Programming Languages* (1988) 88–97.
[126] STATMAN, R., Intuitionistic propositional logic is polynomial-space complete, *Theoret. Comput. Sci.* **9** (1979) 67–72.
[127] STATMAN, R., Number theoretic functions computable by polymorphic programs, in: *Proc. 22nd Ann. IEEE Symp. on Foundations of Computer Science* (1981) 279–282.
[128] STATMAN, R., Completeness, invariance and λ-definability, *J. Symbolic Logic* **47** (1) (1982) 17–26.
[129] STATMAN, R., Equality between functionals, revisited, in: *Harvey Friedman's Research on the Foundations of Mathematics* (North-Holland, Amsterdam, 1985) 331–338.
[130] STATMAN, R., Logical relations and the typed lambda calculus, *Inform. and Control* **65** (1985) 85–97.
[131] STOY, J.E., *Denotational Semantics: The Scott–Strachey Approach to Programming Language Theory* (MIT Press, Cambridge, MA, 1977).
[132] STRACHEY, C., Fundamental concepts in programming languages, Lecture notes, International Summer School in Computer Programming, Copenhagen, 1967.
[133] TAIT, W.W., Intensional interpretation of functionals of finite type, *J. Symbolic Logic* **32** (1967) 198–212.
[134] TAYLOR, P., Recursive domains, indexed category theory and polymorphism, Ph.D. Thesis, Mathematics Dept., Cambridge Univ., 1987.
[135] TROELSTRA, A.S., *Mathematical Investigation of Intuitionistic Arithmetic and Analysis*, Lecture Notes in Mathematics, Vol. 344 (Springer, Berlin, 1973).
[136] TURNER, D.A., Miranda: a non-strict functional language with polymorphic types, in: *IFIP Internat. Conf. on Functional Programming and Computer Architecture*, Lecture Notes in Computer Science, Vol. 201 (Springer, New York, 1985).

[137] US Dept. of Defense, Reference Manual for the Ada Programming Language, GPO 008-000-00354-8, 1980.

[138] VAN DALEN, D.T., The language theory of Automath, Ph.D. Thesis, Technological Univ. Eindhoven, 1980.

[139] WAND, M., Complete type inference for simple objects, in: *Proc. 2nd IEEE Symp. on Logic in Computer Science* (1987) 37–44; Corrigendum in: *Proc. 3rd IEEE Symp. on Logic in Computer Science* (1988) 132.

[140] WAND, M., A simple algorithm and proof for type inference, *Fund. Inform.* **10** (1987) 115–122.

CHAPTER 9

Recursive Applicative Program Schemes

Bruno COURCELLE

Laboratoire d'Informatique, Université Bordeaux I, 351 Cours de la Libération, F-33405 Talence, France

Contents

Introduction 461
1. Preliminary examples 461
2. Basic definitions 463
3. Operational semantics in discrete interpretations 466
4. Operational semantics in continuous interpretations 471
5. Classes of interpretations 475
6. Least fixed point semantics 481
7. Transformations of program schemes 485
8. Historical notes, other types of program schemes and guide to the literature . . . 488
Acknowledgment 490
References 490

HANDBOOK OF THEORETICAL COMPUTER SCIENCE
Edited by J. van Leeuwen

Introduction

Programming languages are very difficult to define and study formally for several different reasons. One reason is that most programming languages are very complex. They must handle many different concepts such as objects of different types, overloading of operators, declarations, and default options, to mention a few.

Another reason is much more fundamental and applies to a very simple language like *pure LISP*, the syntax and semantics of which can be described on one or two pages. As soon as a language is *universal*, i.e., as soon as it can implement every algorithm (or more precisely has the power of Turing machines), important properties such as correctness and termination cannot be decided. This means, roughly speaking, that they cannot be formalized in any useful way.

Program schemes have been introduced in order to overcome (as much as possible) these difficulties.

A programming language is built with different concepts that are related, but somewhat independent:
- data types and data domains
- control structures.

Program schemes formalize the *control structure* of programs, independently of the basic domains involved in the computations. Hence they help to distinguish precisely the control structure from the other aspects. They are also useful for formalizing the top-down construction of programs.

The decision problems for programs can be reformulated in terms of program schemes and then investigated. This makes it possible to understand whether undecidability results arise from the interpretation (for example, the domain of integers the first-order theory of which is undecidable), or from the control structure. From the results of Paterson [61], Friedman [32] and Courcelle [13] one can conclude that undecidability generally arises from the control structure.

A program scheme represents a family of similar programs. Transformations of program schemes represent transformations of the associated programs. The notion of transformation motivates the study of equivalence of program schemes w.r.t. classes of interpretations. There are many types of program schemes. Their multiplicity reflects the multiplicity of control structures used in programming languages and also the multiplicity of abstraction levels at which one can study these control structures.

In this chapter, we give a detailed survey of *recursive applicative program schemes*. We have chosen to present this class because it has a very rich mathematical theory. The last section of this chapter gives an overview of the historical development of the notion of program scheme and suggests further reading on recursive applicative program schemes.

1. Preliminary examples

We present a few examples before introducing the formal definitions.

1.1. EXAMPLE (*factorial function*). Consider the recursive definition

$$\mathbf{Fac}(x) = \mathbf{if}\ x = 0\ \mathbf{then}\ 1\ \mathbf{else}\ \mathbf{Fac}(x-1) * x$$

where $x \in \mathbf{N}$ (and $*$ denotes multiplication). It defines a unique (well-known!) function: $\mathbf{N} \to \mathbf{N}$. If the domain of computation is $\mathbf{Z}$ instead of $\mathbf{N}$ then it defines a partial function: $\mathbf{Z} \to \mathbf{Z}$ (actually the same as when the domain is $\mathbf{N}$ since the computation of $\mathbf{Fac}(x)$ does not terminate for $x < 0$).

1.2. EXAMPLE (*reversal of a word*). Consider now

$$\mathbf{Rev}(x) = \mathbf{if}\ x = \varepsilon\ \mathbf{then}\ \varepsilon\ \mathbf{else}\ \mathbf{Rev}(\mathbf{hd}(x)).\mathbf{tl}(x)$$

where $x \in X^*$ (denoting the set of words over alphabet X), ε denotes the empty word, . denotes the concatenation of two words, $\mathbf{hd}(x)$ denotes the *head*, i.e., the first letter of a nonempty word x, and $\mathbf{tl}(x)$ denotes its *tail*, i.e., the word such that $x = \mathbf{hd}(x).\mathbf{tl}(x)$. It is clear that **Rev** defines the reversal of a word, i.e., the mapping associating the word $a_k a_{k-1} \ldots a_1$ with $x = a_1 a_2 \ldots a_k$, where $a_1, \ldots, a_k \in X$.

1.3. EXAMPLE (*Ackermann's function*). Consider the recursive definition

$$\begin{aligned}\mathbf{Ack}(x, y) = {}&\mathbf{if}\ x = 0\ \mathbf{then}\ y + 1\\ &\mathbf{else\ if}\ y = 0\ \mathbf{then}\ \mathbf{Ack}(x-1, 1)\\ &\mathbf{else}\ \mathbf{Ack}(x-1, \mathbf{Ack}(x, y-1))\end{aligned}$$

where $x, y \in \mathbf{N}$. It defines a total function: $\mathbf{N} \times \mathbf{N} \to \mathbf{N}$. The proof of this function being total is not as immediate as in the two preceding examples.

1.4. EXAMPLE (*homomorphic image of an infinite word*). Less classical than the above examples but meaningful as we shall see in Section 4 is the following recursive definition:

$$\mathbf{Hom}(x) = f(\mathbf{hd}(x)).\mathbf{Hom}(\mathbf{tl}(x))$$

where $x \in X^\omega$ (the set of infinite words over X), $\mathbf{hd}(x)$ is (as in Example 1.2) the leftmost letter of x, and $\mathbf{tl}(x)$ denotes its tail, in this case a word in X^ω. The symbol . again denotes concatenation of two words. In the definition of **Hom**, f is any function: $X \to X^+$ (the set of nonempty words) so that **Hom** is its extension into a function: $X^\omega \to X^\omega$.

These concrete recursive definitions can be written and studied in terms of the following recursive program schemes.

1.5. EXAMPLE (*a recursive program scheme for factorial and reversal*). The expression

$$\varphi(x) = c(p(x), a, b(\varphi(d(x)), e(x)))$$

is called a *recursive program scheme*. As written here, it is purely syntactical. It represents the recursive definition of *factorial* if we take $\mathbf{N}$ as domain of computation and as meaning for a, b, c, d, e and p respectively the constant 1, the multiplication, the conditional (**if** ... **then** ... **else** ...), the decrement function ($\lambda x.x-1$), the identity and the equality to-0 test. We shall say, more shortly, that we take for φ the *interpretation* $\boldsymbol{I}=\langle\mathbf{N}, a_I, b_I, c_I, d_I, e_I, p_I\rangle$ where $a_I, b_I, \ldots$ represent the semantical objects associated with the symbols $a,b,\ldots$, i.e., the constants and functions associated with $a,b,\ldots$ as above.

We shall also say that the recursive definition of factorial in Example 1.1 is an *instance* of the program scheme φ or, more precisely, the *instance of* φ associated with *the interpretation* $\boldsymbol{I}$.

The recursive definition of *reversal* (given in Example 1.2) is the instance of φ associated with the interpretation $\boldsymbol{J}$ the domain of which is X^* and such that

- a_J is ε (the empty word),
- b_J is **.**, the concatenation of words,
- c_J is **if** ... **then** ... **else** ...,
- d_J is **tl**, the tail function,
- e_J is **hd**, the head function,
- p_J is the equality to-ε test.

It is clear that every given recursive definition is an instance of several recursive program schemes.

1.6. EXAMPLE (*a recursive program scheme for Ackermann's function*). We can take

$$\psi(x, y)=c(p(x), f(y), c(p(y), \psi(d(x), a), \psi(d(x), \psi(x, d(y)))))$$

with the interpretation $\boldsymbol{I}'$ consisting of $\boldsymbol{I}$ defined in Example 1.5 and enriched with the increment function (i.e., $\lambda x . x+1$) for function $f_{I'}$.

In this chapter we shall investigate systems of mutually recursive definitions rather than single recursive definitions.

2. Basic definitions

2.1. *Many-sorted algebras*

Many-sorted algebras are treated in detail in the chapter by Wirsing [76] this Handbook; other references are [29, 39]. We mainly review the definitions and notations.

We shall denote by $\mathscr{S}$ a finite set of sorts, by F a finite $\mathscr{S}$-signature, by $\alpha(f)$ the arity (in $\mathscr{S}^*$) of an element f of F, by $\sigma(f)$ its sort (in $\mathscr{S}$), by $\boldsymbol{M}=\langle(M_s)_{s\in\mathscr{S}}, (f_M)_{f\in F}\rangle$ an F-algebra, by $\boldsymbol{M}(F)$ the initial F-algebra and $\boldsymbol{M}(F, X)$ the free F-algebra generated by X where X is a set given with a sort mapping $\sigma: X\rightarrow\mathscr{S}$.

The domain of sort s of $M(F, X)$ is denoted by $M(F, X)_s$ and consists of all well-formed terms written with F, X that are of sort s. When talking of $t\in M(F, X)_s$ we shall refer to X as the set of variables. The pair $(\alpha(f), \sigma(f))$ is written $\alpha(f)\rightarrow\sigma(f)$ and is called the *profile* of f.

An *ordered F-algebra* is an object

$$M=\langle (M_s)_{s\in\mathcal{S}}, (\leqslant_s)_{s\in\mathcal{S}}, (\bot_s)_{s\in\mathcal{S}}, (f_M)_{f\in F}\rangle$$

where $\leqslant_s$ is a partial order on M_s with least element $\bot_s$ for each $s\in\mathcal{S}$. The functions f_M are assumed to be monotone, i.e., to satisfy

$$f_M(d_1,\ldots d_n)\leqslant_s f_M(d'_1,\ldots,d'_n)$$

for all $d_1, d'_1\in M_{s_1},\ldots,d_n, d'_n\in M_{s_n}$ such that $d_1\leqslant_{s_1} d'_1,\ldots,d_n\leqslant_{s_n} d'_n$ where $s=\sigma(f)$ and $s_1\ldots s_n=\alpha(f)$.

We say that M is *continuous* if the partial orders $\leqslant_s$ are *complete*, i.e., if every increasing chain in M_s has a least upper bound (equivalently, if every countable directed subset of M_s has a least upper bound). The least upper bound of a directed subset A of M_s is denoted by **Sup**(A). In the definition of a continuous F-algebra, one also assumes that the functions f_M are *continuous*, i.e., that for all directed subsets A_1 of $M_{s_1},\ldots,A_n$ of M_{s_n}, the directed set $f_M(A_1,\ldots,A_n)=\{f_M(a_1,\ldots,a_n)\mid a_1\in A_1,\ldots, a_n\in A_n\}$ has a least upper bound such that

$$\mathbf{Sup}(f_M(A_1,\ldots,A_n))=f_M(\mathbf{Sup}(A_1),\ldots \mathbf{Sup}(A_n)).$$

The reader will find more details on partially ordered sets considered as semantic domains in the chapters of this Handbook by Mosses [56] and Scott and Gunter [66].

Letting Ω_s be a new constant (i.e., a new nullary symbol) of sort s, one can order the sets $M_\Omega(F,X)_s=_{\text{def}} M(F',X)_s$ for $s\in\mathcal{S}$ by the partial orders $<_s$ defined as the least orders such that $M_\Omega(F,X)=\langle M_\Omega(F,X)_s, (<_s)_{s\in\mathcal{S}}, (\Omega_s)_{s\in\mathcal{S}}, (f_{M(F',X)})_{f\in F}\rangle$ is an ordered F-algebra. (We denote by F' the $\mathcal{S}$-signature $F\cup\{\Omega_s\mid s\in\mathcal{S}\}$). This makes $M_\Omega(F,X)$ to the free *ordered* F-algebra generated by X. There also exists a free *continuous* F-algebra generated by X, denoted by $M_\Omega^\infty(F,X)$.

As mentioned above, the domains of the free F-algebra generated by X, $M(F,X)$, can be concretely constructed as sets of well-formed terms or, equivalently, of finite ordered trees. The domains of the free continuous F-algebra $M_\Omega^\infty(F,X)$ can be concretely constructed as sets of finite and infinite trees. (Finite trees correspond to finite terms in a usual way; infinite trees extend these trees just by allowing infinite paths.) $M_\Omega^\infty(F,X)$ can also be constructed as the *ideal completion* of $M_\Omega(F,X)$. (This notion is presented in Section 5). Infinite trees and their properties relevant to program schemes are investigated in depth in [15]. They are also investigated by Thomas in this Handbook [71], but for a different purpose.

Given any two F-algebras (respectively ordered F-algebras, or continuous F-algebras) M and M', a homomorphism h: $M\to M'$ is a family of mappings $(h_s)_{s\in\mathcal{S}}$, h_s: $M_s\to M'_s$, that preserve the F-operations (respectively map $\bot_s$ to $\bot'_s$ and are monotone, or map $\bot_s$ to $\bot'_s$ and are continuous). The unique homomorphism: $M(F,X)\to M$ (respectively the unique homomorphism of ordered algebras: $M_\Omega(F,X)\to M$, or the unique homomorphism of continuous algebras: $M_\Omega^\infty(F,X)\to M$) that extends a given sort-preserving mapping v: $X\to\bigcup\{M_s\mid s\in\mathcal{S}\}$ is denoted by $h_{M,v}$. If $X=\emptyset$, it is denoted by h_M.

In every F-algebra M, a term t in $M(F, \{x_1,\ldots,x_n\})_s$ defines a mapping

$t_M: M_{s_1} \times \cdots \times M_{s_n} \to M$ (where $s_i = \sigma(x_i)$ for $i = 1, \ldots, n$), such that

$$t_M(d_1, \ldots, d_n) = h_{M,v}(t)$$

where v maps x_i to d_i. (This notation assumes a linear order on the finite set of variables; this order is usually defined by the context).

This notation extends naturally to $t \in \boldsymbol{M}_\Omega(F, \{x_1, \ldots, x_n\})$when $\boldsymbol{M}$ is ordered and to $t \in \boldsymbol{M}_\Omega^\infty(F, \{x_1, \ldots, x_n\})$ when $\boldsymbol{M}$ is continuous.

We denote by X_n the set of variables $\{x_1, \ldots, x_n\}$ linearly ordered in this way. The sort mapping $\sigma: X_n \to \mathscr{S}$ is assumed to be given but is not made explicit in the notations.

2.2. *Recursive applicative program schemes*

Let $\mathscr{S}$ be a finite set of sorts, F a finite $\mathscr{S}$-signature, and Φ another $\mathscr{S}$-signature disjoint from F, $\Phi = \{\varphi_1, \ldots, \varphi_n\}$. A *system of equations over F with set of unknowns* Φ is an N-tuple of equations of the form $\Sigma = \langle \varphi_1(x_{1,1}, \ldots, x_{1,n}) = t_1, \ldots, \varphi_n(x_{n,1}, \ldots, x_{N,n_N}) = t_N \rangle$ where, for each $i = 1, , \ldots, N$, $t_i \in M(F \cup \Phi, \{x_{i,1}, \ldots, x_{i,n_i}\})_{s_i}$, $s_i = \sigma(\varphi_i)$ and $\sigma(x_{i,1}) \ldots \sigma(x_{i,n}) = \alpha(\varphi_i)$. This definition assumes a linear order on Φ. (Usually, Φ is $\{\varphi_1, \ldots, \varphi_N\}$ enumerated in this order.)

A *recursive applicative program scheme* is a pair $S = (\Sigma, t)$ where Σ is as above and $t \in M(F \cup \Phi, X_n)$. We denote by $\alpha(S)$ the sequence $\sigma(x_1) \ldots \sigma(x_n)$ and by $\sigma(S)$ the sort of t; we say that $\alpha(S)$ is the *arity* of S, that $\sigma(S)$ is its *sort* and that $\alpha(S) \to \sigma(S)$ is its profile. We say that $S = (\Sigma, t)$ is a *scheme* for brevity. In such a pair, t is the "main program" and Σ is a set of "auxiliary", mutually recursive procedures.

An F-*interpretation* $\boldsymbol{D}$ is a continuous F-algebra.

2.3. *Equivalent schemes*

Having defined the syntax of schemes and the class of interpretations we shall define in the next two sections the *semantical mapping*, i.e., the mapping associating with a scheme $S = (\Sigma, t)$ of profile $s_1 \ldots s_n \to s$ and with an F-*interpretation* $\boldsymbol{D}$ the *function computed by S in* $\boldsymbol{D}$, denoted by $S_{\boldsymbol{D}}$. It will be a mapping $D_{s_1} \times \cdots \times D_{s_n} \to D_s$.

Several distinct semantical mappings can be defined (call-by-name, call-by-value, call-by-need) as we shall see below. In the context of a fixed semantical mapping, the *equivalence* of two schemes S and S' w.r.t. a class of interpretations $\mathscr{C}$ is defined as follows: S and S' are $\mathscr{C}$-*equivalent*, denoted by $S \equiv_{\mathscr{C}} S'$, iff $\alpha(S) = \alpha(S')$, $\sigma(S) = \sigma(S')$, and $S_{\boldsymbol{D}} = S'_{\boldsymbol{D}}$ for all $\boldsymbol{D} \in \mathscr{C}$.

The need for introducing classes of interpretations and for relativizing equivalences of schemes to them will be explained in Section 5. We shall frequently say that a system Σ is a scheme. In this case the term t (i.e., the "main program") is the left-hand side of the first equation of Σ.

Examples 1.1–1.3 use two sorts: the sort $\boldsymbol{b}$ (for Boolean values) and the sort $\boldsymbol{o}$ (for objects, either integers or words according to the interpretation). The profiles of the symbols $a, b, \ldots$ used in Examples 1.1–1.3 are as follows:

$$a: \to \boldsymbol{o}, \qquad b: \boldsymbol{oo} \to \boldsymbol{o}, \qquad c: \boldsymbol{ooo} \to \boldsymbol{o},$$

$$d, e, f: \boldsymbol{o} \to \boldsymbol{o}, \qquad p: \boldsymbol{o} \to \boldsymbol{b}.$$

The interpretations **I** and **J** used in these examples are not algebras because the functions d_I, d_J, e_J are partial and not total.

We shall deal with this problem in the next two sections where we shall motivate the above notion of an interpretation and define the semantical mapping.

3. Operational semantics in discrete interpretations

Rather than the general notion of an interpretation, we first present a special case: the notion of a discrete interpretation. This notion is sufficient for Examples 1.1–1.3.

3.1. *Partial functions and flat partial orders*

Let $f: D \times \cdots \times D \to D$ be a partial function with n arguments. One can make it into a total function by adding to D a new element $\perp$ (read "bottom") standing for "undefined", and by giving $f(d_1, \ldots, d_n)$ the value $\perp$ when it would be otherwise undefined. So let $\bar{D} = D \cup \{\perp\}$ with $\perp$ not in D. Let $\bar{f}: \bar{D}^n \to \bar{D}$ be such that

$$\bar{f}(d_1, \ldots, d_n) = \begin{cases} f(d_1, \ldots, d_n) & \text{if } d_1, \ldots, d_n \in D \text{ and } f(d_1, \ldots, d_n) \text{ is defined,} \\ \perp & \text{otherwise.} \end{cases}$$

It is convenient to consider $\bar{D}$ as ordered by $\leqslant$ such that

$$d \leqslant d' \text{ iff } d = d' \text{ or } d = \perp \text{ (or both).} \tag{3.1}$$

By reference to the representation of $(\bar{D}, \leqslant, \perp)$ by a Hasse diagram, such an ordered set is called *flat*. These domains are the "primitive domains" of Mosses [56].

It is clear that $(\bar{D}, \leqslant, \perp)$ is complete and $\bar{f}$ is monotone (hence, continuous since every monotone function is continuous in an ordered set where all increasing chains are finite). A monotone function $g: \bar{D}^n \to \bar{D}$ is *total* if $g(d_1, \ldots, d_n) \neq \perp$ for all $d, \ldots, d_n \in D$. It is *strict* if $g(d_1, \ldots, d_n) = \perp$ whenever $d_i = \perp$ for some i. It is of the form $\bar{f}$ for some partial (respectively total) function $f: D^n \to D$ iff it is strict (respectively strict and total).

3.2. *Discrete interpretations*

A *discrete F-interpretation* is an ordered F-algebra of which all partial orders are flat. It follows from a previous remark that such an interpretation is also a continuous F-algebra.

The least element of the domian D_s of sort s of a discrete interpretation $\boldsymbol{D}$ will always be denoted by $\perp_s$. The order $\leqslant_s$ on D_s is thus characterized by (3.1).

Let us define completely the discrete interpretation $\boldsymbol{D}$ where the recursive program schemes associated with Examples 1.1 and 1.3 can be computed.

$$\boldsymbol{D}=\langle D_b, D_o, a_D, b_D, \ldots\rangle;$$
$$D_b=\{\textbf{true}, \textbf{false}, \bot_b\}, \qquad D_o=\mathbf{N}\cup\{\bot_o\};$$
$$a_D=1;$$
$$b_D(x,y)=\begin{cases} x*y & \text{if } x,y\neq\bot,\\ \bot & \text{otherwise;}\end{cases}$$
$$c_D(\bot_b,x,y)=\bot_o,\quad c_D(\textbf{true},x,y)=x,\quad c_D(\textbf{false},x,y)=y \quad \text{for all } x,y\in D_o;$$
$$d_D(\bot_b)=d_D(0)=\bot_b, \qquad d_D(x)=x-1 \quad \text{otherwise;}$$
e_D is the identity;
$$f_D(x)=\begin{cases} x+1 & \text{if } x\neq\bot_o,\\ \bot_o & \text{otherwise;}\end{cases}$$
$$p_D(\bot_o)=\bot_b, \qquad p_D(0)=\textbf{true}, \qquad p_D(x)=\textbf{false} \quad \text{otherwise.}$$

Note that c_D is not strict. (The framework of algebras with partial functions is insufficient to define rigorously an interpretation for φ defined in Example 1.5 so as to obtain **Fac**, the factorial function.)

3.3. *Evaluation by rewriting*

The evaluation of **Fac**(2) can be considered as the result of a sequence of rewritings namely:

$$\begin{aligned}\textbf{Fac}(2) &\to \textbf{if } 2=0 \textbf{ then } 1 \textbf{ else } \textbf{Fac}(2-1)*2\\ &\to \textbf{if false then } 1 \textbf{ else } \textbf{Fac}(2-1)*2\\ &\to \textbf{Fac}(2-1)*2\\ &\to \textbf{Fac}(1)*2\\ &\to \cdots\\ &\to 1*2\\ &\to 2.\end{aligned}$$

The rewritten expressions are built with **Fac** (the “unknown” function symbol), the basic operations ($-, *,$ **if**... **then**... **else**) and denotations for the elements of the domain. The rewriting steps use either the definition of **Fac** or the basic operations. We shall formalize these rewritings and obtain the desired semantical mapping.

Let $\Sigma=\langle\varphi_1(x_{1,1},\ldots)=t_1,\ldots,\varphi_N(x_{N,1},\ldots)=t_N\rangle$ be a system over F. We can consider it as the term rewriting system $\{\varphi_i(x_{i,1},\ldots)\to t_i \mid i\in[N]\}$. This term rewriting system is also denoted by Σ, so that for every set of nullary symbols A a binary relation $\underset{\Sigma}{\to}$ on $M(F\cup\Phi, A)$ is associated with Σ. (We let $\Phi=\{\varphi_1,\ldots,\varphi_N\}$). We refer the reader to the chapter by Dershowitz and Jouannaud [28] for the basic definitions concerning term rewriting systems.

Let $\boldsymbol{D}$ be a discrete F-interpretation. Let us consider every element d of $D=\bigcup\{D_s \mid s\in \mathcal{S}\}$ as a nullary symbol. It is of sort s if it belongs to D_s. Hence $M(F\cup\Phi, D)_s$ is well-defined and is the set of terms of sort s built with the basic function symbols, the unknown function symbols and symbols denoting the elements of $\boldsymbol{D}$. Let $U(\boldsymbol{D})$ be the set of pairs of terms (t,t') in $M(F\cup\{\Omega\},\ X\cup D\cup\{\bot\})$ such that $t=f(t_1,\ldots,t_n)$ for some $f\in F$, where

(1) $t',\ t_i\in X\cup D$ (for all $i=1,\ldots,n$),
(2) t is X-linear, i.e., no variable (of X) occurs twice (or more) in t, and
(3) if t' is a variable x then x occurs in t.

Let $S(\boldsymbol{D})\subseteq U(\boldsymbol{D})$ be the set of all such pairs such that $t_{\boldsymbol{D}}=t'_{\boldsymbol{D}}$ and neither Ω nor $\bot$ occurs in (t,t').

If $\boldsymbol{D}$ is the interpretation defined above, then $S(\boldsymbol{D})$ contains the following pairs (a pair (t,t') is written $t\to t'$):

$$b(2,3)\to 6, \qquad c(\mathbf{true},x,y)\to x,$$
$$c(\mathbf{false},x,y)\to y, \qquad c(\mathbf{true},x,8)\to x.$$

The set $S(\boldsymbol{D})$ will be used as a rewriting system on $M(F\cup\Phi,D)$ making it possible to reduce terms according to operations of $\boldsymbol{D}$. This is a set of *simplification rules* in the sense of Raoult and Vuillemin [63]. We let $\xrightarrow[S(D)]{}$ denote the associated rewriting relation on $M(F\cup\Phi,D)$ and $\xrightarrow[\Sigma,S(D)]{} = \xrightarrow[\Sigma]{}\cup\xrightarrow[S(D)]{}$. The sequence of rewritings $\mathbf{Fac}(2)\xrightarrow{*}2$, taken above as an example, is of the form $t\xrightarrow[\Sigma,S(D)]{*}d$ with $t\in M(F\cup\Phi,D)$ and $d\in D$.

3.1. PROPOSITION. *Let $t\in M(F\cup\Phi,D)_s$. There exists at most one element d in D such that that*

$$t\xrightarrow[\Sigma,S(D)]{*}d \quad \text{and} \quad d\in D_s-\{\bot_s\}.$$

We do not prove this proposition because Proposition 4.4 will state a more general result.

3.4. *The semantical mapping*

By Proposition 3.1, the following is well-defined. For $t\in M(F\cup\Phi,D)_s$ we let

$$t_{\Sigma,D}=\begin{cases} d & \text{if } t\xrightarrow[\Sigma,S(D)]{*}d,\ d\neq\bot_s,\\ \bot_s & \text{if no such } d \text{ exists.}\end{cases}$$

Observe that the value $\bot_s$ is not obtained as the result of a finite computation but only if no finite computation produces a defined (i.e., $\neq\bot_s$) value. The rewriting steps using $S(\boldsymbol{D})$ can only produce defined values or expressions to be evaluated later, since we have excluded $\bot_s$ from the right-hand sides.

A rewriting sequence $t\xrightarrow[\Sigma,S(D)]{*}d$ with $d\neq\bot_s$ is called a *terminated computation*.

If $S=(\Sigma,m)$ is a scheme with $m\in M(F\cup\Phi,\{x_1,\ldots,x_n\})_s$, then the function computed by S in $\boldsymbol{D}$ is $S_{\boldsymbol{D}}: D_{s_1} \times \cdots \times D_{s_n} \to D_s$ (where $s_1 \ldots s_n = \alpha(S)$) defined by

$$S_{\boldsymbol{D}}(d_1, \ldots, d_n) = (m[d_1/x_1, \ldots, d_n/x_n])_{\Sigma, \boldsymbol{D}}. \tag{3.2}$$

Let us also define

$$\varphi_{i\boldsymbol{D}}(d_1, \ldots, d_{n_i}) = (\varphi_i(d_1, \ldots, d_{n_i}))_{\Sigma, \boldsymbol{D}} \tag{3.3}$$

for all $i \in [N]$ ($t[t_1/x_1, \ldots, t_n/x_n]$ denotes the result of the simultaneous substitution of t_i for x_i in t, for all $i = 1, \ldots, n$.) We shall prove below (Theorem 6.2) that $S_{\boldsymbol{D}} = m_{\boldsymbol{D}'}$ where $\boldsymbol{D}'$ is the $(F \cup \Phi)$-algebra consisting of $\boldsymbol{D}$ enriched with $\varphi_{i\boldsymbol{D}}$ as value of φ_i, for all $i = 1, \ldots, n$ and that $(\varphi_{1\boldsymbol{D}}, \ldots, \varphi_{n\boldsymbol{D}})$ is the least solution of Σ in $\boldsymbol{D}$. None of these properties trivially follows from the above definitions

3.5. *Computation rules*

The $\xrightarrow[\Sigma, S(\boldsymbol{D})]{}$ rewriting step can be considered as defining a nondeterministic computation rule in the following sense: given any term t in $M(F \cup \Phi, D)$ considered as a state of computation, every term t' such that $t \xrightarrow[\Sigma, S(\boldsymbol{D})]{} t'$ is a possible next state. If $t' \in D$ then t' is a final state, i.e., a possible result of the computation.

Proposition 3.1 *says that any two (finite) terminated computation sequences yield the same result.* But there may exist infinite computation sequences (yielding no result) as well as (finite) terminated ones.

A *computation rule* is an algorithm R that, given a system Σ over F with set of unknowns Φ, given a discrete F-interpretation $\boldsymbol{D}$ and a term $t \in M(F \cup \Phi, D)$, produces a unique t' such that $t \xrightarrow[\Sigma, S(\boldsymbol{D})]{} t'$, or reports that no such t' exists. We shall then write $t \xrightarrow[\Sigma, S(\boldsymbol{D}), R]{} t'$. By using $t \xrightarrow[\Sigma, S(\boldsymbol{D}), R]{}$ instead of $\xrightarrow[\Sigma, S(\boldsymbol{D})]{}$ one defines $S_{\boldsymbol{D},R}$ and $\varphi_{i\boldsymbol{D},R}$ as $S_{\boldsymbol{D}}$ and $\varphi_{i\boldsymbol{D}}$, by (3.2) and (3.3) above. It is clear that

$$S_{\boldsymbol{D},R} \leqslant S_{\boldsymbol{D}}, \qquad \varphi_{i\boldsymbol{D},R} \leqslant \varphi_{i\boldsymbol{D}}$$

(where the inclusions may be strict). The equality $S_{\boldsymbol{D},R} = m_{\boldsymbol{D}'}$, where $\boldsymbol{D}' = \langle \boldsymbol{D}, (\varphi_{i\boldsymbol{D},R})_{i \in [N]} \rangle$ does not necessarily hold.

Two well-known computation rules are *call-by-name* and *call-by-value*. We define them by using the notion of a *redex*, borrowed from λ-calculus. We first make a few technical assumptions in order to simplify the notations: there is only one sort in $\mathscr{S}$; all functions $f_{\boldsymbol{D}}$ are strict, except the **if**... **then**... **else**... denoted as in the examples of Section 1 by $c(\ldots,\ldots,\ldots)$; we assume that the Boolean values **true** and **false** are represented by objects (for instance 0 and 1 if one deals with integers); they will still be denoted by **true** and **false** for readability.

Let $\Sigma = \langle \varphi_i(x_{i,1}, \ldots, x_{i,n_i}) = t_i;\ i \in [N] \rangle$ be fixed. An *N-redex* is an expression in $M(F \cup \Phi, D)$ in one of the four possible forms listed below (the prefix N stands for "call-by-name"). Together with the syntax of an N-redex we give the result of its *reduction*:

(i) $\varphi_i(s_1, \ldots, s_{n_i})$ for $i \in [N]$, $s_1, \ldots, s_{n_i} \in M(F \cup \Phi, D)$, yielding $t_i[s_1/x_{i,1}, \ldots, s_{n_i}/x_{i,n_i}]$;

(ii) $c(\mathbf{true}, s_1, s_2)$ for $s_1, s_2 \in M(F \cup \Phi, D)$, yielding s_1;

(ii′) $c(\mathbf{false}, s_1, s_2)$ for $s_1, s_2 \in M(F \cup \Phi, D)$, yielding s_2;

(iii) $f(d_1, \ldots, d_k)$ for $f \in F$, $\rho(f) = k$, $d_1, \ldots, d_k \in D - \{\bot\}$, yielding $f_D(d_1, \ldots, d_k)$.

In the first case, the redex reduces by $\xrightarrow[\Sigma]{}$, in the last three cases by $\xrightarrow[S(D)]{}$.

Consider two distinct redices r_1 and r_2 of a term t: either r_1 *is to the left of* r_2 (i.e., t can be written $w_1 r_1 w_2 r_2 w_3$ in prefix Polish notation) or r_2 is to the left of r_1, or r_2 *is inside* r_1 (i.e., t can be written $w_1 r_1 w_2$ and r_1 can be written $w_3 r_2 w_4$) or r_1 is inside r_2. There is no other case; in particular, there is no overlapping, i.e. no critical pair (in the terminology of term rewriting systems; see [28]). Hence, if an expression has N-redices, it has a *leftmost outermost* one, i.e., a unique N-redex having no N-redex to its left and being inside no other N-redex.

We can now define *call-by-name* as the rule which for every t in $M(F \cup \Phi, D)$ selects the leftmost outermost N-redex (if any) and replaces it according to that case of the above-listed four that applies.

The *call-by-value* computation rule is defined as *call-by-name* but now with respect to the class of V-redices, which we are now going define.

A *V-redex* is an expression of the possible forms (ii), (ii′), (iii) as above or of the form (i) with the extra condition that $s_1, \ldots, s_{n_i}$ all belong to $D - \{\bot\}$. The replacements are the same.

3.2. EXAMPLE (*showing that* call-by-name *and* call-by-value *yield different results*). Let

$$\varphi(x, y) = c(x = 0, 1, \varphi(x - 1, \varphi(x, y)))$$

where $x \in \mathbf{N}$. Then consider the computation of $\varphi(1, 0)$ with call-by-name:

$$\begin{aligned} \varphi(1,0) &\to c(1=0, 1, \varphi(1-1, \varphi(1,0))) \\ &\to c(\mathbf{false}, 1, \varphi(1-1, \varphi(1,0))) \\ &\to \varphi(1-1, \varphi(1,0)) \\ &\to c(1-1=0, 1, \varphi((1-1)-1, \varphi(1-1, \varphi(1,0)))) \\ &\to c(0=0, 1, \ldots) \\ &\to c(\mathbf{true}, 1, \ldots) \\ &\to 1. \end{aligned}$$

With call-by-value one obtains

$$\begin{aligned} \varphi(1,0) &\xrightarrow{*} \varphi(1-1, \varphi(1,0)) \quad \text{(as with call-by-name)} \\ &\to \varphi(0, \varphi(1,0)) \\ &\to \varphi(0, c(1=0, 1, \varphi(1-1, \varphi(1,0)))) \\ &\xrightarrow{*} \varphi(0, \varphi(0, \varphi(1,0))) \\ &\to \cdots \end{aligned}$$

and the computation does not terminate since $\varphi(1,0)$ always regenerates itself (by yielding $\varphi(0, \varphi(1,0))$).

It is possible to prove the following results [11, 75]:

$$\varphi_{iD,\text{call-by-value}} \leqslant \varphi_{i,D,\text{call-by-name}} = \varphi_{iD}$$

and the inequality may be strict as shown in Example 3.2.

4. Operational semantics in continuous interpretations

Discrete interpretations are insufficient for the following reasons:

- the product of several discrete interpretations is not a discrete interpretation; this prevents us from considering a tuple of objects as a single one in the appropriate Cartesian product domain;
- the set of monotone functions $D_1 \times \cdots \times D_n \to D$ (where $D_1, \ldots, D_n, D$ are flat partial orders) is a complete partial order but is not flat; this prevents us from defining procedures taking functions as parameters;
- the sets of infinite words cannot be handled in the framework of discrete interpretations; we shall see that it is possible to define recursive program schemes taking infinite words as arguments and producing finite results within a finite number of computation steps.

All these reasons motivate the introduction of a class of interpretations larger than the class of discrete ones. A systematic study of semantic domains can be found in the chapter by Scott and Gunter [66].

4.1. *Interpretations as continuous algebras*

An F-*interpretation* is a continuous F-algebras. In a complete partial order $(D, \leqslant, \bot)$ the distinction between $\bot$ and the other elements of D is not relevant. We must modify the definition of the operational semantics given in Section 3, which was based on this distinction, according to the new situation.

Let Σ be a system over F with set of unknowns Φ, let $\boldsymbol{D}$ be a (continuous) F-interpretation. We shall use $\xrightarrow[\Sigma]{} \subseteq M_\Omega(F \cup \Phi, D) \times M_\Omega(F \cup \Phi, D)$ as in Section 3. Additionally, we shall use the rewriting relation $\xrightarrow[\Omega]{}$ associated with the rewriting rules

$$\varphi_i(x_{i,1}, \ldots, x_{i,n}) \to \Omega_{s_i} \quad \text{for } i \in [N],\ s = \sigma(\varphi_i).$$

Letting $U(\boldsymbol{D})$ be as in Section 3, but also allowing Ω and $\bot$, we let $R(\boldsymbol{D}) \subseteq U(\boldsymbol{D})$ be the set of pairs (t, t') in $U(\boldsymbol{D})$ such that $t_{\boldsymbol{D}} = t'_{\boldsymbol{D}}$. The only difference with $S(\boldsymbol{D})$ is that $\bot_s$ and Ω_s may occur in t and t'. (If $\boldsymbol{D}$ is discrete then $S(\boldsymbol{D}) \subseteq R(\boldsymbol{D})$.) Thus we obtain a rewriting relation $\xrightarrow[R(\boldsymbol{D})]{}$ on $M_\Omega(F \cup \Phi, D)$. We let

$$\xrightarrow[\Sigma, \Omega\, R(\boldsymbol{D})]{} = \xrightarrow[\Sigma]{} \cup \xrightarrow[\Omega]{} \cup \xrightarrow[R(\boldsymbol{D})]{}.$$

We need a few lemmas on these rewriting systems.

4.1. LEMMA. *Let Σ be a system. The rewriting system $\xrightarrow[\Sigma]{}$ is confluent, i.e., for all terms t, s, s', if $t \xrightarrow[\Sigma]{*} s$ and $t \xrightarrow[\Sigma]{*} s'$, then $s \xrightarrow[\Sigma]{*} w$ and $s' \xrightarrow[\Sigma]{*} w$ for some w.*

PROOF. The rewriting system Σ has no critical pair, hence it is (trivially) strongly closed [45, p. 812]. Since it is left-linear, it is strongly confluent [45, Lemma 3.2, p. 813], hence confluent [45, Lemma 2.5, p. 801]. □

4.2. LEMMA. *Let Σ be a system, let $R \subseteq M_\Omega(F, X) \times M_\Omega(F, X)$ be a bilinear rewriting* *system. Then*

$$\xrightarrow[\Sigma,R]{*} = \xrightarrow[\Sigma]{*}\xrightarrow[R]{*} \text{ and } \xrightarrow[\Sigma,\Omega,R]{*} = \xrightarrow[\Sigma]{*}\xrightarrow[\Omega]{*}\xrightarrow[R]{*}.$$

PROOF. The first assertion is Proposition 11 of Raoult and Vuillemin's paper [63, p. 787]. The second one is easily proved with the same techniques. □

4.3. LEMMA. *Let* $\mathbf{D}$ *be a* F-*interpretation. Let* $R_0(\mathbf{D}) = R(\mathbf{D}) \cap M_\Omega(F, D) \times M_\Omega(F, D)$ *and* R *a subset of* $R(\mathbf{D})$. *For every* $t \in M_\Omega(F, D)$, $t \xrightarrow[R_0(\mathbf{D})]{*} t_D$ *and for every* $d \in D$ *if* $t \xrightarrow[R]{*} d$ *then* $t_D = d$.

PROOF. By induction on the structure of t for the first assertion, and by induction on n such that $t \xrightarrow[R]{n} d$ for the second one. □

4.4. PROPOSITION. *Let* $t \in M_\Omega(F \cup \Phi, D)_s$. *The set* $\mathbf{Der}_{\Sigma,D}(t) =_{\mathrm{def}} \{d \in D \mid t \xrightarrow[\Sigma,\Omega,R(\mathbf{D})]{*} d\}$ *is a directed subset of* D_s.

From this proposition, one can define $t_{\Sigma,D} = \mathbf{Sup}(\mathbf{Der}_{\Sigma,D}(t))$. Hence, the definitions of S_D and φ_{iD} follow by (3.2) and (3.3).

PROOF. From Lemma 4.2 it follows that if $t \xrightarrow[\Sigma,\Omega,R(\mathbf{D})]{*} d$ then there exists a t_1 such that $t \xrightarrow[\Sigma,\Omega]{*} t_1 \xrightarrow[R(\mathbf{D})]{*} d$. There exists a unique $t_2 \in M_\Omega(F, D)_s$ such that $t_1 \xrightarrow[\Omega]{*} t_2$ and, by Lemma 4.3, $t_2 \xrightarrow[R(\mathbf{D})]{*} d$. Also by Lemma 4.3, we get $t_{2D} = d$. Hence,

$$\mathbf{Der}_{\Sigma,D}(t) = \{t'_D \mid t' \in M_\Omega(F, D),\ t \xrightarrow[\Sigma,\Omega]{*} t'\}.$$

The term t can be written $w[d_1/x_1, \ldots, d_n/x_n]$ for some $w \in M_\Omega(F, \{x_1, \ldots, x_n\}) = M_\Omega(F, X_n)$ and some $d_1, \ldots, d_n \in D$ (the sorts of $x_1, \ldots, x_n$ being appropriately chosen). Then, for every $t' \in M_\Omega(F, D)$ such that $t \xrightarrow[\Sigma,\Omega]{*} t'$, there exists a $w' \in M_\Omega(F, X_n)$ such that $t' = w'[d_1, \ldots, d_n]$ and $w \xrightarrow[\Sigma,\Omega]{*} w'$. Hence,

$$\mathbf{Der}_{\Sigma,D}(t) = \{w'_D(d_1, \ldots, d_n) \mid w \xrightarrow[\Sigma,\Omega]{*} w'\},$$

where $t = w[d_1, \ldots, d_n]$. The set $L(\Sigma, w) =_{\mathrm{def}} \{w' \mid w \xrightarrow[\Sigma,\Omega]{*}{}'\}$ is directed in $M_\Omega(F, X_n)$ as shown by Nivat (see [59, 43]), and the mapping $w' \mapsto w'_D(d_1, \ldots, d_n)$ (from $M_\Omega(F, X_n)_s$ to D_s) is monotone (this is easy to prove). Hence the image of the former under the latter is precisely $\mathbf{Der}_{\Sigma D}(t)$, and is directed. □

Since $L(\Sigma, w)$ is directed, it has a least upper bound in $M_\Omega^\infty(F, X)$, the continuous algebra of infinite trees. We shall denote it by $T(\Sigma, w)$. It is clear that if $w \in M_\Omega(F \cup \Phi, X_n)_s$, then $T(\Sigma, w) \in M_\Omega^\infty(F, X_n)_s$. Since $\mathbf{M}_\Omega^\infty(F, X_n)$ is the free continuous F-algebra generated by X_n, and by the proof of Proposition 4.4, we have the following corollary.

4.5. COROLLARY. *If* $w \in M_\Omega(F \cup \Phi, X_n)_s$ *and* $d_i \in D_{\sigma(x_i)}$ *for* $i = 1, \ldots, n$ *then* $(w[d_1, \ldots, d_n])_{\Sigma,D} = T(\Sigma, w)_D(d_1, \ldots, d_n)$.

Hence the function S_D defined by a program scheme $S = (\Sigma, m)$ in an interpretation

$\boldsymbol{D}$ can be expressed as $T(\Sigma, m)_{\boldsymbol{D}}$, i.e., as the function in $\boldsymbol{D}$ associated with the (generally infinite) tree $T(\Sigma, m)$.

There are two consequences: the first one is the continuity of $S_{\boldsymbol{D}}$. It follows that $S_{\boldsymbol{D}}$ can be taken as the definition of a new base function and hence that program schemes can be defined in a modular (hierarchical) way. This is interesting for proofs of program correctness in particular. The second consequence is that the semantic function $S \mapsto S_{\boldsymbol{D}}$ can be factorized into two functions namely the function $S \mapsto T(S)$ which only deals with syntactical objects and the function $T \mapsto T_{\boldsymbol{D}}$ which does not depend on S. Hence, we have the following corollary.

4.6. COROLLARY. *Two schemes S and S' are equivalent, i.e. they define the same function in every interpretation, if the associated trees $T(S)$ and $T(S')$ are equal.*

The converse also holds, as we shall see below (Corollary 5.5).

4.2. *Finite maximal elements and terminated computations*

We now consider terminating computations in continuous interpretations. Let $\boldsymbol{D}$ be an F-interpretation.

An element d of D_s is *maximal* if for every $d' \in D_s$, $d \leqslant_s d'$ implies $d = d'$. It is *finite* if for every directed subset A of D_s, $d \leqslant \mathbf{Sup}(\mathrm{A})$ implies $d \leqslant_s d'$ for some $d' \in A$. Hence, a finite maximal element of D_s is a maximal element that cannot be the least upper bound of any directed set A without being in A. It is clear from the definition of $t_{\Sigma,\boldsymbol{D}}$ that the following result holds.

4.7. COROLLARY. *Let $d \neq \perp$ be finite maximal. Then $t_{\Sigma,\boldsymbol{D}} = d$ iff $t \xrightarrow[\Sigma,\Omega,R(\boldsymbol{D})]{*} d$.*

Computation rules can also be defined in continuous interpretations. This is not easy at all, and we refer the reader to the paper by Berry and Lévy [3] on this subject (where all relevant references can be found). Note that in a discrete interpretation, all non-$\perp$ elements are finite maximal. It follows that Corollary 4.7 is close to Proposition 3.1.

We have not yet established that the semantical function defined in Section 3 for discrete interpretations coincides with the one defined here for continuous ones. But this is actually an immediate consequence of the following lemma.

4.8. LEMMA. *Let $\boldsymbol{D}$ be a discrete F-interpretation. Let $t \in M_\Omega(F \cup \Phi, D)$ and $d \in D - \{\perp\}$. Then*

$$t \xrightarrow[S(\boldsymbol{D})]{*} d \ \text{ iff } \ t_{\Omega\boldsymbol{D}} = d \ \text{ iff } \ t_\Omega \xrightarrow[R_0(\boldsymbol{D})]{*} d$$

where t_Ω is the unique element of $M_\Omega(F, D)$ such that $t \xrightarrow[\Omega]{*} t_\Omega$.

PROOF. If $t \xrightarrow[S(\boldsymbol{D})]{*} d$ then $t_\Omega \xrightarrow[S(\boldsymbol{D})]{*} d$ by Lemma 4.2 applied to $R = S(\boldsymbol{D})^{-1}$ and $t_{\Omega\boldsymbol{D}} = d$ by Lemma 4.3 since $S(\boldsymbol{D}) \subseteq R(\boldsymbol{D})$.

Let $t_\Omega = d \neq \perp$. We prove that $t \xrightarrow[S(\boldsymbol{D})]{*} d$ by induction on the structure of t. There are only two cases, either $t = d$ (trivial) or $t = f(t_1, \ldots, t_k)$. Let $d_i = t_{i\boldsymbol{D}}$ for $i = 1, \ldots, k$.

Without loss of generality we can assume that $d_1, \ldots, d_n = \bot$ and that $d_{n+1}, \ldots, d_k \neq \bot$. For $i = n+1, \ldots, k$, one has $t_i \xrightarrow[S(D)]{*} d_i$ by induction. Since $\boldsymbol{D}$ is discrete,

$$f_D(\bot, \ldots, \bot, d_{n+1}, \ldots, d_k) = d \neq \bot$$

implies that $f_D(d'_1, \ldots, d'_n, d_{n+1}, \ldots, d_k) = d$ for all $d'_1, \ldots, d'_n$. Hence the rule $f(x_1, \ldots, x_n, d_{n+1}, \ldots, d_k) \rightarrow d$ is in $S(\boldsymbol{D})$. Hence,

$$\begin{aligned} t = f(t_1, \ldots, t_k) &\xrightarrow[S(D)]{*} f(t_1, \ldots, t_n, d_{n+1}, \ldots, d_k) \\ &\xrightarrow[S(D)]{*} d. \qquad \square \end{aligned}$$

4.9. EXAMPLE. We illustrate Corollary 4.7 by showing that finite computations can deal with infinite words. Let $\mathscr{S} = \{\boldsymbol{i}, \boldsymbol{w}\}$ where $\boldsymbol{i}$ is the sort integer and $\boldsymbol{w}$ is the sort infinite sequence of integers. The signature consists of the following function symbols given with their profiles (infix notation is used):

$$.: \boldsymbol{i}\,\boldsymbol{w} \rightarrow \boldsymbol{w},$$

$$\mathbf{hd}: \boldsymbol{w} \rightarrow \boldsymbol{i}, \qquad \mathbf{tl}: \boldsymbol{w} \rightarrow \boldsymbol{w},$$

$$f: \boldsymbol{i} \rightarrow \boldsymbol{i}, \qquad a: \varepsilon \rightarrow \boldsymbol{i}.$$

We now describe $\boldsymbol{D}$: D_i is the flat domain of integers ($D_i = \mathbf{N} \cup \{\bot\}$) and D_w is the set $(D_i)^{\Omega}$ of infinite sequences of elements of D_i with the componentwise ordering derived from D_i. So the minimal element of D_w is the sequence $(\bot, \bot, \bot, \bot, \ldots)$.

Let us recall from Example 1.4 that $w = \mathbf{hd}_D(w) ._D \mathbf{tl}_D(w)$ for all $w \in D_w$. Here f_D is any mapping: $D_i \rightarrow D_i$, and a_D is any integer.

In the following system Σ, the variables x, y are of sort $\boldsymbol{i}$ and the variables w, u, v are of sort $\boldsymbol{w}$.

$$\Sigma \quad \begin{cases} \theta &= a.\theta, \\ \varphi(x) &= x.\varphi(f(x)), \\ \psi(w) &= f(\mathbf{hd}(w)).\psi(\mathbf{tl}(w)), \\ \mu(u, w) &= \mathbf{hd}(u).\mathbf{hd}(w).\mu(\mathbf{tl}(u), \mathbf{tl}(w)). \end{cases}$$

Hence, in $\boldsymbol{D}$, Σ defines the following objects:

- θ_D, the sequence $(a_D, a_D, a_D, \ldots)$;
- $\varphi_D(x)$, the sequence $(x, f_D(x), f_D^2(x), f_D^3(x), \ldots)$, where x is an integer;
- $\psi_D(w)$, the sequence $(f_D(w_1), f_D(w_2), f_D(w_3), \ldots)$, where $w = (w_1, w_2, w_3, \ldots)$ with $w_1, w_2, w_3, \ldots$ in D_i;
- $\mu_D(u, w)$, the sequence $(u_1, w_1, u_2, w_2, u_3, w_3, \ldots)$, where $u = (u_1, u_2, \ldots)$ with $u_1, u_2, \ldots$ in D_i and w is as above.

None of these sequences can be effectively computed totally. For every $n \geqslant 1$, the nth element of any w in D_w can be expressed as $t_{nD}(w)$ where $t_1 = \mathbf{hd}(w)$, $t_{n+1} = \mathbf{hd}(\mathbf{tl}^n(w))$. Hence $(\Sigma, t_n(\theta))$, $(\Sigma, t_n(\varphi(x)))$, $(\Sigma, t_n(\psi(\varphi(x))))$ and $(\Sigma, t_n(\mu(\theta, \psi(\theta))))$ are schemes of respective profiles $\varepsilon \rightarrow \boldsymbol{i}, \boldsymbol{i} \rightarrow \boldsymbol{i}, \boldsymbol{i} \rightarrow \boldsymbol{i}, \varepsilon \rightarrow \boldsymbol{i}$, i.e., taking values in the domain of integers where, by Corollary 4.7, values can be obtained by finite computations.

We give only one example:

$$t_2(\mu(\theta, \psi(\theta)))$$

$$\mathbf{hd}(\mathbf{tl}(\overline{\mu(\theta, \psi(\theta))})) \underset{1}{\rightarrow} \mathbf{hd}(\overline{\mathbf{tl}(\mathbf{hd}(\theta).\mathbf{hd}(\psi(\theta)).\mu(\mathbf{tl}(\theta), \mathbf{tl}(\psi(\theta))))})$$

$$\underset{2}{\rightarrow} \mathbf{hd}(\overline{\mathbf{hd}(\psi(\theta)).\mu(\mathbf{tl}(\theta), \mathbf{tl}(\psi(\theta)))})$$

$$\underset{3}{\rightarrow} \mathbf{hd}(\overline{\psi(\theta)})$$

$$\underset{4}{\rightarrow} \overline{\mathbf{hd}(f(\mathbf{hd}(\theta)).\psi(\mathbf{tl}(\theta)))}$$

$$\underset{5}{\rightarrow} f(\mathbf{hd}\overline{(\theta)})$$

$$\underset{6}{\rightarrow} f(\overline{\mathbf{hd}(a.\theta)})$$

$$\underset{7}{\rightarrow} \overline{f(a)}$$

$$\underset{8}{\rightarrow} f_D(a_D).$$

At each step we have overlined the redex (in the sense of Section 3). Rewriting steps 3, 5 and 7 use the simplification rule $\mathbf{hd}(x.y) \rightarrow x$. Rewriting step 2 uses the rule $\mathbf{tl}(x.y) \rightarrow y$. Steps 1, 4, 6 use $\underset{\Sigma}{\rightarrow}$ and step 8 uses f_D.

This an example of a *call-by-need* computation (implemented in [44]). Due to space limitations, we have to refer the reader to [3] for more details.

5. Classes of interpretations

Let us recall that two program schemes are equivalent with respect to a class of interpretations $\mathscr{C}$ if they define the same function in every interpretation of $\mathscr{C}$. This equivalence relation is very restrictive if the class of all interpretations or even the class of all discrete interpretations is taken as the reference class. This means that "few" pairs of equivalent program schemes exist or that many "interesting" pairs of equivalent programs cannot be considered as corresponding to pairs of equivalent program schemes.

The introduction of classes of interpretations, specified for instance by algebraic laws to be satisfied by the base operations, brings the possibility of considering some interesting pairs of equivalent programs as associated with pairs of program schemes that are equivalent w.r.t. appropriately chosen classes of interpretations.

5.1. *Universal interpretations*

A *homomorphism* of F-interpretations is a homomorphism of continuous F-algebras. Let $\mathscr{C}$ be a class of interpretations. An interpretation $\boldsymbol{H}$ is *$\mathscr{C}$-initial* if for every $\boldsymbol{D}$ in

$\mathscr{C}$ there exists a homomorphism $h: \boldsymbol{H} \to \boldsymbol{D}$ (we do not require uniqueness; see the remark following Corollary 5.2 below). Let F_X be F augmented with a set X containing countably many nullary symbols of each sort.

For every F-interpretation $\boldsymbol{D}$ and for every assignment $v: X \to D$ (which preserves sorts in an obvious way), we denote by $\boldsymbol{D}_v$ the F_X-interpretation consisting of $\boldsymbol{D}$ augmented with the assignment $v: X \to D$. We denote by $\mathscr{C}_X$ the class of all interpretations of the form $\boldsymbol{D}_v$ for all $\boldsymbol{D}$ in $\mathscr{C}$ and all assignments $v: X \to D$.

An F_X-interpretation $\boldsymbol{H}$ is *$\mathscr{C}$-universal* if it is $\mathscr{C}_X$-initial.

A scheme $S = (\Sigma, m)$ over F (where $m \in M(F \cup \Phi, X_n)$), with profile $s_1 \dots s_n \to s$, can also be considered as a scheme $\bar{S}$ over F_X of arity ε and of sort s. In the former case, S defines a function with n arguments in every F-interpretation $\boldsymbol{D}$. In the latter, $\bar{S}$ defines a value in D_s for every F_X-interpretation $\boldsymbol{D}_v$. Clearly,

$$S_{\boldsymbol{D}}(v(x_1), \dots, v(x_n)) = \bar{S}_{\boldsymbol{D}_v}.$$

The following proposition is an immediate consequence of Corollary 4.5.

5.1. PROPOSITION. *Let S be a scheme over F of profile $s_1 s_2 \dots s_n \to s$, let $\boldsymbol{D}$ and $\boldsymbol{D}'$ be F-interpretations and let $h: \boldsymbol{D} \to \boldsymbol{D}'$ be a homomorphism. Then*

$$S_{\boldsymbol{D}'} \circ (h_{s_1}, \dots, h_{s_n}) = h_s \circ S_{\boldsymbol{D}}.$$

The following corollary is immediate.

5.2. COROLLARY. *Let $\mathscr{C}$ be a class of F-interpretations, let $\boldsymbol{H}$ be $\mathscr{C}$-initial, and let S and S' be two schemes of profile $\varepsilon \to s$ for some s. If $S_{\boldsymbol{H}} = S'_{\boldsymbol{H}}$, then $S \equiv_{\mathscr{C}} S'$.*

In the definition of a $\mathscr{C}$-initial interpretation $\boldsymbol{H}$ we have not required the uniqueness of the homomorphism $h: \boldsymbol{H} \to \boldsymbol{D}$ for $\boldsymbol{D}$ in $\mathscr{C}$. But consider the following commutative diagram:

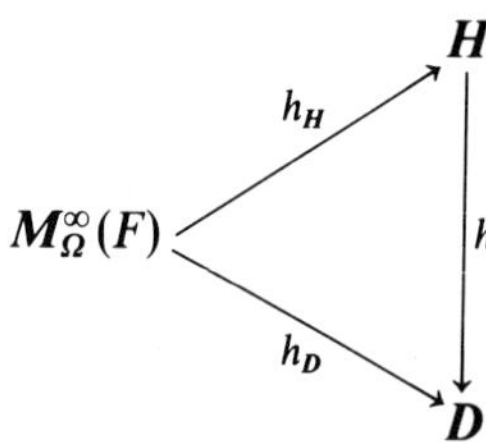

Since h is a homomorphism: $\boldsymbol{H} \to \boldsymbol{D}$, its restriction to $h_H(\boldsymbol{M}^{\infty}_{\Omega}(F))$, that is a subalgebra of $\boldsymbol{H}$, is unique. Note that the proof of Corollary 5.2. deals only with elements of $\boldsymbol{H}$ and $\boldsymbol{D}$ that are images of elements of $\mathbf{M}^{\infty}_{\Omega}(F)$ by h_H and h_D respectively. The following proposition motivates the definition of a $\mathscr{C}$-universal interpretation.

5.3. PROPOSITION. *Let $\mathscr{C}$ be a class of F-interpretations.*

(1) *Let $\boldsymbol{H}$ be a $\mathscr{C}$-universal F_X-interpretation. For every two schemes S and S' of same profile, if $\bar{S}_H = \bar{S}'_H$ then $S \equiv_{\mathscr{C}} S'$.*

(2) *If, furthermore, the restriction* H' *of* H *to the signature* F *belongs to* $\mathscr{C}$, *then* $\bar{S}_H = \bar{S}'_H$ *iff* $S_{H'} = S'_{H'}$ *iff* $S \equiv_{\mathscr{C}} S'$.

PROOF. (1) Let $D \in \mathscr{C}$. By using a renaming of variables if necessary one can assume that $S=(\Sigma, m)$, $S'=(\Sigma', m')$, $m, m' \in M(F \cup \Phi, X_n)$. Let $d_1, \ldots, d_n \in D_{\sigma(x_1)}, \ldots, D_{\sigma(x_n)}$ and let v be any assignment: $X \to D$ such that $v(x_i) = d_i$. Then $S_{D_v} = S_D(d_1, \ldots, d_n)$ and similarly for S'. Since H is $\mathscr{C}_X$-initial, and $\bar{S}_H = \bar{S}'_H$, we have $\bar{S}_{D_v} = \bar{S}'_{D_v}$, hence $S_D(d_1, \ldots, d_n) = S'_D(d_1, \ldots, d_n)$. This holds for all $d_1, \ldots, d_n$, hence $S_D = S'_D$. Since this holds for all D in $\mathscr{C}$, we have $S \equiv_{\mathscr{C}} S'$

(2) Let us assume that $H' \in \mathscr{C}$ where H' is the restriction of H to F. Then $S \equiv_{\mathscr{C}} S'$ implies $S_{H'} = S'_{H'}$. And this implies $\bar{S}_H = S_{H'}(x_{1H}, \ldots, x_{nH})$ and similarly for S'. Hence, we also have $\bar{S}_H = \bar{S}'_H$. □

This proposition shows that the equivalence of two schemes w.r.t. a class of interpretations can be characterized by the equality of the values they compute in a *single* interpretation for a *fixed tuple* of arguments. Here is the first application of this result.

5.4. COROLLARY. *Let* S *and* S' *be two schemes of the same profile. Then* $S \equiv S'$ *iff* $T(S) = T(S')$.

PROOF. Let us apply Proposition 5.3 to the class $\mathscr{C}$ of all F-interpretations and to $H := M^{\infty}_{\Omega}(F, X)$ which is a $\mathscr{C}$-universal interpretation. (This is so since $M^{\infty}_{\Omega}(F, X)$ is the free continuous F-algebra generated by X.) Then

$$\begin{aligned} \bar{S}_H &= T(\bar{S})_H \\ &= T(\bar{S}) \quad \text{(since } T = T_H \text{ for every } T \text{ in } M^{\infty}_{\Omega}(F, X)) \\ &= T(S) \quad \text{(since } L(S) = L(\bar{S})). \end{aligned}$$

The result then follows from Proposition 5.3(2). □

This result is a syntactical characterization of the equivalence of schemes. Courcelle [13, 15] has proved that the problem of deciding the equivalence of schemes is interreducible with the decidability problem of the equivalence of deterministic pushdown automata by using this characterization (see also [35]). The latter problem is still open. Decidable subcases have been given by Rosen [64], Courcelle and Vuillemin [24] and Courcelle and Gallier [19].

Trees of the form $T(S)$ for some scheme S are called *algebraic trees*. They are investigated in [13, 15].

The necessary and sufficient conditions for the equivalence of schemes established in Propositions 5.2–5.4 can be generalized in two directions. The first generalization consists in replacing S by an arbitrary tree T in $M^{\infty}_{\Omega}(F, X_n)$. Hence, for two such trees T and T', one can define $T \equiv_{\mathscr{C}} T'$ iff $T_D = T'_D$ for all $D \in \mathscr{C}$. To extend Proposition 5.3, one can associate with T in $M^{\infty}_{\Omega}(F, X_n)$ the same tree $\bar{T}$, now considered as an element of $M^{\infty}_{\Omega}(F \cup X_n)$. The proof is even simpler than that for program schemes.

This extension is motivated for two reasons: the first one is that the mathematical theory we shall develop is simpler in this way. The second reason is the existence of higher-level program schemes investigated by Damm, Fehr, Indermark, and Gallier [25, 26, 37] for which Corollary 5.4 holds but with nonalgebraic associated trees.

The second generalization consists in rephrasing the above results with $\leqslant$ and $\leqslant_{\mathscr{C}}$ instead of $=$ and $\equiv_{\mathscr{C}}$. Again, the proofs are not more difficult and they also work for trees. Hence, the following results hold, where $T \leqslant_{\mathscr{C}} T'$ means that $T_{\boldsymbol{D}} \leqslant T'_{\boldsymbol{D}}$ for all $\boldsymbol{D} \in \mathscr{C}$:

5.3′. PROPOSITION. *Let $\mathscr{C}$, $\boldsymbol{H}$ and $\boldsymbol{H}'$ be as in Proposition 5.3. For every $T, T' \in M_\Omega^\infty(F, X_n)$, $\bar{T}_H \leqslant \bar{T}'_H$ iff $T_{H'} \leqslant T'_{H'}$ iff $T \leqslant_{\mathscr{C}} T'$.*

5.4′. COROLLARY. *Let S, S' be as in Corollary 5.4. Then $S \leqslant S'$ iff $T(S) \leqslant T(S')$.*

5.2. *Representative interpretations*

Let $\mathscr{C}$ be a class of F-interpretations. A *$\mathscr{C}$-representative interpretation* $\boldsymbol{H}$ is an F_X-interpretation such that for every n and for every $T, T' \in M_\Omega^\infty(F, X_n)$,

$$T \leqslant_{\mathscr{C}} T' \Leftrightarrow T_{\boldsymbol{H}} \leqslant T'_{\boldsymbol{H}}.$$

It follows from Proposition 5.3′ that a $\mathscr{C}$-universal F_X-interpretation belonging to $\mathscr{C}$ is $\mathscr{C}$-representative. The free interpretation $\boldsymbol{M}_\Omega^\infty(F, X)$ is representative w.r.t. the class of all interpretations.

We now examine the existence of $\mathscr{C}$-representative interpretations for general classes $\mathscr{C}$.

5.5. PROPOSITION. *For every class of interpretations $\mathscr{C}$ there exists a $\mathscr{C}$-representative interpretation (not necessarily in $\mathscr{C}$).*

PROOF. Let $\mathscr{C}$ be a class of F-interpretations and let $\mathscr{C}_X$ be the associated class of F_X-interpretations. For every $\boldsymbol{D} \in \mathscr{C}_X$, let $\boldsymbol{D}_0$ be the sub-F_X-algebra $h_{\boldsymbol{D}}(\boldsymbol{M}_\Omega(FUX))$ and let $\boldsymbol{D}_1$ be the least continuous sub-F_X-algebra of $\boldsymbol{D}$. It is clear that the elements of $\boldsymbol{D}_1$ are the least upper bounds in $\boldsymbol{D}$ of the countable directed subsets of $\boldsymbol{D}_0$. Hence, the F_X-algebras $\boldsymbol{D}_1$ for $\boldsymbol{D} \in \mathscr{C}_X$ have a bounded cardinality. It follows that even if $\mathscr{C}_X$ is a proper class (in the set-theoretical sense), one can consider $\mathscr{C}'_X = \{\boldsymbol{D}_1 \mid \boldsymbol{D} \in \mathscr{C}_X\}$ as a set by identifying isomorphic algebras.

Hence one can construct the product algebra $\boldsymbol{H} = \Pi\{\boldsymbol{D}_1 \mid \boldsymbol{D}_1 \in \mathscr{C}'_X\}$ ordered componentwise. It follows that $\boldsymbol{H}$ is a continuous F_X – algebra.

For every $T \in M_\Omega^\infty(F \cup X)$, $T_{\boldsymbol{H}}$ is the family $(T_{\boldsymbol{D}_1})_{\boldsymbol{D}_1 \in \mathscr{C}'_X}$. Hence, for any two trees $T, T' \in M_\Omega^\infty(F \cup X_n)$, $n \geqslant 0$,

$$\begin{aligned} T_{\boldsymbol{H}} \leqslant T'_{\boldsymbol{H}} &\text{ iff } T_{\boldsymbol{D}_1} \leqslant T'_{\boldsymbol{D}_1} \quad \text{for all } \boldsymbol{D}_1 \in \mathscr{C}'_X \\ &\text{ iff } T_{\boldsymbol{D}} \leqslant T'_{\boldsymbol{D}} \quad \text{for all } \boldsymbol{D} \in \mathscr{C}_X \\ &\text{ iff } T_{\boldsymbol{D}'} \leqslant T'_{\boldsymbol{D}'} \quad \text{for all } \boldsymbol{D}' \in \mathscr{C} \end{aligned}$$

(where $\boldsymbol{D}'$ is the restriction of $\boldsymbol{D}$ to the signature F). Hence, $\boldsymbol{H}$ is $\mathscr{C}$-representative. □

In the general case, this result is of no practical interest since comparing T and T' in $\boldsymbol{H}$ is not easier than in all interpretations of $\mathscr{C}$. We now consider an important case where a representative interpretation can be constructed and used.

5.3. *Equational classes of interpretations*

Let R be a possibly infinite subset of $M_\Omega(F,X) \times M_\Omega(F,X)$. Let $\mathscr{C}(R)$ be the class of all F-interpretations $\boldsymbol{D}$ such that $t_{\boldsymbol{D}} \leqslant t'_{\boldsymbol{D}}$ for all $(t,t') \in R$. A class of interpretations $\mathscr{C}$ is equational iff it is of the form $\mathscr{C}(R)$ for some R as above. Examples can be found in [9].

Strictly speaking, an equational class of interpretations should be the class of all interpretations satisfying a given set of equational axioms, i.e., of equalities between finite terms expressing algebraic laws. But since Propositions 5.6 and 5.7 below also work with inequalities between finite terms ($t \leqslant t'$) (and "inequational class" does not sound well), we allow them also as axioms.

For every set R as above, a $\mathscr{C}(R)$-representative interpretation can be constructed. The construction requires several steps, and we first recall the *ideal completion* of a partially ordered set. Let $\boldsymbol{D} = \langle D, \leqslant, \perp \rangle$ be a partially ordered set with least element $\perp$. Let E be the set of nonempty subsets A of D such that

(1) $d_1, d_2 \in A \Rightarrow d_3 \in A$ for some d_3 such that $d_1 \leqslant d_3$, $d_2 \leqslant d_3$ (i.e., A is *directed*);

(2) $d_1 \in A$ and $d_2 \leqslant d_1 \Rightarrow d_2 \in A$ (i.e., A is an *ideal*);

(3) A is countably (or finitely) generated (i.e., A is the least ideal containing some countable or finite set).

Let $\boldsymbol{E} = \langle E, \subseteq, \{\perp\} \rangle$ and $j: D \to E$ be the mapping such that $j(d) = \{d' \in D \mid d' \leqslant d\}$. It can be shown that $\boldsymbol{E}$ is complete, that j is monotone and injective, and that for every monotone mapping $h: \boldsymbol{D} \to \boldsymbol{H}$ where $\boldsymbol{H}$ is a complete partial order there is a unique continuous mapping $\bar{h}: \boldsymbol{E} \to \boldsymbol{H}$ such that $\bar{h} = h \circ j$. This is a universal characterization of $\boldsymbol{E}$ (up to isomorphism) and $\boldsymbol{E}$ is called the ideal completion of $\boldsymbol{D}$. It is denoted by $\boldsymbol{D}^\infty$.

If $\boldsymbol{D}$ is also an ordered F-algebra, then $\boldsymbol{E}$ is a complete F-algebra where $f_E(A_1, \ldots, A_k) = \{d \in \boldsymbol{D} \mid d \leqslant f_D(d_1, \ldots, d_k)$ for some $d_1, \ldots, d_k \in A_1, \ldots, A_k\}$. The universal characterization is similar using homomorphisms of ordered and complete F-algebras (see [23, 52, 78] for more details).

We now start the construction of a $\mathscr{C}(R)$-representative interpretation. A rewriting relation $\underset{R}{\rightarrow}$ on $M_\Omega(F,X)$ is associated with R as usual (see [28]). The relation $(\underset{R}{\rightarrow} \cup <)^*$ is a quasi order $\leqslant^0_R$ on $M_\Omega(F \cup X)$. (Remark that $t \leqslant^0_R t'$ implies $t_{\boldsymbol{D}} \leqslant t'_{\boldsymbol{D}}$ for all $\boldsymbol{D} \in \mathscr{C}(R)$.) Let $\equiv^0_R$ be the associated equivalence relation ($t \equiv^0_R t'$ iff $t \leqslant^0_R t'$ and $t' \leqslant^0_R t$). This makes $\boldsymbol{M}_\Omega(F \cup X)/\equiv^0_R$ into an ordered F-algebra (actually, the initial (F_X, R)-algebra where an (F_X, R)-algebra is an F_X-algebra satisfying the inequalities of R).

Then let $\boldsymbol{H}_R$ be the ideal-completion of $\boldsymbol{M}_\Omega(F \cup X)/\equiv^0_R$, denoted by $[\boldsymbol{M}_\Omega(F \cup X)/\equiv^0_R]^\infty$. It follows that $\boldsymbol{H}_R$ is a continuous (F_X, R)-algebra and, moreover, an initial continuous (F_X, R)-algebra. Hence, $\boldsymbol{H}_R$ is $\mathscr{C}(R)$-universal, belongs to $\mathscr{C}(R)$ and is $\mathscr{C}(R)$-representative. Hence we have proved the following theorem.

5.6. THEOREM. *Let R be a (possibly infinite) set of inequalities. Then $\boldsymbol{H}_R = [\boldsymbol{M}_\Omega(F \cup X)/\equiv^0_R]^\infty$ is $\mathscr{C}(R)$-representative.*

5.7. COROLLARY. (1) *Let* $T, T' \in M_\Omega^\infty(F, X_n)$. *Then* $T \leqslant_{\mathscr{C}(R)} T'$ *iff for every finite* t *in* $M_\Omega(F, X_n)$ *such that* $t \leqslant T$ *there exists* t' *in* $M_\Omega(F, X_n)$ *such that* $t' \leqslant T'$ *and* $t \leqslant_{\mathscr{C}(R)} t'$,

(2) *If* $t, t' \in M_\Omega(F, X)$, *then* $t \leqslant_{\mathscr{C}(R)} t'$ *iff* $t \leqslant_R^0 t'$.

PROOF. Let $T, T' \in M_\Omega^\infty(F, X_n)$. There exist increasing chains $(t_i)_{i\in\mathbf{N}}$ and $(t_i')_{i\in\mathbf{N}}$ in $M_\Omega^\infty(F, X_n)$ such that $T = \mathbf{Sup}_i(t_i)$ and $T' = \mathbf{Sup}_i(t_i')$. Then $T \leqslant_{\mathscr{C}(R)} T'$ iff $\bar{T}_{H_R} \leqslant \bar{T}'_{H_R}$ iff $\mathbf{Sup}_i[t_i] \leqslant \mathbf{Sup}_j[t_j']$ (in H_R) where $[t]$ denotes the equivalence class of t with respect to $\equiv_R^0$. Hence $T \leqslant_{\mathscr{C}(R)} T'$ iff for every $i \in \mathbf{N}$ there exists j such that $t_i \leqslant_R^0 t_j$.

Now consider $t, t' \in M_\Omega(F, X)$: $t \leqslant_{\mathscr{C}(R)} t'$ iff $\bar{t}_{H_R} \leqslant \bar{t}'_{H_R}$ iff $t \leqslant_R^0 t'$, since, for every t in $M_\Omega(F, X)$, $\bar{t}_{H_R} = [t]$. Assertion (2) follows.

Assertion (1) then follows from the first part of the proof. (The remaining technical details are left to the reader.) □

Corollary 5.7 is important for the following reasons: assertion (1) says that the inequality $T \leqslant_{\mathscr{C}(R)} T'$ is expressible in terms of the set of valid inequalities $t \leqslant_{\mathscr{C}(R)} t'$, where t, t' are finite approximations of T and T'. This is not the case for all classes of interpretations, and this property distinguishes the so-called algebraic classes from the other ones (see below.) Furthermore, assertion (2) gives a syntactical characterization of the relation $\leqslant_{\mathscr{C}(R)}$ on finite terms. It is in particular clear that the relation $t \leqslant_{\mathscr{C}(R)} t'$ is recursively enumerable for every finite set R. It is not recursive (hence it is undecidable) in general. Sufficient conditions ensuring decidability have been given by Courcelle [14]. Hence, the combination of assertions (1) and (2) gives a characterization of $T \leqslant_{\mathscr{C}(R)} T'$. This characterization is not decidable in general (even for finite sets R and regular trees T and T'), but it is nevertheless usable for proving or disproving equivalences of program schemes. See [14] for applications, and Example 6.6 below.

5.4. *Algebraic classes of interpretations*

A class of F-interpretations $\mathscr{C}$ is *algebraic* iff for every n and every $T, T' \in M_\Omega^\infty(F, X_n)$,

$$T \leqslant_{\mathscr{C}} T' \iff \forall t \in M_\Omega(F, X_n), \exists t' \in M_\Omega(F, X_n), [t \leqslant T \Rightarrow t' \leqslant T' \text{ and } t \leqslant_{\mathscr{C}} t'].$$

Hence every equational class $\mathscr{C}$ is algebraic but the converse need not be true: observe that this definition only deals with $\leqslant_{\mathscr{C}}$ and not with what is or is not in $\mathscr{C}$.

It has been shown that $\mathscr{D}$, the class of all discrete interpretations, that classes of the form $\mathscr{D}(R) = \mathscr{D} \cap \mathscr{C}(R)$ where R is as in the definition of equational classes, and that classes of the form $\mathscr{D}(\mathscr{A}) =_{\text{def}} \mathscr{D} \cap \mathscr{M}od(\mathscr{A})$ where $\mathscr{M}od(\mathscr{A})$ is the class of all models of a first-order logical theory $\mathscr{A}$ are all algebraic [20]. But these results do not give any syntactical characterization of $t \leqslant_{\mathscr{C}} t'$, $t, t' \in M_\Omega(F, X)$ where $\mathscr{C}$ is any of the above-mentioned classes.

We say that two classes $\mathscr{C}, \mathscr{C}'$ are *equivalent* if $\leqslant_{\mathscr{C}}$ and $\leqslant_{\mathscr{C}'}$ are the same on $M_\Omega^\infty(F, X_n)$ for all $n \geqslant 0$. It is clear that every algebraic class $\mathscr{C}$ is equivalent to the equational class $\mathscr{C}(R)$ where R is the restriction of $\leqslant_{\mathscr{C}}$ to $M_\Omega(F, X) \times M_\Omega(F, X)$. This is a hint to the following theorem [21, 22, 43].

5.8. THEOREM. *For every class of interpretations* $\mathscr{C}$, *the following conditions are* *equivalent*:

(1) $\mathscr{C}$ *is algebraic*;

(2) $\mathscr{C}$ *is equivalent to an equational class*;

(3) *there exists an algebraic* $\mathscr{C}$*-representative interpretation.*

"Algebraic" in condition (3) should be taken in the sense of the theory of partial orders, and comes from the definition by Birkhoff of an algebraic lattice [4]. This result motivates the term "algebraic" in the definition of an algebraic class of interpretations.

We conclude this section by showing a class of interpretations that is not algebraic.

5.9. EXAMPLE (*A nonalgebraic class of interpretations*). Let $F=\{f,g,h\}$ be one-sorted let φ and ψ be defined by the equations

$$\varphi(x)=f(x,\varphi(g(x))), \qquad \psi(x)=f(x,\psi(h(x))).$$

Let $\mathscr{C}$ be the class of all interpretations $\boldsymbol{D}$ such that $\varphi_{\boldsymbol{D}}=\psi_{\boldsymbol{D}}$. We shall prove that $\mathscr{C}$ is not algebraic. It suffices to prove that

$$f(x,f(g(x),\Omega)) \not\leqslant_{\mathscr{C}} t_n \tag{5.1}$$

for all $n\geqslant 0$ where $t_0=\Omega$, $t_n=f(x,t_{n-1}[h(x)/x])$ so that $T(\psi)=\mathbf{Sup}_n(t_n)$.

To do so, we define for $m\geqslant 1$ an F-interpretation $\boldsymbol{D}_m$ with domain $\{\top\}\cup\{t\in M_\Omega(F,X)\mid \mathbf{ht}(t)\leqslant m\}$. The element $\top$ should be read "top" (dual to "bottom") and $\mathbf{ht}(t)$ is the length of a longest path in t from the root to a leaf. We order D_m by

$$d\leqslant d' \text{ iff } d'=\top \text{ or } [d\neq\top,\ d'\neq\top \text{ and } d\leqslant d'],$$

and we define

$$f_{\boldsymbol{D}_m}(d,d')=\begin{cases} f(d,d') & \text{if } d\neq\top,\ d'\neq\top \text{ and } \mathbf{ht}(f(d,d'))\leqslant m,\\ \top & \text{otherwise.}\end{cases}$$

It is then clear that $\boldsymbol{D}_m\in\mathscr{C}$ for all $m\geqslant 1$(observe that $\varphi_{\boldsymbol{D}_m}(d)=\psi_{\boldsymbol{D}_m}(d)=\top$ for all $d\in D_m$), but for every $n\geqslant 0$ there exists an m (actually, any $m>2n$ works) such that $f(x,f(g(x),\Omega))_{\boldsymbol{D}_m}\not\leqslant t_n$. Hence (5.1) has been proved as desired.

6. Least fixed point semantics

Above, semantics has been defined in an operational way: the value of the function computed by a *program*, i.e., a program scheme given with some interpretation, is defined as the result of a finite computation, or as the least upper bound of a set of such results.

From this definition, the only way to prove a property of the function computed by a program is to use inductions on the length of computations. We shall present an alternative characterization, stating that $(\varphi_{1D},\ldots,\varphi_{nD})$ is the least solution of Σ in $\boldsymbol{D}$ (see Sections 3 and 4 for the notations). It yields powerful and handy tools for proving properties of programs.

The main result of this section will be based on the following lemma adapted from the fundamental work of Mezei and Wright [55].

6.1. LEMMA. *Let f be a unary function symbol and Θ be the equation $X=f(X)$. Let $\mathbf{E}$ and $\mathbf{E}'$ be two continuous $\{f\}$-algebras and let $h: \mathbf{E}\to\mathbf{E}'$ be a homomorphism. Then the image under h of the least solution of Θ in $\mathbf{E}$ is its least solution in $\mathbf{E}'$.*

PROOF. The least solution of Θ in $\boldsymbol{E}$ is $x_0=\mathbf{Sup}_i(f^i_E(\bot_E))$. For every i, $h(f^i_E(\bot_E))= f^i_{E'}(h(\bot_E))=f^i_{E'}(\bot_{E'})$. Hence, $h(x_0)=\mathbf{Sup}_i(f^i_{E'}(\bot_{E'}))$, i.e., equal to the least solution of Θ in $\boldsymbol{E}'$. □

Now let $\Sigma=\langle\varphi_i(x_{i,1},\ldots,x_{i,n_i})=t_i;\, i\in[N]\rangle$ be a system over F with set of unknown function symbols $\Phi=\{\varphi_1,\ldots,\varphi_N\}$. Let $T_i=T(\Sigma,\varphi_i(x_{i,1},\ldots,x_{i,n_i}))$ be the (generally infinite) tree associated with φ_i. The tuple $(T_1,\ldots,T_n)$ is the least solution of Σ in $M^\infty_\Omega(F,X)$ (see [59, 15]).

To be more precise, let $E=M^\infty_\Omega(F,\{x_{1,1},\ldots\})\times\cdots\times M^\infty_\Omega(F,\{x_{N,1},\ldots\})$ with the componentwise order associated with $\leqslant$. The system Σ can be considered as a single equation $\Theta=\langle X=f(X)\rangle$ to be solved in $\boldsymbol{E}=\langle E,\, f_E\rangle$ where f_E is the mapping $E\to E$ such that, for $(U_1,\ldots,U_N)\in E$, $f_E(U_1,\ldots,U_N)=(U'_1,\ldots,U'_N)$ and $U'_i= t_i\{U_1/\varphi_1,\ldots,U_N/\varphi_N\}$ (the result of the second-order substitution of U_j for φ_j in t_i, simultaneously for all $j\in[N]$; see [15] for second-order substitutions). The tuple $(T_1,\ldots,T_N)$ is the least solution of Θ in E.

Now let $\boldsymbol{D}$ be any F-interpretation. One can consider Σ as a system of functional equations to be solved in $\boldsymbol{D}$, i.e., as a system for which a function: $D_{\sigma(x_{i,1})}\times\cdots\times D_{\sigma(x_{i,n_1})}\to D_{\sigma(\varphi_i)}$ should be found as a value for φ_i. So let E' be $\Delta_1\times\cdots\times\Delta_N$ where Δ_i is the complete partial order of all continuous functions: $D_{\sigma(x_{i,1})}\times\cdots\times D_{\sigma(x_{i,n_1})}\to D_{\sigma(\varphi_i)}$. Let $f_{E'}: E'\to E'$ be defined as follows. If $(\theta_1,\ldots,\theta_N)\in E'$ then $f_{E'}(\theta_1,\ldots,\theta_N)=(\theta'_1,\ldots,\theta'_N)$ where $\theta'_i=t_{i\boldsymbol{D}[\theta_1,\ldots,\theta_N]}$ for all $i\in[N]$ and $\boldsymbol{D}[\theta_1,\ldots,\theta_N]$ is the $(F\cup\Phi)$-interpretation consisting of $\boldsymbol{D}$ enriched with $(\theta_i)_{i\in[N]}$ where θ_i is the value of φ_i. Hence an n-tuple of functions $(\theta_1,\ldots,\theta_N)\in E'$ is a solution of Σ in $\boldsymbol{D}$ iff it is a solution of Θ in $\boldsymbol{E}'$.

It is not difficult to prove, by an induction on the structure of t, that for all $t\in M_\Omega(F,X)$ and for all $(U_1,\ldots,U_N)\in E$,

$$t\{U_1/\varphi_1,\ldots,U_N/\varphi_N\}_{\boldsymbol{D}}=t_{\boldsymbol{D}[U_{1\boldsymbol{D}},\ldots,U_{N\boldsymbol{D}}]}.$$

It follows that

$$f_E(U_1,\ldots,U_N)_{\boldsymbol{D}}=f_E(U_{1\boldsymbol{D}},\ldots,U_{N\boldsymbol{D}}),$$

i.e., that the application $h: E\to E'$ defined by $h(U_1,\ldots,U_N)=(U_{1\boldsymbol{D}},\ldots,U_{N\boldsymbol{D}})$ is a homomorphism: $\boldsymbol{E}\to\boldsymbol{E}'$. (Some verifications concerning the continuity are left to the reader.)

Hence Lemma 6.1 is applicable and yields the following theorem.

6.2. THEOREM. *Let Σ be a system with unknowns $\{\varphi_1,\ldots,\varphi_N\}$ and $\boldsymbol{D}$ an interpretation. The N-tuple $(\varphi_1,\ldots,\varphi_N)$ is the least solution of Σ in $\boldsymbol{D}$. For every scheme of the form $S=(\Sigma,m)$ one has $S_{\boldsymbol{D}}=m_{\boldsymbol{D}[\varphi_{1\boldsymbol{D}},\ldots,\varphi_{N\boldsymbol{D}}]}$.*

(To obtain the second assertion, it suffices to consider the system $\Sigma' = \Sigma \cup \langle \varphi_0(x_1, \ldots, x_n) = m \rangle$ where φ_0 is a new unknown, and to apply the first assertion to Σ'). In general, a system has several solutions; see [5] for a characterization of all solutions.

Theorem 6.2 has many applications to the theory of program correctness. In the next section we shall give applications to the validation of program transformations. We now derive three proof techniques for proving properties of recursive programs. Since every iterative program can be put in the form of a system of recursive definitions [54], these techniques also apply to programs written in imperative languages.

6.1. *Least and unique fixed-point principles*

Let Σ and $\boldsymbol{D}$ be as in Theorem 6.2. Let $(\theta_1, \ldots, \theta_N)$ be an N-tuple of functions on D having the profiles of $\varphi_1, \ldots, \varphi_N$ respectively. They may be defined by another system Σ' or be defined independently in some other formalism.

LEAST FIXED-POINT PRINCIPLE. *To prove that* $\varphi_{1\boldsymbol{D}} \leqslant \theta_1, \ldots, \varphi_{N\boldsymbol{D}} \leqslant \theta_N$ *it suffices to prove that* $(\theta_1, \ldots, \theta_N)$ *is a solution of* Σ *in* $\boldsymbol{D}$.

Now assume that each function $\varphi_{i\boldsymbol{D}}$ is maximal in $\Delta_i = [D_{s_1} \times \cdots \times D_{s_n} \to D_s]$ (where $s_1 \ldots s_n \to s$ is the profile of φ_i). Then one has the following principle.

UNIQUE FIXED-POINT PRINCIPLE. *To prove that* $\varphi_{1\boldsymbol{D}} = \theta_1, \ldots, \varphi_{N\boldsymbol{D}} = \theta_N$, *it suffices to prove that* $(\theta_1, \ldots, \theta_N)$ *is a solution of* Σ *in* $\boldsymbol{D}$.

The validity of the first principle follows immediately from Theorem 6.2. For the second one, if $\theta_1, \ldots, \theta_n$ is a solution of Σ, then $\varphi_{1\boldsymbol{D}} \leqslant \theta_1, \ldots, \varphi_{N\boldsymbol{D}} \leqslant \theta_N$. Maximality of the mappings $\varphi_{1\boldsymbol{D}}, \ldots, \varphi_{N\boldsymbol{D}}$ implies that the equalities hold.

Maximality of the functions $\varphi_{1\boldsymbol{D}}, \ldots, \varphi_{N\boldsymbol{D}}$ can be proved in the following situations:

(1) when $\boldsymbol{D}$ is discrete and $\varphi_{i\boldsymbol{D}}$ is total (this proof technique has been introduced by McCarthy [53]);

(2) when $\boldsymbol{D} = \boldsymbol{H}_R$ for some set R of equations (see [14] for sufficient conditions ensuring this).

6.2. *Scott's induction principle*

Let Σ and $\boldsymbol{D}$ be as above. Let $P(\theta_1, \ldots, \theta_N)$ be a property of functions $\theta_1 \in \Delta_1, \ldots, \theta_N \in \Delta_N$ where Δ_i is as in the proof of Theorem 6.2. We do not specify any logical calculus in which P is written, we only assume that, for every $(\theta_1, \ldots, \theta_N) \in \Delta_1 \times \cdots \times \Delta_N$, $P(\theta_1, \ldots, \theta_N)$ is either true or false.

We also assume that P is *continuous*, i.e., that, for every N-tuple of increasing sequence $((\theta_1^{(i)})_{i \in \mathbb{N}}, \ldots, (\theta_N^{(i)})_{i \in \mathbb{N}})$ (i.e., $\theta_j^{(i)} \leqslant \theta_j^{(i')}$ for all $j \in [N]$ and all $i \leqslant i'$), we have $P(\mathbf{Sup}_i(\theta_1^{(i)}), \ldots, \mathbf{Sup}_i(\theta_N^{(i)}))$ is true if $P(\theta_1^{(i)}, \ldots, \theta_N^{(i)})$ is true for all $i \geqslant 0$.

With all these hypotheses we can state the following proof rule (usually attributed to

Scott in [24], sometimes also to De Bakker and Scott, in particular by Manna and Vuillemin [51]).

SCOTT INDUCTION PRINCIPLE. *To prove* $P(\varphi_{1D}, \ldots, \varphi_{ND})$ *it suffices to prove that* $P(\bot_{\Delta_1}, \ldots, \bot_{\Delta_N})$ *holds and that, for every* $\theta_1 \in \Delta_1, \ldots, \theta_N \in \Delta_N$, *if* $P(\theta_1, \ldots, \theta_N)$ *holds, then so does* $P(\theta'_1, \ldots, \theta'_N)$ *where* $\theta'_i = t_{iD[\theta_1, \ldots, \theta_N]}$ *for all* $i = 1, \ldots, N$.

Its validity follows from the observation that $\varphi_{iD} = \mathbf{Sup}_j(\theta_i^{(j)})$ where $\theta_i^{(0)} = \bot_{\Delta_i}$ and $\theta_i^{(j+1)} = t_{iD[\theta_1^{(j)}, \ldots, \theta_N^{(j)}]}$ so that, by induction on j, $P(\theta_1^{(j)}, \ldots, \theta_N^{(j)})$ holds for all $j \in \mathbf{N}$. The validity of $P(\varphi_{1D}, \ldots, \varphi_{ND})$ follows from the continuity of P.

We present a last proof method that is more syntactical.

6.3. *Kleene sequences and truncation induction*

Let $S = (\Sigma, m)$ where Σ is as above and $m \in M(F \cup \Phi, X_n)$. The *Kleene sequence of S* is a sequence of elements of $M_\Omega(F, X_n)$, denoted by $(S^{(j)})_{j \geqslant 0}$ *and defined as follows:*

$$S^{(j)} = m\{\varphi_1^{(j)}/\varphi_1, \ldots, \varphi_N^{(j)}/\varphi_N\}$$

where

$$\varphi_i^{(0)} = \Omega_{\sigma(\varphi_i)},$$

$$\varphi_i^{(j+1)} = t_i\{\varphi_1^{(j)}/\varphi_1, \ldots, \varphi_N^{(j)}/\varphi_N\}, \quad \text{for all } i = 1, \ldots, N,$$

It is clear that $\varphi_{iD}^{(j)}$ is equal to $\theta_i^{(j)}$ defined in the proof of validity of Scott's Induction Principle. It follows that $(\mathbf{Sup}_j(\varphi_1^{(j)}), \ldots, \mathbf{Sup}_j(\varphi_N^{(j)}))$ is the least solution of Σ in $\boldsymbol{M}_\Omega^\infty(F, X)$ and that $T(S) = \mathbf{Sup}_j(S^{(j)})$.

Let us now assume that S' is another program scheme, and that $\mathscr{C}$ is a class of interpretations; we have the following principle.

TRUNCATION INDUCTION PRINCIPLE. *To prove that* $S \leqslant_{\mathscr{C}} S'$, *it suffices to prove that* $S^{(j)} \leqslant_{\mathscr{C}} T(S')$ *for all* $j \in \mathbf{N}$.

The proof that $S^{(j)} \leqslant_{\mathscr{C}} T(S')$ can frequently be done by induction on j (possibly with auxiliary inductive hypotheses); a syntactical characterization of $\leqslant_{\mathscr{C}}$ is clearly useful for such a proof (see Corollary 5.7).

6.3. REMARKS (*completeness of the proof principles*). The Truncation Induction Principle is complete in the sense that $S \leqslant_{\mathscr{C}} S'$ iff $S^{(j)} \leqslant_{\mathscr{C}} T(S')$ for all j. (This does *not* mean that a finite proof of the latter condition can be found.) Similarily, the Unique Fixed-point Principle is complete, whenever the functions φ are maximal.

The Least Fixed-point Principle is not complete, since one can have $\varphi_{iD} \leqslant \theta_i$ for all $i \in [N]$ where $(\theta_1, \ldots, \theta_N)$ is *not* a solution of Σ in $\boldsymbol{D}$. Neither is Scott's Principle, since $P(\varphi_{1D}, \ldots, \varphi_{ND})$ may hold without the other conditions.

6.4. EXAMPLE. Let Σ consist of the two equations $\varphi = f(\varphi)$ and $\psi = g(\psi)$. Let also

$R=\{f(g(x))=g(g(f(x))), f(\Omega)=g(\Omega)\}$. We shall prove that $\varphi \equiv_R \psi$. It is easy to prove that for all n

$$f^n(\Omega) \underset{R}{\overset{*}{\longleftrightarrow}} g^{2^n-1}(\Omega),$$

whence $\varphi \leqslant_R \psi$ and $\psi \leqslant_R \varphi$ by truncation induction.

A proof by Scott's induction can also be given as follows, where $\boldsymbol{D}$ is any interpretation in $\mathscr{C}(R)$. Let $P_1(\theta)$ be the property

$$f(\theta) \leqslant \psi_D \quad (\text{i.e., } \forall d \in D,\ f(\theta(d)) \leqslant \psi_D(d)).$$

Scott's induction can be applied to P_1 and the equation defining ψ, giving $f(\psi_D) \leqslant \psi_D$. A second application of Scott's induction to property $P_2(\theta)$, defined as $\theta \leqslant \psi_D$ and to the equation defining φ gives $\varphi_D \leqslant \psi_D$. One finally considers $P_3(\theta)$ defined by

$$\theta \leqslant g(\theta) \wedge g(\theta) \leqslant \varphi_D \wedge g(\theta) \leqslant f(\theta).$$

By Scott's induction applied to the equation defining ψ, this gives $\psi_D \leqslant \varphi_D$, whence the result. The verification of the induction step, i.e. of the implication $P_3(\theta) \Rightarrow P_3(g(\theta))$ uses the following computation:

$$\begin{aligned}
g(\theta) &\leqslant g^2(\theta) && (\text{since } \theta \leqslant g(\theta))\\
&\leqslant g^3(\theta) && (\text{idem})\\
&\leqslant g^2(f(\theta)) && (\text{since } g(\theta) \leqslant f(\theta))\\
&= f(g(\theta)) && (\text{by } R)\\
&\leqslant f(\varphi_D) && (\text{since } g(\theta) \leqslant \varphi_D)\\
&= \varphi_D
\end{aligned}$$

The other verifications are simpler.

(This tricky example is due to R. Milner. It was communicated to the author by J. Vuillemin.)

7. Transformations of program schemes

Program transformations are useful for program development. We refer the reader to the survey papers by Partsch and Steinbruggen [60], and by Pepper [62] on this subject.

Our aim here is to present a few transformation rules applicable to recursive applicative programs. Since iterative programs (written with **for**- and **while**-loops) can be expressed recursively, these rules can be used in particular to transform recursive programs into iterative ones. A major issue is to ensure the equivalence of the obtained programs with the original ones. One can do this by means of ad hoc proofs, but this is not appropriate for a systematic use of such rules. Another method consists in using only tranformation rules that are known to transform programs (or program schemes) into *equivalent* ones: such rules will be called *correct*. Hence, the goal is to determine a set of rules (or of metarules, i.e., of schemes of rules), as large as possible, the correctness of which has been established once and for all. We present a few rules, illustrating the fixed-point approach of Section 6.

7.1. *Equational reasoning on program schemes*

Let Σ be a system over F with unknowns $\varphi_1, \ldots, \varphi_N$. Let R be a set of equations (*not* of inequalities) expressing algebraic laws of the functions denoted by the symbols of F. We shall deal with the following rewriting systems:

$$\xrightarrow[\Sigma,R]{} = \xrightarrow[\Sigma]{} \cup \xrightarrow[R]{} \cup \xrightarrow[R^{-1}]{};$$

$$\xleftrightarrow[\Sigma,R]{} = \xrightarrow[\Sigma]{} \cup \xrightarrow[\Sigma^{-1}]{} \cup \xrightarrow[R]{} \cup \xrightarrow[R^{-1}]{}.$$

If $S = (\Sigma, m)$ and $S' = (\Sigma, m')$ with $m \xrightarrow[\Sigma,R]{*} m'$, it is clear that $S \equiv_R S'$ (where $\equiv_R$ and $\leqslant_R$ stand for $\equiv_{\mathscr{C}(R)}$ and $\leqslant_{\mathscr{C}(R)}$ respectively).

More interesting are the transformations which modify the underlying system. We shall only deal with transformations of *systems* in the remainder of this section.

7.2. *Folding, unfolding, rewriting*

Let Σ and R be as above. Let Σ' be another system with the same unknowns $\varphi_1, \ldots, \varphi_N$. So we have Σ of the form $\langle \varphi_i(\ldots) = t_i;\ i \in [N] \rangle$ and Σ' of the form $\langle \varphi_i(\ldots) = t'_i;\ i \in [N] \rangle$ (we assume that the left-hand sides are exactly the same with the same variables). We say that Σ' is obtained from Σ by *rewriting*, by *unfolding*, by *folding*, by *folding-unfolding* if, respectively,

$t'_i \xleftrightarrow[R]{*} t_i$ for all $i \in [N]$ (denoted by Σ $\mathbf{rewr}_R$ Σ'),

$t_i \xrightarrow[\Sigma,R]{*} t'_i$ for all $i \in [N]$ (denoted by Σ $\mathbf{unf}_R$ Σ'),

$t'_i \xrightarrow[\Sigma,R]{*} t_i$ for all $i \in [N]$ (denoted by Σ $\mathbf{fld}_R$ Σ'),

$t'_i \xleftrightarrow[\Sigma,R]{*} t_i$ for all $i \in [N]$ (denoted by Σ $\mathbf{ufld}_R$ Σ').

We shall compare Σ and Σ' by letting

$$\Sigma \leqslant_R \Sigma' \quad \text{iff} \quad \varphi_{i\Sigma,\boldsymbol{D}} \leqslant \varphi_{i\Sigma',\boldsymbol{D}} \text{ for all } i \in [N] \text{ and all } \boldsymbol{D} \in \mathscr{C}(R);$$

$$\Sigma \equiv_R \Sigma' \quad \text{iff} \quad \varphi_{i\Sigma,\boldsymbol{D}} = \varphi_{i\Sigma',\boldsymbol{D}} \text{ for all } i \in [N] \text{ and all } \boldsymbol{D} \in \mathscr{C}(R);$$

where $\varphi_{i\Sigma,\boldsymbol{D}}$ denotes the function defined by φ_i in $\boldsymbol{D}$ considered as an unknown of Σ and similarly for $\varphi_{i\Sigma',\boldsymbol{D}}$. Then we have the following proposition.

7.1. PROPOSITION. (1) *If* Σ $\mathbf{rewr}_R$ Σ' *or* Σ $\mathbf{unf}_R$ Σ', *then* $\Sigma' \equiv_R \Sigma$.
(2) *If* Σ $\mathbf{fld}_R$ Σ' *or* Σ $\mathbf{ufld}_R$ Σ', *then* $\Sigma' \leqslant_R \Sigma$.

PROOF. If Σ $\mathbf{rewr}_R$ Σ', then every solution of Σ in $\boldsymbol{D}$, where $\boldsymbol{D} \in \mathscr{C}(R)$, is a solution of Σ' and vice versa since, clearly, Σ' $\mathbf{rewr}_R$ Σ. Hence, the least solutions of Σ and Σ' are the same.

If Σ $\mathbf{unf}_R$ Σ', Σ $\mathbf{fld}_R$ Σ' or Σ $\mathbf{unfld}_R$ Σ' then every solution of Σ in $\boldsymbol{D} \in \mathscr{C}(R)$ is a solution of Σ'. Hence the least solution of Σ is a solution of Σ'. It is thus larger than or equal to the least solution of Σ'.

In the case of Σ $\mathbf{unf}_R$ Σ' one also has $\Sigma' \equiv_R \Sigma$ but the proof is quite complicated. (It is a straightforward adaptation of the proof of the corresponding result given in Courcelle

[16, Proposition 6.3].) Let us note that the sets of solutions of Σ and Σ' are not necessarily the same. Only the least solutions are the same. □

Courcelle [14] and Kott [50] have given syntactical conditions on Σ, Σ', R such that Σ $\mathbf{ufld}_R$ Σ', ensuring that $\Sigma' \equiv_R \Sigma$. Since these conditions are quite technical we do not even state them. We rather present another condition on Σ, Σ' such that Σ $\mathbf{ufld}_R$ Σ', ensuring the same thing, but in a much simpler way.

7.3. *Restricted folding-unfolding*

Let Σ, Σ' and R be as in Proposition 7.1. For every $I \subseteq [N]$, we let $\Sigma\uparrow I$ be the set of equations of Σ with a left-hand side $\varphi_i(\ldots.)$ with $i \in I$. We say that Σ' is obtained from Σ by *restricted folding-unfolding* if Σ $\mathbf{ufld}_R$ Σ' and, for some I, $\emptyset \subset I \subset [N]$, the following two conditions hold:

(1) $t'_i = t_i$ for $i \in I$;

(2) $t_i \overset{*}{\underset{\Sigma\uparrow I, R}{\longleftrightarrow}} t'_i$ for $i \in [N] - I$.

We denote this by Σ $\mathbf{rufld}_R$ Σ'.

These conditions mean that the equations of Σ that are used to do the transformation (i.e., the ones of $\Sigma\uparrow I$) are *not* transformed. Hence $\Sigma\uparrow I = \Sigma'\uparrow I$, so that one also has Σ' $\mathbf{rufld}_R$ Σ. Hence, by Proposition 7.1(2), $\Sigma' \leqslant_R \Sigma$ and $\Sigma \leqslant_R \Sigma'$. Hence, $\Sigma \equiv_R \Sigma'$. This proves the following proposition.

7.2. PROPOSITION. *If* Σ $\mathbf{rufld}_R$ Σ', *then* $\Sigma \equiv_R \Sigma'$.

It is not hard to see that Σ and Σ' have the same *sets of solutions* in every interpretation $\boldsymbol{D} \in \mathscr{C}(R)$ if Σ $\mathbf{rufld}_R$ Σ', and not only the same *least* solution as in the cases where Σ $\mathbf{unf}_R$ Σ'.

The reader will find more results on these transformations in [16], in particular, their applications to context-free grammars as well as to program schemes.

We conclude this section with a few examples.

7.3. EXAMPLES. Let $\Sigma = \langle \varphi = f\varphi \rangle$, $\Sigma_1 = \langle \varphi = \varphi \rangle$, $\Sigma_2 = \langle \varphi = ff\varphi \rangle$. (The function symbols f and g are unary, and we omit the parentheses surrounding their arguments). The following facts are clear:

$$\Sigma \, \mathbf{fld} \, \Sigma_1 \quad \text{and} \quad \Sigma_1 < \Sigma,$$

$$\Sigma \, \mathbf{unf} \, \Sigma_2 \quad \text{and} \quad \Sigma_2 \equiv \Sigma \quad (\text{since } \mathbf{Sup}_i \, (f^i \Omega) = \mathbf{Sup}_i \, (f^{2i} \Omega) \text{ in } M_\Omega^\infty(F)).$$

Note that the set of solutions of Σ_2 is strictly larger than the set of solutions of Σ.

Here is a less trivial example proving that $\mathbf{ufld}_R$ is not correct w.r.t. R in general. Let $\Sigma' = \langle \varphi = f\psi,\ \psi = g\psi \rangle$ and $R = \{fgx = gfx\}$. Let $\Sigma'_1 = \langle \varphi = g\varphi,\ \psi = g\psi \rangle$. Then Σ' $\mathbf{ufld}_R$ Σ'_1 since

$$f\psi \underset{\Sigma'}{\longrightarrow} fg\psi \underset{R}{\longrightarrow} gf\psi \underset{\Sigma'^{-1}}{\longrightarrow} g\varphi.$$

But $T(\Sigma'_1, \varphi) <_R T(\Sigma', \varphi)$. (Observe that $g^n\Omega < g^n f\Omega \overset{*}{\underset{R}{\leftrightarrow}} f g^n\Omega$, whence $T(\Sigma'_1, \varphi) \leqslant_R T(\Sigma', \varphi)$; since $f\Omega \leqslant^0_R g^n\Omega$ for no $n \in \mathbf{N}$, the inequality is strict). Hence $\Sigma'_1 <_R \Sigma'$. Let us finally consider $\Sigma'' = \langle \varphi = h(\varphi, \psi),\ \psi = k(\varphi, \varphi) \rangle$ and $\Sigma''_1 = \langle \varphi = h(\varphi, \psi),\ \psi = k(h(\varphi, \psi), h(\varphi, \psi)) \rangle$; then Σ'' **unf** Σ'_1 and, moreover, Σ'' **rufld** Σ''_1. Hence, Σ'' and Σ''_1 have the same sets of solutions and the same least solutions.

8. Historical notes, other types of program schemes and guide to the literature

This section is *not* a complete bibliographical survey but just a commented selection of references concerning other types of program schemes and developments of the theory of recursive applicative ones.

From the basic distinction between *imperative* and *applicative* programs two distinct types of program schemes stem: We first present imperative program schemes since they were introduced first.

8.1. Flowcharts

Imperative program schemes were first introduced by Ianov [47] as a way of representing control structures. Kosaraju [49], followed by Jacob [48], has used them to compare the powers of various control structures.

Less abstract program schemes (modelled out of machine-language programs) have been introduced by Paterson [61], where a lot of undecidability results are established. Decidability results have been obtained by Chandra [12] and Tokura et al. [74] but with some very strict hypotheses making these results of no practical use, especially because they cannot accommodate the notion of a class of interpretations. The basic results can be found in Greibach's book [41].

8.2. Monadic applicative recursive program schemes with fixed conditionals

Polyadic recursive program schemes with *if*... *then*... *else*... considered as a piece of control structure (and not as a base function as we did in this chapter) yield an even more difficult theory than Paterson schemes (see Subsection 8.1), since every Paterson scheme can be translated into such a recursive scheme [54]. Results concerning the translatability of such schemes into imperative ones have been given by Strong and Walker [69, 70] and also by Garland and Luckham [38]. Many results can be found in [41].

By restriction to monadic function symbols and predicates, one gets a class of program schemes close to context-free languages (generalizing Ianov schemes which are closed to finite deterministic automata). Decidability results have been given in [38] and in [2]. The proofs use techniques from language theory. The major result on the subject, essentially closing the theory, is the theorem of Friedman [32] saying that the equivalence problem for monadic recursive schemes with interpreted *if* ... *then*... *else* is interreducible with the (still open) equivalence problem for deterministic pushdown automata. The introduction of constants is considered in [33]. The technique of interpreted value languages in [38] is extended by Courcelle to the case of polyadic schemes by means of context-free graph grammars [17].

All these results focus on decidability questions and yield no result for equivalences w.r.t. classes of interpretations.

8.3. Polyadic recursive applicative program schemes

In this class, investigated in the present chapter, the *if* ... *then* ... *else* ... is considered as a base function and not as a piece of control structure. This apparently small change has actually opened the development of a very rich theory.

The first papers have been published by Courcelle and Vuillemin [24], Courcelle [13], Nivat [59], Vuillemin [75], Guessarian [42], and Courcelle and Nivat [21]. The deep reason for the richness of the theory is the existence of an infinite tree that represents in a semantically meaningful way all computations of a program scheme. (see Proposition 5.3′). Rosen [64] first used this idea for a limited purpose. The computed function is just the function defined by this infinite tree. This point of view has been emphasized by Goguen et al. [40].

Many references have been cited in Sections 1–7; we do not list them here but we draw the reader's attention to the book by Guessarian [43], the extension to nondeterministic schemes by Arnold et al. [1, 57], and the works on operational semantics by Raoult and Vuillemin [63], and by Berry and Levy [3]. In this last work the optimality of computations is considered.

Extensions to higher-type procedures have been investigated in depth in the works of Damm, Fehr and Indermark [25, 26] and of Gallier [37].

8.4. Algebraic theories

A theory of flowcharts has been developped by Elgot, Bloom and Tindell in the framework of *iterative* algebraic theories [30, 31, 6, 7, 8]. A major difference with the preceding approaches is the requirement that, in an interpretation, a system should have a unique solution. This requirement does not pose any difficulty in the free interpretation (consisting of infinite trees), but if one wishes to introduce equations (i.e., to work in an equational class of interpretations), no result comparable to our Theorems 5.7–5.9 can be established so that nearly nothing can be said about the equivalence (or the inequivalence) of programs.

This is not the case for *rational* theories introduced by Wright et al. [77] where an ordering is used and the existence of a *least* fixed point (as opposed to a *unique* fixed point) is imposed. Rational theories have later been developed by Gallier [36]. This line of research has been followed by Schmeck [65] and Ştefănescu [67, 68]. Related works are [58] and [72, 73].

8.5. Applications to the construction and verification of programs

Program transformations have been introduced by Burstall and Darlington [10, 27] and were later investigated by Huet and Lang [46], Kott [50], and Courcelle [14]. See the surveys [60] and [62].

Although program schemes form a good framework for proving properties of programs, there are not that many papers on the subject: see Greibach's book [41], a paper by Gallier [34], and its extension by Courcelle and Deransart [18].

Acknowledgment

I would like to thank Robert Strandh and Kathleen Callaway who helped me to improve the English style of this chapter. This work has been supported by the "Programme de Recherches Coordonnées: Mathématiques et Informatique".

References

[1] ARNOLD, A., P. NAUDIN and M. NIVAT, On the semantics of nondeterministic recursive program schemes, in: M. Nivat and J.C. Reynolds, eds., *Algebraic Methods in Semantics* (Cambridge University Press, 1985) 1–33.

[2] ASHCROFT, E., Z. MANNA and A. PNUELI, Decidable properties of monadic functional schemas, *J. ACM* **20** (1973) 489–499.

[3] BERRY G., and J.J. LEVY, Minimal and optimal computations of recursive programs, *J. ACM* **26** (1979) 148–175.

[4] BIRKHOFF, G., *Lattice Theory* (Amer. Mathematical Soc., Providence, RI, 3rd ed., 1967).

[5] BLOOM, S, All solutions of a system of recursion equations in infinite trees and other contraction theories. *J. Comput. System Sci.* **27** (1983) 225–255.

[6] BLOOM, S. and C. ELGOT, The existence and construction of free iterative theories, *J. Comput. System Sci.* **12** (1976) 305–318.

[7] BLOOM, S., C. ELGOT and J. WRIGHT, Solutions of the iteration equation and extensions of the scalar iteration operation, *SIAM J. Comput.* **9** (1980) 25–45.

[8] BLOOM S., and R. TINDELL, Compatible orderings on the metric theory of trees, *SIAM J. Comput.* **9** (1980) 683–691.

[9] BLOOM S., and R. TINDELL, Varieties of "if-then-else", *SIAM J. Comput.* **12** (1983) 677–707.

[10] BURSTALL, R. and J. DARLINGTON, A transformation system for developing recursive programs, *J. ACM* **24** (1977) 44–67.

[11] CADIOU, J.M., Recursive definitions of partial functions and their computation, Ph.D. Thesis Stanford Univ., Stanford, 1972.

[12] CHANDRA, A., On the decision problems of program schemas with commutative and invertible functions, in: *Proc. ACM Symp. on Principles of Programming Languages* (1973) 235–242.

[13] COURCELLE, B., A representation of trees by languages, *Theoret. Comput. Sci.* **6** (1978) 255–279 and **7** (1978) 25–55.

[14] COURCELLE, B., Infinite trees in normal form and recursive equations having unique solution, *Math. Systems Theory* **13** (1979) 131–180.

[15] COURCELLE, B., Fundamental properties of infinite trees, *Theoret. Comput. Sci.* **25** (1983) 95–169.

[16] COURCELLE, B., Equivalence and transformations of regular systems, *Theoret. Comput. Sci.* **42** (1986) 1–122.

[17] COURCELLE, B., On using context-free graph grammars for analyzing recursive definitions, in: K. Fuchi and L. Kott, eds., *Programming of Future Generation Computers II* (Elsevier, Amsterdam, 1988) 83–122.

[18] COURCELLE, B. and P. DERANSART, Proofs of partial correctness for attribute grammars with applications to recursive procedures and logic programming, *Inform. and Comput.* **78** (1988) 1–55.

[19] COURCELLE, B. and J. GALLIER, Decidable subcases of the equivalence problem for recursive program schemes, *RAIRO Inform Théor. Appl.* **21** (1987) 245–286.

[20] COURCELLE, B. and I. GUESSARIAN, On some classes of interpretations, *J. Comput. System Sci.* **17** (1978) 388–413.

[21] COURCELLE, B. and M. NIVAT, Algebraic families of interpretations, in: *Proc. 17th Ann. IEEE Symp. on Foundations of Computer Science* (1976) 137–146.

[22] COURCELLE, B. and M. NIVAT, The algebraic semantics of recursive program schemes, in: *Proc.*

Mathematical Foundations of Computer Science '78, Lecture Notes in Computer Science, Vol. 64 (Springer, Berlin, 1978) 16–30.

[23] COURCELLE, B. and J.C. RAOULT, Completions of ordered magmas, *Fund. Inform.* **III** (1) (1980) 105–116.

[24] COURCELLE, B. and J. VUILLEMIN, Completeness results for the equivalence of recursive schemes, *J. Comput. System Sci.* **12** (1976) 179–197.

[25] DAMM, W., The IO- and OI-hierarchies, *Theoret. Comput. Sci.* **20** (1982) 95–207.

[26] DAMM, W., E. FEHR and K. INDERMARK, Higher type recursion and self-application as control structures, in: E. Neuhold, ed., *Formal Descriptions of Programming Concepts* (North-Holland, Amsterdam, 1978) 461–487.

[27] DARLINGTON, J. and R. BURSTALL, A system which automatically improves programs, *Acta Inform.* **6** (1976) 41–60.

[28] DERSHOWITZ, N. and J.P. JOUANNAUD, Rewrite systems, in: J. van Leeuwen, ed., *Handbook of Theoretical Computer Science, Vol. B* (North-Holland, Amsterdam, 1990) 243–320.

[29] EHRIG, H. and B. MAHR, *Fundamentals of Algebraic Specifications I* (Springer, Berlin, 1985).

[30] ELGOT, C., Monadic computation and iterative algebraic theories, in: H.E. Rose, ed., *Logic Colloquium 73* (North-Holland, Amsterdam, 1975) 175–230.

[31] ELGOT, C., S. BLOOM and R. TINDELL, The algebraic structure of rooted trees, *J. Comput. System. Sci.* **16** (1978) 362–399.

[32] FRIEDMAN, E., Equivalence problems for deterministic languages and monadic recursion schemes, *J. Comput. System Sci.* **14** (1977) 334–359.

[33] FRIEDMAN, E. and S. GREIBACH, Monadic recursion schemes: the effect of constants, *J. Comput. System Sci.* **18** (1979) 254–266.

[34] GALLIER, J., Nondeterministic flowchart programs with recursive procedures, Part I and II, *Theoret. Comput. Sci.* **13** (1981) 193–223 and 239–270.

[35] GALLIER, J., DPDA's in "atomic normal form" and applications to the equivalence problems, *Theoret. Comput. Sci.* **14** (1981) 155–186; Corrigendum **19** (1982) 229.

[36] GALLIER, J., Recursion closed algebraic theories, *J. Comput. System Sci.* **23** (1981) 69–105.

[37] GALLIER, J., *N*-rational algebras, I: Basic properties and free algebras, II: Varieties and the logic of inequalities *SIAM J. Comput.* **13** (1984) 750–775 and 776–794.

[38] GARLAND, S. and D. LUCKHAM, Program schemes, recursion schemes and formal languages, *J. Comput. System Sci.* **12** (1973) 119–217.

[39] GOGUEN, J., J. THATCHER and E. WAGNER, An initial algebra approach to the specification, correctness, and implementation of abstract data types, in: R. Yeh ed., *Current Trends in Programming Methodology* (Prentice-Hall, Englewood Cliffs, NJ, 1978) 80–149.

[40] GOGUEN, J., J. THATCHER, E. WAGNER and J. WRIGHT, Initial algebra semantics and continuous algebras, *J. ACM* **24** (1977) 68–95.

[41] GREIBACH, S., *Theory of Program Structures: Schemes, Semantics, Verifications*, Lecture Notes in Computer Science, Vol. 36 (Springer, Berlin, 1975).

[42] GUESSARIAN, I., Schemas recursifs polyadiques: équivalences et classes d'interprétations, Doctoral dissertation, 1975, Université de Paris 7.

[43] GUESSARIAN, I., *Algebraic Semantics*, Lecture Notes in Computer Science, Vol. 99 (Springer, Berlin, 1981).

[44] HENDERSON, P. and J. MORRIS, A lazy evaluator, in: *Proc. 3rd ACM Symp. on Principles of Programming Languages* (1976) 95–103.

[45] HUET, G., Confluent reductions: abstract properties and applications to term rewriting systems, *J. ACM* **27** (1980) 797–821.

[46] HUET, G. and B. LANG, Proving and applying program transformations expressed with 2nd order patterns, *Acta Informatica* **11** (19878) 31–55.

[47] IANOV, Y., The logical schemes of algorithms, in: *Problems of Cybernetics I* (Pergamon, Oxford, 1960) 82–140.

[48] JACOB, G., Classes de chartes structurées; quelques méthodes syntaxiques et structurelles, in: *Proc. on Mathematical Foundations of Computer Science* '78 Lecture Notes in Computer Science, Vol. 64 (Springer, Berlin, 1978) 286–297.

[49] KOSARAJU, S., Analysis of structured programs, *J. Comput. System Sci.* **9** (1974) 232–255.

[50] KOTT, L., Unfold/fold program transformations, in same volume as [1] 411–434.

[51] MANNA, Z. and J. VUILLEMIN, Fixpoint approach to the theory of computation, *Comm. ACM* **15** (1972) 528–536.

[52] MARKOWSKY, G. and B. ROSEN, Bases for chain-completed posets, *IBM J. RES. Develop.* **20** (1976) 138–147.

[53] MCCARTHY, J., A basis for a mathematical theory of computation in: P. Braffort and D. Hirschberg, eds., *Computer Programming and Formal Systems* (North-Holland, Amsterdam, 1963) 33–70.

[54] MCCARTHY, J., Towards a mathematical science of computation, in: *Proc. IFIP* (1965) 27–34.

[55] MEZEI, J., and J. WRIGHT, Algebraic automata and context-free sets, *Inform. and Control* **11** (1967) 3–29.

[56] MOSSES, P. Denotational semantics, in: J. van Leeuwen, ed., *Handbook of Theoretical Computer Science, Vol. B* (North-Holland, Amsterdam, 1990) 575–631.

[57] NAUDIN, P., Comparaison et équivalence de sémantiques pour les schémas de programmes non déterministes, *RAIRO Inform. Théor. Appl.* **21** (1987) 59–91.

[58] NELSON, E., Iterative algebras, *Theoret. Comput. Sci.* **25** (1983) 67–94.

[59] NIVAT, M., On the interpretation of recursive polyadic program schemes, in: Symposia Mathematica, Vol. 15 (Academic Press, New York, 1975) 255–281.

[60] PARTSCH, M. and R. STEINBRUGGEN, Program transformation systems, *Computing Surveys* **15** (1983) 199–236.

[61] PATERSON, M., Equivalence problems in a model of computation, Ph.D. Thesis, Univ. of Cambridge, England, 1967.

[62] PEPPER, P., A simple calculus for program transformations (inclusive of induction), *Sci. Comput. Programming* **9** (1987) 221–262.

[63] RAOULT, J.C. and J. VUILLEMIN, Operational and semantic equivalence between recursive programs, *J. ACM* **27** (1980) 772–796.

[64] ROSEN, B., Program equivalence and context-free grammars, *J. Comput. System Sci.* **11**(1975) 358–374.

[65] SCHMECK, H., Algebraic semantics of recursive flowchart schemes, *Inform. and Control* **59** (1983) 108–126.

[66] SCOTT, D. and C. GUNTER, Semantic domains, in: J. van Leeuwen, ed., *Handbook of Theoretical Computer Science, Vol. B* (North-Holland, Amsterdam, 1990) 633–674.

[67] ŞTEFĂNESCU, G., On flowchart theories, Part I: the deterministic case, *J. Comput. System Sci.* **35** (1987) 163–191.

[68] ŞTEFĂNESCU, G., On flowchart theories, Part II: the nondeterministic case, *Theoret. Comput. Sci.* **52** (1987) 307–340.

[69] STRONG, H., Translating recursion equations into flowcharts, *J. Comput. System Sci.* **5** (1971) 254–285.

[70] STRONG, H. and S. WALKER, Characterizations of flowchartable recursions, *J. Comput. System Sci.* **7** (1973) 404–447.

[71] THOMAS, W., Automata on infinite objects, in: J. van Leeuwen, ed., *Handbook of Theoretical Computer Science, Vol. B* (North-Holland, Amsterdam, 1990) 133–191.

[72] TIURYN, J. Fixed points and algebras with infinitely long expressions, *Fund. Inform.* **I** (1978) 107–128; and *Fund. Inform. II* (1979) 317–335.

[73] TIURYN, J., Unique fixed points vs. least fixed points, *Theoret. Comput. Sci.* **12**(1980) 1–26.

[74] TOKURA, N., T. KASAMI and S. FURUTA, Ianov schemes augmented with a pushdown memory, in: *Proc. 15th Ann. IEEE Symp. on switching and Automata Theory* (1974) 84–94.

[75] VUILLEMIN, J., Correct and optimal implementation of recursion in a simple programming language, *J. Comput. System Sci* **9** (1974) 332–334.

[76] WIRSING, M., Algebraic specifications, in: J. van Leeuwen, ed., *Handbook of Theoretical Computer Science, Vol. B* (North-Holland, Amsterdam, 1990) 675–788.

[77] WRIGHT, J., J. THATCHER, E. WAGNER and J. GOGUEN, Rational algebraic theories and fixed point solutions, in: *Proc. 17th Ann. IEEE Symp. on Foundations of Computer Science* (1976) 147–158.

[78] WRIGHT, J.E., WAGNER and J. THATCHER, A uniform approach to inductive posets and inductive closure, *Theoret. Comput. Sci.* **7** (1978) 57–77.

CHAPTER 10

Logic Programming

Krzysztof R. APT

Centre for Mathematics and Computer Science, P.O. Box 4079, 1009 AB Amsterdam, Netherlands,
and
Department of Computer Sciences, University of Texas at Austin, Austin, TX 78712-1188, USA

Contents

1. Introduction 495
2. Syntax and proof theory. 496
3. Semantics 511
4. Computability 523
5. Negative information 531
6. General goals 547
7. Stratified programs 555
8. Related topics 566
Appendix 569
Note 570
Acknowledgment 570
References 571

HANDBOOK OF THEORETICAL COMPUTER SCIENCE
Edited by J. van Leeuwen

1. Introduction

1.1. Background

Some formalisms gain a sudden success and it is not always immediately clear why. Consider the case of logic programming. It was introduced in an article of Kowalski [53] in 1974 and for a long time—in the case of computer science—not much happened. But, sixteen years later, already the Journal of Logic Programming and Annual Conferences on the subject exist and a few hundred of articles on it have been published.

Its success can be attributed to at least two circumstances. First of all, logic programming is closely related to PROLOG. In fact, logic programming constitutes its theoretical framework. This close connection led to the adoption of logic programming as the basis for the influential Japanese Fifth Generation Project. Secondly, in the early eighties a flurry of research on alternative programming styles started and suddenly it turned out that some candidates already existed and even for a considerable time. This led to a renewed interest in logic programming and its extensions.

The power of logic programming stems from two reasons. First, it is an extremely simple formalism. So simple, that some, when confronted with it for the first time, say "Is that all?". Next, it relies on mathematical logic which developed its own methods and techniques and which provides a rigorous mathematical framework. (It should be stated however, that the main basis of logic programming is automatic theorem proving which was developed in a large part by computer scientists.)

The aim of this chapter is to provide a self-contained introduction to the theory of logic programming. In the presentation we try to shed light on the causal dependence between various concepts and notions. Throughout the chapter we attempt to adhere to the notation of Lloyd [64], the book which obviously influenced our presentation. This will hopefully further contribute to the standardization of the notation and terminology in the domain.

1.2. Plan of this paper

We now provide a short description of the content of the chapter. It is hoped that this will facilitate its reading and will allow a better understanding of the structure of its subject.

The aim of Section 2 is to introduce in the fastest possible way the notion of *SLD-resolution* central to the subject of logic programming.

In Section 3 a semantics is introduced with the purpose of establishing soundness of SLD-resolution and several forms of its completeness. Most of these results are collected in the Success Theorem 3.25.

In Section 4 the computability by means of logic programs is investigated. It is among others shown that all recursive functions are computed by logic programs.

SLD-resolution allows us to derive only positive statements. Section 5 deals with the other side of the coin—the derivability of the negative statements. After rejecting the *Closed World Assumption* rule as ineffective, the full effect is directed at an analysis of a stronger but effective rule—the *Negation as Failure* rule and its relation to the construction called *completion of a program*. The final outcome is the Finite Failure Theorem 5.32 dual to the Success Theorem.

After this extensive analysis of how to deal with positive and with negative statements, the mixed statements (so-called *general goals*) are investigated in Section 6. While the resulting form of resolution (called here *SLDNF⁻-resolution*) is sound, the completeness can be obtained only after imposing a number of restrictions, both on the logic programs and the general goals. Finally, in Section 7 we investigate a subclass of general programs, called *stratified programs*, concentrating on their semantics.

The chapter concludes by a short discussion of related topics which are divided into six sections: general programs, alternative approaches, deductive databases, PROLOG, integration of logic and functional programming, and applications in artificial intelligence.

Finally, in the Appendix a short history of the subject is traced.

2. Syntax and proof theory

2.1. First-order languages

Logic programs are simply sets of certain formulas of a first-order language. So to define them, we recall first what a first-order language is, a notion essentially due to G. Frege. By necessity our treatment is reduced to a list of definitions. A reader wishing a more motivated introduction should consult one or more standard books on the subject. Personally, we recommend [70, 92].

A *first-order language* consists of an alphabet and all formulas defined over it. An *alphabet* consists of the following classes of symbols:

- *variables* denoted by x, y, z, v $u, \ldots,$
- *constants* denoted by a, b c, $d, \ldots,$
- *function symbols* denoted by f, g, $.\,, \ldots,$
- *relation symbols* denoted by p, q, $r, \ldots,$
- *propositional constants*, which are **true** and **false**,
- *connectives*, which are $\neg$ (negation), $\vee$ (disjunction), $\wedge$ (conjunction), $\rightarrow$ (implication) and $\leftrightarrow$ (equivalence),
- *quantifiers*, which are $\exists$ (there exists) and $\forall$ (for all),
- *parentheses*, which are (and) and the *comma*, that is: ,.

Thus the sets of connectives, quantifiers and parentheses are fixed. We assume also that the set of variables is infinite and fixed. Those classes of symbols are called *logical symbols*. The other classes of symbols, that is, constants, relation symbols (or just *relations*) and function symbols (or just *functions*), may vary and in particular may be empty. They are called *nonlogical symbols*. Each first-order language is thus determined by its nonlogical symbols.

Each function and relation symbol has a fixed *arity*, that is, the number of arguments. We assume that functions have a positive arity—the rôle of 0-ary functions is played by the constants. In contrast, 0-ary relations are admitted. They are called *propositional symbols*, or simply *propositions*. Note that each alphabet is uniquely determined by its constants, functions and relations.

We now define by induction two classes of strings of symbols over a given alphabet. First we define the class of *terms* as follows:

- a variable is a term,
- a constant is a term,
- if f is an n-ary function and $t_1, \ldots, t_n$ are terms then $f(t_1, \ldots, t_n)$ is a term.

Terms are denoted by s, t, u. Finally, we define the class of *formulas* as follows:

- if p is an n-ary relation and $t_1, \ldots, t_n$ are terms then $p(t_1, \ldots, t_n)$ is a formula (called a *atomic formula*, or just an *atom*),
- **true** and **false** are formulas,
- if F and G are formulas then so are $\neg F$, $(F \wedge G)$, $(F \wedge G)$, $(F \rightarrow G)$ and $(F \leftrightarrow G)$,
- if F is a formula and x is a variable then $\exists x F$ and $\forall x F$ are formulas.

Sometimes we shall write $(G \leftarrow F)$ instead of $(F \rightarrow G)$. Some well known binary functions (like $+$) or relations (like $=$) are usually written in an *infix notation* i.e. between the arguments. Atomic formulas are denoted by A, B and formulas in general by F, G. If F is a quantifier-free formula with variables $x_1, \ldots, x_n$, we write $\exists F$ for $\exists x_1 \ldots \exists x_n F$ and $\forall F$ for $\forall x_1 \ldots \forall x_n F$. Formulas of the form $\forall F$ are called *universal formulas.* A term or formula with no variables is called *ground.*

Given two strings of symbols e_1 and e_2 from the alphabet, we write $e_1 \equiv e_2$ when e_1 and e_2 are identical. Usually these strings will be terms or formulas.

The definition of formulas is rigorous at the expense of excessive use of parentheses. One way to eliminate most of them is by introducing a *binding order* among the connectives and quantifiers. We thus assume that $\neg$, $\exists$ and $\forall$ bind stronger than $\vee$ which in turn binds stronger than $\wedge$ which binds stronger than $\rightarrow$ and $\leftrightarrow$. Also, we assume that $\vee$, $\wedge$, $\rightarrow$ and $\leftrightarrow$ *associate to the right* and omit the outer parentheses. Thus, thanks to the binding order, we can rewrite the formula

$$\forall y \forall x((p(x) \wedge \neg r(y)) \rightarrow (\neg q(x) \vee (A \vee B)))$$

as

$$\forall y \forall x(p(x) \wedge \neg r(y) \rightarrow \neg q(x) \vee (A \vee B))$$

which, thanks to the convention of the association to the right, further simplifies to

$$\forall y \forall x(p(x) \wedge \neg r(y) \rightarrow \neg q(x) \vee A \vee B).$$

Thus completes the definition of a first-order language.

2.2. *Logic programs*

To bar an easy access to newcomers every scientific domain has introduced its own terminology and notation. Logic programming is no exception in this matter but it borrowed most of its terminology from automatic theorem proving. Thus an atom or its negation is called a *literal.* A *positive literal* is just an atom while a *negative literal* is the negation of an atom. Note that **true** and **false** are not atoms.

In turn, a formula of the form

$$\forall(L_1 \vee \cdots \vee L_m)$$

where $L_1, \ldots, L_m$ are literals, is called a *clause.* From now on clauses will be always written in a special form called—yes, you guessed it—a *clausal form*. The above formula in a clausal form is written as

$$A_1, \ldots, A_k \leftarrow B_1, \ldots, B_n$$

where $A_1, \ldots, A_k$ is the list of all positive literals among $L_1, \ldots, L_m$, called *conclusions* and $B_1, \ldots, B_n$ is the list of remaining literals stripped of the negation symbol, called *premises*. Informally, it is to be understood as (A_1 or ... or A_k) if (B_1 and ... and B_n). Thus for example the formula

$$\forall x \forall y (p(x) \vee \neg A \vee \neg q(y) \vee B)$$

looks in clausal form as

$$p(x), B \leftarrow A, q(y).$$

If a clause has only one conclusion ($k=1$), then it is called a *program clause* or a *definite clause*. Its conclusion is then usually called a *head* and the list of its premises a *body*. When the set of premises of a program clause is empty ($n=0$), then we talk of a *unit clause*. They have the form $A \leftarrow$. When the set of conclusions is empty ($k=0$), then we talk of a *goal* or a *negative clause*. They have the form $\leftarrow B_1, \ldots, B_n$. Finally, when both the set of premises and conclusions is empty then we talk of the *empty clause* and denote it by $\square$. It is interpreted as a contradiction.

To understand this interpretation, we are in fact brought to the question of meaning of a formula $L_1 \vee \ldots \vee L_m$ when $m=0$, i.e. of the empty disjunction. Now, the empty disjunction is considered as always false because it asks for an existence of a true disjunct when none of them exists. In contrast, the empty conjunction is considered as always true because it asks for truth of all conjuncts, which holds when none of them exists.

Now, we can define a *logic program* (or just a *program*)—it is a finite nonempty set of program clauses.

Logic programs form a subclass of general logic programs. To define the general programs we first introduce the concept of a *general clause*. It is a construct of the form

$$A_1, \ldots, A_k \leftarrow L_1, \ldots, L_n$$

where $A_1, \ldots, A_k$ are positive literals and $L_1, \ldots, L_n$ are (not necessarily positive) literals. When there is only one conclusion ($k=1$), we talk of a *general program clause*, and when the set of conclusions is empty ($k=0$) we talk of a *general goal*.

A general clause $A_1, \ldots, A_k \leftarrow L_1, \ldots, L_n$ represents the formula

$$\forall (A_1 \vee \ldots \vee A_k \vee \neg L_1 \vee \ldots \vee \neg L_n).$$

Now, a *general logic program* (or just a *general program*) is a finite nonempty set of general program clauses. Note that **true** and **false** are not used to define (general) programs. These formulas will be however needed later, in Subsection 5.5.

With each (general) program P we can uniquely associate a first-order language L_P whose constants, functions and relations are those occurring in P. All considerations concerning a (general) program P refer to the language L_P. In particular, in statements like "Let P be a program and N a goal", N is always assumed to be a goal from L_P.

There are two ways of interpreting a clause $A \leftarrow B_1, \ldots, B_n$. One is: to solve A solve $B_1, \ldots, B_n$. The other is: A is true if $B_1, \ldots, B_n$ are true. The first interpretation is usually called procedural interpretation whereas the second is called declarative interpretation. It is this first interpretation which distinguishes logic programming from first-order logic. We shall discuss this double interpretation in more detail at the end of Section 3.

2.3. *Substitutions*

Consider now a fixed first-order language. In logic programming variables are assigned values by means of a special type of substitutions, called "most general unifiers". Formally, a *substitution* is a finite mapping from variables to terms, and is written as

$$\theta = \{x_1/t_1, \ldots, x_n/t_n\}.$$

Informally, it is to be read: "the variables $x_1, \ldots, x_n$ become (or are *bound to*) $t_1, \ldots, t_n$, respectively".

The notation implies that the variables $x_1, \ldots, x_n$, are different. We also assume that, for $i = 1, \ldots, n$, $x_i \not\equiv t_i$. A pair x_i/t_i is called a *binding*. If all $t_1, \ldots, t_n$ are ground then θ is called *ground*. If θ is a 1-1 and onto mapping from its domain to itself, then θ is called a *renaming*. In other words, θ is a renaming if it is a permutation of the variables from its domain.

Substitutions operate on expressions. By an *expression* we mean a term, a sequence of literals or a clause and denote it by E. For an expression E and a substitution θ, $E\theta$ stands for the result of applying θ to E which is obtained by *simultaneously* replacing each occurrence in E of a variable from the domain of θ by the corresponding term. The resulting expression $E\theta$ is called an *instance* of E. An instance is called *ground* if it contains no variables.

If θ is a renaming then $E\theta$ is called a *variant* of E. Thus, for example, $x < y' + z'$ is a variant of $x < y + z$, since $x < y' + z' \equiv (x < y + z)\{y/y', z/z', y'/y, z'/y\}$, whereas $x < y' + x$ is not.

The following lemma, whose proof we omit, clarifies the concept of a variant and implies that "being a variant of" is a symmetric relation.

2.1. LEMMA. *For all expressions E and F*

E is a variant of F iff E is an instance of F
and F is an instance of E.

Given a program P we denote by ground (P) the set of all ground instances of clauses in P. Note that this set can be infinite. Given an atom A we denote by $[A]$ the set of all its ground instances.

Substitutions can be composed. Given substitutions $\theta = \{x_1/t_1, \ldots, x_n/t_n\}$ and $\eta = \{y_1/s_1, \ldots, y_m/s_m\}$ their *composition* $\theta\eta$ is defined by removing from the set

$$\{x_1/t_1\eta, \ldots, x_n/t_n\eta, y_1/s_1, \ldots, y_m/s_m\}$$

those pairs $x_i/t_i\eta$ for which $x_i \equiv t_i\eta$, as well as those pairs y_i/s_i for which $y_i \in \{x_1,\ldots,x_n\}$.

Thus, for example, when $\theta = \{x/3, y/f(x, 1)\}$ and $\eta = \{x/4\}$ then $\theta\eta = \{x/3, y/f(4, 1)\}$. This definition implies the following simple result.

2.2. LEMMA. *For all substitutions θ, η and γ and an expression E,*
(i) $(E\theta)\eta \equiv E(\theta\eta)$,
(ii) $(\theta\eta)\gamma = \theta(\eta\gamma)$.

This lemma shows that when writing a sequence of substitutions, also in the context of an expression, the parentheses can be omitted. By convention, substitution binds stronger than any connective or quantifier.

We say that a substitution θ is *more general* than a substitution η if for some substitution γ we have $\eta = \theta\gamma$.

2.4. *Unifiers*

Finally, we introduce the notion of unification. Consider two atoms A and B. If for a substitution θ we have $A\theta \equiv B\theta$, then θ is called a *unifier* of A and B and we then say that A and B are *unifiable.* A unifier θ of A and B is called a *most general unifier* (or *mgu* in short) if it is more general than any other unifier of A and B. It is an important fact that if two atoms are unifiable then they have a most general unifier. In fact, we have the following theorem due to Robinson [84].

2.3. THEOREM (Unification Theorem). *There exists an algorithm (called a* unification algorithm) *which for any two atoms produces their most general unifier if they are unifiable and otherwise reports nonexistence of a unifier.*

PROOF. We follow here the presentation of Lassez, Maher and Marriott [62]. We present an algorithm based upon Herbrand's original algorithm [45, p. 148] which deals with solutions of finite sets of term equations. This algorithm was first presented in [71].

Two atoms can unify only if they have the same relation symbol. With two atoms $p(s_1,\ldots,s_n)$ and $p(t_1,\ldots,t_n)$ to be unified we associate a set of equations

$$\{s_1 = t_1,\ldots,s_n = t_n\}.$$

A substitution θ such that $s_1\theta \equiv t_1\theta,\ldots,s_n\theta \equiv t_n\theta$ is called a *unifier* of the set of equations $\{s_1 = t_1,\ldots,s_n = t_n\}$. Thus the set of equations $\{s_1 = t_1,\ldots,s_n = t_n\}$ has the same unifiers as the atoms $p(s_1,\ldots,s_n)$ and $p(t_1,\ldots,t_n)$. Two sets of equations are called *equivalent* if they have the same unifiers.

A (possibly empty) set of equations is called *solved* if it is of the form $\{x_1 = u_1,\ldots,x_n = u_n\}$ where x_i's are distinct variables and none of them occurs in a term u_j.

A solved set of equations $\{x_1 = u_1,\ldots,x_n = u_n\}$ determines the substitution $\{x_1/u_1,\ldots,x_n/u_n\}$. This substitution is a unifier of this set of equations and clearly it is its mgu, that is, it is more general than any other unifier of this set of equations.

Thus to find an mgu of two atoms it suffices to transform the associated set of equations into an equivalent one which is solved. The following algorithm does it if this is possible and otherwise halts with failure.

UNIFICATION ALGORITHM. Nondeterministically choose from the set of equations an equation of a form below and perform the associated action.

(1) $f(s_1, \ldots, s_n) = f(t_1, \ldots, t_n)$: *replace by the equations* $s_1 = t_1, \ldots, s_n = t_n$;

(2) $f(s_1, \ldots, s_n) = g(t_1, \ldots, t_m)$ where $f \not\equiv g$: *halt with failure;*

(3) $x = x$: *delete the equation*;

(4) $t = x$ where t is not a variable: *replace by the equation* $x = t$;

(5) $x = t$ where $x \not\equiv t$ and x has another occurrence in the set of equations: *if* x *appears in* t *then halt with failure, otherwise perform the substitution* $\{x/t\}$ *in every other equation*

The algorithm terminates when no step can be performed or when failure arises. To keep the formulation of the algorithm concise we identified here constants with 0-ary functions. Thus step (1) includes the case $c = c$ for every constant c which leads to deletion of such an equation. Also step (2) includes the case of two constants.

First, observe that for each variable x step (5) can be performed at most once, so this step can be performed only a finite number of times. Subsequent applications (if any) of steps (1) and (4) strictly diminish the total number of occurrences of function symbols on the left-hand side of the equations. This number is not affected by the application of step (3). Moreover, in the absence of step (1), step (3) can be performed only finitely many times. This implies termination.

Next, observe that applications of steps (1), (3) and (4) replace a set of equations by an equivalent one. The same holds in the case of a successful application of step (5) because, for any substitution θ, $x\theta \equiv t\theta$ implies that the substitutions θ and $\{x/t\}\theta$ are identical.

Next, observe that if the algorithm successfully terminates, then by virtue of steps (1), (2) and (4) the left-hand sides of the final equations are variables. Moreover, by virtue of step (5) these variables are distinct and none of them occurs on the right-hand side of an equation. So if the algorithm successfully terminates, it produces a solved set of equations equivalent with the original one.

Finally, observe that if the algorithm halts with failure then the set of equations at the failure step does not have a unifier.

This establishes correctness of the algorithm and concludes the proof of the theorem. □

To illustrate the operation of the above unification algorithm consider the following example.

2.4. EXAMPLE. Consider the following set of equations

$$\{f(x) = f(f(z)),\ g(a, y) = g(a, x)\}.$$

Choosing the first equation, step (1) applies and produces the new equation set

$$\{x = f(z),\ g(a, y) = g(a, x)\}.$$

Choosing the second equation, step (1) applies again and yields

$$\{x = f(z),\ a = a,\ y = x\}.$$

Now by applying step (1) again we get

$$\{x = f(z),\ y = x\}.$$

The only step which can now be applied is step (5). We get

$$\{x = f(z),\ y = f(z)\}.$$

Now no step can be applied and the algorithm successfully terminates.

Call a substitution θ *idempotent* if $\theta\theta = \theta$. Call a unifier of θ of two atoms A and B *relevant* if all variables which appear either in the domain of θ or in the terms from the range of θ also appear in A or B. In Section 2.7 we shall rely on the following observation.

2.5. COROLLARY. *If two atoms are unifiable then they have an mgu which is idempotent and relevant.*

PROOF. The unifier produced by the procedure used in the proof of Theorem 2.3 is of the form $\{x_1/u_1, \ldots, x_n/u_n\}$ where none of the variables x_i occurs in a term u_j, so it is idempotent. Moreover, in the unification algorithm no variables from outside the unified atoms are introduced. Thus the produced mgu is relevant. □

One can prove that idempotent mgu's are relevant but we shall not need this observation in future.

Given a substitution θ denote its domain by $dom(\theta)$ and the set of variables which appear in a term from the range of θ by $r(\theta)$. Given an expression E, denote by $var(E)$ the set of variables which appear in it. The following observation will be needed in Subsection 2.7.

2.6. LEMMA. *Let E be an expression and θ an idempotent substitution. Then*

$$var(E\theta) \cap dom(\theta) = \emptyset.$$

PROOF. It is easy to see that for any substitution θ

$$var(E\theta) \cap dom(\theta) \subseteq r(\theta). \tag{2.1}$$

But for an idempotent substitution θ also

$$dom(\theta) \cap r(\theta) = \emptyset. \tag{2.2}$$

(2.1) and (2.2) imply the claim. □

2.5. Computation process—the SLD-resolution

Logic programs compute through a combination of two mechanisms—replacement and unification. This form of computing boils down to a specific form of theorem proving, called *SLD-resolution.* To better understand this computation process, let us concentrate first on the issue of a replacement in the absence of variables.

Consider for a moment a logic program P in which all clauses are ground. Let $N = \leftarrow A_1, \ldots, A_n$ $(n \geqslant 1)$ be a ground negative clause and suppose that for some i, $1 \leqslant i \leqslant n$ and $k \geqslant 0$, $C = A_i \leftarrow B_1, \ldots, B_k$ is a clause from P. Then

$$N' = \leftarrow A_1, \ldots, A_{i-1}, B_1, \ldots, B_k, A_{i+1}, \ldots, A_n$$

is the result of replacing A_i in N by $B_1, \ldots, B_k$ and is called a *resolvent* of N and C. A_i is called the *selected atom* of N.

Iterating this replacement process we obtain a sequence of resolvents which is called a *derivation.* A derivation can be finite or infinite. If its last clause is empty then we speak of a *refutation* of the original negative clause N. We can then say that, from the assumption that in presence of the program P the clause $N = \leftarrow A_1, \ldots, A_k$ holds, we derived the contradiction, namely the empty clause. This can be viewed as a proof of the negation of N from P.

Assuming for a moment from the reader knowledge of the semantics for first-order logic (which is explained in Subsection 3.1) we note that N stands for $\neg A_1 \vee \cdots \vee \neg A_k$, so its negation stands for $\neg(\neg A_1 \vee \cdots \vee \neg A_k)$ which is semantically equivalent to $A_1 \wedge \cdots \wedge A_k$. Thus a refutation of N can be viewed as a proof of $A_1 \wedge \cdots \wedge A_k$.

If we reverse the arrows in clauses, we can view a program with all clauses ground as a context-free grammar with erasing rules (i.e., rules producing the empty string) and with no start or terminal symbols. Then a refutation of a goal can be viewed as a derivation of the empty string from the word represented by the goal.

An important aspect of logic programs is that they can be used not only to *refute* but also to *compute*—through a repeated use of unification which produces assignments of values to variables. We now explain this process by extending the previous situation to the case of logic programs and negative clauses which can contain variables.

Let P be a logic program and $N = \leftarrow A_1, \ldots, A_n$ be a negative clause. We first redefine the concept of a *resolvent.* Suppose that $C = A \leftarrow B_1, \ldots, B_k$ is a clause from P. If for some i, $1 \leqslant i \leqslant n$, A_i and A unify with an mgu θ, then we call

$$N' = \leftarrow(A_1, \ldots, A_{i-1}, B_1, \ldots, B_k, A_{i+1}, \ldots, A_n)\theta$$

a *resolvent of N and C with the mgu* θ. Thus a resolvent is obtained by performing the following four steps:

(a) select an atom A_i,
(b) try to unify A and A_i,
(c) if (b) succeeds then perform the replacement of A_i by $B_1, \ldots, B_k$ in N,
(d) apply to the resulting clause the mgu θ obtained in (b).

As before, iterating this process of computing a resolvent we obtain a sequence of resolvents called a derivation. But now because of the presence of variables we have to be careful.

By an *SLD-derivation* (we explain the abbreviation SLD in a moment) of $P \cup \{N\}$ we mean a maximal sequence $N_0, N_1, \ldots$ of negative clauses where $N = N_0$, together with a sequence $C_0, C_1, \ldots$ of variants of clauses from P and a sequence $\theta_0, \theta_1, \ldots$ of substitutions such that, for all $i = 0, 1, \ldots,$

(i) N_{i+1} is a resolvent of N_i and C_i with the mgu θ_i,

(ii) C_i does not have a variable in common with $N_0, C_0, \ldots, C_{i-1}$.

The clauses $C_0, C_1, \ldots$ are called the *input clauses* of the derivation. When one of the resolvents N_i is empty then it is the last negative clause of the derivation. Such a derivation is then called an *SLD-refutation*. An SLD-derivation is called *failed* if it is finite and it is not a refutation.

A new element in this definition is the use of variants that satisfy (ii) instead of the original clauses. This condition is called *standardization apart*. Its relevance will be extensively discussed in Section 2.7. The idea is that we do not wish to make the result of the derivation dependent on the choice of variable names. Note for example that $p(x)$ and $p(f(y))$ unify by means of the mgu binding x to $f(y)$. Thus the goal $\leftarrow p(x)$ can be refuted from the program $\{p(f(x))\leftarrow\}$.

The existence of an SLD-refutation of $P\cup\{N\}$ for $N = \leftarrow A_1, \ldots, A_k$ can be viewed as a contradiction. We can then conclude that we proved the negation of N. But N stands for $\forall x_1 \ldots \forall x_s(\neg A_1 \vee \cdots \vee \neg A_k)$, where $x_1, \ldots, x_s$ are all variables appearing in N, so its negation stands for $\neg\forall x_1 \ldots \forall x_s(\neg A_1 \vee \cdots \vee \neg A_k)$ which is semantically equivalent (see Subsection 3.1) to $\exists x_1 \ldots \exists x_s(A_1 \wedge \cdots \wedge A_k)$. Now, an important point is that the sequence of substitutions $\theta_0, \theta_1, \ldots, \theta_m$ performed during the process of the refutation actually provides the bindings for the variables $x_1, \ldots, x_s$. Thus the existence of an SLD-refutation for $P\cup\{N\}$ can be viewed as a proof of the formula $(A_1 \wedge \cdots \wedge A_k)\theta_0 \ldots \theta_m$. We justify this statement in Subsection 3.2.

The restriction of $\theta_0 \ldots \theta_m$ to the variables of N is called a *computed answer substitution* for $P\cup\{N\}$. According to the definition of SLD-derivation, the following two choices are made in each step of constructing a new resolvent:

- choice of the selected atom,
- choice of the input clause whose conclusion unifies with the selected atom.

Now, the first choice is in general dependent on the whole "history" of the derivation up to the current resolvent. Such a history consists of a sequence $N_0, N_1, \ldots, N_{k-1}$ of goals with selected atoms, a goal N_k, a sequence $C_0, C_1, \ldots, C_{k-1}$ of input clauses and a sequence $\theta_0, \theta_1, \ldots, \theta_{k-1}$ of substitutions such that, for all $i=0, \ldots, k-1$, N_{i+1} is a resolvent of N_i and C_i with mgu θ_i where the selected atom of N_i is used in step (a) above. Let now *HIS* stand for the set of all such histories in which the last goal N_k is nonempty.

By a *selection rule* R we now mean a function which, when applied to an element of *HIS* with the last goal $N_k = \leftarrow A_1, \ldots, A_l$, yields an atom A_j, $1 \leqslant j \leqslant l$. Such a general definition allows us to select different atoms in resolvents that occur more than once in the derivation or, in general, in identical resolvents with different histories.

Given a selection rule R, we say that an SLD-derivation of $P\cup\{N\}$ is *via* R if all choices of the selected atoms in the derivation are performed according to R. That is, for each nonempty goal M of this SLD-derivation with a history H, $R(H)$ is the selected atom of M.

Now, SLD stands for Selection rule-driven Linear resolution for Definite clauses.

2.6. *An example*

To the reader overwhelmed with such a long sequence of definitions we offer an example which hopefully clarifies the introduced concepts. We analyze in it the consequences of the choices in (a) and (b).

Consider a simplified version of the 8-puzzle. Assume a 3×3 grid filled with eight moveable tiles. Our goal is to rearrange the tiles so that the blank one is in the middle. We number the fields consecutively as follows:

1	2	3
4	5	6
7	8	9

and represent each legal move as a movement of the "blank" to an adjacent square.

First, we define the relation *adjacent* by providing an exhaustive listing of adjacent squares in ascending order:

$$\text{adjacent}(1,2)\leftarrow, \text{adjacent}(2,3)\leftarrow, \ldots, \text{adjacent}(8,9)\leftarrow, \quad (\textit{horizontal adjacency}),$$
$$\text{adjacent}(1,4)\leftarrow, \text{adjacent}(4,7)\leftarrow, \ldots, \text{adjacent}(6,9)\leftarrow \quad (\textit{vertical adjacency})$$

and using a rule

$$\text{adjacent}(x, y)\leftarrow\text{adjacent}(y, x) \quad (\textit{symmetry}). \tag{a}$$

In total, twenty-four pairs are adjacent. (A more succinct representation would be possible if addition and subtraction functions are available.) Then we define an initial configuration by assuming that the blank is initially, say, on square 1. Thus we have

$$\text{configuration}(1, \text{nil})\leftarrow,$$

where the second argument—here nil—denotes the sequence of squares visited. Finally, we define a legal move by the rule

$$\text{configuration}(x, y.\,l)\leftarrow\text{adjacent}(x, y), \text{configuration}(y, l) \tag{b}$$

where $y.l$ is a list with head y and tail l written in the usual infix notation.

As a goal we choose the negative clause

$$\leftarrow\text{configuration}(5, l)$$

stating that no sequence of visited squares leads to a situation where square 5 is blank. The following represents an SLD-refutation of the goal of length 7.

$\leftarrow$**configuration(5, l)**	(b) $\{l/l_1\}, \{x/5, l/y.l_1\}$
$\leftarrow$**adjacent(5, y)**, configuration(y, l_1)	(a) $\{x/x_1, y/y_1\}, \{x_1/5, y_1/y\}$
$\leftarrow$**adjacent(y, 5)**, configuration(y, l_1)	adjacent(4, 5)$\leftarrow$, $\{y/4\}$
$\leftarrow$**configuration(4, l_1)**	(b) $\{x/x_2, y/y_2, l/l_2\}, \{x_2/4, l_1/y_2.l_2\}$
$\leftarrow$**adjacent(4, y_2)**, configuration(y_2, l_2)	(a) $\{x/x_3, y/y_3\}, \{x_3/4, y_3/y_2\}$
$\leftarrow$**adjacent(y_2, 4)**, configuration(y_2, l_2)	adjacent(1, 4)$\leftarrow$, $\{y_2/1\}$
$\leftarrow$**configuration(1, l_2)**	configuration(1, nil)$\leftarrow$, $\{l_2/\text{nil}\}$
$\square$	

Selected atoms are put in bold. We thus always select the leftmost atom. On the right the input clauses and the mgu's are given. Note that at various places variants of the clauses (a) and (b) are used. The sequence of mgu's performed binds the variable l to 4.1.nil through the consecutive substitutions $\{l/y.l_1\}, \{y/4\}, \{l_1/y_2.l_2\}, \{y_2/1\}, \{l_2/\text{nil}\}$.

This provides the sequence of squares leading to the final configuration. Thus the refutation of the initial goal is constructive in the sense that it provides the value of l for which the formula ← configuration (5, l) does not hold.

Another choice of input clauses can lead to an infinite SLD-derivation. For example, here is a derivation in which we repeatedly use rule (a):

←**configuration(5, l)**	(b) $\{l/l_1\}, \{x/5, l/y.l_1\}$
←**adjacent(5, y)**, configuration(y, l_1)	(a) $\{x/x_1, y/y_1\}, \{x_1/5, y_1/y\}$
←**adjacent(y, 5)**, configuration(y, l_1)	(a) $\{x/x_2, y/y_2\}, \{x_2/y, y_2/5\}$
←**adjacent(5, y)**, configuration(y, l_1)	
...	

Also, another choice of a selection rule can lead to an infinite SLD-derivation. For example, a repeated choice of the rightmost atom and rule (b) leads to an infinite derivation with the goals continuously increasing its length by 1.

2.7. *Properties of SLD-derivations*

In the next sections we shall need the following two lemmas concerning SLD-derivations. Both of them rely on the condition of standardizing apart introduced in Subsection 2.5.

2.7. Lemma. *Let $N_0, N_1, \ldots$ be an SLD-derivation with a sequence $C_0, C_1, \ldots$ of input clauses and a sequence $\theta_0, \theta_1, \ldots$ of mgu's. Suppose that all θ_i's are idempotent and relevant. Then, for all $m \geqslant 0$ and $n > m$,*

(1) $var(N_n) \cap dom(\theta_m) = \emptyset$,

(2) $var(N_n\theta_n) \cap dom(\theta_m) = \emptyset$.

Proof. (1) We prove by induction on i that, for all $i > 0$,

$$var(N_{m+i}) \cap dom(\theta_m) = \emptyset. \tag{2.3}$$

N_{m+1} is of the form $E\theta_m$, so for $i=1$ (2.3) is the consequence of Lemma 2.6. Suppose now that (2.3) holds for some $i > 0$. Since each θ_j is relevant, by the form of N_{j+1}, for all $j \geqslant 0$,

$$var(N_{j+1}) \subseteq var(N_j) \cup var(C_j). \tag{2.4}$$

Since θ_m is relevant,

$$dom(\theta_m) \subseteq var(N_m) \cup var(C_m), \tag{2.5}$$

so using (2.4) m times

$$dom(\theta_m) \subseteq var(N_0) \cup var(C_0) \cup \cdots \cup var(C_m). \tag{2.6}$$

Now

$$\begin{aligned} & var(N_{m+i+1})\cap dom(\theta_m) \\ & \subseteq (var(N_{m+i})\cap dom(\theta_m))\cup(var(C_{m+i})\cap dom(\theta_m)) && \text{(by (2.4) with } j=m+i) \\ & \subseteq var(C_{m+i})\cap(var(N_0)\cup var(C_0)\cup\cdots\cup var(C_m)) && \text{(by (2.3) and (2.6))} \\ & \subseteq \emptyset && \text{(by standardizing apart).} \end{aligned}$$

This proves the induction step and concludes the proof of (1).

(2) It suffices to note that, by assumption on the θ_i's

$$var(N_n\theta_n)\subseteq var(N_n)\cup var(C_n)$$

and use (1), (2.6) and standardizing apart. □

We now show that up to renaming the computed answer substitution of an SLD-derivation does not depend on the choice of variables in the input clauses. To this end we prove a slightly stronger result first, which uses the notion of a resultant of an SLD-derivation.

Given a goal $N=\leftarrow A_1,\ldots,A_k$ we denote by $N^\sim$ the formula $A_1\wedge\cdots\wedge A_k$. Then $\square^\sim$ is the empty conjunction which we identify with **true**. Given an SLD-derivation $N_0,N_1,\ldots$ with a sequenceof mgu's $\theta_0,\theta_1,\ldots$ of length $\geqslant i$, by a *resultant* (*of level i*) we mean the formula

$$N_i^\sim\rightarrow N_0^\sim\theta_0\ldots\theta_{i-1}.$$

Thus the resultant of level 0 is the formula $N_0^\sim\rightarrow N_0^\sim$.

2.8. LEMMA (Variant Lemma) (Lloyd and Shepherdson [66]). *Let $N_0,N_1,\ldots$ and $N'_0,N'_1,\ldots$ be two SLD-derivations of $P\cup\{N\}$ where $N=N_0$ and $N=N'_0$, with the input clauses $C_0,C_1,\ldots$ and $C'_0,C'_1,\ldots$ respectively. Suppose that each C'_i is a variant of C_i and that in each N'_i atoms in the same positions as in N_i are selected. Also, suppose that all mgu's used in the two SLD-derivations are relevant. Then the resultants of these two SLD-derivations are their respective variants.*

PROOF. We prove the claim by induction on the level i of resultants. For $i=0$ there is nothing to prove. Assume the claim holds for some $i\geqslant 0$. Let $\theta_0,\theta_1,\ldots$ be the mgu's of the first SLD-derivation and $\theta'_0,\theta'_1,\ldots$ the mgu's of the second SLD-derivation. By the induction hypothesis

$$Res=N_i^\sim\rightarrow N_0\theta_0\ldots\theta_{i-1}$$

is a variant of

$$Res'=N_i'^\sim\rightarrow N'_0\theta'_0\ldots\theta'_{i-1}.$$

Thus, for a renaming θ with $dom(\theta)\subseteq var(Res')$,

$$Res\equiv Res'\theta. \tag{2.7}$$

By assumption C_i is a variant of C'_i. Thus for a renaming η with $dom(\eta) \subseteq var(C'_i)$

$$C_i \equiv C'_i \eta. \tag{2.8}$$

Given two substitutions σ and ξ with disjoint domains, we denote by $\sigma \cup \xi$ their *union* which is defined in the obvious way. Put now $\gamma = (\theta \cup \eta)\theta_i$. We prove the following four facts:

(1) *γ is well defined.*
(2) *For some σ, $\gamma = \theta'_i \sigma$.*
(3) $N_{i+1} \equiv N'_{i+1}\sigma$.
(4) $N_0 \theta_0 \ldots \theta_i \equiv N'_0 \theta'_0 \ldots \theta'_i \sigma$.

Re (1): We only need to show that the domains of θ and η are disjoint. We first show that

$$var(Res') \cap var(C'_i) = \emptyset. \tag{2.9}$$

By the assumption, $\theta'_0, \ldots, \theta'_{i-1}$ are relevant, so by the same argument as the one used in the previous lemma, but now applied to the ranges of θ'_j instead of their domains, we get, for $j = 0, \ldots, i-1$,

$$r(\theta'_j) \subseteq (N'_0) \cup var(C'_0) \cup \cdots \cup var(C'_{i-1}). \tag{2.10}$$

Also, as in the proof of the previous lemma

$$var(N'_i) \subseteq var(N'_0) \cup var(C'_0) \cup \cdots \cup var(C'_{i-1}). \tag{2.11}$$

Now

$$\begin{aligned} var(Res') &= var(N'_i) \cup var(N'_0 \theta'_0 \ldots \theta'_{i-1}) \\ &\subseteq var(N'_i) \cup var(N'_0) \cup r(\theta'_0) \cup \ldots \cup r(\theta'_{i-1}) \\ &\subseteq var(N'_0) \cup var(C'_0) \cup \ldots var(C'_{i-1}) \quad \text{(by (2.10) and (2.11)} \end{aligned}$$

so (2.9) follows from the standardizing apart. Now note that $dom(\theta) \subseteq var(Res')$ and $dom(\eta) \subseteq var(C'_i)$, so by (2.9) the domains of θ and η are indeed disjoint.

Re (2): Let B' be an atom from C'_i. Then $var(B') \subseteq var(C'_i)$, so by (2.9)

$$var(B') \cap dom(\theta) = \emptyset, \tag{2.12}$$

since $dom(\theta) \subseteq var(Res')$. Similarly, also by (2.9), for an atom A' from N'_i,

$$var(A') \cap dom(\eta) = \emptyset. \tag{2.13}$$

Thus by (2.12), for an atom B' from C'_i,

$$B'(\theta \cup \eta) \equiv B'\eta \tag{2.14}$$

and by (2.13), for an atom A' from N'_i,

$$A'(\theta \cup \eta) \equiv A'\theta. \tag{2.15}$$

Let

$$\begin{aligned} C_i &= B_0 \leftarrow B_1, \ldots, B_k, & N_i &= \leftarrow A_1, \ldots, A_m, \\ C'_i &= B'_0 \leftarrow B'_1, \ldots, B'_k, & N'_i &= \leftarrow A'_1, \ldots, A'_m. \end{aligned}$$

By (2.7) and (2.14), for $j=0,\ldots,k$,

$$B_j \equiv B'_j(\theta \cup \eta) \tag{2.16}$$

and by (2.8) and (2.15), for $j=1,\ldots,m$,

$$A_j \equiv A'_j(\theta \cup \eta). \tag{2.17}$$

Let now A'_l be the selected atom of N'_i. Then A_l is the selected atom of N_i and

$$A_l\theta_i \equiv B_0\theta_i. \tag{2.18}$$

Now

$$\begin{aligned} A'_l\gamma &\equiv A'_l(\theta \cup \eta)\theta_i \\ &\equiv A_l\theta_i && \text{(by (2.17))} \\ &\equiv B_0\theta_i && \text{(by (2.18))} \\ &\equiv B'_0(\theta \cup \eta)\theta_i && \text{(by (2.16))} \\ &\equiv B'_0\gamma, \end{aligned}$$

so γ is a unifier of A'_l and B'_0. Now, since θ_i is an mgu of A'_l and B'_0, for some σ, $\gamma = \theta'_i\sigma$.

Re (3): We have

$$\begin{aligned} N_{i+1} &\equiv \leftarrow (A_1,\ldots,A_{l-1},B_1,\ldots,B_k,A_{l+1},\ldots,A_m)\theta_i \\ &\equiv \leftarrow (A'_1,\ldots,A'_{l-1},B'_1,\ldots,B'_k,A'_{l+1},\ldots,A'_m)(\theta \cup \eta)\theta_i \\ &\qquad\qquad \text{(by (2.16) and (2.17))} \\ &\equiv \leftarrow (A'_1,\ldots,A'_{l-1},B'_1,\ldots,B'_k,A'_{l+1},\ldots,A'_m)\gamma \\ &\equiv \leftarrow (A'_1,\ldots,A'_{l-1},B'_1,\ldots,B'_k,A'_{l+1},\ldots,A'_m)\theta'_i\sigma \quad \text{(by fact (2))} \\ &\equiv N'_{i+1}\sigma \end{aligned}$$

Re (4): We have $dom(\eta) \subseteq var(C'_i)$, so by (2.9)

$$var(N'_0\theta'_0\ldots\theta'_{i-1}) \cap dom(\eta) = \emptyset. \tag{2.19}$$

Now

$$\begin{aligned} N_0\theta_0\ldots\theta_i &\equiv N'_0\theta'_0\ldots\theta'_{i-1}\theta\theta_i && \text{(by (2.7))} \\ &\equiv N'_0\theta'_0\ldots\theta'_{i-1}(\theta \cup \eta)\theta_i && \text{(by (2.19))} \\ &\equiv N'_0\theta'_0\ldots\theta'_{i-1}\gamma \\ &\equiv N'_0\theta'_0\ldots\theta'_{i-1}\theta'_i\sigma && \text{(by the form of } \gamma). \end{aligned}$$

Now, putting facts (3) and (4) together we see that the resultant of level $i+1$ of the first SLD-derivation is an instance of the resultant of level $i+1$ of the second SLD-derivation. By symmetry the resultant of level $i+1$ of the second SLD-derivation is an instance of the resultant of level $i+1$ of the first SLD-derivation. By Lemma 2.1 these resultants are the variants of each other. □

2.9. COROLLARY (Variant Corollary). *Let Φ and Ψ be two SLD-derivations of $P \cup \{N\}$ satisfying the conditions of Lemma 2.8. Suppose that Φ is an SLD-refutation with a* *computed answer substitution θ. Then Ψ is an SLD-refutation with a computed answer substitution η such that $N\theta$ is a variant of $N\eta$.*

PROOF. It suffices to consider resultants of level k of Φ and Ψ, where k is the length of the SLD-refutation Ψ, and apply the previous lemma. □

The above corollary shows that the existence of an SLD-refutation does not depend on the choice of variables in the input clauses.

To be able to use the results of this section we shall assume from now on that *all mgu's used in all SLD-derivations are idempotent and relevant.*

2.8. *Refutation procedures—SLD-trees*

When searching for a refutation of a goal, SLD-derivations are constructed with the aim of generating the empty clause. The totality of these derivations form a *search space.* One way of organizing this search space is by dividing SLD-derivations into categories according to the selection rule used. This brings us to the concept of an SLD-tree.

Let P be a program, N a goal and R a selection rule. The SLD-tree for $P\cup\{N\}$ via R groups all SLD-derivations of $P\cup\{N\}$ via R. Formally the *SLD-tree for $P\cup\{N\}$ via R* is a tree such that

- its branches are SLD-derivations of $P\cup\{N\}$ via R,
- every node N' has exactly one descendant for every clause C of P such that the selected atom A of N' unifies with the head of a variant C' of C. This descendant is a resolvent of N' and C' with A being the selected atom of N'.

We call an SLD-tree *successful* if it contains the empty clause.

The SLD-trees for $P\cup\{N\}$ can differ in size and form.

2.10. EXAMPLE. (Apt and van Emden [4]). Let P be the following program:

1. $\text{path}(x, z) \leftarrow \text{arc}(x, y), \text{path}(y, z)$,
2. $\text{path}(x, x) \leftarrow$,
3. $\text{arc}(b, c) \leftarrow$.

A possible interpretation of P is as follows: arc(x, y) holds if there is an arc from x to y and path(x, y) holds if there is a path from x to y. Figures 1 and 2 show two SLD-trees for $P\cup\{\leftarrow\text{path}(x, c)\}$. The selected atoms are put in bold, used clauses and performed substitutions are indicated. The input clauses at the level i are obtained from the original clauses by adding the subscript "i" to all variables which were used earlier in the derivation. In this way the standardizing apart condition is satisfied. Note that the first tree is finite while the second one is infinite. Both trees contain the empty clause.

2.9. *Bibliographic remarks*

The concepts of unification, resolution and standardization apart were introduced in [84]. Efficient unification algorithms were proposed by Paterson and Wegman [79],

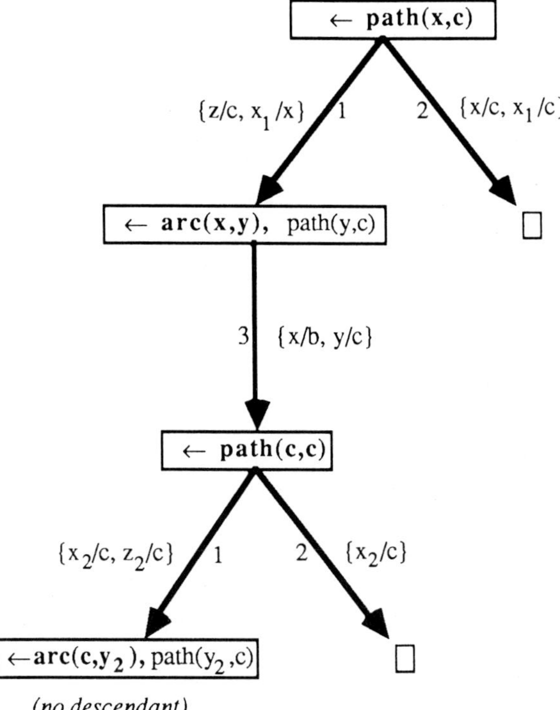

Fig. 1.

and Martelli and Montanari [71]. See also the survey on unification by Siekmann [93].

SLD-resolution is a special case of SL-resolution of Kowalski and Kuehner [57] and was proposed as a basis for programming in Kowalski [53]. The name was first used in [4] where also the notions of a success set and SLD-trees were formally introduced. SLD-trees were informally used in [21] where they were called evaluation trees.

The selection rule was originally required to be a function defined on sequences of atoms. Our formulation follows the suggestion of Shepherdson [88, p. 62]. The proof of Lemma 2.8 differs from the original proof. Corollary 2.9 was independently established in [52].

3. Semantics

3.1. *Semantics for first-order logic*

To understand the *meaning* of a logic program, or a first-order formula in general, we now provide the definition of semantics due to A. Tarski. Again, our treatment is very brief. More extensive discussion of this fundamental issue can be found e.g.

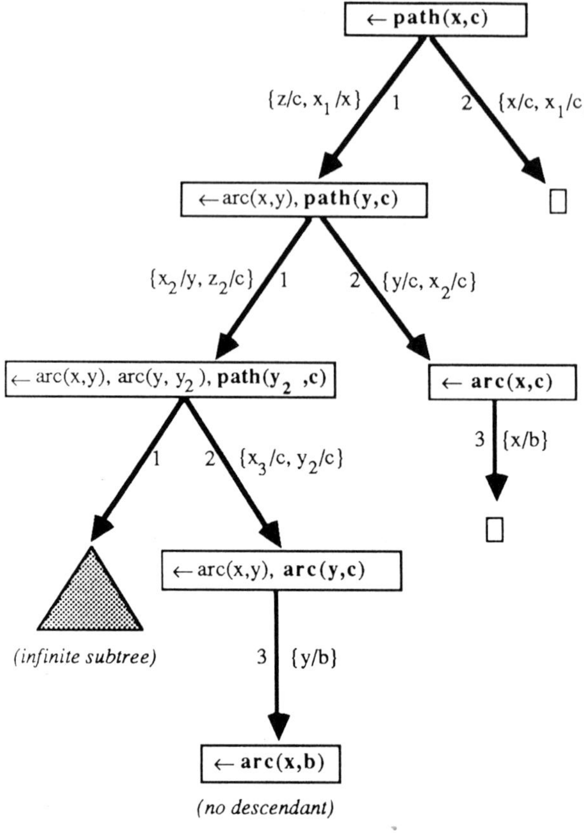

Fig. 2.

in [70], [92]. We begin by defining an interpretation. An *interpretation I* for a first-order language L consists of

- a nonempty set D, called the *domain* of I,
- an assignment for each constant c in L of an element c_I of D,
- an assignment for each n-ary function f in L of a mapping f_I from D^n to D,
- an assignment for each n-ary relation r in L of an n-ary predicate r_I on D, i.e. a subset of D^n.

Our aim is now to define when a formula of L is true in an interpretation for L. To this purpose we first relate terms to elements of the domain of an interpretation. We do this by making use of the notion of a state (or a *variable assignment*). A *state* (*over I*) is simply a function assigning to each variable an element from D.

Given now a state σ, we extend its domain to all terms, that is, we assign to a term t an element $\sigma(t)$ from D proceeding by induction as follows:

- for a constant c we define $\sigma(c)$ as c_I (thus $\sigma(c)$ does not depend on σ),
- if $f(t_1, \ldots, t_n)$ is a term then we define $\sigma(f(t_1, \ldots, t_n))$ as $f_I(\sigma(t_1), \ldots, \sigma(t_n))$, the result

of applying the mapping f_I to the sequence of values associated with the terms $t_1, \ldots, t_n$.

Observe that for a ground term t, $\sigma(t)$ has the same value for all σ.

We can now define a semantics of a formula. Given a formula F we define inductively its truth in a state σ over I, written as $I \models_\sigma F$, as follows:

- if $p(t_1, \ldots, t_n)$ is an atomic formula then

$$I \models_\sigma p(t_1, \ldots, t_n) \text{ iff } (\sigma(t_1), \ldots, \sigma(t_n)) \in p_I,$$

that is, if the sequence of values associated with terms $t_1, \ldots, t_n$ belongs to the predicate p_I,

- $I \models_\sigma$ **true**, not $I \models_\sigma$ **false**,
- if F and G are formulas then

$$\begin{aligned} I \models_\sigma \neg F \quad &\text{iff} \quad \text{not } I \models_\sigma F, \\ I \models_\sigma F \vee G \quad &\text{iff} \quad I \models_\sigma F \text{ or } I \models_\sigma G, \\ I \models_\sigma \forall x F \quad &\text{iff} \quad I \models_{\sigma[x/d]} F \text{ for all } d \in D. \end{aligned}$$

Here $\sigma[x/d]$, for a state σ, an element d of D and a variable x, stands for the state which differs from σ only on the variable x to which it assigns the element d.

This allows us already to define truth of clauses. The truth of other formulas is defined by expressing the remaining connectives and the quantifier $\exists$ in terms of $\neg$, $\vee$ and $\forall$:

$$\begin{aligned} &F \wedge G \quad \text{as} \quad \neg(\neg F \vee \neg G), \\ &F \rightarrow G \quad \text{as} \quad \neg F \vee G, \\ &F \leftrightarrow G \quad \text{as} \quad (F \rightarrow G) \wedge (G \rightarrow F) \quad \text{(and then using the above two definitions)}, \\ &\exists x F \quad \text{as} \quad \neg \forall x \neg F. \end{aligned}$$

Finally, we say that the formula F *is true in the interpretation* I, and write $I \models F$, when for all states σ, $I \models_\sigma F$. Note that $\Box$ as the empty disjunction is false in every interpretation I.

Let now S be a set of formulas. We say that an interpretation I is a *model for* S if every formula from S is true in I. When S has a model, we say that it is *satisfiable* or *consistent*. Otherwise, we say that it is *unsatisfiable* or *inconsistent*. When every interpretation is a model for S, we say that S is *valid*.

Given another set of formulas S' we say that S *semantically implies* S' or S' *is a semantic consequence of* S, if every model of S is also a model of S'. We write then $S \models S'$ and omit the $\{$ and $\}$ brackets if any of these sets has exactly one element. S and S' are *semantically equivalent* if both $S \models S'$ and $S' \models S$ hold.

Several simple facts about semantic consequence and semantic equivalence can be proved and will be used in the sequel. Already in Subsection 2.5 we used the fact that the following formulas are valid:

$$\begin{aligned} &\neg \forall x_1 \ldots \forall x_s F \leftrightarrow \exists x_1 \ldots \exists x_s \neg F, \\ &\neg(A_1 \vee \cdots \vee A_n) \leftrightarrow \neg A_1 \wedge \cdots \wedge \neg A_n, \\ &\neg \neg F \leftrightarrow F. \end{aligned}$$

3.2. Soundness of the SLD-resolution

Recall that, for a goal $N = \leftarrow A_1, \ldots, A_k$, $N^{\sim}$ stands for the formula $A_1 \wedge \cdots \wedge A_k$. Then $\square^{\sim}$ is the empty conjunction, so it is valid. The following lemma is immediate.

3.1. LEMMA. *If M is a resolvent of N and a clause C with an mgu θ then $C \models M^{\sim} \rightarrow N^{\sim}\theta$.*

As a consequence we obtain the following theorem due to Clark [22] justifying the statement made in Subsection 2.5.

3.2. THEOREM (soundness of SLD-resolution). *Let P be a program and $N = \leftarrow A_1, \ldots, A_k$ a goal. Suppose that there exists an SLD-refutation of $P \cup \{N\}$ with the sequence of substitutions $\theta_0, \ldots, \theta_n$. Then $(A_1 \wedge \cdots \wedge A_k)\theta_0 \ldots \theta_n$ is a semantic consequence of P.*

PROOF. Let $N_0, \ldots, N_{n+1}$, with $N_0 = N$ and $N_{n+1} = \square$, be the SLD-refutation in question and let $C_0, \ldots, C_n$ be its input clauses. Applying Lemma 3.1 $n+1$ times we get

$$P \models \square^{\sim} \rightarrow N^{\sim}\theta_0 \ldots \theta_n$$

which implies the claim. □

3.3. COROLLARY. *If there exists an SLD-refutation of $P \cup \{N\}$ then $P \cup \{N\}$ is inconsistent.*

Another straightforward consequence of Lemma 3.1, which will not be used in the sequel, is that all resultants of an SLD-refutation of $P \cup \{N\}$ are semantic consequences of P.

3.4. EXAMPLE. Reconsider now the program P studied in the example in Subsection 2.6 with the goal $\leftarrow$ configuration(5, l). Since we exhibited there an SLD-refutation of $P \cup \{\leftarrow$ configuration(5, l)$\}$, we conclude by the above corollary that $P \cup \{\leftarrow$configuration(5, l)$\}$ is inconsistent, that is, $P \models \exists l$ configuration(5, l). More specifically, by the Soundness Theorem we have $P \models$configuration(5, l)$\theta_0 \ldots \theta_7$ where $\theta_0, \ldots, \theta_7$ is the sequence of performed substitutions. As we saw before, this sequence binds l to 4.1.nil, so we have $P \models$configuration(5, 4.1.nil).

A natural question arises whether a converse of the above corollary or of the Soundness Theorem can be proved, that is, whether certain form of *completeness* of SLD-resolution can be shown. To handle this question we introduce a special class of models of logic programs, called Herbrand models.

3.3. Herbrand models

Let L be a first-order language whose set of constants is not empty. By the *Herbrand universe* U_L for L we mean the set of all ground terms of L. By the *Herbrand base* B_L for

L we mean the set of all ground atoms of L. If L is the first-order language associated with a program P (that is, L is L_P) then we denote U_L and B_L by the U_P and B_P, respectively. Now, by a *Herbrand interpretation* for L we mean an interpretation for L such that

(a) its domain is the Herbrand universe U_L,
(b) each constant in L is assigned to itself,
(c) if f is an n-ary function in L then it is assigned to the mapping from $(U_L)^n$ to U_L defined by assigning the ground term $f(t_1, \ldots, t_n)$ to the sequence $t_1, \ldots, t_n$ of ground terms,
(d) if r is an n-ary relation in L then it is assigned to a set of n-tuples of ground terms.

Thus each Herbrand interpretation for L is uniquely determined by a subset I of the Herbrand base B_L which fixes the assignment of predicates to relation symbols of L by assigning the set $\{(t_1, \ldots, t_n)\colon r(t_1, \ldots, t_n) \in I\}$ to the n-ary relation symbol r. In other words, we can identify Herbrand interpretations for L with (possibly empty) subsets of the Herbrand base B_L. This is what we shall do in the sequel.

To avoid some uninteresting complications we assume from now on that whenever a program P has variables then it also as some constants. This guarantees that its Herbrand base and the set ground (P) are not empty. The case of programs containing variables but no constants is hardly of interest.

With this restriction another uninteresting complication arises when a program uses only propositional symbols. Then its Herbrand universe is empty. To handle this case one can simply drop the condition that a domain of an interpretation is nonempty when L is constant-free and function-free.

By a *Herbrand model* for a set S of sentences we mean a Herbrand interpretation which is a model for S. The following simple lemma shows why Herbrand models naturally arise when studying logic programs.

3.5. LEMMA. *Let S be a set of universal formulas. If S has a model then it has a Herbrand model.*

PROOF. For an interpretation I let $I_H = \{A : A$ is a ground atom and $I \models A\}$ denote the corresponding Herbrand interpretation. A simple induction on the length of the formulas shows that I and I_H satisfy the same quantifier-free ground formulas. From this the lemma follows. □

3.6. COROLLARY. *Let P be a program and N a negative clause. If $P \cup \{N\}$ is consistent then it has a Herbrand model.*

We conclude this section by introducing two often recurring qualifications. A Herbrand model of a set of formulas S is the *least* model of S if it is included in every other Herbrand model of S and it is a *minimal model of* S if no proper subset of it is a Herbrand model of S. The least model is minimal but the converse is not always true (take for example $S = \{A \vee B\}$ with A, B ground atoms).

3.4. *The immediate consequence operator*

To study Herbrand models of programs, following [31], we introduce the *immediate consequence operator* T_p mapping Herbrand interpretations to Herbrand interpretations. For a program P and a Herbrand interpretation I, we put

$$\begin{aligned} A\in T_P(I) \text{ iff } & \text{for some atoms } B_1,\ldots,B_n \\ & A\leftarrow B_1,\ldots,B_n \text{ is a ground}(P) \\ & \text{and } I\models B_1\wedge\cdots\wedge B_n. \end{aligned}$$

Alternatively, for a ground atom A,

$$\begin{aligned} A\in T_P(I) \text{ iff } & \text{for some substitution } \theta \\ & \text{and a clause } B\leftarrow B_1,\ldots,B_n \text{ of } P \\ & \text{we have } A\equiv B\theta \text{ and } I\models(B_1\wedge\cdots\wedge B_n)\theta. \end{aligned}$$

In particular, if $A\leftarrow$ is in P, then every ground instance $A\theta$ of A is in $T_P(I)$ for every I. The following simple observation from [31] relates Herbrand models of P with the operator T_P.

3.7. Lemma. *For a program P and a Herbrand interpretation I, I is a model of P iff $T_P(I)\subseteq I$.*

Proof. First note that I is a model of P iff it is a model of ground(P). Now the latter is true iff, for every clause $A\leftarrow B_1,\ldots,B_n$ in ground(P), $I\models B_1\wedge\cdots\wedge B_n$ implies $I\models A$, i.e. $A\in I$. But this is true iff $T_P(I)\subseteq I$. □

When $T(I)\subseteq I$ holds, I is called a *pre-fixpoint* of T. Thus to study Herbrand models of a program P it suffices to study the pre-fixpoints of its immediate consequence operator T_P. This brings us to a study of operators and their pre-fixpoints in a general setting.

3.5. *Operators and their fixpoints*

Consider now an arbitrary, but fixed, complete lattice (for the definition see e.g. [10]) with the order relation $\subseteq$, the least upper bound operator $\cup$ and the greatest lower bound operator $\cap$. To keep in mind the subsequent applications to logic programs and their interpretations we denote the least element by $\emptyset$, the largest element by B, and the elements of the lattice by I, J, M. Given a set $A=\{I_n: n=0,1,\ldots\}$ of elements, we denote $\bigcup A$ and $\bigcap A$ by $\bigcup_{n=0}^{\infty} I_n$ and $\bigcap_{n=0}^{\infty} I_n$ respectively. Sometimes we rather write $\bigcup_{n<\omega} I_n$ and $\bigcap_{n<\omega} I_n$.

Consider an operator T on the lattice. T is called *monotonic* if, for all I, J, $I\subseteq J$ implies $T(I)\subseteq T(J)$. T is called *finitary* if, for every infinite sequence $I_0\subseteq I_1\subseteq\ldots$,

$$T\left(\bigcup_{n=0}^{\infty} I_n\right)\subseteq\bigcup_{n=0}^{\infty} T(I_n)$$

holds. If T is both monotonic and finitary then it is called *continuous*. A more often used, equivalent definition of continuity is: T is continuous if, for every infinite sequence $I_0 \subseteq I_1 \subseteq \ldots$, it holds that

$$T\left(\bigcup_{n=0}^{\infty} I_n\right) = \bigcup_{n=0}^{\infty} T(I_n).$$

As already mentioned in the previous section, any I such that $T(I) \subseteq I$ is called a *pre-fixpoint of* T. If $T(I) = I$ then I is called a *fixpoint of* T and if $T(I) \supseteq I$ then I is called a *post-fixpoint* of T.

We have the following classical theorem.

3.8. THEOREM (Fixpoint Theorem) (Knaster and Tarski [97]). *A monotonic operator T has a least fixpoint $lfp(T)$ which is also its least pre-fixpoint.*

We now define *powers* of a monotonic operator T. We put

$$T\uparrow 0(I) = I, \qquad T\uparrow(n+1)(I) = T(T\uparrow n(I)), \qquad T\uparrow\omega(I) = \bigcup_{n<\omega} T\uparrow n(I)$$

and abbreviate $T\uparrow\alpha(\emptyset)$ to $T\uparrow\alpha$. Powers of a monotonic operator generalize in a straightforward way to *transfinite powers* $T\uparrow\alpha(I)$ where α is an arbitrary ordinal. We shall not need them in the sequel.

The following well-known fact holds.

3.9. LEMMA. *If T is continuous then $T\uparrow\omega$ is its least pre-fixpoint and its least fixpoint.*

In the next section we apply these observations to the study of Herbrand models.

In Sections 4 and 5 we shall also use largest fixpoints and *downward powers* of monotonic operators. We put for a monotonic operator T

$$T\downarrow 0(I) = I, \qquad T\downarrow(n+1)(I) = T(T\downarrow n(I)), \qquad T\downarrow\omega(I) = \bigcup_{n<\omega} T\downarrow n(I).$$

Downward powers generalize in a straightforward way to *transfinite downward powers* $T\downarrow\alpha(I)$ where α is an arbitrary ordinal. We abbreviate $T\downarrow\alpha(B)$ to $T\downarrow\alpha$.

Note that

$$T\uparrow n(I) \subseteq T\uparrow(n+1)(I)$$

does not necessarily hold, but by monotonicity for all $n \geqslant 0$

$$T\uparrow n \subseteq T\uparrow(n+1)$$

does hold. Analogous statement holds for the downward powers.

The dual theorem to the Fixpoint Theorem 3.8 is the following.

3.10. THEOREM. *A monotonic operator T has a greatest fixpoint $gfp(T)$ which is also its greatest post-fixpoint.*

A monotonic operator T is called *downward continuous* if, for every infinite sequence

$I_0 \supseteq I_1 \supseteq \cdots$, it holds that

$$T\left(\bigcap_{n=0}^{\infty} I_n\right) = \bigcap_{n=0}^{\infty} T(I_n).$$

We have the following well-known lemma.

3.11. LEMMA. *Let T be a monotonic operator. Then for every α we have $T\downarrow\alpha \supseteq gfp(T)$. Moreover, for some α, $T\downarrow\alpha = gfp(T)$. If T is downward continuous then this ordinal is $\leqslant\omega$.*

We denote the smallest ordinal α for which $T\downarrow\alpha = gfp(T)$ by $\|T\downarrow\|$ and call it the *downward closure ordinal of* T or the *closure ordinal* of $T\downarrow$.

3.6. Least Herbrand models

Let us first investigate the properties of the immediate consequence operator. Note that Herbrand interpretations of L with the usual set-theoretic operations from a complete lattice so when studying this operator we can apply the results of the previous section.

3.12. LEMMA. *Let P be a program. Then*
(i) *T_p is finitary,*
(ii) *T_p is monotonic.*

PROOF. (i) Consider an infinite sequence $I_0 \subseteq I_1 \subseteq \cdots$ of Herbrand interpretations and suppose that $A \in T_P(\bigcup_{n=0}^{\infty} I_n)$. Then, for some atoms $B_1, \ldots B_k$, the clause $A \leftarrow B_1, \ldots, B_k$ is in ground(P), and moreover $\bigcup_{n=0}^{\infty} I_n \models B_1 \wedge \cdots \wedge B_k$. But the latter implies that for some I_n, namely the one containing all $B_1, \ldots, B_k$, $I_n \models B_1 \wedge \cdots \wedge B_k$. So $A \in T_P(I_n)$.
(ii) Immediate by definition. □

As an immediate consequence of the above lemma we have the following theorem.

3.13. THEOREM (Characterization Theorem) (Van Emden and Kowalski [31]). *Let P be a program. Then P has a Herbrand model M_P which satisfies the following properties:*
(i) *M_P is the least Herbrand model of P.*
(ii) *M_P is the least pre-fixpoint of T_P.*
(iii) *M_P is the least fixpoint of T_P.*
(iv) *$M_P = T_P\uparrow\omega$.*

PROOF. It suffices to apply Lemma 3.7, Theorem 3.8 and Lemma 3.9. □

By the *success set* of a program P we denote the set of all ground atoms A such that $P \cup \{\leftarrow A\}$ has an SLD-refutation.

3.14. COROLLARY. *The success set of a program P is contained in its least Herbrand model.*

PROOF. By Corollary 3.3 and the above theorem. □

3.7. *Completeness of the SLD-resolution*

We can now return to the problem of completeness. We first prove the converse of Corollary 3.3 that is, the following result due to HILL [46]. The proof is due to APT and VAN EMDEN [4].

3.15. THEOREM (completeness of SLD-resolution). *Let P be a program and N a goal. Suppose $P \cup \{N\}$ is inconsistent. Then there exists an SLD-refutation of $P \cup \{N\}$.*

First we need the following lemma.

3.16. LEMMA. (Substitution Lemma). *Let P be a program, N a goal and θ a substitution. Suppose that there exists an SLD-refutation of $P \cup \{N\theta\}$. Then there exists an SLD-refutation of $P \cup \{N\}$.*

PROOF. We proceed by induction on the length n of the SLD-refutation of $P \cup \{N\theta\}$. By the Variant Corollary 2.9 we can assume that θ does not act on any of the variables appearing in the input clauses of this refutation. Let $N = \leftarrow A_1, \ldots, A_k$.

If $n=1$ then $k=1$ and $A_1\theta$ unifies with a head of a unit input clause. So A_1 unifies with the head of the same clause. This settles the claim.

If $n>1$ then consider the first input clause $B_0 \leftarrow B_1, \ldots, B_m$ of the refutation. For an mgu η we have $A_i\theta\eta \equiv B_0\eta$ where $A_i\theta$ is the selected atom of $N\theta$. Thus, by the assumption on θ, $A_i\theta\eta \equiv B_0\theta\eta$, so A_i and B_0 unify. For some mgu ξ and a substitution γ we have $\theta\eta = \xi\gamma$.

By the assumption on $P \cup \{N\theta\}$ and θ there exists an SLD-refutation of

$$P \cup \{\leftarrow(A_1\theta, \ldots, A_{i-1}\theta, B_1\theta, \ldots, B_m\theta, A_{i+1}\theta, \ldots, A_k\theta)\eta\}$$

of length $n-1$. By the induction hypothesis there exists an SLD-refutation of

$$P \cup \{\leftarrow(A_1, \ldots, A_{i-1}, B_1, \ldots, B_m, A_{i+1}, \ldots, A_k)\xi\}.$$

Consider now an SLD-derivation of $P \cup \{N\}$ in which the first selected atom is A_i and the first input clause is $B_0 \leftarrow B_1, \ldots, B_m$ with the mgu ξ. Its first resolvent is $\leftarrow(A_1, \ldots, A_{-1}, B_1, \ldots, B_m, A_{i+1}, \ldots, A_k)\xi$ which, by the above, settles the claim. □

We now establish the converse of Corollary 3.14.

3.17. LEMMA. *The least Herbrand model of a program P is contained in the success set of P.*

PROOF. We make use of the continuity of the immediate consequence operator T_P which provides an internal structure to M_P. Suppose $A \in M_P$. By the Characterization Theorem 3.13(iv) for some $k>0$, $A \in T_P \uparrow k$. We now prove by induction on k that there exists an SLD-refutation of $P \cup \{\leftarrow A\}$. For $k=1$ the claim is obvious.

If $k>1$, then for some ground atoms $B_1,\dots B_n$ the clause $A\leftarrow B_1,\dots,B_n$ is in ground(P) and $\{B_1,\dots,B_n\}\subseteq T_P\uparrow(k-1)$. By the induction hypothesis, for $i=1,\dots,n$ there exists an SLD-refutation of $P\cup\{\leftarrow B_i\}$. But all B_i are ground so there exists an SLD-refutation of $P\cup\{\leftarrow B_1,\dots,B_n\}$.

Consider now an SLD-derivation of $P\cup\{\leftarrow A\}$ with the first input clause being the one of which $A\leftarrow B_1,\dots,B_n$ is a ground instance. Its first resolvent is a negative clause of which $\leftarrow B_1,\dots,B_n$ is a ground instance. The claim now follows by Lemma 3.16. □

We are now in position to prove the Completeness Theorem.

PROOF OF THEOREM 3.15. Suppose that $N=\leftarrow A_1,\dots,A_n$. M_P is not a model of $P\cup\{N\}$, so N is not true in M_P. Thus, for some substitution θ, $\{A_1\theta,\dots,A_n\theta\}\subseteq M_P$. By Lemma 3.17, for $i=1,\dots,n$ there exists an SLD-refutation of $P\cup\{\leftarrow A_i\theta\}$. But all $A_i\theta$ are ground so there exists an SLD-refutation of $P\cup\{N\theta\}$ and the claim now follows by Lemma 3.16. □

3.8. *Correct answer substitutions*

The Completeness Theorem can be generalized in various ways. We provide here two such generalizations.

First we introduce the following notion. Let P be a program and $N=\leftarrow A_1,\dots,A_n$ a goal. We say that θ is a *correct answer substitution* for $P\cup\{N\}$ if θ acts only on variables appearing in N and $P\models(A_1\wedge\cdots\wedge A_n)\theta$ holds.

Note that if θ is a correct answer substitution for $P\cup\{N\}$ then, for all γ, $P\cup\{N\theta\gamma\}$ is inconsistent. Consequently, $P\cup\{N\}$ is inconsistent as it is equivalent to a weaker statement that, for some γ, $P\cup\{N\gamma\}$ is inconsistent.

The following theorem is a kind of converse of the Soundness Theorem 3.2.

3.18. THEOREM (Clark [22]). *Consider a program P and a goal N. For every correct answer substitution θ for $P\cup\{N\}$ there exists a computed answer substitution σ for $P\cup\{N\}$ such that $N\sigma$ is more general than $N\theta$.*

We present here the proof due to Lloyd [64]. First we need the following strengthening of the Substitution Lemma.

3.19. LEMMA (Lifting Lemma). *Let P be a program, N a goal and θ a substitution. Suppose that there exists an SLD-refutation of $P\cup\{N\theta\}$ with the sequence of mgu's $\theta_0,\dots,\theta_n$. Then there exists an SLD-refutation of $P\cup\{N\}$ with the sequence of mgu's $\theta'_0,\dots,\theta'_n$ such that $\theta'_0\dots\theta'_n$ is more general than $\theta\theta_0\dots\theta_n$.*

PROOF. By a straightforward refinement of the proof of the Substitution Lemma 3.16. □

3.20. LEMMA. *Let P be a program and N a goal. Suppose that θ is a correct answer substitution for $P\cup\{N\}$. Then the empty substitution is a computed answer substitution for $P\cup\{N\theta\}$.*

PROOF. Let $x_1, \ldots, x_n$ be the variables of $N\theta$. Enrich the language of P by adding new constants $a_1, \ldots, a_n$ and let γ be the substitution $\{x_1/a_1, \ldots, x_n/a_n\}$. $P \cup \{N\theta\gamma\}$ is inconsistent, so by the Completeness Theorem 3.15 there exists an SLD-refutation of $P \cup \{N\theta\gamma\}$. By the Variant Corollary 2.9 we can assume that the variables $x_1, \ldots, x_n$ do not appear in the input clauses used in this refutation. But $N\theta\gamma$ is ground, so the answer substitution computed by this refutation is the empty substitution. By textually replacing in this refutation a_i by x_i, for $i = 1, \ldots, n$ we obtain an SLD-refutation of $P \cup \{N\theta\}$ with the empty substitution as the computed answer substitution. □

We are now ready to prove the desired theorem.

PROOF OF THEOREM 3.18. By the above lemma there exists an SLD-refutation of $P \cup \{N\theta\}$ with the empty substitution as the computed answer substitution. Let $\theta_0, \ldots, \theta_n$ be its sequence of mgu's. By the Lifting Lemma 3.19 there exists an SLD-refutation of $P \cup \{N\}$ with a computed answer substitution σ and a sequence of mgu's $\theta'_0, \ldots, \theta'_n$ such that $\theta'_0 \ldots \theta'_n$ is more general than $\theta\theta_0 \ldots \theta_n$.

Then $N\theta'_0 \ldots \theta'_n$ is more general than $N\theta\theta_0 \ldots \theta_n$. But the former goal equals $N\sigma$ whereas the latter equals $N\theta$. □

3.9. *Strong completeness of the SLD-resolution*

Another way to generalize the Completeness Theorem is by taking selection rules into account. We follow here the presentation of Apt and van Emden [4].

3.21. THEOREM (strong completeness of SLD-resolution) (Hill [46]). *Let P be a program and N a goal. Suppose that $P \cup \{N\}$ is inconsistent. Then every SLD-tree with N as root is successful.*

This theorem states that if $P \cup \{N\}$ is inconsistent then there exists an SLD-refutation of $P \cup \{N\}$ via every selection rule.

To prove it we first introduce the following notion. Given a program P we call a goal N *k-refutable*, where $k \geq 1$, if in every SLD-tree with N as root there exists the empty clause with a path length from the root of at most k.

Another straightforward refinement of the proof of Substitution Lemma yields the following.

3.22. LEMMA. *Let P be a program, N a goal and θ a substitution. Suppose that $N\theta$ is k-refutable. Then N is k-refutable.*

The next two lemmas generalize corresponding facts about refuted goals.

3.23. LEMMA. *Let P be a program and let $F_1, \ldots, F_n$ be sequences of atoms. Assume that $F_1, \ldots, F_n$ have no variables in common. If each $\leftarrow F_i$ is k_i-refutable for $i = 1, \ldots, n$ then $\leftarrow F_1 \ldots, F_n$ is $(k_1 + \cdots + k_n)$-refutable.*

PROOF. By straightforward induction on $k_1 + \cdots + k_n$. □

3.24. LEMMA. *If A is in the least Herbrand model of P, then, for some k, $\leftarrow A$ is k-refutable.*

PROOF. By repeating the argument from the proof of Lemma 3.17 using the above lemma with each F_i being a single ground atom. □

We can now prove the strong completeness of SLD-resolution.

PROOF OF THEOREM 3.21. By repeating the argument from the proof of the Completeness Theorem 3.15 using Lemmas 3.24, 3.23 and 3.22. □

Summarizing the results obtained in Sections 3.4, 3.6, 3.7 and the present one, we obtain the following characterizations of the success set.

3.25. THEOREM (Success Theorem). *Consider a program P and a ground atom A. Then the following are equivalent:*
(a) *A is in the success set of P.*
(b) $A \in T_P \uparrow \omega$.
(c) *Every SLD-tree with $\leftarrow A$ as root is successful.*
(d) $P \models A$.

PROOF. First note that, by Corollary 3.6 and the Characterization Theorem 3.13(i), $P \models A$ iff $A \in M_P$. The rest follows by the Characterization Theorem 3.13(iv), Corollary 3.14, Lemma 3.17 and Lemma 3.24. □

The Strong Completeness Theorem shows that when searching for a refutation of a goal any SLD-tree is a complete search space. Of course whether a refutation will be actually found in a successful SLD-tree depends on the tree search algorithm used.

Note that in fact we have proved more.

3.26. THEOREM. *Let P be a program and N a good. If $P \cup \{N\}$ is inconsistent then, for some k, N is k-refutable.*

PROOF. By inspection of the proof of the Strong Completeness Theorem 3.21. □

This indicates that given a program P when searching for a refutation of a goal N it is enough to explore any SLD-tree until a certain depth depending only on N. However, this depth as a function of the goal N is in general not computable. This is an immediate consequence of the results proved in Section 4.

3.10. Procedural versus declarative interpretation

In the last two sections we studied two ways of interpretating the logic programs. They are sometimes referred to as a procedural and declarative interpretation.

Procedural interpretation explains *how* the programs compute, i.e. what is the computational mechanism which underlies the program execution. In the framework of programming languages semantics, it is sometimes referred to as the operational semantics.

On the other hand, *declarative interpretation* provides the meaning of a program, i.e., it attempts to answer the question what semantically follows from the program without analyzing the underlying computational mechanism. In such a way declarative interpretation provides a specification for any underlying computational mechanism, i.e. it explains *what* should be computed by the program. In the framework of programming language semantics, it corresponds with the denotational semantics.

To summarize the above we can say that procedural interpretation is concerned with the *method* whereas declarative interpretation is concerned with the *meaning*. Any form of a completeness theorem can be viewed as a proof of a match between these two interpretations. In practice of course this match can be destroyed when, as explained at the end of the previous subsection, the computational mechanism is supplemented by an incomplete (tree) search algorithm.

3.11. Bibliographic remarks

The name *immediate consequence operator* was introduced in [22]. Gallier [39] presents a different proof of the completeness of the SLD-resolution based on the use of Gentzen systems and indicates how to extend it to obtain a proof of the strong completeness of the SLD-resolution. The strongest completeness result is that of Clark [22], which combines the claims of Theorems 3.18 and 3.21. Lloyd [64] provides a rigorous proof of this theorem.

4. Computability

4.1. Computability versus definability

Once we have defined *how* logic programs compute and analyzed the relation between the proof-theoretic and semantic aspects, let us reflect on the question *what* objects logic programs compute. We show here that logic programs are *computationally complete* in the sense that they have the same computational power as recursive functions.

Assume that the language L has at least one constant, so that the Herbrand universe U_L is not empty. Moreover, assume that L has infinitely many relation symbols in every arity. We say that a program P *computes a predicate* $R \subseteq U_L^n$ *using a relation* r if, for all $t_1, \ldots, t_n \in U_L$,

$$(t_1, \ldots, t_n) \in R \text{ iff there exists an SLD-refutation of } P \cup \{\leftarrow r(t_1, \ldots, t_n)\}.$$

A semantic counterpart of this definition is obtained by saying that a program P *defines a predicate* $R \subseteq U_L^n$ *using a relation* r if, for all $t_1, \ldots, t_n \in U_L$,

$$(t_1, \ldots, t_n) \in R \text{ iff } P \models r(t_1, \ldots, t_n).$$

Both definitions presuppose that $L_P \subseteq L$ and $U_{L_P} = U_L$. We have the following result.

4.1. THEOREM. *Let P be a program, R a predicate and r a relation. Then the following are equivalent*:

(a) *P computes R using r.*
(b) *P defines R using r.*
(c) *For all* $t_1, \ldots, t_n \in U_L$
$$(t_1, \ldots, t_n) \in R \text{ iff } r(t_1, \ldots, t_n) \in M_P.$$

PROOF. By the Success Theorem 3.25 and the Characterization Theorem 3.13. □

Thus the question which predicates are computed by logic programs reduces to the question which predicates are defined over their least Herbrand models.

This question has various answers depending on the form of L. We study here the case when L has finitely many but at least one constant and finitely many but at least one function symbol. Then the Herbrand universe U_L is infinite. The assumption that the set of constants and the set of functions are finite allows us to reverse the question and analyze for a given program P which predicates it computes over its Herbrand universe U_{L_P}. The assumption that in each arity the set of relations is infinite allows us to construct new clauses without syntactic constraints.

4.2. *Enumerability of* U_L

We call a binary predicate R on U_L an *enumeration of* U_L if R defines the successor function on U_L. In other words, R is an enumeration of U_L if we have $U_L = \{f_R^n(u) : n < \omega\}$ where u is some fixed ground term and f_R is a one-one function on U_L defined by $f_R(x) = y$ iff $(x, y) \in R$.

As a first step towards a characterization of predicates computable by logic programs we prove the following result due to Andréka and Németi [1]. Our presentation is based on [12].

4.2. THEOREM (Enumeration Theorem). *There exists a program* successor *which computes an enumeration of* U_L *using a binary relation* succ.

PROOF. The construction of the program *successor* is rather tedious. First we define the enumeration *enum* of U_L which will be computed.

We start by defining inductively the notion of height of a ground term. We put

$$\text{height}(a) = 0 \quad \text{for each constant } a,$$
$$\text{height}(f(t_1, \ldots, t_n)) = \max(\text{height}(t_1), \ldots, \text{height}(t_n)) + 1.$$

Next, we define a well-ordering on all ground terms. To this purpose we first order all constants and all function symbols in some way. We extend this ordering inductively to all ground terms of height $\leq n$ $(n > 0)$ by putting

$$f(s_1, \ldots, s_k) < g(t_1, \ldots, t_m)$$
$$\text{iff } (\text{height}(f(s_1, \ldots, s_k)), f, s_1, \ldots s_k) \prec (\text{height}(g(t_1, \ldots t_m)), g, t_1, \ldots, t_m).$$

Here $\prec$ is a lexicographic ordering obtained from the ordering of natural numbers, ordering of function symbols and the already defined ordering $<$ on ground terms of height $< n$. This extension is compatible with the fragment of $<$ defined so far. By induction, $<$ is defined on all ground terms.

From the following three observations and the assumption about the number of constants and function symbols it follows that $<$ is a well-ordering of type ω:

(a) If height(s) $<$ height(t) then $s<t$.

(b) If height($f(s_1,\ldots,s_k)$)=height($g(t_1,\ldots,t_m)$) and f is smaller than g in the chosen ordering then $f(s_1,\ldots,s_k)<g(t_1,\ldots,t_m)$.

(c) If height($f(s_1,\ldots,s_i,s_{i+1},\ldots,s_k)$)=height($f(s_1,\ldots,s_i,t_{i+1},\ldots,t_k)$) and $s_{i+1}<t_{i+1}$ then $f(s_1,\ldots,s_i,s_{i+1},\ldots,s_k)<f(s_1,\ldots,s_i,t_{i+1},\ldots,t_k)$.

We now define *enum* to be the graph of the $<$-successor function. Note that

(d) if t is the $<$-maximal term of height n then its $<$-successor is the $<$-minimal term of height $n+1$;

(e) otherwise, the $<$-successor of $t=f(t_1,\ldots,t_n)$ is obtained by first locating the rightmost term t_i whose (already defined) $<$-successor t_i' has height smaller than the height of t. Then $f(t_1,\ldots,t_{i-1},t_i',a,\ldots,a,t_n')$ is the $<$-successor of t, where a is the $<$-least constant and t_n' is the $<$-least term s such that height($f(t_1,\ldots,t_{i-1},t_i',a,\ldots,a,s)$)=height($t$).

To compute the relation *enum* we systematically translate its definition into clauses. We proceed by the following steps.

(1) For counting purposes we identify a subset N_L of U_L with the set of natural numbers N. Let f_0 be the smallest function in the chosen ordering. We put

$$N_L=\{\hat{n}: n\in N\}$$

where $\hat{0}=a$ and, for each n, $\widehat{n+1}=f_0(a,\ldots,a,\hat{n})$.

The following program *Nat* computes N_L using a relation *nat*:

$$nat(a)\leftarrow,$$

$$nat(f_0(a,\ldots,a,x))\leftarrow nat(x).$$

In turn, the program S_L obtained by adding to *Nat* the clause

$$s_L(x,f_0(a,\ldots,a,x))\leftarrow nat(x)$$

computes the successor relation on N_L using a relation s_L.

(2) Using the programs *Nat* and S_L the definition of the height function can now be translated into a program *height* with a binary relation h such that

$$height\models h(t,k) \text{ iff } t \text{ is a ground term of height } n, \quad \text{where } k=\hat{n}.$$

(3) Note that $\hat{n}$ is the $<$-minimal term of height n. Thus adding a clause $\min(x,x)\leftarrow nat(x)$ we get a program *minimum* such that

$$minimum\models \min(t,k) \text{ iff } t \text{ is the } <\text{-minimal term of height } n, \quad \text{where } k=\hat{n}.$$

Let now b be the $<$-largest constant and f_1 the largest function in the chosen ordering. Note that the $<$-maximal term of height 0 is b, of height 1 $f_1(b,\ldots,b)$, etc.

Thus adding clauses

$$\max(b, a)\leftarrow,$$

$$\max(f_1(x, \ldots, x), y')\leftarrow\max(x, y), s_L(y, y')$$

we get a program *maximum* such that

$$maximum \models \max(t, k)$$
$$\text{iff } t \text{ is the } <\text{-maximal term of height } n, \quad \text{where } k = \hat{n}.$$

(4) Using the above auxiliary definitions, the program *successor* can now be constructed by translating the statements (d) and (e) into clauses. The details are straightforward though lengthy and we omit them. □

4.3. *Recursive functions*

To characterize the predicates computable by logic programs we need to recall the basic concepts of the recursion theory as developed by S.C. Kleene. We follow here [92].

For brevity denote the sequence $a_1, \ldots, a_n$ by $\bar{a}$. Let, for $i = 1, \ldots, n$, the projection function $\boldsymbol{P}_i^n$ be defined by

$$\boldsymbol{P}_i^n(\bar{a}) = a_i.$$

For a given predicate $R \subseteq N^n$, $\boldsymbol{K}_R$ stands for its characteristic function defined by

$$\boldsymbol{K}_R(\bar{a}) = \begin{cases} 0 & \text{iff } \bar{a} \in R, \\ 1 & \text{iff } \bar{a} \notin R. \end{cases}$$

We define the class of (total) *recursive functions* over N inductively by putting

(R1) the functions $\boldsymbol{P}_i^n$, $+$, $\times$ and $\boldsymbol{K}_<$ are recursive;

(R2) if $g, h_1, \ldots, h_k$ are recursive functions and f is defined by

$$f(\bar{a}) = g(h_1(\bar{a}), \ldots, h_k(\bar{a}))$$

then f is recursive;

(R3) let g be a recursive function such that

$$\forall \bar{a} \exists b\, g(\bar{a}, b) = 0;$$

then the function f defined by

$$f(\bar{a}) = \mu b.\ g(\bar{a}, b) = 0$$

is recursive, where $\mu b.\ R$ stands for the least b such that R holds.

A predicate over N is *recursive* if its characteristic function is recursive. A predicate R is *recursively enumerable* (r.e.) if for some recursive predicate S

$$\bar{a} \in R \text{ iff } \exists b\, (\bar{a}, b) \in S.$$

A predicate R is *r.e. complete* if for every recursively enumerable predicate S there is some recursive function f such that

$$\bar{a} \in S \text{ iff } f(\bar{a}) \in R.$$

R.e. complete predicates are not recursive. It is a well-known fact that there exists a recursively enumerable predicate which is r.e. complete.

In the sequel we shall use various well-known simple results from the theory of recursive functions. We also rely on some standard techniques like coding. This allows us to investigate the complexity of subsets of the Herbrand base B_L as its elements can be coded by natural numbers.

We have the following simple result.

4.3. THEOREM. *For every program P, M_P is recursively enumerable.*

PROOF. By the Characterization Theorem 3.13 (iv) we have $A \in M_P$ iff, for some $k>0$, relation p and $t_1, \ldots, t_n \in U_P$, $A = p(t_1, \ldots, t_n)$ and $p(t_1, \ldots, t_n) \in T_P\uparrow k$. The result now follows by the standard techniques of the recursion theory because the predicate $\{(k, A): A \in T_P\uparrow k\}$ is, after appropriate coding, recursive. □

4.4 Computability of recursive functions

The Herbrand universe U_L does not coincide with natural numbers but thanks to the Enumeration Theorem 4.2 we can make such an identification. This allows us to transfer the notions of the recursion theory from N to U_L.

We now prove the following theorem.

4.4 THEOREM (Computability Theorem) (Andréka and Németi [1]). *For every recursive function f there is a program P which computes the graph of f using a relation p_f.*

PROOF. We assume that each program given here incorporates the program *successor* which uses different relations than those used here. We proceed by induction on the construction of recursive functions.

Re (R1): We can define + in terms of the successor by simply rewriting two well-known axioms of Peano arithmetic as clauses:

$$p_+(x, \hat{0}, x) \leftarrow,$$

$$p_+(x, y, z) \leftarrow succ(y', y),\ succ(z', z),\ p_+(x, y', z').$$

Other functions admit equally straightforward presentations.

Re (R2): Suppose by induction that there exist programs $P_0, \ldots, P_k$ computing the graphs of functions $g, h_1, \ldots, h_k$ using the relations $p_g, p_{h_1}, \ldots, p_{h_k}$ correspondingly. We can assume that $P_0, \ldots, P_k$ have no relations in common, apart from those occurring in *successor*. Then the program $P_0 \cup \cdots \cup P_k$ augmented by the clause

$$p_f(x_1, \ldots, x_l, x_{l+1}) \leftarrow p_{h_1}(x_1, \ldots, x_l, y_1), \ldots, p_{h_k}(x_1, \ldots, x_l, y_k), p_g(y_1, \ldots, y_k, x_{l+1})$$

computes the graph of the function f defined as in (R2).

Re (R3): Let f and g be recursive functions as given in (R3). By induction there exists a program P_g which computes the graph of g using a relation p_g. The program P_f is obtained by adding to P_g the following clauses with a new relation r:

$$p_f(x_1,\ldots,x_k,x_{k+1})\leftarrow p_g(x_1,\ldots,x_{k+1},\hat{0}),\ r(x_1,\ldots,x_{k+1}),$$

$$r(x_1,\ldots,x_k,\hat{0})\leftarrow,$$

$$r(x_1,\ldots,x_k,y)\leftarrow succ(y',y),\ r(x_1,\ldots,x_k,y'),\ p_g(x_1,\ldots,x_k,y',z),\ p_<(\hat{0},z).$$

The intended meaning of $r(x_1,\ldots,x_{k+1})$ is $\forall y(y<x_{k+1}\rightarrow g(x_1,\ldots,x_k,y)>0)$. Note that under this interpretation $r(x_1,\ldots,x_k,y)$ holds and $r(x_1,\ldots,x_k,n+1)$ iff $r(x_1,\ldots,x_k,n)\wedge g(x_1,\ldots,x_k,n)>0$ and this is exactly what the last two clauses express. □

4.5. COROLLARY. *A predicate R on U_L is recursively enumerable iff some program P computes it using a relation r.*

PROOF. (⇒) Suppose that for some recursive predicate S, $\bar{a}\in R$ iff $\exists b\,(\bar{a},b)\in S$. Let P_S be the program computing the characteristic function K_S of S using a relation p_S. Then the program P_S augmented by the clause

$$p_R(x_1,\ldots,x_k)\leftarrow p_S(x_1,\ldots,x_k,y,\hat{0})$$

computes the predicate R using relation p_R.

(⇒) By Theorems 4.1 and 4.3. □

This allows us to prove the converse of the Computability Theorem.

4.6. COROLLARY. *Suppose that a program P computes the graph of a total function using some relation. Then this function is recursive.*

PROOF. A total function is recursive iff its graph is recursively enumerable. □

Also, we can obtain the following characterization of the recursion-theoretic complexity of M_p.

4.7. COROLLARY. *For some program P, M_P is r.e. complete. A fortiori, M_P is not recursive.*

PROOF. Let R be a recursively enumerable, r.e. complete predicate on U_L. By Corollary 4.5 and Theorem 4.1 we have, for all a, $a\in R$ iff $r(a)\in M_P$, where P is a program which computes R using a relation r. This shows that M_P is r.e. complete, as well. □

We conclude this section by mentioning the following strengthening of the Computability Theorem 4.4, which we shall use in the next subsection. Following [12] we call a program P *determinate* if $T_P\uparrow\omega=T_P\downarrow\omega$.

4.8. THEOREM (Blair [12]). *For every recursive function f there is a determinate program P which computes the graph of f using a relation p_f.*

The proof is based on a detailed analysis of the programs constructed in the proof of the Computability Theorem 4.4 and we omit it.

4.5. *Closure ordinals of* $T_P\!\downarrow$

In this subsection we study the downward closure ordinals of the operators T_P for programs P.

We noted in subsection 3.6 that for a program P the operator T_P is continuous. However, T_P does not need to be downward continuous. To see, this consider the following program P:

$$p(f(x)) \leftarrow p(x),$$

$$q(a) \leftarrow p(x).$$

Then for $n \geqslant 1$ we have $T_P\!\downarrow\! n = \{q(a)\} \cup (p(f^k(a)) \colon k \geqslant n\}$, so $T_P\!\downarrow\!\omega = \{q(a)\}$. It follows that $T_P\!\downarrow\!(\omega+1) = \emptyset$, hence $\| T_P\!\downarrow \| = \omega + 1$ and T_P is not downward continuous. Note that, by Lemma 3.11, $gfp(T_P) = T_P\!\downarrow\!(\omega+1) = \emptyset$. This asymmetry is one of the most curious phenomena in the theory of logic programming.

To characterize the downward closure ordinals of the operators T_P we first introduce some definitions. We shall consider well-founded (partial) orderings on natural numbers. For a well-founded ordering R we write $a <_R b$ instead of $(a, b) \in R$ and denote by $dom(R)$ its domain. With each well-founded ordering R we can associate in a standard way an ordinal $\| R \|$ by means of a transfinite induction:

$$\| a \| = \begin{cases} 0 & \text{if } a \text{ is a } <_R\text{-minimal element of } dom(R), \\ sup(\| b \| + 1 \colon b <_R a) & \text{otherwise,} \end{cases}$$

$$\| R \| = sup(\| a \| \colon a \in dom(R)).$$

An ordinal α is called *recursive* if $\alpha = \| R \|$ for some well-founded ordering R which is a recursive predicate. The least nonrecursive ordinal is denoted by ω_1^{ck} (ω_1 of Church and Kleene).

The following theorem characterizes the ordinals $\| T_P\!\downarrow \|$.

4.9. THEOREM (Blair[11]). (i) *For every $\alpha \leqslant \omega_1^{ck}$ there exists a program P such that $\| T_P\!\downarrow \| = \alpha$.*

(ii) *For every program P, $\| T_P\!\downarrow \| \leqslant \omega_1^{ck}$.*

PROOF. (i) It is clear how to construct for any natural number $n \geqslant 0$ a program P such that $\| T_P\!\downarrow \| = n$. Suppose now that $\omega \leqslant \alpha < \omega_1^{ck}$. For some β we have $\alpha = \omega + \beta$. Assume from now on that L has exactly one, unary function symbol f and exactly one constant a. Then U_L coincides with the set of natural numbers. Let R be a recursive well-founded ordering such that $\| R \| = \beta$. Given a relation q we denote by $[q]$ the set of all ground atoms of the form $q(t_1, \ldots, t_n)$.

Let P_1 be the program P from the beginning of this subsection augmented by the clause

$$q(y) \leftarrow p(x).$$

Then $T_{P_1}\downarrow\omega = [q]$ and $T_{P_1}\downarrow\alpha = 0$ for $\alpha > \omega$.

By Theorem 4.8 there exists a determinate program P_2 which computes R using some relation r. We can assume that P_1 and P_2 are disjoint. Then, for any $\alpha \geqslant \omega$,

$$T_{P_2}\downarrow\alpha \cap [r] = R_r, \quad \text{where } R_r = \{r(s,t)\colon (s,t) \in R\}.$$

Let P_3 be the program

$$q(x) \leftarrow r(y,x), q(y)$$

and finally let $P = P_1 \cup P_2 \cup P_3$. Then

$$T_P\downarrow\omega \cap ([q] \cup [r]) = [q] \cup R_r.$$

Thus

$$T_P\downarrow(\omega+1) \cap ([q] \cup [r]) = \{q(s)\colon s \in dom(R),\ \|s\| \geqslant 1\} \cup R_r$$

and more generally, for every γ,

$$T_P\downarrow(\omega+\gamma) \cap ([q] \cup [r]) = \{q(s)\colon s \in dom(R),\ \|s\| \geqslant \gamma\} \cup R_r.$$

Thus, for $\gamma < \beta$,

$$T_P\downarrow(\omega+\gamma) \neq T_P\downarrow(\omega+\gamma+1).$$

Also

$$T_P\downarrow(\omega+\beta) \cap ([q] \cup [r]) = R_r,$$

so

$$T_P\downarrow(\omega+\beta) = T_{P_2}\downarrow(\omega+\beta) = T_{P_2}\downarrow\omega$$

and consequently

$$T_P\downarrow(\alpha+1) = T_P\downarrow\alpha,$$

i.e. $\|T_P\downarrow\| = \alpha$.

The proof that for some program P in fact $\|T_P\downarrow\| = \omega_1^{ck}$ and the proof of (ii) rely on advanced results from recursion theory and are beyond the scope of this paper. □

4.6. *Bibliographic remarks*

There is considerable confusion concerning the actual formulation and origin of the results of the first part of this section. The statement that logic programming has a full power of recursion theory is usually attributed to Tärnlund [98] who showed that Turing machines can be simulated using logic programs. However, in his proof additional function symbols are used and the paper of Andréka and Németi [1] actually appeared earlier as a technical report.

A syntactically stronger form of the Computability Theorem 4.4 in the case when L has exactly one, unary function symbol and exactly one constant was proved in [86]. For such L the Computability Theorem 4.4 is implicitly contained in [94]. Related results were proved in [47, 56, 89, 95]. The last paper discusses all these results in detail. Börger [14] discusses connections between logic programming and computational complexity of various classes of formulas. Fitting [37] studies in detail computability by means of logic programs on domains other than the Herbrand base, in particular integers, words and trees.

That T_P does not need to be downward continuous was originally observed by Andréka and Németi, and Clark. The class of determinate programs is extensively studied in [5], where they are called functional programs.

5. Negative information

5.1. Nonmonotonic reasoning

SLD-resolution is an example of a *sound* method of reasoning because only true facts can be deduced using it. More precisely, we call here a reasoning method "$\vdash$" *sound* if, for all variable-free formulas φ, $P \vdash \varphi$ implies $P \models \varphi$, where $P \vdash \varphi$ denotes that φ can be proved from a program P. And we call "$\vdash$" *weakly sound* if $P \vdash \varphi$ implies consistency of $P \cup \{\varphi\}$. Now, putting (see subsection 2.5) $P \vdash_{\mathrm{SLD}} \exists x_1 \ldots \exists x_s\, (A_1 \wedge \cdots \wedge A_k)$ iff there exists an SLD-refutation of $P \cup \{\leftarrow A_1, \ldots, A_k\}$, we see that "$\vdash_{\mathrm{SLD}}$" is sound by virtue of the Soundness Theorem 3.2.

We call a reasoning method "$\vdash$" *effective* if for any program P the set $\{\varphi : P \vdash \varphi\}$ is recursively enumerable. Now, "$\vdash_{\mathrm{SLD}}$" is easily seen to be effective by using the standard techniques of recursion theory. Effectiveness is a desirable property as it amounts to saying that it is decidable whether an object is a proof of a formula. Ineffective reasoning methods cannot be implemented.

SLD-resolution is also an example of a *monotonic* method of reasoning. We call here a reasoning method "$\vdash$" *monotonic* if, for any two programs P and P',

$$P \vdash \varphi \text{ implies } P \cup P' \vdash \varphi.$$

Otherwise, "$\vdash$" is called *nonmonotonic*. Clearly, if there exists an SLD-refutation of $P \cup \{N\}$ then there also exists an SLD-refutation of $P \cup P' \cup \{N\}$.

However, SLD-resolution is a very restricted form of reasoning, because only positive facts can be deduced using it. This restriction cannot be overcome if soundness or monotonicity is to be maintained. More precisely, the following simple yet crucial observation holds.

5.1. LEMMA. *Let "$\mid\!\sim$" be a reasoning method such that $P \mid\!\sim \neg A$ for some negative ground literal $\neg A$. Then "$\mid\!\sim$" is not sound. Moreover, if "$\mid\!\sim$" is weakly sound then it is not monotonic.*

PROOF. Note that the Herbrand base is a model of P but not a model of $\neg A$. Thus "$\mid\!\sim$" is not sound. Suppose it is monotonic. Then we get $P \cup \{A\} \mid\!\sim \neg A$. But $P \cup \{A\} \cup \{\neg A\}$ is inconsistent, so "$\mid\!\sim$" is not weakly sound. $\square$

However, in some applications it is natural to require that also negative information can be deduced.

5.2. EXAMPLE Consider

$$P = \{element(fire)\leftarrow, element(air)\leftarrow, element(water)\leftarrow, element(earth)\leftarrow, stuff(mud)\leftarrow\}.$$

Then we naturally expect that $\neg element(mud)$, $\neg stuff(fire)$ and similarly with other elements.

By Lemma 5.1 any such extension of SLD-resolution leads to a nonmonotonic reasoning.

5.2. *Closed world assumption*

One natural possibility is to consider here the following rule (or rather metarule):

$$\frac{A \text{ cannot be proved from } P}{\neg A}$$

where A is a ground atom. This rule is usually called the *closed world assumption* (CWA). It was first considered in [83]. The notion of provability referred to in the hypothesis is that in first-order logic. For our purposes it is sufficient to know that it is equivalent here to provability by means of the SLD-resolution.

Given now a program P, consider the set

$$CWA(P) = \{\neg A : A \text{ is a ground atom for which there does not exist an SLD-refutation of } P \cup \{\leftarrow A\}\}.$$

We have the following lemma.

5.3. LEMMA. $\neg A \in CWA(P)$ iff $A \in B_P - M_P$.

PROOF. We have $\neg A \in CWA(P)$ iff A is not in the success set of P. The claim now follows by Corollary 3.14 and Lemma 3.17. □

As an immediate consequence we get this theorem.

5.4. THEOREM (Reiter [83]). *For any program P, $P \cup CWA(P)$ is consistent.*

Thus closed world assumption viewed as a reasoning method is weakly sound. Unfortunately, it is not an effective reasoning method. Namely, we have the following theorem.

5.5. THEOREM. *Assume that L is as in Section 4. Then for some program P the set $CWA(P)$ is not recursively enumerable.*

PROOF. By Corollary 4.7 there exists a program P such that M_P is a recursively enumerable but not recursive subset of U_L. Then, by well-known theorem, $B_P - M_P$, the complement of M_P, is not recursively enumerable. This concludes the proof in view of Lemma 5.3. □

5.3. *Negation as failure rule*

A way out of this dilemma is to adopt some more restrictive forms of unprovability. A natural possibility is to consider $\neg A$ proved when an attempt to prove A using SLD-resolution fails finitely. This leads to the following definitions.

An SLD-tree is *finitely failed* if it is finite and contains no empty clause. Thus all branches of a finitely failed SLD-tree are failed SLD-derivations. Given a program P, its *finite failure set* is the set of all ground atoms A such that there exists a finitely failed SLD-tree with $\leftarrow A$ as root.

We now replace CWA by the following rule:

$$\frac{A \text{ is in the finite failure set of } P}{\neg A}$$

introduced in [21] and called the *negation as failure* rule. (A more appropriate name would be negation as a *finite* failure rule.)

First of all it is useful to note that the negation as failure rule viewed as a reasoning method is weakly sound. Indeed, if A is in the finite failure set of P then by the strong completeness of SLD-resolution (Theorem 3.21) $\neg A$ is in $CWA(P)$, so it suffices to apply Theorem 5.4. Thus by Lemma 5.1 negation as failure is a nonmonotonic form of reasoning. It is also an effective form of reasoning because it is decidable whether a finite tree is a finitely failed SLD-tree.

Finally, observe that using the negation as failure rule we can trivially deduce $\neg$*element*(*mud*) and $\neg$*stuff*(*fire*) from the program P given in Example 5.2.

5.4. *Characterizations of finite failure*

We now provide two characterizations of finite failure, due to Apt and van Emden [4], and Lassez and Maher [61]. We follow here the presentation of [64].

First we introduce the concept of a fair SLD-derivation due to Lassez and Maher [61]. An SLD-derivation is called *fair* if it is either finite or every atom appearing in it is eventually selected. (An atom at the moment of selection will be actually an instance of the original version.) For example, the second derivation given in Subsection 2.6 is not fair as the atom configuration(y, l_1) is never selected in it. An SLD-tree is *fair* if each of its branches is a fair SLD-derivation. A selection rule R is *fair* if all SLD-derivations via R are fair. Thus an SLD-tree is fair if it is via a fair selection rule.

5.6. THEOREM. *Consider a program P and a ground atom A. Then the following are equivalent*:

(a) *A is in the finite failure set of P.*
(b) $A \notin T_P{\downarrow}\omega$.
(c) *Every fair SLD-tree with* $\leftarrow A$ *as root is finitely failed.*

To prove that (a) implies (b) we need two simple lemmas which are counterparts of Lemmas 3.22 and 3.23.

5.7. LEMMA. *Consider a program P, a negative clause N and a substitution θ. If $P\cup\{N\}$ has a finitely failed SLD-tree of depth $\leqslant k$, then so has $P\cup\{N\theta\}$.*

PROOF. By a straightforward induction on k. □

5.8. LEMMA. *Consider a program P and sequences of atoms $F_1,\ldots, F_n$. Assume that $F_1,\ldots, F_n$ have no variables in common. If $P\cup\{\leftarrow F_1,\ldots, F_n\}$ has a finitely failed SLD-tree of depth $\leqslant k$ then so has $P\cup\{\leftarrow F_i\}$ for some $i\in\{1,\ldots, n\}$.*

PROOF. By a simple induction on k using an analogous argument as that in the proof of Lemma 3.23. □

PROOF OF THEOREM 5.6. (a)⇒(b): We prove a stronger claim, namely the following lemma.

5.9. LEMMA. *Suppose $P\cup\{\leftarrow A\}$ has a finitely failed SLD-tree of depth $\leqslant k$. Then $A\notin T_P{\downarrow}k$.*

PROOF. We proceed by induction on k. The claim clearly holds when $k=1$. Assume it holds for $k-1$ and suppose by contradiction that $A\in T_P{\downarrow}k$. Then, for some clause $B\leftarrow B_1,\ldots, B_n$ in P, $A\equiv B\theta$ and $\{B_1\theta,\ldots, B_n\theta\}\subseteq T_P{\downarrow}(k-1)$ for some substitution θ. Thus, for some mgu γ, $A\gamma\equiv B\gamma$ and $\theta=\gamma\sigma$ for some σ. Hence $\leftarrow(B_1,\ldots, B_n)\gamma$ is the root of a finitely failed SLD-tree of depth $\leqslant k-1$. By Lemma 5.7 so is $\leftarrow(B_1,\ldots, B_n)\theta$. Now using Lemma 5.8 with each F_i being a single ground atom we get that, for some i, $1\leqslant i\leqslant n$, the goal $\leftarrow B_i\theta$ is also the root of a finitely failed SLD-tree of depth $\leqslant k-1$. By the induction hypothesis $B_i\theta\notin T_P{\downarrow}(k-1)$ which gives the contradiction. □

To prove that Theorem 5.6(b) implies (c) we need the following lemma.

5.10. LEMMA. *Consider a program P and a goal $\leftarrow A_1,\ldots, A_m$. Suppose there is an infinite fair SLD-derivation $N_0, N_1,\ldots$ with $N_0=\leftarrow A_1,\ldots, A_m$ and the sequence of substitutions $\theta_0, \theta_1,\ldots$. Then for every $k\geqslant 0$ there exists an $n\geqslant 0$ such that*

$$\bigcup_{i=1}^{m} [A_i\theta_0\ldots\theta_n]\subseteq T_P{\downarrow}k.$$

PROOF. We proceed by induction on k. The claim is clearly true if $k=0$. Suppose it holds for $k-1$. Fix $i\in\{1,\ldots, m\}$. By fairness, for some $p\geqslant 0$, the atom $A_i\theta_0\ldots\theta_{p-1}$ is selected in the goal N_p. By the induction hypothesis for some $s\geqslant 0$

$$\bigcup_{j=1}^{q} [B_j\theta_p\ldots\theta_{p+s}]\subseteq T_P{\downarrow}(k-1)$$

holds where N_{p+1} is $\leftarrow B_1,\ldots, B_q$. But

$$[A_i\theta_0\ldots\theta_{p+s}]\subseteq T_P\left(\bigcup_{j=1}^{q}[B_j\theta_p\ldots\theta_{p+s}]\right)$$

so

$$[A_i\theta_0\ldots\theta_{p+s}]\subseteq T_P\downarrow k$$

by the monotonicity of T_P. Thus for each $i\in\{1,\ldots, m\}$ there exists an $n_i\geqslant 0$ such that $[A_i\theta_0\ldots\theta_{n_i}]\subseteq T_P\downarrow k$. Put now $n=\max(n_1,\ldots, n_m)$. □

PROOF OF THEOREM 5.6 (*continued*). (b)⇒(c): Suppose that $A\notin T_P\downarrow\omega$. Consider a fair SLD-tree with $\leftarrow A$ as root. By Lemma 5.10 all of its branches are finite. But this tree does not contain the empty clause. Otherwise, by the Success Theorem 3.25,we would have $A\in T_P\downarrow\omega\subseteq T_P\downarrow\omega$. Thus it is a finitely failed SLD-tree.

(c)⇒(a): Obvious, as for every goal N there is a fair SLD-tree with N as root. □

Equivalence between (a) and (b) is due to Apt and van Emden [4], and between (a) and (c) due to Lassez and Maher [61]. The first equivalence can be seen as a theorem dual to the equivalence between (a) and (b) in the Success Theorem 3.25. The second equivalence can be seen as a counterpart of the equivalence between (a) and (c) in the Success Theorem 3.25 where duality is achieved by restricting the attention to fair SLD-trees.

5.5. *Completion of a program*

Another way of inferring negative information from a logic program is that of using the concept of a completion of a program due to Clark [21].

A program can be seen as a collection of statements of the form "if... then...". This does not allow us to conclude negative facts because only positive conclusions are admitted. But treating the clauses as statements of the form "... iff..." we obtain a stronger interpretation which allows us to draw negative conclusions. In doing so we should exercise some care. For example we wish to interpret the program $\{A\leftarrow B, A\leftarrow C\}$ as $A\leftrightarrow B\vee C$ and not as $(A\leftrightarrow B)\wedge(A\leftrightarrow C)$.

First, assume that "=" is a new binary relation symbol not appearing in P. We write $s\neq t$ as an abbreviation for $\neg(s=t)$. We perform successively the following steps, where $x_1,\ldots, x_n,\ldots$ are new variables.

Step 1: *Remove terms.* Transform each clause $p(t_1,\ldots, t_n)\leftarrow B_1,\ldots, B_m$ of P into

$$p(x_1,\ldots, x_n)\leftarrow(x_1=t_1)\wedge\cdots\wedge(x_n=t_n)\wedge B_1\wedge\cdots\wedge B_m.$$

Step 2: *Introduce existential quantifiers.* Transform each formula $p(x_1,\ldots, x_n)\leftarrow F$ obtained in the previous step into

$$p(x_1,\ldots, x_n)\leftarrow\exists y_1\ldots\exists y_d F,$$

where $y_1,\ldots, y_d$ are the variables of the original clause.

Step 3: *Group similar formulas.* Let

$$p(x_1,\ldots, x_n)\leftarrow F_1,$$

$$\ldots$$

$$p(x_1,\ldots, x_n)\leftarrow F_k$$

be all formulas obtained in the previous step with a relation p on the left-hand side. Replace them by one formula

$$p(x_1,\ldots, x_n)\leftarrow F_1 \vee \cdots \vee F_k.$$

If $F_1 \vee \cdots \vee F_k$ is empty, replace it by **true**.

Step 4: *Handle "undefined" relation symbols.* For each n-ary relation symbol q not appearing in a head of a clause in P add a formula

$$q(x_1,\ldots, x_n)\leftarrow \textbf{false}.$$

Step 5: *Introduce universal quantifiers.* Replace each formula $p(x_1,\ldots, x_n)\leftarrow F$ by

$$\forall x_1 \ldots \forall x_n (p(x_1,\ldots, x_n)\leftarrow F).$$

Step 6: *Introduce equivalence.* In each formula replace "$\leftarrow$" by "$\leftrightarrow$".

We call the intermediate form of P obtained after Step 5 the *IF-definition associated with P* and denote it by $IF(P)$. We call the final form the *IFF-definition associated with P* and denote it by $IFF(P)$. By $ONLY\text{-}IF(P)$ we denote the set of formulas obtained from $IF(P)$ by replacing everywhere "$\leftarrow$" by "$\rightarrow$".

5.11. EXAMPLE. (i) Reconsider the program P from Example 5.2. Then

$$\begin{aligned} IFF(P)=\{&\forall x(element(x)\leftrightarrow x=fire \vee x=air \vee x=water \vee x=earth),\\ &\forall x(stuff(x)\leftrightarrow x=mud)\}. \end{aligned}$$

Note that both $IFF(P)\models \neg stuff(fire)$ and $IFF(P)\models \neg element(mud)$ provided we interpret "$=$" as identity.

(ii) Consider the program

$$\begin{aligned} P=\{&link(a, b)\leftarrow,\ link(b, c)\leftarrow,\\ &connected(u, v)\leftarrow link(u, v),\\ &connected(u, v)\leftarrow link(u, z),\ connected(z, v)\}. \end{aligned}$$

Then

$$\begin{aligned} IFF(P)=\{&\forall x\forall y(link(x, y)\leftrightarrow (x=a \wedge y=b) \vee (x=b \wedge y=c)),\\ &\forall x\forall y(connected(x, y)\leftrightarrow \exists u\exists v((x=u) \wedge (y=v) \wedge link(u, v))\\ &\qquad \vee \exists u\exists v\exists z((x=u) \wedge (y=v) \wedge link(u, z) \wedge connected(z, v)))\}. \end{aligned}$$

It is easy to see that both $IFF(P)\models connected(a, c)$ and $IFF(P)\models \neg connected(a, a)$, provided we interpret "$=$" as identity.

We thus see that negative information can be inferred using the *IFF*-definition provided we interpret the relation symbol "=" properly. The problem of the proper interpretation of "=" is more subtle than it appears. As a first step we extend the interpretation of a first-order language so that "=" is interpreted as identity.

Let I be an interpretation of the first-order language associated with P. We put for any two terms t_1 and t_2 and a state σ over I

$$I \models_\sigma t_1 = t_2 \text{ iff } \sigma(t_1) \text{ and } \sigma(t_2) \text{ are the same elements of the domain of } I.$$

However, this does not yet solve the problem because, even though *mud* and *earth* or a and b are different constants, they still can become equal under some interpretation. To exclude such situations we add to the IFF-definitions the following *free equality axioms* which enforce proper interpretation of "=".

(1) $f(x_1,\ldots,x_n)=f(y_1,\ldots,y_n)\rightarrow x_1=y_1 \wedge \cdots \wedge x_n=y_n$ for each n-ary function f,

(2) $f(x_1,\ldots,x_n)\neq g(y_1,\ldots,y_m)$ for each n-ary function f and m-ary function g such that $f\neq g$,

(3) $x\neq t$ for each variable x and term t such that $x\not\equiv t$ and x occurs in t.

Here, similarly as in the proof of the Unification Theorem 2.3, we identify constants with 0-ary functions. Thus (1) includes $c=c$ for every constant c as a special case, and (2) includes $c\neq d$ for all pairs of distinct constants as a special case.

The resulting interpretation of "=" turns out to be sufficient for our purposes. Observe the striking similarity between the free equality axioms and steps (1), (2) and (5) of the unification algorithm used in the proof of the Unification Theorem 2.3. We shall exploit it in Subsection 5.7.

Given now a program P we denote by *comp*(P) the set of formulas $IFF(P)$ augmented by the free equality axioms. *comp*(P) is called the *completion* of P.

5.6. *Models of completions*

In order to assess the proof-theoretic power of completions, we study their models first. However, in contrast to the case of models of logic programs it is not sufficient to restrict attention here to Herbrand models. This is the content of a proposition we prove at the end of this subsection.

Therefore we shall consider here arbitrary models, but we shall study them by means of a natural generalization of the immediate consequence operator T_P. First, following Jaffar, Lassez and Lloyd [48], we introduce the concept of a *pre-interpretation* for a first-order language L. Its definition is identical to that of an interpretation given in Subsection 3.1 with the exception that the clause explaining the meaning of relations is dropped. We then say that an interpretation I *is based on* J if I is obtained from J by assigning to each n-ary relation r of L an n-ary predicate r_I on the domain of J, that is, by fixing the meaning of the relations of L. Thus each interpretation based on J can be uniquely identified with a set of *generalized atoms*, i.e. objects of the form $r(a_1,\ldots,a_n)$ where r is an n-ary relation of L and $a_1,\ldots,a_n$ are elements of the domain of J. That is what we shall do in the sequel.

We now generalize the operator T_P so that it acts on interpretations based on a given pre-interpretation. To this purpose we first introduce the following useful notation: Fix an interpretation I. Let $A=p(t_1,\ldots,t_n)$ be an atom and let σ be a state over I. Then we denote by $A\sigma$ the generalized atom $p(\sigma(t_1),\ldots,\sigma(t_n))$.

Let now J be a pre-interpretation and let I be an interpretation based on J. For a program P and a generalized atom D, we put

$$D\in T_P^J(I) \quad \text{iff} \quad \text{for some state } \sigma \text{ over } I \text{ and a clause } B\leftarrow B_1,\ldots,B_n \text{ of } P$$
$$\text{we have } D=B\sigma \text{ and } I\models_\sigma B_1\wedge\cdots\wedge B_n.$$

Thus T_P^J maps interpretations based on J to interpretations based on J. The operator T_P^J enjoys several properties similar to those of T_P. We list them in the following lemma omitting the proofs analogous to those of Lemma 3.7 and Lemma 3.12.

5.12. LEMMA. *Let P be a program and J a pre-interpretation. Then*
(i) *T_P^J is finitary.*
(ii) *T_P^J is monotonic.*
(iii) *For an interpretation I based on J, I is a model of P iff $T_P^J(I)\subseteq I$.*

We now wish to prove a similar characterization for models of completions. To this purpose we first note the following.

5.13. LEMMA. *For a program P, P and $IF(P)$ are semantically equivalent.*

PROOF. In Steps 1, 2, 3, 5 each formula is replaced by a semantically equivalent one. In turn, in Step 4 valid formulas are introduced. □

5.14. COROLLARY. *For a program P and a pre-interpretation J, an interpretation I based on J is a model of $IF(P)$ iff $T_P^J(I)\subseteq I$.*

We also have the following theorem.

5.15. THEOREM. *For a program P and a pre-interpretation J, an interpretation I based on J is a model of ONLY-IF(P) iff $T_P^J(I)\supseteq I$.*

To prove it, we first need the following lemma.

5.16. LEMMA. *Let I be an interpretation based on a pre-interpretation J and P a program. Let $\forall x_1\ldots\forall x_n(p(x_1,\ldots,x_n)\rightarrow F)$ be a formula in ONLY-IF(P). Then for every state σ over I*

$$p(x_1,\ldots,x_n)\sigma\in T_P^J(I) \text{ iff } I\models_\sigma F.$$

PROOF. If p does not appear in a head of a clause in P then both sides of the claimed equivalence are necessarily false. Otherwise

$p(x_1, \ldots, x_n)\sigma \in T_P^J(I)$

iff for some state τ over I and some clause $p(t_1, \ldots, t_n) \leftarrow B_1, \ldots, B_m$ of P $I \models_\tau B_1 \wedge \cdots \wedge B_m$ and $\sigma(x_i) = \tau(t_i)$ for $i = 1, \ldots, n$

iff $I \models_\sigma \exists y_1 \ldots \exists y_d((x_1 = t_1) \wedge \cdots \wedge (x_n = t_n) \wedge B_1 \wedge \cdots \wedge B_m)$ for some clause $p(t_1, \ldots, t_n) \leftarrow B_1, \ldots, B_m$ of P with $y_1, \ldots, y_d$ being all its variables

iff $I \models_\sigma F$. □

PROOF OF THEOREM 5.15. We have

I is a model of $ONLY\text{-}IF(P)$

iff for every formula $\forall x_1 \ldots \forall x_n(p(x_1, \ldots, x_n) \rightarrow F)$ in $ONLY\text{-}IF(P)$ and every state σ over I $p(x_1, \ldots, x_n)\sigma \in I$ implies $I \models_\sigma F$

iff (by Lemma 5.16) for every relation p of P and state σ over I $p(x_1, \ldots, x_n)\sigma \in I$ implies $p(x_1, \ldots, x_n)\sigma \in T_P^J(I)$

iff $T_P^J(I) \supseteq I$. □

Combining Corollary 5.14 and Theorem 5.15 we get the following characterization of the models of $IFF(P)$.

5.17. THEOREM. *Let P be a program and J a pre-interpretation. Then an interpretation I based on J is a model of $IFF(P)$ iff $T_P^J(I) = I$.*

PROOF. $IFF(P)$ is semantically equivalent to the set $IF(P) \cup ONLY\text{-}IF(P)$ of formulas. □

Restricting attention to Herbrand interpretations we can now draw some consequences about the completion of P.

5.18. THEOREM. (Apt and Van Emden [4]). *Let P be a program.*

(i) *A Herbrand interpretation I is a model of $comp(P)$ iff $T_P(I) = I$.*
(ii) *$comp(P)$ has a Herbrand model.*
(iii) *For any ground atom A, $comp(P) \cup \{A\}$ has a Herbrand model iff $A \in gfp(T_P)$.*

PROOF. (i) Every Herbrand interpretation is a model of the free equality axioms.
(ii) By (i) and the Characterization Theorem 3.13.
(iii) By (i), Lemma 3.12(ii) and Theorem 3.10. □

Moreover, we have the following observation which brings us to the end of this section.

5.19. THEOREM. *There is a program P and a ground atom A such that $comp(P) \cup \{A\}$ has a model but it has no Herbrand model.*

PROOF. Take the program P considered at the beginning of Subsection 4.5. As $gfp(T_P) = \emptyset$, by Theorem 5.18(iii), $comp(P) \cup \{q(a)\}$ has no Herbrand model. However, $comp(P) \cup \{q(a)\}$ is consistent. Indeed, take as a domain of the interpretation a disjoint union $\mathbb{Z} \cup \mathbb{N}$ of the set of integers and the set of natural numbers. Interpret the constant a as zero in the set $\mathbb{N}$ and f as a successor function, both on the set $\mathbb{Z}$ and the set $\mathbb{N}$. Finally, interpret p as true for all elements of $\mathbb{Z}$ and q true only for the zero of $\mathbb{N}$. The resulting interpretation is a model of $comp(P) \cup \{q(a)\}$. □

In Subsection 5.10 we provide a characterization of the finite failure which provides a more direct proof of the above theorem.

5.7. *Soundness of the negation as failure rule*

Recall that completion of a program was introduced in order to infer negative information from a program. We now relate it to the previously studied way of deducing negative information—that by means of the negation as failure rule. To this purpose we first investigate models of the free equality axioms (Subsection 5.5). Assume a program P and denote these axioms by Eq. As Eq does not refer to relations, it makes sense to say that a pre-interpretation J is a model of Eq. Similarly, it is meaningful to talk about states over a pre-interpretation. For each ground term t denote its value in the domain of J by t_J. We write $J \models_\sigma s = t$ when $\sigma(s)$ equals $\sigma(t)$.

5.20. LEMMA. *Let J be a pre-interpretation which is a model of* Eq. *Then the domain of J contains an isomorphic copy of U_P.*

PROOF. It suffices to show that, for all ground terms s, t, $s_J = t_J$ implies $s \equiv t$. We proceed by induction on the structure of ground terms.

If $s_J = t_J$ then, by axioms (1) and (2), s and t are either the same constants or are respectively of the form $f(s_1, \ldots, s_n)$ and $f(t_1, \ldots, t_n)$. The claim now follows by axiom (1) and the induction hypothesis. □

In the sequel we shall identify this isomorphic copy with U_P. Given a pre-interpretation J let B_J stand for the set of all its generalized atoms. If J is a model of Eq then, by the above lemma, B_J contains an isomorphic copy of the Herbrand base B_P. We identify this copy with B_P.

The following lemma clarifies the relation between the unification and free equality axioms.

5.21. LEMMA (Clark [21]). (a) *If the set $\{s_1, = t_1, \ldots, s_n = t_n\}$ has a unifier then for some of its mgu $\{x_1/u_1, \ldots, x_k/u_k\}$*

$$\text{Eq} \models s_1 = t_1 \wedge \cdots \wedge s_n = t_n \rightarrow x_1 = u_1 \wedge \cdots \wedge x_k = u_k.$$

(b) *If the set $\{s_1 = t_1, \ldots, s_n = t_n\}$ has no unifier then*

$$\text{Eq} \models s_1 = t_1 \wedge \cdots \wedge s_n = t_n \rightarrow \textbf{false}.$$

PROOF. Modify the Unification Algorithm given in the proof of the Unification

Theorem 2.3 as follows. First display each set $\{s_1=t_1,\ldots,s_n=t_n\}$ of equations as a formula $s_1=t_1\wedge\cdots\wedge s_n=t_n$. Then interpret the replacement and deletion steps as operations on these formulas. Interpret the halt with failure action as a replacement of the formula by **false**.

Observe that if ψ is obtained from φ by applying one of the steps of the algorithm then $\mathrm{Eq}\models\varphi\rightarrow\psi$. Indeed, for any x and t, $\varphi_0\wedge x=t\wedge\varphi_1\rightarrow(\varphi_0\wedge\varphi_1)\{x/t\}$ is a valid formula. Other cases are immediate.

The lemma now follows from the correctness of the unification algorithm. $\square$

Given a pre-interpretation J and a state σ over J, call a substitution θ *invariant over a state* σ if, for all x, $\sigma(x)=\sigma(x\theta)$.

5.22. COROLLARY. *Let J be a pre-interpretation which is a model of* Eq. *If for some state σ over J*

$$J\models_\sigma s_1=t_1\wedge\cdots\wedge s_n=t_n,$$

then, for some mgu θ of $\{s_1=t_1,\ldots,s_n=t_n\}$ invariant over σ,

$$J\models(s_1=t_1\wedge\cdots\wedge s_n=t_n)\theta.$$

Call now an interpretation I based on J *good* if, for all conjunctions of atoms F, $I\models_\sigma F$ for some state σ implies $I\models F\theta$ for some substitution θ. Obviously not all interpretations are good. But those of interest to us are. First we need the following two lemmas.

5.23. LEMMA. *Let J be a pre-interpretation which is a model of* Eq. *Let I be based on J. Suppose that I is good. Then $T_P^J(I)$ is good, as well.*

PROOF. Consider a sequence $A_1,\ldots,A_k$ of atoms. The operator T_P^J does not depend on the choice of the names of variables in P. Thus we can assume that each of the variables of P appears in at most one clause of P and none of them appears in $A_1,\ldots,A_k$. Suppose now that $T_P^J(I)\models_\sigma A_1\wedge\cdots\wedge A_k$ for some state σ. By the definition of T_P^J, for each $i=1,\ldots,k$, there exists a clause $B_i\leftarrow B_1^i,\ldots,B_{m_i}^i$ in P and a state τ_i such that $I\models_{\tau_i}B_1^i\wedge\cdots\wedge B_{m_i}^i$ and $A_i\sigma=B_i\tau_i$. Define now a state τ by

$$\tau(x)=\begin{cases}\tau_i(x) & \text{if } x \text{ appears in } B_i\leftarrow B_1^i,\ldots,B_{m_i}^i,\\ \sigma(x) & \text{otherwise.}\end{cases}$$

Then

$$I\models_\tau \bigwedge_{\substack{i=1,\ldots,k\\ j=1,\ldots,m_i}} B_j^i \tag{5.1}$$

and, for each $i=1,\ldots,k$,

$$A_i\tau=B_i\tau. \tag{5.2}$$

By Corollary 5.22 and (5.2) there exists a substitution θ invariant over τ such that, for each $i=1,\ldots,k$,

$$A_i\theta\equiv B_i\theta. \tag{5.3}$$

By the definition of invariance and (5.1)

$$I \models_\tau \bigwedge_{\substack{i=1,\dots,k\\ j=1,\dots,m_i}} B^i_j\theta$$

But I is good, so for some substitution γ

$$I \models_\tau \bigwedge_{\substack{i=1,\dots,k\\ j=1,\dots,m_i}} B^i_j\theta\gamma.$$

We can assume that γ is such that each $B^i_j\theta\gamma$ ground.

Thus by the definition of T^J_P, for each $i=1,\dots,k$, $B_i\theta\gamma \in T^J_P(I)$, i.e. by (5.3) $T^J_P(I) \models (A_1 \wedge \cdots \wedge A_k)\theta\gamma$. This concludes the proof. □

5.24. LEMMA. *Let J be a pre-interpretation which is a model of* Eq. *Let I be based on J. Suppose that I is good. Then $B_p \cap T^J_p(I) = T_p(B_p \cap I)$.*

PROOF. Suppose $A \in B_p \cap T^J_p(I)$. Then, for some state σ over I, $A \equiv B\sigma$ and $I \models_\sigma A_1 \wedge \cdots \wedge A_n$ where $B \leftarrow A_1,\dots, A_n$ is a clause from P. Thus σ when restricted to the variables of B is a ground substitution, say η. We thus have $I \models_\sigma (A_1 \wedge \cdots \wedge A_n)\eta$. But I is good, so for some substitution θ, $I \models (A_1 \wedge \cdots \wedge A_n)\eta\theta$. Thus $B_p \cap I \models (A_1 \wedge \cdots \wedge A_n)\eta\theta$. Moreover $A \equiv B\eta\theta$, so $A \in T_p(B_p \cap I)$.

If now $A \in T_p(B_p \cap I)$ then a fortiori $A \in B_p \cap T^J_p(B_p \cap I)$, so by the monotonicity of T^J_p we have $A \in B_p \cap T^J_p(I)$. □

This lemma states that all ground atoms inferred from I by means of T^J_p can already be inferred by means of T_p, provided I is good.

This brings us to the following important consequences of Lemmas 5.23 and 5.24 which will be also used in Section 6.

5.25. COROLLARY. *Let J be a pre-interpretation which is a model of* Eq.

(i) *For every $n \geqslant 0$, $T^J_P \downarrow n$ is good.*

(ii) *For every $n \geqslant 0$, $B_p \cap T^J_P \downarrow n = T_p \downarrow n$.*

(iii) $B_p \cap T^T_p \downarrow \omega = T_p \downarrow \omega$.

PROOF. We have $T^J_P \downarrow 0 = B_J$. But $B_P \subseteq B_J$, so for all conjunctions of atoms F and all substitutions θ, $B_J \models F\theta$. Thus $T^J_P \downarrow 0$ is good and, by induction using Lemma 5.23, for every $n \geqslant 0$, $T^J_P \downarrow n$ is good.

(ii) We proceed by induction on n. For $n=0$ it is a consequence of the fact that $B_p \subseteq B_J$. Suppose this claim holds for some $n \geqslant 0$. Then

$$\begin{aligned} B_p \cap T^J_p \downarrow (n+1) &= B_p \cap T^J_p(T^J_p \downarrow n) \\ &= T_p(B_p \cap T^J_p \downarrow n) \quad \text{(by (i) and Lemma 5.24)} \\ &= T_p(T_p \downarrow n) \quad \text{(by induction hypothesis)} \\ &= T_p \downarrow (n+1). \end{aligned}$$

This implies the claim for $n+1$.

(iii) Immediate, by (ii). □

Finally, we prove the following lemma which will also be needed in Section 6.

5.26. LEMMA. *Let P be a program and I a model of comp(P). Then $B_P \cap I \subseteq T_P\downarrow\omega$.*

PROOF. I is based on some pre-interpretation J. I is a model of $IFF(P)$, so, by Theorem 5.17, $T_P^J(I)=I$. Thus, by Lemma 3.11, $I \subseteq T_P^J\downarrow\omega$. J is a model of Eq, so by Corollary 5.25(iii) and the above inclusion, the claim follows. □

We can now relate the completion of a program and negation as failure rule.

5.27. THEOREM (soundness of the negation as failure rule) (Clark [21]). *Let P be a program. If A is in the finite failure set of P then comp(P) $\models \neg A$.*

PROOF. Let I be a model of *comp*(P) and suppose that A is in the finite failure set of P. Then, by Theorem 5.6, $A \notin T_p\downarrow\omega$, so by Lemma 5.26 $A \notin B_p \cap I$, i.e. $I \models \neg A$. □

5.8. *Completeness of the negation as failure rule*

We now prove the converse of the above theorem. We follow here essentially the presentation of Lloyd [64] based on a proof due to Wolfram, Maher and Lassez [102]. We first show how to construct models of the free equality axioms.

Let $\mathscr{C}$ be a set of substitutions. We call $\mathscr{C}$ *directed* if

$$\theta, \eta \in \mathscr{C} \Rightarrow \text{there exists a } \gamma \in \mathscr{C} \text{ such that } \theta \leqslant \gamma \text{ and } \eta \leqslant \gamma.$$

Here $\theta \leqslant \gamma$ means that θ is more general than γ. Suppose now that $\mathscr{C}$ is a set of substitutions. For two terms s, t, put

$$s \sim_{\mathscr{C}} t \quad \text{iff} \quad \text{for some } \theta \in \mathscr{C},\ s\theta \equiv t\theta.$$

5.28. LEMMA. *Suppose that $\mathscr{C}$ is a directed set of substitutions. Then $\sim_{\mathscr{C}}$ is an equivalence relation which is a congruence w.r.t. all function symbols. Moreover, the pre-interpretation induced by $\sim_{\mathscr{C}}$ is a model of* Eq.

PROOF. The relation $\sim_{\mathscr{C}}$ is always reflexive and symmetric. By directedness of $\mathscr{C}$ it is also transitive.

Let $[s]$ stand for the equivalence class of term s w.r.t. $\sim_{\mathscr{C}}$. Let f be an n-ary function symbol. If $[s_1]=[t_1],\ldots,[s_n]=[t_n]$ for some terms $s_1, t_1,\ldots, s_n, t_n$, then, by directedness of $\mathscr{C}$, for some $\theta \in \mathscr{C}$,

$$s_1\theta \equiv t_1\theta, \ldots, s_n\theta \equiv t_n\theta.$$

Hence $f(s_1,\ldots,s_n)\theta \equiv f(t_1,\ldots,t_n)\theta$, i.e. $[f(s_1,\ldots,s_n)]=[f(t_1,\ldots,t_n)]$.

Thus the equivalence relation induced by $\sim_{\mathscr{C}}$ is indeed a congruence. This means that $\sim_{\mathscr{C}}$ induces a pre-interpretation of L. That this interpretation is indeed a model of Eq is easy to see, as nonunifiable terms have necessarily different equivalence classes w.r.t. $\sim_{\mathscr{C}}$. □

The essence of the proof of the completeness theorem lies in the following lemma.

5.29. LEMMA. *Consider a program P and a goal N. Suppose there is a nonfailed fair SLD-derivation with N as the initial goal. Then comp(P)$\cup\{\neg N\}$ is consistent.*

PROOF. Let $\Phi = N_0, N_1, \ldots$ with $N_0 = N$ and with the sequence of substitutions $\theta_0, \theta_1, \ldots$ be the *SLD*-derivation in question and let $N = \leftarrow A_1, \ldots, A_s$. Then $\neg N \equiv \exists(A_1 \wedge \cdots \wedge A_s)$. We use this derivation to construct a model of $comp(P) \cup \{\exists(A_1 \wedge \cdots \wedge A_s)\}$. Let $\mathscr{C} = \{\theta_0 \ldots \theta_i\colon i \geqslant 0\}$. Note that $\mathscr{C}$ is directed. By the last lemma the pre-interpretation J induced by $\sim_{\mathscr{C}}$ is a model of Eq. Let $[s]$ denote the equivalence class under $\sim_{\mathscr{C}}$ of a term s.

We now construct an interpretation I based on J by putting

$$I = \{p([t_1], \ldots, [t_n])\colon p(t_1, \ldots, t_n) \text{ appears in a goal from } \Phi\}.$$

We first show that $I \subseteq T_P^J(I)$, i.e. that I is a model of $ONLY\text{–}IF(P)$.

Suppose that $p(t_1, \ldots, t_n)$ appears in a goal N_i of Φ. Since Φ is nonfailed and fair, there exists $j \geqslant i$ such that $p(s_1, \ldots, s_n) \equiv p(t_1, \ldots, t_n)\theta_i \ldots \theta_{j-1}$ is the selected atom in N_j.

In subsection 2.7 we assumed that each mgu θ_l is idempotent and relevant. Thus by Lemma 2.7 for any l, m such that $m > l$, θ_l does not act on the variables from N_m or $N_m\theta_m$. Fix k, $1 \leqslant k \leqslant n$. Thus, since t_k appears in N_i,

$$t_k\theta_l \equiv t_k \quad \text{for } l < i, \tag{5.4}$$

and, since $t_k\theta_i \ldots \theta_j$ appears in $N_j\theta_j$

$$t_k\theta_i \ldots \theta_j\theta_l \equiv t_k\theta_i \ldots \theta_j \quad \text{for } l < j. \tag{5.5}$$

Thus

$$\begin{aligned} t_k\theta_i \ldots \theta_j\theta_0 \ldots \theta_{j-1}\theta_j &\equiv t_k\theta_i \ldots \theta_j\theta_j && \text{(by (5.5) applied } j \text{ times)} \\ &\equiv t_k\theta_i \ldots \theta_j && \text{(by idempotence of } \theta_j\text{)} \\ &\equiv t_k\theta_0 \ldots \theta_j. && \text{(by (5.4) applied } i \text{ times)} \end{aligned} \tag{5.6}$$

Hence for all k, $1 \leqslant k \leqslant n$

$$\begin{aligned} [t_k] &= [t_k\theta_i \ldots \theta_j] && \text{(by (5.6))} \\ &= [s_k\theta_j]. && \text{(by definition of } s_k\text{)} \end{aligned}$$

But by the definition of I we have $p([s_1\theta_j], \ldots, [s_n\theta_j]) \in T_P^J(I)$, so $p([t_1], \ldots, [t_n]) \in T_P^J(I)$, as desired.

Now by Theorem 3.10 and Theorem 5.17 I can be extended to a model of *comp(P)*. By the construction, I is a model of $\exists(A_1 \wedge \cdots \wedge A_s)$, and a fortiori so is its extension. $\square$

We are now in position to prove the desired theorem. It is formulated in a slightly more general form which will be needed in Section 6.

5.30. THEOREM (completeness of the negation as failure rule) (Jaffar, Lassez and Lloyd [48]). *Let P be a program. If, for a goal N, comp(P)$\models$ N, then P$\cup\{N\}$ has a finitely failed SLD-tree.*

PROOF. Assume there is a nonfailed fair SLD-derivation with N as the initial goal. By the last lemma, $comp(P) \cup \{\neg N\}$ is consistent. Thus by contraposition, $comp(P) \models N$ implies that every fair SLD-tree with N as root is finitely failed. Thus $P \cup \{N\}$ has a finitely failed SLD-tree. □

It is perhaps useful to indicate here that, using Lemma 5.29, an alternative proof of the implication (b)⇒(c) in Theorem 5.6 can be given without the use of Lemma 5.10. Indeed, assume there is a nonfailed fair SLD-derivation with $\leftarrow A$ as the initial goal. Then by Lemma 5.29 $comp(P) \cup \{A\}$ has a model. By Lemma 5.26 this implies that $A \in T_P \downarrow \omega$. Thus, by contraposition, $A \notin T_P \downarrow \omega$ implies that every fair SLD-tree with $\leftarrow A$ as root is finitely failed.

5.9. *Equality axioms versus identity*

Clark's [21] original definition of free equality additionally included the following usual equality axioms:

(1) $x = x$,

(2) $x_1 = y_1 \wedge \cdots \wedge x_n = y_n \rightarrow f(x_1, \ldots, x_n) = f(y_1, \ldots, y_n)$ for each function symbol f,

(3) $x_1 = y_1 \wedge \cdots \wedge x_n = y_n \rightarrow (p(x_1, \ldots, x_n) \rightarrow p(y_1, \ldots, y_n))$ for each relation symbol p including $=$.

Denote these axioms by EQ. We did not use EQ at the expense of interpreting equality as identity. Fortunately, both approaches are equivalent as the following well-known theorem (see e.g., [72, p. 80]) shows.

5.31. THEOREM. *Let S be a set of formulas in a first-order language L including $=$. Then for every formula φ*

$$S \models \varphi \text{ iff } S \cup \mathrm{EQ} \models_{+} \varphi,$$

where $\models_{+}$ stands for validity w.r.t. interpretations of L which interpret $=$ in an arbitrary fashion.

PROOF. (⇒): An interpretation of $=$ in a model of EQ is an equivalence relation which is a congruence w.r.t. all function and relation symbols. This implies that every model of EQ is equivalent to (i.e., satisfies the same formulas of) a model in which equality is interpreted as identity. This model has as the domain the equivalence classes of the interpretation of $=$ with the function and relation symbols interpreted in it in a natural way. The proof of the equivalence proceeds by straightforward induction on the structure of the formulas.

(⇐): When $=$ is interpreted as identity, all axioms of EQ became valid. □

5.10. *Summary*

Summarizing the results obtained in Subsections 5.4, 5.7 and 5.8 we obtain the following characterizations of the finite failure.

5.32. THEOREM (Finite Failure Theorem). *Consider a program P and a ground atom A.*

Then the following are equivalent:

(a) *A is in the finite failure set of P.*

(b) $A \notin T_P{\downarrow}\omega$.

(c) *Every fair SLD-tree with* $\leftarrow A$ *as root is finitely failed.*

(d) $comp(P) \models \neg A$.

These results show that the negation as failure rule is a proof-theoretic concept with very natural mathematical properties. Comparing the above theorem with the Success Theorem 3.25, we see a natural duality between the notions of success and finite failure. However, this duality is not complete. By the Characterization Theorem 3.13 and the Success Theorem 3.25, A is in the success set of P iff $A \in lfp(T_P)$. On the other hand, the "dual" statement: A is in the finite failure of P iff $A \notin gfp(T_P)$ does not hold because, as noted in Section 4.5, for certain programs P we have $gfp(T_P) \neq T_P{\downarrow}\omega$.

For any such program P and a ground atom $A \in T_P{\downarrow}\omega - gfp(T_P)$ by the above theorem and Theorem 5.18(iii), $comp(P) \cup \{A\}$ has a model but it has no Herbrand model. This yields a more direct proof of Theorem 5.19.

Clause (d) of the Finite Failure Theorem suggests another possibility of inferring negation. Consider the following rule implicitly studied in [4].

$$\frac{A \text{ is false in all Herbrand models of } comp(P)}{\neg A}$$

Call this rule the *Herbrand rule*. Then the results of this section can be summarized by Fig. 3 from [64, p. 86] assessing the content of Lemma 5.3, Theorem 5.18(iii) and Theorem 5.6.

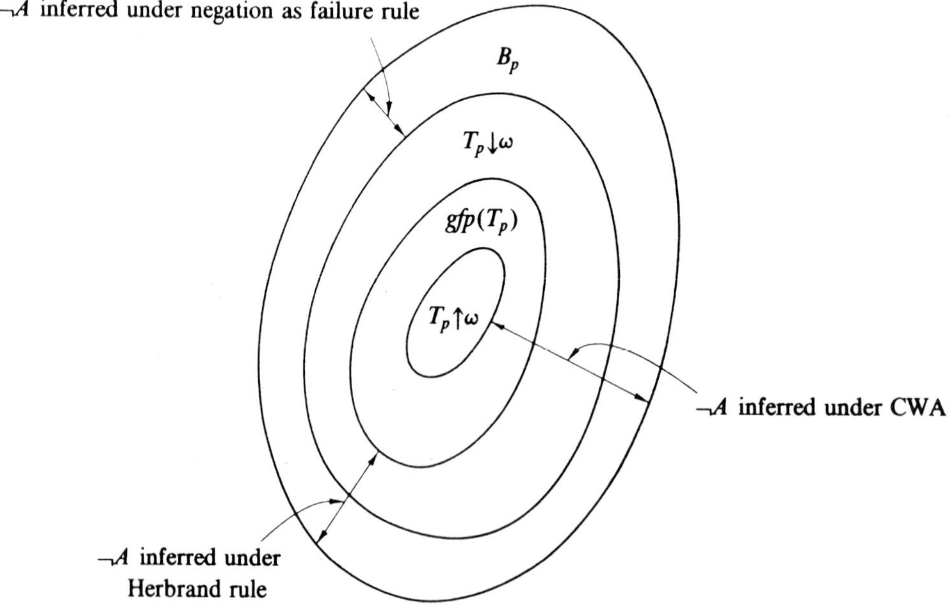

Fig. 3.

5.11. *Bibliographic remarks*

Theorem 5.17 is a straightforward generalization due to Jaffar, Lassez and Lloyd [48] of a special case (Theorem 5.18(a)) proved in [4]. The notion of a finite failure set was introduced in [4].

Lemma 5.20 appears as an exercise in [64, p. 88]. Proofs of Lemma 5.21 and Theorem 5.26 seem to be new. Lemma 5.21 was generalized by Kunen [59] who proved that that Eq is a complete axiomatization for the fragment $L(=)$ of L containing $=$ as the only relation symbol.

Jaffar and Stuckey [49] proved that every program is semantically equivalent to a program P for which $gfp(T_P)=T_P\downarrow\omega$. Maher [69] provided a partial characterization of programs P for which $gfp(T_P)=T_P\downarrow\omega$.

6. General goals

6.1. *SLDNF$^-$-resolution*

When trying to extend the results of Sections 3 and 5 to general programs, we encounter several difficulties. In this paper we examine only a very mild extension of the previous framework, namely the use of logic programs together with *general* goals. This provides some insight into the nature of the new problems.

We have to explain first how general goals are to be refuted. For this purpose we need only to clarify how negative literals are to be resolved. It is natural to use for this purpose the negation as failure rule studied in the previous section. Strictly speaking this rule was defined only for ground atoms, but it can be extended in an obvious way to the nonground case.

This leads us to an extension of the SLD-resolution called SLDNF$^-$-resolution (SLD-resolution with negation as failure rule) introduced in [21]. We added the superscript "$-$" to indicate that it is used here only with nongeneral programs. Formally, we first introduce the notion of a resolvent of a general goal. Let P be a program and $G=\leftarrow L_1,\ldots,L_n$ a general goal. We distinguish two cases. Fix i, $1\leqslant i\leqslant n$.

(a) *Literal L_i is positive.* Suppose that $C=A\leftarrow B_1,\ldots,B_k$ is a clause from P. If L_i and A unify with an mgu θ then

$$\leftarrow(L_1,\ldots,L_{i-1},B_1,\ldots,B_k,L_{i+1},\ldots,L_n)\theta$$

is a *resolvent of* G *and* C *with the mgu* θ.

(b) *Literal L_i is negative*, say $\neg A_i$. Suppose that $P\cup\{\leftarrow A_i\}$ has a finitely failed SLD-tree. Then

$$\leftarrow L_1,\ldots,L_{i-1},L_{i+1},\ldots,L_n$$

is a *resolvent* of G.

L_i is called the *selected literal of* G.

Now, given a program P and a general goal G, by an SLDNF$^-$-*derivation* of $P\cup\{G\}$ we mean a maximal sequence $G_0,G_1,\ldots$ of general goals where $G_0=G$, together with a sequence $C_0, C_1, \ldots$ of variants of clauses from P and a sequence $\theta_0, \theta_1, \ldots$ of substitution such that, for all $i=0, 1, \ldots$,

- if the selected literal in G_i is positive then G_{i+1} is a resolvent of G_i and C_i with the mgu θ_i;
- if the selected literal in G_i is negative then G_{i+1} is a resolvent of G_i, C_i is arbitrary and θ_i is the empty substitution;
- C_i does not have a variable in common with $G_0, C_0, \ldots, C_{i-1}$.

Note that if the selected negative literal $\neg A$ in a general goal G is such that $P \cup \{\leftarrow A\}$ has no finitely failed SLD-tree, then G has no successor in the SLDNF$^-$-derivation. Also note that a successful resolving of a negative literal introduces no variable bindings.

The notions of SLD-refutation, computed answer substitution, selection rule and SLD-trees generalize in an obvious way to the case of SLDNF$^-$-resolution. In particular we can talk of successful and failed SLDNF$^-$-trees.

6.2. *Soundness of the SLDNF$^-$-resolution*

In any soundness or completeness theorem we need to compare the existence of SLDNF$^-$-refutations with some statements referring to semantics of the program under consideration. However, a direct use of the programs is not sufficient here because of the negative literals. For example $P \cup \{\leftarrow \neg A\}$ is always consistent. What we need here is an extension of P which implies some negative information. An obvious candidate is the completion of P, *comp*(P), which was actually introduced by Clark [21] to serve as a meaning of general programs when studying SLDNF-resolution.

After these preparations we can formulate the appropriate soundness theorem, essentially due to Clark [21].

6.1. Theorem (soundness of SLDNF$^-$-resolution). *Let P be a program and $G = \leftarrow L_1, \ldots, L_k$ a general goal. Suppose that there exists an SLDNF$^-$refutation of $P \cup \{G\}$ with the sequence of substitutions $\theta_0, \ldots, \theta_n$. Then $(L_1 \wedge \cdots \wedge L_k)\theta_0 \ldots \theta_n$ is a semantic consequence of comp(P).*

To prove it, we need the following mild generalization of Theorem 5.27, essentially due to Clark [21].

6.2. Lemma. *Consider a program P and an atom A. Suppose there is a finitely failed SLD-tree with $\leftarrow A$ as root. Then comp(P) $\models \neg A$.*

Proof. By Lemma 5.7 there exists an $n_0 \geqslant 1$ such that, for every ground substitution θ, $P \cup \{A\theta\}$ has a finitely failed SLD-tree of depth $\leqslant n_0$. By Lemma 5.9, for every ground substitution θ, $A\theta \notin T_p{\downarrow}n_0$. Suppose now that for some interpretation I based on a pre-interpretation J, $I \models comp(P)$, and moreover, for some state σ, $I \models_\sigma A$. By Theorem 5.17, $T_p^J(I) = I$. Thus, by Lemma 3.11, $I \subseteq T_p^J{\downarrow}n_0$. So we have $T_p^J{\downarrow}n_0 \models_\sigma A$. But by Corollary 5.25(i), $T_P^J\downarrow n_0$ is good, so for some ground substitution θ, $T_P^J\downarrow n_0 \models A\theta$. Now by Corollary 5.25(ii), $A\theta \in T_P^J\downarrow n_0$. This contradicts the former conclusion. □

We can now prove soundness of SLDNF$^-$-resolution.

PROOF OF THEOREM 6.1. Let $A_1, \ldots, A_l$ be the sequence of positive literals of G and $\neg B_1, \ldots, \neg B_m$ the sequence of negative literals of G. If $l=0$ or $m=0$ we disregard the corresponding step in the considerations below.

With each SLDNF$^-$-refutation of $P \cup \{G\}$ we can associate an SLD-refutation of $P \cup \{\leftarrow A_1, \ldots, A_l\}$ obtained by deleting all resolvents arising from the selection of negative literals and by deleting all negative literals in the remaining resolvents. By the soundness of SLD-resolution (Theorem 3.2) and the fact that empty substitutions are used when resolving negative literals,

$$P \models (A_1 \wedge \ldots \wedge A_l)\theta_0 \ldots \theta_n.$$

But $comp(P) \models IF(P)$, so by Lemma 5.13

$$comp(P) \models (A_1 \wedge \ldots \wedge A_l)\theta_0 \ldots \theta_n.$$

Also, by Lemma 6.2, for $i=1, \ldots, m$,

$$comp(P) \models \neg B_i\theta_0 \ldots \theta_{p-1},$$

where $\neg B_i\theta_0 \ldots \theta_{p-1}$ is the selected literal of G_p $(0 \leqslant p \leqslant n)$. Thus

$$comp(P) \models (\neg B_1 \wedge \cdots \wedge \neg B_m)\theta_0 \ldots \theta_n$$

which concludes the proof. □

6.3. COROLLARY. *If there exists an SLDNF$^-$-refutation of $P \cup \{G\}$ then $comp(P) \cup \{G\}$ is inconsistent.*

6.3. Floundering

We now consider the problem of completeness of the SLDNF$^-$-resolution. Unfortunately, even the weakest form of completeness does not hold as the following example shows.

6.4. EXAMPLE. Consider the following program P:

$$p(a) \leftarrow p(a),$$
$$r(b) \leftarrow .$$

Then in every model I of the free equality axioms

$$I \models (\forall x(p(x) \leftrightarrow x=a \wedge p(a))) \rightarrow \neg p(b),$$

so by the definition of completion $comp(P) \models \neg p(b)$, that is, $comp(P) \cup \{\leftarrow \neg p(x)\}$ is inconsistent. However, $P \cup \{\leftarrow p(x)\}$ has no finitely failed SLD-tree, so there is no SLDNF$^-$-refutation of $P \cup \{\leftarrow \neg p(x)\}$.

A natural way out of this dilemma is to impose on SLDNF$^-$-resolution some restrictions. Clearly, the problem is caused here by the use of nonground negative literals. Notice for instance that in the above example $P \cup \{\leftarrow p(b)\}$ has a finitely failed SLD-tree, so there exists an SLDNF$^-$-refutation of $P \cup \{\leftarrow \neg p(b)\}$.

We thus introduce the following restriction. We say that a selection rule is *safe* if it only selects negative literals which are ground. From now on we shall use *only* safe selection rules. But a safe selection rule is not defined on some sequences of literals. This means that certain general goals have no resolvents under a safe selection rule.

We say that an SLDNF$^-$-derivation of $P \cup \{G\}$ via a safe selection rule *flounders* if it is of the form $G_0, \ldots, G_k$ with $G_0 = G$, where G_k contains only nonground negative literals. $P \cup \{G\}$ *flounders* if some SLDNF$^-$-derivation of $P \cup \{G\}$ (via a safe selection rule) flounders.

Obviously, restriction to safe selection rules does not restore completeness of SLDNF$^-$-resolution—a smaller number of selection rules cannot help. But one would hope that a restriction to programs P and general goals G such that $P \cup \{G\}$ does not flounder, does help. Unfortunately, such hopes are vain.

6.5. EXAMPLE. Consider the following program P:

$$r(a) \leftarrow,$$
$$r(b) \leftarrow r(b),$$
$$r(b) \leftarrow q(a),$$
$$q(a) \leftarrow q(a)$$

and the general goal $G = \leftarrow r(x), \neg q(x)$. We now claim that

(i) $P \cup \{G\}$ does not flounder,

(ii) there is no SLDNF$^-$-refutation of $P \cup \{G\}$,

(iii) $comp(P) \cup \{G\}$ is inconsistent.

Both (i) and (ii) are easy to check. To prove (iii), take an interpretation I based on a pre-interpretation J such that $I \models comp(P)$. By Theorem 5.17, $T_P^J(I) = I$. Thus by the form of P the following three facts hold:

(a) $r(a) \in I$,

(b) $q(a) \in I \rightarrow r(b) \in I$,

(c) $q(b) \notin I$.

This means that either $I \models r(a) \wedge \neg q(a)$ or $I \models r(b) \wedge \neg q(b)$ holds, i.e. $I \models \exists x(r(x) \wedge \neg q(x))$, so G is not true in I.

6.4. *Restricted completeness of the SLDNF$^-$-resolution*

Thus to obtain completeness of SLDNF$^-$-resolution, further restrictions are necessary. To this purpose we first introduce the following notions.

Given a program P we define its *dependency graph* D_P by putting for two relations r, q

$$(r, q) \in D_P \text{ iff there is a clause in } P \text{ using } r \text{ in its head and } q \text{ in its body.}$$

We then say that *r refers to q*; *depends on* is the reflexive transitive closure of the relation refers to. Thus a relation does not need to refer to itself, but by reflexivity every relation depends on itself.

Now, given a program P and a general goal G, we say that $P \cup \{G\}$ is *strict* if the relations occurring in positive literals of G depend on different relations than those on which relations occurring in negative literals of G depend. Note that this implies that no relation occurs both in a positive and negative literal of G.

More precisely, given a program P and a set of relations R first put

$$DEP(R) = \{q: \text{some } p \text{ in } R \text{ depends on } q\}.$$

Then $P \cup \{G\}$ is *strict* if

$$DEP(G^+) \cap DEP(G^-) = \emptyset,$$

where G^+ (respectively G^-) stands for the set of relations occurring in positive (respectively negative) literals of G.

Note that for the program P and the general goal G studied in Example 6.5 $P \cup \{G\}$ is not strict.

We now prove the following result established independently by Cavedon and Lloyd [19], and by Apt (unpublished).

6.6. THEOREM (restricted completeness of SLDNF$^-$-resolution). *Let P be a program and G a general goal such that $P \cup \{G\}$ is strict and $P \cup \{G\}$ does not flounder. Suppose comp(P) $\cup \{G\}$ is inconsistent. Then there exists an SLDNF$^-$-refutation of $P \cup \{G\}$.*

In the proof we shall use the following well-known theorem from mathematical logic due to K. Gödel (see e.g. [92]).

6.7. THEOREM (Compactness Theorem). *A set of formulas has a model iff every finite subset of it has a model.*

Using the Compactness Theorem we obtain the following lemma which will be needed in the sequel.

6.8. LEMMA. *Let P be a program. There exists a model N_P of comp(P) such that $B_P \cap N_P = T_P{\downarrow}\omega$.*

PROOF. Let $\{A_1, \ldots, A_n\}$ be a finite set of $T_P{\downarrow}\omega$. By Theorem 5.6, for $i = 1, \ldots, n$, A_i is not in the failure set of P. Thus, by Lemma 5.8, $P \cup \{\leftarrow A_1, \ldots, A_n\}$ does not have a finitely failed SLD-tree. Now, by the completeness of the negation as failure rule (Theorem 5.30), there is a model of *comp(P)* $\cup \{A_1, \ldots, A_n\}$. Thus, by the Compactness Theorem 6.7, *comp(P)* $\cup\, T_P{\downarrow}\omega$ has a model, say N_P. We have $B_P \cap N_P \supseteq T_P{\downarrow}\omega$. Moreover, we have by virtue of Lemma 5.26, $B_P \cap N_P \subseteq T_P{\downarrow}\omega$. □

The model of *comp*(P) constructed in this lemma is in a sense "big". Note that by Theorem 5.6 we have

$$N_P \models \neg A \text{ iff } A \text{ is in the finite failure set of } P.$$

Thus in a sense N_P is "dual" to M_P which is a "small" model of *comp*(P) and for which, by the Characterization Theorem 3.13 and the Success Theorem 3.25,

$$M_P \models A \text{ iff } A \text{ is in the success set of } P.$$

In the proof of Theorem 6.6 we shall use both types of models. But first we need the following simple modification of Lemma 3.9.

6.9. LEMMA. *Let T be a continuous operator on a complete lattice. Suppose that $I \subseteq T(I)$. Then $T\uparrow\omega(I)$ is a fixpoint of T.*

PROOF. Let B be the largest element of the original lattice L. The set $\{J: I \subseteq J \subseteq B\}$ with the operations $\subseteq$, $\cup$ and $\cap$ from L forms a complete lattice with least element I. By assumption on T and I, T is an operator on this lattice and the claim follows by Lemma 3.9. □

Before we apply this lemma, we introduce the following notation. Given two programs P_1 and P_2, we write $P_1 < P_2$ to denote the fact that relations appearing in the heads of clauses from P_2 do not appear in P_1. Informally, when $P_1 < P_2$, then P_1 does not depend on P_2. More formally, we have the following lemma.

6.10. LEMMA. *Let P_1 and P_2 be two programs such that $P_1 < P_2$. Then, for any interpretation I based on a pre-interpretation J and $n \geqslant 1$,*

$$T^J_{P_1}(T^J_{P_2}\uparrow n(I)) = T^J_{P_1}(\emptyset).$$

PROOF. All elements of $T^J_{P_2}\uparrow n(I)$ are of the form $r(t_1, \ldots, t_m)\sigma$ where r appears in a head of a clause from P_2. □

6.11. LEMMA. *Let P_1 and P_2 be two programs such that $P_1 < P_2$. Suppose that I is a model of comp(P_1) based on a pre-interpretation J. Then $T^J_{P_2}\uparrow\omega(I)$ is a model of comp($P_1 \cup P_2$).*

PROOF. By Theorem 5.17 we have $I = T^J_{P_1}(I) \subseteq T^J_{P_1 \cup P_2}(I)$. Moreover, by Lemma 5.12, $T^J_{P_1 \cup P_2}$ is continuous. By Lemma 6.9, $T^J_{P_1 \cup P_2}\uparrow\omega(I)$ is a fixpoint of $T^J_{P_1 \cup P_2}$, so by Theorem 5.17, $T^J_{P_1 \cup P_2}\uparrow\omega(I)$ is a model of *comp*($P_1 \cup P_2$).

On the other hand, using Lemma 6.10 and the fact that $T^J_{P_1}(\emptyset) \subseteq I$, we get by an induction on n

$$T^J_{P_1 \cup P_2}\uparrow n(I) = T^J_{P_2}\uparrow n(I) \quad \text{for } n \geqslant 0.$$

Hence

$$T^J_{P_1 \cup P_2}\uparrow\omega(I) = T^J_{P_2}\uparrow\omega(I). \qquad \Box$$

We can now prove the desired result.

PROOF OF THEOREM 6.6. Let P^+ (respectively P^-) be the set of clauses of P whose heads contain a relation belonging to $DEP(G^+)$ (respectively $DEP(G^-)$). By the assumption of strictness, P^+ and P^- are disjoint. For some set P_0 of clauses

$$P = P_0 \cup P^+ \cup P^-.$$

Note that $P^+ \cup P^- < P_0$. Consider now the interpretation $M_{P^+} \cup_{P^-}$. Note that M_{P^+} and N_{P^-} are disjoint because no relation occurs both in P^+ and P^-. Thus $M_{P^+} \cup N_{P^-}$ is a model of $comp(P^+) \cup comp(P^-)$ i.e. a model of $comp(P^+ \cup P^-)$. This model is based on some pre-interpretation J. By Lemma 6.11, $M = T^J_{P_0}\uparrow\omega(M_{P^+} \cup N_{P^-})$ is a model of $comp(P)$.

By the assumption, $comp(P) \cup \{G\}$ is inconsistent, so, for some state σ,

$$M \models_\sigma A_1 \wedge \cdots \wedge A_l \wedge \neg B_1 \wedge \cdots \wedge \neg B_m$$

where $A_l, \ldots, A_l$ is the sequence of positive literals of G and $\neg B_1, \ldots, \neg B_m$ is the sequence of negative literals of G. If $l=0$ or $m=0$, we disregard the corresponding step in the considerations below.

By the definition of P^+ and P^- and the form of M we have $M_{P^+} \models_\sigma A_1 \wedge \cdots \wedge A_l$ and $N_{P^-} \models_\sigma \neg B_1 \wedge \cdots \wedge \neg B_m$. Thus σ, when restricted to the variables of $A_1 \wedge \cdots \wedge A_l$, is a ground substitution, say θ. By Corollary 3.6 and the Characterization Theorem 3.13(i), θ is a correct answer substitution for $P^+ \cup \{\leftarrow A_1, \ldots, A_l\}$. Applying now Theorem 3.18 we obtain a computed answer substitution γ for $P^+ \cup \{\leftarrow A_1, \ldots, A_l\}$ such that $(\leftarrow A_1, \ldots, A_l)\gamma$ is more general than $(\leftarrow A_1, \ldots, A_l)\theta$. But $(\leftarrow A_1, \ldots, A_l)\theta$ is ground, so in fact γ is more general than θ.

Fix some i, $1 \leqslant i \leqslant m$. By the assumption, $P \cup \{G\}$ does not flounder. Thus if $l=0$ then B_i is ground, so $B_i\sigma$ is a ground atom. If $l>0$ then $B_i\gamma$ is ground, so $B_i\theta$ is ground and consequently $B_i\sigma$ is a ground atom, as well as $B_i\gamma \equiv B_i\sigma$. But $N_P \models_\sigma \neg B_1 \wedge \cdots \wedge \neg B_m$, so $B_i\sigma \in B_P - N_P$. By Lemma 6.8 we now have $B_i\sigma \notin T_P\downarrow\omega$. By Theorem 5.6, $B_i\sigma$ is in the finite failure set of P^-. By the form of P^-, $B_i\sigma$ is in the finite failure set of P.

We thus showed that there exists an SLDNF$^-$-refutation of $P \cup \{G\}$. $\square$

This theorem can be generalized in the same ways as the completeness theorem of SLD-resolution (Theorem 3.15) was. The proofs of these generalizations are straightforward modifications of the above proof and use the generalizations of Theorem 3.15 presented in Subsections 3.8 and 3.9.

6.5. *Allowedness*

Unfortunately, restriction to programs P and general goals G such that $P \cup \{G\}$ does not flounder is not satisfactory as the following theorem shows.

6.12. THEOREM (undecidability of non-floundering). *For some program P it is undecidable whether for a general goal G, $P \cup \{G\}$ does not flounder.*

PROOF. This is a simple consequence of the computability results established in Section 4.4. Let P be a program and $q(x)$ an atom such that the variable x does not appear in P. Note that for any ground atom A there exists an SLD-refutation of $P \cup \{\leftarrow A\}$ iff $P \cup \{\leftarrow A, \neg q(x)\}$ flounders. Indeed, in the SLDNF$^-$-derivations no new negative literals are introduced. By Corollary 3.14 and Lemma 3.17 we thus have

$$A \in M_P \text{ iff } P \cup \{\leftarrow A, \neg q(x)\} \text{ flounders.}$$

But by Corollary 4.7 for some program P, (the complement of) M_P is not recursive. Consequently, it is not decidable whether for such a program P, $P \cup \{\leftarrow A, \neg q(x)\}$ does not flounder. $\square$

A way to solve this problem is by imposing on $P \cup \{G\}$ some syntactic restrictions which imply that $P \cup \{G\}$ does not flounder. To this purpose we introduce the following notion due to Lloyd and Topor [68]. Given a program P and a general goal G, we call $P \cup \{G\}$ *allowed* if the following two conditions are satisfied:

(a) every variable of G appears in a positive literal of G,
(b) every variable of a clause in P appears in the body of this clause.

Note that (a) implies that all negative literals of G are ground if G has no positive literals, and (b) implies that every unit clause in P is ground.

Allowedness is the notion we are looking for as the following theorem shows.

6.13. THEOREM (Lloyd and Topor [68]). *Consider a program P and a general goal G such that $P \cup \{G\}$ is allowed. Then*

(i) *$P \cup \{G\}$ does not flounder,*
(ii) *every computed answer substitution for $P \cup \{G\}$ is ground.*

PROOF. (i) Condition (b) ensures that every general goal appearing in an SLDNF$^-$-derivation satisfies condition (a). Thus $P \cup \{G\}$ does not flounder.

(ii) By the fact that every unit clause in P is ground. $\square$

Property (ii) shows the price we have to pay for ensuring property (i).

Combining Theorems 6.6 and 6.13 we obtain the following conclusion.

6.14. COROLLARY. *Let P be a program and G a general goal such that $P \cup \{G\}$ is strict and allowed. Suppose comp(P) $\cup \{G\}$ is inconsistent. Then there exists an* SLDNF$^-$-*refutation of $P \cup \{G\}$.*

Finally, observe that the definition of allowedness can be weakened a bit by requiring condition (b) to hold only for clauses whose heads contain a relation appearing in $DEP(G^+)$. Indeed, Theorem 6.13 then still holds by virtue of the same argument.

6.6. *Bibliographic remarks*

Usually, the case of programs and *general* goals is not considered separately. Consequently, soundness of the SLDNF$^-$-resolution (Theorem 6.1) is not spelled out separately. The proof of Lemma 6.2 seems to be new. The problem noted in Example 6.4 was first identified in [21]. Example 6.5 seems to be new. The name *floundering* was introduced in [87] but the concept first appeared in [21]. Lemma 6.8 was independently proved in [90]. Theorem 6.12 was independently, but somewhat earlier, proved in Börger [13].

The notion of *strictness* was first introduced in [3] for the case of general programs. The definition adopted here is inspired by Cavedon and Lloyd [19] where a much stronger version of Theorem 6.6 dealing with general programs is proved. The definition of allowedness is a special case of the one introduced in Lloyd and Topor [68] for general programs. Similar, but less general notions were considered in [21, 87, 3].

7. Stratified programs

7.1. Preliminaries

General programs are difficult to analyze because of their irregular behaviour. In this section we study a subclass of general programs obtained by imposing on them some natural syntactic restrictions. Programs from this subclass enjoy several natural properties.

First, we generalize in an obvious way some of the concepts to the case of general programs. To start with, given a general program P we introduce its *immediate consequence operator* T_P by putting for a Herbrand interpretation I

$$\begin{aligned}&A \in T_P(I) \text{ iff for some literals } L_1, \dots, L_n\\&A \leftarrow L_1, \dots, L_n \text{ is in ground}(P)\\&\text{and } I \models L_1 \wedge \cdots \wedge L_n.\end{aligned}$$

Next, given a general program P, we define its *completion* by using the same definition as the one given in Subsection 5.5 but now applied to general clauses instead of clauses. As before, *comp*(P) stands for the completion of P.

Some of the results relating models of P and *comp*(P) to the operator T_P remain valid and will be used in the sequel. We have the following lemma.

7.1. LEMMA. *Let P be a general program and I a Herbrand interpretation.*
(i) *I is a model of P iff $T_P(I) \subseteq I$.*
(ii) *I is a model of comp(P) iff $T_P(I) = I$.*
(iii) *T_P is finitary.*

PROOF. (i) Analogous to the proof of Lemma 3.7.
(ii) Analogous to the proof of Theorem 5.18(i)—all corresponding lemmas remain valid.
(iii) Analogous to the proof of Lemma 3.12(ii). □

Lemma 7.1(iii) remains valid when T_P is considered as an operator on a larger lattice formed by all subsets of $B_{P'}$, where $P \subseteq P'$, as then, for any $J \subseteq B_{P'}$, $T_P(J) = T_P(J \cap B_P)$. We shall use this observation in Subsection 7.4.

It is worthwhile to note that several other results do not generalize to the case of general programs. For example, for the general program $P = \{A \leftarrow \neg B\}$, the associated operator T_P is no longer monotonic, as $T_P(\emptyset) = \{A\}$ and $T_P(\{B\}) = \emptyset$. Thus Lemma 3.12(ii) does not generalize.

The same general program has two minimal models—$\{A\}$ and $\{B\}$ but none of them is the smallest. Thus Theorem 3.13 does not generalize. In turn, completion of the general program $A \leftarrow \neg A$ is inconsistent, so Theorem 5.18(ii) does not generalize either.

We thus see that it is not clear what intended meaning should be associated with a general program. None of the previously available possibilities—the one, semantic, based on M_P and another, proof-theoretic, based on *comp*(P), can be considered.

7.2. *Stratification*

To resolve these difficulties we introduce appropriate syntactic restrictions. Intuitively, we simply disallow a recursion "through negation". To express this idea more precisely we use the notion of a dependency graph introduced in Subsection 6.4. Given a general program P, consider its dependency graph D_P. We call an arc (r, q) from D_P *positive* (respectively *negative*) if there is a general clause in P such that r appears in its head and q appears in a positive (respectively negative) literal of its body. Thus an arc may be both positive and negative.

Following [3, 99] we call a general program *stratified* if its dependency graph does not contain a cycle with a negative arc. An alternative definition of stratified programs is the following: Given a general program P and a relation r, by a *definition of r* (within P) we mean the set of all general clauses of P in whose heads r appears. We call a partition $P = P_1 \cup \cdots \cup P_n$ a *stratification* of P if the following two conditions hold for $i = 1, \ldots, n$:

(i) if a relation appears in a positive literal of a general clause from P_i, then its definition is contained within $\bigcup_{j \leqslant i} P_j$;

(ii) if a relation appears in a negative literal of a general clause from P_i, then its definition is contained within $\bigcup_{j < i} P_j$.

We allow P_1 to be empty. A head of a general clause is viewed here as one of its positive literals. We call each P_i *a stratum*.

Now, both definitions are equivalent as the following lemma shows.

7.2. LEMMA (Apt, Blair and Walker [3]). *A general program P is stratified iff there exists a stratification of P.*

PROOF. If a general program admits some stratifaction then the definition of each relation symbol is contained in some stratum. Assign to each relation the index of the stratum within which it is defined. Then if (p, q) is a positive arc in the dependency graph of P, then the index assigned to q is smaller or equal than that assigned to p, and if (p, q) is a negative arc, then the index assigned to q is strictly smaller than that assigned to p. Thus there are no cycles in the dependency graph through a negative edge.

For the converse, decompose the dependency graph of P into *strongly connected* components each of maximum cardinality, (i.e., such that any two nodes in a component are connected by a cycle). Then the relation "there is an edge from component G to component H" is *well-founded*, since it is finite and contains no cycles. Thus for some n the numbers $1,\ldots,n$ can be assigned to the components so that if there is an edge from G to H, then the number assigned to H is smaller than that assigned to G. Now, let P_i be the subset of the general program P consisting of the definitions of all relations which lie within a component with the number i.

We claim that $P=P_1 \cup \cdots \cup P_n$ is a stratification of P. Indeed, if q is defined within some P_i and refers to r, then r lies in the same component or in a component with a smaller number. In other words, the definition of r is contained in P_j for some $j \leqslant i$. And if this reference is negative, then r lies in a component with a smaller number because, by assumption, there is no cycle through a negative edge. Thus the definition of r is then contained in P_j for some $j<i$. □

This lemma allows us to use both definitions of a stratified general program interchangably.

7.3. EXAMPLE. (i) Consider the general program

$$P=\{p\leftarrow,\ q\leftarrow p,\ r,\ r\leftarrow \neg q\}.$$

Then P is not stratified because the dependency graph of P contains a cycle (q,r), (r,q) with a negative edge.

(ii) Consider the general program $P=\{p\leftarrow, q\leftarrow p, r\leftarrow \neg q\}$. Then P is stratified by $\{p\leftarrow\}\cup\{q\leftarrow p\}\cup\{r\leftarrow\neg q\}$. Also $\{p\leftarrow,\ q\leftarrow p\}\cup\{r\leftarrow\neg q\}$ is a stratification of P.

Thus a general program can be stratified in more than one way.

Of course, it also makes sense to talk about stratification of programs (i.e., general programs "without negation"). By definition, every program is stratified but not every partition of it is a stratification. The following simple lemma relates the notion of stratification to the notation introduced in Subsection 6.4.

7.4. LEMMA. *A partition $P=P_1\cup\cdots\cup P_n$ of a program P is its stratification iff for every $i=1,\ldots,n$ we have $(\bigcup_{j<i}P_j)<P_i$.*

As a first step towards a better understanding of stratified (general) programs, we study in more detail their semantics. In view of Lemma 7.1, to study Herbrand models of a general program P and its completion, it suffices to consider the pre-fixpoints and fixpoints of its immediate consequence operator T_P. However, as just observed, the associated immediate consequence operator T_P does not need to be monotonic. This brings us to the study of nonmonotonic operators and their pre-fixpoints and fixpoints in an abstract setting. We follow here the presentation of [3].

7.3. *Nonmonotonic operators and their fixpoints*

Consider an arbitrary, but fixed, complete lattice and assume the notation used in Subsection 3.5. All operators are considered on this fixed lattice. First we define *cumulative powers* of an operator T. We put

$$T\Uparrow 0(I)=I,\qquad T\Uparrow(n+1)(I)=T(T\Uparrow n(I))\cup T\Uparrow n(I),$$

$$T\Uparrow\omega(I)=\bigcup_{n<\omega}T\Uparrow n(I).$$

Cumulative powers easily relate to the usual powers as clearly for all $\alpha \leqslant \omega$ and I

$$T \Uparrow \alpha(I) = (T \cup Id) \uparrow \alpha(I)$$

where Id is the identity operator, $\cup$ stands for a union of two operators and the powers defined in Subsection 3.5 are now adopted for arbitrary operators.

We have the following lemma.

7.5. LEMMA. *If T is finitary then, for all I, $T \Uparrow \omega(I)$ is a pre-fixpoint of T, i.e.*

$$T(T \Uparrow \omega(I)) \subseteq T \Uparrow \omega(I).$$

PROOF. Since T is finitary,

$$T(T \Uparrow \omega(I)) \subseteq \bigcup_{n=0}^{\infty} T(T \Uparrow n(I)) \subseteq \bigcup_{n=0}^{\infty} T \Uparrow (n+1)(I) \subseteq T \Uparrow \omega(I). \qquad \square$$

We say that an operator T is *growing* if, for all I, J, M,

$$I \subseteq J \subseteq M \subseteq T \Uparrow \omega(I) \text{ implies } T(J) \subseteq T(M).$$

Thus growing is a restricted from of monotonicity.

The following lemma holds.

7.6. LEMMA. *If T is growing then, for all I, $T \Uparrow \omega(I) \subseteq I \cup T(T \Uparrow \omega(I))$.*

PROOF. An easy proof by induction shows that, for all $i \geqslant 0$,

$$T \Uparrow i(I) \subseteq I \cup \bigcup_{n=0}^{\infty} T(T \Uparrow n(I)). \tag{7.1}$$

We now have

$$\begin{aligned} T \Uparrow \omega(I) &= \bigcup_{n=0}^{\infty} T \Uparrow n(I) \\ &\subseteq I \cup \bigcup_{n=0}^{\infty} T(T \Uparrow n(I)) \quad \text{(by (7.1))} \\ &\subseteq I \cup T(T \Uparrow \omega(I)). \quad \text{(by assumption).} \qquad \square \end{aligned}$$

The following corollary generalizes Lemma 6.9 and shows interest in studying finitary and growing operators.

7.7. COROLLARY. *Let T be finitary and growing. Suppose that $I \subseteq T(I)$. Then $T \Uparrow \omega(I)$ is a fixpoint of T.*

PROOF. Since T is growing, $I \subseteq T(I) \subseteq T(T \Uparrow \omega(I))$, so $I \cup T(T \Uparrow \omega(I)) = T(T \Uparrow \omega(I))$ and the claim follows by Lemmas 7.5 and 7.6. $\square$

Next, we study families of operators. Let $T_1, \ldots, T_n$ be operators. We put

$$N_0 = I, \qquad N_1 = T_1 \Uparrow \omega(N_0), \quad \ldots, \quad N_n = T_n \Uparrow \omega(N_{n-1}).$$

Clearly, $N_0 \subseteq N_1 \subseteq \cdots \subseteq N_n$. Of course, all N_i's depend on I and from the context it will be always clear from which one.

Let T stand for the *union* of the operators $T_1, \ldots, T_n$, i.e. for the operator defined by

$$T(X) = \bigcup_{i=1}^{n} T_i(X).$$

We wish to determine under which conditions N_n is a fixpoint of T. To this purpose we introduce the following concept. We call a sequence of operators $T_1, \ldots, T_n$ *local* if, for all I, J,

$$I \subseteq J \subseteq N_n \text{ implies } T_i(J) = T_i(J \cap N_i) \quad \text{for } i = 1, \ldots, n.$$

Informally, locality means that each T_i is determined by its values on the subsets of N_i.

The following two lemmas show interest in studying local sequences of operators.

7.8. LEMMA. *Suppose that the sequence $T_1, \ldots, T_n$ is local and that all T_i's are finitary. Then $T(N_n) \subseteq N_n$.*

PROOF. We have

$$\begin{aligned} T(N_n) &= \bigcup_{i=1}^{n} T_i(N_n) \\ &= \bigcup_{i=1}^{n} T_i(N_i) \quad \text{(by locality)} \\ &\subseteq \bigcup_{i=1}^{n} N_i \qquad \text{(by Lemma 7.5)} \\ &= N_n. \qquad \square \end{aligned}$$

7.9. LEMMA. *Suppose that the sequence $T_1, \ldots, T_n$ is local and that all T_i's are growing. Then $N_n \subseteq I \cup T(N_n)$.*

PROOF. We proceed by induction on n. If $n = 1$, the lemma reduces to Lemma 7.6.

Assume the lemma holds for $n-1$. Then, again by Lemma 7.6,

$$\begin{aligned} N_n &\subseteq N_{n-1} \cup T_n(N_n) \\ &\subseteq I \cup \bigcup_{i=1}^{n-1} T_i(N_{n-1}) \cup T_n(N_n) \quad \text{(by induction hypothesis)} \\ &= I \cup \bigcup_{i=1}^{n-1} T_i(N_n) \cup T(N_n) \qquad \text{(by locality)} \\ &= I \cup T(N_n). \qquad \square \end{aligned}$$

7.10. Corollary. *Suppose that the sequence $T_1, \ldots, T_n$ is local and that all T_i's are finitary and growing. Then $N_n = I \cup T(N_n)$.*

Thus for a local sequence $T_1, \ldots, T_n$ of finitary and growing operators, N_n is a fixpoint of T when $I = \emptyset$.

We now prove that under some assumptions N_n is a minimal pre-fixpoint of T containing I.

7.11. Lemma. *Suppose that the sequence $T_1, \ldots, T_n$ is local and that all T_i's are growing. Suppose $I \subseteq J \subseteq N_n$ and $T(J) \subseteq J$. Then $J = N_n$.*

Proof. We prove by induction on $j = 0, \ldots, n$ that

$$N_j \subseteq J. \tag{7.2}$$

For $j = 0$ it is part of the assumptions. Assume the claim holds for some $j < n$. We now prove by induction on k that

$$T_{j+1} \Uparrow k(N_j) \subseteq J. \tag{7.3}$$

For $k = 0$ this is just (7.2). So assume (7.3) holds for some $k \geqslant 0$. We then have

$$\begin{aligned} T_{j+1} \Uparrow (k+1)(N_j) &\subseteq T_{j+1}(T_{j+1} \Uparrow k(N_j)) \cup J \\ &\subseteq T_{j+1}(J \cap N_{j+1}) \cup J && \text{(by (7.3) and since } T_{j+1} \text{ is growing)} \\ &= T_{j+1}(J) \cup J && \text{(by locality)} \\ &\subseteq J. && \text{(by the assumptions)} \end{aligned}$$

Thus, by induction, for all $k \geqslant 0$ (7.3) holds, so $N_{j+1} \subseteq J$. This proves (7.2) for all $j = 0, \ldots, n$ and concludes the proof. □

Finally, we provide an alternative characterization of N_i. To make it more readable we now assume that $I = \emptyset$. Then, by definition, $N_0 = \emptyset$.

Let now $\boldsymbol{T}_i$ denote the union of $T_1, \ldots, T_i$, i.e. $\boldsymbol{T}_i(X) = T_1(X) \cup \cdots \cup T_i(X)$.

7.12. Lemma. *Suppose that the sequence $T_1, \ldots, T_n$ is local and that all T_i's are finitary and growing. Let*

$$\begin{aligned} \boldsymbol{K}_1 &= \{J : \boldsymbol{T}_1(J) = J,\ T_1(J \cap N_1) \subseteq T_1(J)\}, \\ \boldsymbol{K}_2 &= \{J : \boldsymbol{T}_2(J) = J,\ T_2(J \cap N_2) \subseteq T_2(J),\ N_1 \subseteq J\}, \\ &\ldots \\ \boldsymbol{K}_n &= \{J : \boldsymbol{T}_n(J) = J,\ T_n(J \cap N_n) \subseteq T_n(J),\ N_{n-1} \subseteq J). \end{aligned}$$

Then for $i = 1, \ldots, n$, $\bigcap \boldsymbol{K}_i = N_i$.

Note that each $\boldsymbol{K}_i$ is the collection of all fixpoints of $\boldsymbol{T}_i$ which include N_{i-1}, where additionally the condition $T_i(J \cap N_i) \subseteq T_i(J)$ is required.

PROOF. Fix some i, $1 \leqslant i \leqslant n$. By Corollary 7.10 used for $I = \emptyset$ and $n = i$ and the fact that $N_{i-1} \subseteq N_i$, we conclude that N_i belongs to $\boldsymbol{K}_i$. Thus $\bigcap \boldsymbol{K}_i \subseteq N_i$.

To prove the converse take $J \in \boldsymbol{K}_i$. We prove by induction on k that for $k \geqslant 0$

$$T_i \Uparrow k(N_{i-1}) \subseteq J. \tag{7.4}$$

For $k = 0$ it holds by the definition of $\boldsymbol{K}_i$. Assume this claim holds for some $k \geqslant 0$. Then

$$T_i \Uparrow k(N_{i-1}) \subseteq N_i, \tag{7.5}$$

so by (7.4) and (7.5) and the fact that T_i is growing

$$\begin{aligned} T_i(T_i \Uparrow k(N_{i-1})) &\subseteq T_i(J \cap N_i) \\ &\subseteq T_i(J) \qquad \text{(by definition of } \boldsymbol{K}_i\text{)} \\ &\subseteq \boldsymbol{T}_i(J) \\ &\subseteq J. \qquad \text{(by definition of } \boldsymbol{K}_i\text{).} \end{aligned}$$

Thus the claim holds for $k+1$. This implies $N_i \subseteq J$, so $N_i \subseteq \bigcap \boldsymbol{K}_i$. □

7.4. Semantics of stratified programs

We now apply the results of the previous subsection to provide a semantics for stratified programs. Throughout this section we consider a general program P stratified by $P = P_1 \cup \cdots \cup P_n$. We now define a sequence of Herbrand interpretations by putting

$$M_1 = T_{P_1} \Uparrow \omega(\emptyset), \qquad M_2 = T_{P_2} \Uparrow \omega(M_1), \ \ldots, \ M_n = T_{P_n} \Uparrow \omega(M_{n-1}).$$

Let $M_P = M_n$. Note that M_P depends on the stratification and that for programs P, M_P has already a different meaning. We shall show in the next subsection that these apparent ambiguities in fact do not exist—M_P does not depend on the stratification of P and consequently, by virtue of the Characterization Theorem 3.13, it coincides for programs P with the previous meaning.

We first prove that M_P is a model of P. To this purpose we need the following lemmas.

7.13. LEMMA. *Consider a stratum P_i ($1 \leqslant i \leqslant n$). T_{P_i} considered as an operator on the complete lattice $\{I : I \subseteq B_P\}$ is growing.*

PROOF. Suppose that for some $I \subseteq B_P$, $I \subseteq J \subseteq M \subseteq T_{P_i} \Uparrow \omega(I)$ and let $A \in T_{P_i}(J)$. For some general clause $A \leftarrow L_1, \ldots, L_n$ from ground(P_i) we have $J \models L_1 \wedge \cdots \wedge L_n$. If L_i is positive then also $M \models L_i$. If L_i is negative, say $\neg p(t_1, \ldots, t_k)$, then neither $p(t_1, \ldots, t_k) \in I$ nor p appears in a head of a general clause from P_i because P_i is a stratum. However, for any Herbrand interpretation $N \subseteq B_P$ and a ground atom $r(s_1, \ldots, s_m)$, if $r(s_1, \ldots, s_m) \in T_{P_i} \Uparrow \omega(N)$ then $r(s_1, \ldots, s_m) \in N$ or r appears in a head of general clause from P_i. Thus $p(t_1, \ldots, t_k) \notin T_{P_i} \Uparrow \omega(I)$, so $M \models L_i$, as well. This implies that $A \in T_{P_i}(M)$. □

7.14. LEMMA. *Consider the strata $P_1, \ldots, P_n$. The sequence of operators $T_{P_1}, \ldots, T_{P_n}$ considered on the complete lattice $\{I : I \subseteq B_P\}$ is local.*

PROOF. Choose some $I \subseteq B_P$ and consider the sequence $N_1, \ldots, N_n$ of subsets of B_P defined in the previous subsection. Fix some i, $1 \leqslant i \leqslant n$. Suppose that $p(t_1, \ldots, t_k) \in N_n - N_i$. Then p appears in a head of a general clause from $\bigcup_{j=i+1}^{n} P_j$, so by the definition of stratification p does not appear in a general clause from P_i. Thus $p(t_1, \ldots, t_k) \notin B_{P_i}$. Hence $N_n \cap B_{P_i} \subseteq N_i$ and consequently,

$$N_n \cap B_{P_i} = N_i \cap B_{P_i}, \tag{7.6}$$

since $N_i \subseteq N_n$. Suppose now that $I \subseteq J \subseteq N_n$. We have

$$\begin{aligned} J \cap B_{P_i} &= J \cap N_n \cap B_{P_i} \\ &= J \cap N_i \cap B_{P_i} \quad \text{(by (7.6))}. \end{aligned} \tag{7.7}$$

Thus

$$\begin{aligned} T_{P_i}(J) &= T_{P_i}(J \cap B_{P_i}) && \text{(by definition of } T_{P_i}) \\ &= T_{P_i}(J \cap N_i \cap B_{P_i}) && \text{(by (7.7))} \\ &= T_{P_i}(J \cap N_i). && \text{(by definition of } T_{P_i}). \quad \square \end{aligned}$$

We can now conclude by the following theorem

7.15. THEOREM (Characterization Theorem) (Apt, Blair and Walker [3]). *Let P be a general program stratified by $P = P_1 \cup \cdots \cup P_n$. Then*

(i) *M_P is a Herbrand model of P.*
(ii) *M_P is a minimal Herbrand model of P.*
(iii) *M_P is a Herbrand model of comp(P).*

PROOF. (i) By Lemmas 7.1(i), (iii), 7.13 and 7.14 and Corollary 7.10.
(ii) By Lemmas 7.1(i) and 7.11.
(iii) By Lemmas 7.1(ii), (iii), 7.13 and 7.14 and Corollary 7.10. $\square$.

Finally, we provide an alternative characterization of M_P. To prove the desired theorem we first introduce a notation and prove a lemma.

Given a general program P, let

$$Neg_P = \{A : \text{for some } B \leftarrow L_1, \ldots, L_n \in \text{ground}(P) \text{ and } i,\ 1 \leqslant i \leqslant n,\ L_i = \neg A\}.$$

Thus Neg_P stands for the set of ground instances of atoms whose negation occurs in a hypothesis of general clause from P.

7.16. LEMMA. *Let P be a general program and I, J Herbrand interpretations. Suppose that $I \subseteq J$ and $I \cap Neg_P = J \cap Neg_P$. Then $T_P(I) \subseteq T_P(J)$.*

PROOF. Suppose that $A \in T_P(I)$. For some general clause $A \leftarrow L_1, \ldots, L_n$ from ground(P), we have $I \models L_1 \wedge \cdots \wedge L_n$. If L_i is positive then, by assumption, also $J \models L_i$. If L_i is negative, say $\neg B$, then $B \notin I$, so $B \notin I \cap Neg_P$ and by assumption $B \notin J \cap Neg_P$. But, by definition, $B \in Neg_P$, so $B \notin J$, i.e. $J \models L_i$. This implies that $A \in T_P(J)$. $\square$

Assume now a given stratification $P_1 \cup \cdots \cup P_n$ of P. To shorten the notation let from now on $\boldsymbol{P}_i$ stand for $P_1 \cup \cdots \cup P_i$. Then $P = \boldsymbol{P}_n$. Let M range over the subsets of B_P. Put

$$M(\boldsymbol{P}_1) = \bigcap \{M\colon T_{\boldsymbol{P}_1}(M) = M\},$$
$$M(\boldsymbol{P}_2) = \bigcap \{M\colon T_{\boldsymbol{P}_2}(M) = M,\ M \cap B_{\boldsymbol{P}_1} = M(\boldsymbol{P}_1)\},$$
$$\ldots$$
$$M(\boldsymbol{P}_n) = \bigcap \{M\colon T_{\boldsymbol{P}_n}(M) = M,\ M \cap B_{\boldsymbol{P}_{n-1}} = M(\boldsymbol{P}_{n-1})\}.$$

Note that by Theorem 7.1(ii) each $M(\boldsymbol{P}_i)$ is the intersection of all Herbrand models of $comp(\boldsymbol{P}_i)$ which on the previous Herbrand base $B_{\boldsymbol{P}_{i-1}}$ agree with the previous model $M(\boldsymbol{P}_{i-1})$. In the definition of $M(\boldsymbol{P}_i)$ each $T_{\boldsymbol{P}_i}$ is considered as an operator on the complete lattice B_P. We now prove the following theorem.

7.17. THEOREM (Apt, Blair and Walker [3]). $M_P = M(P)$.

PROOF. We prove by induction that, for $i = 1, \ldots, n$, $M_i = M(\boldsymbol{P}_i)$. This implies the claim since $M_n = M_P$ and $M(\boldsymbol{P}_n) = M(P)$. For $i = 1$ it is a consequence of the Characterization Theorem 3.13 and the fact that $T_{\boldsymbol{P}_1}(M) \subseteq B_{\boldsymbol{P}_1}$.

Suppose the claim holds for some i, $1 \leqslant i < n$. Note that by the Characterization Theorem 7.15(iii) and Lemma 7.1(ii), $T_{\boldsymbol{P}_{i+1}}(M_{i+1}) = M_{i+1}$. Also $M_{i+1} \cap B_{\boldsymbol{P}_i} = M(\boldsymbol{P}_i)$ by the induction hypothesis and the definition of stratification. Thus M_{i+1} is an element of the collection whose intersection is $M(\boldsymbol{P}_{i+1})$. This proves that $M(\boldsymbol{P}_{i+1}) \subseteq M_{i+1}$.

To establish the converse inclusion, take M from the collection whose intersection is $M(\boldsymbol{P}_{i+1})$. Thus

$$T_{\boldsymbol{P}_{i+1}}(M) = M \tag{7.8}$$

and

$$M(\boldsymbol{P}_i) = M \cap B_{\boldsymbol{P}_i}. \tag{7.9}$$

Equation (7.9) implies by the induction hypothesis $M_i = M \cap B_{\boldsymbol{P}_i}$ so

$$M_i \subseteq M. \tag{7.10}$$

Moreover, by the definition of stratification, $M \cap Neg_{P_{i+1}} \subseteq B_{\boldsymbol{P}_i}$, so

$$M \cap Neg_{P_{i+1}} = Neg_{P_{i+1}} \cap B_{\boldsymbol{P}_i}. \tag{7.11}$$

Now

$$\begin{aligned} M \cap M_i \cap Neg_{P_{i+1}} &= M_i \cap Neg_{P_{i+1}} && \text{(by (7.10))} \\ &= M(\boldsymbol{P}_i) \cap Neg_{P_{i+1}} && \text{(by the induction hypothesis)} \\ &= M \cap B_{\boldsymbol{P}_i} \cap Neg_{P_{i+1}} && \text{(by (7.9))} \\ &= M \cap Neg_{P_{i+1}} && \text{(by (7.11)).} \end{aligned}$$

Thus by Lemma 7.16

$$T_{P_{i+1}}(M \cap M_i) \subseteq T_{P_{i+1}}(M). \tag{7.12}$$

We can now apply Lemma 7.12 with $N_j = M_j$ and $T_j = T_{P_j}$. By (7.8), (7.12) and (7.10)

$M \in \boldsymbol{K}_{i+1}$ so $\bigcap \boldsymbol{K}_{i+1} \subseteq M$ and, by Lemma 7.12, $M_{i+1} \subseteq M$. By the choice of M, $M_{i+1} \subseteq M(\boldsymbol{P}_{i+1})$. This concludes the proof of the induction step. □

7.5. *Perfect model semantics*

We now prove that the Herbrand model M_P does not depend on the stratification of P. We follow here the approach of Przymusiński [81]. It is conceptually advantageous to carry out these considerations in a more abstract setting.

Consider a given general program P. Let $<$ be a well-founded ordering on the Herbrand base B_P of P. If $A < B$ then we say that A has a *higher priority than* B.

Let $M, N \subseteq B_P$. We call a Herbrand interpretation N *preferable to* M, and write $N \prec M$, if $N \neq M$ and for every $B \in N - M$ there exists an $A \in M - N$ such that $A < B$. We write $N \preccurlyeq M$ if $N = M$ or $N \prec M$. We call a Herbrand model of P *perfect* if there are no Herbrand models of P preferable to it. Thus a perfect model of P is a $\prec$-minimal Herbrand model of P.

The intuition behind these definitions is the following. N is preferable to M if it is obtained from M by possible adding/removing some atoms and an addition of an atom to N is always compensated by the simultaneous removal from M of an atom of higher priority. This reflects the fact that we are determined to minimize higher-priority atoms even at the cost of adding atoms of lower priority. A model is then perfect if this form of minimization of higher-priority atoms is achieved in it.

The following lemma clarifies the status of perfect models.

7.18. LEMMA. *Let P be a general program and let $<$ be a well-founded ordering on B_P.*

(i) *Every perfect model of P is minimal.*

(ii) *For no two Herbrand interpretations M, N of P, both $M \prec N$ and $N \prec M$.*

PROOF. (i) Immediate, since $N \subseteq M$ implies $N \prec M$.

(ii) Suppose by contradiction that for some Herbrand interpretations M, N of P both $M \prec N$ and $N \prec M$. Then none of them is a subset of the other. Thus $N - M$ is nonempty. Let $A_0 \in N - M$. N is preferable to M, so for some $A_1 \in M - N$, $A_1 < A_0$. But M is preferable to N so, for some $A_2 \in N - M$, $A_2 < A_1$. Continuing in this way we obtain an infinite $<$-descending sequence of ground atoms which contradicts the assumption that $<$ is a well-founded ordering on B_P. □

One can also prove that the relation "N is preferable to M" is a partial order but we shall not need this in the sequel.

Subsequent considerations are carried out for a fixed stratified general program P and a well-founded ordering $<$ on B_P defined by first putting, for two relation symbols p, q,

$$p < q \text{ iff there exists a path from } q \text{ to } p \text{ in } D_P \text{ with a negative arc,}$$

and then putting, for two atoms $A, B \in B_P$,

$$A < B \text{ iff } p < q \text{ where } p \text{ appears in } A \text{ and } q \text{ appears in } B.$$

By the definition of a stratified program, $<$ is a well-founded ordering on B_P. Note that the orientation of $<$ is different than the one suggested by D_P. If $p<q$ then p is defined in a strictly lower stratum than q and all ground atoms containing p are of higher priority than those containing q. Fix from now on a stratification $P_1 \cup \cdots \cup P_n$ of P. Note that $M_P \cap B_{P_i} = M_i$. For a Herbrand interpretation N of L_{P_i}, denote $N \cap B_{P_i}$ by N_i. Note that $N_1 \subseteq N_2 \subseteq \cdots \subseteq N_n$.

7.19. LEMMA. *Let N be a Herbrand model of P. Then for all $i=1,\ldots,n$ we have $M_i \leqslant N_i$.*

PROOF. We proceed by induction on i. Note that $N_i \models P_i$. As P_1 is a program, by the Characterization Theorem 3.13, M_1 is its smallest model. Thus $M_1 \subseteq N_1$, and a fortiori $M_1 \leqslant N_1$.

Suppose the claim holds for some $i \geqslant 1$. Call an element $B \in M_{i+1}$ *regular* if $B \notin N_{i+1}$ implies that, for some $A \in N_{i+1} - M_{i+1}$, $A < B$. To prove that $M_{i+1} \leqslant N_{i+1}$ we need to show that all elements of M_{i+1} are regular.

We have $M_{i+1} = \bigcup_{k=0}^{\infty} T_{P_{i+1}} \Uparrow k(M_i)$. We now prove by induction on k that all elements of $T_{P_{i+1}} \Uparrow k(M_i)$ are regular. To take care of the case $k=0$, consider some $B \in M_i - N_{i+1}$. Then $B \notin N_i$, so, by the induction hypothesis, for some $A \in N_i - M_i$, $A < B$. Moreover, $N_i \subseteq B_{P_i}$, so $A \in B_{P_i}$. But $M_{i+1} \cap B_{P_i} = M_i$, so $A \notin M_{i+1}$. Thus $A \in N_{i+1} - M_{i+1}$ and consequently B is regular.

To take care of the induction step, fix $k \geqslant 0$ and denote $T_{P_{i+1}} \Uparrow k(M_i)$ by M. Assume that all elements of M are regular and consider some $B \in T_{P_{i+1}}(M) - M$. For some general clause $B \leftarrow L_1, \ldots, L_s$ in ground(P_{i+1}), $M \models L_1 \wedge \cdots \wedge L_s$. Let $A_1, \ldots, A_l$ be the positive literals among $L_1, \ldots, L_s$ and let $\neg B_1, \ldots, \neg B_m$ be the negative literals among $L_1, \ldots, L_s$. We have $A_1, \ldots, A_l \in M$ and $B_1, \ldots, B_m \notin M$. Suppose now $B \notin N_{i+1}$. N_{i+1} is a model of P_{i+1}, so either some $A_j \notin N_{i+1}$ or some $B_j \in N_{i+1}$. If some $A_j \notin N_{i+1}$ then $A_j \in M - N_{i+1}$. As A_j is regular, for some $A \in N_{i+1} - M_{i+1}$, $A < A_j$. By the definition of $<$, also $A < B$. If some $B_j \in N_{i+1}$ then $B_j \in N_{i+1} - M$, so $B_j \in N_{i+1} - M_i$. Moreover, by the definition of stratification $B_j \in B_{P_i}$. But $M_{i+1} \cap B_{P_i} = M_i$, so $B_j \notin M_{i+1}$. Thus $B_j \in N_{i+1} - M_{i+1}$. Moreover, by the definition of $<$ we have $B_j < B$.

We thus showed that B is regular. By induction on k we now proved that $M_{i+1} \leqslant N_{i+1}$. Thus by induction on i, we proved the lemma. $\square$

7.20. LEMMA. *Let I, J be Herbrand interpretations for L_P. If for all $i=1,\ldots,n$ we have $I_i \leqslant J_i$, then $I \leqslant J$.*

PROOF. Let $B \in I - J$. For some i, $1 \leqslant i \leqslant n$, we have $B \in I_i - J$. So $B \in B_{P_i}$. But $J_i = J \cap B_{P_i}$, so $B \notin J_i$. Since $I_i \leqslant J_i$, for some $A \in J_i - I_i$, $A < B$. So $A \in B_{P_i}$. But $I_i = I \cap B_{P_i}$, so $A \notin I$. $\square$

This brings us to the main result of this subsection.

7.21. THEOREM (Przymusiński [81]). (i) *For every Herbrand model N of P, $M_P \leqslant N$.*

(ii) *M_P is the unique perfect model of P.*

Proof. (i) By Lemmata 7.19 and 7.20.

(ii) By (i) and Lemma 7.18(ii), M_P is a perfect model of P. By (i) it is also unique. □

Note that (ii) in view of lemma 7.18(i) provides an alternative proof of Theorem 7.15(ii).

7.22. Corollary (Apt, Blair and Walker [3]). *M_P does not depend on the stratification of P.*

Proof. The proof of Theorem 7.12(ii) does not depend on the stratification of M_P. □

Theorems 7.15 and 7.16 show that M_P is a natural model of a stratified program P. However, the most convincing evidence that M_P is indeed natural, is supplied by Theorem 7.22. The notion of a perfect model turns out to be the key concept in assessing the character of M_P.

7.6. *Bibliographic remarks*

Stratified programs form a simple generalization of a class of database queries introduced in [20]. Similar concepts were also introduced in [7] and, in the context of deductive databases, in [76].

The proofs of Theorems 7.17 and 7.21 and of Corollary 7.22 differ from the original ones. The notion of a stratified program was further generalized by Przymusiński [81] to a *locally stratified program.* Lifschitz [63] provides a characterization of the model M_P of a stratified program P using the prioritized circumscription. Other connections between stratification, the model M_P and nonmonotonic reasoning are surveyed in [82]. Apt and Blair [2] analyze the recursion-theoretic complexity of the model M_P.

8. Related topics

Our presentation of logic programming is obviously incomplete. In this section we briefly discuss the subjects we omitted and provide a number of pointers to the literature.

8.1. *General programs*

SLD-resolution and the negation as failure rule was combined by Clark [21] into a more powerful computation mechanism called *SLDNF-resolution* allowing us to refute general goals from general programs. The reader is referred to [65] for a detailed account of SLDNF-resolution.

Shepherdson [90] discusses and compares various approaches to the proof theory and semantics of general programs. The strongest completeness results dealing with the SLDNF-resolution were proved in [19, 60].

8.2. *Alternative approaches*

The approach to logic programming we discussed in this paper is undoubtedly the most widely accepted. However, various alternatives exist and it is worthwhile to point them out.

Proof theory

Fitting [36] proposed on alternative computation mechanism based on a tableau method. Gallier and Raatz [40] introduced a computation mechanism in the form of an interpreter using graph reduction. Brough and Walker [17] studied interpreters with various stopping criteria for function-free programs. Apt, Blair and Walker [3] introduced an interpreter with a loop-checking mechanism and with an ineffective means of handling negative literals. Przymusiński [81] generalized this interpreter to an *SLS-resolution* (Linear resolution with Selection rule for Stratified programs) in which negative literals are resolved in an ineffective way.

Variants of SLD-resolution, called *HLSD*-resolution and *SLD-AL*-resolution were introduced and studied in [76] and [100] respectively.

Semantics

Mycroft [75] suggested to use 3-valued logic (corresponding to the possibilities: provable, refuted and undecidable) to capture the meaning of logic programs. This approach was subsequently studied in detail in [35, 58, 60].

To describe the meaning of general programs Minker [73] proposed the use of minimal models (leading to the *generalized closed world assumption* GCWA), Bidoit and Hull [9] proposed the use of *positivistic models* and Przymusiński [81] introduced the concept of a *perfect model.*

8.3. *Deductive databases*

Deductive databases form an extension of relational databases in which some of the relations are implicitly defined. They can be viewed as logic programs where the *explicitly* defined relations are those defined only by means of unit clauses, whereas the *implicitly* defined relations are those defined by means of non-unit clauses, as well. Moreover, so-called *particularization axioms* are needed to define the intended domain. Additionally, integrity constraints are used to impose a desired meaning on the relations used.

The main difference between deductive databases and logic programming lies in their emphasis on different problems. In deductive databases one studies such issues like query processing (i.e. computation of *all* answers to a given goal), integrity constraint checking, handling of updates (i.e. additions and deletions of ground unit clauses) and processing of negative information.

Recent research concentrates on efficient implementation of recursive queries, i.e. queries about recursively defined relations (see e.g. the survey of Bancilhon and Ramakrishnan [6]), reduction of recursive queries to nonrecursive ones (see e.g. [78]), comparison of expressive power between various query languages (see e.g. [20, 91]), and handling of negative information both in terms of intended semantics (see e.g. [73, 3, 99, 63, 77, 81]) and in terms of query processing, handling of updates and integrity constraint checking (see e.g. [44, 29, 67]).

Earlier research in this area is surveyed in [38] while more recent research is discussed in [51, Section 4] and [74].

8.4. *PROLOG*

PROLOG stands for *pro*gramming in *log*ic. It is a programming language conceived and implemented in the beginning of 1970s by Colmerauer et al. [26]. In its pure form it can be viewed as logic programming with the "left-first" selection rule and with the depth-first strategy for searching the empty node in an SLD-tree. Negation is implemented by means of the negation as failure rule. For efficiency reasons, an important test (the check in step (5) of the Unification Algorithm whether x appears in t—so-called *occur check*) is usually deleted from the unification algorithm and a special control facility (called *cut*) to prune the search tree is introduced. These changes make PROLOG different from logic programming and make it difficult to apply to its study the theoretical results concerning logic programming.

Theoretical study of PROLOG concentrated on efforts to provide a rigorous semantics of it in terms of interpreters explaining the process of SLD-tree traversal (see e.g. [49a]), by means of denotational semantics (see e.g. [34]) or by relating both approaches (see e.g. [28]).

More practical considerations, apart of a study of implementations of PROLOG (see e.g. [18]), led to an investigation of efficient backtracking mechanisms (see e.g. [27]) and of various additions, like metafacilities (see e.g. [13, 96]), modules (see e.g. [41]), control mechanisms (see e.g. [76]) and parallelism (see e.g. Concurrent Prolog of Shapiro [87] and PARLOG of Clark and Gregory [24]).

Good books on PROLOG programming have been written by Bratko [16], and Sterling and Shapiro [96].

8.5. *Integration of logic and functional programming*

Logic or PROLOG programs use relational notation. This makes it awkward to define functions explicitly which have to be rewritten and used as relations. Functional programming is based on the use of functions as primitive objects and shares with logic programming several aspects like the use of recursion as the main control structure and reliance on mathematical logic (especially lambda calculus). Several attempts to combine advantages of both formalisms in one framework originated with the LOGLISP language of Robinson and Siebert [85].

Direct definition of functions by means of equations leads to the problem how in the framework of logic programming equality is to be handled. Solutions to this problem involve the use of *extended unification*, where identity is replaced by equality derivable from axioms defining functions, the use of term rewriting techniques in the form of a *narrowing procedure* and the use of some subset of the standard equality axioms EQ defined in Subsection 5.9 written in a clausal form.

Recent proposals in this area are collected in de Groot and Lindstrom [43] which is a standard reference in this domain. See also [8, 41, 32].

8.6. *Applications in artificial intelligence*

Strictly speaking, logic programming is just a restricted form of automatic theorem proving. Various proposals of extending it to more powerful fragments of certain logics can be seen as attempts to increase its expressive and manipulative power while preserving efficiency. In particular a substantial effort has been made to adapt it to the needs of artificial intelligence. While research in this area is of a much more practical character, we can still single out certain investigations of more theoretical nature.

Use of logic programming as a formalism for knowledge representation and reasoning was advocated by Kowalski [54]. Analysis and implementation of more powerful logics and various forms of reasoning in the framework of logic programming was undertaken by Fariñas del Cerro [33] for modal logic, by Van Emden [30] for quantitative reasoning and by Poole [80] for hypothetical reasoning.

More practical work in this area deals with natural language processing, the original application domain of PROLOG (see e.g. the special issue of the Journal of Logic Programming [50]) and with the use of logic programming and PROLOG for the construction of expert system shells (see e.g. [16, 101].)

Appendix

Short history of the subject

The following is a list of papers and events which have shaped our views of this subject. Obviously, this account of the history of the subject by no means objective (as none is).

1972: A. Colmerauer and R. Kowalski collaborated to develop from resolution theorem proving a programming language.

1973: Colmerauer et al. [26] implemented PROLOG.

1974: Kowalski [53] proposed logic (programming) as a programming language and introduced what is now called SLD-resolution.

1976: Van Emden and Kowalski [31] studied the semantics of logic programs and introduced the ubiquitous immediate consequence operator T_P.

1978: Reiter [83] proposed in the context of deductive databases the *Closed World Assumption* rule as a means of deducing negative information.

1978: Clark [21] introduced the *negation as failure* rule as an effective means of deducing negative information for logic programs and proposed the *completion* of a program, *comp*(*P*), as a description of its meaning.

1979: Kowalski [54] analyzed logic programming as a formalism for knowledge representation and problem solving.

1979: Kowalski [55] investigated logic programming as a formalism for a systematic development of algorithms.

1981: Clark and Gregory [23] proposed a parallel version of logic programming which influenced subsequent language proposals in this area.
1982: Logic programming was chosen as the basis for a new programming language in the Japanese Fifth Generation computer system project.
1982: Apt and Van Emden [4] characterized the SLD-resolution, negation as failure rule and completion of a program by means of the operator T_P and its fixpoints.
1983: In the book [25], edited by K.L. Clark and S.-A. Tärnlund, a number of articles were collected that indicated a wide scope of applications of logic programming and revealed its manipulative and expressive power.
1983: Jaffar, Lassez and Lloyd [48] proved completeness of the negation as failure rule with respect to the completion of a program.
1984: Lloyd [64] gathered in his book several results on logic programming in a single, uniform framework.
1984: A.J. Robinson founded the *Journal of Logic Programming.*
1986: In the book [43] edited by D. de Groot and G. Lindstrom, several approaches aiming at an integration of logic and functional programming were presented.
1986: Apt, Blair and Walker [3] and Van Gelder [99] identified *stratified programs* as a natural subclass of general logic programs and proposed *stratification* as a means of handling negative information.
1986: J. Minker organized the Workshop on Foundations of Deductive Databases and Logic Programming which brought together researchers working in both areas.
1985–1989: M. Fitting and K. Kunen developed in [35, 36, 58, 60] a theory of logic programming based on 3-valued logic.

Note

In this chapter we use the terminology of Lloyd in [64] which differs from that of Lloyd in [65]. In [65] a program is called a definite program and in turn a general program is called a normal program. Similar terminology is used there for goals and general goals.

Acknowledgment

We would like to thank Marc Bezem, Roland Bol, Stephane Grumbach and Jan Willem Klop for detailed comments on the first version of this paper. Also, we profited from discussions with Howard Blair, Lawrence Cavedon, Maarten van Emden, Jean Gallier, Joxan Jaffar, Jean-Louis Lassez, John Lloyd, Michael Maher, Katuscia Palamidessi, Teodor Przymusiński, John Shepherdson, Wayne Snyder and Rodney Topor who commented on the subject of this paper in four languages. Our task was significantly simplified thanks to John Lloyd who collected in [64] most of the results presented here in a single framework. Figure 3 was reproduced with his permission. We would like to thank Eline Meys and Ria Riechelmann-Huis for typing the continuously growing and changing manuscript.

References

[1] ANDRÉKA, H. and I. NÉMETI, The generalized completeness of Horn predicate logic as a programming language, *Acta Cybernet.* **4** (1978) 3–10.

[2] APT, K.R. and H.A. BLAIR, Arithmetic classification of perfect models of stratified programs, in: *Proc. 5th Internat. Conf. on Logic Programming* (MIT Press, Cambridge, MA, 1988) 765–779.

[3] APT, K.R., H.A. BLAIR and A. WALKER, Towards a theory of declarative knowledge, in: J. Minker, ed., *Foundations of Deductive Databases and Logic Programming* (Morgan Kaufmann, Los Altos, CA, 1988).

[4] APT, K.R. and M.H. VAN EMDEN, Contributions to the theory of logic programming, *J. ACM* **29** (3) (1982) 841–862.

[5] AQUILANO, C., R. BARBUTI, P. BOCCHETTI and M. MARTELLI, Negation as failure: Completeness of the query evaluation process for Horn clause programs with recursive definitions, *J. Automat. Reason.* **2** (1986) 155–170.

[6] BANCILHON, F. and R. RAMAKRISHNAN, An amateur's introduction to recursive query processing strategies, in: *Proc. ACM Internat. Conf. on Management of Data* (1986) 16–52.

[7] BARBUTI, R. and M. MARTELLI, Completeness of the SLDNF-resolution for a class of logic programs, in: *Proc. 3rd Internat. Conf. on Logic Programming*, Lecture Notes in Computer Science, Vol. 225 (Springer, Berlin, 1986) 600–614.

[8] BELLIA, M. and G. LEVI, The relation between logic and functional languages, a survey, *J. Logic Programming* **3** (1986) 217–236.

[9] BIDOIT, N. and R. HULL, Positivism versus minimalism in deductive databases, in: *Proc. 5th ACM SIGACT-SIGMOD Symp. on Principles of Database Systems* (1986) 123–132.

[10] BIRKHOFF, G., *Lattice Theory*, American Mathematical Society Colloquium Publications, Vol. 25 (1973).

[11] BLAIR, H.A., The recursion-theoretic complexity of predicate logic as a programming language, *Inform. and Control* **54** (1–2) (1982) 25–47.

[12] BLAIR, H.A., Decidability in the Herbrand base, Manuscript, presented at the Workshop on Foundations of Deductive Databases and Logic Programming, Washington, DC, 1986.

[13] BÖRGER, E., Unsolvable decision problems for PROLOG programs, in: E. Börger, ed., *Computation Theory and Logic*, Lecture Notes in Computer Science, Vol. 270 (Springer, Berlin, 1987) 37–48.

[14] BÖRGER, E., Logic as machine: complexity relations between programs and formulae, in: E. Börger, ed., *Trends in Theoretical Computer Science* (Computer Science Press, Rockville, MD, 1988).

[15] BOWEN, K.A. and R.A. KOWALSKI, Amalgamating language and metalanguage in logic programming, in: K.L. Clark and S.-A. Tärnlung, eds., *Logic Programming* (Academic Press, New York, 1982).

[16] BRATKO, I., *PROLOG Programming for Artificial Intelligence* (Addison Wesley, Reading, MA, 1986).

[17] BROUGH, D. and A. WALKER, Some practical properties of logic programming interpreters, in: *Proc. Japan FGCS84 Conf.* (1984) 149–156.

[18] CAMPBELL, J.A., ed., *Implementations of PROLOG* (Ellis Horwood, Chichester, UK, 1984).

[19] CAVEDON, L. and J. LLOYD, A completeness theorem for SLDNF-resolution, *J. Logic Programming* **7**(4) (1989) 177–193.

[20] CHANDRA, A.K. and D. HAREL, Horn clause queries and generalizations, *J. Logic Programming* **2** (1) (1985) 1–15.

[21] CLARK, K.L., Negation as failure, in: H. Gallaire and J. Minker, eds., *Logic and Data Bases* (Plenum Press, New York, 1978) 293–322.

[22] CLARK, K.L., Predicate logic as a computational formalism, Research Report DOC 79/59, Dept. of Computing, Imperial College, London 1979.

[23] CLARK, K.L. and S. GREGORY, A relational language for parallel programming, in: *Proc. ACM Conf. on Functional Programming Languages and Computer Architecture* (1981) 171–178.

[24] CLARK, K.L. and S. GREGORY, PARLOG: A parallel logic programming language, *ACM Trans. on Programming Languages and Systems* **8** (1) (1986) 1–49.

[25] CLARK, K.L. and S.-A. TÄRNLUND, eds., *Logic Programming* (Academic Press, New York, 1982).

[26] COLMERAUER, A., H. KANOUI, P. ROUSSEL and R. PASERO, Un système de communication

homme-machine en Francais, Tech. Report. Groupe de Recherche en Intelligence Artificielle, Univ. d'Aix-Marseille, 1973.

[27] COX, P.T. and T. PIETRZYKOWSKI, Deduction plans: a basis for intelligent backtracking, in: *IEEE PAMI* **3** (1981) 52–65.

[28] DEBRAY, S.K. and P. MISHRA, Denotational and operational semantics for PROLOG, *J. Logic Programming* **5** (1) (1988) 61–91.

[29] DECKER, H., Integrity enforcement in deductive databases, in: *Proc. 1st Internat. Conf. on Expert Database Systems* (1986).

[30] EMDEN, M.H. VAN, Quantitative deduction and its fixpoint theory, *J. Logic Programming* **3** (1) (1986) 37–53.

[31] EMDEN, M.H. VAN and R.A. KOWALSKI, The semantics of predicate logic as a programming language, *J. ACM* **23** (4) (1976) 733–742.

[32] EMDEN, M.H. VAN and K. YUKAWA, Logic programming with equations, *J. Logic Programming* **4** (4) (1987) 265–288.

[33] FARIÑAS, L., DEL CERRO, MOLOG: A system that extends PROLOG with modal logic, *New Generation Comput.* **4** (1) (1986) 35–50.

[34] FITTING, M., A deterministic PROLOG fixpoint semantics, *J. Logic Programming* **2** (2) (1985) 111–118.

[35] FITTING, M., A Kripke–Kleene semantics for logic programs, *J. Logic Programming* **2** (4) (1985) 295–312.

[36] FITTING, M., Partial models and logic programming, *Theoret. Comput. Sci.* **48** (1986) 229–255.

[37] FITTING, M., *Computability Theory, Semantics, and Logic Programming* (Oxford Univ. Press, New York, 1987).

[38] GALLAIRE, H., J. MINKER and J.M. NICOLAS, Logic and databases: a deductive approach, *ACM Comput. Surveys* **16** (2) (1984) 153–186.

[39] GALLIER, J., *Logic for Computer Science* (Harper & Row, New York, 1986).

[40] GALLIER, and S. RAATZ, A graph-based interpreter for general Horn clauses, *J. Logic Programming* **4** (2) (1987) 119–156.

[41] GALLIER, J. and S. RAATZ, Extending SLD-resolution to equational Horn clauses using E-unification, *J. Logic Programming* **6** (1) (1988) 3–44.

[42] GOGUEN, J.A. and J. MESEGUER, Equality, types, modules and (why not?) generics for logic programming, *J. Logic Programming* **1** (2) (1984) 179–210.

[43] GROOT, D. DE and G. LINDSTROM, eds., *Logic Programming, Functions, Relations and Equations* (Prentice-Hall, Englewood Cliffs, NJ, 1986).

[44] HENSCHEN, L. and H.S. PARK, Compiling the GCWA in indefinite deductive databases, in: J. Minker, ed., *Foundations of Deductive Databases and Logic Programming* (Morgan Kaufmann, Los Altos, CA, 1988).

[45] HERBRAND, J. in: W.D. Goldfarb, ed., *Logical Writings* (Reidel, Dordrecht, 1971).

[46] HILL, R., LUSH-resolution and its completeness, DCL Memo 78, Dept. of Artificial Intelligence, Univ. of Edinburgh, 1974.

[47] ITAI, and J.A. MAKOWSKY, Unification as a complexity measure for logic programming, *J. Logic Programming* **4** (2) (1987) 105–118.

[48] JAFFAR, J., J.-L. LASSEZ and J.W. LLOYD, Completeness of the negation as failure rule, in: *Proc. IJCAI'83* (1983) 500–506.

[49] JAFFAR, J. and P.J. STUCKEY, Canonical logic programs, *J. Logic Programming* **3** (2) (1986) 143–155.

[49a] JONES, N.D. and A. MYCROFT, Stepwise development of operational and denotational semantics for PROLOG, in: *Proc. Internat. Symp. on Logic Programming* (1984) 289–298.

[50] Journal of Logic Programming **4** (1986) *Special Issue on Natural Language and Logic Programming* (MCCORD, M.C., V. DAHL and H. ABRAMSON, guest editors).

[51] KANELLAKIS, P., Elements of relational database theory, in: J. van Leewen, ed., *Handbook of Theoretical Computer Science, Vol. B* (North-Holland, Amsterdam, 1990).

[52] KLOP, J.W. and J.J. CH. MEYER, Toegepaste logica: resolutie logica en epistemische logica, Course Notes, Free University Amsterdam, 1987 in Dutch.

[53] KOWALSKI, R.A., Predicate logic as a programming language, in: *Proc. IFIP'74* (North-Holland, Amsterdam, 1974) 569–574.

[54] KOWALSKI, R.A., *Logic for Problem Solving* (North-Holland, New York, 1979).

[55] KOWALSKI, R.A., Algorithm = logic + control, *Comm. ACM* **22** (7) (1979) 424–435.

[56] KOWALSKI, R.A., The relation between logic programming and logic specification, in: C.A.R. Hoare and J.C. Shepherdson, eds., *Mathematical Logic and Programming Languages* (Prentice-Hall, Englewood Cliffs, NJ, 1985) 11–27.

[57] KOWALSKI, R.A. and D. KUEHNER, Linear resolution with selection function, *Artificial Intelligence* **2** (1971) 227–260.

[58] KUNEN, K., Negation in logic programming, *J. Logic Programming* **4** (4) (1987) 289–308.

[59] KUNEN, K., Answer sets and negation as failure, in: *Proc. 4th Internat. Conf. on Logic Programming* (MIT Press, Cambridge, MA, 1987) 219–228.

[60] KUNEN, K., Signed data dependencies in logic programs, *J. Logic Programming* **7** (4) (1989) 231–245.

[61] LASSEZ, J.-L. and M.J. MAHER, Closures and fairness in the semantics of programming logic, *Theoret. Comput. Sci* **29** (1984) 167–184.

[62] LASSEZ, J.L., M.J. MAHER and K. MARRIOTT, Unification revisited, in: J. Minker, ed., *Foundations of Deductive Databases and Logic Programming* (Morga Kaufmann, Los Altos, CA, 1988).

[63] LIFSCHITZ, V., On the declarative semantics of logic programs with negation, in: J. Minker, ed., *Foundations of Deductive Databases and Logic Programming* (Morgan Kaufmann, Los Altos, CA, 1988).

[64] LLOYD, J.W., *Foundations of Logic Programming* (Springer, Berlin, 1984).

[65] LLOYD, J.W., *Foundations of Logic Programming* (Springer, Berlin, 2nd ed., 1987).

[66] LLOYD, J.W. and J.C. SHEPHERDSON, Partial evaluation in logic programming, Tech. Report CS-87-09, Dept. of Computer Science, Univ. of Bristol, 1987.

[67] LLOYD, J.W., E.A. SONENBERG and R.W. TOPOR, Integrity constraint checking in stratified databases, *J. Logic Programming* **4** (4) (1987) 331–345.

[68] LLOYD, J.W. and R. TOPOR, A basis for deductive databases II, *J. Logic Programming* **3** (1) (1986) 55–67.

[69] MAHER, M., Equivalences of logic programs, in: J. Minker, ed., *Foundations of Deductive Databases and Logic Programming* (Morgan Kaufmann, Los Altos, CA, 1988).

[70] MANIN, Y.I., *A Course in Mathematical Logic* (Springer, New York, 1977).

[71] MARTELLI, A. and U. MONTANARI, An efficient unification algorithm, *ACM Trans. on Programming Languages and Systems* **4** (2) (1982) 258–282.

[72] MENDELSON, E., *Introduction to Mathematical Logic* (Van Nostrand, Princeton, NJ, 2nd ed., 1979).

[73] MINKER, J., On indefinite databases and the closed world assumption, in: D.W. Loveland, ed., *Proc. 6th Conf. on Automated Deduction* Lecture Notes in Computer Science, Vol. 138 (Springer, Berlin, 1982) 292–307.

[74] MINKER, J., Perspectives in deductive databases, *J. Logic Programming* **5** (1) (1988) 33–60.

[75] MYCROFT, A., Logic programs and many-valued logic, in: *Proc. of Symp. on Theoretical Aspects of Computer Science*, Lecture Notes in Computer Science, Vol. 166 (Springer, Berlin, 1984) 274–286.

[76] NAISH, L., *Negation and Control in PROLOG*, Lecture Notes in Computer Science, Vol. 238 (Springer, Berlin, 1986).

[77] NAQVI, S.A., A logic for negation in database systems, in: J. Minker, ed., *Foundations of Deductive Databases and Logic Programming* (Morgan Kaufmann, Los Altos, CA, 1988).

[78] NAUGHTON, J.F. and Y. SAGIV, A decidable class of bounded recursions, in: *Proc. 6th ACM SIGACT-SIGMOD Symp. on Principles of Database Systems* (1987) 227–237.

[79] PATERSON, M.S. and M.N. WEGMAN, Linear unification, *J. Comput. System Sci.* **16** (2) (1978) 158–167.

[80] POOLE, D.L., Default reasoning and diagnosis on theory formation, Tech. Report 86–08, Dept. of Computer Science, Univ. of Waterloo, Waterloo, 1986.

[81] PRZYMUSIŃSKI, T., On the semantics of stratified databases, in: J. Minker, ed., *Foundations of Deductive Databases and Logic Programming* (Morgan Kaufmann, Los Altos, CA, 1988).

[82] PRZYMUSIŃSKI, T., Non-monotonic reasoning vs logic programming: a new perspective, in: Y. Wilks and D. Partridge, eds. *Handbook on the Formal Foundations of A.I.* (Cambridge University Press, Cambridge, in press).

[83] REITER, R., On closed world data bases, in: H. Gallaire and J. Minker, eds., *Logic and Data Bases* (Plenum Press, New York, 1978) 55–76.

[84] ROBINSON, J.A., A machine-oriented logic based on the resolution principle, *J. ACM* **12** (1) (1965) 23–41.

[85] ROBINSON, J.A. and E.E. SIEBERT, LOGLISP: motivation, design and implementation, in: K.L. Clark and S.-A. Tärnlund, eds., *Logic Programming* (Academic Press, New York, 1982) 299–313.

[86] SEBELIK, J. and P. STEPANEK, Horn clause programs for recursive functions, in: K.L. Clark and S.-A. Tärnlund, eds., *Logic Programming* (Academic Press, New York, 1982) 324–340.

[87] SHAPIRO, E.Y., A subset of concurrent PROLOG and its interpreter, Tech. Report TR-003, ICOT, Tokyo, 1983.

[88] SHEPHERDSON, J.C., Negation as failure: a comparison of Clark's completed data base and Reiter's closed world assumption, *J. Logic Programming* **1** (1) (1984) 51–79.

[89] SHEPHERDSON, J.C., Undecidability of Horn clause logic and pure PROLOG, Unpublished manuscript, 1985.

[90] SHEPHERDSON, J.C., Negation in logic programming, in J. Minker, ed. *Foundations of Deductive Databases and Logic Programming* (Morgan Kaufmann, Los Altos, CA, 1988).

[91] SHMUELI, O., Decidability and expressiveness aspects of logic queries, in: *Proc. 6th ACM SIGACT-SIGMOD Symp. on Principles of Database Systems* (1987) 237–249.

[92] SHOENFIELD, J., *Mathematical Logic* (Addison-Wesley, Reading, MA, 1967).

[93] SIEKMANN, J.H., Unification theory, *J. Symbolic Comput.* **7** (1988) 207–274.

[94] SMULLYAN, R.M., *Theory of Formal Systems*, Annals of Mathematical Studies, Vol. 47 (Princeton Univ. Press, Princeton, NJ, 1961).

[95] SONENBERG, E.A. and R. TOPOR, Logic programs and computable functions, Tech. Report 87/5, Dept. of Computer Science, Univ. of Melbourne, 1987.

[96] STERLING, L. and E.Y. SHAPIRO, *The Art of PROLOG* (MIT Press, Cambridge, MA, 1986).

[97] TARSKI, A., A lattice-theoretical fixpoint theorem and its applications, *Pacific J. Math.* **5** (1955) 285–309.

[98] TÄRNLUND, S.-A., Horn clause computability, *BIT* **17** (2) 215–226.

[99] VAN GELDER, A., Negation as failure using tight derivations for general logic programs, in: J. Minker, ed., *Foundations of Deductive Databases and Logic Programming* (Morgan Kaufman, Los Altos, CA, 1988).

[100] VIEILLE, L., A database-complete proof procedure based on SLD-resolution, in: *Proc. 4th Internat. Conf. on Logic Programming* (1987) 74–103.

[101] WALKER, A., Syllog: an approach to PROLOG for non-programmers, in: M. van Caneghem and P.H.D. Warren, eds., *Logic Programming and its Applications* (Ablex, Norwood, NJ, 1986) 32–49.

[102] WOLFRAM, D., M. MAHER and J.L. LASSEZ, A unified treatment of resolution strategies for logic programs, in: *Proc. 2nd Internat. Conf. on Logic Programming* (1984) 263–276.

CHAPTER 11

Denotational Semantics

Peter D. MOSSES

Computer Science Department, Aarhus University, DK-8000 Aarhus C, Denmark

Contents

1. Introduction 577
2. Syntax 578
3. Semantics 585
4. Domains 589
5. Techniques 596
6. Bibliographical notes 623
References 629

HANDBOOK OF THEORETICAL COMPUTER SCIENCE
Edited by J. van Leeuwen

1. Introduction

In programming linguistics, as in the study of natural languages, "syntax" is distinguished from "semantics". The *syntax* of a programming language is concerned only with the *structure* of programs: whether programs are "legal"; the connections and relations between the symbols and phrases that occur in them. *Semantics* deals with what legal programs *mean*: the "behaviour" they produce when executed by computers.

The topic of this chapter, Denotational Semantics, is a framework for the formal description of programming language semantics. The main idea of Denotational Semantics is that each phrase of the language described is given a *denotation*: a mathematical object that represents the contribution of the phrase to the meaning of any complete program in which it occurs. Moreover, the denotation of each phrase is determined just by the denotations of its subphrases.

Thus Denotational Semantics is concerned with giving mathematical *models* for programming languages. Models are *constructed* from given mathematical entities (functions, numbers, tuples, etc.). This is in contrast to the *axiomatic* approach used in other major frameworks, such as Hoare Logic [10] and Structured Operational Semantics [25].

The primary aim of Denotational Semantics is to allow *canonical definitions* of the meanings of programs. A canonical, denotational definition of a programming language documents the *design* of the language. It also establishes a *standard* for implementations of the language—ensuring that each program gives essentially the same results on all implementations that conform to the standard. A denotational definition does *not* specify the techniques to be used in implementations; it may, however, suggest some, and it has been shown feasible to develop implementations systematically from specifications written using the denotational approach. Finally, a denotational definition provides a basis for reasoning about the *correctness* of programs—either directly, or by means of derived proof rules for correctness assertions.

A further aim of Denotational Semantics is to promote *insight* regarding the concepts underlying programming languages. Such insight might help to guide the design of new (and perhaps "better") programming languages.

Currently, most programming language standards documents attempt to define semantics by means of *informal* explanations. This is in contrast to syntax, where formal grammars are routinely used in standards (in preference to informal explanations). However, experience has shown that informal explanations of semantics, even when they are carefully worded, are usually *incomplete* or *inconsistent* (or both), and open to misinterpretation by implementors. They are also an inadequate basis for reasoning about program correctness, and totally unsuitable for generation of implementations. These inherent defects of informal explanations do not afflict denotational definitions (except when definitions are left unfinished, or when their formal status is weakened by excessive use of informal abbreviations and conventions).

This chapter has two purposes. The first of these is to explain the *formalism* used in Denotational Semantics: *abstract syntax*, *semantic functions*, and *semantic domains*. Section 2 relates concrete syntax and abstract syntax. Section 3 considers the nature of semantic functions, and explains the properties of compositionality and full abstractness. Section 4 summarizes the concepts and notation of semantic domains, referring to Gunter and Scott [18] for a detailed presentation of domain theory.

The second purpose of this chapter is to illustrate the major standard *techniques* that are used in denotational descriptions of programming languages: *environments, stores, continuations*, etc. Section 5 explains the relation between these techniques and some fundamental concepts of programming languages, and uses the techniques to give denotational descriptions of many conventional programming constructs.

The bibliographical notes (Section 6) provide references to some significant works on Denotational Semantics.

The reader is expected to be familiar with the basic notions of discrete mathematics (sets, functions, relations, partial orders) and to be prepared to meet a substantial amount of formal notation. Familiarity with programming languages is an advantage, but not essential.

2. Syntax

As mentioned at the beginning, the *syntax* of a programming language is concerned only with the *structure* of programs: which programs are "legal"; what are the connections and relations between the symbols and phrases that occur in them.

There are several kinds of syntax, which we distinguish below. (Readers who are familiar with the distinction between "concrete syntax" and "abstract syntax" may prefer to skip to Section 2.3.)

2.1. Concrete syntax

Concrete syntax treats a language as a set of *strings* over an alphabet of symbols. Concrete syntax is usually specified by a grammar that gives "productions" for generating strings of symbols, using auxiliary "nonterminal" symbols. So-called "regular" grammars are inadequate for specifying syntax of programming languages: "context-free" grammars are required, at least.

DEFINITION. A *context-free grammar* G is a quadruple (N, T, P, s_0) where N is a finite set of *nonterminal symbols*, T is a finite set of *terminal symbols* (disjoint from N), $P \subseteq N \times (N \cup T)^*$ is a finite set of *productions*, and $s_0 \in N$ is the *start symbol*.

(In this section, X^* is the set of strings over X, for any set X; the empty string is indicated by Λ, and string concatenation by juxtaposition. The notation X^* is given a different interpretation when X is a semantic domain, from Section 4 onwards.)

It is common practice to distinguish a *lexical* level and a *phrase* level in concrete syntax. The terminal symbols in the grammar specifying the lexical level are single characters; those in the phrase-level grammar are the *non*terminal symbols of the lexical grammar. Here, let us ignore the distinction between the lexical and phrase levels, for simplicity.

When presenting a grammar, it is enough to list the productions: the sets of nonterminal and terminal symbols are implicit, the start symbol is determined by the first production. We write a production $(a,(x_1 \dots x_n))$ as $a ::= x_1 \dots x_n$. We may also group several productions for the same nonterminal, separating the alternative strings on the right-hand side by "|". (This notation for grammars is essentially the same as so-called BNF.) For later use, a mnemonic name called a *phrase sort* is associated with each nonterminal symbol (we write the phrase sort in parentheses) and occurrences of nonterminal symbols in right-hand sides of productions may be distinguished by subscripts.

An example of a grammar for the concrete syntax of a simple language of expressions is given in Table 1. (The productions for identifiers are omitted, as they are of no interest.)

Table 1
A grammar for concrete syntax

(EXPRESSION)
$E ::= T | E+T | E-T$
(TERM)
$T ::= F \mid T*F$
(FACTOR)
$F ::= I \mid (E)$
(IDENTIFIER)
$I ::=$ *unspecified*

We could define the language of strings generated by a context-free grammar in terms of "derivation steps". For our purposes here, it is more convenient to go straight to the notion of "derivation trees", in which the order of derivation steps is ignored.

DEFINITION. Let L be a set (of *labels*). An *L-labelled tree* t is a pair $(l,(t_1 \dots t_n))$, where $l \in L, n \geqslant 0$, and $t_1 \dots t_n$ is a string of L-labelled trees. We say that t has *label* l and *branches* $t_1, \dots, t_n$.

Let $\mathbf{Tree}_L$ be the set of *finite* L-labelled trees; this is the *least* set that is closed under construction of L-labelled trees.

(Of course, other representations of trees are possible, e.g., as partial functions from "occurrences" to labels.)

DEFINITION. A *derivation tree* according to a grammar $G=(N,T,P,s_0)$ is a finite $(N \cup T)$-labelled tree with label s_0, such that if a node labelled a has branches labelled $x_1, \dots, x_n$, $n \geqslant 0$, then $a \in N$ and $(a,(x_1 \dots x_n)) \in P$, or $a \in T$ and $n=0$.

The set of all derivation trees according to G is denoted by $\mathbf{Tree}_G$.

For notational convenience we identify the tree $(l,(\Lambda))$ with l. This allows us to write, e.g., $(a,(x_1 t x_2))$ to form a derivation tree, where t is a tree and x_1, x_2 are terminal symbols. Figure 1 depicts a derivation tree according to the grammar of Table 1.

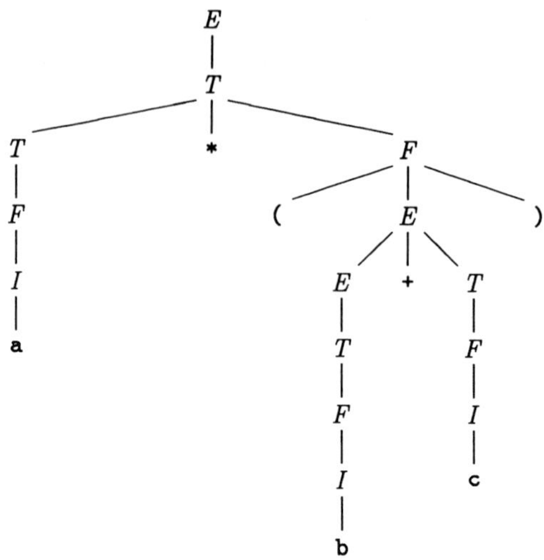

Fig. 1. A derivation tree for concrete syntax.

Now that we know what derivation trees are, let us use them to define the languages generated by grammars:

Definition. For any $t \in \mathbf{Tree}_G$, $\mathbf{frontier}(t) \in T^*$ is defined (inductively) by

$$\mathbf{frontier}(l,(t_1 \ldots t_n)) = \begin{cases} l & \text{if } n=0 \text{ and } l \in T; \\ \mathbf{frontier}(t_1)\ldots\mathbf{frontier}(t_n) & \text{otherwise.} \end{cases}$$

Definition. The *language of strings* $\mathscr{L}(G) \subseteq T^*$ generated by a grammar $G = (N, T, P, s_0)$ is given by

$$\mathscr{L}(G) = \{w \in T^* \mid \exists t \in \mathbf{Tree}_G \colon w = \mathbf{frontier}(t)\}.$$

If any string in $\mathscr{L}(G)$ has more than one derivation tree, then G is said to be *ambiguous*.

Whereas ambiguity seems to be an inescapable feature of natural languages, it is to be avoided in programming languages. For example, there should be no vagueness about whether "a*b+c" is to be read as "a*(b+c)" or as "(a*b)+c", since they should evaluate to different results, in general. (Of course grouping may not matter in some cases, such as "a+b+c".) Moreover, the efficient generation of language parsers from grammars requires special kinds of unambiguous grammars, e.g., satisfying the so-called LALR(1) condition.

Unfortunately, unambiguous grammars tend to be substantially more complex than ambiguous grammars for the same language, and they often require nonterminal symbols and productions that have no relevance to the *essential* phrase structure of the language concerned. For the purposes of semantics, the phrase structure of languages should be as simple as possible, devoid of semantically irrelevant details. Yet there should be no ambiguity in the structure of phrases! Thus we are led to use ambiguous grammars, but to interpret them in such a way that the specified syntactic entities themselves can be unambiguously decomposed. Such a framework is provided by so-called "abstract" syntax.

2.2. *Abstract syntax*

Abstract syntax treats a language as a set of *trees*. The important thing about trees is that, unlike strings, their compositional structure is *inherently* unambiguous: there is only one way of constructing a particular tree out of its (immediate) subtrees.

It is convenient to use derivation trees to represent abstract syntax. Abstract syntax is specified using the same kind of (context-free) grammar that is used for concrete syntax—but now there is no worry about ambiguity. An example of a grammar for abstract syntax is given in Table 2. It gives an appropriate abstract syntax for the language generated by the grammar of Table 1. Notice that the nonterminal symbols T and F, together with the terminal symbols "(" and ")", are not present in the abstract syntax: they were only for grouping in concrete syntax. Also, the various concrete expressions involving operators are collapsed into a single form of abstract expression—at the expense of introducing the nonterminal symbol O. Figure 2 shows a derivation tree according to the grammar of Table 2.

Table 2
A grammar for abstract syntax

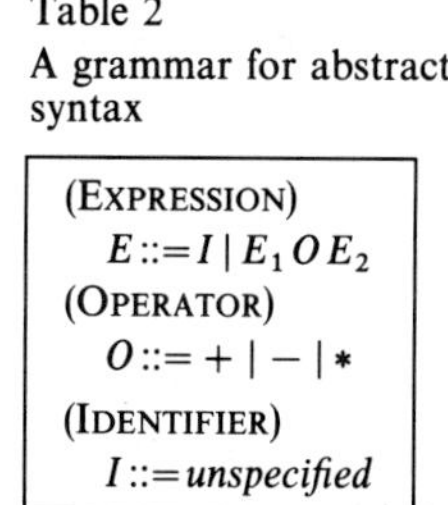

(EXPRESSION)
$E ::= I \mid E_1 \, O \, E_2$
(OPERATOR)
$O ::= + \mid - \mid *$
(IDENTIFIER)
$I ::= \textit{unspecified}$

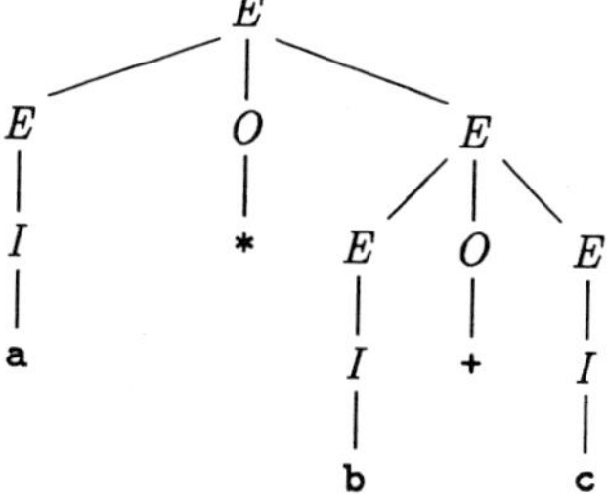

Fig. 2. A derivation tree for abstract syntax.

DEFINITION. The *abstract syntax* defined by a grammar G is $\mathbf{Tree}_G$, the set of derivation trees according to G.

The phrase sorts associated with nonterminal symbols in our grammars (such as EXPRESSION, IDENTIFIER) identify corresponding sets of derivation trees (note that such trees generally occur only as *branches* of trees in $\mathbf{Tree}_G$).

Abstract syntax may be characterized *algebraically*, using the notion of a "signature", as follows.

DEFINITION. Let S be a set (of *sorts*). An *S-sorted signature* Σ is a family of sets $\{\Sigma_{w,s}\}_{w\in S^*, s\in S}$ (of *operators*). A *Σ-algebra* A consists of a family $\{A_s\}_{s\in S}$ of sets (called *carriers*) and for each operator $f\in\Sigma_{s_1\ldots s_n,s}$ a total function $f_A: A_{s_1}\times\cdots\times A_{s_n}\to A_s$.

DEFINITION. A *Σ-homomorphism* $h: A\to B$ (where A and B are Σ-algebras) is a family $\{h_s\}_{s\in S}$ of (total) functions $h_s: A_s\to B_s$ such that for each $f\in\Sigma_{s_1\ldots s_n,s}$ and $a_i\in A_{s_i}$

$$f_B(h_{s_1}(a_1),\ldots,h_{s_n}(a_n))=h_s(f_A(a_1,\ldots,a_n)).$$

The composition $h'\circ h$ of Σ-homomorphisms $h: A\to B$, $h': B\to C$ is the family of functions $\{h'_s\circ h_s\}_{s\in S}$. The identity Σ-homomorphism $\mathbf{id}_A: A\to A$ is the family of identity functions $\{\mathbf{id}_{A_s}\}_{s\in S}$. The Σ-algebras A, B are said to be *isomorphic* when there exist Σ-homomorphisms $h: A\to B$, $h': B\to A$ such that $h'\circ h=\mathbf{id}_A$ and $h\circ h'=\mathbf{id}_B$.

The key concept is that of "initiality":

DEFINITION: A Σ-algebra I is *initial* in a class $\mathbf{C}$ of Σ-algebras iff there is a unique Σ-homomorphism from I to each algebra in $\mathbf{C}$.

2.1. PROPOSITION. *If I and J are both initial in a class $\mathbf{C}$ of Σ-algebras, then I and J are isomorphic.*

PROOF. Let $h: I\to J$, $h': J\to I$ be the unique homomorphisms given by the initiality of I, respectively J. Now $h'\circ h$ and $\mathbf{id}_I$ are both homomorphisms from I to itself; by the initiality of I they must be equal. Similarly, $h\circ h'=\mathbf{id}_J$. $\square$

For each grammar G we define a corresponding signature Σ_G, as follows.

DEFINITION. Let $G=(N, T, P, s_0)$. Then Σ_G is the N-sorted signature with

$$\Sigma_{G,s_1\ldots s_n,s}=\{p\in P\mid p=(s,(u_0s_1\ldots s_nu_n));\ u_0,\ldots,u_n\in T^*\}$$

for each $(s_1\ldots s_n)\in N^*$ and $s\in N$.

By the way, not all signatures can be made into context-free grammars: a signature may have an infinite number of sorts and operators. Notice also that a signature does not have a distinguished "start sort".

Now $\mathbf{Tree}_G$ can be made into a Σ_G-algebra, which we denote by $\mathscr{A}(G)$, as follows. Take the carriers $\mathscr{A}(G)_s$ to be $\mathscr{A}(N, T, P, s)$ for each $s \in N$. (In practice it is convenient to refer to these sets by mnemonic names, associated with nonterminal symbols when grammars are specified.) For each $p \in \Sigma_{G,s_1 \ldots s_n,s}$ with $p=(s,(u_0 s_1 \ldots s_n u_n))$, where $u_1, \ldots, u_n \in T^*$, define a function

$$p_{\mathscr{A}(G)}: \mathscr{A}(G)_{s_1} \times \cdots \times \mathscr{A}(G)_{s_n} \to \mathscr{A}(G)_s$$

by letting, for all $t_i \in \mathscr{A}(G)_{s_i}$, for $i=1, \ldots, n$,

$$p_{\mathscr{A}(G)}(t_1, \ldots, t_n) = (s, (u_0 t_1 \ldots t_n u_n)).$$

2.2. PROPOSITION. *$\mathscr{A}(G)$ is initial in the class of all Σ_G-algebras.*

PROOF. Let A be any Σ_G-algebra. Define $\{h_s: \mathscr{A}(G)_s \to A_s\}_{s \in S}$ inductively, as follows: If $t=(s,(u_0 t_1 \ldots t_n u_n))$ with each $u_i \in T^*$ and each $t_i \in \mathscr{A}(G)_{s_i}$, and $p=(s,(u_0 s_1 \ldots s_n u_n))$, then

$$h_s(t) = p_A(h_{s_1}(t_1), \ldots, h_{s_n}(t_n)).$$

As each $t \in \mathscr{A}(G)_s$ is uniquely decomposable as $(s,(u_0 t_1 \ldots t_n u_n))$, the h_s are well-defined (and they are total since derivation trees are finite). Moreover, it can be seen that this definition is forced by the homomorphic property, so $\{h_s\}_{s \in S}$ is the unique homomorphism from $\mathscr{A}(G)$ to A. Hence $\mathscr{A}(G)$ is initial in the class of all Σ_G-algebras. □

Denotational Semantics defines the semantics of a programming language on the basis of its abstract syntax. The semantics of some concrete syntax may be obtained as well: by giving a function that maps concrete syntax into abstract syntax. Assuming that the concrete grammar is unambiguous, it is enough to map concrete derivation trees into abstract derivation trees. (This map might be neither 1-1 nor onto, in general. Trying to invert it is called "pretty-printing", or "unparsing".)

The specification of the function from concrete to abstract syntax is quite trivial if the grammar for abstract syntax is obtained systematically from that for concrete syntax, just by "unifying" nonterminals and eliminating "chain productions". In fact, the grammar of Table 2 was obtained from that of Table 1 (mainly) in that way; the corresponding map from concrete to abstract syntax may be imagined from the example where the tree in Fig. 1 is mapped to that in Fig. 2.

In general, it is up to the semanticist to *choose* an appropriate abstract syntax for a given language. Different choices may influence the difficulty of specifying the semantics. For instance, consider the rather trivial "language" of binary numerals, with concrete syntax given by the grammar in Table 3. Of course the semantics of binary numerals can be specified for the given syntax; but it turns out (as shown in Section 3) to

Table 3
Concrete syntax for binary numerals

(BINARY-NUMERAL)
$B ::= 0 \mid 1 \mid 0B \mid 1B$

Table 4
Abstract syntax for binary numerals

(BINARY-NUMERAL)
$B ::= 0 \mid 1 \mid B0 \mid B1$

be significantly simpler when the abstract syntax is given by the grammar in Table 4. (This latter grammar is unambiguous, so it could be used for concrete syntax as well as abstract syntax. But in general, the grammars used for abstract syntax are highly ambiguous, e.g., as for expressions in Table 2.)

There do not seem to be any hard and fast rules for choosing grammars for abstract syntax. Usually, one has to compromise between, on the one hand, keeping close to a given grammar for concrete syntax and, on the other hand, facilitating the semantic description.

Note that it is *not* required that the frontiers of the trees generated by the abstract grammar are the strings generated by the given concrete grammar, nor even that the same terminal symbols are used. In fact, some authors prefer to use *disjoint* sets of symbols in concrete and abstract grammars, to avoid altogether any chance of confusion between concrete and abstract syntax. Here, we take the opposite position, and use symbols that make our grammars for abstract syntax strongly suggestive of familiar concrete syntax.

2.3. *Context-sensitive syntax*

The grammars used here for specifying abstract syntax are context-free. But it is well-known that several features of programming languages are *context-sensitive*, and cannot be described by context-free grammars (e.g., that identifiers be declared before they are referred to, and that the "types" of operands match their operators).

In Denotational Semantics, context-sensitive syntax is regarded as a part of semantics, called *static semantics* (because it depends only on the program text, not on the input). For simplicity, let us assume that the static semantics of a program is just a truth value indicating the legality of the program. Then the rest of the semantics of programs can be specified independently of their static semantics—the semantics of programs that are not legal (according to the static semantics) is defined, but irrelevant.

In practice, a proper treatment of static semantics might involve specification of error messages. Also, it may be convenient for a static semantics to yield abstract syntax that reflects context-sensitive disambiguations (for instance, whether occurrences of "+" are arithmetical, or set union), and to define the rest of the semantics on the disambiguated abstract syntax.

Static semantics is not considered further in this chapter. For a study of the semantics of types, see [26].

So much for syntax.

3. Semantics

Consider an entire program in some programming language. What is the nature of its semantics?

First of all let us dismiss any effects that the program might have on human readers, e.g., evoking feelings of admiration or (perhaps more often) disgust. In contrast to philology, programming linguistics is not concerned with subjective qualities at all. The semantics of a program is dependent only on the objective *behaviour* that the program causes (directly) when executed by computers.

Now computers are complex mechanisms, and all kinds of things can be observed to happen when they execute programs: lights flash, disk heads move, electric currents flow in circuits, characters appear on screens or on paper, etc. For programs that are specifically intended to control such physical behaviour, it would be necessary to consider these phenomena in their semantics.

But here, let us restrict our attention to programs whose behaviour is intended to be independent of particular computers. Such programs are typically written in general, high-level programming languages that actually deny the programmer direct control over the details of physical behaviour. The appropriate semantics of these programs is *implementation-independent*, consisting of just those features of program execution that are common to all implementations.

The implementation-independent semantics of a program may typically be modelled mathematically as a *function* (or relation) between *inputs* and *outputs*—where an input or output item might be just a number. The concrete representation of input and output as strings of bits is (usually) implementation-dependent, and hence ignored; likewise the length of time taken for program execution. But *termination* properties are generally implementation-*in*dependent, and should therefore be taken into account in semantics.

Thus the semantics of a program is a mathematical object that models the program's implementation-independent behaviour. The semantics of a programming language consists of the semantics of all its programs.

Actually, some details of semantics are often left *implementation-defined*, e.g., limits on the size of numbers, maximum depth of recursive activation. These are regarded as *parameters* of the semantics; when such parameters are supplied, the implementation-independent semantics of a particular subclass of implementations is obtained.

A *standard* for implementations of a programming language may be established by

(i) specifying the semantics of all programs in the language; and

(ii) specifying a "conformance" relation between semantic objects and implementation behaviours.

Our concern in this chapter is with (i), but let us disgress for a moment to indicate how (ii) might be done.

Assume a correspondence between the inputs and outputs in the semantic model and some physical objects processed by implementations. Let a program and its input be given. If these uniquely determine output (and termination properties), then a conforming implementation, when given the physical representations of the program and input, must produce a representation of the output—computing "for ever" if the semantics specifies non-termination. If, however, there are several possible outputs for a given program and input—i.e., the program is nondeterministic—the implementation need only produce one of them (perhaps not terminating if that is a possibility); the implementation may or may not be nondeterministic itself.

3.1. Denotations

Now back to our main concern: specifying the semantics of programs. The characteristic feature of Denotational Semantics is that one gives semantic objects for *all* phrases—not only for complete programs. The semantic object specified for a phrase is called the *denotation* of the phrase. The idea is that the denotation of each phrase represents the contribution of that phrase to the semantics of any complete program in which it may occur.

The denotations of compound phrases must depend only on the denotations of their subphrases. (Of course, the denotations of basic phrases do not depend on anything.) This is called *compositionality*.

It should be noted that the semantic analyst is free to *choose* the denotations of phrases—subject to compositionality. Sometimes there is a "natural", optimal choice, where phrases have the same denotations whenever they are *interchangeable* (without altering behaviour) in all *complete programs*; then the denotations are called *fully abstract*, and they capture just the "essential" semantics of phrases.

Note that considering interchangeability only in *complete programs* lets the notion of full abstractness refer directly to the *behaviours* of programs, rather than to their denotations. Different choices of which phrases are regarded as complete programs may give different conclusions concerning whether full abstractness obtains.

It is not always easy (or even possible) to find and specify fully abstract denotations, so in practice a compromise is made between simplicity and abstractness.

As an introductory (and quite trivial) example take the binary numerals. An abstract syntax for binary numerals was suggested in Section 2. Now let us extend the syntax to allow "programs" consisting of *signed* binary numerals, see Table 5.

The meanings (i.e., "behaviours") of signed binary numerals are supposed to be integers in **Z**, according to the usual interpretation of binary notation (i.e., the most significant bit is the leftmost), negated if preceded by "$-$". We are free to choose the denotations for unsigned binary numerals B. The natural choice is to let each B denote the obvious natural number in **N**, and such denotations (specified formally in Section 3.2) are indeed fully abstract.

Table 5
Abstract syntax for signed binary numerals

(SIGNED-BINARY-NUMERAL)
$Z ::= B \mid -B$
(BINARY-NUMERAL)
$B ::= \mathsf{0} \mid \mathsf{1} \mid B\mathsf{0} \mid B\mathsf{1}$

Any other choice of denotations is perhaps rather contrived in this simple example, but let us consider an alternative possibility so as to illustrate lack of full abstractness.

We could choose the denotation of B to be a pair $(n, l) \in \mathbf{N}^2$, where n gives the numerical value of B and l gives its *length*. Then the denotation of a signed binary numeral "B" or "$-B$" would be determined just by n. Such denotations can be defined compositionally, but they are not fully abstract: for instance, the phrases "0" and "00" get distinct denotations, yet they can always be interchanged in any signed binary numeral without affecting its meaning.

Now consider the original (concrete) grammar for unsigned binary numerals (Table 4) and regard it as a specification of abstract syntax. With this phrase structure, we are no longer able to take the denotation of B to be just its numerical value: the value of the phrase "1 B" is determined not only by the numerical value of B, but also by the number of its leading zeros. In fact the $(n, l) \in \mathbf{N}^2$ denotations mentioned above turn out to be fully abstract for *this* syntax.

The above example shows that the property of full abstractness can be rather sensitive to the structure of abstract syntax—and thereby casts doubt on its appropriateness as an absolute criterion of the quality of denotational descriptions.

In Denotational Semantics, there is in general a sharp distinction between syntax and semantics, and denotations consist of mathematical objects (such as numbers and functions) that exist completely independently of programming languages. In particular, denotations do not usually incorporate program phrases as components. In fact, it would not conflict with compositionality to let phrases denote even themselves, but such "denotations" tend to have (extremely) poor abstractness.

There are two cases, however, when it *is* desirable to use phrases as denotations:

- *identifiers* usually have to be their own denotations (e.g., in declarations); and
- for languages like LISP where *phrases* can be *computed* the denotation of a phrase essentially corresponds to its abstract syntax (and the benefits of the denotational approach are then questionable since semantic equivalence is just syntactic identity).

3.2. *Semantic functions*

Semantic functions map phrases (of abstract syntax) to their actual denotations. The semantics of a programming language may be specified by defining a semantic function for each sort of phrase.

Recall that abstract syntactic entities have an unambiguous structure. Hence semantic functions may be defined *inductively* by specifying, for each syntactic construct, its denotation in terms of the denotations of its components (if there are any). The conventional way of writing such an inductive definition in Denotational Semantics is as a set of so-called *semantic equations*, with (in general) one semantic equation for each production of the abstract syntax.

Let a, $a_1, \ldots, a_n$ be (possibly subscripted) nonterminal symbols, with associated phrase sorts $s, s_1, \ldots, s_n$. Let $\mathscr{F}_s, \mathscr{F}_{s_1}, \ldots, \mathscr{F}_{s_n}$ be semantic functions mapping phrases of sort $s_{(i)}$ to their denotations (in practice, the semantic functions are usually given mnemonic names when they are introduced). Then the semantic equation for the production "a ::= $u_0 a_1 \dots a_n u_n$" is of the form

$$\mathscr{F}_s[\![u_0 a_1 \dots a_n u_n]\!] = f(\mathscr{F}_{s_1}[\![a_1]\!], \dots, \mathscr{F}_{s_n}[\![a_n]\!]).$$

The way that the denotations of the phrases $a_1, \dots, a_n$ are combined is expressed using whatever notation is available for specifying particular objects—determining a function, written f above.

Note that the emphatic brackets $[\![\ ]\!]$ separate the realm of syntax from that of semantics, which avoids confusion when programming languages contain the same mathematical notations as are used for expressing denotations.

To illustrate the form of semantic equations, let us specify denotations for signed binary numerals (with the abstract syntax given in Table 5). We take for granted the ordinary mathematical notation (0, 1, 2, +, −, ×) for specifying particular integers in **Z** and natural numbers in **N**. The semantic functions ($\mathscr{Z}$ for signed binary numerals, $\mathscr{B}$ for unsigned binary numerals) are defined inductively by the semantic equations given in Table 6.

Table 6
Denotations for signed binary numerals

$\mathscr{Z}$: SIGNED-BINARY-NUMERAL → **Z**
$\mathscr{Z}[\![B]\!] = \mathscr{B}[\![B]\!]$
$\mathscr{Z}[\![-B]\!] = -\mathscr{B}[\![B]\!]$

$\mathscr{B}$: BINARY-NUMERAL → **N**
$\mathscr{B}[\![0]\!] = 0$
$\mathscr{B}[\![1]\!] = 1$
$\mathscr{B}[\![B0]\!] = 2 \times (\mathscr{B}[\![B]\!])$
$\mathscr{B}[\![B1]\!] = (2 \times (\mathscr{B}[\![B]\!])) + 1$

Perhaps the standard interpretation of binary notation is so much taken for granted that we may seem to be merely "stating the obvious" in the semantic equations. But we could just as well have specified alternative interpretations, e.g., by reversing the rôles of "0" and "1", or by making the rightmost bit the most significant.

In effect, the semantic equations *reduce* the semantics of the language described (here, the binary numerals) to that of a "known" language (here, that of ordinary arithmetic). This reduction may also be viewed as a "syntax-directed translation", although it is then essential to bear in mind that phrases are semantically equivalent whenever they are translated to notation that as the same *interpretation*, not merely the same *form*.

An alternative way of specifying semantic functions is to exploit the formulation of abstract syntax as an initial algebra, discussed in Section 2. Recall that the abstract syntax $\mathscr{A}(G)$ specified by a grammar G is a Σ_G-algebra, where Σ_G is the signature corresponding to the productions of G. As $\mathscr{A}(G)$ is the initial Σ_G-algebra (Proposition 2.2), there is a unique Σ_G-homomorphism from $\mathscr{A}(G)$ to any other Σ_G-algebra. So all that is needed is to make the spaces of denotations into a "target" Σ_G-algebra, say D, by defining a function p_G for each $p \in \Sigma_G$, i.e., for each production p of G. Then the semantic functions are given as the components of the unique Σ_G-homomorphism from $\mathscr{A}(G)$ to D.

This approach is known as *Initial Algebra Semantics*. Whereas such an explicit algebraic formulation can be convenient for some purposes, the approach is essentially the same as Denotational Semantics, and it is a simple matter to transform semantic equations into specifications of target algebras—or vice versa—while preserving the semantic functions that are thereby defined.

3.3. Notational conventions

Some abbreviatory techniques are commonly used in semantic equations:

(a) The semantics of a construct may be specified in terms of that of a compound phrase, provided no circularity is introduced into the inductive definition. For instance, we might specify

$$\mathscr{S}[\![\mathsf{if}\ E\ \mathsf{then}\ S]\!] = \mathscr{S}[\![\mathsf{if}\ E\ \mathsf{then}\ S\ \mathsf{else\ skip}]\!]$$

where $\mathscr{S}[\![\mathsf{if}\ E\ \mathsf{then}\ S_1\ \mathsf{else}\ S_2]\!]$ is specified by an ordinary semantic equation. As well as abbreviating the right-hand sides of semantic equations, the use of this technique emphasizes that the syntactic construct is just "syntactic sugar" and does not add anything of (semantic) interest to the language.

(b) There may be several semantic functions for a single phrase sort, say $\mathscr{F}^i : P \to D_i$. This corresponds to a single function $\mathscr{F} : P \to (D_1 \times \cdots \times D_n)$, with the components of denotations being defined separately.

(c) The names of semantic functions may be omitted (when there is no possibility of confusion). In particular, when identifiers are essentially their own denotations, their (injective) semantic function is generally omitted.

These abbreviations have been found to increase the readability of denotational descriptions without jeopardizing their formality.

4. Domains

Appropriate mathematical spaces for the denotations of programming constructs are called (*semantic*) *domains*. *Here*, after a brief introduction to the basic concept of a domain, a summary is given of the notation used for specifying domains and their elements. A thorough explanation of the notation, together with the *theory* of domains, is given by Gunter and Scott [18]. The main *techniques* for choosing domains for use in denotational descriptions of programming languages are demonstrated in Section 5.

4.1. Domain structure

Domains are sets whose elements are partially ordered according to their degree of "definedness". When x is less defined than y in some domain D, we write $x \sqsubseteq_D y$ and say that x *approximates* y in D. (Mention of the domain concerned may be omitted when it is clear from the context.) Every domain D is assumed to have a *least element* $\perp_D$, representing "undefinedness"; moreover there are *limits* $\bigsqcup_n x_n$ for all (countable) increasing sequences $x_0 \sqsubseteq x_1 \sqsubseteq \cdots \sqsubseteq x_n \sqsubseteq \cdots$ (Thus domains are so-called (ω-)*cpos*. Further conditions on domains are imposed by Gunter and Scott [18]; but these conditions need not concern us here, as their primary purpose is to ensure that the class of domains is closed under various constructions.)

For an example, consider the set of *partial functions* from **N** to **N** and, for partial functions f, g, let $f \sqsubseteq g$ iff $\mathbf{graph}(f) \subseteq \mathbf{graph}(g)$. This gives a domain: $\sqsubseteq$ is a partial order corresponding to definedness; the least element $\bot$ is the empty function (but every total function is maximal, so there is no greatest element); and the limit of any increasing sequence of functions is given by the union of the graphs of the functions.

A domain D may be defined simply by specifying its elements and it approximation relation $\sqsubseteq$, as above. But it is tedious to check each time that $\sqsubseteq$ has the required properties—and to define ad hoc notation for identifying elements.

In practice, the domains used in denotational descriptions are generally defined as solutions (up to isomorphism) of *domain equations* involving the standard primitive domains and standard domain constructions. Not only does this ensure that the defined structures really are domains, it also provides us with standard notation for their elements.

The standard *primitive domains* are obtained merely by adding $\bot$ to an unordered (but at most countable) set, of course letting $\bot \sqsubseteq x$ for all x. Domains with such a trivial structure are called *flat*. For example the domain **T** of truth values is obtained by adding $\bot$ to the set {**true**, **false**}, and the domain $\mathbf{N}_\bot$ of natural numbers by adding $\bot$ to **N**.

There are standard domain constructions that correspond closely to well-known set constructions: Cartesian product, disjoint union, function space, and powersets. Of course these domain constructions have to take account of the $\sqsubseteq$ relation, as well as the elements; this leads to several possibilities.

Before proceeding to the details of the standard domain notation, let us consider what *functions* between domains are required. The functions generally needed for the semantics of programming languages are *monotone*, in that they preserve the relation $\sqsubseteq$:

$$x \sqsubseteq y \text{ implies } f(x) \sqsubseteq f(y);$$

and *continuous*, in that they also preserve limits of increasing sequences:

$$x_0 \sqsubseteq x_1 \sqsubseteq \cdots \sqsubseteq x_n \cdots \text{ implies } f\Big(\bigsqcup_n x_n\Big) = \bigsqcup_n f(x_n).$$

Note that we do *not* insist that functions preserve least elements. Those functions f that do satisfy

$$f(\bot) = \bot$$

are called *strict*. Constant functions are nonstrict functions (in general).

Partial functions from **N** to **N** are represented by strict total functions from $\mathbf{N}_\bot$ to $\mathbf{N}_\bot$, the result $\bot$ corresponding to "undefined". Notice, by the way that *all* strict functions on $\mathbf{N}_\perp$ are continuous—but, in practice, we only make use of those that are computable in the usual sense.

The importance of continuity is two-fold:

(i) Let D be a domain. For any continuous function $f: D \to D$ there is at least $x \in D$ such that

$$x = f(x).$$

This x is called the *least fixed point* of f, written $\mathbf{fix}(f)$. It is given by $\bigsqcup_n f^n(\perp)$. Nonmonotone functions on domains need not have fixed points at all; whether monotone (but noncontinuous) functions on domains always have least fixed points depends on the precise structure of domains, but in any case their fixed points are not necessarily obtainable as the limits of countable increasing sequences.

(ii) There exists a nontrivial domain $\mathbf{D}_\infty$ such that

$$\mathbf{D}_\infty \cong \mathbf{D}_\infty \to \mathbf{D}_\infty$$

provided that $\mathbf{D}_\infty \to \mathbf{D}_\infty$ is just the space of *continuous* functions on $\mathbf{D}_\infty$. Domains such as $\mathbf{D}_\infty$ that "contain" their own (continuous) function space are called *reflexive*. If $\mathbf{D}_\infty \to \mathbf{D}_\infty$ were to be the space of *all* functions, $\mathbf{D}_\infty$ would *have* to be the trivial (one-point) domain, by Cantor's Theorem.

Least fixed points of continuous functions provide appropriate denotations for iterative and recursive programming constructs. Reflexive domains are needed for the denotations of constructs that may involve "self-application": procedures with procedure parameters in ALGOL60, functions with dynamic bindings in LISP, assignments of procedures to variables in C, etc. Even when self-application is forbidden (e.g., by type constraints), it may still be simpler to specify denotations as elements of reflexive domains, rather than to introduce infinite families of nonreflexive domains.

The structure of domains described above is further motivated by the fact that domains with continuous functions provide denotations for almost all useful programming constructs. (The only exception seems to be constructs that involve so-called "unbounded nondeterminism", corresponding to infinite sets of implementation-dependent choices.)

4.2. Domain notation

Now for a summary of the notation for specifying domains and their elements, following Gunter and Scott [18]. The (abstract) syntax of the notation is given in Table 7. Some conventions for disambiguating the written representation of the notation, together with some abbreviations for commonly-occurring patterns of notation, are given in Section 4.3.

Let us start with *domain expressions, d*. These may include references to domain *variables w*, whose interpretation is supplied by the context of the domain expression. This context is generally a set of *domain equations* of the form

$$w_1 = d_1, \ldots, w_n = d_n$$

Table 7
Notation for domains and elements

(DOMAIN-EXPRESSIONS)

$$\begin{aligned} d ::= {} & w \mid \mathbf{I} \mid \mathbf{O} \mid \mathbf{T} \mid \mathbf{N}_\perp \mid \\ & d_1 \to d_2 \mid d_1 \circ\!\!\to d_2 \mid \\ & d_1 \times \cdots \times d_n \mid d_1 \otimes \cdots \otimes d_n \mid \\ & d_1 + \cdots + d_n \mid d_1 \oplus \cdots \oplus d_n \mid \\ & d_\perp \mid d^* \mid d^\infty \mid d^\natural \end{aligned}$$

(DOMAIN-VARIABLES)

$w ::= $ *arbitrary symbols*

(EXPRESSIONS)

$$\begin{aligned} e ::= {} & x \mid \perp_d \mid \top \mid \mathbf{true} \mid \mathbf{false} \mid \\ & e_1 =_d e_2 \mid \mathbf{if}\ e_1\ \mathbf{then}\ e_2\ \mathbf{else}\ e_3 \mid 0 \mid \mathbf{succ} \mid \\ & \lambda x \in d.e \mid e_1 e_2 \mid \mathbf{id}_d \mid e_1 \circ e_2 \mid \mathbf{fix}_d \mid \mathbf{strict}_d \mid \\ & (e_1, \ldots, e_n) \mid \langle e_1, \ldots, e_n \rangle \mid \mathbf{on}_i^d \mid \mathbf{smash}_d \mid \\ & [e_1, \ldots, e_n] \mid \mathbf{in}_i^d \mid \mathbf{up}_d \mid \mathbf{down}_d \mid \\ & \{\!| e |\!\} \mid e_1 \uplus e_2 \mid e^\natural \mid \mathbf{ext}_d \end{aligned}$$

(VARIABLES)

$x ::= $ *arbitrary symbols*

where the w_i are distinct and no other variables occur in the d_i. As is shown by Gunter and Scott [18], there is always a "minimal" solution to such a set of equations (up to isomorphism). We need not worry here about the construction of the solution—the equations themselves express all that we really need to know about the defined domains. Note, however, that the simple equation

$$D = D \to D$$

defines D to be the trivial (one-point) domain! Most domain equations that arise in practice do not admit trivial solutions. (Gunter and Scott [18] show how to force non-trivial solutions to $D = D \to D$.)

Element expressions e may include references to element variables x whose domain and interpretation is supplied by the context. Usually the context is just the enclosing (element) expression, but we also allow *auxiliary* definitions of the form

$$x = e \in d.$$

The scope of an auxiliary definition is the entire specification. (Mutually dependent auxiliary definitions may be regarded as abbreviations for independent definitions involving the least fixed point operator.)

Basic domains

- $\perp_d$ denotes the least element of a domain d.
- $\mathbf{I}$ is the one-point domain, consisting only of $\perp_\mathbf{I}$.
- $\mathbf{O}$ is the two-point domain, consisting of $\perp_\mathbf{O}$ and $\top$. (Domains do not usually have greatest elements, so there is no need for a general notation $\top_d$.)

Truth values

- $\mathbf{T}$ is the flat three-point domain of truth values, consisting of $\perp_\mathbf{T}$, **true**, and **false**.

- $e_1 =_d e_2$ tests the equality of e_1 and e_2 in any *flat* domain d. The value is $\bot_{\mathbf{T}}$ if either or both of e_1 and e_2 denote $\bot_d$; otherwise it is **true** or **false**. (The monotonicity and continuity of $=_d$ follow from the flatness of d: equality would not be monotonic on a nonflat domain.)
- **if** e_1 **then** e_2 **else** e_3 requires e_1 to denote an element of $\mathbf{T}$, and e_2, e_3 to denote elements of some domain d. Then it denotes e_2 if e_1 denotes **true**; it denotes e_3 if e_1 denotes **false**; and it denotes $\bot_d$ if e_1 denotes $\bot_{\mathbf{T}}$.

In practice we allow all the usual Boolean functions, extended strictly (in all arguments) to $\mathbf{T}$, and written using infix notation.

Natural numbers

- $\mathbf{N}_\bot$ is the flat domain of natural numbers, consisting of $\bot_{\mathbf{N}_\bot}, 0, 1, \ldots$ (no infinity).
- **succ** denotes the strict extension of the successor function from $\mathbf{N}$ to $\mathbf{N}_\bot$.

In practice we allow all known (computable) functions on the natural numbers, extended strictly in all arguments to $\mathbf{N}_\bot$, and written using infix notation.

Function domains

- $d_1 \rightarrow d_2$ denotes the domain of all continuous functions from the domain denoted by d_1 to the domain denoted by d_2. (Henceforth the tedious "denoted by" is generally omitted.) We have $f \sqsubseteq_{d_1 \rightarrow d_2} g$ iff $f(x) \sqsubseteq_{d_2} g(x)$ for all x in d_1.
- $\lambda x \in d . e$ denotes the (continuous) function f given by defining $f(x) = e$, where x ranges over d. This provides the context for interpreting references to x in e.
- $e_1\, e_2$ denotes the result $f(x)$ of applying the function $f: d \rightarrow d'$ denoted by e_1 to the value $x \in d$ denoted by e_2.
- $\mathbf{id}_d$ denotes the identity function on domain d.
- $e_1 \circ e_2$ denotes the composition of the functions $f_1: d' \rightarrow d''$ and $f_2: d \rightarrow d'$ denoted by e_1 and e_2 respectively, so that for all $x \in d$, $(e_1 \circ e_2)(x) = e_1(e_2(x))$.
- $\mathbf{fix}_d$ denotes the least fixed point operator for domain d, which maps each function f in $d \rightarrow d$ to the least solution x of the equation $x = f(x)$.
- $d_1 \circ\!\!\rightarrow d_2$ denotes the restriction of $d_1 \rightarrow d_2$ to strict functions.
- $\mathbf{strict}_d$ denotes the function that maps each function in $d_1 \rightarrow d_2$ to the corresponding strict function in d, where $d = d_1 \circ\!\!\rightarrow d_2$.

Product domains

- $d_1 \times \cdots \times d_n$ denotes the Cartesian product domain of n-tuples, for any $n \geqslant 2$, generalizing the binary product domain of pairs. We have $(x_1, \ldots, x_n) \sqsubseteq_{d_1 \times \cdots \times d_n} (y_1, \ldots, y_n)$ iff $x_i \sqsubseteq_{d_i} y_i$ for $i = 1, \ldots, n$.
- $(e_1, \ldots, e_n)$ denotes the n-tuple with components $e_1, \ldots, e_n$ for any $n \geqslant 2$.
- $\langle e_1, \ldots, e_n \rangle$ denotes the target tupling of the functions denoted by the e_i, abbreviating $\lambda x \in d . (e_1(x), \ldots, e_n(x))$, where x does not occur in the e_i.
- $\mathbf{on}_i^d$ denotes the projection onto the ith component, mapping $(x_1, \ldots, x_n)$ to x_i, where $d = d_1 \times \cdots \times d_n$.
- $d_1 \otimes \cdots \otimes d_n$ denotes the "smash" product obtained from the Cartesian product by identifying all the n-tuples that have any $\bot$ components. Note that $\otimes$ preserves flatness of domains.

- **smash**$_d$ denotes the function that maps each element of d, where $d=d_1\times\cdots\times d_n$, to the corresponding element of $d_1\otimes\cdots\otimes d_n$, giving $\perp$ if any of the components are $\perp$.

Sum domains

- $d_1+\cdots+d_n$ denotes the "separated" sum domain d whose elements are (distinguished copies of) the elements of the d_i together with a new $\perp_d$. Elements of d originating from different summands d_i are incomparable in d.
- $d_1\oplus\cdots\oplus d_n$ denotes the "coalesced" sum domain d where the $\perp$ elements of (the distinguished copies of) the d_i are identified with $\perp_d$. Note that $\oplus$ preserves flatness of domains.
- $\mathbf{in}_i^d$ denotes the injection function mapping elements of d_i to the corresponding elements of d, where $d=d_1\oplus\cdots\oplus d_n$ and $1\leqslant i\leqslant n$.
- $[e_1,\ldots,e_n]$ denotes the "case analysis" of the functions $f_i: d_i\to d'$ denoted by the e_i, mapping $\mathbf{in}_i(x)$ to the value of $f_i(x)$ for $1\leqslant i\leqslant n$ (but mapping $\perp$ to $\perp$).

Lifted domains

- $d_\perp$ denotes the lifted domain d' obtained by adding a new $\perp_{d'}$ under (a distinguished copy of) d.
- $\mathbf{up}_d$ denotes the function that maps each element of d to the corresponding element of $d_\perp$.
- $\mathbf{down}_d$ denotes the function that maps each element of $d_\perp$ back to the corresponding element of d.

Lists

- d^* denotes the domain of lists of *finite* length, with non-$\perp$ components in d. Note that lists with different lengths are incomparable in $\sqsubseteq$.
- d^∞ denotes the domain of *infinite* lists with components in d. Here, the "empty" list is the infinite list of $\perp$'s. We let $l_1\sqsubseteq_{d^\infty} l_2$ iff every component of l_1 approximates the corresponding component of l_2. Thus the empty list approximates all other lists.

Power domains

- $d^\natural$ denotes the "natural" (convex, Plotkin) powerdomain. Its elements may be imagined as *equivalence classes* of sets, where two sets are equivalent iff this follows from the continuity (and associativity, commutativity and absorption) of the binary union operation. For instance, if $x\sqsubseteq y\sqsubseteq z$, then the sets $\{x, y, z\}$ and $\{x, z\}$ are equivalent; moreover, if $x_0\sqsubseteq x_1\sqsubseteq\cdots\sqsubseteq x_n\sqsubseteq\cdots$ and $X=\{x_n|0\leqslant n\}$, then X is equivalent to $X\cup\{\bigsqcup_n x_n\}$. The other powerdomains (upper, lower) considered by Gunter and Scott [18] are not used in this chapter: they do not accurately reflect the *possibility* of non-termination, as they force sets $X\cup\{\perp\}$ to be equivalent either to $\{\perp\}$ or to X.
- $\{\!|e|\!\}$ denotes the element of $d^\natural$ corresponding to the set $\{x\}$, where e denotes the element $x\in d$.
- $e_1\uplus e_2$ denotes the element of $d^\natural$ corresponding to the union of X_1 and X_2, where e_1 and e_2 denote elements of a powerdomain $d^\natural$ corresponding to the sets X_1 and X_2.

- $e^\natural$ denotes the pointwise extension of e to map $d_1^\natural$ into $d_2^\natural$, where e denotes a function in $d_1 \to d_2$.
- $\mathbf{ext}_d$ extends functions in $d = d_1 \to d_2^\natural$ to $d_1^\natural \to d_2^\natural$.

It is possible to represent the *empty set* by using the domain $\mathbf{O} \oplus d^\natural$ instead of just $d^\natural$; emptiness can be tested for using $[\lambda x \in \mathbf{O}.\, e_1,\ \lambda x \in d^\natural.\, e_2]$. However, there is *no* (continuous) test for *membership* in powerdomains (just as there is no continuous test for equality on nonflat domains).

So much for the basic notation for domains and their elements.

4.3. Notational conventions

When the above notation is written in semantic descriptions, domain expressions in element expressions are generally omitted when they can be deduced from the context. Parentheses are used to indicate grouping, although the following conventions allow some parentheses to be omitted:

- Function domain constructions $\to$ and $\circ\!\!\to$ associate to the right, and have weaker precedence than $+$, $\oplus$, $\times$, and $\otimes$: $D_1 \times D_2 \to D_3 \to D_4$ is grouped as $(D_1 \times D_2) \to (D_3 \to D_4)$;
- application is left-associative, and has higher precedence than the other operators: $f x y$ is grouped as $(f x) y$, and $f \circ g(x)$ is grouped as $f \circ (g(x))$;
- abstraction $\lambda x \in d.\, e$ extends as far as possible: $(\lambda x \in D.f x)$ is grouped as $(\lambda x \in D.(f x))$;
- composition $\circ$ is associative, so its iteration does not need grouping.

(Without these conventions, our semantic descriptions would require an uncomfortable number of parentheses.) Furthermore, when implied unambiguously by the context, the following operations may be omitted:

- isomorphism between w and d, when $w = d$ is a specified domain equation;
- the following isomorphisms (which follow from the definitions of the basic domains and domain constructors):
  - $d_1 + \cdots + d_n \cong (d_1)_\perp \oplus \cdots \oplus (d_n)_\perp$;
  - $\mathbf{O} \cong \mathbf{I}_\perp$;
  - $\mathbf{T} \cong (\mathbf{O} \oplus \mathbf{O})$, mapping **true** to $\mathbf{in}_1(\perp_\mathbf{O})$;
  - $\mathbf{N}_\perp \cong (\mathbf{O} \oplus \mathbf{N}_\perp)$, mapping 0 to $\mathbf{in}_1(\perp_\mathbf{O})$;
  - $d^* \cong (\mathbf{O} \oplus (d \otimes d^*))$;
  - $d^\infty \cong (d \times d^\infty)$;
  - $d^\infty \cong (\mathbf{N}_\perp \circ\!\!\to d)$;
- injections $\mathbf{in}_i : d_i \circ\!\!\to d_i \oplus \cdots \oplus d_n$;
- "bottom extensions" of functions $f : d_i \to d'$ to sum domains: $[\ldots, \perp, f, \perp, \ldots] : d_1 \oplus \cdots \oplus d_n \circ\!\!\to d'$;
- the inclusions of $d_1 \otimes \cdots \otimes d_n$ in $d_1 \times \cdots \times d_n$ and of $d \circ\!\!\to d'$ in $d \to d'$, and the strict inclusion of $d_1 \oplus \cdots \oplus d_n$ in $d_1 + \cdots + d_n$.

Finally, the notation $\lambda(x_1 \in d_1, \ldots, x_n \in d_n).\, e$ abbreviates

$$\lambda x \in d_1 \times \cdots \times d_n.\, (\lambda x_1 \in d_1. \cdots \lambda x_n \in d_n.\, e)(\mathbf{on}_1\, x) \ldots (\mathbf{on}_n\, x)$$

(where x does not occur in e). It denotes the function f defined by $f(x_1, \ldots x_n) = e$.

5. Techniques

The preceding sections introduced all the *formalism* that is needed for specifying denotational descriptions of programming languages: grammars for specifying abstract syntax; domain notation for specifying domains and their elements; and semantic equations for specifying semantic functions mapping syntactic entities to their denotations.

This section gives some examples of denotational descriptions. The main purpose of the examples is to show what *techniques* are available for modelling the fundamental concepts of programming languages (sequential computation, scope rules, local variables, etc.).

Familiarity with these techniques allows the task of specifying a denotational semantics of a language to be factorized into (i) analyzing the language in terms of the fundamental concepts, and (ii) combining the techniques for modelling the concepts involved. Furthermore, the understanding of a given denotational description may be facilitated by recognition of the use of the various techniques.

The programming constructs dealt with in the examples below are, in general, *simplified* versions of constructs to be found in conventional "high-level" programming languages. It is not claimed that the agglomeration of the exemplified constructs would make a particularly elegant and/or practical programming language.

Section 5.1 outlines the semantics of *literals* (numerals, strings, etc.). Then Section 5.2 specifies denotations for arithmetical and Boolean *expressions*, illustrating a simple technique for dealing with "errors". Section 5.3 shows how to specify denotations for *constant declarations*, using "environments" to model scopes. Section 5.4 extends expressions to include *function abstractions*, and gives a denotational description of the *lambda calculus*.

Next, Section 5.5 gives denotations for *variable declarations*, using "stores" and "locations". Then Section 5.6 deals with *statements*, using "direct" semantics; it also explains how the technique of "continuations" can be used to model *jumps*. Section 5.7 describes *procedures* with various *modes* of parameter evaluation.

Section 5.8 distinguishes between the concepts of "batch" and "interactive" *input and output*. Section 5.9 shows how powerdomains can be used to model *nondeterministic programs*. Finally, Section 5.10 introduces "resumptions" and uses them to give denotations for a simple form of *concurrent processes*.

Caveat: In Section 5.2, the denotations of expressions are simply (numerical, etc.) values. But later, they have to be *changed*: in Section 5.3 (to be functions of environments), and again in Section 5.5 (to be functions of stores). Such changes to denotations entail tedious changes to the semantic equations that involve them. This rather unfortunate feature of conventional denotational descriptions stems from the fact that the notation used in the semantic equations has to match the precise domain structure of denotations.

Of course, these changes would be unnecessary if denotations of expressions were to be functions of environments and stores from the start. Although that might be appropriate when giving a denotational description of a complete programming language, it is undesirable in this introduction: the complexity of the denotations of simple constructs would obscure the relation between particular program constructs and the *appropriate* techniques for modelling them.

An alternative approach is to introduce *auxiliary notation* for *combining denotations*. Then when domains of denotations are changed, only the definition of the auxiliary notation requires modification: the semantic equations themselves may be left unchanged. Moreover, the auxiliary notation may be chosen to correspond directly to fundamental concepts, such as "sequencing" and "block structure", so that the semantic equations explicate the fundamental conceptual analysis of the described constructs. Such an approach is presented elsewhere [34]. It would be inappropriate to adopt it here, as it tends to hide the mathematical essence of denotations, and would give a distorted impression of the conventional approach to Denotational Semantics.

5.1. *Literals*

The syntax of a programming language usually includes "literals" (sometimes called "literal constants", or just "constants"). A literal is a symbol (or phrase) that always refers to the same item of data, irrespective of where it occurs. Examples of literals are "true" and "false", numerals, characters, and character strings.

The denotational semantics of literals is fairly straightforward, but somewhat tedious, to specify. We have already seen a simple example: binary numerals (Section 3). So let us skip most of the details here. A skeleton abstract syntax for literals is given in Table 8.

Table 8
Syntax for literals

(LITERAL)
L::= true | false | N | C | CS
(NUMERAL)
N ::= *unspecified*
(CHARACTER)
C ::= *unspecified*
(CHARACTER-STRING)
CS ::= *unspecified*

For the denotations of "true" and "false", we may use the values **true** and **false** of the standard domain **T**. The denotations of numerals should take into account that different implementations generally impose different bounds on the magnitude of numbers, and on the accuracy of "real" numbers. So let the domain of numbers—together with the associated operations—be a *parameter* of the semantics. The same goes for the denotations of characters (the ordering may vary between implementations) and strings (their length may be bounded). Let us leave such parameters as unspecified variables in the semantic description.

For example, let the domains **Num**, **Char**, and **String** be unspecified domain variables together with various variables for elements of, and functions on, these domains; see Table 9. It is straightforward to define the semantic functions introduced in Table 10 in terms of the given elements and functions. The details are omitted here.

Table 9
Domains for literals

Num	=	*unspecified*
zero, one	∈	**Num**
neg	∈	**Num**$\circ\!\!\rightarrow$**Num**
sum, diff	∈	(**Num**$\otimes$**Num**)$\circ\!\!\rightarrow$**Num**
prod, div	∈	(**Num**$\otimes$**Num**)$\circ\!\!\rightarrow$**Num**
Char	=	*unspecified*
ord	∈	**Char**$\circ\!\!\rightarrow$**Num**
chr	∈	**Num**$\circ\!\!\rightarrow$**Char**
String	=	*unspecified*
str	∈	**Char***$\circ\!\!\rightarrow$**String**
chrs	∈	**String**$\circ\!\!\rightarrow$**Char***
V	=	**T**$\oplus$**Num**$\oplus$**Char**$\oplus$**String**

Table 10
Denotations for literals

$\mathscr{L}$: LITERAL→**V**
$\mathscr{N}$: NUMERAL→**Num**
$\mathscr{C}$: CHARACTER→**Char**
$\mathscr{CS}$: CHARACTER-STRING→**String**

By the way, the domains of literal denotations are generally *flat* (and countable). Note in particular that the finite *numerical* approximations to real numbers made by computers should *not* be represented by values related by the *computational* approximation ordering of domains $\sqsubseteq$: once an approximate real number has been computed, further computation does not improve the degree of approximation of *that* number. (Of course, a program may indeed compute a series of approximate numbers, but the numbers are not *necessarily* increasingly good approximations to some particular number.)

5.2. *Expressions*

Expressions in programming languages are constructed using operators and (perhaps) if-then-else from primitive expressions, including literals. Abstract syntax for some typical expressions is given in Table 11. (Further expressions are considered in later sections.)

We take the denotations of expressions to be elements of a domain **EV** that consists of truth values, numbers, etc., representing the result of expression evaluation. The domain **EV** is a so-called "characteristic domain", and its relation to other characteristic domains introduced in later sections can give valuable insight into the essence of

Table 11
Syntax for expressions

(EXPRESSION)
$E ::= L \mid MO\ E_1 \mid E_1\ DO\ E_2 \mid \text{if } E_1 \text{ then } E_2 \text{ else } E_3$
(MONADIC-OPERATOR)
$MO ::= \neg \mid -$
(DYADIC-OPERATOR)
$DO ::= \wedge \mid \vee \mid + \mid - \mid * \mid =$

the described programming language. For now we let **EV** contain the same values as **V**, i.e., the values of literals; later, further expressible values are introduced.

We are now ready to define the denotations of expressions and operators; see Table 12. Note that the notational conventions introduced at the end of Section 4 are much exploited in the semantic equations. For instance, in the equation for if-expressions, there is an application of a function ($\lambda t \in \mathbf{T}. \ldots$) to an argument in **EV**; however, **T** is a summand of **V**, which is isomorphic to **EV**, so the given function, f say, is implicitly extended to $[f, \perp_{\mathbf{Num}\to\mathbf{EV}}, \perp_{\mathbf{Char}\to\mathbf{EV}}, \perp_{\mathbf{String}\to\mathbf{EV}}] \in (\mathbf{T} \oplus \mathbf{Num} \oplus \mathbf{Char} \oplus \mathbf{String}) \to \mathbf{EV}$, and then composed with an isomorphism to give a function in $\mathbf{EV} \to \mathbf{EV}$.

Thus the denotation of an erroneous expression such as "if 42 then ... else ..." is $\perp$. The semantics of such erroneous expressions is actually irrelevant, provided that programs containing them are deemed illegal. More generally, however, it might be better to avoid representing errors by $\perp$, as the *essential* use of $\perp$ (in later sections) is to represent *nontermination.* To do this we would have to introduce special elements for representing errors into *all* domains, and the extra notation for specifying the treatment of errors would be an unwelcome burden in the semantic equations.

Table 12
Denotations for expressions

$\mathbf{EV} = \mathbf{V}$
$\mathcal{E}$: EXPRESSION $\to \mathbf{EV}$
$\mathcal{E}[\![L]\!] = \mathcal{L}[\![L]\!]$
$\mathcal{E}[\![MO\ E_1]\!] = \mathcal{MO}[\![MO]\!](\mathcal{E}[\![E_1]\!])$
$\mathcal{E}[\![E_1\ DO\ E_2]\!] = \mathcal{DO}[\![DO]\!](\mathbf{smash}(\mathcal{E}[\![E_1]\!], \mathcal{E}[\![E_2]\!]))$
$\mathcal{E}[\![\text{if } E_1 \text{ then } E_2 \text{ else } E_3]\!] = (\lambda t \in \mathbf{T}.\ \mathbf{if}\ t\ \mathbf{then}\ \mathcal{E}[\![E_2]\!]\ \mathbf{else}\ \mathcal{E}[\![E_3]\!])$
$(\mathcal{E}[\![E_1]\!])$
$\mathcal{MO}$: MONADIC-OPERATOR $\to (\mathbf{V} \circ\!\!\to \mathbf{V})$
$\mathcal{MO}[\![\neg]\!] = \lambda t \in \mathbf{T}.\ \mathbf{if}\ t\ \mathbf{then\ false\ else\ true}$
$\mathcal{MO}[\![-]\!] = \lambda n \in \mathbf{Num}.\ \mathbf{diff}(\mathbf{zero}, n)$
$\mathcal{DO}$: DYADIC-OPERATOR $\to (\mathbf{V} \otimes \mathbf{V} \circ\!\!\to \mathbf{V})$
$\mathcal{DO}:[\![\wedge]\!] = \lambda(t_1 \in \mathbf{T}, t_2 \in \mathbf{T}).\ \mathbf{if}\ t_1\ \mathbf{then}\ t_2\ \mathbf{else\ false}$
...
$\mathcal{DO}[\![+]\!] = \lambda(n_1 \in \mathbf{Num}, n_2 \in \mathbf{Num}).\ \mathbf{sum}(n_1, n_2)$
$\mathcal{DO}[\![=]\!] = \lambda(v_1 \in \mathbf{V}, v_2 \in \mathbf{V}).(v_1 =_{\mathbf{V}} v_2)$

5.3. *Constant declarations*

Identifiers are symbols used as "tokens" for values. In programming languages, there are various constructs which introduce identifiers and "bind" them to values. It is conventional to refer to the value to which an identifier is bound as the value "denoted" by the identifier, but this terminology is a bit misleading: the *denotation* of an identifier is the identifier itself (or rather, an element of a semantic domain corresponding to the abstract syntax of identifiers).

Let us start with some simple "constant declarations", whose abstract syntax is given in Table 13. The intended effect of the declaration "val $I = E$" is to "bind" I to the value of E. The construct "let CD in E" determines the "scope" of such "bindings": the bindings made by CD are available throughout E—except where overridden by another binding for the same identifier, since "let"s may be nested, giving a "block structure" in expressions. In "CD_1; CD_2", the scope of the bindings introduced by CD_1 includes CD_2. The phrase "rec CD" extends the scope of the declarations in CD to CD itself, making them "mutually recursive".

Table 13
Syntax for constant declarations

(CONSTANT-DECLARATIONS)
$CD ::=$ val $I = E \mid CD_1;\, CD_2 \mid$ rec CD
(EXPRESSION)
$E ::= I \mid$ let CD in E

In the semantics, we write **DV** for the domain that represents the values "denotable" by identifiers. **DV** is a characteristic domain, like **EV**. In real programming languages there are sometimes values that are expressible but not denotable—numbers in ALGOL60, for instance. Less obviously, there may be values that are denotable but not expressible—types in PASCAL, for instance.

"Environments" are used to represent associations between identifiers and denoted values. The domain of environments, together with some basic functions on environments, is defined in Table 14. **Ide** is assumed to be a flat domain corresponding to the abstract phrase sort IDENTIFIER. The element $\top \in \mathbf{O}$ is used to indicate the absence of a denoted value. (To allow the presence of a denoted value to be tested, we would have to lift **DV** to $\mathbf{DV}_\perp$, since the denoted value might be $\perp$.) Notice that **overlay**(e, e') gives precedence to e, whereas **combine**$(e, e') =$ **combine**(e', e) is intended for uniting the bindings of disjoint sets of identifiers.

The result of expression evaluation now depends, in general, on the values bound to the identifiers that occur in it. This dependence is represented by letting the denotation of an expression be a *function* from environments to expressible values—which requires rewriting the semantic equations previously specified for expressions.

Clearly, an appropriate denotation for a constant declaration is a function from environments to environments. But there is a choice to be made: should the resulting environment be the argument environment *extended* by the new bindings, or just the

Table 14
Notation for environments

Env	$=\mathbf{Ide}\to(\mathbf{DV}\oplus\mathbf{O})$
void	$=\lambda I\in\mathbf{Ide}.\ \mathbf{in}_2\top$
	$\in\mathbf{Env}$
bound	$=\lambda I\in\mathbf{Ide}.\ \lambda e\in\mathbf{Env}.\ [\mathbf{id}_{\mathbf{DV}},\bot](e(I))$
	$\in\mathbf{Ide}\to\mathbf{Env}\to\mathbf{DV}$
binding	$=\lambda I\in\mathbf{Ide}.\ \lambda v\in\mathbf{DV}.\ \lambda I'\in\mathbf{Ide}.\ \mathbf{if}\ I=_{\mathbf{Ide}}I'\ \mathbf{then}\ \mathbf{in}_1(v)\ \mathbf{else}\ \mathbf{in}_2(\top)$
	$\in\mathbf{Ide}\to\mathbf{DV}\to\mathbf{Env}$
overlay	$=\lambda(e\in\mathbf{Env},e'\in\mathbf{Env}).\ \lambda I\in\mathbf{Ide}.\ [\mathbf{id}_{\mathbf{DV}},\lambda x\in\mathbf{O}.\ e'(I)](e(I))$
	$\in\mathbf{Env}\times\mathbf{Env}\to\mathbf{Env}$
combine	$=\lambda(e\in\mathbf{Env},\lambda e'\in\mathbf{Env}).\ \lambda I\in\mathbf{Ide}.\ [\lambda d\in\mathbf{DV}.\ [\bot,\lambda x\in\mathbf{O}.\ d],\lambda x\in\mathbf{O}.\ \mathbf{id}_{\mathbf{DV}\oplus\mathbf{O}}]\ (e(I))(e'(I))$
	$\in\mathbf{Env}\times\mathbf{Env}\to\mathbf{Env}$

new bindings by themselves? Let us choose the latter, which gives a bit more flexibility, exploited in later sections.

The denotations of constant declarations and of the related expressions, together with the modified denotations of the previously specified expressions, are defined in Table 15.

The semantics of recursive declarations makes use of $\mathbf{fix}_{\mathbf{Env}}$, which gives the least fixed point of the function in $\mathbf{Env}\to\mathbf{Env}$ to which it is applied. To see that this provides the appropriate denotations, consider $\mathscr{CD}[\![\mathsf{rec\ val}\ I=E]\!](e)$. From the semantic equations we have

$$\begin{aligned}&\mathscr{CD}[\![\mathsf{rec\ val}\ I=E]\!](e)\\&\quad=\mathbf{fix}(\lambda e'\in\mathbf{Env}.\ \mathbf{binding}(I)(\mathscr{E}[\![E]\!](\mathbf{overlay}(e',e)))),\end{aligned}$$

Table 15
Denotations for constant declarations and expressions (modified)

$$\begin{aligned}
&\mathbf{DV}=\mathbf{V}\\
&\mathbf{EV}=\mathbf{V}\\[1ex]
&\mathscr{CD}:\text{CONSTANT-DECLARATIONS}\to\mathbf{Env}\to\mathbf{Env}\\
&\mathscr{CD}[\![\mathsf{val}\ I=E]\!]=\lambda e\in\mathbf{Env}.\ \mathbf{binding}\ I(\mathscr{E}[\![E]\!]e)\\
&\mathscr{CD}[\![CD_1;CD_2]\!]=\lambda e\in\mathbf{Env}.\ (\lambda e_1\in\mathbf{Env}.\ \mathbf{overlay}(\mathscr{CD}[\![CD_2]\!](\mathbf{overlay}(e_1,e)),e_1))\\
&\qquad\qquad(\mathscr{CD}[\![CD_1]\!]e)\\
&\mathscr{CD}[\![\mathsf{rec}\ CD]\!]=\lambda e\in\mathbf{Env}.\ \mathbf{fix}(\lambda e'\in\mathbf{Env}.\ \mathscr{CD}[\![CD]\!](\mathbf{overlay}(e',e)))\\[1ex]
&\mathscr{E}:\text{EXPRESSION}\to\mathbf{Env}\to\mathbf{EV}\\
&\mathscr{E}[\![I]\!]=\lambda e\in\mathbf{Env}.\ \mathbf{bound}\ I\ e\\
&\mathscr{E}[\![\mathsf{let}\ CD\ \mathsf{in}\ E]\!]=\lambda e\in\mathbf{Env}.\ \mathscr{E}[\![E]\!](\mathbf{overlay}(\mathscr{CD}[\![CD]\!]e,e))\\
&\mathscr{E}[\![L]\!]=\lambda e\in\mathbf{Env}.\ \mathscr{L}[\![L]\!]\\
&\mathscr{E}[\![MO\ E_1]\!]=\lambda e\in\mathbf{Env}.\ \mathscr{MO}[\![MO]\!](\mathscr{E}[\![E_1]\!]e)\\
&\mathscr{E}[\![E_1\ DO\ E_2]\!]=\lambda e\in\mathbf{Env}.\ \mathscr{DO}[\![DO]\!](\mathbf{smash}(\mathscr{E}[\![E_1]\!]e,\mathscr{E}[\![E_2]\!]e))\\
&\mathscr{E}[\![\mathsf{if}\ E_1\ \mathsf{then}\ E_2\ \mathsf{else}\ E_3]\!]=\lambda e\in\mathbf{Env}.\ (\lambda t\in\mathbf{T}.\ \mathbf{if}\ t\ \mathbf{then}\ \mathscr{E}[\![E_2]\!]e\ \mathbf{else}\ \mathscr{E}[\![E_3]\!]e)\\
&\qquad\qquad(\mathscr{E}[\![E_1]\!]e)
\end{aligned}$$

i.e. the least $e' \in \mathbf{Env}$ such that

$$e' = \mathbf{binding}(I)(\mathscr{E}[\![E]\!](\mathbf{overlay}(e', e))).$$

Let $v = \mathscr{E}[\![E]\!](\mathbf{overlay}(e', e))$; we have

$$v = \mathscr{E}[\![E]\!]((\mathbf{overlay}(\mathbf{binding}\ I\ v, e))$$

and in fact

$$v = \mathbf{fix}(\lambda v' \in \mathbf{EV}.\ \mathscr{E}[\![E]\!](\mathbf{overlay}(\mathbf{binding}\ I\ v', e))).$$

Notice that a direct circularity in the recursive declarations gives rise to $\bot$ as a denoted value, e.g.,

$$\mathscr{CD}[\![\mathsf{rec\ val}\ I = I]\!] = \mathbf{binding}\ I\ \bot$$

in contrast to a mere "forward reference":

$$\mathscr{CD}[\![\mathsf{rec\ (val}\ I = I';\ \mathsf{val}\ I' = 0)]\!] = \mathbf{overlay}(\mathbf{binding}\ I'\ 0,\ \mathbf{binding}\ I\ 0).$$

The most interesting case is when the sequence of environments e'_n defined by

$$e'_0 = \mathbf{binding}(I)(\mathscr{E}[\![E]\!](\bot)),$$
$$e'_1 = \mathbf{binding}(I)(\mathscr{E}[\![E]\!](\mathbf{overlay}(e'_0, e)),$$
$$\ldots$$
$$e'_{n+1} = \mathbf{binding}(I)(\mathscr{E}[\![E]\!](\mathbf{overlay}(e'_n, e)))$$
$$\ldots$$

is strictly increasing, converging to—but never reaching—the limit point $e' = \bigsqcup_n e'_n$. With the expressions considered so far it is not possible to get such a sequence; but it becomes possible when *function abstractions* are introduced, as in the next section.

5.4. *Function abstractions*

"Functions" in programs resemble mathematical functions: they return values when applied to arguments. In programs, however, the evaluation of arguments may diverge, so it is necessary to take into account not only the relation between argument values and result values, but also the stage at which an argument expression is evaluated: straight away, or when (if ever) the value of the argument is required for calculating the result of the application.

Various programming languages allow functions to be *declared*, i.e., bound to identifiers. Often, functions may also be passed as *arguments* to other functions. But only in a few languages is it possible to *express* functions directly, by means of so-called "abstractions", without necessarily binding them to identifiers. (These languages are generally the so-called "functional programming languages".)

The syntax given in Table 16 allows functions to be expressed by abstractions of the form "fun (val I) E"; we refer to "val I" as the "parameter declaration" of the abstraction (further forms of parameter declaration are introduced later) and to E as the "body". Notice that constant declarations of the form "val I' = fun (val I)E"

Table 16
Syntax for functions and parameter declarations

(EXPRESSION)
$E ::= \text{fun } (PD)\ E \mid E_1\,(E_2)$
(PARAMETER-DECLARATION)
$PD ::= \text{val } I$

resemble "function declarations" in conventional programming languages; recursive references to I' in E are allowed when the declaration is prefixed by "rec".

The phrase "$E_1\,(E_2)$" expresses the application of a function to an argument, with the "actual parameter" E_2 being evaluated *before* the evaluation of the body of the function abstraction is commenced—this "mode" of parameter evaluation is known as "call-by-value". (Functions in programming languages are usually allowed to have lists of parameters; this feature is omitted here, for simplicity.)

There are two distinct possibilities for the scopes of declarations in relation to abstractions, arising from identifiers which occur in the bodies of abstractions, but which refer to outer declarations. With so-called *static* scopes, the scopes of declarations extend into the bodies of an abstraction at the point where the abstraction is introduced, so that the declaration referred to by an identifier is fixed. With *dynamic* scopes, the body of an abstraction is evaluated in the scope of the declarations at each point of application, so that the declaration referred to by an identifier in an abstraction body may vary—and be different from that referred to with static scopes. There is some dispute in the programming community about which of these scope rules is "better". Here, the semantic description of static scopes is illustrated; dynamic scopes are only marginally more complicated to describe.

The domains for use in the semantics of function abstractions are specified in Table 17. Notice that the definitions of **DV** and **EV** supercede the previous definitions. (No changes are needed to the semantic equations for declarations and expressions

Table 17 Denotations for functions and parameter declarations

$$\mathbf{F} = (\mathbf{PV} \circ\!\!\rightarrow \mathbf{FV})_\perp$$
$$\mathbf{PV} = \mathbf{V} \oplus \mathbf{F}$$
$$\mathbf{FV} = \mathbf{V}$$
$$\mathbf{DV} = \mathbf{V} \oplus \mathbf{F}$$
$$\mathbf{EV} = \mathbf{V} \oplus \mathbf{F}$$

$$\mathscr{E}: \text{EXPRESSION} \rightarrow \mathbf{Env} \rightarrow \mathbf{EV}$$
$$\mathscr{E}[\![\text{fun } (PD)\ E]\!] = \lambda e \in \mathbf{Env}.\ (\mathbf{up} \circ \mathbf{strict})(\lambda v \in \mathbf{PV}.\ \mathbf{id}_{\mathbf{FV}}(\mathscr{E}[\![E]\!](\mathbf{overlay}(\mathscr{P}\mathscr{D}[\![PD]\!]v, e))))$$
$$\mathscr{E}[\![E_1\,(E_2)]\!] = \lambda e \in \mathbf{Env}.\ (\mathbf{down} \circ \mathbf{id}_{\mathbf{F}})(\mathscr{E}[\![E_1]\!]e)(\mathscr{E}[\![E_2]\!]e)$$

$$\mathscr{P}\mathscr{D}: \text{PARAMETER-DECLARATIONS} \rightarrow \mathbf{PV} \rightarrow \mathbf{Env}$$
$$\mathscr{P}\mathscr{D}[\![\text{val } I]\!] = \lambda v \in \mathbf{PV}.\ \mathbf{binding}\ I\ v$$

given in Table 12, thanks to our notational conventions about injections and extensions related to sums.)

To model abstractions it is obvious to use functions. The domains consisting of parameter values **PV** and function result values **FV** may be regarded as characteristic domains. Few programming languages allow functions to be returned as results (and some even forbid functions as arguments).

The functions corresponding to the values of abstractions are taken to be *strict*, reflecting value-mode parameter evaluation: $\perp$ represents the nontermination of an evaluation, and the nontermination of an argument evaluation implies the nontermination of the function application. The abstraction values are *lifted* so that an abstraction never evaluates to $\perp$.

Notice that the domain **F** is *reflexive*: it is isomorphic to a domain that (essentially) includes a domain of functions from **F**.

The semantic equations for function abstractions are given in Table 17. Various isomorphisms are left implicit, for instance that between **DV** and **EV**; likewise, some injections and extensions related to sum domains are omitted.

An alternative mode of parameter evaluation is to delay evaluation until the parameter is *used*. This mode is referred to as "call-by-name". (The main difference it makes to the semantics of expressions is that an evaluation which does not terminate with value-mode, *may* terminate when name-mode is used instead.)

Only a few programming languages provide name-mode parameters. Much the same effect, however, can be achieved by passing a (parameterless) abstraction as a parameter, and applying it (to no parameters) wherever the value of the parameter is required.

The main theoretical significance of name-mode abstractions is that they correspond directly to λ-abstractions in the *lambda calculus* of Church (see [4]). Consider the abstract syntax for lambda calculus expressions given in Table 18. The axiom of so-called "β-conversion" of the lambda calculus makes an application "$(\lambda I.\, E)\,(E')$" equivalent to the expression obtained by substituting E' for I in E (with due regard to static scopes of λ-bindings), and this is just E when I does not occur in E.

It is a simple matter to adapt the domains that were used to represent value-mode abstractions, so as to provide a denotational semantics for the lambda calculus. The only necessary changes are to let **FV** include **F**, and to remove the restriction of **F** to strict functions; but let us dispense with the lifting as well, as it is no longer significant. The presence of **V** (in **FV**) ensures that the solution to the domain equations is nontrivial. (The *standard model* for the lambda calculus [18] is obtained by taking

Table 18
Syntax for λ-expressions

(EXPRESSION)
$E ::= (\lambda I.\, E) \mid E_1(E_2) \mid I$
(IDENTIFIER)
$I ::=$ *unspecified*

PV = **FV** = **F**, leaving essentially **F** = **F** → **F**, and the trivial solution has to be avoided another way.)

The denotations for the lambda calculus are specified in Table 19, where for once the injections and extensions related to the sum domain are made explicit (although the isomorphisms between the left- and right-hand sides of the specified domain equations are still omitted).

Table 19
Denotations for λ-expressions

$$\mathbf{F} = \mathbf{PV} \to \mathbf{FV}$$
$$\mathbf{PV} = \mathbf{V} \oplus \mathbf{F}$$
$$\mathbf{FV} = \mathbf{V} \oplus \mathbf{F}$$
$$\mathbf{DV} = \mathbf{V} \oplus \mathbf{F}$$
$$\mathbf{EV} = \mathbf{V} \oplus \mathbf{F}$$

$$\mathscr{E}: \text{EXPRESSION} \to \mathbf{Env} \to \mathbf{EV}$$
$$\mathscr{E}[\![(\lambda I.\, E)]\!] = \lambda e \in \mathbf{Env}.\, \mathbf{in}_2(\lambda v \in \mathbf{PV}.\, \mathscr{E}[\![E]\!](\mathbf{overlay}(\mathbf{binding}\ I\ v, e)))$$
$$\mathscr{E}[\![E_1(E_2)]\!] = \lambda e \in \mathbf{Env}.\, [\bot, \mathbf{id}_{\mathbf{F}}](\mathscr{E}[\![E_1]\!]e)(\mathscr{E}[\![E_2]\!]e)$$

The standard model for the lambda calculus has been extensively studied, and there are some significant theorems about it. Most of these carry over to the denotations defined above. First of all, there is the following theorem that the semantics does indeed model β-conversion.

5.1. PROPOSITION. *For any λ-expressions "$\lambda I.\, E$" and E',*

$$\mathscr{E}[\![(\lambda I.\, E)(E')]\!] = \mathscr{E}[\![[E'/I]E]\!].$$

Here "$[E'/I]E$" is the proper *substitution* of E' for free occurrences of I in E: the identifiers of λ-abstractions in E are assumed (or made) to be different from the free identifiers in E'.

The key to proving the above theorem is the following lemma.

5.2. LEMMA (substitution). *For any λ-expressions E, E', for any identifier I, and for any $e \in \mathbf{Env}$,*

$$\mathscr{E}[\![E]\!](\mathbf{overlay}(\mathbf{binding}(I)(\mathscr{E}[\![E']\!]e), e)) = \mathscr{E}[\![[E'/I]E]\!]e.$$

The following theorem implies that β-reduction is sufficient for symbolic computation of approximations to any desired degree of closeness. Let $\mathscr{A}(E)$ be the set of *approximate normal forms* of E (obtained from E by finite sequences of β-reductions, followed by the replacement of any remaining redexes by an expression "Ω" denoting $\bot$).

5.3. THEOREM (limiting completeness). *For any λ-expression E,*

$$\mathscr{E}[\![E]\!] = \bigsqcup\{\mathscr{E}[\![E']\!] \mid E' \in \mathscr{A}(E)\}.$$

The original proof by Wadsworth [59] involves the introduction of an auxiliary calculus with numerical labels forcing all reduction sequences to terminate. An alternative proof is given by Mosses and Plotkin [36] by introducing an "intermediate" denotational semantics, where denotations are taken to be functions of an argument in the *chain* domain of extended natural numbers (i.e., with ∞): for finite arguments, the intermediate semantics gives approximations, corresponding to the denotations of approximate normal forms; the standard denotations are obtained when the argument is ∞.

5.5. *Variable declarations*

The preceding sections dealt with expressions, constant declarations, and function abstractions. In conventional programming languages, these constructs play a minor rôle in comparison to *statements* (also called "commands"), which operate on "variables". This section deals with the semantics of variables; statements themselves are deferred to the next section.

In programs, variables are entities that provide access to stored data. The *assignment* of a value to a variable as the effect of modifying the stored data, whereas merely inspecting the current value of a variable causes no modification.

This concept of a variable is somewhat different from that of a variable in mathematics. In mathematical *terms*, variables stand for particular unknown values— often, the arguments of functions. These variables do indeed get "assigned" values, e.g., by function application. But the values thus assigned do *not* subsequently vary: a variable refers to the same value throughout the term in which it is used. In fact, mathematical variables correspond closely to *identifiers* in programming languages.

Program variables may be *simple* or *compound*. The latter have component variables that may be assigned values individually; the value of a compound variable depends on the values of its component variables.

Consider the syntax specified in Table 20. The variable declaration "var I: T" determines a "fresh" variable for storing values of the "type" T, and binds I to the variable. Variable declarations are combined by "VD_1, VD_2"; such declarations do not include each other in their scopes (although in our simple example language, it would make no difference if they did, as variable declarations do not refer to identifiers at all). The types "bool", "num" are for declaring simple variables for storing truth values, respectively numbers; the type "T[1..N]" is for declaring compound variables that

Table 20
Syntax for variable declarations and types

(VARIABLE-DECLARATIONS)
VD ::= var I: T | VD_1, VD_2
(TYPE)
T ::= bool | num | T[1 .. N]
(EXPRESSION)
E ::= $E_1[E_2]$

have N independent component variables for storing values of type T. In the expression "$E_1[E_2]$", E_1 is supposed to evaluate to a compound variable v, and E_2 to a positive integer n; then the result is the nth component variable of v.

Types are used for two purposes in programming languages: to facilitate checking that programs are well-formed, prior to execution; and to indicate how much storage to allocate, during execution. Here, we are only concerned with the dynamic semantics of programs, which—in general—does not involve type checking, only storage allocation. (Mitchell [26] provides an extensive study of the semantics of types.)

"Stores" are used to represent associations between *simple* variables and their values. Simple variables are represented by "locations" in stores; their only relevant property is that they can be distinguished from each other. Thus a simple variable identifier gets bound to a location, which in turn gives access to the current value stored in the variable. It is possible for two identifiers to be bound (in the same scope) to the same location: then assignment to the one changes the value of the other. Such identifiers are called "aliases".

Compound variables can be represented by values with variables (ultimately, locations) as components. Whereas assignment to distinct simple variables is independent, distinct compound variables may "share" component variables.

The domain of storable values SV consists of those items of data that can be stored at single locations. It may be considered to be a characteristic domain.

The domain of (states of) stores **S** is defined in Table 21, together with some basic functions on stores. A location mapped to **false** is "free", and a location mapped to **true** is "reserved" but not yet "initialized". Notice that the function "**location**" is left unspecified—it is supposed to select any location that is not reserved in the given state. It is usual to ignore the boundedness of real computer storage in denotational

Table 21
Notation for stores

S	$= \mathbf{Loc} \to (\mathbf{SV} \oplus \mathbf{T})$
Loc	$= \mathbf{O} \oplus \mathbf{Loc}$
empty	$= \lambda l \in \mathbf{Loc}.\ \mathbf{false}$
	$\in \mathbf{S}$
reservation	$= \lambda l \in \mathbf{Loc}.\ \lambda s \in \mathbf{S}.$
	$(\lambda l' \in \mathbf{Loc}.\ \mathbf{if}\ l =_{\mathbf{Loc}} l'\ \mathbf{then\ true\ else}\ s(l'))$
	$\in \mathbf{Loc} \to \mathbf{S} \to \mathbf{S}$
freedom	$= \lambda l \in \mathbf{Loc}.\ \lambda s \in S.$
	$(\lambda l' \in \mathbf{Loc}.\ \mathbf{if}\ l =_{\mathbf{Loc}} l'\ \mathbf{then\ false\ else}\ s(l'))$
	$\in \mathbf{Loc} \to \mathbf{S} \to \mathbf{S}$
store	$= \lambda l \in \mathbf{Loc}.\ \lambda v\ \mathbf{SV}.\ \lambda s \in \mathbf{S}.$
	$(\lambda l' \in \mathbf{Loc}.\ \mathbf{if}\ l =_{\mathbf{Loc}} l'\ \mathbf{then}\ v\ \mathbf{else}\ s(l'))$
	$\in \mathbf{Loc} \to \mathbf{SV} \to \mathbf{S} \to \mathbf{S}$
stored	$= \lambda l \in \mathbf{Loc}.\ \lambda s \in \mathbf{S}.\ [\mathbf{id}_{\mathbf{SV}},\ \bot](s(l))$
	$\in \mathbf{Loc} \to \mathbf{S} \to \mathbf{SV}$
location	$=$ *unspecified*
	$\in \mathbf{S} \to \mathbf{Loc}$
allocation	$= \lambda s \in \mathbf{S}.\ (\lambda l \in \mathbf{Loc}.\ (l,\ \mathbf{reservation}\ l\ s))(\mathbf{location}\ s)$
	$\in \mathbf{S} \to \mathbf{Loc} \times \mathbf{S}$

semantics, so "**location**" may be assumed not to produce $\bot$ (unless applied to a state in which all the locations have somehow been reserved).

Some further notation concerned with compound variables is specified in Table 22. It provides convenient generalizations of the basic functions on stores. **LV** is the domain of all variables; **RV** is the domain of assignable values. (The names of these domains stem from the sides of the assignment statement on which variables and assignable values are used: "left" and "right".) They are considered to be characteristic domains. Usually, as here, **LV** has **Loc** as a summand, and **RV** has **SV** as a summand.

Table 22
Notation for compound variables

$$
\begin{array}{ll}
\mathbf{LV} & = \mathbf{Loc} \oplus \mathbf{LV}^* \\
\mathbf{RV} & = \mathbf{SV} \oplus \mathbf{RV}^* \\
 & \\
\mathbf{allocations} & = \lambda(f \in \mathbf{S} \to \mathbf{LV} \times \mathbf{S}, n \in \mathbf{N}_\bot). \\
 & \quad \mathbf{if}\ n=0\ \mathbf{then}\ \lambda s \in \mathbf{S}.\,(\top, s)\ \mathbf{else} \\
 & \quad (\lambda(l \in \mathbf{LV}, s \in \mathbf{S}).\,(\lambda(l^* \in \mathbf{LV}^*, s' \in \mathbf{S}).\,((l, l^*), s)) \\
 & \qquad\qquad (\mathbf{allocations}(f, n-1)s)) \circ f \\
 & \in (\mathbf{S} \to \mathbf{LV} \times \mathbf{S}) \times \mathbf{N}_\bot \to \mathbf{S} \to \mathbf{LV} \times \mathbf{S} \\
\mathbf{freedoms} & = \lambda(f \in \mathbf{LV} \to \mathbf{S} \to \mathbf{S}, n \in \mathbf{N}_\bot). \\
 & \quad \mathbf{if}\ n=0\ \mathbf{then}\ \lambda l^* \in \mathbf{O}.\,\mathbf{id}_\mathbf{S}\ \mathbf{else} \\
 & \quad \lambda(l \in \mathbf{LV}, l^* \in \mathbf{LV}^*).\,\mathbf{freedoms}(f, n-1)\ l^* \circ f\ l \\
 & \in (\mathbf{LV} \to \mathbf{S} \to \mathbf{S}) \times \mathbf{N}_\bot \to \mathbf{LV} \to \mathbf{S} \to \mathbf{S} \\
\mathbf{component} & = \lambda n \in \mathbf{N}_\bot.\ \mathbf{if}\ n=1\ \mathbf{then}\ \mathbf{on}_1\ \mathbf{else}\ \mathbf{component}\ (n-1) \circ \mathbf{on}_2 \\
 & \in \mathbf{N}_\bot \to \mathbf{LV}^* \to \mathbf{LV} \\
\mathbf{assign} & = [\mathbf{store}, \\
 & \quad [\lambda l \in \mathbf{O}.\,\lambda v \in \mathbf{O}.\,\mathbf{id}_\mathbf{S}, \\
 & \quad\ \lambda(l \in \mathbf{LV}, l^* \in \mathbf{LV}^*).\,\lambda(v \in \mathbf{RV}, v^* \in \mathbf{RV}^*). \\
 & \qquad \mathbf{assign}\ l^*\ v^* \circ \mathbf{assign}\ l\ v]] \\
 & \in \mathbf{LV} \to \mathbf{RV} \to \mathbf{S} \to \mathbf{S} \\
\mathbf{assigned} & = [\mathbf{stored}, \\
 & \quad [\lambda l \in \mathbf{O}.\,\lambda s \in \mathbf{S}.\,(\top, s), \\
 & \quad\ \lambda(l \in \mathbf{LV}, l^* \in \mathbf{LV}^*).\,(\lambda(v \in \mathbf{RV}, s \in \mathbf{S}). \\
 & \qquad\qquad (\lambda(v^* \in \mathbf{RV}^*, s' \in \mathbf{S}).\,((v, v^*), s')) \\
 & \qquad\qquad (\mathbf{assigned}\ l^*\ s)) \circ \mathbf{assigned}\ l]] \\
 & \in \mathbf{LV} \to \mathbf{S} \to \mathbf{RV} \times \mathbf{S}
\end{array}
$$

The denotations of variable declarations and types are given in Table 23. It is convenient to introduce a second semantic function for variable declarations: for specifying that variables are no longer accessible—when exiting the scope of local variable declarations, for instance. Formally, the denotation of a variable declaration VD is the pair $(\mathcal{VD}[\![VD]\!], \mathcal{VU}[\![VD]\!])$.

The appropriate denotations for expressions, declarations, etc., are now functions of stores, as well as environments. Whether expression evaluation should be allowed to affect the store—known as "side-effects"—is controversial: some languages (such as C) actually encourage side-effects in expressions, but allow the order of evaluation of expressions to be specified; others make the order of evaluation of expressions

Table 23
Denotations for variable declarations and types

$$\mathbf{SV} = \mathbf{T} \oplus \mathbf{Num}$$

$$\mathcal{VD}: \text{VARIABLE-DECLARATIONS} \to \mathbf{S} \to (\mathbf{Env} \times \mathbf{S})$$
$$\mathcal{VD}[\![\mathsf{var}\ I\colon T]\!] = (\lambda(l \in \mathbf{LV}, s \in \mathbf{S}).\,(\mathbf{binding}\ I\ l, s)) \circ \mathcal{T}[\![T]\!]$$
$$\mathcal{VD}[\![VD_1, VD_2]\!] = (\lambda(e_1 \in \mathbf{Env}, s_1 \in \mathbf{S}).\,(\lambda(e_2 \in \mathbf{Env}, s_2 \in \mathbf{S}).\,(\mathbf{combine}(e_1, e_2), s_2))\,(\mathcal{VD}[\![VD_2]\!]s_1)) \circ \mathcal{VD}[\![VD_1]\!]$$

$$\mathcal{VU}: \text{VARIABLE-DECLARATIONS} \to \mathbf{Env} \to \mathbf{S} \circ\!\!\to \mathbf{S}$$
$$\mathcal{VU}[\![\mathsf{var}\ I\colon T]\!] = \lambda e \in \mathbf{Env}.\ \mathcal{TU}[\![T]\!](\mathbf{bound}\ I\ e)$$
$$\mathcal{VU}[\![VD_1, VD_2]\!] = \lambda e \in \mathbf{Env}.\ \mathcal{VU}[\![VD_2]\!]e \circ \mathcal{VU}[\![VD_1]\!]e$$

$$\mathcal{T}: \text{TYPE} \to \mathbf{S} \to \mathbf{LV} \times \mathbf{S}$$
$$\mathcal{T}[\![\mathsf{bool}]\!] = \mathbf{allocation}$$
$$\mathcal{T}[\![\mathsf{num}]\!] = \mathbf{allocation}$$
$$\mathcal{T}[\![T[1\,..\,N]]\!] = \mathbf{allocations}(\mathcal{T}[\![T]\!], \mathcal{N}[\![N]\!])$$

$$\mathcal{TU}: \text{TYPE} \to \mathbf{LV} \circ\!\!\to \mathbf{S} \circ\!\!\to \mathbf{S}$$
$$\mathcal{TU}[\![\mathsf{bool}]\!] = \mathbf{freedom}$$
$$\mathcal{TU}[\![\mathsf{num}]\!] = \mathbf{freedom}$$
$$\mathcal{TU}[\![T[1\,..\,N]]\!] = \mathbf{freedoms}(\mathcal{TU}[\![T]\!], \mathcal{N}[\![N]\!])$$

"implementation-dependent", so that the semantics of programs that try to exploit side-effects in expressions becomes nondeterministic. Here, let us forbid side-effects, for simplicity. Thus denotations of expressions may be functions from environment and stores to expressible values—there is no need to return the current store, as it is unchanged.

We must modify the semantic equations for expressions, now that the denotations of expressions take stores as arguments. But first, note that in various contexts, there is an implicit "coercion" when the expression evaluation results in a variable, but the current value of the variable is required. Such contexts include operands of operators and conditions of if-then-else expressions. Very few programming languages insist that the programmer use an explicit operator on a variable in order to obtain its current value.

In practical programming languages, various coercions are allowed. A good example is the coercion from a parameterless function to the result of applying the function, allowed in ALGOL60 and Pascal. Of course, a static semantic analysis could use contextual information to recognize such coercions and replace them by explicit operators. But in general, it is easy enough to deal with coercions directly in the dynamic semantics—although languages like ALGOL68 and ADA allow so many coercions that it may then be preferable to define the dynamic semantics on the basis of an intermediate abstract syntax where the coercions have been made explicit.

It is convenient to introduce a secondary semantic function for expressions $\mathcal{R}$ corresponding to ordinary evaluation followed by coercion (when possible). The modifications to our previous specification are straightforward; the result is shown in Table 24, together with the semantic equation for "$E_1[E_2]$".

Table 24
Denotations for expressions (modified)

$$\mathbf{F} = (\mathbf{PV} \circ\!\!\rightarrow \mathbf{S} \circ\!\!\rightarrow \mathbf{FV})_{\perp}$$
$$\mathbf{PV} = \mathbf{V} \oplus \mathbf{F} \oplus \mathbf{LV}$$
$$\mathbf{FV} = \mathbf{V}$$
$$\mathbf{DV} = \mathbf{V} \oplus \mathbf{F} \oplus \mathbf{LV}$$
$$\mathbf{EV} = \mathbf{V} \oplus \mathbf{F} \oplus \mathbf{LV}$$

$$\mathscr{R}: \text{EXPRESSION} \rightarrow \mathbf{Env} \rightarrow \mathbf{S} \rightarrow \mathbf{RV}$$
$$\mathscr{R}[\![E]\!] = \lambda e \in \mathbf{Env}.\ \lambda s \in \mathbf{S}.\ [\mathbf{id}_{\mathbf{RV}}, \perp, \lambda l \in \mathbf{LV}.\ \mathbf{assigned}\ l\ s](\mathscr{E}[\![E]\!]e\ s)$$

$$\mathscr{E}: \text{EXPRESSION} \rightarrow \mathbf{Env} \rightarrow \mathbf{S} \rightarrow \mathbf{EV}$$
$$\mathscr{E}[\![L]\!] = \lambda e \in \mathbf{Env}.\ \lambda s \in \mathbf{S}.\ \mathscr{L}[\![L]\!]$$
$$\mathscr{E}[\![MO\ E_1]\!] = \lambda e \in \mathbf{Env}.\ \lambda s \in \mathbf{S}.\ \mathscr{MO}[\![MO]\!](\mathscr{R}[\![E_1]\!]e\ s)$$
$$\mathscr{E}[\![E_1\ DO\ E_2]\!] = \lambda e \in \mathbf{Env}.\ \lambda s \in \mathbf{S}.\ \mathscr{DO}[\![DO]\!](\mathbf{smash}(\mathscr{R}[\![E_1]\!]e\ s, \mathscr{R}[\![E_2]\!]e\ s))$$
$$\mathscr{E}[\![\mathsf{if}\ E_1\ \mathsf{then}\ E_2\ \mathsf{else}\ E_3]\!] = \lambda e \in \mathbf{Env}.\ \lambda s \in \mathbf{S}.\ (\lambda t \in \mathbf{T}.\ \mathbf{if}\ t\ \mathbf{then}\ \mathscr{E}[\![E_2]\!]e\ s\ \mathbf{else}\ \mathscr{E}[\![E_3]\!]e\ s)\ (\mathscr{R}[\![E_1]\!]e\ s)$$
$$\mathscr{E}[\![I]\!] = \lambda e \in \mathbf{Env}.\ \lambda s \in \mathbf{S}.\ \mathbf{bound}\ I\ e$$
$$\mathscr{E}[\![\mathsf{let}\ CD\ \mathsf{in}\ E]\!] = \lambda e \in \mathbf{Env}.\ \lambda s \in \mathbf{S}.\ \mathscr{E}[\![E]\!](\mathbf{overlay}(\mathscr{CD}[\![CD]\!]e\ s, e)\ s)$$
$$\mathscr{E}[\![\mathsf{fun}\ (PD)\ E]\!] = \lambda e \in \mathbf{Env}.\ \lambda s \in \mathbf{S}.\ (\mathbf{up} \circ \mathbf{strict})(\lambda v \in \mathbf{PV}.\ \mathbf{id}_{\mathbf{FV}} \circ \mathscr{E}[\![E]\!](\mathbf{overlay}(\mathscr{PD}[\![PD]\!]v, e)))$$
$$\mathscr{E}[\![E_1(E_2)]\!] = \lambda e \in \mathbf{Env}.\ \lambda s \in \mathbf{S}.\ (\mathbf{down} \circ \mathbf{id}_{\mathbf{F}})(\mathscr{E}[\![E_1]\!]e\ s)(\mathscr{E}[\![E_2 e\ s)\ s$$
$$\mathscr{E}[\![E_1[E_2]]\!] = \lambda e \in \mathbf{Env}.\ \lambda s \in \mathbf{S}.\ \mathbf{component}(\mathscr{E}[\![E_1]\!]e\ s, \mathscr{R}[\![E_2]\!]e\ s)$$

$$\mathscr{CD}: \text{CONSTANT-DECLARATIONS} \rightarrow \mathbf{Env} \rightarrow \mathbf{S} \rightarrow \mathbf{Env}$$
$$\mathscr{CD}[\![\mathsf{val}\ I = E]\!] = \lambda e \in \mathbf{Env}.\ \lambda s \in \mathbf{S}.\ \mathbf{binding}\ I\ (\mathscr{E}[\![E]\!]e\ s)$$
$$\mathscr{CD}[\![CD_1;\ CD_2]\!] = \lambda e \in \mathbf{Env}.\ \lambda s \in \mathbf{S}.\ (\lambda e_1 \in \mathbf{Env}.\ \mathbf{overlay}(\mathscr{CD}[\![CD_2]\!](\mathbf{overlay}(e_1, e))\ s, e_1))\ (\mathscr{CD}[\![CD_1]\!]e\ s)$$
$$\mathscr{CD}[\![\mathsf{rec}\ CD]\!] = \lambda e \in \mathbf{Env}.\ \lambda s \in \mathbf{S}.\ \mathbf{fix}(\lambda e' \in \mathbf{Env}.\ \mathscr{CD}[\![CD]\!](\mathbf{overlay}(e', e))\ s)$$

5.6. *Statements*

The statements (or commands) of programming languages include *assignments* of values to variables, and constructs to control the order in which assignments are executed. Some typical syntax for statements is given in Table 25.

In the assignment statement "$E_1 := E_2$", the left-hand side E_1 must evaluate to a variable and E_2 must evaluate to an assignment value. The executions of the statements in "S_1 ; S_2" are sequenced (from left to right!) and "skip" corresponds to an

Table 25
Syntax for statements

(STATEMENTS)
$S ::= E_1 := E_2 \mid S_1;\ S_2 \mid \mathsf{skip} \mid$
$\quad \mathsf{if}\ E\ \mathsf{then}\ S_1 \mid \mathsf{while}\ E\ \mathsf{do}\ S_1 \mid$
$\quad \mathsf{begin}\ VD;\ S_1\ \mathsf{end} \mid$
$\quad \mathsf{stop} \mid I : S_1 \mid \mathsf{goto}\ I$

empty sequence of statements. Conditional execution is provided by "if E then S_1", whereas "while" E do S_1" iterates S_1 as long as E is true. The block "begin VD; S_1 end" limits the scope of the variable declarations in VD to the statements S_1, so that the variables themselves are "local" to the block, and may safely be reused after the execution of S_1—assuming that "pointers" to local variables are not permitted. Let us defer consideration of the remaining statements in Table 25 until later in this section.

The denotational semantics of statements is quite simple: denotations are given by functions, from environments and stores, to stores. The bottom store represents the *nontermination* of statement execution, and the functions are strict in their store argument, reflecting that nontermination cannot be "ignored" by subsequent statements.

We are now ready to define the denotations of statements: see Table 26. Notice that the use of $\mathscr{V}\mathscr{U}$ improves the abstractness of statement denotations: without it, the states produced by statement denotations would depend on the local variables allocated in inner blocks.

Table 26
Denotations for statements (direct)

$$\mathscr{S}: \text{STATEMENTS} \to \mathbf{Env} \to \mathbf{S} \circ\!\!\to \mathbf{S}$$
$$\mathscr{S}[\![E_1 := E_2]\!] = \lambda e \in \mathbf{Env}.\ \lambda s \in \mathbf{S}.\ (\lambda l \in \mathbf{LV}.\ \lambda v \in \mathbf{RV}.\ \mathbf{strict\ assign}\ l\ v\ s)$$
$$(\mathscr{E}[\![E_1]\!]e\ s)(\mathscr{R}[\![E_2]\!]e\ s)$$
$$\mathscr{S}[\![S_1 ; S_2]\!] = \lambda e \in \mathbf{Env}.\ \mathscr{S}[\![S_2]\!]e \circ \mathscr{S}[\![S_1]\!]e$$
$$\mathscr{S}[\![\mathsf{skip}]\!] = \lambda e \in \mathbf{Env}.\ \mathbf{id_S}$$
$$\mathscr{S}[\![\mathsf{if}\ E\ \mathsf{then}\ S_1]\!] = \lambda e \in \mathbf{Env}.\ \lambda s \in \mathbf{S}.\ (\lambda t \in \mathbf{T}.\ \mathbf{if}\ t\ \mathbf{then}\ \mathscr{S}[\![S_1]\!]e\ s\ \mathbf{else}\ s)$$
$$(\mathscr{R}[\![E]\!]e\ s)$$
$$\mathscr{S}[\![\mathsf{while}\ E\ \mathsf{do}\ S_1]\!] = \lambda e \in \mathbf{Env}.\ \mathbf{fix}(\lambda c \in \mathbf{S} \circ\!\!\to \mathbf{S}.\ \lambda s \in \mathbf{S}.$$
$$(\lambda t \in \mathbf{T}.\ \mathbf{if}\ t\ \mathbf{then}\ c(\mathscr{S}[\![S_1]\!]e\ s)\ \mathbf{else}\ s)$$
$$(\mathscr{R}[\![E]\!]e\ s))$$
$$\mathscr{S}[\![\mathsf{begin}\ VD; S_1\ \mathsf{end}]\!] =$$
$$\lambda e \in \mathbf{Env}.\ (\lambda(e' \in \mathbf{Env}, s \in \mathbf{S}).\ \mathscr{V}\mathscr{U}[\![VD]\!](e')(\mathscr{S}[\![S_1]\!](\mathbf{overlay}(e', e))(s)))$$
$$\circ \mathscr{V}\mathscr{D}[\![VD]\!]e$$

The following proposition is a direct consequence of the semantic equations, using the unfolding property of "**fix**".

5.4. PROPOSITION

$$\mathscr{S}[\![\mathsf{while}\ E\ \mathsf{do}\ S_1]\!] = \mathscr{S}[\![\mathsf{if}\ E\ \mathsf{then}\ (S_1 ; \mathsf{while}\ E\ \mathsf{do}\ S_1)]\!].$$

Now let us consider the statement "stop", whose intended effect is that when (if ever) the execution of a statement reaches it, the execution of the enclosing program is terminated—without further changes to the state, just as if control had reached the end of the program normally. We may say that "stop" causes a *jump* to the end of the program. (For now, let programs be simply statements. The semantics of programs is considered further in Section 5.8.)

However, with the denotations for statements used so far, we have (for any statement S_1 and $e \in \mathbf{Env}$):

$$\mathcal{S}[\![S_1\,;\,\textsf{while true do skip}]\!]e = \mathbf{fix}(\mathbf{id}_{\mathbf{C}\to\mathbf{C}}) \circ \mathcal{S}[\![S_1]\!]e$$
$$= \bot_{\mathbf{C}} \circ S[\![S_1]\!]e = \bot_{\mathbf{C}}$$

which is in conflict with the intended equivalence of "stop; while true do skip" to "stop".

In order to deal with "stop", we clearly have to change the denotation of "$S_1\,;\,S_2$". There are two main techniques available for modelling jumps such as "stop": "Flags" and "continuations".

The technique using *flags* is to use a domain of denotations such as $\mathbf{Env} \to \mathbf{S} \circ\!\!\to (\mathbf{S} \oplus \mathbf{S})$. Then a resulting store in (say) the first summand may represent normal termination, and a result in the second summand may represent that "stop" has been executed, so that no further statements are to be executed. Thus we would have

$$\mathcal{S}[\![S_1\,;\,S_2]\!] = [\mathcal{S}[\![S_2]\!]e, \mathbf{in}_2] \circ (\mathcal{S}[\![S_1]\!]e).$$

It is easy to imagine the analogous changes that would be needed to the semantic equations for the other statements, to take account of the two possibilities for resulting stores. (No changes would be needed to the semantic equations for expressions and declarations, as they do not involve statements.)

The alternative technique for dealing with jumps is to let denotations of statements take *continuations* as arguments. The continuation argument represents the semantics of what would be the "rest of the program", if the statement were to terminate normally. In the denotation of each statement, it is specified whether to use the continuation argument, or to ignore it and use a different continuation, such as the empty continuation, which represents a jump to the end of the program. A divergent iterative statement just never gets around to using the argument continuation (and strictness is no longer needed to reflect the preservation of divergence).

In the rest of this section, the use of the continuations technique is illustrated, albeit briefly.

Let the characteristic domains (**DV**, **EV**, etc.) be as usual. The domain of *statement continuations* may be taken to be simply the domain $\mathbf{S} \to \mathbf{S}$ of functions on stores. For uniformity, let *all* denotations be functions of continuations. The continuations of expressions are functions from values to ordinary continuations, those for declarations are functions from environments to continuations, etc. (Auxiliary operations, such as **assign**, could be changed to take continuation arguments as well, if desired.) Such a semantics is called a "continuation semantics"; our previous examples of semantics are called "direct".

Sufficient semantic equations to illustrate the technique of continuations are given in Table 27. (The semantic functions of the continuation semantics are marked with primes to distinguish them from the corresponding direct semantic functions.) Notice the order of composition in the semantic equation for "$S_1\,;\,S_2$": the opposite to that in direct semantics!.

The transformation from direct to continuation semantics is straightforward. It may seem quite obvious that the transformation gives an "equivalent" semantics, but it is nontrivial to prove such results: the relations to be established between the domains of

Table 27
Denotations for statements (continuations)

$\mathbf{C}=\mathbf{S}\to\mathbf{S}$

$\mathscr{E}'$: EXPRESSION$\to\mathbf{Env}\to(\mathbf{EV}\to\mathbf{C})\to\mathbf{C}$
$\mathscr{R}'$: EXPRESSION$\to\mathbf{Env}\to(\mathbf{RV}\to\mathbf{C})\to\mathbf{C}$
$\mathscr{V}\mathscr{D}'$: VARIABLE-DECLARATIONS$\to\mathbf{Env}\to(\mathbf{Env}\to\mathbf{C})\to\mathbf{C}$
...

$\mathscr{S}'$: STATEMENTS$\to\mathbf{Env}\to\mathbf{C}\to\mathbf{C}$
$\mathscr{S}'[\![E_1 := E_2]\!] = \lambda e\in\mathbf{Env}.\ \lambda c\in\mathbf{C}.$
$\qquad \mathscr{E}'[\![E_1]\!]e^*(\lambda l\in\mathbf{LV}.\ \mathscr{R}'[\![E_2]\!]e^*(\lambda v\in\mathbf{RV}.\ c\circ(\mathbf{assign}\ l\ v)))$
$\mathscr{S}'[\![S_1; S_2]\!] = \lambda e\in\mathbf{Env}.\ \lambda c\in\mathbf{C}.\ \mathscr{S}'[\![S_1]\!]e\,(\mathscr{S}'[\![S_2]\!]e\ c)$
$\mathscr{S}'[\![\mathsf{skip}]\!] = \lambda e\in\mathbf{Env}.\ \lambda c\in\mathbf{C}.\ c$
$\mathscr{S}'[\![\mathsf{while}\ E\ \mathsf{do}\ S_1]\!] = \lambda e\in\mathbf{Env}.\ \mathbf{fix}(\lambda g\in\mathbf{C}\to\mathbf{C}.\ \lambda c\in\mathbf{C}.$
$\qquad \mathscr{R}'[\![E]\!]e(\lambda t\in\mathbf{T}.\ \mathbf{if}\ t\ \mathbf{then}\ \mathscr{S}'[\![S_1]\!]e\,(g(c))\ \mathbf{else}\ c))$
$\mathscr{S}'[\![\mathsf{stop}]\!] = \lambda e\in\mathbf{Env}.\ \lambda c\in\mathbf{C}.\ \mathbf{id}_{\mathbf{S}}$
$\mathscr{S}'[\![\mathsf{goto}\ I]\!] = \lambda e\in\mathbf{Env}.\ \lambda c\in\mathbf{C}.\ \mathbf{bound}\ I\ e$

$\mathscr{L}\mathscr{D}'$: STATEMENTS$\to\mathbf{Env}\to\mathbf{C}\to\mathbf{Env}$

the direct and continuation semantics have to be defined recursively, and then shown to be well-defined and "inclusive" [44].

Continuations were originally introduced to model the semantics of general "goto"-statements. Consider again the syntax given in Table 25. An occurrence of a labelled statement "$I: S_1$" may be regarded as a *declaration* that binds I, where the scope of this binding is the smallest enclosing block "begin $VD; S_1$ end".

The execution of "goto I" is intended to jump to the statement labelled by I. It may be seen to consist of

(1) the termination of enclosing statements (including procedure calls) up to the innermost "begin $VD; S_1$ end" that includes the declaration of the label I; then

(2) the execution of those parts of S_1 that follow after the label I; and finally

(3) the normal termination of "begin $VD; S_1$ end", provided that no further jump prevents this.

(Actually, this analysis suggests a direct semantics using flags, where label identifiers are bound to pairs consisting of "activation levels" and direct statement denotations: continuations are not actually *necessary* for the denotational description of "goto"-statements.)

Letting **C** be a summand of **DV**, the value bound to I by "$I: S_1$" is $\mathscr{S}'[\![S_1]\!]e\ c$, where c is the continuation argument of $\mathscr{S}'[\![I: S_1]\!]e$. So assuming that the environment argument e includes this binding, the denotation of the "goto"-statement merely replaces its argument continuation by the continuation bound to I, as specified in Table 27. The declarative component of statement denotations may be expressed by a semantic function $\mathscr{L}\mathscr{D}'$ whose definition involves a fixed point, which reflects that the continuations denoted by label identifiers in a block may be mutually recursive. The details are somewhat tedious; let us omit them here, as unrestricted jumps to labels are not allowed in most modern high-level programming languages.

Note that continuations give possibilities for jumps that are even less "disciplined" than those provided by the "goto" statement: a general continuation need have no relation at all to the context of where it is used!

Continuations have been advocated as a standard technique for modelling programming languages (along with the use of environments and states) in preference to direct semantics. Although the adoption of this policy would give a welcome uniformity in models, it would also make the domains of denotations for simple languages (e.g., the lambda calculus) unnecessarily complex—and, at least in some cases, the introduction of continuations would actually reduce the abstractness of denotations.

The popularity of continuations seems to be partly due to the accompanying notational convenience—especially that the order in which denotations of subphrases occur in semantic equations corresponds to the order in which the phrases are intended to be executed: left to right. (Perhaps direct semantics would be more popular if function application and composition were to be written "backwards".) Another notational virtue of continuations is that "errors" can be handled neatly, by ignoring the continuation argument and using a general error-continuation.

5.7. *Procedure abstractions*

Procedure abstractions are much like function abstractions. The only difference is that the body of a procedure abstraction is a statement, rather than an expression.

By the way, many programming languages do not allow functions to be expressed (or declared) directly: procedures must be used instead. The body of the procedure then includes a special statement that determines the value to be returned (in ALGOL60 and Pascal, this statement looks like an assignment to the procedure identifier!).

Syntax for procedure abstractions is given in Table 28. As with functions, we consider procedures with only a single parameter; but now some more modes of parameter evaluation are introduced.

Table 28
Syntax for procedures

(EXPRESSION)
E::= proc (PD) S_1
(PARAMETER-DECLARATION)
PD::= var I : T | I : T
(STATEMENTS)
S::= $E_1(E_2)$

The procedure abstraction "proc (var I : T) S_1" requires its parameter to evaluate to a variable, and I denotes that variable in the body S_1. This mode of parameter evaluation is usually known as "call-by-reference", but here we refer to it as "variable-mode" parameter evaluation.

The procedure abstraction "proc (I : T) S_1" requires its parameter to be *coercible* to an assignable value; then a local variable is allocated and initialized with the parameter value, and I denotes the variable in the body S_1. This mode of parameter evaluation is usually known as "call-by-value", but it should not be confused with the value-mode parameter evaluation that was considered for function abstractions: that did not involve any local variable allocation. Let us refer to this mode as "copy-mode" parameter evaluation.

The procedure call statement "$E_1(E_2)$" executes the body of the procedure abstraction produced by evaluating E_1, passing the argument obtained by evaluating the parameter E_2.

Note that execution of the procedure body may have an effect on the state, by assignment to a nonlocal variable. With variable-mode parameters, there is also the possibility of modifying the state by assigning to the formal parameter of the abstraction; whereas with copy-mode, such an assignment merely modifies the *local* variable denoted by the parameter identifier. Note also that variable-mode allows two different identifiers to denote the same variable, i.e., "aliasing".

Now for the formal semantics of procedures. The denotations of procedure expressions, parameter declarations, and statements are defined in Table 29.

Table 29
Denotations for procedures

$$\mathbf{P} = (\mathbf{PV} \circ\!\!\rightarrow \mathbf{S} \circ\!\!\rightarrow \mathbf{S})_\perp$$
$$\mathbf{PV} = \mathbf{V} \oplus \mathbf{F} \oplus \mathbf{LV} \oplus \mathbf{P}$$
$$\mathbf{DV} = \mathbf{V} \oplus \mathbf{F} \oplus \mathbf{LV} \oplus \mathbf{P}$$
$$\mathbf{EV} = \mathbf{V} \oplus \mathbf{F} \oplus \mathbf{LV} \oplus \mathbf{P}$$

$$\mathcal{E}: \text{EXPRESSION} \rightarrow \mathbf{Env} \rightarrow \mathbf{S} \rightarrow \mathbf{EV}$$
$$\mathcal{E}[\![\mathsf{proc}\ (PD)\ S]\!] = \lambda e \in \mathbf{Env}.\ \mathbf{up}(\mathbf{strict}\lambda v \in \mathbf{PV}.\ (\lambda(e' \in \mathbf{Env}, s \in \mathbf{S}).\ \mathcal{PU}[\![PD]\!]e'(\mathcal{S}[\![S]\!](\mathbf{overlay}(e', e)))) \circ \mathcal{PD}[\![PD]\!]e)$$

$$\mathcal{PD}: \text{PARAMETER-DECLARATION} \rightarrow \mathbf{PV} \rightarrow \mathbf{S} \rightarrow \mathbf{Env} \times \mathbf{S}$$
$$\mathcal{PD}[\![\mathsf{val}\ I\colon T]\!] = \lambda v \in \mathbf{PV}.\ \lambda s \in \mathbf{S}.\ (\mathbf{binding}\ I\ v, s)$$
$$\mathcal{PD}[\![\mathsf{var}\ I\colon T]\!] = \lambda l \in \mathbf{LV}.\ \lambda s \in \mathbf{S}.\ (\mathbf{binding}\ I\ l, s)$$
$$\mathcal{PD}[\![I\colon T]\!] = \lambda v \in \mathbf{PV}.\ \lambda s \in \mathbf{S}.\ (\lambda v' \in \mathbf{RV}.\ (\lambda(l' \in \mathbf{LV}, s' \in \mathbf{S}).\ (\mathbf{binding}\ I\ l', \mathbf{assign}\ l'\ v'\ s'))\ (\mathcal{T}[\![T]\!]s))\ ([\lambda l \in \mathbf{LV}.\ \mathbf{assigned}\ l\ s, \mathbf{id}_{\mathbf{RV}}, \perp, \perp](v))$$

$$\mathcal{PU}: \text{PARAMETER-DECLARATION} \rightarrow \mathbf{Env} \rightarrow \mathbf{S} \rightarrow \mathbf{S}$$
$$\mathcal{PU}[\![\mathsf{val}\ I]\!] = \lambda e \in \mathbf{Env}.\ \mathbf{id}_{\mathbf{S}}$$
$$\mathcal{PU}[\![\mathsf{var}\ I\colon T]\!] = \lambda e \in \mathbf{Env}.\ \mathbf{id}_{\mathbf{S}}$$
$$\mathcal{PU}[\![I\colon T]\!] = \lambda e \in \mathbf{Env}.\ \mathcal{TU}[\![T]\!](\mathbf{bound}\ I\ e)$$

$$\mathcal{S}: \text{STATEMENTS} \rightarrow \mathbf{Env} \rightarrow \mathbf{S} \circ\!\!\rightarrow \mathbf{S}$$
$$\mathcal{S}[\![E_1(E_2)]\!] = \lambda e \in \mathbf{Env}.\ \lambda s \in \mathbf{S}.\ (\mathbf{down} \circ \mathbf{id}_{\mathbf{P}})(\mathcal{E}[\![E_1]\!]e\ s)(\mathcal{E}[\![E_2]\!]e\ s)\ s$$

The procedure call syntax "$E_1(E_2)$" does not give any indication of the mode of parameter evaluation, so we leave it to the denotation of the parameter declaration to perform any required coercion of the parameter value. An alternative technique is to let the evaluation of the parameter expression E_2 depend on a mode component of the value of the procedure expression E_1.

By the way, the second semantic function for parameter declarations, $\mathscr{PU}$, is analogous to the semantic function $\mathscr{VU}$ for variable declarations, explained in Section 5.5.

5.8. *Programs*

As discussed in Section 3, the semantics of an entire program should be a mathematical representation of the observable behaviour when it is executed by computers (but ignoring implementation-dependent details). Typically, this behaviour involves *streams* of "input" and "output".

By definition, the *input* of a program is the information that is supplied to it by the user; the *output* is the information that the user gets back. However, it is important to take into account not only *what* information is supplied, but also *when* the supply takes place. The main distinction in conventional programming languages is between so-called "batch" and "interactive" input-output.

With *batch* input, all the input to the program is supplied at the start of the program. The input may then be regarded as *stored*, in a "file". Batch output is likewise accumulated in a file, and only given to the user when (if ever) the program terminates.

On the other hand, *interactive* input is provided gradually, as a *stream* of data, while the program is running; the program may have to wait for further input data to be provided before it can proceed. Similarly, interactive output is provided to the user while the program is running, as soon as it has been determined.

Note that interactive input-output allows (later) items of input to *depend* on (earlier) items of output. For instance, input may be stimulated by an output "prompt".

We may regard batch input-output as merely a special case of interactive input-output: the program starts, and then immediately reads and stores the entire input; output is stored until the program is about to terminate, and then the entire output is given to the user.

The *essential* difference between batch and interactive input-output shows up in connection with programs that (on purpose) may run "for ever": batch input-output cannot reflect the semantics of such programs. Familiar examples are traffic-light controllers, operating systems, and screen editors. These programs might, if allowed, read an infinite stream of input, and produce an infinite stream of output. (They might also terminate, in response to particular input—or "spontaneously", when an error occurs.) Moreover, once an item of output has been produced, it cannot be revoked by the program (e.g., the traffic-light controller cannot "undo" the changing of a light).

Consider the abstract syntax for input-output statements and programs specified in Table 30. There is nothing in the given *syntax* that indicates whether the *semantics* of input-output is supposed to be batch or interactive. Let us consider both semantics. We restrict items of input and output to be truth values and numbers, i.e., the same as **SV**.

Table 30
Syntax for programs

(PROGRAM)
$P ::= \mathsf{prog}\ S$
(STATEMENTS)
$S ::= \mathsf{read}\ E \mid \mathsf{write}\ E$

For batch semantics, we may take the representation of streams to be finite lists. The semantic equations for programs, and for read and write statements, are given in Table 31; our previous semantic equations for other statements have to be modified to take account of the extra arguments, but the details are omitted here.

The following proposition confirms that batch output is not observable when program execution does not terminate.

Table 31
Denotations for programs (batch)

$$\mathbf{In} = \mathbf{SV}^*$$
$$\mathbf{Out} = \mathbf{SV}^*$$

$$\mathscr{P} : \text{PROGRAM} \to (\mathbf{In} \circ\!\!\to \mathbf{Out})$$
$$\mathscr{P}[\![\mathsf{prog}\, S_1]\!] = \lambda i \in \mathbf{In}.\ \mathbf{on}_3(\mathscr{S}[\![S_1]\!](\mathbf{void})(\mathbf{empty}, i, \top))$$

$$\mathscr{S} : \text{STATEMENTS} \to \mathbf{Env} \to (\mathbf{S} \otimes \mathbf{In} \otimes \mathbf{Out}) \circ\!\!\to (\mathbf{S} \otimes \mathbf{In} \otimes \mathbf{Out})$$
$$\mathscr{S}[\![\mathsf{read}\, E]\!] = \lambda e \in \mathbf{Env}.\ \lambda(s \in \mathbf{S}, i \in \mathbf{In}, o \in \mathbf{Out}).$$
$$(\lambda l \in \mathbf{Loc}.[\bot, \lambda(v \in \mathbf{SV}, i' \in \mathbf{In}).\mathbf{smash}(\mathbf{store}\ l\ v\ s, i', o)])$$
$$(\mathscr{E}[\![E]\!]e\ s)(i)$$
$$\mathscr{S}[\![\mathsf{write}\, E]\!] = \lambda e \in \mathbf{Env}.\ \lambda(s \in \mathbf{S}, i \in \mathbf{In}, o \in \mathbf{Out}).$$
$$(\lambda v \in \mathbf{SV}.\mathbf{smash}(s, i, \mathbf{extend}\ v\ o))$$
$$(\mathscr{R}[\![E]\!]e\ s)$$

$$\mathbf{extend} = \lambda v \in \mathbf{SV}.\ [\lambda x \in \mathbf{O}.(v, \top),$$
$$\lambda(v' \in \mathbf{SV}, o \in \mathbf{Out}).\ (v', \mathbf{extend}\ v\ o)]$$
$$\in \mathbf{SV} \to \mathbf{Out} \to \mathbf{Out}$$

5.5. PROPOSITION

$$\mathscr{P}[\![\mathsf{prog\ while\ true\ do\ write\ 0}]\!] = \mathscr{P}[\![\mathsf{prog\ while\ true\ do\ skip}]\!] = \bot.$$

The reason for this is that the denotation of the nonterminating while-loop is given by the *least* fixed point of a *strict* function.

Now for interactive input-output semantics for the same language. See Table 32. Let us first change from $\mathbf{SV}^*$ to $\mathbf{SV}^\$$ which represents infinite (and partial) streams. (The only difference between $\mathbf{SW}^\$$ and the standard domain construction $\mathbf{SV}^\infty$ is that the latter allows $\bot$ components to be followed by non-$\bot$ components.) This change by itself would *not* make any substantial difference to the semantics of programs: input-output would still be batch, and the above proposition would still hold.

Table 32
Denotations for programs (interactive, continuations)

$$\mathbf{SV}^{\$} = \mathbf{SV} \otimes \mathbf{SV}^{\$}_{\bot}$$
$$\mathbf{In} = \mathbf{SV}^{\$}$$
$$\mathbf{Out} = \mathbf{SV}^{\$}$$
$$\mathbf{C} = \mathbf{S} \to \mathbf{In} \to \mathbf{Out}$$

$$\mathscr{P}': \text{PROGRAM} \to \mathbf{In} \to \mathbf{Out}$$
$$\mathscr{P}'[\![\mathsf{prog}\ S_1]\!] = \mathscr{S}'[\![S_1]\!](\mathbf{void})(\lambda s \in \mathbf{S}.\ \lambda i \in \mathbf{In}.\ \top)(\mathbf{empty})$$

$$\mathscr{S}': \text{STATEMENTS} \to \mathbf{Env} \to \mathbf{C} \to \mathbf{C}$$
$$\mathscr{S}'[\![\mathsf{read}\ E]\!] = \lambda e \in \mathbf{Env}.\ \lambda c \in \mathbf{C}.$$
$$\mathscr{E}'[\![E]\!]e\ (\lambda l \in \mathbf{Loc}.\ \lambda s \in \mathbf{S}.\ \lambda(v \in \mathbf{SV}, i \in \mathbf{In}).(c \circ \mathbf{store}\ l\ v)\ s\ i))$$
$$\mathscr{S}'[\![\mathsf{write}\ E]\!] = \lambda e \in \mathbf{Env}.\ \lambda c \in \mathbf{C}.$$
$$\mathscr{R}'[\![E]\!]e\ (\lambda v \in \mathbf{SV}.\ \lambda s \in \mathbf{S}.\ \lambda i \in \mathbf{In}.\mathbf{smash}(v, \mathbf{up}(c\ s\ i)))$$

The essential change is to ensure that an item of output becomes incorporated in the program's semantics, irrevocably, as soon as the corresponding "write" statement is executed. There are various ways of achieving this property: in particular, by using continuations. Reverting temporarily to continuation semantics (see Section 5.6) we define the interactive semantics of programs as shown in Table 32.

5.6. Proposition

$$\mathscr{P}'[\![\mathsf{prog\ while\ true\ do\ write\ 0}]\!] \neq \mathscr{P}'[\![\mathsf{prog\ while\ true\ do\ skip}]\!].$$

It is instructive to see how to deal with interactive input-output without using continuations. Consider the domain **IO** defined in Table 33, and let statement denotations be given by functions from environments and stores to **IO**. Each element of **IO** represents a sequence of readings and writings, ending (if at all) with a state. This might not seem particularly abstract, but notice that statement denotations *must* reflect the order in which readings and writing occur, since the semantics of a program in $\mathbf{In} \to \mathbf{Out}$ reveals this information when applied to *partial* inputs.

The semantic equations specified in Table 33 illustrate this technique. The fixed point used in the denotation of "S_1 ; S_2" essentially corresponds to going through the input and output corresponding to S_1 until a final state is reached, and then starting S_2; similarly for programs. It can be shown that interactive output is modelled.

Now consider "piping" the output of one program into the input of another, as expressed by a program construct "$P_1 \mid P_2$". With interactive input-output, both programs can be started simultaneously—but the execution of the second program may have to be suspended to await input that has yet to be output by the first program. The start of the first program could be delayed until the second program actually tries to read from its input (if ever), and then execution could alternate between the two programs, according to the input-output. All these possibilities are expressed by the

Table 33
Denotations for programs (interactive, direct)

$$\mathbf{IO} = \mathbf{S} \oplus (\mathbf{SV} \circ\!\!\rightarrow \mathbf{IO})_\perp \oplus (\mathbf{SV} \otimes \mathbf{IO}_\perp)$$

$$\mathscr{P} : \text{PROGRAM} \rightarrow \mathbf{In} \rightarrow \mathbf{Out}$$
$$\begin{aligned}\mathscr{P}[\![\mathsf{prog}\ S_1]\!] = \mathbf{fix}(&\lambda h \in \mathbf{IO} \rightarrow \mathbf{In} \rightarrow \mathbf{Out}.\\ &[\lambda s \in \mathbf{S}.\, \lambda i \in \mathbf{In}.\, \top,\\ &\ \lambda f \in \mathbf{SV} \rightarrow \mathbf{IO}.\, \lambda(v \in \mathbf{SV}, i \in \mathbf{In}).\, h(f(v))(i),\\ &\ \lambda(v \in \mathbf{SV}, io \in \mathbf{IO}).\, \lambda i \in \mathbf{In}.\, (v, \mathbf{up}(h(io)(i)))])\\ &(\mathscr{S}[\![S_1]\!](\mathbf{void})(\mathbf{empty}))\end{aligned}$$

$$\mathscr{S} : \text{STATEMENTS} \rightarrow \mathbf{Env} \rightarrow \mathbf{S} \circ\!\!\rightarrow \mathbf{IO}$$
$$\begin{aligned}\mathscr{S}[\![E_1 := E_2]\!] &= \lambda e \in \mathbf{Env}.\, \lambda s \in \mathbf{S}.\\ &\quad (\lambda l \in \mathbf{Loc}.\, \lambda v \in \mathbf{SV}.\, \mathbf{in}_1(\mathbf{store}\ l\ v\ s))\\ &\quad (\mathscr{E}[\![E_1]\!] e\ s)(\mathscr{R}[\![E_2]\!] e\ s)\\ \mathscr{S}[\![\mathsf{read}\ E]\!] &= \lambda e \in \mathbf{Env}.\, \lambda s \in \mathbf{S}.\\ &\quad (\lambda l \in \mathbf{Loc}.\, \mathbf{in}_2(\lambda v \in \mathbf{SV}.\, \mathbf{up}(\mathbf{in}_1(\mathbf{store}\ l\ v\ s))))\\ &\quad (\mathscr{E}[\![E_1]\!] e\ s)\\ \mathscr{S}[\![\mathsf{write}\ E]\!] &= \lambda e \in \mathbf{Env}.\, \lambda s \in \mathbf{S}.\\ &\quad (\lambda v \in \mathbf{SV}.\, \mathbf{in}_3(v, \mathbf{up}(\mathbf{in}_1(s))))\\ &\quad (\mathscr{R}[\![E_2]\!] e\ s)\\ \mathscr{S}[\![\mathsf{skip}]\!] &= \lambda e \in \mathbf{Env}.\, \mathbf{id}_{\mathbf{S}}\\ \mathscr{S}[\![S_1; S_2]\!] &= \lambda e \in \mathbf{Env}.\, \lambda s \in \mathbf{S}.\\ &\quad \mathbf{fix}(\lambda g \in \mathbf{IO} \rightarrow \mathbf{IO}.\\ &\qquad [\mathscr{S}[\![S_2]\!] e,\\ &\qquad\ \lambda f \in \mathbf{SV} \rightarrow \mathbf{IO}.\, g \circ f,\\ &\qquad\ \lambda(v \in \mathbf{SV}, io \in \mathbf{IO}).\, (v, \mathbf{up}(g(io)))])\\ &\quad (\mathscr{S}[\![S_1]\!] e\ s)\end{aligned}$$

same semantic equation:

$$\mathscr{P}[\![P_1 \mid P_2]\!] = \mathscr{P}[\![P_2]\!] \circ \mathscr{P}[\![P_1]\!].$$

With batch input-output, the second program does not start until the first one terminates. As with statements, such sequential execution can be modelled by composition of strict functions (the semantic equation for piped programs remains the same, assuming $\mathscr{P}$ is defined as for batch input-output).

5.9. *Nondeterminism*

The final technique illustrated in this chapter is the use of *powerdomains* to model nondeterministic constructs such as "guarded commands" and interleaving.

For our purposes here, it is not necessary to understand the actual structure of power domains. All that we need to know about a powerdomain is that it is equipped with a continuous union operation (associative, commutative, and absorptive), a continuous singleton operation, and that functions on domains can be extended pointwise to powerdomains. (Recall the notation adoption in Section 4. We use only the natural, or convex, power domain; the other power domains do not accurately reflect the possibility of divergence.)

Table 34
Syntax for guarded statements

$$
\begin{array}{|l|}
\hline
(\text{GUARDED-STATEMENTS}) \\
\quad G ::= E \to S_1 \mid G_1 [\,]\, G_2 \\
(\text{STATEMENTS}) \\
\quad S ::= \mathsf{if}\ G\ \mathsf{fi} \mid \mathsf{do}\ G\ \mathsf{od} \\
\hline
\end{array}
$$

Consider the syntax for *guarded statements* given in Table 34. The intention of "$E \to S_1$" is that the statement S_1 is guarded by E and may only be executed if E evaluates to true. So far, this resembles "if E then S_1"; the difference is that guarded statements may be "united" by the construct "G_1 [ ] G_2", whose execution consists of executing precisely one of the guarded statements in G_1 and G_2. Notice that (when E evaluates to a truth value) the guarded statement

$$E \to S_1 [\,]\ \neg E \to S_2$$

expresses a deterministic choice between S_1 and S_2, whereas

$$\mathsf{true} \to S_1\ [\,]\ \mathsf{true} \to S_2$$

expresses a nondeterministic choice.

Both the statements "if G fi" and "do G od" involve the execution of G, when possible. Let us regard the former as equivalent to an empty statement when it is not possible to execute G. With the latter, the execution of G is *repeated*, as many times as possible.

We take the denotations for statements to be functions from environments and stores to elements of the powerdomain $\mathbf{S}^\natural$; these elements represent the nonempty sets of possible states resulting from statement execution (possibly including $\perp$). The denotations of guarded statements are similar, but $\top$ represents the empty set of states. The semantic equations are specified in Table 35. (We do not need to change the denotations of expressions and declarations, which are still deterministic.)

Table 35
Denotations for guarded statements

$$
\begin{array}{|l|}
\hline
\mathcal{G}: \text{GUARDED-STATEMENTS} \to \mathbf{Env} \to \mathbf{S} \circ\!\!\to (\mathbf{O} \oplus \mathbf{S}^\natural) \\
\mathcal{G}[\![E \to S_1]\!] = \lambda e \in \mathbf{Env}.\ \mathbf{strict}\lambda s \in \mathbf{S}. \\
\qquad\qquad (\lambda t \in \mathbf{T}.\ \mathbf{if}\ t\ \mathbf{then}\ \mathbf{in}_2(\mathcal{S}[\![S_1]\!] e\, s)\ \mathbf{else}\ \mathbf{in}_1\ \top)(\mathcal{R}[\![E]\!] e\, s) \\
\mathcal{G}[\![G_1 [\,]\, G_2]\!] = \lambda e \in \mathbf{Env}.\ \mathbf{strict}\lambda s \in \mathbf{S}. \\
\qquad\qquad\quad [\lambda x \in \mathbf{O}.\ \mathbf{id}_{\mathbf{O} \oplus \mathbf{S}^\natural}, \\
\qquad\qquad\quad\ \lambda p_1 \in \mathbf{S}^\natural.\ [\lambda x \in \mathbf{O}.\ \mathbf{in}_2(p_1), \\
\qquad\qquad\qquad\qquad \lambda p_2 \in \mathbf{S}^\natural.\ \mathbf{in}_2(p_1 \uplus p_2)]](\mathcal{G}[\![G_1]\!] e\, s)(\mathcal{G}[\![G_2]\!] e\, s) \\
\\
\mathcal{S}: \text{STATEMENTS} \to \mathbf{Env} \to \mathbf{S} \circ\!\!\to \mathbf{S}^\natural \\
\mathcal{S}[\![\mathsf{if}\ G\ \mathsf{fi}]\!] = \lambda e \in \mathbf{Env}.\ \mathbf{strict}\lambda s \in \mathbf{S}. \\
\qquad\qquad [\lambda x \in \mathbf{O}.\ \{\!|s|\!\},\ \mathbf{id}_{\mathbf{S}^\natural}](\mathcal{G}[\![G]\!] e\, s) \\
\mathcal{S}[\![\mathsf{do}\ G\ \mathsf{od}]\!] = \lambda e \in \mathbf{Env}.\ \mathbf{fix}(\lambda c \in \mathbf{S} \circ\!\!\to \mathbf{S}.\ \mathbf{strict}\lambda s \in \mathbf{S}. \\
\qquad\qquad\qquad\qquad [x \in \mathbf{O}.\ \{\!|s|\!\},\ \mathbf{ext}(c)](\mathcal{G}[\![G]\!] e\, s)) \\
\hline
\end{array}
$$

As an illustration of the semantic equivalence that is induced by the above definitions, consider the two statements S_1, S_2 shown in Table 36. It is obvious that S_2 has the possibility of not terminating; what may be less obvious is that S_1 has precisely the same possibilities, as is expressed in the following theorem.

Table 36
Examples of guarded statements S_1, S_2

x := 0;
y := 0;
do x=0→x := 1
[]x=0→y := y+1
od
	x := 0;
	y := 0;
	do x=0→x := 1
	[]x=0→y := y+1
	[] true→do true→skip od
	od

5.7. PROPOSITION. $\mathscr{S}[\![S_1]\!]=\mathscr{S}[\![S_2]\!]$.

Thus both statements have the possibility of terminating with the variable "y" having *any* (nonnegative) value—or of not terminating. The infinite number of possibilities arises here from the *iteration* of a choice between a *finite* number of possibilities: the possibility of nontermination cannot be eliminated (cf. König's Lemma).

However, one could imagine having a *primitive* statement with an infinite number of possibilities, excluding nontermination. For instance, consider "randomize E", which is supposed to set a variable E to some arbitrary integer. Here we understand "arbitrary" to mean just that the value chosen is completely out of the control of the program—it is implementation-dependent. (Thus a particular implementation might always choose zero, or the successor of the previous choice. Classes of genuinely random implementations could be considered as well.)

It is important to note that our domain of statement denotations above does *not* contain any element that can be used for the denotation of an always terminating "randomize" statement. In fact any attempt to express such a set as

$$\{\!|0|\!\}\uplus\{\!|1|\!\}\uplus\cdots\uplus\{\!|n|\!\}\uplus\cdots$$

as an element of $\mathbf{N}_\perp^\natural$ always ends up by including $\{\!|\perp|\!\}$ as well.

So let us omit further consideration of randomizing statements, and proceed to illustrate a technique known as "resumptions", which is useful for giving a denotational semantics for concurrent processes.

5.10. *Concurrency*

The language constructs considered so far in this chapter come from conventional programming languages, designed to be implemented *sequentially*. Several modern programming languages have constructs for expressing so-called "concurrent processes", and may be implemented on a "distributed system" of computers (or on a single computer that simulates a distributed system). Typically, the processes are executed asynchronously, and they interact by sending messages and making "rendezvous".

In the denotational semantics of concurrent systems, the concurrent execution steps of different processes are usually regarded as "interleaved". Although interleaving is a rather artificial concept when dealing with physically distributed systems (due to the lack of a universal time scale) it is not generally possible to distinguish the possible behaviours of proper concurrent systems from their interleaved counterparts—at least, not unless the observer of the behaviours is distributed too.

The final example of this chapter deals with a very simple form of concurrency: interleaved statements. The syntax of these statements is given in Table 37.

Table 37
Syntax for interleaved statements

(STATEMENTS)
$S ::= S_1 \parallel S_2 \mid \langle S \rangle$

The intention with the statement "$S_1 \parallel S_2$" is that S_1 and S_2 are executed concurrently and asynchronously. If S_1 and S_2 use the same variables, the result of their concurrent execution may depend on the order in which the "steps" of S_1 are executed in relation to those of S_2, i.e., on the interleaving. Let us assume that assignment statements are single, "indivisible" steps of execution, so the state does not change during the evaluation of the left- and right-hand sides. The construct "$\langle S_1 \rangle$" makes the execution of any statement S_1 an indivisible step (sometimes called a "critical region").

Note that when S_1 and S_2 are "independent" (e.g., when they use different variables) an execution of "$S_1 \parallel S_2$" gives the same result as the execution of "$S_1 ; S_2$", or of "$S_2 ; S_1$"; but in general there are other possible results.

Now consider statements

$$S_1 : \mathsf{x} := 1 \qquad S_2 : \mathsf{x} := 0;\ \mathsf{x} := \mathsf{x} + 1.$$

With all our previous denotations for statements, we have $\mathscr{S}[\![S_1]\!] = \mathscr{S}[\![\mathscr{S}_2]\!]$. But when statements include "$S_1 \parallel S_2$", we expect

$$\mathscr{S}[\![S_1 \parallel S_1]\!] \neq \mathscr{S}[\![S_1 \parallel S_2]\!]$$

since the interleaving "$\mathsf{x} := 0; \mathsf{x} := 1; \mathsf{x} := \mathsf{x} + 1$" of S_1 with S_2 sets x to 2, whereas the interleaving of "$\mathsf{x} := 1$" with itself does not have this possibility.

Thus it can be seen that the compositionality of denotational semantics forces $\mathscr{S}[\![S_1]\!] \neq \mathscr{S}[\![S_2]\!]$ when concurrent statements are included. The appropriate denotations for statements are so-called "resumptions", which are rather like segmented ("staccato") continuations. A domain of resumptions is defined in Table 38. The semantic function $\mathscr{S}$ for statements maps environments directly to resumptions, which are themselves functions of stores.

Consider $p = \mathscr{S}[\![S_1]\!] e\, s$. It represents the set of possible results of executing the *first step* of S_1. An element $\mathbf{in}_1(s')$ of this set corresponds to the possibility that there is only

Table 38
Denotations for interleaved statements

$$
\begin{array}{l}
\mathbf{R} = \mathbf{S} \circ\!\!\rightarrow (\mathbf{S} \oplus (\mathbf{R}_\perp \otimes \mathbf{S}))^\natural \\
\\
\mathscr{S}: \text{STATEMENTS} \rightarrow \mathbf{Env} \rightarrow \mathbf{R} \\
\mathscr{S}[\![E_1 := E_2]\!] = \lambda e \in \mathbf{Env}.\ \mathbf{strict} \lambda s \in \mathbf{S}. \\
\qquad (\lambda l \in \mathbf{LV}.\ \lambda v \in \mathbf{RV}.\ \{\!|\mathbf{store}\ l\,v\,s|\!\})(\mathscr{E}[\![E_1]\!]es)(\mathscr{R}[\![E_2]\!]es) \\
\mathscr{S}[\![S_1; S_2]\!] = \lambda e \in \mathbf{Env}.\ \mathbf{fix}(\lambda f \in \mathbf{R} \rightarrow \mathbf{R}.\ \lambda r \in \mathbf{R}. \\
\qquad \mathbf{ext}[\mathscr{S}[\![S_2]\!]e, \\
\qquad\quad \lambda(r' \in \mathbf{R}_\perp, s' \in \mathbf{S}).\ (f(r'), s')] \circ r) \\
\qquad (\mathscr{S}[\![S_1]\!]e) \\
\mathscr{S}[\![\mathsf{skip}]\!] = \lambda e \in \mathbf{Env}.\ \mathbf{strict} \lambda s \in \mathbf{S}.\ \{\!|s|\!\} \\
\mathscr{S}[\![S_1 \| S_2]\!] = \lambda e \in \mathbf{Env}.\ \mathbf{fix}(\lambda g \in (\mathbf{R} \times \mathbf{R}) \rightarrow \mathbf{R}. \\
\qquad \lambda(r_1 \in \mathbf{R}, r_2 \in \mathbf{R}).\ \mathbf{strict} \lambda s \in \mathbf{S}. \\
\qquad\quad (\mathbf{ext}[r_2, \lambda(r'_1 \in \mathbf{R}_\perp, s' \in \mathbf{S}).\ (g(r'_1, r_2), s')](r_1(s))) \uplus \\
\qquad\quad (\mathbf{ext}[r_1, \lambda(r'_2 \in \mathbf{R}_\perp, s' \in \mathbf{S}).\ (g(r_1, r'_2), s')](r_2(s)))) \\
\qquad (\mathscr{S}[\![S_1]\!]e)(\mathscr{S}[\![S_2]\!]e) \\
\mathscr{S}[\![\langle S_1 \rangle]\!] = \lambda e \in \mathbf{Env}.\ \mathbf{fix}(\lambda h \in \mathbf{R} \rightarrow \mathbf{R}.\ \lambda r \in \mathbf{R}. \\
\qquad \mathbf{ext}[\lambda s \in \mathbf{S}.\ \{\!|s|\!\}, \\
\qquad\quad \lambda(r' \in \mathbf{R}, s' \in \mathbf{S}).\ h(r')(s')] \circ r) \\
\qquad (\mathscr{S}[\![S_1]\!]e)
\end{array}
$$

one step, resulting in the state s' (although this "step" might be an indivisible sequence of steps). An element $\mathbf{in}_2(\mathbf{up}\, r, s')$ corresponds to the result of the first step being an *intermediate* state s', together with a resumption r which, when applied to s' (or to some other state) gives the set of possible results from the next step of S_1, and so on.

Resumptions provide adequate denotations for interleaved statements, as the semantic equations in Table 38 show. However, these denotations are not particularly abstract: e.g., we get $\mathscr{S}[\![\mathsf{skip}]\!] \neq \mathscr{S}[\![\mathsf{skip}; \mathsf{skip}]\!]$, even though the two statements are clearly interchangeable in any program. It is currently an open problem to define fully abstract denotations for concurrent interleaved statements (using standard semantic domain constructions).

The technique of resumptions can also be used for expressing denotations of *communicating* concurrent processes (with the "store" component representing pending communications).

We have now finished illustrating the use of the main descriptive techniques of Denotational Semantics: environments, stores, strictness, flags, continuations, power-domains, and resumptions. The various works referenced in the following bibliographical notes provide further illustrations of the use of these techniques, and show how to obtain denotations for many of the constructs to be found in "real" programming languages.

6. Bibliographical notes

This final section refers to some published works on Denotational Semantics and related topics, and indicates their significance.

6.1. Development

The development of Denotational Semantics began with the paper "Towards a formal semantics" [52], written by Christopher Strachey in 1964 for the IFIP Working Conference on *Formal Language Description Languages.* The paper introduces compositionally defined semantic functions that map abstract syntax to "operators" (i.e., functions), and it makes use of the fixed point combinator **Y**, for expressing the denotations of loops. It also introduces (compound) L-values and R-values, in connection with the semantics of assignment and parameter passing. The treatment of identifier bindings follows Landin's approach [21]: identifiers are mapped to bound variables of λ-abstractions.

Strachey's paper "Fundamental concepts of programming languages" [53] provides much of the conceptual analysis of programming languages that underlies their denotational semantics.

The main theoretical problem with Strachey's early work was that, formally, denotations were specified using the type-free lambda calculus, for which there was no known model. In fact, Strachey was merely using λ-abstractions as a convenient way of expressing functions, rather than as a formal calculus. However, the fixed point combinator **Y** was needed (for obtaining a compositional semantics for iterative constructs, for instance). Because **Y** involves self-application it was considered to be "paradoxical": it could be interpreted operationally, but it could not be regarded as expressing a function. By 1969, Dana Scott had become interested in Strachey's ideas. In an exciting collaboration with Strachey, Scott first convinced Strachey to give up the type-free lambda calculus; then he discovered that it did have a model, after all. Soon after that, Scott established the theory of semantic domains, providing adequate foundations for the semantic descriptions that Strachey had been writing.

The original paper on semantic domains by Scott [46] takes domains to be complete lattices (rather than the cpo's used nowadays). Domains have effectively given bases; Cartesian product, (coalesced) sum, and continuous function space are allowed as domain constructors; and solutions of domain equations are found as limits of sequences of embeddings. A domain providing a model for self-application (and hence for the lambda calculus) is given, and a recursively defined domain for the denotations of storable procedures is proposed. (For references to subsequent presentations of domain theory, see [18].)

In a joint paper [48], Scott and Strachey present what is essentially the approach now known as Denotational Semantics (it was called "Mathematical Semantics" until 1976). The paper establishes meta-notation for defining semantic functions, and uses functional notation—rather than the lambda calculus—for specifying denotations. Here, for the first time, denotations are taken to be functions of environments, following a suggestion of Scott. The abstract syntax of finite programs is a set of derivation trees, although it is pointed out that this set could be made into a domain: then semantic functions are continuous, and their existence is guaranteed by the fixed point theorem (see also [47], where partial and infinite programs are considered).

The notion of "characteristic domains" was introduced by Strachey in [54], where characteristic domains are given for ALGOL60 and for a pedagogical language (PAL).

The use of continuations in denotational semantics was proposed by Christopher Wadsworth, and reported in a joint paper with Strachey [55]. The present author was one of the first to exploit the technique, in a denotational description of ALGOL60 [28].

By the mid-1970s, sufficient techniques had been developed for specifying the denotational semantics of any conventional (sequential) programming language. Moreover, John Reynolds [44] and Robert Milne [22] had devised a way of proving the equivalence of denotational descriptions that involve different domains (e.g., direct and continuation semantics for the same language). Wadsworth had shown the relation between the computational and denotational semantics of the lambda calculus [59] (see also [36]). The present author had constructed a prototype "semantics implementation system" (SIS), for generating implementations of programming languages directly from their denotational descriptions [29, 30, 31]. Strachey's inspiration was sorely missed after his untimely death in 1975; but there was confidence that denotational semantics was the best approach to programming language semantics, and that it would be a routine matter to apply it to any real programming language.

Then the increasing interest in *concurrent* systems of processes led to the development of programming languages with nondeterministic constructs. An early treatment by Robin Milner [24] introduced a technique using so-called "oracles", but did not give sufficiently abstract denotations: for instance, nondeterministic choice was not commutative. Then Gordon Plotkin showed how to define powerdomains [40]. The introduction of powerdomains required domains to be cpo's, rather than complete lattices. Moreover, for domains to be closed under powerdomain constructions the cpo's had to be restricted to be so-called SFP objects: limits of sequences of finite cpo's (equivalent to the bifinite cpo's, see [18]). Much of Plotkin's paper is devoted to establishing the SFP framework. Also, the technique of "resumptions" is introduced, and used to define the denotations for some simple parallel programs.

Mike Smyth gave a simple presentation of Plotkin's power domains [49] (and introduced a "weak" powerdomain). Matthew Hennessy and Plotkin together defined a category of "nondeterministic" domains [20], and showed that the Plotkin powerdomain $D^{\natural}$ of a domain D is just the free continuous semilattice generated by D. They also introduced a tensor product for nondeterministic domains, and obtained full abstractness for a simple (although somewhat artificial) parallel programming language. Krzysztof Apt and Plotkin [3] related the Plotkin powerdomain to operational semantics; they showed that Smyth's weak powerdomain (of states) corresponds to Dijkstra's predicate transformers. Plotkin [41] generalized powerdomains to deal with countable nondeterminism. Samson Abramsky has shown [1] that the Plotkin powerdomain gives fully abstract denotations when observable behaviour is characterized by classes of finite experiments. There has also been work on powerdomains using complete metric spaces [2].

Despite all the above works, it is debatable whether the denotational treatment of concurrency is satisfactory. There are difficulties with getting reasonable abstractness of denotations when using resumptions. Moreover, the use of powerdomains gives an unwelcome notational burden. In contrast, Structural Operational Semantics (illustrated in [25]) extends easily from sequential languages to concurrency.

Another problem with the applicability of Denotational Semantics concerns the *pragmatic* aspects of denotational desciptions. For "toy" languages, it is quite a simple matter to "lay the domains on the table" (following [54]), and to give semantic equations that define appropriate (but not necessarily fully abstract) denotations. However, the approach does not scale up easily to "real" programming languages, which (unfortunately) seem to require a large number of complex domains for their denotational semantics. Partly because the semantic equations depend explicitly on the domains of denotations, it can be extremely difficult to comprehend a large denotational description.

A related problem is that it is not feasible to reuse parts of the description of one language (Pascal, say) in the description of another language (MODULA 2, for instance). Analogous problems in software engineering were alleviated by the introduction of "modules". Denotational Semantics has no notation for expressing modules. In fact, if the definitions of the domains of denotations were to be encapsulated in modules, it would not be possible to express denotations using λ-notation in the semantic equations: one would have to use auxiliary operations, defined in the modules, for expressing primitive denotations and for combining denotations. Thus it seems that a high degree of modularity is incompatible with (conventional) denotational semantics.

An aggravating factor, concerning the problem of (writing and reading) large denotational descriptions, may be that the intimate relation between higher-order functions on domains and computational properties is not immediately apparent. (For example, with nonstrict functions, arguments may not need to be evaluated.) It is difficult for the non-specialist to appreciate the abstract denotations of programming constructs.

The effort required to formulate a denotational semantics for a real programming language is reflected by the lack of published denotational descriptions of complete, real programming languages. Efforts have been made for SNOBOL [56], ALGOL60 [28], ALGOL68 [22], Pascal [58], and ADA [11]. In general, these descriptions make some simplifying assumptions about the programming language concerned; they also omit the definitions of various "primitive" functions, and use numerous notational conventions whose formal status is somewhat unclear. (Of course, much the same—and other—criticisms could be made of alternative forms of semantics.)

Hope for the future of denotational semantics lies in the recent popularization of two languages that have been designed with formal (denotational) semantics in mind: Standard ML [19], and Scheme [43]. Although the denotational descriptions of these languages are not used formally as standards for implementations, they do show that it is possible to give complete descriptions of useful languages.

6.2. *Exposition*

There are several expository works that explain the basic notions of Denotational Semantics, and give examples of the prevailing techniques for choosing denotations:

Bob Tennent [57] provides a basic tutorial introduction, containing a semantic description of Reynold's experimental language GEDANKEN and a useful bibliography.

The epic work by Milne and Strachey [23], completed by Milne after Strachey's death, contains careful discussions of many techniques for choosing denotations, including less abstract "nonstandard" denotations. The examples given are related to ALGOL68. It is a valuable reference for further study of Denotational Semantics.

Joe Stoy's book [50] is partly based on Strachey's lectures at Oxford; consequently, scant attention is paid to the syntactic constructs of later programming languages, such as PASCAL. The foreword by Scott gives an detailed appreciation of Strachey and his work.

The book by Mike Gordon [16] takes an engineering approach: it does not explain foundations at all. The techniques illustrated are adequate for the description of most Pascal-like programming languages.

The introductory book on Denotational Semantics by Dave Schmidt [45] includes a rather comprehensive description of domain theory (including powerdomains). The book covers a number of incidental topics, such as semantics-directed compiler generation, and there is a substantial bibliography.

Unfortunately, there is considerable variation in the notation (and notational conventions) used in the works referenced above: almost the only common notational feature is the use of "λ" for function abstraction and juxtaposition for function application! The reader should be prepared to adapt not only to different symbols used for the same constants and operators, but also to different choices of what to regard as primitive and what to define as auxiliary notation. (N.B. the notation presented and used in this chapter is *not* an accepted standard.)

6.3. *Variations*

The approach to semantics presented in this chapter, whose development is sketched above, may be regarded as the main theme of Denotational Semantics: abstract syntax, domains of denotations, semantic functions defined by semantic equations using λ-notation. Some significant variations on this theme are indicated below.

Initial Algebra Semantics

(This approach was sketched in Sections 2 and 3.) Initial Algebra Semantics was developed by Joseph Goguen, Jim Thatcher, Eric Wagner, and Jesse Wright [15]. Although it is formally equivalent to denotational semantics, it has the advantage of making it explicit that abstract syntax is an initial algebra, and that semantic functions are homomorphic. Explicit structural induction proofs in denotational semantics can here be replaced by appeals to initiality. It is easy to extend Initial Algebra Semantics to continuous algebras, so as to allow infinite programs, whereas, with denotational semantics, abstract syntax has to be changed from sets to cpo's. An additional benefit of Initial Algebra Semantics is that one always names the domains of denotations; this seems to encourage the specification of denotations as *compositions*, rather than as applications and abstractions (but this is only a matter of style). Initial Algebra Semantics has not yet been applied to real programming languages.

The French school of Algebraic Semantics [17] has concentrated on the semantics of program schemes, rather than of particular programming languages (see [9]).

OBJ

Goguen and Kamran Parsaye Ghomi show in [14] how the algebraic specification language OBJT (a precursor of OBJ2 [13]) can be used to give modular semantic descriptions of programming languages. Their framework is first-order, and not strictly compositional; but higher-order algebras, which give the power of λ-notation in an algebraic framework [39, 42, 12], could be used instead of OBJ, with similar modularity.

Despite the use of explicit modules, the semantic equations given by Goguen and Parsaye Ghomi are still sensitive to the functionality of denotations. The approach has not been applied to real programming languages.

VDM

The Vienna Development Method, VDM [5], has an elaborate notation, called META-IV, that can be used to give denotational descriptions of programming languages. Although there are quite a few variants of META-IV, these share a substantial, partly standardized auxiliary notation that provides a number of useful "flat" domain constructors (e.g., sets, maps) and declarative and imperative constructs (e.g., let constructions, storage allocation, sequencing, exception handling). However, this auxiliary notation is a supplement to, rather than a replacement for, the λ-notation. The foundations of META-IV have been investigated by Stoy [51] and Brian Monahan [27].

In contrast to Denotational Semantics, VDM avoids the use of higher-order functions and nonstrict abstractions, in order to keep close to the familiar objects of conventional programming. (Andrzej Blikle and Andrzej Tarlecki [7] went even further, and advocated avoidance of reflexive domains.) But as in Denotational Semantics, there are severe problems with large-scale descriptions, due to the lack of modularity. The fact that it has been possible to develop semantic descriptions of real programming languages such as CHILL [8] and ADA [6] in (extended versions of) META-IV is a tribute to the discipline and energy of their authors, rather than evidence of an inherent superiority of META-IV over Denotational Semantics.

Action semantics

Action Semantics [38, 37, 33, 60, 35, 32] is something of a mixture of the denotational, algebraic, and operational approaches to formal semantics. It has been under development since 1977, by the present author and (since 1984) by David Watt. A brief summary of the main features of Action Semantics is given below.

The primary aim is to make it easier to deal with semantic descriptions of "real" programming languages. Factors that have been addressed include modularity (to obtain ease of modification, extension, and reuse) and notation (to improve comprehensibility).

Action Semantics is compositional, just like Denotational Semantics. The essential difference between Action Semantics and Denotational Semantics concerns the entities that are taken as the denotations of program phrases: so-called "actions", rather than functions on semantic domains. Actually, some actions do correspond closely to functions, and are determined purely by the relation between the "information" that they receive and produce. But other actions have a more operational essence: they process information *gradually*, and they may interfere (or collaborate) when put together.

The standard notation for actions is polymorphic, in that actions may be combined without regard to what "kind" of information they process: transient or stored data, bindings, or communications. Furthermore, the different kinds of information are processed independently. This allows the semantic equations for (say) arithmetical expressions to stay the same, even when expressions might be polluted with "side-effects" or communications.

The fundamental concepts of programming languages (as identified by Strachey, and implicit in most denotational descriptions) can be expressed straightforwardly in action notation. The comprehensibility of action semantic descriptions is enhanced by the use of suggestive words, rather than (to programmers) cryptic mathematical symbols: in the usual concrete representation of action notation, for example, the action combinator expressing sequential peformance is written as infixed "**then**". The notation is claimed to be a reasonable "compromise" between previous formal notations and informal English.

The theory of action includes some pleasant algebraic laws. However, the basic understanding of actions is operational, and action equivalence is defined by a (structural) operational semantics [32]. Action notation may also be defined denotationally, and used as auxiliary notation in conventional denotational descriptions as illustrated in [34].

At the time of writing (1989), it is not apparent whether Action Semantics will turn out to be any more palatable than Denotational Semantics to the programming community.

References

[1] ABRAMSKY, S., Experiments, powerdomains, and fully abstract models for applicative multiprogramming, In: *Proc. Internat. Conf. on Foundations of Computation Theory*, Lecture Notes in Computer Science, Vol. 158 (Springer, Berlin, 1983) 1–13.

[2] AMERICA, P., J. DE BAKKER, J.N. KOK and J. RUTTEN, Denotational semantics of a parallel object-oriented language, *Inform. and Comput.* **83**(2) (1989) 152–205.

[3] APT, K.R. and G.D. PLOTKIN, A Cook's tour of countable nondeterminism, in: *Proc. 8th Internat. Coll. on Automata, Languages and Programming*, Lecture Notes in Computer Science, Vol. 115 (Springer, Berlin, 1981) 479–494.

[4] BARENDREGT, H., Functional programming and lambda calculus, in: J. van Leeuwen, ed., *Handbook of Theoretical Computer Science, Vol. B* (North-Holland, Amsterdam, 1990) 321–363.

[5] BJØRNER, D. and C.B. JONES, eds., *Formal Specification and Software Development* (Prentice-Hall, Englewood Cliffs, NJ, 1982).

[6] BJØRNER, D. and O.N. OEST, eds., *Towards a Formal Description of ADA*, Lecture Notes in Computer Science, Vol. 98 (Springer, Berlin, 1980).

[7] BLIKLE, A. and A. TARLECKI, Naive denotational semantics, in: *Proc. 9th IFIP Internat. Congr. on Information Processing* (North-Holland, Amsterdam, 1983) 345–355.

[8] CHILL language definition, recommendation Z200, Study Group XI (CCITT, May 1980).

[9] COURCELLE, B., Recursive applicative program schemes, in: J. van Leeuwen, ed., *Handbook of Theoretical Computer Science, Vol. B* (North-Holland, Amsterdam, 1990) 459–492.

[10] COUSOT, P., Methods and logics for proving programs, in: J. van Leeuwen, ed., *Handbook of Theoretical Computer Science, Vol. B* (North-Holland, Amsterdam, 1990) 841–993.

[11] DONZEAU-GOUGE, V., G. KAHN, B. LANG et al., Formal definition of the ADA programming language (preliminary version), INRIA, Versailles 1980.

[12] DYBJER, P., Domain algebras, in: *Proc. 11th Internat. Coll. on Automata, Languages, and Programming*, Lecture Notes in Computer Science, Vol. 172 (Springer, Berlin, 1984) 138–150.

[13] FUTATSUGI, K., J.A. GOGUEN, J.-P. JOUANNAUD and J. MESEGUER, Principles of OBJ2, in: *Proc. 12th Ann. ACM Symp. on Principles of Programming Languages* (1985) 52–66.

[14] GOGUEN, J.A. and K. PARSAYE-GHOMI, Algebraic denotational semantics using parameterized abstract modules, in: *Proc. Internat. Coll. on Formalization of Programming Concepts*, Lecture Notes in Computer Science, Vol. 107 (Springer, Berlin, 1981) 292–309.

[15] GOGUEN, J.A., J.W. THATCHER, E.G. WAGNER and J.B. WRIGHT, Initial algebra semantics and continuous algebras, *J. ACM* **24**(1977) 68–95.

[16] GORDON, M.J.C., *The Denotational Description of Programming Languages* (Springer, Berlin, 1979).

[17] GUESSARIAN, I., *Algebraic Semantics*, Lecture Notes in Computer Science, Vol. 99 (Springer, Berlin, 1981).

[18] GUNTER, C.A. and D.S. SCOTT, Semantic domains, in: J. van Leeuwen, ed., *Handbook of Theoretical Computer Science, Vol. B* (North-Holland, Amsterdam, 1990).

[19] HARPER, R., D. MACQUEEN and R. MILNER, Standard ML, Report ECS-LFCS-86-2, Computer Science Dept., Univ. of Edinburgh, Edinburgh, 1986.

[20] HENNESSY, M.C.B. and G.D. PLOTKIN, Full abstraction for a simple parallel programming language, in: *Proc. Symp. on Mathematical Foundations of Computer Science*, Lecture Notes in Computer Science, Vol. 74 (Springer, Berlin, 1979) 108–121.

[21] LANDIN, P.J., A formal description of ALGOL60, in: *Proc. IFIP TC2 Working Conf. on Formal Language Description Languages for Computer Programming* (North-Holland, Amsterdam, 1966) 266–294.

[22] MILNE, R.E., The formal semantics of computer languages and their implementations, Ph.D. Thesis, Univ. of Cambridge, 1974; available as Tech. Microfiche TCF–2, Programming Research Group, Univ. of Oxford, 1974.

[23] MILNE, R.E. and C. STRACHEY, *A Theory of Programming Language Semantics* (*Part a, Part b*) (Chapman & Hall, London, 1976).

[24] MILNER, R., Processes: A mathematical model of computing agents, in: *Proc. Logic Colloq. '73* (North-Holland, Amsterdam, 1975) 157–174.

[25] MILNER, R., Operational and algebraic semantics of concurrent processes, in: J. van Leeuwen, ed., *Handbook of Theoretical Computer Science, Vol. B* (North-Holland, Amsterdam, 1990) 1201–1242.

[26] MITCHELL, J.C., Types in programming, in: J. van Leeuwen, ed., *Handbook of Theoretical Computer Science, Vol. B* (North-Holland, Amsterdam, 1990).

[27] MONAHAN, B.Q., A type model for VDM, in: D. Bjørner, C.B. Jones, et al., eds., *VDM–A Formal Method at Work*, Lecture Notes in Computer Science, Vol. 252 (Springer, Berlin, 1987) 210–236.

[28] MOSSES, P.D., The mathematical semantics of ALGOL60, Tech. Monograph PRG-12, Programming Research Group, Univ. of Oxford, 1974.

[29] MOSSES, P.D., Mathematical semantics and compiler generation, Ph.D. Dissertation, Univ. of Oxford, 1975.

[30] MOSSES, P.D., Compiler generation using denotational semantics, in: *Proc. Symp. on Mathematical Foundations of Computer Science*, Lecture Notes in Computer Science, Vol. 45 (Springer, Berlin, 1976) 436–441.

[31] MOSSES, P.D., SIS, Semantics Implementation System: Reference manual and user guide, Tech. Monograph MD–30, Computer Science Dept., Aarhus Univ., 1979. (Note: the system SIS itself is no longer available).

[32] MOSSES, P.D., Action semantics, Draft, Version 8, 1990.

[33] MOSSES, P.D., The modularity of action semantics, Internal Report DAIMI IR–75, Computer Science Dept., Aarhus Univ., 1988.

[34] MOSSES, P.D., A practical introduction to denotational semantics, in: *Formal Description of Programming Concepts*, IFIP State-of-the-Art Report (Springer, Berlin, 1990).

[35] MOSSES, P.D., Unified algebras and action semantics, in: *Proc. 6th Ann. Symp. on Theoretical Aspects of Computer Science*, Lecture Notes in Computer Science, Vol. 349 (Springer, Berlin, 1989) 17–35.
[36] MOSSES, P.D. and G.D. Plotkin, On limiting completeness, *SIAM J. Comput.* **16** (1987) 179–194.
[37] MOSSES, P.D. and D.A. WATT, Pascal: action semantics, Draft, Version 0.3, 1986.
[38] MOSSES, P.D. and D.A. WATT, The use of action semantics, in: *Proc. IFIP TC2 Working Conf. on Formal Description of Programming Concepts III* (North-Holland, Amsterdam, 1987) 135–163.
[39] PARSAYE-GHOMI, K., Higher Order Data Types, Ph.D. Thesis, Computer Science Dept., Univ. of California at Los Angeles, 1981.
[40] PLOTKIN, G.D., A powerdomain construction, *SIAM J. Comput.* **5**(3) (1976) 452–487.
[41] PLOTKIN, G.D., A powerdomain for countable nondeterminism, in: *Proc. 9th Internat. Coll. on Automata, Languages, and Programming*, Lecture Notes in Computer Science, Vol. 140 (Springer, Berlin, 1982) 418–428.
[42] POIGNÉ, A., On semantic algebras, Tech. Report, Informatik II, Universität Dortmund, 1983.
[43] REES, J., W. CLINGER, et al., The revised^3 report on the algorithmic language Scheme, *ACM SIGPLAN Notices* **21**(12) (1986) 37–39.
[44] REYNOLDS, J.C., On the relation between direct and continuation semantics, in: *Proc. 2nd Internat. Coll. on Automata, Languages, and Programming*, Lecture Notes in Computer Science, Vol. 14 (Springer, Berlin, 1974) 157–168.
[45] SCHMIDT, D.A., *Denotational Semantics: A Methodology for Language Development* (Allyn & Bacon, Newton, MA, 1986).
[46] SCOTT, D.S., Outline of a mathematical theory of computation, in: *Proc. 4th Ann. Princeton Conf. on Information Sciences and Systems* (1970) 169–176; a revised and slightly expanded version is Tech. Monograph PRG-2, Programming Research Group, Univ. of Oxford, 1970.
[47] SCOTT, D.S., The lattice of flow diagrams, in: *Proc. Symp. on Semantics of Algorithmic Languages*, Lecture Notes in Mathematics, Vol. 188 (Springer, Berlin, 1971) 311–366.
[48] SCOTT, D.S., and C. STRACHEY, Toward a mathematical semantics for computer languages in: *Proc. Symp. on Computers and Automata*, Microwave Research Institute Symposia Series, Vol. 21 (Polytechnic Institute of Brooklyn, 1971) 19–46.
[49] SMYTH, M.B., Power domains, *J. Comput. System. Sci.* **16** (1978) 23–36.
[50] STOY, J.E., *The Scott–Strachey Approach to Programming Language Theory* (MIT Press, Cambridge, MA, 1977).
[51] STOY, J.E., Mathematical foundations, in D. Bjørner and C.B. Jones, eds., [5, Chapter 3].
[52] STRACHEY, C., Towards a formal semantics, in: *Proc. IFIP TC2 Working Conf. on Formal Language Description Languages for Computer Programming* (North-Holland, Amsterdam, 1966) 198–220.
[53] STRACHEY, C., Fundamental concepts of programming languages, Lecture Notes for a NATO Summer School, Copenhagen; available from Programming Research Group, Univ. of Oxford, 1967.
[54] STRACHEY, C., The varieties of programming language, in: *Proc. Internat. Computing Symp.* (Cini Foundation, Venice, 1972) 222–223; a revised and slightly expanded version is Tech. Monograph PRG-10, Programming Research Group, Univ. of Oxford, 1973.
[55] STRACHEY, C. and C.P. WADSWORTH, Continuations: a mathematical semantics for handling full jumps, Tech. Monograph PRG-11, Programming Research Group, Univ. of Oxford, 1974.
[56] TENNENT, R.D., Mathematical semantics of SNOBOL4, in: *Proc. Ann. ACM Symp. on Principles of Programming Languages* (1973) 95–107.
[57] TENNENT, R.D., The denotational semantics of programming languages, *Comm. ACM* **19** (1976) 437–453.
[58] TENNENT, R.D., A denotational description of the programming language Pascal, Tech. Report, Programming Research Group, Univ. of Oxford, 1978.
[59] WADSWORTH, C.P., The relation between the computational and denotational properties of Scott's D_∞-models of the lambda-calculus, *SIAM J. Comput.* **5** (1976) 488–521.
[60] WATT, D.A., An action semantics of Standard ML, in: *Proc. 3rd Workshop on Mathematical Foundations of Programming Language Semantics*, Lecture Notes in Computer Science, Vol. 298 (Springer, Berlin, 1988) 572–598.

CHAPTER 12

Semantic Domains

C.A. GUNTER

Department of Computer and Information Sciences, University of Pennsylvania, Philadelphia, PA 19104, USA

D.S. SCOTT

Department of Computer Science, Carnegie-Mellon University, Pittsburgh, PA 15213, USA

Contents

1. Introduction 635
2. Recursive definitions of functions 636
3. Effectively presented domains 641
4. Operators and functions 644
5. Powerdomains 653
6. Bifinite domains 660
7. Recursive definitions of domains 663
References 674

HANDBOOK OF THEORETICAL COMPUTER SCIENCE
Edited by J. van Leeuwen

1. Introduction

The theory of domains was established in order to have appropriate spaces on which to define semantic functions for the denotational approach to programming-language semantics. There are two needs: first, there have to be spaces of several different types available, to mirror both the type distinctions in programming languages and also to allow for different kinds of semantical constructs—especially in dealing with languages with side effects; and second, the theory has to account for computability properties of functions—if the theory is to be realistic. The first need is complicated by the fact that types can be both compound (or made up from other types) and recursive (or self-referential), and that a high-level language of types and a suitable semantics of types is required to explain what is going on. The second need is complicated by these complications of the semantical definitions and the fact that it has to be checked that the level of abstraction reached still allows a precise definition of computability.

This degree of abstraction had only partly been served by the state of recursion theory in 1969 when D.S. Scott started working on denotational semantics in collaboration with C. Strachey. In order to fix some mathematical precision, he took over some definitions of recursion theorists such as Kleene, Nerode, Davis, and Platek and gave an approach to a simple type theory of higher-type functionals. It was only after giving an abstract characterization of the spaces obtained (through the construction of bases) that he realized that recursive definitions of types could be accommodated as well—and that the recursive definitions could incorporate function spaces as well. Though it was not the original intention to find semantics of the so-called *untyped λ-calculus*, such a semantics emerged along with many ways of interpreting a very large variety of languages.

A large number of people have made essential contributions to the subsequent developments. They have shown in particular that domain theory is not one monolithic theory, but that there are several different kinds of constructions giving classes of domains appropriate for different mixtures of constructs. The story is, in fact, far from finished even today. In this chapter we will only be able to touch on a few of the possibilities, but we give pointers to the literature. Also, we have attempted to explain the foundations in an elementary way—avoiding heavy prerequisites (such as category theory) but still maintaining some level of abstraction—with the hope that such an introduction will aid the reader in going further into the theory.

The chapter is divided into seven sections. In Section 2 we introduce a simple class of ordered structures and discuss the idea of fixed points of continuous functions as meanings for recursive programs. In Section 3 we discuss computable functions and effective presentations. Section 4 defines some of the operators and functions which are used in semantic definitions and describes their distinguishing characteristics. A special collection of such operators called *powerdomains* are discussed in Section 5. Closure problems with respect to the *convex* powerdomain motivate the introduction of the class of *bifinite* domains which we describe in Section 6. Section 7 deals with the important issue of obtaining fixed points for (certain) operators on *domains*. We illustrate the method by showing how to find domains D satisfying isomorphisms such as $D \cong D \times D \cong D \to D$ and $D \cong \mathbf{N} + (D \to D)$. (Such domains are models of the above-mentioned untyped λ-calculus.)

Many of the proofs for results presented below are only sketched or even omitted. With a few exceptions, the enthusiastic reader should be able to fill in proofs without great difficulty. For the exceptions we provide a warning and a pointer to the literature.

2. Recursive definitions of functions

It is the essential purpose of the theory of domains to study classes of spaces which may be used to give semantics for recursive definitions. In this section we discuss spaces having certain kinds of limits in which a useful fixed point existence theorem holds. We will briefly indicate how this theorem can be used in semantic specification.

2.1. Cpo's and the Fixed Point Theorem

A *partially ordered set* is a set D together with a binary relation $\sqsubseteq$ which is reflexive, anti-symmetric and transitive. We will usually write D for the pair $\langle D, \sqsubseteq \rangle$ and abbreviate the phrase "partially ordered set" with the term "poset". A subset $M \subseteq D$ is *directed* if, for every finite set $u \subseteq M$, there is an upper bound $x \in M$ for u. A poset D is *complete* (and hence a *cpo*) if every directed subset $M \subseteq D$ has a least upper bound $\bigsqcup M$ and there is a least element $\bot_D$ in D. When D is understood from context, the subscript on $\bot_D$ will usually be dropped.

It is not hard to see that *any* finite poset that has a least element is a cpo. The easiest such example is the one-point poset $\mathbf{I}$. Another easy example which will come up later is the poset $\mathbf{O}$ which has two distinct elements $\top$ and $\bot$ with $\bot \sqsubseteq \top$. The *truth value cpo* $\mathbf{T}$ is the poset which has three distinct points, $\bot$, **true**, **false**, where $\bot \sqsubseteq$ **true** and $\bot \sqsubseteq$ **false** (see Fig. 1). To get an example of an infinite cpo, consider the set $\mathbf{N}$ of natural numbers with the discrete ordering (i.e., $n \sqsubseteq m$ if and only if $n = m$). To get a cpo, we need to add a "bottom" element to $\mathbf{N}$. The result is a cpo $\mathbf{N}_\bot$ which is pictured in Fig. 1. This is a rather simple example because it does not have any interesting directed subsets. Consider the ordinal ω; it is not a cpo because it has a directed subset (namely ω itself) which has no least upper bound. To get a cpo, one needs to add a top element to get the cpo $\omega^\top$ pictured in Fig. 1. For a more subtle class of examples of cpo's, let $\mathscr{P}S$ be the set of (all) subsets of a set S. Ordered by ordinary set inclusion, $\mathscr{P}S$ forms a cpo whose least upper bound operation is just set union. As a last example, consider the set $\mathbf{Q}$ of rational numbers with their usual ordering. Of course, $\mathbf{Q}$ lacks the bottom and top elements, but there is another problem which causes $\mathbf{Q}$ to fail to be a cpo: $\mathbf{Q}$ lacks, for example, the square root of 2! However, the unit interval $[0, 1]$ of real numbers *does* form a cpo.

Given cpo's D and E, a function $f: D \to E$ is monotone if $f(x) \sqsubseteq f(y)$ whenever $x \sqsubseteq y$. If f is monotone and $f(\bigsqcup M) = \bigsqcup f(M)$ for every directed M, then f is said to be *continuous*. A function $f: D \to E$ is said to be *strict* if $f(\bot) = \bot$. We will usually write $f: D \circ\!\!\to E$ to indicate that f is strict. If $f, g: D \to E$, then we say that $f \sqsubseteq g$ if and only if $f(x) \sqsubseteq g(x)$ for every $x \in D$. With this ordering, the poset of continuous functions $D \to E$ is

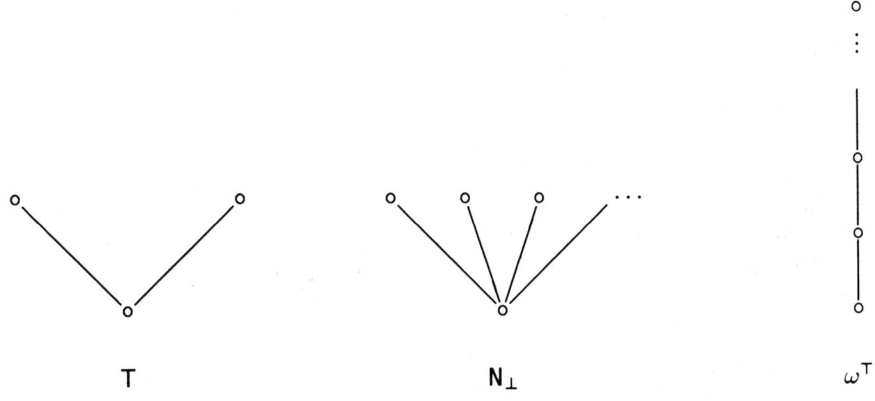

Fig. 1. Examples of cpo's.

itself a cpo. Similarly, the poset of strict continuous functions $D \circ\!\!\rightarrow E$ is also a cpo. (Warning: we use the notation $f: D \rightarrow E$ to indicate that f is a function with domain D and codomain E in the usual set-theoretic sense. On the other hand, $f \in D \rightarrow E$ means that $f: D \rightarrow E$ is continuous. A similar convention applies to $D \circ\!\!\rightarrow E$.)

To get a few examples of continuous functions, note that when $f: D \rightarrow E$ is monotone and D is finite, then f is continuous. In fact, this is true whenever D has no infinite ascending chains. For example, any monotone function $f: \mathbf{N}_\perp \rightarrow E$ is continuous. On the other hand, the function $f: \omega^\top \rightarrow \mathbf{O}$ which sends the elements of ω to $\perp$ and sends $\top$ to $\top$ is monotone, but it is not continuous. Given sets S, T and a function $f: S \rightarrow T$ we define the *extension* of f to be the function $f^*: \mathscr{P}S \rightarrow \mathscr{P}T$ given by taking

$$f^*(X) = \{f(x) \mid x \in X\}$$

for each subset $X \subseteq S$. The function f^* is monotone and, for any collection X_i of subsets of S, we have

$$f^*\left(\bigcup_i X_i\right) = \bigcup_i f^*(X_i).$$

In particular, f^* is continuous. For readers who know a bit about functions on the real numbers, it is worth noting that a function $f: [0, 1] \rightarrow [0, 1]$ on the unit interval may be continuous in the cpo sense without being continuous in the usual sense.

Now, the central theorem may be stated as follows:

2.1. THEOREM (Fixed Point). *If D is a cpo and $f: D \rightarrow D$ is continuous, then there is a point* $\mathbf{fix}(f) \in D$ *such that* $\mathbf{fix}(f) = f(\mathbf{fix}(f))$ *and* $\mathbf{fix}(f) \sqsubseteq x$ *for any* $x \in D$ *such that* $x = f(x)$. *In other words,* $\mathbf{fix}(f)$ *is the least fixed point of* f.

PROOF. Note that $\perp \sqsubseteq f(\perp)$. By an induction on n using the monotonicity of f, it is easy to see that $f^n(\perp) \sqsubseteq f^{n+1}(\perp)$ for every n. Set $\mathbf{fix}(f) = \bigsqcup_n f^n(\perp)$. By the continuity of f, it is easy to see that $\mathbf{fix}(f)$ is a fixed point of f. To see that it is the least such, note that if x is a fixed point of f, then, for each n, $f^n(\bot) \sqsubseteq f^n(x) = x$. □

2.2. *Some applications of the Fixed Point Theorem*

The factorial function

As a first illustration of the use of the Fixed Point Theorem, let us consider how one might define the *factorial function* $\mathbf{fact}: \mathbf{N}_\bot \to \mathbf{N}_\bot$. The usual approach is to say that the factorial function is a strict function which satisfies the following recursive equation for each number n:

$$\mathbf{fact}(n) = \begin{cases} 1 & \text{if } n=0, \\ n * \mathbf{fact}(n-1) & \text{if } n>0, \end{cases}$$

where $*, -: \mathbf{N} \times \mathbf{N} \to \mathbf{N}$ are multiplication and subtraction respectively. But how do we know that there *is* a function **fact** which satisfies this equation? Define a function

$$F: (\mathbf{N}_\bot \circ\!\!\to \mathbf{N}_\bot) \to (\mathbf{N}_\bot \circ\!\!\to \mathbf{N}_\bot)$$

by setting

$$F(f)(n) = \begin{cases} 1 & \text{if } n=0, \\ n * f(n-1) & \text{if } n>0, \\ \bot & \text{if } n=\bot \end{cases}$$

for each $f: \mathbf{N}_\bot \circ\!\!\to \mathbf{N}_\bot$. The definition of F is *not* recursive (F appears only on the left side of the equation) so F certainly exists. Moreover, it is easy to check that F is continuous (but not strict). Hence, by the Fixed Point Theorem, F has a least fixed point $\mathbf{fix}(F)$ and this solution will satisfy the equation for **fact**.

Context-free grammars

One familiar kind of recursion equation is a context-free grammar. Let Σ be an alphabet. One uses context-free grammars to specify subsets of the collection Σ^* of finite sequences of letters from Σ.[1] Here are some easy examples:

(1) $E ::= \varepsilon \mid Ea$

defines the strings of a's (including the empty string ε).

(2) $E ::= a \mid bEb$

defines strings consisting either of the letter a alone or a string of n b's followed by an a followed by n more b's.

(3) $E ::= \varepsilon \mid aa \mid EE$

defines strings of a's of even length.

[1]The superscripted asterisk will be used in three entirely different ways in this chapter. Unfortunately, all of these usages are standard. Fortunately, however, it is usually easy to tell which meaning is correct from context.

We may use the Fixed Point Theorem to provide a precise explanation of the semantics of these grammars. Since the operations $X \mapsto \{\varepsilon\} \cup X\{a\}$, $X \mapsto \{a\} \cup \{b\}X\{b\}$, and $X \mapsto \{\varepsilon\} \cup \{a\}\{a\} \cup XX$ are all continuous in the variable X, it follows from the Fixed Point Theorem that equations such as

(1) $X = \{\varepsilon\} \cup X\{a\}$,
(2) $X = \{a\} \cup \{b\}X\{b\}$,
(3) $X = \{\varepsilon\} \cup \{a\}\{a\} \cup XX$,

corresponding to the three grammars mentioned above, all have least solutions. These solutions are the languages defined by the grammars.

The Schroder–Bernstein Theorem

As a set-theoretic application of the Fixed Point Theorem we offer the proof of the following:

2.2. THEOREM (Schroder–Bernstein). *Let S and T be sets. If $f: S \to T$ and $g: T \to S$ are injections, then there is a bijection $h: S \to T$.*

PROOF. The function $Y \mapsto (T - f^*(S)) \cup f^*(g^*(Y))$ from $\mathscr{P}T$ to $\mathscr{P}T$ is easily seen to be continuous with respect to the inclusion ordering. Hence, by the Fixed Point Theorem, there is a subset

$$Y = (T - f^*(S)) \cup f^*(g^*(Y)).$$

In particular, $T - Y = f^*(S - g^*(Y))$ since

$$\begin{aligned} T - Y &= T - ((T - f^*(S)) \cup f^*(g^*(Y))) \\ &= (T - (T - f^*(S))) \cap (T - (f^*(g^*(Y)))) \\ &= f^*(S) \cap (T - (f^*(g^*(Y)))) \\ &= f^*(S - g^*(Y)). \end{aligned}$$

Now define $h: S \to T$ by

$$h(x) = \begin{cases} y & \text{if } x = g(y) \text{ for some } y \in Y, \\ f(x) & \text{otherwise.} \end{cases}$$

This makes sense because g is an injection. Moreover, h itself is an injection since f and g are injections. To see that it is a surjection, suppose $y \in T$. If $y \in Y$, then $h(g(y)) = y$. If $y \notin Y$, then $y \in f^*(S - g^*(Y))$, so $y = f(x) = h(x)$ for some x. Thus h is a bijection. □

2.3. *Uniformity*

The question naturally arises as to why we take the *least* fixed point in order to get the meaning. In most instances there will be other choices. There are several answers to this question. First of all, it seems intuitively reasonable to take the least defined function satisfying a given recursive equation. But more importantly, taking the least fixed point yields a *canonical* solution. Indeed, it is possible to show that, given a cpo D, the function $\mathbf{fix}_D: (D \to D) \to D$ given by $\mathbf{fix}_D(f) = \bigsqcup_n f^n(\bot)$ is actually *continuous*. But are there other operators like **fix** that could be used? The following definition is helpful.

DEFINITION. A *fixed point operator* F is a class of continuous functions

$$F_D: (D \to D) \to D$$

such that, for each cpo D and continuous function $f: D \to D$, we have $F_D(f) = f(F_D(f))$.

Let us say that a fixed point operator F is *uniform* if, for any pair of continuous functions $f: D \to D$ and $g: E \to E$ and strict continuous function $h: D \circ\!\!\to E$ which makes the following diagram commute

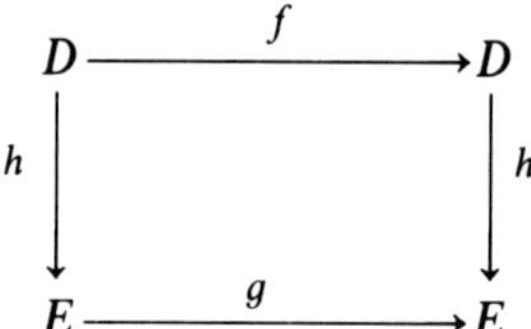

we have $h(F_D(f)) = F_E(g)$. We leave it to the reader to show that **fix** is a uniform fixed point operator. What is less obvious, and somewhat more surprising, is the following theorem.

2.3. THEOREM. **fix** *is the* unique *uniform fixed point operator.*

PROOF. To see why this must be the case, let D be a cpo and suppose $f: D \to D$ is continuous. Then the set

$$D' = \{x \in D \mid x \sqsubseteq \mathbf{fix}(f)\}$$

is a cpo under the order that it inherits from the order on D. In particular, the restriction f' of f to D' has $\mathbf{fix}_D(f)$ as its *unique* fixed point. Now, if $i: D' \to D$ is the inclusion map then the following diagram commutes:

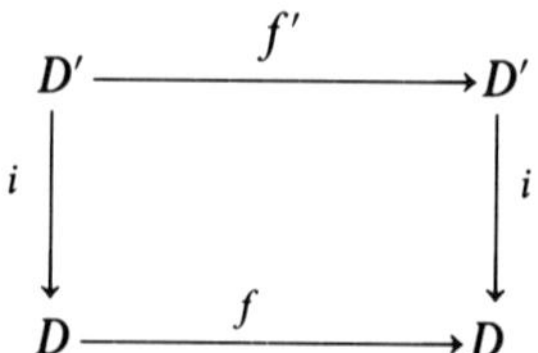

Thus, if F is a uniform fixed point operator, we must have $F_D(f) = F_{D'}(f')$. But $F_{D'}(f')$ is a fixed point of f' and must therefore be equal to $\mathbf{fix}_D(f)$. □

We hope that these results go some distance toward convincing the reader that **fix** is a reasonable operator to use for the semantics of recursively defined functions.

3. Effectively presented domains

There is a significant problem with the full class of cpo's as far as the theory of computation goes. There does not seem to be any reasonable way to define a general notion of *computable function* between cpo's. It is easy to see that these ideas make perfectly good sense for a noteworthy collection of examples. Consider a strict function $f:\mathbf{N}_\perp \circ\!\!\to \mathbf{N}_\perp$. If we take $f(n)=\perp$ to mean that f is *undefined* at n, then f can be viewed as a partial function on $\mathbf{N}$. We wish to have a concept of computability for functions on (some class of) cpo's so that f is computable just in case it corresponds to the usual notion of a partial recursive function. But we must also have a definition that applies to *functionals*, that is, functions which may take functions as arguments or return functions as values. We already encountered a functional earlier when we defined the factorial. To illustrate the point that there is a concept of computability that applies to such operators, consider, for example, a functional $F:(\mathbf{N}_\perp \circ\!\!\to \mathbf{N}_\perp)\circ\!\!\to \mathbf{N}_\perp$ which takes a function $f:\mathbf{N}_\perp \circ\!\!\to \mathbf{N}_\perp$ and computes the value of f on the number 3. The functional F is continuous and it is *intuitively* computable. This intuition comes from the fact that, to compute $F(f)$ on an argument, one needs only know how to compute f on an argument.

Our goal is to define a class of cpo's for which a notion of "finite approximation" makes sense. Let D be a cpo. An element $x \in D$ is *compact* if, whenever M is a directed subset of D and $x \sqsubseteq \bigsqcup M$, there is a point $y \in M$ such that $x \sqsubseteq y$. We let $K(D)$ denote the set of compact elements of D. The cpo D is said to be *algebraic* if, for every $x \in D$, the set $M=\{x_0 \in K(D) \mid x_0 \sqsubseteq x\}$ is directed and $\bigsqcup M = x$. In other words, in an algebraic cpo, each element is a directed limit of its "finite" (compact) approximations. If D is algebraic and $K(D)$ is countable, then we will say that D is a *domain*.

With the exception of the unit interval of real numbers, all of the cpo's we have mentioned so far are domains. The compact elements of the domain $\mathbf{N}_\perp \circ\!\!\to \mathbf{N}_\perp$ are the functions with finite domain of definition, i.e. those continuous functions $f:\mathbf{N}_\perp \circ\!\!\to \mathbf{N}_\perp$ such that $\{n \mid f(n) \neq \perp\}$ is finite. As another example, the collection $\mathscr{P}\mathbf{N}$ of subsets of $\mathbf{N}$, ordered by subset inclusion is a domain whose compact elements are just the finite subsets of $\mathbf{N}$.

One thing which makes domains particularly nice to work with is the way one may describe a continuous function $f:D \to E$ between domains D and E using the compact elements. Let G_f be the set of pairs (x_0, y_0) such that $x_0 \in K(D)$ and $y_0 \in K(E)$ and $y_0 \sqsubseteq f(x_0)$. If $x \in D$, then one may recover from G_f the value of f on x as

$$f(x)=\bigsqcup\{y_0 \mid (x_0, y_0) \in G_f \text{ and } x_0 \sqsubseteq x\}.$$

This allows us to characterize, for example, a continuous function $f:\mathscr{P}\mathbf{N} \to \mathscr{P}\mathbf{N}$ between *uncountable* cpo's with a *countable* set G_f. The significance of this fact for the theory of computability is not hard to see; we will say that the function f is computable just in case G_f is computable (in a sense to be made precise below).

3.1. Normal subposets and projections

Before we give the formal definition of computability for domains and continuous functions, we digress briefly to introduce a useful relation on subposets. Given a poset $\langle A, \sqsubseteq\rangle$ and $x\in A$, let $\downarrow x=\{y\in A \mid y\sqsubseteq x\}$.

DEFINITION. Let A be a poset and suppose $N\subseteq A$. Then N is said to be *normal* in A (and we write $N\lhd A$) if, for every $x\in A$, the set $N\cap\downarrow x$ is directed.

The following lemma lists some useful properties of the relation $\lhd$.

3.1. Lemma. *Let C be a poset with a least element and suppose A and B are subsets of C.*
(1) *If $A\lhd B\lhd C$ then $A\lhd C$.*
(2) *If $A\subseteq B\subseteq C$ and $A\lhd C$ then $A\lhd B$.*
(3) *If $A\lhd C$, then $\bot\in A$.*
(4) *$\langle\mathscr{P}(C),\lhd\rangle$ is a cpo with $\{\bot\}$ as its least element.*

Intuitively, a normal subposet $N\lhd A$ is an "approximation" to A. The notion of normal subposet is closely related to one of the central concepts in the theory of domains. A pair of continuous functions $g: D\to E$ and $f: E\to D$ is said to be an *embedding-projection* pair (g is the embedding and f is the projection) if they satisfy the following

$$f\circ g=\mathbf{id}_D, \qquad g\circ f\sqsubseteq\mathbf{id}_E$$

where $\mathbf{id}_D$ and $\mathbf{id}_E$ are the identity functions on D and E respectively (in future, we drop the subscripts when D and E are clear from context) and composition of functions is defined by $(f\circ g)(x)=f(g(x))$. One can show that each of f and g uniquely determines the other. Hence it makes sense to refer to f as the projection *determined by* g and refer to g as the embedding *determined by* f. There is quite a lot to be said about properties of projections and embeddings and we cannot begin to provide, in the space of this chapter, the full discussion that these concepts deserve (the reader may consult [5, Chapter 0] for this). However, a few observations are essential to what follows. We first provide a simple example.

EXAMPLE. If $f: D\to E$ is a continuous function then there is a strict continuous function $\mathbf{strict}: (D\to E)\to(D\circ\!\!\to E)$ given by

$$\mathbf{strict}(f)(x)=\begin{cases} f(x) & \text{if } x\neq\bot,\\ \bot & \text{if } x=\bot.\end{cases}$$

The function **strict** is a projection whose corresponding embedding is the inclusion map $\mathbf{incl}: (D\circ\!\!\to E)\hookrightarrow(D\to E)$.

In our discussion below we will not try to make much of the distinction between $f: D\circ\!\!\to E$ and $\mathbf{incl}(f): D\to E$ (for example, we may write $\mathbf{id}: D\circ\!\!\to D$ as well as $\mathbf{id}: D\to D$ or even $\mathbf{incl(id)}: D\to D$). From the two equations that define the relationship between a projection and embedding, it is easy to see that a projection is a surjection (i.e. onto) and an embedding is an injection (i.e. one-to-one). Thus one may well think of the image of an embedding $g: D \to E$ as a special kind of sub-cpo of E. We shall be especially interested in the case where an embedding is an *inclusion* as in the case of $D \circ\!\!\to E$ and $D \to E$. Let D be a cpo. We say that a continuous function $p: D \to D$ is a *finitary projection* if $p \circ p = p \sqsubseteq \mathbf{id}$ and $\mathbf{im}(p) = \{p(x) \mid x \in D\}$ is a domain. Note, in particular, that the inclusion map from $\mathbf{im}(p)$ into D is an embedding (which has the corestriction of p to its image as the corresponding projection). It is possible to characterize the basis of $\mathbf{im}(p)$ as follows. as follows.

3.2. LEMMA. *If D is a domain and $p: D \to D$ is a finitary projection, then the set of compact elements of* $\mathbf{im}(p)$ *is just* $\mathbf{im}(p) \cap K(D)$. *Moreover,* $\mathbf{im}(p) \cap K(D) \lhd K(D)$.

Suppose, on the other hand, that $N \lhd K(D)$. Then it is easy to check that the function $p_N: D \to D$ given by

$$P_N(x) = \bigsqcup \{y \in N \mid y \sqsubseteq x\}$$

is a finitary projection. Indeed, the correspondence $N \mapsto p_N$ is inverse to the correspondence $p \mapsto \mathbf{im}(p) \cap K(D)$ and we have the following theorem.

3.3. THEOREM. *For any domain D there is an isomorphism between the cpo of normal substructures of $K(D)$ and the poset* $\mathbf{Fp}(D)$ *of finitary projections on D.*

In particular, if $M \subseteq \mathbf{Fp}(D)$ is directed then $\mathbf{im}(\bigsqcup M)$ is a domain. This is a fact which will be significant later. Indeed, the notions of projection and normal subposet will come up again and again throughout the rest of our discussion.

3.2. Effectively presented domains

Returning now to the topic of computability, we will say that a domain is effectively presented if the ordering on its basis is decidable and it is possible to effectively recognize the finite normal subposets of the basis, according to the following definition.

DEFINITION. Let D be a domain and suppose $d: \mathbf{N} \to K(D)$ is a surjection. Then d is an *effective presentation* of D if

(1) the set $\{(m, n) \mid d_m \sqsubseteq d_n\}$ is effectively decidable, and

(2) for any finite set $u \subseteq \mathbf{N}$, it is decidable whether $\{d_n \mid n \in u\} \lhd K(D)$.

If $\langle D, d\rangle$ and $\langle E, e\rangle$ are effectively presented domains, then a continuous function $f: D \to E$ is said to be *computable* (with respect to d and e) if and only if, for every $n \in \mathbf{N}$, the set $\{(n, m) \mid e_m \sqsubseteq f(d_n)\}$ is recursively enumerable.

Unfortunately, the full class of domains has a serious problem. It is this: there are domains D, E such that the cpo $D \to E$ is *not* a domain (we will return to this topic in Section 6). Since we wish to use $D \to E$ in defining computability at higher types, we need some restriction on domains D and E which will ensure that $D \to E$ is a domain. There are several restrictions which will work. We begin by presenting one which is relatively simple. Another will be discussed later.

DEFINITION. A poset A is said to be *bounded complete* if A has a least element and every bounded subset of A has a least upper bound.

The bounded complete domains are closely related to a more familiar class of cpo's which arise in many places in classical mathematics. A domain D is a (countably based) *algebraic lattice* if every subset of D has a least upper bound. It is not hard to see that a domain D is bounded complete if and only if the cpo $D^{\top}$ which results from adding a new top element to D is an algebraic lattice. The poset $\mathscr{P}\mathbf{N}$ is an example of an algebraic lattice. On the other hand, the bounded complete domain $\mathbf{N}_{\perp} \circ\!\!\rightarrow \mathbf{N}_{\perp}$ lacks a top element and therefore fails to be an algebraic lattice. All of the domains we have discussed so far are bounded complete. In particular, we have the following:

3.4. THEOREM. *If D and E are bounded complete domains, then $D \rightarrow E$ is also a bounded complete domain. Moreover, if D and E have effective presentations, then $D \rightarrow E$ has an effective presentation as well. Similar facts hold for $D \circ\!\!\rightarrow E$.*

PROOF (*sketch*). It is not hard to see that $D \rightarrow E$ is a bounded complete cpo whenever E is. To prove that $D \rightarrow E$ is a domain we must demonstrate its basis. Suppose $N \lhd K(D)$ is finite and $s: N \rightarrow K(E)$ is monotone. Then the function $\mathbf{step}(s): D \rightarrow E$, given by taking $\mathbf{step}(s)(x) = \bigsqcup\{f(y) \mid y \in N \cap \downarrow x\}$, is continuous and compact in the ordering on $D \rightarrow E$. These are called *step functions* and it is possible to show that they form a basis for $D \rightarrow E$. The proof that the poset of step functions has a decidable ordering and finite normal subposets is tedious, but not difficult, using the effective presentations of D and E. The proof of these facts for $D \circ\!\!\rightarrow E$ is essentially the same since the strict step functions form a basis. □

In the remaining sections of this chapter we will discuss a great many operators like $\cdot \rightarrow \cdot$ and $\cdot \circ\!\!\rightarrow \cdot$. We will leave it to the reader to convince himself that all of these operators preserve the property of having an effective presentation. Further discussion of computability on domains may be found in [19, 9]. It is hoped that future research in the theory of domains will provide a general technique which will incorporate computability into the *logic* whereby we reason about the existence of our operators. This will eliminate the need to provide demonstrations of effective presentations. This is a central idea in current investigations but it is beyond the scope of this chapter to discuss it further.

4. Operators and functions

There are a host of operators on domains which are needed for the purposes of semantic definitions. In this section we mention a few of them. An essential technique for building new operators from those which we present here will be introduced below when we discuss solutions of recursive equations.

4.1. *Products*

Given posets D and E, the *product* $D \times E$ is the set of pairs (x, y), where $x \in D$ and $y \in E$. The ordering is coordinatewise, i.e. $(x, y) \sqsubseteq (x', y')$ if and only if $x \sqsubseteq x'$ and $y \sqsubseteq y'$. We define functions $\mathbf{fst}: D \times E \to D$ and $\mathbf{snd}: D \times E \to E$ given by $\mathbf{fst}(x, y) = x$ and $\mathbf{snd}(x, y) = y$. If a subset $L \subseteq D \times E$ is directed, then

$$M = \mathbf{fst}^*(L) = \{x \mid \exists y \in E. (x, y) \in L\},$$
$$N = \mathbf{snd}^*(L) = \{y \mid \exists x \in D. (x, y) \in L\}$$

are directed. In particular, if D and E are cpo's then $\bigsqcup L = (\bigsqcup M, \bigsqcup N)$ and, of course, $\bot_{D \times E} = (\bot_D, \bot_E)$, so $D \times E$ is a cpo. Indeed, if D and E are domains, then $D \times E$ is also a domain with $K(D \times E) = K(D) \times K(E)$. The property of bounded completeness is also preserved by $\times$.

Given cpos D, E, F, one can show that a function f: $D \times E \to F$ is continuous if and only if it is continuous in each of its arguments individually. In other words, f is continuous iff each of the following conditions holds:

(1) For every element $e \in E$, the function $f_1 : D \to F$ given by $x \mapsto f(x, e)$ is continuous;

(2) For every element $d \in D$,the function $f_2 : E \to F$ given by $y \mapsto f(d, y)$ is continuous;

We leave the proof of this equivalence as an exercise for the reader.

It is easy to see that each of the functions **fst** and **snd** is continuous. Moreover, given any cpo F and continuous functions f: $F \to D$ and g:$F \to E$, there is a continuous function $\langle f, g \rangle$:$F \to D \times E$ such that

$$\mathbf{fst} \circ \langle f, g \rangle = f, \qquad \mathbf{snd} \circ \langle f, g \rangle = g,$$

and, for any continuous function h:$F \to D \times E$,

$$\langle \mathbf{fst} \circ h, \mathbf{snd} \circ h \rangle = h.$$

The function $\langle f, g \rangle$ is given by $\langle f, g \rangle(x) = (f(x), g(x))$.

There is another, more pictorial, way of stating these equational properties of the operator $\langle \cdot, \cdot \rangle$ using a commutative diagram. The desired property can be stated in the following manner: given any cpo F and continuous functions f:$F \to D$ and g:$F \to E$, there is a *unique* continuous function $\langle f, g \rangle$ which completes the following diagram:

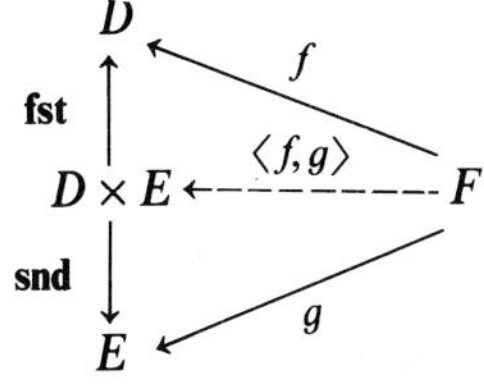

This is referred to as the *universal property* of the operator $\times$. As operators are given below, we will describe the universal properties that they satisfy and these will form the basis of a system of equational reasoning about continuous functions. Virtually all of the functions needed to described the semantics of (a wide variety of) programming languages may be built from those which are used in expressing these universal properties!

Given continuous functions $f: D \to D'$ and $g: E \to E'$, we may define a continuous function $f \times g$ which takes (x, y) to $(f(x), g(y))$ by setting

$$f \times g = \langle f \circ \mathbf{fst}, g \circ \mathbf{snd} \rangle : D \times E \to D' \times E'.$$

It is easy to show that $\mathbf{id}_D \times \mathbf{id}_E = \mathbf{id}_{D \times E}$ and

$$(f \times g) \circ (f' \times g') = (f \circ f') \times (g \circ g').$$

Note that we have "overloaded" the symbol $\times$ so that it works both on pairs of *domains* and pairs of *functions*. This sort of overloading is quite common in mathematics and we will use it often below. In this case (and others to follow) we have an example of what mathematicians call a *functor*.

There is a very important relationship between the operators $\to$ and $\times$. Let D, E and F be cpo's. Then there is a function

$$\mathbf{apply}: ((E \to F) \times E) \to F$$

given by taking $\mathbf{apply}(f, x)$ to be $f(x)$ for any function $f: E \to F$ and element $x \in E$. Indeed, the function **apply** is *continuous*. Also, given a function $f: D \times E \to F$, there is a continuous function

$$\mathbf{curry}(f): D \to (E \to F)$$

given by taking $\mathbf{curry}(f)(x)(y)$ to be $f(x, y)$. Moreover, $\mathbf{curry}(f)$ is the *unique* continuous function which makes the following diagram commute:

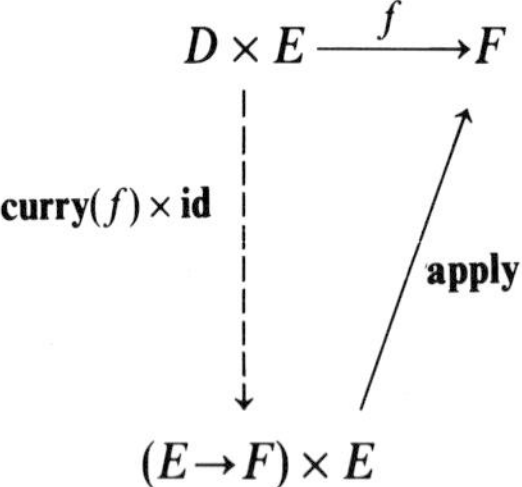

This uniqueness condition is equivalent to the following equation:

$$\mathbf{curry}(\mathbf{apply} \circ (h \times \mathbf{id}_E)) = h \qquad (*)$$

To see this, suppose equation $(*)$ holds and h satisfies $f = \mathbf{apply} \circ (h \times \mathbf{id})$; then

$$\mathbf{curry}(f) = \mathbf{curry}(\mathbf{apply} \circ (h \times \mathbf{id})) = h$$

so the uniqueness condition is satisfied. On the other hand, if $\mathbf{curry}(f)$ is uniquely determined by the diagram above, then equation (∗) follows immediately from the commutativity of the following diagram:

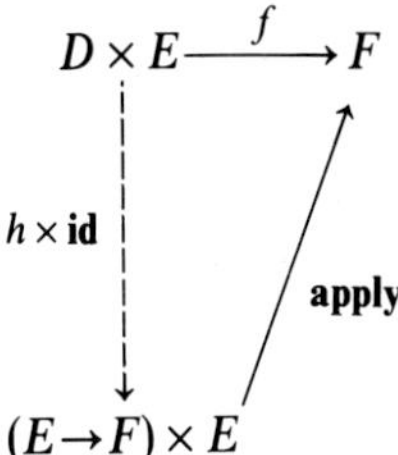

for $f = \mathbf{apply} \circ (h \times \mathbf{id}_E)$.

It is often useful to have a multiary notation for products. We write

$$\times() = \mathbf{I}, \qquad \times(D_1, \ldots, D_n) = \times(D_1, \ldots, D_{n-1}) \times D_n$$

and define projections

$$\mathbf{on}_i : \times(D_1, \ldots, D_n) \to D_i \quad \text{by} \quad \mathbf{on}_i = \mathbf{snd} \circ \mathbf{fst}^{n-i}.$$

Similarly, one defines a multiary version of the pairing operation by taking $\langle\,\rangle$ to be the identity on the one-point domain and defining

$$\langle f_1, \ldots, f_n \rangle = \langle \langle f_1, \ldots, f_{n-1} \rangle, f_n \rangle.$$

These multiary versions of projection and pairing satisfy a universal property similar to the one for the binary product.

4.2. *Church's λ-notation*

If we wish to define a function from, say, natural numbers to natural numbers, we typically do so by describing the action of that function on a generic number x (a *variable*) using previously defined functions. For example, the squaring function f has the action $x \mapsto x * x$ where $*$ is the multiplication function. We may now use f to define other functions: for example, a function g which takes a function $h: \mathbf{N} \to \mathbf{N}$ to $f \circ h$. Continuing in this way we may construct increasingly complex function definitions. However, it is sometimes useful to have a notation for functions which alleviates the necessity of introducing intermediate names. This purpose is served by a terminology known as λ-notation which is originally due to Church.

The idea is this: Instead of introducing a term such as f and describing its action as a function, one simply gives the function a name which is basically a description of what it does with its argument. In the above case one writes $\lambda x.x*x$ for f and $\lambda h.f \circ h$ for g. One can use this notation to define g without introducing f by defining g to be the function $\lambda h.(\lambda x.x*x) \circ h$. The λh at the beginning of this expression says that g is a function which is computed by taking its argument and *substituting* it for the variable h in the expression $(\lambda x.x*x) \circ h$.

The use of the Greek letter λ for the operator which binds variables is primarily a historical accident. Various programming languages incorporate something essentially equivalent to λ-notation using other names. In mathematics textbooks it is common to avoid the use of such notation by assuming conventions about variable names. For example, one may write $x^2 - 2*x$ for the function which takes a real number as an argument and produces as result the square of that number less its double. An expression such as $x^2 + x*y + y^2$ would denote a function which takes two numbers as arguments—that is, the values of x and y—and produces the square of the one number plus the square of the other plus the product of the two. One might therefore provide a name for this function by writing something like: $f(x, y) = x^2 + x*y + y^2$. So f is a function which takes a pair of numbers and produces a number. But what notation should we use for the function g that takes a number n as argument and produces the *function* $n \mapsto x^2 + x*n + n^2$? For example, $g(2)$ is the function $x^2 + 2*x + 4$. It is not hard to see that this is closely related to the function **curry** which we discussed above. Modulo the fact that we defined **curry** for domains above, we might have written $g = \mathbf{curry}(f)$. Or, to define g directly, we would write

$$g = \lambda y.\lambda x.x^2 + x*y + y^2.$$

The definition of f would need to be given differently since f takes a *pair* as an argument. We therefore write:

$$f = \lambda(x, y).x^2 + x*y + y^2.$$

There is no impediment to using this notation to describe higher-order functions as well. For example, $\lambda f.f(3)$ takes a function f and evaluates it on the number 3, and $\lambda f.f \circ f$ takes a function and composes it with itself. But these definitions highlight a very critical issue. Note that both definitions are *ambiguous* as they stand. Does the function $\lambda f.f(3)$ takes, for example, functions from numbers to reals as argument, or does it take a function from numbers to sets of numbers as argument? Either of these would, by itself, make sense. What we need to do is indicate somewhere in the expression the *types* of the variables (and constants if their types are not already understood). So we might write

$$\lambda f : \mathbf{N} \to \mathbf{R}.f(3)$$

for the operator taking a real-valued function as argument and

$$\lambda f : \mathbf{N} \to \mathscr{P}\mathbf{N}.f(3)$$

for the operator taking a $\mathscr{P}\mathbf{N}$-valued function.

So far, what we have said applies to almost any class of spaces and functions where products and an operator like **curry** are defined. But for the purposes of programming semantics, we need a semantic theory that includes the concept of a fixed point. Such fixed points are guaranteed if we stay within the realm of cpo's and continuous functions. But the crucial fact is this: *the process of λ-abstraction preserves continuity.* This is because $\mathbf{curry}(f)$ is continuous whenever f is. We may therefore use the notational tools we have described above with complete freedom and still be sure that recursive definitions using this notation make sense.

Demonstrating that the typed λ-calculus (i.e. the system of notations that we have been describing informally here) is really useful in explaining the semantics of programming languages is not the objective of this chapter. However, one can already see that it provides a considerable latitude for writing function definitions in a simple and mathematically perspicuous manner.

4.3. *Smash products*

In the product $D \times E$ of cpo's D and E, there are elements of the form $(x, \bot)$ and $(\bot, y)$. If $x \neq \bot$ or $y \neq \bot$, then these will be *distinct* members of $D \times E$. In programming semantics, there are occasions when it is desirable to *identify* the pairs $(x, \bot)$ and $(\bot, y)$. For this purpose, there is a collapsed version of the product called the *smash product*. For cpo's D and E, the smash product $D \otimes E$ is the set

$$\{(x, y) \in D \times E \mid x \neq \bot \text{ and } y \neq \bot\} \cup \{\bot_{D \otimes E}\}$$

where $\bot_{D \otimes E}$ is some new element which is not a pair. The ordering on pairs is coordinatewise and we stipulate that $\bot_{D \otimes E} \sqsubseteq z$ for every $z \in D \otimes E$. There is a continuous surjection

$$\mathbf{smash}: D \times E \to D \otimes E$$

given by taking

$$\mathbf{smash}(x, y) = \begin{cases} (x, y) & \text{if } x \neq \bot \text{ and } y \neq \bot, \\ \bot_{D \otimes E} & \text{otherwise.} \end{cases}$$

This function establishes a useful relationship between $D \times E$ and $D \otimes E$. In fact, it is a projection whose corresponding embedding is the function **unsmash**: $D \otimes E \to D \times E$ given by

$$\mathbf{unsmash}(z) = \begin{cases} z & \text{if } z = (x, y) \text{ is a pair,} \\ (\bot, \bot) & \text{if } z = \bot_{D \otimes E}. \end{cases}$$

Let us say that a function $f: D \times E \to F$ is *bistrict* if $f(x, y) = \bot$ whenever $x = \bot$ or $y = \bot$. If $f: D \times E \to F$ is bistrict and continuous, then $g = f \circ \mathbf{unsmash}$ is the unique strict, continuous function which completes the following diagram:

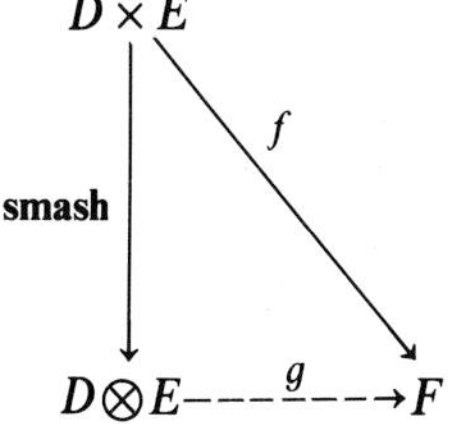

If $f: D \to D'$ and $g: E \to E'$ are strict continuous functions, then

$$f \otimes g = \mathbf{smash} \circ (f \times g) \circ \mathbf{unsmash}$$

is the unique strict, continuous function which completes the following diagram:

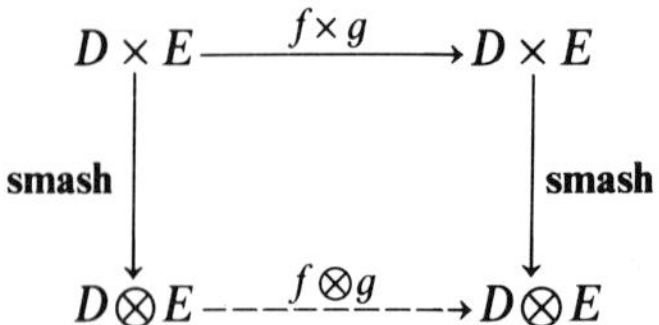

As with the product $\times$ and function space $\rightarrow$, there is a relationship between the smash product $\otimes$ and the strict function space $\circ\!\!\rightarrow$. In particular, there is a strict continuous function **strict_apply** such that for any strict function f there is a unique strict function **strict_curry** such that the following diagram commutes:

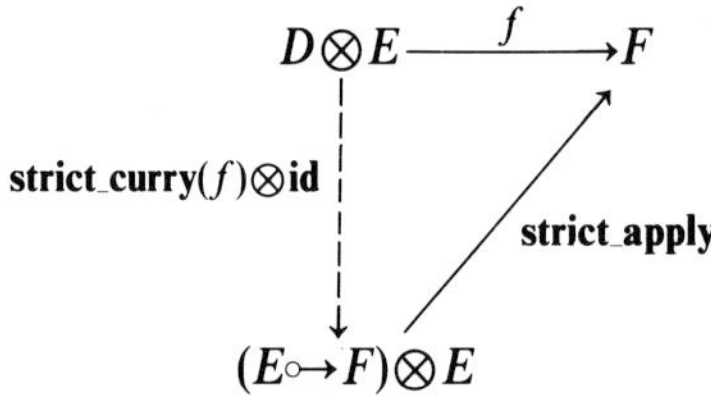

4.4. Sums and lifts

Given cpo's D and E, we define the *coalesced sum* $D \oplus E$ to be the set

$$((D-\{\bot_D\})\times\{0\})\cup((E-\{\bot_E\})\times\{1\})\cup\{\bot_{D\oplus E}\}$$

where $D-\{\bot_D\}$ and $E-\{\bot_E\}$ are the sets D and E with their respective bottom elements removed and $\bot_{D\oplus E}$ is a new element which is not a pair. It is ordered by taking $\bot_{D\oplus E} \sqsubseteq z$ for all $z \in D\oplus E$ and taking $(x, m) \sqsubseteq (y, n)$ if and only if $m=n$ and $x \sqsubseteq y$. There are strict continuous functions **inl**: $D \circ\!\!\rightarrow (D\oplus E)$ and **inr**: $E \circ\!\!\rightarrow (D\oplus E)$ given by taking

$$\mathbf{inl}(x)=\begin{cases}(x,0) & \text{if } x\neq\bot,\\ \bot_{D\oplus E} & \text{if } x=\bot;\end{cases} \qquad \mathbf{inr}(x)=\begin{cases}(x,1) & \text{if } x\neq\bot,\\ \oplus_{D\oplus E} & \text{if } x=\bot.\end{cases}$$

Moreover, if $f: D \circ\!\!\rightarrow F$ and $g: E \circ\!\!\rightarrow F$ are strict continuous functions, then there is a unique strict continuous function $[f, g]$ which completes the following diagram:

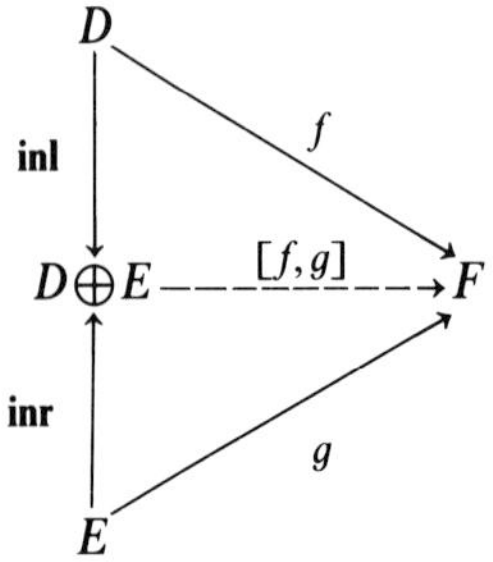

The function $[f, g]$ is given by

$$[f,g](z)=\begin{cases} f(x) & \text{if } z=(x,0),\\ g(y) & \text{if } z=(y,1),\\ \bot & \text{if } z=\bot. \end{cases}$$

Given continuous functions $f: D \circ\!\!\to D'$ and $g: E \circ\!\!\to E'$, we define

$$f \oplus g = [\mathbf{inl} \circ f, \mathbf{inr} \circ g] : D \oplus E \circ\!\!\to D' \oplus E'.$$

As with the product, it is useful to have a multiary notation for the coalesced sum. We define

$$\oplus() = \mathbf{I}, \qquad \oplus(D_1, \ldots, D_n) = \oplus(D_1, \ldots, D_{n-1}) \oplus D_n$$

and

$$\mathbf{in}_i = \mathbf{inr} \circ \mathbf{inl}^{n-i}.$$

One may also define $[f_1, \ldots, f_n]$ and prove a universal property.

Given a cpo D, we define the *lift* of D to be the set $D_\bot = (D \times \{0\}) \cup \{\bot\}$, where $\bot$ is a new element which is not a pair, together with a partial ordering $\sqsubseteq$ which is given by stipulating that $(x, 0) \sqsubseteq (y, 0)$ when $x \sqsubseteq y$ and $\bot \sqsubseteq z$ for every $z \in D_\bot$. In sort, $D_\bot$ is the poset obtained by adding a new bottom to D—see Fig. 2. It is easy to show that $D_\bot$ is a cpo if D is. We define a strict continuous function $\mathbf{down}: D_\bot \circ\!\!\to D$ by

$$\mathbf{down}(z)=\begin{cases} x & \text{if } z=(x,0),\\ \bot_D & \text{otherwise} \end{cases}$$

and a (nonstrict) continuous function $\mathbf{up}: D \to D_\bot$ given by $\mathbf{up}: x \mapsto (x, 0)$. These functions are related by

$$\mathbf{down} \circ \mathbf{up} = \mathbf{id}_D, \qquad \mathbf{up} \circ \mathbf{down} \sqsupseteq \mathbf{id}_{D_\bot}.$$

These inequations are reminiscent of those which we gave for embedding-projection pairs, but the second inequation has $\sqsupseteq$ rather than $\sqsubseteq$. We will discuss such pairs of functions later. Given cpo's D and E and continuous function $f: D \to E$, there is a unique

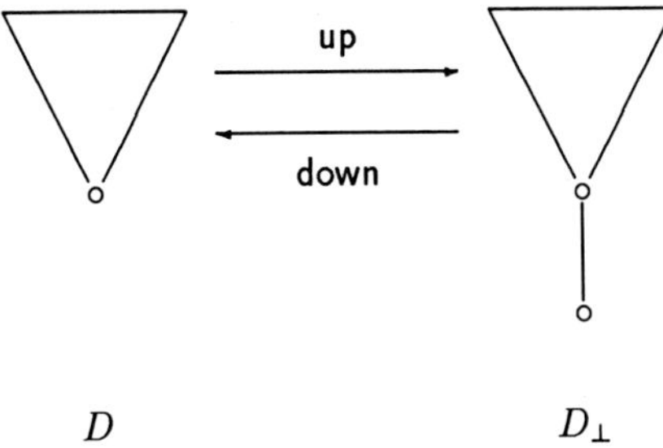

Fig. 2. The lift of a cpo.

strict continuous function $f^\dagger$ which completes the following diagram:

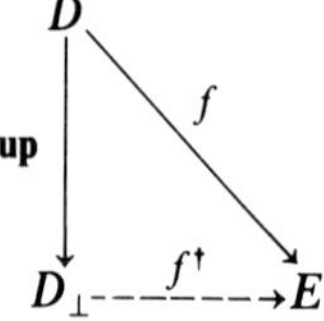

Given a continuous function $f: D \to E$, we define a strict continuous function

$$f_\perp = (\mathbf{up} \circ f)^\dagger : D_\perp \circ\!\!\to E_\perp .$$

Given cpo's D and E, we define the *separated sum* $D + E$ to be the cpo $D_\perp \oplus E_\perp$. By the universal properties for $\oplus$ and $(\cdot)_\perp$, we know that $h = [f^\dagger, g^\dagger]$ is the unique *strict* continuous function which completes the following diagram:

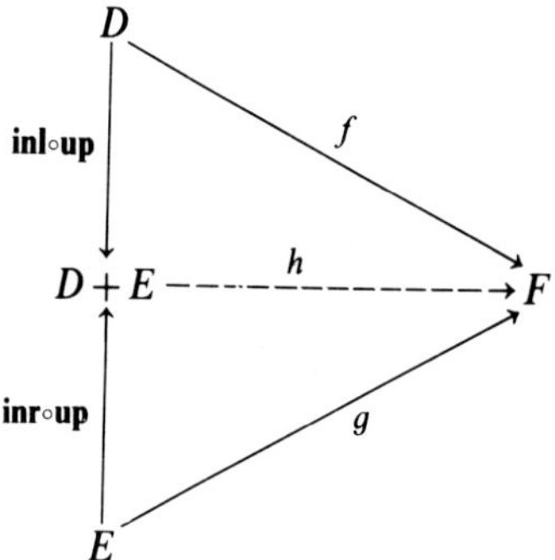

However, h may not be the only *continuous* function which completes the diagram. Given continuous functions $f: D \to D'$ and $g: E \to E'$, we define

$$f + g = f_\perp \oplus g_\perp : D + E \to D' + E' .$$

4.5. *Isomorphisms and closure properties*

There are quite a few interesting relationships between the operators above which are implied by the definitions and commutative diagrams. We list a few of these in the following lemmas.

4.1. LEMMA. *Let D, E and F be cpo's; then*

(1) $D \times E \cong E \times D$,
(2) $(D \times E) \times F \cong D \times (E \times F)$,
(3) $D \to (E \times F) \cong (D \to E) \times (D \to F)$,
(4) $D \to (E \to F) \cong (D \times E) \to F$.

4.2. LEMMA. *Let D, E and F be cpo's; then*

(1) $D \otimes E \cong E \otimes D$,

(2) $(D \otimes E) \otimes F \cong D \otimes (E \otimes F)$,

(3) $(E \oplus F) \circ\!\!\rightarrow D \cong (E \circ\!\!\rightarrow D) \times (E \circ\!\!\rightarrow F)$,

(4) $D \circ\!\!\rightarrow (E \circ\!\!\rightarrow F) \cong (D \otimes E) \circ\!\!\rightarrow F$,

(5) $D \otimes (E \oplus F) \cong (D \otimes E) \oplus (D \otimes E)$

(6) $D_\perp \circ\!\!\rightarrow E \cong D \rightarrow E$.

We already remarked that $D \rightarrow E$ and $D \circ\!\!\rightarrow E$ are bounded complete domains whenever D and E are. It is not difficult to see that similar closure properties will hold for the other operators we have defined in this section.

4.3. LEMMA. *If D and E are bounded complete domains then so are the cpo's* $D \rightarrow E$, $D \circ\!\!\rightarrow E$, $D \times E$, $D \otimes E$, $D + E$, $D \oplus E$, $D_\perp$.

Further discussion of the operators defined in this section and others may be found in [17, 18].

5. Powerdomains

We now turn our attention to another collection of operators on domains. Just as we have defined a computable analog to the *function space*, we will now define a computable analog to the *powerset operation*. Actually, we will produce three such operators. In the domain theory literature these are called *powerdomains*. If D is a domain, we write

- $D^\#$ for the *upper* powerdomain of D,
- $D^\natural$ for the *convex* powerdomain of D, and
- $D^\flat$ for the *lower* powerdomain of D.

The names we use for these operators come from the concepts of upper and lower semicontinuity and the interested reader can consult [22] for a detailed explanation. They commonly appear under other names as well. The convex powerdomain $D^\natural$ was introduced by G. Plotkin [10] and is therefore sometimes referred to as the *Plotkin powerdomain*. The upper powerdomain $D^\#$ was introduced by *M*. Smyth [20] and is sometimes called the *Smyth powerdomain*. For reasons that we will discuss briefly below, this latter power-domain corresponds to the *total correctness* interpretation of programs. Since C.A.R. Hoare has done much to popularize the study of *partial correctness* properties of programs, the remaining powerdomain $D^\flat$—which corresponds to the partial correctness interpretation—sometimes bears his name.

5.1. *Intuition*

There is a basic intuition underlying the powerdomain concept, which can be explained through the concept of *partial information*. To keep things simple, let us assume that we are given a finite poset A and are asked to form the *poset* of finite nonempty subsets of A. As a first guess, one might take the nonempty subsets and order them by subset inclusion. However, this operation ignores the order structure on A! Think of A as a collection of partial descriptions of data elements: $x \sqsubseteq y$ just in case x is a partial description of y. What should it mean for one nonempty subset of A to be a "partial description" of another? There are at least three reasonable philosophies that one might adopt in attempting to answer this question.

Suppose, for example, that I hold a bag of fruit and I wish to give you information about what is in the bag. One such description might be

A fruit in the bag is a yellow fruit or a red fruit.

This description is based on two basic pieces of data: "is a yellow fruit" and "is a red fruit". These are used to *restrict* the kinds of fruit which are in the bag. A more informative description of this kind would provide further restrictions. Consider the following example:

A fruit in the bag is a yellow fruit or a cherry or a strawberry.

It is based on three pieces of data: "is a yellow fruit", "is a cherry" and "is a strawberry". Since these three data provide further restrictions on the contents of the bag (by ruling out the possibility of an apple, for example) it is a more informative statement about the bag's contents. On the other hand,

A fruit in the bag is a yellow fruit or a red fruit or a purple fruit.

is a less informative description because it is more permissive; for instance, it does not rule out the possibility that the bag holds a grape. Now suppose that u, v are subsets of the poset A from the previous paragraph. With this way of seeing things, we should say that u is below v if the restrictions imposed by v are refinements of the restrictions imposed by u: that is, for each $y \in v$, there is an $x \in u$ such that $x \sqsubseteq y$. This is the basic idea behind the *upper powerdomain* of A.

Returning to the bag of fruit analogy, we might view the following as a piece of information about the contents of the bag:

There is some yellow fruit and some red fruit in the bag.

This information is based on two pieces of data: "is a yellow fruit" and "is a red fruit". However, these data are not being used as before. They do not restrict possibilities; instead they offer a *positive* assertion about the contents of the bag. A more informative description of this kind would provide a further enumeration and refinement of the contents:

There is a banana, a cherry and some purple fruit in the bag.

This refined description does not rule out the possibility that the bag holds an apple, but it does ensure that there is a cherry. A statement such as

There is some yellow fruit in the bag

is less informative since it does not mention the presence of red fruit. Now suppose that u, v are subsets of the poset A. With this way of seeing things, we should say that u is below v if the positive assertions provided by u are extended and refined by v: that is, for each $x \in u$, there is a $y \in v$ such that $x \sqsubseteq y$. This is the basic idea behind the *lower powerdomain of A*.

Now, the *convex powerdomain* combines these two forms of information. For example, the assertion

> *If you pull a fruit from the bag, then it must be yellow or a cherry, and you can pull a yellow fruit from the bag and you can pull a cherry from the bag.*

is the combined kind of information. The pair of assertions means that the bag holds some yellow fruit and at least one cherry, but nothing else. A more refined description might be

> *If you pull a fruit from the bag, then it must be a banana or a cherry, and you can pull a banana from the bag and you can pull a cherry from the bag.*

A less refined description might be

> *If you pull a fruit from the bag, then it must be yellow or red, and you can pull a yellow fruit from the bag and you can pull a red fruit from the bag.*

The reader may be curious about what bags of fruit have to do with programming semantics. The powerdomains are used to model nondeterministic computations where one wishes to speak about the set of outcomes of a computation. How one wishes to describe such outcomes will determine which of the three powerdomains is used. We will attempt to illustrate this idea later in this section—when we have given some formal definitions.

5.2. *Formal definitions*

In order to give the definitions of the powerdomains, it is helpful to have a little information about the representation of domains using the concept of a pre-order:

DEFINITION. A *pre-order* is a set A together with a binary relation $\vdash$ which is reflexive and transitive.

It is conventional to think of the relation $a \vdash b$ as indicating that a is "larger" than b (as in mathematical logic, where $\varphi \vdash \psi$ means that the formula ψ follows from the hypothesis φ). Of course, any poset is also a pre-order. On the other hand, a pre-order may fail to be a poset by not satisfying the antisymmetry axiom. In other words, we may have $x \vdash y$ and $y \vdash x$ but $x \neq y$. By identifying elements x, y which satisfy $x \vdash y$ and $y \vdash x$, we obtain an induced partially ordered set from a pre-order (and this is why they are called *pre*-orders). We shall be particularly interested in a special kind of subset of a pre-order.

DEFINITION. An *ideal* over a pre-order $\langle A, \vdash \rangle$ is a subset $s \subseteq A$ such that

(1) if $u \subseteq s$ is finite, then there is an $x \in s$ such that $x \vdash y$ for $y \in u$, and

(2) if $x \in s$ and $x \vdash y$, then $y \in s$.

In short, an ideal is a subset which is directed and downward closed. If $x \in A$ for a pre-order A, then the set

$$\downarrow x = \{y \in A \mid x \vdash y\}$$

is an ideal called the *principal ideal generated by* x. To induce a poset from a pre-order, one can take the poset of principal ideals under set inclusion. The poset of all ideals on a pre-order is somewhat more interesting, as expressed in the following theorem.

5.1. THEOREM. *Given a countable pre-order* $\langle A, \vdash \rangle$, *let* D *be the poset consisting of the ideals over* A, *ordered by set inclusion. If there is an element* $\bot \in A$ *such that* $x \vdash \bot$ *for each* $x \in A$, *then* D *is a domain and* $K(D)$ *is the set of principal ideals over* A.

PROOF. Clearly, the ideals of A form a poset under set inclusion and the principal ideal $\downarrow \bot$ is the least element. To see that this poset is complete, suppose that $M \subseteq D$ is directed and let $x = \bigcup M$. If we can show that x is an ideal, then it is certainly the least upper bound of M in D. To this end, suppose $u \subseteq x$ is finite. Since each element of u must be contained in some element of M, there is a finite collection of ideals $s \subseteq M$ such that $u \subseteq \bigcup s$. Since M is directed, there is an element $y \in M$ such that $z \subseteq y$ for each $z \in s$. Thus $u \subseteq y$ and since y is ideal, there is an element $a \in y$ such that $b \sqsubseteq a$ for each $b \in u$. But $a \in y \subseteq x$, so it follows that x is an ideal.

To see that D is a domain, we show that the set of principal ideals is a basis. Suppose $M \subseteq D$ is directed and $\downarrow a \subseteq \bigcup M$ for some $a \in A$. Then $a \in x$ for some $x \in M$, so $\downarrow a \subseteq x$. Hence $\downarrow a$ is compact in D. Now suppose $x \in D$ and $u \subseteq A$ is a finite collection of elements of A such that $\downarrow a \subseteq x$ for each $a \in u$. Then $u \subseteq x$ and since x is an ideal, there is an element $b \in x$ with $b \vdash a$ for each $a \in u$. Thus $\downarrow a \subseteq \downarrow b$ for each $a \in u$ and it follows that the principal ideals below x form a directed collection. It is obvious that the least upper bound (i.e. union) of that collection is x. Since x was arbitrary, it follows that D is an algebraic cpo with principal ideals of A as its basis. Since A is countable, there are only countably many principal ideals, so D is a domain. □

For any set S, we let $\mathscr{P}_{\mathrm{f}}^*(S)$ be the set of finite nonempty subsets of S. We write $\mathscr{P}_{\mathrm{f}}(S)$ for the set of all finite subsets (including the empty set). Given a poset $\langle A, \sqsubseteq \rangle$, define a pre-ordering $\vdash^{\#}$ on $\mathscr{P}_{\mathrm{f}}^*(A)$ as follows:

$$u \vdash^{\#} v \text{ if and only if } (\forall x \in u)(\exists y \in v).x \sqsupseteq y.$$

Dually, define a pre-ordering $\vdash^{\flat}$ on $\mathscr{P}_{\mathrm{f}}^*(A)$ by

$$u \vdash^{\flat} v \text{ if and only if } (\forall y \in v)(\exists x \in u).x \sqsupseteq y.$$

And define $\vdash^{\natural}$ on $\mathscr{P}_{\mathrm{f}}^*(A)$ by

$$u \vdash^{\natural} v \text{ if and only if } u \vdash^{\#} v \text{ and } u \vdash^{\flat} v.$$

If D is a domain, then let $D^\natural$ be the domain of ideals over $\langle \mathscr{P}_f^*(K(D)), \vdash^\natural \rangle$. We call $D^\natural$ the *convex powerdomain* of D. Similarly, define $D^\#$ and $D^\flat$ to be the domains of ideals over $\langle \mathscr{P}_f^*(K(D)), \vdash^\# \rangle$ and $\langle \mathscr{P}_f^*(K(D)), \vdash^\flat \rangle$ respectively. We call $D^\#$ the *upper powerdomain* of D and $D^\flat$ the *lower powerdomain* of D.

As an example, we compute the lower powerdomain of $\mathbf{N}_\perp$. Since $K(\mathbf{N}_\perp) = \mathbf{N}_\perp$, the lower powerdomain of $\mathbf{N}_\perp$ is the set of ideals over the pre-order $\langle \mathscr{P}_f^*(\mathbf{N}_\perp), \vdash^\flat \rangle$. To see what such an ideal must look like, note first that $u \vdash^\flat u \cup \{\perp\}$ and $u \cup \{\perp\} \vdash^\flat u$ for any $u \in \mathscr{P}_f^*(\mathbf{N}_\perp)$. From this fact it is already possible to see why $\vdash^\flat$ is usually only a *pre-order* and not a poset. Now, if u ad v both contain $\perp$, then $u \vdash^\flat v$ iff $u \supseteq v$. Hence we may identify a ideal $x \in (\mathbf{N}_\perp)^\flat$ with the union $\bigcup x$ of all the elements in x. Thus $(\mathbf{N}_\perp)^\flat$ is isomorphic to the domain $\mathscr{P}\mathbf{N}$ of all subsets of $\mathbf{N}$ under subset inclusion.

Now let us compute the upper powerdomain of $\mathbf{N}_\perp$. Note that if u and v are finite non-empty subsets of $\mathbf{N}_\perp$ and $\perp \in v$, then $u \vdash^\# v$. In particular, any ideal x in $(\mathbf{N}_\perp)^\#$ contains all of the finite subsets v of $\mathbf{N}_\perp$ with $\perp \in v$. So, let us say that a set $u \in \mathscr{P}_f^*(\mathbf{N}_\perp)$ is *nontrivial* if it does not contain $\perp$ and an ideal $x \in (\mathbf{N}_\perp)^\#$ is nontrivial if there is a nontrivial $u \in x$. Now, if u and v are nontrivial, then $u \vdash^\# v$ iff $u \subseteq v$. Therefore, if an ideal x is nontrivial, then it is the principal ideal generated by the intersection of its nontrivial elements! The smaller this set is, the larger is the ideal x. Hence, the nontrivial ideals in the powerdomain (ordered by subset inclusion) correspond to finite subsets of $\mathbf{N}$ (ordered by superset inclusion). If we now throw in the unique trivial ideal, we can see that $(\mathbf{N}_\perp)^\#$ is isomorphic to the domain of sets $\{\mathbf{N}\} \cup \mathscr{P}_f^*(\mathbf{N})$ ordered by superset inclusion.

Finally, let us look at the convex powerdomain of $\mathbf{N}_\perp$. If $u, v \in \mathscr{P}_f^*(\mathbf{N}_\perp)$, then $u \vdash^\natural v$ iff

(1) $\perp \in v$ and $u \supseteq v$, or

(2) $u = v$.

Hence, if x is an ideal and there is a set $u \in x$ with $\perp \notin u$, then x is the principal ideal generated by u. No two distinct principal ideals like this will be comparable. On the other hand, if x is an ideal with $\perp \in u$ for each $u \in x$, then $x \subseteq y$ for an arbitrary ideal y iff $\bigcup x \subseteq \bigcup y$. Thus the convex powerdomain of $\mathbf{N}_\perp$ corresponds to the set of finite, nonempty subsets of $\mathbf{N}$ unioned with the set of arbitrary subsets of $\mathbf{N}_\perp$ that contain $\perp$. The ordering on these sets is like the pre-ordering $\vdash^\natural$ but extended to include infinite sets.

5.3. *Universal and closure properties*

If $s, t \in D^\natural$ then we define a binary operation

$$s \uplus t = \{w \mid u \cup v \vdash^\natural w \text{ for some } u \in s \text{ and } v \in t\}.$$

This set is an ideal and the function $\uplus : D^\natural \times D^\natural \to D^\natural$ is continuous. Similar facts apply when $\uplus$ is defined in this way for $D^\#$ and $D^\flat$. Now, if $x \in D$, define

$$\{\!|x|\!\} = \{u \in \mathscr{P}_f^*(K(D)) \mid \{x_0\} \vdash^\natural u \text{ for some compact } x_0 \sqsubseteq x\}.$$

This forms an ideal and $\{\!|\cdot|\!\} : D \to D^\natural$ is a continuous function. When one replaces $\vdash^\natural$ in this definition by $\vdash^\#$ or $\vdash^\flat$, then similar facts apply. Strictly speaking, we should decorate the symbols $\uplus$ and $\{\!|\cdot|\!\}$ with indices to indicate their types, but this clutters the notation somewhat. Context will determine what is intended.

These three operators $(\cdot)^{\natural}$, $(\cdot)^{\flat}$ and $(\cdot)^{\sharp}$ may not seem to be the most obvious choices for the computable analog of the powerset operator. We will attempt to provide some motivation for choosing them in the remainder of this section. Given the operators $\uplus$ and $\{\!|\cdot|\!\}$, we may say that a point $x \in D$ for a domain D is an "element" of a set s in a powerdomain of D if $\{\!|x|\!\} \uplus s = s$. If s and t lie in a powerdomain of D, then s is a "subset" of t if $s \uplus t = t$. Care must be taken, however, not to confuse "sets" in a powerdomain with sets in the usual sense. The relations of "element" and "subset" described above will have different properties in the three different powerdomains. Moreover, it may be the case that s is a "subset" of t without it being the case that $s \subseteq t$!

To get some idea how the powerdomains are related to the semantics of nondeterministic programs, let us discuss nondeterministic partial functions from $\mathbf{N}$ to $\mathbf{N}$. As we have noted before, there is a correspondence between partial functions from $\mathbf{N}$ to $\mathbf{N}$ and strict functions $f: \mathbf{N}_\perp \circ\!\!\rightarrow \mathbf{N}_\perp$. These may be thought of as the meanings of "deterministic" programs, because the output of a program is uniquely determined by its input (i.e., the meaning is a partial *function*). Suppose, however, that we are dealing with programs which permit some *finite nondeterminism* as discussed in the chapter on denotational semantics (this Handbook). Then we may wish to think of a program as having as its meaning a function $f: \mathbf{N}_\perp \rightarrow P(\mathbf{N}_\perp)$ where P is one of the powerdomains. For example, if a program may give a 1 or a 2 as an output when given a 0 as input, then we will want the meaning f of this program to satisfy $f(0) = \{\!|1|\!\} \uplus \{\!|2|\!\} = \{\!|1, 2|\!\}$. The three different powerdomains reflect three different views of how to relate the various possible program behaviors in the case of divergence. The upper powerdomain identifies program behaviors which *may* diverge. For example, if program P_1 can give output 1 or diverge on any of its inputs, then it will be identified with the program Q which diverges everywhere, since $\{1, \perp\} = \perp = \{\perp\}$ in $(\mathbf{N}_\perp)^{\sharp}$. However, program P_2 which always gives 1 as its output (on inputs other than $\perp$) will *not* have the same meaning as P_1 and $\lambda x.\perp$. On the other hand, if the lower powerdomain is used in the interpretation of these progams, then P_1 and P_2 will be given the same meaning since $\{\!|1, \perp|\!\} = \{\!|1|\!\}$ in $(\mathbf{N}_\perp)^{\flat}$. However, P_1 and P_2 will not have the same meaning as the always divergent program Q since $\{\!|1, \perp|\!\} \neq \perp$ in the lower powerdomain. Finally, in the convex powerdomain, *none* of the programs P_1, P_2, Q have the same meaning since $\{\!|1, \perp|\!\}$, $\{\!|1|\!\}$ and $\{\!|\perp|\!\}$ are all distinct in $(\mathbf{N}_\perp)^{\natural}$.

To derive properties of the powerdomains like those that we discussed in the previous section for the other operators, we need to introduce the concept of a domain with binary operator.

DEFINITION. A *continuous algebra* (*of signature* (2)) is a cpo E together with a continuous binary function $*: E \times E \rightarrow E$. We refer to the following collection of axioms on $*$ as theory $T^{\natural}$:

(1) associativity: $(r*s)*t = r*(s*t)$;
(2) commutativity: $r*s = s*r$;
(3) idempotence: $s*s = s$.

(These are the well-known semilattice axioms.) A *homomorphism* between continuous algebras D and E is a continuous function $f: D \to E$ such that $f(s*t) = f(s)*f(t)$ for all s, $t \in D$.

It is easy to check that, for any domain D, each of the algebras $D^\natural$, $D^\#$ and $D^\flat$ satisfies $T^\natural$. However, $D^\natural$ is the "free" continuous algebra over D which satisfies $T^\natural$, as expressed by the following theorem.

5.2. THEOREM. *Let D be a domain. Suppose $\langle E, * \rangle$ is a continuous algebra which satisfies $T^\natural$. For any continuous $f: D \to E$, there is a unique homomorphism* **ext**$(f): D^\natural \to E$ *which completes the following diagram*:

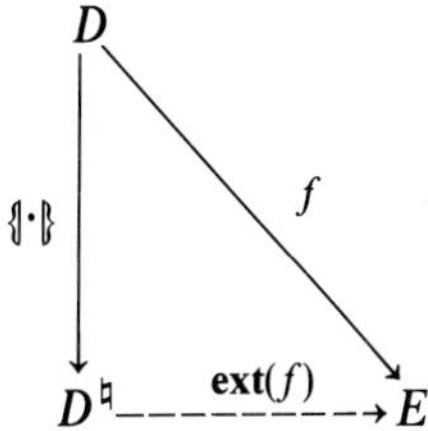

PROOF (*hint*). If $u = \{x_1, \ldots, x_n\} \in s \in D^\natural$, and $\hat{u}$ is the principal ideal generated by u, then define $\mathbf{ext}(f)(\hat{u}) = f(x_1) * \cdots * f(x_n)$. This function has a unique continuous extension to all of $D^\natural$ given by $\mathbf{ext}(f)(s) = \bigsqcup\{\mathbf{ext}(f)(\hat{u}) \mid u \in s\}$.

Now, consider the following axiom:

($4^\#$) $s \uplus t \sqsubseteq s$.

Let $T^\#$ be the set of axioms obtained by adding axiom ($4^\#$) to the axioms in $T^\#$. Similarly, let $T^\flat$ be obtained by adding the axiom

($4^\flat$) $s \sqsubseteq s \uplus t$

to the axioms in $T^\natural$. The point is this: *Theorem 5.2 still holds when $D^\natural$ and $T^\natural$ are replaced by $D^\#$ and $T^\#$ respectively,* or *by $D^\flat$ and $T^\flat$ respectively.*

As was the case with the smash product and lift operators, a diagram like the one in Theorem 5.2 gives rise to another important operation on functions. If $f: D \to E$ is a continuous function, then there is a unique homomorphism $f^\natural$ which completes the following diagram:

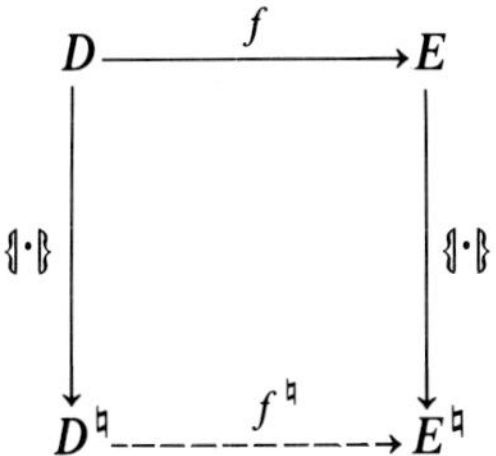

Namely, one defines $f^{\natural} = \mathbf{ext}(\{\!|\cdot|\!\} \circ f)$. Of course, there are functions $f^{\#}$ and $f^{\flat}$ with similar definitions.

Two of the powerdomains preserve the property of bounded completeness.

5.3. LEMMA. *If D is a bounded complete domain then so are $D^{\#}$ and $D^{\flat}$.*

PROOF. We leave for the reader the exercise of showing that a domain D is bounded complete if and only if every finite bounded subset of its basis has a least upper bound. To see that $D^{\flat}$ is bounded complete, just note that, for any pair of sets $u, v \in \mathscr{P}_{\mathrm{f}}^{*}(K(D))$, the ideal generated by their union $u \cup v$ is the least upper bound in $D^{\flat}$ for the ideals generated by u and v. To see that $D^{\#}$ is bounded complete, suppose $u, v, w \in \mathscr{P}_{\mathrm{f}}^{*}(K(D))$ with $w \vdash^{\#} u$ and $w \vdash^{\#} v$. Let w' be the set of elements $z \in K(D)$ such that there are elements $x \in u$ and $y \in v$ and z is the least upper bound of $\{x, y\}$. The set w' is nonempty because $\{u, v\}$ is bounded. Moreover, it is not hard to see that $w \vdash^{\#} w'$ and $w' \vdash^{\#} u$ and $w' \vdash^{\#} v$. Hence the ideal generated by w' is the least upper bound of the ideals generated by u and v. □

6. Bifinite domains

Of the operators that we have discussed so far, only the convex powerdomain $(\cdot)^{\natural}$ does not take bounded complete domains to bounded complete domains. To see this in a simple example, consider the finite poset $\mathbf{T} \times \mathbf{T}$ and the following elements of $\mathscr{P}_{\mathrm{f}}^{*}(\mathbf{T} \times \mathbf{T})$:

$$u = \{\langle \bot, \mathbf{true} \rangle, \langle \bot, \mathbf{false} \rangle\}, \qquad v = \{\langle \mathbf{true}, \bot \rangle, \langle \mathbf{false}, \bot \rangle\},$$
$$u' = \{\langle \mathbf{true}, \mathbf{true} \rangle, \langle \mathbf{false}, \mathbf{false} \rangle\}, \qquad v' = \{\langle \mathbf{true}, \mathbf{false} \rangle, \langle \mathbf{false}, \mathbf{true} \rangle\}.$$

It is not hard to see that u' and v' are *minimal* upper bounds for $\{u, v\}$ with respect to the ordering $\vdash^{\natural}$. Hence no *least* upper bound for $\{u, u'\}$ exists and $(\mathbf{T} \times \mathbf{T})^{\natural}$ is therefore not bounded complete. In this section we introduce a natural class of domains on which *all* of the operators we have discussed above (including the convex powerdomain) are closed. This class is defined as follows.

DEFINITION. Let D be a cpo. Let $\mathscr{M}$ be the set of finitary projections with finite image. Then D is said to be *bifinite* if $\mathscr{M}$ is countable, directed and $\bigsqcup \mathscr{M} = \mathbf{id}$.

The bifinite cpo's are motivated, in part, by considerations from category theory and the definition above is a restatement of their categorical definition. They were first defined by Plotkin [10] (where they are called "**SFP**-objects") and the term "bifinite" is due to P. Taylor. Bifinite domains (and various closely related classes of cpo's) have also been discussed under other names such as "strongly algebraic" [21, 6] and "profinite" [7] domains.

6.1. *Plotkin orders*

As we suggested earlier, the image of a finitary projection $p: D \to D$ on a domain D can be viewed as an approximation to D. A bifinite domain is one which is a directed limit of its finite approximations. But what is this really saying about the structure of D? First of all, it follows from properties of finitary projections that we mentioned earlier that whenever $p: D \to D$ is a finitary projection and $\mathbf{im}(p)$ is finite, then $\mathbf{im}(p) \subseteq K(D)$. From this, together with the fact that the set $\mathcal{M}$ is directed and $\bigsqcup \mathcal{M} = \mathbf{id}$, it is possible to show D is a domain with $\bigcup \{\mathbf{im}(p) \mid p \in \mathcal{M}\}$ as its basis. We may now use the correspondence which we noted in Theorem 3.3 to provide a condition on the basis of a domain which characterizes the domain as being bifinite. Recal that $N \lhd A$ for posets N and A if $N \cap \downarrow x$ is directed for every $x \in A$.

DEFINITION. A poset A is a *Plotkin order* if, for every finite subset $u \subseteq A$, there is a finite set $N \lhd A$ with $u \subseteq N$.

6.1. THEOREM. *The following are equivalent for any cpo D.*

(1) *D is bifinite.*

(2) *D is a domain and K(D) is a Plotkin order.*

To get some idea what a Plotkin order looks like, it helps to have a definition. Given a poset A and a finite set $u \subseteq A$, an upper bound x for u is *minimal* if, for any upper bound y for u, $y \sqsubseteq x$ implies $y = x$. A set v of minimal upper bounds for u is said to be *complete* if, for every upper bound x for u, there is a $y \in v$ with $y \sqsubseteq x$. Now, let A be a Plotkin order and suppose $u \subseteq A$ is finite. Then there is a finite $N \lhd A$ with $u \subseteq N$. The set N must contain a complete set of minimal upper bounds for u (why?). This shows the first fact about Plotkin orders: every finite subset has a complete set of minimal upper bounds. This rules out configurations like the one pictured in Fig. 3(a) where the pair of points indicated by closed circles do not have such a complete set of minimal upper bounds.

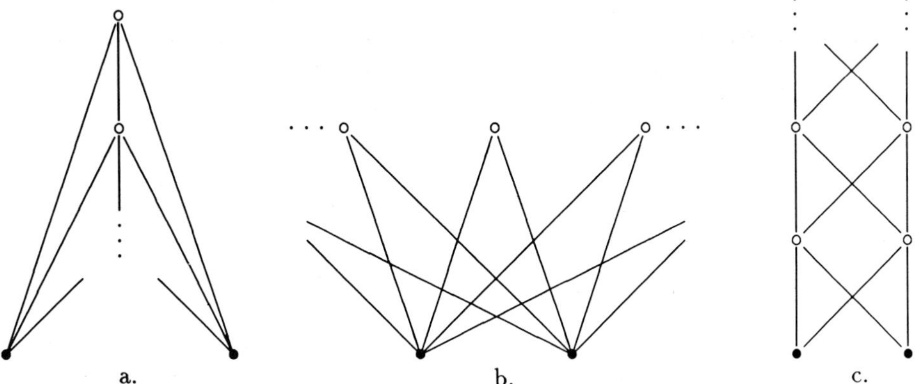

Fig. 3. Posets that are not Plotkin orders.

But the set N is *finite* so we have our second fact: every finite subset must have a *finite* complete set of minimal upper bounds. This rules out configurations like the one pictured in Fig. 3(b) where the pair of points indicated by closed circles has a complete set of minimal upper bounds but not a finite one. However, having finite complete sets of minimal upper bounds for finite subsets is not a sufficient condition for characterizing the Plotkin orders. To see why, let A be a poset which has finite complete sets of minimal upper bounds for finite subsets. If $u \subseteq A$ is finite, let

$$\mathcal{U}(u) = \{x \mid x \text{ is the minimal upper bound for some } v \subseteq u\}.$$

Now, if $u \subseteq N \lhd A$, then $\mathcal{U}(u) \subseteq N$. Hence, $\mathcal{U}^n(u) \subseteq N$ for each n. If N is finite, then there must be an n for which $\mathcal{U}^n(u) = \mathcal{U}^{n+1}(u)$. This is a third fact about Plotkin orders: for each finite $u \subseteq A$, $\mathcal{U}^\infty(u) = \bigcup_n \mathcal{U}^n(u)$ is finite. To see what can go wrong, note that $\mathcal{U}^\infty(u)$ is infinite when u is the pair of points indicated by closed circles in Fig. 3(c).

6.2. *Closure properties*

6.2. PROPOSITION. *A bounded complete domain is bifinite.*

PROOF. Suppose D is bounded complete and $u \subseteq K(D)$ is a finite subset of the basis of D. Let

$$N = \{x \mid x \text{ is the least upper bound of a finite subset of } u\}.$$

Note that N is finite; we claim that $N \lhd K(D)$. Suppose x is the least upper bound of a finite set $v \subseteq K(D)$. Since D is algebraic, there is a directed subset $M \subseteq K(D)$ such that $x = \bigsqcup M$. But the elements of v are compact. Hence, for every $y \in v$, there is a $y' \in M$ with $y \sqsubseteq y'$. Since M is directed, there is some $z \in M$ which is an upper bound for v. Now, $z \sqsubseteq x$ so $x = z$ and x is therefore compact. This shows that $N \subseteq K(D)$. Suppose $v \subseteq N$ is bounded; then the least upper bound of v is the same as the least upper bound of the set $\{x \in u \mid x \sqsubseteq y \text{ for some } y \in v\}$, so the least upper bound of v is in N. Now, if $x \in K(D)$, then $S = (\downarrow x) \cap N$ is bounded. Since S has a least upper bound which, apparently, lies in S, we conclude that S is directed. □

6.3. THEOREM. *If D is bifinite, then the poset* $\mathbf{Fp}(D)$ *of finitary projections on D is an algebraic lattice and the inclusion map i:* $\mathbf{Fp}(D) \hookrightarrow (D \to D)$ *is an embedding.*

PROOF (*sketch*). One uses Theorem 3.3 to show that $\mathbf{Fp}(D)$ is an algebraic lattice. Suppose $f: D \to D$ is continuous. Let

$$S_f = \{x \in K(D) \mid x \sqsubseteq f(x)\}.$$

One can show that there is a least set N_f such that $S_f \subseteq N_f \lhd K(D)$. This set determines a finitary projection p_{N_f} as in the discussion before Theorem 3.3. On the other hand, if $f: D \to D$ is a finitary projection then $N_f = \mathbf{im}(f) \cap K(D)$ and $f = p_{N_f}$. The remaining steps required to verify that $f \mapsto N_f$ is a projection are straightforward. □

6.4. LEMMA. *If D and E are bifinite domains, then so are the cpo's $D\to E$, $D\circ\!\!\to E$, $D\times E$, $D\otimes E$, $D+E$, $D\oplus E$, $D_\perp$, $D^\natural$, $D^\#$ and $D^\flat$.*

PROOF. We will outline proofs for two sample cases. We begin with the function space operator. Suppose $p:D\to D$ and $q:E\to E$ are finitary projections. Given a continuous function $f:D\to E$, define $\Theta(q,p)(f)=q\circ f\circ p$. The function $\Theta(q,p)$ defines a finitary projection on $D\to E$. Moreover, if p and q have finite images, then so does $\Theta(q,p)$. If we let $\mathscr{M}$ be the set of functions $\Theta(q,p)$ such that p and q are finitary projections with finite image, then it is easy to see that $\bigsqcup\mathscr{M}=\mathbf{id}$. Hence $D\to E$ is bifinite. We will encounter the function Θ again in the next section.

To see that $D^\natural$ is bifinite, one shows that the set

$$\mathscr{M}=\{p^\natural \mid p\in \mathbf{Fp}(D) \text{ and } \mathbf{im}(p) \text{ is finite}\}$$

is directed and has the identity as its least upper bound. The functions in $\mathscr{M}$ are themselves finitary projections with finite images, so $D^\natural$ is bifinite. □

One may conclude from this lemma that the bifinite domains have rather robust closure properties. But there is something else about bifinite domains which makes them special. They are the *largest* class of domains which are closed under the operators listed in the Lemma. In fact, there is the following theorem.

6.5. THEOREM. *If D and $D\to D$ are domains, then D is bifinite.*

The theorem is due to Smyth and its proof may be found in [21]. It is carried out by analyzing each of the cases pictured in Fig. 3 and showing that if $D\to D$ is not a domain, then D cannot be bifinite. A similar result for the bounded complete domains can be found in [6].

7. Recursive definitions of domains

Many of the data types that arise in the semantics of computer programming languages may be seen as solutions of *recursive domain equations.* Consider, for example, the equation $T\cong T+T$ (of course, this is an *isomorphism* rather than an *equality*, but let us not make much of this distinction for the moment). How would we go about finding a domain which solves this equation? Suppose we start with the one-point domain $T_0=\mathbf{I}$ as the first approximation to the desired solution. Taking the proof of the Fixed Point Theorem as our guide, we build the domain $T_1=T_0+T_0=\mathbf{I}+\mathbf{I}$ as the second approximation. Now, there is a unique embedding $e_0:T_0\to T_1$, so this gives a precise sense in which T_0 approximates T_1. The next approximation to our solution is the domain $T_2=T_1+T_1$ and again there is an embedding $e_1=e_0+e_0:T_1\to T_2$. If we continue along this path we build a sequence

$$T_0\xrightarrow{e_0}T_1\xrightarrow{e_1}T_2\xrightarrow{e_2}\cdots$$

of approximations to the full simple binary tree. To get a domain, we must add limits for each of the branches. The resulting domain (i.e. the full simple binary tree with the limit points added) is, indeed, a "solution" of $T \cong T+T$. This is all very informal, however; how are we to make this idea mathematically *precise* and, at the same time, sufficiently *general*?

7.1. Solving domain equations with closures

In this section we discuss a technique for solving recursive domain equations by relating domains to functions by the "image" map (**im**) and then using the ideas of the previous section to solve equations. There are two (closely related) ways of doing this which we will illustrate. The first of these is based on the following concept.

DEFINITION. Let D and E be cpo's. A continuous function $r:D \to E$ is a *closure* if there is a continuous function $s:E \to D$ such that $r \circ s = \mathbf{id}$ and $s \circ r \sqsupseteq \mathbf{id}$.

By analogy with the notion of a finitary projection, we will say that a function $r:D \to D$ is a *finitary closure* if $r \circ r = r \sqsupseteq \mathbf{id}$ and $\mathbf{im}(r)$ is a domain. In the event that D is a domain, the requirement that $\mathbf{im}(r)$ be a domain is unnecessary because we have the following lemma.

7.1. LEMMA. *If D is a domain and $r:D \to D$ satisfies the equation $r \circ r = r \sqsupseteq \mathbf{id}$, then $\mathbf{im}(r)$ is a domain.*

The lemma is proved by showing that $\{r(x) \mid x \in K(D)\}$ forms a basis for $\mathbf{im}(r)$. We will say that a domain E is a *closure of D* if it is isomorphic to $\mathbf{im}(r)$ for some finitary closure r on D. We let $\mathbf{Fc}(D)$ be the poset of finitary closures $r:D \to D$.

7.2. LEMMA. *If D is a domain, then $\mathbf{Fc}(D)$ is a cpo.*

DEFINITION. Let us say that an operator F on cpo's is *representable* over a cpo U if and only if there is a continuous function R_F which completes the following diagram (up to isomorphism):

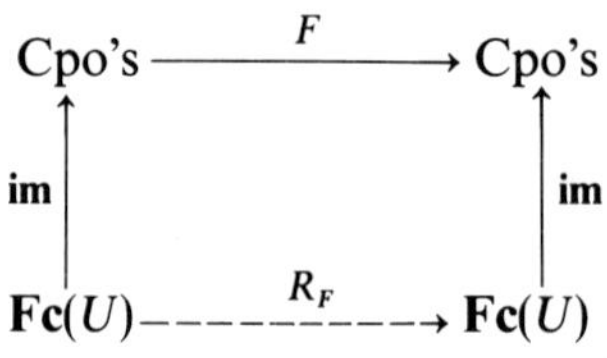

$$\begin{array}{ccc} \text{Cpo's} & \xrightarrow{F} & \text{Cpo's} \\ \uparrow{\scriptstyle \mathbf{im}} & & \uparrow{\scriptstyle \mathbf{im}} \\ \mathbf{Fc}(U) & \dashrightarrow_{R_F} & \mathbf{Fc}(U) \end{array}$$

i.e., $\mathbf{im}(R_F(r)) \cong F(\mathbf{im}(r))$ for every closure r.

This idea extends to multiary operators as well. For example, the function space operator $\cdot \to \cdot$ is representable over a cpo U if there is a continuous function

$$R:\mathbf{Fc}(U) \times \mathbf{Fc}(U) \to \mathbf{Fc}(U)$$

such that, for any $r, s \in \mathbf{Fc}(U)$,

$$\mathbf{im}(R(r,s)) \cong \mathbf{im}(r) \to \mathbf{im}(s).$$

An operator $\langle F_1, \ldots, F_n \rangle$ is defined to be representable if each of the operators F_i is. Note that a composition of representable operators is representable.

7.3. THEOREM. *If an operator F is representable over a domain U, then there is a domain D such that $D \cong F(D)$.*

PROOF. Suppose R_F represents F. By the Fixed Point Theorem, there is an $r \in \mathbf{Fc}(U)$ such that $r = R_F(r)$. Thus $\mathbf{im}(r) = \mathbf{im}(R_F(r)) \cong F(\mathbf{im}(r))$ so $\mathbf{im}(r)$ is the desired doman. $\square$

Now we know how to solve domain equations. For example, to solve $T \cong T + T$ we need to find a domain U and continuous function $f: U \to U$ which represents the operator $F(X) = X + X$. But we are still left with the problem of finding a domain over which such operations may be represented! The next step is to look at a simple structure which can be used to represent several of the operations in which we are interested.

Given sets S and T, let T^S be the set of (all) functions from S into T. If T is a cpo, then T^S is also a cpo under the pointwise ordering. Now, it is not hard to see that the domain equation $X \cong X \times \mathbf{I}^\top$ (where $\mathbf{I}^\top$ is the two-point lattice) has, as one of its solutions, the cpo $(\mathbf{I}^\top)^{\mathbf{N}}$. In fact, this cpo is isomorphic to the algebraic cpo $\mathscr{P}\mathbf{N}$ of subsets of $\mathbf{N}$ which we discussed in the first section. It is particularly interesting because of the following theorem.

7.4. THEOREM. *For any (countably based) algebraic lattice L, there is a closure $r: \mathscr{P}\mathbf{N} \to L$.*

PROOF. Let $l_0, l_1, l_2, \ldots$ be an enumeration of the basis of L. Given $S \subseteq \mathbf{N}$, let $r(S) = \bigsqcup\{l_n \mid n \in S\}$. If $l \in L$, let $s(l) = \{n \mid l_n \sqsubseteq l\}$. We leave for the reader the (easy) demonstration that r, s are continuous with $r \circ s = \mathbf{id}$ and $s \circ r \sqsupseteq \mathbf{id}$. $\square$

Structures such as $\mathscr{P}\mathbf{N}$ are often referred to as *universal domains*, because they have a rich collection of domains as retracts. In the remainder of this section we will discuss two more similar constructions and show how they may be used to provide representations for operators.

Unfortunately, there is no representation for the operator $F(X) = X + X$ over $\mathscr{P}\mathbf{N}$. However, there are some much more interesting operators which *are* representable over $\mathscr{P}\mathbf{N}$. In particular, we have the following lemma.

7.5. LEMMA. *The function space operator is representable over $\mathscr{P}\mathbf{N}$.*

PROOF. Consider the algebraic lattice of continuous functions $\mathscr{P}\mathbf{N} \to \mathscr{P}\mathbf{N}$. By Theorem 7.4, we know that there are continuous functions

$$\Phi_{\to}: \mathscr{P}\mathbf{N} \to (\mathscr{P}\mathbf{N} \to \mathscr{P}\mathbf{N}) \qquad \Psi_{\to}: (\mathscr{P}\mathbf{N} \to \mathscr{P}\mathbf{N}) \to \mathscr{P}\mathbf{N}$$

such that $\Phi_{\rightarrow}\circ\Psi_{\rightarrow}=\mathbf{id}$ and $\Psi_{\rightarrow}\circ\Phi_{\rightarrow}\sqsupseteq\mathbf{id}$. Now, suppose r, $s\in\mathbf{Fc}(\mathscr{P}\mathbf{N})$ (that is, $r\circ r=r\sqsupseteq\mathbf{id}$ and $s\circ s=s\sqsupseteq\mathbf{id}$). Given a continuous function $f:\mathscr{P}\mathbf{N}\rightarrow\mathscr{P}\mathbf{N}$, let $\Theta(s,r)(f)=s\circ f\circ r$ and define

$$R_{\rightarrow}(r,s)=\Psi_{\rightarrow}\circ\Theta(s,r)\circ\Phi_{\rightarrow}.$$

To see that this function is a finitary closure, we take $x\in\mathscr{P}\mathbf{N}$ and compute

$$\begin{aligned}
&(R_{\rightarrow}(r,s)\circ R_{\rightarrow}(r,s))(x)\\
&\quad=(\Psi_{\rightarrow}\circ\Theta(s,r)\circ\Phi_{\rightarrow})(\Psi_{\rightarrow}(s\circ(\Phi_{\rightarrow}(x))\circ r)\\
&\quad=(\Psi_{\rightarrow}\circ\Theta(s,r)\circ\Phi_{\rightarrow}\circ\Psi_{\rightarrow})(s\circ(\Phi_{\rightarrow}(x))\circ r)\\
&\quad=(\Phi_{\rightarrow}\circ\Theta(s,r))(s\circ(\Phi_{\rightarrow}(x))\circ r)\\
&\quad=\Psi_{\rightarrow}((s\circ s)\circ(\Phi_{\rightarrow}(x))\circ(r\circ r))\\
&\quad=\Psi_{\rightarrow}(s\circ(\Phi_{\rightarrow}(x))\circ r)\\
&\quad=R_{\rightarrow}(r,s)(x)
\end{aligned}$$

and

$$R_{\rightarrow}(r,s)(x)=\Psi_{\rightarrow}(s\circ(\Phi_{\rightarrow}(x))\circ r)\sqsupseteq\Psi_{\rightarrow}(\Phi_{\rightarrow}(x))\sqsupseteq x.$$

Thus we have defined a function,

$$R_{\rightarrow}:\mathbf{Fc}(\mathscr{P}\mathbf{N})\times\mathbf{Fc}(\mathscr{P}\mathbf{N})\rightarrow\mathbf{Fc}(\mathscr{P}\mathbf{N})$$

which we now demonstrate to be a representation of the function space operator.

Given r, $s\in\mathbf{Fc}(\mathscr{P}\mathbf{N})$, we must show that there is an isomorphism

$$\mathbf{im}(R(r,s))\cong\mathbf{im}(r)\rightarrow\mathbf{im}(s)$$

for each $r,s\in\mathbf{Fc}(\mathscr{P}\mathbf{N})$. Now, there is an evident isomorphism between continuous functions $f:\mathbf{im}(r)\rightarrow\mathbf{im}(s)$ and continuous functions $g:\mathscr{P}\mathbf{N}\rightarrow\mathscr{P}\mathbf{N}$ such that $g=s\circ g\circ r$. We claim that $\Psi_{\rightarrow}$ cuts down to an isomorphism between such functions and the sets in the image of $R_{\rightarrow}(r,s)$. Since $\Phi_{\rightarrow}\circ\Psi_{\rightarrow}=\mathbf{id}$, we need only show that $(\Psi_{\rightarrow}\circ\Phi_{\rightarrow})(x)=x$ for each $x=R_{\rightarrow}(r,s)(x)$. But if

$$x=\Psi_{\rightarrow}(s\circ(\Phi_{\rightarrow}(x))\circ r)$$

then

$$\begin{aligned}
(\Psi_{\rightarrow}\circ\Phi_{\rightarrow})(x)&=(\Psi_{\rightarrow}\circ\Phi_{\rightarrow}\circ\Psi_{\rightarrow})(s\circ(\Phi_{\rightarrow}(x))\circ r)\\
&=\Psi_{\rightarrow}(s\circ(\Phi_{\rightarrow}(x))\circ r)\\
&=x.
\end{aligned}$$

Hence $\mathbf{im}(R_{\rightarrow}(r,s))\cong\mathbf{im}(r)\rightarrow\mathbf{im}(s)$ and we may conclude that $R_{\rightarrow}$ represents $\rightarrow$ over $\mathscr{P}\mathbf{N}$. □

A similar construction can be carried out for the product operator. Suppose

$$\Phi_{\times}:\mathscr{P}\mathbf{N}\rightarrow(\mathscr{P}\mathbf{N}\times\mathscr{P}\mathbf{N}),\qquad\Psi_{\times}:(\mathscr{P}\mathbf{N}\times\mathscr{P}\mathbf{N})\rightarrow\mathscr{P}\mathbf{N}$$

such that $\Phi_{\times}\circ\Psi_{\times}=\mathbf{id}$ and $\Psi_{\times}\circ\Phi_{\times}\sqsupseteq\mathbf{id}$. For $r,s\in\mathbf{Fp}(\mathscr{P}\mathbf{N})$ define

$$R_{\times}(r,s)=\Psi_{\times}\circ(r\times s)\circ\Phi_{\times}$$

We leave for the reader the demonstration that this makes sense and $R_\times$ represents the product operator.

Suppose that L is an algebraic lattice. Then there are continuous functions

$$\Phi_L : \mathscr{P}\mathbf{N} \to \mathscr{P}\mathbf{N}, \qquad \Psi_L : \mathscr{P}\mathbf{N} \to \mathscr{P}\mathbf{N}$$

such that $\Phi_L \circ \Psi_L = \mathbf{id}$ and $\Psi_L \circ \Phi_L \sqsupseteq \mathbf{id}$. Then the function $R_L(r, s) = \Psi_L \circ \Phi_L$ represents the constant operator $X \mapsto L$, because $\mathbf{im}(\Psi_L \circ \Phi_L) \cong L$. A similar argument can be used to show that a constant operator $X \mapsto D$ is representable over a domain U if and only if D is a closure of U.

7.2. *Modelling the untyped λ-calculus*

It is tempting to try to solve the domain equation $D \cong D \to D$ by the methods just discussed. Unfortunately, the equation $\mathbf{I} \cong \mathbf{I} \to \mathbf{I}$ (corresponding to the fact that on a one-point set there is only one possible self-map) shows that there is no guarantee that the result will be at all interesting. There has to be a way to build in some nontrivial structure that is not wiped out by the fixed point process. Methods are described in [12, 14], but the following, from [13, 15], is more direct and more general.

7.6. LEMMA. *Let U be a nontrivial domain. If the product and function space operators can be represented over U, then there are nontrivial domains D and E such that $E \cong E \times E$ and $D \cong D \to E$.*

PROOF. We can represent $F(X) = U \times X \times X$ over U, so there is a closure A of U such that $A \cong U \times A \times A$. Thus

$$U \times A \cong U \times (U \times A \times A) \cong (U \times A) \times (U \times A).$$

So $E = U \times A$ is nontrivial and $E \cong E \times E$. Now, E is a closure of U, so $G(X) = X \to E$ is representable over U. Hence there is a cpo $D \cong D \to E$. This cpo is nontrivial because E is. □

7.7. THEOREM. *If U is a nontrivial domain which represents products and function spaces, then there is a nontrivial domain D such that $D \cong D \times D \cong D \to D$ and D is the image of a closure on U.*

PROOF. Let D and E be the domains given by Lemma 7.6. Then

$$D \times D \cong (D \to E) \times (D \to E) \cong D \to (E \times E) \cong D \to E \cong D$$

and

$$D \to D \cong D \to (D \to E) \cong (D \times D) \to E \cong D \to E \cong D. \qquad \square$$

We note, in fact, that D will have $\mathscr{P}\mathbf{N}$ itself represented by a closure on U. Hence, to get a nontrivial solution for $D \cong D \to D \cong D \times D$, take U in the theorem to be $\mathscr{P}\mathbf{N}$. What good is such a domain? The answer is that a D satisfying these isomorphisms is a model for a very strong λ-calculus. If we expand the syntax of λ-calculus given in Section 5.3 of the chapter on denotational semantics (this Handbook) to allow pairings, we would have

$$E ::= (\lambda x.E) \mid E_1(E_2) \mid x \mid \mathbf{pair} \mid \mathbf{fst} \mid \mathbf{snd}$$

Now, as pointed out in that same chapter for the type of semantic function defined there, many *different* expressions are mapped into the *same* values. We can say that the model *satisfies* certain equations. In particular, under the isomorphisms obtained in our theorems above, the following equations will be satisfied:

(1) $(\lambda x.E) = (\lambda y.[y/x]E)$ (provided y is not free in E),
(2) $(\lambda x.E)(E') = [E'/x]E$,
(3) $(\lambda x.E(x)) = E$ (provided x is not free in E),
(4) $\mathbf{fst}(\mathbf{pair}(E)(E')) = E$,
(5) $\mathbf{snd}(\mathbf{pair}(E)(E')) = E'$,
(6) $\mathbf{pair}(\mathbf{fst}(E))(\mathbf{snd}(E)) = E$.

In these equations, the third and sixth especially emphasize the isomorphisms $D \cong D \to D$ and $D \cong D \times D$. There are models where $D \to D$ is represented by a closure on D (as is $D \times D$) but where this is not an isomorphism. It follows that the special equations are independent of the others.

In [11] the question is brought up whether we can add to the above equations one relating functional abstraction with pairing. In particular, the following would be interesting:

$$\mathbf{pair}(x)(y) = (\lambda z.\mathbf{pair}(x(z))(y(z))).$$

This equation identifies the primitive pairing with what could be called *pointwise pairing*. This equation is independent from the others, but a model for it can be obtained from the first model by introducing a new pairing and application operation that does things pointwise in a suitable sense. There must be many other kinds of models that relate the functional structure to other constructs as well.

Suppose we have domains that satisfy just the six equations. Then from the primitive operations given, many others can be defined. The operation of λ-abstraction is, to be sure, a variable-binding operator (somewhat like a quantifier), but the others are algebraic in nature. As stated, application is a binary operation, and **pair**, **fst** and **snd** are constants. But we can define binary, ternary, and unary operations such as: $\mathbf{pair}(x)(y)$, $\mathbf{pair}(x)(\mathbf{pair}(y)(z))$, $\mathbf{fst}(x)$, $\mathbf{snd}(y)$, $\mathbf{pair}(\mathbf{snd}(z))(\mathbf{fst}(z))$, and many, many more. In other words, the domain D will become a model of many kinds of algebras.

In general, an *algebra* is a set together with several operations defined on it, taking values in the same set. The simplest situation is to consider finitary operations (i.e., operations taking a fixed finite number of arguments). When giving an algebra, the sequence of arities of the fundamental operations is called the *signature* of the algebra. Thus, a *ring* is often given with just two binary operations (*addition* and *multiplication*) making a signature (2, 2). Now, subtraction is definable in first-order logic from addition, but the definition is not equational. Therefore, it may be better to consider a ring as an algebra of signature (2, 2, 2) with subtraction being taken as primitive. Of, course it is enough to have the minus operation, which is unary. So, a signature (2, 1, 2) is also popular. Strictly speaking, however, different signatures correspond to algebras of different types. Not every algebra of signature (2, 2, 2) is "equivalent" to one of signature (2, 1, 2); rings as algebras have very special properties.

By a *continuous algebra* we mean a domain with various continuous operations singled out. In particular, our λ-calculus model can be considered as a continuous algebra of signature (2, 0, 0, 0, 0, 0). The binary operation is the operation of functional application. Here, 0 indicates a 0-ary operation, which is just a *constant*. We already know the constants **pair**, **fst**, **snd**. The other two popular constants from the literature on λ-calculus are called **S** and **K**. In terms of λ-abstraction they can be defined as follows:

$$\mathbf{S}=(\lambda x.(\lambda y.(\lambda z.x(z)(y(z))))), \qquad \mathbf{K}=(\lambda x.(\lambda y.x))$$

They enjoy many, many equations in the algebra (see, for example, [1]) and, in fact, any equation involving the λ-operator can be rewritten purely algebraically in terms of **S** and **K** and application. Continuous algebras in general have also been a topic of extensive investigation; see [24] and the references there.

We will call an expression in the notation of applicative algebra which has no variables a *combination*. Any combination F defines an n-ary operation: $F(x_1)(x_2)\ldots(x_n)$. What we have been remarking is that the algebras so obtained from combinations can be very rich. In a series of papers [3, 4] Engeler discussed just how rich these algebras can be. A representative result, following Engeler, will be exhibited here.

7.8. THEOREM. *Given a signature* $(s_1, s_2, \ldots, s_n)$, *there are combinations* $F_1, F_2, \ldots, F_n$ *defining operations on D of these arities such that whenever a continuous algebra of this signature is given on a domain A that is a retract of D, then A can be made isomorphic to a subalgebra of this fixed algebra structure on D.*

PROOF. If A is a retract of D, then A can be regarded as a subset of D, and all the continuous operations on A can be naturally extended to continuous operations on D of the same arities. (This does not solve the problem, since the operations on D depend on the choice of A. That is to say, at the start, A is a subalgebra of the wrong algebra on D.) We can call these operations $o_1, o_2, \ldots, o_n$.

We are going to define the representation of A as a subalgebra of D by means of a continuous function $\rho: A \to D$ defined by means of a fixed point equation:

$$\begin{aligned}\rho(a)=\mathbf{pair}(a)&\\ &(\mathbf{pair}(\lambda x_2 \ldots \lambda x_{s_1}.\rho(o_1(a, \mathbf{fst}(x_2), \ldots, \mathbf{fst}(x_{s_1}))))\\ &(\mathbf{pair}(\lambda x_2 \ldots \lambda x_{s_2}.\rho(o_2(a, \mathbf{fst}(x_2), \ldots, \mathbf{fst}(x_{s_2}))))\\ &\vdots\\ &(\mathbf{pair}(\lambda x_2 \ldots \lambda x_{s_n}.\rho(o_n(a, \mathbf{fst}(x_2), \ldots, \mathbf{fst}(x_{s_n}))))\\ &(\mathbf{K}))\ldots).\end{aligned}$$

In this way, we build into ρ the elements from A and the operations as well. The question is how to read off the coded information.

Consider the following combinations:

$$F_1 = \lambda x.\mathbf{fst}(\mathbf{snd}(x))$$
$$F_2 = \lambda x.\mathbf{fst}(\mathbf{snd}(\mathbf{snd}(x)))$$
$$\vdots$$
$$F_n = \lambda x.\mathbf{fst}(\mathbf{snd}(\mathbf{snd}(\ldots\mathbf{snd}(x)))),$$

which have to be rewritten in terms of **S**, **K**, **fst**, and **snd**. We then calculate that

$$F_i(\rho(a_1))(\rho(a_2))\ldots(\rho(a_{s_i})) = \rho(o_i(a_1, a_2, \ldots a_{s_i})).$$

This means that if we consider the algebra $\langle D, F_1, F_2, \ldots, F_n \rangle$, then we can find by means of the definition of ρ any algebra $\langle A, o_1, o_2, \ldots, o_n \rangle$, isomorphic to a subalgebra of the first algebra. □

7.3. *Solving domain equations with projections*

As we mentioned earlier, one slightly bothersome drawback to $\mathscr{P}\mathbf{N}$ as a domain for solving recursive domain equations is the fact that it cannot represent the sum operator $+$. One might try to overcome this problem by using the operator $(\cdot + \cdot)^{\mathsf{T}}$ as a substitute since this *is* representable over $\mathscr{P}\mathbf{N}$. However, the added top element seems unmotivated and gets in the way. It is probably possible to find a cpo which will represent the operators $\times$, $\rightarrow$, $+$. However, for the sake of variety, we will discuss a slightly different method for solving domain equations. Let us say that an operator F on cpo's is *p-representable* over a cpo U if and only if there is a continuous function R_F which completes the following diagram (up to isomorphism):

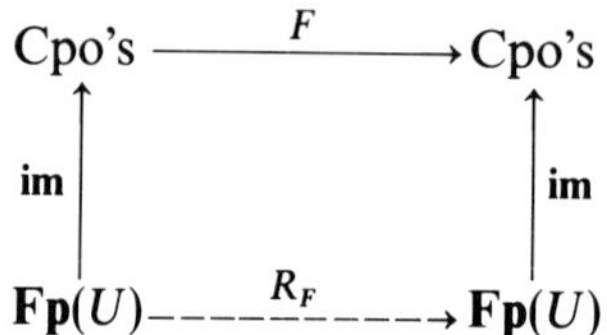

Since there will be no chance of confusion, let us just use the term "representable" for "*p*-representable" for the remainder of this section. Since $\mathbf{Fp}(U)$ is a cpo we can solve domain equations in the same way we did before, *provided we can find domains over which the necessary operators can be represented.*

The construction of a suitable domain is somewhat more involved than was the case for $\mathscr{P}\mathbf{N}$. We begin by describing the basis of a domain $\mathbf{U}$. Let S be the set of rational numbers of the form $n/2^m$ where $0 \leqslant n < 2^m$ and $0 < m$. As the basis $\mathbf{U}_0$ of our domain we take finite (nonempty) unions of half open intervals $[r, t) = \{s \in S \mid r \leqslant s < t\}$. A typical element would look like

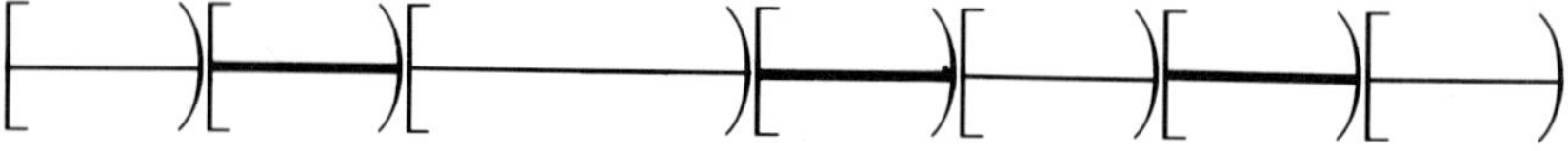

We order these sets by superset so that the interval $[0, 1)$ is the *least* element. There is no

top element under this ordering. If we adjoin the empty set, say $B=\mathbf{U}_0 \cup \{\emptyset\}$, then we get a *Boolean algebra.* (Note that the complement of a finite union of intervals is again one such—unless it is empty.) In particular, any interval contains a proper subinterval, so, as a Boolean algebra, B is *atomless.* But B is countable, and—up to isomorphism—the only countable atomless Boolean algebra is the free one on countably many generators. But this Boolean algebra has the property that every countable Boolean algebra is isomorphic to a subalgebra. Now, suppose A is a countable bounded complete poset. Let B' be the Boolean algebra of subsets of A generated by these subsets of the form $\uparrow x=\{y \in A \mid x \sqsubseteq y\}$ and order this collection by superset so that $\emptyset$ will be its largest element. The map $i: x \mapsto \uparrow x$ is a monotone injection which preserves existing least upper bounds. Moreover, a subset $u \subseteq A$ is bounded just in case $\bigcap_{x \in u} \uparrow x$ is nonempty. Now, if $j: B' \to B$ maps B' isomorphically onto a subalgebra of B, then the composition $j \circ i$ cuts down to an isomorphism between A and a normal subposet $A' \lhd \mathbf{U}_0$. Letting $\mathbf{U}$ be the domain of ideals over $\mathbf{U}_0$ we may now conclude the following theorem.

7.9. THEOREM. *For any bounded complete domain D, there is a projection* $p: \mathbf{U} \to D$.

We can now use this to see that an equation like $X \cong \mathbf{N}_\perp + (X \to X)$ has a solution. The proof that $\to$ is representable over $\mathbf{U}$ is almost identical to the proof we gave above that it is representable over $\mathscr{P}\mathbf{N}$. To get a representation for $+$, take a pair of continuous functions

$$\Phi_+ : \mathbf{U} \to (\mathbf{U}+\mathbf{U}), \qquad \Psi_+ : (\mathbf{U}+\mathbf{U}) \to \mathbf{U}$$

such that $\Phi_+ \circ \Psi_+ = \mathbf{id}$ and $\Psi_+ \circ \Phi_+ \sqsubseteq \mathbf{id}$. Then take

$$R_+(r,s) = \Psi_+ \circ (r+s) \circ \Phi_+.$$

Also, there is a representation $R_{\mathbf{N}_\perp}$ for constant operator $X \mapsto \mathbf{N}_\perp$. Hence the operator $X \mapsto \mathbf{N}_\perp + (X \to X)$ is represented over $\mathbf{U}$ by the function

$$p \mapsto R_+(R_{\mathbf{N}_\perp}(p), R_\to(p,p)).$$

We have, in fact, the following lemma.

7.10. LEMMA. *The following operators are representable over* $\mathbf{U}$: $\to, \circ\!\!\to, \times, \otimes, +, \oplus, (\cdot)_\perp, (\cdot)^\#, (\cdot)^\flat$.

This means that we have solutions over the bounded complete domains for a quite substantial class of recursive equations. More discussion of $\mathbf{U}$ may be found in [16, 17, 18].

7.4. *Representing operators on bifinite domains*

The convex powerdomain $(\cdot)^\natural$ cannot be representable over $\mathbf{U}$ because it does not preserve bounded completeness. We construct a domain over which this operator can be represented as follows. Given a poset A, define $M(A)$ to be the of pairs $(x, u) \in A \times \mathscr{P}_f(A)$ such that $x \sqsubseteq z$ for every $z \in u$. Define a pre-ordering on $M(A)$ by setting $(x, u) \vdash (y, u)$ if and only if there is a $z \in u$ such that $z \sqsubseteq y$. Now, given a domain D, we define D^+ to be the domain of ideals over $\langle M(A), \vdash \rangle$.

7.11. THEOREM. *If D is bifinite, then so is D^+. Moreover, if $D \cong D^+$ and E is any bifinite domain, then there is a projection $p: D \to E$.*

A full proof of the theorem may be found in [7]. We will attempt to offer some hint about how the desired fixed point is obtained. At the first step we take the domain $\mathbf{I} = \{\bot\}$ containing only the single point $\bot$. At the second step, $\mathbf{I}^+$, there are elements $a = (\bot, \{\bot\})$ and $b = (\bot, \emptyset)$ with $b \vdash a$. At the third step, there are five elements

$$(a, \{a\}), (a, \{b\}), (b, \{b\}), (b, \emptyset), (a, \emptyset)$$

which form the partially ordered set $\mathbf{I}^{++}$ pictured in Fig. 4. Note that there is another element $(a, \{a, b\}) \in M(\mathbf{I}^+)$ but this satisfies $(a, \{a\}) \vdash (a, \{a, b\})$ and $(a, \{a, b\}) \vdash (a, \{a\})$, so we have identified these elements in the picture. The next step $\mathbf{I}^{+++}$ has twenty elements (up to equivalence in the sense just mentioned) and it is also pictured in Fig. 4. We leave the task of drawing a picture of $\mathbf{I}^{++++}$ as an exercise for the (zealous) reader. It should be noted that each stage of the construction is *embedded* in the next one by the map $x \mapsto (x, \{x\})$. The closed circles in the figure are intended to give a hint of how this embedding looks. The limit of this process is a domain $\mathbf{V}$.

The technique which we have used to build this domain can be generalized and used for other classes as well [8]. We have the following lemma.

7.12. LEMMA. *The following operators are p-representable over $\mathbf{V}$: $\to, \circ\!\!\to, \times, \otimes, +, \oplus, (\cdot)_\bot, (\cdot)^\#, (\cdot)^\flat, (\cdot)^\natural$.*

As with most of the other operators, to get a representation for $(\cdot)^\natural$, take a pair of continuous functions

$$\Phi_\natural : \mathbf{V} \to \mathbf{V}^\natural, \qquad \Psi_\natural : \mathbf{V}^\natural \to \mathbf{V}$$

such that $\Phi_\natural \circ \Psi_\natural = \mathbf{id}$ and $\Psi_\natural \circ \Phi_\natural \sqsubseteq \mathbf{id}$. Then

$$R_\natural(p) = \Psi_\natural \circ (p^\natural) \circ \Phi_\natural$$

is a representation for the convex powerdomain operator.

We hope that the reader has begun to note a pattern in the way operators are represented. Most of the operators $(\times, \otimes, +, \oplus, (\cdot)_\bot, (\cdot)^\#, (\cdot)^\flat, (\cdot)^\natural)$ may be handled rather straightforwardly using the corresponding action of these operators on functions. Slightly more care must be taken in dealing with the function space and strict function space operators where one must use a function like Θ. The stock of operators that we have defined in this chapter is quite powerful and it can be used for a wide range of denotational specifications. However, the methods that we have used to show facts such as representability (using finitary closures or finitary projections) will apply to a very large class of operators which satisfy certain sufficient conditions.

To understand this phenomenon, one must pass to a more general theory in which such operators are a basic topic of study. This is the theory of *categories*. Many people find it difficult to gain access to the theory of domains when it is described with categorical terminology. On the other hand, it is difficult to explain basic concepts of domain theory without the extremely useful general language of category theory. A good exposition of the relevance of category theory to the theory of semantic domains may be found in [23].

Only a small number of categories of spaces having the properties which we have described above are known to exist. What are the special traits that these categories possess? First of all, they have product and function space functors which satisfy the relationship we described at the beginning of Section 4. This property, known as *Cartesian closure* is a well-known characteristic of categories such as that of sets and functions. But our Cartesian closed categories have not only fixed points for (all) morphisms, but fixed points for many functors as well. It is this latter feature which makes them well adapted to the task of acting as classes of semantic domains. One additional property which makes these categories special is the existence of domains for representing functors.

This is not to say that there are not other categories which will have the desired properties. One particularly interesting example are the stable structures of Berry [2] which we have not discussed here. Interesting new examples of such categories are being uncovered by researchers at the time of the writing of this chapter. The reader will find a few leads to such examples in the published literature, and we expect that many quite different approaches will be put forward in future years.

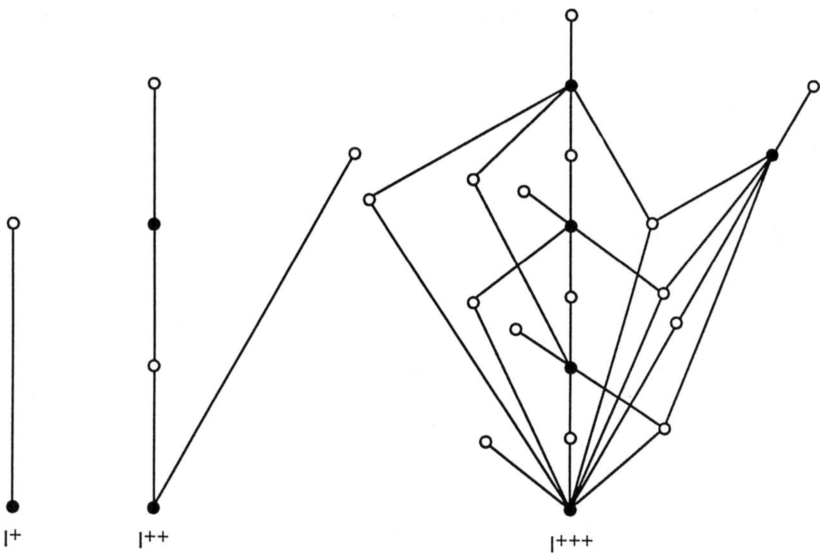

Fig. 4. A domain for representing operators on bifinites.

References

[1] BARENDREGT, H., *The Lambda Calculus: Its syntax and Semantics*, Studies in Logic and the Foundations of Mathematics, Vol. 103 (Elsevier, Amsterdam, revised ed., 1984).

[2] BERRY, G., Stable models of typed λ-calculi, in: *Proc. 5th Internat. Coll. on Automata, Languages and Programming*, Lecture Notes in Computer Science, Vol. 62 (Springer, Berlin, 1978) 72–89.

[3] ENGELER, E., Algebra and combinators, *Algebra Universalis* **13** (1981) 389–392.

[4] ENGELER, E., A combinatory representation of varieties and universal classes. *Algebra Universalis*, to appear.

[5] GIERZ, G., K.H. HOFMANN, K. KEIMEL, J.D. LAWSON, M. MISLOVE and D.S. SCOTT, *A Compendium of Continuous Lattices* (Springer, Berlin, 1980).

[6] GUNTER, C.A., The largest first-order axiomatizable cartesian closed category of domains, in: *Proc. Symp. on Logic in Computer Science* (IEEE Computer Soc. Press, Silver Spring, MD, 1986) 142–148.

[7] GUNTER, C.A., Universal profinite domains, *Inform. and Comput.* **72** (1987) 1–30.

[8] GUNTER, C.A. and A. JUNG, Coherence and consistency in domains, *J. Pure Appl. Algebra* **63** (1990) 49–66.

[9] KAMIMURA, T. and A. TANG, Effectively given spaces, *Theoret. Comput. Sci.* **29** (1984) 155–166.

[10] PLOTKIN, G.D., A powerdomain construction, *SIAM J. Comput.* **5** (1976) 452–487.

[11] RÉVÉSZ, G., Rule-based semantics for an extended lambda-calculus, in: M. Main et al., eds., *Mathematical Foundations of Programming Language Semantics*, Lecture Notes in Computer Science, Vol. 298 (Springer, Berlin, 1987) 43–56.

[12] SCOTT, D.S., Data types as lattices, *SIAM J. Comput.* **5** (1976) 522–587.

[13] SCOTT, D.S., Logic and programming languages, *Comm. ACM* **20** (1976) 634–641.

[14] SCOTT, D.S., The lambda calculus: some models, some philosophy, in: J. Barwise, ed., *The Kleene Symposium* (North-Holland, Amsterdam, 1980) 381–421.

[15] SCOTT, D.S., Relating theories of the lambda calculus, in: J.R. Hindley, ed., *To H.B. Curry: Essays on Combinatory Logic, Lambda Calculus and Formalism* (Academic Press, New York, 1980) 403–450.

[16] SCOTT, D.S., Some ordered sets in computer science, in: I. Rival, ed., *Ordered Sets* (Reidel, Dordrecht, 1981) 677–718.

[17] SCOTT, D.S., Domains for denotational semantics, in: M. Nielsen and E.M. Schmidt, eds., *Proc. 9th Internat. Coll. on Automata, Languages and Programming*, Lecture Notes in Computer Science, Vol. 140 (Springer, Berlin, 1982) 577–613.

[18] SCOTT, D.S., Lectures on a mathematical theory of computation, in: M. Broy and G. Schmidt, eds, *Theoretical Foundations of Programming Methodology*, NATO Advanced Study Institutes Series (Reidel, Dordrecht, 1982) 145–292.

[19] SMYTH, M., Effectively given domains, *Theoret. Comput. Sci.* **5** (1977) 257–274.

[20] SMYTH, M., Power domains, *J. Comput. System Sci.* **16** (1978) 23–36.

[21] SMYTH, M., The largest Cartesian closed category of domains, *Theoret. Comput. Sci.* **27** (1983) 109–119.

[22] SMYTH, M., Power domains and predicate transformers: a topological view, in: J. Diaz, ed., *Proc. 10th Internat. Coll. on Automata, Languages and Programming*, Lecture Notes in Computer Science, Vol. 154 (Springer, Berlin, 1983) 662–676.

[23] SMYTH, M. and G.D. PLOTKIN, The category-theoretic solution of recursive domain equations, *SIAM J. Comput.* **11** (1982) 761–783.

[24] MESEGUER, J., Order completion monads, *Algebra Universalis* **16** (1983) 63–82.

CHAPTER 13

Algebraic Specification

Martin WIRSING

Fakultät für Mathematik und Informatik, Universität Passau, Innstraße 33, B-8390 Passau, FRG

Contents

1. Introduction 677
2. Abstract data types 679
3. Algebraic specifications 690
4. Simple specifications 712
5. Specifications with hidden symbols and constructors 718
6. Structured specifications 737
7 Parameterized specifications 752
8 Implementation 759
9. Specification languages 770
Acknowledgment 780
References 780

HANDBOOK OF THEORETICAL COMPUTER SCIENCE
Edited by J. van Leeuwen

1. Introduction

The proper treatment and choice of the basic data structures is an important and complex part in the process of program construction. In the design phase of the development process, when the basic object sets together with their operations have to be defined, design decisions are made that crucially influence the subsequent development steps and even determine to a considerable extent the structure of the final program. Algebraic specification methods provide techniques for *data abstraction* and the *structured specification, validation* and *analysis of data structures.*

The basic idea of data abstraction seems to have first surfaced in the *class* concept of the SIMULA programming language [46]. The algebraic approach emerged in 1974/75 with the papers [117, 182, 87, 80]. Zilles suggested a new kind of algebra, called "data algebra", based on free algebras, Guttag introduced the important ideas of *consistency* and *sufficient completeness*, and the ADJ-group (formed by J.A. Goguen, J.W. Thatcher, E.G. Wagner, J.B. Wright) gave the first rigorous formalization of abstract data types within standard many-sorted general algebra using initiality.

The basic idea of the algebraic approach consists in describing data structures by just giving the names of the different sets of data, the names of the basic functions and their charactristic properties. The properties are described by formulas (mostly equations) which are invariant under isomorphism. An *abstract data type* is defined to be an isomorphism class of data structures, e.g. an isomorphism class of many-sorted algebras. An *algebraic specification* is a description of one or more such abstract data types. The methodological use of abstract data types in programming is grouping together in one specification all the basic functions that manipulate particular sorts of data, and then "hiding" the representation, in the sense that the *defined data* can only be manipulated by calling the functions provided by the specification [125].

There are three main semantic approaches to algebraic specifications: the *initial algebra approach* advocated in [80, 81], the *terminal algebra approach* advocated in [175], and the *loose approach* in [87, 68, 34]. A fourth approach which is mainly of theoretical interest uses *iso-initial* algebras [27]. Roughly speaking, in the initial algebra of a specification only those terms are identified whose equality can be proven from the axioms; in a terminal algebra only those terms are different whose inequality can be proven from the axioms, in the loose approach all those models are considered which do not contain junk, i.e. all models whose elements can be denoted by ground terms. Initial and terminal algebras of a specification are particularly interesting models without junk.

For the description of large data structures and of systems of data structures it is necessary to compose specifications in a modular way and to build larger specifications from smaller ones. In the simplest form this leads to *hierarchies of specifications* [181]: a hierarchical algebraic specification contains a designated primitive subspecification (which may again be a hierarchical specification). The data of the nonprimitive part of the specification can be viewed as "black boxes" the behaviour of which is given by the effects in the primitive specification. This reflects a basic view in computer science: nonprimitive objects are semantically explained by the effects they have within arbitrary primitive contexts. Another main concept is that of *parameterization* [171]:

parameterized specifications allow one to write specification schemata and avoid rewriting standard parts of specifications any time a new specification is built. They also allow for the design of schematic implementations so that one may concentrate on the implementation of the nonprimitive parts of a specification (cf. e.g. [171, 42]).

Specification languages contain, moreover, a number of specification building operators such as enrichment, renaming, exporting, combination of two specifications and observational abstraction. Historically, the most influential specification language was CLEAR [42, 43]. Other specification languages are LOOK [183, 57], Larch [184] OBJ [63], ACT ONE [50] and ACT TWO [176], ASL [177, 178, 161, 157], PLUSS [65], ASF [15] and OBSCURE [114]. The wide-spectrum languages CIP-L [10], COLD [102] and Extended ML [155] include algebraic specifications as a structuring formalism for specifying data types and programs.

The programming discipline of stepwise refinement advocated by Wirth and Dijkstra suggests that programs be evolved by working gradually by a series of successfully lower-level refinements of the specification toward a specification which is so low-level that it can be regarded as a program. The correctness of the resulting program is ensured provided that each refinement step can be proven correct. Formalization of this approach requires a precise definition of the concept of *refinement*, i.e. of the *implementation of one specification by another*. Several formalizations of the notion of implementation have been suggested (see e.g. [80, 48, 53, 160, 77, 162, 64, 116]. Whatever the chosen formalization is, it must have two fundamental properties: it has to be transitive and monotonic. The former property has been called *vertical composition* and the latter property *horizontal composition* in [73]. In this paper, a very simple notion of implementation will be considered following the approach of [161] where a specification *SP refines* to a specification *SP'* if every model of *SP'* is a model of *SP*. It will be shown that other notions of implementation can be expressed in terms of this simple notion.

This chapter is organized as follows: In Section 2, the basic concepts for the theory of algebraic specifications are reviewed: signatures, terms, many-sorted algebras and computation structures are introduced; the notion of an abstract data type is defined and its basic properties are discussed.

In Section 3, the notion of (simple) algebraic specifications is defined. The three basic semantic approaches (i.e., initial, loose and terminal semantics) are discussed within the framework of total algebras. Four other semantic concepts (error, partial, continuous, and order-sorted algebras), algebras for specifying nondeterminism and concurrency as well as institutions are briefly presented.

In Section 4, the existence of initial and terminal algebras, the lattice structure of the models and the expressive power of initial and terminal algebra specifications is studied.

In Section 5, two additional concepts for algebraic specifications are introduced which are both practically and theoretically essential: hidden symbols and constructors. Hidden functions are necessary to be able to specify equationally all computable data types (Section 5.2). Constructor functions are important for hierarchical (Section 5.4) and structured specifications (cf. Section 6). Lattice and existence theorems for simple specifications are generalized to specifications with hidden symbols and constructors (Section 5.1); it is shown that the expressive power of equational specifications of this kind is complete hyperarithmetical. As a special case of specifications with constructors, hierarchical specifications are introduced and their "observable behaviour" is discussed (Section 5.4).

In Section 6, a general approach to structured specifications is presented. First, a semantic domain for specifications is defined; then a specification-building operation is defined to be a monotonic function over this domain which transforms specifications to specifications (Section 6.1). A number of examples for specification-building operations are given (Sections 6.2–6.5) and their most important properties are summarized in an algebra of specifications (Section 6.6).

In Section 7, parameterized specifications are introduced as specific specification-building operators. Two approaches to parameterization are discussed: the λ-calculus approach (Section 7.1) and the pushout-based approach (Section 7.2). In the former approach, parameterized specifications are considered as specification-building functions whereas in the latter, they are specification morphisms.

In Section 8, the notion of implementing one specification by another is presented. Based on the simple notion of refinement (Section 8.1), other notions known from the literature are defined in terms of this simple notion (Section 8.2). Vertical and horizontal composition properties are studied in Section 8.2 and 8.3. The execution of specifications (without refinement to more concrete versions) using interpretative and compilational techniques is discussed in Section 8.4.

Finally in Section 9, several specification languages are shortly presented including CLEAR, OBJ2, ASL, Larch (Sections 9.1–9.4), as well as some others (Section 9.5).

Several important aspects of algebraic specification will not be treated in the following text; in particular, nothing will be said about the numerous and various applications and connections to other areas; none of the concepts for systematic construction and analysis of specifications and their use in the software development process will be presented (for annotated references on these topics see [138]).

2. Abstract data types

The term *data type* has many informal usages in programming and programming methodology. For example, Gries lists seven interpretations in [85]. However, there is a precise (mathematical) meaning of an *abstract data type* which will be elaborated in the following Sections 2.2–2.4.

The syntactic structure of a data type D is determined by its signature Σ (Section 2.1), i.e. a list of names of different kinds of data (the so-called *sorts*) and lists of notations for distinguished data and operations (the so-called *function symbols*). The semantics of D is based on the notion of *computation structure* (Section 2.2), i.e. a many-sorted algebra A of signature Σ such that all elements of the carrier sets of A can be denoted by terms. An abstract data type abstracts from the concrete representation given by the computation structure A; it is defined as the isomorphism class of a computation structure (Section 2.3). In Section 2.4, the computability of abstract data types is discussed.

2.1. *Signatures and terms*

One of the main characteristics of the algebraic approach to formal specifications is the use of *many-sorted* notions. For any set S, an *S-sorted set* X is a family $\{X_s\}_{s \in S}$ of sets indexed by S.

A *signature* Σ is a pair $\langle S, F \rangle$ where S is a set (of *sorts*) and F is a set (of *function symbols*) such that F is equipped with a mapping *type*: $F \to S^* \times S$. For any $f \in F$, the value *type*(f) is the *type* of f. Constants are represented by nullary function symbols. We write *sorts*(Σ) to denote S, *opns*(Σ) to denote F and $f\colon w \to s$ to denote $f \in F$ with *type*$(f) = w, s$. In the literature, F is often considered as $S^* \times S$-sorted set (cf. [82]). In contrast to the former one, the latter definition allows to use "overloading" of function symbols, e.g. $+ : nat, nat \to nat$ and $+ : int, int \to int$.

2.1.1. EXAMPLE. (1) The basic signature for Boolean values consists of one sort "bool" and two constants "true" and "false".

signature $\Sigma_{BOOL0} \equiv$
  sort *bool*
  functions
    $true\colon \to bool,$
    $false\colon \to bool$
endsignature

(2) The basic signature for natural numbers consists of one sort "nat", one constant "zero" and one unary function symbol "succ" for the successor function.

signature $\Sigma_{NAT0} \equiv$
  sort *nat*
  functions
    $zero\colon \to nat,$
    $succ\colon nat \to nat$
endsignature

(3) A binary function symbol with values in "bool" can be added to Σ_{NAT0} as follows.

signature $\Sigma_{NAT1} \equiv$
  extend $\Sigma_{BOOL0} \cup \Sigma_{NAT0}$ **by**
    function $eq\colon nat, nat \to bool$
endsignature

Here, the expression **extend** $\langle S, F \rangle$ **by sorts** S_1 **functions** $F_1 =_{\text{def}} \langle S \cup S_1, F \cup F_1 \rangle$ denotes the union of the signature $\langle S, F \rangle$ with S_1 and F_1.

(4) A basic signature for sets of natural numbers is the following:

> **signature** $\Sigma_{SETNAT} \equiv$
> **extend** Σ_{NAT1} **by**
> **sort** *set*
> **functions**
> *empty*: $\rightarrow$ *set*,
> *insert* :*nat*, *set* $\rightarrow$ *set*,
> $.\varepsilon.$:*nat*, *set* $\rightarrow$ *bool*
> **endsignature**

The dots to the left and right of the function symbol ε indicate that ε will be used in infix notation.

A signature morphism maps the sorts and function symbols of one signature to the sorts and function symbols of another signature in such a way that the types are preserved. Formally, for any two signatures $\Sigma = \langle S, F \rangle$ and $\Sigma' = \langle S', F' \rangle$ a *signature morphism* $\sigma: \Sigma \rightarrow \Sigma'$ is a pair $\langle \sigma_{sorts}, \sigma_{opns} \rangle$ where $\sigma_{sorts}: S \rightarrow S'$ and $\sigma_{opns}: F \rightarrow F'$ are mappings such that for any $f: w \rightarrow s \in F$, $type(\sigma_{opns}(f)) = \sigma^*(w), \sigma_{sorts}(s)$. Here σ^* denotes the extension of σ_{sorts} to strings of *sorts*; i.e., $\sigma^*(s_1, \ldots, s_n)$ denotes $\sigma_{sorts}(s_1), \ldots, \sigma_{sorts}(s_n)$ for $s_1, \ldots, s_n \in S$. We write $\sigma(s)$ for $\sigma_{sorts}(s)$, $\sigma(w)$ for $\sigma^*(w)$ and $\sigma(f)$ for $\sigma_{opns}(f)$.

The collection of all signatures forms a category, called *Sign*, with the signatures as objects and the signature morphisms as morphisms.

In this chapter, two different kinds of signature morphism will occur: For any two signatures Σ, Σ' with $\Sigma \subseteq \Sigma'$, the *embedding in*: $\Sigma \rightarrow \Sigma'$, where $in(x) =_{\text{def}} x$ for all x, is a signature morphism. Moreover, any mapping between sorts and function symbols induces the following family of signature morphisms. Let $\varphi = \langle \varphi_{srt}, \varphi_{opn} \rangle$ be a pair of mappings $\varphi_{srt}: S_1 \rightarrow S_2, \varphi_{opn}: F_1 \rightarrow F_2$ where S_1, S_2 are sets of sorts and F_1, F_2 are sets of (names of) function symbols; moreover, let $\Sigma = \langle S, F \rangle$ be any signature. Then $\hat{\varphi}: \Sigma \rightarrow \Sigma'$ denotes the signature morphism defind by $\hat{\varphi}(x) =_{\text{def}} \varphi(x)$ if $x \in S_1 \cup F_1$, and x otherwise, where $\Sigma' =_{\text{def}} \langle S', F' \rangle$ is a signature defined by $S' =_{\text{def}} \hat{\varphi}(S)$ and $F' =_{\text{def}} \{\hat{\varphi}(f): \hat{\varphi}^*(w) \rightarrow \hat{\varphi}(s) \mid f: w \rightarrow s \in F\}$. In particular, if $x_1, \ldots, x_n$ are pointwise different sorts and function symbols and φ maps x_i to y_i for $i = 1, \ldots, n$, we write $[x_1 \mapsto y_1, \ldots, x_n \mapsto y_n]$ for $\hat{\varphi}$. If φ is injective, we call $\hat{\varphi}$ a *renaming*; if $\langle S_1, F_1 \rangle \subseteq \Sigma$, we call $\hat{\varphi}$ *extension* of φ.

Every signature $\Sigma = \langle S, F \rangle$ defines a set of syntactically correct expressions which can be formed from free variables and the function symbols of the signature.

Let X be an S-sorted set of free variables (which is disjoint from the constants in F). For every sort $s \in S$ the set $T(\Sigma, X)_s$ *of terms of sort* s (containing elements in X) is the least set containing

(i) every $x \in X_s$ (of sort s) and every nullary operation symbol $f \in F$ with type $\rightarrow s$, and

(ii) every $f(t_1, \ldots, t_n)$ where $f: s_1, \ldots, s_n \rightarrow s$ is an function symbol in F with range s and every t_i $(i = 1, \ldots, n)$ is a term (of sort s_i) in $T(\Sigma, X)_{s_i}$.

Terms without elements of X are called *ground terms* and $T(\Sigma, \emptyset)_s$ is denoted by $T(\Sigma)_s$. If t is a Σ-term, then $FV(t)$ denotes the set of (*free*) *variables* in t, i.e. the set of all elements of X occurring in t. A signature is called *sensible* (cf. [98]) if it admits at least one ground term for each sort. For instance, all signatures in Example 2.1.1 are sensible.

The extension of a signature morphism $\sigma: \Sigma \to \Sigma'$ to a mapping from Σ-terms to Σ'-terms will be denoted by $\sigma^*: T(\Sigma, X) \to T(\Sigma', X)$. *Simultaneous substitution* of the variables $x_1, \ldots, x_n$ by terms $t_1, \ldots, t_n$ (of appropriate sorts) in a term t will be denoted by $t\rho$ where ρ stands for the *substitution* $[t_1/x_1, \ldots, t_n/x_n]$ of the variables $x_1, \ldots, x_n$ by $t_1, \ldots, t_n$.

2.2. *Algebras and computation structures*

A Σ-*algebra* has a carrier set (the elements of a data type) for each sort of Σ and a function on these sets for each function symbol of Σ. We assume that for each sort the carrier set contains at least one element. (This has the advantage that, e.g., the proof system for one-sorted equational logic can easily be generalized to the many-sorted case. On the other hand, the initial algebra does not exist in all cases, a reason for which some authors (e.g., [125, 55]) prefer to admit algebras with empty carrier sets; for a discussion see [76, 77, 143].) A Σ-*homomorphism* maps the "data types" of one Σ-algebra to those of another in such a way that the operations are preserved.

Formally, let $\Sigma = \langle S, F \rangle$ be a signature. A Σ-*algebra* A consists of an S-sorted family of nonempty carrier sets $\{A_s\}_{s \in S}$ (also denoted by A) and of a total function $f^A: A_{s_1} \times \cdots \times A_{s_n} \to A_s$ for each $f: s_1, \ldots, s_n \to s \in F$. A Σ-*homomorphism* $h: A \to B$, where A and B are two Σ-algebras, is a family of maps $\{h_s: A_s \to B_s\}_{s \in S}$ such that for each $f: s_1, \ldots, s_n \to s \in F$ and each $a_1 \in A_{s_1}, \ldots, a_n \in A_{s_n}$, $h_s(f^A(a_1, \ldots, a_n)) = f^B(h_{s_1}(a_1), \ldots, h_{s_n}(a_n))$. A Σ-*isomorphism* is a bijective Σ-homomorphism. The *class of all* Σ-*algebras* is denoted by $Alg(\Sigma)$. Together with Σ-homomorphisms it forms a category which will also be denoted by $Alg(\Sigma)$.

2.2.1. EXAMPLE. Σ_{BOOLO}-algebras are the two-element algebra B, the three-element algebra $B_\perp$ and the unit algebra $B\text{-}U$ where

$$\begin{array}{lll} B_{bool} =_{\text{def}} \{O, L\}, & true^B =_{\text{def}} L, & false^B =_{\text{def}} O, \\ B_{\perp bool} =_{\text{def}} \{O, L, \perp\}, & true^{B_\perp} =_{\text{def}} L, & false^{B_\perp} =_{\text{def}} O, \\ B\text{-}U_{bool} =_{\text{def}} \{1\}, & true^{B\text{-}U} =_{\text{def}} 1, & false^{B\text{-}U} =_{\text{def}} 1. \end{array}$$

Σ_{NATO}-algebras are the algebra N of standard natural numbers and the algebra $N_\perp$ which contains one nonstandard number.

$$\begin{array}{lll} N_{nat} =_{\text{def}} \mathbb{N}, & zero^N =_{\text{def}} 0, & succ^N(n) =_{\text{def}} n+1 \quad \text{for all } n \in \mathbb{N}, \\ N_{\perp nat} =_{\text{def}} \mathbb{N} \cup \{\perp\}, & zero^{N_\perp} =_{\text{def}} 0, & succ^{N_\perp}(n) =_{\text{def}} \begin{cases} n+1 & \text{if } n \in \mathbb{N}, \\ \perp & \text{if } n = \perp. \end{cases} \end{array}$$

There exists a surjective Σ_{BOOLO}-homomorphism h from B onto $B\text{-}U$; h is defined by $h(x) = 1$ for $x \in \{O, L\}$. The canonical embedding i from $\mathbb{N}$ into $\mathbb{N} \cup \{\perp\}$, defined by $i(x) =_{\text{def}} x$ for all $x \in \mathbb{N}$, is a homomorphism from N into $N_\perp$.

Given a Σ'-algebra A' and a signature morphism $\sigma: \Sigma \to \Sigma'$, one can recover the Σ-algebra buried inside A': the σ-*reduct* of A', written $A'|_\sigma$, is the Σ-algebra with the carrier set $(A'|_\sigma)_s =_{\text{def}} A'_{\sigma(s)}$ for each sort $s \in sorts(\Sigma)$, and $f^{A'|_\sigma} =_{\text{def}} \sigma(f)^{A'}$ for each

$f \in opns(\Sigma)$. Similarly, for any Σ'-homomorphism $h': A' \to B'$, where A' and B' are Σ'-algebras, the σ-reduct of h' is the Σ-homomorphism $h'|_\sigma : A'|_\sigma \to B'|_\sigma$ defined by $(h'|_\sigma)_s =_{\text{def}} h'|_{\sigma(s)}$ for all $s \in S$. The mappings $A' \to A'|_\sigma$ and $h' \to h'|_\sigma$ form a functor $\cdot|_\sigma : Alg(\Sigma') \to Alg(\Sigma)$.

If $\Sigma \subset \Sigma'$ and σ is the canonical injection (i.e., $\sigma(x) = x$ for $x \in S \cup F$), the *σ-restriction* of A', written $A'|_\Sigma$, is just the Σ-algebra which is the same as A' except that only carrier sets and functions occur that correspond to sorts and function symbols being not in the image of the canonical injection.

2.2.2. EXAMPLE. The extension of the Σ_{BOOLO} and the Σ_{NATO}-algebras by a binary Boolean function yields the following Σ_{NAT1}-algebras $N\text{-}B$, $N\text{-}B_\perp$, $N\text{-}B2$. The algebras $N\text{-}B$ and $N\text{-}B_\perp$ interpret *eq* as decision function for equality in $\mathbb{N}$ whereas in $N\text{-}B2$ *eq* has an arbitrary interpretation which does not respect the meaning suggested by its name:

$$N\text{-}B|_{\Sigma_{BOOLO}} =_{\text{def}} B,$$

$$N\text{-}B|_{\Sigma_{NATO}} =_{\text{def}} N, \qquad eq^{N\text{-}B}(n,m) =_{\text{def}} \begin{cases} L & \text{if } n=m, \\ O & \text{otherwise,} \end{cases}$$

$$N\text{-}B_\perp|_{\Sigma_{BOOLO}} =_{\text{def}} B_\perp,$$

$$N\text{-}B_\perp|_{\Sigma_{NATO}} =_{\text{def}} N_\perp, \qquad eq^{N\text{-}B_\perp}(n,m) =_{\text{def}} \begin{cases} L & \text{if } n=m,\ n,m \in \mathbb{N}, \\ O & \text{if } n \neq m,\ n,m \in \mathbb{N}, \\ \perp & \text{if } n=\perp \text{ or } m=\perp, \end{cases}$$

$$N\text{-}B2|_{\Sigma_{BOOLO}} =_{\text{def}} B,$$

$$N\text{-}B2|_{\Sigma_{NATO}} =_{\text{def}} N, \qquad eq^{N\text{-}B2}(n,m) =_{\text{def}} \begin{cases} L & \text{if } n \leqslant m, \\ O & \text{otherwise.} \end{cases}$$

Σ_{SETNAT}-algebras can be obtained e.g. as extensions of $N\text{-}B$. The algebra $P0$ extends $N\text{-}B$ by finite sets of natural numbers, the algebra $N0^*$ extends $N\text{-}B$ by finite sequences of natural numbers; in the algebra $P1$ the interpretations of the sort *set* and the function symbols do not correspond to the meaning suggested by the names of these symbols:

$$P0|_{\Sigma_{NAT1}} =_{\text{def}} N\text{-}B,$$

$$P0_{set} =_{\text{def}} \{s \subseteq \mathbb{N} \mid s \text{ finite}\},$$

$$empty^{P0} =_{\text{def}} \emptyset,$$

$$insert^{P0}(n,s) =_{\text{def}} \{n\} \cup s, \qquad n \varepsilon^{P0} s =_{\text{def}} \begin{cases} L & \text{if } n \in s, \\ O & \text{otherwise.} \end{cases}$$

$$P1|_{\Sigma_{NAT1}} =_{\text{def}} N\text{-}B,$$

$$P1_{set} =_{\text{def}} \mathbb{N} \cup \{\perp\},$$

$$empty^{P1} =_{\text{def}} \perp,$$

$$insert^{P1}(n,k) =_{\text{def}} \begin{cases} k & \text{if } n \leqslant k, \\ n & \text{if } n \geqslant k \text{ or } k=\perp, \end{cases} \qquad n \varepsilon^{P1} k =_{\text{def}} \begin{cases} L & \text{if } n \leqslant k, \\ O & \text{otherwise.} \end{cases}$$

$$\begin{aligned}
&NO^*|_{\Sigma NAT1} &&=_{\text{def}} N\text{-}B,\\
&NO^*_{set} &&=_{\text{def}} \mathbb{N}^*,\\
&empty^{NO*} &&=_{\text{def}} \varepsilon \quad \text{(the empty sequence)},\\
&insert^{NO*}(n, w) =_{\text{def}} nw, && n\varepsilon^{NO*}w =_{\text{def}} \begin{cases} L & \text{if } n \text{ occurs in } w,\\ O & \text{otherwise,}\end{cases}
\end{aligned}$$

for all $n \in \mathbb{N}$, $s \subseteq \mathbb{N}$, $k \in \mathbb{N} \cup \{\bot\}$, $w \in \mathbb{N}^*$.

For any signature $\Sigma = \langle S, F \rangle$ and any S-sorted set X, if $T(\Sigma, X)_s$ is nonempty for all $s \in S$, then the so-called *term algebra* (also denoted by $T(\Sigma, X)$) forms a Σ-algebra with carrier set $T(\Sigma, X)_s$ for each $s \in S$, and $f^{T(\Sigma,X)}(t_1, \ldots, t_n) =_{\text{def}} f(t_1, \ldots, t_n)$ for each $f: s_1, \ldots, s_n \to s \in F$ and $t_1 \in T(\Sigma, X)_{s_1}, \ldots, t_n \in T(\Sigma, X)_{s_n}$. If Σ is a sensible signature, then $T(\Sigma) =_{\text{def}} T(\Sigma, \emptyset)$ is an algebra, the so-called *ground term algebra*.

If X is an S-sorted set and A a Σ-algebra, then a map $v: X \to A$ is called *valuation* of X in A or *assignment* of values to the variables of X in A. A valuation is well-defined only if A has nonempty carrier sets for all *sorts* for which A is nonempty. Because of the assumption above of nonempty carrier sets, valuations exist for all S-sorted sets.

The relationship between terms and algebras is given by the notion of *interpretation*. Let $v: X \to A$ be a valuation. The *interpretation* of a term t in A (w.r.t v) is a map $v^*: T(\Sigma, X) \to A$ which is defined as follows:

(i) $v^*(x) =_{\text{def}} v(x)$ for each $x \in X$,

(ii) $v^*(f(t_1, \ldots, t_n)) =_{\text{def}} f^A(v^*(t_1), \ldots, v^*(t_n))$ for each $f: s_1, \ldots, s_n \to s \in F$ and $t_1 \in T(\Sigma, X)_{s_1}, \ldots, t_n \in T(\Sigma, X)_{s_n}$.

The interpretation $v^*: T(\Sigma, X) \to A$ is a Σ-homomorphism which is (for given X, A) the unique Σ-homomorphic extension of v to $T(\Sigma, X)$ (for a proof, see e.g. [82, p. 95]). If t is a ground term, then its interpretation does not depend on the map v; the interpretation is uniquely defined and will be denoted by t^A.

As a consequence, the ground term algebra $T(\Sigma)$ is a distinguished element of the category **Alg**(Σ).

Let K be any category. An object $I \in K$ is called *initial* (object of K) if for all objects $A \in K$ there exists a unique morphism from I to A.

2.2.3. FACT. *Let Σ be a sensible signature. The ground term algebra $T(\Sigma)$ is initial in $Alg(\Sigma)$.*

PROOF. For any Σ-algebra A, $v^*: T(\Sigma) \to A$ is the unique Σ-homomorphism. □

Dually, an object $Z \in K$ is called *terminal* (object of K) if for all objects $A \in K$ there exists a unique morphism from A to Z. The terminal object of $Alg(\Sigma)$ is the trivial *unit algebra* $U(\Sigma)$ where every carrier set consists of exactly one element and the functions are trivially defined.

2.2.4. FACT. *Let Σ be a sensible signature. The unit algebra $U(\Sigma)$ is terminal $Alg(\Sigma)$.*

PROOF. Let $U_s =_{\text{def}} \{u_s\}$ for any $s \in S$. Then, for any Σ-algebra A, the unique Σ-homomorphism $\varphi: A \to U(\Sigma)$ is defined by $\varphi(a) =_{\text{def}} u_s$ for any $a \in A_s$, $s \in S$. □

Initial and terminal algebras are uniquely determined up to isomorphism. For example, the algebra B is isomorphic to $T(\Sigma_{BOOLO})$ and therefore initial in $Alg(\Sigma_{BOOLO})$. Similarly, N is initial in $Alg(\Sigma_{NATO})$ and B-U is terminal in $Alg(\Sigma_{BOOLO})$. All other algebras of Examples 2.2.1 and 2.2.2 are neither initial nor terminal.

A *computation structure* is a many-sorted algebra of which all elements have denotations. The only closed denotations in this setting are the ground terms. Therefore all data of a computation structure are interpretations of ground terms ("generation principle" [9]). Formally, a Σ-algebra is called *reachable* (or *term-generated* [10]) if, for each sort $s \in sorts(\Sigma)$ and each carrier element $a \in A_s$, there exists a term t *with* $a = t^A$, i.e. if $v^* : T(\Sigma) \to A$ is surjective. A reachable Σ-algebra is also called *Σ-computation structure* (or *Σ-data type*). The *class of all Σ-computation structures* is denoted by $Gen(\Sigma)$. Together with the Σ-homomorphisms as morphisms, $Gen(\Sigma)$ forms a category which will be equally denoted by $Gen(\Sigma)$. Obviously, $Gen(\Sigma)$ is empty for nonsensible signatures and nonempty for sensible signatures. To get a further characterization of reachability we need the notion of subalgebra. For any signature Σ and Σ-algebra A, a *Σ-subalgebra* of A is a Σ-algebra B such that for any sort $s \in \Sigma$, $B_s \subseteq A_s$ and for any function symbol $f: s_1, \ldots, s_n \to s \in opns(\Sigma)$ and for all $b_1 \in B_{s_1}, \ldots, b_n \in B_{s_n}$, $f^B(b_1, \ldots, b_n) = f^A(b_1, \ldots, b_n)$. The set of Σ-subalgebras of A is closed under (set-theoretic) intersection. Thus for sensible Σ, every Σ-algebra A contains a *least* (*w.r.t. set inclusion*) *Σ-subalgebra* $R(A)$ which is reachable.

2.2.5. FACT. *Let A be a Σ-algebra. Then the following statements are equivalent*:

(1) *A is reachable*;

(2) *there exists a surjective Σ-homomorphism $h: T(\Sigma) \to A$*;

(3) $A = R(A)$.

PROOF (1)⇒(2): If A is reachable then $v^* : T(\Sigma) \to A$ is a surjective Σ-homomorphism from $T(\Sigma)$ onto A. (2)⇒(3): Let $in : R(A) \to A$ be the canonical embedding of $R(A)$ into A, defined by $in(x) = x$ for all $x \in R(A)$. Then in is a Σ-homomorphism. Hence, the composition $in \circ v^* : T(\Sigma) \to A$ (*of* $v^* : T(\Sigma) \to R(A)$ with in) is a Σ-homomorphism form $T(\Sigma)$ into A. It is surjective since, due to the initiality of $T(\Sigma)$ and assumption (2), the unique Σ-homomorphism from $T(\Sigma)$ into A is surjective. This implies (3). (3)⇒(1): By definition of $R(A)$, $v^* : T(\Sigma) \to R(A)$ is surjective. Because of $A = R(A)$, $v^* : T(\Sigma) \to A$ is surjective and therefore is reachable. □

Homomorphisms between computation structures are uniquely determined by the interpretation of terms.

2.2.6. FACT. *Let A and B be two Σ-computation structures.*

(1) *Let $h: A \to B$ be a map from A to B. Then h is a Σ-homomorphism iff for any ground term $t \in T(\Sigma)$, $h(t^A) = t^B$.*

(2) *There exists at most one Σ-homomorphism from A to B.*

(3) *If $h_1: A \to B$ and $h_2: B \to A$ are two Σ-homomorphisms, then A and B are isomorphic.*

PROOF (1)"⇒": follows by induction on the structure of t using the definition of Σ-homomorphism. "⇐": straightforward, using the reachability of A and B. (2) is a direct consequence of the right-hand side of (1). (3) For any term $t \in T(\Sigma)$, the following holds because of (1): $h_2 \circ h_1(t^A) = t^A$ and $h_1 \circ h_2(t^B) = t^B$. Hence, h_1 and h_2 are bijective Σ-homomorphisms. □

2.3. *Abstract data types*

An abstract data type abstracts away from the concrete representation given by a computation structure. For any sensible signature Σ, an *abstract data type of signature* Σ (or *abstract* Σ-*computation structure*) is an isomorphism class of a Σ-computation structure. The notion of congruence over $T(\Sigma)$ will be the tool for studying the structure of abstract data types.

Let $\Sigma = \langle S, F \rangle$ be a signature and $A \in Alg(\Sigma)$ be an arbitrary Σ-algebra. A Σ-*congruence* on A is an S-sorted equivalence relation $\sim$ which is compatible with all function symbols, i.e. $\sim\, = (\sim_s)_{s \in S}$, and for all $s \in S$, $\sim_s \subseteq A_s \times A_s$ is reflexive, symmetric and transitive and, for any $f: s_1, \ldots, s_n \to s \in opns(\Sigma)$ and all $a_1, b_1 \in A_{s_1}, \ldots, a_n, b_n \in A_{s_n}$, if $a_1 \sim_{s_1} b_1, \ldots,$ and $a_n \sim_{s_n} b_n$ holds, then $f^A(a_1, \ldots, a_n) \sim_s f^A(b_1, \ldots, b_n)$. If A is a term algebra $T(\Sigma, X)$ then the latter condition says that, for all $u_1, v_1 \in T(\Sigma, X)_{s_1}, \ldots, u_n, v_n \in T(\Sigma, X)_{s_n}$ with $u_1 \sim_{s_1} v_1, \ldots, u_n \sim_{s_n} v_n$, the terms $f(u_1, \ldots, u_n)$ and $f(v_1, \ldots, v_n)$ are congruent, i.e. $f(u_1, \ldots, u_n) \sim_s f(v_1, \ldots, v_n)$.

2.3.1. FACT. *If* $\sim$ *is a* Σ-*congruence on* A, *then the quotient* $A/\sim$ *is a well-defined* Σ-*algebra, where* (*for each* $s \in S$) $(A/\sim)_s =_{\text{def}} A_s/\sim_s$ *and for* $f: s_1, \ldots, s_n \to s \in F$, $a_1 \in A_{s_1}, \ldots, a_n \in A_{s_n}$, $f^{A/\sim}([a_1], \ldots, [a_n]) =_{\text{def}} [f^A(a_1, \ldots, a_n)]$.

Let $C(\Sigma)$ be the *set of all* Σ-*congruences* over $T(\Sigma)$. If Σ is sensible, then for any Σ-algebra A one can define an *associated* Σ-*congruence* $\sim^A \in C(\Sigma)$ by

$$t \sim^A t' \quad \Leftrightarrow_{\text{def}} \quad t^A = t'^A$$

for all $t, t' \in T(\Sigma)_s$, $s \in S$. In particular, any Σ-congruence $\sim\, \in C(\Sigma)$ is associated with $T(\Sigma)/\sim$.

2.3.2. FACT. *Let* Σ *be a sensible signature.*

(1) $T(\Sigma)/\sim$ *is a* Σ-*algebra for every* Σ-*congruence* $\sim$ *over* $T(\Sigma)$.

(2) *Every* Σ-*computation structure* A *is isomorphic to* $T(\Sigma)/\sim^A$.

(3) *Let* A *be a* Σ-*computation structure. For any* Σ-*algebra* B, *there exists a* Σ-*homomorphism from* A *to* B *iff* $\sim^A \subseteq \sim^B$.

PROOF. (1) follows from Fact 2.3.1. (2) The mapping $h: A \to T(\Sigma)/\sim^A$, $h(a) =_{\text{def}} [t]$ if $t^A = a$, is a well-defined Σ-isomorphism. (3) The existence of a Σ-homomorphism from A to B is equivalent to $t^A = t'^A \Rightarrow t^B = t'^B$ for all $t \in T(\Sigma)$ (i.e., $\sim^A \subseteq \sim^B$). □

Therefore, every abstract data type of signature Σ corresponds exactly to one Σ-congruence over $T(\Sigma)$ and the existence of a Σ-homomorphism between Σ-computation structures corresponds exactly to the set inclusion of the associated congruences. Hence, according to Fact 2.3.2(3) the following holds.

2.3.3. COROLLARY. *For any two Σ-computation structures A and B, A and B are isomorphic if and only if* $\sim^A = \sim^B$.

The above corollary allows us to use the notation $\sim^D$ for an abstract data type D, i.e. an isomorphism class of computation structures. From now on we identify abstract data types with their associated congruences.

A set M partially ordered w.r.t. a binary relation $\leqslant$ is called an *upper semilattice* if for any two elements $a, b \in M$ a least upper bound $lub\{a, b\} \in M$ exists (i.e., $a \leqslant lub\{a, b\}$, $b \leqslant lub\{a, b\}$ and for all $c \in M$, $a \leqslant c$ and $b \leqslant c$ implies $c \leqslant lub\{a, b\}$). It is *complete* if each subset of M has a least upper bound in M. The notions of lower semilattice and greatest lower bound are defined dually. M is a *(complete) lattice* if it is a (complete) upper and a (complete) lower semilattice. The following lemma is often needed.

2.3.4. LEMMA. *If M is a complete lower semilattice which has a greatest element then M is a complete lattice.*

PROOF. See e.g. [84, p. 24] □

The set $C(\Sigma)$ of all Σ-congruences has a lattice structure.

2.3.5. PROPOSITION. *Let Σ be a sensible signature. The set $C(\Sigma)$ of all Σ-congruences forms a complete lattice w.r.t. set inclusion. Its least element is the identity congruence $\sim^{T(\Sigma)} =_{\text{def}} \{\langle t, t\rangle \mid t \in T(\Sigma)\}$ over $T(\Sigma)$; its greatest element is the universal congruence $\sim^{U(\Sigma)} =_{\text{def}} \{\langle t, t'\rangle \mid t, t' \in T(\Sigma)$ of same sort $s\}$.*

PROOF. It is easy to check that the intersection $\sim^{\cap}$ of any family $\{\sim^j\}_{j \in J}$ of Σ-congruences over $T(\Sigma)$ is itself a Σ-congruence; $\sim^{\cap}$ is defined for all $t, t' \in T(\Sigma)$ by

$$t \sim^{\cap} t' \quad \Leftrightarrow \quad \text{for all } j \in J: t \sim^j t'.$$

Thus $C(\Sigma)$ forms a complete lower semilattice. Obviously, the universal congruence $\sim^{U(\Sigma)}$ is the greatest element. Therefore according to Lemma 2.3.4, $C(\Sigma)$ forms a complete lattice. The identity congruence $\sim^{\Sigma}$ is the intersection of all congruences of $C(\Sigma)$ and therefore the least element. □

The least element $\sim^{T(\Sigma)}$ of $C(\Sigma)$ corresponds exactly to the initial algebra $T(\Sigma)$ of $Alg(\Sigma)$ (and $Gen(\Sigma)$). The greatest element of $C(\Sigma)$ is $\sim^{U(\Sigma)}$ which corresponds exactly to the terminal algebra $U(\Sigma)$ of $Alg(\Sigma)$ (and $Gen(\Sigma)$). More generally, initial and terminal computation structures can be characterized as follows:

2.3.6. LEMMA. *Let Σ be a signature.*

(1) *For any class K of Σ-algebras, a Σ-computation structure $I \in K$ is initial in K iff for any two ground terms $t, t' \in T(\Sigma)$, $t \sim^I t' \Leftrightarrow$ for all $A \in K$: $t \sim^A t'$.*

(2) *For any class K of Σ-computation structures, a Σ-computation structure $Z \in K$ is terminal in K iff for any two ground terms $t, t' \in T(\Sigma)$, $t \sim^Z t' \Leftrightarrow$ there exists $A \in K: t \sim^A t'$.*

PROOF (1) "⇒" follows from the definition of initiality by Fact 2.3.2(3). ("⇐"): By Fact 2.3.2(3) there exists a Σ-homomorphism from I into any algebra of K. The proof of (2) is analogous to the proof of (1). □

2.4. *Computability of abstract data types*

The computability of an abstract data type is a measure for the complexity of this data type; it is also a measure for the adequacy of specification methods. In this section, the notion of computability will be introduced (following [24]) via the notion of coordinatization. This notion derives from the work of [147, 123]. Further interesting results can be found in [173].

Let $\Sigma = \langle S, F \rangle$ be a signature. A Σ-algebra is said to be a *number algebra* if its carrier sets are recursive subsets of the set ℕ of natural numbers. It is called *recursive* if its functions are recursive. A *coordinatization* $\langle C, \alpha \rangle$ of a Σ-algebra A is an epimorphism $\alpha: C \to A$ from a number algebra C of signature Σ onto A. Thus A is isomorphic to $C/\equiv_\alpha$ where for all $n, m \in C_s$, $s \in S$, $n \equiv_\alpha m$ is defined by

$$n \equiv_\alpha m \quad \text{iff} \quad \alpha(n) = \alpha(m) \text{ in } A_s.$$

A coordinatization is called *reachable* if its number algebra is reachable.

The complexity of Σ-algebras will be classified according to the complexity of their coordinatizations. A coordinatization $\langle C, \alpha \rangle$ is called *computable, semicomputable* or *co-semicomputable* if its number algebra C is recursive and for all $s \in S$ the relations $\equiv_\alpha$ are recursive, recursively enumerable or co-recursively enumerable respectively.

Then a Σ-algebra A is said to be *computable* (*semicomputable, co-semicomputable*) if there exists a computable (semicomputable, co-semicomputable respectively) coordinatization of A.

The three computability notions are characterized by the existence of certain epimorphisms; the composition of an epimorphism with an isomorphism yields an epimorphism again. Therefore, the following essential property holds.

2.4.1. FACT. *Computability, semicomputability and co-semicomputability are invariant under isomorphism.*

Thus the three notions qualify as abstract semantic properties for data types. An abstract Σ-data type D is called *computable, semicomputable or co-semicomputable* if there exists a Σ-algebra A in D that is computable, semicomputable or co-semicomputable respectively. By Fact 2.4.1, if one Σ-algebra represents D and is computable, then all representing algebras of D are computable.

In particular, any computable data type is represented by a recursive algebra of numbers whose domains are either finite or the set of natural numbers:

2.4.2. FACT (Representation Lemma). *Any computable Σ-algebra is isomorphic to a* *recursive number algebra R each of whose carrier sets R_s ($s \in S$) is the set $\mathbb{N}$ of natural numbers or the set $\mathbb{N}_m$ of the first m natural numbers.*

PROOF (see e.g. [24]). Let the Σ-algebra A be computable w.r.t. the coordinatization $\langle C, \alpha \rangle$. For each $s \in S$, define the recursive set $B_s \subseteq C_s$ by

$$x \in B_s \quad \text{iff} \quad x \in C_s \wedge \forall y < x. (y \in C_s \Rightarrow \neg(y \equiv_\alpha x)).$$

Then for each $s \in S$, B_s is infinite if and only if A_s is infinite. It is easy to see that the carrier sets $(B_s)_{s \in S}$ can be extended to a Σ-algebra B which is isomorphic to A. Obviously, B is isomorphic to a number algebra R with the required properties. □

Many of the above results can also be stated using the notion of term algebra instead of natural numbers. Consider a fixed injective listing of the set of all ground terms by natural numbers, i.e. a bijective coordinatization $v: C_\Sigma \rightarrow T(\Sigma)$, where C_Σ is a fixed recursive number algebra. For any sensible signature Σ, such a bijective coordinatization always exists. The interpretation function $\cdot^{C_\Sigma}: T(\Sigma) \rightarrow C_\Sigma$ is the inverse of v. It is often called "Gödel numbering" (cf. e.g. [125]).

Now, a Σ-congruence $\sim \subseteq T(\Sigma) \times T(\Sigma)$ is called *recursive* (*recursively enumerable, co-recursively enumerable*) if its inverse image under v is so; i.e. if the relation $\sim^v$ defined by

$$n \sim^v m \quad \text{iff} \quad v(n) \sim v(m) \quad \text{for all } n, m \in (C_\Sigma)_s, s \in S,$$

is recursive, recursively enumerable, or co-recursively enumerable respectively. Equivalently, $\sim^v$ can be defined as follows:

$$t^{C_\Sigma} \sim^v t'^{C_\Sigma} \quad \text{iff} \quad t \sim t' \quad \text{for all } t, t' \in T(\Sigma)_s, s \in S.$$

Obviously, the relation $\sim^v$ is a congruence on C_Σ. For the computability of $\sim$, only the computability of $\sim^v$ has to be considered. For every abstract data type D, the "Gödelization" C_Σ of $T(\Sigma)$ induces a canonical coordinatization $\gamma: C_\Sigma \rightarrow T(\Sigma)/\sim^D$ with $\gamma(n) =_{\text{def}} v(n)^D$ for all $n \in (C_\Sigma)_s$, $s \in S$. The congruence $\equiv_\gamma$ induced by γ is exactly the relation $(\sim^D)^v$. For all $n, m \in (C_\Sigma)_s, s \in S$, the following holds:

$$n \equiv_\gamma m \quad \text{iff} \quad v(n) \sim^D v(m) \quad \text{iff} \quad n (\sim^D)^v m. \tag{2.1}$$

For relating two coordinatizations $\langle C_1, \alpha_1 \rangle$ and $\langle C_2, \alpha_2 \rangle$ of a Σ-algebra A, the notion of recursive reduction is introduced: $\alpha_1: C_1 \rightarrow A$ is said to *recursively reduce* to $\alpha_2: C_2 \rightarrow A$ if there exists an S-sorted recursive mapping $f: C_1 \rightarrow C_2$ such that $\alpha_1 = \alpha_2 \circ f$.

2.4.3. LEMMA. *Let $\langle C_1, \alpha_1 \rangle$ and $\langle C_2, \alpha_2 \rangle$ be two coordinatizations of a Σ-algebra A such that α_1 recursively reduces to α_2. If α_2 is computable (semicomputable, or co-semicomputable, respectively), then so is α_1.*

2.4.4. THEOREM. *Let D be an abstract Σ-data type with associated Σ-congruence $\sim^D$. Then the following properties are equivalent:*

(1) *D is computable, semicomputable, or co-semicomputable respectively.*
(2) *$\sim^D$ is recursive, recursively enumerable, or co-recursively enumerable respectively.*

PROOF. Let $\gamma: C_\Sigma \to T(\Sigma)/\sim^D$ be the canonical coordinatization of the representant $T(\Sigma)/\sim^D$ of D as defined above. (2)⇒(1): If $\sim^D$ is recursive (recursively enumerable, or co-recursively enumerable), then condition (2.1) above implies that D is so.

(1)⇒(2): According to Fact 2.4.1 the existence of coordinatizations is invariant under isomorphism. Therefore, consider w.l.o.g. the representant $T(\Sigma)/\sim^D$ of D and *a* coordinatization $\beta: C \to T(\Sigma)/\sim^D$ of this algebra. The coordinatization $\gamma: C_\Sigma \to T(\Sigma)/\sim^D$ as defined above recursively reduces to $\langle C, \beta\rangle$: because of the uniqueness of the interpretation homomorphism we have

$$\gamma(n) = \nu(n)^D = \beta(\nu(n)^C) \quad \text{for all } n \in (C_\Sigma)_s,\ s \in S;$$

furthermore, the mapping $\cdot^C \circ \nu: C_\Sigma \to C$ is recursive.

Thus by Lemma 2.4.3, if D is computable (semicomputable, or co-semicomputable) via $\langle C, \beta\rangle$, then it is so via $\langle C_\Sigma, \gamma\rangle$ and therefore (2.1) implies that $\sim^D$ is so. □

Using structured specifications as in Section 5 and 6, it is possible to define computation structures which are more complex than semicomputable and co-semicomputable.

A relation $P \subseteq \mathbb{N} \times \mathbb{N}$ is *arithmetical* [152, p. 336] if it has an explicit definition

$$P(n, m) \quad \Leftrightarrow \quad Q_1 x_1 \ldots Q_k x_k . R(n, m, x_1, \ldots, x_k)$$

where R is a recursive relation, each Q_i is a quantifier $\forall$ or $\exists$, and each x_i is a variable for a natural number. In particular, each recursively enumerable relation P has an explicit definition of the form $P(n, m) \Leftrightarrow \exists x_1 . R(n, m, x_1)$; each co-recursively enumerable relation P has a definition of the form $P(n, m) \Leftrightarrow \forall x_1 . R(n, m, x_1)$ where R is a recursive relation (cf. [152, p. 66]).

P is called Π_1^1 if it is defined by

$$P(n, m) \quad \Leftrightarrow \quad \forall \alpha \exists y . R(n, m, \alpha, y);$$

it is called Σ_1^1 if it is defined by

$$P(n, m) \quad \Leftrightarrow \quad \exists \alpha \forall y . R(n, m, \alpha, y)$$

where α is a variable for total functions from $\mathbb{N}$ to $\mathbb{N}$, y is a variable for a natural number and R is recursive. P is *hyperarithmetical* iff it is both Π_1^1 and Σ_1^1. A function f is *arithmetical* (*hyperarithmetical*) iff its graph G_f is so (cf. [152, p. 382]).

Now, a Σ-algebra A is *arithmetical* (*hyperarithmetical* respectively) if it has a coordinatization (C, α) such that C and $\equiv_\alpha$ are arithmetical (hyperarithmetical, respectively). Then an abstract data type D is called *arithmetical* (*hyperarithmetical* respectively) if there exists an arithmetical (hyperarithmetical respectively) algebra in D.

3. Algebraic specifications

Algebraic specifications are an approach to describe abstract data types in an implementation-independent way. The basic idea is to specify an abstract data type by its signature and characteristic properties. Their properties are expressed in a *many-*

sorted logical formalism. Usually, this is a restricted first-order logic such as equational logic (cf. e.g. [82]) or Horn clause equational logic (cf. e.g. [171]) but also other formalisms such as full equational first-order logic (cf. e.g. [181, 120]) or higher-order equational logic [130] may be chosen. Abstract data types are isomorphism classes of computation structures. Thus mathematically, the theory of algebraic specifications is the *theory of isomorphism classes of reachable many-sorted algebras* (Section 3.1).

There are two main methodical approaches to the specification of abstract data types which can roughly be described as follows: either one starts with a data type D as given and then aims at finding an appropriate algebraic specification for D, or one starts with an algebraic specification SP (e.g., for some informally given software or hardware problem) and then tries to develop a more detailed specification SP', a so-called implementation of SP, which specifies an abstract data type. In this paper, mainly the latter approach will be elaborated, for the former see e.g. [82].

In Section 3.2, syntax and semantics of *simple algebraic specifications* are introduced; terminal and initial algebra semantics are defined as specific model classes of such specifications and characterized by their proof-theoretic properties. The total algebra semantics as presented in this paper encounters difficulties if data types with partial functions have to be specified. In Section 3.3 some approaches are presented which try to overcome this problem: *error algebras, partial algebras, continuous algebras* and *order-sorted algebras* (Section 3.3.1–3.3.4). Algebraic approaches for specifying *non-determinism* and *concurrency* are surveyed in Section 3.3.5. Each of these semantic approaches can be combined with the logical formalisms mentioned above. This leads to an enormous growth of different possible specification mechanisms. The concept of *institutions* introduced by [43], abstracts from (many of) these differences and gives a unified view of logical formalisms (Section 3.3.6).

3.1. *Formulas and theories*

Properties of many-sorted algebras will be expressed by many sorted first-order equational formulas.

Let $\Sigma=\langle S,F\rangle$ be a signature. A *Σ-equation* has the form $t=_s t'$ where $t, t' \in T(\Sigma, X)_s$ are terms of a sort $s\in S$. (First-order) *Σ-formulas* are built from Σ-equations using the connectives $\neg$, $\wedge$ and the typed quantifier $\forall$. The set $WFF(\Sigma)$ of Σ-formulas is the least set satisfying the following properties:

(i) every Σ-equation is in $WFF(\Sigma)$;

(ii) if $G, H \in WFF(\Sigma)$ then $\neg G, (G\wedge H)\in WFF(\Sigma)$;

(iii) if $x\in X_s$ and $G\in WFF(\Sigma)$ then $(\forall x{:}s.G)\in WFF(\Sigma)$. The variable x is said to be bound by the universal quantifier $\forall$ in $(\forall x{:}s.G)$.

Σ-equations are also called *atomic formulas.* Further logical operators such as $\vee$, $\Rightarrow$ and $\exists$ are as usual considered as abbreviations: $(G\vee H)=_{\mathrm{def}} \neg(\neg G\wedge \neg H)$, $(G\Rightarrow H)=_{\mathrm{def}}$ $(\neg G\vee H)$, $(G\Leftrightarrow H)=_{\mathrm{def}} ((G\Rightarrow H)\wedge(H\Rightarrow G))$, $(\exists x{:}s.G)=_{\mathrm{def}} \neg(\forall x{:}s.\neg G)$.

Superfluous brackets will be omitted in the usual way; several quantifications of the form $Qx_1{:}s_1,\ldots,Qx_n{:}s_n$ (with $Q\in\{\forall,\exists\}$) will be abbreviated by $Qx_1{:}s_1,\ldots,x_n{:}s_n$. The set of *bound variables* occurring in a formula G will be denoted by $BV(G)$ and the set of all free variables of G (i.e. those variables which occur in G but in a position which is not bound) will be denoted by $FV(G)$. A Σ-formula without free variables is called *Σ-sentence.* A Σ-formula without variables at all is called *ground.* For example, the Σ_{NAT1}-equation $eq(zero, zero) = true$ is a sentence which is ground; the Σ_{NAT0}-formula $\forall x{:}nat. \forall y{:}nat.\ succ(x) = succ(y) \Rightarrow x = y$ is a Σ_{NAT0}-sentence which is not ground. The latter formula has a further particular form: a *Horn clause* of signature Σ is a Σ-formula of the form $(G_1 \wedge \cdots \wedge G_m) \Rightarrow G$ where $G_1, \ldots, G_m, G$ are atomic Σ-formulas and $m \geqslant 0$. Since here all atomic formulas are Σ-equations we speak equivalently of *conditional Σ-equations.*

For any Σ-algebra A, valuation $v: X \to A$ and Σ-formula G the relation *A satisfies G w.r.t v*, written $A, v \models G$, is inductively defined as follows:

(i) $A, v \models t =_s t'$ iff $v^*(t) = v^*(t')$;

(ii) $A, v \models \neg G$ iff $(A, v \models G)$ does not hold;

(iii) $A, v \models (G \wedge H)$ iff $(A, v \models G)$ and $(A, v \models H)$;

(iv) $A, v \models \forall x{:}s.G$ iff $(A, v_x \models G)$ for all valuations $v_x: X \to A$ with $v_x(y) = v(y)$ for all $y \neq x$.

The Σ-algebra A *satisfies* G, written $A \models G$, if $A, v \models G$ holds for all valuations. Notice that according to this definition any Σ-algebra with an empty carrier set would satisfy all Σ-formulas.

The *universal closure* of a Σ-formula G is the Σ-sentence $\forall x_1{:}s_1, \ldots, x_n{:}s_n. G$ where $\{x_1{:}s_1, \ldots, x_n{:}s_n\}$ is the set $FV(G)$ of all free variables of G. Obviously, a Σ-algebra A satisfies G if and only if it satisfies the universal closure of G. A Σ-formula G is *valid* in a class K of Σ-algebras, written $K \models G$, if each $A \in K$ satisfies G. The satisfaction relation is closed under isomorphism: if the Σ-algebra B is isomorphic to A then B satisfies a Σ-formula G if and only if A satisfies it. The isomorphism class of a Σ-algebra A is denoted by $[A]$. Hence a Σ-formula G is valid in $[A]$ if and only if A satisfies G, and a Σ-congruence $\sim\, \in C(\Sigma)$ *satisfies* G if and only if G holds in all Σ-computation structures $\sim$ is associated with.

With every class K of Σ-algebras one may associate the set $Th(K)$ of all Σ-formulas which are valid in K. $Th(K)$ is called the *theory of* K. We write $Th_{EQ}(K)$ for the *equational theory of* K, i.e. for the set of all Σ-equations which are valid in K, and similarly, $Th_{CEQ}(K)$ for the set of all conditional Σ-equations valid in K. On the other hand, with every signature Σ and every set E of Σ-formulas one may associate the class $Alg(\Sigma, E)$ of all Σ-algebras which satisfy all formulas from E, i.e. $Alg(\Sigma, E) =_{\text{def}} \{A \in Alg(\Sigma) \mid A \models G \text{ for all } G \in E\}$. By $Gen(\Sigma, E)$ we denote the class of all Σ-computation structures which satisfy all formulas from E, i.e. $Gen(\Sigma, E) =_{\text{def}} \{A \in Gen(\Sigma) \mid A \models G$ for all $G \in E\}$ and by $C(\Sigma, E)$ we denote the set of all abstract data types associated with $Gen(\Sigma, E)$, i.e. $\{\sim^A \mid A \in Gen(\Sigma, E)\}$. In all cases, E is called the set of *axioms* of $Alg(\Sigma, E)$, $Gen(\Sigma, E)$ and $C(\Sigma, E)$ respectively. More generally, a set E of Σ-formulas is an *axiomatization* of a class K of Σ-algebras (or Σ-computation structures or abstract Σ-data types, respectively) if $K = \{A \in S(\Sigma, E) \mid A \models G \text{ for all } G \in E\}$, $S = Alg$ ($S = Gen$ or $S = C$ respectively).

Axiomatizations are not unique. Obviously, the inclusions $E \subseteq Th(Alg(\Sigma, E))$ and $E \subseteq Th(Gen(\Sigma, E))$ hold for all sets E of Σ-formulas; E and the theory $Th(Alg(\Sigma, E))$ ($Th(Gen(\Sigma, E))$ resp.) are both axiomatizations of $Alg(\Sigma, E)$ ($Gen(\Sigma, E)$ resp.), but for example if E is finite, E is different from $Th(Alg(\Sigma, E))$ and $Th(Gen(\Sigma, E))$. A smaller class of algebras satisfies more formulas than a larger class of algebras, and dually for formulas, a larger set of formulas holds in fewer algebras than a smaller set of formulas: for any two classes K and K' of Σ-algebras closed under isomorphism and any two sets of Σ-formulas E and E',

$$K \subseteq K' \text{ iff } Th(K') \subseteq Th(K); \tag{3.1}$$

$$E \subseteq E' \text{ iff } Alg(\Sigma, E') \subseteq Alg(\Sigma, E) \text{ iff } Gen(\Sigma, E') \subseteq Gen(\Sigma, E) \text{ iff } C(\Sigma, E') \subseteq C(\Sigma, E). \tag{3.2}$$

Since computation structures are particular algebras, we have

$$E \subseteq Th(Alg(\Sigma, E)) \subseteq Th(Gen(\Sigma, E))$$

for all sets E of Σ-formulas. In most cases, both inclusions are strict.

One important property of classes of Σ-algebras (i.e. of $Alg(\Sigma, E)$) is that their associated theories can be generated by a *formal system*: there exists a set of *logical and nonlogical axioms* and a set of (finitary) *inference rules* defining a binary relation $E \vdash G$ which holds if and only if the Σ-formula G is *derivable* from E using the logical and nonlogical axioms and the inference rules.

As nonlogical axioms, the reflexivity, symmetry, transitivity and the substitution property of $=$ are considered:

(i) *Reflexivity*: $\forall x{:}s.\, x =_s x$ for all $s \in S$;

(ii) *Symmetry*: $\forall x,y{:}s.\, x =_s y \Rightarrow y =_s x$ for all $s \in S$;

(iii) *Transitivity*: $\forall x,y,z{:}s.\, x =_s y \wedge y =_s z \Rightarrow x =_s z$ for all $s \in S$;

(iv) *Substitutivity*: $\forall x_1, y_1{:}s_1, \ldots, x_n, y_n{:}s_n.$

$$x_1 =_{s_1} y_1 \wedge \cdots \wedge x_n =_{s_n} y_n \quad \Rightarrow \quad f(x_1, \ldots, x_n) = f(y_1, \ldots, y_n)$$

for all $f: s_1, \ldots, s_n \to s \in F$.

There are several choices for the logical axioms and inference rules (cf. e.g. [8, pp. 34–40]). The important result is Gödel's Completeness Theorem.

3.1.1. THEOREM. *A first-order Σ-formula G is derivable from a set E of Σ-sentences if and only if it is valid in $Alg(\Sigma, E)$; i.e. $E \vdash G$ if and only if $Alg(\Sigma, E) \models G$.*

PROOF. See e.g. [8, pp. 36 or 39]. □

Similarly, for equational and conditional equational formulas there exist formal systems computing the theories $Th_{EQ}(Alg(\Sigma, E))$ and $Th_{CEQ}(Alg(\Sigma, E))$ via binary relations $\vdash_{EQ}$ and $\vdash_{CEQ}$.

The inference rules for Σ-equations consist of two rules: Let H be (quantifier-free) Σ-Horn clause, G_1, G_2, G (possibly empty) conjunctions of Σ-equations and $u, u', v, v' \in T(\Sigma, X)$.

(v) *Substitution rule*: If H is derivable, $x \in X_s$ and $u \in T(\Sigma, X)_s$, then is $H[u/x]$ derivable, too.

(vi) *Cut rule*: If $G_1 \wedge u' = v' \wedge G_2 \Rightarrow u = v$ is derivable and $G \Rightarrow u' = v'$ is derivable, then so is $G_1 \wedge G \wedge G_2 \Rightarrow u = v$.

If a Σ-equation G is derivable from E using the axioms (i)–(iv) and the inference rules (v) and (vi), we write $E \vdash_{EQ} G$. This formal system is sound and complete for sensible signatures.

3.1.2. THEOREM (soundness and completeness of the many-sorted equational calculus). *Let Σ be a sensible signature and let E be a set of conditional Σ-equations. A Σ-equation G is derivable from E (i.e., $E \vdash_{EQ} G$) if and only if $Alg(\Sigma, E) \vdash G$.*

PROOF (*sketch*) (for a full proof see e.g. [141, p. 57]). For the soundness it suffices to check the validity of the logical axioms and inference rules (i)–(vi). This holds because of the assumption of sensible signatures (cf. [98, 76, 77, 143]).

To prove completeness, one considers the following quotient $T(\Sigma, X)/\sim^E$ of the term algebra $T(\Sigma, X)$: X is a *sorts*(Σ)-sorted set of free variables with infinitely many variables of each sort and $\sim$ is the congruence on $T(\Sigma, X)$ defined by $t \sim^E t'$ if and only if $E \vdash_{EQ} t = t'$, for all $t, t' \in T(\Sigma, X)$. The quotient $T(\Sigma, X)/\sim^E$ is a free algebra in $Alg(\Sigma, E)$, i.e. for all Σ-equations $t = t'$, $T(\Sigma, X)/\sim^E \models t = t'$ if and only if $Alg(\Sigma, E) \models t = t'$. This implies the completeness of $\vdash_{EQ}$. □

A sound and complete proof system for equations which includes also nonsensible signatures is given in [55]. The system above is sound and complete w.r.t. *equations*, but *not w.r.t. conditonal equations*. Using the technique of Skolemization it is possible to derive also Horn clauses with this inference system.

3.1.3. COROLLARY. *Let $\Sigma = \langle S, F \rangle$ and let E be a set of Σ-Horn clauses, $H =_{\text{def}} (G_1 \Rightarrow G)$ a Σ-Horn clause and $FV(G) = \{x_1, \ldots, x_m\}$. Fix new constants $c_1, \ldots, c_m$ with type $(c_i) =$ type(x_i) for $i = 1, \ldots, m$ and let $\Sigma' = \langle S, F \cup \{c_1, \ldots, c_m\} \rangle$. Then the following holds:*

$$Alg(\Sigma, E) \models H \quad \textit{iff} \quad E \cup G_1[c_1/x_1, \ldots, c_m/x_m] \vdash_{EQ} G[c_1/x_1, \ldots, c_m/x_m].$$

where $\vdash_{EQ}$ denotes the proof system for Σ'-equations.

PROOF (*sketch*). By Theorem 3.1.2, it is sufficient to show that

$$Alg(\Sigma, E) \models H \quad \text{iff} \quad Alg(\Sigma', E \cup G_1[c_1/x_1, \ldots, c_m/x_m]) \models G[c_1/x_1, \ldots, c_m/x_m].$$

For a proof of this equivalence see e.g. [141]. □

There exists also a sound and complete proof system for Horn clauses which does not use Skolem constants; such a system is given in [163] for the one-sorted case.

The situation is different for classes of computation structures. According to Gödel's Incompleteness Theorem for arithmetic there does not exist a complete formal system for the set $Th(N)$ of all first-order formulas that are true in the standard model N of natural numbers. The reason is that $Th(N)$ is not recursively enumerable, whereas for any formal system the set of all derivable formulas is recursively enumerable (cf. e.g. [8, p. 547]). However, if one admits *semiformal* systems, that is, systems (possibly) containing infinitary rules, one gets also a soundness and completeness theorem.

Let E be a set of Σ-formulas. The theory of $Gen(\Sigma, E)$, $Th(Gen(\Sigma, E))$, is called the *inductive theory* (of (Σ, E)). Its derivation relation $\vdash_I$ is defined by adding for every Σ-formula G containing a free variable x of sort $s \in S$ the following induction rule II_s to the rules of the derivation relation $\vdash$:

(II_s) Infinite Induction:

If $G[t/x]$ is derivable for all $t \in T(\Sigma)_s$ then so is $\forall x{:}s.G$.

The logical and nonlogical axioms remain the same. A Σ-formula G is called *ω-derivable from E* if $E \vdash_I G$.

3.1.4. THEOREM. (Soundness and completeness of the inductive calculus). *Let E be a set of first-order Σ-sentences. A Σ-formula G is ω-derivable from E (i.e., $E \vdash_I G$) if and only if $Gen(\Sigma, E) \models G$.*

PROOF (*sketch*). The proof is a generalization of the ω-Completeness Theorem in [45, p. 81].

Soundness ("⇒") of the induction rule (II_s) for any s in the sorts of Σ is obvious from the definition of $Gen(\Sigma, E)$.

Completeness ("⇐") can be shown analogously to the proof for ω-logic in [45] by applying the "omitting type theorem": let $S = \{s_1, \ldots, s_n\}, n > 0$, and let for every sort $s \in S$ $(t_{i,s})_{i \in \mathbb{N}}$ be an enumeration of the ground terms of sort s. Then we consider the following infinite set $\Phi(x_{s_1}, \ldots, x_{s_n})$ of Σ-formulas in the free variables $x_{s_1}, \ldots, x_{s_n}$ of sorts $s_1, \ldots, s_n$:

$$\Phi(x_{s_1}, \ldots, x_{s_n}) =_{\text{def}} \bigcup_{s \in S} \{x_s \neq t_{1,s}, x_s \neq t_{2,s}, \ldots, x_s \neq t_{n,s}, \ldots\}$$

Now if $\langle \Sigma, E \rangle$ is consistent (i.e., $E \vdash_I G'$ does not hold for all Σ-sentences G'), then one can prove analogously to [45] that the first-order theory $Th(Alg(\Sigma, E'))$ where $E' =_{\text{def}} \{\Sigma\text{-sentence } G \mid E \vdash_I G\}$ locally omits the set $\Phi(x_{s_1}, \ldots, x_{s_n})$ and therefore that according to the omitting type theorem $Gen(\Sigma, E')$ has a model. Hence, also the inductive theory of $Th(Gen(\Sigma, E))$ has a model in $Gen(\Sigma, E)$.

Now, from this result it is easy to deduce the exact formulation of "⇐": Let G be a Σ-sentence such that $E \vdash_I G$ does not hold. Then $\langle \Sigma, E \cup \{\neg G\} \rangle$ is consistent. According to the above result there exists a model in $Gen(\Sigma, E \cup \{\neg G\})$. Hence $Gen(\Sigma, E) \models G$ does not hold. □

As a simple but illustrative example for the power of the induction rule consider $\langle \Sigma, E \rangle$ where Σ consists of one sort s and one constant c and E is $\exists x{:}s.x \neq c$. Then $T(\Sigma)_s = \{c\}$ and $Gen(\Sigma, E)$ is empty. On the other hand, using the induction rule one can deduce a contradiction:

$\vdash_I c = c$ (by reflexivity of =). Then

$\vdash_I (x = c)[t/x]$ for all $t \in T(\Sigma) = \{c\}$ (since $(x = c)[c/x]$ is $c = c$). Hence

$\vdash_I \forall x \in s : x = c$ (by the induction rule II_s)

which is first-order equivalent to the negation $\neg\exists x{:}s.x \neq c$ of the axiom. Thus $E \wedge \neg E$ is ω-derivable and therefore any Σ-formula is ω-derivable from E which implies that $Gen(\Sigma, E)$ is empty.

By adding the infinitary induction rules $\{II_s\}_{s\in S}$ to the equational and the conditional equational calculus one gets sound and complete calculi $\vdash_{I,EQ}$ and $\vdash_{I,CEQ}$ for the inductive theories $Th_{EQ}(Gen(\Sigma, E))$ and $Th_{CEQ}(Gen(\Sigma, E))$. We state this for the equational calculus.

3.1.5. THEOREM (soundness and completeness of the equational inductive theory). *Let E be a set of conditional Σ-equations. A Σ-equation G is ω-derivable from E (i.e.* $E\vdash_{I,EQ}G$*) if and only if,* $Gen(\Sigma, E)\models G$.

PROOF. (see e.g. [141]). The proof is similar to the proof of Theorem 3.1.2. Using the quotient $T(\Sigma)/\sim^E$ of the ground term algebra $T(\Sigma)$ (where $\sim^E = \{u=v \mid E\vdash_{EQ} u=v, u, v\in T(\Sigma)\}$) one proves that for all ground terms $u, v\in T(\Sigma)$, $Gen(\Sigma, E)\models u=v$ if and only if $E\vdash_{EQ} u=v$. Hence, if an arbitrary Σ-equation $t=t'$ is valid in $Gen(\Sigma, E)$ then, for each ground instance $u=v$ (with $u=_{\text{def}} t[u_1/x_1, \ldots, u_n/x_n]$ and $v=t'[u_1/x_1, \ldots, u_n/x_n]$, $\{x_1, \ldots, x_n\}=FV(t)\cup FV(t')$, $u_i, v_i\in T(\Sigma)$ for $i=1, \ldots, n$), there exists a proof by $\vdash_{EQ}$. Applying once the infinitary induction rule yields a proof for $t=t'$ by $\vdash_I$. □

As a consequence of the proof, a Horn sentence H of the form $\forall x_1, \ldots x_n. G_1 \Rightarrow G$ holds in $Gen(\Sigma, E)$ if, for all ground substitutions $\sigma=[t_1/x_1, \ldots, t_n/x_n]$ (where each $t_i\in T(\Sigma)$ is a ground term of the same sort as x_i, $i=1, \ldots, n$), $E\cup\{G_1\sigma\}\vdash_{EQ} G\sigma$. Initial models are characterized by a weaker rule: if $Gen(\Sigma, E)$ admits an initial model I, then H holds in I if and only if for all ground substitutions σ $E\vdash_{EQ} G_1\sigma$ implies $E\vdash_{EQ} G\sigma$ (see [141, Section 4.3] and Lemma 4.1.3). Inductive theories with quantifier-free axioms (such as equational or conditional equational axioms) are equivalent to inductive theories with ground formulas as axioms. The axioms can be replaced by the (infinite) set of their ground instances. Moreover, in such theories the proof of a ground equation does not need induction rules.

3.1.6. FACT. *Let E be a set of quantifier-free Σ-formulas.*

(1) $Gen(\Sigma, E)=Gen(\Sigma, E_{ground})$ *where*

$$E_{ground} =_{\text{def}} \{G[t_1/x_1, \ldots, t_n/x_n] \mid G\in E, FV(G)=\{x_1{:}s_1, \ldots, x_n{:}s_n\}, t_1\in T(\Sigma)_{s_1}, \ldots, t_n\in T(\Sigma)_{s_n}\};$$

(2) *For all ground Σ-terms* $u, v\in T(\Sigma)$: $Gen(\Sigma, E)\models u=v$ *iff* $E\vdash u=v$.

PROOF (1) Obvious, since a Σ-computation structure satisfies a quantifier-free formula G if and only if it satisfies all ground instantiations of G. (2) follows directly from (1) and Theorem 3.1.2 using the equivalence

$$Alg(\Sigma, E)\models u=v \text{ iff } Gen(\Sigma, E)\models u=v$$

which holds for all *ground* equations $u=v\in WFF(\Sigma)$. □

3.2. *Algebraic specifications and their semantics*

Following [95], a specification is a formal documentation for a data type D which guarantees properties of implementations of D for use in proving the correctness of programs over D. Hence, a *first-order axiomatic specification* $[\Sigma, E]$ consists of a signature and a set E of first-order axioms which describe the required properties. The semantics of an axiomatic specification $[\Sigma, E]$ is given by the signature Σ and the class $Alg(\Sigma, E)$ of all first-order models of $[\Sigma, E]$. This idea of axiomatizing implementations is suited to a representation-free view of data types, but without special algebraic assumptions it does not support a method which defines (only) abstract data types. If an axiomatic specification $[\Sigma, E]$ admits one model having a carrier set with infinitely many elements, then, according to the Löwenheim–Skolem Theorem (cf. e.g. [164, 67, p. 79]), it admits models of every infinite cardinality; thus most models in $Alg(\Sigma, E)$ are not reachable and do not have anything to do with abstract data types.

For pure *algebraic* specifications, therefore, the class $Gen(\Sigma, E)$ of all computation structures in $Alg(\Sigma, E)$ is chosen as a basis of the semantics. Three main approaches can be distinguished: the *initial algebra approach*, the *terminal* (*or final*) *algebra approach*, which is related to the behavioral semantics, and the *loose approach*.

In the loose approach, the specification is thought of as a contract possibly open to interpretation by a number of different abstract data types; hence, in this approach, which is appropriate for program verification and program development, the class $Gen(\Sigma, E)$ of all Σ-computation structures satisfying the axioms E is regarded as semantics. However, to be able to define an abstract data type exactly by some axiomatic specification, it is often necessary to have a semantic mechanism M which chooses, uniquely up to isomorphism, a computation structure $M(\Sigma, E)$ from $Gen(\Sigma, E)$. In the initial algebra approach, the initial algebra and, in the terminal algebra approach, the terminal algebra is taken for M. (For the behavioral approach, see Section 5.4 and 6.5.)

A (*simple*) *algebraic specification* $SP = \langle \Sigma, E \rangle$ consists of a signature Σ and a set E of Σ-formulas. If E consists of (conditional) Σ-equations (Σ-sentences, resp.) we speak of (*conditional*) *equational specification* (first-order specification, resp.).

The signature of $SP = \langle \Sigma, E \rangle$ is Σ: $sig(SP) =_{\text{def}} \Sigma$.

According to the three approaches three semantic functions are defined:

(i) $Mod(SP) =_{\text{def}} Gen(\Sigma, E)$, the loose semantics;

(ii) $I(SP) =_{\text{def}} \{I \in Gen(\Sigma, E) \mid I \text{ is initial in } Gen(\Sigma, E)\}$, the initial algebra semantics;[1]

(iii) $Z(SP) =_{\text{def}} \{Z \in Gen(\Sigma, E) \mid Z \text{ is terminal in } Gen(\Sigma, E)\}$, the terminal algebra semantics.[1]

Hence for all three approaches, the semantics of a specification is *abstract*, i.e. it is a class of computation structures which is closed under isomorphism. Any such class can be partitioned into a set of isomorphism classes; the 1–1 correspondence between congruences and isomorphism classes then induces two equivalent semantic descriptions: the set $Mod_{\cong}(SP) =_{\text{def}} \{[A] \mid A \in Mod(SP)\}$ of *isomorphism classes of models* and the set $C(SP)$ of *associated congruences*. A fourth possibility is to consider the theory of the models of SP as the semantics of SP.

3.2.1. FACT (equivalence of semantics). *Let* $SP = \langle \Sigma, E \rangle$ *and* $SP' = \langle \Sigma, E' \rangle$ *be two*

[1] Usually, $Alg(\Sigma, E)$ is considered instead of $Gen(\Sigma, E)$ (cf. e.g. [82, 175]); however, for any conditional equational specification, the initial algebras of $Gen(\Sigma, E)$ and $Alg(\Sigma, E)$ are the same. A similar result holds for terminal algebras.

simple specifications with the same signature. Then the following assertions are equivalent:

(1) $Mod(SP) \subseteq Mod(SP')$;
(2) $Mod_{\cong}(SP) \subseteq Mod_{\cong}(SP')$;
(3) $C(SP) \subseteq C(SP')$;
(4) $Th(Mod(SP)) \supseteq Th(Mod(SP'))$.

PROOF (1)$\Leftrightarrow$(2) holds because of the closure of $Mod(SP)$ under isomorphism. (2)$\Leftrightarrow$(3) is Corollary 2.3.3. (3)$\Leftrightarrow$(4): assertions (3.1) and (3.2) in Section 3.1. □

Hence for comparing two specifications it is sufficient to consider their signature and their classes of models. A specification $SP' = \langle \Sigma', E' \rangle$ is called a *refinement* of the specification $SP = \langle \Sigma, E \rangle$, written $SP' \subseteq SP$, if $\Sigma = \Sigma'$ and $Mod(SP') \subseteq Mod(SP)$; SP and SP' are *equivalent* if $SP' \subseteq SP$ and $SP \subseteq SP'$, i.e. if $\Sigma = \Sigma'$ and $Mod(SP) = Mod(SP')$.

Initial and terminal computation structures do not always exist and, if they exist, they are often not isomorphic (see Section 4.1). But if they are isomorphic for a specification SP, then $Mod(SP)$ consists of exactly one isomorphism class of computation structures, i.e. one abstract data type. A specification SP is called *monomorphic* if initial and terminal models exist and are isomorphic; it is called *satisfiable* if there exists at least one model (i.e. $Mod(SP) \neq \emptyset$).

The characterization of initial and terminal computation structures (Lemma 2.3.6) can be reformulated as follows.

3.2.2. LEMMA. *Let K be a class of Σ-algebras.*

(1) *A Σ-computation structure $I \in K$ is initial in K if and only if for all ground Σ-terms $t, t' \in T(\Sigma)$ the following condition holds*:

$$I \models t = t' \Leftrightarrow \text{for all } A \in K\text{: } A \models t = t'.$$

(2) *A Σ-computation structure $Z \in K$ is terminal in K if and only if for all ground Σ-terms $t, t' \in T(\Sigma)$ the following condition holds*:

$$Z \models t = t' \Leftrightarrow \text{there exists } A \in K\text{: } A \models t = t'.$$

As a consequence of this lemma and Fact 3.1.6(2), an equation holds in an initial model of a specification SP if and only if this equality can be "computed" using the axioms of SP, whereas in a terminal model only those (ground) equations are false whose negation can be proven.

3.2.3. COROLLARY. *Let $SP = \langle \Sigma, E \rangle$ be a specification with a set E of quantifier-free axioms.*

(1) *A Σ-computation structure $I \in Mod(SP)$ is initial in $Mod(SP)$ if and only if for all ground Σ-terms $t, t' \in T(\Sigma)$ the following condition holds*:

$$I \models t = t' \Leftrightarrow t \sim^I t' \Leftrightarrow E \vdash t = t'.$$

(2) *A Σ-computation structure $Z \in Mod(SP)$ is terminal in $Mod(SP)$ if and only if for all ground Σ-terms $t, t' \in T(\Sigma)$ the following condition holds*:

$$Z \models t \neq t' \;\Leftrightarrow\; \neg(t \sim^Z t') \;\Leftrightarrow\; E \vdash \neg(t = t').$$

PROOF. Lemma 3.2.2 and Fact 3.1.6(2). □

In the following we use the notation

spec $SP \equiv$ **signature** Σ **axioms** E **endspec**

to denote a specification $SP = \langle \Sigma, E \rangle$. Larger specifications will be constructed using the operator "**extend . by .**". If $SP_1 = \langle\langle S_1, F_1 \rangle, E_1 \rangle$, then the expression

spec $SP \equiv$ **extend** SP_1 **by sorts** S **functions** F **axioms** E **endspec**

denotes the specification $\langle \Sigma, E_1 \cup E \rangle$ with $\Sigma = \langle S_1 \cup S, F_1 \cup F \rangle$. This specification extends SP_1 by the sorts S, the function symbols F and the axioms E. In the examples the brackets "{" and "}" around sets of sorts, function symbols and axioms are omitted. Moreover, if $SP_2 = \langle\langle S_2, F_2 \rangle, E_2 \rangle$ then $SP_1 + SP_2 =_{\text{def}} \langle\langle S_1 \cup S_2, F_1 \cup F_2 \rangle, E_1 \cup E_2 \rangle$ denotes the union of SP_1 and SP_2.

3.2.4. EXAMPLE. (1) The following specification *BOOL0* consists of the signature Σ_{BOOL0} of truth values (of Example 2.1) and no axioms.

spec $BOOL0 \equiv$ **signature** Σ_{BOOL0} **endspec**

Mod(*BOOL*0) consists of two abstract data types: the abstract data type of the term algebra $T(\Sigma_{BOOL0})$ of Σ_{BOOL0} and the abstract data type of the unit algebra $U(\Sigma_{BOOL0})$ of Σ_{BOOL0}. The carrier set of $T(\Sigma_{BOOL0})$ has exactly two elements corresponding to the two truth values. According to Facts 2.2.3 and 2.2.4, $T(\Sigma_{BOOL0})$ is an initial and $U(\Sigma_{BOOL0})$ is a terminal model of *BOOL*0, i.e. $Mod(BOOL0) = I(BOOL0) \cup Z(BOOL0)$. By adding the inequality $true \neq false$ one obtains a monomorphic specification *BOOLM* with the abstract data type of $T(\Sigma_{BOOL0})$ as model.

spec $BOOLM \equiv$ **signature** Σ_{BOOL} **axioms** $true \neq false$ **endspec**

Thus $Mod(BOOLM) = I(BOOLM) = Z(BOOLM) = I(BOOL0)$.

The following specification *BOOL* extends *BOOLM* by the usual Boolean operations *not*, *and* and *or*, written in infix notation:

spec $BOOL \equiv$
  extend *BOOLM* **by**
    functions $not: bool \to bool$,
      $.and.,.or.: bool, bool \to bool$
    axioms $not(true) = false$, $not(false) = true$,
      $true \; and \; x = x$, $false \; and \; x = false$,
      $x \; or \; y = not(not(x) \; and \; not(y))$
endspec

The specification $BOOL$ is monomorphic; the restriction $M|_{\Sigma_{BOOL0}}$ of any model M of $BOOL$ is a model of $BOOLM$, i.e., $M|_{\Sigma_{BOOL0}} \cong T(\Sigma_{BOOL0})$.

(2) Similarly to $BOOL0$, the signature Σ_{NAT0} induces a specification the initial element of which is the abstract data type of the standard model N (cf. Example 2.2.1) of natural numbers:

spec $NAT0 \equiv$ **signature** Σ_{NAT0} **endspec**

$Mod(NAT0)$ is the class $Gen(\Sigma_{NAT0}, \emptyset)$ of all Σ_{NAT0}-computation structures; $I(NAT0)$ is the abstract data type of $T(\Sigma_{NAT0})$ or equivalently of N; $Z(NAT0)$ is the isomorphism class of the unit algebra of Σ_{NAT0}. By adding two of the Peano axioms one obtains a monomorphic specification $NATM$ of N.

spec $NATM \equiv$
  signature Σ_{NAT0}
  axioms $succ(x) = succ(y) \Rightarrow x = y$,
    $succ(x) \neq zero$
endspec

(3) A monomorphic specification of the algebra N-B (cf. Example 2.2.2) can be obtained by an appropriate enrichment of $BOOLM$ and $NAT0$ (it is not necessary to take $NATM$!).

spec $NAT1 \equiv$
  extend $BOOLM + NAT0$ **by**
    function $eq: nat, nat \rightarrow bool$
    axioms $eq(succ(x), succ(y)) = eq(x, y)$,
      $eq(zero, succ(x)) = false$,
      $eq(succ(x), zero) = false$,
      $eq(x, x) = true$
endspec

The two axioms of $NATM$ are ω-derivable in $NAT1$. Hence, $Mod(NAT1) = I(NAT1) = Z(NAT1) = [N\text{-}B]$.

(4) The following specification $LOOSE$-SET has the computation structure $P0$ of finite sets of natural numbers (cf. Example 2.2.2) as terminal algebra.

spec $LOOSE\text{-}SET \equiv$
  extend $NAT1$ **by**
    sort set
    functions $empty: \rightarrow set$,
      $insert: nat, set \rightarrow set$,
      $.\varepsilon.: nat, set \rightarrow bool$
    axioms $(x \,\varepsilon\, empty) = false$,
      $(x \,\varepsilon\, insert(x, s)) = true$,
      $eq(x, y) = false \Rightarrow (x \,\varepsilon\, insert(y, s)) = (x \,\varepsilon\, s)$
endspec

The initial element of $LOOSE$-SET is the abstract data type of $N0^*$ (cf. Example 2.2.2), the computation structure of all finite sequences of natural numbers:

$$I(LOOSE\text{-}SET)=[N0^*], \qquad Z(LOOSE\text{-}SET)=[P0].$$

Mod(LOOSE-SET) contains infinitely many abstract data types; in fact, the set of isomorphism classes of *Mod(LOOSE-SET)* is not countable. To see this, one may consider partitions of the carrier set $P0_{set}$ of $P0$. Let $p: P0_{set} \rightarrow \{0, 1\}$ be a total function; define an equivalence relation $\sim^p$ on the terms of $\Sigma_{SETNAT} \backslash \{\varepsilon\}$ which identifies two terms t and t' of sort *set* if and only if t and t' represent the same finite set, say s, and $p(s)=1$ holds, i.e. for all $t, t' \in T(\Sigma_{SETNAT} \backslash \{\varepsilon\})_{set}$

$$t \sim^p t' \quad \text{iff} \quad t \equiv t' \text{ or } (P0 \models t = t' \text{ and } p(t^{P0})=1).$$

The relation $\sim^p$ can be extended to a congruence of *LOOSE-SET* by defining "$n\varepsilon t = b$" (for $n \in T(\Sigma_{NAT1})$, $t \in T(\Sigma_{SETNAT})_{set}$, $b \in \{true, false\}$) according to the axioms of *LOOSE-SET*. There exist uncountably many partitions $p: P0_{set} \rightarrow \{0, 1\}$. Hence, *Mod(LOOSE-SET)* is uncountable.

LOOSE-SET can be made to a monomorphic specification by adding two axioms:

spec *SETNAT* ≡
  extend *LOOSE-SET* **by**
    axioms $insert(x, insert(x, s)) = insert(x, s)$,
      $insert(x, insert(y, s)) = insert(y, insert(x, s))$
endspec

By these axioms all terms of sort *set* are identified which contain the same elements of sort *nat*:

$$Mod(SETNAT)=I(SETNAT)=Z(SETNAT)=[P0].$$

Initial algebra specifications need often (cf. Examples 3.2.4(1)–(3)), but not always (cf. Example 3.2.4(4)), less axioms than terminal algebra specifications and loose specifications. One advantage of initial algebra specifications is that they are true (conditional) equational specifications, whereas terminal algebra specifications and monomorphic loose specifications need additional inequality axioms or additional semantic assumptions. For example, Bergstra and Tucker [23] propose to take the coarsest congruence $\sim^Z$ which is different from the unit congruence $\sim^{U(\Sigma)}$; Wand [175] considers extensions of a fixed Σ_0-algebra B with $\Sigma_0 \subseteq \Sigma$, i.e. the terminal element of the class $\{A \in Alg(\Sigma, E) \mid A|_{\Sigma_0} = B\}$. In this paper, we will often assume that a specification *SP* is an enrichment of *BOOLM*; i.e. $true \neq false$ is the only inequation of *SP*, all other axioms are (conditional) equational formulas.

3.3. *Other semantic concepts*

Total algebras provide a simple and elegant basis for the semantics of abstract data types. However, many authors do not consider them appropriate for the description of error situations and nonterminating algorithms. Consider the following specification of binary trees:

spec $TREENAT \equiv$
extend $NAT1$ **by**
sort *tree*
functions *empty* $: \to tree$,
node $: tree, nat, tree \to tree$,
left, *right* $: tree \to tree$,
label $: tree \to nat$,
isempty $: tree \to bool$

axioms $isempty(empty) = true$,
$isempty(node(t_1, n, t_2)) = false$,
$left(node(t_1, n, t_2)) = t_1$,
$right(node(t_1, n, t_2)) = t_2$,
$label(node(t_1 \cdot n, t_2)) = n$
endspec

The specification $TREENAT$ has an initial algebra I_{TREE}, but this algebra does not represent the computation structure of labeled trees over natural numbers: for example, the carrier set of $(I_{TREE})_{nat}$ is not protected, i.e., it is not isomorphic to the standard natural numbers; it contains elements denoted by *label*(*empty*), *label*(*left*(*empty*)), *label*(*right*(*empty*)), *label*(*left*(*left*(*empty*))), ... which are different from any term of the form $succ^n(zero)$, $n \in \mathbb{N}$, denoting a standard natural number; similarly, $(I_{TREE})_{bool}$ does not have two elements, but it consists of infinitely many different elements.

Intuitively, terms such as *label*(*empty*) or *left*(*empty*) denote incorrect or erroneous calls of the function symbols *label* and *left*. In the following, six concepts will be shortly presented. Four of them, error algebras, continuous algebras, partial algebras and order-sorted algebras, are designed for overcoming this problem; the fifth tackles the problem of specifying nondeterminism and concurrency, whereas the sixth one, institutions, aims at abstracting from all these particular algebraic concepts and the corresponding logical formalisms.

3.3.1. *Error algebras*

Error algebras originate from [71]. The main idea is to introduce for every sort one "error constant" and a so-called "ok-predicate" which tells when an element is not error.

Let $SP = \langle \Sigma, E \rangle$ with $\Sigma = \langle S, F \rangle$ be a given equational specification containing an equational specification of truth values as subspecification. For instance, let $BOOL1$ be the specification $BOOL$ of Example 3.2.4(1) with the axiom $true \neq false$ removed. Then in [82] a specification $SP_e = \langle \Sigma_e, E_e \rangle$ is defined with $\Sigma_e = \langle S, F_e \rangle$. First, F_e differs from F by the addition, for each $s \in S$, of

$error_s : \to s$ "error constants",
$ok_s : s \to bool$ "ok-predicates",
$ife_s : bool, s, s \to s$ "if-then-else",
$ifok_{w,s} : s_1, \ldots, s_n, s \to s$ for each $u = v \in E$ with $u, v \in T(\Sigma, \{x_1{:}s_1, \ldots, x_n{:}s_n\})_s$, $w = s_1, \ldots, s_n$ "equations for ok-elements".

The ok-predicates are strict; for any function symbol f, representing a partial function, the domain of f is described by a Boolean term c. Using the "*ifok*"-function symbols it is possible to restrict the range of the universal quantifiers to "non-error" elements. E_e contains the following equations:

$$ife_s(true, x, x') = x, \qquad ife_s(false, x, x') = x',$$

$$ife_s(error_{bool}, x, x') = error_s \quad \text{for each } s \in S,$$

$$ok_s(error_s) = false \quad \text{for each } s \in S,$$

$$ok_s(f(x_1, \ldots, x_n)) = (ok_{s_1}(x_1) \text{ and } \ldots \text{ and } ok_{s_n}(x_n) \text{ and } c(x_1, \ldots, x_n)) \\ \text{where } c \in T(\Sigma, \{x_1, \ldots, x_n\})_{bool},$$

$$f(x_1, \ldots, x_{i-1}, error_{s_i}, x_{i+1}, \ldots, x_n) = error_s \\ \text{for each } f: s_1, \ldots, s_n \to s \in F \text{ and } i = 1, \ldots, n,$$

$$ifok_{w,s}(x_1, \ldots, x_n, x) = ife_s(ok_{s_1}(x_1) \text{ and } \ldots \text{ and } ok_{s_n}(x_n), x, error_s),$$

$$ifok_{w,s}(x_1, \ldots, x_n, u) = ifok_{w,s}(x_1, \ldots, x_n, v) \\ \text{for each } u = v \in E \text{ with } u, v \in T(\Sigma, \{x_1{:}s_1, \ldots, x_n{:}s_n\}),\ w = s_1, \ldots, s_n.$$

3.3.1. THEOREM (Goguen et al. [82]). *Let $SP = \langle\langle S, F\rangle, E\rangle$ be a specification containing BOOLI as subspecification, and define $\langle\langle S, F_e\rangle, E_e\rangle$ as above. Let I denote an initial algebra of SP and assume that SP protects BOOLI, i.e. $I|_{sig(BOOLI)}$ is an initial model of BOOLI. The "error-extension" of I is defined as follows: $A_s =_{\text{def}} I_s \cup \{error_s\}$ for $s \in S$, and for each $w = s_1, \ldots, s_n \in S^*$, $f: w \to s \in F$, $a, a' \in A_s$, $a_1 \in A_{s_1}, \ldots, a_n \in A_{s_n}$, $b \in A_{bool} = \{true^I, false^I, error_{bool}\}$, let*

$$f^A(a_1, \ldots, a_n) =_{\text{def}} \begin{cases} error_s & \text{if any } a_i = error_{s_i}, \\ f^I(a_1, \ldots, a_n) & \text{otherwise,} \end{cases}$$

$$ok_s^A(a) =_{\text{def}} \begin{cases} false & \text{if } a = error_s, \\ true & \text{otherwise,} \end{cases}$$

$$ife_s(b, a, a') =_{\text{def}} \begin{cases} a & \text{if } b = true, \\ a' & \text{if } b = false, \\ error_s & \text{if } b = error_{bool}, \end{cases}$$

$$ifok_{w,s}(a_1, \ldots, a_n, a) =_{\text{def}} \begin{cases} error_s & \text{if any } a_i = error_{s_i}, \\ a & \text{otherwise.} \end{cases}$$

Then A is an initial $\langle\langle S, F_e\rangle, E_e\rangle$-algebra.

3.3.2. EXAMPLE. An "error specification" of binary trees over natural numbers can be given based on an "error specification" NAT_e of the natural numbers which includes an "error specification" of truth values. In order to obtain shorter specifications, the scheme of Theorem 3.3.1 is not exactly applied: the predicates ok_{tree} are defined only for *empty*, $error_{tree}$ and *node*; using the axioms, the definitions of $ok_s(f(x))$ can be derived for $f \in \{isempty, label, left, right\}$. Moreover, due to strictness of the functions, the introduction of the *ifok*-symbols is not necessary; instead, any axiom $u = v$ is replaced by

$$u = ife_s(ok_{s_1}(x_1) \text{ and } \ldots \text{ and } ok_{s_n}(x_n), v, error_s)$$

if $FV(u) \backslash FV(v) = \{x_1{:}s_1, \ldots, x_n{:}s_n\}$.

spec $TREENAT_e \equiv$
extend NAT_e **by**
sort *tree*
functions $empty, error_{tree}: \rightarrow tree,$
$node: tree, nat, tree \rightarrow tree,$
$left, right: tree \rightarrow tree,$
$label: tree \rightarrow nat,$
$isempty, ok_{tree}: tree \rightarrow bool,$
$ife_{tree}: bool, tree, tree \rightarrow tree$
axioms $isempty(error_{tree}) = error_{bool},$
$isempty(empty) = true,$
$isempty(node(t_1, n, t_2))$
$= ife_{bool}(ok_{tree}(t_1) \text{ and } ok_{nat}(n) \text{ and } ok_{tree}(t_2), false, error_{bool}),$

$node(error_{tree}, n, t) = error_{tree},$
$node(t_1, error_{nat}, t_2) = error_{tree},$
$node(t, n, error_{tree}) = error_{tree},$

$label(empty) = error_{nat},$
$label(node(t_1, n, t_2))$
$= ife_{nat}(ok_{tree}(t_1) \text{ and } ok_{nat}(n) \text{ and } ok_{tree}(t_2), n, error_{nat}),$
$label(error_{tree}) = error_{nat},$

$left(empty) = error_{tree},$
$left(node(t_1, n, t_2))$
$= ife_{tree}(ok_{tree}(t_1) \text{ and } ok_{nat}(n) \text{ and } ok_{tree}(t_2), t_1, error_{tree}),$
$left(error_{tree}) = error_{tree},$

$right(empty) = error_{tree},$
$right(node(t_1, n, t_2))$
$= ife_{tree}(ok_{tree}(t_1) \text{ and } ok_{nat}(n) \text{ and } ok_{tree}(t_2), t_2, error_{tree}),$
$right(error_{tree}) = error_{tree},$

$ife_{tree}(true, t_1, t_2) = t_1,$
$ife_{tree}(false, t_1, t_2) = t_2,$
$ife_{tree}(error_{bool}, t_1, t_2) = error_{tree},$

$ok_{tree}(empty) = true,$
$ok_{tree}(node(t_1, n, t_2)) = ok_{tree}(t_1) \text{ and } ok_{nat}(n) \text{ and } ok_{tree}(t_2),$
$ok_{tree}(error_{tree}) = false$

endspec

The specification $TREENAT_e$ is monomorphic if NAT_e is monomorphic. In this case, the following algebra $Tree_e$ of signature $sig(TREENAT_e)$ is the (up to isomorphism) unique model of $TREENAT_e$:

$$(Tree_e)|_{\Sigma_{NAT_e}} =_{\text{def}} \text{“the strict standard algebra over } \mathbb{N} \cup \{\bot\} \text{ and } \{O, L, \bot\}\text{”},$$
$$(Tree_e)_{tree} =_{\text{def}} T(\Sigma_{NATO} \cup \langle\{tree\}, \{empty, node\}\rangle) \cup \{\bot\}.$$

The functions of $Tree_e$ are defined in the obvious way, e.g. for each $t \in Tree_{tree}$

$$label^{Tree}(t) =_{def} \begin{cases} \perp & \text{if } t = empty \text{ or } t = \perp, \\ n & \text{if } t = node(t_1, n, t_2). \end{cases}$$

As this example shows, the explicit axiomatization of error elements leads to illegible specifications in which normal cases and erroneous cases are mixed together. To overcome this problem, Goguen [71] encodes error propagation in the algebras. Bidoit [28] provides a formalism that handles error states by means of declarations; this approach allows one also to declare error-recovery cases which do not propagate erroneous values. Similarly, in the approach of Bidoit et al. [30] recovery can be explicitly introduced.

In [72], Goguen suggests considering error propagation as a special case of coercion and overloading in the framework of order-sorted algebra (cf. Section 3.3.4). This idea has been pursued and simplified by several authors (cf. e.g. [70, 145, 26, 69]. A central aspect in these papers is the explicit characterization of ok-values instead of erroneous ones, and the distinction between safe function symbols that cannot introduce erroneous values and unsafe ones.

3.3.2. Partial algebras

Partial algebras allow for the use of partial functions. Given a signature $\Sigma = \langle S, F \rangle$, a partial Σ-algebra A is defined just as a total Σ-algebra, only the interpretations f^A of the function symbols $f \in F$ may be partial functions. As a consequence, the notions of Σ-homomorphism and of Σ-formula have to be adapted. For homomorphisms there are several possible choices (cf. e.g. [83]). The most convenient notion—used in [149, 38], and many others—seems to be the following: Let A and B be partial Σ-algebras. A function $h: A \to B$ is called a *total Σ-homomorphism* if h is a total function and for each $f: s_1, \ldots, s_n \to s$, for all $a_1 \in A_{s_1}, \ldots, a_n \in A_{s_n}$:

$$f^A(a_1, \ldots, a_n) \text{ defined } \Rightarrow f^B(h(a_1), \ldots, h(a_n)) \text{ defined and}$$
$$h(f^A(a_1, \ldots, a_n)) = f^B(h(a_1), \ldots, h(a_n)).$$

Σ-formulas must be able to express whether a term is defined. Hence, atomic Σ-formulas consist of Σ-equations (as before) and of definedness formulas $D_s(t)$ where $s \in S$, $t \in T(\Sigma, X)_s$ and D_s is the so-called definedness predicate. A partial Σ-algebra A satisfies a definedness formula $D_s(t)$ if the interpretation of t in A is defined. For Σ-equations one may find two different interpretations in the literature: the existential equality $t =^e t'$, which holds only if the interpretations of t and t' are both defined and equal, and the strong equality $t = t'$, which holds if the interpretations of t and t' are both defined and equal or if both interpretations are undefined. Quantification ranges only over defined values. Formally, we have for any partial Σ-algebra A and any total valuation $v: X \to A$:

$$A, v \models D(t) \quad \text{iff } v^*(t) \text{ is defined,}$$
$$A, v \models t =^e t' \quad \text{iff } v^*(t) \text{ and } v^*(t') \text{ are defined and } v^*(t) = v^*(t'),$$
$$A, v \models t = t' \quad \text{iff } v^*(t) = v^*(t') \text{ or } (v^*(t) \text{ and } v^*(t') \text{ are undefined}),$$

$A, v \models \neg G, G \wedge H$ as usual,

$A, v \models \forall x{:}s.G$ iff $A, v \models G$ for all total valuations $v_x\colon X \to A$ with $v_x(y) = v(y)$ for $y \neq x$.

Existential and strong equality are first-order equivalent: for all $t, t' \in T(\Sigma, X)_s$, $s \in S$:

$$\models t =^{e} t' \;\Leftrightarrow\; (t = t' \wedge D(t) \wedge D(t')),$$

$$\models t = t' \;\Leftrightarrow\; (t =^{e} t' \vee (\neg D(t) \vee \neg D(t'))).$$

Moreover, existential equality allows us to derive the definedness predicate:

$$\models D(t) \;\Leftrightarrow\; t =^{e} t \quad \text{for all } t \in T(\Sigma, X)_s,\ s \in S.$$

The existential equality (induced by conditional atomic axioms) has the advantage of being recursively enumerable; i.e. the set of all pairs $\{\langle t, t'\rangle \mid t, t' \in T(\Sigma), E \vdash t =^{e} t'\}$ is recursively enumerable. On the other hand, strong equality is used in recursion theory for defining the equality between recursive functions [152]. For algebraic specifications, the following form of conditional atomic axioms is convenient [36]:

$$u_1 =^{e} v_1 \wedge \cdots \wedge u_n =^{e} v_n \;\Rightarrow\; at$$

where $at \in \{u_{n+1} = v_{n+1},\ D(u_{n+1})\}$, $n \geq 0$, u_i, $v_i \in T(\Sigma, X)_{s_i}$, $i = 1, \ldots, n+1$.

As an example we consider the specification *TREENAT* as a partial algebra specification. Then it has a partial initial algebra $I1$ which represents exactly binary trees over natural numbers:

$$I1_{tree} =_{\text{def}} T(\langle\{tree, nat\}, \{empty, node\}\rangle, \mathbb{N})_{tree},$$

$$empty^{I1} =_{\text{def}} empty,$$

$$node^{I1}(t_1, n, t_2) =_{\text{def}} node(t_1, n, t_2),$$

$$isempty(t) =_{\text{def}} \begin{cases} true & \text{if } t \equiv empty, \\ false & \text{otherwise,} \end{cases}$$

$$label^{I1}(t) =_{\text{def}} \begin{cases} undefined & \text{if } t \equiv empty, \\ n & \text{if } t \equiv node(t_1, n, t_2), \end{cases}$$

$$left^{I1}(t) =_{\text{def}} \begin{cases} undefined & \text{if } t \equiv empty, \\ t_1 & \text{if } t \equiv node(t_1, n, t_2), \end{cases}$$

$$right^{I1}(t) =_{\text{def}} \begin{cases} undefined & \text{if } t \equiv empty, \\ t_2 & \text{if } t \equiv node(t_1, n, t_2), \end{cases}$$

for all t, t_1, $t_2 \in I1_{tree}$, $n \in \mathbb{N}$.

The only partial functions in $I1$ are $label^{I1}$, $left^{I1}$, $right^{I1}$; all other functions are total. In particular, the functions in B and N are all total functions. For instance, for the constants *true* and *false* one may conclude this as follows: Because of the axiom $true \neq false$, strong equality implies that at least one of the constants *true* and *false* must be defined. If $D(true)$ holds, then, because of the axiom $not(false) = true$, $D(false)$ holds as well. Similarly, if $D(false)$ holds, then, because of $not(false) = true$, $D(true)$ holds as well.

Using the Boolean functions *eq* and *isempty* one may deduce that in all partial algebras which are models of *TREENAT* all terms $succ^n(zero)$, $n \in \mathbb{N}$, *empty* and all terms $node(t_1, m, t_2)$, for defined t_1, m, t_2, are defined.

Specifications such as *TREENAT* are *positive* specifications. The axioms imply that certain terms have to be defined, but terms cannot be forced to be undefined. By allowing negated definedness formulas in the conclusions of the axioms, it is possible to specify the undefinedness of terms. For example, by adding the three axioms

$$\neg D(label(empty)), \qquad \neg D(left(empty)), \qquad \neg D(right(empty))$$

to the specification *TREENAT* one obtains a monomorphic specification of binary trees.

The mathematical theory of partial algebras is described in [41]. The initial algebra approach to partial algebraic specifications is developed in [151]. There the axioms of specifications have the form of Horn clauses built from existential equations alone. The approach as presented above has been studied e.g. in [38, 181, 36]. A necessary and sufficient characterization for the existence of initial algebras is given in [2].

In [40] the notion of partial algebras is generalized to a framework where the strictness of the definedness predicate is not required. This leads to an observational approach to specifications as developed in [92, 93] (cf. Section 6.5).

3.3.3. Continuous algebras

Another possibility for the description of nontotal functions, stimulated by the ideas of denotational semantics [167], is to use monotonic or continuous algebras. The carriers of such algebras are partially ordered sets rather than just sets, and the operators are required to be monotonic (in monotonic algebras) or continuous (in continuous algebras). Moreover, in continuous algebras the carriers are required to be *complete* partial orderings. The axioms which are appropriate for specifying classes of monotonic or continuous alegebras are built from inequations of the form $t \leqslant t'$; an equation can be defined by a conjunction of two such inequations:

$$t = t' \;\Leftrightarrow\; t \leqslant t' \wedge t' \leqslant t.$$

Semantically, the relation symbol "$\leqslant$" repesents the partial ordering of the algebra, i.e. it is reflexive, transitive, antisymmetric and compatible with the (interpretations of the) function symbols. The specification of classes of monotonic algebras is similar to error algebras. In many examples, every sort *s* contains a constant $\perp_s$ which is interpreted as the least element of the carrier set associated with *s*. This can be axiomatized by the inequation $\forall x{:}s.\ \perp_s \leqslant x$.

Then often all other axioms are equations. For example, one may give an initial algebra specification of ordered Boolean values with least element as follows:

spec $BOOL_\perp \equiv$
  sort *bool*
  functions *true*, *false*, $\perp : \to bool$,
  $not : bool \to bool$

axioms $\perp \leqslant b$,
$not(\perp)=\perp$,
$not(true)=false$,
$not(false)=true$

endspec

The carrier set of any initial algebra I of $BOOL_{\perp}$ has the form

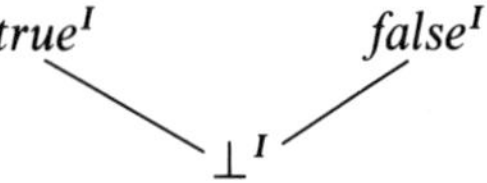

Any terminal algebra is trivial having a one-point carrier set.

A monotonic specification of trees of natural numbers is similar to $TREENAT_e$ in Example 3.3.2. The main difference is that the ok-predicates as defined in Subsection 3.3.1 are not monotonic and therefore have to be replaced by a monotonic specification (i.e., $ok_{tree}(\perp_{tree})=\perp_{tree}$ instead of $ok_{tree}(\perp_{tree})=false$). Thus specifications of partial functions or equivalently of strict functions (i.e., functions f with $f(\ldots,\perp,\ldots)=\perp$) where quantification has to distinguish between "$\perp$" and the "non-$\perp$" elements are often not very elegant. Monotonic and continuous specifications are better suited for describing so-called nonflat domains. For example, consider the following specification $NAT_{\leqslant}$:

spec $NAT_{\leqslant} \equiv$
sort *nat*
functions $zero, \perp: \rightarrow nat$,
$succ: nat \rightarrow nat$
axioms $\perp \leqslant x$,
$zero \leqslant succ(zero)$
endspec

If $NAT_{\leqslant}$ is considered as a monotonic specification, then the order structure of its initial algebra has the form

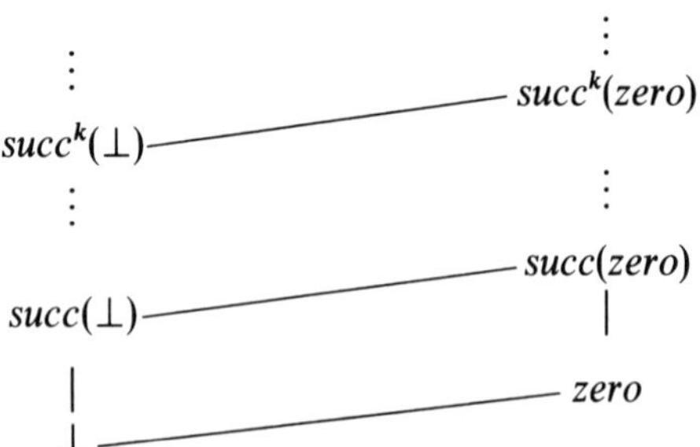

In the case of continuous specifications, any initial algebra contains two elements more, the least upper bounds of the chains $\{succ^k(\perp) \mid k \in \mathbb{N}\}$ and $\{succ^k(zero) \mid k \in \mathbb{N}\}$

where, because of continuity requirements, the former is smaller than the latter. These least upper bounds cannot be denoted by finite terms from $T(\Sigma)$. There are three approaches to deal with this problem: Möller [129] considers *inductively generated* models, i.e., algebras which are isomorphic to the ideal completion of a quotient of the term algebra. Any element of an inductively generated algebra is the least upper bound of a chain (of interpretations of) terms in $T(\Sigma)$. Thus such algebras correspond to algebraic cpo's. In [170] continuous specifications may contain an infinitary operator denoting the least upper bound of a chain of elements. As a consequence, models of specifications are reachable algebras as usual (but induction on the structure of terms involves transfinite methods). A third possibility is to consider reachable algebras which are ordered-embedded in continuous algebras. This technique is used by Möller [130] for the specification of higher-order functions.

The algebraization of continuous algebras has been initiated by M. Nivat and his school (cf. e.g. [133, 86]). The initial algebra approach to continuous algebras is discussed in [81], an interesting application to the semantics of programming languages can be found in [174]. The loose approach to monotonic and continuous algebraic specification is developed by in [129].

3.3.4. *Order-sorted algebras*

Order-sorted algebras improve the expressivity of many-sorted algebras by adding a quasi-ordering relation "$<$" on sorts, not on elements as in Subsection 3.3.3. The standard interpretation of "$<$" is the subset relation: for any two sorts s, s' of a signature Σ, $s<s'$ holds in a Σ-algebra A if $A_s \subseteq A_{s'}$. Due to this ordering on sorts, function symbols may be *polymorphic*: the same function symbol may have two or more functionalities, but it has only one denotation [166].

In particular, an *order-sorted signature* Σ consists of a set S of sorts, an S-sorted family F of function symbols and a set O of subset declarations.

Then an *order-sorted Σ-algebra* consists of interpretations A_s and f^A of the sorts and function symbols of Σ such that, for any $s, s' \in S$,

(1) A_s is a set;

(2) if $s<s' \in O$, then $A_s \subseteq A_{s'}$,

(3) $f^A: D_f^A \to \bigcup\{A_s \mid s \in S\}$ is a total function from the domain D_f^A of f into the union of the carriers of A such that

$$\text{if } (f: s_1, \ldots, s_n \to s) \in F \text{ and } a_i \in A_{s_i} \text{ for } i=1, \ldots, n,$$
$$\text{then } (a_1, \ldots, a_n) \in D_f^A \text{ and } f^A(a_1, \ldots, a_n) \in A_s.$$

From this definition it follows that a function symbol f refers to exactly one semantic object (i.e. no overloading), even if different types are associated with f.

An *order-sorted Σ-homomorphism* $\varphi: A \to B$ is defined as total homomorphism for partial algebras: the homomorphism condition $\varphi(f^A(a_1, \ldots, a_n)) = f^B(\varphi(a_1), \ldots, \varphi(a_n))$ must hold only for $(a_1, \ldots, a_n) \in D_f^A$; moreover, definedness is preserved, i.e. $\varphi(D_f^A) \subseteq D_f^B$.

With these definitions the initial algebra approach extends to the order-sorted case: the particular, initial algebras exist for all equational specifications with sensible signatures; a sound and complete proof system can be given (cf. e.g. [166]).

In the order-sorted approach, trees of natural numbers can be defined as follows:

spec $TREENAT_0 \equiv$
extend $NAT1$ **by**
sorts *tree, empty_tree, nonempty_tree,*
empty_tree < *tree, nonempty_tree* < *tree*
functions *empty* $: \to$ *empty_tree,*
node : *tree, nat, tree* $\to$ *nonempty_tree,*
left, right : *nonempty_tree* $\to$ *tree,*
label : *nonempty_tree* $\to$ *nat*
axioms $left(node(t_1, n, t)) = t_1$,
$right(node(t_1, n, t_2)) = t_2$,
$label(node(t_1, n, t_2)) = n$
endspec

The sorts *empty_tree* and *nonempty_*tree are declared as subsorts of tree. The important point of the example is that with the subsort *nonempty_tree* the correct domains of the selectors *left, right* and *label* can be specified. The initial models of $TREENAT_0$ exactly describe the isomorphism class of finite binary trees with natural numbers as labels.

Therefore, order-sorted specification introduces a degree of partiality in the framework of many-sorted total algebras: in order-sorted specifications, those partial functions can be specified whose domain coincides with (the carriers of) a declared subsort. However, the main feature of order-sorted specifications is their correspondence with object-oriented programming: e.g., the notions of inheritance and of polymorphism in the object-oriented language Eiffel [126] are almost exact counterparts to subsorting and polymorphism in the order-sorted approach (cf. also [79]). An algebraic specification language including subsorting is OBJ2 [63].

Order-sorted predicate logic has been introduced by [134] where it is argued that order-sorted logic is a natural logical language for expressing mathematics. In the field of algebraic specifications, order-sorted algebra originated with Goguen [72]. Gogolla [69] improved and simplified the approach of Goguen [72] and studied methods for error handling with subsorts. The approach as presented above is developed in [166]. This includes order-sorted rewriting and unification as well as hierarchical specifications. In [78] order-sorted logic is extended to include axioms of the form $at_1 \wedge \cdots \wedge at_n \Rightarrow at_{n+1}$ where the atomic formulas as equations or arbitrary predicates. Smolka [165] extended this approach further by a notion of polymorphism which combines parametric polymorphism with the order-sorted polymorphism as defined above.

3.3.5. *Nondeterminism and process algebras*

There are two different ways for describing nondeterministic systems by algebraic specifications: either a nondeterministic function $f: s_1 \to s_2$ is associated to its characteristic Boolean-valued function $f': s_1, s_2 \to bool$ or it is treated directly by introducing nondeterminism in the semantics. The former approach is dealt with by means of partial or total algebra semantics [37, 168]. For the latter Lescanne [115] converted nondeterministic functions into functions whose arguments and results are sets of values; Hesselink [94], Nipkow [132] and Hußmann [100] used multi-algebras as semantics whereas the algebras in [131] form distributive lattices. Then Hußmann [100] and Mosses [131] specified data types using inclusion relations as basic atomic formulas (instead of equations). This provides another approach to order-sortedness [131] but leads to nonconfluent term rewriting systems. In [100] also the soundness and completeness of term rewriting w.r.t. nondeterministic semantics is studied.

Concurrent systems are inherently nondeterministic, so their specification involves all the issues mentioned above. Similar to the approach with Boolean-valued characteristic functions, Broy and Wirsing [39], and Astesiano et al. [4] specified the transition relation of concurrent systems using partial algebra semantics. Equational specifications are given for CCS [128] and ACP [18, 19] but with a different aim: for CCS based on a transition system semantics, equational theorems are derived, whereas for ACP the equational axioms are given and then different semantic models are analyzed. Kaplan and Pnueli [107] combined the ACP-approach with hierarchical specification and implementation (cf. Sections 5.4 and 8.2).

In [56] projection spaces are proposed as models for LOTOS-like specifications ([101], cf. also Section 9.5). In [32, 33] streams and action structures as semantic basis for specifying distributed and concurrent systems are discussed, whereas Kröger [113] used structures and formulas based on temporal logic.

3.3.6. *Institutions*

The preceding sections show that there is a variety of different approaches to the basic concepts of algebraic specifications. Signatures may be many-sorted, order-sorted, polymorphic or higher-order; several notions of algebra ranging from total and partial algebras to order-sorted ones have been presented; the axioms of a specification may be equations, inequations, Horn clauses, first-order formulas etc. There is no reason to view any of these logical systems as superior to the other. The choice must depend on the particular area of application, the personal taste and the available tools. The notion of an *institution* provides a tool for unifying the different approaches to specification by formalizing the concept of a logical system for writing specifications. Any property which can be proven for arbitrary institutions, holds for all approaches to algebraic specification; specification languages which are defined independently of a particular institution can be universally used for building structured specifications in any algebraic framework.

An institution consists of a collection of *signatures* together with, for any signature Σ, a set of *Σ-sentences*, a collection of *Σ-models* and a *satisfaction relation* between Σ-sentences and Σ-models. The signatures are arbitrary abstract objects, not necessarily the many-sorted signatures used in the previous sections. The only semantic requirement is that when the signatures are changed, the induced translation on sentences and models preserve the satisfaction relation. Formally, an *institution INS* consists of

- a category *Sig* (of signatures) (e.g., *Sign*, just after Example 2.2.1 is such a category),
- a functor $Sen: Sig \to Set$ (where *Set* is the category of all sets with mappings as morphisms); $Sen(\Sigma)$ gives, for any signature Σ, the set of all Σ-sentences and for any signature morphism $\sigma: \Sigma \to \Sigma'$ the function $Sen(\sigma): Sen(\Sigma) \to Sen(\Sigma')$ translating Σ-sentences to Σ'-sentences;

- a functor $Mod: Sig \to Cat^{op}$ (where Cat is the category of all categories with functors as morphisms and Cat^{op} is Cat with its morphisms reversed); $Mod(\Sigma)$ gives for any signature Σ the category of Σ-models and for any signature morphism $\sigma: \Sigma \to \Sigma'$ the σ-reduct functor $Mod(\sigma): Mod(\Sigma') \to Mod(\Sigma)$ translating in the reverse direction Σ'-models to Σ-models;
- a satisfaction relation $\models \subseteq |Mod(\Sigma)| \times Sen(\Sigma)$ for each signature Σ (where $|Mod(\Sigma)|$ denotes the objects of $Mod(\Sigma)$) such that for any signature morphism $\sigma: \Sigma \to \Sigma'$ the translations $Mod(\sigma)$ of models and $Sen(\sigma)$ of sentences preserve the satisfaction relation; i.e., for any $G \in Sen(\Sigma)$ and $M' \in |Mod(\Sigma')|$,

$$M' \models Sen(\sigma)(G) \quad \text{iff} \quad Mod(\sigma)(M') \models G.$$

Note that Cat denotes the category of "large" categories; for a discussion of the foundations of this category see e.g. [118].

As an example consider the institution GEQ of ground equational specifications with Σ-algebras, not only Σ-computation structures, as models:

- Sig is the category $Sign$ of signatures with signatures as objects and signature morphisms as morphisms (see Section 2.1).
- For any signature Σ, $Sen(\Sigma)$ is the set of all ground Σ-equations; for any signature morphism $\sigma: \Sigma \to \Sigma'$, $Sen(\sigma)$ maps any ground equation $t=t'$ to the ground Σ'-equation $\sigma^*(t)=\sigma^*(t')$ where σ^* denotes the extension of σ to terms.
- For any signature Σ, $Mod(\Sigma)$ is the category $Alg(\Sigma)$ with Σ-algebras as objects and Σ-homomorphisms as morphisms; for any signature morphism $\sigma: \Sigma \to \Sigma'$, $Mod(\sigma)$ is the σ-reduct functor $\cdot|_\sigma: Alg(\Sigma') \to Alg(\Sigma)$ (see Section 2.2).
- For any signature Σ, $\models_\Sigma$ is the satisfaction relation as defined in Section 3.1.

The notion of institution was introduced by Goguen and Burstall [73] under the name of "language". It is based on to the work of Barwise [7] on abstract model theory which intends to generalize basic results in model theory. In [75], an extensive study of the categorical theory of institutions is studied and applications to specification programming are discussed.

CLEAR [43] is the first specification language basing its semantics on institutions. In [157] a smooth institution-independent semantics for the specification language ASL is developed and put in a general setting. In these papers institutions satisfy some additional requirements which allow to define operators for constructing specifications: e.g., in [43] it is assumed that institutions are *liberal*; i.e., forgetful functors induced by theory morphisms must have left-adjoints; Sannella and Tarlecki [157] require the category Sig to be finitely co-complete and that Mod preserves pushouts and initial objects.

4. Simple specifications

In this section, the model classes of simple specifications of the form $\langle \Sigma, E \rangle$ are analyzed (Section 4.1) and the expressive power of such specifications is studied (Section 4.2). In particular, criteria for the existence of initial and terminal models are given, and it is shown under which conditions the set of isomorphism classes of models forms a complete lattice. For technical convenience, the congruence semantics for specifications is used for this analysis, since the congruences associated with the models of a specification form a set which is naturally ordered by the inclusion relation.

By expressive power we mean the complexity of initial and terminal algebras. It will be shown that initial algebras of conditional equational specifications are semi-computable, whereas terminal algebras are co-semicomputable. Hence, the models of monomorphic specifications are computable. On the other hand, it will be argued that not all computable Σ-algebras admit monomorphic simple specifications of the form $\langle \Sigma, E \rangle$ with a finite set E. The question of the (subrecursive) computational complexity of specifications is not addressed in this section, it is studied e.g. in [5].

4.1. *Lattice and existence theorems*

Let $SP = \langle \Sigma, E \rangle$ be a conditional equational specification with signature $\Sigma = \langle S, F \rangle$. The abstract data types of a specification can be partially ordered according to the existence of homomorphisms between two representatives or equivalently by the set inclusion relation of the associated congruences (cf. Fact 2.3.2(3)).

4.1.1. THEOREM. *Let $SP = \langle \Sigma, E \rangle$ be a conditional equational specification with a sensible signature $\Sigma = \langle S, F \rangle$. Then the set $C(SP)$ of congruences (associated with the models of) SP is a complete sublattice of $C(\Sigma)$. The congruence associated with the initial models of SP is the least element and the unit congruence is the greatest element of $C(SP)$.*

PROOF. As in the proof of Proposition 2.3.5, first note that the intersection of any set of elements of $C(SP)$ belongs to $C(SP)$. In particular, the congruence $\sim^I$ (defined by $t \sim^I t'$ iff $E \vdash t = t'$, for all $t, t' \in T(\Sigma)$) belongs to $C(SP)$ and is initial in $C(SP)$. The unit congruence $\sim^{U(\Sigma)}$ belongs trivially to $C(SP)$; it is the greatest element of $C(SP)$. Thus according to Lemma 2.3.4, $C(SP)$ forms a complete lattice. Note, however, that the least upper bounds in $C(SP)$ need not coincide with those in $C(\Sigma)$. □

(Infinitary) conditional equations together with simple inequations of the form $u \neq v$ are the most general classes of Σ-axioms whose associated model classes are closed under intersection. In order to show that axiom sets containing true disjunctions do not admit initial models, the notion of maximality (cf. [148]) is introduced. A Σ-formula G is called *maximal* in SP if

- either G is valid in SP (i.e., $Mod(SP) \models G$),
- or $\neg G$ is valid in SP (i.e., $Mod(SP) \models \neg G$).

SP is called *maximal w.r.t a sort* $s \in S$, if all ground equations $u = v$ with $u, v \in T(\Sigma)_s$ are maximal. Hence, a monomorphic specification is maximal w.r.t. all sorts.

4.1.2. FACT. (1) *If a specification SP contains a ground axiom of the form $u_1 = v_1 \vee u_2 = v_2$, where neither equation is maximal, then SP does not admit an initial model.*

(2) *Let SP be a specification containing a ground axiom of the form $u \neq v$ with u,*

$v \in T(\Sigma)_s$ for some $s \in S$. If there exists a ground term $t \in T(\Sigma)_s$ such that neither equation $u=t$ nor $v=t$ is maximal in SP, then SP does not admit a terminal model.

PROOF. This follows from the definition of maximality and from the characterization of initial and terminal Σ-computation structures in Lemma 3.2.2. □

In order to be able to obtain nontrivial terminal algebras it is assumed in the remainder of this section that *SP* contains the specification *BOOLM* (Example 3.2.4(1)) as a subspecification, i.e., *SP* contains the axiom *true* $\neq$ *false*. It is not assumed that *SP* is hierarchical w.r.t. *BOOLM* (for this see Section 5.4).

4.1.3. LEMMA. *Let $SP = \langle \Sigma, E\rangle$ be a specification with inequations and conditional equations as axioms; moreover let SP be maximal w.r.t. the sorts of the inequations and the sorts of all premises of conditional equations. Then there exists an equivalent specification $SP' = \langle \Sigma, E'\rangle$ the axioms of which consist of ground equations and ground inequations, such that SP' is maximal w.r.t. the sorts of the inequations.*

PROOF. According to Fact 3.1.6 there exists a set $E'' =_{\text{def}} E_0 \cup E_1$ of ground inequations E_0 and ground conditional equations E_1 such that *SP* and $\langle \Sigma, E''\rangle$ are equivalent. Consider the specification $SP' =_{\text{def}} \langle \Sigma, E'\rangle$ with

$$\begin{aligned} E' =_{\text{def}} & E_0 \\ & \cup \{u=v \mid u, v \in T(\Sigma)_s,\ E \vdash u=v \text{ and } s \text{ is a maximal sort}\} \\ & \cup \{u \neq v \mid u, v \in T(\Sigma)_s,\ E \vdash u \neq v \text{ and } s \text{ is a maximal sort}\} \\ & \cup \{u=v \mid (u_1=v_1 \wedge \ldots \wedge u_n=v_n \Rightarrow u=v) \in E_1 \text{ and } E \vdash u_i=v_i \\ & \qquad \text{for all } i=1,\ldots,n\}. \end{aligned}$$

Obviously, $E \vdash G$ for all $G \in E'$.

On the other hand, let $G =_{\text{def}} (u_1=v_1 \wedge \ldots \wedge u_n=v_n \Rightarrow u=v) \in E_1$. Since G is a ground instance of some axiom of E, each premise $u_i=v_i$ of G, $i=1,\ldots,n$, is maximal in E and therefore by construction either $u_i=v_i$ is in E' or $u_i \neq v_i$ is in E'. We distinguish two cases: first, if for all $i=1,\ldots,n$, $u_i=v_i$ are in E', then, by definition of E', the conclusion $u=v$ is in E' as well; thus $E' \vdash G$. Second, if for some $i_0 \in \{1,\ldots,n\}$, $u_{i_0} \neq v_{i_0} \in E'$, then the negation of the premise $u_{i_0}=v_{i_0}$ is derivable in E' and thus $E' \vdash G$. Hence for all $G \in E_1$, in both cases $E' \vdash G$ holds.

Together with $E_0 \subseteq E'$, this implies that $E' \vdash G$ for all $G \in E''$ and hence $E' \vdash G$ for all $G \in E$ because of the equivalence of *SP* and $\langle \Sigma, E''\rangle$. □

4.1.4. THEOREM. *Let $SP = \langle \Sigma, E\rangle$ be a satisfiable specification such that $E =_{\text{def}} E_0 \cup E_1$ is the union of a set E_0 of inequations and a set E_1 of equations, and, moreover, SP is maximal w.r.t. the sorts of the inequations of E_0. Then the set $C(SP)$ forms a complete lattice. The initial congruence of SP is the smallest and the terminal congruence of SP is the greatest element of $C(SP)$.*

PROOF. It is easy to see that $C(SP)$ forms a complete lower semilattice with the initial abstract data type as smallest element. (Define the intersection of any set of congruence in $C(SP)$ as in the proof of Proposition 2.3.5. Obviously, this intersection preserves the validity of equations and inequations; therefore it is in $C(SP)$). According to Lemma 2.3.4, it remains to prove that $C(SP)$ admits a greatest element $\sim^Z$. Define for all $t, t' \in T(\Sigma)_s$, $s \in S$,

$$t \sim^Z t' \text{ iff there exists } \sim_1, \ldots, \sim_{n-1} \in C(SP) \text{ and } t_0, \ldots, t_n \in T(\Sigma)_s \text{ such that } t_i \sim_i t_{i+1} \text{ for } i=0, \ldots, n-1,\ t \equiv t_0 \text{ and } t' \equiv t_n.$$

Obviously, $\sim^Z$ is a Σ-congruence. We prove that $\sim^Z$ is the terminal congruence of $SP' = \langle \Sigma, E' \rangle$ where E' is the set of ground instances of the axioms E. Since SP' is equivalent to SP (see Fact 3.1.6), $\sim^Z$ is the terminal congruence of SP.

***Case* 1:** *Z is a model of SP. Proof*: Let $e \in E'$ be a axiom of SP'. If $e \equiv (u = v)$ is an equation, then for all $\sim \in C(SP) = C(SP')$, $u \sim v$ holds and therefore $u \sim^Z v$.

Otherwise, e is an inequation $(u \neq v)$ of a maximal sort s. In order to obtain a contradiction suppose that $u \sim^Z v$ holds. Then there would exist $\sim_1, \ldots, \sim_{n-1} \in C(SP)$ and $t_0, \ldots, t_n \in T(\Sigma)_s$ such that $t_0 \equiv u$, $t_n \equiv v$ and $t_i \sim_i t_{i+1}$ for each $i = 0, \ldots, n-1$. Since SP is maximal w.r.t. s, $SP \models t_i = t_{i+1}$ for $i = 0, \ldots, n-1$, and therefore $SP \models u = v$ would hold, which is a contradiction to the satisfiability of SP. Hence $u \not\sim^Z v$.

***Case* 2:** *Z is terminal in SP*. By definition, $\sim \subseteq \sim^Z$ holds for each congruence $\sim \in C(SP)$. □

4.1.5. COROLLARY. *Let $SP = \langle \Sigma, E \rangle$ be a specification containing BOOLM such that $E \backslash \{true \neq false\}$ is equational and SP is maximal w.r.t. bool. Then the set $C(SP)$ forms a complete lattice w.r.t. set inclusion. The initial congruence of SP is the smallest and the terminal congruence of SP is the greatest element of $C(SP)$.*

4.1.6. COROLLARY. *Let $SP = \langle \Sigma, E \rangle$ be a satisfiable specification with inequations and conditional equations as axioms such that SP is maximal w.r.t. the sorts of the inequations and of the premises of the conditional equations of E. Then the set $C(SP)$ forms a complete lattice w.r.t. set inclusion. The initial congruence of SP is the smallest and the terminal abstract data type of SP is the greatest element of $C(\Sigma)$.*

PROOF. According to Lemma 4.1.3, there exists an equivalent specification SP' with ground equations and ground inequations as axioms such that SP' is maximal w.r.t. all sorts of the inequations. Apply Theorem 4.1.4 to SP'. □

4.2. Expressive power of simple specifications

The complexity of specification methods can be measured by the expressive power of the specified algebras. In the following it will be shown, that the initial algebras of (first-order) specifications are semicomputable while the terminal algebras are co-semi-computable; as a consequence, all models of monomorphic (conditional) equational specifications are computable. On the other hand, finite simple equational Σ-specifications describe not all computable Σ-algebras: there exists a computable Σ-computation structure which is *not* the initial algebra of any simple equational Σ-specification with a finite number of axioms. Hence, in order to get complete specification methods it is necessary to introduce further features, the so-called hidden sorts and hidden functions which will be studied in Section 5. Most of the results of this section can be found in [23, 24].

In the following, all signatures are assumed to be finite; i.e., the set of sorts and the set of all function symbols of any signature is assumed to be finite.

4.2.1. THEOREM. *Let $SP=\langle\Sigma, E\rangle$ be a satisfiable specification, and E be recursively enumerable set of first-order axioms.*

(1) *If the initial abstract type $I(SP)$ of SP exists, then $I(SP)$ is semicomputable.*

(2) *If the terminal abstract type $Z(SP)$ of SP exists, then $Z(SP)$ is co-semicomputable.*

(3) *If SP is monomorphic, then all models of SP are computable.*

PROOF. According to Corollary 3.2.3, the initial congruence $\sim^I$ and the *complement* of the terminal congruence $\sim^Z$ of SP can be specified as follows:

$$t \sim^I t' \Leftrightarrow E \vdash t=t' \quad \text{and} \quad \neg(t \sim^Z t') \Leftrightarrow E \vdash \neg(t=t'),$$

for all ground terms $t, t' \in T(\Sigma)_s$, $s \in S$. The recursive enumerability of the provability relation $\vdash$ implies assertions (1) and (2) of the theorem. If SP is monomorphic, then $\sim^I$ and $\sim^Z$ coincide; i.e., the unique congruence associated with the models and its complement are recursively enumerable. Hence, all models are computable. □

Also the converse of Theorem 4.2.1 holds.

4.2.2. FACT. *Let A be a Σ-computation structure.*

(1) *If A is semicomputable, then it has an initial algebra specification $\langle\Sigma, E_I\rangle$ where E_I is a recursively enumerable set of ground equations.*

(2) *If A is co-semicomputable, then it has a terminal algebra specification $\langle\Sigma, E_Z\rangle$ where E_Z is a recursively enumerable set of ground inequations.*

(3) *If A is computable, then it has a monomorphic specification $\langle\Sigma, E_M\rangle$ where E_M is a recursive set of ground equations and ground inequations.*

PROOF. Define

$$E_I =_{\text{def}} \{t=t' \mid A \models t=t',\ t,\ t' \in T(\Sigma)\},$$
$$E_Z =_{\text{def}} \{\neg(t=t') \mid A \models t \neq t', t,\ t' \in T(\Sigma)_s,\ s \in S\},$$
$$E_M =_{\text{def}} E_I \cup E_Z. \quad \square$$

Therefore, the different specification methods have different expressive power. Initial and terminal algebra specifications are mutually incomparable, but are both more expressive than monomorphic specifications. This is also shown by the following two examples (which have a finite number of axioms).

4.2.3. EXAMPLE. (1) *Combinatory logic*: Combinatory logic can be axiomatized by three equations over a signature consisting of three constants K, S, I and a binary application function (see [23], for details see [6]).

spec $CL \equiv$
  sort cl
  functions $K, S, I: \to cl,$
    $.\bullet.: cl, cl \to cl$
  axioms $(K \bullet x) \bullet y = x,$
    $((S \bullet x) \bullet y) \bullet z = (x \bullet z) \bullet (y \bullet z),$
    $(I \bullet x) = x$
endspec

The initial algebra of this specification is known to be the "term model for combinatory logic" which is semicomputable and has, obviously, an equational initial algebra specification. But it is not computable because combinatory logic is a formalism strong enough to define all recursive functions.

(2) *Primitive recursive functions*: A specification of the set of all primitive recursive functions can be given by enriching the monomorphic specification $NAT1$ of natural numbers by a sort *fun* for the (primitive recursive) functionals. The semantics of functionals is defined by an *eval*-operator which applies an n-ary functional to an n-tuple of natural numbers (see [179]).

spec $PREC \equiv$
  extend $NAT1$ **by**
    sort *fun*
    functions
      $zero'$ $: \to fun$ "zero functional"
      $succ'$ $: \to fun,$ "successor functional"
      p_i^n $: \to fun,\ 1 \leq i \leq n,$ "ith projection"
      $comp^n$ $: fun^{(m+1)} \to fun,\ 1 \leq m,\ 0 \leq n$ "composition"
      R $: fun, fun \to fun,$ "primitive recursion"
      $eval$ $: fun, (nat)^n \to nat,\ n \geq 0$
    axioms
      $\forall\, nat\ x,\ nat^n\ d,\ fun\ g, g_1, \ldots g_m, f, h:$
      $eval(zero') = zero \wedge$
      $eval(succ', x) = succ(x) \wedge$
      $eval(p_i^n, d) = d_i \wedge$
      $eval(comp^n(g, g_1, \ldots g_m), d) = eval(g, eval(g_1, d), \ldots eval(g_m, d)) \wedge$
      $eval(R(f, h), d, zero) = eval(f, d) \wedge$
      $eval(R(f, h), d, succ(x)) = eval(h, eval(R(f, h), d, x), d, x)$
endspec

Notice, that this specification is infinite (more precisely, its sets of function symbols and of axioms are recursive) but due to its schematic nature it can easily be changed into a finite one (cf. [21]). In *PREC*, no (nontrivial) equations are required between terms of sort *fun*. Hence, the carrier set of sort *fun* of any initial model is isomorphic to the set of terms $T(\Sigma_{PREC})_{fun}$. For the terminal models Z, the equality in *fun* is characterized by

$$Z \models f = g \text{ iff } PREC \models \forall d{:}nat^n.eval(f, d) = eval(g, d);$$

i.e., two functionals are equal in Z if they are extensionally equivalent. As is well-known, the extensional equality between primitive recursive functionals is co-recursively enumerable but not recursively enumerable and therefore Z is not computable. According to Theorem 4.2.1(1), any initial algebra of a simple specification is semi-computable. Thus neither a finite nor a recursively enumerable simple initial algebra specification of Z exists.

If—as usual—only a finite set of axioms is allowed, then simple equational specifications are not complete for the specification of computable algebras. Majster [121] gave the first example of a computable computation structure which does not have a simple equational initial algebra specification. Other examples are given in [122, 172, 171]. The following example and theorem stem from [24].

4.2.4. THEOREM. *Let Σ_{NAT2} be Σ_{NATO} enriched by a unary function symbol $.**2: nat \to nat$. The Σ_{NAT2}-computation structure $N2$ of standard natural numbers enriched by the usual squaring function does not possess an initial algebra specification $\langle \Sigma, E \rangle$, where E is a finite set of equations. More exactly, $N2$ is defined as follows: $N2|_{\Sigma_{NATO}} =_{\mathrm{def}} N$ (cf. Example 2.2.1), $(n**2)^{N2} =_{\mathrm{def}} n*n$ for all $n \in \mathbb{N}$.*

The proof is given in [24].

Conditional equational specifications are more powerful than equational specifications. Thatcher et al. [171] gave the first example for a computation structure which can be specified as initial algebra of a (finite) conditional equational specification but which does not possess a finite equational specification. In [20] the same property is shown for a computation structure *SOI* of "sets of integers". *SOI* is similar to *P*0 of Example 2.2.2 but instead of containing Boolean functions *eq* and ε it computes the cardinality function $card: set \to nat$.

In [24] there is also an example of a computable computation structure which does not posses an initial algebra specification $\langle \Sigma, E \rangle$ where E is a finite set of *conditional equations.*

5. Specifications with hidden symbols and constructors

In this section the specification method will be extended by two additional features, constructors and hidden parts of the signature.

Constructors are introduced because of methodological reasons: they allow us to determine normal forms for terms and classify the valid structural induction rules for a specification. Simple specifications are examples of specifications with constructors where all function symbols are constructors. Another important example are hierarchical specifications which will be discussed in Section 5.4. Hidden sorts and hidden functions are local definitions in specifications. They facilitate the writing of axioms and, moreover, are introduced to overcome the lack of expressive power of simple specifications (cf. Section 4.2).

Syntax and semantics of specifications with hidden constructors are presented in Section 5.1. In Section 5.2 the structure of the model classes of such specifications is studied. The key property is the notion of *sufficient completeness*. If a specification *SP* with constructors and hidden symbols satisfies this condition, then (under similar assumptions as for Corollary 4.1.6) its class of models forms a complete lattice. The expressive power (Section 5.3) of sufficiently complete specifications is also similar to the one of simple specifications. The only difference is that due to the availability of hidden symbols, all computable data structures are specifiable by finite monomorphic equational specifications (Theorem 5.3.6). In contrast to this, the expressive power of nonsufficiently complete specifications is much greater: the set of monomorphic equational specifications with constructors and hidden symbols is complete hyper-arithmetical (Corollary 5.3.11).

5.1. *Syntax and semantics*

Formally, a *specification SP with hidden symbols and constructors* consists of a quadruple $\langle \Sigma, H\Sigma, E, Cons \rangle$ where

- $\Sigma = \langle S, F \rangle$ is the *signature* of *SP*, also called *export* or *visible signature*;
- $H\Sigma = \langle HS, HF \rangle$, *the extended signature* of *SP*, is a signature containing Σ (i.e., $S \subseteq HS$ and $F \subseteq HF$); the elements of $HS \backslash S$ are called *hidden sorts* and those of $HF \backslash F$ are called *hidden function symbols*;
- $E \subseteq WFF(H\Sigma)$ is the set of *axioms* of *SP*;
- $Cons \subseteq HF$, the *constructor* set of *SP*, is a set of function symbols such that the constructors of the visible sorts are visible function symbols, i.e. $\bigcup\{Cons_s \mid s \in S\} \subseteq F$. The *constructor signature* $\langle HS, Cons \rangle$ is denoted by $H\Sigma_{Cons}$.

A special case of this definition occurs if all function symbols are constructors (i.e., $Cons = HF$) and $H\Sigma$ is different from Σ; then *SP* is called *specification with hidden symbols*. If there are no hidden symbols (i.e., $\Sigma = H\Sigma$), then *SP* is called *specification with constructors*. Any simple specification $\langle \Sigma, E \rangle$ can be seen as specification $\langle \Sigma, \Sigma, E, F \rangle$. In the following, unless explicitly stated, all specifications are considered as specifications with constructors and hidden symbols. A specification *SP* is said to have a *sensible signature* if each sort $s \in HS$ contains at least one ground constructor term, i.e. if for each $s \in HS$, $T(\langle HS, Cons \rangle)_s \neq \emptyset$.

For the definition of the semantics of *SP*, the notion of reachability has to be slightly generalized: an $H\Sigma$-algebra A is called *reachable w.r.t. Cons* if for each sort $s \in HS$ and each carrier element $a \in A_s$ there exists a ground constructor term $t \in T(\langle HS, Cons \rangle)_s$ with $a = t^A$. The class of all $H\Sigma$-algebras which are reachable w.r.t. *Cons* will be denoted by $Gen_{Cons}(H\Sigma)$ and the subclass of these algebras which are also models of $\langle H\Sigma, E \rangle$ will be denoted by $Gen_{Cons}(\langle H\Sigma, E \rangle)$. Thus any $H\Sigma$-algebra B in this class is isomorphic to the following extension $T(B)$ of a quotient of $T(H\Sigma_{Cons})$:

$$T(B)_s =_{\text{def}} (T(\langle HS, Cons \rangle)/\sim_B)_s \qquad \text{for each } s \in HS,$$
$$f^{T(B)}([x_1], \ldots, [x_k]) =_{\text{def}} [f^B(x_1, \ldots, x_k)] \quad \text{for each } f: s_1, \ldots, s_k \to s \in HF$$

where for any ground term $t \in T(\langle HS, Cons \rangle)$, $[t]$ denotes the congruence class of t w.r.t. $\sim_B$ in $T(B)$.

The semantics of $SP = \langle \Sigma, H\Sigma, E, Cons \rangle$ is given by all Σ-computation structures which are Σ-restrictions of those models of $\langle H\Sigma, E \rangle$ whose carrier sets are reachable w.r.t. *Cons*:

$$Mod(SP) =_{\text{def}} \{A \in Gen(\Sigma) \mid \text{there exists a } B \in Gen_{Cons}(\langle H\Sigma, E \rangle) \text{ s.t. } B|_\Sigma = A\};$$
$$I(SP) =_{\text{def}} \{A \in Mod(SP) \mid A \text{ is initial in } Mod(SP)\};$$
$$Z(SP) =_{\text{def}} \{A \in Mod(SP) \mid A \text{ is terminal in } Mod(SP)\}.$$

As in Section 3.2, if $Mod(SP)$ does not admit an initial (terminal respectively) model, then $I(SP)$ ($Z(SP)$ respectively) is empty. Let $\Sigma = \langle S, F \rangle$ and $H\Sigma = \langle HS, HF \rangle$. The definition of $Mod(SP)$ needs some comments.

First, if *Cons* contains hidden function symbols, then it may happen that for each $B \in Gen_{Cons}(\langle H\Sigma, E \rangle)$ the Σ-restriction $B|_\Sigma$ of B is not reachable w.r.t. the visible function symbols F, i.e. $B|_\Sigma \notin Gen(\Sigma)$, and therefore $Mod(SP)$ is empty. The syntactic condition $\bigcup \{Cons_s \mid s \in S\} \subseteq F$ excludes this unsuitable case and ensures that, for each $B \in Gen_{Cons}(\langle H\Sigma, E \rangle)$, $B|_\Sigma$ is a Σ-computation structure.

Second, it may happen, that $Gen_{Cons}(\langle H\Sigma, E \rangle)$ is empty even if $\langle H\Sigma, E \rangle$ has models; in this case, none of the models of $\langle H\Sigma, E \rangle$ is reachable w.r.t. *Cons*. Another particular case occurs when $Gen_{Cons}(\langle H\Sigma, E \rangle)$ consists of exactly one isomorphism class $\sim^{SP}$ whereas $\langle H\Sigma, E \rangle$ is not monomorphic; then the complexity of $\sim^{SP}$ may be complete hyperarithmetical (see below). Hence, it will be important to have a criterion which ensures the equivalence of $Gen(\langle H\Sigma, E \rangle)$ with $Gen_{Cons}(\langle H\Sigma, E \rangle)$.

In the following we use the notation

> **spec** $SP \equiv$
>   **export** Σ **from**
>   **sorts** HS
>   **constructors** *Cons*
>   **functions** $HF \backslash Cons$
>   **axioms** E
> **endspec**

to denote a specification $SP = \langle \Sigma, H\Sigma, E, Cons \rangle$. If it is obvious from the context, only the identifiers of the elements of Σ will be given in the export list.

As in Section 3.2, large specifications will be constructed using the operator "**extend.by.**". Let in the following expression $SP_1 = \langle \Sigma_1, H\Sigma_1, E_1, Cons_1 \rangle$, where $\Sigma_1 = \langle S_1, F_1 \rangle$ and $H\Sigma_1 = \langle HS_1, HF_1 \rangle$, be a specification which is extended by the set S of new sorts, the sets F and *Cons* of new function symbols and the set of axioms E. Thus it is required that the symbols occurring in S, F and *Cons* are disjoint from those in SP_1, i.e. $HS_1 \cap S = \emptyset$ and $HF_1 \cap (F \cup Cons) = \emptyset$, and that in *Cons*, F and E only visible symbols from SP_1 are used, i.e. $sorts(F) \cup sorts(Cons) \subseteq S_1 \cup S$, $sorts(E) \subseteq S_1 \cup S$ and $opns(E) \subseteq F_1 \cup F \cup Cons$. Then the expression

> **spec** $SP \equiv$
>   **export** Σ **from**
>   **extend** SP_1 **by**
>   **sorts** S
>   **constructors** *Cons*
>   **functions** F
>   **axioms** E
> **endspec**

denotes the specification

$$SP =_{\text{def}} \langle \Sigma, \langle S \cup HS_1, Cons \cup F \cup HF_1 \rangle, E \cup E_1, Cons \cup Cons_1 \rangle.$$

In order to guarantee the well-definedness of SP, we moreover assume that Σ forms a signature which is built from the visible part Σ_1 of SP_1 and the new symbols, i.e. $\Sigma \subseteq \langle S_1 \cup S, F_1 \cup Cons \cup F \rangle$, and that all constructors of visible sorts are visible, i.e. $\{c \in Cons_1 \cup Cons \mid range(c) \in sorts(\Sigma)\} \subseteq opns(\Sigma)$.

5.1.1. EXAMPLES. (1) *Truth-values*: The following specification $BOOL$ of truth values uses constructors but no hidden symbols; it is and monomorphic.

spec $BOOL \equiv$
  export *bool, true, false, not, and or* **from**
    sort *bool*
    constructors $true, false: \rightarrow bool$
    functions $not: bool \rightarrow bool$,
      $.and.,.or.: bool, bool \rightarrow bool$
    axioms $true \neq false$,
      $not(true) = false$,
      $not(false) = true$,
      $true\ and\ x = x$,
      $false\ and\ x = false$,
      $x\ or\ y = not(not(x)\ and\ not(y))$
endspec

(2) *Natural numbers*: The following specification NAT of natural numbers uses also constructors but no hidden symbols; it is monomorphic.

spec $NAT \equiv$
  export $sig(BOOL)$, *nat, zero, succ*, $+$, $*$, *eq* **from**
    extend $BOOL$ **by**
      sort *nat*
      constructors $zero: \rightarrow nat$,
        $succ: nat \rightarrow nat$
      functions $.+.,.*.: nat, nat \rightarrow nat$,
        $eq: nat, nat \rightarrow bool$
      axioms $zero + y = y$,
        $succ(x) + y = succ(x + y)$,
        $zero * y = zero$,
        $succ(x) * y = y + (x * y)$,
        $eq(zero, zero) = true$,
        $eq(zero, succ(y)) = false$,
        $eq(succ(x), zero) = false$,
        $eq(succ(x), succ(y)) = eq(x, y)$
endspec

NAT can be used to give a monomorphic specification $NAT2$ with hidden symbols of the computation structure $N2$ of Theorem 4.2.4 (which cannot be specified by simple specifications).

spec $NAT2 \equiv$
  export Σ_{NAT2} **from**
    extend NAT **by**
      function $.**2: nat \rightarrow nat$
      axioms $x**2 = x*x$
endspec

The models of $NAT2$ consist exactly of the isomorphism class of $N2$.

(3) *Finite sets with cardinality*: The following specification *C-SET* describes set-like structures which admit a cardinality operation (cf. Example 3.2.4(4)). The function symbols *if* and ε are used as auxiliary symbols for the definition of the cardinality operation.

spec *C-SET* $\equiv$
  export *nat, zero, succ, set empty, insert, card* **from**
    extend NAT **by**
      sort *set*
      constructors $empty: \rightarrow set$,
        $insert: nat, set \rightarrow set$
      functions $.\varepsilon.: nat, set \rightarrow bool$,
        $card: set \rightarrow nat$,
        $if: bool, nat, nat \rightarrow nat$
      axioms $(x\ \varepsilon\ empty) = false$,
        $(x\ \varepsilon\ insert(y, s)) = (eq(x, y)\ or\ x\ \varepsilon\ s)$,
        $card(empty) = zero$,
        $card(insert(x, s)) = if(x\ \varepsilon\ s, card(s), succ(card(s)))$,
        $if(true, x, y) = x$,
        $if(false, x, y) = y$
endspec

The specification *C-SET* is loose. The algebra *SOI* [20] which describes finite sets with a cardinality operation is a terminal algebra of *C-SET*. This algebra does not admit a (finite) equational initial algebra specification without hidden symbols. By deleting the inequation $true \neq false$ and by adding two axioms $(insert(x, insert(x, s)) = insert(x, s))$ and $(insert(x, insert(y, s)) = insert(y, insert(x, s)))$ to *C-SET*, one obtains an equational initial algebra specification of *SOI* with hidden symbols. Replacing *if* by a condition yields a conditional equational specification of *SOI* without hidden symbols [109].

5.2. *Lattice and existence theorems*

The following definition of sufficient completeness is a refinement of Guttag's original definition [87]. A specification *SP* is called *sufficiently complete* w.r.t. a set $F' \subseteq HF$ of function symbols if for each ground term $t \in T(H\Sigma)$ there exists a ground constructor term $c \in T(\langle HS, F' \rangle)$ such that $E \vdash t = c$; *SP* is called *sufficiently complete* if it is so w.r.t. the set *Cons* of constructors. All specifications in Example 5.1.1 are sufficiently complete.

5.2.1. FACT. *Let SP be a sufficiently complete specification.*

(1) $Gen(\langle H\Sigma, E\rangle) = Gen_{Cons}(\langle H\Sigma, E\rangle)$.

(2) *A Σ-congruence $\sim^I \in C(SP)$ is initial in $C(SP)$ if and only if for all ground Σ-terms $t,t' \in T(\Sigma)$ the following condition holds:*

$$t \sim^I t' \Leftrightarrow E \vdash (t = t').$$

(3) *A Σ-congruence $\sim^Z \in C(SP)$ is terminal in $C(SP)$ if and only if for all ground Σ-terms $t, t' \in T(\Sigma)$ the following condition holds:*

$$\neg(t \sim^Z t') \Leftrightarrow E \vdash \neg(t = t').$$

PROOF. (1) follows from the definition of sufficient completeness.

(2) Let $\sim^I$ be an initial element in $C(SP)$. Then, according to Lemma 3.2.1, for all $t, t' \in T(\Sigma)$

$$\begin{aligned} t \sim^I t' &\Leftrightarrow \text{for all } \sim\, \in C(SP) \colon t \sim t' \quad (\text{because of } t, t' \in T(\Sigma)) \\ &\Leftrightarrow \text{for all } A \in Gen_{Cons}(\langle H\Sigma, E\rangle) \colon A \vdash t = t' \\ &\qquad (\text{because of sufficient completeness}) \\ &\Leftrightarrow \text{for all } A \in Gen(\langle H\Sigma, E\rangle) \colon A \vdash t = t' \\ &\Leftrightarrow E \vdash t = t'. \end{aligned}$$

(3) The proof is analogous to the proof of (2). □

For sufficiently complete specifications a lattice theorem can be directly derived from Corollary 4.1.6.

5.2.2. THEOREM. *Let $SP = \langle \Sigma, H\Sigma, E, Cons\rangle$ be a satisfiable and sufficiently complete specification with inequations and conditional equations as axioms such that the simple specification $\langle H\Sigma, E\rangle$ is maximal w.r.t. the sorts of the inequations and of the premises of the conditional equations of E. Then the set $C(SP)$ of congruences of SP forms a complete lattice w.r.t. set inclusion. The initial congruence of SP is the smallest and the terminal congruence of SP is the greatest element of $C(SP)$.*

PROOF. According to Corollary 4.1.6, $C(\langle H\Sigma, E\rangle)$ forms a complete lattice. Because of the sufficient completeness, each $H\Sigma$-algebra B associated with $\sim^B \in C(\langle H\Sigma, E\rangle)$ is in $Gen_{Cons}(\langle H\Sigma, E\rangle)$ and therefore its Σ-restriction $B|_\Sigma$ is a model of SP. Then it is easy to see that the Σ-restrictions of intersections and unions of congruences in $C(\langle H\Sigma, E\rangle)$ are exactly the intersection and unions of the corresponding models of SP. Thus, $C(SP)$ forms a complete lattice as well. □

5.2.3. COROLLARY. *Let $SP = \langle \Sigma, H\Sigma, E, Cons\rangle$ be a satisfiable and sufficiently complete specification containing BOOL such that $E \backslash \{true \neq false\}$ is equational. Then the set $Mod(SP)$ of abstract models of SP forms a complete lattice w.r.t. set inclusion. The initial abstract data type of SP is the smallest and the terminal abstract data type of SP is the greatest element of $Mod(SP)$.*

PROOF. The sufficient completeness implies that for each ground term b of sort *bool* $E \vdash b = true$ or $E \vdash b = false$ holds. Since SP is satisfiable, exactly one of the two possibilities holds. This clearly implies that SP is maximal w.r.t. *bool*. By applying Theorem 5.2.2 one obtains the assertion of the corollary. □

For example, the specification $C\text{-}SET$ of Example 5.1.1 is sufficiently complete and satisfiable. Therefore, the congruences of $C\text{-}SET$ form a complete lattice.

5.3. *Expressive power of specifications with hidden symbols and constructors*

Specifications with hidden symbols and constructors have considerably more expressive power than simple specifications. The three specification methods initial, terminal and monomorphic are adequate for semicomputable, co-semicomputable and computable abstract data types. If only sufficiently complete specifications are considered, then all three methods are complete. This means that every semicomputable data type can be defined by a sufficiently complete finite initial algebra specification; analogous theorems hold for co-semicomputable and computable data types. For nonsufficiently complete specifications the methods are even complete w.r.t. hyperarithmetical abstract data types.

In order to prove these assertions, we show first that due to the use of hidden symbols, every specification is equivalent to a specification where all axioms are quantifier-free. Then this specification can be transformed into a specification where all but one of the axioms are equations. The only inequation is $true \neq false$. The essential idea for this transformation is that, using the specification $BOOL$ (Example 5.1.1) of truth values, logic can be internalized, i.e. all formulas are coded as Boolean terms (for a similar approach see [141]).

5.3.1. LEMMA. *Each specification* $SP = \langle \Sigma, \langle HS, HF \rangle, E, Cons \rangle$ *is equivalent to a specification* $SP_1 = \langle \Sigma, \langle HS, HF \cup F_1 \rangle, E_1, Cons \rangle$ *with quantifier-free axioms such that*

(i) $|E_1| = |E|$, *and*

(ii) $|F_1| =$ "*the number of occurrences of existential quantifiers in the prenex normal forms of the axioms in E*".

PROOF. Each axiom $e \in E$ *is first-order equivalent to a formula* e_1 of the same signature in prenex normal form; i.e., e_1 has the form $Q_1 x_1 : s_1 \ldots Q_m x_m : s_m . e'_1$ where $Q_i \in \{\forall, \exists\}$ for $i = 1, \ldots, m$ and where e'_1 is quantifier-free (cf. e.g. [164]).

By introducing so-called Skolem functions one can eliminate the existential quantifiers as follows: Let $h_1, \ldots, h_l \in \{1, \ldots, m\}$ be the indices of the universal quantifications $\forall x_{h_1} : s_{h_1}, \ldots, \forall x_{h_l} : s_{h_l}$ in e_1; let $j_1, \ldots, j_{l'}$ be the indices of the existential quantifications in e_1 and let for $i = 1, \ldots, l'$ $\{\forall x_{h_1} : s_{h_1}, \ldots, \forall x_{h_{l(i)}} : s_{h_{l(i)}}\}$ be the set of all universal quantifiers which occur to the left of $\exists x_{j_i} : s_{j_i}$ in e_1. Then consider the "Skolem normal form"

$$e_2 \equiv \forall x_{h_1} : s_{h_1} \ldots \forall x_{h_l} : s_{h_l} . e'_1 [f_i(x_{h_1}, \ldots, x_{h_{l(i)}}) / x_i \text{ for } i = 1, \ldots, l'],$$

where for $i = 1, \ldots, l'$, $f_i : s_{h_1}, \ldots, s_{h_{l(i)}} \to s_{j_i}$ is a new function symbol, the so-called Skolem function symbol. The formula e_2 is equivalent to e_1 in the following sense:

for any algebra $A \in Gen_{Cons}(\langle HS, HF\rangle)$,

$$A \models e_1 \text{ iff there exists } B \in Gen_{Cons}(\langle HS, HF \cup \{f_1, \ldots, f_{l'}\}\rangle)$$
$$\text{such that } B|_{\langle HS,HF\rangle} = A \text{ and } B \models e_2.$$

Therefore, by choosing F_1 to be the union of all Skolem function symbols for axioms in E, and E_1 to be the set of all Skolem normal forms of the axioms in E, one obtains the desired result. □

5.3.2. THEOREM. *For every specification $SP = \langle \Sigma, H\Sigma, E, Cons\rangle$ there exists an equivalent specification $SP' = \langle \Sigma, H\Sigma', E', Cons'\rangle$ with the same export signature such that*

- *E' consists of a set of equations and one ground inequation; $|E'|$ is finite if and only if $|H\Sigma| + |E|$ is finite;*
- *$H\Sigma'$ contains at most one additional (hidden) sort and at most N more additional (hidden) function symbols where $N = 2*HS + 5 +$* |occurrences of existential quantifiers in prenex normal forms of *E*|;
- *Cons' contains at most two additional constants.*

PROOF. We assume w.l.o.g. that *SP* does not contain any Boolean symbols, i.e. $H\Sigma \cap sig(BOOL) = \emptyset$. (Otherwise, it suffices in the following proof to introduce new names instead of those in *sig*(*BOOL*), for *BOOL* see Example 5.1.1(1).) According to Lemma 5.3.1, the axioms in E can be transformed into prenex form and existential quantifiers can be replaced by hidden Skolem function symbols. The resulting specification $SP_1 = \langle \Sigma, H\Sigma_1, E_1, Cons\rangle$ is equivalent to *SP* and contains only quantifier-free axioms.

Then *BOOL* is added and, for each sort $s \in HS$, two hidden function symbols eq_s: $s,s \to bool$ and if_s: $bool, s,s \to s$ are introduced. Each quantifier-free axiom G of E_1 is replaced by an equation $G^* = true$. G^* is inductively defined as follows:

$$(u = v)^* =_{\text{def}} eq_s(u, v) \quad \text{if } u, v \in T(H\Sigma_1, X)_s,\ s \in HS,$$
$$(\neg G)^* =_{\text{def}} not(G^*), \qquad (G \wedge H)^* =_{\text{def}} G^* \textit{ and } H^*.$$

Moreover, for any nonlogical, equational axiom $\forall x_1{:}s_1, \ldots, x_n{:}s_n.\, G$ (i.e., reflexivity, symmetry, transitivity and substitutivity, cf. Section 3.1), the translation $G^* = true$ is introduced as an additional axiom. The following axioms define the function symbol *if* and ensure that for each $s \in HS$, eq_s represents the equality of terms:

$$if_s(true, x, y) = x, \quad if_s(false, x, y) = y, \quad if_s(eq_s(x, y), x, y) = y \quad \text{for all } s \in HS.$$

Then the resulting specification SP' is equivalent to SP, i.e. $Mod(SP) = Mod(SP')$ holds. □

Note that, in general, this translation yields an equational specification which is not sufficiently complete. For example, an equation "$u = v$" is translated into $eq(u, v) = true$; nothing is said about the falsity of Boolean terms. However for certain monomorphic specifications, the translated specification is sufficiently complete.

5.3.3. FACT. *Let $SP = \langle \Sigma, H\Sigma, E, Cons\rangle$ be a satisfiable specification with quantifier-free axioms such that $\langle H\Sigma, E\rangle$ is monomorphic. Then SP and its equational translation SP′ are sufficiently complete.*

PROOF. Let t be any ground term in $T(H\Sigma)$. Since SP is satisfiable there exists a model $A \in Gen_{Cons}(\langle H\Sigma, E\rangle)$ of SP; thus there exists a constructor term $c \in T(\langle HS, Cons\rangle)$ such that $A \models t = c$. Due to the monomorphicity of $\langle H\Sigma, E\rangle$, $t = c$ is maximal and therefore $Gen(\langle H\Sigma, E\rangle) \models t = c$. Then Fact 3.1.6(2) implies $E \vdash t = c$ and thus the sufficient completeness property of SP. Now consider the equational translation $SP' = \langle \Sigma, H\Sigma', E', Cons'\rangle$. Since SP has only quantifier-free axioms, $H\Sigma'$ does not contain Skolem functions; it consists of the union of $H\Sigma$ with the signature of $BOOL$ and the function symbols if_s and eq_s for $s \in HS$. The constructors of SP' are the constructors of SP together with the constructors *true* and *false* of $BOOL$.

The proof of sufficient completeness is by induction on the structure of $T(H\Sigma')$:

1: $t \in T(H\Sigma)_s$, $s \in HS$. Since SP is sufficiently complete, there exists a $c \in T(\langle HS, Cons\rangle)$ such that $E \vdash t = c$ holds. By construction of SP', $E' \vdash eq_s(t, c) = true$ holds. Then, by definition of if, we have $E' \vdash t = if_s(eq_s(t, c), t, c) = c$.

2: $t \equiv eq_s(u, v)$, $u, v \in T(H\Sigma')_s$, $s \in HS$. By the induction hypothesis there exist u', $v' \in T(HS \cup \{bool\}, Cons \cup \{true, false\})_s$ with $E' \vdash u = u'$ and $E' \vdash v = v'$. Because of $s \neq bool$, u' and v' are terms in $T(HS, Cons) \subseteq T(H\Sigma)$. Due to the monomorphicity of $\langle H\Sigma, E\rangle$ either $E \vdash u' = v'$ or $E \vdash u' \neq v'$. Thus, by construction, $E' \vdash eq_s(u', v') = true$ or $E' \vdash eq_s(u', v') = false$ holds. Therefore, $E' \vdash eq_s(u, v) = true$ or $E' \vdash eq_s(u, v) = false$ holds.

3: $t \equiv if_s(b, u, v)$, $b \in T(H\Sigma')_{bool}$, $u, v \in T(H\Sigma')_s$. By the induction hypothesis there exist $b', u', v' \in T(\langle HS \cup \{bool\}, Cons \cup \{true, false\}\rangle)$ with $E' \vdash b = b'$, $E' \vdash u = u'$ and $E' \vdash v = v'$. Because b' is of sort $bool$, b' is either *true* or *false*. Hence by definition of if, $E' \vdash if_s(b, u, v) = z$ with $z \in \{u', v'\} \subseteq T(\langle HS, Cons\rangle)$.

4: For $t \in \{not(b), b \text{ and } b', b \text{ or } b'\}$ it suffices to apply the induction hypothesis. □

5.3.4. THEOREM. *Let $SP = \langle \Sigma, H\Sigma, E, Cons\rangle$ be a sufficiently complete and satisfiable specification, such that E is a recursively enumerable set of first-order axioms.*

(1) *If the initial abstract type I(SP) of SP exists, then I(SP) is semicomputable.*
(2) *If the terminal abstract type Z(SP) of SP exists, then Z(SP) is co-semicomputable.*
(3) *If SP is monomorphic, then all models of SP are computable.*

PROOF. The proof is analogous to the proof of Theorem 4.2.1 because of the characterization of initial and terminal algebras by Fact 5.2.1(2) and 5.2.1(3). □

The following lemma shows that each recursive number algebra (cf. Section 2.4) has a sufficiently complete and monomorphic specification with hidden symbols.

5.3.5. LEMMA. (1) *Let C be a one-sorted recursive number algebra with carrier set $\mathbb{N}$ and signature $\Sigma = \langle \{nat\}, F\rangle$. Then C has a sufficiently complete and monomorphic* *specification of the form*

> **export** Σ **from**
> **extend** $NAT1$ **by**
> **functions** $F \cup F'$
> **axioms** E

where $NAT1$ *is defined in Example* 3.4.2(3), F' *is a set of auxiliary function symbols and E is a set of equations. If F is finite, so are F′ and E.*

(2) *Let* $\Sigma = \langle\{s\}, F\rangle$ *be a one-sorted signature and let* $F_N =_{\text{def}} \{f_N: nat^k \to nat \mid f: s^k \to s \in F\}$ *be the set F indexed by N over nat. There exists a sufficiently complete and monomorphic specification* SP_N *of the form*

> **extend** $NAT1$ **by**
> **functions** $F_N \cup F'$
> **axioms** E_N

where F' *is a set of auxiliary function symbols and* E_N *is a set of equations such that the recursive number algebra* C_Σ *associated with* $T(\Sigma)$ *(see Section* 2.4) *is embedded in any model A of* SP_N; *i.e.* $C_\Sigma \cong A|_{in}$ *where* $in: \Sigma \to sig(SP_N)$ *is defined by* $in(s) =_{\text{def}} nat$, $in(f) =_{\text{def}} f_N$ *for each* $f \in F$. *If F is finite, so are F′ and* E_N.

PROOF. *Proof of* (1): The algebra C consists of the carrier set $\mathbb{N}$ and of recursive functions $f^C: \mathbb{N}^k \to \mathbb{N}$ (for every $f: nat^k \to nat \in F$). We distinguish two cases.

Case 1: f^C is primitive recursive. Then f^C is either the constant 0, the successor function or a projection, or it can be defined by composition and primitive recursion from 0, successor and projections. Thus it has an equational specification which, in the case of the primitive recursion scheme, has the form

$$f(zero, x_1, \ldots, x_{k-1}) = g(x_1, \ldots, x_{k-1}),$$
$$f(succ(x), x_1, \ldots, x_{k-1}) = h(x, x_1, \ldots, x_{k-1}, f(x, x_1, \ldots, x_{k-1})). \tag{5.1}$$

where $g: nat^{k-1} \to nat$, $h: nat^{k+1} \to nat$ are auxiliary function symbols. By induction hypothesis, g and h admit primitive recursive equational specifications without involving f.

Case 2. f^C is recursive (but not primitive recursive). Then the graph of f^C

$$graph(f^C) = \{(x_1, \ldots, x_k, f^C(x_1, \ldots, x_k)) \mid x_1, \ldots, x_k \in \mathbb{N}\}$$

is recursively enumerable. Since every r.e. set has a primitive recursive enumeration, let $h_1^C, \ldots, h_k^C, g^C: \mathbb{N} \to \mathbb{N}$ be primitive recursive functions enumerating $graph(f^C)$, i.e.

$$graph(f^C) = \{(h_1^C(x), \ldots, h_k^C(x), g^C(x)) \mid x \in \mathbb{N}\}.$$

Thus f^C has an equational specification of the form

$$f(h_1(x), \ldots, h_k(x)) = g(x) \tag{5.2}$$

where $h_1, \ldots, h_k, g: nat \to nat$ are auxiliary function symbols which (by (1)) admit primitive recursive equational specifications. Hence, each f^C has a sufficiently complete equational definition w.r.t. the constructors *zero*, *succ* of natural numbers.

For every function symbol $f \in F$, let E_f be the set of equations defining the recursive function f^C including the set of equational definitions of the auxiliary functions and let F_f the set of auxiliary function symbols occurring in E_f. Moreover, assume that the names of the auxiliary function symbols are chosen in a consistent way; i.e., for any $h \in F_f \cap F_g$ $(f, g \in F)$, the equational specifications for h in E_f and E_g coincide.

Then we define the specification of C to the Σ-export specification of the extension SP of the monomorphic specification $NAT1$ of Boolean values and natural numbers by the equational specifications of all $f \in F$:

$$SP_C =_{\text{def}} \textbf{export } \Sigma \textbf{ from } SP \quad \text{where}$$

$$\begin{aligned} SP =_{\text{def}} & \textbf{extend } NAT1 \textbf{ by} \\ & \qquad \textbf{functions } \bigcup \{F_f \mid f \in F\} \\ & \qquad \textbf{axioms } \bigcup \{E_f \mid f \in F\} \end{aligned}$$

SP is sufficiently complete w.r.t. $NAT1$. Since $NAT1$ is monomorphic, SP is also monomorphic and therefore $Mod(SP_C) \cong [C]$ holds.

Proof of (2): C_Σ is a one-sorted recursive number algebra; thus it suffices to apply (1) and to take the appropriate specification SP (see the proof of (1)) for SP_N. □

5.3.6. THEOREM. *Let Σ be a finite signature and let A be a Σ-computation structure.*

(1) *If A is semicomputable, then it has a sufficiently complete initial algebra specification $SP = \langle \Sigma, H\Sigma, E, Cons \rangle$ such that $H\Sigma$ is finite and E is a* finite *set of equations.*

(2) *If A is co-semicomputable, then it has a sufficiently complete terminal algebra specification $SP = \langle \Sigma, H\Sigma, E, Cons \rangle$ such that $H\Sigma$ is finite and $E \backslash \{true \neq false\}$ is a* finite *set of conditional equations.*

(3) *If A is computable, then it has a sufficiently complete and monomorphic specification $SP = \langle \Sigma, H\Sigma, E, Cons \rangle$ such that $H\Sigma$ is finite and $E \backslash \{true \neq false\}$ is a* finite *set of equations.*

PROOF. Let A be a semicomputable or co-semicomputable Σ-computation structure. According to Fact 2.4.1, the computability notions are invariant under isomorphism. Therefore we assume w.l.o.g. that $A = T(\Sigma)/\sim$; i.e., A is the quotient of $T(\Sigma)$ by a Σ-congruence relation $\sim \subseteq T(\Sigma) \times T(\Sigma)$.

Moreover, we assume w.l.o.g. that Σ has the form $\langle \{s\}, F \rangle$, i.e. Σ has only one sort, and that $T(\Sigma)$ is infinite (otherwise, the specification of A is trivial; e.g. it is sufficient to introduce a Boolean function symbol eq: $s, s \to bool$ which specifies all equations and inequations that hold in the finite algebra $T(\Sigma)/\sim$). Let (C_Σ, ν) be the canonical bijective coordinatization of $T(\Sigma)$. According to Lemma 5.3.5(2), C_Σ can be embedded in the sufficiently complete and monomorphic specification

$$SP_N =_{\text{def}} \textbf{extend } NAT1 \textbf{ by functions } F_N \cup F' \textbf{ axioms } E_N$$

where $F_N =_{\text{def}} \{f_N : nat^k \to nat \mid f : s^k \to s \in F\}$ is the set F indexed by N, F' is a set of auxiliary function symbols and E_N is a set of equations.

Proof of (1): Let $A = T(\Sigma)/\sim$ be semicomputable. Then $\sim$ is recursively enumerable; i.e., the relation $\sim^\nu$ defined by

$$i \sim^\nu j \text{ iff } \nu(i) \sim \nu(j) \quad \text{for } i,j \in \mathbb{N}$$

is recursively enumerable. Thus one can choose primitive recursive functions $g^A, h^A: \mathbb{N} \to \mathbb{N}$ to enumerate it so that $\sim^\nu = \{(g^A(x), h^A(x)) \mid x \in \mathbb{N}\}$. Hence by the definition of $\sim^\nu$,

$$\{\nu(g^A(x)) = \nu(h^A(x)) \mid x \in \mathbb{N}\}$$

is the set of ground equations which hold in A. Let $\underline{\nu}$: *nat*$\to s$ and g,h: *nat*$\to$*nat* be function symbols corresponding to ν, g^A, h^A. According to (the proof of) Lemma 5.3.5, g^A and h^A admit a finite set $E_g \cup E_h$ of defining equations. In $E_g \cup E_h$ there may be auxiliary function symbols from a set $F_g \cup F_h$ whose equational specification is also given by $E_g \cup E_h$.

The epimorphism ν can be equationally defined by the following set E_ν of equations:

$$\underline{\nu}(c_N) =_{\text{def}} c \quad \text{for all constants } c \in F,$$

$$\underline{\nu}(f_N(x_1, \ldots, x_k)) =_{\text{def}} f(\underline{\nu}(x_1), \ldots, \underline{\nu}(x_k)) \quad \text{for all } f: s^k \to s \in F.$$

Then the specification

$SP =_{\text{def}}$ **export** Σ **from**
  extend SP_N **by**
    sort s
    constructors F
    functions $F_g \cup F_h \cup \{\underline{\nu}: nat \to s\}$
    axioms $E_g \cup E_h \cup E_\nu \cup \{\forall x: nat.\ \underline{\nu}(g(x)) = \underline{\nu}(h(x))\}$

is a sufficiently complete initial algebra specification of A: the sufficient completeness follows from the surjectivity of ν, the properties of SP_N and the sufficient completeness of the primitive recursive definitions of g^A and h^A; it is an initial algebra specification of A since the last axiom specifies exactly the ground equations which hold in A. No other equations were specified w.r.t. the sort s; due to the monomorphicity of the extension of SP_N by g and h, these are the only equations of sort s, which hold in SP. Thus for all $t, t' \in T(\Sigma)_s$ we have

$$Mod(SP) \models t = t' \text{ iff } A \models t = t'.$$

According to Lemma 3.2.2, A is an initial model of SP.

Proof of (2): Let $A = T(\Sigma)/\sim$ be co-semicomputable. Then the complement $\not\sim$ of $\sim$ is recursively enumerable; i.e., the relation

$$i \not\sim^\nu j \text{ iff } \nu(i) \not\sim \nu(j) \quad \text{for } i,j \in \mathbb{N}$$

is recursively enumerable. Thus one can choose primitive recursive functions $g^A, h^A: \mathbb{N} \to \mathbb{N}$ so that

$$\{\nu(g^A(x)) \neq \nu(h^A(x)) \mid x \in \mathbb{N}\} \tag{5.3}$$

is the set of ground inequations which hold in A.

As in the proof of (1) g^A, h^A and v can be defined by a set of equations $E_g \cup E_h \cup E_v$ using a set $F_g \cup F_h$ of auxiliary function symbols. The inequations (5.3) can be expressed by the following conditional equation:

$$e_{\not\sim} =_{\mathrm{def}} \forall x{:}\, nat.\ \underline{v}(g(x)) = \underline{v}(h(x)) \Rightarrow true = false$$

Then the specification

export Σ **from**
  extend SP_N **by**
    sort s
    constructors F
    functions $F_g \cup F_h \cup \{\underline{v}{:}\, nat \rightarrow s\}$
    axioms $E_g \cup E_h \cup E_v \cup \{e_{\not\sim}\}$

is a sufficiently complete terminal algebra specification of A: The sufficient completeness follows as in the proof of (1). Because of the axiom $e_{\not\sim}$ we have, for all $t,t' \in T(\Sigma)$,

$$Mod(SP) \models t \neq t' \ \text{ iff } \ A \models t \neq t'.$$

According to Lemma 3.2.2(2), *A is a terminal model of SP*.

Proof of (3): Let $A = T(\Sigma)/\sim$ be a computable Σ-computation structure. Then the Σ-congruence $\sim$ is recursive and thus $\sim^v$ and $\not\sim^v$ are recursively enumerable. Let, according to Lemma 5.3.5, $f, g, h, l{:}\, nat \rightarrow nat$ be function symbols with sets E_f, E_g, E_h, E_l resp. of primitive recursive specifications defining the sets of function symbols F_f, F_g, F_h, F_l resp., such that (according to $E_f \cup E_g \cup E_h \cup E_l$) the interpretations of f, g, h, l over the natural numbers enumerate $\sim^v$ and $\not\sim^v$. Then we specify a Boolean function symbol $eq_s{:}\, s, s \rightarrow bool$ which characterizes exactly the congruence $\sim$. The specification

$SP =_{\mathrm{def}}$ **export** Σ **from**
  extend SP_N **by**
    sort s
    constructors F
    functions $F_f \cup F_g \cup F_h \cup F_l \cup \{\underline{v}{:}\, nat \rightarrow s,\ eq_s{:}\, s,s \rightarrow bool\}$
    axioms $E_f \cup E_g \cup E_h \cup E_l \cup E_v$
      $\cup \{eq_s(\underline{v}(f(x)), \underline{v}(g(x))) = true,$
      $eq_s(\underline{v}(h(x)), \underline{v}(l(x))) = false\}$

is a monomorphic and sufficiently complete equational specification. The models of SP consist exactly of the isomorphism class of A. □

The following corollary shows that simple specifications with hidden symbols as considered in the literature e.g. in [23, 24] are enough to specify semicomputable, co-semicomputable and computable data types.

5.3.7. COROLLARY. *Let A be a Σ-computation structure, such that Σ is finite.*

(1) *If A is semicomputable, then there exists a finite simple specification $\langle \Sigma_1, E_1 \rangle$ with*

$\Sigma \subseteq \Sigma_1$ such that the Σ-reduct $I|_\Sigma$ of the initial model I of $\langle \Sigma_1, E_1 \rangle$ is isomorphic to A,

(2) *If A is co-semicomputable, then there exists a finite simple specification $\langle \Sigma_1, E_1 \rangle$ with $\Sigma \subseteq \Sigma_1$ such that the Σ-reduct $Z|_\Sigma$ of the terminal model Z of $\langle \Sigma_1, E_1 \rangle$ is isomorphic to A,*

(3) *If A is computable, then there exists a monomorphic finite simple specification $\langle \Sigma_1, E_1 \rangle$ with $\Sigma \subseteq \Sigma_1$ such that the Σ-reduct $B|_\Sigma$ of any model B of $\langle \Sigma_1, E_1 \rangle$ is isomorphic to A.*

In (1) *and* (3) *E_1 consists of a finite set of equations together with one inequation; in* (2) *E_1 consists of one inequation, one conditional equation and a finite set of equations.*

PROOF. Consider the specifications $SP = \langle \Sigma, H\Sigma, E, Cons \rangle$ of Theorem 5.3.6. Choose $\Sigma_1 =_{\text{def}} H\Sigma$ and $E_1 =_{\text{def}} E$. Because of the sufficient completeness of SP, $Mod(\langle \Sigma_1, E_1 \rangle)|_\Sigma = Mod(SP)$ holds and therefore $\langle \Sigma_1, E_1 \rangle$ has the required properties. □

In [24] a stronger result is shown: computable and co-semicomputable computation structures can be specified with hidden functions but without the use of hidden sorts. For semicomputable computation structures this is still an open question. Therefore, the expressive power of sufficiently complete specifications with constructors is the same as the one of simple specifications with hidden symbols. This situation changes substantially for those specifications with constructors which are not sufficiently complete and admit, however, initial and terminal models (see also [13]).

5.3.8. LEMMA. *Any arithmetical Σ-computation structure has a monomorphic specification $SP = \langle \Sigma, H\Sigma, E, Cons \rangle$ where, except for true $\neq$ false, all axioms are equations. If Σ is finite, so are $H\Sigma$, E and Cons.*

PROOF. Let, w.l.o.g., Σ be one-sorted, i.e. $\Sigma = \langle \{s\}, F \rangle$ for some sort s and some set of function symbols F, and assume that $T(\Sigma)$ is infinite. Let A be an arithmetical Σ-computation structure and assume w.l.o.g. that A is the quotient of $T(\Sigma)$ by some Σ-congruence $\sim$, i.e. $A = T(\Sigma)/\sim$.

According to Lemma 5.3.5(2), the recursive number algebra C_Σ associated with $T(\Sigma)$ can be embedded in the sufficiently complete and monomorphic specification SP_N. Since A is arithmetical, the congruence $\sim^\nu$ associated with the fixed numbering $\nu: C_\Sigma \to T(\Sigma)$ has an explicit definition of the form

$$x \sim^\nu y \Leftrightarrow Q_1 z_1 \ldots Q_n z_n . R_{\sim^\nu}(x, y, z_1, \ldots, z_n)$$

where for $i = 1, \ldots, n$, Q_i are quantifiers, z_i are variables for natural numbers, and $R_{\sim^\nu}$ is a recursive relation. According to Lemma 5.3.5(1), there exists a set E' of recursive definitions specifying $R_{\sim^\nu}$ using a set of auxiliary function symbols F'. Thus, similarly to the proof of Theorem 5.3.6, the following specification is a monomorphic (but not equational) specification of the abstract data type of A.

$SP =_{\text{def}}$ **export** Σ **from**
  extend SP_N **by**
    sort s
    constructors F
    functions $F' \cup \{\underline{v}: nat \to s\}$
    axioms $E' \cup \{\underline{v}(x) = \underline{v}(y) \Leftrightarrow Q_1 z_1 \ldots Q_n z_n . R_{\sim^v}(x, y, z_1, \ldots, z_n)\}$

Applying the construction of Theorem 5.3.2 yields the desired result. □

An application of a theorem of *Spector* yields the following generalization of Lemma 5.3.8 replacing arithmetical by hyperarithmetical.

5.3.9. THEOREM. *Any hyperarithmetical Σ-computation structure (where Σ is finite) has a finite monomorphic specification where, except for true $\neq$ false, all axioms are equations.*

PROOF. According to [152, p. 425], for every hyperarithmetical relation g there exists another hyperarithmetical relation h, such that the pair $\langle g, h\rangle$ is implicitly definable by an arithmetical relation r, which, due to Lemma 5.3.8, has a monomorphic specification. "Implicitly definable" means that there is a first-order formula over the natural numbers involving (apart from logical symbols $\forall$, $\neg$, $\wedge$) only the functions 0, *successor*, *addition*, *multiplication*, and relations r and g, h (where g, h are free variables) such that the graph of $\langle g, h\rangle$ is the only solution of this formula. Lemma 5.3.8 and the construction of the Theorem 5.3.2 then imply the existence of the required specification. □

Conversely, hyperarithmeticity is an upper bound for the complexity of finite, first-order specifications.

5.3.10. THEOREM. *Every initial and every terminal algebra of a finite, first-order specification is hyperarithmetical.*

PROOF. Let, w.l.o.g., $SP = \langle \Sigma, H\Sigma, E, Cons\rangle$ be a one-sorted specification where $\Sigma = \langle \{s_1\}, F\rangle$, $F = \{f_1, \ldots, f_n\}$, $n \geqslant 1$, $H\Sigma = \langle HS, HF\rangle$, $HS = \{s_1, \ldots, s_l\}$, $HF = \{f_1, \ldots, f_{n'}\}$, $l \geqslant 1$, $n' \geqslant n$ and $E = \{e_1, \ldots, e_m\}$ is a set of sentences. Let $\Sigma_{Cons} =_{\text{def}} \langle \{s_1\}, F \cap Cons\rangle$ be the constructor signature of SP.

(1) Let A be an initial model of SP. Without loss of generality it is assumed that the carrier set A_{s_1} of A is the quotient $T(\Sigma_{Cons})/\sim$ of the constructor term algebra by an $H\Sigma$-congruence $\sim$. We show that the recursive number algebra $C_{\Sigma_{Cons}}$ (associated with $T(\Sigma_{Cons})$) can be extended by hyperarithmetical functions to a Σ-algebra C such that the quotient $C/\sim^v$ of C by the congruence $\sim^v$ (associated with $\sim$) is isomorphic to A and that $\sim^v$ is hyperarithmetical. Therefore A is hyperarithmetical.

First, consider the canonical coordinatization $(C_{\Sigma_{Cons}}, v)$ of $T(\Sigma_{Cons})$. $C_{\Sigma_{Cons}}$ is a recursive number algebra; hence the interpretation function $\cdot^C : T(\Sigma_{Cons}) \to \mathbb{N}$ can be chosen to be bijective and the interpretation $f^C : \mathbb{N}^k \to \mathbb{N}$ of any of the constructor function symbols $f: s_1^k \to s_1 \in Cons$ can be chosen to be recursive. Let E_{Cons} be the set of recursive definitions of all $f \in Cons$.

Second, for any non-constructor symbol $f: s_1^k \to s_1 \in F \backslash Cons$, let

$$f^C(x_1, \ldots, x_k) =_{\text{def}} y \text{ iff } f(v(x_1), \ldots, v(x_k)) \sim v(y) \text{ and } \forall z: v(y) \sim v(z) \Rightarrow y \leqslant z.$$

Since A is initial, $\sim$ is characterized by

$$t \sim t' \Leftrightarrow Gen_{H\Sigma_{Cons}}(\langle H\Sigma, E\rangle) \models t = t' \quad \text{for all } t, t' \in T(\Sigma).$$

Thus, the graph g_f of f^C and its complement have explicit definitions of the form

$$\begin{aligned} g_f(\vec{x}, y) \Leftrightarrow\ & \forall f_1, \ldots, f_n \forall eq_{s_1}, \ldots, eq_{s_l}. \\ & [(E_{Cons} \wedge \text{“}(eq_s)_{s \in HS} \text{ is } H\Sigma\text{-congruence”} \wedge e'_1 \wedge \cdots \wedge e'_m \\ & \quad \Rightarrow (eq_{s_1}(f(\vec{x}), y) \wedge \forall z.(eq_{s_1}(y, z) \Rightarrow y \leqslant z)] \end{aligned}$$

where for $i = 1, \ldots, m$, e'_i denotes the result of the substitution of any equation $t =_{s_j} t'$ in e'_i by $eq_{s_j}(t, t')$ where $j \in \{1, \ldots, l\}$, and

$$\neg g_f(\vec{x}, y) \Leftrightarrow \exists z. g_f(\vec{x}, y) \wedge y \neq z.$$

The usual quantifier negation and quantifier contraction rules (cf. [152, p. 375]) imply that f^C is hyperarithmetical.

Third, $\sim^v$ can be shown to be hyperarithmetical as follows. It has the definition

$$\begin{aligned} x \sim^v y \Leftrightarrow\ & \forall f_1, \ldots, f_n \forall eq_{s_1}, \ldots, eq_{s_l}. \\ & [(E_{Cons} \wedge \text{“}(eq_s)_{s \in HS} \text{ is } H\Sigma\text{-congruence”} \wedge e'_1 \wedge \cdots \wedge e'_m \\ & \Rightarrow eq_{s_1}(x, y)]; \end{aligned}$$

its complement can be defined using the least number z_x to which the number x is equivalent (i.e., $z_x =_{\text{def}}$ “*least* z such that $x \sim^v z$”):

$$\begin{aligned} x \not\sim^v y \Leftrightarrow\ & \forall f_1, \ldots, f_n \forall eq_{s_1}, \ldots, eq_{s_l}. \\ & [(E_{Cons} \wedge \text{“}(eq_s)_{s \in HS} \text{ is } H\Sigma\text{-congruence”} \wedge e'_1 \wedge \cdots \wedge e'_m) \\ & \Rightarrow \exists z_x \exists z_y. eq_{s_1}(x, z_x) \wedge eq_{s_1}(y, z_y) \wedge z_x \neq z_y \wedge \\ & \quad \forall z.(eq_{s_1}(x, z) \Rightarrow z_x \leqslant z) \wedge \forall z.(eq_{s_1}(y, z) \Rightarrow z_y \leqslant z)]. \end{aligned}$$

As for f^C, the usual quantifier negation and quantifier contraction rules imply that $\sim^v$ is hyperarithmetical.

It is easy to see that $C/\sim^v$ is isomorphic to A.

(2) A similar argument using

$$\begin{aligned} \neg g_f(\vec{x}, y) \Leftrightarrow\ & \forall f_1, \ldots, f_n \forall eq_{s_1}, \ldots, eq_{s_l}. \\ & [(E_{Cons} \wedge \text{“}(eq_s)_{s \in HS} \text{ is } H\Sigma\text{-congruence”} \wedge e'_1 \wedge \cdots \wedge e'_m) \\ & \wedge eq_{s_1}(f(\vec{x}), y) \wedge \forall z.(eq_{s_1}(y, z) \Rightarrow y \leqslant z) \Rightarrow \forall u, v. eq_{s_1}(u, v)] \end{aligned}$$

works for terminal algebras. □

5.3.11. COROLLARY. *The set of specifications with constructors and hidden symbols is complete hyperarithmetical, i.e. for every Σ-computation structure A the following two* *statements are equivalent:*

(1) *A is hyperarithmetical.*

(2) *A is, up to isomorphism, the only model of a finite monomorphic specification.*

5.4. *Hierarchical specifications*

Constructing specifications in hierarchies is one of the simplest structuring concepts: a hierarchical specification contains a designated "primitive" subspecification (which may again be a hierarchical specification) that can be understood, analyzed and implemented on its own, i.e. without using any information about the overall specification (cf. [87, 88]). On the other hand, an extension of some subspecification can be viewed as a "black box" the behavior of which is given by the effects in the primitive specification; in other words, nonprimitive objects are semantically explained by the effects they have in the primitive specification ("visible" or "observable" behaviors). Following [181], a (simple) *hierarchical specification SP* consists of a triple $\langle \Sigma, E, P\rangle$ where $\langle \Sigma, E\rangle$ is a simple specification and $P = \langle \Sigma_P, E_P\rangle$ is a subspecification of $\langle \Sigma, E\rangle$; i.e., $\Sigma_P \subseteq \Sigma$ and $E_P \subseteq E$.

P is called the primitive specification of SP. A term $t \in T(\Sigma, X)_s$, $s \in S$, is called *primitive* if $t \in T(\Sigma_P, X)_s$; it is called *of primitive sort* if $s \in S_P$. The elements of $S \backslash S_P$ and $F \backslash F_P$ are called *nonprimitive* sorts and function symbols respectively. This definition describes hierarchies with just two layers, the primitive and the nonprimitive part. It is easy to extend these notions and all the following results to several layers.

Semantically, a *model A of a hierarchical specification* $SP = \langle \Sigma, E, P\rangle$ is a model of the simple specification $\langle \Sigma, E\rangle$ (i.e., A is a reachable Σ-algebra which satisfies E) which, in addition, is an expansion of a model of P (i.e., the Σ_P-restriction $A|_{\Sigma_P}$ of A is a model of P). The set of all isomorphism classes of models of SP or, equivalently, the set of all Σ-congruences which are associated with models of SP is (as usual) denoted by $Mod(SP)$.

In other words, a Σ-algebra A is a model of SP if it satisfies E, the constructors of the primitive sorts are the function symbols of P and the constructors of the nonprimitive sorts are all function symbols of $F \backslash F_P$ with range in a nonprimitive sort.

5.4.1. FACT. *Let* $SP = \langle \Sigma, E, P\rangle$ *be a hierarchical specification. Then SP is equivalent to the specification* $\langle \Sigma, \Sigma, E, Cons\rangle$ *with constructors where* $Cons =_{\text{def}} \{f : w \rightarrow s \in F \mid s \in F_P \Rightarrow f \in F_P\}$.

As a consequence, a hierarchical specification SP is called *sufficiently complete* if it is so w.r.t. *Cons*. Since *Cons* contains all function symbols with range in a nonprimitive sort, the sufficient completeness of SP is equivalent to the following property: for all ground terms $t \in T(\Sigma)_s$, $s \in S_p$, of primitive sort there exists a primitive term $p \in T(\Sigma_P)_s$ such that $E \vdash t = p$. (Sufficient conditions for proving this property can be found in [88].)

As for specifications with constructors, a sufficiently complete hierarchical specification is equivalent to a simple specification.

5.4.2. FACT. *If $SP=\langle\Sigma, E, P\rangle$ is a sufficiently complete hierarchical specification, then SP is equivalent to $\langle\Sigma, E\rangle$.*

PROOF. Fact 5.2.1(1). □

The assumption that primitive sorts are observable suggests the following definition. Two Σ-algebras A and B are called *behaviorally equivalent* (on Σ_P) if, for any terms $t, t' \in T(\Sigma)$ of primitive sort, $A \models t=t'$ iff $B \models t=t'$.

5.4.3. FACT. *Let $SP=\langle\Sigma, E, P\rangle$ be a hierarchical specification such that the primitive specification P is monomorphic. For any two models A and B of SP, if there exists a Σ-homomorphism between them, then they are behaviorally equivalent.*

PROOF. Let A, B be models of SP such that there exists a Σ-homomorphism from A to B Then $\sim^A \subseteq \sim^B$ holds for the associated congruences. Thus, for the behavioral equivalence, it is sufficient to show that for any two ground terms, t and t' of a primitive sort, $B \models t=t'$ implies $A \models t=t'$. So let $t, t' \in T(\Sigma)$ of primitive sort such that $B \models t=t'$. Since A is a model of a hierarchical specification, its primitive carrier sets are reachable w.r.t the primitive function symbols; thus there exist primitive terms p and $p' \in T(\Sigma_P)$ with $A \models t=p$ and $A \models t'=p'$. The existence of a Σ-homomorphism from A to B implies $B \models t=p$ and $B \models t'=p'$; thus, because of $B \models t=t'$, we have $B \models p=p'$. Since the Σ_P-restrictions $A|_{\Sigma_P}$ and $B|_{\Sigma_P}$ of A and B are models of the monomorphic specification P, $A \models p=p'$ and therefore $A \models t=t'$ holds as well. □

The above fact says that for the existence of initial and/or terminal models of a hierarchical specification, all models of the specification have to be behaviorally equivalent. Sufficient completeness is a sufficient condition for this property.

5.4.4. FACT. *If $SP=\langle\Sigma, E, P\rangle$ is sufficiently complete and P monomorphic, then all models of SP are behaviorally equivalent.*

PROOF. Let $t, t' \in T(\Sigma)_s$, $s \in S_P$, be two ground terms of primitive sort. We show that the equation $t=t'$ is maximal in $Gen(\Sigma, E)$ which implies the behavioral equivalence: Due to the sufficient completeness, there exist primitive terms $p, p' \in T(\Sigma_P)$ with $E \vdash t=p$ and $E \vdash t'=p'$; therefore, $Gen(\Sigma, E) \models t=t' \Leftrightarrow p=p'$. Because of the monomorphicity of P, the equation $p=p'$ is maximal in P; hence $t=t'$ is maximal in $Gen(\Sigma, E)$. □

The satisfiability for a hierarchical specification is ensured by another property: a hierarchical specification SP is called *hierarchy-consistent* if, for any two ground primitive terms $p, p' \in T(\Sigma_P)$, $E \vdash p=p'$ implies $E_P \vdash p=p'$. Thus for monomorphic P, SP has a model if and only if SP is hierarchy-consistent.

5.4.5. THEOREM. *Let $SP=\langle\Sigma, E, P\rangle$ be a hierarchy-consistent and sufficiently complete hierarchical specification with monomorphic primitive specification P such that the nonprimitive axioms are conditional equations with premises of primitive sort. Then the* *set $C(SP)$ of Σ-congruences of SP forms a nonempty complete lattice w.r.t. set inclusion. The initial congruence of SP is the smallest and the terminal abstract data type of SP is the greatest element of this lattice.*

PROOF. Due to the monomorphicity of P, SP is equivalent to $\langle \Sigma, E' \rangle$ where

$$E' = (E \backslash E_P) \cup \{t = t' \mid P \models t = t',\ t, t' \in T(\Sigma_P)\}$$
$$\cup \{\neg(t = t') \mid P \models \neg(t = t'),\ t, t' \in T(\Sigma_P)_s,\ s \in S_P\}.$$

The inequations of E' are maximal because of the monomorphicity of P, the premises of the conditional equations of $E \backslash E_P$ are maximal due to the sufficient completeness of SP and the monomorphicity of P. Thus it suffices to apply Corollary 4.1.6 and Fact 5.4.2. □

All models of SP are behaviorally equivalent and the terminal models have an important property: they are fully abstract [127, 146]. A Σ-algebra A is called *fully abstract* if for all ground terms t and t' of the same sort $s \in S$,

$$A \models t = t' \text{ iff for all contexts } c \in T(\Sigma, \{x{:}s\})_{s_1},\ s_1 \in S_P, \text{ of primitive sort:}$$
$$A \models c[t/x] = c[t'/x].$$

5.4.6. THEOREM. *Under the assumption of Theorem 5.4.5, any terminal model Z of SP is fully abstract. Z is characterized as follows: for any two ground terms $t, t' \in T(\Sigma)_s$, $s \in S$, $Z \models t = t'$ iff $\forall c \in T(\Sigma, \{x{:}s\})_{s_1}$, $s_1 \in sorts(P)$: $E \vdash c[t/x] = c[t'/x]$.*

PROOF. It is easy (but tedious) to verify that Z is a well-defined Σ-algebra. We prove that Z is a model of SP which is terminal. Note that for any $\sim \in C(SP)$, for any $t,t' \in T(\Sigma)$, if $t \sim t'$ then, for any context c (on the sort of t and t'), $c[t/x] \sim c[t'/x]$. Hence, because of the behavioral equivalence of all models (Fact 5.4.4), $Mod(SP) \models c[t/x] = c[t'/x]$ holds for all contexts c of primitive sort which implies $Z \models t = t'$. Thus $\sim \subseteq \sim^Z$, i.e. $\sim^Z$ is coarser than $\sim$.

According to Lemma 4.1.3, E' (see the proof of Theorem 5.4.5) and therefore E is equivalent to a set of ground inequations E_1 of primitive sort and of ground equations E_0. Because of $\sim^A \subseteq \sim^Z$ for all models A of SP, Z satisfies the equations E_0. It remains to prove that Z satisfies the inequations E_1 and that $Z|_{\Sigma_P}$ is reachable w.r.t. Σ_P. This follows from the fact that for any two terms $t, t' \in T(\Sigma)$ of a primitive sort, $Z \models t = t'$ iff $E \vdash t = t'$, and that SP is sufficiently complete and hierarchy-consistent. □

This characterization of the terminal algebra by contexts is the basis for the behavioral semantics. For these semantics, not all contexts but only those contexts which are observable are considered in e.g. [156].

The expressive power of hierarchical specifications can be deduced from the results in Sections 4 and 5.

If a hierarchical specification is sufficiently complete, then it is equivalent to a simple specification (see Fact 5.4.2). Thus one has the following results.

5.4.7. THEOREM. *Let $SP=\langle\Sigma, E, P\rangle$ be a hierarchical specification which is satisfiable and sufficiently complete. Let E be a recursively enumerable set of first-order axioms.*

(1) *If the initial abstract type I(SP) of SP exists, then I(SP) is semicomputable.*
(2) *If the terminal abstract type Z(SP) of SP exists, then Z(SP) is co-semicomputable.*
(3) *If SP is monomorphic, then all models of SP are computable.*

PROOF. It suffices to apply Theorem 4.2.1. □

If one admits infinite sets of axioms then, according to Fact 4.2.2, it can be shown that semicomputable algebras can be specified by sufficiently complete initial algebra specifications, co-semicomputable algebras by sufficiently complete terminal algebra specifications, and computable algebras by sufficiently complete monomorphic specifications. But if only finite sets of axioms are considered then there exist algebras which cannot be specified without hidden functions. Nonsufficiently complete hierarchical specifications are comparable to nonsufficiently complete specifications with constructors. Therefore, they are complete hyperarithmetical, but for the specification of hyperarithmetical algebras hidden functions are necessary (cf. Corollary 5.3.11 and [13]).

6. Structured specifications

For large algebraic specifications it is convenient to design specifications in a structured fashion by combining and modifying smaller specifications. This supports a modular decomposition into specification of manageable size and helps to master the complexity originating from a large number of function symbols and axioms.

Hierarchical specifications provide a particular way for the stepwise construction of specifications by adding nonprimitive symbols and their characteristics to already defined primitive specifications. This is not the only way for building specifications. Specifications with hidden symbols and constructors allow one to determine which symbols are "exported" by a specification and which function symbols are used to construct data. In general, any specification language determines a set of constructs which can be used to build specifications in a structured manner. In the following, a general approach for describing syntax and semantics of specification languages is presented.

In Section 6.1, a *model semantics* is given for any set of *specification expressions* which is generated by *specification-building operations.* Examples of *specification-building functions are given in Sections* 6.2–6.5. In Section 6.6, the properties of some specifically important specification-building functions are summarized by an *algebra of specifications.* As a consequence, normal forms for specification expressions can be derived.

The examples are chosen either because they are operations in some specification languages or because they are needed in connection with the concepts of implementation of specifications.

6.1. *Semantics of structured specifications*

The (abstract) syntax for writing specifications in some language L can be seen as being given by a set *Spec_expr* of *specification expressions* which are generated by a set of *specification-building operations.* Any such operation, given a list of argument specification expressions, yields a result specification expression. Then any specification expression is a term built by applying a number of specification-building operations to constant specification expressions. Thus, in an abstract setting, any specification language can be considered as a computation structure with specifications as objects (in the carrier set for specification expressions) and specification-building operators as functions.

For the semantics, it is appropriate to consider two different approaches on two different levels of abstraction (cf. e.g. [50, 178]):

(1) a *presentation* semantics which associates with each specification expression a normal form, the so-called *presentation*, and
(2) a *model semantics* which associates with each specification expression its class of models.

The presentation semantics is adequate for the formal manipulation of specification expressions whereas the model semantics determines the abstract mathematical semantics of specifications. Hence, usually, the presentation semantics is finer than the model semantics: any two specification expressions which are presentation equivalent (i.e., which have the same normal form) have the same class of models, but the converse is not always true.

As shown in Section 3.2, for simple axiomatic and algebraic specifications there are several equivalent possible model semantics: one may denote a simple specification by the class of its models, by the set of isomorphism classes of models, by the set of its congruences and by the theory of its class of models. For the model semantics of more general specification languages it is often not possible to consider only computation structures or congruence classes; also nonreachable algebras have to be admitted as models.

One reason is that if specifications are built in several steps, then the constructors are often determined at a late stage of the design process. In the beginning of this process the models are arbitrary, not necessarily reachable algebras. However, it is required that the classes of models be closed under isomorphism. Thus in the following, a model semantics from specification expressions will be given where a specification expression is considered as a finite syntactic object which describes a certain signature (the signature of the data types) and a class of algebras which satisfy the specification. Moreover, due to a normal form theorem (6.6.2), it will be shown how presentation semantics and theory semantics can be associated with (a subset of) the specification expressions.

First, the class *Spec* representing all possible model semantics for specifications will be constructed. We assume the class *Sign* of all signatures to be given (cf. Section 2.1). The collection *Spec* of all specifications consists of all pairs $\langle \Sigma, C \rangle$ of a signature Σ together with a class C of Σ-algebras which is closed under isomorphism (i.e., for all A, $B \in Alg(\Sigma)$, $A \in C$ and $A \cong B$ implies $B \in C$):

$$Spec =_{\text{def}} \{\langle \Sigma, C \rangle \mid \Sigma \in Sign \text{ and } C \subseteq Alg(\Sigma) \text{ is closed under isomorphism}\}.$$

On *Spec* we define two projection functions: the left projection

$$sig: Spec \to Sign, \qquad sig(\langle \Sigma, C \rangle) =_{\text{def}} \Sigma \quad \text{for all } \langle \Sigma, C \rangle \in Spec,$$

assigns to each element of *Spec* its signature; the right projection

$$Mod: Spec \to \{C \subseteq Alg(\Sigma) \mid \Sigma \in Sign\},$$
$$Mod(\langle \Sigma, C \rangle) =_{\text{def}} C \quad \text{for all } \langle \Sigma, C \rangle \in Spec,$$

assigns to each element of *Spec* its class of models (the so-called *model class*).

By adding a least element $\perp$ to *Spec*, the refinement relation $\subseteq$ (defined by $\langle \Sigma, C \rangle \subseteq \langle \Sigma', C' \rangle$ iff $\Sigma = \Sigma'$ and $C \subseteq C'$; cf. Section 3.2) induces a complete partial order on model classes:

$$Spec_\perp =_{\text{def}} Spec \cup \{\perp\},$$
$$\langle \Sigma, C \rangle \sqsubseteq \langle \Sigma', C' \rangle \text{ iff } \langle \Sigma, C \rangle \supseteq \langle \Sigma', C' \rangle \quad \text{for all } \langle \Sigma, C \rangle, \langle \Sigma', C' \rangle \in Spec,$$
$$\perp \sqsubseteq \langle \Sigma, C \rangle \quad \text{for all } \langle \Sigma, C \rangle \in Spec.$$

The element $\perp$ will be used in connection with parameterized specifications (cf. Section 7.1). $Spec_\perp$ forms a category with embedding morphisms induced by the inclusion relation $\sqsubseteq$, $\perp$ is the (trivial) initial object of $Spec_\perp$. The collection of all elements of *Spec* with signature Σ is denoted by $Spec(\Sigma)$:

$$Spec(\Sigma) =_{\text{def}} \{\langle \Sigma, C \rangle \mid C \subseteq Alg(\Sigma) \wedge C \text{ is closed under isomorphism}\}.$$

Specification-building operations are used to build specifications in a structured manner and to exploit this structure as a guide in their use and understanding [42]. Semantically, a specification-building operation is a function on classes of algebras. It maps classes of models (of the argument specifications) to the class of models (over a signature which must be determined as well) of the result specification. There is only one assumption which is made about these operations: specification-building operations are required to be strict and monotonic with respect to the inclusion of classes of algebras. Intuitively, more restrictive argument specifications yield a more restrictive result.

Formally, a *specification-building function* $f: Spec_\perp \to Spec_\perp$ is a strict, monotonic mapping form $Spec_\perp$ to $Spec_\perp$, i.e. f satisfies the following two properties:

(1) *monotonicity*: $sp \sqsubseteq sp'$ implies $f(sp) \sqsubseteq f(sp')$ for all $sp \in Spec_\perp$;

(2) *strictness*: $f(\perp) = \perp$.

Due to the strictness of the functions in the examples of Sections 6.2–6.5, only the values of the function applications to arguments in the domain are given; all other results are $\perp$. In the following, we denote by $[Spec_\perp \to Spec_\perp]$ the domain of all specification-building functions which is pointwise ordered: for all $f, f' \in [Spec_\perp \to Spec_\perp]$,

$$f \sqsubseteq f' \text{ iff for all } sp \in Spec_\perp\text{: } f(sp) \sqsubseteq f'(sp).$$

6.1.1. FACT. (1) *The model class inclusion relation* $\sqsubseteq$ *is a complete partial order on* $Spec_\perp$.

(2) *The ordering relation* $\sqsubseteq$ *is a complete partial order on the class* $[Spec_\perp \to Spec_\perp]$ *of all strict, monotonic functions on* $Spec_\perp$.

PROOF. Obvious. □

The inclusion relation $\supseteq$ will be used in Section 8.1 for the definition of the implementation relation $\rightsquigarrow$. Hence the vertical composition property of $\rightsquigarrow$ (Fact 8.1.1(1)) follows from the transitivity of $\supseteq$; the horizontal composition property follows from the monotonicity of the specification-building functions (Fact 8.1.1(2)).

Using $Spec_\perp$ and $[Spec_\perp \to Spec_\perp]$ one can give a semantics to any set *Spec_expr* of specification expressions by defining a denotational semantic function (cf.[167, 185]):

$$[\![\ ]\!]: Spec\_expr \to Env \to Spec_\perp$$

which associates an element of *Spec* with any specification expression *SP* depending on the environment $Env =_{\text{def}} [Var \to Spec_\perp]$. Any specification-building operation $f: Spec\_expr \to Spec\_expr$ is denoted by a specification-building function $[\![f]\!]: Spec_\perp \to Spec_\perp$; then, given an environment $env \in Env$ and a specification expression *SP*, the semantics of $f(SP)$ w.r.t. *env* is defined by

$$[\![f(SP)]\!](env) =_{\text{def}} [\![f]\!]([\![SP]\!](env)).$$

On the other hand, any specification-building function $f: Spec_\perp \to Spec_\perp$ induces a specification-building operation $\mathbf{f}$: $Spec\_expr \to Spec\_expr$ (written in boldface) in a canonical way by defining

$$[\![\mathbf{f}]\!] =_{\text{def}} f.$$

In the following, we call a specification expression *SP* a *specification* if it does not contain any free identifiers for specifications. In this case, we write $sig(SP)$ and $Mod(SP)$ for $sig([\![SP]\!])$ and $Mod([\![SP]\!])$.

6.2. *Structured specifications without hidden functions*

In this section, the four specification-building operations $\langle .,. \rangle$ *reach*, *translate* and $\cup$ will be presented which are suited to define structured specifications without hidden functions. For example, the specification building constructs of Larch (Section 9.4) can be expressed in that way.

There are two operations for constructing simple specifications.

For any signature Σ and any set E of Σ-formulas, the *basic specification* $[\Sigma, E]$ denotes the pair $\langle \Sigma, Alg(\Sigma, E) \rangle$ in *Spec*, i.e.

$$sig([\Sigma, E]) =_{\text{def}} \Sigma, \qquad Mod([\Sigma, E]) =_{\text{def}} \{A \in Alg(\Sigma) \mid A \models E\}.$$

Thus, the models of $\langle \Sigma, E \rangle$ are all Σ-algebras which satisfy E.

In order to get the subclass of reachable algebras, we introduce the operation $reach_{Cons}: Spec_\perp \to Spec_\perp$. For any signature $\Sigma = \langle S, F \rangle$ with $Cons \subseteq F$ and any class C of Σ-algebras,

$$sig(reach_{Cons}(\langle \Sigma, C \rangle)) =_{\text{def}} \Sigma,$$
$$Mod(reach_{Cons}(\langle \Sigma, C \rangle)) =_{\text{def}} \{A \in C \mid A \text{ is reachable with } Cons\}.$$

Here, the notion of reachability (which is defined below) is a generalization of the notion of Section 2.2. Notice that $reach_{Cons}$ is a partial operation which requires the domain to contain all symbols of *Cons*.

A Σ-algebra A is called reachable with *Cons* if every carrier set $s \in S_{Cons}$ (where $S_{Cons} =_{\text{def}} \{range(f) \mid f \in Cons\}$) is reachable with functions of *Cons* and elements of carrier sets not in S_{Cons}. In other words, A is *reachable with Cons* if, for each sort $s \in S_{Cons}$ and each carrier element $a \in A_s$, there exists a term $t \in T(\langle S, Cons\rangle, X_{S \backslash S_{Cons}})_s$ with variables in sorts of $S \backslash S_{Cons}$ and an assignment $v: X_{S \backslash S_{Cons}} \to A$ such that $v^*(t) = a$. The qualification "with *Cons*" is omitted if $S_{Cons} = S$ and $Cons = F$.

Thus any simple specification $\langle \Sigma, E\rangle$ as in Section 3.2 (with $\Sigma = \langle S, F\rangle$) can be expressed using *[.,.]* and *reach*:

$$\langle \Sigma, E\rangle = \mathbf{reach}_F([\Sigma, E]).$$

The following fact states some basic properties of *[.,.]* and *reach*.

6.2.1. FACT. *For all sp, sp′ ∈ Spec, sets C, C′ of function symbols, signature Σ and sets E, E′ of Σ-formulas*:

(B) *if* $E \vdash e'$ *for all* $e' \in E'$ *then* $[\Sigma, E] \subseteq [\Sigma, E']$;

(R1) $reach_C(reach_{C'}(sp)) = reach_{C'}(reach_C(sp))$;

(R2) $C \subseteq C' \subseteq opns(sig(sp)) \Rightarrow reach_C(reach_{C'}(sp)) = reach_C(sp)$.

PROOF. The proof follows directly from the definitions of *[.,.]* and *reach* if both sides of the equations are well-defined specification expressions, i.e. the semantics is different from $\bot$. If one of the sides is undefined then the other one is so as well. □

In order to combine two classes of algebras, the binary operation "$\cup$" is introduced. Given two elements $sp_1, sp_2 \in Spec$ with the same signature Σ (i.e., $sig(sp_1) = sig(sp_2) = \Sigma$), the union of sp_1 and sp_2 is defined as follows [157]:

$$sig(sp_1 \cup sp_2) =_{\mathbf{def}} \Sigma, \qquad Mod(sp_1 \cup sp_2) =_{\mathbf{def}} Mod(sp_1) \cap Mod(sp_2).$$

The union is commutative, associative, idempotent and commutes with *reach*; the union of basic specifications is equivalent to the union of their axioms:

6.2.2. FACT. *For any sp, sp_1, sp_2 ∈ Spec, set Cons of function symbols, signature Σ and sets E_1, E_2 of Σ-formulas*:

(U1) $sp \cup sp = sp$;

(U2) $sp_1 \cup sp_2 = sp_2 \cup sp_1$;

(U3) $sp \cup (sp_1 \cup sp_2) = (sp \cup sp_1) \cup sp_2$;

(U4) $[\Sigma, E_1] \cup [\Sigma, E_2] = [\Sigma, E_1 \cup E_2]$;

(U5) $reach_{Cons}(sp) \cup sp_1 = reach_{Cons}(sp \cup sp_1)$.

PROOF. Obvious from the definitions. □

Any signature morphism $\sigma: \Sigma \to \Sigma'$ induces a specification-building function *translate . with .* which translates any $\langle \Sigma, C\rangle \in Spec$ into an element of $Spec(\Sigma')$:

$$sig(translate\langle \Sigma, C\rangle \text{ with } \sigma) =_{\mathbf{def}} \Sigma',$$
$$Mod(translate\langle \Sigma, C\rangle \text{ with } \sigma) =_{\mathbf{def}} \{A \in Alg(\Sigma') \mid A|_\sigma \in C\},$$

where $A|_\sigma$ is the σ-reduct of A. This function will be used for the definition of parameter passing in Section 7.1.

Translations compose and can be inductively defined w.r.t. *[.,.]*, *reach* and $\cup$.

6.2.3. FACT. *For any sp*, $sp_1, sp_2 \in Spec$ *with signature* Σ, *set E of* Σ*-formulas, set C of function symbols, signatures* Σ', Σ'', *signature morphisms* $\sigma: \Sigma \to \Sigma'$, $\sigma': \Sigma' \to \Sigma''$:

(T1) $translate[\Sigma, E]\ with\ \sigma = [\Sigma', \sigma^*(E)]$;
(T2) $translate\ reach_C(sp)\ with\ \sigma \subseteq reach_{\sigma(C)}(translate\ sp\ with\ \sigma)$, *the equality holds if* σ *is injective*;
(T3) $translate\ (sp_1 \cup sp_2)\ with\ \sigma = (translate\ sp_1\ with\ \sigma) \cup (translate\ sp_2\ with\ \sigma)$,
(T4) $translate\ (translate\ sp\ with\ \sigma)\ with\ \sigma' = translate\ sp\ with\ \sigma' \circ \sigma$.

The proof easily follows from the definitions.

(T1) states the satisfaction condition for institutions (cf. Section 3.3.6). (T2) and (T3) state the distributivity of *translate* over *reach* and $\cup$. (T4) states the compositionality of *translate* based on the composition of the signature morphisms.

Renaming can be expressed by the *translate* operation (for any bijective ρ):

$$rename\ \langle \Sigma, C \rangle\ via\ \rho =_{\mathrm{def}} translate\ \langle \Sigma, C \rangle\ with\ \rho.$$

Specifications with different signatures can be combined by taking their translations to the union of the signatures: consider any two sp_1, $sp_2 \in Spec$ with signatures $sig(sp_1) = \Sigma_1$ and $sig(sp_2) = \Sigma_2$, and let $in_k: \Sigma_k \to \Sigma_1 \cup \Sigma_2$ be the canonical embeddings defined by $in_k(x) = x$ for $x \in \Sigma_k$, $k = 1, 2$. Then the *sum* $sp_1 + sp_2$ of sp_1 and sp_2 [161] can be defined by

$$sp_1 + sp_2 =_{\mathrm{def}} (translate\ sp_1\ with\ in_1) \cup (translate\ sp_2\ with\ in_2).$$

Notice that $sp_1 + sp_2$ does not take the disjoint union of $Mod(sp_1)$ *and* $Mod(sp_2)$ as model class. This has the advantage that common subparts of sp_1 and sp_2 such as, for example, the Boolean values or the standard natural numbers are not duplicated. On the other hand, the concept can yield an unsatisfiable specification if in sp_1 and sp_2 some sorts or function symbols have the same names but different semantical meaning.

6.2.4. FACT. *The functions reach*, $\cup$, *translate and therefore also rename and* $+$ *are specification-building.*

6.3. *Constructions*

The following functions *restrict*, *derive from* and therefore also the derived operations *export*, *rename*, can be understood as extensions of functions ranging over algebras to functions ranging over classes of algebras. Similarly as in [158], a specification-building function $f: Spec_\perp \to Spec_\perp$ is called a *construction* if it has a definition of the form

$$sig(f(\langle \Sigma, C \rangle)) = \Sigma', \qquad Mod(f(\langle \Sigma, C \rangle)) = \{\varphi(A) \in Alg(\Sigma') \mid A \in C\}$$

where the transformation of the signatures is independent of the model classes (i.e., $sig(f(\langle \Sigma, C \rangle)) = sig(f(\langle \Sigma, C' \rangle))$ for all $C, C' \subseteq Alg(\Sigma)$) and where φ: $Alg(\Sigma) \to Alg(\Sigma')$ is a function on algebras.

6.3.1. FACT. *Constructions are specification-building functions which preserve the* *satisfiability of specifications, i.e. for any construction* $f: Spec_\perp \to Spec_\perp$ *and any* $C \subseteq Alg(\Sigma)$ *with* $\langle \Sigma, C\rangle \in dom(f)$ $C \neq \emptyset \Rightarrow Mod(f(\langle \Sigma, C\rangle)) \neq \emptyset$.

A first example of a construction can be derived from the last subalgebra $R(A)$ of an algebra A. Given a subset $S' \subseteq S$ of sorts of a signature $\Sigma = \langle S, F\rangle$ and a Σ-algebra A, there exists exactly one Σ-subalgebra $R_{S'}(A)$ of A with carriers of sorts not in S' the same as in A and which is reachable with the function symbols with range in S' and the elements of carrier sets not in S', i.e. $R_{S'}(A)_s = A_s$ for all $s \in S \backslash S'$ and $R_{S'}(A)$ is reachable with $\bigcup \{A_s \mid s \in S\backslash S'\} \cup Cons$ where $Cons =_{\text{def}} \{f \in F \mid range(f) \in S'\}$.

This induces a specification-building function, *restrict*$_{S'}$: $Spec_\perp \to Spec_\perp$, which will be important in connection with the notion of implementation of specifications. It is defined for all signatures which contain S'. For any signature $\Sigma = \langle S, F\rangle$, any set $S' \subseteq S$ of sorts and any $C \subseteq Alg(\Sigma)$,

$$sig(restrict_{S'}(\langle \Sigma, C\rangle)) =_{\text{def}} \Sigma,$$

$$Mod(restrict_{S'}(\langle \Sigma, C\rangle)) =_{\text{def}} \{R_{S'}(A) \in Alg(\Sigma) \mid A \in C\}.$$

The operation *derive from* is the basis for renaming specifications and exporting subsignatures. For any signature Σ', any signature morphism $\sigma: \Sigma \to \Sigma'$ and any class C' of Σ'-algebras it is defined as follows:

$$sig(derive\ from \langle \Sigma', C'\rangle by\ \sigma) =_{\text{def}} \Sigma,$$

$$Mod(derive\ from \langle \Sigma', C'\rangle by\ \sigma) =_{\text{def}} [\{A|_\sigma \in Alg(\Sigma) \mid A \in C'\}],$$

where $A|_\sigma$ is the σ-reduct of A and $[\{A|_\sigma \in Alg(\Sigma) | A \in C'\}]$ denotes the closure under isomorphism (cf. Section 2.2).

Let $\Sigma_0 \subseteq \Sigma$ be a subsignature of Σ and $\rho: \Sigma \to \Sigma'$ a bijective renaming of Σ into Σ'. Then the inverse $\rho^{-1}: \Sigma' \to \Sigma$ and the canonical injection $in: \Sigma_0 \to \Sigma$, $in(x) =_{\text{def}} x$, are signature morphisms and *export* and *renaming* can be expressed in terms of *derive* for all $\langle \Sigma, C\rangle \in Spec$:

$$export\ \Sigma_0\ from\ \langle \Sigma, C\rangle =_{\text{def}} derive\ from\ \langle \Sigma, C\rangle by\ in,$$

$$rename\ \langle \Sigma, C\rangle\ via\ \rho = derive\ from\ \langle \Sigma, C\rangle\ by\ \rho^{-1}.$$

More generally, the functions *derive* and *translate* are dual to each other in the following sense.

6.3.2. FACT. *Let* $sp, sp_1 \in Spec$ *with signatures* Σ *and* Σ_1, *resp., and let* $\rho: \Sigma_1 \to \Sigma$ *be a signature morphism. Then the following holds*:

$$derive\ from\ sp\ by\ \rho \subseteq sp_1\ \text{ iff }\ sp \subseteq translate\ sp_1\ with\ \rho.$$

PROOF. ("$\Rightarrow$"): Let $A \in Mod(sp)$. Then $A|_\rho \in Mod(derive\ from\ sp\ by\ \rho)$. By hypothesis, $A|_\rho \in Mod(sp_1)$. Hence $A \in Mod(translate\ sp_1\ with\ \rho)$.

("$\Leftarrow$"): Let $A \in Mod(derive\ from\ sp\ by\ \rho)$, i.e., $\exists B \in Mod(sp)$: $A \cong B|_\rho$. By hypothesis, $B \in Mod(translate\ sp_1\ with\ \rho)$ holds, i.e. $B|_\rho \in Mod(sp_1)$. Hence $A \in Mod(sp_1)$. $\square$

Export satisfies the following basic properties.

6.3.3. FACT. *Let* $sp, sp_1, sp_2 \in Spec$, *C a set of function symbols*, Σ, Σ_1, Σ_2 *signatures*, $\sigma: \Sigma \to \Sigma_2$ *a signature morphism*:

(E1) $C \subseteq opns(\Sigma) \Rightarrow reach_C(export\ \Sigma\ from\ sp) = export\ \Sigma\ from\ reach_C(sp)$;

(E2) $\Sigma = sig(sp_1) \cap sig(sp_2)$
$\Rightarrow (export\ \Sigma\ from\ sp_1) \cup (export\ \Sigma\ from\ sp_2) = export\ \Sigma\ from\ (sp_1 + sp_2)$;

(E3) $\Sigma \subseteq \Sigma_1 \subseteq sig(sp) \Rightarrow export\ \Sigma\ from (export\ \Sigma_1\ from\ sp) = export\ \Sigma\ from\ sp$;

(E4) $export\ sig(sp)\ from\ sp = sp$;

(αh) $\Sigma_1 \subseteq \Sigma = sig(sp) \wedge injective(\sigma)$
$\Rightarrow export\ \Sigma_1\ from\ sp = export\ \Sigma_1\ from\ (translate\ sp\ with\ \sigma)$.

PROOF. (E1) is obvious, for (E2)–(E4) and (αh) see [14]. □

(E1) and (E2) show that the *export* operation commutes with *reach* and $\cup$. (E3)–(E4) are self-evident, the preconditions of (E1)–(E3) ensure the well-definedness of the specification expressions. (αh) shows the renameability of hidden symbols; the condition $\sigma|_{\Sigma_1} = id$ ensures that no visible symbols are renamed; then, due to the injectivity of σ, no name clashes between hidden and visible symbols can occur.

Using *export* and *reach* it is possible to express any specification $SP = \langle \Sigma, H\Sigma, E, Cons \rangle$ with constructors and hidden symbols (see Section 5.1):

$$SP = export\ \Sigma\ from\ reach_{Cons}([H\Sigma, E]).$$

Moreover, *derive* allows one to define the notion of a specification morphism: for any two elements $\langle \Sigma, C \rangle$ and $\langle \Sigma', C' \rangle \in Spec$, a *specification morphism* from $\langle \Sigma, C \rangle$ to $\langle \Sigma', C' \rangle$ is a signature morphism $\sigma: \Sigma \to \Sigma'$ such that

$$derive\ from\ \langle \Sigma', C' \rangle\ by\ \sigma \subseteq \langle \Sigma, C \rangle.$$

Thus the set *SPEC* of specification expressions forms a category where, for $SP, SP' \in SPEC$, $\sigma: SP \to SP'$ is a morphism (also called *specification morphism*) if σ is a specification morphism from $[\![SP]\!]$ to $[\![SP']\!]$. In the literature, specification morphisms are often defined on the level of theories (see e.g. [55]), e.g. for simple equational specifications $\langle \Sigma, E \rangle$, $\langle \Sigma', E' \rangle$, an (equational) specification morphism from $\langle \Sigma, E \rangle$ to $\langle \Sigma', E' \rangle$ is a signature morphism $\sigma: \Sigma \to \Sigma'$ such that for any $e \in E$ the translated formula $\sigma^*(e)$ is derivable from E', i.e. $E' \vdash_{EQ} \sigma^*(e)$.

6.4. Extensions

In Sections 3.2 and 5.1, a construct **extend**.**by**. was introduced for extending specifications in a syntactic way. On the semantic level four different specification-building functions will be considered: enrichments and free, terminal and reachable extensions.

For any signature $\Sigma = \langle S, F \rangle$, any set of sorts S', any $(S \cup S')$-sorted set of function symbols F', any set of conditional $\Sigma' =_{def} \langle S \cup S', F \cup F' \rangle$-*equations* E' and for any class C of Σ-algebras, the semantics of the **extend**.**freely by**. operation is defined as follows:

$$sig(extend\ \langle\Sigma, C\rangle\ freely\ by\ S', F', E') =_{\text{def}} \Sigma',$$
$$Mod(extend\ \langle\Sigma, C\rangle\ freely\ by\ S', F', E') =_{\text{def}} \{F_{S',F',E'}(A) \in Alg(\Sigma') \mid A \in C\},$$

where $F_{S',F',E'}(A)$ denotes the (isomorphism class of) free extensions of A w.r.t. S', F', E'.

Notice that, in general, the free extension does not preserve all Σ-formulas which are valid in C, but it does preserve valid ground equations. Hence, it is a specification morphism for equational specifications but it is not a specification morphism in the sense of Section 6.3. The initial algebra construction is a special case of free extension where $\Sigma = \langle\emptyset, \emptyset\rangle$ is the empty signature; also, quotients are special cases with $S' = F' = \emptyset$:

$$data\ sorts\ S\ functions\ F\ equations\ E =_{\text{def}} extend\ \emptyset\ freely\ by\ S, F, E,$$
$$quotient\ \langle\Sigma, C\rangle\ by\ E =_{\text{def}} extend\ \langle\Sigma, C\rangle\ freely\ by\ \emptyset, \emptyset, E.$$

Terminal extensions are defined similar to free extensions; with the same notations as above (in particular, $\Sigma' =_{\text{def}} \langle S \cup S', F \cup F'\rangle$) the semantics of the **extend . finally by .** operation is given as follows:

$$sig(extend\ \langle\Sigma, C\rangle\ finally\ by\ S', F', E') =_{\text{def}} \Sigma',$$
$$Mod(extend\ \langle\Sigma, C\rangle\ finally\ by\ S', F', E')$$
$$=_{\text{def}} \{A \in Alg(\Sigma') \mid \exists B \in C\colon A \cong Z_{S',F',E'}(B)\},$$

where $Z_{S',F',E'}(B)$ denotes the terminal extension of B w.r.t. S', F', E', i.e. a terminal algebra of the class $\{A \in Alg(\Sigma') \mid A|_\Sigma \cong B \wedge A \models E'\}$.

Note that terminal extensions do not always exist; in contrast to all other specification-building functions above, the domain of the terminal extension functions cannot be characterized only by signature, it depends also on the set of equations E'. A sufficient condition for the existence of a terminal extension of an arbitrary Σ-algebra B can be derived from the sufficient completeness criterion for hierarchical specifications. Let $\underline{B} =_{\text{def}} \{\underline{b} \mid b \in B\}$ be a set of constants representing the elements of B such that $\underline{b}^B = b$ for all $b \in B$, and let

$$Diag(B) =_{\text{def}} \{G \mid B \models G,\ G \equiv (u = v) \text{ or } G \equiv (u \neq v),\ u, v \in T(\langle S, F \cup \underline{B}\rangle)\}$$

be the *diagram* of $\underline{B}$. Then a terminal extension of $\underline{B}$ exists if the hierarchical specification $\langle \Sigma' \cup \underline{B}, E' \cup Diag(B), \langle \Sigma \cup \underline{B}, Diag(B)\rangle\rangle$ is sufficiently complete and hierarchy-consistent (see Theorem 5.4.5). For further criteria see [175, 64].

Extend operations for loose specifications can be modeled using "+" and "reach". For any signature $\Sigma = \langle S, F\rangle$, any set of sorts S', any $(S \cup S')$-sorted set F' of function symbols and any set E' of Σ'-sentences, where $\Sigma' = \langle S \cup S', F \cup F'\rangle$, **enrichment** and **reachable extension** of any $\langle\Sigma, C\rangle \in Spec$ by S', F', E' are defined as follows:

$$enrich\ \langle\Sigma, C\rangle\ by\ S', F', E' =_{\text{def}} \langle\Sigma, C\rangle + [\Sigma', E'],$$
$$extend\ \langle\Sigma, C\rangle\ reachably\ by\ S', F', E' =_{\text{def}} reach_{F'_{Cons}}(enrich\ \langle\Sigma, C\rangle\ by\ S', F', E')$$

where $F'_{Cons} =_{\text{def}} \{f \in F' \mid range(f) \in S'\}$ is the set of all new function symbols with range in a new sort. Hence any model A of the enrichment of $\langle\Sigma, C\rangle$ by S', F', E' satisfies the axioms E' and its Σ-reduct is in C; any model A of the reachable extension is, moreover, reachable on the new sorts with the new function symbols from F' the ranges of which are in S', i.e. $A \in Mod(extend(\langle\Sigma, C\rangle \; reachably \; by \; S', F', E'))$ if and only if $A \in Alg(\Sigma')$, $A|_\Sigma \in C$ and $A \models E'$ and A is reachable with F'_{Cons}. Note that the reachability requirement is void if no new sorts are introduced (i.e. if $S' = \emptyset$); in this case, the reachable extension and enrichment are equivalent:

$$extend \; \langle\Sigma, C\rangle \; reachably \; by \; \emptyset, F', E' = enrich \; \langle\Sigma, C\rangle \; by \; \emptyset, F', E'$$

6.4.1. FACT. *The extension functions enrich and extend .αby. for α∈(freely, finally, reachably} are specification-building functions; extend.freelyby. and extend.finally by. are constructions.*

Any *hierarchical* specification (see Section 5.4) can be expressed using the reachable extension operation:

$\langle\Sigma', E\cup E', \langle\Sigma, E\rangle\rangle$

$= reach_{F\cup F'}(reach_F([\Sigma, E]) + [\Sigma', E'])$ (by definition of hierarchical specification)

$= reach_{F'_{Cons}}(reach_F([\Sigma, E]) + [\Sigma', E'])$ $(*)$

$= extend \; \langle\Sigma, E\rangle \; reachably \; by \; S', F', E'$ (by definition of *extend reachably*)

The equation $(*)$ follows from the fact that reachability on the sorts S of Σ is expressed by the primitive function symbols F; the function symbols in $F' \backslash F'_{Cons}$ do not contribute to the generation of data in $\langle\Sigma, E\rangle$.

The hierarchical extension is not equivalent to the syntactic extension **extend.by.** of simple specifications as defined in Section 3.2 which in general does not preserve the semantics of $\langle\Sigma, E\rangle$:

extend $\langle\Sigma, E\rangle$ **by sorts** S' **functions** F' **axioms** E'

$= \mathbf{reach}_{F\cup F'}(\langle\Sigma, E\rangle + \langle\Sigma', E'\rangle)$

$= \mathbf{reach}_{F\cup F'}(\langle\Sigma', E\cup E'\rangle)$.

6.4.2. EXAMPLE. The difference between the different kinds of extensions can be (trivially) illustrated by extending the specification *BOOLM* (Example 3.2.4(1)) by a third constant *error*. The syntactic extension of *BOOLM* is defined by

spec $BOOLM_e \equiv$ **extend** *BOOLM* **by function** *error*: →*bool* **endspec**

This specification has three nonisomorphic model classes which can be represented by the following three models $I, \mathcal{M}_t$ and $\mathcal{M}_f$:

(1) I is the term model with carrier set $I_{bool} =_{def} \{true, false, error\}$;

(2) $\mathcal{M}_t$ has carrier set $\{true, false\}$ with $error^{\mathcal{M}_t} =_{def} true$;

(3) $\mathcal{M}_f$ has carrier set $\{true, false\}$ with $error^{\mathcal{M}_f} =_{def} false$.

The term model I is the initial model of $BOOLM_e$; its isomorphism class is the model class of the free extension of *BOOLM* by *error*. $BOOLM_e$ does not admit a terminal model, hence the terminal extension of *BOOLM* by error is undefined. The models $\mathcal{M}_t$

and $\mathcal{M}_f$ are representatives of the (isomorphism classes of) models of the reachable extension of *BOOLM*; this extension does not admit *I* as a model since the carrier set {*true*, *false*} is not preserved by *I*. The enrichment has the same class of models as the reachable extension. Enrichment and reachable extension differ for the following specification.

> **spec** *ELEM* ≡
>   **enrich** *BOOL* **by**
>   **sort** *elem*
>   **function** $eq: elem, elem \to bool$
>   **axiom** $x = y \Leftrightarrow eq(x, y) = true$
> **endspec**

Any *sig*(*ELEM*)-algebra *A* satisfying the axiom and containing a model of *BOOL* (Example 5.1.1) as *sig*(*BOOL*)-reduct is a model of *ELEM* whereas the reachable extension does not admit any model since the signature is not sensible (cf. Section 2.2).

Note that because of the constructor set {*true*, *false*} of sort *bool* the axiom above implies "$x \neq y \Leftrightarrow eq(x, y) = false$".

The example above also shows that, in general, extensions do not protect the extended specifications; the carrier set of the free extensions is different from any carrier set of a model of *BOOLM*, the terminal extension does not exist. On the other hand, the *sig*(*NAT*1)-reduct of any model of *LOOSE-SET* (Example 3.2.4(4)) is a model of *NAT*1 and any such model of *NAT*1 can be extended to a model of *LOOSE-SET*. This situation is summarized by the following definition: For any $sp, sp' \in Spec$ with $sig(sp) \subseteq sig(sp')$, sp' is called *persistent extension* of *sp* if

$$sp = export\ sig(sp)\ from\ sp'.$$

A specification-building function $f: Spec_\perp \to Spec_\perp$ is called *persistent* if, for any actual parameter *sp* of *f*, *f*(*sp*) is a persistent extension of *sp*. If *f* is a construction, then for persistency it is enough to consider algebras instead of classes of algebras.

6.4.3. FACT. *For any construction* $f: Spec_\perp \to Spec_\perp$ *associated with* $\varphi: Alg(\Sigma) \to Alg(\Sigma')$ *and extending* Σ *(i.e.* $\Sigma \subseteq \Sigma'$*), the following conditions are equivalent:*

(1) *f is persistent;*
(2) *for any* $A \in Alg(\Sigma)$ *such that* $\varphi(A)$ *is defined,* $\varphi(A)|_\Sigma \cong A$.

PROOF. (1) ⇒ (2) by definition of persistency; (2) ⇒ (1) $Mod(f(\langle \Sigma, C \rangle)) = \{\varphi(A)|_\Sigma \mid A \in C\} = C$. □

Note that Condition (2) is the usual definition of persistency (cf. e.g. [55]). Moreover, any persistent extension *SP'* of a specification *SP* is a *conservative extension*, i.e. for any *sig*(*SP*)-formula *G*, $Mod(SP') \models G$ if and only if $Mod(SP) \models G$ (see e.g. [164]).

A free extension is a specification morphism in the sense of Section 6.3 if it is persistent. Necessary and sufficient criteria for persistency of free extensions can be found in [139, 140].

In many cases, enrichments by functions (without introducing new sorts) are equivalent to free, terminal and reachable extensions as shown in the following fact.

6.4.4. FACT. *Let* $SP' =_{\text{def}}$ **extend** SP **by** $\emptyset, F, E$ *be a persistent and sufficiently complete extension and let* $Mod(SP) \subseteq Gen(sig(SP))$. *Then*

$$Mod(SP') = Mod(\textbf{extend } SP \textbf{ reachably by } \emptyset, F, E).$$

If E is a set of equations, then also $Mod(SP') = Mod(\textbf{extend } SP\ \alpha \textbf{ by } \emptyset, F, E)$ *holds where* $\alpha \in \{\mathit{freely}, \mathit{finally}\}$.

PROOF. If the extension is sufficiently complete then any computation structure is extended in a unique way independently of the special kind of extension. Persistency guarantees that any model of SP can be extended. □

6.5. Observational abstraction

The last important concept is *observational abstraction* [156]. Every equivalence relation $\equiv\ \subseteq Alg(\Sigma) \times Alg(\Sigma)$ which is closed under isomorphism determines a specification-building function $\alpha_{\equiv}: Spec_{\perp} \to Spec_{\perp}$ where for any class C of Σ-algebras,

$$sig(\alpha_{\equiv}(\langle \Sigma, C \rangle)) =_{\text{def}} \Sigma,$$
$$Mod(\alpha_{\equiv}(\langle \Sigma, C \rangle)) =_{\text{def}} \{A \in Alg(\Sigma) \mid \exists A' \in C: A \equiv A'\}.$$

An example is *behavioral abstraction* w.r.t. a set $OBS \subseteq S$ of observable sorts (cf. Section 5.4), which will be denoted by *behavior* $\langle \Sigma, C \rangle$ *w.r.t. OBS*. Let

$$\Sigma_{OBS} =_{\text{def}} \langle OBS, \{f \in F \mid type(f) \in OBS^* \times OBS\}\rangle$$

be the subsignature of Σ associated with OBS:

$$sig(behavior\ \langle \Sigma, C \rangle\ w.r.t.\ OBS) =_{\text{def}} \Sigma,$$
$$Mod(behavior\ \langle \Sigma, E \rangle\ w.r.t.\ OBS) =_{\text{def}} \{A \in Alg(\Sigma) \mid \exists A' \in C: A \text{ and } A' \text{ are behaviorally equivalent on } \Sigma_{OBS}\}.$$

Thus all computations of *behavior* $\langle \Sigma, C \rangle$ *w.r.t. OBS* have the same observable results. As a consequence, the models of a hierarchical specification $\langle \Sigma, E, P \rangle$ are contained in the models of **behavior** $\langle \Sigma, E \rangle$ **w.r.t.** *sorts*(P), i.e.

$$\langle \Sigma, E, P \rangle \subseteq \textbf{behavior } \langle \Sigma, E \rangle \textbf{ w.r.t. } sorts(P).$$

To give an example, behavioral abstraction of the specification *SETNAT* of finite sets of natural numbers (cf. Example 3.2.4(4)) admits all those algebras as models which satisfy the axioms of *LOOSE-SET*:

$$\textit{LOOSE-SET} \subseteq \textbf{behavior } \textit{SETNAT} \textbf{ w.r.t. } \{bool\}.$$

6.5.1. FACT. *Observational and behavioral abstraction are specification-building functions.*

Observational abstraction is transitive and is preserved by the *derive*-operation.

6.5.2. FACT. *Let* $sp \in Spec$, $\equiv$ *a* $sig(sp)$*-equivalence relation*, $\equiv'$ *a* Σ'*-equivalence relation and* $\sigma : \Sigma' \to sig(sp)$ *a signature morphism. Define for all* $A, A' \in Alg(sig(sp))$, $A\ \sigma(\equiv')\ A'$ *iff* $A|_\sigma \equiv' A'|_\sigma$.

(1) $\alpha_\equiv$ *is transitive, i.e.* $\alpha_\equiv(\alpha_\equiv(sp)) = \alpha_\equiv(sp)$,

(2) *derive from* $\alpha_{\sigma(\equiv')}(sp)$ *by* $\sigma \subseteq \alpha_{\equiv'}$(*derive from sp by* σ).

PROOF. (1) holds since $\equiv$ is an equivalence relation. (2) follows from the definition of $\sigma(\equiv')$. □

Consider any construction $\kappa: Spec(\Sigma) \to Spec(\Sigma')$ where the transformation of signatures is given by a signature morphism $\sigma: \Sigma \to \Sigma'$. It preserves behavioral abstraction if it satisfies the following properties [158]: κ (defined by Σ' and φ) is called *observably complete* w.r.t. OBS if for any ground term $t \in T(\Sigma')_{\sigma(OBS)}$ there exists a term $t' \in T(\Sigma)_{OBS}$ such that, for all $A \in Alg(\Sigma)$, $\varphi(A) \models t = \sigma(t')$; it is *observably persistent* if, for all terms $t_1, t_2 \in T(\Sigma)_{OBS}$ and any $A \in Alg(\Sigma)$, $\varphi(A) \models \sigma(t_1) = \sigma(t_2)$ iff $A \models t_1 = t_2$.

6.5.3. LEMMA. *For any signature morphism* $\sigma: \Sigma \to \Sigma'$ *which is injective on sorts and any construction* κ: $Spec(\Sigma) \to Spec(\Sigma')$, *if* κ *is observably complete and observably persistent w.r.t. OBS, then*

$$\kappa(\textit{behavior } \langle \Sigma, C\rangle \textit{ w.r.t. } OBS) \subseteq \textit{behavior } \kappa(\langle \Sigma, C\rangle) \textit{ w.r.t. } \sigma(OBS).$$

PROOF. Assume that $A_1, A_2 \in Alg(\Sigma)$ are behaviorally equivalent w.r.t. OBS; we prove that $\varphi(A_1)$ and $\varphi(A_2)$ are behaviorally equivalent w.r.t. $\sigma(OBS)$, i.e. for any $t'_1, t'_2 \in T(\Sigma')_{\sigma(OBS)}$, $\varphi(A_1) \models t'_1 = t'_2$ iff $\varphi(A_2) \models t'_1 = t'_2$.

("$\Rightarrow$"): Because of observable completeness, let $t_1, t_2 \in T(\Sigma)_{OBS}$ be such that, for all $A_1 \in Alg(\Sigma)$, $\varphi(A_1) \models t'_1 = \sigma(t_1)$ and $\varphi(A_1) \models t'_2 = \sigma(t_2)$. Since σ is injective on OBS, t_1 and t_2 are of the same sort. Because of $\varphi(A_1) \models \sigma(t_1) = \sigma(t_2)$, $A_1 \models t_1 = t_2$ holds by observable persistency and so $A_2 \models t'_2 = t_2$. This implies $\varphi(A_2) \models \sigma(t_1) = \sigma(t_2)$ and therefore $\varphi(A_2) \models t'_1 = t'_2$.

("$\Leftarrow$"): By symmetry. □

Further properties of behavioral abstraction can be found in [156].

6.6. *An algebra of structured specifications*

The specification-building operations induce a language for building structured specifications. In this language, specification expressions can be formed by means of the various specification building operations starting with axiomatic specifications of the form $[\Sigma, E]$ (cf. Section 3.2). In order to modify and restructure such specification expressions one may use identities which can be proven due to the semantics of the operators.

The following specification BSA (for "Basic Specification Algebra") algebraically describes a specification language (which is formed by the subset $\{[\cdot,\cdot]$, *reach*, $+$, *translate, export*} of the specification-building operations). It is hierarchically based on two specifications *SIGNATURE* and *FORMULA*. *SIGNATURE* is a specification of the computation structure of finite signatures and signature morphisms (see [14]) with two sorts *signature* and *signature morphism* and operations for manipulating signatures (such as $\cap$, $\cup$, $\subseteq$) and signature morphisms (such as natural embeddings in Σ, Boolean functions for determining surjectivity and injectivity) of signature morphisms. *FORMULA* is a specification of the algebra of formulas used in the language. Following Section 3.2 this can be e.g. equational formulas, conditional equations or first-order formulas. The axioms of *BSA* summarize the properties of the specification-building functions stated in Sections 6.2–6.3.

spec $BSA \equiv$
extend *SIGNATURE, FORMULA* **by**
sort *spec*
constructors $[.,.]$: *signature, set formula*$\rightarrow$*spec*,
$reach_{.}.$: *set funct, spec*$\rightarrow$*spec*,
$.\cup.$: *spec, spec*$\rightarrow$*spec*,
export . from .: *signature, spec*$\rightarrow$*spec*
translate . by .: *spec, signature morphism*$\rightarrow$*spec*,
function *sig* : *spec*$\rightarrow$*signature*
axioms $\forall\, sp, sp_1, sp_2$: *spec*, $\Sigma, \Sigma_1, \Sigma', \Sigma''$: *signature* C, C': *set funct*, E_1, E_2: *set formula*, $\sigma: \Sigma \rightarrow \Sigma'$, $\sigma': \Sigma' \rightarrow \Sigma''$: *signature morphism*.

(U1) $sp \cup sp = sp$,
(U2) $sp_1 \cup sp_2 = sp_2 \cup sp_1$,
(U3) $sp \cup (sp_1 \cup sp_2) = (sp \cup sp_1) \cup sp_2$,
(U4) $[\Sigma, E_1] \cup [\Sigma, E_2] = [\Sigma, E_1 \cup E_2]$,
(U5) $reach_C(sp) \cup sp_1 = reach_C(sp \cup sp_1)$,

(R1) $reach_C(reach_{C'}(sp)) = reach_{C'}(reach_C(sp))$,
(R2) $C \subseteq C' \subseteq opns(sig(sp)) \Rightarrow reach_C(reach_{C'}(sp)) = reach_C(sp)$,

(E1) $C \subseteq opns(\Sigma) \Rightarrow$
$reach_C(export\ \Sigma\ from\ sp) = export\ \Sigma\ from\ reach_C(sp)$,
(E2) $\Sigma = sig(sp_1) \cap sig(sp_2) \Rightarrow$
$(export\ \Sigma\ from\ sp_1) \cup (export\ \Sigma\ from\ sp_2) = export\ \Sigma\ from\ (sp_1 + sp_2)$,
(E3) $\Sigma \subseteq \Sigma_1 \subseteq sig(sp) \Rightarrow$
$export\ \Sigma\ from\ (export\ \Sigma_1\ from\ sp) = export\ \Sigma\ from\ sp$,
(E4) $export\ sig(sp)\ from\ sp = sp$,

(T1) $translate\ [\Sigma, E]\ with\ \sigma = [\Sigma', \sigma^*(E)]$,
(T2) $injective(\sigma)\ translate\ reach_C(sp)\ with\ \sigma = reach_{\sigma(C)}(translate\ sp\ with\ \sigma)$,
(T3) $translate\ (sp_1 \cup sp_2)\ with\ \sigma = (translate\ sp_1\ with\ \sigma) \cup (translate\ sp_2\ with\ \sigma)$,
(T4) $translate\ (export\ \Sigma\ from\ sp)\ with\ \sigma = export\ \Sigma'\ from\ (translate\ sp\ with\ \hat{\sigma})$,
(T5) $translate\ (translate\ sp\ with\ \sigma)\ with\ \sigma' = translate\ sp\ with\ \sigma' \circ \sigma$,

(αh) $\Sigma_1 \subseteq \Sigma = sig(sp) \wedge \sigma|_{\Sigma_1} = id \wedge injective(\sigma) \Rightarrow$
$export\ \Sigma_1\ from\ sp = export\ \Sigma_1\ from\ (translate\ sp\ with\ \sigma)$,

(S1) $\Sigma \subseteq sig(E) \Rightarrow sig([\Sigma, E]) = \Sigma$,
(S2) $C \subseteq opns(sig(sp)) \Rightarrow sig(reach_C(sp)) = sig(sp)$,
(S3) $sig(sp_1) = sig(sp_2) \Rightarrow sig(sp_1 \cup sp_2) = sig(sp_1)$,
(S4) $\Sigma \subseteq sig(sp) \Rightarrow sig(export\ \Sigma\ from\ sp) = \Sigma$,
(S5) $\Sigma = sig(sp) \Rightarrow sig(translate\ sp\ with\ \sigma) = \Sigma'$

endspec

(U1)–(U5), (R1), (R2), (E1)–(E4), (T1)–(T4), (αh) can be found in the previous sections. (T4) shows that *translate* commutes with *export* if the translation via σ is disjoint from the hidden symbols in *sp*. Here $\hat{\sigma}: sig(sp) \to \Sigma' \cup (sig(sp) \setminus \Sigma)$ denotes the extension of σ to $sig(sp)$ defined by $\hat{\sigma}(x) = \sigma(x)$ if $x \in \Sigma$, and x otherwise. (S1)–(S5) are the natural identities for signatures.

6.6.1. FACT. *The specification-building operations satisfy the axioms of BSA.*

As a consequence, the congruence $\sim_{\text{sem}}$ induced by the semantics of the specification-building operations is associated with a model of *BSA*. The operation $sig: spec \to signature$ is the only operation with range in a primitive sort. Hence, the terminal algebra of *BSA* identifies two specification expressions if and only if they have the same signature. Thus, the terminal congruence is strictly coarser than $\sim_{\text{sem}}$. A deep analysis of several models (of a similar specification of a subsignature) of *BSA* can be found in [14]. The specification *BSA* can be used to derive the following normal form theorem for specification expressions (cf. also [14]).

6.6.2. THEOREM. *For all ground specification expressions $sp \in T(\Sigma_{BSA})_{spec}$ containing only injective signature morphisms there exists a specification expression*

$$sp' \equiv export\ \Sigma'\ from\ (reach_{C_1}(\ldots reach_{C_n}([\Sigma'', E''])\ldots))$$

where Σ', Σ'' are signatures, E'' is a set of Σ''-formulas and, for $i = 1, \ldots, n$, C_i are sets of function symbols such that $BSA \vdash_I sp = sp'$.

PROOF. By structural induction on the form of *sp*: Let *sp* be a well-defined expression. We write $reach_\Gamma(sp)$ for $reach_{C_1}(\ldots reach_{C_n}(sp))$ if $\Gamma = \{C_1, \ldots, C_n\}$ is a powerset of function symbols.

Case 1: $sp \equiv [\Sigma, E]$ has the required form.

Case 2: $sp \equiv reach_C(sp_1)$. By the induction hypothesis, sp_1 has the form $export\ \Sigma\ from\ reach_\Gamma([\Sigma'', E''])$. Then apply (E1).

Case 3: $sp \equiv export\ \Sigma\ from\ sp_1$. Apply (E3) to the induction hypothesis.

Case 4: $sp \equiv sp_1 \cup sp_2$. The well-definedness of *sp* guarantees that sp_1 and sp_2 have the same signature Σ. Then, by induction hypothesis,

$$sp_1 \equiv export\ \Sigma\ from\ reach_{\Gamma_1}([\Sigma_1, E_1]) \quad \text{and}$$
$$sp_2 \equiv export\ \Sigma\ from\ reach_{\Gamma_2}([\Sigma_2, E_2]);$$

because of (αh) and the commutativity (T2) of *translate* with *reach*, one may assume that the hidden symbols of Σ_1 and Σ_2 are disjoint. Thus applying (E2), *sp* is equivalent to

$$export\ \Sigma\ from\ (reach_{\Gamma_1}([\Sigma_1, E_1]) + reach_{\Gamma_2}([\Sigma_2, E_2])).$$

Several applications of (T2), (T1), (U5) and (U4) show that this expression is equivalent to

$$export\ \Sigma\ from\ (reach_{\Gamma_1 \cup \Gamma_2}([\Sigma_1 \cup \Sigma_2,\ E_1 \cup E_2])).$$

Case 5: $sp \equiv translate\ sp_1\ by\ \sigma : \Sigma \to \Sigma'$ with σ injective. By induction hypothesis, sp_1 has the form

$$export\ \Sigma\ from\ reach_\Gamma([\Sigma'', E'']).$$

By (T4) *sp* is equivalent to

$$export\ \Sigma'\ from\ translate\ reach_\Gamma([\Sigma'', E''])\ by\ \hat{\sigma}.$$

Applying (T2) several times and then (T1) yields the required result. □

This normal form theorem associates a presentation semantics with each specification expression in $T(\Sigma_{BSA})_{spec}$ (provided the occurring signature morphisms are injective). In most practical cases, it is equivalent to a specification with hidden symbols and constructors (cf. Section 5).

A first normal form theorem was given by Ehrig et al. [58] for *derive, free* and *reachable extensions.* In [60], refinement relations (i.e. inclusions, cf. Sections 3.2 and 8.1) between specification expressions are systematically derived for many of the specification-building functions of Sections 6.2–6.5.

7. Parameterized specifications

Parameterization is the process of encapsulating a piece of software and abstracting from some names (or more generally from some subexpressions) occuring in it in order to replace them in other contexts by different actual parameters. The most common form (in programming) is parameterization of a piece of code by procedural abstraction. The semantics of such a procedure is a function from states to states.

For algebraic specifications one may distinguish two main approaches to parameterization according to the available syntax: parameterization of a flat specification *SP* means considering a subspecification *SP*0 of *SP* as parameter part; in this case, a parameterized specification is a pair $\langle SP0, SP\rangle$ of specifications such that *SP* is an extension of *SP*0. Parameterization of structured specifications is done by means of *λ-abstraction* (Section 7.1): a parameterized specification is considered as a mapping taking specifications as arguments and giving a specification as result. Parameterization by extension of flat specifications is a particular case of functional abstraction based on the *pushout construction* (Section 7.2). It has been introduced by Thatcher et al. [171] and studied e.g. in [49, 51, 55] for the initial algebra approach, in [64] for the terminal algebra approach and in [180] for the loose approach. The general approach has been initiated by Burstall and Goguen [43] and further studied in [161, 178, 157].

7.1. *The typed λ-calculus approach*

In the λ-calculus approach, parameterized specifications are considered as specification-building functions which are written as λ-expressions. Unlike the classical λ-calculus, one does not allow arbitrary arguments to be substituted for the formal parameters of λ-terms. Instead of having λ-terms of the form $\lambda x.SP$, one considers terms of the form $\lambda x{:}R.SP$, where R is a parameter restriction. Syntactically, a *(structured) parameterized specification PSP* is written as a λ-expression,

$$\lambda X: SP_{\text{par}} \centerdot SP_{\text{body}},$$

where X is an identifier for specifications, SP_{par} is the *formal parameter requirement specification* (*expression*) and SP_{body} is the *target specification* (*expression*). Both expressions are built using specification-building operations and may contain the identifier X. The variable X is *bound* by λ in *PSP*. In general, *PSP* may contain other free and bound variables.

Semantically, a parameterized specification expression is considered as function from classes of $sig([\![SP_{\text{par}}]\!])$-models to $sig([\![SP_{\text{body}}]\!])$-models which depends on the environment *Env* (if *PSP* contains free identifiers); i.e., the semantic interpretation $[\![PSP]\!]$ of *PSP* has type $Env \to Spec_\perp \to Spec_\perp$. For any environment $env \in Env$ and any specification $sp \in Spec_\perp$, the semantics of the λ-expression *PSP* is defined by

$$[\![\lambda X: SP_{\text{par}} . SP_{\text{body}}]\!](env)(sp) =_{\text{def}} \begin{cases} [\![SP_{\text{body}}]\!](env[X \mapsto sp]) & \text{if } \perp \neq sp \subseteq [\![SP_{\text{par}}]\!](env[X \mapsto sp]), \\ \perp & \text{otherwise.} \end{cases}$$

If the formal parameter requirement specification SP_{par} does not contain X, then *PSP* denotes a specification-building function and function application is equivalent to β-reduction.

7.1.1. FACT. *Let* $PSP \equiv \lambda X: SP_{\text{par}} \centerdot SP_{\text{body}}$ *be a parameterized specification expression, where X is not free in* SP_{par}.

(1) *For any environment* $env \in Env$, $[\![PSP]\!](env): Spec_\perp \to Spec_\perp$ *is a specification-building function, i.e. it is strict and monotonic.*

(2) *Function application is equivalent to conditional β-reduction: for any specification expression SP with* $[\![SP]\!](env) \subseteq [\![SP_{par}]\!](env)$,

$$[\![(\lambda X: SP_{\text{par}} . SP_{\text{body}})(SP)]\!](env) =_{\text{def}} [\![SP_{\text{body}}[SP/X]]\!](env).$$

In particular, if SP_{body} *does not contain any other free identifier except X, then for any specification expression SP with* $SP \subseteq SP_{\text{par}}$,

$$(\lambda X: SP_{\text{par}} \centerdot SP_{\text{body}})(SP) = SP_{\text{body}}[SP/X].$$

The proof of (1) follows by induction on the structure of SP_{body}. (2) is a direct consequence of the definition of the semantics of *PSP* and the strictness of the specification-building operations. For the theory of the λ-calculus with parameter restrictions, see [184].

7.1.2. EXAMPLE. (1) The specification *LOOSE-SET* of Example 3.2.4(4) can be parameterized by replacing the natural numbers by arbitrary data on which an equivalence relation is defined. The formal parameter requirement specification *ELEM* extends *BOOL* by a sort *elem* and a relation *eq* describing the equality between elements of sort *elem*.

> **spec** *ELEM* ≡
>   **enrich** *BOOL* **by**
>   **sort** *elem*
>   **function** $eq: elem, elem \rightarrow bool$
>   **axiom** $x = y \Leftrightarrow eq(x, y) = true$
> **endspec**

The target specification expression $LOOSE\text{-}SET_{ELEM}$ is just $LOOSE\text{-}SET[NAT/X, nat/elem]$:

> **spec** $LOOSE\text{-}SET_{ELEM}$ ≡
>   **extend** *X* **reachably by**
>   **sort** *set*
>   **function** $empty: \rightarrow set$,
> $insert: elem, set \rightarrow set$,
> $.\varepsilon.: elem, set \rightarrow bool$
>   **axioms** $(x\,\varepsilon\, empty) = false$,
> $(x\,\varepsilon\, insert(x, s)) = true$,
> $eq(x, y) = false \Rightarrow (x\,\varepsilon\, insert(y, s)) = (x\,\varepsilon\, s)$
> **endspec**

Then a parameterized specification of finite sets can be defined as follows:

> **funct** $PLOOSE\text{-}SET \equiv \lambda X: ELEM.LOOSE\text{-}SET_{ELEM}$ **endfunct**

According to the restrictive definition of conditional β-application above, *PLOOSE-SET* can be applied only to actual parameters with signature *sig*(*ELEM*). A slight change in the definition of *ELEM* remedies this problem partly:

> **spec** $ELEM1 \equiv$ **translate** *ELEM* **with** $in: sig(ELEM) \rightarrow sig(X)$ **endspec**,

> **funct** $PLOOSE\text{-}SET1 \equiv \lambda X: ELEM1.\ LOOSE\text{-}SET_{ELEM}$ **endfunct**

*PLOOSE-SET*1 is applicable to all specifications whose signature contains *sig*(*ELEM*); however, the specification *NAT* is not a correct actual parameter because of the different names of the sort of *NAT* and *ELEM*.

(2) Similarly, one may define a parameterized specification of trees over arbitrary data.

> **funct** $PTREE \equiv \lambda X: ELEM0.\ TREE_{ELEM}$ **endfunct**

where

spec $ELEM0 \equiv [\langle\{elem\},\emptyset\rangle, \emptyset]$**endspec**,

spec $TREE_{ELEM} \equiv$
  extend X **reachably by**
    sort *tree*
    function $empty\colon \to tree$,
      $node\colon tree, elem, tree \to tree$,
      $left, right\colon tree \to tree$,
      $label\colon tree \to elem$,
    axioms $left(node(t_1, n, t_2)) = t_1$,
      $right(node(t_1, n, t_2)) = t_2$,
      $label(node(t_1, n, t_2)) = n$
endspec

Note that because of the total algebra semantics for specifications, the function symbols *left*, *right* and *label* represent *total* functions, even if intuitively the terms *label*(*empty*), *right*(*empty*) and *left*(*empty*) denote "error"-values. This problem can be avoided choosing a different category of algebras such as partial algebras (cf. e.g. [38]) or order-sorted algebras (cf. e.g. [166]).

(3) A parameterized specification, called *COPY*, which produces a specification containing two copies of its actual parameter can be written as follows:

funct $COPY \equiv \lambda X\colon SP_{par}\, X +$ **derive from** X **by**
  $copy\colon sig(X)' \to sig(X)$**endfunct**

where for any signature $\Sigma = \langle S, F\rangle$, Σ' denotes the "primed" signature associated with Σ, i.e. $\Sigma' = \langle S', F'\rangle$ with $S' = \{s' \mid s \in S\}$ and $F' = \{f' \mid f \in F\}$, and where $copy\colon \Sigma' \to \Sigma$ is the signature morphism defined by

$$copy(s') = s \quad \text{for } s' \in S', \qquad copy(f') = f \quad \text{for } f' \in F'.$$

Then $COPY(NAT0)$ has the sorts *nat* and *nat'* and function symbols $zero\colon \to nat$, $zero'\colon \to nat'$, $succ\colon nat \to nat$, $succ'\colon nat' \to nat'$.

Application of an actual parameter SP to a parameterized specification λX: $SP_{par} . SP_{body}$ as defined above requires in particular that the signature of the formal parameter (requirement) SP_{par} is exactly the same as the signature of SP. But this is almost never satisfied as e.g. the Examples 7.1.2(1) and (2) show: $NAT \subseteq ELEM$ does not hold. In order to be able to perform parameter passing for arbitrary actual parameters, it is necessary to rename requirement and target specification appropriately.

A signature morphism $\rho\colon sig(SP_{par}) \to sig(SP)$ is called a *parameter-passing morphism* if it is a specification morphism, i.e. if

derive from SP **by** $\rho \subseteq SP_{par}$

or, equivalently, if

$SP \subseteq$ (**translate** SP_{par} **by** ρ)

(cf. Section 6.3). Now, the function application takes two arguments, the actual parameter SP and a signature morphism ρ, and is defined as follows:

$$(\lambda X: SP_{par} \, . \, SP_{body})(SP, \rho)$$
$$=_{def} (\lambda X: \textbf{translate } SP_{par} \textbf{ with } \rho|_{sig(SP_{par})} . (SP_{body})_\rho)(SP)$$

where $(SP_{body})_\rho =_{def} SP_{body}[\rho(x)/x]$ for all $x \in dom(\rho)$ denotes the specification expression SP_{body} with all sorts and function symbols x of the domain of ρ renamed to $\rho(x)$ and where $\rho|_{sig(SP_{par})}: sig(SP_{par}) \rightarrow sig(SP)$ denotes the restriction of ρ to $sig(SP_{par})$. (For a more general approach to parameterization using the theory of dependent types see [186].) Hence, the function application is well-defined only if the restriction $\rho|_{sig(SP_{par})}$ of ρ is a parameter-passing morphism. In particular, ρ may rename also parts of SP_{body} in order to avoid name clashes with SP.

7.1.3. EXAMPLE. The parameterized specifications *PLOOSE-SET*, *PLOOSE-SET*1 and *PTREE* (see Example 7.1.2) can be applied to the specification *NAT*1 (see Example 3.2.4) via the signature morphism ρ defined by $\rho(elem) =_{def} nat$ and $\rho(x) =_{def} x$ if $x \neq nat$. $\rho|_{sig(ELEM)}$, $\rho|_{sig(ELEM1)}$ and $\rho|_{sig(ELEM0)}$ are obviously parameter-passing morphisms Thus (cf. Example 3.2.4),

$$PLOOSE\text{-}SET(NAT1, \rho) = LOOSE\text{-}SET.$$

On the other hand, consider a signature morphism ρ_2 which associates *eq* with a Boolean function symbol, *less*: *nat*, *nat*$\rightarrow$*bool*, denoting a strict partial ordering "$<$" (defined in an extension *NAT*2) of *NAT*1. Then $\rho_2: sig(ELEM) \rightarrow sig(NAT2)$ is a signature morphism but *not* a parameter-passing morphism (since $x < x$ does not hold in *NAT*2). Hence, *PLOOSE-SET*$(NAT2, \rho_2)$ is undefined.

In order to build sets of sets, one may apply *PLOOSE-SET to LOOSE-SET* with a signature morphism ρ_3 renaming the sort *set* of *PLOOSE-SET into* "*set_of_set*".

$$\textbf{spec } SET\_OF\_LOOSESET \equiv PLOOSE\text{-}SET\,(LOOSE\text{-}SET, \rho_3)\,\textbf{endspec}$$

where ρ_3 is defined by $\rho_3(elem) =_{def} set$, $\rho_3(set) =_{def} set\_of\_set$, $\rho_3(x) =_{def} x.set\_of\_set$ for all $x \in sig(PLOOSE\text{-}SET) \backslash sig(ELEM)$. Hence, e.g., $\rho_3(empty) =_{def} empty.set\_of\_set$ holds.

7.2. *The pushout approach*

In many examples, the parameter requirement specification SP_{par} is a (possibly renamed) subspecification of the target specification (more exactly of $SP_{body}[SP_{par}/X]$). This specific case is the basis for the following widely accepted approach to parameterization (cf. e.g. [49, 51, 43, 64, 55]) which is based on the pushout construction in a suitable category of specifications. In this approach, a parameterized specification $PSP = \lambda X: SP_{par} . SP_{body}$ is considered as a specification morphism $P: SP_{par} \rightarrow SP_{res}$ from the actual parameter specification SP_{par} to the result specification $SP_{res} =_{def} SP_{body}[SP_{par}/X]$. Usually, P is assumed to be an inclusion. Then for any given actual parameter SP and any parameter passing morphism $\rho: SP_{par} \rightarrow SP$, the result $PSP_{po}(SP, \rho)$ of applying P to SP using ρ is defined to be the *pushout* object P' of P and

ρ, that is,

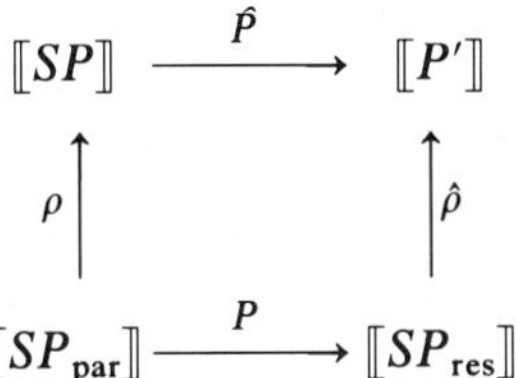

is a pushout in the category $Spec_{\perp}$.

This can be expressed using specification-building operations as follows [158]:

$$PSP_{po}(SP, \rho) = (\textbf{translate } SP \textbf{ with } \hat{P}) \cup (\textbf{translate}(\lambda X : SP_{par} \,.\, SP_{body})(SP_{par}) \textbf{ with } \hat{\rho}).$$

Hence the pushout-based approach may be viewed as a high-level user-oriented mechanism built on top of the β-reduction mechanism.

In the following, the (usual) case where the specification morphism is an inclusion will be studied. For simplicity, it is always assumed that X is the only free variable in SP_{par} and that X and the symbols of SP_{par} are the only free variables in SP_{body}.

A *parameterized extension PSP* is a particular parameterized specification $(\lambda X : SP_{par} \,.\, SP_{body})$ where the formal parameter requirement SP_{par} is a flat specification and where the target specification $SP_{body}[SP_{par}/X]$ is an extension of SP_{par}, i.e., SP_{par} has the form $[\Sigma_{par}, E_{par}]$ with $\Sigma_{par} = \langle S_{par}, F_{par} \rangle$ and SP_{body} has the form **extend** X α **by sorts** S' **functions** F' **axioms** E' such that $\alpha \in \{\emptyset, \textbf{reachably}\}$. The new symbols are assumed to be disjoint from the parameter symbols, i.e. $S_{par} \cap S' = \emptyset$ and $F_{par} \cap F' = \emptyset$. We write $PSP^{\emptyset}$ (PSP^{r}, resp.) to denote the parameterized extension by $\alpha = \emptyset$ (**reachably**, resp.). Each choice of α represents a particular approach to the parameterization of *simple* specifications. The case $\alpha = \emptyset$ corresponds to the *syntax* of parameterized simple specifications since both the formal parameter SP_{par} and the result specification $SP^{\emptyset}_{body}[SP_{par}/X]$ have the form of a simple specification (cf. the definition of **extend**.**by**. in Section 3.2). The other case corresponds to the *semantics* for parameterized loose specifications, for instance, to the reachable extension of the parameter. The following facts and theorems study the relationship between the two parameter-passing mechanisms and between the syntactic extension and the semantic extension.

The *pushout construction for the syntactic extension* can be defined by extending the parameter passing morphism in the following canonical way [55]: Let $\langle \Sigma, E \rangle$ with signature $\Sigma = \langle S, F \rangle$ be a simple specification, $\rho : \Sigma_{par} \to \Sigma$ be a parameter-passing morphism. The canonical extension of ρ is the signature morphism $\hat{\rho} : sig(SP_{res}) \to \hat{\Sigma}$ (where $sig(SP_{res}) = \langle S_{par} \cup S', F_{par} \cup F' \rangle$, $\hat{\Sigma} =_{def} \langle S \uplus S', F \uplus \hat{\rho}(F') \rangle$ with $\uplus$ denoting the disjoint union) induced by $\hat{\rho}(x) =_{def} \rho(x)$ for $x \in S_{par} \cup F_{par}$, and x if $x \in S \cup F$. Then the result of the pushout construction is defined by

$$PSP^{\emptyset}_{po}(SP, \rho) =_{def} \textbf{reach}_{F \cup \hat{\rho}(F')}([\hat{\Sigma}, E \cup \hat{\rho}^{*}(E')]).$$

7.2.1. FACT. *λ-Application and pushout construction are equivalent for syntactic extensions, i.e. for each PSP, ρ and* simple *specification SP*, $PSP^{\emptyset}_{po}(SP, \rho) = PSP^{\emptyset}(SP, \hat{\rho})$.

The proof is obvious from the definitions of parameter passing mechanisms.

For reachable extensions, the model class $PSP^{r}_{po}(SP,\rho)$ of the pushout construction is equivalent to $SP+$ **translate** SP_{res} **with** $\hat{\rho}$ since the specification morphism $\hat{\rho}$ is an inclusion. As a consequence, it is contained in the model class of λ-application; the model classes coincide if the passing morphism is injective. Moreover, the syntactic extension is equivalent with the reachable extension if the latter is sufficiently complete.

7.2.2. THEOREM. *Let SP be an arbitrary specification with signature* $\Sigma=\langle S,F\rangle$ *and let* PSP, ρ, *and an arbitrary specification SP be defined as above.*

(1) *For reachable extensions, the pushout construction is stronger than λ-application; both constructions are equivalent if the parameter passing morphism is injective, i.e.*

$$PSP^{r}_{po}(SP,\rho)\subseteq PSP^{r}(SP,\hat{\rho});$$

$$PSP^{r}_{po}(SP,\rho)=PSP^{r}(SP,\rho)\quad \text{if } \rho \text{ is injective.}$$

(2) *For any* simple *specification* $SP=\langle\Sigma, E\rangle$, *the reachable extension* $PSP^{r}(SP,\hat{\rho})$ *is a hierarchical specification with SP as primitive specification. If it is sufficiently complete, then it is equivalent to the syntactic extension* $PSP^{\emptyset}(SP,\hat{\rho})$.

PROOF. By application of the axioms of *BSA* (see Section 4.3).

(1) $PSP^{r}_{po}(SP,\rho)$

(def. of pushout)

$=SP+$ **translate** SP_{res} **with** $\hat{\rho}$

(def. of SP_{res})

$=SP+$ **translate** $reach_{F'_{Cons}}(SP_{par}+[sig(SP_{res}),E'])$ **with** $\hat{\rho}$

(def. of $+$, (U5))

$=SP+$ **translate** $(SP_{par}+reach_{F'_{Cons}}([sig(SP_{res}),E'])$ **with** $\hat{\rho}$

(def. of $+$, (T3))

$=SP+$ **translate** SP_{par} **with** $\hat{\rho}$
$\quad+$ **translate** $reach_{F'_{Cons}}([sig(SP_{res}), E'])$ **with** $\hat{\rho}$

(ρ is specification morphism)

$=SP+$ **translate** $reach_{F'_{Cons}}([sig(SP_{res}), E'])$ **with** $\hat{\rho}$

(remark on (T2))

$\subseteq SP+\mathbf{reach}_{\hat{\rho}(F'_{Cons})}([\hat{\Sigma},\hat{\rho}^*(E')])$

(def. of $+$, (U5))

$=\mathbf{reach}_{\hat{\rho}(F'_{Cons})}(SP+[\hat{\Sigma},\hat{\rho}^*(E')])$

(def. of λ-application)

$=PSP^{r}(SP,\hat{\rho})$.

If ρ is injective, then $\hat{\rho}$ is so. Hence, because of (T2), the inclusion is an equivalence.

(2) $PSP^{\emptyset}(SP,\hat{\rho})$

(by def.)

$=\mathbf{reach}_{F\cup\hat{\rho}(F'_{Cons})}([\Sigma,E]+[\Sigma',\hat{\rho}^*(E')])$.

$$PSP^{\mathrm{r}}(SP,\hat{\rho})$$

(by def.)

$$= \mathbf{reach}_{\hat{\rho}(F'_{Cons})}(\mathbf{reach}_F([\Sigma,E]) + [\Sigma',\hat{\rho}^*(E')])$$

$(range(F) \cap range(\hat{\rho}(F'_{Cons}) = \emptyset)$

$$= \mathbf{reach}_{F\cup\hat{\rho}(F'_{Cons})}(\mathbf{reach}_F([\Sigma,E] + [\Sigma',\hat{\rho}^*(E')]).$$

Hence $PSP^{\mathrm{r}}(SP,\hat{\rho})$ is a hierarchical specification where SP is the primitive part. According to Fact 5.4.2, $PSP^{\mathrm{r}}(SP,\hat{\rho})$ and $PSP^{\emptyset}(SP,\hat{\rho})$ are equivalent if $PSP^{\mathrm{r}}(SP,\hat{\rho})$ is sufficiently complete w.r.t. F. □

In the initial algebra approach, a parameterized specification can be considered as a free extension of the parameter requirement specification, whereas in the terminal algebra approach, it is considered as a final extension (cf. Section 6.4). The former approach has been studied e.g. in [171, 54, 137], the latter in [175, 64]. In both cases, syntax (i.e., the syntactic extension) and semantics (i.e., the free or final extension) coincide only if the extension is persistent. Sufficient (proof-theoretic) conditions for the persistency are given in [139, 140].

Composition of parameterized specifications is a generalization of the parameter-passing mechanism defined above: given two parameterized specifications $PSP = \lambda X{:}\, SP_{\mathrm{par}}.SP_{\mathrm{body}}$, $PSP1 = \lambda Y{:}\, SP1_{\mathrm{par}}.SP1_{\mathrm{body}}$ and a parameter-passing morphism $\rho{:}\, SP_{\mathrm{par}} \to SP1_{\mathrm{res}}$(where $SP1_{\mathrm{res}} =_{\mathrm{def}} SP1_{\mathrm{body}}[SP1_{\mathrm{par}}/X]$), the composition of PSP and $PSP1$ via ρ is defined to be the application of PSP to SP_{res} via ρ, i.e. $PSP \circ_\rho PSP1 =_{\mathrm{def}} \lambda Y{:}\, SP1_{\mathrm{par}}.PSP(SP1_{\mathrm{res}}, \rho)$. Such compositions are studied in [49, 54, 55] for the initial algebra approach, in [64] for the final algebra approach and in [180] for the loose approach.

The computability of parameterized extensions is studied in [16, 17].

8. Implementation

There are two different approaches for implementing specifications: either the specification is executed as it stands using *interpretative* or *compilational* techniques, or using some program development method a (more efficient) program is developed which implements the specification. The former approach is discussed in Section 8.4. For the latter approach, proceeding from a specification to a program means making a series of design decisions. These include decisions concerning the concrete representation of abstractly defined data types, decisions about how to compute abstractly specified functions (choice of algorithm) and decisions which select between the various possibilities which the high-level specification leaves open [96]. A formalization of this approach requires a precise definition of the concept of correct transformation of more abstract specification into a more concrete one, i.e. of the correct *implementation of one specification by another*.

Depending on the semantic approach to algebraic specification, there have been a number of different interesting approaches in the literature to capture adequately this notion (see e.g. [80, 82, 73, 48, 49, 53, 52, 160, 77, 162, 64, 116, 135, 12, 35, 156, 158]. In the following, the simple notion of *refinement* will be chosen which seems appropriate for the loose approach to specifications (Section 8.1). This notion originates from [161] and was elaborated in [158]. Many of the other notions can and will be defined and explained in terms of this simple notion (Section 8.2). *Parameterized implementations* will be discussed in Section 8.3.

8.1. *Implementation by refinement*

A specification SP1 incorporates more design decisions than a specification *SP* if *SP*1 has the same signature as *SP* and all models of *SP*1 are models of *SP* but possibly some of the models of *SP* are excluded in *SP*1. Formally, *SP*1 *is an implementation of SP* (written $SP \rightsquigarrow SP1$) if $sig(SP1) = sig(SP)$ and $Mod(SP1) \subseteq Mod(SP)$. This notion can be extended to parameterized specifications: Given two parameterized specifications $P1 \equiv \lambda X: SP_{\text{par}}.SP1_{\text{body}}$ and $P \equiv \lambda X: SP_{\text{par}}.SP_{\text{body}}$ with the same formal parameters, then *P*1 is an *implementation of P* (written $P \rightsquigarrow P1$) if, for any actual parameter SP_A with $SP_A \subseteq SP_{\text{par}}$, $P(SP_A) \rightsquigarrow P1(SP_A)$.

Thus also for parameterized specifications, it is required that the implementing specification has the same signature as the implemented specification, i.e.

$$sig(SP_{\text{body}}[SP_{\text{par}}/X]) = sig(SP1_{\text{body}}[SP_{\text{par}}/X]).$$

A direct consequence of the definition is that implementations can be composed *vertically* and *horizontally*. Implementations can be *vertically composed* if the implementation relation is transitive (i.e., $SP \rightsquigarrow SP1$ and $SP1 \rightsquigarrow SP2$ implies $SP \rightsquigarrow SP2$) and they can be *horizontally* composed if the specification-building operations preserve implementations (i.e., $P \rightsquigarrow P1$ and $SP \rightsquigarrow SP1$ implies $P(SP) \rightsquigarrow P1(SP1)$) [73].

8.1.1. FACT (vertical and horizontal composition). *Let SP, SP1, SP2 be specifications and P, P1 be parameterized specifications.*

(1) *If* $SP \rightsquigarrow SP1$ *and* $SP1 \rightsquigarrow SP2$ *then* $SP \rightsquigarrow SP2$.

(2) *If* $SP \rightsquigarrow SP1$, $P \rightsquigarrow P1$ *and SP is an actual parameter of P* (*i.e.*, $[\![P(SP)]\!] \neq \bot$), *then* $P(SP) \rightsquigarrow P'(SP1)$.

PROOF. (1) follows from the transitivity of "$\subseteq$", (2) from the monotonicity of *P* and (1). □

The development of a program from a specification consists of a series of refinement steps $SP_0 \rightsquigarrow SP_1 \rightsquigarrow \cdots \rightsquigarrow SP_n$ where SP_0 is the original high-level specification and SP_n is a program. Vertical composability guarantees the correctness of SP_n w.r.t. SP_0. This considers each of the specifications $SP_0, \ldots, SP_n$ as a single entity. If, however, we decompose any of them using a parameterized specification, say $SP_k = P(SP)$, then the horizontal composability guarantees that any "local" implementations of *SP* by some *SP*1 or of *P* by some *P*1 gives a correct "global" implementation of SP_k. Hence horizontal and vertical composability together guarantee that the results of sub-developments can be performed independently and that their results can always be combined to give a correct specification expression.

In the following, some standard examples for implementations are given.

8.1.2. EXAMPLES (*implementations*). (1) *Truth-values by natural numbers.* The monomorphic specification *BOOLM* (Example 3.2.4(1)) of truth values can be implemented by natural numbers as follows. In a first step, the specification *NATM* (Example 3.2.4(2)) is enriched by new constants *true* and *false* and the sort *nat* is renamed into *bool*:

> **spec** $NATB \equiv$
>   **rename**
>   **enrich** $NATM$ **by**
>   **functions** $true, false: \rightarrow nat$
>   **axioms** $true = zero, false = succ(zero)$
>   **by** $[nat \mapsto bool]$
> **endspec**

For the second step, there are two possibilities:

(i) A quotient is formed (in order to obtain a two-element carrier set) and then the function symbol *succ* is "forgotten":

> **spec** $BOOLM\_by\_NATM\_i \equiv$
>   **export** $sig(BOOLM)$ **from**
>   **quotient** $NATB$ **by** $succ(succ(x)) = succ(x)$
> **endspec**

(ii) The other way round, one may first "forget" the successor function and then "restrict" the carrier set to (the interpretations of) *true* and *false*:

> **spec** $BOOLM\_by\_NATM\_ii \equiv$
> $\textbf{restrict}_{bool}$
>   **export** $sig(BOOLM)$ **from** $NATB$
> **endspec**

Obviously, we have

$$BOOLM \rightsquigarrow BOOLM\_by\_NATM\_i \quad \text{and}$$

$$BOOLM \rightsquigarrow BOOLM\_by\_NATM\_ii$$

Moreover, note that *NAT*1 which contains *BOOLM* as a subspecification is trivially an implementation of *BOOLM*.

$$BOOLM \rightsquigarrow \textbf{export } sig(BOOLM) \textbf{ from } NAT1.$$

(2) *Natural numbers by integers.* The specification *INT* is an implementation of *NAT*0 if the function symbol *pred* is "forgotten" and the carrier set of any model is restricted to the subalgebra generated by *zero* and *succ*:

$$NATM \rightsquigarrow NATM\_by\_INT$$

where

spec $INT \equiv$ **data sorts** *int*
  functions $zero: \rightarrow int$,
    $succ, pred: int \rightarrow int$
  axioms $pred(succ(x)) = x$,
    $succ(pred(x)) = x$
endspec

spec $NATM\_by\_INT \equiv$
  restrict$_{nat}$
    export $sig(NATM)$ **from**
      rename INT **by** $[int \mapsto nat]$
endspec

(3) *Sets by trees.* Sets can be implemented by trees viewed as linear lists as follows. The parameter specification $ELEM0$ of $TREE$ (see Section 7.1) is enriched to $ELEM$, and the body of $TREE$ is enriched by two new function symbols *insert* and ε. Here inserting an element x in a tree t means to construct a new tree t' with x as root and t as right subtree:

funct $TREE\_S \equiv \lambda X: ELEM.$
  rename
    enrich $TREE(X) + BOOL$ **by**
      functions $insert: elem, tree \rightarrow tree$,
        $.\varepsilon.: elem, tree \rightarrow bool$
      axioms $insert(x, t) = node(empty, x, t)$,
        $x \,\varepsilon\, empty = false$,
        $x \,\varepsilon\, node(t_1, x, t_2) = true$,
        $x \neq y \Rightarrow x \,\varepsilon\, node(t_1, y, t_2) = (x \,\varepsilon\, t_1)$ *or* $(x \,\varepsilon\, t_2)$
  by $[tree \mapsto set]$
endfunct

Then the following specification SET_by_TREE is an implementation of SET:

SET_by_TREE

where

funct $SET\_by\_TREE \equiv \lambda X: ELEM$
  quotient
    restrict$_{set}$
      export $sig(SET(X))$ **from** $TREE\_S(X)$
  by $insert(x, insert(x, s)) = insert(x, s)$,
    $insert(x, insert(y, s)) = insert(y, insert(x, s))$
endfunct

To see this, let A be a specification with $A \subseteq ELEM$ and consider any model B of $TREE\_S(A)$. The carrier set B_{set} of B consists of all finite binary trees with labels in B_{elem}.

Applying the operation **export** *sig*($SET(A)$) **from**$\cdots$ to $TREE\_S(A)$ yields the reduct $B|_{sig(SET(A))}$ of B, where only those trees in B_{set} are reachable which are interpretations of a term of the form

$$insert(x_1, insert(x_2, \ldots, insert(x_n, empty)\ldots)), \quad n \geqslant 0 \tag{8.1}$$

(which are by definition equal to

$$node(empty, x_1, node(empty, x_2, \ldots, node(empty, x_n, empty)\ldots)), \; n \geqslant 0)$$

The operation $\mathbf{restrict}_{set}$ yields the smallest *sig*($SET(A)$)-subalgebra $R_{set}(B|_{sig(SET(A))})$ of $B|_{sig(SET(A))}$ where all elements which are not (interpretations of terms) of the form (8.1) are excluded. Finally, the quotient of this subalgebra is a *sig*($SET(A)$)-algebra which is reachable w.r.t. the sort *set* and which satisfies the axioms of $SET(A)$.

(4) *Behaviors of sets by trees.* There is a second way, how trees can be viewed as implementations of sets. If one considers the specification *LOOSE-SET* (Example 3.2.4) which specifies just the "observable behavior" of *SETNAT*, then *TREES_S* applied to *NAT* is a direct implementation of *LOOSE-SET*.

$$LOOSE\text{-}SET \rightsquigarrow TREE\_S(NAT1, \rho).$$

where ρ: *sig*(*ELEM*)$\rightarrow$*sig*(*NAT*1) is induced by $\rho(elem) = nat$. By using (cf. Section 6.5)

$$\textbf{behavior}\ SETNAT\ \textbf{w.r.t.}\ \{bool\} \rightsquigarrow LOOSE\text{-}SET,$$

the vertical composition property implies that the behavior of *SETNAT* is implemented by $TREE\_S(NAT, \rho)$, i.e.,

$$\textbf{behavior}\ SETNAT\ \textbf{w.r.t.}\ \{bool\} \rightsquigarrow TREE\_S(NAT1, \rho).$$

8.2. *Other notions of implementation*

As the examples in Example 8.1.2. suggest, the construction of an implementation of a specification *SP* proceeds in three steps: First, a specification SP' is chosen as the basic specification. Second, SP' is enriched to SP'' by definitions of the sorts and function symbols of *SP* written in terms of the sorts and operations of SP'. Third, some specification-building operators are applied to SP'' and/or *SP* in order to show that SP'' is a correct implementation of *SP*.

For instance in Example 8.1.2(1), *NATM* was chosen as a basic specification for implementing *BOOLM*. Then *NATM* was enriched to *NATB* by definitions of *true*, *false* and by renaming of *nat* in *bool* in order to establish the implementation of the sort *bool*. In a third step, two other specification-building operators (export and quotient, export and restrict) were applied to the implementing specification *NATB* in order to show the correctness of the implementation w.r.t. *BOOLM*. In Example 8.1.2(4) the last step of the implementation of sets by trees consisted of applying the behavior operator to the *implemented* specification *SETNAT*; in this case the functions in $TREE\_S(NAT, \rho)$ "behave like" the functions in *SETNAT* instead of satisfying exactly their definitions.

More generally, constructing an implementation for an element $sp \in Spec$ consists of choosing an element $sp' \in Spec$ and finding specification-building functions α and κ such that $\alpha(sp) \rightsquigarrow \kappa(sp')$. The following definition summarizes (cf. [158]) these observations: sp' is said to **implement** sp **w.r.t.** $\alpha: Spec(sig(sp)) \rightarrow Spec(\Sigma)$ **via** $\kappa: Spec(sig(sp')) \rightarrow Spec(\Sigma)$, written $sp^{\alpha} \rightsquigarrow_{\kappa} sp'$ if $\kappa(sp') \subseteq \alpha(sp)$. In other words, $sp^{\alpha} \rightsquigarrow_{\kappa} sp'$ if κ transforms sp' in such a way that every model of $\kappa(sp')$ is a model of $\alpha(sp)$. For specification expressions SP, SP' and specification-building operations $\boldsymbol{\kappa}$, $\boldsymbol{\alpha}$ we write $SP^{\boldsymbol{\alpha}} \rightsquigarrow_{\boldsymbol{\kappa}} SP'$ if $[\![SP]\!]^{\alpha} \rightsquigarrow_{\kappa} [\![SP']\!]$. Moreover, we do not distinguish between α and $\boldsymbol{\alpha}$ (κ and $\boldsymbol{\kappa}$ resp.) if the distinction between syntax (specification expression) and semantics is clear from the context.

For practical purposes, this notion is too general but it can be used to represent the following more specific concepts of implementation found in the literature.

($\kappa\alpha$) *Constructor and abstractor implementations:* In this concept, developed in [158], κ is a construction and α an observational abstraction (see Sections 6.3 and 6.5): the specification SP' is called an *abstractor implementation* of SP w.r.t. α via κ if α is observational abstraction, κ is a construction, and $SP^{\alpha} \rightsquigarrow_{\kappa} SP'$. It is called a *constructor implementation*, written $SP \rightsquigarrow_{\kappa} SP'$ if, moreover, α is the identity (i.e., κ is constructor and $\kappa(SP') \subseteq SP$).

(RI) *Synthesize-Restrict-Identify implementation:* This notion is a special case of constructor implementation which is often used in practice; it is a composition of the specification-building operations **extend, derive, restrict** and **quotient** (in that order). Let

$$\begin{aligned}
\mathbf{Ext}_{S,F,E} &=_{\text{def}} \lambda X.\, \textbf{extend } X \textbf{ freely by } S,\ F,\ E,\\
\mathbf{D}_{\sigma} &=_{\text{def}} \lambda X.\, \textbf{derive from } X \textbf{ by } \sigma,\\
\mathbf{R}_{S} &=_{\text{def}} \lambda X.\, \textbf{restrict}_S\ X,\\
\mathbf{Q}_{E} &=_{\text{def}} \lambda X.\, \textbf{quotient } X \textbf{ by } E.
\end{aligned}$$

Then SP' is called a *synthesize-restrict-identify implementation* (via the extension Δ, the signature morphism σ and the set E of equations), written $SP \rightsquigarrow_{E,\sigma,\Delta} SP'$, if $SP \rightsquigarrow_{\kappa} SP'$ such that $\kappa = \mathbf{Q}_E \circ \mathbf{R}_{sorts(sig(SP))} \circ \mathbf{D}_{\sigma} \circ \mathbf{Ext}_{\Delta}$ and $\mathbf{Ext}_{\Delta}$ is a persistent extension [53] (cf. also [82, 48]). In [120, 35] the extend-step is a reachable conservative extension. Moreover, [120] requires an (explicit) representation relation instead of the restrict step.

(Beh) *Implementation by behavioral abstraction*: In this notion, every model of the implementing specification is first extended in such a way that its $sig(SP)$-reduct is behaviorally equivalent to a model of the implemented specification SP. The specification SP' is called a *behavioral implementation* of SP via the extension Δ, the signature morphism σ and the set OBS of observable sorts, written

$$SP^{\,OBS} \rightsquigarrow_{\sigma\Delta} SP',$$

if $\mathbf{Beh}_{OBS}(SP) \rightsquigarrow_{\mathbf{D}_{\sigma} \circ \mathbf{Ext}_{\Delta}} SP'$ and $\mathbf{Ext}_{\Delta}$ is a persistent extension, where

$$\mathbf{Beh}_{OBS} =_{\text{def}} \lambda X.\, \textbf{behavior } X \textbf{ w.r.t. } OBS$$

(cf. [77, 35, 162]). An axiomatic approach to behavioral implementation is developed [92].

8.2.1. EXAMPLE. Examples 8.1.2(1)–(2) of truth values, and natural numbers are constructor implementations following the Synthesize-Restrict-Identify schema. For instance, the implementation of *BOOLM* by *BOOLM_by_NATM_ii* is equivalent to a constructor implementation $BOOLM \rightsquigarrow_{\kappa} NATM$ where κ has the form

$\mathbf{Q}_E \circ \mathbf{D}_{in} \circ \mathbf{Ext}_\Delta$ (recall that **export** is a specific **derive** and that for sufficiently complete extensions without introducing new sorts the free extension is equivalent to the syntactic **extend**).

The implementation of *SETNAT* by *LOOSE-SET* (Example 8.1.2(4)) is equivalent to an implementation by behavioral abstraction: $SETNAT^{bool} \rightsquigarrow LOOSE\text{-}SET$.

8.2.2. THEOREM (vertical composition). *Let SP, SP′, SP″ be specifications and* $\alpha, \alpha', \kappa, \kappa'$ *be specification-building operations. If* $SP\ {}^{\alpha}\!\!\rightsquigarrow_{\kappa} SP'$ *and* $SP'\ {}^{\alpha'}\!\!\rightsquigarrow_{\kappa'} SP''$, *then* $SP\ {}^{\alpha}\!\!\rightsquigarrow_{\kappa\circ\kappa'} SP''$, *provided that* α *is transitive (i.e.* $\alpha(\alpha(SP)) \subseteq \alpha(SP)$*) and* κ *commutes with* α *and* α' *(i.e.,* $\kappa(\alpha'(SP')) \subseteq \alpha(\kappa(SP'))$*).*

PROOF. By definition, $\kappa'(SP'') \subseteq \alpha'(SP')$. Then $\kappa(\kappa'(SP'')) \subseteq \kappa(\alpha'(SP'))$ follows from the monotonicity of κ. By vertical composability of $\rightsquigarrow$, it suffices to show that $\kappa(\alpha'(SP')) \subseteq \alpha(SP)$: since κ commutes with α, α' and α is transitive, we have $\kappa(\alpha'(SP')) \subseteq \alpha(\kappa(SP')) \subseteq \alpha(\alpha(SP)) \subseteq \alpha(SP)$. □

The assumptions of the vertical composition theorem (8.2.2) can be relaxed if α and α' are observational abstractions. It holds for constructor implementations without any additional assumptions; for abstractor implementations it is necessary that κ preserves observational equivalences. If κ is a construction, this condition can be stated on the level of algebras: the construction κ *preserves observational equivalences* α and α' w.r.t. the specification SP' if for any two algebras A_1 and $A_2 \in Alg(sig(SP'))$ $A_1 \equiv_{\alpha'} A_2$ and $A_2 \in Mod(SP')$ implies $\kappa(A_1) \equiv_\alpha \kappa(A_2)$.

8.2.3. COROLLARY. *Let SP, SP′, SP″ be specifications, and* α, α' *be observational abstractions.*

(1) *If* $SP \rightsquigarrow_{\kappa} SP'$ *and* $SP' \rightsquigarrow_{\kappa'} SP''$, *then* $SP \rightsquigarrow_{\kappa\circ\kappa'} SP''$.

(2) If $SP\ {}^{\alpha}\!\!\rightsquigarrow_{\kappa} SP'$ *and* $SP'\ {}^{\alpha'}\!\!\rightsquigarrow_{\kappa'} SP''$, *then* $SP\ {}^{\alpha}\!\!\rightsquigarrow_{\kappa\circ\kappa'} SP''$ *provided* κ *preserves observational equivalences.*

PROOF. (1) $\alpha = \alpha' = id$ implies $\alpha \circ \alpha = id$ and $\alpha \circ \kappa = \kappa \circ \alpha$. (2) Any observational abstraction is transitive. □

The requirement of preserving observational equivalences is similar to the requirement in [162] that constructors in implementation cells be *stable*.

8.2.4. EXAMPLE. The implementation of *SETNAT* by *TREE_S*($NAT1, \rho$) (Example 8.1.2(4)) is a trivial example of vertical composition: *LOOSE-SET* is a behavioral implementation of *SETNAT* which is implemented by *TREE*($NAT1, \rho$) via a Synthesize-Restrict-Identify step; obviously, behavioral abstraction is transitive, κ is the identity which preserves observational equivalence.

For implementations using the Synthesize-Restrict-Identify approach or the behavioral construction, vertical composition requires a specific order for the constructions used in the result implementation of SP by SP'' which is not respected by the vertical composition theorem above.

However, because of the required persistency of the extensions, behavioral implementations can be vertically composed if the extension preserves equivalence.

8.2.5. THEOREM (vertical composition for behavioral implementations). (1) *If* $SP^{OBS} \rightsquigarrow_{\mathbf{D}_\sigma \circ \mathbf{Ext}_\Delta} SP'$ *and* $SP'^{OBS'} \rightsquigarrow_{\mathbf{D}_{\sigma'} \circ \mathbf{Ext}_{\Delta'}} SP''$ *are implementations by behavioral abstraction, then* $SP^{OBS} \rightsquigarrow_{\mathbf{D}_{\sigma \circ \sigma'} \circ \mathbf{Ext}_{\sigma'(\Delta) \cup \Delta'}} SP''$ *is an implementation by behavioral abstraction provided that* $\mathbf{Ext}_\Delta$ *preserves observational equivalence (i.e. for all* A, $A' \in Alg(sig(SP'))$, $A \equiv_{OBS'} A'$ *implies* $\mathbf{Ext}_\Delta(A) \equiv_{\sigma(OBS)} \mathbf{Ext}_\Delta(A')$).

(2) *If* $SP^{OBS} \rightsquigarrow_{\mathbf{D}_\sigma} SP'$, $SP'^{OBS'} \rightsquigarrow_{\mathbf{D}_{\sigma'}} SP''$ *and* $\sigma(OBS) \subseteq OBS'$, *then* $SP^{OBS} \rightsquigarrow_{\mathbf{D}_{\sigma \circ \sigma'}} SP''$.

PROOF. (1) The first implementation relation splits into two parts:

$$SP^{OBS} \rightsquigarrow_{\mathbf{D}_\sigma} \mathbf{Ext}_\Delta(SP') \quad \text{and} \quad \mathbf{Ext}_\Delta(SP')^{\sigma(OBS)} \rightsquigarrow_{\mathbf{Ext}_\Delta} SP'.$$

Then, according to Corollary 8.2.3, $SP^{OBS} \rightsquigarrow_\kappa SP''$ is an implementation via $\kappa = \mathbf{D}_\sigma \circ \mathbf{Ext}_\Delta \circ \mathbf{D}_{\sigma'} \circ \mathbf{Ext}_{\Delta'}$, provided that

$$\mathbf{D}_\sigma(\mathbf{Beh}_{\sigma(OBS)}(SP')) \subseteq \mathbf{Beh}_{OBS}(\mathbf{D}_\sigma(SP'))$$

and $\mathbf{Ext}_\Delta$ preserves observational equivalence. The former inclusion holds according to Fact 6.5.2. Moreover, if there are no name clashes between Δ and the range of σ', then $\mathbf{Ext}_\Delta \circ \mathbf{D}_{\sigma'} = \mathbf{D}_{\sigma'} \circ \mathbf{Ext}_{\sigma'(\Delta)}$ holds because of the persistency of $\mathbf{Ext}_\Delta$. This implies $SP^{OBS} \rightsquigarrow_{\kappa'} SP''$ where $\kappa' =_{\text{def}} \mathbf{D}_{\sigma \circ \sigma'} \circ \mathbf{Ext}_{\sigma'(\Delta) \cup \Delta'}$.

(2) follows directly from (1). □

Synthesize-Restrict-Identify implementations can be vertically composed if the synthesize step is trivial.

8.2.6. THEOREM. (vertical composition for Restrict-Identify). *If* $SP \rightsquigarrow_{RI} SP'$ and $SP' \rightsquigarrow_{RI'} SP''$ are implemetations via $RI = \mathbf{Q}_E \circ \mathbf{D}_\sigma$ and $RI' = \mathbf{Q}_{E'} \circ \mathbf{D}_{\sigma'}$ *then* $SP \rightsquigarrow_{RI''} SP''$ *is an implementation via* $RI'' = \mathbf{Q}_{E''} \circ \mathbf{D}_{\sigma \circ \sigma''}$ where $E'' = E' \cup \sim^E|_\sigma$.

PROOF. Let SP, SP', SP'' be specifications with signatures Σ, Σ' and Σ'', respectively, and let $A \in Mod(SP)$. Because of $SP \rightsquigarrow_{RI} SP'$, there exists a unique surjective Σ'-homomorphism $\varphi: A^- \to A^-/\sim^E \in Mod(SP')$ where $A^- = R_{sorts(\Sigma')}(A|_\sigma)$ is the reachable Σ'-subalgebra of $A|_\sigma$. Since $\cdot|_{\sigma'}$ is a functor, $\varphi|_{\sigma'}: A^-|_{\sigma'} \to (A^-/\sim^E)|_{\sigma'}$ is a Σ''-homomorphism and its $sorts(\Sigma'')$-restriction is a Σ''-homomorphism from $R_{sorts(\Sigma'')}(A^-|_\sigma)$ to $B =_{\text{def}} R_{sorts(\Sigma'')}((A^-/\sim^E))|_{\sigma'})$.

Because of $SP' \rightsquigarrow_{RI'} SP''$ and $A^-/\sim^E \in Mod(SP')$, there exists a Σ''-homomorphism Ψ from B to $B/\sim^{E'} \in Mod(SP'')$. Thus $\Psi \circ \varphi|_{\sigma'}$ is a Σ''-homomorphism from

$$R_{sorts(\Sigma'')}(A|_{\sigma' \circ \sigma}) = R_{sorts(\Sigma'')}((R_{sorts(\Sigma')}(A|_\sigma))|_{\sigma'})$$

to $B/\sim^{E'}$. Therefore, $SP \rightsquigarrow_{RI''} SP'$ where $RI'' = \mathbf{Q}_{E' \cup (\sim^E|_{\sigma'})} \circ \mathbf{D}_{\sigma' \circ \sigma}$. □

The following simple example shows the problems for vertical composition with nontrivial synthesize step.

8.2.7. EXAMPLE (Ehrig et al. [53]). Consider the following implementations of $BOOLM$ and $NATM$:

$$\begin{aligned}&\textbf{spec } NATBOOL \equiv \textbf{extend } NATM \textbf{ freely by } \Delta_{BOOL} \textbf{ endspec} \quad \text{where}\\ &\Delta_{BOOL} = \langle\ \textbf{sorts } bool,\\ &\qquad \textbf{functions } c1: nat \to bool,\\ &\qquad\qquad true, false: \to bool\\ &\qquad \textbf{equations } true = c1(zero),\\ &\qquad\qquad false = c1(succ(zero)),\\ &\qquad\qquad c1(succ(x)) = c1(succ(succ(x)))\rangle\end{aligned}$$

$$\begin{aligned}&\textbf{spec } INTNAT \equiv \textbf{extend } INT \textbf{ freely by } \Delta_{NAT} \textbf{ endspec} \quad \text{where}\\ &\Delta_{NAT} = \langle\ \textbf{sorts } nat,\\ &\qquad \textbf{functions } c2: int \to nat,\\ &\qquad\qquad zero': \to nat,\\ &\qquad\qquad succ': nat \to nat\\ &\qquad \textbf{equations } zero' = c2(zero),\\ &\qquad\qquad succ'(c2(x)) = c2(succ(x))\rangle\end{aligned}$$

The specification $NATBOOL$ is an implementation of $BOOLM$ via $\mathbf{D}_{sig(BOOLM)}$ and $INTNAT$ is an implementation of $NATM$ via $\mathbf{R}_{sig(NATM)} \circ \mathbf{D}_\sigma$ where

$$\sigma: sig(NAT0) \to sig(INTNAT), \qquad \sigma(zero) = zero', \qquad \sigma(succ) = succ'.$$

But the extension $INTBOOL$ of INT by the union of the above extensions

$$\textbf{spec } INTBOOL \equiv \quad \textbf{extend } INT \textbf{ freely by } \hat{\sigma}(\Delta_{BOOL}) \cup \Delta_{NAT} \textbf{ endspec}$$

is not an implementation of $BOOL$ (via $\mathbf{D}_{sig(BOOLM)}$), since

$$\begin{aligned}true &= c1(zero') = c1(c2(zero)) = c1(c2(succ(pred(zero))))\\ &= c1(succ'(c2(pred(zero)))) = c1(succ'(succ'(c2(pred(zero)))))\\ &= c1(succ'(c2(succ(pred(zero))))) = c1(succ'(c2(zero)))\\ &= c1(succ'(zero')) = false.\end{aligned}$$

Notice that the extensions $NATBOOL$ of $NATM$ and $INTNAT$ of INT are not observably sufficiently complete w.r.t. observations in the new sorts *bool* and *nat* respectively. Therefore, they cannot be used for implementations w.r.t. observational abstraction.

Proof-theoretic conditions for implementations have been studied among others in [136, 25, 92]. Behavioral implementations of concurrent systems are studied in [107].

8.3. *Parameterized constructor and abstractor implementations*

The vertical composition theorem for the nonparameterized case generalizes directly to the parameterized case as shown in the following theorem.

8.3.1. THEOREM (vertical composition). *For any parameterized specifications P, P', P'', if $P \,{}^{\alpha}\!\!\rightsquigarrow_{\kappa} P'$ and $P' \,{}^{\alpha'}\!\!\rightsquigarrow_{\kappa'} P''$ then $P \,{}^{\alpha}\!\!\rightsquigarrow_{\kappa\circ\kappa'} P''$, provided that α is transitive and κ commutes with α and α'.*

The proof is a pointwise application of the vertical composition for the non-parameterized case.

Similarly to Theorem 8.2.2, the horizontal composition theorem needs a commutation property w.r.t. abstraction α and α'.

8.3.2. THEOREM (horizontal composition). *If* $P\ {}^{\alpha}\!\!\leadsto_{\kappa} P'$ *is an implementation of parameterized specifications w.r.t.* α *via* κ *and if* $SP\ {}^{\alpha'}\!\!\leadsto_{\kappa'} SP'$ *is an implementation w.r.t.* α' *via* κ', *then* $P(SP)\ {}^{\alpha}\!\!\leadsto_{\kappa} P'(\kappa'(SP'))$ *provided that* α *is transitive,* $\alpha'(SP)$ *is an actual parameter of* P *(i.e.,* $\alpha'(SP) \subseteq SP_{\text{par}}$*) and* P *or* $\kappa \circ P'$ *commutes with* α *and* α' *(i.e.,* $P(\alpha'(SP)) \subseteq \alpha(P(SP))$ *or* $P'(\alpha'(SP)) \subseteq \alpha(\kappa(P'(SP)))$*).*

PROOF. Since $SP\ {}^{\alpha'}\!\!\leadsto_{\kappa'} SP'$, by monotonicity of $\kappa \circ P'$, $\kappa(P'(\alpha'(SP))) \supseteq \kappa(P'(\kappa'(SP')))$.

If P commutes with α and α', then $P(\alpha'(SP)) \subseteq \alpha(P(SP))$. This implies $\alpha(P(\alpha'(SP))) \subseteq \alpha(P(SP))$ since α is transitive. Because of $P\ {}^{\alpha}\!\!\leadsto_{\kappa} P'$, $\kappa(P'(\alpha'(SP))) \subseteq \alpha(P(\alpha'(SP)))$ holds and therefore also $\kappa(P'(\kappa'(SP'))) \subseteq \alpha(P(SP))$.

If $\kappa \circ P'$ commutes with α and α', then $\kappa(P'(\alpha'(SP))) \subseteq \alpha(\kappa(P'(SP)))$ and therefore by transitivity, $\kappa(P'(\kappa'(SP'))) \subseteq \alpha(\kappa(P'(SP)))$. Because of $P\ {}^{\alpha}\!\!\leadsto_{\kappa} P'$ and the closure property of α, one can conclude $\alpha(\kappa(P'(SP))) \subseteq \alpha(\alpha(P(SP))) \subseteq \alpha(P(SP))$. $\square$

However, if SP is implemented via a construction, or if P or $\kappa \circ P'$ are observably complete and persistent, then the horizontal composition holds as follows.

8.3.3. COROLLARY. (1) *Let* P *or* $\kappa \circ P'$ *be a construction which is observably persistent and observably complete w.r.t.* OBS. *Then* $P\ {}^{OBS}\!\!\leadsto_{\kappa} P'$ *and* $SP\ {}^{OBS}\!\!\leadsto_{\kappa'} SP'$ *implies* $P(SP)\ {}^{OBS}\!\!\leadsto_{\kappa} P'(\kappa'(SP'))$.

(2) *If* $SP \leadsto_{\kappa'} SP'$ *is an implementation via the construction* κ' *and* $P\ {}^{\alpha}\!\!\leadsto_{\kappa} P'$, *then* $P(SP)\ {}^{\alpha}\!\!\leadsto_{\kappa} P'(\kappa'(SP'))$.

PROOF. (1) According to Lemma 6.5.3, P (or $\kappa \circ P'$, resp.) commutes with behavioral abstraction if P (or $\kappa \circ P'$, resp.) is observably persistent and observably complete.

(2) Because of $P(SP) \subseteq \alpha(P(SP))$, P commutes with α and the identity *id*. $\square$

Although this is not horizontal composition as formulated in [73], it is adequate for stepwise specification development. As Sannella and Tarlecki [158] argue, it guarantees that in the case of a specification formed by applying a parameterized specification P to a specification SP, the developments of P and SP may proceed independently and the results be successfully combined (provided the observational equivalences are preserved). In particular for constructor implementations, if $P \leadsto_{\kappa_1} P_1 \leadsto_{\kappa_2} \cdots \leadsto_{\kappa_n} P_n$ and $SP \leadsto_{\mu_1} SP_1 \leadsto \cdots \leadsto_{\mu_m} SP_m$ then $P(SP) \leadsto_{\kappa_1 \circ \cdots \circ \kappa_n} P_n(\mu_1 \circ \cdots \circ \mu_m(SP_m))$. Now, if P_n is itself a construction, then *$P(SP)$ is implemented by SP_m* via the construction $\kappa_1 \circ \cdots \circ \kappa_n \circ P_n \circ \mu_1 \circ \cdots \circ \mu_m$. For simplifying this expression, identities as proved in Sections 6.2–6.6 and the vertical composition theorems of Section 8.2 can be applied.

Parameterized implementations have been studied in [52, 160, 12] for the Forget-Restrict-Identify-approach, in [116] for an approach similar to constructions, and in [77, 64, 93] for the behavioral approach.

8.4. Executable specifications

Instead of implementing a specification by more and more concrete versions which finally lead to a program one is often interested in executing specifications directly. In this case, a specification language $\mathscr{L}$ is considered as a programming language; any specification (declaration) *SP* corresponds to a module declaration in $\mathscr{L}$; executing *SP* means being able to prove certain kinds of formulas of the theory of *SP* by machine.

There are three main candidates for such formulas:

(i) Given a term $t \in T(sig(SP))$, find its value, i.e. find a term $n \in T(sig(SP))$ in normal form such that the equation $t=n$ is a theorem of *SP*.

(ii) Given a set of equations $E_1 = \{t_j = t'_j \mid j=1, \ldots, m\}$, find substitutions σ for the variables occurring in E_1 that solve these equations, i.e. find one (or all possible) substitutions σ (cf. Section 2.1) such that each equation $t_j\sigma = t'_j\sigma, j=1, \ldots, n$, is is a theorem of *SP*.

(iii) Given an arbitrary sentence, decide whether it is provable or not.

Case (i) corresponds to the execution of programs in a procedural or functional programming language, case (ii) is similar to the execution of logic programs (cf. [1]), and case (iii) is general theorem proving.

The main approaches for solving case (i) are *interpretational* techniques using *rewriting* and *compilational* techniques. Given a (simple) equational specification with a set $E = \{u_i = v_i \mid i=1, \ldots, n\}$ of axioms, one considers, instead of E, a set R of *directed* equations; for instance, one may choose for R the equations of E directed from left to right, i.e. $R = (u_i \rightarrow v_i \mid i=1, \ldots, n\}$. Then R is used to compute by replacing subterms in a given term or formula, as long as possible. When the computations always terminate in a unique normal form for equal terms (i.e., R is terminating and confluent), the system serves to solve the word problem for the equational theory of E (for details see [47]). If R is not confluent or not terminating, then the word problem for E may be undecidable. However, any actual computation of a normal form n for a term t yields a theorem $t=n$ of *SP* and there exist techniques based on the Knuth–Bendix completion procedure [112] for constructing equivalent confluent and terminating term rewriting systems. Rewriting techniques can also be used in the presence of Horn clauses and arbitrary first-order sentences as axioms. Both cases are studied e.g. in [141], for the latter either first-order formulas are coded by equations (cf. Section 5.3), or rewriting is replaced by resolution techniques (cf. [1], for references see also [47]). Recently, compilers for algebraic specifications have been developed. As target languages appear LISP [106], Pascal [67] and abstract machine code [111].

Sets of equations as in case (ii) can be solved using interpretative and compilational techniques. For the latter, the main approach consists in translating conditional equational specifications to a logic programming language such as PROLOG (for a survey see [15]). The interpretative techniques are mainly based on the above-mentioned *completion* procedures and on a general *unification* method, the *narrowing* method: given a (confluent) rewrite system R, a term t narrows to a term s if t contains a (nonvariable) subterm p that unifies, via the substitution σ, with the left-hand side of a rule $u \rightarrow v$ in R (i.e., $p\sigma = u\sigma$) and s is built from $t\sigma$ by the rule $u \rightarrow v$, i.e. $s = t\sigma[v\sigma/p\sigma]$ (for details[2] and references see [47], for conditional equations see [99, 105, 141]).

The narrowing method allows one to prove theorems which hold in $Alg(\Sigma, E)$, i.e. in all first-order models of a specification with axioms E. For proving theorems which hold (only) in the *inductive* theory of a specification (i.e. in $Gen(\Sigma, E)$, cf. Section 3.1), classical theorem proving provides explicit induction rules such as *structural induction* (cf. [9]), while *inductive completion* (often called *induction-less induction*) tries to get rid of induction steps by switching to *proofs of consistency* [97, 103, 108, 125]. Here the proof of inductive theorems $T_1, \ldots, T_n$ is reduced to showing that the underlying specification, say SP, enriched by $T_1, \ldots, T_n$, is a conservative extension of SP. In turn, this property is reduced to specific confluence and termination conditions on $SP \cup \{T_1, \ldots, T_n\}$. A proof-theoretical approach to inductive validity, which generalizes the traditional methods of structured and computational induction to Horn clauses with equality, is described in [142].

Case (iii) requires automated theorem proving for arbitrary first-order sentences (cf. e.g. [59, 44]). In spite of many advances in this field, a general automatic theorem prover seems to be unachievable. For practical purposes (at least by the end of the 1980s), it seems necessary either to restrict the class of formulas syntactically to equations or to Horn clauses (or to aim at intelligent proof checkers which are designed with the right balance between user guidance and automatic mechanisms similar to the *LCF* theorem-proving system [144] or the Edinburgh Logical Framework [91]).

The proof methods presented above are mainly suitable for simple specifications or else for specifications with constructors (as certain forms of narrowing [99] or inductive completion). For executing modular specifications, there exist two main techniques: either the modular structure is flattened to a normal form (cf. Section 6.6) and then techniques for nonmodular specifications are applied (cf. e.g. [15, 66]), or proof systems are developed which directly handle specification expressions (cf. e.g. [154, 156]).

9. Specification languages

In this section, the specification languages CLEAR, OBJ2, ASL and the specification part of Larch are presented. Some other specification languages including LOOK, PLUSS, ACT, ASF, OBSCURE and the specification parts of CIP-L, COLD and Extended ML are shortly discussed.

9.1. Clear

The specification language CLEAR developed by Burstall and Goguen was the first language designed for constructing algebraic specifications in a structured way. The seminal paper on "putting theories together" [42], in which CLEAR was informally presented, had considerable influence on the design of several other specification

[2] More exactly, σ is the most general unifier of p and u; for technical reasons, it is required that the variables in R have been renamed so that they are distinct from those in t.

languages such as OBJ, ASL or OBSCURE as well as on the presentation of the specification-building operations in this paper (cf. Section 6.3).

A specification in CLEAR describes a *theory* (rather than a class of models), usually presented using positive conditional equations as axioms. CLEAR provides *operators for combining* small(er) *theories* to make larg(er) theories. Parameterization in CLEAR is done via *theory procedures*; each such procedure can be applied to various different theories (the actual parameters) to produce new theories which are enrichments of the actual parameters.

A formal category-theoretic semantics of CLEAR is given in [43] based on the notion of institution (as presented in [74], cf. Section 3.3.6); in [153] a more concrete set-theoretic semantics is developed which (as in this paper) associates a model class semantics with each theory-building operation. As explained in the following, the semantics of CLEAR is *loose*, but uses so-called *data constraints* which allow one to construct free extensions; thus initial algebra semantics is also built-in.

The operators of CLEAR are the following:

(1) Basic specifications are written as **theory sorts** S **opns** F **eqns** E **endtheory** which is equivalent to $[\langle S, F\rangle, E]$.

(2) *Any specification* T can be given a name N by **const** $N = T$. Later, this name is kept to identify subspecifications, so-called *subtheories*. An example for a trivial specification is

$$\textbf{const } TRIV \equiv \textbf{theory sorts } elem[\textbf{opns } \emptyset \textbf{ eqns } \emptyset]^3 \textbf{ endtheory}$$

which has all algebras in $Alg(sig(TRIV))$ as models.

(3) A specification can be extended in two ways, the "simple" *enrich* (see Section 6.4) and the *data-enrich* which is defined as follows:

$$\textbf{enrich } T \textbf{ by data } S', F', E' \textbf{ enden } =_{\text{def}} \textbf{extend } T \textbf{ freely by } S', F', E'$$

For simplicity, it is assumed that S' and F' are disjoint from S and F. The subtheories of these enrichments are the subtheories of T. A simple example is the following specification of the (isomorphism class) of the standard model of natural numbers:

const $NAT \equiv$
  [**enrich** $\emptyset$ **by**][3]
  data sort *nat*
    opns $0\colon \to nat$,
        $succ\colon nat \to nat$
    [**eqns** $\emptyset$][3]
  enden

This free extension declares the function symbols 0 and *succ* as constructors of the sort *nat*. Enriching NAT by another function symbol does not change this constructor set. For instance, any model of the specification

$$\textbf{const } NAT_a \equiv \ \textbf{enrich } NAT \textbf{ by opns } a\colon \to nat \textbf{ enden}$$

[3]These (trivial) parts of the expression can be omitted.

is isomorphic to one of the structures N_k defined by $(N_k)_{nat} =_{\text{def}} \mathbb{N}$, $a^{N_k} =_{\text{def}} k$ for some $k \in \mathbb{N}$.

(4) Two specifications can be combined by an operator which constructs an amalgamated union; i.e., informally it builds the union of the subtheories together with the disjoint union of the other symbols. Let T_i, $i=1, 2$, *be two specifications with signature* $\Sigma_i = \langle S_i, F_i \rangle$ and (sets of) subtheories B_i with signatures $\Sigma_{B_i} = \langle S_{B_i}, F_{B_i} \rangle$. Define $\Sigma =_{\text{def}} \langle S_{B_1} \cup S_{B_2} \uplus (S_1 \backslash S_{B_1}) \uplus (S_2 \backslash S_{B_2}), F_{B_1} \cup F_{B_2} \uplus (F_1 \backslash F_{B_1}) \uplus (F_2 \backslash F_{B_2}) \rangle$ to be the amalgamated union of Σ_1 and Σ_2 and in_i: $\Sigma_i \to \Sigma$, $i=1, 2$, to be the associated inclusion morphisms. Then the combination of T_1 and T_2 is defined by

$$T_1 +_{\text{CLEAR}} T_2 =_{\text{def}} (\textbf{translate } T_1 \textbf{ by } in_1) \uplus (\textbf{translate } T_2 \textbf{ by } in_2).$$

The union $B_1 \cup B_2$ is the set of subtheories of $T_1 +_{\text{CLEAR}} T_2$.

(5) The **derive** operation is used to "forget" sorts and operators of a specification possibly renaming the ones remaining. Let T_i, $i=1, 2$ be two specifications with signatues Σ_i and subtheories B_i and let $\sigma: \Sigma_1 \to \Sigma_2$ be a signature morphism such that $B_1 \subseteq B_2$, σ is the identity on B_1 and σ is a specification morphism from T_1 to T_2. The derive operation produces a specification with signature Σ_1 and subtheories B_1 and all models which are derived from T_2 via σ:

$$\textbf{derive } T_1 \textbf{ from } T_2 \textbf{ by } \sigma =_{\text{def}} \textbf{derive from } T_2 \textbf{ by } \sigma.$$

This definition is slightly different from the original definition (for a discussion see [153, pp. 61–62]).

(6) A parameterized specification is called *theory procedure* and has the form

$$\textbf{proc } P(X: SP_{\text{par}}) = SP_{\text{body}}$$

where X is the formal parameter and SP_{par} the so-called *metatheory*. Semantically, a theory procedure is defined as a specification morphism

$$P: SP_{\text{par}} \to SP_{\text{body}}[SP_{\text{par}}/X]$$

Procedure application is the pushout construction: for any actual parameter A and specification morphism $\sigma: SP_{\text{par}} \to A$ (which is the identity on subtheories),

$$P(A[\sigma]) =_{\text{def}} P_{\text{po}}(A, \sigma)$$

is the pushout of σ and the specification morphism P. Subtheories of the procedure application are the subtheories of A and the subtheories of P.

For example, finite sets over arbitrary elements can be specified as follows:

```
proc Set(X: TRIV) =
  enrich X by
    data sort set
        opns empty: → set,
             add: elem, set → set
        eqns add(x, add(x, s)) = add(x, s),
             add(x, add(y, s)) = add(y, add(x, s))
  enden
```

Then the application *Set*(*NAT*[*elem* **is** *nat*]) gives the theory of finite sets of natural numbers; for a specification *BOOL* of the truth values, *Set*(*BOOL*[*elem* **is** *bool*]) gives the theory of finite sets of truth values. Combining the both specifications via

$$Set(NAT[elem\ \mathbf{is}\ nat]) +_{\mathrm{CLEAR}} Set(BOOL[elem\ \mathbf{is}\ bool])$$

yields two different copies of sort *set*, one for sets of natural numbers and one for sets of truth values.

9.2. *OBJ2*

The language OBJ2 developed by Goguen, Meseguer, Jouannaud and Futatsugi [63] supports order-sorted initial algebra specifications. Precursors of OBJ2 are OBJ0 (developed by Goguen), OBJT (developed by Goguen and Tardo), and OBJ1 (developed by Goguen, Meseguer and Plaisted). There are two kinds of specification expressions (called module expressions) in OBJ2, *theories* and *objects*. The former have a loose semantics, the latter denote initial algebras or (in the parameterized case) free extensions. One of the main features of OBJ2 is the introduction of subsorts. This supports the systematic treatment of partial operations (in a total algebra framework), multiple inheritance and error handling. Parameterization and specification-building operations of OBJ2 are inspired by CLEAR, the main difference is that OBJ2 has (only) a renaming operation instead of *derive*, thus OBJ2 has no facility to control the visibility of symbols. On the other hand, OBJ provides three kinds of import mechanisms which define different restrictions on preserved properties of the imported modules.

The mathematical semantics (i.e., the institution) of OBJ2 is based on the theory of order-sorted algebras as presented in [166] (cf. Section 3.4.4). The semantics of specification expressions is based on the notion of *data theories* [74].

In the following, the main features of OBJ2 are shortly presented.

(1) OBJ2's basic entity is the *object* which is a (possibly parameterized) specification describing a free extension of the parameter part. Similarly, a *theory* is a (possibly parameterized) specification describing an arbitrary extension of the parameter part. An object expression has the form

$$\begin{array}{l}\mathbf{obj}\ N[X :: SP_{\mathrm{par}}]\ \mathbf{is}\\ \quad \mathbf{sorts}\ S.\,\mathbf{op}\ f_1.\ \ldots\mathbf{op}\ f_k.\mathbf{var}\ v_1.\ldots\ \mathbf{var}\ v_m.\ \mathbf{eq}\ e_1.\ldots\ \mathbf{eq}\ e_n.\,\mathbf{jbo}\end{array}$$

and denotes the parameterized specification

$$\begin{array}{l}\mathbf{funct}\ N \equiv \lambda X : SP_{\mathrm{par}}.\\ \quad \mathbf{extend}\ X\ \mathbf{freely\ by}\ S,\ \{f_1,\ldots,f_k\},\ \forall v_1,\ldots \forall v_m.\, e_1 \wedge \cdots \wedge e_n\ \mathbf{endfunct}\end{array}$$

Theory expressions are written analogously with "**th**" and "**endth**" instead of "**obj**" and "**jbo**"; for the semantics, one takes the **enrich.by.** operation of CLEAR instead of the free extension. A simple example for a (nonparameterized) theory is the following specification of *TRIV* in OBJ2; this can be used to describe finite trees over arbitrary data:

th *TRIV* **is sort** *elem* endth,

obj *TREE*0[*X*:: *TRIV*] **is**

sort *tree*.
op *empty*: →*tree*.
op *node*: *tree*, *elem*, *tree*→*tree*.
jbo.

(2) Modules can import other modules in three different ways, by "**using**", "**protecting**" and "**extending**". Syntactically, a qualification such as "**using** *M*" is given as an item in the list of attributes of a module, say *N*; semantically, the attribute "**protecting** *M*" requires that *N* is persistent w.r.t. *M*, whereas "using *M*" is similar to the syntactic extension operation (see Section 3.2), it just copies *M* into the body of *N*. "**Extending**" denotes a sufficient condition for ensuring that for model *A* of *M* there exists a model *B* of *N* such that *A* is a subalgebra of $B|_{\mathrm{sig}(M)}$. For example, consider the following extension of a monomorphic specification *NAT* of natural numbers. The theory module

th NAT_a^{p} **is protecting** *NAT*. **op** *a*: → *nat* **endth**

is equivalent to the CLEAR-specification NAT_a (cf. Section 9.1), it protects the standard model of natural numbers. On the other hand, the object expression

obj NAT_a^{u} **is using** *NAT*. **op** *a*: →*nat* **jbo**

denotes the isomorphism class of the term algebra $T(\langle\{nat\}, \{0, a, succ\}\rangle)$. Replacing **using** by **extending** yields the same result whereas the replacement by **protecting** is not correct since the free extension of *NAT* is not persistent.

(3) *Subsorts* are introduced by the keyword **subsorts** followed by assertions of the form $s < s'$ which require that in any model the carrier set of sorts s is included in the one of sort s'. As an example, one may extend the module for *BOOL* by the operations *isempty*, *left*, *label* and *right*; since *left*, *right* and *label* are well-defined only on nonempty trees, a subsort *ne_tree* of *tree* is introduced.

obj *TREE* [*X*:: *TRIV*] **is**
protecting *BOOL*.
sorts *tree*, *ne_tree*.
subsorts *ne_tree* < *tree*.
op *empty*: →*tree*.
op *node*: *tree*, *elem*, *tree*→*ne_tree*.
op *isempty*: *tree*→*bool*.
ops *left*, *right*: *ne_tree*→*tree*.
op *label*: *ne_tree*→*elem*.
var *x*: *elem*.
vars t_1, t_2: *tree*.
eq: $isempty(empty) = true$.
eq: $isempty(node(t_1, x, t_2)) = false$.
eq: $label(t_1, x, t_2) = x$.

eq: $left(node(t_1, x, t_2)) = t_1$.
eq: $right(node(t_1, x, t_2)) = t_2$.
jbo

(4) OBJ2 provides two other constructs for building structured modules, parameter application and renaming. The latter is expressed by an operator **.*.** which is defined exactly as the *rename*-operation in Section 6.2; i.e., for any module M and bijective signature morphism ρ, $M * \rho =_{\text{def}}$ **translate** M **with** ρ. Parameter application is defined by the pushout construction; syntactically, a construct, called **view**, is introduced: for any module identifiers N, A, P and a signature morphism ρ such that A denotes the module SP_A, P denotes the parameterized module $P \equiv \lambda X : SP_{\text{par}} . SP_{\text{body}}$ and ρ: $sig(SP_{\text{par}}) \rightarrow sig(SP_A)$ is a parameter-passing morphism,

view N **of** A **as** P **is** $\rho =_{\text{def}} [\![P]\!]_{\text{po}}(SP_A, \rho)$.

The mapping ρ can be more general than a signature morphism; it can map function symbols of the formal signature to terms in the actual signature. An example for a view is the specification of finite trees over natural numbers:

view *TREENAT* **of** *NAT* **as** *TREE* **is**
sort *elem* **to** *nat*
endview

9.3. ASL

The specification language ASL developed by Sannella, Tarlecki and the author [177, 161, 157] is designed as a kernel language for constructing structured specifications: ASL provides only very basic specification-building operations which can then be used to define higher-level operations. For example, the specification language PLUSS [65], the metalanguage for the (draft) formal definition of ADA [3] and Extended ML [155] have been defined in terms of ASL-expressions. Another main feature of ASL is that specification expressions form only one particular type of ASL-expressions; many other kinds of expressions such as signatures, terms and formulas which are necessary for the constuction of specification expression are first-class objects in ASL [178].

A specification in ASL describes a class of models, usually presented with arbitrary first-order equational formulas as axioms. Although inspired by CLEAR, the specification-building operations of ASL do not contain a free extension operator; instead, ASL offers an operation for abstracting w.r.t. behavioral equivalence. Parameterization in ASL is λ-abstraction.

A set-theoretic denotational semantics of ASL is given in [178], whereas in [157], an institution-independent category-theoretic semantics is defined.

Main aspects of ASL are the following:

(1) ASL has five specification-building operations consisting of *[.,.]*, **reach**, +, **derive from** and **behavior** which correspond exactly to the specification-building functions *[.,.]*, *reach*, +, *derive from* and *behavior* as defined in Sections 6.2–6.6.

Sannella, Tarlecki [157] introduce also *translate* which enable them to express the sum operation "+" by the union "$\cup$". Moreover, they use a generalized version of the reachability operator.

(2) ASL is a strongly typed higher-order functional language; parameterization of specifications is the same as for any type of ASL-expressions: typed λ-abstraction with parameter restriction as presented in Section 7.1.

9.4. *Larch*

The Larch family of specification languages is developed within a project at MIT and DEC System Research Center which started with the work reported in [87]. A language description and many examples are given in [89].

Each Larch specification has components written in two languages: one designed for a specific programming language, the so-called *Larch interface language*, and another common to all programming languages, the so-called *Larch shared language*; the first component is an implementation of the second (in the sense of Section 8).

The shared language is an algebraic specification language, where any specification, called **trait**, denotes a theory (or equivalently a class of models). As e.g. CLEAR, it has mechanisms for building one specification from another (**generated-by**, **partitioned-by**, **imports**, **includes**, **renames**, **assumes**); in contrast to other specification languages, it has also mechanisms for inserting checkable redundancy in specifications (**constrains**, **converts**). Interestingly, Larch does not offer any hiding nor export mechanism; the designers argue that due to the presence of the interface language in which implementations are carried out hiding is not necessary in the shared specification language [89, p. 22].

The language constructs of the *Larch Shared Language* are the following:

(1) A simple algebraic specification is given by a set of function symbols and a set of first-order equational formulas as axioms. The sorts are automatically computed from the types of the function symbols:

$$\textbf{introduces } F, \textbf{ constrains } E =_{\text{def}} [\langle sorts(F), F\rangle, E]$$

(2) The model classes of a trait can be *restricted* by the operators **generated-by** and **partitioned-by** describing reachability and extensionality defined by contexts similar to terminal algebras (cf. e.g. Theorem 5.4.6). For each trait T, sort $s \in sorts(sig(T))$ and set F of function symbols of range s,

$$T, s \textit{ generated by } F =_{\text{def}} \textbf{reach}_F(T);$$

for each set of function symbols F' with one parameter of sort s, s **partitioned-by** F' denotes the additional (extensionality) axiom

$$\forall x, y\colon s. \bigwedge_{f \in F'} f(\ldots, x, \ldots) = f(\ldots, y, \ldots) \Rightarrow x = y.$$

The dots above denote the other parameters of f which are also universally quantified variables. For example, natural numbers are specified in Larch by

Nat: **trait**
  introduces
    zero: $\rightarrow$*nat*
    succ: *nat*$\rightarrow$*nat*
  constrains so that
    nat **generated-by** [*zero*, *succ*]
    nat **partitioned-by** [*succ*]
    for all *x*: *nat*
      $\neg(zero = succ(x))$

The standard model N is (up to isomorphism) the only model of the trait *Nat*.

(3) Traits can be extended by two operators: the operator **includes** corresponds to the sum operator of Section 6.2, whereas the operator **import** asserts that the new specification is a conservative extension of the imported one:

$$T_1 \textbf{ includes } T_2 =_{\text{def}} T_1 + T_2,$$

$$T_1 \textbf{ imports } T_2 =_{\text{def}} \begin{cases} T_1 + T_2 & \text{if } T_2 = \textbf{export } sig(T_2) \textbf{ from } (T_1 + T_2), \\ \bot & \text{otherwise.} \end{cases}$$

For example, a transitive relation on a sort *elem* using a specification *BOOL* of truth values can be given as follows.

Transitive: **trait**
  imports *BOOL*
  introduces.*r*.: *elem*, *elem*$\rightarrow$*bool*
  constrains for all x_1, x_2, x_3: *elem*
    $x_1 r x_2 = true \wedge x_2 r x_3 = true \Rightarrow x_1 r x_3 = true$

Then an equivalence relation is defined by including *Transitive* in a trait with reflexivity and symmetry axioms:

Equivalence: **trait**
  includes *Transitive*
  constrains for all x_1, x_2 : *elem*
    $x_1 r x_1 = true$
    $x_1 r x_2 = true \Rightarrow x_2 r x_1 = true$

(4) Renaming of traits can be done using the **with** operator: for each trait T and any two symbols a and b (such that b occurs in T),

$$T \textbf{ with } [a \textbf{ for } b] =_{\text{def}} \textbf{rename } T \textbf{ with } [a/b]$$

(5) Parameterization is not explicitly included but it can be simulated using the operator **assume**. This operator has the same meaning as **include**, but if a trait with an **assume** clause is imported or included in another trait (this corresponds to parameter application), the assumption will have to be discharged. Hence, **includes**

corresponds to **using**, and **imports** corresponds to **protecting** in OBJ2 (cf. Section 9.2). For example, finite sets over an arbitrary sort *elem* can be specified as follows:

Set: **trait**
  assumes *Equivalence*
  introduces
    $empty: \rightarrow set$
    $add: elem, set \rightarrow set$
    $.\varepsilon.: elem, set \rightarrow bool$
  constrains so that
    set **generated-by** [*empty*, *add*]
    set **partitioned-by** [ε]
    for all $s: set,\ x,\ y: elem$
      $x \varepsilon\ empty = false$
      $x \varepsilon\ add(y, s) = (x\ r\ y)\ or\ (x \varepsilon s)$

For any monomorphic specification of *Equivalence*, the trait *Set* specifies exactly the isomorphism class of finite sets, over elements of the carrier set associated with *elem*. Including *Set* in a trait *Natspec* (which extends *Nat* by *Bool* and an equivalence relation *eq*)

NatSet: **trait**
  includes *Set* **with** [*nat* **for** *elem*, *eq* **for** *r*]
  import *Natspec*

is only allowed if the (renamed) trait *Equivalence* is a subset of the theory of *Natspec*, i.e. if

NatSpec $\subseteq$ **translate** *Equivalence* **with** [*nat*/*elem*, *eq*/*r*]

(6) A particular feature of Larch is the facility of stating intended consequences. There are three basic possibilities: to state that a trait is sufficiently complete (using an attribute **converts**) or partially sufficiently complete (using **exempts**), that the new axioms define a conservative extension (using a list of function symbols in **constrains**), and finally that the model class of a trait is contained in the model class of another trait. This assertion corresponds to the notion of implementation discussed in Section 8.1.

9.5. *Further specification languages*

The language LOOK developed by Zilles, Lucas, Ehrig and Thatcher [183] is inspired by CLEAR but contains much simpler specification building operators: *basic specifications*, *sum*, *free extensions* and *translate* which is injective w.r.t. the constrained part of the specifications. Parameterization is simulated using the *translate* operation, no *export* operation is available. LOOK has two levels of formal semantics, an (elegant) presentation semantics and a model class semantics [57]

The language ACT ONE developed by Ehrig, Fey and Hansen [50] and presented in [55] was strongly influenced by LOOK. It is a kernel language based on pure initial algebra semantics. Every specification expression is a parameterized specification which denotes a strongly persistent free functor. Specification-building operations are basic (parameterized) specification, sum and bijective renaming; parameterization is defined via the pushout construction. The ISO-standard LOTOS [101], which has been developed for the formal definition of protocols and services for computer networks, uses ACT ONE for the definition of data expressions.

The language ACT TWO is a successor of ACT ONE designed for the description of modules [176, 62]. In a simplified way, a module can be understood as a constructor implementation $SP \rightsquigarrow_{\mathbf{R}_{S'} \circ \mathbf{D}_\Sigma} SP'$ where the body specification SP' implements the export specification SP via the construction "Export-Restrict" (cf. Section 8.2). More generally, a module specification consists of interfaces for exports and imports with a common parameter part and a body. Export and import are loose parameterized specifications whereas the body is a parameterized initial algebra specification with the import interface as parameter. A module is correct if the body is a parameterized implementation of the specification-building operation which transforms the input interface into the export interface.

The language PLUSS developed by Gaudel and Bidoit [65] focuses on methodical aspects. Specifications for which constructors are defined form a syntactic category called **specif** whereas specifications which are equivalent to basic ones are called **drafts.** The latter can be manipulated by the specification-building functions *sum* and *enrich*, whereas the former can be put together using conservative extensions and restricted to specific models by the *quotient* operation. The semantics of PLUSS is defined in terms of ASL. In order to support better reusability and refinement of specifications, [29] proposes a new approach to the semantics.

The language ASF developed by Bergstra, Heering and Klint is based on a pure initial algebra approach [15]. Axioms have the form of conditional equations. The specification-building operations are basic specifications, *export*, *rename* and *import* of a sequence of specifications (into an incomplete specification) where the latter operation corresponds to the sum of the sequence freely extended by the incomplete specification.

A so-called "origin rule" allows one to determine the "origin" of a sort or function symbol x, i.e. the specification in which x is declared; this avoids numerous occurrences of renamed versions of the same hidden sort or function within one specification expression. ASF has a presentation semantics: every specification expression has, as normal form, a parameterized specification with hidden functions (cf. Section 5) augmented by a set of origin functions. A translation of ASF specifications into the algebra BSA of specifications (as defined in [14] is envisaged, cf. Section 6.6). The language PSF is an extension of ASF by the ACP-calculus for the specification of concurrent systems [124].

The specification language OBSCURE designed by Lehmann and Loeckx defines a set of specification expressions which is generated by specification-building operations for *sum*, *export*, *renaming*, *quotient* and *restriction* [114]. Every specification expression denotes a construction (cf. Section 8.3), i.e. a particular parameterized specification. In order to get the full expressive power of parameterized specifications, OBSCURE contains also operations for expressing requirements on the parameter and on the body of parameterized specifications. Moreover, it contains an operator for the composition of (parameterized) specifications. For OBSCURE two denotational semantic definitions are given: a loose one and a (mainly) nonloose one where every specification denotes a function from algebras to algebras.

In contrast to all languages above, a wide-spectrum language defines not only specification expressions but also allows one to program in an imperative or/and an applicative style.

The (historically) first example for such a language is CIP-L which was designed within the project CIP aiming at the formal development of programs from specifications by transformations [9, 10, 11]. Specifications in CIP-L are based on the hierarchical approach (see Section 5.4). CIP-L provides facilities for parameterization, export, import and renaming of specifications. The semantics of a specification is a (loose) class of partial algebras.

The specification part of the language COLD developed within the Esprit project METEOR is strongly influenced by ASF [102]. It contains similar specification-building operations as well as the concept of "origin". But the semantics of a specification is, as in CIP-L, a loose class of partial algebras [61].

A third wide-spectrum language is Extended ML designed by Sannella and Tarlecki [155]. Extended ML enhances Standard ML by allowing more information in module interfaces (axioms in ML signatures) and less information in module bodies (axioms in place of code in ML structures). The semantics of Extended ML is based on the institution-independent semantics of ASL [157]. The concept of abstractor implementation as presented in this paper is designed by [158, 159] for program development in Extended ML.

Acknowledgment

The author is indebted to Oliver Geupel, Rolf Hennicker, Jacques Loeckx, Peter Mosses, Peter Padawitz, Victor Pollara, Bernhard Reus, Song Qun and Rainer Weber for many comments and helpful suggestions. Thanks go to Stefan Gastinger and, in particular, to Bernhard Reus for carefully and patiently typing the continuously growing and changing manuscript.

This research has been partially sponsored by Esprit Project 432, METEOR.

References

[1] APT, K., Introduction to logic programming, in J. van Leeuwen, ed., *Handbook of Theoretical Computer Science, Vol. B* (North-Holland, Amsterdam, 1990) 493–574.

[2] ASTESIANO, E. and M. CERIOLI, On the existence of initial models for partial (higher-order) conditional specifications, in: J. Diaz and F. Orejas, eds., *Proc. TAPSOFT '89, Vol. 1*, Lecture Notes in Computer Science, Vol. 351 (Springer, Berlin, 1989) 74–88.

[3] ASTESIANO, E., A. GIOVINI, F. MAZZANTI, G. REGGIO and E. ZUCCA, The ADA challenge for new formal semantic techniques, in: *ADA: Managing the Transition, Proc. of the ADA-Europe Internat. Conf.* (Cambridge University Press, Cambridge, 1986).

[4] ASTESIANO, E., G.F. MASCARI, G. REGGIO and M. WIRSING, On the parameterized algebraic specification of concurrent systems, in: H. Ehrig, C. Floyd, M. Nivat and J.W. Thatcher, *Proc. TAPSOFT'85, Vol. 1*, Lecture Notes in Computer Science, Vol. 185 (Springer, Berlin, 1985) 342–358.

[5] ASVELD, P.R.J. and J.V. TUCKER, Complexity theory and the operational structure of algebraic programming systems, *Acta Inform.* **17** (1982) 451–476.

[6] BARENDREGT, H.P., The type free λ-calculus, in: J.K. Barwise, ed., *Handbook of Mathematical Logic*, (North-Holland, Amsterdam, 1977) 1091–1132.

[7] BARWISE, K.J. Axioms for abstract model theory, *Annals of Math. Logic* **7** (1974) 221–265.

[8] BARWISE K.J. *Handbook of Mathematical Logic*, Studies in Logic and the Foundations of Mathematics, Vol. 90 (North-Holland, Amsterdam, 1977).

[9] BAUER, F.L. and H. WÖSSNER, *Algorithmische Sprache und Programmentwicklung* (Springer, Berlin, 1981); English edition: *Algorithmic Language and Program Development* (Springer, Berlin, 1982).

[10] BAUER, F.L., R. BERGHAMMER, M. BROY, W. DOSCH, F. GEISELBRECHTINGER, R. GNATZ, E. HANGEL, W. HESSE, B. KRIEG-BRÜCKNER, A. LAUT, T. MATZNER, B. MÖLLER, F. NICKL, H. PARTSCH, P. PEPPER, K. SAMELSON, M. WIRSING and H. WÖSSNER, *The Munich Project CIP, Vol. 1: The Wide Spectrum Language CIP-L*, Lecture Notes in Computer Science, Vol. 183 (Springer, Berlin, 1985).

[11] BAUER, F.L., M. BROY, W. DOSCH, R. GNATZ, B. KRIEG-BRÜCKNER, A. LAUT, M. LUCKMANN T. MATZNER, B. MÖLLER, H. PARTSCH, P. PEPPER, K. SAMELSON, R. STEINBRÜGGEN, M. WIRSING and H. WÖSSNER, Programming in a wide spectrum language: a collection of examples, *Sci. Comput. Programming* **1** (1981) 73–144.

[12] BEIERLE, C. and A. VOSS, Algebraic specification and implementations in an integrated software development and verification system, Dissertation, Memo SEKI-85-12, FB Informatik, Univ. Kaiserslautern, 1985.

[13] BERGSTRA, J.A., M. BROY, J.V. TUCKER and M. WIRSING, On the power of algebraic specifications, in: J. Gruska and M. Chytil, eds., *Proc. 10th Internat. Symp. on Mathematical Foundations of Computer Science*, Lecture Notes in Computer Science, Vol. 118, (Springer, Berlin, 1981) 193–204.

[14] BERGSTRA, J.A., J. HEERING and P. KLINT, Module algebra, Report CS-R8617, Centrum voor Wiskunde en Informatica Amsterdam, 1986.

[15] BERGSTRA, J.A., J. HEERING, and P. KLINT, *Algebraic Specification* (ACM Press, New York, 1989).

[16] BERGSTRA, J.A. and J.W. KLOP, Algebraic specifications for parameterized data types with minimal parameter and target algebras, Tech. Report IW 183, Math. Centrum, Dept. of Computer Science, Amsterdam, 1981.

[17] BERGSTRA, J.A. and J.W. KLOP, Initial algebra specifications for parameterized data types, Tech. Report IW 186, Math. Centrum, Dept. of Computer Science, Amsterdam, 1981.

[18] BERGSTRA, J.A. and J.W. KLOP, Algebra of communicating processes, in: J.W. de Bakker, M. Hazewinkel and J.K. Lenstra, eds., *CWI Monograph I, Proc. CWI Symp. on Mathematics and Computer Science* (North-Holland, Amsterdam, 1986) 89–138.

[19] BERGSTRA, J.A. and J.W. KLOP, Process algebra: specification and verification in bisimulation semantics, in: M. Hazewinkel, J.K. Lenstra and L.G.L.T. Meertens, eds., *CWI Monograph 4, Proc. CWI Symp on Mathematics and Computer Science II* (North-Holland, Amsterdam, 1986) 61–94.

[20] BERGSTRA, J.A. and J.-J. MEYER, On specifying sets of integers, *Elektron. Informationsverarb. Kybernet.* **20** (1984) 531–541.

[21] BERGSTRA, J.A. and J.V. TUCKER, A natural data type with a finite equational final semantics but no effective equational initial semantics specification, *Bull. EATCS* **11** (1980) 23–33.

[22] BERGSTRA, J.A. and J.V. TUCKER, The completeness of the algebraic specification methods for computable data types, *Inform. and Control* **54** (1982) 186–200.

[23] BERGSTRA, J.A. and J.V. TUCKER, Initial and final algebra semantics for data type specifications: two characterization theorems, *SIAM J. Comput.* **12** (1983) 366–387.

[24] J.A. BERGSTRA and J.V. TUCKER, Algebraic specifications of computable and semicomputable data types, *Theoret. Comput. Sci.* **50** (1987) 137–181.

[25] BERNOT, G., Correctness proof for abstract implementations, *Inform. and Comput.* **80** (1989) 121–151.

[26] BERNOT, G., M. BIDOIT and C. CHOPPY, Abstract data types with exception handling: an initial algebra approach based on the distinction between exceptions and errors, *Theoret. Comput. Sci.* **46** (1986) 13–46.

[27] BERTONI, A., G. MAURI and P. MIGLIOLI, A characterization of abstract data types as model-theoretic invariants, in: H.A. Maurer, ed., *Proc. 6th Internat. Coll. on Automata, Languages and Programming*, Lecture Notes in Computer Science, Vol. 71 (Springer, Berlin, 1979) 26–37.

[28] BIDOIT, M., Algebraic specification of exception handling and error recovery by means of declarations and equations, in: J. Paredans, ed., *Proc 11th Internat. Coll. on Automata, Languages and Programming*, Lecture Notes in Computer Science, Vol. 172 (Springer, Berlin, 1984) 95–108.

[29] BIDOIT, M., The stratified loose approach: A generalization of initial and loose semantics, Rapport de Recherche No. 402, Orsay, France, 1988.

[30] BIDOIT, M., B. BIEBOW, M.C. GAUDEL, C. GRESSE and G. GUIHO, Exception handling: formal specification and systematic program construction, *IEEE Trans. on Software Eng.* **11** (1985) 242–252.

[31] BIRKHOFF, G. and D. LIPSON, Heterogeneous algebras, *J. Combin. Theory* **8** (1970) 115–133.

[32] BROY, M., Specification and top down design of distributed systems, *J. Comput. System Sci.* **34** (1987) 236–264.

[33] BROY, M., Predicative specification for functional programs describing communicating networks, *Inform. Process. Lett.* **25** (1987) 93–101.

[34] BROY, M., W. DOSCH, H. PARTSCH, P. PEPPER and M. WIRSING, Existential quantifiers in abstract data types, in: H.A. Maurer, ed., *Proc. 6th Internat. Coll. Automata, Languages and Programming*, Lecture Notes in Computer Science, Vol. 71 (Springer, Berlin, 1979) 73–87.

[35] BROY, M., B. MÖLLER, P. PEPPER and M. WIRSING, Algebraic implementations preserve program correctness, *Sci. Comput. Programming* **7** (1986) 35–53.

[36] BROY, M., C. PAIR and M. WIRSING, A systematic study of models of abstract data types, *Theoret. Comput. Sci.* **33** (1984) 139–174.

[37] BROY, M. and M. WIRSING, On the algebraic specification of nondeterministic programming languages, in: E. Astesiano and C. Bohm, eds., *Proc. 6th CAAP*, Genoa, Lecture Notes in Computer Science, Vol. 112 (Springer, Berlin, 1981) 162–179.

[38] BROY, M. and M. WIRSING, Partial abstract data types, *Acta Inform.* **18** (1982) 47–64.

[39] BROY, M. and M. WIRSING, On the algebraic specification of finitary infinite communicating sequential process, in: D. Bjørner, ed., *Proc. IFIP Working Conf. on Formal Description of Programming Concepts II* (North-Holland, Amsterdam, 1983).

[40] BROY, M. and M. WIRSING, Generalized heterogenous algebras and partial interpretations, in: G. Ausiello and M. Protasi, ed., *Proc. 8th CAAP*, Lecture Notes in Computer Science, Vol. 159, (Springer, Berlin, 1983) 1–34.

[41] BURMEISTER, P., *A Model Theoretic Oriented Approach to Partial Algebras*, Mathematical Research, Vol. 31 (Akademie-Verlag, Berlin, 1986).

[42] BURSTALL, R.M. and J.A. GOGUEN, Putting theories together to make specifications, in: *Proc. 5th Internat. Joint Conf. on Artificial Intelligence* (1977) 1045–1058.

[43] BURSTALL, R.M. and J.A. GOGUEN, The semantics of CLEAR, a specification language, in: D. Bjørner, ed., *Proc. Advanced Course on Abstract Software Specifications*, Lecture Notes in Computer Science, Vol. 86 (Springer, Berlin, 1980) 292–232.

[44] BUTLER, R., E. LUSK, W. MCCURE, and R. OVERBEEK, Paths to high-performance theorem proving, in: J. Siekmann, ed., *Proc. 8th Conf. on Automated Deduction*, Lecture Notes in Computer Science, Vol. 230 (Springer, Berlin, 1986) 588–597.

[45] CHANG C.C. and H.J. KEISLER, *Model Theory* (North-Holland, Amsterdam, 1977).

[46] DAHL, O.J., B. MYHRHANG and K. NYGAARD, *Common Base Language* (Norsk Reguesentral, Oslo, 1970).

[47] DERSHOWITZ, N. and J.-P. JOUANNAUD, Rewrite systems, in: J. van Leeuwen, ed., *Handbook of Theoretical Computer Science, Vol. B* (North-Holland, Amsterdam, 1990) 243–320.

[48] EHRICH, H.-D., On the realization and implementation, in: J. Gruska and M. Chytil, eds., *Proc. Internat. Symp. on Mathematical Foundations of Computer Science*, Lecture Notes in Computer Science, Vol. 118 (Springer, Berlin, 1981) 271–280.

[49] EHRICH, H.-D., On the theory of specification, implementation, and parameterization of abstract data types, *J. ACM* **29**(1) (1982) 206–277.

[50] EHRIG, H., W. FEY and H. HANSEN, An algebraic specification language with two levels of semantics, Bericht No. 83-03, Fachbereich 20-Informatik, Technische Univ. Berlin, 1983.

[51] EHRIG, H., H.-J. KREOWSKI, J.W. THATCHER, E.G. WAGNER and J.B. WRIGHT, Parameterized data types in algebraic specification languages (short version), in: J. de Bakker and J. van Leeuwen, eds., *Proc. 7th Internat. Coll. on Automata, Languages and Programming*, Lecture Notes in Computer Science, Vol. 85 (Springer, Berlin, 1980) 157–168.

[52] EHRIG, H. and H.-J. KREOWSKI, Parameter passing commutes with implementation of parameterized data types, in: M. Nielsen and E.M. Schmidt, eds., *Proc. 9th Internat. Coll. on Automata, Languages and Programming*, Lecture Notes in Computer Science, Vol. 140 (Springer, Berlin, 1982) 197–211.

[53] EHRIG, H., H.-J. KREOWSKI, B. MAHR and P. PADAWITZ, Algebraic implementation of abstract data types, *Theoret. Comput. Sci.* **20** (1982) 209–263.

[54] EHRIG, H., H.-J. KREOWSKI, J.W. THATCHER, E.G. WAGNER and J.B. WRIGHT, Parameter passing in algebraic specification languages, *Theoret. Comput. Sci.* **28** (1984) 45–81.

[55] EHRIG, H. and B. MAHR, *Fundamentals of Algebraic Specifications I, Equations and Initial Semantics* EATCS Monographs on Theoretical Computer Science, Vol. 6 (Springer, Berlin, 1985).

[56] EHRIG, H., F. PARISI-PRESICCÉ, P. BOEHM, C. RIECKHOFF, D. DIMITROVICI and M. GROSSE-ROHDE, Algebraic data type and process specifications based on projection spaces, Tech. Report 87-8, Tech. Univ. Berlin, FB 20, 1987.

[57] EHRIG, H., J.W. THATCHER, P. LUCAS and S.N. ZILLES, Denotational and initial algebra semantics of the algebraic specification language LOOK, Tech. Report 84-22, TU Berlin, FB 20, 1984.

[58] EHRIG, H., E.G. WAGNER and J.W. THATCHER, Algebraic specifications with generating constraints, in: J. Diaz, ed., *Proc. 10th Internat. Coll. on Automata, Languages and Programming*, Lecture Notes in Computer Science, Vol. 154 (Springer, Berlin, 1983) 188–202.

[59] EISINGER N., What you always wanted to know about clause graph resolution, in: J. Siekmann, ed., *Proc. 8th Conf. on Automated Deduction*, Lecture Notes in Computer Science, Vol. 230 (Springer, Berlin, 1986) 316–336.

[60] FARRES-CASALS, J., Proving correctness of constructor implementations, LFCS-Report Series, ECS-LFCS-89-72, Dept. of Computer Science, Univ. of Edinburgh, 1989.

[61] FEIJS, L.M.G., H.B.M. JONKERS, C.P.J. KOYMANS and G.R. RENARDEL DE LAVALETTE, Formal definition of the design language COLD-K (preliminary edition), Tech. Report METEOR/t7/PRLE/7, Esprit Project 432, 1987.

[62] FEY, W., Pragmatics, concepts, syntax, semantics and correctness notions of ACT TWO, Dissertation, Berlin, 1989.

[63] FUTATSUGI, K., J.A. GOGUEN, J.-P. JOUANNAUD and J. MESEGUER, Principles of OBJ2, in: *Proc. POPL* (1985) 52–66.

[64] GANZINGER, H., Parameterized specifications: parameter passing and implementation with respect to observability, *ACM TOPLAS* **5** (3) (1983) 318–354.

[65] GAUDEL, M.-C. Towards structured algebraic specifications, in: *ESPRIT '85', Status Report of Continuing Work* (North-Holland, Amsterdam, 1986) 493–510.

[66] GESER, A. and H. HUßMANN, Experiences with the RAP-system—a specification interpreter combining term rewriting and resolution, in: B. Robinet and R. Wilhelm, eds., *Proc. ESOP 86*, Lecture Notes in Computer Science, Vol 213 (Springer, Berlin, 1986) 339–350.

[67] GESER, A., H. HUßMANN, A. MÜCK, A compiler for a class of conditional rewrite systems, in: N. Dershowitz and S. Kaplan, eds., *Proc. 1st Workshop on Conditional Rewrite-Systems*. Lecture Notes in Computer Science, Vol. 308 (Springer, Berlin, 1987) 84–90.

[68] GIARRATANA, V., F. GIMONA and U. MONTANARI, Observability concepts in abstract data type specification, in: A. Mazurkiewicz, ed., *Proc. Internat. Symp. on Mathematical Foundations of Computer Science*, Lecture Notes in Computer Science, Vol. 45 (Springer, Berlin, 1976) 567–578.

[69] GOGOLLA, M., On parametric algebraic specifications with clean error handling, in: H. Ehrig, R. Kowalski, G. Levi and U. Montanari, eds., *Proc. TAPSOFT '87*, Lecture Notes in Computer Science, Vol. 249 (Springer, Berlin, 1987) 81–95.

[70] GOGOLLA, M., K. DROSTEN, U. LIPECK and H.D. EHRICH, Algebraic and operational semantics of specifications allowing exceptions and errors, *Theoret. Comput. Sci.* **34** (1984) 289–313.

[71] GOGUEN, J.A., Abstract errors for abstract data types, in: E. Neuhold, ed., *Formal Description of Programming Concepts* (North-Holland, Amsterdam, 1977).

[72] GOGUEN, J.A., Order-sorted algebra, Semantics and Theory of Computation Report 14, Computer Science Dept, UCLA, 1978.

[73] GOGUEN, J.A. and R.M. BURSTALL, Cat, a system for structured elaboration of correct programs from structured specifications, Tech. Report, CSL-118, Computer Science Lab., SRI International, 1980.

[74] GOGUEN, J.A. and R.M. BURSTALL, Introducing institutions, in: E. Clarke and D. Kozen, eds., *Logic of Programs*, Lecture Notes in Computer Science, Vol. 164 (Springer, Berlin, 1984) 221–255.

[75] GOGUEN, J.A. and R.M. BURSTALL, Institutions: abstract model theory for computer science, Report CSLI-85-30, Stanford Univ. 1985.

[76] GOGUEN, J.A. and J. MESEGUER, Completeness of many-sorted equational logic, *ACM SIGPLAN Notices* **16**(7) (1981) 24–32.

[77] GOGUEN, J.A. and J. MESEGUER, Universal realization, persistent interconnection and implementation of abstract modules, in: M. Nielsen and E.M. Schmidt, eds., *Proc. 9th Internat. Coll. on Automata, Languages and Programming*, Lecture Notes in Computer Science, Vol. 140 (Springer, Berlin, 1982) 265–281.

[78] GOGUEN J.A. and J. MESEGUER, Models and equality for logic programming, in: H. Ehrig and U. Montanari, eds., *Proc. TAPSOFT '87*, Lecture Notes in Computer Science, Vol. 250 (Springer, Berlin, 1987) 1–22.

[79] GOGUEN, J.A. and J. MESEGUER, Unifying functional, object-oriented and relational programming with logical semantics, in: B. Shriver and P. Wegner, eds., *Research Directions in Object-Oriented Programming* (MIT Press, Boston, 1987) 417–477.

[80] GOGUEN, J.A., J.W. THATCHER, E.G. WAGNER and J.B. WRIGHT, Abstract data types as initial algebras and the correctness of data representations, in: *Proc. Conf. on Computer Graphics, Pattern Recognition and Data Structures* (1975) 89–93.

[81] GOGUEN, J.A., J.W. THATCHER, E.G. WAGNER and J.B. WRIGHT, Initial algebra semantics and continuous algebras, *J. ACM* **24** (1977) 68–95.

[82] GOGUEN, J.A., J.W. THATCHER and E.G. WAGNER, An initial algebra approach to the specification correctness, and implementation of abstract data types, IBM Research Report RC-6487; also in: R.T. Yeh, ed., *Current Trends in Programming Methodology, Vol. 4: Data Structuring* (Prentice-Hall, Englewood Cliffs, NJ, 1978) 80–149.

[83] GRÄTZER, G., *Universal Algebra* (Van Nostrand, Princeton, NJ, 1968).

[84] GRATZER, G., *General Lattice Theory* (Birkhäuser, Basel, 1978).

[85] GRIES, D., ed., *Programming Methodology* (Springer, Berlin, 1978).

[86] GUESSARIAN, I., *Algebraic Semantics*, Lecture Notes in Computer Science, Vol. 99 (Springer, Berlin, 1981).

[87] GUTTAG, J.V., The specification and application to programming of abstract data types, Ph.D. Thesis, Univ. of Toronto, 1975.

[88] GUTTAG, J.V. and J.J. HORNING, The algebraic specification of abstract data types, *Acta Inform.* **10** (1978) 27–52.

[89] GUTTAG, J.V., J.J. HORNING and J.M. WING, Larch in five easy pieces, Tech. Report, Digital Systems Research Center, Palo Alto, CA, 1985.

[90] GUTTAG, J.V., E. HOROWITZ and D.R. MUSSER, The design of data type specifications, in: R.T. Yeh, ed., *Current Trends in Programming Methodology, Vol. 4: Data Structuring* (Prentice-Hall, Englewood Cliffs, N.J., 1978) 60–79.

[91] HARPER, R.W., F.A. HONSELL and G.D. PLOTKIN, A framework for defining logics, in: *Proc. 2nd IEEE Symp. on Logic in Computer Science* (1987) 194–204.

[92] HENNICKER, R., Observational implementations, in: B. Monien and R. Cori, *Proc. STACS '89*, Lecture Notes in Computer Science, Vol. 349 (Springer, Berlin, 1989) 59–71.

[93] HENNICKER, R., Implementation of parameterized observational specifications, in: J. Diaz and F. Orejas, eds., *Proc. TAPSOFT '89, Vol. 1*, Lecture Notes in Computer Science, Vol. 351 (Springer, Berlin, 1989) 290–305.

[94] HESSELINK, W.H., A mathematical approach to nondeterminism in data types, *ACM TOPLAS* **10** (1988) 87–117.

[95] HOARE, C.A.R., An axiomatic basis for computer programming, *Comm. ACM* **12** (1969) 576–583.

[96] HOARE, C.A.R., Proofs of correctness of data representations, *Acta Inform.* **1** (1972) 271–281.

[97] HUET, G. and J.M. HULLOT, Proofs by induction in equational theories with constructors, *J. Comput. System. Sci.* **25** (1982) 239–266.

[98] HUET, G. and D.C. OPPEN, Equations and rewrite rules: a survey, in: R.-V. Book, ed., *Formal Language Theory: Perspectives and Open Problems* (Academic Press, New York, 1980).

[99] HUßMANN, H., Unification in conditional equational theories, in: B.V. Caviness, ed., *EUROCAL 85 Vol. 2*, Lecture Notes in Computer Science, Vol. 204 (Springer, Berlin, 1985) 543–553.

[100] HUßMANN, H., Nichtdeterministische algebraische Spezifikation, Dissertation, Univ. Passau, Fak. Math. Informatik, 1989.

[101] ISO DIS 8807, ISO/TC 97/SC 21/WG1 FDT/SCC: LOTOS — a formal description technique based on the temporal ordering of observational behaviour, 1986.

[102] JONKERS, H.B.M., An introduction to COLD-K, in: J.A. Bergstra and M. Wirsing, eds., *Algebraic Methods: Theory, Tools and Applications*, Lecture Notes in Computer Science, Vol. 394 (Springer, Berlin, 1989) 139–205.

[103] JOUANNAUD, J.-P. and E. KOUNALIS, Automatic proofs by induction in equational theories with constructors, in: *Proc. IEEE Symp. on Logic in Computer Science* (1986) 358–360.

[104] KAMIN, S., Final data types and their specification, *ACM TOPLAS* **5**(1) (1983) 97–121.

[105] KAPLAN, S., Simplifying conditional term rewriting systems: Unification, termination and confluence, *J. Symbolic Comput.* **4** (1987) 295–334.

[106] KAPLAN, S., A compiler for conditional term rewriting systems, in: P. Lescanne, ed., *2nd Internat. Conf. on Rewriting Techniques and Applications*, Lecture Notes in Computer Science, Vol. 256 (Springer, Berlin, 1987) 25–41.

[107] KAPLAN, S. and A. PNUELI, Specification and implementation of concurrently accessed data structures, in: F.J. Brandenburg, G. Vidal-Naquet and M. Wirsing, eds., *Proc. STACS 87*, Lecture Notes in Computer Science, Vol. 247 (Springer, Berlin, 1987) 220–244.

[108] KAPUR, D. and D.R. MUSSER, Proof by consistency, *Artificial Intelligence* **31** (1987) 125–157.

[109] KLAEREN, H.A., A simple class of algebraic specifications for abstract software modules, in: P. Dembinsky, ed., *Proc. 9th Internat. Symp. on Mathematical Foundations of Computer Science*, Lecture Notes in Computer Science, Vol. 88 (Springer, Berlin, 1980) 362–374.

[110] KLAEREN, H.A., *Algebraische Spezifikation–Eine Einführung*, Lehrbuch Informatik (Springer, Berlin, 1983).

[111] KLAEREN, H.A. and K. INDERMARK, A new technique for compiling recursive function symbols, in: J.A. Bergstra and M. Wirsing, eds., *Algebraic Methods: Theory, Tools and Applications*, Lecture Notes in Computer Science, Vol. 394 (Springer, Berlin, 1989) 69–90.

[112] KNUTH, D.E. and P.B. BENDIX, Simple word problems in universal algebras, in: J. Leech, ed., *Computational Problems in Abstract Algebra* (Pergamon, Oxford, 1970).

[113] KRÖGER, F., Abstract modules: combining algebraic and temporal logic specification means, *Technique et Sci. Inform.* **6** (1987) 559–573.

[114] LEHMANN, T. and J. LOECKX, The specification language of OBSCURE, in: D.T. Sannella and A. Tarlecki, *Recent Trends in Data Type Specification*, Lecture Notes in Computer Science, Vol. 332 (Springer, Berlin, 1988) 131–153.

[115] LESCANNE, P. ed., Modèles non-deterministes de types abstraits, *RAIRO Inform. Théor. Appl.* **16** (1982) 225–244.

[116] LIPECK, U., Ein algebraischer Kalkül für einen strukturierten Entwurf von Datenabstraktionen, Dissertation, Report 148, Univ. Dortmund, 1983.

[117] LISKOV, B. and S. ZILLES, Programming with abstract data types, *ACM SIGPLAN Notices* **9** (1974) 50–59.

[118] MACLANE, S., *Categories for the Working Mathematician* (Springer, Berlin, 1972).

[119] MACQUEEN D.B. and D.T. SANNELLA, Completeness of proof systems for equational specifications, *IEEE Trans. Software Eng.* **11**(5) (1985) 454–560.

[120] MAIBAUM, T.S.E., P.A.S. VELOSO and M.R. SADLER, A theory of abstract data types for program development: Bridging the gap? in: H. Ehrig, C. Floyd, M. Nivat and J. Thatcher, eds., *Proc. TAPSOFT '85, Vol. 2*, Lecture Notes in Computer Science, Vol. 186 (Springer, Berlin, 1985) 214–230.

[121] MAJSTER, M.E., Limits of the "algebraic" specification of abstract data types, *ACM SIGPLAN Notices* **12** (1977) 37–42.

[122] MAJSTER, M.E., Data types, abstract data types and their specification problem, *Theoret. Comput. Sci.* **8** (1979) 89–127.

[123] MAL'CEV, A.I., Constructive algebras, *Russian Math. Surveys* **16** (1961) 77–129.

[124] MAUW, S., An algebraic specification of process algebra including two examples, in: J.A. Bergstra and M. Wirsing, eds., *Algebraic Methods: Theory, Tools and Applications*, Lecture Notes in Computer Science, Vol. 394 (Springer, Berlin, 1989) 507–554.

[125] MESEGUER, J. and J.A. GOGUEN, Initiality, induction and computability, in: M. Nivat and J. Reynolds, eds., *Algebraic Methods in Semantics* (Cambridge Univ. Press, Cambridge, 1985).

[126] MEYER, B. *Object-Oriented Software Construction* (Prentice-Hall, Englewood Cliffs, NJ, 1988).

[127] MILNER, R., Fully abstract models of typed λ-calculi, *Theoret. Comput. Sci.* **4** (1977) 1–22.

[128] MILNER, R., Operational and algebraic semantics of concurrent processes, in: J. van Leeuwen, ed., *Handbook of Theoretical Computer Science, Vol. B* (North-Holland, Amsterdam, 1990) 1201–1242.

[129] MÖLLER, B., On the algebraic specification of infinite objects—Ordered and continuous models of algebraic types, *Acta Inform.* **22** (1985) 537–578.

[130] MÖLLER, B., Higher-order algebraic specification, Habilitationsschrift, Fakultät Mathematik und Informatik, TU München 1987.

[131] MOSSES, P.D., Unified algebras and substitutions, in: *Proc. 4th Ann IEEE Symp. on Logic in Computer Science* (1989) 304–312.

[132] NIPKOW, T., Nondeterministic data types: models and implementations, *Acta Inform.* **22** (1986) 629–661.

[133] NIVAT, M., On the interpretation of recursive polyadic program schemes, in: *Proc. Istituto Nazionale di Alta Matematica, Symposia Mathematica XV* (Academic Press, London, 1975) 255–281.

[134] OBERSCHELP, A., Untersuchungen zur mehrsortigen Quantorenlogik, *Math. Ann.* **145** (1962) 297–333.

[135] OREJAS, F., Characterizing composability of abstract implementations, in: M. Karpinski, ed., *Proc. 11th Coll. on Foundations of Computation Theory*, Lecture Notes in Computer Science, Vol. 158 (Springer, Berlin, 1983) 335–346.

[136] OREJAS, F., A proof system for checking composability of implementations of abstract data types, in: G. Kahn, D.B. MacQueen and G. Plotkin, eds., *Semantics of Data Types*, Lecture Notes in Computer Science, Vol. 173 (Springer, Berlin, 1983) 357–374.

[137] OREJAS, F., A characterization of passing compatibility for parameterized specifications, *Theoret. Comput. Sci.* **51** (1987) 205–214.

[138] OREJAS, F., H. EHRIG, P. PEPPER, H.-D. EHRICH, B. KRIEG-BRÜCKNER, H.-J. KREOWSKI, H. GANZINGER, R.M. BURSTALL, D.T. SANNELLA, E. ASTESIANO, P. LESCANNE, J.-L. REMY, H. KIRCHNER, H. PARTSCH, M.-C. GAUDEL, J.-P. JOUANNAUD, I. KODRATOFF, M. BROY and M. WIRSING, Compass: a comprehensive algebraic approach to system specification and development, Tech. Report 880613, Univ. Bremen, FB 3 Math.-Informatik, 1988.

[139] PADAWITZ, P., Parameter preserving data type specifications, *J. Comput. System Sci.* **34** (1985) 179–209.

[140] PADAWITZ, P., The equational theory of parameterized specifications, *Inform. and Comput.* **76** (1988) 121–137.

[141] PADAWITZ, P., *Computing in Horn Clause Theories*, EATCS Monographs on Theoretical Computer Science, Vol. 16 (Springer, Berlin, 1988).

[142] PADAWITZ, P., Inductive expansion, Tech. Report MIP-8907, Fakultät Mathematik und Informatik, Univ. Passau 1989.

[143] PADAWITZ, P. and M. WIRSING, Completeness of many-sorted equational logic revisited, *Bull. EATCS* **24** (1984) 88–94.

[144] PAULSON, L.C., *Logic and Computation: Interactive Proof with Cambridge LCF* (Cambridge University Press, Cambridge, 1987).

[145] POIGNÉ, A., Another look at parameterization using algebras with subsorts, in: M. Chytil and P. Koubek, eds., *Mathematical Foundations of Computer Science*, Lecture Notes in Computer Science, Vol. 176 (Springer Berlin, 1984) 471–479.

[146] PLOTKIN, G.P., LCF considered as a programming language, *Theoret. Computer. Sci.* **5** (3) (1977) 223–256.

[147] RABIN, M.O., Computable algebra, general theory and the theory of computable fields, *Trans. Amer. Math. Soc.* **98** (1960) 341–360.

[148] RASIOWA, H. and R. SIKORSKI, *The Mathematics of Metamathematics*, Monografia Matematyczne (Warszawa, 1970).

[149] REICHEL, H., Theorie der Aequoide, Dissertation, Humboldt-Univ., Berlin, 1979.

[150] REICHEL, H., Behavioural equivalence—a unifying concept for initial and final specification methods, in: *Proc. 3rd Hungarian Computer Science Conf.* (1981) 27–39.

[151] REICHEL, H., *Initial Computability, Algebraic Specifications, and Partial Algebras* (Clarendon Press, Oxford, 1987)

[152] ROGERS JR., H., *Theory of Recursive Functions and Effective Computability* (McGraw-Hill, New York, 1967).

[153] SANNELLA, D.T., Semantics, implementation and pragmatics of Clear, a program specification language, Ph.D. Thesis, Dept. of Computer Science, Univ. of Edinburgh, 1982.

[154] SANNELLA, D.T. and R.M. BURSTALL, Structured theories in LCF, in: G. Ausiello and M. Protasi, eds., *Proc. 8th CAAP*, Lecture Notes in Computer Science, Vol. 159 (Springer, Berlin, 1983) 377–391.

[155] SANNELLA, D.T. and A. TARLECKI, Program specification and development in Standard ML, in: *Proc. 12th Ann. ACM Symp. on Principles of Programming Languages* (1985) 67–77.

[156] SANNELLA, D.T. and A. TARLECKI, On observational equivalence and algebraic specifications, *J. Comput. System Sci.* **34** (1987) 150–178.

[157] SANNELLA, D.T. and A. TARLECKI, Specifications in an arbitrary institution, *Inform. and Comput.* **76** (1988) 165–210.

[158] SANNELLA, D.T. and A. TARLECKI, Towards formal development of programs from algebraic specifications: implementations revisited, *Acta Inform.* **25** (1988) 233–281.

[159] SANNELLA, D.T. and A. TARLECKI, Towards formal development of ML programs: foundations and methodology, in: J. Diaz and F. Orejas, eds., *Proc. TAPSOFT '89, Vol. 2*, Lecture Notes in Computer Science, Vol. 352 (Springer, Berlin, 1989) 375–389.

[160] SANNELLA, D.T. and M. WIRSING, Implementation of parameterised specifications, Report CSR-103-82, Dept. of Computer Science, Univ. of Edinburgh; extended abstract in: M. Nielsen and E.M. Schmidt, eds., *Proc. 9th Internat. Coll. on Automata, Languages and Programming*, Lecture Notes in Computer Science, Vol. 140 (Springer, Berlin, 1982) 473–488.

[161] SANNELLA, D.T. and M. WIRSING, A kernel language for algebraic specification and implementation, in: M. Karpinski, ed., *Proc. 11th Coll. on Foundations of Computation Theory*, Lecture Notes in Computer Science, Vol. 158 (Springer, Berlin, 1983) 413–427.

[162] SCHOETT, O., Data abstraction and the correctness of modular programming, Ph.D. Thesis, Report CST-42-87, Univ. of Edinburgh, 1986.

[163] SELMAN, A., Completeness of calculi for axiomatically defined classes of algebras, *Algebra Universalis* **2** (1972) 20–32.

[164] SHOENFIELD, J.R., *Mathematical Logic* (Addison-Wesley, Reading, MA, 1967).

[165] SMOLKA, G., Logic programming with polymorphically order-sorted types, in: J. Grabowski, P. Lescanne and W. Wechler, eds., *Mathematical Research, Vol. 49* (Akademie-Verlag, Berlin, 1988) 53–70.

[166] SMOLKA, G., W. NUTT, J.A. GOGUEN and J. MESEGUER, Order-sorted equational computation, SEKI-Report SR-87-14, Univ. Kaiserslautern, 1987; also in: H. Aït-Kaci and M. Nivat, eds., *Resolution of Equations in Algebraic Structures, Vol. 2* (Academic Press, New York, 1989) 299–367.

[167] STOY, J.E. *Denotational Semantics: The Scott–Strachey Approach to Programming Language Theory* (MIT Press, Cambridge, MA, 1977).

[168] SUBRAHMANYAM, P.A., Nondeterminism in abstract data types, in: S. Even and O. Kariv, eds., *Proc. 8th Internat. Coll. on Automata, Languages and Programming*, Lecture Notes in Computer Science, Vol. 115 (Springer, Berlin, 1981) 148–164.

[169] TARLECKI, A., Quasi-varieties in abstract algebraic institutions, *J. Comput. System Sci.* **33** (1986) 333–360.

[170] TARLECKI, A. and M. WIRSING, Continuous abstract data types, *Fund. Inform.* **9** (1986) 95–125.

[171] THATCHER J.W., E.G. WAGNER and J.B. WRIGHT, Data type specification: parameterization and the power of specification techniques, *ACM TOPLAS* **4** (1982) 711–773.

[172] VELOSO, P.A.S., Traversable stacks with fewer errors, *ACM SIGPLAN Notices* **14** (1979) 55–59.

[173] VRANCKEN, J.L.M., The algebraic specification of semicomputable data types, Report P8705, Programming Research Group, Univ. Amsterdam, 1987.

[174] WAGNER, E.G., J.W. THATCHER and J.B. WRIGHT, Programming languages as mathematical objects, in: J. Winkowski, ed., *Mathematical Foundations of Computer Science, Vol. 10*, Lecture Notes in Computer Science, Vol. 64 (Springer, Berlin, 1978) 84–101.

[175] WAND, M., Final algebra semantics and data type extensions, *J. Comput. System Sci.* **19** (1981) 27–44.

[176] WEBER, H. and H. EHRIG, Programming in the large with algebraic module specifications, in: H.-J. Kugler, ed., *Proc. IFIP 10th World Computer Congress* (North-Holland, Amsterdam, 1986) 675–684.

[177] WIRSING, M., Structured algebraic specifications, in: B. Robinet, ed., *Proc. AFCET Symp. on Mathematics for Computer Science* (1982) 93–108.

[178] WIRSING, M., Structured algebraic specifications: A kernel language, *Theoret. Comput. Sci.* **43** (1986) 123–250.

[179] WIRSING, M. and M. BROY, Abstract data types as lattices of finitely generated models, in: P. Dembiński, ed., *Proc. Internat. Symp. on Mathematical Foundations of Computer Science*, Lecture Notes in Computer Science, Vol. 88 (Springer, Berlin, 1980) 673–685.

[180] WIRSING, M. and M. BROY, An analysis of semantic models for algebraic specifications, in: M. Broy and G. Schmidt, eds., *Proc. Marktoberdorf Summer School on Theoretical Foundations of Programming Methodology* (Reidel, Boston, MA, 1982) 351–412.

[181] WIRSING, M., P. PEPPER, H. PARTSCH, W. DOSCH and M. BROY, On hierarchies of abstract data types, *Acta Inform.* **20** (1983) 1–33.

[182] ZILLES, S.N., Algebraic specification of data types, Computation Structures Group Memo 119, Lab. for Computer Science, MIT, 1974.

[183] ZILLES, S.N., P. LUCAS and J.W. THATCHER, A look at algebraic specifications, IBM Res. Report, Yorktown Heights, NY, 1982.

[184] FEIJS, L.M.G., The calculus $\lambda\pi$, in: M. Wirsing and J.A. Bergstra, eds., *Algebraic Methods: Theory, Tools and Applications*, Lecture Notes in Computer Science, Vol. 394 (Spinger, Berling, 1989) 307–328.

[185] MOSSES, P.D., Denotational semantics, in: J. van Leeuwen, ed., *Handbook of Theoretical Computer Science, Vol. B* (North-Holland, Amsterdam, 1990).

[186] STREICHER, T. and M. WIRSING, Dependent types considered necessary for algebraic specification languages, in: K. Jantke and H. Reichel, eds., *Proc. 7th Workshop on Abstract Data Types*, Lecture Notes in Computer Science (Springer, Berlin, 1990).

CHAPTER 14

Logics of Programs

Dexter KOZEN

Department of Computer Science, Cornell University, Ithaca, NY 14853, USA

Jerzy TIURYN

Institute of Mathematics, University of Warsaw, 00-901 Warsaw, PKiN IXp, Poland

Contents

1. Introduction 791
2. Propositional Dynamic Logic 793
3. First-order Dynamic Logic 811
4. Other approaches 830
Acknowledgment 834
References 834

HANDBOOK OF THEORETICAL COMPUTER SCIENCE
Edited by J. van Leeuwen

1. Introduction

Logics of programs are formal systems for reasoning about computer programs. Traditionally, this has meant formalizing correctness specifications and proving rigorously that those specifications are met by a particular program. Other activities fall into this category as well: determining the equivalence of programs, comparing the expressive power of various operators, synthesizing programs from specifications, etc. These activities range from the highly theoretical to the highly practical. Formal systems too numerous to mention have been proposed for these purposes, each with its own peculiarities.

This chapter gives a brief introduction to some of the basic issues in the study of program logics. We have chosen one system, Dynamic Logic, to illustrate these issues. There are many other reasonable choices: Temporal Logic, Algorithmic Logic, etc. We discuss the relationships among these systems where appropriate. By our choice of Dynamic Logic we do not advocate its use over any other system, but we feel that it is an appropriate vehicle for illustrating the concepts we wish to discuss.

Program logics differ from classical logics in that truth is *dynamic* rather than *static*. In classical predicate logic, the truth value of a formula φ is determined by a valuation of its free variables over some structure. The valuation and the truth value of φ it induces are regarded as immutable; there is no formalism relating them to any other valuations or truth values. In program logics, there are explicit syntactic constructs called *programs* to change the values of variables, thereby changing the truth values of formulas. For example, the program $x:=x+1$ over the natural numbers changes the truth value of the formula "x is even". Such changes occur on a metalogical level in classical predicate logic, for example in the Tarski definition of truth of a formula: if $u:\{x, y, \ldots\} \rightarrow \mathcal{N}$ is a valuation of variables over the natural numbers $\mathcal{N}$, then the formula $\exists x\, x^2 = y$ is defined to be true under the valuation u iff there exists an $a \in \mathcal{N}$ such that the formula $x^2 = y$ is true under the valuation $u[x/a]$, where $u[x/a]$ agrees with u everywhere except x, on which it takes the value a. This definition involves a metalogical operation which produces $u[x/a]$ from u for all possible values $a \in \mathcal{N}$. This operation becomes explicit in Dynamic Logic in the form of the program $x:=?$, called a *nondeterministic assignment*. This is a rather unconventional program, since it is not effective; however, it is quite useful as a descriptive tool. A more conventional way to obtain a square root of y, if it exists, would be the program

$$x:=0;\ \textbf{while}\ x^2 < y\ \textbf{do}\ x:=x+1\ \textbf{od}. \tag{1}$$

In Dynamic Logic, programs are first-class objects on a par with formulas, complete with a collection of operators for forming compound programs inductively from a basis of primitive programs. In the simplest version of *DL*, these program operators are $\cup$ (nondeterministic choice), ; (sequential composition), * (iteration), and ? (test). These operators are already sufficient to generate all **while** programs (which over $\mathcal{N}$ are sufficient to compute all partial recursive functions). To discuss the effect of the execution of a program p on the truth of a formula φ, *DL* uses a modal construct $\langle p\rangle\varphi$, which intuitively states: "It is possible to execute p and halt in a state satisfying φ." For example, the first-order formula $\exists x\, x^2 = y$ is equivalent to the *DL* formula

$$\langle x := ? \rangle x^2 = y. \tag{2}$$

In order to instantiate the quantifier effectively, we might replace the nondeterministic assignment inside the $\langle\,\rangle$ with the **while** program (1); the resulting formula would be equivalent to (2) in $\mathcal{N}$.

1.1. States, input/output relations, traces

A *state* is an instantaneous description of reality. In first-order Dynamic Logic, a state is often taken to be a valuation $u: V \to |\mathcal{A}|$ of variables V ranging over a first-order structure $\mathcal{A} = (|\mathcal{A}|, \ldots, f^{\mathcal{A}}, \ldots, R^{\mathcal{A}}, \ldots)$. In practice, variables may be of different types and range over the different sorts of a many-sorted structure, but for simplicity we usually assume that $\mathcal{A}$ is single-sorted. Here $|\mathcal{A}|$ is the carrier of $\mathcal{A}$, $f^{\mathcal{A}}$ represents a typical function of $\mathcal{A}$, and $R^{\mathcal{A}}$ is a typical relation on $\mathcal{A}$.

A program can be viewed as a transformation on states. Given an initial (input) state, the program will go through a series of intermediate states, perhaps eventually halting in a final (output) state. A sequence of states that can be obtained from the execution of a program p starting from a particular input state is called a *trace*.

The traces of p need not be uniquely determined by their start states; i.e., p may be *nondeterministic*. A trace can be infinite, in which case p is said to be *looping* or *diverging*. A finite trace corresponds to a halting computation of p, and in this case we say that p *halts*, *terminates*, or *converges*. The set of all pairs of first and last states of finite traces of p is called the *input/output relation* of p.

In Dynamic Logic, programs are interpreted as input/output relations. *DL* cannot be used to reason about program behavior not manifested in the input/output relation. For this reason it is inadequate for dealing with programs that are not normally supposed to halt, such as operating systems. Other program logics, such as Temporal Logic or Process Logic, use an execution sequence semantics which overcomes this limitation. However, for programs that are supposed to halt, correctness criteria are traditionally given in the form of an *input/output specification*, consisting of a formal relation between the input and output states that the program is supposed to maintain. The input/output relation of a program carries all the information necessary to determine whether the program is correct relative to such a specification.

1.2. Exogenous and endogenous logics

There are two main approaches to modal logics of programs: the *exogenous* approach, exemplified by Dynamic Logic and its precursor, the Partial Correctness Assertions Method (Hoare Logic) [71], and the *endogenous* approach, exemplified by Temporal Logic and its precursor, the Inductive Assertions Method [44]. A logic is *exogenous* if its programs are explicit in the language. Syntactically, a Dynamic Logic program is a well-formed expression built inductively from primitive programs using a small set of program operators. Semantically, a program is interpreted as its input/output relation. The relation denoted by a compound program is determined by the relations denoted by its parts. This aspect of *compositionality* allows proofs by structural induction. In Temporal Logic, the program is fixed and considered part of the structure over which the logic is interpreted. The current location in the program during execution is stored in a special variable for that purpose, called the *program counter*, and is part of the state along with the values of the program variables. Instead of program operators there are temporal operators that describe how the program variables, including the program counter, change with time. Thus Temporal Logic sacrifices compositionality for a less restricted and hence more generally applicable formalism.

1.3. Historical note

The idea of introducing and investigating a formal system dealing with properties of programs in an abstract model was advanced by Thiele [153] and independently by Engeler [38] in the mid 1960s. Research in program verification flourished in the late 1960s and thereafter with the work of many researchers, notably Floyd [44], Hoare [71], and Salwicki [142]. Dynamic Logic, which emphasizes the modal nature of the program/assertion interaction, was introduced by Pratt in [131] (see also [59]) and is a relatively late development.

There are by now a number of books and survey papers treating logics of programs and Dynamic Logic. We refer the reader to [56, 58, 120, 49, 50, 76].

2. Propositional Dynamic Logic

Propositional Dynamic Logic (*PDL*) was first defined by Fischer and Ladner [42, 43]. It plays the same role in Dynamic Logic that classical propositional logic plays in classical predicate logic. It describes the properties of the interaction between programs and propositions that are independent of the domain of computation.

2.1. Basic definitions

2.1.1. Syntax

Syntactically, *PDL* is a blend of propositional logic, modal logic, and the algebra of regular events. It has expressions of two sorts: *programs* $p, q, r, \ldots$ and *propositions* or *formulas* $\varphi, \psi, \ldots$ There are countably many atomic symbols of each sort, as well as a variety of operators for forming compound expressions from simpler ones:

- propositional operators $\vee$, $\neg$
- program operators $\cup$, ;, *
- mixed operators ?, $\langle\,\rangle$

Compound programs and propositions are defined by mutual induction, as follows. If

φ, ψ are propositions and p, q are programs, then

$$\begin{array}{ll} \varphi \vee \psi & \textit{(propositional disjunction)}, \\ \neg\varphi & \textit{(propositional negation)}, \\ \langle p \rangle \varphi & \textit{(modal possibility)} \end{array}$$

are propositions and

$$\begin{array}{ll} p;q & \textit{(sequential composition)}, \\ p \cup q & \textit{(nondeterministic choice)}, \\ p^* & \textit{(iteration)}, \\ \varphi? & \textit{(test)} \end{array}$$

are programs. The intuitive meanings of the less familiar of these constructs are as follows:

$\langle p \rangle \varphi$ = "It is possible to execute p and terminate in a state satisfying φ",
$p;q$ = "Execute p, then execute q",
$p \cup q$ = "Choose either p or q nondeterministically and execute it",
p^* = "Execute p repeatedly a nondeterministically chosen finite number of times",
$\varphi?$ = "Test φ; proceed if true, fail if false".

The set of propositions is denoted Φ and the set of programs is denoted Ψ. We avoid parentheses by assigning precedence to the operators: unary operators, including $\langle p \rangle$, bind tighter than binary ones, and ; binds tighter than $\cup$. Also, under the semantics to be given in the next section, the operators ; and $\cup$ will turn out to be associative, so we may write $p;q;r$ and $p \cup q \cup r$ without ambiguity.

The primitive operators are chosen for their mathematical simplicity. A number of more conventional programming constructs can be defined from them. The propositional operators $\wedge$, $\rightarrow$, $\leftrightarrow$, *false*, and *true* are defined from $\vee$ and $\neg$ in the usual way. In addition:

$$\begin{array}{l}
\mathbf{skip} = \mathit{true}? \\
\mathbf{fail} = \mathit{false}? \\
[p]\varphi = \neg\langle p \rangle \neg\varphi \\
\mathbf{if}\ \varphi_1 \rightarrow p_1 \,[\!]\, \cdots \,[\!]\, \varphi_n \rightarrow p_n\ \mathbf{fi} = \varphi_1?;p_1 \cup \cdots \cup \varphi_n?;p_n \\
\mathbf{do}\ \varphi_1 \rightarrow p_1 \,[\!]\, \cdots \,[\!]\, \varphi_n \rightarrow p_n\ \mathbf{od} = (\varphi_1?;p_1 \cup \cdots \cup \varphi_n?;p_n)^*;\ (\neg\varphi_1 \wedge \cdots \wedge \neg\varphi_n)? \\
\mathbf{if}\ \varphi\ \mathbf{then}\ p\ \mathbf{else}\ q\ \mathbf{fi} = \mathbf{if}\ \varphi \rightarrow p \,[\!]\, \neg\varphi \rightarrow q\ \mathbf{fi} = \varphi?;p \cup \neg\varphi?;q \\
\mathbf{while}\ \varphi\ \mathbf{do}\ p\ \mathbf{od} = \mathbf{do}\ \varphi \rightarrow p\ \mathbf{od} = (\varphi?;p)^*;\neg\varphi? \\
\mathbf{repeat}\ p\ \mathbf{until}\ \varphi = p;\ \mathbf{while}\ \neg\varphi\ \mathbf{do}\ p\ \mathbf{od} = p;(\neg\varphi?;p)^*;\varphi? \\
\{\varphi\}p\{\psi\} = \varphi \rightarrow [p]\psi
\end{array}$$

The propositions $\langle p \rangle \varphi$ and $[p]\varphi$ are read "diamond p φ" and "box p φ" respectively. The latter has the intuitive meaning: "Whenever p terminates, it must do so in a state satisfying φ." Unlike $\langle p\rangle\varphi$, $[p]\varphi$ does not imply that p terminates. Indeed, the formula $[p]$ *false* asserts that no computation of p terminates. For fixed program p, the operator $\langle p\rangle$ behaves like a possibility operator of modal logic (see [73, 20]). The operator $[p]$ is the modal dual of $\langle p\rangle$ and behaves like a modal necessity operator.

The ternary **if-then-else** operator and the binary **while-do** operator are the usual *conditional test* and *while loop* constructs found in conventional programming languages. The constructs **if-**$[\!]$**-fi** and **do-**$[\!]$**-od** are the *alternative guarded command* and *iterative guarded command* constructs respectively (see [51]). The construct $\{\varphi\}p\{\psi\}$ is the classical Hoare partial correctness assertion [71].

2.1.2. Semantics

The formal semantics of *PDL* comes from modal logic. A *Kripke model* is a pair $\mathcal{M}=(S^{\mathcal{M}}, I^{\mathcal{M}})$ where $S^{\mathcal{M}}=\{u,v,\ldots\}$ is an abstract set of *states* and $I^{\mathcal{M}}$ is an *interpretation function.* Each proposition φ is interpreted as a subset $\varphi^{\mathcal{M}}\subseteq S^{\mathcal{M}}$, and each program p is interpreted as binary relation $p^{\mathcal{M}}$ on $S^{\mathcal{M}}$. We may think of $\varphi^{\mathcal{M}}$ as the set of states satisfying the proposition φ, and $p^{\mathcal{M}}$ as the input/output relation of the program p.

The function $I^{\mathcal{M}}$ assigns an arbitrary subset of $S^{\mathcal{M}}$ to each atomic proposition symbol and an arbitrary binary relation on $S^{\mathcal{M}}$ to each atomic program symbol. Compound programs and propositions receive their meanings inductively:

$$\begin{aligned}
(\varphi\vee\psi)^{\mathcal{M}} &= \varphi^{\mathcal{M}}\cup\psi^{\mathcal{M}}, \qquad (\neg\varphi)^{\mathcal{M}}=S^{\mathcal{M}}-\varphi^{\mathcal{M}},\\
(\langle p\rangle\varphi)^{\mathcal{M}} &= p^{\mathcal{M}}\circ\varphi^{\mathcal{M}}=\{u\mid\exists v\,(u,v)\in p^{\mathcal{M}} \text{ and } v\in\varphi^{\mathcal{M}}\},\\
(p;q)^{\mathcal{M}} &= p^{\mathcal{M}}\circ q^{\mathcal{M}}=\{(u,v)\mid\exists w\,(u,w)\in p^{\mathcal{M}} \text{ and } (w,v)\in q^{\mathcal{M}}\},\\
(p\cup q)^{\mathcal{M}} &= p^{\mathcal{M}}\cup q^{\mathcal{M}},\\
(p^*)^{\mathcal{M}} &= \text{the reflexive transitive closure of } p^{\mathcal{M}},\\
(\varphi?)^{\mathcal{M}} &= \{(u,u)\mid u\in\varphi^{\mathcal{M}}\}
\end{aligned}$$

where the reflexive-transitive closure of a binary relation ρ is $\bigcup_{n<\omega}\rho^n$, where

$$\rho^0=\{(u,u)\mid u\in S^{\mathcal{M}}\}, \qquad \rho^{n+1}=\rho\circ\rho^n.$$

The symbol $\circ$ denotes relational composition.

The defined operators inherit their meanings from these definitions:

$$\begin{aligned}
(\varphi\wedge\psi)^{\mathcal{M}} &= \varphi^{\mathcal{M}}\cap\psi^{\mathcal{M}},\\
([p]\varphi)^{\mathcal{M}} &= \{u\mid\forall v\,(u,v)\in p^{\mathcal{M}}\rightarrow v\in\varphi^{\mathcal{M}}\},\\
\mathit{true}^{\mathcal{M}} &= S^{\mathcal{M}}, \qquad \mathit{false}^{\mathcal{M}}=\emptyset,\\
\mathbf{skip}^{\mathcal{M}} &= \{(u,u)\mid u\in S^{\mathcal{M}}\}, \qquad \mathbf{fail}^{\mathcal{M}}=\emptyset.
\end{aligned}$$

The **if-then-else**, **while-do**, and guarded commands also receive meanings from the above definitions.

It can be argued that the input/output relations given by these definitions capture the intuitive operational meanings of these constructs. For example, the relation associated with the program **while** φ **do** p **od** is the set of pairs (u,v) for which there exist states $u_0, u_1, \ldots, u_n$, $n \geqslant 0$, such that $u = u_0$, $v = u_n$, $u_i \in \varphi^{\mathcal{M}}$ and $(u_i, u_{i+1}) \in p^{\mathcal{M}}$ for $0 \leqslant i < n$, and $u_n \in \neg\varphi^{\mathcal{M}}$.

We often write $\mathcal{M}, u \models \varphi$ for $u \in \varphi^{\mathcal{M}}$ and say that u *satisfies* φ *in* $\mathcal{M}$. We may write $u \models \varphi$ when $\mathcal{M}$ is understood. We write $\mathcal{M} \models \varphi$ if $\mathcal{M}, u \models \varphi$ for all $u \in \mathcal{M}$, and we write $\models \varphi$ and say that φ is *valid* if $\mathcal{M} \models \varphi$ for all $\mathcal{M}$. We say that φ is *satisfiable* if $\mathcal{M}, u \models \varphi$ for some $\mathcal{M}, u$. If Σ is a set of propositions, we write $\mathcal{M} \models \Sigma$ if $\mathcal{M} \models \varphi$ for all $\varphi \in \Sigma$. A proposition ψ is said to be a *logical consequence* of Σ if, for all $\mathcal{M}$, $\mathcal{M} \models \psi$ whenever $\mathcal{M} \models \Sigma$, in which case we write $\Sigma \models \psi$. (Note that this is *not* the same as saying that $\mathcal{M}, u \models \psi$ whenever $\mathcal{M}, u \models \Sigma$.) We say that an inference rule $\frac{\Sigma}{\psi}$ is *sound* if ψ is a logical consequence of Σ.

This version of *PDL* is usually called *regular PDL* because of the primitive operators $\cup$, ;, *, which are familiar from the algebra of regular events (see [72]). Programs can be viewed as regular expressions over the atomic programs and tests. In fact, it can be shown that if φ is an atomic proposition symbol, then any two test-free programs p, q are equivalent as regular expressions if and only if the formula $\langle p \rangle \varphi \leftrightarrow \langle q \rangle \varphi$ is valid.

1. EXAMPLE. Let φ be an atomic proposition, let p be an atomic program, and let $\mathcal{M} = (S^{\mathcal{M}}, I^{\mathcal{M}})$ be a Kripke model with

$$S^{\mathcal{M}} = \{u, v, w\}, \qquad \varphi^{\mathcal{M}} = \{u, v\}, \qquad p^{\mathcal{M}} = \{(u, v), (u, w), (v, w), (w, v)\}.$$

Then $u \models \langle p \rangle \neg\varphi \wedge \langle p \rangle \varphi$, but $v \models [p]\neg\varphi$ and $w \models [p]\varphi$. Moreover,

$$\mathcal{M} \models \langle p^* \rangle [(p;p)^*]\varphi \wedge \langle p^* \rangle [(p;p)^*]\neg\varphi.$$

Other semantics besides Kripke semantics have been studied [12, 112, 77, 78, 134].

2.1.3. *Computation sequences*

Let p be a program. A *finite computation sequence* of p is a finite-length string of atomic programs and tests representing a possible sequence of atomic steps that may occur in a halting execution of p. The set of all such sequences is denoted $CS(p)$. We use the word "possible" loosely—$CS(p)$ is determined by the syntax of p alone, and may contain strings that are never executed in any interpretation.

Formally, the set $CS(p)$ is defined by induction on syntax:

$$\begin{aligned}
CS(p) &= \{p\},\ p \text{ an atomic program or test},\\
CS(p;q) &= \{\alpha;\beta \mid \alpha \in CS(p),\ \beta \in CS(q)\},\\
CS(p \cup q) &= CS(p) \cup CS(q),\\
CS(p^*) &= \bigcup_{n \geqslant 0} CS(p^n)
\end{aligned}$$

where $p^0 = \textbf{skip}$ and $p^{n+1} = p;p^n$. For example, if p is an atomic program and φ is an atomic formula, then the program

$$\textbf{while } \varphi \textbf{ do } p \textbf{ od} = (\varphi?;p)^*;\neg\varphi?$$

has as computation sequences all strings of the form

$$\varphi?;p;\varphi?;p;\cdots;\varphi?;p;\mathbf{skip};\neg\varphi?.$$

Note that each finite computation sequence q of a program p is itself a program, and $CS(q)=\{q\}$. Moreover, the following proposition is not difficult to prove by induction on syntax.

2. PROPOSITION. $p^{\mathcal{M}}=\bigcup_{q\in CS(p)}q^{\mathcal{M}}$.

2.2. A deductive system for PDL

The following Hilbert-style axiom system for *PDL* was formulated by Segerberg [145].

AXIOMS OF *PDL*

1. axioms for propositional logic.
2. $\langle p\rangle\varphi\wedge[p]\psi\rightarrow\langle p\rangle(\varphi\wedge\psi)$.
3. $\langle p\rangle(\varphi\vee\psi)\leftrightarrow\langle p\rangle\varphi\vee\langle p\rangle\psi$.
4. $\langle p\cup q\rangle\varphi\leftrightarrow\langle p\rangle\varphi\vee\langle q\rangle\varphi$.
5. $\langle p;q\rangle\varphi\leftrightarrow\langle p\rangle\langle q\rangle\varphi$.
6. $\langle\psi?\rangle\varphi\leftrightarrow\psi\wedge\varphi$.
7. $(\varphi\vee\langle p\rangle\langle p^*\rangle\varphi)\rightarrow\langle p^*\rangle\varphi$.
8. $\langle p^*\rangle\varphi\rightarrow(\varphi\vee\langle p^*\rangle(\neg\varphi\wedge\langle p\rangle\varphi))$.

Axioms 2 and 3 are not particular to *PDL*, but hold in all normal modal systems (see [73, 20]). Axiom 8 is called the *PDL induction axiom*, and is better known in its dual form (Theorem 3(8) below).

RULES OF INFERENCE

1. *Modus ponens*:

$$\frac{\varphi,\quad \varphi\rightarrow\psi}{\psi}.$$

2. *Modal generalization*:

$$\frac{\varphi}{[p]\varphi}.$$

We write $\vdash\varphi$ if the proposition φ is a theorem of this system, and say that φ is *consistent* if not $\vdash\neg\varphi$. A set Σ of propositions is *consistent* if all finite conjunctions of elements of Σ are consistent.

The soundness of these axioms and rules over the Kripke semantics can be established by elementary arguments in relational algebra.

2.3. Basic properties

We list some basic theorems and derived rules of *PDL* that are provable in the deductive system of the previous section.

3. THEOREM. *The following are theorems of PDL:*

(1) *All propositional tautologies.*
(2) $[p](\varphi \rightarrow \psi) \rightarrow ([p]\varphi \rightarrow [p]\psi)$.
(3) $[p](\varphi \wedge \psi) \leftrightarrow [p]\varphi \wedge [p]\psi$.
(4) $[p \cup q]\varphi \leftrightarrow [p]\varphi \wedge [q]\varphi$.
(5) $[p;q]\varphi \leftrightarrow [p][q]\varphi$.
(6) $[\varphi?]\psi \leftrightarrow (\varphi \rightarrow \psi)$.
(7) $[p^*]\varphi \rightarrow \varphi \wedge [p][p^*]\varphi$.
(8) $(\varphi \wedge [p^*](\varphi \rightarrow [p]\varphi)) \rightarrow [p^*]\varphi$.
(9) $\langle p \rangle(\varphi \wedge \psi) \rightarrow \langle p \rangle \varphi \wedge \langle p \rangle \psi$.
(10) $[p]\varphi \vee [p]\psi \rightarrow [p](\varphi \vee \psi)$.
(11) $\langle p^*;p^* \rangle \varphi \leftrightarrow \langle p^* \rangle \varphi$.
(12) $\langle p^{**} \rangle \varphi \leftrightarrow \langle p^* \rangle \varphi$.
(13) $(\varphi \vee \langle p \rangle \langle p^* \rangle \varphi) \leftrightarrow \langle p^* \rangle \varphi$.
(14) $\langle p^* \rangle \varphi \leftrightarrow (\varphi \vee \langle p^* \rangle(\neg\varphi \wedge \langle p \rangle \varphi))$.
(15) $(\varphi \wedge [p^*](\varphi \rightarrow [p]\varphi)) \leftrightarrow [p^*]\varphi$.

4. THEOREM. *The following are sound rules of inference of PDL:*

(1) *Monotonicity of* $\langle p \rangle$:

$$\frac{\varphi \rightarrow \psi}{\langle p \rangle \varphi \rightarrow \langle p \rangle \psi}.$$

(2) *Monotonicity of* $[p]$:

$$\frac{\varphi \rightarrow \psi}{[p]\varphi \rightarrow [p]\psi}.$$

(3) *Reflexive-transitive closure*:

$$\frac{(\varphi \vee \langle p \rangle \psi) \rightarrow \psi}{\langle p^* \rangle \varphi \rightarrow \psi}.$$

(4) *Loop invariance rule*:

$$\frac{\psi \rightarrow [p]\psi}{\psi \rightarrow [p^*]\psi}.$$

(5) *Hoare composition rule*:

$$\frac{\{\varphi\}p\{\sigma\}, \quad \{\sigma\}q\{\psi\}}{\{\varphi\}p;q\{\psi\}}.$$

(6) *Hoare conditional rule*:

$$\frac{\{\varphi \wedge \sigma\}p\{\psi\},\quad \{\neg\varphi \wedge \sigma\}q\{\psi\}}{\{\sigma\}\ \textbf{if}\ \varphi\ \textbf{then}\ p\ \textbf{else}\ q\ \textbf{fi}\{\psi\}}.$$

(7) *Hoare* **while** *rule*:

$$\frac{\{\varphi \wedge \psi\}p\{\psi\}}{\{\psi\}\ \textbf{while}\ \varphi\ \textbf{do}\ p\ \textbf{od}\{\neg\varphi \wedge \psi\}}.$$

The properties of Theorem 3(2–8) are the modal duals of Axioms 2–8. The converses of Theorem 3(9, 10) are *not* valid. They are violated in state u of the model of Example 1. The rules of Theorem 4(1, 2) say that the constructs $\langle p\rangle\varphi$ and $[p]\varphi$ are *monotone* in φ with respect to the ordering of logical implication. These constructs are also monotone and antitone in p respectively, as asserted by the following metatheorem.

5. PROPOSITION. *If* $p^{\mathcal{M}} \subseteq q^{\mathcal{M}}$, *then for all* φ,

(1) $\mathcal{M} \models \langle p\rangle\varphi \rightarrow \langle q\rangle\varphi$.

(2) $\mathcal{M} \models [q]\varphi \rightarrow [p]\varphi$.

These follow from Axiom 4 and Theorem 3(4).

*2.3.1. The * operator, induction, and reflexive-transitive closure*

The iteration operator * is interpreted as the reflexive-transitive closure operator on binary relations. It is the means by which looping is coded in *PDL*. Looping introduces a level of complexity to *PDL* beyond the other operators. Because of it, *PDL* is not compact: the set

$$\{\langle p^*\rangle\varphi\} \cup \{\neg\varphi,\ \neg\langle p\rangle\varphi,\ \neg\langle p^2\rangle\varphi, \ldots\} \tag{3}$$

is finitely satisfiable but not satisfiable. This seems to say that looping is inherently infinitary; it is thus rather surprising that there should be a finitary complete axiomatization.

The dual propositions Axiom 8 and Theorem 3(8) are jointly called the *PDL induction axiom*. Together with Axiom 7, they completely axiomatize the behavior of *. Intuitively, the induction axiom in the form of Theorem 3(8) says: if φ is true initially, and if the truth of φ is preserved by the program p, then φ will be true after any number of iterations of p. It is very similar to the induction axiom of Peano arithmetic:

$$\varphi(0) \wedge \forall n\,(\varphi(n) \rightarrow \varphi(n+1)) \quad \rightarrow \quad \forall n\,\varphi(n).$$

Here $\varphi(0)$ is the basis of the induction and $\forall n(\varphi(n) \rightarrow \varphi(n+1))$ is the induction step, from which the conclusion $\forall n\,\varphi(n)$ may be drawn. In the *PDL* induction axiom, the basis is φ and the induction step is $[p^*](\varphi \rightarrow [p]\varphi)$, from which the conclusion $[p^*]\varphi$ may be drawn.

The induction axiom is closely related to the reflexive-transitive closure rule (Theorem 4(3)).The significance of this rule is best described in terms of its relationship to Axiom 7. This axiom is obtained by substituting $\langle p^*\rangle\varphi$ for ψ in the premise of the reflexive-transitive closure rule. Axiom 7 thus says that $\langle p^*\rangle\varphi$ is a solution of

$$(\varphi \vee \langle p\rangle X)\rightarrow X; \tag{4}$$

the reflexive-transitive closure rule says that it is the *least* solution to (4) (with respect to logical implication) among all *PDL* propositions.

The relationship between the induction axiom (Axiom 8), the reflexive-transitive closure rule (Theorem 4(3)), and the rule of loop invariance (Theorem 4(4)) is summed up in the following proposition. We emphasize that this result is purely proof-theoretic and is independent of the semantics of Subsection 2.1.2.

6. PROPOSITION. *In PDL without the induction axiom, the following axioms and rules are interderivable*:

(i) *the induction axiom* (*Axiom* 8);
(ii) *the loop invariance rule* (*Theorem* 4(4));
(iii) *the reflexive-transitive closure rule* (*Theorem* 4(3)).

PROOF. First we show that the monotonicity rule (Theorem 4(2)) is derivable in PDL without induction. Assuming the premise $\varphi\rightarrow\psi$ and applying modal generalization, we obtain $[p](\varphi\rightarrow\psi)$; the conclusion $[p]\varphi\rightarrow[p]\psi$ then follows from Theorem 3(2) and modus ponens. The dual monotonicity rule (Theorem 4(1)) can be derived from this rule by pure propositional reasoning.

(i)→(ii): Assume the premise of (ii): $\varphi\rightarrow[p]\varphi$. By modal generalization, $[p^*](\varphi\rightarrow [p]\varphi)$, thus

$$\begin{aligned}\varphi &\rightarrow \varphi\wedge[p^*](\varphi\rightarrow[p]\varphi)\\ &\rightarrow [p^*]\varphi.\end{aligned}$$

The first implication is by propositional reasoning, and the second is the box form of the induction axiom (Theorem 3(8)). By transitivity of implication, we obtain $\varphi\rightarrow[p^*]\varphi$, which is the conclusion of (ii).

(ii)→(iii): Dualizing rule (iii) by purely propositional reasoning, we obtain a rule

$$\frac{\psi \ \rightarrow\ \varphi\wedge[p]\psi}{\psi\rightarrow[p^*]\varphi} \tag{5}$$

equipollent with (iii). It thus suffices to derive (5) from (ii). From the premise of (5), we obtain by propositional reasoning the two formulas

$$\psi\rightarrow\varphi, \tag{6}$$

$$\psi\rightarrow[p]\psi. \tag{7}$$

Applying (ii) to (7), we obtain $\psi\rightarrow[p^*]\psi$, which by (6) and monotonicity gives $\psi\rightarrow[p^*]\varphi$, which is the conclusion of (5).

(iii)→(i): By Axiom 3, propositional reasoning, and Axiom 7, we have

$$\begin{aligned}&\varphi\vee\langle p\rangle(\varphi\vee\langle p^*\rangle(\neg\varphi\wedge\langle p\rangle\varphi))\\ &\quad\rightarrow\ \varphi\vee\langle p\rangle\varphi\vee\langle p\rangle\langle p^*\rangle(\neg\varphi\wedge\langle p\rangle\varphi)\\ &\quad\rightarrow\ \varphi\vee(\neg\varphi\wedge\langle p\rangle\varphi)\vee\langle p\rangle\langle p^*\rangle(\neg\varphi\wedge\langle p\rangle\varphi)\\ &\quad\rightarrow\ \varphi\vee\langle p^*\rangle(\neg\varphi\wedge\langle p\rangle\varphi).\end{aligned}$$

By transitivity of implication,

$$\varphi \vee \langle p\rangle(\varphi \vee \langle p^*\rangle(\neg\varphi \wedge \langle p\rangle\varphi)) \rightarrow \varphi \vee \langle p^*\rangle(\neg\varphi \wedge \langle p\rangle\varphi).$$

Applying (iii), we obtain the induction axiom:

$$\langle p^*\rangle\varphi \rightarrow \varphi \vee \langle p^*\rangle(\neg\varphi \wedge \langle p\rangle\varphi). \qquad \square$$

2.3.2. Encoding of Hoare Logic

Dynamic Logic subsumes Hoare Logic. As an illustration, we show how to derive the Hoare **while** rule (Theorem 4(7)) in *PDL*. The other Hoare rules are also derivable.

Assume that the premise

$$\{\varphi \wedge \psi\}p\{\psi\} \;=\; (\varphi \wedge \psi) \rightarrow [p]\psi \tag{8}$$

holds. We wish to derive the conclusion

$$\{\psi\}\ \textbf{while}\ \varphi\ \textbf{do}\ p\ \textbf{od}\{\neg\varphi \wedge \psi\} \;=\; \psi \rightarrow [(\varphi?;p)^*;\neg\varphi?](\neg\varphi \wedge \psi). \tag{9}$$

Using Theorem 3(5, 6) and propositional reasoning, the right-hand side of (8) is equivalent to $\psi \rightarrow [\varphi?;p]\psi$. Applying the loop invariance rule (Theorem 4(4)), we obtain $\psi \rightarrow [(\varphi?;p)^*]\psi$. By the monotonicity of $[(\varphi?;p)^*]$ (Theorem 4(2)) and propositional reasoning,

$$\psi \rightarrow [(\varphi?;p)^*](\neg\varphi \rightarrow \neg\varphi \wedge \psi).$$

Again by Theorem 3(5, 6) and propositional reasoning, we obtain the right-hand side of (9).

2.4. The small model property

The *small model property* comes from modal logic. Although the proof for *PDL* is substantially more complicated than for simpler modal systems, the same proof technique, called *filtration*, applies. The small model property for *PDL* was first proved by Fischer and Ladner [42, 43].

The small model property says that if φ is satisfiable, then it is satisfied at a state in a model with no more than $2^{|\varphi|}$ states, where $|\varphi|$ is the length (number of symbols) of φ. This immediately gives a naive nondeterministic exponential-time algorithm for the satisfiability problem. A deterministic exponential-time algorithm will be obtained in Subsection 2.6.1.

2.4.1. The relation $\prec$ and the Fischer–Ladner closure

Many proofs in simpler modal systems use induction on subformulas. In *PDL*, the situation is complicated by the simultaneous inductive definitions of programs and propositions; it is further complicated by the behavior of the * operator. We must therefore do our induction on another well-founded relation that resembles the subformula relation but is somewhat more intricate.

Let $\prec$ be the smallest transitive relation on $\{0,1\} \times \Phi$ containing the following inequalities:

1. $(0, \varphi), (0, \psi) \prec (0, \varphi \vee \psi)$,
2. $(0, \varphi) \prec (0, \neg\varphi)$,
3. $(0, \varphi), (1, \langle p \rangle \varphi) \prec (0, \langle p \rangle \varphi)$,
4. $(1, \langle p \rangle \varphi), (1, \langle q \rangle \varphi) \prec (1, \langle p \cup q \rangle \varphi)$,
5. $(1, \langle p \rangle \langle q \rangle \varphi), (1, \langle q \rangle \varphi) \prec (1, \langle p;q \rangle \varphi)$,
6. $(1, \langle p \rangle \langle p^* \rangle \varphi) \prec (1, \langle p^* \rangle \varphi)$,
7. $(0, \varphi) \prec (1, \langle \varphi? \rangle \psi)$.

It is not hard to see that the relation $\prec$ is well-founded.

Define $(i, \varphi) \preccurlyeq (j, \psi)$ if either $(i, \varphi) \prec (j, \psi)$ or $(i, \varphi) = (j, \psi)$. The *Fischer–Ladner closure* of φ, denoted $FL(\varphi)$, is the set

$$FL(\varphi) = \{\psi \mid (j, \psi) \preccurlyeq (0, \varphi),\ j \in \{0, 1\}\}.$$

Let $|A|$ denote the cardinality of a set A and let $|\varphi|$ and $|p|$ denote the length (number of symbols) of φ and p respectively. The following lemma is easily established by induction on the well-founded relation $\prec$.

7. LEMMA.

(1) $|\{(i, \psi) \mid (i, \psi) \preccurlyeq (0, \varphi)\}| \leqslant |\varphi|$.

(2) $|\{(i, \psi) \mid (i, \psi) \preccurlyeq (1, \langle p \rangle \varphi)\}| \leqslant |p|$.

It follows immediately from Lemma 7 that $|FL(\varphi)| \leqslant |\varphi|$. We also have the following lemma.

8. LEMMA. *If* $\langle p \rangle \psi \in FL(\varphi)$ *then* $\psi \in FL(\varphi)$.

2.4.2. Filtration

Given a *PDL* proposition φ and a Kripke model $\mathcal{M} = (S^{\mathcal{M}}, I^{\mathcal{M}})$, we define a new model $\mathcal{M}/FL(\varphi) = (S^{\mathcal{M}/FL(\varphi)}, I^{\mathcal{M}/FL(\varphi)})$, called the *filtration of* $\mathcal{M}$ *by* $FL(\varphi)$, as follows. Define a binary relation $\equiv$ on states of $\mathcal{M}$ by

$$u \equiv v \text{ iff } \forall \psi \in FL(\varphi)\,(u \in \psi^{\mathcal{M}} \leftrightarrow v \in \psi^{\mathcal{M}}).$$

In other words, we collapse u and v if they are not distinguishable by any formula of $FL(\varphi)$. Let

$$\begin{aligned} &[u] = \{v \mid v \equiv u\}, \\ &S^{\mathcal{M}/FL(\varphi)} = \{[u] \mid u \in S^{\mathcal{M}}\}, \\ &\varphi^{\mathcal{M}/FL(\varphi)} = \{[u] \mid u \in \varphi^{\mathcal{M}}\}, \qquad \varphi \text{ atomic}, \\ &p^{\mathcal{M}/FL(\varphi)} = \{([u], [v]) \mid (u, v) \in p^{\mathcal{M}}\}, \qquad p \text{ atomic}. \end{aligned}$$

$I^{\mathcal{M}/FL(\varphi)}$ is extended inductively to compound propositions and programs as in Subsection 2.1.2.

The following lemma relates $\mathcal{M}$ and $\mathcal{M}/FL(\varphi)$. Most of the difficulty in the following lemma is in the correct formulation of the induction hypotheses in the statement of the lemma. Once this is done, the proof is a straightforward induction on the well-founded relation $\prec$.

9. LEMMA (filtration lemma).

(1) *For all* $(0, \psi) \preccurlyeq (0, \varphi)$, $u \in \psi^{\mathcal{M}}$ *iff* $[u] \in \psi^{\mathcal{M}/FL(\varphi)}$.

(2) *For all* $(1, \langle p \rangle \psi) \prec (0, \varphi)$,

(a) *if* $(u, v) \in p^{\mathcal{M}}$ *then* $([u], [v]) \in p^{\mathcal{M}/FL(\varphi)}$;

(b) *if* $([u], [v]) \in p^{\mathcal{M}/FL(\varphi)}$ *and* $v \in \psi^{\mathcal{M}}$, *then* $u \in \langle p \rangle \psi^{\mathcal{M}}$.

Using the filtration lemma, we can prove the small model theorem easily.

10. THEOREM (small model theorem). *Let* φ *be a satisfiable formula of PDL. Then* φ *is satisfied in a model with no more than* $2^{|\varphi|}$ *states.*

PROOF. If φ is satisfiable, then there is a model $\mathcal{M}$ and state $u \in \mathcal{M}$ with $u \in \varphi^{\mathcal{M}}$. Let $FL(\varphi)$ be the Fischer–Ladner closure of φ. By the filtration lemma, $[u] \in \varphi^{\mathcal{M}/FL(\varphi)}$. Moreover, $\mathcal{M}/FL(\varphi)$ has no more states than the number of truth assignments to formulas in $FL(\varphi)$, which by Lemma 7(1) is at most $2^{|\varphi|}$. □

2.4.3. Filtration over nonstandard models

We note for future use that the filtration lemma (Lemma 9) actually holds in more generality. In particular, it holds for *nonstandard models* as well as the standard models defined in Subsection 2.1.2. Nonstandard models of *PDL* were first introduced by Parikh [117].

11. DEFINITION. A *nonstandard Kripke model* is any structure $\mathcal{N} = (S^{\mathcal{N}}, I^{\mathcal{N}})$ that is a Kripke model in the sense of Subsection 2.1.2 in every respect, except that $(p^*)^{\mathcal{N}}$ need not be the reflexive-transitive closure of $p^{\mathcal{N}}$, but only a reflexive, transitive relation containing $p^{\mathcal{N}}$ and satisfying the *PDL* induction axiom (Axiom 8).

It is easily checked that all the axioms and rules of Section 2.2 are still valid over nonstandard models.

12. LEMMA (filtration for nonstandard models). *The filtration lemma* (*Lemma* 9) *holds for nonstandard models.*

PROOF. All cases are identical to the corresponding cases of Lemma 9, except case 2(a) for p^*. For standard models $\mathcal{M}$, the proof is straightforward, using the fact that $(p^*)^{\mathcal{M}}$ is the reflexive-transitive closure of $p^{\mathcal{M}}$. This does not hold in nonstandard models in general, so we must depend on the weaker induction axiom. We argue this case explicitly.

Let $\mathcal{N}$ be a nonstandard model, and suppose $(u, v) \in (p^*)^{\mathcal{N}}$. We wish to show that $([u], [v]) \in (p^*)^{\mathcal{N}/FL(\varphi)}$, or equivalently that $u \in E$, where

$$E = \{w \in \mathcal{N} \mid ([w], [v]) \in (p^*)^{\mathcal{N}/FL(\varphi)}\}.$$

Since E is a union of equivalence classes defined by truth assignments to the elements of $FL(\varphi)$, there is a *PDL* formula ψ_E defining E in $\mathcal{N}$; i.e., $E = \psi_E^{\mathcal{N}}$. There is likewise a *PDL* formula $\psi_{[v]}$ defining just the equivalence class $[v]$. Since $[v] \subseteq E$,

$$\mathcal{N} \vDash \psi_{[v]} \rightarrow \psi_E. \tag{10}$$

Also,

$$\mathcal{N} \models \langle p \rangle \psi_E \rightarrow \psi_E, \tag{11}$$

since if $w \in (\langle p \rangle \psi_E)^{\mathcal{N}}$, then there exists a $t \in \psi_E^{\mathcal{N}} = E$ such that $(w, t) \in p^{\mathcal{N}}$; then $([w], [t]) \in p^{\mathcal{N}/FL(\varphi)}$ by the induction hypothesis, and $([t], [v]) \in (p^*)^{\mathcal{N}/FL(\varphi)}$, thus $([w], [v]) \in (p^*)^{\mathcal{N}/FL(\varphi)}$ and therefore $w \in E$.

Combining (10) and (11), we get

$$\mathcal{N} \models (\psi_{[v]} \vee \langle p \rangle \psi_E) \rightarrow \psi_E.$$

Using the reflexive-transitive closure rule (Theorem 4(3)), which by Proposition 6 is equivalent to the induction axiom, we have that

$$\mathcal{N} \models \langle p^* \rangle \psi_{[v]} \rightarrow \psi_E.$$

But $u \in (\langle p^* \rangle \psi_{[v]})^{\mathcal{N}}$ by assumption; therefore $u \in E$. □

2.5. *Deductive completeness*

Completeness of the deductive system of Section 2.2 was first shown independently by Gabbay [46] and Parikh [117]. A particularly easy-to-follow proof is given in [85]. Completeness is also treated in [132, 135, 12, 112]. The completeness proof sketched here is from [81] and is based on the approach of [12, 135].

13. Lemma. *Let Σ, Γ be maximal consistent sets of PDL propositions. Then the following two statements are equivalent*:

(1) *for all $\psi \in \Gamma$, $\langle p \rangle \psi \in \Sigma$;*
(2) *for all $[p]\psi \in \Sigma$, $\psi \in \Gamma$.*

Define a nonstandard model $\mathcal{F}$ by:

$$S^{\mathcal{F}} = \{\text{maximal consistent sets of propositions of } PDL\},$$
$$\varphi^{\mathcal{F}} = \{u \in S^{\mathcal{F}} \mid \varphi \in u\},$$
$$p^{\mathcal{F}} = \{(u, v) \mid \forall \varphi \in v\ \langle p \rangle \varphi \in u\} = \{(u, v) \mid \forall [p]\varphi \in u\ \varphi \in v\}.$$

The two definitions of $p^{\mathcal{F}}$ are equivalent, by Lemma 13.

14. Proposition. *$\mathcal{F}$ is a nonstandard Kripke model in the sense of Definition* 11.

It should be noted that $\mathcal{F}$ is a *universal model* in the sense that it is defined independently of any particular φ.

15. Theorem (completeness of *PDL*). *If $\models \varphi$ then $\vdash \varphi$.*

Proof. Equivalently, we need to show that if φ is consistent, then it is satisfied in a standard Kripke model. If φ is consistent, then by Zorn's Lemma it is contained in a maximal consistent set u, which is a state of the nonstandard model $\mathcal{F}$. By the filtration lemma for nonstandard models (Lemma 12), φ is satisfied at state $[u]$ in the finite standard model $\mathscr{F}/FL(\varphi)$. □

2.5.1. Logical consequences

In classical logics, a completeness theorem of the form of Theorem 15 is easily adapted to handle the relation of logical consequence $\varphi \models \psi$ between formulas, since usually

$$\varphi \models \psi \quad \text{iff} \quad \models \varphi \rightarrow \psi. \tag{12}$$

Unfortunately, (12) fails in *PDL*, as can be seen by taking $\psi = [p]\varphi$. However, the following result allows Theorem 15, as well as Algorithm 17 of Subsection 2.6.1 below, to be extended to handle logical consequence.

16. PROPOSITION. *Let φ, ψ be any PDL formulas. Then*

$$\varphi \models \psi \quad \textit{iff} \quad \models [(p_1 \cup \cdots \cup p_n)^*]\varphi \rightarrow \psi,$$

where $p_1, \ldots, p_n$ are all atomic programs appearing in φ or ψ.

It is shown in [103] that the problem of deciding whether $\Sigma \models \psi$, where Σ is a fixed r.e. set of *PDL* formulas, is Π_1^1-complete.

2.6. Complexity of the satisfiability problem for PDL

2.6.1. A deterministic exponential-time algorithm

The naive algorithm for the satisfiability problem that comes from the small model theorem is double exponential-time in the worst case. Here we develop an algorithm that runs in deterministic time $2^{O(|\varphi|)}$. In fact, the problem is deterministic exponential-time complete (see Subsection 2.6.2), so a significantly more efficient algorithm is impossible. Deterministic exponential-time algorithms were first given by Pratt [134, 137]. The algorithm given here is from [134].

The algorithm constructs the small model $\mathscr{N} = \mathscr{F}/FL(\varphi)$ obtained in the completeness proof of Section 2.5 explicitly. If φ is satisfiable, then it is consistent, by the soundness of the deductive system of Section 2.2; then φ will be satisfied at some state u of $\mathscr{F}$, and hence at the state $[u]$ of $\mathscr{N}$.

We start with the set S of all truth assignments $u: FL(\varphi) \rightarrow \{\mathit{true}, \mathit{false}\}$. By the construction of $\mathscr{N}$, the states of $\mathscr{N}$ are in one-to-one correspondence with the *consistent* truth assignments to $FL(\varphi)$, thus we may consider $S^{\mathscr{N}}$ to be a subset of S; i.e., for all $\psi \in FL(\varphi)$ and $u \in S^{\mathscr{N}}$,

$$u \in \psi^{\mathscr{N}} \quad \leftrightarrow \quad u(\psi) = \mathit{true}. \tag{13}$$

We will approximate $\mathscr{N}$ with a sequence of models $\mathscr{N}_i = (S_i, I_i), i \geq 0$, such that

$$S \supseteq S_0 \supseteq S_1 \supseteq \cdots \supseteq S^{\mathscr{N}},$$

obtained by deleting from S any truth assignments that we can determine to be inconsistent. When we are done, we will be left with the model $\mathscr{N}$.

The interpretations of the primitive formulas and programs in the models $\mathcal{N}_i$ will be defined in the same way for all i:

$$\psi^{\mathcal{N}_i}=\{u\in S_i \mid u(\psi)=true\}, \quad \psi \text{ atomic,} \tag{14}$$

$$p^{\mathcal{N}_i}=\{(u,v)\in S_i^2 \mid u(\langle p\rangle\psi)=true \text{ whenever } v(\psi)=true\}, \quad p \text{ atomic.} \tag{15}$$

17. ALGORITHM
(1) Construct S.
(2) For each $u\in S$, check whether u respects Axioms 1, 4, 5, and 6 of Section 2.2 and Theorem 2.3(13), all of which can be checked locally. For example, to check Axiom 4, which says

$$\langle p\cup q\rangle\psi \leftrightarrow \langle p\rangle\psi \vee \langle q\rangle\psi,$$

check that $u(\langle p\cup q\rangle\psi)=true$ if and only if either $u(\langle p\rangle\psi)=true$ or $u(\langle q\rangle\psi)=true$. Let S_0 be the set of all $u\in S$ passing this test. The model $\mathcal{N}_0$ is defined by (14) and (15) above.
(3) Repeat the following for $i=0,1,2,\ldots$ until no more u are deleted: find a $\preccurlyeq$-minimal $(1,\langle p\rangle\psi)\preccurlyeq(0,\varphi)$ and $u\in S_i$ violating the property

$$u(\langle p\rangle\psi)=true \rightarrow \exists v\,(u,v)\in p^{\mathcal{N}_i} \text{ and } v(\psi)=true. \tag{16}$$

Delete u from S_i to get S_{i+1}. □

The correctness of this algorithm will follow from the following lemma (cf. Lemma 9).

18. LEMMA. *Let $(j,\xi)\preccurlyeq(0,\varphi)$ be such that every $(1,\langle p\rangle\psi)\prec(j,\xi)$ and $u\in S_i$ satisfy (16).*
(1) *For all $(0,\psi)\preccurlyeq(j,\xi)$ and $u\in S_i$, $u(\psi)=true$ iff $u\in\psi^{\mathcal{N}_i}$.*
(2) *For all $(1,\langle p\rangle\psi)\preccurlyeq(j,\xi)$ and $u,v\in S_i$,*
(a) *if $(u,v)\in p^{\mathcal{N}}$ then $(u,v)\in p^{\mathcal{N}_i}$;*
(b) *if $(u,v)\in p^{\mathcal{N}_i}$ and $v(\psi)=true$, then $u(\langle p\rangle\psi)=true$.*

PROOF. By induction on $\prec$. □

Since every $u\in S^{\mathcal{N}}$ passes the test of Step 2 of the algorithm, $S^{\mathcal{N}}\subseteq S_0$; and (13) and Lemma 18(2(a)) imply that no $u\in S^{\mathcal{N}}$ is ever deleted in Step 3 since, for $u\in S^{\mathcal{N}}$,

$$\begin{aligned} u(\langle p\rangle\psi)=true &\rightarrow u\in(\langle p\rangle\psi)^{\mathcal{N}}\\ &\rightarrow \exists v(u,v)\in p^{\mathcal{N}} \text{ and } v\in\psi^{\mathcal{N}}\\ &\rightarrow \exists v(u,v)\in p^{\mathcal{N}_i} \text{ and } v(\psi)=true.\end{aligned}$$

Thus $S^{\mathcal{N}}\subseteq S_i$, $i\geqslant 0$. Moreover, when the algorithm terminates with the model $\mathcal{N}_n$, then by Lemma 18(1), every $u\in S_n$ (viewed as a truth assignment) is satisfiable, since it is satisfied by the state u in the model $\mathcal{N}_n$; thus $S_n\subseteq S^{\mathcal{N}}$. We can now test the satisfiability of φ by checking whether $u(\varphi)=true$ for some $u\in\mathcal{N}_n$.

Algorithm 17 can be programmed to run in exponential time without much difficulty. The efficiency can be further improved by observing that the p in the $\preccurlyeq$-minimal

$(1, \langle p \rangle \psi)$ violating (16) in Step 3 must be either atomic or of the form q^*, because of the preprocessing in Step 2. This follows easily from Lemma 18. We have shown the following theorem.

19. THEOREM. *There is an exponential-time algorithm for deciding whether a given formula of PDL is satisfiable.*

As previously noted, Proposition 16 allows this algorithm to be adapted to test whether one formula is a logical consequence of another.

2.6.2. A lower bound

In [42, 43], it is shown how *PDL* formulas can encode computations of linear-space-bounded alternating Turing machines. It follows from [18] that the satisfiability problem for *PDL* is deterministic exponential-time hard, therefore requires at least deterministic time $2^{\Omega(n)}$ on formulas of size n.

The satisfiability problem for *PDL* is thus exponential-time complete. In contrast, the satisfiability problem for classical propositional logic is NP-complete.

2.7. Variants of PDL

A number of interesting variants are obtained by extending or restricting *PDL* in various ways. In this section we describe some of these variants and review some of the known results concerning relative expressive power, complexity, and proof theory.

2.7.1. Deterministic PDL and while programs

A program p is said to be (*semantically*) *deterministic* in $\mathcal{M}$ if its traces are uniquely determined by their first states. If p is atomic, this is equivalent to the requirement that $p^{\mathcal{M}}$ be a partial function. The class of *deterministic* **while** *programs*, denoted *DWP*, is the class of programs in which

- the operators $\cup$, ?, and * may appear only in the context of the conditional test, **while** loop, **skip**, or **fail**;
- tests in the conditional test and **while** loop are purely propositional (i.e., there is no occurrence of the $\langle\ \rangle$ operator).

The class of *nondeterministic* **while** *programs*, denoted *WP*, is the same, except unconstrained use of the nondeterministic choice construct $\cup$ is allowed. It is easily shown that if p and q are semantically deterministic in $\mathcal{M}$, then so are **if** φ **then** p **else** q **fi** and **while** φ **do** p **od**.

20. DEFINITION. We define *PDL*(*DWP*) to be the syntactically and semantically constrained version of *PDL* in which

- only deterministic **while** programs are allowed;
- every atomic program is semantically deterministic.

(This version of *PDL* is sometimes called *strict deterministic PDL* in the literature.)

If φ is valid in *PDL*, then φ is also valid in *PDL(DWP)*, but not conversely: the formula

$$\langle p\rangle\varphi\rightarrow[p]\varphi \tag{17}$$

is valid in *PDL(DWP)* but not in *PDL*. Also, *PDL(DWP)* is strictly less expressive than *PDL*, since the formula

$$\langle(p\cup q)^*\rangle\varphi \tag{18}$$

is not expressible in *PDL(DWP)* [55].

Unlike *PDL*, the satisfiability problem for *PDL(DWP)* is PSPACE-complete [55].

If the deductive system of Section 2.2 is modified so as to refer only to deterministic **while** programs, and if axiom (17) is included, the resulting system is sound and complete for *PDL(DWP)* [8].

Related results are obtained in [12, 163, 8].

2.7.2. *Rich and poor tests*

Tests $\varphi?$ in *PDL* are defined for arbitrary propositions φ. This is called *rich-test*. This is substantially more power than one would find in a conventional programming language. Indeed, in the first-order version over the natural numbers, rich test allows undecidable problems to be decided in one step. *Poor-test PDL*, on the other hand, can only test atomic propositions.

The exponential-time hardness result described in Subsection 2.6.2 still holds for poor-test *PDL*, since the construction does not require tests. However, it can be shown [129, 11, 14] that rich-test *PDL* is strictly more expressive than poor-test *PDL*, which in turn is strictly more expressive than test-free *PDL*. These results also hold for *PDL(DWP)* (see Subsection 2.7.1). In fact, formula (18) of test-free *PDL* is not expressible in *PDL(DWP)*.

2.7.3. *Context-free programs*

A PDL program can be viewed as a regular expression denoting the set of its computation sequences (Subsection 2.1.3). The set of computation sequences is thus a regular set in the sense of the theory of formal languages and automata (see [72]).

More complex sets of computation sequences can be allowed as programs to obtain more expressive versions of *PDL*. In particular, *context-free PDL*, denoted PDL_{cf}, is obtained by allowing programs to be context-free sets of computation sequences. This corresponds to a syntax allowing recursive calls without parameters. For example, the sets

$$\{p^n;q;r^n \mid n\geqslant 0\}, \tag{19}$$

$$\{p^n;q^n \mid n\geqslant 0\} \tag{20}$$

where p, q, and r are atomic, are programs of PDL_{cf} but not of *PDL*. Programs may be built compositionally using a context-free grammar-like syntax, or represented by pushdown automata accepting sets of computation sequences.

The satisfiability problem for PDL_{cf} was shown undecidable by Ladner (unpublished); the problem was shown to be Π^1_1-complete in [64], even for *PDL* extended with the single context-free program (19). Curiously, *PDL* extended just by (20) is decidable [116]. Related results can be found in [62, 116].

2.7.4. Automata theory and program logics

A *PDL* program represents a regular set of computation sequences. This same regular set could possibly be represented exponentially more succinctly by a finite automaton. The difference between these two representations corresponds roughly to the difference between **while** programs and flowcharts.

Since finite automata are exponentially more succinct in general, the complexity results of Section 2.6 could conceivably fail if finite automata were allowed as programs. Moreover, we must also rework the deductive system of Section 2.2.

However, it turns out that the completeness and exponential-time decidability results of *PDL* are not sensitive to the representation, and still go through in the presence of finite automata as programs, provided the deductive system of Section 2.2 and the techniques of Sections 2.4–2.6 are suitably modified [134, 135, 67].

In very recent years, the automata-theoretic approach to logics of programs has yielded significant insight into propositional logics more powerful than *PDL*, as well as substantial reductions in the complexity of their decision procedures. Especially enlightening are the connections with automata on infinite strings and infinite trees: by viewing a formula as an automaton and a tree-like model as an input to that automaton, the satisfiability problem for a given formula becomes the emptiness problem for a given automaton. Logical questions are thereby transformed into purely automata-theoretic questions.

This connection has prompted renewed inquiry into the complexity of automata on infinite objects, with considerable success [28, 32, 35, 37, 97, 110, 127, 141, 147, 151, 164, 165, 166, 167, 169, 170]. Especially noteworthy in this area is the recent result of Safra [141] involving the complexity of converting a nondeterministic automaton on infinite strings to an equivalent deterministic one. This result has already had a significant impact on the complexity of decision procedures for several logics of programs [28, 35, 36, 141].

2.7.5. Converse

The *converse operator* $^-$ is a program operator which allows a program to be "run backwards":

$$(p^-)^{\mathcal{M}} = \{(s,t) \mid (t,s) \in p^{\mathcal{M}}\}.$$

This operator strictly increases the expressive power of *PDL*, since the formula $\langle p^- \rangle \varphi$ is not expressible without it. More interestingly, the presence of the converse operator implies that the operator $\langle p \rangle$ is *continuous* in the sense that if Φ is any (possibly infinite) family of formulas possessing a join $\bigvee \Phi$, then $\bigvee \langle p \rangle \Phi$ exists and is logically equivalent to $\langle p \rangle \bigvee \Phi$ [160]. In the absence of the converse operator, there exist nonstandard models for which this fails.

The completeness and single-exponential decidability results of Section 2.5 and Subsection 2.6.1 can be extended to *PDL* with converse [117] provided the following two axioms are added:

$$\varphi \rightarrow [p]\langle p^- \rangle \varphi, \qquad \varphi \rightarrow [p^-]\langle p \rangle \varphi.$$

The filtration lemma (Lemma 9) still holds in the presence of $^-$, as does the finite model property.

The complexity of *PDL* with converse and various forms of well-foundedness constructs (see Subsection 2.7.6 below) is studied in [165].

2.7.6. *Well-foundedness and total correctness*

If p is a deterministic program, the formula $\varphi \rightarrow \langle p \rangle \psi$ asserts the total correctness of p with respect to pre- and postconditions φ and ψ respectively. For *nondeterministic* programs, however, this formula does not express total correctness: it asserts that if φ then *there exists* a halting computation sequence of p yielding ψ, whereas we would really like to assert that if φ then *all* computation sequences of p terminate and yield ψ. Unfortunately, this property is not expressible in *PDL*.

The problem is essentially concerned with *well-foundedness*. A program p is said to be *well-founded* at a state u_0 if there exists no infinite sequence $u_0, u_1, u_2, \ldots$ with $(u_i, u_{i+1}) \in p^{\mathcal{M}}$ for all $i \geqslant 0$.

Several very powerful logics have been proposed to deal with this situation. The most powerful is perhaps the propositional μ-calculus [144, 70, 125, 138, 82, 84, 86, 115, 152, 167]. The first propositional logic of this type was given by Pratt [138]. The version of this logic given in [82] is essentially propositional modal logic with a least fixpoint operator μ, which allows syntactic expression of any property that can be formulated as the least fixpoint of a monotone transformation. The well-foundedness of p is expressed

$$\mu x . [p]x \tag{21}$$

in this logic.

Two somewhat weaker ways of capturing well-foundedness without resorting to the full μ-calculus have been studied. One is to add an explicit predicate **wf** for well-foundedness. Another is to add an explicit predicate **halt**, which asserts that all computations of its argument p terminate. These constructs have been investigated in [66, 68, 115, 150, 151, 152] under the various names **loop**, **repeat**, and Δ.

The predicate **halt** can be defined inductively from **wf**, as follows:

- **halt**(p) if p is an atomic program or test,
- **halt**$(p; q) \leftrightarrow$ **halt**$(p) \wedge [p]$**halt**(q),
- **halt**$(p \cup q) \leftrightarrow$ **halt**$(p) \wedge$ **halt**(q),
- **halt**$(p^*) \leftrightarrow [p^*]$**halt**$(p) \wedge$ **wf**(p).

The propositional μ-calculus is strictly more expressive than *PDL* with **wf** [115, 152], which is strictly more expressive than *PDL* with **halt** [68], which is strictly more expressive than *PDL* [150].

The filtration lemma fails for *PDL* with **halt** or **wf** and for the propositional μ-calculus (except under certain strong syntactic restrictions which render formulas like

(21) ineffable [138]). This can be seen by considering the model $\mathcal{M}=(S^{\mathcal{M}}, I^{\mathcal{M}})$ with

$$S^{\mathcal{M}}=\{(i, j)\in\mathcal{N}^2 \mid 0\leqslant j\leqslant i\}\cup\{u\}$$

and atomic program p with

$$I^{\mathcal{M}}(p)=\{((i, j), (i, j-1)) \mid 1\leqslant j\leqslant i\}\cup\{(u, (i, i)) \mid i\in\mathcal{N}\}.$$

The state u satisfies **halt**(p^*) and **wf**(p), but $[u]$ does not satisfy this formula in any finite filtrate. Despite this failure, these logics do satisfy the finite model property [150, 151, 84]. In the presence of the converse operator, the finite model property fails, since the formula

$$\neg\textbf{halt}(p^*)\wedge[p^*]\textbf{halt}(p^{-*}),\quad p \text{ atomic}$$

is satisfiable but has no finite model. All these logics are decidable [150, 151, 86, 167, 165]; and in fact deterministic single-exponential time complete [35, 141].

There is no known complete axiomatization for *PDL* with **halt** or **wf**, or for the propositional μ-calculus. A completeness result for a syntactically restricted version of the μ-calculus which includes the formula (21) for p a deterministic **while** program is given in [82], and a complete infinitary deductive system is given in [84].

2.7.7. *Other work*

Additional topics related to *PDL*, which space does not permit us to discuss in depth, include work on complementation and intersection of programs [65], nonstandard models [12, 13, 77, 78, 134, 117], Dynamic Algebra [77, 78, 79, 134, 135], Process Logic [118, 133, 113, 60, 168], *PDL* with Boolean assignments [1], restricted forms of the consequence problem [119], concurrency [128], game logic [121, 122, 123], and probabilistic programs [28, 40, 69, 80, 83, 124, 140, 164]. We also refer the reader to Harel's survey [58], which covers many of these topics in more detail.

3. First-order Dynamic Logic

In this section we define various forms of first-order Dynamic Logic (*DL*) and discuss their syntax, semantics, proof theory, and expressiveness. The main difference between first-order *DL* and the propositional version discussed in Section 2 is the presence of a first-order structure $\mathcal{A}$, called the *domain of computation*, over which first-order quantification is allowed. States are no longer abstract points, but *valuations* of a set of variables over $\mathcal{A}$; primitive programs are no longer abstract binary relations, but *assignments* of the form $x:=t$, for example where x is a variable and t is a term; and primitive assertions are first-order formulas.

3.1. *Syntax*

Let $L=(\ldots, f, \ldots, R, \ldots)$ be a finite first-order language with equality. Here f and R denote generic function and relation symbols of L respectively. Each function and relation symbol of L comes with a fixed *arity* (number of inputs). Let $V=\{x_0, x_1, \ldots\}$ be a countable set of *individual variables*. The metasymbols $s, t, \ldots$ range over terms of L.

There are several versions of *DL* that we will discuss, depending on the choice of primitive constructs. In general, these logics are similar to the propositional version introduced in Section 2, with the following key exceptions:

- Primitive programs are assignment statements of the form

$$x := t, \tag{22}$$

  for example. Here $x \in V$ and t is a term of L; this form of assignment is called a *simple assignment*.
- Primitive assertions are atomic formulas of L, i.e. formulas of the form $R(t_1, \ldots, t_n)$ where R is an n-ary relation symbol of L and $t_1, \ldots, t_n$ are terms of L.
- In the inductive definition of formulas, we include the clause

$$\text{if } \varphi \text{ is a formula,} \quad \text{then so is } \exists x\varphi, \text{ where } x \in V.$$

Otherwise, compound programs and formulas are formed exactly as in Subsection 2.1.1, using the connectives ; (sequential composition), $\cup$ (nondeterministic choice), * (iteration), ? (test), $\langle\ \rangle$ (modal possibility), $\vee$ (propositional disjunction), and $\neg$ (propositional negation).

The class R of *regular programs* contains all programs formed from simple assignments (22), $\cup$, ;, *, and ?, in which any formula φ appearing in a test φ? must be a quantifier-free first-order formula. For much of the sequel, we will be concerned with **while** *programs* only. The class of *deterministic* **while** *programs*, denoted *DWP*, is the subclass of R in which the program operators $\cup$, ?, and * are constrained to appear only in the forms

$$\begin{aligned}
&\textbf{skip} = \mathit{true}?, \qquad \textbf{fail} = \mathit{false}?,\\
&\textbf{if } \varphi \textbf{ then } p \textbf{ else } q \textbf{ fi} = \varphi?;p \cup \neg\varphi?;q,\\
&\textbf{while } \varphi \textbf{ do } p \textbf{ od} = (\varphi?;p)^*; \neg\varphi?.
\end{aligned}$$

The class of (*nondeterministic*) **while** *programs*, denoted *WP*, is the same, except that we allow unrestricted use of the nondeterministic choice construct $\cup$.

The definitions of R, *WP*, and *DWP* depend on the language L, but to save notation, we do not make this dependence explicit.

3.1.1. Arrays and stacks

We will eventually want to discuss the power of auxiliary data structures such as *arrays* and *stacks*, as well as a powerful assignment statement called the *nondeterministic assignment*. We introduce the syntactic machinery now so that we can give the semantics of these constructs all at once in the next section.

To handle arrays, we include a countable set of *array variables* $V_{\text{array}} = \{F_0, F_1, \ldots\}$. Each array variable has an associated *arity*, or number of inputs, which we do not represent explicitly. We assume that there are countably many variables of each arity

$n \geqslant 0$. In the presence of array variables, we equate V with the nullary array variables; thus $V \subseteq V_{\text{array}}$. The variables in V_{array} of arity n will range over n-ary functions with arguments and values in the domain of computation. In our exposition, elements of the domain of computation play two roles: they are used as *indices* into an array, and as *values* which can be stored in an array. One might equally well introduce a separate sort for array indices; although conceptually simple, this would complicate notation and would give no new insight.

The classes DWP_{array} and WP_{array} of *deterministic* and *nondeterministic* **while** *programs with arrays* are defined similarly to DWP and WP respectively, except that in addition to simple assignments, we allow *array assignments*. These are similar to simple assignments, except that on the left-hand side we allow a term in which the outermost symbol is an array variable: $F(t_1, \ldots, t_m) := t$. Here F is an m-ary array variable and $t_1, \ldots, t_m, t$ are terms, possibly involving other array variables. Note that when $m = 0$, this reduces to the ordinary simple assignment.

To handle *stacks* or *pushdown stores*, we introduce *stack variables* σ and two new atomic programs:

$$push(\sigma, t), \qquad pop(\sigma, y)$$

where t is a term and $y \in V$. Intuitively, σ represents a stack, the *push* operations pushes the current value of t onto the top of the stack σ, and the *pop* operation pops the top value off the top of the stack σ and assigns that value to the variable y. Formally, σ will range over finite strings of elements of the domain of computation. The classes DWP_{pds} and WP_{pds} are obtained by augmenting the classes of deterministic and nondeterministic **while** programs, respectively, with *one* stack variable σ, and allowing the *push* and *pop* operations above as atomic programs in addition to simple assignments. We emphasize that the programs in DWP_{pds} and WP_{pds} may use *only one* stack—the results change dramatically when two or more stacks are allowed.

If we allow both a stack and arrays in deterministic and nondeterministic **while** programs, we obtain the programming languages $DWP_{\text{array+pds}}$ and $WP_{\text{array+pds}}$ respectively.

3.1.2. *Nondeterministic assignment*

The *nondeterministic assignment* $x := ?$ is a device that arises in the study of fairness [5]. It has often been called *random assignment* in the literature, although we prefer the name *nondeterministic assignment*, since it has nothing to do with randomness or probability. Intuitively, it operates by assigning a nondeterministically chosen element of the domain of computation to the variable x. This construct may be considered an extension of the first-order existential quantifier, in the sense that the two formulas

$$\langle x := ? \rangle \varphi, \qquad \exists x\, \varphi$$

are equivalent. However, the nondeterministic assignment is (at least superficially) more powerful, since it may be iterated.

3.1.3. Rich and poor tests

The variants of DL we have discussed may all be described as "open-test" or "poor-test". This means that only quantifier-free tests are allowed in the **if-then-else** and **while-do** (cf. Subsection 2.7.2). One may allow arbitrary DL formulas in these tests to get the "rich-test" versions. These versions are discussed briefly in Subsection 3.7.4.

3.2. Semantics

In this section, we assign meanings to all the syntactic constructs described above.

Let $\mathscr{A}=(|\mathscr{A}|,\ldots,f^{\mathscr{A}},\ldots,R^{\mathscr{A}},\ldots)$ be a first-order structure for the language L. We call $\mathscr{A}$ the *domain of computation*. Here $|\mathscr{A}|$ is a set, called the *carrier* of $\mathscr{A}$, $f^{\mathscr{A}}$ is an n-ary function $f^{\mathscr{A}}:|\mathscr{A}|^n\rightarrow|\mathscr{A}|$ interpreting the n-ary function symbol f of L, and $R^{\mathscr{A}}$ is an n-ary relation $R^{\mathscr{A}}\subseteq|\mathscr{A}|^n$ interpreting the n-ary relation symbol R of L. (The equality symbol $=$ is always interpreted as the identity relation.) We henceforth dispense with the vertical bars and use the notation $\mathscr{A}$ for both the structure and its carrier.

For $n\geqslant 0$, let $(\mathscr{A}^n\rightarrow\mathscr{A})$ denote the set of all functions $\mathscr{A}^n\rightarrow\mathscr{A}$. By convention, we take $(\mathscr{A}^0\rightarrow\mathscr{A})=\mathscr{A}$. Let $\mathscr{A}^*$ denote the set of all finite-length strings over $\mathscr{A}$.

The structure $\mathscr{A}$ determines a Kripke model, which we will also denote by $\mathscr{A}$, as follows. A *valuation* over $\mathscr{A}$ is a function u assigning an n-ary function over $\mathscr{A}$ to each n-ary array variable, and a finite-length string of elements of $\mathscr{A}$ to each stack variable. That is,

$$\begin{aligned}&u(F)\in(\mathscr{A}^n\rightarrow\mathscr{A}) && \text{if } F \text{ is an array variable,}\\ &u(\sigma)\in\mathscr{A}^* && \text{if } \sigma \text{ is a stack variable.}\end{aligned}$$

Under the convention $(\mathscr{A}^0\rightarrow\mathscr{A})=\mathscr{A}$, and assuming that $V\subseteq V_{\text{array}}$, the individual variables (i.e., the nullary array variables) are assigned elements of $\mathscr{A}$ under this definition:

$$u(x)\in\mathscr{A}\quad \text{if } x\in V.$$

The valuation u extends uniquely to terms t by induction:

$$\begin{aligned}&u(f(t_1,\ldots,t_m))=f^{\mathscr{A}}(u(t_1),\ldots,u(t_m)) && \text{if } f \text{ is an } m\text{-ary function symbol,}\\ &u(F(t_1,\ldots,t_m))=u(F)(u(t_1),\ldots,u(t_m)) && \text{if } F \text{ is an } m\text{-ary array variable.}\end{aligned}$$

If x is a variable of any type and a is an object of the same type, we denote by $u[x/a]$ the new valuation obtained from u by changing the value of x to a, and leaving the values of all other variables intact. For example if F is an m-ary array variable and $f:\mathscr{A}^m\rightarrow\mathscr{A}$, then $u[F/f]$ is the new valuation which assigns the same value as u to all stack variables and array variables other than F, and $u[F/f](F)=f$. If $f:\mathscr{A}^m\rightarrow\mathscr{A}$ is an m-ary function and $a_1,\ldots,a_m,a\in\mathscr{A}$, we denote by $f[a_1,\ldots,a_m/a]$ the m-ary function that agrees with f everywhere except input $a_1,\ldots,a_m$, on which it takes the value a. That is,

$$f[a_1,\ldots,a_m/a](b_1,\ldots,b_m)=\begin{cases}a & \text{if } b_i=a_i,\ 1\leqslant i\leqslant m,\\ f(b_1,\ldots,b_m) & \text{otherwise.}\end{cases}$$

21. DEFINITION. We call valuations u and v *finite variants* of each other if

(1) $u(F)=v(F)$ for all but finitely many array variables F;

(2) for all array variables F of positive arity n,

$$u(F)(a_1,\ldots,a_n)=v(F)(a_1,\ldots,a_n)$$

for all but finitely many n-tuples $a_1,\ldots,a_n \in \mathscr{A}^n$; in other words, u and v may differ on only finitely many array variables, and for those F on which they do differ, the functions $u(F)$ and $v(F)$ may differ on only finitely many values;

(3) for all but finitely many stack variables σ, $u(\sigma)=v(\sigma)$.

The relation "is a finite variant of" is an equivalence relation on valuations. Since a halting computation can run for only a finite time and therefore execute only finitely many assignments, it will not be able to cross equivalence class boundaries; i.e., in the binary relation semantics given below, if the pair (u,v) is an input/output pair of the program p, then v is a finite variant of u.

We are now ready to define the *states* of our Kripke model.

22. DEFINITION. Let $a\in\mathscr{A}$. Let w_a be the valuation in which all arraying and individual variables are interpreted as constant functions taking the value a everywhere, and all stacks are empty. A *state* is any valuation that is a finite variant of w_a for some a. The set of states of $\mathscr{A}$ is denoted $S^{\mathscr{A}}$.

It is meaningful, and indeed useful in some contexts, to take as states the set of all valuations. Our purpose in restricting our attention to states as defined in Definition 22 is to avoid highly complex oracles that might compromise the value of the relative expressiveness results below.

As in Subsection 2.1.2, with every program p, we associate a binary relation $p^{\mathscr{A}}\subseteq S^{\mathscr{A}}\times S^{\mathscr{A}}$ (the *input/output relation* of p), and with every formula φ, we associate a set $\varphi^{\mathscr{A}}\subseteq S^{\mathscr{A}}$. The sets $p^{\mathscr{A}}$ and $\varphi^{\mathscr{A}}$ are defined by mutual induction on the structure of p and φ.

For the basis of this inductive definition, we first give the semantics of all assignment statements discussed in Section 3.1.

(1) The array assignment $F(t_1,\ldots,t_m):=t$ is interpreted as the binary relation

$$(F(t_1,\ldots,t_m):=t)^{\mathscr{A}} \;=\; \{(u,u[F/u(F)[u(t_1),\ldots,u(t_m)/u(t)]])\mid u\in S^{\mathscr{A}}\}.$$

In other words, the array assignment has the effect of changing the value of F on input $u(t_1),\ldots,u(t_m)$ to $u(t)$, and leaving the value of F on all other inputs and the values of all other variables intact. For $m=0$, this definition reduces to the following definition of simple assignment:

$$(x:=t)^{\mathscr{A}} \;=\; \{(u,u[x/u(t)])\mid u\in S^{\mathscr{A}}\}.$$

(2) The stack operation $push(\sigma,t)$ is interpreted as the binary relation

$$push(\sigma,t)^{\mathscr{A}} \;=\; \{(u,u[\sigma/(u(t)\,.\,u(\sigma))])\mid u\in S^{\mathscr{A}}\}.$$

In other words, this operation changes the value of σ from $u(\sigma)$ to the string $u(t).u(\sigma)$, the concatenation of the value $u(t)$ with the string $u(\sigma)$.

(3) The stack operation $pop(\sigma, x)$ is interpreted as the binary relation

$$pop(\sigma, t)^{\mathscr{A}} = \{(u, u[\sigma/tail(u(\sigma))][x/head(u(\sigma), u(x))]) \mid u \in S^{\mathscr{A}}\},$$

where

$$tail(a.\alpha) = \alpha, \qquad tail(\varepsilon) = \varepsilon, \qquad head(a.\alpha, b) = a, \qquad head(\varepsilon, b) = b$$

where ε is the null string. In other words, if $u(\sigma) \neq \varepsilon$, this operation changes the value of σ from $u(\sigma)$ to the string obtained by deleting the first element of $u(\sigma)$, and assigns that element to the variable x; if $u(\sigma) = \varepsilon$, then nothing is changed.

(4) The nondeterministic assignment $x := ?$ for $x \in V$ is interpreted as the relation

$$(x{:=}?)^{\mathscr{A}} = \{(u, u[x/a]) \mid u \in S^{\mathscr{A}},\ a \in \mathscr{A}\}.$$

The meanings of the primitive constructs $\cup$ (nondeterministic choice), ; (sequential composition), * (iteration), ? (test), $\langle\ \rangle$ (modal possibility), $\vee$ (propositional disjunction), and $\neg$ (propositional negation) are exactly as in Subsection 2.1.2, as are those of the defined constructs **skip**, **fail**, **if-then-else**, **while-do**, [ ], etc.

We consider the first-order quantifier $\exists$ a defined construct:

$$\exists x\, \phi \leftrightarrow \langle x{:=}?\rangle\varphi.$$

Thus,

$$\begin{aligned}(\exists x\, \varphi)^{\mathscr{A}} &= (\langle x{:=}?\rangle\varphi)^{\mathscr{A}}\\ &= \{(u, u[x/a]) \mid u \in S^{\mathscr{A}}, a \in \mathscr{A}\} \circ \varphi^{\mathscr{A}}\\ &= \{u \mid \exists a \in A\, u[x/a] \in \varphi^{\mathscr{A}}\}.\end{aligned}$$

The universal quantifier is then given by

$$\forall x\, \varphi \leftrightarrow \neg\exists x\, \neg\varphi \leftrightarrow \neg\langle x{:=}?\rangle\neg\varphi \leftrightarrow [x{:=}?]\varphi.$$

Note that for *deterministic* programs p, $p^{\mathscr{A}}$ is single-valued, thus a partial function from states to states. The partiality of $p^{\mathscr{A}}$ arises from the possibility that p may diverge when starting its computation in certain states. For example, (**while** *true* **do skip od**)$^{\mathscr{A}}$ is the empty relation. For nondeterministic programs p, the relation $p^{\mathscr{A}}$ need not be single-valued.

If $\mathscr{A}$ is a structure and u is a state of $\mathscr{A}$, the pair $(\mathscr{A}, u)$ is called an *interpretation*. As in Subsection 2.1.2, we write $\mathscr{A}, u \models \varphi$ for $u \in \varphi^{\mathscr{A}}$ and say that u *satisfies* φ *in* $\mathscr{A}$. We may write $u \models \varphi$ when $\mathscr{A}$ is understood. We write $\mathscr{A} \models \varphi$ if $\mathscr{A}, u \models \varphi$ for all u, and we write $\models \varphi$ and say that φ is *valid* if $\mathscr{A} \models \varphi$ for all $\mathscr{A}$. We say that φ is *satisfiable* if $\mathscr{A}, u \models \varphi$ for some $(\mathscr{A}, u)$. If Σ is a set of propositions, we write $\mathscr{A} \models \Sigma$ if $\mathscr{A} \models \varphi$ for all $\varphi \in \Sigma$. A proposition ψ is said to be a *logical consequence* of Σ if for all structures $\mathscr{A}$, $\mathscr{A} \models \psi$ whenever $\mathscr{A} \models \Sigma$. We say that an inference rule $\frac{\Sigma}{\psi}$ is *sound* if ψ is a logical consequence of Σ.

In particular, $\mathscr{A}, u \models \langle p\rangle\varphi$ iff there exists a computation of p starting in state u and terminating in a state satisfying φ, and $\mathscr{A}, u \models [p]\varphi$ iff every terminating computation of p starting in state u terminates in a state satisfying φ. For a pure first-order formula φ, the metastatement $\mathscr{A}, u \models \varphi$ has the same meaning as in first-order logic (see, e.g., [19]).

As noted in Subsection 2.3.2, *DL* subsumes Hoare Logic [71]. If p is a program and φ and ψ are pure first-order formulas, the classical Hoare partial correctness formula $\{\varphi\}p\{\psi\}$ of program p with respect to *precondition* φ and *postcondition* ψ is expressed $\varphi \rightarrow [p]\psi$. The corresponding *total correctness formula* (for deterministic programs only—see Subsection 2.7.6) is expressed $\varphi \rightarrow \langle p \rangle \psi$.

If K is a given subset of the syntactic constructs introduced in Section 3.1, we refer to the version of Dynamic Logic built from these constructs as *Dynamic Logic over K*, and denote this logic by $DL(K)$. In particular, we henceforth adopt the following abbreviations:

$$DL = DL(WP), \qquad DDL = DL(DWP),$$
$$DL_{\mathbf{array}} = DL(WP_{\mathbf{array}}), \qquad DL_{\mathbf{array+pds}} = DL(WP_{\mathbf{array+pds}}).$$

3.3. Complexity

3.3.1. The validity problem

First, we discuss the complexity of the validity problem for three fragments of Dynamic Logic: full *DL*, partial correctness formulas, and total correctness formulas. These results give us the first estimate of what can be expected from formal proof systems designed for the above three fragments. The results are stated for nondeterministic **while** programs, but remain true for more powerful programming languages.

The first result of this section is due to Harel, Meyer, and Pratt [61]. The proof can also be found in [58, pp. 551–554].

23. THEOREM. (i) *If L contains at least two unary function symbols and a binary function symbol, then the set of valid DL-formulas is a* Π_1^1*-complete set.*

(ii) *If L contains at least one unary function symbol and one binary function symbol, then the set of valid partial correctness formulas*

$$\{\varphi \rightarrow [p]\psi \mid \varphi, \psi \text{ are first-order formulas and } \models \varphi \rightarrow [p]\psi\}$$

is a Π_2^0*-complete set.*

(iii) *For every L, the set of valid total correctness formulas*

$$\{\varphi \rightarrow \langle p \rangle \psi \mid \varphi, \psi \text{ are first-order formulas and } \models \varphi \rightarrow \langle p \rangle \psi\}$$

is in Σ_1^0.

The assumptions on the language L in the above theorem can presumably be further weakened, but the reader should notice that if L contains no function symbols, then the validity problem for *DL* is in Σ_1^0.

It follows from Theorem 23 that there is no sound and complete finitary proof system capable of dealing with either of the two fragments described in (i) and (ii). For total correctness, however, the situation is different. Such a system will be presented in the next section. Although the reader may feel comfortable with Theorem 23(iii), it should be stressed that only very simple computations are captured by valid total correctness formulas. This is explained by the next result.

24. PROPOSITION. *Let $\varphi \rightarrow \langle p \rangle \psi$ be a valid total correctness formula of DL. There exists a constant $k \geqslant 0$ such that for every structure $\mathcal{A}$ and state u, if $\mathcal{A}, u \models \varphi$, then there exists a computation sequence q of p of length at most k such that $\mathcal{A}, u \models \langle q \rangle \psi$.*

PROOF. This is a standard *compactness argument*. Consider the set $CS(p)$ of finite computation sequences of p defined in Subsection 2.1.3. For any finite computation sequence q, there is a first-order formula φ_q giving necessary and sufficient conditions under which q can execute and terminate successfully; i.e., φ_q is logically equivalent to $\langle q \rangle$*true*. The formula φ_q is defined by induction on the length of q, as follows:

- if q is the empty string, then $\varphi_q = true$;
- if $q = (x := t);q'$, then $\varphi_q = \varphi_{q'}[x/t]$, where $\varphi_{q'}[x/t]$ is the result of substituting t for all free occurrences of x in $\varphi_{q'}$, renaming bound variables as necessary to avoid capture;
- if $q = \theta?;q'$, then $\varphi_q = \theta \wedge \varphi_{q'}$.

By Proposition 2, the total correctness formula in the statement of the proposition is equivalent to the infinitary formula

$$\varphi \rightarrow \bigvee_{q \in CS(p;\psi?)} \varphi_q,$$

which by assumption is valid. Hence, by the compactness of first-order logic, there exists a finite subset $F \subseteq CS(p;\psi?)$ such that the finitary formula

$$\varphi \rightarrow \bigvee_{q \in F} \varphi_q$$

is valid. We may therefore take k to be the maximum of the lengths of the elements of F. □

3.3.2. *Spectral complexity*

In this section, we introduce the notion of *spectral complexity* of a programming language. This notion provides a measure of complexity of the halting problem for programs over finite interpretations. The notion of spectral complexity is relatively new in the literature. This name appears for the first time in [156], though the ideas were already present in [159] and [63]. Spectral complexity plays an important role in establishing the expressive power of logics of programs.

In order to introduce the notion of spectral complexity, we shall need some auxiliary definitions.

25. DEFINITION. An interpretation $(\mathcal{A}, u)$ is said to be *Herbrand-like* if the set $\{u(x) \mid x \in V\}$ generates $\mathcal{A}$; i.e., if every $a \in \mathcal{A}$ is $u(t)$ for some term t over V (thus t contains no array variables of positive arity).

26. DEFINITION. A state u in $\mathcal{A}$ is called *initial* if

(1) there exists an $a \in \mathcal{A}$ such that for all array variables F of positive arity n and $a_1, \ldots, a_n \in \mathcal{A}$, $u(F)(a_1, \ldots, a_n) = a$;

(2) for all stack variables σ, $u(\sigma) = \varepsilon$.

Let $m \geqslant 0$. The pair $(\mathscr{A}, u)$ is called an *m-interpretation* if u is an initial state in $\mathscr{A}$ and there exists an $a \in \mathscr{A}$ such that for all $i \geqslant m$, $u(x_i) = a$.

Two interpretations $(\mathscr{A}, u)$ and $(\mathscr{B}, v)$ are said to be *isomorphic* if there exists an isomorphism $h: \mathscr{A} \to \mathscr{B}$ that commutes with the valuations u and v; i.e., for all m-ary array variables F,

$$v(F)(h(a_1), \ldots, h(a_m)) = h(u(F)(a_1, \ldots, a_m))$$

(in particular, $v(x) = h(u(x))$ for individual variables x); and for any stack variable σ,

$$\text{if } u(\sigma) = a_1 a_2 \ldots a_m, \quad \text{then } v(\sigma) = h(a_1)h(a_2) \ldots h(a_m).$$

Let L be a finite first-order language. We henceforth assume that L contains at least one function symbol of positive arity (otherwise, only trivial relations can be computed). The language L is said to be *rich* if it contains a function or relation symbol (other than equality) of arity at least 2, or at least two function or relation symbols of arity 1; otherwise, L is said to be *poor*. Thus, L is poor if it contains exactly one function symbol of arity one and no relation symbols other than equality.

The essential difference between rich and poor languages is that for fixed m, over structures with n elements, a rich language has exponentially (in n) many pairwise nonisomorphic Herbrand-like m-interpretations, whereas poor languages have only polynomially many. For a rich language L and fixed $m \in \omega$, we can encode every finite Herbrand-like m-interpretation $(\mathscr{A}, u)$ by a binary string '$\mathscr{A}, u$'$\in \{0, 1\}^*$ in such a way that the following properties hold:

(1) the length of '$\mathscr{A}, u$' is polynomial in the cardinality of $\mathscr{A}$;

(2) '$\mathscr{A}, u$' = '$\mathscr{B}, v$' iff $(\mathscr{A}, u)$ and $(\mathscr{B}, v)$ are isomorphic;

(3) the set of codes

$$\{`\mathscr{A}, u\text{'} \mid (\mathscr{A}, u) \text{ is a finite Herbrand-like } m\text{-interpretation}\}$$

is in DSPACE($\log n$).

For a poor language, a similar encoding exists, except the length of '$\mathscr{A}, u$' is logarithmic in the cardinality of A, and the logarithmic function of (3) above should be replaced with a linear one. For our purposes, the form of the encoding is not important, only the existence of one; see [156, 159] for details.

Now we are ready to define the notion of a *spectrum* of a programming language. Let K be a programming language over L. For $p \in K$ and $m \in \omega$, the mth *spectrum of* p is the set

$$SP_m(p) = \{`\mathscr{A}, u\text{'} \mid (\mathscr{A}, u) \text{ is a finite Herbrand-like } m\text{-interpretation, and } \mathscr{A}, u \models \langle p \rangle \mathit{true}\}.$$

The *spectrum* of K is the set

$$SP(K) = \{SP_m(p) \mid p \in K,\ m \in \omega\}.$$

The next three definitions connect spectra with complexity classes. Let $C \subseteq 2^{\{0,1\}^*}$ be a complexity class, and let K be a programming language. We say that the *spectral complexity of K is in C* if $SP_m(p) \in C$ for all $p \in K$ and $m \in \omega$; i.e., $SP(K) \subseteq C$. We say that the *spectral complexity of K is at least C* and write $SP(K) \geqslant C$ if for every $X \in C$ and $m \in \omega$, if X is a set of codes of finite Herbrand-like m-interpretations, then $X = SP_m(p)$ for some $p \in K$. We say that the *spectral complexity of K is equal to C and* write $SP(K) \approx C$ if both $SP(K) \subseteq C$ and $SP(K) \geqslant C$.

We conclude this section by establishing the spectral complexity of the programming languages introduced in Section 3.1. The following result is due to Tiuryn and Urzyczyn [159], except (1)(e), which follows from a more general result of [156].

27. THEOREM. (1) *Let L be a rich language.*

(a) $SP(DWP) \subseteq \text{DSPACE}(\log n)$.

(b) $SP(WP) \subseteq \text{NSPACE}(\log n)$.

(c) $SP(WP_{\text{pds}}) = SP(DWP_{\text{pds}}) \approx \text{DTIME}(2^{O(\log n)}) = \text{PTIME}$.

(d) $SP(WP_{\text{array}}) = SP(DWP_{\text{array}}) \approx \text{DSPACE}(2^{O(\log n)}) = \text{PSPACE}$.

(e) $SP(WP_{\text{array+pds}}) = SP(DWP_{\text{array+pds}}) \approx \text{DTIME}(2^{2^{O(\log n)}}) = \text{EXPTIME}$.

(2) *If L contains a relation symbol of positive arity, exactly one unary function symbol, and no other function symbols, then* $\subseteq$ *in* (1)(a) *and* (1)(b) *can be replaced by* $\approx$.

(3) *If L is poor then* (1)(a)–(1)(e) *carry over, provided* $\log n$ *is replaced by n.*

The proofs of Theorem 27(1)(a, b, d) are by mutual simulations of programs and off-line Turing machines. The proofs of Theorem 27(1)(c, e) are by mutual simulation of programs and Cook's auxiliary pushdown automata (see [72]).

3.4. Deductive systems

We will start our discussion of formal deductive systems for first-order DL with the most promising case (cf. Theorem 23(iii)), namely valid total correctness formulas. Here we will consider the programming language R of regular programs. Extensions to handle arrays can be found in [58, pp. 571–575]. Then, by extending the basic formal proof system S (to be introduced below) in various ways, we will obtain:

(1) a relatively complete system $S()$ for partial correctness formulas;

(2) an arithmetically complete system $S_a()$ for full DL; and

(3) an infinitary complete system $S_\infty()$ for full DL.

Some contrasting negative results are contained in [21, 91, 171].

It should be stressed that unlike *syntax-oriented* formal proof systems of Hoare Logic (cf. [27]), the systems presented in this section are built in a style that resembles first-order reasoning.

28. DEFINITION (*the deductive system S*). The deductive system S consists of the following axioms and rules of inference:

(1) all valid first-order formulas;

(2) all axiom and rules of PDL (see Section 2.2);

(3) $\langle x := t \rangle \varphi \leftrightarrow \varphi[x/t]$, where t is a term, φ is a first-order formula, and $\varphi[x/t]$ denotes the result of substituting t for all free occurrences of x in φ, renaming bound variables as necessary to avoid capture.

We write $\vdash_S \varphi$ if φ is a theorem of this system.

The first result, due to Meyer and Halpern [100], establishes the soundness and completeness of S for total correctness formulas.

29. THEOREM. *For any first-order formulas φ, ψ in L and program p,*

$$\models \varphi \rightarrow \langle p \rangle \psi \quad \textit{iff} \quad \vdash_S \varphi \rightarrow \langle p \rangle \psi.$$

The proof is by induction on p.

It follows from Theorem 23(ii) that we cannot have a result similar to Theorem 29 for partial correctness formulas. A way around this difficulty, suggested by Cook [26], is to consider only *expressive* structures. A structure $\mathscr{A}$ for L is said to be *expressive* for a programming language K with respect to first-order assertions if, for every $p \in K$ and for every first-order formula φ in L, there exists a first-order formula ψ in L such that

$$\mathscr{A} \models \psi \leftrightarrow [p]\varphi.$$

Examples of expressive structures for most programming languages are finite structures and *arithmetical structures.* The latter class of structures was introduced by Moschovakis [109] under the same *acceptable structures* (see also [56]). Briefly, a structure $\mathscr{A}$ is *arithmetical* if it contains a first-order definable copy of the standard model of arithmetic $\mathscr{N} = (\omega, 0, 1, +, \cdot, \leqslant)$, and has first-order definable functions allowing coding and decoding of finite sequences of elements of $\mathscr{A}$.

30. DEFINITION (*the deductive system* $S(\,)$). For a structure $\mathscr{A}$, let $S(\mathscr{A})$ be the system S extended by adding as axioms all first-order formulas valid in $\mathscr{A}$.

The next result, essentially due to Cook [26], establishes the soundness and *relative completeness of* $S(\mathscr{A})$ for partial correctness formulas over structures $\mathscr{A}$ expressive for R.

31. THEOREM. *For every expressive structure $\mathscr{A}$ for R and for every partial correctness formula $\varphi \rightarrow [p]\psi$, where φ, ψ are first-order,*

$$\mathscr{A} \models \varphi \rightarrow [p]\psi \quad \textit{iff} \quad \vdash_{S(\mathscr{A})} \varphi \rightarrow [p]\psi.$$

Again, the proof is by induction on p.

If the system $S(\,)$ is further strengthened and restricted to arithmetical structures, then *arithmetical completeness* can be established for full DL.

32. DEFINITION (*the deductive systems S_a and $S_a(\,)$*). Let S_a be the system S extended with the following two proof rules:

- *quantifier generalization*:

$$\frac{\varphi}{\forall x\, \varphi}$$

- *rule of convergence*:

$$\frac{\varphi(n+1)\rightarrow\langle p\rangle\varphi(n)}{\varphi(n)\rightarrow\langle p^*\rangle\varphi(0)}$$

where $p\in R$, $\varphi(n)$ is a first-order formula with a free variable n ranging over (a copy of) $\mathscr{N}$ such that n does not occur in p, and $+1$ and 0 are replaced by a suitable first-order definition. These requirements can be satisfied in arithmetical structures.

For an arithmetical structure $\mathscr{A}$, let $S_a(\mathscr{A})$ be S_a augmented by adding as axioms all first-order formulas valid in $\mathscr{A}$.

The next result, due to Harel [56], asserts the *arithmetical completeness* of *DL*.

33. THEOREM. *For every arithmetical structure $\mathscr{A}$ and for every $\varphi\in DL$,*

$$\mathscr{A}\models\varphi \quad \textit{iff} \quad \vdash_{S_a(\mathscr{A})}\varphi.$$

By Theorem 23(i), there cannot exist a finitary complete proof system for full *DL*. A complete infinitary proof system was proposed by Mirkowska for Algorithmic Logic [106]. We follow the exposition [58] where this system is presented for *DL*.

34. DEFINITION (*the deductive system S_∞*). Let S_∞ be the system S augmented with the following axiom schemes and proof rules:

- $\langle x:=t\rangle\varphi\leftrightarrow\varphi[x/t]$, where $\varphi\in DL$. (The substitution of a term for a variable in a *DL* formula φ has to be defined carefully, due to the possible presence of programs in φ. The notions of *free* and *bound variable* are not as clear as they are in pure first-order logic. See [58, p. 55] for details.)
- $\varphi\leftrightarrow\psi$, where ψ is φ with some program p replaced by $z:=x;p';x:=z$, for some $z\in V$ not appearing in φ or p, and p' is p with all occurrences of x replaced by z.
- In addition, we also take quantifier generalization as in S_a, and

(∞) *infinitary convergence rule*:

$$\frac{\varphi\rightarrow[p^n]\psi,\quad n\in\omega}{\varphi\rightarrow[p^*]\psi}$$

where $\varphi,\psi\in DL$ and $p\in R$.

Observe that the rule (∞) has infinitely many premises. A *proof* in S_∞ is a possibly infinite sequence of *DL* formulas, each one either an instance of an axiom scheme, or following from previous formulas by application of a proof rule.

The next result is due to Mirkowska [106].

35. THEOREM. *For every $\varphi\in DL$, φ is valid iff* $\vdash_{S_\infty}\varphi$.

There are as many proofs of this theorem as there are proofs of infinitary completeness for $L_{\omega_1\omega}$. In fact, every proof for the latter logic transforms into a proof of Theorem 35. Algebraic methods are used in [106], whereas [58] uses Henkin's method for $L_{\omega_1\omega}$.

3.5. *Expressive power*

Let L be a finite first-order language with equality. The subject of study in this section is the relation of *relative expressiveness* between logics $\mathscr{L}$, $\mathscr{L}'$ over L.

36. DEFINITION. We say that $\mathscr{L}'$ is *more expressive than* $\mathscr{L}$ and write $\mathscr{L} \leqslant \mathscr{L}'$ iff for every $\varphi \in \mathscr{L}$, there exists a $\psi \in \mathscr{L}'$ such that $\mathscr{A}, u \models \varphi \leftrightarrow \psi$ for all structures $\mathscr{A}$ and *initial* states u (Definition 26). We write $\mathscr{L} \equiv \mathscr{L}'$ if both $\mathscr{L} \leqslant \mathscr{L}'$ and $\mathscr{L}' \leqslant \mathscr{L}$, and $\mathscr{L} < \mathscr{L}'$ if $\mathscr{L} \leqslant \mathscr{L}'$ but not $\mathscr{L} \equiv \mathscr{L}'$.

The reason for the restriction to *initial* states in Definition 36 is that if $\mathscr{L}$ and $\mathscr{L}'$ have access to different sets of data types, then they may be trivially incomparable for uninteresting reasons, unless we are careful to limit the states on which they are compared. For example, when comparing DL_{pds} and DL_{array}, we had better disallow input states in which the stack variable σ is nonempty; otherwise, since WP_{pds} programs can access σ and WP_{array} cannot, the DDL_{pds} formula

$$\langle pop(\sigma, x); \textbf{if } x = y \textbf{ then skip else fail fi} \rangle true$$

would be trivially equivalent to no formula of DL_{array}. Definition 36 effectively restricts the *input* states of programs to initial states, but still allows noninitial states as intermediate states in a computation. Recall also that stacks and arrays often play an auxiliary role in programs and are not usually used for input/output.

In the definition of $DL(K)$ given in Section 3.1, the programming language K is an explicit parameter; but the first-order language L over which $DL(K)$ and K are taken should be treated as a parameter as well. It turns out that the relation $\leqslant$ of relative expressiveness is sensitive not only to K, but also to L. This second parameter is often ignored in the literature, creating a source of potential misinterpretation of the results.

Let $L_{\omega\omega}$ denote the first-order predicate calculus with equality over the language L, and let $L_{\omega_1\omega}$ denote the infinitary language that allows countable disjunctions in addition to the usual formation rules of $L_{\omega\omega}$.

We start with a result which orders the versions of DL of Section 3.1 linearly.

37. PROPOSITION. *For every first-order language L,*

(i) $L_{\omega\omega} \leqslant DL \leqslant DL_{\text{pds}} \leqslant DL_{\text{array}} \leqslant DL_{\text{array+pds}} \leqslant L_{\omega_1\omega}$.

(ii) $DDL \leqslant DL$ *and* $DDL_* \leqslant DL_*$ *for* $* \in \{\text{pds, array, array+pds}\}$.

All inequalities of the above proposition are easy to establish, except the third one in (i). This inequality will follow from a more general result (Theorem 45) relating $\leqslant$ with spectra and complexity classes.

It is easy to show that the last inequality in (i) is strict for every language L. Henceforth, we shall assume that L contains at least one function symbol of positive arity. Under this assumption, the first inequality in (i) becomes strict, since it is possible to construct an infinite model for L which is uniquely definable in DL up to isomorphism. By the upward Löwenheim–Skolem Theorem (cf. [19]), this is impossible to $L_{\omega\omega}$.

Strictness of the other inequalities in Proposition 37 will be discussed in the remainder of this section. In order to discuss the relationship of these problems to complexity theory, we must introduce some definitions.

For program p and structure $\mathscr{A}$, let $p^{\mathscr{A}}$ be the binary relation associated with p, as defined in Section 3.2.

38. DEFINITION. A program p is said to be *semantically deterministic* if for every $\mathscr{A}$, $p^{\mathscr{A}}$ is a partial function.

The remarks immediately following Definition 36 at the beginning of this section motivate the following definition.

39. DEFINITION. Let $\rho \subseteq S^{\mathscr{A}} \times S^{\mathscr{A}}$ be a binary relation on states of a structure $\mathscr{A}$. The *ground relation of* ρ is the binary relation

$$ground(\rho) = \{(u{\restriction}V, v{\restriction}V) \mid (u, v) \in \rho \text{ and } u \text{ is an initial state of } \mathscr{A}\}.$$

(Here $u{\restriction}V$ denotes the function u restricted to domain V.) If K and K' are programming languages over L, we write $K \leqslant K'$ iff, for every $p \in K$, there exists a $q \in K'$ such that in every structure $\mathscr{A}$ for L, $ground(p^{\mathscr{A}}) = ground(q^{\mathscr{A}})$.

40. DEFINITION. The programming language K is said to be *semantically closed* under the n-ary programming construct c if for all programs $p_1, \ldots, p_n \in K$, there exists a $q \in K$ such that in every structure $\mathscr{A}$, $c(p_1, \ldots, p_n)^{\mathscr{A}} = q^{\mathscr{A}}$.

For example, "K is semantically closed under sequential composition" means that for every $p, p' \in K$, there exists a $q \in K$ such that $q^{\mathscr{A}} = (p;p')^{\mathscr{A}}$ in all structures $\mathscr{A}$.

Note that K need not contain the construct c; however, if it does, then it is trivially semantically closed under c.

We now define a useful programming construct **run-until**, which works as follows. For $p, q \in K$ and first-order formula φ, the program

run p **until** $\langle q \rangle \varphi$

runs p, but after each atomic step of p, it runs q from the current state and tests whether q would halt in a state satisfying φ. Depending on the outcome, it takes the following actions:

- if q diverges, then the **run-until** statement itself diverges;
- if q terminates in a state v satisfying φ, the entire **run-until** statement terminates in state v;
- if q terminates in a state v satisfying $\neg\varphi$, then control is returned to p in state v to perform the next atomic step.

For regular programs, we can define this construct formally by induction on p:

run p **until** $\langle q \rangle \varphi = p;q$ for p an atomic program or test
run $p;p'$ **until** $\langle q \rangle \varphi =$ (**run** p **until** $\langle q \rangle \varphi$);**if** φ **then skip else run** p' **until** $\langle q \rangle \varphi$ **fi**,
run $p \cup p'$ **until** $\langle q \rangle \varphi =$ (**run** p **until** $\langle q \rangle \varphi$)$\cup$(**run** p' **until** $\langle q \rangle \varphi$),
run p^* **until** $\langle q \rangle \varphi = q \cup$(**run** p **until** $\langle q \rangle \varphi$);($\neg\varphi$?;**run** p **run** $\langle q \rangle \varphi$)*.

The construct also makes sense in more general programing languages, and can be defined formally for any program p equivalent to its set $CS(p)$ of computation sequences, in the sense of Theorem 2. The definition gives **run** p **until** $\langle q\rangle\varphi$ in terms of its computation sequences. The definition of **run** r **until** $\langle q\rangle\varphi$ for r a computation sequence is given above, and for more general programs p,

$$CS(\textbf{run } p \textbf{ until } \langle q\rangle\varphi)=\bigcup_{r\in CS(p)} CS(\textbf{run } r \textbf{ until } \langle q\rangle\varphi).$$

Let WP_{2pds} denote the class of all **while** programs with two stacks.

41. DEFINITION. A programming language K is said to be *acceptable* if it satisfies the following conditions:

(1) $K \leqslant WP_{2pds}$;

(2) K contains all simple assignments $x:=t$ and is semantically closed under sequential composition, conditional, and **while** loop, with tests in the last two restricted to quantifier-free formulas of L;

(3) K is semantically closed under variable renaming; that is, if $p\in K$ and π is any permutation of V, then there exists a $q\in K$ such that for all $\mathscr{A}$,

$$q^{\mathscr{A}}=\{(u\circ\pi, v\circ\pi)\mid (u,v)\in p^{\mathscr{A}}\};$$

(4) K is semantically closed under the **run-until** construct, **run** p **until** $\langle q\rangle\varphi$, for p and q semantically deterministic and φ a quantifier-free formula of L.

Definition 41(1) insures that programs of K perform only effective operations (relative to the interpretation of the symbols in L). WP_{2pds} in (1) can be replaced by any programming language of *universal power*: effective definitional schemes [45], recursive procedures with integer counters, **while** programs with integer-indexed arrays, etc. Conditions (2)–(4) say that K has some semantical flexibility, so that certain operations on programs can be performed without leaving the class K. It should be clear that all the programming languages of Section 3.1 are acceptable. The notion of an acceptable programming language was introduced by Lipton [91] and used by many authors, e.g. [22, 158].

42. DEFINITION. An acceptable programming language K is *semi-universal* (cf. [158]) if for every $m\in\omega$ there exists a semantically deterministic program $p\in K$ such that for every Herbrand-like m-interpretation $(\mathscr{A},u)$,

$$\mathscr{A},u\models\langle p\rangle\, \textit{true} \quad \text{iff} \quad (\mathscr{A},u) \text{ is finite.}$$

The essence of this definition is that semi-universal programming languages have enough power to search every submodel generated by the input. The notion of semi-universality captures the concept of programs with *unbounded memory* in [58].

We now proceed to the last definition.

43. DEFINITION. A programming language K is *divergence-closed* if for every $p\in K$ there exists a $q\in K$ and two variables $x,y\in V$ such that for every finite Herbrand interpretation $(\mathscr{A}, u)$ with A having at least two elements,

$$\mathscr{A}, u \models \langle p \rangle\, true \quad \text{iff} \quad \mathscr{A}, u \models \langle q \rangle (x = y),$$
$$\mathscr{A}, u \models [p]\, false \quad \text{iff} \quad \mathscr{A}, u \models \langle q \rangle (x \neq y).$$

Informally, q decides without diverging whether p possibly terminates.

44. PROPOSITION. *The following programming languages are semi-universal and divergence-closed:*

(1) *for every L containing at least one function symbol of positive arity:* $(D)WP_*$ *for* $* \in \{\text{pds, array, array} + \text{pds}\}$ (*cf.* [158]);

(2) *for every L containing exactly one unary function symbol and no other function symbols of positive arity:* $(D)WP$.

It follows from the result of Immerman [172] that WP is divergence closed for L specified in Proposition 44(2).

The next result, due to Tiuryn and Urzyczyn [156, 158], plays a key role in applying the hierarchy results of complexity theory to problems concerning $\leqslant$.

45. THEOREM. *Let* K_1 *and* K_2 *be programming languages over L such that* K_1 *is acceptable and* K_2 *is semi-universal and divergence-closed. Let* $C_1, C_2 \subseteq 2^{\{0,1\}^*}$ *denote families of sets that are closed downward under logarithmic space or linear space reductions, depending on whether L is rich or poor respectively. Let* $SP(K_i) \approx C_i$ *for* $i = 1, 2$. *The following statements are equivalent:*

(1) $DL(K_1) \leqslant DL(K_2)$,

(2) $SP(K_1) \subseteq SP(K_2)$,

(3) $C_1 \subseteq C_2$.

The equivalence of (1) and (2) is proved in [158, Theorem 5]. The equivalence of (2) and (3) is proved in [156, Theorem 3.9].

Combining Theorem 27, Proposition 44, and Theorem 45 with the known hierarchy theorems of complexity theory, we obtain:

46. COROLLARY. (1) *For every L and for every* $* \in \{\text{pds, array, array} + \text{pds}\}$, $DDL_* \equiv DL_*$.

(2) *If L has exactly one unary function symbol and no other function symbols of positive arity and at least one relation symbol of positive arity other than* $=$, *then*

$$DDL \equiv DL \quad \textit{iff} \quad \text{DSPACE}(\log n) = \text{NSPACE}(\log n).$$

(3) *For every L*, $DL_{\text{pds}} \leqslant DL_{\text{array}}$.

(4) *If L is rich, then*

$$DL_{\text{pds}} \equiv DL_{\text{array}} \quad \textit{iff} \quad \text{DTIME}(2^{O(\log n)}) = \text{DSPACE}(2^{O(\log n)})$$

(*i.e., iff* PTIME = PSPACE).

(5) *For every L*, $DL_{\text{pds}} < DL_{\text{array+pds}}$.

(6) *If L is rich, then*

$$DL_{\mathrm{array}} \equiv DL_{\mathrm{array+pds}} \quad \text{iff} \quad \mathrm{DSPACE}(2^{O(\log n)}) = \mathrm{DTIME}(2^{2^{O(\log n)}})$$

(*i.e., iff* PSPACE = EXPTIME).

(7) *The results of* (2), (4), (6) *hold for poor languages when* $\log n$ *is replaced by* n.

A consequence of Corollary 46 is that many questions on relative expressive power of Dynamic Logics are equivalent to well-known difficult open problems of complexity theory. Most of the results of Corollary 46 were proved in [159].

The following two questions are not settled by the method of spectral complexity.

(1) Does $DDL < DL$ hold for every L containing a function symbol of arity at least 2 or at least two function symbols of arity 1?

(2) Does $DL < DL_{\mathrm{pds}}$ hold for every L?

We conclude this section with a full answer to the first question and a partial answer to the second.

47. THEOREM. *If L contains a function symbol of arity at least* 2 *or at least two function symbols of arity* 1, *then* $DDL < DL$.

This result is due to Stolboushkin and Taitslin [149] and independently to Berman, Halpern and Tiuryn [15]. The proof of [149] uses Adian's Theorem [2]. It constructs an infinite model $\mathscr{A}$ with two unary functions f, g and a constant c satisfying the property:

> for every program $p \in DWP$ there exists a constant $k \in \omega$ such that every terminating computation of p in $\mathscr{A}$ takes at most k steps.

This property of $\mathscr{A}$, together with the compactness of first-order logic, imply that the formula

$$\langle z := c; \mathbf{while}\ z \neq x\ \mathbf{do}\ z := f(z) \cup z := g(z)\ \mathbf{od} \rangle true,$$

which expresses a nondeterministic search, is equivalent to no DDL formula.

The proof of [15] uses a purely combinatorial argument to construct a model $\mathscr{A}$ with the above property. In both cases the construction of $\mathscr{A}$ together with a proof that $\mathscr{A}$ has the desired property is the major difficulty of the proof of Theorem 47. Other proofs of Theorem 47 can be found in [162, 75, 157].

A partial answer to (2) is given by the next result.

48. THEOREM. *If L contains a function symbol of arity at least two, then* $DL < DL_{\mathrm{pds}}$.

This result is due to Erimbetov [39] and independently to Tiuryn [155]. The method of proof in both cases is essentially the same, though [155] contains a slightly more general statement. It says that two certain infinite models, constructed from constants, can be distinguished by no formula of a certain fragment $L^{BM}_{\omega_1\omega}$ of the infinitary language $L_{\omega_1\omega}$. The proof uses two techniques: a *pebbling argument* invented by Paterson and

Hewitt [126] and independently by Friedman [45], and the technique of Ehrenfeucht–Fraissé games [31]. The pebbling argument works in this framework only for languages that satisfy the assumption of Theorem 48. The proof is completed by checking that $DL \leqslant L^{BM}_{\omega_1 \omega}$, and that the two models can be distinguished by a formula of DL_{pds}.

A stronger result holds for deterministic programs. It follows from Corollary 46(1) and Theorem 47 that if L contains a function symbol of arity at least two or at least two function symbols of arity one, then $DDL < DDL_{\text{pds}}$. For languages with exactly one unary function symbol and no function symbols of higher arity, the problems of comparing DL to DL_{pds} and DDL to DDL_{pds} reduce to open problems in complexity theory, such as whether the classes DSPACE($\log n$) and PTIME are equal. A challenging open problem not known to be equivalent to an open problem in complexity theory is the question of whether $DL < DL_{\text{pds}}$ for all languages containing no function symbols of arity at least two and at least two function symbols of arity one.

3.6. *Operational vs. axiomatic semantics*

In this section, we survey some results connected with the following question:

> Can the semantics of a programming language be specified by partial correctness formulas with first-order assertions?

This question is of fundamental importance for the so-called *axiomatic semantics*.

49. Definition. Let K be a programming language. For $p, q \in K$, we say that p is *semantically contained in* q and write $p \subseteq q$ iff, for every model $\mathcal{A}$, $p^{\mathcal{A}} \subseteq q^{\mathcal{A}}$.

For definiteness, we choose $K = WP_{2pds}$, **while** programs with two stacks. Consider the two propositions

(1) $p \subseteq q$,

(2) for all first-order formulas φ, ψ in L, if $\varphi \rightarrow [q]\psi$ is valid, then so is $\varphi \rightarrow [p]\psi$.

The implication (1)→(2) follows immediately from Proposition 5. The essence of the question raised above is whether the converse holds. Meyer conjectured that this was indeed the case, and this conjecture was confirmed by Meyer and Halpern [100] and independently by Bergstra, Tiuryn and Tucker [9].

50. Theorem. *Let L be any first-order language except one containing exclusively unary relations and at most one unary function symbol. Then for every $p, q \in WP_{2\text{pds}}$, if for all first-order formulas φ, ψ in L, $\models \varphi \rightarrow [q]\psi$ implies $\models \psi \rightarrow [p]\psi$, then $p \subseteq q$.*

It is noteworthy that Theorem 50 extends with the same proof of [100, 9] to programming languages whose power goes beyond any acceptable programming language; e.g. flowcharts with arbitrary first-order tests, nondeterministic assignments, stacks, arrays, etc. We also remark that the assertions φ, ψ and programs p, q in Theorem 50 are all over the same language L. It is not known whether Theorem 50 extends to a language L consisting of only one unary function symbol and some number of unary relation symbols. The reader is referred to [89] for a simpler proof of Theorem 50 for the special case of **while** programs. However, the result of [89] is weaker than Theorem 50 for **while** programs, since the former allows the assertions φ, ψ to be in an extension of L.

The definitions of this section can easily be relativized to a first-order theory T, i.e., when interpretations are restricted to models of T. Currently, it is not well understood what conditions on T might be sufficient to imply a relativized version of Theorem 50. It is shown in [9] that the relativized version of Theorem 50 fails for as simple a theory as the equational theory

$$\{f(g(x))=g(f(x))=x\}$$

[9, Theorem 5.8], and for as complex a theory as *Complete Number Theory* [9, Theorem 5.10]. A positive result in this direction by Csirmaz [29] shows that Theorem 50 admits relativization to *Peano Arithmetic*, confirming a conjecture posed in [9].

We conclude this section with a result showing that the validity of the partial correctness formula $\varphi \to [q]\psi$ of Theorem 50 can even be proved in Hoare Logic, provided one allows φ and ψ to be in an extension of L. This result is due independently to Leivant [90] and Meyer [99]. The method of proof given in [99] adapts to any programming language other than WP for which there is a sound and relatively complete proof system.

51. THEOREM. *Let L be any first-order language. There exists an extension $L' \supseteq L$ such that for any $p \in WP_{2\text{pds}}$ and $q \in WP$ over L, if $p \nsubseteq q$, then there exist first-order formulas φ, ψ in L' such that $\varphi \to [q]\psi$ is a theorem of Hoare Logic, but $\varphi \to [p]\psi$ is not valid.*

It is not known, even for L satisfying the assumption of Theorem 50, whether the extension of the language L in Theorem 51 is essential.

The reader is also referred to [10] where various schemes for establishing program inclusion are studied.

3.7. *Other programming languages*

For uniformity of exposition, we have concentrated on the language of **while** programs, occasionally augmented with arrays and stacks. Other definitions appearing in the literature (e.g., [56, 58]) may differ slightly from ours, but the reader should have no difficulty establishing that these versions are equivalent in expressive power.

Below we discuss briefly some possible extensions of WP.

3.7.1. *ALGOL-like languages*

One can add to WP recursive procedures without parameters, recursive procedures with individual parameters passed by name, reference, value/result, or other mechanism, or recursive procedures with higher-order procedure parameters. Blocks with local declarations can be added as well. Depending on scope rules (dynamic vs. static) and the features allowed, one easily arrives at a family consisting of thousands of programming languages.

Proof theory for ALGOL-like languages is well developed by now. The reader is referred to [4, 27, 48] for typical results and further references.

The expressive power of logics based on ALGOL-like programming languages has been studied to a lesser extent. A typical result in this area, which follows from [16, 24], is that Dynamic Logic with recursive procedures in which individual parameters are passed by value/result is equivalent to DL_{pds}.

3.7.2. Nondeterministic assignment

The nondeterministic assignment x:=?, introduced in Section 3.1, chooses an element of the domain of computation nondeterministically and assigns it to x. Thus, it is a device representing *unbounded nondeterminism*, as opposed to the *binary nondeterminism* of the nondeterministic choice construct $\cup$. The programming language WP augmented with the nondeterministic assignment is not an acceptable language. Adding nondeterministic assignment to a programming language usually increases expressive power. For more information, the reader is referred to [58].

3.7.3. Auxiliary data types

WP can be augmented with other data types, such as counters, binary stacks, or higher-order arrays and stacks. Data types can be combined.

Proof theory for programming languages with these data types is not sufficiently developed. Their relative expressive power has been studied more intensively. One interesting result, due to Urzyczyn [161], is that adding a binary stack to DWP results in a logic strictly more expressive than DDL. (The corresponding question for nondeterministic **while** programs is open.) It follows from [155] that this logic is strictly weaker than DL_{pds}. An infinite hierarchy of logics over WP with higher-order arrays and stacks is studied in [156].

3.7.4. Tests

In previous sections, we allowed tests to be quantifier-free first-order formulas. One can increase the power of programs by allowing arbitrary first-order formulas as tests. One can even go further and define programs and formulas by simultaneous induction, allowing programs to test arbitrary formulas. This is called a *rich-test* logic of programs. The reader is referred to [58] for more information on this topic.

4. Other approaches

4.1. Nonstandard Dynamic Logic

Nonstandard Dynamic Logic (NDL) was introduced by Andréka, Németi, and Sain in 1979. The reader is referred to [3, 111] for a full exposition and further references. The main idea behind NDL is to allow nonstandard models of time by referring only to first-order properties of time when measuring the length of a computation. The approach described in [3] and further research in Nonstandard Dynamic Logic is concentrated on proving properties of flowcharts, i.e., programs built up of assignments, conditionals and **goto**'s.

Nonstandard Dynamic Logic is well suited to comparing the reasoning power of various program verification methods. This is usually done by providing a model-theoretic characterization of a given method for program verification. To illustrate this approach, we briefly discuss a characterization of Hoare Logic for partial correctness formulas. For the present exposition, we choose a somewhat simpler formalism which still conveys the basic idea of nonstandard time.

Let L be a first-order language. We fix for the remainder of this section a deterministic **while** program p over L in which the **while-do** construct does not occur. (Such a program is called *loop-free*.) Let $\bar{z}=(z_1,\ldots,z_n)$ contain all variables occurring in p, and let $\bar{y}=(y_1,\ldots,y_n)$ be a vector of n distinct individual variables disjoint from $\bar{z}$.

Since p is loop-free, it has only finitely many computation sequences. One can easily define a quantifier-free first-order formula θ_p with all free variable among $\bar{y}$, $\bar{z}$ which defines the input/output relation of p in all structures $\mathscr{A}$ for L, in the sense that the pair of states (u, v) is in $p^{\mathscr{A}}$ if and only if

$$\mathscr{A},\ v[y_1/u(z_1),\ldots,y_n/u(z_n)]\models\theta_p$$

and $u(x)=v(x)$ for all $x\in V-\{z_1,\ldots,z_n\}$.

Let p^+ be the following deterministic **while** program:

$\bar{y}:=\bar{z}$;

p;

while $\bar{z}\neq\bar{y}$ **do** $\bar{y}:=\bar{z}$; p **od**

where $\bar{z}\neq\bar{y}$ stands for $z_1\neq y_1 \vee\cdots\vee z_n\neq y_n$ and $\bar{y}:=\bar{z}$ stands for $y_1:=z_1;\cdots;y_n=z_n$. Thus program p^+ performs p iteratively until p does not change the state.

The remainder of this section is devoted to giving a model-theoretic characterization, using *NDL*, of Hoare's system for proving partial correctness properties of p^+ relative to a given first-order theory T in L. We denote provability in Hoare Logic by $\vdash_{HL}$.

Due to the very specific form of p^+, the Hoare system reduces to the following rule:

$$\frac{\varphi\rightarrow\chi,\quad \chi[\bar{z}/\bar{y}]\wedge\theta_p\rightarrow\chi,\quad \chi[\bar{z}/\bar{y}]\wedge\theta_p\wedge\bar{z}=\bar{y}\rightarrow\psi}{\varphi\rightarrow[p^+]\psi}$$

where φ, χ, ψ are first-order formulas, and no variable of $\bar{y}$ occurs free in χ.

The next series of definitions introduces a variant of *NDL*. A structure $\mathscr{I}$ for the language consisting of a unary function symbol $+1$ (*successor*), a constant symbol 0, and equality is called a *time model* if the following axioms are valid in $\mathscr{I}$:

(1) $x+1=y+1\rightarrow x=y$,

(2) $x+1\neq 0$,

(3) $x\neq 0\rightarrow\exists y\ y+1=x$,

(4) $x\neq x+1+1+\cdots+1$ (n 1s) for any $n=1, 2,\ldots$

Let $\mathscr{A}$ be a structure for L, and let $\mathscr{I}$ be a time model. A function $\rho:\mathscr{I}\rightarrow\mathscr{A}^n$ is called a *run* of p in $\mathscr{A}$ if the following two infinitary formulas are valid in $\mathscr{A}$:

(1) $\bigwedge_{i\in\mathscr{I}}\theta_p[\bar{y}/\rho(i),\bar{z}/\rho(i+1)]$;

(2) for every first-order formula $\varphi(\bar{z})$ in L,

$$\varphi(\rho(0))\wedge\bigwedge_{i\in\mathscr{I}}(\varphi(\rho(i))\rightarrow\varphi(\rho(i+1)))\;\rightarrow\;\bigwedge_{i\in\mathscr{I}}\varphi(\rho(i)).$$

The first formula says that, for $i\in\mathscr{I}$, $\rho(i)$ is the valuation obtained from $\rho(0)$ after i iterations of the program p. The second formula is the induction scheme along the run ρ.

Finally, we say that a partial correctness formula $\varphi\rightarrow[p^+]\psi$ *follows from T in nonstandard time semantics* and write $T\models_{NT}\varphi\rightarrow[p^+]\psi$ if for every model $\mathscr{A}$ of T, time model $\mathscr{I}$, and run ρ of p in $\mathscr{A}$,

$$\mathscr{A}\models\varphi[\bar{z}/\rho(0)]\;\rightarrow\;\bigwedge_{i\in\mathscr{I}}(\rho(i)=\rho(i+1)\rightarrow\psi[\bar{z}/\rho(i)]).$$

The following characterization theorem is due to Csirmaz [30].

52. THEOREM. *For every first-order theory T in L and first-order formulas φ, ψ in L, the following conditions are equivalent*:

(1) $T\vdash_{HL}\varphi\rightarrow[p^+]\psi$;

(2) $T\models_{NT}\varphi\rightarrow[p^+]\psi$.

Other proof methods have been characterized in the same spirit. The reader is referred to [94] for more information on this issue and further references.

4.2. Algorithmic Logic

Algorithmic Logic (*AL*), first defined by Salwicki in 1970 [142], foreshadowed Dynamic Logic in many respects. Research in *AL* has centered on program verification, infinitary completeness, normal forms for programs, recursive procedures with parameters, and data type specification; see [7, 143] for surveys.

The original version of *AL* allowed deterministic **while** programs and formulas built from the constructs

$$p\varphi,\qquad \bigcup p\varphi,\qquad \bigcap p\varphi$$

corresponding in our terminology to

$$\langle p\rangle\varphi,\qquad \langle p^*\rangle\varphi,\qquad \bigwedge_{n\in\omega}\langle p^n\rangle\varphi$$

respectively, where p is a **while** program and φ is a quantifier-free first-order formula. Mirkowska [107, 108] extended *AL* to allow nondeterministic **while** programs and the constructs

$$\nabla p\varphi,\qquad \Delta p\varphi$$

corresponding in our terminology to

$$\langle p\rangle\varphi,\qquad \mathbf{halt}(p)\wedge[p]\varphi\wedge\langle p\rangle\varphi$$

respectively. The latter asserts that all traces of p are finite and terminate in a state satisfying φ. Complete infinitary deductive systems are given for first-order and propositional versions [107, 108].

Constable [23, 25] and Goldblatt [49] present logics similar to AL and DL for reasoning about deterministic **while** programs.

4.3. *Logic of Effective Definitions*

The Logic of Effective Definitions (*LED*), introduced by Tiuryn in 1978 (see [154]), was intended to study notions of computability over abstract models and to provide a universal framework for the study of logics of programs over such models. It consists of first-order logic augmented with new atomic formulas of the form $p=q$, where p and q are *effective definitional schemes* [45]:

if φ_1 **then** t_1
  else if φ_2 **then** t_2
    else if φ_3 **then** t_3
      else if...

where the φ_i are quantifier-free formulas and t_i are terms over a bounded set of variables, and the function $i \mapsto (\varphi_i, t_i)$ is recursive. The formula $p=q$ is defined to be true in state u if both p and q terminate and yield the same value, or neither terminates.

Model theory and infinitary completeness of *LED* are treated in [154].

4.4. *Temporal Logic*

Temporal Logic (TL) is an alternative application of modal logic to program specification and verification. It was first proposed as a useful tool in program verification by Pnueli [130], and has since been developed by many authors in various forms. This topic is surveyed in depth elsewhere in this Handbook [33].

TL differs from DL chiefly in that it is *endogenous*, i.e., programs are not explicit in the language. Every application has a single program associated with it, and the language may contain program-specific statements such as *at*(l), meaning "execution is currently at location l in the program". Models can be a linear sequence of program states (so-called *linear-time* TL), representing the execution sequence of a deterministic program or a possible execution sequence of a nondeterministic or concurrent program; or a tree of program states (so-called *branching-time* TL), representing the space of all possible computation sequences of a nondeterministic or concurrent program.

Modal constructs used in TL include

$\Box\varphi$ "φ holds in all future states",
$\Diamond\varphi$ "φ holds in some future state",
$\bigcirc\varphi$ "φ holds in the next state"

for linear-time logic, as well as constructs for expressing

"for all paths starting from the present state ... ",
"for some path starting from the present state ... "

for branching-time logic.

Temporal Logic is useful in situations where programs are not normally supposed to halt, such as operating systems, and is particularly well-suited to the study of concurrency. Many of the classical program verification methods such as the *intermittent assertions method* are treated quite elegantly in this framework.

Acknowledgment

We would like to thank the following colleagues for their valuable criticism: J. Bergstra, P. van Emde Boas, E.A. Emerson, J. Halpern, D. Harel, M. Karpiński, N. Klarlund, D. McAllester, A. Meyer, R. Parikh, V. Pratt, G. Smith, and M. Vardi. We are especially indebted to David Harel, on whose excellent survey [58] the present work is modeled.

Dexter Kozen has been supported by NSF Grant CCS-8806979; Jerzy Tiuryn has been supported by a grant from the Polish Ministry of Higher Education R.P.I. 09.

References

[1] ABRAHAMSON, K., Decidability and expressiveness of logics of processes, Ph.D. Thesis, Tech. Report 80-08-01, Dept. of Computer Science, Univ. of Washington, Seattle, 1980.

[2] ADIAN, S.I., *The Burnside Problem and Identities in Groups* (Springer, Heidelberg, 1979).

[3] ANDRÉKA, H., I. NÉMETI and I. SAIN, A complete logic for reasoning about programs via Nonstandard Model Theory. Parts I, II, *Theoret. Comput. Sci.* **17** (1982) 193–212 and 259–278.

[4] APT, K.R., Ten years of Hoare's Logic: a survey—part 1, *ACM Trans. on Programming Languages and Systems* **3** (1981) 431–483.

[5] APT, K.R. and G. PLOTKIN, Countable nondeterminism and random assignment, *J. ACM* **33** (1986) 724–767.

[6] BAKKER, J. DE, *Mathematical Theory of Program Correctness* (Prentice-Hall, Englewood Cliffs, NJ, 1980).

[7] BANACHOWSKI, L., A. KRECZMAR, G. MIRKOWSKA, H. RASIOWA and A. SALWICKI, An introduction to Algorithmic Logic: metamathematical investigations in the theory of programs, in: A. Mazurkiewitz and Z. Pawlak, eds., *Mathematical Foundations on Computer Science* (Banach Center Publications, Warsaw, 1977) 7–99.

[8] BEN-ARI, M., J.Y. HALPERN and A. PNUELI, Deterministic Propositional Dynamic Logic: finite models, complexity and completeness, *J. Comput. System Sci.* **25** (1982) 402–417.

[9] BERGSTRA, J.A., J. TIURYN and J.V. TUCKER, Floyd's principle, correctness theories and program equivalence, *Theoret. Comput. Sci.* **17** (1982) 113–149.

[10] BERGSTRA, J.A. and J.W. KLOP, Proving program inclusion using Hoare's Logic, *Theoret. Comput. Sci.* **30** (1984) 1–48.

[11] BERMAN, F., Expressiveness hierarchy for PDL with rich tests, Tech. Report 78-11-01, Dept. of Computer Science, Univ. of Washington, Seattle, 1978.

[12] BERMAN, F., A completeness technique for *D*-axiomatizable semantics, in: *Proc. 11th Ann. ACM Symp. on Theory of Computing* (1979) 160–166.

[13] BERMAN, F., Semantics of looping programs in Propositional Dynamic Logic, *Math. Systems Theory* **15** (1982) 285–294.

[14] BERMAN, F. and M. PATERSON, Propositional Dynamic Logic is weaker without tests, *Theoret. Comput. Sci.* **16** (1981) 321–328.

[15] BERMAN, P., J.Y. HALPERN and J. TIURYN, On the power of nondeterminism in Dynamic Logic, in: M. Nielsen and E.M. Schmidt, eds., *Proc. 9th Internat. Coll. on Automata, Languages and Programming*, Lecture Notes in Computer Science, Vol. 140 (Springer, Berlin, 1982) 48–60.

[16] BROWN, S., D. GRIES and T. SZYMANSKI, Program schemes with pushdown stores, *SIAM J. Comput.* **1** (1972) 242–268.

[17] BURSTALL, R.M., Program proving as hand simulation with a little induction, in: *Proc. IFIP Congr. on Information Processing* (North-Holland, Amsterdam, 1974) 308–312.

[18] CHANDRA, A.K., D. KOZEN and L. STOCKMEYER, Alternation, *J. ACM* **28** (1) (1981) 114–133.

[19] CHANG, C.C. and H.J. KEISLER, *Model Theory* (North-Holland, Amsterdam, 1973).

[20] CHELLAS, B.F., *Modal Logic: an Introduction* (Cambridge Univ. Press, Cambridge, 1980).

[21] CLARKE, E.M., Programming language constructs for which it is impossible to obtain good Hoare axiom systems, *J. ACM* **26** (1979) 129–147.

[22] CLARKE, E.M., S.M. GERMAN and J.Y. HALPERN, Effective axiomatizations of Hoare Logics, *J. ACM* **30** (1983) 612–636.

[23] CONSTABLE, R.L., On the theory of Programming logics, in: *Proc. 9th Ann. ACM Symp. on Theory of Computing* (1977) 269–285.

[24] CONSTABLE, R.L. and D. GRIES, On classes of program schemata, *SIAM J. Comput.* **1** (1972) 66–118.

[25] CONSTABLE, R.L. and M. O'DONNELL, *A Programming Logic* (Winthrop, Cambridge, MA, 1978).

[26] COOK, S.A., Soundness and completeness of an axiom system for program verification, *SIAM J. Comput.* **7** (1978) 70–80.

[27] COUSOT, P., Methods and logics for proving programs, in. J. van Leeuwen, ed., *Handbook of Theoretical Computer Science, Vol. B* (North-Holland, Amsterdam, 1990) 841–993.

[28] COURCOUBETIS, C. and M. YANNAKAKIS, Verifying temporal properties of finite-state probabilistic programs, *Proc. 29th Ann. IEEE Symp. on Foundations of Computer Science* (1988) 338–345.

[29] CSIRMAZ, L., Determinateness of program equivalence over Peano Axioms, *Theoret. Comput. Sci.* **21** (1982) 231–235.

[30] CSIRMAZ, L., A completeness theorem for Dynamic Logic, *Notre Dame J. Formal Logic* **26** (1985) 51–60.

[31] EHRENFEUCHT, A., An application of games in the completeness problem for formalized theories, *Fund. Math.* **49** (1961) 129–141.

[32] EMERSON, E.A., Automata, tableau, and temporal logics, in: R. Parikh, ed., *Proc. Workshop on Logics of Programs*, Lecture Notes in Computer Science, Vol. 193 (Springer, Berlin, 1985) 79–88.

[33] EMERSON, E.A., Temporal and Modal Logic, in: J. van Leeuwen, ed., *Handbook of Theoretical Computer Science, Vol. B* (North-Holland, Amsterdam, 1990) 995–1072.

[34] EMERSON, E.A. and J.Y. HALPERN, Decision procedures and expressiveness in the Temporal Logic of branching time, *J. Comput System Sci.* **30** (1) (1985) 1–24.

[35] EMERSON, E.A. and C. JUTLA, The complexity of tree automata and logics of programs, in: *Proc. 29th Ann. IEEE Symp. on Foundations of Computer Science* (1988) 328–337.

[36] EMERSON, E.A. and C. JUTLA, On simultaneously determinizing and complementing ω-automata, in: *Proc. 4th Ann. IEEE Symp. on Logic in Computer Science* (1989) 333–342.

[37] EMERSON, E.A. and P.A. SISTLA, Deciding full branching-time logic, *Inform. and Control* **61** (1984) 175–201.

[38] ENGELER, E., Algorithmic properties of structures, *Math. Systems Theory* **1** (1967) 183–195.

[39] ERIMBETOV, M.M., On the expressive power of programming logics, in: *Proc. Conf. on Research in Theoretical Programming* (1981) 49–68 (in Russian).

[40] FELDMAN, Y.A., A decidable propositional Dynamic Logic with explicit probabilities, *Inform. and Control* **63** (1984) 11–38.

[41] FELDMAN, Y.A. and D. HAREL, A Probabilistic Dynamic Logic, *J. Comput. System Sci.* **28** (1984) 193–215.

[42] FISCHER, M.J. and R.E. LADNER, Propositional Modal Logic of programs, in: *Proc. 9th ACM Ann. Symp. on Theory of Computing* (1977) 286–294.

[43] FISCHER, M.J. and R.E. LADNER, Propositional Dynamic Logic of regular programs, *J. Comput. System Sci.* **18** (2) (1979) 194–211.

[44] FLOYD, R.W., Assigning meanings to programs, in: *Proc. AMS Symp. on Applied Mathematics 19* (Amer. Mathematical Soc., Providence, RI, 1967) 19–31.

[45] FRIEDMAN, H., Algorithmic procedures, generalized Turing algorithms, and elementary recursion theory, in: R.O. Gandy and C.M.E. Yates, eds., *Logic Colloquium 1969* (North-Holland, Amsterdam, 1971) 361–390.

[46] GABBAY, D., Axiomatizations of logics of programs, Unpublished manuscript, Bar-Ilan Univ., Ramat-Gan, Israel, 1977.

[47] GABBAY, D., A. PNUELI, S. SHELAH and J. STAVI, On the temporal analysis of fairness, in: *Proc. 7th Ann. ACM Symp. on Principles of Programming Languages* (1980) 163–173.

[48] GERMAN, S.M., E.M. CLARKE and J.T. HALPERN, True relative completeness of an axiom system for the language L4, in: *Proc. 1st IEEE Symp. on Logic in Computer Science* (1986) 11–25.

[49] GOLDBLATT, R., *Axiomatising the Logic of Computer Programming*, Lecture Notes in Computer Science, Vol. 130 (Springer, Berlin, 1982).

[50] GOLDBLATT, R., Logics of time and computation, Center for the Study of Language and Information Lecture Notes 7, Stanford Univ., Stanford, CA, 1987.

[51] GRIES, D., *The Science of Programming* (Springer, New York, 1981).

[52] HAJEK, P. and P. KURKA, A second-order Dynamic Logic with array assignments, *Fund. Inform.* **IV** (1981) 919–933.

[53] HALPERN, J.Y., On the expressive power of Dynamic Logic II, Tech. Report TM-204, Laboratory for Computer Science, MIT, Cambridge, MA, 1981.

[54] HALPERN, J.Y., Deterministic Process Logic is elementary, in: *Proc. 23rd Ann. IEEE Symp. on Foundations of Computer Science* (1982) 204–216.

[55] HALPERN, J.Y. and J.H. REIF, The propositional Dynamic Logic of deterministic, well-structured programs, *Theoret. Comput. Sci.* **27** (1983) 127–165.

[56] HAREL, D., *First-Order Dynamic Logic*, Lecture Notes in Computer Science, Vol. 68 (Springer, Berlin, 1979).

[57] HAREL, D., Proving the correctness of regular deterministic programs: a unifying survey using Dynamic Logic, *Theoret. Comput. Sci.* **12** (1980) 61–81.

[58] HAREL, D., Dynamic Logic, in: D.M. Gabbay and F. Guenthner, eds., *Handbook of Philosophical Logic. II: Extensions of Classical Logic* (Reidel, Boston, MA, 1984) 497–604.

[59] HAREL, D. and D. KOZEN, A programming language for the inductive sets, and applications, *Inform. and Control* **63** (1, 2) (1984) 118–139.

[60] HAREL, D., D. KOZEN and R. PARIKH, Process Logic: Expressiveness, decidability, Completeness, in: *Proc. 21st Ann. IEEE Symp. on Foundations of Computer Science* (1980) 129–142; *J. Comput. System Sci.* **25** (2) (1982) 144–170.

[61] HAREL, D., A.R. MEYER and V.R. PRATT, Computability and completeness in logics of programs, in: *Proc. 9th Ann. ACM Symp. on Theory of Computing* (1977) 261–268.

[62] HAREL, D. and M.S. PATERSON, Undecidability of PDL with $L=\{a^{2^i} \mid i \geqslant 0\}$, *J. Comput. System Sci.* **29** (1984) 359–365.

[63] HAREL, D. and D. PELEG, On static logics, dynamic logics, and complexity classes, *Inform. and Control* **60** (1984) 86–102.

[64] HAREL, D., A. PNUELI and J. STAVI, Propositional Dynamic Logic of nonregular programs, *J. Comput. System Sci.* **26** (1983) 222–243.

[65] HAREL, D., A. PNUELI and M. VARDI, Two-dimensional temporal logic and PDL with intersection, Unpublished manuscript, The Weizmann Institute, Rehovot, Israel, 1982.

[66] HAREL, D. and V.R. PRATT, Nondeterminism in logics of programs, in: *Proc. 5th Ann. ACM Symp. on Principles of Programming Languages* (1978) 203–213.

[67] HAREL, D. and R. SHERMAN, Propositional Dynamic Logic of flowcharts, *Inform. and Comput.* **64** (1985) 119–135.

[68] HAREL, D. and R. SHERMAN, Looping vs. repeating in Dynamic Logic, *Inform. and Control* **55** (1982) 175–192.

[69] HART, S., M. SHARIR and A. PNUELI, Termination of probabilistic concurrent programs, in: *Proc. 9th Ann. ACM Symp. on Principles of Programming Languages* (1982) 1–6.

[70] HITCHCOCK, P. and D. PARK, Induction rules and termination proofs, in: M. Nivat, ed., *Proc. 1st Internat. Coll. on Automata, Languages and Programming* (1973) 225–251.

[71] HOARE, C.A.R., An axiomatic basis for computer programming, *Comm. ACM* **12** (1967) 516–580.

[72] HOPCROFT, J.E. and J.D. ULLMAN, *Introduction to Automata Theory, Languages, and Computation* (Addison-Wesley, Reading, MA, 1979).

[73] HUGHES, G.E. and M.J. CRESSWELL, *An Introduction to Modal Logic* (Methuen, London, 1968).

[74] IANOV, Y.I., The logical schemes of algorithms, in: *Problems of Cybernetics 1* (Pergamon, New York, 1960) 82–140.

[75] KFOURY, A.J., Definability by deterministic and nondeterministic programs with applications to first-order dynamic logic, *Inform and Control* **65** (2–3) (1985) 98–121.

[76] KNIJNENBURG, P.M.W. and J. VAN LEEUWEN, On models for propositional dynamic logic, Tech. Report RUU-CS-89-3, Dept. of Computer Science, Univ. of Utrecht, 1989; *Theoret. Comput. Sci.* (1991).

[77] KOZEN, D., A representation theorem for models of *-free PDL, in: J.W. de Bakker and J. van Leeuwen, eds., *Proc. 7th Internat. Coll. on Automata, Languages and Programming*, Lecture Notes in Computer Science, Vol. 85 (Springer, Berlin, 1980) 351–362.

[78] KOZEN, D., On the duality of dynamic algebras and Kripke models, in: E. Engeler, ed., *Proc. Workshop on Logics of Programs*, Lecture Notes in Computer Science, Vol. 125 (Springer, Berlin, 1981) 1–11.

[79] KOZEN, D., On induction vs. *-continuity, in: D. Kozen, ed., *Proc. Workshop on Logics of Programs*, Lecture Notes in Computer Science, Vol. 131 (Springer, Berlin, 1981)167–176.

[80] KOZEN, D., Semantics of probabilistic programs, *J. Comput. System Sci.* **22** (1981) 328–350.

[81] KOZEN, D., Logics of programs, Unpublished lecture notes, Dept. of Computer Science, Univ. of Aarhus, Denmark, 1981.

[82] KOZEN, D., Results on the propositional μ-calculus, in: M. Neilsen and E.M. Schmidt, eds., *Proc. 9th Internat. Coll. on Automata, Languages and Programming*, Lecture Notes in Computer Science, Vol. 140 (Springer, Berlin, 1982) 348–359.

[83] KOZEN, D., A probabilistic PDL, *J. Comput. System Sci.* **30** (2) (1985) 162–178.

[84] KOZEN, D., A finite model theorem for the propositional μ-calculus, *Studia Logica* **47** (3) (1988) 53–61.

[85] KOZEN, D. and R. PARIKH, An elementary proof of the completeness of PDL, *Theoret. Comput. Sci.* **14** (1981) 113–118.

[86] KOZEN, D. and R. PARIKH, A decision procedure for the propositional μ-calculus, in: E. Clarke and D. Kozen, eds., *Proc. Workshop on Logics of Programs*, Lecture Notes in Computer Science, Vol. 164 (Springer, Berlin, 1983) 313–325.

[87] LAMPORT, L., "Sometime" is sometimes "not never", in: *Proc. 7th Ann. ACM Symp. on Principles of Programming Languages* (1980) 174–185.

[88] LEHMANN, D. and S. SHELAH, Reasoning with time and chance, *Inform. and Control* **53** (3) (1982) 165–198.

[89] LEIVANT, D., Logical and mathematical reasoning about imperative programs, in: *Proc. 12th Ann. ACM Symp. on Principles of Programming Languages* (1985) 132–140.

[90] LEIVANT, D., Hoare's Logic captures program semantics (extended summary), Tech. Report, Computer Science Dept., Carnegie-Mellon Univ., Pittsburgh, PA, 1985.

[91] LIPTON, R.J., A necessary and sufficient condition for the existence of Hoare Logics, in: *Proc. 18th Ann. IEEE Symp. on Foundations of Computer Science* (1977) 1–6.

[92] LUCKHAM, D.C., D. PARK, and M. PATERSON, On formalized computer programs, *J. Comput. System Sci.* **4** (1970) 220–249.

[93] MAKOWSKI, J.A., Measuring the expressive power of Dynamic Logics: an application of abstract model theory, in: *Proc. 7th Internat. Coll. on Automata, Languages and Programming*, Lecture Notes in Computer Science, Vol. 80 (Springer, Berlin, 1980) 409–421.

[94] MAKOWSKY, J.A. and I. SAIN, On the equivalence of weak second-order and nonstandard time semantics for various program verification systems, in *Proc. 1st IEEE Symp. on Logic in Computer Science* (1986) 293–300.

[95] MANDERS, K.L. and R.F. DALEY, The complexity of the validity problem for Dynamic Logic, Unpublished manuscript, Univ. of Pittsburgh, Pittsburgh, PA, 1982.

[96] MANNA, Z. and A. PNUELI, Verification of concurrent programs: temporal proof principles, in: D. Kozen, ed., *Proc. Workshop on Logics of Programs*, Lecture Notes in Computer Science, Vol. 131 (Springer, Berlin, 1981) 200–252.

[97] MANNA, Z. and A. PNUELI, Specification and verification of concurrent programs by ∀-automata, in: *Proc. 14th Ann. ACM Symp. on Principles of Programming Languages* (1987) 1–12.

[98] MEYER, A.R., Ten thousand and one logics of programming, *Bull. EATCS* **10** (1980) 11–29.

[99] MEYER, A.R., Floyd–Hoare Logic defines semantics (preliminary version), in: *Proc. IEEE Symp. on Logic in Computer Science* (1986) 44–48.

[100] MEYER, A.R. and J.Y. HALPERN, Axiomatic definitions of programming languages: a theoretical assessment, *J. ACM*, **29** (1982) 555–576.

[101] MEYER, A.R. and J.C. MITCHELL, Termination assertions for recursive programs: completeness and axiomatic definability, Tech. Report TM-214, Lab. for Computer Science, MIT, Camridge, MA, 1982.

[102] MEYER, A.R. and R. PARIKH, Definability in Dynamic Logic, *J. Comput. System Sci.* **23** (1981) 279–298.

[103] MEYER, A.R., R.S. STREETT and G. MIRKOWSKA, The deducibility problem in Propositional Dynamic Logic, in: E. Engeler, ed., *Proc. Workshop on Logics of Programs*, Lecture Notes in Computer Science, Vol. 125, (Springer, Berlin, 1981) 12–22.

[104] MEYER, A.R. and J. TIURYN, A note on equivalences among logics of programs, in: D. Kozen, ed., *Proc. Workshop on Logics of Programs*, Lecture Notes in Computer Science, Vol. 131 (Springer, Berlin, 1981) 282–299.

[105] MEYER, A.R. and K. WINKLMANN, Expressing program looping in Regular Dynamic Logic, *Theoret. Comput. Sci.* **18** (1982) 301–323.

[106] MIRKOWSKA, G., On formalized systems of Algorithmic Logic, *Bull. Acad. Polon. Sci., Ser. Sci. Math. Astron. Phys.* **19** (1971) 421–428.

[107] MIRKOWSKA, G., Algorithmic Logic with nondeterministic programs, *Fund. Inform.* **III** (1980) 45–64.

[108] MIRKOWSKA, G., PAL—Propositional Algorithmic Logic, in: E. Engeler, ed., *Proc. Workshop on Logics of Programs*, Lecture Notes in Computer Science, Vol. 125 (Springer, Berlin, 1981) 23–101; *Fund. Inform.* **IV** (1981) 675–760.

[109] MOSCHOVAKIS, Y.N., *Elementary Induction on Abstract Structures* (North-Holland, Amsterdam, 1974).

[110] MULLER, D.E., A. SAOUDI and P.E. SCHUPP, Weak alternating automata give a simple explanation of why most Temporal and Dynamic Logics are decidable in exponential time, in: *Proc. 3rd Ann. (IEEE) Symp. on Logic in Computer Science*, (1988) 422–427.

[111] NÉMETI, I., Nonstandard Dynamic Logic, in: D. Kozen, ed., *Proc. Workshop on Logics of Programs*, Lecture Notes in Computer Science, Vol. 131 (Springer, Berlin, 1981) 311–348.

[112] NISHIMURA, H., Sequential method in Propositional Dynamic Logic, *Acta Inform.* **12** (1979) 377–400.

[113] NISHIMURA, H., Descriptively complete Process Logic, *Acta Inform.* **14** (1980) 359–369.

[114] NISHIMURA, H., Arithmetical completeness in first-order Dynamic Logic for concurrent programs, *Publ. Res. Inst. Math. Sci.* **17** (1981) 297–309.

[115] NIWINSKI, D., The propositional μ-calculus is more expressive than the Propositional Dynamic Logic of looping, Unpublished manuscript, 1984.

[116] OLSHANSKI, T. and A. PNUELI, There exist decidable context-free Propositional Dynamic Logics, Unpublished manuscript, The Weizmann Institute, Rehovot, Israel, 1981.

[117] PARIKH, R., The completeness of Propositional Dynamic Logic, in: *Proc. 7th Symp. on Mathematical Foundations of Computer Science*, Lecture Notes in Computer Science, Vol. 64 (Springer, Berlin, 1978) 403–415.

[118] PARIKH, R., A decidability result for second-order Process Logic, in: *Proc. 19th Ann. IEEE Symp. on Foundations of Computer Science* (1978) 177–183.

[119] PARIKH, R., Propositional Logics of programs, in: *Proc. 7th Ann. ACM Symp. on Principles of Programming Languages* (1980) 186–192.

[120] PARIKH, R.K, Propositional Dynamic Logics of programs: a survey, in: E. Engeler, ed., *Proc. Workshop on Logics of Programs*, Lecture Notes in Computer Science, Vol. 125 (Springer, Berlin 1981) 102–144.

[121] PARIKH, R., Propositional Logics of programs: new directions, in: M. Karpinski, ed., *Proc. Internat.*

Conf. on Foundations of Computing Theory, Lecture Notes in Computer Science, Vol. 158 (Springer, Berlin, 1983) 347–359.

[122] PARIKH, R., Propositional Game Logic, in: *Proc. 24th Ann. IEEE Symp. on Foundations of Computer Science* (1983) 195–200.

[123] PARIKH, R., The logic of games and its applications, *Ann. Discrete Math.* **24** (1985) 111–140.

[124] PARIKH, R. and A. MAHONEY, A theory of probabilistic programs, in: E. Clarke and D. Kozen, eds., *Proc. Workshop on Logics of Programs*, Lecture Notes in Computer Science, Vol. 164 (Springer, Berlin, 1983) 396–402.

[125] PARK, D., Finiteness is μ-ineffable, *Theoret. Comput. Sci.* **3** (1976) 173–181.

[126] PATERSON, M.S. and C.E. HEWITT, Comparative schematology, in: *Record Project MAC Conf. on Concurrent Systems and Parallel Computation* (ACM, New York, 1970) 119–128.

[127] PÉCUCHET, J.P., On the complementation of Büchi automata, *Theoret. Comput. Sci.* **47** (1986) 95–98.

[128] PELEG, D., Concurrent Dynamic Logic, *J. ACM* **34** (2) (1987) 450–479.

[129] PETERSON, G.L., The power of tests in Propositional Dynamic Logic, Tech. Report 47, Dept. of Computer Science, Univ. of Rochester, 1978.

[130] PNUELI, A., The Temporal Logic of programs, in: *Proc. 18th Ann. IEEE Symp. on Foundations of Computer Science* (1977) 46–57.

[131] PRATT, V.R., Semantical considerations on Floyd–Hoare Logic, in: *Proc. 17th Ann. IEEE Symp. on Foundations of Computer Science* (1976) 109–121.

[132] PRATT, V.R., A practical decision method for Propositional Dynamic Logic, in: *Proc. 10th Ann. ACM Symp. on Theory of Computing* (1978) 326–337.

[133] PRATT, V.R., Process Logic, in: *Proc. 6th Ann. ACM Symp. on Principles of Programming Languages* (1979) 93–100.

[134] PRATT, V.R., Models of program logics, in: *Proc. 20th Ann. IEEE Symp. on Foundations of Computer Science* (1979) 115–122.

[135] PRATT, V.R., Dynamic algebras and the nature of induction, in: *Proc. 12th Ann. ACM Symp. on Theory of Computing*, (1980) 22–28.

[136] PRATT, V.R., Using graphs to understand PDL, in: D. Kozen, ed., *Proc. Workshop on Logics of Programs*, Lecture Notes in Computer Science, Vol. 131. (Springer, Berlin, 1981) 387–396.

[137] PRATT, V.R., A near-optimal method for reasoning about actions, *J. Comput. System Sci.* **20** (2) (1980) 231–254.

[138] PRATT, V.R., A decidable μ-calculus: preliminary report, in: *Proc. 22nd Ann. IEEE Symp. on Foundations of Computer Science* (1981) 421–427.

[139] RABIN, M.O., Decidability of second-order theories and automata on infinite trees, *Trans. Amer. Math. Soc.* **141** (1969) 1–35.

[140] RAMSHAW, L.H., Formalizing the analysis of algorithms, Ph.D. Thesis, Stanford Univ., Stanford, CA, 1981.

[141] SAFRA, S., On the complexity of ω-automata, in: *Proc. 29th Ann. IEEE Symp. on Foundations of Computer Science* (1988) 319–327.

[142] SALWICKI, A., Formalized algorithmic languages, *Bull. Acad. Polon. Sci., Ser. Sci. Math. Astron. Phys.* **18** (1970) 227–232.

[143] SALWICKI, A., Algorithmic Logic: a tool for investigations of programs, in: R.E. Butts and K.J.J. Hintikka, eds., *Logic, Foundations of Mathematics, and Computability Theory* (Reidel, Dordrecht, 1977) 281–295.

[144] SCOTT, D.S. and J.W. DE BAKKER, A theory of programs, Unpublished notes, IBM, Vienna, 1969.

[145] SEGERBERG, K., A completeness theorem in the Modal Logic of Programs (preliminary report), *Not. Amer. Math. Soc.* **24** (6) (1977) A-552.

[146] SISTLA, A.P. and E.M. CLARKE, The complexity of Propositional Linear Temporal Logics, in: *Proc. 14th Ann. ACM Symp. on Theory of Computing* (1982) 159–168.

[147] SISTLA, A.P., M.Y. VARDI and P. WOLPER, The complementation problem for Büchi automata with application to Temporal Logic, *Theoret. Comput. Sci.* **49** (1987) 217–237.

[148] SOKOLOWSKI, S., Programs as term transformers, *Fund. Inform.* **III** (1980) 419–432.

[149] STOLBOUSHKIN, A.P. and M.A. TAITSLIN, Deterministic Dynamic Logic is strictly weaker than Dynamic Logic, *Inform and Control* **57** (1983) 48–55.

[150] STREETT, R.S., Propositional Dynamic Logic of looping and converse, in: *Proc. 13th Ann. ACM Symp. on Theory of Computing* (1981) 375–381.

[151] STREETT, R.S., Propositional Dynamic Logic of looping and converse is elementarily decidable, *Inform. and Control* **54** (1982) 121–141.

[152] STREETT, R.S., Fixpoints and program looping: reductions from the propositional μ-calculus into Propositional Dynamic Logics of looping, in: R. Parikh, ed., *Proc. Workshop on Logics of Programs*, Lecture Notes in Computer Science, Vol. 193 (Springer, Berlin, 1985) 359–372.

[153] THIELE, H., *Wissenschaftstheoretische Untersuchungen in Algorithmischen Sprachen. Theorie der Graphschemata-Kalkule* (Veb. Deutscher Verlag der Wissenschaften, Berlin, DDR, 1966).

[154] TIURYN, J., A survey of the logic of effective definitions, in: E. Engeler, ed., *Proc. Workshop on Logics of Programs*, Lecture Notes in Computer Science, Vol. 125 (Springer, Berlin, 1981) 198–245.

[155] TIURYN, J., Unbounded program memory adds to the expressive power of first-order programming logics, in: *Proc. 22nd Ann. IEEE Symp. on Foundations of Computer Science* (1981) 335–339; *Inform. and Control* **60** (1984) 12–35.

[156] TIURYN, J., Higher-order arrays and stacks in programming: an application of complexity theory to logics of programs, in: J. Gruska et al. eds., *Proc. Mathematical Foundations of Computer Science*, Lecture Notes in Computer Science, Vol. 233 (Springer, Berlin, 1986) 177–198.

[157] TIURYN, J., A simplified proof of DDL < DL, *Inform. and Comput.* **81** (1989) 1–12.

[158] TIURYN, J. and P. URZYCZYN, Remarks on comparing expressive power of logics of programs, in: M.P. Chytil and V. Koubek, eds., *Proc. Mathematical Foundations of Computer Science*, Lecture Notes in Computer Science, Vol. 176 (Springer, Berlin, 1984) 535–543.

[159] TIURYN, J. and P. URZYCZYN, Some relationships between logics of programs and complexity theory, in: *Proc. 24th IEEE Ann. Symp. on Foundations of Computer Science* (1983) 180–184; *Theoret Comput. Sci.* **60** (1988) 83–108.

[160] TRNKOVA, V. and J. REITERMAN, Dynamic algebras which are not Kripke structures, in: P. Dembińsky, ed., *Proc. 9th Symp. on Mathematical Foundations of Computer Science*, Lecture Notes in Computer Science, Vol. 88 (Springer, Berlin, 1980) 528–538.

[161] URZYCZYN, P., Deterministic context-free Dynamic Logic is more expressive than Deterministic Dynamic Logic of regular programs, in: M. Karpinski, ed., *Proc. Conf. on Fundaments of Computing Theory*, Lecture Notes in Computer Science, Vol. 158 (Springer, Berlin, 1983) 496–504.

[162] URZYCZYN, P., Nontrivial definability by flowchart programs, *Inform. and Control* **58** (1983) 59–87.

[163] VALIEV, M.K., Decision complexity of variants of Propositional Dynamic Logic, in: P. Dembińsky, ed. *Proc. 9th Symp. on Mathematical Foundations of Computer Science*, Lecture Notes in Computer Science, Vol. 88 (Springer, Berlin, 1980) 656–664.

[164] VARDI, M.Y., Automatic verification of probabilistic concurrent finite-state programs, in: *Proc. 26th Ann. IEEE Symp. on Foundations of Computer Science* (1985) 327–338.

[165] VARDI, M.Y., The taming of the converse: reasoning about two-way computations, in R. Parikh, ed., *Proc, Workshop on Logics of Programs*, Lecture Notes in Computer Science, Vol. 193 (Springer, Berlin, 1985) 413–424.

[166] VARDI, M.Y., Verification of concurrent programs: the automata-theoretic framework, in: *Proc. IEEE Symp. on Logic in Computer Science* (1987) 167–176.

[167] VARDI, M. and L. STOCKMEYER, Improved upper ad lower bounds for modal logics of programs: preliminary report, in: *Proc, 17th Ann. ACM Symp. on Thery of Computing* (1985) 240–251.

[168] VARDI, M.Y. and P. WOLPER, Yet another process logic, in: E. Clarke and D. Kozen, eds., *Proc, Workshop on Logics of Programs*, Lecture Notes in Computer Science, Vol. 164 (Springer, Berlin, 1983) 501–512.

[169] VARDI, M.Y. and P. WOLPER, Automata-theoretic techniques for modal logics of programs, in: *Proc. 14th Ann. ACM Symp. on Theory of Computing,* (1984) 446–455.

[170] VARDI, M.Y. and P. WOLPER, An automata-theoretic approach to automatic program verification, in: *Proc. IEEE Symp. on Logic in Computer Science* (1986) 332–344.

[171] WAND, M., A new incompleteness result for Hoare's system, *J. ACM* **25** (1978) 168–175.

[172] IMMERMAN, N., Nondeterministic space is closed under complementation, in: *Proc. 3rd IEEE Structure in Complexity Theory Conf.* (1988) 112–115.

CHAPTER 15

Methods and Logics for Proving Programs

Patrick COUSOT

Ecole Polytechnique LIX, 91128 Palaiseau Cedex, France

Contents

1. Introduction 843
2. Logical, set- and order-theoretic notations 849
3. Syntax and semantics of the programming language 851
4. Partial correctness of a command 858
5. Floyd–Naur partial correctness proof method and some equivalent variants . . 859
6. Liveness proof methods 875
7. Hoare logic 883
8. Complements to Hoare logic 931
References 978

HANDBOOK OF THEORETICAL COMPUTER SCIENCE
Edited by J. van Leeuwen

1. Introduction

Formalizing ideas of Floyd [143] and Naur [319] which, in essence, were already present in [175] and [394] (as recalled in [311]), C.A.R. Hoare introduced in October 1969 an axiomatic method for proving that a program is partially correct with respect to a specification ([208], see the genesis and reprint of this paper in [227, pp. 45–58]). This paper introduced or revealed a number of ideas which originated an evolution of programming from arts and crafts to a science. Hoare logic had a very significant impact on program verification and design methods. It was an essential step in the emergence of "structured programming" in the 1970s. It is also an important contribution to the development of formal semantics of programming languages. Understanding that programs can be a subject of mathematical investigations was also crucial in the development of a theory of programming. This is reflected in the fact that reference [208] is one of the most widely cited papers in computing science (see the bibliography of more than 400 references).

1.1. A brief review of Hoare's seminal paper [208]

The *introduction* of [208] claims that "computer programming is an exact science" and calls for the development of a formal system for reasoning about programs.

The first part of [208] is an attempt to axiomatize *computer arithmetic* in two stages: first axioms are given for arithmetic operations on natural numbers which are valid independently of their computer representation (such as "$x+y=y+x$", "$x+0=x$", etc.) and then choices of supplementary axioms are proposed for characterizing various possible implementations. For example, a finite representation of natural numbers ("$\forall x. 0 \leqslant x \leqslant \text{maxint}$") can lead to various possible interpretations of overflow such as: "$\neg\exists x. (x=\text{maxint}+1)$", i.e. program execution should be stopped; "$\text{maxint}+1=\text{maxint}$", i.e. the result of an overflowing operation should be taken as the maximum value represented; or "$\text{maxint}+1=0$", i.e. arithmetic operations should be computed modulo this maximum value "maxint".

The second part of [208] introduces an axiomatic definition of *program execution*. It defines Hoare correctness formulae "$\{P\}\,C\,\{Q\}$" where C is a program, and the precondition P and postcondition Q are logical formulae describing properties of data manipulated by program C. Such a formula "$\{P\}\,C\,\{Q\}$" means that if execution of C is started in any memory state satisfying precondition P and if this execution does terminate, then postcondition Q will be true upon completion (Hoare correctness formulae were originally written "$P\,\{C\}\,Q$" but are now written as "$\{P\}\,C\,\{Q\}$" to emphasize the rôle of assertions P and Q as comments). The main contribution of [208] is the elucidation of a set of axioms and rules of inference which can be used in correctness proofs (termed "partial" since termination is not involved). For example, if X is a variable identifier, E is an expression, and $P[X \leftarrow E]$ is obtained from P by substituting E for all occurrences of X, then "$\{P[X \leftarrow E]\}\; X := E\,\{P\}$" is an axiom for the assignment command "$X := E$". Intuitively, it simply states that what is true of expression E before the assignment is true of X after assignment of E to X. Another example is the sequential composition "$C_1;C_2$" of commands C_1 and C_2. From

"$\{P\}\ C_1\ \{Q\}$" and "$\{Q\}\ C_2\ \{R\}$", one can infer "$\{P\}\ C_1;C_2\ \{R\}$". Intuitively, this rule of inference states that if Q is true after execution of C_1 starting with P true and if R is true after execution of C_2 starting with Q true, then R is true after the sequential execution of C_1 and C_2 starting with P true. This second part of [208] ends with the formal partial correctness proof of a program for computing the Euclidian division of nonnegative integers by successive subtractions.

The third part of [208] is on *general reservations* about this paper. First side-effects are excluded. Then, and most importantly, termination is not considered. Finally, a number of language features (such as procedures and parallelism) are omitted.

Then Hoare [208] discusses *proofs of program correctness*. An axiomatic approach is indispensable for achieving program reliability. The usefulness of program proving is advocated in view of the cost of programming errors and program testing. It is also useful for program documentation and modification and to achieve portability (the machine-dependent part being clearly identified by the use of implementation-dependent axioms). Difficulties such as unreliable specifications or proof complexity are also foreseen: "program proving, certainly at present, will be difficult even for programmers of high calibre; and may be applicable only to quite simple program designs".

Finally, Hoare [208] discusses *formal language definition*. The axioms and rules of inference can be understood as "the ultimate definitive specification of the meaning of the language". The approach is simple. It can cope with the problem of machine dependence by leaving certain aspects of the language undefined and serve as a guide for language design.

Hoare's [208] last words in the *acknowledgements* anticipated the enormous work on Hoare logic: "The formal material presented here has only an expository status and represents only a minute proportion of what remains to be done. It is hoped that many of the fascinating problems involved will be taken up by others".

1.2. Further work by C.A.R. Hoare on Hoare logic

A number of these problems were treated by C.A.R. Hoare himself. A famous and nontrivial proof, that of the program "Find" [207], was later given in [209]. This paper shows a significant move in the use of Hoare logic from a program verification method (i.e. an a posteriori proof of a complete program) to a program design method: "A systematic technique is described for constructing the program proof during the process of coding it". The use of data abstractions to handle complex data structures was introduced in [211] and exemplified in [212]. Other programming language features were covered later: procedures and parameters [210] including recursion [147], jumps and functions [78], [30], and parallel programs [213], [215]. Hoare logic had an important influence on the design of modern programming languages, Pascal notably [405], [229]. However progress was sometimes slower than anticipated by Hoare in [208]. In the case of parallelism for example, synchronization [214] and communication primitives [217], [221] had to be better understood before introducing formal proof methods [218], [407].

1.3. *Further work by C.A.R. Hoare on methods of reasoning about programs*

With regard to program proofs, Hoare's [208] suffers from a number of weaknesses, some of which were corrected later [224, 225, 226, 223]. First of all, termination is not involved. For example "$\{\text{true}\}$ **while** true **do** skip $\{Q\}$" holds for all assertions Q. This can be considered as a regrettable omission (see the "Envoi" to [227, p. 391]) or as an historically fruitful simplification (since, for example, total correctness was not that simple to understand in the presence of nondeterminism [216]). Later work by C.A.R. Hoare insists upon total correctness [218]. Second, the postcondition Q in Hoare correctness formulae "$\{P\}\,C\,\{Q\}$" is a predicate of the final states alone. Adopting postconditions which are predicates of initial and final states makes it easier to specify relations (as later in [224, 225, 226, 223]). To achieve the same effect in Hoare logic, one has to use so-called logical auxiliary variables such as x in "$\{X=x\}\,C\,\{X=x\}$" to express that execution of C leaves the value of the programming variable X unchanged (assuming that x does not appear in C). Third, Hoare correctness formulae "$\{P\}\,C\,\{Q\}$" do not describe the course of computation of the program C, a flaw in the presence of parallelism. For example, "$\{X=x\}\,C\,\{X=x\}$" does not mean that X is not modified during execution of C, so that we do not know whether "$\{X=x\}\,[C \parallel X := 1]\,\{X=1\}$" holds or not when command C is executed in parallel with assignment "$X := 1$". Sequences of intermediate states (more precisely, messages) were later used in [219]. Fourth, Hoare logic imposes a standard form of reasoning about programs which, for example, lacks the flexibility of choosing between positive or contrapositive arguments or of choosing the most adequate form of induction (the only one allowed being structural induction upon the syntactical structure of programs). Fifth, in the logical treatment of the Hoare correctness formulae "$\{P\}\,C\,\{Q\}$", predicates P, Q and programs C are quite separated. For example, predicates can be conjuncted but Hoare correctness formulae cannot. The use of names for designating objects obeys quite different conventions in predicates and programs. Finally, Hoare logic was a very important step in the understanding that "computers are mathematical machines", that "computer programs are mathematical expressions", that "a programming language is a mathematical theory", and that "programming is a mathematical activity" [220]. However, the main emphasis of Hoare in [208] is on the logical and formal aspects of this mathematical activity. It obliterates the informal and often more elegant mathematical proofs which should also have had their useful counterparts in programming. This point of view is similar to that of a mathematician adhering to rigorous and well-understood proof methods without using exclusively formal mathematical logic.

With regard to program semantics, Hoare's [208] was somewhat optimistic: the axiomatic semantics can no longer be considered as universal but as one of the complementary definitions of programming languages [228].

1.4. *Surveys on Hoare logic*

In the 70s and 80s, Hoare logic has been considerably studied. Most results have been reported in numerous surveys [267, 13, 330, 17, 74, 15, 120, 20, 230] and books [284, 109, 275], to cite a few. This chapter is an elementary but rigorous introduction to Hoare logic. We just assume some elementary familiarity with naïve logical and set-theoretical notations, and some very rudimentary practice with the so-called intermediate assertions method of Floyd; nevertheless we shall give the necessary recalls. We hope that this survey will be useful to readers willing to understand the abundant and often very technical literature on Hoare logic. Numerous references are suggested for further study. We apologize to all those researchers who have been misunderstood or not referenced.

1.5. *Summary*

In Section 2 we fix the logical, set- and order-theoretic notations which are used subsequently.

In Section 3 we define the syntax of while-programs, their operational semantics (that is, a set of finite or infinite sequences of configurations representing successive observable states of an abstract machine executing the program for each possible input data), and their relational semantics (directly defining the relation between initial and final configurations of terminating program executions).

In Section 4 we define the partial correctness "$\underline{\{}p\underline{\}}\,\underline{C}\,\underline{\{}q\underline{\}}$" of a command C with respect to a precondition p and postcondition on q to mean that any terminating execution of C starting from an initial state s satisfying p must end in some final state s' satisfying q.

Section 5 is a presentation of the Floyd–Naur partial correctness proof method. After giving a simple introductory example, we formally derive Floyd–Naur's method from the operational semantics using an elementary stepwise induction principle and predicates attached to program points to express invariant properties of programs. This systematic construction of the verification conditions ensures that the method is semantically sound (i.e. correct) and complete (i.e. always applicable). Then we explain a compositional presentation of Floyd–Naur's method inspired by Hoare logic where proofs are given by induction on the syntactical structure of programs. These two approaches are shown to be equivalent in the strong sense that, up to a difference of presentation, they require verification of exactly the same conditions. Some other partial correctness proof methods are shortly reviewed and shown to be also variants of the basic Floyd–Naur method.

In Section 6, we study total correctness (that is, the conjunction of partial correctness, absence of runtime errors, deadlock freedom and termination) and, more generally, liveness proof methods. After giving a short mathematical recall on well-founded relations, well-orderings and ordinals, we review Floyd's well-founded set method and Burstall's intermittent assertion method.

Section 7 is devoted to Hoare logic for while-programs.

In Subsection 7.1, we first start from a semantical point of view (mathematicians would say a naïve set-theoretic approach) relative to a fixed interpretation specified by the operational semantics. Hoare logic is therefore understood as a set of general theorems for proving partial correctness by structural induction on the syntax of commands. This approach is sound (correct) and complete (always usable) with respect to the operational semantics. In practice, proofs can be presented informally using Owicki's proof outlines (that is, by attaching comments to program points) which is equivalent to Flody's method.

In Subsection 7.2, we study Hoare logic from a syntactical point of view (mathematicians would call it a logical approach) where predicates are required to be machine-representable and proofs to be machine-checkable (but not necessarily machine-derivable). Therefore predicates P, Q and correctness formulae "$\{P\}\ C\ \{Q\}$" are now considered as strings of symbols written according to a precise syntax without fixed meaning. Provability is formalized by rewriting rules for deriving valid theorems by successive transformations of given axioms. More precisely, we first define the syntax of first-order predicates P, Q (allowing only to quantify over elements, but not over subsets or functions) and correctness formulae "$\{P\}\ C\ \{Q\}$". Then we introduce Hilbert-like deductive systems (consisting of axioms such as "$\{P[X \leftarrow E]\}\ X := E\{P\}$" and rules of inference such as "from $\{P\}\ C_1\ \{Q\}$ and $\{Q\}\ C_2\ \{R\}$ infer $\{P\}\ C_1; C_2\ \{Q\}$"), and define formal proofs (i.e., finite sequences of formulae, each of which is either an axiom or else follows from earlier formulae by a rule of inference). This leads to Hoare's classical proof system **H** and to Hoare's proof system **H**′ for proof outlines. Finally, we define free and bound variables in predicates so as to precisely specify the syntactical rules of substitution $P[X \leftarrow E]$.

Subsection 7.3 makes the link between the semantical point of view (i.e., the usual mathematical reasoning about *truth* with respect to the interpretation of programs specified by the operational semantics) and the syntactical point of view (where *provability* is understood as manipulating correctness formulae according to the formal rules of Hoare deductive system **H**). For that purpose we define the semantics of predicates P, Q and correctness formulae "$\{P\}\ C\{Q\}$" in accordance with the relational semantics but parametrized by the meaning of basic symbols $+, *, <, \ldots$ which can be left unspecified.

Subsection 7.4 studies the link between truth (the semantical point of view) and provability (the syntactical point of view). Kurt Gödel showed that truth and provability do not necessarily coincide: provable implies true, refutable implies false but some formulae may be undecidable, that is, neither provable nor refutable (using proofs that can be checked mechanically) although they are either true or false. Therefore the question is whether Hoare's formal proof system captures the true partial correctness formulae, only these (soundness) and ideally all of these (completeness).

Soundness is proved in Subsection 7.4.1.

Completeness and incompleteness issues of Hoare logic are discussed in Subsection 7.4.2. It is shown that the formalism of while-programs is more powerful than the formalism of first-order predicates in the sense that, for example, the reflexive transitive closure of a given basic relation is easily defined by a program whereas it is not (in general) first-order definable (except when full arithmetic on natural numbers is available). Incompleteness results for Hoare logic follow since intermediate assertions needed in the correctness proofs of while-loops cannot in general be expressed. The consequence is that, although the logical language may be simple enough for all true facts about data expressible in this language to be machine-derivable, some programs using the same basic symbols may be either unspecifiable or unprovable within this restricted language. It follows that the logical language must be enriched up to the point where arithmetic is included. Hence the logical language can be used to describe its own deductive system; but then, as shown by Gödel, the deductive system is not powerful enough to prove all true facts expressible in the logical language. It follows that we can only prove Cook's relative completeness of Hoare logic assuming that the logical language is expressive enough to specify the intermediate assertions and that all needed mathematical facts about the data of the program are given. (The alternative, which consists in using richer second-order logical systems, is not considered since proofs would then not even be machine-checkable.) Expressiveness can be defined à la Clarke (using Dijkstra's weakest liberal preconditions) or à la Cook (using strongest liberal postconditions), both notions of expressiveness being equivalent. Examples of inexpressive (abacus arithmetic) and expressive (Peano arithmetic) first-order languages are given. It is shown that expressiveness is sufficient to obtain relative completeness but is not necessary. Finally, we study Clark's characterization problem which consists in characterizing which programming languages have a sound and relatively complete Hoare logic. This is not the case with Algol-like or Pascal-like languages (since soundness and relative completeness of Hoare logic would imply the decidability of the halting problem for finite interpretations where variables can only take a finite number of values). The intuitive reason is again that first-order logical languages are less powerful than Algol- or Pascal-like languages which can manipulate a potentially infinite runtime stack.

In Section 8, we briefly consider complements to Hoare logic: data structures (Subsection 8.1), undefinedness due to runtime errors (Subsection 8.3), aliasing and side-effects (Subsection 8.4), block-structured local variables (Subsection 8.5), goto statements (Subsection 8.6), functions and expressions with side-effects (Subsection 8.7), coroutines (Subsection 8.8) and a guide to the literature on examples of program verification (Subsection 8.11) and other logics extending first-order logic with programs (Subsection 8.12). Three topics are treated more extensively in Section 8.

In Subsection 8.2 we consider procedures. First we define the syntax and relational semantics of recursive parameterless procedures. Then we consider partial correctness proofs based upon computation induction (a generalization of Scott induction) which leads to Hoare's recursion rule. Since this one rule is not complete, we consider Park's fixpoint induction, which, using auxiliary variables, can be indirectly transcribed into Hoare's rule of adaptation. Then these rules are generalized for value–result parameters and numerous examples of applications are provided. References to the literature are given for variable parameters and procedures as parameters.

Parallel programs are handled in Subsection 8.9. First we define the syntax and operational semantics of parallel programs with shared variables and await commands. We review a number of proof methods à la Floyd for such parallel programs, using the unifying point of view of Section 5 where proof methods were shown to derive from a single induction principle and differ only in the way of decomposing into local invariants the global invariant on memory states and control states (or auxiliary variables). In particular, we study in detail the Lamport and Owicki–Gries method, as well as various strengthened or weakened versions, its stepwise (à la Floyd) and syntax-directed (à la Hoare) presentations. This proof method is formalized by

Owicki–Gries logic, and a compositional version is given as introduced by Stirling following Jones. We end with a guide to the literature on Hoare logics for communicating sequential processes.

In Subsection 8.10 we consider proof systems for total correctness. Such proof systems cannot be of pure first-order logical character and must incorporate an external well-founded relation. We first consider the Harel proof rule for while-programs with finitely bounded nondeterminism and its arithmetical completeness. Then we explain the use of transfinite ordinals to deal with unbounded nondeterminism, a situation where a program state may have infinitely many possible successor states. This leads to the total correctness of fair parallel programs. Finally we introduce Dijkstra's weakest preconditions calculus for unbounded nondeterminism.

Numerous references to the literature are given in the last section.

1.6 Hints for reading this survey

This chapter can be read in a number of ways. In all cases, proofs should be omitted on first reading. Here are two examples of nonsequential readings:

(a) Readers purely interested in Hoare logic can read as follows: Section 2 (logical, set- and order-theoretic notations), Section 3 (syntax and semantics of the programming language), Section 4 (partial correctness of a command), Subsection 5.1 (an example of partial correctness proof by the Floyd–Naur method), Subsection 5.3 (a Hoare-style presentation of the Floyd–Naur partial correctness proof method by syntax-directed induction), Subsections 7.1.1 (general theorems for proof construction), 7.2.1 (syntax of predicates and correctness formulae), 7.2.2 (deductive systems and formal proofs), 7.2.3 (Hoare's proof system **H**), and 7.2.5 (syntactical rules of substitution). Then the various complements on Hoare logic of Section 8 can be read in any order; readers more interested in the soundness and completeness problems can go on by Subsections 7.1.2 (semantical soundness and completeness), 7.3 (the semantics of Hoare logic), and 7.4 (the link between syntax and semantics: soundness and completeness issues in Hoare logic).

(b) Readers interested in liveness proof methods but willing to ignore logic can read successively Sections 2 (logical, set- and order-theoretic notations), 3 (syntax and semantics of the programming language), 4 (partial correctness of a command), 5 (Floyd–Naur partial correctness proof method and some equivalent variants), 6 (liveness proof methods), 8.9.1 (operational semantics of parallel programs with shared variables), 8.9.2 (proof methods à la Floyd for parallel programs with shared variables), and 8.10.4 (Dijkstra's weakest preconditions calculus).

2. Logical, set- and order-theoretic notations

For terms of logical character we use the following notations: tt denotes *truth*, ff *falsity*, $\neg$ *negation*, $\wedge$ *conjunction*, $\vee$ *inclusive disjunction*, $\Rightarrow$ *logical implication*, $=$ (or $\Leftrightarrow$) *logical equivalence*, $\forall v.\, p$ *universal quantification* (p is true for any v), $\exists v.\, p$ *existential quantification* (there are some v such that p is true), $\exists! v.\, p$ *unique existential quantification* (there is a unique v such that p is true). $\forall v \in E.p$ is an abbreviation for $\forall v.((v \in E) \Rightarrow p)$ and $\exists v \in E.p$ is $\exists v.((v \in E) \wedge p)$. $\forall v_1, \ldots, v_n \in E.p$ is an abbreviation for $\forall v_1 \in E. \ldots \forall v_n \in E.p$. In the same way, $\exists v_1 \ldots, v_n \in E.p$ is an abbreviation for $\exists v_1 \in E. \ldots \exists v_n \in E.p$. We write $(B \rightarrow X \diamond Y)$ to denote X when B is true and Y otherwise.

$\mathbf{Z}$ (respectively $\mathbf{N}, \mathbf{N}^+$) is the set of integers (positive, strictly positive integers). As usual in computer science, $+$ is the addition, $-$ the subtraction, $*$ the product, *div* the division, *mod* the modulus and $**$ the exponentiation of integers. *odd*(x) (respectively *even*(x)) is true if and only if x is an odd (respectively even) integer.

We accept the intuitive concept of a *set* as a collection of objects called *elements* of the set. The notation $e \in E$ means that e is an element of the set E and $e \notin E = \neg(e \in E)$. The empty set is denoted $\emptyset$. $E \subset E'$, $E \subseteq E'$ and $E = E'$ respectively denote *proper inclusion*, *inclusion* and *equality* of sets, and $(E \supset E') = (E' \subset E)$, $(E \supseteq E') = (E' \subseteq E)$ and $(E \neq E') = \neg(E = E')$. We use the set-theoretic operations $\cup$ (*union*), $\cap$ (*intersection*) and $-$ (*difference*). If a set D is fixed then for subsets E of D the *complement* $\neg E$ of E is $(D - E)$. If E is a set then $\mathscr{P}(E)$ (called the *power set* of E) denotes the set of all subsets of E. If $E_0, \ldots, E_{n-1}$ are sets, the *Cartesian product* $E_0 \times \cdots \times E_{n-1}$ is the set of n-tuples $\langle e_0, \ldots, e_{n-1} \rangle$, with $e_i \in E_i$ for $i = 0, \ldots, n-1$. In symbols: $E_0 \times \cdots \times E_{n-1} = \{\langle e_0, \ldots, e_{n-1} \rangle : e_0 \in E_0 \wedge \cdots \wedge e_{n-1} \in E_{n-1}\}$ where ":" reads "for those which satisfy". The *projection* $\langle e_0, \ldots, e_{n-1} \rangle_i$ is the ith component e_i of the n-tuple $\langle e_0, \ldots, e_{n-1} \rangle$. If $n > 1$, the *elimination* $\langle e_0, \ldots, e_{n-1} \rangle_{\sim i}$ is the $(n-1)$-tuple $\langle e_0, \ldots, e_{i-1}, e_{i+1}, \ldots, e_{n-1} \rangle$. If $E_0 = \cdots = E_{n-1} = E$ then $E^n = E_0 \times \cdots \times E_{n-1}$. We define E^0 to be $\{\emptyset\}$ and identify E^1 with E. The cardinality of a set E is denoted $|E|$; hence if E is finite, $|E|$ is the number of elements of E.

Given a positive integer n and a set E, we define an *n-ary relation* r on E as a subset of E^n. n is called the arity of r and is denoted $\#r$. We say that r is *binary* when $\#r = 2$ and often use the infix notation $x\,r\,y$ for $\langle x, y \rangle \in r$. For example we write $x \leqslant y$ to mean that x is less than or equal to y.

If $r, r' \subseteq E \times E$ are binary relations on E then $r \circ r' = \{\langle e, e' \rangle : \exists e'' \in E. \langle e, e'' \rangle \in r \wedge \langle e'', e' \rangle \in r'\}$ is the *product* of r and r'; $r^{-1} = \{\langle e', e \rangle : \langle e, e' \rangle \in r\}$ is the *inverse* of r. If, moreover, $p \subseteq E$ is a subset of E then $p \rceil r = \{\langle e, e' \rangle \in r : e \in p\}$ is the *left restriction* of r to p and $r \lceil p = \{\langle e, e' \rangle \in r : e' \in p\}$ is the *right restriction* of r to p. The equality relation on E, also called the *diagonal* of E^2, is $\delta = \{\langle e, e \rangle : e \in E\}$.

The *power* of a binary relation r on a set E is defined by recurrence as $r^0 = \delta$, $r^{n+1} = r^n \circ r$ for $n \geqslant 0$. The (*strict*) *transitive closure* r^+ of r is $r^+ = \bigcup\{r^n : n > 0\}$ and the *reflexive transitive closure* r^* of r is $r^* = r^0 \cup r^+$.

A *partially ordered set* is a pair $\langle E, \leqslant \rangle$ where E is a nonempty set and $\leqslant$ is a binary relation on E which is *reflexive* ($\forall a \in E. a \leqslant a$), *antisymmetric* ($\forall a, b \in E. (a \leqslant b \wedge b \leqslant a) \Rightarrow (a = b)$) and transitive ($\forall a, b, c \in E. (a \leqslant b \wedge b \leqslant c) \Rightarrow (a \leqslant c)$). Given $P \subseteq E$, $a \in E$ is an *upper bound* of P if $\forall b \in P. b \leqslant a$; a is the *least upper bound* of P (in symbols *lub* P) if a is an upper bound of P and if b is any upper bound of P then $a \leqslant b$. If *lub* P exists, then it is unique. *Lower bounds* and the *glb* (*greatest lower bound*) are defined dually, that is, by replacing $\leqslant$ by its inverse (also called *dual*) $\geqslant$ defined by $(a \geqslant b) = (b \leqslant a)$. A *complete lattice* is a partially ordered set $\langle E, \leqslant \rangle$ such that *lub* P and *glb* P exist for all P, $P \subseteq E$. It follows that E has a greatest element or *supremum* $\top = lub\ E = glb\ \emptyset$ and a least element or *infimum* $\bot = glb\ E = lub\ \emptyset$. $\langle \mathscr{P}(E), \subseteq \rangle$ is a complete lattice such that $lub = \bigcup$, $glb = \bigcap$, $\top = E$ and $\top = \emptyset$.

Given two sets A and B and a binary relation φ on $A \cup B$, we call φ a *function* of A into B (and write $\varphi: A \to B$) if $(\langle a, b\rangle \in \varphi) \Rightarrow (a \in A \wedge b \in B) \wedge (\forall a \in A.\, \exists! b \in B.\, \langle a, b\rangle \in \varphi)$. $A = dom\, \varphi$ is called the *domain* and $B = rng\, \varphi$ is the *range* of φ. We use the functional notation $b = \varphi(a)$ instead of $\langle a, b\rangle \in \varphi$. The set of all functions of A into B will be denoted B^A or more frequently $A \to B$.

We specify a function $\varphi: A \to B$ without giving it a name using the *lambda notation* "$\lambda x: A \to B.\, e$" where the expression e is such that $\forall a \in A.\, \varphi(a) = e(a)$. We write $\lambda v \in A.\, e$ (respectively $\lambda x.\, e$) when B (respectively A and B) can be easily inferred from e. If $\varphi: A \to B$ then $\varphi[a \leftarrow b]$ is the function $\varphi': A \cup \{a\} \to B \cup \{b\}$ such that $\varphi'(a) = b$ and $\forall a' \in A.((a' \neq a) \Rightarrow (\phi'(a') = \varphi(a')))$. A function $\varphi: \{a_1, \ldots, a_n\} \to \{b_1, \ldots, b_n\}$ such that $\varphi(a_i) = b_i$ for $i = 1, \ldots, n$ will be simply written $[a_1 \leftarrow b_1, \ldots, a_n \leftarrow b_n]$ so that $[a_1 \leftarrow b_1, \ldots, a_n \leftarrow b_n](a_i) = b_i$.

A *family* $\gamma = \langle \gamma_i: i \in I\rangle$ of elements of E is a function φ from the *index set* I into the set E, where $\gamma_i = \varphi(i)$. When $I \subseteq \mathbf{N}$, γ is called a *sequence* of elements of E. More precisely, if $I = \emptyset$ then γ is called an *empty sequence* and is written ε. When $I = \mathbf{N}$, it is an *infinite sequence* and we write $\gamma = \gamma_0, \ldots, \gamma_i, \ldots$ When $n \in \mathbf{N}$ and $I = \{i: 0 \leqslant i < n\}$, γ is called a *finite sequence* of *length n* and we write $\gamma = \gamma_0, \ldots, \gamma_{n-1}$. We define $seq^n\, E$ to be the set of sequences of elements of E of length $n \geqslant 0$, $seq^*\, E = \bigcup_{n \geqslant 0} seq^n\, E$, $seq^\omega\, E$ to be the set of infinite sequences of elements of E and $seq\, E = seq^*\, E \cup seq^\omega\, E$. Moreover, the *concatenation* γe *of* $e \in E$ *to the right of* γ is defined by $\varepsilon e = e$, $\gamma e = \gamma_0, \ldots, \gamma_{n-1}, e$ if γ is a finite sequence of length n and $\gamma e = \gamma$ if γ is an infinite sequence.

Let $\langle A, \leqslant\rangle$ and $\langle B, \preccurlyeq\rangle$ be partially ordered sets. $\varphi: A \to B$ is *monotone* (or *increasing*) if $\forall a, b \in E.(a \leqslant b) \Rightarrow (\varphi(a) \preccurlyeq \varphi(b))$. $x \in A$ is a *fixpoint* of $\varphi: A \to A$ if $\varphi(x) = x$. It is a *prefixpoint* if $x \leqslant \varphi(x)$ and a *postfixpoint* if $\varphi(x) \leqslant x$. Let $\langle L, \leqslant\rangle$ be a complete lattice with infimum $\perp$ and $\varphi: L \to L$ be monotone. The set $\{x \in L: \varphi(x) = x\}$ of fixpoints of φ is a (nonempty) complete lattice for $\leqslant$ with infimum $lfp\, \varphi = glb\{x \in L: \varphi(x) \leqslant x\}$ and supremum $gfp\, \varphi = lub\{x \in L: x \leqslant \varphi(x)\}$ [385]. An *increasing chain* is a sequence γ of elements of A such that $\forall i \in dom\, \gamma - \{0\}.\, \gamma_{i-1} \leqslant \gamma_i$. $\varphi: L \to L$ is *upper-continuous* if $lub\{\varphi(\gamma_i): i \in dom\, \gamma\} = \varphi(lub\{\gamma_i: i \in dom\, \gamma\})$ for any increasing chain γ of L. If φ is upper-continuous then $lfp\, \varphi = lub\{\varphi^n(\perp): n \geqslant 0\}$ where $\forall x \in L.\, \varphi^0(x) = x$ and $\forall n \in \mathbf{N}.\, \forall x \in L.\, \varphi^{n+1}(x) = \varphi(\varphi^n(x))$ [252]. In particular, if r is a binary relation r on a set E then r^* is uniquely determined by the facts that $r^* = \delta \cup r \circ r^*$ (or $r^* = \delta \cup r^* \circ r$) and if $x \subseteq E \times E$ is such that $x = \delta \cup r \circ x$ (or $x = \delta \cup x \circ r$) then $r^* \subseteq x$; that is, $r^* = lfp\, \lambda x: E^2 \to E^2.\, \delta \cup r \circ x$ (or $r^* = lfp\, \lambda x: E^2 \to E^2.\, \delta \cup x \circ r$). If $\langle L, \leqslant, \perp\rangle$ is a complete lattice, $\langle P, \preccurlyeq\rangle$ is a poset with infimum $\perp'$, $\alpha: L \to P$ is strict ($\alpha(\perp) = \perp'$), upper-continuous and $\alpha \circ \varphi = \psi \circ \alpha$, and, moreover, $\varphi: L \to L$ and $\psi: P \to P$ are monotone then $\alpha(lfp\, \varphi) = lfp\, \psi$.

3. Syntax and semantics of the programming language

Since Hoare logic is closely bounded to a programming language, we introduce a very simple Pascal-like [405] language. We first define its *syntax*, that, is the set of well-formed programs. We use an *abstract syntax* describing the structure of programs by trees [298] and leave unspecified the *concrete syntax* (where programs are linear strings of tokens to be parsed unambiguously [3]). Then we define the *operational semantics* of the language, that is, the effect of executing syntactically correct programs. Traditionally, one imagines the program running on an abstract machine with primitive instructions [324]. Its execution steps can be idealized by mapping abstract syntactic constructs in the program to a transition relation on configurations [247]. The language is nondeterministic since variables can be assigned random values so that a configuration may have several possible successors. An execution of the program is a finite or infinite sequence of configurations representing successive observable states of the machine during execution. When observation of intermediate configurations is unnecessary, we use a *relational semantics* directly defining the relationship between initial and final configurations of program executions [228]. This relational semantics is later used as a basis for justifying Hoare's logic [208]. This *axiomatic semantics* associates an axiom or a rule of inference with each kind of basic or structured statement of the language which states what we may assert after execution of that statement in terms of what was true beforehand.

Such complementary definitions of programming language semantics [228, 131, 27] are useful for describing the semantics at various levels of details. This is a natural extension of the original idea in [143, 208] that "the specification of proof techniques provides an adequate formal definition of a programming language" where "an" had to be replaced by "one of" because not all methods have the same power of expression.

3.1. Syntax

Basic commands of our programming language are the null command "skip", and the assignment "$X:=E$" of the value of an expression E to a programming variable X or the nondeterministic assignment "$X:=?$" of a random value to X. Commands C_1, C_2 can be composed sequentially "$(C_1;C_2)$" and conditionally "$(B \to C_1 \diamond C_2)$" according to the value of a Boolean expression B. A command C can also be iterated "$(B * C)$" while B holds.

This *abstract syntax* is more formally defined below. For the time being, the sets **Pvar** of programming variables (ranged over by X), **Expr** of expression (ranged over by E) and **Bexp** of Boolean expressions (ranged over by B) are assumed to be given. **Expr** and **Bexp** will be detailed later. The set **Com** of commands (ranged over by C) is the smallest set closed under the given formation rule (similar to a context-free grammar expressed in Backus–Naur Form (BNF, [320]):

(1) DEFINITION (*syntax of commands*)

X: **Pvar** *programming variables,*

E: **Expr** *expressions,*

B: **Bexp** *Boolean expressions,*

C: **Com** *commands*

$$C ::= \text{skip} \mid X:=E \mid X:=? \mid (C_1;C_2) \mid (B \to C_1 \diamond C_2) \mid (B * C).$$

When necessary we omit parentheses and use a Pascal-like *concrete syntax.*

(2) EXAMPLE (*a program to compute* $x ** y$). The program with concrete syntax

```
        Z:=1;
        while Y<>0 do
          if odd (Y) then
(3)         begin Y:=Y-1; Z:=Z*X end,
          else
            begin Y:=Y div 2; X:=X*X end;
```

has the following abstract syntax:

$$(Z{:=}1;(Y{<>}0*(odd(Y)\rightarrow(Y{:=}Y-1;Z{:=}Z*X)\diamond(Y{:=}Y\,div\,2;\ X{:=}X*X)))). \tag{4}$$

The components of a command C are C itself and the commands appearing in C together with their components. Two instances of the same command C' in a command C should be considered as different components of C. Therefore we mark components $C' \in \mathbf{Comp}[\![C]\!]$ of a command C by the Dewey number designating the root of the subtree of C' in the syntactic tree of C. For example, $\mathbf{Comp}[\![((X:=1;X:=1);X:=1)]\!]=\{((X:=1;\ X:=1);\ X:=1)^{\varepsilon},\ (X:=1;\ X:=1)^{0},\ X:=1^{00},\ X:=1^{01},\ X:=1^{1}\}$ so that "$X:=1^{00}$" is the first, "$X:=1^{01}$" the second and "$X:=1^{1}$" the third instance of assignment command "$X:=1$" in the sequence "$((X:=1;\ X:=1);\ X:=1)$". More formally:

(5) DEFINITION (*components of a command*)

(5.1) $\mathbf{Comp}^{\eta}[\![C]\!]=\{C^{\eta}\}$ if C is skip, $X{:=}E$ or $X{:=}?$,

(5.2) $\mathbf{Comp}^{\eta}[\![(C_1;C_2)]\!]=\{(C_1;C_2)^{\eta}\}\cup\mathbf{Comp}^{\eta0}[\![C_1]\!]\cup\mathbf{Comp}^{\eta1}[\![C_2]\!]$,

(5.3) $\mathbf{Comp}^{\eta}[(B\rightarrow C_1\diamond C_2)]=\{(B\rightarrow C_1\diamond C_2)^{\eta}\}\cup\mathbf{Comp}^{\eta0}[\![C_1]\!]\cup\mathbf{Comp}^{\eta1}[\![C_2]\!]$,

(5.4) $\mathbf{Comp}^{\eta}[\![(B*C)]\!]=\{(B*C)^{\eta}\}\cup\mathbf{Comp}^{\eta0}[\![C]\!]$

so that

(5.5) $\mathbf{Comp}[\![C]\!]=\mathbf{Comp}^{\varepsilon}[\![C]\!]$,

(5.6) $\mathbf{Loops}[\![C]\!]=\mathbf{Comp}[\![C]\!]\cap\{(B*C')^{\eta}:\ B\in\mathbf{Bexp}\wedge C'\in\mathbf{Com}\wedge\eta\in seq^{*}\{0,1\}\}$,

(5.7) $\mathbf{Comp}=\{\mathbf{Comp}[\![C]\!]:\ C\in\mathbf{Com}\}$,

(5.8) $\mathbf{Loops}=\{\mathbf{Loops}[\![C]\!]:\ C\in\mathbf{Com}\}$.

(6) EXAMPLE (*components of program* (4))

$(Z{:=}1;\ (Y{<>}0\ *\ (odd(Y)\ \rightarrow\ (Y{:=}Y-1;\ Z{:=}Z*X)\ \diamond\ (Y{:=}Y\ div\ 2;\ X{:=}X*X))))$

C^{1000} ($Y{:=}Y-1$), C^{1001} ($Z{:=}Z*X$), C^{1010} ($Y{:=}Y\ div\ 2$), C^{1011} ($X{:=}X*X$)

C^{100}, C^{101}

C^{10}

C^{0} $\quad$ C^{1}

C

When there is no ambiguity, the exponent η of a component C^η will be omitted and we identify **Comp** with **Com**.

3.2. *Operational semantics*

A *state* (or *valuation*) is a function s with as domain a given set **Var** of variables (including the set **Pvar** of programming variables) and as range a nonempty set D of objects (called *data* in computer science). $s(v)$ is called the *value* of variable v in state s:

(7) $d{:}\ D$ *data*,

(8) $s{:}\ S = \mathbf{Var} \to D$ *states*.

The *operational semantics* of a command C is a model of its execution. Such an execution will be understood as a finite or infinite sequence "$\gamma_0, \ldots, \gamma_n, \ldots$" of *configurations* such that $\langle \gamma_n, \gamma_{n+1} \rangle \in op[\![C]\!]$ and $op[\![C]\!]$ is the operational transition relation. The initial configuration γ_0 has the form $\langle s_0, C \rangle$ where s_0 records the initial values of variables. Configuration γ_n has the form $\langle s_n, C_n \rangle$ where s_n records the current values of variables after execution of n program steps and C_n records the command remaining to be executed. If execution terminates in configuration γ_l then $\gamma_l = s_l$ where s_l records the final values of variables. In particular, $\langle\langle s, C\rangle, \langle s', C'\rangle\rangle \in op[\![C]\!]$ means that one step of execution of C in state s can lead to state s' with C' being the remainder of C to be executed. If $\langle\langle s, C\rangle, s'\rangle \in op[\![C]\!]$ then execution of C in state s can lead to state s' in one step and is then terminated. Hence we have

(9) $\gamma{:}\ \ \Gamma = (S \times \mathbf{Com}) \cup S$ *configurations*,

(10) $op{:}\ \ \mathbf{Com} \to \mathcal{P}(\Gamma \times \Gamma)$ *operational transition relation.*

The definition of the semantics of expressions will be postponed. For the time being, we will assume that the semantics of expressions is given: if $E \in \mathbf{Expr}$ and $s \in S$ then $\underline{E}(s) = I[\![E]\!](s) \in D$ is the value of expression E in state s. In the same way if $B \in \mathbf{Bexp}$ then $I[\![B]\!]$, also written $\underline{B}$, is the set of states s such that B holds in state s:

(11) $I{:}\,\mathbf{Expr} \to (S \to D),\quad \underline{E} = I[\![E]\!]$ *semantics of expressions,*

(12) $I{:}\,\mathbf{Bexp} \to \mathcal{P}(S),\quad \underline{B} = I[\![B]\!]$ *semantics of Boolean expressions.*

The *operational transition relation* is defined by structural induction on the abstract syntax of commands (in the style of [344] but using a direct recursive definition) as follows:

(13) DEFINITION (*operational transition relation*)

(13.1) $op[\![\text{skip}]\!] = \{\langle\langle s, \text{skip}\rangle, s\rangle : s \in S\}$,

(13.2) $op[\![X{:=}E]\!] = \{\langle\langle s, X{:=}E\rangle, s[\![X \leftarrow \underline{E}(s)]\!]\rangle : s \in S\}$,

(13.3) $op[\![X{:=}?]\!] = \{\langle\langle s, X{:=}?\rangle, s[\![X \leftarrow d]\!]\rangle : s \in S \wedge d \in D\}$,

(13.4) $$op[\![C_1;C_2)]\!] = \{\langle\langle s, (C_1';C_2)\rangle, \langle s', (C_1'';C_2)\rangle\rangle: \langle\langle s, C_1'\rangle, \langle s', C_1''\rangle\rangle \in op[\![C_1]\!]\} \cup \{\langle\langle s, (C_1';C_2)\rangle, \langle s', C_2\rangle\rangle: \langle\langle s, C_1'\rangle, s'\rangle \in op[\![C_1]\!]\} \cup op[\![C_2]\!],$$

(13.5) $$op[\![(B\to C_1 \diamond C_2)]\!] = \{\langle\langle s, (B\to C_1\diamond C_2)\rangle, \langle s, C_1\rangle\rangle: s\in \underline{B}\}\cup op[\![C_1]\!] \cup \{\langle\langle s, (B\to C_1\diamond C_2)\rangle, \langle s, C_2\rangle\rangle: s\notin \underline{B}\}\cup op[\![C_2]\!],$$

(13.6) $$op[\![(B * C)]\!] = \{\langle\langle s, (B*C)\rangle, \langle s, (C;(B*C))\rangle\rangle: s\in\underline{B}\} \cup \{\langle\langle s, (C';(B*C))\rangle, \langle s', (C'';(B*C))\rangle\rangle: \langle\langle s, C'\rangle, \langle s', C''\rangle\rangle \in op[\![C]\!]\} \cup \{\langle\langle s, (C'';(B*C))\rangle, \langle s', (B*C)\rangle\rangle: \langle\langle s, C'\rangle, s'\rangle \in op[\![C]\!]\} \cup \{\langle\langle s, (B*C)\rangle, s\rangle: s\notin \underline{B}\}.$$

(14) EXAMPLE (*operational semantics of program* (4)). Using the components (5) of program (4) given in example (2), we can define labels:

(15) $L_1 = C^\varepsilon = C$, $L_2 = C^1$, $L_3 = (C^{10};C^1)$, $L_4 = (C^{100};C^1)$, $L_5 = (C^{1001};C^1)$, $L_6 = (C^{101};C^1)$, $L_7 = (C^{1011};C^1)$, including a final label L_8 represented by the symbol "$\surd$"

so as to name program points as follows:

(16)
$$\begin{array}{llllllll}(Z:=1; & (Y<>0*(odd(Y)\to & (Y:=Y-1; & Z:=Z*X) & \diamond(Y:=Y\ div\ 2; & X:=X*X)))) & & \\ |L_1 & |L_2 & |L_3 & |L_4 & |L_5 & |L_6 & |L_7 & |L_8\end{array}$$

A label L can also be understood as designating the command remaining to be executed when control is at that point L. According to definition (13), the operational transition relation of program (4) is:

(17) $$\begin{aligned} op[\![C]\!] = {} & \{\langle\langle s, L_1\rangle, \langle s[Z\leftarrow 1],\ L_2\rangle\rangle: s\in S\} \\ & \cup\{\langle\langle s, L_2\rangle, \langle s, L_3\rangle\rangle: s(Y)\neq 0\} \\ & \cup\{\langle\langle s, L_3\rangle, \langle s, L_4\rangle\rangle: odd(s(Y))\} \\ & \cup\{\langle\langle s, L_4\rangle, \langle s[Y\leftarrow s(Y)-1], L_5\rangle\rangle: s\in S\} \\ & \cup\{\langle\langle s, L_5\rangle, \langle s[Z\leftarrow s(Z)*s(X)], L_2\rangle\rangle: s\in S\} \\ & \cup\{\langle\langle s, L_3\rangle, \langle s, L_6\rangle\rangle: even(s(Y)\} \\ & \cup\{\langle\langle s, L_6\rangle, \langle s[Y\leftarrow s(Y)\ div\ 2], L_7\rangle\rangle: s\in S\} \\ & \cup\{\langle\langle s, L_7\rangle, \langle s[X\leftarrow s(X)*s(X)], L_2\rangle\rangle: s\in S\} \\ & \cup\{\langle\langle s, L_2\rangle, s\rangle: s(Y)=0\}. \end{aligned}$$

A possible execution sequence of this program is therefore as follows:

$$\langle[X\leftarrow 3, Y\leftarrow 2, Z\leftarrow 0], L_1\rangle, \langle[X\leftarrow 3, Y\leftarrow 2, Z\leftarrow 1], L_2\rangle,$$
$$\langle[X\leftarrow 3, Y\leftarrow 2, Z\leftarrow 1], L_3\rangle, \langle[X\leftarrow 3, Y\leftarrow 2, Z\leftarrow 1], L_6\rangle,$$

$\langle[X\leftarrow 3, Y\leftarrow 1, Z\leftarrow 1], L_7\rangle, \langle[X\leftarrow 9, Y\leftarrow 1, Z\leftarrow 1], L_2\rangle,$
$\langle[X\leftarrow 9, Y\leftarrow 1, Z\leftarrow 1], L_3\rangle, \langle[X\leftarrow 9, Y\leftarrow 1, Z\leftarrow 1], L_4\rangle,$
$\langle[X\leftarrow 9, Y\leftarrow 0, Z\leftarrow 1], L_5\rangle, \langle[X\leftarrow 9, Y\leftarrow 0, Z\leftarrow 9], L_2\rangle,$
$\langle[X\leftarrow 9, Y\leftarrow 0, Z\leftarrow 9]\rangle.$

3.3. *Relational semantics*

The *relational semantics* or interpretation $I[\![C]\!]$ (also noted $\underline{C}$) of a command C is a relation between states such that $\langle s, s'\rangle \in \underline{C}$ if and only if an execution of command C started in initial state s *may* terminate into final state s':

(18) DEFINITION (Hoare & Lauer [228]) (*relational semantics*)

$$I: \mathbf{Com} \to \mathscr{P}(S \times S)$$
$$I[\![C]\!] = \underline{C} = \{\langle s, s'\rangle : \langle\langle s, C\rangle, s'\rangle \in op[\![C]\!]^*\}.$$

The relational semantics can be characterized as follows:

(19) THEOREM (Hoare & Lauer [228], Greif & Meyer [183])

(19.1) $\underline{\text{skip}} = \{\langle s, s\rangle : s \in S\}$,

(19.2) $\underline{X{:=}E} = \{\langle s, s[X \leftarrow \underline{E}(s)]\rangle : s \in S\}$,

(19.3) $\underline{X{:=}?} = \{\langle s, s[X \leftarrow d]\rangle : s \in S \wedge d \in D\}$,

(19.4) $\underline{(C_1;C_2)} = \underline{C_1} \circ \underline{C_2}$,

(19.5) $\underline{(B \to C_1 \diamond C_2)} = (\underline{B}\rceil \underline{C_1}) \cup (\neg \underline{B}\rceil \underline{C_2})$,

(19.6) $\underline{(B * C)} = (\underline{B}\rceil \underline{C})^* \lceil \neg \underline{B}$

(19.7) $\quad = lfp\, \lambda X.(\delta \lceil \neg \underline{B}) \cup ((\underline{B}\rceil \underline{C}) \circ X)$.

(19.8) $\underline{(B * C)}$ *is the unique* $r \subseteq S^2$ *such that*

(19.8a) $r \subseteq (S \times \neg \underline{B})$,

(19.8b) $\forall q \subseteq S.(q\rceil(\underline{B}\rceil\underline{C}) \subseteq S \times q) \Rightarrow (q\rceil r \subseteq S \times q)$,

(19.8c) $((\underline{B}\rceil\underline{C}) \circ r) \subseteq r$,

(19.8d) $(\delta \lceil \neg \underline{B}) \subseteq (\delta \cap r)$.

(20) EXAMPLE (*calculus of the relational semantics of a simple program*). Assume $D = \mathbf{Z}$; the relational semantics of $C = (Y <> 0 * Y := Y - 1)$ is $\underline{C} = \{\langle s, s[Y \leftarrow 0]\rangle : s(Y) \geqslant 0\}$.

To prove this, observe that by (9.7), $\underline{C} = lfp\, F$ where $F(X) = (\delta \lceil \neg \underline{Y <> 0}) \cup ((Y <> 0\rceil \underline{Y{:=}Y{-}1}) \circ X) = \{\langle s, s\rangle : s(Y) = 0\} \cup \{\langle s, s[Y \leftarrow s(Y) - 1]\rangle : s[Y] \neq 0\} \circ X$. Define $X^i = F^i(\emptyset)$. We have $X^0 = \emptyset$, $X^1 = \{\langle s, s[Y \leftarrow 0]\rangle : s(Y) = 0\}$. Assume by induction hypothesis that $X^i = \{\langle s, s[Y \leftarrow 0]\rangle : 0 \leqslant s(Y) \leqslant i - 1\}$. Then $X^{i+1} = F(X^i) = \{\langle s, s\rangle : s(Y) = 0\} \cup \{\langle s, s[Y \leftarrow s(Y) - 1]\rangle : s[Y] \neq 0\} \circ \{\langle s, s[Y \leftarrow 0]\rangle : 0 \leqslant s(Y) \leqslant i - 1\} = \{\langle s, s\rangle : s(Y) = 0\} \cup \{\langle s, s'\rangle : \exists s''. s[Y] \neq 0 \wedge s'' = s[Y \leftarrow s(Y) - 1] \wedge 0 \leqslant s''(Y) \leqslant i - 1 \wedge s' =$

$s''[Y \leftarrow 0]\} = \{\langle s, s\rangle: s(Y)=0\} \cup \{\langle s, s[Y \leftarrow s(Y)-1][Y \leftarrow 0]\rangle: s[Y] \neq 0 \wedge 0 \leqslant s[Y \leftarrow s(Y)-1](Y) \leqslant i-1\} = \{\langle s, s\rangle: s(Y)=0\} \cup \{\langle s, s[Y \leftarrow 0]\rangle: 1 \leqslant s(Y) \leqslant i\} = \{\langle s, s[Y \leftarrow 0]\rangle: 0 \leqslant s(Y) \leqslant i\}$. It follows that $\underline{C} = lfp\ F = \bigcup_{i \geqslant 0} X^i = \{\langle s, s[Y \leftarrow 0]\rangle: \exists i \geqslant 0.\ 0 \leqslant s(Y) \leqslant i\} = \{\langle s, s[Y \leftarrow 0]\rangle: 0 \leqslant s(Y)\}$.

PROOF OF THEOREM **(19)**. For short we write op instead of $op[\![C]\!]$ when C is clear from the context.

(19.1), (19.2): skip and $\underline{X := E}$ are handled the same way as (19.3) below.

(19.3): $\underline{X := ?} = \{\langle s, s'\rangle: \langle\langle s, X := ?\rangle, s'\rangle \in op^*\}$ [by 18] $= \{\langle s, s'\rangle: \langle\langle s, X := ?\rangle, s'\rangle \in \delta \cup op \circ op^*\}$ [since $r^* = \delta \cup r \circ r^*$] $= \{\langle s, s'\rangle: \exists \gamma \in \Gamma.\ \langle\langle s, X := ?\rangle, \gamma\rangle \in op \wedge \langle \gamma, s'\rangle \in op^*\}$ [since $\langle s, X := ?\rangle \neq s'$] $= \{\langle s, s'\rangle: d \in D \wedge \langle s[X \leftarrow d], s'\rangle \in op^*\}$ [by (13.3)] $= \{\langle s, s[X \leftarrow d]\rangle: s \in S \wedge d \in D\}$ [since, by (13), $\langle \gamma'', \gamma'\rangle \in op$ implies $\gamma'' \notin S$ so that $\langle s'', s'\rangle \in op^*$ if and only if $s'' = s' \in S$].

(19.4): $\langle s, s'\rangle \in \underline{(C_1;C_2)} \Leftrightarrow \langle\langle s, (C_1;C_2)\rangle, s'\rangle \in op^*$ [by (18)] $\Leftrightarrow (\exists s''.\ \langle\langle s, C_1\rangle, s''\rangle \in op^* \wedge \langle\langle s'', C_2\rangle, s'\rangle \in op^*)$ [by 13.4] $\Leftrightarrow (\exists s''.\ \langle s, s''\rangle \in \underline{C_1} \wedge \langle s'', s'\rangle \in \underline{C_2})$ [by (18)] $\Leftrightarrow \langle s, s'\rangle \in \underline{C_1} \circ \underline{C_2}$ [by definition of $\circ$].

(19.5): $\underline{(B \rightarrow C_1 \diamond C_2)} = \{\langle s, s'\rangle: \exists \gamma \in \Gamma.\ \langle\langle s, (B \rightarrow C_1 \diamond C_2)\rangle, \gamma\rangle \in op \wedge \langle \gamma, s'\rangle \in op^*\} = \{\langle s, s'\rangle: s \in \underline{B} \wedge \langle\langle s, C_1\rangle, s'\rangle \in op^*\} \cup \{\langle s, s'\rangle: s \notin \underline{B} \wedge \langle\langle s, C_2\rangle \in op^*\} = (\underline{B} \rceil \underline{C_1}) \cup (\neg \underline{B} \rceil \underline{C_2})$ [by (13.5)].

(19.6): $\langle s, s'\rangle \in (\underline{B} \rceil \underline{C})^* \lceil \neg \underline{B}$ if and only if there are $n \geqslant 1$ and $s_1, \ldots, s_n \in S$ such that $s = s_1$ and, for all $i = 1, \ldots, n-1$, $s_i \in \underline{B}$ and $\langle s_i, s_{i+1}\rangle \in \underline{C}$ and $s_n = s' \notin \underline{B}$; that is, if and only if there is an execution sequence of the form "$\langle s_1, (B * C)\rangle, \langle s_1, (C;(B * C))\rangle, \ldots, \langle s_2, (B * C)\rangle, \langle s_2, (C;(B * C))\rangle, \ldots, \langle s_n, (B * C)\rangle, s_n$" with $s = s_1, \ldots, s_{n-1} \in \underline{B}$ and $s_n = s' \notin \underline{B}$; hence, if and only if $\langle s, s'\rangle \in \underline{(B * C)}$.

(19.7): Let $F = \lambda X.\ \delta \cup (\underline{B} \rceil \underline{C}) \circ X$ and $G = \lambda X.\ (\delta \lceil \neg \underline{B}) \cup (\underline{B} \rceil \underline{C}) \circ X$. We have $F^0(\emptyset) \lceil \neg \underline{B} = G^0(\emptyset) = (\delta \lceil \neg \underline{B})$. Assume by induction hypothesis that $F^n(\emptyset) \lceil \neg \underline{B} = G^n(\emptyset)$; then $F^{n+1}(\emptyset) \lceil \neg \underline{B} = F(F^n(\emptyset)) \lceil \neg \underline{B} = (\delta \lceil \neg \underline{B}) \cup ((\underline{B} \rceil \underline{C}) \circ F^n(\emptyset) \lceil \neg \underline{B}) = (\delta \lceil \neg \underline{B}) \cup ((\underline{B} \rceil \underline{C}) \circ G^n(\emptyset)) = G^{n+1}(\emptyset)$. Since $r^* = lfp\ \lambda x.\ \delta \cup r \circ x$, it follows that $\underline{(B * C)} = (\underline{B} \rceil \underline{C})^* \lceil \neg \underline{B} = (lfp\ F) \lceil \neg \underline{B} = (\bigcup_{n \geqslant 0} F^n(\emptyset)) \lceil \neg \underline{B} = \bigcup_{n \geqslant 0} (F^n(\emptyset) \lceil \neg \underline{B}) = \bigcup_{n \geqslant 0} G^n(\emptyset) = lfp\ G$.

(19.8): We first show that $\underline{(B * C)}$ satisfies conditions (a)–(d):

(a) $\underline{(B * C)} = (\underline{B} \rceil \underline{C})^* \lceil \neg \underline{B} \subseteq S \times \neg \underline{B}$.

(b) If $(q \rceil r \subseteq S \times q)$ then, by induction on $n \geqslant 0$, $(q \rceil r^n \subseteq S \times q)$. This is true for $n=0$ since $q \rceil \delta = \delta \lceil q \subseteq S \times q$. Moreover, $q \rceil r^{n+1} = q \rceil (r^n \circ r) = (q \rceil r^n) \circ r \subseteq (S \times q) \circ r = S^2 \circ (q \rceil r) \subseteq S^2 \circ (S \times q) = S \times q$. Whence if $(q \rceil (\underline{B} \rceil \underline{C}) \subseteq S \times q)$ then $(q \rceil (\underline{B} \rceil \underline{C})^n \subseteq S \times q)$ so that $q \rceil \underline{(B * C)} = q \rceil (\underline{B} \rceil \underline{C})^* \lceil \neg \underline{B} \subseteq q \rceil (\underline{B} \rceil \underline{C})^* = q \rceil (\bigcup_{n \geqslant 0} (\underline{B} \rceil \underline{C})^n) = \bigcup_{n \geqslant 0} (q \rceil (\underline{B} \rceil \underline{C})^n) \subseteq \bigcup_{n \geqslant 0} S \times q = S \times q$.

(c) Since $\underline{(B * C)} = (\delta \lceil \neg \underline{B}) \cup ((\underline{B} \rceil \underline{C}) \circ \underline{(B * C)})$, we have $(\underline{B} \rceil \underline{C}) \circ \underline{(B * C)} \subseteq \underline{(B * C)}$.

(d) Since $\underline{(B * C)} = (\delta \lceil \neg \underline{B}) \cup ((\underline{B} \rceil \underline{C}) \circ \underline{(B * C)})$, we have $(\delta \lceil \underline{B}) \subseteq \underline{(B * C)}$, so that $(\delta \lceil \neg \underline{B}) = \delta \cap (\delta \lceil \neg \underline{B}) \subseteq \delta \cap \underline{(B * C)}$.

Assuming that r satisfies conditions (19.8a–d), we show that $r = \underline{(B * C)}$:

(i) $(\delta \lceil \neg \underline{B}) = \delta \cap r \subseteq r$ and $(\underline{B} \rceil \underline{C}) \circ r \subseteq r$, so that r is a postfixpoint of $F = \lambda X.\ (\delta \lceil \neg \underline{B}) \cup ((\underline{B} \rceil \underline{C}) \circ X)$ whence $\underline{(B * C)} = lfp\ F \subseteq r$.

(ii) To show that $r \subseteq \underline{(B * C)}$, we assume that $\langle s, s'\rangle \in r$ and prove that $\langle s, s'\rangle \in (\underline{B} \rceil \underline{C})^* \lceil \neg \underline{B}$. Let $q = \{s_1 \in S: \langle s, s_1\rangle \in (\underline{B} \rceil \underline{C})^*\}$. We have $q \rceil (\underline{B} \rceil \underline{C}) = \{\langle s_1, s_2\rangle \in (\underline{B} \rceil \underline{C}): \langle s, s_1\rangle \in (\underline{B} \rceil \underline{C})^*\} \subseteq \{\langle s_3, s_2\rangle: \langle s, s_2\rangle \in (\underline{B} \rceil \underline{C})^+\} \subseteq \{\langle s_3, s_2\rangle: \langle s, s_2\rangle \in$

$(\underline{B}\rceil\underline{C})^*\}\subseteq S\times q$. By (19.8b), $(q\rceil r\subseteq S\times q)$ so that $s\in q$ and $\langle s,s'\rangle\in r$ imply $s'\in q$ whence $\langle s,s'\rangle\in(\underline{B}\rceil\underline{C})^*$. Moreover, $\langle s,s'\rangle\in r$ and (19.8a) imply $s'\in\neg\underline{B}$, so that $\langle s,s'\rangle\in(\underline{B}\rceil\underline{C})^*\lceil\neg\underline{B}$. $\square$

Observe that the relational semantics (19) of our language **Com** is not "equivalent" to its operational semantics (13) because information about program termination is lost.

(21) Example (*programs with different operational semantics but identical relational semantics*). Assuming $D = \mathbf{Z}$, C = "$Y{:=}0$" and C' = "$(Y{:=}?;\ (Y <> 0 * Y{:=}Y-1))$" have the same relational semantics $\underline{C} = \underline{C'} = \{\langle s, s[Y\leftarrow 0]\rangle : s\in S\}$. We have $\underline{C} = \{\langle s, s[Y\leftarrow 0]\rangle : s\in S\}$ by (19.2) and $\underline{C} = \underline{Y{:=}?}\circ\underline{(Y<>0*Y{:=}Y-1)}$ [by 10.4] $= \{\langle s, s[Y\leftarrow d]\rangle : s\in S\wedge d\in D\}\circ\{\langle s, s[Y\leftarrow 0]\rangle : 0\leqslant s(Y)\}$ [by (19.3) and example (20)] $= \{\langle s, s[Y\leftarrow d]\ [Y\leftarrow 0]\rangle : 0\leqslant d\} = \{\langle s, s[Y\leftarrow 0]\rangle : s\in S\}$. Hence no distinction is made between program C for which termination with $Y = 0$ is guaranteed and program C' for which termination with $Y = 0$ is possible but not guaranteed (with a usual Pascal-like implementation).

It follows that if we choose (19) (or later Hoare logic) as *the* definition of the semantics of **Com** then a faithful implementation of **Com** should ignore the possibility of nontermination as a viable answer whenever termination is possible. This "angelic" nondeterminism of Floyd [144] could be implemented by parallelism of breadth-first search in order to simultaneously examine all possible choices offered by random assignments (we have to assume a finite number of choices) [197]. The "demonic" nondeterminism of Dijkstra [123, 124] is a radically different alternative where results are valid only if the examined program always terminates. Again, a strictly faithful implementation should use depth-first backtracking to guarantee nontermination if this is possible for one choice in random assignments. In practice, nondeterminism is implemented by choosing arbitrarily, or sometimes fairly, one of the alternatives offered by (19). Then angelic and demonic nondeterminism can be understood as describing the best and worst possible situation [216, 236]. In conclusion, (19) is an approximate version of (13) where "details" about termination are deliberately ignored.

4. Partial correctness of a command

An *assertion* is a set of states. A *specification* is a pair $\langle p, q\rangle$ of assertions on states (where p is called the *input specification* or *precondition* and q is the *output specification* or *postcondition*). A command C is said to be *partially correct* with respect to a specification $\langle p, q\rangle$ (written $\underline{\{p\}C\{q\}}$) if any terminating execution of C starting from an initial state s satisfying p must end in some final state s' satisfying q. Stated in terms of the relational semantics (18), this means that $(p\rceil\underline{C})\subseteq(S\times q)$:

(22) Definition (Floyd [143], Naur [319]) (*partial correctness*)

p, q:	$\mathbf{Ass} = \mathscr{P}(S)$	*assertions*,
$\langle p, q\rangle$:	$\mathbf{Spec} = \mathbf{Ass} \times \mathbf{Ass}$	*specifications*,
$\underline{\{p\}}\underline{C}\underline{\{q\}}$:	$\mathbf{Ass} \times \mathbf{Com} \times \mathbf{Ass} \rightarrow \{\mathrm{tt}, \mathrm{ff}\}$	*partial correctness*

$$\underline{\{p\}}\underline{C}\underline{\{q\}} = (p \rceil \underline{C}) \subseteq (S \times q).$$

(23) EXAMPLE (*partial correctness of program* (4)). Assume for program (4) that **Var** is $\{X, Y, Z, x, y\}$ and D is the set $\mathbf{Z}$ of integers. This program is partially correct with respect to the specification $\langle p, q\rangle$ such that

$$p = \{s \in S\colon\ s(X) = s(x) \wedge s(Y) = s(y) \geqslant 0\},$$
$$q = \{s \in S\colon\ s(Z) = s(x) ** s(y)\}.$$

Otherwise stated, if $x, y \geqslant 0$ respectively denote the initial values of programming variables X, Y and if the corresponding execution terminates then the final value of Z is equal to $x ** y$.

Observe that definition (22) of partial correctness is essentially of semantical nature. It is relative to the operational semantics (13) and is defined in terms of naïve set theory. No reference is made to a particular formal logical language for describing the sets of states p and q (called "assertions" for convenience). Therefore we make no difference between a predicate P (such as "$Y \geqslant 0 \wedge Z*(X ** Y) = x ** y$") and the assertion $\underline{P}$ which it denotes (that is, "$\{s \in S\colon s(Y) \geqslant 0 \wedge s(Z)*(s(X) ** s(Y)) = s(x) ** s(y)\}$" in this example). The consequences of restricting assertions p, q to those that can be formally described by first-order predicates will be studied later in Subsection 7.2.

5. Floyd–Naur partial correctness proof method and some equivalent variants

The first method for proving partial correctness was proposed by Floyd [143] and Naur [319]. After giving a simple introductory example, we formally derive Floyd–Naur's method from the operational semantics (13) using an elementary stepwise induction principle and predicates attached to program points to express invariant properties of programs. This systematic construction of the verification conditions ensures that the method is semantically sound (i.e. correct) and complete (i.e. always applicable). Then we introduce another presentation of Floyd–Naur's method inspired by Hoare logic where proofs are given by induction on the syntactical structure of programs. These two approaches are shown to be equivalent in the strong sense that, up to a difference of presentation, they require verification of exactly the same conditions. Some other partial correctness proof methods are shortly reviewed and shown to be also variants of the basic Floyd–Naur method.

5.1 *An example of partial correctness proof by Floyd–Naur's method*

(24) EXAMPLE (*Partial correctness proof of program* (4)). The informal partial correctness proof of program (4) by Floyd–Naur's method first consists in discovering predicates $P_k(X, Y, Z, x, y)$ associated with each label L_k, $k=1,\ldots,8$, which should relate the values X, Y, Z of variables X, Y, Z whenever control is at L_k to the initial values x, y, z of these variables:

(25)
$\{L_1\}$ $\{P_1(X, Y, Z, x, y)=(X=x \wedge Y=y\geqslant 0)\}$
  $Z:=1;$
$\{L_2\}$ $\{P_2(X, Y, Z, x, y)=(Y\geqslant 0 \wedge Z*(X ** Y)=x ** y)\}$
  {–loop invariant–}
  while $Y<>0$ **do**
$\{L_3\}$ $\{P_3(X, Y, Z, x, y)=(Y>0 \wedge Z*(X ** Y)=x ** y)\}$
    if $odd(Y)$ **then**
    begin
$\{L_4\}$ $\{P_4(X, Y, Z, x, y)=(Y>0 \wedge Z*(X ** Y)=x ** y)\}$
      $Y:=Y-1;$
$\{L_5\}$ $\{P_5(Z, Y, Z, x, y)=(Y\geqslant 0 \wedge Z*(X **(Y+1))=x ** y)\}$
      $Z:=Z*X;$
    end
    else
    begin
$\{L_6\}$ $\{P_6(X, Y, Z, x, y)=(even(Y) \wedge Y>0 \wedge Z*(X ** Y)=x ** y)\}$
      $Y:=Y\,div\,2;$
$\{L_7\}$ $\{P_7(X, Y, Z, x, y)=(Y\geqslant 0 \wedge Z*(X **(2*Y))=x ** y)\}$
      $X:=X*X;$
    end;
$\{L_8\}$ $\{P_8(X, Y, Z, x, y)=(Z=x ** y)\}$

Observe that predicates P_k associated with labels L_k, $k=1,\ldots,8$, can be understood as describing a set of states $\underline{P}_k$. For example,

$$\underline{P}_2=\{s\in S\colon s(Y)\geqslant 0 \wedge s(Z)*(s(X) ** s(Y))=s(x) ** s(y)\}.$$

Then it must be shown that these predicates satisfy *verification conditions*, which can be stated informally as follows:

(26)
(ε) $p \Rightarrow P_1(X,Y,Z,x,y)$ where p is the input specification
(i_1) $P_1(X,Y,Z,x,y) \Rightarrow P_2(X,Y,1,x,y)$
(i_2) $[P_2(X,Y,Z,x,y) \wedge Y\neq 0] \Rightarrow P_3(X,Y,Z,x,y)$
(i_3) $[P_2(X,Y,Z,x,y) \wedge Y=0] \Rightarrow P_8(X,Y,Z,x,y)$
(i_4) $[P_3(X,Y,Z,x,y) \wedge odd(Y)] \Rightarrow P_4(X,Y,Z,x,y)$
(i_5) $[P_3(X,Y,Z,x,y) \wedge even(Y)] \Rightarrow P_6(X,Y,Z,x,y)$
(i_6) $P_4(X,Y,Z,x,y) \Rightarrow P_5(X,Y-1,Z,x,y)$
(i_7) $P_5(X,Y,Z,x,y) \Rightarrow P_2(X,Y,Z*X,x,y)$
(i_8) $P_6(X,Y,Z,x,y) \Rightarrow P_7(X,Y\,div\,2,Z,x\,y)$
(i_9) $P_7(X,Y,Z,x,y) \Rightarrow P_2(X*X,Y,Z,x,y)$
(σ) $P_8(X,Y,Z,x,y) \Rightarrow q$ where q is the output specification.

These verification conditions imply that the predicates are *local invariants*, that is, P_k holds whenever control is at point L_k, that is to say, according to (15), when command L_k remains to be executed. In practice it is only necessary to discover loop invariants since other local invariants can be derived from the loop invariants using these verification conditions.

It follows that the *local invariants* on states attached to program points are equivalent to the following *global invariant* on configurations:

$$i = \bigcup_{k=1}^{7} \{\langle s, L_k\rangle : s \in \underline{P}_k\} \cup \underline{P}_8. \tag{27}$$

i is called *invariant* because it is always true during execution:

$$(\{\langle s, C\rangle : s \in p\} \rceil op[\![C]\!]^*) \subseteq i.$$

It follows immediately that $(p \rceil \underline{C}) \subseteq (S \times q)$.

5.2. *The stepwise Floyd–Naur partial correctness proof method*

A partial correctness proof can always be organized in the same way and it can be reduced to the discovery of local invariants which are then shown to satisfy elementary verification conditions corresponding to elementary program steps. To show this, we first introduce an induction principle expressing the essence of invariance proofs. Then we specialize the induction principle for the operational semantics (13) of language **Com**. This consists in representing the global invariant on configurations by local invariants on states attached to program points. Once properties of programs have been chosen to be expressed in this way, Floyd's verification conditions can be derived by calculus from the operational semantics (13). This construction of the verification conditions for local invariants a priori ensures semantical soundness and completeness of the proof method.

5.2.1. *Stepwise induction principle*

Floyd–Naur's method is usually understood as stepwise induction [286]: to prove that some property i of a program is invariant during the course of the computation, it is sufficient to check that i is true when starting the computation and to show that if i is true at one step of the computation, it remains true after the next step. This means that Floyd–Naur's method consists in applying the following lemma to the operational semantics:

(28) LEMMA (Cousot [89]) (stepwise induction principle). *If* $p, p', q \in \mathcal{P}(E)$ *and* $r \in \mathcal{P}(E \times E)$ *then*

$$[(p \rceil r^* \lceil p') \subseteq (E \times q)] \Leftrightarrow [\exists i \in \mathcal{P}(E).\, (p \subseteq i) \wedge (i \rceil r \subseteq E \times i) \wedge (i \cap p' \subseteq q)].$$

PROOF. For $\Rightarrow$, we observe that $i = \{e \in E : \exists e' \in p.\, \langle e', e\rangle \in r^*\}$ satisfies conditions $p \subseteq i$ and $i \rceil r \subseteq E \times i$, whereas $p \rceil r^* \lceil p' \subseteq E \times q$ implies $i \cap p' \subseteq q$.

For $\Leftarrow$, we observe that, by induction on $n \geqslant 0$, $p \subseteq i$ and $i \rceil r \subseteq E \times i$ imply that $p \rceil r^n \subseteq E \times i$ whence $p \rceil r^* \subseteq E \times i$, so that $p \rceil r^* \lceil p' \subseteq E \times (i \cap p') \subseteq E \times q$. □

The Floyd–Naur partial correctness proof method consists in discovering local assertions on states attached to program points which must be shown to satisfy local verification conditions. As shown by example (24), this can be understood as the discovery of a global assertion i upon configurations which is shown to satisfy a *global verification condition* $\text{gvc}[\![C]\!][p,q](i)$ derived from lemma (28):

(29) THEOREM (Keller [247], Pnueli [345], Cousot [89]) (induction principle for Floyd–Naur's stepwise partial correctness proof method)

(29.1) $\quad \underset{\sim}{\{p\}}\underset{\sim}{C}\underset{\sim}{\{q\}} = [\exists i \in \mathcal{P}(\Gamma).\ \text{gvc}[\![C]\!][p,q](i)]$

(29.2) $\quad$ *where* $\text{gvc}[\![C]\!][p,q](i) = (\forall s \in p.\ \langle s, C\rangle \in i) \wedge (i \rceil op[\![C]\!] \subseteq \Gamma \times i) \wedge (i \cap S \subseteq q)$.

PROOF. $\underset{\sim}{\{p\}}\underset{\sim}{C}\underset{\sim}{\{q\}} = (p \rceil \underset{\sim}{C} \subseteq S \times q)$ [by (22)] $= (p \rceil \{\langle s', s\rangle: \langle\langle s', C\rangle, s\rangle \in op[\![C]\!]^*\} \subseteq S \times q)$ [by (18)] $= (\{\langle s, C\rangle: s \in p\} \rceil op[\![C]\!]^* \lceil S \subseteq \Gamma \times q) = (\exists i \in \mathcal{P}(\Gamma).(\{\langle s, C\rangle: s \in p\} \subseteq i) \wedge (i \rceil op[\![C]\!] \subseteq \Gamma \times i) \wedge (i \cap S \subseteq q))$ [by (28)] $= (\exists i \in \mathcal{P}(\Gamma).(\forall s \in p.\ \langle s, C\rangle \in i) \wedge (i \rceil op[\![C]\!] \subseteq \Gamma \times i) \wedge (i \cap S \subseteq q))$. □

Any i satisfying the verification condition $\text{gvc}[\![C]\!][p,q](i)$ is always true during execution, hence is a *global invariant*. Such a global invariant always exists since there is a *strongest global invariant* which implies all others and can be characterized as a fixpoint:

(30) THEOREM (Park [340], Clarke [72], Cousot [89]) (fixpoint characterization of the strongest global invariant). *The strongest global invariant*

(30.1) $\quad I = \{\gamma: \exists s \in p.\ \langle\langle s, C\rangle, \gamma\rangle \in op[\![C]\!]^*\}$

is such that

(30.2) $I = \mathit{lfp}\ \lambda X: \mathcal{P}(\Gamma).\{\langle s, C\rangle: s \in p\} \cup \{\gamma: \exists \gamma'.\langle \gamma', \gamma\rangle \in X \rceil op[\![C]\!]\}$;
(30.3) *if* $\underset{\sim}{\{p\}}\underset{\sim}{C}\underset{\sim}{\{q\}}$ *then* $\text{gvc}[\![C]\!][p,q](I)$ *holds and* $\forall i.\text{gvc}[\![C]\!][p,q](i) \Rightarrow (I \subseteq i)$.

PROOF. $\langle \mathcal{P}(\Gamma), \subseteq, \emptyset\rangle$ is a complete lattice. $\phi = \lambda X: \mathcal{P}(\Gamma \times \Gamma).\delta \cup X \circ op[\![C]\!]$ and $\psi = \lambda X: \mathcal{P}(\Gamma).\{\langle s, C\rangle: s \in p\} \cup \{\gamma: \exists \gamma'.\langle\gamma', \gamma\rangle \in X \rceil op[\![C]\!]\}$ are monotone. $\alpha = \lambda X: \mathcal{P}(\Gamma \times \Gamma).\{\gamma \in \mathcal{P}(\Gamma): \exists s \in p.\langle\langle s, C\rangle, \gamma\rangle \in X\}$ is strict, upper-continuous and $\alpha \circ \phi = \psi \circ \alpha = \lambda X.\{\gamma \in \mathcal{P}(\Gamma): \exists s \in p.\langle\langle s, C\rangle, \gamma\rangle \in \delta \cup X \circ op[\![C]\!]\}$. Therefore $\alpha(\mathit{lfp}\ \phi) = \alpha(op[\![C]\!]^*) = I = \mathit{lfp}\ \psi$.

Obviously, $\{\langle s, C\rangle: s \in p\} \subseteq I$. Moreover $I \rceil op[\![C]\!] = \{\langle \gamma, \gamma'\rangle: \exists s \in p.\langle\langle s, C\rangle, \gamma\rangle \in op[\![C]\!]^* \wedge \langle \gamma, \gamma'\rangle \in op[\![C]\!]\} \subseteq \{\langle\gamma, \gamma'\rangle: \exists s \in p.\langle\langle s, C\rangle, \gamma'\rangle \in op[\![C]\!]^+\} \subseteq \Gamma \times I$. Finally, $\underset{\sim}{\{p\}}\underset{\sim}{C}\underset{\sim}{\{q\}} \Rightarrow (p \rceil \underset{\sim}{C} \subseteq S \times q)$ [by (22)] $\Rightarrow (p \rceil \{\langle s', s\rangle: \langle\langle s', C\rangle, s\rangle \in op[\![C]\!]^*\} \subseteq S \times q)$ [by (18)] $\Rightarrow (I \lceil S \subseteq \Gamma \times q) \Rightarrow (I \cap S \subseteq q)$.

If $(\forall s \in p.\ \langle s, C\rangle \in i) \wedge (i \rceil op[\![C]\!] \subseteq \Gamma \times i)$ then $\psi(i) \subseteq i$ so that $I = \mathit{lfp}\ \psi = \bigcap\{X \in \mathcal{P}(\Gamma): \psi(X) \subseteq X\}$ [by Tarski [385]] $\subseteq i$. □

(31) EXAMPLE (*application of induction principle* (29) *to the correctness proof of program* (4)). Let us go on with example (24) and show that (26) is equivalent to (29):

(a) $\forall s \in p . \langle s, C_1 \rangle \in i$ is equivalent to $p \subseteq \underline{P}_1$.

(b) $i \rceil op[\![C]\!] \subseteq \Gamma \times i$ is equivalent to the conjunction $\{\langle s, L_k \rangle : s \in \underline{P}_k\} \rceil op[\![C]\!] \subseteq \Gamma \times i$ for $k = 1, \ldots, 7$ and $\underline{P}_8 \rceil op[\![C]\!] \subseteq \Gamma \times i$ so that the proof can be done by cases corresponding to each possible transition from configurations $\langle s, L_k \rangle$ for any s satisfying predicate P_k attached to point L_k. Using in each case the definition of $op[\![C]\!]$, we obtain Floyd's simpler verification conditions:

(b₁) For the assignments $X_k := E_k$, $k = 1, 4, 5, 6, 7$, we have $\langle\langle s, L_k \rangle, \langle s', L_{k'} \rangle\rangle \in op[\![C]\!]$ if and only if $k' = \text{succ}(k)$ where $\text{succ} = [1 \leftarrow 2, 4 \leftarrow 5, 5 \leftarrow 2, 6 \leftarrow 7, 7 \leftarrow 2]$. Therefore the corresponding verification conditions are of the following form given by Floyd: $\forall s \in \underline{P}_k . s[X_k \leftarrow \underline{E}_k(s)] \in \underline{P}_{\text{succ}(k)}$.

For example when $k = 4$, we have to prove that $[s(Y) > 0 \wedge s(Z) * (s(X) ** s(Y)) = s(x) ** s(y)] \Rightarrow [s[Y \leftarrow s(Y) - 1] \in \{s' : s'(Y) \geqslant 0 \wedge s'(Z) * (s'(X) ** (s'(Y) + 1)) = s'(x) ** s'(y)\}]$ or equivalently $[s(Y) > 0 \wedge s(Z) * (s(X) ** s(Y)) = s(x) ** s(y)] \Rightarrow [(s(Y) - 1) \geqslant 0 \wedge s(Z) * (s(X) ** ((s(Y) - 1) + 1)) = s(x) ** s(y)]$ which is obvious.

(b₂) For the test $k = 3$, we have $\langle\langle s, L_3 \rangle, \langle s', L_{k'} \rangle\rangle \in op[\![C]\!]$ if and only if $s' = s$, and if $s \in \underline{Y <> 0}$ then $k' = 4$ else $k' = 6$, so that we have to prove that $(\underline{P}_3 \cap \underline{Y <> 0}) \subseteq \underline{P}_4 \wedge (\underline{P}_3 \cap \underline{Y = 0}) \subseteq \underline{P}_6$.

(b₃) In the same way, for the while-loop $k = 2$, we have to prove that $(\underline{P}_2 \cap \underline{odd(Y)}) \subseteq \underline{P}_3 \wedge (\underline{P}_2 \cap \underline{even(Y)}) \subseteq \underline{P}_8$.

(c) $((i \cap S) \subseteq q) = (\underline{P}_8 \subseteq q)$ is equivalent to $\underline{P}_8 = q$.

5.2.2. *Representing a global invariant on configurations by local invariants on states attached to program points*

However, instead of using a single global invariant i on configurations as in (29), Floyd and Naur proposed to use local invariants on states attached to program points (originally, arcs of flowcharts). Such program points $L \in \mathbf{Lab}[\![C]\!]$ for commands $C \in \mathbf{Comp}$ can be understood as labels specifying where control can reside before, during or after executing a step within C. According to the operational semantics (13), $\mathbf{Lab}[\![C]\!]$ can be chosen as the set of control states C' of configurations $\langle s, C' \rangle \in S \times \mathbf{Comp}$ encountered during execution of command C, together with a final label, arbitrarily denoted "$\surd$", corresponding to configurations $\gamma \in S$ for which execution of C is terminated:

(32) DEFINITION (*labels designating program control points*)

(32.1) $\mathbf{Lab}[\![C]\!] = \mathbf{At}[\![C]\!] \cup \mathbf{In}[\![C]\!] \cup \mathbf{After}[\![C]\!]$,

(32.2) $\mathbf{At}[\![C]\!] = [\![C]\!]$,

(32.3) $\mathbf{In}[\![C]\!] = \emptyset$ if C is skip, $X := E$ or $X := ?$,

(32.4) $\mathbf{In}[\![(C_1;C_2)]\!] = \{(C_1';C_2) : C_1' \in \mathbf{In}[\![C_1]\!]\} \cup \mathbf{At}[\![C_2]\!] \cup \mathbf{In}[\![C_2]\!]$,

(32.5) $\mathbf{In}[\![(B \rightarrow C_1 \diamond _2)]\!] = \mathbf{At}[\![C_1]\!] \mathrm{C} \cup \mathbf{In}[\![C_1]\!] \cup \mathbf{At}[\![C_2]\!] \cup \mathbf{In}[\![C_2]\!]$,

(32.6) $\mathbf{In}[\![(B * C_1)]\!] = \{(C';(B * C_1)) : C' \in \mathbf{At}[\![C_1]\!] \cup \mathbf{In}[\![C_1]\!]\}$,

(32.7) $\mathbf{After}[\![C]\!] = \{\surd\}$.

(33) EXAMPLE (*labels of program* (4)). The labels of program (4) have been defined in example (15). We have $\mathbf{At}[\![C^{10}]\!]=\{C^{10}\}$ and $\mathbf{In}[\![C^{10}]\!]=\{C^{100}, C^{1001}, C^{101}, C^{1011}\}$ whereas $\mathbf{At}[\![C^{\varepsilon}]\!]=\{L_1\}$, $\mathbf{In}[\![C^{\varepsilon}]\!]=\{L_2, L_3, L_4, L_5, L_6, L_7\}$ and $\mathbf{After}[\![C^{\varepsilon}]\!]=\{L_8\}$.

The local invariants are assertions on states attached to program points. More formally they can be defined as a function "inv", which maps labels of C to assertions:

(34) DEFINITION (*local invariants*). $\text{inv}: \mathbf{Lab}[\![C]\!] \rightarrow \mathbf{Ass}$.

(35) EXAMPLE (*local invariants for program* (4)). Local invariants for program (4) have been given in (25): $\text{inv}(L_k)=\underline{P}_k$, $k=1,\ldots,8$.

The local invariants $\text{inv}(L)$, $L \in \mathbf{Lab}[\![C]\!]$ can be understood as describing the global invariant $\gamma(\text{inv}) \in \mathscr{P}(\Gamma)$, which is the set of configurations such that when control is at L, the memory state belongs to $\text{inv}(L)$. Reciprocally, a global invariant $i \in \mathscr{P}(\Gamma)$ can be decomposed into local invariants $\alpha(i)(L)$, $L \in \mathbf{Lab}[\![C]\!]$, defined by the fact that when control is at L the only possible memory states s are those for which the configuration $\langle s, L\rangle$ belongs to i (or s belongs to i if $L=\surd$):

DEFINITION (Cousot & Cousot [92]) (*connection between local and global invariants*)
(36) *Concretization function*:

(36.1) $\gamma: (\mathbf{Lab}[\![C]\!] \rightarrow \mathbf{Ass}) \rightarrow \mathscr{P}(\Gamma)$

(36.2) $\gamma(\text{inv}) = \{\langle s, L\rangle : s \in \text{inv}(L) \wedge L \in \mathbf{Lab}[\![C]\!] - \{\surd\}\} \cup \text{inv}(\surd)$.

(37) *Abstraction function*:

(37.1) $\alpha: \mathscr{P}(\Gamma) \rightarrow (\mathbf{Lab}[\![C]\!] \rightarrow \mathbf{Ass})$

(37.2) $\alpha(i)(L) = \{s : \langle s, L\rangle \in i\}$ if $L \in \mathbf{At}[\![C]\!] \cup \mathbf{In}[\![C]\!]$,

(37.3) $\alpha(i)(L) = i \cap S$ if $L \in \mathbf{After}[\![C]\!]$.

Since α is a bijection, the inverse of which is γ, the discovery of a global invariant $i \in \mathscr{P}(\Gamma)$ satisfying verification condition $\text{gvc}[\![C]\!][p, q](i)$ is equivalent to the discovery of local invariants $\text{inv}(L)$, $L \in \mathbf{Lab}[\![C]\!]$, satisfying verification condition $\text{gvc}[\![C]\!][p, q](\gamma(\text{inv}))$. This leads to the construction of the local verification conditions by calculus [92]. This equivalence is of theoretical interest only, since from a practical point of view each local invariant is simpler than the global one and the task of checking $\text{gvc}[\![C]\!][p, q](\gamma(\text{inv}))$ can be decomposed into the verification of more numerous but simpler conditions, one for each local invariant.

5.2.3. *Construction of the verification conditions for local invariants*

To formally derive the local verification conditions from induction principle (29), we first express the operational semantics (13) in an equivalent form using program steps. From a syntactic point of view, the next elementary step $\mathbf{Step}[\![C]\!][L]$, which will be executed when control is at point $L \in \mathbf{At}[\![C]\!] \cup \mathbf{In}[\![C]\!]$ of command $C \in \mathbf{Comp}$, is an atomic command or a test defined by cases as follows (where $n \geqslant 0$):

(38) DEFINITION (*elementary steps within a command*)
(38.1) $\mathbf{Step}[\![C]\!][(\cdots((C';C_1);C_2)\cdots;C_n)]=C'$ if C' is skip, $X := E$ or $X := ?$;
(38.2) $\mathbf{Step}[\![C]\!][(\cdots(((B\rightarrow C'\diamond C'');C_1);C_2)\cdots;C_n)]=B$;
(38.3) $\mathbf{Step}[\![C]\!][(\cdots(((B*C');C_1);C_2)\cdots;C_n)]=B$.

(39) EXAMPLE (*elementary steps of program* (4)). For program C defined by (4) with labels (15), we have $\mathbf{Step}[\![C]\!]=[L_1\leftarrow Z := 1, L_2\leftarrow Y <> 0, L_3\leftarrow odd(Y), L_4\leftarrow Y := Y-1, L_5\leftarrow Z := Z*X, L_6\leftarrow Y := Y\,div\,2, L_7\leftarrow X := X*X]$.

Again from a syntactic point of view, the next label $\mathbf{Succ}[\![C]\!][L]$, which will be reached after execution of an elementary step when control is at point $L\in\mathbf{At}[\![C]\!]\cup\mathbf{In}[\![C]\!]$ of command $C\in\mathbf{Comp}$, can be defined by cases as follows (where $n\geqslant 0$ and $(\cdots(C_1;C_2)\cdots;C_n)$ is the final label $\surd$ for $n=0$):

(40) DEFINITION (*successors of a program control point*)

(40.1) $$\mathbf{Succ}[\![C]\!][(\cdots((C';C_1);C_2)\cdots;C_n)]=(\cdots(C_1;C_2)\cdots;C_n) \text{ if } C' \text{ is skip, } X := E \text{ or } X := ?,$$

(40.2) $$\mathbf{Succ}[\![C]\!][(\cdots(((B\rightarrow C'\diamond C'');C_1);C_2)\cdots;C_n)]=[\mathrm{tt}\leftarrow(\cdots((C';C_1);C_2)\cdots;C_n),\ \mathrm{ff}\leftarrow(\cdots((C'';C_1);C_2)\cdots;C_n)],$$

(40.3) $$\mathbf{Succ}[\![C]\!][(\cdots(((B*C);C_1);C_2)\cdots;C_n)]=[\mathrm{tt}\leftarrow(\cdots(((C;(B*C));C_1);C_2)\cdots;C_n),\ \mathrm{ff}\leftarrow(\cdots(C_1;C_2)\cdots;C_n)].$$

(41) EXAMPLE (*successors of control points of program* (4)). For program C defined by (4) with labels (15), we have $\mathbf{Succ}[\![C]\!]=[L_1\leftarrow L_2, L_2\leftarrow[\mathrm{tt}\leftarrow L_3, \mathrm{ff}\leftarrow L_8], L_3\leftarrow[\mathrm{tt}\leftarrow L_4, \mathrm{ff}\leftarrow L_6], L_4\leftarrow L_5, L_5\leftarrow L_2, L_6\leftarrow L_7, L_7\leftarrow L_2]$.

Now from a semantical point of view, execution of an elementary step $\mathbf{Step}[\![C]\!][L]$ in memory state s can lead to any successor state $s'\in\mathbf{NextS}[\![C]\!]\langle s, L\rangle$ as follows:

(42) DEFINITION (*successor states*)
(42.1) $\mathbf{NextS}[\![C]\!]\langle s, L\rangle=\{s\}$ if $\mathbf{Step}[\![C]\!][L]$ is skip;
(42.2) $\mathbf{NextS}[\![C]\!]\langle s, L\rangle=\{s[X\leftarrow\underline{E}(s)]\}$ if $\mathbf{Step}[\![C]\!][L]$ is $X := E$;
(42.3) $\mathbf{NextS}[\![C]\!]\langle s, L\rangle=\{s[X\leftarrow d]: d\in D\}$ if $\mathbf{Step}[\![C]\!][L]$ is $X := ?$;
(42.4) $\mathbf{NextS}[\![C]\!]\langle s, L\rangle=\{s\}$ if $\mathbf{Step}[\![C]\!][L]$ is B.

Again from a semantical point of view, the next label $\mathbf{NextL}[\![C]\!]\langle s, L\rangle$, which can be reached after execution of an elementary step in configuration $\langle s, L\rangle$ of command $C\in\mathbf{Comp}$, can be defined by cases as follows:

(43) DEFINITION (*successor control point*)
(43.1) $\mathbf{NextL}[\![C]\!]\langle s, L\rangle=\{\mathbf{Succ}[\![C]\!][L]\}$ if $\mathbf{Step}[\![C]\!][L]$ is skip, $X := E$ or $X := ?$;
(43.2) $\mathbf{NextL}[\![C]\!]\langle s, L\rangle=\{\mathbf{Succ}[\![C]\!][L](s\in\underline{B})\}$ if $\mathbf{Step}[\![C]\!][L]$ is B.

The operational semantics (13) can now be given an equivalent stepwise presentation:

(44) LEMMA (stepwise presentation of the operational semantics)

$$op[\![C]\!] = \{\langle\langle s, L\rangle, \text{final}\langle s', L'\rangle\rangle : s \in S \wedge L \in \mathbf{At}[\![C]\!] \cup \mathbf{In}[\![C]\!] \wedge s' \in \mathbf{NextS}[\![C]\!]\langle s, L\rangle \wedge L' \in \mathbf{NextL}[\![C]\!]\langle s, L\rangle\}$$

where $\text{final}\langle s', \surd\rangle = s'$ *and otherwise* $\text{final}\,\gamma = \gamma$.

We have seen that a partial correctness proof of $\{p\}\underline{C}\{q\}$ by Floyd–Naur's method consists in discovering local invariants $\text{inv} \in \mathbf{Lab}[\![C]\!] \to \mathbf{Ass}$ satisfying $\text{gvc}[\![C]\!][p, q](\gamma(\text{inv}))$. This global verification condition is equivalent to a conjunction of simpler local verification conditions as follows:

(45) THEOREM (Naur [319], Floyd [143], Manna [283, 284]) (Floyd–Naur partial correctness proof method with stepwise verification conditions). *A partial correctness proof of* $\{p\}\underline{C}\{q\}$ *by Floyd–Naur's method consists in discovering local invariants* $\text{inv} \in \mathbf{Lab}[\![C]\!] \to \mathbf{Ass}$, *which must be proved to satisfy the following local verification conditions*:

(45.1) $p \subseteq \text{inv}(L)$ *if* $L \in \mathbf{At}[\![C]\!]$,

(45.2) $\text{inv}(L) \subseteq \text{inv}(\mathbf{Succ}[\![C]\!][L])$ *if* $L \in \mathbf{At}[\![C]\!] \cup \mathbf{In}[\![C]\!] \wedge \mathbf{Step}[\![C]\!][L]$ *is* skip,

(45.3) $\text{inv}(L) \subseteq \{s \in S: s[X \leftarrow \underline{E}(s)] \in \text{inv}(\mathbf{Succ}[\![C]\!][L])\}$ *if* $L \in \mathbf{At}[\![C]\!] \cup \mathbf{In}[\![C]\!] \wedge \mathbf{Step}[\![C]\!][L]$ *is* $X := E$,

(45.4) $\{s[X \leftarrow d]: s \in \text{inv}(L) \wedge d \in D\} \subseteq \text{inv}(\mathbf{Succ}[\![C]\!][L])$ *if* $L \in \mathbf{At}[\![C]\!] \cup \mathbf{In}[\![C]\!] \wedge \mathbf{Step}[\![C]\!][L]$ *is* $X := ?$,

(45.5) $(\text{inv}(L) \cap \underline{B}) \subseteq \text{inv}(\mathbf{Succ}[\![C]\!][L](\text{tt}))$ *if* $L \in \mathbf{At}[\![C]\!] \cup \mathbf{In}[\![C]\!] \wedge \mathbf{Step}[\![C]\!][L]$ *is* B,

(45.6) $(\text{inv}(L) \cap \neg\underline{B}) \subseteq \text{inv}(\mathbf{Succ}[\![C]\!][L](\text{ff}))$ *if* $L \in \mathbf{At}[\![C]\!] \cup \mathbf{In}[\![C]\!] \wedge \mathbf{Step}[\![C]\!][L]$ *is* B,

(45.7) $\text{inv}(L) \subseteq q$ *if* $L \in \mathbf{After}[\![C]\!]$.

Verification condition (45.3) for assignment is backward. This name arises out of the fact that the postcondition $\text{inv}(\mathbf{Succ}[\![C]\!][L])$ is back-transformed into the assertion $\{s \in S: s[X \leftarrow \underline{E}(s)] \in \text{inv}(\mathbf{Succ}[\![C]\!][L])\}$ written in terms of the states before assignment. Verification condition (45.4) for random assignment is forward. Verification condition (45.3) can also be given an equivalent forward form [250]:

(45.8) $\{s[X \leftarrow \underline{E}(s)]: s \in \text{inv}(L)\} \subseteq \text{inv}(\textbf{Succ}[\![C]\!][L])$
if $L \in \textbf{At}[\![C]\!] \cup \textbf{In}[\![C]\!] \wedge \textbf{Step}[\![C]\!][L]$ is $X := E$.

PROOF. By (29), we have to show that $\gamma(\text{inv}(L))$ satisfies $\text{gvc}[\![C]\!][p, q](\gamma(\text{inv}(L)))$. We proceed by simplification of $\text{gvc}[\![C]\!][p, q](\gamma(\text{inv}(L)))$ which constructively leads to local verification conditions (45):

First, $(\forall s \in p. \langle s, C\rangle \in \gamma(\text{inv}(L))) \Leftrightarrow (\forall s \in p. C \in \textbf{At}[\![C]\!] \cup \textbf{In}[\![C]\!] \wedge s \in \text{inv}(C))$ [by (36.2)] $\Leftrightarrow (p \subseteq \text{inv}(C))$ [by (32.2)] $\Leftrightarrow (\forall L \in \textbf{At}[\![C]\!]. p \subseteq \text{inv}(L))$ [by (32.2)].

Then, according to (36) and (44), the condition $\gamma(\text{inv})\rceil op[\![C]\!] \subseteq \Gamma \times \gamma(\text{inv})$ can be decomposed into a conjunction of simpler verification conditions, one for each program step:

$$\begin{aligned}
&\gamma(\text{inv})\rceil op[\![C]\!] \subseteq \Gamma \times \gamma(\text{inv}) \\
&\quad \Leftrightarrow \{\langle s, L\rangle: s \in \text{inv}(L) \wedge L \in \textbf{Lab}[\![C]\!] - \{\surd\}\}\rceil op[\![C]\!] \\
&\qquad \subseteq \Gamma \times [\{\langle s, L\rangle: s \in \text{inv}(L) \wedge L \in \textbf{Lab}[\![C]\!] - \{\surd\}\} \cup \text{inv}(\surd)] \\
&\quad \Leftrightarrow \{\langle\langle s, L\rangle, \text{final}\langle s', L'\rangle\rangle: s \in \text{inv}(L) \wedge L \in \textbf{At}[\![C]\!] \cup \textbf{In}[\![C]\!] \\
&\qquad \wedge s' \in \textbf{NextS}[\![C]\!]\langle s, L\rangle \wedge L' \in \textbf{NextL}[\![C]\!]\langle s, L\rangle\} \\
&\qquad \subseteq \Gamma \times [\{\langle s, L\rangle: s \in \text{inv}(L) \wedge L \in \textbf{Lab}[\![C]\!] - \{\surd\}\} \cup \text{inv}(\surd)] \\
&\quad \Leftrightarrow \forall L \in \textbf{At}[\![C]\!] \cup \textbf{In}[\![C]\!]. \forall s \in \text{inv}(L). \\
&\qquad\quad \{\text{final}\langle s', L'\rangle: s' \in \textbf{NextS}[\![C]\!]\langle s, L\rangle \wedge L' \in \textbf{NextL}[\![C]\!]\langle s, L\rangle\} \\
&\qquad\quad \subseteq [\{\langle s'', L''\rangle: s'' \in \text{inv}(L'') \wedge L'' \in \textbf{Lab}[\![C]\!] - \{\surd\}\} \cup \text{inv}(\surd)].
\end{aligned}$$

We go on by cases, according to (38), (40) and (42):

(a) If $\textbf{Step}[\![C]\!][L]$ is $X := E$ (skip and $X := ?$ are handled the same way), then we have to check that

$$\begin{aligned}
&\{\text{final}\ \langle s', L'\rangle: s' \in \{s[X \leftarrow \underline{E}(s)]\} \wedge L' \in \{\textbf{Succ}[\![C]\!][L]\}\} \\
&\subseteq [\{\langle s'', L''\rangle: s'' \in \text{inv}(L'') \wedge L'' \in \textbf{Lab}[\![C]\!] - \{\surd\}\} \cup \text{inv}(\surd)] \\
&\quad \Leftrightarrow \text{final}\langle s[X \leftarrow \underline{E}(s)], \textbf{Succ}[\![C]\!][L]\rangle \\
&\qquad \in [\{\langle s'', L''\rangle: s'' \in \text{inv}(L'') \wedge L'' \in \textbf{Lab}[\![C]\!] - \{\surd\}\} \cup \text{inv}(\surd)] \\
&\quad \Leftrightarrow s[X \leftarrow \underline{E}(s)] \in \text{inv}(\textbf{Succ}[\![C]\!][L]).
\end{aligned}$$

(and $\forall s \in \text{inv}(L). s[X \leftarrow \underline{E}(s)] \in \text{inv}(\textbf{Succ}[\![C]\!][L])$ is obviously equivalent to $\text{inv}(L) \subseteq \{s \in S: s[X \leftarrow \underline{E}(s)] \in \text{inv}(\textbf{Succ}[\![C]\!][L])\}$ and to $\{s[X \leftarrow \underline{E}(s)]: s \in \text{inv}(L)\} \subseteq \text{inv}(\textbf{Succ}[\![C]\!][L])$).

(b) If $\textbf{Step}[\![C]\!][L]$ is B then we have to check that

$$\begin{aligned}
&\{\text{final}\langle s', L'\rangle: s' \in \{s\} \wedge L' \in \{\textbf{Succ}[\![C]\!][L](s \in \underline{B})\}\} \\
&\subseteq [\{\langle s'', L''\rangle: s'' \in \text{inv}(L'') \wedge L'' \in \textbf{Lab}[\![C]\!] - \{\surd\}\} \cup \text{inv}(\surd)] \\
&\quad \Leftrightarrow \text{final}\langle s, \textbf{Succ}[\![C]\!][L](s \in \underline{B})\rangle \\
&\qquad \in [\{\langle s'', L''\rangle: s'' \in \text{inv}(L'') \wedge L'' \in \textbf{Lab}[\![C]\!] - \{\surd\}\} \cup \text{inv}(\surd)] \\
&\quad \Leftrightarrow s \in \text{inv}(\textbf{Succ}[\![C]\!][L](s \in \underline{B})).
\end{aligned}$$

Finally, $(\gamma(\text{inv}) \cap S \subseteq q) \Leftrightarrow (\text{inv}(\surd) \subseteq q)$ [by (36.2)] $\Leftrightarrow (\forall L \in \textbf{After}[\![C]\!]. \text{inv}(L) \subseteq q)$ [by (32.7)]. □

5.2.4. *Semantical soundness and completeness of the stepwise Floyd–Naur partial correctness proof method*

A proof method is *sound* if it cannot lead to mistaken conclusions. It is *complete* if it is always applicable to prove indubitable facts.

(46) THEOREM (De Bakker & Meertens [112]) (soundness and semantical completeness of the stepwise Floyd–Naur method). *The stepwise presentation of the Floyd–Naur partial correctness proof method is semantically sound and complete.*

PROOF. The method is sound since if inv satisfies (45) then, by construction of (45), gvc$[\![C]\!][p, q](\gamma(\text{inv}))$ holds so that $\underline{\{p\}}\underline{C}\underline{\{q\}}$ derives from (29). It is semantically complete since if $\underline{\{p\}}\underline{C}\underline{\{q\}}$ is true then by (29) we know that $I = \{\langle s, C\rangle : s \in p\}\rceil op[\![C]\!]^*$ satisfies gvc$[\![C]\!][p, q](I)$ so that, by construction, (45) holds for inv$=\alpha(I)$. □

We insist upon *semantical* soundness and completeness as in [112] or [287] since (46) is relative to a given semantics of programs (13) and to a representation of invariants by sets, as opposed to the existence of a formal calculus in a given language to prove partial correctness of programs [157, 363].

5.3. *The compositional Floyd–Naur partial correctness proof method*

Hoare [208] introduced the idea (often called *compositionality*) that the specification of a command should be verifiable in terms of the specifications of its components. This means that partial correctness should be proved by induction on the syntax of programs using their relational semantics (19) instead of an induction on the number of transitions using their operational semantics (13). Following Owicki [335], we give a syntax-directed presentation of Floyd–Naur's method without appeal to a formal logic. To do this we associate preconditions and postcondition with commands and introduce structural verification conditions so that a proof of a composite command is composed of the proofs of its constituent parts. Although later this turns out to be redundant, we prove the semantical soundness and completeness of the method since the underlying reasoning constitutes a simple introduction to relative completeness proofs of Hoare logic.

5.3.1. *Preconditions and postconditions of commands*

A partial correctness proof of $\underline{\{p\}}\underline{C}\underline{\{q\}}$ by Floyd–Naur's method consists in discovering a precondition pre(C') and a postcondition post(C') specifying the partial correctness $\underline{\{\text{pre}(C')\}}\,\underline{C'}\,\underline{\{\text{post}(C')\}}$ of each component C' of command C. This includes an invariant linv(C') for each loop C' within C. Formally "pre", "post" and "linv" can be understood as functions which map components of C to assertions:

(47) DEFINITION (*preconditions, postconditions and loop invariants attached to commands*)

(47.1) pre, post : $\mathbf{Comp}[\![C]\!] \to \mathbf{Ass}$;

(47.2) linv : $\mathbf{Loops}[\![C]\!] \to \mathbf{Ass}$.

(48) EXAMPLE (*preconditions, postconditions and loop invariants for program* (4)). For program C defined by (4) with components defined by (5), we can choose

$$\begin{aligned}
&\mathrm{pre}(C^{\varepsilon})=\mathrm{pre}(C^{0})=\underline{P}_1,\\
&\mathrm{post}(C^{0})=\mathrm{pre}(C^{1})=\mathrm{linv}(C^{1})=\mathrm{post}(C^{1001})\\
&\qquad=\mathrm{post}(C^{100})=\mathrm{post}(C^{1011})=\mathrm{post}(C^{101})=\mathrm{post}(C^{10})=\underline{P}_2,\\
&\mathrm{pre}(C^{10})=\underline{P}_3, \qquad \mathrm{pre}(C^{100})=\mathrm{pre}(C^{1000})=\underline{P}_4,\\
&\mathrm{post}(C^{1000})=\mathrm{pre}(C^{1001})=\underline{P}_5, \qquad \mathrm{pre}(C^{101})=\mathrm{pre}(C^{1010})=\underline{P}_6,\\
&\mathrm{post}(C^{1010})=\mathrm{pre}(C^{1011})=\underline{P}_7, \qquad \mathrm{post}(C^{1})=\mathrm{post}(C^{\varepsilon})=\underline{P}_8.
\end{aligned}$$

5.3.2. Compositional verification conditions

Then these assertions should be proved to satisfy the following verification conditions which are defined compositionally that is, by recursion on the syntax of commands:

(49) DEFINITION (Owicki [335]) (*compositional Floyd–Naur partial correctness proof method*). A partial correctness proof of $\{\underline{p}\}\underline{C}\{\underline{q}\}$ by Floyd–Naur's method consists in discovering preconditions, postconditions and loop invariants (47) which must be proved to satisfy the following compositional verification conditions:

(49.1) $p\subseteq \mathrm{pre}(C) \wedge \mathrm{post}(C)\subseteq q.$

For each component $C' \in \mathbf{Comp}[\![C]\!]$ of C:

(49.2) if C' is skip then $\mathrm{pre}(C')\subseteq \mathrm{post}(C')$,

(49.3) if C' is $X := E$
then $\mathrm{pre}(C')\subseteq \{s\in S: s[X\leftarrow \underline{E}(s)] \in \mathrm{post}(C')\}$,

(49.4) if C' is $X := ?$
then $\{s[X\leftarrow d]: s\in \mathrm{pre}(C') \wedge d\in D\}\subseteq \mathrm{post}(C')$,

(49.5) if C' is $(C_1;C_2)$
then $\mathrm{pre}(C')\subseteq \mathrm{pre}(C_1) \wedge \mathrm{post}(C_1)\subseteq \mathrm{pre}(C_2) \wedge \mathrm{post}(C_2)\subseteq \mathrm{post}(C')$,

(49.6) if C' is $(B\rightarrow C_1 \diamond C_2)$
then $(\mathrm{pre}(C')\cap \underline{B})\subseteq \mathrm{pre}(C_1) \wedge (\mathrm{pre}(C')\cap \neg\underline{B})\subseteq \mathrm{pre}(C_2)$
$\wedge\, \mathrm{post}(C_1)\subseteq \mathrm{post}(C') \wedge \mathrm{post}(C_2)\subseteq \mathrm{post}(C')$,

(49.7) if C' is $(B*C_1)$
then $\mathrm{pre}(C')\subseteq \mathrm{linv}(C') \wedge (\mathrm{linv}(C')\cap \underline{B})\subseteq \mathrm{pre}(C_1)$
$\wedge\, \mathrm{post}(C_1)\subseteq \mathrm{linv}(C') \wedge (\mathrm{linv}(C')\cap \neg\underline{B})\subseteq \mathrm{post}(C')$.

Observe that these compositional verification conditions could also have been defined by an attribute grammar [255] using context-free grammar (1) with attributes “pre”, “post” and “linv” so that (49) expresses the relations between these attributes (see [161, 356]).

(50) EXAMPLE (*compositional verification conditions for program* (4)). These verification conditions are given below for program C defined by (4). Some of them, corresponding to assertions attached to the same label, are obviously satisfied:

$$\begin{aligned}&\text{pre}(C^{\varepsilon})\subseteq\text{pre}(C^{0})\wedge\text{post}(C^{0})\subseteq\text{pre}(C^{1})\wedge\text{post}(C^{1})\subseteq\text{post}(C^{\varepsilon})\\&\wedge\text{pre}(C^{100})\subseteq\text{pre}(C^{1000})\wedge\text{post}(C^{1000})\subseteq\text{pre}(C^{1001})\\&\wedge\text{post}(C^{1001})\subseteq\text{post}(C^{100})\subseteq\text{post}(C^{10})\wedge\text{pre}(C^{101})\subseteq\text{pre}(C^{1010})\\&\wedge\text{post}(C^{1010})\subseteq\text{pre}(C^{1011})\wedge\text{post}(C^{1011})\subseteq\text{post}(C^{101})\\&\qquad\qquad\subseteq\text{post}(C^{10})\subseteq\text{linv}(C^{1}).\end{aligned}$$

Moreover, (49.7) distinguishes the precondition of a loop (e.g., $\text{pre}(C^1)=(X=x\wedge Y=y\geqslant 0\wedge Z=1)$) from its invariant (e.g., $\text{linv}(C^1)=(Y\geqslant 0\wedge Z*(X**Y)=x**\text{y})$):

$$\text{pre}(C^1)\subseteq\text{linv}(C^1).$$

The remaining verification conditions correspond to elementary steps of the program. They are set-theoretic interpretations of formulae (26):

$$\begin{array}{ll}(\varepsilon) & p\subseteq\text{pre}(C^{\varepsilon})\\(i_1) & \text{pre}(C^{0})\subseteq\{s\in S: s[Z\leftarrow 1]\in\text{post}(C^{0})\}\\(i_2) & (\text{linv}(C^{1})\cap\{s\in S: s(Y)\neq 0\})\subseteq\text{pre}(C^{10})\\(i_3) & (\text{linv}(C^{1})\cap\neg\{s\in S: s(Y)\neq 0\})\subseteq\text{post}(C^{1})\\(i_4) & (\text{pre}(C^{10})\cap\{s\in S: \mathit{odd}(s(Y))\})\subseteq\text{pre}(C^{100})\\(i_5) & (\text{pre}(C^{10})\cap\neg\{s\in S: \mathit{odd}(s(Y))\})\subseteq\text{pre}(C^{101})\\(i_6) & \text{pre}(C^{1000})\subseteq\{s\in S: s[Y\leftarrow s(Y)-1]\in\text{post}(C^{1000})\}\\(i_7) & \text{pre}(C^{1001})\subseteq\{s\in S: s[Z\leftarrow s(Z)*s(X)]\in\text{post}(C^{1001})\}\\(i_8) & \text{pre}(C^{1010})\subseteq\{s\in S: s[Y\leftarrow s(Y)\ \mathit{div}\ 2]\in\text{post}(C^{1010})\}\\(i_9) & \text{pre}(C^{1011})\subseteq\{s\in S: s[X\leftarrow s(X)*s(X)]\in\text{post}(C^{1011})\}\\(\sigma) & \text{post}(C^{\varepsilon})\subseteq q.\end{array}$$

5.3.3. *Semantical soundness and completeness of the compositional Floyd–Naur partial correctness proof method*

The compositional presentation of Floyd–Naur's proof method is semantically sound and complete:

(51) THEOREM (soundness of the compositional Floyd–Naur proof method). *If verification conditions* (49) *are satisfied for all components C' of C then*

$$\forall C'\in\textbf{Comp}[\![C]\!].\underline{\{}\text{pre}(C')\underline{\}}\underline{C'}\underline{\{}\text{post}(C')\underline{\}}.$$

It follows from (49.1) *that* $\underline{\{}p\underline{\}}\underline{C}\underline{\{}q\underline{\}}$ *holds.*

PROOF. We prove that $\forall C'\in\textbf{Comp}[\![C]\!].\underline{\{}\text{pre}(C')\underline{\}}\underline{C'}\underline{\{}\text{post}(C'\underline{\}}$ or, equivalently, by (22) that $\forall C'\in\textbf{Comp}[\![C]\!].(\text{pre}(C')\rceil\underline{C'})\subseteq(S\times\text{post}(C'))$ by structural induction on C':

If C' is $X:=E$ then by (49.3) we have $\text{pre}(C')\subseteq\{s\in S: s[X\leftarrow\underline{E}(s)]\in\text{post}(C')\}\Leftrightarrow\{s[X\leftarrow\underline{E}(s)]: s\in\text{pre}(C')\}\subseteq\text{post}(C')\Leftrightarrow(\text{pre}(C')\rceil\underline{C'})\subseteq(S\times\text{post}(C'))$ by (19.2). The cases skip and $X:=?$ are handled the same way.

If C' is $(C_1;C_2)$ then $(\text{pre}(C')\rceil\underline{C_1;C_2})=(\text{pre}(C')\rceil\underline{C_1}\circ\underline{C_2})$ [by (19.4)] $=(\text{pre}(C'\rceil\underline{C_1})\circ\underline{C_2}\subseteq(S\times\text{post}(C_1))\circ\underline{C_2}$ [since $(\text{pre}(C_1)\rceil\underline{C_1})\subseteq(S\times\text{post}(C_1))$ by induction hypothesis] $=S^2\circ(\text{post}(C_1)\rceil\underline{C_2})\subseteq S^2\circ(\text{pre}(C_2)\rceil\underline{C_2})$ [by (49.5)] $\subseteq$

$S^2 \circ (S \times \text{post}(C_2))$ [since $(\text{pre}(C_2)\rceil \underline{C_2}) \subseteq (S \times \text{post}(C_2))$ by induction hypothesis] $=$ $S \times \text{post}(C_2) \subseteq S \times \text{post}(C')$ [by (49.5)]. The case $C' = (B \rightarrow C_1 \diamond C_2)$ is handled the same way.

If C' is $(B * C_1)$ then $(\text{pre}(C_1)\rceil \underline{C_1}) \subseteq (S \times \text{post}(\underline{C_1}))$ holds by induction hypothesis which implies $(\text{linv}(C')\rceil(\underline{B}\rceil \underline{C_1})) \subseteq (S \times \text{linv}(C'))$ by (49.7). Also $\text{pre}(C') \subseteq \text{linv}(C')$ and $(\text{linv}(C') \cap \neg \underline{B}) \subseteq \text{post}(C')$ by (49.6), so that $(\text{pre}(C')\rceil(\underline{B}\rceil \underline{C_1})^* \lceil \neg \underline{B}) \subseteq (S \times \text{post}(C'))$ by (28); hence $(\text{pre}(C')\rceil(\underline{B * C_1})) \subseteq (S \times \text{post}(C'))$ by (19.6). □

(52) THEOREM (semantical completeness of the compositional Floyd–Naur proof method). *If* $\underline{\{p\}C\{q\}}$ *holds then there are functions* pre, post *and* linv *verifying conditions* (49) *for all components* C' *of* C.

PROOF. The proof is by structural induction on C:

If C is $X := E$ then $\underline{\{p\}X := E\{q\}} \Rightarrow (p\rceil \underline{X := E}) \subseteq (S \times q)$ [by (22)] $\Rightarrow \{s[X \leftarrow \underline{E}(s)]: s \in p\} \subseteq q$ [by (19.2)] $\Rightarrow (49.1) \wedge (49.3)$ if we let $\text{pre}(X := E) = p$ and $\text{post}(X := E) = q$. The cases skip and $X := ?$ are handled the same way.

If C is $(C_1; C_2)$ then we let $\text{pre}(C) = \text{pre}(C_1) = p$, $\text{post}(C) = \text{post}(C_2) = q$ and $\text{post}(C_1) = \text{pre}(C_2) = \{s: \exists s' \in p . \langle s', s \rangle \in \underline{C_1}\}$. Then (49.1) and (49.5) are satisfied. It remains to show that (49.2), ..., (49.7) hold for all components C' of C. By induction hypothesis, we just have to show that $\underline{\{\text{pre}(C_1)\}C_1\{\text{post}(C_1)\}}$ and $\underline{\{\text{pre}(C_2)\}C_2\{\text{post}(C_1)\}}$. We have $\underline{\{p\}C_1\{\{s: \exists s' \in p . \langle s', s \rangle \in \underline{C_1}\}\}}$ because $(p\rceil \underline{C_1}) = \{\langle s, s' \rangle: s \in p \wedge \langle s, s' \rangle \in \underline{C_1}\} \subseteq \{\langle s'', s' \rangle: \exists s \in p . \langle s, s' \rangle \in \underline{C_1}\} = S \times \{s: \exists s' \in p . \langle s', s \rangle \in \underline{C_1}\}$. Moreover $\underline{\{p\}(C_1;C_2)\{q\}} \Rightarrow (p\rceil \underline{(C_1;C_2)}) \subseteq (S \times q)$ [by (22)] $\Rightarrow ((p\rceil \underline{C_1}) \circ \underline{C_2}) \subseteq (S \times q)$ [by (19.4)] $\Rightarrow (\forall s', s, s'' \in S . (s' \in p \wedge \langle s', s \rangle \in \underline{C_1} \wedge \langle s, s'' \rangle \in \underline{C_2}) \Rightarrow (s'' \in q)) \Rightarrow ((\{s: \exists s' \in p . \langle s', s \rangle \in \underline{C_1}\}\rceil \underline{C_2}) \subseteq (S \times q)) \Rightarrow \underline{\{\text{pre}(C_2)\}C_2\{\text{post}(C_1)\}}$ by (22).

The proof is similar when C is $(B \rightarrow C_1 \diamond C_2)$ choosing $\text{pre}(C') = p$, $\text{pre}(C_1) = p \cap \underline{B}$, $\text{pre}(C_2) = p \cap \neg \underline{B}$ and $\text{post}(C') = \text{post}(C_1) = \text{post}(C_2) = q$.

If C is $(B * C_1)$ then $\underline{\{p\}(B * C_1)\{q\}} \Rightarrow (p\rceil(\underline{B * C_1})) \subseteq (S \times q)$ [by (22)] $\Rightarrow$ $(p\rceil(\underline{B}\rceil \underline{C_1})^* \lceil \neg \underline{B}) \subseteq (S \times q)$ [by (19.6)] $\Rightarrow$ $[\exists i \in \mathbf{Ass} . (p \subseteq i) \wedge (i\rceil(\underline{B}\rceil \underline{C_1}) \subseteq S \times i) \wedge (i \cap \neg \underline{B} \subseteq q)]$ [by (28)]. We now define $\text{pre}((B * C_1)) = p$, $\text{linv}((B * C_1)) = i$, $\text{post}((B * C_1)) = q$, $\text{pre}(C_1) = i \cap \underline{B}$ and post $(C_1) = i$. It follows that (49.1) and (49.7) hold for C. It remains to show that $\underline{\{\text{pre}(C_1)\}C_1\{\text{post}(C_1)\}}$. This immediately follows from the definitions of $\text{pre}(C_1)$ and post (C_1), $(i \cap \underline{B})\rceil \underline{C_1} \subseteq (S \times i)$ and (22). □

5.4. *Equivalence of stepwise and compositional Floyd–Naur partial correctness proofs*

Examples (26) and (50) show that the compositional Floyd–Naur partial correctness proof method introduces some trivially satisfied verification conditions which do not appear in the stepwise version. Apart from this difference in the presentation, the stepwise and compositional Floyd–Naur partial correctness proofs of program (4) are equivalent. This property is general in the sense that a proof using one presentation can always be derived from a proof using the other presentation. Since the assertions are the same in both presentations, (30.3) and (36) imply that preconditions and postconditions in the compositional presentation (hence later in Hoare logic) are local invariants, a fact which is often taken for granted. By (46), this also implies that the syntax-directed presentation is semantically sound and complete, a fact already proved by (51) and (52).

5.4.1. *The compositional presentation of a stepwise Floyd–Naur partial correctness proof*

The precondition pre(C') of a component C' of a command $C \in \mathbf{Com}$ (and loop invariant linv(C') when C' is a loop) can always be chosen as the local invariant inv(L) attached to the label $L = \mathbf{Lpre}[\![C]\!]\,[C']$ designating where control is just before executing that component C'. The same way, the postcondition post(C') of a component C' of a command C can always be chosen as the local invariant inv(L) attached to the label $L = \mathbf{Lpost}[\![C]\!]\,[C']$ designating where control is just after executing that component C'.

(53) DEFINITION (*program points before and after components of a command*). The label just before and after a component C' of a command $C \in \mathbf{Com}$:

(53.1) $\quad \mathbf{Lpre}[\![C]\!] \in \mathbf{Comp}[\![C]\!] \to \mathbf{Lab}[\![C]\!]$,

(53.2) $\quad \mathbf{Lpost}[\![C]\!] \in \mathbf{Comp}[\![C]\!] \to \mathbf{Lap}[\![C]\!]$

is defined by structural induction on C:

(53.3) $\quad \mathbf{Lpre}[\![C]\!][C] = C$,

(53.4) $\quad \mathbf{Lpost}[\![C]\!][C] = \surd$.

For each $C' \in \mathbf{Comp}[\![C]\!] - \{C\}$ when C is $(C_1; C_2)$, $(B \to C_1 \diamond C_2)$ or $(B * C_1)$:

(53.5) $\quad \mathbf{Lpre}[\![(C_1;C_2)]\!]\,[C'] = (\mathbf{Lpre}[\![C_1]\!]\,[C'];C_2) \quad$ if $C' \in \mathbf{Comp}[\![C_1]\!]$,

(53.6) $\quad \mathbf{Lpre}[\![C_1; C_2)]\!][C'] = \mathbf{Lpre}[\![C_2]\!][C'] \quad$ if $C' \in \mathbf{Comp}[\![C_2]\!]$,

(53.7) $\quad \mathbf{Lpost}[\![(C_1; C_2)]\!][C'] = C_2$
$\quad$ if $C' \in \mathbf{Comp}[\![C_1]\!] \wedge \mathbf{Lpost}[\![C_1]\!][C'] = \surd$,

(53.8) $\quad \mathbf{Lpost}[\![(C_1; C_2)]\!][C'] = (\mathbf{Lpost}[\![C_1]\!][C']; C_2)$
$\quad$ if $C' \in \mathbf{Comp}[\![C_1]\!] \wedge \mathbf{Lpost}[\![C_1]\!][C'] \neq \surd$,

(53.9) $\quad \mathbf{Lpost}[\![(C_1; C_2)]\!][C'] = \mathbf{Lpost}[\![C_2]\!][C'] \quad$ if $C' \in \mathbf{Comp}[\![C_2]\!]$,

(53.10) $\quad \mathbf{Lpre}[\![(B \to C_1 \diamond C_2)]\!][C'] = \mathbf{Lpre}[\![C_1]\!][C'] \quad$ if $C' \in \mathbf{Comp}[\![C_1]\!]$,

(53.11) $\quad \mathbf{Lpre}[\![(B \to C_1 \diamond C_2)]\!][C'] = \mathbf{Lpre}[\![C_2]\!][C'] \quad$ if $C' \in \mathbf{Comp}[\![C_2]\!]$,

(53.12) $\quad \mathbf{Lpost}[\![(B \to C_1 \diamond C_2)]\!][C'] = \mathbf{Lpost}[\![C_1]\!][C'] \quad$ if $C' \in \mathbf{Comp}[\![C_1]\!]$,

(53.13) $\quad \mathbf{Lpost}[\![(B \to C_1 \diamond C_2)]\!][C'] = \mathbf{Lpost}[\![C_2]\!][C'] \quad$ if $C' \in \mathbf{Comp}[\![C_2]\!]$,

(53.14) $\quad \mathbf{Lpre}[\![(B * C_1)]\!][C'] = (\mathbf{Lpre}[\![C_1]\!][C']; (B * C_1)) \quad$ if $C' \in \mathbf{Comp}[\![C_1]\!]$,

(53.15) $\quad \mathbf{Lpost}[\![(B * C_1)]\!][C'] = (B * C_1)$
$\quad$ if $C' \in \mathbf{Comp}[\![C_1]\!] \wedge \mathbf{Lpost}[\![C_1]\!]\,[C'] = \surd$,

(53.16) $\textbf{Lpost}[\![(\underline{B}*C_1)]\!]\,[C'] = (\textbf{Lpost}[\![C_1]\!]\,[C']; (B*C_1))$
if $C' \in \textbf{Comp}[\![C_1]\!] \wedge \textbf{Lpost}[\![C_1]\!]\,[C'] \neq \surd$.

(54) THEOREM (compositional presentation of a stepwise proof). *If* inv $\in \textbf{Lab}[\![C]\!] \to \textbf{Ass}$ *satisfies* (45) *then*

(54.1) $\text{pre} = \lambda C' \in \textbf{Comp}[\![C]\!].\ \text{inv}(\textbf{Lpre}[\![C]\!]\,[C'])$,

(54.2) $\text{linv} = \lambda C' \in \textbf{Loops}[\![C]\!]\ \text{inv}(\textbf{Lpre}[\![C]\!][C'])$,

(54.3) $\text{post} = \lambda C' \in \textbf{Comp}[\![C]\!].\ \text{inv}(\textbf{Lpost}[\![C]\!]\,[C'])$

satisfy (49).

5.4.2. *The stepwise presentation of a compositional Floyd–Naur partial correctness proof*

(55) THEOREM (stepwise presentation of a compositional proof). *If* pre, post $\in \textbf{Comp}[\![C]\!] \to \textbf{Ass}$ *and* linv $\in \textbf{Loops}[\![C]\!] \to \textbf{Ass}$ *satisfy* (49) *then* inv $\in \textbf{Labs}[\![C]\!] \to \textbf{Ass}$ *defined as follows by structural induction on C*:

(55.1) $\text{inv}(C) = \text{linv}(C)$ *if* $C \in \textbf{Loops}$,

(55.2) $\text{inv}(C) = \text{pre}(C)$ *if* $C \in \textbf{Comp} - \textbf{Loops}$,

(55.3) $\text{inv}(\surd) = \text{post}(C)$,

(55.4) $\text{inv}((C_1'; C_2)) = \text{inv}(C_1')$ *if* $C = (C_1; C_2) \wedge C_1' \in \textbf{In}[\![C_1]\!]$,

(55.5) $\text{inv}((C'; (B*C_1))) = \text{inv}(C')$ *if* $C = (B*C_1) \wedge C' \in \textbf{At}[\![C_1]\!] \cup \textbf{In}[\![C_1]\!]$

satisfies (45).

5.5. *Variants of the Floyd–Naur partial correctness proof method*

Lemma (29) and hence Floyd–Naur's method has a great number of equivalent variants, each one leading to a different partial correctness proof methodology [92]. For example, Manna [284] uses an invariant i and a output specification q which relate the possible configurations during execution to the initial states of variables. Otherwise stated, one uses relations between the current and initial values of variables instead of assertions upon their current values. More formally, we have:

(56) DEFINITION (Manna [284]) (*relational partial correctness*)

$p: \textbf{Ispec} = \mathscr{P}(S)$ *input specifications,*

$q: \textbf{Ospec} = \mathscr{P}(S \times S)$ *output specifications,*

$\underline{\{p\}}\underline{C}\underline{\langle q\rangle} : \textbf{Ispec} \times \textbf{Com} \times \textbf{Ospec} \to \{\text{tt}, \text{ff}\}$

$\underline{\{p\}}\underline{C}\underline{\langle q\rangle} = (p \rceil \underline{C}) \subseteq q$ *relational partial correctness.*

Induction principle (29) can be rephrased as follows for relational partial correctness:

(57) Theorem (Manna [284], Cousot & Cousot [92]) (stepwise partial correctness relational proofs using invariants)

$$\underline{\{p\}C\langle q\rangle} = [\exists i \in \mathscr{P}(S \times \Gamma).\{\langle s, \langle s, C\rangle\rangle: s \in p\} \subseteq i \wedge \{\langle s, \gamma'\rangle: \exists \gamma.\langle s, \gamma\rangle \in i \wedge \langle \gamma, \gamma'\rangle \in op[\![C]\!]\} \subseteq i \wedge i \cap S^2 \subseteq q].$$

It is also possible to prove relational partial correctness using an invariant i which relates the possible configurations during execution to the final states of the variables:

(58) Theorem (Morris & Wegbreit [313], Cousot & Cousot [92]) (*subgoal induction*)

(58.1) $\underline{\{p\}C\langle q\rangle} = [\exists i \in \mathscr{P}(\Gamma \times S).\{\langle s, s\rangle: s \in S\} \subseteq i$

(58.2) $\wedge \{\langle \gamma, s\rangle: \exists \gamma'.\langle \gamma, \gamma'\rangle \in op[\![C]\!] \wedge \langle \gamma', s\rangle \in i\} \subseteq i$

(58.3) $\wedge \{\langle s, s'\rangle: s \in p \wedge \langle\langle s, C\rangle, s'\rangle \in i\} \subseteq q].$

Assume that $\langle s, s'\rangle \in \underline{C}$ so that execution $\gamma_0, \ldots, \gamma_n$ of command C starting in configuration $\gamma_0 = \langle s, C\rangle$ with $s \in p$, such that $\langle \gamma_{k-1}, \gamma_k\rangle \in op[\![C]\!]$ for $k = 1, \ldots, n$, does terminate in state $\gamma_n = s'$. Then $\langle \gamma_n, s'\rangle \in i$ by (58.1) and by downward induction on $k = n, n-1, \ldots, 0$, $\langle \gamma_k, s'\rangle \in i$ follows from (58.2). In particular, $\langle \gamma_0, s'\rangle \in i$ whence $\langle s, s'\rangle \in q$ by (58.3). It follows that $(p \rceil \underline{C}) \subseteq q$ whence $\underline{\{p\}C\langle q\rangle}$ holds. Semantical completeness follows from the fact that i can always be chosen as $op[\![C]\!]^* \lceil S$.

Semantical soundness and completeness imply that these partial correctness proof methods are all equivalent. This is also true in the stronger sense that the necessary invariants can be derived from one another:

(59) Theorem (Cousot & Cousot [92], Dijkstra [125]) (equivalence of stepwise induction and subgoal induction)

(59.1) *if* i *satisfies* (57) *then* $i' = \{\langle \gamma, s\rangle: \forall s'.\ \langle s', \gamma\rangle \in i \Rightarrow \langle s', s\rangle \in q\}$ *satisfies* (58);

(59.2) *if* i' *satisfies* (58) *then* $i = \{\langle s, \gamma\rangle: \forall s''.\ \langle \gamma, s''\rangle \in i' \Rightarrow \langle s, s''\rangle \in q\}$ *satisfies* (57).

Proof. If i satisfies (57) then $i \cap S^2 \subseteq q$ so that $\forall s \in S.\ \forall s' \in S.\langle s', s\rangle \in i \Rightarrow \langle s', s\rangle \in q$, hence $\{\langle s, s\rangle: s \in S\} \subseteq i'$ [by 33.1]. Moreover $(\langle \gamma, \gamma'\rangle \in op[\![C]\!] \wedge \langle \gamma', s\rangle \in i' \wedge \langle s', \gamma\rangle \in i) \Rightarrow (\langle s', \gamma'\rangle \in i \wedge \langle \gamma', s\rangle \in i')$ [by (57)] $\Rightarrow (\langle s', \gamma'\rangle \in i \wedge (\forall s'.\langle s', \gamma'\rangle \in i \Rightarrow \langle s', s\rangle \in q))$ [by (59.1)] $\Rightarrow \langle s', s\rangle \in q$ so that $\{\langle \gamma, s\rangle: \exists \gamma'.\langle \gamma, \gamma'\rangle \in op[\![C]\!] \wedge \langle \gamma', s\rangle \in i'\} \subseteq i'$. Finally, $(s \in p \wedge \langle\langle s, C\rangle, s'\rangle \in i') \Rightarrow (\langle s, \langle s, C\rangle\rangle \in i \wedge (\forall s''.\langle s'', \langle s, C\rangle\rangle \in i \Rightarrow \langle s'', s'\rangle \in q))$ [by (57) and (59.1)] $\Rightarrow \langle s, s'\rangle \in q$, hence $\{\langle s, s'\rangle: s \in p \wedge \langle\langle s, C\rangle, s'\rangle \in i'\} \subseteq q$.

If i' satisfies (58) then $\{\langle s, s\rangle: s \in S\} \subseteq i'$ so that $i \cap S^2 = \{\langle s, s'\rangle: \forall s''.\ \langle s', s''\rangle \in i' \Rightarrow \langle s, s''\rangle \in q\} \subseteq \{\langle s, s'\rangle: \langle s', s'\rangle \in i' \Rightarrow \langle s, s'\rangle \in q\} = q$. Moreover $(\langle s, \gamma\rangle \in i \wedge \langle \gamma, \gamma'\rangle \in op[\![C]\!] \wedge \langle \gamma', s''\rangle \in i') \Rightarrow ((\forall s'.\langle \gamma, s'\rangle \in i' \Rightarrow \langle s, s'\rangle \in q) \wedge \langle \gamma, s''\rangle \in i')$ [by (59.2) and (58)]

$\Rightarrow \langle s, s''\rangle \in q$ so that $\{\langle s, \gamma'\rangle : \exists\gamma . \langle s, \gamma\rangle \in i \wedge \langle\gamma, \gamma'\rangle \in op[\![C]\!]\} \subseteq i$. Finally, $\forall s \in p . \forall s''$. $\langle\langle s, C\rangle, s''\rangle \in i' \Rightarrow \langle s, s''\rangle \in q$ [by (58)] hence $\{\langle s, \langle s, C\rangle\rangle : s \in p\} \subseteq i$ [by (59.2)]. □

Replacing i by $\neg j$ in (58), we obtain an equivalent relational partial correctness proof method proceeding by reductio ad absurdum:

(60) THEOREM (Cousot & Cousot [92]) (induction principle for contrapositive proofs)

(60.1) $\quad \underline{\{p\}}\underline{C}\underline{\langle q\rangle} = [\exists j \in \mathscr{P}(\Gamma \times S) . p \rceil \neg q \subseteq \{\langle s, s'\rangle : \langle\langle s, C\rangle, s'\rangle \in j\}$

(60.2) $\quad \wedge j \subseteq \{\langle\gamma, s\rangle : \forall\gamma' . \langle\gamma, \gamma'\rangle \in op[\![C]\!] \Rightarrow \langle\gamma', s\rangle \in j\}$

(60.3) $\quad \wedge \{\langle s, s\rangle : s \in S\} \subseteq \neg j]$.

This consists in proving an invariance property by considering the situation where the contrary property should be true and in establishing that this situation is impossible. Assume that $\langle s, s'\rangle \in \underline{C}$ so that execution $\gamma_0, \ldots, \gamma_n$ of command C starting in configuration $\gamma_0 = \langle s, C\rangle$ with $s \in p$, such that $\langle\gamma_{k-1}, \gamma_k\rangle \in op[\![C]\!]$ for $k = 1, \ldots, n$, does terminate in state $\gamma_n = s'$. Assume, by reductio ad absurdum, that $\langle s, s'\rangle \notin q$. Then $\langle\gamma_0, s'\rangle \in j$ by (60.1) and by induction on $k = 1, \ldots, n$, $\langle\gamma_k, s'\rangle \in j$ follows from (60.2). In particular $\langle\gamma_n, s'\rangle \in j$ in contradiction with $\langle\gamma_n, s'\rangle \notin j$ following from (60.3). Semantical completeness follows from the fact that the contrainvariant j can always be chosen as $\neg(op[\![C]\!]^* \lceil S)$.

This may lead to simpler proofs when the "absurd" configuration is much simpler than the "sensible" one (see [397] for an example).

6. Liveness proof methods

Obviously termination is not implied in partial correctness since, for example, if $\underline{\text{true}} = I[\![\text{true}]\!] = S$, then $\underline{\text{true}*\text{skip}} = \emptyset$ so that $\underline{\{p\}}\underline{\text{true}*\text{skip}}\underline{\{q\}}$ is true for all $p, q \in \mathbf{Ass}$. Floyd [143] originally introduced total correctness as the conjunction of partial correctness and termination. Hoare logic has also been extended to cope with termination and more generally with *liveness* properties of programs [5].

We first introduce execution traces generated by the operational semantics (13) so as to define total correctness and prove that it is the conjunction of partial correctness, deadlock freeness and termination. After giving a short mathematical recall on well-founded relations, well-orderings and ordinals, we introduce Floyd's well-founded set method [143] to prove termination of programs. We next consider an extension of this method to prove liveness properties $P \overset{C}{\leadsto} Q$ stipulating that starting from a configuration of P, program C does eventually reach a configuration of Q [258]. Finally we study Burstall's intermittent assertion method [64] for proving total correctness [291, 187] and generalize it to arbitrary liveness properties. After proper generalization, Burstall's method includes Floyd's method [95] and is more flexible since it allows the combination of inductions on various underlying structures of the program (syntax, computation, data, etc.).

6.1. *Execution traces*

We use execution traces to record the successive configurations that can be encountered during a terminating or nonterminating execution of a program. Since programs are nondeterministic, they can have many different possible executions so that we have to use sets of finite or infinite traces. The theory of traces is surveyed by Mazurkiewicz [296].

Let a set **Com** of programs be given, as well as a set S of states so that the set of configurations is $\Gamma=(S\times \mathbf{Com})\cup S$ and an operational semantics $op\in \mathbf{Com}\to\mathscr{P}(\Gamma\times\Gamma)$.

(61) DEFINITION (*execution traces*). The set of *finite complete execution traces of length* $n\in\mathbf{N}^+$ for command $C\in\mathbf{Com}$ starting in configuration $\gamma\in\Gamma$ is

(61.1) $$\begin{aligned}\Sigma^n[\![C]\!]\gamma=\{\sigma\in seq^n\,\Gamma\colon\ &\sigma_0=\gamma\\ &\wedge\forall i\in\{1,\ldots,n-1\}.\langle\sigma_{i-1},\sigma_i\rangle\in op[\![C]\!]\\ &\wedge\forall\gamma\in\Gamma.\ \langle\sigma_{n-1},\gamma\rangle\notin op[\![C]\!]\},\end{aligned}$$

the set of *infinite traces* of execution of command $C\in\mathbf{Com}$ starting in configuration $\gamma\in\Gamma$ is

(61.2) $$\Sigma^\omega[\![C]\!]\gamma=\{\sigma\in seq^\omega\,\Gamma\colon\ \sigma_0=\gamma\wedge\forall i\in\mathbf{N}.\ \langle\sigma_i,\sigma_{i+1}\rangle\in op[\![C]\!]\},$$

so that the set of *finite traces* of execution of command $C\in\mathbf{Com}$ starting in configuration $\gamma\in\Gamma$ is

(61.3) $$\Sigma^*[\![C]\!]\gamma=\bigcup\{\Sigma^n[\![C]\!]\gamma\colon\ n\in\mathbf{N}\},$$

and the set of *traces* of execution of command $C\in\mathbf{Com}$ starting in configuration $\gamma\in\Gamma$ is

(61.4) $$\Sigma[\![C]\!]\gamma=\Sigma^*[\![C]\!]\gamma\cup\Sigma^\omega[\![C]\!]\gamma.$$

Given $p\in\mathbf{Ass}$, we can define the traces of command C starting in a state of $p\in\mathbf{Ass}$ as

(61.5) $$\Sigma^n[\![C]\!](p)=\bigcup\{\Sigma^n[\![C]\!]\langle s,C\rangle\colon\ s\in p\},$$

(61.6) $$\Sigma^*[\![C]\!](p)=\bigcup\{\Sigma^*[\![C]\!]\langle s,C\rangle\colon\ s\in p\},$$

(61.7) $$\Sigma^\omega[\![C]\!](p)=\bigcup\{\Sigma^\omega[\![C]\!]\langle s,C\rangle\colon\ s\in p\},$$

(61.8) $$\Sigma[\![C]\!](p)=\bigcup\{\Sigma[\![C]\!]\langle s,C\rangle\colon\ s\in p\}.$$

These sets of traces can also be given an equational definition, see [115].

6.2. *Total correctness*

Let us recall (22) that a command C is said to be *partially correct* with respect to a specification $\langle p,q\rangle$ (written $\underline{\{p\}C\{q\}}$) if any terminating execution of C starting from an initial state s satisfying p must end in some final state s' satisfying q. C is said to be *totally correct* with respect to $\langle p,q\rangle$ (written $\underline{[p]C[q]}$) if any execution of C starting from an initial state s satisfying p does terminate properly in a final state s' satisfying q. Partial and total correctness can be defined in terms of execution traces as follows:

DEFINITIONS (Floyd [143]) (*partial and total correctness*)

(62) $$\underline{[p]C[q]}:\mathbf{Ass}\times\mathbf{Com}\times\mathbf{Ass}\to\{\mathrm{tt},\mathrm{ff}\}\quad \textit{total correctness}$$
$$\underline{[p]C[q]}=\forall\sigma\in\Sigma[\![C]\!](p).\ \exists n\in\mathbf{N}^+.\ \sigma\in\Sigma^n[\![C]\!](p)\wedge\sigma_{n-1}\in q,$$

(63) $\underline{\{p\}C\{q\}}: \textbf{Ass} \times \textbf{Com} \times \textbf{Ass} \rightarrow \{\text{tt}, \text{ff}\}$ *partial correctness*

$$\underline{\{p\}C\{q\}} = \forall \sigma \in \Sigma[\![C]\!](p).\, \forall i \in dom\, \sigma.\, (\sigma_i \in S) \Rightarrow (\sigma_i \in q).$$

Total correctness as defined by (62) does not necessarily imply partial correctness (63) because definition (62) does not imply that all states $\sigma_i \in S$ belong to q. However, this follows from (13) because final states $s \in S$ have no possible successor, an hypothesis which we subsequently make about the operational semantics:

(64) HYPOTHESIS (final states are blocking states)

$$\forall C \in \textbf{Com}.\, \forall s \in S.\, \forall \gamma \in \Gamma.\, \langle s, \gamma \rangle \notin op[\![C]\!].$$

Observe that (64) $\Rightarrow$ [(62) $\Rightarrow$ (63)]. A command C is said to *terminate* for initial states $s \in p$ if and only if no execution trace starting from configuration $\langle s, C \rangle$ can be infinite:

(65) DEFINITION (*termination*)

$$\tau\underline{[p]C}: \textbf{Ass} \times \textbf{Com} \rightarrow \{\text{tt}, \text{ff}\}$$
$$\tau\underline{[p]C} = \forall \sigma \in \Sigma[\![C]\!](p).\, \exists n \in \textbf{N}^+.\, \sigma \in \Sigma^n[\![C]\!](p).$$

Termination is *proper* or *clean* for final states $\sigma_{n-1} \in S$. Execution may also end with other blocking states $\sigma_{n-1} \in \Gamma - S$. For example, a sequential program may be blocked by a runtime error such as division by zero or a parallel program may be permanently blocked because all processes are delayed at synchronization commands. Execution of a command C starting with initial states $s \in p$ can be *blocked* if and only if it can reach some state σ_{n-1} which is not final and has no possible successor:

(66) DEFINITION (*blocked execution*)

$$\beta\underline{\{p\}C}: \textbf{Ass} \times \textbf{Com} \rightarrow \{\text{tt}, \text{ff}\}$$
$$\beta\underline{\{p\}C} = \exists n \in \textbf{N}^+.\, \exists \sigma \in \Sigma^n[\![C]\!](p).\, \sigma_{n-1} \notin S.$$

When no execution of a command C starting with initial states $s \in p$ can end in a blocking configuration, we say that these executions are *deadlock free*:

(67) DEFINITION (*deadlock freedom*)

$$\neg\beta\underline{\{p\}C} = \forall n \in \textbf{N}^+.\, \forall \sigma \in \Sigma^n[\![C]\!](p).\, \sigma_{n-1} \in S.$$

Under hypothesis (64), total correctness is the conjunction of partial correctness, deadlock freedom and termination:

(68) THEOREM (characterization of total correctness). (64) *implies*

$$\underline{[p]C[q]} = \underline{\{p\}C\{q\}} \wedge \neg\beta\underline{\{p\}C} \wedge \tau\underline{[p]C}.$$

According to (13), final states $s \in S$ are the only possible states with no possible successor, an hypothesis that is sometimes made about the operational semantics:

(69) HYPOTHESIS (final states are the only blocking states)

$$\forall C \in \mathbf{Com}.\ \forall \gamma \in \Gamma.\ (\forall \gamma' \in \Gamma.\ \langle \gamma, \gamma' \rangle \notin op[\![C]\!]) \Rightarrow (\gamma \in S).$$

Under hypothesis (64) and (69), total correctness can be expressed as the conjunction of partial correctness and termination:

(70) THEOREM (Floyd [143]) (characterization of total correctness). (64) *and* (69) *imply*

$$\underline{[p]C[q]} = \underline{\{p\}C\{q\}} \wedge \underline{[p]C[\text{true}]}.$$

PROOF. (69) implies that $\tau\underline{[p]C} = \underline{[p]C[\text{true}]} = \forall \sigma \in \Sigma[\![C]\!](p).\ \exists i \in dom\ \sigma.\ \sigma_i \in S$ where $\underline{\text{true}} = I[\![\text{true}]\!] = S$, and that $\beta\underline{\{p\}C} = \text{ff}$ since no execution can be blocked. By (64) and (69), $\underline{[p]C[q]} = \forall \sigma \in \Sigma[\![C]\!](p).\ \exists i \in dom\ \sigma.\ \sigma_i \in q = \underline{\{p\}C\{q\}} \wedge \underline{[p]C[\text{true}]}$. □

6.3. *Well-founded relations, well-orderings and ordinals*

A relation $\prec$ on a class W is *well-founded* if and only if every subclass of W has a minimal element; that is, $wf(W, \prec) = [\forall E \subseteq W.\ (E \neq \emptyset \Rightarrow \exists y \in E.\ (\neg \exists z \in E.\ z \prec y))]$ is true. If $wf(W, \prec)$ is true then obviously there is no infinite decreasing sequence $x_0 \succ x_1 \succ \cdots$ where $\succ$ is the inverse of $\prec$.

A relation $\prec$ on a class W is a *strict partial ordering* if and only if it is antireflexive and transitive; that is, $spo(W, \prec) = [(\forall x \in W.\ \neg(x \prec x)) \wedge (\forall x, y, z \in W.\ (x \prec y \wedge y \prec z) \Rightarrow (x \prec z))]$ is true. Observe that if $\leqslant$ is a partial ordering on W then $x < y = (x \leqslant y \wedge x \neq y)$ is a strict partial ordering on W whereas if $<$ is a strict partial ordering on W then $x \leqslant y = (x < y \vee x = y)$ is a partial ordering on W. A *linear ordering* on W is a strict partial ordering such that any two different elements of W are comparable: $lo(W, \prec) = spo(W, \prec) \wedge [\forall x, y \in W.((x \neq y) \Rightarrow (x \prec y \vee y \prec x))]$. A relation $\prec$ on a class W is *well-ordered* if and only if it is a well-founded linear ordering on W: $wo(W, \prec) = wf(W, \prec) \wedge lo(W, \prec)$. A *well-order* is a pair $\langle W, \prec \rangle$ such that $wo(W, \prec)$.

To study common properties of well-ordered relations independently of their support class W, mathematicians have introduced a universal well-order called the class **Ord** of *ordinals* ordered by $<$. We say that two well-orderings $\langle W_1, \prec_1 \rangle$ and $\langle W_2, \prec_2 \rangle$ have the same *order type* if there exists a bijection ι from W_1 onto W_2 such that $x \prec_1 y \Leftrightarrow \iota(x) \prec_2 \iota(y)$. An ordinal can be understood as the class of all well-orderings of the same order type. Intuitively **Ord** is the transfinite sequence $0 < 1 < 2 < \cdots < \omega < \omega + 1 < \omega + 2 < \cdots < \omega + \omega = \omega \cdot 2 < \omega \cdot 2 + 1 < \omega \cdot 2 + 2 < \cdots < \omega \cdot 2 + \omega = \omega \cdot 3 < \cdots < \omega \cdot 4 < \cdots < \omega \cdot \omega = \omega^2 < \omega^2 + 1 < \cdots < \omega^2 \cdot \omega = \omega^3 < \cdots < \omega^\omega = 2_\omega < \cdots < \omega^{\omega^\omega} = 3_\omega < \cdots < \varepsilon_0 = \omega^{\omega^{\omega \cdots}}$ (ω times) $= \omega_\omega < \cdots < \omega_{\omega_\omega} < \cdots$ and so on (although what is behind may seems inaccessible, indeed ineffable). An ordinal α is a *limit ordinal* if it is neither 0 nor the successor of an ordinal; that is, if $\beta < \alpha$ then there is an ordinal γ such that $\beta < \gamma < \alpha$. The first limit ordinal ω is the order type of $\mathbf{N}$ well-ordered by $<$. If

$C \subset \mathbf{Ord}$ then *lub* C is the least upper bound of C $[\forall x \in C . x \leqslant lub\ C \wedge \forall a \in \mathbf{Ord}. ((\forall x \in C . x \leqslant a) \Rightarrow (lub\ C \leqslant a))]$ and $lub^+\ C$ is the least strict upper bound of C $[\forall x \in C . x < lub^+\ C \wedge \forall a \in \mathbf{Ord}. ((\forall x \in C . x < a) \Rightarrow (lub^+\ C \leqslant a))]$. A more detailed and rigorous presentation of ordinals can be found in [370].

A well-founded relation $\prec$ on a set W can be embedded into a well-ordered relation on W (using Knuth's topological sorting algorithm [254] when W is finite), hence into an initial segment of the ordinals. Assuming $wf(W, \prec)$, we can do this by the *rank function* $rk_{(W,\prec)}$ (for short $rk_\prec$) defined by $rk_\prec(x) = lub^+\{rk_\prec(y): y \prec x\}$. Minimal objects x of W (with no $y \prec x$) will have rank 0. The objects x of W which are not minimal but which are such that $y \prec x$ only for minimal objects y of W will have rank 1, and so on. One can easily verify by induction on $\prec$ that $rk_\prec(x)$ is an ordinal. Observe also that $(x \prec y) \Rightarrow (rk_\prec(x) < rk_\prec(y))$. We call $rk_\prec(W) = lub^+\{rk_\prec(x): x \in W\}$ the *rank* of $(W, \prec)$.

6.4. *Termination proofs by Floyd's well-founded set method*

To prove termination, we must discover a well-founded (Floyd in [143] proposed well-ordered) relation $\prec$ on a set W and a *variant function* $f: \Gamma \to W$, and show that its value decreases after each program step: $\forall \langle \gamma, \gamma' \rangle \in op[\![C]\!]. f(\gamma') \prec f(\gamma)$.

(71) EXAMPLE (*proof of termination of program* (4)). Program (4) was proved to be partially correct in example (24) using local invariants (25). To prove termination, let W be $\mathbf{N} \times \{L_1, \ldots, L_8\}$ let and $\prec$ be the well-founded relation on W defined by $\langle Y, L \rangle \prec \langle Y', L' \rangle$ if and only if $(0 \leqslant Y < Y') \vee (Y = Y' \wedge L \ll L')$ where $<$ is the usual ordering on natural numbers $0 < 1 < 2 < \cdots$ and $\ll$ is defined by $L_8 \ll L_6 \ll L_4 \ll L_3 \ll L_2 \ll L_7 \ll L_5 \ll L_1$. Let $f: \Gamma \to W$ be defined by $f(\langle L_i, s \rangle) = f_i(s(Y))$ for $i = 1, \ldots, 7$ and $f(s) = f_8(s(Y))$ where $f_i(y) = \langle y, L_i \rangle$. The proof that the value of f decreases after each program step amounts to the following local arguments:

(72)

(i_1) $[P_1(X, Y, Z, x, y) \wedge Z' = 1 \wedge P_2(X, Y, Z', x, y)] \Rightarrow f_2(Y) \prec f_1(Y)$ (since $L_2 \ll L_1$)

(i_2) $[P_2(X, Y, Z, x, y) \wedge Y \neq 0 \wedge P_3(X, Y, Z, x, y)] \Rightarrow f_3(Y) \prec f_2(Y)$ (since $L_3 \ll L_2$)

(i_3) $[P_2(X, Y, Z, x, y) \wedge Y = 0 \wedge P_8(X, Y, Z, x, y)] \Rightarrow f_8(Y) \prec f_2(Y)$ (since $L_8 \ll L_2$)

(i_4) $[P_3(X, Y, Z, x, y) \wedge odd(Y) \wedge P_4(X, Y, Z, x, y)] \Rightarrow f_4(Y) \prec f_3(Y)$ (since $L_4 \ll L_3$)

(i_5) $[P_3(X, Y, Z, x, y) \wedge even(Y) \wedge P_6(X, Y, Z, x, y)] \Rightarrow f_6(Y) \prec f_3(Y)$ (since $L_6 \ll L_3$)

(i_6) $[P_4(X, Y, Z, x, y) \wedge Y' = Y - 1 \wedge P_5(X, Y', Z, x, y)] \Rightarrow f_5(Y') \prec f_4(Y)$ (since $Y' < Y$)

(i_7) $[P_5(X, Y, Z, x, y) \wedge Z' = Z * X \wedge P_2(X, Y, Z', x, y)] \Rightarrow f_2(Y) \prec f_5(Y)$ (since $L_2 \ll L_5$)

(i_8) $[P_6(X, Y, Z, x, y) \wedge Y' = Y\, div\, 2 \wedge P_7(X, Y', Z, x, y)] \Rightarrow f_7(Y') \prec f_6(Y)$ (since $Y' < Y$)

(i_9) $[P_7(X, Y, Z, x, y) \wedge X' = X * X \wedge P_2(X', Y, Z, x, y)] \Rightarrow f_2(Y) \prec f_7(Y)$ (since $L_2 \ll L_7$).

In practice it is only necessary to prove that all program loops terminate since the ordering $\ll$ on labels directly follows from the syntactic structure of the program.

Floyd's method for proving termination is sound because no infinite decreasing sequence $f(\sigma_0) \gg f(\sigma_1) \gg \cdots \gg f(\sigma_i) \gg \cdots$ with $\sigma \in \Sigma^{\omega}[\![C]\!](p)$ is possible, so that execution of the program must sooner or later terminate in a final state $\sigma_{n-1} \in S$ since, by definition of $\Sigma^{n}[\![C]\!](p)$, we have $\forall \gamma \in \Gamma . \langle \sigma_{n-1}, \gamma \rangle \notin op[\![C]\!]$.

For completeness, observe that the distance $n-i-1$ of the current state σ_i to the final state σ_{n-1} of any execution trace σ of a terminating program is finite and that this distance strictly decreases after each program step. Hence we can hope to be always able to define the variant function $f(\sigma_i)$ as being the distance $n-i-1$. When this is true, we say after Dijkstra [124] that the program *strongly terminates*. The nondeterminism of a program C is *finite* if and only if no configuration of C can have infinitely many possible successors: $\forall \gamma \in \Gamma . \exists n \in \mathbf{N} . |\{\gamma' : \langle \gamma, \gamma' \rangle \in op[\![C]\!]\}| \leqslant n$. It is *bounded* when there is an upper bound on the number of possible successors: $\exists n \in \mathbf{N} . \forall \gamma \in \Gamma . |\{\gamma' : \langle \gamma, \gamma' \rangle \in op[\![C]\!]\}| \leqslant n$. When the nondeterminism of a program is finite then it terminates if and only if it strongly terminates. Then termination can always be proved with $(W, \prec)$ chosen as $(\mathbf{N}, <)$. We say that the nondeterminism of a program is *enumerable* if and only if any configuration of C has an enumerable set of possible successors: $\forall \gamma \in \Gamma . |\{\gamma' : \langle \gamma, \gamma' \rangle \in op[\![C]\!]\}| \leqslant |\mathbf{N}| = \omega$. When the nondeterminism of a program is enumerable but not finite, there may be infinitely many execution traces σ starting with the same given initial state $\sigma_0 = \langle s, C \rangle$ with no finite upper bound on the length n of these traces σ. In this case, the program is said to *weakly terminate*. This is the case of the following example:

$$(X <> 0 * (X < 0 \rightarrow (X := ?; (X < 0 \rightarrow X := -X \diamond \text{skip})) \diamond X := X - 1))$$

where X takes its values in $\mathbf{N}$ (i.e. the set D of data in (7) is $\mathbf{N}$). Then f cannot be chosen integer-valued but we can always find a convenient well-founded range $(W, \prec)$ for f.

6.5. *Liveness*

Burstall [64] generalized Floyd's total correctness property into *liveness*. Given a specification $\langle P, Q \rangle \in \mathscr{P}(\Gamma) \times \mathscr{P}(\Gamma \times \Gamma)$, a command C is said to *inevitably lead from P to Q* (written $P \overset{C}{\leadsto} Q$, using a variant of Lamport's notation [258]) if any execution of C starting from an initial configuration γ of P inevitably reaches a configuration γ' such that $\langle \gamma, \gamma' \rangle \in Q$. Liveness can be defined in terms of execution traces as follows:

(73) DEFINITION (Burstall [64], Lamport [258], Alpern & Schneider [5]) (*liveness*)

$$P \overset{C}{\leadsto} Q : \mathscr{P}(\Gamma) \times \mathbf{Com} \times \mathscr{P}(\Gamma \times \Gamma) \rightarrow \{\text{tt}, \text{ff}\}$$

$$P \overset{C}{\leadsto} Q = \forall \gamma \in P . \forall \sigma \in \Sigma[\![C]\!]\gamma . \exists i \in dom\, \sigma . \langle \gamma, \sigma_i \rangle \in Q.$$

In particular, $\underline{[p]} \underline{C} \underline{[q]} = \{\langle s, C \rangle : s \in p\} \overset{C}{\leadsto} \{\langle \gamma, s' \rangle : \gamma \in \Gamma \wedge s' \in q\}$.

6.6. Generalization of Floyd's total correctness proof method to liveness

Floyd's total correctness proof method can be generalized to liveness properties by the following induction principle:

(74) THEOREM (Pnueli [345], Cousot & Cousot [94]) (Floyd's liveness proof method)

$$P \overset{C}{\rightsquigarrow} Q = [\exists \alpha \in \mathbf{Ord}.\, \exists i \in \alpha \to \mathscr{P}(\Gamma \times \Gamma).$$

(74.1) $$(\forall \gamma \in P.\, \exists \beta < \alpha.\, \langle \gamma, \gamma \rangle \in i(\beta))$$

$$\wedge (\forall \beta < \alpha.\, (\beta > 0)$$

(74.2) $$\Rightarrow (\forall \gamma, \gamma' \in \Gamma.\, (\langle \gamma, \gamma' \rangle \in i(\beta)) \Rightarrow (\exists \gamma'' \in \Gamma.\, \langle \gamma', \gamma'' \rangle \in op[\![C]\!])))$$

$$\wedge (\forall \beta < \alpha.\, (\beta > 0)$$

(74.3) $$\Rightarrow (\forall \gamma, \gamma', \gamma'' \in \Gamma.\, (\langle \gamma, \gamma' \rangle \in i(\beta) \wedge \langle \gamma', \gamma'' \rangle \in op[\![C]\!]) \Rightarrow (\exists \beta' < \beta.\, \langle \gamma, \gamma'' \rangle \in i(\beta'))))$$

(74.4) $$\wedge (i(0) \subseteq Q)].$$

PROOF. We prove soundness ($\Rightarrow$) by reductio ad absurdum. Assume that we have found $\alpha \in \mathbf{Ord}$ and $i \in \alpha \to \mathscr{P}(\Gamma \times \Gamma)$ satisfying verification conditions (74), and that there is a configuration $\gamma \in P$ and a trace $\sigma \in \Sigma[\![C]\!]\gamma$ such that $\forall j \in dom\, \sigma.\, \langle \gamma, \sigma_j \rangle \notin Q$. Then there would be an infinite strictly decreasing sequence of ordinals $\beta \in seq^{\omega}\, \mathbf{Ord}$ such that $\forall j \in \mathbf{N}.\, j \in dom\, \sigma \wedge \langle \gamma, \sigma_j \rangle \in i(\beta_j)$, a contradiction since $<$ is well-founded on $\mathbf{Ord}$. We define $\langle \beta_j\colon j \in \mathbf{N} \rangle$ inductively as follows: by (74.1), $\exists \beta_0 < \alpha.\, \langle \gamma, \gamma \rangle \in i(\beta_0)$ whence $\langle \gamma, \sigma_0 \rangle \in i(\beta_0)$ since $\sigma_0 = \gamma$. If $j \in dom\, \sigma \wedge \langle \gamma, \sigma_j \rangle \in i(\beta_j)$ then $\beta_j \neq 0$, since otherwise, by (74.4), $\langle \gamma, \sigma_j \rangle \in i(\beta_j) \subseteq Q$ so that, by (74.2), $\exists \gamma \in \Gamma.\, \langle \sigma_j, \gamma \rangle \in op[\![C]\!]$ so that, by (61.1), $j+1 \in dom\, \sigma$. Moreover, by (61.1), $\langle \sigma_j, \sigma_{j+1} \rangle \in op[\![C]\!]$ so that, by (74.3), $\exists \beta_{j+1} < \beta_j.\, \langle \gamma, \sigma_{j+1} \rangle \in i(\beta_{j+1})$.

To prove completeness ($\Rightarrow$), assume $\forall \gamma \in P.\, \forall \sigma \in \Sigma[\![C]\!]\gamma.\, \exists i \in dom\, \sigma.\, \langle \gamma, \sigma_i \rangle \in Q$. Define $\langle \underline{\gamma}', \gamma' \rangle \ll \langle \underline{\gamma}, \gamma \rangle = [\underline{\gamma} \in P \wedge \exists \sigma \in \Sigma[\![C]\!]\underline{\gamma}.\, \exists i \in dom\, \sigma.(\forall j < i.\, \langle \sigma_0, \sigma_j \rangle \notin Q) \wedge (\langle \sigma_0, \sigma_i \rangle \in Q) \wedge (\underline{\gamma}' = \underline{\gamma} = \sigma_0) \wedge \exists k < i.\, ((\gamma = \sigma_k) \wedge (\gamma' = \sigma_{k+1}))]$.

We prove by reductio ad absurdum that $\ll$ is well-founded on $\Gamma \times \Gamma$. Assume there is an infinite sequence $\langle \underline{\sigma}_0, \sigma_0 \rangle \gg \langle \underline{\sigma}_1, \sigma_1 \rangle \gg \cdots$. If $\underline{\sigma}_0 = \sigma_0$ then we have $(\forall k \in \mathbf{N}.\, \underline{\sigma}_k = \underline{\sigma}_0 \wedge \langle \underline{\sigma}_0, \sigma_k \rangle \notin Q \wedge \langle \sigma_k, \sigma_{k+1} \rangle \in op[\![C]\!])$ so that $\sigma \in \Sigma[\![C]\!]\underline{\sigma}_0$, in contradiction with $\exists i \in dom\, \sigma.\, \langle \gamma, \sigma_i \rangle \in Q$. If $\underline{\sigma}_0 \neq \sigma_0$ then the same reasoning can be done by concatenation of a finite prefix $\langle \underline{\sigma}'_0, \sigma'_0 \rangle \gg \cdots \gg \langle \underline{\sigma}'_k, \sigma'_k \rangle$ such that $[\underline{\sigma}_1 \in P \wedge \sigma' \in \Sigma[\![C]\!]\underline{\sigma}_1 \wedge \exists i \in dom\, \sigma'.\, \exists k < i.\, (\forall j < i.\, \langle \sigma'_0, \sigma'_j \rangle \notin Q) \wedge (\langle \sigma'_0, \sigma'_i \rangle \in Q) \wedge (\underline{\sigma}_0 = \underline{\sigma}_1 = \sigma'_0) \wedge (\sigma_1 = \sigma'_k) \wedge (\sigma_0 = \sigma'_{k+1})]$ to the left of this sequence $\langle \underline{\sigma}_0, \sigma_0 \rangle \gg \langle \underline{\sigma}_1, \sigma_1 \rangle \gg \cdots$.

We choose $\alpha = rk_{\ll}(\Gamma \times \Gamma)$ and $i(\beta) = \{\langle \gamma, \gamma' \rangle \in P \times \Gamma\colon \exists \sigma \in \Sigma[\![C]\!]\gamma.\, \exists i \in dom\, \sigma.(\forall j < i.\, \langle \sigma_0, \sigma_j \rangle \notin Q) \wedge (\langle \sigma_0, \sigma_i \rangle \in Q) \wedge (\gamma = \sigma_0) \wedge (\exists k \leqslant i.\, \gamma' = \sigma_k) \wedge (\beta = rk_{\ll}\langle \gamma, \gamma' \rangle)\}$. Obviously, $(\forall \gamma \in P.\, rk_{\ll}\langle \gamma, \gamma \rangle < \alpha \wedge \langle \gamma, \gamma \rangle \in i(rk_{\ll}\langle \gamma, \gamma \rangle))$ so that (74.1) holds. If $0 < \beta < \alpha$ and $\langle \gamma, \gamma' \rangle \in i(\beta)$ then $rk_{\ll}\langle \gamma, \gamma' \rangle$ is different from 0 so that $\exists \langle \underline{\gamma}, \underline{\gamma}' \rangle \ll \langle \gamma, \gamma' \rangle$, whence $\langle \gamma', \underline{\gamma}' \rangle \in op[\![C]\!]$ and (74.2) is true. If $0 < \beta < \alpha$, $\langle \gamma, \gamma' \rangle \in i(\beta)$ and $\langle \gamma', \gamma'' \rangle \in op[\![C]\!]$, then let $\sigma \in \Sigma[\![C]\!]\gamma$, $i \in dom\, \sigma$ and $k \leqslant i$ be such that $(\forall j < i.\, \langle \sigma_0, \sigma_j \rangle \notin Q) \wedge (\langle \sigma_0, \sigma_i \rangle \in Q) \wedge (\gamma = \sigma_0) \wedge (\gamma' = \sigma_k) \wedge$

$(\beta = rk_{\ll}\langle\gamma,\gamma'\rangle)$. Since $rk_{\ll}\langle\gamma,\gamma'\rangle$ is different from 0, $\exists\langle\underline{\gamma},\underline{\gamma}'\rangle \ll \langle\gamma,\gamma'\rangle$ whence $[\gamma \in P \wedge \exists\sigma' \in \Sigma[\![C]\!]\gamma . \exists i' \in dom\,\sigma' . (\forall j < i' . \langle\sigma'_0, \sigma'_j\rangle \notin Q) \wedge (\langle\sigma'_0, \sigma'_{i'}\rangle \in Q) \wedge (\underline{\gamma} = \gamma = \sigma'_0) \wedge \exists k' < i . ((\gamma' = \sigma'_{k'}) \wedge (\underline{\gamma}' = \sigma'_{k'+1}))]$ so that $\langle\gamma,\gamma'\rangle \notin Q$, hence $\langle\sigma_0,\sigma_k\rangle \notin Q$ so that $k < i$. Moreover, $\exists\sigma'' \in \Sigma[\![C]\!]\gamma$ with $\sigma''_0 = \sigma_0 = \gamma$, $\sigma''_1 = \sigma_1, \ldots, \sigma''_k = \sigma_k = \gamma'$, $\sigma''_{k+1} = \gamma''$. Then $[\exists i'' \in dom\,\sigma'' . \exists k'' < i'' . (\forall j < i'' . \langle\sigma''_0, \sigma''_j\rangle \notin Q) \wedge (\langle\sigma''_0, \sigma''_{i''}\rangle \in Q) \wedge (\gamma = \gamma = \sigma''_0) \wedge (\gamma' = \sigma''_{k''}) \wedge (\gamma'' = \sigma''_{k''+1})]$ with $k'' = k$ so that $\langle\gamma,\gamma''\rangle \ll \langle\gamma,\gamma'\rangle$. It follows that $\beta' = rk_{\ll}\langle\gamma,\gamma''\rangle < rk_{\ll}\langle\gamma,\gamma'\rangle = \beta$ and $\langle\gamma,\gamma'\rangle \in i(\beta'')$ whence (74.3) holds. Finally, if $\langle\gamma,\gamma'\rangle \in i(0)$ and $\langle\gamma,\gamma'\rangle \notin Q$ then $rk_{\ll}\langle\gamma, \gamma'\rangle = 0$ whence there is no $\langle\underline{\gamma},\underline{\gamma}'\rangle \ll \langle\gamma,\gamma'\rangle$ so that $[\forall\sigma \in \Sigma[\![C]\!]\gamma . \forall i \in dom\,\sigma . (\exists j < i . \langle\sigma_0, \sigma_j\rangle \in Q) \vee (\langle\sigma_0,\sigma_i\rangle \notin Q) \vee \forall k < i . ((\gamma \neq \sigma_k) \vee \forall\underline{\gamma}' \in \Gamma . (\underline{\gamma}' \neq \sigma_{k+1}))]$ so that for any $\sigma \in \Sigma[\![C]\!]\gamma$, choosing the least $i \in dom\,\sigma . \langle\sigma_0,\sigma_i\rangle \in Q$, we have $i > 1$ and $\forall k < i . (\gamma \neq \sigma_k)$, a contradiction for $k = 0$. □

6.7. *Burstall's total correctness proof method and its generalization*

Floyd's total correctness proof method [143] is by induction on the structure of computations where computations are understood as empty or as a step followed by a computation. Proofs à la Floyd are elegant for programs which have exactly this linear structure of computations. For example, this is the case for a program computing the size of a list $L ::= \langle\ \rangle | \langle A; \mathrm{tl}(L)\rangle$ (where A is an atom and $\mathrm{tl}(L)$ is a list) since its structure is of the form $\mathrm{size}(L) = (L = \langle\ \rangle \to 0 \diamond 1 + \mathrm{size}(\mathrm{tl}(L)))$. Hoare [208] remarked that programs (at least those written in structured languages with no goto's, etc.) often have a tree-like computation structure similar to their syntactic structure so that correctness proofs are better handled by induction on this syntactic structure. Burstall [64] adopted the point of view that proofs are better handled by induction on the data structures that are manipulated by the program, since the structure of the computations is often similar to that of these data structures. For example, an iterative program using a stack for computing the size of a tree $T ::= \langle\ \rangle | \langle \mathrm{lf}(T); A; \mathrm{rg}(T)\rangle$ (where $\mathrm{lf}(T)$ and $\mathrm{rg}(T)$ are trees) would have a computation structure of the form $\mathrm{size}\ (T) = (L = \langle\ \rangle \to 0 \diamond 1 + \mathrm{size}(\mathrm{lf}(L)) + \mathrm{size}(\mathrm{rg}(L)))$. The following induction principle generates these points of view by considering that arbitrary computation structures can be specified by a well-founded relation $(W, \prec)$ (or from a mathematical point of view $(\mathbf{Ord}, <)$) which basis ultimately corresponds to elementary program steps:

(75) THEOREM (Cousot & Cousot [95]) (Burstall's liveness proof method)

$$P \overset{C}{\leadsto} Q = [\exists\Lambda \in \mathbf{Ord} . \exists\theta \in \Lambda \to \mathscr{P}(\Gamma \times \Gamma) . \exists\alpha \in \mathbf{Ord} . \exists i \in \Lambda \to \alpha \to \mathscr{P}(\Gamma \times \Gamma) .$$

(75.1) $\qquad (\exists\pi \in \Lambda . \theta_\pi \subseteq P\rceil Q)$

(75.2) $\qquad \wedge (\forall\lambda \in \Lambda . \forall\gamma \in \Gamma . \exists\beta < \alpha . \langle\gamma,\gamma\rangle \in i_\lambda(\beta))$

$\qquad \wedge (\forall\lambda \in \Lambda . \forall\beta < \alpha . \forall\gamma, \gamma' \in \Gamma . \langle\gamma,\gamma'\rangle \in i_\lambda(\beta))$

$\qquad\quad \Rightarrow [(\exists\gamma'' \in \Gamma . \langle\gamma',\gamma''\rangle \in op[\![C]\!]$

(75.3) $\qquad\qquad \wedge \forall\gamma'' \in \Gamma . (\langle\gamma',\gamma''\rangle \in op[\![C]\!] \Rightarrow \exists\beta' < \beta . \langle\gamma,\gamma''\rangle \in i_\lambda(\beta')))$

(75.4) $\qquad\qquad \vee (\exists\lambda' < \lambda . \forall\gamma'' \in \Gamma . (\langle\gamma'\ \gamma''\rangle \in \theta_{\lambda'} \Rightarrow \exists\beta' < \beta . \langle\gamma,\gamma''\rangle \in i_\lambda(\beta')))$

(75.5) $\qquad\qquad \vee (\langle\gamma,\gamma'\rangle \in \theta_\lambda)]])]$.

To prove $P \overset{C}{\leadsto} Q$, we must prove $\Gamma \overset{C}{\leadsto} \theta_\lambda$ for $\lambda \in \Lambda$ so that by (75.1) $\Gamma \overset{C}{\leadsto} \theta_\pi = P \overset{C}{\leadsto} Q$ holds. The liveness proofs $\Gamma \overset{C}{\leadsto} \theta_\lambda$ for $\lambda \in \Lambda$ can be done using (75.2), (75.3) and (75.5), hence by Floyd's liveness proof method (74). However, it is better to exhibit a proof showing the recursive structure of the computations. The basis corresponds to elementary program steps (75.3) whereas induction is described by means of lemmata $\theta_{\lambda'}$, $\lambda' < \lambda$, which are first proved to be correct ($\Gamma \overset{C}{\leadsto} \theta_{\lambda'}$) and can then be used in (75.4) for proving θ_λ. More precisely, $\langle \gamma, \gamma' \rangle \in i_\lambda(\beta)$ if and only if starting execution in configuration γ may lead to configuration γ' from which some final configuration $\underline{\gamma}$ such that $\langle \gamma, \underline{\gamma} \rangle \in \theta_\lambda$ will inevitably be reached . To prove this we can either show by (75.3) that a single program step inevitably leads to a state γ'' with the same property ($\langle \gamma, \gamma'' \rangle \in i_\lambda(\beta')$) but closer to the goal (since $\beta' < \beta$), or else use lemma $\theta_{\lambda'}$ which, according to a previous proof ($\lambda' < \lambda$), states that zero or more program steps inevitably lead to a state γ'' with the same property ($\langle \gamma, \gamma'' \rangle \in i_\lambda(\beta')$) but closer to the goal (since $\beta' < \beta$).

7. Hoare logic

Hoare [208] introduced the idea that partial correctness can be proved compositionally, by induction on the syntax of programs. This idea turned out to be of prime importance in other domains, such as denotational semantics where "the values of expressions are determined in such a way that the value of a whole expression depends functionally on the values of its parts—the exact connection being found through the clauses of the syntactical definition of the language" [369], but it had to be slightly modified to take into account context-sensitive properties of languages.

Hoare [208] also introduced the idea that such proofs can be formalized using a formal logic. The first motivation is that "axioms may provide a simple solution to the problem of leaving certain aspects of a language undefined". To illustrate this point of view, Hoare gives the example of addition (+) and multiplication (∗) of natural numbers. These operations can be formalized by a few axioms of which **N** is a model, They can also be given different consistent interpretations corresponding to various possible implementations ⊕ and ⊗ of + and ∗ in a machine where only a finite subset $\{0, \ldots, \text{maxint}\}$ of **N** is representable. These interpretations include modulo arithmetic $(x \oplus y) = (x + y)$ *mod* $(\text{maxint} + 1)$, firm-boundary arithmetic $(x \oplus y) = (x + y > \text{maxint} \rightarrow \text{maxint} \diamond x + y)$, and overflow arithmetic $(x \oplus y) = (x + y > \text{maxint} \rightarrow \text{undefined} \diamond x + y)$. More generally, the idea would be that a program text may have different interpretations (a computer scientist would say computer implementations), but its correctness should be established once for all its possible interpretations (hence in a machine-independent way). This leads to the second motivation of Hoare's axiomatic semantics: "the specification of proof techniques provides an adequate formal definition of a programming language". The idea first appeared in [143] and was illustrated by Hoare and Wirth [229] on a part of Pascal. The trouble with this axiomatic semantics is that nonstandard, hence computer-unimplementable interpretations are not ruled out [48, 399].

7.1. *Hoare logic considered from a semantical point of view*

7.1.1. *General theorems for proof construction*

In Subsection 5.3, we have considered Hoare logic from a semantical point of view, that is to say, with respect to the conventional operational semantics (13). In summary, this essentially consists in the natural extension of the relational semantics (19) of commands from pairs of states to pairs of sets of states (22). This leads to the proof of partial correctness by structural induction on commands using theorem (49) which can also be rephrased as follows:

(76) THEOREM (Hoare [208], Cook [86]) (semantic interpretation of Hoare logic)

(76.1) $\underline{\{p\}\text{skip}\{p\}} = \text{tt}$;
(76.2) $\underline{\{\{s \in S : s[X \leftarrow \underline{E}(s)] \in p\}\}\underline{X} := \underline{E}\{p\}} = \text{tt}$;
(76.3) $\underline{\{p\}\underline{X} := ?\{\{s[X \leftarrow d : s \in p \wedge d \in \underline{D}\}\}} = \text{tt}$;
(76.4) $\underline{\{p\}(\underline{C_1};\underline{C_2})\{q\}} = (\exists i \in \mathbf{Ass}.\, \underline{\{p\}\underline{C_1}\{i\}} \wedge \underline{\{i\}\underline{C_2}\{q\}})$;
(76.5) $\underline{\{p\}(\underline{B} \rightarrow \underline{C_1} \diamond \underline{C_2})\{q\}} = (\underline{\{p \cap \underline{B}\}\underline{C_1}\{q\}} \wedge \underline{\{p \cap \neg \underline{B}\}\underline{C_2}\{q\}})$;
(76.6) $\underline{\{p \cap \underline{B}\}\underline{C}\{p\}} \Rightarrow \underline{\{p\}(\underline{B} * \underline{C})\{p \cap \neg \underline{B}\}}$;
(76.7) $(\exists p', q' \in \mathbf{Ass}.\, (p \subseteq p') \wedge \underline{\{p'\}\underline{C}\{q'\}} \wedge (q \subseteq q')) = \underline{\{p\}\underline{C}\{q\}}$.

In these theorems, the consequence rule (76.7) has been isolated, whereas in (49) it is distributed over all theorems (76.1) to (76.6). The idea is interesting because proofs concerning properties of objects manipulated by the program (which turn out to be always of the form $p \subseteq p'$) are isolated from proofs concerning the computation (sequence of operations) performed by the program. In practice, this separation leads to excessively tedious proofs. The method proposed by Hoare in [208] "of reducing the tedium of formal proofs is to derive general rules for proof construction out of the simple rules accepted as postulates". For example, derived theorems such as "$(p \subseteq \{s \in S : s[X \leftarrow \underline{E}(s)] \in q\}) \Rightarrow \underline{\{p\}\underline{X} := \underline{E}\{q\}}$" or "$(\{s'[X \leftarrow \underline{E}(s')] : s' \in p\} \subseteq q) \Rightarrow \underline{\{p\}\underline{X} := \underline{E}\{q\}}$" are directly applicable, hence often more useful than (76.2). Also the reciprocal of (76.6) is not true (for example, "$\{X = 0\}$ **while** true **do** $X := X + 1$ $\{X = 0 \wedge \neg\text{true}\}$" holds but "$\{X = 0 \wedge \text{true}\}$ $X := X + 1$ $\{X = 0\}$" does not). Hence, (76.6) does not make completely clear the fact that a loop invariant has to be found (most often differing from the precondition and postcondition). In practice, we prefer a more explicit formulation:

(77) COROLLARY (partial correctness proof of while loops)

$$\underline{\{p\}(\underline{B} * \underline{C})\{q\}} = (\exists i \in \mathbf{Ass}.\, (p \subseteq i) \wedge \underline{\{i \cap \underline{B}\}\underline{C}\{i\}} \wedge ((i \cap \neg \underline{B}) \subseteq q)).$$

(78) EXAMPLE (*partial correctness proof of assignments*). Let us derive "$(\{s'[X \leftarrow \underline{E}(s')] : s' \in p\} \subseteq q) \Rightarrow \underline{\{p\}\underline{X} := \underline{E}\{q\}}$" from (76):

(a) $\{s'[X \leftarrow \underline{E}(s')] : s' \in p\} \subseteq q$ by assumption,
(b) $(s \in p) \Rightarrow (s[X \leftarrow \underline{E}(s)] \in q)$ by (a) and set theory,
(c) $p \subseteq \{s \in S : s[X \leftarrow \underline{E}(s)] \in q\}$ by (b) and set theory,
(d) $\underline{\{\{s \in S : s[X \leftarrow \underline{E}(s)] \in q\}\}\underline{X} := \underline{E}\{q\}}$ by (76.2),

(e) $q \subseteq q$ from set theory,
(f) $\underline{\{q\}}\underline{X := E}\underline{\{q\}}$ by (c), (d), (e) and (76.7).

PROOF OF THEOREMS (76) AND (77)

(1) $\underline{\{p\}}\text{skip}\underline{\{p\}} = ((p \rceil \underline{\text{skip}}) \subseteq (S \times p))$ [by (22)] $= ((p \rceil \{\langle s, s\rangle : s \in S\}) \subseteq (S \times p))$ [by (19.1)] $=$ tt.

(2) $\underline{\{\{s \in S: s[X \leftarrow \underline{E}(s)] \in p\}\}}\underline{X := E}\underline{\{p\}} = ((\{s \in S: s[X \leftarrow \underline{E}(s)] \in p\} \rceil \{\langle s, s[X \leftarrow \underline{E}(s)]\rangle : s \in S\}) \subseteq (S \times p))$ [by (22) and (19.2)] $= (\{\langle s, s[X \leftarrow \underline{E}(s)]\rangle : s[X \leftarrow \underline{E}(s)] \in p\} \subseteq \{\langle s, s'\rangle : s' \in p\}) =$ tt.

(3) $\underline{\{p\}}\underline{X := ?}\underline{\{\{s[X \leftarrow d]: s \in p \wedge d \in D\}\}}$ is tt since $(p \rceil \underline{X := ?}) = \{\langle s, s[X \leftarrow d]\rangle : s \in p \wedge d \in D\}$ by (22) and (19.3).

(4) $\underline{\{p\}}\underline{(C_1; C_2)}\underline{\{q\}} = ((\forall s, s', s''. (s \in p \wedge \langle s, s'\rangle \in \underline{C_1} \wedge \langle s', s''\rangle \in \underline{C_2})) \Rightarrow (s'' \in q))$ by (22) and (19.4). This implies that if we let i be $\{s' : \exists s \in p, . \langle s, s'\rangle \in \underline{C_1}\}$ then $(p \rceil \underline{C_1}) \subseteq (S \times i)$ and $(i \rceil \underline{C_2}) \subseteq (S \times q)$ whence $\underline{\{p\}}\underline{C_1}\underline{\{i\}} \wedge \underline{\{i\}}\underline{C_2}\underline{\{q\}}$ holds. Reciprocally, if $i \in$ **Ass** is such that $\underline{\{p\}}\underline{C_1}\underline{\{i\}} \wedge \underline{\{i\}}\underline{C_2}\underline{\{q\}}$ then $(p \rceil \underline{C_1}) \subseteq (S \times i)$ and $(i \rceil \underline{C_2}) \subseteq (S \times q)$ by (22), so that $(p \rceil \underline{C_1; C_2}) = (p \rceil (\underline{C_1} \circ \underline{C_2})) = ((p \rceil \underline{C_1}) \circ \underline{C_2}) \subseteq ((S \times i) \circ \underline{C_2}) = (S^2 \circ (i \rceil \underline{C_2})) \subseteq (S^2 \circ (S^2 \rceil q)) = (S^2 \rceil q)$.

(5) $\underline{\{p\}}\underline{(B \to C_1 \diamond C_2)}\underline{\{q\}} = [((p \cap \underline{B}) \rceil \underline{C_1} \cup (p \cap \neg \underline{B}) \rceil \underline{C_2}) \subseteq (S \times q)] = (\underline{\{p \cap \underline{B}\}}\underline{C_1}\underline{\{q\}} \wedge \underline{\{p \cap \neg \underline{B}\}}\underline{C_2}\underline{\{q\}})$ by (22) and (19.5).

(6) $\underline{\{p\}}\underline{(B * C)}\underline{\{q\}} = ((p \rceil (\underline{B} \rceil \underline{C})^* \lceil \neg \underline{B}) \subseteq S \times q)$ [by (22) and (19.6)] $= (\exists i \in \textbf{Ass}. (p \subseteq i) \wedge (i \rceil (\underline{B} \rceil \underline{C}) \subseteq S \times i) \wedge ((i \cap \neg \underline{B}) \subseteq q))$ [by (28)] $= (\exists i \in \textbf{Ass}.\ (p \subseteq i) \wedge \underline{\{i \cap \underline{B}\}}\underline{C}\underline{\{i\}} \wedge ((i \cap \neg \underline{B}) \subseteq q))$ [by 22)]. It follows that $\underline{\{p \cap \underline{B}\}}\underline{C}\underline{\{p\}} \Rightarrow ((p \subseteq p) \wedge \underline{\{p \cap \underline{B}\}}\underline{C}\underline{\{p\}} \wedge ((p \cap \neg \underline{B}) \subseteq (p \cap \neg \underline{B}))) \Rightarrow \underline{\{p\}}\underline{(B * C)}\underline{\{p \cap \neg \underline{B}\}}$.

(7) $((p \subseteq p') \wedge \underline{\{p'\}}\underline{C}\underline{\{q'\}} \wedge (q \subseteq q')) \Rightarrow ((p \subseteq p') \wedge ((p' \rceil \underline{C}) \subseteq (S \times q')) \wedge (q \subseteq q')) \Rightarrow ((p \rceil \underline{C}) \subseteq (S \times q)) \Rightarrow \underline{\{p\}}\underline{C}\underline{\{q\}}$ by (22). $\square$

7.1.2. *Semantical soundness and completeness*

If we have proved $\underline{\{p\}}\underline{C}\underline{\{q\}}$ by application of theorem (76) to components $C' \in \textbf{Comp}[\![C]\!]$ of C then we conclude, by structural induction, that $\underline{\{p\}}\underline{C}\underline{\{q\}}$ holds. This is called *semantic soundness*. If we can prove by (22) using (13) that $\underline{\{p\}}\underline{C}\underline{\{q\}}$ holds, then this can always be proved by application of theorem (76) to components $C' \in \textbf{Comp}[\![C]\!]$ of C. This is called *semantic completeness*. We use the epithet "semantic" because soundness and completeness are relative to partial correctness as defined by (22) with respect to operational semantics (13), that is, are defined in terms of mathematical structures without reference to a particular logical language for assertions.

THEOREM (semantical soundness and completeness of Hoare logic)
(79) *Hoare's partial correctness proof method* (76) *is semantically sound.*
(80) *Hoare's partial correctness proof method* (76) *is semantically complete.*

PROOF. The proof that $\underline{\{p\}}\underline{C}\underline{\{q\}}$ is provable by (76) is by structural induction on C.

(1) If $\underline{\{p\}}\underline{\text{skip}}\underline{\{q\}}$ holds then $p \subseteq q$ by (2) and (19.1). Also $p \subseteq p$. Therefore the proof is "$\{p\}\text{skip}\{p\}$ by (76.1) whence $\{p\}\text{skip}\{q\}$ by $p \subseteq p$, $p \subseteq q$ and (76.7)".

(2) If $\underline{\{p\}}\underline{X := E}\underline{\{q\}}$ holds then $\{s'[X \leftarrow \underline{E}(s')]: s' \in p\} \subseteq q$ by (22) and (19.2). Also

$p \subseteq \{s \in S\colon s[X \leftarrow \underline{E}(s)] \in \{s'[X \leftarrow \underline{E}(s')]\colon s' \in p\}\}$. Therefore the proof is simply as follows: "$\{\{s \in S\colon s[X \leftarrow \underline{E}(s)] \in \{s'[X \leftarrow \underline{E}(s')]\colon s' \in p\}\}\}\underline{X := E}\{\{s'[X \leftarrow \underline{E}(s')]\colon s' \in p\}\}$ by (76.2) whence $\{p\}\underline{X := E}\{q\}$ by $p \subseteq \{s \in S\colon s[X \leftarrow \underline{E}(s)] \in \{s'[X \leftarrow \underline{E}(s')]\colon s' \in p\}\}$, $\{s'[X \leftarrow \underline{E}(s')]\colon s' \in p\} \subseteq q$ and (76.7)".

(3) If $\{p\}\underline{X := ?}\{q\}$ holds then $\{s[X \leftarrow d]\colon s \in p \wedge d \in D\} \subseteq q$ by (22) and (19.3). Also $p \subseteq p$. Therefore the proof is "$\{p\}\underline{X := ?}\{\{s[X \leftarrow d]\colon s \in p \wedge d \in D\}\}$ by (76.3) whence $\{p\}\underline{X := ?}\{q\}$ by $p \subseteq p$, $\{s[X \leftarrow d]\colon s \in p \wedge d \in D\} \subseteq q$ and (76.7)".

(4) If $\{p\}\underline{C_1; C_2}\{q\}$ holds then by (76.4) there is an assertion $i \in$ **Ass** such that $\{p\}\underline{C_1}\{i\}$ and $\{i\}\underline{C_2}\{q\}$ are true. Hence, by induction hypothesis, $\{p\}\underline{C_1}\{i\}$ and $\{i\}\underline{C_2}\{q\}$ are provable by induction on C_1 and C_2 using theorem (76). Then applying (76.4), we conclude $\{p\}\underline{C_1;C_2}\{q\}$.

(5) If $\{p\}\underline{(B \rightarrow C_1 \diamond C_2)}\{q\}$ holds then by (76.5) $\{p \cap \underline{B}\}\underline{C_1}\{q\}$ and $\{p \cap \neg\underline{B}\}\underline{C_2}\{q\}$ are true, whence provable by induction on C_1 and C_2 using theorem (76). Then we conclude by (76.5).

(6) If $\{p\}\underline{(B * C)}\{q\}$ holds then by (77) there is an assertion $i \in$ **Ass** such that $(p \subseteq i)$, $\{i \cap \underline{B}\}\underline{C}\{i\}$ and $(i \cap \neg\underline{B} \subseteq q)$ are true. Hence, by induction hypothesis, $\{i \cap \underline{B}\}\underline{C}\{i\}$ is provable by induction on C using theorem (76). The proof goes on with "$\{i\}\underline{(B * C)}\{i \cap \neg\underline{B}\}$ by (76), whence $\{p\}\underline{(B * C)}\{q\}$ by $(p \subseteq i)$, $(i \cap \neg\underline{B} \subseteq q)$ and (76.7)". □

7.1.3. Proof outlines

Hoare [208] was aware that a formal proof is often tedious and claimed that "it would be fairly easy to introduce notational conventions which would significantly shorten it". From a practical point of view, it is indeed essential to be able to present proofs by Hoare's method informally together with the program text. Owicki [335] showed that a proof à la Hoare can be presented as a *proof outline*, that is to say, as a proof à la Floyd where local invariants are attached to program points:

(81) DEFINITION (Owicki [335]) (*proof outline*). The proof outline of a proof of $\{p\}\underline{C}\{q\}$ by theorem (76) is the triple pre, post : **Comp**$[\![C]\!] \rightarrow$ **Ass**, linv : **Loops**$[\![C]\!] \rightarrow$ **Ass** such that $\text{pre}(C) = p$ $\text{post}(C) = q$ and by structural induction on $C' \in$ **Comp**$[\![C]\!]$:

(a) If C' is $(B \rightarrow C_1 \diamond C_2)$ then $\text{pre}(C_1) = \text{pre}(C') \cap \underline{B}$, $\text{pre}(C_2) = \text{pre}(C') \cap \neg\underline{B}$, $\text{post}(C_1) = \text{post}(C_1) = \text{post}(C')$,

(b) If C' is $(C_1;C_2)$ then we can only prove $\{\text{pre}(C')\}\underline{(C_1;C_2)}\{\text{post}(C')\}$ by application of (76.4) once and of (76.7) $n \geqslant 0$ times. Therefore we have $\text{pre}(C') = p_1 \subseteq \cdots \subseteq p_n$, $\{p_n\}\underline{C_1}\{i\}$, $\{i\}\underline{C_2}\{q_n\}$, $q_n \subseteq \cdots \subseteq q_1 = \text{post}(C')$. We let $\text{pre}(C_1) = \text{pre}(C')$, $\text{post}(C_1) = \text{pre}(C_2) = i$ and $\text{post}(C_2) = \text{post}(C')$.

(c) If C' is $(B * C_1)$ then we can only prove $\{\text{pre}(C')\}\underline{(B * C_1)}\{\text{post}(C')\}$ by application of (76.6) once and of (76.7) $n \geqslant 0$ times. Therefore we have $\text{pre}(C') = p_1 \subseteq \cdots p_n = i$, $\{i \cap \underline{B}\}\underline{C_1}\{i\}$, $i \cap \neg\underline{B} = q_n \subseteq \cdots \subseteq q_1 = \text{post}(C')$. We let $\text{pre}(C_1) = i \cap \underline{B}$, $\text{post}(C_1) = i$ and $\text{linv}(C') = i$.

The following two theorems show that Hoare's method (76) is equivalent to the syntax-directed presentation of Floyd's method of Subsection 5.3, hence by (55) is semantically equivalent to Floyd's original stepwise proof method; otherwise stated, that assertions in a proof outline are local invariants:

(82) THEOREM (Presentation à la Floyd of a proof by Hoare logic). *A proof outline of* $\underline{\{p\}C\{q\}}$ *by Hoare's method* (76) *satisfies* (49).

PROOF. (49.1) is true by definition (81). If C is skip then $\underline{\{p\}\text{skip}\{q\}}$ can only be proved using (76.1) and (76.7). It follows that $p \subseteq q$ so that (49.2) holds. The proof is similar when C is $X := E$ or $X := ?$. (49.5), (49.6) and (49.7) directly follow from definition (8.1). □

(83) THEOREM (Presentation à la Hoare of a proof by Floyd's method). *If assertions attached to program points are local invariants in the sense of Floyd* (*i.e., satisfy* (4, 5), *or equivalently by* (54) *and* (55), *satisfy* (49)) *then they can be used to prove partial correctness by Hoare's method* (76).

PROOF. The proof is by structural induction on C. If C is $X := E$ then the proof à la Hoare is "$p \subseteq \{s \in S\colon s[X \leftarrow \underline{E}(s)] \in q\}$ [which is true by (81) and (49.3)] $\underline{\{\{s \in S\colon s[X \leftarrow \underline{E}(s)] \in q\}\}X := E\{q\}}$ [by (76.2) and $q \subseteq q$], hence $\underline{\{p\}X := E\{q\}}$ [by (76.7)]". The proofs are similar when C is skip or $X := ?$.

If C is $(C_1;C_2)$ then, by induction hypothesis, there are proofs à la Hoare of $\underline{\{\text{pre}(C_1)\}C_1\{\text{post}(C_1)\}}$ and $\underline{\{\text{pre}(C_2)\}C_2\{\text{post}(C_2)\}}$. It follows from (81) and (76.4) that $p = \text{pre}(C_1)$ and $\text{post}(C_2) = q$. If we let i be $\text{post}(C_1) = \text{pre}(C_2)$ then we can use (76.7) to prove $\underline{\{p\}C_1\{i\}} \wedge \underline{\{i\}C_2\{q\}}$ and conclude $\underline{\{p\}C_1\{i\}}$ by (76.4). The proofs are similar when C is $(B \rightarrow C_1 \diamond C_2)$ or $(B * C_1)$. □

7.2. *Hoare logic considered from a syntactical point of view*

The purpose of mathematical logic is to provide formal languages for describing the structures with which mathematicians work, and to study the methods of proof available to them. Hoare [208] introduced the same idea in computer science: "Computer programming is an exact science in that all the properties of a program and all the consequences of executing it in any given environment can, in principle, be found out from the text of the program itself by means of purely deductive reasoning. Deductive reasoning involves the application of valid rules of inference to sets of valid axioms. It is therefore desirable and interesting to elucidate the axioms and rules of inference which underlie our reasoning about computer programs".

Hoare logic **HL** consists of first-order predicates $P, Q, \ldots$ for describing assertions $p, q, \ldots$ and correctness formulae $\{P\}C\{Q\}$ for describing the partial correctness $\underline{\{p\}C\{q\}}$ of commands C. First we define the set **HL** of Hoare formulae, that is, the syntax of predicates $P, Q, \ldots$ and correctness formulae $\{P\}C\{Q\}$. We also fill in the details of the syntax in definition (1) of expressions and Boolean expressions of the programming language **Com**. This syntactical definition is parametrized by a *basis* $\Sigma = \langle \mathbf{Cte}, \mathbf{Fun}, \mathbf{Rel}, \# \rangle$ (also called *signature, type,*...), that is, sets of constant, function and relation symbols together with their arity.

Then we introduce Hoare's proof system to define inductively which formulae of Hoare logic are *provably true*. The proof system consists of sets of postulates (axioms) and of syntactic rules of formula manipulation by rewriting (rules of inference) in order to logically derive conclusions from hypotheses. These axioms and rules of inference are defined finitistically so that given formal proofs are checkable by algorithmic means (although the proof itself may not be derivable by a machine). This approach is "syntactical" in that the emphasis is upon a formal language for describing assertions about the values taken by the program variables, and upon a formal deductive system for deriving proofs based upon combinatorial manipulations of formal assertions. Proofs are parametrized by a set of axioms, which are supposed to describe properties of the data manipulated by the programs, the details of which we do not want to enter into. This is usual in logic where the primitive notions $(0, +, \ldots)$ of mathematical theories (such as groups, ...) have no fixed meaning. This is also consistent with Hoare's point of view [208] that "another of the great advantages of using an axiomatic approach is that axioms offer a simple and flexible technique for leaving certain aspects of a language *undefined*, for example, range of integers, accuracy of floating point, and choice of overflow technique. This is absolutely essential for standardization purposes, since otherwise the language will be impossible to implement efficiently on different hardware designs. Thus a programing language standard should consist of a set of axioms of universal applicability, together with a choice of supplementary axioms describing the range of choices facing an implementor".

Then we define which mathematical structures can be understood as being models or interpretations of Hoare logic. Otherwise stated, we define which formulae of Hoare logic **HL** are *semantically true* with respect to a relational semantics C of programs $C \in$ **Com.** This programming language semantics (19) itself depends upon the semantics (also called a *model* or *interpretation*) $\langle D, V \rangle$ of the basis $\Sigma = \langle$**Cte**, **Fun**, **Rel**, $\# \rangle$, that is, upon the domain D of data on which programs operate and the exact interpretation $V[\![c]\!]$ of the basic constants $c \in$ **Cte**, $V[\![f]\!]$ of the functions $f \in$ **Fun** and $V[\![r]\!]$ of the relations $r \in$ **Rel** involved in the program and predicates. By leaving this interpretation as a parameter, we define a family of semantics of Hoare logic with respect to a family of possible operational semantics of the programming language.

Gödel showed that *truth* and *provability* do not necessarily coincide: provable implies true, refutable implies false but some formulae may be undecidable, that is, neither provable nor refutable (using proofs that can be checked mechanically) although they are either true or false [374]. Therefore the question is whether Hoare's formal proof system captures the true partial correctness formulae, only these (soundness), and ideally all of these (completeness).

7.2.1. Syntax of predicates and correctness formulae

We have defined the syntax of the programming language **Com** in definition (1). We now define the syntax of the logical language **HL** which will be used to specify partial correctness of programs. Hoare logic is a first-order language allowing only to quantify over elements, but not over subsets or functions (for example "$\forall A \in \mathscr{P}(\mathbf{N}). (0 \in A \wedge \forall x \in A. (x+1) \in A) \Rightarrow (A = \mathbf{N})$" or "$\forall P. \forall x_1. \ldots \forall x_n. \{P\}$ skip $\{P\}$" are not first-order sentences).

In order to specify the syntax of predicates, we consider given disjoint sets of symbols as follows:

(84) DEFINITION (*symbols*)

(84.1)	X, Y:	**Pvar**	*programming variables,*
(84.2)	x, y:	**Lvar**	*logical variables,*
(84.3)	v, u:	**Var** = **Pvar**∪**Lvar**	*variables,*
(84.4)	c:	**Cte**	*constant symbols,*
(84.5)	f:	**Fun**	*function symbols,*
(84.6)	r:	**Rel**	*relation symbols.*

(85) DEFINITION (*arity of function and relation symbols*). $\#: \mathbf{Fun} \cup \mathbf{Rel} \rightarrow \mathbf{N}^+$.

At any moment we shall use only a finite number of variables but assume that we are given a countably infinite supply of these (in the example, we use capital letters for programming variables and lower-case letters for logical variables). The basis $\Sigma = \langle$**Cte**, **Fun**, **Rel**, $\#\rangle$ of the logical language **HL** is assumed to be given but is otherwise left unspecified.

The purpose of the syntactical definition of the logical language **HL** is to define algorithmically which finite strings of symbols (chosen in **Var**∪**Cte**∪**Fun**∪**Rel**∪{(,), =, ¬, ∧, ∨, ⇒, ∃, ∀, ., {, }, skip, :=, ?, ;, →, ◇, ∗}) belong to **HL**; logicians would say "are well-formed formulae". The sets of *terms, atomic formulae, first-order predicates, correctness formulae* and *formulae* are the smallest sets closed under the following formation rules (from which a recognizer is easy to derive, see for example [3]).

DEFINITION (*syntax of formulae*)

T: **Ter** *terms*

(86) $T ::= v \mid c \mid f(T_1, \ldots, T_{\#f})$,

A: **Afo** *atomic formulae*

(87) $A ::= (T_1 = T_2) \mid r(T_1, \ldots, T_{\#r})$,

P, Q: **Pre** *predicates*

(88) $P ::= A \mid \neg P \mid (P_1 \wedge P_2) \mid (P_1 \vee P_2) \mid (P_1 \Rightarrow P_2) \mid \exists x.P \mid \forall x.P$,

H: **Hcf** *Hoare correctness formulae*

(89) $H ::= \{P\}C\{Q\}$,

F: **HL** *formulae of Hoare logic*

(90) $F := P \mid H$.

(in (87), "=" is a relation of arity 2 in infix notation. The reason why we keep it separate from the relations in **Rel** is that its intended interpretation is fixed as the diagonal (equality) relation δ, whereas the interpretations of members of **Rel** can be specified arbitrarily).

In Subsection 3.1 the syntax of expressions was left unspecified. From now on, we assume that **Expr** is included in the set of terms containing only programming variables and that **BExp** is included in the set of *propositions* (i.e., quantifier-free predicates) containing only programming variables:

DEFINITION (*syntax of expressions*)

E: **Expr** *expressions*

(91) $E ::= X \mid c \mid f(E_1, \ldots, E_{\#f}),$

B: **BExp** *Boolean expressions*

(92) $B ::= (E_1 = E_2) \mid r(E_1, \ldots, E_{\#r}) \mid \neg B \mid (B_1 \wedge B_2) \mid (B_1 \vee B_2) \mid (B_1 \Rightarrow B_2).$

Definitions (84) to (90) are justified by the (restrictive) assumption that all we shall ever have to say about programs is expressible by sentences of **HL**. To formally define the *axiomatic semantics* of the programming language **Com**, it only remains to define exactly all we can assert to be true about programs. This consists in partitioning **HL** into $\mathbf{HL}_{\mathrm{tt}}$ (what is truth) and $\mathbf{HL}_{\mathrm{ff}}$ (what is falsity). To do this, logicians have proposed two complementary approaches:

(1) The *semantical point of view* consists in defining an interpretation $I: \mathbf{HL} \rightarrow \{\mathrm{tt}, \mathrm{ff}\}$ with $\mathbf{HL}_{\mathrm{tt}} = \{F \in \mathbf{HL}\colon I[\![F]\!] = \mathrm{tt}\}$ and $\mathbf{HL}_{\mathrm{ff}} = \{F \in \mathbf{HL}\colon I[\![F]\!] = \mathrm{ff}\}$.

(2) The *syntactical point of view* consists in defining which sentences of **HL** are provable to be true (with given limited means, so that (hopefully) proofs can be checked by mechanical computation).

We first start with provability.

7.2.2. *Deductive systems and formal proofs*

The basis of the inductive definition of formal proofs for a logical language **HL** is provided by a set of *axioms*, that is, a set of formulae of **HL** of which the truth is postulated. Since the set of axioms is usually not finite, we use a finite set $\mathrm{AS} = \{\mathrm{AS}_i\colon i \in \Delta_\alpha\}$ of *axiom schemata* AS_i the instances of which are axioms. Otherwise stated, an axiom schema AS_i is a syntactic rule specifying a set $\{A \in \mathbf{HL}\colon \mathrm{IsAxiom}(A, \mathrm{AS}_i)\}$ of axioms by their syntax. This set of axioms is said to be *recursive* because membership is *decidable*, that is, there is a program $\mathrm{IsAxiom}(A, \mathrm{AS}_i)$ (called a recognizer) which, given any formula $A \in \mathbf{HL}$, will always terminate and answer "tt" or "ff" whether or not formula A belongs to the set of axioms generated by the axiom schema AS_i.

The induction step in the definition of formal proofs is provided by a set of *inferences* $\langle F_1, \ldots, F_n, F\rangle \in \mathbf{HL}^{n+1}$ with $n \geq 1$, traditionally written under the form

$$\frac{F_1, \ldots, F_n}{F}$$

which means that if formulae $F_1,\ldots,F_n$ are provable then so is formula F. A further requirement is again that this set of valid inferences should be recursive. Hence the set of inferences is usually specified by a finite set $\mathrm{IR}=\{\mathrm{IR}_i: i\in\Delta_\iota\}$ of syntactical rules (called *rules of inference*) so that there is a program $\mathrm{IsInference}(F_1,\ldots,F_n,F,\mathrm{IR}_i)$ which, given any formulae $F_1,\ldots,F_n,F$ of **HL**, will always terminate and answer "tt" or "ff" whether or not the inference is correct according to one of the inferences generated by the rule IR_i.

The *deductive system* **H** is the (recursive) set of all axioms generated by the axiom schemata and all inferences generated by the rules of inference:

(93) DEFINITION (*deductive system*)

$$\begin{array}{ll}
\mathrm{AS}=\{\mathrm{AS}_i: i\in\Delta_\alpha\} & \textit{axiom schemata,}\\
\mathrm{IsAxiom}(A,\mathrm{AS}_i) & \textit{axiom recognizer,}\\
\mathrm{IR}=\{\mathrm{IR}_i: i\in\Delta_\iota\} & \textit{rules of inference,}\\
\mathrm{IsInference}(F_1,\ldots,F_n,F,\mathrm{IR}_i) & \textit{inference recognizer,}\\
\mathbf{H}=\bigcup\{\{A\in\mathbf{HL}: \mathrm{IsAxiom}(A,\mathrm{AS}_i)\}: i\in\Delta_\alpha\} & \\
\quad\cup\bigcup\{\{\langle F_1,\ldots,F_n,F\rangle\in\mathbf{HL}^{n+1}: & \\
\qquad\mathrm{IsInference}(F_1\ldots,F_n,F,\mathrm{IR}_i)\}: i\in\Delta_\iota\} & \textit{deductive system.}
\end{array}$$

In order to be able to leave unspecified some aspects of the programming language **Com** (such as machine-dependent features) we assume that we are given an additional set $\mathrm{Th}\subseteq\mathbf{HL}$ of axioms, the truth of which is taken for granted. Th can be understood as a *specification* of the data and operations on these data which are used by the programs of **Com**. A *proof* of F from Th in the deductive system **H** is a finite sequence $F_0,\ldots,F_n$ of formulae, with $F_n=F$, each of which is either a member of Th, an axiom of **H**, or else follows from earlier F_i by one of the inferences of **H**:

(94) DEFINITION (*formal proof*)

$$\begin{aligned}
&\mathrm{Proof}(F_0,\ldots,F_n,F,\mathrm{Th},\mathbf{H})\\
&\quad=[F_n=F]\\
&\qquad\wedge[\forall i\in\{0,\ldots,n\}.\,[(F_i\in\mathrm{Th})\vee(\exists k\in\Delta_\alpha.\,\mathrm{IsAxiom}(F_i,\mathrm{AS}_k))\\
&\qquad\qquad\vee(\exists k\in\Delta_\iota.\,\exists m\geqslant 1.\,\exists j_1<i,\ldots,\exists j_m<i.\\
&\qquad\qquad\quad\mathrm{IsInference}(F_{j_1},\ldots,F_{j_m},F_i,\mathrm{IR}_k))]].
\end{aligned}$$

Clearly, from the above specification we can write a simple combinatorial program to check proofs (provided Th is recursive). Using the more traditional notations of logic [37] we say that F *is provable from* $\mathrm{Th}\subseteq\mathbf{HL}$ in **H**, and write $\vdash_{\mathrm{Th}\cup\mathbf{H}}F$, if there is a proof of F from Th in **H**:

(95) DEFINITION (*provability*)

(95.1) $\vdash_{\mathrm{Th}\cup\mathbf{H}}F$ if $F\in\mathrm{Th}$;

(95.2) $\vdash_{\mathrm{Th}\cup\mathbf{H}}F$ if $F\in\mathbf{H}$;

(95.3) $\vdash_{\mathrm{Th}\cup\mathbf{H}}F$ if $\vdash_{\mathrm{Th}\cup\mathbf{H}}F_1,\ldots,\vdash_{\mathrm{Th}\cup\mathbf{H}}F_m$ and $\frac{F_1,\ldots,F_m}{F}\in\mathbf{H}$.

7.2.3. Hoare's proof system **H**

In Hoare's proof system **H** below [208] $P[v \leftarrow T]$ denotes the predicate obtained by simultaneously substituting term T for all free occurrences of variable v in predicate P. If the substitution would cause an identifier in T to become bound (e.g., $\forall x.(f(x)=y)[y \leftarrow x]$) then a suitable replacement of bound identifiers in P must take place before the substitution in order to avoid conflict (e.g., $\forall z.\ (f(z)=y)[y \leftarrow x]$ is $\forall z.\ (f(z)=x)$). Substitution will be defined more rigorously later on in (118).

DEFINITION (*Hoare proof system*). Axiom schemata of **H**:

(96) $\{P\}\text{skip}\{P\}$ *skip axiom,*

(97) $\{P[X \leftarrow E]\}X:=E\{P\}$ *(backward) assignment axiom,*

(98) $\{P\}X:=?\{\exists X.\, P\}$ *random assignment axiom.*

(A schema of axioms of the form $\{P\}\text{skip}\{P\}$ means that $\forall v_1 .\dots \forall v_n.\ \{Q\}\text{skip}\{Q\}$ is true for all formulae $Q \in \mathbf{Pre}$ with free variables $v_1, \dots, v_n$. This is an approximation of the second-order axiom $\forall P. \forall v_1 .\dots \forall v_n.\ \{P\}\text{skip}\{P\}$ where quantification over predicates is mimicked by a recursive set of axioms corresponding to all possible instances Q of quantified predicate P.)

Rules of inference of **H**:

(99) $$\frac{\{P_1\}C_1\{P_2\},\quad \{P_2\}C_2\{P_3\}}{\{P_1\}(C_1;C_2)\{P_3\}}$$ *composition rule,*

(100) $$\frac{\{P \wedge B\}C_1\{Q\},\quad \{P \wedge \neg B\}C_2\{Q\}}{\{P\}(B \rightarrow C_1 \diamond C_2)\{Q\}}$$ *conditional rule,*

(101) $$\frac{\{P \wedge B\}C\{P\}}{\{P\}(B * C)\{P \wedge \neg B\}}$$ *while rule,*

(102) $$\frac{P \Rightarrow P',\quad \{P'\}C\{Q'\},\quad Q' \Rightarrow Q}{\{P\}C\{Q\}}$$ *consequence rule.*

The backward assignment axiom (97) which corresponds to (45.3) can be given an equivalent forward form corresponding to (45.8) as proposed by FLOYD [143]:

(103) $\{P\}X:=E\{\exists X'.\, P[X \leftarrow X'] \wedge X = E[X \leftarrow X']\}$ *(forward) assignment axiom*

(104) EXAMPLE (*formal partial correctness proof of program* (4) *using* **H**)

(a) $\{(Y-1) \geqslant 0 \wedge Z * X * (X ** (Y-1)) = x ** y\}\ \ Y:=Y-1$
$\{Y \geqslant 0 \wedge Z * X * (X ** Y) = x ** y\}$ by (97)

(b) $\{Y \geqslant 0 \wedge Z * X * (X ** Y) = x ** y\}\ \ Z:=Z * X$
$\{Y \geqslant 0 \wedge Z * (X ** Y) = x ** y\}$ by (97)

(c) $\{(Y-1) \geqslant 0 \wedge Z * X * (X ** (Y-1)) = x ** y\}\ \ (Y:=Y-1; Z:=Z * X)$
$\{Y \geqslant 0 \wedge Z * (X ** Y) = x ** y\}$ by (a), (b), (99)

(d) $(Y > 0 \wedge Z * (X ** Y) = x ** y \wedge odd(Y))$
$\Rightarrow\ ((Y-1) \geqslant 0 \wedge Z * X * (X ** (Y-1)) = x ** y)$ from Th

(e) $(Y \geqslant 0 \wedge Z*(X**Y)=x**y) \Rightarrow (Y \geqslant 0 \wedge Z*(X**Y)=x**y)$ from Th

(f) $\{Y>0 \wedge Z*(X**Y)=x**y \wedge odd(Y)\}$ $(Y:=Y-1; Z:=Z*X)$
$\{Y \geqslant 0 \wedge Z*(X**Y)=x**y\}$ by (d), (c), (e), (102)

(g) $\{(Y\ div\ 2) \geqslant 0 \wedge Z*((X*X)**(Y\ div\ 2)=x**y\}$ $Y:=Y\ div\ 2$
$\{Y \geqslant 0 \wedge Z*((X*X)**Y)=x**y\}$ by (97)

(h) $\{Y \geqslant 0 \wedge Z*((X*X)**Y)=x**y\}$ $X:=X*X$
$\{Y \geqslant 0 \wedge Z*(X**Y)=x**y\}$ by (97)

(i) $\{(Y\ div\ 2) \geqslant 0 \wedge Z*((X*X)**(Y\ div\ 2))=x**y\}$
$(Y:=Y\ div\ 2; X:=X*X)$ $\{Y \geqslant 0 \wedge Z*(X**Y)=x**y\}$ by (g), (h), (99)

(j) $(Y>0 \wedge Z*(X**Y)=x**y \wedge \neg odd(Y))$
$\Rightarrow ((Y div\ 2) \geqslant 0 \wedge Z*((X*X)**(Y\ div\ 2))=x**y)$ from Th

(k) $\{Y>0 \wedge Z*(X**Y)=x**y \wedge \neg odd(Y)\}$
$(Y:=Y\ div\ 2; X:=X*X)$ $\{Y \geqslant 0 \wedge Z*(X**Y)=x**y\}$ by (j), (i), (e), (102)

(l) $\{Y>0 \wedge Z*(X**Y)=x**y\}$
$(odd(Y) \rightarrow (Y:=Y-1; Z:=Z*X) \diamond (Y:=Y\ div\ 2; X:=X*X))$
$\{Y \geqslant 0 \wedge Z*(X**Y)=x**y\}$ by (f), (k), (100)

(m) $(Y \geqslant 0 \wedge Z*(X**Y)=x**y \wedge Y <> 0$
$\Rightarrow (Y>0 \wedge Z*(X**Y)=x**y)$ from Th

(n) $\{Y \geqslant 0 \wedge Z*(X**Y)=x**y \wedge Y <> 0\}$
$(odd(Y) \rightarrow (Y:=Y-1; Z*X) \diamond (Y:=Y div\ 2; X:=X*X))$
$\{Y \geqslant 0 \wedge Z*(X**Y)=x**y\}$ by (m), (l), (e), (102)

(o) $\{Y \geqslant 0 \wedge Z*(X**Y)=x**y\}$
$(Y<>0*(odd(Y) \rightarrow (Y:=Y-1; Z:=Z*X) \diamond (Y:=Y\ div\ 2; X:=X*X)))$
$\{Y \geqslant 0 \wedge Z*(X**Y)=x**y \wedge \neg(Y<>0)\}$ by (n), (10)

(p) $\{Y \geqslant 0 \wedge 1*(X**Y)=x**y\}$ $Z:=1$ $\{Y \geqslant 0 \wedge Z*(X**Y)=x**y\}$ by 97)

(q) $\{Y \geqslant 0 \wedge 1*(X**Y)=x**y\}$
$(Z:=1; (Y<>0*(odd(Y) \rightarrow (Y:=Y-1; Z:=Z*X) \diamond (Y:=Y\ div\ 2; X:=X*X))))$
$\{Y \geqslant 0 \wedge Z*(X**Y)=x**y \wedge y \neg(Y<>0)\}$ by (p), (o), (99)

(r) $(X=x \wedge Y=y \geqslant 0) \Rightarrow (Y \geqslant 0 \wedge 1*(X**Y)=x**y)$ from Th

(s) $(Y \geqslant 0 \wedge Z*(X**Y)=x**y \wedge \neg(Y<>0)) \Rightarrow (Z=x**y)$ from Th

(t) $\{X=x \wedge Y=y \geqslant 0\}$ $(Z:=1; (Y<>0*(odd(Y) \rightarrow (Y:=Y-1; Z:=$
$Z*X) \diamond Y:=Y\ div\ 2; X:=X*X))))$ $\{Z=x**y\}$ by (r), (q), (s), (102)

7.2.4. Hoare's proof system **H**′ *for proof outlines*

If the deductive system **H** is useful for reasoning about Hoare logic, formal proofs using this proof system are totally unworkable (as shown by the level of details needed in example (104)). Proof outlines (81), as introduced by Owicki [335], and Owicki and Gries [337], offer a much more useful linear notation for proofs in which the program is given with assertions interleaved at cutpoints. A natural deduction programming logic of proof outlines was first presented in [84]. Hoare's proof outline system **H**′ below is due to Bergtra and Klop [40], and Lifschitz [273]. Further developments can be found in [365].

DEFINITION (*Hoare proof outline system*)

(105) C': **Com**′ *asserted commands*,

$C' ::= \{P_1\}\mathbf{skip}\{P_2\}$
$\quad | \{P_1\}X := E\{P_2\}$
$\quad | \{P_1\}X := ?\{P_2\}$
$\quad | \{P_1\}(C'_1; \{P_2\}C'_2)\{P_3\}$
$\quad | \{P_1\}(B \rightarrow \{P_2\}C'_1 \diamond \{P_3\}C'_2)\{P_4\}$

$$|\{P_1\}(\boldsymbol{B}*\{P_2\}\boldsymbol{C}'\{P_3\})\{P_4\}$$
$$|\ \{P\}\boldsymbol{C}'$$
$$|\ \boldsymbol{C}'\{P\}$$

(106) $\{P\}\mathbf{skip}\{P\}$ *skip axiom,*

(107) $\{P[X \leftarrow E]\}\boldsymbol{X}{:=}\boldsymbol{E}\{P\}$ *assignment axiom,*

(108) $\{P\}\boldsymbol{X}{:=}\boldsymbol{?}\{\exists X.P\}$ *random assignment axiom,*

(109) $$\frac{\{P_1\}\boldsymbol{C}'_1\{P_2\},\quad \{P_2\}\boldsymbol{C}'_2\{P_3\}}{\{P_1\}(\boldsymbol{C}'_1;\{P_2\}\boldsymbol{C}'_2)\{P_3\}}$$ *composition rule,*

(110) $$\frac{\{P \wedge B\}\boldsymbol{C}'_1\{Q\},\quad \{P \wedge \neg B\}\boldsymbol{C}'_2\{Q\}}{\{P\}(\boldsymbol{B}\rightarrow\{P \wedge B\}\boldsymbol{C}'_1 \diamond \{P \wedge \neg B\}\boldsymbol{C}'_2)\{Q\}}$$ *conditional rule,*

(111) $$\frac{\{P \wedge B\}\boldsymbol{C}'\{P\}}{\{P\}(\boldsymbol{B}*\{P \wedge B\}\boldsymbol{C}'\{P\})\{P \wedge \neg B\}}$$ *while rule,*

(112) $$\frac{(P \Rightarrow P'),\ \{P'\}\boldsymbol{C}'\{Q\}}{\{P\}\{P'\}\boldsymbol{C}'\{Q\}}$$ *left consequence rule,*

(113) $$\frac{\{P\}\boldsymbol{C}'\{Q'\},\ (Q' \Rightarrow Q)}{\{P\}\boldsymbol{C}'\{Q'\}\{Q\}}$$ *right consequence rule.*

(114) EXAMPLE (*proof outline of program* (4) *using* $\mathbf{H}'$)

```
{X=x ∧ Y=y≥0}                                                    by (112), (113)
  {Y≥0 ∧ 1*(X**y}                                                by (107), (109)
    (Z:=1;
  {Y≥0 ∧ Z*(X**Y)=x**y}                                          by (111)
    (Y <> 0*
      {Y≥0 ∧ Z*(X**Y)=x**y ∧ Y <> 0}                             by (112)
        {Y>0 ∧ Z*(X**Y)=x**y}                                    by (110)
          (odd(Y)→
            {Y>0 ∧ Z*(X**Y)=x**y ∧ odd(Y)}                       by (112)
              {(Y−1)≥0 ∧ Z*X*(X**(Y−1))=x**y}                    by (107), (109)
                (Y:=Y−1;
              {Y≥0 ∧ Z*X*(X**Y)=x**y}                            by (107)
                Z:=Z*X)
          ◇
          {Y>0 ∧ Z*(X**Y)=x**y ∧ ¬odd(Y)}                        by (113)
            {Y div 2)≥0 ∧ Z*((X*X)**(Y div 2))=x**y}             by (107), (109)
              (Y:=Y div 2;
            {Y≥0 ∧ Z*((X*X)**Y)=x**y}                            by (107)
              X:=X*X)
          )
          (Y≥0 ∧ Z*(X**Y)=x**y}
    ))
  {Y≥0 ∧ Z*(X**Y)=x**y ∧ ¬(Y <> 0)}
{Z=x**y}.
```

7.2.5. Syntactical rules of substitution

Up to now, we have used informal definitions of variables, bound variables and free variables occurring in a predicate or a command, and of substitution $P[v \leftarrow T]$ of a term T for a variable v in a predicate P. We now give the full formal definitions. This subsection can be omitted on first reading and one may go on with Subsection 7.3.

7.2.5.1. Variables appearing in a term, predicate, command or correctness formula. The set of variables appearing in a term, predicate, command or correctness formula is defined by structural induction as follows:

(115) DEFINITION (*variables appearing in a formula*)

(115.1) $Var(c) = \emptyset$,

(115.2) $Var(v) = \{v\}$,

(115.3) $Var(f(T_1, \ldots, T_{\#f})) = \bigcup\{Var(T_i)\colon 1 \leqslant i \leqslant \#f\}$,

(115.4) $Var((T_1 = T_2)) = Var(T_1) \cup Var(T_2)$,

(115.5) $Var(r(T_1, \ldots, T_{\#r})) = \bigcup\{Var(T_i)\colon 1 \leqslant i \leqslant \#r\}$,

(115.6) $Var(\neg P) = Var(P)$,

(115.7) $Var((P_1 \wedge P_2)) = Var((P_1 \vee P_2)) = Var((P_1 \Rightarrow P_2)) = Var(P_1) \cup Var(P_2)$,

(115.8) $Var(\exists v.\, P) = Var(\forall v.P) = \{v\} \cup Var(P)$,

(115.9) $Var(\text{skip}) = \emptyset$,

(115.10) $Var(X := E) = \{X\} \cup Var(E)$,

(115.11) $Var(X := ?) = \{X\}$,

(115.12) $Var((C_1; C_2)) = Var(C_1) \cup Var(C_2)$,

(115.13) $Var((B \rightarrow C_1 \diamond C_2)) = Var(B) \cup Var(C_1) \cup Var(C_2)$,

(115.14) $Var((B * C)) = Var(B) \cup Var(C)$,

(115.15) $Var(\{P\}C\{Q\}) = Var(P) \cup Var(C) \cup Var(Q)$,

(115.16) $Var(F_1, \ldots, F_n) = Var(F_1) \cup \cdots \cup Var(F_n)$.

7.2.5.2. Bound and free variables appearing in a term, predicate, command or correctness formula. In the formula $\exists x.\, ((+(x, y) = 0))$, variable x is "bounded" by $\exists$ whereas y is sort of floating "free". The notions of *bound* and *free variable* can be made more precise as follows:

(116) DEFINITION (*bound variables appearing in a formula*)

(116.1) $Bound(A) = \emptyset$,

(116.2) $Bound(\neg P) = Bound(P)$,

(116.3) $Bound((P_1 \wedge P_2)) = Bound((P_1 \vee P_2)) = Bound((P_1 \Rightarrow P_2))$
$= Bound(P_1) \cup Bound(P_2)$,

(116.4) $Bound(\exists v.\, P) = Bound(\forall v.\, P) = Bound(P) \cup \{v\}$,

(116.5) $Bound(C) = \emptyset$,

(116.6) $Bound(\{P\}C\{Q\}) = Bound(P) \cup Bound(Q)$,

(116.7) $Bound(F_1, \ldots, F_n) = Bound(F_1) \cup \cdots \cup Bound(F_n)$.

(117) DEFINITION (*free variables appearing in a formula*)

(117.1) $Free(A) = Var(A)$,

(117.2) $Free(\neg P) = Free(P)$,

(117.3) $Free((P_1 \wedge P_2)) = Free((P_1 \vee P_2)) = Free((P_1 \Rightarrow P_2))$
$= Free(P_1) \cup Free(P_2)$,

(117.4) $Free(\exists v.\, P) = Free(\forall v.\, P) = Free(P) - \{v\}$,

(117.5) $Free(C) = Var(C)$,

(117.6) $Free(\{P\}C\{Q\}) = Free(P) \cup Free(C) \cup Free(Q)$,

(117.7) $Free(F_1, \ldots, F_n) = Free(F_1) \cup \cdots \cup Free(F_n)$.

7.2.5.3. Formal definition of substitution of a term for a variable in a term or predicate. The substitution $P[v \leftarrow T]$ denotes the result of renaming bounded occurrences of variables in P so that none of them is v or belongs to $Var(T)$ and then replacing all free occurrences of variable v by term T. Substitution can be formally defined as follows:

(118) DEFINITION (*substitution of a term for a variable*)

(118.1) $$v'[v \leftarrow T] = \begin{cases} T & \text{if } v' = v, \\ v' & \text{if } v' \neq v; \end{cases}$$

(118.2) $c[v \leftarrow T] = c$;

(118.3) $f(T_1, \ldots, T_{\#f})[v \leftarrow T] = f(T_1[v \leftarrow T], \ldots, T_{\#f}[v \leftarrow T])$;

(118.4) $(T_1 = T_2)[v \leftarrow T] = (T_1[v \leftarrow T] = T_2[v \leftarrow T])$;

(118.5) $r(T_1, \ldots, T_{\#r})[v \leftarrow T] = r(T_1[v \leftarrow T], \ldots, T_{\#r}[v \leftarrow T])$;

(118.6) $(\neg P)[v \leftarrow T] = \neg(P[v \leftarrow T])$;

(118.7) $(P_1 \wedge P_2)[v \leftarrow T] = (P_1[v \leftarrow T] \wedge P_2[v \leftarrow T])$;

(118.8) $(P_1 \vee P_2)[v \leftarrow T] = (P_1[v \leftarrow T] \vee P_2[v \leftarrow T])$;

(118.9) $(P_1 \Rightarrow P_2)[v \leftarrow T] = (P_1[v \leftarrow T] \Rightarrow P_2[v \leftarrow T])$;

(118.10) $(\exists v'.P)[v \leftarrow T]$

$$= \begin{cases} \exists v'.P & \text{if } v' = v, \\ \exists v'.(P[v \leftarrow T]) & \text{if } v' \neq v \text{ and } v' \notin Var(T), \\ \exists w.(P[v' \leftarrow w])[v \leftarrow T] & \text{where } w \notin \{v\} \cup Var(T) \cup Var(P) \\ & \text{if } v' \neq v \text{ and } v' \in Var(T); \end{cases}$$

(118.11) $(\forall v'.P)[v \leftarrow T]$

$$= \begin{cases} \forall v'.P & \text{if } v' = v, \\ \forall v'.(P[v \leftarrow T]) & \text{if } v' \neq v \text{ and } v' \notin Var(T), \\ \forall w.(P[v' \leftarrow w])[v \leftarrow T] & \text{where } w \notin \{v\} \cup Var(T) \cup Var(P) \\ & \text{if } v' \neq v \text{ and } v' \in Var(T); \end{cases}$$

7.3. *The semantics of Hoare logic*

We now define the semantics of Hoare logic, that is, an interpretation $I: \mathbf{HL} \rightarrow \{\text{tt}, \text{ff}\}$ defining the truth of predicates and correctness formulae with respect to a relational semantics (19) of the programming language **Com**. This programming language semantics depends upon the semantics (also called a *model* or *interpretation*) $\langle D, V \rangle$ of the basis Σ. By leaving this interpretation unspecified, we define a family of semantics of Hoare logic with respect to a family of possible relational semantics of the programming language.

7.3.1. *Semantics of predicates and correctness formulae*

A *model* or *interpretation* $\langle D, V \rangle$ of the basis $\Sigma = \langle \mathbf{Cte}, \mathbf{Fun}, \mathbf{Rel}, \# \rangle$ specifies the semantics of the common part of the programming and logical languages. It consists of a nonempty set D of data and a function V with domain $\mathbf{Cte} \cup \mathbf{Fun} \cup \mathbf{Rel}$ which define the intended meaning of constants, functions and relations:

(119) DEFINITION (*interpretation of symbols*)
(119.1) $V[c] \in D$;
(119.2) $V[f]: D^{\#f} \rightarrow D$;
(119.3) $V[r] \subseteq D^{\#r}$.

Let us also recall that we have defined *states* (or *valuations*) s assigning a value $s(v) \in D$ to variables $v \in \mathbf{Var}$ (8) and $s[v \leftarrow d]$ for the state s' which agrees with s except that $s'(v) = d$:

DEFINITIONS (*states and assignments*)

(120) $s: S = \mathbf{Var} \rightarrow D$ *states*

(121) $s[v \leftarrow d](u) = (v = u \rightarrow d \diamond s(u))$ *assignment.*

These states have been used to remember the values assigned to programming variables during program execution. They will also be used to specify values for free variables in first-order predicates. Remarkably, programming and logical variables can be handled the same way. This is not always possible for more complicated programming languages.

We now define the semantics or interpretations $\underline{T}=I[\![T]\!]$ of terms T, $\underline{P}=I[\![P]\!]$ of predicates P and $\underline{\{P\}}\underline{C}\underline{\{Q\}}=I[\![\{P\}C\{P\}]\!]$ of correctness formulae $\{P\}C\{Q\}$ with respect to a given model $\langle\underline{D}, \underline{V}\rangle$ (and a given state for terms and predicates):

(122) DEFINITION (*interpretation of terms*)

$$I\colon \mathbf{Ter}\rightarrow(S\rightarrow D),\quad \underline{T}=I[\![T]\!]$$

(122.1) $I[\![v]\!](s)=s(v),$

(122.2) $I[\![c]\!](s)=V[\![c]\!],$

(122.3) $I[\![f(T_1,\ldots,T_{\#f})]\!](s)=V[\![f]\!](I[\![T_1]\!](s),\ldots,I[\![T_{\#f}]\!](s)).$

(123) DEFINITION (*interpretation of predicates*)

$$I\colon \mathbf{Pre}\rightarrow\mathbf{Ass},\quad \underline{P}=I[\![P]\!]$$

(123.1) $I[\![(T_1=T_2)]\!]=\{s\in S\colon \langle I[\![(T_1)]\!](s), I[\![T_2]\!](s)\rangle\in\delta\},$

(123.2) $I[\![r(T_1,\ldots,T_{\#r})]\!]=\{s\in S\colon \langle I[\![T_1]\!](s),\ldots,I[\![T_{\#r}]\!](s)\rangle\in V[\![r]\!]\},$

(123.3) $I[\![\neg P]\!]=S-I[\![P]\!],$

(123.4) $I[\![(P_1\wedge P_2)]\!]=I[\![P_1]\!]\cap I[\![P_2]\!],$

(123.5) $I[\![(P_1\vee P_2)]\!]=I[\![P_1]\!]\cup I[\![P_2]\!],$

(123.6) $I[\![(P_1\Rightarrow P_2)]\!]=(S-I[\![P_1]\!])\cup I[\![P_2]\!],$

(123.7) $I[\![\exists v.P]\!]=\{s\in S\colon (\{s[v\leftarrow d]\colon d\in D\}\cap I[\![P]\!])\neq\emptyset\},$

(123.8) $I[\![\forall v.P]\!]=\{s\in S\colon \{s[v\leftarrow d]\colon d\in D\}\subseteq I[\![P]\!]\}.$

(124) DEFINITION (*intepretation of correctness formulae*)

$$I\colon \mathbf{Hcf}\rightarrow\{\mathrm{tt},\mathrm{ff}\}$$

$$I[\![\{P\}C\{Q\}]\!]=\underline{\{I[\![P]\!]\}}\underline{C}\underline{\{I[\![Q]\!]\}}\quad\text{where } \underline{\{p\}}\underline{C}\underline{\{q\}}=(p\,\rceil\,\underline{C})\subseteq(S\times q).$$

Observe that the truth or falsity of a formula $\{P\}C\{Q\}$ just depends upon the model $\langle D, V\rangle$ since the semantics $\underline{C}$ of command C (19) itself depends only on the semantics $\underline{E}$ of expressions E and $\underline{B}$ of Boolean expressions B which is the same as the semantics of terms (with no logical variables) and predicates (with no quantifiers and logical variables). An interpretation of $\{P\}C\{Q\}$ different from (124) is investigated in [7, 158, 8, 9, 10, 99, 100, 232, 323, 361] using a nonstandard transfinite definition of execution traces.

(125) DEFINITION (*interpretation of formulae*)

$$I\colon \mathbf{HL}\rightarrow\{\mathrm{tt},\mathrm{ff}\}$$

$$I[\![F]\!]=(F\in\mathbf{Pre}\rightarrow I[\![F]\!]=S\diamond I[\![F]\!]).$$

(The function I is polymorphic, so that when $P \in \mathbf{Pre}$ and, depending upon the context we have either $I[\![P]\!] \in \mathcal{P}(S)$ (by (123)) or $I[\![P]\!] \in \{\text{tt}, \text{ff}\}$ (by (125))).

(126) EXAMPLE (*proof of a formula with two different interpretations*). Let us consider the basis $\langle\{0, 1\}, \{+\}, \emptyset, \#\rangle$ with $\#(+)=2$. Then $H=\{X=1\}\ (\neg(X=0)*X:=X+1)\ \{X=0\}$ is a formula of **HL**.

A first interpretation would be $\langle \mathbf{N}, V\rangle$ with $V[\![0]\!]=0_{\mathbf{N}}$, $V[\![1]\!]=1_{\mathbf{N}}$, $V[\![+]\!](x, y)=x+_{\mathbf{N}}y$. With this interpretation, formula H is semantically true ($I[\![H]\!]=\text{tt}$) because execution of the program $(\neg(X=0)*X:=X+1)$ starting in a state s such that $s(X)=1$ will never terminate.

A second interpretation would be $\langle\{0_{\mathbf{N}}, 1_{\mathbf{N}}\}, V\rangle$ with $V[\![0]\!]=0_{\mathbf{N}}$, $V[\![1]\!]=1_{\mathbf{N}}$, $V[\![+]\!](x, y)=(x+_{\mathbf{N}}y)\ \mathit{mod}\ 2_{\mathbf{N}}$. With this interpretation, formula F is semantically true because execution of the program $(\neg(X=0)*X:=X+1)$ always terminates in a state s such that $s(X)=0_{\mathbf{N}}$.

Formula H can be proved to be formally correct from tautologies $\text{Th}=\{((\text{true} \wedge \neg(X=0))\Rightarrow\text{true}), (\text{true}\Rightarrow\text{true}), ((X=1)\Rightarrow\text{true}), ((\text{true} \wedge \neg\neg(X=0))\Rightarrow(X=0))\}$ where true denotes truth, e.g., $(x=x)$, as follows:

(a)	$(\text{true} \wedge \neg(X=0))\Rightarrow\text{true}$	by Th
(b)	$\{\text{true}\}X:=X+1\ \{\text{true}\}$	by (97)
(c)	$(\text{true})\Rightarrow(\text{true})$	by Th
(d)	$\{\text{true} \wedge \neg(X=0)\}\ X:=X+1\ \{\text{true}\}$	by (a), (b), (c), (102)
(e)	$\{\text{true}\}\ (\neg(X=0)*X:=X+1)\ \{\text{true} \wedge \neg\neg(X=0)\}$	by (d), (101)
(f)	$(X=1)\Rightarrow\text{true}$	by Th
(g)	$(\text{true} \wedge \neg\neg(X=0))\Rightarrow(X=0)$	by Th
(h)	$\{X=1\}\ (\neg(X=0)*X:=X+1)\ \{X=0\}$	by (f), (e), (g), (102).

7.3.2. Semantics of substitution

In (118) we have defined the substitution of a term for a variable in a predicate which is used in the assignment axiom schema (97). To prove that this axiom schema is sound we shall need a semantical characterization of substitution. This subsection (7.3.2) can be omitted on first reading.

Informally, substitution commutes with interpretation. More precisely, the interpretation $\underline{T[\underline{v}\leftarrow T']}(s)$ of a term T where T' is substituted for v in state s is the interpretation $\underline{T}(s[v\leftarrow\underline{T'}(s)])$ of term T in state $s'=s[v\leftarrow\underline{T'}(s)]$ which agrees with s except that the value $s'(v)$ of variable v is the interpretation of term T' in state s:

(127) LEMMA (semantics of substitution of a term for a variable in a term)

$$\underline{T[\underline{v}\leftarrow T']}(s)=\underline{T}(s[v\leftarrow\underline{T'}(s)]).$$

PROOF. By structural induction on the syntax of terms:

$\underline{v[\underline{v}\leftarrow T']}(s)=\underline{T'}(s)$ [by (118.1)] $=$ $s[v\leftarrow\underline{T'}(s)](v)$ [by (121)] $=$ $\underline{v}(s[v\leftarrow\underline{T'}(s)])$ [by (122.1)].

When $v'\neq v$, $\underline{v'[\underline{v}\leftarrow T']}(s)=\underline{v'}(s)$ [by (118.1)] $=$ $s(v')$ [by (122.1)] $=$ $s[v\leftarrow\underline{T'}(s)](v')$ [by (121)] $=$ $\underline{v'}(s[v\leftarrow\underline{T'}(s)])$ [by (122.1)].

$\underline{c[\underline{v}\leftarrow T']}(s)=\underline{c}(s)$ [by (118.2)] $=$ $V[\![c]\!]$ [by (122.2)] $=$ $\underline{c}(s[v\leftarrow\underline{T'}(s)])$ [by (122.2)].

$\underline{f(T_1\ldots,T_{\#f})[v\leftarrow T']}(s) = \underline{f(T_1[v\leftarrow T'],\ldots,T_{\#f}[v\leftarrow T'])}(s)$ [by (118.3)] $= V[\![f]\!](\underline{T_1[v\leftarrow T']}(s),\ldots,\underline{T_{\#f}[v\leftarrow T']}(s))$ [by (122.3)] $= V[\![f]\!](\underline{T_1}(s[v\leftarrow \underline{T'}(s)]),\ldots,\underline{T_{\#f}}(s[v\leftarrow \underline{T'}(s)]))$ [by induction hypothesis (127)] $= \underline{f(T_1,\ldots,T_{\#f})}(s[v\leftarrow \underline{T'}(s)])$ [by (122.3)]. □

In the same way, substitution of a term for a variable in a predicate can be semantically characterized by the following lemma:

(128) LEMMA (semantics of substitution of a term for a variable in a predicate)

$$\underline{P[v\leftarrow T]} = \{s\in S\colon s[v\leftarrow \underline{T}(s)]\in \underline{P}\}.$$

PROOF. The proof is (almost) by structural induction on the syntax of predicates. The only difficulty is for $(\forall v'.P)[v\leftarrow T]$ when $v'\neq v$ and $v'\in Var(T)$ because $\forall w.P[v'\leftarrow w]$ is not a syntactic component of $\forall v'.P$. However, they have the same shapes, and more variables of T appear in $\forall v'.P$ than in $\forall w.P[v'\leftarrow w]$. Thus we define the height $\eta(P,T)$ of a predicate P with respect to a term T by structural induction as follows: $\eta(A,T)=0$; $\eta(\neg P,T)=1+\eta(P,T)$; $\eta(P_1\wedge P_2,T)=\eta(P_1\vee P_2,T)=\eta(P_1\Rightarrow P_2,T)=1+\max(\eta(P_1,T),\eta(P_2,T))$; $\eta(\exists x.P,T)=\eta(\forall x.P,T)=1+\eta(P,T)+|Var(P)\cap Var(T)|$. For a given term T, the proof is by induction on the height $\eta(P,T)$ of P. This is long but not difficult. Therefore we only treat few typical cases:

$\underline{(T_1=T_2)[v\leftarrow T]} = \underline{(T_1[v\leftarrow T]=T_2[v\leftarrow T])}$ [by (118.4)] $=\{s\in S\colon \underline{T_1[v\leftarrow T]}(s)=\underline{T_2[v\leftarrow T]}(s)\}$ [by (123.1) and definition of δ] $=\{s\in S\colon \underline{T_1}(s[v\leftarrow \underline{T}(s)])=\underline{T_2}(s[v\leftarrow \underline{T}(s)])\}$ [by (127)] $=\{s\in S\colon s[v\leftarrow \underline{T}(s)]\in \underline{(T_1=T_2)}\}$ [since $(\underline{T_1}(s[v\leftarrow \underline{T}(s)])=\underline{T_2}(s[v\leftarrow \underline{T}(s)]))\Leftrightarrow(s[v\leftarrow \underline{T}(s)]\in\{s'\colon \underline{T_1}(s')=\underline{T_2}(s')\})\Leftrightarrow(s[v\in \underline{T}(s)]\in\underline{(T_1=T_2)})$ by (123.1)].

$\underline{(P_1\Rightarrow P_2)[v\leftarrow T]}=\underline{(P_1[v\leftarrow T]\Rightarrow P_2[v\leftarrow T])}$ [by (118.9)] $=(S-\underline{P_1[v\leftarrow T]})\cup \underline{P_2[v\leftarrow T]}$ [by (123.6)] $=(S-\{s\in S\colon s[v\leftarrow \underline{T}(s)]\in \underline{P_1}\})\cup\{s\in S\colon s[v\leftarrow \underline{T}(s)]\in\underline{P_2}\}$ [by induction hypothesis (128)] $=\{s\in S\colon s[v\leftarrow \underline{T}(s)]\notin \underline{P_1}\}\cup\{s\in S\colon s[v\leftarrow \underline{T}(s)]\in \underline{P_2}\}=\{s\in S\colon s[v\leftarrow \underline{T}(s)]\in(S-\underline{P_1})\cup\underline{P_2}\}=\{s\in S\colon s[v\leftarrow \underline{T}(s)]\in\underline{(P_1\Rightarrow P_2)}\}$ [by (123.6)].

$\underline{(\forall v.P)[v\leftarrow T]}=\underline{\forall v.P}$ [by (118.11)] $=\{s\in S\colon \forall d\in D.s[v\leftarrow d]\in\underline{P}\}$ [by (123.8)] $=\{s\in S\colon \forall d\in D.(s[v\leftarrow \underline{T}(s)])[v\leftarrow d]\in \underline{P}\}$ [since $(s[v\leftarrow d'])[v\leftarrow d]=s[v\leftarrow d]$ by (121)] $=\{s\in S\colon s[v\leftarrow \underline{T}(s)]\in \underline{\forall v.P}\}$ [by (123.8)].

If $v'\neq v$ and $v'\notin Var(T)$ then $\underline{(\forall v'.P)[v\leftarrow T]}=\underline{(\forall v'.(P[v\leftarrow T])}$ [by (118.11)] $=\{s\in S\colon \{s[v'\leftarrow d]\colon d\in D\}\subseteq \underline{P[v\leftarrow T]}\}$ [by (123.8)] $=\{s\in S\colon \{s[v'\leftarrow d]\colon d\in D\}\subseteq\{s\in S\colon s[v\leftarrow \underline{T}(s)]\in\underline{P}\}\}$ [by induction hypothesis (128) since $\eta(P,T)<\eta(\forall v'.P,T)$] $=\{s\in S\colon \forall d\in D.((s[v'\leftarrow d])[v\leftarrow \underline{T}(s)]\in\underline{P})\}=\{s\in S\colon \forall d\in D.((s[v\leftarrow \underline{T}(s)])[v'\leftarrow d]\in\underline{P})\}$ [by (121) since $v\neq v'$] $=\{s\in S\colon s[v\leftarrow \underline{T}(s)]\in \underline{(\forall v'.P)}\}$ [by (123.8)].

If $v'\neq v$ and $v'\in Var(T)$ then $\underline{(\forall v'.P)[v\leftarrow T]}=\underline{\forall w.(P[v'\leftarrow w])[v\leftarrow T]}$ [by (118.11)] $=\{s\in S\colon s[v\leftarrow \underline{T}(s)]\in\underline{\forall w.(P[v'\leftarrow w])}\}$ [by induction hypothesis (128) since $v'\in Var(T)$ and $w\notin\{v\}\cup Var(T)\cup Var(P)$ imply $Var((\forall v'.P))\cap Var(T)=(Var(\forall w.(P[v'\leftarrow w]))\cap Var(T))\cup\{v'\}$, whence $\eta(\forall w.(P[v'\leftarrow w]),T)<\eta(\forall v'.P,T)$ since $\eta(P[v'\leftarrow w],T)=\eta(P,T)$ and $v'\notin Var(\forall w.(P[v'\leftarrow w]))\cap Var(T)$] $=\{s\in S\colon s[v\leftarrow \underline{T}(s)]\in\{s'\in S\colon \{s'[w\leftarrow d]\colon d\in D\}\subseteq \underline{P[v'\leftarrow w]}\}\}$ [by (123.8)] $=\{s\in S\colon \forall d\in D.\ s[v\leftarrow \underline{T}(s)][w\leftarrow d]\in\underline{P[v'\leftarrow w]}\}=\{s\in S\colon \forall d\in D.s[v\leftarrow \underline{T}(s)][w\leftarrow d]\in\{s'\in S\colon s'[v'\leftarrow \underline{w}(s')]\in\underline{P}\}\}$ [by induction hypothesis (128) since $\eta(P[v'\leftarrow w],T)<\eta((\forall v'.P),T)$ because $w\notin Var(P)\cup Var(T)$, whence $\eta(P[v'\leftarrow w],$

$T)\leqslant\eta(P, T)] = \{s\in S: \forall d\in D. s[v\leftarrow\underline{T}(s)][w\leftarrow d][v'\leftarrow\underline{w}(s[v\leftarrow\underline{T}(s)][w\leftarrow d])]\in\underline{P}\} = \{s\in S: \forall d\in D. s[v\leftarrow\underline{T}(s)][w\leftarrow d][v\leftarrow d]\in\underline{P}\}$ [by (122.1) and (121)] $= \{s\in S: \forall d\in D. s[v\leftarrow\underline{T}(s)][v'\leftarrow d][w\leftarrow d]\in\underline{P}$ [by (121) since $w\neq v'$] $= \{s\in S: \forall d\in D. s[v\leftarrow\underline{T}(s)][v'\leftarrow d]\in\underline{P}\}$ [since $w\notin Var(P)$] $= \{s\in S: s[v\leftarrow\underline{T}(s)]\in(\underline{\forall v'.P})\}$ [by (123.8)]. □

7.4. *The link between syntax and semantics: soundness and completeness issues in Hoare logic*

In Subsection 7.2 we have defined the language **HL** of Hoare logic and then the provability $\vdash_{\mathbf{Th}\cup\mathbf{H}}F$ of formulae F. In Subsection 7.3 we have defined the semantics of **HL**, that is, the truth $\underline{F}$ of formulae F. We now investigate the relation between these two definitions, that is, the soundness of provability (is a provable formula always true?) and the completeness of provability (is a true formula always provable?). The deductive system **H** is sound for **HL** (provided all theorems in Th are true). The question of completeness is more subtle because this depends upon which class of interpretations I (induced by the semantics (D, V) of the basis Σ) is considered. Hence we can only prove *relative completeness*, a notion first delineated by Wand [399] and Cook [86].

7.4.1. *Soundness of Hoare logic*

Hoare's deductive system **H** is *sound*: if we have proved $\{P\}C\{Q\}$ from Th using **H** then C is partially correct with respect to specification $\langle\underline{P}, \underline{Q}\rangle$ (assuming that all T in Th are true):

(129) THEOREM (Cook [86]) (soundness of Hoare logic)

$$(\forall T\in \mathrm{Th}. I[\![T]\!]=\mathrm{tt})\wedge(\vdash_{\mathbf{Th}\cup\mathbf{H}}\{P\}C\{Q\}) \quad\Rightarrow\quad \underline{\{P\}C\{Q\}}.$$

The proof of (129) shows that Hoare's formal proof system **H** simply consists in applying theorem (76) without the framework of the restricted logical language **HL**. This proof can be done by a theorem prover [379].

PROOF. Assuming $\forall T\in\mathrm{Th}.\ I[\![T]\!]=\mathrm{tt}$ and given a proof $H_0,\ldots,H_n$ of $\{P\}C\{Q\}$, we prove by induction that for all $i=0,\ldots,n$ we have $I[\![H_i]\!]=\mathrm{tt}$, so that in particular $\underline{\{P\}C\{Q\}}$ is true:

(a) If $H_i\in\mathrm{Th}$, then by hypothesis $I[\![H_i]\!]=\mathrm{tt}$.

(b) If H_i is an axiom of **H**, then three cases have to be considered for any given $P\in\mathbf{Pre}$:

(b₁) For a skip axiom (96), $\underline{\{P\}\mathrm{skip}\{P\}}$ obviously holds by (76.1),

(b₂) For a backward assignment axiom (97), we have $\underline{\{P[X\leftarrow E]\}X{:=}E\{P\}} = \underline{\{\{s\in S: s[v\leftarrow\underline{E}(s)]\in\underline{P}\}\}X{:=}E\{P\}}$ [by (128)] which is true by (76.2),

(b₃) For a forward assignment axiom (103), we have $\underline{\exists X'.P[X\leftarrow X']\wedge X=E[X\leftarrow X']} = \{s\in S: (\{s[X'\leftarrow d]: d\in D\}\cap\underline{P[X\leftarrow X']\wedge X=E[X\leftarrow X']}\neq\emptyset\}$ [by (124) and (123.7)] $= \{s\in S: (\{s[X'\leftarrow d]: d\in D\}\cap\{s\in S: s[X\leftarrow s(X')]\in\underline{P}\}\cap\{s\in S: s(X)=\underline{E}(s[X\leftarrow s(X')])\})\neq\emptyset\}$ [by (123.4), (128), (122.1) and (127)] $= \{s\in S: \exists d\in D. s[X\leftarrow d]\in\underline{P}\wedge s(X)= \underline{E}(s[X\leftarrow d])\}$ whence $\underline{\{P\}X{:=}E\{\exists X'.P[X\leftarrow X']\wedge X=E[X\leftarrow X']\}} = \underline{P}\,\rceil\,\underline{X{:=}E}\subseteq S\times\{s\in S: \exists d\in D. s[X\leftarrow d]\in\underline{P}\wedge s(X) = \underline{E}(s[X\leftarrow d])\}$ [by (124)] $= \{s[X\leftarrow\underline{E}(s)]: s\in\underline{P}\}\subseteq\{s\in S: \exists d\in D. s[X\leftarrow d]\in\underline{P}\wedge s(X) = \underline{E}(s[X\leftarrow d])\}$ [by (19.2)] $= \forall s\in\underline{P}.\exists d\in$

$D.\, s[X \leftarrow \underline{E}(s)][X \leftarrow d] \in \underline{P} \wedge s[X \leftarrow \underline{E}(s)](X) = \underline{E}(s[X \leftarrow \underline{E}(s)][X \leftarrow d]) = \forall s \in \underline{P}.\, \exists d \in D.\, s[X \leftarrow d] \in \underline{P} \wedge \underline{E}(s) = \underline{E}(s[X \leftarrow d])$ which is obviously true by choosing $d = \underline{E}(s)$.

(b₄) For a random assignment axiom we have $\underline{\{P\}X := ?\{\exists X.P\}} = \underline{\{P\}X := ?\{\{s \in S\colon (\{s[X \leftarrow d]\colon d \in D\} \cap \underline{P}) \neq \emptyset\}\}}$ [by (123.7)] $= \underline{\{P\}X := ?\{\{s \in S\colon \exists d \in D.\, s[X \leftarrow d] \in \underline{P}\}\}} = \underline{\{P\}X := ?\{\{s'[X \leftarrow d']\colon d' \in D \wedge s' \in \underline{P}\}\}}$ [by (121) since we let $s' = s[X \leftarrow d]$ and $d' = s(X)$] = tt by (76.3).

(c) If H_i follows from an inference of **H**, then four cases have to be considered for any given $P_1, P_2, P, P', Q, Q' \in$ **Pre**:

(c₁) For a composition inference, assuming $\underline{\{P_1\}C_1\{P_2\}}$ and $\underline{\{P_2\}C_2\{P_3\}}$ by induction hypothesis, we have $\underline{\{P_1\}(C_1; C_2)\{P_3\}} = \mathrm{tt}$ by (76.4),

(c₂) For a conditional inference (100), assuming $\underline{\{P \wedge B\}C_1\{Q\}}$ and $\underline{\{P \wedge \neg B\}C_2\{Q\}}$ by induction hypothesis, we have $\underline{\{P \cap B\}C_1\{Q\}}$ and $\underline{\{P \cap \neg B\}C_2\{Q\}}$ by (123.4) and (123.3) hence $\underline{\{P\}(B \rightarrow C_1 \diamond C_2)\{Q\}}$ is true by (76.5),

(c₃) For a while-inference (101), assuming $\underline{\{P \wedge B\}C\{P\}}$, we have $\underline{\{P \cap B\}C\{P\}}$ [by (123.4)] $= \underline{\{P\}(B * C)\{P \cap \neg B\}}$ [by (76.6)] $= \underline{\{P\}(B * C)\{P \wedge \neg B\}}$ [by (123.4) and (123.3)],

(c₄) For a consequence inference(102), assuming $\underline{(P \Rightarrow P')}$, $\underline{\{P'\}C\{Q'\}}$ and $\underline{(Q' \Rightarrow Q)}$ by induction hypothesis, we have $\underline{P} \subseteq \underline{P'}$ and $\underline{Q'} \subseteq \underline{Q}$ by (123.6) hence $\underline{\{P\}C\{Q\}}$ by (76.7). □

7.4.2. Relative completeness of Hoare logic

7.4.2.1. Completeness and incompleteness issues for Hoare logic. Having defined the syntax **Com** of programs [see (1) with later complements (91) for expressions and (92) for Boolean expressions] parametrized by a basis ⟨**Cte**, **Fun**, **Rel**, #⟩ [(84), (85)], we have approached Hoare's ideas on the formal definition of the partial correctness of programs $C \in$ **Com**, in essentially two different ways:

(a) From the semantic point of view of computer scientists (corresponding to the point of view of mathematicians using the normal everyday set-theoretic apparatus, where objects are thought of in terms of their representation, e.g. sets are understood as collections of objects), we have assumed that the semantics $\langle D, V\rangle$ of the basis ⟨**Cte**, **Fun**, **Rel**, #⟩ is given (119), from which we have defined the sets S of states (120), the set $\mathbf{Ass} = \mathscr{P}(S)$ of assertions as well as the semantics $\underline{E} = I[\![E]\!]$ (122) of expressions E (91) and the semantics $\underline{B} = I[\![B]\!]$ (123) of Boolean expressions B (92), whence the operational semantics $op[\![C]\!]$ (13) and then the relational semantics $\underline{C} = I[\![C]\!]$ (18), (19) of programs $C \in$ **Com**. This leads to the introduction of Hoare's partial correctness specifications:

(130) $\mathbf{HS}(I) = \mathbf{Ass} \times \mathbf{Com} \times \mathbf{Ass}$ *Hoare's specifications.*

where the dependence upon the interpretation I is a shorthand for denoting the dependence upon the basis ⟨**Cte**, **Fun**, **Rel**, #⟩ and its semantics $\langle D, V\rangle$. A triple $\langle p, C, q\rangle$ of **HS**(I) should be understood as specifying that p is the output specification of program C where p and q are sets of states specifying all possible combinations of values of the variables. The set **HS**(I) has been partitioned (22) into a subset representing truth (i.e., which programs C are partially correct with respect to which specifications):

(131) $\mathbf{HS}(I)_{tt} = \{\langle p, C, q\rangle : p \in \mathbf{Ass} \wedge C \in \mathbf{Com} \wedge q \in \mathbf{Ass} \wedge \underline{\{p\}}\underline{C}\underline{\{q\}} = \text{tt}\}$

Hoare's valid specifications;

and a subset representing falsity (i.e., all we know not to be true about the partial correctness of programs):

(132) $\mathbf{HS}(I)_{ff} = \{\langle p, C, q\rangle : p \in \mathbf{Ass} \wedge C \in \mathbf{Com} \wedge q \in \mathbf{Ass} \wedge \underline{\{p\}}\underline{C}\underline{\{q\}} = \text{ff}\}$

Hoare's invalid specifications.

Within this framework, we have explained several (equivalent) partial correctness proof methods by decomposition of the proof that $\underline{\{p\}}\underline{C}\underline{\{q\}} = \text{tt}$ into simpler elementary proofs based upon stepwise induction (45) as in [319, 143] or compositional induction (49), (76) as in [208]. In this development, we have already used a formalized language with only vestigial traces of English but without any linguistic constraint (for example, we felt free to use second-order sentences [such as (19.8b)] and regretfully to misuse English).

(b) Then we have defined the language **Hcf** (89) of Hoare logic for describing partial correctness (or incorrectness) of programs. By choosing the particular language **Hcf**, we deliberately limit our means of expression. Hence we have adopted the syntactical point of view of computer scientists (corresponding to the use of specific formal languages by logicians, where objects are thought of in terms of their denotation and rewriting manipulations). In this framework, we can define truth (and falsity) of partial correctness formulae by two disjoint sublanguages $\mathbf{Hcf}_{tt} \subseteq \mathbf{Hcf}$ (and $\mathbf{Hcf}_{ff} \subseteq \mathbf{Hcf}$). We have used two methods to specify the sublanguage $\mathbf{Hcf}_{tt}$, one defining "truth" and the other "provability":

(b₁) Following the semantic development of logic (also called "model theory" [37, part A]), we have defined the semantics of **Hcf**, by means of an interpretation $I: \mathbf{Hcf} \to \{\text{tt}, \text{ff}\}$ (124) which depends upon the semantics $\langle D, V\rangle$ of the basis $\langle \mathbf{Cte}, \mathbf{Fun}, \mathbf{Rel}, \#\rangle$ and induces a partition of the language **Hcf** into true and false sentences:

(133.1) $\mathbf{Hcf}_{tt}(I) = \{H \in \mathbf{Hcf} : I[\![H]\!] = \text{tt}\}$ *Hoare's valid correctness formulae,*

(133.2) $\mathbf{Hcf}_{ff}(I) = \{H \in \mathbf{Hcf} : I[\![H]\!] = \text{ff}\}$ *Hoare' invalid correctness formulae.*

(Instead of writing $I[\![H]\!] = \text{tt}$, logicians would use the *satisfaction relation* $I \models H$. The set $\mathbf{Hcf}_{tt}(I) = \{H \in \mathbf{Hcf} : I \models H\}$ of sentences of **Hcf** true in I would be called the *theory* of I).

(b₂) Following the calculative development of logic (also called "proof theory" [37, part D]), we have presented Hoare deductive system **H** (93) with its axiom schemata (96) to (98) and rules of inference (99) to (102) and defined provability $\vdash_{\mathrm{Th} \cup \mathbf{H}} H$ of formulae $H \in \mathbf{H}$ with respect to a set Th of postulates (95). Therefore we get the subset of provable sentences (contrary to the ordinary situation in logic, we do not define refutable sentences since we are not interested in partial incorrectness $\neg H$):

(134) $\mathbf{Hcf}_{pr}(\mathrm{Th}) = \{H \in \mathbf{Hcf} : \vdash_{\mathrm{Th} \cup \mathbf{H}} H\}$ *Hoare's provable correctness formulae.*

In order to compare syntactical objects Th, $\mathbf{Hcf}_{pr}(\mathrm{Th})$, $\mathbf{Hcf}_{tt}(I)$ with semantic objects I, $\mathbf{HS}_{tt}(I)$, we can map Hoare correctness formulae $H = \{P\}C\{Q\}$ into triples $\gamma(H) = \langle I[\![P]\!], C, I[\![Q]\!]\rangle$ of $\mathbf{HS}(I)$ and assume that the interpretation I is a model of postulates Th, that is to say, that every formula t of Th is true for interpretation $I: \forall t \in \mathrm{Th}.\, I[\![t]\!] = \text{tt}$. Then we get inclusions between these sets (up to the correspondence

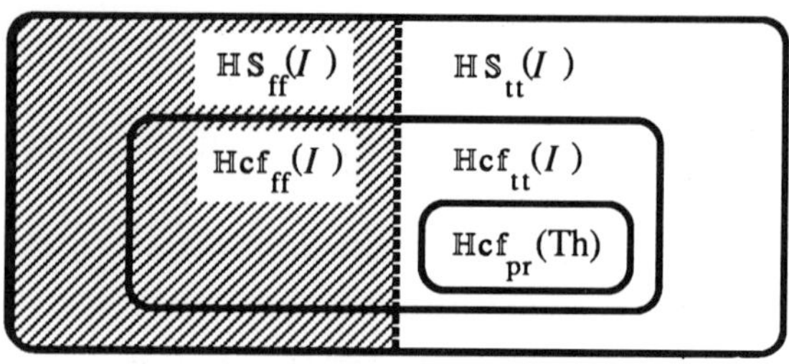

Fig. 1.

γ) as shown in Fig. 1. The fact that $\mathbf{Hcf}_{\mathrm{pr}}(\mathrm{Th}) \subseteq \mathbf{Hcf}_{\mathrm{tt}}(I)$ follows from the soundness theorem (129): every provable formula is true. The inclusions $\gamma(\mathbf{Hcf}_{\mathrm{tt}}(I)) \subseteq \mathbf{HS}(I)_{\mathrm{tt}}$ and $\gamma(\mathbf{Hcf}_{\mathrm{ff}}(I)) \subseteq \mathbf{HS}(I)_{\mathrm{ff}}$ follow from interpretation (123) of correctness formulae. Now, the incompleteness/completeness question is whether these inclusions are strict or not:

(a) $\forall I.\, \mathbf{HS}(I)_{\mathrm{tt}} \subseteq \gamma(\mathbf{Hcf}_{\mathrm{tt}}(I))$: is every true fact about the partial correctness of programs expressible in the restricted language **Hcf**?

(b) $\forall I.\, \mathbf{Hcf}_{\mathrm{tt}}(I) \subseteq \mathbf{Hcf}_{\mathrm{pr}}(\mathrm{Th})$: is every true formula of **Hcf** provable by the deductive system **H**?

In both cases, and without additional hypothesess, the answer is no. Intuitively, the origin of these incompleteness problems is that there exist programs $C \in \mathbf{Com}$ constructing (input-output) relations between objects of the domain D of data that, under the assumption that the programming language **Com** and first-order formulae **Pre** must have the same signature $\langle \mathbf{Cte}, \mathbf{Fun}, \mathbf{Rel}, \# \rangle$, cannot be described by a predicate of **Pre**. To illustrate the phenomenon, we shall use "abacus arithmetic", a limited version of usual arithmetic with 0, $\mathrm{Su}(x) = x + 1$ and $\mathrm{Pr}(x) = x - 1$ (with $0 - 1 = 0$) as only nonlogical symbols.

7.4.2.2. Abacus arithmetic. The purpose of this paragraph is to show that addition cannot be defined by some formula of abacus arithmetic $\mathbf{Pre}_A$ defined by (88) where the basis is $A = \langle \mathbf{Cte}, \mathbf{Fun}, \mathbf{Rel}, \# \rangle$ with $\mathbf{Cte} = \{0\}$, $\mathbf{Fun} = \{\mathrm{Pr}, \mathrm{Su}\}$, $\mathbf{Rel} = \emptyset$, $\#(\mathrm{Pr}) = \#(\mathrm{Su}) = 1$ for interpretation I_A defined by (123) where $D = \mathbf{N}$, $V[\![0]\!] = 0$, $V[\![\mathrm{Pr}]\!](x) = (x = 0 \rightarrow 0 \diamond x -_{\mathbf{N}} 1)$ and $V[\![\mathrm{Su}]\!](x) = x +_{\mathbf{N}} 1$. It is also shown that the *theory* of abacus arithmetic (i.e. $\mathbf{Pre}_{A,\mathrm{tt}}(I_A) = \{P \in \mathbf{Pre}_A : I[\![P]\!] = \mathrm{tt}\}$) is *decidable*, that is, there is an algorithm which given $P \in \mathbf{Pre}_A$ will always terminate and answer "tt" if $P \in \mathbf{Pre}_{A,\mathrm{tt}}(I_A)$ and answer "ff" if this is not true. We will also show that abacus arithmetic has nonstandard interpretations.

7.4.2.2.1. Inexpressibility of addition in abacus arithmetic

(135) Lemma (Enderton [133]) (disjunctive normal form). *A quantifier-free predicate P and its negation $\neg P$ have equivalent* disjunctive normal forms $P^{\vee\wedge}$ *and* $P^{\underline{\vee\wedge}}$ (*i.e.*, $I[\![P]\!] = I[\![P^{\vee\wedge}]\!]$ *and* $I[\![\neg P]\!] = I[\![P^{\underline{\vee\wedge}}]\!]$ *for all interpretations* I) *such that*

$$P^{\vee\wedge} = \bigvee_{i=1,m} \bigwedge_{j=1,n} P_{ij}, \qquad P^{\underline{\vee\wedge}} = \bigvee_{k=1,p} \bigwedge_{l=1,q} \underline{P}_{kl}$$

where the P_{ij} and $\underline{P}_{kl}$ are atomic formulae or negations of atomic formulae.

PROOF. By induction on the syntactical structure of P:

If P is an atomic formula A then $P^{\vee\wedge} = \bigwedge_{i=1,1} \bigwedge_{j=1,1} A$ and $P^{\underline{\vee\wedge}} \bigwedge_{k=1,1} \bigwedge_{l=1,1} \neg A$.

If P is $\neg Q$ then $P^{\vee\wedge} = Q^{\underline{\vee\wedge}}$ and $P^{\underline{\vee\wedge}} = Q^{\vee\wedge}$.

If P is $(Q \vee R)$ then $P^{\vee\wedge} = (Q^{\vee\wedge} \vee R^{\vee\wedge})$ whereas $P^{\underline{\vee\wedge}} = \neg(Q \vee R) = \neg Q \wedge \neg R = (Q^{\underline{\vee\wedge}} \wedge R^{\underline{\vee\wedge}}) = ((\bigvee_{i=1,m} \bigwedge_{j=1,n} \underline{Q}_{ij}) \wedge (\bigvee_{k=1,p} \bigwedge_{l=1,q} \underline{R}_{kl})) = \bigvee_{r=1,n*m} T_r$ with $T_{i+m*(k-1)} = (\bigwedge_{j=1,n} \underline{Q}_{ij} \wedge \bigwedge_{l=1,q} \underline{R}_{kl})$ for $i=1,\ldots, m$; $k=1,\ldots, p$; by using a generalized form of the distributive law $((P_1 \vee P_2) \wedge (P_3 \vee P_4)) = (P_1 \wedge P_3) \vee (P_1 \wedge P_4) \vee (P_2 \wedge P_3) \vee (P_2 \wedge P_4)$.

If P is $(Q \wedge R) = \neg(\neg Q \vee \neg R)$ or P is $(Q \Rightarrow R) = (\neg Q \vee R)$ then we use the previous transformations. □

(136) LEMMA (Enderton [133]) (quantifier elimination). *Any predicate* $P \in \mathbf{Pre}_A$ *is equivalent to a quantifier-free predicate* $P' \in \mathbf{Pre}_A$ *(i.e.,* $I_A[\![P]\!] = I_A[\![P']\!]$*) (with no occurrence of the* Pr *symbol or of terms with two occurrences of the same variable).*

PROOF. Generalizing the simultaneous substitution $P[x \leftarrow t]$ of term t for all occurrences of variable x in predicate P, we can define on the model of (118) the substitution $P[F' \leftarrow F]$ of F for all occurrences of F' in predicate P. We obtain P' from P by repeated application of the following transformations (such that $I_A[\![P]\!] = I_A[\![P']\!]$):

(a) We first eliminate the function symbols Pr from P by repeated transformation of P into

$$P[\mathrm{Pr}(t) \leftarrow x] \wedge (((t=0) \wedge (x=0)) \vee (\neg(t=0) \wedge (\mathrm{Su}(x)=t))),$$

where $x \in \mathbf{Var} - Var(P)$ is a fresh variable, as long as some term $\mathrm{Pr}(t)$ appears within P.

(b) Then we eliminate universal quantifiers $\forall v.\, Q$ from P by repeated transformation of P into $P[\forall v.\, Q \leftarrow \neg(\exists v.\, \neg Q)]$.

(c) Finally we eliminate all subformulae $\exists x.\, Q$ from P, starting from the innermost ones (so that Q can be assumed to be quantifier-free), by repeated application of transformations, in the following order:

(c₁) Q is replaced by its disjunctive normal form $\exists x.\, Q = \exists x.\, Q^{\vee\wedge} = \exists x.\, \bigvee_{i=1,m} \bigwedge_{j=1,n} Q_{ij} = \bigvee_{i=1,m} (\exists x.\, \bigwedge_{j=1,n} Q_{ij})$ so that in the following we can assume that Q is of the form $\bigwedge_{j=1,n} Q_{ij}$ with all Q_{ij} being an atomic formula A or the negation $\neg A$ of an atomic formula A.

(c₂) $\exists x.\, \bigwedge_{i=1,n} Q_i$ where x does not appear in $Q_k, 1 \leqslant k \leqslant n$, is replaced by $Q_k \wedge (\exists x.\, \bigwedge_{i=1,k-1} Q_i \wedge \bigwedge_{j=k+1,n} Q_j)$. Afterwards, using the commutativity of $=$, all Q_i can be assumed to be of the form A or $\neg A$ where A is $(\mathrm{Su}^p(x) = \mathrm{Su}^q(t))$ and t is $0, x$, or another variable $y \neq x$, with $\mathrm{Su}^0(t) = t$ and $\mathrm{Su}^{p+1}(t) = \mathrm{Su}^p(\mathrm{Su}(t))$.

(c₃) All $(\mathrm{Su}^p(x) = \mathrm{Su}^q(x))$ are replaced by $(0=0)$ if $p=q$ and by $\neg(0=0)$ when $p \neq q$, whence no term has two occurrences of the same variable so that in the following we can assume that t is 0 or $y \neq x$:

- If all Q_i are of the form $\neg(\mathrm{Su}^p(x) = \mathrm{Su}^q(t))$ then $\exists x.\, \bigwedge_{i=1,n} Q_i$ asserts that we can find a value of x different from a finite number of given values, so that $\exists x.\, \bigwedge_{i=1,n} Q_i$ is replaced by $(0=0)$ representing truth.
- Else some Q_k is of the form $(\mathrm{Su}^p(x) = t')$ where t' is a term of the form $\mathrm{Su}^r(t)$ with $t=0$ or $t=y \neq x$ so that intuitively x can be chosen as $(t'-p) \geqslant 0$. Therefore, Q_k is replaced

by $(0=0)$ if $p=0$, else by

$$(\neg(t'=0) \wedge \neg(\mathrm{Su}(t')=0) \wedge \cdots \wedge \neg(\mathrm{Su}^{p-1}(t')=0))$$

expressing that $x \geqslant 0$, whereas all terms Q_i, $i=1,\ldots,n$, $i \neq k$, which are of the form $\neg(\mathrm{Su}^q(x)=t_i)$ or $(\mathrm{Su}^q(x)=t_i)$ are respectively replaced by $\neg(\mathrm{Su}^q(t')=\mathrm{Su}^q(t_i))$ or $(\mathrm{Su}^q(t')=\mathrm{Su}^q(t_i))$ since intuitively $x=(t'-p)\geqslant 0$ and $q+x=t_i$ is equivalent to $x=(t'-p)\geqslant 0$ and $q+t'=t_i+p$. Since formula Q no longer contains variable x, $\exists x.\, Q$ can be replaced by Q. □

(137) LEMMA (Enderton [133]) (definability in abacus arithmetic). *A subset* $E \subseteq \mathbf{N}$ *is definable by* $P \in \mathbf{Pre}_A$ *(i.e.,* $\exists v \in \mathbf{Var}.\, E=\{s(v) \in \mathbf{N}: s \in I_A[\![P]\!]\}$*) if and only if it is finite or cofinite (i.e.,* $\mathbf{N}-E$ *is finite).*

PROOF. Let P' be the quantifier-free predicate equivalent to P (136) in disjunctive normal form (135). If $v \notin Var(P')$ then $\forall d \in \mathbf{N}.(s \in I_A[\![P']\!] \Leftrightarrow s[v \leftarrow d] \in I_A[\![P']\!])$ whence $E=\{s(v) \in \mathbf{N}: s \in I_A[\![P']\!]\}=\mathbf{N}$ is cofinite so that in the following we can assume that v occurs free in P'. Then we proceed by induction on the syntactical structure of P'. If P' is an atomic formula A_{ij}, it is of the form $(\mathrm{Su}^p(v)=\mathrm{Su}^q(t))$ where t is 0 or $x \neq v$, whence if t is 0 then $E=\emptyset$ (when $q<p$) or $E=\{q-p\}$ (when $q \geqslant p$) is finite; else t is $x \neq v$ and $E=\{v \in \mathbf{N}: \exists x \geqslant 0.\, p+v=q+x\}=(p \geqslant q \rightarrow \mathbf{N} \diamond \{v \in \mathbf{N}: v \geqslant (q-p)\})$ is cofinite. If P' is a negated atomic formula A_{ij} then E is the complement of a finite or cofinite set, hence it is cofinite or finite. If P' is a conjunction $\bigwedge_{j=1,n} A_{ij}$ of atomic formulae A_{ij} or negations A_{ij} of atomic formulae then E is the intersection of finite or cofinite sets. If all are cofinite then it is cofinite, else it is finite. Finally, if P' is a disjunction $\bigvee_{i=1,m} \bigwedge_{j=1,n} A_{ij}$ then E is the union of finite or cofinite sets, hence it is finite or cofinite. □

(138) LEMMA (Enderton [133]) (inexpressibility of addition in abacus arithmetic). *There is no formula* $P \in \mathbf{Pre}_A$ *equivalent to* $A+B=C$ *(more precisely, such that* $I_A[\![P]\!]=\{s \in S: (s(A)+s(B))=s(C)\}$*).*

PROOF. Otherwise, the set of even naturals would be definable by $\exists k.\,(k+k=v)$ (more precisely $\exists k.\, P[A \leftarrow k][B \leftarrow k][C \leftarrow v]$), in contradiction with (137). □

Let $\mathbf{Com}_A$ be the language (1) with the abacus arithmetic A as basis. Every partial recursive function (such as addition of natural numbers) can be computed by some command $C \in \mathbf{Com}_A$ (cf. [257] and [55, Section 7]). However (138) shows that addition is not expressible in $\mathbf{Pre}_A$. It follows that $\mathbf{Com}_A$ has a greater expressive power than $\mathbf{Pre}_A$. This is the source of incompleteness problems with Hoare logic.

7.4.2.2.2. Decidability of abacus arithmetic

(139) LEMMA (Enderton [133]) (decidability of abacus arithmetic). *The theory* $\mathbf{Pre}_{A,\mathrm{tt}}(I_A)=\{P \in \mathbf{Pre}_A: I[\![P]\!]=\mathrm{tt}\}$ *of abacus arithmetic is decidable.*

PROOF. The algorithm used in the proof of (136) transforms a predicate $P \in \mathbf{Pre}_A$ into an equivalent quantifier-free formula P' with variables $v_1, \ldots, v_n$. Using the same algorithm we can transform $\neg\exists v_1. \ldots \exists v_n. \neg P'$ into a quantifier-free formula P'' with no variables and atoms of the form $(\mathrm{Su}^m(0) = \mathrm{Su}^n(0))$ which are replaced by tt if $m = n$ and ff otherwise. Then the answer is obtained using truth tables ($\mathrm{tt} \wedge \mathrm{tt} = \mathrm{tt}, \mathrm{tt} \wedge \mathrm{ff} = \mathrm{ff}, \ldots$). □

The decidability of $\mathbf{Pre}_{A,\mathrm{tt}}(I_A)$ shows that the fact that true correctness formulae in $\mathbf{Hcf}_{\mathrm{tt}}(I)$ are not provable by $\mathbf{H}$ does not necessarily come from unprovability problems in $\mathbf{Pre}_A$ that would be inherited in Hoare logic through the consequence rule.

7.4.2.2.3. Nonstandard interpretations of abacus arithmetic

(140) LEMMA (Bergstra & Klop [40]) (nonstandard interpretations of abacus arithmetic). *The nonstandard interpretation* I'_A *defined by* $D' = (\{0\} \times \mathbf{N}) \cup (\mathbf{N}^+ \times \mathbf{Z})$, $V'[\![0]\!] = \langle 0, 0\rangle$, $V'[\![Pr]\!](\langle i, x\rangle) = (i = 0 \rightarrow (x = 0 \rightarrow \langle 0, 0\rangle \diamond \langle 0, x -_{\mathrm{N}} 1\rangle) \diamond \langle i, x -_{\mathrm{Z}} 1\rangle)$ *and* $V'[\![\mathrm{Su}]\!](x) = (x = 0 \rightarrow \langle 0, x -_{\mathrm{N}} 1\rangle \diamond \langle i, x -_{\mathrm{Z}} 1\rangle)$ *has the same theory as the standard interpretation* I_A, *i.e., is such that* $\mathbf{Pre}_{A,\mathrm{tt}}(I'_A) = \mathbf{Pre}_{A,\mathrm{tt}}(I_A)$.

This is an extension of the standard naturals: $\langle 0,0\rangle \equiv 0 < \langle 0, 1\rangle \equiv \mathrm{Su}(0) < \langle 0, 2\rangle \equiv \mathrm{Su}^2(0) < \cdots < \langle i, -2\rangle < \langle i, -1\rangle < \langle i, 0\rangle < \langle i, 1\rangle < \langle i, 2\rangle < \cdots < \langle j, -1\rangle < \langle j, 0\rangle < \langle j, 1\rangle < \cdots$ by strictly greater nonstandard infinite numbers, organized by groups isomorphic to integers (however I'_A is not a model of Peano arithmetic with addition and multiplication, because there are too few nonstandard numbers, see [55, Section 17)].

PROOF OF LEMMA (140). If $P' \in \mathbf{Pre}_A$ and $s \in I_A[\![P']\!]$ then obviously $s' = \lambda v. \langle 0, s(v)\rangle \in I'_A[\![P]\!]$. Reciprocally, if $s' \in I'_A[\![P']\!]$ then we define $s \in I_A[\![P]\!]$ where P is equivalent to P', quantifier-free and in disjunctive normal form $\bigvee_{i=1,m} \bigwedge_{j=1,n} P_{ij}$, with the P_{ij} being atoms Q of the form $(\mathrm{Su}^m(0) = \mathrm{Su}^n(0)), (\mathrm{Su}^m(x) = \mathrm{Su}^n(0)), (\mathrm{Su}^m(x) = \mathrm{Su}^n(y))$ with $x \neq y$ or their negation $\neg Q$ (136), (137). One of the $\bigwedge_{j=1,n} P_{ij}$, hence all P_{ij}, must be true in s' and we define s so that all P_{ij} are also true in s. If $x \in V_P = \bigcup_{j=1,n} \mathit{Var}(P_{ij})$ and $s'(x) = \langle k, p\rangle$ then $s(x) = (k = 0 \rightarrow p \diamond (k * \omega) + p - \mu)$, else $s(x) = 0$ where ω is a natural greater than $[\max(\{p : x \in V_P \wedge s'(x) = \langle k, p\rangle\}) - \mu] + \kappa + 1, \mu = \min(\{0\} \cup \{p : x \in V_P \wedge s'(x) = \langle k, p\rangle\})$ and $\kappa \in \mathbf{N}^+$ is greater than all n such that the term $\mathrm{Su}^n(0)$ appears in P. Hence the idea is to map finitely many finite segments of groups of nonstandard numbers into disjoint segments of $\mathbf{N}$ in the same order beyond the finite initial segment used to do the standard computations. Then, the truth of $(\mathrm{Su}^m(0) = \mathrm{Su}^n(0))$ or its negation does not depend upon the states s' and s. If $(\mathrm{Su}^m(x) = \mathrm{Su}^n(0))$ is true in s' then $s'(x) = \langle 0, n - m\rangle$ and $n - m \geqslant 0$ so that it is also true for $s(x) = n - m$. If $\neg(\mathrm{Su}^m(x) = \mathrm{Su}^n(0))$ is true in s' then $s'(x) = \langle k, p\rangle$ with either $k = 0$ and $m + p \neq n$ in which case it is also true for $s(x) = p$, or else $k > 0$ in which case $m + s(x) = m + (k * \omega) + p - \mu > \kappa > n$. If $(\mathrm{Su}^m(x) = \mathrm{Su}^n(y))$ is true in s' then $s'(x) = \langle k, p\rangle$ and $s'(y) = \langle k, q\rangle$ with $m + p = n + q$. Therefore it is also true of $s(x) = (k * \omega) + p - \mu$ and $s(y) = (k * \omega) + q - \mu$. If $\neg(\mathrm{Su}^m(x) = \mathrm{Su}^n(y))$ is true in s' then $s'(x) = \langle k, p\rangle$ and $s'(y) = \langle 1, q\rangle$ with $k = 1$ and $m + p \neq n + q$ in which case $m + s(x) = m + (k * \omega) + p - \mu \neq n + (k * \omega) + q - \mu = n + s(y)$, or else $k \neq 1$ in which case $m + s(x) = n + s(y)$ would imply $\omega \leqslant (k - 1) * \omega = m - n + (q - \mu) - (p - \mu) \leqslant \kappa + (q - \mu) < \omega$. □

No predicate of the standard theory $\mathbf{Pre}_{A,\mathrm{tt}}(I_A)$ can distinguish nonstandard numbers from the standard naturals. Whence partial correctness is unchanged when considering nonstandard interpretations. For example $\{0=0\}C\{X=0\}$ with $C=(\neg(X=0)*X:=\mathrm{Pr}(X))$ is true in the above nonstandard interpretation since C is either started with some value $\langle 0,p\rangle$ of X and terminates with the value $\langle 0,0\rangle$ or else X is initially $\langle i,p\rangle$ and takes successive values $\langle i,p\rangle, \langle i,p-1\rangle, \langle i,p-2\rangle,\ldots$ so that execution of C never terminates [206].

7.4.2.3. Incompleteness results for Hoare logic. We are now in a position to explain the incompleteness results for Hoare logic: (a) the partial correctness of a program C may not be expressible by a formula $\{P\}C\{Q\}$ and, for the same reason that in certain cases first-order languages may be too weak assertion languages, (b) some true formula $\{P\}C\{Q\}$ may not be provable by the deductive system $\mathbf{H}\cup\mathrm{Th}$ where Th is the set of all theorems P true in a given interpretation, and (c) Th may not be axiomatizable by a deduction system $\mathbf{T}$ so that incompleteness problems for the first-order language **Pre** show up in Hoare logic through the consequence rule.

7.4.2.3.1. Unspecifiable partially correct programs. A program based upon abacus arithmetic $\mathbf{Pre}_A$ which computes addition by successive increments is not provable in **H** since, by (138), addition is not expressible in $\mathbf{Pre}_A$. It follows that Hoare logic is incomplete since partial correctness of this program is not expressible in **Hcf**:

(141) Theorem (Bergstra & Tucker [44]) (unspecifiable partially correct programs)

$$\exists I.\,\gamma(\mathbf{Hcf}_{\mathrm{tt}}(I))\neq\mathbf{HS}(I)_{\mathrm{tt}}.$$

Proof. Choosing the basis $A=\langle\{0\},\{\mathrm{Su},\mathrm{Pr}\},\emptyset,\#\rangle$ of abacus arithmetic I_A, $p=\{s\in S: s(X)=s(x)\wedge s(Y)=s(y)\}$, $C=(\neg(X=0)*(Y:=\mathrm{Su}(Y);\,X:=\mathrm{Pr}(X)))$, $q=\{s\in S: s(Y)=s(y)+s(x)\}$ we have $\{p\}C\{q\}$ hence $\langle p,C,q\rangle\in\mathbf{HS}(I)_{\mathrm{tt}}$, whereas by (138) there is no $Q\in\mathbf{Pre}_A$ such that $I_A[\![Q]\!]=q$ whence no $\{P\}C\{Q\}\in\mathbf{Hcf}$ such that $\gamma(\{P\}C\{Q\}) = \langle p,C,q\rangle$. □

The proof shows that "$\{X=x\wedge Y=y\}$ $(\neg(X=0)*(Y:=\mathrm{Su}(Y);\ X:=\mathrm{Pr}(X)))$ $\{Y=y+x\}$" is not expressible in abacus arithmetic $\mathbf{Pre}_A$ because addition cannot be represented in this logic (138). Hence we can enrich the basis A with the addition symbol $+$ to get Presburger arithmetic $\mathrm{PB}=\langle\{0\},\{\mathrm{Su},\mathrm{Pr},+\},\emptyset,\#\rangle$. But then "$\{X=x\wedge Y=y\}$ $(Z:=0;\ (\neg(X=0)*(Z:=Z+Y;\ X:=\mathrm{Pr}(X)))$ $\{Z=x*y\}$" is not expressible because multiplication cannot be represented in this logic [133, Corollary 32G]. Presburger arithmetic has no sound and complete Hoare logic for its while-programs [86, 44] although one can be found when restricting the class of considered programs as in [69]. Hence we must enrich the basis PB with the multiplication symbol $*$ to get Peano arithmetic $\mathrm{PE}=\langle\{0\},\{\mathrm{Su},\mathrm{Pr},+,*\},\emptyset,\#\rangle$ (but then, by Gödel's second incompleteness theorem, Peano arithmetic I_{PE} is no longer first-order axiomatizable). This situation was first described by Harel [197], Meyer and Halpern [300, 301], Bergstra, Tiuryn and Tucker [42] and by Meyer [299]: the partial correctness of a program C on basis Σ can always be expressed by a Hoare correctness formula $\{P\}C\{Q\}$ with predicates $P, Q \notin \mathbf{Pre}_\Sigma$ using extra symbols of Peano arithmetic $\mathrm{PA}=(0, \mathrm{Su}, +, *, <)$ which may not appear in the programs and whose semantics may not be expressed in the language $\mathbf{Pre}_\Sigma$. This is because the functions and relations computed by the program are recursive, hence can be coded in arithmetic [238, Section 4]. But then incompleteness problems in $\mathbf{Pre}_{\mathrm{PA}}$ appear through the consequence rule.

A similar incompleteness argument can be given using the fact that the transitive closure of a first-order expressible binary relation is computable by a while-program but may not be first-order expressible [192, 154]. Then transitive closures [235] or least fixpoints [4] can be turned into logical operators. Such extented first-order logics are studied in [192, 193].

7.4.2.3.2. Unprovable partially correct programs. It may also happen that partial correctness can be specified by a Hoare formula $\{P\}C\{Q\}$, but may not be provable using Hoare's deductive system **H** (96)–(102), even with the help of all true postulates $\mathrm{Th}=\{P \in \mathbf{Pre}\colon I[\![P]\!]=\mathrm{tt}\}$ in the intended interpretation I. As independently shown by Gergely and Szöts [157] and by Wand [399] it might be the case that given $P, Q \in \mathbf{Pre}$ the proof of $\{\underline{P}\}(\underline{B}*\underline{C})\{\underline{Q}\}$ by (77) would involve an intermediate assertion i which cannot be expressed by a first-order predicate: $\neg(\exists P \in \mathbf{Pre}.\ I[\![P]\!]=i)$. In this case, the proof using (77) cannot be carried out into **H** because intermediate (and in particular loop) invariants are not first-order definable. Otherwise stated, intermediate states in the computation of a program (e.g., with recursive procedures) are often much more complicated than the initial and final states involved in paragraph 7.4.2.3.1. Hence the language of first-order logic **Pre** may be too weak to describe precisely enough the set of intermediate states that is computed by a program.

7.4.2.3.2.1. Incompleteness of Hoare logic for an interpretation with decidable first-order theory. A program based upon abacus arithmetic $\mathbf{Pre}_A$ which involves a loop invariant I expressing addition is not provable in **H** since, by (138), the invariant I is not expressible in $\mathbf{Pre}_A$ [40]. It follows that Hoare logic is incomplete:

(142) THEOREM (Gergely & Szöts [157], Wand [399]) (local incompleteness of Hoare logic).

$$\exists I.\ \mathbf{Hcf}_{\mathrm{tt}}(I) \neq \mathbf{Hcf}_{\mathrm{pr}}(\mathrm{Th}).$$

PROOF (Bergstra & Klop [40]). Let us choose abacus arithmetic $\mathbf{Pre}_A$ and define $P=((X=x)\wedge(Y=0))$, $C=(\neg(X=0)*C')$, $C'=(X:=\mathrm{Pr}(X);\ Y:=\mathrm{Su}(Y))$ and $Q=(Y=x)$.

The proof of $\{P\}C\{Q\}$ in **H** must involve the while inference rule (101) with some I such that $\{I \cap \neg(X=0)\}C'\{I\}$, and the consequence rule (102) so that $(P \Rightarrow I)$ and $((I \wedge (X=0)) \Rightarrow Q)$. Then, by the soundness theorem (129), we have $\{\underline{I} \cap \underline{\neg}(\underline{X}\underline{=}\underline{0})\}\underline{C}'\{\underline{I}\}$, whence $\{\langle s, s[X \leftarrow s(X) -_0 1]\ [Y \leftarrow s(Y)+1]\rangle\colon\ s \in \underline{I} \wedge (s(X) \neq 0)\} \subseteq \{\langle s', s''\rangle\colon\ s'' \in \underline{I}\}$ by (19.2), (19.4) and (22) so that by (123) we must have

(a) $\{s \in S\colon (s(X)=s(x)) \wedge (s(Y)=0)\} \subseteq \underline{I}$,

(b) $\{s[X \leftarrow s(X)-1][Y \leftarrow s(Y)+1]\colon s \in \underline{I} \wedge (s(X) \neq 0)\} \subseteq \underline{I}$,

(c) $\{s \in \underline{I}\colon s(X)=0\} \subseteq \{s \in S\colon s(Y)=s(x)\}$.

Let us show that $i=\{s\in S: s(x)=s(Y)+s(X)\}$ is the unique solution $\underline{I}$ of (a), (b), (c). If $s[X\leftarrow 0][Y\leftarrow s(X)+s(Y)]\in\underline{I}$ then $0\leqslant s(X)+s(Y)$ and by (c) we have $s(X)+s(Y)=s(x)$, whence $s\in i$. By (b), we have $s[X\leftarrow n+1][Y\leftarrow s(X)+s(Y)-(n+1)]\in\underline{I}\wedge n+1\leqslant s(X)+s(Y)$ which implies $s[X\leftarrow n][Y\leftarrow s(X)+s(Y)-n]\in\underline{I}\wedge n\leqslant s(X)+s(Y)$. It follows, by induction, that $m\geqslant n\geqslant 0$ implies $(s[X\leftarrow m][Y\leftarrow s(X)+s(Y)-m]\in\underline{I}\wedge m\leqslant s(X)+s(Y))\Rightarrow(s[X\leftarrow n][Y\leftarrow s(X)+s(Y)-n]\in\underline{I}\wedge n\leqslant s(X)+s(Y))\Rightarrow(s\in i)$. When $m=s(X)+s(Y)\geqslant n=s(X)\geqslant 0$, we get $(s[X\leftarrow s(X)+s(Y)][Y\leftarrow 0]\in\underline{I})\Rightarrow(s\in\underline{I})\Rightarrow(s\in i)$. Moreover, if $s\in i$, then $s'=s[X\leftarrow s(X)+s(Y)][Y\leftarrow 0]$ is such that $s'(X)=s(Y)+s(X)=s(x)=s'(x)$ and $s'(Y)=0$, so that by (a) we have $s\in\underline{I}$, whence $(s\in i)\Rightarrow(s\in\underline{I})\Rightarrow(s\in i)$. In conclusion, $i=\underline{I}$.

Now, by (138), an invariant I expressing that $x=(Y+X)$ is not expressible in $\mathbf{Pre}_A$. □

The same way, Abramov [1] proves that an invariant I expressing that $A+B>C$ is needed to prove $\{A>C\}$ $(A>0*(A:=\Pr(A);\ B:=\mathrm{Su}(B)))$ $\{B>C\}$ whereas Abramov [2] shows that I is not expressible in the first-order logic with basis $\langle\{0\},\{\mathrm{Su},\Pr\},\{>\},\#\rangle$.

7.4.2.3.2.2. Incompleteness of Hoare logic for an interpretation with undecidable first-order theory. As above (paragraph 7.4.2.3.1), we could enrich the predicate language **Pre** with Peano arithmetic which would allow more expressive predicates. However, as shown in [86, Section 6, p. 85] this would not solve the completeness problem: if the halting problem is undecidable for interpretation I of **Com** then the set of valid Hoare formulae of the form $\{\text{true}\}C\{\text{false}\}$, $C\in\mathbf{Com}$, is not recursively enumerable, hence cannot be included in the recursively enumerable set of provable Hoare formulae in any formal deductive system Th∪**H**. The rest of this paragraph is devoted to a detailed explanation of the argument. (Another proof of the local incompleteness of Hoare logic (142) not referring to an interpretation I whose first-order theory Th is decidable is given in [270, Theorem 1].)

7.4.2.3.2.2.1. The set of provable Hoare formulae is recursively enumerable. A set E is *recursively enumerable* if there is an algorithm that lists elements of E in some order (of course, since E may be infinite, the list may never be completed but any particular element of E will appear in the list after some finite length of time).

(143) LEMMA (recursive enumerability of provable Hoare formulae). **Hcf** *is recursively enumerable. If* Th *is recursive then* $\mathbf{Hcf}_{\mathrm{pr}}(\mathrm{Th})$ *is recursively enumerable.*

PROOF. Computer scientists know that symbols, finite lists of symbols, finite lists of lists of symbols, etc. can be coded in a machine into an integer in binary representation and that from this representation it is possible to recover the original object. *Gödel numbering* is a similar idea. An odd code $\ulcorner\sigma_i\urcorner$ is associated with the n basic logical or programming symbols σ_i: $\ulcorner=\urcorner=3$, $\ulcorner\Rightarrow\urcorner=5$, $\ulcorner\wedge\urcorner=7$, $\ulcorner\vee\urcorner=9$, $\ulcorner\neg\urcorner=11$, $\ulcorner\forall\urcorner=13$, $\ulcorner\exists\urcorner=15$, $\ulcorner\mathrm{skip}\urcorner=17$, $\ulcorner:=\urcorner=19$, $\ulcorner\Diamond\urcorner=21,\ldots,\ulcorner\sigma_n\urcorner=2n+1$ and with the constant symbols $\ulcorner c_i\urcorner=2(n+1)+10i+1$ where $\mathbf{Cte}\subseteq\{c_i: i\in\mathbf{N}\}$, function symbols $\ulcorner f_i\urcorner=$

$2(n+1)+10i+3$ where $\mathbf{Fun} \subseteq \{f_i : i \in \mathbf{N}\}$, relation symbols $\ulcorner r_i \urcorner = 2(n+1)+10i+5$ where $\mathbf{Rel} \subseteq \{r_i : i \in \mathbf{N}\}$, programming variables $\ulcorner X_i \urcorner = 2(n+1)+10i+7$ where $\mathbf{Pvar} \subseteq \{X_i\colon i \in \mathbf{N}\}$ and logical variables $\ulcorner x_i \urcorner = 2(n+1)+10i+9$ where $\mathbf{Lvar} \subseteq \{x_i\colon i \in \mathbf{N}\}$ are assumed to be enumerable. Then a command or a predicate that is a finite string $\sigma_1, \ldots, \sigma_n$ of symbols will be coded as $\ulcorner \sigma_1, \ldots, \sigma_n \urcorner = 2^{\lceil \sigma_1 \rceil} 3^{\lceil \sigma_2 \rceil} 5^{\lceil \sigma_3 \rceil} \cdots p_n^{\lceil \sigma_n \rceil}$ where p_n is the nth prime number. Then a Hoare correctness formula $\{P\}C\{Q\}$ will be coded as $\ulcorner \{P\}C\{Q\} \urcorner = 2^{\lceil P \rceil} 3^{\lceil C \rceil} 5^{\lceil Q \rceil}$. Then a proof in $\mathbf{H}$, that is, a finite string $F_1, \ldots, F_m$ of predicates or Hoare correctness formulae, will be coded as $\ulcorner F_1, \ldots, F_m \urcorner = 2^{\lceil F_1 \rceil} 3^{\lceil F_2 \rceil} \cdots p_m^{\lceil F_m \rceil}$ where p_m is the mth prime number.

Now, given an integer n, it is possible to decode it. If n is odd then n is the code $\ulcorner \sigma \urcorner$ of a symbol $\sigma = \ulcorner n \urcorner^{-1}$. Else it is even and can be decomposed into its prime factors $n = 2^{n_1} 3^{n_2} \cdots p_m^{n_m}$. The decomposition is unique. If each n_i is odd then n is the code of a finite string of symbols $\ulcorner n \urcorner^{-1} = \ulcorner n_1 \urcorner^{-1} \ulcorner n_2 \urcorner^{-1} \cdots \ulcorner n_m \urcorner^{-1}$. A syntactical recognizer will tell if the string is a syntactically correct command or predicate. Else each n_i can be decomposed into its prime factors. If $m=3$, $\ulcorner n_1 \urcorner^{-1}$ is a predicate P, $\ulcorner n_2 \urcorner^{-1}$ is a command C and $\ulcorner n_3 \urcorner^{-1}$ is a predicate Q, then n is the code of Hoare formula $\{P\}C\{Q\}$. Else it can be checked whether each n_i is the code of a predicate or a Hoare formula $F_i = \ulcorner n_i \urcorner^{-1}$ so that n is the code of a proof $\ulcorner n \urcorner^{-1} = \ulcorner n_1 \urcorner^{-1} \ulcorner n_2 \urcorner^{-1} \cdots \ulcorner n_m \urcorner^{-1}$. Else n is not the Gödel number of a proof, Hoare formula, predicate, command or symbol. Observe that the numbering is injective: different objects have different Gödel numbers; coding and decoding is recursive, that is, can be done by a terminating algorithm as informally described above; and the set of codes is recursive, that is, given any natural number n the algorithm described above always terminates with the object $\ulcorner n \urcorner^{-1}$ coded by n or else answers that n is not a Gödel number.

To do the recursive enumeration of $\mathbf{Hcf}$, we just have to enumerate the natural numbers, for each one we check whether it is the Gödel number of a Hoare formula and then output the corresponding formula. Since all Hoare correctness formulae H have a code $\ulcorner H \urcorner \in \mathbf{N}$ and since no two different formulae can have the same code, no formula H can be omitted in the enumeration.

To do the recursive enumeration of $\mathbf{Hcf}_{\mathrm{pr}}(\mathrm{Th})$, we just have to enumerate the natural numbers, for each one we check whether it is the Gödel number of a proof $F_1, \ldots, F_n$ and then test the validity of the proof using a recognizer to check that F_i is an instance of an axiom scheme or combinatorially check that F_i follows from previous F_j by a rule of inference of $\mathbf{H}$ or we algorithmically check that $F_i \in \mathrm{Th}$ (the algorithm exists since Th is assumed to be recursive). If the proof is correct we output the formula F_n. □

7.4.2.3.2.2.2. The nonhalting problem is not semidecidable for Peano arithmetic. A problem P depending upon data $d \in D$ with a logical answer "yes" or "no" is *decidable* (or *solvable*) (written *Decidable*(P)) if and only if there exists an algorithm (*Decision*(P): $D \to \{\mathrm{tt}, \mathrm{ff}\}$) which when given data d always terminates with output "tt" or "ff" corresponding to the respective answer "yes" or "no" to the problem. A problem is *undecidable* (or *unsolvable*) when no such algorithm exists.

A problem P depending upon data $d \in D$ with a logical answer "yes" or "no" is *semidecidable* (written *Semidecidable*(P)) if and only if there exists an algorithm

(*SemiDecision*(*P*): $D \to \{\text{tt}, \text{ff}\}$) which when given data d always delivers an answer "tt" in a finite amount of time when the answer to the problem is "yes", but may answer "ff" or may be blocked or else may not terminate when the problem for d has answer "no".

The *halting problem* is the problem of deciding whether execution of a command $C \in$ **Com** starting in a given initial state $s \in S$ terminates or not (for the interpretation I where the basis $\langle\{0, 1\}, \{+, *\}, \emptyset, \#\rangle$ has its natural arithmetical interpretation on $\mathbb{N}$).

(144) Lemma (Church [70], Turing [392, 393]) (undecidability of halting problem). *The halting problem is semidecidable but undecidable. The nonhalting problem is not semidecidable.*

Sketch of Proof. Following Hoare and Allison [222], we now briefly sketch a coarse proof for a subset of Pascal. A Pascal program is a finite sequence of characters. It can be represented in Pascal as a text file of arbitrary length. Obviously, we can write a Pascal function $\underline{I}$ of type "**function** $\underline{I}$(**var** F, D: text): Boolean" such that if F is the text of a Pascal function of type "**function** $\underline{F}$(**var** D: text): Boolean" and D is the text of the data of $\underline{F}$ then $\underline{I}(F, D)$ is the result $\underline{F}(D)$ of executing function $\underline{F}$ with data D. $\underline{I}$ is simply a Pascal interpreter written in Pascal but specialized in execution of Boolean functions F with text parameter D.

The semidecision algorithm for the halting problem simply consists in executing F with data D using interpreter $\underline{I}$ and answering "tt" upon termination:

```
function SemiDecisionOfHaltingProblem(var F, D: text);
  var R: Boolean;
begin R := I(F, D); write("tt");
  SemiDecisionOfHaltingProblem := True; end;
```

To show that the halting problem is undecidable, we prove by reductio ad absurdum that we cannot write a termination prover in Pascal that is a function of type "**function** $\underline{T}$(**var** F, D: text): Boolean" such that for all texts F of Pascal functions $\underline{F}$ of type "**function** $\underline{F}$(**var** D: text): Boolean" and all data D of type "text", execution of $\underline{T}$ with data F and D would always terminate and yield a result $\underline{T}(F, D)$ which is "True" if and only if execution of $\underline{I}(F, D)$, i.e. of $\underline{F}$ with data D does terminate. Assuming the existence of such a $\underline{T}$, we let TC be the text: "**function** C(**var** F: text): Boolean; **begin if** $\underline{T}(F, F)$ **then** $C :=$ **not**$\underline{I}(F, F)$ **else** $C :=$ True **end**;". Observe that $\underline{T}(F, F)$ terminates and either $\underline{T}(F, F) =$ True and "$C :=$ **not** $\underline{I}(F, F)$" terminates or $\underline{T}(F, F) =$ False and "$C :=$ True" terminates so that $\underline{T}$(TC, TC) is "True". Then $\underline{I}$(TC, TC) = **if** $\underline{T}$(TC, TC) **then not**$\underline{I}$(TC, TC) **else** True = **not**$\underline{I}$(TC, TC), a contradiction. In conclusion, there is no algorithm by means of which we can test an arbitrary program to determine whether or not it always terminates for given data.

The argument can be rephrased for **Com** using a coding of text files into natural numbers (or, for Turing machines, see rigorous details in [55, Sections 3, 4], [107], [133, Section 3.5], [253, Section 43] or [362, Section 1.9]).

The negation $\neg P$ of a problem P depending upon data $d \in D$ with a logical answer "yes" or "no" is the problem of answering the opposite of P: $\neg P(d) = (P(d) = \text{"yes"} \to$

"no" ◇ "yes"). We have *Decidable(P)* ⇔ [*Semidecidable(P)* ∧ *Semidecidable*(¬*P*)]. ⇒ is obvious. For ⇐ define *Decision(P)(d)* by executing alternatively one step of *SemiDecision(P)(d)* and one step of ¬*SemiDecision*(¬*P*)(*d*) as long as both are not terminated and by terminating *Decision(P)(d)* as soon as one of these fairly interleaved executions of *SemiDecision(P)(d)* and ¬*SemiDecision*(¬*P*)(*d*) is terminated.

If the nonhalting problem were semidecidable then [*Semidecidable*(halting) ∧ *Semidecidable*(¬halting)] would imply *Decidable*(halting), in contradiction with the undecidability of the halting problem. □

7.4.2.3.2.2.3. *The set of valid Hoare formulae for Peano arithmetic is not recursively enumerable*

(145) LEMMA (valid Hoare formulae for Peano arithmetic). *The set* $\mathbf{Hcf}_{tt}(I)$ *is not recursively enumerable for Peano arithmetic (i.e., the interpretation I where the basis* $\langle\{0, 1\}, \{+, *\}, \emptyset, \#\rangle$ *has its natural arithmetical interpretation on* **N**).

PROOF. Assume the contrary. Let $s \in S$ be a state, $Var(C) = X_1, \ldots, X_n$ be the variables of C with initial values $x_1 = s(X_1), \ldots, x_n = s(X)_n) \in \mathbf{N}$ and $P = (X_1 = \mathrm{Su}^{x_1}(0)) \wedge \cdots \wedge (X_n = \mathrm{Su}^{x_n}(0))$ where $\mathrm{Su}^0(0) = 0$ and $\mathrm{Su}^{n+1}(0) = (\mathrm{Su}^n(0) + 1)$. Execution of C in state s never terminates if and only if $I[\![\{P\}C\{\text{false}\}]\!] = \text{tt}$ (where false is $(0=1)$). Hence, execution of C in state s never terminates if and only if the formula $\{P\}C\{\text{false}\}$ is to be found in the recursive enumeration of $\mathbf{Hcf}_{tt}(I)$. It would follow that the nonhalting problem would be semidecidable, in contradiction with (144). □

7.4.2.3.2.2.4. *Incompleteness of Hoare logic for Peano arithmetic*

(146) THEOREM (Cook [86], Apt [13]) (incompleteness of Hoare logic for Peano arithmetic)

$$\exists I . \mathbf{Hcf}_{tt}(I) \neq \mathbf{Hcf}_{pr}(\mathrm{Th}) \quad (\textit{where } \mathrm{Th} \textit{ is recursive and } I \textit{ is Peano arithmetic}).$$

PROOF. $\mathbf{Hcf}_{pr}(\mathrm{Th})$ is recursively enumerable by (143) but $\mathbf{Hcf}_{tt}(I)$ is not by (145), hence these sets are different. □

We say that the interpretation domain *D* is *Herbrand definable* when all elements of *D* can be represented by a term:

(147) DEFINITION (*Herbrand definability*). *D* is *Herbrand definable* if and only if $(\forall d \in D . \exists T \in \mathbf{Ter} . I[\![T]\!] = d)$.

Theorem (146) is also a consequence of (145) combined with the following observation:

(148) THEOREM (Bergstra & Tucker [44], Bergstra & Tiuryn [41]). *If* Th *is recursive,* $|D| = |\mathbf{N}|$, *D is Herbrand definable and* $\mathbf{Hcf}_{tt}(I) = \mathbf{Hcf}_{pr}(\mathrm{Th})$, *then* $\mathbf{Hcf}_{tt}(I)$ *is recursive.*

PROOF. If $C \in \mathbf{Com}$ then for all $n \in \mathbf{N}$ there is $C^n \in \mathbf{Com}-\mathbf{Loops}$ running at most n (assignment or test) steps of C. Indeed, since $|D| = |\mathbf{N}|$, $\mathbf{Lab}[\![C]\!]$ is finite and D is Herbrand definable, there is a bijection η between $\mathbf{Lab}[\![C]\!]$ and some finite subset $\eta(\mathbf{Lab}[\![C]\!]) = \{L_0, \ldots, L_k\}$ of $\mathbf{Ter}$ with $L_0 = \eta(\surd)$ and $L_1 = \eta(C)$. Let $Xc \in \mathbf{Pvar} - Var(C)$ be a fresh variable used as program counter. C^n is $(\cdots((Xc := L_1; I^1); I^2); \cdots I^n)$ where each $I^i = (\neg(Xc = L_0) \rightarrow (S_1; (S_2; \cdots (S_{k-1}; S_k) \cdots)) \diamond \text{skip})$ executes one step of C (unless C is terminated). Each S_j executes the elementary step of C labeled L_j and updates the program counter Xc; therefore S_j is $(Xc = L_j \rightarrow (\mathbf{Step}[\![C]\!][L_j]; Xc := \eta(\mathbf{Succ}[\![C]\!][L_j])) \diamond \text{skip})$ when $\mathbf{Step}[\![C]\!][L]$ is skip, $X := E$ or $X := ?$, and S_j is $(Xc = L_j \rightarrow (B \rightarrow Xc := \eta(\mathbf{Succ}[\![C]\!][L_j]\,(\text{tt})) \diamond Xc := \eta(\mathbf{Succ}[\![C]\!][L_j](\text{ff}))) \diamond \text{skip})$ when $\mathbf{Step}[\![C]\!][L_j]$ is B.

Moreover there is a formula R^n of $\mathbf{Pre}$ such that $\{P\}C^n\{Q\}$ holds if and only if R^n is true. R^n is $(P \Rightarrow wlp(C^n, Q))$ where, by induction on the syntax of $C^n \in \mathbf{Com}-\mathbf{Loops}$, $wlp(\text{skip}, Q) = Q$, $wlp(X := E, Q) = Q[X \leftarrow E]$, $wlp(X := ?, Q) = \forall X . Q$, $wlp((B \rightarrow C_1 \diamond C_2), Q) = ((B \wedge wlp(C_1, Q)) \vee (\neg B \wedge wlp(C_2, Q)))$, $wlp((C_1; C_2), Q) = wlp(C_1, wlp(C_2, Q))$. $wlp(C, Q)$ is explained in more details below, see (151).

If $\mathbf{Hcf}_{\text{tt}}(I) = \mathbf{Hcf}_{\text{pr}}(\text{Th})$ and Th is recursive then $\mathbf{Hcf}_{\text{tt}}(I)$ is recursively enumerable by (143). Moreover $\mathbf{Hcf}_{\text{ff}}(I)$ is also recursively enumerable because $\{P\}C\{Q\} \in \mathbf{Hcf}_{\text{ff}}(I)$ if and only if $R^n \notin \text{Th}$, so that $\mathbf{Hcf}_{\text{ff}}(I)$ can be recursively enumerated by generating all formulae $\{P\}C\{Q\}$ according to syntax (89) and by recursively testing for each of them that $R^n \notin \text{Th}$. We conclude that $\mathbf{Hcf}_{\text{tt}}(I)$ is recursive by running alternatively one step of the algorithms to recursively enumerate $\mathbf{Hcf}_{\text{tt}}(I)$ and $\mathbf{Hcf}_{\text{ff}}(I)$. □

Bergstra, Chmielinska and Tiuryn [38] have shown that the converse of (148) is not true: we may have $\mathbf{Hcf}_{\text{tt}}(I) \neq \mathbf{Hcf}_{\text{pr}}(\text{Th})$ with $\mathbf{Hcf}_{\text{tt}}(I)$ recursive.

7.4.2.3.3. Unprovable valid predicates, mechanical proofs. Following Cook [86], definition (95) of provability $\vdash_{\text{Th} \cup \mathbf{H}} H$ includes the use of a given set Th of postulates which are indispensable in consequence rule (102). Hence Hoare's system **H** can be thought of as being equipped with an infallible oracle answering questions on the validity of first-order predicates. In this way, the reasoning about the programs is separated from the reasoning about the underlying language **Pre** of invariants.

However, Floyd [143] propounded the definition of this set Th of postulates as the set of provable theorems in a formal deductive system. This approach was further advocated by Hoare [208] (with the additional idea that "a programming language standard should consist of a set of axioms of universal applicability, together with a choice from a set of supplementary axioms describing the range of choices facing an implementor"). More precisely, we should define $\text{Th} = \{P \in \mathbf{Pre}: \vdash_{A \cup \mathbf{T}} P\}$ by means of a set of nonlogical axioms A and a deductive system **T**. The set A of nonlogical axioms should be recursive (that is, $P \in A$ should be decidable by a machine) and consistent (A should have at least one *model*, i.e. an interpretation I such that $\forall t \in A . I[\![t]\!] = \text{tt}$). The existence of this deductive system $A \cup \mathbf{T}$ depends upon the class of interpretations I which is considered.

More precisely, by Gödel's completeness theorem of 1930 (see [37, Section A1.4], [55, Section 12], [133, Section 2.5], [238, Section 3] or [253, Section 52]), there is a deductive system **T** such that all (and only) formulae which hold for *all* interpretations

I that are models of some given set of axioms $A \subseteq \mathbf{Pre}$ are provable in $A \cup \mathbf{T}$:

(149) $\{P \in \mathbf{Pre}: \vdash_{A \cup \mathbf{T}} P\} = \{P \in \mathbf{Pre}: \forall I. (\forall t \in A.\ I[\![t]\!] = \mathrm{tt}) \Rightarrow I[\![P]\!] = \mathrm{tt}\}$.

However, when considering *some* given intended interpretation I (for example, Peano arithmetic) or a restricted nonempty class κ of such interpretations I satisfying all axioms of A, it may happen, by Gödel's second incompleteness theorem of 1931 (see [374, Section A.14], [55, Section 16], [133, Section 3.5], [238, Section 9] or [253, Section 44]), that some $P_1 \in \mathbf{Pre}$ is true for I (or all I in κ) but is neither provable nor refutable in $A \cup \mathbf{T}$. By Gödel's 1930 completeness theorem (149), this simply means that P_1 may be true for some interpretations I' and false for other interpretations I''. Hence we should add P_1 or $\neg P_1$ to A in order to eliminate the unintended interpretations I' or I''. But then there is some $P_2 \in \mathbf{Pre}$ which is true for I (or all I in κ) but is neither provable nor refutable in $A \cup \{P_1\} \cup \mathbf{T}$, and so on. This means that unintended interpretations cannot be eliminated when using only first-order concepts [8, 44, 65, 323].

Since program hand-proving is tedious, long and sometimes difficult, the 1970s have lived in hopes of mechanical verification of program correctness [250, 234, 57]. This approach has theoretical limits determined by uncomputability problems: no computer (even ideal ones without size and time limits) can be used to *automatically* prove partial program correctness. This follows from Gödel's 1931 second incompleteness theorem which implies that the theory $\{P \in \mathbf{Pre}: \forall I \in \kappa. I[\![P]\!] = \mathrm{tt}\}$ of a class κ of interpretations satisfying the axioms A (i.e., $\forall I \in \kappa.\ (\forall t \in A. I[\![t]\!] = \mathrm{tt})$) and including Peano arithmetic is not recursive (because otherwise proofs would simply consist in using the terminating algorithm to check that $\forall I \in \kappa. I[\![P]\!] = \mathrm{tt}$). Hence the discovery of the invariant needed in while rule (101) can be partly automated [404, 168, 246] but ultimately requires human intervention. In the same way, the use of consequence rule (102) may also call for such error-prone human interactions. But then, lonesome individuals have to carefully manage large amounts of detailed information produced by machines. This has practical limits discussed in [117]. Experience with a proof editor for interactive proof checking is discussed in [356] and with more ambitious verification environments in [58, 83, 176].

7.4.2.4. Cook's relative completeness of Hoare logic. Cook [86] circumscribed these incompleteness problems by assuming that the set Th of mathematical theorems $P \Rightarrow P'$ which have to be used in consequence rule (102) is given. This corresponds to the common mathematical practice to accept certain notions and structures as basic and work axiomatically from there on, even if we are aware that these notions cannot be completely axiomatized in the restricted language of first-order logic. Cook also assumed that the intermediate invariants needed in composition rule (99) and while rule (101) can be expressed in the first-order language **Pre**. This is called *relative completeness*, which consists in proving that

(150) $$\begin{aligned}&\mathit{Expressive}(\mathbf{Com}, \mathbf{Pre}, op, I)\\ &\quad\Rightarrow \forall C \in \mathbf{Com}. \forall P, Q \in \mathbf{Pre}. (\underline{\{P\}C\{Q\}} \Rightarrow \vdash_{\mathrm{Th} \cup \mathbf{H}} \{P\}C\{Q\})\end{aligned}$$

where $\mathrm{Th} = \{P \in \mathbf{Pre}: I[\![P]\!] = \mathrm{tt}\}$ and $\mathit{Expressive}(\mathbf{Com}, \mathbf{Pre}, op, I)$ is a sufficient (and preferrably necessary) condition implying that intermediate invariants can be expressed in **Pre**. This implies that true Hoare formulae are provable: *Expressive*(**Com**, **Pre**, *op*, *I*) $\Rightarrow$ $\mathbf{Hcf}_{tt}(I) \subseteq \mathbf{Hcf}_{pr}(\mathrm{Th})$.

7.4.2.4.1. Expressiveness à la Clarke. Such a condition *Expressive*(**Com**, **Pre**, *op*, *I*) was first proposed by Cook in [86]; an equivalent one was later proposed by Clarke [71]. Observe that in the proof of $\{P\}(C_1;C_2)\{Q\}$ we can choose an intermediate invariant I such that $\underline{I} = \{s \in S : \forall s' \in S.(\langle s, s' \rangle \in \underline{C_1}) \Rightarrow (s' \in I[\![Q]\!])\}$, whereas in the proof of $\{P\}(B * C)\{Q\}$ we can choose an intermediate invariant I such that $\underline{I} = \{s \in S: \forall s' \in S. \langle (s, s' \rangle \in \underline{B*C}) \Rightarrow (s' \in I[\![Q]\!])\}$. This leads to the following definition.

(151) DEFINITION (Dijkstra [124]) (*weakest liberal precondition*)

$$wlp: \mathscr{P}(S \times S) \times \mathscr{P}(S) \to \mathscr{P}(S)$$
$$wlp(r, q) = \{s \in S: \forall s' \in S. (\langle s, s' \rangle \in r) \Rightarrow (s' \in q)\}.$$

Hence, execution of a command C starting from a state $s \in wlp(\underline{C}, \underline{Q})$ cannot reach a final state not satisfying the predicate Q. The qualifier "liberal" means that non-termination is left as an alternative. The epithet "weakest" means "making as less as possible restrictions on initial states". More precisely, $\underline{P} \subseteq wlp(\underline{C}, \underline{Q})$ is another notation for $\{P\}C\{Q\}$:

(152) LEMMA (Dijkstra [124]) (*definition of partial correctness using wlp*)

$$\underline{\{p\}C\{q\}} \Leftrightarrow p \subseteq wlp(\underline{C}, q).$$

PROOF. $\underline{\{p\}C\{q\}} \Leftrightarrow (p \rceil \underline{C} \subseteq S \times q) \Leftrightarrow [\forall s, s' \in S. (s \in p) \Rightarrow ((\langle s, s' \rangle \in \underline{C}) \Rightarrow (s' \in q))] \Leftrightarrow (p \subseteq wlp(\underline{C}, \underline{Q}))$. $\square$

Termination is obviously not implied since, for example, if execution of command C never terminates then $\underline{C} = \emptyset$ whence $wlp(\underline{C}, q) = S$. Dijkstra's most commonly used predicate transformer *wp* [124] guarantees termination (see Subsection 8.4.8). Possible alternative definitions of *wlp* are studied in [315].

(153) DEFINITION (Clarke [71]) (*expressiveness à la Clarke*). The interpretation I is said to be *expressive* for the languages **Com** and **Pre** if and only if $\forall C \in \mathbf{Com}. \forall Q \in \mathbf{Pre}. \exists P \in \mathbf{Pre}. \underline{P} = wlp(\underline{C}, \underline{Q})$.

The expressiveness requirement rules out the interpretations that have been used in paragraph 7.4.2.3 to prove incompleteness results for Hoare logic. For example, we have the following theorem:

(154) THEOREM (nonexpressiveness of abacus arithmetic). *The standard interpretation* I_A *of abacus arithmetic* $\mathbf{Pre}_A$ *is not expressive for* $\mathbf{Com}_A$.

SKETCH OF PROOF. This follows from the local incompleteness (142) and relative completeness (156) below. One can also prove that for $Q=(Z=0)$ and $C=(\neg(X=0 \wedge Y=0)*((X=0 \rightarrow Y:=Y-1 \diamond X:=X-1); Z:=Z-1))$ we have $wlp(\underline{C}, \underline{Q})=\{s \in S: Z(s)= X(s)+Y(s)\}$ but, by (138), no formula of $\mathbf{Pre}_A$ is equivalent to $(Z=X+Y)$. □

(155) THEOREM (Cook [86]) (expressiveness of Peano arithmetic). *The standard interpretation I_{PE} of Peano arithmetic* $\mathrm{PE}=\langle\{0\},\{\mathrm{Su}, \mathrm{Pr}, + *\},\{<\}, \#\rangle$ *on the domain* $\mathbf{N}$ *of natural numbers is expressive for* $\mathbf{Com}_{\mathrm{PE}}$ *and* $\mathbf{Pre}_{\mathrm{PE}}$.

PROOF. We have, by induction on **Com**, $wlp(\underline{\mathrm{skip}}, \underline{Q})=\underline{Q}$, $wlp(\underline{X:=E}, \underline{Q})= \underline{Q[X \leftarrow E]}$, $wlp(\underline{X:=?}, \underline{Q})=\underline{\forall X. Q}$, $wlp(\underline{(C_1;C_2)}, \underline{Q})=wlp(\underline{C_1}, wlp(\underline{C_2}, \underline{Q}))$, $wlp(\underline{(B \rightarrow C_1 \diamond C_2)}, \underline{Q})=\underline{(B \wedge wlp(C_1, Q)) \vee (\neg B \wedge wlp(C_2, Q))}$. For a while-loop $(B*C)$, the idea is to code the values $\langle s(X_1), \ldots, s(X_k)\rangle$ of free variables $X_1, \ldots, X_k$ of B and C in state s by their Gödel number and then to code the terminating execution traces $\langle s_0, \ldots, s_n\rangle$ of the loop by their Gödel number. To do this we observe that for all $n \in \mathbf{N}$ the coding of any finite sequence of naturals $\langle a_0, \ldots, a_n\rangle \in seq^* \mathbf{N}$ into a natural c and the later decoding of c into the a_i is representable in Peano arithmetic by a predicate $\delta \in \mathbf{Pre}_{\mathrm{PE}}$ with $Var(\delta)=\{c, n, i, a\}$ such that $\forall n \in \mathbf{N}. \forall c \in \mathbf{N}. \exists\langle a_0, \ldots, a_n\rangle \in \mathbf{N}^{n+1}. \forall i \leqslant n. \delta \Leftrightarrow (a=a_i)$ and $\forall n \in \mathbf{N}. \forall\langle a_0, \ldots, a_n\rangle \in \mathbf{N}^{n+1}. \exists c \in \mathbf{N}. \forall i \leqslant n. \delta \Leftrightarrow (a=a_i)$ (see [133, pp. 247–248]). Now $wlp(\underline{(B*C)}, \underline{Q}) = \{s \in S: \exists s' \in S. \exists n \in \mathbf{N}. (\langle s, s'\rangle \in(\underline{B} \lceil \underline{C})^n) \wedge (s' \in \underline{\neg B \wedge Q})\}$ [by (151) and (19.6)] $= \underline{P \wedge \neg B \wedge Q}$ where we have to find a P such that $\underline{P}=\{s \in S: \exists s' \in S. \exists n \in \mathbf{N}. \langle s, s'\rangle \in(\underline{B} \rceil \underline{C})^n\} = \{s \in S: \exists n \in \mathbf{N}. \exists\langle s_0, \ldots, s_n\rangle \in S^{n+1}. (\forall i<n. (s_i \in \underline{B}) \wedge(\langle s_i, s_{i+1}\rangle \in \underline{C})) \wedge(s_n=s)\}$. Assume that X is the only free variable in B and C (generalization consists in coding the values of the variables into an integer); we then have $\underline{P}=\{s \in S: \exists n \in \mathbf{N}. \exists\langle d_0, \ldots, d_n\rangle \in \mathbf{N}^{n+1}. (\forall i<n. (s[X \leftarrow d_i] \in \underline{B}) \wedge(\langle s[X \leftarrow d_i], s[X \leftarrow d_{i+1}]\rangle \in \underline{C})) \wedge(s(X)=d_n)\}$. Let $x \in \mathbf{Lvar}$ be a fresh variable not appearing in B or C. Let, by induction hypothesis, $R \in \mathbf{Pre}_{\mathrm{PE}}$ be such that $\underline{R}=wlp(\underline{C}, \underline{(X=x)})$. We have $\underline{P}=\{s \in S: \exists n \in \mathbf{N}. \exists\langle d_0, \ldots, d_n\rangle \in \mathbf{N}^{n+1}. (\forall i<n. s \in \underline{(B \wedge R)[X \leftarrow d_i][x \leftarrow d_{i+1}]}) \cap \underline{(X=d_n)}\} = \underline{(\exists n. \exists c. (\forall i<n. \exists X. \exists x. \delta[a \leftarrow X] \wedge \delta[i \leftarrow i+1][a \leftarrow x] \wedge B \wedge R) \wedge \delta[i \leftarrow n][a \leftarrow X])}$. □

Bergstra and Tucker [46] have shown that the expressiveness concept is awkward to apply for two-sorted data types. For example, two independent copies of $\mathbf{N}$ can form a two-sorted sorted interpretation that is not expressive.

7.4.2.4.2. Relative completeness of Hoare logic. By the expressiveness requirement on interpretations I, one can always express in **Pre** intermediate invariants which are sufficient to prove the partial correctness of commands C using Hoare logic:

(156) THEOREM (Cook [86]) (relative completeness of Hoare logic)

$$(I \text{ is expressive for } \mathbf{Com} \text{ and } \mathbf{Pre} \wedge \mathrm{Th}=\{P \in \mathbf{Pre}: I[\![P]\!]=\mathrm{tt}\})$$
$$\Rightarrow \forall C \in \mathbf{Com}. \forall P, Q \in \mathbf{Pre}. (\underline{\{P\}C\{Q\}} \Rightarrow \vdash_{\mathbf{Th} \cup \mathbf{H}} \{P\}C\{Q\}).$$

PROOF. By structural induction on formulas $\{P\}C\{Q\}$.

If $\underline{\{P\}\mathrm{skip}\{Q\}}$ then $\underline{P \Rightarrow P}$ is true [by (123.6) and (125)]and so is $\underline{P \Rightarrow Q}$ [by (22), (19.1),

(123.6) and (125)]. It follows by hypothesis that $(P \Rightarrow P)$ and $(P \Rightarrow Q)$ belong to Th. Therefore, the proof of $\{P\}\text{skip}\{Q\}$ consists in applying skip axiom (96) and consequence rule (102).

If $\underline{\{P\}X:=E\{Q\}}$ then $\underline{P} \subseteq \{s \in S: s[X \leftarrow \underline{E}(s)] \in \underline{Q}\}$ [by (22) and (19.2)], whence $\underline{P \Rightarrow Q[X \leftarrow E]}$ is true [by (123.6) and (128)]. Therefore the proof of $\{P\}X:=E\{Q\}$ consists in applying assignment axiom (97) and consequence rule (102).

If $\underline{\{P\}X:=?\{Q\}}$ then $\{s[X \leftarrow d]: s \in \underline{P} \wedge d \in D\} \subseteq \underline{Q}$ [by (22) and (19.3)] whence $\{s: \exists d \in D.\, s[X \leftarrow d] \in \underline{P}\} \subseteq \underline{Q}$ so that $\underline{\exists X.\, P \Rightarrow Q}$ is true [by (123.7), (123.6), and (125)]. The proof of $\{P\}X:=?\{Q\}$ consists in applying random assignment axiom (98) and consequence rule (102).

If $\underline{\{P\}(C_1;C_2)\{Q\}}$ then $\underline{P} \subseteq wlp((\underline{C_1}; \underline{C_2}), \underline{Q})$ [by (152)] $=$ $wlp(\underline{C_1}, wlp(\underline{C_2}, \underline{Q}))$ [by (151) and (19.4)]. By expressiveness, let I and J be such that $\underline{I} = wlp(\underline{C_2}, \underline{Q})$ and $\underline{J} = wlp(\underline{C_1}, \underline{I})$. By induction hypothesis we can prove $\{I\}C_1\{J\}$ and $\{J\}C_2\{Q\}$. Moreover $\underline{P \Rightarrow I}$ and $\underline{P \Rightarrow Q}$ are true so that the proof of $\{P\}(C_1;C_2)\{Q\}$ ends by application of composition rule (99) and consequence rule (102).

If $\underline{\{P\}(B \rightarrow C_1 \diamond C_2)\{Q\}}$ then we can prove $\{P \wedge B\}C_1\{Q\}$ and $\{P \wedge \neg B\}C_2\{Q\}$ [by (22), (19.5), (123.4), (123.3) and induction hypothesis] and conclude by conditional rule (100).

If $\underline{\{P\}(B*C)\{Q\}}$ then let I, J be such that $\underline{I} = wlp((\underline{B*C}), \underline{Q})$ and $\underline{J} = wlp(\underline{C}, \underline{I})$. We have $wlp(\underline{C}, \underline{I}) \subseteq wlp(\underline{C}, \underline{I})$, whence $\underline{\{wlp(C, I)\}C\{I\}}$ [by (152)] so that by induction hypothesis we can prove that $\{J\}C\{I\}$. Moreover, $\underline{I} = \{s \in S: \forall s' \in S.(\langle s, s' \rangle \in (\underline{B*C})) \Rightarrow (s' \in \underline{Q})\}$ [by (151)] $=$ $\{s \in S: \forall s' \in S.(\langle s, s' \rangle \in (\delta \lceil \neg \underline{B}) \cup (\underline{B} \lceil \underline{C}) \circ (\underline{B*C})) \Rightarrow (s' \in \underline{Q})\}$ [by (19.7)] $=$ $\{s \in S: (s \in \neg \underline{B}) \Rightarrow (s \in \underline{Q})\} \cup \{s \in S: \forall s' \in S.(\langle s, s' \rangle \in (\underline{B} \rceil \underline{C}) \circ (\underline{B*C})) \Rightarrow (s' \in \underline{Q})\}$ $=$ $(\underline{\neg B \Rightarrow Q}) \cup \{s \in \underline{B}: \forall s' \in S.(\langle s, s' \rangle \in \underline{C} \circ (\underline{B*C})) \Rightarrow (s' \in \underline{Q})\}$ [by (123)] $=$ $(\underline{\neg B \Rightarrow Q}) \cup (\underline{B} \cap wlp(\underline{C} \circ (\underline{B*C}), \underline{Q}))$ [by (151)] $=$ $(\underline{\neg B \Rightarrow Q}) \cup (\underline{B} \cap wlp(\underline{C}, wlp((\underline{B*C}), \underline{Q})))$ [by (151)]. It follows that $(\underline{I \wedge B}) \Rightarrow \underline{J}$. Whence applying consequence rule (102) with $(I \wedge B) \Rightarrow J$, $\{J\}C\{I\}$ and $I \Rightarrow I$, we can prove $\{I \wedge B\}C\{I\}$ so that $\{I\}(B*C)\{I \wedge \neg B\}$ derives from the while rule (101). Finally $\underline{P \Rightarrow I}$ [by (152)] and $(\underline{I \wedge \neg B}) \Rightarrow \underline{Q}$ so that the proof of $\{P\}(B*C)\{Q\}$ ends by application of consequence rule (102).

Observe that in the above proof, Cook's expressiveness is used only to guarantee *weak expressiveness* [361], that is, that loop invariants for while rule (101) and intermediate invariants for composition rule (99) can be expressed in **Pre**. □

7.4.2.4.3. Expressiveness à la Cook and its equivalence with Clarke's notion of expressiveness. The original notion of *expressiveness* is due to Cook [86] and was expressed in terms of the predicate transformer $slp(\underline{C}, \underline{P})$ (that is, the set of all final states that C can reach when starting in a state satisfying P, such that $\underline{\{p\}C\{slp(C, p)\}}$ and $\underline{\{p\}C\{q\}} \Rightarrow (slp(\underline{C}, p) \subseteq q)$:

(157) DEFINITION (Dijkstra [124]) (*strongest liberal postcondition*)

$$slp(r, p) = \{s \in S: \exists s' \in p.\, \langle s', s \rangle \in r\}.$$

(158) DEFINITION (Cook [86]) (*expressiveness à la Cook*). The interpretation I is said to be *expressive* for the languages **Com** and **Pre** if and only if $\forall C \in \textbf{Com}.\, \exists J \in \textbf{Pre}.\, \underline{J} = slp(\underline{C}, \underline{P})$.

However, expressiveness à la Cook is equivalent to expressiveness à la Clarke. To show that we first observe that the following lemma holds.

(159) LEMMA (relationships between wlp and slp). *$\underline{\{p\}C\{q\}} \Leftrightarrow (slp(\underline{C}, p) \subseteq q)$, $wlp(r, q) = \neg slp(r^{-1}, \neg q)$ and $slp(r, p) = \neg wlp(r^{-1}, \neg p)$.*

PROOF. $\underline{\{p\}C\{q\}} \Leftrightarrow (p \rceil \underline{C} \subseteq S \times q) \Leftrightarrow [\forall s, s' \in S.((s \in p) \wedge (\langle s, s' \rangle \in \underline{C})) \Rightarrow (s' \in q)] \Leftrightarrow (slp(\underline{C}, p) \subseteq q)$. Moreover $\neg slp(r^{-1}, \neg q) = \{s \in S: \neg(\exists s'.s' \in \neg q \wedge \langle s', s \rangle \in r^{-1})\}$ [by (157)] $= \{s \in S: \forall s'.s' \in q \vee \langle s, s' \rangle \notin r\} = wlp(r, q)$ [by (151)]. Finally, $slp(r, p) = \neg\neg slp(r^{-1^{-1}}, \neg\neg p) = \neg wlp(r^{-1}, \neg p)$. □

(160) LEMMA (semantic inversion). *If $\{X_1, \ldots, X_n\} = Free(C)$, $\{x_1, \ldots, x_n\} \cap Free(C) = \emptyset$, $\{x_1, \ldots x_n\} \cap Free(P) = \emptyset$, $\underline{Q} = slp(\underline{C}, \underline{X_1 = x_1} \wedge \cdots \wedge \underline{X_n = x_n})$ and $Q' = (\exists X_1 \ldots . \exists X_n . Q \wedge P)[x_1 \leftarrow X_1] \ldots [x_n \leftarrow X_n]$, then $Q' = slp(C^{-1}, P)$.*

PROOF. We let $n=1$ for simplicity. We first prove that $(\exists s''.s''(X) = s''(x) \wedge \langle s'', s[X \leftarrow d][x \leftarrow s(X)] \rangle \in \underline{C}) \Leftrightarrow (\langle s, s[X \leftarrow d] \rangle \in \underline{C})$. If $\langle s'', s[X \leftarrow d][x \leftarrow s(X)] \rangle \in \underline{C}$ then $s''(y) = s[X \leftarrow d][x \leftarrow s(X)](y)$ for $y \notin Free(C) = \{X\}$ since execution of C does not modify the value of variables not appearing in C. It follows that $s''(y) = s(y)$ for $y \notin \{x, X\}$ and $s''(x) = s(X)$. Whence $s''(X) = s''(x)$ implies $s''(X) = s(X)$, that is, $s''(y) = s(y)$ for $y \neq x$ and $s''(x) = s(X)$ so that $s'' = s[x \leftarrow s(X)]$. It follows that $\langle s[x \leftarrow s(X)], s[X \leftarrow d][x \leftarrow s(X)] \rangle \in \underline{C}$, whence that $\langle s[x \leftarrow d'], s[X \leftarrow d][x \leftarrow d'] \rangle \in \underline{C}$ for all $d' \in D$ since $x \notin Free(C)$. For $d' = s(x)$ we conclude that $\langle s, s[X \leftarrow d] \rangle \in \underline{C}$. Reciprocally, if $\langle s, s[X \leftarrow d] \rangle \in \underline{C}$ then $\langle s[x \leftarrow d'], s[X \leftarrow d][x \leftarrow d'] \rangle \in \underline{C}$ for all $d' \in D$ since $x \notin Free(C)$ so that for $d' = s(X)$ we get $\langle s[x \leftarrow s(X)], s[X \leftarrow d][x \leftarrow s(X)] \rangle \in \underline{C}$, that is, $\langle s'', s[X \leftarrow d][x \leftarrow s(X)] \rangle \in \underline{C}$ with $s'' = s[x \leftarrow s(X)]$ so that $s''(X) = s''(x)$ since $x \neq X$.

$I[\![(\exists X.Q \wedge P)[x \leftarrow X]]\!] = \{s \in S: s[x \leftarrow s(X)] \in I[\![(\exists X.Q \wedge P)]\!]\}$ [by (128), and (122.1)] $= \{s \in S: \exists d \in D. s[x \leftarrow s(X)][X \leftarrow d] \in \underline{Q} \cap \underline{P}\}$ [by (123.7) and (123.4)] $= \{s \in S: \exists d \in D. s[x \leftarrow s(X)][X \leftarrow d] \in slp(\underline{C}, \underline{X = x}) \cap \underline{P}\}$ [by hypothesis of (160)] $= \{s \in S: \exists d \in D. \exists s''. s''(X) = s''(x) \wedge \langle s'', s[X \leftarrow d][x \leftarrow s(X)] \rangle \in \underline{C} \wedge s[x \leftarrow s(X)][X \leftarrow d] \in \underline{P}\}$ [by (157), (123.1), (122.1) and (121) since $X \neq x$] $= \{s \in S: \exists d \in D. \langle s, s[X \leftarrow d] \rangle \in \underline{C} \wedge s[x \leftarrow s(X)][X \leftarrow d] \in \underline{P}\}$ [by the above argument] $= \{s \in S: \exists d \in D. \langle s, s[X \leftarrow d] \rangle \in \underline{C} \wedge s[X \leftarrow d] \in \underline{P}\}$ [since $X \neq x$ implies $s[x \leftarrow s(X)][X \leftarrow d] = s[X \leftarrow d][x \leftarrow s(X)]$ and $x \notin Free(P)$ so that $s[X \leftarrow d][x \leftarrow d'] \in \underline{P}$ implies $s[X \leftarrow d] \in \underline{P}$] $= \{s \in S: \exists s'. \langle s, s' \rangle \in \underline{C} \wedge s' \in \underline{P}\}$ [since $\langle s, s' \rangle \in \underline{C}$ implies $s(y) = s'(y)$ when $y \notin Free(C)$ so that $s' = s[X \leftarrow d]$ where $d = s'(X)$] $= slp(\underline{C}^{-1}, \underline{P})$ [by (157)]. □

(161) THEOREM (Clarke [71], Josko [245], Olderog [327, 328]) (equivalent definitions of expressiveness). *Expressiveness à la Cook is equivalent to expressiveness à la Clarke*: (153) $\Leftrightarrow$ (158).

PROOF. If $\forall C \in \textbf{Com}. \forall P \in \textbf{Pre}. \exists J \in \textbf{Pre}. \underline{J} = slp(\underline{C}, \underline{P})$ then if $\{X_1, \ldots, X_n\} = Free(C)$, $\{x_1, \ldots, x_n\} \cap Free(C) = \emptyset$, $\{x_1, \ldots, x_n\} \cap Free(P) = \emptyset$, $\underline{Q} = slp(\underline{C}, \underline{X_1 = x_1} \wedge \cdots \wedge \underline{X_n = x_n})$ and $Q' = (\forall X_1 \ldots . \forall X_n . Q \Rightarrow \neg P)[x_1 \leftarrow X_1] \ldots [x_n \leftarrow X_n]$, then $\underline{\neg Q'} = \neg slp(\underline{C}^{-1}, \underline{\neg P})$ [by (160)] $= wlp(\underline{C}, \underline{P})$ [by (159)] is expressible in **Pre**.

In the same way, if $\underline{R}=wlp(\underline{C}, \underline{X_1=x_1 \wedge \cdots \wedge X_n=x_n})$ and $R'=(\exists X_1 . \ldots \exists X_n . Q \wedge \neg P)[x_1 \leftarrow X_1] \ldots [x_n \leftarrow X_n]$, then $\underline{\neg R'}=slp(\underline{C}, \underline{P})$. □

7.4.2.4.4. Relative completeness of Hoare logic for arithmetical while-programs and nonstandard interpretations. Hoare logic is relatively complete for while-programs applied to arithmetic:

(162) THEOREM (Cook [86]) (relative completeness of Hoare logic for arithmetical while-programs). $\mathbf{H} \cup \mathrm{Th}(\mathbf{N})$ *is relatively complete for the standard interpretation* I_{PE} *of Peano arithmetic* $\mathrm{PE}=\langle\{0\}, \{\mathrm{Su}, \mathrm{Pr}, +, *\}, \{<\}, \#\rangle$ *on the domain* $\mathbf{N}$ *of natural numbers where* $\mathrm{Th}(\mathbf{N})=\{P \in \mathbf{Pre}_{\mathrm{PE}}: I_{\mathrm{PE}}[\![P]\!]=\mathrm{tt}\}$ *is number theory.*

PROOF. By (155), I_{PE} is expressive for $\mathbf{Com}_{\mathrm{PE}}$ and $\mathbf{Pre}_{\mathrm{PE}}$ so that, by relative completeness (156), the Hoare logic $\mathbf{H} \cup \mathrm{Th}(\mathbf{N})$ is relatively complete for I_{PE}. □

Harel [197] has pointed out that any interpretation I can be expanded to an interpretation with a complete Hoare logic by expanding it to an arithmetical universe (but this expansion may increase the degree of undecidability of the theory of I). A simpler expansion when $\mathbf{Hcf}_{\mathrm{tt}}(I)$ is recursive is proposed in [39].

In (162), the facts about arithmetic that one needs in a program correctness proof are given by the oracle Th(**N**). Bergstra and Tucker [47] use instead Peano's first-order axiomatization of arithmetic (see [253, Section 38], [238, Section 3]). Second-order Peano arithmetic PE^2 over the basis $\mathbf{Cte}=\{0, 1\}$, $\mathbf{Fun}=\{+, *\}$ and $\mathbf{Rel}=\emptyset$ can be formalized by the following axioms (axioms (163.4) to (163.7) are not strictly necessary since addition and multiplication can be defined in the same way as $(x \leqslant y)$ is defined by $(\exists z . (x+z)=y)$):

(163) DEFINITION (*second-order Peano arithmetic* PE^2)
(163.1) $\forall x . \neg(x+1=0)$;
(163.2) $\forall x . \forall y . (x+1=y+1) \Rightarrow (x=y)$;
(163.3) $\forall x . \neg(x=0) \Rightarrow \exists y . (x=y+1)$;
(163.4) $\forall x . (x+0=x)$;
(163.5) $\forall x . \forall y . (x+(y+1))=((x+y)+1)$;
(163.6) $\forall x . (x*0=0)$;
(163.7) $\forall x . \forall y . (x*(y+1))=((x*y)+x)$;
(163.8^2) $\forall P . (P[x \leftarrow 0] \wedge \forall x . P \Rightarrow P[x \leftarrow x+1]) \Rightarrow \forall x . P$.

The last axiom (163.8^2) states that if a property P is true for 0 and is true for the successor $x+1$ of x whenever it is true for x then it is true for all x. Since P ranges over all subsets of **N**, the second-order axiom (163.8^2) describes properties of $|\mathscr{P}(\mathbf{N})|=\aleph_1$ subsets of **N**. To stay in the realm of first-order logic, one can define first-order Peano arithmetic PE^1 which consists of axioms (163.1) to (163.7) plus the axiom scheme (164.8^1):

(164) DEFINITION (*first-order Peano arithmetic* PE^1)

(164.1) $\forall x.\neg(x+1=0)$;

... ...

(164.7) $\forall x.\forall y.(x*(y+1))=((x*y)+x)$;

For all $P \in \mathbf{Pre}_{PE}$.

(164.8^1) $(P[x\leftarrow 0] \wedge \forall x.P \Rightarrow P[x\leftarrow x']) \Rightarrow \forall x.P$.

There are $|\mathbf{N}|=\aleph_0=\omega$ predicates P (the proof uses an enumeration of $\mathbf{Pre}_{PE}$ by Gödel numbers, see (143)) and $|\mathbf{N}|\neq|\mathscr{P}(\mathbf{N})|$, whence ($164.8^1$) describes less subsets of $\mathbf{N}$ than (163.8^2) (the proof that $|\mathbf{N}|\neq|\mathscr{P}(\mathbf{N})|$ is by reductio ad absurdum using a Cantor diagonal argument: if $|\mathscr{P}(\mathbf{N})|=|\mathbf{N}|$ then $\mathscr{P}(\mathbf{N})$ is of the form $\{s_j : j \in \mathbf{N}\}$ where $s_j \subseteq \mathbf{N}$ for all $j \in \mathbf{N}$. Then the set $\{i \in \mathbf{N}: i \notin s_i\}$ would be some element s_k of $\mathscr{P}(\mathbf{N})$, whence a contradiction since either $k \in s_k$ and $s_k=\{i \in \mathbf{N}: i \notin s_i\}$ implies $k \notin s_k$, or $k \notin s_k$ and $s_k=\{i \in \mathbf{N}: i \notin s_i\}$ implies $k \in s_k$). Since PE^1 imposes less constraints on its interpretations than PE^2, PE^1 can have nonstandard interpretations that are disallowed by PE^2. Such nonstandard models of PE^1 ([373], [55, Section 17], or [253, Section 53]) consist of the naturals followed by infinitely many blocks isomorphic to $\mathbf{Z}$: $0, 1, 2, \ldots\ldots\ldots -2'\ -1'\ 0'\ 1'\ 2 \ldots\ldots\ldots -2''\ -1''\ 0''\ 1''\ 2'' \ldots\ldots$ without least or greatest block and between any two blocks there lies a third. It follows that $\mathbf{N}$ is not first-order axiomatizable (although it is by PE^2 since $\mathbf{N}$ is the only model of PE^2) in the sense that there are true facts that can be proved by PE^2 but not by PE^1 (using again the diagonalization argument: if P^i is the predicate with $Free(P^i)=x$ and Gödel number i then Q such that $\forall i \in \mathbf{N}.\ Q[x\leftarrow i]=\neg P^i[x\leftarrow i]$ is not one of them). So why not use second-order logics? Essentially because PE^1 deals with finite sets of integers (as in pure arithmetic) whereas PE^2 deals with infinite sets of integers (as in mathematical analysis) and $\mathscr{P}(\mathbf{N})$ is much more complicated to understand than $\mathbf{N}$ (Cohen [79] proved that there are infinitely many different ways to conceive $\mathscr{P}(\mathbf{N})$ from the same $\mathbf{N}$). In the same way, Hoare logic deals with finite sets of variables and terminating programs, i.e. finite execution traces, and Bergstra and Tucker [47] have shown that Hoare logic for while-programs is essentially first-order: the strongest postcondition calculus can be represented in Peano arithmetic PE^1 (because $slp(\underline{C}, \underline{P})$ can be expressed by a predicate SLP(C, P) of PE^1, see the proof of (155) and (159)) so that Hoare logic over PE^1 is equivalent to PE^1 itself (because $\{P\}C\{Q\}$ is equivalent to (SLP(C, P)$\Rightarrow Q$) by (159)). The comparison of Hoare-style reasoning about programs with reasoning about programs with first-order rendering of predicate transformers is pursued in [269].

7.4.2.4.5. On the unnecessity of expressiveness. Expressiveness is sufficient to obtain relative completeness but it is not necessary: Bergstra and Tucker [43] have shown that Hoare logic can be complete for an inexpressive interpretation I whose first-order theory has some expressive model (i.e., interpretation I' with the same first-order theory $\{P \in \mathbf{Pre}: I'[\![P]\!]=\text{tt}\}$). This point is illustrated by the following theorem:

(165) THEOREM (Bergstra and Tucker [44]) (unnecessity of expressiveness). *Hoare logic* $\mathbf{H}\cup\text{Th}(\mathbf{N})$ *is relatively complete for any model I of Peano arithmetic (such that*

$\forall P \in \mathrm{Th}(\mathbf{N}).I[\![P]\!]=\mathrm{tt}$ *where* $\mathrm{Th}(\mathbf{N})=\{P \in \mathbf{Pre}_{\mathrm{PE}}: I_{\mathrm{PE}}[\![P]\!]=\mathrm{tt}\}$), *but I is not expressive for* $\mathbf{Pre}_{\mathrm{PE}}$ *and* $\mathbf{Com}_{\mathrm{PE}}$ *when I is not the standard model* I_{PE} *of arithmetic.*

PROOF. By (162), the Hoare logic $\mathbf{H} \cup \mathrm{Th}(\mathbf{N})$ is relatively complete for I_{PE}. Since any $P \in \mathbf{Pre}_{\mathrm{PE}}$ is true for the standard interpretation I_{PE} if and only if it is true for the nonstandard interpretation I, $\mathbf{H} \cup \mathrm{Th}(\mathbf{N})$ is also relatively complete for I.

Let $C=(X:=Y;((\neg(X=0)*X:=\mathrm{Pr}(X));X:=Y))$. Execution of C for the nonstandard interpretation I terminates only if the initial value of Y is standard. It follows that $slp(C, S)=\{s \in S: (s(X)=s(Y)) \wedge s(Y) \in \mathbf{N}\}$. Now I is not expressive for $\mathbf{Pre}_{\mathrm{PE}}$ and $\mathbf{Com}_{\mathrm{PE}}$ since otherwise there is a $P \in \mathbf{Pre}_{\mathrm{PE}}$ such that $\underline{P}=slp(C, S)$, so that $\exists Y.P$ is true of X only if X is a standard natural number, in contradiction with the fact that no predicate of $\mathbf{Pre}_{\mathrm{PE}}$ can be used to distinguish among standard and nonstandard numbers. □

Bergstra and Tiuryn [41] have identified and studied two necessary (but not sufficient) conditions that an interpretation I must satisfy if a sound Hoare logic is to be complete for this given I: first, they prove that the first-order theory of I must be *PC-compact*, that is, each asserted program which is true in all models of the theory is true in all models of a finite subset of the theory (if $\mathrm{Th}=\{P \in \mathbf{Pre}: I[\![P]\!]=\mathrm{tt}\}$ then $\forall H \in \mathbf{Hcf}_{\mathrm{tt}}[\![\mathrm{Th}]\!].\exists \mathrm{Th}' \subseteq \mathrm{Th}.(|\mathrm{Th}'| \in \mathbf{N}) \wedge (H \in \mathbf{Hcf}_{\mathrm{tt}}[\![\mathrm{Th}']\!])$ where $\mathbf{Hcf}_{\mathrm{tt}}[\![T]\!]=\{H \in \mathbf{Hcf}: \forall I'.(\forall P \in T.I'[\![P]\!]=\mathrm{tt}) \Rightarrow (I'[\![H]\!]=\mathrm{tt})\}$). Secondly, they prove that the partial correctness theory $\mathbf{Hcf}_{\mathrm{tt}}(I)$ must be decidable relative to its first-order theory Th (as shown in (148)).

From a practical point of view, the incompleteness results about Hoare logic are not restrictive for hand-made proofs, just as Gödel's incompleteness theorems do not prevent mathematicians to make proofs. Only the semantic counterpart of Hoare logic matters and it is complete in the sense of (80). As far as expressiveness is concerned, the limited power of first-order logic can always be overcome using infinitary logics since (76) is expressible in $L\omega_1\omega$ (which allows infinite formulae $\bigwedge \Phi$ and $\bigvee \Phi$ when Φ is a countable set of formulae [37]) as noticed in [136, 137, 33, 33ª] (but then the finitary nature of proofs in ordinary first-order logic $L\omega\omega$ is lost). Also the use of a given theory Th corresponds to the common mathematical practice to accept certain notions and structures as basic and work axiomatically from there on. However, when considering more complicated programming language features, Hoare logic turns out to be incomplete for intrinsic reasons.

7.4.2.5. Clarke's characterization problem. Clarke [71] has shown that some programming languages have no sound and relatively complete Hoare logic. The formal argument is first that if a programming language possesses a relatively complete and sound Hoare logic, then the halting problem for finite interpretations must be decidable, and second that Algol-like [320] or Pascal-like [405] languages have an undecidable halting problem for finite interpretations with $|D| \geqslant 2$. The intuitive reason is that names in predicates $P, Q \in \mathbf{Pre}$ and in commands $C \in \mathbf{Com}$ are used in a similar way: all considered objects, at a given instant of time, are given different names. Hence variables of P, Q and C can be interpreted in exactly the same way by means of states

(120). But when considering Algol- or Pascal-like languages, the naming conventions in P, Q and C are totally different. For example, objects deeply buried in the runtime stack cannot be accessed by their name although they can be modified using procedure calls! Such Algol- or Pascal-like languages are more precisely characterized by the following definition:

(166) DEFINITION (Clarke [71]) (*Clarke languages*). A Clarke language $\mathbf{L}$ is a programming language allowing procedures (with a finite number of local variables and parameters taking a finite number of values, without sharing via aliases) and the following features:

(i) procedures as parameters of procedure calls (without self-application);
(ii) recursion;
(iii) static scoping;
(iv) use of global variables in procedure bodies;
(v) nested internal procedures as parameters of procedure calls.

A Clarke language $\mathbf{L}_j$ is obtained by disallowing feature (j).

The non-existence of Hoare logic for Clarke languages (and other variants of (166), see [71, 270]) introduces the *characterization problem* [74]: what criteria guarantee that a programming language has a sound and relatively complete Hoare logic? First we prove the non-existence of Hoare logics for Clarke languages and next review the literature on the characterization problem.

7.4.2.5.1. Languages with a relatively complete and sound Hoare logic have a decidable halting problem for finite interpretations

(167) LEMMA (Clarke [71]) (decidability of the halting problem...). *If* **Com** *has a sound and relatively complete Hoare logic, then the halting problem must be decidable for all interpretations I on a finite Herbrand definable domain D.*

PROOF. Let some particular finite interpretation I be given. There is a decision procedure to verify that $P \in \mathrm{Th}$, that is, $I[\![P]\!]=\mathrm{tt}$: we just have to check using truth tables that P holds for the finitely many possible combinations of values of the free variables of P. Moreover, since D is finite and Herbrand definable, **Pre** is expressive with respect to **Com** and I: any subset of D can be represented as a finite disjunction of terms representing its elements. Then by the soundness theorem (129) and relative completeness (156) we have $\underline{\{\mathrm{true}\}C\{\mathrm{false}\}} \Leftrightarrow \vdash_{\mathrm{Th}\cup\mathbf{H}} \{\mathrm{true}\}C\{\mathrm{false}\}$ where true $=(x=x)$ and false $=\neg(x=x)$. Since Th is recursive, it follows from (143) that $\mathbf{Hcf}_{\mathrm{pr}}(\mathrm{Th})$ is recursive enumerable, whence so is $\{C\colon \vdash_{\mathrm{Th}\cup\mathbf{H}} \{\mathrm{true}\}C\{\mathrm{false}\}\} = \{C\colon \underline{\{\mathrm{true}\}C\{\mathrm{false}\}}\}$ so that the nonhalting problem is semidecidable. We conclude that the halting problem cannot be undecidable (see (144)). □

7.4.2.5.2. The halting problem for finite interpretations is undecidable for Clarke languages. The halting problem is decidable for while-programs on finite interpretations (we may test for termination (at least theoretically) by watching the execution trace of the program to see if a state is repeated [242]). For recursion one might expect that the program could be viewed as a type of pushdown automaton for which the halting problem is also decidable [85, 243]. However, this is not true for Clarke languages:

(168) LEMMA (Clarke [71], Jones and Muchnick [243]) (undecidability of the halting problem ...). *Clarke languages have an undecidable halting problem for finite interpretations with* $|D| \geqslant 2$.

PROOF. The proofs of Jones and Muchnick [243] (modified in [71]) consist in showing that such languages can be used to simulate queue machines which have an undecidable halting problem. Clarke languages can also simulate the more well-known Turing machines ([392, 393], [134, Section 2], [55, Section 5], [253, Section 41], [362, Section 1.5]). Since, by the *Church Thesis* (i.e., formally unprovable mathematical assertion), all functions intuitively computable algorithmically are computable by Turing machines [70, 251, 393], it follows that all computable functions are programmable in Clarke languages with $|D| \in \mathbf{N}^+ - \{1\}$. Hence, by (144), the halting problem is undecidable. A similar result was previously obtained by Langmaack [265], who showed that the pure procedure mechanism of Algol 60 can simulate any Turing machine.

A Turing machine has a finite number of internal states $Q_0, \ldots, Q_n$. It can read and write symbols chosen in a finite alphabet $S_0, \ldots, S_m$ (containing the blank symbol "–") on a potentially infinite tape marked off into squares by means of a head which can also be moved left or right, one square at a time (see Fig. 2).

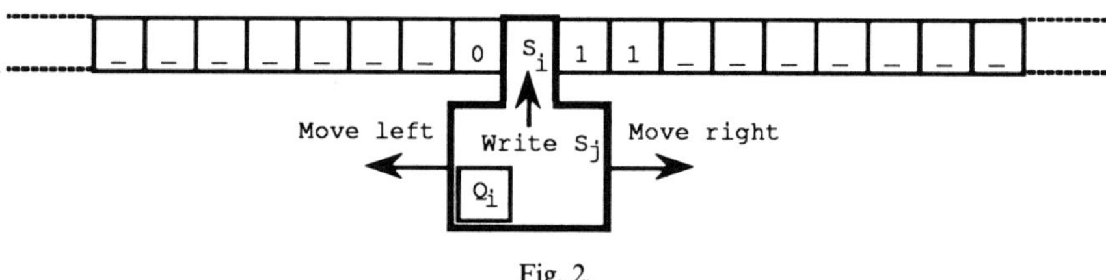

Fig. 2.

An instruction has the form $M(Q_i, S_i, S_j, D, Q_j)$ where D is "Left" or "Right". This instruction is executable if the machine is in the configuration $\langle Q_i, S_i \rangle$, that is, its internal state is Q_i and the symbol scanned under the head is S_i. Its execution consists in overwriting the square under the head by symbol S_j, in moving the head on the tape one square from the present square in the direction indicated by D and in changing the internal state into Q_j. A program consists of a finite number of instructions $M(Q_i, S_i, S_j, D, Q_j)$, $i = 1, \ldots, l$. Its execution consists in repeatedly executing any one of the executable instructions. The execution of the program halts when no instruction is executable. Initially, the tape contains finitely many nonblank symbols.

By induction on the number of steps, it follows that only finitely many squares can be

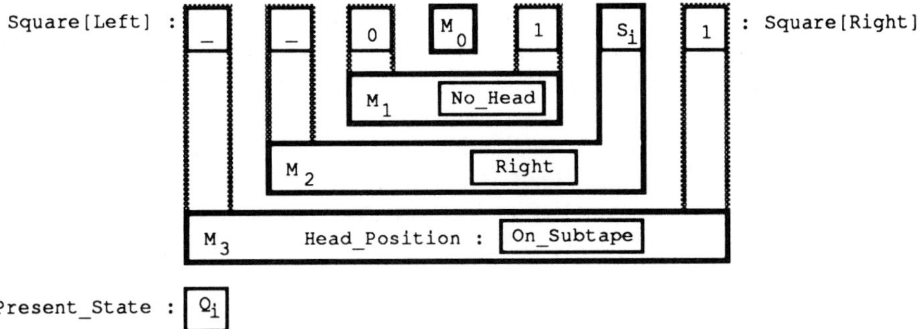

Fig. 3.

nonblank at any time during execution of the program. Therefore, Turing machines can be built up recursively from a finite number of simpler identical machines $M_1, \ldots, M_n$ consisting only of two squares (see Fig. 3). Machine M_0 is empty. The internal state of each machine M_i consists of two squares (Square[Left] and Square[Right]), an indication of whether the head of the Turing machine is on its left square (Head–Position = Left), on its right square (Head–Position = Right), on a square of one of the machines $M_{i-1}, \ldots, M_1$ (Head–Position = On–Subtape), or on a square of one of the machines $M_{i+1}, \ldots, M_n$ (Head–Position = No–Head), and an indication of whether machine M_{i-1} is empty (Is–Empty–Subtape = true). We could have used only Boolean variables as done by the Pascal compiler. To execute the program of the Turing machine, machine M_n has access to the current state Q_i stored in the global variable Present–State (using feature (166(iv)) of Clarke languages). Execution of an instruction of the Turing machine (such as $M(Q_i, S_i, S_j, \text{Left}, Q_j)$ when the head is on the left square of machine M_n (i.e., Head–Position = Left) containing S_i (i.e., Square-[Left] = S_i) may require to extend the tape by one square on side D (D = Left in the example). In this case, machine M_n (currently simulated by procedure Turing–Machine) assigns S_j to its D square, assigns No–Head to its Head–Position, Q_j to Present–State and creates a new machine M_{n+1} (by a recursive call to procedure Turing–Machine, using feature (166(ii)) of Clarke languages). This machine M_{n+1} has two blank squares and Head–Position = Initial–Head–Position = D. M_{n+1} is now in charge of executing the program of the Turing machine. To do this, machine M_{n+1} can ask the cooperation of machine M_n (hence, recursively of machines $M_{n-1}, \ldots, M_1$) using functions and procedures local to M_n and passed to procedure Turing–Machine upon creation of machine M_{n+1} (using features (166(i)) and (166(v)) of Clarke languages). These functions and procedures can be used by M_{n+1} to read (Scanned–Symbol–On–Ends–Of–Subtape) or write (Write–On–Ends–Of–Subtape) the squares of M_n and to read (Head–Position–On–Ends–Of–Subtape) or write (Set–Head–Position–On–Ends–Of–Subtape) the Head–Position of M_n. Procedure M–On–Subtape can be used by machine M_{n+1} to execute an instruction $M(Q_i, S_i, S_j, D, Q_j)$ of the Turing machine when M_{n+1} knows that the head of the Turing machine is not on the squares of machine M_n (by calling Head–Position–On–End–Of–Subtape) so that after execution of this instruction, the head will remain on the subtape represented by machines $M_n, \ldots, M_1$. It follows that in order to simulate the Turing machine, machine M_{n+1} has just to take care of head moves from its squares to those of machine M_n. For example, when the head is on the left square of machine M_{n+1} (i.e., Head–Position = Left) and reads S_i (i.e., Square[Left] = S_i), execution of $M(Q_i, S_i, S_j$, Right, $Q_j)$ consists in writing S_j in this left square, in changing Head–Position to On–Subtape, in changing the Head–Position of machine M_n to Left (by calling Set–Head–Position–On–Ends–Of–Subtape(Left)), and in going to the next state by assignment of Q_j to Present–State. When $n=0$, the Head–Position of machine M_1 is simply changed to Right. Details are given in the following Pascal program (using the static scope execution rule (166(iii)) which states that procedure calls are interpreted in the environment of the procedure's declaration rather than in the environment of the procedure call, thus allowing to access values normally buried deeply in the runtime stack):

```
program Simulate–Turing–Machine;

 const Blank = "–";
 type
   State–Type = 0 .. 107; Symbol–Type = char;
   Head–Position–Type = (Left, Right, On–Subtape, No–Head); Side–Type = Left .. Right;

 function Opposite (D: Side–Type): Side–Type;
   {Opposite(Left) = Right and Opposite(Right) = Left.}
 begin {Opposite}
   case D of
     Left:  Opposite := Right;
     Right: Opposite := Left;
   end;
 end; {Opposite}

 var Present–State: State–Type;
       {Present state of the Turing machine.}
     Stopped: Boolean;
       {True only if the Turing machine must halt (initially false).}
     Configuration–Found: Boolean;
       {To check no invalid configuration (Present–State, scanned symbol) is found.}

 procedure Turing–Machine
   (Initial–Head–Position: Side–Type;
     Is–Empty–Subtape: Boolean;
     function Scanned–Symbol–On–Ends–Of–Subtape(D: Side–Type): Symbol–Type;
     procedure Write–On–Ends–Of–Subtape(D: Side–Type; WS: Symbol–Type);
     function Head–Position–On–Ends–Of–Subtape: Head–Position–Type;
     procedure Set–Head–Position–On–Ends–Of–Subtape (P: Head–Position–Type);
     procedure Dump–Subtape;
     procedure M–On–Subtape(Q: State–Type; S: Symbol–Type;
                 WS: Symbol–Type; D: Side–Type; NQ: State–Type)

 var
   Square: array [Side–Type] of Symbol–Type;
   Head–Position: Head–Position–Type;
```

```
{An infinite tape is represented by its finite nonblank part as a quadruple}
{⟨Square[Left], subtape, Square[Right], Head–Position⟩, where the Head–Position}
{equals D if the head is on "Square [D]" where D is "Left" or "Right", "On Subtape"}
{if the head is on the subtape or else "No–Head" when the head is outside that}
{part of the whole tape. Is–Empty–Subtape is true if and only if the subtape is}
{empty. When the subtape is not empty, it can be manipulated by functions and}
{procedures, similar to the ones explained below for manipulating the tape.}
{A call of "Turing Machine" extends the subtape by two additional blank squares on}
{its left and right ends. The head of the machine is set on one of these}
{additional squares as specified by Initial–Head–Position.}

function Scanned–Symbol–On–Ends–Of–Tape(D: Side–Type): Symbol–Type;
  {Returns the symbol written on the square on the D-end of the tape.}
begin Scanned–Symbol–On–Ends–Of–Tape := Square[D]; end

procedure Write–On–Ends–Of–Tape(D: Side–Type; WS: Symbol–Type);
  {Writes WS on the square on the D-end of the tape.}
begin Square[D] := WS; end;

function Head–Position–On–Ends–Of–Tape: Head–Position–Type;
  {To check if the head of the machine is on the left end of the tape (when the}
  {returned value is Left) or on its right end (when the returned value is Right)}
  {or whether it is on the subtape delimited by the extreme squares (On Subtape)}
  {or if it is outside the part of the ideal infinite tape represented by that}
  {finite tape (No–Head).}
begin Head–Position–On–Ends–Of–Tape := Head–Position; end;

procedure Set–Head–Position–On–Ends–Of–Tape(P: Head–Position–Type);
  {Sets the Head–Position of the tape to P.}
begin Head–Position := P; end;

procedure Dump–Tape;
  {Dump all tape marking the scanned symbol under the head between square brackets.}
  procedure Dump–Square(D: Side–Type);
  begin {Dump–Square}
    if (Head–Position = D) then write ("[") else write (" ");
    if (Square[D] = Blank) then write ("–") else write (Square[D]);
    if (Head–Position = D) then write ("]");
  end; {Dump–Square}
begin Dump–Square(Left); Dump–Subtape; Dump–Square(Right); end;

procedure M(Q: State–Type; S, WS: Symbol–Type; D: Side–Type; NQ: State–Type);
  {Whenever the machine (which is not stopped) comes to state Q (that is, the}
  {instruction labeled Q) while scanning under the head a square where S is }
  {written (set Configuration–Found to true) overwrite this square with WS, move}
  {the head in the direction indicated by D one square from the present square and}
  {proceed to instruction labeled NQ.}
begin {M}
  if (not Configuration–Found) and (not Stopped) and (Present–State = Q) then
  begin
    if (Head–Position = Opposite(D)) then begin
      if (Square[Opposite(D)] = S) then begin
        Configuration–Found := true;
        Square[Opposite(D)] := WS;
```

```
      if Is–Empty–Subtape then Head–Position := D
      else begin {Move the head on the Opposite(D) end of the subtape}
        Set–Head–Position–On–Ends–Of–Subtape (Opposite(D);
        Head–Position := On–Subtape;
      end;
      Present–State := NQ; {Go to next state.}
    end;
  end else if (Head–Position = D) then begin
    if (Square[D] = S) then begin
      Configuration–Found := true;
      Square[D] := WS; Head–Position := No–Head;
      {From now on, the continuation of the simulation of the Turing machine}
      {is delegated to the next call of procedure "Turing–Machine" which}
      {extends the nonempty tape by two new blank squares on its ends and}
      {moves the head of the machine on the square on the D end.}
      Present–State := NQ; {Go to next state.}
      Turing Machine(D false, Scanned–Symbol–On–Ends–Of–Tape,
        Write–On–Ends–Of–Tape, Head–Position–On–Ends–Of–Tape,
        Set–Head–Position–On–Ends–Of–Tape, Dump–Tape, M);
    end;
  end else if (Head–Position = On–Subtape) then begin
    if (Head–Position–On–Ends–Of–Subtape = D)
      and (Scanned–Symbol–On–Ends–Of–Subtape(D) = S) then begin
      Configuration–Found := true;
      Write–On–Ends–Of–Subtape(D, WS);
      {The head leaves the subtape for the D-square}
      Set–Head–Position–On–Ends–Of–Subtape(No–Head); Head–Position := D;
      Present–State := NQ; {Go to next state.}
  end else {The move of the head on the subtape will remain on the subtape.}
    M–On–Subtape (Q, S, WS, D, NQ);
  end;
 end;
end; {M}
begin {Turing–Machine}
  Head–Position := Initial–Head–Position; Square[Left] := Blank; Square[Right] := Blank;
  while (not Stopped) do begin
    {Execute one instruction of the Turing machine}
    Configuration–Found := false;
    {*******************************************************************************}
    {Turing machine of Enderton [134, p. 532]. (This machine computes x+y. The}
    {arguments x and y of the + function are respectively represented by a string of}
    {1s of length x and y. The arguments are separated by a single blank. The tape}
    {is otherwise blank. The head is initially on the leftmost nonblank symbol. The}
    {result is a string of 1s of length x+y.}
    {M(State,  Scan,   Write,  Move,   Next state)}
     M(0,      "1",    "1",    Right,  0);      {Pass over x.}
     M(0,      Blank,  "1",    Right,  1);      {Fill blank square between x and y.}
     M(1,      "1',    "1",    Right,  1);      {Pass over y.}
     M(1,      Blank,  Blank,  Left,   2);      {Move to end of y.}
     M(2,      "1",    Blank,  Left,   3);      {Erase a 1 at end of y.}
     M(3,      "1",    "1",    Left,   3);      {Back up to leftmost 1 of x+y.}
     M(3,      Blank,  Blank,  Right,  4);      {Halt.}
    {*******************************************************************************}
```

```
{Initialize the tape to compute x+y with x=2 and y=3}

  M(100,    Blank, "1",   Left,  101); {Write y.}
  M(101,    Blank, "1",   Left,  102);
  M(102,    Blank, "1",   Left,  103);
  M(103,    Blank, Blank, Left,  104); {Write blank between x and y.}
  M(104,    Blank, "1",   Left,  105); {Write x.}
  M(105,    Blank, "1",   Left,  106);
  M(106,    Blank, Blank, Right  107); {Move head on leftmost 1 of x.}
{*****************************************************************************}
if Present-State=107 then begin
  {Dump the initial tape and start the computation.}
  Dump-Tape; writeln; Present-State:= 0;
end else if (not Configuration-Found) then begin
  {Dump the final tape and halt the computation.}
  Dump-Tape; writeln; Configuration-Found:= true; Stopped:= true;
end;
end;
end; {Turing-Machine}

function Scanned-Symbol-On-Ends-Of-Empty-Tape(D: Side-Type): Symbol-Type;
  begin Scanned-Symbol-On-Ends-Of-Empty-Tape := "?"; end;
procedure Write-On-Ends-Of-Empty-Tape(D: Side-Type; WS: Symbol-Type); begin end;
function Head-Position-On-Ends-Of-Empty-Tape: Head-Position-Type;
  begin Head-Position-On-Ends-Of-Empty-Tape := No-Head; end;
procedure Set-Head-Position-On-Empty-Tape(P: Head-Position-Type); begin end;
procedure Move-On-Empty-Tape(D: Side-Type; Q: State-Type); begin end;
procedure Dump- Empty-Tape; begin end;
procedure M-On-Empty-Tape(Q: State-Tape; S: Symbol-Type;
          WS: Symbol-Type; D: Side-Type; NQ: State-Type); begin end;

begin {Simulate-Turing-Machine}
  Present-State := 100; Stopped := false;
  Turing-Machine(Right, true, Scanned-Symbol-On-Ends-Of-Empty-Tape,
                 Write-On-Ends-Of-Empty-Tape, Head-Position-On-Ends-Of-Empty-Tape,
                 Set-Head-Position-On-Empty-Tape, Dump-Empty-Tape, M-On-Empty Tape);
end. {Simulate-Turing-Machine}
```

Execution of the above program leads to the following initial and final configurations of the Turing machine:

```
- [1] 1 - 1 1 1 - - - - -
- [1] 1 1 1 1 - - - - - -
```

The program can be easily modified to simulate any Turing machine. □

7.4.2.5.3. Languages without sound and relatively complete Hoare logic

(169) THEOREM (Clarke [71]) (non-existence of Hoare logics for Clarke languages). *The Hoare logic for Clarke languages* (166) *is not relatively complete in the class of all expressive interpretations.*

Proof. By (167) and (168) there exists no sound and relatively complete Hoare logic for languages with features (166) since, for finite domains D, **Pre** can be enriched by finitely many constant symbols denoting elements of D so that D is Herbrand definable. □

First-order logic is not expressive for Clarke languages because their control structure is very complex. Leivant and Fernando [270] give another proof of (169) using lambda calculus (for a variant of Clarke languages). They also exhibit a programming language whose control structure is trivial (the language consists of the single program $C=(y:=x;(\neg(y=0)*y:=x+y))$ where $(0,+)$ is a torsion-free abelian group), and yet for which no relatively complete logic exists in the sense of Cook. The idea is that the notion of a torsion-free group $(\forall n\geqslant 1.\ \forall x.\ (x\neq 0)\Rightarrow(n\cdot x\neq 0)$, where $1\cdot x$ is x and $(n+1)\cdot x$ abbreviates $(n\cdot x)+x)$ is not finitely axiomatizable in first-order logic [37, Proposition 2.2] but is completely captured by $\{x\neq 0\}C\{\text{false}\}$. In this case the poverty of the language is precisely what permits certain interpretations to be expressive, interpretations which would not be expressive had the program constructs been used more freely. This result is not in contradiction with relative completeness (156) which holds for **Com** as defined in (1) and (13).

Lipton [274] proved a form of converse to (167) further extended by Clarke, German and Halpern [75], German and Halpern [167], and Urzyczyn [395] who showed that for a deterministic acceptable programming language **Com** with recursion (see the long definition in [75] or [178] and realize that almost all Algol-like programming languages are acceptable [97]), the relative completeness of Hoare logic in the class of expressive and Herbrand-definable interpretations is equivalent to the condition that the halting problem for the programming language must be decidable for finite interpretations. Grabowski [178] proved that the requirement of Herbrand definability can be dropped in the case of partial correctness. The result of Clarke, German and Halpern [75] also holds for total correctness, but Grabowski [179], and Grabowski and Hungar [181] proved that it cannot essentially be strengthened.

7.4.2.6. Nonstandard semantics and logical completeness. Observe that soundness (129) is proved for all interpretations I satisfying all theorems of Th, whereas completeness in the sense of Cook (156) is relative to a given expressive interpretation I satisfying all theorems of Th and not for all interpretations I' with theory Th (i.e., such that Th is exactly the subset of formulae of **Pre** which are true for I'). However, this second understanding fits better with Floyd and Hoare's original idea [143, 208] that Hoare logic defines the semantics of the program where Th is the specification of the operations invoked in the program. It follows that the lack of a general completeness theorem for a sound Hoare logic implies that the operational semantics of the programming language is not the semantics about which the logic is reasoning. This remark has motivated two rather different perspectives on completeness theorems: the first [7, 158, 8, 9, 10, 48, 51, 98, 99, 100, 158, 232, 282, 323, 363] consists in considering "explicit time semantics" or "nonstandard semantics" which can permit of transfinite execution traces, the second [45, 46] consists in considering another notion of completeness called *logical completeness* such that $\mathbf{H}\cup\text{Th}$ is logically complete if and only if any partial correctness formula H which is valid in *all* models of the specification

Th is provable in $\mathbf{H} \cup \mathrm{Th}$ (that is, $\bigcap_I \mathbf{Hcf}_{tt}(I) \subseteq \mathbf{Hcf}_{pr}(\mathrm{Th})$), whereas Csirmaz and Hart [101] only consider finite models. Another variant of the notion of incompleteness is studied in [361]. Soundness and completeness of Hoare logic is studied from an algebraic point of view in [403].

8. Complements to Hoare logic

8.1. Data structures

Assignment axiom (97) is correct for simple variables but cannot be used to handle all data structures. For example, considering one-dimensional arrays, we could deduce $\{1=1\}\ T[T[2]] := 1\ \{T[T[2]]=1\}$ from assignment axiom (97) and since $(T[1]=2 \wedge T[2]=2) \Rightarrow (1=1)$ we deduce that $\{T[1]=2 \wedge T[2]=2\}\ T[T[2]] := 1\ \{T[T[2]]=1\}$ from consequence rule (102) but this is not correct.

One way to handle arrays correctly is to understand the value of an array T as a function $T: \mathit{dom}\ T \rightarrow \mathit{rng}\ T$ where $\mathit{dom}\ T$ is the domain of its indices and $\mathit{rng}\ T$ is the domain of its elements, and to consider assignments to an element as a modification of the whole array [297, 229, 291]. For example, from $\{T=t \wedge (t(2)=2) \Rightarrow (t(1)=1)\}$ we derive that after assignment $T[T[2]] := 1$ we have $\{T=t[t(2)\leftarrow 1] \wedge (t(2)=2) \Rightarrow (t(1)=1)\}$ so that $T[T[2]]=t[t(2)\leftarrow 1](t[t(2)\leftarrow 1](2))=t[t(2)\leftarrow 1](t(2)=2\rightarrow 1 \diamond t(2))=(t(2)=2 \rightarrow t(1) \diamond 1)=1$.

Another way to handle arrays correctly is to understand them as a collection of simple variables with possible aliases [234]. For example, DeBakker [109] suggests the following assignment axiom for subscripted variables:

$$(170) \qquad \{P[T[E_1]\leftarrow E_2]\}\ T[E_1] := E_2\ \{P\}$$

with a refinement of substitution such that

$$(171) \qquad T[I][T[E_1]\leftarrow E_2]=(I=E_1 \rightarrow E_2 \diamond T[I])$$

and the case when an arbitrary expression stands for I is not handled. By way of example, we prove that $\{(T[2]=2) \Rightarrow (T[1]=1)\}\ T[T[2]] := 1\ \{T[T[2]]=1\}$. We have $(T[T[2]]=1)[T[T[2]]\leftarrow 1] = (\exists I.\ T[I]=1 \wedge T[2]=I)[T[T[2]]\leftarrow 1] = (\exists I.\ (I=T[2]\rightarrow 1 \diamond T[I])=1 \wedge (2=T[2]\rightarrow 1 \diamond T[2])=I)$. Now the last formula is implied by $(T[2]=2) \Rightarrow (T[1]=1)$ since if $T[2]=2$ holds then we choose $I=1$ else we choose $I=T[2]$. Thus, by the consequence rule and axiom (170), we get the desired result.

Axioms of assignment applicable to multidimensional arrays or pointers to linked data structures are given in [116, 67, 68, 186, 189, 237, 256, 279, 292, 314, 348, 367]. One can also consult [211, 87, 334, 239, 390, 226] on proving partial correctness properties of programs with user-defined data types and [63, 321] for the special case of linear lists.

8.2. Procedures

First we define the syntax and relational semantics of recursive parameterless procedures. Then we consider partial correctness proofs based upon computation induction (a generalization of Scott induction) which leads to Hoare's recursion rule. Since this rule alone is not complete, we consider Park's fixpoint induction, which, using auxiliary variables, can be indirectly transcribed into Hoare's rule of adaptation. Then these rules are generalized for value–result parameters and numerous examples of application are provided. References to the literature are given for variable parameters and procedures as parameters.

8.2.1. *Recursive parameterless procedures*

8.2.1.1. Syntax and relational semantics of a parameterless procedural language. Let us now consider programs of the form $\text{Pn} :: C_1;C_2$ which consist of a command C_2 calling a single recursive parameterless procedure Pn with body C_1:

(172) DEFINITION (*syntax of a parameterless procedural language*)

(172.1) Pn: **Proc** *procedure names,*

(172.2) Pg: **Prog** *programs,*
$\text{Pg} ::= \text{Pn} :: C_1;C_2$

(172.3) C: **Com** *commands*
$C :: \text{skip} \mid X := E \mid X := ? \mid (C_1;C_2) \mid (B \rightarrow C_1 \diamond C_2) \mid (B * C) \mid \text{Pn}.$

The relational semantics $\underline{\text{Pn}}$ of a procedure Pn is a relation between states such that a call of Pn in state s leads (if execution of such a call does terminate) to a state s' such that $\langle s, s' \rangle \in \underline{\text{Pn}}$. When a command C contains a procedure call, the relational semantics of the command can only be defined if the semantics of the procedure is known. Therefore, we define the semantics $\underline{C}$ of a command C as a function of the semantics r of the procedure $\text{Pn} :: C_1$. The definition of $\underline{C}(r) = I[\![C]\!](r)$ is similar to (19) but for the fact that we must define the effect of a procedure call $\underline{\text{Pn}}(r) = r$ and that of a program: $\underline{\text{Pn} :: C_1;C_2} = \underline{C_2}(r)$ where r is the semantics of procedure Pn. If procedure Pn is not recursive then its semantics is simply $r = \underline{C_1}(\emptyset)$. If it is recursive, the definition is circular since $r = \underline{C_1}(r)$. To avoid paradoxes, we must prove that this equation has a solution and, if it is not unique, we must specify which one is to be considered. To do this, observe that $\mathscr{P}(S \times S)$ is a complete lattice for the partial ordering $\subseteq$ where $\emptyset$ is the infimum, $S \times S$ is the supremum, $\bigcup$ is the least upper bound and $\bigcap$ is the greatest lower bound. Also $\underline{C_1}$ is monotone, i.e. $r \subseteq r' \Rightarrow \underline{C_1}(r) \subseteq \underline{C_1}(r')$. Therefore, according to [385], $\underline{C_1}$ has a least fixed point $\mathit{lfp}\ \underline{C_1} = \bigcap \{r \in \mathscr{P}(S \times S) : \underline{C_1}(r) \subseteq r\}$ such that $\mathit{lfp}\ \underline{C_1} = \underline{C_1}(\mathit{lfp}\ \underline{C_1})$ and $r = \underline{C_1}(r) \Rightarrow (\mathit{lfp}\ \underline{C_1} \subseteq r)$. The semantics of procedure Pn is chosen to be $\mathit{lfp}\ \underline{C_1}$ (this should be justified with respect to an operational semantics as in (19), see [109, Chapter 5]). We get the following definition.

(173) DEFINITION (after Scott & De Bakker [368], De Bakker & De Roever [110]) (*relational semantics*)

(173.1) $I : \textbf{Com} \rightarrow (\mathscr{P}(S \times S) \rightarrow \mathscr{P}(S \times S))$

(173.2) $\underline{\text{skip}}(r)=\{\langle s,s\rangle: s\in S\}$,

(173.3) $\underline{X := E}(r)=\{\langle s, s[X\leftarrow \underline{E}(s)]\rangle: s\in S\}$,

(173.4) $\underline{X := ?}(r)=\{\langle s, s[X\leftarrow d]\rangle: s\in S\wedge d\in D\}$,

(173.5) $\underline{(C_1;C_2)}(r)=\underline{C_1}(r)\circ\underline{C_2}(r)$,

(173.6) $\underline{(B\rightarrow C_1 \diamond C_2)}(r)=(\underline{B}\rceil\underline{C_1}(r))\cup(\neg\underline{B}\rceil\underline{C_2}(r))$,

(173.7) $\underline{(B * C)}(r)=(\underline{B}\rceil\underline{C}(r))^*\lceil\neg\underline{B}$,

(173.8) $\underline{\text{Pn}}(r)=r$,

(173.9) $\underline{\text{Pn} :: C_1;C_2}=\underline{C_2}(\mathit{lfp}\ \underline{C_1})$.

(For a version of (173) taking termination into account, see [206, 108, 119, 343, 281, 27]. A semantics of nondeterministic recursive programs based upon fixpoints of relations on functions rather than fixpoints of functions on relations is presented in [342]).

The recursive programming language (172) with its relational semantics (173) is strictly more powerful than the iterative language (1) with its semantics (19) since, for example, when the data space D is finite, the first may use an unbounded storage space while the second may not (the comparison is pursued in [248, 249]).

8.2.1.2. The recursion rule based upon computational induction. To prove partial correctness, we need, in addition to theorems (76), means of proving $\underline{\{p\}\mathit{lfp}\ C_1\{q\}}$ or, more generally, of proving a property $P(\mathit{lfp}\ F)$ of the least fixpoint $\mathit{lfp}\ F$ of a monotone function F on a complete lattice L:

(174) LEMMA (Scott & De Bakker [368]) (*computation induction*). *If* $\langle L, \leqslant, \bigcup, \bigcap, \bot\rangle$ *is a complete lattice and* $F: L\rightarrow L$ *is monotone then*

$$[P(\bot)\wedge\forall X.\ P(X)\Rightarrow P(F(X))$$
$$\wedge\ \forall\alpha.\ \forall X\in\alpha\rightarrow L.\ (\forall\beta<\alpha.\ P(X_\beta))\Rightarrow P(\textstyle\bigcup_{\beta<\alpha}X_\beta)]\ \Rightarrow\ P(\mathit{lfp}\ F).$$

PROOF. $\mathit{lfp}\ F$ is one of the elements of the transfinite sequence $X_0=\bot,\ldots,X_\alpha=F(X_{\alpha-1})$ for successor ordinals $\alpha,\ldots,X_\alpha=\bigcup_{\beta<\alpha}X_\beta$ for limit ordinals α (see [90]). Therefore we can prove that $P(\mathit{lfp}\ F)$ holds by proving $\forall\alpha.\ P(X_\alpha)$, which, by transfinite induction, is implied by $P(\bot)\wedge\forall X.\ P(X)\Rightarrow P(F(X))\wedge\forall\alpha.\ \forall X.\ (\forall\beta<\alpha.\ P(X_\beta))\Rightarrow P(\bigcup_{\beta<\alpha}X_\beta)$. □

Computation induction is a generalization of *Scott induction* (also called *computational induction* in [286]) which corresponds to the particular case when F is upper-continuous (if so, $\mathit{lfp}\ F=X^\omega$ where $\omega=|\mathbf{N}|$) and P is *admissible* (i.e., $\forall X\in\mathbf{N}\rightarrow L.\ (\forall n\in\mathbf{N}.\ P(X_n))\Rightarrow P(\bigcup_{n\in\mathbf{N}}X_n)$) so that $(P(\bot)\wedge\forall X.\ P(X)\Rightarrow P(F(X))\Rightarrow P(\mathit{lfp}\ F)$. When specialized to the partial correctness proof $\underline{\{p\}\mathit{lfp}\ C_1\{q\}}$ of a recursive procedure $\text{Pn} :: C_1$, computation induction leads to the following theorem, where the assumption $\forall\langle s,s'\rangle\in r.\ \forall v\notin \mathit{Var}(C_1).\ s(v)=s'(v)\wedge\forall d\in D.\ \langle s[v\leftarrow d],\ s'[v\leftarrow d]\rangle\in r$ states that the relational semantics r of C_1 is without side-effects; more precisely, that the variables

v not appearing in C_1 cannot be modified and can have any value during execution of C_1:

(175) THEOREM (partial correctness proof of procedures by computation induction)

$$\begin{aligned}&[\forall r \in \mathscr{P}(S^2).\\ &(\forall \langle s, s'\rangle \in r.\ \forall v \notin Var(C_1).\ s(v)=s'(v) \wedge \forall d \in D.\ \langle s[v\leftarrow d],\ s'[v\leftarrow d]\rangle \in r)\\ &\ \Rightarrow (\underline{\{p\}}r\underline{\{q\}} \Rightarrow \underline{\{p\}}\underline{C_1}(r)\underline{\{q\}})]\\ &\ \Rightarrow\ \underline{\{p\}}\mathit{lfp}\ \underline{C_1}\underline{\{q\}}.\end{aligned}$$

PROOF. We prove $(\forall\langle s, s'\rangle \in \mathit{lfp}\ \underline{C_1}.\ \forall v \notin Var(C_1).\ s(v)=s'(v) \wedge \forall d \in D.\ \langle s[v\leftarrow d],\ s'[v\leftarrow d]\rangle \in \mathit{lfp}\ \underline{C_1} \wedge \underline{\{p\}}\mathit{lfp}\ \underline{C_1}\underline{\{q\}})$ by computation induction (174). $(\forall\langle s, s'\rangle \in \emptyset.\ \forall v \notin Var(C_1).\ s(v)=s'(v) \wedge \forall d \in D.\ \langle s[v\leftarrow d],\ s'[v\leftarrow d]\rangle \in \emptyset \wedge \underline{\{p\}}\emptyset\underline{\{q\}})$ is obviously true. If, by induction hypothesis, $\forall\langle s, s'\rangle \in r.\ \forall v \notin Var(C_1).\ s(v)=s'(v) \wedge \forall d \in D.\ \langle s[v\leftarrow d], s'[v\leftarrow d]\rangle \in r \wedge \underline{\{p\}}r\underline{\{q\}}$ is true, then $\underline{\{p\}}\underline{C_1}(r)\underline{\{q\}}$ holds by hypothesis of theorem (175) and we can prove $\forall\langle s, s'\rangle \in \underline{C_1}(r).\ \forall v \notin Var(C_1).\ s(v)=s'(v) \wedge \forall d \in D.\ \langle s[v\leftarrow d],\ s'[v\leftarrow d]\rangle \in \underline{C_1}(r)$ by structural induction on the syntax (172) of C_1. Finally $\forall\alpha.\ \forall r.(\forall\beta<\alpha.\ (\forall\langle s, s'\rangle \in r_\beta.\ \forall v \notin Var(C_1).\ s(v)=s'(v) \wedge \forall d \in D.\ \langle s[v\leftarrow d],\ s'[v\leftarrow d]\rangle \in r_\beta) \wedge (p\rceil r_\beta \subseteq S \times q)) \Rightarrow ((\forall\langle s, s'\rangle \in \bigcup_{\beta<\alpha} r_\beta.\ \forall v \notin Var(C_1).\ s(v)=s'(v) \wedge \forall d \in D.\ \langle s[v\leftarrow d], s'[v\leftarrow d]\rangle \in \bigcup_{\beta<\alpha} r_\beta) \wedge p\rceil(\bigcup_{\beta<\alpha} r_\beta) \subseteq S \times q)$ is again obviously true. □

Computation induction (175) can be directly translated into Hoare logic by the following *recursion rule* due to Hoare [210] (recursive procedure $\mathrm{Pn} :: C_1$):

(176) $$\frac{\{P\}\mathrm{Pn}\{Q\} \vdash \{P\}C_1\{Q\}}{\{P\}\mathrm{Pn}\{Q\}} \quad \textit{recursion rule.}$$

This rule of inference in the sense of Prawitz [350] means that "$H_1, \ldots, H_n$" is a formal proof of H_n using

$$\mathbf{H} \cup \left\{\frac{H \vdash H'}{H''}\right\}$$

if and only if "$H, H_1, \ldots, H_n$" is a formal proof of H_n (as defined in (94)) using

$$\mathbf{H} \cup \left\{\frac{H'}{H''}\right\}$$

Formally, metarule (176) can be avoided by transforming the proof system into an ordinary one as shown by Apt [13] and by America and De Boer [6].

(177) EXAMPLE (*partial correctness proof by computation induction*). The following program terminates with $X=n$ and $Y=n!$ when initially $X=n\geqslant 0$:

(178)
```
procedure F;
begin
  if X=0 then Y:= 1
  else begin X := X-1; F; X := X+1; Y:= Y*X; end; end;
F;
```

Partial correctness "{true} F $\{Y=X!\}$" can be proved using (176) as follows:

(a)	$\{\text{true}\}\ F\ \{Y=X!\}$	by induction hypothesis
(b)	$\{\text{true} \wedge X=0\}\ Y:=1\ \{Y=X!\}$	by (97), (102)
(c)	$\{Y*X=X!\}\ Y:=Y*X\ \{Y=X!\}$	by (97)
(d)	$\{Y*(X+1)=(X+1)!\}\ X:=X+1\ \{Y*X=X\}$	by (97)
(e)	$\{Y=X!\} \Rightarrow \{Y*(X+1)=(X+1)!\}$	from Th
(f)	$\{Y=X!\}\ X:=X+1\ \{Y*X=X!\}$	by (e), (d), (102)
(g)	$\{\text{true}\}\ X=:X-1\ \{\text{true}\}$	by (97)
(h)	$\{\text{true} \wedge \neg(X=0)\}\ (((X:=X-1;F);X:=X+1);Y:=Y*X)$ $\{Y*X=X!\}$	by (g), (a), (f), (c) (99), (102)
(i)	$\{\text{true}\}\ (X=0\to Y:=1 \diamond (((X:=X-1;F);X:=X-1);Y:=Y*X))$ $\{Y=X!\}$	by (b), (h), (100)
(j)	$\{\text{true}\}\ F\ \{Y=X!\}$	by (a), (i), (175).

Observe that the recursion rule (176) is powerful enough to prove "{true} F $\{Y=X!\}$" because this conclusion (j) can be used as induction hypothesis (a). This is not the case for proving "$\{X=n\}$ F $\{X=n \wedge Y=n!\}$" since then we need induction hypothesis "$\{X=n-1\}$ F $\{X=n-1 \wedge Y=(n-1)!\}$" which cannot be directly derived from the conclusion "$\{X=n\}$ F $\{X=n \wedge Y=n!\}$" using consequence rule (102). Hence, a proof method using (175) or (176) only is not relatively complete [13].

Various proof rules, known as *copy-rule induction*, have been proposed by Gorelick [177], Clarke [71], Langmaack and Olderog [267], Apt [13, 14], Olderog [328, 330, 331], and Trakhtenbrot, Halpern and Meyer [391] to extend recursion rule (176) for higher-order procedure calls.

8.2.1.3. The rule of adaptation based upon fixpoint induction. Since the proof method (175) based upon computation induction is not complete, we come back to the problem of proving a property $P(\mathit{lfp}\ F)$ of the least fixpoint $\mathit{lfp}\ F$ of a monotone function F on a complete lattice L. When $P(\mathit{lfp}\ F)$ is of the form $X \cap \mathit{lfp} F \leqslant Y$, which is the case for $\underline{\{p\}}\mathit{lfp}\ \underline{C_1}\underline{\{q\}}$, we can use fixpoint induction:

(179) LEMMA (Park [340]) (fixpoint induction). *If* $\langle L, \leqslant, \bigcup, \bigcap, \perp\rangle$ *is a complete lattice and* $F: L \to L$ *is monotone, then*

$$(X \cap \mathit{lfp}\ F \leqslant Y) \Leftrightarrow (\exists Z \in L.\ F(Z) \leqslant Z \wedge X \cap Z \leqslant Y).$$

PROOF. For $\Rightarrow$ we can choose $Z = \mathit{lfp}\ F$ so that $F(Z)=Z$ and for $\Leftarrow$ we have $(F(Z) \leqslant Z) \Rightarrow (\mathit{lfp}\ F \leqslant Z)$ since $\mathit{lfp}\ F = \bigcap\{Z: F(Z) \leqslant Z\}$ by [385], whence $X \cap \mathit{lfp} F \leqslant X \cap Z \leqslant Y$. $\square$

When specialized to the partial correctness proof $\underline{\{p\}}\mathit{lfp}\ \underline{C_1}\underline{\{q\}}$ of a recursive procedure $\text{Pn} :: C_1$, fixpoint induction (179) leads to the following proof method (a version of which is used in [88] to establish the partial correctness of clausal programs):

(180) THEOREM (after Park [340], Manna & Pnueli [287]) (partial correctness proof of procedures by fixpoint induction I)

(180.1) $\quad (\exists r \in \mathscr{P}(S^2).\ \underline{C_1}(r) \subseteq r \wedge \underline{\{p\}}r\underline{\{q\}}) \Rightarrow \underline{\{p\}}\mathit{lfp}\ \underline{C_1}\underline{\{q\}}$,

(180.2) $\quad \underline{\{p\}}\mathit{lfp}\, \underline{C_1}\underline{\{q\}} \;\Rightarrow\; (\exists r \in \mathscr{P}(S^2).\ \underline{C_1}(r) \subseteq r \wedge \underline{\{p\}}r\underline{\{q\}}$

$\qquad \wedge\, \forall \langle s, s'\rangle \in r.\ \forall v \notin \mathit{Var}(C).\ (s(v) = s'(v))$

$\qquad \wedge\, \forall d \in D.\ \langle s[v \leftarrow d], s'[v \leftarrow d]\rangle \in r).$

PROOF. We have $(\exists r \in \mathscr{P}(S^2).\ \underline{C_1}(r) \subseteq r \wedge \underline{\{p\}}r\underline{\{q\}}) \Rightarrow (\exists r \in \mathscr{P}(S^2).\ \underline{C_1}(r) \subseteq r \wedge (p \times S) \cap r \subseteq S \times q) \Rightarrow ((p \times S) \cap (\mathit{lfp}\, \underline{C_1}) \subseteq S \times q) \Rightarrow (p \rceil (\mathit{lfp}\, \underline{C_1}) \subseteq S \times q) \Rightarrow \underline{\{p\}}\mathit{lfp}\, \underline{C_1}\underline{\{q\}}$. Reciprocally, if $\underline{\{p\}}\mathit{lfp}\, \underline{C_1}\underline{\{q\}}$ then, obviously, $\underline{C_1}(r) \subseteq r \wedge \underline{\{p\}}r\underline{\{q\}}$ for $r = \mathit{lfp}\, \underline{C_1}$. Moreover, we can prove $P(\mathit{lfp}\, \underline{C_1})$ where $P(r) = (\forall \langle s, s'\rangle \in r.\ \forall v \notin \mathit{Var}(C).\ (s(v) = s'(v)) \wedge \forall d \in D.\ \langle s[v \leftarrow d], s'[v \leftarrow d]\rangle \in r)$ by computation induction (174). $P(\emptyset)$ is obvious. Assuming $P(r)$ we can prove $P(\underline{C_1}(r))$ by structural induction on the syntax of C_1. For example $P(\underline{\mathrm{Pn}}(r)) = P(r)$ [by (173.8)] $=$ tt [by induction hypothesis]. $P(\underline{X} := \underline{E}(r)) = \forall v \notin \{X\} \cup \mathit{Var}(E).\ s(v) = s[X \leftarrow \underline{E}(s)](v) \wedge \forall d \in D.\ \langle s[v \leftarrow d],\ s[X \leftarrow \underline{E}(s)][v \leftarrow d]\rangle \in \{\langle s'', s''[X \leftarrow \underline{E}(s'')]\rangle : s'' \in S\}$ is true, etc. Finally, $\forall \alpha.\ \forall r.\ (\forall \beta < \alpha.\ P(r_\beta)) \Rightarrow P(\bigcup_{\beta < \alpha} r_\beta)$ is obvious. □

(181) EXAMPLE (*partial correctness proof by fixpoint induction* I). Partial correctness "$\{X = n\}\ F\ \{X = n \wedge Y = n!\}$" of program (178) can be proved using (180.1) as follows:

(a)	$r = \{\langle s, s'\rangle : s'(X) = s(X) \wedge s'(Y) = s(X)!\}$	by definition of r
(b)	$\underline{Y} := \underline{1}(r) \subseteq \{\langle s, s[Y \leftarrow 0!]\rangle : s \in S\}$	by (173.2)
(c)	$\underline{X} \underline{=} \underline{0} \rceil \underline{Y} := \underline{1}(r) \subseteq r$	by (b), (a)
(d)	$\underline{X} := \underline{X} \underline{-} \underline{1}(r) \subseteq \{\langle s, s[X \leftarrow s(X) - 1]\rangle : s \in S\}$	by (173.2)
(e)	$\underline{F}(r) \subseteq \{\langle s, s'\rangle : s'(X) = s(X) \wedge s'(Y) = s(X)!\}$	by (173.8), (a)
(f)	$(\underline{X} := \underline{X} \underline{-} \underline{1};\ \underline{F})(r) \subseteq \{\langle s, s'\rangle : s'(X) = s(X) - 1 \wedge s'(Y) = (s(X) - 1)!\}$	by (d), (e), (173.5)
(g)	$\underline{X} := \underline{X} \underline{+} \underline{1}(r) \subseteq \{\langle s, s[X \leftarrow s(X) + 1]\rangle : s \in S\}$	(by (173.2)
(h)	$((\underline{X} := \underline{X} \underline{-} \underline{1}; \underline{F}); \underline{X} := \underline{X} \underline{+} \underline{1})(r) \subseteq \{\langle s, s'\rangle : s'(X) = s(X) \wedge s'(Y) = (s(X) - 1)!\}$	by (f), (g), (173.5)
(i)	$\underline{Y} := \underline{Y} \underline{*} \underline{X}(r) \subseteq \{\langle s, s[Y \leftarrow s(Y) * s(X)]\rangle : s \in S\}$	by (173.2)
(i)	$\neg \underline{X} \underline{=} \underline{0} \rceil (((\underline{X} := \underline{X} \underline{-} \underline{1};\ \underline{F});\ \underline{X} := \underline{X} \underline{+} \underline{1});\ \underline{Y} := \underline{Y} \underline{*} \underline{X})(r) \subseteq r$	by (*f*), (g), (173.5)
(j)	$(\underline{X} \underline{=} \underline{0} \underline{\rightarrow} \underline{Y} := \underline{1} \underline{\diamond} (((\underline{X} := \underline{X} \underline{-} \underline{1};\ \underline{F});\ \underline{X} := \underline{X} \underline{-} \underline{1});\ \underline{Y} := \underline{Y} * \underline{X}))(r) \subseteq r$	by (c), (j), (173.6)
(k)	$\underline{\{}\{s : s(X) = n\}\underline{\}}r\underline{\{}\{s : s(X) = n \wedge s(Y) = n!\}\underline{\}}$	by (a), (22)
(l)	$\underline{\{}\{s : s(X) = n\}\underline{\}}\ \mathit{lfp}\ (\underline{X} \underline{=} \underline{0} \underline{\rightarrow} \underline{Y} := \underline{1} \underline{\diamond} (((\underline{X} := \underline{X} \underline{-} \underline{1};\ \underline{F});$ $\underline{X} := \underline{X} \underline{-} \underline{1});\ \underline{Y} := \underline{Y} \underline{*} \underline{X}))\ \underline{\{}\{s : s(X) = n \wedge s(Y) = n!\}\underline{\}}$	by (j), (k), (178.1).

Theorem (180) is not directly expressible in Hoare logic since $(\exists r \in \mathscr{P}(S^2) \cdot \underline{C_1}(r) \subseteq r \wedge \underline{\{p\}}r\underline{\{q\}})$ does not use only formulae of the style $p' \subseteq q'$ and $\underline{\{p'\}}\underline{C'}\underline{\{q'\}}$. To enforce this, we can let $S' = S^2$, $p' = \{\langle s, s\rangle : s \in S\}$, $\underline{C_1'} = \{\langle\langle \underline{s}, s\rangle, \langle s, s'\rangle\rangle : \langle s, s'\rangle \in \underline{C_1}(r)\}$ and $q' = r$, so that $\underline{\{p'\}}\underline{C_1'}\underline{\{q'\}} = \underline{\{}\{\langle s, s'\rangle : s \in S\}\underline{\}}\{\langle\langle \underline{s}, s\rangle, \langle \underline{s}, s'\rangle\rangle : \langle s, s'\rangle \in \underline{C_1}(r)\}\underline{\{r\}} = (p' \rceil \underline{C_1'} \subseteq S \times q') = (\{\langle\langle \underline{s}, \underline{s}\rangle, \langle \underline{s}, s\rangle\rangle : \langle s, s\rangle \in \underline{C_1}(r)\} \subseteq \{\langle\langle s_0, s_1\rangle, \langle \underline{s}, s\rangle\rangle : \langle \underline{s}, \bar{s}\rangle \in r\}) = \underline{C_1}(r) \subseteq r$. In this translation the relation r between states before and after the procedure call is expressed using predicates upon states but the state space S has been changed into S^2 (as in [288]). To remain in the spirit of traditional Hoare logic, it is better to use logical auxiliary variables not appearing in the program to memorize the value of the programming variables before the procedure call (the importance of these auxiliary variables in correctness proofs for recursive procedures was first realized by Greibach [182] and Gorelick [177] and later thoroughly investigated in [155, 156,

21, 25, 300, 14]. Let $X^{\cup}=\langle X_1,\ldots,X_n\rangle$ be the vector of variables $Var(C_1)=\{X_1,\ldots,X_n\}$ appearing in command C_1. Following Olderog [329], we write $X^{\cup}\|x^{\cup}$ whenever $X^{\cup}=\langle X_1,\ldots,X_n\rangle$, $x^{\cup}=\langle x_1,\ldots,x_m\rangle$, $m=n$ and $\{X_1,\ldots,X_n\}\cap\{x_1,\ldots,x_m\}=\emptyset$ and extend all notations to vectors; for example, $X^{\cup}=x^{\cup}$ means $X_1=x_1\wedge\cdots\wedge X_n=x_m$, $s(X^{\cup})$ stands for $\langle s(X_1),\ldots,s(X_n)\rangle$, $s[X^{\cup}\leftarrow x^{\cup}]$ means $s[X_1\leftarrow x_1]\ldots[X_n\leftarrow x_n]$, $\{X^{\cup}\}$ is $\{X_1,\ldots,X_n\}$, etc. Then we let $p'=\{s\in S\colon\ s(X^{\cup})=s(x^{\cup})\}$ and $q'=\{s'\in S\colon \langle s'[X^{\cup}\leftarrow s'(x^{\cup})],s'\rangle\in r\}$ where $\{X^{\cup}\}=Var(C_1)$ and $X^{\cup}\|x^{\cup}$ so that $\underline{\{}p'\underline{\}}\underline{C_1}(r)\underline{\{}q'\underline{\}}$ implies $\underline{C_1}(r)\subseteq r$. More precisely, we have the following definition.

(182) DEFINITION (Olderog [329]) (*non-interference*)

$$X^{\cup}\|x^{\cup}=[X^{\cup}=\langle X_1,\ldots,X_n\rangle\wedge x^{\cup}=\langle x_1,\ldots,x_m\rangle\wedge m=n \\ \wedge\{X_1,\ldots,X_n\}\cap\{x_1,\ldots,x_m\}=\emptyset].$$

(183) THEOREM (Partial correctness proof of procedures by fixpoint induction II)

$$\begin{aligned}&\exists r\in\mathscr{P}(S^2).\ \underline{C_1}(r)\subseteq r\wedge\underline{\{}p\underline{\}}r\underline{\{}q\underline{\}}\\ &\quad\Leftrightarrow\ \exists p',q'\in\mathscr{P}(S).\ \forall r'\in\mathscr{P}(S^2).\\ &\qquad[(\forall\langle s,s'\rangle\in r'.\ \forall v\notin Var(C_1).\ s(v)=s'(v)\\ &\qquad\quad\wedge\forall d\in D.\ \langle s[v\leftarrow d],s'[v\leftarrow d]\rangle\in r')\\ &\qquad\quad\Rightarrow(\underline{\{}p'\underline{\}}r'\underline{\{}q'\underline{\}}\Rightarrow\underline{\{}p'\underline{\}}\underline{C_1}(r')\underline{\{}q'\underline{\}})]\\ &\qquad\quad\wedge[\forall s\in p.\ \forall d^{\cup}\in D^{\cup}.\ (\forall s'\in S.\ s'[X^{\cup}\leftarrow s(X^{\cup})]\in p'\\ &\qquad\qquad\qquad\qquad\Rightarrow s'[X^{\cup}\leftarrow d^{\cup}]\in q')\\ &\qquad\quad\Rightarrow s[X^{\cup}\leftarrow d^{\cup}]\in q]\end{aligned}$$

where $\{X^{\cup}\}=Var(C_1)$ *and* $X^{\cup}\|x^{\cup}$.

(The assertion $\forall\langle s,s'\rangle\in r'.\ \forall v\notin Var(C_1).\ s(v)=s'(v)\wedge\forall d\in D.\ \langle s[v\leftarrow d],s'[v\leftarrow d]\rangle\in r'$ states that the relational semantics r' of C_1 is without side-effects; more precisely, that in the variables v not appearing in C_1 cannot be modified and can have any value during execution of C_1. We have $\underline{\{}p'\underline{\}}lfp\ \underline{C_1}\underline{\{}q'\underline{\}}$ so that in the assertion $\forall s\in p.\ \forall d^{\cup}\in D^{\cup}.\ (\forall s'\in S.\ s'[X^{\cup}\leftarrow s(X^{\cup})]\in p'\Rightarrow s'[X^{\cup}\leftarrow d^{\cup}]\in q')\Rightarrow s[X^{\cup}\leftarrow d^{\cup}]\in q$, the assumption $\forall s'\in S.\ s'[X^{\cup}\leftarrow s(X^{\cup})]\in p'\Rightarrow s'[X^{\cup}\leftarrow d^{\cup}]\in q'$ states that $\langle p',q'\rangle$ is the specification of the procedure Pn :: C_1. More precisely, any terminating execution of C_1 starting with initial values $s(X^{\cup})$ of the variables $X^{\cup}$ satisfying p' must terminate with final values $d^{\cup}$ of $X^{\cup}$ satisfying q' and this whatever the possible values $s'(v)$ of the variables $v\notin\{X^{\cup}\}$ not appearing in the procedure body C_1 may be. The assertion states that any execution of Pn starting with values $s(X^{\cup})$ of $X^{\cup}$ and terminating with values $d^{\cup}$ of $X^{\cup}$ satisfying specification $\langle p',q'\rangle$ must satisfy the postcondition q whenever the precondition p is satisfied.)

PROOF. For $\Rightarrow$, assume $\underline{C_1}(r)\subseteq r\wedge\underline{\{}p\underline{\}}r\underline{\{}q\underline{\}}$; then $\underline{\{}p\underline{\}}lfp\ \underline{C_1}\underline{\{}q\underline{\}}$ by (180.1) so that, by (180.2), we can assume that $\forall\langle s,s'\rangle\in r.\ \forall v\notin Var(C_1).\ s(v)=s'(v)\wedge\forall d\in D.\ \langle s[v\leftarrow d],$

$s'[v \leftarrow d]\rangle \in r$. Let $p' = \{s \in S : s(X^{\cup}) = s(x^{\cup})\}$ and $q' = \{s' \in S : \langle s'[X^{\cup} \leftarrow s'(x^{\cup})], s'\rangle \in r\}$ where $\{X^{\cup}\} = Var(C_1)$ and $X^{\cup} \| x^{\cup}$.

(A) Assuming $\forall \langle s, s'\rangle \in r'. \forall v \notin Var(C_1). s(v) = s'(v) \wedge \forall d \in D. \langle s[v \leftarrow d], s'[v \leftarrow d]\rangle \in r'$ and $\underline{\{p'\}}r'\underline{\{q'\}}$, we first prove $\underline{\{p'\}}\underline{C_1}(r')\underline{\{q'\}}$.

(a) We first prove that $p' \rceil \underline{C_1}(r') \subseteq p' \rceil \underline{C_1}(r)$. We have $(p' \rceil r' \subseteq S \times q')$ [by (124)] so that $\forall s, s' \in S. (s(X^{\cup}) = s(x^{\cup}) \wedge \langle s, s'\rangle \in r') \Rightarrow \langle s'[X^{\cup} \leftarrow s'(x^{\cup})], s'\rangle \in r$. It follows that if $\langle s, s'\rangle \in r'$ then $\langle s[x^{\cup} \leftarrow s(X^{\cup})], s'[x^{\cup} \leftarrow s(X^{\cup})]\rangle \in r'$ [since $\forall v \notin Var(C_1). \forall d \in D. \langle s[v \leftarrow d], s'[v \leftarrow d]\rangle \in r'$] in which case $s[x^{\cup} \leftarrow s(X^{\cup})](X) = s[x^{\cup} \leftarrow s(X^{\cup})](x^{\cup}) = s(X^{\cup})$ [by (121) since $X^{\cup} \| x^{\cup}$ hence $\{X^{\cup}\} \cap \{x^{\cup}\} = \emptyset$] so that $\langle s'[x^{\cup} \leftarrow s(X^{\cup})] [X^{\cup} \leftarrow s'[x^{\cup} \leftarrow s(X^{\cup})](x^{\cup})], s'[x^{\cup} \leftarrow s(X^{\cup})]\rangle \in r$ and, after simplification by (121), $\langle s'[x^{\cup} \leftarrow s(X^{\cup})][X^{\cup} \leftarrow s(X^{\cup})], s'[x^{\cup} \leftarrow s(X^{\cup})]\rangle \in r$, hence $\langle s, s'\rangle \in r$ since $\forall v \notin Var(C_1). s'(v) = s(v)$. We conclude $r' \subseteq r$, whence by monotony, $\underline{C_1}(r') \subseteq \underline{C_1}(r)$ so that $p' \rceil \underline{C_1}(r') \subseteq p' \rceil \underline{C_1}(r)$.

(b) Then we prove that $\underline{\{p'\}}\underline{C_1}(r)\underline{\{q'\}}$ is true. We have $\underline{\{p'\}}r\underline{\{q'\}} = (p' \rceil r \subseteq S \times q')$[by (124)] $= (\forall \langle s, s'\rangle \in r : s(X^{\cup}) = s(x^{\cup}) \Rightarrow \langle s'[X^{\cup} \leftarrow s'(x^{\cup})], s'\rangle \in r)$ which is true since $\langle s, s'\rangle \in r$ implies $\langle s'[X^{\cup} \leftarrow s(X^{\cup})], s'\rangle \in r$ [since $\forall v \notin Var(C_1). \forall d \in D. \langle s[v \leftarrow d], s'[v \leftarrow d]\rangle \in r$], hence $\langle s'[X^{\cup} \leftarrow s(x^{\cup})], s'\rangle \in r$ [since $s(X^{\cup}) = s(x^{\cup})$]. We conclude $\underline{\{p'\}}\underline{C_1}(r)\underline{\{q'\}}$ since $\underline{C_1}(r) \subseteq r$.

(c) Finally, $\underline{\{p'\}}\underline{C_1}(r')\underline{\{q'\}}$ is true since (b) implies $(p' \rceil \underline{C_1}(r) \subseteq S \times q')$, whence $p' \rceil \underline{C_1}(r') \subseteq S \times q'$ by (a), so that $\underline{\{p'\}}\underline{C_1}(r')\underline{\{q'\}}$ holds.

(B) Let $s \in p$ and $d^{\cup} \in D^{\cup}$ be given. Assuming $(\forall s' \in S. s'[X^{\cup} \leftarrow s(X^{\cup})] \in p' \Rightarrow s'[X^{\cup} \leftarrow d^{\cup}] \in q')$, we prove that $s[X^{\cup} \leftarrow d^{\cup}] \in q$. We have $s[x^{\cup} \leftarrow s(X^{\cup})](X^{\cup}) = s[x^{\cup} \leftarrow s(X^{\cup})](x^{\cup}) = s(X^{\cup})$ [by (121) since $X^{\cup} \| x^{\cup}$, hence $\{X^{\cup}\} \cap \{x^{\cup}\} = \emptyset$], so that $s[X^{\cup} \leftarrow s(X^{\cup})] \in p'$. By definition of p', this implies $s(X^{\cup}) = s[X^{\cup} \leftarrow s(X^{\cup})](X^{\cup}) = s[X^{\cup} \leftarrow s(X^{\cup})](x^{\cup}) = s(x^{\cup})$ [since $X^{\cup} \| x^{\cup}$]. Also $s[X^{\cup} \leftarrow s(X^{\cup})] \in p'$ implies $s[X^{\cup} \leftarrow d^{\cup}] \in q'$, that is, $\langle s[X^{\cup} \leftarrow d^{\cup}] [X^{\cup} \leftarrow s[X^{\cup} \leftarrow d^{\cup}](x^{\cup})], s[X^{\cup} \leftarrow d^{\cup}]\rangle \in r$, after simplification by (121), $\langle s[X^{\cup} \leftarrow s(x^{\cup})], s[X^{\cup} \leftarrow d^{\cup}]\rangle \in r$, hence $\langle s[X^{\cup} \leftarrow s(X^{\cup})], s[X^{\cup} \leftarrow d^{\cup}]\rangle \in r$ [since $s(X^{\cup}) = s(x^{\cup})$], so that $\langle s, s[X^{\cup} \leftarrow d^{\cup}]\rangle \in r$ [by (121)], whence $s[X^{\cup} \leftarrow d^{\cup}] \in q$ [since $s \in p$ and $\underline{\{p\}}r\underline{\{q\}}$].

For $\Leftarrow$, we have $\forall r' \in \mathscr{P}(S^2). (\forall \langle s, s)\rangle \in r'. \forall v \notin Var(C_1). s(v) = s'(v) \wedge \forall d \in D. \langle s[v \leftarrow d], s'[v \leftarrow d]\rangle \in r') \Rightarrow (\underline{\{p'\}}r'\underline{\{q'\}} \Rightarrow \underline{\{p'\}}\underline{C_1}(r')\underline{\{q'\}})$ and (175) which imply $\underline{\{p'\}}\underline{lfp}\ \underline{C_1}\underline{\{q'\}}$. Hence, by (180.2), there exists $r \in \mathscr{P}(S^2)$ such that $\underline{C_1}(r) \subseteq r \wedge \underline{\{p'\}}r\underline{\{q'\}} \wedge \forall \langle s, s'\rangle \in r. \forall x \notin Var(C_1). (s(x) = s'(x)) \wedge \forall d \in D. \langle s[x \leftarrow d], s'[x \leftarrow d]\rangle \in r$.

To prove $\underline{\{p\}}r\underline{\{q\}}$, we assume that $s \in p$ and $\langle s, s'\rangle \in r$ so that $s' = s[X^{\cup} \leftarrow s'(X^{\cup})]$ since $\forall x \notin \{X^{\cup}\}.\ s(x) = s'(x)$. Also $\forall s'' \in S.\ \langle s''[X^{\cup} \leftarrow s(X^{\cup})], s''[X^{\cup} \leftarrow s'(X^{\cup})]\rangle \in r$ since $\forall x \notin \{X^{\cup}\}. \forall d \in D. \langle s[x \leftarrow d], s'[x \leftarrow d]\rangle \in r$. Hence $\underline{\{p'\}}r\underline{\{q'\}}$ implies $\forall s'' \in S. s''[X^{\cup} \leftarrow s(X^{\cup})] \in p' \Rightarrow s''[X^{\cup} \leftarrow s'(X^{\cup})] \in q'$. Then $(\forall s'' \in S. s''[X^{\cup} \leftarrow s(X^{\cup})] \in p' \Rightarrow s''[X^{\cup} \leftarrow d^{\cup}] \in q')$ where $d^{\cup} = s'(X^{\cup})$ implies $s[X^{\cup} \leftarrow d^{\cup}] \in q$, that is, $s[X^{\cup} \leftarrow s'(X^{\cup})] \in q$ so that, in conclusion, $s' \in q$. $\square$

Theorem (183) can be directly transcribed in Hoare logic using recursion rule (176) and the following rule of adaptation (due to Morris [312] and Olderog [329]):

(184) $$\frac{\{P'\}C\{Q'\}}{\{\forall x^{\cup}. (\forall y^{\cup}. (P' \Rightarrow Q'[X^{\cup} \leftarrow x^{\cup}])) \Rightarrow Q[X^{\cup} \leftarrow x^{\cup}]\}C\{Q\}}$$ *rule of adaptation*

where $X^{\cup} \| x^{\cup}$ with $\{x^{\cup}\} \cap Free(P', Q', C, Q) = \emptyset$ and $\{y^{\cup}\} = Free(P', Q') - \{X^{\cup}\}$.

A first version of this rule of adaptation was introduced by Hoare [210] and slightly extended in [234, 138]. Olderog [329] proved that it is sound and relatively complete and can replace the substitution rule I, the invariance rule and the elimination rule of Apt [13]. Although Hoare's adaptation rule is relatively complete, Morris [312] and Olderog [329] showed that it does not give the best possible precondition and proposed a strengthened version. An incorrect version of Hoare's adaptation rule was designed by Guttag, Horning and London [194], and by London, Guttag, Horning and Lampson [278]. The error was pointed out by Gries and Levin [189] who proposed a sound version. A slightly simpler version of the adaptation rule of Gries and Levin [189, 188] was proposed by Martin [294] but Bijlsma, Wiltink and Matthews [49] showed that both versions were equivalent. As pointed out by Olderog [329], and Bijlsma, Wiltink and Matthews [50] the rule of Gries and Levin is relatively incomplete. They proposed a sound and relatively complete variant. Another sound and relatively complete variant was designed by Cartwright and Oppen [67, 68].

(185) EXAMPLE (Apt [13]) (*recursive factorial procedure with global variables*). To prove partial correctness $\{X=n\}\ F\ \{X=n \wedge Y=n!\}$ of procedure F of program (178) we use recursion rule (176). Assuming $\{X=n\}\ F\ \{X=n \wedge Y=n!\}$ by induction hypothesis, we must prove $\{X=n-1\}\ F\ \{X=n-1 \wedge Y=(n-1)!\}$ for the recursive call of F. By the rule of adaptation (184) where $X^{\cup}=\langle X, Y\rangle$, $x^{\cup}=\langle x, y\rangle$ and $y^{\cup}=\langle n\rangle$ we derive $\{\forall x.\,\forall y.\,(\forall n.\,((X=n)\Rightarrow(x=n \wedge y=n!)))\Rightarrow(x=n-1 \wedge y=(n-1)!)\}\ \ F\ \ \{X=n-1 \wedge Y=(n-1)!\}$. By consequence rule (102), it remains to show that $(X=n-1) \Rightarrow (\forall x.\,\forall y.\,(\forall n'.\,((X=n')\Rightarrow(x=n' \wedge y=n'!)))\Rightarrow(x=n-1 \wedge y=(n-1)!))$ which is obviously true. Then the correctness proof of procedure F can be completed as shown by the following proof outline:

```
procedure F;
begin
  {X=n}
  if X=0 then
    {X=n ∧ X=0} Y:= 1 {X=n ∧ Y=n!}
  else begin
    {X=n ∧ X≠0} x:= x−1; {X=n−1}
    F;
    {X=n−1 ∧ Y=(n−1)!} x:= x+1; {X=n ∧ Y=(n−1)!} Y:= Y*X;
  end;
  {X=n ∧ Y=n!}
end;
{X=y}F; {X=y ∧ Y=y!}.
```

For the main call of procedure F it remains to prove $\{X=y\}\ F\ \{X=y \wedge Y=y!\}$. Since $(X=y) \Rightarrow (\forall x'.\,\forall y'.\,(\forall n.\,((X=n)\Rightarrow(x'=n \wedge y'=n!)\Rightarrow(x'=y \wedge y'=y!)))$, this follows from the rule of adaptation (184) and the consequence rule (102).

(186) EXAMPLE (De Bakker and Meertens [112]) (*showing the usefulness of auxiliary variables*). The following program computes $2^{n_0}-1$ when n_0 is positive:

```
procedure F;
begin
  {N=n≥0 ∧ S=s}
  if N>0 then begin
    {N=n>0 ∧ S=s} N:=N−1; {N=n−1≥0 ∧ S=s}
    F;
    {N=n−1≥0 ∧ S=s+2^(n−1)−1} s:= s+1; {N=n−1≥0 ∧ S=s+2^(n−1)}
    F;
    {N=n−1≥0 ∧ S=s+2^(n−1)+2^(n−1)−1} N:= N+1;
  end;
  {N=n≥0 ∧ S=s+2^n−1}
  end;
  {N=n_0≥0 ∧ S=0} F; {N=n_0≥0 ∧ S=2^(n_0)−1}.
```

Its partial correctness proof uses the recursion rule (176) with induction hypothesis $\{N=n\geqslant 0 \wedge S=s\}\ F\ \{N=n\geqslant 0 \wedge S=s+2^n-1\}$ and the adaptation and consequence rules (184) and (102) for the main and recursive calls of F. For example, $\{N=n-1\geqslant 0 \wedge S=s\}\ F\ \{N=n-1\geqslant 0 \wedge S=s+2^{n-1}-1\}$ follows from the fact that $(N=n-1\geqslant 0 \wedge S=s)$ implies $(\forall n', s'.(\forall \underline{n}, \underline{s}.\ ((N=\underline{n}\geqslant 0 \wedge S=\underline{s})\Rightarrow(n'=\underline{n}\geqslant 0 \wedge s'=\underline{s}+2^{\underline{n}}-1))\Rightarrow (n'=n-1\geqslant 0 \wedge s'=s+2^{n-1}-1))$. The proof proposed in [112] uses an infinite pattern of assertions (one for each program step) because in their proof system one cannot use logical variables (n, s) to relate the different values of the program variables (N, S) at different stages of the computation, so that the invariants have to express the current value of N and S in terms of their initial values n_0 and 0 and of a representation of the computation history (see [182, 177, 155, 156, 21, 25, 300, 14]).

8.2.1.4. Hoare-like deductive systems with context-dependent conditions. Hoare deduction systems are Hilbert-style systems [37] where proofs proceed by induction on the syntax of programs. Since this syntax is specified by a context-free grammar [3] it is not directly possible to express context-dependent conditions such as the correspondence between procedure calls and procedure declarations (specifying which procedure bodies are associated with procedure names). In fact, a formula $\{P\}C\{Q\}$ is often relative to declarations of named objects which can be represented by a mapping from names to objects called an *environment* [369]. Therefore, most often, Hoare correctness formulae have the form $\langle \rho \mid \{P\}C\{Q\}\rangle$ where ρ is an environment associating procedure bodies to procedure names (and informally all procedure names occurring free in C are declared in ρ and there are no procedure names occurring free in ρ). The axioms and rules of inference propagate the environment from declarations to calls:

(187.1) $$\frac{\langle [\mathrm{Pn}\leftarrow C_1] \mid \{P\}C_2\{Q\}\rangle}{\{P\}\mathrm{Pn} :: C_1; C_2\{Q\}}$$ *rule of programs,*

(187.2) $$\langle \rho \mid \{P\}\ \mathrm{skip}\ \{P\}\rangle$$ *skip axiom,*

(187.3) $$\frac{\langle \rho \mid \{P\}C_1\{Q\}\rangle,\quad \langle \rho \mid \{Q\}C_2\{R\}\rangle}{\langle \rho \mid \{Q\}(C_1; C_2)\{R\}\rangle}$$ *composition rule,*

(187.4) $$\frac{\langle\rho \mid \{P\}\mathrm{Pn}\{Q\}\rangle \vdash \langle\rho \mid \{P\}\rho(\mathrm{Pn})\{Q\}\rangle}{\langle\rho \mid \{P\}\mathrm{Pn}\{Q\}\rangle}$$ *recursion rule.*

The environment component ρ clearly corresponds to the environment argument in denotational semantics [369, 388] and to synthesized attributes in attribute grammars [255]. It was first introduced in the context of Hoare logic by Clarke [71] and by Apt and De Bakker [22].

8.2.2. Value–result parameters

Let us now consider a procedural language with value–result parameters:

(188) DEFINITION *(syntax of a procedural language with value and/or result parameters)*

(188.1) Pn: **Proc** *procedure names,*

(188.2) Pg: **Prog** *programs,*
$\mathrm{Pg} ::= \mathrm{Pn}(?U, ?!V, !W) :: C_1; C_2$

(188.3) C: **Com** *commands*
$C ::= \mathrm{skip} \mid X := E \mid X := ? \mid (C_1; C_2) \mid (B \rightarrow C_1 \diamond C_2) \mid (B * C) \mid \mathrm{Pn}(E, Y, Z).$

U, V and W are distinct formal parameters of recursive procedure Pn with body C_1 which are respectively passed by value, value–result and result. Therefore a call $\mathrm{Pn}(E, Y, Z)$ with expression E and variables Y and Z as actual parameters consists in creating *new* local variables named U, V, W within C_1, respectively initialized to the value of E, to the value of Y and to any value d of D, then in executing the body C_1 of procedure Pn (which may modify the values of local variables U, V, W as well as that of the global variables (but not the ones named U, V and W which are hidden by their local homonyms)), and finally in assigning the values of the local variables V and W to the variables Y and Z (in that order) and then in destroying the local variables named U, V, W:

(189) DEFINITION (after De Roever [118]) *relational semantics of value and/or result parameter passing)*

(189.1) $\underline{\mathrm{Pn}}(\underline{E}, \underline{Y}, \underline{Z})(r)$
$= \{\langle s, s'[U \leftarrow s(U)][V \leftarrow s(V)][W \leftarrow s(W)][Y \leftarrow s'(V)][Z \leftarrow s'(W)]\rangle:$
$\exists d \in D.\, \langle s[U \leftarrow \underline{E}(s)][V \leftarrow s(Y)][W \leftarrow d], s'\rangle \in r\},$

(189.2) $\underline{\mathrm{Pn}(?U, ?!V, !W) :: C_1; C_2} = \underline{C_2}(\mathit{lfp}\, \underline{C_1}).$

(190) EXAMPLE. Let Pg be $\mathrm{Pn}(?A, ?!B, !C) :: C_1; C_2$ where C_1 is $(B := A + B; (A := A + 1; (C := A + 1; D := C + 1)))$ and C_2 is $(C_3; \mathrm{Pn}(A+1, B, B))$ with $C_3 = (A := 1; (B := 10; C := 100))$. By (173.3) and (173.5), the relational semantics of the procedure body is $\mathit{lfp}\, \underline{C_1} = \underline{C_1}(\emptyset) = \{\langle s, s[A \leftarrow s(A)+1][B \leftarrow s(A)+s(B)][C \leftarrow s(A)+2][D \leftarrow s(A)+3]\rangle: s \in S\}$. By (189.1), the semantics of the procedure call is $\underline{\mathrm{Pn}}(\underline{A+1}, \underline{B}, \underline{B}) = \{\langle s, s'[A \leftarrow s(A)][B \leftarrow s(B)][C \leftarrow s(C)][B \leftarrow s'(B)][B \leftarrow s'(C)]\rangle:\ \exists d \in D.\, \langle s[A \leftarrow \underline{A+1}(s)][B \leftarrow s(B)][C \leftarrow d], s'\rangle \in \mathit{lfp}\, \underline{C_1}\} = \{\langle s, s[D \leftarrow s(A)+4][B \leftarrow s(A)+3]\rangle: s \in S\}$. The semantics of the initialization part of the program body is $\underline{C}_3(lfp\,\underline{C}_1)=\{\langle s, s[A\leftarrow 1][B\leftarrow 10][C\leftarrow 100]\rangle: s\in S\}$ so that the semantics of the program is $\underline{\mathrm{P}}\mathrm{g}=\underline{C}_2(lfp\,\underline{C}_1)=\{\langle s, s[A\leftarrow 1]\ [B\leftarrow 10][C\leftarrow 100][D\leftarrow (s[A\leftarrow 1]\ [B\leftarrow 10][C\leftarrow 100](A))+4][B\leftarrow (s[A\leftarrow 1][B\leftarrow 10]\ [C\leftarrow 100](A))+3]\rangle: s\in S\}=\{\langle s, s[A\leftarrow 1][C\leftarrow 100][D\leftarrow 5][B\leftarrow 4]\rangle: s\in S\}$.

This semantics leads to an analog of theorem (183) and then to the following proof rules (the rule of adaptation (192) could be simplified as in [189, 188, 294, 49] by restricting the use of global variables, the assignments to value parameters, or assuming that value–result and result parameters are different) (recursive procedure $\mathrm{Pn}(?U, ?!V, !W) :: C_1$):

(191) *Recursion rule*:

$$\frac{\{P\}\ \mathbf{Pn(?}U\mathbf{, ?!}V\mathbf{, !}W\mathbf{)}\ \{Q\}\vdash\{P\}C_1\{Q\}}{\{P\}\ \mathbf{Pn(?}U\mathbf{, ?!}V\mathbf{, !}W\mathbf{)}\ \{Q\}},$$

(192) *Rule of adaptation*:

$$\{P'\}\ \mathbf{Pn(?}U\mathbf{, ?!}V\mathbf{, !}W\mathbf{)}\ \{Q'\}$$

$$\begin{array}{l}
\{\forall x^{\cup}.\,\forall “u, “v, “w, “y, “z, v', w'.\\
\ ((U=“u)\wedge(V=“v)\wedge(W=“w)\wedge(Y=“y)\wedge(Z=“z)\wedge\\
\ \ (\forall m^{\cup}.\\
\ \ \ ((\exists ‘w.\,P'[U\leftarrow E[U\leftarrow “u][V\leftarrow “v][W\leftarrow “w][Y\leftarrow “y][Z\leftarrow “z]][V\leftarrow “v][W\leftarrow ‘w][Y\leftarrow “y][Z\leftarrow “z])\\
\ \ \ \ \Rightarrow(\exists u', y', z'.\,Q'[X^{\cup}\leftarrow x^{\cup}][U\leftarrow u'][V\leftarrow v'][W\leftarrow w'][Y\leftarrow y'][Z\leftarrow z'])))))\\
\ \Rightarrow Q[Z\leftarrow w'][Y\leftarrow v'][X^{\cup}\leftarrow x^{\cup}][U\leftarrow “u][V\leftarrow “v][W\leftarrow “w]\}\\
\mathbf{Pn(}E\mathbf{,}\ Y\mathbf{,}\ Z\mathbf{)}\\
\{Q\}
\end{array}$$

where

- $X^{\cup}=Var(E, C_1)-\{U, V, W, Y, Z\}$ is the value of the global variables (not used as actual result parameters and not homonyms of actual or formal parameters) before the call $\mathrm{Pn}(E, Y, Z)$,
- $x^{\cup}$ such that $X^{\cup}\|x^{\cup}$ are fresh variables denoting the value of $X^{\cup}$ after the call $\mathrm{Pn}(E, Y, Z)$,
- $“u$, $“v$, $“w$, $“y$, $“z$ are fresh variables denoting the values of the global variables U, V, W, Y, Z before the call,
- $‘w$ is a fresh variable denoting the undetermined initial value $d\in D$ of formal result parameter W when execution of the procedure body C_1 is started,
- u', v', w' are the values of the local formal parameters U, V and W after execution of the procedure body C_1 and before the result parameters passing,
- $\{m^{\cup}\}=Free(P', Q')-(\{X^{\cup}\}\cup\{U, V, W, Y, Z\})$ is the set of logical variables used for the specification of the procedure body C_1,
- y', z' are the values of the global variables Y and Z after execution of the procedure body C_1 and before the result-parameters passing,
- all fresh variables are distinct and do not appear in $Free(P', Q', C_1, E, Q)\cup\{U, V, W, Y, Z\}$.

When reading rule (192) it must be remembered that substitution is left to right, that is to say, $P[X\leftarrow x][Y\leftarrow y]$ is $P'[Y\leftarrow y]$ where $P'=P[X\leftarrow x]$.

(193) EXAMPLE (Martin [294]) (*once used to show the inconsistency of erroneous procedure-call proof rules*). Assuming $\{P'\}\ \mathrm{Pn}(!Z)\ \{Q'\}$ where $P' = \text{true}$ and $Q' = ((Z=1) \vee (Z=2))$, we can determine P such that $\{P\}\ \mathrm{Pn}(C)\ \{Q\}$ holds with $Q = (C=2)$ by the rule of adaptation (192):

$$
\begin{aligned}
P = \quad & \forall \text{``}w, \text{``}z, \text{``}w.\, ((Z = \text{``}w) \wedge (C = \text{``}z) \\
& \qquad\qquad \wedge ((\exists \text{`}w.\, P'[Z \leftarrow \text{`}w][C \leftarrow \text{``}z]) \\
& \qquad\qquad\quad \Rightarrow (\exists z\text{'}.\, Q'[Z \leftarrow w\text{'}][C \leftarrow z\text{'}]))) \\
& \Rightarrow Q[C \leftarrow w\text{'}][Z \leftarrow \text{``}w] \\
= \quad & \forall \text{``}w, \text{``}z, w\text{'}.\, ((Z = \text{``}w) \wedge (C = \text{``}z) \\
& \qquad\qquad \wedge ((w\text{'} = 1) \vee (w\text{'} = 2))) \\
& \Rightarrow (w\text{'} = 2) \\
= \quad & \text{false.}
\end{aligned}
$$

Martin [294] gives David Gries the credit for this example who used it to demonstrate the inconsistency of a number of procedure-call proof rules. The Euclid proof rule [278], for example gives $P = \text{true}$.

(194) EXAMPLE (Bijlsma, Matthews & Wiltink [49]) (*on the use of auxiliary variables*). Consider a procedure $\mathrm{Pn}(?X, !Z)$ for rounding the real number X to a nearby integer Z (such as $Z := \mathrm{floor}(X)$, $Z := \mathrm{ceil}(X)$ or $Z = \mathrm{round}(X)$). The specification $\{P'\}\mathrm{Pn}\{Q'\}$ where $P' = (m \leqslant X \leqslant m+1)$ and $Q' = (Z = m \vee Z = m+1)$ states that Z is obtained from X by rounding any non-integer X either up or down to an integer, and by using X itself if X happens to be an integer. Now consider the call $\mathrm{Pn}(A, C)$ with postcondition $Q = (C=0)$. The corresponding precondition P such that $\{P\}\ \mathrm{Pn}(A, C)\ \{Q\}$ holds is given by the rule of adaptation (192) as

$$
\begin{aligned}
P = \quad & \forall \text{``}u, \text{``}w, \text{``}z, w\text{'}. \\
& ((X = \text{``}u) \wedge (Z = \text{``}w) \wedge (C = \text{``}z) \wedge \\
& \ (\forall m. \\
& \ \ ((\exists \text{`}w.\, P'[X \leftarrow A[X \leftarrow \text{``}u][Z \leftarrow \text{``}w][C \leftarrow \text{``}z]][Z \leftarrow \text{`}w][C \leftarrow \text{``}z]) \\
& \ \ \ \Rightarrow (\exists u\text{'}, y\text{'}, z\text{'}.\, Q'[X \leftarrow u\text{'}][Z \leftarrow w\text{'}][C \leftarrow z\text{'}])))) \\
& \Rightarrow Q[C \leftarrow w\text{'}][X \leftarrow \text{``}u][Z \leftarrow \text{``}w] \\
= \quad & \forall w\text{'}.\, (\forall m.\, ((m \leqslant A \leqslant m+1) \Rightarrow (w\text{'} = m \vee w\text{'} = m+1))) \Rightarrow (w\text{'} = 0) \\
= \quad & (A = 0).
\end{aligned}
$$

This example was designed by Bijlsma, Matthews and Wiltink [49] to show that the precondition in the rule of [188, Theorem 12.4.1, p. 161] (which would be $P = \text{false}$) is not the weakest that can be inferred solely from the procedure's specification in cases when the specification involves auxiliary logical variables.

(195) EXAMPLE (*homonym formal and actual parameters*). Let Pg be $\mathrm{Pn}(?A, ?!B, !C) ::$ $C_1; C_2$ where C_1 is $(B := A + B; (A := A+1; (C := A+1; D := C+1)))$ and C_2 is $(C_3;$ $\mathrm{Pn}(A+1, B, B))$ with $C_3 = (A := 1; (B := 10; C := 100))$. By (97), the composition rule (99) and the consequence rule (102), we have $\{A=a \wedge B=b \wedge C=c\}\ C_1\ \{A=a+1 \wedge B=a+b \wedge C=a+2 \wedge D=a+3\}$, whence by recursion rule (191) we derive $\{P'\}$ Pn(?A, ?!B, !C) $\{Q'\}$ where P' is $(A=a \wedge B=b \wedge C=c)$ and Q' is $(A=a+1 \wedge B=a+b \wedge C=a+2 \wedge D=a+3)$. Let Q be $(A=a \wedge B=b \wedge C=c \wedge D=d)$. In order to derive the corresponding precondition P by the rule of adaptation (192) we observe that $X^{\cup}=\langle D\rangle$, $m^{\cup}=\langle a,b,c\rangle$, whence

$$\begin{aligned}
P= \quad & (\forall x. \forall \text{``}u, \text{``}v, \text{``}w, \text{``}y, \text{``}z, v', w'.\\
& \quad ((A=\text{``}u) \wedge (B=\text{``}v) \wedge (C=\text{``}w) \wedge (B=\text{``}y) \wedge (B=\text{``}z)\\
& \quad\ \wedge (\forall a, b, c.\\
& \qquad ((\exists \text{`}w.\, P'[A \leftarrow E[A \leftarrow \text{``}u][B \leftarrow \text{``}v][C \leftarrow \text{``}w][B \leftarrow \text{``}y][B \leftarrow \text{``}z]][B \leftarrow \text{``}v]\\
& \qquad\qquad [C \leftarrow \text{`}w][B \leftarrow \text{``}y][B \leftarrow \text{``}z])\\
& \qquad \Rightarrow (\exists u', y', z'.\, Q'[D \leftarrow x][A \leftarrow u'][B \leftarrow v'][C \leftarrow w'][B \leftarrow y'][B \leftarrow z'])))))\\
& \quad \Rightarrow Q[B \leftarrow w'][B \leftarrow v'][D \leftarrow x][A \leftarrow \text{``}u][B \leftarrow \text{``}v][C \leftarrow \text{``}w])\\
= \quad & (A=a=b-3=d-4 \wedge C=c).
\end{aligned}$$

(196) EXAMPLE (*recursive factorial procedure with value-result parameters*). Let us prove the partial correctness of the following program:

```
procedure F(?X, !Y);
begin
  {X=n}
  if X=0 then
    {X=n ∧ X=0} Y:=1
  else begin
    {X=n ∧ X≠0} F(X−1, Y); {X=n ∧ Y=(n−1)!} Y:= Y * X;
  end;
  {X=n ∧ Y=n!)
  end;
{X=y}F(X, Y); {X=y ∧ Y=y!}.
```

To prove partial correctness $\{X=n\}\ F(?X, !Y)\ \{X=n \wedge Y=n!\}$ of procedure F, we use recursion rule (191). Assuming $\{X=n\}\ F(?X, !Y)\{X=n \wedge Y=n!\}$ by induction hypothesis, we must prove $\{X=n \wedge X\neq 0\}\ F(X-1, Y)\ \{X=n \wedge Y=(n-1)!\}$ for the recursive call of F. By consequence rule (102) and rule of adaptation (192) (where $\{X^{\cup}\}=\{x^{\cup}\}=\emptyset$ and $m^{\cup}=n$, $P'=(X=n)$, $Q'=(X=n \wedge Y=n!)$ and $Q=(X=n \wedge Y=(n-1)!)$) we must show that

$$\begin{aligned}
& (X=n \wedge X\neq 0)\\
& \Rightarrow (\forall \text{``}u, \text{``}w, \text{``}z, w'.\\
& \quad ((X=\text{``}u) \wedge (Y=\text{``}w) \wedge (Y=\text{``}z)\\
& \quad \wedge (\forall n.\, ((\exists \text{`}w.\, P'[X \leftarrow (X-1)[X \leftarrow \text{``}u][Y \leftarrow \text{``}w][Y \leftarrow \text{``}z]][Y \leftarrow \text{`}w][Y \leftarrow \text{``}z])\\
& \qquad \Rightarrow (\exists u', z'.\, Q'[X \leftarrow u'][Y \leftarrow w'][Y \leftarrow z']))))\\
& \quad \Rightarrow Q[Y \leftarrow w'][X \leftarrow \text{``}u][Y \leftarrow \text{``}w]),
\end{aligned}$$

that is, after simplification,

$$(X=n \wedge X \neq 0) \Rightarrow (\forall``u, w'.((X=``u) \wedge (w'=(``u-1)!)) \Rightarrow (``u=n \wedge w'=(n-1)!))$$

which is obvious. Then the correctness poof of procedure F can be completed as shown by the above proof outline. Knowing that $\{X=n\}\ F(?X, !Y)\ \{X=n \wedge Y=n!\}$ holds, it remains to prove $\{X=y\}\ F(X, Y)\ \{X=y \wedge Y=y!\}$ for the main call of procedure F. This follows from the rule of adaptation (192) and the consequence rule (102), since $(X=y) \Rightarrow (\forall``u, ``w, w'.((X=``u) \wedge (Y=``w) \wedge (\forall n.((\exists`w.(X=n)[X \leftarrow ``u][Y \leftarrow `w]) \Rightarrow (\exists u'.(X=n \wedge Y=n!)[X \leftarrow u'][Y \leftarrow w'])))) \Rightarrow (X=y \wedge Y=y!)[Y \leftarrow w'][X \leftarrow ``u])$.

8.2.3. Complements on variable parameters and procedures as parameters

Apt's survey [13] on Hoare logic is mainly concerned with procedures and parameter mechanisms. More recent progress concerning procedures have been reported in [267, 330]. In particular, for a treatment of variable (reference, address, ...) parameters, see [210, 234, 138, 366, 109 (Chapter 9), 189, 13, 67, 164, 166, 371]. A separation of procedural abstraction and parameter passing is attempted in [308, 309, 310].

Clarke's theorem (169) [71] put a borderline to sound and relatively complete Hoare logics for programs with procedures. Clarke also claimed that the languages $\mathbf{L}_j$ (as defined in (166)) do have a sound and relatively complete Hoare logic. For $j \neq 4$ these claims were either proved in [71] or later established by Olderog [328]. These languages $\mathbf{L}_j, j \neq 4$, are easy to axiomatize since, for each program, there is a bound on the number of procedure environments (associating a procedure body to a procedure name) that can be reached during execution [330, 105, 106]. It follows that each of the procedure environments can be treated as a separate case. The case of $\mathbf{L}_4$ was more difficult because it can give rise to nonhomomorphic chains of procedure environments that grow arbitrarily long.

(197) EXAMPLE. Consider the following program written in an $\mathbf{L}_4$ subset of Pascal:

```
program L4;

  procedure P(X: integer; procedure Q(Z: integer));
    procedure L(X: integer); begin Q(X-1) end;
  begin if X>0 then P(X-1, L) else Q(X) end;

  procedure M;
    var N: integer;
    procedure R(X: integer); begin writeln(X) end;
  begin write("N= "); readln(N); P(N, R) end; {M}

begin M end.
```

The main procedure M reads the value n of N and calls $P(n, L^n)$ with $L^n = R$ which recursively calls $P(n-1, L^{n-1})$, $P(n-2, L^{n-2}), \ldots, P(1, L^1)$ and $P(0, L^0)$ thus creating a chain of procedures **procedure** $L^{i-1}(X\text{: integer})$; **begin** $L^i(X-1)$ **end;** for $i=n, \ldots, 1$ so that execution of $P(0, L^0)$ consists in calling $L^0(0)$ whence $L^1(-1), L^2(-2), \ldots, L^{n-1}(-n+1)$ and $L^n(-n)$, that is, $R(-n)$ which finally prints $-n$.

The first axiomatizations of Clarkes' language $\mathbf{L}_4$ proposed by Damm and Josko [105, 106] and Olderog [331, 332] used higher-order assertion languages to make assertions about this unbounded number of procedure environments (i.e., state-transformation): all concepts for Hoare logic are lifted from programs transforming states into states to programs transforming state-transformations into state-transformations. This leads to completeness not relative to the first-order theory of the interpretation. German, Clarke and Halpern [164, 166] were the first to propose a relative completeness proof for $\mathbf{L}_4$ with a first-order oracle and Cook's notion of expressiveness [86]. However, the logic extends Hoare logic by allowing quantifiers over first-order variables ($\forall x. H, \ldots$) and other logical connectives ($H \Rightarrow H', \ldots$) to be used on the level of Hoare formulae (an idea going back to Harel, Pnueli and Stavi [199] and also exploited by Sieber [371]), but is relatively complete only for Hoare correctness formulae H. Later Goerdt [173] proposed an indirect axiomatic proof method for $\mathbf{L}_4$ which involves the translation of programs into finitely typed lambda calculus and then the application of Goerdt's Hoare calculus [172]. This calculus was shown to be relatively complete (without Herbrand definability hypothesis) by Goerdt [174]. Hoare logic has also been extended to other types of languages with higher-order concepts in [139, 111, 172, 173, 174, 195, 266, 391].

8.3. *Undefinedness*

When the evaluation of expressions in the programming language can lead to runtime errors, it is necessary to deal with partially defined functions. Hoare logic can be easily extended when the domain of these partial functions is given (for example, in the case of array bounds outside their declared range) or can be easily defined (for example in the case of dangling pointers) as shown by Sites [372], German [162, 163], Coleman and Hugues [80], and Blikle [52]. Contrary to mathematicians, the computer scientists are often confronted with the problem of proving properties about partial objects without the knowledge of their domain. In this case first-order 2-valued logic is inadequate as pointed out by Ashcroft, Clint and Hoare [30] who have discovered an error in [78] for nonterminating functions. Alternatives consist in introducing an extra element $\perp$ in the data domain which, by the axiomatization, is forced to behave like a "divergent" or "undefined" data object [66] or, more generally, in using a 3-valued logic covering undefinedness in program proofs as proposed by Barringer, Cheng and Jones [36] and studied by Hoogewijs in [231].

8.4. *Aliasing and side-effects*

The assignment axiom (97) is incorrect when aliases (i.e., distinct identifiers designating a shared storage location) are allowed. One can hide the source of the problem by prohibiting interference between identifiers in the programming language or in the proofs [357, 360, 293] or consider augmented versions of Hoare logic which allow aliasing between variables [237, 366, 68, 328, 266, 391].

In the same way, the assignment axiom (97), the conditional (100) and the while (101) rule are incorrect when the evaluation of expressions can have side-effects [292].

A number of Hoare-like axiomatizations of expression languages or expressions with side-effects [102, 256, 353, 367] modify Hoare logic by introducing Hoare correctness formulae for expressions $\{P\}E\{Q\}$ where the precondition P and postcondition Q explicitly depend upon a distinguished variable standing for the value returned by the evaluation of expression E. Alternatives consist in considering other logics explicitly referring to the state of computation [292] in using the value of a programming language expression as the underlying primitive [54] or in transforming the program into procedural form [266].

8.5. *Block-structured local variables*

If we extend the syntax (188) of programs with block-structured local variables:

(198) $C ::= (\mathbf{var}\ X; C_1)$,

local variable X is like a bound variable of a predicate (see (116)) so that blocks satisfy the property that systematic replacement of the local variable X by some fresh variable Y preserves the meaning of the block: $(\mathbf{var}\ X; C)$ is equivalent to $(\mathbf{var}\ Y; C[X \leftarrow Y])$, provided that Y does not occur free in C. For example, according to this *static scope* rule, $(\mathbf{var}\ X; ((\mathbf{var}\ X; C_1)\, C_2))$ is equivalent to $(\mathbf{var}\ Y; ((\mathbf{var}\ Z; C_1[X \leftarrow Z]); C_2[X \leftarrow Y]))$ where $Y, Z, \notin Var(C_1, C_2)$. This leads to the following rule [210, 229, 12, 13, 109], where the renaming of X for Y is performed to distinguish between the occurrences of local X in C and possible free occurrences of nonlocal X in P or Q:

(199) $$\frac{\{P\}\ C[X \leftarrow Y]\ \{Q\}}{\{P\}\ (\mathbf{var}\ X; C)\ \{Q\}} \quad \textit{block rule}$$

where $Y \notin Free(P, Q) \wedge ((X = Y) \vee Y \notin Free(C))$.

Observe that rule (199) may necessitate alteration $C[X \leftarrow Y]$ of the program text C, but this can be avoided (naïve solutions such as [131] fail to treat the scope rule properly, see the discussion in [13, 145, 328, 329, 330, 331] for further details). Observe also that rule (199) does not take into account the possibility of running out of new storage locations [391]. There are other difficulties in defining the semantics of such blocks (see [109 (Chapter 6), 391, 196, 304] for a discussion). For example, mapping local variables into a global memory using a stack discipline is an overspecification, since then, for example, $\{\text{true}\}\ ((\mathbf{var}\ X;\ X := 0);\ (\mathbf{var}\ Z;\ Y := Z))\ \{Y = 0\}$ would hold because X and Z are allocated at the same address. Although this phenomenon can be observed in a number of implementations, it is contrary to the specification of block-structured languages where the initial value of local variables is usually *undetermined* so that, as shown by (199), $(\mathbf{var}\ X; C)$ should be equivalent to $(\mathbf{var}\ X; (X := ?; C))$, see [405] for example. However this introduces unbounded nondeterminism (see the corresponding difficulties in Subsection 8.10.2). Another way to cope with uninitialized local variables [13] is to use an extra uninitialized value $\omega \in D$ so that $(\mathbf{var}\ X; C)$ is equivalent to $(\mathbf{var}\ X; (X := \omega; C))$. But then one can prove $\{\text{true}\}\ ((\mathbf{var}\ X;\ Y := X);\ (\mathbf{var}\ X; Z := X))\ \{X = Y\}$, which is not true for most implementations where the initial value of local variables is undetermined. De Bakker [109] introduces simple conditions which guarantee that variables are initialized before being used. One can also force initialization upon declaration with the syntax (**var** $X := E; C_1$) (and use union types to allow for an initial value denoting logical uninitialization).

8.6. *Goto statements*

Goto's with static labels cause no problem with Floyd's partial correctness proof method [143]. For each label L, if $\{P_i,\ i=1,\ldots,n\}$ is the set of preconditions of the statements **goto** L in the program (including the postcondition of the command sequentially preceding the command labeled L) and Q is the precondition of the command labeled L, then Floyd's verification condition is simply $(\bigcup_{i=1,\ldots,n} P_i) \Rightarrow Q$. The difficulty with Hoare logic is to express this verification condition compositionally, by induction on the syntax of programs [326]. Various solutions have been proposed by Clint and Hoare [78], Donahue [131], Wang [401], Kowaltowski [256], De Bruin [114], and Lifschitz [273]. An inconsistency problem with goto rules in [78, 131, 256] was noticed by Arbib and Alagic [28] and by O'Donnell [326]. Scope problems with jumps out of procedures have not been explicitly dealt with (but for weakened forms of procedure escapes, see [145]).

Techniques similar to that used in theorem (169) have also been used to obtain incompleteness results for programming languages that include label variables with retention [71, 74].

8.7. *Functions and expressions (with side-effects)*

Functions are not called in order to change states—the realm that assertions can capture—but to return a value. In the proof rule proposed by Clint and Hoare [78] the result returned by the function call $f(x)$ (in the computer science sense) is simply denoted in predicates by $f(x)$ (in the mathematical sense). This excludes side-effects in functions (since, for example, the mathematical identity $f(x)+f(x)=2f(x)$ is not valid whenever a call to f increments x by 1). Moreover, Ashcroft, Clint and Hoare [30] have noticed that the proof rule of Clint and Hoare [78] yields an inconsistency whenever a defined function fails to halt for some possible argument, even if the value of the function is never computed for that argument, as explained in [326]. Hence the problem is again the one of undefinedness which is not correctly handled in predicates (see Subsection 8.3).

The problem can be circumvented by reduction of expression evaluation to execution of a sequence of assignment statements [111], by transforming the programmer-declared functions into procedures [266], or by modifying Hoare logic by explicitly introducing one or more symbols to denote expression values [102, 256, 353, 366], or by introducing primitive notations to make explicit assertions about the value of programming language expressions [53, 54, 378].

8.8. *Coroutines*

Clint [76] extended Hoare logic to simple coroutines (block-body and single coroutine combination) based upon the semi-coroutine concept of Simula 67

[402, 104]. It was further extended by Pritchard [352] to multiple coroutines and Dahl [103] to multiple dynamic instances of coroutines. They use auxiliary variables to accumulate the computation and communication history in stacks of arbitrary size. Clarke [73] showed that, in the case of simple static coroutines, history variables are useless and that auxiliary variables of bounded size simulating program counters are enough. Clint [77] also argues that history variables, although not necessary, can help for clarity and ease of verification.

Techniques similar to that used in theorem (169) have also been used to obtain incompleteness results for programming languages that include coroutines with local recursive procedures that can access global variables [71, 74].

8.9. *Parallel programs*

The controversy on the usefulness of program verification [117] can hardly be extended to parallel programs because their correctness, which is often very intricate, cannot be checked by nonreproducible and nonexhaustive tests. Therefore, clear programming notations as well as correctness proofs are indispensable, at least when designing the underlying basic algorithms [122, 11].

The evolution of Hoare logic for parallel programs is tightly coupled with the slow emergence of clear notations for expressing process synchronization and communication to which Hoare largely contributed, from shared variables [213, 215, 337, 338, etc.], then monitors [214, 233, 336] and finally synchronous message passing [217, 24, 271, etc.].

We discuss proof methods for parallel programs with shared variables and briefly survey the case of synchronous message passing in Hoare's communicating sequential processes CSP [217].

8.9.1. *Operational semantics of parallel programs with shared variables*

We consider (a simplified version of) parallel programs with shared global variables as introduced by Owicki [335] and by Owicki and Gries [337]:

(200) DEFINITION (*syntax of parallel programs*)

(200.1) Pp: **Papr** *parallel programs*

$$\text{Pp} ::= [C_1 \parallel C_2 \parallel \cdots \parallel C_n] \quad n \geqslant 2,$$

(200.2) C: **Com** *sequential commands*

$$C ::= \text{skip} \mid X := E \mid X := ? \mid (C_1; C_2) \mid (B \rightarrow C_1 \diamond C_2) \mid (B * C) \mid (B \text{ ¿ } C) \mid \surd.$$

Execution of a program "$[C_1 \parallel C_2 \parallel \cdots \parallel C_n]$" consists in executing processes $C_1, C_2, \ldots$ and C_n in parallel. These commands act upon implicitly declared shared global variables. Evaluation of a Boolean expression B, execution of assignments "$X := E$" or "$X := ?$" and execution of await commands "$(B \text{ ¿ } C)$" are *atomic* or *indivisible actions*, that is, no concurrent action can modify the value of the variables involved in this action. When a process attempts to execute an await command "$(B ¿ C)$", it is delayed until the condition B is true. Then the command C is executed as an indivisible action. Evaluation of B is part of the indivisible action so that another process may not change variables so as to make B false after B has been evaluated but before C begins execution. Upon termination of "$(B ¿ C)$", parallel processing continues. If two or more processes are waiting for the same condition B, any one of them may be allowed to proceed when B becomes true, while the others continue waiting. The order in which processes are scheduled is indifferent, for example a *weak fairness hypothesis* would be that no nonterminated, hence permanently enabled process C_k can be indefinitely delayed. A *strong fairness hypothesis* would be that no process can be indefinitely delayed if condition B can be infinitely often evaluated to true while that process is waiting; see [259, 268, 289, 149] for more details on fairness hypotheses. Owicki [335] and Owicki and Gries [337] assume that await commands $(B ¿ C)$ cannot be imbricated since this could lead to deadlocks (as in "[(true ¿ (false ¿ skip)) ∥ skip]"). Execution of a program "$[C_1 \parallel C_2 \parallel \cdots \parallel C_n]$" is terminated when all processes have finished their execution. We use the empty command "$\surd$" to denote termination.

To define the operational semantics of parallel programs, we introduce control states:

(201) Definition (*labels designating control states*). If Pp is $[C_1 \parallel C_2 \parallel \cdots \parallel C_n]$ then

(201.1) $\mathbf{At}[\![(B ¿ C)]\!] = \{(B ¿ C)\}$, $\mathbf{In}[\![(B ¿ C)]\!] = \emptyset$, $\mathbf{After}[\![(B ¿ C)]\!] = \{\surd\}$,

(201.2) $\mathbf{Lab}[\![\mathrm{Pp}]\!] = \mathbf{Lab}[\![C_1]\!] \times \cdots \times \mathbf{Lab}[\![C_n]\!]$,

(201.3) $\mathbf{PpLa} = \bigcup\{\mathbf{Lab}[\![\mathrm{Pp}]\!]: \mathrm{Pp} \in \mathbf{Papr}\}$,

(201.4) $\mathbf{At}[\![\mathrm{Pp}]\!] = \mathbf{At}[\![C_1]\!] \times \cdots \times \mathbf{At}[\![C_n]\!]$,

(201.5) $\mathbf{After}[\![\mathrm{Pp}]\!] = \mathbf{After}[\![C_1]\!] \times \cdots \times \mathbf{After}[\![C_n]\!]$,

(201.6) $\mathbf{In}[\![\mathrm{Pp}]\!] = \mathbf{Lab}[\![\mathrm{Pp}]\!] - \mathbf{At}[\![\mathrm{Pp}]\!] - \mathbf{After}[\![\mathrm{Pp}]\!]$.

The *operational semantics* of parallel programs is defined by interleaved executions of atomic actions as defined by (13.1) to (13.6):

(202) Definition (*operational semantics of parallel programs* (*compositional presentation*)

(202.1) $d: D$ *data*,

(202.2) $s: S = \mathbf{Var} \to D$ *states*,

(202.3) $\gamma: \Gamma = (S \times \mathbf{Com}) \cup (S \times \mathbf{PpLa})$ *configurations*,

(202.4) $op: (\mathbf{Com} \cup \mathbf{Papr}) \to \mathcal{P}(\Gamma \times \Gamma)$ *operational transition relation*,

(202.5) $op[\![(B ¿ C)]\!] = \{\langle\langle s, (B ¿ C)\rangle, s'\rangle: s \in \underline{B} \wedge \langle\langle s, C\rangle, s'\rangle \in op[\![C]\!]^*\}$,

(202.6) $op[\![\surd]\!] = \emptyset$,

(202.7) $op[\![[C_1 \parallel C_2 \parallel \cdots \parallel C_n]]\!]$
$= \{\langle\langle s, L\rangle, \langle s', L[k \leftarrow L'_k]\rangle\rangle : k \in \{1, \ldots, n\}$
$\wedge \langle\langle s, L_k\rangle, \langle s', L'_k\rangle\rangle \in op[\![C_k]\!]$
$\cup \{\langle\langle s, L\rangle, \langle s', L[k \leftarrow \surd]\rangle\rangle : k \in \{1, \ldots, n\} \wedge \langle\langle s, L_k\rangle, s'\rangle \in op[\![C_k]\!]\}$.

(203) EXAMPLE. We illustrate the proof methods for parallel programs on the following program Pp which increments the value of variable X by $a+b$ where a and b are given integer constants:

(204) $[X := X + a \quad \parallel \quad X := X + b].$
$|L_{11} \qquad |L_{12} \quad |L_{21} \qquad |L_{22}$

Its operational semantics is (for simplicity we write $\langle s, \langle L_1, \ldots, L_n\rangle\rangle$ as $\langle s, L_1, \ldots, L_n\rangle$):

$$\begin{aligned} op[\![\mathrm{Pp}]\!] = \; & \{\langle\langle s, L_{11}, L_{21}\rangle, \langle s[X \leftarrow s(X)+a], L_{12}, L_{21}\rangle\rangle : s \in S\} \\ & \cup \{\langle\langle s, L_{11}, L_{22}\rangle, \langle s[X \leftarrow s(X)+a], L_{12}, L_{22}\rangle\rangle : s \in S\} \\ & \cup \{\langle\langle s, L_{11}, L_{21}\rangle, \langle s[X \leftarrow s(X)+b], L_{11}, L_{22}\rangle\rangle : s \in S\} \\ & \cup \{\langle\langle s, L_{12}, L_{21}\rangle, \langle s[X \leftarrow s(X)+b], L_{12}, L_{22}\rangle\rangle : s \in S\} \end{aligned}$$

where $L_{11} = (X := X+1)^0$, $L_{12} = \surd^0$, $L_{21} = (X := X+1)^1$ and $L_{22} = \surd^1$. This program is partially correct $\underline{\{p\}\mathrm{Pp}\{q\}}$ for the specification $p = \{s \in S : s(X) = s(X)\}$ and $q = \{s \in S : s(X) = s(x) + a + b\}$.

By (44), the operational semantics (202) can also be given a stepwise presentation:

(205) LEMMA (operational semantics of parallel programs (stepwise presentation))

(205.1) $op[\![[C_1 \parallel C_2 \parallel \cdots \parallel C_n]]\!]$
$= \{\langle\langle s, L\rangle, \langle s', L[k \leftarrow L'_k]\rangle\rangle : k \in \{1, \ldots, n\} \wedge L_k \notin \mathbf{After}[\![C_k]\!]$
$\wedge s' \in \mathbf{NextS}[\![C_k]\!]\langle s, L_k\rangle \wedge L'_k \in \mathbf{NextL}[\![C_k]\!]\langle s, L_k\rangle\}$

where

(205.2) $\mathbf{Step}[\![C]\!][(\cdots(((B \text{¿} C'); C_2)\cdots; C_n)] = (B \text{¿} C')$,

(205.3) $\mathbf{Succ}[\![C]\!][(\cdots(((B \text{¿} C_1); C_2)\cdots; C_n)] = (\cdots(C_1; C_2)\cdots; C_n)$;

if $\mathbf{Step}[\![C]\!][L]$ *is* $(B \text{¿} C')$ *then*

(205.4) $\mathbf{NextS}[\![C]\!]\langle s, L\rangle = \{s' : s \in \underline{B} \wedge \langle\langle s, C'\rangle, s'\rangle \in op[\![C]\!]^*\}$,

(205.5) $\mathbf{NextL}[\![C]\!]\langle s, L\rangle = \{\mathbf{Succ}[\![C]\!][L] : s \in \underline{B}\}$.

(206) EXAMPLE (*stepwise operational semantics of parallel program* (204)). The stepwise presentation of the operational semantics of program $\mathrm{Pp} = [C_1 \parallel C_2]$ defined in (204) is specified by $C_1 = (X := X + a)$, $C_2 = (X := X + b)$, $\mathbf{At}[\![C_1]\!] = \{L_{11}\}$, $\mathbf{In}[\![C_1]\!] = \emptyset$, $\mathbf{After}[\![C_1]\!] = \{L_{12}\}$, $\mathbf{Lab}[\![C_1]\!] = \{L_{11}, L_{12}\}$, $\mathbf{At}[\![C_2]\!] = \{L_{21}\}$, $\mathbf{In}[\![C_2]\!] = \emptyset$, $\mathbf{After}[\![C_2]\!] = \{L_{22}\}$, $\mathbf{Lab}[\![C_2]\!] = \{L_{21}, L_{22}\}$, $\mathbf{NextS}[\![C_1]\!]\langle s, L_{11}\rangle = \{s[X \leftarrow s(X) + a]\}$,

NextL$[\![C_1]\!]\langle s, L_{11}\rangle = \{L_{12}\}$, **NextS**$[\![C_2]\!]\langle s, L_{21}\rangle = \{s[X \leftarrow s(X)+b]\}$, **NextL**$[\![C_2]\!]$ $\langle s, L_{21}\rangle = \{L_{22}\}$.

8.9.2. Proof methods à la Floyd for parallel programs with shared variables

We follow the style of presentation of Cousot and Cousot in [93]. From a semantical point of view, proofs of partial correctness $\underline{\{p\}\text{Pp}\{q\}}$ consist in discovering local invariants $I \in$ **Linv**$[\![\text{Pp}]\!]$ attached to control points, and in proving that they satisfy local verification conditions $\text{lvc}[\![\text{Pp}]\!][p, q](I)$ expressing that local invariants must remain true after execution of atomic actions. Hence, without complementary hypotheses on the set **Linv**$[\![\text{Pp}]\!]$ of local invariants we can assume that proof methods à la Floyd are of the form

(207) $\quad [\exists I \in \mathbf{Linv}[\![\text{Pp}]\!] . \text{lvc}[\![\text{Pp}]\!][p, q](I)]$.

Up to a connection (α, γ) between local invariants $I \in$ **Linv**$[\![\text{Pp}]\!]$ and global invariants $i \in \mathscr{P}(\Gamma)$:

(208) $\quad \alpha \in \mathscr{P}(\Gamma) \rightarrow \mathbf{Linv}[\![\text{Pp}]\!]$,

(209) $\quad \gamma \in \mathbf{Linv}[\![\text{Pp}]\!] \rightarrow \mathscr{P}(\Gamma)$

this proof method (207) consists in applying induction principle (29.1):

(210) $\quad [\exists i \in \mathscr{P}(\Gamma) [\![\; \text{gvc}[\![\text{Pp}]\!][p, q](i)]$.

This induction principle involves a single global invariant i on configurations (i.e., memory and control states), which is true for initial configurations satisfying p, remains true after each program step and implies q for final configurations:

(211) $\quad \text{gvc}[\![\text{Pp}]\!][p, q](i) =$

(211.1) $\quad\quad (\forall s \in p . \langle s, C_1, \ldots, C_n\rangle \in i)$

(211.2) $\quad\quad \wedge (i\rceil op[\![\text{Pp}]\!] \subseteq \Gamma \times i)$

(211.3) $\quad\quad \wedge (\forall \langle s, \surd, \ldots, \surd\rangle \in i . s \in q)$.

By theorem (29.1) this induction principle (210) is semantically sound ($\Leftarrow$) and complete ($\Rightarrow$):

(212) $\quad \underline{\{p\}\text{Pp}\{q\}} \Leftrightarrow [\exists i \in \mathscr{P}(\Gamma) . \; \text{gvc}[\![\text{Pp}]\!][p, q](i)]$.

The connection $i = \gamma(I)$ and $I = \alpha(i)$ between (207) and (210) expresses the fact that the global invariant i can be decomposed into local assertions I attached to control points. It induces a logical connection between local and global verification conditions:

(213) $\quad \forall I \in \mathbf{Linv}[\![\text{Pp}]\!] . \text{lvc}[\![\text{Pp}]\!][p, q](I) \Rightarrow \text{gvc}[\![\text{Pp}]\!][p, q](\gamma(I))$,

(214) $\quad \forall i \in \mathscr{P}(\Gamma) . \text{gvc}[\![\text{Pp}]\!][p, q](i) \Rightarrow \text{lvc}[\![\text{Pp}]\!][p, q](\alpha(i))$

which, together with (212), ensures the soundness and semantical completeness of (207):

(215) $\quad \underline{\{p\}\text{Pp}\{q\}} \Leftrightarrow [\exists I \in \mathbf{Linv}[\![\text{Pp}]\!] . \text{lvc}[p, q](I)]$.

This approach can also be used to formally construct proof methods by symbolic calculus: Given gvc⟦Pp⟧$[p, q]$, α and γ, we can let lvc⟦Pp⟧$[p, q](I)$ be gvc⟦Pp⟧$[p, q](\gamma(I))$ (so that the proof method is sound by construction) and check semantical completeness by showing that gvc⟦Pp⟧$[p, q](i) \Rightarrow$ lvc⟦Pp⟧$[p, q](\alpha(i)) =$ gvc⟦Pp⟧$[p, q](\gamma(\alpha(i)))$ (which is often obvious because gvc⟦Pp⟧$[p, q]$ is monotone and (α, γ) is a Galois connection so that $\forall i \in \mathscr{P}(\Gamma).\, i \Rightarrow \gamma(\alpha(i))$). For example, this point of view was applied to the design of a partial correctness proof method for CSP programs in [91].

We now explain a few partial correctness proof methods for parallel programs following these guidelines.

8.9.2.1. Using a single global invariant. Ashcroft's partial correctness proof method [29] and the one of Keller [247] (illustrated by Babich in [32]) consist in directly applying induction principle (29.1).

(216) EXAMPLE. To prove partial correctness of (204) by (210), we can use the global invariant

$$\begin{aligned} i = &\{\langle s, L_{11}, L_{21}\rangle : s(X) = s(x)\} \cup \{\langle s, L_{12}, L_{21}\rangle : s(X) = s(x) + a\} \\ &\cup \{\langle s, L_{11}, L_{22}\rangle : s(X) = s(x) + b\} \cup \{\langle s, L_{12}, L_{22}\rangle : s(X) = s(x) + a + b\} \end{aligned}$$

and show that (for all $s, s' \in S$)

$$\begin{aligned} &(s \in p) \Rightarrow (\langle s, L_{11}, L_{21}\rangle \in i), \\ &(\langle s, L_{11}, L_{21}\rangle \in i \wedge s' = s[X \leftarrow s(X) + a]) \Rightarrow (\langle s', L_{12}, L_{21}\rangle \in i), \\ &(\langle s, L_{11}, L_{22}\rangle \in i \wedge s' = s[X \leftarrow s(X) + a]) \Rightarrow (\langle s', L_{12}, L_{22}\rangle \in i), \\ &(\langle s, L_{11}, L_{21}\rangle \in i \wedge s' = s[X \leftarrow s(X) + b]) \Rightarrow (\langle s', L_{11}, L_{22}\rangle \in i), \\ &(\langle s, L_{12}, L_{21}\rangle \in i \wedge s' = s[X \leftarrow s(X) + b]) \Rightarrow (\langle s', L_{12}, L_{22}\rangle \in i), \\ &(\langle s, L_{12}, L_{22}\rangle \in i) \Rightarrow (s \in q). \end{aligned}$$

In the Ascroft–Keller method, $\mathbf{Linv}_{\mathrm{AK}}$⟦Pp⟧ is chosen as $\mathscr{P}(\Gamma)$ and $(\alpha_{\mathrm{AK}}, \gamma_{\mathrm{AK}})$ is identity. According to the operational semantics (205), the verification condition $(i\rceil op$⟦Pp⟧$\subseteq \Gamma \times i)$ of (210) can be decomposed into simpler verification conditions corresponding to each atomic action. Otherwise stated, $(i\rceil op$⟦Pp⟧$\subseteq \Gamma \times i)$ is equivalent to

(217) $$\begin{aligned} &\forall s, s' \in S.\, \forall L \in \mathbf{Lab}[\![\mathrm{Pp}]\!].\, \forall k \in \{1, \ldots, n\}. \\ &\quad (\langle s, L\rangle \in i \wedge L_k \notin \mathbf{After}[\![C_k]\!] \wedge s' \in \mathbf{NextS}[\![C_k]\!]\langle s, L_k\rangle \\ &\quad \wedge L'_k \in \mathbf{NextL}[\![C_k]\!]\langle s, L_k\rangle) \Rightarrow (\langle s', L[k \leftarrow L'_k]\rangle \in i). \end{aligned}$$

This means that starting with i true of $\langle s, L_1, \ldots, L_k, \ldots, L_n\rangle$ and executing any atomic action labeled L_k of any process C_k of the program (that is, a transition in that process C_k from configuration $\langle s, L_k\rangle$ to configuration $\langle s', L'_k\rangle$) leaves i true of $\langle s', L_1, \ldots, L'_k, \ldots, L_n\rangle$.

The difficulty with this method is that for large programs the single global invariant i tends to be unmanageable without being decomposable into simpler assertions.

8.9.2.2. Using an invariant on memory states for each control state. Early attempts towards the decomposition of the global invariant such as [31, 272] involves the transformation of the parallel program into an equivalent nondeterministic one upon which Floyd–Naur's partial correctness proof method is applied. As observed by Keller [247], this simply consists in using induction principle (210) with local invariants on memory states attached to each control state.

(218) Example (*partial correctness proof of program* (204) *by Ashcroft and Manna's method*). To prove partial correctness of parallel program (204) by (220), we can use the local invariants

$$\begin{aligned} I\langle L_{11}, L_{21}\rangle &= \{s \in S: s(X) = s(x)\},\\ I\langle L_{11}, L_{22}\rangle &= \{s \in S: s(X) = s(x) + b\},\\ I\langle L_{12}, L_{21}\rangle &= \{s \in S: s(X) = s(x) + a\},\\ I\langle L_{12}, L_{22}\rangle &= \{s \in S: s(X) = s(x) + a + b\}\end{aligned}$$

and show that (for all $s, s' \in S$):

- initialization (220.1):

$$(s \in p) \Rightarrow (s \in I\langle L_{11}, L_{21}\rangle);$$

- induction (220.2):

$$\begin{aligned} &(s \in I\langle L_{11}, L_{21}\rangle \wedge s' = s[X \leftarrow s(X) + a]) \Rightarrow (s' \in I\langle L_{12}, L_{21}\rangle),\\ &(s \in I\langle L_{11}, L_{22}\rangle \wedge s' = s[X \leftarrow s(X) + a]) \Rightarrow (s' \in I\langle L_{12}, L_{22}\rangle),\\ &(s \in I\langle L_{11}, L_{21}\rangle \wedge s' = s[X \leftarrow s(X) + b]) \Rightarrow (s' \in I\langle L_{11}, L_{22}\rangle),\\ &(s \in I\langle L_{12}, L_{21}\rangle \wedge s' = s[X \leftarrow s(X) + b]) \Rightarrow (s' \in I\langle L_{12}, L_{22}\rangle);\end{aligned}$$

- finalization (220.3):

$$(s \in I\langle L_{12}, L_{22}\rangle) \Rightarrow (s \in q).$$

To formally construct this proof method, we exactly follow the development of Subsection 5.2 for the Floyd–Naur method. Then we introduce local invariants and their connection with the global invariant of (210):

(219) Definition (*connection between local and global invariants*)

(219.1) $\quad \mathbf{Linv}_{\mathbf{AM}}[\![\mathrm{Pp}]\!] = \mathbf{Lab}[\![\mathrm{Pp}]\!] \rightarrow \mathbf{Ass} \quad (\text{where } \mathbf{Ass} = \mathscr{P}(S)),$

(219.2) $\quad \alpha_{\mathbf{AM}}: \mathscr{P}(\Gamma) \rightarrow \mathbf{Linv}[\![\mathrm{Pp}]\!], \quad \alpha(i)(L) = \{s: \langle s, L\rangle \in i\},$

(219.3) $\quad \gamma_{\mathbf{AM}}: \mathbf{Linv}[\![\mathrm{Pp}]\!] \rightarrow \mathscr{P}(\Gamma), \quad \gamma(I) = \{\langle s, L\rangle: L \in \mathbf{Lab}[\![\mathrm{Pp}]\!] \wedge s \in I(L)\}.$

Then we derive local verification conditions by $\mathrm{lvc}[\![\mathrm{Pp}]\!][p, q](I) = \mathrm{gvc}[\![\mathrm{Pp}]\!][p, q]$ $(\gamma_{\mathbf{AM}}(I))$:

(220) Theorem (Ashcroft–Manna partial correctness proof method for parallel programs). *If* Pp *is* $[C_1 \parallel C_2 \parallel \cdots \parallel C_n]$ *then* $\underline{\{p\}}\underline{\text{Pp}}\underline{\{q\}}$ *holds if and only if there exists* $I \in \mathbf{Lab}[\![\text{Pp}]\!] \to \mathbf{Ass}$ *such that*

(220.1) *if* $L \in \mathbf{At}[\![\text{Pp}]\!]$, *then* $p \subseteq I(L)$;

(220.2) *if* $s, s' \in S$, $L \in \mathbf{At}[\![\text{Pp}]\!] \cup \mathbf{In}[\![\text{Pp}]\!]$, $k \in \{1, \ldots, n\}$, *then*

$$(s \in I(L) \wedge L_k \notin \mathbf{After}[\![C_k]\!] \wedge s' \in \mathbf{NextS}[\![C_k]\!]\langle s, L_k\rangle \\ \wedge L'_k \in \mathbf{NextL}[\![C_k]\!]\langle s, L_k\rangle) \Rightarrow (s' \in I(L[k \leftarrow L'_k]));$$

(220.3) *if* $L \in \mathbf{After}[\![\text{Pp}]\!]$, *then* $I(L) \subseteq q$.

Otherwise stated, starting with $I\langle L_1, \ldots, L_k, \ldots, L_n\rangle$ true of s and executing any atomic action labeled L_k of any process C_k of the program (that is, a transition in that process C_k from configuration $\langle s, L_k\rangle$ to configuration $\langle s', L'_k\rangle$) leaves $I\langle L_1, \ldots, L'_k, \ldots, L_n\rangle$ true of s'. These local verification conditions can further be detailed as in (45) for each possible kind of atomic action. In particular, when the atomic action $\mathbf{Step}[\![C_k]\!][L_k]$ labeled L_k is an await command $(B ¿ C)$ with next label $L'_k = \mathbf{Succ}[\![C_k]\!][L_k]$ we must prove $\underline{\{I(L) \cap B\}}\underline{C}\underline{\{I(L[k \leftarrow L'_k])\}}$ to which Floyd's stepwise partial correctness proof method (45) is directly applicable.

As shown by the success of the Floyd–Naur partial correctness proof method, this approach is very well suited for sequential programs. However, for parallel programs $\text{Pp} = [C_1 \parallel C_2 \parallel \cdots \parallel C_n]$ the number of local invariants $I(L)$, $L \in \mathbf{Lab}[\![\text{Pp}]\!]$, grows as the number $|\mathbf{Lab}[\![\text{Pp}]\!]|$ of control states, that is, as the product $|\mathbf{Lab}[\![C_1]\!]| \cdot \mathbf{Lab}[\![C_2]\!]| \cdot \cdots \cdot |\mathbf{Lab}[\![C_n]\!]|$ of the sizes of the processes $C_1, C_2, \ldots, C_n$. Apart from trivial programs, this exponential explosion is rapidly unmanageable.

8.9.2.3. Using an invariant on memory states for each program point. In paragraph 8.9.2.2, the Floyd–Naur partial correctness proof method was generalized to parallel programs by using an invariant on memory states for each control state $\langle L_1, \ldots, L_n\rangle \in \mathbf{Lab}[\![\text{Pp}]\!]$. For sequential programs this is equivalent to the use of an invariant $I(L) \in \mathbf{Ass}$ on memory states for each program point $L \in \mathbf{Prpt}[\![\text{Pp}]\!]$. These two points of view do not coincide for parallel programs. So, as first suggested by Ashcroft [29], the Floyd–Naur proof method can also be generalized to parallel programs using an invariant on memory states for each program point. This consists in the following definitions:

(221) DEFINITION (*connection between local and global invariants*)

(221.1) $\mathbf{Prpt}[\![[C_1 \parallel \cdots \parallel C_n]]\!] = \bigcup\{\mathbf{Lab}[\![C_k]\!] : k \in \{1, \ldots, n\}\}$,

(221.2) $\mathbf{Linv}_{\mathrm{A}}[\![\text{Pp}]\!] = \mathbf{Prpt}[\![\text{Pp}]\!] \to \mathbf{Ass}$,

(221.3) $\alpha_{\mathrm{A}} : \mathscr{P}(\Gamma) \to \mathbf{Linv}_{\mathrm{A}}[\![\text{Pp}]\!]$, $\alpha_{\mathrm{A}}(i)(l) = \{s : \exists L \in \mathbf{Lab}[\![\text{Pp}]\!] . \langle s, L[k \leftarrow l]\rangle \in i\}$,

(221.4) $\gamma_{\mathrm{A}} : \mathbf{Linv}_{\mathrm{A}}[\![\text{Pp}]\!] \to \mathscr{P}(\Gamma)$, $\gamma_{\mathrm{A}}(I) = \{\langle s, L\rangle : \forall j \in \{1, \ldots, n\} . s \in I(L_j)\}$.

The induction step $(\gamma_{\mathrm{A}}(I) \rceil op[\![\text{Pp}]\!] \subseteq \Gamma \times \gamma_{\mathrm{A}}(I))$ of the corresponding verification conditions $\text{lvc}[\![\text{Pp}]\!][p, q](I) = \text{gvc}[\![\text{Pp}]\!][p, q](\gamma_{\mathrm{A}}(I))$ can be, following Owicki and Gries [337], decomposed into a sequential proof and a proof of interference freedom (also called "monotony condition" in [258]). Sequential correctness asserts that executing any atomic action labeled L of any process C_k of the program (that is, a transition of that process C_k from configuration $\langle s, L\rangle$ to configuration $\langle s', L'\rangle$) when starting with $I(L)$ true of s makes $I(L')$ true of s'. Interference freedom asserts that for every label L'' in a different process C_m, starting with both $I(L'')$ and $I(L)$ true of s leaves $I(L'')$ true of s':

(222) THEOREM (incomplete partial correctness proof method for parallel programs). *If* Pp *is* $[C_1 \parallel C_2 \parallel \cdots \parallel C_n]$ *then* $\underline{\{p\}}\underline{\text{Pp}}\underline{\{q\}}$ *holds if there exists* $I \in \textbf{Prpt}[\![\text{Pp}]\!] \rightarrow \textbf{Ass}$ *such that*

(222.1) *if* $k \in \{1, \ldots, n\}$ *and* $L \in \textbf{At}[\![C_k]\!]$, *then* $p \subseteq I(L)$;

(222.2) *if* $s, s' \in S$, $k \in \{1, \ldots, n\}$, $L \in \textbf{At}[\![C_k]\!] \cup \textbf{In}[\![C_k]\!]$ *and* $L' \in \textbf{Lab}[\![C_k]\!]$, *then*

$$[s \in I(L) \wedge s' \in \textbf{NextS}[\![C_k]\!]\langle s, L\rangle \wedge L' \in \textbf{NextL}[\![C_k]\!]\langle s, L\rangle] \Rightarrow s' \in I(L');$$

(222.3) *if* $s, s' \in S$, $k \in \{1, \ldots, n\}$, $L \in \textbf{At}[\![C_k]\!] \cup \textbf{In}[\![C_k]\!]$, $m \in \{1, \ldots, n\} - \{k\}$ *and* $L'' \in \textbf{Lab}[\![C_m]\!]$, *then*

$$[s \in I(L'') \wedge s \in I(L) \wedge s' \in \textbf{NextS}[\![C_k]\!]\langle s, L\rangle] \Rightarrow s' \in I(L'');$$

(222.4) $\bigcap\{I(L): \exists k \in \{1, \ldots, n\}. L \in \textbf{After}[\![C_k]\!]\} \subseteq q$,

but the reciprocal is not true.

(223) EXAMPLE (*partial correctness proof of program* (204) *with method* (222)). Assuming $a=1$ and $b=2$, we can apply (222) to prove partial correctness of parallel program (204). The verification conditions are the following:

- initialization (222.1):

$$p \subseteq I(L_{11}), \qquad p \subseteq I(L_{21});$$

- sequential correctness (222.2):

$$s \in I(L_{11}) \Rightarrow s[X \leftarrow s(X)+1] \in I(L_{12}),$$
$$s \in I(L_{21}) \Rightarrow s[X \leftarrow s(X)+2] \in I(L_{22});$$

- interference freedom (222.3):

$$s \in I(L_{11}) \wedge s \in I(L_{21}) \Rightarrow s[X \leftarrow s(X)+2] \in I(L_{11}),$$
$$s \in I(L_{12}) \wedge s \in I(L_{21}) \Rightarrow s[X \leftarrow s(X)+2] \in I(L_{12}),$$
$$s \in I(L_{21}) \wedge s \in I(L_{11}) \Rightarrow s[X \leftarrow s(X)+1] \in I(L_{21}),$$
$$s \in I(L_{22}) \wedge s \in I(L_{11}) \Rightarrow s[X \leftarrow s(X)+1] \in I(L_{22});$$

- finalization (222.4):

$$I(L_{12}) \cap I(L_{22}) \subseteq q.$$

They are satisfied by the following local invariants:

$$I[L_{11}] = \{s \in S: s(X) = s(x) \vee s(X) = s(x)+2\},$$
$$I[L_{12}] = \{s \in S: s(X) = s(x)+1 \vee s(X) = s(x)+3\},$$
$$I[L_{21}] = \{s \in S: s(X) = s(x) \vee s(X) = s(x)+1\},$$
$$I[L_{22}] = \{s \in S: s(X) = s(x)+2 \vee s(X) = s(x)+3\}.$$

Using example (204) with $a=b=1$, Keller [247], and Owicki and Gries [337] have shown that the corresponding proof method is semantically incomplete:

(224) COUNTEREXAMPLE (*incompleteness of method* (222) *for program* (204)). Formally, the strongest global invariant $\underline{i}$ satisfying the global verification condition gvc⟦Pp⟧$[p, q]$ in (210) for program (204) is given by its fixpoint characterization (30) as

$$\begin{aligned}\underline{i} &= \mathit{lfp}\ \lambda X:\mathscr{P}(\Gamma).\ \{\langle s, C_1,\ldots,C_n\rangle:\ s\in p\}\cup\{\gamma:\ \exists\gamma'.\ \langle\gamma',\gamma\rangle\in X\rceil op[\![\mathrm{Pp}]\!]\}\\ &=\{\langle s, L_{11}, L_{21}\rangle:\ s(X)=s(x)\}\cup\{\langle s, L_{12}, L_{21}\rangle:\ s(X)=s(x)+a\}\\ &\quad\cup\{\langle s, L_{11}, L_{22}\rangle:\ s(X)=s(x+b\}\cup\{\langle s, L_{12}, L_{22}\rangle: s(X)=s(x)+a+b\}.\end{aligned}$$

It follows by monotony, that the strongest local invariants are $\underline{I}=\alpha_{\mathrm{A}}(\underline{i})$, that is to say,

$$\begin{aligned}&\underline{I}(L_{11})=\{s\in S\colon s(X)=s(x)\vee s(X)=s(x)+b\},\\ &\underline{I}(L_{12})=\{s\in S\colon s(X)=s(x)+a\vee s(X)=s(x)+a+b\},\\ &\underline{I}(L_{21})=\{s\in S\colon s(X)=s(x)\vee s(X)=s(x)+a\},\\ &\underline{I}(L_{22})=\{s\in S\colon s(X)=s(x)+b\vee s(X)=s(x)+a+b\}.\end{aligned}$$

When $a=b=1$, they are too weak to satisfy the interference freedom (222.3) and finalization (222.4) verification conditions as given in example (223).

8.9.2.4. Using an invariant on memory and control states for each program point. The use of an invariant on memory states for each program point is incomplete because the relation between memory and control states is lost. This can be avoided by using local invariants $I(L_k)$ attached to program points L_k of each process C_k of the program Pp specifying the relation between the memory state and the control state of other processes:

(225) $$\begin{aligned}&\mathbf{Linv}_{\mathrm{L}}[\![\mathrm{Pp}]\!]\\ &=\{I\in\mathbf{Prpt}[\![\mathrm{Pp}]\!]\to\mathscr{P}(S\times\mathbf{Prpt}[\![\mathrm{Pp}]\!]^{n-1})\colon \forall k\in\{1,\ldots,n\}.\ \forall L_k\in\mathbf{Lab}[\![C_k]\!].\\ &\quad I(L_k)\subseteq S\times\mathbf{Lab}[\![C_1]\!]\times\cdots\times\mathbf{Lab}[\![C_{k-1}]\!]\times\mathbf{Lab}[\![C_{k+1}]\!]\times\cdots\times\mathbf{Lab}[\![C_n]\!]\}.\end{aligned}$$

This way of expressing the global invariant can be extirpated from Newton [325] and is clear in [258]. These local invariants can be written as a proof outline (as in Subsection 7.2.4 and [337]) in which a predicate P_k representing $I(L_k)$ is attached to control point L_k. Control predicates à la Lamport [258, 260] and Lamport and Schneider [264] can be used in P_k to explicitly mention the control state. Such control predicates can, to some extent, be defined in a language-independent way [96].

EXAMPLE (*proof outline for parallel program* (204)). To prove partial correctness of parallel program (204), we can use the following local invariants:

(226) $$\begin{aligned}&I[L_{11}]=\{\langle s, L_{21}\rangle:\ s(X)=s(x)\}\cup\{\langle s, L_{22}\rangle:\ s(X)=s(x)+b\},\\ &I[L_{12}]=\{\langle s, L_{21}\rangle:\ s(X)=s(x)+a\}\cup\{\langle s, L_{22}\rangle:\ s(X)=s(x)+a+b\},\\ &I[L_{21}]=\{\langle s, L_{11}\rangle:\ s(X)=s(x)\}\cup\{\langle s, L_{12}\rangle:\ s(X)=s(x)+a\},\\ &I[L_{22}]=\{\langle s, L_{11}\rangle:\ s(X)=s(x)+b\}\cup\{\langle s, L_{12}\rangle:\ s(X)=s(x)+a+b\}\end{aligned}$$

which can be specified by the following proof outline:

(227) $\{X=x\}$

$[L_{11}$: $\{(\text{at}(L_{21}) \wedge X=x) \vee (\text{at}(L_{22}) \wedge X=x+b)\}$

$\quad \mathbf{X := X+a}$

$\quad L_{12}$: $\{(\text{at}(L_{21}) \wedge X=x+a) \vee (\text{at}(L_{22}) \wedge X=x+a+b)\}$

$\| \; L_{21}$: $\{(\text{at}(L_{11}) \wedge X=x) \vee (\text{at}(L_{12}) \wedge X=x+a)\}$

$\quad \mathbf{X := X+b}$

$\quad L_{22}$: $\{(\text{at}(L_{11}) \wedge X=x+b) \vee (\text{at}(L_{12}) \wedge X=x+a+b)\}]$

$\{X=x+a+b\}$.

This decomposition of the global invariant can be specified by the following connection between local and global invariants (recall that $\langle L_1,\ldots,L_n\rangle_{\sim k}$ is $\langle L_1,\ldots, L_{k-1}, L_{k+1},\ldots,L_n\rangle$):

(228) DEFINITION (*connection between local and global invariants*)

(228.1) $\alpha_L : \mathscr{P}(\Gamma) \to \mathbf{Linv}_L[\![\text{Pp}]\!]$, $\quad \alpha_L(i)(L_k) = \{\langle s, L_{\sim k}\rangle : \langle s, L\rangle \in i\}$,

(228.2) $\gamma_L : \mathbf{Linv}_L[\![\text{Pp}]\!] \to \mathscr{P}(\Gamma)$, $\quad \gamma_L(I) = \{\langle s, L\rangle : \exists k \in \{1,\ldots,n\}.\, \langle s, L_{\sim k}\rangle \in I(L_k)\}$

Observe that (α_L, γ_L) is a bijection between $\mathscr{P}(\Gamma)$ and $\mathbf{Linv}_L[\![\text{Pp}]\!]$ which ensures the soundness and semantical completeness of the derived proof method.

8.9.2.4.1. The strengthened Lamport and Owicki–Gries method. The local verification conditions $\text{lvc}[\![\text{Pp}]\!][p, q](I) = \text{gvc}[\![\text{Pp}]\!][p, q](\gamma_L(I))$ corresponding to the above definition of $\mathbf{Linv}_L[\![\text{Pp}]\!]$ and (α_L, γ_L) were first designed by Cousot and Cousot [93] and later by Lamport [263]. The proof method is similar to that of [258, 337] but for the fact that it is strengthened by allowing the use of the proof outline of the processes not involved in the sequential or interference freedom proof.

(229) EXAMPLE. To prove partial correctness of parallel program (204), we can use local invariants (226) and check the following local verification conditions (for all $s, s' \in S$, $c_1 \in \{L_{11}, L_{12}\}$ and $c_2 \in \{L_{21}, L_{22}\}$):

- *Initialization* (230.1): The initialization step $(\forall s \in p.\, \langle s, C_1,\ldots,C_n\rangle \in \gamma_A(I))$ states that the input specification p implies the invariants attached to entry points of the processes $C_1,\ldots,C_n$ of Pp:

$$s \in p \Rightarrow \langle s, L_{21}\rangle \in I(L_{11}), \qquad s \in p \Rightarrow \langle s, L_{11}\rangle \in I(L_{21}).$$

- *Induction step*: The induction step $(\gamma_A(I) \rceil op[\![\text{Pp}]\!] \subseteq \Gamma \times \gamma_A(I))$ must be checked for all atomic actions of all processes C_k of program Pp, that is, all transitions of that process C_k from configuration $\langle s, L\rangle$ to configuration $\langle s', L'\rangle$. The proof can be decomposed into sequential and interference freedom proofs:

(a) *Sequential correctness* (230.2): To prove sequential correctness we must show that, starting with $I(L)$ true of $\langle s, L_1, \ldots, L_{k-1}, L_{k+1}, \ldots, L_n\rangle$ (as well as the assertions $I(L_j)$ attached to control points L_j of the other processes $C_j, j \in \{1, \ldots, n\} - \{k\}$, true of $\langle s, L_1, \ldots, L_{j-1}, L_{j+1}, \ldots, L_{k-1}, L, L_{k+1}, \ldots, L_n\rangle$), makes $I(L')$ true of $\langle s', L_1, \ldots, L_{k-1}, L_{k+1}, \ldots, L_n\rangle$:

$$\begin{aligned}
&[\langle s, c_2\rangle \in I(L_{11}) \wedge \langle s, L_{11}\rangle \in I(L_{21}) \cup I(L_{22})]\\
&\quad \Rightarrow \langle s[X \leftarrow s(X)+a], c_2\rangle \in I(L_{12}),\\
&[\langle s, c_1\rangle \in I(L_{21}) \wedge \langle s, L_{21}\rangle \in I(L_{11}) \cup I(L_{12})]\\
&\quad \Rightarrow \langle s[X \leftarrow s(X)+b], c_1\rangle \in I(L_{22}).
\end{aligned}$$

(b) *Interference freedom* (230.3): Interference freedom asserts that for every label L'' in a different process C_m, $m \in \{1, \ldots, n\} - \{k\}$, starting with both $I(L)$ true of $\langle s, L_1, \ldots, L_{m-1}, L'', L_{m+1}, \ldots, L_{k-1}, L_{k+1}, \ldots, L_n\rangle$ and $I(L'')$ true of $\langle s, L_1, \ldots, L_{m-1}, L_{m+1}, \ldots, L_{k-1}, L, L_{k+1}, \ldots, L_n\rangle$ (as well as the assertions $I(L_j)$ attached to control points L_j of the other processes C_j, $j \in \{1, \ldots, n\} - \{k, m\}$, true of $\langle s, L_1, \ldots, L_{m-1}, L'', L_{m+1}, \ldots, L_{j-1}, L_{j+1}, \ldots, L_{k-1}, L, L_{k+1}, \ldots, L_n\rangle$), leaves $I(L'')$ true of $\langle s', L_1, \ldots, L_{m-1}, L_{m+1}, \ldots, L_{k-1}, L', L_{k+1}, \ldots, L_n\rangle$:

$$\begin{aligned}
&[\langle s, c_2\rangle \in I(L_{11}) \wedge \langle s, L_{11}\rangle \in I(L_{21})] \Rightarrow \langle s[X \leftarrow s(X)+b], c_2\rangle \in I(L_{11}),\\
&[\langle s, c_2\rangle \in I(L_{12}) \wedge \langle s, L_{12}\rangle \in I(L_{21})] \Rightarrow \langle s[X \leftarrow s(X)+b], c_2\rangle \in I(L_{12}),\\
&[\langle s, c_1\rangle \in I(L_{21}) \wedge \langle s, L_{21}\rangle \in I(L_{11})] \Rightarrow \langle s[X \leftarrow s(X)+a], c_1\rangle \in I(L_{21}),\\
&[\langle s, c_1\rangle \in I(L_{22}) \wedge \langle s, L_{22}\rangle \in I(L_{11})] \Rightarrow \langle s[X \leftarrow s(X)+a], c_1\rangle \in I(L_{22}).
\end{aligned}$$

- *Finalization* (230.4): The final step of the partial correctness proof $(\forall \langle s, [\surd \parallel \cdots \parallel \surd]\rangle \in \gamma_{\mathrm{A}}(I).\, s \in q)$ shows that final states satisfy the output specification q:

$$[\langle s, L_{22}\rangle \in I(L_{12}) \wedge \langle s, L_{12}\rangle \in I(L_{22})] \Rightarrow s \in q.$$

This proof method $[\exists I \in \mathbf{Linv}_{\mathrm{L}}[\![\mathrm{Pp}]\!].\, \mathrm{lvc}_{\mathrm{SLO}}[\![\mathrm{Pp}]\!][p, q](I)]$ directly follows from the choice $\mathrm{lvc}_{\mathrm{SLO}}[\![\mathrm{Pp}]\!][p, q](I) = \mathrm{gvc}[\![\mathrm{Pp}]\!][p, q](\gamma_{\mathrm{L}}(I))$ with $\mathrm{gvc}[\![\mathrm{Pp}]\!][p, q](i)$ defined by (211) and operational semantics (205):

(230) THEOREM (Cousot & Cousot [93]) (strengthened Lamport and Owicki–Gries method). *$\{p\}\underline{\mathrm{Pp}}\{q\}$ where* Pp *is $[C_1 \parallel C_2 \parallel \cdots \parallel C_n]$ is true if and only if their exists $I \in \mathbf{Linv}_{\mathrm{L}}[\![\mathrm{Pp}]\!]$ such that for all $k \in \{1, \ldots, n\}$, $m \in \{1, \ldots, n\} - \{k\}$ and $s, s' \in S$,*

(230.1) *if $l \in \mathbf{At}[\![C_k]\!]$ and $L \in \mathbf{At}[\![\mathrm{Pp}]\!]$, then $s \in p \Rightarrow \langle s, L_{\sim k}\rangle \in I(l)$;*

(230.2) *if $l \in \mathbf{At}[\![C_k]\!] \cup \mathbf{In}[\![C_k]\!]$ and $L \in \mathbf{Lab}[\![\mathrm{Pp}]\!]$, then*

$$\begin{aligned}
&[\langle s, L_{\sim k}\rangle \in I(l) \wedge \forall j \in \{1, \ldots, n\} - \{k\}.\, \langle s, L[k \leftarrow l]\rangle_{\sim j} \in I(L_j)\\
&\wedge s' \in \mathbf{NextS}[\![C_k]\!]\langle s, l\rangle \wedge l' \in \mathbf{NextL}[\![C_k]\!]\langle s, l\rangle] \Rightarrow \langle s', L_{\sim k}\rangle \in I(l');
\end{aligned}$$

(230.3) *if $l \in \mathbf{At}[\![C_k]\!] \cup \mathbf{In}[\![C_k]\!]$, $l'' \in \mathbf{Lab}[\![C_m]\!]$ and $L \in \mathbf{Lab}[\![\mathrm{Pp}]\!]$, then*

$$\begin{aligned}
&[\langle s, L[k \leftarrow l]_{\sim m}\rangle \in I(l'') \wedge \langle s, L[m \leftarrow l'']_{\sim k}\rangle \in I(l) \wedge \forall j \in \{l, \ldots, n\} - \{k, m\}.\\
&\langle s, L[m \leftarrow l''][k \leftarrow l]_{\sim m}\rangle \in I(L_j) \wedge s' \in \mathbf{NextS}[\![C_k]\!]\langle s, l\rangle \wedge l' \in \mathbf{NextL}[\![C_k]\!]\langle s, l\rangle]\\
&\Rightarrow \langle s', L[k \leftarrow l']_{\sim m}\rangle \in I(l'');
\end{aligned}$$

(230.4) *if* $L \in \mathbf{After}[\![\mathrm{Pp}]\!]$, *then* $[\forall k \in \{1, \ldots, n\}. \langle s, L_{\sim k}\rangle \in I(L_k)] \Rightarrow s \in q$.

8.9.2.4.2. Newton's method. Newton's proof method [325] (although designed for quite a different definition of concurrent programs) is a weakened version of Lamport's, and Owicki and Gries's method [258, 337] in the sense that interference freedom simply consists in proving that any local invariant $I(L'')$, $L'' \in \mathbf{Lab}[\![C_m]\!]$, remains true after execution of any atomic action labeled L of any other process C_k, $k \neq m$, (and this without assuming that $I(L)$ holds before executing this atomic action).

(231) EXAMPLE (*partial correctness proof of program* (204) *by Newton's method*). To prove partial correctness of parallel program (204), we can use local invariants (226) and check the following local verification conditions (for all $s, s' \in S$, $c_1 \in \{L_{11}, L_{12}\}$ and $c_2 \in \{L_{21}, L_{22}\}$):

- initialization (232.1):

$$s \in p \Rightarrow \langle s, L_{21}\rangle \in I(L_{11}), \qquad s \in p \Rightarrow \langle s, L_{11}\rangle \in I(L_{21});$$

- sequential correctness (232.2):

$$\langle s, c_2\rangle \in I(L_{11}) \Rightarrow \langle s[X \leftarrow s(X)+a], c_2\rangle \in I(L_{12}),$$
$$\langle s, c_1\rangle \in I(L_{21}) \Rightarrow \langle s[X \leftarrow s(X)+b], c_1\rangle \in I(L_{22});$$

- interference freedom (232.3):

$$\langle s, c_2\rangle \in I(L_{11}) \Rightarrow \langle s[X \leftarrow s(X)+b], c_2\rangle \in I(L_{11}),$$
$$\langle s, c_2\rangle \in I(L_{12}) \Rightarrow \langle s[X \leftarrow s(X)+b], c_2\rangle \in I(L_{12}),$$
$$\langle s, c_1\rangle \in I(L_{21}) \Rightarrow \langle s[X \leftarrow s(X)+a], c_1\rangle \in I(L_{21}),$$
$$\langle s, c_1\rangle \in I(L_{22}) \Rightarrow \langle s[X \leftarrow s(X)+a], c_1\rangle \in I(L_{22});$$

- finalization (232.4):

$$[\langle s, L_{22}\rangle \in I(L_{12}) \wedge \langle s, L_{12}\rangle \in I(L_{22})] \Rightarrow s \in q.$$

This proof method $[\exists I \in \mathbf{Linv}_{\mathrm{L}}[\![\mathrm{Pp}]\!].\ \mathrm{lvc}_{\mathrm{N}}[\![\mathrm{Pp}]\!][p, q](I)]$ is sound (since $\mathrm{lvc}_{\mathrm{N}}[\![\mathrm{Pp}]\!][p, q](I) \Rightarrow \mathrm{lvc}_{\mathrm{SLO}}[\![\mathrm{Pp}]\!][p, q](I) \Rightarrow \mathrm{gvc}[\![\mathrm{Pp}]\!][p, q](\gamma(I)) \Rightarrow \underline{\{p\}\mathrm{Pp}\{q\}}$), and relatively complete (since any $I \in \mathbf{Linv}_{\mathrm{L}}[\![\mathrm{Pp}]\!]$ satisfying $\mathrm{lvc}_{\mathrm{SLO}}[\![\mathrm{Pp}]\!][p, q](I)$ can be strengthened into I' such that $I'(l) = \{\langle s, L_{\sim k}\rangle \in I(l):\ \ \forall j \in \{1, \ldots, n\} - \{k\}. \langle s, L[k \leftarrow 1]\rangle_{\sim j} \in I(L_j)\}$ satisfying $\mathrm{lvc}_{\mathrm{N}}[\![\mathrm{Pp}]\!][p, q](I')$):

(232) THEOREM (Newton's method). $\underline{\{p\}\mathrm{Pp}\{q\}}$ *where* Pp is $[C_1 \parallel C_2 \parallel \cdots \parallel C_n]$ *is true if and only if there exists* $I \in \mathbf{Linv}_{\mathrm{L}}[\![\mathrm{Pp}]\!]$ *such that for all* $k \in \{1, \ldots, n\}$, $m \in \{1, \ldots, n\} - \{k\}$ *and* $s, s' \in S$,

(232.1) *if* $l \in \mathbf{At}[\![C_k]\!]$ *and* $L \in \mathbf{At}[\![\mathrm{Pp}]\!]$, *then* $s \in p \Rightarrow \langle s, L_{\sim k}\rangle \in I(l)$;

(232.2) *if* $l \in \mathbf{At}[\![C_k]\!] \cup \mathbf{In}[\![C_k]\!]$ *and* $L \in \mathbf{Lab}[\![\mathrm{Pp}]\!]$, *then*

$$[\langle s, L_{\sim k}\rangle \in I(l) \wedge s' \in \mathbf{NextS}[\![C_k]\!]\langle s, l\rangle \wedge l' \in \mathbf{NextL}[\![C_k]\!]\langle s, l\rangle]$$
$$\Rightarrow \langle s', L_{\sim k}\rangle \in I(l');$$

(232.3) *if* $l \in \mathbf{At}[\![C_k]\!] \cup \mathbf{In}[\![C_k]\!]$, $l'' \in \mathbf{Lab}[\![C_m]\!]$ *and* $L \in \mathbf{Lab}[\![\mathrm{Pp}]\!]$, *then*

$$[\langle s, L[k \leftarrow l]_{\sim m}\rangle \in I(l'') \wedge s' \in \mathbf{NextS}[\![C_k]\!]\langle s, l\rangle \wedge l' \in \mathbf{NextL}[\![C_k]\!]\langle s, l\rangle] \Rightarrow \langle s', L[k \leftarrow l']_{\sim m}\rangle \in I(L'')$$

(232.4) *if* $L \in \mathbf{After}[\![\mathrm{Pp}]\!]$, *then* $[\forall k \in \{l, \ldots, n\}.\ \langle s, L_{\sim k}\rangle \in I(L_k)] \Rightarrow s \in q$.

Observe that interference freedom (232.3) disappears when considering single-process programs ($n=1$, in which case (232) exactly amounts to Floyd's stepwise partial correctness proof method (45)) or multiprocess programs with assertions about parts of the store such that only operations acting upon separate parts may be performed concurrently (as in [215] or [295] for example).

Although partial correctness proof methods (230) and (232) are both semantically complete, it may be the case that some assertions $I \in \mathbf{Linv}_{\mathrm{L}}[\![\mathrm{Pp}]\!]$ satisfy (230) but cannot be proved to be invariant using (232) without being strengthened.

(233) EXAMPLE (*weak invariants for program* (204)). Parallel program (204) with $a=1$ and $b=2$ is partially correct with respect to specification $\langle p, q\rangle$ such that $p=\{s \in S$: $s(X)=0\}$ and $q=\{s \in S$: $s(X)=3\}$. This can be proved using the verification conditions of (230) given at example (229) and the following invariants:

$$I[L_{11}]=\{\langle s, L_2\rangle: \mathit{even}(s(X))\}, \qquad I[L_{12}]=\{\langle s, L_2\rangle: s(X)=1 \vee s(X)=3\},$$
$$I[L_{21}]=\{\langle s, L_1\rangle: s(X)=0 \vee s(X)=1\},$$
$$I[L_{22}]=\{\langle s, L_1\rangle: s(X)=2 \vee s(X)=3\}.$$

These invariants are too weak to satisfy the verification conditions of (232) given in example (231).

8.9.2.4.3. The lattice of proof methods including Lamport's method. Any proof method $[\exists I \in \mathbf{Linv}_{\mathrm{L}}[\![\mathrm{Pp}]\!].\mathrm{lvc}[\![\mathrm{Pp}]\!][p, q](I)]$ with local verification conditions $\mathrm{lvc}[\![\mathrm{Pp}]\!][p, q]$ such that $\mathrm{lvc}_{\mathrm{N}}[\![\mathrm{Pp}]\!][p, q](I) \Rightarrow \mathrm{lvc}[\![\mathrm{Pp}]\!][p, q](I) \Rightarrow \mathrm{lvc}_{\mathrm{SLO}}[\![\mathrm{Pp}]\!][p, q](I)$ for all $I \in \mathbf{Linv}_{\mathrm{L}}[\![\mathrm{Pp}]\!]$ is sound (since $\mathrm{lvc}[\![\mathrm{Pp}]\!][p, q](I) \Rightarrow \mathrm{lvc}_{\mathrm{SLO}}[\![\mathrm{Pp}]\!][p, q](I) \Rightarrow \mathrm{gvc}[\![\mathrm{Pp}]\!][p, q](\gamma(I)) \Rightarrow \underline{\{p\}}\underline{\mathrm{Pp}}\underline{\{q\}}$), and semantically complete (since $\underline{\{p\}}\underline{\mathrm{Pp}}\underline{\{q\}} \Rightarrow [\exists i \in \mathscr{P}(\Gamma).\mathrm{gvc}[\![\mathrm{Pp}]\!][p, q](i)] \Rightarrow [\exists i \in \mathscr{P}(\Gamma).\mathrm{lvc}_{\mathrm{N}}[\![\mathrm{Pp}]\!][p, q](\alpha(i))] \Rightarrow [\exists I \in \mathbf{Linv}_{\mathrm{L}}[\![\mathrm{Pp}]\!].\ \mathrm{lvc}_{\mathrm{N}}[\![\mathrm{Pp}]\!][p, q](I)] \Rightarrow [\exists I \in \mathbf{Linv}_{\mathrm{L}}[\![\mathrm{Pp}]\!].\ \mathrm{lvc}_{\mathrm{N}}[\![\mathrm{Pp}]\!][p, q](I)]$). This is the case of Lamport's method [258] which is Owicki and Gries's method [337] with control predicates instead of auxiliary variables:

(234) THEOREM (Lamport [258]) (*Lamport's partial correctness proof method*). *$\underline{\{p\}}\underline{\mathrm{Pp}}\underline{\{q\}}$ where* Pp *is* $[C_1 \parallel C_2 \parallel \cdots \parallel C_n]$ *is true if and only if their exists* $I \in \mathbf{Linv}_{\mathrm{L}}[\![\mathrm{Pp}]\!]$ *such that for all* $k \in \{1, \ldots, n\}$, $m \in \{1, \ldots, n\} - \{k\}$ *and* $s, s' \in S$,

(234.1) *if* $l \in \mathbf{At}[\![C_k]\!]$ *and* $L \in \mathbf{At}[\![\mathrm{Pp}]\!]$, *then* $s \in p \Rightarrow \langle s, L_{\sim k}\rangle \in I(l)$;

(234.2) *if* $l \in \mathbf{At}[\![C_k]\!] \cup \mathbf{In}[\![C_k]\!]$ *and* $L \in \mathbf{Lab}[\![\mathrm{Pp}]\!]$, *then*

$$[\langle s, L_{\sim k}\rangle \in I(l) \wedge s' \in \mathbf{NextS}[\![C_k]\!]\langle s, l\rangle \wedge l' \in \mathbf{NextL}[\![C_k]\!]\langle s, l\rangle] \Rightarrow \langle s', L_{\sim k}\rangle \in I(l');$$

(234.3) *if* $l \in \mathbf{At}[\![C_k]\!] \cup \mathbf{In}[\![C_k]\!]$, $l'' \in \mathbf{Lab}[\![C_m]\!]$ *and* $L \in \mathbf{Lab}[\![\mathrm{Pp}]\!]$, *then*

$$[\langle s, L[k \leftarrow l]_{\sim m}\rangle \in I(l'') \wedge \langle s, L[m \leftarrow l'']_{\sim k}\rangle \in I(l) \wedge s' \in \mathbf{NextS}[\![C_k]\!]\langle s, l\rangle \wedge l' \in \mathbf{NextL}[\![C_k]\!]\langle s, l\rangle] \Rightarrow \langle s', L[k \leftarrow l']_{\sim m}\rangle \in I(L'');$$

(234.4) *if* $L \in \mathbf{After}[\![\mathrm{Pp}]\!]$, *then* $[\forall k \in \{1, \ldots, n\}. \langle s, L_{\sim k}\rangle \in I(L_k)] \Rightarrow s \in q$.

(235) EXAMPLE (*partial correctness proof of program* (204) *by Lamport's method*). To prove partial correctness of parallel program (204), we can use local invariants (226) and check the following local verification conditions (for all $s, s' \in S$, $c_1 \in \{L_{11}, L_{12}\}$ and $c_2 \in \{L_{21}, L_{22}\}$):

- Initialization (234.1):

$$s \in p \Rightarrow \langle s, L_{21}\rangle \in I(L_{11}), \qquad s \in p \Rightarrow \langle s, L_{11}\rangle \in I(L_{21});$$

- Sequential correctness (234.2):

$$\langle s, c_2\rangle \in I(L_{11}) \Rightarrow \langle s[X \leftarrow s(X)+a], c_2\rangle \in I(L_{12}),$$
$$\langle s, c_1\rangle \in I(L_{21}) \Rightarrow \langle s[X \leftarrow s(X)+b], c_1\rangle \in I(L_{22});$$

- Interference freedom (234.3):

$$[\langle s, c_2\rangle \in I(L_{11}) \wedge \langle s, L_{11}\rangle \in I(L_{21})] \Rightarrow \langle s[X \leftarrow s(X)+b], c_2\rangle \in I(L_{11}),$$
$$[\langle s, c_2\rangle \in I(L_{12}) \wedge \langle s, L_{12}\rangle \in I(L_{21})] \Rightarrow \langle s[X \leftarrow s(X)+b], c_2\rangle \in I(L_{12}),$$
$$[\langle s, c_1\rangle \in I(L_{21}) \wedge \langle s, L_{21}\rangle \in I(L_{11})] \Rightarrow \langle s[X \leftarrow s(X)+a], c_1\rangle \in I(L_{21}),$$
$$[\langle s, c_1\rangle \in I(L_{22}) \wedge \langle s, L_{22}\rangle \in I(L_{11})] \Rightarrow \langle s[X \leftarrow s(X)+a], c_1\rangle \in I(L_{22});$$

- Finalization (234.4):

$$[\langle s, L_{22}\rangle \in I(L_{12}) \wedge \langle s, L_{12}\rangle \in I(L_{22})] \Rightarrow s \in q.$$

8.9.2.5. Using an invariant on memory states with auxiliary variables for each program point

8.9.2.5.1. A stepwise presentation of the Owicki–Gries method. Owicki and Gries's partial correctness proof method [337] is based upon (234) using *auxiliary variables* (also called "dummy", "phantom", "ghost", "history", "mythical" or "thought" variables) for completeness.

(236) DEFINITION (Owicki and Gries [337]) (*auxiliary variables*). $AV \subseteq \mathbf{Pvar}$ is a set of *auxiliary variables* for a parallel program Pp and a specification $\langle p, q\rangle$ if and only if AV is finite, any $X \in AV$ appears in Pp only in assignments of the form $X := E$ and q does not depend upon some $X \in AV$ (i.e., $\forall X \in AV. \forall s \in S. \forall d \in D. s[X \leftarrow d] \in q$).

We say that Pp is obtained from Pp′ by *elimination of auxiliary variables* AV if Pp can be obtained from Pp′ by deleting all assignments to the variables of AV and subsequently replacing some of the components of the form "(true ¿ $Y := E$)" by "$Y := E$".

We say that p is obtained from $p' \in \mathbf{Ass}$ by *elimination of auxiliary variables* $AV = \{X_1, \ldots, X_n\}$ if and only if $\exists d_n \in D. \ldots \exists d_n \in D. p = \{s[X_1 \leftarrow d_1] \ldots [X_n \leftarrow d_n] : s \in p'\}$.

Auxiliary variables can be used to record the history of execution or indicate which part of a program is currently executing. The Owicki–Gries proof method is sound since the elimination of auxiliary variables does not change the result of program execution. It is semantically complete since auxiliary variables can simulate program counters:

(237) THEOREM (Owicki and Gries [337]) (Owicki–Gries partial correctness proof method). $\underline{\{p\}\mathrm{Pp}\{q\}}$ *if and only if there exists* Pp′ *and* p' *such that* $\underline{\{p'\}\mathrm{Pp}'\{q\}}$ *holds by* (234) *and there exists a set* AV *of auxiliary variables for* Pp′ *and* $\langle p', q\rangle$ *such that* p *and* Pp *are respectively obtained from* p' *and* Pp′ *by elimination of auxiliary variables* AV.

(238) EXAMPLE (*auxiliary variables for the partial correctness proof of program* (204)). The partial correctness of parallel program (204) can be proved using the Owicki–Gries method (237) as shown by the following proof outline:

$$\{X=x \wedge L1=1 \wedge L2=1\}$$
$$[\{(L1=1 \wedge L2=1 \wedge X=x) \vee (L1=1 \wedge L2=2 \wedge X=x+b)\}$$
$$(\mathbf{true} ¿ (X := X+a;\ L1 := 2))$$
$$\{(L1=2 \wedge L2=1 \wedge X=x+a) \vee (L1=2 \wedge L2=2 \wedge X=x+a+b)\}$$
$$\| \ \{(L2=1 \wedge L1=1 \wedge X=x) \vee (L2=1 \wedge L1=2 \wedge X=x+a)\}$$
$$(\mathbf{true} ¿ (X =: X+b;\ L2 := 2))$$
$$\{(L2=2 \wedge L1=1 \wedge X=x+b) \vee (L2=2 \wedge L1=2 \wedge X=x+a+b)\}]$$
$$\{X=x+a+b\}.$$

8.9.2.5.2. On the use of auxiliary variables. As shown by example (238) auxiliary variables can simulate program counters and this ensures the equivalence of the Owicki–Gries method (237) with the Lamport method (234), hence its semantical completeness. Lamport [263] claims that "although dummy variables can represent the control state, the implicit nature of this representation limits their utility". However, contrary to program counters, the values of auxiliary variables are not bounded, whence they can also be used to record computation histories [381]. This introduces additional power. For example, as shown by Apt [14], the invariants can be restricted to recursive predicates. The argumentation of Lamport in favor of control predicates goes on with the claim that "the use of explicit control predicates allows a strengthening of the ordinary Owicki–Gries method that makes it easier to write annotations". Observe however that (237) corresponds to (234) and that a strengthened version corresponding to (230) can be used instead. Then example (233) shows that the usefulness of this strengthened version does not depend upon the use of control predicates or auxiliary variables. To designate which part of a program is currently executing, program counters have the default that their value changes after each execution of an atomic action within that part. For example, introducing extra "skip" commands in a program would change nothing when using auxiliary variables but would introduce additional interference-freedom checks upon control predicates. In return, changes to auxiliary variables are not concomitant with an atomic action. Critical sections (such as "(true ¿ ($X := X + a$; $L1 := 2$))" in example (238)) can be used for assignments but not for Boolean tests. De Roever's suggestion [120] to use a dynamic extension of Apt, Francez and De Roever's "indivisibility" brackets [24] can be very useful in that respect. Finally, the use of program counters would be advantageous for program verifiers since "intelligence" would be needed to introduce auxiliary variables.

8.9.2.5.3. A syntax-directed presentation of the Owicki–Gries method. Following the guidelines of Subsections 5.2, 5.3 and 5.4, we can give a syntax-directed presentation of (237) due to Owicki [335] (see also [14]), where definition of program components is extended by $\mathbf{Comp}^{\eta}[\![[C_1' \parallel \cdots \parallel C_n']]\!] = \bigcup\{\mathbf{Comp}^{\eta i}[\![C_i']\!]: i=1,\ldots,n\}$ and $\mathbf{Comp}^{\eta}[\![(B ¿ C)]\!] = \{(B ¿ C)^{\eta}\} \cup \mathbf{Comp}^{\eta 0}[\![C]\!]$:

(239) Theorem (Owicki [335]) (syntax-directed presentation of the Owicki–Gries proof method) $\{p\}\underline{\mathrm{Pp}}\{q\}$ *if and only if there exists* $\mathrm{Pp}' \in \mathbf{Papr}$, *a set AV of auxiliary variables for* Pp′ *and* $\langle p, q\rangle$ *and preconditions* $\mathrm{pre} \in \mathbf{Comp}[\![\mathrm{Pp}']\!] \to \mathbf{Ass}$ *and postconditions* $\mathrm{post} \in \mathbf{Comp}[\![\mathrm{Pp}']\!] \to \mathbf{Ass}$ *such that*

(239.1) Pp *is obtained from* Pp′ *by elimination of auxiliary variables AV*;

(239.2) $p \subseteq \mathrm{pre}(\mathrm{Pp}') \wedge \mathrm{post}(\mathrm{Pp}') \subseteq q$.

Each component $C \in \mathbf{Comp}[\![\mathrm{Pp}']\!]$ *of program* Pp′ *is sequentially correct*:

(239.3) *if* C *is* skip *then* $\mathrm{pre}(C) \subseteq \mathrm{post}(C)$;

(239.4) *if* C *is* $X := E$, *then* $\mathrm{pre}(C) \subseteq \{s \in S: s[X \leftarrow \underline{E}(s)] \in \mathrm{post}(C)\}$;

(239.5) *if* C *is* $X := ?$, *then* $\{s[X \leftarrow d]: s \in \mathrm{pre}(C) \wedge d \in D\} \subseteq \mathrm{post}(C)$;

(239.6) *if* C *is* $(C_1; C_2)$, *then*

$$\mathrm{pre}(C) \subseteq \mathrm{pre}(C_1) \wedge \mathrm{post}(C_1) \subseteq \mathrm{pre}(C_2) \wedge \mathrm{post}(C_2) \subseteq \mathrm{post}(C);$$

(239.7) *if* C *is* $(B \to C_1 \diamond C_2)$, *then*

$$(\mathrm{pre}(C) \cap \underline{B}) \subseteq \mathrm{pre}(C_1) \wedge (\mathrm{pre}(C) \cap \neg\underline{B}) \subseteq \mathrm{pre}(C_2) \wedge \mathrm{post}(C_1) \subseteq \mathrm{post}(C)$$
$$\wedge\ \mathrm{post}(C_2) \subseteq \mathrm{post}(C);$$

(239.8) *if* C *is* $(B * C_1)$, *then*

$$\mathrm{pre}(C) \subseteq \mathrm{post}(C_1) \wedge (\mathrm{post}(C_1) \cap \underline{B}) \subseteq \mathrm{pre}(C_1) \wedge (\mathrm{post}(C_1) \cap \neg\underline{B}) \subseteq \mathrm{post}(C);$$

(239.9) *if* C *is* $(B ¿ C_1)$, *then*

$$(\mathrm{pre}(C) \cap \underline{B}) \subseteq \mathrm{pre}(C_1) \wedge \mathrm{post}(C_1) \subseteq \mathrm{post}(C).$$

No await "$(B ¿ C)$" assignment "$X := E$" or *"$X := ?$" command* $C_j' \in \mathbf{Comp}[\![C_j]\!]$ *of a process* C_j *of program* Pp *of the form* "$[C_1 \parallel \cdots \parallel C_n]$" *interferes with the proof of components* $C_i' \in \mathbf{Comp}[\![C_i]\!]$ *of other processes* C_i, $i \neq j$:

(239.10) $\underline{\{}\text{pre}(C_i')\cap\text{pre}(C_j')\underline{\}}\,\underline{C}_j'\underline{\{}\text{pre}(C_i')\underline{\}}$;

(239.11) $\underline{\{}\text{post}(C_i')\cap\text{pre}(C_j')\underline{\}}\,\underline{C}_j'\underline{\{}\text{post}(C_i')\underline{\}}$.

8.9.3. *Hoare logics for parallel programs with shared variables*

8.9.3.1. Owicki–Gries logic. After Hoare [213, 215], Owicki and Gries [335, 337, 338] were the first to extend Hoare's paper [208] to parallel programs with shared variables (see also the summary given by Dijkstra [128]). The difficulty is that although (239) is syntax-directed, it is not compositional in the sense of De Roever [120], that is, "the specification of a program should be verifiable in terms of the specification of its syntactic subprograms". More precisely, the sequential proof which is context-free can be expressed in Hoare's style but interference freedom which is context-sensitive cannot. Therefore Owicki and Gries's method **OG** [337] is usually informally presented in Hoare's style [208] adding to **H** the following *rules of inference of* **OG**:

(240) $$\frac{\{P\}\text{Pp}'\{Q\}}{\{P\}\text{Pp}\{Q\}}$$ *auxiliary variables elimination rule*

provided Pp is obtained from Pp′ by elimination of auxiliary variables AV and Q contains no variable of AV.

(241) $$\frac{\{P\}\text{Pp}\{Q\}}{\{P[X\leftarrow T]\}\text{Pp}\{Q\}}$$ *substitution rule*

provided $X \notin Free(\text{Pp}, Q)$.

(242) $$\frac{\{P\wedge B\}C\{Q\}}{\{P\}(B\,¿\,C)\{Q\}}$$ *await rule.*

(243) $$\frac{\{P_1\}C_1\{Q_1\}\cdots\{P_n\}C_n\{Q_n\}}{\{P_1\wedge\cdots\wedge P_n\}\,[C_1\parallel\cdots\parallel C_n]\,\{Q_1\wedge\cdots\wedge Q_n\}}$$ *parallelism rule*

provided $\{P_i\}C_i\{Q_i\}$, $i=1,\ldots,n$, are interference-free.

Rule (240) allows for the elimination of auxiliary variables in the program, whereas substitution rule (241) allows for the elimination of auxiliary variables in the precondition. Rule (243), where interference freedom is defined as in (239.10–11), is not compositional because interference freedom of the whole is directly reduced to that of its atomic parts and not to that of its constituent parts. This proof system of [337] can be extended to a proof outline as in [365]. The relative completeness of **OG** is proved in [14].

8.9.3.2. Stirling compositional logic. Lamport [260] proposed a first compositional version of Hoare logic, named GHL. It essentially consists in axiomatizing (217), with the disadvantages of global invariants. However GHL can be developed in a programming-language-independent way [264, 96]. Another language-independent step toward compositionality was [170].

A most important observation is that a compositional specification of a command should include a first part called *assumption* or *rely-condition* describing the desired behavior of the command and a second part called *commitment* or *guarantee-condition* describing the behavior of the environment [261]. A first step was taken by Francez and Pnueli [153] who introduced statements of the form $\langle\varphi\rangle C\langle\psi\rangle$ meaning that in any execution such that the environment behaves according to assumption φ, it is guaranteed that command C behaves according to ψ. However, this significantly departs from Hoare triples, since φ and ψ are temporal formulae [346]. A further step was taken by Jones [240, 241] who introduced specifications of commands under the form (Rc, Gc): $\{P\}C\{Q\}$ where P is a precondition and Q is a postcondition over states whereas the assumption Rc and the guarantee-condition Gc are sets of pairs of states characterizing the interference of command C with other processes. More precisely, Rc defines the relations which can be assumed to exist between the free variables of command C in states changed by other processes. Gc is a commitment which must be respected by all state transformations of C, thus constraining the interference which may be caused by command C.

An axiomatization **S** à la Hoare [208] and Owicki–Gries [337] of Jones' papers [240, 241] was proposed by Stirling [383, 384]. In order to remain in the realm of first-order logic where states but not pairs of states are considered, each $P \in \mathbf{Pre}$ determines a set of changes which are invariant with respect to it:

$$(244) \qquad \begin{aligned} &\mathbf{Inv}: \mathbf{Pre} \to \mathscr{P}(S \times S) \\ &\mathbf{Inv}[\![P]\!] = \{\langle s, s'\rangle \colon s \in \underline{P} \Rightarrow s' \in \underline{P}\}. \end{aligned}$$

The interpretation is that if $\langle s, s'\rangle \notin \mathbf{Inv}[\![P]\!]$, then the transition from state s to state s' interferes with the truth of P. This is extended to families R of formulae in **Pre**:

$$(245) \qquad \begin{aligned} &\mathbf{Inv}: \mathscr{P}(\mathbf{Pre}) \to \mathscr{P}(S \times S) \\ &\mathbf{Inv}[\![R]\!] = \bigcap\{\mathbf{Inv}[\![P]\!] \colon P \in R\}. \end{aligned}$$

Stirling [384] then defines an invariant implication between sets of formulae:

$$(246) \qquad (R \Rrightarrow G) \Leftrightarrow (\mathbf{Inv}[\![R]\!] \subseteq \mathbf{Inv}[\![G]\!])$$

which is characterized by the following lemma:

(247) LEMMA
(247.1) $(R \Rrightarrow G) \Leftrightarrow (\forall Q \in G. \forall R' \subseteq R. (\forall P \in R'. P \Rightarrow Q) \vee (\forall P \in R - R'. \neg P \Rightarrow \neg Q))$;
(247.2) $R \Rrightarrow R$;
(247.3) $(R \Rrightarrow G \wedge G \Rrightarrow H) \Rightarrow (R \Rrightarrow H)$;
(247.4) $(G \subseteq R) \Rightarrow (R \Rrightarrow G)$;
(247.5) $(P \Leftrightarrow Q \wedge R \Rrightarrow G) \Rightarrow (R \cup \{P\} \Rrightarrow G \cup \{Q\})$;
(247.6) $R \cup \{P, Q\} \Rrightarrow R \cup \{P \wedge Q\}$.

Stirling's proof system **S** [384] consists of **H** for proving properties of subcommands within await commands plus the following axiom schemata and rules of inference:

(248) $(R, G): \{P\}\text{skip}\{P\}$ *skip axiom,*

(249) $$\frac{R \Rrightarrow \{P\}, \quad P \Rightarrow Q[X \leftarrow E], \quad \forall I \in G. (P \wedge I) \Rightarrow I[X \leftarrow E]}{(R, G): \{P\}\ X := E\ \{Q\}}$$ *assignment rule,*

(250) $$\frac{(R, G): \{P_1\}C_1\{P_2\}, \quad (R, G): \{P_2\}C_2\{P_3\}}{(R, G): \{P_1\}(C_1;C_2)\{P_3\}}$$ *composition rule,*

(251) $$\frac{R \Rrightarrow \{P\}, \quad (R, G): \{P \wedge B\}C_1\{Q\}, \quad (R, G): \{P \wedge \neg B\}C_2\{Q\}}{(R, G): \{P\}(B \rightarrow C_1 \diamond C_2)\{Q\}}$$ *conditional rule,*

(252) $$\frac{R \Rrightarrow \{P\}, \quad (R, G): \{P \wedge B\}C\{P\}}{(R, G): \{P\}(B * C)\{P \wedge \neg B\}}$$ *while rule,*

(253) $$\frac{R \Rrightarrow R', \quad P \Rightarrow P', \quad (R', G'): \{P'\}C\{Q'\}, \quad Q' \Rightarrow Q, \quad G' \Rrightarrow G}{(R, G): \{P\}C\{Q\}}$$ *consequence rule,*

(254) $$\frac{R \Rrightarrow \{P\}, \quad \{P \wedge B\}C\{Q\}, \quad \forall I \in G. \{P \wedge B \wedge I\}C\{I\}}{(R, G): \{P\}(B \text{¿} C)\{Q\}}$$ *await rule,*

(255) $$\frac{\begin{array}{c} R_1 \Rrightarrow \{Q_1\}, \quad (R_1, R_2 \cup G): \{P_1\}C_1\{Q_1\}, \\ (R_2, R_1 \cup G): \{P_2\}C_2\{Q_2\}, \quad R_2 \Rrightarrow \{Q_2\} \end{array}}{(R_1 \cup R_2, G): \{P_1 \wedge P_2\}[C_1 \parallel C_2]\{Q_1 \wedge Q_2\}}$$ *parallelism rule*

(256) $$\frac{(R, G): \{P\}\text{Pp}\{Q\}}{\{P\}\text{Pp}\{Q\}}$$ *derelativization rule,*

(257) $$\frac{\{P\}\text{Pp}'\{Q\}}{\{P\}\text{Pp}\{Q\}}$$ *auxiliary variables elimination rule*

provided Pp is obtained from Pp′ by elimination of auxiliary variables AV and Q contains no variable of AV,

(258) $$\frac{\{P\}\text{Pp}\{Q\}}{\{P[X \leftarrow T]\}\text{Pp}\{Q\}}$$ *substitution rule*

provided $X \notin \mathit{Free}(\text{Pp}, Q)$.

(259) EXAMPLE. The partial correctness proof (238) of parallel program (204) is given by the following proof outline:

$$\{P\}\{I_0\}\ [\{I_{11}\}(\text{true}\ ¿\ (X := X+a;\ L1 := 2))\{I_{12}\}$$
$$\|\ \{I_{21}\}(\text{true}\ ¿\ (X := X+b;\ L2 =: 2))\{I_{22}\}]\ \{Q\}$$

where

$$P = (X = x),$$
$$I_0 = (X = x \wedge L1 = 1 \wedge L2 = 1),$$
$$I_{11} = ((L1 = 1 \wedge L2 = 1 \wedge X = x) \vee (L1 = 1 \wedge L2 = 2 \wedge X = x+b)),$$
$$I_{12} = ((L1 = 2 \wedge L2 = 1 \wedge X = x+a) \vee (L1 = 2 \wedge L2 = 2 \wedge X = x+a+b)),$$
$$I_{21} = ((L2 = 1 \wedge L1 = 1 \wedge X = x) \vee (L2 = 1 \wedge L1 = 2 \wedge X = x+a)),$$
$$I_{22} = ((L2 = 2 \wedge L1 = 1 \wedge X = x+b) \vee (L2 = 2 \wedge L1 = 2 \wedge X = x+a+b)),$$
$$Q = (X = x+a+b)$$

can be formalized by **S** as follows:

(a) $\{I_{11}, I_{12}\} \Rightarrow \{I_{11}\}$ by (247.4)
(b) $\{I_{11} \wedge \text{true}\}\ (X := X+a;\ L1 := 2)\ \{I_{12}\}$ by **H**∪Th
(c) $\{I_{11} \wedge \text{true} \wedge I_{21}\}\ (X := X+a;\ L1 := 2)\ \{I_{21}\}$ by **H**∪Th
(d) $\{I_{11} \wedge \text{true} \wedge I_{22}\}\ (X := X+a;\ L1 := 2)\ \{I_{22}\}$ by **H**∪Th
(e) $(\{I_{11}, I_{12}\}, \{I_{21}, I_{22}\}):\{I_{11}\}\ (\text{true}\ ¿\ (X := X+a; L1 := 2))\ \{I_{12}\}$ by (a), (b), (c), (d), (254)
(f) $\{I_{21}, I_{22}\} \Rightarrow \{I_{21}\}$ by (247.4)
(g) $\{I_{21} \wedge \text{true}\}\ (X := X+b;\ L2 := 2)\ \{I_{22}\}$ by **H**∪Th
(h) $\{I_{21} \wedge \text{true} \wedge I_{11}\}\ (X := X+b;\ L2 := 2)\ \{I_{11}\}$ by **H**∪Th
(i) $\{I_{21} \wedge \text{true} \wedge I_{12}\}\ (X := X+b;\ L2 := 2)\ \{I_{12}\}$ by **H**∪Th
(j) $(\{I_{21}, I_{22}\}, \{I_{11}, I_{12}\}):\{I_{11}\}\ (\text{true}\ ¿\ (X := X+b; L2 := 2))\ \{I_{22}\}$ by (f), (g), (h), (i), (254)
(k) $\{I_{11}, I_{12}\} \Rightarrow I_{12}$ by (247.4)
(l) $\{I_{21}, I_{22}\} \Rightarrow I_{22}$ by (247.4)
(m) $(\{I_{11}, I_{12}, I_{21}, I_{22}\}, \emptyset)$:
$\{I_{11} \wedge I_{21}\}\ [(\text{true}\ ¿\ (X := X+a;\ L1 := 2))$
$\|\ (\text{true}\ ¿\ (X := X+b;\ L2 := 2))]\ \{I_{12} \wedge I_{22}\}$ by (k), (e), (j), (l), (255)
(n) $\{I_{11} \wedge I_{21}\}\ [(\text{true}\ ¿\ (X := X+a;\ L1 := 2))$
$\|\ (\text{true}\ ¿\ (X := X+b;\ L2 := 2))]\ \{I_{12} \wedge I_{22}\}$ by (m), (256)
(o) $\{I_0\}\ [(\text{true}\ ¿\ (X := X+a;\ L1 := 2))$
$\|\ (\text{true}\ ¿\ (X := X+b;\ L2 := 2))]\ \{Q\}$ by Th, (n), (102)
(p) $\{I_0\}\ [X := X+a \,\|\, X := X+b]\ \{Q\}$ by (o), (257)
(q) $\{I_0[L1 \leftarrow 1,\ L2 \leftarrow 1]\}\ [X := X+a \,\|\, X := X+b]\ \{Q\}$ by (p), (258)
(r) $\{P\}\ [X := X+a \,\|\, X := X+b]\ \{Q\}$ by Th, (q), (102)

Additional techniques for proving partial or total correctness of parallel programs with shared variables are extensively discussed in number of surveys such as [17, 35, 120, 365].

8.9.4. Hoare logics for communicating sequential processes

Hoare [217, 221] introduced CSP (*Communicating Sequential Processes*), a language for parallel programs with communication via synchronous unbuffered message-passing. A program has the form "$[\text{Pl}_1 :: C_1 \,\|\, \text{Pl}_2 :: C_2 \,\|\, \cdots \,\|\, \text{Pl}_n :: C_n]$" where process labels $\mathrm{Pl}_1, \ldots, \mathrm{Pl}_n$ respectively designate parallel processes $C_1, \ldots, C_n$. Shared variables are disallowed.

Communication between processes Pl_i and Pl_j $(i \neq j)$ is possible if process Pl_i is to execute a send primitive "$\mathrm{Pl}_j\,!\,E$" and process Pl_j is to execute a receive primitive "$\mathrm{Pl}_i\,?\,X$". The first process ready to communicate has to wait as long as the other one is not ready to execute the matching primitive. Their execution is synchronized and it results in the assignment of the value of expression E (depending upon the values of the local variables of Pl_i) to the variable X (which is local to Pl_j). For example, "$\{X=a\}$ $[\mathrm{Pl}_1 :: \mathrm{Pl}_2!X \parallel \mathrm{Pl}_2 :: (\mathrm{Pl}_1?Y;\ \mathrm{Pl}_3!Y) \parallel \mathrm{Pl}_3 :: \mathrm{Pl}_2?Z]\ \{Z=a\}$" is true.

Nondeterminism is introduced via the alternation command "$(B_1; G_1 \rightarrow C_1 \diamond B_2; G_2 \rightarrow C_2 \diamond \cdots \diamond B_n;\ G_n \rightarrow C_n)$" where the guards "$B_k; G_k$", $k=1, \ldots, n$, consist of a Boolean expression B_k followed by a send "$\mathrm{Pl}_j\,!\,E$" or "skip" command G_k. Its execution consists in selecting and executing an arbitrary successful guard $B_k;G_k$ (where B_k evaluates to true and process Pl_j is ready to communicate if G_k is "$\mathrm{Pl}_j\,!\,E$") and then the corresponding alternative C_k. Their is no fairness hypothesis upon the choice between successful guards. For the repetition command "$*(B_1; G_1 \rightarrow C_1 \diamond B_2; G_2 \rightarrow C_2 \diamond \cdots \diamond B_n; G_n \rightarrow C_n)$", this is repeated until all guards "B_k; G_k" fail, that is, B_k evaluates to false or process Pl_j has terminated if G_k is "$\mathrm{Pl}_j\,!\,E$". This is called the distributed termination convention [23].

Cousot and Cousot [91] extended Floyd's proof method to CSP using control predicates. Levin and Gries [271] extended Owicki and Gries's axiomatic method [337] to CSP using global shared auxiliary variables (to simulate control states). In sequential proofs, communication is simply ignored, thus any assertion may be placed after a communication command:

(260) $\{P\}\ \mathrm{Pl}_i\,!\,E\ \{Q\}$ *send rule*,

(261) $\{P\}\ \mathrm{Pl}_j\,?\,X\ \{Q\}$ *receive rule*.

A satisfaction proof (also called cooperation proof) is then provided for any pair of communication commands which validates these assumptions:

(262) $(P \wedge P') \Rightarrow (Q \wedge Q')[X \leftarrow E]$ *satisfaction proof*

when

$$[\cdots \parallel \mathrm{Pl}_i :: \cdots \{P\}\ \mathrm{Pl}_j\,!\,E\ \{Q\} \cdots \parallel \cdots \parallel \mathrm{Pl}_j :: \cdots \{P'\}\ \mathrm{Pl}_i\,?\,X\ \{Q'\} \cdots \parallel \cdots].$$

Not all matching pairs of communication commands can have rendezvous, so that satisfaction proofs for dynamically unmatching pairs can be avoided by a simple static analysis of programs [16, 387]. The use of shared auxiliary variables necessitates interference-freedom proofs, but many trivial ones can be omitted [318]. Apt, Francez and De Roever [24] succeeded in restricting the use of auxiliary variables so that the assertions used in the proof of Pl_i do not contain free variables subject to changes in Pl_j, $j \neq i$. The soundness and relative completeness of their proof system was shown by Apt in [15]. A simplified and more comprehensive presentation is given by Apt in [18]. A restricted and modified version was later introduced by Apt [19] to prove the correctness of distributed termination algorithms à la Francez [148]. Joseph, Moitra and Soundararajan [244] have extended their proof rules for fault-tolerant programs written in a version of CSP and executing on a distributed system whose nodes may fail. However, in these approaches, one cannot deal with the individual processes of a program in isolation from the other processes. The special case of a single process interacting with its environment is considered in [158].

To deal with the individual processes of a program in isolation, Lamport and Schneider [264] reinterpret Hoare's triple $\{P\}C\{Q\}$ so that $P = Q$ is a global invariant during execution of C whereas Brookes [60, 61] introduces a new class of assertions for expressing sets of execution traces. In order to remain faithful to Hoare's interpretation of P as a precondition, Soundararajan [380, 382] and subsequently Zwiers, De Bruin and De Roever [409], and Zwiers, De Roever and Van Emde Boas [410] allowed to reason about hidden variables that correspond to sequences of messages sent and received by each process up to some moment during the execution of that process, an idea going back for example to Dahl [103] for coroutines, to Misra and Chandy [307] for networks of processes, and to Zhou Chao Chen and Hoare [407], Hoare [218, 219], Hehner and Hoare [204], Francez, Lehmann and Pnueli [152], Olderog and Hoare [333], and Fauconnier [140] for CSP.

Similar proof rules have been developed for the ADA™ rendezvous by Gerth [169] and by Gerth and De Roever [171]; for Brinch Hansen's distributed processes [59] by Sobel and Soundararajan [375] and De Roever [121]; for a version of Milner's calculus of communicating systems [306] by Ponse [347]; and for more abstract communication mechanisms named "scripts" by Taubenfeld and Francez [386] and by Francez, Hailpern and Taubenfeld [150]. Hoare logic was also extended for proving partial correctness of parallel logic programming languages [317].

The dynamic creation and destruction of processes is considered in [409, 113, 141].

Additional techniques for proving partial or total correctness of communicating sequential processes with nested parallelism, hiding of communication channels, buffered message passing etc. are extensively discussed in a number of surveys such as [18, 19, 35, 230, 408].

8.10. Total correctness

Hoare logic was originally designed for proving partial correctness but has been extended to cope with termination [288, 401, 376, 197] including the case of recursive procedures [199, 377, 189, 13, 302, 303, 294, 339, 49, 50, 6].

8.10.1. Finitely bounded nondeterminism and arithmetical completeness

In the case of deterministic while-programs (**Com** without random assignment "$X := ?$") we can use Harel's rule [197] where $P(n)$ stands for $P[x \leftarrow n]$ and "x" is an integer-valued logical variable not in $Var(B, C)$:

$$(263) \qquad \frac{P(n+1) \Rightarrow B, \quad [P(n+1)]C[P(n)], \quad P(0) \Rightarrow \neg B}{[\exists n.\ P(n)](B * C)[P(0)]} \qquad \textit{while rule.}$$

Soundness follows from the fact that nontermination would lead to an infinite strictly decreasing sequence of integers values $n, n-1, \ldots$ for the logical variable x. Semantic completeness follows from the remark that if execution of the while-loop does terminate, then after each iteration in the loop body the number of remaining iterations must strictly decrease and so, can always be chosen as the value of the logical variable x.

Observe that we now go beyond first-order logic and consider **N**-logic (also called ω-logic [37]), that is, a two-sorted language with a fixed structure **N**. Since **N** is infinite, **N**-logic is stronger than first-order logic. This is because there cannot be a sound and relatively complete deductive system based on a first-order oracle for total correctness [206]. For the oracle to be a realistic analog of an axiomatic proof system, it should be a uniform recursive enumeration procedure P of the theory $\text{Th} = \{\mathbf{P} \in \mathbf{Pre}\colon I[\![P]\!] = \text{tt}\}$, i.e., the procedure should operate exactly in the same way over interpretations I with their theories Th equal to one another and should be totally sound, i.e., as in (129), sound over all interpretations with theory Th. The argument given in [13] is that if such a procedure P exists, we could prove using P and the relatively complete deductive system that "[true] C [true]" where C is "$(X := 0; (\neg(X = Y) * X := X + 1))$" holds for the standard interpretation I_{PE} of arithmetic which is expressive by (165). Since P is uniform and totally sound, C should be guaranteed to terminate for all initial values of X and Y, but this is not true for the nonstandard interpretations of arithmetic when the initial value of X is a standard natural number and that of Y is a nonstandard one. Moreover, it is shown in [179] that there cannot be a deductive system that is sound and relatively complete for total correctness even if, for acceptable languages (e.g., Pascal-like languages [75]), the deductive system is required to be sound only for expressive interpretations.

It remains to look for classes of interpretations for which total correctness is relatively complete. The idea of Harel [197], called *arithmetical completeness*, consists in extending the interpretation to an arithmetic universe by augmenting it, if necessary, with the natural numbers and additional apparatus for encoding finite sequences into one natural. More precisely, following Grabowski [179] where the set of natural numbers is not primitive but first-order definable in the interpretations involved, an interpretation I is *k-weakly arithmetic* if and only if I is expressive and there exist first-order formulae $N(x)$, $E(x, y)$, $Z(x)$, $Add(x, y, z)$, $Mult(x, y, z)$ with at most k quantifiers and respectively n, $2n$, $2n$, N, $3n$, $3n$ free variables for some n such that E defines an equivalence relation on I^n and formulae N, E, Z, Add and $Mult$ define on the set $\{x\colon I[\![N(x)]\!]\}$ the model M such that the quotient model M/E is isomorphic to the standard model I_{PE} of Peano arithmetic PE with equality $\langle\{0\}, \{\text{Su}, +, *\}, \emptyset, \#\rangle$. Grabowski states in [179] that for every acceptable programming language with recursion and for every $k \in \mathbf{N}$, Hoare logic for total correctness is relatively complete for k-weakly arithmetic interpretations (but not for ∞-weakly arithmetic interpretations). Grabowski proceeds along in [180] with the comparison of arithmetical versus relative completeness.

In conclusion, the proof systems for total correctness cannot be of pure first-order logical character but must incorporate the standard model for Peano arithmetic or an external well-founded relation.

8.10.2. *Unbounded nondeterminism*

While-rule (263) implies the *strong termination* of while-loops, that is (cf. [126]), for each initial state s there is an integer $n(s)$ such that the loop $(B * C)$ is guaranteed to terminate in at most $n(s)$ iterations. As first observed by Back [33[a]] no such bound can exist for the program given in [124, p. 77]:

$$(X <> 0 * (X < 0 \rightarrow (X := ?; (X < 0 \rightarrow X := -X \diamond \text{skip})) \diamond X := X - 1))$$

in which case termination that is not strong is called *weak termination*. In the case of finitely bounded nondeterminism ($\forall \gamma \in \Gamma .\ |\{\gamma' \in \Gamma : \langle \gamma, \gamma' \rangle \in op[\![C]\!]\}| \in \mathbf{N}$), weak termination implies strong termination. However, when termination is weak but not strong, one can use the following rule due to Apt and Plotkin [27] and directly deriving from Floyd's liveness proof method (74), where $P(\alpha)$ stands for $P[x \leftarrow \alpha]$ and "x" is an ordinal-valued logical variable not in $Var(B, C)$:

(264)
$$\frac{(P(\alpha) \wedge \alpha > 0) \Rightarrow B, \quad [P(\alpha) \wedge \alpha > 0]C[\exists \beta < \alpha .\ P(\beta)], \quad P(0) \Rightarrow \neg B}{[\exists \alpha .\ P(\alpha)]\ (B * C)\ [P(0)]}$$

while rule.

The use of ordinal-valued loop counters (as in induction principle (74)) was first advocated in [56] but in fact was already proposed by Floyd [143] and incorporated in Hoare logic by Manna and Pnueli [288] under the form of well-founded sets. Apt and Plotkin [27] have shown that, in the case of countable nondeterminism (such that $|D| = \omega$), rule (264) is sound and complete (provided the assertion language **Pre** contains the sort of countable ordinals including the constant 0 and the order relation $<$ upon ordinals), and that all recursive ordinals are needed.

8.10.3. *Total correctness of fair parallel programs*

8.10.3.1. Fairness hypotheses and unbounded nondeterminism. Total correctness of parallel programs usually depends upon *weak* or *strong fairness hypotheses* [259, 268, 289, 290, 149], that is, assumptions that an action which can be, respectively, permanently or infinitely often executed will eventually be executed. For example, termination of "$X :=$ true; $[(X * \text{skip}) \parallel X := \text{false}]$" is not guaranteed without a weak fairness hypothesis since the first process "$(X * \text{skip})$" can loop for ever if the second process "$X :=$ false" is never activated. Under the weak fairness hypothesis that no nonterminated process can be eternally delayed, the command "$X :=$ false" must eventually be executed so that the while command "$(X * \text{skip})$" terminates and so does the parallel command "$[(X * \text{skip}) \parallel X := \text{false}]$".

We write $P \overset{\text{wf}[C_1 \| \cdots \| C_n]}{\rightsquigarrow} Q$ to state that execution of parallel program $[C_1 \parallel \ldots \parallel C_n]$ under weakly fair execution hypothesis inevitably leads from P to Q. More precisely, this fairness hypothesis is that on all infinite execution traces σ no process k is permanently enabled and never activated:

(265) Definition (*weak fairness hypothesis*)

$$\begin{aligned}&\forall \gamma \in \Gamma .\ \forall \sigma \in \Sigma^{\omega}[\![[C_1 \parallel \cdots \parallel C_n]]\!]\gamma .\ \forall k \in \{l, \ldots, n\}.\\ &\quad \neg(\forall i \geqslant 0 .\ \exists \gamma' \in \Gamma .\ \langle \sigma_i, \gamma' \rangle \in op[\![C_k]\!] \wedge \langle \sigma_i, \sigma_{i+1} \rangle \notin op[\![C_k]\!]).\end{aligned}$$

Observe that in practice the waiting delay is always bounded (say, by the lifetime of the computer) but this bound is unknown. Hence, fairness hypotheses with unbounded waiting delays should be understood as theoretical simplifications of actual scheduling policies. However this introduces unbounded nondeterminism since for example "N" can have any finite final value in the program "$(X := \text{false};\ N := 0);\ [(X * N := N+1) \parallel X := \text{false}]$" which terminates under the weak fairness hypothesis.

8.10.3.2. Failure of the Floyd liveness proof method. Floyd's liveness proof method (74) is not (directly) applicable to prove $P \overset{\text{wf}[C_1\|\cdots\|C_n]}{\rightsquigarrow} Q$, since under fairness hypotheses there may be no variant function decreasing after each program step.

(266) COUNTEREXAMPLE (*failure of the Floyd liveness proof method for weak fairness*). Floyd's liveness proof method (74) is not applicable to prove termination of "$[(X * \text{skip}) \parallel X := \text{false}]$" executed under the weak fairness hypothesis (265). For simplification, the operational semantics of this program C can be defined by $\Gamma = \{a, b\}$ and $op[\![C]\!] = op[\![C_1]\!] \cup op[\![C_2]\!]$ with $op[\![C_1]\!] = \{\langle a, a\rangle\}$ and $op[\![C_2]\!] = \{\langle a, b\rangle\}$ where a is the configuration in which either "$(X * \text{skip})$" or "$X := \text{false}$" is executable and b is the final configuration. Let $P = \{a\}$ and $Q = \{\langle a, b\rangle\}$. Applying (74), we would have some α, i and an infinite chain of ordinals β_k, $k \geqslant 0$, such that $\langle a, a\rangle \in i(\beta_0)$ and $0 < \beta_0 < \alpha$ [by (74.1) and (74.4) since $\langle a, a\rangle \notin Q$] so that assuming by induction hypothesis that $\langle a, a\rangle \in i(\beta_k)$ and $\beta_k > 0$, there exists some β_{k+1} such that $\langle a, a\rangle \in i(\beta_{k+1})$ [by (74.3) since $\langle a, a\rangle \in op[\![C]\!]$] and $0 < \beta_{k+1} < \beta_k$ [by (74.4) since $\langle a, a\rangle \notin Q$]: a contradiction since β_k, $k \geqslant 0$, is strictly decreasing.

The difficulty is due to the fact that some program steps (such as an iteration in the loop "$(X * \text{skip})$" of example (266)) do not directly contribute to termination. However, such inoperative steps contribute indirectly to termination in that their execution brings the scheduler nearer the choice of an operative step. This can be taken into account by coding the scheduler into the program [26] or by requiring the variant function to decrease only for such operative steps [268].

8.10.3.3. The transformational approach. Floyd's total correctness proof method can be generalized to liveness properties of weakly fair parallel programs $[C_1 \parallel \cdots \parallel C_n]$ by application of induction principle (74) to a semantics including a scheduler which ensures that execution of the program is weakly fair:

(267) DEFINITION (Apt & Olderog [26]) (*operational semantics of weakly fair parallel programs*)

$$\Gamma' = (\{1, \ldots, n\} \to \mathbf{N}) \times \Gamma$$

$$op'[\![C_1 \parallel \cdots \parallel C_n]\!]$$

$$= \{\langle\langle p, \gamma\rangle, \langle p', \gamma'\rangle\rangle\colon \exists k \in \{1, \ldots, n\}.$$

$$\langle \gamma, \gamma'\rangle \in op[\![C_k]\!] \wedge ([p_k > 0 \wedge p' <_k = p] \vee [B(p, \gamma) \wedge p' > 0])\}$$

where

$$\begin{aligned}
&(p' <_k = p) \\
&\quad = [(p'_k < p_k) \wedge (\forall j \in \{1, \dots, k-1, k+1, \dots, n\}.\ p'_j = p_j)], \\
&B(p, \gamma) = [\forall k \in \{1, \dots, n\}.\ (\forall \gamma' \in \Gamma.\ \langle \gamma, \gamma' \rangle \notin op[\![C_k]\!] \vee p_k = 0)], \\
&p' > 0 = [\forall k \in \{1, \dots, n\}.\ p_k > 0].
\end{aligned}$$

Execution is organized into rounds, within which each process C_k will be blocked or else will be executed at least one and at most p_k steps so that $p_1, \dots, p_n$ can be interpreted as priorities assigned, respectively, to processes $C_1, \dots, C_n$. An execution step within a round consists in executing a step of some process C_k with a non-zero priority $p_k > 0$. After this step the priority p'_k of that process C_k has strictly decreased while the priority p'_j of other processes C_j is left unchanged (whence $p' <_k = p$). A new round begins when all processes are either blocked or have a zero priority (whence $B(p, \gamma)$ holds), all priorities being strictly positive at the beginning of the next round (whence $p' > 0$). This is weakly but not strongly fair since a process which is almost always enabled but infinitely often disabled may never be activated.

Applying Floyd's liveness induction principle (74) to this transformed semantics, we get the following induction principle:

(268) THEOREM (Apt & Olderog [26]) (liveness proof method for weakly fair parallel programs)

$$P \overset{\mathrm{wf}[C_1 \| \cdots \| C_n]}{\rightsquigarrow} Q$$

$$= [\exists \alpha \in \mathbf{Ord}.\ \exists i \in \alpha \to (\{1, \dots, n\} \to \mathbf{N}) \to \mathscr{P}(\Gamma \times \Gamma).$$

(268.1) $\quad (\forall \gamma \in P.\ \forall p \in (\{1, \dots, n\} \to \mathbf{N}).\ \exists \beta < \alpha.\ \langle \gamma, \gamma \rangle \in i(\beta)(p))$

$\quad \wedge (\forall \gamma, \gamma' \in \Gamma.\ \forall p, p' \in (\{1, \dots, n\} \to \mathbf{N}).\ \forall \beta < \alpha.$

(268.2) $\quad \langle \gamma, \gamma' \rangle \in i(\beta)(p) \Rightarrow \langle \gamma, \gamma' \rangle \in Q$

(268.3) $\quad \vee [(\exists \gamma'' \in \Gamma.\ \exists k \in \{1, \dots, n\}.\ \langle \gamma', \gamma'' \rangle \in op[\![C_k]\!] \wedge (p_k > 0 \vee B(p, \gamma')))$

$\quad \wedge (\forall k \in \{1, \dots, n\}.\ \forall \gamma'' \in \Gamma.$

$\quad [\langle \gamma', \gamma'' \rangle \in op[\![C_k]\!] \wedge ([p_k > 0 \wedge p' <_k = p] \vee [B(p, \gamma') \wedge p' > 0])]$

(268.4) $\quad \Rightarrow [\exists \beta' < \beta.\ \langle \gamma, \gamma'' \rangle \in i(\beta')(p')]])]]$.

(269) EXAMPLE (*termination of a weakly fair parallel program by* (268)). Applying (268) to prove termination of "[(X ∗ skip) ∥ X := false]" with simplified operational semantics defined by $\Gamma = \{a, b\}$, $op[\![C_1]\!] = \{\langle a, a \rangle\}$, $op[\![C_2]\!] = \{\langle a, b \rangle\}$ and specification $P = \{a\}$ and $Q = \{\langle a, b \rangle\}$, we can choose $\alpha = \omega$ and $i = \lambda \beta.\ \lambda p.\ (\beta = 0 \to \{\langle a, b \rangle\} \diamond (\beta = p_1 \to \{\langle a, a \rangle\} \diamond \emptyset))$.

Including the scheduling policy into the induction principle by counting the number of steps executed within each round is often very clumsy.

8.10.3.4. The intermittent well-foundedness approach. Floyd's total correctness proof method can also be generalized to liveness properties of weakly fair parallel programs by the following more elegant induction principle. The variant function β is not assumed to decrease after each computation step. When no such progress is possible, it is sufficient that there exists one permanently enabled process C_k which decreases the rank β when activated. By fairness hypothesis, this is inevitable. Hence, one has only to record the identity k of the next process C_k which will make progress to the computation:

(270) THEOREM (Lehmann, Pnueli & Stavi [268]) (liveness proof method for weakly fair parallel programs)

$$P \overset{\mathrm{wf}[C_1\|\cdots\|C_n]}{\rightsquigarrow} Q$$
$$= [\exists \alpha \in \mathbf{Ord}.\ \exists i \in \alpha \to \{1,\dots,n\} \to \mathscr{P}(\Gamma \times \Gamma).$$

(270.1) $$(\forall \gamma \in P.\ \exists \beta < \alpha.\ \exists k \in \{1,\dots,n\}.\ \langle \gamma, \gamma \rangle \in i(\beta)(k))$$
$$\wedge (\forall \gamma, \gamma' \in \Gamma.\ \forall \beta < \alpha.\ \forall k \in \{1,\dots,n\}.\ \langle \gamma, \gamma' \rangle \in i(\beta)(k)$$

(270.2) $$\Rightarrow \langle \gamma, \gamma' \rangle \in Q$$

(270.3) $$\vee [(\exists \gamma'' \in \Gamma.\ \langle \gamma', \gamma'' \rangle \in op[\![C_k]\!])$$
$$\wedge (\forall j \in \{1,\dots,n\}.\ \forall \gamma'' \in \Gamma.\ \langle \gamma', \gamma'' \rangle \in op[\![C_j]\!]$$

(270.4) $$\Rightarrow [(\exists \beta' < \beta.\ \exists k' \in \{1,\dots,n\}.\ \langle \gamma, \gamma'' \rangle \in i(\beta')(k'))$$

(270.5) $$\vee (j \neq k \wedge \langle \gamma, \gamma'' \rangle \in i(\beta)(k))])])].$$

Starting from any initial state $\gamma \in P$ (270.1), each program step makes progress toward the goal Q (270.2) since all nonfinal intermediate states γ' (which satisfy invariant $i(\beta)(k)$, where β bounds the number of remaining operative steps and k is an enabled operative process) are nonblocking states (270.3), whence must have a successor state γ'' either by an operative step (270.4), in which case γ'' is closer to the goal, or by an inoperative step (270.5), in which case process C_k remains permanently enabled, which, by the weak fairness hypothesis, guarantees a future progress by (270.4).

(271) EXAMPLE (*termination of a weakly fair parallel program by* (270)). Applying (270) to prove termination of "[(X * skip) ‖ X := false]" with simplified operational semantics defined by $\Gamma = \{a, b\}$, $op[\![C_1]\!] = \{\langle a, a \rangle\}$, $op[\![C_2]\!] = \{\langle a, b \rangle\}$ and specification $P = \{a\}$ and $Q = \{\langle a, b \rangle\}$, we can choose $\alpha = 2$, $i(0) = \lambda k.\{\langle a, b \rangle\}$. $i(1)(1) = \emptyset$ and $i(1)(2) = \{\langle a, a \rangle\}$.

Approaches (265) and (270) can be generalized to arbitrary semantics [94] and formalized in the temporal-logic framework [289, 290]. Lehmann, Pnueli and Stavi [268], and Manna and Pnueli [290] consider generalizations of (270) for strong fairness. Grümberg and Francez [190], and Grümberg, Francez and Katz [191] respectively consider weak and strong equifairness, where a permanently or infinitely often enabled process is infinitely often activated with the further requirement for the scheduler to give an equally fair chance to each process in a group of jointly enabled processes. Francez and Kozen [151] present a unifying generalization of these fairness and equifairness notions by parametrization of the enabling and activating conditions.

Apt's [17, 20] and Francez's [149] are surveys of fair parallel program correctness proof methods with numerous references to the literature.

8.10.4. Dijkstra's weakest preconditions calculus

Dijkstra [123, 124] introduced the calculus of weakest preconditions as a generalization of Hoare logic to total correctness as first considered in [143].

(272) DEFINITION (Dijkstra [123]) (*weakest precondition*)

$$wp: \mathbf{Com} \times \mathbf{Ass} \rightarrow \mathbf{Ass}$$
$$wp(C, q) = \bigcup \{p \in \mathbf{Ass}: \underline{[}p\underline{]}\underline{C}\underline{[}q\underline{]}\}.$$

The operational interpretation of the assertion $wp(C, q)$ is that any valid implementation of C, when started in any state satisfying $wp(C, q)$, should lead to a finite computation that ends in a state of q and that it is the weakest such assertion:

(273) THEOREM (Dijkstra [124]) (*characterization of weakest preconditions*)

$$\forall C \in \mathbf{Com}.\ \forall p, q \in \mathbf{Ass}.$$

(273.1) $\underline{[}wp(C, q)\underline{]}\underline{C}\underline{[}q\underline{]}$

(273.2) $\wedge\ \underline{[}p\underline{]}\underline{C}\underline{[}q\underline{]} \Rightarrow p \subseteq wp(C, q).$

Dijkstra axiomatized wp as follows [123] (the continuity condition (274.5) was later introduced in [124, p. 72] to take into account strong termination of finitely bounded nondeterminism [126, 127]):

(274) THEOREM (Dijkstra [123, 124]) (healthiness criteria)

$$\forall C \in \mathbf{Com}.\ \forall p, q \in \mathbf{Ass}.$$

(274.1) $wp(C, \emptyset) = \emptyset$

(274.2) $(p \subseteq q) \Rightarrow (wp(C, p) \subseteq wp(C, q))$

(274.3) $wp(C, p \cap q) = wp(C, p) \cap wp(C, q)$

(274.4) $wp(C, p \cup q) = wp(C, p) \cup wp(C, q)$

(274.5) $\forall s \in S.\ |\{s': \langle s, s'\rangle \in \underline{C}\}| \in \mathbf{N}$

$$\Rightarrow \forall p \in \mathbf{N} \rightarrow \mathbf{Ass}.\ (\forall i \in \mathbf{N}.\ p_i \subseteq p_{i+1}) \Rightarrow \left(wp\left(C, \bigcup_{i \in \mathbf{N}} p_i\right) = \bigcup_{i \in \mathbf{N}} wp(C, p_i)\right).$$

Continuity of wp (274.5) is obviously violated for unbounded nondeterminism, since for example, if we let $D=\mathbf{N}$ and $p_i=\{s\in : s(X)\leqslant i\}$ for $i\geqslant 0$ then $\forall i\in\mathbf{N}.\, p_i\subseteq p_{i+1}$ but $S=wp(X:=?,\ S)=wp(X:=?,\ \bigcup_{i\in\mathbf{N}}p_i)\neq\bigcup_{i\in\mathbf{N}}wp(X:=?,p_i)=\bigcup_{i\in\mathbf{N}}\emptyset=\emptyset$, a contradiction when $|D|\geqslant 1$.

Dijkstra's weakest preconditions form a calculus for the derivation of programs that "turned program development into a calculational activity (and the idea of program correctness into a calculational notion)" [129]. This point of view was extensively developed in [124, 188]. The basis of this calculus is the following theorem (where case (275.7) corresponds to bounded nondeterminism):

(275) THEOREM (Dijkstra [123]) (cf. [406, 216, 72, 200, 27])

(275.1) $wp(\text{skip}, q)=q$,

(275.2) $wp(X:=E, q)=\{s\in S\colon s[X\leftarrow \underline{E}(s)]\in q\}$,

(275.3) $wp(X:=?, q)=\{s\in S\colon \forall d\in D.\, s[X\leftarrow d]\in q\}$,

(275.4) $wp((C_1;C_2), q)=wp(C_1, wp(C_2, q))$,

(275.5) $wp((B\rightarrow C_1 \diamond C_2), q)=(\underline{B}\cap wp(C_1, q))\cup(\neg\underline{B}\cap wp(C_2, q))$,

(275.6) $wp((B*C), q)=\mathit{lfp}\ \lambda X.\,(\underline{B}\cap wp(C, X))\cup(\neg\underline{B}\cap q)$,

(275.7)
$$\begin{aligned}&\forall s\in S.\,|\{s'\colon\langle s,s'\rangle\in\underline{C}\}|\in\mathbf{N}\\&\quad\Rightarrow \forall p\in\mathbf{N}\rightarrow\mathbf{Ass}.\\&\qquad(p_0=\neg\underline{B}\cap q)\wedge(\forall i\in\mathbf{N}.\ p_{i+1}=(\underline{B}\cap wp(C,p_i))\cup p_0)\\&\qquad\Rightarrow\left(wp((B*C),q)=\bigcup_{i\in\mathbf{N}}p_i\right).\end{aligned}$$

The derivation of weakest preconditions can be impractical for loops. Therefore (275.6) and (276.7) are advantageously replaced by the following theorem (using an invariant p and a variant function t as in [143].

(276) THEOREM (Dijkstra [124]) (cf. [33[a], 130])

$$\begin{aligned}&(\exists D.\,\exists W\subseteq D.\,\exists t\in D^S.\\&\ wf(W,\prec)\\&\ \wedge((p\cap\underline{B})\subseteq\{s\in S\colon t(s)\in W\}\\&\quad\wedge(\forall x\in D.\,(p\cap\underline{B}\cap\{s\in S\colon t(s)=x\})\subseteq wp(C, p\cap\{s\in S\colon t(s)\prec x\})))\\&\quad\Rightarrow(p\subseteq wp((B*C),\neg\underline{B}\cap q)).\end{aligned}$$

Dijkstra's weakest precondition calculus has been formalized in a number of ways such as, for example, using the infinitary logics $L\omega_1\omega$ (for finitely bounded nondeterminism) or $L\omega_1\omega_1$ (for unbounded nondeterminism, [33, 33[a]]), linear algebra [280], category theory [398], etc. It can be extended to more language features [119, 305, 200, 396, 189, 109 (Chapter 7), 142, 188, 341, 294, 132, 49, 50, 62, 205], thus losing part of their original simplicity when considering complicated languages. Various generalizations have been introduced in [33, 224, 225, 223, 236, 262, 322, 34].

8.11. *Examples of program verification*

Classical examples of program verification using Floyd–Naur's proof method, Hoare logic or Dijkstra's weakest preconditions calculus are given in [276, 277, 209, 211, 147, 285, 123, 185, 400, 275, 146, 184].

8.12. *Other logics extending first-order logic with programs*

Hoare's idea of extending first-order logic with programs or for program proofs has also been exploited in a number of formal systems such as the *algorithmic logics* of Engeler [135, 136, 137], Salwicki [364], and Rasiowa [355]; the *computational logic* of Boyer and Moore [57, 58]; the *dynamic logic* of Pratt [348], and Harel [197, 198]; the *first-order programming logic* of Cartwright [65, 66]; the *predicative semantics* of Hehner [201, 202], Hoare [219], and Hehner, Gupta and Malton [203]; the *programming logic* of Constable [81, 82], Constable and O'Donnell [84], and Constable, Johnson and Eichenlaub [83]; the *situational calculus* of Manna and Waldinger [292]; the *programming* calculus of Morris [316]; the *specification logic* of Reynolds [359] (see also [389]); the *weakest preconditions calculus* of Dijkstra [124] (see Subsection 8.10.3); the *weakest prespecification* of Hoare and He Jifeng [224, 225], Hoare, He Jifeng and Sanders [226], Hoare, Hayes, He Jifeng, Morgan, Roscoe, Sanders, Sorensen and Sufrin [223] (see also Chapter 14 on "Logics of Programs" by Kozen and Tiuryn and Chapter 16 on "Temporal and Modal Logic" by Emerson in this volume of the Handbook).

References

[1] Abramov, S.V., Remark on the method of intermediate assertions, *Soviet Math. Dokl.* **24**(1) (1981) 91–93.

[2] Abramov, S.V., The nature of the incompleteness of the Hoare system, *Soviet Math. Dokl.* **29**(1) (1984) 83–84.

[3] Aho, A.V., R. Sethi, and J.D. Ullman, *Compilers; Principles, Techniques and Tools* (Addison-Wesley, Reading, MA, 1986).

[4] Aho, A.V. and J.D. Ullman, Universality of data retrieval languages, in: *Conf. Record 6th ACM SIGACT-SIGPLAN Symp. on Principles of Programming Languages* (1979) 110–117.

[5] Alpern, B. and F.B. Schneider, Defining liveness, *Inform. Process. Lett.* **21** (1985) 181–185.

[6] America, P. and F.S. De Boer, Proving total correctness of recursive procedures, *Inform. and Comput.* **84**(2) (1990) 129–162.

[7] Andréka, H. and I. Németi, Completeness of Floyd logic, *Bull. Section of Logic, Wroclaw* **7** (1978) 115–121.

[8] Andréka, H., I. Németi and I. Sain, Completeness problems in verification of programs and program schemes, in: J. Becvàr, ed., *Mathematical Foundations of Computer Science 1979*, Lecture Notes in Computer Science, Vol. 74 (Springer, Berlin, 1979) 208–218.

[9] Andréka, H., I. Németi and I. Sain, A characterization of Floyd-provable programs, Lecture Notes in Computer Science, Vol. 118 (Springer, Berlin, 1981) 162–171.

[10] ANDRÉKA, H., I. NÉMETI and I. SAIN, A complete logic for reasoning about programs via non-standard model theory, Parts I–II, *Theoret. Comput. Sci.* **17** (1982) 193–212 and 259–278.

[11] ANDREWS, G.R., Parallel programs: proofs, principles, and practice, *Comm. Assoc. Comput. Mach.* **24**(3) (1981) 140–146.

[12] APT, K.R., A sound and complete Hoare-like system for a fragment of Pascal, Research Report IW 96/78, Afdeling Informatica, Mathematisch Centrum, Amsterdam, 1978.

[13] APT, K.R., Ten years of Hoare's logic: a survey—Part I, *ACM Trans. Programming Languages and Systems* **3**(4) (1981) 431–483.

[14] APT, K.R., Recursive assertions and parallel programs, *Acta Inform.* **15** (1981) 219–232.

[15] APT, K.R., Formal justification of a proof system for communicating sequential processes, *J. Assoc. Comput. Mach.* **30**(1) (1983) 197–216.

[16] APT, K.R., A static analysis of CSP programs, in: E. Clarke and D. Kozen, eds., *Logics of Programs*, Lecture Notes in Computer Science. Vol. 164 (Springer, Berlin, 1983) 1–17.

[17] APT, K.R., Ten years of Hoare's logic: a survey—Part II: nondeterminism, *Theoret. Comput. Sci.* **28** (1984) 83–109.

[18] APT, K.R., Proving correctness of CSP programs, a tutorial, in: M. Broy, ed., *Control Flow and Dataflow: Concepts of Distributed Programming* (Springer, Berlin, 1985) 441–474.

[19] APT, K.R., Correctness proofs of distributed termination algorithms, in: K.R. Apt, ed., *Logics and Models of Concurrent Systems*, Nato ASI Series, Vol. F13, (Springer, Berlin, 1985) 147–167.

[20] APT, K.R., Proving correctness of concurrent programs: a quick introduction, in: E. Böger, ed., *Trends in Theoretical Computer Science* (Computer Science Press, Rockville, MD, 1988) 305–345.

[21] APT, K.R., J.A. BERGSTRA and L.G.L.T. MEERTENS, Recursive assertions are not enough—or are they?, *Theoret. Comput. Sci.* **8** (1979) 73–87.

[22] APT, K.R. and J.W. DE BAKKER, Semantics and proof theory of PASCAL procedures, in: A. Salomaa and M. Steinby, eds., *Proc. 4th Internat. Coll. on Automata, Languages and Programming*, Lecture Notes in Computer Science, Vol. 52 (Springer, Berlin, 1977) 30–44.

[23] APT, K.R. and N. FRANCEZ, Modeling the distributed termination convention of CSP, *ACM Trans. Programming Languages and Systems* **6**(3) (1984) 370–379.

[24] APT, K.R., N. FRANCEZ and W.P. DE ROEVER, A proof system for communicating sequential processes, *ACM Trans. Programming Languages and Systems* **2**(3) (1980) 359–385.

[25] APT, K.R. and L.G.L.T. MEERTENS, Completeness with finite systems of intermediate assertions for recursive program schemes, *SIAM J. Comput.* **9**(4) (1980) 665–671.

[26] APT, K.R. and E.-R. OLDEROG, Proof rules and transformations dealing with fairness, *Sci. Comput. Programming* **3** (1983) 65–100.

[27] APT, K.R. and G.D. PLOTKIN, Countable nondeterminism and random assignment, *J. Assoc. Comput. Mach.* **33**(4) (1986) 724–767.

[28] ARBIB, M.A. and S. ALAGIC, Proof rules for gotos, *Acta Inform.* **11** (1979) 139–148.

[29] ASHCROFT, E.A., Proving assertions about parallel programs, *J. Comput. System Sci.* **10**(1) (1975) 110–135.

[30] ASHCROFT, E.A., M. CLINT and C.A.R. HOARE, Remarks on "Program proving: jumps and functions by M. Clint and C.A.R. Hoare", *Acta Inform.* **6** (1976) 317–318.

[31] ASHCROFT, E.A. and Z. MANNA, Formalization of properties of parallel programs, *Machine Intelligence* **6** (1970) 17–41.

[32] BABICH, A.F., Proving the total correctness of parallel programs, *IEEE Trans. Software Engrg.* **5**(6) (1979) 558–574.

[33] BACK, R.J.R., *Correctness Preserving Program Refinements: Proof Theory and Applications*, Mathematical Centre Tracts, Vol. 131 (Mathematisch Centrum, Amsterdam, 1980).

[33ᵃ] BACK, R.J.R., Proving total correctness of nondeterministic programs in infinitary logic, *Acta Inform.* **15** (1981) 233–249.

[34] BACK, R.J.R. and J. VON WRIGHT, A lattice-theoretical basis for a specification language, in: J.L.A. van de Snepscheut, ed., *Mathematics of Program Construction*, Lecture Notes in Computer Science, Vol. 375 (Springer, Berlin, 1989) 139–156.

[35] BARRINGER, H., *A Survey of Verification Techniques for Parallel Programs*, Lecture Notes in Computer Science, Vol. 191 (Springer, Berlin, 1985).

[36] BARRINGER, H., J.H. CHENG and C.B. JONES, A logic covering undefinedness in program proofs, *Acta Inform.* **21** (1984) 251–269.

[37] BARWISE, J., An introduction to first-order logic, in: J. Barwise, ed., *Handbook of Mathematical Logic* (North-Holland, Amsterdam, 1978) 5–46.

[38] BERGSTRA, J.A., A. CHMIELINSKA and J. TIURYN, Another incompleteness result for Hoare's logic, *Inform. and Control* **52** (1982) 159–171.

[39] BERGSTRA, J.A., A. CHMIELINSKA and J. TIURYN, Hoare's logic is incomplete when it does not have to be, Lecture Notes in Computer Science, Vol. 131 (Springer, Berlin, 1982) 9–23.

[40] BERGSTRA, J.A. and J.W. KLOP, Proving program inclusion using Hoare's logic, *Theoret. Comput. Sci.* **30** (1984) 1–48.

[41] BERGSTRA, J.A. and J. TIURYN, PC-compactness, a necessary condition for the existence of sound and complete logics for partial correctness, in: E. Clarke and D. Kozen, eds., *Logics of Programs*, Lecture Notes in Computer Science, Vol. 164 (Springer, Berlin, 1983) 45–56.

[42] BERGSTRA, J.A., J. TIURYN, and J.V. TUCKER, Floyd's principle, correctness theories and program equivalence, *Theoret. Comput. Sci.* **17** (1982) 113–149.

[43] BERGSTRA, J.A. and J.V. TUCKER, Algebraically specified programming systems and Hoare's logic, Lecture Notes in Computer Science, Vol. 115 (Springer, Berlin, 1981) 348–362.

[44] BERGSTRA, J.A. and J.V. TUCKER, Some natural structures which fail to possess a sound and decidable Hoare-like logic for their while-programs, *Theoret. Comput. Sci.* **17** (1982) 303–315.

[45] BERGSTRA, J.A. and J.V. TUCKER, Expressiveness and the completeness of Hoare's Logic, *J. Comput. System Sci.* **25**(3) (1982) 267–284.

[46] BERGSTRA, J.A. and J.V. TUCKER, Two theorems about the completeness of Hoare's logic, *Inform. Process. Lett.* **15**(4) (1982) 143–149.

[47] BERGSTRA, J.A. and J.V. TUCKER, Hoare's logic and Peano's arithmetic, *Theoret. Comput. Sci.* **22** (1983) 265–284.

[48] BERGSTRA, J.A. and J.V. TUCKER, The axiomatic semantics of programs based on Hoare's logic, *Acta Inform.* **21** (1984) 293–320.

[49] BIJLSMA, A., J.G. WILTINK and P.A. MATTHEWS, Equivalence of the Gries and Martin proof rules for procedure calls, *Acta Inform.* **23** (1986) 357–360.

[50] BIJLSMA, A., J.G. WILTINK and P.A. MATTHEWS, A sharp proof rule for procedures in wp semantics, *Acta Inform.* **26** (1989) 409–419.

[51] BIRÓ, B., On the complete verification methods, *Bull. Section of Logic, Wroclaw* **10**(2) (1981).

[52] BLIKLE, A., The clean termination of iterative programs, *Acta Inform.* **16** (1981) 199–217.

[53] BOEHM, H.-J., A logic for expressions with side effects, in: *Conf. Record 9th ACM SIGACT-SIGPLAN Symp. on Principles of Programming Languages* (1982) 268–280.

[54] BOEHM, H.-J., Side effects and aliasing can have simple axiomatic descriptions, *ACM Trans. Programming Languages and Systems* **7**(4) (1985) 637–655.

[55] BOOLOS, G.S. and R.C. JEFFREY, *Computability and Logic* (Cambridge Univ. Press, 1974, 1980).

[56] BOOM, H.J., A weaker precondition for loops, *ACM Trans. Programming Languages and Systems* **4**(4) (1982) 668–677.

[57] BOYER, R.S. and J.S. MOORE, *A Computational Logic* (Academic Press, New York, 1979).

[58] BOYER, R.S. and J.S. MOORE, *A Computational Logic Handbook* (Academic Press, New York, 1988).

[59] BRINCH HANSEN, P., Distributed processes: a concurrent programming concept, *Comm. ACM* **21**(11) (1978) 934–941.

[60] BROOKES, S.D., On the axiomatic treatment of concurrency, in: S.D. Brookes, A.W. Roscoe and G. Winskel, eds., *Seminar on Concurrency*, Lecture Notes in Computer Science, Vol. 197 (Springer, Berlin, 1984) 1–34.

[61] BROOKES, S.D., A semantically based proof system for partial correctness and deadlock in CSP, in: *Proc. Symp. on Logic in Computer Science* (1986) 58–65.

[62] BROY, M. and G. NELSON, Can fair choice be added to Dijkstra's calculus?, Research Report MIP-8902, Fakultät für Mathematik und Informatik, Univ. Passau, Fed. Rep. Germany, 1989; also submitted to *ACM Trans. Programming Languages and Systems.*

[63] BURSTALL, R.M., Some techniques for proving correctness of programs which alter data structures, *Machine Intelligence* **7** (1972) 23–50.

[64] BURSTALL, R.M., Program proving as hand simulation with a little induction, in: *Information Processing 74* (North-Holland, Amsterdam, 1974) 308–312.

[65] CARTWRIGHT, R., Non-standard fixed points in first-order logic, in: E. Clarke and D. Kozen, eds., *Logics of Programs*, Lecture Notes in Computer Science, Vol. 164 (Springer, Berlin, 1983) 129–146.

[66] CARTWRIGHT, R., Recursive programs as definitions in first order logic, *SIAM J. Comput.* **13**(2) (1984) 374–408.

[67] CARTWRIGHT, R. and D.C. OPPEN, Unrestricted procedure calls in Hoare's logic, in: *Conf. Record 5th ACM SIGACT-SIGPLAN Symp. on Principles of Programming Languages* (1978) 131–140.

[68] CARTWRIGHT, R. and D.C. OPPEN, The logic of aliasing, *Acta Inform.* **15** (1981) 365–384.

[69] CHERNIAVSKY, J. and S. KAMIN, A complete and consistent Hoare axiomatics for a simple programming language, in: *Conf. Record 4th ACM SIGACT-SIGPLAN Symp. on Principles of Programming Languages* (1977) 131–140; and *J. Assoc. Comp. Mach.* **26** (1979) 119–128.

[70] CHURCH, A., An unsolvable problem of elementary number theory, *Amer. J. Math.* **58** (1936) 345–363.

[71] CLARKE, JR, E.M., Programming language constructs for which it is impossible to obtain good Hoare axiom systems, in: *Conf. Record 4th ACM SIGACT-SIGPLAN Symp. on Principles of Programming Languages* (1977) 10–20; and *J. Assoc. Comput. Mach.* **26**(1) (1979) 129–147.

[72] CLARKE, JR, E.M., Program invariants as fixedpoints, *Computing* **21** (1979) 273–294.

[73] CLARKE, JR, E.M., Proving correctness of coroutines without history variables, *Acta Inform.* **13** (1980) 169–188.

[74] CLARKE, JR, E.M., The characterization problem for Hoare logic, *Phil. Trans. Soc. London A* **312** (1984) 423–440.

[75] CLARKE, JR, E.M., S.M. GERMAN and J.Y. HALPERN, Effective axiomatizations of Hoare Logics, *J. Assoc. Comput. Mach.* **30**(3) (1983) 612–636.

[76] CLINT, M., Program proving: coroutines, *Acta Inform.* **2** (1973) 50–63.

[77] CLINT, M., On the use of history variables, *Acta inform.* **16** (1981) 15–30.

[78] CLINT, M. and C.A.R. HOARE, Program proving: jumps and functions, *Acta Inform.* **1** (1972) 214–224.

[79] COHEN, P.J., *Set Theory and the Continuum Hypothesis* (Benjamin, New York, 1966).

[80] COLEMAN, D. and J.W. HUGUES, The clean termination of Pascal programs, *Acta Inform.* **11** (1979) 195–210.

[81] CONSTABLE, R.L., On the theory of programming logic, in: *Conf. Record 9th Ann. ACM Symp. on Theory of Computing* (1977) 269–285.

[82] CONSTABLE, R.L., Mathematics as programming, in: E. Clarke and D. Kozen, eds., *Logics of Programs*, Lecture Notes in Computer Science, Vol. 164 (Springer, Berlin, 1983) 116–128.

[83] CONSTABLE, R.L., S. JOHNSON and C. EICHENLAUB, *Introduction to the PL/CV2 Programming Logic*, Lecture Notes in Computer Science, Vol. 135 (Springer, Berlin, 1982).

[84] CONSTABLE, R.L. and M.J. O'DONNELL, *A Programming Logic* (Winthrop, Cambridge, MA, 1978).

[85] COOK, S.A., A characterization of pushdown machines in terms of time-bounded computers, *J. Assoc. Comput. Mach* **18**(1) (1971) 4–18.

[86] COOK, S.A., Soundness and completeness of an axiom system for program verification, *SIAM J. Comput.* **7**(1) (1978) 70–90.

[87] COOK, S.A. and D.C. OPPEN, An assertion language for data structures, in: *Conf. Record 2nd ACM SIGACT-SIGPLAN Symp. on Principles of Programming Languages* (1975) 160–166.

[88] COURCELLE, B., Proofs of partial correctness for iterative and recursive computations, in: The Paris Logic Group, ed., *Logic Colloquium '85* (North-Holland, Amsterdam, 1987) 89–110.

[89] COUSOT, P., Semantic foundations of program analysis, in: S.S. Muchnick and N.D. Jones, eds., *Program Flow Analysis: Theory and Practice* (Prentice Hall, Englewood Cliffs, NJ, 1981) 303–342.

[90] COUSOT, P. and R. COUSOT, A constructive version of Tarski's fixpoint theorems, *Pacific J. Math.* **82**(1) (1979) 43–57.

[91] COUSOT, P. and R. COUSOT, Semantic analysis of communicating sequential processes, in: J.W. De Bakker and J. Van Leeuwen, eds., *Proc. 7th Internat. Coll. on Automata, Languages and Programming*, Lecture Notes in Computer Science, Vol. 85 (Springer, Berlin, 1980) 119–133.

[92] COUSOT, P. and R. COUSOT, Induction principles for proving invariance properties of programs, in: D. Néel, ed., *Tools & Notions for Program Construction* (Cambridge Univ. Presss, Cambridge, 1982) 75–119.

[93] COUSOT, P. and R. COUSOT, Invariance proof methods and analysis techniques for parallel programs, in: A.W. Biermann, G. Guiho and Y. Kodratoff, eds., *Automatic Program Construction Techniques* (Macmillan, New York, 1984) 243–272.

[94] COUSOT, P. and R. COUSOT, "A la Floyd" induction principles for proving inevitability properties of programs, in: M. Nivat and J. Reynolds, eds., *Algebraic Methods in Semantics* (Cambridge Univ. Press, Cambridge, 1985) 277–312.

[95] COUSOT, P. and R. COUSOT, Sometime = always + recursion ≡ always; on the equivalence of the intermittent and invariant assertions methods for proving inevitability properties of programs, *Acta Inform.* **24** (1987) 1–31.

[96] COUSOT, P. and R. COUSOT, A language-independent proof of the soundness and completeness of Generalized Hoare Logic, *Inform. and Comput.* **80**(2) (1989) 165–191.

[97] CRASEMANN, CH. and H. LANGMAACK, Characterization of acceptable by Algol-like programming languages, in: E. Clarke and D. Kozen, eds., *Logics of Programs*, Lecture Notes in Computer Science, Vol. 164 (Springer, Berlin, 1983) 129–146.

[98] CSIRMAZ, L., Structure of program runs of nonstandard time, *Acta Cybernet.* **4** (1980) 325–331.

[99] CSIRMAZ, L., On the completeness of proving partial correctness, *Acta Cybernet.* **5** (1981) 181–190.

[100] CSIRMAZ, L., Programs and program verifications in a general setting, *Theoret. Comput. Sci.* **16** (1981) 199–210.

[101] CSIRMAZ, L. and B. HART, Program correctness on finite fields, in: *Proc. Symp. on Logic in Computer Science* (1986) 4–10.

[102] CUNNINGHAM, R.J. and M.E.J. GILFORD, A note on the semantic definition of side effects, *Inform. Process. Lett.* **4**(5) (1976) 118–120.

[103] DAHL, O.-J., An approach to correctness proofs of semi-coroutines, in: A. Blikle, ed., *Proc. Symp. and Summer School on Mathematical Foundations of Computer Science*, Lecture Notes in Computer Science, Vol. 28 (Springer, Berlin, 1975) 157–174.

[104] DAHL, O.J. and K. NYGAARD, SIMULA – An ALGOL-based simulation language, *Comm. ACM* **9** (1966) 671–678.

[105] DAMM, W. and B. JOSKO, A sound and relatively* complete axiomatization of Clarke's language L_4, in: E. Clarke and D. Kozen, eds., *Logics of Programs*, Lecture Notes in Computer Science, Vol. 164 (Springer, Berlin, 1983) 161–175.

[106] DAMM, W. and B. JOSKO, A sound and relatively* complete Hoare-Logic for a language with higher type procedures, *Acta Inform.* **20** (1983) 59–101.

[107] DAVIS, M., Unsolvable problems, in: J. Barwise, ed., *Handbook of Mathematical Logic* (North-Holland, Amsterdam, 1978) 567–594.

[108] DE BAKKER, J.W., Semantics and termination of nondeterministic recursive programs, in: S. Michaelson and R. Milner, eds., *Proc. 3rd Internat. Coll. on Automata, Languages and Programming* (1976) 436–477.

[109] DE BAKKER, J.W., *Mathematical Theory of Program Correctness* (Prentice Hall, Englewood Cliffs, NJ, 1980).

[110] DE BAKKER, J.W. and W.P. DE ROEVER, A calculus for recursive program schemes, in: M. Nivat, ed., *Proc. 1st Internat. Coll. on Automata, Languages and Programming* (1972) 167–196.

[111] DE BAKKER, J.W., J.W. KLOP and J.-J. CH. MEYER, Correctness of programs with function procedures, in: D. Kozen, ed., *Logics of Programs*, Lecture Notes in Computer Science, Vol. 131 (Springer, Berlin, 1982) 94–112.

[112] DE BAKKER, J.W. and L.G.L.T. MEERTENS, On the completeness of the inductive assertion method, *J. Comput. System Sci.* **11**(3) (1975) 323–357.

[113] DE BOER, F.S., A proof rule for process-creation, in: M. Wirsing, ed., *Formal Description of Programming Concepts III* (Proceedings of the IFIP TC2 WG2.2 Working Conference on Formal Description of Programming Concepts, Ebberup, Denmark, 25–28 August 1986) (1987).

[114] DE BRUIN, A., Goto statements: semantics and deduction system, *Acta Inform.* **15** (1981) 385–424.

[115] DE BRUIN, A., On the existence of Cook semantics, *SIAM J. Comput.* **13**(1) (1984) 1–13.

[116] DEMBINSKI, P. and R.L. SCHWARTZ, The pointer type in programming languages: a new approach, in: B. Robinet, ed., *Programmation, Proc, 2nd Internat. Symp. on Programming* (1976) 89–105.

[117] DE MILLO, R.A., R.J. LIPTON and A.J. PERLIS, Social processes and proofs of theorems and programs, *Comm. ACM* **22**(5) (1979) 271–280.

[118] DE ROEVER, W.P., Recursion and parameter mechanisms: an axiomatic approach, in: J. Lœckx, ed., *Proc. 2nd Internat. Coll. on Automata, Languages and Programming*, Lecture Notes in Computer Science, Vol. 14 (Springer, Berlin, 1974) 34–65.

[119] DE ROEVER, W.P., Dijkstra's predicate transformer, non-determinism, recursion and termination, in: *Mathematical Foundations of Computer Science*, Lecture Notes in Computer Science, Vol. 45 (Springer, Berlin, 1976) 472–481.

[120] DE ROEVER, W.P., The quest for compositionality, a survey of assertion-based proof systems for concurrent programs, Part 1: concurrency based on shared variables, in: E.J. Neuhold and C. Chroust, eds., *Formal Models in Programming* (Elsevier Science Publishers B.V., Amsterdam, 1985) 181–205.

[121] DE ROEVER, W.P., The cooperation test: a syntax-directed verification method, in: K.R. Apt, ed., *Logics and Models of Concurrent Systems*, NATO ASI Series, Vol. F 13 (Springer, Berlin, 1985) 213–257.

[122] DIJKSTRA, E.W., A constructive approach to the problem of program correctness, *BIT* **8** (1968) 174–186.

[123] DIJKSTRA, E.W., Guarded commands, nondeterminacy and formal derivation of programs, *Comm. ACM* **18**(8) (1975) 453–457.

[124] DIJKSTRA, E.W., *A Discipline of Programming* (Prentice Hall, Englewood Cliffs, NJ, 1976).

[125] DIJKSTRA, E.W., On subgoal induction, in: *Selected Writings on Computing: A Personal Perspective* (Springer, Berlin, 1982) 223–224.

[126] DIJKSTRA, E.W., On weak and strong termination, in: *Selected Writings on Computing: A Personal Perspective* (Springer, Berlin, 1982) 355–357.

[127] DIJKSTRA, E.W., The equivalence of bounded nondeterminacy and continuity, in: *Selected Writings on Computing: A Personal Perspective* (Springer, Berlin, 1982) 358–359.

[128] DIJKSTRA, E.W., A personal summary of the Gries–Owicki theory, in: *Selected Writings on Computing: A Personal Perspective* (Springer, Berlin, 1982) 188–199.

[129] DIJKSTRA, E.W., Invariance and non-determinacy, *Phil. Trans. Roy. Soc. London A* **312** (1984) 491–499.

[130] DIJKSTRA, E.W. and A.J.M. GASTEREN, A simple fixpoint argument without the restriction to continuity, *Acta Inform.* **23** (1986) 1–7.

[131] DONAHUE, J.E., *Complementary Definitions of Programming Language Semantics*, Lecture Notes in Computer Science, Vol. 42 (Springer, Berlin, 1976).

[132] ELRAD, T. and N. FRANCEZ, A weakest precondition semantics for communicating processes, *Theoret. Comput. Sci.* **29** (1984) 231–250.

[133] ENDERTON, H.B., *A Mathematical Introduction to Logic* (Academic Press, New York, 1972).

[134] ENDERTON, H.B., Elements of recursion theory, in: J. Barwise, ed., *Handbook of Mathematical Logic* (North-Holland, Amsterdam, 1978) 527–566.

[135] ENGELER, E., Algorithmic properties of structures, *Math. Systems Theory* **1** (1967) 183–195.

[136] ENGELER, E., Remarks on the theory of geometrical constructions, in: J. Barwise, ed., *The Syntax and Semantics of Infinitary Languages*, Lecture Notes in Mathematics, Vol. 72 (Springer, Berlin, 1968) 64–76.

[137] ENGELER, E., Algorithmic logic, in: J.W. De Bakker, ed., *Foundations of Computer Science*, Mathematical Center Tracts, Vol. 63 (Mathematisch Centrum, Amsterdam, 1975) 57–85.

[138] ERNST, G.W., Rules of inference for procedure calls, *Acta Inform.* **8** (1977) 145–152.

[139] ERNST, G.W., J.K. NAVLAKHA and W.F. OGDEN, Verification of programs with procedure-type parameters, *Acta Inform.* **18** (1982) 149–169.

[140] FAUCONNIER, H., Sémantique asynchrone et comportements infinis en CSP, *Theoret. Comput. Sci.* **54** (1978) 277–298.

[141] FIX, L. and N. FRANCEZ, Proof rules for dynamic process creation and destruction, Manuscript, 1988.

[142] FLON, L. and N. SUZUKI, The total correctness of parallel programs, *SIAM J. Comput.* **10**(2) (1981) 227–246.

[143] FLOYD, R.W., Assigning meanings to programs, in J.T. Schwartz, ed., *Proc. Symp. in Applied Mathematics 19* (1967) 19–32.

[144] FLOYD, R.W., Nondeterministic algorithms, *J. Assoc. Comput. Mach.* **14**(4) (1967) 636–644.

[145] FOKKINGA, M.M., Axiomatization of declarations and the formal treatment of an escape construct, in: E.J. Neuhold, ed., *Formal Descriptions of Programming Concepts* (North-Holland, Amsterdam, 1978) 221–235.

[146] FOKKINGA, M.M., A correctness proof of sorting by means of formal procedures, *Sci. Comput. Programming* **9** (1987) 263–269.

[147] FOLEY, M. and C.A.R. HOARE, Proof of a recursive program: QUICKSORT, *Comput. J.* **14**(4) (1971) 391–395.

[148] FRANCEZ, N., Distributed termination, *ACM Trans. Programming Languages and Systems* **2**(1) (1980) 42–55.

[149] FRANCEZ, N., *Fairness* (Springer, Berlin, 1986).

[150] FRANCEZ, N., B. HAILPERN and G. TAUBENFELD, Script: a communication abstraction mechanism and its verification, in: K.R. Apt, ed., *Logics and Models of Concurrent Systems*, NATO ASI Series, Vol. F13 (Springer, Berlin, 1985) 169–212.

[151] FRANCEZ, N. and D. KOZEN, Generalized fair termination, in: *Conf. Record 11th ACM SIGACT-SIGPLAN Symp. on Principles of Programming Languages* (1984) 46–53.

[152] FRANCEZ, N., D. LEHMANN and A. PNUELI, A linear history semantics for languages for distributed programming, *Theoret. Comput. Sci.* **32** (1984) 25–46.

[153] FRANCEZ, N. and A. PNUELI, A proof method for cyclic programs, *Acta Inform.* **9** (1978) 133–157.

[154] GAIFMAN, H. and M.Y. VARDI, A simple proof that connectivity of finite graphs is not first-order definable, *Bull. EATCS* (1985) 43–45.

[155] GALLIER, J.H., Semantics and correctness of nondeterministic flowchart programs with recursive procedures, in: G. Ausiello and C. Böhm, eds., *Proc. 5th Internat. Coll. on Automata, Languages and Programming*, Lecture Notes in Computer Science, Vol. 62 (Springer, Berlin, 1978) 252–267.

[156] GALLIER, J.H., Nondeterministic flowchart programs with recursive procedures: semantics and correctness, *Theoret. Comput. Sci.* **13** (1981) Part I: 193–223; Part II: 239–270.

[157] GERGELY, T. and M. SZŐTS, On the incompleteness of proving partial correctness, *Acta Cybernet.* **4**(1) (1978) 45–57.

[158] GERGELY, T. and L. URY, Time models for programming logics, in: B. Dömölki and T. Gergely, eds., *Mathematical Logic in Computer Science*, Colloquia Mathematica Societatis János Bolyai, Vol. 26 (North-Holland, Amsterdam, 1981) 359–427.

[159] GERGELY, T. and L. ÚRY, Specification of program behavior through explicit time considerations, in: S.H. Lavington, ed., *Information Processing '80* (North-Holland, Amsterdam, 1980) 107–111.

[160] GERGELY, T. and L. ÚRY, A theory of interactive programming, *Acta Inform.* **17** (1982) 1–20.

[161] GERHART, S.L., Correctness-preserving program transformations, in: *Conf. Record 2nd ACM SIGACT-SIGPLAN Symp. on Principles of Programming Languages* (1975) 54–66.

[162] GERMAN, S.M., Automatic proofs of the absence of common runtime errors, in: *Conf. Record 5th ACM SIGACT-SIGPLAN Symp. on Principles of Programming Languages* (1978) 105–118.

[163] GERMAN, S.M., Verifying the absence of common runtime errors in computer programs, Report No. STAN-CS-81-866, Dept. of Computer Science, Stanford Univ., 1981.

[164] GERMAN, S.M., E.M. CLARKE, JR and J.Y. HALPERN, Reasoning about procedures as parameters, in: E. Clarke and D. Kozen, eds., *Logics of Programs*, Lecture Notes in Computer Science, Vol. 164 (Springer, Berlin, 1983) 206–220.

[165] GERMAN, S.M., E.M. CLARKE, JR and J.Y. HALPERN, True relative completeness of an axiom system for the language *L4* (abridged), in: *Proc. Symp. on Logic in Computer Science* (1986) 11–25.

[166] GERMAN, S.M., E.M. CLARKE, JR and J.Y. HALPERN, Reasoning about procedures as parameters in the language *L4*, *Inform. and Comput.* **83**(3) (1989) 265–360.

[167] GERMAN, S.M. and J.Y. HALPERN, On the power of the hypothesis of expressiveness, IBM Research Report RJ 4079, 1983.

[168] GERMAN, S.M., and B. WEGBREIT, A synthesizer of inductive assertions, *IEEE Trans. Software Engrg.* **1**(1) (1975) 68–75.

[169] GERTH, R., A sound and complete Hoare axiomatization of the ADA-rendezvous, in: M. Nielsen

and E.M. Schmidt, eds. *Proc. 9th Internat. Coll. on Automata, Languages and Programming*, Lecture Notes in Computer Science, Vol. 140 (Springer, Berlin, 1982) 252–264.

[170] GERTH, R., Transition logic: how to reason about temporal properties in a compositional way, in: *Proc. 16th Ann. ACM Symp. on Theory of Computing* (1983) 39–50.

[171] GERTH, R. and W.P. DE ROEVER, A proof system for concurrent ADA programs, *Sci. Comput. Programming* **4** (1984) 159–204.

[172] GOERDT, A., A Hoare calculus for functions defined by recursion on higher types, in: R. Parikh, ed., *Logics of Programs*, Lecture Notes in Computer Science, Vol. 193 (Springer, Berlin, 1985) 106–117.

[173] GOERDT, A., Hoare logic for lambda-terms as basis of Hoare logic for imperative languages, in: *Proc. Symp. on Logic in Computer Science 1987* (1987) 293–299.

[174] GOERDT, A., Hoare calculi for higher-type control structures and their completeness in the sense of Cook, in: M.P. Chytil, L. Jane and V. Koubek, eds., *Proc 13th Symp. on Mathematical Foundations of Computer Science*, Lecture Notes in Computer Science, Vol. 324 (Springer, Berlin, 1988) 329–338.

[175] GOLDSTINE, H.H. and J. VON NEUMANN, Planning and coding of problems for an electronic computing instrument, Report for U.S. Ord. Dept., in: A. Taub, ed., *Collected Works of J. von Neumann, Vol. 5* (Pergamon, New York, 1965) 80–151.

[176] GOOD, D.I., Mechanical proofs about computer programs, *Phil. Trans. Roy. Soc. London* A **312** (1984) 389–409.

[177] GORELICK, G.A., A complete axiomatic system for proving assertions about recursive and non-recursive procedures, Tech. Report 75, Dept. Computer Science, Univ. of Toronto, Toronto, 1975.

[178] GRABOWSKI, M., On the relative completeness of Hoare logics, in: *Conf. Record 11th ACM SIGACT-SIGPLAN Symp. on Principles of Programming Languages* (1984) 258–261; *Inform. and Control* **66** (1986) 29–44.

[179] GRABOWSKI, M., On the relative incompleteness of logics for total correctness, in: R. Parikh, ed., *Logics of Programs*, Lecture Notes in Computer Science, Vol. 193 (Springer, Berlin, 1985) 118–127.

[180] GRABOWSKI, M., Arithmetical completeness versus relative completeness, *Studia Logica* **XLVII** (3) (1988) 213–220.

[181] GRABOWSKI, M. and X. HUNGAR, On the existence of effective Hoare logics, in: *Proc. Symp. on Logic in Computer Science* (1988).

[182] GREIBACH, S.A., *Theory of Program Structures: Schemes, Semantics, Verification.* Lecture Notes in Computer Science, Vol. 36 (Springer, Berlin, 1975).

[183] GREIF, I. and A.R. MEYER, Specifying the semantics of **while** programs: a tutorial and critique of a paper by Hoare and Lauer, *ACM Trans. Programming Languages and Systems* **3**(4) (1981) 484–507.

[184] GRIBOMONT, P.E., Stepwise refinement and concurrency: a small exercise, in: J.L.A. van de Snepscheut, ed., Lecture Notes in Computer Sciences, Vol. 375 (Springer, Berlin, 1989) 219–238.

[185] GRIES, D., An exercise in proving parallel programs correct, *Comm. ACM* **20**(12) (1977) 921–930.

[186] GRIES, D., The multiple assignment statement, *IEEE Trans. Software Engrg.* **4**(2) (1978) 89–93.

[187] GRIES, D., Is "sometime" ever better than "always"?, *ACM Trans. Programming Languages and Systems* **1**(1979) 258–265.

[188] GRIES, D., *The Science of Programming* (Springer, Berlin, 1981).

[189] GRIES, D. and G.M. LEVIN, Assignment and procedure call proof rules, *ACM Trans. Programming Languages and Systems* **2**(4) (1980) 564–579.

[190] GRÜMBERG, O. and N. FRANCEZ, A complete proof rule for (weak) equifairness, IBM Research Report, RC-9634, T.J. Watson Research Center, 1982.

[191] GRÜMBERG, O., N. FRANCEZ and S. KATZ, A complete proof rule for strong equifair termination, in: E. Clarke and D. Kozen, eds., *Logics of Programs*, Lecture Notes in Computer Science, Vol. 164 (Springer, Berlin, 1983) 257–278.

[192] GUREVICH, Y., Toward logic tailored for computational complexity, in: M.M. Richter, E. Börger, W. Oberschelp, B. Schinzel and W. Thomas, eds., *Computation and Proof Theory*, Lecture Notes in Mathematics, Vol. 1104 (Springer, Berlin, 1984) 175–216.

[193] GUREVICH, Y., Logic and the challenge of computer science, in: E. Börger, ed., *Trends in Theoretical Computer Science* (Computer Science Press, Rockville, MD, 1988) 1–58.

[194] GUTTAG, J.V., J.J. HORNING and R.L. LONDON, A proof rule for Euclid procedures, in: E.J. Neuhold, ed., *Formal Description of Programming Concepts* (North-Holland, Amsterdam, 1978) 211–220.

[195] HALPERN, J.Y., A good Hoare axiom system for an ALGOL-like language, in: *Conf. Record 11th ACM SIGACT-SIGPLAN Symp. on Principles of Programming Languages* (1984) 262–271.

[196] HALPERN, J.Y., A.R. MEYER and B.A. TRAKHTENBROT, The semantics of local storage, or what makes the free-list free? in: *Conf. Record 11th ACM SIGACT-SIGPLAN Symp. on Principles of Programming Languages* (1984) 245–254.

[197] HAREL, D., *First-order Dynamic Logic*, Lecture Notes in Computer Science, Vol. 68 (Springer, Berlin, 1979).

[198] HAREL, D., Proving the correctness of regular deterministic programs: a unifying survey using dynamic logic, *Theoret. Comput. Sci.* **12** (1980) 61–81.

[199] HAREL, D., A. PNUELI, and J. STAVI, A complete axiomatic system for proving deductions about recursive programs, in: *Proc. 9th Ann. ACM Symp. on Theory of Computing* (1977) 249–260.

[200] HEHNER, E.C.R., Do considered od: a contribution to the programming calculus, *Acta Inform.* **11** (1979) 287–304.

[201] HEHNER, E.C.R., *The Logic of Programming* (Prentice Hall, Englewood Cliffs, NJ, 1984).

[202] HEHNER, E.C.R., Predicative programming, *Comm. ACM* **27** (1984) 134–151.

[203] HEHNER, E.C.R., L.E. GUPTA and A.J. MALTON, Predicative methodology, *Acta Inform.* **23** (1986) 487–505.

[204] HEHNER, E.C.R. and C.A.R. HOARE, A more complete model of communicating processes, *Theoret. Comput. Sci.* **26** (1983) 105–120.

[205] HESSELINK, W.H., Predicate-transformer semantics of general recursion, *Acta Inform.* **26** (1989) 309–332.

[206] HITCHCOCK, P. and D.M.R. PARK, Induction rules and proofs of program termination, in: M. Nivat, ed., *Proc. 1st Internat. Coll. on Automata, Languages and Programming* (1973) 225–251.

[207] HOARE, C.A.R., Algorithm 63, Partition; Algorithm 64, Quicksort; Algorithm 65, Find, *Comm. ACM* **4**(7) (1961) 321–322.

[208] HOARE, C.A.R., An axiomatic basis for computer programming, *Comm. ACM* **12**(10) (1969) 576–580, 583; also in: ref. [227] 45–58.

[209] HOARE, C.A.R., Proof of a program: Find, *Comm. ACM* **14**(1) (1971) 39–45; also in: ref. [227] 59–74.

[210] HOARE, C.A.R., Procedures and parameters: an axiomatic approach, in: E. Engeler, ed., *Symp. on Semantics of Algorithmic Languages*, Lecture Notes in Mathematics, Vol. 188 (Springer, Berlin, 1971) 102–116; also in: ref. [227] 75–88.

[211] HOARE, C.A.R., Proof of correctness of data representations, *Acta Inform.* **1** (1972) 271–281; also in: ref. [227] 103–116.

[212] HOARE, C.A.R., Proof of a structured program: "the sieve of Eratosthenes", *Comput. J.* **15**(4) (1972) 321–325; also in: ref [227] 117–132.

[213] HOARE, C.A.R., Towards a theory of parallel programming, in: C.A.R. Hoare and R.H. Perrott, eds., *Operating System Techniques* (Academic Press, New York, 1972) 61–71.

[214] HOARE, C.A.R., Monitors: an operating system structuring concept, *Comm. ACM* **17**(10) (1974) 549–557; also in: ref. [227] 17–191.

[215] HOARE, C.A.R., Parallel programming: an axiomatic approach, *Computer Languages* **1**(2) (1975) 151–160; also in: ref. [227] 245–258.

[216] HOARE, C.A.R., Some properties of predicate transformers, *J. Assoc. Comput. Mach.* **25**(3) (1978) 461–480.

[217] HOARE, C.A.R., Communicating sequential processes, *Comm. ACM* **21**(8) (1978) 666–677; also in: ref. [227] 259–288.

[218] HOARE, C.A.R., A calculus of total correctness for communicating processes, *Sci. Comput. Programming* **1**(1–2) (1981) 49–72.

[219] HOARE, C.A.R., Programs as predicates, *Phil. Trans. Roy. Soc. London A* **312** (1984) 475–489; also in: C.A.R. Hoare and J.C. Shepherdson, eds., *Mathematical Logic and Programming Languages* (Prentice Hall, New York, 1985) 141–154; also in: ref. [227] 333–349.

[220] HOARE, C.A.R., The mathematics of programming, in: S.N. Maheshwari, ed., *Proc. 5th Coll. on Foundations of Software Technology and Theoretical Computer Science*, Lecture Notes in Computer Science, Vol. 206 (Springer, Berlin, 1985) 1–18; also in: ref. [227] 351–370.

[221] HOARE, C.A.R. *Communicating Sequential Processes* (Prentice Hall, New York, 1985).

[222] HOARE, C.A.R. and D.C.S. ALLISON, Incomputability, *Computing Surveys* **4**(3) (1972) 169–178.

[223] HOARE, C.A.R., I.J. HAYES, HE JIFENG, C.C. MORGAN, A.W. ROSCOE, J.W. SANDERS, I.H. SORENSEN and B.A. SUFRIN, Laws of programming, *Comm. ACM* **30**(8) (1987) 672–686.

[224] HOARE, C.A.R. and HE JIFENG, The weakest prespecification, *Fund. Inform.* **9** (1986), Part I: 51–84, Part II: 217–252.

[225] HOARE, C.A.R. and HE JIFENG, The weakest prespecification, *Inform. Process. Lett.* **24**(2) (1987) 127–132.

[226] HOARE, C.A.R., HE JIFENG and J.W. SANDERS, Prespecification in data refinement, *Inform. Process. Lett.* **25**(2) (1987) 71–76.

[227] HOARE, C.A.R. and C.B. JONES, eds., *Essays in Computing Science* (Prentice Hall, New York, 1989).

[228] HOARE, C.A.R. and P. LAUER, Consistent and complementary formal theories of the semantics of programming languages, *Acta Inform.* **3** (1984) 135–155.

[229] HOARE, C.A.R. and N. WIRTH, An axiomatic definition of the programming language PASCAL, *Acta Inform.* **2** (1973) 335–355; also in: ref. [227] 153–169.

[230] HOOMAN, J. and W.-P. DE ROEVER, The quest goes on: a survey of proof systems for partial correctness of CSP, in: J.W. De Bakker, W.-P. De Roever and G. Rozenberg, eds., *Current Trends in Concurrency, Overviews and Tutorials*, Lecture Notes in Computer Science, Vol. 224 (Springer, Berlin, 1989) 343–395.

[231] HOOGEWIJS, A., Partial-predicate logic in computer science, *Acta Inform.* **24** (1987) 381–393.

[232] HORTALÁ-GONZÁLEZ, M.-T. and M. RODRÍGUEZ-ARTALEJO, Hoare's logic for nondeterministic regular programs: a nonstandard completeness theorem, Lecture Notes in Computer Science, Vol. 194 (Springer, Berlin, 1985) 270–280.

[233] HOWARD, J.H., Proving monitors, *Comm. ACM* **19**(5) (1976) 273–279.

[234] IGARASHI, S., R.L. LONDON and D.C. LUCKHAM, Automatic program verification I: a logical basis and its implementation, *Acta Inform.* **4**(1975) 145–182.

[235] IMMERMAN, N., Languages which capture complexity classes, in: *Proc. 15th Ann. ACM Symp. on Theory of Computing* (1983) 347–354.

[236] JACOBS, D. and D. GRIES, General correctness: a unification of partial and total correctness, *Acta Inform.* **22** (1985) 67–83.

[237] JANSSEN, T.M.V. and P. VAN EMDE BOAS, On the proper treatment of referencing, dereferencing and assignment, Lecture Notes in Computer Science, Vol. 52, (Springer, Berlin, 1977) 282–300

[238] JOHNSTONE, P.T., *Notes on Logic and Set Theory* (Cambridge Univ. Press, Cambridge, 1987).

[239] JONES, C.B., *Software Development: A Rigorous Approach* (Prentice Hall, Englewood Cliffs, NJ, 1980).

[240] JONES, C.B., Specification and design of (parallel programs), in: R.E.A. Mason, ed., *Information Processing 83* (Elsevier Science Publishers B.V., Amsterdam, 1983) 321–332.

[241] JONES, C.B., Tentative steps toward a development method for interfering programs, *ACM Trans. Programming Languages and Systems* **5**(4) (1983) 596–619.

[242] JONES, N.D. and S.S. MUCHNICK, Even simple programs are hard to analyze, *J. Assoc. Comput. Mach.* **24**(2) (1977) 338–350.

[243] JONES, N.D. and S.S. MUCHNICK, Complexity of finite memory programs with recursion, *J. Assoc. Comput. Mach.* **25** (1978) 312–321.

[244] JOSEPH, M., A. MOITRA and N. SOUNDARARAJAN, Proof rules for fault tolerant distributed programs, *Sci. Comput. Programming* **8** (1987) 43–67.

[245] JOSKO, B., A note on expressivity definitions in Hoare logic, Schriften zur Informatik und angewandten Mathematik, Rheinisch-Westfalische TH Aachen, Bericht 80, 1983.

[246] KATZ, S. and Z. MANNA, Logical analysis of programs, *Comm. ACM* **19** (1976) 188–206.

[247] KELLER, R.M., Formal verification of parallel programs, *Comm. ACM* **19** (7) (1976) 371–384.

[248] KFOURY, A.J., Definability by programs in first-order structures, *Theoret. Comput. Sci.* **25** (1983) 1–66.

[249] KFOURY, A.J. and P. URZYCZYN, Necessary and sufficient conditions for the universality of programming formalisms, *Acta Inform.* **22** (1985) 347–377.

[250] KING, J., A program verifier, Ph.D. Thesis, Carnegie-Mellon Univ., 1969.

[251] KLEENE, S.C., λ-Definability and recursiveness, *Duke Math. J.* **2** (1936) 340–353.

[252] KLEENE, S.C., *Introduction to Metamathematics* (Van Nostrand, Princeton, NJ, 1952).

[253] KLEENE, S.C., *Mathematical Logic* (Wiley, New York, 1967).

[254] Knuth, D.E., *The Art of Computer Programming, Vol. 1, Fundamental Algorithms* (Addison-Wesley, Reading, MA, 1968).

[255] Knuth, D.E., Semantics of context-free languages, *Math. Systems Theory* **2**(2) (1968) 127–145; correction: *Math. Systems Theory* **5** (1971) 95–96.

[256] Kowaltowski, T., Axiomatic approach to side effects and general jumps, *Acta Inform.* **7** (1977) 357–360.

[257] Lambek, J., How to program an infinite abacus, *Canad. Math. Bull.* **4** (1961) 295–302.

[258] Lamport, L., Proving the correctness of multiprocess programs, *IEEE Trans. Software Engrg.* **3**(2) (1977) 125–143.

[259] Lamport, L., "Sometime" is sometimes "not never", in: *Conf. Record 7th ACM SIGACT-SIGPLAN Symp. on Principles of Programming Languages* (1980) 174–185.

[260] Lamport, L., The "Hoare logic" of concurrent programs, *Acta Inform.* **14** (1980) 31–37.

[261] Lamport, L., Specifying concurrent program modules, *ACM Trans. Programming Languages and Systems* **5**(2) (1983) 190–222.

[262] Lamport, L., *win* and *sin*: predicate transformers for concurrency, *ACM Trans. Programming Languages and Systems* **12**(3) (1990) 396–428.

[263] Lamport, L., Control predicates are better than dummy variables for reasoning about program control, *ACM Trans. Programming Languages and Systems* **10**(2) (1988) 267–281.

[264] Lamport, L. and F.B. Schneider, The "Hoare logic" of CSP, and that all, *ACM Trans. Programming Languages and Systems* **6**(2) (1984) 281–296.

[265] Langmaack, H., On procedures as open subroutines, Part I, *Acta Inform.* **2** (1973) 311–333; Part II, *Acta Inform.* **3** (1974) 227–241.

[266] Langmaack, H., Aspects of programs with finite modes, in: M. Karpinski, ed., *Foundations of Computation Theory*, Lecture Notes in Computer Science, Vol. 158 (Springer, Berlin, 1983) 241–254.

[267] Langmaack, H. and E.R. Olderog, Present-day Hoare-like systems for programming languages with procedures: power, limits and most likely extensions, in: J.W. De Bakker and J. Van Leeuwen, eds., *Proc. 7th Internat. Coll. on Automata, Languages and Programming*, Lecture Notes in Computer Science, Vol. 85 (Springer, Berlin, 1980) 363–373.

[268] Lehmann, D., A. Pnueli and J. Stavi, Impartiality, justice and fairness: the ethics of concurrent termination, in: S. Even and O. Kariv, eds., *Proc. 8th Internat. Coll. on Automata, Languages and Programming*, Lecture Notes in Computer Science, Vol. 115 (Springer, Berlin, 1981) 264–277.

[269] Leivant, D., Partial-correctness theories as first-order theories, in: R.Parikh, ed., *Logics of Programs*, Lecture Notes in Computer Science, Vol. 193 (Springer, Berlin, 1985) 190–195.

[270] Leivant, D. and T. Fernando, Skinny and fleshy failures of relative completeness, in: *Conf. Record 14th ACM SIGACT-SIGPLAN Symp. on Principles of Programming Languages* (1987) 246–252.

[271] Levin, G.M. and D. Gries, A proof technique for communicating sequential processes, *Acta Inform.* **15** (1981) 281–302.

[272] Levitt, K.N., The application of program-proving techniques to the verification of synchronization processes, in: *1972 AFIPS Fall Joint Computer Conf.*, AFIPS Conference Proceedings, Vol. 41 (1972) 33–47.

[273] Lifschitz, V., On verification of programs with goto statements, *Inform. Process. Lett.* **18** (1984) 221–225.

[274] Lipton, R.J., A necessary and sufficient condition for the existence of Hoare Logics, in: *Proc. 18th Ann. IEEE–ACM Symp. on Foundations of Computer Science* (1977) 1–6.

[275] Loeckx, J. and K. Sieber, *The Foundations of Program Verification* (Teubner–Wiley, New York, 1987).

[276] London, R.L., Proof of algorithms: a new kind of certification (certification of algorithm 245, TREESORT 3), *Comm. ACM* **13**(6) (1970) 371–373.

[277] London, R.L., Proving programs correct: some techniques and examples, *BIT* **10** (1970) 168–182.

[278] London, R.L., J.V. Guttag, J.J. Horning, B.W. Lampson, J.G. Mitchell and G.J. Popek, Proof rules for the programming language Euclid, *Acta Inform.* **10** (1978) 1–26.

[279] Luckham, D.C. and N. Suzuki, Verification of arrays, record and pointer operations in Pascal, *ACM Trans. Programming Languages and Systems* **1** (1979) 226–244.

[280] Main, M.G. and D.B. Benson, Functional behavior of nondeterministic programs, in: M. Karpinski,

ed., *Foundations of Computation Theory*, Lecture Notes in Computer Science, Vol. 158 (Springer, Berlin, 1983) 290–301.

[281] MAJSTER-CEDERBAUM, M.E. A simple relation between relational and predicate transformer semantics for nondeterministic programs, *Inform. Process. Lett.* **11**(4, 5) (1980) 190–192.

[282] MAKOWSKY, J.A. and I. SAIN, On the equivalence of weak second-order and nonstandard time semantics for various program verification systems, in: *Proc. Symp. on Logic in Computer Science* (1986) 293–300.

[283] MANNA, Z., The correctness of programs, *J. Comput. System Sci.* **3**(2) (1969) 119–127.

[284] MANNA, Z., Mathematical theory of partial correctness, *J. Comput. System Sci.* **5**(3) (1971) 239–253.

[285] MANNA, Z., *Mathematical Theory of Computation* (McGraw-Hill, New York, 1974).

[286] MANNA, Z., S. NESS and J. VUILLEMIN, Inductive methods for proving properties of programs, *ACM SIGPLAN Notices* **7**(1) (1972) 27–50.

[287] MANNA, Z. and A. PNUELI, Formalization of properties of functional programs, *J. Assoc. Comput. Mach.* **17**(3) (1970) 555–569.

[288] MANNA, Z. and A. PNUELI, Axiomatic approach to total correctness of programs, *Acta Inform.* **3** (1974) 243–264.

[289] MANNA, Z. and A. PNUELI, Adequate proof principles for invariance and liveness properties of concurrent programs, *Sci. Comput. Programming* **4** (1984) 257–289.

[290] MANNA, Z. and A. PNUELI, Completing the temporal picture, in: G. Ausiello, M. Dezani-Ciancaglini and S. Ronchi Della Rocca, eds., *Proc. 16th Internat. Coll. on Automata, Languages and Programming*, Lecture Notes in Computer Science, Vol. 372 (Springer, Berlin, 1989) 534–558.

[291] MANNA, Z. and R. WALDINGER, Is "sometime" sometimes better than "always"?, *Comm. ACM* **21**(2) (1978) 159–172.

[292] MANNA, Z. and R. WALDINGER, Problematic features of programming languages: a situational-calculus approach, *Acta Inform.* **16** (1981) 371–426.

[293] MASON, I.A., Hoare's logic in the LF, LFCS report series ECS-LFCS-87-32, Laboratory for Foundations of Computer Science, Univ. of Edinburgh, 1987.

[294] MARTIN, A.J., A general proof rule for procedures in predicate transformer semantics, *Acta Inform.* **20** (1983) 301–313.

[295] MAZURKIEWICZ, A., Invariants of concurrent programs, in: J. Madey, ed., *Internat. Conf. on Information Processing, IFIP–INFOPOL-76* (1977) 353–372.

[296] MAZURKIEWICZ, A., Basic notions of trace theory, in: J.W. De Bakker, W.-P. De Roever and G. Rozenberg, eds., *Linear Time, Branching Time and Partial Orders in Logics and Models for Concurrency*, Lecture Notes in Computer Science, Vol. 354 (Springer, Berlin, 1989) 285–363.

[297] MCCARTHY, J., Towards a mathematical science of computation, in: C.M. Popplewell, ed., *Information Processing, Proc. IFIP Congress* (1962) 21–28.

[298] MCCARTHY, J., A basis for a mathematical theory of computation, in: P. Braffort and D. Hirschberg, eds., *Computer Programming and Formal Systems* (1963) 33–69.

[299] MEYER, A.R., Floyd–Hoare logic defines semantics: Preliminary version, in: *Proc. Symp. on Logic Computer Science* (1986) 44–48.

[300] MEYER, A.R. and J.Y. HALPERN, Axiomatic definitions of programming languages, a theoretical assessment, in: *Conf. Record 7th ACM SIGACT-SIGLAN Symp. on Principles of Programming Languages* (1980) 203–212; and *J. Assoc. Comput. Mach.* **29** (1982) 555–576.

[301] MEYER, A.R. and J.Y. HALPERN, Axiomatic definitions of programming languages II, in: *Conf. Record 8th ACM SIGACT-SIGPLAN Symp. on Principles of Programming Languages* (1981) 139–148.

[302] MEYER, A.R. and J.C. MITCHELL, Axiomatic definability and completeness for recursive programs, in: *Conf. Record 9th ACM SIGACT-SIGPLAN Symp. on Principles of Programming Languages* (1982) 337–346.

[303] MEYER, A.R. and J.C. MITCHELL, Termination assertions for recursive programs: completeness and axiomatic definability, *Inform. and Control* **56** (1983) 112–138.

[304] MEYER, A.R. and K. SIEBER, Towards fully abstract semantics for local variables: preliminary report, in: *Proc. 15th Ann. AGM SIGACT-SIGPLAN Symp. on Principles of Programming Languages* (1988) 191–203.

[305] Milne, R., Transforming predicate transformers, in: E.J. Neuhold, ed., *Formal Descriptions of Programming Concepts* (North-Holland, Amsterdam, 1978) 31–65.
[306] Milner, A.J.R.G., *A Calculus of Communicating Systems*, Lecture Notes in Computer Science, Vol. 92 (Springer, Berlin, 1980).
[307] Misra, J. and K.M. Chandy, Proofs of networks of processes, *IEEE Trans. Software Engng.* **7**(4) (1981) 417–426.
[308] Morgan, C.C., The specification statement, *ACM Trans. Programming Languages And Systems* **10**(3) (1988) 403–419.
[309] Morgan, C.C., Procedures, parameters and abstraction: separate concerns, *Sci. Comput. Programming* **11** (1988) 17–27.
[310] Morgan, C.C., Data refinement by miracles, *Inform. Process. Lett.* **26** (1988) 243–246.
[311] Morris, F.L. and C.B. Jones, An early program proof by Alan Turing, *Ann. Hist. Comput.* **6**(2) (1984) 139–143.
[312] Morris, J.H., Comments on "procedures and parameters", Manuscript (undated and unpublished).
[313] Morris, J.H. and B. Wegbreit, Subgoal induction, *Comm. ACM* **20**(4) (1977) 209–222.
[314] Morris, J.M., A general axiom of assignment, in: M. Broy and G. Schmidt, eds., *Theoretical Foundations of Programming Methodology* (1982) 25–34.
[315] Morris, J.M., Varieties of weakest liberal preconditions, *Inform. Process. Lett.* **25** (1987) 207–210.
[316] Morris, J.M., A theoretical basis for stepwise refinement and the programming calculus, *Sci. Comput. Programming* **9** (1987) 287–306.
[317] Murakami, M., Proving partial correctness of guarded Horn clauses programs, in: K. Furukawa, H. Tanaka and T. Fujisaki, eds., *Proc. 6th Conf. on Logic Programming*, Lecture Notes in Computer Science, Vol. 315 (Springer, Berlin, 1988) 215–235.
[318] Murtagh, T.P., Redundant proofs of non-interference in Levin–Gries CSP program proofs, *Acta Inform.* **24** (1987) 145–156.
[319] Naur, P., Proof of algorithms by general snapshots, *BIT* **6** (1966) 310–316.
[320] Naur, P., ed., Report on the algorithmic language Algol 60, *Comm. ACM* **3**(5) (1960) 299–314; Revised report on the algorithmic language Algol 60, *Comm. ACM* **6**(1) (1960) 1–17.
[321] Nelson, G., Verifying reachability invariants of linked structures, in: *Conf. Record 10th Ann. ACM SIGACT-SIGPLAN Symp. on Principles of Programming Languages* (1983) 38–47.
[322] Nelson, G., A generalization of Dijkstra's calculus, *ACM Trans. Programming Languages and Systems* **11**(4) (1989) 517–561.
[323] Németi, I., Nonstandard runs of Floyd-provable programs, in: A. Salwicki, ed., *Logics of Programs and their Applications*, Lectures Notes in Computer Science, Vol. 148 (Springer, Berlin, 1980) 186–204.
[324] Neuhold, E.J., The formal description of programming languages, *IBM System J.* **2** (1971) 86–112.
[325] Newton, G., Proving properties of interacting processes, *Acta Inform.* **4** (1975) 117–126.
[326] O'Donnell, M.J., A critique of the foundations of Hoare style programming logic, *Comm. ACM* **25** (12) (1982) 927–935.
[327] Olderog, E.-R., General equivalence of expressivity definitions using strongest postconditions, resp. weakest preconditions, Bericht 8007, Institut für Informatik und praktische Mathematik, Kiel Universität, 1980.
[328] Olderog, E.-R., Sound and complete Hoare-like calculi based on copy rules, *Acta Inform.* **16** (1981) 161–197.
[329] Olderog, E.-R., On the notion of expressiveness and the rule of adaptation, *Theoret. Comput. Sci.* **24** (1983) 337–347.
[330] Olderog, E.-R., Hoare's logic for programs with procedures—what has been achieved?, in: E. Clarke and D. Kozen, eds., *Logics of Programs*, Lecture Notes in Computer Science, Vol. 164 (Springer, Berlin, 1983) 385–395.
[331] Olderog, E.-R., A characterization of Hoare's logic for programs with Pascal-like procedures, in: *Proc. 15th Ann. ACM Symp. on Theory of Computing* (1983) 320–329.
[332] Olderog, E.-R., Correctness of programs with PASCAL-like procedures without global variables, *Theoret. Comput. Sci.* **30** (1984) 49–90.
[333] Olderog, E.-R. and C.A.R. Hoare, Specification-oriented semantics for communicating processes, *Acta Inform.* **23** (1986) 9–66.

[334] OPPEN, D.C. and S.A. COOK, Proving assertions about programs that manipulate data structures, in: *Proc. 7th Ann. ACM Symp. on Theory of Computing* (1975) 107–116.

[335] OWICKI, S.S., Axiomatic proof techniques for parallel programs, Ph.D. Thesis, TR-75-251, Computer Science Dept., Cornell Univ., 1975.

[336] OWICKI, S.S., Verifying concurrent programs with shared data classes, in: E.J. Neuhold, ed., *Formal Description of Programming Concepts* (North-Holland, Amsterdam, 1978) 279–299.

[337] OWICKI, S.S. and D. GRIES, An axiomatic proof techniques for parallel programs I, *Acta Inform.* **6** (1976) 319–340.

[338] OWICKI, S.S. and D. GRIES, Verifying properties of parallel programs: an axiomatic approach, *Comm. ACM* **19**(5) (1976) 279–285.

[339] PANDYA, P. and M. JOSEPH, A structure-directed total correctness proof rule for recursive procedure calls, *Comput. J.* **29** (6) (1986) 531–537.

[340] PARK, D.M.R., Fixpoint induction and proofs program properties, in: B. Meltzer and D. Michie, eds., *Machine Intelligence* **5** (Edinburgh Univ. Press, 1969) 59–78.

[341] PARK, D.M.R., A predicate transformer for weak fair iteration, in *Proc. 6th IBM Symp. on Mathematical Foundations of Computer Science, Logical Aspects of Programs*, (1981) 259–275.

[342] PLAISTED, D.A., The denotational semantics of nondeterministic recursive programs using coherent relations, in *Proc, Symp. on Logic in Computer Science 1986* (1986) 163–174.

[343] PLOTKIN, G.D., A powerdomain construction, *SIAM J. Comput.* **5** (1976) 452–487.

[344] PLOTKIN, G.D., A structural approach to operational semantics, Research Report DAIMI FN-19, Computer Science Dept. Aarhus Univ. Denmark 1981.

[345] PNUELI, A., The temporal logic of programs, in *Proc. 18th Ann. IEEE-ACM Symp. on Foundations of Computer Science* (1977) 46–57.

[346] PNUELI, A., In transition from global to modular temporal reasoning about programs, in: K.R. Apt, ed., *Logics and Models of Concurrent Systems*, Nato ASI Series, Vol. F13 (Springer Berlin, 1985) 123–144.

[347] PONSE, A., Process expressions and Hoare's logic, Research Report CS-R8905, Centrum voor Wiskunde en Informatica, Amsterdam, 1989.

[348] PRATT, V.R., Semantical considerations on Floyd–Hoare logic, in: *Proc. 17th Ann. IEEE-ACM Symp. on Foundations of Computer Science* (1976) 109–121.

[349] PRATT, V.R., Process logic: preliminary report, in: *Conf. Record 6th ACM SIGACT-SIGPLAN Symp. on Principles of Programming Languages* (1979) 93–100.

[350] PRAWITZ, D., *Natural Deduction, A Proof-Theoretic Study* (Almqvist & Wiksell, Stockholm, 1965).

[351] PRESBURGER, M., Über die Vollständigkeit eines gewissen Systems der Arthmetik ganzer Zahlen, in welchem die Addition als einzige Operation hervortritt, *Comptes Rendus du Premier Congrès des Mathématiciens Slaves* (1929) 92–101.

[352] PRITCHARD, P., A proof rule for multiple coroutine systems, *Inform. Process. Lett.* **4**(6) (1976) 141–143.

[353] PRITCHARD, P., Program proving — expression languages, in: *Information Processing 77* (North-Holland, Amsterdam, 1977) 727–731.

[354] RABIN, M.O., Decidable theories in: J. Barwise, ed., *Handbook of Mathematical Logic* (North-Holland, Amsterdam, 1978) 595–629.

[355] RASIOWA, H., Algorithmic logic and its extensions, a survey, in: *Proc. 5th Scandinavian Logic Symp.* (1979) 163–174.

[356] REPS, T. and B. ALPERN, Interactive proof checking, in: *Conf. Record 11th ACM SIGACT-SIGPLAN Symp. on Principles of Programming Languages* (1984) 36–45.

[357] REYNOLDS, J.C., Syntactic control of interference, in: *Conf. Record 5th ACM SIGACT-SIGPLAN Symp. on Principles of Programming Languages* (1978) 39–46.

[358] REYNOLDS, J.C., *The Craft of Programming* (Prentice Hall, Englewood Cliffs, NJ, 1981).

[359] REYNOLDS, J.C., Idealized Algol and its specification logic, in: D. Néel, ed., *Tools & Notions for Program Construction* (Cambridge Univ. Press, Cambridge, 1982).

[360] REYNOLDS, J.C., Syntactic control of interference, Part 2, in: G. Ausiello, M. Dezani-Ciancaglini and S. Ronchi Della Rocca, eds., *Proc. 16th Internat. Coll. on Automata, Languages and Programming*, Lecture Notes in Computer Science, Vol. 372 (Springer, Berlin, 1989) 704–722.

[361] Rodríguez-Artalejo, M., Some questions about expressiveness and relative completeness in Hoare's logic, *Theoret. Comput. Sci.* **39** (1985) 189–206.

[362] Rogers, Jr, H., *Theory of Recursive Functions and Effective Computability* (McGraw-Hill, New York, 1967).

[363] Sain, I., A simple proof for the completeness of Floyd's method, *Theoret. Comput. Sci.* **35** (1985) 345–348.

[364] Salwicki, A., Formalized algorithmic languages, *Bull. de l'Académie Polonaise des Sciences, Série des Sciences Mathématiques, Astronomiques et Physiques* **18**(5) (1970) 227–232.

[365] Schneider, F.B. and G.R. Andrews, Concepts for concurrent programming, in: J.W. De Bakker, W.P. De Roever and G. Rozenberg, eds., *Current Trends in Concurrency, Overviews and Tutorials*, Lecture Notes in Computer Science, Vol. 224 (Springer, Berlin, 1986) 670–716.

[366] Scwartz, R.L., An axiomatic treatment of ALGOL68 routines, in: H.A. Maurer, ed., *Proc. 6th Internat. Coll. on Automata, Languages and Programming*, Lecture Notes in Computer Science, Vol. 71 (Springer, Berlin, 1979) 530–545.

[367] Schwartz, R.L. and D.M. Berry, A semantic view of ALGOL 68, *Computer Languages* **4** (1979) 1–15.

[368] Scott, D.S. and J.W. De Bakker, A theory of programs, Unpublished manuscript, *Seminar on Programming*, IBM Research Center, Vienna, 1969.

[369] Scott, D.S. and C. Strachey, Toward a mathematical semantics for computer languages, in: J. Fox, ed., *Computers and Automata* (Wiley, New York, 1972) 19–46.

[370] Shoenfield, J.R., Axioms of set theory, in: J. Barwise, ed., *Handbook of Mathematical Logic* (North-Holland, Amsterdam, 1978) 321–344.

[371] Sieber, K., A partial correctness logic for procedures, in: R. Parikh, ed., *Logics of Programs*, Lecture Notes in Computer Science, Vol. 193 (Springer, Berlin, 1985) 320–342.

[372] Sites, R.L., Proving that computer programs terminate cleanly, Research Report STAN-CS-74-418, Computer Science Dept., Stanford Univ., 1974.

[373] Skolem, T., Über die Nicht-charakterisierbarkeit der Zahlenreihe mittels endlich oder abzählbar unendlich vieler Aussagen mit ausschliesslich Zahlenvariablen, *Fund. Math.* **23** (1934) 150–161.

[374] Smorynski, C., The incompleteness theorems, in: J. Barwise, ed., *Handbook of Mathematical Logic* (North-Holland, Amsterdam, 1978) 821–865.

[375] Sobel, A.E.K. and N. Soundararajan, A proof system for distributed processes, in: R. Parikh, ed., *Logics of Programs*, Lecture Notes in Computer Science, Vol. 193 (Springer, Berlin, 1985) 343–358.

[376] Sokolowski, S., Axioms for total correctness, *Acta Inform.* **9** (1976) 61–71.

[377] Sokolowski, S., Total correctness for procedures, in: J. Gruska, ed., *Proc. 6th Symp. on Mathematical Foundations of Computer Science*, Lecture Notes in Computer Science, Vol. 53 (Springer, Berlin, 1977) 475–483.

[378] Sokolowski, S., Partial correctness: the term-wise approach, *Sci. Comput. Programming* **4** (1984) 141–157.

[379] Sokolowski, S., Soundness of Hoare's logic: an automated proof using LCF, *ACM Trans. on Programming Languages and Systems* **9**(1) (1987) 100–120.

[380] Soundararajan, N., Correctness proofs of CSP programs, *Theoret. Comput. Sci.* **24** (1983) 131–141.

[381] Soundararajan, N., A proof technique for parallel programs, *Theoret. Comput. Sci.* **31** (1984) 13–29.

[382] Soundararajan, N., Axiomatic semantics of communicating sequential processes, *ACM Trans. on Programming Languages and Systems* **6**(4) (1984) 647–662.

[383] Stirling, C., A compositional reformulation of Owicki–Gries's partial correctness logic for a concurrent while language, in: L. Kott, ed., *Proc. 13th Internat. Coll. on Automata, Languages and Programming*, Lecture Notes in Computer Science, Vol. 226 (Springer, Berlin, 1986) 408–415.

[384] Stirling, C., A generalization of Owicki–Gries's Hoare logic for a concurrent while language, *Theoret. Comput. Sci.* **58** (1988) 347–359.

[385] Tarski, A., A lattice theoretic fixpoint theorem and its applications, *Pacific J. Math.* **25**(2) (1955) 285–309.

[386] Taubenfeld, G. and N. Francez, Proof rules for communication abstraction, in: M. Joseph and R. Shyamasundar, eds., *Foundations of Software Technology and Theoretical Computer Science*, Lecture Notes in Computer Science, Vol. 181 (Springer, Berlin, 1984) 444–465.

[387] TAYLOR, R.N., A general-purpose algorithm for analyzing concurrent programs, *Comm. ACM* **26**(5) (1983) 362–376.

[388] TENNENT, R.D., The denotational semantics of programming languages, *Comm. ACM* **19**(8) (1976) 437–453.

[389] TENNENT, R.D., Semantical analysis of specification logic, in: R. Parikh, ed., *Logics of Programs*, Lecture Notes in Computer Science, Vol. 193 (Springer, Berlin, 1985) 373–386.

[390] TIURYN, J., A simple programming language with data types: semantics and verification, in: R. Parikh, ed., *Logics of Programs*, Lecture Notes in Computer Science, Vol. 193 (Springer, Berlin, 1985) 387–405.

[391] TRAKHTENBROT, B.A., J.Y. HALPERN and A.R. MEYER, From denotational to operational and axiomatic semantics for Algol-like languages: an overview, in: E. Clarke and D. Kozen, eds., *Logics of Programs*, Lecture Notes in Computer Science, Vol. 164 (Springer, Berlin, 1983) 474–500.

[392] TURING, A.M., On computable numbers, with an application to the Entscheidungsproblem, *Proc. London Math. Soc., Ser. 2*, **42** (1936) 230–265; correction, *Ibidem* **43** (1937) 544–546.

[393] TURING, A.M., Computability and λ-definability, *J. Symbolic Logic* **2** (1937) 153–163.

[394] TURING, A.M., On checking a large routine, in: *Report of a Conference on High-speed Automatic Calculating Machines*, University Mathematics Laboratory, Cambridge (1949) 67–69.

[395] URZYCZYN, P., A necessary and sufficient condition in order that a Herbrand interpretation be expressive relative to recursive programs, *Inform. and Control* **56** (1983) 212–219.

[396] VAN LAMSWEERDE, A. and M. SINTZOFF, Formal derivation of strongly correct concurrent programs, *Acta Inform.* **12** (1979) 1–31.

[397] VERJUS, J.-P., On the proof a distributed algorithm, *Inform. Process. Lett.* **25**(3) (1987) 145–147.

[398] WAGNER, E.G., A categorical view of weakest liberal preconditions, Lecture Notes in Computer Science, Vol. 240 (Springer, Berlin, 1986) 198–205.

[399] WAND, M., A new incompleteness result for Hoare's system, *J. Assoc. Comput. Mach.* **25**(1) (1978) 168–175.

[400] WAND, M., *Induction, Recursion and Programming* (North-Holland, Amsterdam, 1980).

[401] WANG, A., An axiomatic basis for proving total correctness of goto-programs, *BIT* **16** (1976) 88–102.

[402] WANG, A. and O.-J. DAHL, Coroutine sequencing in a block structured environment, *BIT* **11** (1971) 425–449.

[403] WECHLER, A., Hoare algebras versus dynamic algebras, in: *Algebra, Combinatorics and Logic in Computer Science, Vol. I, II*, Colloquia Mathematica Societatis János Bolyai, Vol. 42 (North-Holland, Amsterdam, 1986) 835–847.

[404] WEGBREIT, B., The synthesis of loop predicates, *Comm. ACM* **17** (1974) 102–112.

[405] WIRTH, N., The programming language PASCAL, *Acta Inform.* **1**(1) (1971) 35-63.

[406] YEH, R.T., Verification of nondeterministic programs, Technical Report TR-56, Dept. of Computer Sciences, Univ. of Texas, Austin, TX, 1976; revised 1977.

[407] ZHOU CHAO CHEN and C.A.R. HOARE, Partial correctness of communicating sequential processes, in: *Proc. 2nd Internat. Conf. on Distributed Computing Systems* (1981) 1–12.

[408] ZWIERS, J., *Compositionality, Concurrency and Partial Correctness, Proof Theories for Networks of Processes and their Relationship*, Lecture Notes in Computer Science, Vol. 321 (Springer, Berlin, 1989).

[409] ZWIERS, J., A. DE BRUIN and W.P. DE ROEVER, A proof system for partial correctness of dynamic networks of processes, in: E. Clarke and D. Kozen, eds., *Logics of Programs*, Lecture Notes in Computer Science, Vol. 164 (Springer, Berlin, 1983) 513–527.

[410] ZWIERS, J., W.P. DE ROEVER and P. VAN EMDE BOAS, Compositionality and concurrent networks: soundness and completeness of a proof system, in: W. Brauer, ed., *Proc. 12th Internat. Coll. on Automata, Languages and Programming*, Lecture Notes in Computer Science, Vol. 194 (Springer, Berlin, 1985) 509–519.

CHAPTER 16

Temporal and Modal Logic

E. Allen EMERSON

Computer Sciences Department, University of Texas at Austin, Austin, TX 78712, USA

Contents

1. Introduction 997
2. Classification of temporal logics 998
3. The technical framework of Linear Temporal Logic 1000
4. The technical framework of Branching Temporal Logic 1011
5. Concurrent computation: a framework 1017
6. Theoretical aspects of Temporal Logic 1021
7. The application of Temporal Logic to program reasoning 1048
8. Other modal and temporal logics in computer science 1064
Acknowledgment 1067
References 1067

HANDBOOK OF THEORETICAL COMPUTER SCIENCE
Edited by J. van Leeuwen

1. Introduction

The class of Modal Logics was originally developed by philosophers to study different "modes" of truth. For example, the assertion P may be false in the present world, and yet the assertion *possibly* P may be true if there exists an alternate world where P is true. Temporal Logic is a special type of Modal Logic; it provides a formal system for qualitatively describing and reasoning about how the truth values of assertions change over time. In a system of Temporal Logic, various temporal operators or "modalities" are provided to describe and reason about how the truth values of assertions vary with time. Typical temporal operators include *sometimes* P which is true now if there is a future moment at which P becomes true and *always* Q which is true now if Q is true at all future moments.

In a landmark paper [83] Pnueli argued that Temporal Logic could be a useful formalism for specifying and verifying correctness of computer programs, one that is especially appropriate for reasoning about nonterminating or continuously operating concurrent programs such as operating systems and network communication protocols. In an ordinary sequential program, e.g. a program to sort a list of numbers, program correctness can be formulated in terms of a Precondition/Postcondition pair in a formalism such as *Hoare's Logic* because the program's underlying semantics can be viewed as given by a *transformation* from an initial state to a final state. However, for a continuously operating, *reactive* program such as an operating system, its normal behavior is a nonterminating computation which maintains an ongoing interaction with the environment. Since there is no final state, formalisms such as Hoare's Logic which are based on a transformational semantics, are of little use for such nonterminating programs. The operators of temporal logic such as *sometimes* and *always* appear quite appropriate for describing the time-varying behaviour of such programs.

These ideas were subsequently explored and extended by a number of researchers. Now Temporal Logic is an active area of research interest. It has been used or proposed for use in virtually all aspects of concurrent program design, including specification, verification, manual program composition (development), and mechanical program synthesis. In order to support these applications a great deal of mathematical machinery connected with Temporal Logic has been developed. In this survey we focus on this machinery, which is most relevant to theoretical computer science. Some attention is given, however, to motivating applications.

The remainder of this paper is organized as follows: In Section 2 we describe a multiaxis classification of systems of Temporal Logic, in order to give the reader a feel for the large variety of systems possible. Our presentation centers around only a few—those most thoroughly investigated—types of Temporal Logics. In Section 3 we describe the framework of Linear Temporal Logic. In both its propositional and first-order forms, Linear Temporal Logic has been widely employed in the specification and verification of programs. In Section 4 we describe the competing framework of Branching Temporal Logic which has also seen wide use. In Section 5 we describe how Temporal Logic structures can be used to model concurrent programs using non-determinism and fairness. Technical machinery for Temporal reasoning is discussed in Section 6, including decision procedures and axiom systems. Applications of Temporal

Logic are discussed in Section 7, while in the concluding Section 8 other modal and temporal logics in computer science are briefly described.

2. Classification of temporal logics

We can classify most systems of TL (Temporal Logic) used for reasoning about concurrent programs along a number of axes: propositional versus first-order, global versus compositional, branching versus linear, points versus intervals, and past versus future tense. Most research to date has concentrated on global, point-based, discrete-time, future-tense logics; therefore our survey will focus on representative systems of this type. However, to give the reader an idea of the wide range of possibilities in formulating a system of Temporal Logic, we describe the various alternatives in more detail below.

2.1. *Propositional versus first-order*

In a propositional TL, the nontemporal (i.e., nonmodal) portion of the logic is just classical propositional logic. Thus formulae are built up from *atomic propositions*, which intuitively express atomic facts about the underlying state of the concurrent system, *truth-functional connectives*, such as $\wedge$, $\vee$, $\neg$ (representing "and", "or", and "not", respectively), and the temporal operators. Propositional TL corresponds to the most abstract level of reasoning, analogous to classical propositional logic.

The atomic propositions of propositional TL are refined into expressions built up from variables, constants, functions, predicates, and quantifiers, to get *First-order TL*. There are several different types of First-order TLs. We can distinguish between *uninterpreted* First-order TL where we make no assumptions about the special properties of structures considered, and *interpreted* First-order TL where a specific structure (or class of structures) is assumed. In a *fully* interpreted First-order TL, we have a specific domain (e.g. *integer* or *stack*) for each variable, a specific, concrete function over the domain for each function symbol, and so forth, while in a *partially* interpreted First-order TL we might assume a specific domain but, e.g., leave the function symbols uninterpreted. It is also common to distinguish between *local* variables which, by the semantics, are assigned different values in different states, and *global* variables which are assigned a single value which holds globally over all states. Finally, we can choose to impose or not impose various syntactic restrictions on the interaction of quantifiers and temporal operators. An unrestricted syntax will allow, e.g., modal operators within the scope of quantifiers. For example, we have instances of *Barcan's Formula:* $\forall y\ \mathit{always}\,(P(y)) \equiv \mathit{always}\,(\forall y\ P(y))$. Such unrestricted logics tend to be highly undecidable. In contrast we can disallow such quantification over temporal operators to get a restricted first-order TL consisting of essentially propositional TL plus a first-order language for specifying the "atomic" propositions.

2.2. *Global versus compositional*

Most systems of TL proposed to date are *endogenous*. In an endogenous TL, all temporal operators are interpreted in a single universe corresponding to a single concurrent program. Such TLs are suitable for *global* reasoning about a complete, concurrent program. In an *exogenous* TL, the syntax of the temporal operators allows expression of correctness properties concerning several different programs (or program fragments) in the same formula. Such logics facilitate *compositional* (or *modular*) program reasoning: we can verify a complete program by specifying and verifying its constituent subprograms, and then combining them into a complete program together with its proof of correctness, using the proofs of the subprograms as lemmas (cf. [5, 86]).

2.3. *Branching versus linear time*

In defining a system of temporal logic, there are two possible views regarding the underlying nature of time. One is that the course of time is linear: at each moment there is only one possible future moment. The other is that time has a branching, tree-like nature: at each moment, time may split into alternate courses representing different possible futures. Depending upon which view is chosen, we classify a system of temporal logic as either a linear-time logic in which the semantics of the time structure is linear, or a system of branching-time logic based on the semantics corresponding to a branching-time structure. The temporal modalities of a temporal logic system usually reflect the character of time assumed in the semantics. Thus, in a logic of linear time, temporal modalities are provided for describing events along a single time line. In contrast, in a logic of branching time, the modalities reflect the branching nature of time by allowing quantification over possible futures. Both approaches have been applied to program reasoning, and it is a matter of debate as to whether branching or linear time is preferable (cf. [53, 26, 87]).

2.4. *Points versus intervals*

Most temporal logic formalisms developed for program reasoning have been based on temporal operators that are evaluated as true or false of *points* in time. Some formalisms (cf. [100, 74]), however, have temporal operators that are evaluated over *intervals* of time, the claim being that use of intervals greatly simplifies the formulation of certain correctness properties.

The following related issue has to do with the underlying structure of time.

2.5. *Discrete versus continuous*

In most temporal logics used for program reasoning, time is *discrete* where the present moment corresponds to the program's current state and the next moment corresponds to the program's immediate successor state. Thus the temporal structure corresponding to a program execution, a sequence of states, is the nonnegative integers.

However, tense logics interpreted over a *continuous* (or dense) time structure such as the reals (or rationals) have been investigated by philosophers. Their application to reasoning about concurrent programs was proposed in [6] to facilitate the formulation of fully abstract semantics. Such continuous-time logics may also have applications in so-called real-time programs where strict, quantitative performance requirements are placed on programs.

2.6. *Past versus future*

As originally developed by philosophers, temporal modalities were provided for describing the occurrence of events in the past as well as the future. However, in most temporal logics for reasoning about concurrency, only future-tense operators are provided. This appears reasonable since, as a rule, program executions have a definite starting time, and it can be shown that, as a consequence, inclusion of past-tense operators adds no expressive power. Recently, however, it has been advanced that use of the past-tense operators might be useful simply in order to make the formulation of specifications more natural and convenient (cf. [60]). Moreover, past-tense operators appear to play an important role in compositional specification somewhat analogous to that of history variables.

3. The technical framework of Linear Temporal Logic

3.1. *Timelines*

In linear temporal logic the underlying the structure of time is a totally ordered set $(S, <)$. In the sequel we will further assume that the underlying structure of time is isomorphic to the natural numbers with their usual ordering $(\mathbb{N}, <)$. Note that, under our assumption, time

(i) is discrete,

(ii) has an initial moment with no predecessors, and

(iii) is infinite into the future.

We remark that these properties seem quite appropriate in view of our intended application: reasoning about the behavior of ongoing concurrent programs. Property (i) reflects the fact that modern-day computers are discrete, digital devices; property (ii) is appropriate since computation begins at an initial state; and (iii) is appropriate since we develop our formalism for reasoning about ongoing, ideally nonterminating behavior.

Let AP be an underlying set of atomic proposition symbols, which we denote by P, Q, P_1, Q_1, P', Q', etc. We can then formalize our notion of a timeline as a *linear-time structure* $M=(S,x,L)$ where

- S is a set of *states*,
- $x\colon \mathbb{N} \to S$ is an infinite sequence of states, and
- $L\colon S \to \text{PowerSet(AP)}$ is a labelling of each state with the set of atomic propositions in AP true at the state.

We usually employ the more convenient notation $x=(s_0, s_1, s_2, \ldots)=(x(0),\ x(1),$

$x(2),\ldots$) to denote the *timeline* x. Alternative terminology permits us to refer to x as a *fullpath*, or *computation sequence*, or *computation*, or simply as a *path*; the latter could cause confusion in rare instances since we intend the *maximal path* x, not just one of its prefixes, whence the term fullpath; but ordinarily no confusion will result. (Other synonyms include execution, execution sequence, trace, history, and run.)

REMARK. We could convey the same information by associating a truth value to each atomic proposition at each state by defining L as a mapping AP $\rightarrow$ PowerSet(S) which assigns to each atomic proposition the set of states at which it is true. Another equivalent alternative is to use a mapping $L: S \times \mathrm{AP} \rightarrow \{true, false\}$ such that $L(s, P) = true$ iff it is intended that P be true at s. Still another alternative is to have $L: S \rightarrow (\mathrm{AP} \rightarrow \{true, false\})$ so that $L(s)$ is an interpretation of each proposition symbol at state s. In the future, we will use whichever presentation is most convenient for the purpose at hand, assuming the above equivalences to be obvious.

3.2. Propositional Linear Temporal Logic

In this subsection we will define the formal syntax and semantics of Propositional Linear Temporal Logic (PLTL). The basic temporal operators of this system are Fp ("sometime p"; also read as "eventually p"), Gp ("always p"; also read as "henceforth p"), Xp ("nexttime p"), and p U q ("p until q"). Figure 1 illustrates their intuitive meanings. The formulae of this system are built up from atomic propositions, the truth-functional connectives ($\wedge$, $\vee$, $\neg$, etc.) and the above-mentioned temporal operators. This system, or some slight variation thereof, is frequently employed in applications of temporal logic to concurrent programming.

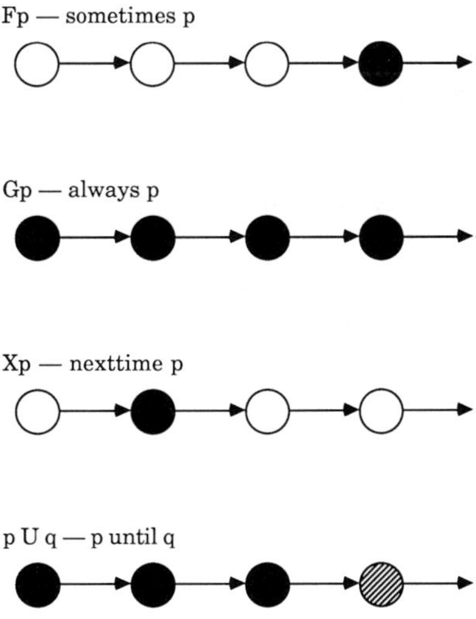

Fig. 1. Intuition for linear-time operators.

3.2.1. Syntax

The set of formulae of Propositional Linear Temporal Logic (PLTL) is the least set of formulae generated by the following rules:

(1) each atomic proposition P is a formula;
(2) if p and q are formulae then $p \wedge q$ and $\neg p$ are formulae;
(3) if p and q are formulae then $p \mathsf{U} q$ and $\mathsf{X}p$ are formulae.

The other formulae can then be introduced as abbreviations in the usual way: For the propositional connectives, $p \vee q$ abbreviates $\neg(\neg p \wedge \neg q)$, $p \Rightarrow q$ abbreviates $\neg p \vee q$, and $p \equiv q$ abbreviates $(p \Rightarrow q) \wedge (q \Rightarrow p)$. The Boolean constant *true* abbreviates $p \vee \neg p$, while *false* abbreviates $\neg$*true*. Then the temporal connective $\mathsf{F}p$ abbreviates (*true* $\mathsf{U}\, p$) and $\mathsf{G}p$ abbreviates $\neg\mathsf{F}\neg p$. It is convenient to also have $\mathsf{F}^{\infty}p$ abbreviate $\mathsf{GF}p$ (infinitely often), $\mathsf{G}^{\infty}p$ abbreviate $\mathsf{FG}p$ ("almost everywhere"), and $(p\,\mathsf{B}\,q)$ ("p before q") abbreviate $\neg((\neg p)\mathsf{U}\, q)$.

REMARK. The above is an *abstract syntax* where we have suppressed detail regarding parenthesization, binding power of operators, and so forth. In practice, we use the following notational conventions, supplemented by auxiliary parentheses as needed: the connectives of highest binding power are the temporal operators F, G, X, U, B, F^{∞}, and G^{∞}; the operator $\neg$ is of next highest binding power, followed by $\wedge$, followed by $\vee$, followed by $\Rightarrow$, followed finally by $\equiv$ as the operator of least binding power.

EXAMPLE. $\neg p_1 \mathsf{U}\, q_1 \wedge r_1 \vee r_2$ means $((\neg(p_1 \mathsf{U}\, q_1)) \wedge r_1) \vee r_2$.

3.2.2. Semantics

We define the semantics of a formula p of PLTL with respect to a linear-time structure $M = (S, x, L)$ as above. We write $M, x \models p$ to mean that "in structure M formula p is true of timeline x." When M is understood we write $x \models p$. The notational convention that $x^i =$ the suffix path $s_i, s_{i+1}, s_{i+2} \ldots$ is used. We define $\models$ inductively on the structure of the formulae:

(1) $x \models P$ iff $P \in L(s_0)$, for atomic proposition P;
(2) $x \models p \wedge q$ iff $x \models p$ and $x \models q$,
$x \models \neg p$ iff it is not the case that $x \models p$;
(3) $x \models (p \mathsf{U}\, q)$ iff $\exists j\, (x^j \models q$ and $\forall k < j\, (x^k \models p))$,
$x \models \mathsf{X}p$ iff $x^1 \models p$.

The modality $(p \mathsf{U}\, q)$, read as "p until q", asserts that q does eventually hold and that p will hold everywhere prior to q. The modality $\mathsf{X}p$, read as "next time p", holds now iff p holds at the next moment.

For conciseness, we took the temporal operator U and X as primitive, and defined the others as abbreviations. However, the other operators are themselves of sufficient independent importance that we also give their formal definitions explicitly.

The modality $\mathsf{F}q$, read as "sometimes q" or "eventually q" and meaning that at some future moment q is true, is formally defined so that

$$x \models \mathsf{F}q \text{ iff } \exists j\, (x^j \models q).$$

The modality $\mathsf{G}q$, read as "always q" or "henceforth q" and meaning that at all future moments q is true, can be formally defined as

$$x \models \mathsf{G}q \text{ iff } \forall j\, (x^j \models q).$$

The modality $(p \,\mathsf{B}\, q)$, read as "p precedes q" or "p before q" and which intuitively means that "if q ever happens in the future, it is strictly preceded by an occurrence of p", has the following formal definition:

$$x \models (p \,\mathsf{B}\, q) \text{ iff } \forall j\, (x^j \models q \text{ implies } \exists k < j\, (x^k \models p)).$$

The modality $\mathsf{F}^{\infty} p$, which is read as "infinitely often p", intuitively means that it is always true that p eventually holds, or in other words that p is true infinitely often, and can be defined formally as

$$x \models \mathsf{F}^{\infty} p \text{ iff } \forall k\, \exists j \geqslant k\, (x^j \models p).$$

The modality $\mathsf{G}^{\infty} p$, which is read as "almost everywhere p" or "almost always p", intuitively means that p holds at all but a finite number of times, and can be defined as

$$x \models \mathsf{G}^{\infty} p \text{ iff } \exists k\, \forall j > k\, (x^j \models p).$$

3.2.3. Basic definitions

We say that PLTL formula p is *satisfiable* iff there exists a linear-time structure $M = (S, x, L)$ such that $M, x \models p$. We say that any such structure defines a *model* of p. We say that p is *valid*, and write $\models p$, iff for all linear-time structures $M = (S, x, L)$ we have $M, x \models p$. Note that p is valid iff $\neg p$ is not satisfiable.

3.2.4. Examples

We have the following examples:

- $p \Rightarrow \mathsf{F}q$ intuitively means that "if p is true now then at some future moment q will be true". This formula is satisfiable, but not valid.
- $\mathsf{G}(p \Rightarrow \mathsf{F}q)$ intuitively means that "whenever p is true, q will be true at some subsequent moment". This formula is also satisfiable, but not valid.
- $\mathsf{G}(p \Rightarrow \mathsf{F}q) \Rightarrow (p \Rightarrow \mathsf{F}q)$ is a valid formula, but its converse only satisfiable.
- $p \wedge \mathsf{G}(p \Rightarrow \mathsf{X}p) \Rightarrow \mathsf{G}p$ means that if p is true now and whenever p is true it is also true at the next moment, then p is always true. This formula is valid, and is a temporal formulation of mathematical induction.
- $(p \,\mathsf{U}\, q) \wedge ((\neg p) \,\mathsf{B}\, q)$ means that p will be true until q eventually holds, and that the first occurrence of q will be preceded by $\neg p$. This formula is unsatisfiable.

Significant validities. The duality between the linear temporal operators are illustrated by the following assertions:

$$\models \mathsf{G}\neg p \equiv \neg \mathsf{F}p, \qquad \models \mathsf{F}\neg p \equiv \neg \mathsf{G}p,$$
$$\models \mathsf{X}\neg p \equiv \neg \mathsf{X}p,$$
$$\models \mathsf{F}^{\infty}\neg p \equiv \neg \mathsf{G}^{\infty} p, \qquad \models \mathsf{G}^{\infty}\neg p \equiv \neg \mathsf{F}^{\infty} p,$$
$$\models ((\neg p) \,\mathsf{U}\, q) \equiv \neg (p \,\mathsf{B}\, q).$$

The following are some important implications between the temporal operators, which cannot be strengthened to equivalences:

$$\models p \Rightarrow \mathsf{F}p, \qquad \models \mathsf{G}p \Rightarrow p,$$
$$\models \mathsf{X}p \Rightarrow \mathsf{F}p, \qquad \models \mathsf{G}p \Rightarrow \mathsf{X}p,$$
$$\models \mathsf{G}p \Rightarrow \mathsf{F}p, \qquad \models \mathsf{G}p \Rightarrow \mathsf{XG}p,$$
$$\models p\,\mathsf{U}\,q \Rightarrow \mathsf{F}q, \qquad \models \mathsf{G}^{\infty}q \Rightarrow \mathsf{F}^{\infty}q.$$

The idempotence of F, G, F^{∞}, and G^{∞} are asserted below:

$$\models \mathsf{FF}p \equiv \mathsf{F}p, \qquad \models \mathsf{F}^{\infty}\mathsf{F}^{\infty} \equiv \mathsf{F}^{\infty}p, \qquad \models \mathsf{GG}p \equiv \mathsf{G}p, \qquad \models \mathsf{G}^{\infty}\mathsf{G}^{\infty}p \equiv \mathsf{G}^{\infty}p.$$

Note that, of course, $\mathsf{XX}p \equiv \mathsf{X}p$ is not valid. We also have that X commutes with F, G, and U:

$$\models \mathsf{XF}p \equiv \mathsf{FX}p, \qquad \models \mathsf{XG}p \equiv \mathsf{GX}p,$$
$$\models ((\mathsf{X}p)\,\mathsf{U}\,(\mathsf{X}q)) \equiv \mathsf{X}(p\,\mathsf{U}\,q).$$

The infinitary modalities F^{∞} and G^{∞} "gobble up" other unitary modalities applied to them:

$$\models \mathsf{F}^{\infty}p \equiv \mathsf{XF}^{\infty}p \equiv \mathsf{FF}^{\infty}p \equiv \mathsf{GF}^{\infty}p \equiv \mathsf{F}^{\infty}\mathsf{F}^{\infty}p \equiv \mathsf{G}^{\infty}\mathsf{F}^{\infty}p,$$
$$\models \mathsf{G}^{\infty}p \equiv \mathsf{XG}^{\infty}p \equiv \mathsf{FG}^{\infty}p \equiv \mathsf{GG}^{\infty}p \equiv \mathsf{F}^{\infty}\mathsf{G}^{\infty}p \equiv \mathsf{G}^{\infty}\mathsf{G}^{\infty}p,$$

(Note that in the above we make use of the abuse of notation that $\models a_1 \equiv \cdots \equiv a_n$ abbreviates the $n-1$ valid equivalences $\models a_1 \equiv a_2, \ldots, \models a_{n-1} \equiv a_n$.) The F, F^{∞} operators have an existential nature, the G, G^{∞} operators a universal nature, while the U operator is universal in its first argument and existential in its second argument. We thus have the following distributivity relations between these temporal operators and the Boolean connectives $\wedge$ and $\vee$:

$$\models \mathsf{F}(p \vee q) \equiv (\mathsf{F}p \vee \mathsf{F}q), \qquad \models \mathsf{F}^{\infty}(p \vee q) \equiv (\mathsf{F}^{\infty}p \vee \mathsf{F}^{\infty}q),$$
$$\models \mathsf{G}(p \wedge q) \equiv (\mathsf{G}p \wedge \mathsf{G}q), \qquad \models \mathsf{G}^{\infty}(p \wedge q) \equiv (\mathsf{G}^{\infty}p \wedge \mathsf{G}^{\infty}q),$$
$$\models ((p \wedge q)\,\mathsf{U}\,r) \equiv ((p\,\mathsf{U}\,r) \wedge (q\,\mathsf{U}\,r)),$$
$$\models (p\,\mathsf{U}\,(q \vee r)) \equiv ((p\,\mathsf{U}\,q) \vee (p\,\mathsf{U}\,r)).$$

Since the X operator refers to a unique next moment, it distributes with all the Boolean connectives:

$$\models \mathsf{X}(p \vee q) \equiv (\mathsf{X}p \vee \mathsf{X}q), \qquad \models \mathsf{X}(p \wedge q) \equiv (\mathsf{X}p \wedge \mathsf{X}q),$$
$$\models \mathsf{X}(p \Rightarrow q) \equiv (\mathsf{X}p \Rightarrow \mathsf{X}q), \qquad \models \mathsf{X}(p \equiv q) \equiv (\mathsf{X}p \equiv \mathsf{X}q).$$

(Note that $\models \mathsf{X}\neg p \equiv \neg \mathsf{X}p$ was given above.)

When we mix operators of universal and existential characters we get the following implications, which again cannot be strengthened to equivalences:

$$\models (\mathsf{G}p \vee \mathsf{G}q) \Rightarrow \mathsf{G}(p \vee q), \qquad \models (\mathsf{G}^{\infty}p \vee \mathsf{G}^{\infty}q) \Rightarrow \mathsf{G}^{\infty}(p \vee q),$$
$$\models \mathsf{F}(p \wedge q) \Rightarrow \mathsf{F}p \wedge \mathsf{F}q, \qquad \models \mathsf{F}^{\infty}(p \wedge q) \Rightarrow (\mathsf{F}^{\infty}p \wedge \mathsf{F}^{\infty}q),$$
$$\models ((p\,\mathsf{U}\,r) \vee (q\,\mathsf{U}\,r)) \Rightarrow ((p \vee q)\,\mathsf{U}\,r),$$
$$\models (p\,\mathsf{U}(q \wedge r)) \Rightarrow ((p\,\mathsf{U}\,q) \wedge (p\,\mathsf{U}\,r)).$$

We next note that the temporal operators below are monotonic in each argument:

$$\models \mathsf{G}(p \Rightarrow q) \Rightarrow (\mathsf{G}p \Rightarrow \mathsf{G}q), \qquad \models \mathsf{G}(p \Rightarrow q) \Rightarrow (\mathsf{F}p \Rightarrow \mathsf{F}q),$$
$$\models \mathsf{G}(p \Rightarrow q) \Rightarrow (\mathsf{X}p \Rightarrow \mathsf{X}q),$$
$$\models \mathsf{G}(p \Rightarrow q) \Rightarrow (\mathsf{F}^{\infty}p \Rightarrow \mathsf{F}^{\infty}q), \qquad \models \mathsf{G}(p \Rightarrow q) \Rightarrow (\mathsf{G}^{\infty}p \Rightarrow \mathsf{G}^{\infty}q),$$
$$\models \mathsf{G}(p \Rightarrow q) \Rightarrow ((p \,\mathsf{U}\, r) \Rightarrow (q \,\mathsf{U}\, r)),$$
$$\models \mathsf{G}(p \Rightarrow q) \Rightarrow ((r \,\mathsf{U}\, p) \Rightarrow (r \,\mathsf{U}\, q)).$$

Finally, we have following important *fixpoint characterizations* of the temporal operators (cf. Section 8.4):

$$\models \mathsf{F}p \equiv p \vee \mathsf{XF}p, \qquad \models \mathsf{G}p \equiv p \wedge \mathsf{XG}p,$$
$$\models (p \,\mathsf{U}\, q) \equiv q \vee (p \wedge \mathsf{X}(p \,\mathsf{U}\, q)), \qquad \models (p \,\mathsf{B}\, q) \equiv \neg q \wedge (p \vee \mathsf{X}(p \,\mathsf{B}\, q)).$$

3.2.5. Minor variants of PLTL

One minor variation is to change the basic temporal operators. There are a number of variants of the until operator $p \,\mathsf{U}\, q$, which is defined as the *strong until*: there does exist a future state where q holds and p holds until then. We could write $p \,\mathsf{U}_{\mathrm{s}}\, q$ or $p \,\mathsf{U}_{\exists}\, q$ to emphasize its strong, existential character. The operator *weak until*, written $p \,\mathsf{U}_{\mathrm{w}}\, q$ (or $p \,\mathsf{U}_{\forall}\, q$), is an alternative. It intuitively means that p holds for as long as q does not, even forever if need be. It is also called the *unless* operator. Its technical definition can be formulated as

$$x \models p \,\mathsf{U}_{\forall}\, q \quad \text{iff} \quad \forall j\,((\forall k \leqslant j\,(x^k \models \neg q)) \text{ implies } x^j \models p)$$

exhibiting its "universal" character. Note that, given the Boolean connectives, each until operator is expressible in terms of the other:

(a) $p \,\mathsf{U}_{\exists} q \equiv p \,\mathsf{U}_{\forall} q \wedge \mathsf{F}q$,
(b) $p \,\mathsf{U}_{\forall} q \equiv p \,\mathsf{U}_{\exists} q \vee \mathsf{G}p \equiv p \,\mathsf{U}_{\exists} q \vee \mathsf{G}(p \wedge \neg q)$.

We also have variations based on the answer to the question: does the future include the present? The future does include the present in our formulation, and is thus called the *reflexive* future. We might instead formulate versions of the temporal operators referring to the *strict* future, i.e., those times strictly greater than the present. A convenient notation for emphasizing the distinction involves use of $>$ or $\geqslant$ as a superscript:

- $\mathsf{F}^{>}p$: $\exists$ a strict future moment when p holds,
- $\mathsf{F}^{\geqslant}p$: $\exists$ a moment, either now or in the future, when p holds,
- $\mathsf{F}^{>}p \equiv \mathsf{XF}^{\geqslant}p$
- $\mathsf{F}^{\geqslant}p \equiv p \vee \mathsf{F}^{>}p$.

Similarly we have the strict always ($\mathsf{G}^{>}p$) in addition to our "ordinary" always ($\mathsf{G}^{\geqslant}p$). The *strict* (strong) until $p \,\mathsf{U}^{>}\, q \equiv \mathsf{X}(p \,\mathsf{U}\, q)$ is of particular interest. Note that *false* $\mathsf{U}^{>}\, q \equiv \mathsf{X}(\textit{false} \,\mathsf{U}\, q) \equiv \mathsf{X}q$. The single modality strict, strong until is enough to define all the other linear-time operators (as shown by Kamp [46]).

REMARK. One other common variation is simply notational. Some authors use $\Box p$ for $\mathsf{G}p$, $\Diamond p$ for $\mathsf{F}p$, and $\bigcirc p$ for $\mathsf{X}p$.

Another minor variation is to change the underlying structure to be any initial segment I of $\mathbb{N}$, possibly a finite one. This seems sensible because we may want to reason about terminating programs as well as nonterminating ones. We then correspondingly alter the meanings of the basic temporal operators, as indicated (informally) below:

- $\mathsf{G}p$: for all subsequent times in I, p holds,
- $\mathsf{F}p$: for some subsequent time in I, p holds,
- $p\,\mathsf{U}\,q$: for some subsequent time in I, q holds, and p holds at all subsequent times until then.

We can also now distinguish two notions of nexttime:

- $\mathsf{X}_{\forall}p$: weak nexttime—if there exists a successor moment then p holds there,
- $\mathsf{X}_{\exists}p$: strong nexttime—there exists a successor moment and p holds there.

Note that each nexttime operator is the dual of the other: $\mathsf{X}_{\exists}p \equiv (\neg \mathsf{X}_{\forall} \neg p$ and $\mathsf{X}_{\forall}p \equiv \neg \mathsf{X}_{\exists} \neg p)$. Also we may use $\mathsf{X}_s p$ for $\mathsf{X}_{\exists}p$ and $\mathsf{X}_w p$ for $\mathsf{X}_{\forall}p$.

REMARK. Without loss of generality, we can restrict our attention to structures where the timeline is $\mathbb{N}$ and still get the effect of finite timelines. This can be done in either of two ways:

(a) Repeat the final state so the finite sequence $s_0 s_1 \ldots s_k$ of states is represented by the infinite sequence $s_0 s_1 \ldots s_k s_k s_k \ldots$ (This is somewhat like adding a self-loop at the end of a finite, directed linear graph.)

(b) Have a proposition P_{GOOD} true for exactly the good (i.e., finite) portion of the timeline.

Adding past-tense temporal operators. As used in computer science, all temporal operators are future-tense; we might use the following suggestive notation and terminology for emphasis:

- $\mathsf{F}^{+}p$: sometime in the future p holds,
- $\mathsf{G}^{+}p$: always in the future p holds,
- $\mathsf{X}^{+}p$: nexttime p holds (note that "next" implicitly implies the future),
- $p\,\mathsf{U}^{+}\,q$: sometime in the future q holds and p holds subsequently until then.

However, as originally studied by philosophers there were past-tense operators as well; we can use the corresponding notation and terminology:

- $\mathsf{F}^{-}p$: sometime in the past p holds,
- $\mathsf{G}^{-}p$: always in the past p holds,
- $\mathsf{X}_{\exists}^{-}p$: lasttime p holds (note that "last" implicitly refers to the past),
- $p\,\mathsf{U}^{-}q$: sometime in the past q holds and p holds previously until then.

When needed for emphasis we use PLTLF for the logic with just future-tense operators, PLTLP for the logic with just past-tense operators, and PLTLB for the logic with both.

For temporal logic using the past-tense operators, given a linear-time structure $M=(S, x, L)$ we interpret formulae over a pair (x, i), where x is the timeline and the natural number index i specifies *where* along the timeline the formula is true. Thus, we write $M, (x, i) \models p$ to mean that "in structure M along timeline x at time i formula p holds true"; when M is understood we write just $(x, i) \models p$. Intuitively, pair (x, i) corresponds to the suffix x^i, which is the forward interval $x[i{:}\infty)$ starting at time i, used in the definition of the future-tense operators. When the past is allowed, the pair (x, i) is needed since formulae can reference positions along the entire timeline, both forward and backward of position i. If we restrict our attention to just the future tense as in the definition of PLTL, we can omit the second component of (x, i)—in effect assuming that $i=0$, and that formulae are interpreted at the beginning of the timeline—and write $x \models p$ for $(x, 0) \models p$.

The technical definitions of the basic past-tense operators are as follows:

$$(x, i) \models p \, \mathsf{U}^- q \text{ iff } \exists j (j \leqslant i \text{ and } (x, j) \models q \text{ and } \forall k (j < k \leqslant i \text{ implies } (x, j) \models p)),$$
$$(x, i) \models \mathsf{X}_\exists^- p \text{ iff } i > 0 \text{ and } (x, i-1) \models p.$$

Note that the lasttime operator is strong, having an existential character, asserting that there is a past moment; thus it is false at time 0. The other past connectives are then introduced as abbreviations as usual: e.g., the weak lasttime $\mathsf{X}_\forall^- p$ for $\neg \mathsf{X}_\exists^- \neg p$, $\mathsf{F}^- p$ for $(true \, \mathsf{U}^- p)$, and $\mathsf{G}^- p$ for $\neg \mathsf{F}^- \neg p$.

For comparison we also present the definitions of some of the basic future-tense operators using the pair (x, i) notation:

$$(x, i) \models (p \, \mathsf{U} \, q) \text{ iff } \exists j (j \geqslant i \text{ and } (x, j) \models q \text{ and } \forall k (i \leqslant k < j \text{ implies } (x, k) \models p)),$$
$$(x, i) \models \mathsf{X} p \text{ iff } (x, i+1) \models p,$$
$$(x, i) \models \mathsf{G} q \text{ iff } \forall j (j \geqslant i \text{ implies } (x, j) \models q),$$
$$(x, i) \models \mathsf{F} q \text{ iff } \exists j (j \geqslant i \text{ and } (x, j) \models q).$$

REMARK. Philosophers used a somewhat different notation. $\mathsf{F}^- p$ was usually written as $\mathsf{P} p$, $\mathsf{G}^- p$ as $\mathsf{H} p$, and $p \, \mathsf{U}^- q$ as $p \, \mathsf{S} \, q$ meaning "p since q". We prefer the present notation due to its more uniform character.

The decision whether to allow i to float or to anchor it at 0 yields different notions of equivalence, satisfiability, and validity. We say that a formula p is *initially satisfiable* provided there exists a linear-time structure $M = (S, x, L)$ such that $M, (x, 0) \models p$. We say that a formula p is *initially valid* provided that for all timeline structures $M = (S, x, L)$ we have $M, (x, 0) \models p$. We say that a formula p is *globally satisfiable* provided that there exists a linear-time structure $M = (S, x, L)$ and time i such that $M, (x, i) \models p$. We say that a formula p is *globally valid* provided that for all linear-time structures $M = (S, x, L)$ and times i we have $M, (x, i) \models p$.

In an almost trivial sense inclusion of the past-tense operators increases the expressive power of our logic: We say that formula p is *globally equivalent* to formula q, and write $p \equiv_g q$, provided that $\forall$ linear structure x, $\forall$ time $i \in \mathbb{N}$ $[(x, i) \models p$ iff $(x, i) \models q]$.

3.1. THEOREM. *As measured with respect to gobal equivalence,* PLTLB *is strictly more expressive than* PLTLF.

PROOF. The formula $\mathsf{F}^- Q$ is not expressible in PLTLF, as can be seen by considering two structures x, x' as depicted below.

$$\overset{Q}{\bullet} \longrightarrow \bullet \sim \sim \sim \sim \rightarrow$$

$$\overset{\neg Q}{\bullet} \longrightarrow \bullet \sim \sim \sim \sim \rightarrow$$

The structures are identical except for their respective state at time 0. At time 1, $\mathsf{F}^- Q$ distinguishes the two structures (i.e., $(x, 1) \models \mathsf{F}^- Q$ and $(x', 1) \not\models \mathsf{F}^- Q$), yet future-tense PLTLF cannot distinguish $(x, 1)$ from $(x', 1)$, since the infinite suffixes beginning at time 1 are identical. □

Yet in the sense that programs begin execution in an initial state, inclusion of the past-tense operators adds no expressive power.

We say that formula p is *initially equivalent* to formula q, and write $p \equiv_i q$, provided that $\forall$ linear structure x, $[(x, 0) \models p \text{ iff } (x, 0) \models q]$.

3.2. Theorem. *As measured with respect to initial equivalence,* PLTLB *is equivalent in expressive power to* PLTLF.

This can be proved using results regarding the theory of linear orderings (cf. [34]). We also note the following relationship between $\equiv_i$ and $\equiv_g$:

3.3. Proposition. $p \equiv_g q$ *iff* $\mathsf{G}p \equiv_i \mathsf{G}q$.

By convention we shall take "satisfiable" to mean "initially satisfiable" and "valid" to mean "initially valid", unless otherwise stated. Intuitively, this makes sense since programs start execution in an initial state. Moreover, whenever we refer to expressive power we are measuring it with respect to initial equivalence, unless otherwise stated. One benefit of comparing expressive power on the basis of initial equivalence is that it suggests we view formulae of PLTL and its variants as defining sets of sequences, i.e. formal languages (see Section 6).

3.3. *First-Order Linear Temporal Logic (FOLTL)*

First-order linear temporal logic (FOLTL) is obtained by taking propositional linear temporal logic (PLTL) and adding to it a first-order language L. That is, in addition to atomic propositions, truth-functional connectives, and temporal operators, we now also have predicates, functions, individual constants, and individual variables, each interpreted over an appropriate domain with the standard Tarskian definition of truth.

Symbols of L

We have a first-order language L over a set of *function* symbols and a set of *predicate* symbols. The zero-ary function symbols comprise the subset of *constant* symbols. Similarly, the zero-ary predicate symbols are known as the *proposition* symbols. Finally, we have a set of individual *variable* symbols. We use the following notations:

- $\varphi, \psi, \ldots,$ etc. for n-ary, $n \geq 1$, predicate symbols,
- $P, Q, \ldots,$ etc. for proposition symbols,
- $f, g, \ldots,$ etc. for n-ary, $n \geq 1$, function symbols,
- $c, d, \ldots,$ etc. for constant symbols, and
- $y, z, \ldots,$ etc. for variable symbols.

We also have the distinguished binary predicate symbol $\approx$, known as the *equality symbol*, which we use in the standard infix fashion. Finally, we have the usual quantifier symbols $\forall$ and $\exists$, denoting universal and existential quantification respectively, which are applied to individual variable symbols, using the usual rules regarding scope of quantifiers, and free and bound variables.

Syntax of $\boldsymbol{L}$

The *terms* of $\boldsymbol{L}$ are defined inductively by the following rules:

(T1) each constant c is a term;

(T2) each variable y is a term;

(T3) if $t_1, \ldots, t_n$ are terms and f is an n-ary function symbol then $f(t_1, \ldots, t_n)$ is a term.

The *atomic formulae* of $\boldsymbol{L}$ are defined by the following rules:

(AF1) each 0-ary predicate symbol (i.e. atomic proposition) is an atomic formula;

(AF2) if $t_1, \ldots, t_n$ are terms and ψ is an n-ary predicate then $\psi(t_1, \ldots, t_n)$ is an atomic formula;

(AF3) if t_1, t_2 are terms then $t_1 \approx t_2$ is also an atomic formula.

Finally, the (compound) formulae of $\boldsymbol{L}$ are defined inductively as follows:

(F1) each atomic formula is a formula;

(F2) if p, q are formulae then $(p \wedge q)$, $\neg p$ are formulae;

(F3) if p is a formula and y is a free variable in p then $\exists y p$ is a formula.

Semantics of $\boldsymbol{L}$

The semantics of $\boldsymbol{L}$ is provided by an interpretation I over some domain D. The interpretation I assigns an appropriate meaning over D to the (nonlogical) symbols of $\boldsymbol{L}$: Essentially, the n-ary predicate symbols are interpreted as concrete, n-ary relations over D, while the n-ary function symbols are interpreted as concrete, n-ary functions on D. (Note that an n-ary relation over D may be viewed as an n-ary function $D^n \to \mathbb{B}$, where $\mathbb{B} = \{true, false\}$ is the distinguished Boolean domain.) More precisely I assigns a meaning to the symbols of $\boldsymbol{L}$ as follows:

- for an n-ary predicate symbol ψ, $n \geqslant 1$, the meaning $I(\psi)$ is a function $D^n \to \mathbb{B}$;
- for a proposition symbol P, the meaning $I(P)$ is an element of $\mathbb{B}$;
- for an n-ary function symbol f, $n \geqslant 1$, the meaning $I(f)$ is a function $D^n \to D$;
- for an individual constant symbol c, the meaning $I(c)$ is an element of D;
- for an individual variable symbol y, the meaning $I(y)$ is an element of D.

The interpretation I is extended to arbitrary terms, inductively

$$I(f(t_1, \ldots, t_n)) = I(f)(I(t_1), \ldots, I(t_n))$$

We now define the meaning of truth under interpretation I of formula p, written $I \models p$. First, for atomic formulae we have

$I \models P$ where P is an atomic proposition iff $I(P) = true$;

$I \models \psi(t_1, \ldots, t_n)$ where ψ is an n-ary predicate and $t_1, \ldots, t_n$ are terms iff $I(\psi)(I(t_1), \ldots, I(t_n)) = true$;

$I \models t_1 \approx t_2$ iff $I(t_1) = I(t_2)$.

Next, for compound formulae we have

$I \models p \wedge q$ iff $I \models p$ and $I \models q$;
$I \models \neg p$ iff it is not the case that $I \models p$;
$I \models \exists y\, p$ where y is a free variable in p
iff there exists some $d \in D$ such that $I[y \leftarrow d] \models p$ where $I[y \leftarrow d]$ is the interpretation identical to I except that y is assigned value d.

Global versus logical symbols

For defining First-Order Linear Temporal Logic (FOLTL), we assume that the set of symbols is divided into two classes, the class of *global* symbols and the class of *local* symbols. Intuitively, each global symbol has the same interpretation over all states; the interpretation of a local symbol may vary, depending on the state at which it is evaluated. We will subsequently assume that all function symbols (and thus all constant symbols) are global, and that all n-ary predicate symbols, for $n \geqslant 1$, are also global. Proposition symbols (i.e. 0-ary predicate symbols) and variable symbols may be local or may be global.

A (*first-order*) *linear-time structure* $M = (S, x, L)$ is defined just as in the propositional case, except that L now associates with each state s an interpretation $L(s)$ of all symbols at s such that, for each global symbol w, $L(s)(w) = L(s')(w)$ for all $s, s' \in S$. Note that the structure M has an underlying domain D, as for $\boldsymbol{L}$. Also, it is sometimes convenient to refer to the global interpretation I associated with M by $I(w) = L(s)(w)$, where w is any global symbol and s is any state of M. (Note that implicitly given with a structure is its *signature* or *similarity type* consisting of the alphabets of all the different kinds of symbols. The signature of the structure is assumed to match that of the language (FOLTL).)

Description of FOLTL

We are now ready to define the language of First-Order Linear Temporal Logic (FOLTL) obtained by adding $\boldsymbol{L}$ to PLTL. First, the terms of FOLTL are those generated by rules (T1–3) for $\boldsymbol{L}$, plus the rule

(T4) if t is a term then $\mathsf{X}t$ is a term (intuitively, denoting the immediate future value of term t).

The atomic formulae of FOLTL are generated by the same rules as for $\boldsymbol{L}$, but now are used in conjunction with the expanded set of rules (T1–4) for terms. Finally, the (compound) formulae of FOLTL are defined inductively using the following rules:

(FOLTL1) each atomic formula is a formula;
(FOLTL2) if p, q are formulae, then so are $p \wedge q$, $\neg p$;
(FOLTL3) if p, q are formulae, then so are $p \mathsf{U} q$, $\mathsf{X}p$;
(FOLTL4) if p is a formula and y is a free variable in p, then $\exists y\, p$ is a formula.

The semantics of FOLTL is provided by a first-order linear-time structure M over a domain D as above. Global interpretation I of M assigns a meaning to each global symbol, while the local interpretations $L(-)$ associated with M assign a meaning to each local symbol.

Since the terms of FOLTL are generated by rules (T1–3) for $\boldsymbol{L}$ plus the rule (T4) above, we extend the meaning function—now denoted by a pair (M, x)—for terms:

- $(M, x)(c) = I(c)$, since all constants are global;
- $(M, x)(y) = I(y)$, where y is a global variable;
- $(M, x)(y) = L(s_0)(y)$, where y is a local variable and $x = (s_0, s_1, s_2, \ldots)$;
- $(M, x)(f(t_1, \ldots, t_n)) = (M, x)(f)((M, x)(t_1), \ldots, (M, x)(t_n))$;
- $(M, x)(\mathrm{X}t) = (M, x^1)(t)$.

Now the extension of $\models$ is routine. For atomic formulae we have

- $M, x \models P$ iff $I \models P$, where P is a global proposition;
- $M, x \models P$ iff $L(s_0)(P) = \mathit{true}$, where P is a local proposition and $x = (s_0, s_1, s_2, \ldots)$;
- $M, x \models \psi(t_1, \ldots, t_n)$ iff $(M, x)(\psi)((M, x)(t_1), \ldots, (M, x)(t_n)) = \mathit{true}$;
- $M, x \models t_1 \approx t_2$ iff $(M, x)(t_1) = (M, x)(t_2)$.

We finish off the semantics of FOLTL with the inductive definition of $\models$ for compound formulae:

- $M, x \models p \wedge q$ iff $M, x \models p$ and $M, x \models q$;
- $M, x \models \neg p$ iff it is not the case that $M, x \models p$;
- $M, x \models (p \,\mathrm{U}\, q)$ iff $\exists j (M, x^j \models q$ and $\forall k < j\, (M, x^k \models p))$;
- $M, x \models Xp$ iff $M, x^1 \models p$;
- $M, x \models \exists y\, p$ where y is a global variable free in p iff there exists some $d \in D$ for which $M[y \leftarrow d], x \models p$, where $M[y \leftarrow d]$ is the structure having global interpretation $I[y \leftarrow d]$ identical to I except y is assigned the value d.

A formula p of FOLTL is *valid* iff for every first-order linear-time structure $M = (S, x, L)$ we have $M, x \models p$. The formula p is *satisfiable* iff there exists $M = (S, x, L)$ such that $M, x \models p$.

REMARK. For notational simplicity we have assumed that $\boldsymbol{L}$ is a one-sorted first-order language. Thus each symbol (function symbol, predicate symbol, etc.) is of the same sort and is interpreted over the single domain D. For certain applications, it is more convenient to assume that $\boldsymbol{L}$ is a multisorted language, where the symbols of $\boldsymbol{L}$ are partitioned into different sets, each of which corresponds to a different domain with different argument positions. The extension to multisorted languages is routine although a bit cumbersome notationally.

4. The technical framework of Branching Temporal Logic

4.1. *Tree-like structures*

In branching-time temporal logics, the underying structure of time is assumed to have a branching tree-like nature where each moment may have many successor moments. The structure of time thus corresponds to an infinite tree. In the sequel, we will further assume that along each path in the tree, the corresponding timeline is isomorphic to $\mathbb{N}$. We do allow a node in the tree to have infinitely many (even uncountably many) successors, while we require each node to have at least one successor. It will turn out that, as far as our branching temporal logics are concerned, such trees are indistinguishable from trees with finite, even bounded, branching. Trees of the latter type have a natural correspondence with the computations of concurrent or nondeterministic programs, as discussed in the next section.

We say that a *temporal structure* $M=(S, R, L)$ where

- S is the set of *states*,
- R is a total *binary relation* $\subseteq S \times S$ (i.e., one where $\forall s \in S\, \exists t \in S\, (s, t) \in R$), and
- $L: S \rightarrow \text{PowerSet(AP)}$ is a labelling which associates with each state s an interpretation $L(s)$ of all atomic proposition symbols at state s.

We may view M as a labelled, directed graph with node set S, arc set R, and node labels given by L. We say (the graph of) M

(a) is *acyclic* provided it contains no directed cycles;

(b) is *tree-like* provided that it is acyclic and each node has at most one R-predecessor (i.e., there is no "merging" of paths); and

(c) is a *tree* provided that it is tree-like and there exists a unique node—called the *root*—from which all other nodes of M are reachable and that has no R-predecessors.

We have not required that (the graph of) M be a tree. However, we may assume, without loss of generality, that it is. We define the structure $M=(\hat{S}, \hat{R}, \hat{L})$, which is called the structure obtained by unwinding M starting at state $s_0 \in S$, where $\hat{S}, \hat{R}$ are respectively the least subsets of $S \times \mathbb{N}$, $\hat{S} \times \hat{S}$ such that

- $(s_0, 0) \in \hat{S}$;
- if $(s, n) \in \hat{S}$, then $\{(t, n+1)$: t is an R-successor of s in $M\} \subseteq \hat{S}$, and $\{((s, n), (t, n+1))$: t is an R-successor of s in $M\} \subseteq \hat{R}$;

and $\hat{L}((s, n))=L(s)$. Then (the graph of) $\hat{M}$ is a tree with root $(s_0, 0)$, and it is easily checked that, for all the branching-time logics we will consider, a formula p holds at s_0 in M iff p holds at $(s_0, 0)$ in $\hat{M}$ (see Fig. 2).

4.2. *Propositional Branching Temporal Logics*

In this section we provide the formal syntax and semantics for two representative systems of propositional branching-time temporal logics. The simpler logic, CTL (Computational Tree Logic) allows basic temporal operators of the form a path quantifier—either A ("for all futures") or E ("for some future")—followed by a single one of the usual linear temporal operators G ("always"), F ("sometime"), X ("nexttime"), or U ("until"). It corresponds to what one might naturally first think of as a branching time logic. CTL is closely related to branching-time logics proposed in [53, 23, 8] and was itself proposed in [12]. However, as we shall see, its syntactic restrictions significantly limit its expressive power. We therefore also consider the much richer language CTL*, which is sometimes referred to informally as full branching-time logic. The logic CTL* extends CTL by allowing basic temporal operators where the path quantifier (A or E) is followed by an arbitrary linear-time formula, allowing Boolean combinations and nestings, over F, G, X, and U. It was proposed as a unifying framework in [26] subsuming both CTL and PLTL, as well as a number of other systems. Related systems of high expressiveness are considered in [1, 108, 115].

Syntax

We now give a formal definition of the syntax of CTL*. We inductively define a class of state formulae (true or false of states) using rules (S1–3) below and a class of path

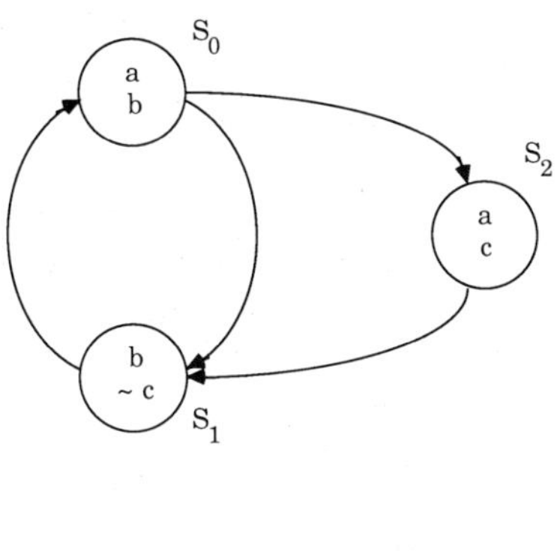

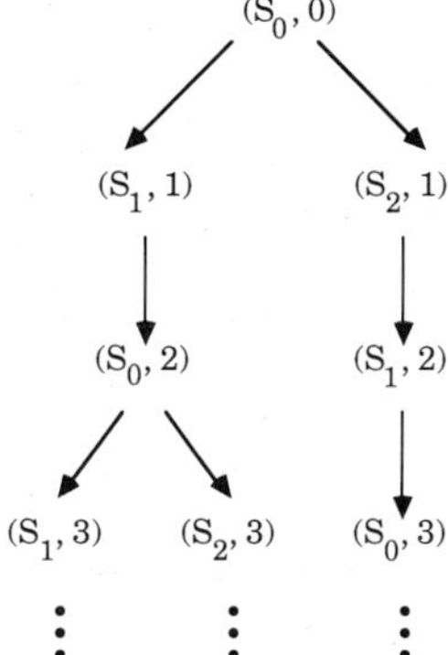

Fig. 2. Unwinding a (finite) structure into an infinite tree.

formulae (true or false of paths) using rules (P1–3) below:

(S1) each atomic proposition P is a state formula;
(S2) if p, q are state formulae then so are $p \wedge q$, $\neg p$;
(S3) if p is a path formula then $\mathsf{E}p$, $\mathsf{A}p$ are state formulae;
(P1) each state formula is also a path formula;
(P2) if p, q are path formulae then so are $p \wedge q$, $\neg p$;
(P3) if p, q are path formulae then so are $\mathsf{X}p$, $p\,\mathsf{U}\,q$.

The set of state formulae generated by the above rules forms the language CTL*. The other connectives can then be introduced as abbreviations in the usual way.

REMARK. We could take the view that $\mathsf{A}p$ abbreviates $\neg\mathsf{E}\neg p$, and give a more terse syntax in terms of just the primitive operators E, $\wedge$, $\neg$, X, and U. However, the present approach makes it easier to give the syntax of the sublanguage CTL below.

The restricted logic CTL is obtained by restricting the syntax to disallow Boolean combinations and nestings of linear-time operators. Formally we replace rules (P1–3) by

(P0) if p, q are state formulae then $\mathsf{X}p$, $p\,\mathsf{U}\,q$ are path formulae.

The set of state formulae generated by rules (S1–3) and (P0) form the language CTL.

The other Boolean connectives are introduced as above, while the other temporal operators are defined as abbreviations as follows: $\mathsf{EF}p$ abbreviates $\mathsf{E}(true\,\mathsf{U}\,p)$, $\mathsf{AG}p$ abbreviates $\neg\mathsf{EF}\neg p$, $\mathsf{AF}p$ abbreviates $\mathsf{A}(true\,\mathsf{U}\,p)$, and $\mathsf{EG}p$ abbreviates $\neg\mathsf{AF}\neg p$. (Note that this definition can be seen to be consistent with that of CTL*.)

Also note that the set of path formulae generated by rules by (P1–3) yields the linear-time PLTL.

Semantics

A formula of CTL* is interpreted with respect to a structure $M=(S, R, L)$ as defined above. A *fullpath* of it is an infinite sequence $s_0, s_1, s_2, \ldots$ of states such that $\forall i\,(s_i, s_{i+1}) \in R$. We use the convention that $x=(s_0, s_1, s_2, \ldots)$ denotes a fullpath, and that x^i denotes the suffix path $(s_i, s_{i+1}, s_{i+2}, \ldots)$. We write $M, s_0 \models p$ (respectively $M, x \models p$) to mean that state formula p (respectively path formula p) is true in structure M at state s_0 (respectively of fullpath x). We define $\models$ inductively as follows:

(S1) $M, s_0 \models p$ iff $p \in L(s_0)$,
(S2) $M, s_0 \models p \wedge q$ iff $M, s_0 \models p$ and $M, s_0 \models q$,
$M, s_0 \models \neg p$ iff not$(M, s_0 \models p)$;
(S3) $M, s_0 \models \mathsf{E}p$ iff $\exists$ fullpath $x=(s_0, s_1, s_2, \ldots)$ in M, $M, x \models p$,
$M, s_0 \models \mathsf{A}p$ iff $\forall$ fullpath $x=(s_0, s_1, s_2, \ldots)$ in M, $M, x \models p$;
(P1) $M, x \models p$ iff $M, s_0 \models p$;
(P2) $M, x \models p \wedge q$ iff $M, x \models p$ and $M, x \models q$,
$M, x \models \neg p$ iff not$(M, x \models \neg p)$;
(P3) $M, x \models p\,\mathsf{U}\,q$ iff $\exists i\,[M, x^i \models q$ and $\forall j\,(j<i$ implies $M, x^j \models p)]$,
$M, x \models \mathsf{X}p$ iff $M, x^1 \models p$.

A formula of CTL is also interpreted using the CTL* semantics, using rule (P3) for path formulae generated by rule (P0).

We say that a state formula p (respectively path formula p) is *valid* provided that for every structure M and every state s (respectively fullpath x) in M we have $M, s \models p$ (respectively $M, x \models p$). A state formula p (respectively path formula p) is *satisfiable* provided that for some structure M and some state s (respectively fullpath x) in M we have $M, s \models p$ (respectively, $M, x \models p$).

Generalized semantics

We can define CTL* and other logics over various generalized notions of structure. For example, we could consider more general structures $M=(S, X, L)$ where S is a set of states and L a labelling of states as usual, while $X \subseteq S^\omega$ is a family of infinite computation sequences (fullpaths) over S. The definition of CTL* semantics carries over directly, with path quantification restricted to paths in X, provided that "a fullpath x in M" is understood to refer to a fullpath x in X.

In the most general case X can be completely arbitrary. However, it is often helpful to impose certain requirements on X (cf. [53, 90, 1, 22]). We say that X is *suffix closed* if, whenever computation $s_0 s_1 s_2 \ldots \in X$, then the suffix $s_1 s_2 \ldots \in X$. Similarly, X is *fusion closed* if, whenever $x_1 s y_1, x_2 s y_2 \in X$, then $x_1 s y_2 \in X$. The idea is that the system should always be able to follow the prefix of one computation and then continue along the suffix $s y_2$ of another computation; thus the computation actually followed is the "fusion" of two others. Both suffix and fusion closure are needed to ensure that the future behavior of a program depends only on the current state and not on how the state is reached.

We may also wish to require that X be *limit closed* meaning that whenever x_1y_1, $x_1x_2y_2$, $x_1x_2x_3y_3, \ldots$ are all elements of X, then the infinite path $x_1x_2x_3\ldots$, which is the limit of the prefixes $x_1, x_1x_2, x_1x_2x_3\ldots$, is also in X. In short, if it is possible to follow a path arbitrarily long, then it can be followed forever. Finally, a set of paths is *R-generable* if there exists a total binary relation R on S such that a sequence $x = s_0s_2s_2\ldots \in X$ iff $\forall i\,(s_i, s_{i+1}) \in R$. It can be shown that X is R-generable iff it is limit closed, fusion closed and suffix closed. Of course, the basic type of structures we ordinarily consider are R-generable, which correspond to the execution of a program under pure nondeterministic scheduling (cf. [22]).

Some such restrictions on the set of paths X are usually needed in order to have the abstract, computation path semantics reflect the behavior of actual concurrent programs. An additional advantage of these restrictions is that they ensure the validity of many commonly accepted principles of temporal reasoning. For example fusion closure is needed to ensure that $\mathsf{EFEF}p \equiv \mathsf{EF}p$. Suffix closure is needed for $\mathsf{EF}p \wedge \neg p \Rightarrow \mathsf{EXEF}p$, and limit closure for $p \wedge \mathsf{AGEX}p \Rightarrow \mathsf{EG}p$. An R-generable structure satisfies all these natural properties.

Another generalization is to define a *multiprocess temporal structure*, which is a refinement of the notion of a branching temporal structure that distinguishes between different processes. Formally, a multiprocess temporal structure $M = (S, \boldsymbol{R}, L)$ where

- S is a set of states,
- $\boldsymbol{R}$ is a finite family $\{R_1, \ldots, R_k\}$ of binary relations R_i on S (intuitively, R_i represents the transitions of process i) such that $R = \bigcup \boldsymbol{R}$ is total (i.e., $\forall s \in S\, \exists t \in S\,(s,t) \in R$),
- L associates with each state an interpretation of symbols at the state.

Just as for a (uniprocess) temporal structure, a multiprocess temporal structure may be viewed as a directed graph with labelled nodes and arcs. Each state is represented by a node that is labelled by the atomic propositions true there, and each transition relation R_i is represented by a set of arcs that are labelled with index i. Since there may be multiple arcs labelled with distinct indices between the same pair of nodes, technically the graph-theoretic representation is a directed multigraph.

The previous formulation of CTL* over uniprocess structures refers only to the atomic formulae labelling the nodes. However, it is straightforward to extend it to include in effect, arc assertions indicating which process performed the transition corresponding to an arc. This extension is needed to formulate the technical definitions of fairness in the next section, so we briefly describe it.

Now, a fullpath $x = (s_0, d_1, s_1, d_2, s_2, \ldots)$ as depicted below

$$\underset{s_0}{\bullet} \xrightarrow{d_1} \underset{s_1}{\bullet} \xrightarrow{d_2} \underset{s_2}{\bullet} \xrightarrow{d_3} \underset{s_3}{\bullet} \xrightarrow{d_4} \cdots$$

is an infinite sequence of states s_i alternating with relation indices d_{i+1} such that $(s_i, s_{i+1}) \in R_{d_{i+1}}$, indicating that process d_{i+1} caused the transition from s_i to s_{i+1}. We also assume that there are distinguished propositions $enabled_1, \ldots, enabled_k$, $executed_1, \ldots\ executed_k$, where intuitively $enabled_j$ is true of a state exactly when process j is enabled, i.e. when a transition by process j is possible, and *excuted*$_j$ is true of a transition when it is performed by process j. Technically, each *enabled*$_j$ is an atomic proposition—and hence a state formula—true of exactly those states in domain R_j:

$$M, s_0 \models enabled_j \text{ iff } s_0 \in \text{domain } R_j = \{s \in S: \exists t \in S\,(s, t) \in R\}$$

while each *executed*$_j$ is an atomic arc assertion—and a path formula such that

$$M, x \models executed_j \text{ iff } d_1 = j.$$

It is worth pointing out that there are alternative formalisms that are essentially equivalent to this notion of a (multiprocess) structure. A *transition system M* is a formalism equivalent to a multiprocess temporal structure consisting of a triple $M = (S, \boldsymbol{R}, L)$ where $\boldsymbol{R}$ is a finite family of transitions $\tau_i: S \to \text{PowerSet}(S)$. To each transition τ_i there is a corresponding relation $R_i = \{(s, t) \in S \times S: t \in \tau_i(s)\}$ and conversely. Similarly, there is a correspondence between multiprocess temporal logic structures and **do-od** programs (cf. [19]). Assume we are given a **do-od** program $\rho = \textbf{do } B_1 \to A_1 \,\square \cdots \square\, B_k \to A_k \textbf{ od}$, where each B_i may be viewed as subset of the state space S and each A_i as a function $S \to S$. Then we may define an equivalent structure $M = (S, \boldsymbol{R}, L)$, where each $R_i = \{(s, t) \in S: s \in B_i \text{ and } t = A_i(s)\}$, and L gives appropriate meanings to the symbols in the program. Conversely, given a structure M, there is a corresponding generalized **do-od** program ρ, where by generalized we mean that each action A_i is allowed to be a relation; viz., it is $\textbf{do } B_1 \to A_1 \,\square \cdots \square\, B_k \to A_k \textbf{ od}$, where each $B_i = \text{domain } R_i = \{s \in S: \exists t \in S\,(s, t) \in R_i\}$ and $A_i = R_i$.

We can define a single type of general structure which subsumes all of those above. We assume an underlying set of symbols, divided into global and local subsets as before and called *state symbols* to emphasize that they are interpreted over states, as well as an additional set of *arc assertion symbols* that are interpreted over transitions $(s, t) \in R$. Typically we think of $L((s, t))$ as the set of indices (or names) of processes which could have performed the transition (s, t). A (*generalized*) *fullpath* is now a sequence of states s_i (alternating with arc assertions d_i) as depicted above.

Now we say that a *general structure* $M = (S, R, X, L)$ where

- S is a set of *states*,
- R is a total *binary relation* $\subseteq S \times S$,
- X is a set of *fullpaths* over S, and
- L is a mapping associating with each state s an interpretation $L(s)$ of all state symbols at s, and with each transition $(s, t) \in R$ an interpretation of each arc assertion at (s, t).

There is no loss of generality due to including R in the definition: for any set of fullpaths X, let $R = \{(s, t) \in S \times S$: there is a fullpath of the form $ystz$ in X, where y is a finite sequence of states and z an infinite sequence of states in $S\}$; then all consecutive pairs of states along paths in X are related by R.

The extensions needed to define CTL* over such a general structure M are straightforward. The semantics of path quantification as specified in rule (S3) carries over directly to the general M, provided that a "fullpath in M" refers to one in X. If d is an arc assertion we have that

$$M, x \models d \text{ iff } d \in L((s_0, s_1)).$$

4.3. *First-Order Branching Temporal Logic*

We can define systems of First-Order Branching Temporal Logic. The syntax is obtained by combining the rules for generating a system of propositional Branching Temporal Logic plus a (multisorted) first-order language. The underlying structure $M=(S,R,L)$ is extended so that it associates with each state s an interpretation $L(s)$ of local and global symbols at state s, including in particular local variables as well as local atomic propositions. The semantics is given by the usual Tarskian definition of truth. Validity and satisfiability are defined in the usual way. The details of the technical formulation are closely analogous to those for first-order linear temporal logic and are omitted here.

5. Concurrent computation: a framework

5.1. *Modelling concurrency by nondeterminism and fairness*

Our treatment of concurrency is the usual one where concurrent execution of a system of processes is modelled by the nondeterministic interleaving of atomic actions of the individual processes. The semantics of a concurrent program is thus given by a computation tree: a concurrent program starting in a given state may follow any one of a (possibly infinite) number of different computation paths in the tree (i.e., sequences of execution states) corresponding to the different sequences of nondeterministic choices the program might make. Alternatively, the semantics can be given simply by the set of all possible execution sequences, ignoring that they can be organized into a tree, for each possible starting state.

We remark that it is always possible to model concurrency by nondeterminism, since by picking a sufficiently fine level of granularity for the atomic actions to be interleaved, any behavior that could be produced by true concurrency (i.e., true simultaneity of action) can be simulated by interleaving. In practice, it is helpful to use as coarse a granularity as possible, since it reduces the number of interleavings that must be considered.

There is one additional consideration in the modelling of concurrency by nondeterminism. This is the fundamental notion of *fair scheduling assumptions*, commonly called *fairness*, for short.

In a truly concurrent system, implemented physically, it would be reasonable to assume that each sequential process P_i of a concurrent program $P_1 \| \cdots \| P_n$ is assigned to its own physical processor. Depending on the relative rates of speed at which the physical processors ran, we would expect that the corresponding nondeterministic choices modelling this concurrent system would favor the faster processes more often. For a very simple example, consider a system $P_1 \| P_2$ with just two processes. If each process ran on its own physical processor, and the processor ran at approximately equal speeds, we would expect the corresponding sequence of interleavings of steps of the individual processes to be of the form:

$P_1P_2P_1P_2P_1P_2\ldots$ or $P_2P_1P_2P_1P_2P_1\ldots$ or perhaps
$P_1P_1P_2P_1P_2P_2P_1P_1P_2\ldots$

where, for each i, after i steps in all have been executed, roughly $i/2$ steps of each individual process has been executed. If processor 1 ran, say, three times faster than processor 2 we would expect corresponding interleavings such as

$$P_1P_1P_1P_2P_1P_1P_1P_2P_1P_1P_1P_2\ldots$$

where steps of process P_1 occur about three times more often than steps of process P_2.

Now, on the other hand, we would not expect to see a sequence of actions such as $P_1P_1P_1P_1\ldots$, where process P_1 is always chosen while process P_2 is never chosen. This would be *unfair* to process P_2. Under the assumption that each processor is always running at some positive, finite speed, regardless of how the relative ratios of the processor's speed might vary, we would thus expect to see *fair* sequences of interleavings where each process is executed infinitely often. This notion of fair scheduling thereby corresponds to the reasonable and very weak assumption that each process makes some progress. In the sequel, we shall assume that the nondeterministic choices of which process is next to execute a step are such that the resulting infinite sequence is fair.

For the present we let the above notion of fairness—that each process be executed infinitely often—suffice; actually, however, there are a number of technically distinct refinements of this notion (see, for example, the book by Francez [32] as well as [1, 33, 34, 53, 57, 85, 94, 60, 27]). Some of these will be described subsequently.

Thus to model the semantics of concurrency accurately we need fairness assumptions in addition to the computation sequences generated by nondeterministic interleaving of the execution of individual processes.

We remark on an advantage afforded by fairness assumptions. By the principle of separation of concerns, we should distinguish the issue of correctness of a program, from concerns with its efficiency or performance. Correctness is a qualitative sort of property. To say that we are concerned that a program be totally correct means that we wish to establish that it does eventually terminate meeting a certain postcondition. Establishing just when it terminates is a quantitative sort of property that is distinct from the qualitative notion of eventually terminating. Temporal Logic is especially appropriate for such qualitative reasoning. Moreover, fairness assumptions facilitate such qualitative reasoning. Since fairness corresponds to the very weak qualitative notion that each process is running at some finite positive speed, programs proved correct under a fair scheduling assumption will be correct no matter what the rates are at which the processors actually run.

We very briefly summarize the preceding discussion by saying that, for our purposes, concurrency = nondeterminism + fairness. Somewhat less pithily but more precisely and completely, we can say that a concurrent program amounts to a global state transition system, with global state space essentially the Cartesian product of the state spaces of the individual sequential processes and transitions corresponding to the atomic actions of the individual sequential processes, plus a fairness constraint and a starting condition. The behavior of a concurrent program is then described in terms of the trees (or simply sets) containing all the computation sequences of the global state transition system which meet the fair scheduling constraint and starting condition.

5.2. *Abstract model of concurrent computation*

With the preceding motivation, we are now ready to describe our abstract model of concurrent computation.

An *abstract concurrent program* is a triple $(M, \varphi_{\mathrm{START}}, \Phi)$ where M is a (multiprocess) temporal structure, φ_{START} is an atomic proposition corresponding to a distinguished set of starting states in M, Φ is a fair scheduling constraint which we, for convenience, take to be specified in linear Temporal Logic.

Among possible fairness constraints, the following are very common ones:

(1) *Unconditional fairness* (also known as *impartiality*): an infinite sequence is impartial iff every process is executed infinitely often during the computation, which is expressed by $\Phi = \bigwedge_{i=1}^{k} \mathsf{F}^{\infty} executed_i$.

(2) *Weak fairness* (also known as *justice*): an infinite computation sequence is *weakly fair* iff every process enabled almost everywhere is executed infinitely often, which is expressed by $\Phi = \bigwedge_{i=1}^{k} (\mathsf{G}^{\infty} enabled_i \Rightarrow \mathsf{F}^{\infty} executed_i)$.

(3) *Strong fairness* (also known simply as *fairness*): an infinite computation sequence is strongly fair iff every process enabled infinitely often is executed infinitely often, which is expressed by $\Phi = \bigwedge_{i=1}^{k} (\mathsf{F}^{\infty} enabled_i \Rightarrow \mathsf{F}^{\infty} executed_i)$.

5.3. *Concrete models of concurrent computation*

Different concrete models of concurrent computation can be obtained from our abstract model by refining it in various ways. These include:

(i) providing structure for the global state space,

(ii) defining (classes of) instructions which each process can execute to manipulate the state space, and

(iii) providing concrete domains for the global state space.

We now describe some concrete models of concurrent computation.

Concrete models of parallel computation based on shared variables

Here, we consider parallel programs of the form $P_1 \| P_2 \cdots \| P_k$ consisting of a finite, fixed set of sequential processes $P_1, \ldots, P_k$ running together in parallel. There is also an underlying set of *variables* $v_1, \ldots, v_m$ assuming values in a domain D that are *shared* among the processes in order to provide for interprocess communication and coordination. Thus, the global state set S consists of tuples of the form

$$(l_1, \ldots, l_k, v_1, \ldots, v_m) \in \times_{h=1}^{k} \mathrm{LOC}(P_h) \times \times_{h'=1}^{m} D_{h'},$$

where each process P_i has an associated set $\mathrm{LOC}(P_i) = \{l_i^1, \ldots, l_i^{n_i}\}$ of locations. Each process P_i is described by a transition diagram with nodes labelled by locations. Alternatively, a process can be described by an equivalent text. Associated with each arc (l, l') there is an *instruction* I which may be executed by process P_i whenever process P_i is selected for execution and the current global state has the location of P_i at l. The instruction I is presented as a guarded command $B \rightarrow A$, where guard B is a predicate over the variables $\bar{v}$ and action A is an assignment $\bar{u} := \bar{e}$ of a tuple of expressions to the corresponding tuple of variables.

It is possible to make further refinements of the model. By imposing appropriate restrictions on the way instructions can access (i.e., read) and manipulate (i.e., write) the data, we can get models ranging from those that can perform "test-and-set" instructions, which permit a read followed by a write in a single atomic operation on a variable, to models that only permit an atomic read or an atomic write of a variable.

We might also wish to impose restrictions on which processes are allowed which kind of access to which variables. One such rule is that each variable v is "owned" by some one unique process P (think of v as being in the "local" memory of process p); then, each process can read any variable in the system, while only the process which owns a variable can write into it. This specialization is referred to as the *distributed shared-variables* model.

Still, another refinement is to specify a specific domain for the variables, say $\mathbb{N}$ = the natural numbers. Yet another is to specify the type of instructions (e.g., "copy the value of variable y into variable z"). They can be combined to get a completely concrete program with instructions such as "load the value of variable z into variable y and decrement by the natural number 1."

Concrete models of parallel computation based on message passing

This model is similar to the previous one. However, each process has its own set of local variables $y_1, \ldots, y_n$ that cannot be accessed by other processes. All interprocess communication is effected by message passing primitives similar to those of CSP [44]; processes communicate via *channels*, which are essentially message buffers of length 0. The communication primitives are

- $B;e!\alpha$: send the value of expression e along channel α, provided that guard predicate B is enabled and there is a corresponding receive command ready;
- $B;v?\alpha$: receive a value along channel α and store it in variable v, provided that the guard predicate B is enabled and there is a corresponding send command ready.

As in CSP, we assume that message transmission occurs as a single, synchronous event, with sender and receiver simultaneously executing the send, respectively receive primitive.

REMARK. For programs in one of the above concrete frameworks, we use atomic propositions such as $\mathrm{at}l_i^j$ to indicate that, in the present state, process i is at location l_j.

5.4. Connecting the concurrent computation framework with temporal logic

For an abstract concurrent program $(M, \varphi_{\mathrm{START}}, \Phi)$ and Temporal Logic formula p we write $(M, \varphi_{\mathrm{START}}, \Phi) \models p$ and read it precisely (and a bit long-windedly) as "for program M with starting condition φ_{START} and fair scheduling constraint Φ, formula p holds true"; the technical definition is as follows:

(i) in the *linear-time* framework: $(M, \varphi_{\mathrm{START}}, \Phi) \models p$ iff $\forall x$ in M such that $M, x \models \varphi_{\mathrm{START}}$ and $M, x \models \Phi$, we have $M, x \models p$;

(ii) in the *braching-time* framework: $(M, \varphi_{\mathrm{START}}, \Phi) \models p$ iff $\forall s$ in M such that $M, s \models \varphi_{\mathrm{START}}$ we have $M, s \models p_\Phi$, where p_Φ is the branching-time formula obtained from p by relativizing all path quantifications to scheduling constraint Φ; i.e., by replacing (starting at the innermost subformulae and working outward) each subformula $\mathsf{A}q$ by $\mathsf{A}(\Phi \Rightarrow q)$ and $\mathsf{E}q$ by $\mathsf{E}(\Phi \wedge q)$.

6. Theoretical aspects of Temporal Logic

In this section we discuss the work that has been done in the computing science community on the more purely theoretical aspects of Temporal Logic. This work has tended to focus on decidability, complexity, axiomatizability, and expressiveness issues. Decidability and complexity refer to natural decision problems associated with a system of Temporal Logic including (i) satisfiability—given a formula, does there exist a structure that is a model of the formula?, (ii) validity—given a formula, is it true that every structure is a model of the formula?, and (iii) model checking—given a formula together with a particular finite structure, is the structure a model of the formula? (Note that a formula is valid iff its negation is not satisfiable, so satisfiability and validity are, in effect, equivalent problems.) Axiomatizability refers to the question of the existence of deductive systems for proving all the valid formulae of a system of Temporal Logic, and the investigation of their soundness and completeness properties. Expressiveness concerns what correctness properties can and cannot be formulated in a given logic. The bulk of theoretical work has thus been to analyze, classify, and compare various systems of Temporal Logic with respect to these criteria, and to study the tradeoffs between them. We remark that these issues are not only of intrinsic interest, but are also significant due to their implications for mechanical reasoning applications.

6.1. Expressiveness

6.1.1. Linear time expressiveness

It turns out that PLTL has intimate connections with formal language theory. This connection was first articulated in the literature by Wolper who argued in [119] that PLTL "is not sufficiently expressive":

6.1. THEOREM. *The property* G_2Q, *meaning that "at all even times* (0, 2, 4, 6, ... etc.), Q *is true", is not expressible in* PLTL.

To remedy this shortcoming Wolper [119] suggested the use of an extended logic based on grammar operators; for example, the grammar

$$V_0 \rightarrow Q; true; V_0$$

defines the set of models of G_2Q. This relation with formal languages is discussed in more detail subsequently.

Quantified PLTL. Another way to extend PLTL is to allow quantification over atomic propositions (cf. [118, 102]). The syntax of PLTL is augmented by the formation rule

> if p is a formula and Q is an atomic proposition occurring free in p, then $\exists Q\, p$ is the formula too.

The semantics of $\exists Q\, p$ is given by

$$M, x \models \exists Q\, p \text{ iff there exists a linear structure } M' = (S, x, L') \text{ such that } M', x \models p \text{ where } M = (S, x, L) \text{ and } L' \text{ differs from } L \text{ in at most the truth value of } Q.$$

The formula $\exists Q\, p$ thus represents existential quantification over Q; since, under the interpretation M, Q may be viewed as defining an infinite sequence of truth values, one for every state s along x, this is a type of *second-order* quantification. We use $\forall Q\, p$ to abbreviate $\neg \exists Q \neg p$; of course, it means universal quantification over Q.

The extended temporal operator $\mathsf{G}_2 Q$ can be defined in QPLTL.

$$\mathsf{G}_2 Q \equiv_l \exists Q' (Q' \wedge X \neg Q' \wedge \mathsf{G}(Q' \Leftrightarrow XXQ') \wedge \mathsf{G}(Q' \Rightarrow Q)).$$

It can be shown that QPLTL coincides in expressive power with a number of formalisms from language theory including the just discussed grammar operators of [119].

6.1.2. *Monadic theories of linear ordering*

The First-Order Language of Linear Order (FOLLO) is the formal system corresponding to the "right-hand-side" of the definitions of the basic temporal operators of PLTL. A formula of FOLLO is interpreted over a linear-time structure $(S, <)$; for our purposes, as usual we only consider $(\mathbb{N}, <)$ or $(I, <)$ where I is an initial segment of $\mathbb{N}$. The language of linear order is built from the following symbols:

- P, Q, etc., denoting monadic (one-argument) predicate symbols (and intuitively corresponding to atomic propositions);
- t, u, etc., denoting individual variables (and intuitively ranging over moments of time in $\mathbb{N}$); and
- $<$: the distinguished *less-than* symbol (representing the temporal ordering).

A linear-time structure $M = (S, x, L)$ is then defined just as for PLTL; note that L may be viewed as assigning to each monadic predicate symbol in AP the set of times at which it is true. The formulae of FOLLO are those generated by the following rules:

(LO0) if t, u are individual variables, then $t < u$ is a formula;

(LO1) if P is a monadic predicate symbol and t is an individual variable, then $P(t)$ is a formula;

(LO2) if p, q are formulae, then so are $p \wedge q$, $p \vee q$, $\neg p$;

(LO3) if p is a formula and t is a free individual variable in p then $\exists t p$ is a formula.

The Second-Order Language of Linear Order (SOLLO) is obtained by using the following rule:

(LO4) if Q is a monadic predicate symbol not in AP that appears free in formula p, then $\exists Q\, p$ is a formula.

The semantics of SOLLO and FOLLO are defined in the obvious way. The results described below indicate how the expressive powers of the variants of PLTL and the theories of linear ordering compare.

6.2. THEOREM. *As measured with respect to initial equivalence, the relative expressive power of these linear-time formalisms is*

$$\text{PLTLF} \equiv_i \text{PLTLB} \equiv_i \text{FOLLO} <_i \text{SOLLO} \equiv_i \text{QPLTLB} \equiv_i \text{QPLTLF}.$$

For the sake of thoroughness, we include the following theorem.

6.3. THEOREM. *As measured with respect to global equivalence, the relative expressive power of these linear-time formalisms is as depicted below, where any two logics not connected by a chain of $\equiv_g$'s and $<_g$'s are of incomparable expressive power:*

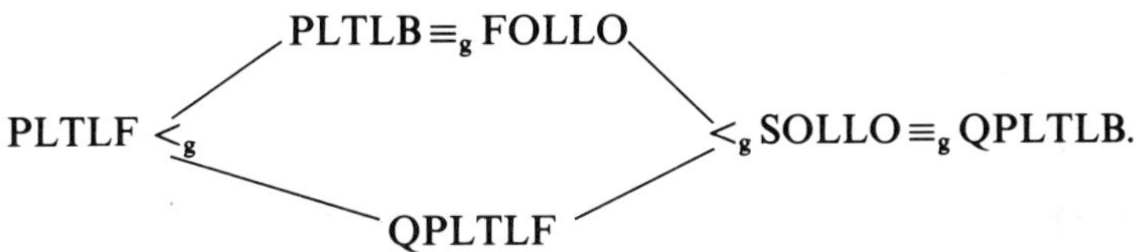

Note that QPLTLB (respectively QPLTLF) denotes the version of PLTL with quantification over auxiliary propositions having both past- and future-tense operators (respectively future-tense temporal operators only).

6.1.3. Regular languages and PLTL

There is an intimate relationship between languages definable in (variants and extensions of) PLTL, the monadic theories of linear ordering, and the regular languages. We will first consider languages of finite strings, and then languages of infinite strings. In the sequel let Σ be a finite alphabet. For simplicity, we further assume $\Sigma = \text{PowerSet}(AP)$ for some set of atomic propositions AP. Moreover, we assume that the empty string λ is excluded so that the languages of finite strings are subsets of Σ^+, rather than Σ^*.

Languages of finite strings. Before presenting the results we briefly review regular-expression notations and certain concept concerning finite-state automata. The reader is referred to [112] for more details. There are several types of regular-expression notations:

(a) the *restricted regular expressions* which are those built up from the alphabet symbols σ, for each $\sigma \in \Sigma$, and $\cdot$, $\cup$, and $*$, denoting "concatenation", "union", and "Kleene (or star) closure" respectively;

(b) the *general regular expressions* which are those built up from the alphabet symbols σ for each $\sigma \in \Sigma$ and $\cdot$, $\cap$, $\neg$, $\cup$, $*$ denoting "concatenation", "intersection", "complementation (with respect to Σ^*)", "union", and "Kleene (or star) closure" respectively;

(c) the *star-free regular expressions* are those general regular expressions with no occurrence of $*$.

The restricted regular expressions are equivalent in expressive power to the general regular expressions: however, the star-free regular expressions are strictly less expressive.

A finite-state automaton $M = (Q, \Sigma, \delta, q_0, F)$ is said to be *counter-free* iff it is *not* the case that there exist distinct states $q_0, \ldots, q_{k-1} \in Q$, $k \geqslant 1$, and a word $w \in \Sigma^+$ such that $q_{i+1} \bmod k \in \delta(q_i, w)$. A language L is said to be *noncounting* iff it is accepted by some counter-free finite-state automaton. Intuitively, a counter-free automaton cannot count modulo n for any $n \geqslant 2$. It is also known that the noncounting languages coincide with those expressible by star-free regular expressions. We now have the following results.

6.4. Theorem. *The following are equivalent conditions on languages L of finite strings:*
(a) $L \subseteq \Sigma^+$ *is definable in* PLTL.
(b) $L \subseteq \Sigma^+$ *is definable in* FOLLO.
(c) $L \subseteq \Sigma^+$ *is definable by a star-free regular expression.*
(d) $L \subseteq \Sigma^+$ *is definable by a counter-free finite-state automaton.*

This result thus accounts for why Wolper's property $\mathsf{G}_2 P$ is not expressible in PLTL, for it requires counting modulo 2. The following result also suggests why his regular grammar operators suffice.

6.5. Theorem. *The following are equivalent conditions for languages L of finite strings:*
(a) $L \subseteq \Sigma^+$ *is definable in* QPLTL.
(b) $L \subseteq \Sigma^+$ *is definable* SOLLO.
(c) $L \subseteq \Sigma^+$ *is definable by a regular expression.*
(d) $L \subseteq \Sigma^+$ *is definable by a finite-state automaton.*

The equivalence of conditions (b), (c), and (d) was established using lengthy and difficult arguments in the monograph of McNaughton and Pappert [71]. The equivalence of conditions (a) and (b) in Theorem 6.4 was established in Kamp [46], while for Theorem 6.5 it was established in [60]. Direct translations between PLTL and star-free regular expressions were given in [123].

Remark. Since we have past-tense operators, it is natural to think of history variables. If $x = (s_0, s_1, s_2, \ldots)$ is a computation then the most general history variable h would be that which at time j has accumulated the complete history $s_0 \ldots s_j$ up to (and including) time j. The expressive power of a language with history variables depends on the type of predicates we may apply to the history variables. One natural type of history predicate is of the form $[\alpha]_H$ where α is a (star-free) regular expression, with semantics given by

$(x, i) \models [\alpha]_H$ iff $s_0 \ldots s_i$ considered as a string over $\Sigma = \text{PowerSet(AP)}$ is in the language over Σ denoted by α.

These history variables will be helpful in describing canonical forms for languages of infinite strings subsequently.

Languages of infinite strings. In extending the notion of regular language to encompass languages of infinite strings, the principal concern is how to finitely describe an infinite string. For finite-state automata this is done using an extended notion of acceptance involving repeating a designated set of states infinitely (See [112]). The framework of regular expressions can be similarly extended, in one of two ways:

(i) by adding an infinite repetition operator ω: if α is an (ordinary) regular expression, then α^ω represents all strings of the form $a_1a_2a_3\ldots$, where each $a_i \in \alpha$;

(ii) by adding a limit operator lim: if α is an ordinary regular expression, then $\lim \alpha$ consists of all those strings in Σ^ω which have infinitely many (distinct) finite prefixes in α.

We now have the two results below which follow from an assemblage of results in the literature (cf. [46, 60, 112]).

6.6. THEOREM. *The following are equivalent conditions for languages L of infinite strings:*

(a) $L \subseteq \Sigma^\omega$ *is definable in* QPLTL.
(b) $L \subseteq \Sigma^\omega$ *is definable in* SOLLO.
(c_0) $L \subseteq \Sigma^\omega$ *is definable by an ω-regular expression, i.e. an expression of the form* $\bigcup_{i=1}^m \alpha_i\beta_i^\omega$, *where* α_i, β_i *are regular expressions.*
(c_1) $L \subseteq \Sigma^\omega$ *is definable by an ω-limit regular expression, i.e. an expression of the form* $\bigcup_{i=1}^m \alpha_i \lim \beta_i$, *where* α_i, β_i *are regular expressions.*
(c_2) $L \subseteq \Sigma^\omega$ *is representable as* $\bigcup_{i=1}^m (\lim \alpha_i \cap \neg \lim \beta_i)$, *where* α_i, β_i *are regular expressions.*
(c_3) $L \subseteq \Sigma^\omega$ *is expressible as* $\bigvee_{i=1}^m (\mathsf{F}^\infty[\alpha_i]_H \wedge \neg \mathsf{F}^\infty[\beta_i]_H)$, *where* α_i, β_i *are regular expressions.*

For the case of the star-free ω-languages we have the following result.

6.7. THEOREM. *The following are equivalent conditions for languages L of infinite strings:*

(a) $L \subseteq \Sigma^\omega$ *is definable in* PLTL.
(b) $L \subseteq \Sigma^\omega$ *is definable in* FOLLO.
(c_1) $L \subseteq \Sigma^\omega$ *is definable by an ω-regular expression, i.e. an expression of the form* $\bigcup_{i=1}^m \alpha_i \lim \beta_i$, *where* α_i, β_i *are star-free regular expressions.*
(c_2) $L \subseteq \Sigma^\omega$ *is representable as* $\bigcup_{i=1}^m (\lim \alpha_i \cap \neg \lim \beta_i)$, *where* α_i, β_i *are star-free regular expressions.*
(c_3) $L \subseteq \Sigma^\omega$ *is expressible in the form* $\bigvee_{i=1}^m (\mathsf{F}^\infty[\alpha_i]_H \wedge \neg \mathsf{F}^\infty[\beta_i]_H)$, *where* α_i, β_i *are star-free regular expressions.*

Result 6.7(c_0), analogous to Result 6.6(c_0), was intentionally omitted—because it does not hold as noted in [112]. It is not the case that $\bigcup_{i=1}^m \alpha_i\beta_i^\omega$, where α_i, β_i are star-free regular expressions, must itself denote a star-free regular set. For example, consider the language $L = (00 \cup 1)^\omega$. L is expressible as a union of $\alpha_i\beta_i^\omega$: take $m = 1$, $\alpha_i = \lambda$, $\beta_1 = 00 \cup 1$. But L, which consists intuitively of exactly those strings for which there is an even number of 0s between every consecutive pair of 1s, is no definable in FOLLO, nor is it star-free regular.

Remark. One significant issue we do not address here in any detail—and which is not very thoroughy studied in the literature—is that of *succinctness*. Here we refer to how long or short a formula is needed to capture a given correctness property. Two formalisms may have the same raw expressive power, but one may be much more succinct than the other. For example, while FOLLO and PLTL have the same raw expressive power, it is known that FOLLO can be significantly (nonelementarily) more succinct than PLTL (cf. [73]).

6.1.4. Branching time expressiveness

Analogy with the linear temporal framework suggests several formalisms for describing infinite trees that might be compared with branching temporal logic. Among these are: finite-state automata on infinite trees, the monadic second-order theory of many successors (S*n*S), and the monadic second-order theory of partial orders. However, not nearly so much is known about the comparison with related formalisms in the branching-time case.

One difficulty is that, technically, the types of branching objects considered differ. Branching Temporal Logic is interpreted over structures which are, in effect, trees with nodes of infinite outdegree, whereas, e.g., tree automata take input trees of fixed finite outdegree. Another difficulty is that the logics, such as CTL*, as ordinarily considered, do not distinguish between, e.g., left and right successor nodes, whereas the tree automata can.

To facilitate a technical comparison, we therefore restrict our attention to (a) structures corresponding to infinite binary trees and (b) tree automata with a "symmetric" transition function that does not distinguish, e.g., left from right. Then we have the following result from [29] comparing logics augmented with existential quantification over atomic propositions with tree automata.

6.8. Theorem. (i) EQCTL* *is exactly as expressive as symmetric pairs automata on infinite trees.*

(ii) EQCTL *is exactly as expressive as symmetric Büchi automaton infinite trees.*

Here, EQCTL* consists of the formula $\exists Q_1 \ldots \exists Q_m f$, where f is a CTL* formula and the Q_i are atomic propositions appearing in f. The semantics is that, given a structure $M=(S,R,L)$, $M,s \models \exists Q_1 \ldots \exists Q_m f$ iff there exists a structure $M'=(S,R,L')$ such that $M',s \models f$ and L' differs from L at most in the truth assignments to each Q_i, $1 \leqslant i \leqslant m$. Similarly, EQCTL consists of formulae $\exists Q_1 \ldots Q_m f$, where f is a CTL formula.

A related result is the following one from [35].

6.9. Theorem. CTL* *is exactly as expressive as the monadic second-order theory of two successors with set quantification restricted to infinite paths, over infinite binary trees.*

REMARK. By augmenting CTL* with arc assertions which allow it to distinguish outgoing arc i from $i+1$, the result extends to infinite n-ary trees, $n>2$. By taking $n=1$, the result specializes to the "expressive completeness" result of Kamp [46] that PLTL is equivalent in expressive power to FOLLO (our Theorem 6.7(a, b)).

While less is known about comparisons of BTLs (Branching-Time Logics) against external "yardsticks", a great deal is known about comparisons of BTLs against each other. This contrasts with the reversed situation for LTLs (Linear-Time Logics). Perhaps this reflects the much greater degree of "freedom" due to the multiplicity of alternative futures found in the BTL framework.

It is useful to define the notion of a *basic modality* of a BTL. This is a formula of the form $\mathsf{A}p$ or the form $\mathsf{E}p$, where p is a pure linear-time formula (containing no path quantifiers.) Then a formula of a logic may be seen as being built up by combining basic modalities using Boolean connectives and nesting. For example, $\mathsf{EF}P$ is a CTL basic modality; so is $\mathsf{AF}Q$. $\mathsf{EFAF}Q$ is formula of CTL (but not a basic modality) obtained by nesting $\mathsf{AF}Q$ within $\mathsf{EF}P$ (more precisely, by substituting $\mathsf{AF}Q$ for P within $\mathsf{EF}P$). $\mathsf{E}(\mathsf{F}P \wedge \mathsf{F}Q)$ is a basic modality of CTL*, but not a basic modality nor a formula of CTL.

A large number of sublanguages of CTL* can be defined by controlling the way the linear-time operators combine using Boolean connectives and nesting of operators in the basic modalities of the language. For instance, we use $B(\mathsf{F}, \mathsf{X}, \mathsf{U})$ to indicate the language where only a single linear-time operator X, F, or U can follow a path quantifier, and $B(\mathsf{F}, \mathsf{X}, \mathsf{U}, \wedge, \neg)$ to indicate the language where Boolean combinations of these linear operators are allowed, but not nesting of the linear operators. Thus formula $\mathsf{E}(\mathsf{F}p \wedge \mathsf{G}q)$ is in the language $B(\mathsf{F}, \mathsf{X}, \mathsf{U}, \wedge, \neg)$ but not in $B(\mathsf{X}, \mathsf{F}, \mathsf{U})$.

The diagram in Fig. 3 shows how some of these logics compare in expressive power. The notation $L_1 < L_2$ means that L_1 is strictly less expressive than L_2, which holds provided that

(a) $\forall$ formula p of L_1, $\exists$ a formula q of L_2 such that $\forall$ structure M and $\forall$ state s in M, $M, s \models p$ iff $M, s \models q$, and

(b) the converse of (a) does not hold,

$$
\begin{array}{c}
\text{CTL*} \\
6\ \vee \\
\mathrm{B}(\mathrm{X},\mathrm{F},\mathrm{U},\overset{\infty}{\mathrm{F}},\wedge,\neg) \\
5\ \vee \\
\mathrm{B}(\mathrm{X},\mathrm{F},\mathrm{U},\overset{\infty}{\mathrm{F}}) \\
4\ \vee \\
\mathrm{B}(\mathrm{X},\mathrm{F},\mathrm{U},\wedge,\neg) \\
3\ \equiv \\
\mathrm{B}(\mathrm{X},\mathrm{F},\mathrm{U}) \\
2\ \vee \\
\mathrm{B}(\mathrm{X},\mathrm{F},\wedge,\neg) \\
1\ \vee \\
\mathrm{B}(\mathrm{X},\mathrm{F}) \\
0\ \vee \\
\mathrm{B}(\mathrm{F})
\end{array}
$$

Fig. 3. Hierarchy of branching-time logics.

while $L_1 \equiv L_2$ means that L_1 and L_2 are equivalent in expressive power, and $L_1 \leqslant L_2$ means $L_1 < L_2$ or $L_1 \equiv L_2$.

Most of the logics shown are known from the literature. $B(\mathsf{F})$ is the branching-time logic of Lamport [53], having basic modalities of the form A or E followed by F or G. The logic $B(\mathsf{X}, \mathsf{F})$, which has basic modalities of the form A or E followed by X, F, or G, was originally proposed in [8] as the logic UB. The logic $B(\mathsf{X}, \mathsf{F}, \mathsf{U})$ is of course CTL. The logic $B(\mathsf{X}, \mathsf{F}, \mathsf{U}, \mathsf{F}^{\infty}, \wedge, \neg)$ is essentially the logic proposed in [23]; its infinitary modalities F^{∞} and G^{∞} permit specification of fairness properties.

We now give some rough, high-level intuition underlying these results. Semantic containment along each edge follows directly from syntactic containment in all cases, except edges 2 and 4, which follow given the semantic equivalence of edge 3 (discussed below).

The X operator (obviously) cannot be expressed in terms of the F operator, which accounts for edge 0: $B(\mathsf{F}) < B(\mathsf{F}, \mathsf{X})$. Similarly, the U operator cannot be expressed in terms of X, F and Boolean connectives. This was known "classically" (cf. [46]), and accounts for edge 2: $B(\mathsf{X}, \mathsf{F}, \wedge, \neg) < B(\mathsf{X}, \mathsf{F}, \mathsf{U})$.

To establish the equivalence of edge 3, we need to provide a translation of $B(\mathsf{X}, \mathsf{F}, \mathsf{U}, \wedge, \neg)$ into $B(\mathsf{X}, \mathsf{F}, \mathsf{U})$. The basic idea behind this translation can be understood by noting that $\mathsf{E}(\mathsf{F}P \wedge \mathsf{F}Q) \equiv \mathsf{EF}(P \wedge \mathsf{EF}Q) \vee \mathsf{EF}(Q \wedge \mathsf{EF}P)$. However, it is a bit more subtle than that; the ability to do the translation in all cases depends on the presence of the until (U) operator (cf. edge 1). The following validities, two of which concern the until, can be used to inductively translate each $B(\mathsf{X}, \mathsf{F}, \mathsf{U}, \wedge, \neg)$ formula into an equivalent $B(\mathsf{X}, \mathsf{F}, \mathsf{U})$ formula:

$$\begin{aligned} &\mathsf{E}((p_1 \,\mathsf{U}\, q_1) \wedge (p_2 \,\mathsf{U}\, q_2)) \\ &\quad \equiv \mathsf{E}((p_1 \wedge p_2) \,\mathsf{U}\, (q_1 \wedge \mathsf{E}(p_2 \,\mathsf{U}\, q_2))) \vee \mathsf{E}((p_2 \wedge p_1) \,\mathsf{U}\, (q_2 \wedge \mathsf{E}(p_1 \,\mathsf{U}\, q_1))), \\ &\mathsf{E}(\neg(p \,\mathsf{U}\, q)) \equiv \mathsf{E}((\neg q \wedge \neg p) \,\mathsf{U}\, (q \wedge p)) \vee \mathsf{EG}\neg q, \\ &\mathsf{E}(\neg \mathsf{X}p) \equiv \mathsf{EX}\neg p. \end{aligned}$$

$\mathsf{EF}^{\infty}Q$, a $B(\mathsf{X}, \mathsf{F}, \mathsf{U}, \mathsf{F}^{\infty})$ formula, is not expressible in $B(\mathsf{X}, \mathsf{F}, \mathsf{U})$ accounting for the strict containment on arc 4. This is probably the most significant result, for it basically says that correctness under fairness assumptions cannot be expressed in a BTL with a simple set of modalities. For example, the property that P eventually becomes true along all fair computations (fair inevitability of P) is of the form $\mathsf{A}(\mathsf{F}^{\infty}Q \Rightarrow \mathsf{F}P)$ for even a (very) simple fairness constraint like $\mathsf{F}^{\infty}Q$. Neither it, nor its dual $\mathsf{E}(\mathsf{F}^{\infty}Q \wedge \mathsf{G}P)$, is expressible in $B(\mathsf{X}, \mathsf{F}, \mathsf{U})$, since by taking P to be *true* the dual becomes $\mathsf{EF}^{\infty}Q$.

The inexpressibility of $\mathsf{EF}^{\infty}Q$ was established in [23], using recursion-theoretic arguments to show that the predicate transformer associated with $\mathsf{EF}^{\infty}Q$ is Σ_1^1-complete while the predicate transformers for $B(\mathsf{X}, \mathsf{F}, \mathsf{U})$ are arithmetical. The underlying intuition is that $\mathsf{EF}^{\infty}Q$ uses second-order quantification in an essential way to assert that there exists a sequence of nodes in the computation tree where Q holds. Another version of this inexpressiveness result was established by Lamport [53] in a somewhat different technical framework. Still another proof of this result was given by Emerson and Halpern [26]. The type of inductive, combinatorial proof used is paradigmatic of the proofs of many inexpressiveness results for TL, so we describe the main idea here.

6.10. THEOREM. $\mathsf{EF}^{\infty}Q$ *is not expressible in* $B(\mathsf{X}, \mathsf{F}, \mathsf{U})$

PROOF (*idea*). We inductively define two sequences $M_1, M_2, M_3, \ldots$ and N_1, N_2, $N_3, \ldots$ of structures as shown in Fig. 4. It is plain that for all i,

$$M_i, s_i \models \mathsf{EF}^{\infty}Q \quad \text{and} \quad N_i, s_i \models \neg\mathsf{EF}^{\infty}Q \tag{6.1}$$

Thus $\mathsf{EF}^{\infty}Q$ distinguishes between the two sequences. However, we can show by an inductive argument that each formula of $B(\mathsf{X}, \mathsf{F}, \mathsf{U})$ is "confused" by the two sequences, in that

$$M_i, s_i \models p \text{ iff } N_i, s_i \models p \quad \text{for all } i \geqslant \text{the length of } p. \tag{6.2}$$

If some formula p of $B(\mathsf{X}, \mathsf{F}, \mathsf{U})$ were equivalent to $\mathsf{EF}^{\infty}Q$, we would then have, for i equalling the length of p, that

$$M, S_i \models p \quad \text{and} \quad N, s_i \models \neg p \qquad \text{by virtue of (6.1)}$$

and also that

$$N, s_i \models p, \quad \text{by virtue of (6.2), a contradiction.} \qquad \square$$

The strict containment along the rest of the edges follow from these inexpressiveness results: $\mathsf{E}(\mathsf{F}P \wedge \mathsf{G}Q)$ is not expressible in $B(\mathsf{X}, \mathsf{F})$, for edge 1. $\mathsf{E}(\mathsf{F}^{\infty}P_1 \wedge \mathsf{F}^{\infty}P_2)$ is not expressible in $B(\mathsf{X}, \mathsf{F}, \mathsf{U}, \mathsf{F}^{\infty})$, for edge 5. $\mathsf{A}(\mathsf{F}(P \wedge \mathsf{X}P))$ is not expressible in

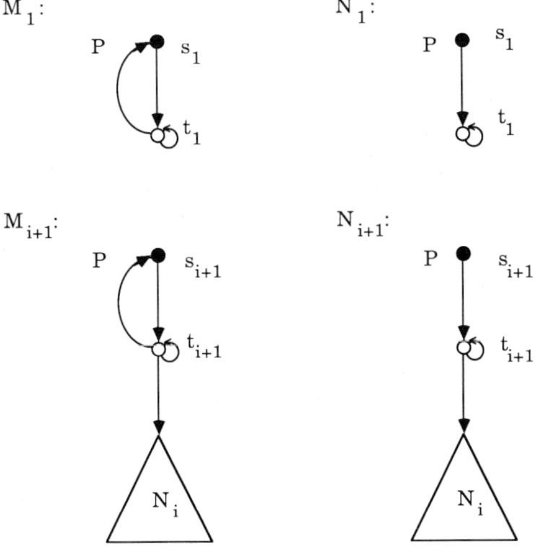

Fig. 4.

$B(\mathsf{X}, \mathsf{F}, \mathsf{U}, \mathsf{F}^{\infty}, \wedge, \neg)$, for edge 6. The proofs are along the lines of the theorem above for $\mathsf{EF}^{\infty}Q$.

It is also possible to compare branching- with linear-time logics. When a linear-time formula is interpreted over a program, there is usually an implicit universal quantification over all possible computations. This suggests that when given a linear-time language L, which is of course a set of path formulae, we convert it into a branching-time language by prefixing each path formula by the universal path quantifier A. We thus get the corresponding branching language $\mathrm{BL}(L) = \{\mathsf{A}p: p \in L\}$. Figure 5 shows how various branching and linear logics compare. Not surprisingly, the major limitation of linear time is its inability to express existential path quantification (cf. [53, 26]).

6.11. THEOREM. *The formula* $\mathsf{EF}P$ *is not expressible in any* BL(–) *logic.*

6.2. Decision procedures for Propositional Temporal Logics

In this section we discuss algorithms for testing if a given formula p_0 in a system of Propositional TL is satisfiable. The usual approach to developing such algorithms is to first establish the *small model property* for the logic: if a formula is satisfiable, then it is satisfiable in a "small" finite model, where "small" means of size bounded by some function, say, f, of the length of the input formula. This immediately yields a decision procedure for the logic. Guess a small structure M as a candidate model of given formula p_0; then check that M is indeed a model of p_0. This check can be done by exhaustive search, since M is finite, and can often be done efficiently.

An elegant technique for establishing the small model property is through use of the *quotient construction*, also called—in classical modal logic—*filtration*, where an equivalence relation of small finite index is defined on states. Then equivalent states are identified to collapse a possibly infinite model to a small finite one.

An example of a quotient construction is its application to yield a decision procedure for Propositional Dynamic Logic of [31], discussed in [50]. There, the equivalence relation is defined so that, in essence, two states are equivalent when they agree (i.e., have the same truth value) on all subformulae of the formula p_0 being tested for satisfiability. This yields a decision procedure of nondeterministic exponential-time complexity, calculated as follows: The total complexity is the time to guess a small candidate model plus the time to check that it is indeed a model. The candidate model can be guessed in time polynomial in its size which is exponential in the length of p_0, since for a formula of length n there are about n subformulae and 2^n equivalence classes.

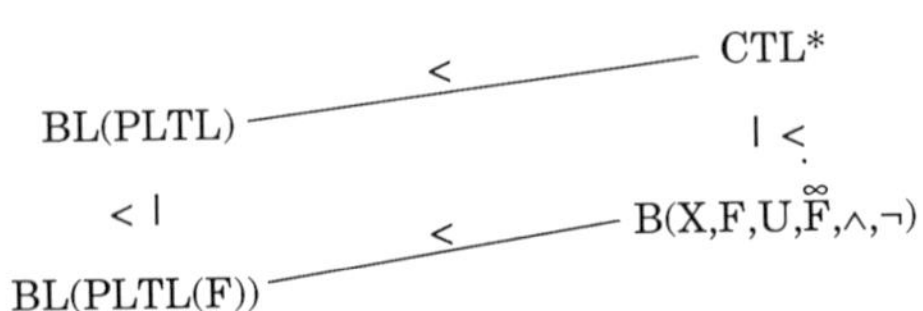

Fig. 5. Comparing linear with branching time.

And it turns out that checking that the candidate model is a genuine model can be done in polynomial time.

Of course the deterministic time complexity of the above algorithm is double exponential. The complexity can be improved through use of the tableau construction.

A *tableau* for formula p_0 is a finite directed graph with nodes labelled by subformulae associated with p_0 that, in effect, encodes all potential models of p_0. In particular, as in the case of Propositional Dynamic Logic, the tableau contains as a subgraph the quotient structure corresponding to any model of p_0. The tableau can be constructed, and then tested for consistency to see if it contains a genuine quotient model. Such testing can often be done efficiently. In the case of Propositional Dynamic Logic, the tableau is of size exponential in the formula length, while the testing can be done in deterministic polynomial time in the tableau size, yielding a deterministic single exponential time decision procedure.

For some logics, no matter how we define a finite index equivalence relation on states, the quotient construction yields a quotient structure that is not a model. However, for many logics, the quotient structure still provides useful information. It can be viewed as a "pseudo-model" that can be unwound into a genuine, yet still small, model. The tableau construction, moreover, can still be used to perform a systematic search for a pseudo-model, to be unwound into a genuine model.

We remark that the tableau construction is a rather general one, which applies to many logics. Tableau-based decision procedures for various logics are given in [90, 8, 7, 118, 119, 43]. See also the excellent survey by Wolper [120]. In the sequel we describe a tableau based decision procedure for CTL formulae, along the lines of [24, 25]. The following definitions and terminology are needed.

We assume that the candidate formula p_0 is in *positive normal form*, obtained by pushing negations inward as far as possible using de Morgan's laws ($\neg(p \vee q) \equiv \neg p \wedge \neg q$, $\neg(p \wedge q) \equiv \neg p \vee \neg q$) and dualities ($\neg \mathsf{AG}p \equiv \mathsf{EF}\neg p$, $\neg\mathsf{A}[p \mathsf{U} q] \equiv \mathsf{E}[\neg p \mathsf{B} q]$, etc.). This at most doubles the length of the formula, and results in only atomic propositions being negated. We write $\sim p$ for the formula in positive normal form equivalent to $\neg p$. The *closure* of p_0, $\mathrm{cl}(p_0)$, is the least set of subformulae such that

- each subformulae of p_0, including p_0 itself, is a member of $\mathrm{cl}(p_0)$;
- if $\mathsf{EF}q$, $\mathsf{EG}q$, $\mathsf{E}[p \mathsf{U} q]$, or $\mathsf{E}[p \mathsf{B} q] \in \mathrm{cl}(p_0)$ then, respectively, $\mathsf{EXEF}q$, $\mathsf{EXEG}q$, $\mathsf{EXE}[p\mathsf{U}q]$, or $\mathsf{EXE}[p \mathsf{B} q] \in \mathrm{cl}(p_0)$;
- if $\mathsf{AF}q$, $\mathsf{AG}q$, $\mathsf{A}[p \mathsf{U} q]$, or $\mathsf{A}[p \mathsf{B} q] \in \mathrm{cl}(p_0)$ then, respectively, $\mathsf{AXAF}q$, $\mathsf{AXAG}q$, $\mathsf{AXA}[p \mathsf{U} q]$, or $\mathsf{AXA}[p \mathsf{B} q] \in \mathrm{cl}(p_0)$;

The *extended closure* of p_0, $\mathrm{ecl}(p_0) = \mathrm{cl}(p_0) \cup \{\sim p\colon p \in \mathrm{cl}(p_0)\}$. Note that $\mathrm{card}(\mathrm{ecl}(p_0)) = O(\mathrm{length}(p_0))$.

At this point we give the technical definitions for the quotient construction, as they are needed in the proof of the small model theorem of CTL. We also show the quotient construction by itself is inadequate for getting a small model theorem for CTL.

Let $M = (S, R, L)$ be a model of p_0, let H be a set of formulae, and let $\equiv_H$ be an equivalence relation on S induced by agreement on the formulae in H, i.e. $s \equiv_H t$ whenever $\forall q \in H$, $M, s \models q$ iff $M, t \models q$. We use $[s]$ to denote the equivalence class $\{t\colon t \equiv_H s\}$ of s. Then the quotient structure of M by $\equiv_H$, $M/{\equiv_H}$, is (S', R', L') where

$S' = \{[s]: s \in S\}$, $R' = \{([s], [t]): (s, t) \in R\}$, and $L'([s]) = L(s) \cap H$. Ordinarily, we take $H = \text{ecl}(p_0)$.

However, as the following theorem shows, no way of defining the equivalence relation for the quotient construction preserves modelhood.

6.12. THEOREM. *For every set H of* (CTL) *formulae, the quotient construction does not preserve modelhood for the formula* AFP. *In particular, there is a model M of* AFP *such that for every finite set H, $M/\equiv_H$ is not a model for* AFP.

PROOF (*idea*). Note the structure shown in Fig. 6(a) is a model of AFP. But however the quotient relation collapses the structure, two distinct states s_i and s_j will be identified, resulting in a cycle in the quotient structure along which P is always false, as suggested in Fig. 6(b). Hence AFP does not hold along the cycle. □

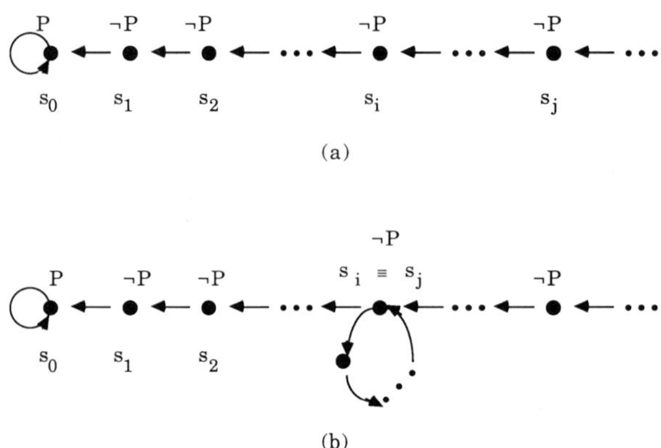

Fig. 6. The structure in diagram (a) is a model of AFP. Its quotient structure in diagram (b) is not.

We now proceed with the technical development needed. To simplify the exposition, we assume that the candidate formula p_0 is of the form $p_1 \wedge$ AGEX *true*, syntactically reflecting the semantic requirement that each state in a structure have a successor state.

We say that a formula is *elementary* provided that it is a proposition, the negation of a proposition, or has main connective AX or EX. Any other formula is *nonelementary*. Each nonelementary formula may be viewed as either a conjunctive formula $\alpha \equiv \alpha_1 \wedge \alpha_2$ or a disjunctive formula $\beta \equiv \beta_1 \vee \beta_2$. Clearly, $f \wedge g$ is an α formula and $f \vee g$ is a β formula. A modal formula may be classified as α or β based on its fixpoint characterization (cf. Section 8.4); e.g., EF$p = p \vee$ EXEFp is a β formula and AG$p =$

$p \wedge \mathsf{AXAG}p$ is an α formula. The following table summarizes the classification:

$\alpha = p \wedge q$	$\alpha_1 = p$	$\alpha_2 = q$
$\alpha = \mathsf{A}[p\,\mathsf{B}\,q]$	$\alpha_1 = \sim q$	$\alpha_2 = p \vee \mathsf{AXA}[p\,\mathsf{B}\,q]$
$\alpha = \mathsf{E}[p\,\mathsf{B}\,q]$	$\alpha_1 = \sim q$	$\alpha_2 = p \vee \mathsf{EXE}[p\,\mathsf{B}\,q]$
$\alpha = \mathsf{AG}q$	$\alpha_1 = q$	$\alpha_2 = \mathsf{AXAG}q$
$\alpha = \mathsf{EG}q$	$\alpha_1 = q$	$\alpha_2 = \mathsf{EXEG}q$
$\beta = p \vee q$	$\beta_1 = p$	$\beta_2 = q$
$\beta = \mathsf{A}[p\,\mathsf{U}\,q]$	$\beta_1 = q$	$\beta_2 = p \wedge \mathsf{AXA}[p\,\mathsf{U}\,q]$
$\beta = \mathsf{E}[p\,\mathsf{U}\,q]$	$\beta_1 = q$	$\beta_2 = p \wedge \mathsf{EXE}[p\,\mathsf{U}\,q]$
$\beta = \mathsf{AF}q$	$\beta_1 = q$	$\beta_2 = \mathsf{AXAF}q$
$\beta = \mathsf{EF}q$	$\beta_1 = q$	$\beta_2 = \mathsf{EXEF}q$

A formula of the form $\mathsf{A}[p\,\mathsf{U}\,q]$ or $\mathsf{E}[p\,\mathsf{U}\,q]$ is an *eventuality* formula. An eventuality makes a promise that something will happen. This promise must be *fulfilled*. The eventuality $\mathsf{A}[p\,\mathsf{U}\,q](\mathsf{E}[p\,\mathsf{U}\,q])$ is fulfilled for s in M provided that, for every (respectively for some) path starting at s, there exists a finite prefix of the path in M whose last state is labelled with q and all of whose other states are labelled with p. Since $\mathsf{AF}q$ and $\mathsf{EF}q$ are special cases of $\mathsf{A}[p\,\mathsf{U}\,q]$ and $\mathsf{E}[p\,\mathsf{U}\,q]$ respectively, they are also eventualities. In contrast, $\mathsf{A}[p\,\mathsf{B}\,q]$, $\mathsf{E}[p\,\mathsf{B}\,q]$, and their special cases $\mathsf{AG}q$ and $\mathsf{EG}q$, are *invariance* formulae. An invariance property asserts that whatever happens to occur (if anything) will meet certain conditions (cf. Subsection 7.1).

We say that a *prestructure* M is a triple (S, R, L) just like a structure, except that the binary relation R is not required to be total. An *interior* node of a prestructure is one with at least one successor. A *frontier* node is one with no successors.

It is helpful to associate certain consistency requirements on the labelling of a (pre)structure:

Propositional Consistency Rules:

(PC0) $\sim p \in L(s)$ implies $p \notin L(s)$;
(PC1) $\alpha \in L(s)$ implies $\alpha_1 \in L(s)$ and $\alpha_2 \in L(s)$;
(PC2) $\beta \in L(s)$ implies $\beta_1 \in L(s)$ or $\beta_2 \in L(s)$.

Local Consistency Rules:

(LC0) $\mathsf{AX}p \in L(s)$ implies, $\forall$ successor t of s, $p \in L(t)$;
(LC1) $\mathsf{EX}p \in L(s)$ implies, $\exists$ successor t of s, $p \in L(t)$.

A *fragment* is a prestructure whose graph is a dag (directed acyclic graph) such that all of its nodes satisfy (PC0–2) and (LC0) above, and all of its interior nodes satisfy (LC1) above.

A *Hintikka structure* (*for* p_0) is a structure $M = (S, R, L)$ (with $p_0 \in L(s)$ for some $s \in S$) which meets the following conditions:

(1) the propositional consistency rules (PC0–2),
(2) the local consistency rules (LC0–1), and
(3) each eventuality is fulfilled.

6.13. PROPOSITION. *If structure $M = (S, R, L)$ defines a model of p_0 and each s is labelled with exactly the formula in* ecl(p_0) *true at s, then M is a Hintikka structure for p_0. Conversely, a Hintikka structure for p_0 defines a model of p_0.*

If M is a Hintikka structure, then for each node s of M and each eventuality r in ecl(p_0) such that $M, s \models r$, there is a fragment (call it DAG$[s, r]$) which certifies fulfillment of r at s in M. What is the nature of this fragment? It has s as its *root*, i.e. the node from which all other nodes in DAG$[s, r]$ are reachable. If r is of the form $\mathsf{AF}q$, then DAG$[s, \mathsf{AF}q]$ is obtained by taking node s and all nodes along all paths emanating from s up to and including the first state where q is true. The resulting subgraph is indeed a dag all of whose frontier nodes are labelled with q. If r were of the form $\mathsf{A}[p \,\mathsf{U}\, q]$, DAG$[s, \mathsf{A}[p \,\mathsf{U}\, q]]$ would be the same, except that its interior nodes are all labelled with p. In the case of DAG$[s, \mathsf{EF}q]$, take a shortest path leading from node s to a node labelled with q, and then add sufficient successors to ensure that (LC1) holds of each interior node on the path. In the case of DAG$[s, \mathsf{E}[p \,\mathsf{U}\, q]]$, the only change is that p labels each interior node on the path.

In a Hintikka structure M for p_0, each fulfilling fragment DAG$[s, r]$ for each eventuality r, is "cleanly embedded" in M. If we collapse M by applying a finite-index quotient construction, the resulting quotient structure is not, in general, a model because cycles are introduced into such fragments. However, there is still a fragment, call it DAG$'[s, r]$, "contained" in the quotient structure of M. It is simply no longer cleanly embedded. Technically, we say prestructure $M_1 = (S_1, R_1, L_1)$ is *contained* in prestructure $M_2 = (S_2, R_2, L_2)$ whenever $S_1 \subseteq S_2$, $R_1 \subseteq R_2$, and $L_1 = L_2|S_1$, the labelling L_2 restricted to S_1. We say that M_1 is *cleanly embedded* in M_2 provided M_1 is contained in M_2, and also every interior node of M_1 has the same set of successors in M_1 as in M_2.

A *pseudo-Hintikka structure* (*for* p_0) is a structure $M = (S, R, L)$ (with $p_0 \in L(s)$ for some $s \in S$) which meets the following conditions:

(1) the propositional consistency rules (PC0–2),

(2) the local consistency rules (LC0–1), and

(3) each eventuality is *pseudo-fulfilled* in the following sense:

(3)(a) $\mathsf{AF}q \in L(s)$ (resp. $\mathsf{A}[p \,\mathsf{U}\, q] \in L(s)$) implies there is a finite fragment—called DAG$[s, \mathsf{AF}q]$ (resp. DAG$[s, \mathsf{A}[p \,\mathsf{U}\, q]]$)—rooted at s contained in M such that for *all* frontier nodes t of the fragment, $q \in L(t)$ (resp. and for all interior nodes u of the fragment, $p \in L(u)$);

(3)(b) $\mathsf{EF}q \in L(s)$ (resp. $\mathsf{E}[p \,\mathsf{U}\, q] \in L(s)$) implies there is a finite fragment—called DAG$[s, \mathsf{EF}q]$ (resp. DAG$[s, \mathsf{E}[p \,\mathsf{U}\, q]]$)—rooted at s contained in M such that for *some* frontier node t of the fragment, $q \in L(t)$ (resp. and for all interior nodes u of the fragment, $p \in L(u)$).

6.14. THEOREM (Small Model Theorem for CTL). *Let p_0 be a* CTL *formula of length n. Then the following are equivalent*:

(a) *p_0 is satisfiable.*

(b) *p_0 has an infinite tree model with finite branching bounded by* O(n).

(c) *p_0 has a finite model of size* $\leqslant \exp(n)$.

(d) *p_0 has a finite pseudo-Hintikka structure of size* $\leqslant \exp(n)$.

PROOF (*sketch*). We show that (a)$\Rightarrow$(b)$\Rightarrow$(d)$\Rightarrow$(c)$\Rightarrow$(a).

(a)⇒(b): Suppose $M, s \models p_0$. Then, as described in Subsection 5.1, M can be unwound into an infinite tree model M_1, with root state s_1 a copy of s. It is possible that M_1 has infinite branching at some states, so (if needed) we chop out spurious successor states to get a bounded branching subtree M_2 of M_1 such that still $M_2, s_1 \models p_0$. We proceed down M_1 level-by-level deleting all but n successors of each state. The key idea is that for each formula $\mathsf{EX}q \in L(s)$, where s is a retained node on the current level, we keep a successor t of s of least q-rank, where the q-rank(s) is defined as the length of the shortest path from s fulfilling q if q is of the form $\mathsf{EF}r$ or $\mathsf{E}[p \mathsf{U} r]$, and is defined as 0 if q is of any other form. This will ensure that each eventuality of the form $\mathsf{EF}r$ or $\mathsf{E}[p \mathsf{U} r]$ is fulfilled in the tree model M_2. Moreover, since there are at most O(n) formulae of the form $\mathsf{EX}q$ in ecl(p_0), the branching at each state of M_2 is bounded by O(n).

(b)⇒(d): Let M be a bounded branching infinite tree model with root s_0 such that M, $s_0 \models p_0$. We claim that the quotient structure $M' = M/\equiv_{\mathrm{ecl}(p_0)}$ is a pseudo-Hintikka structure. It suffices to show that for each state $[s]$ of M', and each eventuality r in the label of $[s]$ there is a finite fragment contained in M' certifying pseudo-fulfillment of r. We sketch the argument in the case $r = \mathsf{AF}q$. The argument for other types of eventuality is similar.

So suppose $\mathsf{AF}q$ appears in the label of $[s]$. By definition of the quotient construction, in the original structure M, $\mathsf{AF}q$ is true at state s, and thus there exists a finite fragment DAG$[s, \mathsf{AF}q]$ with root s cleanly embedded in M. Extract (a copy of) the fragment DAG$[s, \mathsf{AF}q]$. Chop out states with duplicate labels. Given two states s, s' with the same label, let the deeper state replace the shallower, where the depth of a state is the length of the longest path from the state back to the root s_0. This ensures that after the more shallow node has been chopped out, the resulting graph is still a dag, and moreover, a fragment. Since we can chop out any pair of duplicates, the final fragment, call it DAG$'[[s], \mathsf{AF}q]$, has at most a single occurrence of each label. Therefore (a copy of) DAG$'[[s], \mathsf{AF}q]$ is contained in the quotient structure M'. It follows that M' is a pseudo-Hintikka model as desired.

(d)⇒(c): Let $M = (S, R, L)$ be a pseudo-Hintikka model for p_0. For simplicity we identify a state s with its label $L(s)$. Then for each state s and each eventuality $q \in s$, there is a fragment DAG$[s, q]$ contained in M certifying fulfillment of q. We show how to splice together copies of the DAGs, in effect unwinding M, to obtain a Hintikka model for p_0.

For each state s and each eventuality q, we construct a dag rooted at s, DAGG$[s, q]$. If $q \in s$ then DAGG$[s, q]$ = DAG$[s, q]$; otherwise DAGG$[s, q]$ is taken to be the subgraph consisting of s plus a sufficient set of successors to ensure that local consistency rules (LC0–1) are met.

We now take (a single copy of) each DAGG$[s, q]$ and arrange them in a matrix as shown in Fig. 7, the rows range over eventualities $q_1, \ldots, q_m$ and the columns range over the states $s_1, \ldots, s_N$ in the tableau. Now each frontier node s in row i is replaced by the copy of s that is the root of DAGG$[s, q_{i+1}]$ in row $i+1$. Note that each fullpath through the resulting structure goes through each row infinitely often. As a consequence, the resulting graph defines a model of p_0, as can be verified by induction on the structure of formulae. The essential point is that each eventuality q_i is fulfilled along each fullpath where needed, at least by the time the fullpath has gone through row i.

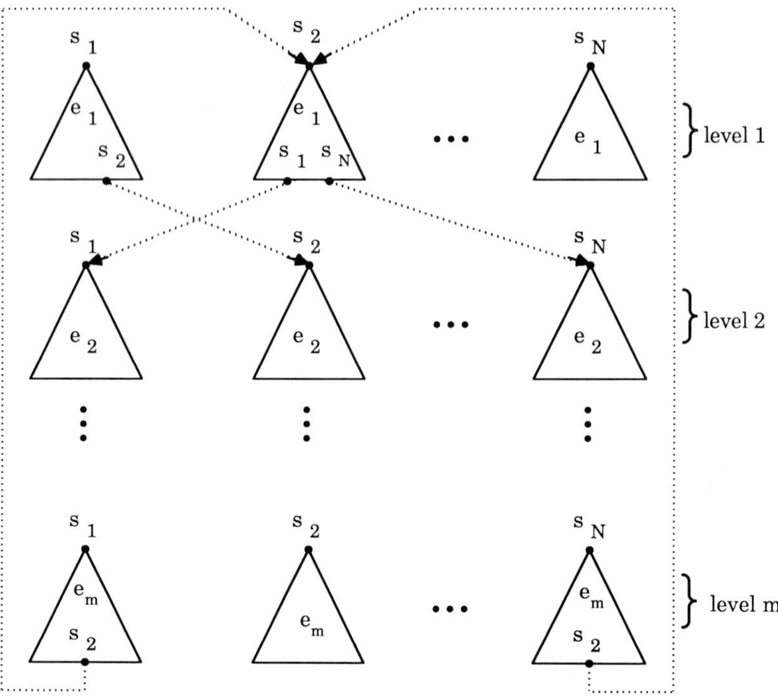

Fig. 7. The diagram above shows an array of $m \times N$ dags; the ijth entry is (a copy of) DAGG$[s_j, e_i]$. Drawing an arc $s \cdots > t$ indicates that node s is to be replaced by node t. This illustrates how to construct a finite model for p_0 from the DAGGs for its eventualities.

The cyclic model consists of $m \cdot N$ DAGGs, each consisting of N nodes. It is thus of size $m \cdot N^2$ nodes, where the number of eventualities $m \leqslant n$ and the number of tableau nodes $N \leqslant 2^n$, and n is the length of p_0. We can chop out duplicate nodes with the same label within a row, using an argument based on the depth of a node like that used above in the proof of (b)⇒(d), to get a model of size $m \cdot N = \exp(n)$.

(c)⇒(a) is immediate. □

We now describe the tableau-based decision procedure for CTL. Let p_0 be the candidate CTL formula which is to be tested for satisfiability. We proceed as follows:

(1) Build an initial *tableau* $T = (S, R, L)$ for p_0, which encodes potential pseudo-Hintikka structures for p_0. Let S be the collection of all maximal, propositionally consistent subsets s of ecl(p_0), where by maximal we mean that for every formula $p \in$ ecl(p_0), either p or $\sim p \in s$, while propositionally consistent refers to rules (PC0–2) above. Let $R \subseteq S \times S$ be defined so that $(s, t) \in R$ unless $\mathsf{AX}p \in s$ and $\sim p \in t$, for some formula $\mathsf{AX}p \in$ ecl(p_0). Let $L(s) = s$. Note that the tableau as initially constructed meets all propositional consistency rules (PC0–2) and local consistency rule (LC0).

(2) Test the tableau for consistency and pseudo-fulfillment of eventualities, by

repeatedly applying the following deletion rules until no more nodes in the tableau can be deleted:

- Delete any state s such that eventuality $r \in L(s)$ and there does *not* exist a fragment DAG$[s, r]$ rooted at s contained in the tableau which certifies pseudo-fulfillment of r.
- Delete any state which has no successors.
- Delete any state which violates (LC1).

Note that this portion of the algorithm must terminate, since there are only a finite number of nodes in the tableau.

(3) Let T' be the final tableau. If there exists a state s' in T' with $p_0 \in L(s')$ then return "YES, p_0 is satisfiable"; If not, then return "NO, p_0 is unsatisfiable".

To test the tableau for the existence of the appropriate fragments to certify fulfillment of eventualities, we can use a ranking procedure. For an $\mathsf{A}[p \,\mathsf{U}\, q]$ eventuality, initially assign rank 1 to all nodes labelled with q and rank ∞ to all other nodes. Then, for each node s and each formula r such that $\mathsf{EX}r$ is in the label of s, define $\text{SUCC}_r(s) = \{s' \colon s'$ is a successor of s in the tableau with $r \in$ label of $s'\}$ and compute $\text{rank}(\text{SUCC}_r(s)) = \min\{\text{rank}(s') \colon s' \in \text{SUCC}_r(s)\}$. Now for each node s of $\text{rank} = \infty$ such that $p \in L(s)$, let $\text{rank}(s) = 1 + \max\{\text{rank}(\text{SUCC}_r(s)) \colon \mathsf{EX}r \in L(s)\}$. Repeatedly apply the above ranking rules until stabilization. A node has finite rank iff $\mathsf{A}[p \,\mathsf{U}\, q]$ is fulfilled at it in the tableau. Testing for fulfillment of an $\mathsf{AF}q$ is a special case of the above, ignoring the formula p. To test for fulfillment of $\mathsf{E}[p \,\mathsf{U}\, q]$, use a procedure like the above, but compute $\text{rank}(s) = 1 + \min\{\text{rank}(\text{SUCC}_r(s)) \colon \mathsf{EX}r \in L(s)\}$. Testing for fulfillment of $\mathsf{EF}q$ is again a special case, where the formula p is ignored.

6.15. THEOREM. *The problem of testing satisfiability for* CTL *is complete for deterministic exponential time.*

PROOF (*idea*). The above algorithm can be shown to run in deterministic exponential time in the length of the input formula, since the size of the tableau is, in general, exponential in the formula size, and the tableau can be constructed and tested for containment of a pseudo-Hintikka structure in time polynomial in its size. This establishes the upper bound. The lower bound follows by a reduction from alternating, polynomial-space bounded Turing machines, similar to that used to establish exponential time hardness for Propositional Dynamic Logic (see [50]). □

The above formulation of the CTL decision procedure is sometimes known as the *maximal model* approach, since the nodes in the initial tableau are maximal, propositionally consistent sets of formulae and we put in as many arcs as possible. One drawback is that its average-case complexity is as bad as its worst-case complexity, since it always constructs the exponential-size collection of maximal, propositionally consistent sets of formulae.

An alternative approach is to build the initial tableau incrementally, which in practice often results in a significant decrease in the size and time required to construct it. The tableau construction will now begin with a bipartite graph $T' = (C, D, R_{CD}, R_{DC}, L)$ where nodes in C are referred to as states while nodes in D are known as prestates;

$R_{CD} \subseteq C \times D$ and $R_{DC} \subseteq D \times C$. The labels of the states will be *sparsely downward closed* set of formulae in ecl(p_0), i.e., sets which satisfy (PC0), (PC1), and

(PC2′): $\beta \in L(s)$ implies either $\beta_1 \in L(s)$ or $\beta_2 \in L(s)$.

Initially, let C be the empty set, D a single prestate d labelled with p_0.

repeat
  let e be a frontier node of T'
  if e is a prestate d **then**
    let $c_1, \ldots, c_k$ be states whose labels comprise all the sparsely downward closed supersets of $L(d)$
    add $c_1, \ldots, c_k$ as R_{DC}-successors of d in T'
    Note: if any c_i has the same label as another state c' already in T' then identify c_i and c (i.e. delete c_i and draw an R_{DC}-arc from d to c').
  if e is a state c labelled with next-time formulae $\mathsf{AX}p_1, \ldots, \mathsf{AX}p_l$, $\mathsf{EX}q_1, \ldots, \mathsf{EX}q_k$
    then create prestates $d_1, \ldots, d_k$ labelled with sets respectively $\{p_1, \ldots, p_l, q_1\}, \ldots, \{p_1, \ldots, p_l, q_k\}$ and add them as R_{CD}-successors to c in T'
    Note: if any d_i has the same label as another prestate d' already in T', then identify d_i and d' as above.
until all notes in T' have at least one successor

Now the tableau $T = (C, R, L|C)$ where C is the set of states in T' above and $R = R_{CD} \circ R_{DC}$; $L|C$ is the labelling L restricted to C. Then the remainder of the decision procedure described previously can be applied to this new tableau constructed incrementally.

REMARK. It is possible to construct the original type of tableau incrementally. Let the initial prestate be labelled with $p_0 \vee \sim p_0$ and use maximal, propositionally consistent sets for the labels of states.

The decision procedure for CTL also yields a deterministic exponential-time decision procedure for PLTL, as expressed by the following theorem.

6.16. THEOREM. *Let p_0 be a* PLTL *formula in positive normal form. Let p_1 be the* CTL *formula obtained from p_0 by replacing each temporal operator* $\mathsf{F}, \mathsf{G}, \mathsf{X}, \mathsf{U}, \mathsf{B}$ *by* $\mathsf{AF}, \mathsf{AG}, \mathsf{AX}, \mathsf{AU}, \mathsf{AB}$, *respectively. Then p_0 is satisfiable iff p_1 is satisfiable.*

We can in fact do better for PLTL and various fragments of it. The following results on the complexity of deciding linear time are due to Sistla and Clarke [104].

6.17. THEOREM. *The problem of testing satisfiability for* PLTL *is* PSPACE-*complete.*

PROOF (*idea*). To establish membership in PSPACE, we design a nondeterministic algorithm that, given an input formula p_0, guesses a satisfying path through the tableau for p_0 which defines a linear model of size exp(n), where n = length(p_0). This path can be guessed and verified to be a model in only O(n) space, since the algorithm need only remember the label of the current and next state along the path, and the point where the path loops back, in order to check that eventualities are fulfilled. PSPACE-hardness can be established by a generic reduction from polynomial-space Turing machines. □

For the sublanguage of PLTL restricted to allowing only the F operator (and its dual G) and denoted PLTL(F), further improvement is still possible. We first establish the following somewhat surprising, result.

6.18. THEOREM. (Linear-Size Model Theorem for PLTL(F)). *If a* PLTL(F) *formula* p_0 *of length n is satisfiable, then it has a finite linear model of size* O(n).

PROOF (*idea*). The important insight is that truth of a PLTL(F) formula only depends on the set of successor states, and not on their order or arrangement. Now suppose p_0 is satisfiable. Let $x = s_0, s_1, s_2, \ldots$ be a model of p_0. Then there exist i and j such that $i < j$ and $s_i = s_j$ and the set of states appearing infinitely often along x equals $\{s_i, \ldots, s_{j-1}\}$. Let x' be the linear structure obtained be deleting all states of index greater than $j-1$ and making s_i the successor of s_{j-1}. It is readily checked that $x' \models p_0$. Moreover, since the order of successor states does not matter we can, in general, delete many states while preserving the truth of p_0 in the resulting linear structure. We need only retain, in the "loop" from state s_i to s_{j-1} and back, a single state labelled q for each formula Fq that appears in the label of some state in the loop. The other states in the loop can be deleted, reducing its size to at most n states. We also need to ensure that each Fq that appears somewhere in the "stem" from s_0 to s_{i-1} is fulfilled by a q labelling some subsequent state. The other states in the stem can be deleted reducing the size of the stem to at most n states. The final structure x'' is still a model of p_0, and is of size at most $2n$ states. □

6.19. THEOREM. *The problem of testing satisfiability for* PLTL(F) *is* NP-*complete.*

PROOF (*idea*). Membership in NP follows using the Linear-Size Model Theorem. An algorithm can be designed that, given a formula of length n, guesses a candidate model of size O(n) and then checks that it is indeed a model in time O(n^2). NP-hardness follows since the logic subsumes Propositional Logic. □

Finally, it can be shown that the complexity of testing satisfiability of the very expressive branching-time logic CTL* has an upper bound of deterministic double exponential time, by means of a quite elaborate reduction to the nonemptiness problem for finite-state automata on infinite trees (see Section 6.5). A lower bound of deterministic double exponential time has also been established by a reduction from alternating exponential-space Turing machines in [129]. (Note that by double exponential we mean exp(exp(n)), where exp(n) is a function c^n for some $c > 1$.) Thus we have the following theorem (cf. [125]).

6.20. THEOREM. *The problem of testing satisfiability for* CTL* *is complete for deterministic double exponential time.*

6.3. Deductive systems

A deductive system for a temporal logic consists of a set of axiom schemes and inference rules. A formula p is said to be *provable*, written $\vdash p$, if there exists a finite sequence of formulae, ending with p such that each formula is an instance of an axiom or follows from previous formulae by application of one of the inference rules. A deductive system is said to be *sound* if every provable formula is valid. It is said to be *complete* if every valid formula is provable.

Consider the following axioms and rules of inference:

Axiom Schemes:

(Ax1) All validities of propositional logic;
(Ax2) $\mathsf{EF}p \equiv \mathsf{E}[\mathit{true}\ \mathsf{U}\ p]$;
(Ax2b) $\mathsf{AG}p \equiv \neg\mathsf{EF}\neg p$;
(Ax3) $\mathsf{AF}p \equiv \mathsf{A}[\mathit{true}\ \mathsf{U}\ p]$;
(Ax3b) $\mathsf{EG}p \equiv \neg\mathsf{AF}\neg p$;
(Ax4) $\mathsf{EX}(p \vee q) \equiv \mathsf{EX}p \vee \mathsf{EX}q$;
(Ax5) $\mathsf{AX}p \equiv \neg\mathsf{EX}\neg p$;
(Ax6) $\mathsf{E}(p\ \mathsf{U}\ q) \equiv q \vee (p \wedge \mathsf{EXE}(p\ \mathsf{U}\ q))$;
(Ax7) $\mathsf{A}(p\ \mathsf{U}\ q) \equiv q \vee (p \wedge \mathsf{AXA}(p\ \mathsf{U}\ q))$;
(Ax8) $\mathsf{EX}\ \mathit{true} \wedge \mathsf{AX}\mathit{true}$;
(Ax9) $\mathsf{AG}(r \Rightarrow (\neg q \wedge \mathsf{EX}r)) \Rightarrow (r \Rightarrow \neg\mathsf{A}(p\ \mathsf{U}\ q))$;
(Ax9b) $\mathsf{AG}(r \Rightarrow (\neg q \wedge \mathsf{EX}r)) \Rightarrow (r \Rightarrow \neg\mathsf{AF}q))$;
(Ax10) $\mathsf{AG}(r \Rightarrow (\neg q \wedge (p \Rightarrow \mathsf{AX}r))) \Rightarrow (r \Rightarrow \neg\mathsf{E}(p\ \mathsf{U}\ q))$;
(Ax10b) $\mathsf{AG}(r \Rightarrow (\neg q \wedge \mathsf{AX}r)) \Rightarrow (r \Rightarrow \neg\mathsf{EF}q))$;
(Ax11) $\mathsf{AG}(p \Rightarrow q) \Rightarrow (\mathsf{EX}p \Rightarrow \mathsf{EX}q)$.

Rules of Inference:

(R1) if $\vdash p$ then $\vdash \mathsf{AG}p$ (generalization);
(R2) if $\vdash p$ and $\vdash p \Rightarrow q$ then $\vdash q$ (modus ponens).

This deductive system for CTL is easily seen to be sound. We can also establish the following theorem (cf. [25, 8]).

6.21. THEOREM. *The above deductive system for* CTL *is complete.*

PROOF (*sketch*). Suppose p_0 is valid. Then $\sim p_0$ is unsatisfiable. We apply the above tableau-based decision procedure to $\sim p_0$. All nodes whose label includes $\sim p_0$ will be eliminated. In the sequel, we use the following notation and terminology. We use $\bigwedge s$ to denote the conjunction of all formulae labelling node s. We also write $p \in s$ for $p \in L(s)$, and we say that formula p is *consistent* provided that not $\vdash \sim p$.

***Claim* 1**: *If node s is deleted, then* $\vdash \sim(\bigwedge s)$.

Assuming the claim, we will show that $\vdash p_0$. We will use the formulae below, whose validity can be established by propositional reasoning:

$$\vdash q \equiv \bigvee\{\bigwedge s\text{: } s \text{ is a node in the tableau and } q \in s\} \quad \text{for each formula } q \in \text{ecl}(p_0)$$
$$\equiv \bigvee\{\bigwedge s\text{: } s \text{ is a node in the tableau and } q \in s \text{ and } \bigwedge s \text{ is consistent}\},$$

$$\vdash \mathit{true} \equiv \bigvee\{\bigwedge s\text{: } s \text{ is a node in the tableau}\}$$
$$\equiv \bigvee\{\bigwedge s\text{: } s \text{ is a node in the tableau and } \bigwedge s \text{ is consistent}\}.$$

Thus $\vdash \sim p_0 \equiv \bigvee\{\bigwedge s\text{: } s \text{ is a node in the tableau and } \sim p_0 \in s\}$. Because $\sim p_0$ is unsatisfiable, the decision procedure will delete each node s containing $\sim p_0$ in its label. By Claim 1 above, for each such node s that is eliminated, $\vdash \sim\bigwedge s$. Thus we get $\vdash \sim\sim p_0$ and also $\vdash p_0$.

Before proving Claim 1, we establish this one.

***Claim* 2:** *If* $(s,t) \notin R$ *as originally constructed then* $\bigwedge s \wedge \mathsf{EX} \bigwedge t$ is inconsistent.

Proof: Suppose $(s,t) \notin R$. Then, for some formula $\mathsf{AX}p$, $\mathsf{AX}p \in s$ and $\sim p \in t$. Thus we can prove the following:
(a) $\vdash \bigwedge s \Rightarrow \mathsf{AX}p$ (since $\mathsf{AX}p \in s$),
(b) $\vdash \bigwedge t \Rightarrow \sim p$ (since $\sim p \in t$),
(c) $\vdash \mathsf{AG}(\bigwedge t \Rightarrow \sim p)$ (generalization rule),
(d) $\vdash \mathsf{EX} \bigwedge t \Rightarrow \mathsf{EX} \neg p$ ((Ax11): monotonicity of EX operator),
(e) $\vdash (\bigwedge s \wedge \mathsf{EX} \bigwedge t) \Rightarrow \mathsf{AX}p \wedge \mathsf{EX} \sim p$ (lines a, d and propositional reasoning),
(f) $\vdash (\bigwedge s \wedge \mathsf{EX} \bigwedge t) \Rightarrow \mathit{false}$ ((Ax5) and def. AX operator),
(g) $\vdash \sim(\bigwedge s \wedge \mathsf{EX} \bigwedge t)$ (propositional reasoning).
Thus we have established that $\bigwedge s \wedge \mathsf{EX} \bigwedge t$ is inconsistent, thereby completing the proof of Claim 2.

We are now ready to give the proof of Claim 1. We argue by induction on when a node is deleted that if node s is deleted then $\vdash \sim\bigwedge s$.

Case 1: if $\bigwedge s$ is consistent, then s is not deleted on account of having no successors. To see this, we note that we can prove

$$\vdash \bigwedge s \equiv \bigwedge s \wedge \mathsf{EX}\mathit{true}$$
$$\equiv \bigwedge s \wedge \mathsf{EX}(\bigvee\{\bigwedge t\text{: } \bigwedge t \text{ is consistent}\})$$
$$\equiv \bigwedge s(\bigvee\{\mathsf{EX} \bigwedge t\text{: } \bigwedge t \text{ is consistent}\})$$
$$\equiv \bigvee\{\bigwedge s \wedge \mathsf{EX} \bigwedge t\text{: } \bigwedge t \text{ is consistent}\}.$$

Thus if $\bigwedge s$ is consistent, $\bigwedge s \wedge \mathsf{EX} \bigwedge t$ is consistent for some t. By Claim 1 above, $(s,t) \in R$ in the original tableau. By induction hypothesis, node t is not eliminated. Thus $(s,t) \in R$ in the current tableau, and node s is not eliminated due to having no successors.

Case 2: node s is eliminated on account of $\mathsf{EX}q \in s$, but s has no successor t with $q \in t$. This is established using an argument like that in Case 1.

Case 3: node s is deleted on account of $\mathsf{EF}q \in s$, which is not fulfilled (ranked) at s. Let $V = \{t\text{: } \mathsf{EF}q \in t \text{ but not fulfilled}\} \cup \{t\text{: } \mathsf{EF}q \notin t\}$. Note that node $s \in V$. Moreover, the complement of V is the set $\{t\text{: } \mathsf{EF}q \in t \text{ and fulfilled}\}$.

Let $r = \bigvee\{\bigwedge t\text{: } t \in V\}$. We claim that $\vdash r \Rightarrow (\neg q \wedge \mathsf{AX}r)$. It is clear that $\vdash r \Rightarrow \neg q$, because $\neg q \in t$ for each $t \in V$ and $\vdash \bigwedge t \Rightarrow \neg q$. We must now show that $\vdash r \Rightarrow \mathsf{AX}r$. It suffices to show that, for each $t \in V$, $\vdash \bigwedge t \Rightarrow \mathsf{AX}r$. Suppose not; then $\exists t \in V$, $\bigwedge t \wedge \mathsf{EX} \sim r$ is consistent. Since $\neg r = \bigvee \{\bigwedge t' \colon t' \notin V\}$, $\exists t \in V \exists t' \notin V$, $\bigwedge t \wedge \mathsf{EX} \bigwedge t'$ is consistent. By Claim 2 above, $(t, t') \in R$ as originally constructed, and since $\bigwedge t$ and $\bigwedge t'$ are each consistent, neither is eliminated by induction hypothesis. So $(t, t') \in R$ in the current tableau. Since $t' \notin V$, $\mathsf{EF}q \in t'$ and is ranked. But by virtue of the arc (t, t') in the tableau, t should also be ranked for $\mathsf{EF}q$, a contradiction to t being a member of V. Thus $\vdash r \Rightarrow \mathsf{AX}r$.

By generalization, $\vdash \mathsf{AG}(r \Rightarrow \mathsf{AX}r)$ and, by the induction axiom for EF and modus ponens, $\vdash r \Rightarrow \neg \mathsf{EF}q$. Now $\vdash \bigwedge s \Rightarrow r$, by definition of r (as the disjunction of formulae for each state in V, which includes node s). However, we assumed $\mathsf{EF}q \in s$, which of course means that $\vdash \bigwedge s \Rightarrow \mathsf{EF}q$. Thus $\vdash \bigwedge s \Rightarrow \mathit{false}$, so that $\bigwedge s$ is inconsistent.

The proofs for the other cases for eventualities $\mathsf{E}(p \mathsf{U} q)$, $\mathsf{AF}q$, and $\mathsf{A}(p \mathsf{U} q)$ are similar to that for Case 3. □

6.4. *Model checking*

The model checking problem (roughly) is: given a finite structure M and a Propositional TL formula p, does M define a model of p? For any Propositional TL, the model checking problem is decidable since if needed, we can do an exhaustive search through the paths of the finite input structure. The problem has important applications to mechanical verification of finite-state concurrent systems (see Section 7.3). The significant issues from the theoretical standpoint are to analyze and classify logics with respect to the complexity of model checking. For some logics, which have adequate expressive power to capture certain important correctness properties, we can develop very efficient algorithms for model checking. Other logics cannot be model-checked so efficiently.

We say "roughly" because there is some potential ambiguity in the above definition. What system of TL is the formula p from? In particular, is it branching- or linear-time? Also, what does it mean for a structure M to be a model of a formula p? From the definition of satisfiability for a formula p_0 of branching-time logic, a state formula, it seems that we should say that a structure M is a model of a formula p_0 provided it contains a state s such that $M, s \models p_0$. From the technical definition of satisfiability for a formula p_0 of linear-time logic, it appears we should say that a structure M is a model of a formula p_0 provided it contains a fullpath x such that $M, x \models p_0$. However, the number of fullpaths can be exponential in the size of a finite structure M. It thus seems that the complexity of model checking for linear time could be very high, since in effect an examination of all paths through the structure could be required.

To overcome these difficulties, we therefore formalize the model-checking problem as follows: The *Branching-Time Logic Model Checking Problem* (*BMCP*) formulated for Propositional Branching-Time Logic BTL is: Given a finite structure $M = (S, R, L)$ and a BTL formula p, determine for each state s in S whether $M, s \models p$ and, if so, label s with p. The *Linear-Time Logic Model Checking Problem* (*LMPC*) for Propositional Linear-Time Logic LTL can be similarly formulated as follows: given a finite structure $M = (S, R, L)$ and an LTL formula p, determine for each state in S whether there is a fullpath satisfying p starting at s and, if so, label s with $\mathsf{E}p$.

This definition of LMCP may, at first glance, appear to be incorrectly formulated because it defines truth of linear-time formulae in terms of states. However, one should note that there is a fullpath in finite structure M satisfying linear-time formula p iff there is such a fullpath starting at some state s of M. It thus suffices to solve LMCP and then scan the states to see if one is labelled with $\mathsf{E}p$. We can also handle the applications-oriented convention that a linear-time formula p is true of a structure (representing a concurrent program) iff it is true of all (initial) paths in the structure, because p is true of all paths in the structure iff $\mathsf{A}p$ holds at all states of the structure. Since $\mathsf{A}p \equiv \neg\mathsf{E}\neg p$, by solving LMCP and then scanning all (initial) states to check whether $\mathsf{A}p$ holds, we get a solution to the applications formulation.

We now analyze the complexity of model-checking linear time. The next three results are from [104].

6.22. LEMMA. *The model checking problem for* PLTL *is polynomial-time reducible (transformable) to the satisfiability problem for* PLTL.

PROOF (*sketch*). The key idea is that we can readily encode the organization of a given finite structure into a PLTL formula. Suppose $M=(S, R, L)$ is a finite structure and p_0 a PLTL formula, over an underlying set of atomic propositions AP. Let AP′ be an extension of AP obtained by including a new, "fresh" atomic proposition Q_s for each state $s \in S$. The local organization of M at each state s is captured by the formula

$$q_s = Q_s \Rightarrow \left(\bigwedge_{P \in L(s)} P \wedge \bigwedge_{P \notin L(s)} \neg P \wedge \bigvee_{(s,t) \in R} \mathsf{X} Q_t \right)$$

while the formula below asserts that the above local organization prevails globally:

$$q' = \mathsf{G}\left(\left(\sum_{s \in S} Q_s = 1 \right) \wedge \bigwedge_{s \in S} q_s \right)$$

and means, in more detail, that exactly one Q_s is true at each time and that the corresponding q_s holds.

Claim: *There exists a fullpath* x_1 *in* M *such that* $M, x_1 \models p_0$ *iff* $q' \wedge p_0$ *is satisfiable.*

The $\rightarrow$-direction is clear: annotate M with propositions from AP′. The path x_1 so annotated is a model of $q' \wedge p_0$.

The $\leftarrow$-direction can be seen as follows: Suppose $M', x \models q' \wedge p_0$. The $x = u_0, u_1, u_2, \ldots$ matches the organization of M in that, for each i, (a) with state u_i we associate a state s of M—the unique one such that $M', u_i \models Q_s$—that satisfies the same atomic propositions in AP as does s (call it $s(u_i)$); and (b) the successor u_{i+1} along x of u_i is associated with a state $t = s(u_{i+1})$ of M which is a successor of s in M. Thus, the path $x_1 = s(u_0), s(u_1), s(u_2), \ldots$ in M is such that $M, x_1 \models p_0$. □

6.23. THEOREM. *The model checking problem for* PLTL *is* PSPACE-*complete.*

PROOF (*idea*). Membership in PSPACE follows from the preceding lemma and the theorem establishing that satisfiability is in PSPACE. PSPACE-hardness follows by a generic reduction from PSPACE Turing machines. □

REMARK. The above PSPACE-completeness result holds for PLTL(F, X), the sublanguage of PLTL obtained by restricting the temporal operators to just X, F, and its dual G. It also holds for PLTL(U), the sublanguage of PLTL obtained by restricting the temporal operators to just U and its dual B.

6.24. THEOREM. *The problem of model checking for* PLTL(F) *is* NP-*complete.*

PROOF (*idea*). To establish membership in NP, we design a nondeterministic algorithm that guesses a finite path in the input structure M leading to a strongly connected component such that any unwinding of the component prefixed by some finite path comprises a candidate model of the input formula p_0. To check that it is indeed a model, evaluate each subformula of each state of the candidate model, which can be done in polynomial time. NP-hardness follows by a reduction from 3-SAT. □

We now turn to model checking for branching-time logic. First we have the following theorem from [121].

6.25. THEOREM. *The model checking problem for* CTL *is in deterministic polynomial time.*

This result is somewhat surprising since CTL seems somehow more complicated than the linear-time logic PLTL. Because of such seemingly unexpected complexity results, the question of the complexity of model-checking has been an issue in the branching- versus linear-time debate. Branching time, as represented by CTL, appears to be more efficient than linear time, but at the cost of potentially valuable expressive power, associated with, for example, fairness.

However, the real issue for model checking is not branching versus linear time, but simply what are the basic modalities of the branching-time logic to be used. Recall that the basic modalities of a branching-time logic are those of the form Ap or Ep, where p is a "pure" linear-time formula containing no path quantifiers itself. Then we have the following result of [27].

6.26. THEOREM. *Given any model-checking algorithm for a linear logic* LTL, *there is a model-checking algorithm for the corresponding branching logic* BTL, *whose basic modalities are defined by the* LTL, *of the same order of complexity.*

PROOF (*idea*). Simply evaluate nested branching-time formulae Ep or Ap by recursive descent. For example, to model-check EFAGP, recursively model-check AGP, then label every state labelled with AGP with a fresh proposition Q and model-check EFQ. □

For example, CTL* can be reduced to PLTL since the basic modalities of CTL* are of the form A or E followed by a PLTL formula. As a consequence we get the following corollary (cf. [13]).

6.27. COROLLARY. *The model checking problem for* CTL* *is* PSPACE-*complete.*

Thus the increased expressive power of the basic modalities of CTL* incurs a significant complexity penalty. However, it can be shown that basic modalities for reasoning under fairness assumptions do not cause complexity difficulties for model checking. These matters are discussed further in Section 7.

6.5. *Automata on infinite objects*

There has been a resurgence of interest in finite-state automata on infinite objects, due to their close connection to TL. They provide an important alternative approach to developing decision procedures for testing satisfiability for propositional temporal logics. For linear-time temporal logics, the tableau for formula p_0 can be viewed as defining a finite automaton on infinite strings that essentially accepts a string iff it defines a model of the formula p_0. The satisfiability problem for linear logics is thus reduced to the emptiness problem of finite automata on infinite strings. In a related but somewhat more involved fashion, the satisfiability problem for branching-time logics can be reduced to the nonemptiness problem for finite automata on infinite trees.

For some logics, the only known decision procedures of elementary time complexity (i.e., of time complexity bounded by the composition of a fixed number of exponential functions) are obtained by reductions to finite automata on infinite trees. The use of automata transfers some difficult combinatorics onto the automata-theoretic machinery. Investigations into such automata-theoretic decision procedures is an active area of research interest.

We first outline the automata-theoretic approach for linear time. As suggested by Theorem 6.16, the tableau construction of CTL can be specialized, essentially by dropping the path quantifiers to define a tableau construction for PLTL. The extended closure of a PLTL formula p_0, ecl(p_0), is defined as for CTL, remembering that, in a linear structure, $\mathsf{E}p \equiv \mathsf{A}p \equiv p$. The notions of maximal and propositionally consistent subsets of ecl(p_0) are also defined analogously. The (initial) tableau for p_0 is then a structure $T=(S, R, L)$, where S is the set of maximal, propositionally consistent subsets of ecl(p_0), i.e. states; $R \subseteq S \times S$ consists of the transitions (s, t) defined by the rule $(s, t) \in R$ exactly when $\forall$ formula $\mathsf{X}p \in \mathrm{ecl}(p_0)$, $\mathsf{X}p \in s$ iff $p \in t$; and $L(s) = s$ for each $s \in S$.

We may view the tableau for PLTL formula p_0 as defining the transition diagram of a nondeterministic finite-state automaton $\mathscr{A}$ which accepts the set of infinite strings over alphabet $\Sigma = \mathrm{PowerSet(AP)}$ that are models of p_0, by letting the arc (u, v) be labelled with AtomicPropositions(v), i.e., the set of atomic propositions in v. Technically, $\mathscr{A}$ is a tuple of the form $(S \cup \{s_0\}, \Sigma, \delta, s_0, -)$ where $s_0 \notin S$ is a unique start state, δ is defined so that $\delta(s_0, a) = \{\text{states } s \in S\text{: } p_0 \in s \text{ and AtomicPropositions}(s) = a\}$ for each $a \in \Sigma$, $\delta(s, a) = \{\text{states } t \in S\text{: } (s, t) \in R \text{ and AtomicPropositions}(s) = a\}$. The acceptance condition is defined below. A *run* r of $\mathscr{A}$ on input $x = a_1 a_2 a_3 \ldots \in \Sigma^\omega$ is an infinite sequence of states $s_0 s_1 s_2 \ldots$ such that $\forall i \geqslant 0$ $\delta(s_i, a_{i+1}) \supseteq \{s_{i+1}\}$. Note that $\forall i \geqslant 1$

AtomicPropositions(s_i) $= a_i$. Any run of $\mathscr{A}$ would correspond to a model of p_0, in that $\forall i \geq 1, x^i \models \wedge$ {formulae p: $p \in s_i$}, except that eventualities might not be fulfilled. To check fulfillment, we can easily define acceptance in terms of complemented pairs (cf. [112]). If ecl(p_0) has m eventualities $(p_1 \mathsf{U} q_1), \ldots, (p_m \mathsf{U} q_m)$, we let $\mathscr{A}$ have m pairs (RED_i, GREEN_i) of lights. Each time a state containing $(p_i \mathsf{U} q_i)$ is entered, flash RED_i; each time a state containing q_i is entered, flash GREEN_i. A run r is accepted iff, for each $i \in [1{:}m]$, there are infinitely many RED_i flashes implies there are infinitely many GREEN_i flashes iff every eventuality is fulfilled iff the input string x is a model of p_0.

We can convert $\mathscr{A}$ into an equivalent nondeterministic Büchi automaton $\mathscr{A}_1$, where acceptance is defined in terms of a single GREEN light flashing infinitely often. We need some terminology. We say that the eventuality $(p \mathsf{U} q)$ is *pending* at state s of run r provided that $(p \mathsf{U} q) \in s$ and $q \notin s$. Observe that run r of $\mathscr{A}$ on input x corresponds to a model of p_0 iff not($\exists$ eventuality $(p \mathsf{U} q) \in$ ecl(p_0), $(p \mathsf{U} q)$ is pending almost everywhere along r) iff $\forall$ eventuality $(p \mathsf{U} q) \in$ ecl(p_0), $(p \mathsf{U} q)$ is not pending infinitey often along r. The Büchi automaton $\mathscr{A}_1$ is then obtained from $\mathscr{A}$ augmenting the state with an $(m+1)$valued counter. The counter is incremented from i to $i+1$ mod$(m+1)$ when the ith eventuality $(p_i \mathsf{U} q_i)$ is next seen to be not pending along the run r. When the counter is reset to 0, flash GREEN and set the counter to 1. (If $m=0$, flash GREEN in every state.) Now observe that there are infinitely many GREEN flashes iff $\forall i \in [1{:}m]$ $(p_i \mathsf{U} q_i)$ is not pending infinitely often iff every pending eventuality is eventually fulfilled iff the input string x defines a model of p_0. Moreover, $\mathscr{A}_1$ still has $\exp(|p_0|) \times \mathrm{O}(|p_0|) = \exp(|p_0|)$ states.

Similarly, the tableau construction for a branching-time logic with relatively simple modalities such as CTL can be viewed as defining a Büchi tree automaton that, in essence, accepts all models of a candidate formula p_0. (More precisely, every tree accepted by the automaton is a model of p_0, and if p_0 is satisfiable there is some tree accepted by the automaton.) General automata-theoretic techniques for reasoning about a number of relatively simple logics, including CTL, using Büchi tree automata have been described by Vardi and Wolper [116].

For branching-time logics with richer modalities such as CTL*, the tableau construction is not directly applicable. Instead, the problem reduces to constructing a tree automaton for the branching-time modalities (such as $\mathsf{A}p$) in terms of the string automaton for the corresponding linear-time formula (such as p). This tree automaton will in general involve a more complicated acceptance condition such as pairs or complemented pairs, rather than the simple Büchi condition. Somewhat surprisingly, the only known way to build the tree automaton involves difficult combinatorial arguments and/or appeals to powerful automata-theoretic results such as McNaughton's construction [70] for determinizing automata on infinite strings.

The principal difficulty manifests itself with just the simple modality $\mathsf{A}p$. The naive approach of building the string automaton for p and then running it down all paths to get a tree automaton for $\mathsf{A}p$ will not work. The string automaton for p must be determinized first. To see this, consider two paths xy and xz in the tree which start off with the same common prefix x but eventually separate to follow two different infinite suffixes y or z. It is possible that p holds along both paths, but, in order for the nondeterministic automaton to accept it might have to "guess" while reading a particular symbol of x whether it will eventually read the suffix y or the suffix z. The state it guesses for y is in general different from the state it guesses for z. Consequently, no single run of a tree automaton based on a nondeterministic string automaton can lead to acceptance along all paths.

For a CTL* formula of length n, use of classical automata-theoretic results yields an automaton of size triple exponential in n. (Note that by triple exponential we mean exp(exp(exp(n))), etc.) The large size reflects the exponential cost to build the string automaton as described above for a linear-time formula p plus the double exponential cost of McNaughton's construction to determinize it. Nonemptiness of the automaton can be tested in exponential time to give a decision procedure of deterministic time complexity quadruple exponential in n. In [29] Emerson and Sistla showed that, due to the special structure of the string automata derived from linear temporal logic formulae, such string automata could be determinized with only single exponential blow-up. This reduced the complexity of the CTL* decision procedure to triple exponential. Further improvement is possible as described below.

The size of a tree automaton is measured in terms of two parameters: the number of states and the number of pairs in the acceptance condition. A careful analysis of the tree automaton constructions in temporal decision procedures shows that the number of pairs is logarithmic in the number of states, and for CTL* we get an automaton with double exponential states and single exponential pairs. An algorithm in [125] shows how to test nonemptiness in time polynomial in the number of states, while exponential in the number of pairs. For CTL* this yields a decision procedure of deterministic double exponential time complexity, matching the lower bound of [129].

One drawback to the use of automata is that, due to the delicate combinatorial constructions involved, there is usually no clear relationship between the structure of the automaton and the syntax of the candidate formula. An additional drawback is that in such cases the automata-theoretic approach provides no aid in finding sound and complete axiomatizations. For example, the existence of an explicit, sound and complete axiomatization for CTL* has been an open question for some time. (Note that we refer here to an axiomatization for its validities over the usual semantics generated by a binary relation; interestingly, for certain nonstandard semantics, complete axiomatizations are known (cf. [1, 58]).)

However, there are certain definite advantages to the automata-theoretic approach. First, it does provide the only known elementary time decision procedures for some logics. Secondly, automata can provide a general, uniform framework encompassing temporal reasoning (cf. [113, 117, 114]). Automata themselves have been proposed as a potentially useful specification language. Automata, moreover, bear an obvious relation to temporal structures, abstract concurrent programs, etc. This makes it possible to account for various types of temporal reasoning applications such as program synthesis and mechanical verification of finite-state programs in a conceptually uniform fashion. Verification systems based on automata have also been developed (cf. [51]).

We note that not only has the field of TL benefitted from automata theory, but the converse holds as well. For example, the tableau concept for the branching-time logic CTL, particularly the state/prestate formulation, suggests a very helpful notion of the transition diagram for a tree automaton (cf. [124]). This has made it possible to apply tableau-theoretic techniques to automata, resulting in more efficient algorithms for testing nonemptiness of automata, which in turn can be used to get more efficient decision procedures for satisfiability of TLs (cf. [125]). Still another improved nonemptiness algorithm, motivated by program synthesis applications is given in [128]. New types of automata on infinite objects have also been proposed to facilitate reasoning in TLs (cf. [108, 129, 66]). A particularly important advance in automata theory motivated by TL is Safra's construction [98] for determinizing an automaton on infinite strings with only a single exponential blow-up, without regard to any special structure possessed by the automaton. Not only is Safra's construction an exponential improvement over McNaughton's construction, but it is conceptually much more simple and elegant. In this way we see that not only can TL sometimes benefit from adopting the automata-theoretic viewpoint, but also, conversely and even synergistically, the study of automata on infinite objects has been advanced by work motivated by and using the techniques of TL.

7. The application of Temporal Logic to program reasoning

Temporal Logic has been suggested as a formalism especially appropriate to reasoning about ongoing concurrent programs, such as operating systems, which have a *reactive* nature, as explained below (cf. [88]).

We can identify two different classes of programs (also referred to as systems). One class consists of those ordinarily described as "sequential" programs. Examples include a program to sort a list, programs to implement a graph algorithm as discussed in, say the chapter on graph algorithms (see Volume A of this Handbook, Chapter 10), and programs to perform a scientific calculation. What these programs have in common is that they normally terminate. Moreover, their behavior has the following pattern: they initially accept some input, perform some computation, and then terminate yielding final output. For all such systems, correctness can be expressed in terms of a precondition/postcondition pair in a formalism such as Hoare's Logic or Dijkstra's weakest preconditions, because the systems' underlying semantics can be viewed as a *transformation* from initial states to final states, or from postconditions to preconditions.

The other class of programs consists of those which are continuously operating, or, ideally, nonterminating. Examples include operating systems, network communication protocols, and air traffic control systems. For a continuously operating program its normal behavior is an arbitrarily long, possibly nonterminating computation, which maintains an ongoing interaction with the environment. Such programs can be described as *reactive* systems. The key point concerning such systems is that they maintain an ongoing interaction with the environment, where intermediate outputs of the program can influence subsequent intermediate inputs to the program. Reactive systems thus subsume many programs labelled as concurrent, parallel, or distributed, as well as process control programs. Since there is in general no final state, formalisms such as Hoare's Logic, which are based on an initial-state–final-state semantics, are of little use for such reactive programs. The operators of temporal logic such as *sometimes* and *always* appear quite appropriate for describing the time-varying behavior of such programs.

What is the relationship between concurrency and reactivity? They are in some sense independent. There are transformational programs that are implemented to exploit parallel architectures (usually, to speed up processing, allowing the output to be obtained more quickly). A reactive system could also be implemented on a sequential architecture.

On the other hand, it can be recommended that, in general, concurrent programs should be viewed as reactive systems. In a concurrent program consisting of two or more processes running in parallel, each process is generally maintaining an ongoing interaction with its environment, which usually includes one or more of the other processes. If we take the compositional viewpoint, where the meaning of the whole is defined in terms of the meaning of its parts, then the entire system should be viewed in the same fashion as its components, and the view of any system is a reactive one. Even if we are not working in a compositional framework, the reactive view of the system as a whole seems a most natural one in light of the ongoing behavior of its components. Thus, in the sequel when we refer to a concurrent program, we mean a reactive, concurrent system.

There are two main schools of thought regarding the application of TL to reasoning about concurrent programs. The first might be characterized as "proof-theoretic". The basic idea is to manually compose a program and a proof of its correctness using a formal deductive system, consisting of axioms and inference rules, for an appropriate temporal specification language. The second might be characterized as "model-theoretic". The idea here is to use decision procedures that manipulate the underlying temporal models corresponding to programs and specifications to automate the tasks of program construction and verification. We subsequently outline the approach of each of these two schools. First, however, we discuss the types of correctness properties of practical interest for concurrent programs and their specification in TL.

7.1. *Correctness properties of concurrent programs*

There are large number of correctness properties that we might wish to specify for a concurrent program. These correctness properties usually fall into two broad classes (cf. [83, 77]). One class is that of "safety" properties also known as "invariance" properties. Intuitively, a safety property asserts that "nothing bad happens". The other class consists of the "liveness' properties also referred to as "eventuality" properties or "progress" properties. Roughly speaking, a liveness property asserts that "something good will happen". These intuitive descriptions of safety and liveness are made more precise below, following [88].

A *safety* property states that each finite prefix of a (possibly infinite) computation meets some requirement. Safety properties thus are those that are (initially) equivalent to a formula of the form $\mathrm{G}p$ for some past formula p. The past formula describes the condition required of finite prefixes, while the G operator ensures that p holds of all finite prefixes. Note that this formal definition of safety requires that always "nothing bad has happened yet", consistent with the intuitive characterization of [77] mentioned above.

Any formula built up from past formulae, the propositional connectives $\wedge$ and $\vee$, and the future temporal operators G and U_w can be shown to express a safety property. For example,

$$(p \, U_w \, q) \equiv_i G(G^- p \vee F^-(q \vee X^- G^- p)).$$

A number of concrete examples of safety properties can be given. The *partial correctness* of a program ith respect to a precondition φ and postcondition ψ, which stipulates that if program execution begins in a state satisfying φ, then if it terminates the final state satisfies ψ, is expressed by

$$\mathrm{at}l_0 \wedge \varphi \;\Rightarrow G(\mathrm{at}l_h \Rightarrow \psi)$$

where the program's start label is l_0 and its halt label is l_h. (Note that this formula is initially equivalent to

$$G(F^-(\neg(\mathrm{at}l_0 \wedge \varphi) \wedge X_w^- \, \mathit{false})) \vee G(\mathrm{at}l_h \Rightarrow \psi))$$

thereby demonstrating that it is a safety property according to the technical definition.)

Other safety properties include *global invariance* of assertion p which is simply expressed by Gp. To capture *local invariance*, which means that p holds whenever control is at location l, we write $G(\mathrm{at}l \Rightarrow p)$.

The requirement of *mutual exclusion* for a two-process solution to the critical-section problem can be written

$$G(\neg(\mathrm{atCS}_1 \wedge \mathrm{atCS}_2))$$

where atCS_i indicates that control of process i is at its critical section.

Another very important property for concurrent programs is *freedom from deadlock*. A concurrent program is *deadlocked* if no process is enabled to proceed. The formula $G(\mathrm{enabled}_1 \vee \cdots \vee \mathrm{enabled}_m)$ captures freedom from deadlock for a concurrent program with m processes.

Liveness properties are in some sense dual to safety properties, requiring that some finite prefix property hold a certain number of times.

The *basic liveness* properties are technically defined to be those (initially) expressible in the form Fp, $F^\infty p$, or $G^\infty p$, where p is a past formula required to hold for some, for infinitely many, or for all but a finite number, respectively of the finite prefixes of a computation. It is interesting to note that $(p \, U_s \, q) \equiv_i F(q \wedge X_w^- G^- p)$ for any past formulae p and q, thus showing the strong until to be a basic liveness property, even though it is not immediately obvious that it can be expressed in the required form. Also note that $Fp \equiv_i GF(F^- p) \equiv_i FG(F^- p)$ and is technically redundant, even though we find it more convenient to keep Fp separated out. A more serious redundancy is that, by our definition, each safety property is a basic liveness property, since $Gp \equiv_i GF(G^- p)$ for any past formula p.

If we wish to avoid this redundancy, we can first define an *invincible* past formula p to be such that every finite sequence x has a finite extension x' with $(x', \mathrm{length}(x')) \models p$ (i.e., with p holding at the last state of x').

We then define the *pure liveness* properties to be those initially equivalent to one of the formulae $\mathsf{F}p$, $\mathsf{GF}p$, $\mathsf{FG}p$ for some invincible past formula p. Note that any satisfiable state formula p is an invincible past formula, so that the pure liveness formulae still include a broad range of properties. However, $(p\,\mathsf{U}_s\,q)$ is not a pure liveness property, because while $(p\,\mathsf{U}_s\,q) \equiv_i \mathsf{F}(q \wedge \mathsf{X}^-\mathsf{G}^-p)$, the formula $q \wedge \mathsf{X}^-\mathsf{F}^-p$ is not invincible. It is expressible as the conjunction of a safety property and a pure liveness property: $(p\,\mathsf{U}_s\,q) \equiv_i (p\,\mathsf{U}_w\,q) \wedge \mathsf{F}q$.

Note that if p is a pure livenesss property, then it has the following characteristic: every finite sequence x can be extended to a finite or infinite sequence x' such that $(x', 0) \models p$. This corresponds to the intuitive characterization of liveness that "something good will happen" of [77].

Further work on syntactic and semantic characterizations of safety and liveness properties is given in [2, 103].

One important generic liveness property has the form

$$\mathsf{G}(p \Rightarrow \mathsf{F}q)$$

for past formulae p and q, and is called *temporal implication* (cf. [83, 53]). Many specific correctness properties are instances of temporal implication, as described below.

An *intermittent assertion* is expressed by

$$\mathsf{G}((\mathrm{at}l \wedge \varphi) \;\Rightarrow\; \mathsf{F}(\mathrm{at}l' \wedge \varphi'))$$

meaning that whenever φ is true at location l, then φ' will eventually be true at location l' (cf. [11, 68]). An important special type of intermittent assertion is *total correctness* of a program with respect to a precondition φ and postcondition ψ. It is expressed by

$$\mathrm{at}l_0 \wedge \varphi \;\Rightarrow\; \mathsf{F}(\mathrm{at}l_{\mathrm{h}} \wedge \psi)$$

which indicates that if the program starts in a state satisfying φ, then it halts in a state satisfying ψ.

The property of *guaranteed accessibility* for a process in a solution to the mutual exclusion problem to enter its critical section, once it has indicated that it wishes to do so is expressed by

$$\mathsf{G}(\mathrm{atTry}_i \Rightarrow \mathsf{F}\mathrm{atCS}_i)$$

where atTry_i and atCS_i indicate that process i is in its Trying section or Critical section respectively. This property is sometimes referred to as *absence of individual starvation for process i*. General guaranteed accessibility is of the form

$$\mathsf{G}(\mathrm{at}l \Rightarrow \mathsf{F}\mathrm{at}l').$$

Still another property expressible in this way is *responsiveness*. Consider a system consisting of a resource controller that monitors access to a shared resource by competing user processes. We would like to ensure that each request for access eventually leads to a response in the form of a granting of access. This is captured by an assertion of the form $\mathsf{G}(\mathrm{req}_i \Rightarrow \mathsf{F}\mathrm{grant}_i)$ where req_i and grant_i are predicates indicating that a request by process is made or a grant of access to process i is given respectively.

The fairness properties discussed in Section 5 are also liveness properties.

A final general type of correctness property is informally known as the *precedence* properties. These properties have to do with temporal ordering, precedence, or priority of events. We shall not give a formal definition but instead illustrate the class by several examples.

To express *absence of unsolicited response* as in the resource controller example above, where we want a grant_i to be issued only if preceded by a req_i, we can write

$$\neg\text{grant}_i \;\Rightarrow\; (\neg\text{grant}_i \,\mathsf{U}_{\mathrm{w}}\, \text{req}_i).$$

Alternatively, we can write ($\text{req}_i \,\mathsf{B}\, \text{grant}_i$), where we recall that the precedes operator $(p \,\mathsf{B}\, q)$ asserts that the first occurrence of q, if any, is strictly preceded by an occurrence of p.

The important property of First-In First-Out (FIFO) responsiveness can be written in a straightforward but slightly imprecise fashion as

$$(\text{req}_i \,\mathsf{B}\, \text{req}_j) \;\Rightarrow\; (\text{grant}_i \,\mathsf{B}\, \text{grant}_j).$$

A more accurate expression is

$$(\text{req}_i \wedge \neg\text{req}_j \wedge \neg\text{grant}_j) \;\Rightarrow\; ((\neg\text{grant}_j) \,\mathsf{U}_{\mathrm{w}}\, \text{grant}_i)$$

where we rely on the assumption that once a request has been made, it is not withdrawn before it has been granted. Hence, $\text{req}_i \wedge \neg\text{req}_j$ implies that process i's request preceded that of process j.

It is interesting to note the importance of correctly formalizing our intuitive understanding of the problem in the formal specification language. An important application where this issue arises is the specification of correct behavior for a message buffer. Such buffers are often used in distributed systems based on message passing, where one process transmits messages to another process via an intermediate, asynchronous buffer that temporarily stores messages in transit.

We assume that the buffer has an input channel x and output channel y. It also has unbounded storage capacity and is assumed to operate according to FIFO discipline. We want to specify that the log of input/output transactions for the buffer is correct, viz. that the sequence of messages output on channel y equals the sequence of messages input on channel x.

An important limitation of PLTL and related formalisms was established by Sistla et al. [105] which shows that an unbounded FIFO buffer cannot be specified in PLTL. Essentially, the problem is that any particular formula p of PLTL is of a fixed size and corresponds to a bounded-size finite-state automaton, while the buffer can hold an arbitrarily large sequence of messages, thereby permitting the finite automaton to become "confused". Moreover, the problem is not alleviated by extending the formalism to be pure (i.e., uninterpreted) FOLTL (cf. [47]).

However, as noted in [105], there exist partially interpreted FOLTLs which make it possible to capture correct behavior for a message buffer. One such logic provides history variables that accumulate the string of all previous states along with a prefix predicate ($\leqslant$) on these histories. The safety portion of the specification is given by $\mathsf{G}(y \leqslant x)$ which asserts that the sequence of messages output is always a prefix of the sequence of messages input. The liveness requirement is expressed by $\forall z\, \mathsf{G}(x=z \Rightarrow \mathsf{F}(y=z))$ which ensures that whatever sequence appears along the input channel is eventually replicated along the output channel.

The essential feature of the above specification based on histories is the ability to, in effect, associate a unique sequence number with each message, thereby ensuring that all messages are distinct. Using in(m) to indicate that message m is placed on input channel x and out(m) for the placement of message on output channel y, we have the following alternative specification in the style of [47]: The formula

$$\forall m\ \mathsf{G}(\mathrm{in}(m)\ \mathsf{B}\ \mathrm{out}(m))$$

specifies that any message output must have been previously input.

The formula

$$\forall m\, \forall m'\ \mathsf{G}(\mathrm{in}(m) \wedge \mathsf{XF}\mathrm{in}(m') \Rightarrow \mathsf{F}(\mathrm{out}(m) \wedge \mathsf{XF}\mathrm{out}(m')))$$

asserts that FIFO discipline is maintained, i.e. messages are output in the same order they were input.

The liveness requirement is expressed by

$$\forall m\ \mathsf{G}(\mathrm{in}(m) \Rightarrow \mathsf{F}\mathrm{out}(m))$$

while the assumption of message uniqueness is captured by

$$\forall m\, \forall m'\ \mathsf{G}((\mathrm{in}(m) \wedge \mathsf{XF}\mathrm{in}(m')) \Rightarrow (m \neq m')).$$

Note that the requirement of message uniqueness is essential for the correctness of the specification. Without it, a computation with, e.g., the same message output twice for each input message would be permitted.

Recently, Wolper [121] has provided additional insight into the power of logical formalisms for specifying message buffers. First, he pointed out that PLTL is a priori inadequate for specifying message buffers when the underlying data domain is infinite, since each PLTL formula is finite. However, he goes on to show that PLTL is nonetheless adequate for specifying message buffer protocols that satisfy the *data-independence* criterion, which requires that the behavior of the protocol does not depend on the value or content of a message. While it is in general undecidable whether a protocol is data-independent, a simple syntactic check of the protocol, if positive, ensures data-independence. This amounts to checking that the only possible operations performed on message contents are reading from channels to variables, writing from variables to channels, and copying between variables.

It is shown in [121] that it is enough for data-independent buffer protocols to assert correctness over a three-symbol message alphabet $\Sigma = \{m_1, m_2, m_3\}$, so that the input is of the form $m_3^* m_1 m_3^* m_2 m_3^\omega$ iff the output is of the form $m_3^* m_1 m_3^* m_2 m_3^\omega$. This matching of output to input can be expressed in PLTL, using propositions in_m_i and out_m_i (assumed to be exclusive and exhaustive), $1 \leqslant i \leqslant 3$, to indicate the appearance of message m_i on the input channel and on the output channel respectively, as

$$((\text{in\_}m_3\ \mathsf{U}\,(\text{in\_}m_1 \wedge \mathsf{X}(\text{in\_}m_3\ \mathsf{U}\,(\text{in\_}m_2 \wedge \mathsf{XG}\,\text{in\_}m_3))))$$
$$\Rightarrow (\text{out\_}m_3\ \mathsf{U}\,(\text{out\_}m_1 \wedge \mathsf{X}(\text{out\_}m_3\ \mathsf{U}\,(\text{out\_}m_2 \wedge \mathsf{XG}\,\text{out\_}m_3))))$$
$$\wedge \bigwedge_{i=1,\ldots,3} (\text{out\_}m_i\ \mathsf{B}\ \text{in\_}m_i).$$

Intuitively this works because it ensures that each pair of distinct input messages are transmitted through to the output correctly; since the buffer is assumed to be oblivious to the message contents, the only way it can ensure such correct transmission for the three-symbol alphabet is to transmit correctly over any alphabet, including those with distinct messages.

The reader may have noticed that the above example specifications were given in linear TL. If we wished to express them in branching TL we would merely need to prefix each assertion by the universal path quantifier. The reason linear TL sufficed was that, above, we were mainly interested in properties holding of all computations of a concurrent program. If we want to express lower bounds on nondeterminism and/or concurrency, we need the ability to use existential path quantification, provided only by branching-time logic. Such lower bounds are helpful in applications such as program synthesis. Moeover, branching time makes it possible to distinguish between *inevitability* of predicate P, which is captured by $\mathsf{AF}P$, and *potentiality* of predicate P, which is captured by $\mathsf{EF}P$. It also ensures that our specification logic is closed under semantic negation, so that we can express, for example, not only absence of deadlock along all futures but also the possibility of deadlock along some future (cf. [53, 26, 87]).

7.2. *Verification of concurrent programs: proof-theoretic approach*

A great deal of work has been done investigating the proof-theoretic approach to verification of concurrent programs using TL (cf. e.g. [84, 61, 63, 64, 53, 36, 77, 54, 99]). Typically, one tries to prove, by hand, that a given program meets a certain TL specification using various axioms and inference rules for the system of TL. A drawback of this approach is that proof construction is often a difficult and tedious task, with many details that require considerable effort and ingenuity to organize in an intellectually manageable fashion. The advantage is that human intuition can provide useful guidance that would be unavailable in a (purely) mechanical verification system. It should also be noted that the emphasis of this work has been to develop axioms, rules, and techniques that are useful in practice, as demonstrated on example programs, as opposed to metatheoretic justifications of proof systems.

A proof system in the LTL framework has been given by Manna and Pnueli [64] consisting of three parts:

(i) A *general* part for reasoning about temporal formulae valid over all interpretations. This includes PLTL and FOLTL;

(ii) A *domain* part for reasoning about variables and data structures over specific domains, such as the natural numbers, trees, lists, etc.; and

(iii) A *program* part specialized to program reasoning.

This system is referred to as a *global* system, since it is intended for reasoning about a program *as a whole*. In this survey, we focus on some useful proof rules from the program part, applicable to broad classes of properties. The reader is referred to [63, 88] for more detail.

The rules are presented in the form

$$\frac{\begin{array}{c} A_1 \\ \vdots \\ A_n \end{array}}{B}$$

where $A_1, \ldots, A_n$ are premises and B is the conclusion. The meaning is that if all the premises are shown to hold for a program then the conclusion is also true for the program.

The following *invariance* rule (INVAR) is adequate for proving most safety properties. Let φ be an assertion:

$$\text{(INVAR)}\ \frac{\begin{array}{c} \varphi \\ \mathsf{G}(\varphi \Rightarrow \mathsf{X}\varphi) \end{array}}{\mathsf{G}\varphi}$$

Note that this rule really has the form of an induction rule. The first premise, the basis, ensures that φ holds initially. The second premise, the induction step, states that whenever φ holds, it also holds at the following moment. The conclusion is thus that φ always holds.

To perform the induction step, we must show that φ is preserved across all atomic actions of the program. In practice this can often be determined by inspection, considering only the potentially falsifying transitions and ignoring those which obviously cannot make φ *false*.

As an example, we now verify safety for Peterson's solution [81] to the mutual exclusion problem shown in Fig. 8. Each process has a noncritical section (l_0, m_0

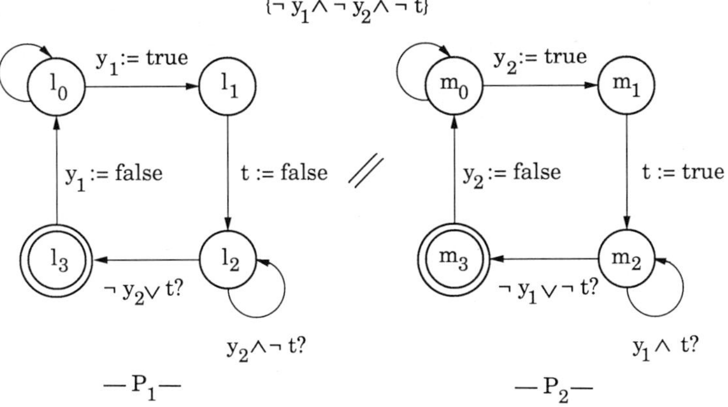

Fig. 8.

respectively) in which it idles unless it needs access to its critical section (l_3, m_3, respectively), signalled by entry into its trying region (l_1 and l_2, m_1 and m_2 respectively) Presence in the critical sections should be mutually exclusive. The safety property we wish to establish is thus that the system never reaches a state where both processes are in their respective critical sections at the same time: $\mathsf{G}(\neg(\mathrm{at}l_3 \wedge \mathrm{at}m_3))$.

It is helpful to establish several preliminary invariances. We use the notation $\mathrm{at}l_{1\ldots3}$ to abbreviate $\mathrm{at}l_1 \vee \mathrm{at}l_2 \vee \mathrm{at}l_3$:

$$\begin{array}{ll} \mathsf{G}\varphi_1, & \varphi_1\colon y_1 \equiv \mathrm{at}l_{1\ldots3}, \\ \mathsf{G}\psi_1, & \psi_1\colon y_2 \equiv \mathrm{at}m_{1\ldots3}, \\ \mathsf{G}\varphi_2, & \varphi_2\colon \mathrm{at}l_3 \wedge \mathrm{at}m_2 \Rightarrow t, \\ \mathsf{G}\psi_2, & \psi_2\colon \mathrm{at}m_3 \wedge \mathrm{at}l_2 \Rightarrow t, \\ \mathsf{G}\varphi, & \varphi\colon \neg(\mathrm{at}l_3 \wedge \mathrm{at}m_3). \end{array}$$

φ_1 plainly holds initially. Only transitions of process P_1 can affect it. Transitions $l_0 \rightarrow l_1$ leaves it true. Each of the other transitions of P_1 preserve its truth also, since y_1 is true whenever P_1 is at l_1, l_2, or l_3, and false when P_1 is at l_0. Thus $\mathsf{G}\varphi_1$ is established.

A similar argument proves $\mathsf{G}\psi_1$.

φ_2 is vacuously true initially. The only potentially falsifying transitions for φ_2 are

(i) $l_1 \rightarrow l_2$ ensures $\mathrm{at}l_3$ is *false*, so φ_2 is preserved;

(ii) $l_2 \rightarrow l_3$ while $\mathrm{at}m_2$ which is enabled only when $\neg y_2 \vee t$ holds; since y_2 is *true*, by virtue of ψ_1 and $\mathrm{at}m_2$, it must be that t is *true* both before and after the transition: hence φ_2 is preserved.

(iii) $m_1 \rightarrow m_2$ makes t *true*, so that ψ_2 is again preserved.

Thus $\mathsf{G}\varphi_2$ is established.

A similar argument establishes ψ_2.

Now, to prove $\mathsf{G}\varphi$, we first note that φ hold initially. The only potentially falsifying transitions are in fact never enabled:

(i) $l_2 \rightarrow l_3$ by process P_1 while process P_2 is at m_3—By ψ_2, t is *false* and by ψ_1, y_2 holds. Since the enabling condition for the transition is $\neg y_2 \vee t$, the transition is never enabled.

(ii) $m_2 \rightarrow m_3$ by process P_2 while process P_1 is at l_3 which is similarly shown to be impossible.

Thus $\mathsf{G}\varphi$ (i.e., $\mathsf{G}(\neg(\mathrm{at}l_3 \wedge \mathrm{at}m_3))$) is established.

We have the following liveness rule (LIVE), which is adequate for establishing eventualities based on a single step of a *helpful* process P_k, assuming weak fairness. Here we have formulae φ and ψ, and write $\mathsf{X}_k p$ for $\mathrm{enabled}_k \Rightarrow (\mathrm{executed}_k \Rightarrow \mathsf{X}p)$, which means that the next execution of a step of process P_k will establish p. The rule is

$$\text{(LIVE)}\ \frac{\begin{array}{c} \mathsf{G}(\varphi \Rightarrow \mathsf{X}(\varphi \vee \psi)) \\ \mathsf{G}(\varphi \Rightarrow \mathsf{X}_k\psi) \\ \mathsf{G}(\varphi \Rightarrow \psi \vee \mathrm{enabled}_k) \end{array}}{\mathsf{G}(\varphi \Rightarrow \mathsf{F}\psi)}$$

Often several invocations of (LIVE) must be linked together to prove an eventuality. We thus have the following rule (CHAIN):

$$\text{(CHAIN)}\ \frac{\mathsf{G}\left(\varphi_i \Rightarrow \mathsf{F}\left(\bigvee_{j<i} \varphi_j \vee \psi\right)\right)}{\mathsf{G}\left(\bigvee_{i \leqslant k} \varphi_i \Rightarrow \mathsf{F}\psi\right)}$$

In many cases the rule (CHAIN) is adequate, in particular for finite-state concurrent programs. In some instances however, no a priori bound on the number of intermediate assertions φ_i can be given. We therefore use an assertion $\varphi(a)$ with parameter a ranging over a given well-founded set $(W, <)$, which is a set W partially ordered by $<$ having no infinite decreasing sequence $a_1 > a_2 > a_3 > \cdots$. Note that this rule, (WELL), generalizes the (CHAIN) rule, since we can take W to be the interval $[1{:}k]$ with the usual ordering and $\varphi(i) = \varphi_i$.

$$\text{(WELL)}\ \frac{\mathsf{G}(\varphi(a) \Rightarrow \mathsf{F}(\exists b < a\,(\varphi(b) \vee \psi)))}{\mathsf{G}((\exists a\, \varphi(a)) \Rightarrow \mathsf{F}\psi)}$$

We illustrate the application of the (CHAIN) rule on Peterson's [81] algorithm for mutual exclusion. We wish to prove guaranteed accessibility:

$$\mathsf{G}(\mathrm{at}l_1 \Rightarrow \mathsf{F}\mathrm{at}l_3)$$

(which is sometimes also called absence of starvation for process P_1), indicating that whenever process 1 wants to enter its critical section, it will eventually be admitted.

We define the following assertions:

ψ: $\mathrm{at}l_3$

φ_1: $\mathrm{at}l_2 \wedge \mathrm{at}m_2 \wedge t$ $\qquad \varphi_2$: $\mathrm{at}l_2 \wedge \mathrm{at}m_1$

φ_3: $\mathrm{at}l_2 \wedge \mathrm{at}m_0$ $\qquad \varphi_4$: $\mathrm{at}l_2 \wedge \mathrm{at}m_3$

φ_5: $\mathrm{at}l_2 \wedge \mathrm{at}m_2 \wedge \neg t$ $\qquad \varphi_6$: $\mathrm{at}l_1$

and establishing the corresponding temporal implication by an application of the (LIVE) rule in order to meet the hypothesis of the (CHAIN) rule:

$\mathsf{G}(\varphi_6 \Rightarrow \mathsf{F}(\varphi_5 \vee \varphi_4 \vee \varphi_3 \vee \varphi_2))$, using helpful process P_1,
$\mathsf{G}(\varphi_5 \Rightarrow \mathsf{F}\varphi_4)$, using helpful process P_2,
$\mathsf{G}(\varphi_4 \Rightarrow \mathsf{F}\varphi_3)$, using helpful process P_2,
$\mathsf{G}(\varphi_3 \Rightarrow \mathsf{F}(\varphi_2 \vee \psi))$, using helpful process P_1,
$\mathsf{G}(\varphi_2 \Rightarrow \mathsf{F}(\varphi_1 \vee \psi))$, using helpful process P_2,
$\mathsf{G}(\varphi_1 \Rightarrow \mathsf{F}\psi)$, using helpful process P_1.

The (CHAIN) rule now yields $G(\varphi_6 \Rightarrow F\psi)$, i.e., $G(\mathrm{at}l_1 \Rightarrow F\mathrm{at}l_3)$ as desired. The argument can be summarized in a *proof lattice* as depected in Fig. 9 (cf. [77, 63].

7.3. *Mechanical synthesis of concurrent programs from Temporal Logic specifications*

One ambitious but promising possibility is that of automatically synthesizing concurrent programs from high-level specifications expressed in Temporal Logic. Here one deals with the *synchronization skeleton* of the program, which is an abstraction of the actual program where detail irrelevant to synchronization is suppressed. For example, in the synchronization skeleton for a solution to the critical section problem, each process's critical section may be viewed as a single node since the internal structure of the critical section is unimportant. Most solutions to synchronization problems in the literature are in fact given as synchronization skeletons. Because synchronization skeletons are in general finite state, a propositional version of Temporal Logic suffices to specify their properties.

The synthesis method exploits the small model property of the Propositional TL. It uses a decision procedure so that, given a TL formula p, it will decide whether p is satisfiable or unsatisfiable. If p is satisfiable, a finite model of p is constructed. In this application, unsatisfiability of p means that the specification is inconsistent (and must

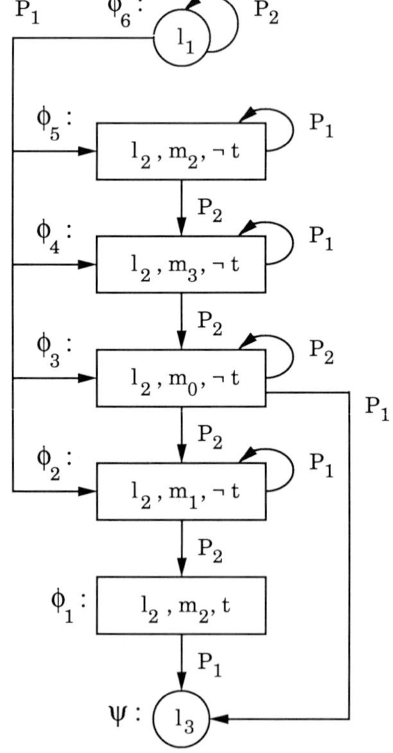

Fig. 9. Proof lattice.

be reformulated). If the formula p is satisfiable, then the specification it expresses is consistent. A model for p with a finite number of states is constructed by the decision procedure. The synchronization skeleton of a program meeting the specification can be read from this model. The small model property ensures that any program whose synchronization properties can be expressed in the TL can be realized by a system of concurrently running processes, each of which is a finite-state machine.

One suitable logic is the branching-time logic CTL. It has been used to specify and to synthesize, e.g., a starvation-free solution to the mutual exclusion problem (cf. [24]). Consider two processes P_1 and P_2, where each process is always in one of three regions of code: NCS_i—the NonCritical Section, TRY_i—the TRYing Secton, or CS_i—the Critical Section, which it cycles through, in order, repeatedly. When it is in region NCS_i, process P_i performs "noncritical" computations which can proceed in parallel with computations by the process P_j. At certain times, however, P_i may need to perform certain "critical" computations in the region CS_i. Thus, P_i remains in NCS_i as long as it has not yet decided to attempt critical section entry. When and if it decides to make this attempt, it moves into the region TRY_i. From there it enters CS_i as soon as possible, provided that the mutual exclusion constraint $\neg(\mathrm{atCS}_1 \wedge \mathrm{atCS}_2)$ is not violated. It remains in CS_i as long as necessary to perform its "critical" computations and then re-enters NCS_i.

It is assumed that only transitions between different regions of sequential code are recorded. Moves entirely within the same region are not considered in specifying synchronization. Moreover, the programs are running in a shared-memory environment with test-and-set primitives. The behavior of the system can be specified using the formulae listed below:

(1) start state:

$$\mathrm{atNCS}_1 \wedge \mathrm{atNCS}_2;$$

(2) mutual exclusion:

$$\mathsf{AG}(\neg(\mathrm{atCS}_1 \wedge \mathrm{atCS}_2));$$

(3) absence of starvation for P_i $(i=1,2)$:

$$\mathsf{AG}(\mathrm{atTRY}_i \Rightarrow \mathsf{AF}\mathrm{atCS}_i)$$

plus some additional formulae to formally specify the information regarding the model of concurrent computation which was informally communicated in the above narrative. The global state transition diagram of a program meeting the conjunction of the above specification, obtained by applying the synthesis method outlined, is shown in Fig. 10. Solutions to other well known synchronization problems such as readers-writers and dining philophers can also be synthesized.

A closely related synthesis method for CSP programs based on the use of a decision procedure for PLTL was given in [69]. In the recent study [128] a method for synthesizing an individual component of a reactive system from a specification in (essentially) CTL* is described. Earlier informal efforts toward synthesis of concurrent programs from TL-like formalisms include [55, 95].

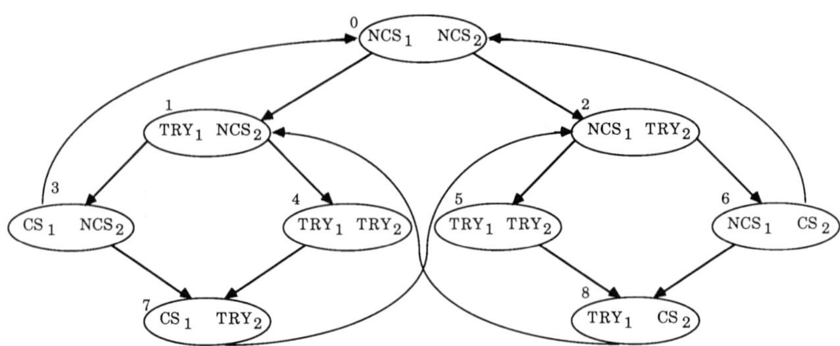

Fig. 10. Global state transition graph synthesized for the two-process mutual exclusion problem.

There are a number of advantages to this type of automatic program synthesis method. It obviates the need to compose a program as well as the need to construct a correctness proof. Moreover, since it is algorithmic rather than heuristic in nature, it is both sound and complete. It is sound in that any program produced as a solution does in fact meet the specification. It is complete in that if the specification is satisfiable, a solution will be generated.

A drawback of this method is, of course, the (at least) exponential complexity of the decision procedure. Is this an insurmountable barrier to the development of this method into a practical software tool? Recall that, while deciding satisfiability of propositional formulae requires exponential time in the worst case using the best known algorithms, the average-case performance appears to be substantially better, and working automatic theorem provers and program verifiers are a reality. Similarly, the performance in practice of the decision procedure used by the synthesis method may be substantially better than the potentially exponential-time worst case (see [28]). Furthermore, synchronization skeletons are generally small. It therefore seems conceivable that this approach may, in the long run, turn out to be useful in practical applications.

7.4. *Automatic verification of finite-state concurrent systems*

The global state transition graph of a finite state concurrent system may be viewed as a finite temporal logic structure, and a model-checking algorithm (cf. Section 6.3) can be applied to determine whether the structure is a model of a specification expressed as a formula in an appropriately chosen system of Propositional TL. In other words, the model-checking algorithm is used to determine whether a given finite-state program meets a particular correctness specification. Provided that the model-checking algorithm is efficient, this approach is potentially of wide applicability since a large class of concurrent programming problems have finite-state solutions, and the interesting properties of many such systems can be specified in a Propositional TL. For example, many network communication protocols can be modelled as a finite-state system.

The basic idea behind this mechanical model-checking approach to verification of finite-state systems is to make brute-force graph reachability analysis efficient and expressive through the use of TL as an assertion language. Of course, research in protocol verification has attempted to exploit the fact that protocols are frequently finite-state, making exhaustive graph reachability analysis possible. The advantage offered by model checking seems to be that it provides greater flexibility in formulating specifications through the use of TL as a single, uniform assertion language that can express a wide variety of correctness properties. This makes it possible to reason about, e.g., both safety and liveness properties with equal facility.

Historically, Pnueli [83] showed that the problem of deciding truth of a linear temporal formula over a finite structure was decidable. However, Pnueli's decision procedure was nonelementary, and the problem is PSPACE-complete in general (Theorem 6.17). The term "model checking" was coined by Clarke and Emerson [12], who gave an efficient (polynomial-time) model-checking algorithm for the branching-time logic CTL, and first proposed that it could be used as the basis of a practical automatic verification technique. At roughly the same time, Queille and Sifakis [93] gave a model-checking algorithm for a similar branching-time logic, but did not analyze its complexity.

To illustrate how model-checking algorithms work, we now describe a simple model-checking algorithm for CTL. Note that it is similar to the global flow analysis algorithms used in compiler optimization. Assume that $M=(S, R, L)$ is a finite structure and p_0 is a CTL formula. The goal is to determine at which states s of M we have $M, s \models p_0$. The algorithm is designed to operate in stages: the first stage processes all subformulae of p_0 of length 1, the second stage processes all subformulae of p_0 of length 2, and so on. At the end of the ith stage, each state will be labelled with the set of all subformulae of length $\leqslant i$ that are true at the state. To perform the labelling at stage i, information gathered in earlier stages is used. For example, subformula $q \wedge r$ should be placed in the label of a state s precisely when q and r are both already in the label of s. For the modal subformula $\mathsf{A}[q \,\mathsf{U}\, r]$, information from the successor states of s, as well as state s itself, is used. Since $\mathsf{A}[q \,\mathsf{U}\, r] \equiv q \vee \mathsf{AXA}[q \,\mathsf{U}\, r]$, $\mathsf{A}[q \,\mathsf{U}\, r]$ is initially added to the label of each state already labelled with r. Then satisfaction of $\mathsf{A}[q \,\mathsf{U}\, r]$ is propagated outward, by repeatedly adding $\mathsf{A}[q \,\mathsf{U}\, r]$ to the label of each state labelled by q and having $\mathsf{A}[q \,\mathsf{U}\, r]$ in the label of all successors.

Let

$$(\mathsf{A}[q \,\mathsf{U}\, r])^1 = r, \qquad (\mathsf{A}[q \,\mathsf{U}\, r])^{j+1} = r \vee (q \wedge \mathsf{AX}(\mathsf{A}[q \,\mathsf{U}\, r])^j).$$

It can be shown that $M, s \models (\mathsf{A}[q \,\mathsf{U}\, r])^j$ iff $M, x \models \mathsf{A}[q \,\mathsf{U}\, r]$ and along every path starting at s, r holds within distance j. Thus, states where $(\mathsf{A}[q \,\mathsf{U}\, r])^1$ holds are found first, then states where $(\mathsf{A}[q \,\mathsf{U}\, r])^2$ holds, etc. If $\mathsf{A}[q \,\mathsf{U}\, r]$ holds, then $(\mathsf{A}(q \,\mathsf{U}\, r])^{\mathrm{card}(S)}$ must hold since all loop-free paths in M are of length $\leqslant \mathrm{card}(S)$. Thus if, after card($S$) steps of propagating outward, $\mathsf{A}[q \,\mathsf{U}\, r]$ has still not been found to hold at state s, then $\mathsf{A}[q \,\mathsf{U}\, r]$ is false at s. Satisfaction of the other CTL modality $\mathsf{E}[p \,\mathsf{U}\, q]$ propagates outward in the same fashion.

This version of the algorithm can be naively implemented to run in time linear in the length of p_0 and quadratic in the size of structure M. A more clever version of the algorithm can be implemented to run in time linear in the length of the input formula p and the size of M (cf. [13]):

```
for i=1 to length(p_0)
  for each subformula p of p_0 of length i
    case on the form of p
      p=P, an atomic proposition /* nothing to do */
      p=q∧r: for each s∈S
                if q∈L(s) and r∈L(s)
                then add q∧r to L(s)
              end
      p=¬q:   for each s∈S
                if q∉L(s)
                then add ¬q to L(s)
              end
      p=EXq:  for each s∈S
                if (for some successor t of s, q∈L(t))
                then add EXq to L(s)
              end
      p=A[q U r]: for each s∈S
                    if r∈L(s)
                    then add A[q U r] to L(s)
                  end
                  for j=1 to card(S)
                    for each s∈S
                      if q∈L(s) and (for each successor t of s, A[q U r]∈L(t))
                      then add A[q U r] to L(s)
                    end
                  end
      p=E[q U r]: for each s∈S
                    if r∈L(s)
                    then add E[q U r] to L(s)
                  end
                  for j=1 to card(S)
                    for each s∈S
                      if q∈L(s) and (for some successor t of s, E[q U r]∈L(t))
                      then add E[q U r] to L(s)
                    end
                  end
    end of case
  end
end
```

One limitation of the logic CTL is, of course, that it cannot express correctness under fair scheduling assumptions. However, the extended logic FairCTL described in [27] can express correctness under fairness (cf. [93]). A FairCTL specification (p_0, Φ_0) consists of a functional assertion p_0, which is a state formula, and an underlying fairness assumption Φ_0, which is a pure path formula. The functional assertion p_0 is expressed in essentially CTL syntax with basic modalities of the form either A_Φ ("for all *fair* paths") or E_Φ ("for some *fair* path") followed by one of the linear-time operators F, G, X, or U. The path quantifiers range over paths meeting the fairness constraint Φ_0, which is a Boolean combination of the infinitary linear time operators F^∞ ("infinitely often") and G^∞ ("almost always"), applied to propositional arguments. We can then view a subformula such as $\mathsf{A}_\Phi \mathsf{F}P$ of functional assertion p_0 as an abbreviation for the CTL* formula $\mathsf{A}[\Phi_0 \Rightarrow \mathsf{F}P]$. Similarly, $\mathsf{E}_\Phi \mathsf{G}P$ abbreviates $\mathsf{E}[\Phi_0 \wedge \mathsf{G}P]$. In this way FairCTL inherits its semantics from CTL*. Provided that Φ_0 is in the canonical form

$$\bigvee_{i=1}^{n} \bigwedge_{j=1}^{n_i} (\mathsf{F}^\infty p_{ij} \vee \mathsf{G}^\infty q_{ij}),$$

the model-checking problem for FairCTL can be solved in time linear in the input structure size and small polynomial in the specification size.

Nevertheless, there are still correctness properties that one might like to describe that are not expressible within FairCTL, although they are describable in CTL* or even PLTL. The PSPACE-completeness of these latter logics, on first hearing, would seem to be a serious drawback. Lichtenstein and Pnueli [59] noted however that model checking is a problem with two input parameters, the structure and the specification, and then proceeded to develop a model-checking algorithm for PLTL of complexity exponential in the length of the specification but only linear in the size of the structure. They argued that since specifications are generally quite short while the structures representing programs are usually quite large, the exponential complexity in the specification size can be discounted. In practice, the dominating factor in the complexity should thus be the linear growth in the structure size.

It is worth pointing out that model checking, despite (because of?) its simplicity, is one approach to automatic verification that really seems to be useful in practice. It has been used to verify a large variety of finite-state concurrrent programs. These programs range from examples in the academic literature on concurrency to large-scale network communication protocols. For instance, a solution to the mutual exclusion problem given in [77] and proved correct there manually using Linear TL is actually finite-state. It was mechanically verified using the CTL model-checking algorithm as described in [13]. Model checking is also applicable to the design of VLSI hardware and asynchronous circuits: Clarke has developed an efficient implementation of the CTL model checker along with various pieces of support software, which together forms the EMC (Extended Model Checker) system at Carnegie-Mellon University. In [127] the use of the EMC system resulted in the detection of a previously unknown error in a circuit for a self-timed queue element published in the text [72]. Other applications to the design of sequential circuits are discussed in [9, 10, 20], and as well as the overview article [15]. Finally, model checking is applicable to large-scale network communication protocols. Indeed, one project in France [101] has bought dedicated hardware to use for model-checking network protocols. Finite-state systems with on the order of 10^5 states (and arcs) can currently be handled.

Despite the above practical successes, a potentially serious drawback to the entire model-checking approach is that the size of the global state transition graph grows exponentially with the number of processes. Recent work in [14, 106, 15] suggests that it may be possible to avoid this exponential blow-up in some cases for concurrent systems with many "copies" of the same process, although this is not possible in general (cf. [3]). Other work on reducing the size of the state graph based on hierarchical specification and hiding of states at lower levels of abstraction is presented in [127].

8. Other modal and temporal logics in computer science

8.1. *Classical Modal Logic*

The class of Modal Logics was originally developed by philosophers to study different "modes" of truth. Such modes include possibility, necessity, obligation, knowledge, belief, and perception. Among the most important modes of truth are what "must be" true (necessity) and what "may be" true (possibility). For example, the assertion P may be false in the present world, and yet the assertion *possibly* P true if there exists another world where P is true. The assertion *necessarily* P is true provided that P is true in all worlds.

Thus we have the well known *possible worlds* semantics of Kripke, where the truth value of a modal assertion at a world depends on the truth value(s) of its subassertion(s) at other possible worlds. This is formalized in terms of a *Kripke structure* $M=(S, R, L)$ consisting of an underlying set S of possible *worlds*, also called *states*, an accessibility relation $R \subseteq S \times S$ between worlds, and a labelling L which provides an interpretation of primitive (i.e., nonmodal) assertions at each world. The technical definitions are such that *possibly* P is true at world s iff P is true at some world accessible from s, and *necessarily* P is true at world s iff P is true at all worlds accessible from s.

As we have seen, Temporal Logic is a particular kind of Modal Logic that has been specialized for reasoning about program behavior. Temporal Logic provides a much richer set of modalities, varying in how their truth value depends on which argument(s) hold(s) at which worlds, with the accessibility relation corresponding to the evolution of a concurrent system over time.

8.2. *Propositional Dynamic Logic*

An alternative development of a Modal Logic framework for program reasoning is represented by Dynamic Logic, originally proposed by Pratt [89] in the first-order version, specialized to the propositional version by Fischer and Ladner [31], and, in general studied intensively by Harel [41] and others. (Detailed treatments of Dynamic Logic can be found in [50, 42].) The basic modalities of Propositional Dynamic Logic (PDL) are of the form $\langle\alpha\rangle p$ where α is a regular expression over "atomic programs" and p is a formula. The intuitive meaning is that there exists an execution of program α leading from the present state to a state where p holds. PDL may be viewed as a Propositional Branching-Time Logic, with basic modalities of the form $\mathsf{E}\beta$, where β is a regular expression over atomic propositions (node labels) and atomic programs (arc labels), and which means that there exists a path (a sequence of alternating states and arcs) starting at the present state that matches the regular expression β. A variety of extensions of PDL have been proposed in order to increase its expressive power. One is of particular interest to temporal logicians, viz. that with a repetition construct referred to as PDL + repeat or Δ-PDL. In Temporal Logic terms, its basic modalities are of the form $\mathsf{E}\beta$, where β is an ω-regular expression such as $\alpha\gamma^{\omega}$. The ω iteration operator corresponds to the *repeat* operator Δ. Δ-PDL strictly subsumes CTL* in expressive power, and is thus able to express modalities such as $\mathsf{AF}p$ that cannot be expressed in ordinary PDL. Historically, Δ-PDL is important for reasons beyond its high expressive power. It is with Δ-PDL that automata-theoretic techniques for testing satisfiability were pioneered by Streett [108].

8.3. *Probabilistic logics*

Various probabilistic Temporal Logics have been proposed. For instance, Lehmann and Shelah [58] describe a logic, TC, with essentially the same syntax as CTL*, but where $\mathsf{A}p$ means: "for almost all paths (i.e., for a set of paths of measure 1) p holds", and $\mathsf{E}p$ means: "for significantly many paths (i.e., for a set of paths of positive measure) p holds". They give a deterministic double-exponential-time decision procedure and a sound and complete axiomatization for it. Interestingly, the logic TC has the same axiomatization as the logic MPL (Modal Logic Process) of Abrahamson [1]. MPL has essentially the same syntax as CTL* but interprets it over more abstract structures, where the set of paths is not required to be generated by a binary relation. A probabilistic version of CTL is considered in [43].

8.4. *Fixpoint logics*

Temporal operators can be characterized in terms of extremal fixpoints of monotonic functionals. Let $M=(S, R, L)$ be a structure. We use PRED(S) to denote the lattice of total predicates over state set S, where each predicate is identified with the set of states which make it true and the ordering on predicates is set inclusion. Then a formula p defines a member of PRED(S), $\{s\in S: M, s\models p\}$, and if it contains an atomic proposition Q, e.g. $p(Q)$, then it defines a function PRED(S)$\rightarrow$PRED(S) where the value of $p(Q)$ varies as Q varies. Let τ: PRED(S)$\rightarrow$PRED(S) be given; then τ is said to be

- *monotonic* provided $P\subseteq Q$ implies $\tau(P)\subseteq\tau(Q)$;
- $\cup$-*continuous* provided that $P_1\subseteq P_2\subseteq P_3\subseteq\cdots$ implies $\tau(\bigcup_i P_i)=\bigcup_i\tau(P_i)$.
- $\cap$-*continuous* provided that $P_1\supseteq P_2\supseteq P_3\supseteq\cdots$ implies $\tau(\bigcap_i P_i)=\bigcap_i\tau(P_i)$.

A predicate P is said to be a *fixpoint* of functional τ if $P=\tau(P)$. The theorem of Tarski–Knaster [110] ensures that a monotonic functional τ: PRED(S)$\rightarrow$PRED(S) always has a least fixpoint, $\mu Z.\tau(Z)=\bigcap\{Y:\tau(Y)=Y\}$, and a greatest fixpoint $\nu Z.(Z)=\bigcup\{Y: \tau(Y)=Y\}$. Whenever τ is $\cup$-continuous then $\mu Z.\tau(Z)=\bigcup_i\tau^i(\mathit{false})$, and whenever τ is $\cap$-continuous then $\nu Z.(Z)=\bigcap_i \tau^i(true)$. (Note that $\tau^2(false)=\tau(\tau(false))$, *etc.*)

For example, below are shown the fixpoint characterizations for certain CTL modalities.

$$\mathsf{EF}P=\mu Z.P\vee \mathsf{EX}Z, \qquad \mathsf{AG}P=\nu Z.P\wedge \mathsf{AX}Z,$$
$$\mathsf{AF}P=\mu Z.P\vee \mathsf{AX}Z, \qquad \mathsf{EG}P=\nu Z.P\wedge \mathsf{EX}Z.$$

Intuitively, the properties characterized as least fixpoints correspond to eventualities, while those characterized as greatest fixpoints are invariance properties.

Assume for simplicity that each state in the underlying structure has a finite number of successors. Then each of the above functionals is both $\cup$-continuous and $\cap$-continuous in the argument Z, and we can readily establish the correctness for the above characterizations. For example, to show that $\mathsf{EF}P=\mu Z.\tau(Z)$, with $\tau(Z)=P\vee \mathsf{EX}Z$, it suffices to show that for each i (ranging over $\mathbb{N}$), $\tau^i(false)=\{$states s in M: there exists a path of length $\leqslant i$ in M from state s to some state t such that $M, t\models P\}$.

These fixpoint characterizations are used in the model-checking algorithm of Section 7 and in the tableau-based decision procedure of Section 6.

We can define an entire logic built up from atomic proposition constants $P, Q, \ldots$, atomic proposition variables $Y, Z, \ldots$, Boolean connectives $\wedge$, $\vee$, $\neg$, the next-time operators EX, AX, and the least fixpoint μ and greatest fixpoint ν operators. We require that each formula be syntactically monotone, meaning that fixpoint formulae such as $\mu Z.\tau(Z)$ (or $\nu X.\tau(Z)$) are legal only when Z appears under an even number of negations within τ. The semantics is given in the obvious way suggested above.

Essentially this system was dubbed the "the Propositional Mu-Calculus" by Kozen [48]. This Mu-Calculus has very considerable expressive power. It can encode (and in fact subsumes) CTL, FairCTL, CTL*, and, interpreted over multiprocess structures, also PDL and PDL + *repeat*. In practical terms it also allows expression of extended modalities such as P is true at all even moments along all futures, which is captured by $\nu Z.P\wedge \mathsf{AXAX}Z$. Related systems were considered in [23, 91]. Other proposals for formalisms based on fixpoints can be found in, e.g., [4, 79, 97, 19, 80].

8.5. *Knowledge*

Recently there has been interest in the development of modal and temporal logics for reasoning about the states of knowledge in reactive systems. Knowledge can be especially important in the realm of distributed systems where processes are geographically dispersed and, at any given moment, possess only incomplete knowledge regarding the status of other processes in the system. Indeed, in many informal instances of reasoning about the behavior of distributed systems, it is a natural metaphor to refer to what a process *knows*. Logics of knowledge represent an effort to provide a formal basis for such reasoning.

A number of systems have been proposed (cf. [39, 56, 52, 21]). Typical modalities include K_iP which means that "process i knows p" and $\mathsf{C}p$ which means that "p is common knowledge" in the sense that "all processes know p, all processes know that all processes know p, all processes know that all processes know that all processes know p,..." These modalities of knowledge can be combined in various ways with temporal operators to permit reasoning about distributed systems. Certain semantic constraints, expresssed as axioms (e.g. $\mathsf{K}_i\mathsf{X}p \Rightarrow \mathsf{X}\mathsf{K}_i p$), are usually required. Subtle interactions between the syntax of the logic and the assumptions made regarding the model of distributed computation can lead to widely varying complexities for the decision problems of the resulting logics (cf. [126]). In general, this seems a promising area for future research. We refer the reader to the excellent survey by Halpern [38] for an in-depth treatment.

Acknowledgment

We would like to thank Amir Pnueli for very helpful comments.

This work was supported in part by US NSF Grant CCR8511354, ONR Contract N00014-86-K-0763 and Netherlands ZWO Grant nf-3/nfb 62-500.

References

[1] ABRAHAMSON, K., Decidability and expressiveness of logics of processes, Ph.D. Thesis, Univ. of Washington, Seattle, 1980.

[2] ALPERN, B. and F. SCHNEIDER, Defining safety and liveness, Tech. Report, Cornell Univ., Ithaca, NY, 1985.

[3] APT, K. and D. KOZEN, Limits for automatic verification of finite state systems, *Inform. Process. Lett.* **22**(6) (1986) 307–309.

[4] BAKKER, J. DE and D. SCOTT, A theory of programs, Unpublished notes, IBM Seminar, Vienna, 1969.

[5] BARRINGER, H., R. KUIPER and A. PNUELI, Now you may compose Temporal Logic specifications, in: *Proc. 16th Ann. ACM Symp. on Theory of Computing* (1984) 51–63.

[6] BARRINGER, H., R. KUIPER and A. PNUELI, A really abstract concurrent model and its temporal logic, in: *Proc. 13th Ann. ACM Symp. on Principles of Programming Languages* (1986) 173–183.

[7] BEN-ARI, M., J.Y. HALPERN and A. PNUELI, Deterministic Propositional Dynamic Logic: finite models, complexity, and completeness, *J. Comput. System Sci.* **25** (1982) 402–417.

[8] BEN-ARI, M., A. PNUELI and Z. MANNA, The Temporal Logic of Branching Time, in: *Proc. 8th Ann. Symp. on Principles of Programming Languages* (1981) 164–176; journal version, *Acta Inform.* **20** (1983) 207–226.

[9] BROWNE, M., E.M. CLARKE and D. DILL, Checking the correctness of sequential circuits, in: *Proc. IEEE Internat. Conf. on Computer Design* (1985) 545–548.

[10] BROWNE, M., E.M. CLARKE, D. DILL and B. MISHRA, Automatic verification of sequential circuits using temporal logic, *IEEE Trans. Comput.* **35**(12) (1986) 1035–1044.

[11] BURSTALL, R., Program proving considered as hand simulation plus induction, in: *Proc IFIP Congr. on Information Processing* (1974) 308–312.

[12] CLARKE, E.M. and E.A. EMERSON, Design and synthesis of synchronization skeletons using Branching Time Temporal Logic, in: *Proc. Workshop on Logics of Programs*, Lecture Notes in Computer Science, Vol. 131 (Springer, Berlin, 1981) 52–71.

[13] CLARKE, E.M., E.A. EMERSON and A.P. SISTLA, Automatic verification of finite state concurrent system using temporal logic, in: *Proc. 10th Ann. ACM Symp. on Principles of Programming Languages* (1983); journal version, *ACM Trans. on Programming Languages and Systems* **8**(2) (1986) 244–263.

[14] CLARKE, E.M., O. GRUMBERG and M.C. BROWNE, Reasoning about networks with many identical finite state processes, in: *Proc. 5th Ann. ACM Symp. on Principles of Distributed Computing* (1986) 240–248.

[15] CLARKE, E.M. and O. GRUMBERG, Avoiding the state explosion problem in temporal model checking algorithms, in: *Proc. 6th Ann. ACM Symp. on Principles of Distributed Computing* (1987) 294–303.

[16] CLARKE, E.M. and O. GRUMBERG, Research on automatic verification of finite state concurrent systems, *Ann. Rev. Comput. Sci.* **2** (1987) 269–290.

[17] CLARKE, E.M. and B. MISHRA, Automatic verification of asynchronous circuits, in: *Proc. Workshop on Logics of Programs*, Lecture Notes in Computer Science, Vol. 164 (Springer, Berlin, 1984) 101–115.

[18] COURCOUBETIS, C., M.Y. VARDI and P.L. WOLPER, Reasoning about fair concurrent programs, in: *Proc. 18th Ann. ACM Symp. on Theory of Computing* (1986) 283–294.

[19] DIJKSTRA, E.W., *A Discipline of Programming* (Prentice-Hall, Englewood Cliffs, NJ, 1976).

[20] DILL, D. and E.M. CLARKE, Automatic verification of asynchronous circuits using Temporal Logic, *IEEE Proc.* **133** (1986) 276–282.

[21] DWORK, C. and Y. MOSES, Knowledge and common knowledge in a Byzantine environment I: crash failures, in: J.Y. Halpern, ed., *Proc. 1st Conf. on Theoretical Aspects of Reasoning about Knowledge* (1986) 149–170.

[22] EMERSON, E.A., Alternative semantics for temporal logics, *Theoret. Comput. Sci.* **26** (1983) 121–130.

[23] EMERSON, E.A. and E.M. CLARKE, Characterizing correctness properties of parallel programs as fixpoints, in: *Proc. 7th Internat. Coll. on Automata, Languages and Programming*, Lecture Notes in Computer Science, Vol. 85 (Springer, Berlin, 1981) 169–181.

[24] EMERSON, E.A. and E.M. CLARKE, Using Branching Time Temporal Logic to synthesize synchronization skeletons, *Sci. Comput. Programming* **2** (1982) 241–266.

[25] EMERSON, E.A. and J.Y. HALPERN, Decision procedures and expressiveness in the Temporal Logic of Branching Time, *J. Comput. System Sci.* **30**(1) (1985) 1–24.

[26] EMERSON, E.A. and J.Y. HALPERN, "Sometimes" and "Not Never" revisited: on Branching versus Linear Time Temporal Logic, *J. ACM* **33**(1) (1986) 151–178.

[27] EMERSON, E.A. and C.L. LEI, Modalities for model checking: branching time strikes back, in: *Proc. 12th Ann. ACM Symp. on Principles of Programming Languages* (1985) 84–96; journal version, *Sci. Comput. Programming* **8** (1987) 275–306.

[28] EMERSON, E.A., T.H. SADLER and J. SRINIVASAN, Efficient temporal reasoning, in: *Proc. 16th Ann. ACM Symp. on Principles of Programming Languages* (1989) 166–178.

[29] EMERSON, E.A. and A.P. SISTLA, Deciding full Branching Time Logic, *Inform. and Control* **61**(3) (1984) 175–201.

[30] ENDERTON, H.B., *A Mathematical Introduction to Logic* (Academic Press, New York, 1972).

[31] FISCHER, M.J. and R.E. LADNER, Propositional Dynamic Logic of regular programs, *J. Comput. System Sci.* **18** (1979) 194–211.

[32] FRANCEZ, N., *Fairness* (Springer, New York, 1986).

[33] FRANCEZ, N., and D. KOZEN, Generalized fair termination, in: *Proc. 11th Ann. ACM Symp. on Principles of Programming Languages* (1984) 46–53.

[34] GABBAY, D., A. PNUELI, S. SHELAH and J. STAVI, On the temporal analysis of fairness, in: *Proc. 7th Ann. ACM Symp. on Principles of Programming Languages* (1980) 163–173.

[35] HAFER, T., and W. THOMAS, Computation tree logic CTL* and path quantifiers in the monadic theory of the binary tree, in: *Proc. 14th Internat. Coll. on Automata, Languages and Programming*, Lecture Notes in Computer Science, Vol. 267 (Springer, Berlin, 1987) 269–279.

[36] HAILPERN, B.T., *Verifying Concurrent Processes Using Temporal Logic*, Lecture Notes in Computer Science, Vol. 129 (Springer, Berlin, 1982).

[37] HAILPERN, B.T. and S.S. OWICKI, Verifying network protocols using Temporal Logic, in: *Proc. Trends and Applications 1980: Computer Network Protocols* (1980) 18–28.

[38] HALPERN, J.Y., Using reasoning about knowledge to analyze distributed systems, *Ann. Rev. Comput. Sci.* **2** (1987) 37–68.

[39] HALPERN, J.Y. and Y. MOSES, Knowledge and common knowledge in a distributed environment, in: *Proc. 3rd Ann. ACM Symp. on Principles of Distributed Computing* (1984) 50–61.

[40] HALPERN, J.Y. and Y. SHOHAM, A Propositional Modal Logic of time intervals, in: *Proc. IEEE Symp. on Logic in Computer Science* (1986) 279–292.

[41] HAREL, D., Dynamic logic: axiomatics and expressive power, Ph.D. Thesis, Lab. for Computer Science, MIT, 1979; also available as Lecture Notes in Computer Science, Vol. 68 (Springer, Berlin, 1979).

[42] HAREL, D., Dynamic logic, in: D. Gabbay and F. Guenthner, eds., *Handbook of Philosophical Logic Vol. II: Extensions of Classical Logic* (Reidel, Boston, MA, 1984) 497–604.

[43] HART, S. and M. SHARIR, Probabilistic Temporal Logics for finite and bounded models, in: *Proc. 16th Ann. ACM Symp. on Theory of Computing* (1984) 1–13.

[44] HOARE, C.A.R., Communicating sequential processes, *Comm. ACM* **21**(8) (1978) 666–676.

[45] HOSSLEY, R. and C. RACKOFF, The emptiness problem for automata on infinite trees, in: *Proc. 13th Ann. IEEE Symp. on Switching and Automata Theory* (1972) 121–124.

[46] KAMP, H., Tense logic and the theory of linear order, Ph.D. Dissertation, UCLA, Los Angeles, 1968.

[47] KOYMANS, R., Specifying message buffers requires extending Temporal Logic, in: *Proc. 6th Ann. ACM Symp. on Principles of Distributed Computing* (1987) 191–204.

[48] KOZEN, D., Results on the propositional μ-Calculus, *Theoret. Comput. Sci.* **27** (1983) 333–354.

[49] KOZEN, D. and R. PARIKH, An elementary proof of completeness for PDL, *Theoret. Comput. Sci.* **14** (1981) 113–118.

[50] KOZEN, D. and J. TIURYN, Logics of programs, in: J. van Leeuwen, ed., *Handbook of Theoretical Computer Science, Vol. B* (North-Holland, Amsterdam, 1990) 789–840.

[51] KURSHAN, R.P., Testing containment of omega regular languages, Tech. Report 1121-861010-33-TM, AT&T Bell Labs, Murray Hill, NJ, 1986.

[52] LADNER, R. and J. REIF, The logic of distributed protocols, in: J. Halpern, ed., *Proc. Conf. on Theoretical Aspects of Reasoning about Knowledge* (1986) 207–222.

[53] LAMPORT, L., Sometimes is sometimes "Not never"—on the Temporal Logic of programs, in: *Proc. 7th Ann. ACM Symp. on Principles of Programming Languages* (1980) 174–185.

[54] LAMPORT, L., What good is Temporal Logic? in: *Proc. IFIP Congr. on Information Processing* (1983) 657–667.

[55] LAVENTHAL, M., Synthesis of synchronization code for data abstractions, Ph.D. Thesis, MIT, Cambridge, MA, 1978.

[56] LEHMANN, D., Knowledge, common knowledge, and related puzzles, in: *Proc. 3rd Ann. ACM Symp. on Principles of Distributed Computing* (1984) 62–67.

[57] LEHMANN, D., A. PNUELI and J. STAVI, Impartiality, justice and fairness: the ethics of concurrent termination, in: *Proc. 8th Internat. Coll. on Automata, Languages and Programming*, Lecture Notes in Computer Science, Vol. 115 (Springer, Berlin, 1981) 264–277.

[58] LEHMANN, D. and S. SHELAH, Reasoning about time and chance, *Inform. and Control* **53**(3) (1982) 165–198.

[59] LICHTENSTEIN, O. and A. PNUELI, Checking that finite state concurrent programs satisfy their linear specification, in: *Proc. 12th Ann. ACM Symp. on Principles of Programming Languages* (1985) 97–107.

[60] LICHTENSTEIN, O., A. PNUELI and L. ZUCK, The glory of the past, in: *Proc. Conf. on Logics of Programs*, Lecture Notes in Computer Science, Vol. 193 (Springer, Berlin, 1985) 196–218.

[61] MANNA, Z. and A. PNUELI, Verification of concurrent programs: temporal proof principles, in: D. Kozen, ed., *Proc. Workshop on Logics of Programs*, Lecture Notes in Computer Science, Vol. 131 (Springer, Berlin, 1982) 200–252.

[62] MANNA, Z. and A. PNUELI, Verification of concurrent programs: the temporal framework, in: R.S. Boyer and J.S. Moore, eds., *The Correctness Problem in Computer Science* (Academic Press, London, 1982) 215–273.

[63] MANNA, Z. and A. PNUELI, Verification of concurrent programs: a temporal proof system, in: *Proc. 4th School on Advanced Programming* (1982) 163–255.

[64] MANNA, Z. and A. PNUELI, How to cook a proof system for your pet language, in: *Proc. 10th Ann. ACM Symp. on Principles of Programming Languages* (1983) 141–154.

[65] MANNA, Z. and A. PNUELI, Adequate proof principles for invariance and liveness properties of concurrent programs, *Sci. Comput. Programming* **4**(3) (1984) 257–290.

[66] MANNA, Z. and A. PNUELI, Specification and verification of concurrent programs by $\forall$-automata (extended abstract), in: *Proc. 14th Ann. ACM Symp. on Principles of Programming Languages* (1987) 1–12.

[67] MANNA, Z. and A. PNUELI, A hierarchy of temporal properties, in: *Proc. 6th Ann. ACM Symp. on Principles of Distributed Computing* (1987) 205.

[68] MANNA, Z. and R. WALDINGER, Is "sometimes" sometimes better than "always"?: intermittent assertions in proving program correctness, *Comm. ACM* **21**(2) (1978) 159–172.

[69] Manna, Z. and P.L. Wolper, Synthesis of communicating processes from temporal logic specifications, *ACM Trans. Programming Languages and Systems* **6**(1) (1984) 68–93.

[70] McNaughton, R., Testing and generating infinite sequences by a finite automaton, *Inform. and Control* **9** (1966) 521–530.

[71] McNaughton, R. and S. Pappert, *Counter-Free Automata* (MIT Press, Cambridge, MA, 1962).

[72] Mead, C. and L. Conway, *Introduction to VLSI Systems* (Addison-Wesley, Reading, MA, 1980).

[73] Meyer, A.R., Weak monadic second order theory of one successor is not elementarily recursive, in: *Proc. Logic Colloquium*, Lecture Notes in Mathematics, Vol. 453 (Springer, Berlin, 1975) 132–154.

[74] Moszkowski, B., Reasoning about digital circuits, Ph.D. Thesis, Stanford Univ., 1983.

[75] Muller, D.E., Infinite sequences and finite machines, in: *Proc. 4th Ann. IEEE Symp. on Switching Theory and Logical Design* (1963) 3–16.

[76] Nguyen, V., A. Demers, S. Owicki and D. Gries, A modal and temporal proof system for networks of processes, *Distrib. Comput.* **1**(1) (1986) 7–25.

[77] Owicki, S.S. and L. Lamport, Proving liveness properties of concurrent programs, *ACM Trans. on Programming Languages and Systems* **4**(3) (1982) 455–495.

[78] Parikh, R., A decidability result for second order process logic, in: *Proc. 19th Ann. IEEE Symp. on Foundations of Computer Science* (1978) 177–183.

[79] Park, D., Fixpoint induction and proof of program semantics, in: B. Meltzer and D. Michie, eds., *Machine Intelligence, Vol. 5* (Edinburgh Univ. Press, Edinburgh, 1970) 59–78.

[80] Park, D., On the semantics of fair parallelism, in: *Abstract Software Specification*, Lecture Notes in Computer Science, Vol. 86 (Springer, Berlin, 1980) 504–524.

[81] Peterson, G.L., Myths about the mutual exclusion problem, *Inform. Process. Lett.* **12**(3) (1981) 115–116.

[82] Pinter, S. and P.L. Wolper, A Temporal Logic for reasoning about partially ordered computations, in: *Proc. 3rd Ann. ACM Symp. on Principles of Distributed Computing* (1984) 28–37.

[83] Pnueli, A., The Temporal Logic of programs, in: *Proc. 18th Ann. IEEE Symp. on Foundations of Computer Science* (1977) 46–57.

[84] Pnueli, A., The temporal semantics of concurrent programs, *Theoret. Comput. Sci.* **13** (1981) 45–60.

[85] Pnueli, A., On the extremely fair termination of probabilistic algorithms, in: *Proc. 15th Ann. ACM Symp. on Theory of Computing* (1983) 278–290.

[86] Pnueli, A., In transition from global to modular reasoning about concurrent programs, in: K.R. Apt, ed., *Logics and Models of Concurrent Systems* (Springer, Berlin, 1984).

[87] Pnueli, A., Linear and branching structures in the semantics and logics of reactive systems, in: *Proc. 12th Internat. Coll. on Automata, Languages and Programming* (Springer, Berlin, 1985) 15–32.

[88] Pnueli, A., Application of temporal logic to the specification and verification of reactive systems: a survey of current trends, in: J.W. de Bakker, W.P. de Roever and G. Rozenberg, eds., *Current Trends in Concurrency: Overviews and Tutorials*, Lecture Notes in Computer Science, Vol. 224 (Springer, Berlin, 1986).

[89] Pratt, V., Semantical considerations on Floyd–Hoare Logic, in: *Proc. 17th Ann. IEEE Symp. on Foundations of Computer Science* (1976) 109–121.

[90] Pratt, V., Models of program logics, in: *Proc. 20th Ann. IEEE Symp. on Foundations of Computer Science* (1979) 115–122.

[91] Pratt, V., A decidable μ-calculus, in: *Proc. 22nd Ann. IEEE Symp. on Foundations of Computer Science* (1981) 421–427.

[92] Prior, A., *Past, Present, and Future* (Clarendon Press, Oxford, 1967).

[93] Queille, J.P. and J. Sifakis, Specification and verification of concurrent programs in CESAR, in: *Proc. 5th Internat. Symp. on Programming*, Lecture Notes in Computer Science, Vol. 137 (Springer, Berlin, 1982) 195–220.

[94] Queille, J.P. and J. Sifakis, Fairness and related properties in transition systems, *Acta Inform.* **19** (1983) 195–220.

[95] Ramarithram, K. and R. Keller, Specification and synthesis of synchronizers, in: *Proc. 9th Internat. Conf. on Parallel Processing* (1980) 311–321.

[96] Rescher, N. and A. Urquhart, *Temporal Logic* (Springer, Wien, 1971).

[97] ROEVER, W.P. DE, *Recursive Program Schemes: Semantics and Proof Theory*, Mathematical Centre Tracts, Vol. 70 (Centre for Mathematics and Computer Science, Amsterdam, 1976).
[98] SAFRA, S., On the complexity of omega-automata, in: *Proc. 29th Ann. IEEE Symp. on Foundations of Computer Science* (1988) 319–327.
[99] SCHWARTZ, R. and P. MELLIAR-SMITH, From state machines to Temporal Logic: specification methods for protocol standards, *IEEE Trans. on Comm.* **30**(12) 2486–2496.
[100] SCHWARTZ, R., P. MELLIAR-SMITH and F. VOGT, An interval logic for higher-level temporal reasoning, in: *Proc. 2nd Ann. ACM Symp. on Principles of Distributed Computing* (1983) 173–186.
[101] SIFAKIS, J., Personal communication, 1987.
[102] SISTLA, A.P., Theoretical issues in the design of distributed and concurrent systems, Ph.D. Thesis, Harvard Univ., Cambridge, MA, 1983.
[103] SISTLA, A.P., Characterization of safety and liveness properties in Temporal Logic, in: *Proc. 4th Ann. ACM Symp. on Principles of Distributed Computing* (1985) 39–48.
[104] SISTLA, A.P. and E.M. CLARKE, The complexity of Propositional Linear Temporal Logic, *J. ACM*, **32**(3) (1985) 733–749.
[105] SISTLA, A.P., E.M. CLARKE, N. FRANCEZ and A.R. MEYER, Can message buffers be axiomatized in Temporal Logic?, *Inform. and Control* **63**(1, 2) (1984) 88–112.
[106] SISTLA, A.P. and S.M. GERMAN, Reasoning about many processes, in: *Proc. IEEE Symp. on Logic in Computer Science* (1987) 138–152.
[107] SISTLA, A.P., M.Y. VARDI and P.L. WOLPER, The complementation problem for Büchi automata with applications to Temporal Logic, *Theoret. Comput. Sci.* **49** (1987) 217–237.
[108] STREETT, R., Propositional Dynamic Logic of looping and converse, *Inform. and Control* **54** (1982) 121–141. (Full version: Propositional Dynamic Logic of looping and converse, Ph.D. Thesis, Lab for Computer Science, MIT, Cambridge, MA, 1981).
[109] STREETT, R. and E.A. EMERSON, The Propositional μ-Calculus is elementary, in: *Proc. 11th Internat. Coll. on Automata, Languages and Programming*, Lecture Notes in Computer Science, Vol. 172 (Springer, Berlin, 1984) 465–472; journal version, An automata theoretic decision procedure for the Propositional μ-Calculus, *Inform. and Comput.* **81**(3) (1989) 249–264.
[110] TARKSI, A., A lattice-theoretical fixpoint theorem and its applications, *Pacific J. Math.* **55** (1955) 285–309.
[111] THOMAS, W., Star-free regular sets of omega-sequences, *Inform. and Control* **42** (1979) 148–156.
[112] THOMAS, W., Automata on infinite objects, in: J. van Leeuwen, ed., *Handbook of Theoretical Computer Science, Vol. B* (North-Holland, Amsterdam, 1990) 133–191.
[113] VARDI, M., The taming of converse: reasoning about two-way computations, in: *Proc. Workshop on Logics of Programs*, Lecture Notes in Computer Science, Vol. 193 (Springer, Berlin, 1985) 413–424.
[114] VARDI, M., Verification of concurrent programs: the automata-theoretic framework, in: *Proc. IEEE Symp. on Logic in Computer Science* (1987) 167–176.
[115] VARDI, M. and P. WOLPER, Yet another process logic, in: *Proc. Workshop on Logics of Programs*, Lecture Notes in Computer Science, Vol. 164 (Springer, Berlin, 1984) 501–512.
[116] VARDI, M. and P. WOLPER, Automata theoretic techniques for modal logics of programs, in: *Proc. 16th Ann. ACM Symp. on Theory of Computing* (1984) 446–456; journal version, *J. Comput. System. Sci.* **32** (1986) 183–221.
[117] VARDI, M. and P. WOLPER, An automata-theoretic approach to automatic program verification, in: *Proc. IEEE Symp. on Logic in Computer Science* (1986) 332–344.
[118] WOLPER, P., Synthesis of communicating processes from Temporal Logic specifications, Ph.D. Thesis, Stanford Univ., Palo Alto, CA, 1982.
[119] WOLPER, P., Temporal Logic can be more expressive, *Inform. and Control* **56**(1, 2) (1983) 72–93.
[120] WOLPER, P., The tableau method for Temporal Logic: an overview, *Logique et Anal.* **28** (1985) 119–136.
[121] WOLPER, P., Expressing interesting properties of programs in Propositional Temporal Logic, in: *Proc. 13th Ann. ACM Symp. on Principles of Programming Languages* (1986) 184–193.
[122] WOLPER, P., On the relation of programs and computations to models of Temporal Logic, Manuscript, Univ. of Liege, Belgium, 1988.

[123] ZUCK, L., Past temporal logic, Ph.D. Dissertation, Weizmann Institute, Rehovot, Israel, 1986.

[124] EMERSON, E.A., Automata, tableaux, and temporal logic, in: R. Parikh, ed., *Proc. Conf. on Logics of Programs*, Lecture Notes in Computer Science, Vol. 193 (Springer, Berlin, 1985) 79–88.

[125] EMERSON, E.A. and C.S. JUTLA, The complexity of free automata and logics of programs, in: *Proc. 29th Ann. IEEE-CS Symp. on Foundations of Computer Science* (1988) 328–337.

[126] HALPERN, J.Y. and M.Y. VARDI, The complexity of reasoning about knowledge and time (extended abstract), in: *Proc. 18th Ann. ACM Symp. on Theory of Computing* (1986) 304–315.

[127] MISHRA, B. and E.M. CLARKE, Hierarchical verification of asynchronous circuits using temporal logic, *Theoret. Comput. Sci.* **38** (1985) 269–291.

[128] PNUELI, A. and R. ROSNER, On the synthesis of a reactive module, in: *Proc. 16th Ann. ACM Symp. on the Principles of Programming Languages* (1988) 179–190.

[129] VARDI, M. and L. STOCKMEYER, Improved upper and lower bounds for modal logics of programs, in: *Proc. 7th Ann. ACM Symp. on Theory of Computing* (1985) 240–251.

CHAPTER 17

Elements of Relational Database Theory

Paris C. KANELLAKIS

Department of Computer Science, Brown University, P.O. Box 1910, Providence, RI 02912, USA

Contents

1. Introduction 1075
2. The relational data model 1079
3. Dependencies and database scheme design 1112
4. Queries and database logic programs 1121
5. Discussion: other issues in relational database theory 1140
6. Conclusions 1144
Acknowledgment 1144
References 1144

HANDBOOK OF THEORETICAL COMPUTER SCIENCE
Edited by J. van Leeuwen

1. Introduction

The goal of this chapter is to provide a systematic and unifying introduction to relational database theory, including some of the recent developments in database logic programming. The first part covers the two basic components of the relational data model: its specification component, that is, the database scheme with dependencies, and its operational component, that is, the relational algebra query language. The choice of basic constructs, for specifying the semantically meaningful databases and for querying them, is justified through an in-depth investigation of their properties. Some important research themes are reviewed in this context: the analysis of the hypergraph syntax of a database scheme and the extensions of the query language using deduction or universal relation assumptions. The subsequent parts of the chapter are structured around the two fundamental concepts illustrated in the first part, namely dependencies and queries. The main themes of dependency theory are implication problems and applications to database scheme design. Queries are classified in a variety of ways, with emphasis on the connections between the expressibility of query languages, finite model theory and logic programming. The theory of queries is very much related to research on database logic programs, which are an elegant formalism for the study of the principles of knowledge base systems. The optimization of such programs involves both techniques developed for the relational data model and new methods for analyzing recursive definitions. The exposition closes with a discussion of how relational database theory deals with the problems of complex objects, incomplete information and database updates.

1.1. Some motivation and historical remarks

The practical need for efficient manipulation of large amounts of structured information and the basic insight that "data should be treated as an integrated resource, independent of application programs" led to the development of database management as an important area of computer science research. Database research has had a major impact on software systems and computer science in general; it has provided one of the few paradigms of man–machine interaction which is both of a very high level, akin to programming in logic, and computationally efficient. *Database theory* grew as the theory corresponding to and directly influencing a number of *database management system* (DBMS) implementations.

Database technology presents the theoretical community with challenging research questions, which can be classified into three broad categories: problems of *relational database theory*, problems of *transaction processing* and problems of *access methods.* The unity of research in these dissimilar areas has been provided by the database management systems themselves, which are large integrated software systems addressing issues in all three areas. Of the three facets of database theory, relational theory is the one more closely identified with the field. It is also the subject of our presentation and it may be viewed as an important application of mathematical logic to computer science problems. The emphasis of the other two facets of database theory is on algorithms and data structures; and these are outside the scope of our exposition.

For a view of the subject in its entirety we refer the reader to [280], an advanced textbook on databases, and to [182], an introductory-level one. Relational database theory is the topic of [201], a monograph which can be used to fill in many of the proofs omitted here. Transaction processing, e.g., *database concurrency control* and *recovery*, has many connections to the theory of operating systems and to distributed algorithms. We recommend two recent monographs in this area: [231] for the theoretical development and [57] for a more applied treatment. The last branch of this threefold classification, the study of access methods, is the area closest to the theory of data structures and is best reviewed in that context. Its distinguishing characteristic is an emphasis on large volumes of structured data residing in secondary storage; see [280].

The pioneering work of E.F. Codd, in the early 1970s, offered both the theoretical and practical communities an elegant, well-motivated and appropriately restricted computational paradigm for database management: *the relational data model.* Much of the systems effort in the 1970s was devoted to efficient implementations of this approach; a largely successful endeavor, which has made relational DBMSs a database technology standard. For historical perspectives on the relational data model and on the interplay of theory and practice we recommend [79, 267, 279].

The relational data model is the cornerstone of relational database theory, which was to a large extent developed by analyzing and enriching this original kernel. In this framework, a *database* is a set of (finite) relations interpreting a set of nonlogical relation symbols, the *database scheme.* Relations are manipulated using the "procedural" *relational algebra* query language or its equivalent "declarative" *relational calculus.* The duality of algebra and calculus is based on an algebraization of first-order definitions. The addition of *recursion* (via fixpoints or equivalently via deduction) to first-order definitions leads to more expressive query languages, which are related to logic programming. Database query programs have been studied extensively. Their optimization based on "compile-time" transformations and efficient "run-time" evaluation is crucial, if one is to map a high-level programming paradigm into a computationally feasible one.

The querying capability of the relational data model (i.e., the definition of functions from databases to databases) is complemented by a capability for specification of "the legal" databases (i.e., the definition of sets of databases). This allows constraining the databases under consideration and, consequently, expressing more knowledge about the objects represented. Codd defined *functional dependencies* as a useful device for this purpose. *Dependencies*, in general, are semantically meaningful and syntactically restricted sentences of the predicate calculus that must be satisfied by any "legal" database. Their presence remedies some of the semantic poverty of relations, e.g., with pure relations one has trouble representing the fact that some relationships are one-to-one or one-to-many. Studies of the decision problem and other computational properties of dependencies have been motivated by questions of good database scheme design. Interestingly, they have also contributed to basic research in mathematical logic.

Theoretical research on the relational data model itself attained a certain degree of maturity in the late 1970s and early 1980s. The efficiency and robustness provided by the relational DBMSs of the 1970s opened the way to more ambitious goals. Current research on logic and databases aims at developing logic programming into a database tool of comparable efficiency. Here the term *knowledge base* is appropriate given the potential new applications. This has brought the logic programming and database communities into closer contact and has stimulated new research directions for theoretical computer science. For a historical perspective on the relationship between the two fields we recommend [123, 124, 207, 215, 216].

In order to highlight the unity of the subject, we structure our presentation around the two basic concepts of dependencies and queries. This allows us to introduce the current work on logic and databases as the natural evolution of the first investigations into the relational data model.

For more details on dependencies and queries, we recommend the surveys [117, 65] respectively. We stress the contributions of database theory to mathematical logic and, in particular, to finite model theory; see [140, 141]. Space limitations do not allow a detailed examination of many interesting questions. Fortunately, we can refer the reader to a number of additional recent surveys: [294] for dependency theory, [168] for relational algebra optimization, [32] for database logic program evaluation, [172] for queries and parallel computation, and [163] for queries and the expressibility of logics.

Despite its elegance, relational theory does not give satisfactory answers to many database problems. This is, in part, the price one has to pay for the simplicity of the relational data model. For example, it is often easier to define hierarchical structures directly, than indirectly through dependencies. In order to represent and manipulate *complex objects* one can extend the data type relation; this is usually done by introducing *set* and *tuple constructors*. We discuss some of the work on complex objects and recommend [155] for a recent survey of this area, as well as [156] for further readings on *semantic data models*. We also discuss two other issues of practical significance: querying *incomplete information* databases and *updating* a database, i.e., coping with uncertainty and with dynamic change. In both cases relational database theory offers interesting solutions.

Relational formalisms for DBMSs have the advantage of being "value-oriented" [279]. That is, a clear separation is achieved between the logical and the physical description of the data and access is specified at the logical level by value. This is also in the spirit of logic programming. At present, there is a growing interest in alternative "object-oriented" approaches, e.g., see [29].

"Object-oriented" approaches emphasize the organization of the data into *inheritance* hierarchies of meaningful *abstract data types*: *objects that encapsulate both the data and the operations for accessing it.* There is a fair amount of ongoing experimentation in object-oriented DBMSs. Some of this work realizes earlier semantic data model proposals. Some of it may be viewed as design of programming languages with types that *persist* beyond the scope of programs. What is the right formal model for the object-oriented framework is an interesting open question. Its solution will (most probably) combine elements of relational database theory and of the theory of abstract data types.

1.2. *A roadmap of the contents*

A brief description of the material is now in order.

A detailed overview of the relational data model is contained in Section 2. After some basic definitions, in Section 2.1, we attempt to clarify the two choices made: of query language and dependencies. Section 2.2 is a three-part justification of relational algebra based on (1) its equivalence to the relational calculus, (2) its low computational complexity, and (3) its potential for query optimization via the homomorphism technique. The three-part argument for functional dependencies, in Section 2.3, has analogous goals and is based on (1) the elegant axiomatization of these special sentences, (2) their polynomial-time computational properties, and (3) the preservation of many of these properties when other semantically meaningful statements are added. The attribute notation is a particular feature of relational database theory; the database scheme is a hypergraph on a set of attributes. In Section 2.4, we illustrate some of the advantages of this notation. Hypergraph acyclicity is a simple means to test syntactic conditions on the database scheme, which implies a host of interesting semantic properties for the database. We take a critical look at the limitations of the relational data model in Section 2.5. We describe two extensions to the basic framework: deductive and universal relation data models. We argue why these extensions are related and comment on the wealth of analysis about them.

Dependency theory is the subject of Section 3. The classification of dependencies in Section 3.1 focuses on some important classes of statements and on the main computational question: testing for dependency implication. This problem is closely related to the decision problem in logic. In Section 3.2, we examine database scheme design: from (1) defining schemes with desirable semantic properties such as independence, to (2) listing normal form schemes that have these properties and are also constructible.

In deductive and universal relation data models, dependencies assume (indirectly) some of the functionality of query languages. In Section 4 we come back to the study of extensions of the relational data model, but through a direct investigation of the expressive power of query language primitives. The classification of queries in Section 4.1 proceeds: (1) along the lines of computational complexity, and (2) based on the expressibility of logics over finite structures, fixpoint logics in particular. An important and robust class of queries are the fixpoint queries, which correspond to the functions expressible in many database logic programming formalisms. Their analysis is at the heart of the topical interest in knowledge bases. In Section 4.2, we identify some of the emerging themes in this area: (1) logic programs without function symbols but which may have negation (we call them recursive rules), (2) optimization of recursive rules via the stage function technique, and (3) basic "top-down" versus "bottom-up" tradeoffs in recursive rule evaluation. We close our discussion of queries, in Section 4.3, with references to the work on complex object data models.

Section 5 is a discussion of how relational database theory deals with incomplete information and updates. These are topics that have attracted a fair amount of attention. Unfortunately, they do not seem to fit in as tidy a framework as queries and dependencies. These two topics, together with the open questions of query and dependency theory, account for much of the ongoing research in relational database theory by the end of the 1980s.

References to the literature are given in the text and there are additional bibliographic comments at the ends of sections. As with any presentation of this type, the choice of material has to be selective and the list of references is by no means comprehensive.

2. The relational data model

2.1. Relational algebra and functional dependencies

We start with some definitions and notation that are basic in database theory. We then introduce the syntax and the semantics of relational algebra and of functional dependencies.

Let $\mathscr{U}$ be a countably infinite set of *attributes*. We denote attributes using capital letters from the beginning of the alphabet, $A, B, C, A_1, \ldots,$ and capital letters from the end of the alphabet, $X, Y, Z, X_1, \ldots,$ to denote sets of attributes. We usually do not distinguish between attribute A and attribute set $\{A\}$. Notation XY is a convenient shorthand for the union of attribute sets X and Y. Thus, ABC denotes the attribute set $\{A, B, C\}$.

Every attribute A has an associated set of *values* $\Delta[A]$, called A's *domain. The domain* is the set of values Δ, the union of the domains of all the attributes. Δ is countably infinite and disjoint from $\mathscr{U}$. We denote values using small letters, $a, b, c, a_1, \ldots,$ from the beginning of the alphabet. Throughout our exposition we fix $\mathscr{U}$ and Δ, moreover, we assume that $\Delta = \Delta[A]$ for each A in $\mathscr{U}$. This is for notational simplicity only; under some weak technical conditions, the theory can be developed for any $\Delta[A]$'s (see Remark 2.1.4).

The following definitions highlight the clean distinction made in the relational data model between *relation schemes* and *relations* (the latter are sometimes called *relation instances* or *states* in the literature). This distinction parallels the one in mathematical logic between the syntactic concept of *vocabulary of relation symbols* and the semantic concept of *structure over this vocabulary.*

2.1.1. DEFINITION. The *universe* U is a finite subset of $\mathscr{U}$. A *relation scheme* R is a subset of U. A *database scheme* D over U is a set of relation schemes with union U.

2.1.2. DEFINITION. Let D be a database scheme over U, R a relation scheme and X a subset of U. An *X-tuple* t is a mapping from X into Δ, such that each A in X is mapped to an element of $\Delta[A]$. A *relation* r over R is a finite set of R-tuples. A *database* d over D is a set of relations; one relation over each relation scheme of D.

We omit the qualifications (over ...) in the above definitions when they are clear from the context; also we use tuple instead of X-tuple if X is understood. We use $D, D_1, \ldots,$ for database schemes and $R, R_1, \ldots,$ for relation schemes. We denote tuples by $t, t_1, \ldots,$ relations by $r, r_1, \ldots,$ and databases by $d, d_1, \ldots$; finally, we use the notation $\alpha(r)$ for the relation scheme of relation r and $\alpha(d)$ for the database scheme of database d. A relation r over R is represented in figures by a "table" with "columns" named with the attributes of R and with the tuples as "rows".

Note that in database theory we define a relation (a database) to be a finite set. There are situations where it is useful to adopt a more liberal view and allow *unrestricted relations* (*unrestricted databases*), i.e., both finite and infinite sets of tuples. In these cases, we will explicitly use the term unrestricted.

2.1.3. REMARK. Attribute notation is the norm in database theory literature and is quite useful. In Section 2.4, we will see some of its advantages. A relation scheme R such that $|R|=n$ for some integer $n \geqslant 0$, and a relation symbol in logic with *arity n*, are very similar concepts. The only minor difference is whether to use attributes or numbers to name "columns" of relations. The choice is largely a matter of convenience and there is the obvious translation based on some arbitrary ordering of the attributes. Hence, notation R is the only symbol used here in an overloaded fashion, which is always disambiguated by the context. We use R either as a relation scheme, or as a relation symbol, or as a relation variable.

We first define the operational component of the relational data model. This is the *relational algebra query language* proposed by Codd in [75]. The programs of this language are algebraic expressions, which denote mappings between databases called *queries*. The language is based on a small number of primitive operations on relations, which we now describe as follows. In clause (1) we have the *scheme restrictions* required of the *argument* relation schemes and the scheme of the *result* relation; in clause (2) we have the semantics of each operation. It is easy to see that these operations are well defined on relations, i.e., the result is a relation provided the scheme restrictions are satisfied.

Let t be an X-tuple and Z a subset of X, then the *projection* of t on Z, denoted $t[Z]$, is the restriction of the mapping t on Z.

PROJECTION. $\pi_X(r)$ is the projection of r on X.
(1) $X \subseteq \alpha(r)$ and $\alpha(\pi_X(r))=X$.
(2) $\pi_X(r)=\{t[X]: t \in r\}$.

NATURAL JOIN. $r_1 \bowtie r_2$ is the (natural) join of r_1 and r_2.
(1) no scheme restriction and $\alpha(r_1 \bowtie r_2)=\alpha(r_1) \cup \alpha(r_2)$.
(2) $r_1 \bowtie r_2=\{t: t$ is an $(\alpha(r_1) \cup \alpha(r_2))$-tuple such that $t[\alpha(r_1)] \in r_1$ and $t[\alpha(r_2)] \in r_2\}$.

UNION. $r_1 \cup r_2$ is the union of r_1 and r_2.
(1) $\alpha(r_1)=\alpha(r_2)$ and $\alpha(r_1 \cup r_2)=\alpha(r_1)$.
(2) $r_1 \cup r_2=\{t: t \in r_1$ or $t \in r_2\}$.

DIFFERENCE. r_1-r_2 is the difference of r_1 minus r_2.
(1) $\alpha(r_1)=\alpha(r_2)$ and $\alpha(r_1-r_2)=\alpha(r_1)$.
(2) $r_1-r_2=\{t: t \in r_1$ and $t \notin r_2\}$.

SELECTION. $\varsigma_{A=B}(r)$ is the selection on r by $A=B$.
(1) $A, B \in \alpha(r)$ and $\alpha(\varsigma_{A=B}(r))=\alpha(r)$.
(2) $\varsigma_{A=B}(r)=\{t: t \in r$ and $t[A]=t[B]\}$.

RENAMING. $\varrho_{B|A}(r)$ is the renaming in r of A into B.

(1) $A \in \alpha(r)$, $B \notin \alpha(r)$ and $\alpha(\varrho_{B|A}(r)) = (\alpha(r) - \{A\}) \cup \{B\}$.

(2) $\varrho_{B|A}(r) = \{t\colon \text{for some } t' \in r,\ t[B] = t'[A] \text{ and } t[C] = t'[C] \text{ when } C \neq B\}$.

A feature that we have omitted from our relational algebra is *constants*, that is, symbols representing the elements of the domain. The presence of constants does not affect the essentials of database theory and they can always be added with a certain degree of notational difficulty, see [66, 67]. We use query languages without constants throughout our exposition; except in Section 4.2.3 where we study algebraic properties of the selections on constants, $\varsigma_{A=a}$. (The symbol ς is used for selection, since the more familiar symbol σ will be used to denote a dependency.)

2.1.4. REMARK. Note that renaming allows the introduction of attributes not in $\alpha(r)$; this is important for the algebraization theorem of the next section. We could have defined the operations without scheme restrictions by making the result empty if the restrictions are violated (see [64]). The conventions we have chosen instead are quite common in the literature and do not affect any of the theorems. In the scheme restriction for renaming, by our assumption about the domain, $\Delta = \Delta[A] = \Delta[B]$. In general, a technical condition is used for the set $\mathscr{U}$ of attributes (see [158]): "For each attribute A, there is an infinite sequence of attributes $A_1, A_2, \ldots$, with the same attribute domain as A. These are the attributes A can be renamed into".

2.1.5. DEFINITION. Let $D = \{R_1, \ldots, R_m\}$ be a database scheme, then the *relational algebra expressions over D* are the expressions E generated by the following grammar, subject to the scheme restrictions:

$$E := R_1 \mid \cdots \mid R_m \mid \pi_X(E) \mid (E \bowtie E) \mid (E \cup E) \mid (E - E) \mid \varsigma_{A=B}(E) \mid \varrho_{B|A}\ \ (E).$$

It is clear that one may represent a relational algebra expression over D as a tree, i.e., the parse tree of its generation. Each internal node of this tree is labeled by a relational operation and each leaf by a relation scheme of D over universe U. For each expression there is an associated relation scheme $\alpha(E)$, easily determined from the scheme restrictions. (Note that, because of renaming, there might be attributes in $\alpha(E)$ that are not in U). The symbols of a relational expression are syntactic symbols, e.g., R_i is a relation variable. These symbols are interpreted in the following intuitive way.

A relational algebra expression E over D denotes a function from databases over D to databases over $\{\alpha(E)\}$. This (total) function is defined via a "bottom-up" evaluation of the *parse tree* of E. On *input* database d over D associate: (1) with each leaf of the tree with label R, the relation r over R of d, and (2) with each internal node of the tree, the result of its label operation with argument relations the relations associated with its successor nodes. The *output* $E(d)$ is the database of one relation associated with the tree's root.

If d consists of one relation r, we use the abbreviation $r' = E(r)$ for $\{r'\} = E(d) = E(\{r\})$. Let us illustrate Definitions 2.1.1, 2.1.2 and 2.1.5 with an example.

2.1.6. EXAMPLE. Consider database scheme $D = \{R\}$, where the universe U is the same as the relation scheme $R = AB$. One can think of relation r over R (the only relation in our database d over D) as a directed graph. This graph has no isolated nodes, its arcs are the tuples of r and its nodes are the unary relation (i.e., a set) $\pi_A(r) \cup \pi_B(r)$. Let E, E_1, E_2, and E_3 be the relational algebra expressions:

$$E = R, \qquad E_1 = \varrho_{B|A}(\varrho_{C|B}(R)), \qquad E_2 = (R \bowtie \varrho_{B|A}(\varrho_{C|B}(R))),$$

$$E_3 = (R \cup \varrho_{B|C}(\pi_{AC}(R \bowtie \varrho_{B|A}(\varrho_{C|B}(R))))).$$

Note that R is used as a relational variable and all these expressions are *subexpressions* of E_3, i.e., their parse trees are subtrees of its parse tree.

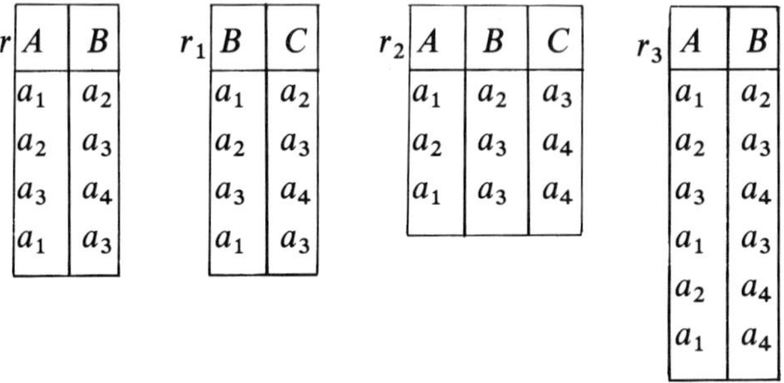

r: A	B
a_1	a_2
a_2	a_3
a_3	a_4
a_1	a_3

r_1: B	C
a_1	a_2
a_2	a_3
a_3	a_4
a_1	a_3

r_2: A	B	C
a_1	a_2	a_3
a_2	a_3	a_4
a_1	a_3	a_4

r_3: A	B
a_1	a_2
a_2	a_3
a_3	a_4
a_1	a_3
a_2	a_4
a_1	a_4

Fig. 1. The relations of Example 2.1.6.

Let r be as in Fig. 1, then we have $r = E(r)$. Also, $r_i = E_i(r)$, for $i = 1, 2, 3$, the other relations in Fig. 1. Note how renaming just manipulates the relation scheme, how join works and how duplicates are eliminated when projection or union are performed. To see the intuitive meaning of these expressions let us examine $r_3 = E_3(r)$. For any relation r, the relation r_3 consists of those pairs of nodes $\langle a, b \rangle$ of the graph represented by r such that there is a directed path of one or two arcs from a to b.

2.1.7. DEFINITION. Let E_1, E_2 be relational algebra expressions over D such that $\alpha(E_1) = \alpha(E_2)$. We say that E_1 is *contained* in E_2, denoted $E_1 \subseteq_e E_2$, if for all databases d over D we have that every tuple in $E_1(d)$ is also in $E_2(d)$. We say that E_1 and E_2 are *equivalent*, denoted $E_1 \equiv_e E_2$, if $E_1 \subseteq_e E_2$ and $E_2 \subseteq_e E_1$ hold.

It is possible to define other relational operations, besides projection, join, union, difference, selection and renaming. For instance, one can define *intersection* as the obvious set-theoretic intersection on relations with the same scheme. However, join on relations with the same scheme does exactly that. For a large repertoire of relational operations, e.g., *Cartesian-product, equi-join, division,* we can just use an equivalent expression built from the basic set of operations we chose. Another such example is *select* with conditions that are propositional (i.e., $\vee$, $\wedge$, $\neg$,) combinations of atoms of the form $A = B$. For detailed expositions of relational algebra we refer to any of the standard database textbooks, e.g., [201, 280].

Relational operations and relational algebra expressions can be defined as above (verbatim) for unrestricted relations. This allows us to define, as in Definition 2.1.7, *unrestricted containment* and *unrestricted equivalence*, which we will carefully distinguish from *containment* and *equivalence*. Testing for expression containment and equivalence are central computational problems for expression optimization. The unrestricted notions imply the finite ones (since for them d ranges over finite and infinite databases), but they are not necessarily the same. Let us see why.

Testing for containment is *co-recursively enumerable* (co-r.e.), by going through all possible inputs (i.e., all finite d's) and simultaneously checking for noncontainment via the bottom-up evaluation. But testing unrestricted containment is *recursively enumerable* (r.e.), by reducing it to the validity of a sentence of the *first-order* (f.o.) *predicate calculus with equality* [105] and using the Gödel Completeness Theorem. For this reduction one can use the correspondence between relational algebra expressions and f.o. formulas from Section 2.2.1. Thus, equality of the finite and unrestricted notions would imply decidability; the problems would be r.e. and co-r.e. and therefore *recursive*.

This equality is not true in general because unrestricted containment and equivalence of relational expressions are *undecidable*. This can be shown to follow from the undecidability of validity of f.o. sentences. For the undecidability of unrestricted containment of special subclasses of relational expressions see [158, 150]. Containment and equivalence of relational expressions are also undecidable; this can be shown using the undecidability of validity over finite structures of [276].

For many cases of interest, we have equality of the finite and unrestricted notions and thus *decidability*. For instance, one such case is for the expressions built using only projection and join, the class of *project-join* expressions. Database theory has developed around properties of project-join expressions. Here is an example, which highlights some important project-join containment properties.

2.1.8. EXAMPLE. The operation of join is *associative* and *commutative*. Therefore, if $r_1, \ldots, r_m$ are relations respectively over $R_1, \ldots, R_m$ then the join $r_1 \bowtie \cdots \bowtie r_m$ is unambiguously defined. One property of this join is that $\pi_{R_j}(\bowtie\{r_i: 1 \leqslant i \leqslant m\}) \subseteq r_j$ for $1 \leqslant j \leqslant m$.

Let $D = \{R_1, \ldots, R_m\}$ be a database scheme, whose relation schemes have union R. If r is a relation over R, we call the set of relations $\{\pi_{R_1}(r), \ldots, \pi_{R_m}(r)\}$ a (full) *decomposition* of r, and we denote it by $\pi_D(r)$. We use $\bowtie\pi_D(r)$ for $\bowtie\{\pi_{R_i}(r): 1 \leqslant i \leqslant m\}$. An important property that holds for decompositions is that for all r, $r \subseteq \bowtie\pi_D(r)$.

If $r = \bowtie\pi_D(r)$ holds then r has a *lossless decomposition* over D, else it has a *lossy decomposition*. The idea here is that by decomposing relation r into its projections we lose no information when the decomposition is lossless, because then we can get the original relation back by joining the projections. The importance of lossless decompositions was identified in [13, 241].

In relational algebra expression notation, we have the following containments, i.e., these are *identities*: (1) $R \subseteq_e \bowtie\pi_D(R)$, see above, and (2) $\bowtie\pi_D(\bowtie\pi_D(R)) \equiv_e \bowtie\pi_D(R)$, idempotence.

We now turn to the specification component of the relational data model, the functional dependency (fd for short). This type of constraint can be thought of as a generalization of "keys-of-records". It was originally identified by Codd in [76]. A tuple of a relation represents a relationship among certain values, but by itself it does not provide any information about the nature of this relationship. For example, it would be useful to know that only relations representing functional relationships are acceptable. This is exactly what the addition of fd's accomplishes.

2.1.9. DEFINITION. Let R be a relation scheme and X, Y be subsets of R. An expression of the form $X \to Y$ is called a *functional dependency over* R (fd over R). The fd $\sigma = X \to Y$ is satisfied by relation r over R, when, for all tuples t_1, t_2 in r, if $t_1[X]=t_2[X]$ then $t_1[Y]=t_2[Y]$. Relation r satisfies a set Σ of fd's if r satisfies all the fd's in Σ.

A set Σ of fd's over R can be viewed as a semantic specification of a set of databases over database scheme $\{R\}$. That is, Σ specifies the set of databases satisfying Σ. Relations that do not satisfy some σ in Σ cannot be "possible worlds" of the application being modeled. We denote this specification as the pair $\langle \{R\}, \Sigma \rangle$. This basic specification mechanism is developed in the theory of *dependencies*, through extensions and variations of fd's. As we shall see, fd's can be represented as sentences of the f.o. predicate calculus with equality. Moreover, they have many interesting and qualitatively different representations.

2.1.10. EXAMPLE. Let R be ABC. Consider the relation r over $\alpha(r)=R$ in Fig. 2. It is easy to verify that r satisfies the set of fd's $\Sigma = \{A \to B,\ B \to C,\ C \to B\}$ and that it does not satisfy the fd's $B \to A$, $BC \to A$. To see an intuitive application of this example: think of A as EMPLOYEE, B as DEPARTMENT and C as MANAGER in a COMPANY database. Every employee works in a unique department, every department has a unique manager and every manager directs a unique department.

Two general concepts have emerged from the definitions and examples. A mapping between databases can be described by an expression of a query language. For these expressions testing containment is a fundamental problem. A set of databases can be specified by a database scheme coupled with dependencies. In the last example of this section, let us indicate how *containment equations* can be used instead of dependencies. This theme is developed in [305], which is an in-depth study of project-join expression containments.

2.1.11. EXAMPLE. The pair $\langle \{ABC\}, \{A \to B\} \rangle$ specifies the set of relations as follows: $FD = \{r : r$ over ABC and satisfying $A \to B\}$. Using containment we can also specify a set of relations $CN = \{r : r$ over ABC and satisfying $E(r) \subseteq E'(r)\}$, where E, E' are the following relational algebra expressions over $R = ABC$ with $\alpha(E) = \alpha(E') = BB'$:

$$E = \pi_{BB'}(\pi_{AB}(R) \bowtie \pi_{AB'}(\varrho_{B'|B}(R))), \qquad E' = \varsigma_{B=B'}(E).$$

A simple argument suffices to show that the two sets of relations FD and CN are identical. This technique can be generalized to any set of fd's.

r	A	B	C
	a	b	c
	a'	b	c
	a''	b'	c'

Fig. 2. The database of Example 2.1.10.

2.1.12. ADDITIONAL BIBLIOGRAPHIC COMMENTS. The definitions of the relational data model and the insight that they can be used as a foundation of database management are due to Codd [75–79]. The best known early pioneer of the idea of relations as databases is Childs [72]; see [279] where some of the prehistory of the field is investigated.

Relational algebra, as indicated in [158], can be studied as a *cylindric algebra.* There has been a great deal of work on the subject of cylindric algebras, much of it by Tarski and his school, see [150] for a standard exposition. Even in this well studied mathematical setting, the problems and the solutions of database theory are often novel. This is because of the practical motivation and the emphasis on computational complexity and on finite structures. Database theory has also contributed directly to the theory of cylindrical algebras, e.g., the *cylindric lattices* of [83].

2.2. *Why relational algebra?*

Our goal in this section is to demonstrate that the choice of relation as the only data type and of *manipulating this data type only via the relational algebra operations* is by no means ad hoc.

2.2.1. *Algebra = calculus*

The relational algebra query language is a typical "procedural" programming formalism. The basic observation of Codd [75] is that this "procedural" language has a very natural "declarative" counterpart, the *relational calculus query language.*

For the purposes of this presentation it suffices to describe the relational calculus in an informal but intuitive fashion. As we mentioned in Remark 2.1.3, relation scheme R and relation symbol R are similar concepts, modulo an ordering of the attributes. Under this simple translation the database scheme $\{R_1, \ldots, R_m\}$ becomes a vocabulary of relation symbols. We further assume that $\mathcal{V}$ is a countably infinite set of *variables,* disjoint from the attribute set $\mathcal{U}$ and the value set Δ (we denote variables using $x, y, z, x_1, \ldots$). For a detailed treatment of the relational calculus we refer to [201], which uses relation schemes and to [280], which uses relation symbols.

The *formulas* of the f.o. predicate calculus with equality over vocabulary $\{R_1, \ldots, R_m\}$ are built out of *atomic formulas* of the form $x = y$ and $R_i(x_1 \ldots x_{n_i})$, $1 \leqslant i \leqslant m$ (where R_i has arity $n_i \geqslant 0$) using the *propositional connectives* $\vee$, $\wedge$, $\neg$ and *quantifiers* $\forall x$, $\exists x$. *Free* and *bound* variables are defined in the standard fashion [105]; *sentences* are formulas without free variables. Let us denote by $\varphi(x_1, \ldots, x_n)$ a f.o. formula with exactly n *distinct* free variables $x_1, \ldots, x_n$.

There is a connection at the definition level between database theory and finite model theory. If d is a database then the set of values that occur in d is finite; let us call it δ. (To be technically consistent with mathematical logic if d is empty $\delta=\{a\}$ for some fixed value a. In fact, we could develop the theory using as the domain δ some superset of the values occurring in d, see Section 4.1.) Thus, given database d over $\{R_1,\ldots,R_m\}$ we have a finite f.o. relational structure $\langle\delta,d\rangle$.

Let $\varphi(x_1,\ldots,x_n)$ be a formula as above, let d be a database whose relations are *interpreting* the relation symbols $\{R_1,\ldots,R_m\}$, let δ be as above and let $a_1,\ldots,a_n$ be values from δ *interpreting* the n variables $x_1,\ldots,x_n$ respectively. We can define *satisfaction* of $\varphi(a_1,\ldots,a_n)$ by $\langle\delta,d\rangle$ in the standard f.o. fashion [105]; it is denoted by $\langle\delta,d\rangle\models\varphi(a_1,\ldots,a_n)$.

2.2.1. DEFINITION. A *relational calculus expression* F over $\{R_1,\ldots,R_m\}$, with $\alpha(F)=R$, is an expression of the form $\{R,x_1,\ldots,x_n\colon\varphi(x_1,\ldots,x_n)\}$, where R is a relation symbol of arity $n\geqslant 0$ and φ is a f.o. formula with exactly n distinct free variables. F defines a function from databases over $\{R_1,\ldots,R_m\}$ to databases over $\{R\}$ as follows: on input database d, with set of values δ, the output $F(d)$ is the database consisting of the relation $\{a_1\ldots a_n\colon\langle\delta,d\rangle\models\varphi(a_1,\ldots,a_n)\}$.

Query language expressions, in general, are used to denote functions of some *type*: from databases over $\{R_1,\ldots,R_m\}$ to databases over $\{R\}$. Using Definition 2.1.7 we can define containment and equivalence for expressions of the same type, e.g., $E\equiv_e F$ for a relational algebra expression E and a relational calculus expression F. Hence, when we mention expression equivalence or containment, we will imply that the types of the expressions are the same. Similar comments apply to unrestricted equivalence and containment.

2.2.2. THEOREM. *Every relational algebra expression can be translated, in time polynomial in its size, into an equivalent relational calculus expression. Every relational calculus expression can be translated, in time polynomial in its size, into an equivalent relational algebra expression.*

The proof of this theorem, known as Codd's Theorem [75, 77], is by structural induction on algebra and calculus expressions. The arguments are simple, but the theorem has great practical significance. The efficient translation of calculus into algebra reveals a "procedural" evaluation for the function defined "declaratively" by a calculus expression. The expressive power of other query languages can be assessed, based on whether they can express or not all the calculus or algebra queries. We will use the term PTIME *language equivalent* for the two conditions of Theorem 2.2.2. So, *relational algebra and relational calculus are* PTIME *language equivalent.*

For an example, the relational expression E_3 from Example 2.1.6 is equivalent and unrestricted equivalent to the relational calculus expression $\{R',x,y\colon R(xy)\vee\exists z(R(xz)\wedge R(zy))\}$.

Let d be an unrestricted database. The satisfaction $\langle\Delta,d\rangle\models\varphi(a_1,\ldots,a_n)$ is defined as in the finite case, only Δ is used instead of δ and $a_1,\ldots,a_n$ are from Δ. If we keep the syntax of relational calculus, but use Δ instead of δ for the semantics, we have *the relational calculus with unrestricted domain.* This is a query language for unrestricted databases, because the outputs could be infinite even for finite inputs. To see this consider $\{R, x\colon \neg R(x)\}$, where if the input is finite, the output is infinite because negation is with respect to Δ.

2.2.3. REMARK. For the relational calculus with unrestricted domain there is a theorem analogous to Theorem 2.2.2: *The relational calculus with unrestricted domain and the relational algebra with "complement" are* PTIME *language unrestricted equivalent.* In this case we have to extend relational algebra on unrestricted databases with the operation of *complement*: $\neg(r) = \{t\colon t \notin r\}$, where $\alpha(\neg(r)) = \alpha(r)$. Both these query languages are over unrestricted databases, so we use unrestricted equivalence.

There is one flaw with the relational calculus. For some expression F it might be possible to find a database input d such that the output is different in the relational calculus from the relational calculus with unrestricted domain. In these cases F is called *domain-dependent* and using Δ instead of δ matters, e.g., in $F = \{R, x\colon \neg R(x)\}$. If no such databases d exist, then F is called *domain-independent.* Domain-independent formulas have been proposed as a class of "well-formed" relational calculus expressions. The rationale here is that when one evaluates a domain-dependent formula, one usually computes all of δ; this can be wasteful even if feasible, and should be avoided if possible.

Unfortunately, the class of domain-independent formulas is not recursive [99, 288]. Fortunately, they have recursive subclasses for which Theorem 2.2.2 can be refined. One example, are the *safe formulas* of [280]. For these one can show: *The relational calculus, the safe relational calculus and the relational algebra are all* PTIME *language equivalent* [201, 280]. Note that domain-independent formulas are defined semantically, whereas safe formulas are defined syntactically. Another interesting fact is that *given a relational calculus expression F and a database d, it is decidable if the output is the same in the relational calculus and in the relational calculus with unrestricted domain,* see [27].

Codd's Theorem can be specialized to subsets of the relational calculus and algebra. The *positive existential calculus* of [67] consists of the relational calculus expressions built with $\exists$, $\vee$, $\wedge$, $=$ but without $\forall$, $\neg$. The relational algebra without difference consists of the relational algebra expressions built using all the defining operations of the algebra, except for difference.

2.2.4. THEOREM. *The positive existential relational calculus and the relational algebra without difference are* PTIME *language equivalent.*

2.2.5. ADDITIONAL BIBLIOGRAPHIC COMMENTS. Kuhns in [184] was among the first to propose using the predicate calculus as a database query language. Codd's Theorem originally appeared in [75, 77] for an algebra and a calculus with constants, comparators ($\leqslant$), and with variables in the calculus ranging over tuples. We presented a pure version of these languages (without constants or $\leqslant$) and a domain as opposed to a tuple relational calculus.

Let us comment on the relationship between Codd's Theorem and cylindric algebras.

A version of this theorem was known early on in the logic community, e.g., [73, 274]. It is used to show Tarski's Algebraization Theorem, see [150]. Tarski's Algebraization Theorem provides a more general "algebraization" of the f.o. predicate calculus with equality, because the data types manipulated are not restricted to be relations and the properties of the operations are specified equationally. However, it was only after Codd's seminal papers that the practical implications of this work became clear.

Theorem 2.2.2 should be properly viewed as one instance of a theorem scheme. For example, an analogous theorem holds for a relational algebra with aggregate operations [178], e.g., with operations max, min, etc. Similar theorems can be shown for other data models, see [1, 92, 187]. In the area of complex object data models, algebraization has been a central theme, see [155] for a survey and [234, 157, 188, 286] for recent developments in the area. Algebra = calculus theorems can be shown for subsets of the predicate calculus as in Theorem 2.2.4. For instance, an algebra for the f.o. predicate calculus without equality can be found in [64].

The notion of domain-independent formulas is due to Fagin in [111] (the definition used here is a variation of the original one). We refer the reader to [111] for a review of a number of syntactically defined sets of formulas that can be used instead of the safe formulas of [280]. A recent analysis of domain independence can be found in [285].

2.2.2. LOGSPACE data complexity

We assume some familiarity with complexity classes for sequential computation such as the classes LOGSPACE, NLOGSPACE, PTIME, NP (NPTIME), PHIER, PSPACE, EXPTIME, with the class NC for parallel computation, and with the notion of logspace-completeness in a complexity class; see the surveys [125, 80]. All these classes are properly contained in EXPTIME. The following containments ($\subseteq$) are conjectured to be proper ($\subset$).

$$\begin{aligned}\text{LOGSPACE} \subseteq \text{NLOGSPACE} \subseteq \text{NC} \subseteq \text{PTIME} \subseteq \text{NPTIME}\\ \subseteq \text{PHIER} \subseteq \text{PSPACE} \subset \text{EXPTIME}.\end{aligned}$$

Fagin's work in [106] was crucial in establishing the first link between computational complexity and finite model theory (and thus database theory) through a logical characterization of NP. We will return to this subject in Section 4, where we examine a whole spectrum of possible query languages; see also the survey [163]. In this subsection, we introduce the basic definition of data complexity from [67] and we examine the data complexity of relational algebra queries.

Query language expressions denote queries, i.e., functions from databases to databases. What is the computational complexity of these functions? Let us assume that a database d, represented in some standardized binary encoding, has size $|d|$. This size typically dominates, by many orders of magnitude, the size of a query expression and is therefore the asymptotic parameter of interest. In other words, we may assume that the query expression E has size bounded by some constant, since we can exhaustively analyze it before we apply it to the database.

2.2.6. DEFINITION. The *data complexity* of the query denoted by expression E is the computational complexity of testing membership in the set $\{\langle t, d\rangle$: tuple t is in $E(d)\}$.

We say that a query is in a computational complexity class if its membership problem is in this class; and that it is logspace-complete in this class if its membership problem is.

The following observation complements the effective translation of Theorem 2.2.2, by emphasizing that the evaluation can be performed efficiently with respect to the database. Given the potential large size of the database, PTIME algorithms are the only "reasonable" ones. Among them, NC algorithms are "preferable" because they facilitate the use of parallel processing.

2.2.7. THEOREM. *Relational algebra or calculus queries are in* LOGSPACE.

Note that LOGSPACE$\subseteq$NC, so this theorem states that there is a lot of potential parallelism in relational algebra queries. Much of the work on access methods realizes this promise, e.g., very efficient join algorithms for external storage. To prove Theorem 2.2.7, one uses the relational operation semantics and the fact that the width of the tables processed is fixed. The data complexity for many query languages, both declarative and procedural, is determined in [291].

There is a price to pay for the low data complexity. This is the limited expressive power of relational algebra or calculus. Let r be a relation representing an undirected graph and $Trans(r)$ its transitive closure. The *transitive closure* query maps r to $Trans(r)$ for all r.

2.2.8. THEOREM. *The transitive closure query is not expressible in relational algebra or calculus.*

For unrestricted relations this theorem is still true and has an easy proof using the Compactness Theorem of first-order logic [105]. Unfortunately, compactness and other useful properties fail when attention is restricted to finite structures only [140, 141]. Theorem 2.2.8 was first shown in [107], in a stronger form using Ehrenfeucht–Fraisse game techniques. The stronger form is: *it is not possible to characterize connected graphs using a monadic existential second-order sentence over finite structures* (i.e., the second-order quantifiers are over monadic relations, they form a prefix of the sentence and they are all existential). Other proofs appear in [16, 67, 119, 121, 160].

2.2.9. ADDITIONAL BIBLIOGRAPHIC COMMENTS. Theorem 2.2.8 has been generalized to certain classes of queries, [42, 84]. Ajtai and Gurevich have recently announced that it holds for all "unbounded Datalog programs" (see Section 4.2.2). The difficulty here is the finiteness assumption, which requires the development of tools to replace the Compactness Theorem of first-order logic.

Data complexity is not the only measure of interest. If we assume that the query expression is part of the input then we have the notion of *expression complexity*. Vardi has shown in [291] that expression complexity is typically one exponential higher than data complexity. The expression complexity of project-join expressions is further investigated in [81, 153, 205].

From these investigations one can infer that relational operations are less structured than common integer operations, exponentiation included. This is because, for relational algebra expressions, intermediate results of a computation can be much larger than both inputs and outputs. Circuits of relational operations are compared to Boolean circuits in [304].

2.2.3. Query optimization and homomorphisms

The procedural nature of a query language facilitates the "compile-time" optimization of programs. Query optimization has been used to implement the relational data model in a reasonably efficient fashion, e.g., [23, 268, 302]. In these implementations, a relational algebra expression is algorithmically transformed into an equivalent expression that is better by some measure of complexity. The subject of query optimization flourished in the 1970s and is still an active area of research; see [168] for a survey.

Two optimization ideas occur most often in the literature. We summarize them in the two bullets below. The rationale for these is as follows. The join has been identified as the most expensive relational operation and its cost grows with the size of the argument relations. Thus, it is preferable to apply it on relations after these have been reduced in size through selection. Also, it pays to perform as few joins as possible.

- The first optimization idea is the algebraic rewriting of expressions, with emphasis on the *propagation of selections before the joins* [16]. An early use of this idea appears in [230]. A companion problem, usually solved by dynamic programming, is to find the *best order of join evaluations* in an expression $\bowtie\{r_i: 1 \leqslant i \leqslant m\}$.
- The second idea is the *minimization of the total number of joins.*

We will defer the first idea to the context of more powerful query languages in Section 4.2.3. Here, we focus on the second one, which reveals the importance of *homomorphism* techniques for expression containment and database theory in general. Let us proceed with the formalization of the second question.

The set of variables $\mathscr{V}$ can be partitioned into disjoint subsets, one for each attribute in $\mathscr{U}$. Each attribute A corresponds to $\mathscr{V}[A] = \{x, x_1, x_2, \ldots\}$; x, the lexicographically first one of these variables, is called *distinguished* and the others *nondistinguished.* We also extend the notion of a *tuple* from a mapping into Δ to a mapping into $\Delta \cup \mathscr{V}$.

2.2.10. Definition. A *tableau* T over relation scheme R is a set of R-tuples such that if tuple t is in T then $t[A]$ is in $\mathscr{V}[A]$. The *target relation scheme* $\alpha(T)$ of T is the set $\{A$: for some t in T, $t[A]$ is distinguished$\}$. The *summary* t_T of T is the $\alpha(T)$-tuple of its distinguished variables.

For examples of tableaux see Fig. 3 in Example 2.3.12. These examples appear in Section 2.3, in order to emphasize the intimate connections between tableaux and dependencies.

There are two ways of thinking about tableaux:

(1) The first way is to consider them as relations over R with domain $\mathscr{V}$. In this case we can think of $T_1 \subseteq T_2$ as a set inclusion between two relations over the same scheme R.

(2) The second way of thinking about tableaux is as expressions of the *tableau query language.*

A tableau T denotes a (total) function from relations over R to relations over $\alpha(T)$ as follows: A *valuation* h is a mapping from $\mathscr{V}$ into Δ; valuation h can be extended to tuples componentwise. On input relation r over R, the output $T(r)$ is a relation over $\alpha(T)$, where

$$T(r)=\{h(t_T): h \text{ is a valuation such that } h(t) \text{ is in } r \text{ for each tuple } t \text{ of } T\}.$$

As expressions, tableaux define mappings from single-relation databases to single-relation databases. Expression equivalence and containment in this case are denoted by $T_2 \subseteq_e T_1$ and $T_2 \equiv_e T_1$, where we must have $\alpha(T_1)=\alpha(T_2)$.

A subtle point is that $\subseteq$ and $\subseteq_e$ are different concepts. In fact, if $T_1 \subseteq T_2$ and $\alpha(T_1)=\alpha(T_2)$ then $T_2 \subseteq_e T_1$; this is no coincidence.

A *homomorphism* h from T_1 to T_2 is a mapping from $\mathscr{V}$ into $\mathscr{V}$ such that

(1) $h(x)=x$ for x distinguished, and

(2) $h(T_1)\subseteq T_2$ for $h(T_1)=\{h(t): t$ a tuple in T_1, h extending componentwise$\}$.

For example, if $T_1 \subseteq T_2$, then there is a homomorphism defined by the identity mapping from T_1 to itself as part of T_2. Note the similarity of definition between a valuation and a homomorphism; it is easy to see that a homomorphism composed with a valuation is a valuation. Using composition and both interpretations of tableaux, one can show the following simple but central fact in database theory.

2.2.11. THEOREM. *For two tableaux T_1, T_2 with the same summary we have that $T_2 \subseteq_e T_1$ iff there is a homomorphism from T_1 to T_2.*

The homomorphism technique was first developed for *conjunctive queries* in [70]. Conjunctive queries are a superset of tableau queries; see Remark 2.2.15 below. The basic results for tableau optimization (Theorems 2.2.11–2.2.13) are from [14, 15]. The following two theorems illustrate the possible optimizations of relational expressions, using the tableau as a tool. To illustrate the expressive power of the tableau language we consider project-join relational expressions. There are tableau queries that cannot be expressed as project-join ones; see [305] for the precise expressive power of tableaux.

2.2.12. THEOREM. *For every project-join expression over $\{R\}$ there is an equivalent tableau T over R constructed recursively as follows:*

(1) *if $E=R$ then T is the tableau with one R-tuple of all distinguished variables,*

(2) *if $E=\pi_X(E_1)$ and T_1 is the tableau of E_1 then T is obtained from T_1 by changing each distinguished symbol x in $\mathscr{V}[A]$ such that $A \notin X$ into a new nondistinguished symbol, and*

(3) *If $E=(E_1 \bowtie E_2)$ and T_1, T_2 are the tableaux of E_1, E_2 then T is the union of the sets of tuples of T_1, T_2.*

2.2.13. THEOREM. *Each tableau T has a minimal subset of its tuples $T_{\min}$ equivalent to T. This $T_{\min}$ is unique up to renaming of the nondistinguished symbols. If T is the tableau of the project-join expression E then the $T_{\min}$ above is the tableau of a project-join expression equivalent to E with a minimum number of joins.*

Tableau minimization is NP-complete. This follows from the results of [70]. Tighter bounds, e.g., for project-join expressions, are derived in [14, 15] as well as syntactic constraints on tableaux under which the problem can be solved in PTIME. Thus, in general, to minimize the number of joins, one might have to examine all homomorphisms from T to T; this is not as bad as it sounds because T is an expression and not a database.

Theorem 2.2.11 can be extended from a single tableau to sets of tableaux, all with the same summary. The output for such a set of tableaux is the union of the outputs of the single tableau queries. The following theorem is from [260].

2.2.14. THEOREM. $\{T_{21}, \ldots, T_{2m}\} \subseteq_e \{T_{11}, \ldots, T_{1n}\}$ *iff for each* T_{2j}, $1 \leqslant j \leqslant m$, *there exists a* T_{1i}, $1 \leqslant i \leqslant n$, *such that* $T_{2j} \subseteq_e T_{1i}$.

2.2.15. REMARK. As expressions, tableaux have typed variables and a single relation input. We say that they are *typed and untagged*, where tags would correspond to multiple relation inputs. The above development can be carried out for untyped variables and many relation inputs; such an expression is called an *untyped and tagged tableau* and is essentially the same as a conjunctive query of [70]. The typed untagged case is particularly important in the theory of dependencies. On the other hand, unions of untyped tagged tableaux express the positive existential queries (see Theorem 2.2.4). This easily follows from putting positive existential formulas in disjunctive normal form. A consequence is that there is an algorithm for testing positive existential query containment.

Removing the typing does not affect any of Theorems 2.2.11–2.2.14. Untyped tableau queries T' are monotone, that is, for all relations r_1, r_2 we have that $r_1 \subseteq r_2$ entails $T'(r_1) \subseteq T'(r_2)$. In addition, for a tableau T and for all relations r we have that $r \subseteq T(r)$; because there is always a homomorphism from T to the identity tableau (one tuple of all distinguished variables). Typing does, however, account for some additional special structure. A good source for the many properties that result from typing is [115].

2.2.16. ADDITIONAL BIBLIOGRAPHIC COMMENTS. The homomorphism technique is related to *subsumption* techniques, which have been used since the early 1970s to determine the equivalence of logic programs. The algorithmic analysis was developed as part of database theory. We recommend [199, 254] for discussions of this technique for logic programming applications.

Tableaux can be extended to include constants and other features such as inequalities [179]. They can be defined over databases that satisfy dependencies. This rich topic was initiated in [14, 15] for fd's. Containment in the presence of another common type of dependencies, *inclusion* dependencies (see Section 3), has been examined in [169, 173].

2.3. Why functional dependencies?

The study of dependency theory began with the introduction of fd's in [76] and grew into a rich topic, interesting in its own right. In this section we present the fundamentals of dependency theory using fd's and some of their extensions, and we defer generalizations and applications to Section 3. We would like to emphasize that this choice is not only for reasons of exposition. Fd's are the most relevant dependencies from a practical standpoint. Also, most of the fundamental concepts were introduced and are best illustrated in this restricted context.

Throughout this section we assume one relation scheme with fd's defined on it.

2.3.1. Dependency implication and its axiomatization

The following definition of implication applies to dependencies in general.

2.3.1. DEFINITION. Let Σ be a set of dependencies and σ a single dependency. We say that Σ *implies* σ, $\Sigma \models \sigma$, if every unrestricted database that satisfies Σ also satisfies σ; Σ *finitely implies* σ, $\Sigma \models_f \sigma$, if every database that satisfies Σ also satisfies σ.

The importance of finite implication in database theory first became apparent in Bernstein's work [54]. As we shall see below, finite implication and implication coincide for fd's, as they often do in dependency theory. Although finite implication is the relevant notion from a practical standpoint, implication is also important because it is closely related to unsatisfiability of logical sentences.

2.3.2. DEFINITION. Two sets of dependencies Σ, Σ' are *equivalent* when they are satisfied by the same set of databases. In this case we say that Σ is a *cover* for Σ' and vice versa. If Σ is a cover of Σ' and it is also a subset of Σ', then we say it is a *contained cover*.

Clearly, Σ' is redundant if it has a contained cover Σ that is a proper subset of it, because Σ is a more economical specification equally expressive as Σ'.

To come back to finite implication, it is easy to see that Σ and Σ' are equivalent iff $\Sigma \models_f \sigma'$ for all σ' in Σ' and $\Sigma' \models_f \sigma$ for all σ in Σ. For fd's we can use $\models$ instead of $\models_f$, since these are the same. Testing if Σ' is redundant can be similarly reduced to finite implication.

2.3.3. EXAMPLE. As was observed by Nicolas [227], fd's can be represented as sentences of the f.o. predicate calculus with equality. Let us demonstrate how this is done for universe ABC and the fd $\sigma = A \rightarrow B$. The vocabulary in the predicate calculus will be $\{R\}$, where the arity of R is 3 and A corresponds to the first argument of relation symbol R etc. The fd σ is expressed by the sentence

$$\varphi_\sigma = \forall x \forall y \forall z \forall y_1 \forall z_1 (R(xyz) \wedge R(xy_1z_1)) \Rightarrow (y = y_1)$$

It follows from the definition of fd's that the set of finite relational structures satisfying φ_σ and the set of relations satisfying σ are the same. This is also true for unrestricted relations. Similar arguments show that any set of fd's can be expressed by a set of sentences of the f.o. predicate calculus with equality. Note, however, that the database notation for fd's is often preferable, because it is less cumbersome to use and it intuitively captures the meaning of fd's.

The identification of a dependency σ with a sentence φ_σ is true for fd's and for many other dependencies. One of its consequences is the reduction of dependency (finite) implication to the (finite) unsatisfiability problem of f.o. logic. Let $\Sigma=\{\sigma_1,\ldots,\sigma_k\}$, then $\Sigma\models_{(f)}\sigma$ iff we have that the sentence $\varphi_{\sigma_1}\wedge\cdots\wedge\varphi_{\sigma_k}\wedge\neg\varphi_\sigma$ is (finitely) unsatisfiable.

Recall that a sentence is (finitely) unsatisfiable if it has no (finite) models. Also, unsatisfiability for the f.o. predicate calculus with equality is r.e., by the Gödel Completeness Theorem, and finite unsatisfiability is co-r.e., by enumerating and testing all finite structures. From unsatisfiability we can infer finite unsatisfiability, from finite satisfiability we can infer satisfiability and from implication we can infer finite implication (but the converses do not always hold). From this discussion it follows that if a set of dependencies is identified with a set of sentences, for which satisfiability and finite satisfiability coincide, then dependency implication and finite implication coincide and are decidable. This happens to be the case for fd's.

2.3.4. THEOREM. *For fd's finite implication and implication are the same and decidable.*

The following argument for this theorem is somewhat of an overkill, since the theorem can be shown without an excursion into satisfiability. However, it is quite instructive since it may be used for many nontrivial extensions of fd's.

Let us look at the structure of the sentence $\varphi_{\sigma_1}\wedge\cdots\wedge\varphi_{\sigma_k}\wedge\neg\varphi_\sigma$ in the fd case. This can be written as a $\exists^*\forall^*$-sentence, that is, a sentence in prenex normal form whose quantifier prefix consists of a string of $\exists$s followed by a string of $\forall$s. This is known as a sentence of the *initially extended Bernays–Schönfinkel class*, for which satisfiability and finite satisfiability coincide [101].

Implication is a semantic notion, which is commonly studied using the syntactic device of a *formal system* or *axiomatization*. An axiomatization (for a set of dependencies in general) consists of *axiom schemes* and *inference rules*. A *derivation* of a dependency σ from a set of dependencies Σ is a sequence of dependencies $\sigma_1,\ldots,\sigma_n$ such that $\sigma_n=\sigma$ and each σ_i in the sequence is either a member of Σ, or an instance of an axiom scheme, or follows from preceding σ_j's in the sequence via an instance of an inference rule. For an example, see formal system FD below.

We denote the existence of a derivation of σ from Σ by $\Sigma\vdash\sigma$. We say that a formal system $\vdash$ ($\vdash_f$) is *sound* for (finite) implication if $\Sigma\vdash\sigma$ ($\Sigma\vdash_f\sigma$) entails $\Sigma\models\sigma$ ($\Sigma\models_f\sigma$); and it is *complete* if $\Sigma\models\sigma$ ($\Sigma\models_f\sigma$) entails $\Sigma\vdash\sigma$ ($\Sigma\vdash_f\sigma$).

Formal systems for fd implication and finite implication were first studied by Armstrong [21]. His original system was slightly different from the sound and complete system FD, which we use here. FD consists of one axiom scheme and two inference rules. If Σ, σ are dependencies over R then any subset of R can be substituted for X, Y and Z. It is also an example of a *k-ary* system for $k=2$, because each inference rule has at most two antecedents:

- **FDr** reflexivity axiom scheme: $\vdash X\rightarrow\emptyset$;
- **FDt** transitivity inference rule: $X\rightarrow Y$ *and* $Y\rightarrow Z\ \vdash\ X\rightarrow Z$;
- **FDa** augmentation inference rule: $X\rightarrow Y\ \vdash\ XZ\rightarrow YZ$.

2.3.5. THEOREM. *The system* FD *is sound and complete for (finite) implication of fd's.*

Let us comment on the proof. Soundness of a formal system for implication entails soundness for finite implication (the converse is not true for all dependencies) and it is usually straightforward to show. Completeness for finite implication entails completeness for implication (the converse is not true for all dependencies) and it is typically the harder property to show. For this direction one usually examines the combinatorics of derivations and, if σ is not derivable from Σ, then one constructs a counterexample database satisfying Σ and falsifying σ. See [201, 280, 294] for standard expositions of the counterexample construction.

A closer examination of the FD system reveals that, to derive σ from Σ, one need only restrict attention to derivations with attributes in Σ and σ. This observation and Theorem 2.3.5 are an alternative proof of Theorem 2.3.4.

Let us close this subsection with some additional properties of fd's. Fd's are domain-independent sentences, as defined in Section 2.2.1. This is important, because testing for their satisfaction does not depend on whether Δ or δ is used.

Let Σ be a set of dependencies from some class of dependencies Σ^*. The *closure* of Σ over Σ^* are all dependencies in this class implied by Σ: it is denoted by Σ^+. We say that an unrestricted relation is an *Armstrong* relation if it satisfies all dependencies in Σ^+ and *simultaneously* falsifies all those in $\Sigma^* - \Sigma^+$. An interesting property for the set of fd's over a fixed universe is that there exists a finite Armstrong relation for fd's.

2.3.6. ADDITIONAL BIBLIOGRAPHIC COMMENTS. The first complete and sound axiomatization for fd's is from [21], but rules for fd's and some of their properties appeared early on in [97]. The concept of Armstrong relation is due to Fagin from [111]; research on these relations is surveyed in [110]. Their structure for fd's is examined in [37] and their existence for more general statements is investigated in [111]. They have been used as a tool for "database design by example" in [210].

The expressive power of fd's as a specification mechanism is examined in [154, 126]. For other "global" dependency questions different from implication, such as the existence of Armstrong relations, see [289].

2.3.2. PTIME computational and algebraic properties

The computational complexity of fd implication was considered by Beeri and Bernstein in [35], who demonstrated that implication can be performed optimally in linear time. This required a more detailed analysis than the decidability property, which follows from the equality of implication and finite implication.

Let Σ be a set of fd's over the universe, A, B attributes and X, Y sets of attributes from the universe. It is not hard to see that every fd $X \rightarrow Y$ is equivalent to the set of fd's $\{X \rightarrow B: B \in Y\}$. Let us define *closure*$(X, \Sigma) = \{A: \Sigma \models X \rightarrow A\}$. Given Σ, X, Y, if we can compute *closure*(X, Σ) then, clearly, we can decide the fd implication $\Sigma \models X \rightarrow Y$ in the same amount of time. The following theorem and algorithm are due to Beeri and Bernstein [35]. The algorithm can be made to run in linear time with the appropriate data structures.

2.3.7. THEOREM. (*Finite*) *implication of fd's is in linear time.*

```
procedure closure(X, Σ)
  ATRLIST := X; FDLIST := Σ;
  repeat
    erase all occurrences of attributes in ATRLIST
    from the left-hand sides of fd's in FDLIST;
    for each ∅→Y on FDLIST do
      ATRLIST := ATRLIST ∪ Y
  until no new attributes are added to ATRLIST in repeat loop
  return closure := ATRLIST
end.
```

Extensive use of this algorithm has been made in database scheme design. From Definition 2.3.2, recall the notions of *cover* and *contained cover* for fd's. Such covers are *minimum* if they contain the minimum number of fd's possible. As shown in [200], it is possible to compute minimum covers of fd's in quadratic time. An algorithm for minimum contained covers is presented in [35], where it is also shown that this computational task is NP-complete.

2.3.8. THEOREM. *Minimum covers of fd's can be computed in* PTIME *and minimum contained covers in* NP; *but deciding if there is a contained cover with less than k fd's, for a given k, is* NP-*complete.*

In the previous subsection we established that every fd σ (every set of fd's Σ) can be associated with a sentence φ_σ (a set of sentences φ_Σ). From the relationship of dependency (finite) implication and (finite) unsatisfiability it follows that $\Sigma \models_{(f)} \sigma$ iff $\varphi_\Sigma \models_{(f)} \varphi_\sigma$, where the second $\models_{(f)}$ is (finite) implication for sentences of the f.o. predicate calculus with equality. This is not the only relationship with mathematical logic; fd's have a number of elegant algebraic properties.

THE PROBLEM OF DEPENDENCY IMPLICATION FOR TWO-TUPLE RELATIONS. The counterexample relation in the proof of Theorem 2.3.5, which was used for showing completeness, may be chosen to have only two tuples. Thus, for fd's one can show that $\Sigma \models_{(f)} \sigma$ iff $\Sigma \models_2 \sigma$, where $\models_2$ is dependency implication over two-tuple relations.

THE PROBLEM OF IMPLICATION FOR PROPOSITIONAL HORN CLAUSES. One need not go to the f.o. predicate calculus with equality to represent fd's: it suffices to consider propositional Horn clauses. Without loss of generality, we have fd's with single attribute right-hand sides.

Now translate attribute A into a propositional constant A, translate fd $\sigma = A_1 \ldots A_n \to A$ into a propositional Horn clause $prop_\sigma = A_1 \wedge \cdots \wedge A_n \Rightarrow A$, and a set of fd's Σ into the obvious set of propositional Horn clauses $prop_\Sigma$. One can show that $\Sigma \models_{(f)} \sigma$ iff $prop_\Sigma \models prop_\sigma$, where the second $\models$ is propositional sentence implication. The converse of this reduction also holds, i.e., propositional Horn clause implication is immediately reducible to fd implication. Thus, Horn clause implication can be decided in linear time.

THE GENERATOR PROBLEM FOR FINITELY PRESENTED ALGEBRAS. Let Γ be a finite set of *generators* (function symbols of arity 0) and O a finite set of *operators* (function symbols of arity >0). A (ground) *term* τ is built out of $\Gamma \cup O$ in the standard f.o. fashion and a (ground) *equation eq* is a statement of the form $\tau = \tau'$. Equations are thus sentences of the f.o. predicate calculus with equality over vocabulary $\Gamma \cup O$ and they are satisfied by structures over this vocabulary, called *algebras*, according to the semantics of the calculus. If eq_Σ is a finite set of equations and eq_σ an equation then the implication $eq_\Sigma \models eq_\sigma$ is defined as in the calculus.

The uniform word problem for finitely presented algebras is the problem of deciding $eq_\Sigma \models eq_\sigma$, given a finite set of equations eq_Σ and an equation eq_σ.

The generator problem for finitely presented algebras is defined as follows: on input of (1) a finite set of equations eq_Σ over $\Gamma \cup O$, (2) an element γ of Γ, and (3) a subset Γ' of Γ, decide if there exists a term τ over $\Gamma' \cup O$ such that $eq_\Sigma \models \gamma = \tau$. If the answer is yes we denote this by *Generator*(eq_Σ, γ, Γ'). Both the uniform word problem and the generator problem are shown to be in PTIME in [183]. For the generator problem the algorithm of [183] is a generalization of the [35] algorithm for fd closure and was independently derived.

To see the relationship of fd closure with the generator problem, translate each attribute A of the universe into a generator A. Translate each fd $\sigma = A_1 \ldots A_n \to A$ into eq_σ, that is, the equation $A = f_\sigma(A_1 \ldots A_n)$, and a set of fd's into the obvious finite set of equations eq_Σ. The universe has become the set of generators, and the function symbols for the fd's in Σ (one per fd) form the set of operators. One can show that $\Sigma \models_{(f)} X \to A$ iff *Generator*(eq_Σ, A, X).

THE UNIFORM WORD PROBLEM FOR Γ-SEMILATTICES. A Γ-semilattice is an algebra $\langle \Lambda, \odot, \Gamma \rangle$, where Λ is a nonempty set, the *carrier* of the algebra, $\odot$ is a binary associative commutative idempotent operation on the carrier, and Γ is a finite set of named elements of the carrier. It is a finite Γ-semilattice if the carrier is finite. A (ground) term τ built out of operator $\odot$ and generators Γ is interpreted over a Γ-semilattice in the natural way; the meaning of τ is the element of the carrier computed by interpreting $\odot$ as the binary operation and Γ as the corresponding named elements of the carrier. A (ground) equation *leq* has the form $\tau = \tau'$ and is satisfied by a Γ-semilattice if τ and τ' have the same meaning in this semilattice. We say that a set of equations leq_Σ (finitely) implies equation leq_σ, denoted $leq_\Sigma \models^{l}_{(f)} leq_\sigma$, if every (finite) Γ-semilattice that satisfies every equation in leq_Σ also satisfies equation leq_σ. Note that this is a uniform word problem. But the algebra is no longer finitely presented, because the presentation consists of leq_Σ together with associativity commutativity and idempotence axioms for $\odot$. These properties are expressible using equations with variables, i.e., *nonground* equations.

To see the relationship with fd implication, translate each attribute A of the universe into a generator A, translate each fd $\sigma = A_1 \ldots A_n \to A$ into leq_σ which is the equation $A_1 \odot \cdots \odot A_n \odot A = A_1 \odot \cdots \odot A_n$, and a set of fd's Σ into the obvious set of equations

leq_Σ. One can show that $\Sigma \models_{(f)} \sigma$ iff $leq_\Sigma \models^l_{(f)} leq_\sigma$. It is also true that finite implication and implication are the same over Γ-semilattices.

We have seen six qualitatively different formulations of fd implication, which we summarize below. In all these statements finite implication and implication coincide.

2.3.9. THEOREM. *Let Σ be a set of fd's and σ be fd $X \rightarrow A$; then $\Sigma \models_{(f)} \sigma$ iff $\Sigma \models_2 \sigma$ iff $\varphi_\Sigma \models_{(f)} \varphi_\sigma$ iff $prop_\Sigma \models prop_\sigma$ iff Generator(eq_Σ, A, X) iff $leq_\Sigma \models^l_{(f)} leq_\sigma$.*

2.3.10. ADDITIONAL BIBLIOGRAPHIC COMMENTS. The connection of fd's, two-tuple implication and propositional logic was first identified by Fagin. It has been extended to the class of functional and *multivalued* (see Section 2.3.3) dependencies in [255].

The connection of fd's and generator problems was established in [86], where it was also extended to the class of functional and inclusion dependencies (see Section 3), in two ways. One of these extensions holds for implication and the other for functional and *unary inclusion* (see Section 2.3.3) dependency finite implication.

The connection of fd's and Γ-semilattices is part of the "folklore" of database theory. It has been extended to the class of *partition* dependencies and Γ-lattices in [89]. A Γ-lattice $\langle \Lambda, \odot, \oplus, \Gamma \rangle$ has two operations $\odot$ and $\oplus$, such that $\langle \Lambda, \odot, \oplus \rangle$ is a *lattice* and Γ are named elements of Λ. A partition dependency is defined as an equation between ground terms over $\odot$, $\oplus$ and elements of Γ. A partition dependency can be interpreted as a statement about relations over the universe and it generalizes the functional dependency. As shown in [89], because of the lattice semantics, there are certain partition dependencies which are not expressible using f.o. sentences; despite this, partition dependency implication is in PTIME and is the same as finite implication.

2.3.3. *Functionality, decomposition and inclusion*

Decomposing a relation r over R into a set of relations over $D = \{R_1, \ldots, R_m\}$ with $R = U = \bigcup_{i=1}^m R_i$ (i.e., computing $\pi_D(r) = \{\pi_{R_1}(r), \ldots, \pi_{R_m}(r)\}$ as in Example 2.1.8) is a technique commonly used in storing, querying and updating data. One can think of "the world" as being relation r and of $\pi_D(r)$ as its convenient representation.

The set of "possible worlds" is specified by $\langle \{U\}, \Sigma \rangle$, where Σ is a set of dependencies over U. This simplifying framework is known as the *pure universal relation assumption*, where U is the universe, r is a universal relation satisfying Σ, and $\langle \{U\}, \Sigma \rangle$ denotes the set of universal relations satisfying Σ.

A (*full*) *decomposition* π_D is thus a function from $\langle \{U\}, \Sigma \rangle$ into the set of databases over D. The least one could require of this function is that it be *injective*, i.e., one-to-one. This would guarantee that it has a left inverse, called the *reconstruction* function, from the range of π_D onto $\langle \{U\}, \Sigma \rangle$. The existence of the reconstruction function allows a decomposition without loss of information, since "the parts can be put together again to form the original whole". This is called the *representation principle* in [36], an early and good survey of database scheme design. A desirable candidate for the reconstruction function is the join of the relations over D. Note that *join reconstruction* is a desirable feature, but it does not always follow from injectiveness [290]; additional conditions might be required.

Decomposition π_D is *lossless* if for all r in $\langle\{U\},\Sigma\rangle$ we have that $r=\bowtie\pi_D(r)$. If a decomposition is lossless then it is easy to see that it is injective and that join reconstruction is possible. One of the first results in database theory involved conditions for lossless decompositions into two parts, in the presence of one fd [97, 149].

By taking $\bigcup_{i=1}^{m} R_i = R \subseteq U$ ($\subseteq$ instead of $=$), it is possible to express *embedded decomposition* properties of the universal relation (instead of only full ones). To describe embedded lossless decompositions we add assertions to our dependency language of the form for all r, $\pi_R(r)=\bowtie\pi_D(r)$ (note that $\pi_R(r)\subseteq\bowtie\pi_D(r)$ is a tautology). These are typed statements, in the sense that they only involve comparisons of values of the same attribute; this is a similarity with fd's.

Statements about embedded lossless decompositions are natural integrity constraints. The importance for database scheme design of two-part full lossless decompositions, called multivalued dependencies, and of two-part embedded lossless decompositions, called embedded multivalued dependencies, was identified early on in [96, 108, 309]. Multipart lossless decompositions, called join and embedded join dependencies, were first defined and studied in [13, 241].

Inclusion dependencies, first identified in [60], are another useful class of statements about relational databases. Here we concentrate on unary inclusion dependencies, the simplest (and most common) inclusion dependencies. For example, they can be used to express referential integrity and ISA constraints. To describe them it suffices to add assertions to our dependency language of the form for all r, $\pi_A(r)\subseteq\pi_B(r)$. They are unlike fd's, because they are untyped. Also, unlike fd's, they can be used as constraints between two different relations in a more general multirelational setting.

In the following definition we summarize the additions, motivated by decomposition and inclusion properties, to our repertoire of dependencies.

2.3.11. DEFINITION. Let D be a database scheme whose $m \geqslant 2$ relation schemes have union $R\subseteq U$, where U is the universe. *Embedded join dependency* (ejd) $\bowtie[D]$ is satisfied by relation r over U if $\pi_R(r)=\bowtie\pi_D(r)$. It is called a *join dependency* (jd) if $R=U$; an *embedded multivalued dependency* (emvd) if $m=2$; a *multivalued dependency* (mvd) if $R=U$ and $m=2$. The mvd $\bowtie[\{XY, XZ\}]$ with $Z=U-XY$ is denoted by $X\twoheadrightarrow Y$. *Unary inclusion dependency* (uind) $A\subseteq B$ is satisfied by relation r over U if $\pi_A(r)\subseteq\pi_B(r)$.

Tableau notation (recall Section 2.2.3) is often used in the context of fd's and ejd's. Every ejd $\sigma=\bowtie[D]$ is associated with a tableau T_σ, which is the one equivalent to the expression $\bowtie\pi_D(U)$ (see Theorem 2.2.12). So its summary is an R-tuple of distinguished variables, where $R\subseteq U$. Every fd $\sigma=X\rightarrow Y$ is associated with a tableau T_σ, which is the one associated with the mvd $X\twoheadrightarrow Y$. Note that for both fd's and jd's the summary of the associated tableau is a U-tuple of distinguished variables.

Clearly, every mvd, jd and emvd is an ejd. Also, every mvd is an emvd and a jd. The following example illustrates most of these definitions.

2.3.12. EXAMPLE. Let $U=ABCA'$ and let fd $AA'\rightarrow C$, ejd $\bowtie[\{AB, AC\}]$ and uind $C\subseteq B$ be the dependencies specifying the "legal" universal relations. The relation r in

r	A	B	C	A′
	a	b	c	a'
	a	b'	c	a'
	a	b	c'	a''
	a	b'	c'	a''
	a_1	c	c'	a''
	a_2	c'	c'	a''

summary	x	y	z	x'
T_{fd}	x	y_1	z	x'
	x	y	z_1	x'

summary	x	y	z	
T_{emv}	x	y	z_1	x'_1
	x	y_1	z	x'_2

Fig. 3. The relation and tableaux of Example 2.3.12.

Fig. 3 satisfies the three dependencies. The two tableaux of Fig. 3 correspond to the fd $AA' \to C$ and to the emvd $\bowtie[\{AB, AC\}]$ respectively. The first tableau also corresponds to the mvd $AA' \twoheadrightarrow C$ or $\bowtie[\{AA'C, AA'B\}]$.

One may think of this situation as an INTERESTS universal relation, where A stands for NAME, B for JOB, C for SPORT and A' for SEASON. The fd asserts that SEASON and NAME functionally determine SPORT, (each person does one sport each season). The ejd is an emvd, which can be interpreted as follows: every value of NAME is associated with a set of values of JOB and a set of values of SPORT and these two sets are independent of each other. That is, if tuples abc, ab_1c_1 are in the projection of the universal relation on NAME JOB SPORT then tuple abc_1 is also in this projection. Intuitively, NAME embedded multivalued determines JOB and SPORT, which are independent of each other. The uind asserts that every SPORT is someone's JOB (there are professional athletes in each sport).

It is possible to express these dependencies using sentences of the f.o. predicate calculus, with equality for the fd and without equality for the emvd and uind. We list these sentences. One should note the presence of the $\exists$ quantifiers for the emvd and uind; this is an important difference from the fd.

$$\text{fd:}\quad \forall x \forall y_1 \forall z \forall x' \forall y \forall z_1 (R(xy_1zx') \wedge R(xyz_1x')) \Rightarrow (z = z_1),$$

$$\text{emvd:}\quad \forall x \forall y \forall z_1 \forall x'_1 \forall y_1 \forall z \forall x'_2 (R(xyz_1x'_1) \wedge R(xy_1zx'_2)) \Rightarrow \exists x'_3 R(xyzx'_3),$$

$$\text{uind:}\quad \forall x_1 \forall x_2 \forall x_3 \forall x_4 R(x_1x_2x_3x_4) \Rightarrow \exists y_1 \exists y_2 \exists y_3 R(y_1x_3y_2y_3).$$

2.3.13. Remark. Of course, where one draws the line for dependencies is an educated, but somewhat arbitrary choice. As we shall see in Section 3, fd's and ejd's are typical cases of the larger class of *embedded implicational dependencies* (eid's), which were identified by a number of researchers independently as the natural closure of fd's and ejd's [51, 52, 111, 305]. Eid's are unirelational and typed, but can be generalized to the multirelational and untyped case; this closure is referred to generically as *embedded dependencies* in [117], and this is where we will draw the line in Section 3. Inclusion dependencies (ind's) are among the most studied embedded dependencies that are not eid's; uind's are special ind's that are also not eid's. As sentences of the f.o. predicate calculus with equality, embedded dependencies have quantifier prefix $\forall^*\exists^*$. They are called *full dependencies* if the quantifier prefix is only $\forall^*$. The full eid's are called *full implicational dependencies* (fid's); fd's and jd's are fid's.

Here we focus on ejd's, fd's and uind's in order to stress the existence of PTIME computational properties. For these statements implication can be expressed as an unsatisfiability problem of a f.o. sentence, whose quantifier prefix is $\exists^*\forall^*\exists^*$. (The argument is similar as in the fd case, only the last existential quantifiers must be added). For fd's and jd's this sentence has quantifier prefix $\exists^*\forall^*$. Thus finite implication and implication coincide and are decidable for fd's and jd's; this need not be the situation in the presence of either ejd's or uind's.

We use N for the input size to a dependency implication problem. Let us first examine testing a decomposition for losslessness, given a set of fd's. It is easy to show that full decomposition π_D is lossless iff the jd $\bowtie[D]$ is implied by the given set of fd's. Also, here finite implication and implication coincide. The first algorithm for this problem used $O(N^4)$ time and was described in [13]. It was improved in [197, 100]; the best current bound from [100] is based on the operation of congruence closure and is $O(N^2 \log^2 N/\log k)$ time and $O(N^2 k)$ space, for k a parameter with $1 \leqslant k \leqslant N$. The proof also applies to an embedded decomposition, i.e., an ejd as the implied statement. The results were extended in [173] to finite implication (in $O(N^3)$ time) and implication (same bounds as [100]) of an ejd by a set of fd's and uind's. As shown in [60] finite implication and implication differ in the presence of fd's and uind's.

2.3.14. THEOREM. *Finite implication and implication of an ejd by a set of fd's and uind's do not coincide, but are both in* PTIME.

In this theorem, it is important that the domains of attributes be unbounded, since two-tuple relations no longer characterize the implication problems. If some domain has only two values then implication of a jd by a set of fd's is NP-complete [171].

An important technique grew out of the algorithm of [13] for testing lossless joins. This algorithm was extended in [203] to antecedent statements being fd's and jd's and named the *chase*. The chase can be further extended into a semidecision procedure for embedded dependency implication and an exponential decision procedure for full dependency implication, see [51, 52] for the general setting. In its most general form it is similar to resolution with paramodulation and it uses a correspondence between tableaux and embedded dependencies. For the ind chase we refer to [169]. But we should note that equational forms of reasoning seem more natural for ind problems [86, 87, 82, 218]. Let us describe the chase for jd's and fd's from [203].

In [203] it is also shown that the *chase* procedure below is Church–Rosser, i.e., its result is unaffected by nondeterministic choices in its steps. Note that, procedure *chase*(Σ, σ) is a decision procedure, since it always terminates. From its description one can see that no new symbols are generated during the computation and this fact accounts for the termination. It runs in exponential time.

procedure *chase*(Σ, σ)
input Σ a set of fd's and jd's, σ an fd or jd;
  $T := T_\sigma$, where T_σ is the tableau associated with σ; *success* := *false*;
  repeat
    if σ' is an fd in Σ such that relation $T \not\models \sigma'$ because $t[A] \neq t'[A]$
    then replace all occurrences of one of $t[A], t'[A]$ by the other,
      keep the distinguished symbol if one is such
      or else keep the lowest subscript nondistinguished symbol
    if σ' is a jd in Σ such that relation $T \not\models \sigma'$ because $T \neq \bowtie \pi_D(T)$
    then replace relation T by relation $\bowtie \pi_D(T)$
  until no further modifications of T are possible;
  if σ is fd $X \rightarrow Y$
  then *success* := *true* if all nondistinguished vars for Y in T_σ became
    distinguished in T;
  if σ is jd $\bowtie[D]$
  then *success* := *true* if the summary of T_σ is a tuple of T;
output if *success* = *true* **then** $\Sigma \models_{(f)} \sigma$ **else** $\Sigma \not\models_{(f)} \sigma$;
end.

Using properties of the chase it is possible to show a number of positive results. These we summarize in Theorem 2.3.15, which complements Theorem 2.3.14. The simultaneous presence of fd's and uind's makes finite implication and implication different, but they are still both decidable and sometimes efficiently so. We state the theorem using the more general embedded implicational dependencies (instead of ejd's and fd's) and full implicational dependencies (instead of jd's and fd's); see Remark 2.3.13. The results for ejd's, jd's and fd's only are from [203, 205, 292], for eid's and fid's only are from [287], and the addition of uind's is from [173].

2.3.15. THEOREM. (1) *Implication and finite implication of an eid or uind from a set of fid's and uind's do not coincide, but are both in* EXPTIME.

(2) *Implication and finite implication of an mvd, fd or uind from a set of fid's and uind's do not coincide, but are both in* PTIME.

Unfortunately, the precise complexity of the ejd implication problems is still open. The only known lower bounds are NP-hardness ones [47, 48, 50, 205]. For a discussion of the open questions see [117]. For example, it was shown in [118] that *testing whether a set of mvd's implies a jd is* NP-*hard*.

One way of understanding implication is through the development of complete and sound formal systems for it, e.g., the system FD for fd's. Such formal systems are often nontrivial to construct, particularly in the presence of jd's, see [48, 50, 265]. Also, unlike for fd's, no complete and sound k-ary system (for some fixed k) might exist. For example, it was shown in [60] that this is the case for many fd and ind implication problems, including fd and uind finite implication. This is also a difficulty for emvd implication, see [235, 259].

In the restricted world of mvd's, fd's and uind's, (i.e., the only ejd's allowed are mvd's), the situation is almost ideal. All implication problems are efficiently solvable and have sound and complete formal systems that are simple.

The first sound and complete formal system for fd's and mvd's appeared in [38]; here we give a slightly modified version MFD from [294]. Mvd and fd implication was solved in $O(N^4)$ in [34] and these bounds were improved in [122, 146, 249]; the best current bound is $O(N \log N)$ from [122]. This computational behavior is due in large part to the algebraic properties of what is called in [38] a *dependency basis*. The results were extended in [173] to also include uind's, with an overhead of $O(N^3)$ for finite implication and of $O(N \log N)$ for implication.

We summarize the known results in Fig. 4. Namely, the complexities of the implication problems together with sound and complete formal systems for these problems. We have already seen system FD = {FDr, FDa, FDt}. To these rules we add a number of rules and use the following convention in Fig. 4: MD = {MDc, MDa, MDd}, MFD = FD ∪ MD ∪ {MFDt, MFDi}, etc. We also note when finite and unrestricted notions are equal. All the rules that follow are 2-ary ones. The only exception is the last set of "cycle rules", where we have one cycle rule for each odd positive integer k.

- **MDc** complement axion scheme: $\vdash X \twoheadrightarrow U - X$;
- **MDa** augmentation inference rule: $X \twoheadrightarrow Y \vdash XZ \twoheadrightarrow YZ$;
- **MDd** difference inference rule, where $Y \cap Z = \emptyset$:

$$X \twoheadrightarrow Y \textit{ and } Z \twoheadrightarrow Y_1 \vdash X \twoheadrightarrow Y - Y_1;$$

- **MFDt** translation inference rule: $X \rightarrow Y \vdash X \twoheadrightarrow Y$;
- **MFDi** intersection inference rule, where $Y \cap Z = \emptyset$:

$$X \twoheadrightarrow Y \textit{ and } Z \rightarrow Y_1 \vdash X \rightarrow Y \cap Y_1;$$

- **UDr** reflexivity axiom scheme: $\vdash A \subseteq A$;
- **UDt** transitivity inference rule: $A \subseteq B \textit{ and } B \subseteq C \vdash A \subseteq C$;
- **UFDe** $\emptyset$ interaction inference rule: $\emptyset \rightarrow A \textit{ and } B \subseteq A \vdash A \subseteq B \textit{ and } \emptyset \rightarrow B$;
- **CD** one cycle inference rule for each odd positive integer k:

$$A_0 \rightarrow A_1 \textit{ and } A_2 \subseteq A_1, \ldots, A_{k-1} \rightarrow A_k \textit{ and } A_0 \subseteq A_k$$
$$\vdash A_1 \rightarrow A_0 \textit{ and } A_1 \subseteq A_2, \ldots, A_k \rightarrow A_{k-1} \textit{ and } A_k \subseteq A_0.$$

The efficient solution for mvd, fd and uind problems provides a convenient pause in our excursion into dependency theory. The next logical step is the investigation of emvd's, which motivated many of the generalizations to be examined in Section 3. The implication problems for emvd's are the major open questions in dependency theory. No upper or lower bounds are known for these problems beyond the trivial ones (even the NP-hardness bounds for ejd's do not apply to emvd's).

2.3.16. OPEN PROBLEM. Are emvd implication and finite implication decidable?

Σ, σ	$\models$	$\models_f$	$\vdash$	$\vdash_f$
fd	O(N)	$=\models$	FD	$=\vdash$
mvd	O($N \log N$)	$=\models$	MD	$=\vdash$
uind	O(N)	$=\models$	UD	$=\vdash$
mvd, fd	O($N \log N$)	$=\models$	MFD	$=\vdash$
mvd, uind	O($N \log N$)	$=\models$	MUD	$=\vdash$
fd, uind	O(N)	O(N^3)	UFD	CUFD
mvd, fd, uind	O($N \log N$)	O(N^3)	UMFD	CUMFD

Fig. 4. Mvd, fd, uind implication problems.

2.4. *On hypergraphs and the syntax of database schemes*

With a database scheme D or a jd $\bowtie[D]$ or a decomposition π_D we can associate a *hypergraph.* Hypergraphs, like undirected graphs, consist of a finite set of *nodes* (in this case, the attributes in D) and a finite set of sets of nodes or *edges* (in this case the relation schemes of D), where we assume that each node belongs to some edge. The many applications of hypergraph theory to database theory illustrate the use of attribute notation.

Most graph notions, such as paths or connectivity, immediately carry over to hypergraphs. Unlike graphs, however, hypergraphs have a number of inequivalent notions of "acyclicity". The *acyclicity* we examine here is the first one proposed in the context of database theory [39]. It is also known as α-acyclicity, to distinguish it from other variants examined in [112]. There is a large number of combinatorial characterizations of acyclicity [40, 114], many of which have applications in database theory. Here we use the operational definition based on the GYO reduction, independently identified in [132] and [307].

2.4.1. DEFINITION. Database scheme D, jd $\bowtie[D]$, decomposition π_D and hypergraph D are *acyclic* if hypergraph D can be reduced to the empty set by some sequence of applications of the following two rules: (1) if an edge is a subset of another edge then delete it, and (2) if a node belongs to precisely one edge then delete it.

Testing Definition 2.4.1 is clearly in PTIME. As shown in [273], this can be done optimally in linear time. An example of an acyclic hypergraph is given pictorially in Fig. 5. Note that a subset of its edges form a cyclic hypergraph; this behavior is somewhat counterintuitive and does not occur for γ-acyclicity [112]. Despite such anomalies, acyclic database schemes are well motivated from a semantic point of view, e.g., see [193, 176].

An attractive feature of acyclicity, from a computational point of view, is that a number of NP-hard questions involving jd's or decompositions can be resolved in PTIME, provided the input is acyclic. Following the exposition of [117], we present three such questions. The first is about jd's, the second about decompositions, and the final one about database schemes and the pure universal relation assumption.

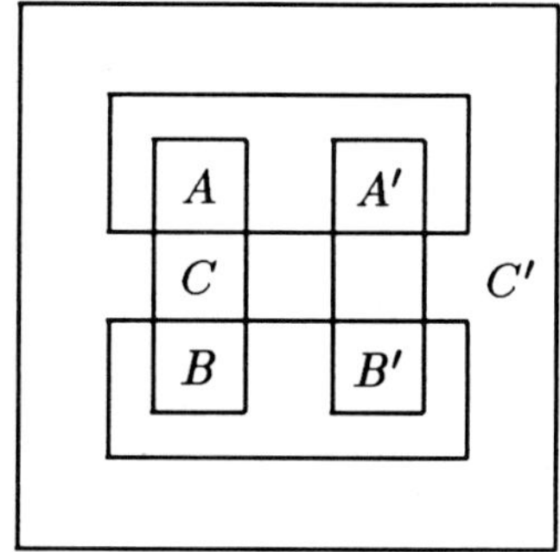

Fig. 5. An acyclic hypergraph.

A jd is acyclic iff it is equivalent to (i.e., it implies and is implied by) some set of mvd's. This is one of the characterizations of acyclicity from [114]. From [40] it follows that, given an acyclic jd, an equivalent set of mvd's can be found in PTIME. The converse, given a set of mvd's, find an equivalent jd or report that none exists, was shown in [130] also to be in PTIME. There is an interesting consequence. Recall that testing implication of a jd by a set of mvd's is NP-hard [118], but testing implication of an acyclic jd by a set of mvd's is in PTIME. To see this, use the [114] characterization, the PTIME construction [40] and mvd implication. In summary, we have the following theorem.

2.4.2. THEOREM. (1) *A jd is acyclic iff there exists a set of mvd's equivalent to it.*

(2) *Given an acyclic jd, an equivalent set of mvd's can be constructed in* PTIME.

(3) *Given a set of mvd's, finding an equivalent jd or that none exists is in* PTIME.

(4) *Implication of an acyclic jd from a set of mvd's is in* PTIME.

As shown in [290] about decompositions, injectiveness (see Section 2.3.3) does not always entail that the reconstruction function is the join. This is the case when the dependencies are jd's and fd's and the decomposition is in two relation schemes [53, 90]. In fact, the results of [53] are for the more general case of fid's (see Remark 2.3.13) and acyclic decompositions. In [173] it is shown that uind's can be added without changing the theorem.

2.4.3. THEOREM. *Let the possible universal relations $\langle\{U\}, \Sigma\rangle$ be determined by a set Σ of fid's and uind's over U, and let the decomposition π_D be acyclic; then:*

(1) *if π_D is injective, the reconstruction function is the join, and*

(2) *testing if π_D is injective is equivalent to testing $\Sigma \models_f \bowtie[D]$ and is in* PTIME.

We say that a database $d=\{r_1,\ldots,r_m\}$ over D is *join-consistent* or is the *projection of a universal relation* if there is a relation r such that $\pi_D(r)=d=\{r_1,\ldots,r_m\}$. In other words, the pure universal relation assumption states that all databases over D are join-consistent. This is only a simplifying assumption. Not all databases are join-consistent, as one can see from simple examples. Another problem is that, given a d over D, testing whether it is join-consistent is NP-complete, see [153].

A less restrictive assumption about a database $d=\{r_1,\ldots,r_m\}$ over D is that it is *pairwise-consistent*. Namely, every two relations r_i, r_j over R_i, R_j respectively agree on their common attributes $R_i \cap R_j$, i.e., $\pi_{R_i \cap R_j}(r_i)=\pi_{R_i \cap R_j}(r_j)$ for $1 \leqslant i, j \leqslant m$. An equivalent way of defining pairwise consistency is requiring every database $\{r_i, r_j\}$, $1 \leqslant i, j \leqslant m$, to be join-consistent. Every join-consistent database is pairwise-consistent, but not conversely. Pairwise consistency can be expressed using inclusion dependencies. It is easy to see that pairwise consistency can be tested in PTIME. From [40] we have the following characterization of acyclicity.

2.4.4. THEOREM. (1) *If database scheme D is acyclic, then any database over D that is pairwise-consistent is also join-consistent.*

(2) *If database scheme D is not acyclic, then there is a pairwise-consistent database over D which is not join-consistent.*

(3) *Given database d over acyclic D, testing whether d is join-consistent is in* PTIME.

2.4.5. ADDITIONAL BIBLIOGRAPHIC COMMENTS. Reference [303] contains a large number of problems solvable in PTIME in the presence of acyclicity. Acyclicity is a very useful tool for query optimization. This has led to an investigation of query evaluation based on the *semijoin* operation, e.g., [55, 56, 127, 128, 129, 257]. Acyclicity in the presence of fd's is the topic of many studies, e.g., [190, 245, 246, 256]. Various forms of acyclicity defined in [112] were further examined in [26, 93, 137]. Finally, acyclicity has been used in database scheme design and in the context of universal relation data models.

2.5. On logic and the semantics of databases

We have, until now, used one basic paradigm: that the database is a finite f.o. relational structure. This *model-theoretic* approach is not the only one possible. In fact, a *proof-theoretic* approach might be more appropriate for issues beyond the capabilities of the relational data model, such as the problems of incomplete information and updates that we address in Section 5.

Let us re-examine our basic premise that the database *is* the semantics. Instead, we can view databases as *f.o. theories* of a very special form, called *extended relational theories* by Reiter in [240]. A f.o. theory is a set of sentences of the f.o. predicate calculus with equality, over some vocabulary of constant and relation symbols. For the relational data model this point of view is equivalent to our original premise. An advantage of databases as structures is the natural development of the rich repertoire of algebraic techniques that we have already seen. On the other hand, a fresh point of view may lead to other important data models.

We discuss two such types of data models: deductive and universal relation data models. Our exposition highlights the fact that at a fundamental level these data models share many ideas, even if they have been developed separately.

The study of data model based on f.o. theories is known as the area of *deductive databases.* Research in this area was greatly influenced by an early collection of papers on the subject [123] and many of the original advances are surveyed in [124]. Recently, there has been much work on databases and logic programming that is related to deductive databases, see [216]. There is an intimate connection between deductive databases and the study of query languages beyond relational algebra, in particular, database logic programs. This has been demonstrated in [68] and is emphasized in [216]. We therefore chose to examine these subjects together, in some detail, in Section 4.

In this section we set up the f.o. theory framework for the relational data model. Modifications of this framework lead to nontrivial extensions, both from the point of view of query language expressive power and in terms of representation power for incompletely specified data. We also focus on the *weak universal relation assumption* and its relationship to deductive databases. Given the expressive power of the f.o. theory point of view, the historical choice of the terminology "weak" is rather unfortunate (its justification is that, as an assumption, it is weaker than the pure universal relation assumption).

2.5.1. Deductive data models and first-order theories

Consider a vocabulary consisting of relation symbols $\{R_1, \ldots, R_m\}$ and of a countably infinite set of constant symbols $\mathscr{C}$. This set $\mathscr{C}$ and the set of values Δ are put into one-to-one and onto correspondence, i.e., a in $\mathscr{C}$ corresponds to a in Δ. A *finite f.o. theory* θ is a finite set of sentences of the f.o. predicate calculus with equality over this vocabulary.

Consider a standard proof system for the f.o. predicate calculus with equality [105]. The consistency of a theory is the proof-theoretic property that not all sentences are provable using the theory and the standard proof system. It is well known that consistency is the proof-theoretic analog of satisfiability: a finite f.o. theory is *consistent* iff it is satisfiable [105]. So, let F be an expression of the relational calculus as in Definition 2.2.1: $\{R, x_1, \ldots, x_n: \varphi(x_1, \ldots, x_n)\}$. Then the new framework results from defining $F(\theta)$ as follows:

$$F(\theta) = \begin{cases} \{a_1 \ldots a_n: a_i \text{ appears in } \theta,\ 1 \leqslant i \leqslant n,\ \theta \models_f \varphi(a_1, \ldots, a_n)\} \\ \quad \text{if } \theta \text{ consistent,} \\ \emptyset \quad \text{if } \theta \text{ inconsistent.} \end{cases}$$

Note that $\models_f$ is finite implication of sentences, and not satisfaction by a finite model as in Definition 2.2.1. This approach is best exemplified by the work of Reiter, e.g. [238–240]. It has been called *proof-theoretic* because, under certain sufficient conditions for the sentences, finite implication $\models_f$ may be replaced by *provability* $\vdash_f$.

Thus, the tuples of $F(\theta)$ correspond to all the variable assignments verifying φ (or the facts) that we can prove from θ in first-order logic. This conservative approach towards what is true, i.e., *facts are true only if they can be proven*, is known as *the closed world assumption* [238]. Now let us try to relate this new approach to the relational data model.

If d is a database, we will construct a finite f.o. theory θ_d called an *extended relational theory*. This θ_d is constructed from d using the union of four sets of axioms AT, UN, DO, CO. We use shorthand $\vec{x}$ for $x_1, \ldots, x_n$ and $\vec{a}$ for $a_1, \ldots, a_n$, where n is determined from the context and equality $\vec{x} = \vec{a}$ is componentwise.

- **AT**: a sentence $R(\vec{a})$ for each R-tuple $\vec{a}$ in d.
- **UN**: a sentence $a \neq a'$ for each pair of distinct values a, a' occurring in d.
- **DO**: a sentence $\forall x(x=a_1 \vee \cdots \vee x=a_k)$ if the values occurring in d are exactly $\{a_1, \ldots, a_k\}$.
- **CO**: a sentence $\forall \vec{x}(R(\vec{x})) \Rightarrow (\vec{x} = \vec{a}_1 \vee \ldots \vee \vec{x} = \vec{a}_j)$ for each symbol R, where the R-tuples in d are exactly $\{\vec{a}_1, \ldots, \vec{a}_j\}$. If $j=0$, this sentence is $\forall \vec{x}(\neg R(\vec{x}))$.

The axioms AT (atomic facts) just express the tuples as ground atoms; the axioms UN (uniqueness) just express that different symbols are different objects; and the axioms DO (domain closure) and CO (completion) guarantee that θ_d defines a unique model d, modulo automorphisms on $\mathscr{C}$.

If d is a database and Σ a set of dependencies then $F(\theta_d \cup \Sigma)$ has the following two properties (Definition 2.2.1 is used for $F(d)$ and ER stands for extended relational theory).

- **ER1**: $F(\theta_d \cup \Sigma) = F(\theta_d) = F(d)$ if d satisfies Σ.
- **ER2**: $F(\theta_d \cup \Sigma) = \emptyset$ if d does not satisfy Σ.

Note how in these properties the problem of testing whether d satisfies Σ is separated from computing $F(d)$. The facts that can be proven from $\theta_d \cup \Sigma$ are the same as the result of querying the database according to Definition 2.2.1.

2.5.1. Remarks. If we assume that $\Sigma = \emptyset$ and that only a subset of the UN axioms are present, then we have a situation of incomplete information. There is a bound on the database domain and the tuples of the relations. However, not all constant symbols need represent distinct values. The uniqueness of the model of θ_d is lost, i.e., there might be many "possible worlds". In this case, the definition of $F(\theta_d)$ above has been used to provide semantics for querying incomplete information databases, see [295, 240].

One way to model incomplete information is to omit some of the axioms of an extended relational theory. In Remark 2.5.1 we deleted some of the UN axioms. An interesting situation arises when we have dependencies and we delete some of the CO axioms.

Let us assume that the vocabulary of relation symbols is $\{R_1, \ldots, R_m\}$. We are given an extended relational theory θ_d^c, *without the* CO *axioms for some of the relation symbols*, and a set of dependencies Σ over this vocabulary. Sentences $\theta_d^c \cup \Sigma$ form a *deductive database*. The theory θ_d^c represents the known facts about the application; Σ can be used to derive other true facts as follows.

2.5.2. Definition. If d is a database and $\theta = \theta_d^c \cup \Sigma$ is a deductive database and R is a relation symbol in its vocabulary then R denotes the query

$$R[d, \Sigma] = \begin{cases} \{a_1 \ldots a_n : a_i \text{ appears in } \theta, 1 \leqslant i \leqslant n, \theta \models_f R(a_1, \ldots, a_n)\} \\ \quad \text{if } \theta \text{ consistent,} \\ \emptyset \quad \text{if } \theta \text{ inconsistent.} \end{cases}$$

Note that this definition gives us a data model. Relations (extended relational theories) are manipulated via a query language implicitly defined by $\models_f$. The language makes little distinction between dependency satisfaction and finite implication. For example, a calculus query $F(d)$ can be expressed using a deductive data model by incorporating F's definition into Σ, provided that the set of dependencies allowed is sufficiently expressive.

2.5.3. EXAMPLE. Let R be a binary relation symbol and θ_r^c be an extended relational theory over R without the CO axiom for R, where r is a directed graph without isolated nodes. *Trans*(r) is the transitive closure of r, which is not expressible in relational calculus, see Theorem 2.2.8. It is not hard to see that $R[r, \{\sigma\}] = \mathit{Trans}(r)$, where

$$\sigma = \forall x \forall y \forall z (R(xz) \wedge R(zy)) \Rightarrow R(xy).$$

Deductive data models are parameterized by the possible syntax of the dependencies of Σ. In Example 2.5.3 the dependencies belong to the language of *universally closed Horn clauses*. The query language, implictly defined by $\models_f$ for such Horn clauses, is known as *Datalog* [68, 207] and we will examine it in Section 4 in some detail. We will also find universally closed Horn clauses as natural dependencies, called *rule dependencies*, that generalize jd's in Section 3. This Horn form of Σ is sufficient to guarantee that, the semantics based on Definition 2.5.2 are the same as the minimal Herbrand model semantics for the deductive database seen as a logic program, see [18, 20, 216].

Syntactic classes for Σ can be used to define query languages that are more expressive than relational calculus; see Example 2.5.3 and Theorem 2.2.8. We believe, however, that the study of these formalisms is best conducted as a study of query languages. This allows for a clean separation between issues of querying and of checking integrity of the database. We thus refer the reader to Section 4, where we study expressibility in various logics as opposed to the properties of deduction.

2.5.2. Universal relation data models and first-order theories

One of the original motivations for introducing the relational data model was to free the programmer from the need to procedurally specify how the data should be accessed; the so-called "navigation problem". When programming at the level of abstraction of relational algebra, the programmer need not worry about the access paths within the data structures that implement relations; the so-called "physical navigation problem". Nevertheless, "logical navigation" is still necessary. Even in the declarative relational calculus one has to specify how to connect various relations (but with less effort than in the algebra). Universal relation data models emerged through the work of many researchers, whose goal was to further simplify "logical navigation".

The ideal goal is to use the attribute notation and the hypergraph structure of database schemes, as a query language that is more declarative than relational calculus. Instead of using a relational calculus expression F (from databases over D to databases over $\{\alpha(F)\}$), just use $\alpha(F)$. The hope is that the database scheme structure and the dependencies have enough information for the system to automatically choose some F by using $\alpha(F)$. This of course involves thoughtful database scheme design and might work only for some expressions F.

We have already seen *the pure universal relation assumption*, first proposed as a simplifying assumption about database scheme design. The set of possible worlds are specified as $\langle\{U\}, \Sigma\rangle$ and only databases that are decompositions of such *universal* relations are considered. A potential problem is that many databases are not decompositions of any universal relation. Even testing whether there exists a universal relation r such that the given d is a decomposition of r is NP-complete [153].

These difficulties have led to the formulation of an alternative assumption, the *weak universal relation assumption.* This assumption, instead of the existence of a universal assumption, postulates the existence of a *weak universal relation.* A very good exposition of the subject can be found in [206]. In our definitions we limit Σ to the full dependencies of [117] (see also Section 3). For the embedded ones we refer to [206].

2.5.4. Definition. Let $d=\{r_1,\ldots,r_m\}$ be a database over $D=\{R_1,\ldots,R_m\}$ with universe U and Σ be a set of full dependencies. A relation r over U is a *weak universal relation* for d w.r.t. Σ if (1) r satisfies Σ, and (2) $r_i \subseteq \pi_{R_i}(r)$, $1 \leqslant i \leqslant m$. Each subset X of U denotes the query

$$X[d,\Sigma]=\bigcap\{\pi_X(r): r \text{ is a weak universal relation for } d \text{ w.r.t. } \Sigma\}.$$

Note that this definition gives us data models, which are parametrized by the class of dependencies of which Σ is a subset, e.g., fd's or jd's together with fd's etc. Relations are manipulated and a query language is provided: queries are specified implicitly from the attribute set X, the dependencies Σ and the intersection condition.

Let us determine the precise relationship of universal relation data models with deductive data models. The vocabulary we use has one relation symbol X for each subset X of U and the constant symbols $\mathscr{C}$ from the previous subsection. We build a finite f.o. theory θ_d^{u}, as the union of four sets of sentences AT, UN, IN, CN. The first two sets AT, UN are the same as for extended relational theory θ_d (see Section 5.2.1). The sentences IN express that for each X the relation over X is a subset of the projection on X of the weak universal relation. The sentences CN express the converse that for each X the projection on X of the weak universal relation is a subset of the relation on X. Let us illustrate these conditions by an example, where $U=ABC$ and $X=AB$.

- **IN**: for $U=ABC$, $X=AB$ include the sentence $\forall x\forall y\exists z X(xy) \Rightarrow U(xyz)$.
- **CN**: for $U=ABC$, $X=AB$ include the sentence $\forall x\forall y\forall z U(xyz) \Rightarrow X(xy)$.

An important observation from [206] is that if we consider theory θ_d^{u} instead of θ_d^{c} in Definition 2.5.2 then we get the $X[d, \Sigma]$ of Definition 2.5.4. Namely, if $\theta = \theta_d^{u} \cup \Sigma$ then

$$X[d,\Sigma]=\begin{cases}\{a_1 \dots a_n\colon a_i \text{ appears in } \theta, 1 \leqslant i \leqslant n, \theta \models_f X(a_1,\dots,a_n)\} \\ \quad \text{if } \theta \text{ consistent}, \\ \emptyset \quad \text{if } \theta \text{ inconsistent}.\end{cases}$$

One difference from deductive databases is the availability of a rather large vocabulary of relation symbols X, namely all the subsets of U. This is to facilitate "logical navigation". Another difference is the special emphasis in universal data models on Σ's consisting of fd's. With fd's, query evaluation has a distinctly different flavor than other deductive database query evaluation techniques, e.g., the ones for Σ consisting of rule dependencies.

The two obvious question arise from Defiition 2.5.4.

(1) How does one test *consistency*, in this case the existence of some weak universal relation of d w.r.t. Σ?

(2) How does one *evaluate* $X[d, \Sigma]$?

Since the two characterizations we have are not procedural, there could be an infinite number of candidate weak universal relations of which we need to compute the intersection. Fortunately, there is a naive but procedural method to answer these questions. The equivalence of the following method with Definition 2.5.4 is shown in [206, 250]:

(i) Construct a *representative universal relation* for database d; this consists of the tuples of d padded with uniquely occurring special symbols called *nulls* that turn these tuples into U-tuples.

(ii) This representative universal relation is viewed as a tableau and is chased by the dependencies in Σ, using a generalization of the chase from Section 2.3.3. Since Σ consists of full dependencies, this procedure terminates. If two non-nulls are equated then d, Σ are inconsistent, else they are consistent.

(iii) If d, Σ are consistent then $X[d, \Sigma]$ is the projection on X of the result of the chase, restricted to tuples without nulls.

A consequence is that query evaluation becomes dependency implication, where the representative universal relation is viewed as the dependency inferred. The following theorem is from [152] for fd's and from [134] for other dependencies. For embedded dependencies, the chase may not terminate and the various questions become undecidable.

2.5.5. THEOREM. *Under the weak universal assumption, testing consistency of d and Σ and evaluating $X[d, \Sigma]$ when Σ is a set of fd's are both in* PTIME. *They are* EXPTIME-*complete, when Σ is a set of full dependencies.*

Constructing and chasing the representative universal relation is not particularly efficient. There has been a great deal of work on better procedural query evaluation methods, e.g., [133, 204, 206, 212, 250–253, 303, 304]. The goal of these efforts is, given

Σ, X and D, to find a relational algebra expression E over D such that for all databases d over D we have $E(d) = X[d, \Sigma]$. This would reduce the query in the universal data model to a relational algebra query. Such an E need not always exist. If an E exists, we call Σ, X, D *algebraic*.

It is easy to see that the transitive closure query of Example 2.5.3 can be computed in a universal relation data model: one just needs $U = AB$ and a single full dependency; so this case is not algebraic. There are algebraic cases where E exists and can be computed efficiently, for example if Σ is one jd [250, 303]. For this case, with X extended to a positive existential query instead of just a projection, see [304].

The algebraic cases (where the difference relational operation is excluded) are characterized in [206] by a property of the chase procedure, i.e., termination in a fixed number of steps for any d. This property is called *uniform boundedness* of the chase. There are instances where uniform boundedness can be decided: if Σ is a set of fid's without equality and $X = U$, see [253].

Let us close this section with yet another application of dependency implication. We would like to compare the pure and the weak universal relation assumptions. One desirable condition is that if the pure assumption holds then we get the same result by projecting from the universal relation as we do by following the semantics of Definition 2.5.4. Formally we want that, for all r satisfying Σ and such that $d = \pi_D(r)$, we have $\pi_X(r) = X[d, \Sigma]$. In [206] this is shown to be true iff $\Sigma \models_f \sigma$, where σ is a *projected-join dependency* (pjd) and relation r satisfies σ if $\pi_X(\bowtie \pi_D(r)) = \pi_X(r)$.

2.5.6. Additional Bibliographic Comments. Weak universal relations have also been used to model incomplete information [298]. The weak universal relation assumption can be modified to capture the fine distinctions between values present but unknown, and values whose presence is even unknown [24, 89]. Semantics based on set partitions are proposed for the weak universal relation assumption in [89]; this shows the close connection between universal relation models and the deductive data model of [271].

3. Dependencies and database scheme design

3.1. Dependency classification

In this section we try to classify the many dependencies that have been examined in the literature. We follow the classification framework of embedded dependencies from [117]. We highlight the importance of three subclasses of these dependencies: embedded implicational dependencies [51, 52, 111, 305], inclusion dependencies [60], and rule dependencies [68, 124]. We outline some of the decidability and complexity bounds on their implication problems.

3.1.1. Definition. An *atom* is a formula of the form $R(z_1, \ldots, z_j)$, a *relational atom*, or is a formula of the form $z_1 = z_2$, an *equality atom*, where $z_1, \ldots, z_j$ are not necessarily distinct variables and R is a relation symbol of arity $j \geqslant 0$. Let $\varphi(x_1, \ldots, x_n)$ and $\psi(y_1, \ldots, y_m)$ be two nonempty conjunctions of atoms such that $x_1, \ldots, x_n$, $n \geqslant 1$, and

$y_1, \ldots, y_m, m \geqslant 1$, are the distinct variables appearing in these conjunctions respectively. An *embedded dependency* is a sentence of the f.o. predicate calculus with equality of the following form:

$$\forall x_1 \ldots \forall x_n \varphi(x_1, \ldots, x_n) \Rightarrow \exists z_1 \ldots \exists z_k \psi(y_1, \ldots, y_m).$$
$$\text{where } \{z_1 \ldots z_k\} = \{y_1 \ldots y_m\} - \{x_1 \ldots x_n\}.$$

A set of embedded dependencies Σ is satisfied by a database d if d satisfies each sentence in Σ, where the relation r_1 of d interprets R_1 etc. Note that embedded dependencies are satisfied by a database with empty relations and that each embedded dependency is a domain-independent sentence. Also a set of dependencies Σ can be written as a sentence in prenex normal form with quantifier prefix $\forall^* \exists^*$. In the definition above, conjunction φ is called the *body* and conjunction ψ the *head* of an embedded dependency. Without loss of generality, "the equality symbol need only occur in the head ψ and only between variables that also appear in the body φ"; we will hence assume that this well-formedness condition about equality is true. There are five common restrictions on embedded dependencies that give us five classes of dependencies:

(1) the *full* are those without $\exists$ quantifiers;
(2) the *unirelational* are those with one relation symbol only;
(3) the 1-*head* are those with a single atom in the head;
(4) the *tuple-generating* are those without the equality symbol;
(5) the *equality-generating* are full, 1-head, with an equality atom as head.

For full dependencies, being 1-head is not a constraining assumption, since each full dependency can be replaced by an equivalent set of full 1-head dependencies. One can always replace an embedded dependency by a set of tuple-generating dependencies and equality-generating dependencies, see [287]. Within this five-fold classification, three special subclasses of dependencies have been identified for their practical significance.

UNIRELATIONAL TYPING AND EID'S. Let R be the relational symbol for the unirelational case. Partition the sets of variables into sets corresponding to the arguments of R. A unirelational dependency is *typed* if

(1) its equality atoms are between two variables both from a set corresponding to some argument of R, and

(2) its relational atoms have in each argument of R a variable from the set corresponding to that argument.

The embedded dependencies that are typed unirelational are called *embedded implicational dependencies* (eid's). They were identified, independently and in a variety of formalisms, as the natural closure of many sets of dependencies [51, 52, 111, 305]. The full, 1-head eid's are called *full implicational dependencies* (fid's). In Section 2.3.3 we have already encountered *functional, join, multivalued dependencies* (fd's, jd's, mvd's), all examples of fid's. Other examples of 1-head eid's that we have encountered in Section 2.3.3 are *embedded join, embedded multivalued dependencies* (ejd's, emvd's). All these examples, with the exception of fd's, are tuple-generating. The tuple-generating, 1-head eid's are called *embedded template dependencies* (etd's) [248]. Each etd can be described by a tableau, where its tuples describe the body and its summary describe the single head atom with the existentially quantified variables of the head being those distinguished variables missing from the summary. Note that, for dependency satisfaction, we are not only interested in the tableau mapping, but in the database being closed under this mapping. When the summary of the tableau has all the distinguished variables then we have *full template dependencies* (ftd's); we have encountered such fid's without = in Section 2.5.2. At the end of Section 2.5.2 we also saw a sublclass of etd's containing the ejd's: these are the *projected-join dependencies* (pjd's) of [206]. Finally, the fd's with $|X|=|Y|=1$ are called *unary* fd's.

MULTIRELATIONAL CONSTRAINTS AND IND'S. *Inclusion dependencies* (ind's) are all embedded dependencies such that the head is one relational atom with no multiple occurrences of variables and the body is one relational atom with no multiple occurrences of variables. For example, $\forall x \forall y \forall z R_1(xyz) \Rightarrow \exists z' R_2(yxz')$ is an ind. Ind's, defined in [60], are very useful in describing multirelational constraint's, e.g., "referential integrity constraints". Note that ind's are tuple-generating, 1-head, but may involve more than one relation symbol and may be untyped. We have already encountered *unary inclusion dependencies* (uind's) for a single relation in Section 2.3.3. Uind's may be defined, in the obvious way, for may relation symbols. The results of Section 2.3. extend to the multirelational case as long as uind's are the only multirelational statements.

There is notational simplicity in denoting single-relation uind's as $A \subseteq B$. To extend this notation for ind's one must be careful and sequences instead of sets of attributes must be used. If R_1, R_2 are relation schemes, $\langle X \rangle$ a sequence of k distinct attributes of R_1 and $\langle Y \rangle$ a sequence of k distinct attributes of R_2, then $R_1\langle X \rangle \subseteq R_2\langle Y \rangle$ is a ind, and every ind can be represented in this fashion. For the previous example, this notation would give us $R_1\langle AB \rangle \subseteq R_2\langle BA \rangle$ where $R_1=ABC$ and $R_2=ABA'$. If $k=1$, we have uind's, $k=2$ *binary* ind's etc.

DEDUCTION AND RULE DEPENDENCIES. The full, 1-head, tuple-generating dependencies are called *rule dependencies* or rules for short. We have already mentioned in Sections 2.5.1–2 how rules may be used for querying in deductive and universal relation data models. They play an important part in the area of database logic programs [68, 124]. The analysis and optimization of these programs use many techniques from dependency theory, e.g., [88, 254]. A rule may be represented by an untyped tagged tableau with a summary that is full and can have repetitions of variables (see Remark 2.2.15). Ftd's are a special case of rules; for example, the results of [253] for ftd's apply to unirelational and typed database logic programs.

We have already defined the problems of finite implication and implication of dependencies in Definition 2.3.1. For each class of dependencies we have a special case of these implication problems. For example, an instance of eid implication consists of a set of eid's Σ and an eid σ; the question is whether $\Sigma \models \sigma$ and all these eid's are over one relation symbol R (but there is no a priori bound on the arity of R).

As we described in Section 2.3.3 (finite) implication for a class of embedded dependencies can be identified with (finite) unsatisfiability for a class of sentences in prenex normal form with quantifier prefix $\exists^*\forall^*\exists^*$. For full dependencies this quantifier prefix becomes $\exists^*\forall^*$ and, as a consequence, finite implication and implication are the same and decidable. In fact, this decision problem is in EXPTIME by a chase decision procedure for full dependencies. One corollary is that rules can be checked algorithmically for certain redundancies. For embedded dependencies the situation is qualitatively different: finite implication and implication need not coincide.

The typing in eid's entails a large degree of structure, e.g., algebraic properties such as the *faithfulness under direct product of* [111]. Eid implication can also be axiomatized [305, 52]. There is an elegant semidecision procedure for eid implication that generalizes the fd-jd chase [51]. Unfortunately, finite implication and implication are different even for etd's and, as shown independently in [49, 69], they are undecidable. These undecidability bounds were strengthened to pjd's independently in [142, 293]. The following two theorems summarize our current understanding about eid implication problems. The first is from [69] and the second from [142, 293].

3.1.2. THEOREM. *Finite implication and implication of fid's coincide and are* EXPTIME-*complete.*

3.1.3. THEOREM. *Finite implication and implication of eid's do not coincide and are both undecidable even for pjd's.*

The first result about ind's was that, despite the fact that ind's are not full, finite implication and implication coincide. Theorem 3.1.4 is from [60], where ind implication is also axiomatized.

3.1.4. THEOREM. *Finite implication and implication of ind's coincide and are* PSPACE-*complete.*

The implication problems for ind's alone and for ind's together with fd's have a qualitatively different flavor from the eid problems. For example, as illustrated in the axiomatizations of [218, 86, 87], equational formal systems are better suited for studying ind and fd implication than is the chase. The chase is still useful, but generally hard to reason about. A careful analysis of the chase in [169] suffices to show that implication of an eid from a set of ind's is PSPACE-complete. A sufficient condition for the termination of the chase is *ind acyclicity* [266]. Finite implication and implication of fid's and acyclic ind's coincide; but there are exponential lower bounds for acyclic ind's with fd's [86] and even for acyclic ind's alone implication is NP-complete [87].

As we have seen in Section 2.3.3, uind's and fid's interact in an interesting fashion, [173]. Fid and uind finite implication and implication differ, but are both decidable (Theorem 2.3.15). For fd's, mvd's and uind's all decision questions are in PTIME (Fig. 4).

Unfortunately, the good properties of ind's alone, or uind's and fid's, or acyclic ind's and fid's do not extend to the general case. The following theorem was shown independently [71, 218].

3.1.5. THEOREM. *Finite implication and implication do not coincide and are both undecidable even for unary fd's and binary ind's.*

The relevance of undecidability results, and of other complexity lower bounds, depends on whether arbitrary instances of an implication problem correspond to "real world" situations, see [264, 266] for studies of this issue. Pairwise consistency (see Section 2.4) is a natural universal relation assumption that is expressible using ind's across relations. Let the only dependencies within relations be unary fd's. The *ufd-graph* consists of a node for each attribute of each relation scheme and of an arc $\langle R.A, R.B\rangle$ for each unary fd $A \rightarrow B$ of R. Unary fd's and pairwise consistency interact in a surprinsingly nontrivial fashion. The following is from [86, 87].

3.1.6. THEOREM. *Implication of unary fd's in the presence of pairwise consistency is undecidable. If the ufd-graph is acyclic, finite implication and implication of ufd's in the presence of pairwise consistency coincide and are decidable.*

There are a number of technical implication problems that remain open, particularly for finite implication of eid's with ind's. For example decidability is open for

(1) finite implication of an eid from a set of ind's [169],

(2) finite implication of uind's, fd's and emvd's [173], or

(3) finite implication of ufd's in the presence of pairwise consistency—Theorem 3.1.6 is for implication.

However, the remaining outstanding question is emvd (finite) implication (2.3.16). Dependency theory is a well developed subject, which matured in the late 1970s and early 1980s. Much of the current attention is directed towards rule implication problems, because of their relationship to database logic program optimization.

3.1.7. ADDITIONAL BIBLIOGRAPHIC COMMENTS. There are some classes of dependencies that do not fit in the embedded dependency framework, e.g. in [290], arbitrary f.o. sentences are used as dependencies; the embedded dependencies are generalized further in [83], the afunctional dependencies of [95] are not embedded ones, and the partition dependencies of [89] are not f.o. expressible (see 2.3.10).

The concept of eid's generalized and unified much of the early work on dependency theory, e.g., [126, 138, 213, 227, 233, 235, 248, 259]. See [117] for the history and the relationship between these eid subclasses. Recent results on special ind and fd implication problems can be found in [61, 82, 86, 87, 169, 173, 190].

3.2. *Database scheme design*

The principal issue in database scheme design can be summarized in the following fashion: replace a specification $\langle\{U\}, \Sigma\rangle$ of universal relations r satisfying dependencies Σ over U by a *good* specification $\langle D, \Sigma'\rangle$ of multirelational databases $d = \{r_1, \ldots, r_m\}$ satisfying dependencies Σ' over $D = \{R_1, \ldots, R_m\}$, where $U = \bigcup_{i=1}^{m} R_i$. The goal is to replace r by d, where $d = \pi_D(r)$ and π_D is a decomposition (see Section 2.3.3). The intended use of this decomposition is twofold, both for querying and updating the database. We use the notational conventions of this paragraph throughout this section.

Database scheme design is more of an art than a science. However, certain general principles and crisp mathematical questions have emerged as a result of this endeavor. There are a number of standard expositions [182, 201, 280] and an early survey [36] to which we refer the reader for design-rule motivations such as the elimination of *update anomalies*. Instead we focus on two topics:

(1) what is a *good* decomposition π_D of $\langle \{U\}, \Sigma \rangle$, and

(2) what are the most common *normal forms* for $\langle D, \Sigma' \rangle$.

In (1) we assume that Σ consists of a set of full dependencies. Theorems 3.2.1, 3.2.2 and Definition 3.2.3 can be extended to embedded dependencies verbatim if we also allow unrestricted relations. In (2) we limit ourselves further to fd's and jd's, since these are the dependencies commonly examined in the context of normalization.

3.2.1. Independent schemes

The least one can require of a decomposition is that it be *injective*, i.e., one-to-one. In Section 2.3.3 we have already encountered this requirement, called the *representation principle*. It is also useful that the reconstruction operator, whose existence is implied by injectiveness, be the join. Although the representation principle does not imply join reconstruction, this is so in important special cases, e.g., Theorem 2.4.3.

Let us first assume the pure universal relation assumption and the desirability of injectiveness and join reconstruction. Theorem 3.2.1, from [46, 202], reduces these requirements to a dependency implication problem. There is an analogous semantic statement if we assume the weak universal relation assumption. Theorem 3.2.2 from [206] (see also end of Section 2.5.2): let r be a universal relation satisfying Σ and $d = \pi_D(r)$; then we want the query $U[d, \Sigma]$ to give us back r. Both theorems highlight the importance of having the join dependency $\bowtie[D]$ as part of the design requirements.

3.2.1. THEOREM. *Let Σ be a set of full dependencies over U. Decomposition π_D is injective over $\langle \{U\}, \Sigma \rangle$ with join as the reconstruction operation iff $\Sigma \models \bowtie[D]$.*

3.2.2. THEOREM. *Let Σ be a set of full dependencies over U and π_D a decomposition. For all universal relations r satisfying Σ we have that $r = U[\pi_D(r), \Sigma]$ iff $\Sigma \models \bowtie[D]$.*

Decomposing a universal relation into parts is certainly useful from a storage minimization point of view. But it may be wasteful from a querying standpoint. This is an instance of a classical time–space tradeoff. Under the pure universal relation assumption, we might have to perform the expensive reconstruction of the universal relation in order to answer a query. Under the weak universal relation assumption, we might have to do the same for the representative universal relation; much of the research on querying weak universal relations has focused on avoiding representative relation based evaluation.

Let $\{r_1, r_2\}$ be a decomposition of r and $\{r'_1, r'_2\}$ a decomposition of r', where r, r' are universal relations satisfying Σ. Another desirable property of decompositions is that the components r_1, r_2, r'_1, r'_2 do not depend on each other; in the sense that $\{r_1, r'_2\}$ and $\{r'_1, r_2\}$ are also decompositions of universal relations that satisfy the given dependencies. This condition, called the *separation principle*, would facilitate integrity and therefore database updating by limiting integrity checks to the parts instead of the whole. Unfortunately, under the pure universal assumption there are conditions on the whole that one cannot circumvent, e.g., that the parts are projections of a single relation.

Let us formalize the separation principle under the two universal relation assumptions. The sets of databases over D called PGSAT and WGSAT consist of those databases that are decompositions of universal relations satisfying Σ, under the pure and weak assumptions respectively. The G in PGSAT, WGSAT stands for global.

$$\text{PGSAT}=\{d:\exists r\models\Sigma,\ \forall r'\in d \text{ we have } r'=\pi_{R'}(r)\},$$
$$\text{WGSAT}=\{d:\exists r\models\Sigma,\ \forall r'\in d \text{ we have } r'\subseteq\pi_{R'}(r)\}.$$

Inverting the quantification order we get larger sets PLSAT and WLSAT, where L stands for local. In these sets, every relation is a piece of some decomposition of a universal relation satisfying Σ and these relations are put together into databases with only one additional guarantee. The only additional guarantee about the databases is made in the second argument of the intersection. For the pure assumption we require join consistency. For the weak assumption the analogous statement is always true:

$$\text{PLSAT}=\{d:\forall r'\in d, \exists r\models\Sigma \text{ and } r'=\pi_{R'}(r)\}$$
$$\cap\{d:\exists r, \forall r'\in d \text{ we have } r'=\pi_{R'}(r)\},$$
$$\text{WGSAT}=\{d:\forall r'\in d, \exists r\models\Sigma \text{ and } r'\subseteq\pi_{R'}(r)\}$$
$$\cap\{d:\exists r, \forall r'\in d \text{ we have } r'\subseteq\pi_{R'}(r)\}.$$

We clearly have PGSAT$\subseteq$PLSAT and WGSAT$\subseteq$WLSAT. The separation principle can be formalized as the *surjectiveness* conditions PGSAT$=$PLSAT and WGSAT$=$ WLSAT respectively.

3.2.3. Definition. Given $\langle\{U\},\Sigma\rangle$ where Σ is a set of full dependencies and D a database scheme over U, *D is independent* under the pure (weak) universal relation assumption if (1) $\Sigma\models\bowtie[D]$, and (2) PGSAT$=$PLSAT (WGSAT$=$WLSAT).

Independence was first proposed in [241] for Σ a set of fd's and for D having two relation schemes. It was extended to many relation schemes in [46, 202] and generalized further to full dependencies in [290]. In all this work the pure assumption is postulated, also injectiveness and surjectiveness are shown equivalent to $\Sigma\models\bowtie[D]$ and surjectiveness. Independence under the weak assumption is from [135].

Let us now limit attention to Σ's that are sets of fd's. An important computational notion for testing separation is that "the relation schemes of D must embed a cover for Σ". We say that D *embeds a cover* for Σ if there is *dependency preservation* as follows: Let $D=\{R_1,\ldots,R_m\}$ and Σ^+ be the closure of Σ under fd-implication, then $\pi_{R_i}(\Sigma^+)$, $1\leqslant i\leqslant m$, are the fd's of Σ^+ with attributes exclusively from R_i. Dependency preservation occurs when Σ and $\bigcup_{i=1}^m\pi_{R_i}(\Sigma^+)$ are equivalent sets of fd's. It is easy to see that dependency preservation is decidable, but it is more involved to show that it is decidable efficiently for fd's. The efficient dependency preservation test of [41]

(Theorem 3.2.4) can be used to determine independence under the pure assumption (Theorem 3.2.5). This has been extended by [135] to an independence test under the weak assumption (Theorem 3.2.5).

3.2.4. THEOREM. *Testing if a database scheme embeds a cover for a set of fd's is in* PTIME.

3.2.5. THEOREM. *Testing if a database scheme and a set of fd's are independent is in* PTIME, *under both the pure and the weak universal relation assumptions.*

3.2.6. ADDITIONAL BIBLIOGRAPHIC COMMENTS. There has been a fair amount of work on querying independent schemes under the weak assumption, e.g., [25, 62, 63, 135, 165, 252].

Injectiveness and $\Sigma \models \bowtie[D]$ can be used in lieu of each other provided we also require surjectiveness [290]. For a comparison of these two and other conditions (which are inequivalent without surjectiveness) we refer to [22, 43, 201].

3.2.2. Normal form schemes

Most of design has dealt with fd's, mvd's and the more general jd's; we thus limit Σ here to a set of fd's and jd's. We can use the chase algorithm of Section 2.3.3 to construct Σ^+, the closure of Σ with respect to fd-jd implication.

The designer is faced with the task of replacing a specification $\langle\{U\}, \Sigma\rangle$ by the multiple specifications $\langle\{R_i\}, \Sigma_i\rangle$, $1 \leqslant i \leqslant m$, where $D = \{R_1, \ldots, R_m\}$ is a database scheme over U and Σ_i is as follows: $\Sigma_i = \pi_{R_i}(\Sigma^+)$, that is, Σ_i consists of those fd's and jd's implied by Σ whose attributes are from R_i, $1 \leqslant i \leqslant m$. Note that Σ_i is closed under fd-jd implication.

A number of *normal forms* have been proposed for $\langle\{R_i\}, \Sigma_i\rangle$, $1 \leqslant i \leqslant m$. These are conditions on each individual $\langle\{R_i\}, \Sigma_i\rangle$. Normal forms guarantee less "redundancy" and fewer "update anomalies" [36]. In the previous subsection we outlined some properties that would make a decomposition into m normal forms desirable. One would like that for these normal forms the decomposition π_D is lossless. In addition, if possible, one would like that D is independent with respect to Σ.

The *first normal form* 1-NF corresponds to our definition of relation, so it is satisfied by any $\langle\{R_i\}, \Sigma_i\rangle$. Non-1-NF databases have relations whose values are not indivisible, e.g., values could be sets, tuples, relations etc. Non-1-NF databases have been studied as part of complex object data models (Section 4.3). For further normal forms we need some preliminary definitions.

A most important concept for database scheme design is that of a *key* of R_i: "A subset X of R_i is a key if $X \to R_i$ is in Σ_i and no proper subset of X has this property". Keys capture the essential relationships between attributes. The maintenance of key integrity constraints is supported by most database systems. Σ_i(*keys*) is the set of fd's $X \to R_i$ in Σ_i, where X is a key.

A *nontrivial* dependency is one that does not hold in all databases, i.e., it is not implied by the empty set of dependencies. $X \to Y$ over R is trivial iff Y is a subset of X. $X \twoheadrightarrow Y$ over R is trivial iff either Y is a subset of X or the union of Y and X is R. We can now define *third normal form* (3-NF), from [75]. 2-NF is a weaker normal form than is subsumed by 3-NF [75, 280], so we omit it in our presentation.

Let Σ_i be a set of fd's over R_i, closed under fd implication.

3-NF: $\langle\{R_i\}, \Sigma_i\rangle$ is in 3-NF if, for each nontrivial fd $X \to A$ in Σ_i, we have that either X contains a key of R_i or that A is an attribute of a key of R_i.

Given a $\langle\{U\}, \Sigma\rangle$, where Σ is a set of fd's, it is possible to construct $\langle\{R_i\}, \Sigma_i\rangle$, $1 \leqslant i \leqslant m$, all in 3-NF, such that the decomposition is independent under the pure assumption (by losslessness and dependency preservation). For Σ a set of fd's, dependency preserving 3-NF's were first constructed in [54], in PTIME, using a method called synthesis. It is possible to also satisfy the lossless join condition, again in PTIME, [58, 228].

Most database systems support integrity maintainance for keys. Thus, if in a normal form we limit Σ_i to the dependencies implied by Σ_i(keys), integrity maintenance will be easier to implement. This is one reason for trying to remove the last condition of 3-NF: "A is an attribute of a key of R_i".

The *Boyce–Codd normal form* (BC-NF) was first defined in [77] for fd's. We present two equivalent definitions from [108] of BC-NF for fd's and jd's. From these definitions it is clear that BC-NF implies 3-NF. The *fourth normal form* (4-NF) is from [108], for which we also provide two equivalent definitions. The *project-join normal form* (PJ-NF) is the strongest normal form for fd's and jd's from [109].

Let Σ_i be a set of fd's and jd's over R_i closed under fd-jd implication.

- **BC-NF:** $\langle\{R_i\}, \Sigma_i\rangle$ is in BC-NF if for each nontrivial fd $X \to Y$ in Σ_i we have that X contains a key of R_i.
- **4-NF:** $\langle\{R_i\}, \Sigma_i\rangle$ is in 4-NF if for each nontrivial mvd $X \twoheadrightarrow Y$ in Σ_i we have that X contains a key of R_i.

The following two equivalent definitions of BC-NF and 4-NF naturally lead to the definition of PJ-NF.

Let Σ_i be a set of fd's and jd's over R_i closed under fd-jd implication.

- **BC-NF:** $\langle\{R_i\}, \Sigma_i\rangle$ is in BC-NF if for each fd $\sigma \in \Sigma_i$ we have $\Sigma_i(\text{keys}) \models \sigma$.
- **4-NF:** $\langle\{R_i\}, \Sigma_i\rangle$ is in 4-NF if for each mvd $\sigma \in \Sigma_i$ we have $\Sigma_i(\text{keys}) \models \sigma$.
- **PJ-NF:** $\langle\{R_i\}, \Sigma_i\rangle$ is in PJ-NF if for each jd $\sigma \in \Sigma_i$ we have $\Sigma_i(\text{keys}) \models \sigma$

For PJ-NF⇒4-NF⇒BC-NF⇒3-NF⇒2-NF⇒1-NF, see [108, 109]. Moreover, using dependency implication and successive lossless decomposition into two parts, it is possible to transform $\langle\{U\}, \Sigma\rangle$ losslessly into $\langle\{R_i\}, \Sigma_i\rangle$, $1 \leqslant i \leqslant m$, which can be in any of these normal forms.

Unfortunately, even BC-NF is sometimes too strong a condition. Testing whether a given scheme is in BC-NF is NP-hard [35]. This implies that constructing such normal forms can be inefficient in the size of Σ. Things are even worse when independence is sought, because dependencies are not necessarily preserved. There are sets of fd's Σ such that no database scheme D exists which is in BC-NF and independent with respect to Σ, under the pure assumption.

Is there an ultimate normal form? The definition of PJ-NF was further generalized in [109] to the *domain-key normal form.* The semantic definition of domain-key normal form is analogous to the one for PJ-NF, where Σ_i might include arbitrary constraints, including bounds on the domain sizes. In [109], this semantic definition was shown to correspond to a formalization of the absence of update anomalies, and thus represent the ultimate normal form. PJ-NF retains the advantage of being constructible via successive lossless decompositions. The generality of domain-key normal form is at the expense of constructibility.

3.2.7. ADDITIONAL BIBLIOGRAPHIC COMMENTS. Third normal form is extended in [194] to eliminate certain redundancies across relations. Database design has been combined with various types of acyclicity, e.g., [130, 176, 308]. A normal form for nested relations appears in [229].

Testing for BC-NF is not the only natural NP-hard problem that arises in scheme design with fd's. Finding a minimum-size key of a relation scheme with fd's is NP-complete [198].

4. Queries and database logic programs

4.1. *Query classification*

The study of mappings from databases to databases, or *queries*, is essential in order to understand the limitations and the possible extensions of the relational data model.

We have already established the central role of relational algebra or calculus queries, also referred to as f.o. queries. It is important to distinguish between the queries themselves as mappings and the relational algebra expressions $E(\cdot)$ or calculus expressions $F(\cdot)$, which are programs denoting these mappings. We have also investigated tableau expressions $T(\cdot)$, which define a proper subset of the f.o. queries; see [67]. On the other hand, we have seen that the transitive closure query is not an f.o. query. Interestingly, using a deductive or a universal relation data model it is possible to express transitive closure via a program $R[\cdot, \Sigma]$ or $X[\cdot, \Sigma]$, where Σ is a set of rule dependencies.

In this section we present a systematic view of the classes of queries that are most common in database theory. For this we follow the general classification of [65].

4.1.1. *The computable queries and their data complexity*

The first natural question in this context is to delimit the queries that one can conceivably compute. For this let us think of the database d over D as a finite relational structure $(\bar{\delta}, r_1, \ldots, r_m)$, where $\bar{\delta}$ is a finite set containing the set of values δ appearing in the database. (This is a more liberal view of what constitutes the database domain than the use of δ in Section 2.2.2). We are interested in computable mappings between finite relational structures.

Computability for unrestricted relational structures has been extensively studied in mathematical logic, e.g., [221] but the emphasis has been on infinite models. The finite case is qualitatively different, even if the syntactic concepts are the same, see [140, 141]. In fact, the study of database queries has directly contributed to our understanding of finite models.

The class of *computable queries* is defined by Chandra and Harel in [66]. It captures three basic intuitions. First, that the query has to be a partial recursive function f from finite relational structures to finite relational structures. Second, that for each $d=(\bar{\delta}, r_1, \ldots, r_m)$, the values appearing in $f(d)$ are in $\bar{\delta}$. Finally that, isomorphic inputs should be mapped to isomorphic outputs. The third restriction is because relations are unordered collections of tuples; therefore the answer to a query should not be affected by the implementation details of how this set is stored, i.e., which is the first tuple etc. More specifically, any automorphism on the input structure should leave the output unchanged; this has also been proposed in [16, 28, 232]. The isomorphism requirement is the natural generalization of the automorphism requirement. Let us be more precise.

Consider two databases over D, say $d=(\bar{\delta}, r_1, \ldots, r_m)$ and $d'=(\bar{\delta}', r'_1, \ldots, r'_m)$. These databases are *isomorphic* if there is a bijection $h: \bar{\delta} \to \bar{\delta}'$ that extends componentwise to tuples and is such that for all i, $1 \leqslant i \leqslant m$, $t \in r_i \Rightarrow h(t) \in r'_i$ and $t' \in r'_i \Rightarrow h^{-1}(t') \in r_i$. The *isomorphism* h is denoted by $d \leftrightarrow^h d'$, and is called an *automorphism* if $d=d'$.

4.1.1. Definition. Let D be a database scheme and R a relation scheme such that $|R|=n$. A *computable query* from D to $\{R\}$ is a function f, which, on input database $d=(\bar{\delta}, r_1, \ldots, r_m)$ over D, has output database $f(d)$ over $\{R\}$ such that

(1) f is partial recursive,

(2) $f(d)=(\bar{\delta}, r)$, where $r \subseteq \bar{\delta}^n$,

(3) if $d \leftrightarrow^h d'$ then $f(d) \leftrightarrow^h f(d')$.

It is easy to see that f.o. queries are computable and that the transitive closure query is computable. Therefore, not all computable queries are f.o. ones. A programming language is proposed in [66] for the computable queries. This is an extension of relational algebra, where the crucial new features are a *while* looping construct and variables that can be assigned relations of any arity. A different programming language with the same expressive power is proposed in [7]. Unbounded arity variables are replaced there by the ability to create new values. Necessary and sufficient conditions on programming language constructs in order to express the computable queries is an interesting research issue.

Computable queries can be classified according to their data complexity (see Definition 2.2.6). Thus, class QPTIME consists of the computable queries with PTIME data complexity. Note that there is a difference between all queries with PTIME data complexity and QPTIME queries; the latter must satisfy the isomorphism condition of Definition 4.1.1. Similarly, we have computable queries for other complexity classes (i.e., QLOGSPACE, ...). It is shown in [67] that, under some weak technical condition (complexity class closure under many-one logspace reducibility), the various computable query classes defined this way are related to each other as the complexity classes. For example, QLOGSPACE = QPTIME iff LOGSPACE = PTIME.

As we saw in Theorem 2.2.7, f.o. queries are contained in QLOGSPACE. There are, however, simple queries in QLOGSPACE that are not f.o. queries, e.g., computing the parity of the database domain [67]. Thus, we have the following inclusions, where the containments ($\subseteq$) are strongly conjectured to be proper ($\subset$):

$$\text{F.O.} \subset \text{QLOGSPACE} \subseteq \text{QNLOGSPACE} \subseteq \text{QNC} \subseteq \text{QPTIME}$$
$$\subseteq \text{QNPTIME} \subseteq \text{QPHIER} \subseteq \text{QPSPACE} \subset \text{QEXPTIME} \subset \text{COMPUTABLE}.$$

There are close connections between *second-order* logic, interpreted over finite structures, and this query classification. As is shown in [106], *the queries in* QNPTIME *are precisely the queries expressible by existential second-order formulas over finite structures* (these formulas consist of a prefix of existential second-order quantifiers followed by an f.o. formula). A closely related result was independently shown in [170]. The connection to second-order logic was extended in [272]: *the queries in* QPHIER *are precisely the queries expressible by second-order formulas over finite structures.* For other connections to second-order logic we refer to [191].

A natural (infinitary, as opposed to second-order) extension of the f.o. predicate calculus with equality is fixpoint logic, also known as the μ-calculus [221]. This was used as a query language in [67] to define a class of computable queries, the *fixpoint queries* FP. Fixpoint queries are an important extension of f.o. queries. They have been studied extensively through universal and deductive data models and, more recently, through database logic programs. QPTIME properly contains FP, since computing the parity of the database domain is not in FP. There are fixpoint queries that are logspace-complete in PTIME and thus most probably not in QNC. This is an important distinction from f.o. queries, since much of the efficiency of the relational data model is due to the large degree of potential parallelism inherent in f.o. queries (Theorem 2.2.7).

The relationship between complexity classes and various logics has been investigated in detail; see the survey [163]. Many of the results in the area assume the ability to test for $<$, which is a special relation that linearly orders the database domain. This change in the rules of the game is certainly possible, since strings are used to represent values and strings are usually equipped with lexicographic order. Also, the ability to test for $<$ can add to the expressive power of the language. The presence of $<$ has allowed the comparison of a variety of extensions of f.o. logic (e.g., especially when the resulting data complexity is in PTIME [162]) by identifying expressibility with computational complexity. On the other hand, $<$ is not a particularly intuitive programming construct.

Most of the analysis of fixpoint queries, within database theory, does not assume the presence of $<$. However, its addition has an interesting effect on expressibility: *the queries with* PTIME *data complexity are precisely the queries expressible by fixpoint formulas over finite structures with* $<$. This was shown independently in [161, 262, 291]. In the proofs, all PTIME computations are simulated using $<$ (not only the computations satisfying condition (3) of Definition 4.1.1). An interesting open issue is to design a language for exactly QPTIME.

4.1.2. ADDITIONAL BIBLIOGRAPHIC COMMENTS. The data complexity of many query languages (both procedural and declarative) is determined in [291]. Two more measures are defined in [291], expression complexity and expression + data complexity. They correspond respectively to asymptotic growth in the program only and in both program + data. These measures are typically one exponential higher than data complexity.

The finer structure of f.o. queries is examined in [67], where a query hierarchy is built based on the alternation of first-order quantifiers in a calculus expression $F(\cdot)$.

A large repertoire of programming language constructs can be added to relational algebra and the expressive power of the resulting query languages is examined in [64]. Various *looping* constructs, in particular, lead to interesting query classes, see [67, 64, 65, 7, 8]. *Nondeterminate* queries that are relations, as opposed to functions, are studied in [7, 8], as part of an investigation of update languages. Programming formalisms for fixpoint queries are also examined in [148, 189].

4.1.2. *The fixpoint queries*

Transitive closure can be expressed by the addition to relational calculus of least fixpoint equations, see [16]. If R_2 is interpreted as a directed graph then this graph's transitive closure is the least, under subset ordering, interpretation of R_1 that satisfies the following equation among query expressions: $R_1(xy) \equiv R_2(xy) \vee \exists z(R_2(xz) \wedge R_1(zy))$. The right-hand side of this equation is a f.o. formula that is monotone and positive in R_1. In general, let $\varphi(\vec{x}; R_1)$ be an f.o. formula, where $\vec{x}$ is its vector of distinct free variables of size $n_1 \geqslant 0$ and R_1 is a relation symbol of arity n_1. This formula $\varphi(\vec{x}; R_1)$ might have other relation symbols occurring in it, but we highlight R_1 because it will be on the left-hand side of the equation, (i.e., it will be a variable).

Formula $\varphi(\vec{x}; R_1)$ is *monotone* in R_1 when, given any database d over all the relation symbols of $\varphi(\vec{x}; R_1)$ except R_1, for all r, r' relations over R_1, $r \subseteq r'$ entails $\{\vec{c}\colon \varphi(\vec{c}; r_1)\} \subseteq \{\vec{c}\colon \varphi(\vec{c}; r')\}$ where the $\vec{c}$ are vectors of size n_1 of values from the domains of d, r, r'.

A sufficient condition for the monotonicity of $\varphi(\vec{x}; R_1)$ in R_1 is that every occurrence of R_1 in this formula is under an even number of negations. In that case we say that $\varphi(\vec{x}; R_1)$ is *positive* in R_1. Every f.o. formula monotone in R_1 (viewed as a query expression) is unrestricted equivalent to some f.o. formula positive in R_1. This is known as Lyndon's Lemma in model theory. Thus, the syntactic condition of positivity, which is easy to test, matches the semantic condition of monotonicity for unrestricted databases. Unfortunately, the transformation from monotone to positive is not effective because, testing for monotonicity is undecidable in both the unrestricted and finite cases [140]. A surprising fact for finite structures is that f.o. formulas monotone in R_1 are more expressive (viewed as query expressions) than positive ones. The following theorem is from [17].

4.1.3. THEOREM. *For finite structures, there is a f.o. formula monotone in R_1 that is not equivalent to any f.o. formula positive in R_1.*

For formulas $\varphi(\vec{x}; R_1)$ monotone in R_1, the equation $R_1 \equiv \varphi(\vec{x}; R_1)$ has a least fixpoint solution on any database d over all the relation symbols of $\varphi(\vec{x}; R_1)$ except R_1.

This follows from the Tarski–Knaster Theorem. The *least fixpoint of* $\varphi(\vec{x}; R_1)$ *on* d, denoted r_∞, can be be evaluated in an unbounded but finite number of stages (by the finiteness of d) as follows:

$$r_0 = \emptyset, \qquad r_{i+1} = \{\vec{c} : \varphi(\vec{c}; r_i)\}, \qquad r_\infty = \bigcup_{i \geqslant 0} r_i.$$

If $\varphi(\vec{x}; R_1)$ is neither positive nor monotone in R_1, it is still possible to define a solution for the equation $R_1 \equiv \varphi(\vec{x}; R_1)$ using inflationary semantics. The *inflationary (or inductive) fixpoint of* $\varphi(\vec{x}; R_1)$ *on* d, denoted r_∞, can be evaluated in an unbounded but finite number of stages as follows:

$$r_0 = \emptyset, \qquad r_{i+1} = r_i \cup \{\vec{c} : \varphi(\vec{c}; r_i)\}, \qquad r_\infty = \bigcup_{i \geqslant 0} r_i.$$

The inflationary fixpoint of $\varphi(\vec{x}; R_1)$ on d is the same as the least fixpoint of $(\varphi(\vec{x}; R_1) \vee R_1(\vec{x}))$ on d. Also, the inflationary and the least fixpoint on d coincide if $\varphi(\vec{x}; R_1)$ is monotone. In both fixpoint constructions above it is easy to see that $r_i \subseteq r_{i+1}$, $i \geqslant 0$.

The *fixpoint queries* FP are defined in [67] using fixpoint formulas. Informally, fixpoint formulas are obtained recursively using the f.o. constructors $\exists$, $\forall$, $\vee$, $\wedge$, $\neg$ as well as a fixpoint constructor μ. The constructor μ binds a relation symbol R_1 appearing free in a fixpoint formula and is applicable only when R_1 appears positively in this formula, i.e., under an even number of negations. The semantics of $\mu R_1 \psi(\vec{x}; R_1)$ on d are the same as the least fixpoint of $\psi(\vec{x}; R_1)$ on d (where, recursively, $\psi(\vec{x}; R_1)$ is a fixpoint formula). We refer to [67] for the detailed definitions. We omit them here because by Theorem 4.1.5 below one application of μ suffices.

Given Theorem 4.1.3, one might consider the set of fixpoint formulas, formed under the positivity restrictions on the applicability of μ, as a somewhat arbitrary choice. By keeping the least fixpoint semantics but making the applicability of μ more liberal (μ applicable to all monotone formulas) one can define a set of queries FM. Of course, in this case it is undecidable to test whether μ is applicable. By adopting the inflationary semantics one can remove all restrictions on the applicability of μ and define a set of queries FI. It follows from the definitions that FP $\subseteq$ FM $\subseteq$ FI. The following theorem from [143] testifies to the robustness of FP; we stress the finiteness because the theorem is not true for unrestricted databases.

4.1.4. THEOREM. *For finite structures*, FP = FM = FI.

In [161], Immerman showed another surprising result, also not true for unrestricted databases: *the complement, with respect to the database domain, of the least fixpoint of an f.o. formula on a database can be expressed as the least fixpoint of some other formula on this database.* In fact, fixpoint formulas have equivalent normal forms where the μ constructor is applied only once, [143], see also [161]. This normal form has been simplified in [8], where it is shown that the $\forall$ constructor is unnecessary. The fixpoint queries can be expressed taking inflationary fixpoints of *existential* f.o. formulas and their projections. Thus, we need to consider only inflationary fixpoints of f.o. formulas with an $\exists^*$ quantifier prefix and $\neg$ limited to the matrix; see also [141, Remark Section 6].

We summarize the above discussion in Theorem 4.1.5 using a formalism that is closer to database logic programs. We use a system of equations for which we define simultaneous least and inflationary fixpoints on databases. These equations are mutually recursive. We could have eliminated the mutual recursion and have a single equation, as is done with fixpoint formulas. This elimination imposes certain technical restrictions on the vocabulary in [67] and is why $\neq$ is used in [67]. The use of a system has two advantages: it is close to the notation used for database logic programs and it leads to the definition of equation width.

Let us therefore consider a system of equations $R_i \equiv \varphi_i(\vec{x}_i; R_1, \ldots, R_m)$, $1 \leqslant i \leqslant m$, where each $\varphi_i(\vec{x}_i; R_1, \ldots, R_m)$ is an f.o. formula, $\vec{x}_i$ is its vector of n_i distinct free variables and R_i is a relation symbol of arity $n_i \geqslant 0$. The *equation width* is the maximum n_i for $1 \leqslant i \leqslant m$. The notation emphasizes the fact that the left-hand sides are the relation variables (they need not occur in each right-hand side). A database d interprets all the relation symbols occurring in the right-hand sides except for $R_1, \ldots, R_m$. The inflationary fixpoint of a system of equations on d is defined analogously to the inflationary fixpoint of one equation on d; only a sequence of databases $d_0, \ldots, d_\infty$ must be used instead of a sequence of relations $r_0, \ldots, r_\infty$ [68]. If the $\varphi_i(\vec{x}_i; R_1, \ldots, R_m)$, $1 \leqslant i \leqslant m$, are monotone in $R_1, \ldots, R_m$ the same can be done for the least fixpoint on d.

4.1.5. THEOREM. *For finite structures, each fixpoint query can be expressed as the projection on R_1 of the inflationary fixpoint of $R_i \equiv \varphi_i(\vec{x}_i; R_1, \ldots, R_m)$, $1 \leqslant i \leqslant m$, where each $\varphi_i(\vec{x}_i; R_1, \ldots, R_m)$ in this system of equations is an existential f.o. formula.*

4.1.6. EXAMPLE. First, let us illustrate inflationary fixpoints of systems of equations using a system that fails to compute the complement of the transitive closure.

$$R_1(xy) \equiv \neg R_2(xy),$$
$$R_2(xy) \equiv \exists z R_3(xy) \vee (R_3(xz) \wedge R_2(zy)).$$

Given a database d over $\{R_3\}$ (the input graph), let r_∞, r'_∞ be the projections on R_1, R_2 of the inflationary fixpoint on d. Then $r_0 = r'_0 = \emptyset$ and $r'_1 =$ (the input graph) and $r_1 =$ (all possible pairs of nodes). In fact, $r'_\infty =$ (the transitive closure of the input graph) and $r_\infty =$ (all possible pairs of nodes).

The following system (from [8]) succeeds in computing the complement of the transitive closure, given an input graph as a database over $\{R_5\}$. The semantics used are inflationary.

$$R_1(xy) \equiv \exists x' \exists y' \neg R_4(xy) \wedge R_3(x'y') \wedge \neg R_2(x'y').$$
$$R_2(xy) \equiv \exists x' \exists y' \exists z' R_4(xy) \wedge R_4(x'z') \wedge R_4(z'y') \wedge \neg R_4(x'y').$$
$$R_3(xy) \equiv R_4(xy).$$
$$R_4(xy) \equiv \exists z R_5(xy) \vee (R_5(xz) \wedge R_4(zy)).$$

The idea is that, as the transitive closure is inserted in R_4 by the successive stages of the fixpoint construction, it is also copied in R_2 and R_3. The copying is one stage behind the transitive closure construction in R_4 and its last stage is copied in R_3 but not R_2. By finiteness there is a last stage! Only then are tuples inserted in R_1 that have not been inserted in R_4.

4.1.7. ADDITIONAL BIBLIOGRAPHIC COMMENTS. The equational formalism used in Theorem 4.1.5 is identical with the language Datalog$^\neg$ defined in [181, 8]. The existence and uniqueness of least fixpoints of existential f.o. formulas is studied in [181] from the point of view of computational complexity.

The fixpoint queries are further classified in [67] according to *fixpoint width.* Although some progress has been made in understanding the properties of fixpoint width [119], there are still many open questions. The fixpoint width of [67] differs from the equation width used above and in [42, 98, 221]; the difference is in the treatment of mutual recursion. For some recent results on equation width, see [10, 102].

4.2. Database logic programs

The semantics, the optimization and the evaluation of database logic programs are some of the most active research topics in database theory today. This activity is bound to render obsolete any attempted survey of the area. There are, however, strong ties to other more mature topics, such as the theory of dependencies and the theory of queries. We outline the state of the field by emphasizing these connections.

In database logic programs, Datalog$^\neg$ and the fixpoint queries play the role that relational calculus and the f.o. queries play in the relational data model. Datalog, a sublanguage of Datalog$^\neg$, has received most of the attention in terms of program optimization and evaluation. This is analogous to the importance of positive existential programs in the relational data model. Most of this work has direct applications to the development of new efficient knowledge base systems.

4.2.1. From Datalog to Datalog$^\neg$

DATALOG AND DATALOG$^\neg$ SYNTAX. The vocabulary of a Datalog$^{(\neg)}$ program H is the set of relation symbols occurring in H. It is partitioned into the intensional database symbols (IDB's) $\{R_1, \ldots, R_m\}$ and the extensional database symbols (EDB's) $D = \{R_{m+1}, \ldots, R_k\}$ (we also call the EDB's D the set of *input relation symbols*). The IDB symbol R_1 is called the *output relation symbol.*

H consists of Datalog$^{(\neg)}$ *rules*, where every rule has a *head* and a *body*. Only IDB's occur in the heads of rules and every IDB occurs in the head of some rule.

Each Datalog$^\neg$ rule is of the form $R(\vec{x})\text{:-}\varphi$.

(1) The head $R(\vec{x})$ consists of an IDB R of arity $n \geqslant 0$ and of a vector $\vec{x}$ of variables. Without loss of generality we can take $\vec{x} = xy\ldots$ the vector of the first n distinct variables, in some standard ordering of the variables (i.e., heads are normalized).

(2) The body φ is a list of equality atoms, relational atoms and negations of such atoms. This list is terminated by a point (.) and the atoms are separated by commas (,). The variables in $\vec{x}$ must occur in φ.

Datalog rules are Datalog$^\neg$ rules without any negated atoms in their bodies.

DATALOG AND DATALOG$^\neg$ FIXPOINT SEMANTICS. Given Datalog$^{(\neg)}$ program H, construct the following formula for each IDB R_i, $1 \leqslant i \leqslant m$: let $\varphi_i(\vec{x}; R_1, \ldots, R_m)$ be the disjunction of the bodies of all rules with head $R_i(\vec{x})$, where each body is a conjunction of its list of atoms prefixed by existential quantifiers for all variables except $\vec{x}$. That is, we have changed in the bodies comma (,) into "and" ($\wedge$), point (.) into "or" ($\vee$), and have existentially quantified all the variables not in the head.

We form the system $S(H)$ of equations: $R_i(\vec{x}) \equiv \varphi_i(\vec{x}; R_1, \ldots, R_m)$, $1 \leqslant i \leqslant m$. Let d be a database over input symbols D, i.e., over the vocabulary except $\{R_1, \ldots, R_m\}$. If H is a Datalog$^{\neg}$ program, then $S(H)$ has an inflationary fixpoint on d (see Section 4.1.2); moreover, all φ_i's are f.o. existential. If H is a Datalog program then the R_i's appear only positively in the right-hand sides of the equations and $S(H)$ has a least fixpoint on d (see Section 4.1.2), which by monotonicity is the same as the inflationary fixpoint.

On input database d over D, the output of H is the projection on R_1 of the inflationary fixpoint (or the least fixpoint in the case of Datalog) defined by $S(H)$ on d.

It is clear that Datalog queries are Datalog$^{\neg}$ queries, which in turn are fixpoint queries. We have assigned fixpoint semantics to Datalog and Datalog$^{\neg}$ programs. Such fixpoints semantics can be computed using the standard Tarski construction (see Section 4.1.2). This construction is also known as *naive bottom-up evaluation* and can be viewed as an "operational semantics" for database logic programs.

Note the similarity of Datalog notation with the Prolog programming language syntax and the differences in their "operational semantics". The Datalog rules (negation is excluded here) are Horn clauses [105], when one views (:–) as ($\Leftarrow$), (,) as ($\wedge$), and (.) as ($\vee$). Recall that there are no negations in the bodies of Datalog rules. In the Horn clause case there is a fundamental fact from logic programming [18, 20]: that *least fixpoint semantics coincide with minimum model semantics*. So let us present the Datalog minimum model semantics.

DATALOG MINIMUM MODEL SEMANTICS. Given a Datalog program H and a database d over D, the *minimum model* $\mathcal{M}(d, H)$ is defined as follows. The *Herbrand base* of d and H is the set of all possible *ground* relational atoms (i.e., relational atoms with values substituted for the variables) that can be constructed using the vocabulary of H and the values in d. The Herbrand base is finite since d and H are finite.

$\mathcal{M}(d, H)$ is the least, under set inclusion, subset of the Herbrand base containing d and satisfying each rule of H interpreted as a universally closed Horn sentence. That is, in each rule (:–) is changed into ($\Leftarrow$) and comma (,) into ($\wedge$) and *all* the variables of the resulting Horn clause are universally quantified.

On input database d over D the output of H is the projection on R_1 of $\mathcal{M}(d, H)$.

4.2.1. THEOREM. *The least fixpoint and the minimum model semantics of a Datalog program coincide.*

It is clear that rule dependencies are Datalog rules. There are only minor differences in conventions of notation; e.g., in Datalog rules instead of having repeated variables in the head we use $=$ in the body. That is the only nontrivial use of $=$ that we make in Datalog and can be eliminated with repeated variables. We have avoided using impure features in Datalog, such as $\neq$ [68] or constants in the rules. This is in keeping with our definition of relational calculus without constants and with our use of systems of equations to express mutual recursion. It is simple to add constants to the rules. As long as these constants are added to the Herbrand base, all the definitions generalize in the straightforward fashion. If the only negations are inequalities in the bodies, we have a sublanguage of Datalog$^\neg$ called Datalog$^{\neq}$.

Datalog queries are a proper subset of the fixpoint queries, because of the absence of negation. They are incomparable with f.o. queries, because of the ability to express transitive closure in Datalog, but they have many interesting properties. They contain the positive existential queries, for which the homomorphism technique was originally developed. On the other hand, it is easy to see the Datalog$^\neg$ queries are exactly the fixpoint queries (see Theorem 4.1.5).

Querying in deductive and universal relation data models with rule dependencies can be reformulated as querying with Datalog. The chase may now be viewed as an *evaluation algorithm*. Consider the input database d as a set of untyped tagged tableaux that are chased using the rules of program H as tuple-generating rule dependencies (see [134]). The result of the chase is the minimum model $\mathcal{M}(d, H)$. Note that, in the construction of least fixpoints in Datalog and of inflationary fixpoints in Datalog$^\neg$, a particular order of rule applications is used. A fine point is that, by the Church–Rosser property of the chase, the order of rule applications is immaterial for Datalog. However, it is significant for Datalog$^\neg$. Finally, as we shall see, another use of the chase algorithm is for the *optimization* of Datalog programs.

4.2.2. EXAMPLE. Let us rewrite the systems of Example 4.1.6 in Datalog$^\neg$ notation. The first system becomes the following set of rules. Note that the second and third rule form a Datalog program for the transitive closure.

$$R_1(xy) :\!- \neg R_2(xy).$$
$$R_2(xy) :\!- R_3(xz),\ R_2(xy).$$
$$R_2(xy) :\!- R_3(xy).$$

The second system becomes the following set of rules.

$$R_1(xy) :\!- \neg R_4(xy),\ R_3(x'y'),\ \neg R_2(x'y').$$
$$R_2(xy) :\!- R_4(xy),\ R_4(x'z'),\ R_4(z'y'),\ \neg R_4(x'y').$$
$$R_3(xy) :\!- R_4(xy).$$
$$R_4(xy) :\!- R_5(xz),\ R_4(zy).$$
$$R_4(xy) :\!- R_5(xy).$$

Both sets of rules in this example are what is known as *stratified*. Their IDB's can be partitioned into a *linearly ordered set of strata*, such that

(1) a positive atom $R'(\vec{z})$ is in the body of a rule with head $R(\vec{x})$ iff the stratum of R' is less or equal than the stratum of R, and

(2) a negative atom $\neg R'(\vec{z})$ is in the body of a rule with head $R(\vec{x})$ iff the stratum of R' is strictly less than the stratum of R.

In Example 4.2.2, the ordered strata for the first set of rules are $\langle\{R_2\}, \{R_1\}\rangle$ and for the second set of rules $\langle\{R_4, R_3\}, \{R_2\}, \{R_1\}\rangle$ or $\langle\{R_4\}, \{R_3, R_2\}, \{R_1\}\rangle$ could be possible ordered strata.

Stratified sets of rules can be assigned least fixpoint semantics [68] or equivalent minimum model semantics [19] by evaluating the strata bottom-up and constructing the least fixpoint for a lower stratum before going on to a higher one. The particular linear order chosen is immaterial and, in addition, within each stratum a Church–Rosser property applies for the evaluation of rules just as in Datalog. Thus, the stratified evaluation of certain Datalog$^{\neg}$ programs can be used to define the class of *stratified* queries. Note that, in Example 4.2.2, both sets of rules express the complement of the transitive closure when they are evaluated in a stratified way. Under the inflationary semantics, only the second set of rules computes the complement of the transitive closure.

Stratified queries have been proposed as a natural and implementable means of introducing negation into database logic programs, see [19, 68, 222, 283]. They properly contain both Datalog and f.o. queries. Operationally they are related to Clark's *negation as failure* principle for general logic programs [74].

A difficulty with this aproach is that, the number of strata in a stratified program is bounded. Using the inflationary semantics of Datalog$^{\neg}$, it is possible to simulate an unbounded number of strata. As shown in [91, 180], the stratified queries are a proper subset of the fixpoint queries. Let P.E. be the positive existential queries, S-DATALOG the stratified queries, etc.

4.2.3. THEOREM. *For finite databases,* DATALOG *and* F.O. *are incomparable and we have*

$$\begin{aligned}\text{P.E.} &= (\text{DATALOG} \cap \text{F.O.}) \subset (\text{DATALOG} \cup \text{F.O.})\\ &\subset \text{S-DATALOG} \subset \text{S-DATALOG}^{\neg} = \text{FP} \subset \text{QPTIME}.\end{aligned}$$

This theorem summarizes our understanding of the expressibility of negation. One point in the theorem must be further clarified. It is easy to see that P.E. $\subseteq$ (DATALOG$\cap$F.O.). For unrestricted databases P.E. = (DATALOG$\cap$F.O.) by a compactness argument. It has been recently claimed by Ajtai and Gurevich that this also holds in the finite case (see [84] for special cases).

4.2.4. ADDITIONAL BIBLIOGRAPHIC COMMENTS. The introduction of negation into logic programs has been a major topic of research. We have limited our exposition to negation in database queries and stressed the properties of negation in finite model theory. We refer to [18, 216] for more information on negation and minimal models. We refer to [284] for some recent results on minimum model semantics that also deal with inflationary fixpoints.

4.2.2. Query optimization and stage functions

We have defined minimum model and equivalent, least fixpoint semantics for Datalog programs (programs for short in this subsection). An alternative and useful way of presenting the construction of least fixpoint semantics is via derivation trees.

A ground atom t is in the minimum model $\mathcal{M}(d, H)$ iff there is a derivation tree for t from d and H. A derivation tree for t is a tree, where each node of the tree is labeled by a ground atom such that

(1) each leaf is labeled by a tuple of d;

(2) for each internal node there is a rule of H whose variables can be instantiated so that the head is the label of that node and the body is the set of labels of its children; and

(3) the root is labeled by t.

The depth of a derivation tree is the length of its longest path and its size is the number of nodes. Derivation trees are descriptions of computations of special alternating Turing machines associated with each database logic program [282, 172]. They are particular to Datalog and are commonly used to analyze logic programs.

4.2.5. DEFINITION. Let $H(d)$ be the output of program H on input database d. The *stage function* of H is $\xi(n, H) = \max\{\xi(d, H): |d| \leqslant n$ and $|d|$ is the size of d where $\xi(d, H) = \min\{i$: for each t in $H(d)$ there is a derivation tree of depth $\leqslant i\}$.

Stage functions are a major topic in [221], where they are defined for unrestricted databases. The name "stage function" is used because $\xi(d, H)$ is the first iteration of the naive bottom-up evaluation by which all the output has been constructed. After this iteration, the output does not change, even if other ground atoms are added to the minimum model. Stage functions can also be defined for Datalog$^{\neg}$ programs, using this intuition about iterations and the construction of inflationary fixpoints.

Recall that, by convention, the input symbols are the EDB's and the output symbol is an IDB. If H is a program let program $\bar{H}$ have the same set of rules, but all the vocabulary as output symbols. We use the term *program stage function* of H for $\xi(n, \bar{H})$. If a H has only a single IDB then we have that $\xi(n, \bar{H}) = \xi(n, \bar{H})$.

The distinction between IDB's and EDB's is somewhat artificial and is based on the fact that the IDB's are initialized to $\emptyset$ in the iterative construction of fixpoints. It is possible to define *uniform* least and inflationary fixpoint semantics by initializing the IDB's as arbitrary relations. Then the vocabulary of H, the output and the input symbols are all the same. Under these uniform semantics, we have the *uniform stage function* of H, denoted $\bar{\xi}(n, \bar{H})$. An interesting special case are *single-rule programs* or *sirups*: they consist of a single rule and are given the uniform least fixpoint semantics [172]. So they have a single IDB and the uniform stage function.

Let us relate the stage functions of a program H. The *equation width* of H is the largest arity of an IDB in H. It is easy to show that for all n, $\xi(n, H) \leqslant \xi(n, \bar{H}) \leqslant \bar{\xi}(n, \bar{H})$ and $\bar{\xi}(n, \bar{H}) = O(n^k)$, where k is the equation width of H. Programs with equation width 1 are called monadic, with width 2 binary, etc. Note that programs can have arbitrary arity EDB's occurring in the bodies of the rules.

4.2.6. EXAMPLE. There are some other classes of Datalog programs, that have been investigated. One example is *linear* programs: in the bodies of the rules of these programs there can be at most one IDB occurrence. Another example are *chain* programs [282]: their syntax greatly resembles that of context-free grammars. Recall that the semantics for sirups are uniform (i.e., all relation symbols are initialized to arbitrary nonempty relations).

Program H_0 is a linear, chain, binary sirup known as the "same-generation" program:

$$R_1(xy) \text{:-} R_2(xz),\ R_1(zz'),\ R_3(z'y).$$

Program H_1 is a nonlinear, chain, binary sirup known as the "ancestor" program:

$$R_1(xy) \text{:-} R_1(xz), R_1(zy).$$

Note the difference of the ancestor program from the uniformly interpreted sirup H':

$$R_1(xy) \text{:-} R_1(xz), R_2(zy).$$

H_1 expresses the same query as the following program consisting of two rules:

$$R_1(xy) \text{:-} R_1(xz), R_2(zy). \qquad R_1(xy) \text{:-} R_2(xy).$$

Program H_2 is a nonlinear, monadic sirup that is not a chain. It expresses the "path accessibility problem". This is the prototypical logspace-complete problem in PTIME discovered by Cook:

$$R_1(x) \text{:-} R_1(y), R_1(z), R_2(xyz).$$

Program H_3 is a sirup whose uniform stage function is O(1):

$$R_1(xy) \text{:-} R_2(x), R_1(zy).$$

The conventions on output and uniformity lead to many notions of containment for two programs with the same input and output symbls. The containment that we have previously encountered for other query languages has a direct analog for Datalog programs, called *containment*. We use the term *program containment* when the output symbols are all symbols and the term *uniform containment* under the uniform semantics. The first basic observation is that *all the kinds of containment are the same as their unrestricted containment counterparts.* It is a special feature of Datalog that containments and unrestricted containments coincide and are therefore co-r.e. The second observation is that *uniform containment* $\Rightarrow$ *program containment* $\Rightarrow$ *containment*, but the converses need not hold.

Using context-free language theory, one can show that program containment and containment are undecidable (Π_1^0-complete) even for linear, binary, chain programs [85, 269]. Using regular language theory, even for non-chain programs, one can show that all the kinds of containment are decidable for monadic programs [85]. The important connection of program rules and rule dependencies becomes clear when we consider uniform containment. It was first noted in [88] for full template dependencies (ftd's) and in [254] for all rules that uniform containment is the same as rule dependency implication and therefore EXPTIME-complete. Therefore, the chase is useful as a Datalog program optimization technique. The many possible notions of equivalence between logic programs are investigated in [199].

4.2.7. THEOREM. *Program H is uniformly contained in program H′ iff the set of rule dependencies H is implied by the set H′.*

The presence of recursion in the language generates new questions, beyond containments. The most fundamental of these questions is whether recursion is bounded. As shown in [120] undecidablity of recursion boundedness can be translated into undecidability for most other questions concerning recursion. Interestingly, the first bounded recursion definitions and uses appeared for universal models [206] and deductive data models [217].

4.2.8. DEFINITION. A program H is *bounded* if $\xi(n, H)=\mathrm{O}(1)$, it is *program bounded* if $\xi(n, \bar{H})=\mathrm{O}(1)$, and it is *uniformly bounded* if $\bar{\xi}(n, \bar{H})=\mathrm{O}(1)$.

Program H_3 in Example 4.2.6 is uniformly bounded. Clearly, *uniform boundedness* $\Rightarrow$ *program boundedness* $\Rightarrow$ *boundedness*, but the converses need not hold. From the definition one can see that undecidability of uniform boundedness automatically translates into undecidability for program boundedness, etc. Also, that decidability of boundedness implies decidability of all the other kinds.

Boundedness, and all the other varieties as well, imply that the query expressed by the program is a positive existential query and therefore an f.o. query. It is also true that uniform unboundedness, and all the other varieties as well, imply that the query expressed by the program is not a positive existential query [226]. If we allowed unrestricted databases then, by a compactness argument, unboundedness implies non-f.o. expressibility. For databases this has been recently claimed by Ajtai and Gurevich. It is true because of Datalog's special features. For Datalog$^{\neg}$ the situation is different, as Kolaitis has observed: there is an unbounded Datalog$^{\neg}$ program which is f.o. expressible over finite structures.

Another example of the special character of Datalog from [88] is a *gap* theorem: *uniform unboundedness implies* $\bar{\xi}(n, \bar{H})=\Omega(\log n)$.

Boundedness is decidable for various subclasses of Datalog programs. This was first demonstrated in [164, 224] for subclasses of linear programs with PTIME decision procedures. For chain programs, testing for boundedness becomes testing for context-free language finiteness. For full template dependencies, uniform boundedness is shown decidable in [253]. In this case (without loss of generality [115]) the program is a sirup and uniform boundedness is NP-hard. Boundedness is shown decidable for monadic programs in [85]; regular language theory is used to analyze the decidability and the complexity of all monadic boundedness problems. Other decidable cases appear in [226, 297].

Unfortunately, boundedness is an undecidable property for Datalog programs in general. This was shown for uniform boundedness of linear, single-IDB, 7-ary programs and for program boundedness of linear, single-IDB, 4-ary programs in [120]. Even boundedness of linear single-IDB, binary programs is undecidable [297].

Uniformity is not the only assumption under which the complexities of noncontainment and boundedness differ. Whereas program boundedness is r.e. (Σ_1^0-complete) [120], boundedness is even harder (Σ_2^0-complete), [85]. This is because of the possible final projection. Thus, we have the following theorem.

4.2.9. THEOREM. *Uniform boundedness is undecidable, even for linear, single-*IDB, *7-ary programs. Program boundedness is undecidable, even for linear, single-*IDB, *binary programs. Boundedness is decidable for monadic programs, for chain programs and for ftd programs.*

For the case of a single IDB and a single recursive rule, there has been some recent progress. This case is slightly more general than sirups, because of the possibility of nonrecursive initialization rules. If the recursive rule is linear then it generalizes both [164, 224]. Recently, undecidability has been claimed (by Abiteboul) if the recursive rule is nonlinear, and decidability has been claimed (by Vardi) if the recursive rule is linear. Sirup uniform boundedness is still open; an NP-hardness lower bound is known [172].

A definite decidability–undecidability boundary is emerging, even if many questions remain unresolved. The complexity bounds are still not tight for the monadic case. The effects of adding constants to programs are only partly understood, [297]. This is also true about the effects of equalities and inequalities in the rule bodies, e.g., monadic, single-IDB boundedness is undecidable for Datalog$^{\neq}$, but uniform boundedness in this case is open [120]. Undecidability proofs to date use the fact that rules may be *disconnected*, see [85]; for the connected case there are only NP-hardness lower bounds [297].

4.2.10. OPEN PROBLEM. In which case is boundedness decidable? In particular, what is its status for sirups and for connected programs?

Unbounded stage functions have also been analyzed. Some Datalog queries are logspace-complete in PTIME. They are, in a worst-case sense, inherently sequential and are not in QNC, unless NC = PTIME. If testing for linear order is not a primitive of the query language, then it is possible to derive unconditional lower bounds for these queries. For example, the following theorem follows from [160].

4.2.11. THEOREM. *Any Datalog$^{\neg}$ program H expressing "path system accessibility" has* $\xi(n, H) = \Omega(\log^k n)$ *for each fixed* $k \geqslant 0$.

On the other hand, queries in QNC have a lot of inherent parallelism. For example, linear programs express such queries; this is one reason why so much attention has been given to linear program evaluation. Determining if a program expresses a query in QNC is undecidable [282, 120], but can be decided in special cases, e.g., for chain sirups [12].

Sufficient conditions for detecting implicit parallelism have been proposed. A semantic condition on the size of derivation trees proposed in [282] has the interesting property that (although it is not effectively testable) if it is true for a given program H, then H can be transformed automatically into an equivalent H' which is naively bottom-up evaluated in $O(\log n)$ stages and is thus in QNC.

4.2.12. THEOREM. *Let H be a program such that every t in $H(d)$ has a polynomial-size derivation tree. Then H can be transformed into an equivalent H' such that $\xi(n, H') = O(\log n)$.*

4.2.13. ADDITIONAL BIBLIOGRAHIC COMMENTS. Membership in NC, given the condition of Theorem 4.2.12, follows from [244]. The more interesting consequence is that H can be transformed into H' such that $\xi(n, H') = O(\log n)$; this represents a dynamic version of the parallel algorithm technique proposed in [214]. Theorem 4.2.12 is extended in [172].

An interesting application is the existence of a nonlinear chain program expressing a QNC query that is not expressible by any linear chain program [282] or by any linear program [10].

It is possible to use the techniques of parallel computational complexity to analyze not only database logic programs, but logic programs in general. These programs contain f.o. terms (not only variables and constants) and the primitive operation for their evaluation is unification [242]. Term unification is in linear time [236], but is also logspace-complete in PTIME [103, 306] and only restricted cases are known to be in NC [104].

4.2.3. Query evaluation and selection propagation

Up to this point we have outlined much of relational database theory without making use of constants and, in particular, of the selection operation $\varsigma_{A=a}$ from relational algebra with constants. We now examine such selections, in order to discuss one query optimization technique that is of great practical significance.

As we mentioned in Section 2.2.3, "performing the selections before the joins" is a recurrent theme of relational algebra optimization. It is, in principle, always possible to perform *selection propagation* for an f.o. query. This is typically done by rewriting the parse tree of a relational expression and propagating the selections to its leaves. It is usually accompanied by a dynamic programming analysis of the order of join computations. Query evaluation is commonly performed in two phases:

(1) a "compile-time" phase, when selection is propagated to the leaves and the order of joins is planned, and

(2) a "run-time" phase, when the database is accessed.

Of course there is a whole spectrum of possibilities, within this "compile-time" and "run-time" classification.

Selection propagation into logic programs is one of the fundamental issues in the implementation of database logic programs, because of its impact on efficiency. In this more general setting the query language is Datalog$^{\neg}$ (usually Datalog) with constants, instead of the less expressive relational algebra with constants. The importance of this issue and the first partial solutions for database logic programs were proposed in [278, 151].

We refer to [32] for a survey and performance comparison of the various evaluation methods, which use selection propagation. As in the f.o. case, query evaluation can be divided into "compile-time" and "run-time" phases. However, many proposals have been made, where these phases are interleaved. It is also typical to use *seminaive* bottom-up evaluation instead of naive bottom-up evaluation, where in the seminaive case the Tarski construction is performed in such a way that the work at iteration i is not duplicated by the work at iteration $i+1$.

We illustrate the formal setting of Datalog with constants using a simple example; syntax and semantics are intuitive generalizations of the constant-free case.

4.2.14. EXAMPLE. Consider the program with constant a,

$$R_1(x) :- R_2(ax).$$
$$R_2(xy) :- R_2(xz), R_3(zy).$$
$$R_2(xy) :- R_3(xy).$$

The input is a direct graph represented by a relation over R_3, and the output (over R_1) is the set of nodes x, such that $\langle ax \rangle$ is in the transitive closure of the input. A bottom-up evaluation of this program corresponds to a computation of all-pairs reachability and then a restriction to the nodes reachable from a. The following program is equivalent (R_2 is no longer needed) but its bottom-up evaluation corresponds to a single-source reachability computation from source a.

$$R_1(x) :- R_1(z), R_3(zx)$$
$$R_1(x) :- R_3(ax).$$

Naive or seminaive bottom-up evaluation are ways of constructing the output to a query that do not take advantage of selections until the minimum model has been constructed. On the contrary, the typical Prolog interpreter would produce the output in "top-down" fashion by reasoning backwards from the selection. Between these two extremes, there is a whole spectrum of evaluation stategies. These are based on a trade-off between "bottom-up" and "top-down" evaluation. In Example 4.2.14, the second program is produced from the first program via "top-down" reasoning about their equivalence, then it is evaluated via a "bottom-up" method.

As illustrated in [32], many other possibilities exist, depending on which optimizations are applied and on whether they are performed at "compile-time" or "run-time". Many of the algorithms are expressed in algebraic terms, i.e., as relational algebra expressions together with *while* constructs. In some cases, arithmetic operations are used to control the number of iterations.

The first sufficient conditions for commuting recursion, in the form of a least fixpoint operator, and selection were introduced by Aho and Ullman in [16]. This approach does not consider other possibilities of equivalence transformations on the program, beyond a straightforward rewriting of the selection operation into the recursion. The magic-set strategy of [31] adopts a more general view of selection propagations. Assume that a relation is defined via a Datalog program and that a selection is applied to this relation (in Prolog terminology, a relation is defined using a Prolog program without function symbols and a variable in the goal is bound to a constant). The magic-set strategy propagates the information about the selection by computing *magic* sets of values, which are then used to prune useless rule applications in a "bottom-up" method. The magic sets themselves are computed using rules that are less costly than the original rules.

The following example illustrates the magic-set strategy, as a program optimization applied at "compile-time". The "same-generation" program is translated into a different, but equivalent, program. This is accomplished by adding atoms to the bodies of already existing rules (typically with binary IDB's). These atoms are of the form *magic*(x) and are computed using less costly rules (typically with monadic IDB's). For the details of the transformation we refer to [31] and for an interpretation of it using quotients of context-free languages to [42].

4.2.15. EXAMPLE. Consider the following version of the "same generation" program,

$$R_1(x) \text{:-} R_2(ax)$$
$$R_2(xy) \text{:-} R_3(xy), R_4(zy)$$
$$R_2(xy) \text{:-} R_3(xz), R_2(zz'), R_4(z'y).$$

Think of $R_3(xz)$ as x is-child-of z and of $R_4(z'y)$ as z' is-parent-of y. The $R_2(xy)$ stands for x belongs-in-same-generation-as y and R_1 contains all those x in a's generation. The magic-set transformation produces an equivalent program, where the new monadic IDB *magic* restricts the application of the old rules to whenever x is instantiated as an ancestor of a.

$$R_1(x) \text{:-} R_2(ax)$$
$$R_2(xy) \text{:-} R_3(xz), R_4(zy), \mathit{magic}(x).$$
$$R_2(xy) \text{:-} R_3(xz), R_2(zz'), R_4(z'y), \mathit{magic}(x).$$
$$\mathit{magic}(a) \text{:-}$$
$$\mathit{magic}(x) \text{:-} \mathit{magic}(y), R_3(yx),$$

Evaluation based on counting path lengths of ground tuples has been proposed in [31] and is refined in [247], where it is combined with magic sets. One restriction is that input relations must be acyclic directed graphs. This use of *counting* is a departure from the pure Datalog context, but it is justified by the considerable gain in performance of *magic counting* [247, 211]. Even specific programs, such as the program of Example 4.2.15, are challenging computational questions when the evaluation is allowed to use counters [145] and the inputs are arbitrary relations with cycles.

An important issue in this area is the development of a framework in which the plethora of evaluation methods can be analyzed and compared. The *capture rule* framework of [278] allows the combination of these methods, provided they have a minimum degree of modularity. A systematic treatment of the methods is attempted in [45].

The propagation of selections is formalized in [42] as the problem of *finding an equivalent program of smaller equation width.* This formulation allows boundedness to be used as an analytical technique. It also indicates some fundamental differences from selection propagation for f.o. queries.

For *chain* Datalog program H, selection propagation corresponds to a regularity condition [42] on the language $L(H)$. This $L(H)$ is generated from the rules of H, without the variables, as context-free grammar productions taking the EDB's as terminals and the IDB's as nonterminals. If the regularity condition holds, as in the program of Example 4.2.14, then it is possible to decrease the equation width. If the regularity condition does not hold, as for example in the program of Example 4.2.15, it is not possible to decrease the equation width, even if some heuristic improvements are possible. The analysis of [225] may be viewed as defining regularity conditions for *non-chain* programs. The regularity condition of [225] allows for a clean comparison between a variety of evaluation methods.

"Top-down" evaluation strategies are closer to the spirit of Prolog interpreters. Their analysis presents many nontrivial computational questions, even when the only function symbols are constants. The simplest question here is termination, which is trivially true for "bottom-up" methods. The price to pay for the clever "top-down" propagation of selection information is the possibility of nontermination for these methods. In some cases it is possible to analyze the program and decide termination [11, 219, 281].

Much of the query evaluation literature is concerned with linear database logic programs. One reason for this is that queries defined this way are in QNC and consequently have a large amount of potential parallelism. An important question is how to efficiently use many processors to valuate such queries in parallel. The core combinatorial problem is *source–sink reachability in directed graphs* [172, 279]. This problem is known to be in NC. For undirected graphs, a linear number of processors suffice in order to achieve $O(\log^2 n)$ parallel time on most parallel computation models. (This makes the processor–time product $O(n \log^2 n)$, which is close to the $O(n)$ product for the one-processor case). For directed graphs, the best processor–parallel-time product for NC algorithms is $O(n^k)$, identical with the sequential time bounds for n by n matrix multiplication. (There is an increase from the $O(n)$ product for the one-processor case).

4.2.16. Open Question. Are there processor-efficient NC algorithms for source–sink directed reachability?

4.2.17. Additional Bibliographic Comments. For some new, interesting, general program transformations we refer to [237, 261].

We refer the reader to [32] for an extensive bibliography of evaluation methods. Here we only present a sample from a currently active area of research. It is important to stress that many of the evaluation methods form integral parts of new logic database systems, e.g. [220, 44, 277].

A question related to the termination of various evaluation methods is that of safety for database logic programs. This is analogous to safety for relational calculus, only now fixpoints add a new dimension to the problem [177, 258].

4.3. Query languages and complex object data models

An important aspect of the relational data model and of its logic programming extensions is their simplicity. Relations are the only data type. They are the common denominator, which facilitates the comparison and the analysis of many query formalisms. Semantics are introduced using dependencies, which are (as we have seen) closely related to queries. Despite the elegance of the theory, relations are flat structures and the representation of hierarchical database structures via relations is sometimes forced. It is often simpler to deal explicitly with more complex data types than to express them indirectly using dependencies.

There has been wide interest in hierarchical database structures. Of particular interest are those built using finite *tuple* and *set* constructors. These constructors are the basis for defining a large repertoire of data types. Instances of these data types are manipulated using query languages that are extensions of relational algebra and calculus.

The first proposal to generalize the relational data model by removing the 1-NF assumption was made by Makinouchi [208]. Algebras have been proposed for nested (i.e., non-1-NF) relations, e.g., [1, 2, 167, 187, 243, 263, 275]. These are usually based on the two operations NEST and UNNEST introduced by [167]. An early higher-order calculus (but no algebra) is contained in [166]. First-order calculi can be found in [1, 187, 243]. A good survey of the area is [155].

The expressive power of relational algebra can be significantly extended by the addition of a *powerset operator* (first considered in [187]). It is shown in [1] that transitive closure can be expressed by an algebra extended with the powerset. It is interesting to consider looping constructs, least fixpoint constructs etc. in the context of nested relation algebras, see [144]. Most of these enrichments of nested relation algebras turn out to have the same expressive power as the algebra with the powerset operator in [1]. As shown in [234], if the inputs and outputs are 1-NF relations, if there is no powerset operation or looping construct etc., and if the intermediate results are nested relations then, the expressive power of many of these formalisms is still that of relational algebra.

Algebra = calculus theorems are shown in both [187, 1]. The calculus of [187] has many similarities with recursive rules, whereas the one in [1] is a more intuitive extension of the relational calculus. The two settings are somewhat different in terms of their data types: both have *set, tuple and union types*, and [187] also allows *recursive types* (but only for inputs). Also, *object identifiers* are present in [187]; these can be thought of as pointers to values. Both settings are generalized in [6].

Another basic idea that has been explored recently is to use levels of nesting of the powerset to classify queries. In this way, queries can be classified into hierarchies going beyond PHIER-queries. See [157, 188] for the detailed formulation and results.

Much recent attention has been devoted to complex object models, with logic-based query languages, e.g., [1, 30, 59, 187]. In this context the *tuple* constructor leads to first-order terms, which can be manipulated using term unification or matching (see Comments 4.2.13). This is a significant step towards full integration of databases and logic programming. On the other hand, adding the *set* constructor leads to nontrivial extensions of conventional logic programming.

Incorporating *sets* into database logic programs is a promising research direction [4, 30, 44, 185, 186, 270]. The solution proposed in [270] makes use of the difference between data and expression complexity. For a comparison of the various proposed solutions we refer to [186].

An important distinction that can be made in complex object languages is whether the language can be *statically type checked* or not. For example, this is possible in [1, 4, 187, 188], but not in [30]. The key is the separation of scheme and instance, which is present in database languages but not in logic programming languages like Prolog.

Interestingly, it is possible to have a simple generalization of most complex object models with a language that

(1) has set, tuple, union, and recursive type for both input and output,

(2) can be statically type checked, and

(3) has full computational capabilities.

An example is the language IQL of [6], which is based on recursive rules and object identifiers.

Complex object data models, semantic data models [156], and the more recent *object-oriented data models* are all efforts to circumvent the semantic limitations of relations. A variety of new applications require languages that are much closer to full programming languages than are the current database query languages [29]. The most challenging open question in this area is the development of an elegant, mathematical model to serve as a foundation for the object-oriented database experimental efforts.

5. Discussion: other issues in relational database theory

There are database problems orthogonal to the issues addressed in the theories of dependencies and queries. We close our presentation with two such issues: *incomplete information* and *database updates.*

These are important in a wider context. A theory of incomplete information should be a theory of "knowledge", with many formal connections to modal logics. Updates are an example of dealing with "dynamic change", a problem that has motivated much of the research in the area of the logics of programs. Here, we limit ourselves to formulations and solutions that have a distinct database character.

5.1. The problems of incomplete information

The implicit assumption of the relational data model is that the database is a completely determined finite structure. Reiter's work on first-order theories as databases makes explicit the implicit assumptions about the relational data model [238, 239, 240].

To deal with incomplete information, we need representations of sets of "possible worlds", i.e., the identification we have been making of a database with a single "possible world" must be relaxed. Consequently, the syntax itself of the database must be used as a specification. This was done in Section 2.5 with extended relational theories for the complete information case, with deductive and universal relation models, and with relaxing the uniqueness axioms (Remark 2.5.1).

In terms of what is a "possible world", there is a fundamental distinction between *open* and the *closed* extensions of relations. Open extensions are best exemplified by the weak universal relation assumption, i.e., we know some tuples completely and any containing universal relation is a "possible world". A closed extension is a *conservative* open extension, i.e., it is an open extension that consists only of tuples whose existence we can infer. A detailed comparison of the two approaches, based on computational complexity, is made in [296].

The first-order logical framework can be augmented to express facts about the uncertainty itself. This has been done through the addition of *modal* operators, e.g., for possibility or certainty, see [192, 195, 196].

For the rest of this section, we focus on various algebraic mechanisms for modeling uncertainty. The most commonly used is the mechanism of *null values*. Their presence changes the basic data type of relation as follows.

A *Codd table* [78] is the result of replacing some occurrences of values in a relation by *distinct* variables, called *null values*. A *condition* φ is a propositional ($\vee, \wedge, \neg$) combination of equality atoms $x=y$ or $x=c$, where x, y are variables and c is a value. Conditions may be associated with Codd table T in two ways:

(1) a *global condition* φ_T is associated with the whole of T, and

(2) a *local condition* φ_t is associated with a tuple t of T (if no association is specified the default is the trivial true condition $x=x$).

A *conditional table* $\mathscr{T}$ is a Codd table T with associated global and local conditions, see [159]. One can visualize a conditional table as the rows of an untyped tableau with constants, together with a column of conditions.

A valuation h is a function from variables to values, that naturally extends to tuples and Codd tables by defining $h(c)=c$ for each value c. A condition is satisfied or falsified by h in the obvious way. A conditional table $\mathscr{T}$ (with Codd table T) represents the set of relations:

$$\mathscr{T}_{\text{rel}}=\{r\text{: there is a } h\models\varphi_T \text{ and } r \text{ consists of the tuples } h(t) \text{ for which } h\models\varphi_t\}.$$

Why are conditional tables an interesting representation? Codd tables are a simple way of generalizing relations. Between them and conditional tables there is a large repertoire of representations [5]. For example, the first-order theories in [295] can be represented using a Codd table and a conjunctive global condition. In [136] global conditions are used to describe dependency satisfaction. The chase may be applied to conditional tables [136, 201]. The most important reason is that there is a closure theorem for conditional tables, which allows simple and semantically meaningful querying of incomplete information databases. (One has to expand Codd tables with local conditions in order to realize this closure [159, 3]). According to these semantics, the output is a representation of *all the possibilities* in the following fashion. *If $\mathscr{T}$ is a conditional table* (*without global condition*) *and E a relational algebra expression, then from E and $\mathscr{T}$ it is possible to construct another $\mathscr{T}'$ conditional table* (*without global condition*) *such that:* $\mathscr{T}'_{\text{rel}}=E(\mathscr{T}_{\text{rel}})=\{E(r)\text{: } r\in\mathscr{T}_{\text{rel}}\}$.

There is a large volume of work on querying incomplete information databases. The goal of most of this research has been to determine the correct semantics of applying a program to an incomplete information database. Closure conditions are one desirable alternative. They often lead to variants of the relational operations with which to process null values. For example, one common problem is how to extend the $\bowtie$ operation to relations with null values. We refer to [159] for many more references on the subject.

Many of the proposed algorithms focus on *the certain* tuples, i.e., tuples contained in all "possible worlds", as opposed to describing all possibilities. We encountered this *conservative* approach, when we examined querying in deductive and universal relation data models. When querying for the certain tuples, an algorithm is *sound* if it produces certain tuples only and it is *complete* if it produces all the certain tuples.

Unfortunately, it is often the case that completeness implies an exponential increase in data complexity. This was first shown in [295] and further investigated in [5]. Thus, the sound algorithms in [139, 240, 295] trade completeness for efficiency. There is one important special case, where things are ideal. This was identified in [295, 159] for positive existential queries: *To derive the certain answers to a second-order positive query on a Codd table with a satisfiable conjunctive global condition it suffices to*

(1) *incorporate the equality atoms in the table,*

(2) *ignore the inequality atoms, and*

(3) *proceed as if the table were a relation.*

The null values we have been discussing are "values present but unknown", sometimes constrained through conditions. Other kinds of null values, e.g., "values whose presence is unknown", have been studied in [310].

Many ideas from mathematical logic have been proposed as modeling tools for uncertainty (e.g., three-valued logic) but their applicability is hard to assess. In summary, for the representation and querying of incomplete information databases, there are many partial solutions but no satisfactory full answer. It seems that the further away we move from the relational data model the fewer analytical and algebraic tools are available and the more we have to rely on general-purpose theorem-proving heuristic techniques.

5.2. *The problems of database updates*

Database manipulation languages have primitives for both querying and updating relations. The characteristic that distinguishes database updating from updating the state in more general programming languages is the simplicity of the changes made. Tuples are only *inserted, deleted* or values *modified* in the one underlying relation data type. It is possible to study these operations in the context of both algebra-like and calculus-like query languages. These investigations span language definition [7, 8], program optimization [9], dependencies [299] and incomplete information [3].

A first-order theory approach to updates was initiated in [116] and elaborated in [113]. The semantics of *insert, delete* for first-order theories is formulated in terms of sets of these theories. The broader logical framework and novel notions, such as equivalence that is preserved under common updates or *equivalence forever*, illustrate new challenges to database theory. Algorithmic problems in this setting are resolved in [300]; see [301] for an overview.

Although broader than the practical use of updates in many database systems, the above concepts are more specialized than a direct use of some form of dynamic logic [147]. For two recent applications of dynamic logic to database logic programs see [209, 223].

We close our exposition with the view *update problem.* This is a problem that combines incomplete information with update issues. It is a challenging problem in general, which has been elegantly formalized in relational database theory.

VIEW UPDATES. Individual users often want to deal only with part of the information of the database. This is why database systems provide the *view* facility. The user would like to query and update the database view in blissful ignorance of the underlying database.

In database systems, a view is generally implemented by storing its definition, i.e., some program in a query language supported by the system. Thus, querying is done by composing the query on the view with the view definition. Sometimes things are simplified further by precomputing and storing the view as a relation, called *materializing* the view. Updating a view is much more challenging. A tuple insertion in or deletion from the view may make the underlying database inconsistent ("no possible world") or ambiguous ("many possible worlds").

A first attempt at resolving the difficulties of view updates was made in [94]. An elegant semantic solution was proposed in [33], which is quite independent of the relational data model and stated in terms of mappings.

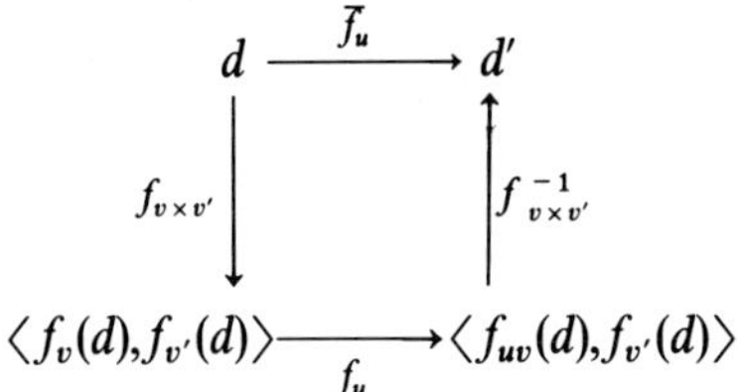

Fig. 6. View update semantics.

A *view* f_v and an *update* on this view f_u are both mappings of the appropriate types (from the database scheme to the view scheme and from the view scheme to itself respectively). How can f_u be translated into a database update $\bar{f}_u$ (of type from the database scheme to itself)? The basic insight of [33] here is that of a *complementary view* $f_{v'}$ of f_v, where the mapping from databases d to databases $\langle f_v(d), f_{v'}(d)\rangle$ must be injective. This mapping is denoted by $f_{v\times v'}$. *A view can have many complementary views; choosing one that must remain constant assigns unambiguous semantics to a view update.*

This is because the database can be reconstructed from the updated view and the constant complement (recall the theory of decompositions from Section 3.2.1). Once a complementary view is chosen, translating under this constant complement amounts to finding a database d' such that

(1) $f_v(d') = f_u(f_v(d)) = f_{uv}(d)$, and

(2) $f_{v'}(d') = f_{v'}(d)$.

If such a d' exists then, by injectiveness, it is unique. The update mapping f_u is translatable iff such a d' exists for any d (the independent two-part decompositions from Section 3.2.1 are a sufficient condition for translatability). If f_u is translatable then its translation is $\bar{f}_u = f^{-1}_{v\times v'} \circ f_{uv\times v'}$.

References [33, 90] describe additional properties of this approach, in particular for families of updates closed under composition. The semantics of [33] are depicted pictorially in the commuting diagram of Fig. 6. They are a clarifying (even if initial) step in resolving a challenging problem. Their complexity, for the simplest possible set of dependencies (fd's) and views (projections), is examined in [90]. In many cases the computational complexity of a full implementation of view updates is prohibitive. Much of the recent work, e.g., [131, 174, 175, 300, 301], has concentrated on feasible algorithmic approximations.

6. Conclusions

Queries and dependencies are major themes which we have used to present relational database theory. A number of problems remain unresolved. Among the crisp and perhaps hard mathematical problems that have emerged we have highlighted dependency implication (2.3.16), recursive rule boundedness (4.2.10) and parallel computation (4.2.16) questions.

In the area of logic and databases, we are only beginning to understand the possible uses of negation and the large variety of evaluation methods. Much remains to be done on the subjects of complex objects, incomplete information, and database updates. Finally, developing the foundations for object-oriented database languages will (most likely) involve new conceptual contributions that use the theory of abstract data types.

Acknowledgment

We would like to thank the many researchers who provided helpful, detailed comments to earlier versions of this paper. In particular, we would like to thank: Serge Abiteboul, Ed Chan, Claude Delobel, Ron Fagin, Marc Gyssens, Michael Kifer, Gabi Kuper, Jack Minker, Jeff Naughton, Jan Paredaens, Oded Shmueli, Dirk Tahon, Allen Van Gelder, Dirk Van Gucht, Victor Vianu, and Mihalis Yannakakis. We would also like to acknowledge Ashok Chandra, Stavros Cosmadakis, Phokion Kolaitis, and Moshe Vardi for many clarifying discussions on the fine points of the theory of queries. Support for the writing of this chapter was provided partly by NSF Grant IRI-8617344 and partly by an Alfred P. Sloan Foundation Fellowship.

References

[1] ABITEBOUL, S. and C. BEERI, On the manipulation of complex objects, in: *Proc. Internat. Workshop on Theory and Applications of Nested Relations and Complex Objects* (1987).

[2] ABITEBOUL, S. and N. BIDOIT, Non first normal form relations to represent hierarchically organized data, in: *Proc. 3rd ACM Symp. on Principles of Database Systems* (1984) 191–200.

[3] ABITEBOUL, S. and G. GRAHNE, Update semantics for incomplete databases, in: *Proc. 11th Conf. on Very Large Databases* (1985) 1–12.

[4] ABITEBOUL, S. and S. GRUMBACH, COL: A logic based language for complex objects, in: *Proc. 1st Internat. Conf. on Extending Database Technology* (1988) 271–293.

[5] ABITEBOUL, S., P.C. KANELLAKIS and S. GRAHNE, On the representation and querying of sets of possible worlds, in: *Proc. ACM SIGMOD* (1987) 10–19.

[6] ABITEBOUL, S. and P.C. KANELLAKIS, Object identity as a query language primitive, in: *Proc. ACM SIGMOD* (1989).

[7] ABITEBOUL, S. and V. VIANU, A transaction language complete for database update and specification, in: *Proc. 6th ACM Symp. on Principles of Database Systems* (1987) 260–268.

[8] ABITEBOUL, S. and V. VIANU, Procedural and declarative database update languages, in: *Proc. 7th ACM Symp. on Principles of Database Systems* (1988) 240–250.

[9] ABITEBOUL, S. and V. VIANU, Equivalence and optimization of relational transactions, *J. ACM* **35**(1) (1988) 130–145.

[10] AFRATI, F. and S.S. COSMADAKIS, Expressiveness of restricted recursive queries, in: *Proc. 21st Ann. ACM Symp. on Theory of Computing* (1989) 113–126.

[11] AFRATI, F., C.H. PAPADIMITRIOU, G. PAPAGEORGIOU, A. ROUSSOU, Y. SAGIV and J.D. ULLMAN, Convergence of sideways query evaluation, in: *Proc. 5th ACM Symp. on Principles of Database Systems* (1986) 24–30.

[12] AFRATI, F. and C.H. PAPADIMITRIOU, The parallel complexity of simple chain queries, in: *Proc. 6th ACM Symp. on Principles of Database Systems* (1987) 210–213.

[13] AHO, A.V., C. BEERI and J.D. ULLMAN, The theory of joins in relational databases, *ACM Trans. Database Systems* **4**(3) (1979) 297–314.

[14] AHO, A.V., Y. SAGIV and J.D. ULLMAN, Equivalence of relational expressions, *SIAM J. Comput.* **8**(2) (1979) 218–246.

[15] AHO, A.V., Y. SAGIV and J.D. ULLMAN, Efficient optimization of a class of relational expressions, *ACM Trans. Database Systems* **4**(4) (1979) 435–454.

[16] AHO, A.V. and J.D. ULLMAN, Universality of data retrieval languages, in: *Proc. 6th Ann. ACM Symp. on Principles of Programming Languages* (1979) 110–117.

[17] AJTAI, M. and Y. GUREVICH, Monotone versus positive, *J. ACM* **34**(4) (1987) 1004–1015.

[18] APT, K., Logic programming, in: J. van Leeuwen, ed., *Handbook of Theoretical Computer Science, Vol. B* (North-Holland, Amsterdam, 1990) 493–574.

[19] APT, K., H. BLAIR and A. WALKER, Towards a theory of declarative knowledge, in: J. Minker, ed., *Foundations of Deductive Databases and Logic Programming* (Morgan-Kaufman, Los Altos, CA, 1988) 89–148.

[20] APT, K. and M. VAN EMDEN, Contributions to the theory of logic programming, *J. ACM* **29**(3) (1982) 841–862.

[21] ARMSTRONG, W.W., Dependency structures of database relationships, in: *Proc. IFIP '74* (1974) 580–583.

[22] ARORA, A.K. and C.R. CARLSON, The information preserving properties of relational data base transformations, in: *Proc. 4th Conf. on Very Large Databases* (1978) 352–359.

[23] ASTRAHAN, M.M., ET AL., System R: a relational approach to data management, *ACM Trans. Database Systems* **1**(2) (1976) 97–137.

[24] ATZENI, P. and M.C. DE BERNARDIS, A new basis for the work instance model, in: *Proc. 6th ACM Symp. on Principles of Database Systems* (1987) 79–86.

[25] ATZENI, P. and E.P.F. CHAN, Efficient query answering in the representative instance approach, in: *Proc. 4th ACM Symp. on Principles of Databases Systems* (1985) 181–188.

[26] AUSIELLO, G., A. D'ATRI and M. MOSCARINI, Chordality properties on graphs and minimal conceptual connections in semantic data models, in: *Proc. 4th ACM Symp. on Principles of Database Systems* (1985) 164–170.

[27] AYLAMAZYAN, A.K., M.M. GIGULA, A.P. STOLBOUSKIN and G.F. SCHWARTZ, Reduction of the relational model with infinite domain to the case of finite domains, *Dokl. Akad. Nauk. SSSR* **286**(2) (1986) 308–311.

[28] BANCILHON, F., On the completeness of query languages for relational data bases, in: *Proc. 7th Symp. on Mathematical Foundations of Computer Science*, Lecture Notes in Computer Science, Vol. 64 (Springer, Berlin, 1978) 112–123.

[29] BANCILHON, F., Object-oriented database systems, in: *Proc. 7th ACM Symp. on Principles of Database Systems* (1988) 152–162.

[30] BANCILHON, F. and S. KHOSHAFIAN, A calculus for complex objects, in: *Proc. 5th ACM Symp. on Principles of Database Systems* (1986) 53–59.

[31] BANCILHON, F., D. MAIER, Y. SAGIV and J.D. ULLMAN, Magic sets and other strange ways to implement logic programs, in: *Proc. 5th ACM Symp. on Principles of Database Systems* (1986) 1–15.

[32] BANCILHON, F. and R. RAMAKRISHNAN, An amateur's introduction to recursive query processing, in: *Proc. ACM SIGMOD* (1986) 16–52; also in: J. Minker, ed., *Foundations of Deductive Databases and Logic Programming* (Morgan-Kaufmann, Los Altos, CA, 1988) 439–518.

[33] BANCILHON, F. and N. SPYRATOS, Update semantics of relational views, *ACM Trans. Database Systems* **6**(4) (1981) 557–575.

[34] BEERI, C., On the membership problem for functional and multivalued dependencies in relational databases, *ACM Trans. Databases Systems* **5**(3) (1980) 241–259.

[35] BEERI, C. and P.A. BERNSTEIN, Computational problems related to the design of normal form relational schemes, *ACM Trans. Database Systems* **4**(1) (1979) 30–59.

[36] BEERI, C., P.A. BERNSTEIN and N. GOODMAN, A sophisticate's introduction to database normalization theory, in: *Proc. 4th Conf. on Very Large Databases* (1978) 113–124.

[37] BEERI, C., M. DOWD, R. FAGIN and R. STATMAN, On the structure of Armstrong relations for functional dependencies, *J. ACM* **31**(1) (1984) 30–46.

[38] BEERI, C., R. FAGIN and J.H. HOWARD, A complete axiomatization for functional and multivalued dependencies in database relations, in: *Proc. ACM SIGMOD* (1977) 47–61.

[39] BEERI, C., R. FAGIN, D. MAIER, A.O. MENDELZON, J.D. ULLMAN and M. YANNAKAKIS, Properties of acyclic database schemes, in: *Proc. 13th Ann. ACM Symp. on Theory of Computing* (1981) 355–362.

[40] BEERI, C., R. FAGIN, D. MAIER and M. YANNAKAKIS, On the desirability of acyclic database schemes, *J. ACM* **30**(3) (1983) 479–513.

[41] BEERI, C. and P. HONEYMAN, Preserving functional dependencies, *SIAM J. Comput.* **10**(3) (1981) 647–656.

[42] BEERI, C., P.C. KANELLAKIS, F. BANCILHON and R. RAMAKRISHNAN, Bounds on the propagation of selection into logic programs, in: *Proc. 6th ACM Symp. on Principles of Database Systems* (1987) 214–227.

[43] BEERI, C., A.O. MENDELZON, Y. SAGIV and J.D. ULLMAN, Equivalence of relational database schemes, *SIAM J. Comput.* **10**(2) (1981) 352–370.

[44] BEERI, C., S. NAQVI, R. RAMAKRISHNAN, O. SHMUELI and S. TSUR, Sets and negation in a logic database language (LDL1) in: *Proc. 6th ACM Symp. on Principles of Database Systems* (1987) 21–37.

[45] BEERI, C. and R. RAMAKRISHNAN, On the power of magic, in: *Proc. 6th ACM Symp. on Principles of Database Systems* (1987) 269–283.

[46] BEERI, C. and J. RISSANEN, Faithful representation of relational database schemes, IBM Res. Report RJ2722, IBM Research Lab., San Jose, CA, 1980.

[47] BEERI, C. and M.Y. VARDI, On the complexity of testing implications of data dependencies, Res. Report, Hebrew Univ., Jerusalem, 1980.

[48] BEERI, C. and M.Y. VARDI, Formal systems for join dependencies, Res. Report, Hebrew Univ., Jerusalem, 1981.

[49] BEERI, C. and M.Y. VARDI, The implication problem for data dependencies, in: *Proc. 8th Internat. Coll. on Automata, Languages and Programming*, Lecture Notes in Computer Science, Vol. 115 (Springer, Berlin, 1981) 73–85.

[50] BEERI C. and M.Y. VARDI, On the properties of join dependencies, in: H. Gallaire, J. Minker and J.-M. Nicolas, eds., *Advances in Database Theory, Vol. 1* (Plenum, New York, 1981) 25–72.

[51] BEERI, C. and M.Y. VARDI, A proof procedure for data dependencies, *J. ACM* **31**(4) (1984) 718–741.

[52] BEERI, C. and M.Y. VARDI, Formal systems for tuple and equality generating dependencies, *SIAM J. Comput.* **13**(1) (1984) 76–98.

[53] BEERI, C. and M.Y. VARDI, On acyclic database decompositions, *Inform. and Control* **61**(2) (1984) 75–84.

[54] BERNSTEIN, P.A., Synthesizing third normal form relations from functional dependencies, *ACM Trans. Databases Systems* **1**(4) (1976) 277–298.

[55] BERNSTEIN, P.A. and D.W. CHIU, Using semi-joins to solve relational queries, *J. ACM* **28**(1) (1981) 25–40.

[56] BERNSTEIN, P.A. and N. GOODMAN, The power of natural semi-joins, *SIAM J. Comput.* **10** (1981) 751–771.

[57] BERNSTEIN, P.A., V. HADZILACOS and N. GOODMAN, *Concurrency Control and Recovery in Database Systems* (Addison-Wesley, Reading, MA, 1987).

[58] BISKUP, J., U. DAYAL and P.A. BERNSTEIN, Synthesizing independent database schemas, in: *Proc. ACM SIGMOD* (1979) 143–152.

[59] BUNEMAN, P., S. DAVIDSON and A. WATTERS, A semantics for complex objects and approximate queries, in: *Proc. 7th ACM Symp. on Principles of Database Systems* (1988) 302—314.

[60] CASANOVA, M.A., R. FAGIN and C.H. PAPADIMITRIOU, Inclusion dependencies and their interaction with functional dependencies, *J. Comput. System Sci.* **28**(1) (1984) 29–59.

[61] CASANOVA, V. and V.M.P. VIDAL, Towards a sound view integration methodology, in: *Proc. 2nd ACM Symp. on Principles of Database Systems* (1983) 36–47.

[62] CHAN, E.P. and H.J. HERNANDEZ, Independence reducible database schemes, in: *Proc. 7th ACM Symp. on Principles of Database Systems* (1988) 163–173.

[63] CHAN, E.P. and A.O. MENDELZON, Answering queries on the embedded complete database schemes, *J. ACM* **34**(2) (1987) 349–375.

[64] CHANDRA, A.K., Programming primitives for database languages, in: *Proc. 8th Ann. ACM Symp. on Principles of Programming Languages* (1981) 50–62.

[65] CHANDRA, A.K., Theory of database queries, in: *Proc. 7th ACM Symp. on Principles of Database Systems* (1988) 1–9.

[66] CHANDRA, A.K. and D. HAREL, Computable queries for relational data bases, *J. Comput. System. Sci.* **21**(2) (1980) 156–178.

[67] CHANDRA, A.K., D. HAREL, Structure and complexity of relational queries, *J. Comput. System Sci.* **25**(1) (1982) 99–128.

[68] CHANDRA, A.K. and D. HAREL, Horn clause queries and generalizations, *J. Logic Programming* **2**(1) (1985) 1–15.

[69] CHANDRA, A.K., H.R. LEWIS and J.A. MAKOWSKY, Embedded implicational dependencies and their inference problem, in: *Proc. 13th Ann. ACM Symp. on Theory of Computing* (1981) 342–352.

[70] CHANDRA, A.K. and P.M. MERLIN, Optimal implementation of conjunctive queries in relational data bases, in: *Proc. 9th Ann. ACM Symp. on Theory of Computing* (1977) 77–90.

[71] CHANDRA, A.K. and M.Y. VARDI, The implication problem for functional and inclusion dependencies is undecidable, *SIAM J. Comput.* **14**(3) (1985) 671–677.

[72] CHILDS, D.L., Feasibility of a set-theoretical data structure—a general structure based on a reconstituted definition of relation, in: *Proc. IFIP'68* (1968) 162–172.

[73] CHIN, L.H. and A. TARSKI, Remarks on projective algebras, abstract, *Bull. Amer. Math. Soc.* **54** (1948) 80–81.

[74] CLARK, K.L., Negation as failure, in: H. Gallaire and J. Minker, eds., *Logic and Databases* (Plenum, New York, 1978) 293–322.

[75] CODD, E.F., A relational model of data for large shared data banks, *Comm. ACM* **13**(6) (1970) 377–387.

[76] CODD, E.F., Further normalization of the data base relational model, in: R. Rustin, ed., *Data Base Systems* (Prentice-Hall, Englewood Cliffs, NJ, 1972) 33–64.

[77] CODD, E.F., Relational completeness of database sublanguages, in: R. Rustin, ed., *Data Base Systems* (Prentice-Hall, Englewood Cliffs, NJ, 1972) 65–98.

[78] CODD, E.F., Extending the data base relational model to capture more meaning, *ACM Trans. on Database Systems* **4**(4) (1979) 397–434.

[79] CODD, E.F., Relational databases: a practical foundation for productivity, *Comm. ACM* **25**(2) (1982) 102–117.

[80] COOK, S.A., A taxonomy of problems with fast parallel algorithms, *Inform. and Control* **64** (1985) 2–22.

[81] COSMADAKIS, S.S., The complexity of evaluating relational queries, *Inform. and Control* **58** (1983) 101–112.

[82] COSMADAKIS, S.S., Equational theories and database constraints, Ph.D. Dissertation, Res. Report TR-346, Lab. for Computer Science, MIT, Cambridge, MA, 1985.

[83] COSMADAKIS, S.S., Database theory and cylindric lattices, in: *Proc. 27th Ann. IEEE Symp. on Foundations of Computer Science* (1987) 411–420.

[84] COSMADAKIS, S.S., On the first order expressibility of recursive queries, in: *Proc. 8th ACM Symp. on Principles of Database Systems* (1989) 311–324.

[85] COSMADAKIS, S.S., H. GAIFMAN, P.C. KANELLAKIS, and M.Y. VARDI, Decidable optimization

problems for database logic programs, in: *Proc. 20th Ann. ACM Symp. on Theory of Computing* (1988) 477–490.

[86] COSMADAKIS, S.S. and P.C. KANELLAKIS, Equational theories and database constraints, in: *Proc. 17th Ann. ACM Symp. on Theory of Computing* (1985) 273–284.

[87] COSMADAKIS, S.S., and P.C. KANELLAKIS, Functional and inclusion dependencies: a graph theoretic approach, in: P.C. Kanellakis and F. Preparata, eds., *Advances in Computing Research, Vol. 3* (JAI Press, Greenwich, CT, 1986) 164–185.

[88] COSMADAKIS, S.S. and P.C. KANELLAKIS, Parallel evaluation of recursive rule queries, in: *Proc. 5th ACM Symp. on Principles of Database Systems* (1986) 280–293.

[89] COSMADAKIS, S.S., P.C. KANELLAKIS and S. SPYRATOS, Partition semantics for relations, *J. Comput. System Sci.* **32**(2) (1986) 280–293.

[90] COSMADAKIS, S.S. and C.H. PAPADIMITRIOU, Updates of relational views, *J. ACM* **31**(4) (1984) 742–760.

[91] DAHLHAUS, E., Skolem normal forms concerning the least fixpoint, in: E. Börger, ed., *Computation Theory and Logic*, Lecture Notes in Computer Science, Vol. 270 (Springer, Berlin, 1987) 101–106.

[92] DAHLHAUS, E. and J.A. MAKOWSKY, Computable directory queries, in: *Proc. 11th CAAP '86*, Lecture Notes in Computer Science, Vol. 214 (Springer, Berlin, 1986) 254–265.

[93] D'ATRI, A. and M. MOSCARINI, Recognition algorithms and design methodologies for acyclic database schemes, in: P.C. Kanellakis and F. Preparata, eds., *Advances in Computing Research, Vol. 3* (JAI Press, Greenwich, CT, 1986) 164–185.

[94] DAYAL, U. and P.A. BERNSTEIN, On the correct translation of update operations on relational views, *ACM Trans. Database Systems* **8**(3) (1982) 381–416.

[95] DE BRA, P. and J. PAREDAENS, Conditional dependencies for horizontal decompositions, in: *Proc. 10th Internat. Coll. on Automata, Languages and Programming*, Lecture Notes in Computer Science, Vol. 154 (Springer, Berlin, 1983) 67–82.

[96] DELOBEL, C., Normalization and hierarchical dependencies in the relational data model, *ACM Trans. Database Systems* **3**(3) (1978) 201–222.

[97] DELOBEL, C. and R.C. CASEY, Decomposition of a database and the theory of Boolean switching functions, *IBM J. Res. Develop.* **17**(5) (1972) 370–386.

[98] DE ROUGEMONT, M., Second-order and inductive definability on finite structures, *Z. Math. Logik* **33** (1987) 47–63.

[99] DIPAOLA, R.A., The recursive unsolvability of the decision problem for a class of definite formulas, *J. ACM* **16**(2) (1969) 324–327.

[100] DOWNEY, P.J., R. SETHI and R.E. TARJAN, Variations on the common subexpression problem, *J. ACM* **27**(4) (1980) 758–771.

[101] DREBEN, B.S. and W.D. GOLDFARB, *The Decision Problem: Solvable Classes of Qualificational Formulas* (Addison-Wesley, Reading, MA, 1979).

[102] DUBLISH, P. and S.N. MAHESHWARI, Expressibility of bounded-arity fixed-point query hierarchies, in: *Proc. 8th ACM Symp. on Principles of Database Systems* (1989) 324–336.

[103] DWORK, C., P.C. KANELLAKIS and J. MITCHELL, On the sequential nature of unification, *J. Logic Programming* **1**(1) (1984) 35–50.

[104] DWORK, C., P.C. KANELLAKIS and L. STOCKMEYER, Parallel algorithms for term matching, *SIAM J. of Comput.* **17**(4) (1988) 711–731.

[105] ENDERTON, H.B., *A Mathematical Introduction to Logic* (Academic Press, New York, 1972).

[106] FAGIN, R., Generalized first-order spectra and polynomial-time recognizable sets, in: R. Karp, ed., *Complexity of Computation*, SIAM–AMS Proceedings, Vol. 7 (1974) 43–73.

[107] FAGIN, R., Monadic generalized spectra, *Z. Math. Logik* **21** (1975) 89–96.

[108] FAGIN, R., Multivalued dependencies and a new normal form for relational databases, *ACM Trans. Database Systems* **2**(3) (1977) 262–278.

[109] FAGIN, R., A normal form for relational databases that is based on domains and keys, *ACM Trans. Database Systems* **6** (3) (1981) 387–415.

[110] FAGIN, R., Armstrong Databases, in: *Proc. 7th IBM Symp. on Mathematical Foundations of Computer Science* (1982); also, IBM Res. Report RJ3440, IBM Res. Lab., San Jose, CA, 1982.

[111] FAGIN, R., Horn clauses and databases dependencies, *J. ACM* **29**(4) (1982) 952–985.

[112] FAGIN, R., Degrees of acyclicity for hypergraphs an relational database schemes, *J. ACM* **30**(3) (1983) 514–550.

[113] FAGIN, R., G. KUPER, J.D. ULLMAN and M.Y. VARDI, Updating logical databases, in: P.C. Kanellakis and F. Preparata, eds., *Advances in Computing Research, Vol. 3* (JAI Press, Greenwich, CT, 1986) 1–18.

[114] FAGIN, R., A.O. MENDELZON and J.D. ULLMAN, A simplified universal relational assumption and its properties, *ACM Trans. Database Systems* **7**(3) (1982) 343–360.

[115] FAGIN, R., J.D. ULLMAN, D. MAIER and M. YANNAKAKIS, Tools for template dependencies, *SIAM J. Comput.* **12** (1983) 36–59.

[116] FAGIN, R., J.D. ULLMAN and M.Y. VARDI, On the semantics of updates in database, in: *Proc. 2nd ACM Symp. on Principles of Database Systems* (1983) 352–365.

[117] FAGIN, R. and M.Y. VARDI, The theory of data dependencies: a survey in: M. Anshel and W. Gewirtz, eds., *Mathematics of Information Processing*, Symp. in Appl. Math., Vol. 34 (1986) 19–72.

[118] FISCHER, P.C. and D.M. TSOU, Whether a set of multivalued dependencies implies a join dependencies is NP-hard, Res. Report, Vanderbilt Univ., Nashville, TN, 1981.

[119] GAIFMAN, H., On local and nonlocal properties, in: J. Sterne, ed., *Proc. Logic Colloquium* (1981) 105–132.

[120] GAIFMAN, H., H. MAIRSON, Y. SAGIV and M.Y. VARDI, Undecidable optimization problems for database logic progams, in: *Proc. 2nd IEEE Symp. on Logic in Computer Science* (1987) 106–115.

[121] GAIFMAN, H. and M.Y. VARDI, A simple proof that connectivity of finite graphs is not first-order definable, *Bull. EATCS* **26** (1985) 43–45.

[122] GALIL, Z., An almost linear-time algorithm for computing a dependency basis in a relational database, *J. ACM* **29**(1) (1982) 96–102.

[123] GALLAIRE, H. and J. MINKER, eds., *Logic and Databases* (Plenum, New York, 1978).

[124] GALLAIRE, H., J. MINKER and J.-M. NICOLAS, Logic and databases: a deductive approach, *ACM Comput. Surveys* **16**(2) (1984) 153–185.

[125] GAREY, M.R. and D.S. JOHNSON, *Computers and Intractability: A Guide to the Theory of NP-Completeness* (Freeman, San Francisco, CA, 1979).

[126] GINSBURG, S. and S.M. ZAIDDAN, Properties of functional dependency families, *J. ACM* **29**(4) (1982) 678–698.

[127] GOODMAN, N. and O. SHMUELI, Tree queries: a simple class of queries, *ACM Trans. Database Systems* **7**(4) (1982) 653–677.

[128] GOODMAN, and O. SHMUELI, Syntactic characterization of tree database schemas, *J. ACM* **30**(4) (1983) 767–786.

[129] GOODMAN, N. and O. SHMUELI, The tree projection theorem and relational query processing, *J. Comput. System Sci.* **28**(1) (1984) 60–79.

[130] GOODMAN, N. and Y.C. TAY, Synthesizing fourth normal form relations from multivalued dependencies, Res. Report, Harvard Univ., Cambridge, MA, 1983.

[131] GOTTLOB, G., P. PAOLINI and R. ZICARI, Properties and update semantics of consistent views, *ACM Trans. Database Systems* **13**(4) (1988) 486–524.

[132] GRAHAM, M.H., On the universal relation, Ph.D. Dissertation Univ. of Toronto, Toronto, 1979.

[133] GRAHAM, M.H., Functions in database, *ACM Trans. Database Systems* **8**(1) (1983) 81–109.

[134] GRAHAM, M.H., A.O. MENDELZON and M.Y. VARDI, Notions of dependency satisfaction, *J. ACM* **33**(1) (1986) 105–129.

[135] GRAHAM, M.H. and M. YANNAKAKIS, Independent database schemes, *J. Comput. System Sci.* **28**(1) (1984) 121–141.

[136] GRAHNE, G., Dependency satisfaction in databases with incomplete information, in: *Proc. 10th Conf. on Very Large Databases* (1984).

[137] GRAHNE, G. and K.-J. RÄIHÄ, Characterizations for acyclic database schemes, in: P.C. Kanellakis and F. Preparata, eds., *Advances in Computing Research, Vol. 3* (JAI Press, Greenwich, CT, 1986) 19–42.

[138] GRANT, J. and B.E. JACOBS, On the family of generalized dependency constraints, *J. ACM* **29**(4) (1982) 986–997.

[139] GRANT, J. and J. MINKER, Answering queries in indefinite databases and the null value problem, in: P.C. Kanellakis and F. Preparata, eds., *Advances in Computing Research, Vol. 3* (JAI Press, Greenwich, CT, 1986) 19–42.

[140] GUREVICH, Y., Toward a logic tailored for computational complexity, in: M.M. Richter et al., eds., *Computation and Proof Theory*, Lecture Notes in Mathematics, Vol. 1104 (Springer, Berlin, 1984) 175–216.

[141] GUREVICH, Y., Logic and the challenge of computer science, in: E. Börger, ed., *Trends in Theoretical Computer Science* (Computer Science Press, Rockville, MD, 1988) 1–57.

[142] GUREVICH, Y. and H.R. LEWIS, The inference problem for template dependencies, in: *Proc. 1st ACM Symp. on Principles of Database Systems* (1982) 221–229.

[143] GUREVICH, Y. and S. SHELAH, Fixed-point extensions of first-order logic, *Annals of Pure Appl. Logic* **32** (1986) 265–280; also in: *Proc. 26th Ann. IEEE Symp. on Foundations of Computer Science* (1985) 346–353.

[144] GYSSENS, M. and D. VAN GUCHT, The powerset algebra as a result of adding programming constructs to the nested relational algebra, in: *Proc. ACM SIGMOD* (1988) 225–232.

[145] HADDAD, R. and J. NAUGHTON, Counting methods for cyclic relations, in: *Proc. 7th ACM Symp. on Principles of Database Systems* (1988) 333–340.

[146] HAGIHARA, K., M. ITO, K. TANIGUCHI and T. KASAMI, Decision problems for multivalued dependencies in relational databases, *SIAM J. Comput.* **8**(2) (1979) 247–264.

[147] HAREL, D., *First-Order Dynamic Logic*, Lecture Notes in Computer Science, Vol. 68 (Springer, Berlin, 1979).

[148] HAREL, D. and D. KOZEN, A programming language for the inductive sets and applications, *Inform. and Control* **63** (1984) 118–139.

[149] HEATH, I.J., Unacceptable file operations in a relational data base, in: *Proc. ACM SIGFIDET Workshop on Data Description, Access, and Control* (1971).

[150] HENKIN, L., J.D. MONK and A. TARSKI, *Cylindric Algebras, Part I* (North-Holland, Amsterdam, 1971); *Part II* (North-Holland, Amsterdam, 1985).

[151] HENSCHEN, L.J. and S.A. NAQVI, On compiling queries in recursive first-order databases, *J. ACM* **31**(1) (1984) 47–85.

[152] HONEYMAN, P., Testing satisfaction of functional dependencies, *J. ACM* **29**(3) (1982) 668–677.

[153] HONEYMAN, P., R.E. LADNER and M. YANNAKAKIS, Testing the universal instance assumption, *Inform. Process. Lett.* **10**(1) (1980) 14–19.

[154] HULL, R., Finitely specifiable implicational dependency families, *J. ACM* **31**(2) (1984) 210–226.

[155] HULL, R., A survey of theoretic research on typed complex database objects, in: J. Paredaens, ed., *Databases* (Academic Press, New York, 1987) 193–256.

[156] HULL, R. and R. KING, Semantic database modeling: survey, applications, and research issues, *ACM Comput. Surveys* **19** (1987) 201–260

[157] HULL, R. and J. SU, On the expressive power of database queries with intermediate types, in: *Proc. 7th ACM Symp. on Principles of Databases Systems* (1988) 39–51.

[158] IMIELINSKI, T. and W. LIPSKI, The relational model of data and cylindric algebras, *J. Comput. System Sci.* **28**(1) (1984) 80–102.

[159] IMIELINSKI, T. and W. LIPSKI, Incomplete information in relational databases, *J. ACM* **31**(4) (1984) 761–791.

[160] IMMERMAN, N., Number of quantifiers is better than number of tape cells, *J. Comput. System Sci.* **22**(3) (1981) 65–72.

[161] IMMERMAN, N., Relational queries computable in polynomial time, *Inform. and Control* **68** (1986) 86–104.

[162] IMMERMAN, N., Languages which capture complexity classes, *SIAM J. Comput.* **16**(4) (1987) 760–778.

[163] IMMERMAN, N., Expressibility as a complexity measure: results and directions, Res. Report DCS-TR-538, Yale Univ., New Haven, CT, 1987.

[164] IOANNIDIS, Y.E., A time bound on the materialization of some recursively defined views, in: *Proc. 11th Conf. on Very Large Databases* (1985) 219–226.

[165] ITO, M., M. IWASAKI and T. KASAMI, Some results on the representative instance in relational databases, *SIAM J. Comput.* **14**(2) (1985) 334–354.

[166] JACOBS, B.E., On database logic, *J. ACM* **29**(2) (1982) 310–332.

[167] JAESCHKE, G. and H.-J. SCHEK, Remarks on the algebra on non first normal form relations, in: *Proc. 1st ACM Symp. on Principles of Database Systems* (1982) 124–138.

[168] JARKE, M. and J. KOCH, Query optimization in database systems, *ACM Comput. Surveys* **16** (1984) 111–152.

[169] JOHNSON, D.S. and A. KLUG, Testing containment of conjunctive queries under functional and inclusion dependencies, *J. Comput. System Sci.* **28**(1) (1984) 167–189.

[170] JONES, N.G. and A.L. SELMAN, Turing machines and the spectra of first-order sentences, *J. Symbolic Logic* **39** (1974) 139–150.

[171] KANELLAKIS, P.C., On the computational complexity of cardinality constraints in relational databases, *Inform. Process. Lett.* **11**(2) (1980) 98–101.

[172] KANELLAKIS, P.C., Logic programming and parallel complexity, in: J. Minker, ed., *Foundations of Deductive Databases and Logic Programming* (Morgan-Kaufman, Los Altos, CA, 1988) 547–586.

[173] KANELLAKIS, P.C., S.S. COSMADAKIS and M.Y. VARDI, Unary inclusion dependencies have polynomial time inference problems, in: *Proc. 15th Ann. ACM Symp. on Theory of Computing* (1983) 264–277.

[174] KELLER, A., Algorithms for translating view updates to database updates for views involving selections projections and joins, in: *Proc. 4th ACM Symp. on Principles of Database Systems* (1985) 154–163.

[175] KELLER, A. and J.D. ULLMAN, On complementary and independent mappings, in: *Proc. ACM SIGMOD* (1984).

[176] KIFER, M. and C. BEERI, An integrated approach to logical design of relational database schemes, *ACM Trans. Database Systems* **11**(2) (1986) 134–158.

[177] KIFER, M., R. RAMAKRISHNAN and A. SILBERSCHATZ, An axiomatic approach to deciding query safety in deductive databases, in: *Proc. 7th ACM Symp. Princples of Database Systems* (1988) 52–60.

[178] KLUG, A., Equivalence of relational algebra and relational calculus query languages having aggregate functions, *J. ACM* **29**(3) (1982) 699–717.

[179] KLUG, A., On conjunctive queries containing inequalities, *J. ACM* **35**(1) (1988) 146–160.

[180] KOLAITIS, P.G., The expressive power of stratified logic programs, Manuscript, 1987; submitted to *Inform. and Comput.*

[181] KOLAITIS, P.G. and C.H. PAPADIMITRIOU, Why not negation by fixpoint?, in: *Proc. 7th ACM Symp. on Principles of Database Systems* (1987) 231–239.

[182] KORTH, H.F. and A. SILBERSCHATZ, *Database System Concepts* (McGraw-Hill, New York, 1986).

[183] KOZEN, D., Complexity of finitely presented algebras, in: *Proc. 9th Ann. ACM Symp. on Theory of Computing* (1977) 164–177.

[184] KUHNS, J.L., Answering questions by computer: a logical study, Res. Report, RM-5428-PR, Rand Corp., Santa Monica, CA, 1967.

[185] KUPER, G.M. Logic programming with sets, in: *Proc. 6th ACM Symp. on Principles of Database Systems* (1987) 11–20.

[186] KUPER, G.M., On the expressive power of logic programming languages with sets, in: *Proc. 7th ACM Symp. on Principles of Database Systems* (1988) 10–14.

[187] KUPER, G. and M.Y. VARDI, A new approach to database logic, in: *Proc. 3rd ACM Symp. on Principles of Database Systems* (1984) 86–96.

[188] KUPER, G. and M.Y. VARDI, On the complexity of queries in the logical database model, in: *Proc. 2nd Internat. Conf. on Database Theory* (1988) 267–280.

[189] LAKSHMANAN, V.S. and A.O. MENDELZON, Inductive pebble games and the inductive power of Datalog, in: *Proc. 8th ACM Symp. on Principles of Database Systems* (1989) 301–311.

[190] LAVER, K., A.O. MENDELZON and M.H. GRAHAM, Functional dependencies on cyclic database schemes, in: *Proc. ACM SIGMOD* (1983) 79–91.

[191] LEIVANT, D., Descriptive characterizations of computational complexity, Res. Report, Carnegie-Mellon Univ., 1988.

[192] LEVESQUE, H.J., Foundations of a functional approach to knowledge representation, *Artificial Intelligence* **23** (1984) 155–212.

[193] LIEN, E., On the equivalence of database models, *J. ACM* **29**(2) (1982) 333–363.

[194] LING, T., F. TOMPA and T. KAMEDA, An improved third normal form for relational databases, *ACM Trans. Databases Systems* **6**(2) (1981) 326–346.

[195] LIPSKI, W., On semantic issues connected with incomplete information databases, *ACM Trans. on Database Systems* **4**(3) (1979) 262–296.
[196] LIPSKI, W., On databases with incomplete information, *J. ACM* **28**(1) (1981) 41–70.
[197] LIU, L. and A. DEMERS, An algorithm for testing the lossless join property in relational databases, *Inform. Process Lett.* **11**(2) (1980) 73–76.
[198] LUCCHESI, C.L. and S.L. OSBORN, Candidate keys for relations, *J. Comput. System Sci.* **17**(2) (1978) 270–279.
[199] MAHER, M.J., Equivalences of logic programs, in: J. Minker, ed., *Foundations of Deductive Databases and Logic Programming* (Morgan-Kaufmann, Los Altos, CA, 1988) 627–658.
[200] MAIER, D., Minimum covers in the relational database model, *J. ACM* **27**(4) (1980) 664–674.
[201] MAIER, D., *The Theory of Relational Databases* (Computer Science Press, Rockville, MD, 1983).
[202] MAIER, D., A.O. MENDELZON, F. SADRI and J. D. ULLMAN, Adequacy of decompositions of relational databases, *J. Comput. System Sci.* **21**(3) (1980) 368–379.
[203] MAIER, D., A.O. MENDELZON and Y. SAGIV, Testing implications of data dependencies, *ACM Trans. Database Systems* **4**(4) (1979) 455–469.
[204] MAIER, D., D. ROZENSHTEIN and D.S. WARREN, Window functions, in: P.C. Kanellakis and F. Preparata, eds., *Advances in Computing Research, Vol. 3* (JAI Press, Greenwich, CT, 1986) 213–246.
[205] MAIER, D., Y. SAGIV and M. YANNAKAKIS, On the complexity of testing implications of functional and join dependencies, *J. ACM* **28**(4) (1981) 680–695.
[206] MAIER, D., J.D. ULLMAN and M.Y. VARDI, On the foundations of the universal relation model, *ACM Trans. Database Systems* **9**(2) (1984) 283–308.
[207] MAIER, D. and D.S. WARREN, *Computing with Logic: Logic Programming with Prolog* (Benjamin/ Cummings, Menlo Park, CA, 1988).
[208] MAKINOUCHI, A., A consideration of normal form of not-necessarily-normalized relations in the relational data model, in: *Proc. 3rd Conf. on Very Large Databases* (1977) 447–453.
[209] MANCHANDA, S. and D.S. WARREN, A logic-based language for database updates, in: J. Minker, ed., *Foundations of Deductive Databases and Logic Programming* (Morgan-Kaufmann, Los Altos, CA, 363–394.
[210] MANNILA, H. and K.-J. RÄIHÄ, Small Armstrong relations for database design in: *Proc. 4th ACM Symp. on Principles of Database Systems* (1985) 245–250.
[211] MARCHETTI-SPACCAMELA, A., A. PELAGGI and D. SACCA, Worst-case complexity analysis of methods for logic query implementation, in: *Proc. 6th ACM Symp. on Principles of Database Systems* (1987) 294–301.
[212] MENDELZON, A.O., Database states and their tableaux, *ACM Trans. Database Systems* **9**(2) (1984) 264–282.
[213] MENDELZON, A.O. and D. MAIER, Generalized mutual dependencies and the decomposition of data base relations, in: *Proc. 5th Conf. on Very Large Databases* (1979) 75–82.
[214] MILLER, G.L. and J.H. REIF, Parallel tree contraction and its applications, in: *Proc. 26th Ann IEEE Symp. on Foundations of Computer Science* (1985) 478–489.
[215] MINKER, J., Perspectives in deductive databases, *J. Logic Programming* **5**(1) (1988) 33–60.
[216] MINKER, J., ed., *Foundations of Deductive Databases and Logic Programming* (Morgan-Kaufmann, Los Altos, CA, 1988).
[217] MINKER, J. and N. NICOLAS, On recursive axioms in relational databases, *Inform. Systems* **8** (1982) 1–13.
[218] MITCHELL, J.C., The implication problem for functional and inclusion dependencies, *Inform. and Control* **56**(3) (1983) 154–173.
[219] MORRIS, K., An algorithm for ordering subgoals in NAIL! in: *Proc. 7th ACM Symp. on Principles of Database Systems* (1988) 82–88.
[220] MORRIS, K., J.D. ULLMAN and A. VAN GELDER, Design overview of the NAIL! system, in: *Proc. 3rd Internat. Conf. on Logic Programming*, Lecture Notes in Computer Science, Vol. 225 (Springer, Berlin, 1986) 554–568.
[221] MOSCHOVAKIS, Y.N., *Elementary Induction on Abstract Structures* (North-Holland, Amsterdam, 1974).
[222] NAQVI, S., A logic for negation in database system, in: *Proc. Workshop on Logic Databases* (1986).

[223] NAQVI, S. and R. KRISHNAMURTHY, Database updates in logic programming, in: *Proc. 7th ACM Symp. on Principles of Database Systems* (1988).

[224] NAUGHTON, J.F., Data independent recursion in deductive databases, in: *Proc. 5th ACM Symp. on Principles of Databases Systems* (1986) 267–279.

[225] NAUGHTON, J.F., One-sided recursions, in: *Proc. 6th ACM Symp. on Principles of Database Systems* (1987) 340–348.

[226] NAUGHTON, J.F. and Y. SAGIV, A decidable class of bounded recursions, in: *Proc. 6th ACM Symp. on Principles of Database Systems* (1987) 227–236.

[227] NICOLAS, J.-M., First order logic formalization for functional, multivalued, and mutual dependencies, in: *Proc. ACM SIGMOD* (1978) 40–46.

[228] OSBORN, S.L., Normal forms for relational databases, Ph.D. Dissertation, Res. Report, Univ. of Waterloo, Waterloo, 1977.

[229] OZSOYOGLOU, Z.M. and L.-Y. YUAN, A new normal form for nested relations, *ACM Trans. Databases Systems* **12**(1) (1987) 111–136.

[230] PALERMO, F.P., A database search problem in: J.T. Tou, ed., *Information Systems COINS IV* (Plenum, New York, 1974).

[231] PAPADIMITRIOU, C.H., *The Theory of Concurrency Control* (Computer Science Press, Rockville, MD, 1986).

[232] PAREDAENS, J., On the expressive power of the relational algebra, *Inform. Process. Lett.* **7**(2) (1978).

[233] PAREDAENS, J. and D. JANSSENS, Decompositions of relations: a comprehensive approach, in: H. Gallaire, J. Minker and J-M. Nicolas, eds., *Advances in Data Base Theory, Vol. 1* (Plenum, New York, 1981 73–100.

[234] PAREDAENS, J. and D. VAN GUCHT, Possibilities and limitations of using flat operators in nested algebra expressions, in: *Proc. 7th ACM Symp. on Principles of Database Systems* (1988) 29–38.

[235] PARKER, D.S. and K. PARSAYE-GHOMI, Inference involving embedded multivalued dependencies and transitive dependencies, in: *Proc. ACM SIGMOD* (1980) 52–57.

[236] PATERSON, M.S. and M.N. WEGMAN, Linear unification, *J. Comput. System Sci.* **16** (1978) 158–167.

[237] RAMAKRISHNAN, R., Y. SAGIV, J.D. ULLMAN and M.Y. VARDI, Proof-tree transformation theorems and their applications, in: *Proc. 8th ACM Symp. on Principles of Database Systems* (1989) 172–182.

[238] REITER, R., On closed world data bases, in: H. Gallaire and J. Minker, eds., *Logic and Databases* (Plenum, New York, 1978) 55–76.

[239] REITER, R., Towards a logical reconstruction of relational database theory, in: M.L. Brodie, J.L. Mylopoulos and J.W. Schmidt, eds., *On Conceptual Modeling* (Springer, New York, 1984) 163–189.

[240] REITER, R., A sound and sometimes complete query evaluation algorithm for relational databases with null values, *J. ACM* **33**(2) (1986) 349–370.

[241] RISSANEN, J., Independent components of relations, *ACM Trans. Database Systems* **2**(4) (1977) 317–325.

[242] ROBINSON, J.A., A machine oriented logic based on the resolution principle, *J. ACM* **12**(1) (1965) 23–41.

[243] ROTH, M.A., H.F. KORTH and A. SILBERSCHATZ, Theory of non-first-normal-form relational databases, Res. Report TR-84-36, Univ. of Texas, Austin, TX, 1986.

[244] RUZZO, W.L., Tree-size bounded alternation, *J. Comput. System Sci.* **21**(2) (1980) 218–235.

[245] SACCA, D., Closures of database hypergraphs, *J. ACM* **32**(4) (1985) 774–803.

[246] SACCA, D., F. MANFREDI and A. MECCHIA, Properties of database schemata with functional dependencies, in: P.C. Kanellakis and F. Preparata, eds., *Advances in Computing Research, Vol. 3* (JAI Press, Greenwich, CT, 1986) 105–137.

[247] SACCA, D. and C. ZANIOLO, On the implementation of a simple class of logic queries for databases, in: *Proc. 5th ACM Symp. on Principles of Database Systems* (1986) 16–23.

[248] SADRI, U.F. and J.D. ULLMAN, Template dependencies: a large class of dependencies in relational database and their complete axiomatization, *J. ACM* **29**(2) (1981) 363–372.

[249] SAGIV, Y., An algorithm for inferring multivalued dependencies with an application to propositional logic, *J. ACM* **27**(2) (1980) 250–262.

[250] SAGIV, Y., Can we use the universal assumption without using nulls?, in: *Proc. ACM SIGMOD* (1981) 108–120.

[251] SAGIV, Y., A characterization of globally consistent database and their correct access paths, *ACM Trans. Database Systems* **8**(2) (1983) 266–286.

[252] SAGIV, Y., Evaluation of queries in independent database schemes, Res. Report, Hebrew Univ., Jerusalem, 1984.

[253] SAGIV, Y., On computing restricted projections of representative instances, in: *Proc. 4th ACM Symp. on Principles of Database Systems* (1985) 171–180.

[254] SAGIV, Y., Optimizing Datalog programs, in: J. Minker, ed., *Foundations of Deductive Databases and Logic Programming* (Morgan-Kaufmann, Los Altos, CA, 1988) 659–698.

[255] SAGIV, Y., C. DELOBEL, D.S. PARKER and R. FAGIN, An equivalence between relational database dependencies and a fragment of propositional logic, *J. ACM* **28**(3) (1981) 435–453.

[256] SAGIV, Y. and O. SHMUELI, On finite FD-acyclicity, in: *Proc. 5th ACM Symp. on Principles of Database Systems* (1986) 173–182.

[257] SAGIV, Y. and O. SHMUELI, The equivalence of solving queries and producing tree projections, in: *Proc. 5th ACM Symp. on Principles of Database Systems* (1986) 160–172.

[258] SAGIV, Y. and M.Y. VARDI, Safety of Datalog queries over infinite database, in: *Proc. 8th ACM Symp. on Principles of Database Systems* (1989) 160–172.

[259] SAGIV, Y. and S. WALECKA, Subset dependencies and a completeness result for a subclass of embedded multivalued dependencies, *J. ACM* **29**(1) (1982) 103–117.

[260] SAGIV, Y. and M. YANNAKAKIS, Equivalence among expressions with the union and difference operators, *J. ACM* **27**(4) (1980) 633–655.

[261] SARAIYA, Y., Linearising nonlinear recursions in polynomial time, in: *Proc. 8th ACM Symp. on Principles of Database Systems* (1989) 182–190.

[262] SAZONOV, V., A logical approach to the problem of "P = NP?", in: *Proc. Mathematical Foundations of Computer Science '80*, Lecture Notes in Computer Science, Vol. 88 (Springer, Berlin, 1980) 562–575.

[263] SCHECK, H.-J. and M. SCHOLL, The relational model with relation-valued attributes, *Inform. Systems* (1986).

[264] SCIORE, E., Real-world MVDs, in: *Proc. ACM SIGMOD* (1981) 121–132.

[265] SCIORE, E., A complete axiomatization of full join dependencies, *J. ACM* **29**(2) (1982) 373–393.

[266] SCIORE, E., Comparing the universal instance and relational data models, in: P.C. Kanellakis and F. Preparata, eds., *Advances in Computing Research, Vol. 3* (JAI Press, Greenwich, CT, 1986) 139–163.

[267] SELINGER, P., Chickens and eggs—The interrelations of systems and theory, in: *Proc. 6th ACM Symp. on Principles of Database Systems* (1987) 250–253.

[268] SELINGER, P., M.M. ASTRAHAN, D.D. CHAMBERLIN, R.A. LORIE and T.G. PRICE, Access path selection in a relational database management system, in: *Proc. ACM SIGMOD* (1979) 23–34.

[269] SHMUELI, O., Decidability and expressiveness aspects of logic queries, in: *Proc. 6th ACM Symp. on Principles of Database Systems* (1987) 237–249.

[270] SHMUELI, O., S. TSUR and C. ZANIOLO, Rewriting of rules containing set terms in a logic data language (LDL), in: *Proc. 7th ACM Symp. on Principles of Database Systems* (1988) 15–28.

[271] SPYRATOS, N., The partition model: a deductive database model, *ACM Trans Database Systems* **12**(1) (1987) 1–37.

[272] STOCKMEYER, L., The polynomial-time hierarchy, *Theoret. Comput. Sci.* **3** (1977) 1–22.

[273] TARJAN, R.E. and M. YANNAKAKIS, Simple linear-time algorithms to test chordality of graphs, test acyclicity of hypergraphs, and selectively reduce acyclic hypergraphs, *SIAM J. Comput.* **13**(3) (1984) 566–579.

[274] TARSKI, A. and F.B. THOMPSON, Some general properties of cylindric algebras (abstract), *Bull. Amer. Math. Soc.* **58** (1952) 65.

[275] THOMAS, S.J. and P.C. FISCHER, Nested relational structures, in: P.C. Kanellakis and F. Preparata, eds., *Advances in Computing Research, Vol. 3* (JAI Press, Greenwich, CT, 1986) 269–307.

[276] TRAHTENBROT, B.A., Impossibility of an algorithm for the decision problem in finite classes, *Dokl. Akad. Nauk USSR* **70** (1950) 569–572.

[277] TSUR, S. and C. ZANIOLO, LDL: A logic-based data-language, in: *Proc. 12th Conf. on Very Large Databases* (1986).

[278] ULLMAN, J.D., Implementation of logical query languages for databases, *ACM Trans. Databases Systems* **10**(3) (1985) 289–321.

[279] ULLMAN, J.D., Database theory—past and future, in: *Proc. 6th ACM Symp. on Principles of Database Systems* (1987) 1–10.

[280] ULLMAN, J.D., *Principles of Database and Knowledge Base Systems: Volume I* (Computer Science Press, Rockville, MD, 1988).

[281] ULLMAN, J. and M. VARDI, The complexity of ordering subgoals, in: *Proc. 7th ACM Symp. on Principles of Database Systems* (1988) 74–81.

[282] ULLMAN, J.D. and A. VAN GELDER, Parallel complexity of logical query programs, *Algorithmica* **3**(1) (1988) 5–42.

[283] VAN GELDER, A., Negation as failure using tight derivations for general logic programs, in: *Proc. 3rd IEEE Symp. on Logic Programming* (1986) 127–139.

[284] VAN GELDER, A., The alternating fixpoint of logic programs with negation, in: *Proc. 8th ACM Symp. on Principles of Database Systems* (1989) 1–11.

[285] VAN GELDER, A. and R. TOPOR, Safety and correct translation of relational calculus formulas, in: *Proc. 6th ACM Symp. on Principles of Database Systems* (1987) 313–328.

[286] VAN GUCHT, D., On the expressive power of the extended relational algebra for the unnormalized relational model, in: *Proc. 6th ACM Symp. on Principles of Database Systems* (1987) 302–312.

[287] VARDI, M.Y., The implication problem for data dependencies in relational databases, Ph.D. Dissertation, Res. Report, Hebrew Univ., Jerusalem, 1981.

[288] VARDI, M.Y., The decision problem for database dependencies, *Inform. Process. Lett.* **12**(5) (1981) 251–254.

[289] VARDI, M.Y., Global decision problems for relational databases, in: *Proc. 22nd Ann. IEEE Symp. on Foundations of Computer Science* (1981) 198–202.

[290] VARDI, M.Y., On decomposition of relational databases, in: *Proc. 23rd Ann. IEEE Symp. on Foundations of Computer Science* (1982) 176–185.

[291] VARDI, M.Y., The complexity of relational query languages, in: *Proc. 14th Ann. ACM Symp. on Theory of Computing* (1982) 137–146.

[292] VARDI, M.Y., Inferring multivalued dependencies from functional and join dependencies, *Acta Inform.* **19** (1983) 305–324.

[293] VARDI, M.Y., The implication and finite implication problems for typed template dependencies, *J. Comput. System Sci.* **28**(1) (1984) 3–28.

[294] VARDI, M.Y., Fundamentals of dependency theory, IBM Res. Report RJ4858 (IBM Research Lab., San Jose, CA, 1985).

[295] VARDI, M.Y., Querying logical databases, *J. Comput. System Sci.* **32**(2) (1986).

[296] VARDI, M.Y., On the integrity of databases with incomplete information, in: *Proc. 5th ACM Symp. on Principles of Database Systems* (1986) 252–266.

[297] VARDI, M.Y., Decidability and undecidability results for boundedness of linear recursive queries, in: *Proc. 7th ACM Symp. on Principles of Databases Systems* (1988) 341–351.

[298] VASSILIOU, Y., A formal treatment of imperfect information in data management, Ph.D. Dissertation, Res. Report CSRG-TR-123, Univ. of Toronto, Toronto, 1980.

[299] VIANU, V., Dynamic functional dependencies and database aging, *J. ACM* **34**(1) (1987) 28–59.

[300] WINSLETT, M., A model-theoretic approach to updating logical databases, in: *Proc. 5th ACM Symp. on Principles of Database Systems* (1986) 224–234.

[301] WINSLETT, M., A framework for comparison of update semantics, in: *Proc. 7th ACM Symp. on Principles of Database Systems* (1988) 315–324.

[302] WONG, E. and K. YOUSSEFI, Decomposition—a strategy for query processing, *ACM Trans. Database Systems* **1**(3) (1976) 223–241.

[303] YANNAKAKIS, M., Algorithms for acyclic database schemes, in: *Proc. 7th Conf. on Very Large Databases* (1981) 82–94.

[304] YANNAKAKIS, M., Querying weak instances, in: P.C. Kanellakis and F. Preparata, eds., *Advances in Computing Research, Vol. 3* (JAI Press, Greenwich, CT, 1986) 185–212.

[305] YANNAKAKIS, M. and C. PAPADIMITRIOU, Algebraic dependencies, *J. Comput. System Sci.* **25**(2) (1982) 3–41.

[306] YASUURA, H., On the parallel complexity of unification, Res. Report ER-83-01, Yajima Lab., 1983.

[307] YU, C.T. and M.Z. OZSOYOGLU, An algorithm for tree-query membership of a distributed query, in: *Proc. IEEE COMPSAC* (1979) 306–312.

[308] YUAN, L.Y. and M.Z. OZSOYOGLU, Logical design of relational databases schemes, in: *Proc. 7th ACM Symp. on Principles of Database Systems* (1987) 38–47.
[309] ZANIOLO, C., Analysis and design of relational schemata for database systems, Ph.D. Dissertation, Res. Report ENG-7669, Univ. of California at Los Angeles, 1976.
[310] ZANIOLO, C., Database relations with null values, *J. Comput. System Sci.* **28**(1) (1984) 142–166.

CHAPTER 18

Distributed Computing: Models and Methods

Leslie LAMPORT

Systems Research Center, Digital Equipment Corporation,
130 Lytton Avenue, Palo Alto, CA 94301, USA

Nancy LYNCH

Laboratory for Computer Science, MIT,
545 Technology Square, Cambridge MA 02139, USA

Contents

1. What is distributed computing 1159
2. Models of distributed systems 1160
3. Reasoning about distributed algorithms 1167
4. Some typical distributed algorithms 1178
References 1196

HANDBOOK OF THEORETICAL COMPUTER SCIENCE
Edited by J. van Leeuwen

1. What is distributed computing?

In the term *distributed computing*, the word *distributed* means spread out across space. Thus, distributed computing is an activity performed on a spatially distributed system. Although one usually speaks of a distributed system, it is more accurate to speak of a distributed *view* of a system. A hardware designer views an ordinary sequential computer as a distributed system, since its components are spread across several circuit boards, while a Pascal programmer views the same computer as nondistributed. An important problem in distributed computing is to provide a user with a nondistributed view of a distributed system—for example, to implement a distributed file system that allows the client programmer to ignore the physical location of his data.

We use the term *model* to denote a view or abstract representation of a distributed system. We will describe and discuss models informally, although we do present formal methods that can be used to reason about them.

The models of computation generally considered to be distributed are *process models*, in which computational activity is represented as the concurrent execution of sequential processes. Other models, such as Petri nets [79], are usually not studied under the title of distributed computing, even though they may be used to model spatially distributed systems. We therefore restrict our attention to process models.

Different process models are distinguished by the mechanism employed for interprocess communication. The process models that are most obviously distributed are ones in which processes communicate by *message passing*–a process sends a message by adding it to a message queue, and another process receives the message by removing it from the queue. These models vary in such details as the length of the message queues and how long a delay may occur between when a message is sent and when it can be received. There are two significant assumptions embodied in message–passing models:

- message passing represents the dominant cost of executing an algorithm;
- a process can continue to operate correctly despite the failure of other processes.

The first assumption distinguishes the use of message passing in distributed computing from its use as a synchronization mechanism in nondistributed concurrent computing. The second assumption characterizes the important subfield of fault-tolerant computing. Some degree of fault tolerance is required of most real distributed systems, but one often studies distributed algorithms that are not fault tolerant, leaving other mechanisms (such as interrupting the algorithm) to cope with failures.

Other process models are considered to be distributed if their interprocess communication mechanisms can be implemented efficiently enough by message passing, where efficiency is measured by the message-passing costs incurred in achieving a reasonable degree of fault tolerance. Algorithms exist for implementing virtually any process model by a message-passing model with any desired degree of fault tolerance. Whether an implementation is efficient enough, and what constitutes a reasonable degree of fault tolerance are matters of judgement, so there is no consensus on what models are distributed.

2. Models of distributed systems

2.1. *Message-passing models*

2.1.1. *Taxonomy*

A wide variety of message-passing models can be used to represent distributed systems. They can be classified by the assumptions made about four separate concerns: network topology, synchrony, failure, and message buffering. Different models do not necessarily represent different systems; they may be different views of the same system. An algorithm for implementing (or simulating) one model with another provides a mechanism for implementing one view of a system with a lower-level view. The entire goal of system design is to implement a simple and powerful user-level view with the lower-level view provided by the hardware.

Network topology The network topology describes which processes can send messages directly to which other processes. The topology is described by a *communication graph* whose nodes are the processes, and where an arc from process i to process j denotes that i can send messages directly to j. Most models assume an undirected graph, where an arc joining two processes means that each can send messages to the other. However, one can also consider directed graph models in which there can be an arc from i to j without one from j to i, so i can send messages to j but not vice versa. We use the term *link* to denote an arc in the communication graph; a message sent directly from one process to another is said to be sent over the link joining the two processes.

In some models, each process is assumed to know the complete set of processes, and in others a process is assumed to have only partial knowledge—usually the identity of its immediate neighbors. The simplest models, embodying the strongest assumptions, are ones with a completely connected communication graph, where each nonfaulty process knows about and can send messages directly to every other nonfaulty process. Routing algorithms are used to implement such a model with a weaker one.

Synchrony In the following discussion, all synchrony conditions are assumed to apply only in the absence of failure. Failure assumptions are treated separately below.

A *completely asynchronous* model is one with no concept of real time. It is assumed that messages are eventually delivered and processes eventually respond, but no assumption is made about how long it may take.

Other models introduce the concept of time and assume known upper bounds on message transmission time and process response time. For simplicity, in our examples we will use the simplest form of this assumption, that a message generated in response to an event at any time t (such as the receipt of another message) arrives at its destination by time $t+\delta$, where δ is a known constant.

Processes need some form of real-time clock to take advantage of this assumption. The simplest type of clock is a *timer*, which measures elapsed time; the instantaneous values of different processes' timers are independent of one another. Timers are used to detect failure, the assumption made above implying that a failure must have occurred if the reply to a message is not received within 2δ seconds of the sending of that message.

Some models make the stronger assumption that processes have synchronized clocks that run at approximately the correct rate of one second of clock time per second of real time. The simplest such assumption, which we use in our discussion, is that at each instant, the clocks of any two processes differ by at most ε for some known constant ε. Algorithms can use synchronized clocks to reduce the number of messages that need to be sent. For example, if a process is supposed to send a message at a known time t, then the receiving process knows that there must have been a failure if the message did not arrive by approximately time $t+\delta+\varepsilon$ on its clock—the δ due to delivery time and the ε due to the difference between the two processes' clocks. Thus, one can test for failure by sending a single message rather than the query and response required with only timers. It appears to be a fundamental property of distributed systems that algorithms which depend upon synchronized clocks incur a delay proportional to the bound on clock differences (taken to be ε in our discussion).

Given a bound on the ratio of the running rates of any two processes' timers, and the assumed bound on message and processing delays, algorithms exist for constructing synchronized clocks from timers. These algorithms are discussed later.

The most strongly synchronous model is one in which the entire computation proceeds in a sequence of distinct rounds. At each round, every process sends messages, possibly to every other process, based upon the messages that it received in previous rounds. Thus, the processes act like processors in a single synchronous computer. This model is easily simulated using synchronized clocks by letting each round begin $\delta+\varepsilon$ seconds after the preceding one.

Failure In message-passing models, one can consider both process failures and communication failures. It is commonly assumed that communication failure can result only in lost messages, although duplication of messages is sometimes allowed. Models in which incorrect messages may be delivered are seldom studied because it is believed that in practice, the use of redundant information (checksums) allows the system to detect garbled messages and discard them.

Models may allow transient errors that destroy individual messages, or they may consider only failures of individual links. A link failure may cause all messages sent over the link to be lost or, in models with timers or clocks, a failed link may deliver messages too late. Since algorithms that use timers or clocks usually discard late messages, there is little use in distinguishing between late and lost messages. Of particular concern in considering link failures is whether or not one considers the possibility of *network partition*, where the communication graph becomes disconnected, making it impossible for some pairs of nodes to communicate with each other.

The weakest assumption made about process failure is that failure of one process cannot affect communication over a link joining two other processes, but any other behavior by the failed process is possible. Such models are said to allow *Byzantine* failure.

More restrictive models permit only *omission* failures, in which a faulty process fails to send some messages. (Since late messages are usually discarded, failures that cause a process to send messages too late can be considered omission failures.)

The most restrictive models allow only *halting* failures, in which a failed process does nothing. In the subclass of *fail-step* models, other processes know when a process has failed [75].

In addition to the actual failure mode, some models make assumptions about how a failed process may be restarted. Models that allow only halting failures often assume some form of *stable storage* that is not affected by a failure. A failed process is restarted with its stable storage in the same state as before the failure and with every other part of its state restored to some initial values.

Failure models are problematic because it is difficult to determine how accurately they describe the behavior of real systems. It seems to be a widely held view among implementers of distributed systems that message loss and link failure adequately represent intercomputer communication failures. Whether or not a particular model of process failure is suitable depends upon the degree of reliability one requires of the system. There is general agreement that halting failure represents the most common type of computer failure—the familiar "system crash". It seems to provide a suitable model when only modest reliability is required. Omisson faults, caused by unusual demand slowing down a computer's response time, should probably be considered when greater reliability is required. When extremely high reliability is required—especially when failure of the entire system could be life threatening—it seems necessary to assume Byzantine failures.

As we describe later, algorithms that tolerate Byzantine failures are more costly than ones that tolerate only more restricted failures. Less costly algorithms can be achieved by strengthening Byzantine failure models to allow *digital signatures* [26]. It is assumed that given an arbitrary data item D, any nonfaulty process i can generate a digital signature $S(i, D)$ such that any other process can determine whether a particular value v equals $S(i, D)$ for a given D, but no other process can generate $S(i, D)$. Although digital signatures are a cryptographic concept, in practical fault-tolerant algorithms they are implemented with redundancy. It is believed that, by the careful use of redundancy, the assumption made about digital signatures can be achieved with high enough probability to allow the use of the model even when extremely high reliability is required.

Message Buffering In message-passing models, there is a delay between when a message is sent and when it is received. Such a delay implies that there is some form of message buffering. Models may assume either finite or infinite buffers. With finite buffers, any link may contain only a fixed maximum number of messages that have been sent over that link but not yet received. When the link's buffer is full, attempts to send an additional message over the link either fail and produce some error response to the sending process or else cause the sending process to wait until there is room in the buffer. With infinite buffers, there may be arbitrarily many unreceived messages in a link's buffer, and the sender can always send another message over the link. Although any real system has a finite capacity, this capacity may be large enough to make infinite buffering a reasonable abstraction.

If a link's buffer can hold more than one message, it is possible for messages to be received in a different order than they were sent. Models with *FIFO* (first-in first-out) buffering assume that messages that are not lost are always received in the same order in which they were sent. Many algorithms for asynchronous systems work only under the assumption of FIFO buffering. In most algorithms for systems with timers or synchronized clocks, a process does not send a message to another process until it knows that the previous message to that process has either been delivered or lost, so FIFO buffering need not be assumed. At the lowest level, real distributed systems usually provide FIFO buffering. This need not be the case at higher levels, where messages may be routed to their destinations along multiple possible paths. However, if it is not provided by the underlying communication mechanism, FIFO buffering can be implemented by numbering the messages.

2.1.2. Measuring complexity

There are two basic complexity measures for distributed algorithms; time and message complexity. The time complexity of an algorithm measures the time needed both for message transmission and for computation within the processes. However, computations performed by individual processes are traditionally ignored, only message-passing time being counted. This is a reasonable approximation for current computer networks in which message delivery time is usually several milliseconds or more, while computer operations are measured in microseconds. However, a millisecond is only a thousand microseconds, and a practical algorithm should not perform millions of extra calculations to save a few messages. Moreover, the large difference between message delivery time and processing time should not be taken for granted. Although it takes much longer for electromagnetic signals to travel within a processor than between processors in a spatially distributed system, current processing speed is limited primarily by circuit delays rather than transmission delays. With current technology, the high cost of sending a message is an artifact of the way systems are designed, since electrical signals can travel a kilometer in a few microseconds.

The usual measure of message-passing time for an algorithm is the length of the longest chain of messages that occurs before the algorithm terminates, where each message in the chain except the first is generated by the receipt of the previous one. For completely asynchronous models, where no assumptions are made about message delivery times, this seems to be the only reasonable way to measure worst-case message-passing time; for synchronous models that operate in rounds, it is just the number of rounds. The measure can be refined to take account of more precise timing assumptions—for example, if transmission delays are different for different links. Of course, processing time should be included in the time complexity if it is significant.

The most common measure of message complexity is the total number of messages transmitted. If messages contain on the order of a few hundred bits or more, then the total number of bits sent might be a better measure of the cost than the number of messages. In many algorithms, a process brodcasts the same message to n other processes. Depending upon the implementation details of the system, such a broadcast might cost as much as sending n separate messages, or it might cost no more than sending a single message.

Tradeoffs between time and message complexity are often possible. The minimal time algorithm is usually simple, with more complex algorithms saving messages, but taking longer to terminate. It is often possible to "improve" algorithms by reducing their message complexity at the expense of their time complexity. However, many distributed systems contain few enough processes that an algorithm with a message complexity proportional to the square of the number of processes is quite practical and is often better than a more complicated one that uses fewer messages but takes longer.

As with sequential algorithms, there is also the question of whether to measure worst-case or average behavior—for example, whether to measure the maximum number of messages that can be sent or the expected number (in the sense of probability theory). When high reliability is required, worst-case behavior is usually the appropriate measure. In other cases, the average cost may be more important. Average costs have been derived mainly for probabilistic algorithms, in which processes make random choices.

2.2. *Other models*

Other models of concurrent systems are usually described in terms of language constructs for interprocess communication. This can lead to the confusion of underlying concepts (what one says) with language issues (how one says it), but we know of no simple alternative for classifying the standard models.

2.2.1. *Shared variables*

In the earliest models of concurrency, processes communicate through *global shared variables*—program variables that can be read and written by all processes. Initially, the shared variables were accessed by the ordinary program operations of expression evaluation and assignment; later variations included synchronization primitives such as semaphores [28] and monitors [47] to control access to shared variables. Global shared-variable models provide a natural representation of multiprocessing on a single computer with one or more processors connected to a central shared memory.

The most natural and efficient way to implement global shared variables with message passing is to have each shared variable maintained by a single process. That process can access the variable locally; it requires two messages for another process to read or write the variable. A read requires a query and a response with the value; a write requires sending the new value and receiving an acknowledgement that the operation was done—the acknowledgement is required because the correctness of shared-variable algorithms depends upon the assumption that a write is completed before the next operation is begun.

Such an implementation of global shared variables is not at all fault-tolerant, since failure of the process holding the variable blocks the progress of any other process that accesses it. A fault-tolerant implementation must maintain multiple copies of the variable at different processes, which requires much more message passing. Hence, global shared-variable models are not generally considered to be distributed.

A more restrictive class of models permits interprocess communication only through *local shared variables*, which are shared variables that are "owned" by individual processes. A local shared variable can be read by multiple processes, but it can be written only by the process that owns it. Reading a variable owned by a failed process is assumed to return some default value.

2.2.2. *Synchronous communication*

Synchronous communication was introduced by Hoare in his Communicating Sequential Processes (CSP) language [48]. In CSP, process i sends a value v to process j by executing the *output command* $j!v$; process j receives that value, assigning it to variable x, by executing the *input command* $i?x$. Unlike the case of ordinary message passing, the input and output commands are executed synchronously. Execution of a $j!v$ operation is delayed until process i is ready to execute an $i?x$ operation, and vice versa. Thus, a CSP communication operation waits until a corresponding communication operation can be executed in another process.

There is an obvious way to implement sychronous communication with message passing. Process i begins execution of a $j!v$ command by sending a message to j with the value v; when process j is ready to execute the corresponding $i?x$ command, it sends an acknowledgement message to i and proceeds to its next operation. Process i can continue its execution when it receives the acknowledgement.

Many concurrent algorithms require that a process be prepared to communicate with any one of several processes, but actually communicate with only one of them before doing some further processing. With synchronous communication primitives, this means that a process must be prepared to execute any one of a set of input and/or output commands. If each process could be waiting for an arbitrary set of communication commands, then deciding which communications should occur could require a complicated distributed algorithm. For example, consider a network of three processes, each of which is ready to communicate with either one of the other two. Any pair of them can execute their corresponding communication actions, but only one pair may do so, and deciding upon that pair requires a distributed algorithm. To get around this difficulty, CSP allows a process to wait for an arbitrary set of input commands, but it may not be waiting for any other communication if it is ready to perform an output command. The choice of which communication to perform can then be made within a process, so each communication action requires only two messages.

Although CSP allows an efficient implementation with message passing, it does not permit fault-tolerant algorithms. A process i that is waiting to execute a $j!v$ command cannot continue unless process j executes a corresponding $i?x$ command. The failure of process j therefore halts the execution of process i. (This could be avoided if i could wait to communicate with any one of several processes, which CSP prohibits.) Despite this difficulty, CSP is often considered a distributed model.

Closely related to synchronous communication is the *remote procedure call* or *rendezvous*. A remote procedure call is executed just like an ordinary procedure call, except the procedure is executed in another process. It can be implemented with two messages: one to send the arguments of the call in one message and another to return the result. Halting and omission failures can be handled by having the procedure call return an error result or raise an exception if no response to the first message is received. Remote procedure call is currently the most widely used language construct for implementing distributed systems without explicit message-passing operations.

2.3. *Fundamental concepts*

The theory of sequential computing rests upon fundamental concepts of computability that are independent of any particular computational model. If there are any such fundamental formal concepts underlying distributed computing, they have yet to be developed. At present, the field seems to consist of a collection of largely unrelated results about individual models. Nevertheless, one can make some informal observations that seem to be important.

Underlying almost all models of concurrent systems is the assumption that an execution consists of a set of discrete events, each affecting only part of the system's state. Events are grouped into processes, each process being a more or less completely sequenced set of events sharing some common locality in terms of what part of the state they affect. For a collection of autonomous processes to act as a coherent system, the processes must be synchronized.

From the original work on concurrent process synchronization, two distinct classes of synchronization problem emerged: contention and cooperation. The archetypical contention problem is the *mutual exclusion* problem, in which each process has a critical section and processes must be synchronized so that no two of them execute their critical section at the same time [27]. As originally stated, this problem includes the requirement that a process be allowed to halt when not executing its critical section or its synchronization protocol. With this requirement, solutions are possible in shared-variable models but not in asynchronous message-passing models, which require that a process receive a message from every other process before it can enter its critical section. However, the mutual exclusion problem without this requirement has been studied in asynchronous message-passing systems.

The classic problem in cooperation is the *bounded buffer* problem, in which an unbounded sequence of values are transmitted in order from a sender process to a receiver process, using a fixed-length array of registers as a buffer. The receiver must wait when the buffer is empty, and the sender must wait when the buffer is full. This problem is best viewed as a symmetrical one, in which the sender generates filled buffer elements for use by the receiver and the receiver generates empty buffer elements for use by the sender.

The fundamental difference between these two forms of synchronization is that in contention problems a process must be able to make unlimited progress even if other processes fail to progress, while in cooperation problems the progress of one process depends upon the progress of another. For example, in the mutual exclusion problem, a process may enter its critical section an unlimited number of times while other processes are not requesting entrance, but in the bounded buffer problem, after the producer has filled the buffer it cannot proceed until the consumer creates an empty buffer element.

Problems of contention and cooperation appear in all models of concurrency. A class of problems that has arisen in the study of message-passsing models is that of *global consistency*. For example, in a distributed banking system, one would like all branches of the bank to have a consistent view of the balance of any single account. In general, one would like to describe a distributed system in terms of its current global state. The global consistency problem is to ensure that all processes have a consistent view of the state. In the banking example, the amount of money currently in each account is part of the state.

To define a global state, there must be a total ordering of all transactions—to determine if there is enough money in my account for a withdrawal request to be granted, one must know if a deposit action occurred before or after the request. In an asynchronous message-passing model, there is no natural total ordering of events, only the partial ordering among events defined by letting event a precede event b if there is information flow permitting a to affect b. The definition of a global state requires completing the partial ordering of events, defined by the causality relation, to a total ordering. Achieving global consistency can be reduced to the problem of guaranteeing that all processes choose the same total ordering of events, thereby having the same definition of the global system state. One method of achieving this common total ordering is through the use of *logical clocks* [52]. A logical clock is a counter maintained by each process with the property that if event a precedes event b, then the time of event a precedes the time of event b, where the time of an event is measured on the logical clock of the process at which the event occurred. Logical clocks are implemented by attaching a *timestamp*, containing the current value of the sender's logical clock, to each message.

Because there is no unique definition of a global state in a message-passing model, it is sometimes mistakenly argued that one should not use the global state in reasoning about such models. The absence of a unique definition of the global state does not mean that we cannot reason in terms of an arbitrarily chosen definition. The method of reasoning we describe below, which involves reasoning about the state of a system, is useful for all concurrent models, including message-passing ones.

Another way of viewing the global consistency problem is in terms of knowledge. The problem exists because it is impossible for a process to know the current global state, since the concurrent activity of the processes can render its knowledge obsolete. It is rather natural to think about distributed algorithms in terms of what each process knows, and reasoning about the limitations on a process's knowledge forms the basis for proofs of many of the impossibility results decribed below. However, only recently has there been an attempt to perform this reasoning within formal theories of knowledge [45]. These theories of knowledge provide a promising approach to a fundamental theory of distributed processing, but, at this writing, it is too early to know how successful they will prove to be.

3. Reasoning about distributed algorithms

Concurrent algorithms can be deceptive; an algorithm that looks simple may be quite complex, allowing unanticipated behavior. Rigorous reasoning is necessary to determine if an algorithm does what it is supposed to, and rigorous reasoning requires a formal foundation.

Here, we discuss *verification*—proving properties of concurrent algorithms. In verification, the properties to be proved are stated in terms of the algorithm itself—that is, in terms of the algorithm's variables and actions. The related field of *specification*, in which the properties to be satisfied are expressed in higher-level, implementation-independent terms, is considered briefly in Section 3.6. Specification methods must deal with the subtle question of what it means for a lower-level algorithm to implement a higher-level description. This question does not arise in the verification methods that we discuss, since the description of the algorithm and the properties to be proved are expressed in terms of the same objects.

3.1. *A system as a set of behaviors*

We have already seen that there are a wide variety of computational models of concurrent systems. However, they can almost all be described in terms of a single formal model, which forms the basis for our discussion of verification. In this model, we represent a concurrent system by a triple consisting of a set $\boldsymbol{S}$ of *states*, a set $\boldsymbol{A}$ of *actions*, and a set Σ of *behaviors*, each behavior being a finite or infinite sequence of the form

$$s_0 \xrightarrow{\alpha_1} s_1 \xrightarrow{\alpha_2} s_2 \cdots \tag{1}$$

where each s_i is a state and each α_i is an action. (If the sequence is finite, then it ends with a state s_n.) A state describes the complete instantaneous state of the system, an action is a system operation that is taken to be indivisible, and a behavior represents an execution of the system whose ith action α_i takes the system from state s_{i-1} to state s_i. The set Σ represents the set of all possible system executions.

Most verification methods regard a behavior as either a sequence of states or a sequence of actions. Having states and actions in a behavior allows our discussion to apply to both approaches.

To reason about a system, one must first describe the triple $\boldsymbol{S}, \boldsymbol{A}, \Sigma$ that represents it—for example, by a program in some programming language. Properties of the system are expressed by assertions about the set Σ. Here are three examples to indicate, very informally, how this is done.

Mutual exclusion: For every state s_i of every behavior of Σ, in s_i there is at most one process in its critical section. (For a state to be a complete description of the instantaneous state of the system, it must describe which processes are in their critical sections.)

Lockout-freedom: (This property asserts that a process that wants to enter its critical section eventually does so.) For every behavior of the form (1) in Σ and every $i \geqslant 0$, if s_i is a state in which a process is requesting entry to its critical section, then there is some $j > i$ such that s_j is a state in which that process is in its critical section.

Bounded message delay: If α_i is the action of sending a message, s_{i-1} is a state in which the time is T, and s_j is a state in which the time is greater than $T + \delta$, then there is a k with $i < k < j$ such that α_k is the action of receiving that message. (This assumes that the current time is part of the state.)

3.2. *Safety and liveness*

Any model is an abstraction that represents only some aspects of the system, and the choice of model restricts the class of properties one can reason about. Most formal reasoning about concurrent systems has been aimed at proving two kinds of properties: *safety* and *liveness*. Intuitively a safety property asserts that something bad does not happen, and a liveness property asserts that something good eventually does happen.

In sequential computing, the most commonly studied safety property is partial correctness—if the program is started with correct input, then it does not terminate with the wrong answer, and the most commonly studied liveness property is termination—the program eventually terminates. A richer variety of safety and liveness properties are studied in concurrent computing; for example, mutual exclusion and bounded message delay are safety properties and lockout-freedom is a liveness property.

There are other classes of properties besides safety and liveness that are of interest—for example, the assertion that there is a 0.99 probability that the transmission delay is less than δ is neither a safety nor a liveness property. However, safety and liveness are the major classes of properties for which there are well developed methods of formal reasoning, so we will restrict our attention to them.

A safety or liveness property is an assertion about an individual behavior. It is satisfied by the system if it is true for all behaviors in Σ. A safety property is one that is false for a behavior if and only if it is false for some finite initial prefix of the behavior. (Intuitively, if something bad happens, then it happens after some finite number of actions.) A liveness property is one in which any finite behavior can be extended to a finite or infinite behavior (not necessarily a behavior of the program) that satisfies the property [4]. (Intuitively, after any finite portion of the behavior, it must still be possible for a good thing to happen.)

3.3. *Describing a system*

To give a formal description of a system, one must define the sets of states $\boldsymbol{S}$, actions A, and behaviors Σ. A state is defined to be as assignment of values to some set of variables, where the variables may include ordinary program variables, message buffers, "program counters", and whatever else is needed to describe completely the instantaneous state of the computation. The set of actions is usually explicitly enumerated—for example, it may include all actions of the form *i sends m to j* for particular processes i and j and a particular message m. Actions represent internal operations of the system as well as input and output operations.

There are two general approaches to describing the set Σ. They may be called the *constructive* and *axiomatic* approaches, though we shall see that these names are misleading. In the constructive approach, one describes Σ by a program, where Σ is defined to be the set of all possible behaviors obtained by executing the program. The program may be written in a conventional programing language, or in terms of a formal model such as *I/O automata* [63] or Unity [22]. In the axiomatic approach, one describes Σ by a set of axioms, where Σ is defined to be the set of all sequences of the form of formula (1) that satisfy the axioms. The axioms may be written in a formal system—some form of temporal logic [35, 77] being a currently popular choice—or in a less formal mathematical notation.

Axiomatic descriptions lead directly to a method of reasoning. If $\mathscr{S}$ is the set of axioms that describe Σ, and C is a property expressed in the same formal system as $\mathscr{S}$, then the system satisfies C if and only if the formula $\mathscr{S} \vdash C$ is valid. On the other hand, constructive descriptions are often more convenient than axiomatic ones, since programming languages are designed especially for describing computations while formal systems are usually chosen for their logical properties.

In a constructive description, one specifies the possible state transitions $s \xrightarrow{\alpha} t$ caused by each action α of $\boldsymbol{A}$. A behavior of the form (1) is in Σ only if (i) each transition $s_{i-1} \xrightarrow{\alpha_i} s_i$ is a possible state transition of α_i, and (ii) it is either infinite or it terminates in a state in which no further action is possible.

Formally, one defines a relation $\Gamma(\alpha)$ on $\boldsymbol{S}$ for each action α of $\boldsymbol{A}$, where $(s, t) \in \Gamma(\alpha)$ if and only if executing the action α in state s can produce state t. The action α is said to be *enabled* in state s if there exists some state t with $(s, t) \in \Gamma(\alpha)$. For example, the operation *send m to j* in process i's code is represented by the action α such that (s, t) is in $\Gamma(\alpha)$ if and only if s is a state in which control in process i is at operation α and t is the same as s except with m added to the queue of messages from i to j and with control in process i at the next operation after α; this action is enabled if and only if control is at the operation and the message queue is not full. The behavior (1) is in Σ only if (i) $(s_{i-1}, s_i) \in \Gamma(\alpha_i)$ for all i, and (ii) the sequence is either infinite or ends in a state s_n in which no action is enabled. Observe that condition (i) is a safety property.

In this definition, we include in Σ behaviors that start in any arbitrary state, including intermediate states one expects to encounter only in the middle of a computation and states that cannot occur in any computation. The properties one proves are of the form: if a behavior starts in a certain initial state, then ... For example, the set Σ for a mutual exclusion algorithm includes behaviors starting with several processes in their critical section. It is customary to include in the description a set of valid initial states, and to include in Σ only those behaviors starting in such a state. However, we find it more convenient not to assume any preferred starting states because, as we shall see, when proving liveness properties one must reason about the system's behavior starting from a point in the middle of the computation.

In addition to satisfying the two conditions above, sequences in Σ are usually required to satisfy some kind of fairness condition. For example, one may require that the sequence contain infinitely many actions from every process unless a point is reached after which no further actions of the process are enabled. This condition is expressed more formally by requiring that for every process k, either infinitely many of the α_i are actions of k or else there is some n such that no action of k is enabled in any state s_i with $i > n$. Fairness conditions are liveness properties.

In practice, fairness conditions do not affect the safety properties of a system. This means that if all behaviors in the set Σ described by a program satisfy a safety property C, then all behaviors satisfying only condition (i), with no fairness requirement, also satisfy C. Intuitively, safety properties are assertions about any arbitrarily long finite portion of the behavior, while liveness properties restrict only the infinite behavior. One can easily devise fairness conditions that affect safety properties—for example, the fairness requirement that every process executes infinitely many actions implies the safety property that no process ever reaches a halting state. However, such fairness conditions are unnatural and are never assumed in practice.

For reasoning about safety properties, one can therefore ignore condition (ii) and fairness conditions and consider only the relations $\Gamma(\alpha)$ defined by the actions. (In fact, (ii) really is a fairness condition.) Conversely, formal models that do not include fairness conditions are suitable only for studying safety properties, not liveness properties.

One can express conditions (i) and (ii) and the fairness conditions in a suitably chosen formal system. Expressing them in this way provides an axiomatic semantics for constructive descriptions, meaning that every program description in the form of a program can be translated into a collection of axioms. Thus, constructive descriptions can be viewed as a special class of axiomatic ones. In particular, we can adopt the simple approach to formal reasoning in which a program satisfies a property C if and only if $\mathscr{S}\vdash C$ is valid, where $\mathscr{S}$ is the translation of the program as a set of axioms. While this approach provides a formal definition of what it means for a program to satisfy a property, it does not necessarily provide a practical method for reasoning about programs because the axioms derived from conditions (i) and (ii) may be too complicated.

3.4. *Assertional reasoning*

Verifying that a system satisfies a property C means showing that every behavior satisfying the definition of system behaviors also satisfies C. The obvious way of doing this is to reason directly about sequences, using either a temporal logic or direct mathematical reasoning about sequences. The problem with such an approach is that concurrent systems can exhibit a wide variety of possible behaviors. Reasoning directly about behaviors can become quite complex, with many different cases to consider. It is not clear if there are satisfactory methods for coping with this complexity.

Assertional methods attempt to overcome this difficulty by reducing the problem of reasoning about concurrent systems to that of reasoning separately about each individual action. In an assertional method, attention is concentrated on the states. A behavior is considered to be a (finite or infinite) sequence of states $s_0 \rightarrow s_1 \rightarrow s_2 \rightarrow \cdots$ and properties are expressed in terms of *state predicates*—boolean-valued functions on the set of states. Safety and liveness properties are handled by separate techniques.

3.4.1. *Simple safety properties*

It is convenient to introduce a bit of temporal logic to express properties. We interpret a state predicate P as an assertion about behaviors by defining P to be true for a behavior if and only if it is true for the first state of the behavior. We define $\Box P$ to be the assertion that is true for a behavior if and only if P is true for all states in the behavior, so $\Box P$ asserts that P is "always" true.

Traditional assertional methods prove safety properties of the form $P \Rightarrow \Box Q$ for state predicates P and Q. Most safety properties that have been considered are of this form, with P being the predicate asserting that program control is at the beginning and all program variables have their correct initial values. For example, partial correctness is expressed by letting Q be the predicate asserting that if control is at the end then the variables have the correct final values, and mutual exclusion is expressed by letting Q be the predicate asserting that no two processes are in their critical sections. Proving such a property means showing that a certain class of states in S, namely the states in which Q is false, do not appear in any behaviors in Σ that begin in a state with P true.

We say that a state predicate I is an *invariant* of a system if no action in A can make I false. More formally, I is an invariant if and only if, for every action α in A and every pair (s, t) in $\Gamma(\alpha)$, if $I(s)$ is true then $I(t)$ is true. A simple induction argument shows that if I is an invariant then $I \Rightarrow \Box I$ is true for every behavior in Σ. (What we call an invariant is also called a *stable property*, and the term "invariant" is often used to mean a stable property that is true of the initial state.)

In assertional methods, one proves $P \Rightarrow \Box Q$ by finding a predicate I such that (i) I is an invariant, (ii) P implies I, and (iii) I implies Q. Since the invariance of I means that $I \Rightarrow \Box I$ is true for every sequence in Σ, it follows easily from (ii) and (iii) that $P \Rightarrow \Box Q$ is true for every sequence in Σ.

```
variables x, y: boolean;
cobegin loop α: ⟨await x=y⟩;
             β: ⟨critical section⟩;
             γ: ⟨x := ¬x⟩
        end loop
□
        loop λ: ⟨await y≠x⟩;
             μ: ⟨critical section⟩;
             ν: ⟨y := ¬y⟩
        end loop
coend
```

Fig. 1. A simple synchronization protocol.

As a simple example, consider the two-process program in Fig. 1, where each process cycles repeatedly through a loop composed of three statements, the angle brackets enclosing atomic actions. This program describes a common hardware synchronization protocol that ensures that the two processes alternately execute their critical sections. (For simplicity the critical sections are represented by atomic actions.) We prove that this algorithm guarantees mutual exclusion, which means that control is not at the *critical section* statements in both processes at the same time.[1] Mutual exclusion is expressed formally as the requirement $P \Rightarrow \Box Q$, where the predicates P and Q are defined by

$$P \equiv at(\alpha) \wedge at(\lambda), \qquad Q \equiv \neg(at(\beta) \wedge at(\mu));$$

where $at(\alpha)$ is the predicate asserting that control in the first process is at statement α, and the other "*at*" predicates are similarly defined.

[1]This protocol does not solve the original mutual exclusion problem because one process cannot progress if the other halts.

The invariant I used to prove this property is defined by

$$I \equiv ((at(\beta) \vee at(\gamma)) \Rightarrow (x = y)) \wedge ((at(\mu) \vee at(\nu)) \Rightarrow (x \neq y)).$$

If the critical sections do not change x or y, then executing any atomic action of the program starting with I true leaves I true, so I is an invariant. It is also easy to check that $P \Rightarrow I$ and $I \Rightarrow Q$, which imply $P \Rightarrow \Box Q$.

The method of proving safety properties of the form $P \Rightarrow \Box Q$ can be generalized to prove properties of the form $P \wedge \Box R \Rightarrow \Box Q$ for predicates P, Q, and R. Such properties are used in proving liveness properties. We say that a predicate I is *invariant under the constraint* R if any action executed in a state with $I \wedge R$ true leaves I true or makes R false. If I is invariant under the constraint R, then $I \wedge \Box R \Rightarrow \Box I$ is true for every behavior in Σ. One can therefore prove $P \wedge \Box R \Rightarrow \Box Q$ by finding a predicate I such that (i) I is an invariant under the constraint R, (ii) P implies I, and (iii) I implies Q. Thus, the ordinary assertional method for proving $P \Rightarrow \Box Q$ is extended to prove properties of the form $P \wedge \Box R \Rightarrow \Box Q$ by replacing invariance with invariance under the constraint R.

The hard part of an assertional proof is constructing I and verifying that it is an invariant (or an invariant under a constraint). The predicate I can be quite complicated, and finding it can be difficult. However, proving that it is an invariant is reduced to reasoning separately about each individual action.

Experience has indicated that this reduction is usually simpler and more illuminating than reasoning directly about the behaviors for proving safety properties that are easily expressed in the form $P \Rightarrow \Box Q$. However, reasoning about behaviors has been more successful for proving properties that are not easily expressed in this form. It is usually the case that safety properties one proves about a particular algorithm are of the form $P \Rightarrow \Box Q$, while general properties one proves about classes of algorithms are not.

Because the invariant I can be complicated, one wants to decompose it and further decompose the proof of its invariance. This is done by the *Owicki–Gries method* [65], in which the invariant is written as a program annotation with predicates attached to program control points. In this method, I is the conjunction of predicates of the form: "If program control is at this point, then the attached predicate is true." The decomposition of the invariance proof is based upon the following principle: if I is an invariant and I' is invariant under the constraint I then $I \wedge I'$ is an invariant.

A number of variations of the Owicki–Gries method have been proposed, usually for the purpose of handling particular styles of interprocess communication [5, 57]. These methods are usually described in terms of proof rules—the individual steps one goes through in proving invariance—without explicitly mentioning I or the underlying concept of invariance. This has tended to obscure their simple common foundation.

3.4.2. *Liveness properties*

If P and Q are predicates, then $P \leadsto Q$ is defined to be true if, whenever a state is reached in which P is true, then eventually a state will be reached in which Q is true. More precisely, $P \leadsto Q$ is true for the sequence (1) if, for every n, if $P(s_n)$ is true then there exists an $m \geqslant n$ such that $Q(s_m)$ is true. Most liveness properties that one wishes to prove about systems are expressible in the form $P \leadsto Q$. For example, termination is expressed by letting P assert that the program is in its starting state and letting Q assert that the program has terminated; lockout-freedom is expressed by letting P assert that some process k is requesting entry to its critical section and letting Q assert that k is in its critical section.

The basic method of proving liveness properties is by a counting argument, using a *well-founded* set—one with a partial ordering relation $\succ$ such that there are no infinite chains of the form $e_i \succ e_2 \succ \cdots$ Suppose we construct a function w from the set of states to a well-founded set with the following property: if the system is in a state s in which $Q(s)$ is false, then it must eventually reach a state t in which either $Q(t)$ is true or $w(s) \succ w(t)$. Since the value of w cannot decrease forever, this implies that Q must eventually become true.

To prove $P \leadsto Q$, we construct such a function w and prove that it has the required property—namely, that its value must keep decreasing unless Q becomes true. In this proof, we may assume the truth of any predicate R such that $P \Rightarrow \Box R$ is true for all behaviors in Σ. This is a generalization of the usual method for proving termination of a loop in a sequential program, in which w decreases with each iteration of the loop and R asserts that the loop invariant[2] is true if control is at the start of the loop.

One still needs some way of proving that w must decrease unless Q becomes true, assuming the truth of a predicate R that satisfies $P \Rightarrow \Box R$. The simplest approach is to prove that each action in $\boldsymbol{A}$ either decreases the value of w or else makes Q true—in other words, that for every action α and every $(s, t) \in \Gamma(\alpha)$, $R(s) \wedge \neg Q(s)$ implies $w(s) \succ w(t) \vee Q(t) \vee \neg R(t)$.

This approach works only if the validity of the property $P \leadsto Q$ does not depend upon any fairness assumptions. To see how it can be generalized to handle fairness, consider the simple fairness assumption that if an action is continuously enabled, then it must eventually be executed—in other words, for every behavior (1) and every $n > 0$, if α is enabled in all states s_i with $i \geqslant n$, then $\alpha = \alpha_i$ for some $i > n$. Under this assumption, it suffices to show that every action either leaves the value of w unchanged or else decreases it, and that there is at least one action α whose execution decreases w, where α remains enabled until it is executed. Again, this need be proved only under the assumption that Q remains false and R remains true, where R is a predicate satisfying $P \Rightarrow \Box R$.

The problem with this approach is that the precise rules for reasoning depend upon the type of fairness assumptions. An alternative approach uses the single framework of temporal logic to reason about any kind of fairness conditions. We have already written the liveness property to be proved ($P \leadsto Q$) and the safety properties used in its proof (properties of the form $P \Rightarrow \Box R$) as temporal logic formulas. The fairness conditions are also expressible as a collection of temporal logic formulas. Logically, all that must be done is to prove, using the rules of temporal logic, that the fairness conditions and the safety properties imply the desired liveness property. The problem is to decompose this proof into a series of simple steps.

The decomposition is based upon the following observation. Let $\mathscr{A}$ be a well-founded set of predicates. Suppose that, using safety properties of the form $P \Rightarrow \Box R$, for

[2] A loop invariant is not an invariant according to our definition, since it asserts only what must be true when control is at a certain point, saying nothing about what must be true at the preceding control point.

every predicate A in $\mathscr{A}$ we can prove that

$$A \leadsto (Q \vee \exists A' \in \mathscr{A} : A \succ A')$$

The well-foundedness of $\mathscr{A}$ then implies that Q must eventually become true. This decomposition is indicated by a *proof lattice*[3] consisting of Q and the elements of $\mathscr{A}$ connected by lines, where downward lines from A to $A_1, \ldots, A_n$ denotes the assertion $A \leadsto A_1 \vee \cdots \vee A_n$.

An argument using a proof lattice $\mathscr{A}$ of predicates is completely equivalent to a counting argument using a function w with values in a well-founded set; either type of argument is easily translated into the other. These counting arguments work well for proving liveness properties that do not depend upon fairness assumptions. When fairness is required, it is convenient to use more general proof lattices containing arbitrary temporal logic formulas, not just predicates.

To illustrate the use of such proof lattices, we consider the mutual exclusion algorithm of Fig. 2. For simplicity, the noncritical sections have been eliminated and the critical sections are represented by atomic actions, which are assumed not to modify x or y. Under the fairness assumption that a continuously enabled action must eventually be executed, this algorithm guarantees that the first process eventually enters its critical section. (However, the second process might remain forever in its **while** loop.) The proof that the algorithm satisfies the liveness property $at(\alpha) \leadsto at(\gamma)$ uses the proof lattice of Fig. 3. The individual $\leadsto$ relations represented by the lattice are numbered and are explained below.

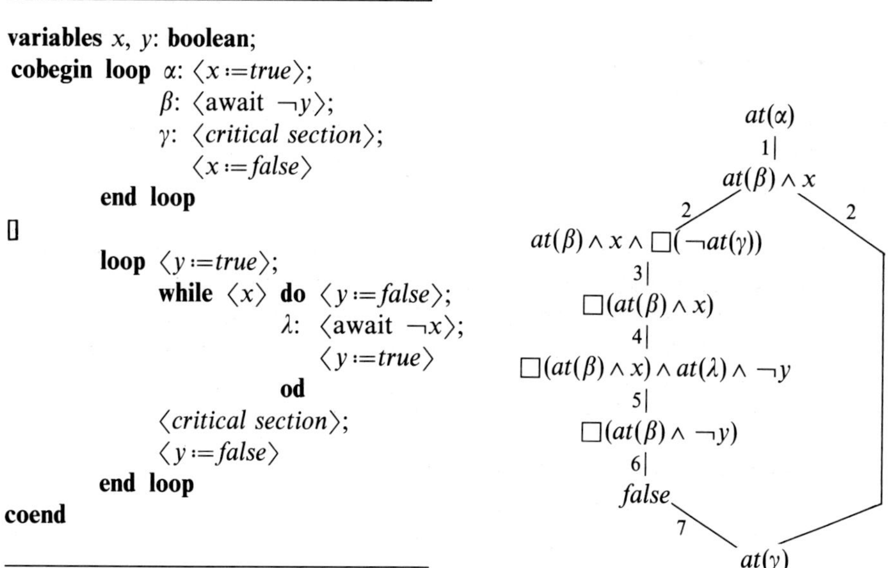
variables x, y: **boolean**;
cobegin loop α: $\langle x := true \rangle$;
β: $\langle$await $\neg y\rangle$;
γ: $\langle$*critical section*$\rangle$;
$\langle x := false \rangle$
end loop
□
loop $\langle y := true \rangle$;
while $\langle x \rangle$ **do** $\langle y := false \rangle$;
λ: $\langle$await $\neg x\rangle$;
$\langle y := true \rangle$
od
$\langle$*critical section*$\rangle$;
$\langle y := false \rangle$
end loop
coend
$at(\alpha)$
1
$at(\beta) \wedge x$
2
2
$at(\beta) \wedge x \wedge \Box(\neg at(\gamma))$
3
$\Box(at(\beta) \wedge x)$
4
$\Box(at(\beta) \wedge x) \wedge at(\lambda) \wedge \neg y$
5
$\Box(at(\beta) \wedge \neg y)$
6
false
7
$at(\gamma)$

Fig. 2. A simple mutual exclusion algorithm. Fig. 3. Proof lattice for mutual exclusion algorithm.

[3]The term "proof lattice" is used even though $\mathscr{A}$ need not be a lattice.

(1) $at(\alpha) \leadsto (at(\beta) \wedge x)$ follows from the fairness assumption, since action α is enabled when $at(\alpha)$ is true.

(2) This is an instance of the temporal logic tautology

$$P \leadsto (Q \vee (P \wedge \Box \neg Q))$$

which is valid because Q either eventually becomes true or else remains forever false. (We are using linear-time temporal logic [35, Section 2.3].)

(3) This $\leadsto$ relation is actually an implication, asserting that if the first process is at statement β with x true and never reaches γ, then it must remain forever at β with x true. This implication is of the form $(P \wedge \Box R) \Rightarrow \Box Q$ and is proved by finding an invariant under the constraint R, as explained in Section 3.4.1.

(4) If x remains true forever, then the fairness assumption implies that control in the second process must eventually reach λ with y false. A formal proof of this assertion would use another proof lattice in which each $\leadsto$ relation represents a single step of the second process.

(5) This is another property of the form $(P \wedge \Box R) \Rightarrow \Box Q$, proved by finding an invariant under the constraint R.

(6) Action β is enabled when $at(\beta) \wedge \neg y$ holds, so by the fairness assumption, $\Box(at(\beta) \wedge \neg y)$ implies that β must eventually be executed, making $at(\beta)$ false. Since $\Box at(\beta)$ asserts that $at(\beta)$ is never false, this is a contradiction.

(7) *false* implies anything.

The proof lattice formalizes a simple style of intuitive reasoning. Further examples of the use of proof lattices can be found in [66].

Temporal logic appears to be the best method for proving liveness properties that depend upon fairness assumptions. There seems little reason to use less formal methods for reasoning about behaviors, since such reasoning can be expressed compactly and precisely with temporal logic. However, the verification of liveness properties has received less attention than the verification of safety properties, and any conclusions we draw about the best approach to verifying liveness properties must be tentative.

3.5. *Deriving algorithms*

We have discussed methods for reasoning about algorithms, without regard to how the algorithms are developed. There is increasing interest in methods for deriving correct algorithms. What is meant exactly by "deriving" an algorithm varies. It may consist of simply developing the correctness proof along with the algorithm. Such an approach, based upon assertional methods and the Unity language, is taken by Chandy and Misra [22]. At the other extreme are approaches in which the program is derived automatically from a formal specification [35, Section 7.3].

An appealing approach to the development of correct algorithms is by program transformation. One starts with a simple algorithm whose correctness is obvious, and transforms it by a series of refinement steps, where each step yields an equivalent program. Perhaps the most elegant instance of this approach is Milner's Calculus of Communicating Systems (CCS) [64], where refinement steps are based upon simple algebraic laws. However, the simplicity and elegance of CCS break down in the presence of fairness, so CCS is not well suited for developing algorithms whose correctness depends upon fairness.

Methods for deriving concurrent algorithms are comparatively new and have thus far had only limited success. Automatic methods can derive only simple, finite-state algorithms. While informal methods can often provide elegant *post hoc* derivations of existing algorithms, it is not clear how good they are at deriving new algorithms. Finding efficient algorithms—whether efficiency is judged by theoretical complexity measures or by implementation in a real system—is still an art rather than a science. We still need to verify algorithms independently of how they are developed.

3.6. Specification

To determine whether an algorithm is correct, we need a precise specification of the problem it purports to solve. In the classical theory of computation, a problem is specified by describing the correct output as a function of the input. Such an input/output function is inadequate for specifying a problem in concurrency, which may involve a complex interaction of the system and its environment.

As discussed above, a behavior of a concurrent system is usually modeled as a sequence of states and/or actions. A specification of a system—that is, a specification of what the system is supposed to do—consists of the set of all behaviors considered to be corrrect. Another approach, taken by CCS [64], is to model a concurrent system as a tree of possible actions, where branching represents nondeterminism. The specification is then a single tree rather than a set of sequences.

With any specification method, there arises the question of exactly what it means for a particular system to implement a specification. This is a very subtle question. Details that are insignificant for sequential programs may determine whether or not it is even possible to implement a specification of a concurrent system. Some of the issues that must be addressed are the following:

- No system can function properly in the face of completely arbitrary behavior by the environment. How can an implementation specify appropriate constraints on the environment (for example, that the environment not change the program's local variables) without "illegally" constraining the environment (for example, by preventing it from generating any input)?
- The granularity of action of the specification is usually much coarser than that of the implementation—for example, sending a message may be a single specification action, while executing each computer instruction is a separate implementation action. What does it mean to implement a single specification action by a set of lower-level actions?
- The granularity of data in the specification may be coarser than in the implementation—for example, messages versus computer words. What does it mean to implement one data structure with another?

Space does not permit a description of proposed specification methods and how they have addressed (or failed to address) these issues. We can only refer the reader to a small selection from the extensive literature on specification [50, 55, 63, 76].

4. Some typical distributed algorithms

In this section, we discuss some of the most significant algorithms and impossibility results in this area. We restrict our attention to four major categories of results: shared-variable algorithms, distributed consensus algorithms, distributed network algorithms, and concurrency control. Although we are neglecting many interesting topics, these four areas provide a representative picture of distributed computing.

In early work, algorithms were presented rather informally, without formal models or rigorous correctness proofs. The lack of rigor led to errors, including the publication of incorrect algorithms. The development of formal models and proof techniques such as those discussed in Section 3, as well as a generally higher standard of rigor, has made such errors less common. However, algorithms are still published with inadequate correctness proofs, and synchronization errors are still a major cause of "crashes" in computer systems.

4.1. Shared variable algorithms

Shared variable algorithms represent the beginnings of distributed computing theory, and many of the ideas that are important elsewhere in the area first appear here. Today, programming languages provide powerful synchronization primitives and multiprocess computers provide special instructions to simplify their implementation, so the early synchronization algorithms are seldom used. However, higher-level contention and cooperation problems still exist, and these early algorithms provide insight into these problems.

4.1.1. Mutual exclusion

The prototypical contention problem is that of *mutual exclusion.* Dijkstra [27] presents a mutual exclusion algorithm which uses indivisible read and write operations on shared variables. In addition to ensuring mutual exclusion, the algorithm ensures the liveness property that some process eventually enters its critical section if there are any contending processes. Lockout-freedom is not guaranteed; the system might grant the resource repeatedly to the same process, excluding another process forever. This algorithm is significant because prior to its discovery, it was not even clear that the problem could be solved.

Dijkstra's algorithm inspired a succession of additional solutions to the mutual exclusion problem. Some of this work improves upon his algorithm by adding the requirement that the solution be fair to individual processes. Fairness can take several forms. The strongest condition usually stated is FIFO (first-in first-out), while the weakest is lockout-freedom. There are intermediate possibilities: there might be an upper bound on the number of times one process can be bypassed by another while it is waiting for the resource ("bounded waiting"), or the time for a process to obtain the resource might be bounded in terms of its own step time. (These last two conditions are very different: the former is an egalitarian condition which tends to cause all processes to move at the same speed, while the latter tends to allow faster processes to move ahead of slower processes.) The work on mutual exclusion includes a collection of algorithms satisfying these various fairness conditions.

An interesting example of a mutual exclusion algorithm is Lamport's "bakery algorithm" [51], so called because it is based on the processes choosing numbers, much as customers do in bakery. The bakery algorithm was the first FIFO solution, and it was the first solution to use only local shared variables (see Section 2.2.1). It also has the fault-tolerance property that if a process stops during its protocol, and its local shared variables subsequently revert to their initial values, then the rest of the system continues correctly without it. This property permits a distributed implementation that tolerates halting failures.

The most important property of the bakery algorithm is that it was the first algorithm to implement mutual exclusion without assuming lower-level mutual exclusion of read and write accesses to shared variables. Accesses to shared variables may occur concurrently, where reads that occur concurrently with writes are permitted to return arbitrary values. Concurrent reading and writing is discussed in Section 4.1.4.

Peterson and Fischer [70] contribute a complexity-theory perspective to the mutual exclusion area. They describe a collection of algorithms which include strong fairness and resiliency properties, and which also keep the size of the shared variables small. Of particular interest is their "tournament algorithm" which builds an n-process mutual exclusion algorithm from a binary tree of 2-process mutual exclusion algorithms. They also describe a useful way to prove bounds on time complexity for asynchronous parallel algorithms: assuming upper bounds on the time for certain primitive occurrences (such as process step time and time during which a process holds the resource), they infer upper bounds on the time for occurrences of interest (such as the time for a requesting process to obtain the resource). Their method can be used to obtain reasonable complexity bounds, not only for mutual exclusion algorithms, but also for most other types of asynchronous algorithms.

The development of many different fairness and resiliency conditions, and of many complex algorithms, gave rise to the need for rigorous ways of reasoning about them. Burns et al. [17] introduce formal models for shared-variable algorithms, and use the models not only to describe new memory-efficient algorithms, but also to prove impossibility results and complexity lower bounds. The upper and lower bound results in [17] are for the amount of shared memory required to achieve mutual exclusion with various fairness properties. The particular model assumed there allows for a powerful sort of access to shared memory, via indivisible "test and set" (combined read and write) operations. Even so, Burns and his coauthors are able to prove that $\Omega(n)$ different values of shared memory are required to guarantee fair mutual exclusion. More precisely, guaranteeing freedom from lockout requires at least $n/2$ values, while guaranteeing bounded waiting requires at least n values.

The lower bound proofs in [17] are based on the limitations of "local knowledge" in a distributed system. Since processes' actions depend only on their local knowledge, processes must act in the same way in all computations that look identical to them. The proofs assume that the shared memory has fewer values than the claimed minimum and derive a contradiction. They do this by describing a collection of related computations and then using the limitation on shared memory size and the pigeonhole principle to conclude that some of these computations must look identical to certain processes. But among these computations are some for which the problem specification requires the processes to act in different ways, yielding a contradiction. The method used here—proving that actions based on local knowledge can force two processes to act the same when they should act differently—is the fundamental method for deriving lower bounds and other impossibility results for distributed algorithms.

The lower bound results in [17] apply only to deterministic algorithms—that is, algorithms in which the actions of each process are uniquely determined by its local knowledge. Recently, randomized algorithms, in which processes are permitted to toss fair coins to decide between possible actions, have emerged as an alternative to deterministic algorithms. A randomized algorithm can be thought of as a strategy for "playing a game" against an "adversary", who is usually assumed to have control over the inputs to the algorithm and the sequence in which the processes take steps. In choosing its own moves, the adversary may use knowledge of previous moves. A randomized algorithm should, with very high probability, perform correctly against any allowable adversary.

One of the earliest examples of such a randomized algorithm was developed by Rabin [72] as a way of circumventing the limitations proved in [17]. The shared memory used by Rabin's algorithm has only $O(\log n)$ values, in contrast to the $\Omega(n)$ lower bound for deterministic algorithms. Rabin's algorithm is also simpler than the known deterministic mutual exclusion algorithms that use $O(n)$-valued shared memory. A disadvantage is that Rabin's algorithm is not solving exactly the same problem—it is not absolutely guaranteed to grant the resource to every requesting process. Rather, it does so with probability that grows with the amount of time the process waits. Still, in some situations, the advantages of simplicity and improved performance may outweigh the small probability of failure.

The mutual exclusion problem has also been studied in message-passing models. The first such solution was in [52], where it was presented as a simple application of the use of logical clocks to totally order system events (see Section 2.3). Mutual exclusion was reduced to the global consistency problem of getting all processes to have a consistent view of the queue of waiting processes. More recently, several algorithms have been devised which attempt to limit the number of messages required to solve the problem. A generalization to k-exclusion, in which up to k processes can be in their critical section at the same time has also been studied.

The reader can consult the book by Raynal [74] for more information and more pointers into the extensive literature on mutual exclusion.

4.1.2 Other contention problems

The *dining philosophers* problem [29] is an important resource allocation problem in which each process ("philosopher") requires a specific set of resources ("forks"). In the traditional statement of the problem, the philosophers are arranged in a circle, with a fork between each pair of philosophers. To eat, each philosopher must have both adjacent forks. Dijkstra's solution is based on variables (semaphores) shared by all processes, and thus is best suited for use within a single computer.

One way to restrict access to the shared variables is by associating each variable with a resource, and allowing only the processes that require that resource to access the variable. This arrangement suggests solutions in which processes simply visit all their resources, attempting to acquire them one at a time. Such a solution permits deadlock, where processes obtain some resources and then wait forever for resources held by other processes. In the circle of dining philosophers, deadlock arises if each one first obtains his left fork and then waits for his right fork.

The traditional dining philosophers problem is symmetrical if processes are identical and deterministic, and all variables are initialized in the same way. If processes take steps in round-robin order, the system configuration is symmetrical after every round. This implies that if any process ever obtained all of its needed resources, then every process would, which is impossible. Hence, there can be no such completely symmetric algorithm. The key to most solutions to this problem is their method for breaking symmetry.

There are several ways of breaking symmetry. First, there can be a single "token" that is held by one process, or circulated around the ring. To resolve conflict, the process with the token relinquishes its resources in exchange for a guarantee that it can have them when they next become available. Second, alternate processes in an even-sized ring can attempt to obtain their left or right resources first; this strategy can be used not only to avoid deadlock, but also to guarantee a small upper bound on waiting time for each process. Third, Chandy and Misra [21] describe a scheme in which each resource has a priority list, describing which processes have stronger claims on the resource. These priorities are established dynamically, depending on the demands for the resources. Although the processes are identical, the initial configuration of the algorithm is asymmetric: it includes a set of priority lists that cannot induce cycles among waiting processes. The rules used in [21] to modify the priority lists preserve acyclicity, and so deadlock is avoided.

Finally, Rabin and Lehmann [73] describe a simple randomized algorithm, that uses local random choices to break symmetry. Each process chooses randomly whether to try to obtain its left or right fork first. In either case, the process waits until it obtains its first fork, but only tests once to see if its second fork is available. If it is not, the process relinquishes its first fork and starts over with another random choice. This strategy guarantees that, with probability 1, the system continues to make progress.

These symmetry-breaking techniques avoid deadlock and ensure that the system makes progress. They provide a variety of fairness and performance guarantees.

4.1.3. Cooperation problems

For shared-variable models, cooperation problems have received less attention than contention problems. The only cooperation problems that have been studied at any length are *producer–consumer* problems, in which processes produce results that are used as input by other processes. The simplest producer–consumer problem is the bounded buffer problem (Section 2.3). A very general class of producer–consumer problems involves the simulation of a class of Petri nets known as marked graphs [24], where each node in the graph represents a process and each token represents a value. An example of this class is the problem of passing a token around a ring of processes, where the token can be used to control access to some resource.

An interesting problem that combines aspects of both contention and cooperation is concurrent garbage collection, in which a "collector" process running asynchronously with a "mutator" process must identify items in the data structure that are no longer accessible by the mutator and add those items to a "free list". This is basically a producer–consumer problem, with the collector producing free-list items and the mutator consuming them. However, the problem also involves contention because the mutator changes the data structure while the collector is examining it.

The reason cooperation problems have not been studied as extensively as contention problems is probably that they are easier to solve. For example, in concurrent garbage collection algorithms, it is the contention or access to the data structure rather than the cooperative use of the free list that poses the challenge. However, there is one important property that is harder to achieve in cooperation problems than in contention problems—namely, *self-stabilization*. An algorithm is said to be self-stabilizing if, when started in any arbitrary state, it eventually reaches a state in which it operates normally [30]. For example, a self-stabilizing token-passing algorithm can be started in a state having any number of tokens and will eventually reach a state with just one token that is being passed around. It is generally easy to devise self-stabilizing contention problems because processes go through a "home" state in which they are reinitialized—for example, a process in the dining philosophers problem eventually reaches a state in which it is not holding or requesting any forks—and the whole algorithm is reinitialized when every process has reached its home state. On the other hand, cooperation problems do not have such a home state. For example, the symmetry in the bounded buffer problem means that an empty buffer and a full buffer are symmetric situations, and neither of them can be considered a "home" state. Dijkstra's self-stabilizing token-passing algorithms [30] are currently the only published self-stabilizing cooperation algorithms.

Self-stabilization is an important fault-tolerance property, since it permits an algorithm to recover from any transient failure. This property has not received the attention it deserves.

4.1.4. *Concurrent readers and writers*

With the exception of the bakery algorithm, all of the work we have described so far assumes that processes access shared memory using primitive operations (usually read and write operations), each of which is executed indivisibly. The ability to implement multiple processors with a single integrated circuit has rekindled interest in shared memory models that do not assume indivisibility of reads and writes. Rather, they assume that operations on a shared varible have duration, that reads and writes that do not overlap behave as if they were indivisible, but that reads and writes that overlap can yield less predictable results [54]. The bakery algorithm assumes *safe* shared variables—ones in which a read that is concurrent with a write can return an arbitrary value from the domain of possible values for the variable. Another possible assumption is a *regular* shared variable, in which a read that overlaps a write is guaranteed to return either the old value or the one being written; however, two successive reads that overlap the same write may obtain first the new value then the old one. A still stronger assumption is an *atomic* shared variable, which behaves as if each read and each write occurred at some fixed time within its interval.

Using safe, regular, or atomic shared variables, it is possible to simulate shared variables having indivisible operations, so that algorithms designed for the stronger models can be applied in the weaker models. This work has evolved from the traditional readers-writers algorithms based on mutual exclusion [25], through nontraditional algorithms that allow concurrent reading and writing [69], to more recent algorithms for implementing one class of shared variable with a weaker class [54, 18, 14].

Recently, Herlihy [46] has considered atomic shared variables that support operations other than reads and writes. He has shown that read-write atomic variables cannot be used to implement more powerful atomic shared variables such as those supporting test-and-set operations. He has also shown that other types of atomic variables are "universal", in the sense that they can be used to implement atomic shared variables of arbitrary types. Herlihy's impossibility proof proceeds by showing that atomic read-write shared variables cannot be used to solve a version of the distributed consensus problem discussed in the following subsection.

4.2. *Distributed consensus*

Achieving global consistency requires that processes reach some form of agreement. Problems of reaching agreement in a message-passing model are called *distributed consensus* problems. There are many such problems, including agreeing (exactly or approximately) on values from some domain, synchronizing actions of different processes, and synchronizing software clocks. Distributed consensus problems arise in areas as diverse as real-time process-control systems (where agreement might be needed on the values read by replicated sensors) and distributed database systems (where agreement might be needed on whether or not to accept the results of a transaction). Since global consistency is what makes a collection of processes into a single system, distributed consensus algorithms are ubiquitous in distributed systems.

Consensus problems are generally easy to solve if there are no failures; in this case, proceses can exchange information reliably about their local states, and thereby achieve a common view of the global state of the system. The problem is considerably harder, however, when failures are considered. Consensus algorithms have been presented for almost all the classes of failure described in Section 2.1.1.

Distributed consensus problems have been a popular subject for theoretical research recently, because they have simple mathematical formulations and are surprisingly challenging. They also provide a convenient vehicle for comparing the power of models that make different assumptions about time and failures.

4.2.1. *The two-generals problem*

Probably the first distributed consensus problem to appear in the literature is the "two-generals problem" [44], in which two processes must reach agreement when there is a possibility of lost messages. The problem is phrased as that of two generals, who communicate by messages, having to agree upon whether or not to attack a target. The following argument can be formalized to show that the problem is unsolvable when messages may be lost. Reaching at least one of the two possible decisions, say the decision to attack, requires the successful arrival of at least one message. Consider a scenario Σ in which the fewest delivered messages that will result in agreement to attack are delivered, and let Σ' be the same scenario as Σ except that the last message delivered in scenario Σ is lost in Σ', and any other messages that might later be sent are also lost. Supose this last message is from general A to general B. General A sees the same messages in two scenarios, so he must decide to attack. However, the minimality assumption of Σ implies that B cannot also decide to attack in scenario Σ', so he must make a different decision. Hence, the problem is unsolvable.

4.2.2. *Agreement on a value*

The agreement problem requires that processes agree upon a value. Communication is assumed to be reliable, but processes are subject to failures (either halting, omission, or Byzantine). Each of the processes begins the algorithm with an input value. After the algorithm has completed, each process is to decide upon an output value. There are two constraints on the solution: (a) (*agreement*) all nonfaulty processes must agree on the output, and (b) (*validity*) if all nonfaulty processes begin with the same input value, that value must be the output value of all nonfaulty processes. For the case of Byzantine faults, this problem has been called the *Byzantine generals problem.* (Other, equivalent formulations of the problem have also been used.)

In the absence of failures, this problem is easy to solve: processes could simply exchange their values, and each could decide upon the majority value. The following example shows the kinds of difficulties that can occur, when failures are considered. Consider three processes, A, B and C. Suppose that A and B begin with input 0 and1 respectively. Suppose that C is a Byzantine faulty processor, which acts toward A as if C were nonfaulty and started with 0, but as if B were faulty. At the same time, C acts toward B as if C were nonfaulty and started with 1, but as if A were faulty. Since A's view of execution is consistent with A and C being nonfaulty and starting with the same input 0, A is required to decide 0. Analogously, B is required to decide 1. But this means that A and B have been made to disagree, violating the agreement requirement of the problem. This example can be elaborated into a proof of the impossibility of reaching agreement among $3t$ processes if t processes might be faulty.

The problem of reaching agreement on a value was studied by Pease, Shostak and Lamport [67, 56] in a model with Byzantine failures and computation performed in a sequence of rounds. (They also described the implementation of rounds with synchronized clocks.) Besides containing the impossibility proof described in the last paragraph, these papers also contain two subtle algorithms. The first is a recursive algorithm that requires $3t+1$ processes and tolerates Byzantine faults. The second requires only $t+1$ processes, but assumes digital signatures (Section 2.1.1). Both algorithms assume a completely connected network.

Dolev [32] considers the same problem in an arbitrary network graph. For t Byzantine failures, he shows how to implement an algorithm similar to that of [56], provided that the network is at least $(2t+1)$-connected (and has at least $2t+1$ processes). He also proves a matching lower bound.

A series of results, starting with [38] and culminating in [34], shows that any synchronous algorithm for reaching agreement on a value, in the presence of t failures—even the simple halting failures—requires at least $t+1$ rounds of message exchange in the worst case. As usual, these arguments are based on the limitations caused by local knowledge in distributed algorithms; by assuming fewer rounds, a “chain” of computations is constructed that leads to a contradiction. In the first computation in the chain, nonfaulty processes are constrained by the problem statement to decide 0, while in the last computation in the chain, nonfaulty processes are constrained to decide 1. Further, any two consecutive computations in the chain share a nonfaulty process to which the two computations look the same; this process therefore reaches the same decision in both computations. Hence, all nonfaulty processes decide upon the same value in every computation in the chain, which yields the required contradiction.

Dwork and Moses [34] provide explicit, intuitive definitions for the “knowledge” that individual processes have at any time during the execution of an algorithm. Their problem statements, algorithms, and lower bound proofs are based on these definitions. This work suggests that formal models and logics of knowledge may provide useful high-level ways of reasoning about distributed algorithms.

Bracha [15] is able to circumvent the $t+1$ lower bound on rounds with a randomized algorithm; his solution uses only $O(\log n)$ rounds, but requires cryptographic techniques that rest on special assumptions. More recently, Feldman and Micali [36] have improved Bracha’s upper bound to a constant. Chor and Coan [23] give another randomized algorithm that requires $O(t/\log n)$ rounds, but does not require any special assumptions.

The consensus algorithms mentioned above all assume a synchronous model of computation. Fischer, Lynch and Paterson [40] study the problem of reaching agreement on a value in a completely asynchronous mode. They obtain a surprising fundamental impossibility result: if there is the possibility of even one simple halting failure, then an asynchronous system of deterministic processes cannot guarantee agreement. This result suggests that while asynchronous models are simple, general, and popular, they are too weak for studying fault tolerance.

The impossibility result is proved by first showng that any asynchronous consensus protocol that works correctly in the absence of faults must have a reachable configuration C in which there is a single “decider” process i—one that is capable, on its own, of causing either of two different decisions to be reached. If this protocol is also required to tolerate a single process failure, then, starting from C, all the processes except i must be able to reach a decision. But, this decision will conflict with one of the possible decisions process i might reach on its own. (Herlihy used a similar technique to prove the impossibility result mentioned in the previous subsection.)

There are several ways to cope with the limitation described in [40]. One can simply add some synchrony assumptions—the weakest ones commonly used are timers and bounded message delay. Alternatively, one can use an asynchronous deterministic algorithm, but attempt to reduce the probability that a failure will upset correct behavior. This approach is sometimes used in practice when only modest reliability is needed, but there has been no rigorous attempt to analyze the reliability of the resulting systems.

Another possibility is to use randomized rather than deterministic algorithms. For example, Ben-Or [12] gives a randomized algorithm for reaching agreement on a value in the completely asynchronous model, allowing Byzantine faults. The algorithm never permits disagreement or violates the validity condition; however, instead of guaranteeing eventual termination, it guarantees only termination with probability 1.

A good survey of the early work in this area appears in [37].

4.2.3. *Other consensus problems*

Other distributed consensus problems have been studied under the assumption that processes can be faulty but communication is reliable. One such problem is that of reaching approximate, rather than exact, agreement on a value. Each process begins with an initial (infinite-precision) real value, and must eventually decide on a real value subject to: (a) (agreement) all nonfaulty processes' decisions must agree to within ε, and (b) (*validity*) the decision value for any nonfaulty process must be within ε of the range of the initial values of the nonfaulty processes. Processes are permitted to send real values in messages.

Although the problems of exact and approximate agreement seem to be quite similar, reaching approximate agreement is considerably easier; in particular, there are simple deterministic algorithms for approximate agreement in asynchronous models—even in the presence of Byzantine faults. It seems almost paradoxical that deterministic processes can reach agreement on real values to within any predetermined ε, but they cannot reach exact agreement on a single bit.

Another consensus problem is achieving simultaneous action by distributed processes, in a model with timers in which all messages take exactly the same (known) time for delivery. This problem, sometimes called the "distributed firing squad problem", yields results very similar to those for agreement on a value. In fact, for the case of Byzantine faults, a general transformation converts any algorithm for agreement to an algorithm for simultaneous action. The firing squad algorithm is obtained by running many instances of the agreement algorithm, each deciding whether the processes should fire at a particular time. The first instance that reaches a positive decision triggers the simultaneous firing action.

In this transformation, many instances of a Byzantine agreement algorithm are executed concurrently. Those instances that are not actually carrying out any interesting computation can be implemented in a trivial way by letting all of their messages be special "null" messages that are not actually sent. This trick of sending a message by not sending a message is also used in [53] to give fault-tolerant distributed simulations of centralized algorithms.

Another consensus problem is establishing and maintaining synchronized local clocks in a distributed system. It is closely related to both of the preceding problems (reaching approximate agreement and achieving simultaneous action), since it may be viewed as simultaneously reaching approximate agreement on a clock value, or as reaching exact agreement on a clock value at approximately the same instant. The problem is one of implementing synchronized clocks using timers that run at approximately the same rate, usually assuming initial synchronization of the clocks. However, it is generally described in terms of maintaining the synchronization (to within ε) of the processes' clocks despite a small, varying difference in their clock rates.

Clock synchronization is difficult to achieve in the presence of faulty processes, Many algorithms to solve this problem have been suggested, analyzed, and compared in the literature, and some have been used in implementing systems. In most algorithms for maintaining synchronization among clocks that are initially synchronized, a new round is begun when the clocks reach predetermined values. In each round, processes exchange information about their clock values and use the information to adjust their own clocks. Synchronization algorithms that do not assume the clocks to be initially synchronized use other methods, since they cannot depend upon the clocks to determine when the first round should begin.

Lower bounds and impossibility results have also been proved for clock synchronization problems. Of particular interest is the result of Dolev, Halpern and Strong [33] showing that clock synchronization problems cannot be solved for $3t$ processes if t of them can exhibit Byzantine failures.

This impossibility result is reminiscent of the impossibility result described earlier for agreement on a value, where the problem cannot be solved with $3t$ processes in the presence of t Byzantine failures. In fact, a $3t$ versus t impossibility result also holds for many other consensus problems under Byzantine failures, including approximate agreement and simultaneous action. Furthermore, all of these problems are unsolvable in network graphs having less than $(2t+1)$-connectivity. These impossibility results do not apply if authentication is used.

Since all of these bounds are tight, it is apparent that there must be a common reason for the many similar results. Fischer, Lynch and Merritt [39] tie together this large collection of impossibility results with a common proof technique. We illustrate this technique by proving the 3 versus 1 impossibility result for reaching agreement on a value. Assume for the sake of obtaining a contradiction that there is such a solution for the system $\mathscr{T}$ consisting of the three processes A, B, and C arranged in a triangle. Let $\mathscr{H}$ be a new system, consisting of two copies of each of A, B and C, in the hexagonal arrangement shown in Fig. 4. Note that system $\mathscr{H}$ looks locally like the original system $\mathscr{T}$.

Let Σ be a computation of $\mathscr{H}$ that results if $\mathscr{H}$ is run with each of its six processes

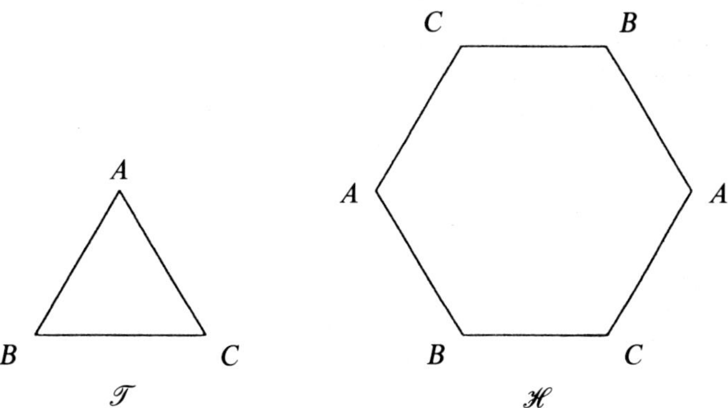

Fig. 4. The systems $\mathscr{T}$ and $\mathscr{H}$.

behaving exactly like the corresponding nonfaulty process of $\mathscr{T}$. Consider any pair of nonfaulty processes in $\mathscr{H}$, say the upper right-hand copies of A and B. There is a computation Σ' of $\mathscr{T}$, with C faulty, in which A and B receive the same inputs as they do in the computation Σ. By our assumption, A and B agree on the same value in Σ'. Since the copies of A and B have the same view in Σ as their namesakes do in Σ', they must also agree on the same value. Moreover, if A and B have the same input value, then that is their output value. Since this proof works for any pair of adjacent processes in $\mathscr{H}$, this shows that, in any computation of $\mathscr{H}$, (a) all processes must agree upon the same output value, and (b) if adjacent processes have the same input value then that is their output value. Letting the upper right-hand copies of A and B have input values of 1 and the lower left-hand copies of A and B have input values of 0, this implies that the output of every process must equal both 0 and 1, which is the required contradiction. Other impossibility results are proved similarly, using slightly more complicated systems for $\mathscr{H}$.

4.2.4. *The distributed commit problem*

The *transaction commit* problem for distributed databases is the problem of reaching agreement, among the nodes that have participated in a transaction, about whether to *commit* or *abort* the transaction. (We say more about transactions in Section 4.4.) The requirements are: (a) (*agreement*) all nonfaulty processes' decisions must agree, and (b) (*validity*) (i) if any process's initial value is "abort", then the decision must be "abort", and (ii) if all processes' initial values are "commit" and no failure occurs, then the decision must be "commit". The problem has traditionally been studied under the assumption of halting failures and the loss of individual messages. The impossibility result of [40] implies that the commit problem cannot be solved in the completely asynchronous model, for even a single faulty process—even with reliable communication. The impossibility result for the two-generals problem implies that the commit problem cannot be solved if messages can be lost, even if message delays are otherwise bounded and processes are reliable and have synchronized clocks.

Most commit protocols, such as the popular *two-phase commit* algorithm, have a failure window—a period during the computation when a single halting failure can prevent termination. Using the assumptions that the processes have synchronized clocks and there is a known upper bound on message delivery time, one can construct a commit protocol that has no failure window from a synchronous algorithm for reaching agreement on a value. However, the synchronous model does not permit communication failure, so the loss of a message must be considered to be a failure of either the sending or receiving process. The *three-phase commit* protocol of Skeen [77] is another commit protocol without a failure window; it assumes reliable message delivery and detectable failures.

4.3. *Network algorithms*

We now describe a class of algorithms for message-passing models, which we call network algorithms, in which the behavior of the algorithm depends strongly on the network topology. Most of these algorithms are designed to solve problems arising in communication in computer networks. They usually assume a completely asynchronous, failure-free model. Most of them can be divided into two categories, which we call *static* and *dynamic*. Static algorithms are assumed to operate in fixed networks and to start with all their inputs available at the beginning; dynamic algorithms also operate in fixed networks but receive some of their inputs interactively. Another way of viewing the distinction is that static algorithms are based upon unchanging information in the initial states of the processes, while dynamic algorithms use changing information from the changing state of the application processes. A network problem can have both static and dynamic versions, but the two versions are usually treated separately in the literature. We also consider some algorithms designed to operate in changing networks, and some algorithms designed to ensure reliable message delivery over a single unreliable link.

4.3.1. Static algorithms

Route-determination algorithms. In communication networks, it is often important for processes that have local information about the speed, bandwidth, and other costs of message transmission to their immediate network neighbors, to determine good routes through the network for communicating with distant processes. If such routes are to be determined infrequently, it may be useful to consider the static problem in which the local information is assumed to be available initially and fixed during execution of the route-determination algorithm.

Different applications require different notions of what constitutes a "good" set of routes through the network. For example, if the routes are used primarily for broadcasting a single message to all other processes, unnecessary message duplication can be avoided by establishing a spanning tree of the network. If a weight is associated with each link in the network to represent the cost of sending a message over that link, a minimum-weight spanning tree (MST) can be used to minimize the total cost of the broadcast.

Gallager, Humblet and Spira [43] present an efficient distributed algorithm for finding a minimum-weight spanning tree in a network with n nodes and e edges. The algorithm is based upon the following two observations: if all edge weights are distinct, then the MST is unique; and the minimum-weight external edge of any subtree of the MST is in the MST. The algorithm grows the MST by coalescing fragments until the complete MST is formed. Initially, each node is a fragment, and a fragment coalesces with the one at the other end of its minimum-weight external edge.

The main achievement of this algorithm is to keep the number of messages small. Each time a fragment with f nodes computes its minimum-weight external edge, $O(f)$ messages are required. Naively coalescing fragments could lead to as many as $\Omega(n^2)$ messages. By using a priority scheme to determine when fragments are permitted to coalesce, this algorithm generates only $O(n \log n + e)$ messages.

Although the basic idea is simple, the algorithm itself is quite complicated. Certain simple "high-level" tasks, such as determining a fragment's minimum-weight external edge, are implemented as a series of separate steps occurring at different processes. The steps implementing different high-level tasks interleave in complicated ways. The correctness of the algorithm is not obvious; in fact, only recently have careful correctness proofs appeared [42, 81]. While these proofs use techniques based upon those described in Section 3, they are lengthy and difficult to check. In general, network algorithms are typically longer and harder to understand than the other types of distributed algorithms we are considering, and rigorous correctness proofs are seldom given.

Many other network algorithms are also designed to minimize the number of messages sent. While message complexity is easy to define and amenable to clean upper and lower bound results, time bounds may be more important in practice. However, there have so far been few upper and lower time bounds derived for network problems.

Other route-determination algorithms have been proposed for finding MST's in a directed graph of processes [49] and for determining other routing structures, such as the set of shortest paths between all pairs of nodes and breadth-first and depth-first spanning trees. Also, a basic lower bound of $\Omega(e)$ has been proved for the number of messages required to implement broadcast in an arbitrary network [10].

Leader election. In this problem, a network of identical processes must choose a "leader" from among themselves. The processes are assumed to be indistinguishable, except that they may possess unique identifiers. The difficulty lies in breaking the symmetry. Solutions can be used to implement a fault-tolerant token-passing algorithm; if the token is lost, the leader-election algorithm is invoked to decide which process should possess the token.

Peterson [68] has devised a leader-election algorithm for a completely asynchronous ring of processes with unidirectional communication; it uses at most $O(n \log n)$ messages in the worst case. On the other hand, Frederickson and Lynch [41] have shown that at least $\Omega(n \log n)$ messages are required in the worst case, even in a ring having synchronous and bidirectional communication.

These results would characterize the message complexity in the important special case of a ring of processes but for an interesting technicality. The Frederickson–Lynch lower bound assumes that the algorithm uses process identifiers only in order comparisons, but not in counting or more general arithmetic operations. Almost all published election algorithms satisfy this assumption. The lower bound also holds for more general uses of identifiers if, for each ring size, the algorithm satisfies a uniform time bound, independent of the process identifiers. Without this technical assumption, the problem can be solved with only $O(n)$ messages by an algorithm taking an unbounded amount of time [4, 80]. Although unlikely to be of practical use, this algorithm provides an interesting extreme time–message tradeoff.

The election problem has been solved under many different assumptions: the network can be a ring, a complete graph, or a general graph; the graph can be directed or undirected; the processes might have unique identifiers or be identical; the individual processes might or might not know the size or shape of the network; the algorithm can be deterministic or randomized; communication can be synchronous, asynchronous, or partially synchronous; and failures might or might not be allowed. The problem has provided an opportunity to study a single problem under many different assumptions, but no general principles have yet emerged.

Other problems. Other static problems include the computation of functions, such as the median and other order statistics, where the inputs are initially distributed. Attiya et al. [8] and Abrahamson et al. [1] have obtained especially interesting upper and lower bound results—many surprisingly tight—about the number of messages required to compute functions in a ring.

4.3.2. *Dynamic algorithms*

Distributed termination. In this problem, each process is either active or inactive. Only an active process can send a message, and an inactive process can become active only by receiving a message. A termination detection algorithm detects when no further process activity is possible—that is, when all processes are simultaneously inactive and no messages are in transit.

This problem was first solved by Dijkstra [31] for the special case in which the application program is a "diffusing computation"—one where all activity originates from and returns to one controlling process. Other researchers have addressed the problem of detecting termination in CSP programs; because CSP programs admit the possibility of deadlock as well as normal termination, these algorithms must also recognize deadlock. Termination can also be detected using global snapshot algorithms, discussed later in this section.

Distributed deadlock detection. Here it is assumed that processes request resources and release them, and there is some mechanism for granting resources to requesting processes. However, resources may be granted in such a way that deadlock results—for example, in the dining philosophers problem, each philosopher may have requested both forks and received only his right fork, so the system is deadlocked because no one can obtain his left fork. A deadlock detection algorithm detects such a situation, so appropriate corrective action can be taken—usually forcing some processes to relinquish resources already granted.

The simplest instance of the problem, in which each process is waiting for a single resource held by another process, is solved by detecting cycles of the form "*A* is waiting for a resource held by *B*, who is waiting for a resource ... held by *A*." Straightforward cycle-detection algorithms can be applied, but they may not be efficient. A more complicated solution is required if process requests have a more interesting structure, such as "any one of a set of resources" or "any two from set *S* and any one from set *T*". In such cases, the problem may involve detecting a knot or other graph structure, instead of a cycle.

A difficulty in designing distributed deadlock detection algorithms is avoiding the detection of "false deadlocks". Consider a ring of processes each of which occasionally requests a resource held by the next, but in which there is no deadlock. An algorithm that simply checks the status of all processes in some order could happen to observe every process when it is waiting for a resource and incorrectly decide that there is deadlock. The algorithm of Chandy, Misra, and Haas [19] is a typical algorithm that does not detect false deadlocks.

Global snapshots. The global state of a distributed system consists of the state of each process and the messages on each transmission line. A global snapshot algorithm attempts to determine the global state of a system. A trivial algorithm would instantaneously "freeze" the execution of the system and determine the state at its leisure, but such an algorithm is seldom feasible. Moreover, as explained in Section 2.3, determining the global state requires knowing the complete temporal ordering of events, which may be impossible. Therefore, a global snapshot algorithm is required only to determine a global state that is consistent with the known temporal ordering of events. This is sufficient for most purposes. This problem was studied by Chandy and Lamport [20], who presented a simple global snapshot algorithm.

A global snapshot algorithm can be used whenever one wants global information about a distributed system. In a distributed banking system, such an algorithm can determine the total amount of money in the bank without halting other banking transactions. Similarly, one can use a global snapshot algorithm to checkpoint a system for failure recovery without halting the system.

A general class of applications is detecting when an invariant property holds. Recall that an invariant is a property of the state which, once it holds, will continue to hold in all subsequent states. (Invariants of distributed systems are often called "stable properties".) Distributed termination and deadlock are invariants. If an invariant holds in the consistent state observed by a global snapshot algorithm, then it also holds in all global states reached by the system after the snapshot algorithm terminates. Thus, one can detect termination, deadlock, or any other invariant property by obtaining a consistent global snapshot and checking it for that property.

A global snapshot algorithm can transform an algorithm for solving a static network problem to one that solves the dynamic version of the same problem. For example, the static version of the deadlock detection problem, in which the set of resources held and requests pending never changes, is easier to solve than the dynamic version because there is no possibility of detecting false deadlocks. The harder dynamic version can be solved by taking a global snapshot, then running an algorithm for the static problem on the state determined by that snapshot. It is not necessary to collect the global snapshot information in one place; the static deadlock detection can be done with a distributed algorithm. This strategy is used in a deadlock detection algorithm by Bracha and Toueg [16].

Synchronizers Many simple algorithms have been designed for strongly synchronous networks—ones in which the entire computation proceeds in a series of rounds. A *network synchronizer* is a program designed to convert such an algorithm to one that can run in a completely asynchronous network. Awerbuch [9] has designed a collection of network synchronizers, varying in their message and time complexity. They have been used to produce asynchronous algorithms that are more efficient than previously known ones for breadth-first search and the determination of maximum flows and shortest paths.

The simplest of the synchronizers transforms an algorithm for a synchronous network into an asynchronous algorithm that has approximately the same execution time. This seems to imply that any problem can be solved just as quickly in asynchronous networks as in synchronous networks. However, Arjomandi, Fischer and Lynch [6] showed that there are some problems whose solution requires much more time (greater by a multiplicative factor of the network diameter) in an asynchronous than in a synchronous network. A typical problem is for all nodes to perform a sequence of outputs, in such a way that every node does its ith output before any node does its $(i+1)$st. A synchronous system can perform r such output rounds in time r, but an asynchronous system requires extra time for communication between all the nodes in between each pair of rounds.

4.3.3. Changing networks

The algorithms discussed so far in this subsection are designed to operate in communication networks that are fixed while the algorithm is being executed. Algorithms for the same problems are also required for the harder case where network links may fail and recover during execution—that is, for *changing networks.*

One can translate any algorithm for a fixed but arbitrary network into one that works for a changing network as follows. The nodes run the fixed-network algorithm as long as the network does not appear to change. Whenever a node detects a change, it stops executing the old instance of the fixed-network algorithm and begins a new instance, this time on the changed network. Thus, there can be may instances of the fixed-network algorithm executing simultaneously. The different instances are distinguished by means of "instance identifiers" attached to the messages.

It is not difficult to implement this approach using an unbounded number of instance identifiers, each chosen to be larger than the previous one used by the node. Afek, Awerbuch and Gafni [2] have developed a method that requires only a finite number of identifiers. However, simply bounding the number of instance identifiers is of little practical significance, since practical bounds on an unbounded number of identifiers are easy to find. For example, with 64-bit identifiers, a system that chooses ten per second and was started at the beginning of the universe would not run out of identifiers for several billion more years. However, through a transient error, a node might choose too large an identifier, causing the system to run out of identifiers billions of years too soon—perhaps within a few seconds. A self-stabilizing algorithm using a finite number of identifiers would be quite useful, but we know of no such algorithm.

4.3.4. Link protocols

Links, joining nodes in a network, are implemented with one or more *physical channel,* each delivering low-level messages called "packets". Packet delivery is not necessarily reliable. A *link protocol* is used to implement reliable message communication using unreliable physical channels. Of course, it is necessary to make some assumptions about the types of failures permitted for the physical channels. For example, channels might be assumed to lose and reorder messages, but not to duplicate or fabricate them. In addition, some liveness assumption on the physical channel is needed to ensure that messages are eventually delivered; a common assumption is that if infinitely many packets are sent, then eventually some message will be delivered.

The Alternating Bit Protocol is a link protocol that assumes the physical channel may lose packets but cannot reorder them. When a sender wishes to transmit a message, it assembles a packet consisting of the message and a single-bit "header", and transmits this packet repeatedly on the physical channel. Upon receipt of the packet, the receiver sends the header bit back to the sender. When the sender receives a header bit that is the same as the one it is currently transmitting, it knows that its current message has been received and switches to the next message, using the opposite header bit.

This protocol does not work if the physical channels can reorder messages. In fact, Lynch, Mansour and Fekete [60] showed that no protocol with bounded headers can work over non-FIFO physical channels if the best-case number of packets required to deliver each message must be bounded. Attiya et al. [7] complete the picture by showing that this latter assumption is necessary—that there is a (not very practical) protocol using bounded headers if the best-case number of packets required to deliver one message is permitted to grow without bound.

Baratz and Siegel [11] developed link protocols that tolerate "crashes" of the participating nodes, with loss of information in the nodes' states. Their algorithm requires the node at each end of the link to have one bit of "stable memory" that survives crashes. It is shown in [60] that this bit of stable memory is necessary.

Aho et al. [3] have studied the basic capabilities of finite-state link protocols.

4.4. Concurrency control in databases

A database consists of a collection of items that are individually read and written by the operations of programs called *transactions.* A *concurrency control algorithm* executes each transaction so it either acts like an atomic action, with no intervening steps of other transactions, or aborts and does nothing. This condition, called *serializability*, ensures that the system acts as if all transactions that are not aborted are executed in some serial order. This order must be consistent with the order in which any externally visible actions of the transactions occur. The serializability condition for databases is very similar to the atomic condition discussed earlier for shared variables.

By making transactions appear atomic, concurrency control makes the system easier to understand. For this reason, atomic transactions have been proposed as a basic construct in distributed programming languages and systems such as Argus [58] and Camelot [78].

Allowing transactions to be aborted permits more efficient concurrency control algorithms. An algorithm can make scheduling decisions that lead to faster execution but may produce a nonserializable execution; it aborts any transaction whose execution would not appear atomic. It is sometimes useful to abort a transaction for reasons other than maintaining serializability. The transaction might be running very slowly and holding needed resouces, or the person who submitted the transaction could change his mind and want it aborted.

The simplest concurrency control algorithm is one that actually runs the transactions serially, one at a time. However, such an algorithm is not satisfactory because it eliminates the possibility of concurrent execution of transactions, even if they access disjoint sets of data items.

4.4.1. Techniques

Hundreds of papers have been written about concurrency control algorithms, and many techniques have been proposed. We discuss only the two most popular ones: locking and timestamps. We refer the reader to the textbook by Bernstein, Hadzilacos and Goodman [13] for a more complete survey of concurrency control algorithms and an exposition of some of the underlying theory, and to Lynch et al. [61, 62] for a general theory of concurrency control algorithms.

Locking. The concurrency control method used most often in commercial systems is *locking.* A locking algorithm requires a transaction to obtain a *lock* on each data item before accessing it, preventing conflicting operations on the item by different transactions. There are usually two kinds of locks: exclusive locks that enable the owner to read or write the item, and shared locks that enable the owner only to read it. Several transactions can hold shared locks on the same item, but a transaction cannot have an exclusive lock while any other transaction holds either kind of lock on that item.

In a classic paper, Eswaran et al. [59] showed that serializability is guaranteed by *two-phase* locking, in which a transaction does not acquire any new locks after releasing a lock—for example, if it requests all locks at the beginning and releases them all at the end. However, if locks are acquired one at a time, deadlock is possible in which each member of some set of transactions is waiting for a lock held by another member of the set. Such deadlock must be detected, and the deadlock must be broken by aborting one or more waiting transactions. The effects of any aborted transactions must be undone; this may require saving the original values of all data items that have been modified by transactions which have not yet completed.

The notions of shared and exclusive locks can be generalized to other kind of locks, depending on the semantics of the operations on the database—in particular, on which operations commute. These other classes of locks lead to more general and efficient concurrency control mechanisms than ones based only on shared and exclusive locks.

Timestamps. Instead of using locks, some algorithms use timestamps (described in Section 2.3) to control access to data items. A timestamp is assigned to each transaction, and the algorithm ensures that transactions not aborted are executed as if they were run serially in the order specified by their timestamps. This serial execution order is obtained if operations to the same item are performed in timestamp order, where the timestamp of an operation is defined to be that of the transaction to which it belongs. One way to implement this condition is not to execute an operation on an item until all operations to that item with smaller timestamps have been executed. In a distributed database, waiting until no operation with an earlier timestamp can arrive may be expensive. Alternatively, one can abort a transaction if it tries to access a data item that has already been accessed by a transaction with a later timestamp.

So far, we have assumed that only a single version of the item is maintained. Additional flexibility can be achieved by keeping several earlier versions as well, since it is no longer necessary to abort a transaction when it accesses an item that has already been accessed by transactions with later timestamps. For example, if the transaction is just reading the item, the serial order can be preserved by reading an earlier version. Some of these earlier versions might be needed anyway to restore the item's value if a transaction is aborted.

While timestamps seem to offer some advantages over locking, almost all existing database systems use locking. This may be at least partly due to timestamp algorithms being more complicated than the commonly used locking methods.

4.4.2. Distribution issues

In distributed systems, items can be located at multiple sites. A concurrency control algorithm must guarantee that all sites affected by a transaction agree on whether or not it is aborted. This agreement is obtained by a commit protocol (Section 4.2.4).

Copies of an item may be kept at several sites, to increase its availability in the event of site failure or to make reading the item more efficient. For transactions to appear atomic, they must provide the appearance of accessing a single copy. One method of ensuring this is to require each operation to be performed on some subset of the copies—a read using the most recent value from among the copies it reads. Atomicity is guaranteed if the transactions are serialized and the sets of copies of any item accessed by any two operations have at least one element in common—for example, if each read reads two copies and each write writes all but one copy.

4.4.3. Nested transactions

The concept of a transaction has been generalized to *nested* transactions, which are transactions that can invoke subtransactions as well as execute operations on items. The nesting is described as a tree; each transaction is the parent of the subtransactions it invokes, and an added root transaction serves as the parent of all top-level transactions. Serializability is generalized to the requirement that for every node in the transaction tree, all its children appear to run serially. Algorithms based upon locking and timestamps have been devised for implementing this more general condition.

With nested transactions, failures can be handled by aborting a subtransaction without aborting its parent. The parent is informed that its child has aborted and can take corrective action. Nested transactions appear as a fundamental concept in the Argus distributed programming language [58] and in the Camelot system [78].

The framework presented in [61] is general enough for modeling nested transactions as well as ordinary single-level transactions.

References

[1] ABRAHAMSON, K., A. ADLER, L. HIGHAM and D. KIRKPATRICK, Probabilistic solitude verification on a ring, in: *Proc. 5th Ann. ACM Symp. on Principles of Distributed Computing* (1986) 161–173.

[2] AFEK, Y., B. AWERBUCH and E. GAFNI, Applying static network protocols to dynamic networks, in: *Proc. 28th Ann. IEEE Symp. on Foundations of Computer Science* (1987) 358–370.

[3] AHO, A.V., J. ULLMAN, A. WYNER and M. YANNAKAKIS, Bounds on the size and transmission rate of communication protocols, *Comput. Math. Appl.* **8**(3) (1982) 205–214.

[4] ALPERN, B. and F.B. SCHNEIDER, Defining liveness, *Inform. Process. Lett.* **21**(4) (1985) 181–185.

[5] APT, K.R., N. FRANCEZ and W.P. DE ROEVER, A proof system for communicating sequential processes, *ACM Trans. on Programming Languages and Systems* **2**(3) (1980) 359–385.

[6] ARJOMANDI, E., M.J. FISCHER and N.A. LYNCH, Efficiency of synchronous versus asynchronous distributed systems, *J. ACM* **30**(3) (1983) 449–456.

[7] ATTIYA, H., M. FISCHER, D. WANG and L. ZUCK, Reliable communication over an unreliable channel, unpublished manuscript.

[8] ATTIYA, H., M. SNIR and M.K. WARMUTH, Computing on an annonymous ring, *J. ACM* **35**(4) (1988) 845–875.

[9] AWERBUCH, B., Complexity of network synchronization, *J. ACM* **32**(4) (1985) 804–823.

[10] AWERBUCH, B., O. GOLDREICH, D. PELEG and R. VAINISH, On the message complexity of broadcast: basic lower bound, Tech. Memo TM-365, MIT Laboratory for Computer Science, 1988.

[11] BARATZ, A.E. and A. SEGALL, Reliable link initialization procedures, *IEEE Trans. Comm.* **36**(2) (1988) 144–152.

[12] BEN-OR, M. Another advantage of free choice: completely asynchronous agreement protocols, in: *Proc. 2nd Ann. ACM Symp. on Principles of Distributed Computing* (1983) 27–30.

[13] BERNSTEIN, P.A., V. HADZILACOS and N. GOODMAN, *Concurrency Control and Recovery in Database Systems* (Addison-Wesley, Reading, MA, 1987).

[14] BLOOM, B., Constructing two-writer atomic registers, *IEEE Trans. Comput.* **37**(12) (1988) 1506–1514.

[15] BRACHA, G., An O(log *n*) expected rounds randomized Byzantine generals algorithm, in: *Proc. 17th Ann. ACM Symp. on Theory of Computing* (1985) 316–326.

[16] BRACHA, G. and S. TOUEG, Distributed deadlock detection, *Distributed Comput.* **2**(3) (1987) 127–138.

[17] BURNS, J.E., P. JACKSON, N.A. LYNCH, M.J. FISCHER and G.L. PETERSON, Data requirements for implementation of *n*-process mutual exclusion using a single shared variable *J. ACM* **29**(1) (1982) 183–205.

[18] BURNS, J.E. and G. L. PETERSON, Constructing multireader atomic values from non-atomic values, in: *Proc. 6th Ann. ACM Symp. on Principles of Distributed Computing* (1987) 222–231.

[19] CHANDY, K.M., L.M. HAAS and J. MISRA, Distributed deadlock detection, *ACM Trans. Comput. Systems* **1**(2) (1983) 144–156.

[20] CHANDY, K.M. and L. LAMPORT, Distributed snapshots: determining global states of a distributed system, *ACM Trans. Comput. Systems* **3**(1) (1985) 63–75.

[21] CHANDY, M. and J. MISRA, The drinking philosophers problem, *ACM Trans. on Programming Languages and Systems* **6**(4) (1984) 632–646.

[22] CHANDY, K.M. and J. MISRA, *Parallel Program Design* (Addison-Wesley, Reading, MA 1988).

[23] COAN, B.A. and B. CHOR, A simple and efficient randomized Byzantine agreement algorithm, in: *Proc. 4th Symp. on Reliability in Distributed Software and Database Systems* (1984) 98–106.

[24] COMMONER, F., A.W. HOLT, S. EVEN and A. PNUELI. Marked directed graphs, *J. Comput. System Sci.* **5**(6) (1971) 511–523.

[25] COURTOIS, P.J., F. HEYMANS and D.L. PARNAS, Concurrent control with "readers" and "writers", *Comm. ACM* **14**(10) (1971) 667–668.

[26] DIFFIE, W. and M.E. HELLMAN, Privacy and authentication: an introduction to cryptography, *Proc. IEEE* **67**(3) (1979) 397–427.

[27] DIJKSTRA, E.W., Solution of a problem in concurrent programming control, *Comm. ACM* **8**(9) (1965) 569.

[28] DIJKSTRA, E.W., The structure of the "THE" multiprogramming system, *Comm. ACM* **11**(5) (1968) 341–346.

[29] DIJKSTRA, W., Hierarchical ordering of sequential processes, *Acta Inform.* **1** (1971) 115–138.

[30] DIJKSTRA, E.W., Self-stabilizing systems in spite of distributed control, *Comm. ACM* **17**(11) (1974) 643–644.

[31] DIJKSTRA, E.W. and C.S. SCHOLTEN, Termination detection for diffusing computations, *Inform. Process. Lett.* **11**(1) (1980) 1–4.

[32] DOLEV, D., The Byzantine generals strike again, *J. Algorithms* **3**(1) (1982) 14–30.

[33] DOLEV, D., J. HALPERN and H.R. STRONG, On the possibility and impossibility of achieving clock synchronization, in: *Proc. 16th Ann. ACM Symp. on Theory of Computing* (1984) 504–511.

[34] DWORK, C. and Y. MOSES, Knowledge and common knowledge in a Byzantine environment I: crash

failures (extended abstract), in: J. Halpern, ed., *Proc. Conf. on Theoretical Aspects of Reasoning About Knowledge* (1986) 149–170.

[35] EMERSON, E.A., Temporal and modal logic, in: J. van Leeuwen, ed., *Handbook of Theoretical Computer Science, Vol. B* (North-Holland, Amsterdam, 1990) 995–1072.

[36] FELDMAN, P. and S. MICALI, Optimal algorithms for Byzantine agreement, in: *Proc. 20th Ann. ACM Symp. on Theory of Computing* (1988) 148–161.

[37] FISCHER, M.J., The consensus problem in unreliable distributed systems (a brief survey), in: M. Karpinski, ed., *Proc. Internat. Conf. on Foundations of Computation Theory*, Lecture Notes in Computer Science, Vol. 158 (Springer, Berlin, 1983) 127–140.

[38] FISCHER, M.J. and N.A. LYNCH, A lower bound for the time to assure interactive consistency, *Inform. Process. Lett.* **14**(4) (1981) 183–186.

[39] FISCHER, M.J., N.A. LYNCH and M. MERRITT, Easy impossibility proofs for distributed consensus problems, *Distributed Comput.* **1**(1) (1986) 26–39.

[40] FISCHER, M.J., N. LYNCH and M.S. PATERSON, Impossibility of distributed consensus with one faulty process, *J. ACM* **32**(2) (1985) 374–382.

[41] FREDERICKSON, G.N. and N.A. LYNCH, Electing a leader in a synchronous ring, *J. ACM* **34**(1) (1987) 98–115.

[42] GAFNI, E. and C.T. CHOU, Understanding and verifying distributed algorithms using stratified decomposition, in: *Proc. 7th Ann. ACM Symp. on Principles of Distributed Computing* (1988) 44–65.

[43] GALLAGER, R.G., P.A. HUMBLET and P.M. SPIRA, A distributed algorithm for minimum-weight spanning trees, *ACM Trans. on Programming Languages and Systems* **5**(1) (1983) 66–77.

[44] GRAY, J.N., Notes on database operating systems, in: R. Bayer, R.M. Graham, and G. Seegmuller, eds., *Operating Systems: An Advanced Course*, Lecture Notes in Computer Science, Vol. 60. (Springer, Berlin, 1978) 393–481.

[45] HALPERN, J.Y. and Y. MOSES, Knowledge and common knowledge in a distributed environment. in: *Proc. 3rd Ann. ACM Symp. on Principles of Distributed Computing* (1984) 50–61.

[46] HERLIHY, M.P., Impossibility and universality results for wait-free synchronization, in: *Proc. 7th Ann. ACM Symp. on Principles of Distributed Computing* (1988) 276–290.

[47] HOARE, C.A.R., Monitors: an operating system structuring concept, *Comm. ACM* **17**(10) (1974) 549–557.

[48] HOARE, C.A.R., Communicating sequential processes, *Comm. ACM* **21**(8) (1978) 666–677.

[49] HUMBLET, P.A., A distributed algorithm for minimum weight directed spanning trees, *IEEE Trans. Comm.* **31**(6) (1983) 756–762.

[50] LAM, S.S. and A.U. SHANKAR, Protocol verification via projections, *IEEE Trans. on Software Engrg.* **10**(4) (1984) 325–342.

[51] LAMPORT, L., A new solution of Dijkstra's concurrent programming problem, *Comm. ACM* **17**(8) (1974) 453–455.

[52] LAMPORT, L., Time, clocks, and the ordering of events in a distributed system, *Comm. ACM* **21**(7) (1978) 558—565.

[53] LAMPORT, L., Using time instead of timeout for fault-tolerant distributed systems, *ACM Trans. on Programming Languages and Systems* **6**(2) (1984) 254–280.

[54] LAMPORT, L., On interprocess communication, *Distributed Comput.* **1**(2) (1986) 77–101.

[55] LAMPORT, L., A simple approach to specifying concurrent systems, *Comm. ACM* **32**(1) (1989) 32–45.

[56] LAMPORT, L., R. SHOSTAK and M. PEASE, The Byzantine generals problem, *ACM Trans. on Programming Languages and Systems* **4**(3) (1982) 382–401.

[57] LEVIN, G.M. and D. GRIES, A proof technique for communicating sequential processes, *Acta Inform.* **15** (3) (1981) 281–302.

[58] LISKOV, B., The Argus language and system, in: M. Paul and H.J. Siegert, eds., *Distributed Systems: Methods and Tools for Specification*, Lecture Notes in Computer Science, Vol. 190 (Springer, Berlin, 1985) Chapter 7, 343–430.

[59] LORIE,R.A., K.P. ESWARAN, J.N. GRAY and I.L. TRAIGER, The notions of consistency and predicate locks in a database system, *Comm. ACM* **19**(11) (1976) 624–633.

[60] LYNCH, N.A., Y. MANSOUR and A. FEKETE, The data link layer: two impossibility results, in: *Proc. 7th Ann. ACM Symp. on Principles of Distributed Computation* (1988) 149–170; also MIT Tech. Memo MIT/LCS/TM-355, Lab. for Computer Science, 1988.

[61] LYNCH, N.A., M. MERRITT, W. WEIHL and A. FEKETE, A theory of atomic transactions, in: M. Gyssens, J. Paredaens and D. Van Gucht, eds., *Proc. 2nd Internat. Conf. on Database Theory*, Lecture Notes in Computer Science, Vol. 326 (Springer, Berlin, 1988) 41–71.

[62] LYNCH, N.A., M. MERRITT, W. WEIHL and A. FEKETE, *Atomic Transactions* (Morgan Kaufmann, Los Altos, CA, in preparation).

[63] LYNCH, N. and M. TUTTLE, Hierarchical correctness proofs for distributed algorithms, in: *Proc. 6th Ann. ACM Symp. on Principles of Distributed Computing* (1987) 137–151.

[64] MILNER, R., *A Calculus of Communicating Systems*, Lecture Notes in Computer Science, Vol. 92 (Springer, Berlin, 1980).

[65] OWICKI, S. and D. GRIES, An axiomatic proof technique for parallel programs I, *Acta Inform.* **6**(4) (1976): 319–340.

[66] OWICKI, S. and L. LAMPORT, Proving liveness properties of concurrent programs, *ACM Trans. on Programming Languages and Systems* **4**(3) (1982): 455–495.

[67] PEASE, M., R. SHOSTAK and L. LAMPORT, Reaching agreement in the presence of faults, *J. ACM* **27**(2) (1980): 228–234.

[68] PETERSON, G.L., An $O(n \log n)$ unidirectional algorithm for the circular extremal problem, *ACM Trans. on Programming Languages and Systems* **4**(4) (1982): 758–762.

[69] PETERSON, G.L., Concurrent reading while writing, *ACM Trans. on Programming Languages and Systems*, **5**(1) (1983) 46–55.

[70] PETERSON, G. and M. FISCHER, Economical solutions for the critical section problem in a distributed system, in: *Proc. 9th Ann. ACM Symp. on Theory of Computing*, (1977) 91–97.

[71] PNUELI, A., The temporal logic of programs, in: *Proc. 18th Ann. IEEE Symp. on the Foundations of Computer Science* (1977) 46–57.

[72] RABIN, M.O., N-process mutual exclusion with bounded waiting by $4 \log n$ valued shared variable, *J. Comput. System Sci.* **25**(1) (1982) 66–75.

[73] RABIN, M and D. LEHMANN, On the advantages of free choice: a symmetric and fully distributed solution to the dining philosophers problem, in: *Proc. 8th Ann. ACM Symp. on Principles of Programming Languages* (1981) 133–138.

[74] RAYNAL, M., *Algorithms for Mutual Exclusion* (MIT Press, Cambridge, MA, 1986).

[75] SCHNEIDER, F.B. and G.R. ANDREWS, Concepts for concurrent programming, in: J.W. de Bakker, W.-P. de Roever and G. Rozenberg, eds., *Current Trends in Concurrency*, Lecture Notes in Computer Science Vol. 222 (Springer, Berlin, 1986) 669–716.

[76] SCHWARTZ, R.L. and P.M. MELLIAR-SMITH, From state machines to temporal logic: specification methods for protocol standards, *IEEE Trans. Comm.* **30**(12) (1982): 2486–2496.

[77] SKEEN, M.D., Crash recovery in a distributed database system, Ph.D. Thesis, Univ. of California, Berkeley, 1982.

[78] SPECTOR, A.Z., D. THOMPSON, R.F. PAUSCH, J.L. EPPINGER, D. DUCHAMP, R. DRAVES, D.S. DANIELS, and J.J. BLOACH, Camelot: a distributed transaction facility for Mach and the Internet–an interim report, Tech. Report CMU-CS-87-129, Carnegie–Mellon Univ., 1987.

[79] THIAGARAJAN, P.S., Some aspects of net theory, in B.T. Denvir, W.T. Harwood, M.I. Jackson and M.J. Wray, eds., *The Analysis of Concurrent Systems*, Lecture Notes in Computer Science, Vol. 266 (Springer, Berlin, 1985) 26–54.

[80] VITÁNYI, P., Distributed elections in an Archimedean ring of processors, in: *Proc. 16th Ann. ACM Symp. on Theory of Computing* (1984) 542–547.

[81] WELCH, J., L. LAMPORT and N. LYNCH, A lattice-structured proof of a minimum spanning tree algorithm, in: *Proc. 7th Ann. ACM Symp. on Principles of Distributed Computing* (1988) 28–43.

CHAPTER 19

Operational and Algebraic Semantics of Concurrent Processes

Robin MILNER

Computer Science Department, University of Edinburgh, The King's Buildings, Mayfield Road, Edinburgh, UK EH9 3JZ

Contents

1. Introduction 1203
2. The basic language 1206
3. Strong congruence of processes 1215
4. Observation congruence of processes 1223
5. Analysis of bisimulation equivalences 1234
6. Confluent processes 1238
7. Sources 1240
Note added in proof 1241
References 1241

HANDBOOK OF THEORETICAL COMPUTER SCIENCE
Edited by J. van Leeuwen

1. Introduction

This chapter is concerned with a particular class of mathematical models for concurrent communicating processes, namely the class of algebraic calculi. The view taken is that processes are not mathematical objects whose nature is universally agreed upon, and that therefore we cannot undertake this study by starting from a well-understood domain of entities known as processes. Rather we adopt the approach by which the lambda calculus was first introduced. We begin by defining a small language whose constructions reflect simple operational ideas; the meaning of these constructions is presented by means of structured operational semantics [30]. But although we take the liberty of calling the closed terms of our language *processes*, we do not claim to have defined the semantic notion of process until we have introduced a suitable congruence relation over these terms; then a (semantic) process is understood to be a congruence class of terms.

This congruence relation is therefore interpreted as equality of processes. Much of the chapter is taken up with developing the algebra of this equality, in particular with the existence and uniqueness of solutions of equations which define processes recursively. The essence of the congruence is that it is built upon the idea of *observing* a process, which we take to mean communicating with it; the idea is that processes are equal iff they are indistinguishable in any experiment based upon observation.

After establishing the congruence and developing its algebra in Sections 2–4, in Section 5 it is demonstrated that observation equivalence is characterised by a simple modal logic. Finally, in Section 6 a brief study appears of the ideas of *confluence* and *determinacy* of processes.

The theory presented here is in contrast with that of Petri nets; see [32] for example. The first difference is that the idea of *causal independence* (called *concurrency* by net theorists) is not here taken as fundamental; in our discipline of observation, causality is not directly observable and therefore the concurrent occurrence of two observable actions is not distinguished from their occurrence in arbitrary sequence. This intentional lack of distinction is strongly questioned by many, including particularly the Petri net community. The *event structures* introduced by Nielsen, Plotkin and Winskel [27] provide an important model in which these different views can be studied in the same framework; a recent paper by Boudol and Castellani carries this unifying approach further [6]. The second difference from net theory is that the combinational structure of process expressions, reflecting perhaps the physical structure of a machine which performs a process, plays a prominent part in our theory. The relative advantages of the two approaches are not discussed, and the discussion is indeed difficult; the aim here is to pursue the algebraic theory in its own terms as far as space permits.

There are many different approaches to the algebraic treatment of processes. The main distinctions among them are the constructions by which processes are assembled, the method by which process expressions are endowed with meaning, and the notion of equivalence among process expressions. A detailed and structured comparison of these approaches is not possible without a detailed presentation of each, since there is no established framework in which they can all be developed together. The main body of this chapter therefore consists of a single approach; the author hopes thereby to equip the reader to study any of the approaches to greater depth. But to place the chapter in perspective, the remainder of this section is devoted to a survey of the characteristics of the main algebraic approaches, and some remarks about alternative approaches. At the end of Sections 2 and 4 some detailed comparison with the other algebraic approaches is given.

Petri's net theory [32] was the first significant theory of concurrent systems. It is not algebraic in character, but is a generalisation of the classical theory of automata which was studied in depth in the 1950s [33]. In classical automata theory, an automaton assumes (or "is in") a single *state* at each moment; the theory considers the nature and structure of the *transitions* which take automata from one state to another. By contrast, net theory introduces a notion which may be called a *partitioned* state; the nodes of a net may be interpreted as *conditions* which may or may not hold of a system, and a (global) state consists of the simultaneous holding of many conditions. Thus the activity of a subsystem may be represented by the changing of holdings of some, but not all, conditions. These changes occur via transitions; a *transition*—in this refined model—is now defined by the *pre*-conditions which must hold before it occurs, and the *post*-conditions which will hold after it occurs. The development and literature of net theory is extensive; it achieves a sophisticated classification of nets, and many analytical techniques which pertain to certain classes of nets.

An early approach which is more algebraic is COSY, originated by Lauer and Campbell [11]. This work places greater emphasis upon the structure of a concurrent system, as built from a set of components which perform concurrently. (Indeed, the term "algebraic", applied to a theory of concurrency, usually refers to the way in which processes are built from smaller processes within that theory; the operators of the algebra are just the process constructors.) A COSY *program* (which may here be understood to mean "system description") consists of a collection of *paths*, each representing a sequentially acting process; each path is defind by a *regular expression* (which is one means by which a sequential automaton is described in classical automata theory), and the paths interact via the synchronisation of certain transitions on one path with similarly named transitions in other paths. The semantics of COSY is closely related to net theory; see the recent textbook [18]. Though the presentation is algebraic, the analytical methods place greater emphasis on concepts from net theory than upon algebraic (e.g. equational) techniques.

The algebraic emphasis was increased in the author's *Calculus of Communicating Systems*, known as CCS [23]. The main step taken, in this respect, was to treat *all* process-building constructions, whether sequential (to do with flow of control) or parallel (to do with the way concurrently acting processes are assembled into a system), as operators of a single algebraic theory. In this theory, the equations explain how the behaviour of a composite system unfolds. Independently of this, Hoare had designed and published his concurrent programming language *Communicating Sequential Processes*, known as CSP [16]. These two developments—an algebra and a programming language—agreed in the essential matter of the primitive means of interaction between two processes, namely the *synchronisation* of the *sending* of a datum by one agent with its *receipt* by another.

There is now a family of algebraic approaches, which can be characterised by (1) a thorough-going use of equations and inequalities among process expressions, and by (2) an acceptance of synchronised communication (rather than, say, the use of a shared memory) as the primitive means of interaction among system components. The principal members of this family are: Hoare's CSP [7, 17], a theory developed from the programming language in conjunction with Brookes and Roscoe; Hennessy's *Algebraic Theory of Processes*, called here ATP, which he developed with de Nicola [26, 13, 14]; the Process Algebra ACP developed by Bergstra and Klop [5] and by Baeten [3]; Meije and its relatives, developed by Berry, Boudol and de Simone [2, 34]; and the author's CCS [25], developed with the help of Hennessy and Park. In addition, Milne [22] has designed an algebraic method *Circal* specifically for VLSI design.

These approaches, while agreeing to place the emphasis upon algebra, have significant differences. Deferring details to later sections, we comment here upon some differences in general method. CSP and ATP give semantics to expressions by presenting an abstract (set-theoretic) model of processes. The model is more or less the same in each case, though differently presented; it is discussed at the end of Section 4. The informal idea behind it is that one should equate two processes if (a) they can execute exactly the same sequences of actions, and (b) for each such sequence they have the same possibility of entering a deadlocked state after executing it. CCS on the other hand presents the meaning of processes operationally, as mentioned at the very beginning of this section, with a view to maintaining a firm hold upon the dynamic intuitions which often guide computing theory; it then defines a notion of equivalence—*bisimilarity*—based upon this operational meaning. This is a much finer equivalence than that of CSP or ATP, but the latter equivalence can also be characterised operationally; Hennessy [14] presents this characterisation via a concrete notion of *testing*. But there are many different notions of testing, with varying power to discriminate between processes; in a striking paper Abramsky introduces a family of testing notions, of which the strongest exactly characterises bisimilarity [1].

Bergstra et al. [5] developed a family of process algebras, collectively called ACP, in which many equivalences and many different constructions are studied within a single framework. A significant feature of this work is the smooth treatment of sequential composition of processes; though sequential composition is definable in terms of (parallel) Composition in the present calculus—which is why it is here omitted from the basic constructors, for the sake of parsimony—its presence as a basic constructor leads to a pleasant algebra. The emphasis in ACP is more upon algebra, less upon operational semantics, and in more recent papers they have extensively studied the effect—including consistency—of many combinations of equational laws.

Although this chapter confines itself to algebraic approaches, some other prominent approaches to modelling concurrent processes will now be briefly outlined. An all-embracing theory is not in sight, and even if it emerges we shall still need to choose different special theories of concurrency for different applications.

First, many authors emphasise the use of *temporal logic* in specifying and verifying systems; a recent exposition is by Pnueli [31]. (See also Chapter 16 on temporal modal logic in this Handbook.) Temporal logic is a particular modal logic, whose modalities are time concepts like *always*, *eventually* etc. Clarke et al. have found temporal logic a good basis for machine-assisted reasoning about concurrent systems [11].

Second, an important line of work (stemming from Hoare) is that of axiomatic systems for proofs about imperative concurrent programs; the assertions of these proof systems are variants of the well-known "Hoare triple" $\{p\}C\{q\}$, meaning that if property p holds before command C is executed then property q will hold when and if C terminates. A formal system of this kind is given by Owicki and Gries [28]. A related but different approach is adopted by Chandy and Misra [10]; they give a methodology for developing proven programs from specifications which are expressed by means of assertions such as the above.

Third, various domain-theoretic models have been explored. An early study is in a paper already cited [27]; more recently, Broy [8] has employed to good effect a simpler domain consisting of finite and infinite streams.

Fourth, much work has been devoted to the difficult problem of *modular* design and verification of concurrent systems. In a modular methodology one must be able to specify the behaviour of a subsystem not completely, but under the assumption that its neighbour subsystems will behave in certain ways. Barringer et al. have tackled this problem from the temporal logic standpoint [4]. Other prominent work on modularity is by Lamport [20] and Jones [19]. Also, Stirling has incorporated Jones' insights into an axiomatic system which refines that of Owicki and Gries mentioned above, allowing modular proofs of a wider class of parallel programs [35].

We have identified each of the above strands of research by some distinguishing feature, but the spectrum of concurrency research is more complex than this selection of strands may suggest. There are other strands, and much good work is of interest for the very reason that it weaves some of the strands together. In the absence of anything like a complete classification of this burgeoning subject, it is possible here only to sample, not to summarise, the field.

2. The basic language

We begin by defining the *expressions* of our process language. Having done this we shall present its operational semantics as a *Labelled Transition System*. This leads to the notion of a *derivation tree*, which records the successive transitions or actions which may be performed by given process. We introduce the notion of *sort*, which provides a convenient means of classifying processes; a sort is a set of labels, and we attribute a sort L to a process if L contains all the labels which occur in the derivation tree of the process. This leads to the notion of *flow graph*, a means of depicting the structure of a process as composed from subprocesses which perform concurrently and may communicate with one another. We then present a richer language, derived from the basic language, in which the passage of data values between processes can be conveniently expressed.

The section ends with a discussion of some of the alternative algebraic operators which have been proposed for process languages.

2.1. *Syntax and notation*

We presuppose an infinite set $\mathscr{A}$ of *names* $a, b, c, \ldots$. Then $\bar{\mathscr{A}}$ is the set of *co-names* $\bar{a}, \bar{b}, \bar{c}, \ldots$; $\mathscr{A}$ and $\bar{\mathscr{A}}$ are disjoint and are in bijection via $(\bar{\ })$; we declare $\bar{\bar{a}} = a$.

$\mathscr{L}=\mathscr{A}\cup\bar{\mathscr{A}}$ is the set of *labels*; we shall use l, l' to range over $\mathscr{L}$, and K, L to range over subsets of $\mathscr{L}$, and we define $\bar{L}=\{\bar{l}: l\in L\}$. Labels will identify the actions which may be performed by a process. We also introduce a distinguished *silent action* $\tau\notin\mathscr{L}$; this special action is considered to be unobservable. We set $Act=\mathscr{L}\cup\{\tau\}$, and we shall use α, β to range over Act.

A function $f: \mathscr{L}\to\mathscr{L}$ is called a *relabelling function* provided $f(\bar{l})=\overline{f(l)}$; we extend f to Act by decreeing that $f(\tau)=\tau$. We shall often write f in the form $(l'_1/l_1, \ldots, l'_n/l_n)$ when $f(l_i)=l'_i$, $f(\bar{l}_i)=\bar{l}'_i$ and otherwise f is the identity function; here the names of l_i are distinct.

We presuppose a countable set $\mathscr{X}$ of *process variables* $X, Y, \ldots$. We shall often use I to stand for an arbitrary (possibly infinite) indexing set. Then the set $\mathscr{E}$ of *process expressions* $E, F, \ldots$ (also called terms) is the smallest set including $\mathscr{X}$ and the following expressions—when E, E_i are already in $\mathscr{E}$:

$\alpha.E$, a *Prefix* ($\alpha\in Act$),

$\Sigma_{i\in I}E_i$, a *Summation*,

$E_0|E_1$, a *Composition*,

$E\backslash L$, a *Restriction* ($L\subseteq\mathscr{L}$),

$E[f]$, a *Relabelling* (f a relabelling function),

$\mathbf{fix}_j\{X_i=E_i: i\in I\}$, a *Recursion* ($j\in I$).

In the final form (Recursion) the variables X_i are *bound* variables. For any expression E, its *free* variables $fv(E)$ are those which occur unbound in E. We say E is *closed* if $fv(E)=\emptyset$; in this case we say E is a *process*, and we shall use P, Q, R to range over the processes $\mathscr{P}$.

Roughly, the meaning of the process constructions is as follows. $\alpha.P$ is the process which performs α and then behaves as P; $\Sigma_{i\in I}P_i$ is the process which may behave as any of P_i; $P_0|P_1$ represents P_0 and P_1 performing concurrently, with possible communication; $P\backslash L$ behaves like P but with actions in $L\cup\bar{L}$ prohibited; $P[f]$ behaves like P but with the actions relabelled by f; finally, $\mathbf{fix}_j\{X_i=E_i: i\in I\}$ is the jth component of a distinguished "solution" of the recursive process equations $X_i=E_i$ ($i\in I$).

We shall usually be concerned only with finite Summations $\Sigma_{i\in I}P_i$, and indeed confine ourselves to the cases $I=\emptyset$ and $I=\{0, 1\}$. When $I=\emptyset$ we write $\mathbf{0}$, which we call *Inaction*; it is the process which can do nothing. When $I=\{0, 1\}$ we write P_0+P_1.

We shall often write $\tilde{E}$ for $\{E_i: i\in I\}$ when I is understood, and write $\tilde{X}=\tilde{E}$ for an I-indexed set of equations. Other abbreviations are $\Sigma\tilde{E}$ for a Summation and $\mathbf{fix}_j\tilde{X}\tilde{E}$ for a Recursion. Also $\mathbf{fix}\tilde{X}\tilde{E}$ means $\{\mathbf{fix}_j\tilde{X}\tilde{E}: j\in I\}$. For the simultaneous substitution of E_i for free occurrences of X_i in E we write $E\{\tilde{E}/\tilde{X}\}$, assuming that bound variables are changed where necessary to avoid clashes.

2.2. *Operational semantics*

A *Labelled Transition System* (LTS) is of the form $(S, A, \{\xrightarrow{a}: a\in A\})$ where S is a set of *states*, A is a set of *actions* and each $\xrightarrow{a}$ is a subset of $S\times S$, called an *action relation* over S. We call $s\xrightarrow{a}s'$ an *action* of $s\in S$; in this case (a, s') is a *successor* of s and s' is an *a-successor* (or just *successor*) of s. For $t=a_1 \ldots a_n \in A^*$ we define $\xrightarrow{t} = \xrightarrow{a_1} \cdots \xrightarrow{a_n}$, and we call $s \xrightarrow{t} s'$ a *transition* of s; then (t, s') is a *derivative* of s and s' is a *t-derivative* (or just *derivative*) of s.

We shall define the LTS $(\mathscr{E}, Act, \{\xrightarrow{\alpha} : \alpha \in Act\})$. The action relations $\xrightarrow{\alpha}$ are defined to be the smallest which obey the following rules, in which the action below the line is to be inferred from those above the line:

$$\text{ACT:} \quad \frac{}{\alpha.E \xrightarrow{\alpha} E} \qquad\qquad \text{SUM}_j\text{:} \quad \frac{E_j \xrightarrow{\alpha} E'_j}{\sum_{i\in I} E_i \xrightarrow{\alpha} E'_j} (j \in I)$$

$$\text{COM}_0\text{:} \quad \frac{E_0 \xrightarrow{\alpha} E'_0}{E_0|E_1 \xrightarrow{\alpha} E'_0|E_1} \qquad\qquad \text{COM}_1\text{:} \quad \frac{E_1 \xrightarrow{\alpha} E'_1}{E_0|E_1 \xrightarrow{\alpha} E_0|E'_1}$$

$$\text{COM}_2\text{:} \quad \frac{E_0 \xrightarrow{l} E'_0 \qquad E_1 \xrightarrow{\bar{l}} E'_1}{E_0|E_1 \xrightarrow{\tau} E'_0|E'_1}$$

$$\text{RES:} \quad \frac{E \xrightarrow{\alpha} E'}{E\backslash L \xrightarrow{\alpha} E'\backslash L} (\alpha \notin L \cup \bar{L}) \qquad\qquad \text{REL:} \quad \frac{E \xrightarrow{\alpha} E'}{E[f] \xrightarrow{f(\alpha)} E'[f]}$$

$$\text{REC}_j\text{:} \quad \frac{E_j\{\mathbf{fix}\,\tilde{X}\tilde{E}/\tilde{X}\} \xrightarrow{\alpha} E'_j}{\mathbf{fix}_j\tilde{X}\tilde{E} \xrightarrow{\alpha} E'_j} (j \in I)$$

In most cases these rules are easily seen to realise the rough description of the process constructions given in Section 2.1. For Summation, note that the first action of $\sum_{i\in I} E_i$ determines which alternative E_j is selected, the others being discarded. For Composition, the rules COM_0 and COM_1 permit the concurrent performance of E_0 and E_1; the rule COM_2 permits a synchronising communication between E_0 and E_1 whenever they may perform complementary actions.

The rule REC_j states that the actions of (the jth component of) the distinguished "solution" of $\tilde{X} = \tilde{E}$ are just those inferrable by unwindng the Recursion. In effect, this rule *defines* the distinguished solution, though as we see later there are other solutions in general.

For simplicity we sometimes use an alternative formulation of Recursion. Instead of the **fix** construction we introduce *process constants* $A, B, \ldots$ into the language. We then admit a set $\tilde{A} \stackrel{\text{def}}{=} \tilde{P}$ of *defining equations* for the constants; of course, $\tilde{P}$ may also contain the constants $\tilde{A}$. Then in place of the rules REC_j we have a simple rule

$$\text{REC}'\text{:} \quad \frac{P \xrightarrow{\alpha} P'}{A \xrightarrow{\alpha} P'} \quad \text{whenever } A \stackrel{\text{def}}{=} P.$$

We shall often need to prove properties with the logical form "if $E \xrightarrow{\alpha} E'$ then ...". In the absence of Recursion such properties can be proved by induction on the structure of E, but this in infeasible in general because, with either Recursion rule, an action for an expression is inferred from an action of a "larger" expression. So we use instead a stronger form of induction, namely upon the depth of the inference by which $E \xrightarrow{\alpha} E'$

is inferred. We call this *action induction*; it is a sound principle just because inference trees—i.e., compound inferences via the rules—are well-founded (in fact finite).

In what follows we shall omit many parentheses in process expressions, by declaring that Restriction and Relabelling have strongest binding power, then Prefix, then Composition, then Summation. Thus, for example,

$$P|\tau.Q\backslash a+R[f] \quad \text{means} \quad (P|(\tau.(Q\backslash a)))+(R[f]).$$

This also illustrates that $Q\backslash l$ means $Q\backslash\{l\}$.

2.3. *Derivation trees and transition graphs*

A *derivation tree* (of E) is of the form

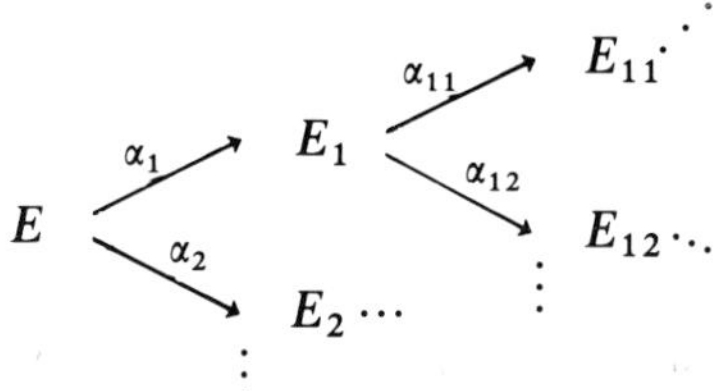

where the outgoing arcs from each non-leaf node are all the actions of the expression at that node. The tree can be infinite; it is *total* if the expressions at leaf nodes have no actions, otherwise it is *partial*.

The following shows that the derivatives of a process are themselves processes, so that the derivation tree of a process contains no free variables.

2.3.1. PROPOSITION. *If* $E \xrightarrow{\alpha} E'$ *and* $fv(E)=\emptyset$ *then* $fv(E')=\emptyset$.

PROOF. An easy action induction. □

2.3.2. EXAMPLE. A simple partial derivation tree is the following:

$$((a.E+b.\mathbf{0})|\bar{a}.F)\backslash a \;\begin{cases} \xrightarrow{\tau} (E|F)\backslash a \\ \xrightarrow{b} (\mathbf{0}|\bar{a}.F)\backslash a \end{cases}$$

It may be checked that all actions of the root expression are present, but to extend the tree, E and F must be made explicit.

2.3.3. EXAMPLE. Now consider the Recursive definition

$$A \stackrel{\text{def}}{=} ((a.(b.A+c.\mathbf{0})+b.\mathbf{0})\,|\,\bar{a}.0)\backslash a$$

whose right-hand side arises from setting $E\equiv b.A+c.\mathbf{0}$, $F\equiv\mathbf{0}$ in Example 2.3.2. Then, using REC′, the (infinite) total derivation tree of A can be developed:

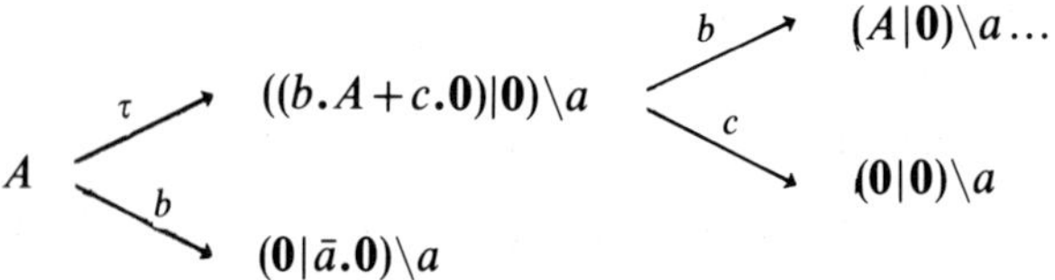

Now it will follow from our theory of equality of processes that

$$(A|\mathbf{0})\backslash a = A, \qquad (\mathbf{0}|\bar{a}.\mathbf{0})\backslash a = (\bar{a}.\mathbf{0})\backslash a = \mathbf{0}$$

and hence the behaviour of A can be represented by the transition graph

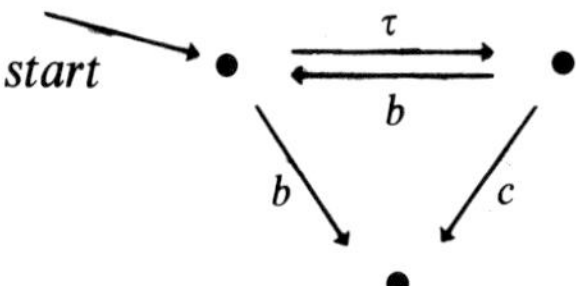

Moreover, this transition graph also unfolds into the derivation tree of B, where

$$B \stackrel{\text{def}}{=} \tau.(b.B + c.\mathbf{0}) + b.\mathbf{0}$$

We shall not treat transition graphs formally. We mention them here for comparison with the classical theory of automata, where the meaning of an automaton (transition graph) is the set of finite strings which it accepts. By contrast, we pay more attention to the branching structure of derivation trees.

2.4. Sorts

2.4.1. Definition. *P has sort $L \subseteq \mathscr{L}$*, or $P : L$, if every label l in the total derivation tree of P is a member of L.

For example, $A:\{b, c\}$ in Example 2.3.3. The essential property of sorts is the following simple consequence of the definition.

2.4.2. Proposition. (1) *If $P: K$ and $K \subseteq L$ then $P: L$.*
(2) *$P: L$ iff, whenever $P \xrightarrow{\alpha} P'$, then $\alpha \in L \cup \{\tau\}$ and $P' : L$.*

It is also obvious that if $P: L_i$ for each $i \in I$, then $P: \bigcap_{i \in I} L_i$, so every process has a minimum sort. In general, this minimum sort cannot be effectively computed. However, sorts can be assigned to process expressions by simple syntactic rules; for this purpose we first assume that to each variable X and each constant A are assigned sorts $\mathscr{L}(X)$ and $\mathscr{L}(A)$.

2.4.3. Definition. The (*distinguished*) sort $\mathscr{L}(E)$ of each E is given by the following rules:

$$\mathscr{L}(\alpha.E)=\mathscr{L}(E)\cup\{\alpha\}-\{\tau\}, \qquad \mathscr{L}(\Sigma_{i\in I}E_i)=\bigcup_{i\in I}\mathscr{L}(E_i),$$
$$\mathscr{L}(E_0|E_1)=\mathscr{L}(E_0)\cup\mathscr{L}(E_1), \qquad \mathscr{L}(E\backslash L)=\mathscr{L}(E)-(L\cup\bar{L}),$$
$$\mathscr{L}(E[f])=f(\mathscr{L}(E)), \qquad \mathscr{L}(\mathbf{fix}_j\tilde{X}\tilde{E})=\mathscr{L}(X_j)$$

with the proviso that, in the last case $\mathscr{L}(E_i)\subseteq\mathscr{L}(X_i)$ for each $i\in I$; also for each defining equation $A \stackrel{\text{def}}{=} P$ we require $\mathscr{L}(P)\subseteq\mathscr{L}(A)$.

Note that $\mathscr{L}(E)$ is not always minimum; for example $(a.b.\mathbf{0})\backslash a$: $\emptyset$, while $\mathscr{L}((a.b.\mathbf{0})\backslash a)=\{b\}$. The following proposition supports the definition:

2.4.4. PROPOSITION. P: $\mathscr{L}(P)$.

PROOF. A simple action induction, using the characterisation of Proposition 2.4.2(2). □

The importance of sorts is that certain equations will hold only under an assumption about sorts; for example $P\backslash b=P$ if $b, \bar{b}\notin\mathscr{L}(P)$.

2.5. *Flow graphs*

We introduce flow graphs to represent the compositional structure of processes in which several subprocesses are performing concurrently; they record the *spatial* structure of processes, in contrast to derivation trees or transition graphs which record their *temporal* structure.

2.5.1. DEFINITION. A *flow graph* is a graph in which
(1) each *node* is labelled by a process (a generator);
(2) each node labelled P possesses a *port* with *internal label* l, for each $l\in\mathscr{L}(P)$;
(3) each arc joins two ports of different nodes;
(4) each port may possess an *external label* $\in\mathscr{L}$;
(5) two ports joined by an arc either possess no external labels, or posses complementary external labels.

The *sort* $\mathscr{L}(G)$ of a flow graph G is its set of external port labels.

2.5.2. EXAMPLE. Let P_1:$\{a,b\}$, P_2:$\{b,c\}$, P_3:$\{\bar{b},d\}$. Then the processes P_1, P_2 and P_3 by themselves, and the process

$$P\equiv((P_1\,|\,P_2[\bar{a}/b,b/c])\,|\,P_3)\backslash b$$

are depicted by the four flow graphs in Fig. 1. Notice that the flow graph for P_1 by itself has internal and external labels equal. But the internal and external labels of a port may differ if a Relabelling is involved, and the external label may be absent if a Restriction is involved. Notice also that the sort $\mathscr{L}(G)=\{a,\bar{a},d\}$ of the flow graph G depicting P is exactly the sort $\mathscr{L}(P)$ defined by Definition 2.4.3.

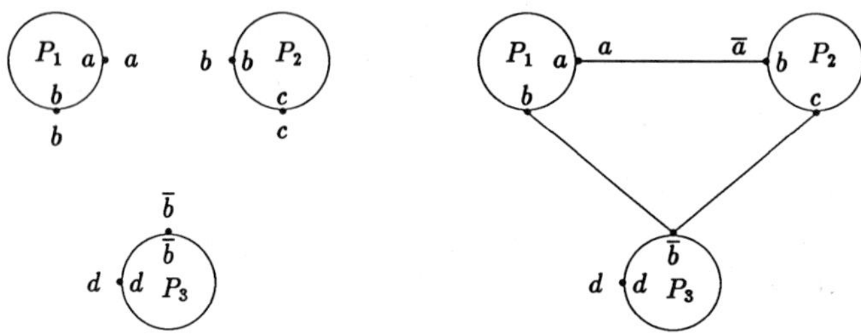

Fig. 1.

Flow graphs are indispensable as a tool for understanding complex parallel processes, though we shall not require them in our semantic treatment. They are formal objects, and indeed the operations of Composition, Restriction and Relabelling—which we call the *static constructors*—can be interpreted very simply over flow graphs in the following way:

- $G_0|G_1$ is obtained by joining each port in G_0 with external label l to every port in G_1 with external label $\bar{l}$;
- $G\backslash L$ is obtained from G by deleting all external labels in $L\cup\bar{L}$;
- $G[f]$ is obtained from G by replacing each external label l in G by $f(l)$.

It can be checked that G in Example 2.5.2 is built in this way. Many identities hold in this algebra of flow graphs. Indeed, the *static laws* (involving the static constructors), which we prove in the next section for processes, also hold in flow graph algebra. Further, it can even be shown—at least under the condition that the relabelling functions are one-to-one when restricted to the sorts of their operands—that the static laws are *complete* for flow graph algebra. Thus flow graphs are the free algebra for the static laws; our process semantics can therefore be regarded as the unique morphism from process algebra to flow graph algebra induced by the depiction of the generators.

2.6. *The extended language*

For many applications it is important to represent the passage of data values between processes. So first assume that V is the set of data values (we do not trouble to consider different types of value), and that there are *value variables* x, y,... and *value expressions* e built from constants and standard functions (e.g. arithmetic or Boolean operations) over V.

To represent input of a arbitrary value at a port with name a, we introduce the construction

$$ax.E$$

in which x is a *bound* value variable. We now consider a not as a single name, but standing for a family of names $\{a_v : v\in V\}$. Similarly we introduce the construction

$$\bar{a}e.E$$

to represent output of the value of e at the port with co-name $\bar{a}$; again $\bar{a}$ now stands for the family $\{\bar{a}_v : v \in V\}$. Next, we allow V-indexed families of process constants, $\{A_v : v \in V\}$, and introduce defining equations of the form

$$A(x) \stackrel{\text{def}}{=} P$$

where x will normally occur free in P; this admits processes which are parametric upon V. Finally, it is convenient to allow a conditional construction

$$\textsf{if } e \textsf{ then } E_1 \textsf{ else } E_2$$

2.6.1. EXAMPLE. Consider a process $A : \{a, \bar{b}\}$ which alternately inputs an integer v at a, and outputs at $\bar{b}$ the least even integer $\geqslant v$. A can be defined by:

$$A \stackrel{\text{def}}{=} ax.A'(x),$$
$$A'(x) \stackrel{\text{def}}{=} \textsf{if even}(x) \textsf{ then } \bar{b}x.A \textsf{ else } \bar{b}(x+1).A$$

In translating expressions of the extended language into the basic language, we confine our attention to expressions which contain no free value variables $x, y, \ldots$. (If E contains x free, then it can be considered as a family of expressions $E\{v/x\}$, one for each value constant v.)

The translation is defined by induction upon the structure of expressions. In stating it, we need only consider value expressions e which contain no value variables $x, y, \ldots$. This is because the translation of an input Prefix "$a(x)$." will replace x by a Summation over all values $v \in V$, thus eliminating all occurrences of the bound variable x. We can therefore treat each value expression e as identical with the constant value v to which it evaluates.

For each expression E without free value variables, its translated form $\hat{E}$ is given as follows:

E	$\hat{E}$
$ax.F$	$\sum_{v \in V} a_v.\widehat{F\{v/x\}}$
$\bar{a}e.F$	$\bar{a}_e.\hat{F}$
$\textsf{if } e \textsf{ then } E_1 \textsf{ else } E_2$	$\begin{cases} \hat{E}_1 & \text{if } e = \mathit{true} \\ \hat{E}_2 & \text{if } e = \mathit{false} \end{cases}$
$A(e)$	A_e

(translation distributes over the other constructions). Thus the translation of Example 2.6.1 is given by

$$A \stackrel{\text{def}}{=} \textstyle\sum_{v \in V} a_v.A'_v,$$

$$A'_v \stackrel{\text{def}}{=} \begin{cases} \bar{b}_v.A & \text{(if } v \text{ is even)} \\ \bar{b}_{v+1}.A & \text{(if } v \text{ is odd)} \end{cases} \quad \text{for each } v \in V.$$

The translation is possible just because, in the basic language, both Summations and Recursions have been indexed by arbitrary sets I. Most of our theory applies to this infinitary language; it is only when we consider complete axioms for finite-state processes that we wish to confine ourselves to the finitary language in which the indexing sets are kept finite.

2.7. *Alternative constructors*

There is nothing definitive about the set of constructors presented here; it will add to the reader's perspective to see a few alternatives.

Hoare's CSP has an alternative form of Composition which allows many (not just two) agents to synchronise upon a single action; moreover, CSP does not employ complementary actions. In the present framework, his constructor is best presented as ${}_K\|_L$, a binary operator dependent upon two sorts K and L. In the construction $P\,{}_K\|_L\,Q$, the idea is that P can independently perform any action not in L, and Q can independently perform any action not in K, but $P\,{}_K\|_L\,Q$ can perform any action in $K \cap L$ provided that P and Q perform it simultaneously. Thus, for example if $P \xrightarrow{a} P'$, $Q \xrightarrow{a} Q'$ and $R \xrightarrow{a} R'$ then

$$P\,{}_a\|_a\,Q\,{}_a\|_a\,R \xrightarrow{a} P'\,{}_a\|_a Q'\,{}_a\|_a R'$$

(where we have written a for the subscript $\{a\}$). Note that no τ action is formed by this synchronisation.

Hoare also replaces Restriction by an operator which he calls *Hiding*; we write it as P/L. Instead of preventing any action in L (and its complementary action), the Hiding operator $/L$ converts this action into a τ action (in terms of the understanding given in this paper). Thus, continuing the above example we have

$$(P_a\|_a Q_a\|_a R)/a \xrightarrow{\tau} (P'_a\|_a Q'_a\|_a R')/a.$$

These constructors have the advantage of allowing multiway synchronisation, but the composition depends upon explicitly supplied sorts. The two forms of composition have different merits, but this is not of primary importance since they are amenable to the same semantic treatment.

Some systems, in particular ACP and CSP, introduce sequential composition—which is a generalisation of Prefix—as a basic constructor. Thus $P;Q$ is an agent which behaves like P until P terminates properly (i.e. not deadlocked), and then behaves like Q. For this to work, it is necessary to distinguish proper termination from deadlock. CSP does this by introducing a distinguished action $\surd$, which is the last action of a properly terminating agent; ACP on the other hand introduces a special agent δ, standing for deadlock.

A pleasant feature of ACP is that it treats actions as a distinguished class of agents—the atomic or indivisible ones. ACP defines communication first by a binary operator | upon actions, satisfying the laws

$$a|b = b|a, \qquad (a|b)|c = a|(b|c), \qquad \delta|a = \delta.$$

This "action algebra" can be chosen arbitrarily subject to the conditions above, and the choice determines which actions may be synchronised—$a|b=\delta$ means that a and b cannot be synchronised. The constructor | is then extended to nonatomic agents; in general $P|Q$ *forces* P and Q to synchronise their first actions. (There is another parallel constructor $P \| Q$ which allows P's and Q's first actions to be independent.) ACP has the Restriction operator; it also has the silent action τ, and a Hiding operator which converts a given set of actions to τ.

Meije [2] and SCCS (a synchronous version of CCS; see [25]) keep actions and agents distinct, but allow multiway communication by taking the algebra of actions to be a commutative monoid, in which the binary operation $\times$ means synchronisation, and τ is the unit. They even allow the commutative monoid to be a commutative *group*, in which the inverse is our familiar complementation of actions—the difference being that *any* pair of actions whatever may be synchronised (in contrast with ACP). A noteworthy achievement of the French group is the result of Simone [34] that all constructors with a certain from of operational definition are definable in terms of a basic few, both in Meije and in SCCS. There is no such result yet for the other calculi.

Milne [22] rings yet another change in synchronisation, in his Circal; he allows actions to be, not arbitrary elements of a monoid, but *sets* of elementary actions. Each of the various forms of synchronisation has advantages; it is a little unfortunate that no clear best alternative has emerged among them.

3. Strong congruence of processes

3.1. Discussion

In this section we seek an appropriate equivalence relation over processes, which gives no special status to the silent τ-action. The following section will present a weaker (larger) equivalence which reflects the idea that the τ-action should indeed be silent, i.e. unobservable; but for the stronger equivalence of this section we shall already be able to derive several useful algebraic laws.

Perhaps the most obvious equivalence of processes is one which requires merely that they should possess the same transitions. More exactly, we might declare P and Q to be equivalent just when, for all $t \in Act^*$, $P \xrightarrow{t}$ iff $Q \xrightarrow{t}$; here we use $P \xrightarrow{t}$ to mean that $P \xrightarrow{t} P'$ for some P'.

But consideration of deadlock leads us to reject this proposal. For P and Q would be equivalent if they have the following total derivation trees:

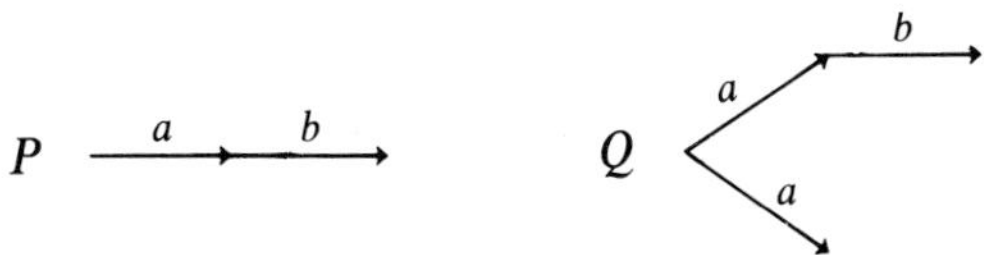

But in this case, after performing a, P will always be able to perform b while Q may not. Thus, in an "environment" which *demands b* after a, P will not deadlock while Q may. So apparently our proposed equivalence is too large.

On the other hand we may be *too* restrictive; we may take P and Q to be equivalent just when their derivation trees are isomorphic. This would deny the equivalence of

P and Q with trees

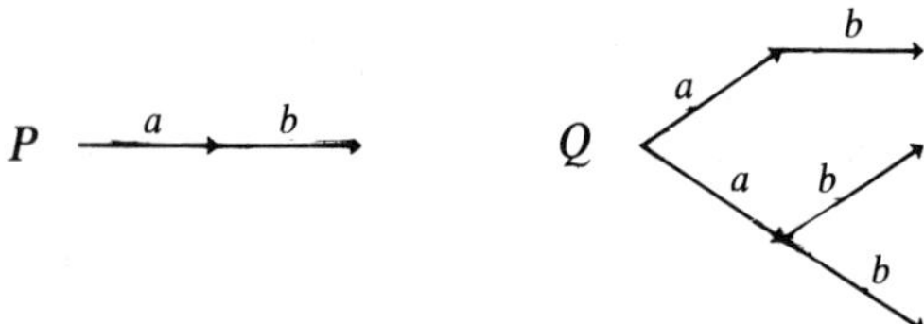

even though, at each stage, the same actions are possible.

We therefore seek an intermediate notion, with the following property:

P and Q are equivalent
iff, for all $\alpha \in Act$ each α-success of P is equivalent
to some α-successor of Q, and conversely. (3.1)

We shall see that such an equivalence exists; moreover it is indeed a congruence, and has many useful algebraic properties.

3.2. *Strong bisimulation*

3.2.1. DEFINITION. Let $\mathscr{F}$ be the function over binary relations $\mathscr{S} \subseteq \mathscr{P} \times \mathscr{P}$ defined as follows: $\langle P, Q \rangle \in \mathscr{F}(\mathscr{S})$ iff, for all $\alpha \in Act$,

(i) whenever $P \xrightarrow{\alpha} P'$ then, for some $Q', Q \xrightarrow{\alpha} Q'$ and $\langle P', Q' \rangle \in \mathscr{S}$;
(ii) whenever $Q \xrightarrow{\alpha} Q'$ then, for some P', $P \xrightarrow{\alpha} P'$ and $\langle P', Q' \rangle \in \mathscr{S}$.

It is easy to see that an equivalence relation $\sim$ satisfies (3.1) iff it is a fixed point of $\mathscr{F}$—i.e. $\sim = \mathscr{F}(\sim)$. But we must ensure that such a relation exists.

3.2.2. DEFINITION. $\mathscr{S} \subseteq \mathscr{P} \times \mathscr{P}$ is a *strong bisimulation* if $\mathscr{S} \subseteq \mathscr{F}(\mathscr{S})$.

3.2.3 PROPOSITION. *$\mathscr{F}$ is monotone; that is, $\mathscr{S} \subseteq \mathscr{S}'$ implies $\mathscr{F}(\mathscr{S}) \subseteq \mathscr{F}(\mathscr{S}')$.*

3.2.4. DEFINITION. Converse and composition of relations are defined as follows:

$$\mathscr{S}^{-1} = \{\langle Q, P \rangle : P\,\mathscr{S}\,Q\},$$
$$\mathscr{S}\mathscr{S}' = \{\langle P, R \rangle : \text{for some } Q,\ P\,\mathscr{S}\,Q \text{ and } Q\,\mathscr{S}'\,R\}.$$

3.2.5. PROPOSITION. *If $\mathscr{S}_i$ $(i \in I)$ are strong bisimulations, then so is $\bigcup_{i \in I} \mathscr{S}_i$.*

PROOF An easy application of Proposition 3.2.3. □

3.2.6. PROPOSITION. *If $\mathscr{S}$ and $\mathscr{S}'$ are strong bisimulations, then so are* $\mathrm{Id}_{\mathscr{P}}$, $\mathscr{S}^{-1}$ *and* $\mathscr{S}\mathscr{S}^{1}$.

PROOF Immediate from Definition 3.2.1. □

We are now ready to define $\sim$, our *strong congruence* relation.

3.2.7. DEFINITION. $\sim\ = \bigcup\{\mathcal{S} : \mathcal{S}$ is a strong bisimulation$\}$.

3.2.8. PROPOSITION. $\sim$ *is the largest strong bisimulation, and is also an equivalence relation.*

PROOF. The first part follows from Proposition 3.2.5; for the second part, reflexivity, symmetry and transitivity follow from Proposition 3.2.6. □

Now we can show that $\sim$ is the largest equivalence satisfying (3.1).

3.2.9. PROPOSITION. $\sim\ = \mathcal{F}(\sim)$, *and* $\sim$ *is the largest fixed point of* $\mathcal{F}$.

PROOF From Proposition 3.2.8 we have $\sim\ \subseteq \mathcal{F}(\sim)$. Hence $\mathcal{F}(\sim) \subseteq \mathcal{F}(\mathcal{F}(\sim))$ by monotonicity, so $\mathcal{F}(\sim)$ is a strong bisimulation and, by Proposition 3.2.8, is included in $\sim$; hence $\sim\ = \mathcal{F}(\sim)$. But all fixed points of $\mathcal{F}$ are strong bisimulations, so the largest strong bisimulation—being a fixed point—is also the largest fixed point. □

The rest of this section is organised as follows. First, we shall derive several equational laws for $\sim$. The first group consists of simple properties of Summation; the second group consists of the static laws mentioned in Section 2; the third group is a single theorem, the *expansion law*, which relates the *static constructors* (Composition, Restriction and Relabelling) to the *dynamic constructors* (Prefix and Summation), and which yields several simple corollaries.

Next we introduce the notion of strong bisimulation *up to* $\sim$, which provides a useful proof technique. We also extend $\sim$ from processes to process expressions. This allows us to discharge our obligation to prove that $\sim$ is preserved by all the process constructors; that is, that it *is* a congruence.

Finally, we show that under a simple condition a set of equations $\tilde{X} \sim \tilde{E}$ possesses a unique solution up to strong congruence.

3.3. *Equational properties of strong congruence*

Our first result is simple, namely that $(\mathcal{P}/\sim, +, \mathbf{0})$ is a monoid with absorption.

3.3.1. PROPOSITION (monoid laws)

(1) $P+Q \sim Q+P$, (2) $P+(Q+R) \sim (P+Q)+R$,
(3) $P+P \sim P$, (4) $P+\mathbf{0} \sim P$.

PROOF. Consider (1) only (the others are similar). Let $P+Q \xrightarrow{\alpha} R$; then this action must be inferred either from $P \xrightarrow{\alpha} R$ by SUM_0 or from $Q \xrightarrow{\alpha} R$ by SUM_1; in either case the opposite rule yields also $Q+P\xrightarrow{\alpha}R$. Hence both sides have identical α-successors, so $P+Q\sim Q+P$. □

For strong congruence, we find no other laws satisfied by the dynamic constructors Prefix and Summation. One possibility would be the distributive law $\alpha.(P+Q)\sim\alpha.P+\alpha.Q$; however, we have implicitly rejected this law. For if it were true then with the help of Proposition 3.3.1(4) we could deduce $a.b.\mathbf{0}\sim a.b.\mathbf{0}+a.\mathbf{0}$, and we have rejected this equivalence in Section 3.1 because of deadlock considerations. In fact the reader can easily check that property (3.1) denies this equivalence.

However, when we weaken the equivalence in the next section to allow for the unobservability of τ, then we shall find a few laws for the τ-action.

We now proceed to deduce the static laws for Composition, Restriction and Relabelling. We call these the static constructors because they are preserved by transition; for example if $(P|Q)\backslash L\xrightarrow{\alpha}R$ then R must be of the form $(P'|Q')\backslash L$. It follows that these constructors represent spatial rather than temporal structure, so it is natural to find that they can be interpreted over flow graphs.

The static laws depend sometimes on a sort condition; we consider an example. In general we do not have $(P|Q)\backslash a\sim P\backslash a\,|\,Q\backslash a$. The reader may check that this fails in the flow graph interpretation, when for example $P:\{a,b\}$ and $Q:\{\bar{a}\}$. It also fails in the process interpretation; this can be seen by taking $P\equiv a.b.\mathbf{0}$ and $Q\equiv\bar{a}.\mathbf{0}$, for then the left-hand side is equivalent to $\tau.b.\mathbf{0}$ and the right-hand side to $\mathbf{0}$. Intuitively a Restriction may only be distributed over a Composition when it does not affect a possible "communication link" between the components; this is reflected in the side condition on (7) below.

3.3.2. PROPOSITION (static laws)

(1) $P|Q\sim Q|P$.
(2) $P|(Q|R)\sim(P|Q)|R$.
(3) $P|\mathbf{0}\sim P$.
(4) $P\backslash L\sim P$ *if* $\mathscr{L}(P)\cap(L\cup\bar{L})=\emptyset$.
(5) $P\backslash K\backslash L\sim P\backslash(K\cup L)$.
(6) $P[f]\backslash L\sim P\backslash f^{-1}(L)[f]$.
(7) $(P|Q)\backslash L\sim P\backslash L\,|\,Q\backslash L$ *if* $\overline{\mathscr{L}(P)}\cap\mathscr{L}(Q)\cap(L\cup\bar{L})=\emptyset$.
(8) $P[\mathrm{Id}_{\mathscr{L}}]\sim P$.
(9) $P[f]\sim P[f']$ *if* $f\restriction\mathscr{L}(P)\sim f'\restriction\mathscr{L}(P)$.
(10) $P[f][f']\sim P[f'\circ f]$.
(11) $(P|Q)[f]\sim P[f]\,|\,Q[f]$ *if* $f\restriction(L\cup\bar{L})$ *is one-to-one, where* $L=\mathscr{L}(P|Q)$.

PROOF. All the laws may be proved by exhibiting appropriate strong bisimulations. For (1), for example, we must prove that

$$\mathscr{S}=\{\langle P_1|P_2,P_2|P_1\rangle\colon P_1,P_2\in\mathscr{P}\}.$$

is a strong bisimulation. So let $P_1|P_2\xrightarrow{\alpha}Q$; it is enough to find R such that $P_2|P_1\xrightarrow{\alpha}R$ and $\langle Q,R\rangle\in\mathscr{S}$ (with a symmetric argument). But $P_1|P_2\xrightarrow{\alpha}Q$ must be inferred from COM_0, COM_1 or COM_2, so $Q\equiv P_1'|P_2'$ where *either* $P_1\xrightarrow{\alpha}P_1'$ and $P_2\equiv P_2'$, *or* $P_1\equiv P_1'$

and $P_2 \xrightarrow{\alpha} P'_2$, *or* $\alpha = \tau$ and $P_1 \xrightarrow{l} P'_1$ and $P_2 \xrightarrow{\bar{l}} P'_2$. In each case we have $P_2 | P_1 \xrightarrow{\alpha} R \equiv P'_2 | P'_1$, and $\langle Q, R \rangle \in \mathcal{S}$ as required.

Proofs of the other laws differ only in detail, not in principle. □

Some of the static laws are more readily understood in terms of simple special cases. Often, for example, a relabelling function $f = (l'_1/l_1, \ldots, l'_n/l_n)$ has l_i and l'_i all distinct, and in this case $f = (l'_n/l_n) \circ \cdots \circ (l'_1/l_1)$; hence $P[f] = P[l'_1/l_1] \cdots [l'_n/l_n]$ by Proposition 3.3.2(10). So it is often enough to deal with the simple relabelling functions of the form l'/l. A few simple consequences of Proposition 3.3.2 are as follows.

3.3.3. COROLLARY. (1) *$P[l'/l] \sim P$ if $l, \bar{l} \notin \mathcal{L}(P)$.*

(2) *$P \backslash l \sim P[l'/l] \backslash l'$ if $l', \bar{l}' \notin \mathcal{L}(P)$.*

(3) *$P \backslash L[l'/l] \sim P[l'/l] \backslash L$ if $l, l' \notin L \cup \bar{L}$.*

These consequences show that Restriction and Relabelling behave very much like binding and syntactic substitution. But Relabelling is a constructor (operator) of the calculus, not a metasyntactic operation; it could not be treated as the latter, since—as Proposition 3.3.2(7) shows—it does not always distribute over Composition, but only under a side condition on sorts.

In many applications, a set of processes $P_1, \ldots, P_n$ are connected in some way represented by a construction $\mathscr{C}[P_1, \ldots, P_n]$, where $\mathscr{C}[\,]$ involves just the static constructors. So this construction can be depicted as a flowgraph with $P_1, \ldots, P_n$ labelling the nodes. For semantic analysis it turns out very often (in fact, whenever only one-to-one relabelling functions are used) that static constructions $\mathscr{C}[P_1, \ldots, P_n]$ can be transformed by the static laws into the form

$$(P_1[f_1] | \cdots | P_n[f_n]) \backslash L$$

which we call a *standard concurrent form*. The next theorem indicates how the successors of such a form are derived from the successors of its components P_i. In fact, the theorem—and its corollaries—cover everything that we normally need to know about the interaction between the static and dynamic constructions.

3.3.4. PROPOSITION (the expansion law). *Let $P \equiv (P_1[f_1] | \cdots | P_n[f_n]) \backslash L$, $n \geqslant 1$. Then*

$$\begin{aligned} P \sim \textstyle\sum \{ & f_i(\alpha).(P_1[f_1] | \cdots | P'_i[f_i] | \cdots | P_n[f_n]) \backslash L: \\ & P_i \xrightarrow{\alpha} P'_i,\ i \leqslant n,\ f_i(\alpha) \notin L \cup \bar{L} \} \\ + \textstyle\sum \{ & \tau.(P_1[f_1] | \cdots | P'_i[f_i] | \cdots | P'_j[f_j] | \cdots | P_n[f_n]) \backslash L: \\ & P_i \xrightarrow{l_1} P'_i,\ P_j \xrightarrow{l_2} P'_j,\ i < j \leqslant n,\ f_i(l_1) = \overline{f_j(l_2)} \}. \end{aligned}$$

REMARK. The first Summation corresponds to actions performed by a single component of P, while the second Summation corresponds to communications performed by pairs of components.

PROOF (*outline*). It is quite easy to prove first, by induction upon n, that if

$Q \equiv Q_1 | \cdots | Q_n$ then

$$Q \sim \Sigma\{\alpha.(Q_1 | \cdots | Q_i' | \cdots | Q_n) \colon i \leqslant n,\ Q_i \xrightarrow{\alpha} Q_i'\} \\ + \Sigma\{\tau.(Q_1 | \cdots | Q_i' | \cdots | Q_j' | \cdots | Q_n) \colon i < j \leqslant n,\ Q_i \xrightarrow{l} Q_i',\ Q_j \xrightarrow{\bar{l}} Q_j'\}.$$

The general case is then obtained by setting $Q_i \equiv P_i[f_i]$ and applying the Restriction $\backslash L$ to both sides. □

Several useful corollaries of this theorem arise from setting $n=1$, and either $S_i = \mathrm{Id}_{\mathscr{L}}$ or $L = \emptyset$. They show how Restriction and Relabelling "distribute" over Prefix and Summation.

3.3.5. COROLLARY

(1) $$(\alpha.P)\backslash L \sim \begin{cases} \mathbf{0} & \textit{if } \alpha \in L \cup \bar{L}, \\ \alpha.P\backslash L & \textit{otherwise.} \end{cases}$$

(2) $(\alpha.P)[f] \sim f(\alpha).P[f]$.

(3) $(P+Q)\backslash L \sim P\backslash L + Q\backslash L$.

(4) $(P+Q)[f] \sim P[f] + Q[f]$.

3.4. Substitutivity of strong congruence

We must demonstrate that $\sim$ is indeed a congruence relation, i.e. that it is substitutive everywhere. First, we refine the proof technique of strong bisimulation a little. We know that to show $P \sim Q$ it is enough to show $P \mathscr{S} Q$ for some strong bisimulation $\mathscr{S}$. But for this purpose, we need only a slightly weaker property of $\mathscr{S}$.

3.4.1. DEFINITION. $\mathscr{S}$ is a *strong bisimulation up to* $\sim$ if $\sim\mathscr{S}\sim$ is a strong bisimulation.

3.4.2. PROPOSITION. *If $\mathscr{S}$ is a strong bisimulation up to $\sim$ then $\mathscr{S} \subseteq \sim$.*

PROOF. Let $P \mathscr{S} Q$. Then clearly $P \sim\mathscr{S}\sim Q$, so $P \sim Q$ since $\sim\mathscr{S}\sim\ \subseteq\ \sim$. □

The next result shows that a property of $\mathscr{S}$ slightly weaker than $\mathscr{S} \subseteq \mathscr{F}(\mathscr{S})$ is enough to show that $\mathscr{S}$ is a strong bisimulation up to $\sim$ and hence that $S \subseteq\ \sim$.

3.4.3. PROPOSITION. *If $\mathscr{S} \subseteq \mathscr{F}(\sim\mathscr{S}\sim)$ then $\mathscr{S} \subseteq\ \sim$.*

PROOF. It is enough to show that $\sim\mathscr{S}\sim\ \subseteq \mathscr{F}(\sim\mathscr{S}\sim)$ from the hypothesis. For this purpose, let $P \sim\mathscr{S}\sim Q$, i.e., $P \sim P_1\ \mathscr{S}\ Q_1 \sim Q$. Let $P \xrightarrow{\alpha} P'$; we wish to find Q' so that $Q \xrightarrow{\alpha} Q'$ and $P' \sim\mathscr{S}\sim Q'$. This is done by filling in the bottom row of the following diagram from left to right, starting with the given P':

$$\begin{array}{ccccccc} P & \sim & P_1 & \mathscr{S} & Q_1 & \sim & Q \\ \alpha\downarrow & & \downarrow\alpha & & \downarrow\alpha & & \downarrow\alpha \\ P' & \sim & P_1' & \sim\ \mathscr{S}\ \sim & Q_1' & \sim & Q' \end{array}$$

This—with a symmetric argument—completes the proof. □

The first proposition about substitutivity is the following.

3.4.4. PROPOSITION. *Let* $P_1 \sim P_2$. *Then*

(1) $\alpha.P_1 \sim \alpha.P_2$.

(2) $P_1 + Q \sim P_2 + Q$.

(3) $P_1 | Q \sim P_2 | Q$.

(4) $P_1 \backslash L \sim P_2 \backslash L$.

(5) $P_1[f] \sim P_2[f]$.

PROOF. (1) and (2) are easily established by working directly with property (3.1) in Section 3.1. For (3), it is enough to show that

$$S = \{\langle P_1 | Q, P_2 | Q\rangle : P_1 \sim P_2\}$$

is a strong bisimulation. The details are routine, and (4) and (5) can be established by a similar method. □

This tells us that $\sim$ is substitutive everywhere except under Recursion. For the latter, we first extend $\sim$ to expressions.

3.4.5. DEFINITION. Let E and F have free variables $\tilde{X}$. Then $E \sim F$ if, for all P, $E\{\tilde{P}/\tilde{X}\} \sim F\{\tilde{P}/\tilde{X}\}$.

3.4.6. PROPOSITION. *Let* $\tilde{E} \sim \tilde{F}$. *Then* $\mathbf{fix}\, \tilde{X}\tilde{E} \sim \mathbf{fix}\, \tilde{X}\tilde{F}$.

PROOF. It will be enough to assume that $\tilde{E}, \tilde{F}$ have free variables $\tilde{X}$ (at most), since by use of Definition 3.4.5 we can then extend the result to the general case in which $\mathbf{fix}\, \tilde{X}\tilde{E}$ contains (other) free variables. Now let

$$\tilde{P} \equiv \mathbf{fix}\, \tilde{X}\tilde{E}, \qquad \tilde{Q} \equiv \mathbf{fix}\, \tilde{X}\tilde{F}.$$

Then it will be enough to show that

$$S = \{\langle G\{\tilde{P}/\tilde{X}\}, G\{\tilde{Q}/\tilde{X}\}\rangle : fv(G) \subseteq \tilde{X}\}$$

is a strong bisimulation up to $\sim$: for the result follows by taking $G \equiv X_i$ for each i, where $\tilde{X} = \{X_i : i \in I\}$.

For this purpose, it is enough to show that if $G\{\tilde{P}/\tilde{X}\} \xrightarrow{\alpha} P'$ then, for some Q', $G\{\tilde{Q}/\tilde{X}\} \xrightarrow{\alpha} Q'$ and $P' \mathcal{S}{\sim}\, Q'$. This property can be established by action induction, by case analysis on the form of G. The interesting case is when $G \equiv X_i$; then we have $G\{\tilde{P}/\tilde{X}\} \equiv P_i \xrightarrow{\alpha} P'$, so $E_i\{\tilde{P}/\tilde{X}\} \xrightarrow{\alpha} P'$ by a shorter inference, so by induction $E_i\{\tilde{Q}/\tilde{X}\} \xrightarrow{\alpha} Q''$ with $P' \mathcal{S}{\sim}\, Q''$; hence because $E_i \sim F_i$, $F_i\{\tilde{Q}/\tilde{X}\} \xrightarrow{\alpha} Q'$ with $Q'' \sim Q'$; hence finally $G\{\tilde{Q}/\tilde{X}\} \equiv Q_i \xrightarrow{\alpha} Q'$ also, and $P' \mathcal{S}{\sim}\, Q'$. □

This completes our proof of the substitutivity of $\sim$, and allows us to treat it as an equality relation. It remains to see the behaviour of fixed points with respect to this relation.

3.5. Unique fixed points up to strong congruence

The first result about fixed points is that the distinguished "solution" of $\tilde{X}=\tilde{E}$ is indeed a solution, up to $\sim$.

3.5.1. PROPOSITION. $\mathbf{fix}\,\tilde{X}\tilde{E}\sim\tilde{E}\{\mathbf{fix}\tilde{X}\tilde{E}/\tilde{X}\}$.

PROOF. By REC, clearly $\mathbf{fix}_j\tilde{X}\tilde{E}$ and $E_j\{\mathbf{fix}\tilde{X}\tilde{E}/\tilde{X}\}$ have identical successors for each j. □

Now, in general, solutions of $\tilde{X}\sim\tilde{E}$ cannot be unique; the equations $\tilde{X}=\tilde{X}$ have *all* processes as solutions. But a simple constraint ensures uniqueness.

3.5.2. DEFINITION. X is *weakly guarded* in E if each free occurrence of X is within some subexpression $\alpha.F$ of E. E is *weakly guarded* if all variables are weakly guarded in E.

Note that X is weakly guarded in $\tau.X$. In the next section we need a stronger notion of guardedness, which requires the guard to be a member of $\mathcal{L}$, to ensure uniqueness of fixed points for the weaker notion of equality developed there.

3.5.3. PROPOSITION. *Let the variables $\tilde{X}$ be weakly guarded in $\tilde{E}$, and let $\tilde{F}\sim\tilde{E}\{\tilde{F}/\tilde{X}\}$. Then $\tilde{F}\sim\mathbf{fix}\tilde{X}\tilde{E}$.*

PROOF. Using Definition 3.4.5, it will be enough to consider the case in which $fv(\tilde{E})\subseteq\tilde{X}$, so that $\tilde{E}$ are weakly guarded expressions. Then it will be enough to prove that if $\tilde{P}\sim\tilde{E}\{\tilde{P}/\tilde{X}\}$ and $\tilde{Q}\sim\tilde{E}\{\tilde{Q}/\tilde{X}\}$, then $\tilde{P}\sim\tilde{Q}$. For this, it will again be enough to show that

$$\mathcal{S}=\{\langle G\{\tilde{P}/\tilde{X}\}, G\{\tilde{Q}/\tilde{X}\}\rangle\colon fv(G)\subseteq\tilde{X}\}\cup\mathrm{Id}_{\mathcal{P}}$$

is a strong bisimulation up to $\sim$; the result follows by taking $G\equiv X_i$ for each i.

To prove this, we need to show that if $G\{\tilde{P}/\tilde{X}\}\xrightarrow{\alpha}P'$ then, for some Q', $G\{\tilde{Q}/\tilde{X}\}\xrightarrow{\alpha}Q'$ and $P'\sim\mathcal{S}\sim Q'$. As in Proposition 3.4.6, this can be established by action induction, and again the interesting case is when $G\equiv X_i$; the details are rather similar to Proposition 3.4.6, and we leave the reader to fill them in. □

The outcome of this section is that strong congruence provides a tractable notion of equality for processes, and allows many nontrivial equalities to be derived. But it is deficient in a vital respect; it treats τ on the same basis as all other actions, and properties which we would expect to hold if τ is unobservable, such as $\alpha.\tau.P=\alpha.P$, do not hold if "$=$" is taken to mean strong congruence. The next section, which follows the lines of the present section in many ways, will remove this defect by widening the notion of equality.

To conclude the sections we should remark that, even without giving τ special treatment, there are other interesting notions of equivalence. The first, and simplest, is *strong trace equivalence*; another is *strong failures equivalence*. We shall not describe these; their weak versions are more interesting, and are described at the end of the next section. From that description the reader should have no difficulty in defining the strong versions.

4. Observation congruence of processes

This section is devoted mainly to a congruence relation over processes, based upon observation of their behaviour. Many different congruences and equivalences have been studied, and new ones continue to emerge. The one presented here is one of the finest—i.e. it equates fewer processes than most others. At the end of the section we examine some of the others, distinguishing among them by suitably chosen examples.

4.1. Observation equivalence

What does it mean for τ to be silent, or unobservable? A first answer might be that two processes should be eqivalent if they become strongly congruent when the τ-actions are excised from their derivation trees. Under this proposal we would equate $\mathbf{P} \equiv a.\mathbf{0}+\tau.b.\mathbf{0}$ with $Q \equiv a.\mathbf{0}+b.\mathbf{0}$:

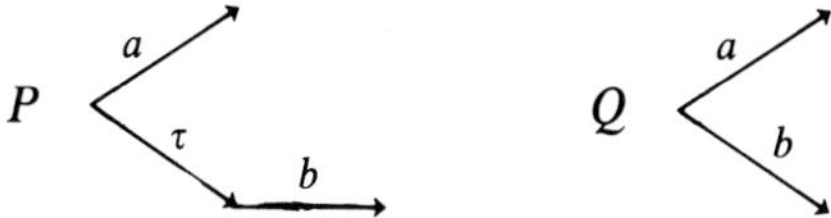

But this leads to difficulty. Since τ is to be unobservable, which we take to mean uncontrollable by the environment, P can perform τ autonomously and thus forego its ability to perform a; Q however preserves this ability. So, in an environment which demands a, Q will not deadlock, while P may. So τ, though unobservable directly, can affect the observability of visible actions.

We shall capture this indirect effect by requiring that, for P and Q to be observation equivalent, each action by P must be matched by a sequence of actions by Q with the same visible content, and conversely. The following definitions make this precise.

4.1.1. DEFINITION. Let $t=\alpha_1 \ldots \alpha_n \in Act^*$. Then

(1) $\xrightarrow{t} \stackrel{\text{def}}{=} \xrightarrow{\alpha_1} \cdots \xrightarrow{\alpha_n}$;

(2) $\hat{t} \in Act^*$ is the result of removing all τ's from t;

(3) $\stackrel{t}{\Longrightarrow} \stackrel{\text{def}}{=} (\xrightarrow{\tau})^* \xrightarrow{\alpha_1} (\xrightarrow{\tau})^* \ldots (\xrightarrow{\tau})^* \xrightarrow{\alpha_n} (\xrightarrow{\tau})^*$.

Note that $\hat{\tau}=\varepsilon$, the empty sequence, and $\stackrel{\varepsilon}{\Rightarrow}=(\xrightarrow{\tau})^*$. Also $\stackrel{\hat{t}}{\Rightarrow}$represents all sequences of actions with the same visible content as $\xrightarrow{t}$. We shall write $\Rightarrow$ for $\stackrel{\varepsilon}{\Rightarrow}$.

4.1.2. DEFINITION. For $t \in Act^*$, E' is a *t-descendant* of E if $E \stackrel{\hat{t}}{\Rightarrow} E'$.

We therefore seek an equivalence with the following property:

> P and Q are equivalent
> iff, for all $\alpha \in Act$, each α-successor of P is equivalent to some
> α-descendant of Q, and conversely. (4.1)

We arrive at this, which we call *observation equivalence*, in exactly the same way as we arrived at strong congruence.

4.2. Bisimulation

4.2.1. DEFINITION.[1] Let $\mathscr{G}$ be the function over binary relations $\mathscr{S} \subseteq \mathscr{P} \times \mathscr{P}$ defined as follows: $\langle P, Q \rangle \in \mathscr{G}(\mathscr{S})$ iff, for all $\alpha \in Act$,
(i) whenever $P \overset{\alpha}{\Rightarrow} P'$ then, for some Q', $Q \overset{\hat{\alpha}}{\Rightarrow} Q'$ and $\langle P', Q' \rangle \in \mathscr{S}$;
(ii) whenever $Q \overset{\alpha}{\Rightarrow} Q'$ then, for some P', $P \overset{\hat{\alpha}}{\Rightarrow} P'$ and $\langle P', Q' \rangle \in \mathscr{S}$.

So observation equivalence will be a fixed point of $\mathscr{G}$.

4.2.2. DEFINITION. $\mathscr{S} \subseteq \mathscr{P} \times \mathscr{P}$ is a (*weak*) *bisimulation* if $\mathscr{S} \subseteq \mathscr{G}(\mathscr{S})$

4.2.3. PROPOSITION. *$\mathscr{G}$ is monotone.*

4.2.4. PROPOSITION. *If $\mathscr{S}_i$ $(i \in I)$ are bisimulations, so is $\bigcup_{i \in I} \mathscr{S}_i$*

4.2.5. PROPOSITION. *If $\mathscr{S}, \mathscr{S}'$ are bisimulations, so are $\mathrm{Id}_{\mathscr{P}}, \mathscr{S}^{-1}$ and $\mathscr{S}\mathscr{S}'$*

We now define $\approx$, *observation equivalence* or *bisimilarity*.

4.2.6. DEFINITION. $\approx \; = \bigcup \{\mathscr{S} : \mathscr{S}$ is a bisimulation$\}$

4.2.7. PROPOSITION. *$\approx$ is the largest bisimulation and it is an equivalence.*

4.2.8. PROPOSITION. $\approx \; = \mathscr{G}(\approx)$, *and* $\approx$ is the largest fixed point of $\mathscr{G}$.

Thus $\approx$ is the largest relation satisfying (4.1). To establish $P \approx Q$, it is clearly enough to find a bisimulation $\mathscr{S}$ such that $P \mathscr{S} Q$, since $\mathscr{S} \subseteq \approx$. This is the most frequently useful technique for showing $P \approx Q$; however, there is another characterisation of $\approx$ which is intuitively satisfying since it is expressed solely in terms of $\overset{s}{\Rightarrow}$ $(s \in \mathscr{L}^*)$ rather than in terms of $\overset{\alpha}{\Rightarrow}$ and $\overset{\hat{\alpha}}{\Rightarrow}$.

4.2.9. PROPOSITION. *$P \approx Q$ iff, for all $s \in \mathscr{L}^*$,*
(i) *whenever $P \overset{s}{\Rightarrow} P'$ then, for some Q', $Q \overset{s}{\Rightarrow} Q'$ and $P' \approx Q'$;*
(ii) *whenever $Q \overset{s}{\Rightarrow} Q'$ then, for some P', $P \overset{s}{\Rightarrow} P'$ and $P' \approx Q'$.*
Moreover, $\approx$ is the largest relation with this property.

[1] See "Note added in proof" at the end of this chapter.

4.2.10. PROPOSITION. *$P \sim Q$ implies $P \approx Q$.*

PROOF. It follows directly from the definitions that every strong bisimulation is also a bisimulation. □

4.2.11. PROPOSITION. *$P \approx \tau.P$.*

PROOF. It is easy to show that $\{\langle P, \tau.P\rangle\} \cup \mathrm{Id}_{\mathscr{P}}$ is a bisimulation. □

This result shows us that observation equivalence is not a congruence. For it is not preserved by Summation; we have $b.\mathbf{0} \approx \tau.b.\mathbf{0}$, but if $a \neq b$ it is easily seen that $a.\mathbf{0}+b.\mathbf{0} \not\approx a.\mathbf{0}+\tau.b.\mathbf{0}$, as we would expect from the discussion at the beginning of Section 4.1. However, observation equivalence is preserved by the static constructors:

4.2.12. PROPOSITION. *If $P \approx Q$ then $P|R \approx Q|R$, $P\backslash L \approx Q\backslash L$ and $P[f] \approx Q[f]$.*

PROOF. For composition, it is enough to show that $\{\langle P|R, Q|R\rangle : P \approx Q,\ R \in \mathscr{P}\}$ is a bisimulation; a similar argument also works in the other cases. □

Later we shall see that the Prefix constructor even strengthens the relation.

For further work, we also need to deal with the more general notion of bisimulation *up to* $\approx$, analogous to strong bisimulation up to $\sim$.

4.2.13. DEFINITION. $\mathscr{S}$ is a *bisimulation up to* $\approx$ if $\approx\mathscr{S}\approx$ is a bisimulation.

4.2.14. PROPOSITION. *If $\mathscr{S}$ is a bisimulation up to $\approx$ then $\mathscr{S} \subseteq \approx$.*

4.2.15. PROPOSITION. *If $\mathscr{S} \subseteq \mathscr{G}(\approx\mathscr{S}\approx)$ then $\mathscr{S} \subseteq \approx$.*

Finally we extend $\approx$ to expressions in the obvious way.

4.2.16. DEFINITION. Let $fv(E), fv(F) \subseteq \tilde{X}$. Then $E \approx F$ if, for all $\tilde{P}$, $E\{\tilde{P}/\tilde{X}\} \approx F\{\tilde{P}/\tilde{X}\}$.

4.3. Observation congruence

We must now tackle the difficulty that $\approx$ is not a congruence. We look for a congruence which is as close to $\approx$ as possible. The definition is as follows, and we use the equality symbol "=" for the congruence relation since it is our final choice of congruence over $\mathscr{P}$.

4.3.1. DEFINITION. $P = Q$, i.e. P and Q are *observation congruent*, if for all $\alpha \in Act$

(i) whenever $P \overset{\alpha}{\Rightarrow} P'$ then, for some Q', $Q \overset{\alpha}{\Rightarrow} Q'$ and $P' \approx Q'$;

(ii) whenever $Q \overset{\alpha}{\Rightarrow} Q'$ then, for some P', $P \overset{\alpha}{\Rightarrow} P'$ and $P' \approx Q'$.

Note that this is a direct definition in terms of $\approx$, and that the clauses differ from those for $\approx$ only in the use of $\overset{\alpha}{\Rightarrow}$ in place of $\overset{\hat{\alpha}}{\Rightarrow}$; thus each action of P or Q must be matched by *at least* one action of the other. But this only applies to the initial actions.

We immediately extend equality to expressions in the obvious way.

4.3.2. DEFINITION. Let $fv(E)$, $fv(F) \subseteq \tilde{X}$; then $E=F$ if, for all $\tilde{P}$, $E\{\tilde{P}/\tilde{X}\}=F\{\tilde{P}/\tilde{X}\}$.

We can give a simple alternative characterisation of equality as follows.

4.3.3. PROPOSITION. *$P=Q$ iff, for all R, $P+R\approx Q+R$.*

PROOF ($\Rightarrow$): It is easy to show that $\{\langle P+R, Q+R\rangle\}\cup\approx$ is a bisimulation.

($\Leftarrow$): Suppose $P\neq Q$. Then, for example, there are α and P' such that $P\overset{\alpha}{\Rightarrow}P'$ but, whenever $Q\overset{\alpha}{\Rightarrow}Q'$, then $P'\not\approx Q'$. Choose $R\equiv l.\mathbf{0}$, $l\notin\mathscr{L}(P)\cup\mathscr{L}(Q)$. Clearly, $P+R\overset{\alpha}{\Rightarrow}P'$; it remains to show that whenever $Q+R\overset{\hat{\alpha}}{\Rightarrow}Q'$ then $P'\not\approx Q'$. If $\alpha=\tau$ and $Q'\equiv Q+R$ then $P'\not\approx Q'$ since $Q+R\overset{l}{\rightarrow}\mathbf{0}$ while P cannot perform l; otherwise $Q+R\overset{\alpha}{\Rightarrow}Q'$, and since $R\overset{\alpha}{\Rightarrow}Q'$ is impossible (because $\alpha\neq l$) we must have $Q\overset{\alpha}{\Rightarrow}Q'$, so $P'\not\approx Q'$ by assumption. □

4.3.4. PROPOSITION. *$P\sim Q$ implies $P=Q$ implies $P\approx Q$.*

PROOF. For the first part, use property (3.1) of strong congruence, together with Definition 4.3.1 of equality and Proposition 4.2.10. For the second part, $P=Q$ implies $P+\mathbf{0}\approx Q+\mathbf{0}$ by Proposition 4.3.3; now use $P+\mathbf{0}\sim P$. □

4.3.5. PROPOSITION. *Observation congruence ($=$) is an equivalence relation.*

PROOF. Use Proposition 4.3.3. □

4.3.6. PROPOSITION. *$P\approx Q$ implies $\alpha.P=\alpha.Q$.*

PROOF. Direct from Definition 4.3.1. □

Thus the Prefix constructor not only preserves $\approx$, but strengthens it to equality. We are now ready to show that equality is a congruence.

4.3.7. PROPOSITION. *If $P=Q$ then $\alpha.P=\alpha.Q$, $P+R=Q+R$, $P|R=Q|R$, $P\backslash L=Q\backslash L$ and $P[f]=Q[f]$.*

PROOF. The first follows from Proposition 4.3.6, and the second from Proposition 4.3.3 using associativity of Summation. The remainder follow directly from Definition 4.3.1 of equality together with Proposition 4.2.12. □

It remains to show that Recursion preserves equality.

4.3.8. PROPOSITION. *If $\tilde{E}=\tilde{F}$ then $\mathbf{fix}\tilde{X}\tilde{E}=\mathbf{fix}\tilde{X}\tilde{F}$.*

PROOF. It is enough to treat the case in which $fv(\tilde{E}), fv(\tilde{F})\subseteq\tilde{X}$; the general case then follows from Definition 4.3.2. So let $\tilde{P}\equiv\mathbf{fix}\tilde{X}\tilde{E}$ and $\tilde{Q}\equiv\mathbf{fix}\tilde{X}\tilde{F}$. We proceed, as in

Proposition 3.4.6, to show that

$$\mathcal{S} = \{\langle G\{\tilde{P}/\tilde{X}\},\ G\{\tilde{Q}/\tilde{X}\}\rangle : fv(G) \subseteq \tilde{X}\}$$

is a bisimulation up to $\approx$. In fact, we need something slightly stronger (since this would only show $\tilde{P} \approx \tilde{Q}$, by setting $G \equiv X_i$ for each i); we shall prove that

$$\begin{aligned}&\text{if } G\{\tilde{P}/\tilde{X}\} \stackrel{\alpha}{\Rightarrow} P' \\ &\text{then, for some } Q',\ G\{\tilde{Q}/\tilde{X}\} \stackrel{\alpha}{\Rightarrow} Q' \text{ and } P' \mathcal{S} \approx Q'.\end{aligned} \tag{4.2}$$

This clearly makes $\mathcal{S}$ a bisimulation up to $\approx$, and by setting $G \equiv X_i$ for each i we get $\tilde{P} = \tilde{Q}$ by using the definition of equality (Definition 4.3.1).

Just as in Proposition 3.4.6, we prove (4.2) by action induction; the details are only slightly more refined than in Proposition 3.4.6. □

We now turn to two further results which show how close the relations of bisimilarity and equality are to each other.

4.3.9. DEFINITION. P is *stable* if $P \stackrel{\tau}{\to} P'$ is impossible for all P'.

4.3.10. PROPOSITION. *If $P \approx Q$ and both are stable then $P = Q$.*

PROOF. An easy consequence of Definition 4.3.1. □

The second of the two results is rather surprising.

4.3.11. PROPOSITION. *$P \approx Q$ iff ($P = Q$ or $P = \tau.Q$ or $\tau.P = Q$).*

PROOF. ($\Leftarrow$): Easy, by Propositions 4.2.11 and 4.3.4.

($\Rightarrow$): We distinguish three cases:

(1) $P \stackrel{\tau}{\to} P^* \approx Q$ for some P^*;

(2) $Q \stackrel{\tau}{\to} Q^* \approx P$ for some Q^*;

(3) Neither (1) nor (2) obtains.

In cases (1) and (2) we prove $P = \tau.Q$ and $\tau.P = Q$ respectively, while in case (3) we prove $P = Q$. □

The results which we have obtained in this section so far provide a rich repertoire of tools for proving bisimilarity or equality. The most powerful of these is to prove $P \approx Q$ by exhibiting a bisimulation containing the pair $\langle P, Q\rangle$; often this can be strengthened to an equality by stability (Proposition 4.3.10) or by applying the Prefix constructor (Proposition 4.3.6).

In the course of proofs we can freely use the static laws (Proposition 3.3.2) and the expansion law (Proposition 3.3.4) and their corollaries and also the monoid properties of Summation (Proposition 3.3.1), since strong congruence is stronger than equality. To enrich the algebra of equality, we now extend the four laws of Proposition 3.3.1 by three more concerning both Prefix and Summation; they can be regarded as absorption laws due to the unobservability of τ, and we call these seven laws together the *dynamic laws*. It is remarkable that our algebraic laws can be classified into three classes; the dynamic laws, the static laws, and the single expansion law which captures the interaction between the dynamic and static constructors.

4.3.12. PROPOSITION (τ-laws). (1) $\alpha.\tau.P=\alpha.P$.

(2) $P+\tau.P=\tau.P$.

(3) $\alpha.(P+\tau.Q)+\alpha.Q=\alpha.(P+\tau.Q)$.

PROOF. Easy application of the definition of equality (Definition 4.3.1). □

Later we shall see that the dynamic laws are complete for equality of sequential processes, i.e. those not involving the static constructors.

4.4. *Unique fixed points up to equality*

In Section 3.5 we saw that equations $\tilde{X}\sim\tilde{E}$ possess a unique solution up to $\sim$, provided that the variables $\tilde{X}$ are weakly guarded in $\tilde{E}$. We now ask when the equations $\tilde{X}=\tilde{E}$ possess a unique solution up to $=$. Immediately we see that weak guarding is not a strong enough condition; for if we consider the equation $X=\tau.X$, then we see that any process of the form $\tau.P$ is a solution, since, by Proposition 4.3.12(1), $\tau.P=\tau.\tau.P$. Similarly, we can show that $X=\tau.X+a.\mathbf{0}$ has $\tau.(P+a.\mathbf{0})$ as solution for any P. This leads us to strengthen the guarding condition as follows.

4.4.1. DEFINITION. X is (*strongly*) *guarded* in E if every free occurrence of X lies within a subexpression $l.F$ of E.

But we now meet another difficulty. In the presence of the static constructors, it is not enough to demand that X be (strongly) guarded in E to ensure that $X=E$ has a unique solution. Consider $X=(\bar{a}.\mathbf{0}|a.X)\backslash a$. This simplifies to $X=\tau.(X\backslash a)$, and again $\tau.P$ will always be a solution provided that $a,\bar{a}\notin\mathscr{L}(P)$. So we look for a further constraint.

4.4.2. DEFINITION. X is *sequential* in E if no free occurrence of X lies within a static constructor in E.

Thus, for example, X is sequential in $a.(P|Q)\backslash b+\tau.X$ (though not guarded), but not in $a.(X\backslash a)$.

For the rest of this subsection we shall be concerned often with expressions G such that $fv(G)\subseteq\tilde{X}$, and we shall often write $G(\tilde{P})$ for $G\{\tilde{P}/\tilde{X}\}$. We first need a lemma, which states the under a certain condition G's successors are independent of $\tilde{P}$.

4.4.3. LEMMA. *Let G have free variables $\subseteq\tilde{X}$, all guarded and sequential in G, and let $G(\tilde{P})\overset{\alpha}{\to}P'$. Then for some H with $fv(H)\subseteq\tilde{X}$, $G\overset{\alpha}{\to}H$, and $P'\equiv H(\tilde{P})$; also, for any $\tilde{Q}$,*

$G(\tilde{Q})\xrightarrow{\alpha}H(\tilde{Q})$. Moreover, the variables $\tilde{X}$ are sequential in H, and if $\alpha=\tau$ they are also guarded in H.

PROOF. By action induction. $\square$

Now we are ready to prove the main result.

4.4.4. PROPOSITION. *Let the variables $\tilde{X}$ be guarded and sequential in $\tilde{E}$, and let $\tilde{F}=\tilde{E}\{\tilde{F}/\tilde{X}\}$. Then $\tilde{F}=\mathbf{fix}\tilde{X}\tilde{E}$.*

PROOF. By Definition 4.3.2 it will be enough to assume $fv(\tilde{E})\subseteq\tilde{X}$. Then it will be enough to prove that if $\tilde{P}=\tilde{E}(\tilde{P})$ and $\tilde{Q}=\tilde{E}(\tilde{Q})$ then $\tilde{P}=\tilde{Q}$. For this we shall show that

$$\mathcal{S}=\{\langle G(\tilde{P}),G(\tilde{Q})\rangle : fv(G)\subseteq\tilde{X},\ \tilde{X}\text{ sequential in }G\}\cup\mathrm{Id}_{\mathcal{P}}$$

is a bisimulation up to $\approx$. Actually we prove the stronger fact that

$$\begin{array}{ll}\text{if } G(\tilde{P})\overset{\alpha}{\Rightarrow}P' & \\ \text{then, for some } Q',\ G(\tilde{Q})\overset{\alpha}{\Rightarrow}Q' \text{ and } P'\approx\mathcal{S}\approx Q'. & (4.3)\end{array}$$

This makes $\mathcal{S}$ a bisimulation up to $\approx$, and gives us $\tilde{P}=\tilde{Q}$ by setting $G\equiv X_i$ for each i.

We first prove something like (4.3) for silent actions:

$$\begin{array}{ll}\text{if } G(\tilde{P})\Rightarrow P' & \\ \text{then, for some } Q',\ G(\tilde{Q})\Rightarrow Q' \text{ and } P'\approx\mathcal{S}\approx Q'; & (4.4)\end{array}$$

so let $G(\tilde{P})\Rightarrow P'$. Now $G(\tilde{P})=G(\tilde{E}(\tilde{P}))$, so $G(\tilde{E}(\tilde{P}))\Rightarrow P''\approx P'$. But $\tilde{X}$ are guarded and sequential in G, so by Lemma 4.4.3 repeatedly, $\tilde{X}$ are guarded and sequential in some G' where

$$P''\equiv G'(\tilde{P}) \quad\text{and}\quad G(\tilde{E}(\tilde{Q}))\Rightarrow G'(\tilde{Q})$$

and since $G(\tilde{Q})=G(\tilde{E}(\tilde{Q}))$ we also have $G(\tilde{Q})\Rightarrow Q'\approx G'(\tilde{Q})$ which establishes (4.4)

We now prove (4.3). Let $G(\tilde{P})\overset{\alpha}{\Rightarrow}P'$, so that

$$G(\tilde{E}(\tilde{P}))\ \Rightarrow\xrightarrow{\alpha}\Rightarrow\ P''\approx P'.$$

Now by Lemma 4.4.3 repeatedly, taking us through $\Rightarrow\xrightarrow{\alpha}$, there is a G' with free sequential variables $\tilde{X}$ such that this derivation is of form

$$G(\tilde{E}(\tilde{P}))\ \Rightarrow\xrightarrow{\alpha}\ G'(\tilde{P})\ \Rightarrow\ P''\approx P'$$

and also

$$G(\tilde{E}(\tilde{Q}))\ \Rightarrow\xrightarrow{\alpha}\ G'(\tilde{Q})$$

and by (4.4) this derivation can be continued by

$$G'(\tilde{Q})\Rightarrow Q'' \quad\text{with } P''\approx\mathcal{S}\approx Q''.$$

Finally, since $G(\tilde{E}(\tilde{Q}))=G(\tilde{Q})$, we deduce

$$G(\tilde{Q}) \stackrel{\alpha}{\Rightarrow} Q' \approx Q''.$$

So $P' \approx \mathscr{S} \approx Q'$ and the proof is complete. □

This result is important in practice. Often we have a specification $\mathbf{fix}\tilde{X}\tilde{E}$ where $\tilde{X}$ are guarded and sequential in $\tilde{E}$, and we wish to prove a highly concurrent system $\tilde{P}$ equal to the specification by proving $\tilde{P}=\tilde{E}(\tilde{P})$.

4.5. *Completeness of equational laws*

We now turn to the question of whether we have "enough" equational laws. We have alluded in Section 2.5 to one positive result; under a mild constraint the static laws of Proposition 3.3.2 are complete for the algebra of flow graphs, which is evidence that we have "enough" laws concerning the static constructors alone. Here we consider mainly the dynamic constructors and recursion; we shall only be able to summarise the results, due to space limitation.

One first result is a negative one. The mixture of Recursion and Composition (with the other constructors) is enough to allow us effectively to translate the ith Turing machine, in any enumeration, to a process TM_i such that $\mathrm{TM}_i \sim \mathbf{fix}X(\tau.X)$ iff the machine diverges starting from blank tape, and hence the set of valid strong congruences is not recursively enumerable, so cannot be effectively axiomatised. This translation is best carried out by first defining a process representing a stack and then modelling a Turing machine by a pair of stacks, and only employs finitely indexed Summations and Recursions. A similar result holds for equality in place of strong congruence. We therefore look at subsets of $\mathscr{P}$; for the remainder of this section we assume first that all Summations and Recursions are finitely indexed.

4.5.1. DEFINITION. E is *finite* if it contains no Recursion; E is *sequential* if it contains no Composition, Restriction or Relabelling. □

4.5.2. PROPOSITION. *The expansion law* (*Proposition* 3.3.4) *and the monoid laws for Summation* (*Proposition* 3.3.1) *are complete for strong congruence of finite processes.*

PROOF (*outline*). In the absence of Recursion, the expansion law can be used to convert every finite process to a strongly congruent sequential process. This can be placed in the standard form $\Sigma\alpha_i.P_i$ where each P_i is also in standard form, using Proposition 3.3.1(4) to remove redundant occurrences of $\mathbf{0}$. Then it can be shown that the other monoid laws are enough to prove all valid strong congruences between standard forms. □

4.5.3. PROPOSITION. *The expansion law, the monoid laws and the τ-laws* (*Proposition* 4.3.12) *are complete for equality of finite processes.*

PROOF (*outline*). We convert each process to standard form $P \equiv \Sigma\alpha_i.P_i$ as in Proposition 4.5.2. Then, by Proposition 4.3.12(2) and (3) we can convert P to a "full" standard form in which, whenever $P \stackrel{\alpha}{\Rightarrow} P'$, $\alpha.P$ is a summand of P—i.e. $P \stackrel{\alpha}{\rightarrow} P'$. Finally, it can be shown that the monoid laws with Proposition 4.3.12(1) are enough to prove all valid equations between full standard forms. □

Now we turn to processes which are sequential but not finite. There are three laws for Recursion, which are easily proved sound, and whose purpose is to convert all Recursions to those in which the expressions $\tilde{E}$ are weakly or strongly guarded. For simplicity we state these only for single Recursions $\mathbf{fix}XE$.

4.5.4. PROPOSITION. $\mathbf{fix}X(X+E)\sim\mathbf{fix}XE$.

4.5.5. PROPOSITION. (1) $\mathbf{fix}X(\tau.X+E)=\mathbf{fix}X(\tau.E)$.
(2) $\mathbf{fix}X(\tau.(X+E)+F)=\mathbf{fix}X(\tau.X+E+F)$.

The point of these laws is that, once Recursions are weakly or strongly guarded, the appropriate unique fixed point rule (Proposition 3.5.3 or 4.4.4) can be applied.

4.5.6. PROPOSITION. *The monoid laws, together with the recursion laws for strong congruence* (*Propositions* 3.5.1, 3.5.3 *and* 4.5.4) *are complete for strong congruence of sequential processes.*

The proof of this result is quite detailed and we omit it here; we also omit the proof of our final completeness result, the following.

4.5.7. PROPOSITION. *The monoid laws, the τ-laws and the Recursion laws for equality* (*Proposition* 3.5.1, 4.5.4, 4.5.5 *and* 4.4.4) *are complete for equality of sequential processes.*

The diagram in Fig. 2 summarises the completeness results.

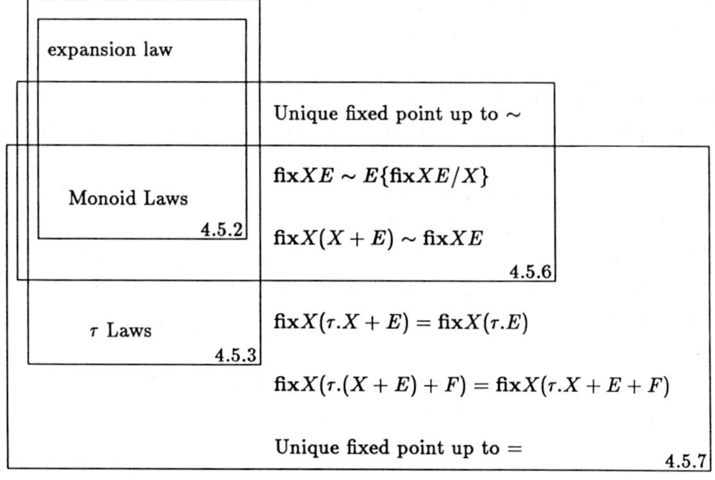

Fig. 2.

4.6. *Alternative notions of equivalence*

Hitherto we have studied two main equivalences: strong and weak bisimilarity. (Actually observation congruence is a third, but closely allied to weak bisimilarity.) We shall now look at two further equivalences. They both abstract from internal (τ) actions like weak bisimilarity, but they are both much more generous; they equate more agents.

Our first equivalence is perhaps the simplest of all: *trace equivalence*. It equates two agents if and only if they can perform exactly the same sequences of observable actions.

4.6.1. DEFINITION. P and Q are (*weakly*) *trace equivalent*, written $P \approx_1 Q$, if for all $s \in \mathscr{L}^*$.

$$P \stackrel{s}{\Rightarrow} \text{ if and only if } Q \stackrel{s}{\Rightarrow}.$$

4.6.2. PROPOSITION. *If* $P \approx Q$ *then* $P \approx_1 Q$.

The main disadvantage of trace equivalence is that it does not tell us anything about deadlock; for example, $a.\mathbf{0}+\tau.\mathbf{0} \approx_1 a.\mathbf{0}$ even though the left-hand side can reach a state in which a and all other observable actions are impossible—i.e. a deadlock—while the right-hand side cannot avoid doing a. But if we are content with such a coarse equivalence then we can exploit several pleasant algebraic properties. First, $\approx_1$ is a congruence relation.

4.6.3. PROPOSITION. $\approx_1$ *is preserved by all the basic combinators. That is, if* $P \approx_1 Q$ *and* $P_i \approx_1 Q_i$ *then* $\alpha.P \approx_1 \alpha.Q$, $\sum_i P_i \approx_1 \sum_i Q_i$, $P_0|P_1 \approx_1 Q_0|Q_1$, $P\backslash L \approx_1 Q\backslash L$ *and* $P[f] \approx_1 Q[f]$. *Also if* $\tilde{E} \approx_1 \tilde{F}$ *then* $\mathbf{fix}\tilde{X}\tilde{E} \approx_1 \mathbf{fix}\tilde{X}\tilde{F}$.

Thus, in contrast with $\approx$, we do not have to refine $\approx_1$ to obtain a congruence. Second, there is a unique solution theorem for $\approx_1$ which is false in general for observation congruence.

4.6.4. PROPOSITION. (1) $\alpha.(P+Q) \approx_1 \alpha.P+\alpha.Q$.
(2) $(P+Q)|R \approx_1 P|R+Q|R$.
(3) $P \approx_1 \tau.P$.

The first two of these laws state that Prefix and Composition both distribute over Summation, up to $\approx_1$. They are very useful in proof; distribution is a powerful device, and trace equivalence is correspondingly a very tractable theory. (It is, of course, the main equivalence studied in classical automata theory.)

The fact that trace equivalence equates a deadlocking agent to one which does not deadlock indicates that it is too weak in general. In some cases, however, it will be adequate. In particular we shall see in Proposition 6.2.6 that, for agents which are known to be confluent, it agrees completely with bisimilarity.

An interesting equivalence which lies between trace equivalence and bisimilarity is *failures equivalence*. It is important, because it appears to be the weakest equivalence which never equates a deadlocking agent with one which does not deadlock. We shall describe it briefly here; for a more detailed presentation the reader should consult Hoare [17]. Matthew Hennessy, with Rocco de Nicola, has given an elegant alternative characterisation of this equivalence called *testing equivalence*, in which two agents are equivalent just when they pass the same tests of a certain simple nature [26, 14].

4.6.5. DEFINITION. A *failure* is a pair (s, L), where $s \in \mathscr{L}^*$ is called a *trace* and $L \subseteq \mathscr{L}$ is a set of labels. The failure (s, L) is said to belong to an agent P if there exists P' such that

- $P \overset{s}{\Rightarrow} P'$;
- $P' \overset{\tau}{\nrightarrow}$, i.e. P' is stable;
- For all $l \in L$, $P'' \overset{l}{\nrightarrow}$.

Thus, if (s, L) belongs to P it means that P can perform s and thereby reach a state in which no further action (not even τ) is possible if the environment will only permit actions in L. As a simple example, consider $A \overset{\text{def}}{=} a.(b.c.\mathbf{0}+b.d.\mathbf{0})$. Then A has the following failures:

- (ε, L) for all L such that $a \notin L$;
- (a, L) for all L such that $b \notin L$;
- (ab, L) for all L such that $\{c, d\} \nsubseteq L$;
- (abc, L) and (abd, L) for all L.

Most important here are the failures of the form (ab, L); if the environment wishes to *ensure* that A can continue after b then it must permit both c and d, since it cannot predict which will be chosen.

4.6.6. DEFINITION. Two agents P and Q are said to be *failures-equivalent*, written $P \approx_{\mathrm{f}} Q$, if they possess exactly the same failures.

It is easy to see that failures equivalence implies trace equivalence, since a trace is part of a failure. That they are not the same equivalence can be seen from the following example: Let $P \equiv a.(b.\mathbf{0}+c.\mathbf{0})$ and $Q \equiv a.b.\mathbf{0}+a.c.\mathbf{0}$, where $b \neq c$. Then $P \approx_1 Q$; on the other hand $P \not\approx_{\mathrm{f}} Q$, since Q has the failure $(a, \{b\})$ but P does not.

The contrast with bisimilarity can best be seen by considering the agents $A \overset{\text{def}}{=} a.(b.c.\mathbf{0}+b.d.\mathbf{0})$ and $B \overset{\text{def}}{=} a.b.c.\mathbf{0}+a.b.d.\mathbf{0}$. It turns out that their failures are the same, so $A \approx_{\mathrm{f}} B$; but it is easy to check that $A \not\approx B$. Intuitively, we may say that bisimilarity is more sensitive than failures equivalence to the branching structure of an agent. Indeed, this is emphasised by the following law for failures equivalence:

$$\alpha.(\beta.P+\beta.Q) \approx_{\mathrm{f}} \alpha.\beta.P+\alpha.\beta.Q$$

which does not hold for bisimilarity. This is a constrained form of distributive law; it may be compared with the unconstrained distributive law, Proposition 4.6.4(1), which holds for trace equivalence but not for failures equivalence.

Failures equivalence has a pleasant algebra, and has also an associated partial order (based upon inclusion, rather than equality, of the failure-sets of two agents) which makes it very useful as a means of specification. Bisimilarity, on the other hand, benefits from the simple proof technique of finding bisimulations. Both have a good intuitive character. As we have presented them, it is unfortunately not quite true to say that

$P \approx Q$ implies $P \approx_f Q$, because failures equivalence is more discriminating with respect to divergent behaviour (infinite internal action). But it is possible to refine bisimilarity slightly in this respect, at the same time defining an associated partial order [36]; then the implication does indeed hold—both for the equivalences and for their associated partial orders. Moreover, bisimilarity—or its associated congruence—then appears to be the strongest equivalence based upon observable actions that one can reasonably demand; it is this characteristic which led us to give it primary place.

5. Analysis of bisimulation equivalences

5.1. Hierarchies of equivalence

Both strong congruence and bisimilarity are largest fixed points of a function $\mathscr{H}$ over binary relations, which can be defined for any Labelled Transition System $(\mathscr{P}, A, \{\xrightarrow{a}: a \in A\})$ as follows.

5.1.1. DEFINITION. For any binary relation $\mathscr{S}$ over $\mathscr{P}$, $\langle P, Q\rangle \in \mathscr{H}(\mathscr{S})$ iff, for all $a \in A$,
(i) whenever $P \xrightarrow{a} P'$ then, for some $Q', Q \xrightarrow{a} Q'$ and $P' \mathscr{S} Q'$;
(ii) whenever $Q \xrightarrow{a} Q'$ then, for some $P', P \xrightarrow{a} P'$ and $P' \mathscr{S} Q'$.

Here the set $\mathscr{P}$ need have no structure—except that provided by the action relations $\xrightarrow{a}$, If we take $\mathscr{P}$ to be the processes of our calculus, then strong congruence arises as the largest fixed point of $\mathscr{H}$ when $A = Act^*$ and $\xrightarrow{t}$ is as defined in Definition 4.1.1(1) for $t \in Act^*$; bisimilarity arises by taking $A = \mathscr{L}^*$ and using the relations $\xRightarrow{s} (s \in \mathscr{L}^*)$ as defined in Definition 4.1.1(3). Actually we used $\xrightarrow{\alpha}$ $(\alpha \in Act)$ in defining strong congruence, but it is easy to see that using $\xrightarrow{t}$ does not alter the congruence; similarly—as hinted by Proposition 4.2.9—the action relations $\xRightarrow{s}$ can be seen to yield bisimilarity. By choosing action relations indexed by sequences we get a more interesting hierarchy, defined for our arbitrary LTS as follows.

5.1.2. DEFINITION. The relations $\simeq_\kappa$ over $\mathscr{P}$, for each ordinal κ, are defined as follows:
(1) $\simeq_0 \; = \mathscr{P} \times \mathscr{P}$;
(2) $\simeq_{\kappa+1} = \mathscr{H}(\simeq_\kappa)$;
(3) $\simeq_\lambda = \bigcap_{\kappa<\lambda} \simeq_\kappa$, for a limit ordinal λ.

We use $\mathscr{O}$ for the ordinals, and define in general:

5.1.3. DEFINITION. $\simeq \; = \bigcap_{\kappa \in \mathscr{O}} \simeq_\kappa$.

The following is a straightforward, partly standard, result.

5.1.4. Proposition. (1) *If* $\lambda > \kappa$ *then* $\simeq_\lambda \subseteq \simeq_\kappa$.
(2) $\simeq_\kappa$ *and* $\simeq$ *are equivalence relations.*
(3) $\simeq$ *is the largest fixed point of* $\mathscr{H}$

Thus both strong congruence and bisimilarity of processes are limits of decreasing sequences of equivalence relations. The following results about $\simeq_1$ are interesting:

5.1.5. PROPOSITION. (1) *$P \sim_1 Q$ iff, for all $t \in Act^*$, ($P \xrightarrow{t}$ iff $Q \xrightarrow{t}$).*
(2) *$P \approx_1 Q$ iff, for all $s \in \mathscr{L}^*$, ($P \overset{s}{\Rightarrow}$ iff $Q \overset{s}{\Rightarrow}$).*

Thus, at the first level of the hierarchies which approximate strong congruence and bisimilarity, P and Q are equivalent just when they can perform the same sequences of *actions* in the first case, and of *visible actions* in the second case. These are the *strong* and *weak trace equivalences* mentioned at the end of Section 4. We already know that $\sim$ and $\approx$ are smaller than $\sim_1$ and $\approx_1$; now we shall see that the hierarchies are properly decreasing at least as far as $\omega + 1$.

5.1.6. PROPOSITION. *The sequence $\sim_0, \sim_1, \ldots, \sim_\omega$ is properly decreasing.*

PROOF (*outline*). For $i < \omega$, let P_i and Q_i be as follows:

$$P_0 \equiv b.\mathbf{0}, \qquad Q_0 \equiv c.\mathbf{0},$$
$$P_{i+1} \equiv a.(P_i + Q_i), \qquad Q_{i+1} \equiv a.P_i + a.Q_i,$$

where $b \neq c$.

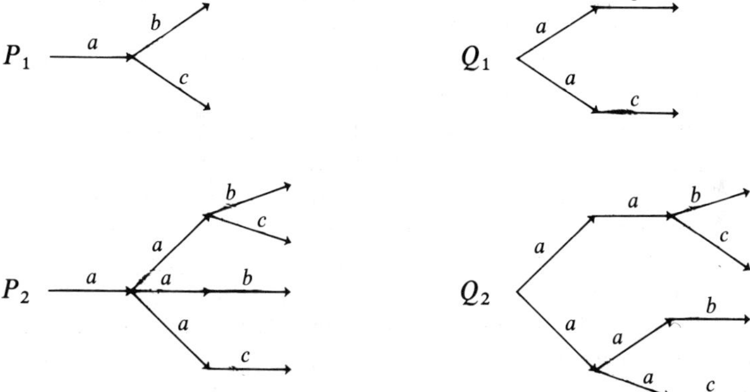

Then by induction it is easy to prove

$$P_i \sim_i Q_i \sim_i P_i + Q_i,$$
$$P_i \not\sim_{i+1} Q_i \not\sim_{i+1} P_i + Q_i \not\sim_{i+1} P_i$$

which implies the stated result. □

5.1.7. PROPOSITION. *The sequence $\approx_0, \approx_1, \ldots, \approx_\omega$ is properly decreasing.*

PROOF. The same proof suffices, since $\sim_\kappa$ and $\approx_\kappa$ coincide in the absence of τ-actions. □

We now show that the sequences decrease still further if we allow infinite sums.

5.1.8. PROPOSITION. *$\sim_{\omega+1} \neq \sim_\omega$ and $\approx_{\omega+1} \neq \approx_\omega$.*

PROOF. Using P_i and Q_i as above, let R_i and R_ω be as follows:

$$\begin{aligned}
R_0 &\equiv \textstyle\sum_{0\leqslant i<\omega} P_i,\\
R_1 &\equiv Q_0 + \textstyle\sum_{1\leqslant i<\omega} P_i,\\
R_2 &\equiv Q_0 + Q_1 + \textstyle\sum_{2\leqslant i<\omega} P_i,\\
&\dots\\
R_\omega &\equiv Q_0 + Q_1 + Q_2 + \cdots
\end{aligned}$$

Then set $R \equiv \sum_{i<\omega} . R_i$ and $R' \equiv R + a.R_\omega$:

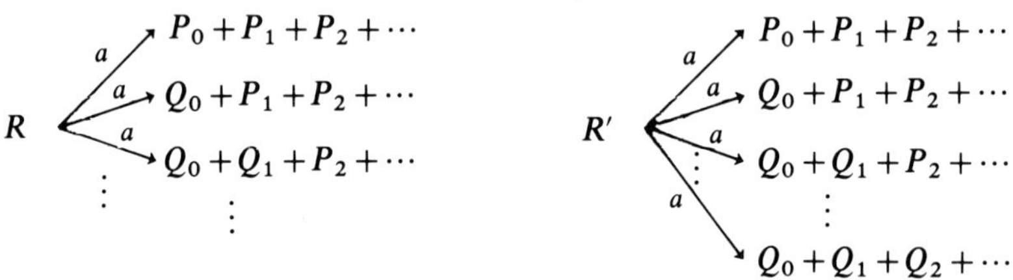

First, it follows from the analysis in Proposition 5.1.6 that $R_\omega \sim_i R_i$ but $R_\omega \not\sim_{i+1} R_i$. We now show that $R \sim_i R'$ for each $i < \omega$, and hence $R \sim_\omega R'$. For clearly all derivatives of R are exactly matched by R', and all derivatives of R' are exactly matched by R except $R' \xrightarrow{\alpha} R_\omega$. But we can choose $R \xrightarrow{\alpha} R_i$ to match this up to $\sim_i$, and this is enough to ensure $R \sim_i R'$.

On the other hand there is no a-successor of R which matches the a-successor R_ω of R' up to $\sim_\omega$, since $R_\omega \not\sim_\omega R_i$; hence $R \not\sim_{\omega+1} R'$.

We conclude that $\sim_{\omega+1} \neq \sim_\omega$, and also $\approx_{\omega+1} \neq \approx_\omega$ since $\sim_\kappa$ and $\approx_\kappa$ coincide in the absence of τ. □

It is interesting to note that this example has infinite branching but only finite paths. It can be converted to an example with only finite branching but infinite paths (involving τ), to show that $\approx_{\omega+1} \neq \approx_\omega$. But we shall show that $\sim_{\omega+1}$ and $\sim_\omega$, and also $\approx_{\omega+1}$ and $\approx_\omega$, coincide on finite processes.

5.2. *Logical characterization of the hierarchy*

The distinction between P_2 and Q_2 in Proposition 5.1.6 can be described by the following property enjoyed by Q_2 but not by P_2:

> an a-action is *possible* such that after any further a-action it is *necessary* that both a b-action and a c-action are *possible*. (5.1)

This triple alternation of *possibility* and *necessity* seems to be what is needed to express a property which will distinguish P_2 from Q_2, with respect to $\sim_3$. We therefore introduce the following logical language $\mathscr{PL}$ to characterise the members of our arbitrary LTS.

5.2.1. DEFINITION. $\mathscr{PL}$ is the smallest class of formulae such that
(1) if $F \in \mathscr{PL}$ then $\neg F \in \mathscr{PL}$;
(2) if $F_i \in \mathscr{PL}$ for each i in an indexing set I, then $\bigwedge_{i\in I} F_i \in \mathscr{PL}$;
(3) if $F \in \mathscr{PL}$ and $a \in A$, then $\langle a \rangle F \in \mathscr{PL}$.

$\mathscr{PL}$ is nonempty, since (2) ensures that the empty conjunction $\bigwedge_{i\in\emptyset} F_i$, which we write as TRUE, is present. We also define FALSE $\equiv \neg$TRUE, $\bigvee_{i\in I} F_i \equiv \neg \bigvee_{i\in I} \neg F_i$, and $[a] F \equiv \neg \langle a \rangle \neg F$. Property (5.1) will be formalised as

$$\langle a \rangle [a] (\langle b \rangle \text{TRUE} \wedge \langle c \rangle \text{TRUE}). \tag{5.2}$$

In fact, to ensure that this captures the meaning intended by (5.1), we interpret $\mathscr{PL}$ by the *satisfaction relation* $\models \subseteq \mathscr{P} \times \mathscr{PL}$ as follows.

5.2.2. DEFINITION. (1) $P \models \neg F$ iff not $P \models F$.
(2) $P \models \bigwedge_{i\in I} F_i$ iff, for all $i \in I$, $P \models F_i$.
(3) $P \models \langle a \rangle F$ iff, for some $P', P \xrightarrow{a} P' \models F$.

Thus for P_2 and Q_2 in Proposition 5.1.6 and the formula F at (5.2) we have $P_2 \not\models F$, $Q_2 \models F$.

5.2.3. DEFINITION. (1) depth$(\neg F)$ = depth(F).
(2) depth$(\bigwedge_{i\in I} F_i) = \sup_{i\in I}(F_i))$.
(3) depth$(\langle a \rangle F)$ = depth$(F) + 1$.

Then we define the levels of $\mathscr{PL}$ as follows.

5.2.4. DEFINITION. $\mathscr{PL}_\kappa = \{F\colon \text{depth}(F) \leqslant \kappa\}$ for $\kappa \in \mathcal{O}$

Now we are ready for the main result, that each level in the equivalence hierarchy is exactly characterised by the corresponding level in $\mathscr{PL}$.

5.2.5. PROPOSITION. (1) *$P \simeq_\kappa Q$ iff, for every $F \in \mathscr{PL}_\kappa$, ($P \models F$ iff $Q \models F$).*
(2) *$P \approx Q$ iff, for every $F \in \mathscr{PL}$, ($P \models F$ iff $Q \models F$).*

PROOF (*outline*). (2) follows directly from (1) since $\mathscr{PL} = \bigcap_\kappa \mathscr{PL}_\kappa$. The proof of (1) is by induction on κ. The crucial step is in proving (1) for $\kappa+1$; assuming $P \not\simeq_{\kappa+1} Q$ we have to find $F \in \mathscr{PL}_{\kappa+1}$ such that $P \models F$ and $Q \not\models F$. By assumption, there is w.l.o.g. an action $P \xrightarrow{a} P'$ such that whenever $Q \xrightarrow{a} Q'$ then $P' \not\simeq_\kappa Q'$. So let $\{Q_i\colon i \in I\}$ be the set of Q' such that $Q \xrightarrow{a} Q'$; then (with the help of negation) there exists an $F_i \in \mathscr{PL}_\kappa$ for each $i \in I$, such that $P' \models F_i$ and $Q_i \not\models F_i$. Take $F \equiv \langle a \rangle \bigwedge_{i\in I} F_i$; clearly, $P \models F$ and $Q \not\models F$.

The remainder of the proof, including the inductive step for limit ordinals, is routine. □

The only part of this proof which requires a possibly infinite conjunction of formulae is the part we have detailed. From this we learn that for finite processes, since the set $\{Q_i\colon i \in I\}$ in the proof must be finite, the finitary language $\mathscr{PL}^{\text{fin}}$ (i.e., only finite conjunctions) is enough. But every formula in this subset has finite depth, so $\mathscr{PL}_{\omega}^{\mathrm{fin}} = \mathscr{PL}^{\mathrm{fin}}$; hence we deduce the following proposition.

5.2.6. PROPOSITION. *On finite processes,* $\sim_{\omega}$ *coincides with* $\sim$, *and* $\approx_{\omega}$ *coincides with* $\approx$.

6. Confluent processes

6.1. Determinism

We would like to define a notion of determination for processes, and to identify a set of constructions which preserve determinism. A benefit from this would be a sub-language of processes which can be seen to be deterministic from their syntactic form, and which admit special proof techniques for establishing equivalence and other properties.

A natural first attempt at a definition of determinism might be as follows: P is deterministic if, whenever $P \overset{s}{\Rightarrow} P'$ and $P \overset{s}{\Rightarrow} P''$, then $P' \approx P''$. This would immediately yield that whenever $P \Rightarrow P'$, then $P \approx P'$; an advantage would be that, in checking whether some $\mathscr{S}$ is a bisimulation up to $\approx$, only a subset of the pairs in $\mathscr{S}$ would need to be tested.

We certainly hope that our eventual definition will imply the above property, but the property as it stands is too weak to be preserved by simple constructions. For let $P \equiv a.\mathbf{0} + b.\mathbf{0}$ and $Q \equiv \bar{a}.\mathbf{0}$; both clearly possess the property, but (if $a \neq b$) $(P|Q)\backslash a = \tau.\mathbf{0}$ does not. We therefore seek something stronger. Since we do wish to allow this kind of restricted composition, it appears that summation should only be admitted when each alternative action does not preclude the later occurrence of any of the others; this means that $P \equiv a.\mathbf{0} + b.\mathbf{0}$ should be excluded, but $a.b.\mathbf{0} + b.a.\mathbf{0}$ should be admitted.

6.2. Confluence

6.2.1. DEFINITION. The set $\mathscr{C}$ of *confluent processes* is the largest set closed under successor and such that, for all $P \in \mathscr{C}$ and derivatives P', P'' as shown, Q can be found to complete each of the following diagrams ($l \neq l'$):

$$\begin{array}{ccc} P & \overset{\tau}{\rightarrow} & P' \\ \Downarrow & & \Downarrow \\ P'' & \Rightarrow\approx & Q \end{array} \qquad \begin{array}{ccc} P & \overset{\tau}{\rightarrow} & P' \\ {\scriptstyle l}\Downarrow & & {\scriptstyle l}\Downarrow \\ P'' & \Rightarrow\approx & Q \end{array}$$

$$\begin{array}{ccc} P & \overset{l}{\rightarrow} & P' \\ {\scriptstyle l'}\Downarrow & & {\scriptstyle l'}\Downarrow \\ P'' & \overset{l}{\Rightarrow}\approx & Q \end{array} \qquad \begin{array}{ccc} P & \overset{l}{\rightarrow} & P' \\ {\scriptstyle l}\Downarrow & & \Downarrow \\ P'' & \Rightarrow\approx & Q \end{array}$$

To justify the definition, we note that there is a *largest* such set—namely the union of all sets satisfying the conditions. We present the definition in this form since it shows just how each (single) action of P should "commute up to $\approx$" with a sequence of other actions. Note however that the last diagram, involving two possibly different l-descendants, only requires common τ-descendants (up to $\approx$). There is another way of characterising confluence which involves the idea of the *excess* of $r \in \mathcal{L}^*$ over $s \in \mathcal{L}^*$; this excess is gained by working through r from left to right deleting any member of s–but respecting multiplicity of ocurrence. Thus the excess of $r = aba$ over $s = bacb$ is a, while the excess of s over r is cb.

6.2.2. DEFINITION. r/s, the *excess of r over s*, is defined as follows:

$$\varepsilon/s = \varepsilon, \qquad (lr)/s = \begin{cases} l(r/s) & \text{if } l \notin s, \\ r/(s/l) & \text{if } l \in s. \end{cases}$$

6.2.3. PROPOSITION. *P is confluent iff, for all r, s* $\in \mathcal{L}^*$, *the following diagram can be completed in the sense of* 6.2.1:

$$\begin{array}{ccc} P & \stackrel{r}{\Rightarrow} & P' \\ {\scriptstyle s}\Downarrow & & {\scriptstyle s/r}\Downarrow \\ P'' & \stackrel{r/s}{\Rightarrow}\approx & Q \end{array}$$

The essential properties of confluence are captured by three results relating it to bisimilarity. First, we wish to know that it is really a property of observation equivalence classes.

6.2.4. PROPOSITION. *If* $P \approx Q$ *and P is confluent then so is Q.*

Next, we wish to know that silent actions have "no effect" up to $\approx$ for confluent processes.

6.2.5. PROPOSITION. *If P is confluent and* $P \stackrel{\tau}{\to} P'$ *then* $P \approx P'$.

Finally, we find that bisimilarity of confluent processes reduces to accepting the same sequences of visible actions.

6.2.6. PROPOSITION. *If P and Q are confluent then* $P \approx Q$ *iff* $P \approx_1 Q$

These results indicate that confluent processes are very well behaved.

6.3. *Constructions preserving confluence*

Let us first see that Summation and Composition fail to preserve confluence in general. For the first, clearly $a.\mathbf{0}$ and $b.\mathbf{0}$ are confluent, but $a.\mathbf{0} + b.\mathbf{0}$ is not (see the third diagram of Definition of 6.2.1). For the second, we need only observe that $a.\mathbf{0}|\bar{a}.\mathbf{0}$ is not confluent, since it is equal to $a.\bar{a}.\mathbf{0} + \bar{a}.a.\mathbf{0} + \tau.\mathbf{0}$. The other constructors do preserve confluence, so we look for constrained forms of Summation and Composition.

We first observe that $P|Q$ may not be confluent, even if P and Q are, if $\mathcal{L}(P) \cap \mathcal{L}(Q) \neq \emptyset$; an example is $a.b.\mathbf{0}|a.\mathbf{0}$, for which we can check that the fourth condition of Definition 6.2.1 fails. Next, $P|Q$ may not be confluent if $\mathscr{L}(P)\cap\overline{\mathscr{L}(Q)}\neq\emptyset$; the example in the previous paragraph suffices. We are led to the following definition.

6.3.1. DEFINITION. $P\parallel_L Q \stackrel{\text{def}}{=} (P|Q)\backslash L$ is a *Confluent Composition* if $\mathscr{L}(P)\cap\mathscr{L}(Q)=\emptyset$ and $\mathscr{L}(P)\cap\overline{\mathscr{L}(Q)}\subseteq L\cup\bar{L}$.

For example, $a.b.\mathbf{0}\parallel_a c.\bar{a}.\mathbf{0}$ is a Confluent Composition, and the reader may easily verify that it is confluent. We now turn to a mixture of Prefix and Summation.

6.3.2. DEFINITION. $(\alpha_1|\alpha_2|\cdots|\alpha_n).P \stackrel{\text{def}}{=} \Sigma_{1\leqslant i\leqslant n}\alpha_i.(\alpha_1|\cdots|\alpha_{i-1}|\alpha_{i+1}|\cdots|\alpha_n).P$ is a *Confluent Sum.*

The effect of this construction is to perform $\alpha_1,\ldots,\alpha_n$ in any order before entering P. A simple example is $(a|b).\mathbf{0}\equiv a.b.\mathbf{0}+b.a.\mathbf{0}$. Note that the Prefix constructor $\alpha.P$ is a special case with $n=1$.

6.3.3. DEFINITION. E is in *confluent form* if it is built using only Confluent Sum, Confluent Composition, Restriction, Relabelling and Recursion.

We now come to the main result of this section.

6.3.4. PROPOSITION. *Every process in confluent form is confluent.*

Space limitation forbids us to give the proof of this result, which is rather detailed especially in the case of Confluent Composition. Its importance is that many practical examples of systems can be built in confluent form, and are thus amenable to especially simple analysis based upon Propositions 6.2.5 and 6.2.6. There are indeed weaker, i.e. more general, notions of confluent form which cover a wider class of applications.

In general, however, we are forced to admit that even if a complete system is confluent, and may be proved so by ad hoc analysis, it is often impossible or unnatural to build it in confluent form. One can find other examples of this general phenomenon in programming; for instance, a whole program may be always-terminating, but it may be most inconvenient to build it by adhering to a discipline—such as primitive recursion—which makes this property manifest from the form of the program.

7. Sources

In this chapter we have accumulated the main results of a particular theory of communicating systems, and we have omitted or only outlined several of the proofs. The main ideas were first presented in [23]. The most important new ingredient after this was the notion of bisimulation discovered by David Park [29]; though the observation equivalence of [23] coincides with that based on bisimulation in most interesting cases, the conceptual framework of bisimulation is distinctly superior, and its proof technique is distinctly more tractable.

The notion of infinite Summation, and its use in deriving the extended language of Section 2.6, was introduced in [24], which also contains some more detailed proofs. The completeness results for finite processes, Propositions 4.5.2 and 4.5.3, were first reported in [15]. Matthew Hennessy provided the idea for the simple proof of Proposition 4.5.3 which we have outlined; Proposition 4.3.11 is also due to him. The hierarchies of equivalence and the logical characterisations of Section 5 were also first treated in [15]. The study of confluence was begun in [23], greatly improved here with the help of Park's bisimulation.

Note added in proof

In the author's book [25], $\xrightarrow{\alpha}$ was used instead of $\stackrel{\alpha}{\Rightarrow}$ in the antecedent of each clause in Definition 4.2.1: (i) whenever $P\xrightarrow{\alpha}P' \ldots$; (ii) whenever $Q\xrightarrow{\alpha}Q' \ldots$. This yields a different function $\mathcal{G}$, but makes no difference to the notion of weak bisimulation. The form with $\xrightarrow{\alpha}$ is better for some practical purposes, but needs more care in handling the notion of bisimulation up to $\approx$ (here Definition 4.2.13). In fact, the corresponding Definition 5.8 of [25] is wrong, because it fails to support Proposition 5.6 of [25] (which corresponds to Proposition 4.2.15 here). The book will be corrected in reprinting.

References

[1] ABRAMSKY, S., Observation equivalence as a testing equivalence, *Theoret. Comput. Sci.* **53** (1987) 225–241.

[2] AUSTRY, D. and G. BOUDOL, Algèbre de processus et synchronisation, *Theoret. Comput. Sci.* **30**(1) (1984) 91–131.

[3] BAETEN, J.C.M., *Process Algebra* (Kluwer, Deventer, 1986) (in Dutch; English translation in preparation).

[4] BARRINGER, H., R. KUIPER and A. PNUELI, Now you may compose temporal logic specifications, in: *Proc 16th Ann. ACM Symp. on Theory of Computing* (1984) 51–63.

[5] BERGSTRA, J.A., and J.W. KLOP, Algebra of communicating processes with abstraction, *Theoret. Comput. Sci.* **37**(1) (1985) 77–121.

[6] BOUDOL, G. and I. CASTELLANI, On the semantics of concurrency: partial orders and transition systems, in: *Proc. TAPSOFT'87*, Lecture Notes in Computer Science, Vol. 249 (Springer, Berlin, 1987) 123–137.

[7] BROOKES, S.D., C.A.R. HOARE and A.W. ROSCOE, A theory of communication sequential processes, *J. ACM* **31**(3) (1984) 560–599.

[8] BROY, M., Fixed point theory for communication and concurrency: in: D. Bjørner, ed., *Formal Description of Programming Concepts II* (North-Holland, Amsterdam, 1983) 125–148.

[9] CAMPBELL, R. and A. HABERMANN, The specification of process synchronisation by path expressions, in: *Proc. Internat. Symp. on Operating Systems*, Lecture Notes in Computer Science, Vol. 16 (Springer, Berlin, 1974) 89–102.

[10] CHANDY, K.M. and J. MISRA, *Parallel Program Design: a Foundation* (Addison-Wesley, Reading, MA, 1988).

[11] CLARKE, E.M., E.A. EMERSON and A.P. SISTLA, Automatic verification of finite-state concurrent systems using temporal logic specifications, *ACM TOPLAS* **8**(2) (1986) 244–263.

[12] DE NICOLA, R., Testing equivalence and fully abstract models for communicating processes Ph.D. Thesis, Report CST-36-85, Computer Science Dept., Univ. of Edinburgh, 1985.

[13] Hennessy, M., Acceptance trees, *J. ACM* **32**(4) (1985 896–928.

[14] Hennessy, M., *Algebraic Theory of Processes* (MIT Press, Cambridge, MA, 1988).

[15] Hennessy, M. and R. Milner, Algebraic laws for nondeterminism and concurrency, *J. ACM* **32**(1) (1985) 137–161.

[16] Hoare, C.A.R., Communicating sequential processes, *Comm. ACM* **21**(8) (1978) 666–677.

[17] Hoare, C.A.R., *Communicating Sequential Processes* (Prentice-Hall, Englewood Cliffs, NJ, 1985).

[18] Janicki, R. and P.E. Lauer, *Specification and Analysis of Concurrent Systems: the COSY Approach*, EATCS Monographs on Theoretical Computer Science (Springer, Berlin, to appear).

[19] Jones, C.B., Specification and design of (parallel) programs, in: *Proc. IFIP 9th World Computer Congr. on Information Processing* (1983) 321–332.

[20] Lamport, L., Specifying concurrent program modules, *ACM TOPLAS* **6**(2) (1983) 190–222.

[21] Lauer, P.E. and R.H. Campbell, Formal semantics of a class of high-level primitives for co-ordinating concurrent processes, *Acta Inform.* **5** (1975) 297–332.

[22] Milne, G., Circal and the representation of communication, concurrency and time, *ACM TOPLAS* **7** (1985) 290–298.

[23] Milner, R., *A Calculus of Communicating Systems*, Lecture Notes in Computer Science, Vol. 92 (Springer, Berlin, 1980).

[24] Milner, R., Calculi for synchrony and asynchrony, *Theoret. Comput. Sci.* **25** (1983) 267–310.

[25] Milner, R., *Communication and Concurrency* (Prentice-Hall, Englewood Cliffs, NJ, 1989).

[26] Nicola, R. de and M. Hennessy, Testing equivalences for processes, *Theoret. Comput. Sci.* **34**(1) (1984) 83–134.

[27] Nielsen, M., G. Plotkin and G. Winskel, Petri nets, event structures and domains, *Theoret. Comput. Sci.* **13** (1981) 85–108.

[28] Owicki, S. and D. Gries, An axiomatic proof technique for parallel programs I, *Acta Inform.* **6**(1) (1976) 319–340.

[29] Park, D., Concurrency and automata on infinite sequences, in: *Theoretic Computer Science*, Lecture Notes in Computer Science, Vol. 104 (Springer, Berlin, 1981) 167–183.

[30] Plotkin, G., A structural approach to operational semantics, Report DAIMI FN-19, Computer Science Dept., Aarhus Univ., 1981.

[31] Pnueli, A., Temporal logic, in: C.A.R. Hoare, ed., *Proc. Univ. of Texas Year of Programming, Vol 1, Concurrent Programming* (Addison-Wesley, Reading, MA, 1989).

[32] Reisig, W., *Petri Nets*, EATCS Monographs on Theoretical Computer Science, Vol. 4 (Springer, Berlin, 1983).

[33] Shannon, C.E. and J. McCarthy, eds., *Automata Studies* (Princeton Univ. Press, Princeton, NJ, 1956).

[34] Simone, R. de, Higher-level synchronising devices in MEIJE-SCCS, *Theoret. Comput. Sci.* **37**(3) (1985) 245–267.

[35] Stirling, C.P., A generalisation of Owicki–Gries' Hoare logic for a concurrent while language, *Theoret. Comput. Sci.* **58** (1988) 347–360.

[36] Walker, D.J., Bisimulation and divergence, *Inform. and Comput.*, to appear.

SUBJECT INDEX

abort 1188, 1194
  of completion procedure 296, 298, 301
absence of individual starvation 1051
absence of unsolicited response 1052
abstract concurrent program 1019
abstract data type 686
abstract syntax (*see also* syntax, abstract) 1002
abstraction 326
  behavioral 748
  λ- 752
  lambda 369, 375
  observational 748
  type 432
abstraction function 864
abstractness, full 586
abstractor implementation 764
AC (*see* associative-commutative axioms)
acceptable meaning function 417
acceptable structure 821
acceptance condition (*see also* recognizable) 136
accepting pair 149
access methods 1075
AC-completion (*see* completion, extended, associative-commutative)
Ackermann's function 277
ACP 1205
ACT ONE 778
ACT TWO 779
action 1168, 1207
  silent 1207
  visible 1223
action relation 1207
action semantics 628
acyclic structure 1012
adequacy of rules 256
adherence 163
admissible logical relation 418
adversary 1180
agreement 1183
  approximate 1186
  Byzantine 1184
AL (*see also* Algorithmic Logic) 832
algebra 205, 260, 582, 668
  absolutely free (*see* algebra, term)
  arithmetical 690
  combinatory 388
  computable 266, 688
  constructor 264, 266
  continuous 464, 636, 707
  co-semicomputable 688
  cylindric 1085
  Dynamic 811
  error 702
  finitely presented 305, 1097
  free 262, 279
  ground term 262, 263
  heterogeneous (*see* algebra, many-sorted)
  hyperarithmetical 690
  initial 262, 264
  initial ~ semantics 589, 627
  many-sorted 205, 305, 463
  normal-form 263, 264
  number 688
  ordered 464
  order-sorted 305, 709
  partial 705
  process 710
  quotient 252, 262, 279
  recursive number 688
  relational 1081
  semicomputable 688
  Σ- 682
  sub 685
  term 262, 279, 684
  typed lambda 388
  unit 684
algebraic calculus 1203
algebraic cpo 641
algebraic (logical) relation 428
algebraic ω-language 162
algebraic power series 126
algebraic semantics (*see also* semantics) 260
  of concurrent processes 1201
algebraic series 45
algebraic specification 697
algebraic theory 489
algebraic tree 477
algorithm
  bakery 1179

distributed 1159
dynamic 1189, 1191
Knuth, Morris and Pratt's 33
Markov 114, 245, 308
McNaughton and Yamada's 47
Moore's 45
Morris and Pratt's 32
polynomial 227
randomized 1180
shared variable 1178
static 1189, 1192
synchronizer 1192
unification 282, 501
algorithm(s)
axiomatic description of 1169
constructive description of 1169
deriving (distributed) 1176
Algorithmic Logic (AL) 832
$Alg(\Sigma)$ 682
$Alg(\Sigma, E)$ 692
aliasing 946
almost always 1003
almost everywhere 1003
alpha conversion (*see* literal similarity)
α-equivalent 377
alphabet 4
commutative 47
partially commutative 48
alphabetic morphism 83
alternating automaton 146, 177
Alternating Bit Protocol 1193
always 1003
ambiguity 65, 126
degree of 126
inherent 65
(in rational expressions) 18
intrinsic 87
strong 87
anomaly, update 1119
aperiodic set 22
application 326, 369, 375
of equation 252
of rule 252
type 433
applicative structure 410
extensional 384
in a category 411
Kripke 410
typed 382
apply(f, x) 646
approximate agreement 1186
arc assertion 1015
Argus 1194, 1196
arithmetic
abacus 904
Peano 909, 913, 917, 920, 921
Presburger 48, 282, 908
arithmetical 690
arithmetical completeness 822
arithmetical structure 821
arity of operator 249
Armstrong relation 1095
array assignment 813, 815
arrow (of a category) 397
ASF 779
ASL 775
assertion 858
arc 1015
inductive 792
intermittent 834, 1051
assertional reasoning 1171
assignment 373, 684, 852, 897
array 813, 815
nondeterministic 791, 813, 816, 830
random 813
simple 812, 815
type 373
assignment axiom 892, 894, 931
assignment of value 260
associated Σ-congruence 686
associative-commutative axioms 257, 259, 265, 282, 299
associative-commutative completion (*see also* completion) 299, 305
associativity 282
assumption
closed world 532, 1107
pure universal relation 1110
weak universal relation 1110
asynchronous model 1160
at predicate 1172
atom 497
generalized 537
atomic 1183, 1194
atomic action 949, 1194
atomic formula 691
ATP 1205
attribute 1079
authentication 1187
automata theory 1, 809
automaton 1045
alternating 146, 177
asynchronous 48
Büchi 136, 168, 1046
complete 5
counter-free 4
decomposition of 49
definite 51

deterministic 5
finite 1, 5
∀- 161
I/O 1169
local 28
minimal 9
Muller 148, 169
multitape 46
n-local universal 27
nondeterministic (*see* automaton, finite)
normalized 4
on infinite object 1045
on infinite string 1045
on infinite tree 1045
on ω-words 136
probabilistic 45
Rabin 149, 169
random access 50
reversible 50
Streett 169
subtree 180
symmetric pairs 1026
synchronizing 46
tree (*see also* tree automaton) 167
two-way 46
unambiguous 47
with a distance function 21
with multiplicities 45
auxiliary definition 592
axiom 890, 891
assignment 892, 894, 931
associative-commutative 257, 259, 265, 282, 299
equational 256, 261
for *PDL* 797
skip 892, 894, 940
axiom schema 890
axioms (in specification) 699
axiomatic description of algorithms 1169
axiomatic semantics 828, 890
axiomatic specification 697
axiomatizability (*see also* deductive systems) 1021
equational 261

bag (*see* multiset)
bakery algorithm 1179
Barcan's formula 998
base type (in $\lambda^{\rightarrow}$) 374
basic liveness property 1050
basic modality 1027
basic specification 740
basis 351, 887
finite (*see* equational theory, finitely based)

before 1003
behavior 585, 748, 1168
behavioral abstraction 748
behavioral implementation 764
behaviorally equivalent 735
β-conversion 327
bi-infinite word 147
binary relation 251, 266
binding 326
dynamic 357
static 358
binding occurrence 376
Birkhoff's Completeness Theorem 262, 263
bisimilarity 1205
bisimulation
strong 1216
weak 1224
blocked execution 877
Boolean 331
Boolean circuit 67
Boolean ring 258, 282, 304
Boolean semiring 45
Borel hierarchy 153
bottom 341
bottom-up evaluation 1128
bottom-up tree automaton 167
bound variable 375
bounded buffer 1166, 1181, 1182
bounded waiting 1178
boundedness (program, uniform) 1133
box operator 794
Boyce–Codd normal form (BC-NF) 1120
branching time 833, 999
Branching-Time Temporal Logic 1003, 1012
breadth-first search 1192
breadth-first spanning tree 1190
breaking symmetry 1181
British Museum method 282
broadcast 1189, 1190
Büchi automaton 136, 1046
Büchi tree automaton 168
Büchi-recognizable 136, 169
Byzantine agreement 1184
Byzantine failure 1161
Byzantine generals 1184

calculus, relational 1085
Calculus of Communicating Systems (CCS) 156, 1176, 1204
call-by-name 470
call-by-need 475
call-by-value 470
Camelot 1194, 1196
cancellation law 23

Cantor topology 152
capture rule framework 1137
carrier (*see also* universe) 582, 814
Cartesian closed category 396, 403
Cartesian closure 401, 646, 673
Cartesian product 261, 342
category 209, 308
  Cartesian closed 396, 403
  definition of a 396
  well-pointed 409
category theory 261
causal independence 1203
Cayley graph 97
ccc-relation (between Cartesian closed categories) 429
CCS (Calculus of Communicating Systems) 156, 1176, 1204
chain 178
chain argument 1185
chain logic 184
chain rule 1131
changing networks 1189, 1193
channel, physical 1193
characteristic series 107
characterization problem 922, 929
chase procedure 1101
checkpoint 1192
choice, nondeterministic 794
Chomsky hierarchy 110
Chomsky normal form 65
Church's λ-notation 647
Church's Thesis 266, 924
Church–Rosser property 324, 339, 379
  of binary relation 266
  of rewrite system (*see* confluence of rewriting)
  weak 424
CIP-L 780
circuit
  Boolean 67
  sequential 3
class of interpretations 475
  algebraic 480
  equational 479
Classical Modal Logic 1064
class-rewriting system 257, 264, 278, 289, 291
clausal form 498
clause 304, 497
  definite 498
  general program 498
  Horn 306, 692
cleanly embedded 1034
CLEAR 770
cliff, in proof 289, 289, 303
clock 1160
  logical 1167, 1180
clock synchronization 1183, 1186
closed term 328
closed world assumption 532, 1107
closure 1031
  Cartesian 401, 646, 673
  congruence 252, 301
  finitary 664
  Fischer–Ladner 802
  reflexive 251
  reflexive-transitive 251, 253
  rewrite 252
  symmetric 251
  (topological) 153
  transitive 251, 850, 1089
closure ordinal 518
coalesced sum (of cpo's) 650
Cobham's theorem 39
Codd table 1141
code 47
code, semaphore 51
codomain (of a morphism) 397
coercion 609
Coffee Can problem 245, 253, 257, 260
coherence 268
COLD 780
colimit 231
combinator 328, 387
  fixed point 329
  serial 359
  super 359
combinatory algebra 388
combinatory logic 245, 256, 288
combinatory model 387
combinatory term 356
command 852, 932
  asserted, 923
commit 1188, 1196
common ancestor relation 267
common descendent relation 267
common knowledge 1066
communicating processes 1203
Communicating Sequential Processes (CSP) 968, 1165, 1204
communication, multiway 1215
communication graph 1160
commutation
  of relations 278
  quasi-~ of relations 278
commutativity 257, 259, 283, 289, 291, 299
compact element (of a cpo) 641
compactness 799, 818
Compactness theorem 551
compatibility 269

compatible 338
compatible graph property 237
complement (of a relation) 1087
complementary view 1143
complete 1040
  observably 749
  sufficiently 306, 722
complete lattice 341, 850
complete partial order (cpo) 372, 393–394, 464, 636
  directed subset of 393
complete partially ordered set (*see also* cpo) 372, 393–394, 464, 636
completeness
  arithmetical 822, 970, 971
  computational (of logic programs) 523
  deductive 804
  for $\lambda^{\rightarrow}$ 382
  infinitary 822
  of equational laws 1230
  of *PDL* 804
  of rewrite system 256
  of SLD-resolution 519
  of SLD resolution, strong 521
  of the negation-as-failure rule 544
  of unifiers 282
  relative 902, 915, 917, 920
  semantical 868, 871, 885
  with empty types 386
  without empty types 385
completion
  abstract 292
  extended 298, 303
  extended, associative-commutative 299, 304, 305
  extended, commutative 299
  ordered 300, 305, 307
  unfailing (*see* completion, ordered)
  see also completion procedure
completion of a program 537
completion procedure 247, 292
  abortion of 296, 298, 301
  correctness 298
  soundness 293
completion sequence 293, 297
  failure 295, 298
  fair 297, 300, 302
  infinite 293
  success 293, 298
complex object 1138
complexity 1021
  data 1088
  expression 1089
  message 1163
  of *DL* 817
  of *PDL* 805
  spectral 818
  time 1163
component 853
composed
  horizontally 760
  vertically 760
compositeness, criterion for critical pairs 288
composition 332, 1207
  confluent 1240
  of substitutions 250
  parallel 1205
compositional 999
compositional semantics 586
compositionality 793
computability 523
computable (*see also* semicomputable) 688
computable function (on a domain) 643
computable function on modest sets 389
computable query 1122
computation
  concurrent 1017
  diffusing 1191
  domain of 811, 814
  terminated 468
computation rule 469
computation sequence 796, 1001
computation structure 685
  Σ- 685
computation tree logic 181
computed answer substitution 504
concatenation 851
concretization, function 864
concurrency 1017, 1048, 1049, 1203
concurrency control 1194
  database 1194
concurrent computation 1017
concurrent processes 1201
concurrent program 160
  abstract 1019
  correctness 1049
concurrent reading and writing 1179, 1183
concurrent system 1017, 1019, 1048
  finite state 1042, 1060
conditional equational specification 697
conditional rewriting 306
conditional Σ-equations 692
conditional table 1141
cone (of context-free languages) 84
  principal 84
  rational 84
  substitution-closed 85
configuration 854

forbidden 222
confluence 379
local 424
of binary relations 267
of binary relations, local 267, 286, 290, 307
of binary relations, strong 267
(of processes) 1238
of rewriting 255, 256, 258, 266, 285
of rewriting, strong 288
confluent composition 1240
confluent form 1240
confluent grammar 236
confluent sum 1240
congruence 139, 221, 251, 262, 338
associated Σ- 686
co-recursively enumerable 689
finite 221
locally finite 221
observation 1225
permutative 257, 302
recursive 689
recursively enumerable 689
Σ- 686
strong 1217
congruence class (*see also* rewrite system, congruence class) 252, 257, 285, 295
finiteness 282, 286
congruence closure (*see* closure, congruence)
conjecture, Ehrenfeucht 120
connectedness criterion for critical pairs 288
connectivity 1187
consensus 1183
conservative extension 295, 747
consistency 797, 821
of specification 264
proof by (*see* proof by consistency)
consistent 513
constant 249
constant symbol (in $\lambda^{\rightarrow}$) 374
construction 742
constructive description of algorithms 1169
constructor 249, 264, 719
dynamic 1217
free 264, 265
static 1217
constructor implementation 764
contained 1034
containment 1082
contention problems 1166
context 250, 251
context-free grammar 62, 162, 215
context-free graph 182, 238
context-free graph-grammar 236
context-free group 96
context-free language 45, 62, 110, 215
deterministic 67
inherently ambiguous 65
context-free ω-language 162
context-free set of graphs 211
context-free set of hypergraphs 211
contexts 8
context-sensitive language 110
continuation 612
continuation semantics 612
continuous algebra 658
continuous function 394, 590
(map) 341
of several arguments 394
(on a cpo) 395, 636
(on a cpo), computable 643
continuous function hierarchy 395
continuous functional 1065
contradiction 305
control point 863
control structure 461
convergence of rewrite system (*see* rewrite system, convergent)
converse 809
conversion 326
α- 329
β- 327
convertible 328
cooperation problems 1166
coordinatization 688
co-recursively enumerable 689
coroutine 948
correct answer substitution 520
correctness 577
of concurrent programs 161, 1049
partial 795, 817
partial 858, 877, 884, 892, 916, 934, 937, 953
relational partial 873
total 817, 877, 970, 972
correctness formula 889, 898
co-semicomputable 688
COSY 1204
counting evaluation 1137
counting monadic second-order logic 237
cpo (*see also* complete partially ordered set) 636
algebraic 641
bifinite 660
truth value 636, 637
criterion for critical pairs (*see* critical pair, criterion)
critical pair 286, 292, 293
criterion 288, 290, 298, 303
persisting 298

Critical Pair Lemma 286, 288, 290, 298, 307, 309
  Extended 289, 300
  Ordered 290, 302
critical peak (*see* peak in proof, critical)
critical section 1172
cross product (*see* Cartesian product)
cross-section theorem 46
CSP (*see also* Communicating Sequential Processes) 1165, 1191, 1204
curry(f) 646
Curry–Howard isomorphism 439
Curry type inference rule 444
Curry typing 444
currying 327
cylindric algebra 1085
cylindric lattice 1085

data complexity 1088
data independence 1053
data model
  deductive 1107
  object-oriented 1077
  semantic 1077
  universal relation 1109
  value-oriented 1077
data type 685
  abstract 686
database 1079
  deductive 567
  distributed 1183, 1188
  extensional 1127
  incomplete information 1140
  intensional 1127
  unrestricted 1080
database concurrency control 1076, 1194
database logic program 1127
database management system (DBMS) 1075
database recovery 1076
database scheme 1079
  independent 1117
  normal form 1119
database scheme design 1116
database theory 1075
  relational 1075
database update 1142
Datalog 1127
$\text{Datalog}^{\neg}$ 1127
$\text{Datalog}^{\neq}$ 1129
dawg 10
DDL 817
deadlock 1050
  false 1191
deadlock detection 1191
deadlock freedom 877
decidability 890, 904, 906, 911
decidability of ground reducibility (*see* reducibility, ground)
decidability of validity (*see also* word problem) 247, 255, 259, 293, 296
decidable theory 146, 180, 202
decision procedure 1030
  for validity 247, 255, 259, 293, 296, 301
decomposition 1083
  embedded 1099
  full 1083
  lossless 1083
  lossy 1083
deduction 247, 304
deductive completeness 804
deductive data model 1107
deductive database 567
deductive system 891
  for *DL* 820
  for Temporal Logic 1040
definability 523
definable graph 234
definable hypergraph 234
definable property 198
definable set 198
defined function 264
δ-rule 347, 348
demodulation (*see* rewriting)
denotation 577
denotational semantics 341, 575
dense 40
dependency 1076
  contained cover 1093
  cover 1093
  embedded 1113
  embedded implicational 1113
  embedded join 1099
  embedded multivalued 1099
  embedded template 1113
  finite implication 1093
  full 1113
  full implicational 1113
  full template 1114
  functional 1084
  implication 1093
  inclusion 1114
  join 1099
  minimum cover 1096
  multivalued 1099
  partition 1098
  preservation 1118
  projected-join 1114
  rule 1114

unary inclusion 1099
dependency axiomatization 1093
dependency basis 1103
dependency graph 551
depth-first spanning tree 1190
derivability relation 253
derivable (*see also* ω-derivable from E) 693
derivation 252, 253, 503
leftmost 62
SLD- 503
SLDNF- 497
(with a context-free grammar) 62
derivation tree 62, 110, 1131, 1209
derivative (of a regular expression) 47
derive from 743
derived operation 206
deriving (distributed) algorithms 1176
determinacy (of games) 173
determinism (of processes) 1238
deterministic Büchi automaton 147
deterministic finite automaton (*see* automaton)
deterministic pushdown automaton (dpda) 67
deterministic tree automaton 167
Dewey decimal notation 250
diagonal 850
diagram 745
diamond lemma 267
diamond operator 794
diamond property 379
difference (of relations) 1080
diffusing computation 1191
digital signature 1162
Dining Philosophers 1180
Diophantine equation, undecidability of solvability 282, 284, 285
directed 341
directed equation (*see* rewrite rule)
directed set 479, 636
directed subset (of a cpo) 393–394
directed-complete predicate 430
discrete convergence 123
disjunctive normal form example 270, 276
distributed algorithm 1159
distributed computing 1159
distributed consensus 1183
distributed database 1183, 1188
divergence of completion procedure (*see* completion sequence, infinite)
divergence-closed 825
divisibility order (*see* ordering on terms, simplification)
DL (*see also* Dynamic Logic) 811
domain 589, 641
bifinite 660
bounded complete 644
effectively presented 643
of a morphism 397
semantic 641
domain equation 591
recursive 663
solving a 665
domain-independent formula 1087
domain of computation 811, 814
domain part (of proof system) 1054
domain-key normal form 1121
domain-ordered set of values 393
do-od program 1016
dot-depth 50, 158
dpda (*see* deterministic pushdown automaton)
DWP 812
Dynamic Algebra 811
dynamic algorithm 1189, 1191
dynamic laws 1228
Dynamic Logic (*DL*) 811
Nonstandard 830

edge 197
label of 5
of an automaton 5
effective definitional schemes 833
effective presentation (of a domain) 643
Ehrenfeucht conjecture 120
election (of a leader) 1190
elementary formula 1032
elementary step 865
embedded decomposition 1099
embedded dependency 1113
embedded implicational dependency (eid) 1113
embedded join dependency (ejd) 1099
embedded multivalued dependency (emvd) 1099
embedded template dependency (etd) 1113
embedding 681
homeomorphic 274
embedding-projection pair 642
emptiness problem
for Büchi automata 138
for Büchi tree automata 171
for Rabin tree automata 172
empty word 4, 106
enabled 1170
encompassment ordering (*see* ordering on terms, encompassment)
endogenous 792, 833
endogenous logic 999
enrichment 745
entropic groupoid 291, 302
enumerable

co-recursively 689
recursively 689
environment 383, 600
logically related 417
satisfying type assignment 383
environment model 384
equality axioms 545
free 537
equality set (of two morphisms) 119
equation 251, 691
application of 252
between lambda terms 376, 378
conditional 306
conditional Σ- 692
Diophantine (*see* Diophantine equation)
directed (*see* rewrite rule)
domain (*see* domain equation)
extended 386
orientable 293
permutative (*see* congruence, permutative)
persisting 293, 298
recursive domain 663
satisfiability (*see* satisfiability of equation)
semantic 587
Σ- 691
solution of 261, 280, 282
unorientable 295, 298
equation width 1126
equational axiom 256, 261
equational axiomatizability 261
equational class of interpretations 479
equational laws, completeness of 1230
equational model (*see* algebra)
equational proof (*see* proof, equational)
equational set 214
equational soundness 413
equational specification 207, 697
equational specification
conditional 697
equational step 252
equational theorem-proving 247
equational theory 247, 251, 255, 256, 263, 279, 292, 692
finitely based 256, 257, 261
equations
polynomial system of 214
regular system of 232
system of 465
system of (defining a context-free language) 66
equivalence 1082
failures 1233
(of grammars) 62
rational 83
strong failures 1223
strong trace 1223
testing 1233
unrestricted 1083
weak trace 1232
equivalence class 420
finiteness (*see* congruence class, finiteness)
equivalence forever 1142
equivalence problem (for parallel rewriting systems) 121
equivalence relation 381
logical partial 419
(over processes) 1215
partial 391, 421
equivalent, behaviorally 735
equivalent classes of interpretations 480
equivalent program schemes 465
erasure function 443
error 599
error algebra 702
error term 266, 306
evaluation 1135
bottom-up 1136
counting 1137
magic-counting 1137
magic-set 1136
naive bottom-up 1128
top-down 1136
event structure 1203
eventuality 1033, 1049
eventually 1002
exclusive lock 1195
exclusive-or normal form 258, 304
execution
blocked 877
program 843
sequence 792, 796
existential type 441
exogenous 792
exogenous logic 999
expansion law 1217
export 720, 743
expression 499, 852, 890, 948
graph 203
(in a process language) 1207
ω-regular 1025
process 1207
project-join 1083
rational (*see* rational expression)
regular (*see also* rational expression) 45, 1204
relational algebra 1081
relational calculus 1086
expression complexity 1089
expressive structure 821

expressiveness 916, 918, 919, 921, 1021
linear-time 1031
relative 823
extend
by 699
finally by 745
freely by 744
reachably by 745
extended closure 1031
extended equation 386
Extended ML 780
extended rational expression (*see* rational expression, extended)
extended relational theory 1108
extended signature 719
extension
conservative 747
of rule (*see* rewrite rule, extended)
parameterized 757
persistent 747
reachable 745
extensional applicative structure 384
extensional collapse 421
extensional database (EDB) 1127
extensional equality of functions 371

factor (of a word) 25
recurrent 41
factoring 304
fail-stop model 1162
failure 1233
Byzantine 1161
halting 1161
of completion (*see* completion sequence, failure)
omission 1161
failure function 32
failure window 1188
fair parallel program 972
fair scheduling assumption 1017
fairness 972, 1017, 1170
of completion (*see* completion sequence, fair)
strong 1019
unconditional 1019
weak 972, 1019
fairness condition 156
false 348
false 331
false deadlock 1191
family of languages 84
fault-tolerance computing 1159
Fibonacci representation 72
Fibonacci sequence 43
field 261
real closed 282
FIFO (First-In-First-Out) 1052, 1162
filtration 802, 1030
for nonstandard models 803
finite antichain property 272
finite automaton (*see* automaton)
finite failure set 533
Finite Failure Theorem 545
finite first-order theory 1107
finite graph property 237
finite power property 20
finite process 1230
finite-state concurrent system 1042, 1060
finite-state program 161
probabilistic 161
finite-state strategy 175
finitely presented algebra 1097
firing squad 1186
first normal form (1-NF) 1119
first-order axiomatic specification 697
first-order language 496
semantics of 511
first-order linear-time structure 1010
first-order logic 198
(in connection with ω-languages) 156
first-order property 198
first-order theory 202
finite 1107
Fischer–Ladner closure 802
fixed point (*see also* fixpoint, least fixed point) 1217
double ~ theorem 334
initial 231
second ~ theorem 336
theorem 329, 637
unique 1222
fixed point combinator 329
fixed point operator 164, 396, 640
uniform 640
fixpoint (*see also* fixed point) 517, 851, 1065
inflationary 1125
least 1125
(of a substitution) 37
fixpoint characterization (of temporal operators) 1005, 1066
fixpoint logics 1065
fixpoint query 1124
Fixpoint theorem (*see also* Fixed Point theorem) 517
fixpoint width 1127
flag 612
flat partial order 466
floundering 549
flow analysis 1061

flow graph 1211
flowchart 488
Floyd–Naur proof method (*see* proof, Floyd–Naur method)
fold-unfold program transformation 486
∀-automata 161
forbidden configuration 222
forbidden minor 223
formal language 107
formal parameter requirement specification 753
formal power series 107
formal series 45
formal system 693
formula 1085
  atomic 691, 889, 898
  Barcan's 998
  correctness 889, 898
  domain-independent 1087
  elementary 1032
  first-order 304
  first-order, unsatisfiability 304
  ground 692
  invincible 1050
  monotone 1124
  nonelementary 1032
  path 1012
  positive 1124
  safe 1087
  Σ- 691
  state 1012
formulas-as-types 439
fourth normal form (4-NF) 1120
fragment 1033, 1034
freedom from deadlock (*see also* deadlock freedom) 1050
frontier 165
  outer 166
frontier node 1033
fulfill 1033
full decomposition 1083
full dependency 1113
full implicational dependency (fid) 1113
full template dependency (ftd) 1114
fullpath 1001, 1014
fully abstract 586, 736
fully invariant 251
function 590, 948
  computable (on a domain) 643
  continuous 590
  monotone 590
  monotonic (on a partial order) 394
  polymorphic 433
  strict 590
function (on a cpo) 636
  continuous 636
  monotone 636
  strict 394, 636
function hierarchy
  continuous 398
  set-theoretic 398
function space 342
function symbols 249
  associative-commutative 278, 299
  (*see also* associative-commutative axioms)
function type 370
functional 641
functional dependency (fd) 1084
functional program 323
functional programming 321
functional reflexivity 251, 282
functor 646
fusion closed 1014
future-tense 1000, 1006

Gale–Stewart game 174
game
  determinacy of 173
  Gale–Stewart 174
  regular 175
game logic 811
Gandy hull 421
garbage collection 1182
general part (of proof system) 1054
general product (Π) type 451
general structure 1014, 1016
general sum (Σ) type 451
generality of substitution 250
generality of unifier 280, 282
generalization, least 279
generated by 776
generating function 70
generator 84
generator problem 1097
$Gen(\Sigma, E)$ 692
glb (*see* greatest lower bound)
global condition 1141
global consistency 1166, 1180, 1183
global element
  of a category 409
  of a Kripke model 412
global invariance 1050
global reasoning 999
global shared variable 1164
global snapshot 1192
global state 1167, 1192
globally equivalent 1007
globally satisfiable 1007
globally valid 1007

goal 498
  general 498
goto 948
grammar 109
  ambiguous 580
  confluent 236
  context-free 62, 162, 215, 578
  hypergraph 211
  matrix 112
  NTS- 69
  phrase-structure 109
  reduced 65
  regular tree- 212
  tree 265
  unambiguous 126
grammar form 122
grammatical family 122
graph
  Cayley 97
  communication 1160
  context-free 182, 238
  context-free set of 211
  definable 234
  dependency 551
  flow 1211
  Hamiltonian 200
  marked 1181
  partial k-tree 224
  perfect 201
  planar 202
  series-parallel 211
  transition 1210
graph expression 203
graph-grammar, context-free 236
graph operation 203
graph property 198
  compatible 237
  finite 237
graph reduction 359
graph rewriting (*see also* rewriting, graph) 193, 308
greatest lower bound (glb) 850
  of terms 279
Greibach algebraic ω-language 163
Greibach language 85
Greibach normal form (*see also* normal form, double Greibach) 65, 70
ground 692
ground instance 499
ground reducibility (*see* reducibility, ground ~ of terms)
ground relation 824
ground term 681
group 260
  context-free 96
  free 50, 63
  Hotz 68
  presentation of a 96
  solvable 46
group theory 261
  Abelian 300, 301, 302
  commutative (*see* group theory, Abelian)
  fragment 294, 295
group-free 159
growth function 117
growth matrix 118
guaranteed accessibility 1051
guarded command
  alternative 795
  iterative 795

halt operator 810
halting failure 1161
halting problem 912, 923, 924
Hamiltonian graph 200
healthiness criteria 976
helpful process 1056
henceforth 1003
Henkin model 382
Herbrand base 514
Herbrand definability 913
Herbrand interpretation 515
Herbrand model 515
Herbrand universe 262, 514
Herbrand's theorem 304
Hercules 271
hidden 719
hiding operator 1214
hierarchical specification 734
hierarchy
  Borel 153
  Chomsky 110
  continuous function 395, 398
  recursive function ~ 390
  recursive function ~ on modest sets 389
  set-theoretic function 398
hierarchy-consistent 735
Higman's lemma 274
Hilbert's Tenth Problem 284
Hintikka structure 1033
history (of logic programming) 569
history variables 1024, 1052
Hoare Logic 798, 799, 801, 817, 829, 831, 883, 884, 887, 929, 965, 968
  semantics 897
  soundness 901
hom set 397
homomorphic image 261

homomorphism 261, 270, 382, 464, 582, 658, 1091
  between applicative structures 383
  Σ- 682
  total Σ- 705
  unique 262
horizontally composed 760
Horn clause 692
Hotz group 68
Hydra 271, 279
hyperarithmetical 690
hyperedge 204
  replacement 211
hypergraph 204, 309, 1104
  context-free set of 211
  definable 234
  width of a 206
hypergraph acyclicity 1104
hypergraph expression 205
hypergraph grammar 211
hypergraph operation 204
hypergraph rewriting rule 210

I/O automaton 1169
ideal 656
  polynomial 304
  principal 656
ideal completion 479
idempotency 258, 282
idempotent 10
identity (for rational expressions) 12
  equational (*see* equation)
if-then-else 261, 331, 348
immediate consequence operator 516
impartiality 1019
implementation 585, 760
  abstractor 764
  behavioral 764
  constructor 764
  model with standard observables 426
  of 760
  synthesize-restrict-identify 764
implementation of completion 298
implementation of initial algebra semantics 263
implicit parallelism 1134
implicit typing 444
impredicative polymorphism 433
inclusion dependency (ind) 1114
incomplete information database 1140
incompleteness 908, 909, 910, 913, 957
inconsistency 264, 265, 303
independent database schemes 1117
index of a set 22
induction
  computational 933, 934
  fixpoint 935
  inductionless 265, 302
  principle 861
  Scott 933
  structural 265
  subgoal 874
  transfinite 254, 271
  well-founded 254
induction axiom (of *PDL*) 797, 799, 800
induction principle
  Scott's 484
  truncation 484
inductionless induction (*see* proof by consistency)
inductive assertion 792
inductive set of predicates 216
inductive theory 263, 265, 303, 695
inductively generated 709
inevitability 1054
inexpressibility 904, 906
inference (of automata) 50
inference problem 443
  type 443
inference rule 251, 292, 306
  collapse 293, 299, 301
  compose 293, 299
  deduce 293, 299, 301
  delete 293, 299, 301
  extend 299
  orient 293, 299
  simplify 293, 299, 301
inference rules of *PDL* 797
infinitary completeness 822
infinite string 1045
infinite tree 464
infinite word 462
infinitely often 1003
inflationary fixpoint 1125
initial 684
initial algebra (*see also* algebra, initial) 262, 264
initial algebra semantics (*see also* semantics) 262, 589, 627
initial fixed point 231
initial functions 331
initially equivalent 1008
initially satisfiable 1007
initially valid 1007
input/output relation 792, 795, 815
input/output specification 792
instance 499
  of term 250
instance relation on typings 446

institution 711
intensional database (IDB) 1127
interchange lemma 76
interference freedom 959
interior node 1033
intermittent assertion 834, 1051
interpretation 512, 684, 795, 814, 816, 897, 898, 1009
  algebraic 481
  declarative 523
  discrete 466
  global 1010
  Herbrand 515
  Herbrand-like 818
  local 1010
  m- 819
  nonstandard 907
  of term 270, 274
  pre 537
  procedural 522
  representative 478
  universal 476
invariance 1033, 1049
invariance rule 1055
invariant 1172, 1192
  global 862, 954, 955, 958
  local 863, 864, 954, 955, 958
  loop 868
invariant under a constraint 1173
inverse 850
  of binary relation 251
invincible formula 1050
irreducibility (*see* reducibility, ground- of term)
irreducible term (*see* normal form)
isomorphism 261, 382, 582
  between applicative structures 382
  Curry–Howard 439
iteration 72
iteration lemma 10
  of Bader and Moura 73
  of Bar-Hillel et al. 73
  of Ogden 73
iteration operator 794
iterative pair 72
  degenerated 78
  strict 83

join consistency 1105
join dependency (jd) 1099
join of relations 1080
join reconstruction 1098
joinability relation 267
justice 1019
justification
  of proof step 252, 297

KADS 304
KB 304
k-exclusion 1180
key 1119
Kleene sequence 484
Kleene's theorem 14
knowledge 1066, 1167, 1185
  common 1066
  local 1179
knowledge base 1077
Knuth, Morris and Pratt's algorithm 33
Knuth–Bendix completion procedure (*see* completion procedure)
König's lemma 275
k-recognizable set of integers 36
Kripke
  applicative structure 410
  lambda model 410
  logical relation 429
Kripke model 413, 795
  nonstandard 803
Kripke structure 161, 181, 1064
Krohn–Rhodes theorem 49
Kruskal's Tree Theorem 273, 276, 278
Kuratowski's theorem 222

label 863, 950, 1207
labelled transition system 1206
λ- (*see also under* lambda-)
λ-abstraction 647, 752
lambda abstraction 369, 375
lambda algebra, typed 388
λ-calculus 325, 647
  call-by-name 360
  lazy 360
  models of the 648, 667
  partial 360
  polymorphic (second-order) 353
  untyped 635, 648, 667
lambda calculus (*see also* λ-calculus) 245, 259, 267, 325, 604
  call-by-name 360
  lazy 360
  partial 360
  polymorphic 433
  second-order 433
  typed 388
λ-definable 331
λδ-calculus 348
lambda-lifting 359

λ-notation 647
lambda notation 851
λ-term 327
  typed 375, 424
λYδ-calculus 354
language 107
  Clarke 923, 929
  context-free 45, 62, 110
  context-sensitive 110
  deterministic context-free 67
  Dyck 62
  formal 107
  Goldstine 64
  Greibach 85
  inherently ambiguous context-free 65
  lazy 341
  Lukasiewicz 63
  of completely parenthesized expressions 63
  of linear order 1022
  of palindromes 64
  poor 819
  recursively enumerable 110
  regular 110
  rich 819
  right-linear 110
  semi-universal 825
  symmetric 64
Larch 776
lattice 261, 280
  complete 341, 687
  cylindric 1085
  proof 1175
law
  cancellation 23
  dynamic 1228
  expansion 1217
  Leibniz's 251
  recursion 1231
  static 1218
lazy language 341
lazy reduction 341
leader 1190
leader election 1190
least fixed point principle 483
least fixed point semantics 481
least fixpoint 1125
least Σ-subalgebra 685
least upper bound (lub) 850
  of terms 280
LED (*see also* Logic of Effective Definitions) 833
left-linearity, of rewrite rules 253, 265, 270, 288, 289, 300
Leibniz's law 251
lemma
  Critical Pair 286, 288, 290, 298, 307, 309
  diamond 267
  Higman's 274
  interchange 76
  iteration (*see also* iteration lemma) 10
  König's 275
  narrowing 284
  Newman's 267
  Parallel Moves 288
  pumping (*see* iteration lemma)
  Variant 507
  Zorn's 276
let 347
let declaration (in ML) 437
letrec 347
lexicographic ordering 274, 281, 297
lexicographic path ordering (*see* path ordering, lexicographic)
lift 651
limit of completion sequence 293, 298
limit closed 1015
linear order, language of 1022
linear rule 1131
Linear Temporal Logic 1000
  First-Order 1008
linear time 833, 999
linear-time expressiveness 1021
linear-time structure 1000
linear-time Temporal Logic 159
link 1160, 1193
LISP 357
literal 497
literal similarity 250
liveness 875, 880, 1169
liveness property 156, 1049
liveness rule 1056
local characterization 1236
local condition 1141
local confluence 424
local invariance 1050
local knowledge 1179
local set 11
local shared variable 1164
locally testable set 25
lock
  exclusive 1195
  shared 1195
locking 1195
lockout-freedom 1168
logic
  Algorithmic (*AL*) 832
  chain 184
  Classical Modal 1064

combinatory 245, 256, 288
computation tree 181
Dynamic (DL) 811
endogenous 999
exogenous 999
first-order 198
first-order (in connection with ω-languages) 156
fixpoint 1065
game 811
Hoare (*see also* Hoare Logic) 883, 884, 887, 929, 965, 968
linear-time Temporal 159
modal 791, 793, 797
Modal 997
monadic second-order 143, 199, 237
Nonstandard Dynamic (NDL) 830
of Effective Definitions (LED) 833
of programs 791, 809
Owicki–Gries 965
path 184
process 181
Process 811
Propositional Dynamic (PDL) 793
Stirling compositional 965
Temporal (TL, *see also* Temporal Logic) 793, 833, 997, 1206
logic program 498
database 1127
logical clock 1167, 1180
logical consequence 796, 805, 816
logical partial equivalence relation 419
logical partial function 419
logical relation 415
admissible 418
algebraic 428
logically related environments 417
logics of programs, modal 180
LOOK 778
loop invariance rule 798, 800
loop operator 810
lossless decomposition 1083
lossy decomposition 1083
L-system 116
lub (*see* least upper bound)

machine, SECD 358
magic-counting evaluation 1137
magic-set evaluation 1136
many-sorted algebra 205
marked graph 1181
marking scheme in completion procedure 298
Markov algorithm 114, 245, 308
Markov chain 3
matching of terms 250, 259, 282, 284
associative 259
associative-commutative 259, 299
commutative 259
materialized view 1143
matrix grammar 112
maximal 713
maximal model 1037
maximum flow 1192
McNaughton and Yamada's algorithm 47
McNaughton's theorem 149
meaning function 417
acceptable 417
Meije 1205
message buffer 1052
message complexity 1163
message passing 1020, 1159
mgu (most general unifier, *see also* unifier, most general) 500
minimal upper bounds 661
complete set of 661
minimalization 332
minimum model 1128
minor 223
minor-closed 223
m-interpretation 819
minimum spanning tree 1189
ML, Extended 780
ML programming language 437
Mod 697, 712, 739
modal generalization 797
modal logic 791, 793, 797
Modal Logic 997, 1064
Classical 1064
modal logics of programs (decision problem) 181
modality, basic 1027
model (*see also* data model) 513, 697, 897, 1159
asynchronous 1160
combinatory 387
distributed shared-variables 1020
environment 384
equational (*see* algebra)
equational standard (*see* algebra, initial)
fail-stop 1162
Henkin 382
Herbrand 515
Kripke 413, 795
Kripke lambda 410
maximal 1037
minimum 1128
nonstandard 803
nontrivial 386
observational equivalence of 426

  of $\lambda^{\rightarrow}$ 382
  of the lambda calculus 341, 345
  perfect 564
  process 1159
  recursion-theoretic 388
  synchronous 1161
  time 831
  universal 804
model checking 161
model checking problem 1042
  branching-time logic 1042
  linear-time logic 1042
model class 739
model theorem
  linear-size 1039
  small 1034
model-theoretic 1049
modest set 371, 389
modular design 1206
modulo system (*see* rewriting, congruence class)
modus ponens 797
monadic second-order logic 143, 199
  counting 237
monadic second-order property 199
monadic second-order theory of two successors 1026
monadic theories (of linear ordering) 1022
monadic theory 203
monitor 1164
monogenicity 114
monoid 5, 105, 282
  aperiodic 22
  commutative 105
  free 4, 105
  reset 49
  syntactic 8
  transition 7
monoid laws 1217
monoids, wreath product of 49
monomorphic 698
monotone 851
monotone formula 1124
monotonic function (on partial order) 394
monotonic functional 1065
monotonic reasoning method 531
monotonicity (*see* rewrite relation)
Moore's algorithm 45
morphism 4, 106
  alphabetic 83
  arrow of a category 397
  parameter-passing 755
  signature 681
  specification 744
  syntactic 8
Morris and Pratt's algorithm 32
most general unifier (mgu) (*see also* unifier, most general) 500
μ-calculus 810
mu-calculus 1066
Muller acceptance 148, 169
Muller automaton 148
Muller tree automaton 169
multiplicative dependence 39
multiplicity (in rational expression) 18
multiprocess temporal structure 1015
multiset (bag) 264, 268, 269, 275, 280, 297
multiset ordering (*see* ordering, multiset)
multiset path ordering (*see* path ordering, multiset)
multitape finite automaton 46
multivalued dependency (mvd) 1099
μ-term 185
mutual exclusion 1050, 1059, 1166, 1178

naive bottom-up evaluation 1128
narrowing 284, 303, 307
narrowing lemma 285
naturally extended (*see* operator, strict)
NDL (*see also* Nonstandard Dynamic Logic) 830
necessity 1236
negation-as-failure rule 533
negative information 531
nested relation 1139
nested transaction 1196
network partition 1161
network topology 1160
Newman's lemma (*see* diamond lemma)
nexttime 1002
nexttime operator 833
Nivat's theorem 46
n-local universal automaton 27
no confusion (*see* consistency, of specification)
Noetherian relation (*see* relation, terminating)
no junk (*see* sufficient completeness, of specification)
nondeterminism 247, 254, 283, 710, 880
  bounded 970
  for modelling concurrency 1017
  unbounded 971, 972
nondeterministic assignment 791, 813, 816, 830
nondeterministic choice 794, 830
nonelementary formula 1032
nongenerator 87
nonhalting problem 911
nonmonotonic reasoning method 531
nonprimitive 734
Nonstandard Dynamic Logic (*NDL*) 830

nonstandard model 803, 811
nontermination 604, 616
 of completion (*see* completion sequence, infinite)
 of rewriting (*see* termination, of rewriting)
nontrivial model 386
normal form 267, 338, 378, 379, 1119
 Boyce–Codd 1119
 Chomsky 65
 disjunctive 270, 276
 domain-key 1120
 double Greibach 66
 exclusive-or 258, 304
 first 1119
 fourth 1120
 Greibach 65, 70
 has a 338
 (in partially commutative monoid) 48
 is a 338
 (of a context-free grammar) 65
 of ground term 263
 of proof 297
 of term 247, 252, 253, 254, 256, 259, 265, 269
 of term, uniqueness 247, 253, 256
 positive 1031
 project-join 1120
 third 1120
normal form algebra (*see* algebra, normal form)
normal form database schemes 1119
normalizability relation 253
normalization 422
 strong 444
 unique 253, 263
NP 226
NP-complete 227
NTS-grammar 69
null value 1141
number algebra 688
Number Theory 829
numerals 330, 331
 Church's 330

OBJ 263
OBJ2 773
object 1077
 complex 1138
 (of a category) 397
object identifier 1139
object-oriented data model 1077
OBSCURE 779
observably complete 749
observably persistent 749
observation congruence 1225
observation equivalence 1224
observational abstraction 748
observationally equivalent models 426
occur check 281
occurrence (*also* position) 250
 binding 376
ω-automaton 809
ω-derivable from E 695
ω-Kleene closure 163
ω-language 136
 algebraic 162
 context-free 162
 Greibach algebraic 163
 regular 138
 star-free 158
ω-regular 138
ω-regular expression 1025
ω-word 136
omega complete 263
omission failure 1161
operation, fundamental 260
operational semantics (*see also* semantics) 263, 288, 337, 466, 828, 854, 855, 858, 949, 950, 951, 973
 of concurrent processes 1201
operator 582
 arity of 249
 box 794
 continuous 516
 converse 809
 diamond 794
 finitary 516
 growing 558
 halt 810
 hiding 1214
 immediate consequence 516
 iteration 794, 799
 loop 810
 mixed 793
 monotonic 516
 nexttime 833
 nonmonotonic 557
 program 793
 propositional 793
 repeat 810
 star 794, 799
 strict 261
 temporal 997
 test 794
 wf 810
operators, representing 664
order
 complete partial (*see also* cpo) 636
 partial 636
 Plotkin 661

ordered completion (*see* completion, ordered)
ordered pair 331
ordering
  linear 878
  multiset 275, 281, 297
  on proofs 297
  partial 850
  partial, well-founded 254
ordering on terms 270
  encompassment 250, 253, 278, 278, 292, 297, 301
  Knuth–Bendix 277
  path (*see* path ordering)
  polynomial 271, 274, 278
  reduction 270, 292, 297, 298, 301
  rewrite 251, 270
  rewrite, total 271, 290
  simplification 274, 290
  simplification, complete 290, 301
  well-founded 270
order-sorted algebra 305, 709
ordinal 271, 277, 878
ordinal number 147
overlap, between rules 286, 293, 299
Owicki–Gries logic 965
Owicki–Gries method 1173
Owicki–Gries proof method 958, 963

P 227
packet 1193
pair
  accepting 149
  iterative (*see also* iterative pair) 72
  ordered 331
pairwise consistency 1106
palindromes, language of 64
Parallel Moves lemma 288
parallel program (*see* program, parallel)
parallel rewriting (*see also* rewriting) 288
parameter
  procedure 945
  value-result 941
  variable 945
parameter passing morphism 755
parameterized extension 757
parameterized specification 752
  structured 753
paramodulation 247, 282, 307
parse tree (*see* derivation tree)
partial algebra 705
partial correctness (*see also* correctness, partial) 795, 817, 1050, 1169
partial enumeration function 389
partial equivalence relation (per) 391
partial function 466
partial k-tree 224
partial order 393, 636
  complete (*see also* cpo) 636
  flat 466
partially commutative alphabet 48
partially ordered set (*see also* poset) 636
partition dependency 1098
past-tense 1000, 1006
path 166, 179, 1204
  end 5
  label 5
  length 5
  origin 5
  successful 5
path formula 1012
path logic 184
path ordering 275
  lexicographic 277, 290, 291
  lexicographic, decidability 277
  multiset 276, 278
  multiset, complexity 277
  recursive 277, 295
pattern 32
  of proof 296
  unavoidable 278
pda (*see* pushdown automaton)
PDL (*see also* Propositional Dynamic Logic) 793
  axioms for 797
  completeness of 804
  complexity of 805
  context-free 808
  Δ 810
  deterministic 807
  inference rules of 797
  regular 796
  satisfiability problem for 805
  strict deterministic 807
  with Boolean assignments 811
peak
  in derivation 267, 268
  in proof 286, 289, 296, 297
  in proof, critical 286, 289, 297
Peano arithmetic (*see also* arithmetic, Peano) 271, 799, 829
perfect graph 201
perfect model 564
persistent, observably 749
persistent extension 747
Peterson's solution 1055
Petri net 48, 1159, 1181, 1203
  in connection with ω-words 155, 165
phrase-structure grammar 109

physical channel 1193
piecewise testable set 28
pigeonhole principle 272, 1179
PLA (*see* programmable logic array)
place (*see also* occurrence, position) 250
planar graph 202
Plotkin order 661
PLUSS 779
point (of a category) 409
polymorphic composition 432
polymorphic function 432
polymorphic identity 433
polymorphic lambda-calculus 433, 353
polymorphism 434
  impredicative 434
  predicative 434
polynomial 107
  real 274
polynomial algorithm 227
polynomial system of equations 214
port 1211
poset (partially ordered set) 636
  bounded complete 644
  complete (*see also* cpo) 636
  Plotkin order 661
position (in term) 250
positive formula 1124
positive normal form 1031
possibility 1236
possible worlds (in Kripke lambda model) 410
possibly 1064
Post canonical system 113
Post correspondence problem 119
Post system 113
postcondition 817, 868
post-fixpoint 517
potentiality 1054
power series 107
  algebraic 126
  formal 107
  rational 123
power set (of a relation) 1139
powerdomain
  convex 653, 657
  Hoare 653, 657
  lower 653, 657
  Plotkin 653, 657
  Smyth 653, 657
  upper 653, 657
precedence 275, 1052
precondition 817, 868
  weakest 976
  weakest liberal 916
predicate 216
  directed-complete 430
  first-order 889, 898
predicate calculus
  first-order 247, 304
  first-order with equality 285, 305
predicative polymorphism 433
prefix 1207
  of a word 25
pre-fixpoint 517
pre-interpretation 537
pre-order 655
Presburger arithmetic (*see also* arithmetic, Presburger) 48, 282, 908
prestructure 1033
pre-term 380
  of $\lambda^{\rightarrow,\Pi}$ 380
  of $\lambda^{\rightarrow}$ 380
primitive 734
primitive recursion 332
primitive recursivity 271
principal type scheme 352
principal typing 443
priorities of rules 307
probabilistic finite automaton 45
probabilistic finite-state program 161
probabilistic logics 1065
probabilistic program 811
procedure 931, 937, 941
process (*see also* processes) 1166, 1201
  finite 1230
  helpful 1056
  sequential 1230
  stable 1227
process algebra 710, 1205
process control 324
process expression 1207
process logic 181
Process Logic 811
process model 1159
processes (*see also* process, concurrent processes)
  communicating 1203
  concurrent (*see also* semantics of -) 1203
  confluent 1238
producer-consumer problem 1181
production (*see* rewriting rule)
program 498
  abstract concurrent 1019
  allowed 554
  closed term of observable type 426
  completion of a 537
  concurrent 160
  context-free 808

database logic 1127
do-od 1016
fair parallel 972
finite-state 161
functional 323
logic 498
nondeterministic 812
parallel 949, 950, 951, 952
probabilistic 811
probabilistic finite-state 161
reactive 1048
regular 812
single-rule 1131
stratified 556
weakly fair parallel 973
while 812
with arrays 813
with stacks 813
program annotation 1173
program boundedness 1133
program counter 793
program execution 843
program logics 809
program part (of proof system) 1054
program reasoning 1048
program scheme
equivalent 465
monadic recursive applicative 488
polyadic recursive applicative 489
recursive 186
recursive applicative 465
program stage function 1131
program transformation 485
fold-unfold 486
programmable logic array (PLA) 45
programming language
acceptable 825
ML 437
rewrite-based 247, 266, 307
semi-universal 825
programming language semantics 263
programming, functional 321
programs, modal logics of 181
project-join expression 1083
project-join normal form (PJ-NF) 1120
projected-join dependency (pjd) 1114
projection 146, 642
finitary 643
of a relation 1080
Prolog 282, 568
proof
by consistency 265, 302
by contradiction 304
contrapositive 875
direct 248, 255
equational 252
equational, normal-form (*see* normal form, of proof)
equational, rewrite 254, 296, 297
Floyd–Naur method, compositional 868, 871
Floyd–Naur method, stepwise 859, 861, 866, 871, 881, 973
formal 891
Lamport method 958
liveness 881, 882, 974, 975
outline 886, 893
Owicki–Gries method 958, 963
satisfaction 969
termination, Floyd's well-founded set method 879
proof lattice 1058, 1175
proof method (*see* proof)
proof normalization theorem 297
proof rules 1173
proof simplification 296
proof system 376, 1054
proof theory 292
proof-theoretic 1049
property
Church–Rosser (*see* Church–Rosser property)
definable 198
finite power 20
first-order 198
monadic second-order 199
(*see also* graph property)
propositional calculus (*see* Boolean ring)
Propositional Dynamic Logic (*see also PDL*) 793, 1064
protocol
alternating bit 1193
three-phase commit 1188
two-phase commit 1188
provability 888, 891
provability relation 252
proving programs (*see also* proof) 843
pseudo-fulfilled 1034
pseudo-Hintikka structure 1034
PSF 779
pumping lemma (*see* iteration lemma)
pure liveness property 1051
pure universal relation assumption 1110
pushdown automaton (pda) 67
deterministic 67
(as a rewriting system) 111
(over ω-words) 155, 164
pushdown tree automaton 186
pushout 209, 756

quasi-commutation (of relations) 278
quasi-order 479
quasi-ordering 250
quasi-reducibility (*see* reducibility, ground- of a term)
quasi-simplification ordering (*see* ordering on terms, simplification)
query 1075
query evaluation 1135
query language 1075
query optimization 1090
quote 358
quotient 745
quotient
  applicative structure 420
  Henkin model 382
quotient algebra 252, 262, 279
quotient construction 1030

R-generable 1015
Rabin acceptance 147, 169
Rabin automaton 149
Rabin index 155, 185
Rabin tree automaton 169
Rabin's tree theorem 180
Rabin-recognizable 169
Ramsey's theorem 141, 273
random assignment 813
randomized algorithm 1180
rank (*see* arity, of operator)
rank function 879
ranked alphabet 166
rational (*see also* regular)
rational expression 12
  extended 21
  unambiguous 18
rational function 47
rational ω-language 138
rational ω-relation 138
rational operation 11
rational power series 123
rational relation 13
rational set 13
rational theory 489
rational transduction 82
rational tree 282
reach 740
reachable 685, 719
reachable extension 745
reaching agreement 1183
reactive program 1048
reactive system 175
readers-writers problem 1183
reasoning
  backward 247
  forward 247
  program 1048
reasoning method
  monotonic 531
  nonmonotonic 531
recognizable (*also* acceptance condition) 136
  Büchi 136, 169
  Muller 148, 169
  1- 154
  Rabin 149, 169
  Streett 169
  2- 154
recognizable set 15, 221
  in a monoid 8
recognizable set of numbers (*see also* *k*-recognizable set) 35, 52
recovery, database 1076
recursion 1076, 1207
recursion laws 1231
recursion-theoretic model 388
recursive 689, 890
recursive definition 462
recursive enumerability 263
recursive function 332, 526
recursive function hierarchy 390
  on modest sets 389
recursive number algebra 688
recursive path ordering (*see* path ordering, recursive)
recursive program scheme 186
recursive type 1139
recursively enumerable 689, 910, 913
recursively enumerable language 110
redex 252, 469
  outermost 266
reducibility
  ground ~ of term 264
  ground ~ of term, complexity 265
  ground ~ of term, decidability 265
  of term, inductive (*see* reducibility, ground ~ of a term)
reducibility relation 253
reduct, σ- 682
reduction 338, 378–379
  eager 341
  graph 359
  lazy 341
  leftmost 340
  normal 340
  one-step 379
  weak 356
reduction machine 323
reduction ordering (*see* ordering on terms, re-

duction)
reduction rule 378, 469
reduction strategy 340
referential integrity 1099
referential transparency 324
refinement 698
reflexive 344
reflexive future 1005
refutation 503
regular (*see also* rational) 1182
regular expression 45, 138, 1023, 1204
regular game 175
regular language 110
regular ω-language 138
regular program 812
regular system of equations 232
regular tree 172, 233
regular tree-grammar 212
regular tree language 166
regulated rewriting 112
relabelling 1207
relation 1079
  action 1207
  admissible logical 423
  algebraic (logical) 428
  Armstrong 1095
  binary 251, 266
  binary monotonic (*see also* rewrite relation) 251
  ccc- (between Cartesian closed categories) 429
  common ancestor 267
  common descendant 267
  complement of a 1087
  derivability 253
  difference of 1080
  equivalence (over processes) 1215
  ground 824
  input/output 792, 795, 815
  inverse of binary 251
  join 1080
  joinability 267
  logical 416
  logical (*see* logical relation)
  nested 1139
  Noetherian 254, 267
  normalizability 253
  power set of 1139
  projection of 1080
  provability 252
  reducibility 253
  rewrite 251
  satisfaction 692, 903, 1237
  terminating 254, 267
  transition 854
  universal (*see* universal relation)
  unrestricted 1080
  well-founded 878
relational algebra 1081
relational algebra expression 1081
relational calculus 1085
relational calculus expression 1086
relational database theory 1075
relative completeness (*see* completeness, relative)
reliability 1162
remote procedure call (RPC) 1165
renaming 499, 681, 743
renaming (of a relation) 1081
renaming of variables (*see* literal similarity)
rendezvous 1165
repeat operator 810
replacement of equals 251
replacement of subterm 250
representation, Fibonacci 72
representation independence 425
representation principle 1098
representative universal relation 1111
reset monoid 49
residual, left 17
resolution 247, 304
  SLD- 503
  SLDNF- 547
resolvent 503
response, absence of unsolicited 1052
responsiveness 1051
restrict 743
  σ- 683
restriction 850, 1207
resultant 507
resumption 622
retract 344
REVE 304
rewrite relation 251
rewrite rule 252
  conditional 306
  extended 289, 299
  persisting 293
  reduction of 293
rewrite system 245, 252
  associative-commutative 257, 271, 278, 292
  canonical 253, 291, 293, 298, 301
  Church–Rosser (*see* confluence of rewriting)
  complete (*see* completeness of rewrite system)
  conditional 306
  confluent (*see* confluence of rewriting)
  congruence class 257, 264, 278, 285, 289
  convergent 247, 253, 256, 260, 262, 284, 285, 286, 289, 291, 292, 307

decidability of confluence 288, 307
finite 256
fully-simplified (*see* rewrite system, reduced)
ground 253, 270
ground convergent 256, 263, 264, 265, 291, 303
monadic (*see* rewrite system, string)
multisorted 305
nonambiguous (*see* rewrite system, nonoverlapping)
nonoverlapping 286
normalizing 253, 254, 266, 287
ordered 259, 290, 300
order-sorted 305
orthogonal 288, 291
priority 307
reduced 253, 291, 300
reduced, uniqueness 291, 298
regular (*see* rewrite system, orthogonal)
string 253, 266, 270, 286
terminating (*see* termination of rewriting)
uniquely normalizing 253, 260
rewriting
decidability of 307
extended 258, 289, 298
graph 308
of graphs 193
parallel 288
regulated 112
typed (*see* rewrite system, multisorted)
rewriting modulo a congruence (*see* rewrite system, congruence class)
rewriting rule 468
rewriting rule (production) 109
rewriting system (*see also* rewrite system) 109
parallel 116
term 468
right-linear language 110
ring
Boolean 258, 282, 304
commutative 261
road coloring problem 46
round 1161
route-determination 1189
RRL 304
rule 1127
assignment 967
auxiliary variables elimination 965, 967
await 965, 967
block 947
composition 892, 894, 940, 967
conditional 892, 894, 967
consequence 892, 894, 967
derelativization 967
of adaptation 938, 942
of inference 891, 892
parallelism 965, 967
program 940
receive 969
recursion 934, 940, 942
rewrite (*see* rewrite rule)
send 969
substitution 965, 967
while 892, 894, 967, 970, 972
rule body 1127
rule dependency 1114
rule head 1127
run-until 824

safe 1182
safe formula 1087
safety 1049, 1169
safety property 156
satisfaction
of a formula by a model 385
of a type assignment 385
satisfaction relation 692, 903, 1237
satisfiability 1021
for modal logics of programs 181
satisfiability of equation 261, 262, 279, 307
undecidability 284
satisfiability problem (for *PDL*) 805
satisfiable 698, 1003, 1007, 1011, 1014
saturation 8
of ω-languages 139
SCCS 1215
SCHEME 358
scheme, database 1079
Schützenberger's theorem 23
Schützenberger's representation theorem 124
SECD machine 358
second-order lambda-calculus 353, 433
second-order logic 201
counting monadic 237
monadic 143, 199
second-order theory of two successors, monadic 1026
Segerberg axioms 797
selection of relations 1080
selection propagation 1135
selection rule 504
selector 249
self-stabilization 1182
semantic containment 828
semantic data model 1077
semantic domain 641
semantic equation 587
semantic function 587

semantically closed 824
semantically deterministic 807, 824
semantics 260, 445, 585, 795, 814
  action 628
  algebraic 260
  axiomatic 828, 890
  compositional 586
  continuation 612
  denotational 341, 575
  direct 612
  execution sequence 792, 796
  initial algebra 262, 589, 627
  least fixed point 481
  of Hoare Logic 897
  of recursively defined functions 638
  of stratified program 561
  of types, simple 445
  operational (*see* operational semantics)
  relational 856, 858, 932, 941
  simple ~ on types 445
  static 584
semantics (for first-order languages) 511
semantics (of concurrent processes)
  algebraic 1201
  operational 1201
semantics of types, simple 445
semaphore 1164
semaphore code 51
semicomputable (*see also* co-semicomputable) 688
semidecidability 911
  of satisfiability 282
semidecision procedure, for validity 247, 298, 301
semigroup 256, 257
semijoin 1106
semilattice 1097
  upper 687
semiring 12, 106
  Boolean 45
  idempotent 12
  tropical 20
semi-Thue system (*see also* rewrite system, string) 210
semi-universal language 825
sensible 681, 719
separated sum (of cpo's) 652
separation principle 1118
sequence 851
  completion (*see* completion sequence)
  Fibonacci 43
  Kleene 484
  Thue 120
  Thue–Morse 38
sequential calculus 143
sequential circuit 3
sequential composition 794
sequential process 1230
sequentiality 288
serializability 1194
series
  algebraic 45
  characteristic 107
  formal 45
  power (*see* power series)
series-parallel graph 211
set
  aperiodic 22
  complete partially ordered (*see* cpo)
  context-free ~ of graphs 211
  context-free ~ of hypergraphs 211
  definable 198
  directed 479, 636
  domain-ordered 393
  equality (of two morphisms) 119
  equational 214
  finite failure 533
  hom 397
  index of a 22
  inductive ~ of predicates 216
  local 11
  locally testable 25
  magic 1136
  modest 371, 389
  multi (*see* multiset)
  of words recognized by an automaton 5
  partially ordered (*see also* poset) 636
  piecewise testable 28
  power (of a relation) 1139
  rational 13
  recognizable 15, 221
  recognizable (in a monoid) 8
  recognizable ~ of integers 36
  regular (*see* rational set)
  S-sorted 680
  star-free 22
  success 518
  test 265
  underlying (*see* universe)
  well-founded 1174
set of words recognized by an automaton 5
set type 1139
set-theoretic function hierarchy 384
shared lock 1195
shared variable 1019, 1164
  local 1164
shared variable algorithm 1178
shortest paths 1192

side-effect 946
sig 739
σ-reduct 682
σ-restriction 683
Σ-algebra 682
Σ-computation structure 685
Σ-congruence 686
Σ-equation 691
Σ-formula 691
Σ-homomorphism 682
Σ-isomorphism 682
Σ-sentence 692
Σ-subalgebra 685
Sign 681
signature 463, 582, 669, 680, 887, 1010
  extended 719
  for $\lambda^{\rightarrow}$ 374
  visible 719
  (*see also* vocabulary) 249
signature morphism 681
similarity type 1010
simple algebraic specification 697
simple assignment 812, 815
simple semantics of types 445
simplification rule 468
simultaneous substitution 682
single-rule program (sirup) 1131
skip axiom 892, 894, 940
Skolem constant 291, 302
Skolemization 291, 302, 304
Skolem–Mahler–Lech theorem 125
SLD-derivation 503
  fair 533
SLD-resolution 503
SLD-tree 510
SLDNF-resolution 547
small cancellation theory 305
small model property 801, 1030
small model theorem 803, 1034
smash product 649
snapshot (global) 1191
sofic system 52
solution, of equation 261, 280, 282
solved form 280
sometimes 1002
S1S (second-order theory of one successor) 144
sort 249, 582, 680, 1210
sound 1040
soundness 413, 796, 797, 816
  equational 413
  of completion (*see* completion procedure, soundness)
  of Hoare Logic 901
  of rules 256
  of SLD-resolution 514
  of SLDNF-resolution 548
  of the negation-as-failure rule 543
  semantical 868, 870, 885
  type 414
source 203
source-sink reachability 1138
spanning tree
  breadth-first 1190
  depth-first 1190
  minimum 1189
sparsely downward closed 1038
Spec 738
Spec-expr 740
specification 792, 858, 891, 902, 1167, 1177
  algebraic 697
  axiomatic 697
  basic 740
  conditional equational 697
  equational 207, 697
  first-order axiomatic 697
  formal parameter requirement 753
  hierarchical 734
  input/output 792
  of data types, algebraic 248
  parameterized 752
  simple algebraic 697
  structured parameterized 753
  target 753
specification expression 740
specification morphism 744
specification with hidden symbols 719
specification-building function 739
spectral complexity 818
spectrum 819
square-free word 71
S-sorted set 680
stable process 1227
stable property 1172, 1192
stable storage 1162
stack example 248, 252, 253, 256, 262, 263, 266, 288, 303, 305
stack operations 813
stage function 1131
standard completion (*see* completion)
standard concurrent form 1219
standards 585
star operator 794, 799
star-free 157, 158
star-free set 22
star-height 19
  extended 22
starvation, absence of individual 1051
state 512, 792, 795, 815, 854, 897, 1168, 1204

initial 5, 818
terminal 5
state formula 1012
state predicate 1171
static algorithm 1189, 1192
static laws 1218
step
equational 252
rewrite 254
store 607
strategy 174
finite-state 175
for rewriting, leftmost-outermost 288
for rewriting, optimal 288
for rewriting, parallel-outermost 288
stratification 556, 1129
stratified program 556
semantics of 561
stream 325, 616, 1206
Streett acceptance 169
strict function (on a cpo, *see also* function, strict) 394
strict future 1005
strict operator (*see* operator, strict)
strictness analysis 360
string, infinite (*see also* word) 1045
string-matching 33
string rewriting (*see* rewrite system, string)
strong fairness 1019
strong normalization 422
strongest postcondition, liberal 918
structural induction (*see* induction, structural)
structure 196, 1000, 1010, 1012, 1015, 1016
acceptable 821
acyclic 1012
event 1203
first-order linear-time 1010
general 1014, 1016
Hintikka 1033
Kripke 1064
linear-time 1000
multiprocess temporal 1015
pseudo-Hintikka 1034
temporal 1012
tree 1012
tree-like 1012
S2S (second-order theory of two successors) 178
subalgebra 261, 685
least Σ- 685
Σ- 682
subgoal induction 874
subject reduction theorem 352
subposet, normal 642
subquotient (of a set) 391
subset construction 6
subsort 305, 710
substitution 37, 84, 108, 499, 682, 895, 896, 899, 900
computed answer 504
correct answer 520
finite 108
of terms 250
simultaneous 682
syntactic 85
uniform 113
substitution instance (of a type assignment) 445
substitutivity 1220
subsumption 250, 279, 282, 289
subsumption relation (on typings) 443
subterm 250
proper 250
subterm ordering 250, 272, 274
subtree 166
subtree automaton 180
subword 28
success, of completion (*see* completion sequence, success)
success set 518
succinctness 1026
sufficient completeness of specification 249, 264, 306
sufficiently complete 306, 722
suffix closed 1014
suffix transducer 34
suffix tree 51
summation 1207
supercombinator 359
superposition (*see* overlap)
superposition test 286, 292
support 107
symmetry, breaking 1181
synchronization 1204
multiway 1214
synchronization problems 1166
synchronization skeleton 1058
synchronizer 1192
synchronizing word 46
synchronous model 1161
syndetic 40
syntactic monoid 8, 143
syntactic morphism 8
syntax 248, 374, 578
abstract 581, 851, 852, 1002
concrete 578, 851
contextsensitive 584
synthesis 1058
synthesize-restrict-identify implementation 764
System F 433

system of equations 465
defining a contextfree language 66
polynomial 214
regular 232

table
Codd 1141
conditional 1141
tableau 1031, 1090
tableau summary 1090
tables, parallel rewriting with 118
target specification 753
temporal implication 1051
Temporal Logic (*TL*) 793, 833, 997, 1171, 1206
Branching-Time 999, 1012
extended 160
First-Order Branching 1017
First-Order Linear 1008
Linear-Time 159
Linear 999, 1001
Propositional Branching 1012
Propositional Linear 1001
temporal operator 997
temporal structure 1012
multiprocess 1015
term 889, 898
closed 328
combinatory 356
first order 249
flat 299
ground 250, 256, 263
ground 681
ground ~ constructor 249, 264
instance of 250
interpretation of 270, 274
irreducible (*see* normal form)
irreducible (*see* normal form of term)
lambda- (*see also* lambda-term) 327
matching (*see* matching of terms)
monadic 250, 253
μ- 185
term algebra 262, 279, 684
term constant (in $\lambda^{\rightarrow}$) 374
term rewriting system (*see also* rewrite system) 468
term-generated 685
terminal 684
terminated computation 468
termination 877, 879, 974, 975, 1169, 1191
of binary relation 254, 267
of rewriting 254, 256, 258, 266, 269
of unification 281
strong 880
weak 880
termination proof, Floyd's well-founded set method 879
term-rewriting system (*see* rewrite system)
test 794
poor 808, 814
rich 808, 814, 830
test operator 794
test set 265
test-and-set 1179
testing equivalence 1233
theorem
Birkhoff's Completeness 262, 263
Cobham's 39
Compactness 551
cross-section 46
double fixed point 334
Finite Failure 545
Fixed Point 329, 637
Fixpoint 517
Herbrand's 304
Kleene's 14
Krohn–Rhodes 49
Kruskal's Tree 273, 276, 278
Kuratowski's 222
linear size model 1039
McNaughton's 149
Nivat's 46
proof normalization 297
Rabin's tree 180
Ramsey's 141, 273
Schützenberger's 23
Schützenberger's representation 124
second fixed point 336
Skolem–Mahler–Lech 125
small model 803, 1034
subject reduction 352
Unification 500
Zielonka's 48
theorem-proving
equational 247
first-order 304
inductive 302, 305
theory 692, 903, 904
algebraic 489
automata 1, 809
category 261
database 1075
decidable 146, 181, 202
equational (*see also* equational theory) 692
extended relational 1108
finite first-order 1107
first-order 202
group (*see* group theory)
inductive 263, 265, 303, 695

  monadic 203
  monadic second-order ~ of two successors 1026
  proof 292
  rational 489
  relational database 1075
  second-order ~ of one successor (S1S) (*see also* WS1S) 144
  second-order ~ of two successors (S2S) (*see also* WS2S) 178
  small cancellation 305
  typed lambda 378
  undecidable 183, 202
  weak monadic 147
third normal form (3-NF) 1120
Thomson construction 47
three-phase commit protocol 1188
Thue sequence 120
Thue–Morse sequence 38
time 999
  continuous 999
  discrete 999
time complexity 1163
time model 831
timeline 1001
timer 1160
timestamp 1167, 1195
TL (*see also* Temporal Logic) 793, 833
top 341
top-down evaluation 1136
top-down tree automaton 167
topological dynamics 52
topology, Cantor 152
total correctness (*see also* correctness, total) 817, 1051
total Σ-homomorphism 705
tournament 1179
trace 792, 876, 1233
trace equivalence 1232
transaction 1194
  nested 1196
transaction processing 1075
transducer, suffix 34
transduction (*see also* rational relation) 82
transduction, rational 82
transfinite induction (*see* induction, transfinite)
transformation rule
  check 281, 282
  coalesce 281
  conflict 281
  decompose 280, 283, 284
  delete 280
  eliminate 281, 283, 284
  imitate 283, 284
  merge 281
  mutate 283, 284
  splice 283
transition array 6
transition graph 1210
transition monoid 7
transition network 3
transition relation 854
transition system 1016
  labeled 1206
transitive closure 251, 850, 1089
translate ... with 741
tree 165, 200
  algebraic 477
  automaton (*see also* tree automaton) 167, 168, 169
  binary 166
  concatenation 166
  derivation 62, 110, 1131, 1209
  infinite 464
  language 166
  rational 282
  regular 233
  SLD- 510
  spanning (*see* spanning tree)
  suffix 51
tree automaton 167
  bottom-up 167
  Büchi 168
  deterministic 167
  Muller 169
  pushdown 186
  Rabin 169
  top-down 167
tree language, regular 166
tree model property 181
tree structure 1012
tree-decomposition 224
tree-like structure 1012
tree-width 224
trie 7
true 348
true 331
truth (*see also* validity) 888
tuple 1079
tuple type 1139
Turing machine 245, 266, 924
Turing machine (over ω-words) 155, 165
two-generals problem 1183
two-phase commit protocol 1188
two-way automaton 146
two-way finite automaton 46
typable 352
type (*see also* sort) 249, 350

base (in $\lambda^{\rightarrow}$) 374
existential 441
function 370
general product (Π) 451
general sum (Σ) 451
simple semantics of 445
tuple 1139
union 1139
universes of 435
type 680
type abstraction 432
type application 433
type assignment 373
type constant (in $\lambda^{\rightarrow}$) 374
type frame (for $\lambda^{\rightarrow}$) 383
type inference problem 443
in ML 448
type inference rule, Curry 444
type scheme, principal 352
type soundness 385
type system 365
type-closed property (of terms) 424
typed applicative structure 382
typed lambda algebra 388
typed λ-calculus 753
typed lambda calculus 388
typed lambda term 375, 424
typed lambda theory 378
typed rewriting (*see also* rewriting) 305
type-free 326
type:type assumption 434
typing
Curry 444
implicit 444
principal 445
typing axiom 384
typing context (*see also* type assignment) 373

ultimately periodic infinite word 43
unambiguous context-free grammar 65
unambiguous grammar 126
unary inclusion dependency (uind) 1099
unbounded nondeterminism 830
unconditional fairness 1019
undecidability 911, 912
undecidable theory 183, 202
undefinedness 946
underlying set (*see* universe)
unfailing completion (*see* completion, ordered)
unifiability of terms 262, 279
second-order 282
unification 446
associative-commutative 282, 299
order-sorted 306
semantic 282
syntactic 279
unification algorithm 501
complete 282
unification theorem 500
unifier 500
most general (mgu) 280, 281, 282, 286, 289, 290, 500
uniform boundedness 1133
uniform stage function 1131
uniform word problem 1097
uniformization 184
union (of relations) 1080
union of rewrite systems, disjoint 278, 289
union type 1139
unique fixed point 1222
unique normalization 253, 263
unit algebra 684
Unity 1169
universal closure 692
universal model 802
universal property 646
universal relation
pure assumption 1110
weak assumption 1110
universal relation data model 1109
universality 1183
universe (*see also* carrier) 260
Herbrand 262, 514
universes of types 435
unless 1005
unrestricted containment 1083
unrestricted database 1080
unrestricted equivalence 1083
unrestricted relation 1080
until 1002
strong 1005
weak 1005
update, database 1142
update anomaly 1119
upper-continuous 851
upper semilattice 687

valid 692, 1003, 1007, 1011, 1014
validity 1003
in equational theories 261
in equational theory (*see* decision procedure for validity)
in initial model 263
validity problem 805, 817
valley, in proof (*see* proof, equational, rewrite)
valuation 344, 684, 791, 811, 814, 1091
value 260
assignable 608

denotable 600
expressible 599
parameter 604
result 604
storable 607
value-oriented data model 1077
variable 249, 327, 681
auxiliary 962, 963
bound 326, 375, 691, 895
free 326, 681, 896
local 947
shared 949, 1164
variable renaming (*see* literal similarity)
variant 499
Variant lemma 507
variety 261
variety (of monoids) 49
verification 1167
verification (of concurrent programs) 1054, 1060
verification condition 862, 866, 869
vertically composed 760
view
complementary 1143
materialized 1143
view update problem 1143
visible signature 719
vocabulary 249
finite 272

weak Church–Rosser property 424
weak fairness 1019
weak fairness hypothesis 972
weak monadic theory 147
weak universal relation assumption 1110
weakest precondition 976
liberal 916
weakly definable 178, 184
weakly fair parallel program 973
well-founded induction (*see* induction, well-founded)
well-founded ordering (*see* ordering, partial, well-founded)
well-founded relation 878
well-founded set 1174
well-foundedness 810
well-order 878
well-pointed category 409
well-quasi-ordering 272
wf operator 810
while program 807, 812
width (of a hypergraph) 206
word 4
bi-infinite 147
empty 4, 106
factor of a 25, 41
generalized 165
infinite 462
length of 4
ω 136
prefix of a 25
square-free 71
synchronizing 46
ultimately periodic infinite 43
word problem 247, 256, 257, 263, 266, 302, 305
of finitely generated groups 96
WP 807, 812
wreath product of monoids 49
WS1S (weak S1S) 147
WS2S (weak S2S) 178

Zielonka's theorem 48
Zorn's lemma 276